CHEMICAL FUNCTIONALIZATION OF CARBON NANOMATERIALS

CHEMISTRY AND APPLICATIONS

CHEMICAL FUNCTIONALIZATION OF CARBON NANOMATERIALS

CHEMISTRY AND APPLICATIONS

Edited by

Vijay Kumar Thakur
Washington State University, Pullman, USA

Manju Kumari Thakur
Himachal Pradesh University, India

CRC Press
Taylor & Francis Group
Boca Raton London New York

CRC Press is an imprint of the
Taylor & Francis Group, an **informa** business

CRC Press
Taylor & Francis Group
6000 Broken Sound Parkway NW, Suite 300
Boca Raton, FL 33487-2742

First issued in paperback 2017

© 2016 by Taylor & Francis Group, LLC
CRC Press is an imprint of Taylor & Francis Group, an Informa business

No claim to original U.S. Government works

ISBN 13: 978-1-138-89457-0 (pbk)
ISBN 13: 978-1-4822-5394-8 (hbk)

Library of Congress Cataloging-in-Publication Data

Chemical functionalization of carbon nanomaterials : chemistry and applications / edited by Vijay Kumar Thakur and Manju Kumari Thakur.
     pages cm
Includes bibliographical references and index.
     ISBN 978-1-4822-5394-8 (hardcover : alk. paper) 1. Carbon nanofibers. 2. Carbon nanotubes. 3. Supramolecular chemistry. I. Thakur, Vijay Kumar, 1981- editor. II. Thakur, Manju Kumari, editor.

TA455.C3C44 2016
547'.1226--dc23   2015008962

Visit the Taylor & Francis Web site at
http://www.taylorandfrancis.com

and the CRC Press Web site at

To my parents and teachers, who helped me become what I am today.

Vijay Kumar Thakur

Contents

Preface

Carbon-based nanomaterials are rapidly emerging as one of the most fascinating materials in the twenty-first century. Carbon is a very important element in the periodic table that essentially forms the basis of life on Earth. Elemental carbon exists in two natural allotropes, diamond and graphite, having sp^3- and sp^2-hybridized carbon atoms, respectively. Both allotropes of carbon represented the carbon family for a long time, until the discovery of fullerene in 1985. The serendipitous discovery of fullerene marked the beginning of an era of synthetic carbon allotropes from naturally occurring diamond and graphite. Soon after the discovery of 0D fullerene, the synthetic carbon family has been graced by the addition of quasi-1D carbon nanotubes (CNTs), whose discovery in 1991 created a boom in the scientific world. Their large length (up to several microns) and small diameter (a few nanometers) result in a large aspect ratio. They can be seen as the nearly 1D form of fullerenes. Therefore, these materials are expected to possess additional interesting electronic, mechanical, and molecular properties. More recently, the discovery of graphene made it a flagship material harbingering the age of nanotechnology. The arrangement of carbon atoms in each crystal is the fundamental difference between these various structures. These revolutionary allotropes of carbon represent an attractive research field in the nanomaterials sector with important economic potential. The most popular nanomaterials of the carbon family to date are fullerenes, CNTs, and graphenes, with dimensions ranging from 0.5 to 100 nm. These carbon-based nanomaterials possess unique and novel properties, such as remarkable mechanical strength, electrical conductivity, and optical, chemical, and thermal properties due to their unique and intriguing size. The prime advantages of the carbon nanomaterials include high surface-area-to-volume ratio and unique thermal, optical, mechanical, and electrical properties to name a few. The characteristic structures of carbon-based nanomaterials promote them to interact with organic molecules through covalent and noncovalent bonds. Noncovalent interaction can be named as hydrogen bonding, π–cation interaction, π–π stacking, π–anion electrostatic forces, hydrophobic interactions, and van der Waals forces. Carbon allotropes have attracted scientists, and materials that consist of conjugated π-bond systems are topologically confined objects in zero, one, two, or three dimensions.

In spite of their numerous advantages, carbon-based nanomaterials also suffer from a few drawbacks. One of the biggest disadvantages is their low solubility in nearly all solvents and the lack of bonding caused by different forces, such as the attractive van der Waals interaction. Solubility is a vital property for processability as it concerns purification. Another significant drawback is the lack of compatibility/bonding/inertness toward other materials. To overcome the disadvantages of these carbon-based nanomaterials, surface modification through chemical functionalization is an imperative technique capable of overcoming the inherent advantages of the various carbon nanomaterials. Functionalizing carbon-based nanomaterials with an element of interest facilitated a lot to conquer their demerits and also augment their activity in many cases. Functionalization of carbon-based nanomaterials can be achieved by different techniques, and the two most practiced methodologies include covalent functionalization and noncovalent functionalization. Given their practical implications for commercial use, the surface chemistry of carbon nanomaterials is a topic of huge interest, and strong efforts are being dedicated in developing novel methods for the modification and qualitative and quantitative characterization of the surface functional groups.

Thus, given the immense advantages of carbon nanomaterials, this book primarily focuses on their chemical functionalization using different techniques. Several critical issues and suggestions for future work are comprehensively discussed with the hope to provide a deep insight into the state of the art of functionalized carbon nanomaterials. We thank Leong Li-Ming (acquisitions editor) and the publisher (CRC Press and Taylor & Francis Group) for the invaluable help in the organization of the editing process.

Finally, we thank our parents for their continuous encouragement and support.

Vijay Kumar Thakur, PhD
Washington State University
Pullman, Washington

Manju Kumari Thakur, MSc, MPhil, PhD
Himachal Pradesh University
Shimla, India

Editors

Vijay Kumar Thakur, PhD, has been working as research faculty (staff scientist) in the School of Mechanical and Materials Engineering at Washington State University, Pullman, Washington, since September 2013. He was formerly a research scientist at Temasek Laboratories at Nanyang Technological University, Singapore, and a visiting research fellow at the Department of Chemical and Materials Engineering at Lunghwa University of Science and Technology (LHU), Taiwan. His research interests include the synthesis and processing of biobased polymers, nanomaterials, polymer micro-/nanocomposites, nanoelectronic materials, novel high dielectric constant materials, electrochromic materials for energy storage, green synthesis of nanomaterials, and surface functionalization of polymers/nanomaterials. He completed his postdoctorate in materials science and engineering at Iowa State University and earned a PhD in polymer chemistry (2009) at the National Institute of Technology. In his academic career, he has published more than 77 SCI journal research articles in the field of polymers/materials science and holds one U.S. patent. He has also published 15 books and 33 book chapters on the advanced state of the art of polymers/materials science with numerous publishers. He is an editorial board member of several international journals, including *Advanced Chemistry Letters, Lignocelluloses, Drug Inventions Today, International Journal of Energy Engineering,* and *Journal of Textile Science and Engineering,* and he is also a member of scientific bodies around the world. In addition to being on the editorial board of journals, he also serves as the guest editor of *Journal of Nanomaterials, International Journal of Polymer Science,* and *Journal of Chemistry.*

Manju Kumari Thakur, MSc, MPhil, PhD, is an assistant professor of chemistry at the Division of Chemistry, Government Degree College Sarkaghat, Himachal Pradesh University, Shimla, India, since June 2010. She earned a BSc in chemistry, botany, and zoology; an MSc and an MPhil in organic chemistry; and a PhD in polymer chemistry at the Chemistry Department at Himachal Pradesh University, Shimla, India. She has a rich experience in the field of organic chemistry, biopolymers, composites/nanocomposites, hydrogels, applications of hydrogels in the removal of toxic heavy metal ions, drug delivery, etc. She has published more than 30 research papers in several international journals, coauthored 2 books, and also published 28 book chapters in the field of polymeric materials.

Contributors

E.C. Abdullah
Malaysia–Japan International Institute of Technology
Universiti Teknologi Malaysia
Kuala Lumpur, Malaysia

Sindhu Karthika Ammini
Department of Pediatrics
Oman Medical Complex
Ibri, Sultanate of Oman

I.V. Antonova
A.V. Rzhanov Institute of Semiconductor Physics SB RAS
Novosibirsk, Russia

S. Ganesh Babu
Department of Chemistry
National Institute of Technology, Tiruchirappalli
Tiruchirappalli, India

and

Nano Fusion Technology Research Lab
Division of Frontier Fibers
Institute for Fiber Engineering
Interdisciplinary Cluster for Cutting Edge Research
National University Corporation
Shinshu University
Nagano, Japan

S.S. Barkade
Department of Chemical Engineering
University Institute of Chemical Technology
North Maharashtra University
Jalgaon, India

J. Sadhik Basha
Faculty of Process Operations Technology
International Maritime College Oman
Sohar, Sultanate of Oman

Sukumar Basu
Department of Electronics and Telecommunication
  Engineering
IC Design and Fabrication Centre
Jadavpur University
Kolkata, India

B.A. Bhanvase
Department of Chemical Engineering
Laxminarayan Institute of Technology
Nagpur, India

A.E. Burakov
Department of Equipment and Technologies of Nanoproduct
  Manufacture
Tambov State Technical University
Tambov, Russia

Diego Cazorla-Amorós
Department of Inorganic Chemistry and Materials Institute
Universidad de Alicante
Alicante, Spain

Saurabh Chaudhury
Department of Electrical Engineering
National Institute of Technology, Silchar
Silchar, India

Zhongfang Chen
Department of Chemistry
Institute for Functional Nanomaterials
University of Puerto Rico
San Juan, Puerto Rico

Suman Chowdhury
Department of Physics
University of Calcutta
Kolkata, India

Matthew T. Cole
Electrical Engineering Division
Department of Engineering
Cambridge University
Cambridge, United Kingdom

Clare Collins
Electrical Engineering Division
Department of Engineering
Cambridge University
Cambridge, United Kingdom

Ritwika Das
Department of Physics
University of Calcutta
Kolkata, India

M.P. Deosarkar
Department of Chemical Engineering
Vishwakarma Institute of Technology
Pune, India

B. Dewangan
Department of Plastics and Polymer Technology
Laxminarayan Institute of Technology
Nagpur, India

Shivani Dhall
Department of Physics
National Institute of Technology, Kurukshetra
Kurukshetra, India

Ana M. Díez-Pascual
Department of Analytical Chemistry, Physical Chemistry
   and Chemical Engineering
Alcalá University
Madrid, Spain

Siqi Ding
School of Civil Engineering
Dalian University of Technology
Dalian, Liaoning, People's Republic of China

Deyang Du
Department of Physics
Southeast University
Nanjing, Jiangsu, People's Republic of China

Gary J. Ellis
Department of Polymer Physics, Elastomers and Energy
   Applications
Institute of Polymer Science and Technology
Madrid, Spain

Joaquim C.G. Esteves da Silva
Faculty of Sciences
Department of Chemistry and Biochemistry
University of Porto
Porto, Portugal

Xin Fang
Department of Macromolecular Science
and
Laboratory of Advanced Materials
State Key Laboratory of Molecular Engineering
   of Polymers
Fudan University
Yangpu, Shaghai, People's Republic of China

G.R. Gajare
Department of Chemical Engineering
Sinhgad College of Engineering
Pune, India

Signorino Galvagno
Department of Electronic Engineering, Industrial Chemistry
   and Engineering
University of Messina
Messina, Italy

P. Ganesan
Department of Mechanical Engineering
University of Malaya
Kuala Lumpur, Malaysia

Haili Gao
College of Materials Engineering
Fujian Agriculture and Forestry University
Fuzhou, Fujian, People's Republic of China

Liehui Ge
Department of Materials Science and NanoEngineering
Rice University
Houston, Texas

P.R. Gogate
Department of Chemical Engineering
Institute of Chemical Technology
Mumbai, India

P.S. Goh
Faculty of Petroleum and Renewable Energy Engineering
Advanced Membrane Technology Research Centre
Universiti Teknologi Malaysia
Skudai, Malaysia

Cristina Gómez-Aleixandre
Department of Surfaces, Coatings and Molecular Astrophysics
Materials Science Institute of Madrid
Spanish National Research Council
Madrid, Spain

Marián A. Gómez-Fatou
Department of Polymer Physics, Elastomers and Energy
   Applications
Institute of Polymer Science and Technology
Madrid, Spain

Jian Ru Gong
CAS Key Laboratory for Nanosystem and Hierarchical
   Fabrication
National Center for Nanoscience and Technology
Beijing, People's Republic of China

Carolina González-Gaitán
Materials Institute
Universidad de Alicante
Alicante, Spain

P. González-García
Postgraduate and Research Direction
Center for Engineering and Industrial Development
Querétaro, México

M. Gopiraman
Nano Fusion Technology Research Lab
Division of Frontier Fibers
Institute for Fiber Engineering
Interdisciplinary Cluster for Cutting Edge Research
National University Corporation
Shinshu University
Ueda, Japan

Baoguo Han
School of Civil Engineering
Dalian University of Technology
Dalian, Liaoning, People's Republic of China

Surajit Kumar Hazra
Department of Physics and Materials Science
Jaypee University of Information Technology
Waknaghat, India

Daniela Iannazzo
Department of Electronic Engineering, Industrial Chemistry
    and Engineering
University of Messina
Messina, Italy

A.F. Ismail
Faculty of Petroleum and Renewable Energy Engineering
Advanced Membrane Technology Research Centre
Universiti Teknologi Malaysia
Skudai, Malaysia

N.K. Jain
Pharmaceutical Nanotechnology Research Laboratory
ISF College of Pharmacy
Moga, India

and

Pharmaceutics Research Laboratory
Department of Pharmaceutical Sciences
Dr. H. S. Gour University
Sagar, India

Debnarayan Jana
Department of Physics
University of Calcutta
Kolkata, India

N.S. Jayakumar
Department of Chemical Engineering
University of Malaya
Kuala Lumpur, Malaysia

Peng Jin
School of Materials Science and Engineering
Hebei University of Technology
Tianjin, People's Republic of China

R. Karvembu
Department of Chemistry
National Institute of Technology
Tiruchirappalli, India

Marianna V. Kharlamova
Faculty of Physics
University of Vienna
Vienna, Austria

and

Department of Materials Science
Lomonosov Moscow State University
Moscow, Russia

I.S. Kim
Nano Fusion Technology Research Lab
Division of Frontier Fibers
Institute for Fiber Engineering
Interdisciplinary Cluster for Cutting Edge Research
National University Corporation
Shinshu University
Ueda, Japan

Antigoni E. Koletti
Laboratory of Analytical Chemistry
Department of Chemistry
Aristotle University of Thessaloniki
Thessaloniki, Greece

Naoki Komatsu
Department of Chemistry
Shiga University of Medical Science
Otsu, Japan

N.E. Kornienko
Faculty of Physics
Department of Theoretical Physics
Taras Shevchenko National University of Kyiv
Kyiv, Ukraine

A.E. Kucherova
Department of Equipment and Technologies of Nanoproduct
    Manufacture
Tambov State Technical University
Tambov, Russia

João M.M. Leitão
Faculty of Pharmacy
University of Coimbra
Coimbra, Portugal

Chi Li
Display Research Centre
School of Electronic Science and Engineering
Southeast University
Nanjing, Jiangsu, People's Republic of China

Lizhao Liu
School of Science
and
Key Laboratory of Materials Modification by Laser, Ion and
    Electron Beams
Dalian University of Technology
Ministry of Education
Dalian, Liaoning, People's Republic of China

Xin Lu
Department of Macromolecular Science
and
Laboratory of Advanced Materials
State Key Laboratory of Molecular Engineering
    of Polymers
Fudan University
Yangpu, Shaghai, People's Republic of China

Xiaoguang Luo
Department of Physics
Southeast University
Nanjing, Jiangsu, People's Republic of China

Carlos Marco
Department of Polymer Physics, Elastomers and Energy
    Applications
Institute of Polymer Science and Technology
Madrid, Spain

Gerardo Martínez
Department of Polymer Physics, Elastomers and Energy
    Applications
Institute of Polymer Science and Technology
Madrid, Spain

A.L. Martínez-Hernández
Division of Postgraduate Studies and Research
Technological Institute of Queretaro
and
Center of Applied Physics and Advanced Technology
National Autonomous University of Mexico
Querétaro, México

Neelesh Kumar Mehra
Pharmaceutical Nanotechnology Research Laboratory
ISF College of Pharmacy
Moga, India

and

Pharmaceutics Research Laboratory
Department of Pharmaceutical Sciences
Dr. H. S. Gour University
Sagar, India

William I. Milne
Electrical Engineering Division
Department of Engineering
Cambridge University
Cambridge, United Kingdom

Satyendra Mishra
Department of Chemical Engineering
University Institute of Chemical Technology
North Maharashtra University
Jalgaon, India

Kyungsun Moon
Department of Physics
Yonsei University
Seoul, South Korea

Omid Moradi
Department of Chemistry
Shahr-e-Qods Branch
Islamic Azad University
Tehran, Iran

Emilia Morallón
Department of Physical Chemistry and Materials Institute
Universidad de Alicante
Alicante, Spain

N.M. Mubarak
Department of Chemical Engineering
University of Malaya
and
Department of Chemical and Petroleum Engineering
UCSI University
Kuala Lumpur, Malaysia

Shabbir Muhammad
Department of Physics
College of Science
King Khalid University
Abha, Kingdom of Saudi Arabia

Roberto Muñoz
Department of Surfaces, Coatings and Molecular Astrophysics
and
Department of Materials for Information Technologies
Materials Science Institute of Madrid
Spanish National Research Council
Madrid, Spain

J.B. Naik
Department of Chemical Engineering
University Institute of Chemical Technology
North Maharashtra University
Jalgaon, India

Masayoshi Nakano
Department of Materials Engineering Science
Graduate School of Engineering Science
Osaka University
Osaka, Japan

Palash Nath
Department of Physics
University of Calcutta
Kolkata, India

A.P. Naumenko
Faculty of Physics
Department of Experimental Physics
Taras Shevchenko National University of Kyiv
Kyiv, Ukraine

F. Navarro-Pardo
Division of Postgraduate Studies and Research
Technological Institute of Queretaro
and
Center of Applied Physics and Advanced Technology
National Autonomous University of Mexico
Querétaro, México

and

National Institute of Scientific Research
University of Quebec
Varennes, Québec, Canada

N.A. Nebogatikova
A.V. Rzhanov Institute of Semiconductor Physics SB RAS
Novosibirsk, Russia

Toktam Nezakati
Division of Surgery and Interventional Science
Centre for Nanotechnology and Regenerative Medicine
University College London
London, United Kingdom

B.C. Ng
Faculty of Petroleum and Renewable Energy Engineering
Advanced Membrane Technology Research Centre
Universiti Teknologi Malaysia
Skudai, Malaysia

Jinping Ou
School of Civil Engineering
Dalian University of Technology
Dalian, Liaoning, People's Republic of China

and

School of Civil Engineering
Harbin Institute of Technology
Harbin, Heilongjiang, People's Republic of China

Richard Parmee
Electrical Engineering Division
Department of Engineering
Cambridge University
Cambridge, United Kingdom

Huisheng Peng
Department of Macromolecular Science
and
Laboratory of Advanced Materials
State Key Laboratory of Molecular Engineering
of Polymers
Fudan University
Yangpu, Shaghai, People's Republic of China

D.V. Pinjari
Department of Chemical Engineering
Institute of Chemical Technology
Mumbai, India

Alessandro Pistone
Department of Electronic Engineering, Industrial Chemistry
and Engineering
University of Messina
Messina, Italy

Reza Pourazizi
Faculty of New Sciences and Technologies
Composites Research Laboratory
University of Tehran
Tehran, Iran

Raghavan Prasanth
Department of Materials Science and NanoEngineering
Rice University
Houston, Texas

V.Ya. Prinz
A.V. Rzhanov Institute of Semiconductor Physics SB RAS
Novosibirsk, Russia

A. Rubavathy Jaya Priya
Department of Physical Chemistry
School of Chemical Sciences
University of Madras
Chennai, India

Teng Qiu
Department of Physics
Southeast University
Nanjing, Jiangsu, People's Republic of China

Roham Rafiee
Faculty of New Sciences and Technologies
Composites Research Laboratory
University of Tehran
Tehran, Iran

Jun-Won Rhim
School of Physics
Korea Institute for Advanced Study
and
Department of Physics
Yonsei University
Seoul, South Korea

I.V. Romantsova
Department of Equipment and Technologies of Nanoproduct Manufacture
Tambov State Technical University
Tambov, Russia

Debesh Ranjan Roy
Department of Applied Physics
S. V. National Institute of Technology
Surat, India

Ramiro Ruiz-Rosas
Materials Institute
Universidad de Alicante
Alicante, Spain

Hamidreza Sadegh
Department of Chemistry
Islamic Azad University
Tehran, Iran

J.N. Sahu
Department of Petroleum and Chemical Engineering
Institut Teknologi Brunei
Tungku Gadong, Brunei

Horacio J. Salavagione
Department of Polymer Physics, Elastomers and Energy
    Applications
Institute of Polymer Science and Technology
Madrid, Spain

Victoria F. Samanidou
Laboratory of Analytical Chemistry
Department of Chemistry
Aristotle University of Thessaloniki
Thessaloniki, Greece

Dirtha Sanyal
Radioactive Ion Beam Facility Group
Variable Energy Cyclotron Centre
Kolkata, India

Alexander M. Seifalian
Division of Surgery and Interventional Science
Centre for Nanotechnology and Regenerative Medicine
University College London
and
Royal Free London NHS Foundation Trust Hospital
London, United Kingdom

Esha V. Shah
Department of Applied Physics
S. V. National Institute of Technology
Surat, India

Ramin Shahryari-ghoshekandi
Department of Chemistry
Islamic Azad University
Tehran, Iran

Himani Sharma
Excitonics and Nanostructure Laboratory
Department of Electrical and Computer Engineering
University of Alberta
Edmonton, Alberta, Canada

Jae Ho Shin
Department of Chemistry
Kwangwoon University
Seoul, South Korea

A.K. Shukla
Thin Film Laboratory
Department of Physics
Indian Institute of Technology, Delhi
New Delhi, India

Geoffrey S. Simate
School of Chemical and Metallurgical Engineering
University of the Witwatersrand, Johannesburg
Johannesburg, Gauteng, South Africa

Eliana F.C. Simões
Faculty of Sciences
Department of Chemistry and Biochemistry
University of Porto
Porto, Portugal

Sanjeet Kumar Sinha
Department of Electrical Engineering
National Institute of Technology, Silchar
Silchar, India

S.H. Sonawane
Department of Chemical Engineering
National Institute of Technology, Warangal
Warangal, India

Kapil Sood
School of Physics and Material Science
Thapar University
Patiala, India

M.V. Strikha
V.E. Lashkariov Institute of Semiconductor Physics
National Academy of Sciences of Ukraine
Kyiv, Ukraine

Abu Bakar Sulong
Faculty of Engineering and Built Environment
Department of Mechanical and Materials Engineering
Universiti Kebangsaan Malaysia
Bangi, Malaysia

Shengwei Sun
School of Civil Engineering
Harbin Institute of Technology
Harbin, Heilongjiang, People's Republic of China

Xuemei Sun
Department of Macromolecular Science
and
Laboratory of Advanced Materials
State Key Laboratory of Molecular Engineering
of Polymers
Fudan University
Yangpu, Shaghai, People's Republic of China

Aaron Tan
Division of Surgery and Interventional Science
Centre for Nanotechnology and Regenerative Medicine
and
Medical School
University College London
London, United Kingdom

Soon Huat Tan
School of Chemical Engineering
Universiti Sains Malaysia
Nibong Tebal, Malaysia

Chengchun Tang
School of Materials Science and Engineering
Hebei University of Technology
Tianjin, People's Republic of China

Manju Kumari Thakur
Division of Chemistry
Government Degree College, Sarkaghat
Himachal Pradesh University
Shimla, India

Vijay Kumar Thakur
School of Mechanical and Materials Engineering
Washington State University
Pullman, Washington

A.G. Tkachev
Department of Equipment and Technologies of Nanoproduct
Manufacture
Tambov State Technical University
Tambov, Russia

V.D. Vankar
Thin Film Laboratory
Department of Physics
Indian Institute of Technology
New Delhi, India

C. Velasco-Santos
Division of Postgraduate Studies and Research
Technological Institute of Queretaro
and
Center of Applied Physics and Advanced Technology
National Autonomous University of Mexico
Querétaro, México

J.R. Wong
Department of Chemical and Petroleum Engineering
UCSI University
Kuala Lumpur, Malaysia

Qian Wen Yeang
School of Chemical Engineering
Universiti Sains Malaysia
Nibong Tebal, Malaysia

Shabi Abbas Zaidi
Department of Chemistry
Kwangwoon University
Seoul, South Korea

Liqing Zhang
School of Civil Engineering
Dalian University of Technology
Dalian, Liaoning, People's Republic of China

Jijun Zhao
Key Laboratory of Materials Modification by Laser, Ion and
    Electron Beams
Dalian University of Technology
Ministry of Education
Dalian, Liaoning, People's Republic of China

Li Zhao
Department of Chemistry
Shiga University of Medical Science
Otsu, Japan

SECTION I

CHEMICAL FUNCTIONALIZATION OF CARBON NANOMATERIALS

Structure and Chemistry

CHAPTER 1

CONTENTS

Carbon Allotropes and Fascinated Nanostructures

The High-Impact Engineering Materials of the Millennium

Raghavan Prasanth, Sindhu Karthika Ammini, Liehui Ge, Manju Kumari Thakur, and Vijay Kumar Thakur

1.1 INTRODUCTION

The concepts of nanotechnology or nanomaterials that seeded nanotechnology were first discussed in 1959 by the renowned physicist Richard Feynman in his talk *There's Plenty of Room at the Bottom*, in which he described the possibility of synthesis via direct manipulation of atoms. The term *nanotechnology* was first used by Norio Taniguchi in 1974, though it was not widely known. Inspired by Feynman's concepts, K. Eric Drexler independently used the term *nanotechnology* in his 1986 book *Engines of Creation: The Coming Era of Nanotechnology*, which proposed the idea of a nanoscale *assembler* that would be able to build a copy of itself and of other items of arbitrary complexity with atomic control. Also in 1986, Drexler cofounded the *Foresight Institute* (with which Drexler is no longer affiliated) to help increase public awareness and understanding of nanotechnology concepts and implications. Thus, the emergence of nanotechnology as a field in the 1980s occurred through convergence of Drexler's theoretical and public work, which developed and popularized a conceptual framework for nanotechnology and high-visibility experimental advances that drew additional wide-scale attention to the prospects of atomic control of matter. For example, the invention of the scanning tunneling microscope in 1981 provided unprecedented visualization of individual atoms and bonds and was successfully used to manipulate individual atoms in 1989. The microscope's developers Gerd Binnig and Heinrich Rohrer at IBM Zurich Research Laboratory received a Nobel Prize in Physics in 1986 [1].

Nanotechnology is the engineering of functional systems at the molecular scale. This covers both current work and concepts that are more advanced. In its original sense, nanotechnology refers to the projected ability to construct items from the bottom up, using techniques and tools being developed today to make complete, high-performance products. One nanometer (nm) is one billionth, or 10^{-9}, of a meter. By comparison, typical carbon–carbon bond lengths, or the spacing between these atoms in a molecule, are in the range 0.12–0.15 nm, and a DNA double helix has a diameter around 2 nm. On the other hand, the smallest cellular life forms, the bacteria of the genus *Mycoplasma*, are around 200 nm in length. By convention, nanotechnology is taken as the scale range 1–100 nm following the definition used by the National Nanotechnology Initiative in the United States. The lower limit is set by the size of atoms (hydrogen has the smallest atoms, which are approximately a quarter of 1 nm diameter) since nanotechnology must build its devices from atoms and molecules. The upper limit is more or less arbitrary but is around the size that phenomena not observed in larger structures start to become apparent and can be made use of in the nanodevice. These new phenomena make nanotechnology distinct from devices that are merely miniaturized versions of an equivalent macroscopic device; such devices are on a larger scale and come under the description of microtechnology [2]. To put that scale in another context, the comparative size of a nanometer to a meter is the same as that of a marble to the size of the Earth. Or another way of putting it: a nanometer is the amount an average man's beard grows in the time it takes him to raise the razor to his face [3]. Two main approaches are used in nanotechnology. In the bottom-up approach, materials and devices are built from molecular components that assemble themselves schematically by principles of molecular recognition. In the top-down approach, nano-objects are constructed from larger entities without atomic-level control [4]. Areas of physics such as nanoelectronics, nanomechanics, nanophotonics, and nanoionics have evolved during the last few decades to provide a basic scientific foundation of nanotechnology.

Carbon is the sixth element of the periodic table. Each carbon atom has six electrons that occupy $1s^2$, $2s^2$, and $2p^2$ atomic orbital. It can hybridize in sp (e.g., C_2H_2), sp^2 (e.g., graphite), or sp^3 (e.g., CH_4) forms. This property is unique to carbon in its particular group in periodic table. With an atomic number of 6, carbon is the fourth most abundant element in the universe by mass after hydrogen, helium, and oxygen. It forms more compounds than any other element, with almost 10 million pure organic compounds. Abundance, together with the unique diversity of organic compounds and their unusual polymer-forming ability at the temperatures commonly encountered on Earth, makes the element the chemical basis of all known life. Discoveries of very stable nanometer-size sp^2 carbon-bonded materials such as fullerenes [5], nanodiamonds, carbon nanotubes (CNTs) [6], and graphene [7] have stimulated the research in this field.

Nanoscience and nanotechnology highlighted around carbon since the opening point mainly due the extensive study of materials started from fullerenes to benchmarked graphene. The production of microtubes of carbon in 1960 by Bacon [8] was one of the main discoveries in carbon cage structures. The era of carbon materials started from the discovery of fullerenes [9] and then continued through CNTs by Iijima and Ichihashi [6] followed by large-scale synthesis of CNTs by Ebbesen and Ajayan [10]. These nanostructures constitute a bridge between molecules to bulk systems of carbon. Many approaches are there to classify carbon nanostructures; the whole range of dimensionalities is fitted in these structures beginning from 0D (fullerenes, diamond) to 3D structures (nanocrystalline diamond, fullerite), including 1D CNTs and 2D graphene [7]. Within this group of *new carbon* materials, texture on a nanometer scale based on the preferred orientation of anisotropic hexagonal layers plays an important role in their properties. Some of the *new* graphitic materials contain nanostructural units within a complex hierarchical structure such as carbon fibers consisting of CNTs in their cores. Because of its valency, carbon can form many allotropes in which well-known forms are diamond and graphite. The classifications, synthesis, and applications will be discussed in the following sections. While the history of synthetic graphite begins in the nineteenth century [11], artificial diamonds were not synthesized until the middle of the twentieth century. Since then, both graphite- and diamond-related groups of carbon materials have experienced several waves of renewed interest in scientific communities when new types of materials or synthesis techniques had been discovered. Within the graphite-based group, new materials (*new carbon*), such as carbon fibers, glass-like carbons, and pyrolytic carbons, were developed in the early 1960s and found broad industrial applications [12]. The most significant relatively recent application of this class of carbon material is probably lithium-ion rechargeable batteries that use nanostructured carbon anodes, which have made possible portable electronic devices.

A new era in carbon materials began with the discovery of the buckminsterfullerene family (buckyballs) at Rice University, USA, in the mid-1980s [9] followed by the discovery of fullerene nanotubules (buckytubes) [13]. Table 1.1 summarizes the milestones of carbon nanomaterials and related subjects. The discovery of these structures motivated the scientific community to set in motion a new worldwide research boom that seems still to be growing. As mentioned earlier, fullerene nanotubules and graphite-based materials are inherently connected, and researchers who produced carbon filaments had been unknowingly growing nanotubes by arc discharge process for synthesizing fullerene decades before Iijima's publication [13].

TABLE 1.1 Discoveries of Carbon Nanomaterials and Related Subjects

Year	Event	Reference
1952	Hollow graphitic carbon fibers that are 50 nm in diameter.	[14]
1960	Production of carbon tubes with a graphite layer structure.	[8]
1966	A large hollow cage molecule is suggested.	[15]
1970	The soccer-ball-shaped C_{60} molecule is suggested.	[16]
1973	Prediction that C_{60} would be stable.	[17]
1976	Growth of nanometer-scale carbon fibers.	[18]
1979	Idea of a space elevator using *continuous pseudo-1D diamond crystal*.	[19]
1980	First observation of nanotubes.	[20]
1983	Synthesis of dodecahedrene ($C_{20}H_{20}$).	[21]
1984	Report on preferential stability of even-atom carbon clusters, especially of a 60-carbon species.	[22]
1985	Discovery of C_{60}, fullerenes.	[9]
1990	Fullerene synthesis in large quantities.	[23]
1991	Discovery of MWNTs.	[13]
1991	Prediction of hyperfullerenes.	[24]
1992	Observation of carbon onions.	[25]
1993	Discovery of SWNTs.	[6]
1994	Fullerenes found in meteorites.	[26]
1996	Ropes of SWNTs.	[27]
1997	First CNT single-electron transistors (operating at low temperature).	[28]
1997	The first suggestion of using CNTs as optical antennas.	[29]
1998	CVD synthesis of aligned nanotube films.	[30]
1998	The first CNT FETs are demonstrated.	[31]
2000	The first demonstration proving that bending of CNT changes their resistance.	[32]
2001	Nanotube single crystals.	[33]
2001	Report on a technique for separating semiconducting and metallic CNTs.	[34]
2002	MWNTs demonstrated to be the fastest known oscillators (>50 GHz).	[35]
2003	NEC announced the stable fabrication technology of CNT transistors.	[36]
2004	Nature published a photo of an individual 4 cm long SWNT.	[37]
2005	Demonstrated high-definition 10 cm flat-screen made using CNTs.	[38]
2005	Nanotube sheet synthesized with dimensions 5 × 100 cm.	[39]
2006	Gadget invented by researchers in Rice University for sorting CNTs by size and electrical properties.	[40]
2006	Nanotubes used as a scaffold for damaged nerve regeneration.	[41]
2009	Nanotubes incorporated in virus battery.	[42]
2012	IBM creates a 9 nm CNT transistor that outperforms silicon.	[43]
2013	Researchers build a CNT computer.	[38]

1.2 CLASSIFICATIONS OF CARBON NANOMATERIALS

1.2.1 Allotropes of Carbon

Carbon is well known to form distinct solid-state allotropes with diverse structures and properties ranging from sp³-hybridized diamond to sp²-hybridized graphite. Historically, chemists have known only two allotropes, pure forms, of carbon: graphite, a greasy, electrically conducting black substance, and diamond, crystal clear, electrically insulating, and harder than any other solid. But they have constantly theorized about other possible carbon allotropes. Mainly there are eight allotropes of carbon: (1) diamond, (2) graphite, (3) lonsdaleite, (4) C_{60} (buckminsterfullerene or buckyball), (5) C_{540}, (6) C_{70}, (7) amorphous carbon, and (8) CNT or buckytube. Until the 1960s when *new carbon* materials were synthesized, only two allotropic forms of carbon were known, graphite and diamond, including their polymorphous modifications. Recently only *amorphous carbon* is considered as a third allotrope of carbon. Graphite is other most common allotropes of carbon, the most thermodynamically stable form of carbon at room temperature. Therefore, it is used in thermochemistry as the standard state for describing the heat of formation of carbon compounds. Graphite consists of a layered 2D structure where each layer possesses a hexagonal honeycomb structure of sp^2-bonded carbon atoms with a C–C bond length of 1.42 Å. These thick single-atom layers (i.e., graphene layers) interact via noncovalent van der Waals forces with an interlayer spacing of 3.35 Å. The weak interlayer bonding in graphite implies that single graphene layers can be exfoliated via mechanical or chemical methods as will be outlined in detail later. Graphene is often viewed as the 2D building block of other sp^2-hybridized carbon nanomaterials in that it can be conceptually rolled or distorted to form CNTs and fullerenes. Graphite is an electrical conductor and is applicable in electronics. Graphite conducts electricity, due to delocalization of the π-bond electrons above and below the planes of the carbon atoms. These electrons are free to move, so are capable to conduct electricity. However, the electricity is only conducted along the plane of the layers. In diamond, all four outer electrons of each carbon atom are localized between the atoms in covalent bonding. The movement of electrons is constrained and diamond does not conduct an electric current. In graphite, each carbon atom uses only three of its four outer energy level electrons in covalently bonding to three other carbon atoms in a plane. Each carbon atom contributes one electron to a delocalized system of electrons that is also a part of the chemical bonding. The delocalized electrons are free to move throughout the plane. So graphite conducts electricity along the planes of carbon atoms. Amorphous carbon is the carbon that does not have any crystalline structure. As with all glassy materials, some short-range order can be observed, but there is no long-range pattern of atomic positions. While completely amorphous carbon can be produced, most of them contain microscopic crystals of graphite-like or even diamond-like carbon.

There are many ways to classify the carbon materials based on the state, structure, dimension, etc. In principle, there are different approaches that can be used to classify carbon nanostructures; however, the appropriate classification scheme depends on the field of application of the nanostructures. Mostly the carbon nanomaterials are classified on the basis of its dimensionality. By this classification, the entire range of dimensionalities is represented in the nanocarbon world, beginning with 0D structures (fullerenes, diamond clusters) and including 1D (nanotubes), 2D (graphene), and 3D (nanocrystalline diamond, fullerite) structures. In a different approach, the classification based on the scale of characteristic sizes of the nanomaterials and this scheme of classification naturally allow more consideration of complicated hierarchical structures of carbon materials such as carbon fibers and carbon polyhedral particles. Also, classification based on different shapes and spatial arrangements of elemental structural units of carbon-caged nanostructures provides a very vivid and useful picture of the numerous forms of carbon structures at the nanoscale [44]. Regarding the last approach, the spatial distribution of penta- and hexa-rings within structures also can provide a basis for classification [45].

In terms of a more fundamental basis for the classification of carbon nanostructures, it would be logical to develop a classification scheme based on existing carbon allotropes that is inherently connected with the nature of bonding in carbon materials. Ironically, there is no consensus on how many carbon allotropes/forms are defined at present. From time to time publications appear proposing new crystalline forms or allotropic modifications of carbon. Whether fullerenes or carbynes are considered as new carbon allotropes depends to a large extent on the corresponding scientific [46,47]. Sometimes the *fullerene community* appears to ignore the carbynes, which were discovered in the 1960s, and similarly, the *carbyne community* does not classify fullerenes as an allotrope [46]. Table 1.2 shows the properties of carbon allotropes and its schematic for classifying existing carbon forms based on the types of chemical bonds in carbon, with each valence state corresponding to a certain form of a simple substance. As discussed earlier, elemental carbon exists in three bonding states corresponding to sp^3, sp^2, and sp hybridization of the atomic orbitals, and the corresponding three carbon allotropes with an integer degree of carbon bond hybridization are diamond, graphite, and carbine [48]. All other forms of carbon are classified based on the transitional forms as *mixed* short-range-order carbon forms and *intermediate* carbon forms with a noninteger degree of carbon bond hybridization, sp^n. The *mixed* short-range-order carbon forms such as diamond-like carbon, vitreous carbon, soot, and carbon blacks have more or less arranged carbon atoms of different hybridization states. Numerous hypothetical structures like graphynes and *superdiamond* also

TABLE 1.2 Types of Geometrical Characteristics of Selected Carbon Entities and Their
Simplest Assemblies Observed at the Nanoscale

Schematic View	Entity	Characteristic Size
	Fullerene	Smallest, C_{20} Most abundant C_{60}
	Carbon onion	Outer diameter 10 nm–1 μm Inner diameter 0.7–1 nm
	SWNT	Diameter 1–10 nm Length 1 nm–10 μm
	MWNT	Length 10 nm–1 μm Outer diameter 2–30 nm

(Continued)

TABLE 1.2 (Continued) Types of Geometrical Characteristics of Selected Carbon Entities and Their Simplest Assemblies Observed at the Nanoscale

Schematic View	Entity	Characteristic Size
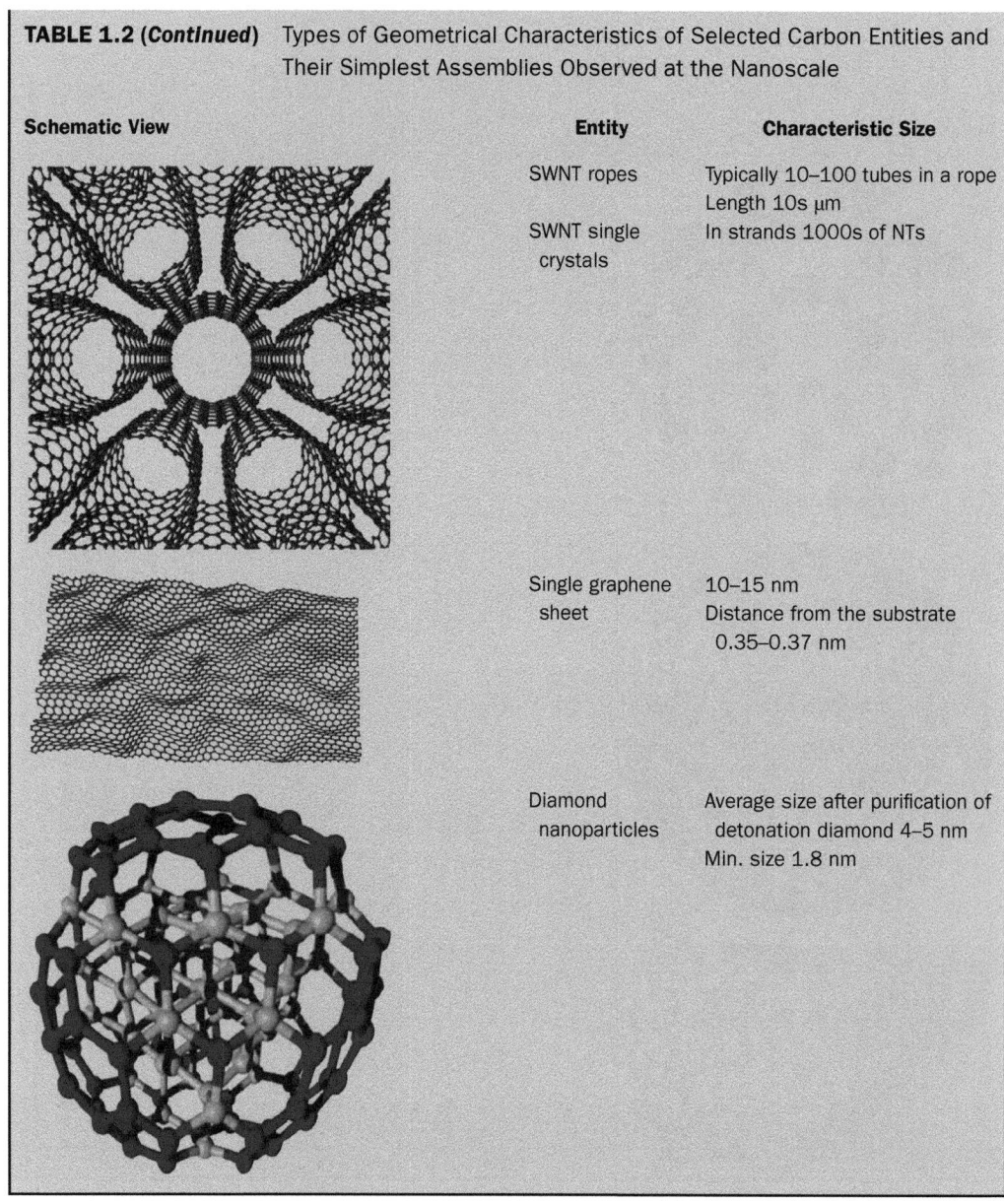	SWNT ropes	Typically 10–100 tubes in a rope Length 10s μm
	SWNT single crystals	In strands 1000s of NTs
	Single graphene sheet	10–15 nm Distance from the substrate 0.35–0.37 nm
	Diamond nanoparticles	Average size after purification of detonation diamond 4–5 nm Min. size 1.8 nm

come under the category *mixed* short-range-order carbon forms. In *intermediate* carbon forms with a noninteger degree of carbon bond hybridization, sp^n have many subgroups depending on the value of integer "n" in sp^n carbon bond hybridization. When the value of n in sp^n carbon bond hybridization $1 < n < 2$ includes various monocyclic carbon structures. Similarly when $2 < n < 3$, the intermediate carbon forms are comprised of closed-shell carbon structures such as fullerenes (the degree of hybridization in C_{60} is ~2.28), carbon onions and CNTs, and hypothetical tori.

1.3 CARBON AT NANOSCALE: ONE-DIMENSIONAL CARBON MATERIALS

1.3.1 Buckminsterfullerene or Buckyballs

Buckminsterfullerene (buckyball) is a spherical fullerene molecule with the formula C_{60}. It has a cage-like fused-ring structure (truncated icosahedron) that resembles a soccer ball, made of 20 hexagons and 12 pentagons, with a carbon atom at each vertex of each polygon and a bond along each polygon edge. It was first generated in 1985 by Harold Kroto, James R. Heath, Sean O'Brien, Robert Curl, and Richard Smalley at Rice University [9]. Kroto, Curl, and Smalley were awarded the 1996 Nobel Prize in Chemistry for their roles in the discovery of buckminsterfullerene and the related class of molecules, the fullerenes. The name is a reference to Buckminster Fuller, as C_{60} resembles his trademark geodesic domes. Buckminsterfullerene is the most common naturally occurring fullerene molecule, as it can be found in small quantities in soot [49,50]. Buckminsterfullerene is one of the largest objects to have been shown to exhibit wave–particle duality, as stated in the theory every object exhibits this behavior [51]. Its discovery led to the exploration of a new field of chemistry, involving the study of fullerenes.

Theoretical predictions of buckyball molecules appeared in the early 1970s. Osawa, a Japanese chemist [16], wrote of a molecule that would be made of 80 carbon atoms, a stable, spherical molecule in the shape of a soccer ball, but they went largely unnoticed. About a decade later, Orville L. Chapman, an organic chemist at the University of California, Los Angeles (UCLA), thought on the same idea of soccer-ball-shaped carbon 60 molecules. When he was spending summer in Germany on a Humboldt fellowship, Chapman raised a question in mind on making a new form of carbon and initiated an effort to synthesize this odd form of carbon; however, he failed. The concept of this carbon form has become famous as the buckyball, more formally known as buckminsterfullerene, the third allotrope of carbon, which has become the focus of research in chemistry laboratories everywhere. While Chapman and his colleagues, including Robert L. Whetten and Francois Diederich, were working at UCLA to synthesize carbon 60, Richard Smalley and his team at Rice University were busy with a totally unrelated subject, the study of atomic clusters. They used a device of their own invention in which laser energy was used to vaporize a sample of an element, which was then blown into a mass spectrometer by a stream of helium for analysis. Richard Smalley and his team at Rice University had no interest in carbon materials until Harry Kroto, a visitor from the University of Sussex in England, suggested that their device could cast light on the nature of the carbon clusters detected in interstellar space, and then Richard Smalley ran the experiment. While running the experiment, they found an unusual concentration of large, even-numbered carbon clusters, with an unusual abundance of C_{60}, and they developed a technique to isolate these substances. They used laser vaporization of a suitable target to produce clusters of atoms. Another interesting fact is that, at the same time, astrophysicists were working along with spectroscopists to study infrared emissions from giant red carbon stars [52]. Smalley and his team were able to use a laser vaporization technique to create carbon clusters that could potentially emit infrared at the same wavelength as had been emitted by the red carbon star [53]. Hence, the inspiration came to Smalley and his team to use the laser technique on graphite to create the first fullerene molecule. Now it is widely known that the researchers at Rice University independently reinvented the idea of a spherical carbon allotrope called buckminsterfullerene. Later, two physicists, Donald Huffman of the University of Arizona and Wolfgang Kraetschmer of the Max Planck Institute for Nuclear Physics in Germany, reported C_{70} and larger fullerenes, and it opened up a new area of research on fullerenes and the family of ball-shaped carbon allotropes frequently called as buckyballs. Using laser evaporation of graphite, they found C_n clusters (where n > 20 and even) of which the most common were C_{60} and C_{70}. In the early 1970s, the chemistry of unsaturated carbon configurations was studied by a group at the University of Sussex, led by Harry Kroto and David Walton. The discovery of buckyballs was surprising, as the scientists aimed the experiment at producing carbon plasmas to replicate and characterize unidentified interstellar matter. Mass spectrometry analysis of the product indicated the formation of spheroidal carbon molecules. The experimental evidence, a strong peak at 720 atomic mass units, indicated that a carbon molecule with 60 carbon atoms was forming, but provided no structural information. The research group concluded after reactivity experiments that the most likely structure was a spheroidal molecule. The idea was quickly rationalized as the basis of an icosahedral symmetry closed-cage structure. Kroto mentioned geodesic dome structures of the noted futurist and inventor Buckminster Fuller as influences in the naming of this particular substance as buckminsterfullerene. In 1996, Robert Curl, Harold Kroto, and Richard Smalley were awarded Nobel Prize in Chemistry for the discovery of C_{60}, the buckyballs.

In 1990, W. Kraetschmer and D. R. Huffman developed a simple and efficient method of producing fullerenes in gram and even kilogram amounts, which has boosted the fullerene research. In this technique, carbon soot is produced from two high-purity graphite electrodes by igniting an arc discharge between them in an inert atmosphere (helium gas). Alternatively, soot is produced by laser ablation of graphite or pyrolysis of aromatic hydrocarbons. Fullerenes are extracted from the soot using a multistep procedure. First, the soot is dissolved in appropriate organic solvents. This step yields a solution containing up to 75% of C_{60}, as well as other fullerenes. These fractions are separated using chromatography [54]. Generally, the fullerenes are dissolved in hydrocarbon or halogenated hydrocarbon and separated using alumina columns [55].

1.3.1.1 Band structure and superconductivity

In 1991, Haddon et al. [56] found that intercalation of alkali metal atoms in solid C_{60} leads to metallic behavior. In 1991, it was revealed that potassium-doped C_{60} becomes superconducting at 18 K [57,58]. This was the highest transition temperature for a molecular superconductor. Since then, superconductivity has been reported in fullerene doped with various other alkali metals [59,60]. It has been shown that the superconducting transition temperature in alkaline-metal-doped fullerene increases with the unit-cell volume [61,62]. As cesium forms the largest alkali ion, cesium-doped fullerene is an important material in this family. Recently, superconductivity at 38 K has been reported in bulk Cs_3C_{60}, [63] but only under applied pressure. The highest superconducting transition temperature of 33 K at ambient pressure is reported for Cs_2RbC_{60} [64]. The increase of transition temperature with the unit-cell volume had been believed to be evidence for the BCS mechanism of C_{60} solid superconductivity, because inter-C_{60} separation can be related to an increase in the density of states on the Fermi level, $N(\varepsilon_F)$. Therefore, there have been many efforts to increase the interfullerene separation, in particular, intercalating neutral molecules into the A_3C_{60} lattice to increase the interfullerene spacing while the valence of C_{60} is kept unchanged. However, this ammoniation technique has revealed a new aspect of fullerene intercalation compounds: the Mott transition and the correlation between the orientation/orbital order of C_{60} molecules and the magnetic structure [64].

The C_{60} molecule is composed of solid, weakly bound molecules. The fullerites are therefore molecular solids, in which the molecular properties still survive. The discrete levels of a free C_{60} molecule are only weakly broadened in the solid, which leads to a set of essentially nonoverlapping bands with a narrow width of about 0.5 eV [57]. For an undoped C_{60} solid, the fivefold h_u band is the HOMO level, and the threefold t_{1u} band is the empty LUMO level, and this system is a band insulator. But when the C_{60} solid is doped with metal atoms, the metal atoms give electrons to the t_{1u} band or the upper threefold t_{1g} band [65]. This partial electron occupation of the band may lead to metallic behavior. However, A_4C_{60} is an insulator, although the t_{1u} band is only partially filled and it should be a metal according to the band theory [66]. This unpredicted behavior may be explained by the Jahn–Teller effect, where spontaneous deformations of high-symmetry molecules induce the splitting of degenerate levels to gain the electronic energy. The Jahn–Teller type of electron–phonon interaction is strong enough in C_{60} solids to destroy the band picture for particular valence states [67]. A narrowband or strongly correlated electronic system and degenerated ground states are important points to understand in explaining superconductivity in fullerene solids. When the interelectron repulsion U is greater than the bandwidth, an insulating localized electron ground state is produced in the simple Mott–Hubbard model. This explains the absence of superconductivity at ambient pressure in cesium-doped C_{60} solids [63]. Electron-correlation-driven localization of the t_{1u} electrons exceeds the critical value, leading to the Mott insulator. The application of high pressure decreases the interfullerene spacing; therefore, cesium-doped C_{60} solids become metallic and superconducting. A fully developed theory of C_{60} solid superconductivity is still lacking, but it has been widely accepted that strong electronic correlations and the Jahn–Teller electron–phonon coupling [68] produce local electron pairings that show a high transition temperature close to the insulator–metal transition [69].

1.3.2 Nanodiamonds

Diamond is one well-known allotrope of carbon. It is a metastable form of carbon that possesses a 3D cubic lattice with a lattice constant of 3.57 Å and C–C bond length of 1.54 Å. Diamond is used widely in both industrial applications and jewelry due to its hardness and high dispersion of light. Diamond is the hardest known natural mineral. Each carbon atom in a diamond is covalently bonded to four other carbons in a tetrahedron. These tetrahedrons form a 3D network of six-membered carbon rings, in the chair conformation, allowing for zero bond angle strain. This stable network of covalent bonds and hexagonal rings is the reason that diamond is so incredibly strong. Mixed states are also possible and form the basis of amorphous carbon, diamond-like carbon, and nanocrystalline diamond. Diamond was synthesized from graphite by high-pressure/high-temperature methods in the1950s, and low-pressure chemical vapor deposition (CVD) of diamond polycrystalline films had been developed at the beginning of the 1960s. Nanoscale diamond particles were first produced by detonation in the USSR in the 1960s [70], but they remained essentially unknown to the rest of the world until the end of the 1980s [71]. The area of the CVD of diamond films experienced several shifts of scientific and funding activity, with the last peak taking place in the United States in the mid-1990s. In the past few years, the interest in diamond thin films has increased as research activities related to the booming nanotechnology have grown worldwide and new synthesis methods of nanocrystalline diamond films have been discovered [72,73]. In the beginning of the 1960s, diamond powder was synthesized via shock wave compression induced by solid explosive detonation of carbon materials (graphite, carbon black) mixed with metal powder (Ni, Cu, Al, Co) placed in a capsule (which was destroyed after the process) by DuPont de Nemours and Co., a leader in explosive technology previously applied to other materials. The produced polycrystalline diamond particles of micron size varied from 1 to 60 μm (trade name Mypolex™). However, the prepared diamond particles consist of nanometer-sized diamond grains, 1–50 nm. For a long time, this synthesized diamond material has been used for high-precision polishing applications.

Another approach for producing diamond powder by a more effective means with a reusable detonation capsule is the conversion of carbon-containing compounds into diamond during firing of explosives in hermetic tanks [74], and this type of nanodiamond, known as ultradispersed diamond (UDD) or detonation diamond, is much less known and has been described in a recently published book by Vereschagin [75]. This method was initiated in Russia in the early 1960s soon after DuPont's work on shock wave synthesis and was a very active area of research in the 1980s, where it was studied independently by different groups of researchers. Publications in this area at that time were very scarce, with some reports appearing decades after the actual discoveries had been made [74]. The first work on nanodiamond produced by detonation in the United States was published in 1987, where the method of synthesis was described [71]. In 1983 the NPO "ALTAI" was founded in Russia, the first industrial company to commercialize the process of detonation diamond production in bulk quantities (tons of the products per year). According to a USSR government report (1989) on UDD production, it was planned to increase UDD production by up to 250 million carats per year [75]. At the present time, the production of detonation diamond by "ALTAI" is limited. Currently, there are several commercial centers in the world producing UDD, particularly in Russia (e.g., the *Diamond Center* in St. Petersburg), Ukraine (e.g., Alit), Belorussia, Germany, Japan, and China. A center for the production of UDD is being organized in India.

Then, beginning in the late 1990s, a number of important breakthroughs led to wider interest in these particles, which are now known as nanodiamonds. First, colloidal suspensions of individual diamond particles with diameters of 4–5 nm (*single-digit* nanodiamonds) became available [76]. Second, researchers started to use fluorescent nanodiamonds as a nontoxic alternative to semiconductor quantum dots for biomedical imaging [77,78]. Third, nanoscale magnetic sensors based on nanodiamonds were developed. Fourth, the chemical reactivity of the surface of nanodiamonds allowed a variety of wet [79,80] and gas [81] chemistry techniques to be employed to tailor the properties of nanodiamonds for use in composites and also for other applications [82–86] such as attaching drugs and biomolecules [87–89]. Fifth, new environmentally benign purification techniques were developed, and these allowed high-purity nanodiamond powders with controlled surface chemistry to be produced in large volumes at a low cost [90,91]. Finally, nanodiamond was found to be less toxic than other carbon nanoparticles [92–94] and, as a result, is currently being considered for applications in biomedical imaging, drug delivery, and other areas in medicine.

1.3.3 Carbon Nanotubes

The first report on CNTs came in 1952 by Radushkevich and Lukyanovich as hollow graphitic fibers and became the limelight in the last couple of decades due to its interesting electronic, mechanical and thermal properties [13]. In 1991, Sumio Iijima scientifically reported CNTs, long, thin cylinders of carbon, as a by-product of fullerene synthesis. Remarkable progresses have been developed on this area in more than two decades, and still the topic is hot as far as research in nanoscience is concerned. These are large macromolecules that are unique for their size, shape, and remarkable physical properties. CNT is considered as an allotrope of carbon that has a cylindrical and tubular structure; however, it is different from other allotropes of carbon like graphite, diamond, and fullerene. These are 1D structures having very high aspect ratio (length-to-diameter ratio), which opened up a new area of study in nanotechnology [95,96].

They can be thought of as a sheet of graphite (a hexagonal lattice of carbon) rolled into a cylinder-shape structure with a diameter that is of the order of a few nanometers, while their length can be of the order of several millimeters and length-to-diameter ratio of up to 132,000,000:1 [97], significantly larger than any other material. These intriguing structures have sparked much excitement in recent years and a large amount of research has been dedicated to their understanding. These cylindrical carbon molecules have unusual properties, which are valuable for nanotechnology, electronics, optics, and other fields of materials science and technology. In particular, owing to their extraordinary thermal conductivity and mechanical and electrical properties, CNTs find applications as additives to various structural materials. For instance, nanotubes form a tiny portion of the material(s) in some (primarily carbon fiber) baseball bats, golf clubs, or car parts [98].

CNTs, tubular carbon molecules, are basically members of the fullerene structural family, but while the fullerene's molecules form a spherical shape, nanotubes are cylindrical structures with the ends covered by half a fullerene molecule. Their name is derived from their long, hollow structure with the walls formed by one-atom-thick sheets of carbon, called graphene. These sheets are rolled at specific and discrete (chiral) angles, and the combination of the rolling angle and radius decides the nanotube properties, for example, whether the individual nanotube shell is a metal or semiconductor (Figure 1.1). Based on the chirality, the nanotubes are classified as

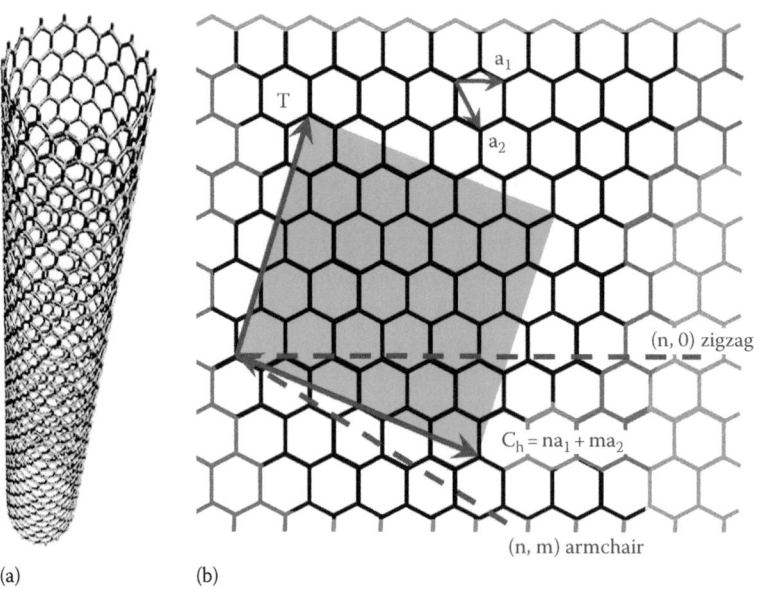

(a)   (b)

FIGURE 1.1
Schematic illustration of (a) CNT and (b) its unit vectors. The (n, m) nanotube naming scheme can be thought of as a vector (C_h) in an infinite graphene sheet that describes how to *roll up* the graphene sheet to make the nanotube. T denotes the tube axis, and a_1 and a_2 are the unit vectors of graphene in real space.

metallic, nonmetallic, or semiconducting in nature. The way the graphene sheet is wrapped can be represented by a pair of indices (n, m), where n and m denote the number of unit vectors along the two directions of a hexagonal crystal lattice of graphene. The chirality is defined by the chiral vector, $C_h = na_1 + ma_2$. CNTs have a zigzag structure for m = 0 and armchair structure for n = m, and if both conditions are not satisfied, it is classified as chiral [99]. The chirality greatly affects the electronic properties of CNTs. For a given (n, m), if (2n + m) is a multiple of 3, it is considered metallic, otherwise semiconductor.

Several types of nanotubes exist; but they can be divided into two main categories: (1) single-walled carbon nanotubes (SWNTs) (this can be envisioned as a cylinder composed of a rolled-up graphene sheet around a central hollow core) and (2) multiwalled carbon nanotubes (MWNTs) (this consists of two or more graphene layers holding together with van der Waals forces between adjacent layers and folded as hollow cylinders) [100]. The form of nanotubes is identified by a sequence of two numbers, the first one of which represents the number of carbon atoms around the tube, while the second identifies an offset of where the nanotube wraps around to. The chemical bonding of CNTs is sp^2 bonding similar to graphite. Based on the rolling angle of the graphite sheet against the tube axis, CNTs can have three kinds of structures: (1) armchair, (2) zigzag, and (3) chiral. Individual nanotubes naturally align themselves into *ropes* held together by van der Waals forces, more specifically π-stacking. Applied quantum chemistry, specifically orbital hybridization, best describes chemical bonding in nanotubes. The chemical bonding between carbon atoms inside nanotubes is always of the sp^2 type, similar to those found in graphite, and provides them their unique strength. These bonds, which are stronger than the sp^3 bonds found in alkanes and diamond, provide nanotubes their unique strength. Moreover, due to the chemical bonding, they align themselves into ropes held together by van der Waals force and can merge together under high pressure, trading some sp^2 bonds to sp^3 and producing very strong wires of nanometric lateral dimension.

The physical properties of nanotubes make them potentially useful in nanometric scale electronic and mechanical applications. Currently, the physical properties are still being discovered and disputed. Nanotubes have a very broad range of electronic, thermal, and structural properties that change depending on the different kinds of nanotube (defined by its diameter, length, and chirality, or twist). In terms of mechanical strength, CNTs are considered to have high tensile strength and elastic modulus, making them strong and stiff materials. The physical properties of CNTs are compared to other allotropes of carbon and illustrated in Table 1.2. Considerable research efforts are going on modifying CNT properties such as chirality, purity, length, and surface properties for binding with various materials [30,99–103]. Based on their characteristic properties, CNTs are identified as ideal materials for energy storage, conductive adhesive, inks and grease, reinforcing fillers, catalyst supports, and many other advanced applications.

1.3.3.1 Single-walled carbon nanotubes

SWNTs are nanometer-diameter cylinders consisting of a single graphene sheet wrapped up to form a tube. Since their discovery in the early 1990s, there has been intense activity exploring the electrical properties of these systems and their potential applications in electronics. Experiments and theory have shown that these tubes can be either metals or semiconductors, and their electrical properties can rival, or even exceed, the best metals or semiconductors known. The first studies on metallic tubes were done in 1997 [28,104] and the first on semiconducting tubes in 1998 [104]. In the intervening 5 years, a large number of groups have constructed and measured nanotube devices, and most major universities and industrial laboratories now have at least one group studying their properties.

Most SWNTs have a diameter close to 1 nm, with a tube length that can be millions of times longer. The structure of a SWNT can be theorized by covering a one-atom-thick layer of graphite called graphene into a seamless cylinder. The way the graphene sheet is wrapped is represented by a pair of indices (n, m). The integers n and m denote the number of unit vectors along two directions in the honeycomb crystal lattice of graphene. If m = 0, the nanotubes are called zigzag nanotubes, and if n = m, the nanotubes are called armchair nanotubes. Otherwise, they are called chiral. The diameter of an ideal nanotube can be calculated from its (n, m) indices as follows:

$$d = \frac{a}{\pi}\sqrt{n^2 + nm + m^2} = 78.3\sqrt{\left(n+m\right)^2 - nm}\,\text{pm}$$

where a = 0.246 nm.

SWNTs are an important variety of CNTs because most of their properties change significantly with the (n, m) values, and this dependence is nonmonotonic (see the Kataura plot). In particular, their bandgap can vary from zero to about 2 eV and their electrical conductivity can show metallic or semiconducting behavior. SWNTs are likely candidates for miniaturizing electronics. The most basic building block of these systems is the electric wire, and SWNTs with diameters of an order of a nanometer can be excellent conductors [105,106]. One useful application of SWNTs is in the development of the first intermolecular field-effect transistors (FETs). The first intermolecular logic gate using SWCNT FETs was made in 2001 [107]. A logic gate requires both a p-FET and an n-FET. Because SWNTs are p-FETs when exposed to oxygen and n-FETs otherwise, it is possible to protect half of an SWNT from oxygen exposure, while exposing the other half to oxygen. This results in a single SWNT that acts as a *not* logic gate with both p- and n-type FETs within the same molecule. SWNTs are dropping precipitously in price, from around \$1500/g as of 2000 to retail prices of around \$50/g of as-produced 40%–60% by weight SWNTs as of March 2010.

The SWNTs show remarkable electrical properties. Its band structure is quite unusual and is called a zero-bandgap semiconductor since it is metallic in some directions and semiconducting in others. In an SWNT, the momentum of the electrons moving around the circumference of the tube can be quantified. This quantization results in tubes that are either 1D metals or semiconductors depending on how the allowed momentum states compare to the preferred directions for conduction. The tube acts as a 1D metal with a Fermi velocity vf = 8×10^5 m/s comparable to typical metals. If the axis is chosen differently, the tube acts as a 1D semiconductor, with a bandgap between the filled hole states and the empty electron states. The bandgap is predicted to be E_g = 0.9 eV/d, where d is the diameter of the tube. Nanotubes can therefore be either metals or semiconductors, depending on how the tube is rolled up. This remarkable theoretical prediction has been verified using a number of measurement techniques.

1.3.3.2 Multiwalled carbon nanotubes

MWNTs consist of multiple rolled layers (concentric tubes) of graphene. There are two models that can be used to describe the structures of MWNTs: the Russian doll model and the parchment model. In the Russian doll model, sheets of graphite are arranged in concentric cylinders, SWNT within a larger SWNT. In the parchment model, a single sheet of graphite is rolled in around itself, resembling a scroll of parchment or a rolled newspaper. The interlayer distance in MWNTs is close to the distance between graphene layers in graphite, approximately 3.4 Å. The Russian doll structure is observed more commonly. Its individual shells can be described as SWNTs, which can be metallic or semiconducting. Because of statistical probability and restrictions on the relative diameters of the individual tubes, one of the shells, and thus the whole MWNT, is usually a zero-gap metal.

Double-walled carbon nanotubes (DWNTs) form a special class of nanotubes because their morphology and properties are similar to those of SWNT but their resistance to chemicals is significantly improved. This is especially important when functionalization is required (this means grafting of chemical functions at the surface of the nanotubes) to add new properties to the CNT. In the case of SWNT, covalent functionalization will break some C=C double bonds, leaving *holes* in the structure on the nanotube and, thus, modifying both its mechanical and electrical properties. In the case of DWNT, only the outer wall is modified. DWNT synthesis on the gram scale was first proposed in 2003 [108], by the CCVD technique, from the selective reduction of oxide solutions in methane and hydrogen. The telescopic motion ability of inner shells [109] and their unique mechanical properties [110] will permit the use of MWNTs as main movable arms in nanomechanical devices. A retraction force that occurs during the telescopic motion is caused by the Lennard-Jones interaction between shells, and its value is about 1.5 nN [111].

1.3.3.3 Synthesis of carbon nanotubes

CNTs in bulk can be produced by arc discharge, laser ablation, gas-phase catalytic growth of carbon monoxide, and CVD techniques [30,99–103]. The main techniques for the synthesis of CNTs are the arc discharge method, CVD, laser ablation, etc. Most of these processes take place in a vacuum or with process gases. CVD growth of CNTs can occur in vacuum or at atmospheric pressure. Large quantities of nanotubes can be synthesized by these methods; advances in catalysis and continuous growth are making CNTs more commercially feasible.

1.3.3.4 Arc discharge

Carbon arc discharge method, initially used for producing C_{60} fullerenes, is the most common and perhaps easiest way to produce CNTs, as it is rather simple. However, it is a technique that produces a complex mixture of components and requires further purification—to separate the CNTs from the soot and the residual catalytic metals present in the crude product. Iijima first observed the growth of CNTs during a carbon arc discharge by using a current of 100 amps that was envisioned to produce fullerenes [13]. However, the first macroscopic production of CNTs was made in 1992 by Ebbesen and Ajayan at NEC's Fundamental Research Laboratory [10]. The method used was the same as what Iijima used for the CNT production. This method creates CNTs through arc vaporization of two carbon rods placed end to end, separated by approximately 1 mm, in an enclosure that is usually filled with inert gas at low pressure. During this process, the carbon contained in the negative electrode sublimates because of the high discharge temperatures. Recent investigations have shown that it is also possible to create CNTs with the arc method in liquid nitrogen. A direct current of 50–100 A, driven by a potential difference of approximately 20 V, creates a high discharge temperature between the two electrodes. The discharge vaporizes the surface of one of the carbon electrodes and forms a small rod-shaped deposit on the other electrode. Producing CNTs in high yield depends on the uniformity of the plasma arc and the temperature of the deposit forming on the carbon electrode. The yield for this method is up to 30% by weight and it yields both SWNT and MWNT with lengths of up to 50 μm.

1.3.3.5 Laser ablation

In 1996, CNTs were first synthesized using a dual-pulsed laser and achieved high yields of about 70 wt% and purity. In laser ablation, a pulsed laser vaporizes a graphite target in a high-temperature reactor while an inert gas is bled into the chamber. During the process, graphite rods with a 50:50 catalyst mixture of cobalt and nickel will vaporize at 1200°C in flowing argon, followed by heat treatment in a vacuum at 1000°C to remove the C_{60} and other fullerenes. The initial laser vaporization pulse was followed by a second pulse, to vaporize the target more uniformly. The use of two successive laser pulses minimizes the amount of carbon deposited as soot. The second laser pulse breaks up the larger particles ablated by the first one and feeds them into the growing nanotube structure. Nanotubes grow on the cooler surfaces of the reactor as the vaporized carbon condenses. A water-cooled surface may be included in the system to collect the nanotubes. The material produced by this method appears as a mat of *ropes*, 10–20 nm in diameter and up to 100 μm or more in length. Each rope is found to consist primarily of a bundle of SWNTs, aligned along a common axis. By varying the growth temperature, the catalyst composition, and other process parameters, the average nanotube diameter and size distribution can be varied.

This process was developed by Dr. Richard Smalley and coworkers at Rice University, USA, who, at the time of the discovery of CNTs, were blasting metals with a laser to produce various metal molecules. When they heard of the existence of nanotubes, they replaced the metals with graphite to create MWNTs [112]. Later that year, the team used a composite of graphite and metal catalyst particles (the best yield was from a cobalt and nickel mixture) to synthesize SWNTs [113].

Arc discharge and laser vaporization are currently the principal methods for obtaining small quantities of high-quality CNTs. However, the laser vaporization method is more expensive than either arc discharge or CVD. There are few more drawbacks involved with arc discharge and laser vaporization methods. First, both the methods produce CNTs by evaporating the carbon source, so it has been unclear how to scale up production to the industrial level using these approaches. The second issue relates to the fact that vaporization methods grow CNTs in highly tangled forms, mixed with unwanted forms of carbon and/or metal species. The CNTs thus produced are difficult to purify, manipulate, and assemble for building nanotube device architectures for practical applications. The laser ablation method yields around 70% and produces primarily SWNTs with a controllable diameter determined by the reaction temperature [114].

1.3.3.6 Thermal plasma and plasma torch

CNTs can also be synthesized by a thermal plasma method. It was first invented in 2000 at Institut National de la Recherche Scientifique (INRS), in Varennes, Canada, by Olivier Smiljanic. In this method, the aim is to reproduce the conditions prevailing in the arc discharge and laser ablation approaches, but a carbon-containing gas is used instead of graphite vapors to supply the carbon necessary for the production of SWNT. Doing so, the growth of SWNT is more efficient (decomposing a carbon-containing gas can be 10 times less energy consuming than graphite vaporization). It is also continuous and occurs at low cost. To produce a continuous process, a gas mixture composed of argon, ethylene, and ferrocene is introduced into a microwave plasma torch, where it is atomized by the atmospheric pressure plasma, which has the form of an intense *flame*. The fumes created by the flame are found to contain SWNT, metallic and carbon nanoparticles, and amorphous carbon [115]. The induction thermal plasma method can produce up to 2 g of nanotube material per minute, which is higher than the arc discharge or the laser ablation methods.

Another way to produce SWNTs with a plasma torch is to use the induction thermal plasma method, implemented in 2005 by groups from the University of Sherbrooke and the National Research Council of Canada. The method is similar to arc discharge in that both use ionized gas to reach the high temperature necessary to vaporize carbon-containing substances and the metal catalysts necessary for the ensuing nanotube growth. The thermal plasma is induced by high-frequency oscillating currents in a coil and is maintained in flowing inert gas. Typically, a feedstock of carbon black and metal catalyst particles is fed into the plasma and then cooled down to form SWNTs. Different SWNT diameter distributions can be synthesized.

1.3.3.7 Chemical vapor deposition

CVD is the most widely used method for the production of CNTs [116]. CVD of hydrocarbons over a metal catalyst is a classical method that has been used to produce various carbon materials such as carbon fibers and filaments for over 20 years. Large amounts of CNTs can be formed by catalytic CVD of acetylene over cobalt and iron catalysts supported on silica or zeolite. The carbon deposition activity seems to relate to the cobalt content of the catalyst, whereas the CNTs' selectivity seems to be a function of the pH in catalyst preparation. Fullerenes and bundles of SWNTs were also found among the MWNTs produced on the carbon/zeolite catalyst. The method of catalytic vapor phase deposition of carbon was first reported in 1952 [117,118], but it was not until 1993 [119] that CNTs were formed by this process. In 2007, researchers at the University of Cincinnati (UC) developed a process to grow aligned CNT arrays of length 18 mm on a FirstNano ET3000 CNT growth system [120].

During CVD, a substrate is prepared with a layer of metal catalyst particles, most commonly nickel, cobalt [121], iron, or a combination [122]. The metal nanoparticles are mixed with a catalyst support such as MgO or Al_2O_3 to increase the surface area for higher yield of the catalytic reaction of the carbon feedstock with the metal particles. One issue in this synthesis route is the removal of the catalyst support via an acid treatment, which sometimes could destroy the original structure of the CNTs. However, alternative catalyst supports that are soluble in water have proven effective for nanotube growth [123]. The metal nanoparticles can be produced by reduction of oxides or oxide solid solutions. The diameters of the nanotubes that are to be grown are related to the size of the metal particles. This can be controlled by patterned (or masked) deposition of the metal, annealing, or by plasma etching of a metal layer. The substrate is heated to approximately 700°C. To initiate the growth of nanotubes, two gases are bled into the reactor: a process gas (such as ammonia, nitrogen, or hydrogen) and a carbon-containing gas (such as acetylene, ethylene, ethanol, or methane). Nanotubes grow at the sites of the metal catalyst; the carbon-containing gas is broken apart at the surface of the catalyst particle, and the carbon is transported to the edges of the particle, where it forms the nanotubes. This mechanism is still being studied [124]. The catalyst particles can stay at the tips of the growing nanotube during growth or remain at the nanotube base, depending on the adhesion between the catalyst particle and the substrate [125]. Thermal catalytic decomposition of hydrocarbon has become an active area of research and can be a promising route for the bulk production of CNTs. Fluidized bed reactor is the most widely used reactor for CNT preparation. Scale-up of the reactor is the major challenge [126].

Some researchers are experimenting with the formation of CNTs from ethylene. Supported catalysts such as iron, cobalt, and nickel, containing either a single metal or a mixture of metals, seem to induce the growth of isolated SWNTs or SWNT bundles in the ethylene atmosphere. The production of SWNTs, as well as DWNTs, on molybdenum and molybdenum–iron alloy catalysts has also been demonstrated. CVD of carbon within the pores of a thin alumina template with or without a nickel catalyst has been achieved. Ethylene was used with reaction temperatures of 545°C for nickel-catalyzed CVD and 900°C for an uncatalyzed process. The resultant carbon nanostructures have open ends, with no caps. Methane has also been used as a carbon source. In particular it has been used to obtain *nanotube chips* containing isolated SWNTs at controlled locations. High yields of SWNTs have been obtained by catalytic decomposition of a H_2/CH_4 mixture over well-dispersed metal particles such as cobalt, nickel, and iron on magnesium oxide at 1000°C. The synthesis of composite powders containing well-dispersed CNTs can be achieved by selective reduction in a H_2/CH_4 atmosphere of oxide solid solutions between a nonreducible oxide such as Al_2O_3 or $MgAl_2O_4$ and one or more transition metal oxides. The reduction produces very small transition metal particles at a temperature of usually >800°C. The decomposition of CH_4 over the freshly formed nanoparticles prevents their further growth and thus results in a very high proportion of SWNTs and fewer MWNTs.

If plasma is generated by the application of a strong electric field during growth (plasma-enhanced CVD), then the nanotube growth will follow the direction of the electric field [30]. By adjusting the geometry of the reactor, it is possible to synthesize vertically aligned CNTs (i.e., perpendicular to the substrate), a morphology that has been of interest to researchers interested in electron emission from nanotubes. Without the plasma, the resulting nanotubes are often randomly oriented. Under certain reaction conditions, even in the absence of plasma, closely spaced nanotubes will maintain a vertical growth direction resulting in a dense array of tubes resembling a carpet or forest.

Of the various means for nanotube synthesis, CVD shows the most promise for industrial-scale deposition, because of its price/unit ratio and because CVD is capable of growing nanotubes directly on a desired substrate, whereas the nanotubes must be collected in other growth techniques. The growth sites are controllable by careful deposition of the catalyst [127]. In 2007, a team from Meijo University demonstrated a high-efficiency CVD technique for growing CNTs from camphor [128]. Researchers at Rice University, until recently led by the late Richard Smalley, have concentrated upon finding methods to produce large, pure amounts of particular types of nanotubes. Their approach grows long fibers from many small seeds cut from a single nanotube; all of the resulting fibers were found to be of the same diameter as the original nanotube and are expected to be of the same type as the original nanotube [129].

1.4 CARBON AT NANOSCALE: TWO-DIMENSIONAL CARBON MATERIALS

1.4.1 Graphene

Graphene is the most shining star on the horizon of materials science and technology of the current decade. Graphene is pure carbon in the form of a very thin, nearly transparent sheet, one atom thick, and the term *graphene* first appeared in 1987 [130] to describe single sheets of graphite as a constituent of graphite intercalation compounds. Graphene—a single atomic layer of graphite—consists of a 2D honeycomb structure of sp^2-bonded carbon atoms. *Graphene* is a combination of graphite and the suffix -ene, named by Hanns-Peter Boehm [131], who described single-layer carbon foils in 1962 [132]. Graphene is an allotrope of carbon in the structure of a plane sp^2-hybridized atom with a bond length of 0.142 nm. It is a single, tightly packed nearly transparent sheet thin layer of pure carbon atoms that are bonded together in a hexagonal honeycomb lattice. It is remarkably strong for its very low weight (100 times stronger than steel), and it conducts heat and electricity with great efficiency. Graphene is a crystalline allotrope of carbon with 2D properties. In graphene, carbon atoms are densely packed in a regular sp^2-bonded atomic-scale hexagonal pattern. Graphene can be described as a one-atom-thick layer of graphite. It is the basic structural element of other allotropes, including graphite, charcoal, CNTs, and fullerenes and so called the mother of all carbonaceous materials. It can also be considered as an indefinitely large aromatic molecule, the limiting case of the family of flat polycyclic aromatic hydrocarbons. This 2D material is strictly the pure form of carbon, and it exhibits high crystal and electronic quality. Graphene also exhibits exceptionally high physical properties, is the strongest and lightest material discovered, and has the highest thermal and electrical conductivity. The strength of the graphene is reported as 100 times the tensile strength of steel. Furthermore, carbon is an excellent conductor of electricity; graphene does this very efficiently with electrical current densities 1,000,000 times that of copper. The material has high thermal conductivity too, and because it is a 2D material that is almost transparent, it interacts in other interesting and useful ways with light and with other materials.

When graphene was first reliably produced in the lab in 2004 by Geim and Novoselov, single-atom-thick crystallites were extracted from bulk graphite by lifting graphene layers from graphite with adhesive tape and then transferred onto a silicon wafer. This structure is bringing new properties that are opening up many new opportunities for the one of the oldest elements known to man. Graphene, unlike CNTs, has edges that can react chemically. These exposed carbon molecules have special reactivity, as do any imperfections in the graphene sheets. Because of its 2D structure and the lateral availability of the carbon, graphene is now known to be the most reactive form of carbon.

Graphene's stability is due to a tightly packed, periodic array of carbon atoms and an sp^2 orbital hybridization—a combination of orbitals p_x and p_y that constitute the σ-bond. Graphene has three σ-bonds and one π-bond. The final p_z electron makes up the π-bond and is key to the half-filled band that permits free-moving electrons. It is discovered that carbon atoms at the edge of graphene sheets have special chemical reactivity and graphene has the highest ratio of edgy carbons.

1.4.2 Graphene Oxide

Graphite oxide was first prepared by Oxford chemist Benjamin C. Brodie in 1859, by treating graphite with a mixture of potassium chlorate and fuming nitric acid. Graphite oxide sheets, now called graphene oxide (GO), is the product of chemical exfoliation of graphite that has been known for more than a century. It is typically synthesized by reacting graphite powders with strong oxidizing agents such as $KMnO_4$ in concentrated sulfuric acid. The oxidation of graphite breaks up the extended 2D conjugation of the stacked graphene sheets into nanoscale graphitic sp^2 domains surrounded by disordered, highly oxidized sp^3 domains as well as defects of carbon vacancies. The resulting GO sheets are derivatized by carboxylic acid at the edges and phenol, hydroxyl, and epoxide groups mainly at the basal plane. Therefore, the sheets can readily exfoliate to form a stable, light-brown-colored, single-layer suspension in water. This severe functionalization of the conjugated network renders GO sheets insulating. However, conductivity may be partially restored conveniently by thermal or chemical treatment, producing chemically modified graphene sheets. The structure and properties of graphite oxide depend on particular synthesis method and degree of oxidation.

GO, a nonconductive hydrophilic carbon material, attracted a great deal of interest in recent times for a variety of reasons. The tremendous potential of GO and its functionalized derivatives make this an ideal candidate for many applications that include energy storage, catalysis, and composites, and hence there is extensive research being carried out in this area worldwide. In the present study, it is proposed to synthesize GO via a modified method. From a fundamental perspective, graphene and derivatives of graphene are also ideal templates for verifying many quantum mechanical phenomena. The design and preparation of advanced functional materials based on 3D architectures of graphene and GO matrices with selective incorporation of metal and metal oxide nanolayers/clusters for possible applications will be taken up.

1.4.3 Graphene Story

The discovery of the wonder material of the decade graphene had already been studied theoretically in 1947 by Wallace [133] as a textbook example for calculations in solid-state physics. The unusual electronic structure and properties of the material were predicted. In 1956, McClure [134] wrote the wave equation for excitations, and its similarity to the Dirac equation was discussed in 1984 by Semenoff [135] and DiVincenzo [136]. It was a big surprise to the scientific community when Andre Geim, Konstantin Novoselov, and their collaborators from the University of Manchester (UK) and the Institute for Microelectronics Technology in Chernogolovka (Russia) presented their discovery on graphene structures and the experimental results. In 2004, Novoselov et al. [7] first published the fabrication, identification, and atomic force microscopy (AFM) characterization of graphene in science. Andrei Geim and Konstantin Novoselov, two Russian-émigré scientists at the University of Manchester, were simply playing about with flakes of layered graphite in an attempt to investigate the electrical properties of the thinnest layer of graphite. In extracting the thin layers of graphite from its crystal, they used a simple but effective mechanical exfoliation method with the help of sticky Scotch tape, exactly the same technique first suggested and tried by R. Ruoff's group [137], who were, however, not able to identify any monolayers. With the Scotch tape, they tried to make thinner flakes by repeatedly sticking and peeling off further layers from the original cleaved flakes of few atom thick and realized they could get the thinnest of one-atom-thick layer, the most possible layers. By using an optical method (AFM), they were able to identify the fragments made up of a single layer and graphene, a material with unique and immensely interesting properties, which promises to transform the future, bringing the Nobel Prize in Physics to Novoselov and Geim in 2010. Playing about with Scotch tape in the lab of course sounds a jokey thing to do, but soon the funny play turned into a deadly serious game of a great scientific discovery that would have been impossible if not for the well-prepared minds of Geim and Novoselov. In 2010, when the two scientists, Novoselov and Geim, won their joint Nobel Prize for the groundbreaking discovery, the Nobel committee made a point of citing the *playfulness* that was one of the hallmarks of the way they have worked together. After receiving the Noble Prize, Novoselov said, "A playful idea is perfect to start things but then you need a really good scientific intuition that your playful experiment will lead to something, or it will stay as a joke forever. Joking for a week or two is the right way to go, but you don't want to make your whole research into a joke."

1.5 CONCLUSION

Different carbon allotropes are discussed with special attention on its size, structure, and properties. A wide range of potential applications are found for these little gems. They find a respectful place in tribology, drug delivery, bioimaging, electronics and energy industry, and engineering and aviation materials. Among other allotropes of carbon, nanotubes and graphene have the simplest chemical composition and atomic bonding configuration; however, nanotubes exhibit the most extreme diversity and richness among nanomaterials in structures and structure–property relations. Due to the exceptional electrical conductivity, CNTs have emerged as a very promising new class of electronic materials. Both metallic and semiconducting CNTs are found to possess electrical characteristics that compare favorably to the best electronic materials available. Because of the biocompatibility and nontoxicity, CNTs, nanodiamonds, and buckyballs became promising materials for cancer treatment and drug delivery. Graphene is the real gem in the materials science and technology of the current decade and is the strongest and lightest material ever discovered. They have higher electrical conductivity and thermal conductivity (10 times higher) than copper. These nanostructured carbon materials, especially graphite, nanotubes, and graphene, offer great promise in numerous electrode applications in energy devices due to their high surface area and area-specific capacitance and medical applications due to the biocompatibility. The wide range of potential applications for nanodiamonds will continue to drive research in this field forward. Better understanding of the structure and surface chemistry of these nanomaterials will lead to greater control over their properties and also help to the mass production volumes. The search for simple but effective new ways to make these nanostructures will also continue, and any increase in supply will certainly lead to new applications.

ACKNOWLEDGMENTS

The authors wish to thank the members in Prof. Pulickel M. Ajayan's research group at the Department of Mechanical Engineering and Nanoscience, Rice University, Houston, USA, and Prof. Michael R. Kessler's group at the School of Mechanical and Materials Engineering, Washington State University, Pullman, USA.

REFERENCES

1. Binnig G., Rohrer H. (1986) Scanning tunneling microscopy, *IBM J. Res. Develop.* 30:4.
2. Prasad S. K. (2008) *Modern Concepts in Nanotechnology*, pp. 31–32, Discovery Publishing House, New Delhi, India.
3. Kahn, J. (2006) Nanotechnology, *Natl. Geogr.* June:98–119.
4. Nishiguchi, K. et al. (2006) *Appl. Phys. Lett.* 88:183–191.
5. Guo T., Jin C. M., Smalley R. E. (1991) Doping bucky—Formation and properties of boron-doped buckminsterfullerene, *J. Phys. Chem.* 95:4948–4950.
6. Iijima S., Ichihashi T. (1993) Single-shell carbon nanotubes of 1-nm diameter, *Nature* 363:603–605.
7. Novoselov K. S., Geim A. K., Morozov S. V., Jiang D., Zhang Y., Dubonos S. V., Grigorieva I. V., Firsov A. A. (2004) Electric field effect in atomically thin carbon films, *Science* 306:666–669.
8. Bacon R. (1960) Growth, structure and properties of graphite whiskers, *J. Appl. Phys.* 31:283–290.
9. Kroto H. W., Heath J. R., O'Brien S. C., Curl R. F., Smalley R. E. (1985) C_{60}: Buckminsterfullerene, *Nature* 318:162–163.
10. Ebbesen T. W., Ajayan P. M. (1992) Large-scale synthesis of carbon nanotubes, *Nature* 358:220–222.
11. Collin G. (2000) On the History of Technical Carbon, *CFI-Ceramic Forum Intern.* 77:28–35.

12. Inagaki M. (2000) *New Carbons—Control of Structure and Functions*, Elsevier, London, U.K.
13. Iijima S. (1991) Helical microtubules of graphitic carbon, *Nature* 354:56–58.
14. Monthioux M., Kuznetsov V. L. (2006) Who should be given the credit for the discovery of carbon nanotubes? *Carbon* 44:1621–1623.
15. Jones D. E. H. (1982) Hollow molecules, *In The inventions of Daedalus: A compendium of plausible schemes*, (6th Ed.), pp. 118–119, Oxford: W. H. Freeman.
16. Osawa E. (1970) Superaromaticity, *Kagaku (Chemistry)* 25:854–863.
17. GBochvar D. A., Galperin E. G. (1973) Huckel (4N + 2) rule and some polycondensed systems, *Dokl. Akad. Nauk, SSSR* 209:610–612. (English translation, (1973) *Proc. Acad. Sci. USSR*, 209:239–241).
18. Oberlin A., Endo M., Koyama T. (1976) Filamentous growth of carbon through benzene decomposition, *J. Cryst. Growth* 3:335.
19. 1D Diamond Crystal—A continuous pseudo-one dimensional diamond crystal—Maybe a nanotube? Retrieved October 21, 2006, http://www.technovelgy.com/ct/content.asp?Bnum=699.
20. Iijima S. (1980) High resolution electron microscopy of some carbonaceous materials, *J. Microscopy* 119:99–111.
21. Paquette L. A., Ternansky R. J., Balogh D. W., Kentgen G. (1983) Total synthesis of dodecahedron, *J. Am. Chem. Soc.* 105:5446–5545.
22. Rohlfing E. A., Cox D. M., Kaldor A. (1984) Production and characterization of supersonic carbon cluster beams, *J. Chem. Phys.* 81:3322–3330.
23. Kraetschmer W., Lamb L. D., Fostiropoulos K., Huffman D. R. (1990) Solid C-60—A new form of carbon, *Nature* 347:354–358.
24. Curl R. F., Smalley R. E. (1991) Fullerenes, *Sci. Am.* 265:54–63.
25. Ugarte D. (1992) Curling and closure of graphitic networks under electron-beam irradiation, *Nature* 359:707–709.
26. Bethune D. S., Kiang C. H., DeVries M. S., Gorman G., Savoy R., Beyers R. (1993) Cobalt-catalysed growth of carbon nanotubes with single-atomic-layer walls, *Nature* 365:605–607.
27. Thess A., Lee R., Nikolaev P., Dai H., Petit P., Robert J., Xu C. et al. (1996) Crystalline ropes of metallic carbon nanotubes, *Science* 273:483–487.
28. Bockrath M., Cobden D. H., McEuen P. L., Chopra N. G., Zettl A., Thess A., Smalley R. E. (1997) Single-electron transport in ropes of carbon nanotubes, *Science* 275:1922–1925.
29. Patent US6700550—Optical antenna array for harmonic generation, mixing and signal amplification—Google Patents. Retrieved January 30, 2013, Google.com.
30. Ren Z. F., Huang Z. P., Xu J. W., Wang J. H., Bush P., Siegal M. P., Provencio P. N. (1998) Synthesis of large arrays of well-aligned carbon nanotubes on glass, *Science* 282:1105–1107.
31. Tans S. J., Verschueren A. R. M., Dekker C. (1998) Room-temperature transistor based on a single carbon nanotube, *Nature* 393:49–52.
32. Tombler, T. W., Zhou, C., Alexseyev, L., Kong, J., Dai, H., Liu, L., Jayanthi, C. S., Tang, M., Wu, S. Y. (2000) Reversible electromechanical characteristics of carbon nanotubes under local-probe manipulation, *Nature* 405:769–772.
33. Schilttler R. R., Seo J. W., Gimzewski J. K., Durkan C., Saifullah M. S. M., Welland M. E. (2001) Single crystals of single-walled carbon nanotubes formed by self-assembly, *Science* 292:1136–1139.
34. Collins P. G., Arnold M. S., Avouris P. (2001) Engineering carbon nanotubes and nanotube circuits using electrical breakdown, *Science* 292:706–709.

35. Zheng Q., Jiang Q. (2002) Multiwalled carbon nanotubes as gigahertz oscillators, *Phys. Rev. Lett.* 88:045503.
36. Tests verify carbon nanotube enable ultra high performance transistor (Press release). NEC. September 19, 2003. Retrieved October 21, 2006, http://www.nec.co.jp/press/en/0309/1901.html.
37. Zheng, L. X., O'connell M. J., Doorn S. K., Liao X. Z., Zhao Y. H., Akhadov E. A. et al. (2004) Ultralong single-wall carbon nanotubes, *Nat. Mater.* 3:673–676.
38. Shulaker M. M., Hills G., Patil N., Wei H., Chen H. Y., Wong H. S. P., Mitra S. (2013) Carbon nanotubes used in computer and TV screens, *Nature* 501:526–530.
39. Carbon-nanotube fabric measures up. Nanotechweb.org. Retrieved August 18, 2005, http://nanotechweb.org/cws/article/tech/22915.
40. Gadget sorts nanotubes by size. Retrieved June 27, 2006, http://www.newscientist.com/article/dn9419-gadget-sorts-nanotubes-by-size.html#.U-LOkPldWBw.
41. Optic nerve regrown with a nanofibre scaffold. Retrieved March 13, 2006, http://www.newscientisttech.com/channel/tech/nanotechnology/dn8840-optic-nerve-regrown-with-a-nanofibre-scaffold-.html.
42. New virus-built battery could power cars, electronic devices. Retrieved April 2, 2009, http://newsoffice.mit.edu/2009/virus-battery-0402.
43. Franklin A. D., Luisier M., Han S. J., Tulevski G., Breslin C. M., Gignac L., Lundstrom M. S., Haensch W. (2012) 10 nm Carbon nanotube transistor, *Nano Lett.* 12:758–762.
44. Osawa E., Yoshida M., Fujita M. (1994) Shape and fantasy of Fullerenes, *MRS Bull.* 19:33.
45. Cataldo F. (2002) The impact of fullerene-like concept in carbon black science, *Carbon* 40:157–162.
46. Kudryavtsev Y. P., Evsyukov S. E., Guseva M. B., Babaev V. G., Khvostov V. V. (1993) Carbyne-the third allotropic form of carbon, *Russ. Chem. Bull.* 42:399–413.
47. Lagow R. J., Kampa J. J., Wei H. C., Battle S. L., Genge J. W., Laude D. A., Harper C. J. et al. (1995) Synthesis of linear acetylenic carbon: The "sp" carbon allotrope, *Science* 267:362–367.
48. Heinmann R. B., Evsyukov S. E., Koga, Y. (1997) Carbon allotropes: A suggested classification scheme based on valance orbital hybridization, *Carbon* 35:1654–1658.
49. Howard J. B., McKinnon J. T., Makarovsky Y., Lafleur A. L., Johnson M. E. (1991) Fullerenes C_{60} and C_{70} in flames, *Nature* 352:139–141.
50. Howard J. B., Lafleur A. L., Makarovsky Y., Mitra S., Pope C. J., Yadav T. K. (1992) Fullerenes synthesis in combustion, *Carbon* 30:1183.
51. Markus A., Olaf N., Julian V. A., Claudia K., Gerbrand V. D. Z., Anton Z. (1999) Wave-particle duality of C_{60}, *Nature* 401:680–682.
52. Leger A., d'Hendecourt L., Verstraete I., Schmidt W. (1987) Remarkable candidates for the carrier of the diffuse interstellar bands—C_{60}, *Astr. Astrophys.* 203:145.
53. Dietz T. G., Duncan M. A., Powers D. E., Smalley R. E. (1981) Laser production of supersonic metal cluster beams, *J. Chem. Phys.* 74:6511–6512.
54. Katz E. A. (2006) Fullerene thin films as photovoltaic material, in Soga T., ed., *Nanostructured Materials for Solar Energy Conversion*, Elsevier, Oxford, U.K., pp. 361–443.

55. Shriver, A. (2010) *Inorganic Chemistry* (5th edn.), W. H. Freeman and Company, New York.

56. Haddon R. C., Hebard A. F., Rosseinsky M. J., Murphy D. W., Duclos S. J., Lyons K. B., Miller, B. et al. (1991) Conducting films of C_{60} and C_{70} by alkali-metal doping, *Nature* 350:320–322.

57. Gunnarsson, O. (1997) Superconductivity in fullerides, *Rev. Mod. Phys.* 69:575–606.

58. Hebard A. F., Rosseinsky M. J., Haddon R. C., Murphy D. W., Glarum S. H., Palstra, T. T. M., Ramirez A. P., Kortan A. R. (1991) Superconductivity at 18 K in potassium-doped C_{60}, *Nature* 350:600–601.

59. Rosseinsky M., Ramirez A., Glarum S., Murphy D., Haddon R., Hebard A., Palstra T., Kortan A., Zahurak S., Makhija A. (1991) Superconductivity at 28 K in $Rb_x C_{60}$, *Phys. Rev. Lett.* 66:2830–2832.

60. Chen C. C., Kelty S. P., Lieber C. M. (1991) $(Rb_x K_{1-x})_3 C_{60}$ Superconductors: Formation of a continuous series of solid solutions, *Science* 253:886–888.

61. Zhou O., Zhu Q., Fischer J. E., Coustel N., Vaughan G. B. M., Heiney P. A., McCauley J. P., Smith A. B. (1992) Compressibility of $M_3 C_{60}$ fullerene superconductors: Relation between Tc and lattice parameter, *Science* 255:833–835.

62. Brown C. M., Takenobu T., Kordatos K., Prassides K., Iwasa Y., Tanigaki K. (1999) Pressure dependence of superconductivity in the $Na_2 Rb_{0.5} Cs_{0.5} C_{60}$ fulleride, *Phys. Rev. B* 59:4439–4444.

63. Ganin A. Y., Takabayashi Y., Khimyak Y. Z., Margadonna S., Tamai A., Rosseinsky M. J., Prassides K. (2008) Bulk superconductivity at 38 K in a molecular system, *Nature* 7:367–371.

64. Tanigaki K., Ebbesen T. W., Saito S., Mizuki J., Tsai J. S., Kubo Y., Kuroshima S. (1991) Superconductivity at 33 K in $Cs_x Rb_y C_{60}$, *Nature* 352:222–223.

65. Steven E., Mark P. (1991) Electronic structure of crystalline $K_6 C_{60}$, *Phys. Rev. Lett.* 67:1610–1612.

66. Steven E., Mark P. (1993) Electronic structure of superconducting $Ba_6 C_{60}$, *Phys. Rev. B* 47:14657–14659.

67. Iwasa Y., Takenobu T. (2003) Superconductivity, Mott Hubbard states, and molecular orbital order in intercalated fullerides, *J. Phys. Condens. Matter* 15:R495–R519.

68. Han J., Gunnarsson O., Crespi V. (2003) Strong superconductivity with local Jahn-Teller phonons in C_{60} solids, *Phys. Rev. Lett.* 90:167006.

69. Capone M., Fabrizio M., Castellani C., Tosatti E. (2002) Strongly correlated super-conductivity, *Science* 296:2364–2366.

70. Danilenko V. V. (2004) On the history of the discovery of nanodiamond synthesis, *Phys. Solid State* 46:595–599.

71. Greiner N. R., Phillips D. S., Johnson J. D., Volk F. (1998) Diamonds in detonation soot, *Nature* 333:440–442.

72. Gruen D. M. (1999) Nanocrystalline diamond films, *Annu. Rev. Mater. Sci.* 29:211–259.

73. Gogotsi Y., Welz S., Ersoy D. A., McNallan M. J. (2001) Conversion of silicon carbide to crystalline diamond-structured carbon at ambient pressure, *Nature* 411:283–287.

74. Volkov K. V., Danilenko V. V., Elin V. I. (1990) Synthesis of diamond from the carbon in the detonation products of explosives, *Fizika Gorenia i Vzriva* (Russian) 26:123–125.

75. Vereschagin A. L. (2001) *Detonation Nanodiamonds*, Altai State Technical University, Barnaul, Russian Federation, in Russian.

76. Ozawa M., Inaguma M., Takahashi M., Kataoka F., Krueger A., Ōsawa E. (2007) Preparation and behavior of brownish, clear nanodiamond colloids, *Adv. Mater.* 19:1201–1206.

77. Chang Y. R., Lee H. Y., Chen K., Chang C. C., Tsai D. S., Fu C. C., Lim T. S. et al. (2008) Mass production and dynamic imaging of fluorescent nanodiamonds, *Nat. Nanotechnol.* 3:284–288.

78. Mochalin V. N., Gogotsi Y. (2009) Wet chemistry route to hydrophobic blue fluorescent nanodiamond, *J. Am. Chem. Soc.* 131:4594–4595.

79. Maze J. R., Stanwix P. L., Hodges J. S., Hong S., Taylor J. M., Cappellaro P., Jiang L. et al. (2008) Nanoscale magnetic sensing with an individual electronic spin in diamond, *Nature* 455:644–647.

80. Krueger A. (2008) Diamond nanoparticles: Jewels for chemistry and physics, *Adv. Mater.* 20:2445–2449.

81. Spitsyn B. V., Davidson J. L., Gradoboev M. N., Galushko T. B., Serebryakova N. V., Karpukhina T. A., Melnik N. N. (2006) Inroad to modification of detonation nanodiamond, *Diam. Relat. Mater.* 15:296–299.

82. Behler K. D., Stravato A., Mochalin V., Korneva G., Yushin G., Gogotsi Y. (2009) Nanodiamond–polymer composite fibers and coatings, *ACS Nano* 3:363–369.

83. Zhang Q., Mochalin V. N., Neitzel I., Knoke I. Y., Han J., Klug C. A., Zhou J. G., Lelkes P. I., Gogotsi Y. (2011) Fluorescent PLLA–nanodiamond composites for bone tissue engineering, *Biomaterials* 32:87–94.

84. Wang D. H., Tan L. S., Huang H. J., Dai L. M., Osawa E. (2009) *In-situ* nanocomposite synthesis: Arylcarbonylation and grafting of primary diamond nanoparticles with a poly(ether–ketone) in polyphosphoric acid, *Macromolecules* 42:114–124.

85. Cheng J. L., He J. P., Li C. X., Yang Y. L. (2008) Facile approach to functionalize nanodiamond particles with V-shaped polymer brushes, *Chem. Mater.* 20:4224–4230.

86. Mochalin, V. N., Neitzel I., Etzold B. J. M., Peterson A., Palmese G., Gogotsi Y. (2001) Covalent incorporation of aminated nanodiamond into an epoxy polymer network, *ACS Nano* 5:7494–7502.

87. Shimkunas R. A., Robinson E., Lam R., Lu S., Xu X., Zhang X. Q., Huang H., Osawa E., Ho D. (2009) Nanodiamond–insulin complexes as pH-dependent protein delivery vehicles, *Biomaterials* 30:5720–5728.

88. Purtov K. V., Petunin A. I., Burov A. E., Puzyr A. P., Bondar V. S. (2010) Nanodiamonds as carriers for address delivery of biologically active substances, *Nanoscale Res. Lett.* 5:631–636.

89. Alhaddad A., Adam M. P., Botsoa J., Dantelle G., Perruchas S., Gacoin T., Mansuy C. et al. (2011) Nanodiamond as a vector for siRNA delivery to Ewing sarcoma cells, *Small* 7:3087–3095.

90. Osswald S., Yushin G., Mochalin V., Kucheyev S. O., Gogotsi Y. (2006) Control of sp^2/sp^3 carbon ratio and surface chemistry of nanodiamond powders by selective oxidation in air, *J. Am. Chem. Soc.* 128:11635–11642.

91. Shenderova O., Koscheev A., Zaripov N., Petrov I., Skryabin Y., Detkov P., Turner S., Tendeloo G. V. (2011) Surface chemistry and properties of ozone-purified detonation nanodiamonds, *J. Phys. Chem. C* 115:9827–9837.

92. Schrand A. M., Johnson J., Dai L., Hussain S. M., Schlager J. J., Zhu L., Hong Y., Osawa E. (2009) Nanomaterials, in Webster T. J., ed., *Safety of Nanoparticles: From Manufacturing to Medical Applications*, Nanostructure science and technology, cytotoxicity and genotoxicity of carbon, pp. 159–187, Springer, New York.

93. Schrand A. M., Hens S. A. C., Shenderova O. A. (2009) Nanodiamond particles: Properties and perspectives for bioapplications, *Crit. Rev. Solid State Mater. Sci.* 34:18–74.

94. Schrand A. M., Huang H., Carlson C., Schlager J. J., Sawa O. E., Hussain S. M., Dai L. (2007) Are diamond nanoparticles cytotoxic? *J. Phys. Chem. B* 111:2–7.

95. Lin Y., Taylor S., Li H., Fernando, K. A. S., Qu L., Wang W., Gu L., Zhou B., Sun Y. P. (2004) Advances toward bioapplications of carbon nanotubes, *J. Mater. Chem.* 14:527–541.

96. Moniruzzaman M., Winey K. I. (2006) Polymer nanocomposites containing carbon nanotubes, *Macromolecules* 39:5194–5205.

97. Wang X., Li Q., Xie J., Jin Z., Wang J., Li Y., Jiang K., Fan S. (2009) Fabrication of ultralong and electrically uniform single-walled carbon nanotubes on clean substrates, *Nano Lett.* 9:3137–3141.

98. Gullapalli S., Wong M. S. (2011) Nanotechnology: A guide to nano-objects, *Chem. Eng. Prog.* 107:28–32.

99. Thostenson E. T., Ren Z. F., Chou T. W. (2001) Advances in the science and technology of carbon nanotubes and their composites, *Compos. Sci. Technol.* 61:1899–1912.

100. Bethune D. S., Johnson R. D., Salem R. J., de Varies M. S., Yannoni C. S. (1993) Atoms in carbon cages: The structure and properties of endohedral fullerenes, *Nature* 366:123–128.

101. Journet C., Maser W. K., Bernier P., Loiseau A., de la Chapelle M. L., Lefrant S., Deniard P., Lee R., Fischer J. E. (1997) Large scale production of single walled carbon nanotubes by the electric arc technique, *Nature* 388:756–758.

102. Rinzler A. G., Liu J., Dai H., Nikolaev P., Huffman C. B., Rodriguez-Macias F. J., Boul P. J. et al. (1998) Large-scale purification of single-wall carbon nanotubes: Process, product and characterization, *Appl. Phys. A Mater. Sci. Process.* 67:29–37.

103. Nikolaev P., Bronikowski M. J., Bradley R. K., Fohmund F., Colbert D. T., Smith K. A., Smalley R. E. (1999) Gas-phase catalytic growth of single-walled carbon nanotubes from carbon monoxide, *Chem. Phys. Lett.* 313:91–97.

104. Tans S. J., Devoret M. H., Dai H., Thess A., Smalley R. E., Georliga L. J., Dekker C. (1997) Individual single-wall carbon nanotubes as quantum wires, *Nature* 386:474–477.

105. Mintmire J. W., Dunlap B. I., White C. T. (1992) Are fullerene tubules metallic? *Phys. Rev. Lett.* 68:631–634.

106. Dekker C. (1999) Carbon nanotubes as molecular quantum wires, *Phys. Today* 52:22–28.

107. Martel R., Derycke V., Lavoie C., Appenzeller J., Chan K., Tersoff, J., Avouris Ph. (2001) Ambipolar electrical transport in semiconducting single-wall carbon nanotubes, *Phys. Rev. Lett.* 87:256805–256808.

108. Flahaut E., Bacsa R., Peigney A., Laurent C. (2003) Gram-scale CCVD synthesis of double-walled carbon nanotubes, *Chem. Commun.* 12:1442–1443.

109. Zettl J. A. (2000) Low-friction nanoscale linear bearing realized from multiwall carbon nanotubes, *Science* 289:602–604.

110. Treacy M. M. J., Ebbesen T. W., Gibson J. M. (1996) Exceptionally high Young's modulus observed for individual carbon nanotubes, *Nature* 381:678–680.

111. Zavalniuk V., Marchenko S. (2011) Theoretical analysis of telescopic oscillations in multi-walled carbon nanotubes, *Low Temp. Phys.* 37:337–342.

112. Guo T., Nikolaev P., Rinzler A. G., Tomanek D., Colbert D. T., Smalley R. E. (1995) Self-assembly of tubular fullerenes, *J. Phys. Chem.* 99:10694–10697.

113. Guo T., Nikolaev P., Thess A., Colbert D., Smalley R. E. (1995) Catalytic growth of single-walled nanotubes by laser vaporization, *Chem. Phys. Lett.* 243:49–54.
114. Collins P. G., Avouris P. (2000) Nanotubes for electronics, *Sci. Am.* 283:67–69.
115. Smiljanic O., Stansfield B. L., Dodelet J. P., Serventi A., Désilets S. (2002) Gas-phase synthesis of SWNT by an atmospheric pressure plasma jet, *Chem. Phys. Lett.* 356:189–193.
116. Kumar M., Ando Y. (2010) Chemical vapor deposition of carbon nanotubes: A review on growth mechanism and mass production, *J. Nanosci. Nanotechnol.* 10:3739–3758.
117. Radushkevich L. V. (1952) The structure of carbon formed by the thermal decomposition of carbon monoxide on the iron touch, *J. Phys. Chem.* 26:88–95.
118. Walker Jr. P. L., Rakszawski J. F., Imperial G. R. (1959) Carbon formation from carbon monoxide-hydrogen mixtures over iron catalysts. I. Properties of carbon formed, *J. Phys. Chem.* 63:133–140.
119. José-Yacamán M., Miki-Yoshida M., Rendón L., Santiesteban J. G. (1993) Catalytic growth of carbon microtubules with fullerene structure, *Appl. Phys. Lett.* 62:657–659.
120. Beckman W. (2007) UC Researchers Shatter World Records with Length of Carbon Nanotube Arrays. University of Cincinnati, Cincinnati, OH, http://www.uc.edu/News/NR.aspx?ID=5700.
121. Inami N., Ambri M. M., Shikoh E., Fujiwara A. (2007) Synthesis-condition dependence of carbon nanotube growth by alcohol catalytic chemical vapor deposition method, *Sci. Technol. Adv. Mater.* 8:292–295.
122. Ishigami N., Ago H., Imamoto K., Tsuji M., Iakoubovskii K., Minami N. (2008) Crystal plane dependent growth of aligned single-walled carbon nanotubes on sapphire, *J. Am. Chem. Soc.* 130:9918–9924.
123. Eftekhari A., Jafarkhani P., Moztarzadeh F. (2006) High-yield synthesis of carbon nanotubes using a water-soluble catalyst support in catalytic chemical vapor deposition, *Carbon* 44:1343–1345.
124. Naha S., Ishwar K. P. (2008) A model for catalytic growth of carbon nanotubes, *J. Phys. D Appl. Phys.* 41:065304.
125. Banerjee S., Naha S., Ishwar K. P. (2008) Molecular simulation of the carbon nanotube growth mode during catalytic synthesis, *Appl. Phys. Lett.* 92:233121.
126. Pinilla J. L., Moliner R., Suelves I., Lazaro M., Echegoyen Y., Palacios J. (2007) Production of hydrogen and carbon nanofibers by thermal decomposition of methane using metal catalysts in a fluidized bed reactor, *Int. J. Hydrogen Energ.* 32:4821–4829.
127. Neupane S., Lastres M., Chiarella M., Li W. Z., Su Q., Du G. H. (2012) Synthesis and field emission properties of vertically aligned carbon nanotube arrays on copper, *Carbon* 50:2641–2650.
128. Kumar M., Ando Y. (2007) Carbon nanotubes from camphor: An environment-friendly nanotechnology, *J. Phys. Confer. Ser.* 61:643–646.
129. Smalley R. E., Li Y., Moore, V. C., Price B. K., Colorado R., Schmidt H. K., Hauge R. H., Barron A. R., Tour J. M. (2006) Single wall carbon nanotube amplification: En route to a type-specific growth mechanism, *J. Am. Chem. Soc.* 128:15824–15829.
130. Mouras S., Hamm A., Djurado D., Cousseins J. C. (1987) Synthesis of first stage graphite intercalation compounds with fluorides, *Revue de Chimie Minerale* 24:572–582.

131. Boehm H. P., Setton R., Stumpp E. (1994) Nomenclature and terminology of graphite intercalation compounds, *Pure Appl. Chem.* 66:1893–1901.

132. Boehm H. P., Clauss A., Fischer G. O., Hofmann U. (1962) Das Adsorptionsverhalten sehr dünner Kohlenstoffolien, *Zeitschrift für anorganische und allgemeine Chemie* (in German) 316:119–127.

133. Wallace P. R. (1947) The band theory of graphite, *Phys. Rev.* 71:622–634.

134. McClure J. W. (1956) Diamagnetism of graphite, *Phys. Rev.* 104:666–671.

135. Semenoff G. W. (1984) Condensed-matter simulation of a three-dimensional anomaly, *Phys. Rev. Lett.* 53:2449–2452.

136. DiVincenzo D. P., Mele E. J. (1984) Self-consistent effective-mass theory for intralayer screening in graphite intercalation compounds, *Phys. Rev. B* 29:1685–1694.

137. Lu X. K., Yu M. F., Huang H., Ruoff R. S. (1999) Tailoring graphite with the goal of achieving single sheets, *Nanotechnology* 10:269–272.

CHAPTER 2

CONTENTS

Different Functionalization Methods of Carbon-Based Nanomaterials

2

Toktam Nezakati, Aaron Tan, and Alexander M. Seifalian

2.1 INTRODUCTION

Carbon-based nanomaterials (CNMs) possess special and novel properties in their specific structure with at least one dimension in nanometer size. High surface-to-volume ratio and unique thermal, optical, mechanical, and electrical properties can be named as their main advantages. The characteristic structures of CNMs enable them to interact with organic molecules through covalent and noncovalent bonds. Noncovalent interaction can be named as hydrogen bonding, π–cation, π–π stacking, π–anion electrostatic forces, hydrophobic interactions, and van der Waals forces. Carbon allotropes have attracted scientists, and materials that consist of conjugated π-bond systems are topologically confined objects in zero, one, two, or three dimensions. CNMs have found a wide range of applications in different sample preparation technologies. In this chapter, we will consider a broad overview on the different functionalizations on CNMs, including graphene, carbon nanotube (CNT), fullerene, and nanodiamond (ND), as well as their functionalized forms. We will focus on CNM chemistry and focus on functionalized addends and different modifications that could be applied to CNT, fullerene, graphene, and ND-based electronic and optical devices and biomaterial products production.

2.2 COVALENT FUNCTIONALIZATION

Covalent bonds occur between organic chemicals and CNMs if both of the chemicals and CNMs have certain functional groups such as $-COOH$, $-OH$, and $-NH_2$ (Huang et al. 2002; Piao et al. 2009). These covalent bonds can be depicted by spectroscopic studies with Fourier transform infrared spectroscopy, x-ray photoelectron spectroscopy (XPS), and nuclear magnetic resonance techniques (Piao et al. 2009). Compared to the noncovalent interactions such as π–π stacking, π–cation interaction, π–anion interaction, and hydrophobic interaction, the attachment of organic molecules to CNMs through a covalent bond can resist any desorption and is much stronger. Thus, covalent functionalization of CNMs has been broadly utilized to form a variety of nanostructures with excellent physical and chemical properties (Banerjee and Wong 2002; Riggs et al. 2000). Functionalization of CNMs with covalent functionalization is normally achieved through reactions such as free-radical chemistry (Ying et al. 2003), amidation (Chen and Mitra 2008), esterification (Kim et al. 2004), carboxylation (Wang et al. 2005), fluorination (Mickelson and Huffman 1998), diazonium (Bahr and Tour 2001), Bingel reaction (Coleman et al. 2003), and composite formation (Daniel et al. 2007; Shen et al. 2007).

Carbon nanomaterials can be chemically functionalized through aryl diazonium salts, benzoyl peroxide (BPO), styrene, nitrenes, carbenes, and aryne cycloaddition. In the following, we will explain differently these methods with their application.

One of the moderately reactive species applied to covalently functionalize carbon nanomaterial is aryl diazonium salts. The functionalization reaction is to proceed via a free-radical mechanism.

CNM substrates originating from various chemical processes have different reactivity toward aryl diazonium salts. Chemically converted graphene, which is based on reduced graphene oxide (GO), illustrates very high reactivity to diazonium salts, with the basal plane and the edges becoming functionalized (Lomeda et al. 2008). Epitaxial graphene grown from silicon carbide (SiC) is efficiently surface grafted using diazonium salts (Bekyarova et al. 2009). Micromechanical exfoliated graphene samples were functionalized by nitrophenyldiazonium salt (4-nitrophenyldiazonium [4-NPD]) tetrafluoroborate as depicted in Figure 2.1a. The diazonium reaction with graphite or mechanically exfoliated graphene presents preferentially edge reactivity (Lim et al. 2010; Sun et al. 2010). Soluble graphene could be attained via the exfoliation from

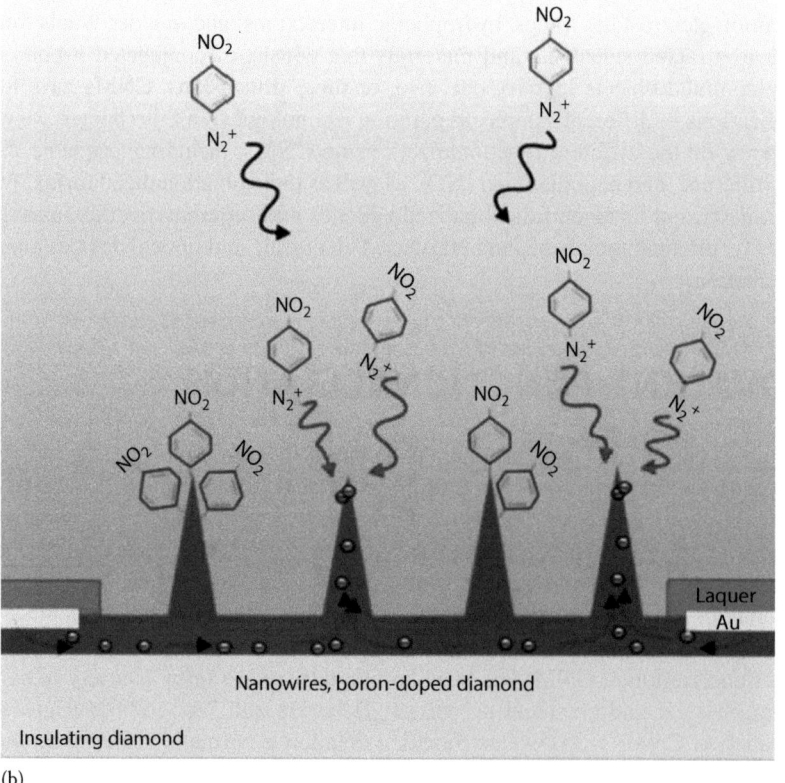

(a)

(b)

FIGURE 2.1

(a) The reaction between graphene and nitrophenyl diazonium salts. (From Bekyarova, E. et al., *J. Am. Chem. Soc.*, 131(4), 1336, 2009.) (b) Phenyl linker molecules are preferentially attached to tips of wires due to the electrochemical attachment schema. (From Nebel, C.E. et al., *Diamond Relat. Mater.*, 18(5–8), 910, 2009.)

edge of functionalized graphite. Another functionalization chemistry similar to diazonium is azide. Nitrenes are considered as reactive intermediates and formed from photolysis or thermolysis of azides (Liu and Yan 2011). Covalent reactivity of CNMs are affected by topological and strain defects (Boukhvalov and Katsnelson 2008, 2009). It was studied that less reactive domain is located in perfect basal plane than defective and strained graphene spots (Boukhvalov and Katsnelson 2008, 2009).

Hummer's method of oxidizing the graphite ($KMnO_4$, $NaNO_3$, H_2SO_4) represents a noncontrollable method to functionalize covalently the basal plane of graphene. This is due to GO cut into small pieces with some holes on its surface. GO and GO nanoribbons with less defects are optimized using the new methods recently (Higginbotham et al. 2010; Marcano et al. 2010). This optimized method increases the amount of $KMnO_4$ and excludes the $NaNO_3$ and uses a 9:1 mixture of H_2SO_4/H_3PO_4.

The electrical measurement proposes that the basal plane of the GO yield in the optimized reaction is much higher and less disrupted than when using Hummer's method.

2.2.1 Free-Radical Functionalization

Covalent chemical functionalization on carbon nanomaterial can be carried out by organic compounds such as aryl diazonium salts, BPO, styrene, nitrenes, carbenes, and aryne cycloaddition. In the following, we will explain differently these methods with their application. The functionalization, properties, derivatization, and application techniques of CNMs are also described. Aryl diazonium salts, as an organic compound, used for CNMs covalently functionalization. The functionalization reaction proceeds via a free-radical mechanism. Synthetic carbon allotropes as nanomaterial substrates arising from different chemical processes have diverse reactivity toward aryl diazonium salts.

2.2.1.1 Diazonium salts

Aryl diazonium cations are extensively used to covalently functionalize CNM substrates that exhibit a broad range of composition, from sp^2 graphene to sp^3 diamond and ND (Jayasundara et al. 2013). This organic compound is one of the methods to covalently functionalize conductive and semiconductive materials (Pinson and Podvorica 2005).

The functionalization reaction proceeds through a free-radical mechanism. The functionalized material acts as an electron transfer agent that donates an electron to the aryl diazonium ion to form an aryl radical after the removal of nitrogen gas. The high reactivity of diazonium salts, coupled with the concerted N_2 split, is believed to be responsible for the generation of aryl radicals within the vicinity of the substrate (Pinson and Podvorica 2005). Rather than diffusing into the surrounding medium, this close proximity promotes the reaction between the substrate with the radical. Due to the rapid decomposition of diazonium compounds, they should be utilized immediately after synthesis/preparation (Sinitskii et al. 2010).

Since chemical or electrochemical oxidation may impair the CNM surface, a new method was developed, which involved the electrochemical reduction of a phenyldiazonium derivative. Surface modification process involves the formation of a diazonium radical, followed by the formation of a covalent bond to the glassy carbon (GC) electrode given by

$$R\text{-}C_6H_4N_2^+ + e^- \rightarrow N_2 + R\text{-}C_6H_4^{\bullet}$$

$$GC + R\text{-}C_6H_4^{\bullet} \rightarrow GC\text{-}C_6H_4\text{-}R$$

where R is a parasubstituent such as carboxy and nitro (Downard et al. 1995; Liu and Mccreery 1995). Immobilization of glucose oxidase (Bourdillon et al. 1992) and control of protein adsorption via appropriate utilization of R groups can be conducted using these modified electrodes (Downard et al. 1995). It has been recently reported that nucleophilic substitution reaction with substituted phenyldiazonium ions can be used to chemically modify polyaniline films (Liu and Freund 1996). Additionally, the corresponding aniline can be used to synthesize large numbers of substituted phenyldiazonium salts, which have been electrochemically studied in the last decades (Elofson and Gadallah 1969).

Graphene is a chemically inert material and stable in air up to 200°C (Chen et al. 2011); accordingly, to be able to break its sp^2 structural bonds to begin any functionalization reaction, high energy is required. Hot tungsten filament is used to generate atomic hydrogen to reduce the graphite to graphene (Savchenko 2009).

Moreover, single graphene sheets are tremendously (or remarkably) conductive, where the distinctive properties of mobility of charge carriers and the electrical conductivity can reach to 200,000 cm^2/(V s) (Bolotin et al. 2008) up to 6,000 S/m (Du et al. 2008), respectively.

One of the vast majorities of research on the properties of graphene sheets such as electrical conductivity has focused on the noncovalent mixing methods between graphene sheets and polymers. A broad range of Electrical Conductivity (EC) values of these composite materials has been disclosed from 10^{-14} to 10^4 S/cm (Bekyarova et al. 2009; Kim et al. 2010; Liang et al. 2009; Stankovich et al. 2006). In this case, in the noncovalent mixing method, the EC of a single graphene sheet cannot be handled, and the graphene composites are unsettled upon outside stimulants like vibrations, temperature, and pressure.

Graphene nanomaterial can be functionalized with aryl diazonium salts to cause change in the electronic properties of graphene.

Preparation of single-layered and multilayered graphene sheets was done via exfoliation of bulk graphene on silicon wafers. This was followed by annealing and treatment with diazonium salt, rendering only the top layer and edges to be reactive.

Epitaxial graphene grown on SiC substrate was treated with 4-NPD salt molecule, representing the covalent attachment of the aryl group to graphene. The reaction is due to the spontaneous electron transfer from the graphene layer and its substrate to the diazonium salt. XPS was conducted to detect covalent bond formation. Cyclic voltammetry was done to measure the surface coverage of nitrophenyl group. This was shown to be around 1×10^{15} molecules/cm^2, as opposed to the theoretical value of around 8×10^{15} molecules/cm^2 for a close-packed single layer of nitrophenyl groups that are vertically aligned. More detailed studies on the reactivity of 4-NPD salt with graphene and the electron transfer chemistry were conducted by Strano and coworkers (Sharma et al. 2010).

Functionalization could also be represented by Raman D/G ratio measurements. The molecule of 4-NPD salt's Raman results shows that the reactivity of single-layered graphene is 10 times more than the bilayered or multilayered graphene. Due to the polarization dependence of the disorder D, the peak intensity is expected to be twice as high as in the edge compared to the bulk of single-layer graphene, in terms of reactivity (Park and Yan 2013).

It has been shown that covalent attachment of aryl groups to the basal plane of graphene can increase its conductivity (Huang et al. 2011). The difference in conductivity could be attributed to increased concentration of reagents and longer reaction time, and the degree of functionalization in yielded graphene is reported higher. However, there are conflicting reports that suggest that aryl molecule groups act as defects at lower degrees of modifications, thereby decreasing the carrier mobility by reducing the mean free path and/or opening a band gap. In comparison, more modification on graphene hole density originated from rising electron transfer, resulting in the aryl molecule groups. This will result into being more conductive, since it is large for decreasing carrier mobility.

The introduction of a band gap in epitaxial graphene by the utilization of (p-nitrophenyl) diazonium tetrafluoroborate was proposed by Haddon et al. (Niyogi et al. 2010). The covalent attachment of the aryl group to the planar graphene will cause the alteration of sp^2 carbon to sp^3 carbon. The conjugation length of delocalized carbon lattice will change due to this conversion. The functionalized band gap reading of epitaxial graphene is 0.36 eV using angle-resolved photoemission spectroscopy technique. The solubility of pristine graphene was improved via diazonium functionalization without the need for an external stabilizer (Liu and Mccreery 1995; Sun et al. 2010). Precipitation and aggregation of pristine graphene would occur in the absence of a surfactant. Functionalized aryl diazonium salts on graphene result in solubility of 0.01–0.02 mg/mL in Dimethyl Formamide (DMF) (Liu and Mccreery 1995; Sun et al. 2010), which corresponds to the solubility of graphene when using tetrabutylammonium hydroxide (Li et al. 2008). Functionalized graphene solubility in chloroform was inspected up to 27 µg/mL by Hirsch et al. Meanwhile, in situ polymerization of p-t-butylphenyldiazonium tetrafluoroborate or p-sulfonylphenyldiazonium was conducted on graphite powder (Englert et al. 2011; Huang et al. 2011). It was thus postulated that the increase in conductivity of the functionalized graphene in comparison with pristine graphene was due to the charge transfer effect of nitrophenyl groups.

Hence, there is an unmet need for the development of technologies to modify the EC of graphene composites. The high conductivity of graphene is due to its fully conjugated structure. Consequently, modification of graphene conductivity can be modulated via the controlled destruction of the conjugation system.

Modification of graphene sheets can be done by three different mono- and bifunctional precursors using covalent or π–π stacking mechanisms. Depending on the amount of monoaryl diazonium salts (MDSs) used, the ECs of graphene papers that are covalently modified by MDS can be controlled over 5 orders of magnitude. Stronger, cross-linked structures were seen in graphene papers that were constructed with bifunctional aryl diazonium salts. Modification with bipyrene terminal molecular wire (BPMW) showed swollen netlike structures but decreased Tensile Strength (TSs) because of the twisted structure of BPMW. Increased UV absorption at around 300 nm was also observed in BPMW modification of graphene. Raman scattering, fluorescence spectroscopy, and x-ray diffraction were also done on the modified graphene papers. Potential applications include use in electronics and material science for these highly porous controlled EC graphene papers. Partially destructed conjugation single graphene sheets are currently under investigation. Aryl reaction to the Single Wall Carbon Nanotube (SWCNT) sidewall with aryl diazonium compounds can yield porphyrin–SWCNT covalent composites with short rigid phenylene spacer (Liu et al. 2012).

Through the aryl addition reaction to the SWCNT sidewall with adopting aryl diazonium compounds, the covalent composite of porphyrin–SWCNT with short, stiff phenylene spacer will result (Cheng and Adronov 2006; Guo et al. 2006; He et al. 2010; Umeyama et al. 2007, 2011). These direct addition reactions from aryl diazonium compounds to the SWCNT sidewalls are convenient and promote easy attachment of functional groups.

Free-based tetraphenylporphyrin composite and SWCNTs applying in-situ-created diazonium porphyrin compounds were proposed by Chen and Zhang et al. (H_2P-DSWCNT) (Guo et al. 2006). In the steady-state fluorescence spectrum, an efficiency of >90% in response to the reference of porphyrin is observed. Photoinduced ET from $1H_2P^*$ to SWCNT was a likely explanation of the observed rising in the fluorescence quenching with raised solvent polarity (Giordani et al. 2009; Guo et al. 2006). In the measurements of time-resolved nanosecond transient absorption, the formation of the charge-separated state with a decay lifetime of nanosecond in DMF was observed. The effective intramolecular quenching of H_2P-DSWCNT is due to the vigorous electronic coupling between SWCNT and conjugated π-systems of the porphyrin facilitated via a thorough interaction of bond as a result of the short, stiff phenylene spacer (Re = 5.8 Å). This is in differentiation to the considered fluorescence quenching of ZnP-E-SWCNT and ZnP-PSWCNT (Arai et al. 2009; Das et al. 2012).

A N_2 purged glove box is used to conduct the reduction of diazonium salts via the application of a fixed potential of −0.05 V (vs. Ag/Ag$^+$) for 2s (Figure 2.1b). Sonication of nitrophenyl-modified diamond surfaces with acetonitrile and acetone is then performed. Electrochemical reduction of 4-nitrophenyl groups ($-C_6H_5NO_2$) is done in 0.1M KCl solution of EtOH–H_2O solvent, forming aminophenyl groups ($-C_6H_5NH_2$) (Nebel et al. 2009). Putting into use a fixed potential of −50 mV (vs. Ag/Ag$^+$) for 2s is necessary for the bonding of nitrophenyl molecules from diazonium salts. By 50% charge reduction, during the bonding process, the nitro- to aminophenyl conversion is observed, which is related to the bonding on a steady diamond surface. The intensely dissimilar attachment scheme can be due to the phenyl bonding to the wire tips (Nebel et al. 2009). The reduction of the diazonium salt BF_4^-, $^+N_2$-C_6H_4-$CH(CH_3)Br$ is done in a simple method, that is, by using the spontaneous electroless chemical method at the surface of ND. The modified fluorescent nanodiamond (fND) (fND-Br) as microinitiators can be considered to surface-confined atom transfer radical polymerization (ATRP) of tert-butyl methacrylate (tBMA). Hydrolysis of these hybrids is then conducted to obtain fluorescent nanodiamond-poly(methacrylic acid) nanoparticles. These peptide-functionalized hairy fND particles (fND-PtBMA) are then activated using 1-ethyl-3-(3-dimethylaminopropyl) carbodiimide/N-hydroxysuccinimide for the purposes of covalent attachment of bovine serum albumin (Dahoumane et al. 2009). A versatile method can be used to couple aryl molecules on the surfaces of diamonds via Suzuki coupling steps and aryl diazonium salt derivatization. This method has a higher specificity compared to photochemical coupling methods utilizing alkyl linkers. A simple wet-solution chemistry can be used to couple phenyl boronic acid on intrinsic hydrogenated diamond for reactions that involve bromophenyl as the synthon.

These surface processing methods would work well on high pressure/temperature, or detonation of ND powder, mainly due to the fact that the spontaneous derivatization does not require photo-chemical or electrical activation and indicates the method of coupling aryl molecules. Diamond with hydrogen termination will cause the spontaneous coupling of aryl diazonium salts, but not on oxygen-terminated diamonds. Distinguished observation between hydrogen terminated and oxygen terminated can be due to the antithetic electron affinities and conductivities, which will cause the energy barrier in interfacial charge transfer on the surfaces (Zhong et al. 2008).

2.2.1.2 Benzoyl peroxide

Benzoyl is an organic compound with the formula C_6H_5CO-. The structural formula of two benzoyl groups bridged by a peroxide is $[C_6H_5C(O)]_2O_2$. In addition to diazonium salts, BPO can also generate phenyl radicals, which happens by using hydrogen peroxide and benzoyl chloride by considering that the oxygen–oxygen bond in peroxides is not strong enough; therefore, BPO will undergo homolysis (symmetrical fission) (Brodie 2006), building free radicals:

$$[C_6H_5C(O)]_2O_2 \rightarrow 2C_6H_5CO_2^\bullet$$

BPOs are obtained under reaction of prepared organic peroxide by treating benzoyl chloride with barium peroxide (Brodie 2006), which is shown as follows:

$$2C_6H_5C(O)Cl + BaO_2 \rightarrow [C_6H_5C(O)]_2O_2 + BaCl_2$$

The homolysis is usually caused by heating. BPO-initiated in situ functionalization of carbon nano-materials shows a simple means of generating reactive sites on the surface of CNTs by employing free-radical interactions with the surrounding polymer matrix (Mcintosh et al. 2007).

Immersing the mono-/multilayer graphene flakes on silicon wafer in a solution of BPO and treating under intense laser irradiation will generate the functionalized graphene. Drastic decrease in electrical conductivity and rise in the hole doping level are distinctive properties of this reaction (Liu et al. 2009). A hot electron from photoexcited graphene is accepted by the surface-adsorbed BPO. This then decomposes benzoyloxyl radicals and benzoate and is subsequently converted to a phenyl radical via the removal of CO_2 (Figure 2.2a). It was observed that there was a higher reactivity (around 14 times) for single-layer graphene, compared to double-layer graphene. This was due to the single-layer graphene being more reactive at the interface of induced corrugation and due to the absence of interlayer π stacking or direct substrate contact (Liu et al. 2009).

Covalent functionalization of graphene with residual oxygen-containing functional groups was conducted by Hsiao et al. They prepared the graphene/maleic anhydride-poly(oxyalkylene)amines (MA-POA2000) by free-radical grafting and direct modification (Hsiao et al. 2010). Figure 2.2b illustrates an overview of the covalent functional procedure. In order to prepare the free-radical graphene/MAPOA2000 (F-graphene/MA-POA2000), 200 mg of graphene was placed in 100 mL of tetrahydrofuran (THF) via shear mixing for 1 h and was sonicated for 15 min. Initiation of the free-radical reaction was done by BPO at 70°C for 1 h. Ten milliliters of THF was used to dissolve MA-POA2000 and was then slowly added to the graphene suspension and refluxed at 70°C for 8 h (Hsiao et al. 2010).

A chemical method to construct effective aspect ratio of graphene nanosheets (GNSs) using an in situ radical addition between GNS and polymer matrix was described by Zhang et al. This method allows the blending of unsaturated poly-(vinylidene fluoride-trifluorethylene) (P(VDF-TrFE)), named P(VDF-TrFEDB), for double bonds, with surface-modified GNS and BPO as a free-radical initiator. Free radicals formed from the breakdown of BPO would add to the double bonds on the P(VDF-TrFE-DB) chain at high temperatures, which would occur between the polymer backbone and double bonds on GNS (Wen et al. 2013). Increased compatibility and reduction of the dielectric loss via the induction by the interface between them are mainly attributed to the chemical bonds formed between the P(VDF-TrFE-DB) matrix and GNS. The increase in threshold composition and reduction in conductivity and dielectric loss of the resultant composites at high GNS loading content are due to the defect caused by the addition reaction from the polymeric radicals onto GNS (Wen et al. 2013).

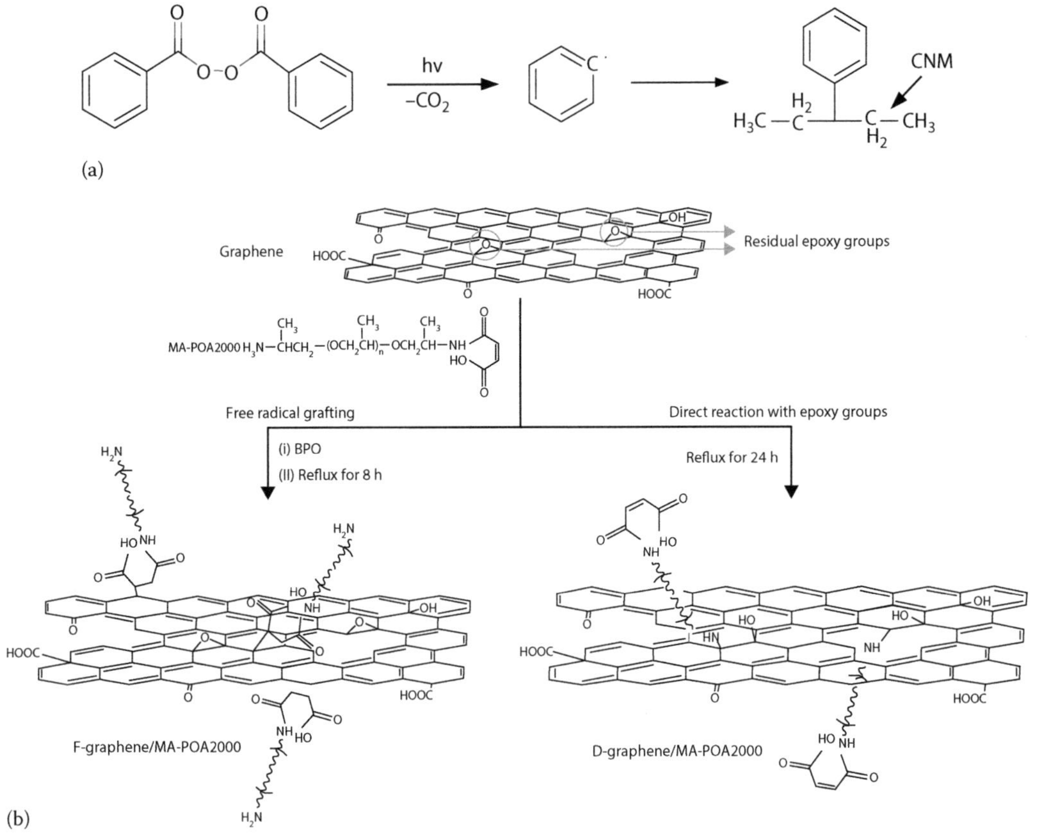

FIGURE 2.2
(a) Reaction between graphene and absorbed benzoyl peroxide. (From Liu, H. et al., *J. Am. Chem. Soc.*, 131(47), 17099, 2009.) (b) The preparation of F-graphene/MA-POA2000 and D-graphene/MA-POA2000. (From Hsiao, M.-C. et al., *ACS Appl. Mater. Interf.*, 2(11), 3092, 2010.)

The thermal decomposition of four diacyl peroxides (benzoyl, 4-methoxybenzoyl, phthaloyl, and trifluoroacetyl) in the presence of SWCNTs results to the functionalized SWCNT with the BPO (Engel et al. 2008). Also, phenyl-modified CNT material produced by the thermal dissociation of BPO is utilized for the fabrication of water-soluble tubes (Liang et al. 2006).

Studies have shown that sulfonation of phenylated SWCNT can be done by reacting it with sulfuric acid and by dispersing them in oleum (Billups and Liang 2007; Liang et al. 2006). The –H on phenyl groups would be substituted with $-SO_3H$ at a position para to their attachment to the SWCNT. These phenylsulfonated SWCNTs are water soluble and can be used in a wide range of industries including electronics, energy, and pharmaceutical (Kharisov et al. 2009). It was reported by Barrera et al. that high temperature and high shear would cause the breakdown of BPO, thereby generating free radicals. This facilitates the covalent attachment of SWCNT to the polypropylene matrix (Mcintosh et al. 2007).

Different end groups like fluorine, bromine, methoxy, phenyl, and nitro groups are capable of synthesizing a series of electron acceptors based on o-xylenyl C_{60} bis-adduct (OXCBA). The derivatives functionalized, 2,3-dimethylnaphthalene and three o-xylene, at the phenyl ring constituents, fluorine, methoxy, and nitro groups, proceed with N-bromosuccinimide in the presence of BPO in 1,2-dichloroethane under reflux and form ingredients for reaction with C_{60} (Kim et al. 2012). Distinctive properties, structural and electrical, will change due to the changes of end groups in OXCBA derivatives. Tuning the hydrophobicity of OXCBA derivatives could influence by changing the end groups, and this will result in systematic control of the miscibility and interfacial interaction with poly(3-hexylthiophene) (P3HT) (Kim et al. 2012).

It is also possible to form a uniform polystyrene/poly(divinyl benzene) (PS/PDVB) micrometer-sized composite particles. This begins with a single-step swelling process at room temperature of the PS template particles with dibutyl phthalate (a swelling solvent) droplets containing DVB and BPO and is followed by annealing with the polymerization of DVB within the swollen PS template particles. If the annealing temperature is controlled, it is possible to control the crystallinity, particle size and size distribution, composition, magnetic properties, and surface area (Snovski et al. 2011).

On the ND surface, the multistep organic transformations could be resulted with compile the essential functional groups over covalent bonds. This reaction will follow the radical reaction without purification of commercial BPO, including 25% water (Takimoto et al. 2010).

2.2.1.3 Styrene

Styrene is an aromatic organic compound with the formula C_8H_8. It comprises a benzene ring with a vinyl group attached to it. Annealing of Fe^{2+}-treated porous PS microspheres using a two-step seed emulsion polymerization reaction was done to produce high-specific-surface-area magnetic porous carbon microspheres (MPCMSs). Sulfonation of the microspheres was then carried out. Loading of Fe^{2+} was done by ion exchange, and annealing at 250°C under ambient atmosphere was done for 1 h, to yield PS-250 composite. MPCMS-500 was annealed at 500°C, while MPCMS-800 was annealed at 800°C (Zhou et al. 2014). This study highlights the potential for carbon spheres to be used as Fenton catalyst for environmental pollution cleanup. Irradiation of graphite with high-intensity ultrasound gave interesting radical chemistry, as reported by Suslick and Xu (2011). Exfoliation of graphite and polymerization onto graphene flakes were achieved via sonication for 2 h at 0°C under Ar. It has been proposed that the functionalization of graphene and PS occurs due to the generation of radicals during the sonication process. The Raman D band seen is attributed to the conversion of sp^2 to sp^3 carbon, showing that covalent bond formation has occurred. PS-functionalized graphene is easily dissolved in organic solvents without precipitation.

2.2.2 Amidation

The amidation reaction of GOs via the addition of ethylenediamine can prove challenging when taking into consideration the solubility of GOs. The partial hydrophobicity and hydrophilicity of GOs do not facilitate dispersion in nonpolar organic solvents, compared to CNTs, which can be dispersed in toluene for the purposes of amidation. Hence, pyridine was selected (rather than toluene) as a polar organic solvent for amidation. Modification of thionyl chloride chemistry during the functionalization of GNS facilitated amidation without the loss of aqueous dispersibility. This is in contrast to other studies where positively charged graphene sheets have low aqueous dispersibility. Layer-by-layer (LBL) self-assembly between carboxylic acid–functionalized and amine-functionalized graphene sheets is responsible for the ability to control light transmittance and film thickness (Kotal et al. 2013). Figure 2.3 depicts a three-step process used to construct a carbon nanofiber (CNF)–Polyanaline (PANI) composite by attaching PANI onto isocyanate-functionalized CNF via an amide group. Amide-functionalized CNF (TCNF; Step No-II) was obtained from the covalent functionalization of toluene diisocyanate (TDI) to carboxylated CNF (XCNF). A urea derivative on its surface was then obtained from the reaction with excess aniline. The aniline monomer was then oxidatively polymerized in the presence of ammonium peroxydisulfate (Step No-IIIa). PANIg-TCNF was then finally formed after the reaction with a urea derivative.

Due to its high reactivity with the carboxyl group of XCNF in the absence of catalyst, TDI is used as the modifying agent to yield dense PANI on the TCNF surface. The formation of a π-conjugated system is due to the phenyl-substituted amide linkage in connecting PANI and CNF. Effective transfer of Faradic charge in the CNF network is seen in this highly conducting network. A decreased degradation of structural conformation of PANI with repetitive charge–discharge reaction might occur due to the prohibition of the accumulation of charge. Hence, facilitating charge transfer between the TCNF–PANI composite and lowering the overall resistance would improve its rate capability with regard to fast charge–discharge kinetics and rapid ion diffusion (Kotal et al. 2013).

The influence of functional groups on desorption hysteresis was studied on multiwalled carbon nanotubes (MWCNTs). An investigation of adsorption–desorption of phenols, nitrobenzenes, and anilines on five MWCNTs with varying degrees of surface oxidation was conducted. In the case of phenols, nitrobenzenes, and 4-nitroaniline MWCNTs, there was an absence of desorption hysteresis. On the other hand, in the case of aniline and 4-methylaniline MWCNTs, significant desorption hysteresis was seen (although not on unoxidized MWCNT). It was also observed that the amidation reaction of amino group of anilines with groups containing oxygen (i.e., carboxyl or lactonic groups)

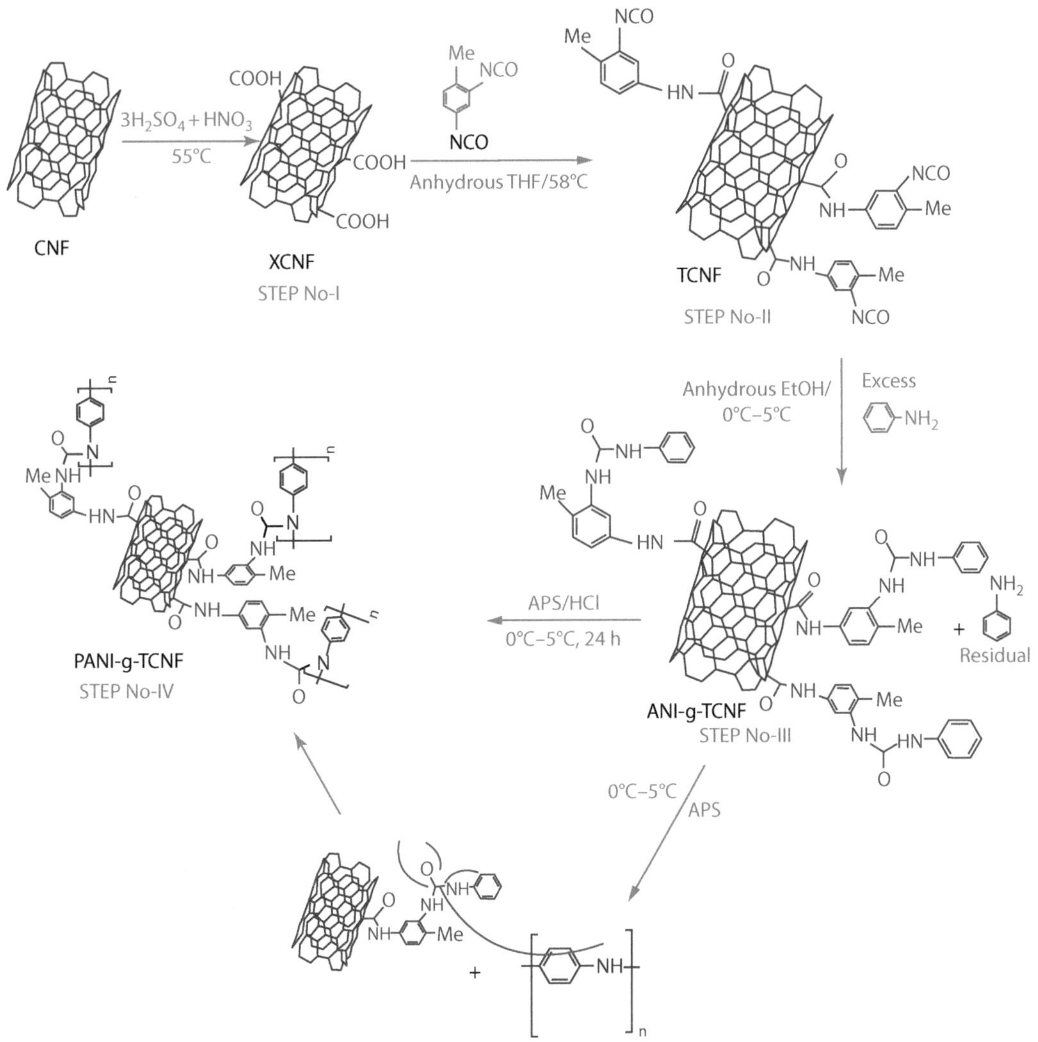

FIGURE 2.3
Schematic diagram for the preparation of a PANI-g-TCNF composite by functionalizing XCNF with TDI through amidation followed by reaction with excess aniline to form a urea derivative and residual aniline, which was subsequently polymerized and grafted with a urea derivative. (From Kotal, M. et al., *ACS Appl. Mater. Interf.*, 5, 8374, 2013.)

resulted in the formation of an amide bond. Hence, we postulate that the desorption hysteresis may be due to the immobilization of organic compounds on the surface of CNTs as a consequence of the irreversible chemical/reaction binding. The process of chemical immobilization is selective and is dependent on the surface groups containing oxygen of CNTs (Perreault et al. 2014; Wu et al. 2013).

2.2.3 Esterification

Esters are chemical compounds derived from carboxylic acids and consist of a carbonyl and an ether linkage. Usually, ester is a derivation of an inorganic or organic acid replacing one −OH (hydroxyl) group by an −O-alkyl (alkoxy) group, generally from carboxylic acids and alcohols.

Reversible addition fragmentation transfer (RAFT) as one of the controlled radical polymerization methods uses hydrogel, with a cross-linking agent, to prepare graphene/poly (acrylic acid) (graphene/PAA). The resultant product via an esterification reaction would be a covalently coupled RAFT agent onto graphene surface (Liu et al. 2014a); this leads to a thermally stable and pH-sensitive hydrogel graphene/PPA.

To improve the limitation of UV light response and fast generation/recombination of electron-hole pair properties of TiO$_2$ nanoparticles, the interface formation between TiO$_2$ and graphene sheets

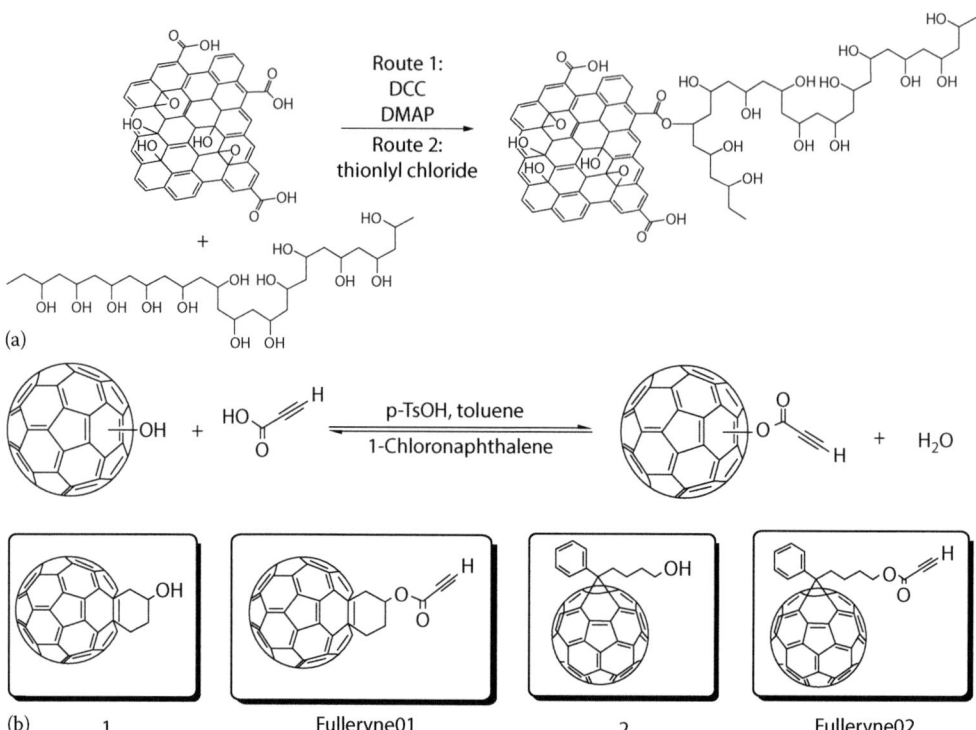

FIGURE 2.4
(a) Schematic illustration of the esterification of graphite oxide with PVA. (From Salavagione, H.J. et al., *Macromolecules*, 42(17), 6331, 2009.) (b) Synthesis of fullerene 01 and fullerene 02 by Fisher esterification. (From Zhang, W.-B. et al., Polymer, 52(19), 4221, 2011.)

could result in a homogeneous attachment of TiO_2 nanoparticles on graphene sheets. To obtain this, esterification reaction will take place between hydroxyl groups of TiO_2 and graphene carboxylic groups. The reduction of Cr (VI) to Cr (III) in sunlight results in a developed photocatalytic decrease for 5% graphene compositing in TiO_2 (Zhang et al. 2012).

Esterification of carboxylic groups in GO with hydroxyl groups in polyvinyl alcohol (PVA) can be considered in the synthesis of GO–polymer composite sheets (Figure 2.4a) (Salavagione et al. 2009). The synthesis of GO–polymer composite sheets can be made by esterification of carboxylic groups in GO with hydroxyl groups in PVA. Two strategies are involved, the first is directly esterifying GO and the second involves the acyl chloride derivative (GOCl). The final compounds, GO-es-PVA and GOCl-es-PVA, are soluble in both water and DMSO in the presence of heat, similar to PVA and PVACNT. Much attention has recently been focused on PVA-functionalized GO, especially in relation to its characterization and isolation. Nonreacted GO is removed via centrifugation, thereby leaving only the PVA-functionalized GO. To ensure the esterification reactions are complete, comprehensive spectroscopic analysis was conducted (Salavagione et al. 2009).

Six different molecular weight PVAs functionalized the GO by esterification reaction. Esterification reaction uses the carboxylic groups of GO and hydroxyl groups of PVA and will result in covalently functionalized f-(PVA) GO. This functionalization will lead to improvements of Young's modulus and tensile strength (Cano et al. 2013).

Wallace et al. covalently attached polycaprolactone chains to free carboxyl groups on graphene sheets via an esterification reaction. A stable anhydrous dimethylformamide dispersion of a highly reduced form of graphene is used to obtain peripheral ester linkages in the reaction. This yields enhanced homogeneity, resulting in increased Young's modulus, tensile strength, and electrical conductivity (about 14 times) compared to pristine polymer at less than 10% graphene content (Sayyar et al. 2013). The esterification process of GO is similar to CNT (Huang et al. 2002). Since the MWCNTs were well dispersed in host nematic-liquid crystal (N-LCs) when the nanotubes are functionalized with phenylcyclohexyl mesogenic moieties, the functionalized LC-MWCNTs, synthesized by Steglich esterification, result in good dispersion stability in ethanol for 3 weeks. Therefore, the composite of LC-MWCNTs in the presence of N-LCs shows improvement in miscibility for concentrations up to 5 wt% due to the substantial affinity between the LC moieties on MWNTs in the presence of N-LCs. Improvement in miscibility allows for alignment of LCMWNTs with introduction of a magnetic field. Therefore, alignment and dispersion of CNTs in LC matrix will result in the functionalization of CNTs with the presence of liquid crystal moiety (Yoo et al. 2013).

To improve the alkyne-functionalized fullerene, one type of esterification called Fisher esterification can be used, adopting 1-chloronaphthalene as a solvent to form and extend the fullerene derivatives. Consequently the design will boost the solubility of fullerene derivatives and in order to complete the reaction, it allows a steady azeotropic distillation for the elimination of water to drive the reaction with yields >90%. (Zhang et al. 2011). Fulleryne01 (Figure 2.4b), a C_{60} propiolate, has been shown to possess *click* reactivity in reactions with azide-functionalized polymer in our previous reports (Bhargava et al. 2006; Zhang et al. 2010a). Considering that only a yield of 17% was observed, its practical usage was substantially impaired due to this low availability. Utilizing 1-chloronaphthalene as the solvent and a custom-made reactor, Fisher reported, for the very first time, improved synthesis of Fulleryne01 with up to 92% yield. The synthesis of Fulleryne02 (Figure 2.4b), a derivative of phenyl-C61-butyric methyl ester (PCBM), can be proceeded via this method. Modular construction of advanced fullerene materials using building blocks via the *click* chemistry approach can thus serve as a viable and efficient method of synthesis. The impact of covalent bonds of fullerene and the hydroxyl groups on the properties of photovoltaic potential can be studied using two new $[C_{60}]$ methanofullerenes consisting of hydroxyl groups into a fullerene dimeric derivative with the same fullerene unit. It has been shown that compatibility between the donor material P3HT and methanofullerenes is low. Indeed, poor photovoltaic performance was seen in the new compounds, compared with [6,6]-PCBM. Compared to monofullerene derivatives, dimeric fullerene derivatives can give rise to improvements in photovoltaic properties. Hence, it could be seen that the performance of fullerene-based solar cells is affected by the covalent aggregation of fullerene (Zhang et al. 2011).

Two new $[C_{60}]$ methanofullerenes consisting of hydroxyl groups, incorporating a fullerene dimeric derivative and with the same fullerene unit, can be incorporated to study the impact of covalent linkage of fullerene and the hydroxyl groups on the properties of photovoltaic fullerene derivatives. The hydroxyl groups will show low compatibility between the methanofullerenes and the donor material P3HT, and new compounds will show poor photovoltaic performance compared with [6,6]-PCBM. Nonetheless, dimeric fullerene derivatives can show improvements in photovoltaic properties than monofullerene derivatives, and this will suggest that the covalent aggregation of fullerene could extremely influence the conduct of fullerene-based solar cells (Ge et al. 2012).

2.2.4 Fluorination

Fluorine is a highly electronegative element, and the more electronegativity will lowering the bond length. Fluorine is well known in compounds within the chemical industry, along with pharmaceuticals, agrochemicals, and devices. Its existence is considered to improve bioavailability, stability, and lipophilicity. The fluorinated material advantage was performed during the *Manhattan Project* (manipulation of UF6). Fluorine promotes thermal stability enhancement (C–F 107 kcal/mol) and increases lipophilicity to increase bioavailability, resembling enzyme substrate (comparable in size to H, 1.47 vs. 1.20 Å) and isoelectronic effects to −O and −OH. The increase in electronegativity of fluorine regularly changes because of the effect of metabolic transformations by chemical reactivity (Su 2008).

Thermal fluorination of MWCNTs was achieved to enhance the hydrogen storage capacity, based on three applications. First, MWCNT surfaces were changed via thermal fluorination to make a route to store hydrogen molecules inside the MWCNTs. This surface analysis enhanced the MWCNT number irregularity via attacking fluorine radicals. Second, through thermal fluorination of the pore volume and surface area, the hydrogen adsorption site number will enhance. Finally, the induced fluorine groups increased capacity of hydrogen storage within attraction effects on the electron in the hydrogen molecules because of the huge fluorine electronegativity (Im et al. 2011a).

Fluorination of SWCNT was studied at 25°C under 1 atm pure fluorine gas. In such situation, the resultant C–F bonding is much weaker than for samples fluorinated at 280°C. If at low temperature fluorination is achieved, fluorine atoms could be detached from the structure by vacuum or slow heating until 300°C with no significant casualty on the tubes. Eventually, through thermal defluorination, the arising sample can be refluorinated alike the pristine tubes (Chamssedine et al. 2011).

Thermal fluorination of MWCNTs was performed at different temperatures (100°C–1000°C) to study the development of the reaction temperature. The high performance of the electrode of NO gas sensor depends on the temperature of fluorination, which could be considered into three different regions, distincted at 400°C and 1000°C. Considering the first temperature domain, the introduction of fluorine functional groups onto MWCNTs reflected the differing direction in electrical resistance change in comparison with classical p-type MWCNTs. For the second temperature domain, the introduction of the fluorine functional groups was weakened by inducing fluorinated carbon gases, resulting in the disintegration of MWCNTs and the improvement of classical p-type gas sensor conduct. In the third temperature region above 1000°C, a repaired carbon structure was shown, also showing bent nanotubes achieved from loss by fluorination and after repair because of the high temperature. The gas sensing showed improvement by thermal fluorination, subsequently causing electrophilic attraction; in humid condition, this improves the hydrophobicity and, for NO gases, provides adsorption sites. Therefore, gas sensor performance is enhanced and achieved by thermal fluorination of MWCNTs (Im et al. 2011b).

The characteristics of poly(dimethylsiloxane) (PDMS)-based fouling-release coatings filled with pristine and fluorinated MWCNTs are announced. The fluorination of MWCNTs was induced from a three-step method, consisting of oxidation in an acid mixture, reduction by diisobutylaluminum hydride, and silanization by (heptadecafluoro-1,1,2,2-tetrahydrodecyl) triethoxysilane. Atomic force microscopy and critical surface energy determination were investigated to characterize the surface chemistry and structure of the coatings. Both types of nanotubes showed some diversity in surface properties of the coatings. The mechanical property characterization showed no changes on both tensile strength and modulus of the pristine and fluorinated MWCNTs. The existence of the fluorinated MWCNTs on the surface of the coatings enhanced the fouling-release properties by decreasing the pseudobarnacle adhesion strength by 67% in comparison to the unfilled sample. The results raised a 47% weakening in the adhesion strength of pseudobarnacles on the coatings filled with the intact MWCNTs (Irani et al. 2013).

Synthesis of easy, low-cost, and effective fluorinated graphene with tunable C/F atomic ratio ($R_{C/F}$) has been introduced by the reaction between dispersed GO and hydrofluoric acid. The study showed that by grafting fluorine onto the basal plane of graphene, it is easy to adjust the $R_{C/F}$ by controlling the reaction conditions. The morphology of as-synthesized fluorinated graphene is like a sheet with thickness of one to two layers and an adjustable energy band gap from 1.82 to 2.99 eV, applicable in optoelectronic and photonic devices (Wang et al. 2012).

For the fluorinated CNFs, the electrochemical properties have been studied by measuring the galvanostatic discharges. By using the static method, electrochemical performances of the resulting materials can be compared and great improvements can also be shown. For the material using the direct path, the discharge potential shows 2.27 V and rises to 2.4 V for static path achieved from double gas discharge (Ahmad et al. 2013).

By annealing ND in fluorine gas, fluorinated nanodiamond (F-ND) was obtained, which, for homogenous treatment of ND particles, the fluorine gas induces periodically. Based on the results, ND particle surfaces chemically adsorbed the fluorine atoms, and finally on the surface of the F-ND, carbon fluorine bond formed and 6.4% fluorine detected. The agglomerated ND particles are separated by fluorination, and eventually, the agglomerated ND particle size reduces to tens of nm. Pristine ND stability was lower than F-ND stability in ethanol and water. The anodic electrophoretic deposition of the F-ND particles providing that the F-ND particles were negatively charged (Huang et al. 2012).

2.2.5 Carboxylation

Another chemical reaction in chemistry is called carboxylation, which is known as one of the common methods of functionalization for carbon materials. In the field of biomedicine, one of the methods to introduce peptides to polymers is performed in the presence of carboxylic groups.

Through cooperation with oxidizing inorganic acids, it will build dangling bonds on the surface of carbon materials that are continuously oxidized to hydroxyl (–OH), carbonyl (–CO), and carboxyl (–COOH) functional groups.

The biodurability of single-wall CNTs is dependent on the surface functional groups of CNTs. The characterization of single-wall CNTs with carboxylic groups is able to go through the 90-day degradation in an in vitro assay of simulation of phagolysosome, which results in decreasing length and aggregation of ultrafine solid carbonaceous debris. The same approach also shows that the unfunctionalized ozone-treated and aryl-sulfonated tubes will not degrade under the same condition.

The designated chemistry aspect of acid carboxylation introduces –COOH surface functional groups and results in collateral catastrophe to the tubular graphenic backbone that brings points of attack for more oxidative degradation. The outcome will lead to the important use of application of functionalized CNTs where biodegradability will advance safety or bring function (Liu et al. 2010b).

Multiwalled CNTs carboxylated to varying weight percentages were added to polysulfone membranes. The degree to which the CNTs were retained within the membrane during membrane production, operation, and cleaning was examined as a function of the extent of CNT carboxylation. The effects of CNTs on polymeric membranes such as increases in tensile strength, changes in surface hydrophilicity, and changes in membrane permeability were evaluated as a function of CNT carboxylation, which was found to be coupled to CNT retention within the membrane. It was found that CNTs functionalized to a higher degree form more homogeneous polymer solutions, which leads to greater improvements in the aforementioned membrane characteristics. However, CNTs functionalized to a higher degree were also found to more readily leave the membrane during immersion precipitation and membrane cleaning. Therefore, a balance was discovered between the benefits associated with increased dispersibility and hydrophilicity, and the disadvantages associated with decreased retention, increased leaching, and decreased strength of CNTs with greater carboxylation (De Lannoy et al. 2013).

The carboxylation effect of CNT on Young's modulus is conducted by applying a molecular dynamics (MD) method. The interatomic interaction in SWCNT and MWCNT with various amounts of attached carboxylic groups is modelled by a force field. Based on the results, the simulations of MD indicated the increase in –COOH groups on CNT causing the decrease of the Young modulus; the amount of decrease is lower in SWCNT compared to MWCNT (Coto et al. 2011).

Carboxyl (–COOH) covalent functionalization of nitrogen (N)- and boron (B)-doped graphene and CNT is investigated in another study (Al-Aqtash and Vasiliev 2011). The results indicated that the carboxylic group binding energy decreases with N-doping and increases with B-doping on defective CNTs and graphenes, which shows that the surface reactivity of the CNMs will differ based on the substituted doping (Al-Aqtash and Vasiliev 2011).

In the other study, a mesoporous graphene nanoball (MGB) is developed through a precursor-assisted chemical vapor deposition process from a polymer/metal precursor solution. Mesoporosity diameters of 4.27 nm and surface area of 508 m^2/g are presented in MGB; p-doped MGB also indicated a conductivity of 6.5 S/cm (Lee et al. 2013). The supercapacitors based on MGB indicated good capacitance of 206 F/g, which is a great achievement for supercapacitors using the MGB electrodes, indicating their potential for different applications (Lee et al. 2013).

The efficiency of P3HT–fullerene (C_{60}) bulk heterojunction photovoltaic cells improved by applying functionalized NDs into the photoactive layer (Lau et al. 2013). A C_{60} composite synthesized through microwave induced functionalization. The carboxylated ND indicated a 53% enhancement in short-circuit current density J_{sc} in comparison with C_{60}, which indicated both electron transport efficient, ND, and electron acceptor, C_{60} are advantages here. Based on the results, the functionalized NDs with C_{60} show improvement in performance on the polymer photovoltaic cells (Lau et al. 2013).

2.2.6 CNM Functionalization with Carbenes

Two successfully used carbene precursors to functionalize CNMs are diazirines (Ismaili et al. 2011) and chloroform (Chua et al. 2012). Chloroform can be transferred to the surface of the CNMs and form dichlorocarbene by a phase transfer catalyst (Hine 1950). A three-member heterocyclic ring called diazirine contains sp^3 azo group bonded to carbon. It releases nitrogen due to irradiation and heating and decomposes like azides and provides electron-deficient carbene components. In the absence of the reaction paths, the highly reactive carbene is prone to cycloaddition reaction for high yield of C=C (Brunner et al. 1980). One of the main diazirine compounds is 3-aryl-3(trifluoromethyl) diazirine; this is due to the absence of carbene intermolecular rearrangement (Brunner et al. 1980).

The sidewall functionalization of soluble HiPco SWCNTs by addition of dichlorocarbene is reported. The dichlorocarbene-functionalized SWCNTs [(s-SWCNT)CCl_2] in organic solvents such as THF and dichlorobenzene could retain the solubility. The dichlorocarbene functionalization degree was in the range of 12% and 23% by using various amounts of the dichlorocarbene precursor. The sidewall carbon atoms of SWCNTs saturated due to the addition of dichlorocarbene. Then delocalized partial double bonds are replaced with cyclopropanes, and there is drastic change in the SWCNT optical spectrum (Hu et al. 2003; Silva-Tapia et al. 2012).

Pristine C_{60} possesses a poor solubility in common solvents, which cause the usage limitation in many applications; this can be overcome by organic functionalization of the fullerene. C_{60} is subject to nucleophilic, radical cycloadditions as well as to carbene additions, due to the high electron deficiency of the polyene nature. Thus, derivatives of fullerene and their synthesis have been obtained through direct functionalization of C_{60} in the final stage, which allowed highly soluble C_{60} derivative preparation and also increased the scope of fullerene applications (Constant et al. 2014; Lopez et al. 2011).

Photoaffinity developed labels for proteins (Korshunova et al. 2000); diazirines have found applications in the synthesis of functional materials and surface modification. Microdiamond monocrystalline powder is used to create a robust covalent gold nanoparticle (AuNP)–diamond hybrid material with photochemical generation of a reactive carbene-modified AuNP (Ismaili and Workentin 2011).

2.2.7 CNM Functionalization via Aryne Cycloaddition

Arynes are an uncharged reactive species forming an aromatic ring, and by the abstraction of two orthosubstituents on aromatic hydrocarbon, arenes are formed. The reaction of Diels–Alder (DA) in the presence of dienes is known as a common reaction of arynes (Lu et al. 2004, 2010; Meier et al. 1998). Other common methods are to apply the functional groups that allow a direct reaction to potential initiator moiety attachment, such as the attachment of 1,3-dipolar, DA, and cycloaddition nitrene on CNTs or graphene sp^2 carbon (Ménard-Moyon et al. 2006; Munirasu et al. 2010; Nebhani and Barner-Kowollik 2010; Sarkar et al. 2011; Zydziak et al. 2011, 2013). First, $Cr(CO)_6$ in 1,4-dioxane is reacted with 3 mg of SWCNT, and the resulting chromium-SWCNTs were added to 2,3 dimethoxy-1,3-butadiene **1** in THF solution and sonicated for 5 min, followed by heating at 50°C and 1.3 GPa (Figure 2.5a, path A). Pressure was released after 60 h and visible light radiated to the mixture to induce decomplexation of chromium (Ménard-Moyon et al. 2006).

By centrifugation, CNTs are separated and, with organic solvent, washed for several times. Accordingly, a sample of SWCNTs was treated as earlier but without diene **1** (Figure 2.5a, path B) (Ménard-Moyon et al. 2006). Figure 2.5b illustrates the functionalization of C_{60} in two-step processes, the cyclopentadienyl-functionalized polyethylene glycol (PEG) monomethyl ether (3) used. First, to form the tosylated PEG monomethyl ether (2), PEG monomethyl ether (1) is tosylated. To form the cyclopentadienyl-functionalized PEG monomethyl ether (3), this is followed by tosyl group nucleophilic substitution with cyclopentadiene. Finally, for C_{60} functionalization in the absence of catalyst at ambient temperature in 5 min, the cyclopentadienyl-functionalized PEG monomethyl ether (3) is used (Nebhani and Barner-Kowollik 2010).

A cross-linkable octa-acrylate polyhedral oligomeric silsesquioxane (POSS) is coated on MWCNT, through the combination of ATRP with DA cycloadditions. Through esterification reaction from furfuryl alcohol and 2-bromoisobutyryl bromide, the furfuryl-2-bromoisobutyrate is synthesized. Then, via DA reaction, between furfuryl-2-bromoisobutyrate and MWCNTs, the MWCNT-based initiators of ATRP are synthesized. To initiate the ATRP reaction of the octa-acrylate POSS on the surface of MWCNT, the MWCNT-based initiators are used. To provide a composite with low dielectric loss and high dielectric permittivity, polyvinylidene fluoride is compounded by MWCNT coated with POSS (Zhang et al. 2014).

In another study, graphene functionalization through aryne cycloaddition reported which graphene sheets were treated with 1,2-(trimethylsilyl) phenyl triate and CsF (Zhong et al. 2010).

The functionalization degree was determined as 1 per 17 carbons from the 36.5% weight loss measured by Thermal Gravimetric Analysis (TGA), or 1 per 16 carbons from the integration of the F1 and C1 peaks by XPS.

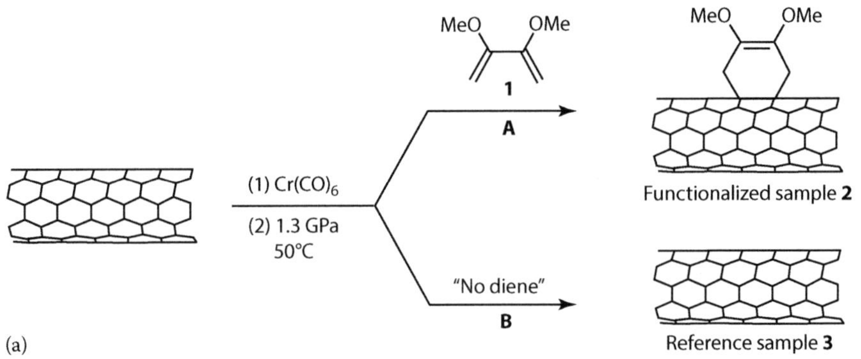

FIGURE 2.5
(a) Cycloaddition. (From Ménard-Moyon, C. et al., *J. Am. Chem. Soc.*, 128(46), 14764, 2006.)
(b) Functionalization of fullerenes (4) to via cyclopentadienyl functionalized PEG (3) synthesized
from tosylated PEG (2), which in turn is obtained via reaction of PEG (1) with p-toluenesulfonyl chloride. (From Nebhani, L. and Barner-Kowollik, C., *Macromol. Rapid. Commun.*, 31(14), 1298, 2010.)

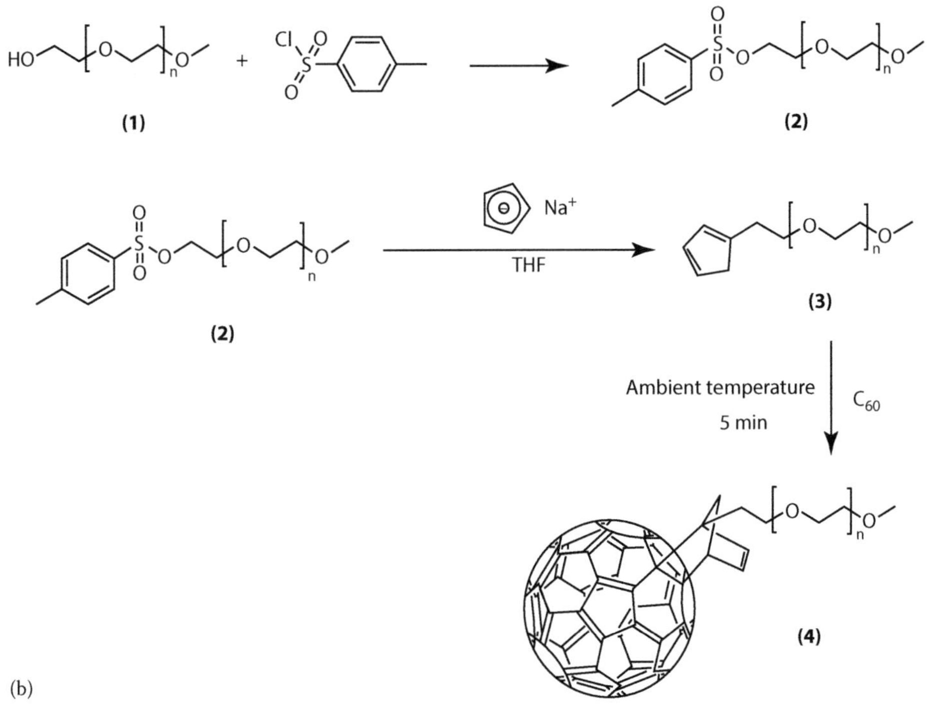

2.3 NONCOVALENT FUNCTIONALIZATION

Noncovalent CNM functionalization involves the adsorption or intercalation of molecules or atom charge transfer (Chan et al. 2013). Noncovalent functionalization is mostly used to modify work function and carrier concentration of the material (Choudhury et al. 2010; Gunes et al. 2010; Zhang et al. 2010b). This doping method moderates the concentrations and does not change much of the CNM chemical properties or band structure (i.e., band gap opening). In solution-based processes, noncovalent modification of CNM is used to improve CNM solubility (Choi et al. 2010; Kim et al. 2011; Liu et al. 2010a). Noncovalent functionalization, without changing the electronic structure, improves the solubility (Malig et al. 2013). The π−π interaction between graphene sheets results in the formation of multilayers. In nature, pristine graphene sheets are hydrophobic, and then they aren't soluble in polar solvents. Thus, it makes functionalization of graphene sheets important for their future applications (Georgakilas et al. 2012). To improve the solubility of graphene in common solvents, and prevent stacking, noncovalent functionalization with different organic compounds is essential. Noncovalent functionalization with π-interactions provides functional group attachment

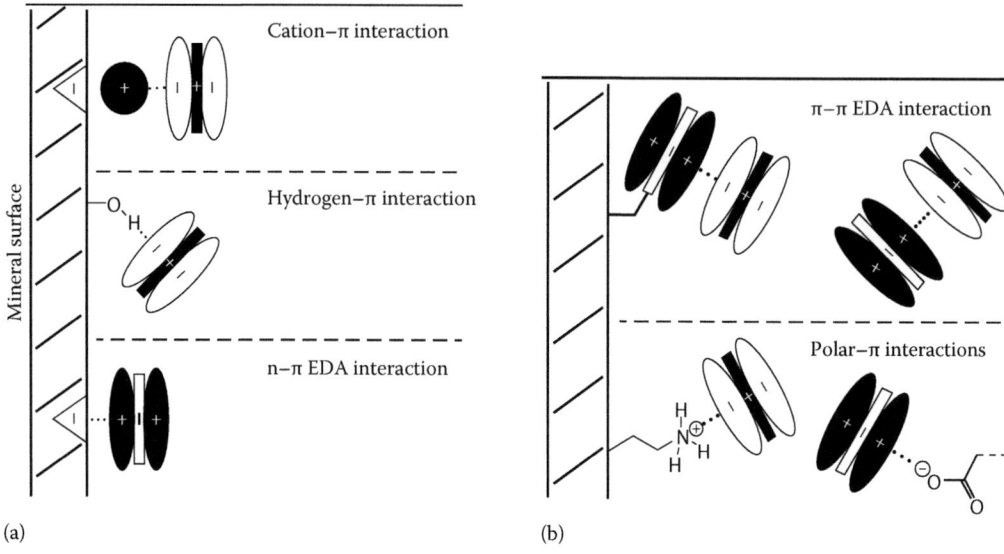

FIGURE 2.6
Schematic representation of aromatic interactions involving the π-system (indicated by doted-lines). Quadrupoles of π-donors and π-acceptors (a) Interactions with mineral surface sites. Exchangeable cations in cation–π interactions are drawn without potential hydration shell. (b) Intermolecular interactions. Polar–π interactions shown are organic cation–π (left) and n–π EDA/anion-π (right) interactions. Triangles represent permanent charge at mineral surface. (From Keiluweit, M. and Kleber, M., *Environ. Sci. Technol.*, 43(10), 3421, 2009.)

to CNM without disturbing the electronic characteristic of material, therefore making it an attractive synthetic method, especially for graphenes and CNTs (Karousis et al. 2010; Tasis et al. 2006).

In recent years, the noncovalent surface modification of CNTs, fullerenes, and graphenes by macromolecules or surfactants to provide stable suspensions of individually dispersed CNMs was broadly used in the preparation of organic solutions. A detailed discussion on the noncovalent modification of CNT sidewalls by a variety of chemical species is presented in the following paragraphs (Georgakilas et al. 2012). Figure 2.6 represents the aromatic interactions involving the π-system, which are shown by dotted lines. Quadruples of π-donors and π-acceptors are shown. Interactions with mineral surface sites are shown in Figure 2.6a; exchangeable cations in cation–π interactions are drawn without potential hydration shell and exhibited. Figure 2.6b illustrates the intermolecular interactions. Polar–π interactions illustrated are organic n–π electron donor–acceptor (EDA)/ anion–π (right) and cation–π (left) interactions (Keiluweit and Kleber 2009). Triangles represent permanent charge at mineral surface (Keiluweit and Kleber 2009).

2.3.1 CNM–Ligand Noncovalent Interactions

The structure and electronic properties of CNM, including their interaction with molecules and atoms, have been investigated broadly. Noncovalent intermolecular interactions involving π-networks are fundamental to the stabilization of proteins, enzyme–drug complexes, DNA–protein complexes, organic supramolecules, and functional nanomaterials (Hong et al. 2001a; Meyer et al. 2003). These interactions involving π-networks are most related to the field of fabrication and CNM structure of nanodevices; the subtle changes in the electronic characteristics of their π-systems could lead to dramatic effects in the characteristic, structure, and properties of the nanomaterial (Hong et al. 2001b; Lee et al. 2009; Singh et al. 2007). In recent years, extensive investigations have been conducted to comprehend the nature of π-interactions including the H–π interaction, π–π interaction, cation–π interaction, anion–π interaction, and nonpolar gas–π interaction (Kim et al. 2000; Riley et al. 2010). These π-interactions are playing the main role in graphene- and CNT-based sensors and devices (Hong et al. 2005). Extensive investigations have been conducted regarding the energetic significance of π-interactions. The π-interaction strength is determined by the combination of effects of attractive forces (such as dispersive, electrostatic, and inductive interactions) and repulsive forces. These effects illustrate the characteristic differences in directionality, physical origin, and magnitude (Morishita et al. 2010).

2.3.2 π–Cation Interaction

A noncovalent interaction between a surface of an aromatic π–donor and cation is called cation–π interaction (Figure 2.6).

The interaction between two neutral molecules tends to be much weaker when compared to neutral molecule and ion interaction, especially when the neutral molecule is easily polarizable (Bowen 1991; MoroKuma. K et al. 1977). A noncovalent π–cation interaction results when a neutral π-network interacts with closed-shell cation (Mahadevi and Sastry 2013).

Recently, there is controversy regarding the microscopic mechanism of π-bond electron-rich CNMs, such as CNT, fullerene, and graphene, and dispersion in ionic liquids (Zhao and Hu 2013). Graphene surface and alkali metal ion interaction is done in the presence and absence of an applied electric field perpendicular to the surface conducted.

Density functional theory (DFT) is used to interpret the adsorption properties. Based on the results, electric field forced the positively charged ion near to the graphene, where the charge transfer between the electron-rich graphene surface and alkali metal cations increases. The excess electrons cause a negative charge on the alkali metal ion at a certain level of electric field (Nie and Li 2013).

Through a stacked membrane composed of GO sheets, which is called a GO membrane, the selective transmembrane transport properties of alkaline and alkali earth cations are investigated. The ion transport procedure of thermodynamics declares the composition between the cation interactions and the thermal motions generated with the GO sheets resulting in temperature to penetration behavior variation to cations such as Ba^{2+}, Ca^{2+}, K^+, and Mg^{2+}. The investigated graphene and metal atom interactions are quantified based on the plane-wave-basis-set DFT approach. The selective transportation of ion transmembrane mechanism is found to be consistent with π–cation interactions in biological systems.

In the π–cation interaction, the desolation effect of ions causes the selectivity of GO membrane toward the ions' penetration, and cations are considered with the sp^2 clusters of GO membrane. GO membrane can be a promising candidate in the membrane separation through selectivity toward different ions (Sun et al. 2014).

2.3.3 π–π Interaction

π–π EDA interactions (or π–π stacking) are noncovalent interactions between π–acceptor and –donor molecules (Figure 2.6); these interaction forces could be as high as 17 kJ/mol (Salonen et al. 2011).

The capability of CNMs as a spacer to form electrochemically functionalized nanostructures is investigated onto electrodes in a manageable manner through LBL chemistry. Positively charged methylimidazolium-functionalized MWCNTs and methylene green were used as examples of electroactive components and electrochemically useful components for the assembly, respectively. Electrochemical studies illustrate that the assembled nanostructures provide great electrochemical properties and electrocatalytic activity toward the oxidation of nicotinamide adenine dinucleotide. Therefore, it could be used as bioelectronic devices such as transducers. This could be illustrated by using an alcohol dehydrogenase–based electrochemical biosensor and glucose dehydrogenase–based glucose/O_2 biofuel cell as typical examples. This investigation provides a simplified path to the manageable formation of electrochemically functionalized CNMs, which can be used for the development of molecular biodevices such as biofuel cells and biosensors (Wang et al. 2011).

2.3.4 π–Anion Interaction

In recent years, aromatic ring and anion interactions had shown more attraction in addition to hydrogen bonding and other π-interactions, such as π–cation and π–π stacking.

π–Anion interaction illustrated the attraction of anion to the centroid of the aromatic ring, while the weak σ-interaction indicated that anion is placed over the periphery of the aromatic ring (Wang and Wang 2013).

2.3.5 Polymer Composites

Researchers have attempted to incorporate CNMs into polymer composites, but most fell short of the desired results (He et al. 2007; Pötschke et al. 2002; Wang et al. 2013). Several reasons could cause for this dissatisfaction.

CNMs are inherently inert and difficult to disperse in polymer matrixes, and due to their high aspect ratio and strong intersheet and intertube van der Waals interaction, they can be easily agglomerated and entangled. The weak interfacial adhesion between CNMs and polymers also restricts the application of CNMs as reinforcement.

A homogeneous dispersion and proper interfacial adhesion between CNMs and a polymer matrix must be obtained for CNMs to be effective nanofillers in polymer composites. Surface modifications of CNMs by acid treatment and covalent functionalization were conducted to improve the interfacial adhesion and dispersion (Li et al. 2007; Zeng et al. 2006). Moreover, these modifications usually require sophisticated control of the reaction conditions and negatively impact many of the desirable properties of the CNMs due molecular structure damaging of the CNMs and the aspect ratio reduction.

Electric properties can be implemented with pyrene derivatives to maintain the CNM structure.

Mechanical strength of the $nylon_{66}$ (PA_{66})/MWCNT composite can be improved by the use of interfacial agents such as pyrene derivatives. Pyrene derivatives are noncovalently functionalized with MWCNTs by physisorption, which are mixed with PA_{66} using a twin extruder.

PA_{66} is covalently bonded with pyrene derivatives, such as 1-pyrenebutyric chloride (PBC), and adsorbed onto MWCNTs, between the amine end groups in PA_{66} and acyl chloride groups in PBC through a condensation reaction. Fabrication of PA_{66} composite matrix with PBC-adsorbed MWCNTs indicated the best interfacial adhesion and the highest level of MWCNT dispersion in PA_{66} matrix. Due to the increase in composite interfacial adhesion, the mechanical strength of the composite increases too. The dispersion of MWCNTs in PA_{66}/MWCNT–PBC composite stems from the high interfacial adhesion energy between PBC-PA_{66} molecules and PA_{66} matrix formed on MWCNTs during melt extrusion (Choi et al. 2014).

CNT dispersants are effective for high-melting polymer/CNT composites; maleimide polymers (MIPs) are synthesized using N-substituted maleimide to impart physical adsorption on the CNT surfaces and to achieve stable CNT-dispersed solutions and high heat resistance. The MIPs indicated good solubility and strong physical adsorption on different CNT surfaces in a broad range of organic solvents and acted as great CNT dispersants (Morishita et al. 2010).

The high-quality pristine graphene and metal nanostructure combination is highly expected for various applications. The combinational process of noncovalent functionalization of high-quality pristine graphene with poly(amidoamine) (PAMAM) dendrimer and the homogenous attachment of metal nanoparticles on graphene surface provided a method for large-scale production of metal nanoparticle and pristine graphene hybrid composite conductive polymer. By the absence of solvents, stable functionalized graphene nanofluid by PAMAM is provided.

Homogenous dispersion and intrinsic structure are presented in graphene sheets.

By thermal reduction method of reducing agents and using PAMAM as stabilizing homogenous dispersion of silver NP (AgNP), graphene hybrid can be obtained.

The hybrid is used as nanofiller to produce epoxy-based conductive composites for electrical interconnects. This is due to the combined effects of the high electrical conductivity of pristine graphene and enhanced contacts between the fillers by low-temperature sintering of AgNPs and at low percolation threshold (Liu et al. 2014b).

Flexible multifunctional 1-pyrenecarboxylic acid (PCA)–graphene–PDMS hybrid structures are fabricated by transferring PCA-functionalized graphene film onto PDMS substrates. The noncovalent π-interaction attachment of PCA on graphene enables to produce a range of functionalities that do not exist in pristine graphene, without sacrificing the conducting nature of graphene. Promising optical properties are indicated in hybrid structures. This is in the form of UV absorption/suppression and photodetection, as well as molecular sensing properties, which are applicable in molecular detectors and pressure sensors (An et al. 2011).

Graphene adhesive properties become related to the nanoscale regime due to van der Waals interactions (Reuven et al. 2012). One of the tools for creating the anisotropic nanopatterned systems is provided by polymer self-assembly through graphene-mediated noncovalent interactions. By unzipping MWCNTs, the supramolecular self-assembly of biofunctionally modified poly(2-methoxystyrene) on graphene nanoribbons is provided. The glycol-modified polymer promoted to self-assemble into structured nanopatterns with preserved bioactivity in this method. Van der Waals interaction enhancement and charge transfer association from polymer to graphene are attributed by self-assembly. These observations indicate that the assembly yields a prospective path to novel CNM systems (Reuven et al. 2012).

REFERENCES

Ahmad, Y. et al. 2013. Enhanced performances in primary lithium batteries of fluorinated carbon nanofibers through static fluorination. *Electrochimica Acta*, 114, 142–151. Available at: http://linkinghub.elsevier.com/retrieve/pii/S0013468613019075 [Accessed May 11, 2014].

Al-Aqtash, N. and Vasiliev, I. 2011. Ab initio study of boron- and nitrogen-doped graphene and carbon nanotubes functionalized with carboxyl groups. *The Journal of Physical Chemistry C*, 115(38), 18500–18510. Available at: http://pubs.acs.org/doi/abs/10.1021/jp206196k.

An, X. et al. 2011. Optical and sensing properties of 1-pyrenecarboxylic acid-functionalized graphene films laminated on polydimethylsiloxane membranes. *ACS Nano*, 5(2), 1003–1011.

Arai, T. et al. 2009. Zinc porphyrins covalently bound to the side walls of single-walled carbon nanotubes via flexible bonds: Photoinduced electron transfer in polar solvent. *The Journal of Physical Chemistry C*, 113(32), 14493–14499. Available at: http://pubs.acs.org/doi/abs/10.1021/jp904193f.

Bahr, J.L. and Tour, J.M. 2001. Highly functionalized carbon nanotubes using in situ generated diazonium compounds. *Chemical Materials*, (14), 3823–3824.

Banerjee, S. and Wong, S.S. 2002. Synthesis and characterization of carbon nanotube–nanocrystal heterostructures. *Nano Letters*, 2(3), 195–200. Available at: http://pubs.acs.org/doi/abs/10.1021/nl015651n.

Bekyarova, E. et al. 2009. Chemical modification of epitaxial graphene: Spontaneous grafting of aryl groups. *Journal of the American Chemical Society*, 131(4), 1336–1337. Available at: http://www.ncbi.nlm.nih.gov/pubmed/19173656.

Bhargava, P. et al. 2006. Self-assembled polystyrene-block-poly (ethylene oxide) micelle morphologies in solution. *Macromolecules*, 39(14), 4880–4888.

Billup, E. and Liang, F. 2007. Water-soluble single-wall carbon nanotubes as a platform technology for biomedical applications. USPTO Patent Number US2007/0110658, May 17, Application Number 11/516426, Filed September 6, 2006.

Bolotin, K.I. et al. 2008. Ultrahigh electron mobility in suspended graphene. *Solid State Communications*, 146(9–10), 351–355. Available at: http://linkinghub.elsevier.com/retrieve/pii/S0038109808001178 [Accessed April 28, 2014].

Boukhvalov, D.W. and Katsnelson, M.I. 2008. Chemical functionalization of graphene with defects. *Nano Letters*, 8(12), 4373–4379. Available at: http://www.ncbi.nlm.nih.gov/pubmed/19367969.

Boukhvalov, D.W. and Katsnelson, M.I. 2009. Enhancement of chemical activity in corrugated graphene. *The Journal of Physical Chemistry C*, 113(32), 14176–14178. Available at: http://pubs.acs.org/doi/abs/10.1021/jp905702e.

Bourdillon, C. et al. 1992. Immobilization of glucose oxidase on a carbon surface derivatized by electrochemical reduction of diazonium salts. *Journal of Electroanalytical Chemistry*, 336(1–2), 113–123. Available at: http://linkinghub.elsevier.com/retrieve/pii/0022072892802667.

Bowen, R.D. 1991. Ion-neutral complexes. *Accounts of Chemical Research*, 24(12), 364–371. Available at: http://pubs.acs.org/doi/abs/10.1021/ar00012a002.

Brodie, V.B.C. 2006. Ueber die Bildung der Hyperoxyde organischer Siiureradicale. *European Journal of Organic Chemistry*, 108, 79–83.

Brunner, J., Senn, H., and Richards, F.M. 1980. 3-Trifluoromethyl-3-phenyldiazirine. A new carbene generating group for photolabeling reagents. *Journal of Biological Chemistry*, 255, 3313–3318.

Cano, M. et al. 2013. Improving the mechanical properties of graphene oxide based materials by covalent attachment of polymer chains. *Carbon*, 52, 363–371. Available at: http://linkinghub.elsevier.com/retrieve/pii/S000862231200783X [Accessed May 2, 2014].

Chamssedine, F. et al. 2011. Fluorination of single walled carbon nanotubes at low temperature: Towards the reversible fluorine storage into carbon nanotubes. *Journal of Fluorine Chemistry*, 132(12), 1072–1078. Available at: http://linkinghub. elsevier.com/retrieve/pii/S0022113911001989 [Accessed May 18, 2014].

Chan, C.K. et al. 2013. Electrochemically driven covalent functionalization of graphene from fluorinated aryl iodonium salts. *The Journal of Physical Chemistry C*, 117(23), 12038–12044. Available at: http://pubs.acs.org/doi/abs/10.1021/jp311519j.

Chen, S. et al. 2011. Oxidation resistance of graphene-coated Cu and Cu/Ni alloy. *ACS Nano*, 5(2), 1321–1327. Available at: http://www.ncbi.nlm.nih.gov/pubmed/21275384.

Chen, Y. and Mitra, S. 2008. Fast microwave-assisted purification, functionalization and dispersion of multi-walled carbon nanotubes. *Journal of Nanoscience and Nanotechnology*, 8, 5770–5775.

Cheng, F. and Adronov, A. 2006. Suzuki coupling reactions for the surface functionalization of single-walled carbon nanotubes. *Chemistry of Materials*, 18(23), 5389–5391. Available at: http://pubs.acs.org/doi/abs/10.1021/cm061736j.

Choi, E.-Y. et al. 2010. Noncovalent functionalization of graphene with end-functional polymers. *Journal of Materials Chemistry*, 20, 1907–1912.

Choi, E.Y., Roh, S.C., and Kim, C.K. 2014. Noncovalent functionalization of multi-walled carbon nanotubes with pyrene-linked nylon66 for high performance nylon66/multi-walled carbon nanotube composites. *Carbon*, 72, 160–168. Available at: http://linkinghub. elsevier.com/retrieve/pii/S0008622314001201 [Accessed May 18, 2014].

Choudhury, D. et al. 2010. XPS evidence for molecular charge-transfer doping of graphene. *Chemical Physics Letters*, 497(1–3), 66–69. Available at: http://linkinghub. elsevier.com/retrieve/pii/S0009261410010390 [Accessed May 14, 2014].

Chua, C.K., Ambrosi, A., and Pumera, M. 2012. Introducing dichlorocarbene in graphene. *Chemical Communications (Cambridge, England)*, 48(43), 5376–5378. Available at: http://www.ncbi.nlm.nih.gov/pubmed/22527304 [Accessed May 18, 2014].

Coleman, K.S. et al. 2003. Functionalization of single-walled carbon nanotubes via the Bingel reaction. *Journal of the American Chemical Society*, 125(29), 8722–8723. Available at: http://www.ncbi.nlm.nih.gov/pubmed/12862456.

Constant, C. et al. 2014. Orthogonal functionalization of a fullerene building block through copper-catalyzed alkyne–azide and thiol–maleimide click reactions. *Tetrahedron*, 70(18), 3023–3029. Available at: http://linkinghub.elsevier.com/retrieve/pii/S0040402014003019 [Accessed May 18, 2014].

Coto, B. et al. 2011. Molecular dynamics study of the influence of functionalization on the elastic properties of single and multiwall carbon nanotubes. *Computational Materials Science*, 50(12), 3417–3424. Available at: http://linkinghub.elsevier. com/retrieve/pii/S0927025611003958 [Accessed May 14, 2014].

Dahoumane, S.A. et al. 2009. Protein-functionalized hairy diamond nanoparticles. *Langmuir : The ACS Journal of Surfaces and Colloids*, 25(17), 9633–9638. Available at: http://www.ncbi.nlm.nih.gov/pubmed/19634873 [Accessed May 11, 2014].

Daniel, S. et al. 2007. A review of DNA functionalized/grafted carbon nanotubes and their characterization. *Sensors and Actuators B: Chemical*, 122(2), 672–682. Available at: http://linkinghub.elsevier.com/retrieve/pii/S0925400506004527 [Accessed May 6, 2014].

Das, S.K. et al. 2012. Functionalization of diameter-sorted semiconductive SWCNTs with photosensitizing porphyrins: Syntheses and photoinduced electron transfer. *Chemistry (Weinheim an der Bergstrasse, Germany)*, 18(36), 11388–11398. Available at: http://www.ncbi.nlm.nih.gov/pubmed/22807374 [Accessed May 10, 2014].

De Lannoy, C.-F., Soyer, E., and Wiesner, M.R. 2013. Optimizing carbon nanotube-reinforced polysulfone ultrafiltration membranes through carboxylic acid functionalization. *Journal of Membrane Science*, 447, 395–402. Available at: http://linkinghub.elsevier.com/retrieve/pii/S0376738813005899 [Accessed May 13, 2014].

Downard, A.J., Roddick, A.D., and Bond, A.M. 1995. Covalent modification of carbon electrodes for voltammetric differentiation of dopamine and ascorbic acid. *Analytica Chimica Acta*, 317(1–3), 303–310. Available at: http://linkinghub.elsevier.com/retrieve/pii/0003267095003975.

Du, X. et al. 2008. Approaching ballistic transport in suspended graphene. *Nature Nanotechnology*, 3(8), 491–495. Available at: http://www.ncbi.nlm.nih.gov/pubmed/18685637 [Accessed May 2, 2014].

Elofson, R.M. and Gadallah, F.F. 1969. Substituent effects in the polarography of aromatic diazonium salts. *The Journal of Organic Chemistry*, 34, 854–857.

Engel, P.S. et al. 2008. Reaction of single-walled carbon nanotubes with organic peroxides. *The Journal of Physical Chemistry C*, 112(3), 695–700.

Englert, J.M. et al. 2011. Covalent bulk functionalization of graphene. *Nature Chemistry*, 3(April), 279–286.

Ge, J. et al. 2012. Photovoltaic properties of dimeric methanofullerenes containing hydroxyl groups. *Chemical Physics Letters*, 535, 100–105. Available at: http://linkinghub.elsevier.com/retrieve/pii/S0009261412003879 [Accessed May 17, 2014].

Georgakilas, V. et al. 2012. Functionalization of graphene: Covalent and non-covalent approaches, derivatives and applications. *Chemical Reviews*, 112(11), 6156–6214. Available at: http://www.ncbi.nlm.nih.gov/pubmed/23009634.

Giordani, S. et al. 2009. Multifunctional hybrid materials composed of [60]fullerene-based functionalized-single-walled carbon nanotubes. *Carbon*, 47(3), 578–588. Available at: http://linkinghub.elsevier.com/retrieve/pii/S0008622308005769 [Accessed April 30, 2014].

Gunes, F. et al. 2010. Layer-by-layer doping of few-layer graphene film. *ACS Nano*, 4(8), 4595–4600.

Guo, Z. et al. 2006. Covalently porphyrin-functionalized single-walled carbon nanotubes: A novel photoactive and optical limiting donor? Acceptor nanohybrid. *Journal of Materials Chemistry*, 16(29), 3021. Available at: http://xlink.rsc.org/?DOI=b602349e [Accessed May 10, 2014].

He, L. et al. 2010. Meso-meso linked diporphyrin functionalized single-walled carbon nanotubes. *Journal of Photochemistry and Photobiology A: Chemistry*, 216(1), 15–23. Available at: http://linkinghub.elsevier.com/retrieve/pii/S101060301000362X [Accessed May 10, 2014].

He, X. et al. 2007. Preparation of a carbon nanotube/carbon fiber multi-scale reinforcement by grafting multi-walled carbon nanotubes onto the fibers. *Carbon*, 45(13), 2559–2563. Available at: http://linkinghub.elsevier.com/retrieve/pii/S000862230700406X [Accessed April 28, 2014].

Higginbotham, A.L. et al. 2010. Lower-defect graphene oxide nanoribbons from multi-walled carbon nanotubes. *ACS Nano*, 4(4), 2059–2069.

Hine, J. 1950. Carbon dichloride as an intermediate in the basic hydrolysis of chloroform. A mechanism for substitution reactions at a saturated carbon atom. *Journal of the American Chemical Society*, 72(6), 2438–2445.

Hong, B.H. et al., 2001a. Self-assembled arrays of organic nanotubes with infinitely long one-dimensional H-bond chains. *Journal of American Chemical Society*, 123, 10748–10749.

Hong, B.H. et al., 2001b. Ultrathin single-crystalline silver nanowire arrays formed in an ambient solution phase. *Science*, 294, 348–351.

Hong, B.H. et al. 2005. Extracting subnanometer single shells from ultralong multiwalled carbon nanotubes. *Proceedings of the National Academy of Sciences of the United States of America*, 102(40), 14155–14158. Available at: http://www.pubmedcentral.nih.gov/articlerender.fcgi?artid=1242308&tool=pmcentrez&rendertype=abstract.

Hsiao, M.-C. et al. 2010. Preparation of covalently functionalized graphene using residual oxygen-containing functional groups. *ACS Applied Materials and Interfaces*, 2(11), 3092–3099. Available at: http://www.ncbi.nlm.nih.gov/pubmed/20949901 [Accessed May 4, 2014].

Hu, H. et al. 2003. Sidewall functionalization of single-walled carbon nanotubes by addition of dichlorocarbene. *Journal of the American Chemical Society*, 125(48), 14893–14900. Available at: http://www.ncbi.nlm.nih.gov/pubmed/14640666.

Huang, H. et al. 2012. Improvement of suspension stability and electrophoresis of nanodiamond powder by fluorination. *Applied Surface Science*, 258(8), 4079–4084. Available at: http://linkinghub.elsevier.com/retrieve/pii/S0169433211020216 [Accessed May 18, 2014].

Huang, P. et al. 2011. Graphene covalently binding aryl groups: Conductivity increases rather than decreases. *ACS Nano*, 5(10), 7945–7949. Available at: http://www.ncbi.nlm.nih.gov/pubmed/21923180.

Huang, W. et al. 2002. Attaching proteins to carbon nanotubes via diimide-activated amidation. *Nano Letters*, 2(4), 311–314.

Im, J.S. et al. 2011a. Effect of thermal fluorination on the hydrogen storage capacity of multi-walled carbon nanotubes. *International Journal of Hydrogen Energy*, 36(2), 1560–1567. Available at: http://linkinghub.elsevier.com/retrieve/pii/S0360319910021038 [Accessed May 18, 2014].

Im, J.S. et al. 2011b. Thermal fluorination effects on carbon nanotubes for preparation of a high-performance gas sensor. *Carbon*, 49(7), 2235–2244. Available at: http://linkinghub.elsevier.com/retrieve/pii/S0008622311000820 [Accessed May 18, 2014].

Irani, F., Jannesari, A., and Bastani, S. 2013. Effect of fluorination of multiwalled carbon nanotubes (MWCNTs) on the surface properties of fouling-release silicone/MWCNTs coatings. *Progress in Organic Coatings*, 76(2–3), 375–383. Available at: http://linkinghub.elsevier.com/retrieve/pii/S0300944012002950 [Accessed May 10, 2014].

Ismaili, H. et al. 2011. Light-activated covalent formation of gold nanoparticle-graphene and gold nanoparticle-glass composites. *Langmuir: The ACS Journal of Surfaces and Colloids*, 27(21), 13261–13268. Available at: http://www.ncbi.nlm.nih.gov/pubmed/21928860.

Ismaili, H. and Workentin, M.S. 2011. Covalent diamond-gold nanojewel hybrids via photochemically generated carbenes. *Chemical Communications (Cambridge, England)*, 47(27), 7788–7790. Available at: http://www.ncbi.nlm.nih.gov/pubmed/21637867 [Accessed May 2, 2014].

Jayasundara, D.R., Cullen, R.J., and Colavita, P.E. 2013. In situ and real time characterization of spontaneous grafting of aryldiazonium salts at carbon surfaces. *Chemistry of Materials*, 25(7), 1144–1152. Available at: http://pubs.acs.org/doi/abs/10.1021/cm4007537.

Karousis, N., Tagmatarchis, N., and Tasis, D. 2010. Current progress on the chemical modification of carbon nanotubes. *Chemical Reviews*, 110(9), 5366–5397. Available at: http://www.ncbi.nlm.nih.gov/pubmed/20545303.

Keiluweit, M. and Kleber, M. 2009. Molecular-level interactions in soils and sediments: The role of aromatic π-systems. *Environmental Science and Technology*, 43(10), 3421–3429.

Kharisov, B.I. et al. 2009. Recent advances on the soluble carbon nanotubes. *Industrial and Engineering Chemistry Research*, 48(2), 572–590. Available at: http://pubs.acs.org/doi/abs/10.1021/ie800694f.

Kim, H., Miura, Y., and Macosko, C.W. 2010. Graphene/polyurethane nanocomposites for improved gas barrier and electrical conductivity. *Chemistry of Materials*, 22(11), 3441–3450. Available at: http://pubs.acs.org/doi/abs/10.1021/cm100477v [Accessed May 6, 2014].

Kim, J.E. et al. 2011. Graphene oxide liquid crystals. *Angewandte Chemie (International ed. in English)*, 50(13), 3043–3047. Available at: http://www.ncbi.nlm.nih.gov/pubmed/21404395 [Accessed May 2, 2014].

Kim, K. et al. 2012. Effects of solubilizing group modification in fullerene bis-adducts on normal and inverted type polymer solar cells. *Chemistry of Materials*, 24, 2373–2381.

Kim, K.S., Tarakeshwar, P., and Lee, J.Y. 2000. Molecular clusters of pi-systems: Theoretical studies of structures, spectra, and origin of interaction energies. *Chemical Reviews*, 100(11), 4145–4186. Available at: http://www.ncbi.nlm.nih.gov/pubmed/11749343.

Kim, W. et al. 2004. Functionalization of shortened SWCNTs using esterification. *Bulletin of Korean Chemical Society*, 25(9), 1301–1302.

Korshunova, G.A. et al. 2000. Photoactivatable reagents based on aryl (trifluoromethyl) diazirines: Synthesis and application for studying nucleic acid–protein interactions. *Molecular Biology*, 34(6), 823–839.

Kotal, M., Thakur, A.K., and Bhowmick, A.K. 2013. Polyaniline–carbon nanofiber composite by a chemical grafting approach and its supercapacitor application. *ACS Applied Materials and Interfaces*, 5, 8374–8386.

Lau, X.C., Desai, C., and Mitra, S. 2013. Functionalized nanodiamond as a charge transporter in organic solar cells. *Solar Energy*, 91, 204–211. Available at: http://linkinghub.elsevier.com/retrieve/pii/S0038092X13000418 [Accessed May 18, 2014].

Lee, J.-S. et al. 2013. Chemical vapor deposition of mesoporous graphene nanoballs for supercapacitor. *ACS Nano*, 7(7), 6047–6055. Available at: http://www.ncbi.nlm.nih.gov/pubmed/23782238.

Lee, J.Y. et al. 2009. Near-field focusing and magnification through self-assembled nanoscale spherical lenses. *Nature*, 460, 498–501.

Li, L. et al. 2007. Structure and crystallization behavior of nylon 66/multi-walled carbon nanotube nanocomposites at low carbon nanotube contents. *Polymer*, 48(12), 3452–3460. Available at: http://linkinghub.elsevier.com/retrieve/pii/S0032386107003631 [Accessed May 18, 2014].

Li, X. et al. 2008. Highly conducting graphene sheets and Langmuir-Blodgett films. *Nature Nanotechnology*, 3(9), 538–542. Available at: http://www.ncbi.nlm.nih.gov/pubmed/18772914 [Accessed May 8, 2014].

Liang, F. et al. 2006. Highly exfoliated water-soluble single-walled carbon nanotubes. *Chemistry of Materials*, 18(6), 1520–1524. Available at: http://pubs.acs.org/doi/abs/10.1021/cm0526967.

Liang, J. et al. 2009. Electromagnetic interference shielding of graphene/epoxy composites. *Carbon*, 47(3), 922–925. Available at: http://linkinghub.elsevier.com/retrieve/pii/S0008622308007148 [Accessed May 10, 2014].

Lim, H. et al. 2010. Spatially resolved spontaneous reactivity of diazonium salt on edge and basal plane of graphene without surfactant and its doping effect. *Langmuir: The ACS Journal of Surfaces and Colloids*, 26(14), 12278–12284. Available at: http://www.ncbi.nlm.nih.gov/pubmed/20536169 [Accessed May 5, 2014].

Liu, G. and Freund, M.S. 1996. Nucleophilic substitution reactions of polyaniline with substituted benzenediazonium ions: A facile method for controlling the surface chemistry of conducting polymers. *Chemistry of Materials*, 4756(16), 1164–1166.

Liu, H. et al. 2009. Photochemical reactivity of graphene. *Journal of the American Chemical Society*, 131(47), 17099–17101. Available at: http://www.ncbi.nlm.nih.gov/pubmed/19902927.

Liu, J. et al. 2012. Using molecular level modification to tune the conductivity of graphene papers. *The Journal of Physical Chemistry C*, 116, 17939–17946.

Liu, J. et al., 2014a. RAFT controlled synthesis of graphene/polymer hydrogel with enhanced mechanical property for pH-controlled drug release. *European Polymer Journal*, 50, 9–17. Available at: http://linkinghub.elsevier.com/retrieve/pii/S0014305713005077 [Accessed May 9, 2014].

Liu, K. et al. 2014b. Noncovalently functionalized pristine graphene/metal nanoparticle hybrid for conductive composites. *Composites Science and Technology*, 94, 1–7. Available at: http://linkinghub.elsevier.com/retrieve/pii/S0266353814000074 [Accessed May 18, 2014].

Liu, L.-H. and Yan, M. 2011. Functionalization of pristine graphene with perfluorophenyl azides. *Journal of Materials Chemistry*, 21(10), 3273. Available at: http://xlink.rsc.org/?DOI=c0jm02765k [Accessed May 5, 2014].

Liu, S. et al. 2010a. Stable aqueous dispersion of graphene nanosheets: Noncovalent functionalization by a polymeric reducing agent and their subsequent decoration with Ag nanoparticles for enzymeless hydrogen peroxide detection. *Macromolecules*, 43(23), 10078–10083. Available at: http://pubs.acs.org/doi/abs/10.1021/ma102230m [Accessed May 16, 2014].

Liu, X., Hurt, R.H., and Kane, A.B., 2010b. Biodurability of single-walled carbon nanotubes depends on surface functionalization. *Carbon*, 48(7), 1961–1969. Available at: http://www.pubmedcentral.nih.gov/articlerender.fcgi?artid=2844903&tool=pmcentrez&rendertype=abstract [Accessed May 6, 2014].

Liu, Y. and Mccreery, R.L. 1995. Reactions of organic monolayers on carbon surfaces observed with unenhanced Raman spectroscopy. *Journal of American Chemistry Society*, 117(18), 11254–11259.

Lomeda, J.R. et al. 2008. Diazonium functionalization of surfactant-wrapped chemically converted graphene sheets. *Journal of the American Chemical Society*, 130(48), 16201–16206. Available at: http://www.ncbi.nlm.nih.gov/pubmed/18998637.

Lopez, A.M., Mateo-Alonso, A., and Prato, M. 2011. Materials chemistry of fullerene C60 derivatives. *Journal of Materials Chemistry*, 21, 1305.

Lu, X. et al. 2004. Addition of benzyne to Gd@C82. *Chemistry of Materials*, 16(6), 953–955.

Lu, X. et al. 2010. Nitrated benzyne derivatives of La@C 82: Addition of NO2 and its positional directing effect on the subsequent addition of benzynes. *Angewandte Chemie*, 122(3), 604–607. Available at: http://doi.wiley.com/10.1002/ange.200905024 [Accessed May 15, 2014].

Mahadevi, A.S. and Sastry, G.N. 2013. Cation-π interaction: Its role and relevance in chemistry, biology, and material science. *Chemical Reviews*, 113(3), 2100–2138. Available at: http://www.ncbi.nlm.nih.gov/pubmed/23145968.

Malig, J., Jux, N., and Guldi, D.M. 2013. Toward multifunctional wet chemically functionalized graphene-integration of oligomeric, molecular, and particulate building blocks that reveal photoactivity and redox activity. *Accounts of Chemical Research*, 46(1), 53–64. Available at: http://www.ncbi.nlm.nih.gov/pubmed/22916796.

Marcano, D.C. et al. 2010. Improved synthesis of graphene oxide. *ACS Nano*, 4(8), 4806–4814. Available at: http://www.ncbi.nlm.nih.gov/pubmed/21286599.

McIntosh, D., Khabashesku, V.N., and Barrera, E.V. 2007. Benzoyl peroxide initiated in situ functionalization, processing, and mechanical properties of single-walled carbon nanotube-polypropylene composite fibers. *The Journal of Physical Chemistry*, 111(4), 1592–1600.

Meier, M.S. et al. 1998. Benzyne adds across a closed 5–6 ring fusion in C 70: Evidence for bond delocalization in fullerenes. *Journal of the American Chemical Society*, 7863(14), 2337–2342.

Ménard-Moyon, C. et al. 2006. Functionalization of single-wall carbon nanotubes by tandem high-pressure/Cr(CO)6 activation of Diels-Alder cycloaddition. *Journal of the American Chemical Society*, 128(46), 14764–14765. Available at: http://www.ncbi.nlm.nih.gov/pubmed/17105260.

Meyer, E.A., Castellano, R.K., and Diederich, F. 2003. Interactions with arenes interactions with aromatic rings in chemical and biological recognition. *Angewandte Chemie International Edition, England*, 42(11), 1210–1250.

Mickelson, E.T. and Huffman, C.B. 1998. Fluorination of single-wall carbon nanotubes. *Chemical Physics Letters*, 296(1–2), 188–194.

Morishita, T. et al. 2010. Noncovalent functionalization of carbon nanotubes with maleimide polymers applicable to high-melting polymer-based composites. *Carbon*, 48(8), 2308–2316. Available at: http://linkinghub.elsevier.com/retrieve/pii/S0008622310001715 [Accessed May 18, 2014].

MoroKuma K. et al. 1977. Why do molecules interact? The origin of electron donor-acceptor complexes, hydrogen bonding, and proton affinity. *Accounts of Chemical Research*, 10, 294–300.

Munirasu, S. et al. 2010. Functionalization of carbon materials using the Diels-Alder reaction. *Macromolecular Rapid Communications*, 31(6), 574–579. Available at: http://www.ncbi.nlm.nih.gov/pubmed/21590945 [Accessed May 15, 2014].

Nebel, C.E. et al. 2009. Diamond nano-wires, a new approach towards next generation electrochemical gene sensor platforms. *Diamond and Related Materials*, 18(5–8), 910–917. Available at: http://linkinghub.elsevier.com/retrieve/pii/S092596350800561X [Accessed May 11, 2014].

Nebhani, L. and Barner-Kowollik, C. 2010. Functionalization of fullerenes with cyclopentadienyl and anthracenyl capped polymeric building blocks via Diels-Alder chemistry. *Macromolecular Rapid Communications*, 31(14), 1298–1305. Available at: http://www.ncbi.nlm.nih.gov/pubmed/21567528 [Accessed May 15, 2014].

Nie, J.C. and Li, H. 2013. Some unexpected behavior of the adsorption of alkali metal ions onto the graphene surface under the effect of external electric field. *The Journal of Physical Chemistry C*, 117(41), 21509–21515.

Niyogi, S. et al. 2010. Spectroscopy of covalently functionalized graphene. *Nano Letters*, 10(10), 4061–4066. Available at: http://www.ncbi.nlm.nih.gov/pubmed/20738114 [Accessed May 10, 2014].

Park, J. and Yan, M. 2013. Covalent functionalization of graphene with reactive intermediates. *Accounts of Chemical Research*, 46, 181–189.

Perreault, F., Tousley, M.E., and Elimelech, M. 2014. Thin-film composite polyamide membranes functionalized with biocidal graphene oxide nanosheets. *Environmental Science and Technology Letters*, 1(1), 71–76. Available at: http://pubs.acs.org/doi/abs/10.1021/ez4001356.

Piao, L. et al. 2009. The adsorption of L-phenylalanine on oxidized single-walled carbon nanotubes. *Journal of Nanoscience and Nanotechnology*, 9(2), 1394–1399. Available at: http://www.ncbi.nlm.nih.gov/pubmed/19441532.

Pinson, J. and Podvorica, F. 2005. Attachment of organic layers to conductive or semiconductive surfaces by reduction of diazonium salts. *Chemical Society Reviews*, 34(5), 429–439. Available at: http://www.ncbi.nlm.nih.gov/pubmed/15852155 [Accessed May 7, 2014].

Pötschke, P., Fornes, T.D., and Paul, D.R. 2002. Rheological behavior of multiwalled carbon nanotube/polycarbonate composites. *Polymer*, 43(11), 3247–3255. Available at: http://linkinghub.elsevier.com/retrieve/pii/S0032386102001519.

Reuven, D.G. et al. 2012. Self-assembly of biofunctional polymer on graphene nanoribbons. *ACS Nano*, 6(2), 1011–1017. Available at: http://www.ncbi.nlm.nih.gov/pubmed/22239759.

Riggs, J.E. et al. 2000. Strong luminescence of solubilized carbon nanotubes. *Journal of the American Chemical Society*, 122(14), 5879–5880.

Riley, K.E. et al. 2010. Stabilization and structure calculations for noncovalent interactions in extended molecular systems based on wave function and density functional theories. *Chemical Reviews*, 110(9), 5023–5063. Available at: http://www.ncbi.nlm.nih.gov/pubmed/20486691.

Salavagione, H.J., Gómez, M.A., and Martínez, G. 2009. Polymeric modification of graphene through esterification of graphite oxide and poly(vinyl alcohol). *Macromolecules*, 42(17), 6331–6334. Available at: http://pubs.acs.org/doi/abs/10.1021/ma900845w [Accessed May 11, 2014].

Salonen, L.M., Ellermann, M., and Diederich, F. 2011. Aromatic rings in chemical and biological recognition: Energetics and structures. *Angewandte Chemie (International ed. in English)*, 50(21), 4808–4842. Available at: http://www.ncbi.nlm.nih.gov/pubmed/21538733 [Accessed May 28, 2014].

Sarkar, S. et al. 2011. Diels-Alder chemistry of graphite and graphene: Graphene as diene and dienophile. *Journal of the American Chemical Society*, 133(10), 3324–3327. Available at: http://www.ncbi.nlm.nih.gov/pubmed/21341649.

Savchenko, A. 2009. Transforming graphene. *Science*, 323(January), 589–590.

Sayyar, S. et al. 2013. Covalently linked biocompatible graphene/polycaprolactone composites for tissue engineering. *Carbon*, 52, 296–304. Available at: http://linkinghub.elsevier.com/retrieve/pii/S0008622312007683 [Accessed August 6, 2013].

Sharma, R. et al. 2010. Anomalously large reactivity of single graphene layers and edges toward electron transfer chemistries. *Nano Letters*, 10(2), 398–405. Available at: http://www.ncbi.nlm.nih.gov/pubmed/20055430 [Accessed May 4, 2014].

Shen, J. et al. 2007. Thermo-physical properties of epoxy nanocomposites reinforced with amino-functionalized multi-walled carbon nanotubes. *Composites Part A: Applied Science and Manufacturing*, 38(5), 1331–1336. Available at: http://linkinghub.elsevier.com/retrieve/pii/S1359835X06003010 [Accessed May 13, 2014].

Silva-Tapia, A.B., García-Carmona, X., and Radovic, L.R. 2012. Similarities and differences in O2 chemisorption on graphene nanoribbon vs. carbon nanotube. *Carbon*, 50(3), 1152–1162. Available at: http://linkinghub.elsevier.com/retrieve/pii/S000862231100858X [Accessed May 1, 2014].

Singh, N.J. et al. 2007. De novo design approach based on nanorecognition toward development of functional molecules/materials and nanosensors/nanodevices. *Pure and Applied Chemistry*, 79, 1057–1075.

Sinitskii, A. et al. 2010. Kinetics of diazonium functionalization of chemically converted graphene nanoribbons. *ACS Nano*, 4(4), 1949–1954. Available at: http://www.ncbi.nlm.nih.gov/pubmed/20345149.

Snovski, R., Grinblat, J., and Margel, S. 2011. Synthesis and characterization of magnetic poly(divinyl benzene)/Fe3O4, C/Fe3O4/Fe, and C/Fe onionlike fullerene micrometer-sized particles with a narrow size distribution. *Langmuir: The ACS Journal of Surfaces and Colloids*, 27(17), 11071–11080. Available at: http://www.ncbi.nlm.nih.gov/pubmed/21806045.

Stankovich, S. et al. 2006. Graphene-based composite materials. *Nature*, 442(7100), 282–286. Available at: http://www.ncbi.nlm.nih.gov/pubmed/16855586 [Accessed September 20, 2013].

Su, S. 2008. Fluorination of organic compounds. *Baran Group Meeting*.

Sun, P. et al. 2014. Selective trans-membrane transport of alkali and alkaline earth cations through graphene oxide membranes based on cation-π interactions. *ACS Nano*, 8(1), 850–859. Available at: http://www.ncbi.nlm.nih.gov/pubmed/24401025.

Sun, Z. et al. 2010. Soluble graphene through edge-selective functionalization. *Nano Research*, 3(2), 117–125. Available at: http://link.springer.com/10.1007/s12274-010-1016-2 [Accessed May 5, 2014].

Takimoto, T. et al. 2010. Preparation of fluorescent diamond nanoparticles stably dispersed under a physiological environment through multistep organic transformations. *Chemistry of Materials*, 22(11), 3462–3471. Available at: http://pubs.acs.org/doi/abs/10.1021/cm100566v [Accessed May 12, 2014].

Tasis, D. et al. 2006. Chemistry of carbon nanotubes. *Chemical Reviews*, 106(3), 1105–1136. Available at: http://www.ncbi.nlm.nih.gov/pubmed/16522018.

Umeyama, T. et al. 2007. Electrophoretic deposition of single-walled carbon nanotubes covalently modified with bulky porphyrins on nanostructured SnO2 electrodes for photoelectrochemical devices. *The Journal of Physical Chemistry C*, 111(30), 11484–11493.

Umeyama, T. et al. 2011. Effects of fullerene encapsulation on structure and photophysical properties of porphyrin-linked single-walled carbon nanotubes. *Chemical Communications (Cambridge, England)*, 47(42), 11781–11783. Available at: http://www.ncbi.nlm.nih.gov/pubmed/21935543 [Accessed May 10, 2014].

Wang, D.-X. and Wang, M.-X. 2013. Anion-π interactions: Generality, binding strength, and structure. *Journal of the American Chemical Society*, 135(2), 892–897. Available at: http://www.ncbi.nlm.nih.gov/pubmed/23244296.

Wang, X. et al. 2011. Graphene as a spacer to layer-by-layer assemble electrochemically functionalized nanostructures for molecular bioelectronic devices. *Langmuir: The ACS Journal of Surfaces and Colloids*, 27(17), 11180–11186. Available at: http://www.ncbi.nlm.nih.gov/pubmed/21793577.

Wang, X. et al. 2013. Effect of carbon nanotube length on thermal, electrical and mechanical properties of CNT/bismaleimide composites. *Carbon*, 53, 145–152. Available at: http://linkinghub.elsevier.com/retrieve/pii/S0008622312008548 [Accessed May 6, 2014].

Wang, Y., Iqbal, Z., and Malhotra, S.V. 2005. Functionalization of carbon nanotubes with amines and enzymes. *Chemical Physics Letters*, 402(1–3), 96–101. Available at: http://linkinghub.elsevier.com/retrieve/pii/S0009261404018846 [Accessed May 13, 2014].

Wang, Z. et al. 2012. Synthesis of fluorinated graphene with tunable degree of fluorination. *Carbon*, 50(15), 5403–5410. Available at: http://linkinghub.elsevier.com/retrieve/pii/S0008622312006185 [Accessed May 5, 2014].

Wen, F. et al. 2013. Chemical bonding-induced low dielectric loss and low conductivity in high-K poly(vinylidenefluoride-trifluorethylene)/graphene nanosheets nanocomposites. *ACS Applied Materials and Interfaces*, 5(19), 9411–9420. Available at: http://www.ncbi.nlm.nih.gov/pubmed/24016800.

Wu, W. et al. 2013. Influence of functional groups on desorption of organic compounds from carbon nanotubes into water: Insight into desorption hysteresis. *Environmental Science and Technology*, 47(15), 8373–8382. Available at: http://www.ncbi.nlm.nih.gov/pubmed/23848495.

Xu, H. and Suslick, K.S. 2011. Sonochemical preparation of functionalized graphenes. *Journal of the American Chemical Society*, 133(24), 9148–9151. Available at: http://www.ncbi.nlm.nih.gov/pubmed/21604712.

Ying, Y. et al. 2003. Functionalization of carbon nanotubes by free radicals. *Organic Letters*, 5(9), 1471–1473. Available at: http://www.ncbi.nlm.nih.gov/pubmed/12713301.

Yoo, H.J. et al. 2013. Dispersion and magnetic field-induced alignment of functionalized carbon nanotubes in liquid crystals. *Synthetic Metals*, 181, 10–17. Available at: http://linkinghub.elsevier.com/retrieve/pii/S0379677913003494 [Accessed May 17, 2014].

Zeng, H. et al. 2006. In situ polymerization approach to multiwalled carbon nanotubes-reinforced nylon 1010 composites: Mechanical properties and crystallization behavior. *Polymer*, 47(1), 113–122. Available at: http://linkinghub.elsevier.com/retrieve/pii/S0032386105016393 [Accessed May 6, 2014].

Zhang, K., Kemp, K.C., and Chandra, V. 2012. Homogeneous anchoring of TiO2 nanoparticles on graphene sheets for waste water treatment. *Materials Letters*, 81, 127–130. Available at: http://linkinghub.elsevier.com/retrieve/pii/S0167577X12006416 [Accessed May 7, 2014].

Zhang W. et al. 2010a. Soft fullerene materials: click chemistry and supramolecular assemblies. *Polymer Preprints*, 51(1), 59.

Zhang, W. et al. 2014. Controlled dielectric properties of polymer composites from coating multiwalled carbon nanotubes with octa-acrylate silsesquioxane through Diels–Alder cycloaddition and atom transfer radical polymerization. *Industrial and Engineering Chemistry Research*, 53(16), 6699–6707. Available at: http://pubs.acs.org/doi/abs/10.1021/ie404204g.

Zhang, W.-B. et al. 2011. Improved synthesis of fullerynes by Fisher esterification for modular and efficient construction of fullerene polymers with high fullerene functionality. *Polymer*, 52(19), 4221–4226. Available at: http://linkinghub.elsevier.com/retrieve/pii/S003238611100601X [Accessed May 14, 2014].

Zhang, Y.-H. et al. 2010b. Tuning the electronic structure and transport properties of graphene by noncovalent functionalization: Effects of organic donor, acceptor and metal atoms. *Nanotechnology*, 21, 065201.

Zhao, Y. and Hu, Z. 2013. Graphene in ionic liquids: Collective van der Waals interaction and hindrance of self-assembly pathway. *The Journal of Physical Chemistry B*, 117(36), 10540–10547. Available at: http://www.ncbi.nlm.nih.gov/pubmed/23957744.

Zhong, X. et al. 2010. Aryne cycloaddition: Highly efficient chemical modification of graphene. *Chemical Communications (Cambridge, England)*, 46(39), 7340–7342. Available at: http://www.ncbi.nlm.nih.gov/pubmed/20820532 [Accessed May 15, 2014].

Zhong, Y.L. et al. 2008. Suzuki coupling of aryl organics on diamond. *Chemistry of Materials*, 20(9), 3137–3144. Available at: http://pubs.acs.org/doi/abs/10.1021/cm703686w.

Zhou, L. et al. 2014. Preparation and characterization of magnetic porous carbon microspheres for removal of methylene blue by a heterogeneous Fenton reaction. *ACS Applied Materials and Interfaces*, 6(10), 7275–7285. Available at: http://www.ncbi.nlm.nih.gov/pubmed/24731240.

Zydziak, N. et al. 2011. One-step functionalization of single-walled carbon nanotubes (SWCNTs) with cyclopentadienyl-capped macromolecules via diels à alder chemistry. *Macromolecules*, 44(9), 3374–3380.

Zydziak, N., Yameen, B., and Barner-Kowollik, C. 2013. Diels–Alder reactions for carbon material synthesis and surface functionalization. *Polymer Chemistry*, 4(15), 4072. Available at: http://xlink.rsc.org/?DOI=c3py00232b [Accessed May 15, 2014].

CHAPTER 3

CONTENTS

Role of Top and Interlayer Metal Nanoparticle Grafting on CNTs

Improved Raman Scattering and Electron Emission Investigations

3

Himani Sharma, A.K. Shukla, and V.D. Vankar

3.1 INTRODUCTION

Carbon is a very important element in the periodic table that essentially forms the basis of life on Earth. Elemental carbon exists in two natural allotropes, diamond and graphite, having sp^3- and sp^2-hybridized carbon atoms, respectively. Both these allotropes of carbon represented the carbon family for a long time, until the discovery of fullerene in 1985.[1,2]

The serendipitous discovery of fullerene, also known as buckminsterfullerene, by Kroto and Smalley[2] marked the beginning of an era of synthetic carbon allotropes from the naturally occurring diamond and graphite. Soon after the discovery of 0D fullerene, the synthetic carbon family has been graced by the addition of quasi-1D carbon nanotubes (CNTs), whose discovery in 1991 by Iijima[3] created a boom in the scientific world. Their large length (up to several microns) and small diameter (a few nanometers) result in a large aspect ratio. They can be seen as the nearly 1D form of fullerenes. Therefore, these materials are expected to possess additional interesting electronic, mechanical, and molecular properties. Especially in the beginning, all theoretical studies on CNTs focused on the influence of the nearly 1D structure on molecular and electronic properties. More recently, the discovery of graphene by Novoselov and Geim[4] made it a flagship material harbingering the age of nanotechnology. Graphene is also advertised as a mother of all synthetic carbon allotropes with its 2D structure.[5] The arrangement of carbon atoms in each crystal is the fundamental difference between these various structures.[6]

These revolutionary allotropes of carbon represent an attractive research field in the nanomaterials sector with important economic potential. The most popular nanomaterials of the carbon family to date are fullerene, CNTs, and graphene with dimensions ranging from 0.5 to 100 nm. These nanostructures exhibit remarkable electrical, optical mechanical, chemical, and thermal properties due to their unique and intriguing size and structure that make them a potential candidate for various applications.

3.2 CARBON NANOSTRUCTURES: A CHRONOLOGICAL ASSESSMENT

Carbon nanostructures (CNs) are regarded as artificially composed structures with the nanometer size. Their properties are the subject of both theoretical and experimental investigation.[6] The CNs span an astounding range of extremes, considering that they are all merely structural formations of the same element. The various CNs are fullerene,[7] CNTs,[8] graphene, graphene nanoflakes (GNFs),[5,9] etc. The chronological presentation of CNs is shown in Figure 3.1.[6]

3.2.1 Carbon Nanotubes

CNTs can be described as the seamless cylinder of a rolled-up hexagonal network of carbon atoms, which is capped with half a fullerene molecule at the end. The uniqueness of the nanotube arises from its structure and the inherent subtlety in the structure, which is the helicity (local symmetry) in the arrangement of the carbon atoms in hexagonal arrays on their surface honeycomb lattices. The helicity, along with the diameter (which determines the size of the repeating structural unit), introduces significant changes in the electronic density of states and hence provides a unique electronic character for the nanotubes. This novel electronic property creates a range of fascinating electronic device applications and has been the subject of discussion.

The other factor of importance that determines the uniqueness in physical properties is topology, or the closed nature of individual nanotube shells; when individual layers are closed on to themselves, certain aspects of the anisotropic properties of graphite disappear, making the structure

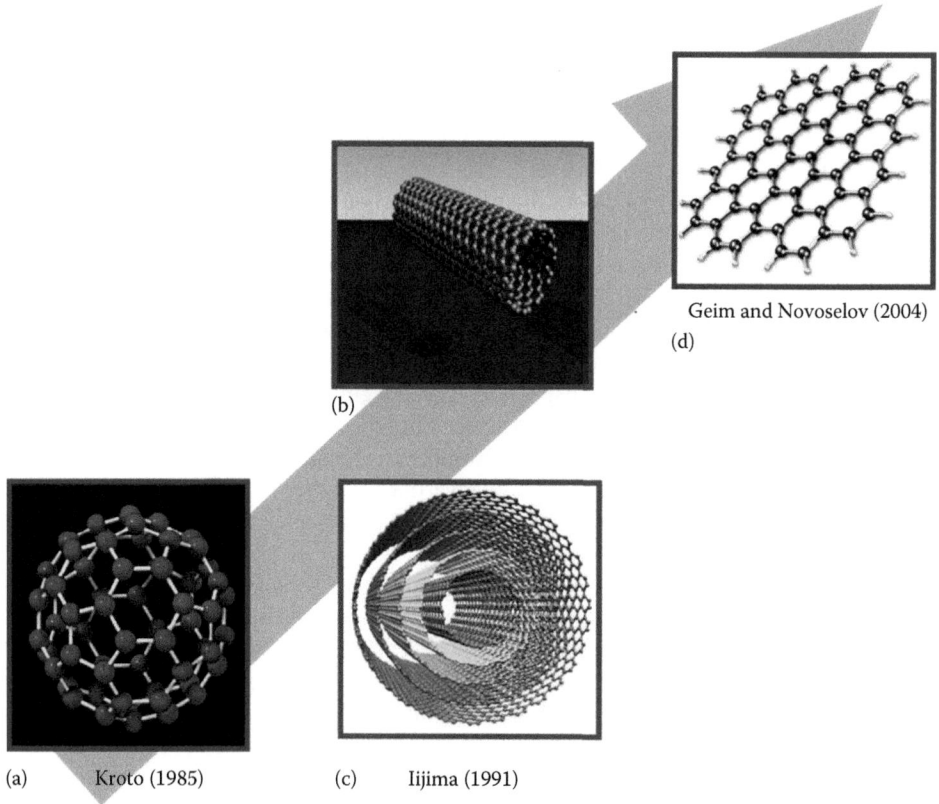

Geim and Novoselov (2004)

(d)

(b)

(a)   Kroto (1985)   (c)   Iijima (1991)

FIGURE 3.1
Synthetic allotropes of carbon that include (a) fullerene, (b) SWCNTs, (c) MWCNTs, and (d) graphene.

remarkably different from graphite. The combination of size, structure, and topology endows nanotubes with important structural and electrical properties combined with special surface properties, and the applications based on these properties form the central topic of this chapter. In addition to the helical lattice structure and closed topology, topological defects in nanotubes (five-member Stone–Wales defects near the tube ends, aiding in their closure), akin to those found in the fullerene structures, result in local perturbations to their electronic structure, for example, the ends or caps of the nanotubes are more metallic than the cylinders, due to the concentration of pentagonal defects.[10,11] These defects also enhance the reactivity of tube ends, giving the possibility of opening the tubes, functionalizing the tube ends, and filling the tubes with foreign substances.[12–16] Among the nanotubes, two varieties, which differ in the arrangement of their graphene cylinders, share the limelight, single-walled CNTs (SWCNTs) and multiwalled CNTs (MWCNTs).

SWCNTs can be considered as long, wrapped graphene sheets with an aspect ratio (length to diameter ratio) of about 1000 so they can be considered as nearly 1D structures with a typical diameter of about 1.4 nm that is similar to a buckyball.[17] They have a tendency to form in bundles that are parallel in contact and consist of tens to hundreds of nanotubes.[18] CNTs are generally metallic in nature, but SWCNTs can be both metallic and semiconducting depending upon their chirality.

MWCNTs are closed graphite tubules rolled like a graphite sheet with a higher degree of structural perfection. They are made of concentric cylinders placed around a common central hollow. The diameters of MWCNTs usually range between 2 and 50 nm, and the interlayer spacing (d) between sheets is about 0.34 nm.[19] This interlayer spacing in MWNTs is slightly larger than the single-crystal graphite value (0.335 nm) since in these tubes there is a severe geometrical constraint when forming the concentric seamless cylinders while maintaining the graphite spacing between them.[20] The 3D structural correlation that prevails in single-crystal graphite (ABAB stacking) is lost in the nanotubes, and the layers are rotationally disordered with respect to each other.

3.2.2 Graphene/Graphene Nanoflakes

Graphene is a rising star on the horizons of material science and condensed matter physics.[4,5] Graphene is the monolayer of carbon atoms that is tightly packed into a 2D honeycomb lattice. It is the basic building block for graphitic materials of all other dimensionalities and is projected as the mother of all carbon allotropes.[5] It can be wrapped up into 0D fullerenes, rolled into 1D nanotubes, or stacked into 3D graphite. Graphene possesses higher structure perfection due to continuous 2D stacking of carbon atoms with a nearest neighbor distance of 1.42 Å. The crystal perfection in graphene causes the ballistic transport of charge carriers at submicrometer distances.[4] Thus, it exhibits a metallic character with zero bandgap and is often termed as zero-gap semiconductor or zero-overlap semimetal.

The electronic properties of graphene change with the number of graphitic layers and the relative position of atoms in adjacent layers (stacking order). The limit of thickness at which graphene can still be considered as such can be determined by the rapid change in the electronic structure as the number of layers increase. As the number of layers increases, the band structure becomes more complicated as several charge carriers appear.[21] Therefore, three different types of pseudo-2D crystals are distinguished: graphene, double, and few-layer (3 to <10) graphene. More than 10 layers of graphene can be referred to as multilayer graphene or GNFs, depending on the edge thickness. In GNFs, high density of open edges and high surface area further increase their chemical stability.[22] The GNF structure comprises unfolded graphene layers. The edges of CNF may also act as sharp tips for field emission purpose.[23]

Thicker structures can be considered as thin films or slabs of graphite.[5] The stacking order or disorder has been found to dramatically change the electronic properties of multilayer graphene, introducing Dirac fermions due to symmetry breaking, even in graphite crystals to form graphene stacks.[24,25] All these nanostructures (SWCNTs, MWCNTs, graphene, and GNFs) of the carbon family are referred to as CNs. These CNs exhibit different physical properties due to difference in the arrangement of carbon atoms in a particular structure. The properties of these CNs are provided in Table 3.1.[26,27]

TABLE 3.1 Properties of Carbon Nanostructures

Properties	SWCNTs	MWCNTs	Graphene Nanoflakes
Average diameter and thickness (nm)	1.2–1.4 nm	10–50 nm	2–30 nm
Maximum current density (A/cm^2)	10^{13}	10^9	10^7
Bandgap (eV)	0 eV (metallic) 0.5 (semiconducting)	0	0 (can be modified)
Work function (eV)	5.05	4.9	4.8
Young's modulus (TPa)	1	1.28	—
Tensile strength (GPa)	100	150	
Density (g/cm^3)	1.33	2.5	3.6
Thermal conductivity (W/cmK)	58	30	28

Source: Sharma, H., Growth, structure and electron emission characteristics of carbon nanostructures synthesized by microwave plasma enhanced chemical vapour deposition process, PhD dissertation, Indian Institute of Technology, Delhi, India, 2012.

3.3 PROCESSING OF CARBON NANOSTRUCTURES

The state-of-the-art CN production encompasses numerous methods and new routes are continuously being developed. Essentially, nanostructures are formed in the same way, but the process that causes the formation differs. The various methods used for the processing of CNs are arc discharge, laser ablation, and chemical vapor deposition (CVD).[28,29]

Out of these methods, the most attractive and commercially used is the CVD method.[30] As compared to arc discharge and laser ablation methods, CVD is a simple and economic technique for synthesizing CNs at low temperature and ambient pressure. The crystalline quality of CNs grown by the arc discharge and laser ablation methods is superior to the CVD-grown CNs. However, in yield and purity, CVD overcomes the arc and laser methods.[31] A brief overview of these techniques is mentioned in Table 3.2. However, CVD provides better control over CNs in comparison to other techniques. It is versatile in the sense that it offers harnessing plenty of hydrocarbons in any state (solid, liquid, or gas), enables the use of various substrates, and allows CNT growth in a variety of forms, such as powder, thin or thick films, aligned or entangled, straight or coiled nanotubes, or a desired architecture of nanotubes on predefined sites of a patterned substrate. It also offers better control on the growth parameters. The relative ease with which one can set up a CVD system also makes it the most promising route for the mass production of CNs. In order to use CNs in novel devices, it is necessary to produce these materials with a high crystallinity in a large scale at economic costs.

3.3.1 Chemical Vapor Deposition

CVD is a chemical process in which volatile precursors are used to provide a carbon feed source to a catalyst particle at elevated temperatures (350°C–1000°C). In a CVD reaction, the catalyst particles can reside in free space, the so-called floating catalysts, or they can reside on a substrate (supported catalysts). In such process, carbon saturates on the catalyst particles and precipitates in the form of a CN and is discussed in much detail later.

Depending on the energy sources, various CVD-based methods are used for the growth of CNs. Few of them are thermal CVD, plasma-enhanced CVD (PECVD), and hybrid laser-assisted thermal CVD (LCVD).[23,32,33] These various forms of CVD in wide use make it the most common route by which CNs are formed.

It has been shown that CVD is amenable for CNTs and other nanostructures, suitable for fabrication of electronic devices, sensors, field emitters, and other applications where controlled growth over masked areas is needed for further processing.[34,35] Out of all CVD techniques, PECVD has been investigated for its ability to produce vertically aligned nanotubes. PECVD uses energy to

TABLE 3.2 Comparison of Various Growth Techniques to Form Carbon Nanostructures

Methods	Arc Discharge	Chemical Vapor Deposition	Laser Ablation
Pioneer	Iijima (1991)	Yacaman et al. (1993)	Smalley (1995)
Setup	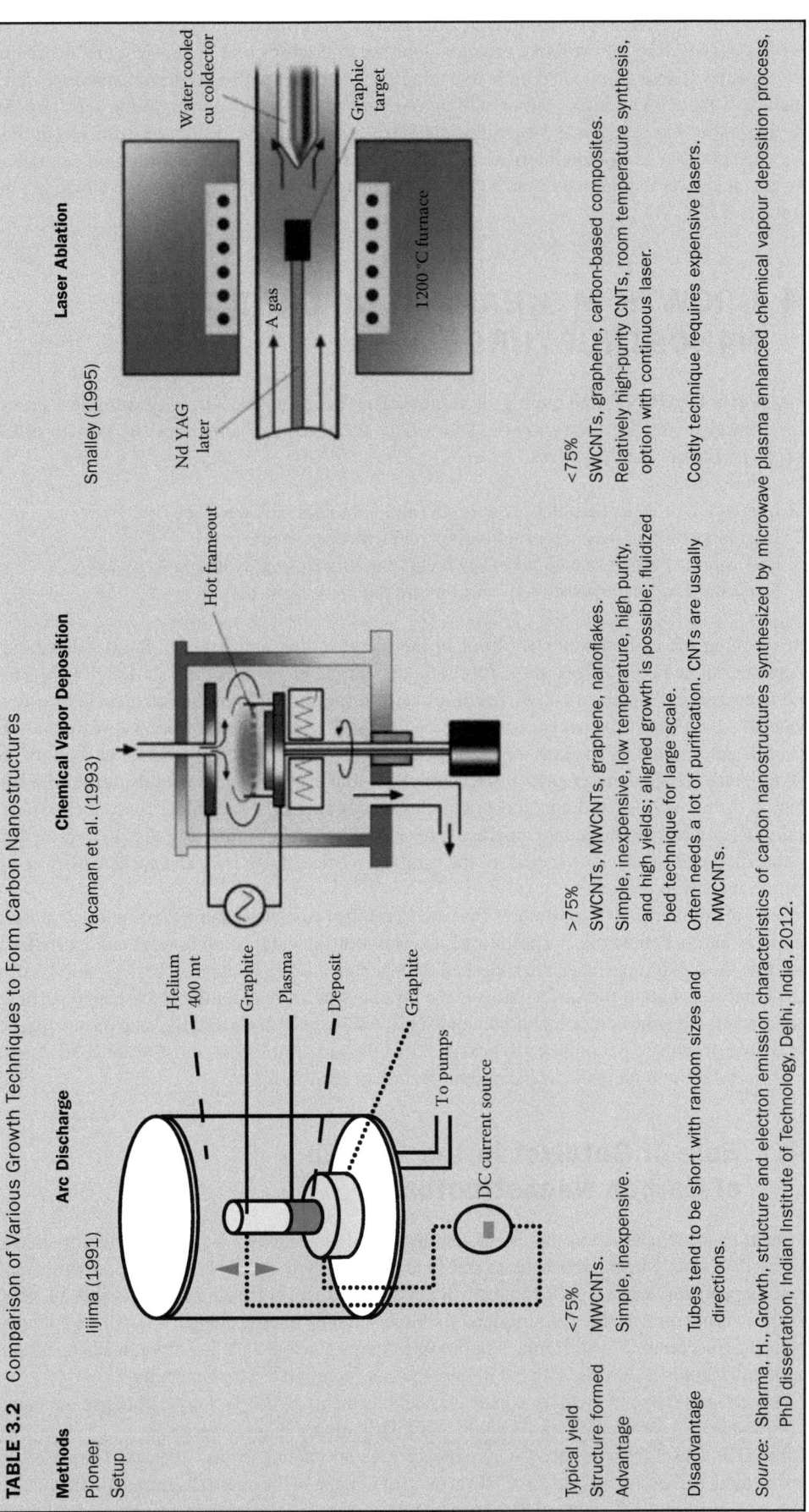		
Typical yield	<75%	>75%	<75%
Structure formed	MWCNTs.	SWCNTs, MWCNTs, graphene, nanoflakes.	SWCNTs, graphene, carbon-based composites.
Advantage	Simple, inexpensive.	Simple, inexpensive, low temperature, high purity, and high yields; aligned growth is possible; fluidized bed technique for large scale.	Relatively high-purity CNTs, room temperature synthesis, option with continuous laser.
Disadvantage	Tubes tend to be short with random sizes and directions.	Often needs a lot of purification. CNTs are usually MWCNTs.	Costly technique requires expensive lasers.

Source: Sharma, H., Growth, structure and electron emission characteristics of carbon nanostructures synthesized by microwave plasma enhanced chemical vapour deposition process, PhD dissertation, Indian Institute of Technology, Delhi, India, 2012.

generate a glow discharge (plasma) in which the energy is transferred into a gas mixture. This transforms the gas mixture into reactive radicals, ions, neutral atoms and molecules, and other highly excited species. These atomic and molecular fragments interact with a substrate, and depending on the nature of these interactions, either etching or deposition processes occur at the substrate. Since the formation of the reactive and energetic species in the gas phase occurs by collision in the gas phase, the substrate can be maintained at a low temperature. Hence, film formation can occur on substrates at a lower temperature than is possible in the conventional CVD process, which is a major advantage of PECVD.

3.4 GROWTH MECHANISM OF CARBON NANOSTRUCTURES

The growth of CNs (CNTs and multilayer graphene) has been topical. Various groups have proposed different mechanisms for their growth. The widely accepted mechanism for the growth of CNTs proposed by Baker et al.[36] involves

1. Interaction of hydrocarbon precursor gas and hot metal nanoparticles
2. Decomposition of precursor into carbon and hydrogen species
3. Diffusion of carbon in and out of the metal particle causing hydrogen to fly away
4. Nucleation and incorporation of carbon into the growing structure

After reaching the carbon-solubility limit in the metal at that temperature, the dissolved carbon precipitates out and crystallizes in the form of a cylindrical network having no dangling bonds and hence is energetically stable. The diffusion of carbon species through the nanoparticle on the hot surface on which hydrocarbons occurs to the cooler rear faces at which carbon is precipitated from solution is determined as the key step in this mechanism. The driving force for carbon diffusion is the temperature gradient created in the particle by the exothermic decomposition of the hydrocarbon at the hot surface and endothermic deposition of carbon at the colder rear faces, which are initially in contact with the support surface. This allows the hollow structure (CNT) to grow having the same diameter as the cross section of the catalytic particle. Both bulk and surface diffusion play an important role in growth.

The growth mechanism postulates that metal catalyst particles can be floating or supported on graphite or another substrate.[37] The interaction between the catalyst and metal nanoparticles presumes that the catalyst particles are spherical or pear shaped.[38] In that case, the deposition will take place on only one half of the surface (this is the lower curvature side for the pear-shaped particles).[38] Depending on the adhesion of metal nanoparticles with the substrate and the size of nanoparticles, tip and base growth of nanotubes are possible. For the growth of CNs, most of the CVD methods employ acetylene and methane as the carbon precursor feedstock.

3.4.1 Role of Catalyst in the Growth of Carbon Nanostructures

The metal nanoparticles act as nanoscale templates and hence play a significant role for nucleation and growth of CNs. Metal nanoparticles stabilize the growth of the CNTs against premature closure caused by pentagon formation on the growth edge. This effect is based on the fact that a C atom at a hexagon corner on the edge has significantly larger binding energy compared with that of an atom at the pentagon corner. This difference in binding energies, about 1 eV per atom, leads to easier dissolution of C atoms if they are located in pentagons as compared with hexagons.[39]

The size and shape of metal nanoparticles defines the final morphology and hence the diameter of CNs. Figure 3.2 shows the size variation and distribution of nanoparticles.

Iron (Fe), nickel (Ni), or cobalt (Co) nanoparticles are often used as catalysts. The rationale for choosing these metals as catalysts for CVD growth of CNs lie in the phase diagrams for the metals and carbon. Solubility and diffusivity of carbon in metal lattice are two essential parameters in determining

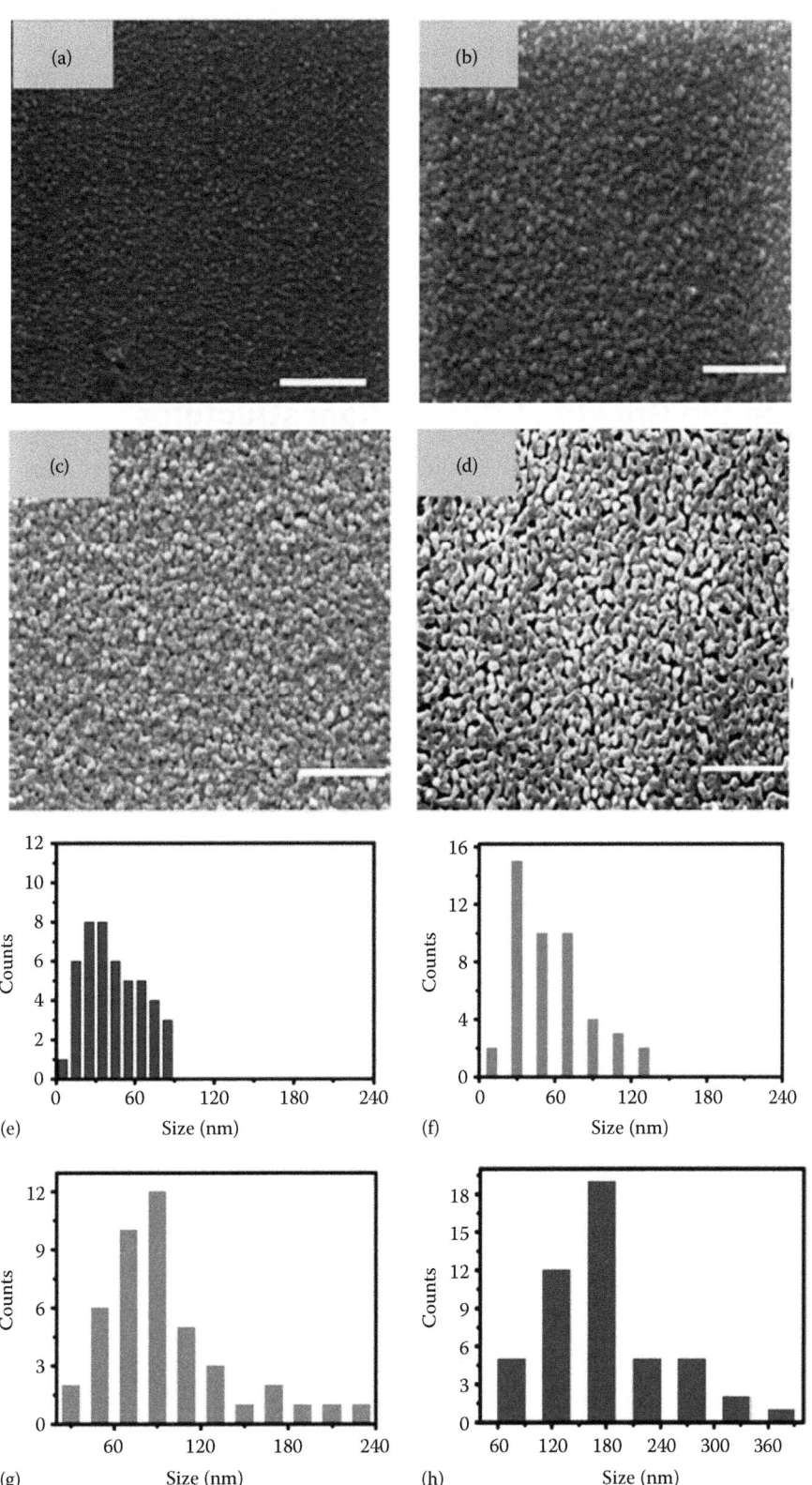

FIGURE 3.2
SEM micrographs and statistical distribution of Fe nanoparticles with catalyst thickness: (a and e) 5 nm, (b and f) 10 nm, (c and g) 20 nm, and (d and h) 50 nm. The scale bar corresponds to 1 μm. (Reprinted from *Mater. Chem. Phys.*, 137, Sharma, H., Shukla, A.K., and Vankar, V.D., Influence of Fe nanoparticles diameters on the structure and electron emission studies of carbon nanotubes and multilayer graphene, 802, Copyright 2013, with permission from Elsevier.)

the tube growth. At high temperatures, carbon has a finite solubility with these metals that leads to the formation of metal carbon solid-state solutions and therefore to the aforementioned growth. According to the binary phase diagrams of carbon and metal, the eutectic point (lowest melting point) and the carbon content at the point for various kinds of carbon–metal alloy are as follows: C–Ni (1326°C, 8.0 wt%), C–Fe (1153°C, 4.2 wt%), and C–Co (1320°C, 2.6 wt%). Carbon atoms diffuse in the metal lattices through the interstitial sites and through vacancies depending on the sufficient temperature to activate them.[40] Interstitial diffusion proceeds faster along specific crystallographic planes. Further diffusion is faster through the surface than through bulk.[41] On the other hand, diffusion of carbon in the carbide phase (Ni_3C and Fe_3C) is known to be a slower process than diffusion of carbon in pure metal. Here, iron (Fe) is used as catalyst and its thickness is varied between 5 and 50 nm.

3.4.2 Effect of Catalyst Nanoparticle Diameters on the Growth of Carbon Nanostructures

The size of the nanoparticles varies from 10 to 180 nm in different films, as seen in Figure 3.2a through c. However, in Figure 3.2d, nanoclusters of ~250–300 nm are formed, owing to the large thickness of the Fe film. A distribution in the density and size of nanoparticles can be seen in Figure 3.2e through h. Scanning electron microscope images of CNs, grown on different-sized Fe nanoparticles, are shown in Figure 3.3. Figure 3.3a shows the dense growth

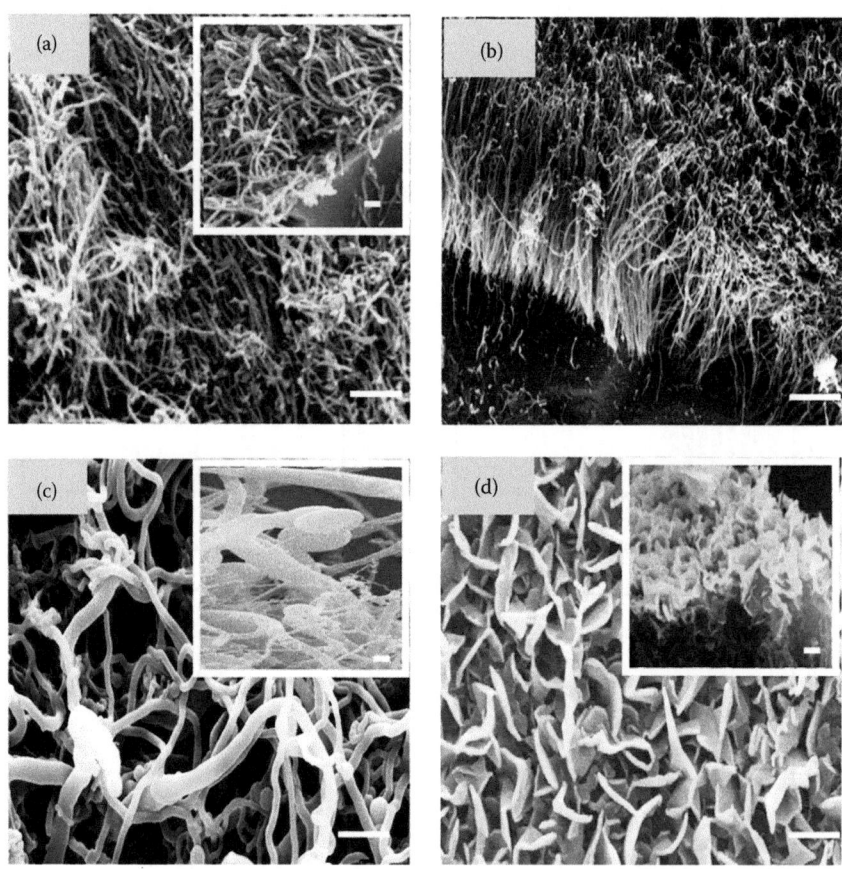

FIGURE 3.3
SEM micrographs showing the variation in the morphology of CNs of (a) a 5 nm sample, inset image showing the cross-sectional view of CNTs; (b) 10 nm sample showing vertically aligned CNTs; (c) 20 nm sample with noodle-shaped CNTs, inset image showing the cross-sectional view of the sample; and (d) 50 nm sample showing the formation of graphene flakes, inset image showing the vertically aligned graphene flakes. The scale bars correspond to 1 mm. (Reprinted from *Mater. Chem. Phys.*, 137, Sharma, H., Shukla, A.K., and Vankar, V.D., Influence of Fe nanoparticles diameters on the structure and electron emission studies of carbon nanotubes and multilayer graphene, 802, Copyright 2013, with permission from Elsevier.)

of CNTs, depending upon the distribution of Fe nanoparticles. However, vertically aligned growth of CNTs can be seen in Figure 3.3b. Figure 3.3c shows the entangled growth of CNTs with an increase in Fe nanoparticle diameters. A 2D growth is seen in Figure 3.3d that corresponds to Multi layered graphene (MLG) structures and is attributed to the condensation of carbon vapors to form a sheetlike morphology. It is seen that as the size and shape of the particles vary, morphology and hence microstructure are also different.[42] This is further confirmed by high-resolution transmission electron microscope (HRTEM), shown in Figure 3.4. Figure 3.4a through d show the variation in the number of walls and diameters of CNs and the size of Fe nanoparticles. A nanoparticle remaining at the CNT base provides the curvature for budding tubes. Cylindrical nanoparticles are able to form CNTs with straight walls, whereas on conical nanoparticles, conical CNT nuclei tend to form leading to the growth of *bamboo-like* CNTs. The formation of bamboo-like structures in the CVD process may be attributed to the periodic closure and subsequent nucleation of successive conical layers.[43,44] A schematic illustration for the growth mechanism is shown in Figure 3.5. Depending on the shape and size of the catalyst particles and on the distribution of more favorable nucleation sites at the metal/graphitic interface, it is reasonably assumed that the growth of the graphitic monolayers proceeds via island nucleation and coalescence, according to the Volmer–Weber model.[40] Thus, depending on the size of catalyst particles, different CNs such as SWCNTs, MWCNTs, nanoflakes, and MLGs are grown.

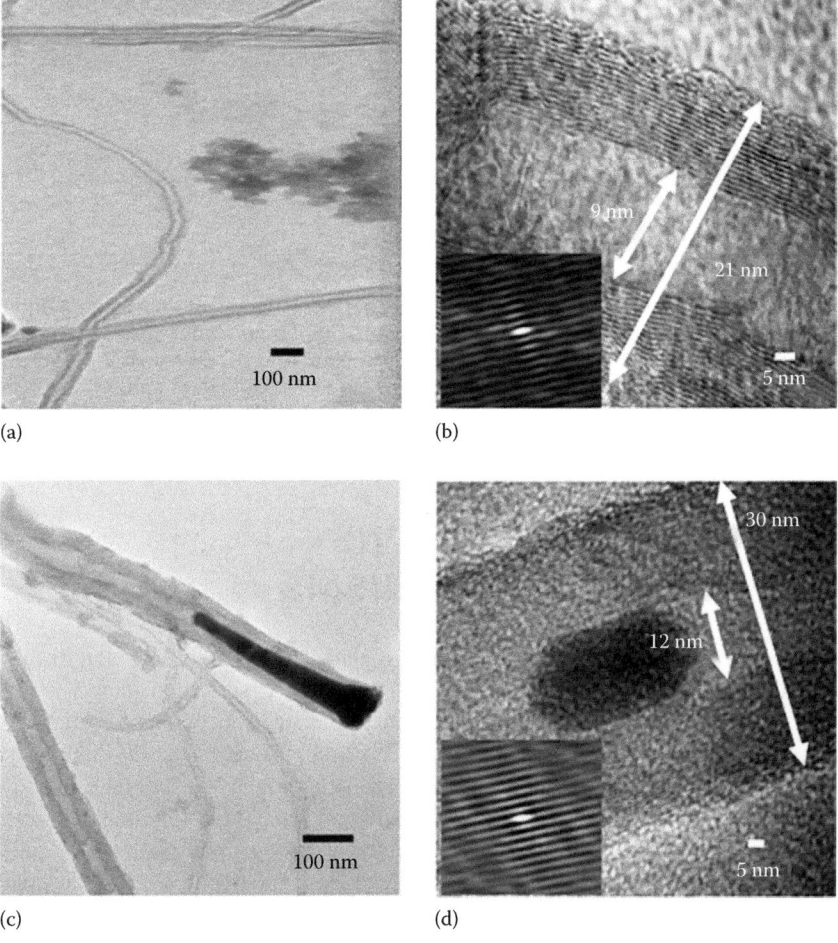

(a)   (b)

(c)   (d)

FIGURE 3.4
HRTEM micrographs and corresponding lattice planes of CNs: (a and b) 5 nm and (c and d) 10 nm. The inset images are the autocorrelated images showing the lattice planes for each sample.

(Continued)

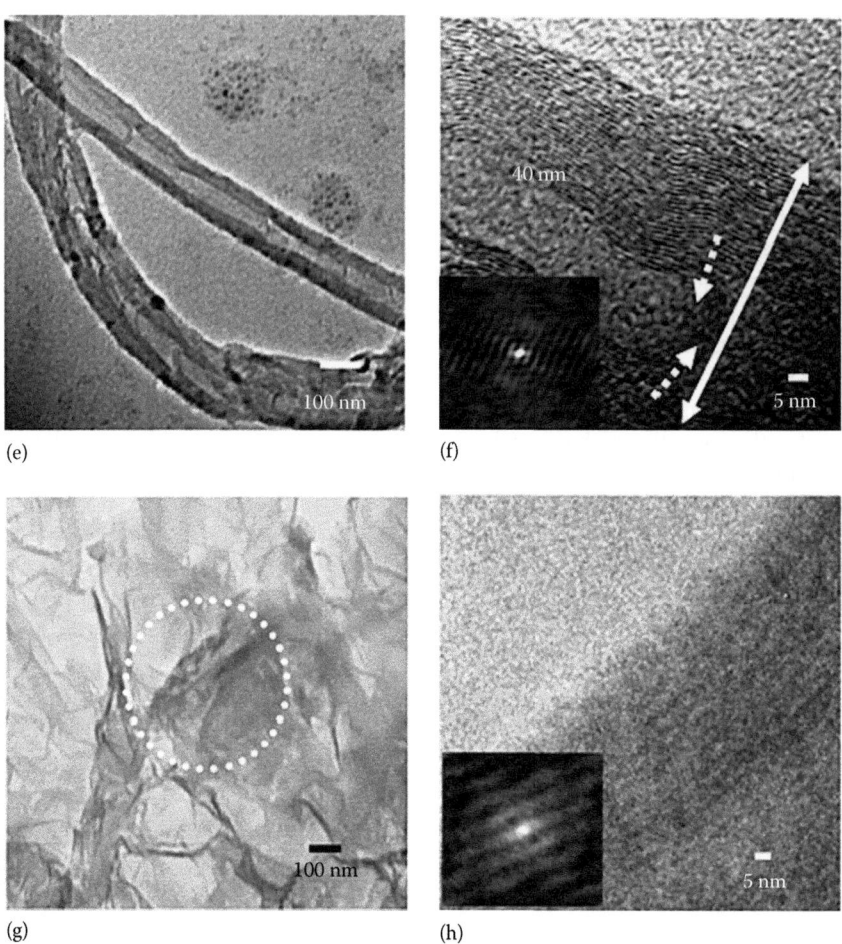

(e)　　(f)

(g)　　(h)

FIGURE 3.4 (*Continued*)
HRTEM micrographs and corresponding lattice planes of CNs: (e and f) 20 nm, and (g and h) 50 nm. The inset images are the autocorrelated images showing the lattice planes for each sample. (Reprinted from *Mater. Chem. Phys.*, 137, Sharma, H., Shukla, A.K., and Vankar, V.D., Influence of Fe nanoparticles diameters on the structure and electron emission studies of carbon nanotubes and multilayer graphene, 802, Copyright 2013, with permission from Elsevier.)

3.5 POTENTIAL APPLICATIONS AND RELATED MECHANISMS

CNs are considered as best contenders in the fields of nanoelectronic, sensing, and display devices. The right combination of properties—nanometer size diameter, structural integrity, high electrical conductivity, and chemical stability—make them applicable for these applications. Flat panel displays are one of the important applications of CNs and particularly CNTs that have got most attention. Samsung (FED lab) has demonstrated several generations of prototypes, including a 9 in. red–blue–green color display that can reproduce moving images.[45] Nanotube-based lamps are similar to displays in comprising a nanotube-coated surface opposing a phosphor-coated substrate, but they are less technically challenging and require much less investment. High-performance prototypes seem suitable for early commercialization, having a lifetime of 8000 h, the high efficiency (for green phosphors) of environmentally problematic mercury-based fluorescent lamps, and the luminance required for large stadium-style displays. Other than emission CNs are explored for gas and molecular sensors. The first realized major commercial application of MWNTs is their use

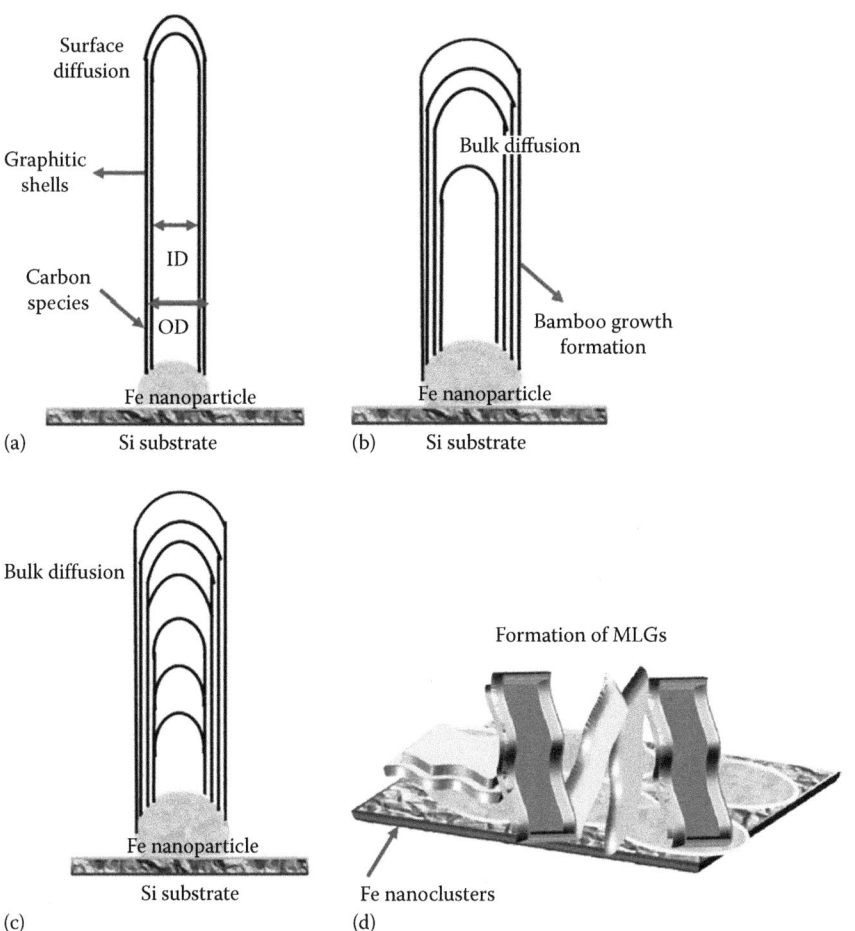

FIGURE 3.5
Schematic model showing the growth of CNs: (a) hollow CNTs, (b) hollow and bamboo-shaped CNTs, (c) bamboo-shaped CNTs, and (d) GNFs, dependent on the size of Fe nanoparticles/nanoclusters. (Reprinted from *Mater. Chem. Phys.*, 137, Sharma, H., Shukla, A.K., and Vankar, V.D., Influence of Fe nanoparticles diameters on the structure and electron emission studies of carbon nanotubes and multilayer graphene, 802, Copyright 2013, with permission from Elsevier.)

as electrically conducting components in polymer composites. These applications can further be improved by modifying the surface of CNs.

3.5.1 Electron Emission

Carbon-based nanomaterials are potential contenders for electron emission applications that can be explored for flat panel displays and vacuum microwave power amplifier devices. The high aspect ratio, extremely small curvature, high field enhancement factor, high stability, and good electrical conductivity make them one of the exclusive materials for these applications.[46,47]

Electron emission is a quantum mechanical tunneling phenomenon where the emission of electrons occurs from the surface of the material to vacuum by an external applied electric field (Figure 3.6). This phenomenon occurs in high electric fields of the order 10^7–10^8 V/cm. The high electric field narrows the potential barrier at the emitter–vacuum interface sufficiently for the electrons to have a significant probability of tunneling from the emitter surface to vacuum. For electric field more than 10^3 V/μm, emission of electrons via the tunneling process is observed. It is caused by field ionization of trapped electrons into the conduction band or by electrons tunneling from the metal Fermi energy into the insulator/semiconductor conduction band.[6]

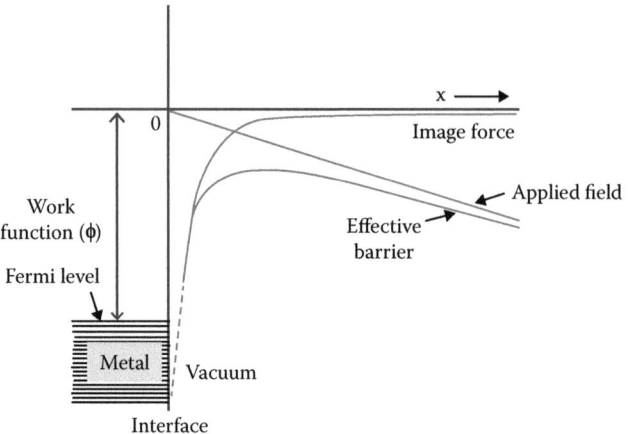

FIGURE 3.6
Potential energy diagram illustrating the effect of an external electric field on the energy barrier of electrons at the metal surface.

The electron emission characteristics of CNs are given by the Fowler–Nordheim (F-N) theory, which is given as[42,48]

$$J_M \approx \lambda_M a \phi^{-1} \gamma_C^2 E_M^2 \exp\left[\frac{-v_F b \phi^{3/2}}{\gamma_C E_M}\right] v_E \tag{3.1}$$

where
  J_M is the macroscopic current density
  E_M is the macroscopic field
  λ_M is a macroscopic preexponential correction factor and is the correction factor
  The values of constant are a = 1.54×10^{-6} A eV V^{-2} and b = 6.83×10^7 eV$^{3/2}$ V cm^{-1}
  γ_C is a macroscopic field enhancement factor and is given by

$$\gamma_C = \frac{\sigma b \phi^{3/2}}{S^{fit}} \tag{3.2}$$

  σ is the slope correction factor, with the value of 0.95 (taken on orthodox emission)
  S^{fit} is the slope of the line fitted to the F-N plot
  γ_C is calculated by considering the work function of CNTs as 5 eV[49]

The F-N plot is obtained by plotting ln (J_M/E_M^2) versus $1/E_M$. The distance between cathode and anode is 250 μm.

The emission of electrons from carbon-based nanostructures is explained on the basis of the F-N model. CNs exhibit excellent electron emission properties as the electrons are emitted from the tips that are of nanometer dimension. The tunneling of electrons from carbon-based structures is accounted in terms of the field enhancement factor (β), which is an important parameter for electron emission investigations. The value of β depends on the shape and also on the height and diameter of emitter. The typical value for β is calculated from the negative slope of the F-N plot. β is also related with the geometrical enhancement that is given in terms of (h/r) for carbon-based emitters. CNs with high aspect ratios has increased value of β that improves the emission properties of CNTs. CNTs with an average bundle length less than 10 μm have a small β factor. From Equation 3.2, it is seen that β is related with the work function.

The electronic work function, which is the energy difference between the Fermi levels and the vacuum level, is the parameter of a material that determines the electron emission characteristics.

Work function is an intrinsic property of a material and cannot be changed. However, modifications can be performed in a material by various ways that affect the density of states near the Fermi level to improve the electron emission properties and hence β. As metal nanoparticle adsorbs on CNTs, bands are split around the Fermi level due to the mixing between π-orbitals of CNTs and d-orbitals of metal nanoparticles resulting in Fermi level shifting. Thus, modification occurs by increase in the density of states that results in the increases of tunneling currents. The aligned MWNTs were found to have a larger density of states at the Fermi level and a slightly lower work function than the random MWNTs, which is attributed to the difference in the electronic state between the tip and the sidewall of the CNTs. For various device applications, the performance of an emitter is also discussed in terms of the turn-on field (E_{on}) and (E_{th}) threshold field. E_{on} and E_{th} are the macroscopic fields required to produce $J_M = 10\ \mu A/cm^2$ and $1\ mA/cm^2$, respectively. The lower the values of E_{on} and E_{th}, the better is the electron emission behavior of an emitter.

The electron emission of CNs depends upon various factors such as morphology of CNs, aspect ratio, work function, CN–substrate interaction, and catalyst effect. Few of them are discussed in this chapter.

3.5.2 Surface Enhanced Raman Scattering

As the name suggests, surface enhanced Raman spectroscopy (SERS) provides the same information that normal Raman spectroscopy does, but with a greatly enhanced signal. It is a surface-sensitive technique that enhances Raman scattering by molecules adsorbed on rough metal surfaces. The enhancement factor can be as much as 10^{10}–10^{11}, which means the technique may detect single molecules.[50–53]

Before getting into the detailed discussions on SERS mechanisms, it is essential to move into the domain of plasmons, which are the collective oscillations of the electrons. In metal nanoparticles, the excitation is confined to the near-surface region and is termed as surface plasmons. Nanoparticles are small clusters with a diameter of about 10–100 nm. As the diameter of a nanoparticle is of the order of the penetration depth of electromagnetic waves, the excitation light is able to penetrate the particle. The field inside the particle shifts the conduction electrons collectively with respect to the fixed positive charge of the lattice ions. The electrons build up a charge on the surface at one side of the particle. The attraction of this negative charge and the positive charge of the remaining lattice ions on the opposite side results in a restoring force. If the frequency of the excitation light field is in resonance with the frequency of this collective oscillation, even a small exciting field leads to a strong oscillation. The resonance frequency is mainly determined by the strength of the restoring force. The properties of these excitations depend strongly on particle geometry and interparticle coupling and give rise to a variety of effects, such as frequency-dependent absorption and scattering or near-field enhancement. For example, the optical properties of molecules (or other light emitters like quantum dots) can be strongly modified upon their electromagnetic interaction with particle plasmons, which might be beneficial for novel-sensing applications.

After having an essence of plasmons, SERS mechanisms can be discussed. The enhanced Raman scattering from molecules on metals (or vice versa) can be realized in two ways: (1) electromagnetic SERS and (2) chemical SERS.

Electromagnetic SERS is governed by the resonant interactions between the optical field and surface plasmons in the metal. The incident laser light is in resonance with the plasmons due to dipolar oscillations of conduction electrons. Thus, the coherent interaction of the incoming electric field with the dipolar field leads to a redistribution of electric field intensities around the metal nanoparticles.[54] A molecule (e.g., CNTs) attached to the nanoparticles experiences an enhanced intensity.[55]

The enhancement in the intensity depends upon the shape, size, and roughness of the metal nanoparticles. Au, Ag, and Cu are the best materials for SERS as they fulfill the resonance condition in the visible and near infrared.

In chemical SERS, electronic levels in the adsorbed molecule become shifted or broadened because of charge transfer between the molecule and metal.

3.6 ROLE OF INTERLAYER IN THE STRUCTURE AND ELECTRON EMISSION PROPERTIES OF CARBON NANOSTRUCTURES

The catalytic activity of metals depends on the carbon–metal interaction. The ability of a metal to form new compounds is related to its capability of forming and breaking bonds that depends on the electronic structure of a particular metal. In order to address this issue, the thermodynamics of carbon–metal interaction have been studied.[56] The carbon–metal interaction involves the reaction of carburization as follows:

$$zM + C_xH_y \rightarrow M_zC_x + \left(\frac{1}{2}\right)yH_2 \tag{3.3}$$

where

  M denotes a metal
  C_xH_y is hydrocarbon precursor gas
  M_zC_x corresponds to the particular metal carbide

The Gibbs free energy (ΔG) for the reaction is given as

$$\Delta G = \Delta H - T\Delta S \tag{3.4}$$

where

  ΔH is the enthalpy of the reaction
  T is the reaction temperature
  ΔS corresponds to the entropy of the system

On considering the pure metal catalyst, the enthalpy of the particular reaction is the difference in the enthalpy of formation of the metal carbide and carbon source. Other than Fe, Ni, and Co, which are referred to as typical catalysts for the growth of CNs, few atypical catalysts such as aluminum (Al), indium (In), platinum (Pt), titanium (Ti), palladium (Pd), chromium (Cr), and silver (Ag) have also been used for the growth of CNs.[56–59] The overlapping of the d-orbital of these metals and p-orbital of carbon play an important role in the growth of CNs using various metals.

  However, the high temperatures associated with the CVD technique might produce an undesired interaction of the metallic catalyst with the Si substrate, which could deteriorate the catalytic efficiency.[60] The formation of iron silicides has been presented in some studies. In order of avoid silicide formation, an interlayer is introduced between the Si substrate and the catalyst.[57,61] Various materials such as Al, Ag, Cr, Mo, Pt, Ti, TiN, and W have been used as interlayers in order to avoid the hindrance in catalysis and thus exploring the CNs for fabrication of electronic devices.[57,58,62–64] The interlined material should have the following properties[65]:

  1. The interlaying material should minimize the transport rate between the catalyst and the substrate. It should constitute the traffic barrier between the catalyst and the substrate.
  2. The interlaying material should be thermodynamically stable.
  3. The interlaying material should have a low contact resistance with the catalyst and the substrate and must adhere well with both the layers.

Figure 3.6 shows the difference in CNTs after using Ti as an interlayer. Few defective walls are seen in CNTs without Ti as an interlayer (Figure 3.7a). However, CNTs with Ti as an interlayer shows much improved and less defective growth (Figure 3.7b). The autocorrelated images of CNTs without and with a Ti interlayer are shown in Figure 3.7c and d, respectively. The schematic display of Ti-interlined CNTs is seen in Figure 3.7e. Using Ti as an interlayer facilitates the growth of CNTs by providing more nucleation sites along with Fe. Therefore, denser growth is achieved in comparison to Fe-catalyzed CNTs. These Ti-interlined CNTs result in enhanced emission in comparison to Fe-catalyzed CNTs, shown in Figure 3.8. Figure 3.8a shows a schematic where Ti-interlined CNTs act as cathode. The distance

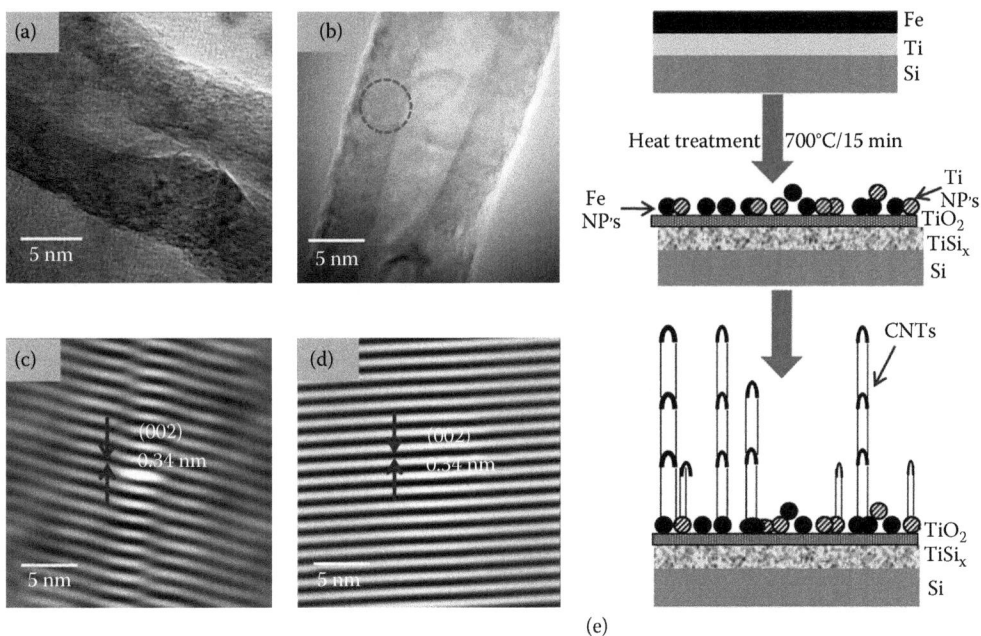

FIGURE 3.7
HRTEM images representing (a) MWCNTs with defects and kinks without a Ti interlayer. (b) Very few defects are seen in Ti-interlined MWCNTs. (c and d) Autocorrelated images corresponding to Figure 3.7a,b, respectively. Defects can be seen in the HRTEM image of Fe-catalyzed CNTs, which shows the resolved (002) lattice planes corresponding to graphene. However, Ti-interlined CNTs are less defective as compared to Fe CNTs. (e) Schematic model showing the growth of MWCNTs using Ti as an interlayer. The Ti acts as a barrier layer and also helps in the growth of CNTs along with Fe. NPs: nanoparticles. (Reprinted from *Mater. Chem. Phys.*, 137, Sharma, H., Shukla, A.K., and Vankar, V.D., Influence of Fe nanoparticles diameters on the structure and electron emission studies of carbon nanotubes and multilayer graphene, 802, Copyright 2013, with permission from Elsevier.)

between cathode and anode was roughly 250 nm. Depending on the thickness of the Ti interlayer, electron emission was explained on the basis of the double-layer and direct emission method.[57]

3.7 SURFACE MODIFICATION OF CARBON NANOSTRUCTURES

Till now the structure, properties, and growth of CNs have been discussed. It is now well established that CNTs and other CNs have so remarkable electronic and structural properties that can be used as active building blocks for a large variety of nanoscale devices. To really disclose the true technological potential of CNs in different areas of nanotechnology, they have been incorporated in a variety of materials with even more attractive properties and possible applications. Due to their large active surface area, good conductivity, and thermal stability, a large variety of entities such as metals, semiconductors, metal oxides, polymers, and biomolecules have been interfaced with CNTs.[66–70] The surface and structure of CNs can be modified by various ways. Among them, the following two methods are discussed.

3.7.1 Surface Modification by Grafting Nanoparticles over Carbon Nanostructures

The doping or coating methods have been known as practically viable approaches to tailor the structural, electrical, and optical properties of CNs.[71–73] Thus, modification in the structure and properties of CNs by various materials results in the formation of hybrid structures.[74] These hybrid materials made of metals (or metal oxides) and CNs have been attracting much attention due to modification in

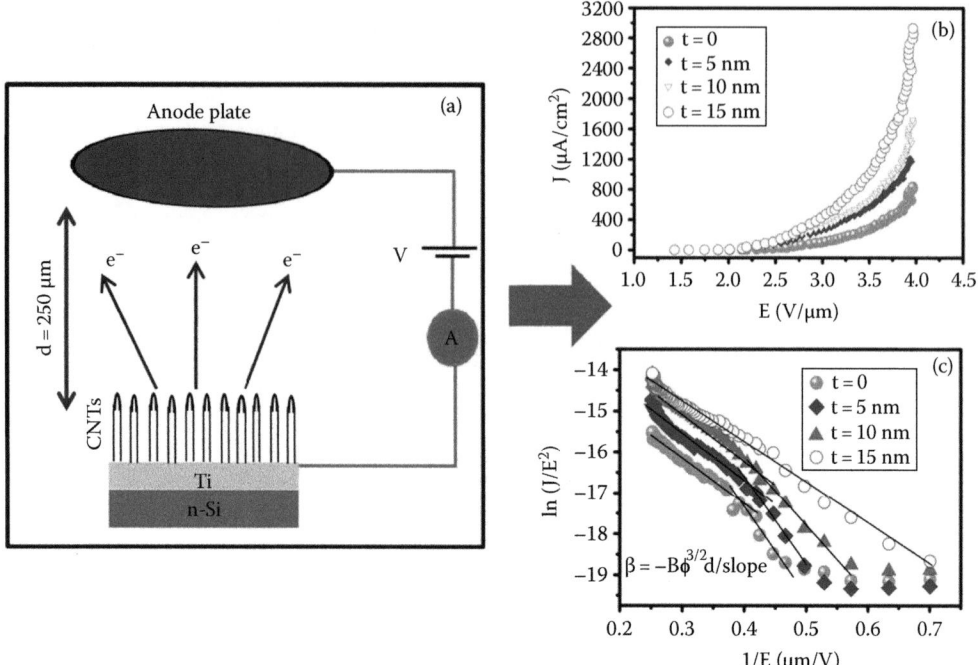

FIGURE 3.8

(a) Schematic diagram of the field emission from Ti-interlined MWCNTs. d is the distance between the cathode and the anode. (b) The macroscopic emission current density (J_M) as a function of the applied macroscopic electric field (E_M) from MWCNTs using Ti interlayers with thicknesses t = 0, 5, 10, and 15 nm with a constant thickness of Fe. (c) F-N plot for the same samples. From the slope of the plot, the field enhancement factor is calculated. (Reprinted from *Mater. Chem. Phys.*, 137, Sharma, H., Shukla, A.K., and Vankar, V.D., Influence of Fe nanoparticles diameters on the structure and electron emission studies of carbon nanotubes and multilayer graphene, 802, Copyright 2013, with permission from Elsevier.)

their electronic structure through a charge transfer and orbital hybridization. The metals passivate the defect sites on the CNT surface, thereby stabilizing CNT mechanically and chemically.[67]

The coating of metal onto a CNT surface involves first the adsorption of metal atoms onto the surface, followed by diffusion of these adatoms across the surface until nucleation of islands occurs when the diffusing adatoms form a stable nucleus. After the formation of a stable nucleus at a nucleation center, the next incoming adatoms can either attach to an existing nucleus or diffuse on the surface until they encounter another adatom to form a new stable nucleus. The number of nucleation centers N_c is given by

$$N_c \propto \left(\frac{F}{D_0} \right)^{1/3} \tag{3.5}$$

where

    F is the incoming vapor flux
    D_0 the diffusion constant

The diffusion constant on the substrate is given by

$$D_0 = v\exp\left(\frac{-E_{diff}}{kT} \right) \tag{3.6}$$

where

    v is a prefactor for diffusion
    E_{diff} is the diffusion energy that can be empirically described as $E_{diff} \sim E_b/4$; E_b is the binding
        energy of an adatom to the substrate[75]

Therefore, the density of nucleation centers depends on the interaction between the metallic atoms and the CNT surface: a higher interaction between adatoms and the substrate will lead to a higher nucleation density. He et al. carried out the studies on the coating geometry of different metals (Fe, Au, Pd, Ni, and Ti) and investigated that these metals interact differently with SWCNTs. The overlapping between the orbital modifies the density of states of CNs so that these hybrid materials can be used as a potential materials for various applications such as field emission and localized surface plasmon resonance (LSPR).

3.7.1.1 Grafting of Ti nanoparticles and enhanced electron emission properties

It is studied that Ti, Ni, and Pd have low metal–SWNT interfacial energies and high diffusion barriers that are responsible for forming continuous or quasicontinuous layers on the SWNT surface. In contrast, Au has small diffusion barriers and poor SWNT surface wetting; thus, they tend to aggregate and form large clusters. The binding energy between Fe and SWNTs is large, and due to the large cohesion energy and poor wetting, Fe may form isolated clusters. The interaction between CNTs and various metals also involves the interaction between the $p\pi$-orbitals of carbon and the d-orbitals of various metals.

Ti film of thickness 2–5 nm was coated over CNTs to observe the changes in the electron emission properties (Figure 3.9).[67] The average diameter of Ti nanoparticles is 5–10 nm as seen in Figure 3.9b. It is seen that Ti nanoparticles adsorb on the defective walls to passivate the defects. The schematic illustration for the same is seen in Figure 3.9c, where Ti nanoparticles are adsorb on the sidewalls. The Ti nanoparticle–grafted CNTs were tested for electron emission studies. It was found that Ti modifies the local work function of CNTs and provides more density of sides. As a result, electron emission properties of CNTs are enhanced. The E_{on} value of Ti nanoparticle–grafted CNTs were found to be ~0.8 V/μm, as compared to pristine CNTs (~1.8 V/μm) and γ_C—the factor reported was ~1.14×10^4, corresponding to enhancement electron emission properties of CNTs.[67]

3.7.1.2 Grafting of Au nanoparticles and surface-enhanced Raman scattering

In the work related to modification of CNTs, these unique structures were grafted with the Au thin film of 5 nm using inert gas evaporation at two different deposition pressures, 4×10^{-3} and 4×10^{-4} Torr, respectively.[74] It was found that using different inert gas pressures leads to variation in size of Au nanoparticles owing to their different mean free paths. Figure 3.10 shows the attachment of Au nanoparticle–grafted CNTs. It is seen from the HRTEM images that Au nanoparticles are physically adsorbed on the sidewalls of CNTs (Figure 3.10a). The schematic display of Au-grafted CNTs is shown in Figure 3.10b. Micro-Raman scattering was recorded for Au-grafted CNTs, using two different laser wavelengths, 514.5 and 1064 nm, respectively (Figure 3.10c and d). Various Raman modes are seen in the Raman spectrum. The peak position at 1355 cm^{-1} corresponds to the D mode, confirming transverse optical (TO) vibrations of carbon atoms.[76] This mode appears due to defect and disordering in the CNs. Another mode that corresponds to the graphitic in-plane vibration of carbon atoms (longitude optical [LO] mode) that is present at 1580 cm^{-1} is known as the G mode.[61,77] The variation in these modes is assimilated to the structural, electronic, and optical properties of CNTs and graphene.

It is seen in Figure 3.10c that the intensity of Au-grafted CNTs enhances in comparison to the pristine CNTs. However, Au nanoparticles grafted at 4×10^{-4} Torr results in more intense spectra in comparison to that of Au nanoparticles grafted at 4×10^{-3} Torr. The enhancement in Raman scattering for Au-grafted CNTs is attributed to the local surface plasmons exhibited by the Au nanoparticles on impinging laser light. Electric field enhancement at the surface of Au nanoparticles due to LSPR results in enhanced Raman scattering, also known as SERS.

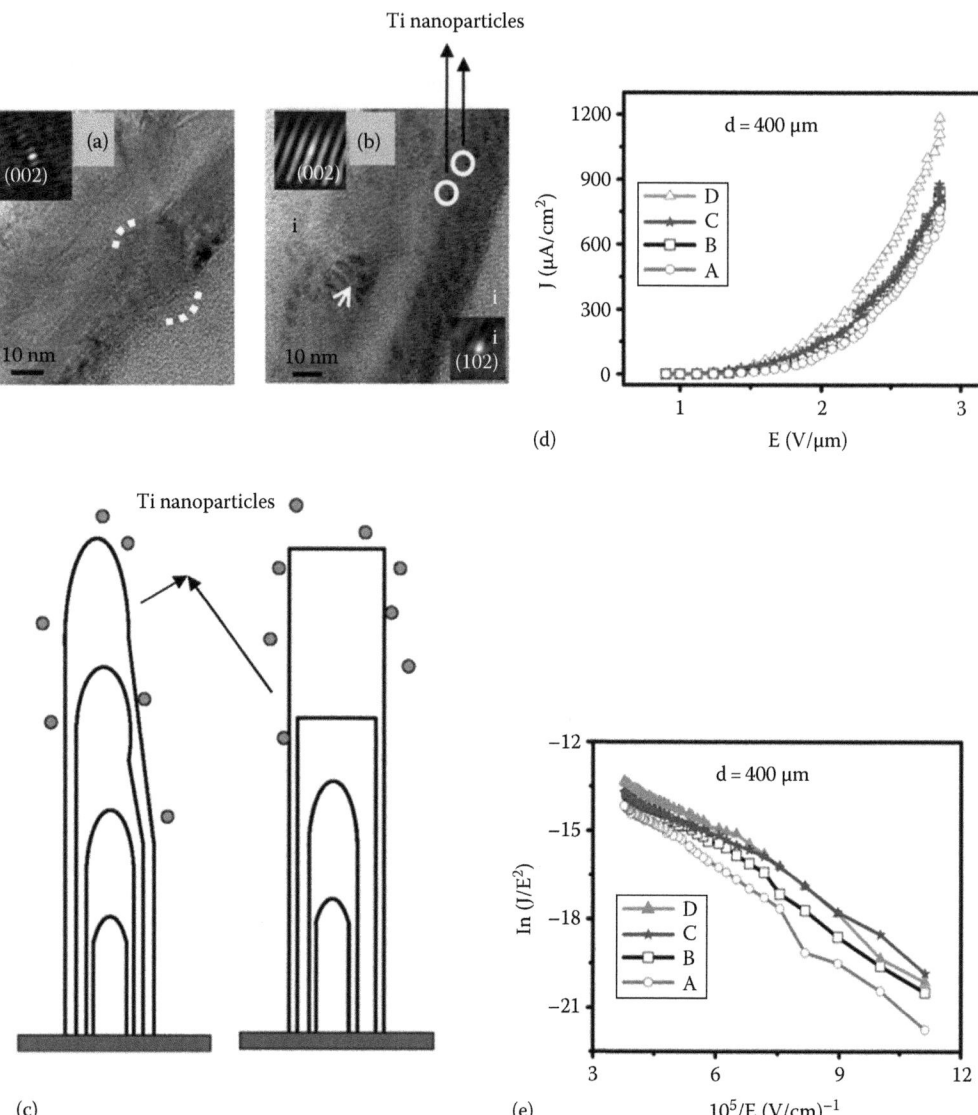

FIGURE 3.9
HRTEM images showing (a) defect planes in bare MWCNTs by a dotted line and (b) Ti-coated MWCNTs, where Ti nanoparticles have been encircled; the arrow shows the Ti planes. Inset images showing (a) (002) planes of pristine MWCNT; (b) (i) (002) planes of Ti-coated MWCNT, with d = 0.34 nm; and (ii) (102) planes of Ti, with d = 0.23 nm; (c) a schematic presentation of Ti nanoparticles adhering to the CNT walls and passivation of defects. Here, d is the interplanar distance; (d, e) J versus E plot for samples using different thicknesses of Ti film at the distance of 400 μm (b) corresponding F-N plot. d is the distance between the cathode and anode. The turn-on and threshold field values are taken at 10 μA/cm² and 1 mA/cm². (Reprinted from *Mater. Chem. Phys.*, 137, Sharma, H., Shukla, A.K., and Vankar, V.D., Influence of Fe nanoparticles diameters on the structure and electron emission studies of carbon nanotubes and multilayer graphene, 802, Copyright 2013, with permission from Elsevier.)

The enhancement in Raman scattering depends on the following factors: magnitude of the field coupled to MWCNTs, size of Au nanoparticles (rAu), distance from CNTs (dAu–MWCNT), the dielectric constant of Au, and incident field strength.

The enhanced scattering for Au-grafted CNTs at 4×10^{-4} Torr is due to the bigger size of Au nanoparticles and small interparticle distance between them in comparison to 4×10^{-3} Torr. Furthermore, no SERS phenomenon was observed for the samples tested for 1064 nm laser wavelength (Figure 3.10d).

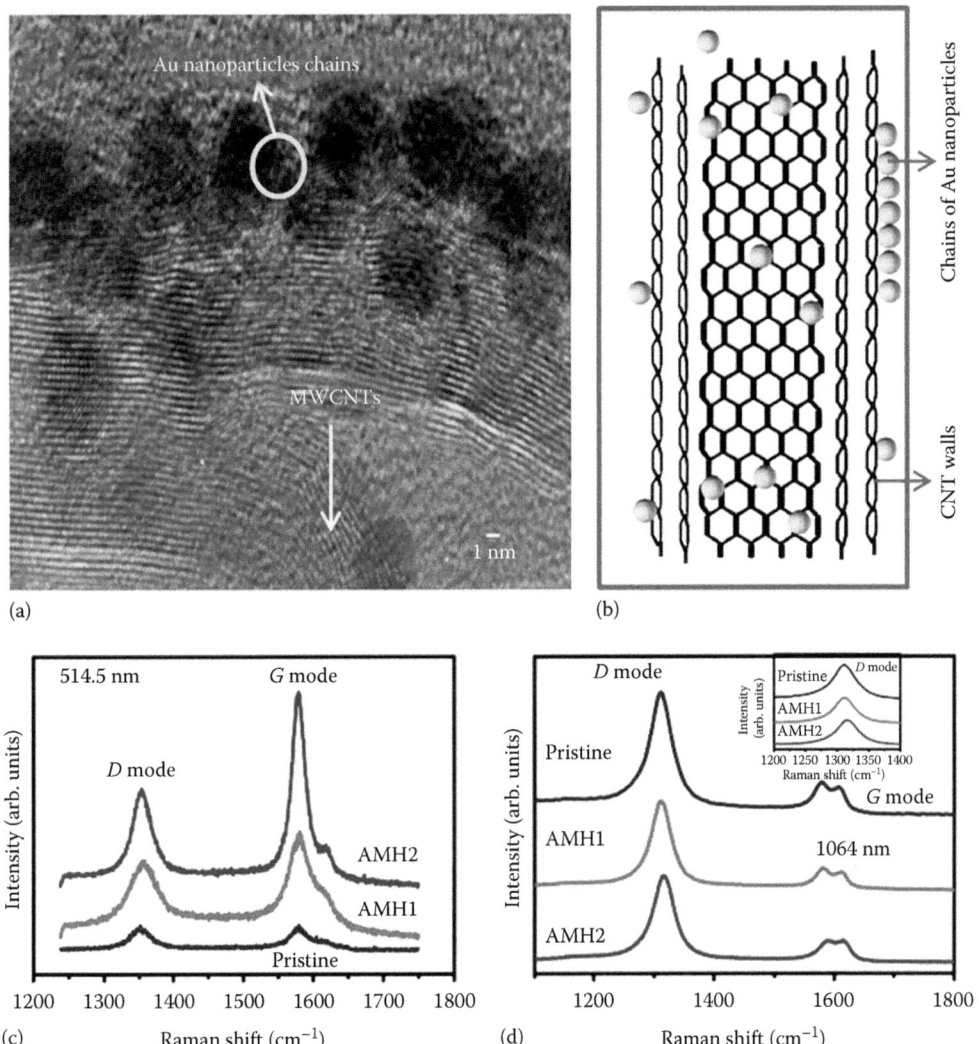

FIGURE 3.10
(a) HRTEM micrograph showing the adsorption of Au nanoparticles on MWCNT defective edges and (b) schematic model showing the piling up of Au nanoparticles on MWCNTs. (c) SERS studies showing the increase in the intensity of the G mode for samples deposited at different inert gas pressures as compared to pristine samples using 514.5 nm excitation wavelength, and (d) micro-Raman spectra showing the dispersion in the D mode for samples deposited at different inert gas pressures as compared to pristine samples using 1064 nm excitation wavelength. (Reprinted from *Mater. Chem. Phys.*, 137, Sharma, H., Shukla, A.K., and Vankar, V.D., Influence of Fe nanoparticles diameters on the structure and electron emission studies of carbon nanotubes and multilayer graphene, 802, Copyright 2013, with permission from Elsevier.)

3.8 SUMMARY

The growth, structure, and electron emission characteristics of CNs have been found to be affected by the thicknesses of the Fe (catalyst) film. It is observed that depending on the Fe thickness, the morphology of the CNs changes from vertically aligned CNTs to entangled CNTs to MLGs for 5, 10, 20, and 50 nm, respectively. The number of graphitic shells and hence the diameter of CNs increase for large-size Fe nanoparticles. The growth and electron emission characteristics of CNTs are affected by the Ti interlayer thickness. It is observed that Ti, depending on its thickness, hinders the formation of iron silicide and also interacts with Fe to provide more nucleation sites for CNT growth. Electron emission characteristics of Ti-interlined CNTs are found to be superior as compared to the CNTs without a Ti interlayer. Electron emission is explained on the basis of the double-barrier model for low thicknesses of Ti. In another set of studies, the enhanced electron emission properties and crystallinity of Ti–CN hybrids were studied. It was observed that Ti-grafted CNs have low work function and higher field enhancement factor in comparison to those pristine CNTs. The effect of Au nanoparticles of different sizes decorated over MWCNT was investigated in the form of Au–MWCNT hybrids correlated with each other.

ACKNOWLEDGMENT

The financial support of the Ministry of Information Technology (MIT), Government of India, is gratefully acknowledged.

REFERENCES

1. Hirsch, A. 2010. The era of carbon allotropes. *Nature Materials* 9:868–871.
2. Kroto, H. W., Heath, J. R., O'Brien, S. C., Curl, R. F., and Smalley, R. E. 1985. C60: Buckminsterfullerene. *Nature* 318:162–163.
3. Iijima, S. 1991. Helical microtubules of graphitic carbon. *Nature* 354:56–58.
4. Novoselov, K. S., Geim, A. K., Morozov, S. V. et al. 2004. Electric field effect in atomically thin carbon films. *Science* 306:666–669.
5. Geim, A. K. and Novoselov, K. S. 2007. The rise of graphene. *Nature Materials* 6:183–191.
6. Sharma, H. 2012. Growth, structure and electron emission characteristics of carbon nanostructures synthesized by microwave plasma enhanced chemical vapour deposition process. PhD dissertation, Indian Institute of Technology, Delhi, India.
7. Sumio, I. 1980. Direct observation of the tetrahedral bonding in graphitized carbon black by high resolution electron microscopy. *Journal of Crystal Growth* 50:675–683.
8. Wang, X., Li, Q., Xie, J. et al. 2009. Fabrication of ultralong and electrically uniform single-walled carbon nanotubes on clean substrates. *Nano Letters* 9:3137–3141.
9. Wu, Y., Yang, B., Zong, B. et al. 2004. Carbon nanowalls and related materials. *Journal of Materials Chemistry* 14:469–477.
10. Ajayan, P. M., Ichihashi, T., and Iijima, S. 1993. Distribution of pentagons and shapes in carbon nano-tubes and nano-particles. *Chemical Physics Letters* 202:384–388.
11. Carroll, D. L., Redlich, P., Ajayan, P. M. et al. 1997. Electronic structure and localized states at carbon nanotube tips. *Physical Review Letters* 78:2811–2814.
12. Ebbesen, T. W., Ajayan, P. M., Hiura, H., and Tanigaki, K. 1994. Purification of nanotubes. *Nature* 367:519.
13. Liu, J., Rinzler, A. G., Dai, H. et al. 1998. Fullerene pipes. *Science* 280:1253–1256.
14. Ajayan, P. M. and Iijima, S. 1993. Capillarity-induced filling of carbon nanotubes. *Nature* 361:333–334.
15. Tsang, S. C., Chen, Y. K., Harris, P. J. F., and Green, M. L. H. 1994. A simple chemical method of opening and filling carbon nanotubes. *Nature* 372:159–162.
16. Dujardin, E., Ebbesen, T. W., Hiura, H., and Tanigaki, K. 1994. Capillarity and wetting of carbon nanotubes. *Science* 265:1850–1852.
17. Hongjie, D. 2002. Carbon nanotubes: Opportunities and challenges. *Surface Science* 500:218–241.
18. Zhao, Q., Gan, Z., and Zhuang, Q. 2002. Electrochemical sensors based on carbon nanotubes. *Electroanalysis* 14:1609–1613.
19. Paradise, M. and Goswami, T. 2007. Carbon nanotubes—Production and industrial applications. *Materials and Design* 28:1477–1489.

20. Ajayan, P. M. 1999. Nanotubes from carbon. *Chemical Reviews* 99:1787–1800.

21. Partoens, B. and Peeters, F. M. 2006. From graphene to graphite: Electronic structure around the K point. *Physical Review B* 74:075404.

22. Shiji, K., Hiramatsu, M., Enomoto, A. et al. 2005. Vertical growth of carbon nanowalls using rf plasma-enhanced chemical vapor deposition. *Diamond and Related Materials* 14:831–834.

23. Srivastava, S. K., Shukla, A. K., Vankar, V. D., and Kumar, V. 2005. Growth, structure and field emission characteristics of petal like carbon nano-structured thin films. *Thin Solid Films* 492:124–130.

24. Luk'yanchuk, I. A. and Kopelevich, Y. 2006. Dirac and normal fermions in graphite and graphene: Implications of the quantum hall effect. *Physical Review Letters* 97:256801.

25. Latil, S., Meunier, V., and Henrard, L. 2007. Massless fermions in multilayer graphitic systems with misoriented layers: Ab initio calculations and experimental fingerprints. *Physical Review B* 76:201402.

26. Valentin, N. P. 2004. Carbon nanotubes: Properties and application. *Materials Science and Engineering: R: Reports* 43:61–102.

27. Shiraishi, M. and Ata, M. 2001. Work function of carbon nanotubes. *Carbon* 39:1913–1917.

28. Saito, Y., Nakahira, T., and Uemura, S. 2003. Growth conditions of double-walled carbon nanotubes in arc discharge. *The Journal of Physical Chemistry B* 107:931–934.

29. Scott, C. D., Arepalli, S., Nikolaev, P., and Smalley, R. E. 2001. Growth mechanisms for single-wall carbon nanotubes in a laser-ablation process. *Applied Physics A: Materials Science and Processing* 72:573–580.

30. Hart, A. J., Boskovic, B. O., Chuang, A. T. H. et al. 2006. Uniform and selective CVD growth of carbon nanotubes and nanofibres on arbitrarily microstructured silicon surfaces. *Nanotechnology* 17:1397.

31. Kumar, M. and Ando, Y. 2010. Chemical vapor deposition of carbon nanotubes: A review on growth mechanism and mass production. *Journal of Nanoscience and Nanotechnology* 10:3739–3758.

32. Handuja, S., Srivastava, P., and Vankar, V. 2010. On the growth and microstructure of carbon nanotubes grown by thermal chemical vapor deposition. *Nanoscale Research Letters* 5:1211–1216.

33. Shiokawa, T., Zhang, B. P., Suzuki, M., Kobayashi, T., and Ishibashi, K. 2005. In *Microprocesses and Nanotechnology Conference, 2005 International*, October 25–28, 2005, pp. 92–93.

34. Kong, J., Soh, H. T., Cassell, A. M., Quate, C. F., and Dai, H. 1998. Synthesis of individual single-walled carbon nanotubes on patterned silicon wafers. *Nature* 395:878–881.

35. Delzeit, L., Chen, B., Cassell, A. et al. 2001. Multilayered metal catalysts for controlling the density of single-walled carbon nanotube growth. *Chemical Physics Letters* 348:368–374.

36. Baker, R. T. K. 1989. Catalytic growth of carbon filaments. *Carbon* 27:315–323.

37. Kiang, C. H., Endo, M., Ajayan, P. M., Dresselhaus, G., and Dresselhaus, M. S. 1998. Size effects in carbon nanotubes. *Physical Review Letters* 81:1869–1872.

38. Sinnott, S. B., Andrews, R., Qian, D. et al. 1999. Model of carbon nanotube growth through chemical vapor deposition. *Chemical Physics Letters* 315:25–30.

39. Louchev, O. A., Laude, T., Sato, Y., and Kanda, H. 2003. Diffusion-controlled kinetics of carbon nanotube forest growth by chemical vapor deposition. *The Journal of Chemical Physics* 118:7622–7634.

40. Ducati, C., Alexandrou, I., Chhowalla, M., Robertson, J., and Amaratunga, G. A. J. 2004. The role of the catalytic particle in the growth of carbon nanotubes by plasma enhanced chemical vapor deposition. *Journal of Applied Physics* 95:6387–6391.
41. Holstein, W. L. 1995. The roles of ordinary and soret diffusion in the metal-catalyzed formation of filamentous carbon. *Journal of Catalysis* 152:42–51.
42. Sharma, H., Shukla, A. K., and Vankar, V. D. 2013. Influence of Fe nanoparticles diameters on the structure and electron emission studies of carbon nanotubes and multilayer graphene. *Materials Chemistry and Physics* 137:802–810.
43. Zhang, X. X., Li, Z. Q., Wen, G. H. et al. 2001. Microstructure and growth of bamboo-shaped carbon nanotubes. *Chemical Physics Letters* 333:509–514.
44. Lee, C. J. and Park, J. 2000. Growth model of bamboo-shaped carbon nanotubes by thermal chemical vapor deposition. *Applied Physics Letters* 77:3397–3399.
45. Baughman, R. H., Zakhidov, A. A., and de Heer, W. A. 2002. Carbon nanotubes—The route toward applications. *Science* 297:787–792.
46. Murakami, H., Hirakawa, M., Tanaka, C., and Yamakawa, H. 2000. Field emission from well-aligned, patterned, carbon nanotube emitters. *Applied Physics Letters* 76:1776–1778.
47. Cheng, Y. and Zhou, O. 2003. Electron field emission from carbon nanotubes. *Comptes Rendus Physique* 4:1021–1033.
48. Richard, G. F. 2012. Extraction of emission parameters for large-area field emitters, using a technically complete Fowler–Nordheim-type equation. *Nanotechnology* 23:095706.
49. Bonard, J. 1998. Field emission from single-wall carbon nanotube films. *Applied Physics Letters* 73:918.
50. Blackie, E. J., Ru, E. C. L., and Etchegoin, P. G. 2009. Single-molecule surface-enhanced raman spectroscopy of nonresonant molecules. *Journal of the American Chemical Society* 131:14466–14472.
51. Le Ru, E. C., Blackie, E., Meyer, M., and Etchegoin, P. G. 2007. Surface enhanced Raman scattering enhancement factors: A comprehensive study. *The Journal of Physical Chemistry C* 111:13794–13803.
52. Nie, S. and Emory, S. R. 1997. Probing single molecules and single nanoparticles by surface-enhanced Raman scattering. *Science* 275:1102–1106.
53. Le Ru, E. C., Meyer, M., and Etchegoin, P. G. 2006. Proof of single-molecule sensitivity in surface enhanced Raman scattering (SERS) by means of a two-analyte technique. *The Journal of Physical Chemistry B* 110:1944–1948.
54. Kneipp, K. 2007. Surface-enhanced Raman scattering. *Physics Today* 60(11):40–46.
55. Subramaniam, C., Sreeprasad, T. S., Pradeep, T. et al. 2007. Visible fluorescence induced by the metal semiconductor transition in composites of carbon nanotubes with noble metal nanoparticles. *Physical Review Letters* 99:167404.
56. Esconjauregui, S., Whelan, C. M., and Maex, K. 2009. The reasons why metals catalyze the nucleation and growth of carbon nanotubes and other carbon nanomorphologies. *Carbon* 47:659–669.
57. Sharma, H., Shukla, A. K., and Vankar, V. D. 2011. Effect of titanium interlayer on the microstructure and electron emission characteristics of multiwalled carbon nanotubes. *Journal of Applied Physics* 110:033726.
58. Kaushik, V., Sharma, H., Girdhar, P., Shukla, A. K., and Vankar, V. D. 2011. Structural modification and enhanced electron emission from multiwalled carbon nanotubes grown on Ag/Fe catalysts coated Si-substrates. *Materials Chemistry and Physics* 130:986–992.

59. Chhoker, S., Vinayak, S., Shukla, A. K., and Vankar, V. D. 2011. Field emission studies of carbon nanostructures synthesised over Ni–Cr catalyst layer. *Journal of Experimental Nanoscience* 6:374–388.
60. de los Arcos, T., Vonau, F., Garnier, M. G. et al. 2002. Influence of iron—Silicon interaction on the growth of carbon nanotubes produced by chemical vapor deposition. *Applied Physics Letters* 80:2383–2385.
61. Sharma, H., Shukla, A. K., and Vankar, V. D. 2012. Structural modifications and enhanced Raman scattering from multiwalled carbon nanotubes grown on titanium coated silicon single crystals. *Thin Solid Films* 520:1902–1908.
62. Kabir, M. S., Morjan, R. E., Nerushev, O. A. et al. 2005. Plasma-enhanced chemical vapour deposition growth of carbon nanotubes on different metal underlayers. *Nanotechnology* 16:458.
63. Merkulov, V. I., Lowndes, D. H., Wei, Y. Y., Eres, G., and Voelkl, E. 2000. Patterned growth of individual and multiple vertically aligned carbon nanofibers. *Applied Physics Letters* 76:3555–3557.
64. Xuhui, S., Ke, L., Raymond, W. et al. 2010. The effect of catalysts and underlayer metals on the properties of PECVD-grown carbon nanostructures. *Nanotechnology* 21:045201.
65. Nicolet, M. A. and Bartur, M. 1981. Diffusion barriers in layered contact structures. *Journal of Vacuum Science and Technology* 19:786–793.
66. Georgakilas, V., Gournis, D., Tzitzios, V. et al. 2007. Decorating carbon nanotubes with metal or semiconductor nanoparticles. *Journal of Materials Chemistry* 17:2679–2694.
67. Sharma, H., Kaushik, V., Girdhar, P. et al. 2010. Enhanced electron emission from titanium coated multiwalled carbon nanotubes. *Thin Solid Films* 518:6915–6920.
68. Wildgoose, G. G., Banks, C. E., and Compton, R. G. 2006. Metal nanoparticles and related materials supported on carbon nanotubes: Methods and applications. *Small* 2:182–193.
69. Giulianini, M., Waclawik, E. R., Bell, J. M. et al. 2009. Poly(3-hexyl-thiophene) coil-wrapped single wall carbon nanotube investigated by scanning tunneling spectroscopy. *Applied Physics Letters* 95:143116–143113.
70. Lin, Y., Taylor, S., Li, H. et al. 2004. Advances toward bioapplications of carbon nanotubes. *Journal of Materials Chemistry* 14:527–541.
71. Fischer, J. E. 2002. Chemical doping of single-wall carbon nanotubes. *Accounts of Chemical Research* 35:1079–1086.
72. Charlier, J. C., Terrones, M., Baxendale, M. et al. 2002. Enhanced electron field emission in B-doped carbon nanotubes. *Nano Letters* 2:1191–1195.
73. Kim, D. S., Lee, S. M., Scholz, R. et al. 2008. Synthesis and optical properties of ZnO and carbon nanotube based coaxial heterostructures. *Applied Physics Letters* 93:103108.
74. Sharma, H., Agarwal, D. C., Shukla, A. K., Avasthi, D. K., and Vankar, V. D. 2013. Surface-enhanced Raman scattering and fluorescence emission of gold nanoparticle–multiwalled carbon nanotube hybrids. *Journal of Raman Spectroscpoy* 44:12–20.
75. Zhang, Y., Franklin, N. W., Chen, R. J., and Dai, H. 2000. Metal coating on suspended carbon nanotubes and its implication to metal–tube interaction. *Chemical Physics Letters* 331:35–41.
76. Dresselhaus, M. S., Dresselhaus, G., Saito, R., and Jorio, A. 2005. Raman spectroscopy of carbon nanotubes. *Physics Reports* 409:47–99.
77. Dresselhaus, M. S., Jorio, A., Hofmann, M., Dresselhaus, G., and Saito, R. 2010. Perspectives on carbon nanotubes and graphene Raman spectroscopy. *Nano Letters* 10:751–758.

CHAPTER 4

CONTENTS

Overview on the Functionalization of Carbon Nanotubes

4

N.M. Mubarak, J.N. Sahu, J.R. Wong, N.S. Jayakumar, P. Ganesan, and E.C. Abdullah

4.1 INTRODUCTION

Carbon nanotubes (CNTs) are found to have wide applications and great potential usage in various fields such as nanobiotechnology, material science, and chemistry due to their astonishing structural, electrical, and mechanical properties (Agüí et al. 2008, Balasubramanian and Burghard 2006, Mita et al. 2007). However, difficulty is faced while applying CNTs for applications due to their low solubility in nearly all solvent and bundling between CNT tubules caused by the attractive van der Waals interaction force among the nanotubes. Solubility is a vital property for process ability as it concerns with the purification (Li et al. 2007). Thus, surface modification of CNTs should be made through the functionalization process in order to compensate the disadvantages of CNTs. Functionalization of CNTs is made through selective chemical reaction that depends on the functional group, which is required to attach to the surface of CNTs. With the presence of functional groups on CNTs, the solubility of CNTs in aqueous media will be improved depending on the selected functionalization method (Li et al. 2007, Salzmann et al. 2007). Nevertheless, researchers have shown that severe structure destruction of multiwalled carbon nanotubes (MWCNTs) could occur during functionalization, a damage including shortening or breaking of MWCNT length and decamping at the end of nanotubes. These structural damages could vary the variable properties of CNTs or even cause the loss of certain properties of nanotubes. The covalent functionalization of CNTs allows functional groups to be attached to the tube end or sidewalls. Covalent sidewall functionalization forms sp³ carbon sites on the tubes, which disrupts band-to-band transitions of π-electrons and may cause loss of properties of CNTs, such as their high conductivity and remarkable mechanical properties. Noncovalent functionalization is particularly attractive in comparison with covalent functionalization. There are three advantages: (1) it allows attachment of chemical handles without affecting the structure and electronic network of the tubes, (2) it utilizes a simple preparation procedure, and (3) it is applicable to as-produced full-length CNTs. The solubility of CNTs in a solvent depends on the type and concentration of the molecules adsorbed on the tube surface. CNT composites with a variety of noncovalent wrapping agents are described in the following three sections.

4.2 FUNCTIONALIZATION OF CNTs

CNTs have various diameter and length size distributions. Most of them depend on their structures, which depend on the electronic properties from metallic to semiconducting. The impurity of CNTs decreases the overall yield of usable material and will interfere with most of the desired properties of CNTs. Besides that, in all the forms of CNTs, they are difficult to disperse and dissolve in the in water or organic compounds as resistance to wetting. Another difficulty is to make composites of insoluble nanotubes with other materials (Hirsch and Vostrowsky 2005).

To overcome these problems, many researchers develop a new technique called functionalization of nanotubes, that is, the attachment of *chemical functionalities*. This functionalization can improve the solubility and processibility. In other words, it can produce a unique property that is combined with other types of materials (Ausman et al. 2000). Functionalization of CNTs with other chemical groups on the sidewall is attempting to modify the properties required for an application in hand. For example, chemical modification of the sidewalls may improve the adhesion characteristics of CNTs in a host polymer matrix to make functional composites (Chen et al. 1998). Modification of CNTs encouraged greater use of their intrinsic properties, as well as the capability to modify these properties. Noncovalent methods preserve the pristine CNT structure, while covalent modification introduces structural perturbations. A brief summary of functionalization of CNTs is discussed as follows.

4.2.1 Covalent Functionalization

In general, the term *covalent functionalization* comes from covalent bond. The covalent bond itself can be defined as a form of chemical bonding that is characterized by the sharing of pairs of electrons between atoms. The stable balance of attractive and repulsive forces between atoms when they share electrons is known as covalent bonding (Georgakilas et al. 2002). Covalent bonding commonly has many types of interactions, including σ-bonding, π-bonding, metal-to-metal bonding, agnostic reaction, and three-center two-electron bonds (Georgakilas et al. 2002). Covalent bonding does not necessarily require the two atoms be of the same elements. However, the covalency will have greatest results in the same element (Hauke and Hirsch 2010, Sinnott 2002). In this method, CNTs are functionalized via not reversible attachment of functional groups on the tips and/or sidewalls of nanotubes. This type of surface modification has been achieved by three different approaches that are thermally activated chemistry, photochemical functionalization, and electrochemical modification (Balasubramanian and Burghard 2005). However, this study only concerns on the means of thermally activated chemistry to avoid confusion of the scope of this research. Covalent functionalization can be subcategorized in several discrete groups listed in the following section. Unlike the noncovalent method, covalent functionalization on single-walled CNTs (SWCNTs) will perturb the electronic structure and cause an irreversible loss of double bonds within the nanotubes. These variations of structure may affect the conductive property, inhibiting the further CNT application. In the case of MWCNTs, new properties can be added to their structure without altering their inner structure of the nanotubes as well as their electronic structure.

In addition, covalent functionalization of CNTs allows the functional groups to be attached to the tube end or sidewalls. The loss of properties of CNTs, such as conductivity and mechanical properties, can be the cause of the formation of sp^3 carbon sites on the tubes. It was disturbing band-to-band transition of electrons. However, the covalent sidewall will significantly expand the utility of the nanotube structure by increasing the solubility of the tubes to organic solvent and provide anchoring sites on the tube wall to improve the interaction on the nanotubes with other materials (Kuzmany et al. 2004). The chemical oxidation of SWCNTs can be described by several processes: (1) oxidative attack of existing active sites such as sidewall functional groups or Stone–Wales defects, (2) generation of additional active sites, (3) generation of a vacancy in the graphene layer, and (4) consumption of the graphene sidewall around the vacancy or nanotube end. The consumption process can occur along the circumferential or axial direction of the nanotube and is the only process that affects length. Circumferential consumption will induce cutting of the nanotube and appears to preferentially leave an exposed armchair edge, while axial consumption will induce shortening of the nanotube (Ziegler et al. 2005).

SWCNTs posses two distinct regions of differing reactivities toward covalent chemical modification. The presence of five-membered rings at the caps leads to a relatively higher reactivity at these points compared to the reactivity of fullerene (Basiuk et al. 2004). By comparison, functionalization of the sidewall comprising the regular graphene framework is more difficult to accomplish. In general, addition reactions to the partial carbon–carbon double bonds cause the transformation of sp^2- into sp^3-hybridized carbon atoms, which is associated with a change from a trigonal-planar local bonding geometry to a tetrahedral geometry. This process is energetically more favorable at the caps due to their pronounced curvature in two dimensions, in marked contrast to the sidewall with its comparatively low curvature in only one dimension. On the other hand, the nonzero curvature makes the sidewall more reactive than a planar graphene sheet. In the same way, the binding energy of atoms or functional groups on the sidewall should increase with decreasing tube diameter. This tendency is supported by theoretical studies, as has been reported, for instance, for the bonding of alkyl radicals to the sidewall of SWCNT (Mylvaganam and Zhang 2004). On the contrary, the concave curvature of the inner surface of the nanotube imparts a very low reactivity toward addition reaction (Chen et al. 2004). CNTs have been proposed as nanocontainers for reactive gas atoms, analogous to fullerenes encapsulating nitrogen atoms.

In reality, however, nanotubes are not ideal structures but rather contain defects formed during synthesis. Typically 1%–3% of the carbon atoms of a nanotube are located at the defect site (Niyogi et al. 2002). A frequently encountered type of defect is the so-called Stone–Wales defect, which is comprised of two pairs of five-membered and seven-membered rings. A Stone–Wales defect leads to a local deformation of the graphitic sidewall and thereby introduces an increased curvature in this region. The strongest curvature exhibits at the interface between the two five-membered rings; as a result of this curvature, addition reactions are most favored at the carbon–carbon double bond in these positions (Balasubramanian and Burghard 2005). The presence of carboxyl groups leads to the reduction of van der Waals interaction between the CNTs, which strongly facilitates the separation of nanotubes into individual tubes. Additionally, the attachment of a suitable group renders the tube soluble in aqueous or organic solvent, opening the possibility of further modification through subsequent solution-based chemistry (Shiral Fernando et al. 2004).

In fact, not all covalent functionalizations of CNTs give a good result. Some barrier appears in developing and studying CNTs due to insolubility of CNTs in aqueous and organic solvent. The difficulties of covalent functionalities of CNTs occur when the structural changes develop in the orbital hybridization from sp^2 to sp^3. This behavior can give an undesirable deterioration in their optical, electrical, and mechanical properties. To overcome this behavior, the noncovalent functionalization of CNTs was developed. This method gives very less effect on the electrical structure of their curled graphene sheets. This is making a great interest due to promising practical uses in various applications.

4.2.2 Oxidation and Carboxyl-Based Couplings

The most common techniques for functionalization are via the liquid-phase or gas-phase oxidation of nanotubes, introducing carboxylic groups and some other oxygen-bearing functionalities such as carbonyl, hydroxyl, nitro, and ester groups into the tubes. This oxidation process involves ultrasonic treatment of CNTs in the mixture of concentrated nitric acid and sulfuric acid with a 1:3 volume-to-volume ratio (Balasubramanian and Burghard 2005). Such extreme condition leads to the opening of tube caps and the formation of holes in the sidewalls; then an oxidative is etched along the walls leading to the release of carbon dioxide. In addition, considerable side functionalization also takes place mainly because of the large aspect ratio of the CNT. As a result, products for this treatment are nanotube fragments with length in the range of 100–300 nm whose sidewall and ends are featured with high density of several oxygen-containing groups, mainly carboxyl group (Chen et al. 1998). An alternative way for this method is performed by refluxing the CNTs in concentrated nitric acid, or using oxygen, ozone, or air as oxidant at elevated temperatures, or the combination of nitric acid and air oxidation. Furthermore, the chemical modification is then limited mostly to the opening of the tube caps and formation of functional groups at defect sites along the sidewalls. CNTs functionalized in this manner basically retain their pristine electronic and mechanical properties. Any modification of the C–C bonding structure away from sp^2 bonding configuration will induce an impurity state in the gap region and significantly change the electronic states and conducting properties of the CNT (Balasubramanian and Burghard 2006, Zhao et al. 2004).

In general, oxidation treatment will introduce defects on the surface of the nanotube, oxidize the CNTs leading *hole doping*, and produce impurity states at the Fermi level of CNTs. Moreover, this treatment tends to purify the raw material by removal of impurities; oxidation can be used to etch or cut, shorten, and open the CNTs. The shortening and cutting of nanotube via oxidation often depend on the rate and extent of the reaction, which gives rise to a new length distribution. As a result, the cutting mechanism will leave the CNTs with open and oxygenated ends (Dillon et al. 1999). The oxidative stability depends on the tube's diameter as well as the production process responsible for the tube's dimension. A study has shown that smaller-diameter tubes are more rapidly air oxidized than larger-diameter tubes with the aid of the resonance-enhanced Raman radial breathing mode (Zhou et al. 2001). It is also proven in another study that smaller-diameter SWCNTs produced by the high-pressure carbon monoxide conversion process are more reactive toward ozone than larger-diameter SWCNTs generated by mixed concentrated nitric acid and sulfuric acid solution (Bahr and Tour 2002). For industrial applications, acid-treated CNTs often assembled on a number of surfaces such as highly oriented pyrolytic graphite, silver, and silicon.

To minimize the problem of poor matrix–SWCNT connectivity and phase segregation in the polymer–SWCNT composite, assembling the acid-treated, oxidized, and negatively charged SWCNTs layer by layer with the positively charged polymer (for instance, polyethyleneimine polyelectrolyte) is recommended. As a result, a nanometer-scale composite with 50 wt.% SWCNT loading can be achieved after subsequent chemical cross-linking. The attachment of appropriate functional groups on CNTs renders their solubility in aqueous or organic solvent and, meanwhile, increases the possibility of further modifications through subsequent solution-based chemistry. For instance, carboxyl functional group present in the CNTs is useful for further treatment by enabling the covalent coupling of molecules through the creation of amide and ester bonds with other particles (enzyme, nucleic acids, metal complexes, or semiconductor and metal nanoparticles) as shown in Figure 4.1. This characteristic is vital in this study as the immobilization of cellulase enzyme on functionalized MWCNTs requires the formation of the amide bond between these two components. Moreover, the presence of carboxyl groups reduces the van der Waals interactions between CNT molecules that enhance the separation nanotube bundles into singular tube.

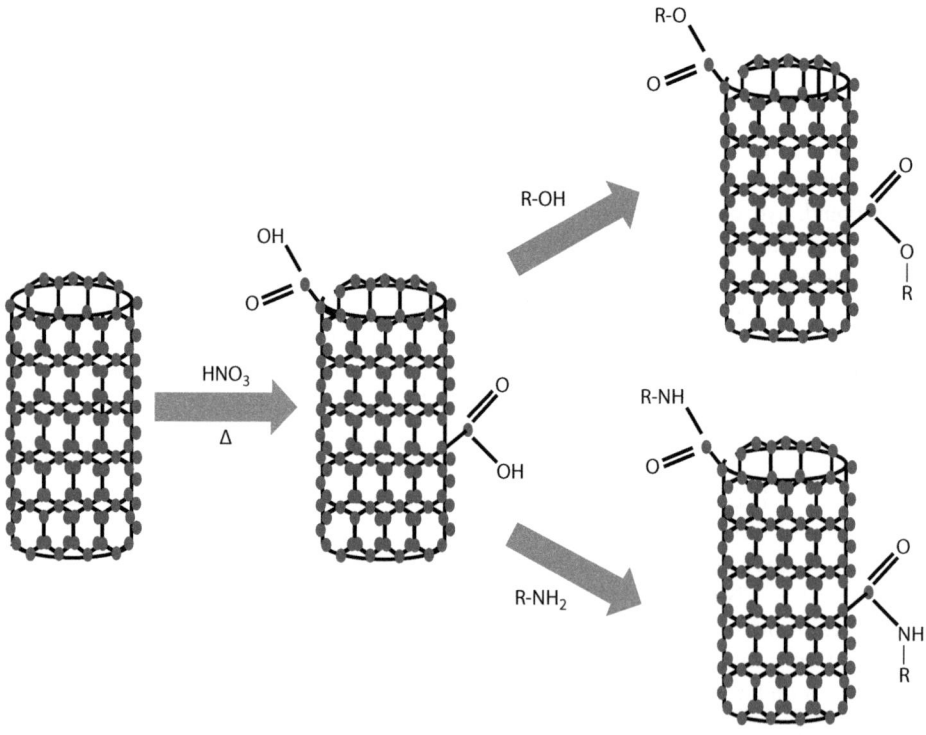

FIGURE 4.1
Surface modification of CNTs through thermal oxidation followed by subsequent amidization or esterification of carboxyl groups. (Adapted from Balasubramanian, K. and Burghard, M., *Small*, 1(2), 180, 2005.)

4.2.3 Defect Functionalization: Sidewall Addition Reaction

This method is correlated to the previous method described earlier as it involves a two-step functionalization including the introduction of carboxyl group via oxidation, followed by the formation of ester or amide linkage (Jiang et al. 2004), which is shown in Figure 4.2 in the previous section. As compared to the acid treatment, the influence on mechanical and electronic properties of CNTs is relatively weak (Jiang et al. 2004). In general, this method allows direct coupling of functional groups onto the carbon framework of the nanotubes. This direct coupling effect greatly enhances the solubility of CNTs. Under this category of functionalization, it can be further allocated into three subcategories with different sidewall addition reactions such as amidation, esterification, and thiolation (Hirsch and Vostrowsky 2005). In the amidation process, carboxamides can be formed via carboxylic acid chlorides and allow for the decoration of oxidized tubes with aryl amines, aliphatic amines, peptides, amino acid derivatives, and amino-group-substituted dendrimers as nucleophiles. The carboxylic groups can be activated by conversion into acyl chloride groups with thionyl chloride, and the acyl chlorides formed can be transformed to carboxamides by amidation as shown in Figure 4.2. For instance, the conversion of the acid functionality of SWCNTs to the amide of octadecyl amine will lead to the shortened and soluble SWCNTs.

In addition, the acyl chloride–functionalized SWCNTs (SWCNT–COCl) are susceptible to react with other nucleophiles such as alcohol (for instance, octadecyl alcohol) for the formation of soluble ester-functionalized SWCNTs. Besides, attempts had been made for the preparation of soluble polymer-bound and dendritic ester-functionalized SWCNTs achieved by attaching poly(vinyl acetate-co-vinyl alcohol) and hydrophilic and lipophilic dendron-type benzyl alcohols, respectively, to the SWCNT-COCl. These functional groups could be detached under basic and acidic hydrolysis conditions, and therefore additional evidence for the nature of the attachment was provided (Riggs et al. 2000). Several types of SWCNT–ester are illustrated in Figure 4.3. In the thiolation process, the thiol group is directly attached to the nanotubes via subsequent carboxylation with sonication, reduction with the aids of sodium borohydride ($NaBH_4$), and chlorination with thionyl

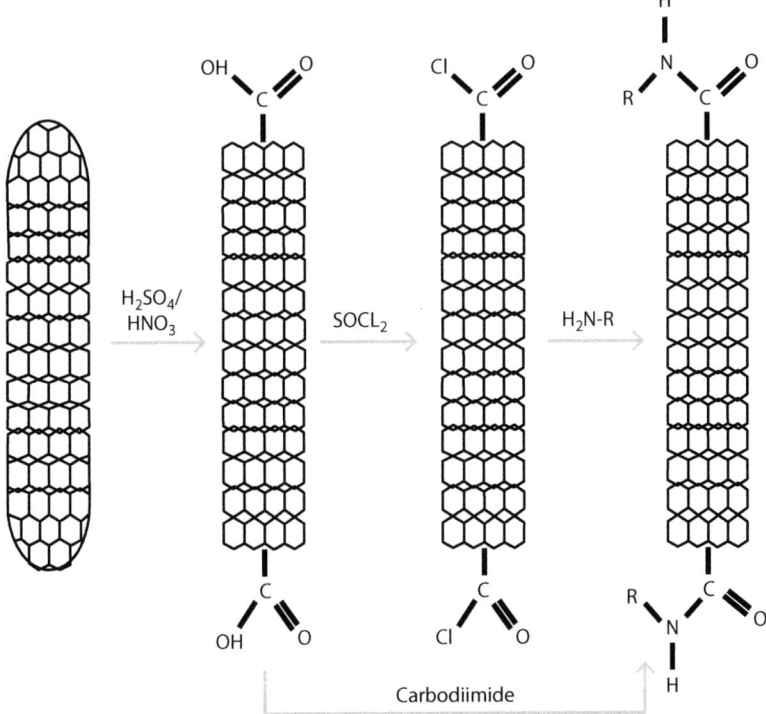

FIGURE 4.2
Schematic representation of oxidative etching of SWCNTs followed by treatment with thionyl chloride and subsequent amidation. (With kind permission from Springer Science+Business Media: *Functional Molecular Nanostructures*, Functionalization of carbon nanotubes, 2005, pp. 193–237, Hirsch, A. and Vostrowsky, O., (ed.) Schlüter, A.D., Springer, Berlin, Germany.)

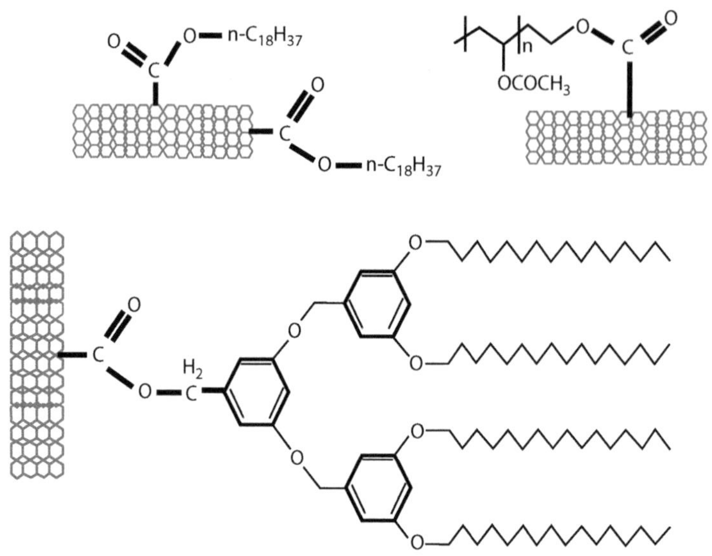

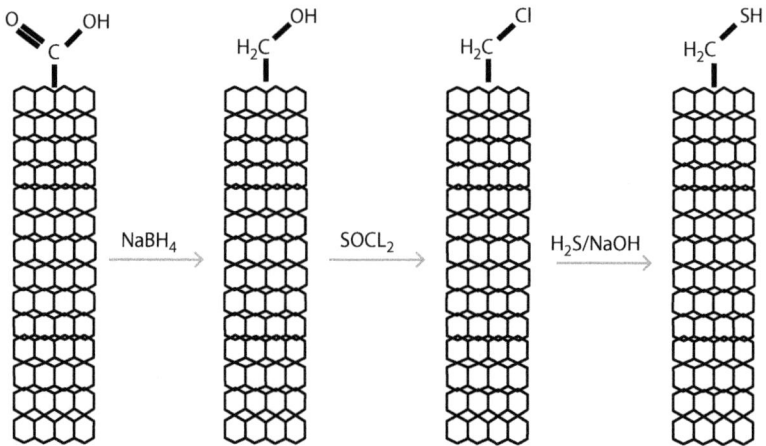

FIGURE 4.3
Schematic representation of SWCNT–ester. (With kind permission from Springer Science+Business Media: *Functional Molecular Nanostructures*, Functionalization of carbon nanotubes, 2005, pp. 193–237, Hirsch, A. and Vostrowsky, O., (ed.) Schlüter, A.D., Springer, Berlin, Germany.)

FIGURE 4.4
Schematic representation of thiolation of CNTs. (With kind permission from Springer Science+Business Media: *Functional Molecular Nanostructures*, Functionalization of carbon nanotubes, 2005, pp. 193–237, Hirsch, A. and Vostrowsky, O., (ed.) Schlüter, A.D., Springer, Berlin, Germany.)

chloride ($SOCl_2$), followed by thiolation with the mixture of sodium sulfite (Na_2S) and sodium hydroxide (NaOH) to the open end of the CNT as shown in Figure 4.4.

Apart from the three processes discussed earlier, there are other reactions that can be categorized into the sidewall addition reaction such as halogenations and hydrogenation; however, for ease of discussion, those reactions are broken down into the subsequent different sections. An overview of certain addition reactions on the sidewall of CNTs is listed in Figure 4.5.

4.2.4 Hydrogenation of Carbon Nanotubes

Generally, hydrogenation of CNTs was the first reaction sequence that is based on the reduction of CNTs with alkali metal in liquid ammonia, which is also known as the Birch reduction (Guldi 2010). The most common approach for the hydrogenation of CNTs is through the lithium

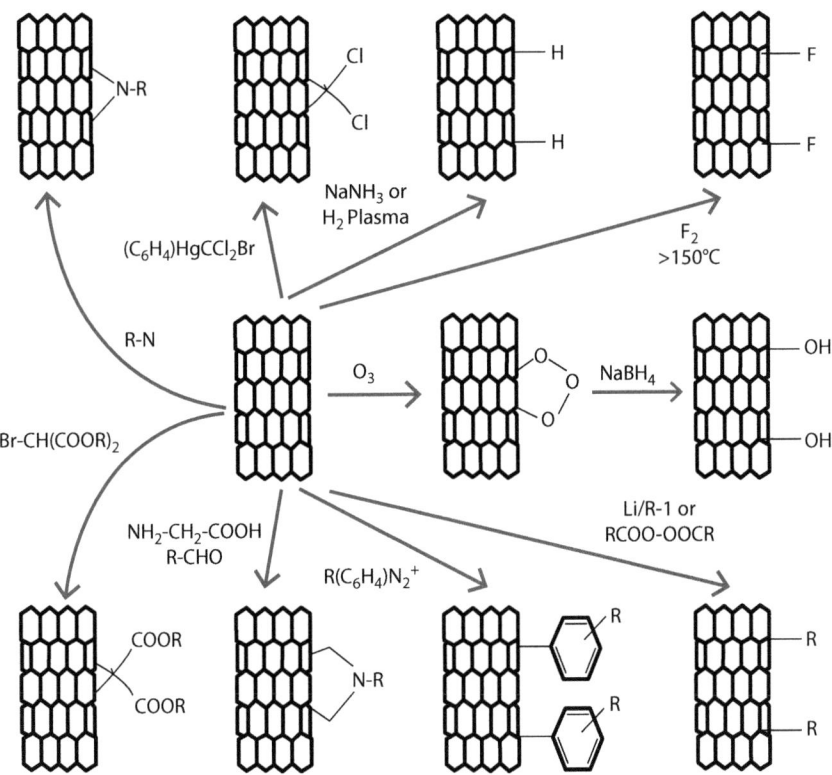

FIGURE 4.5
Overview of functionalization of the nanotube sidewall with addition reactions. (Adapted from Balasubramanian, K. and Burghard, M., *Anal. Bioanal. Chem.*, 385(3), 452, 2006.)

dispersed in diaminoethane. Typically, 200 mg of carbon and 100 mg of Li were mixed in a glove box, and the reaction tube was then attached to a vacuum line; after that, dried NH_3 (20 mL) was condensed to the reaction mixture from a Na/NH_3 solution; such process can be considered as a modified Birch reduction method. Besides, the applied method also similar to the reactions of C_8H graphite intercalated compound with weak protic acids in tetrahydrofuran suspension, which give rise to the formation of partially reduced graphite with the composition of C_8H (Bergbreiter 1985). Based on the investigation, it shows that the protonation of reduced Li would tend to also reduce the SWCNT intermediates with methanol leading to the formation of hydrogenated SWCNTs (Krueger 2008). No obvious preference for the reaction of metallic and small-diameter tubes was observed during the hydrogenation process, which is in stark contrast to the alkylation reactions. Past research also pointed out that the degree of functionalization is considerably lower than that observed for the corresponding alkylation (Jung et al. 2006). The resulting products of the dissolved metal reductions are covalently bonded CH derivatives, and they are principally different from the absorbed H_2 derivatives used for hydrogen storage (Jones and Bekkedahl). Moreover, the strongly bound hydrogen releases from the materials can be achieved at around 500°C.

The attachment of hydrogen atom onto nanotube is always an exothermic reaction. It is essential to take note that the hydrogenated derivatives were thermally stable up to 450°C, while above this temperature, a characteristic decomposition would start to take place (Mickelson et al. 1998). Generally, hydrogen bound to SWCNTs should not be released until it reaches 500°C, which indicates a robust attachment. It is reported that the organic hydrocarbon with the formula of CnHnS was estimated to be stable up to the radius of an (8,8) nanotube, with binding energies proportional to 1/R. Throughout the calculation, zigzag nanotubes were found more likely to be hydrogenated than armchair tubes with equal radii due to the differences in their electronic structures. Hydrogenation occurred even on the inner tubes of MWCNTs, as revealed by the overall corrugation and its chemical composition. Despite the fact that graphite has a lower reactivity as compared to CNTs due to the curvature, turbostratic graphite flakes and powder would result in a C/H ratio of 4:1 upon hydrogenation (Yildirim et al. 2001). A novel access route–hydrogenated SWCNT is provided via the use of high-boiling polyamines as hydrogenation reagents as polyamine-based hydrogenated CNTs possess both efficient and thermally reversible properties (Marshall et al. 2006).

4.2.5 Halogenation on Carbon Nanotube

In this category, halogen like fluorine, chloride, and bromide is attached to the sidewall of CNTs by a process called halogenation. Mickelson et al. (1998) reported the first nondestructive and extensive controlled sidewall fluorinations of SWCNTs using elemental fluorine as the fluorinating agent. The highest degree of fluorination could be achieved with a CNT–fluoride composition by using iodine pentafluoride, IF_5. However, the fluorinated CNT exhibits significant changes in its spectroscopic properties that are confirmed by Raman scattering and UV absorption, and it showed that there was an electronic perturbation that was not found in the previous oxidized CNTs. The electrical conductivity of the fluorinated SWCNTs differed dramatically from the pristine SWCNT where untreated SWCNTs were reported to possess a resistance of 10–15 Ω, whereas fluorinated SWCNTs had a resistance of more than 20 MΩ (Hirsch and Vostrowsky 2005). This fluorination may result in modifying the electronic properties of the tube to be either metallic or semiconducting depending on the method of fluorine application. The major drawback of this mode of functionalization is the high degree of fluorine atom addition, reflecting a greater number of tube defects and damage. However, this feature made the fluorinated CNT become highly soluble and open new avenues of solution-phase chemistry. Once fluorinated, the CNTs can serve as a staging point for a wide variety of further sidewall chemical functionalizations (Gu et al. 2002).

The second attempt that has been made in this category is the chlorination process in which the origin of the idea was to substitute carbon atoms within the hexagon network of nanotubes with silicon atoms, thus gaining an sp^3-like stable defect center as a trapping site for the chemisorption of other atoms or molecules. Then these defects can be readily modified to silicon halide derivatives by different routes or the silicon halide complex directly incorporated (Fagan et al. 2003). The halogen atoms are coupled to the nanotube lattice via electrolytical evolution of chloride onto an anode made from CNTs on foil. Moreover, other oxygen-bearing groups such as carboxyl and hydroxyl are formed at the same time enhancing the solvation of the CNTs in alcohol or water. The chlorinated CNTs can be further converted to triphenylmethyl lithium or sodium amide to add the corresponding functional groups. The last attempt for halogenation is through bromination where the low susceptibility of CNTs to bromination was mainly utilized as a means of purification for MWCNTs contaminated by other carbon particles (Fagan et al. 2003, 2004).

4.2.6 Substitution Reaction on Fluorinated Carbon Nanotubes

For the past few decades, the direct sidewall addition of elemental fluorine was one of the first efficacious alternations of sp^2 carbon framework. Different pathways of functionalization leading to the fluorinated CNTs have been reviewed and developed, including the plasma-based functionalization (Holzinger et al. 2001). The reaction involved is applicable only for fluorinated nanotubes. Fluorinated nanotubes have become a widely used starting material for further chemical transformation steps; throughout this method, the fluorine atom within the nanotube is replaced via nucleophilic substitution reaction, therefore enabling a flexible approach in providing the sidewalls with other types of functional groups as illustrated in Figure 4.6. By using amine (such as n-butylamine) or bifunctional amines (such as polyethyleneimines and alkoxysilanes), a facial introduction of functionality on the basis of sidewall derivatives is possible. A study has shown that the reaction of fluorinated HiPCO SWCNTs with ω-amino acids could lead to water-soluble SWCNT derivatives. Furthermore, the solubility of those fluorinated CNTs is controlled by the length of the hydrocarbon chain of the amino acids and is highly dependent on the pH value of the solution (Ashcroft et al. 2006).

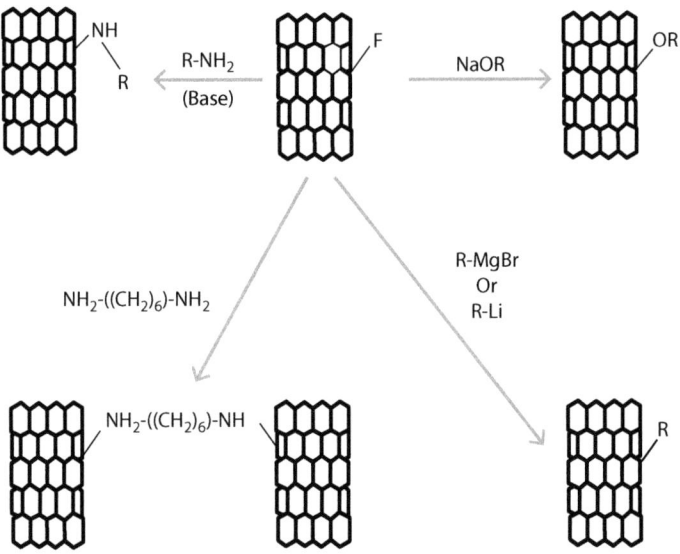

FIGURE 4.6
Functionalization of the sidewall through nucleophilic substitution reaction in fluorinated
nanotubes. (Adapted from Balasubramanian, K. and Burghard, M., *Anal. Bioanal. Chem.*, 385(3),
452, 2006.)

A study has shown that about 15% of the carbon atoms of the sidewall have been successfully
bearing a functional group when chemicals like alcohols, amines, nucleophiles, alkyl lithium compounds, and Grignard reagents are employed in the functionalization (Fagan et al. 2003). These
reactions are thought to proceed via a concentrated, allylic displacement mechanism, as backside
SN_2 attack is not possible and the stability of a cationic SN_1 intermediate is uncertain. An extension of this displacement capability to functional group–tolerant reagents will further the utility of
this chemistry. In addition, this bifunctional derivative demonstrates improved dispersibility in solvents like water and dimethylformamide. Another promising route to fluorinated CNTs is achieved
through the coupling of free thiol or thiophene group to gold nanoparticles, and an investigation
is made to the corresponding conjugates through the use of atomic force microscopy (AFM) and
scanning tunneling microscopy (STM). Based on the results, it is shown that both derivatives yield
a similar degree of functionalization, but the only difference is the distribution of the functional
groups. Thus, a statement can be made that the AFM and the complexation of chemical marker
nanoparticles are an astonishing method for the determination of the presence of functionalization
on the SWCNT sidewall, whereas STM provides excellent information regarding the location of
substituents (Guldi 2010).

4.2.7 Addition of Radicals

The experimental proof for a covalent sidewall functionalization of SWCNTs via radical species
was given after 1 year from the theoretical prediction. Generally, those reactive radical species can be generated from various precursors either photophysical or thermally route (Gergely
et al. 2007). The addition of perfluorinated alkyl radicals, attained by photoinduction from heptadecafluorooctyl iodide, to SWCNTs would yield a perfluorooctyl-derivatized CNTs. Aryl or
alkyl peroxides are commonly used as radical initiators. In this method, a thermal cleavage of
lauroyl and benzoyl peroxides leads to the generation of highly reactive alkyl and phenyl radicals
that would attack the SWCNTs, leadings to sidewall derivative materials that have been characterized by a broad variety of spectroscopic methods (Yokoi et al. 2005). The combination of
the benzoyl peroxide as a radical starter and alkyl iodines permits the covalent attachment of a
huge variety of different functionalities to SWCNTs. Moreover, this concept is also applicable
for grafting polymers onto the sidewalls of MWCNTs. For example, treated poly(oxyalkylene)
amines, equipped with a terminal maleic anhydride, with the aid of benzoyl peroxide, lead to the
intermediate formation of a free radical species that is chemically grafted onto the sidewall of

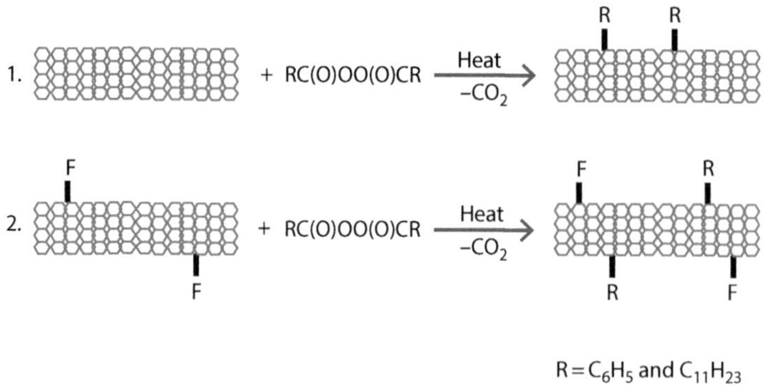

$R = C_6H_5$ and $C_{11}H_{23}$

FIGURE 4.7
Functionalization of SWCNTs and fluorinated F-SWCNTs with benzoyl ($R = C_6H_5$) and lauroyl ($R = C_{11}H_{23}$) peroxides. (Adapted from Ni, B. and Sinnott, S.B., *Phys. Rev. B*, 61(24), R16343, 2000.)

the dispersed MWCNTs (Oh et al. 2008). The concept of radical-based functionalization can be further extended to the generation of highly functional SWCNT building blocks throughout the reaction between SWCNTs with glutaric or succinic acid acyl peroxides. A study has shown that a 90% excess yield of sidewall carboxylic acid–functionalized SWCNT materials can be produced by grinding SWCNTs with succinic acyl peroxide and heating at a corresponding power at a temperature of 100°C for approximately 4 min (Dyke and Tour 2003). In the cases of DWCNTs, the thermal or ultrasonic decomposition of the broad variety of azobis compounds could lead to sidewall-derivatized DWCNTs exhibiting heterofunctionalities such as hydroxyl and carboxyl acid moieties. These functional groups on the sidewall of DWCNTs serve as a reactive site to ease the covalent coupling effect of DWCNT with other materials such as platinum or rhodium nanoparticles (Dyke and Tour 2003). An alternative way for the addition of radicals is made through the reaction of organic peroxide free radical with pristine CNTs or fluorinated CNTs as shown in Figure 4.7. Throughout the solution-phase UV/Vis–NIR spectra, it indicates that there is complete loss of the van Hove absorption band structures, after the functionalized SWCNT with the free radical. This is mainly due to the disruption of graphene π-bonded electronic structure of the sidewalls of the nanotubes (Peng et al. 2003). In addition, radical-based functionalization process sequences can be used to attach well-defined polymers onto CNTs resulting in sidewall composite materials. For instance, polystyrene-based copolymers and homopolymers can be grafted onto SWCNTs via a radical coupling reaction involving polymer-centered radicals with the aids of loss of stable free radical nitroxide capping agents (Nakamura et al. 2007). The resulting polymer–SWCNT composites possess high solubility in a variety of organic solvents. In other approaches, the radicals generated are added to the chain end of MWCNTs to covalently functionalized polymer–MWCNT composite materials.

4.2.8 Addition of Nucleophilic Carbenes

The first functionalization of CNT sidewalls with a carbon-based substituent has been reported by Haddon since 1998 (Hu et al. 2003). Dichlorocarbene can be generated in situ from chloroform with potassium hydroxide or from phenyl (bromodichloromethyl) mercury covalently attached to the soluble SWCNTs. Each covalently bound imidazolidine addend bears a positive charge, whereas the other one negative charge/addend would be transferred to the delocalized tube surface as shown in Figure 4.8. Most of the derivative nanotubes were highly soluble in a solvent like dimethyl sulfoxide, which permits the ease of separation of unreacted and insufficiently functionalized CNTs. In general, this electrophilic carbene addition method only gave a low functionalization degree with about 2%; however, the degree of functionalization can be further increased up to 23% with the use of a soluble alkyl chain–functionalized HiPCO SWCNT as a starting material. The strength of the interband electronic van Hove transitions in this adduct is suppressed due to the presence of sp³-hybridized carbon atoms within the sidewalls, based upon the conversion of metallic SWCNTs into semiconducting in the additional

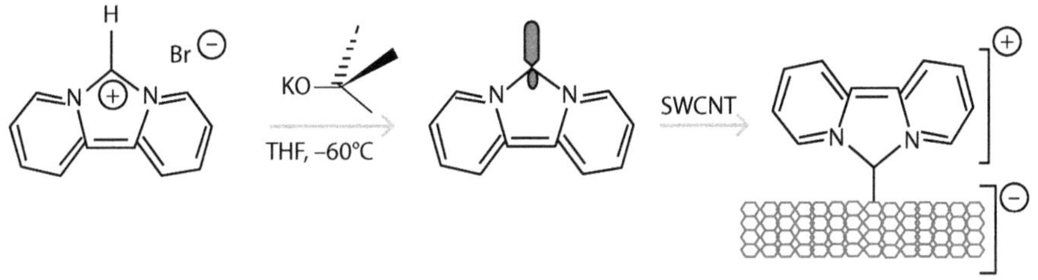

FIGURE 4.8
Sidewall functionalization by the addition of nucleophilic dipyridyl imidazolidine to the electrophilic SWCNT π-system. (With kind permission from Springer Science+Business Media: *Functional Molecular Nanostructures*, Functionalization of carbon nanotubes, 2005, pp. 193–237, Hirsch, A. and Vostrowsky, O., (ed.) Schlüter, A.D., Springer, Berlin, Germany.)

steps (Holzinger et al. 2003). In the early study, it has been shown that the SWCNT sidewall could also be attacked by the nucleophilic carbene. Fagan et al. (2003) reported that a method for the production of zwitterionic polyadducts is reported via the reaction of an electrophilic SWCNT π-system with the nucleophilic dipyridyl imidazolidine with the electrophilic. In comparison to the proposed cyclopropane structure in the case of dichlorocarbene, the addition of nucleophilic dipyridyl imidazolidine would lead to a zwitterionic reaction product because of the astonishing stability of the resultant aromatic 14π-perimeter of the imidazole addend (Moghaddam et al. 2004). Although the gain in solubility of these carbon derivatives could also be accomplished with the use of other chemical decorations of the nanotubes, the doping that occurs on the surface of the tubes in this method offers a new way to modify tube properties, thus leading to a controlled intervention into the CNT's electronic properties. In addition, Tagmatarchis et al. (2002) reported that an interesting approach has been made in the year 2002, that is, the functionalization of SWCNTs through electrophilic addition of $CHCl_3$ in the presence of $AlCl_3$. Then, hydroxyl-functionalized SWCNTs were obtained from hydrolysis of the coproduced labile chlorinated intermediate species. Next, the coupling with propionyl chloride would lead to the formation of the corresponding ester SWCNT derivatives. These functionalized SWCNTs can be used as the starting materials to graft polystyrene from the surface throughout atom transfer radical polymerization (ATRP). Typically, these ATRP initiators are attached to the SWCNTs by esterification of 2-chlorophyll chloride with the surface-bound hydroxyl functionalities (Yinghuai et al. 2005). This functionalization method does not allow the identification of the structure of modified tubes but also offers an improved solubility in organic solvent. Moreover, subsequent polymerization with styrene can be made on these functionalized SWCNTs to yield styrene-grafted SWCNT composite materials. The concept of the Friedel–Crafts-type electrophilic addition to the SWCNT sidewall has been generalized to develop a microwave-based functionalization sequence that allows the use of a broad variety of alkyl halides for the (Usrey et al. 2005). Besides, the polyacylation of SWCNTs under Friedel–Crafts conditions is probable by the reaction of sodium chloride with acyl chloride and aluminum chloride at high temperatures. It is essential to take note that the acylated materials (e.g., the use of perfluorinated acyl chlorides leads to the formation of perfluoroalkyl-decorated SWCNTs) produced are highly soluble in organic solvent (Wang et al. 2007). In addition, sidewall acylation of MWCNTs with various four-substituted benzoic acid derivatives can be achieved by direct Friedel–Crafts acylation in mild phosphorous pentoxide/polyphosphoric acid. The resulting functionalized MWCNTs take the form of bundles with average diameters of 40–70 nm, depending upon the polarity of the surface functionality introduced. Moreover, thermal and dispensability properties of the functionalized MWCNTs are greatly influenced by the nature of the group attached to the surface of nanotubes (Vigolo et al. 2009).

4.2.9 Other Approaches for Covalent Functionalization

The rapid growth of the biotechnology industry has led to the continued growth of methodology for the functionalization of CNTs. There are several rising-up procedures for the CNT surface modification such as electrochemical functionalization with the use of aryl diazonium through

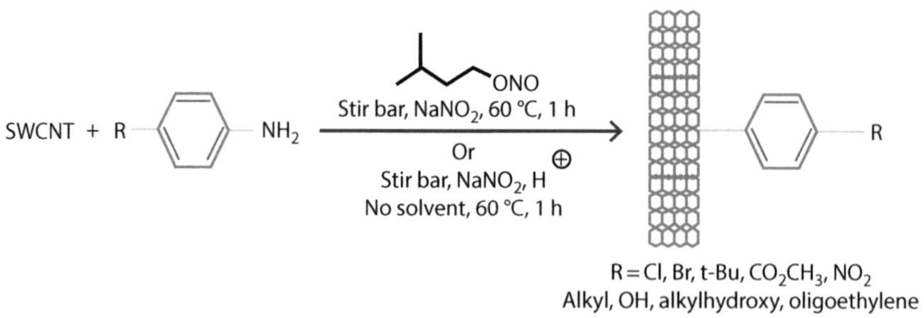

FIGURE 4.9
Schematic representation of solvent-free functionalization of the nanotube. (Adapted from Lalonde, J. and Margolin, A.: Immobilization of enzymes, *Enzyme Catalysis in Organic Synthesis: A Comprehensive Handbook*, 2nd edn., Wiley Press, New York, pp. 163–184, 2002. Copyright Wiley-VCH Verlag GmbH & Co. KGaA. Reprinted with permission.)

cathodic and anodic couplings. The functionalization is achieved when the corresponding reactive aryl radicals are generated from the diazonium salts via one-electron reduction as shown in Figure 4.9. At the end of the reaction, the resulting tubes remained unbundled throughout their entire lengths and it was incapable of reroping (Lalonde and Margolin 2002). This method would synthesize a very high degree of functionalization, up to one in nine carbons on the nanotube having an organic addend. Furthermore, researches have been made for the functionalization cycloaddition onto the nanotube such as carbene (e.g., dichlorocarbene) addition, cycloaddition of alkoxycarbonyl nitrenes obtained from azides, nucleophilic cyclopropanation, and 1,3-dipolar addition of azomethine ylides (Ni and Sinnott 2000). In the cases of Diels–Alder cycloadditions of o-quinodimethane to the sidewall of CNTs, it is most often obtained via the aromatic stabilization of the transition state and reaction product. In this reaction mechanism, the double bonds of SWCNTs act as dienophiles (Carlsson 2006). The rate of the Diels–Alder addition reaction is improved by the use of electron-withdrawing substituents on the monoene (the *dienophile*), with fluorinated SWCNTs used as starting materials (Ballesteros et al. 2007). Generally, these SWCNTs with activated sidewalls have been thermally reacted with 2,3-dimethyl-1, 3-butadiene, anthracene, and 2-trimethylsiloxy-1, 3-butadiene yielding highly functionalized SWCNT derivatives (Ménard-Moyon et al. 2006).

The concept of nucleophilic sidewall addition can even extend to other substance classes that contain heteroatoms. In a recent study, it has been shown that in situ generated n-propylamine can be covalently attached to the sidewalls of HiPCO SWCNTs through a direct nucleophilic addition reaction (Fantini et al. 2007). The reaction of n-propylamine with n-butyllithium could yield the corresponding amide, which would attack the SWCNTs in an attempt to generate negatively charged SWCNT derivatives that are consequently reoxidized by air into the corresponding neutral amino-functionalized SWCNT derivatives (Louie 2001). These reaction product exhibits a significantly improved solubility in nearly all the organic solvents, and the first concentration-dependent studies imply that the functionalization degree is correlated to the concentration of amide used for the sequence of functionalization (Charlier 2002). Research has also discovered a convenient route for sidewall functionalization of CNTs by reacting aryl diazonium salts with SWCNTs in an electrochemical reaction, with the aid of bucky paper as the working electrode. By varying the acyl diazonium salts involved in the reaction, a broad range of highly functionalized, well-soluble SWCNT adducts became feasible (Bahr et al. 2001). In this method, the corresponding reactive aryl radicals are synthesized from the diazonium salts throughout the one-electron reduction. Afterward, this concept was adopted for a solvent-based thermally induced reaction with the site generation of the diazonium compound by reacting the isoamyl nitrile with the aniline derivatives (Bahr and Tour 2001).

4.3 NONCOVALENT FUNCTIONALIZATION

The noncovalent functionalization process has an advantage over covalent functionalization due to its ability of not altering the system of the sidewall of the nanotubes. Hence, the final structural properties of the CNTs after functionalization won't be affected yet the solubility will increase drastically (Chen et al. 2003). Furthermore, through this method, the CNTs are functionalized by aromatic compounds, surfactants, and polymers (Shim et al. 2002). This method type is based on hydrophobic, van der Waals, or π–π-interactions. The major advantage of this method is that it would not perturb the electronic structure of CNTs, particularly SWCNTs. Apart from the electronic structure, the chemical structure of the π-network of CNTs is also not disrupted except for the CNT's length that shortens due to sonication involved in this modification process, but most of the physical properties of CNTs are well preserved in this approach. However, functionalization by noncovalent means is difficult to control and difficult to characterize. It involves the use of the weak force and/or bond between molecules, which is hard to calculate. Besides, especially in solution, other interactions could occur and eventually lead to the replacement of the molecule in the nanotube by a solvent molecule. This method involves the use of aromatic compounds, surfactants, and other reagents to carry out the functionalization. For instance, surfactant is used to exfoliate the bundles of SWCNTs as well as ionic liquids in order to promote CNT manipulation and further reactions, the use of polymers in wrapping DNA molecules, and the attachment of protein to CNTs by noncovalent interaction by the use of biosensor application (Jiang et al. 2004) and endohedral fullerenes, which is the trapping of molecules inside the CNT structure via noncovalent functionalization (Tasis et al. 2006). Chen et al. (2001) produced a new method of noncovalent functionalization through immobilization of biological molecules in the walls of CNTs to enhance the production of an advanced biosensor material based on the component of pyrene. On the other hand, polymers act as an excellent wrapping material due to the π–π-stacking and van der Waals interactions between the polymer molecules and the surface of CNTs (Balasubramanian and Burghard 2006). Due to this success of wrapping, surfactant polymers have been assigned to functionalize CNTs as well. On the same note, introduction of surfactants on the surface of CNTs also helps CNTs to get rid of the weak van der Waals forces by replacing it with electrostatic forces (Hu et al. 2009). The deprivation of the usage of surfactants for the functionalization of CNTs is that it is harmful and toxic for biological applications. Hence, rather than using it thoroughly for every biological applications, the usage is limited for the biomedical applications only (Rey et al. 2006, Sinha and Yeow 2005).

The nature of the polyaromatic graphitic surface of CNTs has allowed it to easily bind with aromatic molecules via π–π-stacking (Chen et al. 2002). By taking these advantages, CNTs can be easily functionalized with molecules like purine derivatives, aromatic DNA base unit (Moon et al. 2008), fluorescein (Nakayama-Ratchford et al. 2007), and porphyrin derivatives (Guldi et al. 2005) via noncovalent functionalization. In biological applications, an ideal noncovalent functionalization coating on CNTs should meet the following criteria that are ecofriendly (materials used are nontoxic and biocompatible) and strong with a stable coating where it is high enough to prevent detachment from the CNT surface in biological solutions (for instance, serum that contains high protein and salt contents), and the amphiphilic coating molecules should possess very low CMC values so that the CNT coating remains stable after the removal of a large portion of excess coating molecules from the CNT suspension (Liu et al. 2009). In addition, the coating molecules should comprise functional groups that are available for bioconjugation with molecules like antibodies to create functionalized CNTs coupled with a different functional group for specific applications. Figure 4.10 shows the irreversible absorbing of pyrenebutanoic acid, succinimidyl ester onto the sidewall of SWCNTs via π-stacking. The anchored pyrene moieties on SWCNTs are highly stable against desorption in aqueous solutions. Thus, it eventually leads to the functionalization of SWCNTs with succinimidyl ester groups that are specifically reactive to nucleophilic substitution by amine groups in the protein, resulting in the formation of an amide bond (Hirsch and Vostrowsky 2005).

In addition, a study has been made on the noncovalent interaction between CNTs and six-membered ring molecules such as cyclohexane, benzene, and 2,3-dichloro-5,6-dicyano-1,4-benzoquinone with the aid of first-principles calculations (Balavoine et al. 1999). From the analysis research, it is demonstrated that hybridization between the nanotube valence bands and the benzoquinone molecular level transform the semiconductor tube into a metallic one. Besides, the coupling of π-electrons between tubes and aromatic molecules was observed, which indicates that noncovalent functionalization of CNTs by aromatic molecules might be effective for the contribution of the electronic properties of CNTs. It was suggested that immobilization

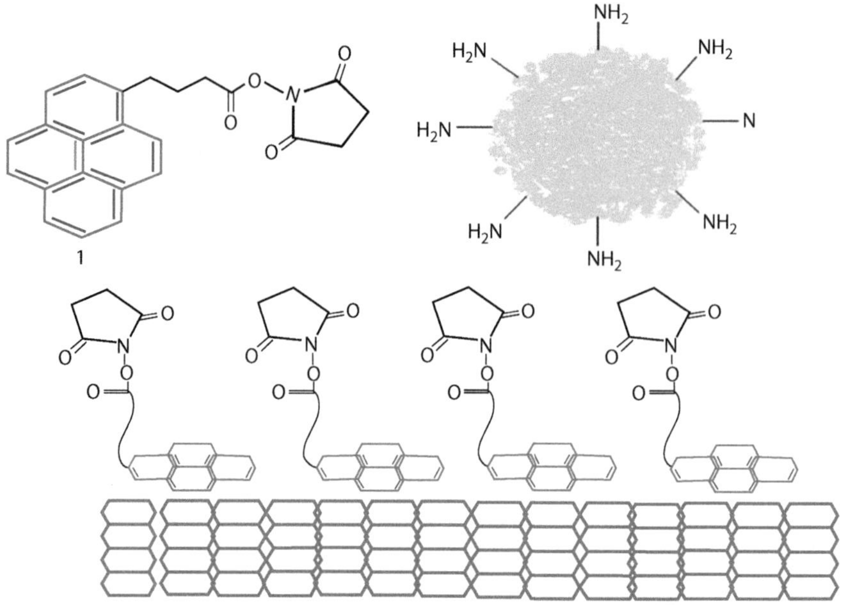

FIGURE 4.10
1-Pyrenebutanoic acid succinimidyl ester irreversibly adsorbed onto the sidewall of an SWCNT via p-stacking. Amino groups of a protein react with the anchored succinimidyl ester to form amide bonds for protein immobilization. (Adapted from Chen, R.J. et al., *Proc. Natl. Acad. Sci.*, 100(9), 4984, 2003.)

is strong and physical and does not require covalent bonding; and taking the example illustrated in Figure 4.1, the succinimidyl group of functionalized SWCNTs could nucleophilically substitute the primary or secondary amino groups from proteins such as streptavidin or ferritin, which allowed the immobilization of the biopolymers on the tubes (Sun et al. 2001).

Furthermore, polymer coating of CNTs can also be achieved via noncovalent functionalization with purified MWCNTs or SWCNTs. From the obtained results, it is demonstrated that this is a high-yield and nondestructive method for the purification of MWCNTs (Negra et al. 2003). The polymer will host selectively suspended nanotubes relative to impurities, giving a 91% pure nanotube material after removal by filtration. The wrapping of SWCNTs with polymers carrying polar side chains such as polystyrene sulfonate or polyvinylpyrrolidone would lead to a stable solution of the corresponding SWCNT/polymer complexes in water. The wrapping of polymer ropes around the tube lattice of MWCNTs occurs in a well-ordered periodic fashion. On the other hand, at low loading fractions, the polymer intercalated between the nanotubes would lead to an unraveling of ropes and cause a significant decrease of the interactions between the individual SWCNTs (Murphy et al. 2002).

4.4 SUMMARY AND PROSPECTIVE

The unique properties of CNTs make it desirable for different applications, particularly in the separation process such as wastewater treatment and protein separation drug delivery. Due to its hydrophobicity properties, CNTs applied in wastewater treatment require functionalization, such as changing some of the graphite properties to make nanotubes soluble in different media or attaching different groups or even inorganic particles for future utilization of modified nanotubes. The functionalization process is very crucial in the modification of the adsorbent surface in the adsorption process of organic compounds, especially in increasing the hydrophilic properties of adsorbents such as CNTs in water and increasing the availability of functional groups such as carboxylic, hydroxyl, carbonyl, and lactonic. Chemical modification of the adsorbent surface is the most common method used throughout the literature due to its ease in preparation. Hence, modification of CNTs creates various applications in present and future applications of nanotechnology.

ACKNOWLEDGMENT

This research is financially supported by the University of Malaya, Ministry of Higher Education, High Impact Research program (UM.C/HIR/MOHE/ENG/20).

REFERENCES

Agüí, L., Yáñez-Sedeño, P., and Pingarrón, J. M. 2008. Role of carbon nanotubes in electroanalytical chemistry: A review. *Analytica Chimica Acta*, 622(1), 11–47.

Ashcroft, J. M. et al. 2006. Functionalization of individual ultra-short single-walled carbon nanotubes. *Nanotechnology*, 17(20), 5033.

Ausman, K. D. et al. 2000. Organic solvent dispersions of single-walled carbon nanotubes: Toward solutions of pristine nanotubes. *The Journal of Physical Chemistry B*, 104(38), 8911–8915.

Bahr, J. L. and Tour, J. M. 2001. Highly functionalized carbon nanotubes using in situ generated diazonium compounds. *Chemistry of Materials*, 13(11), 3823–3824.

Bahr, J. L. and Tour, J. M. 2002. Covalent chemistry of single-wall carbon nanotubes. *Journal of Materials Chemistry*, 12(7), 1952–1958.

Bahr, J. L. et al. 2001. Functionalization of carbon nanotubes by electrochemical reduction of aryl diazonium salts: A bucky paper electrode. *Journal of the American Chemical Society*, 123(27), 6536–6542.

Balasubramanian, K. and Burghard, M. 2005. Chemically functionalized carbon nanotubes. *Small*, 1(2), 180–192.

Balasubramanian, K. and Burghard, M. 2006. Biosensors based on carbon nanotubes. *Analytical and Bioanalytical Chemistry*, 385(3), 452–468.

Balavoine, F. et al. 1999. Helical crystallization of proteins on carbon nanotubes: A first step towards the development of new biosensors. *Angewandte Chemie International Edition*, 38(13/14), 1912–1915.

Ballesteros, B. et al. 2007. Synthesis, characterization and photophysical properties of a SWNT-phthalocyanine hybrid. *Chemical Communications*, (28), 2950–2952.

Basiuk, E. V. et al. 2004. Direct solvent-free amination of closed-cap carbon nanotubes: A link to fullerene chemistry. *Nano Letters*, 4(5), 863–866.

Bergbreiter, D. E. 1985. Alkali metal organometallic polymers: Preparation and applications in polymer and reagent syntheses. *Metal-Containing Polymeric Systems*, J. E. Sheats et al. (eds.). Springer, Berlin, Germany, pp. 405–424.

Carlsson, J. M. 2006. Curvature and chirality dependence of the properties of point defects in nanotubes. *Physica Status Solidi (B)*, 243(13), 3452–3457.

Charlier, J.-C. 2002. Defects in carbon nanotubes. *Accounts of Chemical Research*, 35(12), 1063–1069.

Chen, J. et al. 1998. Solution properties of single-walled carbon nanotubes. *Science*, 282(5386), 95–98.

Chen, J. et al. 2002. Noncovalent engineering of carbon nanotube surfaces by rigid, functional conjugated polymers. *Journal of the American Chemical Society*, 124(31), 9034–9035.

Chen, R. J. et al. 2001. Noncovalent sidewall functionalization of single-walled carbon nanotubes for protein immobilization. *Journal of the American Chemical Society*, 123(16), 3838–3839.

Chen, R. J. et al. 2003. Noncovalent functionalization of carbon nanotubes for highly specific electronic biosensors. *Proceedings of the National Academy of Sciences*, 100(9), 4984–4989.

Chen, Z. et al. 2004. Side-wall opening of single-walled carbon nanotubes (SWCNTs) by chemical modification: A critical theoretical study. *Angewandte Chemie*, 116(12), 1578–1580.

Dillon, A. C. et al. 1999. A simple and complete purification of single-walled carbon nanotube materials. *Advanced Materials*, 11(16), 1354–1358.

Dyke, C. A. and Tour, J. M. 2003. Solvent-free functionalization of carbon nanotubes. *Journal of the American Chemical Society*, 125(5), 1156–1157.

Fagan, S. B. et al. 2003. Functionalization of carbon nanotubes through the chemical binding of atoms and molecules. *Physical Review B*, 67(3), 033405.

Fagan, S. B. et al. 2004. Substitutional Si doping in deformed carbon nanotubes. *Nano Letters*, 4(5), 975–977.

Fantini, C., Usrey, M., and Strano, M. 2007. Investigation of electronic and vibrational properties of single-walled carbon nanotubes functionalized with diazonium salts. *The Journal of Physical Chemistry C*, 111(48), 17941–17946.

Georgakilas, V. et al. 2002. Organic functionalization of carbon nanotubes. *Journal of the American Chemical Society*, 124(5), 760–761.

Gergely, A. et al. 2007. Modification of multi-walled carbon nanotubes by Diels-Alder and Sandmeyer reactions. *Journal of Nanoscience and Nanotechnology*, 7(8), 2795–2807.

Gu, Z. et al. 2002. Cutting single-wall carbon nanotubes through fluorination. *Nano Letters*, 2(9), 1009–1013.

Guldi, D. M. 2010. *Carbon Nanotubes and Related Structures: Synthesis, Characterization, Functionalization, and Applications*. John Wiley and Sons, Chichester, U.K.

Guldi, D. M. et al. 2005. Novel photoactive single-walled carbon nanotube-porphyrin polymer wraps: Efficient and long-lived intracomplex charge separation. *Advanced Materials*, 17(7), 871–875.

Hauke, F. and Hirsch, A. 2010. Covalent functionalization of carbon nanotubes. *Carbon Nanotubes and Related Structures: Synthesis, Characterization, Functionalization, and Applications*, D. M. Guldi and N. Martín (eds.). John Wiley and Sons, Chichester, U.K., pp. 135–198.

Hirsch, A. and Vostrowsky, O. 2005. Functionalization of carbon nanotubes. *Functional Molecular Nanostructures*, A. D. Schlüter (ed.). Springer, Berlin, Germany, pp. 193–237.

Holzinger, M. et al. 2001. Sidewall functionalization of carbon nanotubes. *Angewandte Chemie International Edition*, 40(21), 4002–4005.

Holzinger, M., Abraham, J., Whelan, P., Graupner, R., Ley, L., Hennrich, F., Kappes, M., and Hirsch, A. 2003. Functionalization of single-walled carbon nanotubes with (R-)oxycarbonyl nitrenes, *Journal of the American Chemical Society*, 125, 8566–8580.

Hu, C. Y. et al. 2009. Non-covalent functionalization of carbon nanotubes with surfactants and polymers. *Journal of the Chinese Chemical Society*, 56(2), 234–239.

Hu, H. et al. 2003. Sidewall functionalization of single-walled carbon nanotubes by addition of dichlorocarbene. *Journal of the American Chemical Society*, 125(48), 14893–14900.

Jiang, K. et al. 2004. Protein immobilization on carbon nanotubes via a two-step process of diimide-activated amidation. *Journal of Materials Chemistry*, 14(1), 37–39.

Jones, A. D. K. and Bekkedahl, T. 1997. Storage of hydrogen in single-walled carbon nanotubes. *Nature*, 386, 377.

Jung, A. et al. 2006. Quantitative determination of oxidative defects on single walled carbon nanotubes. *Physica Status Solidi (B)*, 243(13), 3217–3220.

Krueger, A. 2008. New carbon materials: Biological applications of functionalized nanodiamond materials. *Chemistry—A European Journal*, 14(5), 1382–1390.

Kuzmany, H. et al. 2004. Functionalization of carbon nanotubes. *Synthetic Metals*, 141(1), 113–122.

Lalonde, J. and Margolin, A. 2002. Immobilization of enzymes. (eds.). *Enzyme Catalysis in Organic Synthesis: A Comprehensive Handbook*, 2nd edn. K. Drauz and H. Waldmann Wiley Press, New York, pp. 163–184.

Li, C.-C. et al. 2007. A new and acid-exclusive method for dispersing carbon multi-walled nanotubes in aqueous suspensions. *Colloids and Surfaces A: Physicochemical and Engineering Aspects*, 297(1), 275–281.

Liu, Z. et al. 2009. Carbon nanotubes in biology and medicine: In vitro and in vivo detection, imaging and drug delivery. *Nano Research*, 2(2), 85–120.

Louie, S. G. 2001. Electronic properties, junctions, and defects of carbon nanotubes. M. S. Dresselhaus, G. Dresselhaus, and Ph. Avouris (eds.). *Carbon Nanotubes*, Springer, Heidelberg, Germany, pp. 113–145.

Marshall, M. W., Popa-Nita, S., and Shapter, J. G. 2006. Measurement of functionalised carbon nanotube carboxylic acid groups using a simple chemical process. *Carbon*, 44(7), 1137–1141.

Ménard-Moyon, C. et al. 2006. Functionalization of single-wall carbon nanotubes by tandem high-pressure/Cr (CO) 6 activation of diels-alder cycloaddition. *Journal of the American Chemical Society*, 128(46), 14764–14765.

Mickelson, E. et al. 1998. Fluorination of single-wall carbon nanotubes. *Chemical Physics Letters*, 296(1), 188–194.

Mita, D. et al. 2007. Enzymatic determination of BPA by means of tyrosinase immobilized on different carbon carriers. *Biosensors and Bioelectronics*, 23(1), 60–65.

Moghaddam, M. J. et al. 2004. Highly efficient binding of DNA on the sidewalls and tips of carbon nanotubes using photochemistry. *Nano Letters*, 4(1), 89–93.

Moon, H. K. et al. 2008. Effect of nucleases on the cellular internalization of fluorescent labeled DNA-functionalized single-walled carbon nanotubes. *Nano Research*, 1(4), 351–360.

Murphy, R. et al. 2002. High-yield, nondestructive purification and quantification method for multiwalled carbon nanotubes. *The Journal of Physical Chemistry B*, 106(12), 3087–3091.

Mylvaganam, K. and Zhang, L. 2004. Nanotube functionalization and polymer grafting: An ab initio study. *The Journal of Physical Chemistry B*, 108(39), 15009–15012.

Nakamura, T. et al. 2007. Chemical modification of single-walled carbon nanotubes with sulfur-containing functionalities. *Diamond and Related Materials*, 16(4), 1091–1094.

Nakayama-Ratchford, N. et al. 2007. Noncovalent functionalization of carbon nanotubes by fluorescein-polyethylene glycol: Supramolecular conjugates with pH-dependent absorbance and fluorescence. *Journal of the American Chemical Society*, 129(9), 2448–2449.

Negra, F. D., Meneghetti, M., and Menna, E. 2003. Microwave-assisted synthesis of a soluble single wall carbon nanotube derivative. *Fullerenes, Nanotubes and Carbon Nanostructures*, 11(1), 25–34.

Ni, B. and Sinnott, S. B. 2000. Chemical functionalization of carbon nanotubes through energetic radical collisions. *Physical Review B*, 61(24), R16343.

Niyogi, S. et al. 2002. Chemistry of single-walled carbon nanotubes. *Accounts of Chemical Research*, 35(12), 1105–1113.

Oh, S. B. et al. 2008. Facile covalent attachment of well-defined poly (t-butyl acrylate) on carbon nanotubes via radical addition reaction. *Journal of Nanoscience and Nanotechnology*, 8(9), 4598–4602.

Peng, H. et al. 2003. Sidewall functionalization of single-walled carbon nanotubes with organic peroxides. *Chemical Communications*, (3), 362–363.

Rey, D. A., Batt, C. A., and Miller, J. C. 2006. Carbon nanotubes in biomedical applications. *Nanotechnol Law Business*, 3, 263.

Riggs, J. E. et al. 2000. Strong luminescence of solubilized carbon nanotubes. *Journal of the American Chemical Society*, 122(24), 5879–5880.

Salzmann, C. G. et al. 2007. The role of carboxylated carbonaceous fragments in the functionalization and spectroscopy of a single-walled carbon-nanotube material. *Advanced Materials*, 19(6), 883–887.

Shim, M. et al. 2002. Functionalization of carbon nanotubes for biocompatibility and biomolecular recognition. *Nano Letters*, 2(4), 285–288.

Shiral Fernando, K., Lin, Y., and Sun, Y.-P. 2004. High aqueous solubility of functionalized single-walled carbon nanotubes. *Langmuir*, 20(11), 4777–4778.

Sinha, N. and Yeow, J. T.-W. 2005. Carbon nanotubes for biomedical applications. *IEEE Transactions on NanoBioscience*, 4(2), 180–195.

Sinnott, S. B. 2002. Chemical functionalization of carbon nanotubes. *Journal of Nanoscience and Nanotechnology*, 2(2), 113–123.

Sun, Y., Wilson, S. R., and Schuster, D. I. 2001. High dissolution and strong light emission of carbon nanotubes in aromatic amine solvents. *Journal of the American Chemical Society*, 123(22), 5348–5349.

Tagmatarchis, N. et al. 2002. Sidewall functionalization of single-walled carbon nanotubes through electrophilic addition. *Chemical Communications*, (18), 2010–2011.

Tasis, D. et al. 2006. Chemistry of carbon nanotubes. *Chemical Reviews*, 106(3), 1105–1136.

Usrey, M. L., Lippmann, E. S., and Strano, M. S. 2005. Evidence for a two-step mechanism in electronically selective single-walled carbon nanotube reactions. *Journal of the American Chemical Society*, 127(46), 16129–16135.

Vigolo, B. et al. 2009. Evidence of sidewall covalent functionalization of single-walled carbon nanotubes and its advantages for composite processing. *Carbon*, 47(2), 411–419.

Wang, G.-J. et al. 2007. Synthesis of water-soluble single-walled carbon nanotubes by RAFT polymerization. *Polymer*, 48(3), 728–733.

Yildirim, T., Gülseren, O., and Ciraci, S. 2001. Exohydrogenated single-wall carbon nano-tubes. *Physical Review B*, 64(7), 075404.

Yinghuai, Z. et al. 2005. Substituted carborane-appended water-soluble single-wall carbon nanotubes: New approach to boron neutron capture therapy drug delivery. *Journal of the American Chemical Society*, 127(27), 9875–9880.

Yokoi, T. et al. 2005. Chemical modification of carbon nanotubes with organic hydrazines. *Carbon*, 43(14), 2869–2874.

Zhao, J. et al. 2004. Electronic properties of carbon nanotubes with covalent sidewall functionalization. *The Journal of Physical Chemistry B*, 108(14), 4227–4230.

Zhou, W. et al. 2001. Structural characterization and diameter-dependent oxidative stability of single wall carbon nanotubes synthesized by the catalytic decomposition of CO. *Chemical Physics Letters*, 350(1), 6–14.

Ziegler, K. J. et al. 2005. Controlled oxidative cutting of single-walled carbon nanotubes. *Journal of the American Chemical Society*, 127(5), 1541–1547.

CHAPTER 5

CONTENTS

Collective Nature of Chemical Bonds in Fullerenes and Fullerites C$_{60}$

Vibrational Resonances, Vibrational–Electronic Interactions, and Anomalous Enhancement of Bands in the Vibrational Spectra of Nanofilms (Results of Vibrational Spectroscopy and Quantum Chemical Calculations)

5

N.E. Kornienko and A.P. Naumenko

5.1 INTRODUCTION

Physical–chemical properties of most material media are largely determined by chemical bonds (CBs) between atoms in molecules and condensed matter, including solid and liquid state, as well as a variety of nanomaterials (NMs). For the development of new ideas about the collective nature of CBs in large molecules such as C$_{60}$, as well as polymeric and biological macromolecules and condensed matter, some physics–chemical facts, the nature of which still remains little studied, are essential. Here, above all, the significant differences of CB in small molecules and condensed matter should be noted. Most of the molecules, including the robust carbon nanotubes (CNTs) and important biological macromolecules (DNA and RNA), are typically composed of atoms of 1, 2, and 3 periods of the Mendeleev periodic table (PT) of elements. But a huge number of practically important structural and functional materials consist mainly of atoms 4, 5, and 6 from PT periods. For a better understanding of these differences in CB in simple molecules, carbon NMs, and metals and metal alloys, let us consider the power and energy characteristics of CB in these objects in more detail.

It is well known that the CB of atoms in molecules can be characterized by force constants K, which are determined by the frequencies of normal vibrations ν. For two atomic molecules K = $\mu(2\pi c\nu)^2$, where $\mu = m_1m_2/(m_1 + m_2)$—reduced mass and c is the light velocity. For example, using the table data on vibrational frequencies ν for alkali metal dimers [1], it is easy to see that K-values decrease with the increase of period number n, to which the alkali metal atom belongs. In particular, upon transition from Li_2 to Cs_2, the force constant K decreases 3.68 times. Similar regularities are observed for halogen molecules X_2 (X=F, Cl, Br, I). In this series, the value of K decreases 2.67 times. That is, the normal CB for heavy atoms is significantly weakened.

At the same time, the empirical relationship $\mathbf{Q_m \approx RT_D(n - 1)}$ for the molar melting heats for metals Qm was established quite a long time ago [2], where R is the universal gas constant and TD is the Debye temperature associated with the boundary frequencies of phonons $\mathbf{f_{max}}$ $(\mathbf{kT_D = hf_{max}})$. It is essential that the melting heat per atom in $\mathbf{Q_1 = Q_m/N_A}$ ($\mathbf{N_A}$—Avogadro's number), uniquely corresponding to the energy of cCB in the metal, increases for higher periods $\mathbf{n}$ of PT. This differs radically condensed matter and NM on the properties of simple molecules. It should be noted that the analyzed empirical formula has no relation to the gas medium, and by the universal gas constant R and the Boltzmann constant k ($\mathbf{R = kN_A}$), the interrelation of energy and temperature is performed: $\mathbf{Q_1 = kT_D(n - 1) = hf_{max}(n - 1)}$. The trend of increasing the CB energy with the increasing number of the period n is well confirmed for Group II of PT (metals from Be to Ba). Using tabular data [1] it is easy to see that for metals Be, Mg, Ca, Sr, and Ba, the dependence $\mathbf{Q_1/hf_{max}(n)}$ is almost linear in the transition from Ba to Be, which increases by 7.73 times. The same relationship holds not only for metals but remains valid for molecular crystals F_2, Cl_2, Br_2, and I_2. In this case, the value $\mathbf{Q_1/hf_{max}}$ is increased to more than 22 times upon transition from F_2 to I_2. These examples for simple standard molecules and solids convincingly demonstrate the existence of two well-defined regularities of the weakening of CB in molecules and their strengthening in condensed matter in the transition from light atoms to the atoms of higher periods n in the PT of elements [3].

For condensed matter unlike small molecules, qualitatively new regularities appear, which are associated with collective-wave properties of their vibrational and electronic excitations. The collective properties of vibrational states occur at the level of CB. In a spatially distributed matter, even a weak wave nonlinearity can lead to significant effects [4–12] due to the accumulation of nonlinear wave interactions. This wave nonlinearity differs significantly from local nonlinearities in small systems, where nonlinear effects are usually mild.

Nonlinear wave processes are well known in laser physics and nonlinear optics, but to date these ideas are not transferred to the fundamental quantum physics and materials science. To date, the nonlinear interaction of the vibrational modes (NIVM) in nanostructures, in particular in typical carbon materials (fullerenes, nanotubes, graphene, onion-like carbon), is insufficiently studied. We point out that the general regularities of the spatial dynamics of nonlinear resonant wave interactions have been studied previously in detail in many publications [4–12]. For a theoretical description of nonlinear wave resonance processes, there is a well-developed formalism of coupled systems of nonlinear Maxwell equations for interacting waves of different frequencies and the equations for the density matrix of the material media.

In the case of nonlinear resonant interaction of thermally excited low-frequency vibrational modes, the higher vibrational states (overtones and tones total) are generated. They approach the electronic states (ESs) and interact effectively with them, leading to their changes. This is particularly important in semiconductors, which include fullerites C_{60}. In this case self-consistent with the vibrational excitation change, ES promotes abnormal growth of vibrational nonlinearity, which leads to the phenomenon of strong vibrational-electronic interaction (VEI) [13–21]. It should be noted that the efficiency of nonlinear wave processes is significantly increased in terms of multiple vibrational resonances, which are well developed in fullerenes C_{60} [13–15]. It promotes the transformation of the vibrational energy of the thermally excited low-frequency Hg vibrations (1,2) of the C_{60} molecule ($\nu \approx 270$–340 cm^{-1}) to the middle frequency range (700–800 cm^{-1}), where the Raman-active Hg vibrations (3,4) exist, and further to the high-frequency region of vibrations Flu(4), Ag(2), Hg(8) ($\nu \approx 1429$, 1469 and 1576 cm^{-1}). Strong VEIs are common to many substances and are manifested in the anomalous enhancement (up to 104 times) of the absorption bands of higher vibrational modes [18,19] and the appearance of new electronic absorption bands in the oscillations [17–19,21].

Naturally, multiple vibrational resonances lead to increased VEI. Strong VEIs lead to a change of ES, which is manifested in the anomalous increase of the band intensities in the vibrational spectra, as well as the induction of essentially new electronic bands (EBs) in the bandgap of C$_{60}$. Therefore, fullerenes and fullerites C$_{60}$ are good model objects for the study of the collective nature of CB and chemical functionalization of carbon NMs. The investigations of fullerenes and molecular crystals (fullerites) and polymerized forms are of great interest for solving both fundamental and applied tasks. This problem is dedicated to extensive literature, including a number of reviews and books [22–29]. Both extensive research and application of carbon materials are connected with the ability of carbon atoms to form various types of CBs, which leads to a very wide range of changes in the optical, mechanical, and electromagnetic properties of the corresponding materials, so that the carbon is the basis of life on Earth. The existence of different phases of fullerenes (monomers, dimers, 1D and 2D polymers) in a small energetic range (~0.01 eV/atom) is one of the peculiarities of fullerenes as an allotropic form of carbon [24]. It is also noteworthy a manifestation of the collective properties of the CBs in C$_{60}$ when they interact with metal atoms [30–35].

In the literature devoted to the study of fundamental physical properties of different types of carbon materials, including fullerenes, nanotubes, onion-like carbon, and other spectral studies posses the significant place [36–59], as they provide the most complete information about the still little studied nonlinear quantum properties of complex molecules and condensed matter, including C$_{60}$ and its polymeric forms. Despite the huge number of publications devoted to the study and application of different types of fullerenes, a number of their properties and empirically derived regularities do not have a proper understanding at a fundamental level. It should be noted that currently sufficiently developed methods of quantum chemical calculations (QCCs) are based on first principles, but their accuracy is significantly worse than the capabilities of modern spectroscopic methods. Therefore, the results of spectroscopic methods, in particular the study of Raman spectra and IR absorption of micro- and nanofilms of C$_{60}$ compared with the results of QCC, allow receiving a new physical–chemical information promoting the further development of quantum physics.

In connection with this the anomalies in changing of the vibrational frequencies and intensities of vibrational bands (VBs) in the Raman and IR spectra of fullerene C$_{60}$ are considered in detail. Both experimental results of other authors and the results of our experiments are widely used. An important place in our study covers a comparison of the observed regularities with the results of QCC. The relationship of the examined spectral anomalies for C$_{60}$ with the set of the sequential vibrational resonances Hg(1)+Hg(2)=Hg(3), Hg(1)+Ag(1)=Hg(4), 2Hg(3)=Hg(7), Hg(3)+Hg(4)≈Ag(2), 2Hg(4)≈Hg(7) is studied [13–15,30–35,57–62]. Therefore, we consider a new phenomenon of vibrational resonance instability of C$_{60}$ molecules and compare the various QCC methods [63]. In general, the present work aimed to establish new regularities peculiar to molecules with a large number of normal vibrations and multiple vibrational resonances. This should contribute to a strategic pathway analysis of biological macromolecules and convergence of physics, chemistry, and biology.

In nanoparticles, a considerable proportion of atoms close to the surface and, for fullerenes, the proportion of surface atoms increase to 100%, which leads to high anharmonicity of the vibrations and the nonlinearity for nanostructured media in the vibrational region. In particular, for C$_{60}$ nanofilms on the silicon substrate with 60 nm thickness, the generation of the second and third harmonics of the exciting laser radiation has been observed [25]. Even in the nonresonant region (1064 nm), the measured value of the third-order nonlinear susceptibility $\chi^{(3)} = 2 \cdot 10^{-10}$ units CGSE puts the fullerenes on one of the first places among the nonlinear materials on a molecular basis. Therefore, we carried out the study of nonlinear quantum regularities on the example of nanostructures, where they appear quite clearly.

The problem of collective properties of CB related to bridging the existing gap between the real physical–chemical materials science and fundamental physics. Materials science, biology, medicine, and partially chemistry actually refer to a higher hierarchical level, which is essentially described in detail by modern science. Essentially, the problem of collective CB boils down to deepen and broaden the relationship of classical and quantum physics. As we know the trivial limiting transition h → 0, according to the de Broglie relation $\lambda = h/p$, where λ is the wavelength and p is the mechanical impulse which actually neglects the wave properties of matter ($\lambda = 0$). We consider that a strategic development path here corresponds to W. Heisenberg ideas "… that even such inherently linear theory as quantum theory, will be replaced by a non-linear" [64]. This idea was expressed by them in 1967, probably under the influence of great success in the study of nonlinear wave phenomena in the initial stage of development of laser physics, quantum electronics, and nonlinear optics [4–7].

Our key idea in the development of new ideas about the collective CBs is that they are consistent with the collective vibrational modes. And largely practically classical vibrational motion of the atoms is consistent with the actual quantum behavior of the electron subsystem by higher vibrational states (overtones and sum tones) that are approaching the ES and effectively interact with them. It should be noted that it is very difficult to take into account NIVM and VEI directly in the Schrödinger equation. Therefore at the initial stage of development of nonlinear quantum physics (NQPh), we transfer the ideas laser physics, nonlinear optics, and resonance spectroscopy to materials science and condensed matter physics. According to the results of our work on the theory of nonlinear resonant wave interactions [8–12], the self-consistent solutions of the coupled system of wave equations for a number of waves, and the equations for the density matrix of the environment in which these waves interact are considered. In this case a comparison of the results of spectral experiments with the results of the nonlinear wave theory and QCC for fullerenes C_{60} plays an important role. In particular, the development of ideas about collective CB based on QCC for complexes C_{60} with metal atoms [30–35]. In general, it is a significant development of optical–mechanical analogy with the inclusion of nonlinear wave interactions of vibrational modes (VM) and binding vibrational-electronic states. Fundamental importance to developing concept is to establish a nontrivial relationship of classical and quantum physics, one of the clearest manifestations of which is CB.

Due to the fact that the key role of NIVM and VEI is ignored in many studies of carbon NMs, the aim of this work is the study of ES changes under the vibrational inducing of essentially new EBs in the vibrational field of fullerenes C_{60} and appropriate enhancement of active and *silent* VBs in the Raman and IR absorption spectra of C_{60} nanofilms [47–53,57–62]. A comparison of the intensities of the VBs in the Raman and IR spectra for nanofilms of C_{60} with thickness at 150–250 nm and microfilms with thickness at 1–2 μm has been done. In the Raman spectra of revealed nanofilms, the enhancement of the active bands Hg (1÷8) by 2÷7 times in comparison with a slight weakening of the Ag(1,2) band for microfilms has been observed. Even greater enhancement has been observed in the Raman spectra for nanofilms inactive in icosahedral symmetry I_h vibrations, Gg, u, Hu, $T_{1,2g}$, T_{2u} etc. Nonmonotonic dependence of the intensities of the observed EB from the thickness of the nanofilms and reduction of enhancement of the VBs under the polymerization of C_{60} nanofilms C_{60} through N_2H_4 have been established, which further confirms the collective nature of CB. Anomalous enhancement of EP by two to three orders of magnitude and anomalous enhancement of the VBs under the influence of intense laser radiation is shown [47–53].

The collective nature of CB for complexes $C_{60}M$ with metal atoms (M = Fe, Sn) has been studied on the basis of the QCC analysis of electron density transfer from the metal atoms and its distribution on the C_{60} molecule. Collective properties of CB are manifested wherein relatively weak bonds with metal atoms C_{60} (length bonds, the CN ~ 2 Å, and low frequencies of the corresponding vibrations, ~100 cm^{-1}) due to a redistribution of electron density lead to a 1%–2% reduction of all single and double bonds in the C_{60} molecule, despite the Coulomb repulsion of the charge distribution (*quantum chemical compression* of the molecules). An even greater degree of collective properties is characterized by a dynamic change in CB charges and electronic polarizabilities during vibration, resulting in a change in intensities of the band's active vibrations and induced activity in a number of *silent* vibrations.

The obtained results explain the significant differences of observed intensities of VBs with the results of KHS that points the way to improve them. The developed concept deepens the understanding of the role of the vibrational resonances on the nonlinear interaction of vibrational modes and vibrational-electronic interactions in the formation of CBs and their modifications in the physics of condensed matter and nanostructures. In general, C_{60} can be good model objects for the development of NQPh and the unification of physics and chemistry similar to how a hydrogen atom was a model object for the formation of quantum physics at the initial stage of its development.

5.2 THE EXPERIMENTAL PROCEDURE AND NUMERICAL PROCESSING OF SPECTRAL DATA

Fullerite C_{60} films were deposited on substrates of crystalline silicon Si(001) with thickness' 0.5 mm by hot or cold deposition. The fullerite C_{60} polycrystalline microfilms with thinness 0,6; 1,2 и 2 μm and nanofilms with thinness 150–200 nm have been investigated. The Raman spectra were studied using automated spectrometers Horiba Jobin Yvon T64000 and DFS-24 with the cooled photomultiplier and a photon counting system. Raman spectra were excited by Ar^+ laser with a wavelengths

$\lambda_L = 514.5$ and 488 nm. To reduce the influence of photopolymerization of C$_{60}$, the cylindrical focusing of the laser radiation (spot size of 0.3×2.5 mm^2) was used, laser intensity was ~2 W/cm^2. With sharp focusing of laser excitation λ_L up to a diameter of ~2 μm intensity of the exciting radiation exceeds 10 kW/cm^2, what allows us to study the impact of light radiation on the Raman spectra and the VEI. Preliminary laser irradiation of C$_{60}$ films has been studied additionally.

IR spectra of C60 have been studied by Fourier spectrometer Nicolet NEXUS-470. It should be noted that in the region 400–8000 cm^{-1}, one to three interference maxima have been observed due to the interference in fullerene thin film. Frequent oscillations caused by interference in a silicon plate have been removed by numerical smoothing of IR spectra. To correct account the Fresnel losses at the input and output surfaces of the samples, first the transmission spectra T (ν) have been recorded. Further the optical density spectra $D(\nu) = -\ln [T(\nu)/TF(\nu)]$ have been calculated, where TF(ν)—determined by spectrum $T(\nu)$ the amount of attenuation of the transmitted radiation due to reflection losses and scattering by boundaries of the sample at intervals of maximum transparency. The absorption coefficients $\alpha(\nu) = D(\nu)/d$, where d—thickness testing fullerite sample were determined from absorbance spectra. Spectral resolution of Raman and IR spectra was better than 0.5 cm^{-1}.

In order to increase the signal/noise ratio, the numerical analysis of the spectra was carried out with optimal numerical smoothing. In the numerical decomposition of vibrational and electronic bands on the individual spectral components the method of least squares with the variation of the frequency positions of the components, their intensity, shape and half-width have been used. Because of the existence of high-level fluctuations for broadband background compared with vibrational bands for it, more smoothing has been used. Vibrational bands are numerically allocated for broadband background in Raman and IR spectra by polynomial approximation using the required number of control points in the experimental fields of intensity minima. For the correct comparison of the intensities of various vibrational bands of C$_{60}$ microfilms and nanofilms, the normalization of dedicated lanes on the intensities of the strongest Raman and IR lines of Ag(2) 1469 cm^{-1} and F1u(1) 527 cm^{-1} has been conducted. For C$_{60}$ nanofilms together with these lines, the line $\nu_{Si} = 520$ cm^{-1} of crystalline silicon substrate is observed, which can be used to classify the C$_{60}$ nanofilms by thickness d. With decreasing of d, the intensity of ν_{Si} increases, and for microfilms C$_{60}$ with thickness about 1.2–2 μm, this line as a rule does not appear.

For the numerical decomposition of the vibrational and significantly broader electronic bands on the individual elementary components, the least-squares method with a variation of the frequency position of the spectral components, their intensity, shape, and half-width have been used. Number of valid components of the spectral components of each band are determined by the number of negative minima calculated from the experimental data of the second derivatives after their optimal smoothing. Correctness of numerical decomposition analyzed bands on individual components has been controlled by agreement of the second derivatives of the experimental and theoretical forms of their contours.

5.3 ANOMALIES IN THE VIBRATIONAL SPECTRA OF C$_{60}$

5.3.1 General Information about the Vibrations of C$_{60}$ and Spectral Features of Fullerites

By the number of degrees of freedom of isolated C$_{60}$ molecule is (3N-6) except the translations and rotations and 174 intramolecular vibrations according to group-theoretical analysis are grouped by 46 normal vibrational modes: 4T1u + 2Ag + 8Hg + 3T1g + 4F2g + 6Gg + Au + 4T1u + 5T2u + 6Gu + 7Hu. Here the indices g and u denote the symmetric and asymmetric vibrational modes relative to the center of symmetry. Because of the high symmetry of the C$_{60}$ molecule, the majority of its vibrations have a significant degeneration: 15 fivefold degenerate modes of the Hg (1 ÷ 8) and Hu (1 ÷ 7), 12 fourfold degenerate modes of Gg (1 ÷ 6) and Gu (1 ÷ 6), and 16 triply degenerate vibrations of T1g (1 ÷ 3), T2g (1 ÷ 4), T1u (1 ÷ 4), and T2u (1 ÷ 5). Only three vibrations of Ag(1,2) and Au are nondegenerated. According to the selection rules for the symmetry group of the icosahedron I$_h$, four vibrations of T1u (i) with frequencies of 526, 575, 1182, and 1429 cm^{-1} are active in the IR absorption spectrum and 10 vibrations, two fully symmetric modes of Ag(1) at 496 cm^{-1} and Ag(2) at 1469 cm^{-1}, and eight vibrations of Hg (k) with the Raman shifts 272, 433, 709, 776, 1101, 1252, 1425, and 1577 cm^{-1}. It should be noted that the lowest-frequency Hg mode (1) corresponds to the transformation of the spheroid to an ellipsoid of revolution (*pumpkin* mode), and nondegenerate modes of Ag(1) and Ag(2) refer to the symmetric breathing mode and tangential resizing pentagons of the C_{60} molecule, respectively.

The molecular crystal forces between C_{60} molecules are much weaker than the covalent forces between atoms in the individual molecules. This is consistent with the small quantities of crystalline splitting observed in fullerite C_{60} at low temperatures [41–43]. Note that the use of pure isotopes $^{12}C_{60}$ and $^{13}C_{60}$ allows sharing of crystalline and isotope splitting [42]. The weakness of the intermolecular interaction is further supported by the fact that the highest frequencies of the intermolecular vibrations is ~50 cm^{-1} [65,66], much lower than the lowest frequency of 272 cm^{-1} for intramolecular vibrations of Hg(1). This is in agreement with the good correlation observed in crystals of C_{60} vibration frequencies with the values obtained as a result of QCC for isolated molecules. Therefore, in most cases, the interpretation of the spectra of fullerites in the notation of free vibrations of the C_{60} molecule is used.

The high sensitivity of the vibrational spectra of C_{60} to the action of relatively weak disturbances led to the lowering of the symmetry associated with an extremely high degree of degeneration. As a result of such effects the degeneracy is lifted, resulting in a splitting of the corresponding vibrational state or ES. We performed a detailed analysis of the extensive literature on the spectroscopy of the fullerenes C_{60} in order to identify and organize a set of observed anomalies that are a little explained in terms of generally accepted scientific ideas. At the same time we pay much attention to the agreement of the observed spectral anomalies with the results of consistent QCC coming from the first principles. We assume that the latter leads to important effects of VEIs that significantly affect the properties of the studied systems. This allows detecting new regularities appearing in complex molecules and the corresponding molecular media as a result of multiple vibrational resonances and possible nonlinear interaction of vibrational modes. We estimate the changes in the positions of the VBs in the Raman and IR absorption spectra of fullerenes C_{60} at different physical and chemical influences.

Variations of the position of the VBs in the Raman spectra and IR absorption spectra of fullerenes C_{60} at different physicochemical impacts were analyzed. The effects of temperature, external pressure, doping of fullerite C_{60} with alkaline and earth element atoms, interaction with organic compounds, various types of polymerization, the effect of resonant laser radiation, and ion irradiation by neutron implantations have been examined. Both changes in the frequency of the VBs and their half widths and intensities have been studied, in this study the case of nonresonant excitation of the Raman spectra and the resonant excitation of cases in different parts of the lowest ES. It should be noted that crystalline C_{60} is a semiconductor with a bandgap of ~1.5–1.7 eV, but there are Frenkel excitons with energies of 1.55 and 1.83 eV [24]. Thus the band edge absorption band corresponds to the emission with wavelength of ~730–830 nm, and the exciton resonances are realized for emissions 800 and 677.5 nm.

Therefore we first analyze the frequencies and half widths of the VBs of fullerene C_{60} and then turn to the changes in the intensities of the Raman and IR spectra. In many cases, a comparison of the spectral data with the results of QCC has been done for strong detection of the observed anomalies.

5.3.2 Change of Frequencies and Half Width of Vibrational Bands

Let's compare the frequencies of the fundamental vibrations of fullerene C_{60} observed in the Raman (Ag(1,2) and Hg(1 ÷ 8)) and IR spectra (T1u (1 ÷ 4)) with the results of the first-principles calculations. Figure 5.1a shows the difference $\Delta\nu$ of calculated and experimentally observed frequencies of active modes of C_{60}. In this case calculations of frequencies of the fullerene C_{60} were carried out using the program Gaussian03 Hartree–Fock (HF) basis 6-31G*. It is evident that the values $\Delta\nu$ are quite significant and dramatically increase in a high-frequency region, reaching very large values for spectroscopy of ~200 cm^{-1}. It should be noted that $\Delta\nu$ values for the various types of vibrations are grouped in a narrow range that characterizes the logical behavior of such deviations, and they are well approximated by part of the Lorentzian function, as shown in Figure 5.1a.

Note that the development of methods of QCCs, including using different bases, was aimed at reducing the differences, but without using scaling factors frequency, differences still remain.

The inset in Figure 5.1a shows the ratio of experimental and calculated frequencies ν(expt.)/ν(calc.). This ratio varies within a wide range and cannot be accounted for by a permanent scaling factor

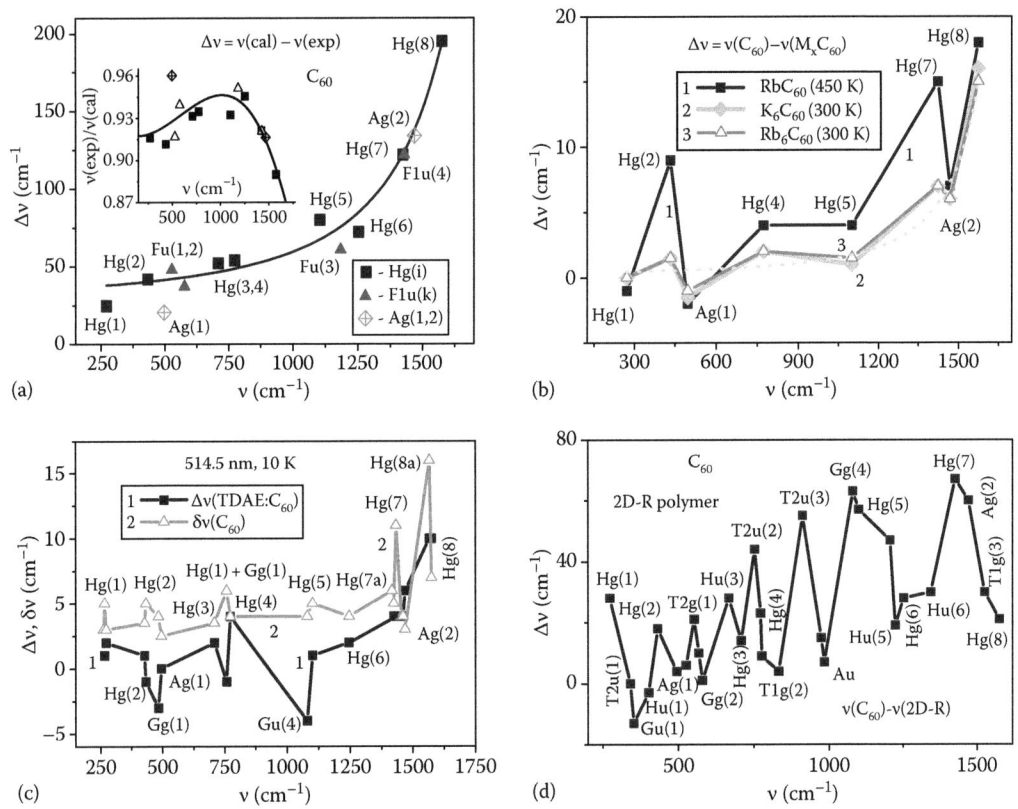

FIGURE 5.1

Softening of observed frequencies of Raman and IR-active modes of fullerene C$_{60}$ (a) and the entire set of active and spectral dependences of the frequency differences of Raman-active modes of fullerene C$_{60}$ in complexes with potassium and rubidium, according to References [39,67] (b) and in combination with C$_{60}$ tetra (dimethylamino) ethylene (TDAE:C$_{60}$) [40] and pristine C$_{60}$ (curve 1), as well as an increase in the half widths of the vibrational bands in the Raman spectra δν C$_{60}$ single crystal according to [40] (curve 2) (c), and low-frequency shifts of active and *silent* vibrations C$_{60}$ in 2D rhombohedral phase (2D-R) according to References [68,69] compared to the nonpolymerized fullerite C$_{60}$ (d).

(~0.90–0.95), as it is done in many papers. In general, the results shown in Figure 5.1a characterize the existence of systemic factors that are not taken into account in modern *ab initio* quantum calculations.

The sharp increase of the frequency difference $\Delta\nu$ precisely in the high-frequency region indicates the important role of higher vibrational states. They are approaching the low-energy ES and lead to effects of VEIs. It this case a significant change of the ES is possible, as well as CB changes, which is manifested in changes in the frequencies and intensities of the VBs. That is why we analyze the variation of the frequencies and intensities of the VBs in order to obtain additional information about the VEI. Naturally in this case a consistent account in the calculation of total tones and overtones is required that is not currently being done. Moreover, most of the calculations are carried out in the adiabatic approximation, which makes consistent inclusion of the VEI.

The real significance of the detected systematic differences between experiment and theory is confirmed by a similar *softening* of frequencies C$_{60}$ by doping with alkali metal atoms, the interaction of fullerenes with organic molecules, as well as the effect of temperature, as illustrated in Figure 5.1b through d. Figure 5.1b shows the difference between the vibration frequencies observed in the Raman spectra of pure C$_{60}$ and in composites with potassium and rubidium according to numerical data [39,40,67]. It is essential that for composites K$_6$C$_{60}$ and Rb$_6$C$_{60}$ and Cs$_6$C$_{60}$, the difference between the frequencies $\Delta\nu$ is also located near the wing of the Lorentzian as shown by the dotted line in Figure 5.1b. Significantly, in this case, the observed values are compared only with vibration frequencies, which are not related to the methods of theoretical calculations. Consideration of C$_{60}$ doped with alkali metal atoms is of practical interest in connection with the discovery of their superconductivity at critical temperatures (T_C = 19–33 K) [25]. Softening of the vibration frequencies in the compounds of C_{60} with alkali metals clearly indicates the importance of the changes in the electron subsystem.

Similar changes of the frequency differences of the vibrations $\Delta\nu$ compared to pure C_{60} are observed under the binding of fullerenes C_{60} to tetra(dimethylamino)ethylene (TDAE:C_{60}) [40], which is illustrated in Figure 5.1c (see curve 1). In this case for a number of active and *silent* vibrations of Gg(1), Gu(4)-negative values $\Delta\nu$ are manifested, indicating a strengthening of the vibrations simultaneously with the *softening* of high-frequency oscillations. It is clear that the formation of chemical complexes is also associated with a change in the ES.

Figure 5.1a also shows the systemic increase of observed half widths $\delta\nu$ of the VBs in the Raman spectrum of C_{60} single crystal at 10 K, which is particularly evident for a number of split components of the most high-frequency oscillations of Hg(7) and Hg(8). The half width of the VBs associated with the relaxation rate of vibrational excitationsis known, which is associated with a nonlinear parametric decay data of vibrational excitations on the lower-frequency vibrational modes. Thus, the increase in half widths of the high-frequency VBs $\delta\nu$ is ultimately associated with the increasing role of the vibrational nonlinearity. This also supports the idea of the important role of the nonlinear interaction of vibrational modes, which leads to abnormal strengthening of VEI.

A significant increase in the vibrational nonlinearity has been observed in composites Sm_xC_{60}, where x = 1,2,3 according to the Raman spectra at 514.5 nm excitation [70]. This is confirmed by a strong increase in the half widths $\delta\nu$ of VBs of Hg with increasing degree of doping C_{60} by Sm. In particular, in Sm_3C_{60} the bands Hg(1) and Hg(7) become broader, respectively, by 6.2 and 26.6 times, which is associated with increased nonlinear relaxation processes for C_{60} complexes with metals. Simultaneously, a narrow band of full symmetrical vibrations of Ag(1,2) are broadened by 1.2–1.8 times, indicating their lower role in relaxation processes.

Among the 46 normal vibrations of fullerene C_{60}, 14 vibrations are active in the vibrational spectra (10 in the Raman spectrum and 4 in IR absorption spectrum), and 32 vibrations are inactive *silent* ones. We carried out a careful analysis and systematization of the set of *silent* vibrations of the C_{60} molecule, the results of which will be discussed later. It is essential that at high pressures and polymerization of fullerite C_{60} in the spectrum except for active modes, many inactive icosahedral symmetry vibrations are observed. The behavior of C_{60} at high pressures is of interest in connection with their easier transformation into diamond compared to graphite. This requires a pressure of 20 GPa at room temperature and the need to convert the graphite to a pressure of 50 GPa at a temperature of about 900 K [26]. At high pressure P on pure C_{60} or their composites with metals, strong changes in their vibrational frequency are observed.

Systematization and refinement of frequencies of *silent* vibrations allowed to calculate the correct differences between the frequencies of unpolymerized and polymerized various forms of F_{ul}lerite C60, which is illustrated in Figure 5.1d. It shows the results of processing the experimental data tabulated in [68,69], the collected vibrational frequencies in the Raman spectra for the 2D polymerized rhombohedral phase (2D-R). Figure 5.1d shows a good tendency to softening of high-frequency vibrations under polymerization. Moreover, for a number of high-frequency vibrations of Hg(7), Gg(4) and T2u(3) $\Delta\nu$ values approach the values of 60–70 cm^{-1}, which is substantially greater than those seen at Figure 5.1b and c. This shows a significant change in the structure of C_{60} molecules in the polymerized phases. A few smaller, low-frequency shifts of vibration frequencies are compared to the initial C_{60} observed in the tetragonal phase C_{60} (2D-T) [71] and the 1D polymer orthorhombic (1D-O) according to [72]. But there is also a tendency of *softening* of frequencies of such high-frequency vibrations of Hg(7) and Ag(2). The presence of negative values $\Delta\nu$ for a number of vibrations id shown in Figure 5.1d, and Figure 5.1b indicates increased CBs. An observed new regularity of oscillatory changes in $\Delta\nu$ values with increasing frequency ν in Figure 5.2a indicates the complexity associated with nonlinear interactions of vibrational modes and vibrational-electronic interactions.

To determine the frequency of inactive vibrations in the vibrational spectra of C_{60}, a lot of effort was made. We considered not only the data of IR and Raman spectra but also inelastic neutron scattering, which allows all types of vibrations of C_{60} regardless of their symmetry (see, e.g., [37]). The spectral observation of inactive vibrations is possible in the case of resonance Raman and gain Raman scattering, when the selection rules change and the Jahn–Teller effect manifests. Inelastic neutron scattering [22,73,74] and electron energy loss [75,76] spectra exhibit all the vibrational modes regardless of their symmetry, which would be easier for the observation and identification of inactive

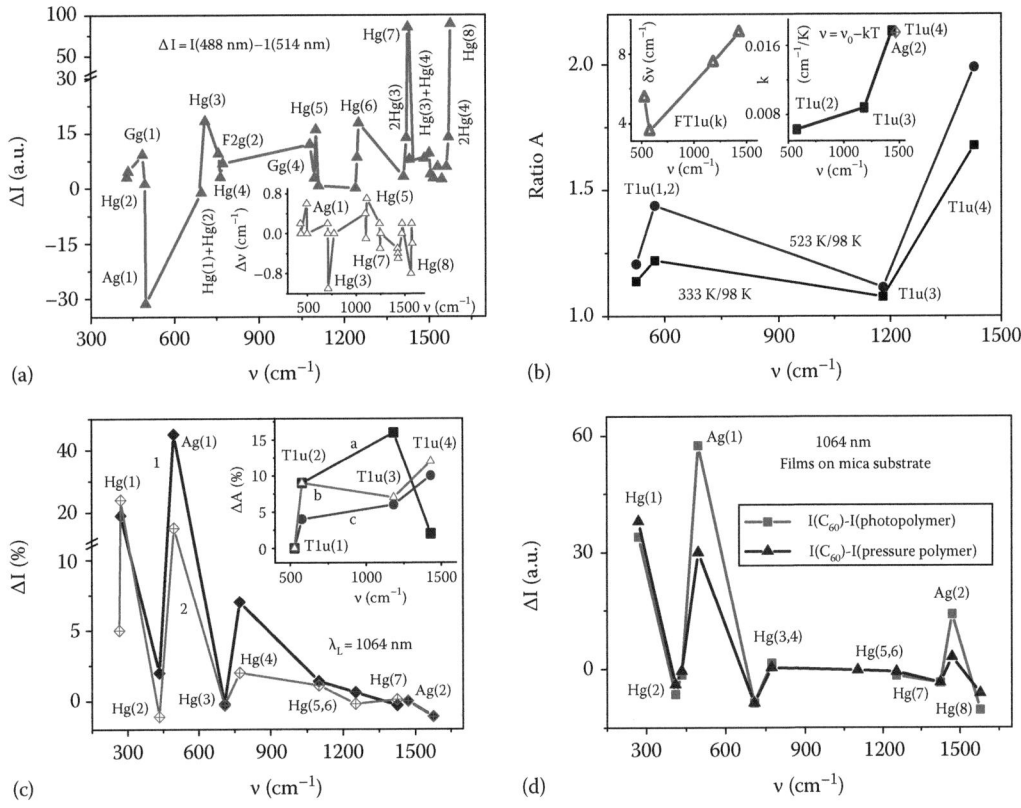

FIGURE 5.2
Changes in the intensities of the vibrational bands and their frequencies (inset) in the Raman spectra of C₆₀ film when the exciting radiation wavelength changes (514 nm → 488 nm) under results [41] (a) and the ratio of the intensities of the T1u (k) in the IR spectra of C₆₀ at change in temperature (98 K → 333 K → 523 K) according to [90], as well as an increase in the half-width δν high-frequency IR bands (left inset), and temperature coefficients k reduce the frequency bands T1u (2–4) and Ag(2) (right inset) according to results [90] and [91], respectively (b). Variations of the intensities of the vibrational bands in the main Raman spectrum at nonresonant excitation laser of 1064 nm by numerical data [92–94] and IR-type bands T1u under numerical data [92,95–97] (inset) (c) and also a decrease in the intensities of the vibrational bands in the Raman spectra of C₆₀ films on substrates of mica at photopolymerization and polymerization pressure on the results of [98] (d).

vibrational modes. However, these methods are less accurate than the methods of optical spectroscopy and do not facilitate accurate identification of the totality of the vibrational modes of C₆₀.

The significant role in establishing real frequency of *silent* vibrations belongs to their spectral manifestations due to symmetry constraint violations. Lowering of the symmetry and activation of previously *silent* vibrations of C₆₀ are possible due to intermolecular interactions and the crystal field in the presence of ¹³C isotopes; the influence of defects in molecules or crystals, in particular the dislocations and faces of nanocrystallite interaction with other metal atoms or molecules; and polymerization. Significant reduction in C₆₀ symmetry takes place when ¹³C isotope atoms are injected. The natural isotopic abundance of ¹³C is 1.11%, so about half of the C₆₀ molecules have low symmetry and the majority of previously *silent* vibrations should appear in the IR and Raman spectra. However, according to QCC for isotope substituted molecules 13CC₅₉ [15,31,32,35], intensities permitted in this case *silent* vibrations in most cases also a minor, which also makes it difficult to interpret uniquely inactive C₆₀ vibrations.

QCC results and the degrees of degeneracy vibrations are useful for identifying the most low-frequency "silent" vibrations of T2u(1), Gu(1), and Hu(1), which in the vibrational spectra clearly correspond to frequencies 343.5, 355.5, and 404.5 cm⁻¹. However, even in well-known publications [26–28], the incorrect assignment of these vibrations is used, which is not consistent with QCC. However, even after the establishment of a correct identification of *silent* vibrations in the survey [22], the authors of [77] devoted to the analysis of the accuracy of various methods QCC, for comparison with experimental data were chosen [36], which does not lead to the right conclusions.

Here it should be noted that the assignment of the calculated vibrations of C_{60} by symmetry types can be accomplished using the Gaussian03 program, as was done in [78]. This example shows how important the problem of the correct identification of the true values of *silent* vibrations of the C_{60} molecule. We point out that for the higher frequency of *silent* vibrations, the order of vibrational modes in some cases depends on the QCC method (e.g., HF or density functional), which makes the problem of classifying the experimentally observed VBs even more complex.

Next, we briefly consider the problem of the influence of crystal field symmetry. For high-temperature (T > 260 K) face-centered cubic structure of fullerite C_{60} with T_h symmetry, the degeneracy is removed for modes with maximum degeneracy, Gg,u → Ag,u + Tg,u and Hg,u → Eg,u + Tg,u [22,79]. While vibrations of F1g, F2g, and Gg become active in Raman scattering, asymmetric modes such as Tu (including types associated with vibrations of Gu, Hu, and T2u) become active in the IR spectra. Because of the weakness of van der Waals forces in the molecular crystal C_{60}, the intensities of allowed additional bands in the vibrational spectra of F1g, F2g, and Gg types in the Raman spectra are quite low. This complicates the selection of *silent* vibrations among a large number of second-order lines, the intensities of which often exceed the weak first-order line for inactive vibrations in icosahedral symmetry.

For example, in the Raman spectrum at 5 K, noticeable intensities had additional bands of 485, 757.3, 1078.6, and 1481.2 cm^{-1} [41]. In this case the interpretation of the band at 485 cm^{-1} is that the vibration of Gg(1) is not in doubt, because this vibration with a distinct fine structure previously has been observed at 2 K [42]. The authors of [22] assigned the band at 757 cm^{-1} to the vibration of T2g(2), while in[80], it was assigned to Gu(3) vibration, and authors [36,37] described it as the sum tone of Hg(1) + Gg(1). For the exact set of frequencies of Gu(1) and Hu(1) at 355.5 и 404.5 cm^{-1}, it is easy to see that the band is at 757 cm^{-1} taking into account the possible anharmonic shift that corresponds exactly to the total tone of Gu(1) + Hu(1). Thus, the considered band is in very good agreement with the calculated frequencies of two sum frequencies and should not be attributed to the *silent* vibrations. In this regard we choose the frequency at 712 cm^{-1} for T2g(2) vibrations according to the recent work [14], explaining it according to reliable data on IR absorption [13]. Further, the band 1079 cm^{-1} refers to the *silent* vibration of Gg(4) [22]. Furthermore, according to our analysis considered, the frequency corresponds to the sum tone of T2u(1) + Gu(2), active in the Raman spectrum. Moreover, the frequency of 1079 cm^{-1} corresponds to the correct correlation between calculated and experimental values for a range of vibration frequencies of Gg (i), where i = 1 ÷ 6. As a result we assign the frequency 1079 cm^{-1} to the *silent* vibration Gg(4), which is imposed on the total active tone of T2u(1) + Gu(2), which leads to the observed increase in the intensity of the resulting bands. Band at 1481 cm^{-1}, located on the high-frequency wing of the most intense narrow Raman band Ag(2) in [22], refers to the *silent* vibration of Gg(6), and according to [38], it corresponds to the inactive vibration of T1g(3) or to the sum tone of Hg(1) + T2g(3). In [38], the band refers to the sum tone of Hg(3) + Hg(4). According to the results of one of the authors of this chapter, this band, no doubt, refers to the sum tone of Hg(3) + Hg(4), as it is connected with the whole system of successive vibrational resonances [13–15,30–35,57–62], as will also be discussed in the next section. For *silent* vibration of Gg(6), a higher frequency (1502 cm^{-1}) is chosen, which corresponds to the observed IR band [80].

In a multiplier unit cell, such as in a simple cubic lattice at the low-temperature phase (below 249 K), when the unit cell contains four molecules in the vibrational spectra, the additional spectral bands appear as a result of the folding of the phonon branches, including relevant Davydov splitting [13,41,114]. As a result of the reduced symmetry, the dipole IR-active vibrations T1u type can be observed in the Raman spectra, and the vibrations of Ag(1) and Ag(2), on the contrary, can be observed in the IR absorption spectra [13,14], what will be considered later.

A significant amount of vibrations of the C_{60} molecule inactive for icosahedral symmetry I_h is manifested in the vibrational spectra after polymerization of fullerites. It should be noted that the photopolymerization of C_{60} films may occur at the study of the Raman spectra in the case of the laser photons with energies greater than 1.8 eV. In particular, for the orthorhombic structure of polymers C_{60} due to the lowering of symmetry up to D_{2h}, the vibrations such as Hg are split into nondegenerate vibrations such 2Ag + B1g + B2G + B3g allowed in the Raman spectra [81]. Other modes split similarly, for example, Gu → Au + B1u + B2u + B3u, T1,2g → B1g + B2g + B3g, and T2u → B1u + B2u + B3u. We point out that the majority of the vibrations with D_{2h} symmetry except Au are allowed in the vibrational spectra [82]. However, the significant changes in vibration frequencies of fullerite at high pressures and the changing their order prevent a reliable identification of *silent* vibrations of C_{60}. The spectra of inelastic neutron scattering [1,24,25] and electron energy loss [26,27] exhibit all the vibrational modes regardless of their symmetry, which would be easier for the observation and identification of inactive vibrational modes. However, these methods are less accurate than the methods of optical spectroscopy, which does not facilitate accurate identification of the totality of the vibrational modes of C$_{60}$.

The problem of *silent* vibrations of highly symmetric fullerene C$_{60}$ molecule is not complete. According to results of main works where the identification of all *silent* vibrations has been done [22,36,37,80,83], anomalously large variations in the values of their frequencies greater than a several hundred cm^{-1} exist. As a result, even when discarding the extreme values of vibration frequencies, they still remain abnormally large (60–140 cm^{-1}), needing further analysis.

Briefly discussed here are a number of factors necessary to consider the phonon dispersion of the vibrational modes $\nu(k)$, as confirmed by some differences in the frequencies of the VBs in the Raman and IR spectra [14]. This is especially important for the second-order bands, for which the vibrational states of the Brillouin zone boundary are well manifested, and where the density of vibrational states can reach the maximum values. Furthermore, it should be noted that second-order bands do not always have a large half width, as suggested in [22]. This is due to the fact that a high density of vibrational states in the region of large wave vectors can lead to sharp peaks in the observed spectra. The narrow lines in the spectra of the second and higher orders may be associated with broadband group synchronisms [84,85], which occur when implementing the addition of frequencies within certain phonon bands. We observed such effects in the second-order Raman spectra of some crystals [86,87]. In addition, the sharp emission peaks can be formed as a result of the new phenomenon of concentration of energy in the nonlinear interaction of waves [88]. We hope that a deeper and more reliable analysis of the vibrational modes of the set of all different types of fullerenes should lead to a better understanding of the physics of quantum chemical phenomena and improvement of *ab initio* calculations of complex molecules, including biologically active molecules.

5.3.3 Changes in the Intensities of the Vibrational Bands with Temperature and Exciting Radiation Frequency Changes

As is known, during spectral studies, the change of the intensity of the VBs in many cases is more important than the change of frequency. In particular, the variations in the bond intensities ν_{OH} of stretching OH vibrations characterize the change in strength of hydrogen bonds. In this regard, we focus here on the change of the intensities of the VBs in the Raman and IR spectra of C$_{60}$ as the frequency of the exciting laser radiation or temperature changes. Figure 5.4a shows the changes in the intensities of the VBs in the Raman spectrum of the fullerite C$_{60}$ at 5 K and the change of the excitation laser wavelength from 514.5 to 488 nm according to [44]. We remark here that at room temperature, for fullerite C$_{60}$, usually the differences in spectra at these wavelength intervals are insignificant [89]. Due to the fact that the intensities of the laser radiation are different for different frequencies, for a correct comparison of the intensities of the Raman bands were prenormalization spectra on the intensity of bands Hg(6), which was taken as 10 a.u. Figure 5.2a shows that changes in the intensities I(488 nm) − I(514 nm) for all VBs (observed in [41]) have a distinct tendency of increasing values ΔI with the increasing of frequency vibrations. It should be mentioned that according to the tabulated data [41], the frequencies of the vibrations ($\Delta \nu = \nu$ (488 nm) − ν (514 nm)) is sufficiently small (not exceeding 1 cm^{-1}), which demonstrates the inset in Figure 5.2a. Very large changes in the intensities of the high-frequency fundamental vibrations of Hg(7,8) of C$_{60}$ indicate strong changes in the electronic polarizabilities of the molecules of the corresponding vibrations ($I \sim d\alpha/dq^2$). This confirms our concept of vibration-induced significant changes of the electron subsystem.

Figure 5.2b shows the ratio of intensities of the main IR absorption bands T1u (1 ÷ 4) at temperatures of 333, 523, and 98 K according to the tabulated data [90]. And for IR-active modes of C$_{60}$, the significant changes in the intensity of the high-frequency bands of T1u(4) at1429 cm^{-1} are also observed, due to the change of the dipole moments during vibration. Even more clearly considered high trend changes in the properties of C$_{60}$ are found for the half widths of the IR bands $\delta\nu$ T1u (1 ÷ 4), and temperature coefficients k reduce the frequency of the IR bands $\nu = \nu_0 - kT$, as demonstrated in Figure 5.2b. We investigated the IR transmission spectra for a C$_{60}$ film with a thickness of 2 μm on crystalline Si(100), and values of k were taken from [61]. It is interesting to note that the value of the temperature coefficient k for the most intensive band Ag(2) in Raman spectra of fullerite C_{60} determined from [91] almost exactly coincides with the value k for the closest frequency IR mode T1u(4).

It is expedient to compare changes in the intensities of the VBs in the Raman spectra in the case of resonant excitation of nonresonant excitation. Figure 5.2c shows abnormally large variations in the intensities of low-frequency bands in the Raman spectra under laser excitation at 1064 nm. Here are the differences between the normalized intensities of the bands according to the tabular data of different authors [92–94]. The observed differences may be associated with the differences in the properties of fullerenes C_{60} under study and the impact of different powers of the exciting radiation, which will be analyzed further in the succeeding text. Significantly, in the case of long-wavelength excitation, the variations in intensity of Hg(1) and Ag(1) bands may exceed 20%–40% of the strongest bands of Ag(2). It should be noted that this regularity has an opposite trend as those changes in the properties for maximum high-frequency vibrations of C_{60} discussed earlier.

Figure 5.2c shows similar situation changes in the intensities of IR absorption bands T1u (1 ÷ 4) of pure fullerene C_{60} in the works of various authors. Shown here difference normalized by the strongest IR band T1u(1) higher-frequency IR absorption bands ΔA according ethyl tabular data [22,92,95–97]. It is evident that there can be observed age values of ΔA with increasing frequency, reaching 10% and more drastic changes in the midrange.

The existence of abnormally large variations in the intensity of low-frequency Raman bands excited by 1064 nm nonresonant radiation confirmed by the reduced intensities of the bands Hg(1) and Ag(1) during the polymerization of C_{60} films. The calculated differences in observed intensities of Raman bands for pristine C_{60} films and films after photopolymerization and polymerization by pressure (data from [98]) are shown in Figure 5.2d. It is seen that in this case the changes in the intensities of low-frequency bands Hg(1) and Ag(1) substantially more than an hour to bands the Ag(2). This shows that the normalization of the bands in Figure 5.2c is not essential, although it does not allow to detect changes in the intensity of bands of Ag(2). Detecting large similarities Figure 5.2c and d suggests that the observed anomalous variation of intensities of low-frequency Raman bands of C_{60} may be associated with an uncontrolled change of the ES under laser irradiation, which will be analyzed further later.

It is illustrated on Figure 5.2d, where the calculated by us differences in observed intensities of Raman bands for pristine C_{60} films and films after photopolymerization and polymerization by pressure (data from [98]) are shown. It is seen that in this case the changes in the intensities of low-frequency bands Hg(1) and Ag(1) substantially more than an hour to bands the Ag(2). This shows that the normalization of the bands in Figure 5.2c is not essential, although it does not allow to detect changes in the intensity of bands of Ag(2). Detecting large similarities Figure 5.2c and d suggests that the observed anomalous variation of intensities of low-frequency Raman bands of C_{60} may be associated with an uncontrolled change of the ES under laser irradiation, which will be analyzed further later.

5.4 ON THE NATURE OF STRONG CHANGES IN THE FREQUENCIES AND INTENSITIES OF THE VIBRATIONAL BANDS IN THE RAMAN AND IR ABSORPTION SPECTRA OF C_{60}

5.4.1 Influence of the Vibrational Resonances on the Intensity of the Vibrational Bands

Let us first consider the influence of temperature and interaction of fullerene C_{60} with TDAE on the intensities of VBs in the Raman spectra. For this we use tabular data [40.89] for the intensities of the Raman bands of fullerite C_{60} under temperature change from 10 K to room, as well as complex TDAE: C_{60} at 10K in the case of resonant laser excitation 514.5 nm according to the tabulated data [40,89]. For a correct comparison of the intensities of all observed bands in the vibrational spectra at different conditions of the experiment, we carried out their normalization to the intensity of the strongest line in the Raman spectrum of Ag(2) with a frequency of 1469 cm^{-1}, which was

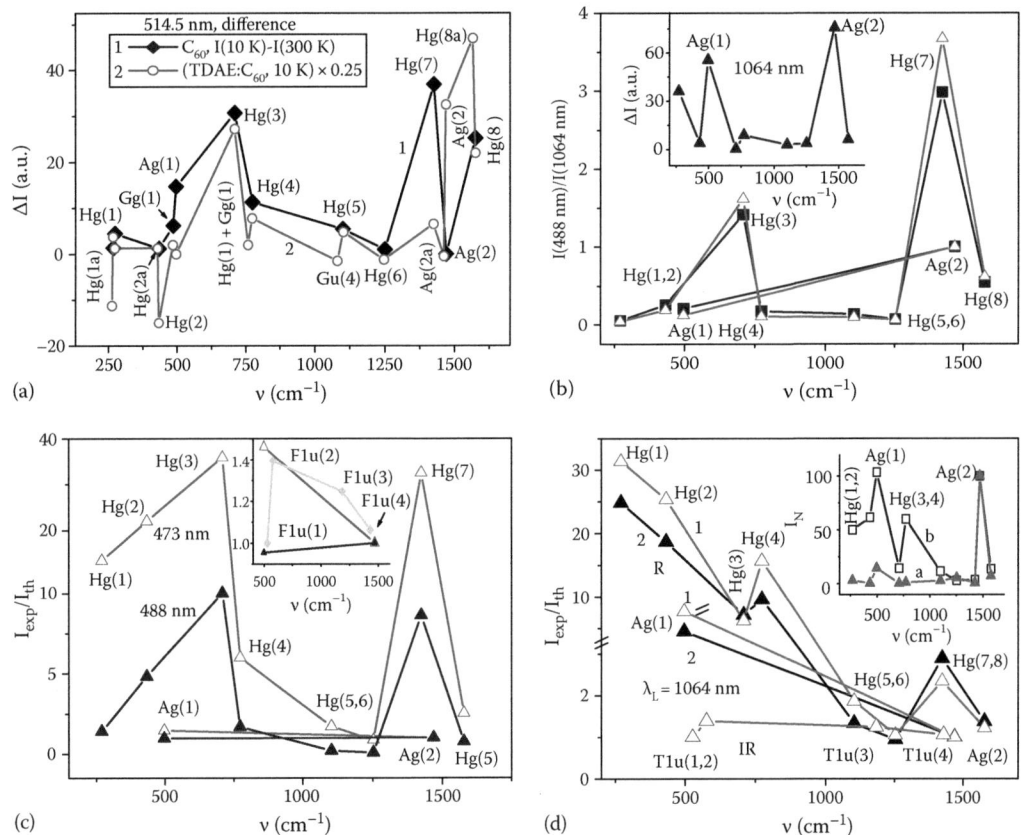

FIGURE 5.3

Spectral dependences of the differences of the intensities of vibrational bands in the Raman spectra of fullerite C$_{60}$ at 10 K [40] and 293 K [89] (1), as well as a decrease in the intensities of the single-crystal TDAE:C$_{60}$ at 10 K compared with pristine C$_{60}$ at 10 K according to numerical data [40] (2) (a), and the ratio of the band intensities for Raman spectra of fullerite C$_{60}$ at 488 nm excitation according to our measurements and 1064 nm excitation according to the tabulated data ([92], filled symbols and [93], open symbols) (b) and the weakening of vibrational bands under neutron irradiated of polycrystalline C$_{60}$ according to the results of [99] (inset). (c) The ratio of intensities of experimentally observed vibrational bands in the Raman and IR (inset) spectra of C$_{60}$ excited by 488 nm (our measurements) and 473 nm [102] and calculated intensities of vibrations of C$_{60}$; (d) the ratio of the intensities of the experimental and calculated normalized vibrational bands in the Raman spectra in the long-wavelength excitation and the IR spectrum of C$_{60}$.

taken as 100 a.u. Figure 5.3 shows the difference between the observed intensities ΔI of VBs in the Raman spectra of C$_{60}$ at 10 and 293 K (see dependence 1) as well as the difference ΔI between the Raman spectra of a single crystal of C$_{60}$ and composite TDAE:C$_{60}$ (curve 2). Characteristically, with increasing temperature, as well as the interaction of C$_{60}$ with organic compounds, the intensity of the majority of bands in the Raman spectra is reduced. Moreover, the greatest changes in intensity reaching 35%–45% of the intensity of the band Ag(2) are observed for vibrations of Hg(3) and Hg(7).

This is due to a nonlinear vibrational resonance of 2Hg(3) $\approx$ Hg(7) [13–15,30–35,57–62] mentioned earlier. The calculated frequency of the overtone 2Hg(3) equal to 1418 cm^{-1} in fact coincides with the band 1417 cm^{-1} observed in Hg(7) [40]. However, its intensity was only ~4.4% of the maximum intensity of the band Hg(7). At the same time in the [41], the intensity of the line 2Hg(3) with the frequency 1418.3 cm^{-1} was more than 50% of the most intense components of the 1424 cm^{-1} band Hg(7). These redundant variations of the given band intensity clearly indicate its nonlinear nature. Good observed high-frequency line 1481 cm^{-1} closed to the strongest band Ag(2) at 1469 cm^{-1} refers to the sum tone of Hg(3) + Hg(4) with the estimated frequency at 1482 cm^{-1} [14,38]. Nonlinear generation of high-frequency vibrational modes of ~Hg(7) and Ag(2) with a resonant interaction of low-frequency modes Hg(1) + Hg(2) $\approx$ Hg(3), Hg(1) Ag + (1) $\approx$ Hg(4) promotes the conversion of thermal energy from low-frequency to the high-frequency region in References [13–15].

It is essential that the nonequilibrium excitation of higher vibrational modes can be realized in a cascaded manner as a result of the resonant generation of overtones and sum tones of Hg(1) + Hg(2) = Hg(3), Hg(1) Ag + (1) = Hg(4), Hg(3) + Hg(4) = Ag(2), 2Hg(3) ≈ Hg(7), 2Hg(4) ≈ Hg(8), which was first identified in References [13,14,60,62]. Existence of the NIVM of fullerites and the possibility of the sum mode generation Hg(1) + Hg(2) and Hg(1) + Ag(1), which is resonant with Hg(3) and Hg(4), supports the observation of high-frequency components at 297 and 501 cm^{-1} for Raman bands Hg(1) and Ag(1) [13,14], the corresponding parallel developing processes of different frequency generation of Hg(3,4) − Hg(2). It is important that authors of [41] at 5 K clearly observe the considered number of bands of sum tones and overtones, Hg(1) + Hg(2), Hg(1) + Ag(2), 2Hg(3) ≈ Hg(7), Hg(3) + Hg(4) ≈ Ag(2), and 2Hg(4) close to Hg(8), which confirms the results of our earlier works [13,14] and is consistent with the important role of the nonlinear interaction of vibrational modes and changes of the condition of electronic vibration from heavy-electron interactions. Maximum resonant nonlinear interaction of vibrational modes increases the efficiency of these processes.

Figure 5.3a shows that at a temperature 10 K, the fivefold degenerate bands Hg(1,2,5,8) contain two spectral components due to the partial removal of the degeneracy of Hg → Eg + Tg [22,79] as a result of lowering of the symmetry of the molecule position in the crystal of cubic symmetry. Furthermore, at a low temperature the band of the sum tone of Hg(1) + Gg(1) at 758 cm^{-1} occurs in the spectrum [27], which is only 15 cm^{-1} different from the frequency of the fundamental vibration Hg(4). In this case, the sum tone of Hg(1) Ag + (1) [13,14] should appear in the spectrum with estimated frequency of 768 cm^{-1}, which is only 4.5 cm^{-1} different from the line maximum Hg(4) and comparable to its half width.

Thus, besides the tendency of strengthening change frequencies of VBs and their half widths in the high-frequency region of the spectrum in the earlier examples, new characteristic regularity of the maximum change in the intensities of the Raman bands for resonantly interacting vibrational modes of Hg(3,7) appeared. This demonstrates the importance of the nonlinear interaction of vibrational modes. This regularity is confirmed by comparing the intensities of the VBs in the Raman spectra under resonant excitation at 488 nm and nonresonant long-wavelength excitation with Nd:YAG laser at 1064 nm, as illustrated in Figure 5.3b. We used the results of our studies of the Raman spectra of the C_{60} microfilm with thickness of about 2 μm on crystalline Si(100) at 488 nm excitation and the tabular results of References [92,93] in the case of excitation by 1064 nm. In this case, the observed regularity of increasing intensities of resonantly coupled vibrational modes Hg(3,7) is even more pronounced. As for the breathing mode for Ag(1), as well as a number of weak modes of Hg (i), there is a weakening of the intensities of VBs in the excitation of 488 nm compared to 1064 nm nonresonant excitation.

The inset in Figure 5.3b shows the decrease of the intensity of the fundamental VBs g and Ag in the Raman spectrum of powdered crystalline C_{60} sample under neutron irradiation at a dose of $1.32 \cdot 10^{15}$ n/cm^2 compared to the initial sample in the case of excitation by Nd:YAG laser at 1064 nm according to the results [99]. In this case, the maximum changes for the bands of Ag(1) and Ag(2) may be associated with the resonant interaction of 3Ag(1) ≈ Ag(2) [100]. Such peaks can be identified when comparing the Raman spectra (excitation 488 nm) single crystals and thin films of C_{60} according to [101]. In the films of C_{60} as compared with the crystal, the intensity of the bands Hg(3) and Hg(7), Ag(2), and for the very low-frequency vibrations Hg(1) minimizes.

Thus, besides the significant changes of ESs, the important role of nonlinear resonant vibrational modes of interaction that leads to the generation of higher vibrational states has been proved. In semiconductors, including fullerite C_{60}, they approach its ESs, interact with them, and modify them. As a result, it leads to increased VEIs, which are a key element for the vibrational modes' relationship and changes of CB. Thus, with strong VEI the collective properties of the vibrational modes are transferred to the CB in condensed matter, where they would also become collective. And this important fundamental regularity should play a significant role in chemical functionalization.

A good indicator of strengthening intermolecular interactions in fullerites is the value of the Davydov splitting of VBs [13,14]. Strengthening of Davydov splitting for the surface areas of fullerite found in the IR spectra of diffuse reflection is also further evidenced by the nonlinear nature of the observed phenomena [14]. It is due to the fact that fluctuations in the surface region of the atoms have a higher anharmonicity, which is associated with enhanced nonlinear susceptibilities near the surface, which opens the way to explore the mechanisms of formation of the surface tension. These examples show the importance of nonlinear resonant interaction of vibrational modes of molecules with a large number of normal vibrations.

Very good maxima of resonance vibrations of Hg(3) and Hg(7) are shown in the comparison, experimentally observed with calculated intensities in the vibrational spectra of C$_{60}$. In Figure 5.3c the relationship of experimental and calculated intensities of the Raman-active bands Hg (1 ÷ 8) and Ag(1,2) is shown as well as the IR bands T1u (1 ÷ 4) (inset). Here we use our measurements of IR and Raman spectra of C$_{60}$ under laser excitation at 488 nm and data for laser excitation at 473 nm according to [102] as well as calculated intensities of active VBs by the HF (basis 6-31G*) method. Previously all of the Raman and IR spectra were normalized to the intensities of the strongest bands Ag(2) and T1u(1). It can be seen that the relationship I_{exp}/I_{th} for both excitation wavelengths reaches maxima just for resonance vibrations of Hg(3) and Hg(7).

The inset in Figure 5.3c in enlarged scale shows the relationship I_{exp}/I_{th} for IR-active modes T1u (2 ÷ 4) and Ag(1). It is seen that for vibration Ag(1) at 488 nm excitation, the value I_{exp}/I_{th} is a bit less than 1, while at 473 nm, the intensity of this band increases. IR vibration differences between theory and experiment do not exceed 40%, which is significantly less than for the bands Hg(3,7). These facts show that modern QCC methods better describe the activity of vibrations of C$_{60}$ in the IR spectra than the changes in the electronic polarizabilities during the vibrations responsible for the intensities of VBs in the Raman spectra.

Even in the case of nonresonant excitation of Raman spectra by 1064 nm laser line there are significant differences in the behavior of relations I_{exp}/I_{th}, as illustrated in Figure 5.3d. Here in the inset, the normalized Ag(2) band calculated (spectrum a) and experimental (spectrum b) intensities of the VBs in the Raman scattering are illustrated. Clearly, under long-wavelength excitation at 1064 nm, there is an experimentally observed significant increase of the intensities of low-frequency bands Ag(1) and Hg(1.2), as well as Hg(3.4). As a result the ratio I_{exp}/I_{th} for bands Hg (i) greatly increased at low-frequency region and bands Ag(1) reach the values about ~30, which is similar to a giant wing of the Rayleigh line close to the exciting laser line. Here the curves 1 correspond to the experimental tabulated data [92], and the curves of 2—[93]. But in these dependences small maxima are observed for vibrations Hg(4) and Hg(7). The IR spectra of the differences between observed and calculated values of the band intensities are essentially less, which in an enlarged scale are shown in the inset in Figure 5.3c.

Thus, in the case of nonresonant excitation, there is a more pronounced increase of the electronic polarizabilities for low-frequency vibrations of C$_{60}$, and in the case of resonant excitation, more high-frequency vibrations. In the latter case the nonlinear resonant interaction of vibrational modes associated with the observation of vibrations maxima for Hg(3) and Hg(7) is clearly manifested (see Figure 5.3a and b).

5.4.2 Resonance-Vibrational Instability of C$_{60}$ Molecules

To study the particular sensitivity of the highly symmetrical structure of the fullerene C$_{60}$ to small perturbations, the comparison of vibrations calculated with the use of the program Gaussian03 HF (basis 6-31G*) of high- and low-symmetry molecules C$_{60}$ and ^{13}CC$_{59}$ has been fulfilled. The difference between the frequencies $\Delta\nu$ of the corresponding vibrations of C$_{60}$ and ^{13}CC$_{59}$ is shown in Figure 5.4. The absolute values $\Delta\nu$ reach ±50 cm^{-1}, and the relative values $\Delta\nu/\nu$ in many cases exceed 4%, and for some vibrations it reaches 6%. Characteristically, in a number of isotopically substituted molecule, the frequencies of some vibrations do not decrease, but rather increase, despite the mass increasing and all CBs practically unchanged. In particular, it is observed for inactive C$_{60}$ vibrations of Gg(2), Hu(3), Gu(3), T2g(3), Au, etc., with $\Delta\nu < 0$ (see Figure 5.4a). In most cases, the extreme values of $\Delta\nu < 0$ are realized for the *silent* vibrations. The maximum vibration softening mitigation in isotopically substituted ^{13}CC$_{59}$ molecule for which $\Delta\nu > 0$ is attained for vibrations Gu(2), Gg(3), Gg(4) T1u(3), Hg(6), T2g(4), Gg, u(6) (see Figure 5.4a). Thus, for the majority of the *silent* vibrations of the fullerene C$_{60}$ is an abnormally high sensitivity of their states to small perturbations highly symmetrical C$_{60}$ structure.

This explains the very large scatter of the selectable frequencies of *silent* vibrations, what was discussed at the end of Section 5.2.2. Really both of C$_{60}$ and 13CC$_{59}$ molecules are present in comparable amounts and for which the corresponding *silent* vibrations differ by ±(10–59) cm^{-1}. In this case, without understanding the high sensitivity of perturbations of the properties of C$_{60}$, even the correct method of selecting *silent* vibrations can lead to significant differences in outcomes.

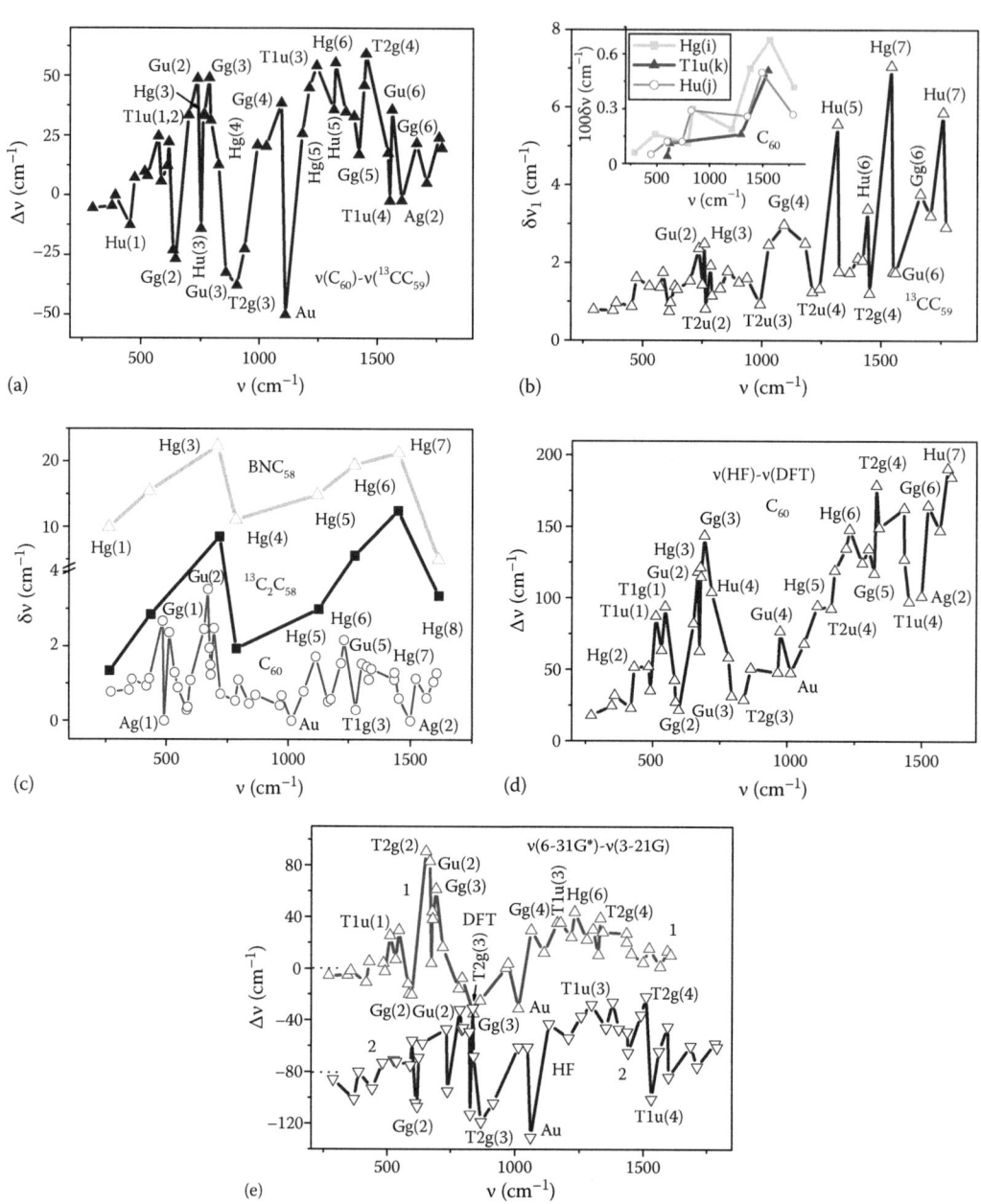

FIGURE 5.4

A frequency difference of all vibrations of fullerenes C_{60} and $^{13}CC_{59}$ (a) and the magnitude of the frequency splitting $\delta\nu$ for all degenerate in C_{60} vibrations of $^{13}CC_{59}$ (b), calculated by HF, as well as value $\delta\nu$ for C_{60} molecule calculated by the DFT (B3LYP) method, and molecules $^{13}C_2C_{58}$ and BNC_{58} (HF method) (c), and the differences between the vibration frequencies of C_{60} calculated using the HF and DFT methods (d), and the differences between the frequencies of vibrations calculated using DFT (1) and HF methods (2 shift at -80 cm^{-1}) in the bases 6-31G* and 3-21G [77] (e).

In Figure 5.4, there are two characteristic maxima in the average spectral behavior of the quantity $\Delta\nu$: in the low-frequency region, about 700 cm^{-1} and near 1330 cm^{-1} at higher frequencies. Characteristically, the dipole-active modes of T1u(1.2) and T1u(3.4) also play a significant role in the formation of this maximum instability. The similarity between these maxima and the maxima of the intensity changes of VBs of the fullerene C_{60}, found by us and shown in Figure 5.3a through d, allows a deeper understanding of the mechanisms of the particular instability of complex molecules. They are associated with the sequential occurrence of vibrational resonances in molecules with a large number of normal vibrations, which significantly enhance the nonlinear interaction of vibrational modes and lead to increased VEI. This defines a new quality of complex molecules and condensed matter, which can play a significant role in the chemical functionalization of carbon materials, as well as understanding the true mechanisms of biologically active molecules including proteins and DNA.

Simultaneously with the shift of the vibration frequency in isotopically substituted 13CC$_{59}$ molecule as a result of the symmetry, lowering the removal of degeneracy and the corresponding splitting of the vibrational bands occur. The total for all splitting of degenerate vibrations of fullerene is shown in Figure 5.4b. It is evident that there is also a well-expressed increasing trend in the splitting $\delta\nu_1$ in the high-frequency region, which also indicates manifestations KEV. Note that the maxima $\delta\nu_1$ are also associated with a number of *silent* vibrations of Gg(4,6) and Hu(5,6,7). It also describes the development of the instability of the molecules. However, the maximum values of splitting $\delta\nu_1$ in the low- and high-frequency part of the spectrum are achieved for active resonant vibrations of Hg(3,7).

It is interesting to point out that the beginnings of instability are manifested in the C$_{60}$ molecule. According to the QCC using the HF method, even the C$_{60}$ molecule manifested its light asymmetry and splitting of degenerate vibrations with a very small quantity $\delta\nu \sim 10^{-3}$ cm^{-1}. This is illustrated in the inset in Figure 5.4b, which shows a very small dependence of the spectral splitting of both active-type vibrations Hg (i) and T1u(k) and for *silent* vibrations, for example, Hu (j). Naturally, these splittings ($\sim 1-6 \cdot 10^{-3}$ cm^{-1}) are associated with disappearing small asymmetry of molecules C$_{60}$ and are difficult to detect experimentally. It is very important that the values of these splitting naturally increase with increasing frequency. It characterizes their relationship with the nonrandomness and VEI.

It is characteristic that in Figure 5.4a and b, the sharply oscillating dependences of the frequency shifts $\Delta\nu$ and splitting $\delta\nu_1$ of the vibrations appear, which is similar to an oscillating spectral dependence $\Delta\nu$ under polymerization of fullerites (see Figure 5.1d) actually observed experimentally. This confirms the validity of the observed *spectral oscillations*. The correlation of splitting $\delta\nu$ for different types of vibrations of C$_{60}$ should be noted (see inset Figure 5.4b), as well as achieving maximum values $\Delta\nu$ for vibrations of different types (T1u(4), Hu(6)) in the vicinity of resonant vibration Hg(7). Probably the observed oscillations are associated with the corresponding *wavelike* changes in the ES of the fullerene C$_{60}$, indicating a partial accounting VEI by modern QCC.

It should be noted that the density functional method (DFT) when taking into account electron correlation and the phenomenon of resonant vibrational instability (RVI) in fullerene C$_{60}$ exist even more so. This is confirmed by the large splitting of degenerate vibrations, as demonstrated by the lower curve in Figure 5.4c. It shows the splitting of all degenerate modes of the fullerene C$_{60}$ except nondegenerate vibrations Ag(1,2) and Au. For many vibrations the values of the splitting $\delta\nu_1$ reach about 1.5–3.5 cm^{-1}. Most importantly, the existence of two maxima values of splitting $\delta\nu_1$ undoubtedly connects with the influence of vibrational resonances. In particular the main maxima $\delta\nu_1$ may be associated with the vibrational resonances Gu(2) + T1u(1) $\approx$ Hg(6).

Besides the molecule ^{13}CC$_{59}$, QCC was conducted using the HF method for molecules ^{13}C$_2$C$_{58}$ with two isotope atoms ^{13}C, as well as for isoelectronic substitution BNC$_{58}$ [15]. For fullerenes ^{13}C$_2$C$_{58}$ and BNC$_{58}$, spectral dependences of the splitting $\delta\nu_1$ for vibrations Hg(k) are shown in Figure 5.4c. Here it clearly reveals two maximum values $\delta\nu_1$ for vibrations Hg(3) and Hg(7), where the vibration splitting $\delta\nu_1$ reaches values of 8.5–12.5 and 22.4–21.3 cm^{-1}. This is due to nonlinear resonant coupled 2Hg(3) $\approx$ Hg(7) [13–15].

According to our concept a critical role in the phenomenon of RVI plays the NIVM. But it is not taken into account in QCC where only the system of independent normal vibrations is considered. In small molecules with a few degrees of freedom, normal modes with high accuracy are independent. However, in large molecules due to the small spacing between the frequencies of vibrations, real anharmonicity leads to the relationship of closed normal vibrations. In this regard, the standard QCC partially account the interaction of vibrations, even without explicit consideration of the NIVM and disregarding the vibrations of the second and higher orders. Therefore, the maxima $\delta\nu$ for resonantly coupled modes Hg(3,7) are observed.

In connection with this the appearance of RVI can be expected even in QCCs of symmetric C$_{60}$ molecules. For this purpose, we compared the results of calculations of vibrational frequencies on the basis of HF and DFT methods, using different bases. Figure 5.4d shows the different frequencies of normal vibrations of C$_{60}$ $\Delta\nu = \nu$ (HF) $- \nu$ (DFT), calculated using the HF (basis 6-31G*) and B3LYP (basis 3-21G) methods. It is seen that the calculated differences $\Delta\nu$ increase in a high-frequency region and reaches the values of 150–200 cm^{-1}. This shows that the DFT method is more correct to examine actual softening of vibration frequencies and electron correlations are associated with VEI. In Figure

5.4d excepting the VEI effect, two peaks in the value $\Delta\nu$ appear in the region of low-frequency vibrations Hg(3), Gu(2), and Gg(3) and high-frequency vibrations Hg(6), T2g(4), Gg(6), Hu(7), etc. This may indicate an implicit influence on vibrational resonances on the results of QCC.

These conclusions are confirmed by the comparison of different variants KHR, the results of which are published in the detailed work [77]. Figure 5.4e shows the difference between the vibration frequencies $\Delta\nu$, calculated using the B3LYP method and basis 6-31G* and 3-21G (upper curve). The lower curve corresponds to the vibration difference between the frequencies calculated in the same bases using the HF method. For convenience, the latter dependence is shifted by -80 cm^{-1}. The existence of two well-defined maxima envelopes for all frequencies convincingly confirms the influence of the considered vibrational resonances on the results of QCC for fullerenes C$_{60}$. Thus in the vicinity of the low-frequency peak in the case of the DFT method, the magnitudes $\Delta\nu$ may exceed 80 cm^{-1}, and the relative difference $\Delta\nu/\nu$ reaches 14%, indicating the importance of even partial influence on the effects under study.

Moreover, the nonrandomness of the observed differences indicates the repetition of the details of the two curves in Figure 5.4d. In particular, in the high vibrational maximum a local increases $\Delta\nu$ repeated for the vibrations of T1u(3), Hg(6), and T2g(4) and in the low-frequency region—for the vibrations of Gu(2) and Gg(3). And this correspondence is saved despite an overall decrease in the values of all frequencies when the DFT method was used due to a more correct account of vibration softening. Very good agreement for vibrations in the two QCC methods observed for negative minima $\Delta\nu$, for example, for vibrations Au, Gg(2), T2g(3), etc.

Thus, in the current *ab initio* QCC the vibrational resonances are well manifested in the values of vibrational frequency shifts $\Delta\nu$ and spectral splitting $\delta\nu$ and essential worse effect on the intensities of the bands in the vibrational spectra, although the influence of vibrational resonances on the experimentally observed intensities is very significant.

5.5 STUDY OF NEW ELECTRONIC STATES IN THE REGION OF VIBRATIONAL BANDS OF FULLERITE C$_{60}$

5.5.1 Review of Studies on the Observation of Broadband Background in the Vibrational Spectra of Fullerite C$_{60}$

In the Raman and IR spectra of nanostructures, as a rule there is broadband background (BB). It is essential that BB is observed in the Raman spectra, even for nonresonant radiation, when cannot be excited photoluminescence. For example, in [103], it was reported that exited by line at 785 nm of a diode laser Raman spectrum of C$_{60}$ has BB with broad main maxima in the field of vibrations Hg(2), Ag(1), and Hg(7), Ag(2), Hg(8). The BB maxima of the same vibrations of C$_{60}$ appear for composites Sm$_X$C$_{60}$ where x $\geq$ 2, at 514.5 nm excitation [70]. Moreover, with an increase doping level (x = 3–6), the background in Raman spectra of Sm$_X$C$_{60}$ significantly enhanced. These facts show that the intensity of BB inextricably connected as with active vibrational modes of the fullerene C$_{60}$, so simultaneously with the change of the ES.

Excited by 780 nm Raman spectrum of the composite fullerene with cuban (C$_{60}\cdot$C$_8$H$_8$), a strong background with a maximum abound 200 cm^{-1} is observed, which extends up to the strong line of Ag(2) [104]. Moreover, such BB is observed not only at room temperature but at 83 K. Significantly, in the region of maximum BB, the intensities of the lines Hg(1) and Ag(1) greatly increase and have approximately twice intensity of the Ag(2). It also confirms the connection BB with vibrational modes and the changes of the ES, since the intensities of the Raman lines are proportional to changes in electronic polarizabilities under vibrations (I ~ d2α/dq2). Low-frequency BB with an maximum abound 200 cm^{-1} was also observed in the Raman spectrum of the endohedral fullerene Li@C$_{60}$ under more long-wavelength excitation at 1064 nm [105]. At the same excitation in the spectrum of the fullerite C$_{60}$ film, BB appears peaking at 800–1000 cm^{-1} when the polymerization resulted by external pressure [98].

The important fact observed in the Raman spectra BB is not associated with photoluminescence confirmed by the observation of a similar background in the IR spectra. Thus, in [89], BB is present in the IR spectra of fullerenes C$_{60}$ and C$_{70}$. In the IR spectrum of fullerenes C$_{60}$ observed background with maxima at 750 and 1050 cm^{-1}, that is, in the region of vibrations in Hg(3,4) and Hg(5). Even greater intensity BB was recorded in both low- and high-frequency regions of the IR spectra of chlorine–fullerenes—C$_{60}$Cl$_{4}$, C$_{60}$Cl$_{28}$, and C$_{60}$Cl$_{30}$ [106]. In the known paper [38] in the Raman spectra of pure fullerite C$_{60}$ at 488 nm excitation, even at 20 K, the number of maxima of BB in the areas of CHF ~ 500, 1300–1400, 1900–2450, and 3300 cm^{-1}. Note that these peaks are also associated with the active vibrations of fullerite C$_{60}$, including strong dipole-active modes T1u(1–4) and its overtones. At a temperature of 523 K, the BB increases and dramatically enhances in the high-frequency region $\nu > 2300$ cm^{-1} [38]. This is analogous to the change of the vibrational frequencies and intensities of the bands that were analyzed in Section 5.2 and confirms the concept of developing strong VEI.

Previously, we have established the electronic nature of the BB in the vibrational spectra and proved the importance of strong KEV for inducing ES fundamentally new in the bandgap semiconductors and dielectrics [17–19,21]. In the Raman spectra the induced EB abnormally strongly increased during the irradiation of fullerite C$_{60}$ by electrons [21], neutrons [99], and ions D^{+} [107] C^{2+} [108], Au [109], and others, as well as at high pressure [110,111]. In this section, we first consider the difference of the new EB in the spectra of micro- and nanofilms of fullerite C$_{60}$ and extraction of VBs on the background of abnormally broad EB and also study the changes of EB under polymerization of fullerenes C$_{60}$ by diamine. Next a less broad local EP arising in the field of the strongest vibrational modes will be considered, as well as increasing induced of EP with increasing laser intensity and nonmonotonic dependence of the intensities of the EP from thickness of nanofilms.

5.5.2 Appearance of New Broad Electronic Bands in the Vibrational Spectra of Fullerene C$_{60}$ Films

The examples observed by us an abnormally wide EB in the Raman spectra (RS) of fullerite C$_{60}$ shown in Figure 5.5c. Figure 5.5b demonstrates the overview RS of micro- and nanofilms of C$_{60}$, respectively. In the studied microfilm with thickness of 2 μm, a wide EB well approaching by Gaussian $I_E = I_0 + A_{exp} [-(\nu - \nu_0)^2/2\delta\nu)^2]$ with a little regular background with intensity I_0. The support experimental points using for this approximation are shown in Figure 5.5 by light triangles. The obtained values of all constants are also shown in Figure 5.5. It should be noted that the half width of the analyzed EB $\delta\nu = 1650$ cm^{-1} and its high frequency $\nu_0 = 1870$ cm^{-1} are comparable in magnitude, which characterize the broadband of observed EB that is significantly wider than the VBs. These are very broad EB with maxima in the high-frequency region occurring after exposing C$_{60}$ by Au ions (200 MeV) [109], as well as the polymerization by pressure [98]. The relationship of induced EB at a dose of 10^{13} ions Au ions/cm^2 with vibrational states confirmed by decrease of more than four times absorption for bands Fu (1–4) and a 20%–50% increase in half width of the high-frequency bands Fu(3,4) [109].

For RS of nanofilms of C$_{60}$ the other behavior of BB is usually observed: the intensities of broad EB decrease with increasing f, as shown in Figure 5.5b. In this case we can apply their approximation by one or two exponents: $I_E = I_0 + A \exp(-\nu/\nu_1) + B \exp(-\nu/\nu_2)$. Characteristically, for hot depositing method of nanofilms, the BB in the high-frequency region is almost two orders weaker of magnitude (see spectrum 1 at Figure 5.5b) than at the cold depositing process (upper spectrum 2). The obtained typical values of all constants for the spectra 1 and 2 are shown in Figure 5.5b. We point out that at the cold deposition process the permanent background component I_0 greatly increased. Similar EB detected in the IR spectra of C$_{60}$ films. Later considered VBs will always be allocated at the observed broad EB, as shown in Figure 5.5b.

For C$_{60}$ nanofilms together with the breathing radial mode Ag(1) at 497 cm^{-1}, the line of crystalline silicon substrate $\nu_{Si} = 520$ cm^{-1} is observed, which is shown in detail in the inset Figure 5.5b. The intensity of this line in the future we will use for the classification of C$_{60}$ nanofilms by thickness—with increasing of nanofilm thickness d band intensity Ag(1) increases and ν_{Si} lines of silicon decreases (compare spectra 1 and 2 in the inset). In microfilms of C$_{60}$ 1.2–2 μm thick line, ν_{Si} does not manifest (see Figure 5.5).

Change of observed EB after polymerization of nanofilms by hydrazine N$_2$H$_4$ is illustrated in Figure 5.5c (see spectra of 1p and 2p). The broad EB enhanced in films obtained by hot deposition

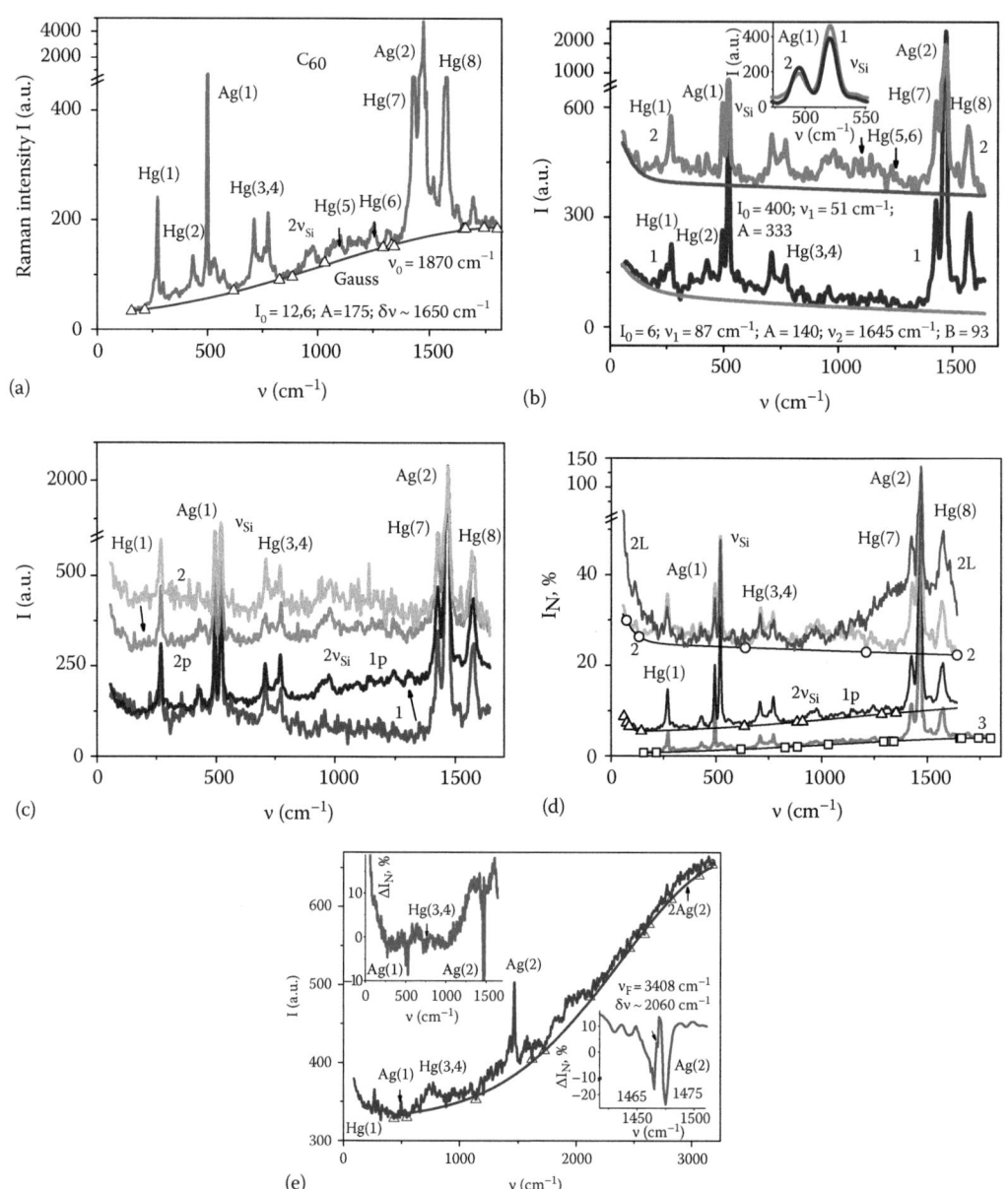

FIGURE 5.5

Decomposition of broad EB in the Raman spectra of fullerite C_{60} microfilms with thickness about 2 μm (a) and nanofilm thickness about ~200 nm in hot (1) and cold (2) deposition method (b), and change in EP after polymerization of nanofilms by diamine (spectra 1p, 2p) (c), as well as a comparison of broad EB to the normalized Raman spectra of various nanofilms of fullerene C_{60} (1p, 2, and 2L) and microfilm 2 μm (3) (d); the enhancement of EP in C_{60} microfilm at high laser intensity (3L) (e). The insets show the marked lines of the breathing vibration Ag (1) and a crystalline silicon substrate νSi for C_{60} nanofilms (b) and the general appearance of difference spectra 2L and 2 in (d) and its fragment in the field of Ag(2) line (e).

(range 1p on Figure 5.5c) and weakened for films of cold deposition (2p) can be seen. But the most importantly is that their intensity increases with frequency ν the same manner as intensity of t EB in RS of C_{60} microfilm that indicates the VEI enhancement in the high-frequency range. It should be mentioned that after polymerization of NF using N_2H_4 in spectra 1p and 2p the broadband in the vicinity of vibrational overtone $2\nu_{Si} \sim 900$–1000 cm^{-1} well observed. This is due to polymerization of C_{60} and the collectivization of the vibrational modes and increasing VEI. It is interesting to note that in the RS of microfilm in Figure 5.5a at the complete absence of line 520 cm^{-1} band, the overtone $2\nu_{Si}$ is often clearly seen, which should be due to resonance of second-order vibrations of fullerene C_{60} and the silicon substrate as well as the influence of the VEI.

For a correct comparison of the intensities of the new EB in various nano- and microfilms of the C$_{60}$ Raman spectra were normalized on the intensity of the strongest line Ag(2) at 1469 cm^{-1}, which illustrated by Figure 5.5d. It is evident that in microfilm (d = 2 μm, spectrum 3) broad EP is much weaker than in nanofilms even hot method of their preparation (range 1p). In a irradiation of nanofilms of C$_{60}$ as a low-frequency part of EP (an analogue of the Rayleigh line wing) may increase and their high-frequency part near the most intense Raman bands Ag(2) and Hg(7,8), as shown on the spectrum 2L in Figure 5.5d. It is seen that in the latter case the intensities of local EB in the vicinity of vibrations Ag(2), Hg(7,8) are greatly increased, which will be considered in more detail in the next section. It should be noted that according to Figure 5.5d in nanofilms 1p and 2 (especially with additional laser irradiation), together with the increasing of BB, the intensities of the bands Hg(7,8) greatly increased too, which is an additional proof of developed concept of strong VEI and inducing of new ES in the bandgap.

A general view of the difference between spectra 2L and 2 is shown in Figure 5.5d upper inset Figure 5.5e. It is seen that with increasing of the intensity of local EB in low- and high-frequency area, there is weakening of most of the VBs, including Ag(1,2), Hg(1,3,4), and ν_{Si}. On the other insert in Figure 5.5e in enlarge scale the difference spectra under consideration 2L + 2 in the normalized bands Ag(2) are shown. It turns out that with additional laser irradiation band Ag(2) is not extended, which would have to be at the photopolymerization, and suddenly narrows. This can be explained by the influence of the nonlinear interaction of vibrational modes, which leads to the appearing of a central sharp peak in the line shape of Ag(2) of the difference spectrum. It is known that the nonlinear wave interaction as a rule leads to a narrowing of the spectrum of the generated radiation that proves the connection with EP-induced vibrational modes.

At strong focusing of the exciting radiation in λ_L in the fullerite microfilm with thickness of about 1.2 μm, a wide total EB is enhanced by more than 100 times and its maximum is much higher than the intensity of the strongest line Ag(2) at 1469 cm^{-1}, which is shown in Figure 5.5 (spectrum 3L). At the time the maximum of EB moves from the most intense bands Ag(2), Hg(7,8) in the Raman spectrum to their overtones that characterize the increasing of interaction of atomic and electronic subsystems (violation of the adiabatic approximation) and is a strong evidence of the developed nonlinear vibrational-electronic concept.

In conclusion, we note that the observation of the low- and high-frequency EB in nano- and microfilm of fullerites C$_{60}$ as shown in Figure 5.5a and b like an abnormally large difference between observed and calculated intensities of the VBs at the nonresonant and resonant excitation of the Raman spectra (see Figure 5.3c and d). It can be assumed that this analogy has a deep physical meaning associated with two types of induced electron states: (1) at low frequencies, that is, near the frequency of the exciting laser radiation λ_L, and (2) in the electronic high-frequency bands of vibrations Ag(2), Hg(7,8) or their harmonics. Weaker influence of laser radiation on a substance in the case of nonresonant excitation or less effective spatial accumulation of nonlinear wave processes in nanofilms leads to the emergence of low-frequency EB (analogue of the Rayleigh line wing). Resonance radiation has a stronger effect on fullerite C$_{60}$ and induces the appearance of a high-energy EB. This is particularly evident with the additional laser treatment of substance (range on Figure 5.5d, 2L) and increasing the laser intensity when the new EB is abnormally amplified and shifted to the overtone bands of Ag(2), Hg(7,8) (see Figure 5.5e). Thus, the set of characteristic properties of vibration-induced new EP prove their nonlinear nature and strong relationship with EB and collective CBs.

5.5.3 Local Induced Electronic Bands and the Influence of Laser Radiation on Vibrational–Electronic Interaction

According to Figure 5.5a through c in the vicinity of numerous vibrations apart from very broad EB with half widths more than $\delta\nu = 10^3$ cm^{-1}, the less broad *islet* EBs occur. Moreover these additional EB may be more intensive than main bands. For example, in the vicinity of VBs Hg(3,4) with Raman shifts 709 and 773 cm^{-1}, local EBs are stronger than in the vicinity of the strongest Hg(1) band near 270 cm^{-1} (see Figure 5.5a and b). The subtraction of the local EB in the region of Hg (3,4), Ag(2), Hg(7,8) vibrations for polymerized by diamine NF of C_{60}$ is illustrated by Figure 5.6a and b. In Figure 5.6 EB approximated with cubic polynomial using two support points in the low- and high-frequency field of considered fragment of RS of fullerite C$_{60}$ nanofilms. One can see that half widths

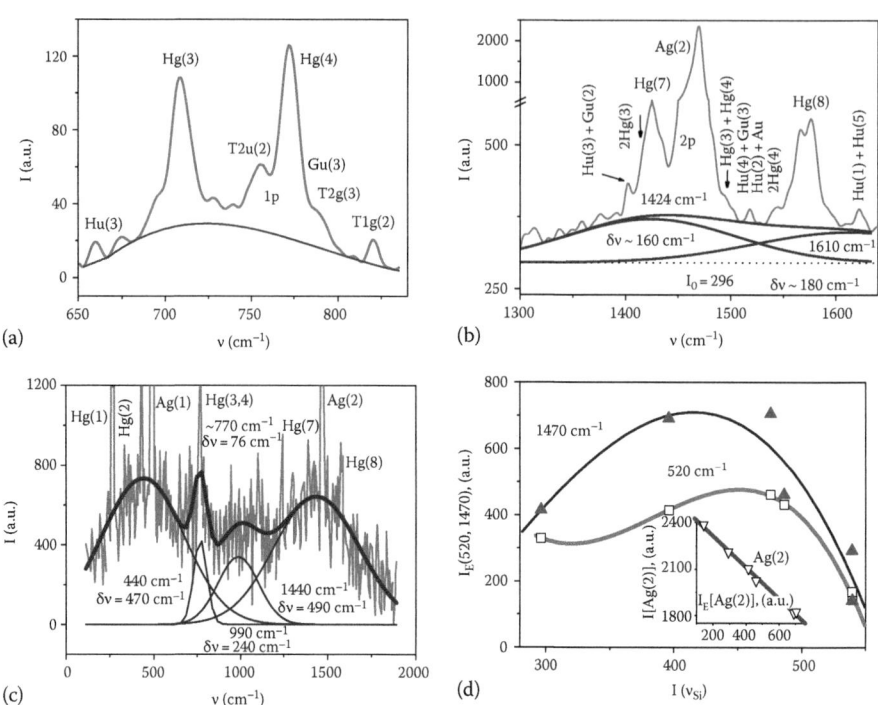

FIGURE 5.6
Subtraction of the wide EB in Raman spectra of C_{60} nanofilms with under hot (a) and could
(b) deposition process (a) and polymerization by diamine in the vicinity of vibrational bands Hg
(3,4) (a) and Ag(2), Hg (7,8) (b), as well as the results of numerical extraction of wide and local EB
in RS of fullerite C_{60} under excitation by 785 nm according to [103] (c) and nonmonotonic dependences of the intensities I_E of EB in the vicinity of Ag(2) and ν_{Si} bands from the intensity of the line
ν_{Si} = 520 cm^{-1}, associated uniquely with the thickness d of nanofilms (d). The inset shows the
decrease of the peak intensity of the band Ag(2) with increasing EB intensity I_E in the vicinity of
current vibration.

$\delta\nu$ of local EB are about 100–200 cm^{-1} and that is substantially less than the half width of EB shown
in Figure 5.5a,c and b, but considerably less than the half widths of the VBs. Here, in particular, are
seen overtones 2Hg(3), 2Hg(4) and the overall tone of Hg(3) + Hg(3), as discussed in Section 5.2.

Localization of *islet* EB in the region of vibrations that can be clearly seen by coincidence, the
local EB at 1424 cm^{-1} with a vibration Hg(7) in Figure 5.6b, confirms their relationship with the
vibrations and VEI. Probably the coincidence of the band Hg(7) with IR-active vibration F1u(4) at
1429 cm^{-1} plays an important role here. Peak intensities of considered local EB are approaching to
high narrow VBs of C_{60}, and the integral intensity of EB is significantly more.

Similar local EB appeared in RS of fullerites C_{60} at a pressure of 36–40 GPa at room temperature [111]. In this case the local EB arise in the fields of vibrations Ag(1), Hg(2,3), and Ag(2),
Hg(7,8). In this high-frequency vibrations dramatically shifted from 1520–1600 cm^{-1} to the region
1660–1750 cm^{-1}, which is also associated with changes in EB and strong VEI. When the temperature
rises to 1085 K, the appearance of a local EB occurs at a lower pressure about 10 times [110]. This is
due to the fact that the higher thermal excitation of interacting vibrational modes leads to an increase
in the efficiency of nonlinear interactions. Therefore, change in properties of matter under pressure
may be reduced. The inertia of process of EB occurrence of vibrations Ag(2), Hg(8) (~8 min) [110] in
our opinion is connected with the need to accumulate the effects of nonlinear resonant interaction of
thermally excited vibrational modes with the generation of higher vibrational states approaching ES.

Similar local EB with maxima in the low-frequency region of vibrations of C_{60} Ag(1), F1u(1,2),
Hg(3) and high-frequency vibrations Ag(2), Hg(8) has occurred in the case of neutron irradiation
with a dose of $1.32 \cdot 10^{15}$ n/cm^2 [99]. Thus spectral intensity of abnormally strong background is
more than three times higher than the peak intensity of the band Ag(2). Moreover with the EB
enhancing the intensities of VBs are greatly reduced, as well as under the influence of additional
laser radiation on fullerite C_{60} (see inset in Figure 5.5c), which is also associated with the specific
vibrational–electron interaction. Under ion irradiation new VC usually appears in the region of VBs

Ag(2), Hg(7,8) [107,108]. So when irradiated with deuterons with energy 5 keV, the induction of new EP occurs at a dose of $5 \cdot 10^{15}$ D$^+$/cm^2 [107]; and under ion C^{2+} irradiation with an energy about 7 MeV, the local EB occurred at a dose of 10^{15} ions/cm^2 [108].

The numerical results of numerical extraction of wide EB in the field of vibrations Hg(1,2), Ag(1), and Hg(7,8), Ag(2) are shown in Figure 5.6c, as well as a narrow and weak local EB in the region of Hg(3,4) and Hg(5) vibrations in the Raman spectrum of C$_{60}$ using a diode laser excitation at 785 nm in accordance with data of [103]. In this case, the effect of the phonons with lesser energy leads to a smaller half-width-inducible EB ($\delta\nu \approx 440$–490 cm^{-1}). Wherein instead of one large EB shown in Figure 5,a,e with the value of $\delta\nu \approx 1500$–2000 cm^{-1} there are two EP: low- and high-frequency with $\delta\nu \approx 470$–490 cm^{-1}. Subtracted local EBs in the region of vibrations Hg(3,4) and Hg(5) have lower intensities and half widths.

Nonmonotonic dependence of the intensity I_E of the broad EB thickness d of nanofilms has been established, as shown in Figure 5.6d. Here are the dependences of I_E at 520 and 1470 cm^{-1} from the intensity of line $\nu_{Si} = 520$ cm^{-1} of the crystalline silicon substrate, which is uniquely related to the thickness d of nanofilms. Figure 5.6d shows the experimental points approached by cubic polynomials. There is a sharp decrease of the intensities I_E for thin films (at high-intensity line ν_{Si}) due to insufficient spatial accumulation of nonlinear wave processes. The existence of the nanofilm optimal thickness, in which the intensities I_E of the new EB reach a maximum, is connected with the decrease of the vibrational nonlinearity with increasing nanofilm thickness d (in the region of lower values of the intensity of 520 cm^{-1} line). This clearly demonstrates the important role of NIVM and considered phenomena just for nanostructures and confirms the nonlinear nature of the new EB. The role of strong VEI also confirmed by linear attenuation Ag(2) line intensity with increasing of the intensity I_E of broad EB of this band (1470 cm^{-1}) is shown in the inset Figure 5.6d. In strengthening EB the intensities of VBs decrease, which is also confirmed in the region of bands Hg(3,4). Earlier, the strong decrease of the VBs at the increasing of BB have been observed under neutron irradiation [99], as well as the polymerization of C$_{60}$ [98], but that does not have a physical explanation.

5.6 ENHANCING THE VIBRATIONAL BANDS IN THE RAMAN AND IR SPECTRA OF C$_{60}$ NANOFILMS

5.6.1 Enhancement of Active Vibrational Bands in the Raman Spectra of C$_{60}$ Nanofilms

A comparison of the intensities of various vibrational bands in the Raman and IR spectra of C$_{60}$ microfilms and nanofilms was conducted with the normalization of dedicated lanes on the intensities of the strongest lines of Ag(2) at 1469cm^{-1} and F1u(1) at 527cm^{-1}, ($I_N = I_j/I_{max}$). A comparison of normalized vibrational bands of C$_{60}$ nano- and microfilms was allocated on background of broad electronic bands in RS, shown in Figure 5.7a through c. Consider first the change of the intensities of the Hg(7,8) bands in the RS, where local electronic bands begin to be induced by a laser mileradiation of fullerene C$_{60}$ and under increasing external pressure. Figure 5.7 demonstrate the comparison of the high-frequency vibrational bands Hg(7,8) and Ag(2) for C$_{60}$ nanofilms fabricated by the hot deposition method (spectrum 1) and microfilm with thickness 2 μm (range 3). It can be seen that in nanofilms, the fundamental bands Hg(7,8), and the set of sum frequencies, for example, Hu(3) + Gu(3), Hu(2) + Au et al, and "silent" vibration Gg(6) significantly enhanced. In particular, the band Hg(7) is amplified more than two times, and the band Gg(6) and second order bands amplified even more. We also note that the band Ag(2), on which the considered fragments of spectra were normalized, in nanofilms significantly broadened.

Figure 5.7b shows a comparison of allocated and normalized vibrational bands of micro- and nanofilms of C$_{60}$ bands Hg(3,4), including the nanofilms polymerized by diamine. Importance of the analyses of changes in the intensities of these bands connects with the fact that the bands Hg(3,4) play a central role in the series of vibrational resonances. It is seen that in C$_{60}$ nanofilms, the bands Hg(3,4) are enhanced more than twice compared to the microfilms. Intense "silent" vibrations T2 (2,3), T2u(2), Gu(2), Gg(3), etc. enhance even more. Here should be noted the weakening of the band Hg(3) and increasing of Hg(4) at the polymerization of nanofilms 1 and 2 by diamine N$_2$H$_4$ (compare spectra 1p and 1 Figure 5.7b).

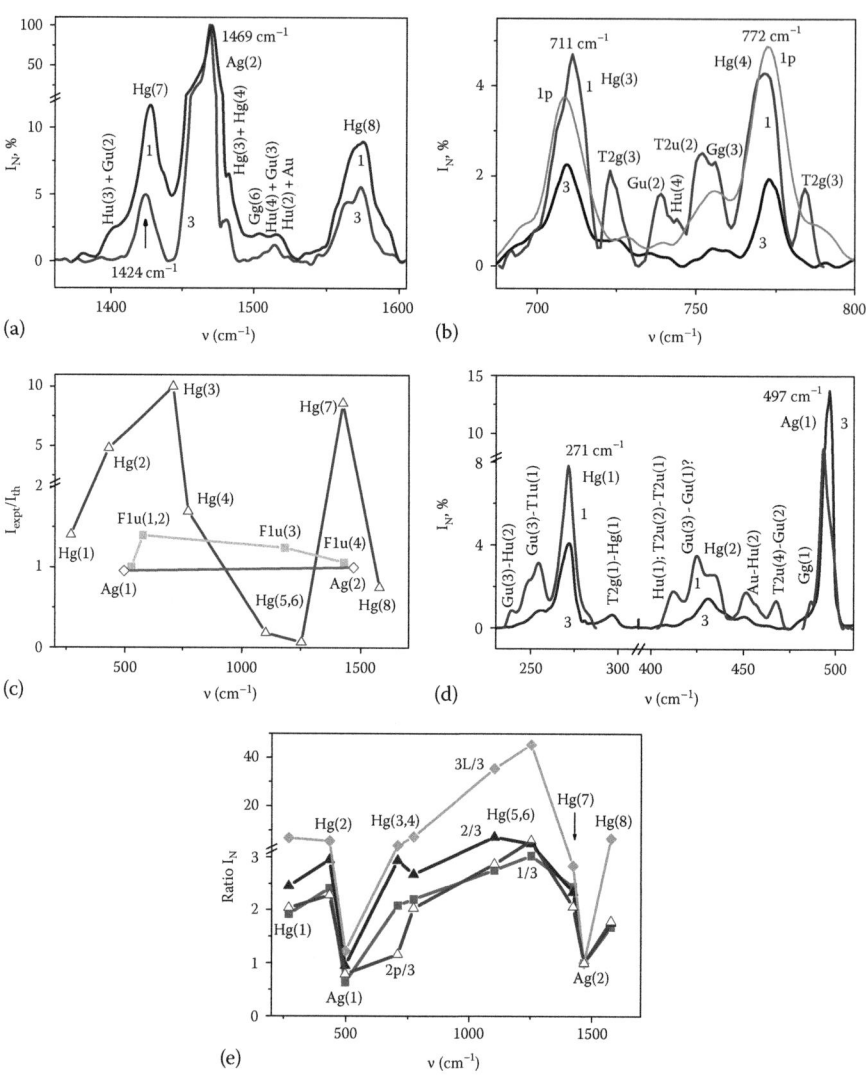

FIGURE 5.7

The comparison of the normalized to line Ag(2) RS fragments of nanofilms fabricated by hot deposition (1) and polymerized by diamine (1p) (1p) and the microfilm C_{60} 2 µm thick (3) in the vicinity of high-frequency bands Hg(7,8) (a) and Hg(3,4) (b), and in low-frequency field of bands Hg (1,2) and Ag (1) (d), and the relationship of normalized intensities of the observed and calculated active bands Hg(1-8) in RS (open symbols) and T1u (1-4) in IR spectrum (filled symbols), (c) and reinforcement bands of active vibration Hg (1-8) in nanofilms C_{60} preparing by hot(1) and cold (2) deposition and polymerized by diamine films (2p) and microfilm at an elevated intensity of the exciting radiation (3L) compared to the initial thickness of 2 µm microfilm (e).

It should be pointed out that according to the results of QCC based on Hartree–Fock method, the intensity of the vibration Hg(4) in the RS is more than 5 times greater than the vibration Hg(3), and this difference is aggravated further when density functional method is used. However, in the Raman spectra of micro- and nanofilms, the inverse ratio of the intensities of considered bands is observed: here the intensities of band Hg(3) are slightly larger than for bands Hg(4). It has not yet received an adequate explanation as very low intensity of the observed bands Hg(5,6), which, compared with the calculation weakened more than an order of magnitude, as illustrated Figure 5.7c.

It is important that after polymerization of nanofilms with N_2H_4 the band Hg(4) increases, and Hg(3) is reduced, that is, the intensities of the bands Hg(3,4) change in approaching to the results obtained by QCC. This fact allows to understand the true causes of the differences between the results of spectral experiments and that of QCC. Bottom line here is that after polymerization of nanofilms, their total vibrational nonlinearity decreases. This is confirmed by a significant weakening of the majority of bands of "silent" vibrations in the polymerized nanofilms, which are virtually absent in microfilms. In fact, it establishes the connection between the discussed spectral anomalies and the nonlinear interaction of the vibrational modes and strong VEI, that is, not taking into account in QCC.

Changes in the intensities of the most low-frequency vibrational bands of Hg(1,2) and Ag(1) in the RS of fullerenes C$_{60}$ are shown in Figure 5.7d. It is evident that bands Hg(1,2) in nanofilms enhanced about two times, and the bands of "silent" vibrations Gg(1) and Hu(1) and a plurality of difference tones specified by Figure 5.7d are amplified significantly higher. The difference frequencies Gu(3)-T1u(1), T2u(2)-T2u(1), Au-Hu(2), and T2u(4)—Gu(2) with the participation of "silent" vibrations are most enhanced. It is interesting to note that the difference tone of asymmetrical "silent" vibrations mainly manifested in the low-frequency region of the Raman spectra of C$_{60}$ fullerite. In particular, the "silent" vibrations T2u(2) – T2u(1) and Gu(3) – Gu(1) are very close to the fundamental vibrations Hu(1) and Hg(2), that is, significantly expands the set of possible vibrational resonances.

The possibility of the manifestation of difference frequencies is associated with participation in these most low-frequency inactive vibrations T2u(1), Gu(1), etc., and their thermal excitation at room temperature. Spectral component 425 cm^{-1} in the vicinity of the band Hg(2) can be due both to splitting of the vibrations Hg(2) in the crystal field, and difference tone Gu(3) – Gu(1). It is essential that the most intense in the low frequency spectrum of C$_{60}$ microfilm line Ag(1) at 497 cm^{-1} shifted to 493 cm^{-1} in nanofilms, as well as using the 473 nm radiation [102]. At the same time the band Ag(1) is weakened together with band Ag(2), as shown in the inset Figure 5.6d. Thus, in nanofilms, C$_{60}$ obtained a large enhancement of the intensities of active in RS bands Hg(i), where i = 1 ÷ 8, with the relative weakening of the most intense in the RS bands Ag(1,2).

The main part of obtained results on enhancement of the vibrational bands in the RS of C$_{60}$ nanofilms compared to microfilms is summarized in Figure 5.7. Here are shown the ratio of the maximum of bands of active vibrations Hg(1-8) in different studied nanofilms and microfilm with thickness of 2 μm. Spectra according to the numbering adopted earlier, the relationships t1/3 relate to a hot method for preparing nanofilms, and 2/3 relate to the cold one; relations 2p/3 relate to polymerized films 2, and 3L/3 relate to a large laser radiation exposure on the substance under strong focusing of the exciting radiation λ_L = 488 nm. It is seen that substantial enhancement of the vibrational bands is achieved for vibrations Hg(1,2), Hg(3,4), and Hg(7,8). The greatest enhancement of vibrational bands, reaching 36–46 times, is observed for the bands Hg(5,6) and is realized with high intensity of the exciting radiation λ_L even for microfilms (d = 1,2 μm, the relationship 3L/3). For polymerized with N$_2$H$_4$ nanofilms the enhancement of most bands is weakened (see the ratio 2p/3), which is associated with a decrease of vibrational nonlinearity and weakening of VEI.

These results open the way to understanding the significant differences between those observed in the RS intensities of vibrational bands and results of quantum chemical calculations (QCC), as shown in Figure 5.7c. For fully symmetric vibrations of C$_{60}$ fullerenes results of QCC quite well correlated with the experiment. For IR active modes T1u(2.3), a slight increase of absorption comparable with the calculation is observed. Abnormally large differences of experimental and calculated results are observed for a number of bands Hg (j). In particular, experimental bands Hg(3) and Hg(7) are 8–10 times stronger as compared with the results of QCC that were eventually due to the influence of the vibrational resonance 2Hg(3) = Hg(7). Conversely, the observed bands Hg(5,6), are weakened by 5.5 and 14.5 times respectively compared with the theory (see Figure 5.7c). It is essential that the maxima of the ratio of the intensities of the vibrational bands I_{expt}/I_{th} for vibrations Hg(3) and Hg(7) on Figure 5.7c consistent with increasing of the intensities of these bands in nanofilms C$_{60}$ in Figure 5.7e. Since the growth of the intensities of the vibrational bands in nanofilms is associated with NIVM and VEI, the obtained differences VEI with experiment may be caused by these phenomena. This important conclusion is confirmed by the previous analysis of changes in the intensities of the bands Hg(3,4) under the polymerization. Further researches in this direction should explain the reasons for the anomalously large differences in the intensities of the bands Hg(5,6) given by experiment and theory.

5.6.2 Abnormal Enhancement of "Silent" Vibrations in Raman Spectra of C$_{60}$ Nanofilms

The anomalously strong growth of the band intensities for inactive for icosahedral symmetry I$_h$ vibrations Gg,u, Hu, T$_{1,2}$ g, T$_2$u and Au is observed in the Raman spectra of C60 Nanofilms. Examples of manifestations of inactive vibrations in the high- and low-frequency region in RS are

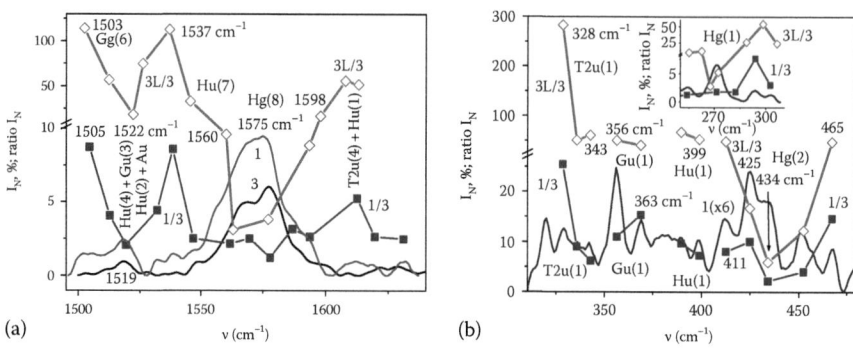

FIGURE 5.8

Enhancement fragments of normalized Raman spectra (RS) of nanofilms C$_{60}$ in the region of the band Hg(8) (a) and the low-frequency *silent* vibrations and F2u(1), Gu(1), and Hu(1) and active vibrations Hg(2) and Hg(1) (inset) (b) and enhancement number of *silent* and active modes in the RS of nanofilms (1/3) and microfilms at an elevated intensity of the exciting radiation at 488 nm (3L/3) compared with the original microfilm.

shown in Figure 5.8a and b. Some of the identified spectral component in the region of vibrations Hg(8) and Hg(1), Hg(2) are shown in these figures, taking into account the possibility of splitting and showing the main vibration frequency. In the field of bands Hg(8) the most high-frequency inactive vibrations Gg(6) and Hu(7) appear. Let us indicate here that the maximum of the band at 1519 cm^{-1} can be treated as one of the components of the splitting vibrations Gg(6) and to the sum tones Hu(4) + Gu(3), Hu(2) + Au, consistent with the selection rules [22]. Similarly, the side components in the field of active vibration bands Hg(8) and Hg(1,2) may relate to splitting components of these vibrations or to the tones of the second or higher order.

Because of the large number of the observed vibrational bands shown in Figure 8, b, as well as in Figure 5.7b and d, as the intensities greatly increased as a result of nonlinear processes VEIV, these vibrational bands do sufficient overlap. In these circumstances, to correctly determine the gain for individual vibrations, a numerical spectral decomposition of complex lines in the selected spectra of C$_{60}$ nano- and microfilm onto separate components has been conducted. The obtained results were used to correctly determine the coefficients of enhancement of vibrational bands in nanofilms C$_{60}$ in comparison with microfilms. Thus, calculated spectral dependences of the enhancement of active and "silent" vibrational bands in the Raman spectra of C$_{60}$ nanofilms compared with microfilm of 2 μm for considered fragments of spectra are shown in Figure 5.8b.

Here curves 1/3 correspond to the enhancement of the vibrational bands in nanofilms and 3L/3 to the enhancement of the vibrational bands in C$_{60}$ microfilm with a thickness of 1.2 microns and under a strong focusing of the exciting radiation 488 nm and power about 1 mW to 2 μm diameter.

In general, it can be stated that the side spectral components are enhanced to a greater degree than the central components (see Figure 5.8a and b). This relates to both active and inactive vibrational bands. For example, for the active band Hg(8) in nanofilm 1 under strong focusing of the exciting radiation, the central component is enhanced by three times, and the side ones enhanced by 10–17 times. This regularity is evident for the inactive bands Gg(6) too. In nanofilm 1, the band maximum at 1519 cm^{-1} is enhanced 2.1 times, and the side parts of the band enhanced 8.8 times. Under a strong laser irradiation on microfilm C$_{60}$ the central part of the band Gg(6) is enhanced 18.2 times, and the side components 112 times.

Figure 5.8. Enhancement fragments of normalized Raman spectra (RS) of nanofilms C$_{60}$ in the region of the band Hg(8) (a) and the low-frequency "silent" vibrations and F2u(1), Gu(1), Hu(1) and active vibrations Hg(2) and Hg(1) (inset) (b) and enhancement number of "silent" and active modes in the RS of nanofilms (1/3), and microfilms at an elevated intensity of the exciting radiation at 488 nm (3L/3) compared with the original microfilm.

In the low-frequency region bands of inactive vibrations, T2u(1), Gu(1), and Hu(1) are observed between the vibrations Hg(1) and Hg(2), as shown in Figure 5.8b. Pointed positions of these "silent" vibrations are consistent with the results of QCC that is easy to install, taking into account various degenerations of these vibrations. We note that in known publications [36–38,77], the assignments of these vibrations are not consistent with the data of QCC. As in the field of vibration Hg(8), the low-frequency bands Hg(1) and Hg(2) manifest regularity of high enhancement of side spectral components. For example, for the band Hg(2) in nanofilm#1, the central component is enhanced 2.2 times, and the side parts 10–15 times. In case of powerful exposure of laser radiation the center of the band Hg(2) is enhanced up to 6 times, and side components 45–48 times.

The inset in Figure 5.8b shows a fragment of RS in the field of the lowest-frequency band Hg(1) and the enhancement of individual spectral components in the nanofilm and microfilm in the case of strong focusing of the laser radiation. In the latter case, the center of band Hg(1) is enhanced 2.8 times, and side components 12–26 times. For high-frequency satellites of band Hg(1) corresponding to the difference tone Hg(3) – Hg(2) [13,14], the enhancement by ~53 times is achieved.

For the lowest-frequency "silent" vibrations, T2u(1), Gu(1), and Hu(1), the observed spectral components in nanofilms C$_{60}$ are enhanced by 5–25 times (see Figure 5.8b), and under a strong laser irradiation—by 50–300 times even in microfilm. The observed trend of increasing enhancement of bands of inactive vibrations in the low-frequency region should be noted. This correlates with the increase of intensity-induced electronic bands in the initial nanofilms, as shown in Figure 5.5b and d, and the increasing low-frequency bands Hg(1) and Ag(1) with increasing low-frequency background in the Raman spectrum of the composite C$_{60}$:C$_8$H$_8$ [104].

Thus, strong active bands of vibrations Hg(i) in the Raman spectra of fullerene C$_{60}$ nanofilms with thickness of 150–250 nm are enhanced 2–7 times, and weaker bands Hg(5,6), as well as active side components of the bands can be enhanced approximately 35–50 times compared to microfilms. Anomalously strong increase in the intensities of the vibrational bands in Raman and IR spectra are observed for a number of "silent" vibrations, as well as a number of sum and difference tones with a high degree of participation of inactive vibrations. In many cases, these bands in the vibrational spectra of C$_{60}$ nanofilms, for example, the band of "silent" vibration Gu(1) on Figure 5.8b, are close to intensity of the active bands, but their enhancement reaches 50–300 times compared with the thickness of 1–2 μm.

5.7 CONCEPT OF THE COLLECTIVE CHEMICAL BONDS AND THEIR MANIFESTATION IN THE INTERACTION FULLERENES C$_{60}$ WITH METAL ATOMS

5.7.1 Development of Ideas about the Collective Properties of Chemical Bonds

(106) The idea of the collective properties of CBs in the condensed matter has bee suggested by one of the authors when the problem of hydration of metal cations in aqueous solutions was analyzed [112]. Moreover a correlation hydration energies of various cations (including multiply) was found as well as anions with shared ionization energies of corresponding atoms. On this basis, the physical nature of the phenomenon of hydration of ions in aqueous electrolyte solutions was established, which associated with a substantial rearrangement of the electron shell of the ions and the compensation of their charges.

(107) It has been found the recovery of metal cations incorporated into network of hydrogen bonds to neutral atoms as a result of transfer of electron density from water. As a result of the modification of quantum-chemical bonds the ion charges actually distributed over a considerable volume of water, with the creation of the "charge of the atmosphere." Clearly, this leads to a significant reduction of energy of electrostatic fields around ions. The ions are reduced to states close to neutral atoms, but they are not identical, since recovered atoms are not free, and associated with the liquid. Essentially, that the compensation charge of cations performed with the participation of many electrons from the totality of the surrounding water molecules, which determines the collective properties of CB modified cations with water molecules. Thus, the hydrated ions in aqueous solutions are special state of matter where the missing valence electrons are compensated by cations bound electrons of water and anion charges—due to the displacement of protons.

(108) It should be noted that in classical work P. Debye and E. Hückel, 1923 [113] introduces the notion of *charge atmosphere* around the ions in solution, but it had the opposite sign, and it was about

the statistical distribution of oppositely charged ions in an electrostatic field of a considered ion. The problem of collective CBs is essentially quantum and a classical statistical approach does not apply to it.

In the future, the problem of collective CB has been evolved by the way of development of ideas about the vibrational instability of condensed matter [18,114]. The essence of this instability is in the fact that under excitation of vibrational states with energy ~1%–10% of the binding energy as a result of VEI, the stability of the ES, providing cCB in substance, is lost and their changing starts.

The collective nature of the majority of CB in molecular, covalent, and ionic matter, as well as metals and alloys, is associated with a significant difference in CB in condensed matter, macromolecules, and nanoparticles from CB in simple molecules that have already been considered in the introduction. Moreover, it was substantiated in the previous sections using a wide range of results of the vibrational spectra of various fullerites. And the key role of vibrational resonances and the nonlinear interaction of vibrational modes with the generation of higher vibrational states, which leads to strong VEI and CB changes, has been proven. As a result, collective CBs were associated with a deeper relationship of classical and quantum physics. As is well known the vibrational modes in condensed matter are always delocalized and have collective properties; in particular, they are characterized by a wave vector. As a result, the electron–vibrational coupling of CB in condensed matter and NMs also has collective properties.

For the purpose of better understanding of the problem of chemical functionalization and polymerization of fullerites C_{60}, as well as improving the physicochemical characteristics of nanostructured theoretical research of formation of metal atom CBs with fullerenes C_{60}, experimental studies of the vibrational spectra at the interaction of C_{60} with a range of metals 4, 5, and 6 periods in the PT of elements have been done [15,31,115–120]. The complexes of C_{60} with metal atoms are good model systems for QCCs and demonstration of the characteristic peculiarities of collective CB.

5.7.2 Calculating the Spatial Redistribution of Charges in the C_{60} Molecule When Interacting with Metal Atoms

Relatively large sizes of the C_{60} molecules, their high symmetry, and delocalization of the normal vibrations allow to theoretically demonstrate new collective properties of CB, what is more difficult to do for crystals. Using the Gaussian program, detailed theoretical studies of CB for metal atoms Sn, Fe, and Ti C60 molecules have been performed [15,31], which is of interest for fundamental science, including chemical functionalization of carbon materials, and for modern materials science. For comparisons the QCC were performed to study the interaction of C_{60} with an additional carbon atom. The complexes and $C_{60}M$ и $C_{60}MC_{60}$, where M is the metal atom, have been studied. At the same time to solve the Schrödinger equation, the HF method was used. In contrast to QCC for geometric configuration of the complexes and electronic binding energies that are being made in many papers, vibrational frequencies and intensities of the VBs in the IR absorption and Raman scattering were calculated. The charge transfer from the metal atoms to the number of electrophilic molecule of fullerene C_{60} and redistribution of electrical charges within the molecule C_{60} have been studied in detail.

It is shown that the metal atoms M are attached to fullerene C_{60} by using one of the double bond between C60 hexagons or by single bonds between pentagons and hexagons. By careful QCC it was found that the electron density of the metal atoms is not only transferred to the nearest carbon atoms C_1 but distributed on a significant part of the C_{60} molecule, which is illustrated in Figure 5.9a and b [30,31]. Here the red color corresponds to a negative charge and green to positive. We note that in the case of formation of CB of fullerene with an additional carbon atom (left in Figure 5.9), the charge q = 0.45e, where e is the electron charge that is transferred to fullerene. When the molecule C_{60} interacts with atoms Sn, Ti, and Fe, the charges q = 0.977e; 1.086e and 1.415e are transferred to fullerene, respectively. In the case of interaction of Ti atoms with two C_{60} molecules, as shown in Figure 5.9b, the charge q = 2.059e is transferred to two molecules of C_{60}, which is approximately two times greater than for the complex $C_{60}Ti$.

It is essential that at interaction with the metal atoms, the transferred charge is more evenly distributed over the entire surface of the spherical C_{60} molecules than when interacting with the C atom (see Figure 5.9b). It should be noted that the relationship of the fullerene C_{60} with the carbon atom is stronger than with metal atoms, but the transferred charge is smaller and localizes on the smaller part of a fullerene molecule. In our opinion, in the case of interaction with metal atoms, the d-states of metals play a significant role. An alternating position of positive and negative charges on the carbon atoms of

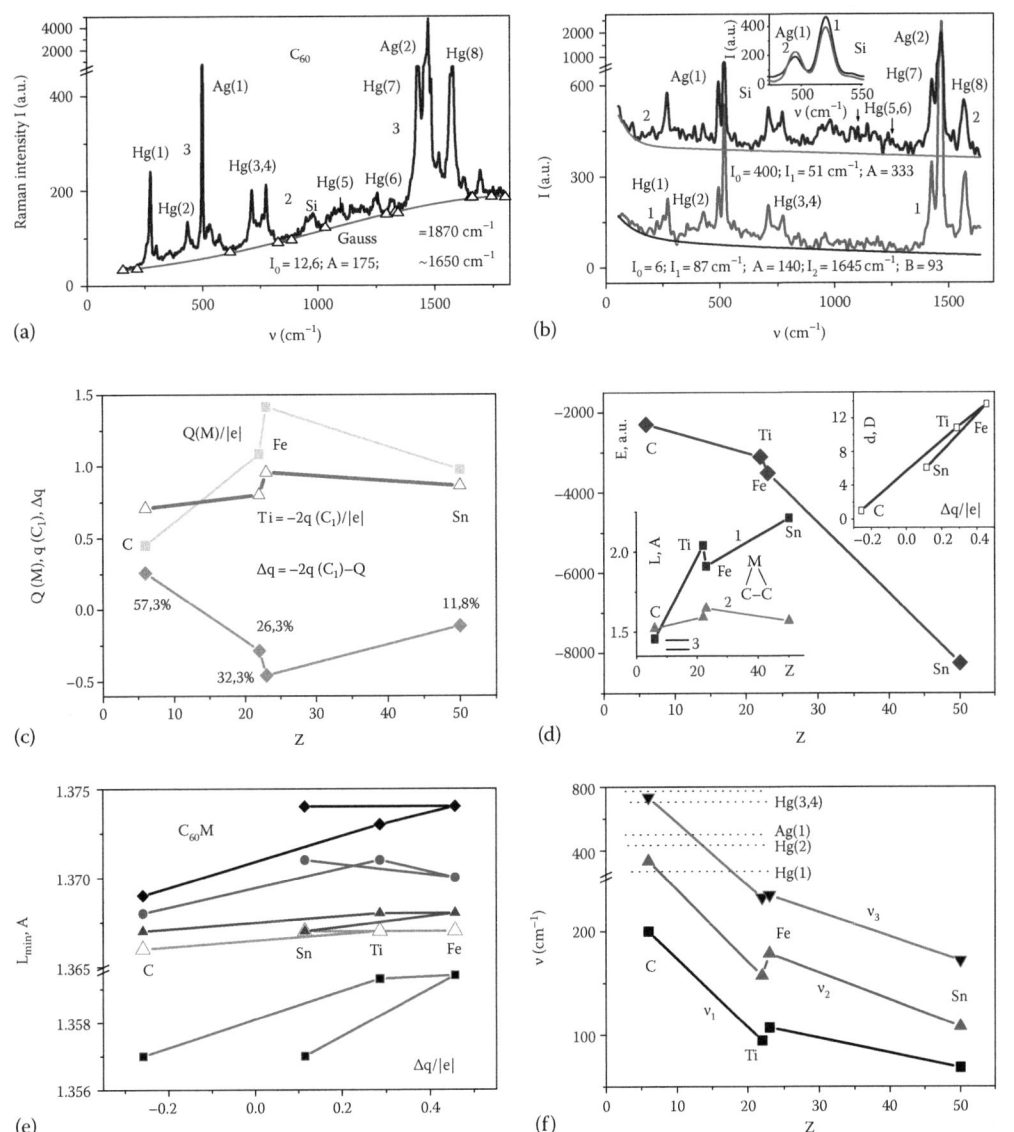

FIGURE 5.9
Charge redistribution in the complexes C$_{60}$ with C atoms (q = 0.45 e) (left) and Sn (q = 0.977 e) (a) and in the complex S$_{60}$TiS$_{60}$ (q = 2,059 e) (b) according to the results of QCC (q > 0 green; q < 0 red color) and the dependences of atomic charges of M and C$_1$ complexes C$_{60}$M and distributed on the C$_{60}$ molecule charge Δq [30,31] (c), and total binding energies E(Z) and bond lengths L(Z) of the C–M (1) and C–C (2) bonds on the atomic number Z of the atom M and the correlation of the dipole moment d of complex and distributed charge Δq (inset) (d), and the dependence of the lengths of the shortest five double bonds C=C in complexes C$_{60}$M on the magnitude of electron charge distribution Δq (e) and local vibration frequency ν$_{1,2,3}$ of atoms MC$_2$ on Z [30] (f).

the fullerene adjacent to collective CB should be pointed: Two close C$_1$ atoms are negatively charged, four of their neighbors C$_2$ are positively charged, and the next eight atoms C$_3$ are again negative. Note that in the case of CB of the fullerene C$_{60}$ with atoms of Ti and Fe only, 4 C atoms are positively charged, with atom Sn 8 atoms, and if C$_{60}$ bonding with C 24 atoms are positively charged and 18 atoms have no charges. It is characteristic that according to Figure 5.9a and b at interaction with the carbon atom, there is an even distribution of positive charge on the fullerene surface, as in the case of CB with a metal atom on the surface of the fullerene, and the negative charge is more evenly distributed (red).

Consider in more detail the distribution of charges, the bond lengths between neighboring atoms, and other characteristics of the complexes of C$_{60}$ with metals. In Figure 5.9c the dependences of charges are localized on the atoms C, Ti, Fe, and Sn, as well as the neighboring two atoms C$_1$, on the atomic number of the attached atom Z [30,31]. We should point out here that all the charges we normalize on the absolute magnitude of the electron charge $|e|$. Because of the large electrophilicity of the C_{60} molecule, all M atoms are positively charged ($Q > 0$), and the atoms C_1 are negatively charged ($q(C1) < 0$). When interacting C_{60} with atoms C, Ti, Fe, and Sn charges on atom C_1 are $-0.354 |e|$, $-0.4 |e|$, $-0.479 |e|$, and $-0.431 |e|$, respectively. Figure 5.9c demonstrates the amount of charge Δq, distributed over the surface of the C_{60} molecule. From the law of a charge storage $\Delta q + 2q(C1) + Q = 0$, it follows that $\Delta q = -2q(C_1) - Q$, and the absolute value of the distributed charge $|\Delta q|$ is the difference of the absolute values of the charges on the atoms M and $2C_1$. It is important that during the interaction of the fullerene C_{60} with carbon, the charge Δq is positive, while at CB with metal atom, charge Δq is negative. Distributed over the surface of the C_{60}, the charge varies between 12% and 57% of the value of the atomic charges M.

The dependence of the total binding energy E (Z) for a series of complexes $C_{60}M$ is shown in Figure 5.9d [30]. It is seen that the maximum $|E|$ is reached for Sn, although in this case the length of the C–M bond is the highest and equals 2.21 A (curve 1 in the inset). For C_{60} complexes with Fe and Sn atoms, the bond lengths C_1–C_1 are, respectively, 1.646 and 1.567 A (curve 2 in Figure 5.9d). They are considerably higher than the standard length of single (1.391 A) and double (1.463 A) bonds in C_{60}, marked by three horizontal bars in the lower inset in Figure 5.9d. This is due to the electrostatic attraction of the charges $q(C_1)$. And even in the CB of the fullerene C_{60} with an additional C atom as a result of breach of the neutrality of C atoms, the bond length C_1–C_1 is greater than the length of a single C–C bond in the C_{60} molecule. However, despite the mutual attraction of the charges Q (M) and $2q(C1)$, the length of C–M is longer.

It was established that the magnitudes of the dipole moments of complexes d are correlated with the charge values Δq and Q (M), reaching for the metal atoms 10 Debye and more, as illustrated in the upper inset in Figure 5.9d. When C_{60} connects with atoms Sn, Fe, and Ti, the excess electron density Δq is distributed over 50–54 C atoms. It is shown that the charges on the atoms of the fullerene molecules in complexes $C_{60}M$ excluding C_1 atoms with the maximum charges range from 0 to 0,056e, and in an average, one atom charge q1 $\approx$ (0,002–0,007) e <0 is localized. The maximum in the electron density distribution for the complexes $C_{60}M$ is shifted toward larger negative charges in a series of Sn, Ti, and Fe (up to $-0,005 |e|$) and reached to 12–14 atoms.

5.7.3 Quantum Chemical Compression of the Fullerene C_{60} and Collective Mechanisms for Strengthening of Chemical Bonds

Despite the strong Coulomb repulsion between the distributed charges of the same sign, for collective CBs of metal atoms with C_{60} molecules, the shortening of lengths of most of the double bonds by an average of 1.3%, while single -0.75% is characteristic. The dependences for shortest 5 shortest C=C L_{min} bonds of $C_{60}M$ on Δq values are shown in Figure 5.9 [30]. In complexes $C_{60}Sn$ and $C_{60}S$, the minimum values of the lengths (C = 1.357 A) are reduced by more than 2%. Similarly, single C–C bonds are reduced. As a result, the accession of additional atoms to fullerene C_{60} leads to abnormally strong quantum chemical compression of C_{60}. From a comparison given in Figure 5.9d and e, one can see the nontrivial relationship of total bond energies E for complexes $C_{60}M$ and shortening of the bonds in C_{60}. Thus, collective CBs are characterized not only by local connections C–M but also the strengthening of all the bonds in the C_{60} molecule, which should increase the strength of fullerites.

Besides the delocalized normal vibrations of the fullerene C_{60} for complexes $C_{60}M$, local deformation $\nu_{1,2}$ and ν_3 stretching vibrations involving groups of atoms C_2M are characteristic. Dependences of the frequencies of these vibrations on the Z are shown in Figure 5.9f [30]. It is evident that for atoms Ti, Fe, and Sn, the frequencies $\nu_{1,2,3}$ are significantly below the minimum frequencies of vibrations Hg(1.2) of the fullerene C_{60}. And for easier bounded atoms C, the frequency ν_3 is about Hg(3.4) and the frequency of ν_2 vibration is close to Hg(1.2). According to the length of the C–M bond (1.52 A) and high-frequency ν_3 C atoms, coupling with C_{60} is strong enough, in spite of the small binding energy of the complex E (Figure 5.9d). Conversely, for complex $C_{60}Sn$, the value of E is maximum, and the length of the C–Sn is large and the frequencies of local vibrations $\nu_{1,2,3}$ are low enough, which characterizes weaker local bonds. These apparent contradictions associated with the collective nature of CB—the weaker bonds of metal atoms with C_{60}—lead to a significant strengthening of the many other bonds as a result of redistribution of the electron density and the effects of VEI. Collective properties of CB manifested in the fact that relatively weak bonds of atoms Sn, Fe, and Ti with C$_{60}$ (great lengths about 1.9–2.2 A and low frequency of vibrations) due to a redistribution of electron density lead to a significant (1%–2%) reduction of all bonds in molecules C$_{60}$ (*quantum chemical compression* of molecules), despite the Coulomb repulsion of the distributed charges. The existence of such compression of fullerenes when interacting with metals was confirmed by x-ray diffraction patterns. In this x-ray peaks are shifted to larger angles, indicating a decrease in the distance between the C$_{60}$ molecules.

Ultimately, the collective properties of CB in macromolecules and condensed matter are associated with cooperative properties of the majority of elementary excitations (vibrational, electronic) and strong vibrational–electron interaction (violation of the adiabatic approximation). To a large extent the collective properties of CB are associated with changes in the intensities of the VBs and other characteristics of the vibrational modes. Figure 5.10a shows the ratio of the intensities of the VBs for the complex C$_{60}$Sn and fullerene C$_{60}$ obtained by QCC (curves 1 and 1 IR) [31]. It is very important that by calculating the vibrations of Hg (1 ÷ 8) in the complex C$_{60}$Sn, the well-defined peaks have been obtained for resonance vibrations of Hg(3,7). According to the results presented in Figures 5.3c and 5.7c for such vibrations of Hg(3,7), the maximum differences of experimental and theoretical values of the intensities of the VBs in the Raman spectra have been obtained.

Carried out QCC for C$_{60}$ complexes with metals and study the characteristic properties of collective CB showed that the growth of the intensities of the Hg(3,7) in the composite C$_{60}$Sn are associated with the transfer of electron density from the tin atom Sn and its delocalization over the entire molecule C$_{60}$. Figure 5.10 also shows the ratio of the experimental and calculated intensities of Hg-type bands in the Raman spectrum of the composite C$_{60}$Sn (curve 2). It is evident that a number of actually high intensities are observed compared with the results of QCC. This is due to the

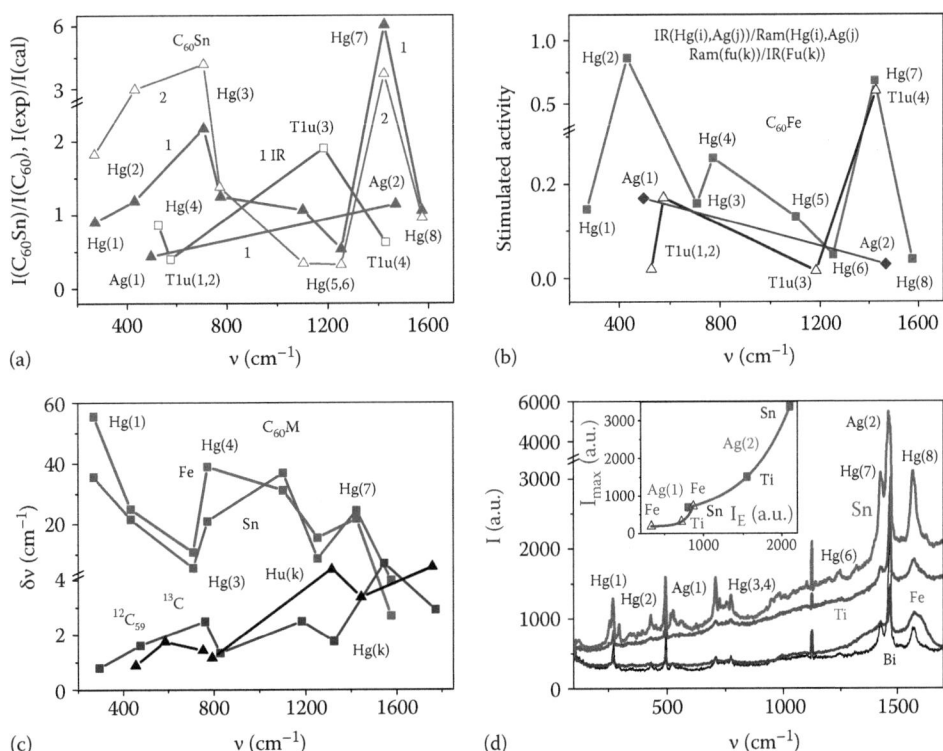

FIGURE 5.10
The relationships of calculated intensities of vibrational bands C$_{60}$Sn and C$_{60}$ [31] (1) and the ratio of the experimental and calculated intensities of the composite C$_{60}$Sn in the Raman spectrum (2) (a), the calculated simulated activity of vibrations of Ag (1.2) and Hg (1 ÷ 8) in the IR spectrum and T1u (1 ÷ 4) in the Raman spectrum C$_{60}$Fe (b), the calculated spectral splittings of δ𝜈 for active modes Hg (1–8) in the complexes C$_{60}$M (M=Fe, Sn), and their comparison with the magnitude of the splitting for induced bands Hu (1–7) in molecules ^{12}C$_{59}$^{13}C [15,31] (c), as well as an increase in the intensity of the new EB in RS of fullerite C$_{60}$ under their interaction with atoms of Bi, Fe, Ti, and Sn [21] (d).

fact that modern methods of QCC clearly not taken into account the NIVM. However, the carried out researches have convincingly demonstrated that the increase in the intensities of the Hg(3,7) and the weakening of Hg(5,6) as in complexes $C_{60}M$, and in initial C_{60} associated with changes in the collective ES due to strong VEI. The role of VEI abnormally increases as a result of the collective properties of B and nonlinear interaction of the vibrations.

5.7.4 Spectral Splittings of Vibrations, Induction of Activities of *Silent* Vibrations, and Additional Electronic States in the Composites of the Fullerenes C_{60} with Metals

In addition to the previously mentioned characteristics the collective properties of CB are largely determined by a dynamic change of charges and electronic polarizabilities when atom vibrate, which is manifested in the intensities of the IR absorption and Raman scattering bands, as well as inducing activities of *silent* vibrations, frequency splitting of degenerate vibrations and their displacements relative frequencies of the fullerene C_{60} vibrations [15,31]. A complete characterization of collective CBl inextricably connected with a detailed analysis of the vibrational modes, including improved opportunities for implementation of vibrational resonances in the splitting of the vibration frequencies. In particular, metal complexes with fullerites the intensity of bands of the inactive C_{60} modes greatly increase, so they can exhibit catalytic properties.

Induced activities of vibrations Ag(1.2) and Hg(1 ÷ 8) in the IR spectrum of the complex $C_{60}Fe$ and vibrations T1u(1 ÷ 4) in the Raman spectrum of $C_{60}Fe$ is shown in Figure 5.10b. This induced activities are normalized to the activity of the same bands in the spectra: IR(Hg (i), Ag(1,2))/Ram (Hg (i), Ag (1,2)) and Ram (T1u (k))/IR (T1u (k)). It is seen that vibrations Hg(2), Hg(4), Hg(7), and T1u(4) must be observed in additional spectra induced by the interaction with Fe atom. As a result, the intensities of the *silent* in C_{60} vibrations may approach the intensities of active modes. Similar induced activities appear to truly *silent* vibrations T1, 2g; T2u; Gg, u; Hu. For example, the activity of bands Hu(5,7) for the isotopically substituted fullerene $^{13}C^{12}C_{59}$ approaching to the weakest active bands in the Raman spectra. However, for C_{60} complexes with metal atoms the calculated intensities of induced bands of *silent* vibrations can be abnormally large. Moreover, in the Raman spectra they are usually larger than in the IR spectra, that is also connected with precision of modern QCC and requires special additional study.

The spectral splitting of $\delta\nu$ and frequency shifts $\Delta\nu$ of active in the vibrational spectra vibrations Hg(1 ÷ 8) and T1u(1 ÷ 4) as a result of lowering of the symmetry have been studied, as illustrated Figure 5.10c [15,30,31]. Here's a comparison of the values of splitting for active in the Raman spectrum vibrations Hg(1 ÷ 8) and *silent* vibrations Hu(1 ÷ 7) in the presence of ^{13}C isotope of the fullerene C_{60} and interaction with atoms of Sn and Fe. The presence of Isotopes ^{13}C leads to splitting $\delta\nu = 1 ÷ 7$ cm^{-1}, and in the complexes $C_{60}M$ the splitting $\delta\nu$ reaches 20–50 cm^{-1} and more. We point out that the molecule $^{13}C^{12}C_{59}$ splitting for active modes Hg (k) compared with splitting for *silent* vibrations Hu (1 ÷ 7) in C_{60}. Frequency splitting of degenerate vibrations significantly greater half widths of the same bands for fullerene C_{60}, that greatly facilitates the conditions for the sequence of resonances for vibrational modes. This role can perform and isotopically, which may explain the violation of the natural isotope distribution for the number of elements in living organisms. The increasing role of vibrational resonances of low-frequency gain and promotes intermolecular vibrations in the 50–180 cm^{-1} by laser or electron irradiation of C60 [57]. Other aspects of the collective cholesterol solids were analyzed in [121]. Problem of multiple vibrational resonances also analyzed in [122] when interacting of water with porous silicon. Thus, multiple vibrational resonances and strong VEI can play an important role in a variety of phenomena.

For values $\delta\nu$ according Figure 5.10c the maxima can be observed for resonant vibrations, which is similar Figure 5.4c. However, there are appear more complex regularities due to the fact that instead of increasing quantities $\delta\nu$ in the high-frequency region from the observed abnormal growth in the low frequency area and maximum values achieved for fluctuation $\delta\nu$ Hg(1). This must be due to the existence of low- and high-frequency additional ES. The new law is confirmed experimentally observable splitting of $\delta\nu = 41.6$ cm^{-1} band for Hg(1) in the Raman spectrum of the composite S60Sn, while for the band Hg(3) $\delta\nu = 13.8$ cm^{-1} [31].

In the high-frequency region in Figure 5.10c the visible vibrations frequency shifts Hg(7,8) of complex C$_{60}$M toward lower frequencies are visibly seen. It is shown that the softening of vibrational frequencies of the composites fullerenes with metals can reach 20 cm^{-1}. Significant growth of splittings δν and intensities of VBs for complexes C$_{60}$MC$_{60}$ compared with C$_{60}$M also confirms the idea of the delocalization of the ES.

Previously, it has been found experimentally the significant enhancement of Ag(2) band in the Raman spectrum of C$_{60}$ and its low-frequency shift of 1469 → 1461 cm^{-1} when interacting of fullerite C$_{60}$ with a thin film of Sn [21]. There were also observed changes in RS of C$_{60}$ under ion irradiation of Ti, Fe and in contact with Bi and In nanofilms [21]. The metal covering with thickness of 5–10 nm and more has bigger contact with the top or bottom surfaces of thick C$_{60}$films with thickness about 1.2 μm. For all considered cases, the interaction of C$_{60}$ with metals, as well as their laser or high-energy electrons (1.8 MeV) irradiation are repeated with slight variations four main band spectral components Ag(2) 1459.6–1461.4; 1464–1465; 1466–1467 and 1469.4–1470.2 cm^{-1}. Line 1468–1470 cm^{-1} refers to the vibrations of C$_{60}$ in the ground state, the line about 1464–1466 cm^{-1} is usually attributed to fluctuations in the dimer [123] or molecules in the excited triplet state [124]. Line near 1458–1460 and 1447 cm^{-1} associated respectively with the linear polymer and polymerized tetragonal phase [125]. These results confirm the substantial change in the properties of fullerite at interaction with metal atoms.

The greatest changes in the RS of fullerites after interaction with metal atoms consist in a strong increasing of BB, which corresponds to the appearance of new broad EB in the bandgap of fullerite C$_{60}$ and, observed together with them VBs, as shown in Figure 5.10d [21]. In interaction between Bi and In with the intensities of new EP in the vicinity of Ag(2) was respectively around 33% and 40% of the peak intensity of this band. Note that the initial samples of fullerite C$_{60}$ the intensity of background in RS was less than 5%. And under the interaction of C$_{60}$ with Sn atoms the relative intensity of EP increased to 63%. Like most of the earlier considered cases the intensity of EP increases to high-frequency region, which is associated with the phenomenon of strong VEI.

Very high relative intensity of the new EP was observed upon irradiation of fullerite C$_{60}$ by high-energy ions of Ti and Fe (140 keV, 1013 ions/cm^2). In the case of Ti-the intensity of induced EP exceeds 9% the peak intensity of bands Ag(2) and for the Fe ions—on 15%. As a result of strong VEI close to new EP the intensity of Raman bands Hg(7,8) (see Figure 5.10d) significantly increases, which also confirms the change of the ES. Electronic nature of the BB in RS associated with electrophilic C$_{60}$ molecules and transfer of electron density from the metal atoms, as evidenced by an increase in the intensities of the joint electronic and VBs. These illustrated in inset Figure 5.10d, where the dependences of peak intensities of the full symmetric vibrations Ag(1,2) on the intensity of the electronic background in I$_E$ in the region of these bands are shown. It is seen that the intensity of the VBs Ag (1,2) sharply increases with increasing of the intensity of EP in the vicinity of considered vibrations.

The proposed concept is consistent with the significant increase of EP under irradiation of fullerite C$_{60}$ by high-energy electrons (1.8 MeV) with a dose of 100–2500 Mrad [21]. In this case, the maximum of the ES is achieved not only in the region of bands Ag(2), Hg(7,8), but increases dramatically near exciting laser line. As for C$_{60}$ nanofilms it shows the possibility of inducing two types of ES. As for strong laser excitation (see Figure 5.5) the rate of new EPO increases strongly at high dose electron irradiation (~2500 Mrad). In this case, the intensity of the induced EP almost three times higher than the peak intensity of the band Ag(2) and under intense laser irradiation new EP intensified even more. These results confirm the developing concept of strong VEI, vibrational induction of new ES and the collective nature of CB.

5.8 DISCUSSION AND CONCLUSIONS

This work is aimed at broadening and deepening the representations about CB of complex molecules and condensed matter, including crystalline solids and amorphous and liquid media. Carbon materials, because of their wide dissemination of knowledge, and in particular molecular and polymerized fullerenes C$_{60}$, including a variety of chemically modified carbon materials (composites with metals, C$_{60}$, grafted with different functional groups), are good model systems for the development of new concepts of collective CBs. This problem is related to the further development of the initial quantum concepts and significant expansion of the optical–mechanical analogy. Known regularities of nonlinear interaction of laser radiation with high effective harmonic generation of high and sum frequencies we transfer to the field of materials science. Due to abnormally high vibrational nonlinearity of condensed matter, as well as large molecules, they are characterized by the development of nonlinear resonant interaction of vibrational modes with the generation of higher vibrational states that are approaching the ES and interact strongly with them.

Strong VEIs lead to changes in the ES and CB. Thus, collective CBs are associated with the NIVM and VEI and are a good example of a nontrivial relationship of classical and quantum physics. On an example of collective CB, we first pay attention on the necessity of the development of new nonlinear quantum concepts.

The conclusions on closely related peculiarities of collective CBs in fullerenes and fullerites C_{60} and the development of new nonlinear quantum concepts, including vibrational resonances, strong VEI and strengthening bands in the vibrational spectra are done in the following:

1. The development of nonlinear quantum concepts based on a large array of experimental data, which for a long time did not have a proper understanding at a fundamental level. Here we should indicate the radical differences of CB in simple molecules and condensed matter, and the majority are described in Sections 5.2 and 5.3 with spectral anomalies relating to changes in the frequencies and intensities of the VBs in the Raman and IR absorption spectra of different fullerites C_{60}.
2. The importance of nonlinear resonant interaction of vibrational modes of $2Hg(3) \approx Hg(7)$ and others is confirmed by maximal splitting of resonant VBs of Hg(3) and Hg(7) with isotopic or isoelectronic substitution in molecules $^{13}CC_{59}$, $^{13}C_2C_{58}$, and BNC_{58} [15] and abnormal increasing of intensities of active and *silent* vibrations in the Raman spectra of nanofilms (NFs) with a thickness of 150–250 nm, compared with microfilms with thickness of 1–2 μm.
3. The maximum enhancement of the intensities of resonant bands of Hg(3) and Hg(7) at the resonant excitation Raman scattering has been detected, as well as maximum attenuation of these bands with increasing temperature and the interaction of the fullerene C_{60} with organic compounds. It is important to note that the neglect of the nonlinear interaction of vibrations in modern QCC leads to the maximum ratio of the observed and theoretical values of the intensities I_{expt}/I_{th} especially for resonant vibrations of Hg(3,7). Moreover, there is an increase in the I_{expt}/I_{th} ratios up to values 35–200 with increasing frequency and intensity of the exciting laser radiation, which confirms the key role of nonlinear interactions of vibrations and strong VEI. Convincingly the role of the nonlinear interaction of vibrations exhibits altered frequencies and intensities of the VBs in the Raman spectra of carbon materials with increasing intensity of the exciting laser radiation [126,127].
4. The phenomenon of RVI of the fullerene C_{60} has been found, which is manifested in abnormally large $\Delta\nu \sim \pm 50$ cm^{-1} shifts of the vibrational frequencies ($\Delta\nu/\nu \approx 4\%$–6%) when the mass of the fullerene changes about 0.14% in isotopically substituted molecule $^{13}CC_{59}$. It is characteristic that maxima of these instabilities are correlated with resonance modes of Hg(3,7). It is shown that in the modern QCC methods, the vibrational resonances manifest as in the differences of vibrational frequencies calculated using DFT or HF methods and various bases. The values of $\Delta\nu/\nu$ can exceed 10%, which complicates the selection of a true set of *silent* vibrations of the fullerene C_{60}, as well as the choice of the optimal basis in QCC. It was shown for the first time that for molecules with a large number of normal vibrations, the real anharmonicity leads to a relationship of close vibrations.
5. Strong VEI and changing of the ES are confirmed by the appearance of principally new electronic bands (EP) observed in the Raman scattering spectra and IR absorption spectra of C_{60} in the field of VBs. It should distinguish low- and high-frequency vibration-induced EP. EPs close to exciting laser line and the area of low-frequency vibrations of Hg(1,2) and Ag(1) are observed in NF fullerite C_{60}. Their existence is also confirmed by the maximum variations of the intensities of the Hg(1,2) and Ag(1) in the Raman spectra at nonresonant excitation. Broadband EP with half widths $\delta\nu$ greater than 10^3 cm^{-1} is observed in the microfilm of fullerites C_{60}. The integrated intensity of new EP can be several orders of magnitude greater than the integrated intensities of the Raman bands. Their electronic nature proved a strong bond with the intensities of the VBs and the strong increase in the intensities of EPO doping C_{60} metals, as well as irradiation with high-energy electrons.

6. In nanofilms the intensity of new EB is significantly higher than in microfilms. The existence of optimal thicknesses of NP of C$_{60}$, for which the EP intensity reaches a maximum, confirms their nonlinear nature. For thin NP the EP intensities became weaker as a result of insufficient spatial accumulation of nonlinear wave processes, and in thicker films of C$_{60}$ vibration, nonlinearity weakened. Abnormal enhancement of the intensities of the majority of the VBs in NF is associated with the change of the ES and gain broad EB. It is proved that in NF when the EP intensities increase, the intensity of Ag(2) decreases, which confirms the connection of EB and vibrations. The enhancement EB in the region of the vibrations of Hg(7,8) and Ag(2) for polymerization of NP using N$_2$H$_4$ is shown. When exposed to intense laser radiation on microfilm C$_{60}$, new EB significantly enhanced and shifted to the field of overtones 2Ag(2), 2Hg(7,8), which characterizes the increased interaction of atomic and electronic subsystems (violation of the adiabatic approximation).

7. The observation of local EB with half width $\delta\nu \approx 100\text{--}200$ cm^{-1} in the areas of vibrations of Hg(3) and Hg(7) associated with the vibrational resonances of Hg(1) + Hg(2) = Hg(3), 2Hg(3) = Hg(7), is strong evidence of the developed nonlinear vibrational-electronic concept. There is an observed increase in the intensity of low- and high-frequency EA with additional laser irradiation NP. A strong increase of the irradiation intensities EB of microfilms of C$_{60}$ by ions Ti and Fe has been demonstrated, as well as the formation of composites with atoms Sn, In, and Bi. In this case, a significant change in the ES is confirmed by a noticeable increase in the intensity of VBs of Hg(7,8). Spectral intensity of these EBs can be compared and even exceed the intensity of the strongest peak Ag(2) in the Raman spectra. Redistribution of electron density between the metal atoms and the fullerene C$_{60}$, according to the results of the QCC, confirms the electronic nature of the new EB.

8. QCC for fullerene complexes C$_{60}$M and C$_{60}$MC$_{60}$ with metal atoms (M=Sn, Ti, Fe, Sn) were performed. The transfer of electron density from the metal atom and its distribution in almost all the atoms C$_{60}$ molecules were studied in detail. For comparison the calculations of the coupling C$_{60}$ with an additional carbon atom when the charge redistribution is carried out on a smaller part of the fullerene have been provided. It was established that collective CBs of fullerene C$_{60}$ with metal atoms are linked with a strong redistribution of the electron density and high total binding energy, bond lengths C–M ~ 2A, and small forces of local CB in the set of atoms MC$_2$. This is confirmed by the low frequencies of bending and stretching vibrations in the group MC$_2$.

9. A characteristic feature of the collective properties of fullerenes CB with metal atoms is a reduction of all single and double bonds in the bonded C$_{60}$ molecule by 1%–2% compared to the isolated molecule C$_{60}$. This indicates a quantum chemical molecule compression, despite the Coulomb repulsion of distributed negative charges. This result was confirmed by x-ray diffraction. In addition to these collective CB properties, a dynamic change of the charges and electronic polarizabilities during atomic vibrations is largely determined, resulting in a change in intensity of the IR absorption bands and Raman scattering and induced activities of *silent* vibrations.

10. On the base of QCC for the complex C$_{60}$Sn, it was found that for the ratio of the intensities of the VBs of the complex and initial C$_{60}$ fullerene, the maxima are equal to 2 and 6 for resonant vibrations of Hg(3,7). Thus, we have shown that large values of the I$_{expt}$/I$_{th}$ ratios for experimentally observed and calculated intensities of the VBs Hg(3,7) as in complexes C$_{60}$M and in initial C$_{60}$ fullerite are associated with changes in the collective ESs due to strong VEI that are poorly accounted in modern methods of QCC. Frequency splitting of degenerate vibrations in the complexes C$_{60}$M reaches values of 10–40 cm^{-1} and more, making it easier to fulfill the conditions of vibrational resonances and leads to an increase of the NIVM and the manifestation of the effects of strong VEI.

Thus, by the NIVM and strong VEI, the collective properties of the vibrational motion influence variations in the quantum properties of the electron subsystem and CB, and they also inevitably become collective. This nontrivial relationship between classical and quantum regularities should play a significant role in the chemical functionalization, showing the way to the optimal choice, merging with other fullerenes or carbon material functional groups to achieve the desired properties. Comparison of the results of vibrational spectroscopy and QCC shows the limits of the past and points the way to improve them. One might think that the established regularities are characteristic for other nanostructured materials. In conclusion, the developed nonlinear quantum ideology and the carried out researches of spectral features in Raman spectra of fullerites can contribute to receive more information about the dynamics of the nonlinear interaction of vibrational modes, as well as electron–vibrational dynamics and changes in the properties of CB that needs further study.

REFERENCES

1. I.S. Grigoreva, E.Z. Meylihova, eds., *Physical Quantities: A Handbook*, Energoatomizdat, Moscow, Russia, 1991, 1232 p. (in Russian).
2. A.R. Ubbelohde, *The Molten State of Matter*, Wiley, New York, 1978.
3. N.E. Kornienko, A.N. Kornienko, Fundamental differences of chemical bonds in molecules and condensed matter, *Proceedings of the International Conference on Modern Problems of Condensed Matter Physics*, Kyiv, Ukraine, October 10–13, 2012, pp. 97–99.
4. N. Blombergen, *Nonlinear Optics*, W.A. Benjamin, Inc., New-York, 1965.
5. F. Zernike, J. Midwinter, *Applied Nonlinear Optics*, Wiley-Interscience Publication, New York, 1973.
6. V.S. Butylkin, A.E. Kaplan, Y.G. Hronopulo, E.I. Yakubovich, *Resonant Interaction Light with Matter*, Nauka, Moscow, Russia, 1977, 222pp. (in Russian).
7. Y.R. Shen, *The Principles of Nonlinear Optics*, Wiley-Interscience Publication, New York, 1987.
8. N.E. Kornienko, Spatial Evolution of Wave Amplitudes, Stability of Solutions and Bifurcation in the Processes of the Stimulated Emission of Sum and Difference Frequencies Under the Two-Photon Resonance, *Quantum Electronics*, **12**(8), 1595–1601, 1985 (in Russian).
9. N.E. Kornienko, Bifurcation and limit the effectiveness of four-photon resonance generation of sum and difference frequencies under biharmonic pumping. *Optics and Spectroscopy*, **60**(1), 186–188, 1986 (in Russian).
10. N.E. Kornienko, M.F. Kornienko, A.P. Naumenko, A.M. Fedorchenko, About four-parametric frequency conversion in the two-photon absorption of the signal and pump, *Optics and Spectroscopy*, **60**(3), 650–654, 1986 (in Russian).
11. N.E. Kornienko, Five-wave approximation in the theory of stimulated light scattering, *Ukraine Journal of Physics*, **47**(5), 435–440, 2002 (in Russian).
12. N.E. Kornienko, S.I. Mihnitsky, Wave enlightenment in matter and high-effective generation under stimulated Raman scattering on polaritons, *Ukraine Journal of Physics*, **47**(8), 726–737, 2002 (in Russian).
13. M.E. Kornienko, M.P. Kulish, S.A. Alekseev, O.P. Dmitrenko, J.L. Pavlenko, Fine structure of bands in vibrational spectra of fullerite C_{60}, *Ukraine Journal of Physics*, **55**(6), 732–738, 2010.
14. N.E. Kornienko, N.P. Kulish, S.A. Alekseev, O.P. Dmitrenko, E.L. Pavlenko, Fine Band Structure of the Vibrational Spectra of Fullerite C_{60} and Enhancement of Intermolecular Interaction in High_Temperature Phase, *Optics and Spectroscopy*, **109**(5), 742–752, 2010.
15. N.E. Kornienko, V.A. Brusentsov, E.L. Pavlenko, Carbon nanoparticles in condensed media, Collected scientific papers, Minsk, Belarus, 2013, pp. 264–270 (in Russian).
16. N.E. Kornienko, About relationship between heats of f melting of crystals and the optical phonons energies, *Bulletin of the University of Kiev*, Physics and Mathematics Series, (4), 466–476, 2004; (3), 520–534, 2005 (in Russian).

17. N.E. Kornienko, The effects of strong phonon-electron interaction 1. Funding of the new type of electronic bands *Bulletin of the University of Kiev*, Physics and Mathematics Series, (3), 489–499, 2006; (3), 248–256, 2008 (in Russian).

18. N.E. Kornienko, V.I. Grygoruk, A.N. Kornienko, Phenomenon of Nonlinear Compression and Vibrationaly Instability of Condensed Matter (Development Foundations Nonlinear Quantum Physics). In "Actual problems of physics Solid State". *Collected Dokladov Mezhdunarodnaya Scientific Conference on Actual Problems of Physics Solid State*, Minsk, Belarus, 2011, vol. 1, pp. 26–28 (in Russian).

19. N.E. Kornienko, The development of nonlinear vibrational-electronic concepts of solutions and physical and chemical bases of living. 1. The nature of the solubility and eutectics, the change in vibrational and electronic states and energy conversion in solutions and living organisms. *Physics of the Alive. 16*, 5–22, 2008; On the development of nonlinear quantum macro- and nonlinear wave model of the energy channels of living organisms (the nature of the Chinese meridians), **17**(1), 5–43, 2009; **17**(2), 5–39, 2009 (in Russian).

20. N.E. Kornienko, N.L. Sheiko, A.N. Kornienko, T.Y. Nikolaenko, Properties of quasiliquid water film in the ice promelting range. 1. Temperature dependences of water nanofilm thickness and viscoelastic properties of polycrystalline ice, *Ukraine Journal of Physics,* **58**(2), 151–162, 2013.

21. N.E. Kornienko, O.P. Dmitrenko, M.P. Kulish, V.V. Strelchuk, Carbon nanoparticles in condensed media, Collection of scientific articles, Minsk, Belarus, 2013, pp. 251–256 (in Russian).

22. J. Menéndez, J.B. Page, *Vibrational Spectroscopy of C$_{60}$*, Light Scattering in Solids Series, vol. 8, Springer-Verlag, Berlin, Germany, 2000.

23. M.S. Dresselhaus, G. Dresselhaus, P.C. Eklund, *Science of Fullerenes and Carbon Nanotubes*, Academic Press, San Diego, CA, 1996.

24. T.L. Makarova, Electrical and optical properties of pristine and polymerized fullerenes (Review), *Physics and Technology of Semiconductors*, **35**(3), 257–293, 2001 (in Russian).

25. A.V. Eletskii, B.M. Smirnov, Fullerenes, *Uspekhi Physics Science*, **163**(2), 33–60, 1993 (in Russian).

26. A.V. Eletskii, B.M. Smirnov, Fullerenes and carbon structures, *Uspekhi Physics Science*, **165**(9), 977–1008, 1995 (in Russian).

27. P.C. Eklund, ed., *Fullerene Polymers and Fullerene Polymer Composites*, Springer Series in Materials Science, Springer, Berlin, Germany, 2000.

28. A. Krueger, *Carbon Materials and Nanotechnology*, Wiley-VCH Verlag GmbH, Weinheim, Germany, 2010.

29. M.S. Amer, *Raman Spectroscopy, Fullerenes and Nanotechnology*, RSC Nanoscience and Nanotechnology No.13, Royal Society of Chemistry, London, U.K., 2010.

30. N.E. Kornienko, V.A. Brusentsov, E.L. Pavlenko, N.P. Kulish, Fullerenes and nanostructures in condensed matter, Collected articles, Minsk, Belarus, 2011, pp. 377–383 (in Russian).

31. N.E. Kornienko, N.P. Kulish, E.L. Pavlenko, V.A. Brusentsov, V.V. Strelchuk, *International Scientific Conference on Actual Problems of Physics Solid State*, Collected reports, Minsk, Belarus, 2011, vol. 1, pp. 118–120 (in Russian).

32. N.E. Kornienko, N.P. Kulish, E.L. Pavlenko, V.A. Brusentsov, Collective properties chemical bonds of atoms Sn, Fe, Ti with fullerenes C$_{60}$, *Materials 51 Mezhdunarodnoy Conference on Actual Problems of Strength*, Kharkov, Ukraine, May 16–20, 2011, p. 109 (in Russian).

33. N.E. Kornienko, N.P. Kulish, E.L. Pavlenko, V.A. Brusentsov, V.V. Strelchuk, Collective chemical bonding of metal atoms with C_{60} and quantum-chemical compression molecules, *Proceedings of the International Conference on Modern Problems of Condensed Matter Physics*, Kyiv, Ukraine, October 10–13, 2012, pp. 28–30, c. 102–104 (in Russian).

34. N. Korniyenko, N. Kulish, O. Dmytrenko, O. Pavlenko, V. Brusentsov, V. Strelchuk, Collective chemical bonds of atoms of metals with fullerene molecules C_{60}: Theory and experiment, *Materials XIII International Conference on Physics and Technology of Thin Films and Nanosystems*, Ivano-Frankivsk, Ukraine, May 16–21, 2011, vol. 2, p. 31.

35. N.E. Kornienko, N.P. Kulish, E.L. Pavlenko, V.A. Brusentsov, V.V. Strelchuk, Collective nature chemical bonds and quantum-chemical phenomenon of compression molecules, *III International Scientific Conference on Nanostructural Materials—2012: Russia—Ukraine—Belarus*, St. Petersburg, FL, November 19–22, 2012, p. 79 (in Russian).

36. K.A. Wang, A.M. Rao, P.C. Eklund, M.S. Dresselhaus, G. Dresselhaus, *Physical Review B*, **48**, 11375–11380, 1993.

37. M.C. Martin, X. Du, J. Kwon, L. Mihaly, *Physical Review B*, **50**(11), 173–183, 1994-I.

38. Z.-H. Dong, P. Zhou, J.M. Holden, P.C. Eklund, M.S. Dresselhaus, G. Dresselhaus, *Physical Review B*, **48**, 2862–2865, 1993.

39. P.C. Eklund, P. Zhou, K.-A. Wang, G. Dresselhaus, M.S. Dresselhaus, *Journal of Physics and Chemistry of Solids*, **53**, 1391, 1992.

40. K. Pokhodnia, J. Demsar, A. Omerzu, D. Mihailovic, *Physical Review B*, **55**(6), 3757–3762, 1997.

41. A.V. Peschanskii, A.Yu. Glamazda, V.I. Fomin, V.A. Karachevtsev, Raman scattering in non-polymerized and photo-polymerized C60 films at 5K, *Journal of Low Temperature Physics*, **38**(9), 1077–1087, 2012 (in Russian).

42. P.J. Horoyski, M.L.W. Thewalt, T.R. Anthony, Raman Fine Structure in Crystalline C60: The Effects of Merohedral Disorder, Isotopic Substitution, and Crystal Field, *Physical Review Letters*, **74**, 194–197, 1995.

43. C.C. Homes, P.J. Horoyski, M.L.W. Thewalt, B.H. Clayman, T.R. Anthony, Effect of isotopic disorder of the Fu modes in crystalline C_{60}, *Physical Review B*, **52**, 16892–16900, 1995.

44. N.E. Kornienko, A.P. Naumenko, Structure of vibrational bands of the first and second-order Raman spectra of carbon nanotubes and their anharmonicity of vibrations, in V.S. Gorelick, ed., *Raman—80 Years of Research*, pp. 206–218, Physical Institute of Sciences RAN, Moscow, Russia, 2008, 604pp. (in Russian).

45. A.P. Naumenko, N.E. Kornienko, V.M. Yashchuk, V.N. Bliznyuk, S. Singamaneni, Phonon-like light scattering in polycrystalline carbon structures, *Ukraine Journal of Physics*, **57**(2), 197–207, 2012.

46. A.P. Naumenko, N.E. Korniyenko, V.N. Yaschuk, S. Singamaneni, V.N. Bliznyuk, Raman spectroscopy of carbon nanostructures: Nonlinear effects and anharmonicity, in C. Kumar, ed., *Raman Spectroscopy for Nanomaterials Characterization* (Chapter 7), Springer-Verlag, Berlin, Germany, 2012, DOI: 10.1007/978-3-642-20620-7_7.

47. N.E. Kornienko, A.P. Naumenko, Strong vibration-electron interactions and vibration band enhancement in vibrational spectra of C_{60} nanofilms and singlewalled carbon nanotubes, *Proceedings of the International Conference on Nanomaterials: Applications and Properties*, vol. 2, no. 1, 01001(5pp.), 2013, 03NCNN40-1-5.

48. N.E. Kornienko, A.P. Naumenko, Strong vibration-electron interactions and increased intensities of the bands in the vibrational spectra of C$_{60}$ nanofilms, *Proceedings of the VI International Scientific Conference on Topical Problems of Solid State Physics*, Minsk, Belarus, October 15–18, vol. 2, pp. 7–9, 2013 (in Russian).

49. M. Kornienko, A. Naumenko, The vibrational band enhancements for active and "silent" vibrations in the Raman and IR spectra of the fullerene C$_{60}$ nanofilms, *Ukraine Journal of Physics*, **59**(3), 339–346, 2014.

50. N.E. Kornienko, A.P. Naumenko, Enhancement of active and "silent" vibrational bands in Raman and IR spectra of C$_{60}$ nanofilms, *XXI International School-Seminar of Galyna Puchkovska "Spectroscopy of Molecules and Crystals" (XXI ISSSMC)*, Book of Abstracts, Berehove, Ukraine, September 22–29, 2013, pp. 200–201.

51. N.E. Kornienko, A.P. Naumenko, Strong electron-vibration interactions and induced electronic states in nano- and microfilms of fullerene C$_{60}$, *XXI International School-Seminar of Galyna Puchkovska "Spectroscopy of Molecules and Crystals" (XXI ISSSMC)*, Book of Abstracts, Berehove, Ukraine, September 22–29, 2013, pp. 202–203.

52. N.E. Kornienko, A.P. Naumenko, On the origin of anomalous changes in the vibrational band intensities in nanofilms C$_{60}$, *International Research and Practice Conference: Nanotechnology and Nanomaterials (NANO-2013)*, Bukovel, Ukraine, August 25–September 1, 2013, Abstract Book, p. 358.

53. N.E. Kornienko, A.P. Naumenko, Strengthening of the vibrational bands in the Raman and IR spectra of C60 and nanofilms strong vibrational-electronic interactions, *IV International Conference on Nanoscale Systems: Structure, Properties, Technology*, Kyiv, Ukraine, November 19–22, 2013, г., c. 332 (in Russian).

54. N.E. Kornienko, A.P. Naumenko, Changing the properties of carbon nanotubes and fullerite C$_{60}$ under laser radiation, *Materials 51th International Conference on Actual Problems of Strength*, Kharkov, Ukraine, May 16–20, 2011, p. 108 (in Russian).

55. N.E. Kornienko, A.D. Rud, Anomalies of vibrational anharmonicity and internal self-compression onion-like carbon nanostructures, *Proceedings of the IX International Conference on Electronics and Applied Physics*, Kyiv, Ukraine, October 23–26, 2013, pp. 63–64.

56. N.E. Kornienko, A.D. Rud, Formation of liquid and solid nanodiamonds in onion-like carbon structures, Abstracts of the *4th International Samsonovskoy Conference on Materials Science of Refractory Compounds*, Kiev, Ukraine, May 21–23, 2014.

57. N.E. Kornienko, N.P. Kulish, E.L. Pavlenko, V.V. Strelchuk, O.P. Dmitrenko, Change C$_{60}$ structure under the influence of the metal atoms, the electron and laser irradiation, *Bulletin of the Tambov University*, Natural and Technical Sciences Series, **15**(3), 951–952, 2010 (in Russian).

58. N.E. Kornienko, N.P. Kulish, E.L. Pavlenko, V.V. Strelchuk, S.A. Alekseev, O.P. Dmitrenko, Multiple vibrational resonances in C60, the vibration-electron interactions and investigation of the mechanisms of polymerization, *Abstracts XXIV Congress on Spectroscopy*, Moscow, Russia, February 28–March 5, 2010, vol. 1, pp. 181–182 (in Russian).

59. N.E. Kornienko, N.P. Kulish, O.L. Pavlenko, O.P. Dmytrenko, Observation of multiple vibration resonances in Raman spectrums of fullerite C$_{60}$ under electron irradiation, *AIP Conference Proceedings*, Melville, New York; *XXII International Conference on Raman Spectroscopy*, Boston, MA, August 8–13, 2010, pp. 455–456.

60. M.E. Korniyenko, M.P. Kulish, O.L. Pavlenko, S.A. Alekseev, O.P. Dmytrenko, M.M. Bilyi, Multiply resonances in Raman spectra of fullerite C_{60}, Davydov splitting and polymerization mechanisms, *XIX International School-Seminar on Spectroscopy of Molecules and Crystals*, Beregovoe, Ukraine, September 19–27, 2009, Abstracts, pp. 120–122.

61. N.E. Kornienko, N.P. Kulish, E.L. Pavlenko, V.V. Strelchuk, Multitude of resonances in the vibrational spectra of C_{60} and vibrational-electronic interactions, *Proceedings of the II International Scientific Conference on Nanostructured Materials 2010: Belarus-Russia-Ukraine, NANO 2010*, Kiev, Ukraine, October 19–22, 2010, p. 484 (in Russian).

62. M.Ye. Korniyenko, M.P. Kulish, O.L. Pavlenko, S.A. Alekseev, O.P. Dmytrenko, Multiple vibration resonances in fullerite C_{60}, vibration-electron interaction and polymerization mechanism, *Proceedings of the IV International Conference on Electronics and Applied Physics*, Kyiv, Ukraine, October 21–24, 2009, pp. 65–66.

63. N.E. Kornienko, Resonant vibrational-electronic instability of C_{60}: Results of experiments and quantum-chemical calculations, *Materials of the 55th International Conference on Actual Problems of Strength*, Kharkiv, Ukraine, June 9–13, 2014, p. 37 (in Russian).

64. W. Heisenberg, Nonlinear problems in physics, *Physics Today*, **20**, 27–37, 1967.

65. J.D. Axe, S.C. Moss, D.A. Neumann, in H. Ehrenreich, F. Spaepen, eds., *Solid State Physics*, Academic Press, New York, vol. 48, p. 149, 1994.

66. P.J. Horoyski, M.L.W. Thewalt, T.R. Anthony, Raman Fine Structure in Crystalline C60: The Effects of Merohedral Disorder, Isotopic Substitution, and Crystal Field, *Physical Review B*, **52**(10), R6951–R6954, 1995.

67. J. Winter, H. Kuzmany, Face-centered-cubic to orthorombic phase transition in single-crystal RbC60 analyzed by Raman scattering, *Physical Review B*, **52**, 7115, 1995.

68. K.P. Meletov, G.A. Kourouklis, Pressure- and temperature-induced transformations in crystalline polymers of C_{60}, *Journal of Experimental and Theoretical Physics*, **115**(4), 706–722, 2012.

69. K.P. Meletov, J. Arvanitidis, G.A. Kourouklis, K. Prassides, Y. Iwasa, Structural stability of the rhombohedral 2D polymeric phase of C60 studied by in situ Raman scattering at pressures up to 30 GPa, *Chemical Physics Letters*, **357**, 307, 2002.

70. Y. Wang, X. Cao, H.U. Han, G. Lan, Raman spectra of samarium–fullerene intercalation compounds, *Journal of Physics and Chemistry of Solids*, **63**, 2053–2056, 2002.

71. K.P. Meletov, J. Arvanitidis, S. Assimopoulos, G.A. Kourouklis, B. Sundqvist, *Journal of Experimental and Theoretical Physics*, **95**(4), 736–747, 2002.

72. A.M. Rao, P.C. Eklund, J.-L. Hodeau, L. Marque, M. Nunez-Regueiro, Infrared and Raman studies of pressure-polymerized C_{60}s, *Physical Review B*, **55**(7), 4766–4773, 1997-I.

73. J.R.D. Copley, D.A. Neumann, W.A. Kamitakahara, *Canadian Journal of Physics*, **73**, 763, 1995.

74. C. Coulombeau, H. Jobic, C.J. Carlile, S.M. Bennington, C. Fabre, A. Rassat, On the vibrational spectrum of C_{60} measured by neutron inelastic scattering. Fullerene Science and Technology, *Fullerene Science and Technology*, **2**, 247, 1994.

75. V. Schettino, P.R. Salvi, R. Bini, G. Cardini, On the vibrational assignment of fullerene C60, *Journal of Chemical Physics*, **101**, 11079, 1994.

76. J.M. Auerhammer, T. Kim, M. Knupfer, M.S. Golden, J. Fink, N. Tagmatarchis, K. Prassides, Vibrational and electronic excitations of (C59N)(2), *Solid State Communications*, **177**, 697–701, 2001.

77. R.-H. Xie, G.W. Bryant, L. Jensen, J. Zhao, V.H. Smith, Jr., *Journal of Chemical Physics*, **118**(19), 8621–8635, 2003.

78. E.V. Butyrskaya, S.A. Zapryagaev, Computer simulation of the infrared spectra of endohedral metallofullerenes Li$_2$C$_{60}$ and Na$_2$C$_{60}$, *Solid State Physics*, **51**(3), 613–618, 2009 (in Russian).

79. G. Dresselhaus, M.S. Dresselhaus, P.C. Eklund, Symmetry for lattice modes in C60 and alcali-metal-doped C$_{60}$, *Physical Review B*, **45**(12), 6923–6930, 1992-II.

80. V. Schettino, M. Pagliai, L. Ciabini, G. Cardini, The vibrational spectrum of fullerene C$_{60}$, *Journal of Physical Chemistry A*, **105**, 11192–11196, 2001.

81. J. Winter, H. Kuzmany, P.A. Person, P. Jacobson, B. Sundquist, Charge transfer in alcali-metal-doped polymeric fullerenes, *Physical Review B*, **54**(24), 17486–17491, 1995.

82. L.M. Sverdlov, M.A. Kovner, E.P. Kraynov, *Vibrational Spectra of Polyatomic Molecules*, Nauka, Moscow, Russia, 559pp., 1970 (in Russian).

83. C.H. Choi, M. Kertesz, L. Mihaly, Vibrational Assignment of All 46 Fundamentals of C60 and C60(6-): Scaled Quantum Mechanical Results Performed in Redundant Internal Coordinates and Compared to Experiments, *Journal of Physical Chemistry A*, **104**, 102, 2000.

84. S.G. Dolinchuk, N.E. Kornienko, V.I. Zadorozhnii, Noncritical vectorial phase matchings in nonlinear optics of crystals and infrared up-conversion, *Infrared Physics Technology*, **35**(7), 881–895, 1994.

85. V.I. Zadorozhnii, N.E. Kornienko, T.S. Sidenko, Broadband tunable and multiple noncritical phase matchings in LBO crystal, *Functional Materials*, **12**(N1), 91–97, 2005.

86. Н.Е. Корниенко, В.И. Задорожный, С.Ю. Кутовой, Т.С. Сиденко, The concept of vector group and a multiple of noncritical phase synchronism for the processes of frequency conversion in nonlinear crystals, *International Scientific Conference on Actual Problems of Physics Solid State*, Collected reports, Minsk, Belarus, 2005, vol. 2, pp. 253–255 (in Russian).

87. N.E. Kornienko, V.I. Zadorozhnii, S.Yu. Kutovyi, New Trends in Physics of Nonlinear Group Syncronisms in the Phonon Field of Condenced Matter, *International Conference Problems of Theoretical Physics*, Kyiv, Ukraine, 2012, pp. 89–90.

88. Н.Е. Корниенко, *International Scientific Conference on Actual Problems of Physics Solid State*, Collected reports, Minsk, Belarus, 2007, vol. 3, pp. 204–206 (in Russian).

89. R. Meilunas, R.P.H. Chang, Infrared and Raman spectra of C60 and C70 solid films at room temperature, *Journal of Applied Physics*, **70**(9), 5128–5130, 1991.

90. S. Iglesias-Groth, F. Cataldo, A. Manchado, Infrared spectroscopy and integrated molar absorptivity of C$_{60}$ and C$_{70}$ fullerenes at extreme temperatures, *Monthly Notices of Royal Astronomical Society*, **413**(1), 213–222, 2011.

91. K.P. Meletov, E. Liarokapis, J. Arvanitidis, K. Papagelis, G.A. Kourouklis, S. Ves, Softening of phonon modes in C60 crystals induced by laser irradiation: Thermal effects, *Journal of Experimental and Theoretical Physics*, **114**(5), 1785–1794, 1998.

92. B. Chase, N. Herron, E. Holler, Vibrational spectroscopy of fullerenes (C60 and C70). Temperature dependant studies, *Journal of Physical Chemistry*, **96**, 4262, 1992.

93. P.R. Birkett, H.W. Kroto, R. Taylor, D.R.M. Walton, R.I. Grose, P.J. Hendra, P.W. Fowler, The Raman spectra of C$_{60}$Br$_6$, C$_{60}$Br$_8$ and C$_{60}$Br$_{24}$, *Chemical Physics Letters*, **205**, 399, 1993.

94. K. Lynch, C. Tanke, F. Menzel, W. Brockner, P. Scharf, E. Stumpp, FT-Raman Spectroscopic Studies of C60 and C70 Subsequent to Chromatographic Separation Including Solvent Effects, *Journal of Physical Chemistry*, **99**, 7985, 1995.

95. K.-J. Fu, W.L. Karney, O.L. Chapman, S.-M. Huang, R.B. Kaner, F. Diederich, K. Holczer, R.L. Whetten, Giant vibrational resonances in A_6C_{60} compounds, *Physical Review B*, **46**, 1937, 1992.

96. M.C. Martin, D. Koller, L. Mihaly, *In situ* infrared transmission study of Rb- and K-doped fullerenes, *Physical Review B*, **47**, 14607, 1993.

97. J. Onoe, K. Takeuchi, *In situ* high-resolution infrared spectroscopy of a photopolymerized C_{60} film, *Physical Review B*, **54**, 6167, 1996.

98. T. Wagberg, P. Jacobsson, Comparative Raman study of photopolymerized and pressure-polymerized C_{60} films, *Physical Review B*, **60**(7), 4535–4538, 1999-I.

99. T. Braun, H. Rausch, J. Mink, Raman spectroscopy of the effect of reactor neutron irradiation on the structure of polycrystalline C60, *Carbon*, **43**, 855–894, 2005.

100. O.N. Bubel, S.A. Vyrko, E.F. Kislyakov, N.A. Poklonsky, Full-symmetrical vibrations of fullerene C_{60}, *JETP Letters*, **71**(12), 741–744, 2000 (in Russian).

101. P.M. Rafailov, C. Thomsen, A. Bassil, K. Komatsu, W. Bacsa, Inelastic light scattering of hydrogen containing open-cage fullerene ATOCF, *Physica Status Solidi*, **242**(12), R106–R108, 2005.

102. L.V. Baran, Fullerenes and nanostructures in condensed matter, In *Collected articles "Fullerenes and Nanostructures in Condensed Matter"*, Minsk, Belarus, 2013, pp. 302–307 (in Russian).

103. R. Poloni, D. Machon, M.V. Fernandez-Serra, S. Le Floch, S. Pascarelli, G. Montagnac, H. Cardon, A. San-Miguel1, High-pressure stability of Cs_6C_{60}, *Physical Review B*, **77**, 125413(6), 2008.

104. R.C. Haddon A.F. Hebard, M.J. Rosseinsky, D.W. Murphy, S.J. Duclos, K.B. Lyons, B. Miller, J.M. et al., Conducting Films of C_{60} and C_{70} by Alkali Metal Doping, *Nature*, **350**, 320, 1991.

105. A. Gromov, N. Krawez, A. Lassesson, D.I. Ostrovskii, E.E.B. Campbell, Optical properties of Endohedral Li@C60, *Current Applied Physics*, 2, 51–55, 2002.

106. P.A. Troshin, A. Łapinski, A. Bogucki, M. Połomska, R.N. Lyubovskaya, Structure of amorphized C60 films studied by Raman spectroscopy and X-ray photoelectron spectroscopy, *Carbon*, **44**, 2770–2777, 2006.

107. R. Ookawa, K. Takahiro, K. Kawatsura, F. Nishiyama, S. Yamamoto, H. Naramoto, Structure of amorphized C60 films studied by Raman spectroscopy and X-ray photoelectron spectroscopy, *Nuclear Instruments and Methods in Physics Research B*, **206**, 175–178, 2003.

108. K. Narumi, H. Naramoto, Modification of C60 thin films by ion irradiation, *Surface and Coatings Technology*, **158–159**, 364–367, 2002.

109. G.-H. Jeonga, T. Okadaa, T. Hirataa, R. Hatakeyamaa, K. Tohji, Fullerene negative ion irradiation toward double-walled carbon nanotubes using low energy magnetized plasma, *Thin Solid Films*, **464–465**, 299–303, 2004.

110. A. Dzwilewski, A. Talyzin, G. Bromiley, S. Dub, L. Dubrovinsky, Characterization of phases synthesized close to the boundary of C60 collapse at high temperature high pressure conditions, *Diamond and Related Materials*, **16**, 550–1556, 2007.

111. R. Poloni, G. Aquilanti, P. Toulemonde, S. Pascarelli, S. Le Floch, D. Machon, D. Martinez-Blanco, G. Morard, A. San-Miguel, High-pressure phase transition in Rb_6C_{60}, *Physical Review B*, **77**, 2054339(1–8), 2008.

112. N.E. Kornienko, Quantum regularities in aqueous solutions of electrolytes. 1. Nature of solubility of substances in water and hydration of ions *Bulletin of the University of Kiev, Physics and Mathematics Series*, (2), 438–451, 2006 (in Russian).

113. P. Debye, Selected works, Articles 1909–1965, L., Nauka, 1987, 163–202 (in Russian).

114. N.E. Kornienko, V.I. Grygoruk, A.N. Kornienko, Oscillatory instability structure solids: Maxima of stability, types of bonds in metals and compress atoms, *Bulletin of the Tambov University*, Natural and Technical Sciences Series, **15**(3), 953–954, 2010 (in Russian).

115. O.L. Pavlenko, O.P. Dmytrenko, M.P. Kulish, M.Ye. Korniyenko, Yu.I. Prylutskyy, M.M. Belyi, O.D. Rud, E.M. Shpilevsky, Polymerization of fullerene C$_{60}$ films under doping by indium atoms, *Fullerenes, Nanotubes and Carbon Nanostructures*, **19**, 1–5, 2011.

116. O.P. Dmytrenko, O.L. Pavlenko, M.P. Kulish, M.A. Zabolotnyi, M.E. Kornienko, V.A. Brusentsov, V.M. Rubii, E.M. Shpilevskyi, Component hybridization in thin granulated C$_{60}$-Cu nanocomposite films, *Ukraine Journal of Physics*, **56**(8), 828–837, 2011.

117. M.E. Kornienko, H.P. Kulish, O.L. Pavlenko, V.V. Strelchuk, O.P. Dmytrenko, Changes in the structure of fullerite C$_{60}$ under influence of metal atoms, electron an laser irradiation. *Bulletin of the University of Tambov, Series Natural and Technical Sciences*, **15**, 291–306, 2011.

118. O.L. Pavlenko, M.P. Kulish, O.P. Dmytrenko, M.E. Kornienko, A.I. Momot, V.A. Brusentsov, Ju.E. Grabovsky, E.M. Shpilevskyi, V.V. Strelchuk, A.D. Rud, V.M. Tkach, Chemical interaction and polymerization in films of C60-Sn, *Nanosystems, Nanomaterials, Nanotechnologies*, **9**(2), 291–306, 2011.

119. Pavlenko O.L., Dmytrenko O.P., Kulish M.P., Kornienko M.E., Zabolotnyi M.A. et al., Radiation damages and polymerization of fullerite C$_{60}$ films under irradiation with Ti ions. *Questions of Atomic Science and Technics*, **2**, 22–27, 2011.

120. O.L. Pavlenko, O.P. Dmytrenko, M.P. Kulish, V.A. Brusentsov, M.E. Kornienko, V.V. Strelchuk, E.M. Shpilevskyi, Fullerenes and nanostructures in condensed matter, Collected articles, Minsk, Belarus, 2011, pp. 92–97 (in Russian).

121. N.E. Kornienko, A.N. Kornienko, Collective nature of the chemical bonds in solids bodies, *Proceedings of the XVIII International Conference on Physics of Strength and Plasticity Materials*, Samara, Russia, July 1–3, 2012, p. 29 (in Russian).

122. M.E. Kornienko, V.I. Grygoruk, V.A. Makara, A.M. Korniyenko, V.B. Shevchenko, Spectral research of water and multiple resonances in nano- and mesoporous silica, *Bulletin of the University of Kiev*, Physics and Mathematics Series, (2), 275–278, 2012 (in Russian).

123. A.V. Talyzin, A. Dzwilewski, T. Wagberg, Temperature dependence of C 60 Raman spectra up to 840 K, *Solid State Communications*, **140**, 178–181, 2006.

124. G. Chambers, A.B. Dalton, L.M. Evans, H.J. Byrne, Observation and identification of the molecular triplet in C60 thin films, *Chemical Physics Letters*, **345**, 361–366, 2001.

125. L.G. Grechko, N.E. Kornienko, V.I. Zadorozhnii, A.M. Fedorchenko, Nonlinear quasiphase synchronism in the processes of parametric conversion of optical frequencies in resonant media, *Optics and Spectroscopy*, **55**, 209, 1983 (in Russian).

126. P.H. Tan, Y.M. Deng, Q. Zhao, Temperature-dependent Raman spectra and anomalous Raman phenomenon of highly oriented pyrolytic graphite, *Physical Review*, **58**, 5435, 1998.

127. P. Corio, P.S. Santos, M.A. Pimenta, M.S. Dresselhaus, Evolution of the molecular structure of metallic and semiconducting carbon nanotubes under laser irradiation, *Chemical Physics Letters*, **360**, 557, 2002.

CHAPTER 6

CONTENTS

Carbon Material Supported Nanostructures in Catalysis

6

S. Ganesh Babu, M. Gopiraman,
R. Karvembu, and I.S. Kim

6.1 INTRODUCTION

The world of carbon played only on the well-known modifications, graphite and diamond, until the 1980s. The scenario was completely changed by the discovery of the molecular carbon allotropes, namely, fullerenes, carbon nanotubes (CNTs), graphene, carbon nanohorns, and carbon onions. Because of their exceptional behaviors, researchers turned straight away into the exploration and modification of these new groups of materials. Owing to their versatility in terms of textural/ surface properties as well as morphology, chemical inertness under harsh conditions, thermal stability, and mechanical resistance, nanostructured carbonaceous materials are recognized as potential resources in various applications [1,2]. Each of these carbon-based materials has its own unique properties. To adduce, graphene, a 2D sp^2-hybridized carbon material (CM), has a single-atom-thick sheet of carbon atoms arranged in a honeycomb pattern. Undoubtedly, it is the world's thinnest, strongest, and stiffest material, along with extraordinary conducting behavior toward both heat and electricity. It showed promises to be a potential material in many emerging applications because of its very huge surface area and other exclusive physical and chemical properties. The theoretical surface area of monolayer graphene was found to be 2620 m^2/g [3,4]. Similarly, CNTs also posses unique structural, mechanical, optical, thermal, and electrical properties and hence are widely utilized in various fields [5]. Both single-walled carbon nanotubes (SWCNTs) and multiwalled carbon nanotubes (MWCNTs) can be used either as isolated nanostructures or in the form of macroscopic assemblies, such as CNT papers and films [6], yarns [7], fibers [8], and forests [9]. Each form has its own applications in different fields. Likewise, fullerene is another allotropic form of carbon that is in the form of a hollow sphere, entirely composed of carbon. Since 1990, the practical applicability of fullerenes and their derivatives, especially in the field of medicinal chemistry, has been studied thoroughly [10].

Even though these immaculate nanostructured carbonaceous materials have specific properties, the insolubility and inertness toward many chemicals limit their usage under many circumstances. Functionalizing these materials with an element of interest facilitated a lot to conquer this demerit and also augment its activity in many cases. Functionalization of carbon-based nanostructures can be achieved by two different methodologies, namely, covalent functionalization and noncovalent functionalization. In the case of covalent functionalization, there is a new covalent bond formation between the carbon (usually at the termini of the carbon framework) and inward atom/molecule [11]. Introduction of heteroatoms (P, O, N, and/or S), anchoring of inorganic metal complexes, and grafting of nanocomposites are being accomplished by covalent functionalization. Unfortunately, covalent functionalization disturbed π-conjugation exhibited in the carbon skeleton, and hence, it resulted in the reduced optical, electrical, and thermal properties that mainly depend on the extended π-conjugation. But for the incorporation of metal nanoparticles (MNPs) and embedding the biological moieties (such as DNA and RNA), noncovalent functionalization is the better approach one can think of. In the case of noncovalent functionalization, the foreign moiety may cohere onto the CM by means of van der Waals' forces or π–π interaction or physical adsorption [12].

Many researchers compared the activity of functionalized CMs with the original one and recognized that functionalized material showed good activity in almost all the cases. This chapter furnishes a clear outlook about the functionalization of nanostructured CMs and its versatile application in catalysis and applicability in energy and environmental problems.

6.2 FUNCTIONALIZATION

6.2.1 Covalent Functionalization

Introduction of heteroatom, inorganic complexes, or nanocomposites onto the CMs can be achieved through covalent functionalization. Covalent functionalization mainly changes the hybridization of one or more carbons in the network from sp^2 to sp^3 [13]. This kind of functionalization also causes structural alterations (defective sites) in the CMs and hence is called as defect functionalization [14]. Such functionalization improves the dispersion of CMs in solvents and polymers (Table 6.1) [11]. In some cases, this covalent functionalization of CMs improves the solubility, increases the electron mobility, enhances the electrocatalytic activity, introduces the hydrophilicity, and so on.

6.2.1.1 Doping of heteroatoms

Because of the enhancement in the physical and (electro)chemical properties of CMs, many attempts are made to introduce heteroatoms onto the carbon framework in recent years. For instance, the acid-treated MWCNTs are completely stable and highly dispersed in water more than 100 days at room temperature [15]. Similarly, Panchakarla et al. reported that covalent functionalization of double-walled carbon nanotubes (DWNTs) by aliphatic amide function made it soluble in nonpolar solvents [16]. Likewise, doping of heteroatoms (e.g., N, B, P, and O) to the carbon network of graphene introduces the electrocatalytic active sites and increases the surface hydrophilicity of the material, increasing the electrical conductivity [17].

6.2.1.1.1 Introduction of nitrogen

Among the different heteroatoms, the more vastly used heteroatom is nitrogen, which was demonstrated in many nitrogen-doped CMs, namely, N-doped CNTs [18], colloidal graphene quantum

TABLE 6.1 Common SWCNT Sidewall Functionalization Methodologies and Degree of Solubility

Methodology	Addend	Degree of Functionalization	Highest Solubility
Diazonium	Aryl	1 addend in every 10 carbons in sodium dodecyl sulfate (SDS)/ water and 1 addend in every 25 carbons in organic solvent or neat	0.8 mg/mL in N,N-dimethylformamide (DMF)
Diazonium in oleum	Aryl	1 addend in every 20 carbons	0.25 mg/mL in H_2O
Fluorination	Fluorine	1 addend in every 2 carbons	1 mg/mL in 2-propanol
Radical chemistry	Pyrrolidine	1 addend in every 100 carbons	50 mg/mL in $CHCl_3$
Azomethine ylides or nitrene	Aziridine	1 addend in every 50 carbons	1.2 mg/mL in DMSO
Bingel reaction	Cyclopropane	1 addend in every 50 carbons	Not given

Source: Dyke, C.A. and Tour, J.M., *J. Phys. Chem. A*, 108, 11151, 2004.

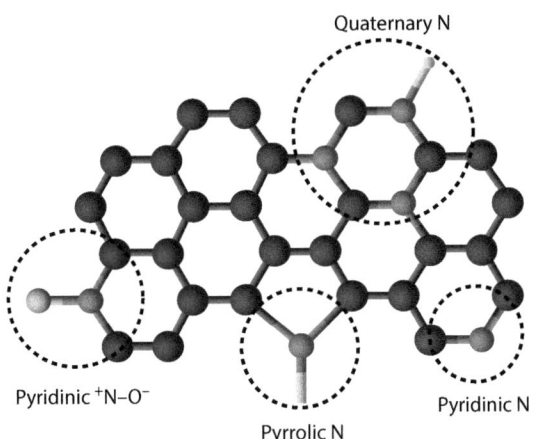

FIGURE 6.1
Configuration of nitrogen atoms in N-doped graphene.

dots (GQDs) [19], and carbon submicrometer spheres [20]. Introduction of nitrogen to CMs can be classified into two categories: The first one is replacement of existing carbon atom with the nitrogen atom and the other class is functionalization of terminal carbons of CMs by nitrogen-containing compounds. Doping of nitrogen to CMs, especially to graphene nanosheets (GNSs), is getting attraction because of its ability to replace the Pt-based electrocatalyst without compromise in activity particularly toward oxygen reduction reaction (ORR). Nitrogen can exist in four different forms over the graphene sheet, namely, quaternary N (or graphitic N), pyridinic N, pyridinic $^+$N–O$^-$, and pyrrolic N (Figure 6.1). This kind of nitrogen doping causes defects in carbon lattice, and among these nitrogen types, pyridinic N and quaternary N are sp^2 hybridized, which can contribute one p-electron to the π system, and the pyrrolic N is sp^3 hybridized, which contributes two p-electrons [21].

N-doped graphene was prepared by different physicochemical methods such as chemical vapor deposition (CVD) [4,22], segregation growth [23], solvothermal [24], arc discharge [25,26], thermal treatment [27,28], and plasma treatment [29]. CVD is a well-known versatile tool that synthesizes CMs such as graphene [30], CNTs [31], and carbon nanofibers (CNFs) [32]. Recently, it is widely used for the bulk production of nitrogen-doped carbon nanotubes (NCNTs) [33] as well as N-graphene [34]. The major advantage of CVD is that the nitrogen doping can be varied either by altering the flow rate [4] or by changing the ratio between carbon and nitrogen sources [34]. Rao et al. reported the preparation of N-graphene by arc discharge method in the presence of pyridine vapor or NH$_3$ [35]. Based on the preparation technique and convenience, different nitrogen precursors were used for the doping of nitrogen to the carbon framework. Some of the commonly used nitrogen sources are NH$_3$ [4], acetonitrile [36], pyridine [37], Li$_3$N [26], melamine [38], N$_2$ plasma [39], N$_2$H$_4$ [40], NH$_3$ plasma [41], urea [42], and so on. Different atomic percentages of nitrogen-doped CMs obtained by changing the synthesis methodology and nitrogen precursor are described in Table 6.2.

NCNTs were prepared by pyrolysis of acetonitrile over a cobalt catalyst [43]. As in N-doped graphene, NCNTs also showed nitrogen in different configurations (pyridinic N, pyrrole N, quaternary N, and pyridine-*N*-oxide), which was confirmed by x-ray photoelectron spectroscopic (XPS) studies. The catalytic performance of the as-prepared NCNTs was tested toward the ORR and found to have one order of magnitude lower activity than that of a commercial Pt/C catalyst.

Apart from the replacement of carbon atom by nitrogen atom, there is another process used to introduce nitrogen to CMs, which forms a C–N bond between nitrogen-containing compounds and the CMs. For illustration, [60]fulleroazepines were synthesized by treating [60]fullerene with N-substituted-2-amino biaryls in the presence of palladium catalyst (Scheme 6.1) [44]. The simultaneous formation of C–N and C–C bonds followed by C–H bond activation was achieved successfully.

[60]Fullerenes showed a positive response against HIVs [45], but the major drawback in [60] fullerenes is its insoluble nature. This lack of solubility can be overcome by functionalizing [60]

TABLE 6.2 Nitrogen Doping Methods and Nitrogen Concentration on Graphene

Synthesis Method	Precursors	N Content (%)
CVD	CH_4/NH_3	1.2–8.9
	$NH_3/CH_4/H_2/Ar$	4
	Acetonitrile	~9
	Pyridine	~2.4
Segregation growth	Nitrogen-containing boron layer	0.3–2.9
Solvothermal	Li_3N/CCl_4	4.5
	$N_3C_3Cl_3/Li_3N/CCl_4$	16.4
Arc discharge	Pyridine/H_2/He	0.6
	NH_3/H_2/He	1
	Pyridine/He	1.4
Thermal treatment	N^+ ion-irradiated graphene, NH_3	1.1
	NH_3/Ar	2.0–2.8
	NH_3	6.7–10.78
	Melamine	7.1–10.1
Plasma treatment	N_2 plasma	8.5
N_2H_4 treatment	N_2H_4, NH_3	4.01–5.21
	N_2H_4	1.04

Source: Wang, H. et al., *ACS Catal.*, 2, 781, 2012.

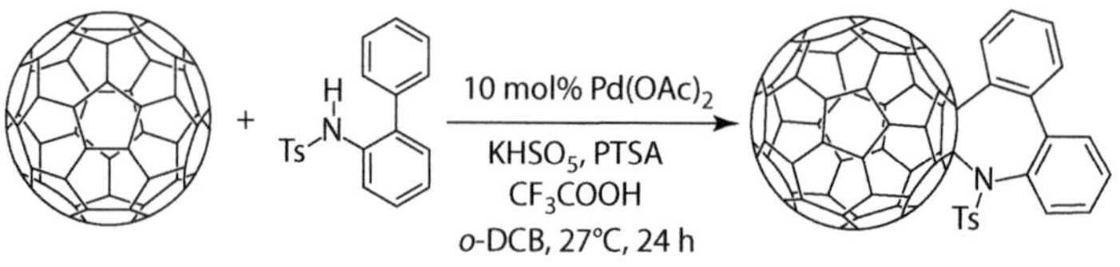

SCHEME 6.1
Synthesis of [60]fulleroazepines.

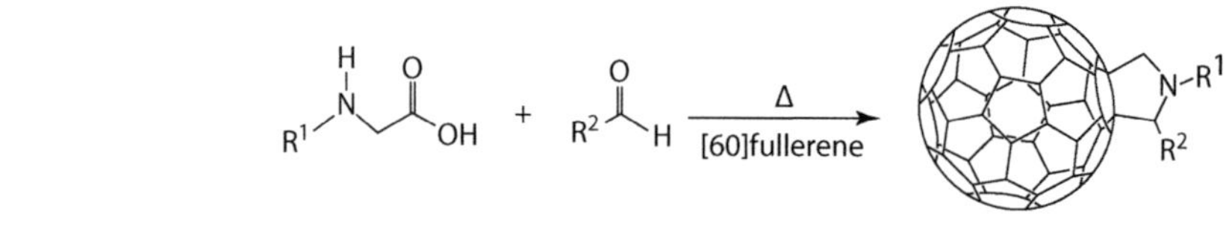

SCHEME 6.2
(3 + 2) Cycloaddition of azomethine ylides with [60]fullerene.

fullerenes with any nitrogen-containing compounds. One of the frontier strategies for the functionalization of [60]fullerenes is the [3 + 2] cycloaddition reaction with azomethine ylides [46,47]. It was easily achieved by treating N-substituted glycines and aldehydes with [60]fullerenes (Scheme 6.2).

Similarly, CNTs were also functionalized with nitrogen compounds to enhance their utility. Azidothymidine (AZT) is a type of antiretroviral drug used for the treatment of HIV/AIDS infection, which was loaded over MWCNTs by the photoetching technique (Scheme 6.3) and exclusively used for the in situ synthesis of DNA molecule on the side walls of well-aligned MWCNTs [48].

6.2.1.1.2 Introduction of sulfur

As like nitrogen doping, introduction of sulfur to the carbon network also finds much attention in recent years because of its utility in various fields such as fuel cells, solar cells, and electrochemical applications. Ultrasound-assisted sulfur-doped graphene oxide (GO) was prepared using

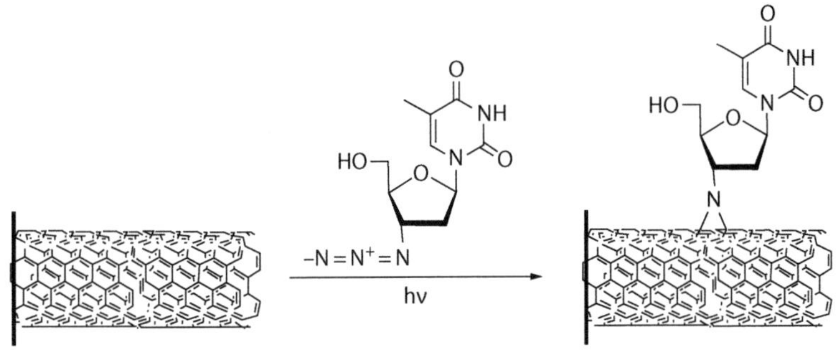

SCHEME 6.3
Photoetching route for the synthesis of AZT-doped well-aligned MWCNTs.

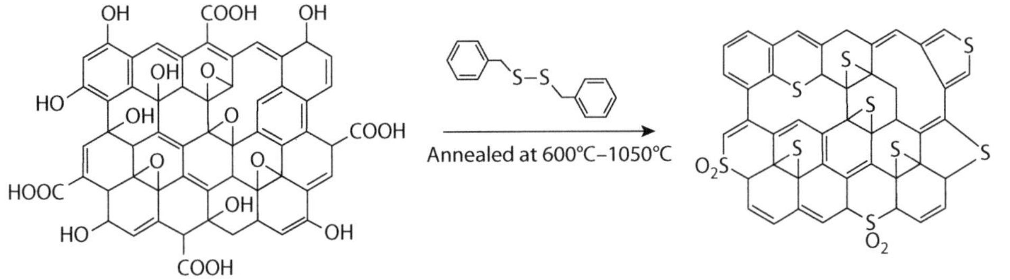

SCHEME 6.4
Schematic illustration for the preparation of S-graphene using BDS.

benzyl disulfide (BDS) and annealed at high temperature under argon atmosphere, as illustrated in Scheme 6.4 [49]. The formation of C–S–C and C–SO$_x$–C bonds (x = 2, 3, and 4) was confirmed by XPS analysis.

Because of the enhanced synergetic effect, simultaneous doping of nitrogen and sulfur gained more attention rather than doping sulfur alone to the CMs. Thiourea was used as a single nitrogen and sulfur precursor source for the doping of N and S to graphene sheets by annealing GO with thiourea. The as-prepared N,S-doped graphene showed a direct four-electron reaction pathway, high onset potential, high current density, and high stability [50].

Synergistically enhanced activity was observed with nitrogen and sulfur dual-doped mesoporous graphene as an electrocatalyst for the ORR [51]. Similarly, the one-pot hydrothermal approach was developed for the preparation of nitrogen and sulfur co-doped graphene frameworks by using ammonium thiocyanate as a single precursor source [52]. Pyrrolic/graphitic-N structures were found to be dominant while doping nitrogen and sulfur to the graphene layers using pyrimidine and thiophene as precursors for nitrogen and sulfur, respectively. The catalytic activity of S- and N-doped GO was found to be superior to that of mono N-doped carbon nanomaterials [53]. Moreover, the dual-doped catalyst showed an excellent methanol tolerance and the durability of this catalyst was relatively higher than the commercial Pt/C catalysts. Likewise, a porous 3D N and S co-doped graphene (3D-NGS) with a high sulfur content of 87.6 wt.% was achieved by a one-pot solution method [54]. Sulfur was well dispersed within the 3D framework and the as-fabricated 3D-NGS composite showed excellent rate competence and durability. Most of these doping processes are high-temperature methods and/or carried out under drastic conditions. Guo et al. demonstrated a novel route to synthesize N and S co-doped graphene under ambient conditions (37°C) by microbial respiration of sulfate-reducing bacteria (SRB) [55]. The SRB not only introduced N and S to the graphene layer but also are involved in the reduction of GO.

6.2.1.1.3 Introduction of oxygen

Unlike the heteroatoms such as nitrogen and sulfur, oxygen is unable to replace the carbon atom in carbon-based materials because of its different hybridization configuration. But introduction of oxygen functional groups (–C–OH, –CHO, –C=O, –COOH, and C–O–C) onto the carbon framework

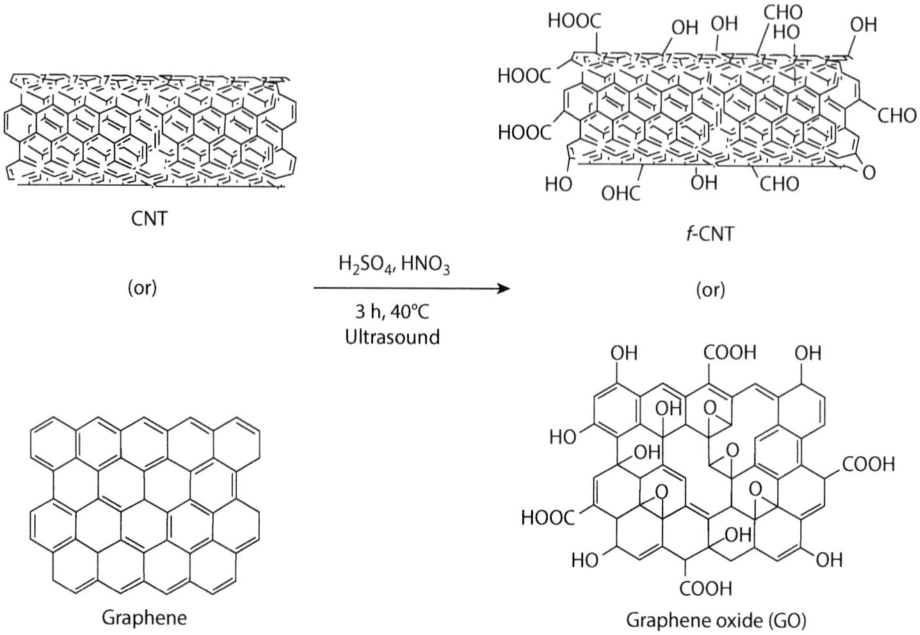

SCHEME 6.5
Introduction of oxygen functionalities to CNT and graphene.

is much easier than introducing N and/or S to it. Simple acid treatment to graphene or CNTs is even enough to introduce oxygen functionalities. The mineral acids oxidize the terminal carbons, leading to the formation of oxygen functional groups at the edges of CMs. Introduction of oxygenated functionalities made the CMs hydrophilic, hence it is able to completely disperse in aqueous medium. Hummers and Offeman developed a simple methodology for the preparation of graphitic oxide from graphite using $KMnO_4$, $NaNO_3$, conc. H_2SO_4, and H_2O_2, which not only introduced oxygen functional groups but also exfoliated the layers [56]. GNSs were chemically treated with conc. H_2SO_4 and conc. HNO_3 (3:1) and kept in ultrasonic irradiation to introduce oxygen functionalities [57]. A similar trend was also followed for the preparation of functionalized CNTs (Scheme 6.5) [58]. The presence of these oxygen functionalities enhances the hydrophilicity of the CMs, and hence, the loading of MNPs and metal oxides onto the surface is very easy. But this functionalization retards the electron mobility of the CMs.

6.2.1.2 Anchoring of metal complexes

Because of the chemical inertness, very high mechanical strength, electrical conductivity, dispersive power, and high surface area, CMs were emerging as solid supports for transition or noble metal catalysts. This covalent anchoring can be accomplished either by reaction of the metal complexes with the carbon support surface groups or mediated through a spacer that is previously grafted on the carbon support. In the first case, the inorganic metal complex can form an axial coordination bond with the functional groups accessible at the surface of CMs, whereas in the second case, a bifunctional molecule (spacer) binds the carbon support and the transition metal complex. The presence of oxygen-rich carbon surface facilitates the anchoring of inorganic metal complexes. Catalytically active functional carbon nanomaterial was synthesized by treating polymer-functionalized CNTs with $CuCl_2$ [59]. Polymer functionalization was carried out via amidation of carboxylated CNTs (Scheme 6.6).

The immobilization of inorganic metal complexes on CMs makes the catalyst as heterogeneous so that it can be reused for several cycles (which hold back the demerits of homogeneous catalysis). For illustration, a very good recyclability was demonstrated for the pyrrolidine diphosphine ligand–based chiral rhodium complex, which was covalently anchored onto functionalized CNTs (Scheme 6.7). Negligible amount of leaching of active site along with no loss in activity and enantioselectivity was observed for the reused catalyst [60].

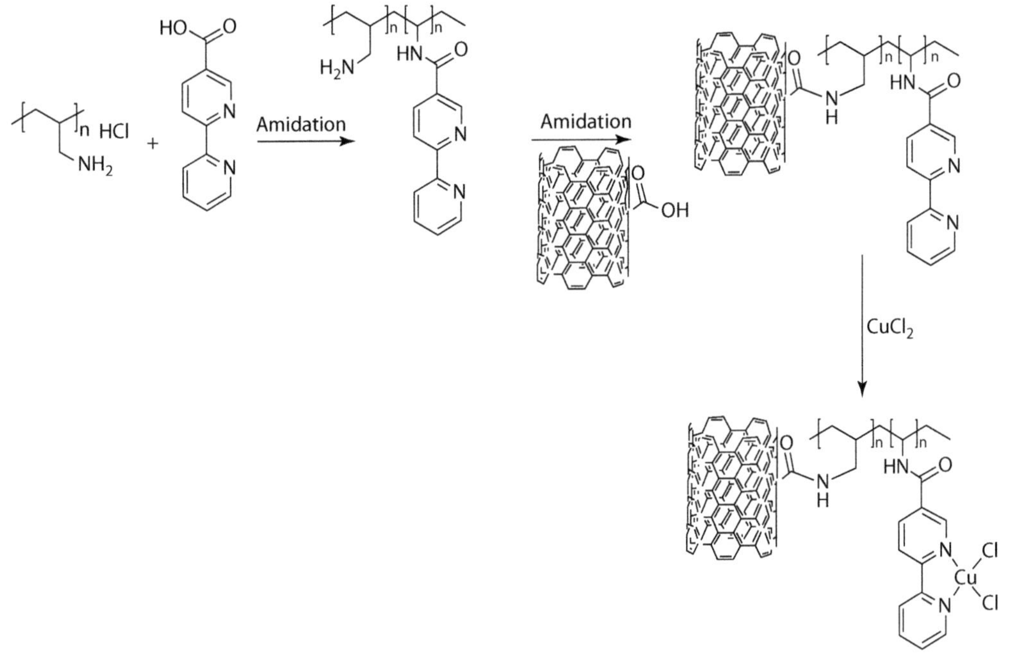

SCHEME 6.6
Anchoring of Cu complex on SWCNT.

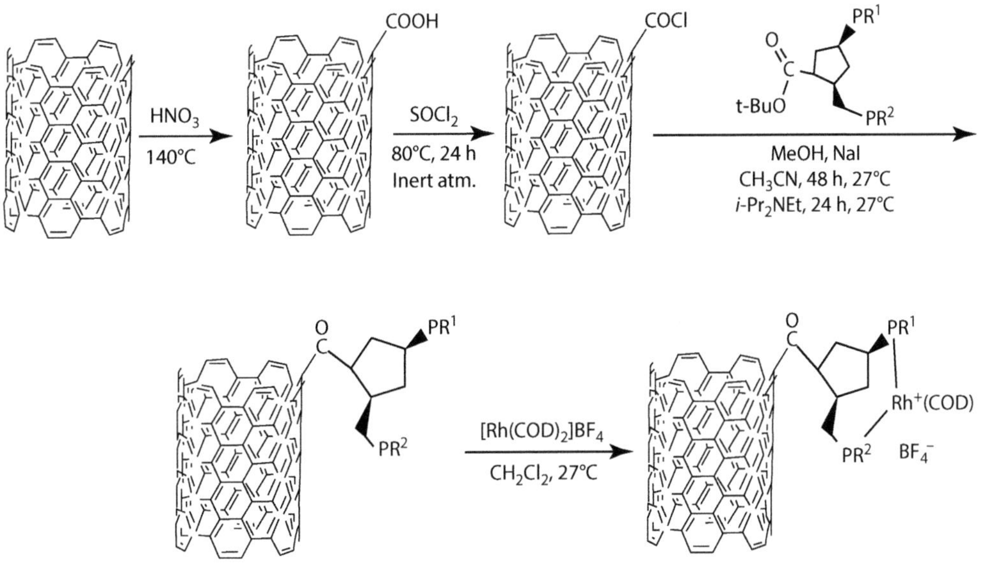

SCHEME 6.7
Covalent anchoring of chiral Rh complex onto CNT.

Vaska's complex, $IrCl(CO)[P(C_6H_5)_3]_2$, was treated with raw and oxygen-functionalized CNTs to get Ir coordinated nanotubes (Figure 6.2). In the former case (with raw CNTs), the coordination takes place through η^2-coordination process (Figure 6.2a), while in the latter case (with f-CNTs), the coordination is through the surface-bound oxygen atoms, which formed a hexacoordinate Ir complex (Figure 6.2b) [61]. In the same way, Wilkinson's catalyst was functionalized with CNTs to improve the solubility and also to enhance the optical property. Authors claim that Rh coordinated through the oxygen atoms and formed a hexacoordinate structure around the Rh atom. This Rh-functionalized CNTs are soluble in DMSO (>150 mg/L) and also to an extent in DMF and tetrahydrofuran (THF) [62]. Ruthenium triphenylphosphine complex was attached to carboxyl-functionalized CNTs (Figure 6.2c). Similarly, vanadyl(IV) Schiff base complexes were covalently anchored onto mercapto-modified CNTs (Figure 6.2d) [63]. Several metal acetylacetonates ($M(acac)_x$) were grafted onto

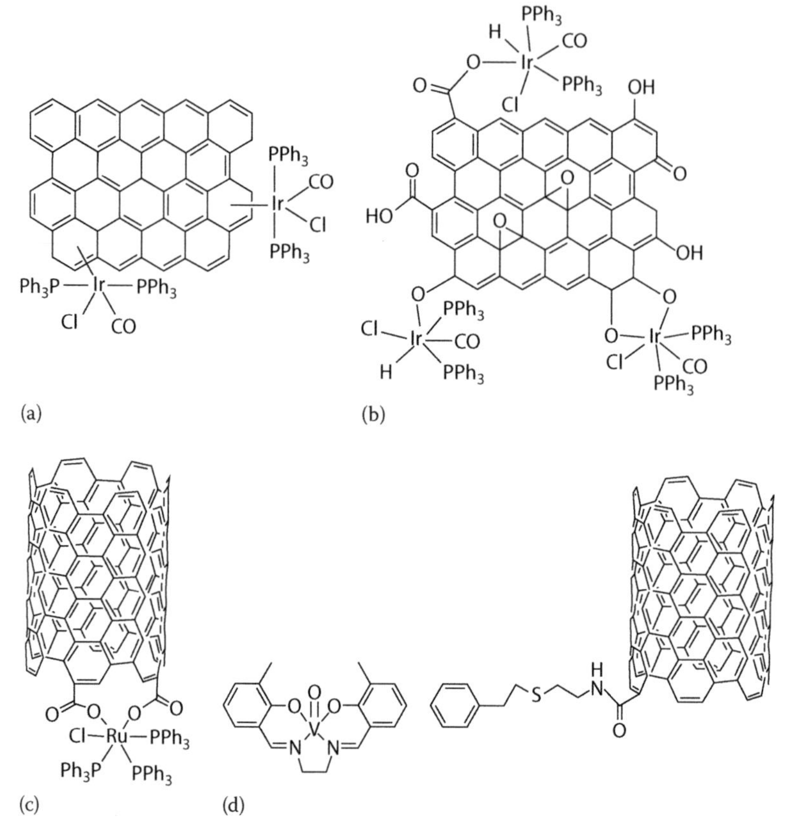

FIGURE 6.2
Structure of Ir complex anchored on (a) raw and (b) oxidized graphene and structure of CNTs anchored on (c) Ru complex and (d) V(IV) Schiff base complex.

the amine-functionalized CMs (such as activated carbon and CNTs), which were prepared by treating CM with air and nitric acid [64–67].

6.2.1.3 Grafting of nanocomposites

Nanocomposites were grafted over the CMs in order to extend the utility of them. Anchoring of organic chains to the surface of CMs can be achieved in two ways, namely, *grafting to* and *grafting from*. *Grafting to* is a process in which the end-functionalized polymer forms a bond to the reactive surface groups on the CMs, whereas in the case of *grafting from* process, immobilization of the initiators (monomer) onto the surface of CMs is carried out and then in situ surface polymerization is performed to generate the tethered polymer chains. In other words, *grafting from* is a process where the polymer nanocomposites were grown on the surface of CMs [68]. Polyaniline (PANI) was grafted either over the carboxyl-functionalized MWCNT surface by simple in situ polymerization in an emulsion system [69] or over the 4-aminobenzoyl-functionalized MWCNTs (Scheme 6.8) [70].

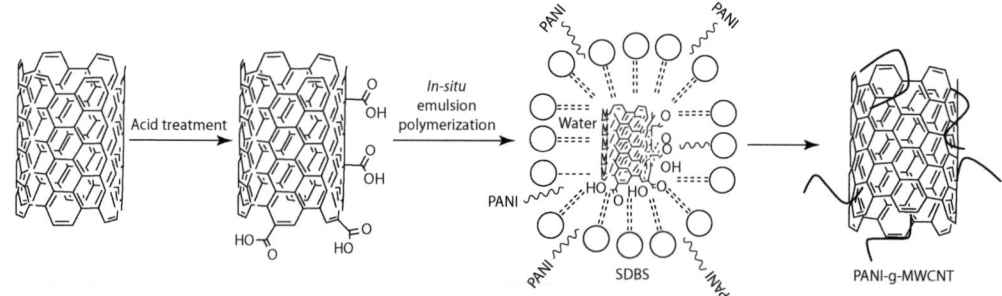

SCHEME 6.8
Grafting of PANI onto carboxyl-functionalized MWCNTs.

Further, it was blended with polyetherimide (PEI) using *N*-methyl-2-pyrrolidone as solvent. The formed PANI-g-MWCNT/PEI nanocomposite showed excellent conducting behavior, higher than pristine MWNTs/PEI composite. Similarly, MWCNTs/polyimide (PI) films were prepared from MWCNTs and PI by simple mechanical stirring. Relatively good dispersion of MWCNT in the prepared nanocomposite was investigated by microscopic studies [71].

6.2.2 Noncovalent Functionalization

Indeed, CMs are highly hydrophobic in nature and chemically inert under normal conditions, so they cannot be dissolved in any organic solvents [72]. Therefore, these materials have also been extensively used as a support for active metal catalyst in heterogeneous catalysis [73]. Generally, metal catalysts are attached onto CMs via noncovalent functionalization of MNPs in order to improve the activity and efficiency of these metal catalysts. Noncovalent functionalization (or physical adsorption) involves weak interactions (π-interactions), which cause no change on the basal plane structure of CMs (as in the case of CNTs and graphene sheets only) and their electronic properties being largely retained [3]. Moreover, the conductive nature of these CMs obviously assists a homogeneous dispersion of MNP catalysts on the CMs [74]. Apart from MNP catalysts, some polymeric materials are also attached on CMs via noncovalent functionalization method for energy conversion. In some cases, the heteroatom (such as N, S, and O)-doped CMs have also been used to support MNPs to enhance the ORR activity and stability due to the reduced aggregation of MNPs on the CM supports [75–77]. In fact, the noncovalent functionalization of CMs by MNPs can be transformed to a very active catalyst through the interactions between the active metal clusters and carbon vacancies. There are two main methods for the noncovalent functionalization of CMs by MNPs: (1) wet synthesis and (2) dry synthesis.

6.2.2.1 Wet impregnation

CMs including CNTs and GNSs are an important class of nanomaterials that are receiving greater attention. These materials are easy to produce by various well-known methods including those based on wet chemistry [78]. The CM-based composites have found important applications as high-performance battery electrodes, gas sensors, catalysts, or potential hydrogen storage matrices, due to their high conductivity and ultimately large surface-to-volume ratio [79–82]. Hence, many researchers' proposed efficient methods for the preparation of MNP/CM composites by the wet synthesis method.

Jibin et al. [83] demonstrated a noncovalent DNA decoration of GO and reduced graphene oxide (RGO) by the wet synthesis method (Scheme 6.1). The GO composites (AuNP–DNA–GO and AuNP–DNA–RGO) were prepared by incubating DNA–GO (or DNA–RGO) with an excess amount of phosphine-protected gold nanoparticles (Au NPs; 6 nm diameter) in TBE buffer supplemented with NaCl solution for 12 h. To remove unbound Au NPs and isolate the AuNP–DNA–GO (or AuNP–DNA–RGO) assemblies, the reaction mixture was loaded on a 2% agarose gel and run in TBE solution at 10 V/cm for 30 min. The AuNP–DNA–GO or AuNP–DNA–RGO conjugates were collected by rinsing the corresponding gel wells with running buffer. The recovered composites were directly examined by atomic force microscopy (AFM) and UV–Vis absorbance spectroscopy (Scheme 6.9).

Hee et al. [84] reported an extraordinary result related to the reduction of metal ions in the presence of CNTs. The author confirmed that Pt and Au NPs could be formed in the presence of CNTs, from metal precursors without any reducing agent. They proposed that a direct redox reaction occurs between the metal ions and CNTs. The MNPs formed spontaneously and exclusively on the walls of SWCNTs right after the immersion of SWCNTs in a solution of $HAuCl_4$ (Au^{3+}) or of Na_2PtCl_4 (Pt^{2+}) for 3 min. They concluded that this metal deposition without using the reducing agent was effective only for Au and Pt. Other metal cations such as Ag^+, Ni^{2+}, and Cu^{2+} could not be reduced in the same way because of their lower redox potentials.

Minati et al. [85] reported a new microwave irradiation method to functionalize the MWCNTs with oxygen functional groups (carboxyl, carbonyl, or hydroxyl) without the use of strong acids such as HNO_3/H_2SO_4 or ultrasonication. The oxygen functional groups serve as

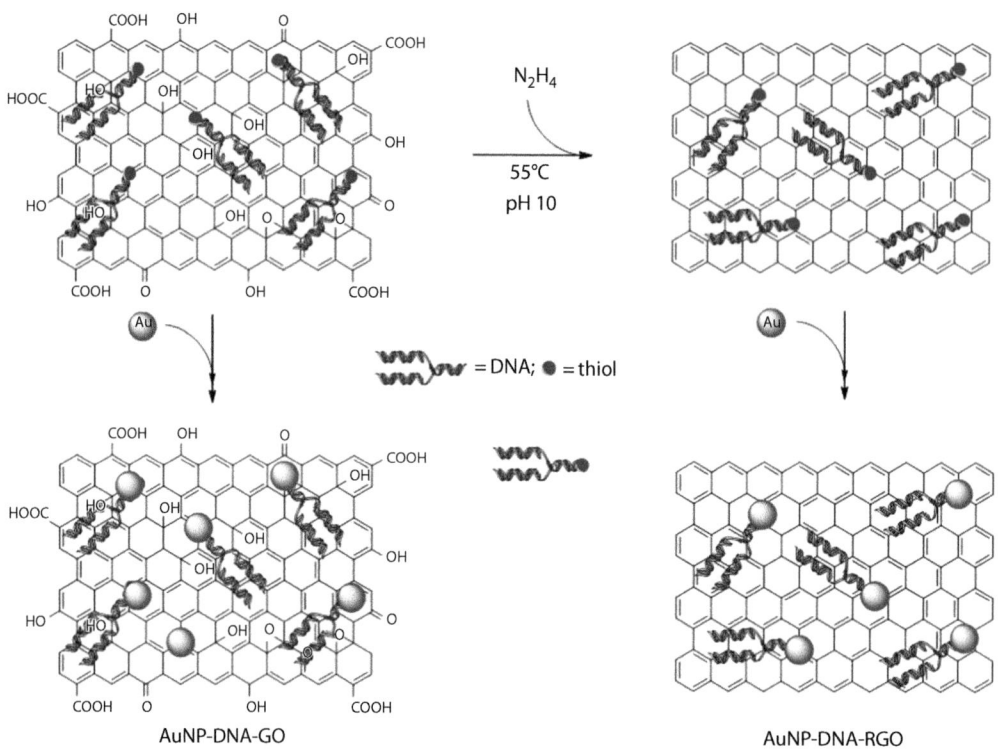

SCHEME 6.9
DNA coating and aqueous dispersion of GO and RGO, which were then used as 2D bionanointer-faces for homogeneous assembly of metal–carbon heteronanostructures. (From Jinbin, L. et al., *J. Mater. Chem.*, 20, 900, 2010.)

effective anchoring centers for the reduction of the metal ions present in solution. MWCNTs were dispersed into a water solution containing HAuCl$_4$, using ethylene glycol (EG) as reducing agent. After microwave irradiation, AuNPs uniformly dispersed on the CNT surfaces were formed. Similarly, Makala et al. [86] have treated SWCNTs with diluted HNO$_3$ to create oxygenated functions on the SWCNT walls. Chloroauric acid (HAuCl$_4$) and EG were used as the Au sources and reducing agent, respectively. The final material contained about 10% of AuNPs with a mean diameter of about 1–2 nm.

Bhalchandra et al. [87] developed a new and simple microwave treatment method for the preparation of RhNPs supported on MWCNTs (RhNPs/MWCNTs). Rh/MWNT catalysts have been prepared by direct mixing of acidified (microwave treatment in acid mixture) MWCNTs with tridecylamine-capped rhodium NPs followed by microwave treatment for 1 min.

CNT-supported metallic NPs including Pt, Rh, and bimetallic Pd–Rh were prepared by wet synthesis–based microemulsion method [88]. For synthesizing the CNT-supported metallic nanocatalysts, the water-in-hexane microemulsion method was employed. The microemulsion-templated synthesis was carried out at room temperature using relevant metal ions in the aqueous phase. Pure H$_2$ gas at a pressure of >1 atm was bubbled through the aqueous phase to reduce the metal ions in the water core of the microemulsion. The system was stirred vigorously during the reduction stage in the presence of the carboxylic acid–functionalized MWCNTs. MNPs formed in the water core of the microemulsion can be effectively transferred to the surfaces of the functionalized MWCNTs. After hydrogen reduction, the CNT-supported MNPs precipitated at the bottom of the flask were separated, washed with ethanol several times, and then dried in an oven for catalytic applications.

RGO was used as a support to anchor semiconductor and MNPs by Kamat et al. [89]. The AuNPs/RGO and TiO$_2$NPs/RGO composites were prepared by wet synthesis method (Scheme 6.10). In this method, a dilute solution of NaBH$_4$ was added to RGO solution followed by the controlled addition of HAuCl$_4$ dropwise under stirring followed by calcinations at higher temperature. The composite materials were used for various applications such as catalysis, light energy conversion,

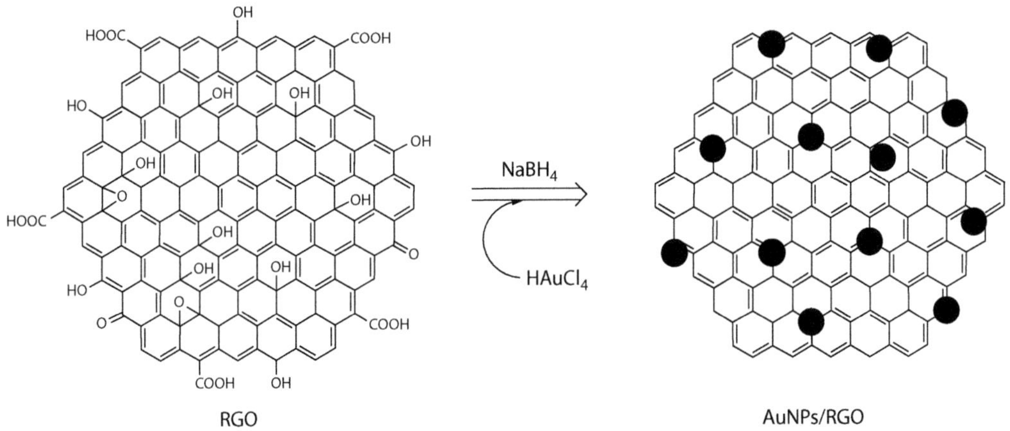

SCHEME 6.10
Pictorial illustration of attaching metal and metal oxide NPs to RGO sheets. (From Byunghoon, Y. et al., *J. Phys. Chem. C*, 113, 1520, 2009.)

and fuel cells. Similarly, Xiang-Rong and coworkers [90] have reported a rapid, convenient, and environmentally benign method for the fabrication of MNP/MWCNT composites. NPs of Pd, Rh, and Ru were deposited onto *f*-MWCNTs (acid-treated MWCNTs) through a simple hydrogen reduction of metal–β-diketone precursors in supercritical carbon dioxide.

In spite of this effectiveness, this wet synthesis method often suffers from several drawbacks. Particularly, in order to obtain a homogeneous distribution and very good adhesion of MNPs on CMs, many factors such as the type of organic solvents, concentration of metal precursors, reducing agents, deposition time, and temperature need to be controlled very carefully; therefore, a wet synthesis approach is very limited [79,91].

6.2.2.2 Dry synthesis

Recently, the dry synthesis method has been receiving a great deal of attention due its simplicity, better adhesion, and advantages of least parameters to control. The drawbacks for the synthesis of MNPs/CMs found in the wet synthesis method have been resolved by the dry synthesis method. Since the CMs are highly hydrophobic in nature, prior to the decoration of MNPs via dry synthesis, the creation of the oxygen functional groups (C–OH, C–O–C, C=O, and COOH) on CMs is important [92]. In fact, the presence of oxygen functional groups can play a bridging role between the MNPs and support [93]. However, in case of some CMs such as activated carbon, CNFs, and carbon black, the creation of the oxygen functional groups is very difficult.

Lin et al. [94] described a rapid, solventless method for the decoration of CNTs (multi- and single) with MNPs. In this method, straightforward two-step process involves the dry mixing of a precursor metal salt with CNTs followed by heating in an inert atmosphere. Interestingly, the procedure is scalable and generally applicable to various other carbon substrates (e.g., CNFs, expanded graphite, CNTs, and carbon black) and many metal salts (e.g., Ag, Au, Co, Ni, and Pd acetates). The AgNP-decorated CNTs have been reported as a model system, and the composites were prepared under various mixing techniques, metal loading levels, thermal treatment temperatures, and nanotube oxidative acid treatments. They found that the AgNPs is strongly attached on the surface of the CNTs.

Copper(II) oxide nanoparticles (CuONPs) were successfully decorated on *f*-MWCNTs using copper acetate precursor by a very simple *dry synthesis* method [58]. In this method, $Cu(OAc)_2$ was added into *f*-MWCNTs and mixed well by a mortar and pestle. The homogeneous mixture of *f*-MWCNTs and $Cu(OAc)_2$ was obtained in 10–15 min. Then the mixture was calcinated under argon atmosphere at 350°C for 3 h in a muffle furnace. The microscopic and spectroscopic results showed that the CuONPs are homogeneously dispersed and strongly attached on the MWCNTs. Advantage of CuO/MWCNT nanocatalyst is the recovery of pure MWCNTs from the used nanocatalyst by the simple acid treatment.

Lu et al. [95] have demonstrated a facile dry decoration of GO sheets with aerosol Ag nanocrystals from an arc plasma source using an electrostatic force directed assembly technique at room temperature. In this method, the aerosol silver nanocrystals were produced through physical vaporization of a solid precursor material (Ag wire, 99.999% purity) using a mini-arc plasma source sustained between a tungsten cathode and a graphite anode.

Recently, metallic RuNPs were successfully decorated on GNSs for the very first time by the dry synthesis method [57]. Initially, the bi- and few-layered GNSs were obtained from graphene nanoplatelets (GNPs) via a solution-phase exfoliation (SPE) method by sonicating the GNPs in N-methylpyrrolidone solvent for 12 h at 4°C. Then the resultant GNSs were chemically treated with concentrated H_2SO_4 and HNO_3. The resultant f-GNSs were used for the decoration of RuNPs. Ru(acac)$_3$ was added into f-GNSs and mixed well by a mortar and pestle under ambient conditions. The homogeneous mixture of f-GNSs and Ru(acac)$_3$ was obtained within 10–15 min. The impregnated Ru(acac)$_3$ was thermally decomposed into metallic RuNPs by calcination at 300°C for 3 h under argon atmosphere. Figure 6.3 shows a schematic illustration of the procedure for the preparation of GNS-RuNPs. After catalytic application, the GNS-RuNPs were separated out from the reaction mixture and analyzed by various spectroscopic and microscopic techniques; the results revealed that the GNS-RuNPs are physically as well as chemically stable. Owing to the high stability of used catalyst (u-GNS-RuNPs), it was further applied in various catalytic reactions. RuO$_2$ nanorod hybrid GNS (u-GNS-RuO$_2$NRs) was prepared from u-GNS-RuNPs by simple calcination. Similarly, by increasing calcination temperature, RuO$_2$ nanorods (RuNRs) were decorated on GNP support for the very first time by the simple *dry synthesis* method. In this method, Ru(acac)$_3$ was added into f-GNPs and mixed well by a mortar and pestle until the homogeneous mixture of f-MWCNTs and Ru(acac)$_3$ was obtained. Then the mixture was calcined under nitrogen atmosphere at 500°C for 1 h in a muffle furnace.

Ultrafine ruthenium oxide nanoparticles (RuO$_2$NPs) with an average diameter of 1.3 nm were anchored on GNPs using Ru(acac)$_3$ precursor utilizing the dry synthesis method [96]. Before decorating RuO$_2$NPs, GNPs were chemically functionalized with various oxygen functional groups by simple acid treatment. The authors realized the merit of this method by observing the transmission electron microscopic (TEM) images of the GNPs-RuO$_2$NPs, which showed an excellent attachment of RuO$_2$NPs on GNPs with Ru loading of 2.68 wt%, as confirmed by scanning electron microscopy-energy dispersive X-ray spectroscopy (SEM-EDS). The XPS and XRD of GNPs-RuO$_2$NPs revealed that the chemical state of Ru on GNPs was +4.

Recently, a huge surface area RuO$_2$NPs/SWCNT catalyst was prepared using Ru(acac)$_3$ precursor using the straightforward *dry synthesis* method [97]. The surface area of RuO$_2$/SWCNT was found to be 415.77 m²/g. Interestingly, the size of the RuO$_2$NPs was about 0.9 nm (Figure 6.4). TEM images and Raman spectrum of the resultant material (RuO$_2$/SWCNT) revealed an excellent adhesion and homogeneous dispersion of RuO$_2$NPs on anchoring sites of the SWCNTs. The SEM-EDS result showed that the weight percentage of Ru in RuO$_2$/SWCNT was 13.79%. Chemical state of Ru in RuO$_2$/SWCNT was +4, as confirmed by XPS and XRD analyses.

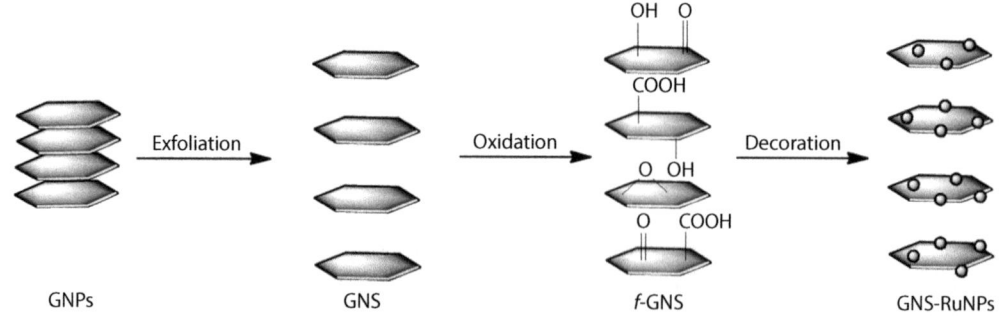

FIGURE 6.3
Schematic illustration for the preparation of GNS-RuNPs. (From Gopiraman, M. et al., *J. Phys. Chem. C*, 117, 23582, 2013.)

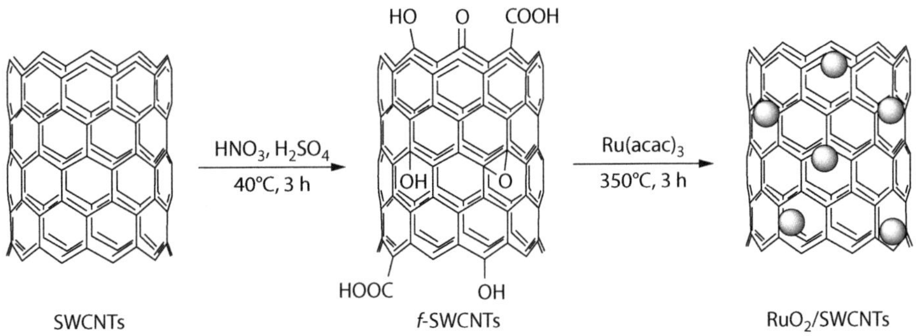

FIGURE 6.4
Schematic illustration for the preparation of RuO$_2$/SWCNTs. (From Heck, R.F. and Nollwy, J.P., *J. Org. Chem.*, 37, 2320, 1972.)

6.3 APPLICATIONS

6.3.1 For Organic Transformations

Catalytic organic reactions play a significant role in chemical industries for the synthesis of organic building blocks, natural products, pharmaceuticals, and agricultural derivatives [98,99]. There are several catalytic systems reported for various organic transformations. Among them, MNP/CM composites have shown higher activity. The outstanding catalytic activity of MNPs/CMs is mainly due to very less aggregation of MNPs, effective dispersion in various solvents, and larger surface area of the MNPs/CMs [73]. Particularly, they have provided the advantages of high atom efficiency, simplified isolation of product, and easy recovery and recyclability of the catalysts [100]. Moreover, the decoration of MNPs onto the CM support has shown more versatility in carrying out the highly selective catalytic processes [101].

Heck and Suzuki–Miyaura reactions are significant methods to produce carbon–carbon (C–C) bonds via cross coupling of aryl halides with olefins or arylboronic acids, which are key synthetic steps in the production of agrochemicals, polymers, and pharmaceuticals [102]. Certainly, Pd-catalyzed olefination of aryl halides (Heck–Mizoroki reaction) is one of the most powerful tools for the construction of C–C bonds [103]. In fact, the Pd-based catalytic systems are highly efficient and they generally offer excellent product yields with good selectivity. Recently, Gil et al. [104] prepared PdNP-supported graphite oxide and used as a catalyst for the Suzuki–Miyaura coupling reaction with higher activity than the commercial Pd–C catalyst. Similarly, Cano et al. [105] prepared CNT-supported PdNPs and used as a catalyst for the Heck and Suzuki coupling reactions. They have found that the prepared CNTs–PdNPs showed very high conversions and good recyclability. Yoon and coworkers [106] described an unusual high catalytic activity of the CNT-supported bimetallic Pd/Rh NPs for the hydrogenation of anthracene. Similarly, Ioni and coworkers [107] found that PdNPs decorated on carbon support such as GO showed higher activity for the Mizoroki–Heck reaction.

Nitrogen-containing heterocycles are prevalent structural motifs in various fields such as biological, pharmaceutical, and material sciences [108]. Particularly, imidazole derivatives are efficient antibacterial, antimalarial, antiviral, antimycobacterial, and antifungal compounds [109,110]. Gopiraman and coworkers have prepared CuO/MWCNT composite using Cu(OAc)$_2$ precursor by a very simple *dry synthesis* method and used such composite as a heterogeneous nanocatalyst for the N-arylation of imidazole (Scheme 6.11) [58]. They found that the catalytic system requires a very low amount

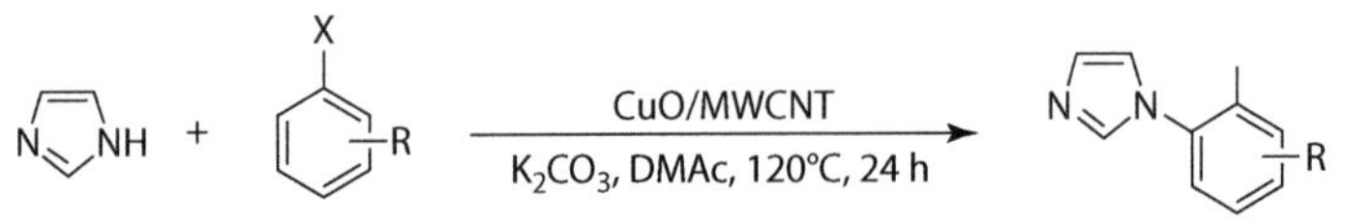

SCHEME 6.11
CuO/MWCNT-catalyzed N-arylation of imidazole with various aryl halides. (From Gopiraman, M. et al., *Carbon*, 62, 135, 2013.)

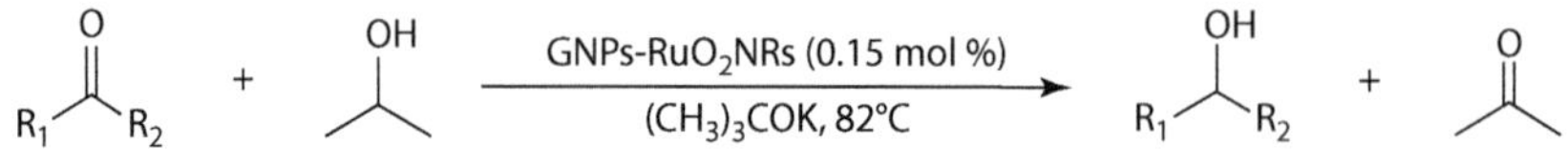

SCHEME 6.12
GNP-RuO$_2$NR-catalyzed transfer hydrogenation of carbonyl compounds. (From Gopiraman, M. et al., *Catal. Sci. Technol.*, 3, 1485, 2013.)

(5 mg, 0.98 mol% of Cu) of CuO/MWCNT catalyst; this is the lowest amount of Cu used for this reaction till date. The catalysis is heterogeneous in nature and the catalyst is reusable. Interestingly, after the catalytic reaction, *f*-MWCNTs were successfully separated out from the used CuO/MWCNT.

It was found that GNP-supported RuO$_2$NRs is an effective and reusable heterogeneous catalyst for the transfer hydrogenation of aromatic aldehydes and ketones in good yield with excellent selectivity (Scheme 6.12) [111]. The catalyst exhibited excellent catalytic activity, chemoselectivity, stability, and reusability in the reduction of aldehydes and ketones. Similarly, Baoqiang et al. [112] have investigated the transfer hydrogenation of alkynes, alkenes, and nitro aromatics using graphene-supported PtNPs.

Pina et al. [113] investigated the activity of gold-based mono- and bimetallic catalysts in the oxidation of tertiary amines to afford the corresponding *N*-oxides. They found that the Au/C catalyst showed an excellent catalytic activity for the oxidation of tertiary amines to the corresponding *N*-oxides; however, other catalysts, namely, Rh/C, Pt/C, AuRh/C, and AuPt/C, are less effective and often require an alkali as a promoter to improve the activity of the catalysts.

Most of the industrially important carbonyl compounds such as aldehydes and ketones are prepared via catalytic organic transformation [114], in which a selective aerial oxidation of primary and secondary alcohols is the prime route. For this purpose, RuNPs were successfully decorated on GNSs for the very first time using a very simple dry synthesis method [57]. The resultant composite (GNS-RuNPs) was used as a nanocatalyst for the aerial oxidation of alcohols after being optimized. The scope of the catalytic system was extended with various aliphatic, aromatic, alicyclic, benzylic, allylic, amino, and heterocyclic alcohols (Scheme 6.13). They found that 0.036 mol% (5 mg) of GNS-RuNPs is enough for the aerial oxidation of alcohols, the lowest amount of catalyst among so far reported. Owing to the high stability of used catalyst (*u*-GNS-RuNPs), it was further applied in transfer hydrogenation after suitable modifications. RuO$_2$ nanorod hybrid GNSs (*u*-GNS-RuO$_2$NRs) were obtained from *u*-GNS-RuNPs by simple calcination. The catalytic activity of *u*-GNS-RuO$_2$NRs toward the transfer hydrogenation of various aromatic, alicyclic, and heterocyclic ketones was found to be excellent. Finally, authors concluded that the GNS-RuNP catalyst is highly efficient, tunable, and versatile.

Aliphatic and aromatic *tert*-amine oxides (amine *N*-oxides) are essential and key components in the formulation of several cosmetic products as well as in biomedical applications [115]. Particularly, *N*-oxides of aromatic amines are extensively used as protecting groups, auxiliary agents, and oxidants in various organic reactions. They are often used as potential cytoximes for the treatment of solid tumors and also as ligands for the preparation of useful transition metal complexes. Ultrafine RuO$_2$NPs with an average diameter of 1.3 nm were anchored on GNPs. The resultant material (GNPs-RuO$_2$NPs) was used as a heterogeneous catalyst for the *N*-oxidation of tertiary amines for the first time [96]. After the optimization of reaction conditions for *N*-oxidation of triethylamine, the scope of the reaction was extended with various aliphatic, alicyclic, and aromatic tertiary amines (Scheme 6.14). The GNPs-RuO$_2$NPs showed an excellent catalytic activity in terms of yields even at a very low amount of Ru catalyst (0.13 mol%). It was found that the GNPs-RuO$_2$NPs were heterogeneous in nature, chemically as well as physically very stable, and reused for up to five times.

The transition metal–catalyzed C–C cross coupling reaction is a key step in the synthesis of organic building blocks, natural products, pharmaceuticals, and agricultural derivatives [116,117].

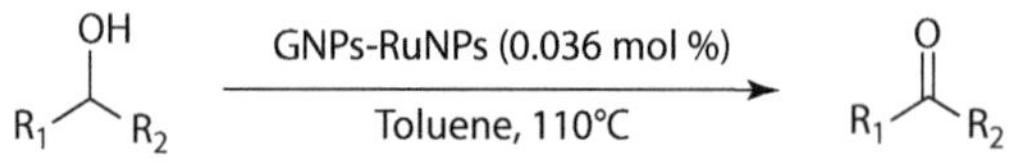

SCHEME 6.13
GNS-RuNP-catalyzed oxidation of alcohols. (From Gopiraman, M. et al., *J. Phys. Chem. C*, 117, 23582, 2013.)

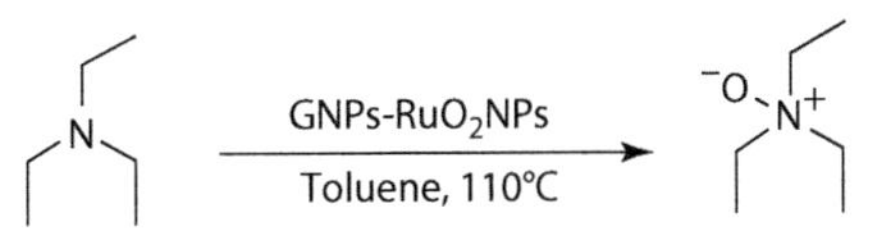

SCHEME 6.14
N-oxidation of tertiary amines catalyzed by GNPs-RuO$_2$NPs. (From Gopiraman, M. et al., *Catal. Sci. Technol.*, 4, 2099, 2014.)

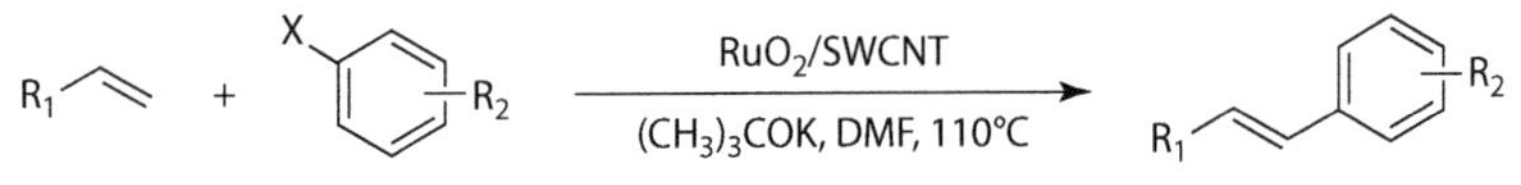

SCHEME 6.15
Substrate scope of the RuO$_2$/SWCNT-catalyzed Heck-type olefination of aryl halides. (From Gopiraman, M. et al., *ACS Catal.*, 4, 2118, 2014.)

A highly efficient SWCNT-supported RuO$_2$NP-based catalytic system for the Heck olefination of aryl halides was developed [97]. The authors found that the substrate scope of the reaction could be efficiently carried out with as low as 0.9 mol% of the supported RuO$_2$ catalyst over a wide range of substrates in moderate to excellent yields with a good TON and TOF values. Interestingly, the iodoarenes, less reactive bromo- and chloroarenes can also be effectively olefinated using this catalytic system (Scheme 6.15). The RuO$_2$/SWCNT is the most active Ru-based heterogeneous catalyst for the Heck olefination of aryl halides among those reported so far. The RuO$_2$/SWCNT is highly regioselective and chemoselective for the Heck olefination reaction. Heterogeneity and reusability of RuO$_2$/SWCNT were found to be good. Finally, it was concluded that the simple synthesis and excellent activity make RuO$_2$/SWCNT an alternate choice to the existing Ru-based catalysts for Heck coupling reactions.

6.3.2 For Environmental Applications

Catalytic applications of carbonaceous materials are not only restricted to organic transformations but also extended to solve environmental and energy-related problems. From the environmental point of view, making pollution-free natural resources (water, air, and land) is an important issue. Among the pollution-related problems, treatment of sewage water is the foremost concern of many chemists, because highly colored effluents (e.g., dyes such as methylene blue, methyl orange, rhodamine b [RhB], acridine orange, crystal violet, and Congo red) from textile, food, leather, and pharmaceutical industries cause severe environmental problems [118]. A recent survey declares that about 280,000 t dyes are released into the environment per annum by the textile industry [119]. Some of the well-established techniques for the degradation of (in)organic pollutants, namely, coagulation, reverse osmosis, membrane filtration, oxidation, ion exchange, complexometric methods, and biological methods, under aerobic and anaerobic circumstances suffer with the disadvantages like requirement of special instruments and/or speciality chemicals [120]. Hence, photocatalytic degradation of organic pollutants is a unique technique for wastewater treatment, which is free from these drawbacks [121]. The key factor in photodegradation is the prevention of recombination of electron–hole pairs of semiconducting photocatalyst materials, which slow down the catalytic performance. Graphene, which is having very high electron mobility, has the ability to prevent the recombination process and hence widely used as a solid support for semiconducting materials. For example, BaCrO$_4$ is a semiconducting material that can excite the electron from the valence band (VB) to the conduction band (CB) even under visible light illumination. But poor photocatalytic performance was experienced with BaCrO$_4$ due to its fast charge recombination. But in graphene-supported BaCrO$_4$, graphene served as an electron collector and transporter and hence enhanced the photocatalytic degradation activity [122].

Similarly, the Au–RGO nanocomposite was prepared by reducing chloroauric acid (HAuCl$_4$) over the RGO surface and was used as an effective visible light–driven photocatalyst for the degradation of RhB [123]. CdS/Al$_2$O$_3$/GO and CdS/ZnO/GO nanocomposites were prepared by the single-step hydrothermal process and were found to be an efficient photocatalyst for the (1) degradation of organic dye (methyl orange), (2) reduction of Cr(VI), and (3) also the release of H$_2$ from water.

The presence of GNSs lengthens the lifetime of the photogenerated charge carriers [124]. Li and coworkers found that loading of RGO narrows down the band gap energy of TiO_2 (P25), which clearly attributed to the chemical bonding between P25 and RGO. Authors demonstrated an enhanced photocatalytic activity of TiO_2–RGO nanocomposite for the degradation of methylene blue [125]. Recently, a review listed out the progress of graphene-based photocatalyst for wastewater treatment [126]. Palladium NPs were successfully decorated over hierarchical carbon structures (CNT/foam) and employed for the reductive dechlorination of harmful carbon tetrachloride [127].

6.3.3 For Energy Applications

Heteroatom (N, S, B, O, P, etc.)-doped, inorganic complex–anchored, and nanocomposite-grafted carbonaceous materials showed excellent physical and electrochemical properties and hence drawn tremendous attention in the past decade. Especially these heteroatom-doped CMs have been exclusively studied for the replacement of Pt-based catalysts in fuel cell applications. The key process in fuel cells is ORR, which is commercially performed using the Pt/C catalyst. But the deactivation of catalyst through CO poisoning and high cost of Pt restrict the practical applicability of the catalyst. To overcome these issues, many researchers proposed less cost and recyclable catalyst for ORR. Among them, nitrogen-doped graphene nanocatalyst gained major attention as a metal-free catalyst and also because of its comparable or better efficacy and exceptional durability. For instance, Shao et al. reported nitrogen-doped graphene as a metal-free catalyst for ORR in fuel cells and also showed high catalytic activity toward oxygen evolution reaction (OER). The authors pictured out the potential application of nitrogen-doped graphene as a bifunctional electrocatalyst for both ORR and OER [42].

The high doping level of nitrogen (10.1 atomic%) was achieved in graphene sheets by the catalyst-free thermal annealing approach and was used as metal-free catalyst for ORR. Authors concluded that the pyridine-like nitrogen component in nitrogen-doped graphene determines the electrocatalytic activity toward ORR but not the nitrogen content [30]. In concurrence with this, only 2.8% nitrogen-loaded graphene, which was prepared by the solvothermal method at 900°C, showed very high catalytic activity toward ORR under alkaline conditions, and also the catalyst illustrated excellent durability [31]. In addition to its electrocatalytic activity toward ORR, nitrogen-doped graphene can also be used as an efficient catalyst for H_2O_2 reduction. The activity of nitrogen-doped graphene is better than the activity of perfect graphene. Further, the catalyst exhibited remarkable durability and selectivity [128].

Like in fuel cells, carbonaceous materials establish its imprint in solar cells as well. In 2012, researchers are able to achieve 8.6% for a prototype solar cell consisting of a wafer of silicon coated with a layer of graphene doped with trifluoromethanesulfonyl-amide (TFSA). Zhu et al. improved the efficiency of perovskite solar cells from 8.81% to 10.15% by the insertion of an ultrathin GQDs layer between perovskite and TiO_2. The authors concluded that the improved cell efficiency was due to the much faster electron extraction in the presence of GQDs (90–106 ps) than without their presence (260–307 ps). Furthermore, they highlighted that GQDs can behave as an ultrafast electron tunnel for optoelectronic devices [129]. Recently, another research group claimed that graphene-based solar cell by ditching the silicon all together led the efficiency to 15.6%. Moreover, graphene-based metal oxide nanocomposites have the potential to throw a novel pathway for the development of low-cost solar cells [130].

Similarly, carbon-based catalytic systems were studied in detail for the splitting of water to generate H_2 fuel, because H_2 is recorded as one of the ultimate clean fuels. Discharge of H_2 from water using a semiconducting photocatalyst and solar energy is a promising and attractive approach. Gao et al. reported the ultrasonic-assisted synthesis procedure for the preparation of TiO_2–GO composites and demonstrated the synergetic effect of GO for H_2 production. In addition, GO–TiO_2 (3 wt.% GO) showed the highest H_2 generation rate (305.6 μmol/h), which was about 13 times higher than those of TiO_2 microsphere [131]. A facile solvothermal technique was developed for the synthesis of TiO_2 (P25)–RGO composites. The resultant microstructures were used for the evolution of H_2 through water splitting [132]. The recent reviews illustrated the complete utility of graphene-based materials/photocatalysts for the production of H_2 either by light-driven water splitting or by means of photoelectrochemical (PEC) technology [132,133].

Carbon-based rechargeable batteries attract many researchers, particularly after the commercialization of the Li-ion battery. Morita et al. successfully demonstrated the practical applicability of layered boron–carbon–nitride (BC_2N) as a negative electrode matrix for rechargeable lithium batteries [134]. Like other heteroatom-doped carbon-based materials such as BC_2N, nitrogen-doped graphene was also expected to have enhanced Li battery properties. As anticipated, the nitrogen-doped graphene layer, which was prepared by the liquid precursor–based CVD technique, showed better activity in lithium battery application. To study the reversible Li-ion intercalation properties, the nitrogen-doped graphene layer was grown directly on Cu current collectors. The authors reported that the activity of nitrogen-doped graphene was almost double as compared to the activity of pristine graphene [39].

Ultracapacitors (UCs) have the ability to both store and release electrical energy. It is mainly based on the electrostatic interactions between ions in the electrolyte and electrodes. Almost all the carbon nanomaterials such as porous CMs [135], CNTs [136–138], and graphene [139–141] have been used as UC electrodes due to their large surface areas and high conductivities. Even then the UC capacitance is not up to the expected level in many cases. Recently, nitrogen-doped graphene was used as UC electrodes to resolve the insufficient capacitance issue whose capacitances ($\sim$280 F/$g_{electrode}$) were about four times larger than those of original graphene-based counterparts. Furthermore, excellent cycle life (>200,000), high power capability, and compatibility with flexible substrates were also observed [142].

Doping of nitrogen to the CMs clearly showed n-type electron doping behavior and can be utilized for potential applications including field-effect transistors (FETs) [143,144]. For example, nitrogen-doped graphene obtained by NH_3 annealing after N^+-ion irradiation of graphene was fabricated for FETs. The process was monitored by the source–drain conductance and back-gate voltage (G_{sd}–V_g) curves [29]. The Dirac point position moved from positive V_g to negative V_g, which suggested the transition of graphene from p-type to n-type after annealing in NH_3. Likewise, concurrent segregation technique was employed for the growth of nitrogen-doped graphene over a sandwiched Ni(C)/B(N)/SiO_2/Si substrate at high temperature. To illustrate the effect of nitrogen doping toward back-gated FETs, both nitrogen-doped graphene and pristine graphene were fabricated separately on 300 nm SiO_2/Si substrates and used for FETs. The authors found that nitrogen-doped graphene displayed an astonishing n-type behavior with an effective band gap of 0.16 eV. It is noteworthy to mention that nitrogen doping could tune the electrical properties of graphene [25].

6.4 CONCLUSIONS

CMs are highly attractive for a wide range of potential applications. In order to tune these materials toward various applications, surface functionalization of these materials is highly essential. Various covalent and noncovalent functionalization methods have been developed for preparing CM-supported nanostructured materials with appropriate structures and properties. Owing to the various sensitive applications of these surface functionalized materials, they have opened a new field of nanotechnology. In this chapter, we have summarized the recent progress in the research on covalent and noncovalent functionalization of CMs such as CNTs, carbon fibers, graphene and fullerene, fabrication methods for constructing the functionalized CMs into organic transformations, energy conversion, and environmental applications.

REFERENCES

1. Serp, P. Surface inorganic chemistry and metal-based catalysis, In *Comprehensive Inorganic Chemistry II*, 2nd edn, J. Reedijk and K. Poeppelmeier, eds., Elsevier, Amsterdam, the Netherlands, 2013, Vol. 7, pp. 323–369.
2. Reinoso, F.R., Escribano, A.S. Carbon as catalyst support, In *Carbon Materials for Catalysis*, P. Serp and J.L. Figueiredo, eds., John Wiley and Sons, Inc., New York, 2009, pp. 131–155.
3. Georgakilas, V., Otyepka, M., Bourlinos, A.B., Chandra, V., Kim, N., Kemp, K.C., Hobza, P., Zboril, R., Kim, K.S. 2012. Functionalization of graphene: Covalent and noncovalent approaches, derivatives and applications. *Chem. Rev.* 112:6156–6214.

4. Wei, D., Liu, Y., Wang, Y., Zhang, H., Huang, L., Yu, G. 2009. Synthesis of N-doped graphene by chemical vapor deposition and its electrical properties. *Nano Lett.* 9:1752–1758.

5. Thess, A., Lee, R., Nikolaev, P., Dai, H., Petit, P., Robert, J., Xu, C. et al. 1996. Crystalline ropes of metallic carbon nanotubes. *Science* 273:483–487.

6. Spitalsky, Z., Aggelopoulos, C., Tsoukleri, G., Tsakiroglou, C., Parthenios, J., Georga, S., Krontiras, C., Tasis, D., Papagelis, K., Galiotis, C. 2009. The effect of oxidation treatment on the properties of multi-walled carbon nanotube thin films. *Mater. Sci. Eng. B Adv.* 165:135–138.

7. Jiang, K.L., Li, Q.Q., Fan, S.S. 2002. Nanotechnology: Spinning continuous carbon nanotube yarns—Carbon nanotubes weave their way into a range of imaginative macroscopic applications. *Nature* 419:801.

8. Li, Z.F., Luo, G.H., Wei, F., Huang, Y. 2006. Microstructure of carbon nanotubes/ PET conductive composites fibers and their properties. *Compos. Sci. Technol.* 66:1022–1029.

9. Dassios, K.G., Galiotis, C. 2012. Polymer-nanotube interaction in MWCNT/ poly(vinyl alcohol) composite mats. *Carbon* 50:4291–4294.

10. Kratschmer, W., Lamb, L.D., Fostiropoulos, K., Huffman, D.R. 1990. Solid C60: A new form of carbon. *Nature* 347:354–358.

11. Dyke, C.A., Tour, J.M. 2004. Covalent functionalization of single-walled carbon nanotubes for materials applications. *J. Phys. Chem. A* 108:11151–11159.

12. Bilalis, P., Katsigiannopoulos, D., Avgeropoulos, A., Sakellariou, G. 2014. Non-covalent functionalization of carbon nanotubes with polymers. *RSC Adv.* 4:2911–2934.

13. Schnorr, J.M., van der Zwaag, D., Walish, J.J., Weizmann, Y., Swager, T.M. 2013. Sensory arrays of covalently functionalized single-walled carbon nanotubes for explosive detection. *Adv. Funct. Mater.* 23:5285–5291.

14. Zhang, Y., Ge, J., Wang, L., Wang, D., Ding, F., Tao, X., Chen, W. 2013. No high performance oxygen reduction reaction. *Sci. Rep.* 3:2771–2778.

15. Feng, J., Sui, J., Cai, W., Gao, Z. 2008. Microstructure and mechanical properties of carboxylated carbon nanotubes/poly(L-lactic acid) composite. *J. Compos. Mater.* 42:1587–1595.

16. Panchakarla, L.S., Govindaraj, A. 2008. Covalent and non-covalent functionalization and solubilization of double-walled carbon nanotubes in nonpolar and aqueous media. *J. Chem. Sci.* 120:607–611.

17. Xue, Y.H., Liu, J., Chen, H., Wang, R.G., Li, D.Q., Qu, J., Dai, L.M. 2012. Nitrogen-doped graphene foams as metal-free counter electrodes in high-performance dye-sensitized solar cells. *Angew. Chem. Int. Ed.* 51:12124–12127.

18. Gong, K., Du, F., Xia, Z., Durstock, M., Dai, L. 2009. Nitrogen-doped carbon nanotube arrays with high electrocatalytic activity for oxygen reduction. *Science* 323:760–764.

19. Li, Q., Zhang, S., Dai, L., Li, L.S. 2012. Nitrogen-doped colloidal graphene quantum dots and their size-dependent electrocatalytic activity for the oxygen reduction reaction. *J. Am. Chem. Soc.* 134:18932–18935.

20. Ai, K., Liu, Y., Ruan, C., Lu, L., Lu, G. 2013. sp^2 C-dominant N-doped carbon submicrometer spheres with a tunable size: A versatile platform for highly efficient oxygen-reduction catalysts. *Adv. Mater.* 25:998–1003.

21. Wang, H., Maiyalagan, T., Wang, X. 2012. Review on recent progress in nitrogen-doped graphene: Synthesis, characterization, and its potential applications. *ACS Catal.* 2:781–794.

22. Qu, L., Liu, Y., Baek, J.B., Dai, L. 2010. Nitrogen-doped graphene as efficient metal-free electrocatalyst for oxygen reduction in fuel cells. *ACS Nano* 4:1321–1326.

23. Zhang, C., Fu, L., Liu, N., Liu, M., Wang, Y., Liu, Z. 2011. Synthesis of nitrogen-doped graphene using embedded carbon and nitrogen sources. *Adv. Mater.* 23:1020–1024.

24. Deng, D., Pan, X., Yu, L., Cui, Y., Jiang, Y., Qi, J., Li, W.X. et al. 2011. Toward N-doped graphene *via* solvothermal synthesis. *Chem. Mater.* 23:1188–1193.

25. Panchakarla, L.S., Subrahmanyam, K.S., Saha, S.K., Govindaraj, A., Krishnamurthy, H.R., Waghmare, U.V., Rao, C.N.R. 2009. Synthesis, structure, and properties of boron- and nitrogen-doped graphene. *Adv. Mater.* 21:4726–4730.

26. Ghosh, A., Late, D.J., Panchakarla, L.S., Govindaraj, A., Rao, C.N.R. 2009. NO_2 and humidity sensing characteristics of few-layer graphenes. *J. Exp. Nanosci.* 4:313–322.

27. Guo, B., Liu, Q., Chen, E., Zhu, H., Fang, L., Gong, J.R. 2010. Controllable N-doping of graphene. *Nano Lett.* 10:4975–4980.

28. Geng, D., Chen, Y., Chen, Y., Li, Y., Li, R., Sun, X., Ye, S., Knights, S. 2011. High oxygen-reduction activity and durability of nitrogen-doped graphene. *Energy Environ. Sci.* 4:760–764.

29. Shao, Y., Zhang, S., Engelhard, M.H., Li, G., Shao, G., Wang, Y., Liu, J., Aksay, I.A., Lin, Y. 2010. Nitrogen-doped graphene and its electrochemical applications. *J. Mater. Chem.* 20:7491–7496.

30. Reina, A., Jia, X.T., Ho, J., Nezich, D., Son, H., Bulovic, V., Dresselhaus, M.S., Kong, J. 2009. Large area, few-layer graphene films on arbitrary substrates by chemical vapor deposition. *Nano Lett.* 9:30–35.

31. Zhang, J., Zou, H., Qing, Q., Yang, Y., Li, Q., Liu, Z., Guo, X., Du, Z. 2003. Effect of chemical oxidation on the structure of single-walled carbon nanotubes. *J. Phys. Chem. B* 107:3712–3718.

32. Che, G., Lakshmi, B.B., Martin, C.R., Fisher, E.R., Ruoff, R.S. 1998. Chemical vapor deposition based synthesis of carbon nanotubes and nanofibers using a template method. *Chem. Mater.* 10:260–267.

33. Maldonado, S., Morin, S., Stevenson, K.J. 2006. Structure, composition, and chemical reactivity of carbon nanotubes by selective nitrogen doping. *Carbon* 44:1429–1437.

34. Luo, Z., Lim, S., Tian, Z., Shang, J., Lai, L., MacDonald, B., Fu, C., Shen, Z., Yu, T., Lin, J. 2011. Pyridinic N doped graphene: Synthesis, electronic structure, and electrocatalytic property. *J. Mater. Chem.* 21:8038–8044.

35. Subrahmanyam, K.S., Panchakarla, L.S., Govindaraj, A., Rao, C.N.R. 2009. Simple method of preparing graphene flakes by an arc-discharge method. *J. Phys. Chem. C* 113:4257–4259.

36. Reddy, A.L.M., Srivastava, A., Gowda, S.R., Gullapalli, H., Dubey, M., Ajayan, P.M. 2010. Synthesis of nitrogen-doped graphene films for lithium battery application. *ACS Nano* 4:6337–6342.

37. Jin, Z., Yao, J., Kittrell, C., Tour, J.M. 2011. Large-scale growth and characterizations of nitrogen-doped monolayer graphene sheets. *ACS Nano* 5:4112–4117.

38. Sheng, Z.H., Shao, L., Chen, J.J., Bao, W.J., Wang, F.B., Xia, X.H. 2011. Catalyst-free synthesis of nitrogen-doped graphene *via* thermal annealing graphite oxide with melamine and its excellent electrocatalysis. *ACS Nano* 5:4350–4358.

39. Imran, J.R., Rajalakshmi, N., Ramaprabhu, S. 2010. Nitrogen doped graphene nanoplatelets as catalyst support for oxygen reduction reaction in proton exchange membrane fuel cell. *J. Mater. Chem.* 20:7114–7117.

40. Wang, D.W., Gentle, I.R., Lu, G.Q. 2010. Enhanced electrochemical sensitivity of PtRh electrodes coated with nitrogen-doped graphene. *Electrochem. Commun.* 12:1423–1427.

41. Lin, Y.C., Lin, C.Y., Chiu, P.W. 2010. Controllable graphene N-doping with ammonia plasma. *Appl. Phys. Lett.* 96:133110–133113.

42. Xiong, C., Wei, Z., Hu, B., Chen, S., Li, L., Guo, L., Ding, W., Liu, X., Ji, W., Wang, X. 2012. Nitrogen-doped carbon nanotubes as catalysts for oxygen reduction reaction. *J. Power Sources* 215:216–220.

43. Kundu, S., Nagaiah, T.C., Xia, W., Wang, Y., Dommele, S.V., Bitter, J.H., Santa, M., Grundmeier, G., Bron, M., Schuhmann, W., Muhler, M. 2009. Electrocatalytic activity and stability of nitrogen-containing carbon nanotubes in the oxygen reduction reaction. *J. Phys. Chem. C* 113:14302–14310.

44. Rajeshkumar, V., Chan, F.W., Chuang, S.C. 2012. Palladium-catalyzed and hybrid acids-assisted synthesis of [60] fulleroazepines in one pot under mild conditions: Annulation of n-sulfonyl-2-aminobiaryls with [60] fullerene through sequential C–H bond activation, C–C and C–N bond formation. *Adv. Synth. Catal.* 354:2473–2483.

45. Bosi, S., Ros, T.D., Spalluto, G., Prato, M. 2003. Fullerene derivatives: An attractive tool for biological applications. *Eur. J. Med. Chem.* 38:913–923.

46. Maggini, M., Scorrano, G., Prato, M. 1993. Addition of azomethine ylides to C60: Synthesis, characterization, and functionalization of fullerene pyrrolidines. *J. Am. Chem. Soc.* 115:9798–9799.

47. Prato, M., Maggini, M. 1998. Fulleropyrrolidines: A family of full-fledged fullerene derivatives. *Acc. Chem. Res.* 31:519–526.

48. Moghaddam, M.J., Taylor, S., Gao, M., Huang, S., Dai, L., McCall, M.J. 2004. Highly efficient binding of DNA on the sidewalls and tips of carbon nanotubes using photochemistry. *Nano Lett.* 4:89–93.

49. Yang, Z., Yao, Z., Li, G., Fang, G., Nie, H., Liu, Z., Zhou, X., Chen, X., Huang, S. 2012. Sulfur-doped graphene as an efficient metal-free cathode catalyst for oxygen reduction. *ACS Nano* 6:205–211.

50. Wang, X., Wang, J., Wang, D., Dou, S., Ma, Z., Wu, J., Tao, L., Shen, A., Ouyang, C., Liu, Q., Wang, S. 2014. One-pot synthesis of nitrogen and sulfur co-doped graphene as efficient metal-free electrocatalysts for the oxygen reduction reaction. *Chem. Commun.* 50:4839–4842.

51. Liang, J., Jiao, Y., Jaroniec, M., Qiao, S.Z. 2012. Sulfur and nitrogen dual-doped mesoporous graphene electrocatalyst for oxygen reduction with synergistically enhanced performance. *Angew. Chem. Int. Ed.* 51:11496–11500.

52. Su, Y., Zhang, Y., Zhuang, X., Li, S., Wu, D., Zhang, F., Feng, X. 2009. Low-temperature synthesis of nitrogen/sulfur co-doped three-dimensional graphene frameworks as efficient metal-free electrocatalyst for oxygen reduction reaction. *Carbon* 62:296–301.

53. Xu, J., Dong, G., Jin, C., Huang, M., Guan, L. 2013. Sulfur and nitrogen co-doped, few-layered graphene oxide as a highly efficient electrocatalyst for the oxygen-reduction reaction. *ChemSusChem* 6:493–499.

54. Wang, C., Su, K., Wan, W., Guo, H., Zhou, H., Chen, J., Zhang, X., Huang, Y. 2014. High sulfur loading composite wrapped by 3D nitrogen-doped graphene as a cathode material for lithium–sulfur batteries. *J. Mater. Chem. A* 2:5018–5023.

55. Guo, P., Xiao, F., Liu, Q., Liu, H., Guo, Y., Gong, J.R., Wang, S., Liu, Y. 2013. One-pot microbial method to synthesize dual-doped graphene and its use as high-performance electrocatalyst. *Sci. Rep.* 3:3499.

56. Hummers, W.S., Offeman, R.E. 1958. Preparation of graphitic oxide. *J. Am. Chem. Soc.* 80:1339.

57. Gopiraman, M., Babu, S.G., Khatri, Z., Kai, W., Kim, Y.A., Endo, M., Karvembu, R., Kim, I.S. 2013. Dry synthesis of easily tunable nano ruthenium supported on graphene: Novel nanocatalysts for aerial oxidation of alcohols and transfer hydrogenation of ketones. *J. Phys. Chem. C* 117:23582–23596.

58. Gopiraman, M., Babu, S.G., Khatri, Z., Kai, W., Kim, Y.A., Endo, M., Karvembu, R., Kim, I.S. 2013. An efficient, reusable copper-oxide/CNT catalyst for N-arylation of imidazole. *Carbon* 62:135–148.

59. Bailey, M.M., Heddleston, J.M., Davis, J., Staymates, J.L., Walker, A.R.H. 2014. Functionalized, carbon nanotube material for the catalytic degradation of organophosphate nerve agents. *Nano Res.* 7:390–398.

60. Gheorghiu, C.C., Machado, B.F., de Lecea, C.S.M., Gouygou, M., Martinez, M.C.R., Serp, P. 2014. Chiral rhodium complexes covalently anchored on carbon nanotubes for enantioselective hydrogenation. *Dalton Trans.* 43:7455–7463.

61. Banerjee, S., Wong, S.S. 2002. Functionalization of carbon nanotubes with a metal-containing molecular complex. *Nano Lett.* 2:49–53.

62. Oliveira, P., Ramos, A.M., Fonseca, I., do Rego, A.B., Vital, J. 2005. Oxidation of limonene over carbon anchored transition metal Schiff base complexes: Effect of the linking agent. *Catal. Today* 102–103:67–77.

63. Baleizao, C., Gigante, B., Garcia, H., Corma, A. 2004. Vanadyl salen complexes covalently anchored to single-wall carbon nanotubes as heterogeneous catalysts for the cyanosilylation of aldehydes. *J. Catal.* 221:77–84.

64. Silva, A.R., Figueiredo, J.L., Freire, C., de Castro, B. 2005. Copper(II) acetylacetonate anchored onto an activated carbon as a heterogeneous catalyst for the aziridination of styrene. *Catal. Today* 102–103:154–159.

65. Valente, A., do Rego, A.M.B., Reis, M.J., Silva, I.F., Ramos, A.M., Vital, J. 2001. Oxidation of pinane using transition metal acetylacetonate complexes immobilized on modified activated carbon. *Appl. Catal. A: Gen.* 207:221–228.

66. Silva, A.R., Martins, M., Freitas, M.M.A., Figueiredo, J.L., Freire, C., de Castro, B. 2004. Anchoring of copper(II) acetylacetonate onto an activated carbon functionalized with a triamine. *Eur. J. Inorg. Chem.* 2004:2027–2035.

67. Jarrais, B., Silva, A.R., Freire, C. 2005. Anchoring of vanadyl acetylacetonate onto amine-functionalized activated carbons: Catalytic activity in the epoxidation of an allylic alcohol. *Eur. J. Inorg. Chem.* 2005:4582–4589.

68. Matrab, T., Chancolon, J., Lhermite, M.M., Rouzaud, J., Deniau, G., Boudou, J., Chehimi, M.M., Delamar, M. 2006. Atom transfer radical polymerization (ATRP) initiated by aryl diazonium salts: A new route for surface modification of multiwalled carbon nanotubes by tethered polymer chains. *Colloids Surf. A Physicochem. Eng. Asp.* 287:217–221.

69. Li, X., Du, Z., Zhang, C., Li, H., Zou, W. 2013. Preparation of polyaniline grafted multiwalled carbon nanotubes and conductive application in polyetherimide. *Polym. Adv. Technol.* 24:151–156.

70. Jeon, I.Y., Kang, S.W., Tan, L.S., Baek, J.B. 2014. Two-Step direct arylation for synthesis of naphthalenediimide-based conjugated polymer. *J. Polym. Sci. Part A: Polym. Chem.* 52:1401–1407.

71. Ogasawara, T., Ishida, Y., Ishikawa, T., Yokota, R. 2004. Characterization of multi-walled carbon nanotube/phenylethynyl terminated polyimide composites. *Composite, Part A* 35:67–74.

72. Sanjib, B., Lawrence, T.D. 2009. A novel approach to create a highly ordered monolayer film of graphene nanosheets at the liquid–liquid interface. *Nano Lett.* 9:167–172.

73. Machadoa, B.F., Serp, P. 2012. Graphene-based materials for catalysis. *Catal. Sci. Technol.* 2:54–75.

74. Yan, J., Wei, T., Shao, B., Fan, Z., Qian, W., Zhang, M., Wei, F. 2010. Preparation of a graphene nanosheet/polyaniline composite with high specific capacitance. *Carbon* 48:487–493.

75. Kou, R., Shao, Y.Y., Wang, D.H., Engelhard, M.H., Kwak, J.H., Wang, J., Viswanathan, V.V. et al. 2009. Enhanced activity and stability of Pt catalysts on functionalized graphene sheets for electrocatalytic oxygen reduction. *Electrochem. Commun.* 11:954–957.

76. Jafri, R.I., Rajalakshmi, N., Ramaprabhu, S. 2010. Nitrogen doped graphene nanoplatelets as catalyst support for oxygen reduction reaction in proton exchange membrane fuel cell. *J. Mater. Chem.* 20:7114–7117.

77. Dong, L.F., Gari, R.R.S., Li, Z., Craig, M.M., Hou, S.F. 2010. Graphene-supported platinum and platinum-ruthenium nanoparticles with high electrocatalytic activity for methanol and ethanol oxidation. *Carbon* 48:781–787.

78. Dai, L. 2012. Functionalization of graphene for efficient energy conversion and storage. *Acc. Chem. Res.* 46:31–42.

79. Xiaozhu, Z., Xiao, H., Xiaoying, Q., Shixin, W., Can, X., Freddy, Y.C.B., Qingyu, Y., Peng, C., Hua, Z. 2009. In situ synthesis of metal nanoparticles on single-layer graphene oxide and reduced graphene oxide surfaces. *J. Phys. Chem. C* 113:10842–10846.

80. Bhattacharya, A., Bhattacharya, S., Majumder, C., Das, G.P. 2010. Transition-metal decoration enhanced room-temperature hydrogen storage in a defect-modulated graphene sheet. *J. Phys. Chem. C* 114:10297–10301.

81. Hector, A.B., Jie, M., Zunfeng, L., Randall, M.S., Zhenan, B., Yongsheng, C. 2008. Evaluation of solution-processed reduced graphene oxide films as transparent conductors. *ACS Nano* 3:463–470.

82. Omid, A. 2010. Graphene nanomesh by ZnO nanorod photocatalysts. *ACS Nano* 4:4174–4180.

83. Jinbin, L., Yulin, L., Yueming, L., Jinghong, L., Zhaoxiang, D. 2010. Noncovalent DNA decorations of graphene oxide and reduced graphene oxide toward water-soluble metal–carbon hybrid nanostructures *via* self-assembly. *J. Mater. Chem.* 20:900–906.

84. Hee, C.C., Moonsub, S., Sarunya, B., Hongjie, D. 2002. Spontaneous reduction of metal ions on the sidewalls of carbon nanotubes. *J. Am. Chem. Soc.* 124:9058–9059.

85. Minati, L., Antonini, V., Dalla Serra, M., Speranza, G. 2012. Multifunctional branched gold–carbon nanotube hybrid for cell imaging and drug delivery. *Langmuir* 28:15900–15906.

86. Makala, S.R., Saurabh, A., Nikki, B., Ganapathiraman, R. 2006. Microwave-assisted single-step functionalization and in situ derivatization of carbon nanotubes with gold nanoparticles. *Chem. Mater.* 18:1390–1393.

87. Bhalchandra, A.K., Suman, S., Shivappa, B.H., Vijayamohanan, K.P. 2008. Highly selective catalytic hydrogenation of arenes using rhodium nanoparticles supported on multiwalled carbon nanotubes. *J. Phys. Chem. C* 112:13317–13319.

88. Byunghoon, Y., Horng-Bin, P., Chien, M.W. 2009. Relative catalytic activities of carbon nanotube-supported metallic nanoparticles for room-temperature hydrogenation of benzene. *J. Phys. Chem. C* 113:1520–1525.

89. Kamat, P.V. 2010. Graphene-based nanoarchitectures anchoring semiconductor and metal nanoparticles on a two-dimensional carbon support. *J. Phys. Chem. Lett.* 1:520–527.

90. Xiang-Rong, Y., Yuehe, L., Chongming, W., Mark, H.E., Yong, W., Chien, M.W. 2004. Supercritical fluid synthesis and characterization of catalytic metal nanoparticles on carbon nanotubes. *J. Mater. Chem.* 14:908–913.

91. Yang, J., Tian, C., Wang, L., Fu, H. 2001. An effective strategy for small-sized and highly-dispersed palladium nanoparticles supported on graphene with excellent performance for formic acid oxidation. *J. Mater. Chem.* 21:3384–3390.

92. Fan, X., Peng, W., Li, J., Li, X., Wang, S., Zhang, G., Zhang, F. 2008. Deoxygenation of exfoliated graphite oxide under alkaline conditions: A green route to graphene preparation. *Adv. Mater.* 20:4490–4493.

93. Kuila, T., Bhadra, S., Yao, D., Kim, N.H., Bose, S., Lee, J.H. 2010. Recent advances in graphene based polymer composites. *Prog. Polym. Sci.* 35:1350–1375.

94. Lin, Y., Watson, K.A., Fallbach, M.J., Ghose, S., Smith, J.G., Delozier, D.M., Cao, W., Crooks, R.E., Connell, J.W. 2009. Rapid, solventless, bulk preparation of metal nanoparticle-decorated carbon nanotubes. *ACS Nano* 3:871–884.

95. Lu, G., Mao, S., Park, S., Ruoff, R.S., Chen, J. 2009. Facile, noncovalent decoration of graphene oxide sheets with nanocrystals. *Nano Res.* 2:192–200.

96. Gopiraman, M., Bang, H., Babu, S.G., Wei, K., Karvembu, R., Kim, I.S. 2014. Catalytic *N*-oxidation of tertiary amines on RuO_2NPs anchored graphene nanoplatelets. *Catal. Sci. Technol.* 4:2099–2106.

97. Gopiraman, M., Karvembu, R., Kim, I.S. 2014. Highly active, selective and reusable RuO_2/SWCNT catalyst for Heck olefination of aryl halides. *ACS Catal.* 4:2118–2129.

98. Diederich, F., Stang, P.J. 1998. *Metal-Catalyzed Cross-Coupling Reactions*. Wiley-VCH, Weinheim, Germany.

99. Baur, J.A., Sinclair, D.A. 2006. Therapeutic potential of resveratrol: The in vivo evidence. *Nat. Rev. Drug Discov.* 5:493–506.

100. Schaetz, A., Zeltner, M., Stark, W.J. 2012. Carbon modifications and surfaces for catalytic organic transformations. *ACS Catal.* 2:1267–1284.

101. Nethravathi, C., Anumol, E.A., Rajamathi, M., Ravishankar, N. 2011. Highly dispersed ultrafine Pt and PtRu nanoparticles on graphene: Formation mechanism and electrocatalytic activity. *Nanoscale* 3:569–571.

102. Heck R.F., Nollwy J.P. 1972. Palladium-catalyzed vinylic hydrogen substitution reactions with aryl, benzyl, and styryl halides. *J. Org. Chem.* 37:2320–2322.

103. Gaudin, J.M. 2000. Synthesis and organoleptic properties of p-menthane lactones. *Tetrahedron Lett.* 56:4769–4776.

104. Gil, M.S., Luigi, R., Peter, S., Willi, B., Rolf, M. 2009. Palladium nanoparticles on graphite oxide and its functionalized graphene derivatives as highly active catalysts for the Suzuki–Miyaura coupling reaction. *J. Am. Chem. Soc.* 131:8262–8270.

105. Cano, M., Benito, A., Maser, W.K., Urriolabeitia, E.P. 2011. One-step microwave synthesis of palladium–carbon nanotube hybrids with improved catalytic performance. *Carbon* 49:652–658.

106. Yoon, B., Wai, C.M. 2005. Microemulsion-templated synthesis of carbon nanotube-supported Pd and Rh nanoparticles for catalytic applications. *J. Am. Chem. Soc.* 127:17174–17175.

107. Ioni, Y.V., Lyubimov, S.E., Davankov, V.A., Gubin, S.P. 2013. The use of palladium nanoparticles supported on graphene oxide in the Mizoroki-Heck reaction. *Russ. J. Inorg. Chem.* 58:392–394.

108. Gungor, T., Fouquet, A., Teulon, J.M., Provost, D., Cazes, M., Cloarec, A. 1992. Cardiotonic agents–Synthesis and cardiovascular properties of novel 2-arylbenzimidazoles and azabenzimidazoles. *J. Med. Chem.* 35:4455–4463.

109. Leeson, P.D., Springthorpe, B. 2007. The influence of drug-like concepts on decision-making in medicinal chemistry. *Nat. Rev. Drug. Discov.* 6:881–890.

110. Zhang, W., Ramamoorthy, Y., Kilicarslan, T., Nolte, H., Tyndale, R.F., Sellers, E.M. 2002. Inhibition of cytochrome P450 by antifungal imidazole derivatives. *Drug. Metab. Dispos.* 30:314–318.

111. Gopiraman, M., Babu, S.G., Khatri, Z., Kai, W., Endo, M., Karvembu, R., Kim, I.S. 2013. Facile and homogeneous decoration of RuO_2 nanorods on graphene nanoplatelets for transfer hydrogenation of carbonyl compounds. *Catal. Sci. Technol.* 3:1485–1489.

112. Baoqiang, S., Lei, H., Tingting, Y., Xueqin, C., Hongwei, G. 2012. Highly-dispersed ultrafine Pt nanoparticles on graphene as effective hydrogenation catalysts. *RSC Adv.* 2:5520–5523.

113. Pina, C.D., Falletta, E., Rossi, M. 2007. Selective oxidation of tertiary amines on gold catalysts. *Top. Catal.* 44:325–329.

114. Punniyamurthy, T., Velusamy, S., Iqbal, J. 2005. Recent advances in transition metal catalyzed oxidation of organic substrates with molecular oxygen. *Chem. Rev.* 105: 2329–2364.

115. Campeau, L.C., Stuart, D.R., Leclerc, J.P., Laperle, M.B., Villemure, E., Sun, H.Y., Lasserre, S., Guimond, N., Lecavallier, M., Fagnou, K. 2009. Palladium-catalyzed direct arylation of azine and azole N-oxides: Reaction development, scope and applications in synthesis. *J. Am. Chem. Soc.* 131:3291–3306.

116. Vanrheenen, V., Cha, D.Y., Hartley, W.M. 1978. Catalytic osmium tetroxide oxidation of olefins: Cis-1,2-cyclohexanediol. *Org. Synth.* 6:342–347.

117. Babu, S.G., Neelakandeswari, N., Dharmaraj, N., Jackson, S.D., Karvembu, R. 2013. Copper(II) oxide on aluminosilicate mediated Heck coupling of styrene with aryl halides in water. *RSC Adv.* 3:7774–7781.

118. Patil, P.S., Shedbalekar, U.U., Kalyani, D.C., Jadhav, J.P. 2008. Biodegradation of reactive blue 59 by isolated bacterial consortium PMB11. *J. Ind. Microbiol. Biotechnol.* 35:1181–1190.

119. Jin, X.C., Liu, G.Q., Xu, Z.H., Tao, W.Y. 2007. Decolorization of a dye industry effluent by *Aspergillus fumigatus* XC6. *Appl. Microbiol. Biotechnol.* 74:239–243.

120. Slokar, Y.M., Marechal, A.M.L. 1998. Methods of decoloration of textile wastewaters. *Dyes Pigm.* 37:335–356.

121. Kepa, U., Mazanek, E.S., Stepniak, L. 2008. The use of the advanced oxidation process in the ozone + hydrogen peroxide system for the removal of cyanide from water. *Desalination* 223:187–193.

122. Gawande, S.B., Thakare, S.R. 2013. Synthesis of visible light active graphene-modified $BaCrO_4$ nanocomposite photocatalyst. *Int. Nano Lett.* 3:1–8.

123. Xiong, Z., Zhang, L.L., Ma, J., Zhao, X.S. 2010. Photocatalytic degradation of dyes over graphene–gold nanocomposites under visible light irradiation. *Chem. Commun.* 46:6099–6101.

124. Khan, Z., Chetia, T.R., Vardhaman, A.K., Barpuzary, D., Sastri, C.V., Qureshi, M. 2012. Visible light assisted photocatalytic hydrogen generation and organic dye degradation by CdS–metal oxide hybrids in presence of graphene oxide. *RSC Adv.* 2:12122–12128.

125. Zhang, H., Lv, X., Li, Y., Wang, Y., Li, J. 2010. P25-graphene composite as a high performance photocatalyst. *ACS Nano* 4:380–386.

126. Lina, Z., Wallera, G.H., Liua, Y., Liua, M., Wong, C. 2013. 3D nitrogen-doped graphene prepared by pyrolysis of graphene oxide with polypyrrole for electrocatalysis of oxygen reduction reaction. *Nano Energy* 2:241–248.

127. Vijwani, H., Agrawal, A., Mukhopadhyay, S.M. 2012. Dechlorination of environmental contaminants using a hybrid nanocatalyst: Palladium nanoparticles supported on hierarchical carbon nanostructures. *J. Nanotechnol.* 2012:1–9.

128. Xie, G., Zhang, K., Guo, B., Liu, Q., Fang, L., Gong, J.R. 2013. Graphene-based materials for hydrogen generation from light-driven water splitting. *Adv. Mater.* 25:3820–3839.
129. Zhu, Z., Ma, J., Wang, Z., Mu, C., Fan, Z., Du, L., Bai, Y., Fan, L., Yan, H., Phillips, D.L., Yangs, S. 2014. Efficiency enhancement of perovskite solar cells through fast electron extraction: The role of graphene quantum dots. *J. Am. Chem. Soc.* 136:3760–3763.
130. Wang, J.T.W., Ball, J.M., Barea, E.M., Abate, A., Webber, J.A.A., Huang, J., Saliba, M., Sero, I.M., Bisquert, J., Snaith, H.J., Nicholas, R.J. 2014. Low-temperature processed electron collection layers of graphene/TiO_2 nanocomposites in thin film perovskite solar cells. *Nano Lett.* 14:724–730.
131. Gao, P., Sun, D.D. 2013. Ultrasonic preparation of hierarchical graphene-oxide/ TiO_2 composite microspheres for efficient photocatalytic hydrogen production. *Chem.–Asian J.* 8:2779–2786.
132. Cheng, P., Yang, Z., Wang, H., Cheng, W., Chen, M., Shangguan, W., Ding, G. 2012. TiO_2-graphene nanocomposites for photocatalytic hydrogen production from splitting water. *Int. J. Hydrogen Energy* 37:2224–2230.
133. Xiang, Q., Yu, J. 2013. Graphene-based photocatalysts for hydrogen generation. *J. Phys. Chem. Lett.* 4:753–759.
134. Morita, M., Hanada, T., Tsutsumi, H., Matsuda, Y., Kawaguchi, M. 1992. Layeredstructure BC_2N as a negative electrode matrix for rechargeable lithium batteries. *J. Electrochem. Soc.* 139:1227–1230.
135. Chmiola, J., Yushin, G., Gogotsi, Y., Portet, C., Simon, P., Taberna, P.L. 2006. Anomalous increase in carbon capacitance at pore sizes less than 1 nanometer. *Science* 313:1760–1763.
136. An, K.H., Kim, W.S., Park, Y.S., Moon, J.M., Bae, D.J., Lim, S.C., Lee, Y.S., Lee, Y.H. 2001. Electrochemical properties of high-power supercapacitors using single-walled carbon nanotube electrodes. *Adv. Funct. Mater.* 11:387–392.
137. Yoon, B.J., Jeong, S.H., Lee, K.H., Kim, S.H.S., Park, G.C., Han, H.J. 2004. Electrical properties of electrical double layer capacitors with integrated carbon nanotube electrodes. *Chem. Phys. Lett.* 388:170–174.
138. Futaba, D.N., Hata, K., Yamada, T., Hiraoka, T., Hayamizu, Y., Kakudate, Y., Tanaike, O., Hatori, H., Yumura, M., Iijima, S. 2006. Shape-engineerable and highly densely packed single-walled carbon nanotubes and their application as super-capacitor electrodes. *Nat. Mater.* 5:987–994.
139. Stoller, M.D., Park, S.J., Zhu, Y.W., An, J.H., Ruoff, R.S. 2008. Synthesis of nitrogen-doped graphene films for lithium battery application. *Nano Lett.* 8:3498–3502.
140. Wang, Y., Shi, Z., Huang, Y., Ma, Y., Wang, C., Chen, M., Chen, Y. 2009. Supercapacitor devices based on graphene materials. *J. Phys. Chem. C* 113:13103–13107.
141. Wang, D.W., Li, F., Wu, Z.S., Ren, W., Cheng, H.M. 2009. Electrochemical interfacial capacitance in multilayer graphene sheets: Dependence on number of stacking layers. *Electrochem. Commun.* 11:1729–1732.
142. Jeong, H.M., Lee, J.W., Shin, W.H., Choi, Y.J., Shin, H.J., Kang, J.K., Choi, J.W. 2011. Nitrogen-doped graphene for high-performance ultracapacitors and the importance of nitrogen-doped sites at basal planes. *Nano Lett.* 11:2472–2477.
143. Li, X.L., Wang, X.R., Zhang, L., Lee, S.W., Dai, H. 2008. Chemically derived, ultra-smooth graphene nanoribbon semiconductors. *Science* 319:1229–1232.
144. Wang, X.R., Ouyang, Y.J., Li, X.L., Wang, H.L., Guo, J., Dai, H. 2008. Room-temperature all-semiconducting sub-10-nm graphene nanoribbon field-effect transistors. *Phys. Rev. Lett.* 100:206803–206804.

CHAPTER 7

CONTENTS

Synthesis and Electronic Properties of Single-Walled Carbon Nanotubes Filled with Inorganic Compounds and Metals

7

Marianna V. Kharlamova

7.1 INTRODUCTION

The single-walled carbon nanotubes (SWCNTs) are known for their exceptional physical, chemical, mechanical, and structural properties.[1] This makes them prospective components of nanoscale and molecular electronic devices. For instance, applications of SWCNTs in the fields of nanoelectronics,[2–5] bioelectronics,[6,7] and thin-film flexible electronics[8–11] have been already realized. Although sufficient progress has been made in constructing nanotube-based elements of nanoelectronics,[10,12,13] full potential of SWCNTs is still not implemented in practical applications. The limiting factor is the presence of nanotubes of both conductivity types (metallic and semiconducting) in the typical synthesized samples, which causes inhomogeneity of their electronic properties.

Several approaches have been developed to control the electronic properties of SWCNTs and obtain metallicity-sorted SWCNTs. They can be divided into three groups. The first group includes the methods of selective synthesis of SWCNTs with a defined conductivity type. For instance, the selective synthesis of semiconducting SWCNTs and even single-chiral nanotubes was succeeded.[14–16] The second group includes the methods of separation of synthesized SWCNTs. To date, a large variety of chemical and physical approaches were established to separate SWCNTs by structural parameters (length[17–21] and diameter[22–26]), conductivity type,[27–37] and even chiral angle.[38–42] The third group includes the methods of chemical modification of SWCNTs, such as covalent attaching of different functional groups to the outer surface of nanotubes,[43–50] noncovalent functionalization and wrapping of SWCNTs with molecules,[51–56] substitution of carbon atoms of SWCNT walls by other atoms,[57–63] intercalation of SWCNT bundles,[64–73] and filling of the SWCNT internal channels. In contrast to the methods from the first and second groups, which allow synthesis and sorting of SWCNTs with predefined properties, the methods of chemical functionalization of SWCNTs open a way to tune their electronic properties.

The filling of SWCNTs implies the encapsulation of substances into the internal channels of nanotubes. Starting from 1998,[74,75] the topic attracted considerable interest, and a large variety of substances, such as fullerenes[76–83] and their derivatives,[84–92] metallocenes[93–101] and other molecules,[102–106] metals,[107–117] nonmetals,[118–120] metal oxides,[121–124] hydroxides,[125] metal halogenides,[117,126–157] and metal chalcogenides,[117,158–166] were introduced into the SWCNT channels. A strong experimental basis was developed for the filling of SWCNTs with different substances in a large yield.[167–173] The possibility of the encapsulation of substances with appropriate physical and chemical properties inside the SWCNTs allows tailoring their electronic structure and controlling the doping level.[169] This makes the filling of the SWCNT channels a very promising method for the directional modification of the electronic properties of nanotubes.

The aim of this chapter is to compare the modification of the electronic properties of SWCNTs upon filling their channels with different substances and elucidate the correlation between the chemical nature of the encapsulated substances and their influence on the SWCNT properties.

This chapter focuses on the investigation of SWCNTs filled with a large number of inorganic compounds (metal halogenides and metal chalcogenides) and simple substances (metals), which possess different chemical and physical properties. The first section is dedicated to the description of the procedures used to fill the SWCNTs with these substances. The second section summarizes the results of the investigation of filled SWCNTs. At the beginning, the filling ratio and crystallinity of the inserted substances are analyzed. The structure of the encapsulated substances is considered, and its correlation with the chemical nature of the incorporated compounds is analyzed on the basis of the high-resolution transmission electron microscopy (HRTEM) data. After that, the electronic properties of the filled SWCNTs are discussed. The results of the Raman spectroscopy, optical absorption spectroscopy, x-ray photoelectron spectroscopy (XPS), and near-edge x-ray absorption fine structure spectroscopy (NEXAFS) investigations are presented and analyzed in order to determine the electronic structure of the filled SWCNTs. The local interactions including the formation of chemical bonds between the introduced compounds and the nanotube walls are explored. Finally, the correlation between the chemical nature of the encapsulated substances and their influence on the SWCNT electronic properties is discussed, and the schemes of the modification of the electronic structure of nanotubes upon their filling with metal halogenides, metal chalcogenides, and metals are presented.

7.2 SYNTHESIS OF FILLED SWCNTs

The filling of the SWCNT internal channels can be performed by different methods. They include approaches using the gas phase, liquid phase (melts and solutions), plasma, and chemical reactions. These methods were reviewed in detail in previous reports.[167,168,170–173] Here, it should be concluded that the choice of the filling method depends on the substance that is desired to introduce inside the SWCNTs.

The inorganic compounds, such as metal halogenides and chalcogenides, which are under consideration in this chapter, were the most often encapsulated inside the SWCNT channels by the liquid method using melts, because this approach was shown to be the most effective for achieving high filling degrees of nanotubes.[167,168,170,171] The first stage of all filling procedures was the opening of nanotube ends. The typical conditions for 1.4 nm diameter nanotubes were the annealing in air at 500°C for 30 min. The open-ended SWCNTs were then mixed with a corresponding compound in a specified molar ratio. The obtained mixtures were evacuated in quartz ampoules at high vacuum and sealed. The ampoules were heated up to the temperature of 100°C above the melting point of the compound. This temperature was maintained for several hours (typically 6–10 h), and then the samples were slowly cooled down. Such synthesis procedure allowed filling the nanotubes in a large yield and therefore achieving sufficient modifications of their electronic properties, which were investigated in depth for the SWCNTs filled with a variety of metal halogenides (AgCl, AgBr, AgI,[128] $MnCl_2$, $MnBr_2$,[133] $FeCl_2$, $FeBr_2$, FeI_2,[130] $CoBr_2$,[131] CuCl, CuBr, CuI,[129] CuI,[127,137] $NiCl_2$, $NiBr_2$,[132] $ZnCl_2$, $ZnBr_2$, ZnI_2,[134] $CdCl_2$, $CdBr_2$, CdI_2,[135] $ErCl_3$,[126] $TbCl_3$,[136] $TmCl_3$,[117] and SnF_2[138]) and metal chalcogenides (GaSe,[158] GaTe, SnS,[159] SnSe, SnTe,[160] Bi_2Se_3,[159] and Bi_2Te_3[117]). Table 7.1 summarizes the temperatures of synthesis of nanotubes filled with metal halogenides and metal chalcogenides.

Because metals have high melting temperatures, they can not be introduced inside the nanotubes by the melt method. The encapsulation of metals into the SWCNTs was performed by a two-step approach, including the filling of the nanotubes with metal salt with subsequent thermal treatment for the decomposition of the salt to pure metal. Preopened SWCNTs were put into saturated solution of metal nitrate ($AgNO_3$[107,110,112,113,117] or $Cu(NO_3)_2$[112,114]). This mixture was sonicated and kept for several days. After that, the nanotubes were filtrated and dried. The subsequent annealing of the obtained samples in air (in case of silver) or in hydrogen (in case of copper) atmosphere at 300–500°C led to the formation of metallic nanoparticles inside the SWCNT channels. The inclusion of multistep washing procedures into the synthesis process allowed obtaining clean samples, which was important for their further spectroscopic investigations.

TABLE 7.1 Temperatures of Synthesis of the SWCNTs Filled with Metal Halogenides and Metal Chalcogenides

Sample	$T_{synthesis}$ (°C)	Sample	$T_{synthesis}$ (°C)
Filling of SWCNTs with metal halogenides			
$MnCl_2$@SWCNT	750	$ZnCl_2$@SWCNT	400
$MnBr_2$@SWCNT	798	$ZnBr_2$@SWCNT	494
$FeCl_2$@SWCNT	774	ZnI_2@SWCNT	546
$FeBr_2$@SWCNT	784	$CdCl_2$@SWCNT	668
FeI_2@SWCNT	687	$CdBr_2$@SWCNT	669
$CoCl_2$@SWCNT	840	CdI_2@SWCNT	488
$CoBr_2$@SWCNT	778	AgCl@SWCNT	555
CoI_2@SWCNT	615	AgBr@SWCNT	530
$NiCl_2$@SWCNT	1101	AgI@SWCNT	660
$NiBr_2$@SWCNT	1063	CuCl@SWCNT	530
$TbCl_3$@SWCNT	688	CuBr@SWCNT	600
$TmCl_3$@SWCNT	924	CuI@SWCNT	705
Filling of SWCNTs with metal chalcogenides			
GaSe@SWCNT	1060	SnTe@SWCNT	907
GaTe@SWCNT	924	Bi_2Se_3@SWCNT	806
SnS@SWCNT	980	Bi_2Te_3@SWCNT	686
SnSe@SWCNT	961		

Note: The data from reports [117,128–136,158–160] were taken into consideration for preparation of the table.

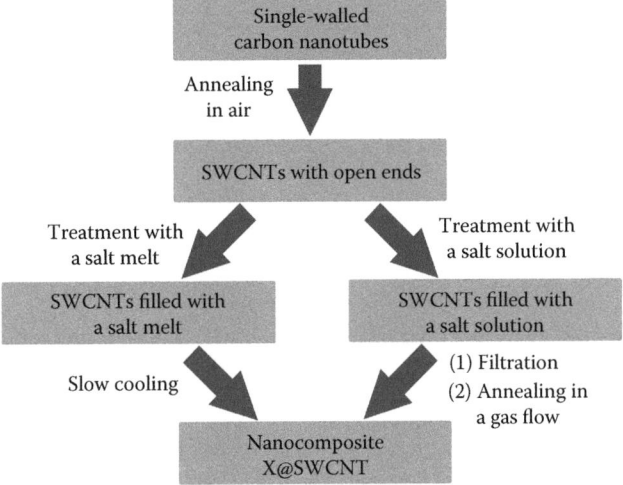

FIGURE 7.1
The scheme of the filling of SWCNTs with inorganic compounds (metal halogenides and metal chalcogenides) and metals.

Figure 7.1 demonstrates the scheme of the filling of SWCNTs with inorganic compounds (metal halogenides and metal chalcogenides) and metals.

7.3 INVESTIGATION OF FILLED SWCNTs

The properties of the filled SWCNTs were investigated by complementary experimental methods. The filling of the nanotube channels and crystallization of the encapsulated substances were examined by HRTEM. The electronic properties of the filled SWCNTs were studied on a qualitative level (the detection of charge transfer) by optical absorption spectroscopy and Raman spectroscopy and on a quantitative level (the determination of the Fermi level shift) by XPS. The chemical bonding between the SWCNTs and inserted substances was explored by NEXAFS.

7.3.1 Filling Ratio and Crystallinity of Encapsulated Substances

HRTEM allows visualization of individual atoms inside the SWCNTs, and thus, this is the most representative and direct method to prove the SWCNT filling. The HRTEM data obtained for SWCNTs filled with metal halogenides, metal chalcogenides, and metals were used to estimate the filling degree of the nanotube channels and crystallinity of the encapsulated substances and to find the correlation between these values and the chemical nature of the filler. Table 7.2 summarizes the estimated filling degrees and crystallinity of the inserted metal halogenides (MHal$_n$, M = Mn, Fe, Co, Ni, Cu, Zn, Ag, Cd, Tb, Tm, Hal = Cl, Br, I) and metal chalcogenides (GaX, SnX, Bi$_2$X$_3$, X = S, Se, Te). The data from reports[117,128–136,158–160] were taken into consideration for the preparation of the table.

The filling degree (i.e., the ratio between filled nanotube length and total nanotube length on the HRTEM micrographs) is in the range from 30% to 90% for the metal halogenide-filled SWCNTs.

TABLE 7.2 Estimated Filling Degrees and Crystallinity of Metal Halogenides and Metal Chalcogenides Encapsulated into the SWCNTs (from the HRTEM Data)

Sample	Filling Degree (%)	Crystallinity	Sample	Filling Degree (%)	Crystallinity
The SWCNTs filled with metal halogenides					
MnCl$_2$@SWCNT	30	Amorphous	ZnCl$_2$@SWCNT	30	Amorphous
MnBr$_2$@SWCNT	50	Amorphous + crystalline	ZnBr$_2$@SWCNT	70	Amorphous + crystalline
FeCl$_2$@SWCNT	50	Amorphous + crystalline	ZnI$_2$@SWCNT	90	Crystalline
FeBr$_2$@SWCNT	60	Crystalline	CdCl$_2$@SWCNT	60	Amorphous + crystalline
FeI$_2$@SWCNT	70	Crystalline	CdBr$_2$@SWCNT	60	Amorphous + crystalline
CoCl$_2$@SWCNT	40	Amorphous	CdI$_2$@SWCNT	80	Crystalline
CoBr$_2$@SWCNT	50	Amorphous + crystalline	AgCl@SWCNT	50	Amorphous
CoI$_2$@SWCNT	80	Crystalline	AgBr@SWCNT	60	Amorphous + crystalline
NiCl$_2$@SWCNT	40	Amorphous	AgI@SWCNT	90	Crystalline
NiBr$_2$@SWCNT	60	Crystalline	CuCl@SWCNT	40	Amorphous
TbCl$_3$@SWCNT	80	Crystalline	CuBr@SWCNT	60	Amorphous + crystalline
TmCl$_3$@SWCNT	70	Crystalline	CuI@SWCNT	90	Crystalline
The SWCNTs filled with metal chalcogenides					
GaSe@SWCNT	60	Amorphous + crystalline	SnS@SWCNT	40	Amorphous
GaTe@SWCNT	80	Crystalline	SnSe@SWCNT	50	Amorphous + crystalline
Bi$_2$Se$_3$@SWCNT	50	Amorphous + crystalline	SnTe@SWCNT	80	Crystalline
Bi$_2$Te$_3$@SWCNT	70	Crystalline			

Note: The data from reports [117,128–136,158–160] were taken into consideration for preparation of the table.

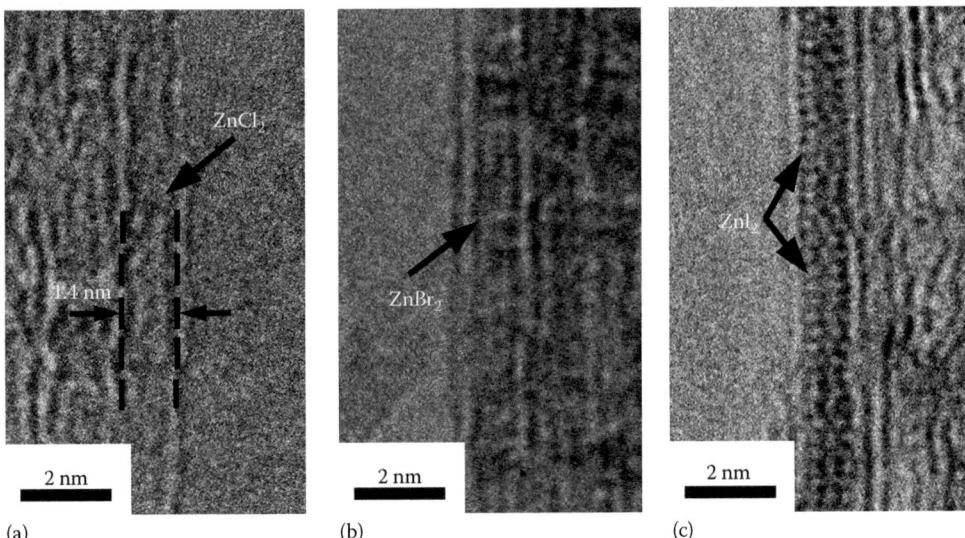

FIGURE 7.2
The HRTEM micrographs of the SWCNTs filled with zinc chloride (a), zinc bromide (b), and zinc iodide (c). (Modified from Kharlamova, M.V. et al., *Eur. Phys. J. B*, 85, 34, 2012.)

These compounds form amorphous nanoparticles or nanocrystals with well-ordered structure inside the nanotubes. The filling degree and crystallinity depend on the type of metal halogenide. For halogenides of 3d metals (Mn, Fe, Co, Ni, Cu, Zn) and 4d metals (Ag, Cd), the filling degree increases in line with chloride–bromide–iodide. In case of metal chlorides, the value is 30%–50%; metal bromides, 50%–70%; and metal iodides, 70%–90%. The crystallinity also increases from metal chloride to metal bromide to metal iodide. The metal chlorides form amorphous particles or mixtures of amorphous and crystallized phases inside the SWCNTs, as metal bromides. The filling of SWCNTs with metal iodides leads to the formation of 1D nanocrystals with well-ordered structure.

Figure 7.2 shows the HRTEM micrographs of the SWCNTs filled with zinc chloride, bromide, and iodide.[134] All of them demonstrate atoms of the inserted salts inside the SWCNT channels. In case of $ZnCl_2$, atoms are disordered and form amorphous nanoparticles (Figure 7.2a), whereas $ZnBr_2$ is partly crystallized (Figure 7.2b). The micrograph of the SWCNTs filled with ZnI_2 demonstrates contrast elements (corresponding to individual atoms), which are placed periodically along the SWCNTs axis, that is, 1D nanocrystal presences inside the nanotube (Figure 7.2c). The filling degree of the SWCNTs increases from 30% for zinc chloride to 70% for zinc bromide to 90% for zinc iodide.

The tendency of the increase in the crystallinity from metal chloride to metal bromide to metal iodide coincides with the increase in halogen anion radius ($r_{(Cl^-)} = 0.181$ nm, $r_{(Br^-)} = 0.196$ nm, $r_{(I^-)} = 0.220$ nm). In case of metal chloride, the diameter of an embedded nanoparticle is smaller than the diameter of an internal channel of nanotube. As a result, atoms of salt turn out to be movable inside the channel, which impedes the stabilization of crystal structure. In contrast, the diameter of a nanoparticle of embedded metal iodide is comparable with the diameter of the SWCNT channel, and it leads to the stabilization of crystal structure.[128,129,167,168] Besides, electron affinity decreases in line with Cl–Br–I (from 3.617 [Cl] to 3.365 [Br] to 3.059 [I]), and this parameter can also influence the stability of nanocrystals inside the nanotube channels.[129]

The radius of metal cation also influences the filling degree of SWCNTs with metal halogenides and crystallinity of the introduced salts. Both these parameters increase in case of chlorides of rare-earth metals (Tb and Tm). For these salts, the filling degrees of SWCNTs amount to 70%–80%, and they form nanocrystals inside the channels, whereas the encapsulated chlorides of 3d metals are amorphous or form mixtures of amorphous and crystalline phases (Table 7.2). Indeed, the radii of cations of rare-earth metals are 0.107 nm (Tb^{3+}) and 0.102 nm (Tm^{3+}), which are significantly larger than the radii of 3d metal cations (from 0.069 nm [Ni^{2+}] to 0.091 nm [Mn^{2+}]).

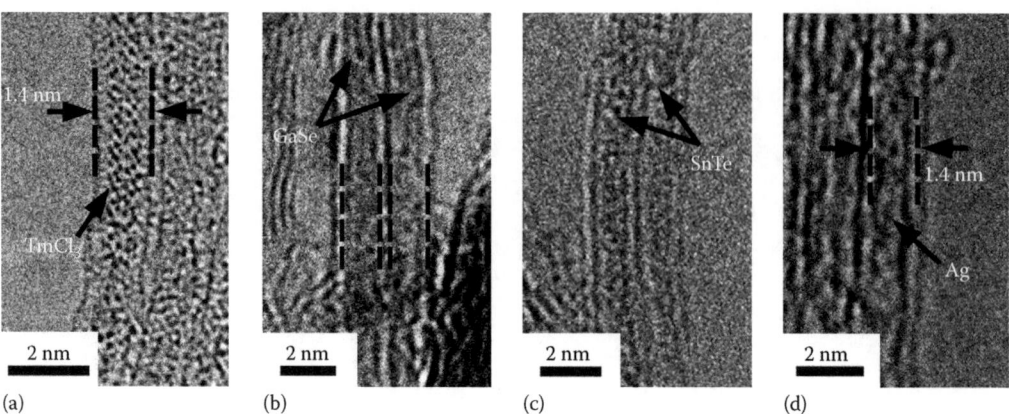

FIGURE 7.3
The HRTEM micrographs of the SWCNTs filled with thulium chloride (a), gallium selenide (b), tin telluride (c), and silver (d). (Modified from Kharlamova, M.V. and Niu, J.J., *Appl. Phys. A*, 109, 25, 2012; Kharlamova, M.V. et al., *Appl. Phys. A*, 112, 297, 2013; Kharlamova, M.V., *JETP Lett.*, 98, 272, 2013; Yashina, L.V. et al., *J. Phys. Chem. C*, 115, 3578, 2011.)

Consequently, in case of rare-earth metal halogenides, the stabilization of nanocrystals inside the SWCNT channels is already observed for chlorides. Figure 7.3a shows the HRTEM micrograph of the SWCNTs filled with thulium chloride.[117]

The filling degrees of nanotubes with metal chalcogenides depend on the chemical nature of compounds and are in the range from 40% to 80%. The filling degree of SWCNTs increases with an increase in chalcogen radius ($r_{(S^{2-})} = 0.184$ nm, $r_{(Se^{2-})} = 0.198$ nm, $r_{(Te^{2-})} = 0.221$ nm). In case of bismuth and gallium selenides and tellurides, the value amounts to 50%–60% and 70%–80%, respectively. For tin chalcogenides, the filling degree increases from 40% to 50% to 80% in the line SnS–SnSe–SnTe. The maximal filling degree of the SWCNTs is observed for GaTe and SnTe (80%) (Table 7.2).

The encapsulated compounds form amorphous nanoparticles or nanocrystals with well-ordered structure. Their crystallinity increases with an increase in chalcogen radius. For instance, tin sulfide forms amorphous nanoparticles, whereas tin selenide forms a mixture of amorphous and crystalline phases. The filling of SWCNTs with tin telluride leads to the formation of 1D nanocrystals (Table 7.2).

The tendency of the increase in the filling degree and crystallinity with the increase in the radius of chalcogen is analogous to the effect observed for metal halogenides, discussed earlier. In case of metal sulfide and selenide, the diameter of nanoparticles is smaller than the diameter of the internal channels of SWCNTs. As a result, the atoms of the introduced compound are movable inside the channels, and it complicates the formation of well-ordered nanocrystals. In contrast, in case of metal telluride, the diameter of nanoparticles is comparable with the diameter of the SWCNTs, which results in the stabilization of nanocrystals. Figure 7.3b and c shows the HRTEM micrographs of the SWCNTs filled with gallium selenide[158] and tin telluride.[160] One may recognize atoms of the encapsulated compounds inside the nanotube channels, which are periodically placed along the SWCNT axis, that is, nanocrystals with well-ordered structure are formed inside the nanotubes.

The HRTEM data of metal-filled SWCNTs prove the successful incorporation of silver and copper inside the SWCNT channels. Figure 7.3d demonstrates the HRTEM micrograph of the nanotube filled with silver, where atoms of metals are recognized inside the SWCNTs.[112]

The estimated filling degrees of SWCNTs with metals amount up to 30%. These values are smaller than the ones for metal halogenide– and metal chalcogenide–filled SWCNTs, because the solution filling method does not allow achieving the filling degrees larger than 30%–40%.[167,168,170,171]

Also, it should be noted that the HRTEM micrographs do not demonstrate metallic nanoparticles on the outer surface of nanotubes, which is important for the investigation of the modified electronic properties of filled SWCNTs.

7.3.2 Electronic Properties of Filled SWCNTs

7.3.2.1 Detection of charge transfer in filled SWCNTs

The detection of charge transfer in the filled SWCNTs was performed by optical absorption spectroscopy and Raman spectroscopy.

Figure 7.4 demonstrates the optical absorption spectra of the pristine SWCNTs and nanotubes filled with zinc iodide,[134] gallium telluride,[159] and bismuth telluride. There are several peaks in the spectrum of the pristine nanotubes at energies of 0.64, 1.20, and 1.72 eV, which correspond to electronic transitions between van Hove singularities (vHs) of semiconducting and metallic nanotubes. Taking into consideration the Kataura plot, for 1.4–1.6 nm diameter SWCNTs, the peak at the energy of 0.6–0.8 eV corresponds to the electronic transitions between the first vHs of semiconducting nanotubes E_{11}^S, the peak at the energy of 1.0–1.4 eV belongs to the electronic transitions between the second vHs of semiconducting tubes (E_{22}^S), and the peak at the energy of 1.7–2.0 eV can be assigned to the electronic transitions between the first vHs of metallic nanotubes (E_{11}^M).[174] Thus, the energy difference between the corresponding first and second vHs in the valence band and the conduction band of semiconducting SWCNTs amounts to 0.64 and 1.20 eV, accordingly. Assuming that the Fermi level is positioned in the middle between the corresponding vHs, the distance from the Fermi level to the first and second vHs of semiconducting SWCNTs equals to 0.32 and 0.60 eV, respectively. The energy difference between the first vHs of metallic tubes amounts to 1.72 eV; therefore, the distance from the Fermi level to this singularity is 0.86 eV.

The spectrum of the ZnI_2-filled SWCNTs demonstrates significant changes (Figure 7.4). The suppression of the peak corresponding to the electronic transitions between the first vHs of semiconducting SWCNTs is observed. It is evidence of the charge transfer in the filled nanotubes. If there is a charge transfer from the SWCNT walls, the Fermi level of SWCNTs downshifts, and the first vHs in the valence band of semiconducting nanotubes are emptied. This is the case of acceptor doping of SWCNTs. If there is a charge transfer to the SWCNT walls, the Fermi level of SWCNTs upshifts, and the first vHs in the conduction band of semiconducting nanotubes are filled. This is the case of donor doping of SWCNTs. Thus, the optical absorption spectroscopy data give an evidence of the charge transfer in the filled nanotubes. However, these two cases could not be separated by this method. The optical absorption spectra of SWCNTs filled with other metal halogenides were reported, and the disappearance of the E_{11}^S peak was observed for SWCNTs filled with halogenides of silver,[128] iron,[130] cobalt,[131] nickel,[132] copper,[129] zinc,[134] terbium,[136] and cadmium.[135] This effect is a sign of significant modification of electronic structure of SWCNTs upon their filling.

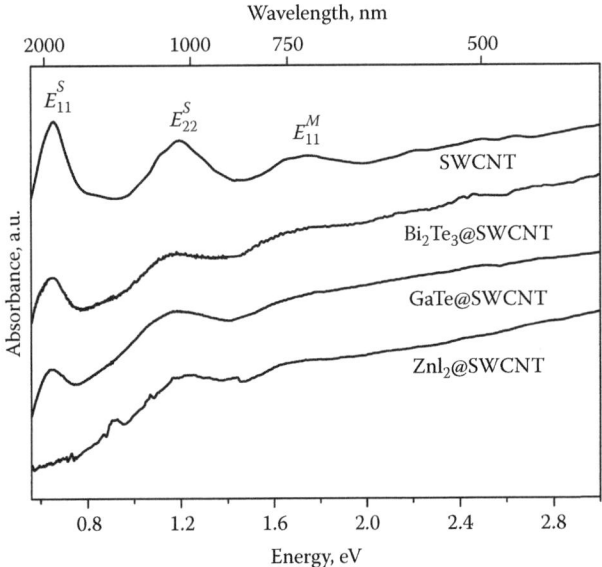

FIGURE 7.4
The optical absorption spectra of the pristine SWCNTs and nanotubes filled with zinc iodide,[134] gallium telluride,[159] and bismuth telluride.

The spectra of gallium telluride– and bismuth telluride–filled SWCNTs do not show significant changes as compared with the spectrum of the pristine tubes (Figure 7.4). In particular, the E_{11}^S peak is not suppressed, as it was observed in the spectra of metal halogenide–filled nanotubes. It means that even if the charge transfer takes place in the filled SWCNTs, the Fermi level stays above the first vHs in the valence band of semiconducting SWCNTs (in case of a charge transfer from the SWCNT walls and lowering the Fermi level) or below the first vHs in the conduction band of semiconducting SWCNTs (in case of a charge transfer to the SWCNT walls and increasing the Fermi level). The same results were also reported for SWCNTs filled with GaSe,[158] Bi$_2$Se$_3$,[159] SnS,[159] and SnTe.[160] However, the confident conclusion about the charge transfer in metal chalcogenide–filled SWCNTs cannot be drawn from the optical absorption spectroscopy data.

Further investigation of the electronic structure of the filled SWCNTs on a qualitative level was carried out by Raman spectroscopy. The Raman spectrum of SWCNTs contains two main regions[175]: a radial breathing mode (RBM) at frequencies below 200 cm^{-1}, which belongs to synchronous radial vibrations of carbon atoms (A$_{1g}$ symmetry), and a G-mode at frequencies of 1500–1600 cm^{-1}, which is assigned to C–C bond vibrations (A, E$_1$, E$_2$ symmetries[176]). The G-band includes two most intense components[177]: the G$^-$-mode at lower frequencies (about 1540 cm^{-1}) and the G$^+$-mode at higher frequencies (about 1590 cm^{-1}). They belong to longitudinal optical (LO) phonon (LO corresponds to the displacement of carbon atoms along the tube axis) in metallic (G$^-$) and semiconducting (G$^+$) SWCNTs.[178] The G$^+$-mode shows a shoulder on the lower-frequency side, which originates from the transversal optical (TO) phonon (TO corresponds to the circumferential displacement of the atoms) in semiconducting SWCNTs.[178] It should be noted that the G-bands of semiconducting and metallic SWCNTs have different shapes. The G-mode of semiconducting tubes is a narrow peak with a Lorentzian shape, whereas the G-band of metallic tubes has the broad profile of a Breit–Wigner–Fano (BWF) function due to phonon–plasmon coupling in the presence of conduction electrons.[175,179–181]

Figure 7.5 compares the RBM- and G-bands of Raman spectra of the pristine nanotubes and the SWCNTs filled with cadmium chloride,[135] gallium selenide,[158] tin sulfide,[159] and silver[112] that were acquired at laser energies of 1.96 eV (λ_{ex} = 633 nm) and 1.58 eV (λ_{ex} = 785 nm).

According to the Kataura plot, the lasers of different energies excite electronic transitions between vHs of metallic and semiconducting SWCNTs of different diameters. The laser with an energy of 1.96 eV (λ_{ex} = 633 nm) excites the E_{11}^M electronic transitions of 1.4 nm metallic tubes and the E_{33}^S electronic transitions of 1.6 nm semiconducting SWCNTs.[174] The RBM-band of Raman spectrum of the pristine nanotubes acquired at 1.96 eV (λ_{ex} = 633 nm) (Figure 7.5a) contains two peaks at 156 and 172 cm^{-1}. Taking into consideration the fact that the positions of the RBM peaks are inversely proportional to the SWCNT diameter by the equation $\omega_{RBM} = 227/d_t\sqrt{1 + Cd_t^2}$, where C = 0.05786 nm^{-2},[182] these peaks correspond to semiconducting SWCNTs with a diameter of 1.5 nm and metallic nanotubes with a diameter of 1.4 nm, accordingly. The G-band of the spectrum (Figure 7.5b) is a broad asymmetric peak, which is characteristic of metallic nanotubes. It contains three components positioned at 1546, 1564, and 1592 cm^{-1} (Table 7.3). The possible chiralities of exciting SWCNTs are (20,1), (19,3), (17,4), (16,6), (15,8), and (14,10) for SWCNTs with a diameter of 1.5 nm and (18,0), (17,2), (16,4), (13,7), (14,5), and (10,10) for SWCNTs with a diameter of 1.4 nm.[128,129]

According to the Kataura plot, the laser with an energy of 1.58 eV (λ_{ex} = 785 nm) excites the E_{11}^M electronic transitions of metallic nanotubes with a diameter in the range of 1.4–1.5 nm.[174] The RBM-band of Raman spectrum of the pristine SWCNTs contains two peaks at 161 and 171 cm^{-1} (Figure 7.5c), which correspond to SWCNTs with diameters of 1.50 and 1.40 nm, respectively. The G-band of the spectrum (Figure 7.5d) is a broad asymmetric peak that includes three components at 1552, 1568, and 1591 cm^{-1} (Table 7.3). The possible chiralities of exciting SWCNTs are (15,6), (14,8), and (11,11) for nanotubes with a diameter of 1.5 nm and (18,0), (17,2), and (16,4) for SWCNTs with a diameter of 1.4 nm.[128,129]

The RBM- and G-bands of Raman spectra of the filled SWCNTs demonstrate changes, which depend on the chemical nature of the encapsulated compound and laser excitation energy.

The RBM-band of Raman spectrum of the CdCl$_2$-filled SWCNTs acquired at a laser energy of 1.58 eV (λ_{ex} = 785 nm) has only a little upshift of peaks by 1 cm^{-1} as compared to the one of the pristine SWCNTs (Figure 7.5c), whereas the spectrum of the filled SWCNTs acquired at a laser energy of 1.96 eV (λ_{ex} = 633 nm) demonstrates the shift of peaks by 4–11 cm^{-1} toward higher

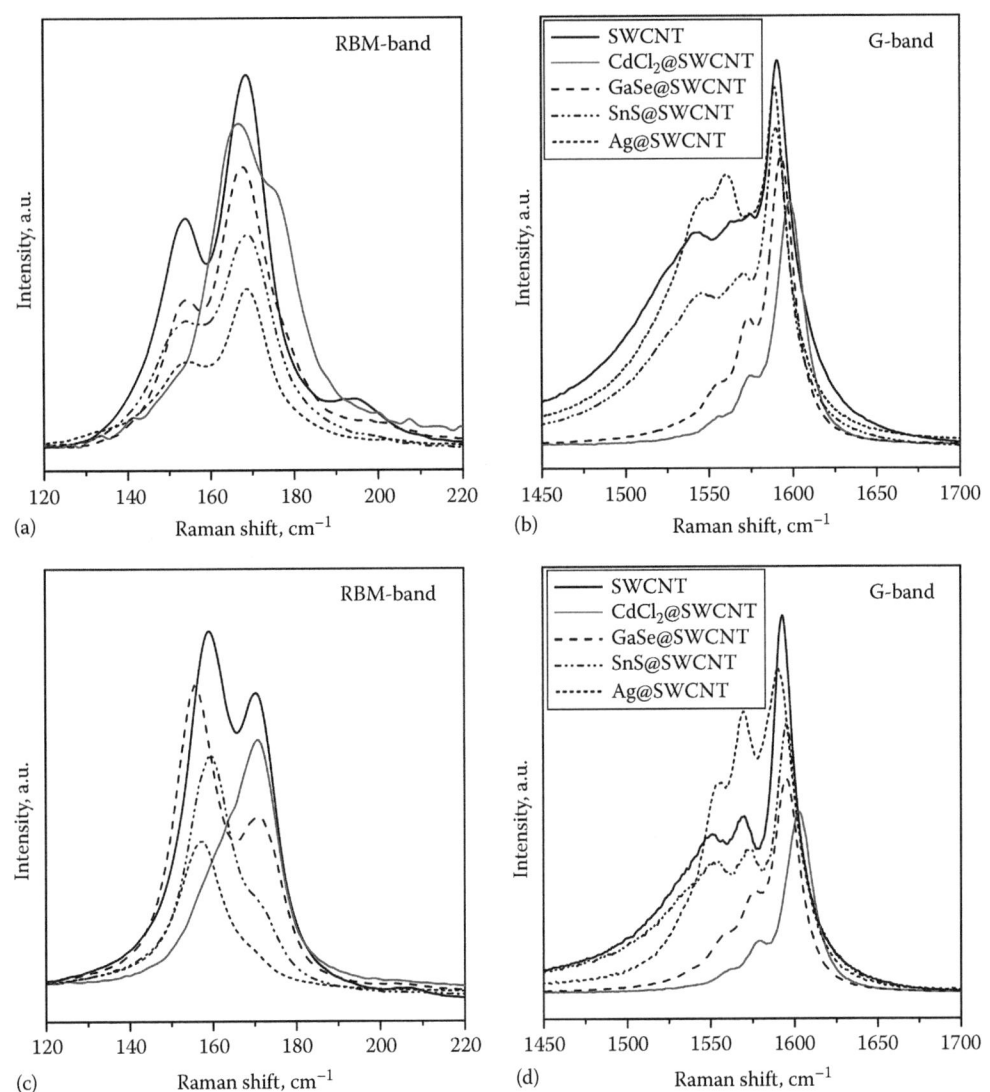

FIGURE 7.5
The RBM- and G-bands of Raman spectra of the pristine nanotubes and SWCNTs filled with cadmium chloride,[135] gallium selenide,[158] tin sulfide,[159] and silver[112] that were acquired at laser energies of 1.96 eV (λ_{ex} = 633 nm) (a, b) and 1.58 eV (λ_{ex} = 785 nm) (c, d).

frequencies (Figure 7.5a). Also, the profile of the RBM-band of Raman spectra of the filled nanotubes acquired at both laser energies is significantly modified as compared to the one of the pristine SWCNTs, because the relative intensities of the RBM peaks change drastically. It should be noted that the profile of the RBM-band of Raman spectrum of the CdCl$_2$-filled SWCNTs obtained at a laser energy of 1.58 eV (λ_{ex} = 785 nm) is similar to the profile of RBM-band of Raman spectrum of the pristine SWCNTs acquired at a laser energy of 1.96 eV (λ_{ex} = 633 nm) (Figure 7.5a and c). In both spectra, the peak of 1.4 nm metallic nanotubes has the maximal intensity. It may testify to changes in resonance excitation energy of metallic nanotubes with a diameter of 1.5 nm upon their filling. As a result, laser with an energy of 1.58 eV (λ_{ex} = 785 nm) detects smaller diameter nanotubes (1.40 nm).

The G-band of Raman spectra of the CdCl$_2$-filled SWCNTs acquired at laser energies of 1.96 eV (λ_{ex} = 633 nm) and 1.58 eV (λ_{ex} = 785 nm) demonstrates an upshift by 6–11 cm^{-1} (Figure 7.5b and d, Table 7.3). This is caused by the change in the C–C bond energy and the electronic structure of nanotubes upon their filling due to the charge transfer between the SWCNT walls and encapsulated compound. This explanation is in agreement with the modification of the RBM-band of the spectra, discussed earlier. The same trend was observed in the Raman spectra of SWCNTs

TABLE 7.3 Positions of the RBM- and G-Lines in the Raman Spectra of the Pristine SWCNTs, CdCl$_2$@SWCNT, GaSe@SWCNT, SnS@SWCNT, and Ag@SWCNT Samples Acquired at Laser Excitation Energies of 1.96 eV (λ_{ex} = 633 nm) and 1.58 eV (λ_{ex} = 785 nm)

Sample	E$_{ex}$ (eV)	RBM-Band (cm^{-1})		G-Band (cm^{-1})		
				G$^-$	G$_{TO}^+$	G$_{LO}^+$
SWCNT	1.96	156	172	1546	1564	1592
CdCl$_2$@SWCNT		167 (+11)	176 (+4)	1556 (+10)	1575 (+11)	1598 (+6)
GaSe@SWCNT		155 (−1)	170 (−2)	1556 (+10)	1573 (+9)	1594 (+2)
SnS@SWCNT		157 (+1)	172	1547 (+1)	1566 (+2)	1591 (−1)
Ag@SWCNT		155 (−1)	172	1544 (−2)	1561 (−3)	1588 (−4)
SWCNT	1.58	161	171	1552	1568	1591
CdCl$_2$@SWCNT		162 (+1)	172 (+1)	1561 (+9)	1579 (+11)	1602 (+11)
GaSe@SWCNT		158 (−3)	172 (+1)	1558 (+6)	1576 (+8)	1594 (+3)
SnS@SWCNT		162 (+1)	172 (+1)	1553 (+1)	1570 (+2)	1592 (+1)
Ag@SWCNT		160 (−1)	169 (−2)	1555 (+3)	1570 (+2)	1590 (−1)

Note: The shifts of peak positions in comparison to the ones of the pristine SWCNTs are given in parentheses. The data from reports [112,135,158,159] are included in the table.

filled with halogenides of manganese,[133] iron,[130] cobalt,[131] nickel,[132] copper,[129] zinc,[134] terbium,[136] thulium,[117] silver,[128] and cadmium,[135] as well as chromium oxide,[110] organic molecules,[105] and simple chalcogens.[119]

The G-band of Raman spectra of the CdCl$_2$-filled SWCNTs acquired at laser energies of 1.58 eV (λ_{ex} = 785 nm) and 1.96 eV (λ_{ex} = 633 nm) is also characterized by a change in the profile from metallic to semiconducting shape (Figure 7.5b and d). This may be connected with changes in resonance excitation conditions of nanotubes or band gap opening in the electronic structure of SWCNTs upon their filling. It should be noted that this effect often accompanies the upshift of the G-band, and it was reported for all aforementioned nanocomposites.

Thus, the observed modifications of Raman spectra of the SWCNTs filled with metal halogenides testify to a significant alteration of the electronic properties of nanotubes as a result of the charge transfer between the SWCNT walls and inserted compounds.

The Raman spectra of metal chalcogenide–filled SWCNTs demonstrate changes as compared to the spectra of the pristine SWCNTs that depend on the chemical nature of the compound and laser excitation energy. The RBM- and G-bands of Raman spectra of GaSe- and SnS-filled SWCNTs acquired at laser energies of 1.96 eV (λ_{ex} = 633 nm) and 1.58 eV (λ_{ex} = 785 nm) are presented in Figure 7.5. The RBM-band of Raman spectra of the filled SWCNTs demonstrates only slight shifts of peaks (Table 7.3). At the same time, the profile of the RBM-band stays almost unchanged in these spectra (Figure 7.5a and c). This is in contrast with the case that was observed for metal halogenide–filled SWCNTs, where the relative intensities of the RBM peaks significantly changed. This is evidence that the conditions of resonance excitation of nanotubes are not affected by their filling.

The changes in the G-bands of Raman spectra of the filled nanotubes acquired at laser energies of 1.96 eV (λ_{ex} = 633 nm) and 1.58 eV (λ_{ex} = 785 nm) are different for GaSe and SnS (Figure 7.5b and d). The G-band of the GaSe-filled SWCNTs acquired at both laser energies shows significant changes, which are similar to those observed for metal halogenide–filled tubes. All three components of the G-band are shifted as compared to the spectrum of the pristine SWCNTs. For spectra acquired at 1.96 eV (λ_{ex} = 633 nm) and 1.58 eV (λ_{ex} = 785 nm), the metallic component G$^-$ is upshifted by 10 and 6 cm^{-1}, respectively; the semiconducting component G$_{TO}^+$ is shifted by +9 and +8 cm^{-1}, respectively; and semiconducting component G$_{LO}^+$ is shifted by +2 and +3 cm^{-1}, respectively (Table 7.3). The shift of the components is assigned to changes in the electronic structure of the filled SWCNTs as a result of the charge transfer between the nanotubes and introduced compound. The larger shift of the metallic component may testify to a stronger influence of the embedded gallium selenide on the electronic properties of metallic tubes than semiconducting SWCNTs. Besides the shifts of the peaks of the G-band, the spectra of the GaSe-filled SWCNTs acquired at both laser excitation energies demonstrate the change in the G-band profile from metallic to semi-conducting shape, which additionally proves significant modification of the electronic structure of gallium selenide–filled nanotubes.

The G-band of the SnS@SWCNT sample is not modified in comparison with the spectrum of the pristine SWCNTs, except of minor shifts of the peaks by 1–2 cm^{-1} (Table 7.3). Taking into consideration the HRTEM data, discussed earlier, that testified the filling of SWCNT channels with tin sulfide, one may assume that the encapsulation of this chemical compound into the SWCNTs does not lead to a noticeable alteration of the electronic properties of nanotubes. The similar results were also reported for tin telluride,[160] bismuth selenide,[159] and bismuth telluride[117] embedded into the SWCNT channels.

The changes observed in Raman spectra of metal-filled SWCNTs are different from those observed for metal halogenide– and metal chalcogenide–filled nanotubes. The Raman spectra of the Ag-filled SWCNTs are presented in Figure 7.5.

The RBM-bands of the spectra of the Ag@SWCNT sample (Figure 7.5a and c) show only slight modifications as compared to the spectra of the pristine SWCNTs, notably a small downshift of the RBM peaks by 1–2 cm^{-1} (Table 7.3). The G-bands of Raman spectra of the silver-filled nanotubes demonstrate more significant changes (Figure 7.5b and d). Firstly, the shift of the G-band components is observed. It reaches 2–4 cm^{-1} for the G-band peaks in the Raman spectrum acquired at a laser energy of 1.96 eV ($\lambda_{ex} = 633$ nm) and 1–3 cm^{-1} for the G-peaks in the spectrum acquired at 1.58 eV ($\lambda_{ex} = 785$ nm) (Table 7.3). This tendency is assigned to modifications of the electronic structure of the filled SWCNTs due to the charge transfer between the SWCNT walls and encapsulated silver nanoparticles. Secondly, the G-band of Raman spectra of the filled SWCNTs demonstrates an alteration of the intensity ratios of the peaks. An increase in the relative intensity of the peak of metallic nanotubes (G$^-$) as compared to the peaks of semiconducting SWCNTs (G$_{TO}^+$ and G$_{LO}^+$) takes place. This effect is an evidence of the increase of the content of metallic nanotubes in the sample as a result of the transition of semiconducting SWCNTs to a metallic state. Such transition occurs due to the charge transfer from the embedded metallic nanoparticles to the nanotube walls, that is, donor doping of SWCNTs. The same trend was observed in Raman spectra of SWCNTs filled with other metals, such as copper.[112,114] It should be noted that this tendency is the opposite of the one that took place for metal halogenide–filled nanotubes, discussed earlier, where the profile of the G-band changed from metallic to semiconducting shape as a result of the band gap opening in the electronic structure of metallic nanotubes and their transition to a semiconducting state.

Thus, the Raman spectroscopy data clearly reflect differences in the influence of substances of different chemical nature on the electronic properties of SWCNTs.

7.3.2.2 Ascertainment of local interactions in filled SWCNTs

To detect the local interactions (chemical bonding) in the filled SWCNTs, the samples were investigated by NEXAFS. Figure 7.6 compares the C 1s x-ray absorption spectra of the pristine SWCNTs and nanotubes filled with nickel bromide,[132] and gallium selenide.[158]

The spectrum of the pristine SWCNTs contains two main peaks: the π*-resonance at an energy of 285.4 eV, which corresponds to electron transitions to the π*-band of SWCNTs, and the σ*-resonance at an energy of 291.7 eV, which belongs to electron transitions to the σ*-band of SWCNTs.[183]

The spectrum of the nickel bromide–filled SWCNTs demonstrates the peaks of the π*- and σ*-resonances. Additionally, a new peak shifted by about 1.35 eV appears at the low-energy side of the π*-resonance (labeled A in Figure 7.6). The origin of this peak is connected with an emergence of new unoccupied level in the conduction band of SWCNTs due to the hybridization of the π-orbitals of carbon with 3d orbitals of nickel, that is, the formation of chemical bonds between the SWCNT walls and encapsulated nickel bromide.

The appearance of this peak was also observed in the C 1s NEXAFS spectra of SWCNTs filled with nickel chloride,[132] erbium chloride,[126] halogenides of iron,[130] zinc,[134] silver,[128] copper,[129] and cadmium.[135] It should be noted that despite the fact that the new localized level is formed due to the hybridization of the π-orbitals of carbon and different d orbitals of metals (3d for halogenides of Ni, Fe, Zn, and Cu, 4d for halogenides of Cd and Ag, and 5d for ErCl$_3$), the position of this new level differs only slightly for different metal halogenides.

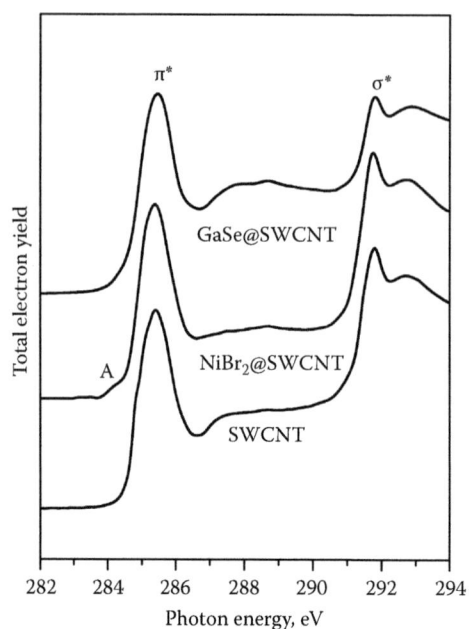

FIGURE 7.6
The C 1s NEXAFS spectra of the pristine SWCNTs and nanotubes filled with nickel bromide,[132] and gallium selenide.[158]

Thus, the chemical bonds are formed in the metal halogenide–filled SWCNTs. In this case, the charge transfer between the SWCNT walls and the formed localized levels can take place.[129] Besides, the charge transfer is a result of doping of SWCNTs by the encapsulated compounds due to the differences in the work functions of nanotubes and metal halogenides, and it leads to the equalization of their Fermi levels. Most probably, both these effects contribute to the appearance of the changes in optical absorption spectra and Raman spectra of the metal halogenide–filled SWCNTs, described earlier.

In contrast to the C 1s NEXAFS spectra of metal halogenide–filled SWCNTs, the spectra of metal chalcogenide–filled nanotubes do not demonstrate significant changes as compared with the spectrum of the pristine SWCNTs. Figure 7.6 shows the spectrum of gallium selenide–filled SWCNTs. The new peak at the low-energy side of the π^*-resonance does not appear in this spectrum. This testifies that the filling of nanotubes with GaSe does not lead to the formation of chemical bonds between the inserted compound and nanotube walls.

7.3.2.3 Determination of the fermi level shift in filled SWCNTs

To directly confirm the Fermi level shift with corresponding change in the work function for the filled SWCNTs, the work function measurements and investigation of the valence band of metal halogenide–filled SWCNTs were performed.

Figure 7.7 shows the secondary electron cutoff and valence band spectra of the pristine SWCNTs and nanotubes filled with zinc bromide.[134] The secondary electron cutoff spectra of the pristine and filled SWCNTs (Figure 7.7a) are sharp peaks with maxima at kinetic energies of 4.9 and 5.6 eV, respectively. The electron work function corresponds to the value of the kinetic energy at the half maximum of the peak. The work functions for the pristine and $ZnBr_2$-filled SWCNTs are 4.8 and 5.1 eV, accordingly.[134] Because the change in the work function is directly connected with an alteration of the Fermi level, the shift of the Fermi level of the filled SWCNTs can be calculated as the difference between the work functions of the pristine and filled SWCNTs. This value is −0.3 eV for the $ZnBr_2$-filled SWCNTs and is comparable with the one obtained for copper halogenide–filled nanotubes.[129]

The valence band spectrum of the zinc bromide–filled SWCNTs demonstrates the shift of the π-resonance peak (corresponding to the photoelectron emission from the π-band of nanotubes[184,185]) toward the Fermi level as compared to the SWCNT spectrum (Figure 7.7b). The π-resonance is

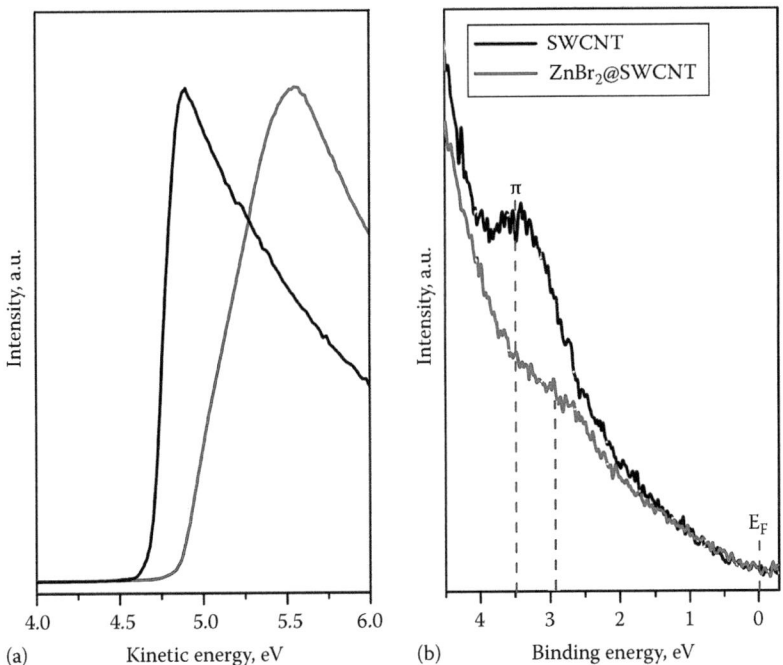

FIGURE 7.7
The secondary electron cutoff (a) and valence band spectra (b) of the pristine SWCNTs and
nanotubes filled with zinc bromide. (Modified from Kharlamova, M.V. et al., *Eur. Phys. J. B*, 85, 34,
2012.)

located at a binding energy of 3.22 eV for the pristine SWCNTs and at 2.96 eV for the ZnBr$_2$@
SWCNT sample. This effect is caused by the change in the work function of SWCNTs upon their
filling. The difference in the positions of the π-resonance for the pristine and filled SWCNTs is
equal to the Fermi level shift. The value is -0.26 eV, and it is in agreement with the value obtained
by the work function measurements. The downshift of the π-resonance was also observed in the
valence band spectra of copper halogenide–filled SWCNTs,[129] which testified the downshift of the
Fermi level of SWCNTs due to acceptor doping.

The shift of the Fermi level of SWCNTs upon their filling leads to changes in the C 1s XPS
spectra, too. Moreover, these changes depend on the chemical nature of the substances, introduced
into the SWCNT channels. Figure 7.8 shows the C 1s XPS spectra of the pristine SWCNTs and
nanotubes filled with cadmium chloride,[135] gallium selenide,[158] bismuth telluride,[117] tin telluride,[160]
and silver.[117]

The spectrum of the pristine SWCNTs is a narrow peak, which is fitted with one component
centered at a binding energy of 284.60 eV (Figure 7.8a). The spectrum of the CdCl$_2$-filled SWCNTs
is fitted with three components, as it is shown in Figure 7.8b. The parameters of component I are
similar to those of the pristine nanotubes (Table 7.4). Taking into account the fact that the sample
consists of filled and unfilled SWCNTs, this component is attributed to the unfilled SWCNTs in
the CdCl$_2$@SWCNT sample. Components II and III belong to the filled nanotubes. Component II
is shifted toward lower binding energies by 0.36 eV as compared to component I (Table 7.4). The
origin of this component is connected with the change in the work function of SWCNTs upon their
filling due to a downshift of the Fermi level, which leads to corresponding shift of all peaks of the
filled nanotubes to lower binding energies. This effect is caused by the charge transfer from the
SWCNT walls to the encapsulated cadmium chloride, that is, acceptor doping of nanotubes, which
is a result of the difference in the work functions of the SWCNTs and compound and leads to the
equalization of their Fermi levels. The appearance of the additional component at the lower-energy
side from component I was also observed in the C 1s XPS spectra of nanotubes filled with cadmium bromide and iodide[135] and halogenides of silver,[128] cobalt,[131] manganese,[133] zinc,[134] copper,[129]
terbium,[136] and thulium.[117] Thus, the XPS data for the metal halogenide–filled SWCNTs confirm
the results of optical absorption spectroscopy and Raman spectroscopy. The origin of less intensive
and broader component III that shifted toward higher energies as compared to component I is not

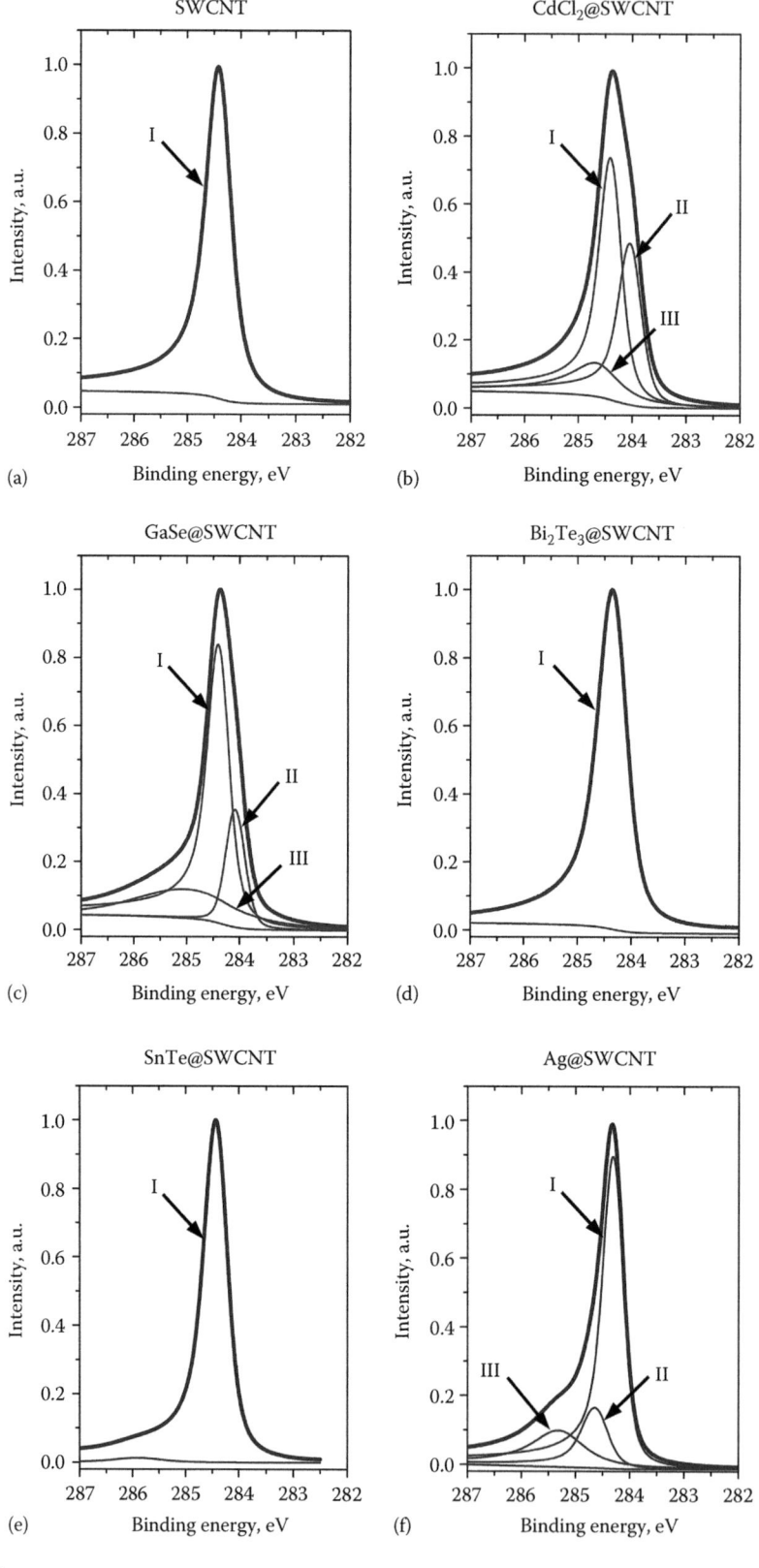

FIGURE 7.8

The C 1s XPS spectra of the pristine SWCNTs (a) and nanotubes filled with cadmium chloride (b), gallium selenide (c), bismuth telluride (d), tin telluride (e), and silver (f). (Modified from Kharlamova, M.V. et al., *Appl. Phys. A*, 112, 297, 2013; Kharlamova, M.V. et al., *J. Mater. Sci.*, 48, 8412, 2013; Kharlamova, M.V., *JETP Lett.*, 98, 272, 2013; Yashina, L.V. et al., *J. Phys. Chem. C*, 115, 3578, 2011.)

TABLE 7.4 Results of the Fitting of the C 1s XPS Spectra of the Pristine SWCNTs, $CdCl_2$@ SWCNT, GaSe@SWCNT, Bi_2Te_3@SWCNT, SnTe@SWCNT, and Ag@SWCNT Samples

Sample	Component	Assignment	Relative Intensity	Binding Energy (eV)
SWCNT	I	SWCNTs	1.000	284.38
$CdCl_2$@SWCNT	I	SWCNTs	0.554	284.38
	II	Doped SWCNTs	0.370	284.02 (−0.36)
	III	Local interactions	0.076	284.63 (+0.25)
GaSe@SWCNT	I	SWCNTs	0.621	284.38
	II	Doped SWCNTs	0.297	284.10 (−0.28)
	III	Local interactions	0.082	284.80 (+0.42)
Bi_2Te_3@SWCNT	I	SWCNTs + doped SWCNTs	1.000	284.33
SnTe@SWCNT	I	SWCNTs + doped SWCNTs	1.000	284.41
Ag@SWCNT	I	SWCNTs	0.764	284.29
	II	Doped SWCNTs	0.147	284.62 (+0.33)
	III	Local interactions	0.089	285.27 (+0.98)

Note: The shifts of peak positions of components II and III in comparison to the ones of component I are given in parentheses. The data from reports [117,135,158,160] are included in the table.

clear. One of explanations is local interactions between carbon atoms of nanotubes and atoms of the encapsulated compounds.[128,129]

The spectrum of the GaSe-filled SWCNTs shown in Figure 7.8c is also fitted with three components. As for metal halogenide–filled nanotubes, the origin of the second component at lower binding energies than the component of the unfilled SWCNTs is caused by the charge transfer from nanotube walls to the encapsulated compound, which is a result of acceptor doping of SWCNTs due to the difference in the work functions of gallium selenide and nanotubes. This conclusion is in agreement with the Raman spectroscopy data. It should be noted that the shift value of 0.28 eV is smaller than the energy difference between the Fermi level and the first vHs in the valence band of semiconducting SWCNTs (0.32 eV) (obtained from the optical absorption spectroscopy data, discussed earlier). It means that the shifted Fermi level stays still above this singularity. As a result, the E_{11}^S electronic transitions between the first vHs in the valence band and the conduction band of semiconducting SWCNTs are not suppressed, and the E_{11}^S peak is observed in the optical absorption spectrum of the gallium selenide–filled nanotubes. The same effect was also remarked for gallium telluride–filled SWCNTs,[159] but this tendency differs from the one observed for metal halogenide–filled SWCNTs (described earlier), where stronger doping of SWCNTs by the inserted salts caused the shift of the Fermi level below the first vHs in the valence band of semiconducting SWCNTs, and the E_{11}^S peak disappeared in the optical absorption spectra.

The C 1s XPS spectra of the nanotubes filled with bismuth and tin chalcogenides are fitted with only one component. Figure 7.8d and e shows the spectra of the Bi_2Te_3@SWCNT and SnTe@ SWCNT samples. The parameters of this component are similar to those of the pristine nanotubes (Table 7.4). Taking into consideration the fact that the HRTEM data prove the filling of the SWCNT channels with bismuth and tin chalcogenides, one may conclude that component I in the spectra of nanocomposites belongs to the filled SWCNTs, and the absence of other components (observed in the spectra of SWCNTs filled with metal halogenides and gallium chalcogenides) testifies that the incorporation of bismuth telluride and tin telluride into the SWCNT channels does not lead to the modification of the electronic structure of nanotubes. This conclusion is in agreement with the optical absorption spectroscopy and Raman spectroscopy data. The same results were also reported for Bi_2Se_3- and SnS-filled SWCNTs.[159]

The C 1s XPS spectrum of the nanotubes filled with silver is presented in Figure 7.8f. The spectrum is fitted with three components, as the spectra of metal halogenide–filled SWCNTs. The difference is that in the spectrum of metal-filled nanotubes, the second component arises at higher binding energies than the component of the unfilled SWCNTs (Table 7.4). It means that the filling of nanotubes with silver leads to the change in the work function of SWCNTs due to an upshift of the Fermi level. The shift value approximately equal to the energy difference between the maxima of components I and II, and it amounts to 0.33 eV. This effect is caused by the charge transfer from the incorporated metallic nanoparticles to the SWCNT walls, that is, donor doping of SWCNTs. This conclusion is in agreement with the Raman spectroscopy data, discussed earlier, where Raman spectra of the silver-filled SWCNTs demonstrated the shift of the G-band peaks and the change in the G-band profile to a metallic shape. The same effect was observed for copper-filled nanotubes.[112,114]

7.3.2.4 General scheme of influence of different fillers on electronic structure of SWCNTs

Based on the results of the investigation of the filled SWCNTs by optical absorption spectroscopy, Raman spectroscopy, x-ray absorption spectroscopy, and XPS, the correlation between the chemical nature of the encapsulated substances and their influence on the electronic properties of SWCNTs was found.

According to the data described in the previous sections of this chapter, the compounds and simple substances inserted into the SWCNTs can be divided into four groups by the character of their influence on the electronic properties of nanotubes. Every group includes substances with similar chemical nature and electronic structure. Figure 7.9 presents the schemes of the modification of the electronic structure of SWCNTs upon their filling with metal halogenides, metal chalcogenides, and metals.

The first group includes metal halogenides, which are wide band gap semiconductors and insulators and have larger work function than the value of the pristine SWCNTs. The filling of

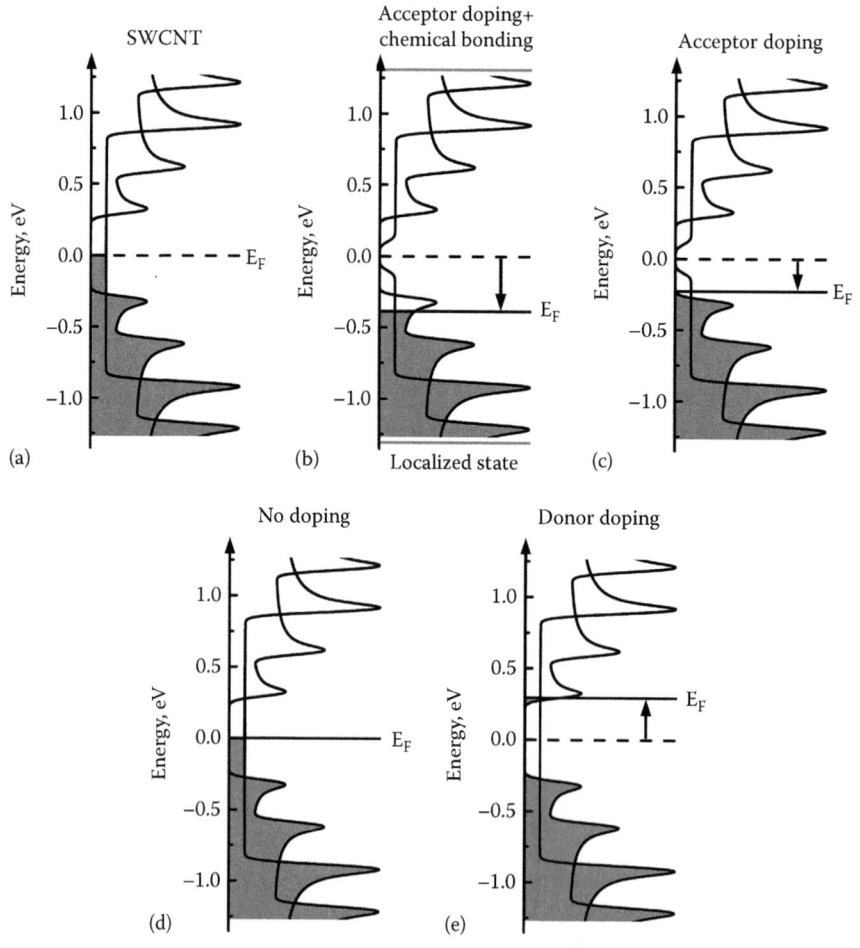

FIGURE 7.9
The schemes of the modification of the electronic structure of SWCNTs (a) upon their filling with metal halogenides (b), gallium chalcogenides (c), bismuth and tin chalcogenides (d), and metals (e).

nanotubes with these substances leads to substantial modifications of the intrinsic electronic structure of SWCNTs (Figure 7.9a). The downshift of the Fermi level of SWCNTs due to the charge transfer from the nanotube walls to the encapsulated salts is observed, that is, acceptor doping of SWCNTs takes place. Moreover, the formation of localized states in the band structure of SWCNTs occurs, which is a result of chemical bonding between atoms of the inserted compounds and atoms of the nanotubes (Figure 7.9b).

The second group includes gallium chalcogenides, which are semiconductors and have larger work function than the corresponding value of the pristine SWCNTs. The incorporation of these substances into the nanotube channels leads to acceptor doping of SWCNTs, which is accompanied by lowering their Fermi level as a result of the charge transfer from the nanotubes to the embedded compounds (Figure 7.9c).

The third group includes bismuth and tin chalcogenides, which are narrow band gap semiconductors and have the comparable work function as the value of the pristine SWCNTs. The encapsulation of these compounds into the nanotube channels does not result in modifications of the electronic structure of SWCNTs (Figure 7.9d).

The fourth group includes metals, which have smaller work function than the corresponding value of the pristine SWCNTs. The filling of nanotube channels with these substances leads to donor doping of SWCNTs, which is accompanied by the upshift of their Fermi level due to the charge transfer from the incorporated metal nanoparticles to the SWCNT walls (Figure 7.9e).

Thus, the electronic structure of SWCNTs can be modified in a controllable way via the encapsulation of substances of different chemical nature into the nanotube channels.

7.4 CONCLUSIONS

To conclude, in this chapter, the synthesis and electronic properties of SWCNTs filled with inorganic compounds (metal halogenides and metal chalcogenides) and metals were considered. The filling procedures of SWCNTs were described. The structure of the encapsulated substances was analyzed, and its correlation with the chemical nature of the compounds was explored. It was shown that the filling degree of SWCNTs and crystallinity of the inserted substances increase with the increase in halogen and chalcogen radius, and the values of $4f$ metal chlorides are larger than the ones of $3d$ and $4d$ metal chlorides. The influence of substances of different chemical nature on the electronic properties of SWCNTs was investigated. The possibility of achieving both acceptor and donor doping by the encapsulation of substances with appropriate chemical and physical properties inside the SWCNTs was demonstrated. It was shown that the incorporation of metal halogenides into SWCNTs causes the charge transfer from the nanotube walls to the encapsulated salts, that is, acceptor doping of SWCNTs, and the formation of chemical bonds between atoms of the inserted compounds and atoms of nanotubes. The filling of SWCNTs with gallium chalcogenides also leads to acceptor doping of SWCNTs, but the chemical bonds are not formed. The inserted bismuth and tin chalcogenides do not influence the electronic structure of SWCNTs. The encapsulation of metals into the SWCNTs leads to the charge transfer from the incorporated metal nanoparticles to the SWCNT walls, that is, donor doping of nanotubes. The summarized data demonstrate that the electronic structure of SWCNTs can be tailored in an ambipolar manner via the incorporation of appropriate substances into their channels, which makes the filling of SWCNTs a promising approach to realize high-performance nanotube-based electronics.

ACKNOWLEDGMENTS

The author thanks Dr. L.V. Yashina and Prof. A.V. Lukashin for the given possibility to perform scientific work at the Department of Materials Science, Lomonosov Moscow State University, and for their support, interest in the present work, and fruitful discussions. The author also thanks Prof. A.V. Krestinin (Institute of Problems of Chemical Physics, RAS, Chernogolovka, Russia) and Prof. Yu. Gogotsi (Drexel University, United States) for providing the SWCNTs; Dr. J.J. Niu (Drexel University, United States), Dr. A.V. Egorov (Lomonosov Moscow State University),

Prof. N.A. Kiselev (Institute of Crystallography, RAS, Moscow), and Dr. D.I. Petukhov (Lomonosov Moscow State University) for the HRTEM measurements of samples; and Dr. L.V. Yashina, Dr. M.M. Brzhezinskaya (BESSY, Germany), Dr. A.A. Eliseev (Lomonosov Moscow State University), and Dr. V.S. Neudachina (OJSC "Giredmet," Moscow) for their help with the investigation of samples by NEXAFS and XPS.

REFERENCES

1. Saito, R., Dresselhaus, G., Dresselhaus, M.S. 1998. *Physical Properties of Carbon Nanotubes*. London, U.K.: Imperial College Press.
2. McEuen, P.L., Fuhrer, M.S., Park, H.K. 2002. Single-walled carbon nanotube electronics. *IEEE Trans Nanotechnol* 1:78–85.
3. Javey, A., Guo, J., Farmer, D.B. et al. 2004. Self-aligned ballistic molecular transistors and electrically parallel nanotube arrays. *Nano Lett* 4:1319–1322.
4. Tans, S.J., Verschueren, A.R.M., Dekker, C. 1998. Room-temperature transistor based on a single carbon nanotube. *Nature* 393:49–52.
5. Avouris, P., Chen, Z.H., Perebeinos, V. 2007. Carbon-based electronics. *Nat Nanotechnol* 2:605–615.
6. Huang, Y.X., Sudibya, H.G., Fu, D.L. et al. 2009. Label-free detection of ATP release from living astrocytes with high temporal resolution using carbon nanotube network. *Biosens Bioelectron* 24:2716–2720.
7. Huang, Y.X., Palkar, P.V., Li, L.J., Zhang, H., Chen, P. 2010. Integrating carbon nanotubes and lipid bilayer for biosensing. *Biosens Bioelectron* 25:1834–1837.
8. Bradley, K., Gabriel, J.C.P., Gruner, G. 2003. Flexible nanotube electronics. *Nano Lett* 3:1353–1355.
9. Artukovic, E., Kaempgen, M., Hecht, D.S., Roth, S., Gruner, G. 2005. Transparent and flexible carbon nanotube transistors. *Nano Lett* 5:757–760.
10. Cao, Q., Kim, H.S., Pimparkar, N. et al. 2008. Medium-scale carbon nanotube thin-film integrated circuits on flexible plastic substrates. *Nature* 454:495–500.
11. Wang, C., Chien, J.C., Takei, K. et al. 2012. Extremely bendable, high-performance integrated circuits using semiconducting carbon nanotube networks for digital, analog, and radio-frequency applications. *Nano Lett* 12:1527–1533.
12. Bachtold, A., Hadley, P., Nakanishi, T., Dekker, C. 2001. Logic circuits with carbon nanotube transistors. *Science* 294:1317–1320.
13. Chen, Z.H., Appenzeller, J., Lin, Y.M. et al. 2006. An integrated logic circuit assembled on a single carbon nanotube. *Science* 311:1735–1735.
14. Bachilo, S.M., Balzano, L., Herrera, J.E. et al. 2003. Narrow (n,m)-distribution of single-walled carbon nanotubes grown using a solid supported catalyst. *J Am Chem Soc* 125:11186–11187.
15. Ding, L., Tselev, A., Wang, J.Y. et al. 2009. Selective growth of well-aligned semiconducting single-walled carbon nanotubes. *Nano Lett* 9:800–805.
16. Li, Y.M., Mann, D., Rolandi, M. et al. 2004. Preferential growth of semiconducting single-walled carbon nanotubes by a plasma enhanced CVD method. *Nano Lett* 4:317–321.
17. Duesberg, G.S., Burghard, M., Muster, J., Philipp, G., Roth, S. 1998. Separation of carbon nanotubes by size exclusion chromatography. *Chem Commun* 435–436.
18. Huang, X.Y., Mclean, R.S., Zheng, M. 2005. High-resolution length sorting and purification of DNA-wrapped carbon nanotubes by size-exclusion chromatography. *Anal Chem* 77:6225–6228.

19. Xu, X.Y., Ray, R., Gu, Y.L. et al. 2004. Electrophoretic analysis and purification of fluorescent single-walled carbon nanotube fragments. *J Am Chem Soc* 126:12736–12737.
20. Yang, Y.L., Xie, L.M., Chen, Z. et al. 2005. Purification and length separation of single-walled carbon nanotubes using chromatographic method. *Synthet Met* 155:455–460.
21. Ziegler, K.J., Schmidt, D.J., Rauwald, U. et al. 2005. Length-dependent extraction of single-walled carbon nanotubes. *Nano Lett* 5:2355–2359.
22. Arnold, M.S., Stupp, S.I., Hersam, M.C. 2005. Enrichment of single-walled carbon nanotubes by diameter in density gradients. *Nano Lett* 5:713–718.
23. Borowiak-Palen, E., Pichler, T., Liu, X. et al. 2002. Reduced diameter distribution of single-wall carbon nanotubes by selective oxidation. *Chem Phys Lett* 363:567–572.
24. Menna, E., Della Negra, F., Dalla Fontana, M., Meneghetti, M. 2003. Selectivity of chemical oxidation attack of single-wall carbon nanotubes in solution. *Phys Rev B* 68:193412.
25. Ortiz-Acevedo, A., Xie, H., Zorbas, V. et al. 2005. Diameter-selective solubilization of single-walled carbon nanotubes by reversible cyclic peptides. *J Am Chem Soc* 127:9512–9517.
26. Vetcher, A.A., Srinivasan, S., Vetcher, I.A. et al. 2006. Fractionation of SWNT/nucleic acid complexes by agarose gel electrophoresis. *Nanotechnology* 17:4263–4269.
27. An, L., Fu, Q.A., Lu, C.G., Liu, J. 2004. A simple chemical route to selectively eliminate metallic carbon nanotubes in nanotube network devices. *J Am Chem Soc* 126:10520–10521.
28. Chattopadhyay, D., Galeska, I., Papadimitrakopoulos, F. 2003. A route for bulk separation of semiconducting from metallic single-wall carbon nanotubes. *J Am Chem Soc* 125:3370–3375.
29. Chen, Z.H., Du, X., Du, M.H. et al. 2003. Bulk separative enrichment in metallic or semiconducting single-walled carbon nanotubes. *Nano Lett* 3:1245–1249.
30. Collins, P.C., Arnold, M.S., Avouris, P. 2001. Engineering carbon nanotubes and nanotube circuits using electrical breakdown. *Science* 292:706–709.
31. Hassanien, A., Tokumoto, M., Umek, P. et al. 2005. Selective etching of metallic single-wall carbon nanotubes with hydrogen plasma. *Nanotechnology* 16:278–281.
32. Krupke, R., Hennrich, F., von Lohneysen, H., Kappes, M.M. 2003. Separation of metallic from semiconducting single-walled carbon nanotubes. *Science* 301:344–347.
33. Lutz, T., Donovan, K.J. 2005. Macroscopic scale separation of metallic and semiconducting nanotubes by dielectrophoresis. *Carbon* 43:2508–2513.
34. Maeda, Y., Kimura, S., Kanda, M. et al. 2005. Large-scale separation of metallic and semiconducting single-walled carbon nanotubes. *J Am Chem Soc* 127:10287–10290.
35. Maeda, Y., Kanda, M., Hashimoto, M. et al. 2006. Dispersion and separation of small-diameter single-walled carbon nanotubes. *J Am Chem Soc* 128:12239–12242.
36. Menard-Moyon, C., Izard, N., Doris, E., Mioskowski, C. 2006. Separation of semiconducting from metallic carbon nanotubes by selective functionalization with azomethine ylides. *J Am Chem Soc* 128:6552–6553.
37. Miyata, Y., Maniwa, Y., Kataura, H. 2006. Selective oxidation of semiconducting single-wall carbon nanotubes by hydrogen peroxide. *J Phys Chem B* 110:25–29.
38. Arnold, M.S., Green, A.A., Hulvat, J.F., Stupp, S.I., Hersam, M.C. 2006. Sorting carbon nanotubes by electronic structure using density differentiation. *Nat Nanotechnol* 1:60–65.

39. Lustig, S.R., Jagota, A., Khripin, C., Zheng, M. 2005. Theory of structure-based carbon nanotube separations by ion-exchange chromatography of DNA/CNT hybrids. *J Phys Chem B* 109:2559–2566.

40. Strano, M.S., Zheng, M., Jagota, A. et al. 2004. Understanding the nature of the DNA-assisted separation of single-walled carbon nanotubes using fluorescence and Raman spectroscopy. *Nano Lett* 4:543–550.

41. Zheng, M., Jagota, A., Strano, M.S. et al. 2003. Structure-based carbon nanotube sorting by sequence-dependent DNA assembly. *Science* 302:1545–1548.

42. Zheng, M., Jagota, A., Semke, E.D. et al. 2003. DNA-assisted dispersion and separation of carbon nanotubes. *Nat Mater* 2:338–342.

43. Alvaro, M., Atienzar, P., de la Cruz, P. et al. 2004. Sidewall functionalization of single-walled carbon nanotubes with nitrile imines. Electron transfer from the substituent to the carbon nanotube. *J Phys Chem B* 108:12691–12697.

44. Callegari, A., Marcaccio, M., Paolucci, D. et al. 2003. Anion recognition by functionalized single wall carbon nanotubes. *Chem Commun* 2576–2577.

45. Coleman, K.S., Chakraborty, A.K., Bailey, S.R., Sloan, J., Alexander, M. 2007. Iodination of single-walled carbon nanotubes. *Chem Mater* 19:1076–1081.

46. Kawasaki, S., Komatsu, K., Okino, F., Touhara, H., Kataura, H. 2004. Fluorination of open- and closed-end single-walled carbon nanotubes. *Phys Chem Chem Phys* 6:1769–1772.

47. Marcoux, P.R., Hapiot, P., Batail, P., Pinson, J. 2004. Electrochemical functionalization of nanotube films: Growth of aryl chains on single-walled carbon nanotubes. *New J Chem* 28:302–307.

48. Strano, M.S., Dyke, C.A., Usrey, M.L. et al. 2003. Electronic structure control of single-walled carbon nanotube functionalization. *Science* 301:1519–1522.

49. Wang, Y.B., Iqbal, Z., Mitra, S. 2005. Microwave-induced rapid chemical functionalization of single-walled carbon nanotubes. *Carbon* 43:1015–1020.

50. Li, J.B., Huang, Y.X., Chen, P., Chan-Park, M.B. 2013. In situ charge-transfer-induced transition from metallic to semiconducting single-walled carbon nanotubes. *Chem Mater* 25:4464–4470.

51. Cha, M., Jung, S., Cha, M.H. et al. 2009. Reversible metal-semiconductor transition of ssDNA-decorated single-walled carbon nanotubes. *Nano Lett* 9:1345–1349.

52. Fernando, K.A.S., Lin, Y., Wang, W. et al. 2004. Diminished band-gap transitions of single-walled carbon nanotubes in complexation with aromatic molecules. *J Am Chem Soc* 126:10234–10235.

53. Matarredona, O., Rhoads, H., Li, Z.R. et al. 2003. Dispersion of single-walled carbon nanotubes in aqueous solutions of the anionic surfactant NaDDBS. *J Phys Chem B* 107:13357–13367.

54. Moore, V.C., Strano, M.S., Haroz, E.H. et al. 2003. Individually suspended single-walled carbon nanotubes in various surfactants. *Nano Lett* 3:1379–1382.

55. Nakashima, N., Tomonari, Y., Murakami, H. 2002. Water-soluble single-walled carbon nanotubes via noncovalent sidewall-functionalization with a pyrene-carrying ammonium ion. *Chem Lett* 638–639.

56. Shin, H.J., Kim, S.M., Yoon, S.M. et al. 2008. Tailoring electronic structures of carbon nanotubes by solvent with electron-donating and -withdrawing groups. *J Am Chem Soc* 130:2062–2066.

57. Ayala, P., Grueneis, A., Gemming, T. et al. 2007. Tailoring N-doped single and double wall carbon nanotubes from a nondiluted carbon/nitrogen feedstock. *J Phys Chem C* 111:2879–2884.

58. Ayala, P., Plank, W., Gruneis, A. et al. 2008. A one step approach to B-doped single-walled carbon nanotubes. *J Mater Chem* 18:5676–5681.

59. Borowiak-Palen, E., Pichler, T., Graff, A. et al. 2004. Synthesis and electronic properties of B-doped single wall carbon nanotubes. *Carbon* 42:1123–1126.

60. Elias, A.L., Ayala, P., Zamudio, A. et al. 2010. Spectroscopic characterization of N-doped single-walled carbon nanotube strands: An X-ray photoelectron spectroscopy and Raman study. *J Nanosci Nanotechnol* 10:3959–3964.

61. Krstic, V., Rikken, G.L.J.A., Bernier, P., Roth, S., Glerup, M. 2007. Nitrogen doping of metallic single-walled carbon nanotubes: n-type conduction and dipole scattering. *Europhys Lett* 77:37001.

62. McGuire, K., Gothard, N., Gai, P.L. et al. 2005. Synthesis and Raman characterization of boron-doped single-walled carbon nanotubes. *Carbon* 43:219–227.

63. Wiltshire, J.G., Li, L.J., Herz, L.M. et al. 2005. Chirality-dependent boron-mediated growth of nitrogen-doped single-walled carbon nanotubes. *Phys Rev B* 72:205431.

64. Graupner, R., Abraham, J., Vencelova, A. et al. 2003. Doping of single-walled carbon nanotube bundles by Bronsted acids. *Phys Chem Chem Phys* 5:5472–5476.

65. Kim, K.K., Bae, J.J., Park, H.K. et al. 2008. Fermi level engineering of single-walled carbon nanotubes by $AuCl_3$ doping. *J Am Chem Soc* 130:12757–12761.

66. Kramberger, C., Rauf, H., Knupfer, M. et al. 2009. Potassium-intercalated single-wall carbon nanotube bundles: Archetypes for semiconductor/metal hybrid systems. *Phys Rev B* 79:195442.

67. Kukovecz, A., Pichler, T., Pfeiffer, R., Kuzmany, H. 2002. Diameter selective charge transfer in p- and n-doped single wall carbon nanotubes synthesized by the HiPCO method. *Chem Commun* 1730–1731.

68. Lee, R.S., Kim, H.J., Fischer, J.E. et al. 2000. Transport properties of a potassium-doped single-wall carbon nanotube rope. *Phys Rev B* 61:4526–4529.

69. Liu, X., Pichler, T., Knupfer, M., Fink, J. 2003. Electronic and optical properties of alkali-metal-intercalated single-wall carbon nanotubes. *Phys Rev B* 67:125403.

70. Liu, X., Pichler, T., Knupfer, M., Fink, J., Kataura, H. 2004. Electronic properties of $FeCl_3$-intercalated single-wall carbon nanotubes. *Phys Rev B* 70:245435.

71. Rauf, H., Pichler, T., Knupfer, M., Fink, J., Kataura, H. 2004. Transition from a Tomonaga-Luttinger liquid to a Fermi liquid in potassium-intercalated bundles of single-wall carbon nanotubes. *Phys Rev Lett* 93:096805.

72. Shimoda, H., Gao, B., Tang, X.P. et al. 2002. Lithium intercalation into opened single-wall carbon nanotubes: Storage capacity and electronic properties. *Phys Rev Lett* 88:015502.

73. Suzuki, S., Bower, C., Watanabe, Y., Zhou, O. 2000. Work functions and valence band states of pristine and Cs-intercalated single-walled carbon nanotube bundles. *Appl Phys Lett* 76:4007–4009.

74. Sloan, J., Hammer, J., Zwiefka-Sibley, M., Green, M.L.H. 1998. The opening and filling of single walled carbon nanotubes (SWTs). *Chem Commun* 347–348.

75. Smith, B.W., Monthioux, M., Luzzi, D.E. 1998. Encapsulated C_{60} in carbon nanotubes. *Nature* 396:323–324.

76. Burteaux, B., Claye, A., Smith, B.W. et al. 1999. Abundance of encapsulated C_{60} in single-wall carbon nanotubes. *Chem Phys Lett* 310:21–24.

77. Jeong, G.H., Hirata, T., Hatakeyama, R., Tohji, K., Motomiya, K. 2002. C_{60} encapsulation inside single-walled carbon nanotubes using alkali-fullerene plasma method. *Carbon* 40:2247–2253.

78. Kataura, H., Maniwa, Y., Kodama, T. et al. 2001. High-yield fullerene encapsulation in single-wall carbon nanotubes. *Synthet Met* 121:1195–1196.

79. Kataura, H., Maniwa, Y., Abe, M. et al. 2002. Optical properties of fullerene and non-fullerene peapods. *Appl Phys A* 74:349–354.

80. Khlobystov, A.N., Britz, D.A., Wang, J.W. et al. 2004. Low temperature assembly of fullerene arrays in single-walled carbon nanotubes using supercritical fluids. *J Mater Chem* 14:2852–2857.

81. Sloan, J., Dunin-Borkowski, R.E., Hutchison, J.L. et al. 2000. The size distribution, imaging and obstructing properties of C_{60} and higher fullerenes formed within arc-grown single walled carbon nanotubes. *Chem Phys Lett* 316:191–198.

82. Smith, B.W., Monthioux, M., Luzzi, D.E. 1999. Carbon nanotube encapsulated fullerenes: A unique class of hybrid materials. *Chem Phys Lett* 315:31–36.

83. Yudasaka, M., Ajima, K., Suenaga, K. et al. 2003. Nano-extraction and nano-condensation for C_{60} incorporation into single-wall carbon nanotubes in liquid phases. *Chem Phys Lett* 380:42–46.

84. Britz, D.A., Khlobystov, A.N., Wang, J.W. et al. 2004. Selective host-guest interaction of single-walled carbon nanotubes with functionalised fullerenes. *Chem Commun* (2):176–177.

85. Chamberlain, T.W., Camenisch, A., Champness, N.R. et al. 2007. Toward controlled spacing in one-dimensional molecular chains: Alkyl-chain-functionalized fullerenes in carbon nanotubes. *J Am Chem Soc* 129:8609–8614.

86. Chamberlain, T.W., Champness, N.R., Schroder, M., Khlobystov, A.N. 2011. A piggyback ride for transition metals: Encapsulation of exohedral metallofullerenes in carbon nanotubes. *Chem A-Eur J* 17:668–674.

87. Gimenez-Lopez, M.D., Chuvilin, A., Kaiser, U., Khlobystov, A.N. 2011. Functionalised endohedral fullerenes in single-walled carbon nanotubes. *Chem Commun* 47:2116–2118.

88. Okazaki, T., Suenaga, K., Hirahara, K. et al. 2002. Electronic and geometric structures of metallofullerene peapods. *Physica B* 323:97–99.

89. Okazaki, T., Shimada, T., Suenaga, K. et al. 2003. Electronic properties of $Gd@C_{82}$ metallofullerene peapods: $(Gd@C_{82})_n@SWNTs$. *Appl Phys A* 76:475–478.

90. Smith, B.W., Luzzi, D.E., Achiba, Y. 2000. Tumbling atoms and evidence for charge transfer in $La_2@C_{80}@SWNT$. *Chem Phys Lett* 331:137–142.

91. Suenaga, K., Okazaki, T., Wang, C.R. et al. 2003. Direct imaging of $Sc_2@C_{84}$ molecules encapsulated inside single-wall carbon nanotubes by high resolution electron microscopy with atomic sensitivity. *Phys Rev Lett* 90:055506.

92. Suenaga, K., Taniguchi, R., Shimada, T. et al. 2003. Evidence for the intramolecular motion of Gd atoms in a $Gd_2@C_{92}$ nanopeapod. *Nano Lett* 3:1395–1398.

93. Guan, L.H., Shi, Z.J., Li, M.X., Gu, Z.N. 2005. Ferrocene-filled single-walled carbon nanotubes. *Carbon* 43:2780–2785.

94. Li, L.J., Khlobystov, A.N., Wiltshire, J.G., Briggs, G.A.D., Nicholas, R.J. 2005. Diameter-selective encapsulation of metallocenes in single-walled carbon nanotubes. *Nat Mater* 4:481–485.

95. Liu, X.J., Kuzmany, H., Ayala, P. et al. 2012. Selective enhancement of photoluminescence in filled single-walled carbon nanotubes. *Adv Funct Mat* 22:3202–3208.

96. Plank, W., Pfeiffer, R., Schaman, C. et al. 2010. Electronic structure of carbon nanotubes with ultrahigh curvature. *ACS Nano* 4:4515–4522.

97. Shiozawa, H., Pichler, T., Kramberger, C. et al. 2008. Fine tuning the charge transfer in carbon nanotubes via the interconversion of encapsulated molecules. *Phys Rev B* 77:153402.

98. Shiozawa, H., Pichler, T., Gruneis, A. et al. 2008. A catalytic reaction inside a single-walled carbon nanotube. *Adv Mater* 20:1443–1449.

99. Shiozawa, H., Pichler, T., Kramberger, C. et al. 2009. Screening the missing electron: Nanochemistry in action. *Phys Rev Lett* 102:046804.
100. Shiozawa, H., Kramberger, C., Rummeli, M. et al. 2009. Electronic properties of single-walled carbon nanotubes encapsulating a cerium organometallic compound. *Phys Status Solidi B* 246:2626–2630.
101. Kharlamova, M.V., Sauer, M., Saito, T. et al. 2013. Inner tube growth properties and electronic structure of ferrocene-filled large diameter single-walled carbon nanotubes. *Phys Status Solidi B* 250:2575–2580.
102. Morgan, D.A., Sloan, J., Green, M.L.H. 2002. Direct imaging of o-carborane molecules within single walled carbon nanotubes. *Chem Commun* 2442–2443.
103. Shiozawa, H., Silva, S.R.P., Liu, Z. et al. 2010. Low-temperature growth of single-wall carbon nanotubes inside nano test tubes. *Phys Status Solidi B* 247:2730–2733.
104. Shiozawa, H., Kramberger, C., Pfeiffer, R. et al. 2010. Catalyst and chirality dependent growth of carbon nanotubes determined through nano-test tube chemistry. *Adv Mater* 22:3685–3689.
105. Takenobu, T., Takano, T., Shiraishi, M. et al. 2003. Stable and controlled amphoteric doping by encapsulation of organic molecules inside carbon nanotubes. *Nat Mater* 2:683–688.
106. Yanagi, K., Miyata, Y., Kataura, H. 2006. Highly stabilized beta-carotene in carbon nanotubes. *Adv Mater* 18:437–441.
107. Borowiak-Palen, E., Ruemmeli, M.H., Gemming, T. et al. 2006. Silver filled single-wall carbon nanotubes - Synthesis, structural and electronic properties. *Nanotechnology* 17:2415–2419.
108. Borowiak-Palen, E., Mendoza, E., Bachmatiuk, A. et al. 2006. Iron filled single-wall carbon nanotubes—A novel ferromagnetic medium. *Chem Phys Lett* 421:129–133.
109. Chamberlain, T.W., Zoberbier, T., Biskupek, J. et al. 2012. Formation of uncapped nanometre-sized metal particles by decomposition of metal carbonyls in carbon nanotubes. *Chem Sci* 3:1919–1924.
110. Corio, P., Santos, A.P., Santos, P.S. et al. 2004. Characterization of single wall carbon nanotubes filled with silver and with chromium compounds. *Chem Phys Lett* 383:475–480.
111. Govindaraj, A., Satishkumar, B.C., Nath, M., Rao, C.N.R. 2000. Metal nanowires and intercalated metal layers in single-walled carbon nanotube bundles. *Chem Mater* 12:202–205.
112. Kharlamova, M.V., Niu, J.J. 2012. Comparison of metallic silver and copper doping effects on single-walled carbon nanotubes. *Appl Phys A* 109:25–29.
113. Kharlamova, M.V., Niu, J.J. 2012. Donor doping of single-walled carbon nanotubes by filling of channels with silver. *J Exp Theor Phys* 115:485–491.
114. Kharlamova, M.V., Niu, J.J. 2012. New method of the directional modification of the electronic structure of single-walled carbon nanotubes by filling channels with metallic copper from a liquid phase. *JETP Lett* 95:314–319.
115. Kitaura, R., Nakanishi, R., Saito, T. et al. 2009. High-yield synthesis of ultrathin metal nanowires in carbon nanotubes. *Angew Chem Int Ed* 48:8298–8302.
116. Loebick, C.Z., Majewska, M., Ren, F., Haller, G.L., Pfefferle, L.D. 2010. Fabrication of discrete nanosized cobalt particles encapsulated inside single-walled carbon nanotubes. *J Phys Chem C* 114:11092–11097.
117. Kharlamova, M.V., Yashina, L.V., Lukashin, A.V. 2013. Comparison of modification of electronic properties of single-walled carbon nanotubes filled with metal halogenide, chalcogenide, and pure metal. *Appl Phys A* 112:297–304.

118. Chancolon, J., Archaimbault, F., Pineau, A., Bonnamy, S. 2006. Filling of carbon nanotubes with selenium by vapor phase process. *J Nanosci Nanotechnol* 6:82–86.

119. Chernysheva, M.V., Kiseleva, E.A., Verbitskii, N.I. et al. 2008. The electronic properties of SWNTs intercalated by electron acceptors. *Physica E* 40:2283–2288.

120. Fan, X., Dickey, E.C., Eklund, P.C. et al. 2000. Atomic arrangement of iodine atoms inside single-walled carbon nanotubes. *Phys Rev Lett* 84:4621–4624.

121. Bajpai, A., Gorantla, S., Loffler, M. et al. 2012. The filling of carbon nanotubes with magnetoelectric Cr_2O_3. *Carbon* 50:1706–1709.

122. Costa, P.M.F.J., Sloan, J., Rutherford, T., Green, M.L.H. 2005. Encapsulation of Re_xO_y clusters within single-walled carbon nanotubes and their in tubulo reduction and sintering to Re metal. *Chem Mater* 17:6579–6582.

123. Hulman, M., Kuzmany, H., Costa, P.M.F.J., Friedrichs, S., Green, M.L.H. 2004. Light-induced instability of PbO-filled single-wall carbon nanotubes. *Appl Phys Lett* 85:2068–2070.

124. Mittal, J., Monthioux, M., Allouche, H., Stephan, O. 2001. Room temperature filling of single-wall carbon nanotubes with chromium oxide in open air. *Chem Phys Lett* 339:311–318.

125. Thamavaranukup, N., Hoppe, H.A., Ruiz-Gonzalez, L. et al. 2004. Single-walled carbon nanotubes filled with M OH (M = K, Cs) and then washed and refilled with clusters and molecules. *Chem Commun* (15):1686–1687.

126. Ayala, P., Kitaura, R., Nakanishi, R. et al. 2011. Templating rare-earth hybridization via ultrahigh vacuum annealing of $ErCl_3$ nanowires inside carbon nanotubes. *Phys Rev B* 83:085407.

127. Chernysheva, M.V., Eliseev, A.A., Lukashin, A.V. et al. 2007. Filling of single-walled carbon nanotubes by CuI nanocrystals via capillary technique. *Physica E* 37:62–65.

128. Eliseev, A.A., Yashina, L.V., Brzhezinskaya, M.M. et al. 2010. Structure and electronic properties of AgX (X = Cl, Br, I)-intercalated single-walled carbon nanotubes. *Carbon* 48:2708–2721.

129. Eliseev, A.A., Yashina, L.V., Verbitskiy, N.I. et al. 2012. Interaction between single walled carbon nanotube and 1D crystal in CuX@SWCNT (X = Cl, Br, I) nanostructures. *Carbon* 50:4021–4039.

130. Kharlamova, M.V., Brzhezinskaya, M., Vinogradov, A. et al. 2009. The forming and properties of one-dimensional $FeHal_2$ (Hal = Cl, Br, I) nanocrystals in channels of single-walled carbon nanotubes. *Russ Nanotechnol* 4:77–87.

131. Kharlamova, M.V., Eliseev, A.A., Yashina, L.V. et al. 2010. Study of the electronic structure of single-walled carbon nanotubes filled with cobalt bromide. *JETP Lett* 91:196–200.

132. Kharlamova, M.V., Yashina, L.V., Eliseev, A.A. et al. 2012. Single-walled carbon nanotubes filled with nickel halogenides: Atomic structure and doping effect. *Phys Status Solidi B* 249:2328–2332.

133. Kharlamova, M.V., Eliseev, A.A., Yashina, L.V., Lukashin, A.V., Tretyakov, Y.D. 2012. Synthesis of nanocomposites on basis of single-walled carbon nanotubes intercalated by manganese halogenides. *J Phys: Conf Ser* 345:012034.

134. Kharlamova, M.V., Yashina, L.V., Volykhov, A.A. et al. 2012. Acceptor doping of single-walled carbon nanotubes by encapsulation of zinc halogenides. *Eur Phys J B* 85:34.

135. Kharlamova, M.V., Yashina, L.V., Lukashin, A.V. 2013. Charge transfer in single-walled carbon nanotubes filled with cadmium halogenides. *J Mater Sci* 48:8412–8419.

136. Kharlamova, M.V. 2013. Comparison of influence of incorporated 3d-, 4d-and 4f-metal chlorides on electronic properties of single-walled carbon nanotubes. *Appl Phys A* 111:725–731.

137. Kumskov, A.S., Zhigalina, V.G., Chuvilin, A.L. et al. 2012. The structure of 1D and 3D CuI nanocrystals grown within 1.5–2.5 nm single wall carbon nanotubes obtained by catalyzed chemical vapor deposition. *Carbon* 50:4696–4704.

138. Zakalyukin, R.M., Mavrin, B.N., Dem'yanets, L.N., Kiselev, N.A. 2008. Synthesis and characterization of single-walled carbon nanotubes filled with the superionic material SnF_2. *Carbon* 46:1574–1578.

139. Bendall, J.S., Ilie, A., Welland, M.E., Sloan, J., Green, M.L.H. 2006. Thermal stability and reactivity of metal halide filled single-walled carbon nanotubes. *J Phys Chem B* 110:6569–6573.

140. Brown, G., Bailey, S.R., Sloan, J. et al. 2001. Electron beam induced in situ clusterisation of 1D $ZrCl_4$ chains within single-walled carbon nanotubes. *Chem Commun* 845–846.

141. Brown, G., Bailey, S.R., Novotny, M. et al. 2003. High yield incorporation and washing properties of halides incorporated into single walled carbon nanotubes. *Appl Phys A* 76:457–462.

142. Flahaut, E., Sloan, J., Coleman, K., Green, M. 2001. Synthesis of 1D P-block halide crystals within single walled carbon nanotubes. *AIP Conf Proc* 59:283–286.

143. Flahaut, E., Sloan, J., Friedrichs, S. et al. 2006. Crystallization of 2H and 4H PbI_2 in carbon nanotubes of varying diameters and morphologies. *Chem Mater* 18:2059–2069.

144. Friedrichs, S., Falke, U., Green, M.L.H. 2005. Phase separation of LaI_3 inside single-walled carbon nanotubes. *Chemphyschem* 6:300–305.

145. Hutchison, J.L., Sloan, J., Kirkland, A.I., Green, M.L.H., Green, M.L.H. 2004. Growing and characterizing one-dimensional crystals within single-walled carbon nanotubes. *J Electr Microsc* 53:101–106.

146. Hutchison, J.L., Grobert, N., Zakalyukin, R.M. et al. 2008. The behaviour of 1D CuI crystal@SWNT nanocomposite under electron irradiation. *AIP Conf Proc* 999:79–92.

147. Kirkland, A.I., Meyer, M.R., Sloan, J., Hutchison, J.L. 2005. Structure determination of atomically controlled crystal architectures grown within single wall carbon nanotubes. *Microsc Microanal* 11:401–409.

148. Kiselev, N.A., Zakalyukin, R.M., Zhigalina, O.M. et al. 2008. The structure of 1D CuI crystals inside SWNTs. *J Microsc -Oxford* 232:335–342.

149. Zakalyukin, R.M., Demyanets, L.N., Kiselev, N.A., Kumskov, A.S., Kislov, M.B., Krestinin, A.V., Hutchison, J.L. One-dimensional SnF_2 single crystals in the inner channels of single-wall carbon nanotubes: I. Preparation and basic characterization. *Crystallogr Rep* 55:507–512.

150. Kiselev, N.A., Kumskov, A.S., Zakalyukin, R.M. et al. 2012. The structure of nanocomposite 1D cationic conductor crystal@SWNT. *J Microsc* 246:309–321.

151. Philp, E., Sloan, J., Kirkland, A.I. et al. 2003. An encapsulated helical one-dimensional cobalt iodide nanostructure. *Nat Mater* 2:788–791.

152. Sloan, J., Wright, D.M., Woo, H.G. et al. 1999. Capillarity and silver nanowire formation observed in single walled carbon nanotubes. *Chem Commun* 699–700.

153. Sloan, J., Novotny, M.C., Bailey, S.R. et al. 2000. Two layer 4:4 co-ordinated KI crystals grown within single walled carbon nanotubes. *Chem Phys Lett* 329:61–65.

154. Sloan, J., Kirkland, A.I., Hutchison, J.L., Green, M.L.H. 2002. Integral atomic layer architectures of 1D crystals inserted into single walled carbon nanotubes. *Chem Commun* 1319–1332.

155. Sloan, J., Friedrichs, S., Meyer, R.R. et al. 2002. Structural changes induced in nanocrystals of binary compounds confined within single walled carbon nanotubes: A brief review. *Inorg Chim Acta* 330:1–12.

156. Sloan, J., Grosvenor, S.J., Friedrichs, S. et al. 2002. A one-dimensional BaI_2 chain with five- and six-coordination, formed within a single-walled carbon nanotube. *Angew Chem Int Ed* 41:1156–1159.

157. Sloan, J., Kirkland, A.I., Hutchison, J.L., Green, M.L.H. 2003. Aspects of crystal growth within carbon nanotubes. *C R Phys* 4:1063–1074.

158. Kharlamova, M.V. 2013. Novel approach to tailoring the electronic properties of single-walled carbon nanotubes by the encapsulation of high-melting gallium selenide using a single-step process. *JETP Lett* 98:272–277.

159. Kharlamova, M.V. 2014. Comparative analysis of electronic properties of tin, gallium and bismuth chalcogenide-filled single-walled carbon nanotubes. *J Mater Sci* 49:8402–8411.

160. Yashina, L.V., Eliseev, A.A., Kharlamova, M.V. et al. 2011. Growth and characterization of one-dimensional SnTe crystals within the single-walled carbon nanotube channels. *J Phys Chem C* 115:3578–3586.

161. Carter, R., Sloan, J., Kirkland, A.I. et al. 2006. Correlation of structural and electronic properties in a new low-dimensional form of mercury telluride. *Phys Rev Lett* 96:215501.

162. Eliseev, A.A., Chernysheva, M.V., Verbitskii, N.I. et al. 2009. Chemical reactions within single-walled carbon nanotube channels. *Chem Mater* 21:5001–5003.

163. Kumskov, A.S., Eliseev, A.A., Freitag, B., Kiselev, N.A. 2012. HRTEM of 1DSnTe@SWNT nanocomposite located on thin layers of graphite. *J Microsc* 248:117–119.

164. Li, L.J., Lin, T.W., Doig, J. et al. 2006. Crystal-encapsulation-induced band-structure change in single-walled carbon nanotubes: Photoluminescence and Raman spectra. *Phys Rev B* 74:245418.

165. Wang, Z.Y., Li, H., Liu, Z. et al. 2010. Mixed low-dimensional nanomaterial: 2D ultranarrow MoS_2 inorganic nanoribbons encapsulated in quasi-1D carbon nanotubes. *J Am Chem Soc* 132:13840–13847.

166. Carter, R., Suyetin, M., Lister, S. et al. 2014. Band gap expansion, shear inversion phase change behaviour and low-voltage induced crystal oscillation in low-dimensional tin selenide crystals. *Dalton Trans* 43:7391–7399.

167. Eliseev, A., Yashina, L., Kharlamova, M., Kiselev, N. 2011. One-dimensional crystals inside single-walled carbon nanotubes: Growth, structure and electronic properties. In *Electronic Properties of Carbon Nanotubes*, ed. J.M. Marulanda, pp. 127–156. Rijeka, Croatia: InTech.

168. Eliseev, A.A., Kharlamova, M.V., Chernysheva, M.V. et al. 2009. Preparation and properties of single-walled nanotubes filled with inorganic compounds. *Russ Chem Rev* 78:833–854.

169. Kharlamova, M.V. 2013. Electronic properties of pristine and modified single-walled carbon nanotubes. *Phys-Usp* 56:1047–1073.

170. Monthioux, M. 2002. Filling single-wall carbon nanotubes. *Carbon* 40:1809–1823.

171. Monthioux, M., Flahaut, E., Cleuziou, J.P. 2006. Hybrid carbon nanotubes: Strategy, progress, and perspectives. *J Mater Res* 21:2774–2793.

172. Eletskii, A.V. 2000. Endohedral structures. *Uspekhi Fizicheskikh Nauk* 170:113–142.

173. Eletskii, A.V. 2004. Sorption properties of carbon nanostructures. *Phys-Usp* 47:1119–1154.

174. Kataura, H., Kumazawa, Y., Maniwa, Y. et al. 1999. Optical properties of single-wall carbon nanotubes. *Synthet Met* 103:2555–2558.

175. Dresselhaus, M.S., Dresselhaus, G., Jorio, A., Souza, A.G., Saito, R. 2002. Raman spectroscopy on isolated single wall carbon nanotubes. *Carbon* 40:2043–2061.

176. Dresselhaus, M.S., Eklund, P.C. 2000. Phonons in carbon nanotubes. *Adv Phys* 49:705–814.

177. Jorio, A., Pimenta, M.A., Souza, A.G. et al. 2003. Characterizing carbon nanotube samples with resonance Raman scattering. *New J Phys* 5:139.

178. Fouquet, M., Telg, H., Maultzsch, J. et al. 2009. Longitudinal optical phonons in metallic and semiconducting carbon nanotubes. *Phys Rev Lett* 102:075501.

179. Brown, S.D.M., Corio, P., Marucci, A. et al. 2000. Anti-stokes Raman spectra of single-walled carbon nanotubes. *Phys Rev B* 61:R5137–R5140.

180. Brown, S.D.M., Jorio, A., Corio, P. et al. 2001. Origin of the Breit-Wigner-Fano line-shape of the tangential G-band feature of metallic carbon nanotubes. *Phys Rev B* 63:155414.

181. Shim, M., Ozel, T., Gaur, A., Wang, C.J. 2006. Insights on charge transfer doping and intrinsic phonon line shape of carbon nanotubes by simple polymer adsorption. *J Am Chem Soc* 128:7522–7530.

182. Araujo, P.T., Maciel, I.O., Pesce, P.B.C. et al. 2008. Nature of the constant factor in the relation between radial breathing mode frequency and tube diameter for single-wall carbon nanotubes. *Phys Rev B* 77:241403(R).

183. Kramberger, C., Rauf, H., Shiozawa, H. et al. 2007. Unraveling van Hove singularities in x-ray absorption response of single-wall carbon nanotubes. *Phys Rev B* 75:235437.

184. Ishii, H., Kataura, H., Shiozawa, H. et al. 2003. Direct observation of Tomonaga–Luttinger-liquid state in carbon nanotubes at low temperatures. *Nature* 426:540–544.

185. Ayala, P., Miyata, Y., De Blauwe, K. et al. 2009. Disentanglement of the electronic properties of metallicity-selected single-walled carbon nanotubes. *Phys Rev B* 80:205427.

CHAPTER 8

CONTENTS

Chemical Functionalization of Endohedral Metallofullerenes

Changes Caused from the Outside

8

Peng Jin, Chengchun Tang, and Zhongfang Chen

8.1 INTRODUCTION

Endohedral metallofullerenes (EMFs) are novel hybrid molecules enclosing a variety of metal ions or metal-containing clusters in various fullerene cages.[1] They have unique core–shell structures and fascinating properties that are never expected for their parent fullerenes. An intrinsic character of EMFs is the existence of charge transfer from the incarcerated metals to the outer carbon frameworks. The negatively charged cages exhibit stabilities that are mostly different from those of the neutral ones, and many labile fullerenes that violate the isolated pentagon rule (IPR)[2] are therefore successfully achieved in their endohedral forms. To date, the encaged metals mainly originate from Group III (Sc, Y, and lanthanides) on the periodic table, with minority from Groups I, II, and IV.

EMFs can be classified in terms of the inner composition. Two classical types, namely, mono-metallofullerenes (mono-EMFs) such as La@C_{82}[3] and dimetallofullerenes (di-EMFs) such as La$_2$@C_{80},[4] are first isolated and simply differentiated based on the metal number. The major isomer of La@C_{82} bears a $C_{2v}(9)$-C_{82} cage[5] with the La atom located close to a hexagon ring along the C_2 axis of the cage (Figure 8.1).[6] La$_2$@C_{80} has two La atoms trapped in a $I_h(7)$-C_{80} cage, and the whole molecule adopts D_{2h} symmetry.[7] The two La atoms can freely rotate inside the cage at room temperature.[8] The formal charge transferred from each metal to the cage is $3e$, and their electronic configurations can be denoted as La^{3+}@C_{82}^{3-} and (La^{3+})$_2$@C_{80}^{6-}, respectively.

In 1999, trimetallic nitride template (TNT) Sc$_3$N@I_h(7)-C_{80} was accidentally discovered due to N$_2$ leakage into the arc discharge chamber.[9] It has a planar and rotational Sc$_3$N moiety, which formally transfers $6e$ to the cage, resulting in a (Sc^{3+})$_3$N^{3-}@C_{80}^{6-} electronic structure (Figure 8.1). Since then, nitride clusterfullerenes (NCFs) have received tremendous attention because of their surprisingly high yields. Thus far, besides the metal nitride clusters, many complex species have been successfully encased, including metal carbides (M$_{2,3,4}$C$_2$, e.g., Sc$_2$C$_2$@C$_{84}$),[10] hydrogenated metal carbides (Sc$_3$CH, e.g., Sc$_3$CH@C$_{80}$),[11] metal nitrogen carbides (M$_{1,3}$NC, e.g., Sc$_3$NC@C$_{80}$),[12] metal oxides (Sc$_{2,4}$O$_{1,2,3}$, e.g., Sc$_4$O$_2$@C$_{80}$),[13] and metal sulfides (M$_2$S, e.g., Sc$_2$S@C$_{82}$).[14]

In recent years, the chemical functionalization of EMFs has become a very active research area mainly because (1) it can help characterize the exact molecular structures of EMFs with X-ray crystallography, since functionalized derivatives are easier to crystallize than their pristine forms; (2) it is an effective means to finely modulate the electronic and magnetic properties of

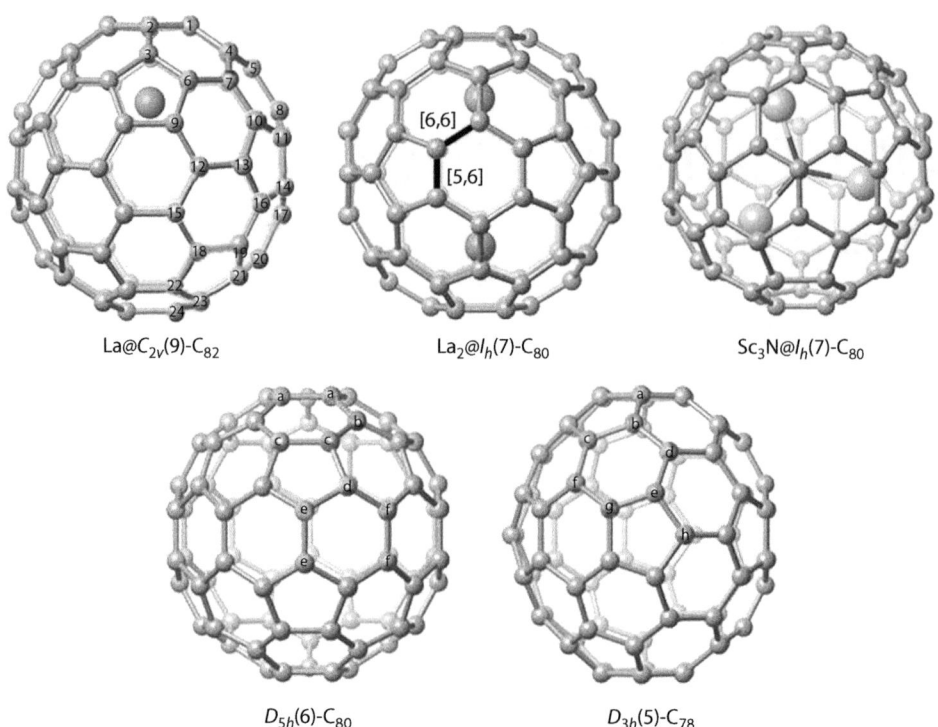

La@C_{2v}(9)-C_{82} La$_2$@I_h(7)-C_{80} Sc$_3$N@I_h(7)-C_{80}

D_{5h}(6)-C_{80} D_{3h}(5)-C_{78}

FIGURE 8.1
Structures of some EMFs and fullerene cages.

EMFs to generate new functional materials with multiple potential applications; (3) it can improve the solubility of EMFs to obtain water-soluble derivatives to be used in biomedicines; and (4) the position and motion of encaged metals can be regulated by exohedral functionalization, which may find application in the design of nanodevices. In this chapter, we present an exhaustive survey of important research progress in the chemical functionalization of EMFs. Typical functionalization reactions including Diels–Alder reaction, 1,3-dipolar cycloaddition, Bingel–Hirsch reaction, carbene addition reactions, silylation reaction, photochemical reaction, free-radical reaction, noncovalent complexation, and other miscellaneous reactions are thoroughly summarized and reviewed. Finally, the potential applications of functionalized EMFs are addressed.

8.2 FUNCTIONALIZATION REACTIONS OF ENDOHEDRAL METALLOFULLERENES

8.2.1 Disilylation

Disilylation is the first chemical reaction to functionalize EMFs. In 1995, Akasaka et al. reported the photochemical reaction between La@C_{2v}(9)-C_{82} and 1,1,2,2,-tetrakis(2,4,6-trimethylphenyl)-1,2-disilrane[(Mes$_2$Si$_2$)$_2$CH$_2$; Mes = 2,4,6-trimethylphenyl] in toluene solution (Figure 8.2).[15] Mass spectrum confirmed the formation of 1:1 La@C_{82}(Mes$_2$Si$_2$)$_2$CH$_2$ monoadduct. They obtained similar results when using 1,1,2,2,-tetrakis(2,6-dimethylphenyl)-1,2-disilrane as the reactant. New sets of octet lines in the EPR spectrum for La@C_{82}(Mes$_2$Si$_2$)$_2$CH$_2$ imply the coexistence of several regioisomers. The low regioselectivity was later explained by the favorable singly occupied molecular orbital (SOMO) distribution of La@C_{82}.[16] The hyperfine coupling constants (hfccs) (1.7 and 1.8 G for two major isomers) of the adducts are much larger than that of La@C_{82} due to the electron donation from the disilirane moiety to the carbon cage, which suppresses the charge transfer (otherwise 3e) from the La atom. The disilylation can also proceed by thermal treatment at 80°C. With high electron-accepting ability, M@C_{82}(M = Y,[17] Ce,[18] Pr,[19] and Gd[20]) can react with disilirane both

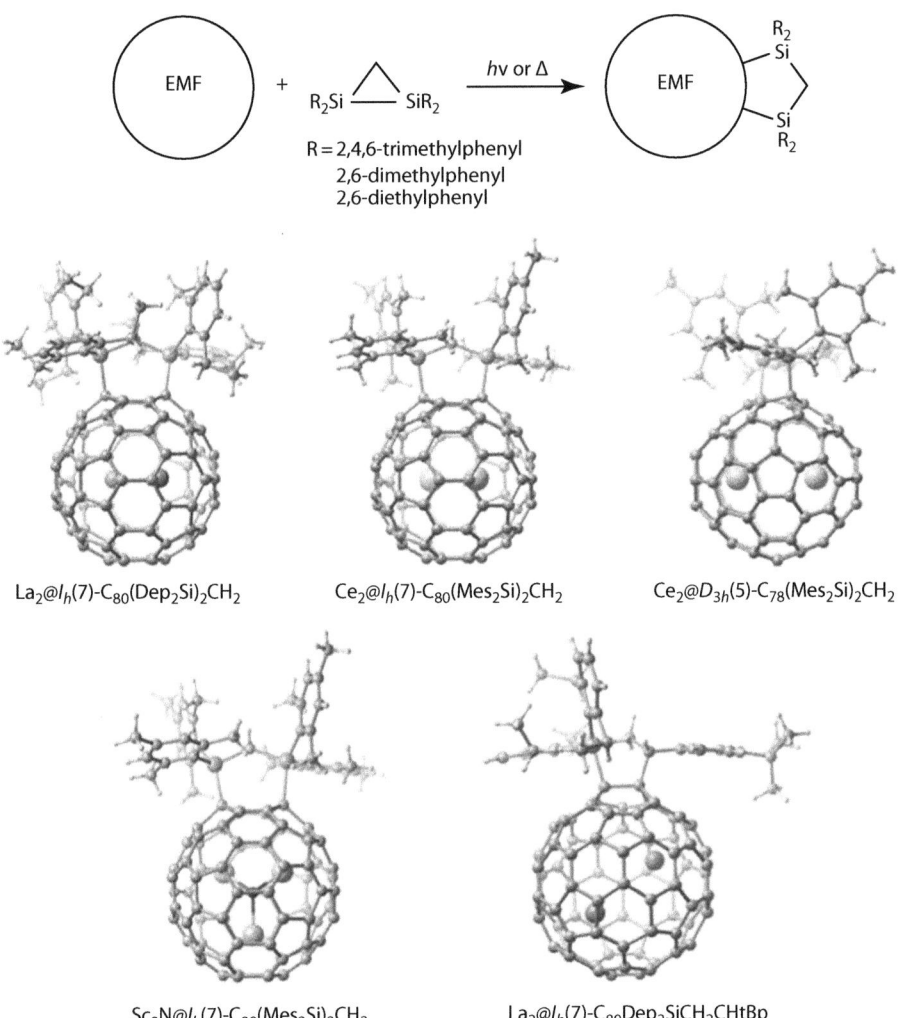

La$_2$@I_h(7)-C$_{80}$(Dep$_2$Si)$_2$CH$_2$ Ce$_2$@I_h(7)-C$_{80}$(Mes$_2$Si)$_2$CH$_2$ Ce$_2$@D_{3h}(5)-C$_{78}$(Mes$_2$Si)$_2$CH$_2$

Sc$_3$N@I_h(7)-C$_{80}$(Mes$_2$Si)$_2$CH$_2$ La$_2$@I_h(7)-C$_{80}$Dep$_2$SiCH$_2$CHtBp

FIGURE 8.2

Disilylation of EMFs. (Product structures from Wakahara, T. et al., *Chem. Commun.*, 2680, 2007; Yamada, M. et al., *J. Am. Chem. Soc.*, 127, 14570, 2005; Yamada, M. et al., *J. Am. Chem. Soc.*, 132, 17953, 2010; Yamada, M. et al., *Chem. Commun.*, 558, 2008; Wakahara, T. et al., *J. Am. Chem. Soc.*, 128, 9919, 2006.)

photochemically and thermally. In contrast, La@C_s(6)-C$_{82}$ failed to give any detectable signal in neither electron spin resonance (ESR) nor mass spectrum.[21]

Likewise, the photochemical reaction of La@C$_{82}$ with 1,1,2,2-tetrakis(2,6-diethylphenyl)-1,2-digermirane [(Dep$_2$Ge)$_2$CH$_2$; Dep = 2,6-diethylphenyl] afforded at least three isomeric mono-adducts.[16,22] Notably, the addition can proceed thermally, even at 20°C. The facile reaction is rationalized by considering that digermirane has a lower oxidation potential than disilirane.

The reactivity of EMFs can be tuned by altering their charge states. For M@C$_{82}$ (M = Y, La, and Ce), their monocations reacted readily with disilirane to afford 1:1 adducts at room temperature in the dark, whereas no product was obtained for their monoanions either thermally (80°C) or photochemically (<400 nm).[23]

The (Mes$_2$Si$_2$)$_2$CH$_2$ derivatives of La$_2$@I_h(7)-C$_{80}$ and Sc$_2$@C$_{84}$(III)(actually Sc$_2$C$_2$@ C_{3v}(8)-C$_{82}$[24,25]) were also synthesized.[26] La$_2$@C$_{80}$ reacts with disilirane both photochemically and thermally, whereas Sc$_2$C$_2$@C_{3v}(8)-C$_{82}$ is only reactive under photoirradiation mainly due to its more negative reduction potential than that of La$_2$@C$_{80}$(−0.97 vs. −0.31 V). Density functional theory (DFT) calculations revealed that the additions of one or two disiliranes onto the La$_2$@C$_{80}$ cage energetically favor a 1,4-postion.[27] The appreciable charge transfer from the disiliranes to La$_2$@C$_{80}$ leads to an electronic configuration of (La$_2$@C$_{80}$)$^{0.9(1.8)−}$[(H$_2$Si)$_2$CH$_2$]$_{1(2)}^{0.9(1.8)+}$. The additional electrons enter the La–La σ-bonding orbital and shorten the metal–metal distance. The addition pattern is supported by the recent X-ray crystallographic analysis on $La_2@C_{80}(Dep_2Si)_2CH_2$, in which each La atom resides toward a hexagon at the equator of the cage (Figure 8.2).[28] Consistent with the theoretical calculations,[27] a large broadening of the ^{139}La NMR linewidth was observed when temperature increased from 183 to 308 K, indicating that the two La atoms perform a 2D hopping motion between two sites along the equator. For the $Ce_2@C_{80}(Mes_2Si)_2CH_2$ first synthesized in 2005, X-ray data and NMR measurement show that the two Ce atoms are even fixed in the equator plane (Figure 8.2).[29] Such controllable metal motion by chemical functionalization may help designing novel molecular devices. In addition, the photochemical and thermal reaction of $Ce_2@ D_{5h}(6)-C_{80}$ with disilirane was also reported.[30]

On the other hand, the thermal carbosilylation of $La_2@I_h-C_{80}$ with silirane (silacyclopropane) afforded two diastereomers of the carbosilylated derivatives $La_2@I_h-C_{80}Dep_2SiCH_2CHtBp$ (tBp = 4-$tert$-butylphenyl).[31] There are two types of double bonds available on the I_h-C_{80} cage (Figure 8.1). One is the [5,6]-ring junction shared by one pentagon and one hexagon, and the other is the [6,6]-ring junction between two hexagons, with the former featuring a symmetric plane. X-ray crystallographic analysis revealed that the addition occurs at the [5,6]-ring junction with the silacyclopentane ring exhibiting an envelope conformation (Figure 8.2). The La atoms are disordered inside the cage and may circuit along a band of 10 contiguous hexagons, a scenario which is supported by the variable-temperature ^{139}La NMR spectra. The adducts exhibit intermediate electrochemical properties between those of the pristine $La_2@I_h-C_{80}$ and bis-silylated $La_2@I_h-C_{80}(Dep_2Si)_2CH_2$.[28] Thus, the electronic properties of EMFs can be finely tuned by gradually changing the number of silyl groups.

$Ce_2@D_{3h}(5)-C_{78}$ and its bis-silylated derivative were synthesized and fully characterized.[32] The X-ray crystallographic analysis shows that the two Ce atoms are disordered over two locations (Ce–Ce distance of 4.036 Å for the major components) and face toward the hexagonal rings on the C_3 axis of the cage (Figure 8.2). The comparative ^{13}C NMR measurements reveal that the Ce atoms are situated at the equator in the pristine $Ce_2@C_{78}$ but more tightly localized in the monoadduct.

The encased species plays a critical role in the chemical reactivity of EMFs. Although holding the same cage and electronic structure ($I_h-C_{80}^{6-}$) as $La_2@C_{80}$, $Sc_3N@I_h-C_{80}$ only reacts with disilirane under photoirradiation to afford a bis-silylated product $Sc_3N@C_{80}(Mes_2Si)_2CH_2$ because of its higher lowest unoccupied molecular orbital (LUMO) level and more negative first reduction potential than $La_2@C_{80}$ (−1.22 vs. −0.31 V).[33] NMR spectroscopy and X-ray crystallographic analysis revealed that the cycloadduct has 1,2- and 1,4-isomers (Figure 8.2), and the former isomerizes to the latter by heating.[34] The free motion of Sc_3N cluster in the pristine $Sc_3N@C_{80}$[9] is seriously prohibited and may be restricted in the plane perpendicular to the cage equator. Charge is transferred from the disilirane to the EMF molecule and result in an electronic configuration of $(Sc_3N@C_{80})^{1.2-}[(Mes_2Si)_2CH_2]^{1.2+}$.

To summarize, classical EMFs react with nucleophilic disilirane both photochemically and thermally due to their high electron affinities, whereas clusterfullerenes only undergo photochemical reactions.

8.2.2 Diels–Alder Cycloaddition

The Diels–Alder reaction refers to the [4 + 2] cycloaddition between a conjugated diene and an alkene (commonly termed as dienophile) to form a cyclohexene system.

In 2002, Dorn et al. first reported the thermal reaction of $Sc_3N@I_h(7)-C_{80}$ with ^{13}C-labeled 6,7-dimethoxyisochroman-3-one to afford a $Sc_3N@C_{80}C_8H_6(OCH_3)_2$ cycloadduct (Figure 8.3).[35] ^{13}C NMR spectrum and X-ray crystal analysis confirmed that the reaction took place at a [5,6] bond with an elongated C–C bond distance (1.626 Å).[36] The Sc_3N cluster adopts a planar geometry and locates far from the addend. The same Diels–Alder reaction for $Gd_3N@C_{80}$ led to the formation of the first isolated NCF bisadduct in low yield.[37]

The room-temperature reaction of $La@C_{2v}(9)-C_{82}$ with excess cyclopentadiene (Cp) in toluene gave rise to 1:1 adduct in a low yield (44%) due to the fast retro-reaction (Figure 8.3).[38] When 1,2,3,4,5-pentamethylcyclopentadiene(Cp*) was chosen as the reactant, however, the $La@C_{82}Cp*$ monoadduct showed higher stability with much lower decomposition rate.[39] X-ray crystallographic analysis and recent DFT calculations[40] confirmed that the two additions occur at the C21–C23 position (Figure 8.3). The stability of $La@C_{82}Cp*$ is mainly due to the long-range dispersion interactions.[40] The high regioselectivity can be understood by the large pyramidalization angle

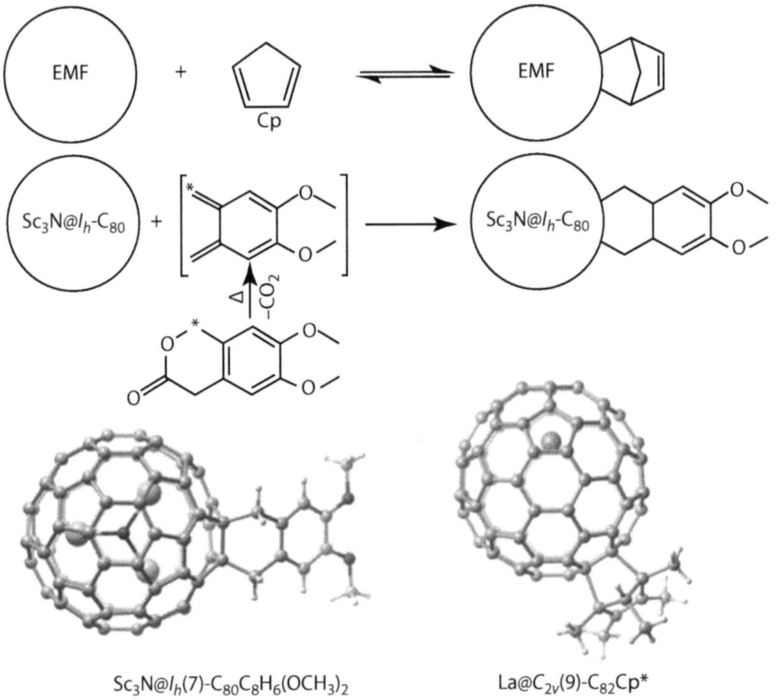

FIGURE 8.3
Diels–Alder reactions of EMFs. (Product structures from Lee, H.M. et al., *J. Am. Chem. Soc.*, 124, 3494, 2002; Maeda, Y. et al., *Chem. Eur. J.*, 16, 2193, 2010.)

(measured from the π-orbital axis vector [POAV][41]) and the positive charge density of C21 and C23, as well as large thermal stability of the corresponding adduct. The formation of La@C_{82}Cp* proceeds *via* a concerted mechanism[42] and is both kinetically and thermodynamically favored. Interestingly, La@C_{82}Cp* assumes two orientations in the crystal structure, and 60% of the adduct form a dimer. The bisadduct of La@C_{82} with Cp* and adamantylidene (Ad) was also achieved.

By means of DFT calculations, Swart, Solà, and coworkers systematically studied the Diels–Alder reaction of 1,3-*cis*-butadiene with X@C_{78} (X = Sc_3N, Y_3N, Ti_2C_2)[43–45] and X@C_{80}(X = La_2, Y_3, Sc_3N, Y_3N, Gd_3N, Lu_3N, Sc_3C_2, Sc_3CH, Sc_3NC, Sc_4C_2, Sc_4O_2, Sc_4O_3).[46,47] Compared with the hollow cages, EMFs exhibit less reactivity toward butadiene mainly due to formal charge transfer from the metal cluster to the cage, which leads to higher-lying LUMOs. The LUMO alteration and structural deformation of the carbon framework result in certain types of unreactive bonds are activated, and some are deactivated. For all the I_h-C_{80}-based EMFs except Y_3@I_h-C_{80}, the [5,6] adduct is thermodynamically favored than [6,6] one, and they coexist in many cases. The trapped metal cluster significantly affects the reactivity of EMFs, and larger size generally leads to less regioselectivity. The D_{5h}(6)-C_{80} counterparts exhibit much higher reactivity.[46,48]

Significantly, TNT EMFs can be quickly and efficiently isolated from hollow fullerenes and classical EMFs by taking advantaging of their different chemical reactivities in the Diels–Alder reactions.[49]

8.2.3 Prato Addition

The 1,3-dipolar cycloaddition reaction between (metallo)fullerenes and azomethine ylides, namely, Prato reaction, was first reported in 1993 and has been widely employed to prepare various (metallo) fulleropyrrolidines (Figure 8.4).[50]

In 2004, Akasaka and coworkers synthesized a monoadduct and a bisadduct of La@C_{2v}(9)-C_{82} metallofulleropyrrolidines by heating a toluene solution of La@C_{82} with excess *N*-methylglycine and paraformaldehyde.[51] The addition is rather efficient and up to 99% of La@C_{82} has been converted. EPR and UV–vis–NIR absorption spectra showed that the electronic structures of the pyrrolidine cycloadducts, especially the bisadduct, are different from that of the pristine La@C_{82}.

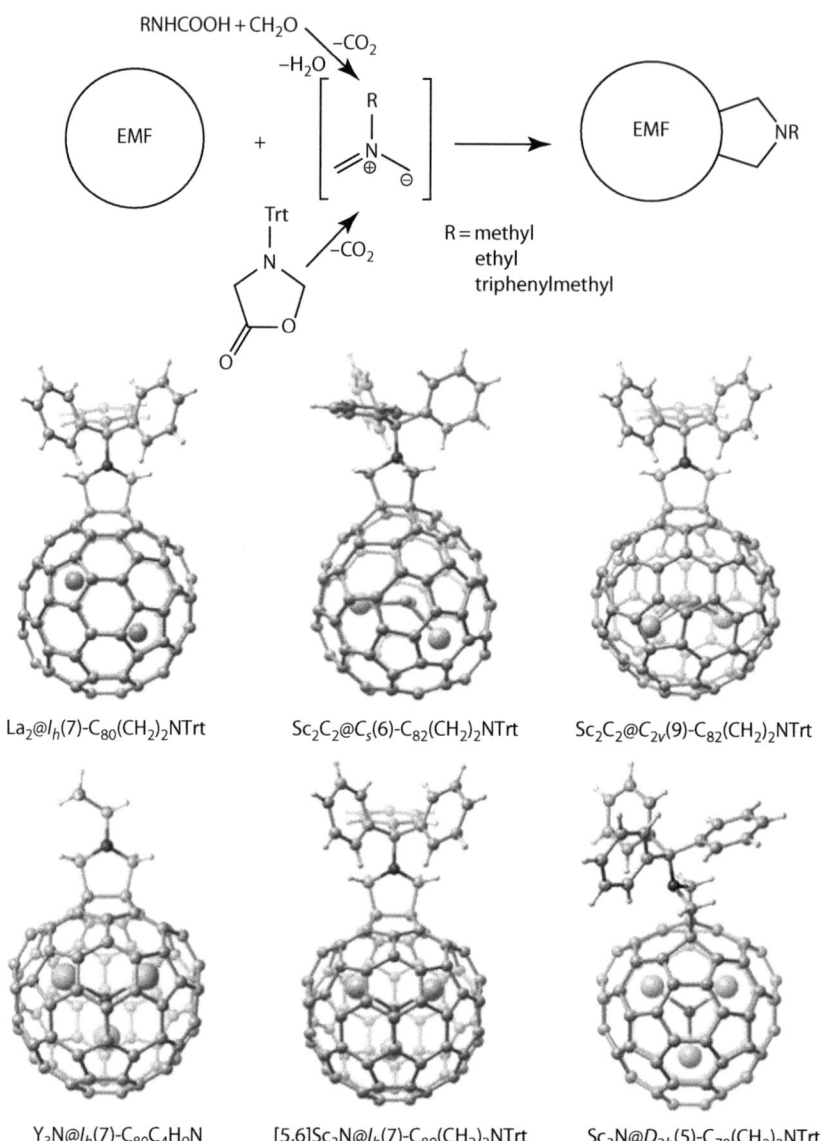

FIGURE 8.4
Prato reactions of EMFs. (Product structures from Yamada, M. et al., *J. Am. Chem. Soc.*, 128, 1402, 2006; Echegoyen, L. et al., *Chem. Commun.*, 2653, 2006; Cai, T. et al., *J. Am. Chem. Soc.*, 128, 6486, 2006; Cai, T. et al., *J. Am. Chem. Soc.*, 129, 10795, 2007; Lu, X. et al., *J. Am. Chem. Soc.*, 133, 19553, 2011; Lu, X. et al., *Angew. Chem. Int. Ed.*, 51, 5889, 2012.)

In the same year, Gu et al. found that azomethine ylides obtained from the reaction between sarcosine and aldehydes with different substituted groups can react with $M@C_{82}(M = Y, Gd)$ to achieve different number of pyrrolidine rings on the cages.[52,53]

$M_2@I_h(7)\text{-}C_{80}(CH_2)_2NTrt$ (M = La, Ce; Trt = triphenylmethyl) pyrrolidinodimetallofullerenes were synthesized by the thermal reaction of $M_2@C_{80}$ with *N*-triphenylmethyl-5-oxazolidinone in toluene (Figure 8.4).[54,55] Both the [6,6] and [5,6] regioisomers were obtained with different relative ratio (4:1 and 1:1 for M = La and Ce, respectively). X-ray data of the two [6,6]-closed adducts (Figure 8.4) show slanted metal atoms inside the cage (La–La, 3.8231 Å; Ce–Ce, 3.8998 Å). In contrast, paramagnetic NMR spectra suggested the two Ce atoms are collinear with the pyrrolidine group in the [5,6] adduct. Thus, the metal position can be finely tuned by addition pattern. Theoretical calculations and NMR spectra suggest restricted metal motion upon the exohedral modification.

The first TNT pyrrolidinometallofullerenes were achieved by heating *o*-dichlorobenzene (*o*-DCB) solutions containing $M_3N@I_h\text{-}C_{80}(M = Sc$ and Er) and excess formaldehyde and *N*-methylglycine.[56,57] ^1H, ^{13}C NMR, and heteronuclear multiple quantum coherence (HMQC) spectra of the ^{13}C-enriched Sc$_3$N@C$_{80}$ adduct suggest that the reaction may take place at the [5,6] double bond. The first pyrrolidine monoadduct of Y$_3$N@I_h-C$_{80}$ was later reported.[58] A comparative study on the Prato reaction of M$_3$N@C$_{80}$ (M = Sc, Y, Er)[59] reveals that, in contrast to the [5,6] addition pattern in the Diels–Alder and pyrrolidine products of Sc$_3$N@I_h-C$_{80}$,[35,36,56,57] Y$_3$N@C$_{80}$ and Er$_3$N@C$_{80}$ afforded both [5,6] and [6,6] ones. The observed reversible and irreversible electrochemical cathodic behaviors for the [5,6] and [6,6] monoadducts, respectively, render electrochemistry a practical tool to characterize the two regioisomers. In line with the experiments,[59] the [6,6] and [5,6] adducts of M$_3$N@C$_{80}$(M = Sc, Y) were proposed as the kinetic and thermodynamic products, respectively.[60] The [6,6] adduct of Y$_3$N@C$_{80}$ can isomerize to the thermodynamically stable [5,6] isomer and exhibit a *dancing* pyrrolidine on the cage surface. The detailed structure of the [5,6] Y$_3$N@C$_{80}$(C$_4$H$_9$N) was confirmed by the X-ray data (Figure 8.4).[61] The Y$_3$N cluster slightly deviates from planarity inside the cage due to its large size. Two Y atoms are located nearby two [5,6] junctions and straddle the addition site, whereas the third one is far from the addend.

The 1,3-dipolar reaction of Sc$_3$N@I_h-C$_{80}$ with *N*-triphenylmethyl-5-oxazolidinone yielded two monoadducts and a bisadduct.[62] ^{1}H and ^{13}C NMR, COSY, HMQC, and UV–vis spectra suggested that the addition occurred at both the [5,6] and [6,6] junctions. X-ray data unambiguously confirmed their structures with disordered Sc atoms far from the addend. The [6,6] adduct can be converted to the [5,6] one (Figure 8.4) by heating its refluxing chlorobenzene solution, indicating that the former and the latter are kinetically and thermodynamically controlled products, respectively. A pirouette-like interconversion mechanism was proposed, which involves the formation of an intermediate zwitterionic species from the [6,6] adduct, followed by the rotation of the CH$_2$-cage bond to form the [5,6] adduct. The same cycloaddition to Sc$_3$N@D_{5h}(6)-C$_{80}$ is much faster and gives rise to two mono- and two bis-tritylpyrrolidino adducts.[48] ^{1}H NMR and correlation spectroscopy (COSY) spectra suggest that one monoadduct is an asymmetric product at a [6,6] bond (a–b, b–c, or d–f, Figure 8.1), whereas the other is thermodynamically stable and corresponds to a symmetrical addition at a [6,6] junction (e–e). However, recent DFT calculations[63] point that two energetically the most favored addition sites are [5,6] bonds a–a and c–c.

Recently, Aroua and Yamakoshi reported the Prato reaction of M$_3$N@I_h-C$_{80}$(M = Sc, Y, Gd, Lu) and the formation of fulleropyrrolidinebis(carboxylic acid) derivatives.[64] High-performance liquid chromatography (HPLC), NMR, and vis–NIR spectra confirmed the initial formation of kinetic [6,6] monoadducts, which further rearranged to the thermodynamically stable [5,6] adducts. Larger clusters lead to lower conversion rates, and the thermal rearrangements are even reversible for Y$_3$N@C$_{80}$ and Gd$_3$N@C$_{80}$. The [6,6] adduct of Gd$_3$N@C$_{80}$ was successfully functionalized through a coupling reaction to afford a highly water-soluble derivative.

Sc$_3$N@D_{3h}(5)-C$_{78}$ exhibits much higher reactivity than Sc$_3$N@I_h-C$_{80}$ and affords two mono- and one di-*N*-tritylpyrrolidino adducts.[65] D_{3h}(5)-C$_{78}$ has 8 types of carbons and 13 sets of C–C bonds (Figure 8.1). For the monoadducts, ^{1}H and ^{13}C NMR spectra as well as DFT calculations indicate that they are [6,6] regioisomers (c–f and b–d), as previously suggested based on the bond order analysis.[66] X-ray crystallographic analysis unambiguously confirmed addition pattern of the c–f isomer (Figure 8.4). The inner Sc$_3$N cluster is restricted in a horizontal plane of the cage and has no direct interaction with the addition site.

The Prato reactions of mixed-metal NCFs were reported. In 2007, Wang et al. investigated the regioselectivity of Sc$_x$Gd$_{3-x}$N@I_h-C$_{80}$(x = 0–3) in the reaction with formaldehyde and *N*-ethylglycine.[67] According to the HPLC profile, [5,6] adduct dominates in the Sc$_3$N@C$_{80}$ case, whereas the yield of [6,6] adduct gradually increases from Sc$_2$GdN@C$_{80}$ to ScGd$_2$N@C$_{80}$ and finally becomes the major isomer for Gd$_3$N@C$_{80}$. Similar trend was found for Sc$_{3-x}$Y$_x$N@I_h-C$_{80}$(x = 0–3).[68] Larger metal cluster introduces more stress and thus result in more deformed C$_{80}$ cage. Although the [5,6] isomer holds the lowest energy in most cases, the relative stability of the two regioisomers is dramatically decreased and even inversed with increasing cluster size. These studies, together with the previous reports,[56–60,64,67] underscore the critical size effect of inner species on the regioselectivity of EMFs.[69]

A paramagnetic Sc$_3$C$_2$@I_h-C$_{80}$ fulleropyrrolidine monoadduct was recently synthesized, and ^{13}C NMR and HMQC spectra suggest that its addition occurs at the [5,6]-ring junction.[70] FTIR spectra and DFT calculations reveal that the trifoliate-shaped Sc$_3$C$_2$ cluster deforms and its symmetry reduces from C_{3v} to C_{2v} upon the exohedral addition. The Sc hfccs of the pyrrolidine monoadduct (8.602 G for one nucleus, 4.822 G for two nuclei) obviously differs from that of the pristine Sc$_3$C$_2$@

C_{80} (6.256 G for three nuclei). The results can be understood in terms of their different spin density distributions: in contrast to the homogeneous distribution on the three Sc atoms in pristine $Sc_3C_2@C_{80}$, a higher spin density resides on the Sc atom far from the pyrrolidine addend. Afterward, nine $Sc_3C_2@C_{80}$ fulleropyrrolidine bisadducts were prepared, with the second pyrrolidine addend most likely attacking a [5,6]-ring junction.[71] The spin densities on the Sc nuclei become completely unequal, and their distributions can be finely manipulated by altering the second reaction sites. The controllable paramagnetism induced by exohedral functionalization is significant for the design of novel molecular devices.

The reaction of $Sc_2C_2@C_s(6)$-C_{82} with N-triphenylmethyl-5-oxazolidinone afforded one monoadduct.[72] X-ray data revealed that the addition selectively occurs at a [6,6] junction far from the butterfly-shaped Sc_2C_2 cluster due to the favorable LUMO distribution at the site (Figure 8.4). The same 1,3-dipolar reaction for $Sc_2C_2@C_{2v}(9)$-C_{82} affords three monoadduct isomers, A, B, and C, with a relative ratio of 2:1.5:1.[73] According to the LUMO distributions, their reaction sites are suggested to be the [5,6], [6,6], and [6,6] junctions, respectively, and the addition pattern of isomer A was confirmed by X-ray analysis (Figure 8.4). The chemical modification drastically altered the electronic and electrochemical properties of $Sc_2C_2@C_{82}$. Most of the redox potentials of $Sc_2C_2@C_s(6)$-C_{82} are cathodically shifted by 0.1–0.3 V, whereas the UV–vis–NIR spectra lines are blue shifted. The electrochemical bandgap of $Sc_2C_2@C_{2v}(9)$-C_{82} (0.99 V) is markedly enlarged upon exohedral functionalization (A, 1.40 V; B, 1.23 V; C, 1.19 V).

8.2.4 Bingel–Hirsch Reaction

In a typical Bingel–Hirsch reaction, carbanion is first generated by the dehydrogenation of bromomalonate with 1,8-diazabicyclo[5.4.0]undec-7-ene (DBU) and then attacks fullerene cages to form cyclopropanated derivatives (Figure 8.5).[74]

The Bingel–Hirsch reaction of La@$C_{2v}(9)$-C_{82} with diethyl bromomalonate in the presence of DBU afforded five La@$C_{82}CBr(COOC_2H_5)_2$ monoadducts; four of them are diamagnetic and ESR silent.[75,76] NMR spectra suggest that all the ESR-inactive monoadducts are singly bonded. For the most abundant isomer, X-ray analysis confirmed the addition at the C23 site (Figure 8.1), which has the highest positive charge and large local strain in the underivatized La@C_{82}, very reactive toward nucleophilic attack (Figure 8.5). Several positively charged carbon atoms far from the metal are suggested as the possible reaction sites for other single-bonded adducts. For the paramagnetic isomer, NMR and UV–vis–NIR spectra suggest that it may have open-cage structure.

By using malonate under elevated temperature, a La@$C_{82}[CH(COOC_2H_5)_2]_2$ bisadduct was prepared.[77] X-ray data reveal that the reaction selectively occurs at the adjacent C21 and C23 sites. The La atom occupies several sites inside the cage, with the most abundant one deviating from the C_2 axis and far from the reaction site. The bisadduct has a pair of enantiomers, which couple to form a dimer in the single crystal with long intercage bond (1.638 Å, Figure 8.5). This is the first unambiguous evidence for EMFs dimerization.

In 2010, Echegoyen et al. prepared several Bingel–Hirsch derivatives of $M_3N@I_h(7)$-C_{80}(M = Sc, Lu) by the reaction with 2-bromodiethyl malonate, 2-bromo-1,3-dipyrrolidin-1-ylpropane-1,3-dionate bromide, and 9-bromo fluorine in the presence of DBU or NaH.[78] ¹H NMR spectra point that the monoadducts likely have a [6,6] addition pattern, and the products were afterward suggested to form under kinetic control by theoretical calculations.[79]

The room-temperature cyclopropanation reaction of $Y_3N@I_h$-C_{80} with excess diethyl bromomalonate in the presence of DBU afforded monoadduct $Y_3N@C_{80}C(COOC_2H_5)_2$ with addition occurring at the [6,6] bond.[58] In contrast, $Sc_3N@C_{80}$ does not react at all under the same conditions. $Y_3N@I_h$-$C_{80}C(COOCH_2Ph)_2$ was synthesized later, and X-ray structure showed that it is a fulleroid with the addition occurring at the [6,6] ring junction close to one Y atom (Figure 8.5).[80] The inner Y_3N cluster is disordered and assumes a planar geometry within the spacious cavity (for the major occupancy). Theoretical calculations indicate a hampered cluster rotation due to the exohedral modification.

A comparative study on the Bingel reactions of $Gd_3N@C_{2n}(2n = 80, 82, 84, 88)$ was conducted.[81,82] Under the same conditions, $Gd_3N@C_{80}$ afforded both a mono- and a biadduct as well as several multiadducts, whereas three and one monoadducts were obtained for $Gd_3N@C_s(39663)$-C_{82}

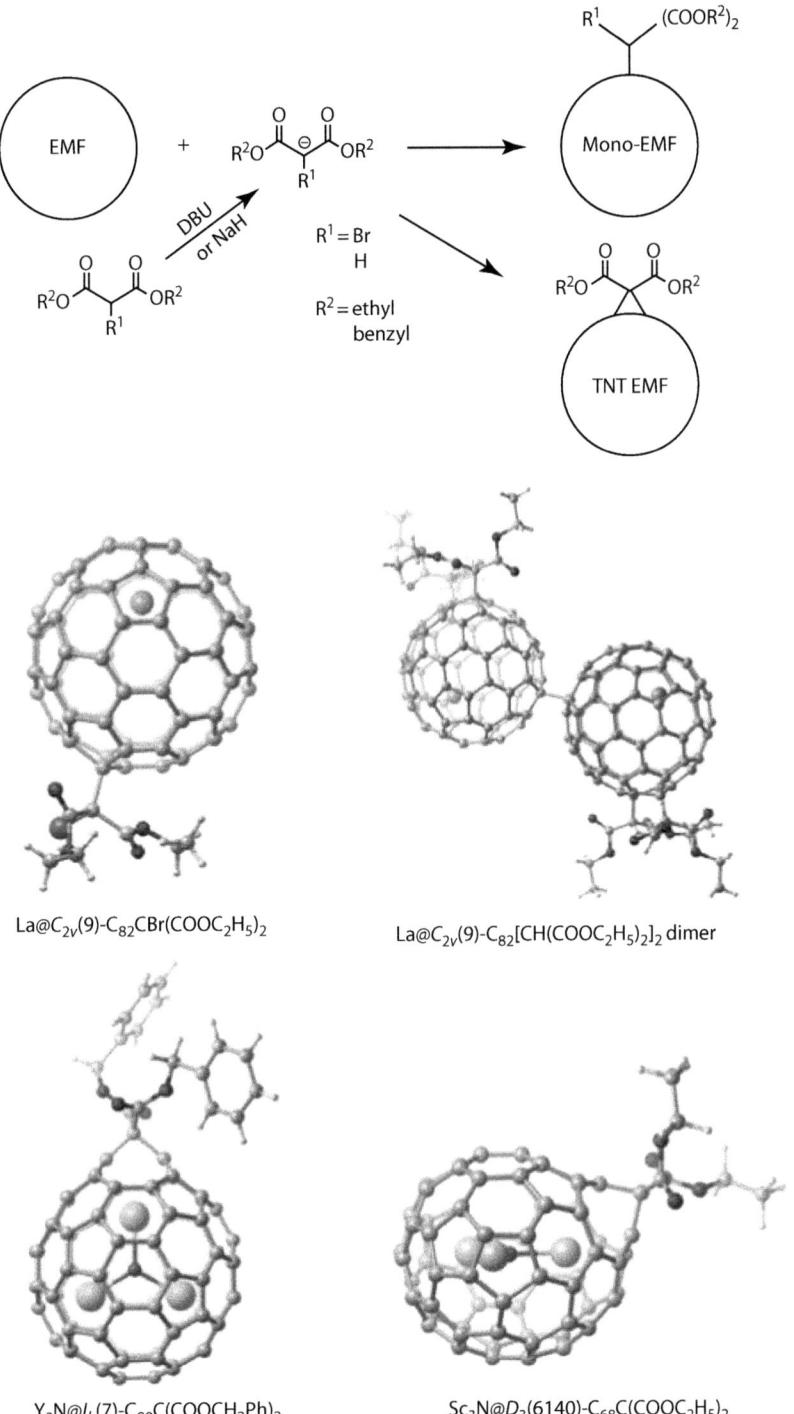

La@C_{2v}(9)-C_{82}CBr(COOC$_2$H$_5$)$_2$

La@C_{2v}(9)-C_{82}[CH(COOC$_2$H$_5$)$_2$]$_2$ dimer

Y$_3$N@I_h(7)-C_{80}C(COOCH$_2$Ph)$_2$

Sc$_3$N@D_3(6140)-C_{68}C(COOC$_2$H$_5$)$_2$

FIGURE 8.5

Bingel–Hirsch reactions of EMFs. (Product structures from Feng, L. et al., *J. Am. Chem. Soc.*, 127, 17136, 2005; Feng, L. et al., *J. Am. Chem. Soc.*, 128, 5990, 2006; Alegret, N. et al., *Chem. Eur. J.*, 19, 5061, 2013; Lukoyanova, O. et al., *J. Am. Chem. Soc.*, 129, 10423, 2007.)

and Gd$_3$N@C_s(51365)-C_{84}, respectively, and Gd$_3$N@C_{88} is totally unreactive. The reactivity decreases with increasing cage size.

Sc$_3$N@D_{3h}(5)-C_{78} reacts with diethyl bromomalonate in the presence of DBU to afford a mono-adduct and a bisadduct as main products.[83] According to the ^{1}H and ^{13}C NMR spectra, the mono- and bisadduct have mirror and C_{2v} symmetry, respectively. Theoretical calculations and NMR spectrum point that both the adducts feature closed cyclopropyl structure. The characteristic peak of Sc$_3$N@

C_{78} at 614 nm in the UV–vis adsorption spectra disappeared in the two adducts, suggesting dramatic change of the electronic structure. The addition site of the second addend was proposed based on the LUMO distribution.

The Bingel–Hirsch reaction of the non-IPR $Sc_3N@D_3(6140)$-C_{68} was first reported in 2008 and yielded $Sc_3N@C_{68}[C(COOC_2H_5)_2]_n$ (n = 1–5) diethyl malonate adducts.[84] ^1H and ^{13}C NMR spectra suggest that the monoadduct has an asymmetric structure with reduced molecular symmetry. NMR, UV–vis spectra, and DFT calculations indicate that the addition occurs at a [6,6] bond adjacent to the fused pentagons (Figure 8.5) and leads to an open-cage structure. The addition was kinetically controlled.[79] However, its exact structure remains unclear and still needs further investigation. Note that the Bingel–Hirsch reactions of non-IPR $Gd_3N@C_s(39663)$-C_{82} and $Gd_3N@C_s(51365)$-C_{84} are also theoretically suggested to occur at the [6,6] bonds near the pentalene motif and yield fulleroids.[85] Thus, different non-IPR TNT EMFs may have analogous reaction paths in this reaction.

Although $TiSc_2N@C_{80}$ is isoelectronic to $Sc_3N@C_{80}^-$ radical and different from neutral $Sc_3N@C_{80}$ in terms of only one metal atom, its reaction with diethyl bromomalonate in the presence of DBU yields two monoadducts, which are ESR-silent and likely singly bonded.[86]

Significantly, one can utilize many otherwise completely insoluble EMFs by synthesizing their soluble Bingel derivatives and *turn the waste into treasure*. For example, the Bingel adduct $Gd@C_{60}[C(COOCH_2CH_3)_2]_{10}$ was synthesized *via* multiple cycloaddition of bromomalonates to $Gd@C_{60}$ and further hydrolyzed to water-soluble $Gd@C_{60}[C(COOH)_2]_{10}$.[87] The latter exhibits properties comparable to commercially available Gd(III) chelate–based magnetic resonance imaging (MRI) contrast agents.[88]

8.2.5 Carbene Addition

In 2002, Gu et al. reported the reaction between $Tb@C_{82}$ and aryldiazoacetates with the aid of a copper(I) catalyst $Cu(MeCN)_4PF_6$.[89] Multiadducts $Tb@C_{82}[CH_2(C_6H_5)CO_2CH_3]_n$ (n = 1–6) were observed in the mass spectra, and their similar XPS spectra as that of pristine $Tb@C_{82}$ suggest that the Tb atom is in a trivalent state. The exohedral modification changes the electronic structure of the carbon cage and lead to disappearance of characteristic bands of $Tb@C_{82}$ in the adsorption spectrum.

For years, Akasaka and coworkers continuously conducted a series of photochemical reactions between 2-adamantane-2,3-[3H]-diazirine (AdN_2) and $M@C_{82}$ (M = Sc, Y, La, Ce, Gd) (Figure 8.6).[90–95] Although holding different metals, all $M@C_{2v}(9)$-C_{82} (M = Y, La, Ce, Gd) afforded two adducts, suggesting their similar chemical reactivity and regioselectivity. The X-ray analysis on the major (Figure 8.6) and minor isomer clarified that the electrophilic Ad group is located at the C1–C2 and C2–C3 bonds (Figure 8.1), respectively; both of which are opened. The major isomer isomerizes to the thermodynamically more stable minor isomer under thermal treatment.[93] For the pristine $M@C_{82}$, the off-centered metal results in strong metal–cage interactions and large charge density and local strain at the reaction sites, especially the C2 atom. In $Sc@C_{2v}(9)$-C_{82}, however, the small ionic radius of Sc^{3+} leads to much close metal–cage contact and facilitates electron backdonation from the cage. The strong Sc-cage bonding interaction results that both C1 and C2 atoms exhibit high charge density and the Ad group bonds to the adjacent carbons after attacking C1 or C2 and thus forms four adducts in total.[94] The most abundant isomer of $Sc@C_{82}$ is thus the one with Ad addition to the C1–C2 bond and also bears an open-cage structure.

Two $La@C_s(6)$-C_{82}Ad monoadducts were synthesized by the photochemical reaction.[95] Analysis based on local stain, charge distribution, and relative energies suggested that the additions are highly regioselective and may take place at the carbon sites very close to the metal.

The photochemical reaction of divalent $Yb@C_{2v}(3)$-C_{80} with AdN_2 afforded only one monoadduct.[96] X-ray analysis shows that the Ad group attacks a [5,6] junction in the most curved region with an open-cage geometry (Figure 8.6). The addition leads to much flexible metal compared to that of the pristine $Yb@C_{80}$. The similar reaction of $Yb@C_2(13)$-C_{84} afforded three monoadducts with likely fulleroid structures.[97] X-ray analysis on the most abundant isomer found that the addition site is a [5,6] bond (Figure 8.6). In contrast to the flexible metal in $Yb@C_2(13)$-C_{84}, Yb atom resides steadily under a hexagon far from the addend. In both cases, the regioselectivity is explained by the release of local strain rather than the charge density distribution.

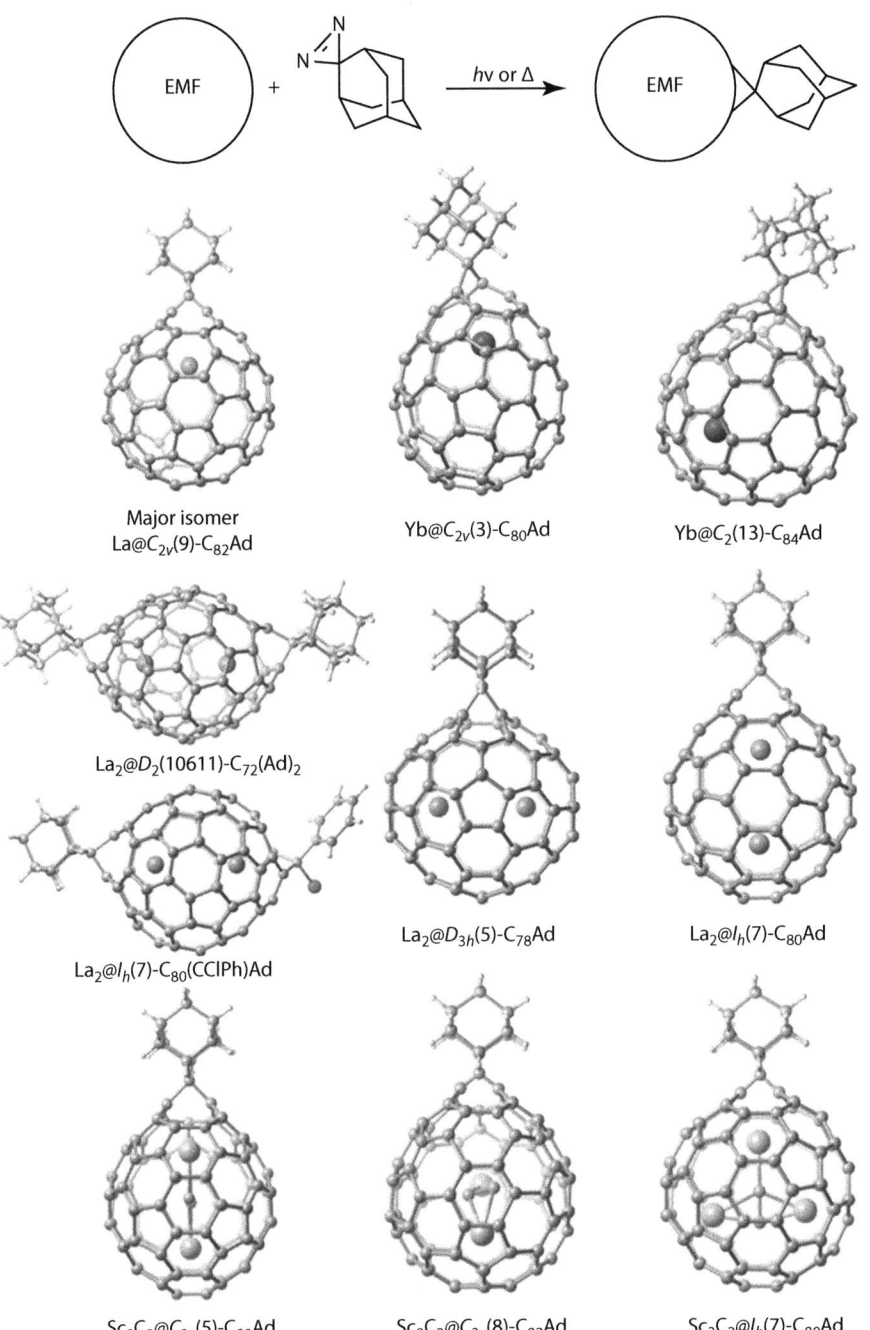

FIGURE 8.6
Carbene addition reactions between EMFs and 2-adamantane-2,3-[3H]-diazirine (AdN$_2$). (Product structures from Iiduka, Y. et al., *Angew. Chem. Int. Ed.*, 46, 5562, 2007; Maeda, Y. et al., *J. Am. Chem. Soc.*, 126, 6858, 2004; Xie, Y. et al., *Angew. Chem. Int. Ed.*, 52, 5142, 2013; Zhang, W. et al., *J. Am. Chem. Soc.*, 135, 12730, 2013; Lu, X. et al., *Angew. Chem. Int. Ed.*, 47, 8642, 2008; Cao, B. et al., *J. Am. Chem. Soc.*, 130, 983, 2008; Yamada, M. et al., *J. Am. Chem. Soc.*, 130, 1171, 2008; Kurihara, H. et al., *J. Am. Chem. Soc.*, 133, 2382, 2011; Iiduka, Y. et al., *J. Am. Chem. Soc.*, 127, 12500, 2005; Ishitsuka, M.O. et al., *J. Am. Chem. Soc.*, 133, 7128, 2011.)

The Ad addition to di-EMFs such as M$_2$@C$_{2n}$ (M = La, Ce; 2n = 72, 78, and 80) were also pursued.[98–101] For La$_2$@C$_{72}$, six stable monoadducts were successfully isolated and characterized, and they exhibit similar UV–vis–NIR spectra as that of the pristine La$_2$@C$_{72}$, suggesting their open-cage structures.[98] X-ray data of the three most abundant isomers confirmed that their cage isomer is the non-IPR D_2(10611)-C$_{72}$ with two pairs of fused pentagons at the two poles (vide infra).

Consistent with the 82 lines in ^{13}C NMR spectrum (for the one in highest yield), Ad group does not attack the [5,5] bonds to form otherwise symmetric structure, but favors the [5,6] junctions in the fused-pentagon regions. The La atoms reside close to the [5,5] bonds, which are significantly stabilized by the strong metal–cage interactions. The highest charge density is found for those carbons adjacent to the [5,5] junctions; they are thus very reactive toward the electrophilic Ad group. Since $La_2@C_{72}$ has two pairs of fused pentagons, a variety of biscarbene adducts $La_2@C_{72}(Ad)_2$ were further synthesized.[99] The bisadducts exhibit similar adsorption spectra but different redox potentials compared with those of $La_2@C_{72}$ and the most abundant $La_2@C_{72}Ad$. Thus, Ad additions can tune the electrochemical properties of $La_2@C_{72}$, while keeping its electronic structure almost intact. For the most abundant $La_2@C_{72}(Ad)_2$ isomer, X-ray data show that the Ad groups are added to the [5,6] bonds in the fused-pentagon regions, corresponding to that of the most abundant monoadduct (Figure 8.6).[98] The whole molecule thus holds C_2 symmetry with an open-cage structure. Its larger La–La distance (4.3 Å) than that of the monoadducts (4.171–4.178 Å) indicates substantial release of the metal–metal repulsion. The absence of multiadducts suggests that the fused-pentagon regions at the two poles are responsible for the high reactivity of $La_2@C_{72}$.

$La_2@D_{3h}(5)\text{-}C_{78}$ reacts with AdN_2 both thermally and photochemically.[100] The photochemical reaction proceeds very fast and affords four monoadducts and at least eight bis- as well as trisadducts in a minute. In contrast, the thermal reaction is slow and exhibits low regioselectivity with more than seven monoadducts. ^{13}C NMR spectrum and X-ray analysis of one major $La_2@C_{78}Ad$ isomer show that it has C_s symmetry with the two La atoms located along the C_3 axis (major occupation, La–La distance: 4.081 Å, Figure 8.6). The Ad addition selectively occurred at a [5,6] bond, which is parallel to the C_3 axis and perpendicular to the σ_h plane of the cage, and results in an open-cage structure. Several cage carbons on the polar and equator regions are suggested as the possible addition sites for other monoadducts due to their large local strains and high charge densities.

The photochemical reaction of $M_2@C_{80}$ (M = La, Ce) with excess AdN_2 in toluene afforded two monoadducts in 80% yield.[101] ^{13}C NMR spectra show 44 lines for the carbon cage, indicating that the $M_2@C_{80}Ad$ adducts have a [6,6]-open structure. X-ray diffraction (XRD) data of $La_2@C_{80}Ad$ confirmed its addition pattern (C–C distance: 2.166 Å, Figure 8.6). The two La atoms are collinear with the spiro carbon, and their separation of 4.031 Å is longer than that of the pristine $La_2@C_{80}$ by 0.2–0.3 Å. Similar structure for $Ce_2@C_{80}(Ad)$ was proposed based on the paramagnetic ^{13}C NMR spectrum. The addition-induced bond cleavage considerably expands the cage and reduces the metal–metal repulsion. DFT calculations found a minimum on the electrostatic potential map inside $[C_{80}(Ad)]^{6-}$. Accordingly, the free 3D rotation of the La atoms in $La_2@C_{80}$ is obviously hindered in its Ad derivative.

The rapid rotation of spherical EMFs in the crystal lattice prohibits direct XRD structural characterization. Their carbene derivatives are frequently employed due to their higher feasibility to crystallize than their pristine forms. By this means, several previously assumed classical EMFs such as $Sc_2C_{82}(I)$, $Sc_2C_{84}(III)$, and Sc_3C_{82} were convincingly clarified to be actually metal carbide clusterfullerenes $Sc_2C_2@C_{2v}(5)\text{-}C_{80}$,[102,103] $Sc_2C_2@C_{3v}(8)\text{-}C_{82}$[25], and $Sc_3C_2@I_h(7)\text{-}C_{80}$,[104] respectively (Figure 8.6).

The photochemical reaction of $La@C_{2v}(9)\text{-}C_{82}$ with phenylchlorodiazirine ($PhClCN_2$) afforded a $La@C_{82}(CClPh)$ carbene monoadduct.[105] The addition was suggested to occur at the C1–C2 site and afford a pair of isoenergetic diastereomers differing only in the direction of chlorine and phenyl groups. The photochemical reaction of $La_2@C_{80}$ with $PhClCN_2$ in toluene afforded a monoadduct $La_2@C_{80}(CClPh)$, which then reacts with AdN_2 to generate three bis-functionalized $La_2@C_{80}(CClPh)Ad$ derivatives.[106] For $La_2@C_{80}(CClPh)$, NMR and X-ray crystallographic analysis reveal its [6,6]-open-cage structure with metals collinear to the spiro carbon. Consistent with the theoretical predictions based on the local strain and charge density, X-ray data unambiguously confirm that the Ad group in one bisadduct connects with a [6,6] junction at the opposite side from the CClPh group and leads to local open-cage structure (Figure 8.6). The doubly opened geometry markedly expands the inner cavity and leads to an extremely long La–La distance of 4.16 Å.

$La_2@C_{80}$ glycoconjugates were synthesized by the room-temperature reaction with glycosylidene carbene generated *in situ* from a diazirine precursor.[107] NMR suggests the yield of two diastereomers of [6,6]-open monoadducts. Compared with $La_2@C_{80}Ad$, the adduct is more soluble in organic solvents probably due to the introduction of polarity with sugar-like structure.

8.2.6 Radical Reaction

In the radical reactions, paramagnetic and diamagnetic EMFs are generally attached by an odd and even number of singly bonded addends, respectively, to achieve a closed-shell configuration.

In 1995, Suzuki et al. reported the first radical addition of EMFs by conducting the reaction between La@C_{82} with excess diphenyldiazomethane (Ph$_2$CN$_2$) in a toluene solution at 60°C.[108] The mass spectrum showed several peaks for the La@C_{82}(CPh$_2$)$_n$ ($n = 1$–3) adducts. The four new octets in the EPR spectrum may correspond to four monoadducts, whereas the broad signals (in the derivative of a Gaussian line) are ascribed to bis- and triadducts.

La@C_{2v}(9)-C_{82} can undergo a radical coupling reaction with toluene in the presence of N-triphenylmethyl-5-oxazolidinone (serves as initiator) under thermal condition to afford four ESR-silent La@C_{82}(CH$_2$C$_6$H$_5$) derivatives.[109] The reaction occurs even upon photoirradiation or in sunlight without the initiator and can be extended to La@C_s-C_{82} and Ce@C_{2v}-C_{82}. Clearly, toluene is very reactive toward paramagnetic EMFs. In addition, substituted benzenes such as p-xylene, o-xylene, p-$tert$-butyltoluene, and $\alpha,\alpha,2,4$-tetrachlorotoluene can also be employed as the reagents. The photochemical reaction of La@C_{2v}-C_{82} with $\alpha,\alpha,2,4$-tetrachlorotoluene in 1,2-dichlorobenzene (o-DCB) solution affords at least eight adducts. In sharp contrast to previous M@C_{82} monoadducts, X-ray data show that one major La@C_{82}(CHClC$_6$H$_3$Cl$_2$) isomer features trichlorobenzyl group singly connecting with the C10 site (Figure 8.7), which exhibits large spin density in the pristine La@C_{82}. Moreover, metallofulleropyrrolidine derivatives La@C_{2v}-C_{82}(C$_2$H$_4$NCPh$_3$) were also obtained by the thermal reaction with N-triphenylmethyl-5-oxazolidinone in benzene solution. For comparison, the EPR intensity of La@C_{82} decreases much faster in toluene than in benzene, indicating more favorable addition of benzyl radical than azomethine ylide. Interestingly, pristine La@C_{2v}-C_{82} in a yield as high as 96% can be recovered from these radical adducts by heating in the presence of a radical trapping reagent such as 2,2,6,6-tetramethylpiperidine-1-oxyl (TEMPO).[110]

A dibenzyl derivative of Sc$_3$N@I_h-C_{80} was synthesized in high yield via its reaction with photochemically generated benzyl radicals.[111] ^{1}H and ^{13}C NMR spectra as well as DFT calculations suggested that the Sc$_3$N@C_{80}(CH$_2$C$_6$H$_5$)$_2$ adduct possesses C_2 symmetry and stems from a 1,4-addition, which was unambiguously confirmed by X-ray crystallographic analysis (Figure 8.7). Although disordered, the Sc$_3$N cluster keeps planar inside the cage. The adduct exhibits a maximum at 898 nm in the UV–vis adsorption spectrum, in contrast to that of pristine Sc$_3$N@I_h-C_{80}(735 nm). Similar spectrum change was also observed for Lu$_3$N@I_h-C_{80}(CH$_2$C$_6$H$_5$)$_2$ obtained by the same reaction. Clearly, the benzylic substituents significantly modify the electronic structure of these two TNT EMFs.

By adding excess ^{13}C-labeled diethyl malonate and Mn(OAc)$_3 \cdot$2H$_2$O to Sc$_3$N@C_{80} in chlorobenzene, Dorn et al. synthesized two monoadducts Sc$_3$N@C_{80}[^{13}C(COOC$_2$H$_5$)$_2$] and Sc$_3$N@C_{80}[^{13}CHCOOC$_2$H$_5$], as well as multiadducts with up to octa-methano addends.[112] The two monoadducts were suggested to be [6,6]-open methanofullerides by the combined NMR and UV measurements as well as theoretical calculations. These are the first thermodynamically stable [6,6] adducts of Sc$_3$N@I_h-C_{80}. By the same methodology, Lu$_3$N@C_{80} analogues with up to deca-addends were obtained.

Shinohara et al. first synthesized La@C_{82}(C$_8$F$_{17}$)$_2$ by irradiating a toluene solution containing La@C_s(6)-C_{82} and perfluorooctyl iodide and isolated the products using a fluorous-phase partitioning approach.[113] The obtained seven isomers have paramagnetic nature, which impedes further structural characterization by NMR spectroscopy. X-band EPR spectra show that their isotropic hfccs are obviously different from that of La@C_{82}. In addition, the onsets of their adsorption spectra are all blueshifted with respect to that of La@C_{82}, indicating their larger highest occupied molecular orbital (HOMO)–LUMO gaps.

The electron-withdrawing CF$_3$ radical has been widely used for the functionalization of (metallo)fullerenes. The first trifluoromethylation reaction of EMFs was reported in 2005.[114] In the experiment, Y@C_{82}-enriched extract was mixed with silver trifluoroacetate (CF$_3$COOAg) and heated to 400°C for 10 h (Figure 8.7). Mass spectrum shows that, in addition to the main product Y@C_{82}(CF$_3$)$_5$, Y@C_{82}(CF$_3$)$_n$($n = 1, 3$) and Y$_2$@C_{80}(CF$_3$)$_n$($n = 1, 3$) were also obtained. According to the 2D ^{19}F NMR spectroscopic analysis and DFT calculations, possible structures for the two isolated Y@C_{82}(CF$_3$)$_5$ isomers favor a chain of four 1,4-C_6(CF$_3$)$_2$ edge-sharing hexagons on the Y@C_{2v}-C_{82} cage surface. The Gd@C_{82} analogues were achieved using similar procedure.[115]

In contrast to previous reports on the different reactivity between Sc$_3$N@I_h-C_{80} and Sc$_3$N@D_{5h}-C_{80}, both of them react readily with CF$_3$ radicals at almost the same rate at 520°C.[116] DFT

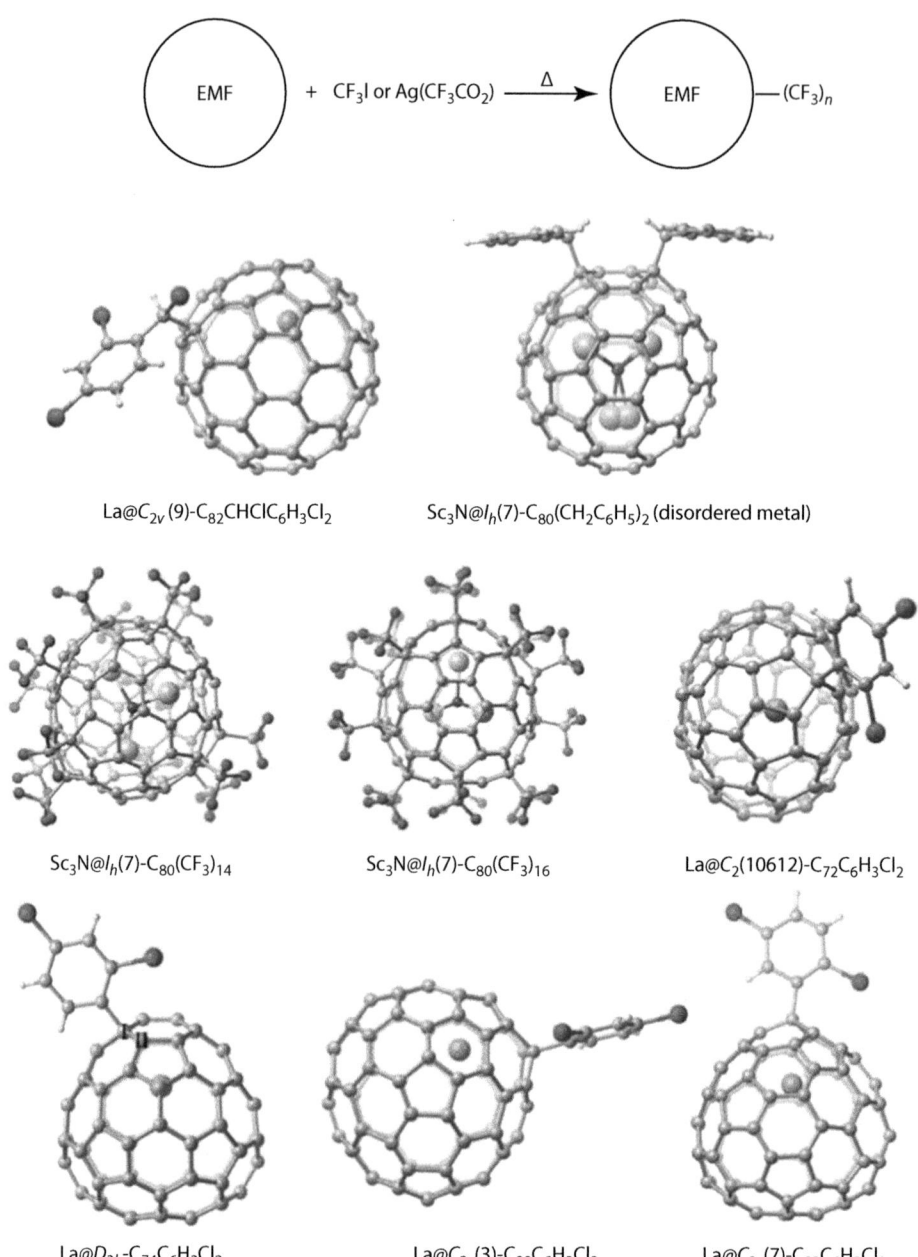

FIGURE 8.7
Radical reactions of EMFs. (Product structures from Takano, Y. et al., *J. Am. Chem. Soc.*, 130, 16224, 2008; Shu, C. et al., *J. Am. Chem. Soc.*, 130, 17755, 2008; Shustova, N. et al., *J. Am. Chem. Soc.*, 131, 17630, 2009; Nikawa, H. et al., *J. Am. Chem. Soc.*, 127, 9684, 2005; Lu, X. et al., *Angew. Chem. Int. Ed.*, 50, 6356, 2011; Nikawa, H. et al., *J. Am. Chem. Soc.*, 131, 10950, 2009; Akasaka, T. et al., *Angew. Chem. Int. Ed.*, 49, 9715, 2010; Wakahara, T. et al. *J. Am. Chem. Soc.*, 128, 14228, 2006.)

calculations suggested that the Sc atoms in the fluoroalkylated derivatives energetically favor to bond with those cage carbons that are *para* to the cage sp^3-C(CF$_3$) atoms. Four isomers for Sc$_3$N@ I_h-C$_{80}$(CF$_3$)$_{14,16}$ (two for each) were also synthesized.[117,118] All the adducts feature multiple CF$_3$ groups at [6,6,6] sites, which hold low pyramidal angle and are generally less reactive. The stability of [6,6,6] addition pattern is considered due to the preferred formation of isolated aromatic regions and their coordination with the inner metals. The Sc$_3$N cluster is distorted upon the exohedral substitutions, and its geometry highly depends on the addition pattern. X-ray data show that the attachment of multiple CF$_3$ groups considerably deformed the C$_{80}$ cage and led to a nonspherical outer carbon framework with restricted inner metal motion (Figure 8.7). In addition, some efforts have been devoted to the different charged states of the $Sc_3N@C_{80}(CF_3)_n$ ($n = 2, 10, 12$) derivatives and disclose diverse spin density distribution and metal motion.[119–121]

By the reaction with 1,2,4-trichlorobenzene, missing metallofullerenes such as $La@C_{2n}$ ($2n = 72$, 74, 80, 82) have been successfully isolated and characterized as their dichlorophenyl derivatives.[122–126]

Two series (I, II) of isomers for $La@C_{74}(C_6H_3Cl_2)$ were obtained; each consists of three members differing in the substitution pattern (2,4-, 2,5-, or 3,4-) of the dichlorophenyl group.[122,123] X-ray data show that their cage is D_{3h}-C_{74}, the only IPR isomer of C_{74} (Figure 8.7). Categories I and II correspond to C-I and C-II sites, respectively. The introduction of $C_6H_3Cl_2$ group leads to locally deformed cage with the metal slightly deviating from the σ_h plane.[127] Notably, $La@C_{74}(C_6H_3Cl_2)$ is the first EMF derivative featuring a single bond connecting the EMF and addend.

Three $La@C_{72}(C_6H_3Cl_2)$ derivatives were obtained, and DFT calculations together with X-ray crystal analysis confirmed their non-IPR $C_2(10612)$-C_{72} framework (Figure 8.7), and the addition occurs at a carbon site adjacent to the [5,5] bond.[124] On the basis of $La@C_{72}(C_6H_3Cl_2)$, the first chiral EMFs was achieved by its copper-catalyzed Prato reaction with a N-metalated azomethine ylide.[128]

The reaction of $La@C_{80}$ with 1,2,4-trichlorobenzene afforded three $La@C_{80}(C_6H_3Cl_2)$ isomers.[125] X-ray crystallographic analysis reveal that they are based on an unprecedented $C_{2v}(3)$-C_{80} cage (Figure 8.7). Although $La@C_{2v}(3)$-C_{80} is energetically less stable than $La@C_{2v}(5)$-C_{80}, it has a high radical character with small ionization potential (IP) and electron affinity (EA), favorable for the reaction with $C_6H_3Cl_2$ radical. Similar phenomenon was found for $La@C_{82}(C_6H_3Cl_2)$; its two isomers are based on a $C_{3v}(7)$-C_{82} cage, albeit $La@C_{3v}(7)$-C_{82} is 16.1 kcal mol^{-1} higher in energy than $La@C_{3v}(8)$-C_{82}.[126] Its dichlorophenyl group is linked to a [6,6,6] carbon on the C_3 axis *via* a single bond (Figure 8.7). Clearly, exohedral modification changes the relative stability of their derivatives.

In all the cases, the La atom is close to the addition site, which thus features both high spin density and local stain and is thus rather reactive toward the dichlorophenyl radical.

8.2.7 Benzyne Cycloaddition

The first [2 + 2] cycloaddition to EMFs is the reaction of $Gd@C_{82}$ with benzyne generated from the diazotization of anthranilic acid with isoamyl nitrite *in situ*.[129] Two main monoadducts exhibit distinct adsorption spectra from that of $Gd@C_{82}$. Clearly, the electronic structure has been changed upon exohedral modification.

By the similar procedure, the cycloaddition of benzyne to $La@C_{2v}(9)$-C_{82} and $Sc_3N@I_h(7)$-C_{80} was conducted.[130,131] For $La@C_{82}$, mass spectra suggest that up to ten benzene groups have been successfully attached onto the cage. There also exist peaks ascribed to one NO_2 group singly bonded to each adduct, leading to the observed absence of ESR signal.[130] $La@C_{82}(C_6H_4)_2NO_2$ was isolated as the first EMF trisadduct. X-ray analysis reveals that it contains both singly and cyclically bonded addends: the NO_2 group is added at the C18 site *via* a single N–C bond, and the two benzenes are attached to [5,6] bonds *para* to C18 and form closed cyclobutene rings (Figure 8.8). The La atom is positioned far from the three substituents. The carbon site for NO_2 attack has both high spin density and large POAV value, very reactive toward radical attack. The attached NO_2 affects the subsequent addition pattern of benzynes by imparting high local strain to the two carbon sites at the *para*-positions. The trisadduct has a UV–vis–NIR spectrum dramatically different from that of the parent $La@C_{82}$.

In contrast, no nitrated derivatives were obtained for $Sc_3N@I_h$-C_{80}.[131] Mass spectrum showed that two stable monoadducts were successfully obtained. ^1H NMR spectra and X-ray analysis confirmed that they are symmetric [5,6] and unsymmetrical [6,6] regioisomers, with the planar Sc_3N cluster far from the benzene addend (Figure 8.8). Similar to the previous report,[116] the Sc atoms in the [5,6] adduct reside nearby two cage bonds that are *para* to the sp^3 carbons. The [6,6] adduct exhibits reversible cathodic and anodic behavior. The first reduction potentials of the two adducts are shifted to less negative than that of pristine of $Sc_3N@I_h$-C_{80}, suggesting their low-lying LUMOs.

8.2.8 Noncovalent Functionalization

In 2002, Yang et al. first prepared the host–guest complex of $Dy@C_{82}$ and *p-tert*-butylcalix[8]arene (Figure 8.9a) and found that it was more stable than the C_{60} analogue once formed.[132]

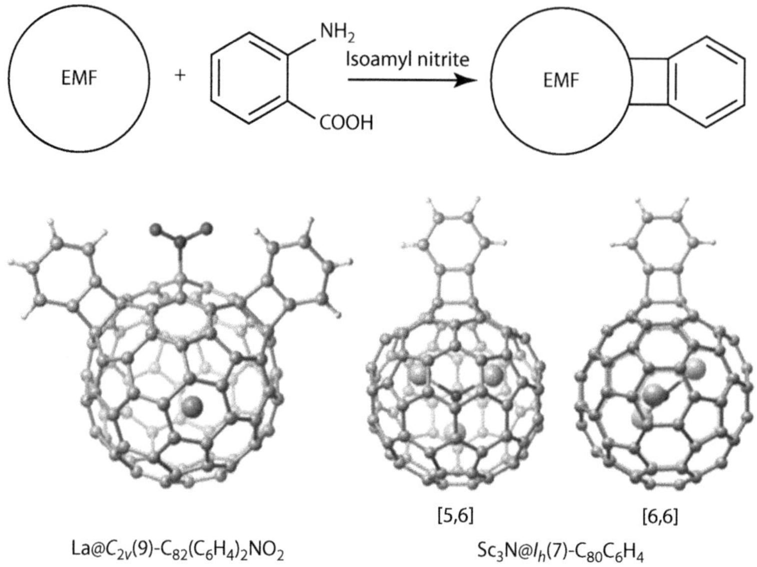

FIGURE 8.8

[2 + 2] cycloaddition of benzyne to EMFs. (Product structures from Lu, X. et al., *Angew. Chem. Int. Ed.*, 49, 594, 2010; Li, F.-F. et al., *J. Am. Chem. Soc.*, 133, 1563, 2011.)

In 2006, Akasaka et al. reported the 1:1 complexes of La@C_{2v}-C_{82} with various azacrown ethers, namely, 1,4,7,10,13,16-hexaazacyclooctadecane; 1,4,7,10,13,16-hexamethyl-1,4,7,10,13,16-hexaazacyclooctadecane (Figure 8.9b); and mono-aza-18-crown-6 ether (Figure 8.9c)[133] as well as unsaturated thiacrown ethers of various ring size (15-, 18-, 21-, and 24-membered, Figure 8.9d) in solution.[134] The formation of such host–guest supramolecular systems is accompanied by electron-transfer process. The size matching between the host and guest molecules plays an important role in their complexation strength, and the 21-membered thiacrown ether can perfectly include La@C_{82}.

The complexation of La@C_{82} with organic donor molecules such as *N*,*N*,*N*′,*N*′-tetramethyl-*p*-phenylenediamine (TMPD) (Figure 8.9e) in nitrobenzene solution exhibits reversible intermolecular electron transfer from the latter to the former.[135] Such spin-site exchange systems show thermochromic and solvatochromic characteristics and may find applications in optoelectronic and magnetic field. Similarly, the ground-state electron-transfer interaction between La$_2$@C_{80} and various organic donors such as TMPD leads to formation of stable 1:1 complexes.[136] The binding of [La$_2$@C_{80}]$^{.-}$/ [TMPD]$^{.+}$ radical pair is reversible and becomes more stable at low temperature and in solvent with high permittivity.

Porphyrin is often used to construct donor–acceptor systems with EMFs. Recently, Akasaka and Guldi et al. reported that C_{60}, La@C_{82}Py, and La$_2$@C_{80}Py (Py = pyridyl functional group) can form 1:1 hybrid with zinc tetraphenylporphyrin (ZnP) (binding constants of ca. 10^4), whereas pristine La@C_{82} failed.[137] Transient absorption spectra showed the formation of fullerene$^{.-}$ ZnP$^{.+}$ radical ion pairs due to electron transfer upon photoexcitation at 420 nm (ZnP excitation). They found the fastest charge separation for the La@C_{82} complex, whereas the slowest charge recombination for the La$_2$@C_{80} complex due to the lowest reduction potential of La@C_{82} and the metal-localized SOMO in La$_2$@C_{80}^-, respectively.

La@C_{2v}-C_{82} can be trapped by an isophthaloyl-bridged porphyrin dimer.[138] Notably, the complexation does not affect the spectroscopic behaviors such as the EPR and adsorption spectra of La@C_{82} or the emission bands of the porphyrin dimer. It is therefore suggested that the complexation only results from noncovalent interactions between the porphyrin rings as well as that between nitrogen atoms and La@C_{82}, without any intracomplex electronic interaction.

An inclusion complex of La@C_{82} and cyclodimeric copper(II) porphyrin *cyclo*-[P$_{Cu}$]$_2$: *cyclo*-[P$_{Cu}$]$_2$ ⊃ La@C_{82} was recently reported.[139] It can transform to *cage*-[P$_{Cu}$]$_2$ ⊃ La@C_{82} by ring-closing olefin metathesis of its side-chain olefinic termini (Figure 8.9f). Simultaneously, the spin coupling of the complex changes from ferromagnetic to ferrimagnetic. DFT calculations suggest that *cyclo*- and *cage*-[P$_{Cu}$]$_2$ ⊃ La@C_{82} energetically favor quartet and doublet spin states, respectively.

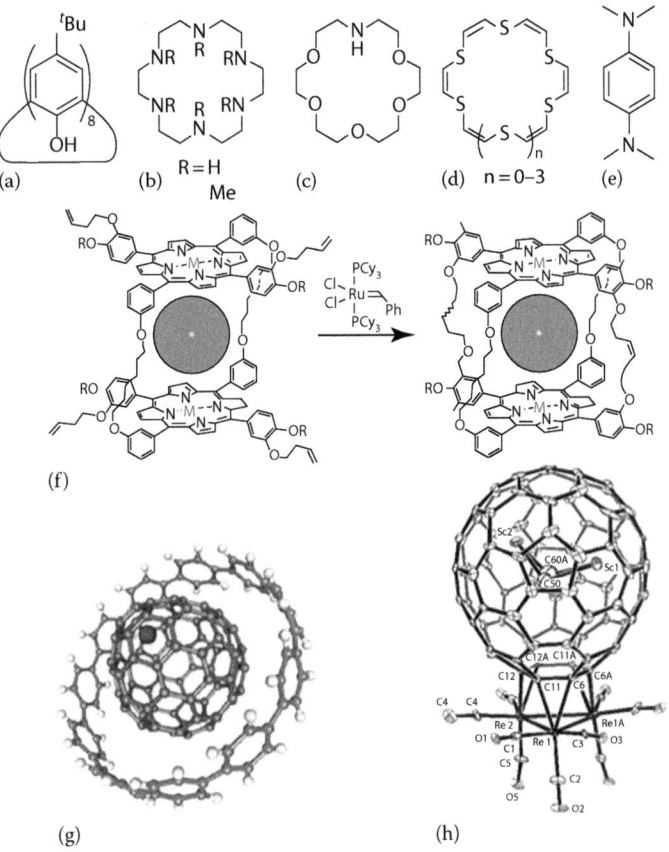

FIGURE 8.9
(a–e) Organic molecules used to construct supramolecular complexes with EMFs.
(f) *Cyclo*-[P_{Cu}]$_2$ ⊃ La@C_{82} and *cage*-[P_{Cu}]$_2$ ⊃ La@C_{82}, (g) Complexation between EMF and CPP,
(h) [(μ-H)$_3$Re$_3$(CO)$_9$-(η^2, η^2, η^2-Sc$_2$C$_2$@C_{3v}(8)-C$_{82}$)]. (From Hajjaj, F. et al., *J. Am. Chem. Soc.*, 133,
9290, 2011; Nakanishi, Y. et al., *Angew. Chem. Int. Ed.*, 53, 3102, 2014; Chen, C.-H. et al.,
Angew. Chem. Int. Ed., 51, 13046, 2012.)

A novel nonchromatographic approach to selectively extract a given EMF from a metallofullerene mixture in solution was realized by the size-selective complexation of EMFs with extended π-conjugated systems such as cycloparaphenylene (CPP).[140] To host-specify EMF molecules, the nanoring size of CPP should be carefully screened. For example, ^{1}H NMR, UV spectroscopic titration, and fluorescence quenching experiments confirmed that [11]CPP can strongly capture C_{82}-based EMFs (such as M$_x$@C_{2v}-C_{82}(M = Gd, Tm, Lu, x = 1 or 2) with large binding constants (Figure 8.9g), whereas [12]CPP did not show any encapsulation affinity. Clearly, the selectivity is only dependent on the cage size and irrespective of the inner species.

Recently, Sc$_2$C$_2$@C_{3v}(8)-C$_{82}$ was treated with [(μ-H)$_3$Re$_3$(CO)$_{11}$-(NCMe)] in chlorobenzene to afford an air-stable complex [(μ-H)$_3$Re$_3$(CO)$_9$-(η^2, η^2, η^2-Sc$_2$C$_2$@C_{3v}(8)-C$_{82}$)].[141] The XRD data show that the addition is highly regioselective, and it occurred at a sumanene-type hexagon, with the Re$_3$ triangle nearly parallel to the protruding hexagon plane (Figure 8.9h). Very recently, the analogous complexes for Sc$_2$@C_{3v}(8)C$_{82}$, Sc$_2$C$_2$@C_{2v}(5)-C$_{80}$, and Sc$_2$O@C_s(6)-C$_{82}$ were also synthesized.[142] In contrast to the pristine EMFs, these derivatives are readily isolated by using HPLC. Their facile decomplexation *via* carbonylation promises an attractive method to purify unresolved EMFs.

8.2.9 Miscellaneous Reactions

Water-soluble EMFs derivatives are desirable for the practical application in biomedicines. In addition to the Gd@C_{60},[87,88] many efforts are devoted to Gd@C_{82}. For example, thermal reaction between nucleophilic glycine methyl/ethyl esters and Gd@C_{82} are reported to afford noncycloadducts with up to four/eight substituents on the cage surface.[143]

By treating a toluene solution of Dy@C_{2v}(9)-C_{82} with dimethyl acetylenedicarboxylate and triphenylphosphine (PPh$_3$) at ambient temperature for 10 h, Yang et al. obtained a monoadduct in high yield.[144] In the reaction, PPh$_3$ first reacted with an acetylene carbon to generate a zwitterion intermediate, which then attacked the fullerene surface in a [2 + 1] fashion. Single crystal data show that the addition site is a [6,6] double bond (C1–C2 site in Figure 8.1), which is opened (2.153 Å) upon the functionalization.

The encapsulation of Sc$_3$N remarkably suppresses the reactivity of C_{80} toward hydrogen or fluorine addition due to the cluster-cage electron transfer.[145] Even so, stable hydrogenated derivatives such as Sc$_3$N@C_{80}H$_{52}$ are still expected provided that there are enough bonding interactions between Sc atoms and hydrogen-free carbons on the cage. Compared with the planar geometry in Sc$_3$N@C_{80}, the metal cluster favors a pyramidal shape in the hydrogenated derivatives.

Electrosynthesis is a very promising route for EMF functionalization. Recently, [Lu$_3$N@I_h-C_{80}]$^{2-}$ dianion was generated by controlled potential electrolysis and reacted with excess electrophile PhCHBr$_2$ to afford a methano-adduct.[146] The ^{1}H NMR and UV–vis–NIR spectra of the obtained Lu$_3$N@I_h-C_{80}(CHC$_6$H$_5$) suggest that it most likely has a [6,6]-open structure. This is further supported by the observed irreversible cathodic electrochemical behavior for this derivative. In contrast to [Lu$_3$N@I_h-C_{80}]$^{2-}$, [Sc$_3$N@I_h-C_{80}]$^{2-}$ is completely unreactive under identical conditions, probably due to less cage contribution to HOMO.[146]

The reaction of a substituted tetrazine with Sc$_3$C$_2$@I_h-C_{80} unexpectedly yielded an open-cage bisfulleroid derivative featuring a doubly bridged 14-membered ring.[147] NMR measurements and theoretical simulations suggested that the addition occurred at the [5,6] junction. The EPR spectra exhibit two sets of hfccs at 6.73 G (one Sc nucleus) and 4.00 G (two Sc nuclei). Thus, although the addend can induce the cage expansion and trap one Sc atom in the formed bisfulleroid bulge, the other two Sc atoms can still rotate around the cage to become equivalent.

8.3 POTENTIAL APPLICATIONS

The potential applications of EMFs and their derivatives in various fields have been well documented and recently summarized by us and others.[1,148–150] One attractive aspect is the employment of water-soluble Gd-based EMFs derivatives for high-efficiency MRI contrast reagents.[151] For example, the polyhydroxyl Gd@C_{82}(OH)$_{40}$ exhibits a longitudinal relaxivity r_1 value (81 mM^{-1} s^{-1}) much higher than the commercially available Gd-DTPA(DTPA = diethylenetriaminepentaacetic acid) (3.9 mM^{-1} s^{-1}).[152] In addition, the Ho-based EMFs such as ^{166}Ho$_x$@C_{82}(OH)$_y$ can serve as radiotracers and therapeutic radiopharmaceuticals.[153] Gd@C_{82}(OH)$_{22}$ nanoparticles show antitumor activity and can be used in cancer therapy.[154] Herein, we only selectively highlight the related functionalization reactions in the photovoltaics field.

Stable electron donor–acceptor conjugates of La@C_{2v}-C_{82}/La$_2$@I_h-C_{80} with π-extended tetrathiafulvalene (exTTF) were achieved by the 1,3-dipolar cycloaddition of exTTF-containing azomethine ylides (Figure 8.10a) to exclusively afford [5,6] metallofulleropyrrolidines.[155,156] Transient absorption spectra showed that they are excited into their singlet states upon 387 nm photoexcitation, which induced a fast intramolecular electron transfer from the exTTF to EMFs and resulted in a stable EMF$^{\cdot-}$–pyrrolidine–exTTF$^{\cdot+}$ radical ion pair state. Notably, the La@C_{82} ion pair exhibits a rather slow charge recombination: 2.4 ± 0.5 ns in THF and 1.1 ± 0.3 ns in cyclohexyl isonitrile.

The electron donor–acceptor conjugate Ce$_2$@I_h-C_{80}-ZnP formed *via* a [2 + 1] cycloaddition showed intriguing charge transfer chemistry.[157] Although linked by a flexible 2-oxyethyl butyrate spacer (Figure 8.10c), the donor and acceptor moieties exhibit strong interactions. Steady-state fluorescence and transient absorption measurements reveal the formation of (Ce$_2$@I_h-C_{80})$^{\cdot-}$-(ZnP)$^{\cdot+}$ and (Ce$_2$@I_h-C_{80})$^{\cdot+}$-(ZnP)$^{\cdot-}$ in nonpolar and polar solvents, respectively. The similar solvent-dependent behavior was also reported for the corresponding conjugate of La$_2$@C_{80}, whereas the Sc$_3$N@C_{80} analogue remains the same Sc$_3$N@$C_{80}$$^{\cdot-}$-ZnP$^{\cdot+}$ pair state regardless of the solvent.[158] On the contrary, when La$_2$@C_{80} was linked with strong electron acceptor such as 11,11,12,12-tetracyano-9,10-anthra-*p*-quinodimethane (TCAQ) by the Prato reaction, an La$_2$@$C_{80}$$^{\cdot+}$-TCAQ$^{\cdot-}$ ion pair formed in both nonpolar and polar media (Figure 8.10b).[159]

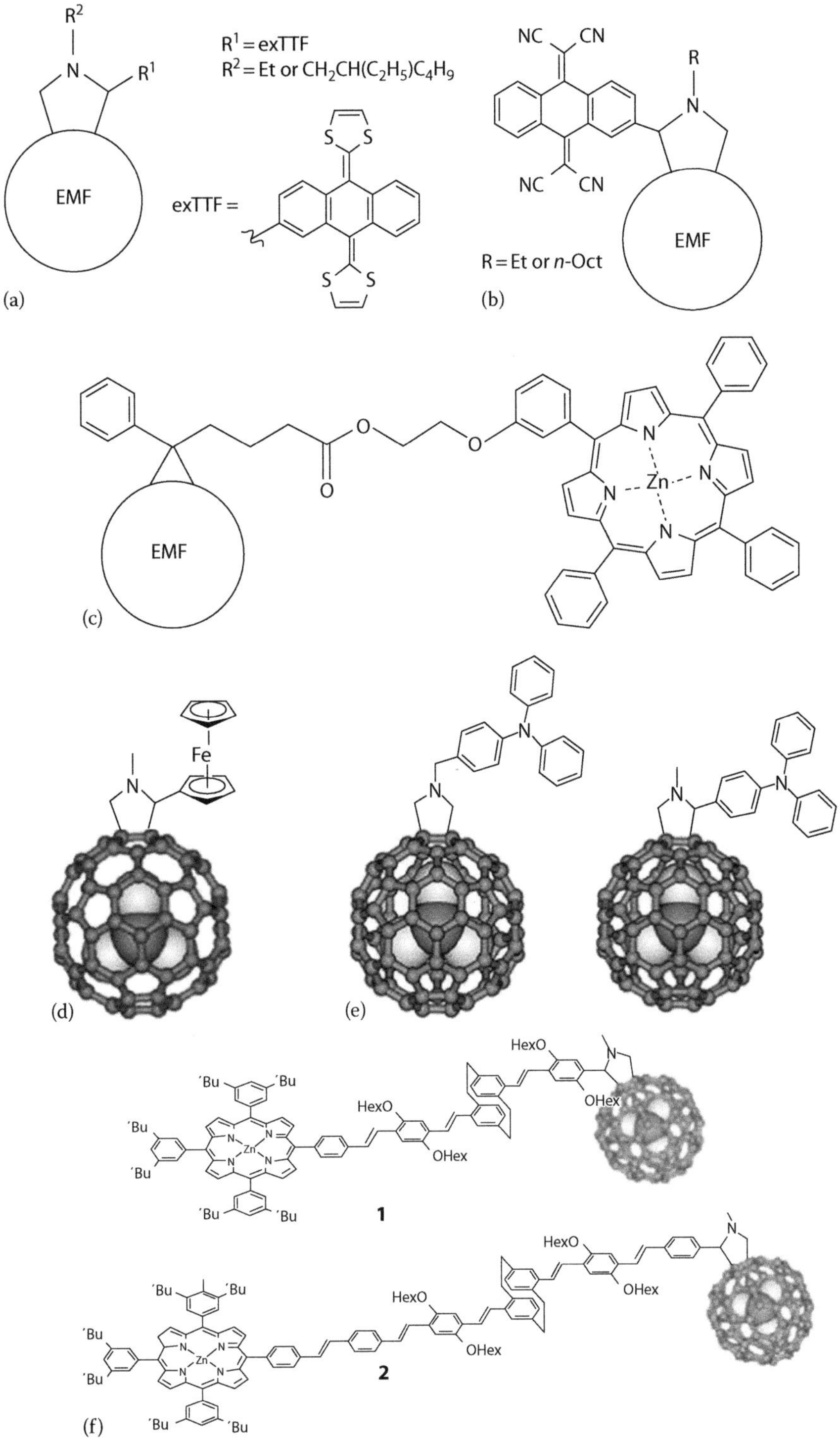

FIGURE 8.10

Some EMF-based donor–acceptor conjugates. (a) EMF–pyrrolidine–exTTF, (b) EMF–pyrrolidine–TCAQ, (c) EMF–ZnP conjugate linked by the 2-oxyethyl butyrate spacer, (d) $Sc_3N@C_{80}$-pyrrolidine-ferrocene (from Pinzón, J.R. et al., *Angew. Chem. Int. Ed.*, 47, 4173, 2008), (e) two $Sc_3N@$ C_{80}-pyrrolidine-TPA dyads with the donor on different positions of pyrrolidine ring (adapted from Pinzón, J.R. et al., *J. Am. Chem. Soc.*, 131, 7727, 2009), and (f) $Sc_3N@C_{80}$-pyrrolidine-ZnP dyads with donor-to-acceptor distances of 32.74 Å (**1**) and 45.94 Å (**2**). (From Wolfrum, S. et al., *Chem. Commun.*, 47, 2270, 2011.)

The thermal reaction of $Sc_3N@I_h$-C_{80} with ferrocene (Fc) carboxaldehyde and sarcosine in a toluene/o-DCB solution afforded a ferrocenyl pyrrolidine adduct, which was suggested as a [5,6] regioisomer (Figure 8.10d).[160] Upon photoexcitation, the time-resolved transient absorption spectroscopy showed that the $Sc_3N@C_{80}$ centered singlet excited state decays rapidly (lifetime: 5 ps) and confirmed the formation of metastable radical ion pair state $(Sc_3N@C_{80})^{\cdot-}$—pyrrolidine—$(Fe(Cp)_2)^{\cdot+}$ (lifetime: 128 and 84 ps in CS_2 and o-DCB, respectively).

A series of $M_3N@C_{80}$ (M = Sc, Y)-based dyads containing Fc, exTTF, or phthalocyanine (Pc) derivatives as the donor units were further synthesized via 1,3-dipolar cycloaddition of azomethine ylides or Bingel–Hirsch reactions.[161] Electrochemical studies suggested the addition patterns as [5,6] and [6,6] regioisomers in the cases of $Sc_3N@C_{80}$ and $Y_3N@C_{80}$, respectively. These donor–acceptor systems hold higher HOMO level but similar LUMO level as that of pristine $M_3N@C_{80}$ and thus exhibit smaller electrochemical gaps than the latter. The photophysical study on a ferrocenyl–$Sc_3N@C_{80}$-fulleropyrrolidine dyad confirmed the formation of radical ion pair state due to photoinduced electron transfer.

The large donor–acceptor separation plays a crucial role in the charge separation dynamics. Echegoyen et al. synthesized two isomeric [5,6] triphenylamine (TPA)-$Sc_3N@I_h$-C_{80} electron donor–acceptor conjugates by 1,3-dipolar cycloaddition reactions (Figure 8.10e).[162] They found a longer-time $TPA^{\cdot+}$–$Sc_3N@C_{80}{}^{\cdot-}$ charge-separated state and better thermal stability when the donor is linked to the N atom than to the 2-position of the pyrrolidine ring. Recently, Martín, Echegoyen, and Guldi et al. further prepared two $Sc_3N@C_{80}$-ZnP dyads (center-to-center distances: 32.74 Å for **1** and 45.94 Å for **2**) with long-range charge transfer between $Sc_3N@C_{80}$ and photoexcited ZnP (Figure 8.10f).[163] The resultant $(Sc_3N@C_{80})^{\cdot-}$ $(ZnP)^{\cdot+}$ radical ion pair states exhibit long lifetimes (1.0 and 1.2 μs for **1** and **2**, respectively). This renders $Sc_3N@C_{80}$-ZnP very promising materials for the design of artificial photosynthetic system.

Recently, a linear $Lu_3N@I_h$-C_{80}-PDI (PDI = 1,6,7,12-tetrachloro-3,4,9,10-perylenediimide) electron donor–acceptor conjugate was synthesized.[164] The Lu_3N cluster introduces a considerable electron nuclear hyperfine coupling, which leads to the efficient intersystem crossing from singlet radical ion pair state to the triplet one, and the latter exhibits extreme longer lifetime (ca. 1000 times) than the former.

8.4 CONCLUDING REMARKS

In summary, chemical functionalization endows EMFs with tunable and flexible properties and renders them rather versatile materials for various applications.

Numerous experimental and theoretical efforts have disclosed the following aspects. The reactivity and regioselectivity of EMFs are jointly affected by the species and size of the inner metal cluster, outer cage size and isomer, as well as the reaction type. The reactivity of EMFs is tunable upon oxidation and reduction. Regarding the regioselectivity, however, none of geometric or electronic parameters can independently predict the exact addition sites; in this sense, the recently discovered close relationship between aromaticity and thermodynamic stability of EMFs derivatives is notable.[165,166] For the non-IPR EMFs, their [5,5] bonds are remarkably stabilized by the internal metals and result that most of the exohedral additions only occur at the adjacent cage carbons. Sometimes, for a given functionalization reaction, the thermodynamic and kinetic products are different, and their interconversion is controlled by temperature. The metals affect the addition pattern, which will in turn regulate the position, orientation, and motion of the internal metals. The cluster-cage-addend interplay determines the final structure and property of one EMF derivative.

We believe that, with the rapid advance of fullerene chemistry, exohedral modifications involving new cages, metal clusters, and functionalization reactions will continuously appear in the near future and bring fantastic changes to the EMFs world.

ACKNOWLEDGMENTS

Support in China by NSFC (21103224), and in USA by Department of Defense (Grant W911NF-12-1-0083) and partially by NSF (Grant EPS-1010094) is gratefully acknowledged.

REFERENCES

1. For a recent review, see Popov, A., Yang, S., and Dunsch, L. 2013. Endohedral fullerenes. *Chem. Rev.* 113:5989–6113.
2. Kroto, H. W. 1987. The stability of the fullerenes C_n, with n = 24, 28, 32, 36, 50, 60 and 70. *Nature* 329:529–531.
3. Chai, Y., Guo, T., Jin, C. M. et al. 1991. Fullerenes with metals inside. *J. Phys. Chem.* 95:7564–7568.
4. Alvarez, M. M., Gillan, E. G., Holczer, K., Kaner, R. B., Min, K. S., and Whetten, R. L. 1991. La_2C_{80}: A soluble dimetallofullerene. *J. Phys. Chem.* 95:10561–10563.
5. Fowler, P. W. and Manolopoulos, D. E. 1995. *An Atlas of Fullerenes*, Clarendon, Oxford, U.K.
6. Akasaka, T., Wakahara, T., Nagase, S. et al. 2000. $La@C_{82}$ anion. An unusually stable metallofullerene. *J. Am. Chem. Soc.* 122:9316–9317.
7. Akasaka, T., Nagase, S., Kobayashi, K. et al. 1997. ^{13}C and ^{139}La NMR studies of $La_2@C_{80}$: First evidence for circular motion of metal atoms in endohedral dimetallofullerene. *Angew. Chem. Int. Ed. Engl.* 36:1643–1645.
8. Kobayashi, K., Nagase, S., and Akasaka, T. 1996. Endohedral dimetallofullerenes $Sc_2@C_{84}$ and $La_2@C_{80}$. Are the metal atoms still inside the fullerene cages? *Chem. Phys. Lett.* 261:502–506.
9. Stevenson, S., Rice, G., Glass, T. et al. 1999. Small-bandgap endohedral metallofullerenes in high yield and purity. *Nature* 401:55–57.
10. Wang, C.-R., Kai, T., Tomiyama, T. et al. 2001. A scandium carbide endohedral metallofullerene: $(Sc_2C_2)@C_{84}$. *Angew. Chem. Int. Ed.* 40:397–399.
11. Krause, M., Ziegs, F., Popov, A. A., and Dunsch, L. 2007. Entrapped bonded hydrogen in a fullerene: The five-atom cluster Sc_3CH in C_{80}. *ChemPhysChem* 8:537–540.
12. Wang, T.-S., Feng, L., Wu, J.-Y. et al. 2010. Planar quinary cluster inside a fullerene cage: Synthesis and structural characterizations of $Sc_3NC@C_{80}$-I_h. *J. Am. Chem. Soc.* 132:16362–16364.
13. Stevenson, S., Mackey, M. A., Stuart, M. A. et al. 2008. A distorted tetrahedral metal oxide cluster inside an icosahedral carbon cage. Synthesis, isolation, and structural characterization of $Sc_4(\mu_3\text{-}O)_2@I_h\text{-}C_{80}$. *J. Am. Chem. Soc.* 130:11844–11845.
14. Dunsch, L., Yang, S., Zhang, L., Svitova, A., Oswald, S., and Popov, A. A. 2010. Metal sulfide in a C_{82} fullerene cage: A new form of endohedral clusterfullerenes. *J. Am. Chem. Soc.* 132:5413–5421.
15. Akasaka, T., Kato, T., Kobayashi, K. et al. 1995. Exohedral adducts of $La@C_{82}$. *Nature* 374:600–601.
16. Kato, T., Akasaka, T., Kobayashi, K. et al. 1997. Chemical reactivities of endohedral metallofullerenes. *J. Phys. Chem. Solids* 58:1779–1783.
17. Yamada, M., Feng, L., Wakahara, T. et al. 2005. Synthesis and characterization of exohedrally silylated $M@C_{82}$ (M = Y and La). *J. Phys. Chem. B* 109:6049–6051.
18. Wakahara, T., Kobayashi, J. i., Yamada, M. et al. 2004. Characterization of $Ce@C_{82}$ and its anion. *J. Am. Chem. Soc.* 126:4883–4887.
19. Akasaka, T., Okubo, S., Kondo, M. et al. 2000. Isolation and characterization of two $Pr@C_{82}$ isomers. *Chem. Phys. Lett.* 319:153–156.
20. Akasaka, T., Nagase, S., Kobayashi, K. et al. 1995. Exohedral derivatization of an endohedral metallofullerene $Gd@C_{82}$. *J. Chem. Soc., Chem. Commun.* 1343–1344.
21. Kato, T., Akasaka, T., Kobayashi, K. et al. 1996. ESR study on the reactivity of two isomers of LaC_{82} with disilirane. *Appl. Magn. Reson.* 11:293–300.

22. Akasaka, T., Kato, T., Nagase, S. et al. 1996. Chemical derivatization of endohedral metallofullerene La@C_{82} with digermirane. *Tetrahedron* 52:5015–5020.

23. Maeda, Y., Miyashita, J., Hasegawa, T. et al. 2005. Chemical reactivities of the cation and anion of M@C_{82} (M = Y, La, and Ce). *J. Am. Chem. Soc.* 127:2143–2146.

24. Iiduka, Y., Wakahara, T., Nakajima, K. et al. 2006. ^{13}C NMR spectroscopic study of scandium dimetallofullerene, Sc_2@C_{84} vs. Sc_2C_2@C_{82}. *Chem. Commun.* 2057–2059.

25. Iiduka, Y., Wakahara, T., Nakajima, K. et al. 2007. Experimental and theoretical studies of the scandium carbide endohedral metallofullerene Sc_2C_2@C_{82} and its carbene derivative. *Angew. Chem. Int. Ed.* 46:5562–5564.

26. Akasaka, T., Nagase, S., Kobayashi, K. et al. 1995. Synthesis of the first adducts of the dimetallofullerenes La_2@C_{80} and Sc_2@C_{84} by addition of a disilirane. *Angew. Chem. Int. Ed. Engl.* 34:2139–2141.

27. Kobayashi, K., Nagase, S., Maeda, Y., Wakahara, T., and Akasaka, T. 2003. La_2@C_{80}: Is the circular motion of two La atoms controllable by exohedral addition? *Chem. Phys. Lett.* 374:562–566.

28. Wakahara, T., Yamada, M., Takahashi, S. et al. 2007. Two-dimensional hopping motion of encapsulated La atoms in silylated La_2@C_{80}. *Chem. Commun.* 2680–2682.

29. Yamada, M., Nakahodo, T., Wakahara, T. et al. 2005. Positional control of encapsulated atoms inside a fullerene cage by exohedral addition. *J. Am. Chem. Soc.* 127:14570–14571.

30. Yamada, M., Mizorogi, N., Tsuchiya, T., Akasaka, T., and Nagase, S. 2009. Synthesis and characterization of the D_{5h} isomer of the endohedral dimetallofullerene Ce_2@C_{80}: Two-dimensional circulation of encapsulated metal atoms inside a fullerene cage. *Chem. Eur. J.* 15:9486–9493.

31. Yamada, M., Minowa, M., Sato, S. et al. 2010. Thermal carbosilylation of endohedral dimetallofullerene La_2@I_h-C_{80} with silirane. *J. Am. Chem. Soc.* 132:17953–17960.

32. Yamada, M., Wakahara, T., Tsuchiya, T. et al. 2008. Location of the metal atoms in Ce_2@C_{78} and its bis-silylated derivative. *Chem. Commun.* 558–560.

33. Iiduka, Y., Ikenaga, O., Sakuraba, A. et al. 2005. Chemical reactivity of Sc_3N@C_{80} and La_2@C_{80}. *J. Am. Chem. Soc.* 127:9956–9957.

34. Wakahara, T., Iiduka, Y., Ikenaga, O. et al. 2006. Characterization of the bis-silylated endofullerene Sc_3N@C_{80}. *J. Am. Chem. Soc.* 128:9919–9925.

35. Iezzi, E. B., Duchamp, J. C., Harich, K. et al. 2002. A symmetric derivative of the trimetallic nitride endohedral metallofullerene, Sc_3N@C_{80}. *J. Am. Chem. Soc.* 124:524–525.

36. Lee, H. M., Olmstead, M. M., Iezzi, E., Duchamp, J. C., Dorn, H. C., and Balch, A. L. 2002. Crystallographic characterization and structural analysis of the first organic functionalization product of the endohedral fullerene Sc_3N@C_{80}. *J. Am. Chem. Soc.* 124:3494–3495.

37. Stevenson, S., Stephen, R. R., Amos, T. M., Cadorette, V. R., Reid, J. E., and Phillips, J. P. 2005. Synthesis and purification of a metallic nitride fullerene bisadduct: Exploring the reactivity of Gd_3N@C_{80}. *J. Am. Chem. Soc.* 127:12776–12777.

38. Maeda, Y., Miyashita, J., Hasegawa, T. et al. 2005. Reversible and regioselective reaction of La@C_{82} with cyclopentadiene. *J. Am. Chem. Soc.* 127:12190–12191.

39. Maeda, Y., Sato, S., Inada, K. et al. 2010. Regioselective exohedral functionalization of La@C_{82} and its 1,2,3,4,5-pentamethylcyclopentadiene and adamantylidene adducts. *Chem. Eur. J.* 16:2193–2197.

40. Garcia-Borràs, M., Luis, J. M., Swart, M., and Solà, M. 2013. Diels–Alder and retro-Diels–Alder cycloadditions of (1,2,3,4,5-pentamethyl)cyclopentadiene to La@C_{2v}-C_{82}: Regioselectivity and product stability. *Chem. Eur. J.* 19:4468–4479.

41. Haddon, R. C. 1988. π-electrons in three dimensions. *Acc. Chem. Res.* 21:243–249.

42. Sato, S., Maeda, Y., Guo, J.-D. et al. 2013. Mechanistic study of the Diels–Alder reaction of paramagnetic endohedral metallofullerene: Reaction of La@C_{82} with 1,2,3,4,5-pentamethylcyclopentadiene. *J. Am. Chem. Soc.* 135:5582–5587.

43. Osuna, S., Swart, M., Campanera, J. M., Poblet, J. M., and Solà, M. 2008. Chemical reactivity of D_{3h} C_{78} (metallo)fullerene: Regioselectivity changes induced by Sc_3N encapsulation. *J. Am. Chem. Soc.* 130:6206–6214.

44. Osuna, S., Swart, M., and Solà, M. 2009. The Diels-Alder reaction on endohedral $Y_3N@$ C_{78}: The importance of the fullerene strain energy. *J. Am. Chem. Soc.* 131:129–139.

45. Garcia-Borràs, M., Osuna, S., Luis, J. M., Swart, M., and Solà, M. 2012. The exo-hedral Diels–Alder reactivity of the titanium carbide endohedral metallofullerene $Ti_2C_2@D_{3h}$-C_{78}: Comparison with D_{3h}-C_{78} and $M_3N@D_{3h}$-C_{78} (M = Sc and Y) reactivity. *Chem. Eur. J.* 18:7141–7154.

46. Osuna, S., Valencia, R., Rodríguez-Fortea, A., Swart, M., Solà, M., and Poblet, J. M. 2012. Full exploration of the Diels–Alder cycloaddition on metallofullerenes $M_3N@$ C_{80} (M = Sc, Lu, Gd): The D_{5h} versus I_h isomer and the influence of the metal cluster. *Chem. Eur. J.* 18:8944–8956.

47. Garcia-Borràs, M., Osuna, S., Luis, J. M., Swart, M., and Solà, M. 2013. A complete guide on the influence of metal clusters in the Diels–Alder regioselectivity of I_h-C_{80} endohedral metallofullerenes. *Chem. Eur. J.* 19:14931–14940.

48. Cai, T., Xu, L., Anderson, M. R. et al. 2006. Structure and enhanced reactivity rates of the D_{5h} $Sc_3N@C_{80}$ and $Lu_3N@C_{80}$ metallofullerene isomers: The importance of the pyracylene motif. *J. Am. Chem. Soc.* 128:8581–8589.

49. Ge, Z., Duchamp, J. C., Cai, T., Gibson, H. W., and Dorn, H. C. 2005. Purification of endohedral trimetallic nitride fullerenes in a single, facile step. *J. Am. Chem. Soc.* 127:16292–16298.

50. Maggini, M., Scorrano, G., and Prato, M. 1993. Addition of azomethine ylides to C_{60}: Synthesis, characterization, and functionalization of fullerene pyrrolidines. *J. Am. Chem. Soc.* 115:9798–9799.

51. Cao, B., Wakahara, T., Maeda, Y. et al. 2004. Lanthanum endohedral metallofulleropyr-rolidines: Synthesis, isolation, and EPR characterization. *Chem. Eur. J.* 10:716–720.

52. Lu, X., He, X., Feng, L., Shi, Z., and Gu, Z. 2004. Synthesis of pyrrolidine ring-fused metallofullerene derivatives. *Tetrahedron* 60:3713–3716.

53. Feng, L., Lu, X., He, X., Shi, Z., and Gu, Z. 2004. Reactions of endohedral metallofullerenes with azomethine ylides: An efficient route toward metallofuller-ene-pyrrolidines. *Inorg. Chem. Commun.* 7:1010–1013.

54. Yamada, M., Wakahara, T., Nakahodo, T. et al. 2006. Synthesis and structural characterization of endohedral pyrrolidinodimetallofullerene: $La_2@C_{80}(CH_2)_2NTrt$. *J. Am. Chem. Soc.* 128:1402–1403.

55. Yamada, M., Okamura, M., Sato, S. et al. 2009. Two regioisomers of endohedral pyrrolidinodimetallofullerenes $M_2@I_h$-$C_{80}(CH_2)_2NTrt$ (M = La, Ce; Trt = trityl): Control of metal atom positions by addition positions. *Chem. Eur. J.* 15:10533–10542.

56. Cai, T., Ge, Z., Iezzi, E. B. et al. 2005. Synthesis and characterization of the first trimetallic nitride templated pyrrolidino endohedral metallofullerenes. *Chem. Commun.* 3594–3596.

57. Cardona, C. M., Kitaygorodskiy, A., Ortiz, A., Herranz, M. Á., and Echegoyen, L. 2005. The first fulleropyrrolidine derivative of $Sc_3N@C_{80}$: Pronounced chemical shift differences of the geminal protons on the pyrrolidine ring. *J. Org. Chem.* 70:5092–5097.

58. Cardona, C. M., Kitaygorodskiy, A., and Echegoyen, L. 2005. Trimetallic nitride endohedral metallofullerenes: Reactivity dictated by the encapsulated metal cluster. *J. Am. Chem. Soc.* 127:10448–10453.

59. Cardona, C. M., Elliott, B., and Echegoyen, L. 2006. Unexpected chemical and electrochemical properties of $M_3N@C_{80}$ (M = Sc, Y, Er). *J. Am. Chem. Soc.* 128:6480–6485.

60. Rodríguez-Fortea, A., Campanera, J. M., Cardona, C. M., Echegoyen, L., and Poblet, J. M. 2006. Dancing on a fullerene surface: Isomerization of $Y_3N@(N$-ethylpyrrolidino-$C_{80})$ from the 6,6 to the 5,6 regioisomer. *Angew. Chem. Int. Ed.* 45:8176–8180.

61. Echegoyen, L., Chancellor, C. J., Cardona, C. M. et al. 2006. X-Ray crystallographic and EPR spectroscopic characterization of a pyrrolidine adduct of $Y_3N@C_{80}$. *Chem. Commun.* 2653–2655.

62. Cai, T., Slebodnick, C., Xu, L. et al. 2006. A pirouette on a metallofullerene sphere: Interconversion of isomers of N-tritylpyrrolidino I_h $Sc_3N@C_{80}$. *J. Am. Chem. Soc.* 128:6486–6492.

63. Osuna, S., Rodríguez-Fortea, A., Poblet, J. M., Solà, M., and Swart, M. 2012. Product formation in the Prato reaction on $Sc_3N@D_{5h}$-C_{80}: Preference for [5,6]-bonds, and not pyracylenic bonds. *Chem. Commun.* 48:2486–2488.

64. Aroua, S. and Yamakoshi, Y. 2012. Prato reaction of $M_3N@I_h$-C_{80} (M = Sc, Lu, Y, Gd) with reversible isomerization. *J. Am. Chem. Soc.* 134:20242–20245.

65. Cai, T., Xu, L., Gibson, H. W. et al. 2007. $Sc_3N@C_{78}$: Encapsulated cluster regiocontrol of adduct docking on an ellipsoidal metallofullerene sphere. *J. Am. Chem. Soc.* 129:10795–10800.

66. Campanera, J. M., Bo, C., and Poblet, J. M. 2006. Exohedral reactivity of trimetallic nitride template (TNT) endohedral metallofullerenes. *J. Org. Chem.* 71:46–54.

67. Chen, N., Zhang, E., Tan, K., Wang, C.-R., and Lu, X. 2007. Size effect of encaged clusters on the exohedral chemistry of endohedral fullerenes: A case study on the pyrrolidino reaction of $Sc_xGd_{3-x}N@C_{80}$ (x = 0 – 3). *Org. Lett.* 9:2011–2013.

68. Chen, N., Fan, L. Z., Tan, K. et al. 2007. Comparative spectroscopic and reactivity studies of $Sc_{3-x}Y_xN@C_{80}$ (x = 0–3). *J. Phys. Chem. C* 111:11823–11828.

69. Yang, S. F., Popov, A. A., and Dunsch, L. 2008. Carbon pyramidalization in fullerene cages induced by the endohedral cluster: Non-scandium mixed metal nitride clusterfullerenes. *Angew. Chem. Int. Ed.* 47:8196–8200.

70. Wang, T., Wu, J., Xu, W. et al. 2010. Spin divergence induced by exohedral modification: ESR study of $Sc_3C_2@C_{80}$ fulleropyrrolidine. *Angew. Chem. Int. Ed.* 49:1786–1789.

71. Wang, T., Wu, J., Feng, Y. et al. 2012. Preparation and ESR study of $Sc_3C_2@C_{80}$ bisaddition fulleropyrrolidines. *Dalton Trans.* 41:2567–2570.

72. Lu, X., Nakajima, K., Iiduka, Y. et al. 2011. Structural elucidation and regioselective functionalization of an unexplored carbide cluster metallofullerene $Sc_2C_2@C_s(6)$-C_{82}. *J. Am. Chem. Soc.* 133:19553–19558.

73. Lu, X., Nakajima, K., Iiduka, Y. et al. 2012. The long-believed $Sc_2@C_{2v}(17)$-C_{84} is actually $Sc_2C_2@C_{2v}(9)$-C_{82}: Unambiguous structure assignment and chemical functionalization. *Angew. Chem. Int. Ed.* 51:5889–5892.

74. (a) Bingel, C. 1993. Cyclopropylation of fullerenes. *Chem. Ber.* 126:1957–1959. (b) Hirsch, A., Lamparth, I., and Karfunkel, H. R. 1994. Fullerene chemistry in three dimensions: Isolation of seven regioisomeric bisadducts and chiral trisadducts of C_{60} and di(ethoxycarbonyl)methylene. *Angew. Chem. Int. Ed.* 33:437–438. (c) Hirsch, A., Lamparth, I., Grösser, T., and Karfunkel, H. R. 1994. Regiochemistry of multiple additions to the fullerene core: Synthesis of a T_h-symmetric hexakisadduct of C_{60} with bis(ethoxycarbonyl)methylene. *J. Am. Chem. Soc.* 116:9385–9386.

75. Feng, L., Nakahodo, T., Wakahara, T. et al. 2005. A singly bonded derivative of endohedral metallofullerene: $La@C_{82}CBr(COOC_2H_5)_2$. *J. Am. Chem. Soc.* 127:17136–17137.

76. Feng, L., Wakahara, T., Nakahodo, T. et al. 2006. The Bingel monoadducts of La@C$_{82}$: Synthesis, characterization, and electrochemistry. *Chem. Eur. J.* 12:5578–5586.

77. Feng, L., Tsuchiya, T., Wakahara, T. et al. 2006. Synthesis and characterization of a bisadduct of La@C$_{82}$. *J. Am. Chem. Soc.* 128:5990–5991.

78. Pinzón, J. R., Zuo, T., and Echegoyen, L. 2010. Synthesis and electrochemical studies of Bingel–Hirsch derivatives of M$_3$N@I$_h$-C$_{80}$ (M = Sc, Lu). *Chem. Eur. J.* 16:4864–4869.

79. Alegret, N., Rodríguez-Fortea, R., and Poblet, J. M. 2013. Bingel–Hirsch addition on endohedral metallofullerenes: Kinetic versus thermodynamic control. *Chem. Eur. J.* 19:5061–5069.

80. Lukoyanova, O., Cardona, C. M., Rivera, J. et al. 2007. "Open rather than closed" malonate methano-Fullerene derivatives. The formation of methanofulleroid adducts of Y$_3$N@C$_{80}$. *J. Am. Chem. Soc.* 129:10423–10430.

81. Chaur, M. N., Melin, F., Athans, A. J. et al. 2008. The influence of cage size on the reactivity of trimetallic nitride metallofullerenes: A mono- and bis-methanoadduct of Gd$_3$N@C$_{80}$ and a monoadduct of Gd$_3$N@C$_{84}$. *Chem. Commun.* 2665–2667.

82. Alegret, N., Chaur, M. N., Santos, E., Rodríguez-Fortea, A., Echegoyen, L., and Poblet, J. M. 2010. Bingel–Hirsch reactions on non-IPR Gd$_3$N@C$_{2n}$ (2n = 82 and 84). *J. Org. Chem.* 75:8299–8302.

83. Cai, T., Xu, L., Shu, C. et al. 2008. Selective formation of a symmetric Sc$_3$N@C$_{78}$ bisadduct: Adduct docking controlled by an internal trimetallic nitride cluster. *J. Am. Chem. Soc.* 130:2136–2137.

84. Cai, T., Xu, L., Shu, C., Reid, J. E., Gibson, H. W., and Dorn, H. C. 2008. Synthesis and characterization of a non-IPR fullerene derivative: Sc$_3$N@C$_{68}$[C(COOC$_2$H$_5$)$_2$]. *J. Phys. Chem. C* 112:19203–19208.

85. Alegret, N., Salvadó, P., Rodríguez-Fortea, A., and Poblet, J. M. 2013. Bingel–Hirsch addition on non-isolated-pentagon-rule Gd$_3$N@C$_{2n}$(2n = 82 and 84) metallofullerenes: Products under kinetic control. *J. Org. Chem.* 78:9986–9990.

86. Yang, S., Chen, C., Li, X., Wei, T., Liu, F., and Wang, S. 2013. Bingel–Hirsch monoadducts of TiSc$_2$N@I$_h$-C$_{80}$ *versus* Sc$_3$N@I$_h$-C$_{80}$: Reactivity improvement *via* internal metal atom substitution. *Chem. Commun.* 49:10844–10846.

87. Bolskar, R. D., Benedetto, A. F., Husebo, L. O. et al. 2003. First soluble M@C$_{60}$ derivatives provide enhanced access to metallofullerenes and permit in vivo evaluation of Gd@C$_{60}$[C(COOH)$_2$]$_{10}$ as a MRI contrast agent. *J. Am. Chem. Soc.* 125:5471–5478.

88. Sitharaman, B., Bolskar, R. D., Rusakova, I., and Wilson, L. J. 2004. Gd@C$_{60}$[C(COOH)$_2$]$_{10}$ and Gd@C$_{60}$(OH)$_x$: Nanoscale aggregation studies of two metallofullerene MRI contrast agents in aqueous solution. *Nano Lett.* 4:2373–2378.

89. Feng, L., Zhang, X., Yu, Z., Wang, J., and Gu, Z. 2002. Chemical modification of Tb@C$_{82}$ by copper(I)-catalyzed cycloadditions. *Chem. Mater.* 14:4021–4022.

90. Maeda, Y., Matsunaga, Y., Wakahara, T. et al. 2004. Isolation and characterization of a carbene derivative of La@C$_{82}$. *J. Am. Chem. Soc.* 126:6858–6859.

91. Akasaka, T., Kono, T., Takematsu, Y. et al. 2008. Does Gd@C$_{82}$ have an anomalous endohedral structure? Synthesis and single crystal X-ray structure of the carbene adduct. *J. Am. Chem. Soc.* 130:12840–12841.

92. Lu, X., Nikawa, H., Feng, L. et al. 2009. Location of the yttrium atom in Y@C$_{82}$ and its influence on the reactivity of cage carbons. *J. Am. Chem. Soc.* 131:12066–12067.

93. Takano, Y., Aoyagi, M., Yamada, M. et al. 2009. Anisotropic magnetic behavior of anionic Ce@C$_{82}$ carbene adducts. *J. Am. Chem. Soc.* 131:9340–9346.

94. Hachiya, M., Nikawa, H., Mizorogi, N., Tsuchiya, T., Lu, X., and Akasaka, T. 2012. Exceptional chemical properties of Sc@C$_{2v}$(9)-C$_{82}$ probed with adamantylidene carbene. *J. Am. Chem. Soc.* 134:15550–15555.

95. Akasaka, T., Kono, T., Matsunaga, Y. et al. 2008. Isolation and characterization of carbene derivatives of La@C$_{82}$(C$_s$). *J. Phys. Chem. A* 112:1294–1297.

96. Xie, Y., Suzuki, M., Cai, W. et al. 2013. Highly regioselective addition of adamantylidene carbene to Yb@C$_{2v}$(3)-C$_{80}$ to afford the first derivative of divalent metallofullerenes. *Angew. Chem. Int. Ed.* 52:5142–5145.

97. Zhang, W., Suzuki, M., Xie, Y. et al. 2013. Molecular structure and chemical property of a divalent metallofullerene Yb@C$_2$(13)-C$_{84}$. *J. Am. Chem. Soc.* 135:12730–12735.

98. Lu, X., Nikawa, H., Nakahodo, T. et al. 2008. Chemical understanding of a non-IPR metallofullerene: Stabilization of encaged metals on fused-pentagon bonds in La$_2$@C$_{72}$. *J. Am. Chem. Soc.* 130:9129–9136.

99. Lu, X., Nikawa, H., Tsuchiya, T. et al. 2008. Bis-carbene adducts of non-IPR La$_2$@C$_{72}$: Localization of high reactivity around fused pentagons and electrochemical properties. *Angew. Chem. Int. Ed.* 47:8642–8645.

100. Cao, B., Nikawa, H., Nakahodo, T. et al. 2008. Addition of adamantylidene to La$_2$@C$_{78}$: Isolation and single-crystal X-ray structural determination of the monoadducts. *J. Am. Chem. Soc.* 130:983–989.

101. Yamada, M., Someya, C., Wakahara, T. et al. 2008. Metal atoms collinear with the spiro carbon of 6,6-open adducts, M$_2$@C$_{80}$(Ad) (M = La and Ce, Ad = adamantylidene). *J. Am. Chem. Soc.* 130:1171–1176.

102. Kurihara, H., Lu, X., Iiduka, Y. et al. 2011. Sc$_2$C$_2$@C$_{80}$ rather than Sc$_2$@C$_{82}$: Templated formation of unexpected C$_{2v}$(5)-C$_{80}$ and temperature-dependent dynamic motion of internal Sc$_2$C$_2$ cluster. *J. Am. Chem. Soc.* 133:2382–2385.

103. Kurihara, H., Lu, X., and Iiduka, Y. 2012. Chemical understanding of carbide cluster metallofullerenes: A case study on Sc$_2$C$_2$@C$_{2v}$(5)–C$_{80}$ with complete X-ray crystallographic characterizations. *J. Am. Chem. Soc.* 134:3139–3144.

104. Iiduka, Y., Wakahara, T., Nakahodo, T. et al. 2005. Structural determination of metallofullerene Sc$_3$C$_{82}$ revisited: A surprising finding. *J. Am. Chem. Soc.* 127:12500–12501.

105. Ishitsuka, M. O., Enoki, H., Tsuchiya, T. et al. 2010. Chemical modification of endohedral metallofullerene La@C$_{82}$ with 3-chloro-3-phenyldiazirine. *Phosphorus Sulfur Silicon Relat. Elem.* 185:1124–1130.

106. Ishitsuka, M. O., Sano, S., Enoki, H. et al. 2011. Regioselective bis-functionalization of endohedral dimetallofullerene, La$_2$@C$_{80}$: Extremal La–La distance. *J. Am. Chem. Soc.* 133:7128–7134.

107. Yamada, M., Someya, C. I., Nakahodo, T., Maeda, Y., Tsuchiya, T., and Akasaka, T. 2011. Synthesis of endohedral metallofullerene glycoconjugates by carbene addition. *Molecules* 16:9495–9504.

108. Suzuki, T., Maruyama, Y., Kato, T. et al. 1995. Chemical reactivity of a metallofullerene: EPR study of diphenylmethano-La@C$_{82}$ radicals. *J. Am. Chem. Soc.* 117:9606–9607.

109. Takano, Y., Yomogida, A., Nikawa, H. et al. 2008. Radical coupling reaction of paramagnetic endohedral metallofullerene La@C$_{82}$. *J. Am. Chem. Soc.* 130:16224–16230.

110. Takano, Y., Ishitsuka, M. O., Tsuchiya, T., Akasaka, T., Kato, T., and Nagase, S. 2010. Retro-reaction of singly bonded La@C$_{82}$ derivatives. *Chem. Commun.* 46:8035–8036.

111. Shu, C., Slebodnick, C., Xu, L. et al. 2008. Highly regioselective derivatization of trimetallic nitride templated endohedral metallofullerenes via a facile photochemical reaction. *J. Am. Chem. Soc.* 130:17755–17760.

112. Shu, C., Cai, T., Xu, L. et al. 2007. Manganese(III)-catalyzed free radical reactions on trimetallic nitride endohedral metallofullerenes. *J. Am. Chem. Soc.* 129:15710–15717.

113. Tagmatarchis, N., Taninaka, A., and Shinohara, H. 2002. Production and EPR characterization of exohedrally perfluoroalkylated paramagnetic lanthanum metallofullerenes: (La@C$_{82}$)-(C$_8$F$_{17}$)$_2$. *Chem. Phys. Lett.* 355:226–232.

114. Kareev, I. E., Lebedkin, S. F., Bubnov, V. P. et al. 2005. Trifluoromethylated endohedral metallofullerenes: Synthesis and characterization of Y@C$_{82}$(CF$_3$)$_5$. *Angew. Chem. Int. Ed.* 44:1846–1849.

115. Kareev, I. E., Bubnov, V. P., and Yagubskii, E. B. 2008. Endohedral gadolinium-containing metallofullerenes in the trifluoromethylation reaction. *Russ. Chem. Bull. Int. Ed.* 57:1486–1491.

116. Shustova, N. B., Popov, A. A., Mackey, M. A. et al. 2007. Radical trifluoromethylation of Sc$_3$N@C$_{80}$. *J. Am. Chem. Soc.* 129:11676–11677.

117. Shustova, N., Chen, Y. S., Mackey, M. A. et al. 2009. Sc$_3$N@(C$_{80}$-I_h(7))(CF$_3$)$_{14}$ and Sc$_3$N@(C$_{80}$-I_h(7))(CF$_3$)$_{16}$. endohedral metallofullerene derivatives with exohedral addends on four and eight triple-hexagon junctions. Does the Sc$_3$N cluster control the addition pattern or vice versa? *J. Am. Chem. Soc.* 131:17630–17637.

118. Yang, S., Chen, C., Lanskikh, M. A., Tamm, N. B., Kemnitz, E., and Troyanov, S. I. 2011. New isomers of trifluoromethylated derivatives of metal nitride cluster fullerene: Sc$_3$N@C$_{80}$(CF$_3$)$_n$ (n = 14 and 16). *Chem. Asian J.* 6:505–509.

119. Popov, A. A., Shustova, N. B., Svitova, A. L. et al. 2010. Redox-tuning endohedral fullerene spin states: From the dication to the trianion radical of Sc$_3$N@C$_{80}$(CF$_3$)$_2$ in five reversible single-electron steps. *Chem. Eur. J.* 16:4721–4724.

120. Popov, A. A. and Dunsch, L. 2011. Charge controlled changes in the cluster and spin dynamics of Sc$_3$N@C$_{80}$(CF$_3$)$_2$: The flexible spin density distribution and its impact on ESR spectra. *Phys. Chem. Chem. Phys.* 13:8977–8984.

121. Shustova, N. B., Peryshkov, D. V., Kuvychko, I. V. et al. 2011. Poly(perfluoroalkylation) of metallic nitride fullerenes reveals addition-pattern guidelines: Synthesis and characterization of a family of Sc$_3$N@C$_{80}$(CF$_3$)$_n$ (n = 2−16) and their radical anions. *J. Am. Chem. Soc.* 133:2672–2690.

122. Nikawa, H., Kikuchi, T., Wakahara, T. et al. 2005. Missing metallofullerene La@C$_{74}$. *J. Am. Chem. Soc.* 127:9684–9685.

123. Lu, X., Nikawa, H., Kikuchi, T. et al. 2011. Radical derivatives of insoluble La@C$_{74}$: X-ray structures, metal positions, and isomerization. *Angew. Chem. Int. Ed.* 50:6356–6359.

124. Wakahara, T., Nikawa, H., Kikuchi, T. et al. 2006. La@C$_{72}$ having a non-IPR carbon cage. *J. Am. Chem. Soc.* 128:14228–14229.

125. Nikawa, H., Yamada, T., Cao, B. et al. 2009. Missing metallofullerene with C$_{80}$ cage. *J. Am. Chem. Soc.* 131:10950–10954.

126. Akasaka, T., Lu, X., Kuga, H. et al. 2010. Dichlorophenyl derivatives of La@C$_{3v}$(7)-C$_{82}$: Endohedral metal induced localization of pyramidalization and spin on a triple-hexagon junction. *Angew. Chem. Int. Ed.* 49:9715–9719.

127. Tang, C., Deng, K., Tan, W. et al. 2007. Influence of a dichlophenyl group on the geometric structure, electronic properties, and static linear polarizability of La@C$_{74}$. *Phys. Rev. A* 76:013201.

128. Sawai, K., Takano, Y., Izquierdo, M. et al. 2011. Enantioselective synthesis of endohedral metallofullerenes. *J. Am. Chem. Soc.* 133:17746–17752.

129. Lu, X., Xu, J., He, X., Shi, Z., and Gu, Z. 2004. Addition of benzyne to Gd@C$_{82}$. *Chem. Mater.* 16:953–955.

130. Lu, X., Nikawa, H., Tsuchiya, T. et al. 2010. Nitrated benzyne derivatives of La@C$_{82}$: Addition of NO$_2$ and its positional directing effect on the subsequent addition of benzynes. *Angew. Chem. Int. Ed.* 49:594–597.

131. Li, F.-F., Pinzón, J. R., Mercado, B. Q., Olmstead, M. M., Balch, A. L., and Echegoyen, L. 2011. [2+2] cycloaddition reaction to Sc$_3$N@I_h-C$_{80}$. The formation of very stable [5,6]- and [6,6]-adducts. *J. Am. Chem. Soc.* 133:1563–1571.

132. Yang, S. and Yang, S. 2002. Preparation and film formation behavior of the supramolecular complex of the endohedral metallofullerene Dy@C$_{82}$ with calix[8]arene. *Langmuir* 18:8488–8495.

133. Tsuchiya, T., Sato, K., Kurihara, H. et al. 2006. Host–guest complexation of endohedral metallofullerene with azacrown ether and its application. *J. Am. Chem. Soc.* 128:6699–6703.

134. Tsuchiya, T., Kurihara, H., Sato, K. et al. 2006. Supramolecular complexes of La@C$_{82}$ with unsaturated thiacrown ethers. *Chem. Commun.* 3585–3587.

135. Tsuchiya, T., Sato, K., Kurihara, H. et al. 2006. Spin-site exchange system constructed from endohedral metallofullerenes and organic donors. *J. Am. Chem. Soc.* 128:14418–14419.

136. Tsuchiya, T., Wielopolski, M., Sakuma, N. et al. 2011. Stable radical anions inside fullerene cages: Formation of reversible electron transfer systems. *J. Am. Chem. Soc.* 133:13280–13283.

137. Tsuchiya, T., Rudolf, M., Wolfrum, S. et al. 2013. Coordinative interactions between porphyrins and C$_{60}$, La@C$_{82}$, and La$_2$@C$_{80}$. *Chem. Eur. J.* 19:558–565.

138. Pagona, G., Economopoulos, S. P., Aono, T., Miyata, Y., Shinohara H., and Tagmatarchis, N. 2010. Molecular recognition of La@C$_{82}$ endohedral metallofullerene by an isophthaloyl-bridged porphyrin dimer. *Tetrahedron Lett.* 51:5896–5899.

139. Hajjaj, F., Tashiro, K., Nikawa, H. et al. 2011. Ferromagnetic spin coupling between endohedral metallofullerene La@C$_{82}$ and a cyclodimeric copper porphyrin upon inclusion. *J. Am. Chem. Soc.* 133:9290–9292.

140. Nakanishi, Y., Omachi, H., Matsuura, S. et al. 2014. Size-selective complexation and extraction of endohedral metallofullerenes with cycloparaphenylene. *Angew. Chem. Int. Ed.* 53:3102–3106.

141. Chen, C.-H., Yeh, W.-Y., Liu, Y.-H., and Lee, G.-H. 2012. [(μ–H)$_3$Re$_3$(CO)$_9$(η^2,η^2,η^2-Sc$_2$C$_2$@C$_{3v}$(8)-C$_{82}$)]: Face-capping cluster complex of an endohedral fullerene. *Angew. Chem. Int. Ed.* 51:13046–13049.

142. Chen, C.-H., Lin, D.-Y., and Yeh, W.-Y. 2014. Regiospecific coordination of Re$_3$ clusters with the sumanene type hexagons on endohedral metallofullerenes and higher fullerenes that provides an efficient separation method. *Chem. Eur. J.* 20:5768–5775.

143. Lu, X., Zhou, X., Shi, Z., and Gu, Z. 2004. Nucleophilic addition of glycine esters to Gd@C$_{82}$. *Inorg. Chim. Acta* 357:2397–2400.

144. Li, X., Fan, L., Liu, D. et al. 2007. Synthesis of a Dy@C$_{82}$ derivative bearing a single phosphorus substituent via a zwitterion approach. *J. Am. Chem. Soc.* 129:10636–10637.

145. Campanera, J. M., Heggie, M. I., and Taylor, R. 2005. Analysis of polyaddition levels in *i*-Sc$_3$NC$_{80}$. *J. Phys. Chem. B* 109:4024–4031.

146. Li, F.-F., Rodríguez-Fortea, A., Poblet, J. M., and Echegoyen, L. 2011. Reactivity of metallic nitride endohedral metallofullerene anions: Electrochemical synthesis of a Lu$_3$N@I$_h$-C$_{80}$ derivative. *J. Am. Chem. Soc.* 133:2760–2765.

147. Kurihara, H., Iiduka, Y., Rubin, Y. et al. 2012. Unexpected formation of a Sc$_3$C$_2$@C$_{80}$ bisfulleroid derivative. *J. Am. Chem. Soc.* 134:4092–4095.

148. Lu, X., Feng, L., Akasaka, T., and Nagase, S. 2012. Current status and future developments of endohedral metallofullerenes. *Chem. Soc. Rev.* 41:7723–7760.

149. Cong, H., Yu, B., Akasaka, T., and Lu, X. 2013. Endohedral metallofullerenes: An unconventional core–shell coordination union. *Coord. Chem. Rev.* 257:2880–2898.

150. Jin, P., Tang, C., and Chen, Z. 2014. Carbon atoms trapped in cages: Metal carbide clusterfullerenes. *Coord. Chem. Rev.* 270–271:89–111.

151. Ghiassi, K. B., Olmstead, M. M., and Balch, A. L. 2014. Gadolinium-containing endohedral fullerenes: Structures and function as magnetic resonance imaging (MRI) agents. *Dalton Trans.* 43:7346–7358.

152. Mikawa, M., Kato, H., Okumura, M. et al. 2001. Paramagnetic water-soluble metallofullerenes having the highest relaxivity for MRI contrast agents. *Bioconjug. Chem.* 12:510–514.

153. Cagle, D. W., Kennel, S. J., Mirzadeh, S., Alford, J. M., and Wilson, L. J. 1999. In vivo studies of fullerene-based materials using endohedral metallofullerene radiotracers. *Proc. Natl. Acad. Sci. USA.* 96:5182–5187.

154. Meng, J., Liang, X., Chen X., and Zhao, Y. 2013. Biological characterizations of [Gd@C$_{82}$(OH)$_{22}$]$_n$ nanoparticles as fullerene derivatives for cancer therapy. *Integr. Biol.* 5:43–47.

155. Takano, Y., Obuchi, S., Mizorogi, N. et al. 2012. Stabilizing ion and radical ion pair states in a paramagnetic endohedral metallofullerene/π-extended tetrathiafulvalene conjugate. *J. Am. Chem. Soc.* 134:16103–16106.

156. Takano, Y., Herranz, M. Á., Martín, N. et al. 2010. Donor-acceptor conjugates of lanthanum endohedral metallofullerene and π-extended tetrathiafulvalene. *J. Am. Chem. Soc.* 132:8048–8055.

157. Guldi, D. M., Feng, L., Radhakrishnan, S. G. et al. 2010. A molecular Ce$_2$@I_h-C$_{80}$ switch—unprecedented oxidative pathway in photoinduced charge transfer reactivity. *J. Am. Chem. Soc.* 132:9078–9086.

158. Feng, L., Radhakrishnan, S. G., Mizorogi, N. et al. 2011. Synthesis and charge-transfer chemistry of La$_2$@I_h-C$_{80}$/Sc$_3$N@I_h-C$_{80}$-zinc porphyrin conjugates: Impact of endohedral cluster. *J. Am. Chem. Soc.* 133:7608–7618.

159. Takano, Y., Obuchi, S., Mizorogi, N. et al. 2012. An endohedral metallofullerene as a pure electron donor: Intramolecular electron transfer in donor–acceptor conjugates of La$_2$@C$_{80}$ and 11,11,12,12-tetracyano-9,10-anthra-p-quinodimethane (TCAQ). *J. Am. Chem. Soc.* 134:19401–19408.

160. Pinzón, J. R., Plonska-Brzezinska, M. E., Cardona, C. M. et al. 2008. Sc$_3$N@C$_{80}$-ferrocene electron-donor/acceptor conjugates as promising materials for photovoltaic applications. *Angew. Chem. Int. Ed.* 47:4173–4176.

161. Pinzón, J. R., Cardona, C. M., Herranz, M. Á. et al. 2009. Metal nitride cluster fullerene M$_3$N@C$_{80}$ (M = Y, Sc) based dyads: Synthesis, and electrochemical, theoretical and photophysical studies. *Chem. Eur. J.* 15:864–877.

162. Pinzón, J. R., Gasca, D. C., Sankaranarayanan, S. G. et al. 2009. Photoinduced charge transfer and electrochemical properties of triphenylamine I_h-Sc$_3$N@C$_{80}$ donor–acceptor conjugates. *J. Am. Chem. Soc.* 131:7727–7734.

163. Wolfrum, S., Pinzón, J. R., Molina-Ontoria, A. et al. 2011. Utilization of Sc$_3$N@C$_{80}$ in long-range charge transfer reactions. *Chem. Commun.* 47:2270–2272.

164. Rudolf, M., Feng, L., Slanina, Z., Akasaka, T., Nagase, S., and Guldi, D. M. 2013. A metallofullerene electron donor that powers an efficient spin flip in a linear electron donor–acceptor conjugate. *J. Am. Chem. Soc.* 135:11165–11174.

165. Garcia-Borràs, M., Osuna. S., Swart, M., Luis, J. M., Echegoyen, L., and Solà, M. 2013. Aromaticity as the driving force for the stability of non-IPR endohedral metallofullerene Bingel–Hirsch adducts. *Chem. Commun.* 49:8767–8769.

166. Garcia-Borràs, M., Osuna, S., Luis, J. M., Swart, M., and Solà, M. 2014. The role of aromaticity in determining the molecular structure and reactivity of (endohedral metallo)fullerenes. *Chem. Soc. Rev.* 43:5089–5105.

CHAPTER 9

CONTENTS

Electrochemical Methods to Functionalize Carbon Materials

9

Carolina González-Gaitán, Ramiro Ruiz-Rosas,
Emilia Morallón, and Diego Cazorla-Amorós

9.1 INTRODUCTION

Surface chemistry plays a relevant role in the physicochemical properties of carbon materials. It is highly influenced by the presence of heteroatoms (mainly oxygen and nitrogen, but also phosphorus, boron, and sulfur) that may form different surface functionalities. They can be found naturally on the carbon surface, but they can also be generated during the carbon material preparation or by subsequent treatments. The occurrence of diverse surface functionalities over the carbon surface governs their reactivity, their chemical and physical stability, and their structure. Along with their porous structure, these properties dictate the potential use of carbon materials in adsorption (Boehm 2008), catalysis (Bandosz 2008), energy storage (Radovic 2010), sensors (McCreery 2008), and biomedical (Bianco et al. 2005; Liu et al. 2007) applications. This is especially manifested in the case of the promising nanostructured carbon materials discovered in the last two decades but is also essential for classic materials like graphite, carbon blacks, and activated carbons.

Given the practical implications of the surface chemistry on the commercial use of carbon-based materials, the surface chemistry of carbon materials is a topic of huge interest, and a strong effort is being dedicated in developing methods for the modification and qualitative and quantitative characterization of the surface functional groups. Regarding surface chemistry characterization, pioneer research by Walker and Boehm (Vastola et al. 1964; Boehm 1966; Bansal et al. 1970) was followed in the 1970s and the 1980s and further expanded in the last two decades, being the majority of those works devoted to the characterization of the most frequently found surface oxygen and nitrogen functionalities. Some examples of the abundant literature are the following: Román-Martínez et al. (1993), Boehm (1994), Pels et al. (1995), Stańczyk et al. (1995), Zielke et al. (1996), Biniak et al. (1997), De la Puente et al. (1997), Zhang and Ritter (1997), Figueiredo et al. (1999), Kuznetsova et al. (2001), Salame and Bandosz (2001), Boehm (2002), Raymundo-Piñero et al. (2002), Vix-Guterl et al. (2004), Zhou et al. (2007), Gorgulho et al. (2008), Karousis et al. (2010), and Kundu et al. (2010). Thanks to these valuable published results and the development and spreading of new analytical tools, it is possible to find nowadays a whole body of technical literature that deals with the measurement and characterization of surface functional groups using a variety of experimental techniques and methods.

Rather than trying to discuss exhaustively the surface chemistry of carbon materials, this chapter is focused on the modification of surface chemistry, but dedicating more effort to electrochemical methods. If interested in highly detailed and/or general-oriented discussion, the reader can easily find several excellent reviews devoted to the study of surface chemistry in carbon materials (Leon y Leon and Radovic 1992; Bandosz and Ania 2006; Bandosz 2008; Boehm 2008; Radovic 2010). Figure 9.1 presents a brief scheme of the most frequent functionalization techniques available for the modification of surface chemistry, along with a few examples of their use, that are going to be detailed in the following sections.

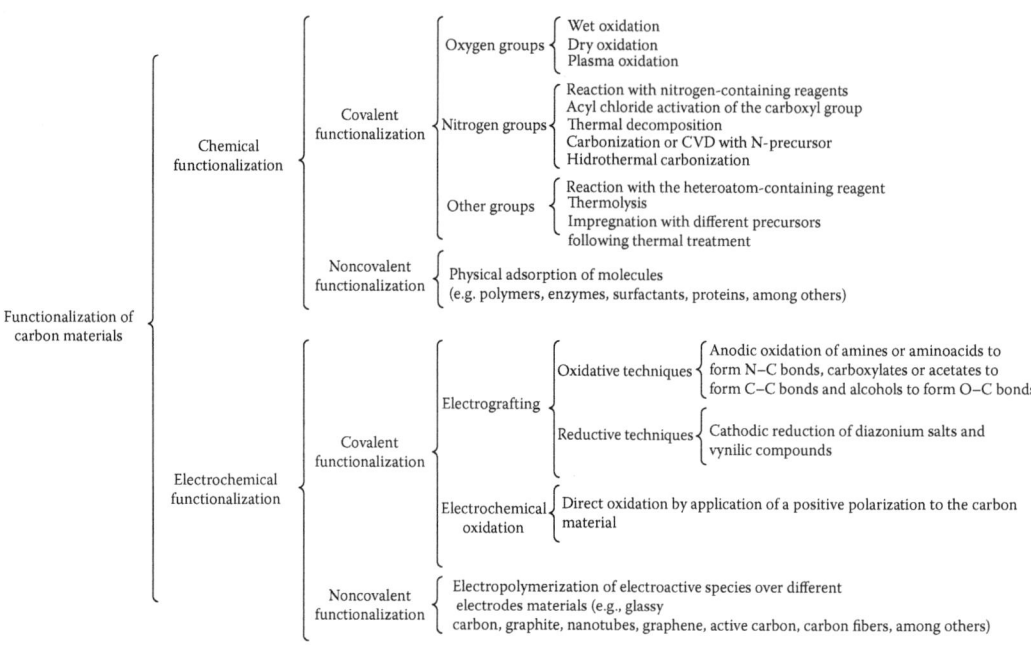

FIGURE 9.1
Scheme of the functionalization techniques that will be reviewed in this chapter.

In order to keep the scope of the review at a convenient level, most of the chemical functionalization methods will be briefly explained, as the text is devoted to electrochemical functionalization techniques. However, that section on chemical functionalization is necessary to understand the differences between both approaches. Before exploring the chemical or electrochemical reaction pathways, we will start with a short section on reactivity of carbon materials, which is convenient to understand the functionalization of carbon materials.

9.2 REACTIVITY OF CARBON SURFACE

Most, if not all, carbon materials that will be reviewed in this chapter have a predominant sp^2 bonding type in carbon atoms, which means that they are mainly composed of single or stacked graphene sheets; functionalization of allotropes like diamond is out of the scope of this work. Independently of the arrangement and curvature of the graphene layers, all carbon materials have active sites where covalent bonds can be formed between external molecules and the carbon surface forming a new functional group or grafting a new molecule. Marked differences have been assumed for long time when considering the reactivity of C atoms with strong covalent carbon–carbon bonds located inside the graphene layer, that is, basal plane sites, and those lying either in zigzag or armchair environment at the edges of the layer, that is, edge sites (Abrahamson 1973; Kinoshita 1988; Walker et al. 1991). The nature of the active sites is still a matter of controversy, although recent studies seem to confirm that carbine-type sites at armchair edges and carbene-type at zigzag edges are responsible for the reactivity of graphene and carbon nanotubes. (Radovic and Bockrath 2005; Radovic 2009; Silva-Tapia et al. 2012). Since these carbon atoms with unsatisfied valence found on edge sites are more reactive than those in the basal plane, the ratio of basal plane to edge atoms, which is directly related to the size of the graphene layer, is a good index for the possibilities of forming covalently bonded functional groups. Nevertheless, the basal plane is far to be inactive. The high π electron density of the basal plane increases the dispersive adsorption potential of graphene (Radovic et al. 1997), allowing noncovalent functionalization which will be later discussed, and provides for some basicity to the carbon surface (Leon y Leon et al. 1992). Moreover, epoxide groups in the basal plane can be formed by oxygen spillover from adsorbed O$_2$ on carbene-type edge sites (Radovic et al. 2011), opening the door to sidewall functionalization on carbon nanotubes CNTs.

The strain induced in C–C bonds by the pronounced curvature that may exist in the graphene layers has been also proposed as highly affecting the reactivity of the basal plane, especially in CNTs (Hirsch 2002; Lehman et al. 2011). The effect of curvature in the reactivity of C–C bond of CNT in oxygen chemisorption has been theoretically demonstrated for armchair-type CNTs (Silva-Tapia et al. 2012). The curvature can also be related to the presence of pentagons or heptagons in the hexagonal network, and the bonds in these polygons have been theoretically identified as reactive sites for oxidation in graphene and single-walled CNTs (Fujimori et al. 2012). Yao confirmed that pentagonal or heptagonal defects in multiwalled CNTs act as initiation sites for their oxidation (Yao et al. 2011).

The enhanced reactivity of curved graphene layers can be also found in carbon materials obtained by template nanocasting methods. Since the structure of these carbon materials is the negative replica of the nanosized channels of the template, they are composed by highly curved and defective stacked graphene layers, which are far more reactive to oxidation and electrooxidation than the pyrolytic carbon obtained without using a porous template in identical synthesis conditions (Nishihara et al. 2009; Berenguer et al. 2013).

It should be taken into consideration that the initial differences in reactivity between edge and basal plane atoms can be overcome by carefully selecting both the adequate material and functionalization method. For example, Dongil et al. (2011) studied the generation of oxygen functionalities in high surface area graphite and in carbon nanofibers, which have rather different edge-to-basal plane site ratios. They carried out the functionalization in both materials using a traditional wet oxidation method with concentrated HNO_3 and plasma oxidation with oxygen. The plasma oxidation was able to oxidize both edge and basal plane sites, as it was previously shown (Paredes et al. 2000), while HNO_3 oxidation mainly oxidizes edge sites. In terms of efficiency, it was found that oxidation of the high surface area graphite was more effective than that of carbon nanofibers, being the highest when combined with the wet oxidation method, because of the higher edge-to-basal plane site ratio.

Therefore, a high amount of edge sites can be seen as an advantage when designing surface functionalization of carbon materials. A relative abundance of active sites occurs in the case of highly disordered and porous carbon materials, like activated carbons, but it is also found in ordered, yet again highly porous, nanocasted carbons (Berenguer et al. 2013). For these porous materials, the practical limit for profiting of such abundance of active sites comes from diffusional limitations that render most of the active sites unavailable for attaching functional groups. For highly ordered carbon materials, like graphene or CNTs, edge sites can be found as defects generated during the most frequently used synthesis method of graphene through graphene oxide (Hirsch et al. 2013) or can be produced by oxidizing treatments in the highly curved tips of CNTs, where strains weaken the C–C bonds (Hirsch 2002). However, an excessive generation of active sites for functionalization will strongly modify the intrinsic electronic, optical, and mechanical properties of nanostructured carbon materials (Niyogi et al. 2002; Zhao et al. 2004). For instance, a large density of functional groups and defects cause a decrease in the conductivity of functionalized graphene sheets, such that significant ohmic losses will occur during their use as electrodes (Punckt et al. 2010).

9.3 CHEMICAL FUNCTIONALIZATION OF CARBON MATERIALS

9.3.1 Noncovalent Functionalization Methods

Noncovalent functionalization is the most preferred route when, for a given application, it is essential to maintain the extended π-conjugation of nanostructured carbon materials (Kuzmany et al. 2004; Simmons et al. 2009; Georgakilas et al. 2012). In this case, polymers (Chen et al. 2002; Baskaran et al. 2005), surfactants (Hu et al. 2009), enzymes (Shim et al. 2002), proteins (Chen et al. 2001b), and amino-containing molecules (Chen et al. 2003), among others, have been used. The interactions between the molecule and the carbon surface arise between the π-system of the carbon structure and ligands with hydrogen, cation, and anion or π-systems in their structure, and they have been profusely studied on the basis of their energetic and geometrical significance (Georgakilas et al. 2012).

Many examples can be found in molecules containing large π-systems because they are strongly adsorbed at the surface of the carbon material (Hu et al. 2009; Georgakilas et al. 2012).

The strength of noncovalent functionalization results from the simultaneous effect of electrostatic, dispersive, inductive, and exchange repulsion forces and highly relies on the presence of an extended π-system in the carbon structure. Consequently, this functionalization is especially suited for those materials in which an ordered structure and a largely exposed aromatic surface are found, as CNTs (Chen et al. 2001b; Simmons et al. 2009) and graphene (Bai et al. 2009; Kuila et al. 2012). It is extensively used for producing materials for biological and sensor applications (Chen et al. 2003; Agüí et al. 2008). These noncovalent modification techniques retain the desired properties of these carbon materials, and in addition, the presence of the new molecules at the surface facilitates their processability even in aqueous solution. In the case of highly porous materials, there are few examples (Godino-Salido et al. 2014). For these porous materials, the blockage of their porosity when adsorbing large molecules imposes important limitations to the use of noncovalent functionalization.

9.3.2 Covalent Functionalization Methods

Covalent functionalization by chemical methods has been strongly developed in order to tailor the properties of the carbon material by the addition of certain heteroatoms or molecules. The incorporation of heteroatoms like oxygen, nitrogen, sulfur, and phosphorus (Datsyuk et al. 2008; Stein et al. 2009; Ayala et al. 2010; Yang et al. 2012a,b; Hirsch et al. 2013; Johns and Hersam 2013; Shen and Fan 2013; Kiciński et al. 2014) is not difficult compared to grafting of other organic molecules in which the yield of the process is generally low and is mainly applicable to carbon nanostructures like CNTs, carbon nanofibers, and graphene (Bahr and Tour 2001; Ying et al. 2003; Tagmatarchis and Prato 2004; Li et al. 2005; Wang et al. 2005; Peng and Wong 2009; Quintana et al. 2010, 2013; Kuila et al. 2012). As a consequence, a large number of papers have been published on modification of surface chemistry by introduction of heteroatoms since they may provide clear improvements in different application areas. For example, they can be anchorage sites for catalysts or they can be catalysts themselves, they can provide pseudocapacitance, they can increase the wettability and the affinity toward polar molecules, they can be reaction sites for further carbon material modification, and they can improve the oxidation resistance of the material.

The intense research carried out for tailoring the functionalities was initially done with classic carbon materials, like graphites, carbon blacks, and activated carbons, and it has been successfully applied to new carbon forms. Precedents for the functionalization of graphene and CNTs can be found in studies of the oxidation of carbon black, activated carbon, activated carbon fibers and graphites (Donnet 1968; Fitzer and Weiss 1987; Moreno-Castilla et al. 2000), chemical modification of carbon electrodes (Elliott and Murray 1976), the generation of nitrogen groups on the surface of activated carbons (Stöhr et al. 1991; Jansen and van Bekkum 1994) and activated carbon fibers (Delamar et al. 1997), immobilization of enzymes (Osborn et al. 1982) or even porphyrins over glassy carbon (Willman et al. 1980) and activated carbons (Silva et al. 2002), functionalization of carbon blacks by grafted polymers (Tsubokawa 1992), etc. These studies are an example of the large amount of literature available about functionalization and grafting of carbon materials, which are only partially covered in more recent reviews (Stein et al. 2009; Bhatnagar et al. 2013; Figueiredo 2013). Some reviews about functionalization of CNTs (Bahr and Tour 2002; Banerjee and Wong 2002; Hirsch 2002; Kuzmany et al. 2004; Balasubramanian and Burghard 2005; Zeng 2008; Peng and Wong 2009; Karousis et al. 2010; Johns and Hersam 2013) and graphene (Boukhvalov and Katsnelson 2008; Englert et al. 2011; Salavagione et al. 2011; Chen et al. 2012; Georgakilas et al. 2012; Kuila et al. 2012; Lu et al. 2012; Pantelides et al. 2012; Dai 2013; Hirsch et al. 2013; Johns and Hersam 2013; Quintana et al. 2013) can be found nowadays. A general conclusion that can be deduced regarding the functionalization of carbon materials with heteroatoms by chemical methods is the low selectivity toward given functional groups.

In the following sections, a brief summary of the most usual treatments for covalent functionalization with oxygen-, nitrogen-, sulfur-, and phosphorus-containing groups will be presented in order to show a general view of this important area of research. It will also be useful for a better understanding of the electrochemical functionalization techniques that will be reviewed later in this chapter.

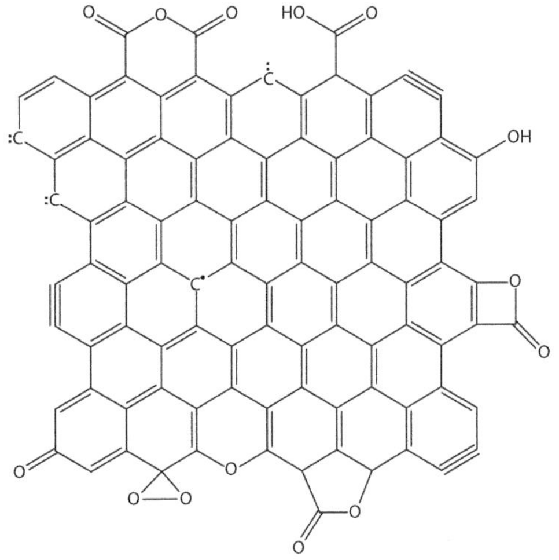

FIGURE 9.2
The most frequent oxygen functionalities found on carbon surface. (Modified from Fanning, P.E.
and Vannice, M.A., *Carbon*, 31(5), 721, 1993; Radovic, L.R., Surface chemical and electrochemical properties of carbons, in *Carbons for Electrochemical Energy Storage and Conversion Systems*,
F. Beguin and E. Frackowiak, eds., Taylor & Francis Group, Boca Raton, FL, pp. 163–219, 2010.)

9.3.2.1 Oxygen functionalization

Oxygen functionalities are inherent to any carbon surface exposed to the atmosphere. Traditionally,
surface oxygen groups (SOGs) are divided into two groups regarding their acidic or basic nature
(Boehm 1966, 1994; Donnet 1968). In such classification, carboxylic, anhydrides, and lactones
groups are regarded as acid, whereas phenols, quinones, carbonyls, and ethers are considered as
slightly basic groups. Figure 9.2 shows a scheme of a graphene layer with the most habitual SOGs,
and it also considers additional issues of the surface chemistry that are relevant to carbon materials,
like the presence of radicals (Fanning and Vannice 1993; Radovic 2010).

There are two traditional routes for generation of SOGs: mainly wet oxidation (Linares-Solano
et al. 1990; Otake and Jenkins 1993; Vinke et al. 1994; Ago et al. 1999; Moreno-Castilla et al.
2000), where the carbon material is contacted with a solution of an oxidizing agent, and dry oxidation, where the carbon surface is exposed to an oxidizing gas, usually air, at mild temperatures
(Polovina et al. 1997; Ago et al. 1999; Wang et al. 2004; Bandosz and Ania 2006). Plasma oxidation is also available for such purpose, although its use is less frequent (Esumi et al. 1987; García
1998; Ago et al. 1999; Domingo-García et al. 2000; Paredes et al. 2000; Tang et al. 2007). The
wet oxidation route is preferred for the formation of acid groups, with the air oxidation generating
higher amounts of basic and neutral groups. In any case, the selectivity of these methods is low. All
these kinds of groups can be decomposed into CO and CO_2 under heating (Román-Martínez et al.
1993), making available active sites that could be reoxidized when putting again the carbon material
in contact with the oxygen of atmosphere. Since these groups decompose at different intervals of
temperatures (Figueiredo et al. 1999), subsequent heat treatments in inert or reducing atmospheres
can be employed for modulating to some extent the nature of SOGs (Román-Martínez et al. 1993).

9.3.2.2 Nitrogen functionalization

Nitrogen is one of the most frequent heteroatoms that can be found in the surface of carbon
materials. In general, nitrogen can be found either bonded to one (amine group) or two carbon
atoms (pyridine and pyrrole groups) and also substituting one carbon atom in the graphene layer
(quaternary nitrogen) (Biniak et al. 1997; Raymundo-Piñero et al. 2002; Bellafont et al. 2014). There
still exists some controversy about the chemical nature of some of these functionalities, especially
regarding quaternary N. The position of the N heteroatom within the graphene layer also mediates
the properties of these groups and may produce interesting local structural changes in the graphene
layer. For example, pyridine and pyrrole groups inside the graphene layer involve the appearance

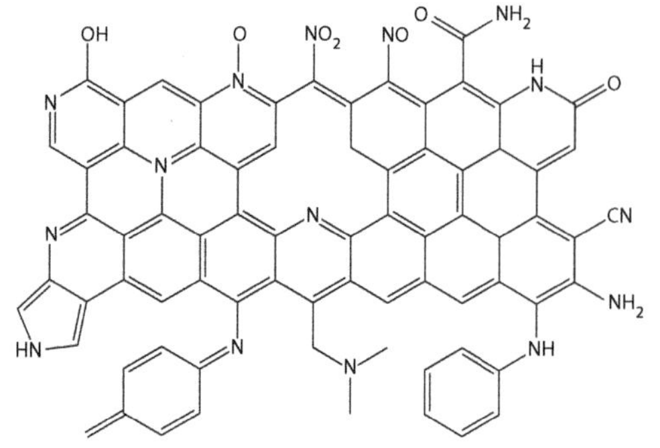

FIGURE 9.3
Nitrogen functionalities found on carbon surface.

of a vacancy, and quaternary nitrogen being in a valley position (close to the edge of the graphene layer) is more stable than in a center position (Yamada et al. 2014). Additional nitrogen-containing functional groups appear if oxygen is taken into consideration (i.e., pyridone groups). A scheme with the different possibilities for N functionalities is included in Figure 9.3.

Each of these functionalities contributes to the physicochemical properties of carbon materials like surface acidity or basicity, wettability, and electrical conductivity, among others (Van Dommele et al. 2008; Mabena et al. 2011). This can be beneficial in many applications like CO_2 capture (Hao et al. 2010), environmental remediation (Tanada et al. 1999), energy storage (Wood et al. 2014), replacement of noble metal as catalyst in fuel cells (Wu et al. 2011) and other catalyzed systems, or biomedical applications (Ramanathan et al. 2005).

Nitrogen functionalities can be produced onto the surface of carbon materials by several methods (Boehm 1994; Raymundo-Piñero et al. 2002, 2003; Wang et al. 2005; White et al. 2009; Shen and Fan 2013), being the most frequent:

- Reaction with nitrogen-containing reagents, either in gas phase or liquid phase, that is, ammonia, urea, and NO
- Converting carboxyl functionalities into amides by acyl chloride activation of the carboxyl group after $SOCl_2$ treatment
- Thermal decomposition of a nitrogen-containing precursor or polymer (melamine, polyacrylonitrile, polypyrrole, polyaniline [PANI], etc.) in the presence of the carbon material
- Carbonization or chemical vapor deposition (CVD) using a nitrogen-containing precursor followed, if necessary, by chemical or physical activation for porosity development
- Hydrothermal carbonization of nitrogen-containing biomass precursors

The temperature has a critical effect in the selectivity of the reaction. In general, treatments at high temperatures promote the formation of quaternary N, pyridine and pyrrole groups, and decomposition of less stable lactams, amines, and imines.

Most of those treatments can be used for functionalization of CNTs (Wang et al. 2005; Ayala et al. 2010) or graphene (Georgakilas et al. 2012), and for both materials, the amide bridge plays a key role in the functionalization by polymer, enzymes, and proteins.

9.3.2.3 Sulfur functionalization

Sulfur can be naturally found in coal, although most of it is not chemically bonded to the carbon surface. Nevertheless, different types of sulfur functionalities can be attached to the carbon surface and can be classified according to the number of carbon atoms bonded to them as sulfides or sulfoxides (two carbon atoms) and thiols or thioquinones (one carbon atom). Disulfides can be also found. The most common S-functionalities are thiols and sulfoxides (You et al. 2014), and each of them provides different properties to the carbon material. In general, the presence of sulfur confers them a higher chemical stability, and it is a catalytic center for many reactions.

These properties can be profited in heterogeneous catalysis, adsorption processes, and conversion and storage of energy (Kiciński et al. 2014).

Functionalization of an activated carbon surface with sulfur is possible by contacting it with sulfur-containing reagents such as H_2S, CS_2, or SO_2 (Bandosz and Ania 2006). Further studies of functionalization by chemical methods achieved this kind of modification in CNTs (Tripathi et al. 2011; Shi et al. 2013; Zhou et al. 2013), carbon nanospheres (You et al. 2014), graphene (Yang et al. 2012b; Su et al. 2013; Wang et al. 2013), porous carbons (Kiciński et al. 2014), and carbon aerogels (Wohlgemuth et al. 2012), among others.

9.3.2.4 Phosphorus functionalization

The doping of carbon materials with phosphorus has been studied for a long time for inhibition of the $C-O_2$ reaction (McKee 1981; McKee et al. 1984a), since it is important for oxidation protection (Wu and Radovic 2006; Rosas et al. 2012) and fire retardant (Bandosz 2008).

Phosphorus groups can be directly attached to carbon forming C–P bonds or through oxygen atoms by C–O–P bonds. It is found in the form of phosphines, phosphonates, phosphates, and polyphosphates (Rosas et al. 2009). The most frequent method for phosphorus functionalization is phosphoric acid activation of a carbon precursor, usually a lignocellulosic precursor. Polymerization of a carbon precursor in the presence of oxoacids of phosphorus generates a highly developed porous structure rich in phosphate and polyphosphate bridges in their carbon matrix, being partially responsible for the development of pore structure (Jagtoyen and Derbyshire 1998; Strelko Jr et al. 1999; Puziy et al. 2002; Rosas et al. 2009, 2010). Impregnation of carbons with organophosphorus compounds, H_3PO_4, $POCl_3$, acid phosphates, and metal phosphates, followed by thermal treatment at mild temperatures, is also effective for protecting them against corrosion, thanks to the phosphorus species formed in the carbon surface (Wu and Radovic 2006). Hydrothermal treatments have also been recently applied for successfully grafting P groups in the carbon surface (Wu et al. 2012).

Phosphorous functional groups result in an effective modification of electrical properties and chemical reactivity of carbon materials, giving rise to numerous applications that are being developed recently, especially in catalysis (Bedia et al. 2010; Rosas et al. 2010, 2012) due to their strong acid properties. Phosphorous has also emerged as a new possibility to modify carbon materials for energy storage applications due to the similar electronic configuration of nitrogen. Functionalization of CNTs for catalysis application has also been reported by oxidation, impregnation with $(NH_4)_3PO_4$ as phosphorus precursor, and calcination up to 550°C for 6 h (Zhang et al. 2008). Strategies such as thermolysis have been used to achieve P-doped graphite layers and nanotubes, showing their potential application for oxygen reduction reaction as a possibility of metal-free catalyst (Liu et al. 2011; Zhou et al. 2014).

9.3.2.5 Other surface functional groups

It is possible to functionalize the surface of carbon materials with other, but less frequent, heteroatoms. For instance, boron substitution is possible in carbon materials resulting in a significant modification of their properties without noticeable structural changes, even though the inserted amount is very limited. These boron species have been used to protect composite carbon materials from oxidation (Allardice and Walker 1970; McKee et al. 1984b) and to improve hydrogen physisorption (Chung et al. 2008). It also constitutes a p-type doping element with electrocatalytic activity when inserted into CNTs (Han et al. 1999; Yang et al. 2011; Zhao et al. 2013) and graphene (Panchakarla et al. 2009). Its effect on accelerating the development of graphitic stacking (Han et al. 1999) and on improving the electrode performance in lithium-ion rechargeable batteries has also been reported (Dahn et al. 1992). A relevant example for functionalization with this element is boron substitution in diamond crystals that results in materials with great interest as electrode in electrochemical applications (Spitsyn et al. 1981; Fryda et al. 1999).

It is possible to functionalize carbon materials with halogens like chlorine (Stacy et al. 1965; Tobias and Soffer 1985; Barthos et al. 2005; Li et al. 2011), bromine (Gupta et al. 1995; Hutson et al. 2007; Barpanda et al. 2011), iodine (Rai Puri and Bansal 1966; Hill and Marsh 1968; Yao et al. 2012; Poh et al. 2013), or fluorine (Mickelson et al. 1998; Touhara and Okino 2000; Osuna et al. 2010; Robinson et al. 2010; Jung et al. 2011).

9.4 ELECTROCHEMICAL FUNCTIONALIZATION OF CARBON MATERIALS

Electrochemical methods are used for surface modification of materials because they have several advantages compared to chemical methods (Bagotsky 2006; Berenguer et al. 2012): (1) the procedures are simple and can be immediately interrupted, (2) they can be run at room temperature and atmospheric pressure and using extremely small volumes, (3) the reaction conditions can be very precisely reproduced, and (4) the treatments have higher sensitivity and selectivity.

Several electrochemical techniques can be used for functionalization of carbon materials in a selective and controlled mode: potentiostatic methods where the electrode is kept at constant potential, potentiodynamic methods that are based in a sweep potential with time, and galvanostatic methods where the current is controlled and kept constant during the process (Bard et al. 2001; Pletcher et al. 2010).

Electrochemical approaches can be used for both covalent and noncovalent modifications of carbon materials (Downard 2000; Balasubramanian and Burghard 2005; Zeng 2008; Ma et al. 2010; Chan et al. 2013). We will comment briefly about noncovalent methods, since we will dedicate most of the section to covalent modification of carbon materials.

9.4.1 Noncovalent Strategies

Electrochemistry can be used for a controlled growth of polymer thin films over the carbon surface, which is a representative example for this section. Either positive or negative polarization of the surface in the aforementioned conditions (galvanostatic, potentiostatic, potentiodynamic) will be used depending on the conditions required for the functionalization (Gao et al. 2000; Chen et al. 2001a). Noncovalent attachment of polymeric films over carbon material has been studied with different polymers, although it cannot be discarded that some covalent attachment of monomers can occur through surface functionalities already existing in the carbon material (Bleda-Martínez et al. 2007). As an example of the literature existing on this topic, deposition of PANI thin films in activated carbon fibers was studied by chemical and electrochemical processes resulting in materials with higher capacitance than the pristine fibers for supercapacitor applications (Salinas-Torres et al. 2013). By carefully selecting the experimental conditions, a thin film of PANI is obtained within the microporosity of the fibers, which allowed a strong interaction of PANI to the carbon surface avoiding its loss upon cycling. In addition, the developed thin film reduces the problems derived from polymer volume changes when submitted to continuous charge and discharge cycles (Salinas-Torres et al. 2012, 2013).

9.4.2 Covalent Functionalization Methods

9.4.2.1 Electrografting

Electrografting refers to an electrochemical reaction that leads to a covalent attachment of an organic layer to a conductive substrate (Bélanger and Pinson 2011). These methods result in the incorporation of a variety of functional groups to the carbon structure through the selection of specific organic molecules and have many possibilities and applications in different fields. Electrografting can be achieved by performing oxidation and also reduction reactions; both have different advantages that have to be considered depending on the envisaged modification, for example, in some cases, it is necessary to avoid oxidative conditions because they can lead to undesired oxidation of the carbon material, which can be acceptable in other cases (Baranton and Bélanger 2005).

9.4.2.1.1 Reductive techniques

Those methods that use electrochemical functionalization of carbon surfaces through negative polarization can be considered as reductive techniques. There are different approaches for this purpose, including reduction of diazonium salts and vinylic compounds, among others (Bélanger and Pinson 2011).

The reduction of diazonium salts is the most representative and highly studied reductive technique. An aryl radical is produced by reduction of an aryl diazonium salt, which reacts then with the carbon surface producing a covalent bond between a carbon atom and the aryl group (Delamar et al. 1992). As the diazonium salts are not stable, it is necessary to prepare them at the moment in which the reaction is done. Thus, the diazonium salt is synthesized from an aryl amine in the same cell where the electrochemical treatment is performed, with the advantage of being generated very close to the electrode surface where the reduction is going to occur. In addition, the reduction is more likely to occur on an unmodified surface, creating just a thin layer on the electrode surface (McCreery 2008). The modification can be done in different media: aprotic media like acetonitrile are the most common (Baranton and Bélanger 2005); acid aqueous solution (Bélanger and Pinson 2011) and ionic liquids have also been used (Price et al. 2005; Fukushima and Aida 2007).

Many studies have been done with different carbon materials using diazonium salts: glassy carbon (Delamar et al. 1992; Saby et al. 1997; Ortiz et al. 1998; Baranton and Bélanger 2005), CNTs (Bahr et al. 2001; Price and Tour 2006; Fukushima and Aida 2007; Jousselme et al. 2008; Santos et al. 2008), highly ordered pyrolytic graphite (HOPG; Allongue et al. 1997), and carbon fibers (Delamar et al. 1997), among others.

Electrografting using a variety of aryl diazonium radicals with different heteroatoms or functional groups (Br, Cl, NO_2, COOH, SO_3H) has been studied. From these studies, it seems that the solvent used does not have a very important effect since diazonium salts in acetonitrile and in situ radicals prepared in aqueous solutions originate grafted molecules with comparable results; as a consequence, in situ radical generation in aqueous solution is a good alternative for grafting molecules that are insoluble in nonaqueous solvents (Baranton and Bélanger 2005). Single-walled CNTs can also be functionalized using diazonium salts that produce different degrees of functionalization depending on the reactivity of the salt and the carbon surface, since the presence of defects and the size of the tubes have a significant influence (Bahr et al. 2001).

Direct electrografting of acrylates in aqueous solutions also shows a different approach to reductive techniques. The molecular structure of the monomer is crucial in this method because it can allow or inhibit the electrografting. The proposed mechanism includes the diffusion of micelles of monomers in the carbon vicinities and reorganization as a bilayer when the applied potential produces the reduction of double bonds in the layer and, consequently, forms a covalently grafted species. This process can be followed by polymerization of the monomer (Gabriel et al. 2010).

Functionalization of multiwalled CNTs with polyacrylonitrile has been studied using electrochemical process with negative polarization, resulting in a new material that improves the solubility of the pristine CNTs in specific solvents. The process has to be performed under very strong conditions in organic media, so that this is a disadvantage compared to the techniques in aqueous media (Petrov et al. 2004).

9.4.2.1.2 Oxidative techniques

Oxidative techniques are the second alternative to graft different organic molecules in the carbon surface. Oxidative electrografting of amines, carboxylates, and alcohols has been studied for this purpose, and it gives different possibilities of functionalization depending on the precursor used (Bélanger and Pinson 2011).

Oxidation of amines (aliphatic or aromatics) for grafting on carbon surfaces has been studied by cyclic voltammetry and chronoamperometry. The process is a one-electron oxidation of the amine group that gives a radical, which then is attached to the surface by a C–N bond giving different properties depending on the nature of the organic amine used. Primary, secondary, and tertiary amines have different reactivities, being the primary amines the most reactive and the tertiary amines the less one; in some cases, they even do not react due to steric effects (Deinhammer et al. 1994).

An interesting and illustrative example of the potentiality of this method is the modification of glassy carbon surfaces using 4-aminobenzoic acid (4-ABA; Yang et al. 2006; Holm et al. 2007), 4-aminobenzenesulfonic acid (4-ABSA; Li et al. 2004), and 4-aminobenzylphosphonic acid (4-ABPA; Yang et al. 2005) in aqueous solution that shows that the grafting of species with a terminal carboxyl, sulfonic, and phosphonic groups, respectively, is possible. The process can be done in ethanol and aqueous solutions, and it occurs after an irreversible oxidation process in which a thin film of the amino acid is formed at the electrode surface blocking further oxidation processes. These modified electrodes are useful for sensor applications due to their acidic character (Liu et al. 2000; Li et al. 2004; Yang et al. 2006). In the case of 4-ABPA, two reactions may occur. A one-electron amine oxidation where the grafting is via carbon–nitrogen bond, and a Kolbe-like reaction where HPO_3 is formed leading to a carbon–carbon bond between the organic molecule (i.e., aminobenzene) and the carbon surface. Each possibility can be controlled using a proper window potential. To achieve C–N bond, it is necessary to have a potential of 0.75 V (vs. Ag/AgCl), with C–C bonding requiring a potential of 0.90 V (vs. Ag/AgCl). This shows that electrochemical techniques have the possibility for controlling the selectivity of the process depending on the required functionalization (Yang et al. 2005).

Linking of nitroxyl radicals was studied in graphite felt electrodes via amine bridges by anodic oxidation of amine compounds in different solvents. It was found that aqueous solutions lead to higher amounts of grafted species at the electrode surface (Geneste et al. 2002; Geneste and Moinet 2005).

Oxidation of different nitro-containing carboxylates, alcohols, and salts has been done in order to graft different groups to graphite felts using aqueous solutions at different pHs. The results showed that the functionalization of graphite felts by anodic oxidation leads to chemically stable bonding and that can be used in indirect electrolysis (Geneste et al. 2002). Studies of grafting with 1,3-propanediol were successfully performed by cyclic voltammetry using a $LiClO_4$ aqueous electrolyte in a five-scan process in which the alcohol-grafted product was then used as an intermediate for further modifications of the grafted group (Verma et al. 2011).

9.4.2.2 Electrochemical oxidation of carbon materials

Electrochemical oxidation of carbon materials refers to the creation of SOGs by application of a positive polarization. The electrochemical generation of SOGs in carbon materials when submitted to positive polarization was demonstrated long time ago (Kinoshita and Bett 1973a), and the possibility of titrating the amount of oxygen inserted by cyclic voltammetry was proposed by Kinoshita in the 1970s (Kinoshita and Bett 1973b). However, those studies were initially oriented to understand the oxidation process of carbon materials in order to extend the lifetime of industrial carbon electrodes (McBreen et al. 1981; Stonehart 1984; Antonucci et al. 1988), rather than modifying the surface chemistry by the formation of oxygen functionalities. Reinforcement of carbon fiber composites was one of the first applications in which electrooxidation was employed for this purpose (Donnet and Ehrburger 1977; Fitzer and Weiss 1987). The adhesion between the carbon fibers and the polymer matrix was improved when oxygen functionalities were electrochemically bonded to the surface of the carbon fibers. Yue et al. electrochemically treated carbon fibers as anode using 1% wt KNO_3 aqueous electrolyte and using specific charges between 133 and 10,600 C/g. The generation of carboxyl and ester groups occurs together with the development of narrow microporosity. The authors pointed out that although the generation of microporosity would worsen the adhesion for the preparation of composites, these materials could be useful for adsorption-based applications (Yue et al. 1999). Like carbon fibers, most carbon materials can be electrooxidized when submitted to anodic treatment. Commercial F-400 activated carbon was electrochemically functionalized in 0.5 M Na_2SO_4 aqueous electrolyte inside a fixed bed column using platinum as electrode for the generation of surface acid groups that enhanced the phenol adsorption (Mehta and Flora 1997). BPL activated carbon was electrooxidized by anodic treatment in 0.5 M K_2SO_4 aqueous electrolyte at two constant currents using different treatment times, and the possibility of obtaining a low and controlled oxidation of the carbon surface was observed (Barton et al. 1997). Single-walled CNTs can be electrooxidized using HCl or KCl as electrolytes when working at potentials higher than 1.1 V versus Ag/AgCl at potentiostatic conditions (Rafailov et al. 2008). Vertically aligned CNTs were electrochemically oxidized as a purification treatment and for enhancing the accessibility of the surface of the nanotubes (Ye et al. 2006). An example of the versatility of electrochemical treatments is found in this work, where the electrooxidized CNTs were later used as support of platinum nanoparticles that were deposited by electroreduction. The electrochemical oxidation of graphite in anodic configuration was achieved in acid media (Metrot and Fischer 1981), with graphite oxide as the main product when aqueous solutions of diluted HNO_3 are used (Sorokina et al. 2001).

TABLE 9.1 Porosity and Surface Chemistry of Some Porous Carbons Studied

Sample	S_{BET}, m²/g	$V_{DR}(N_2)$, cm³/g	$V_{DR}(CO_2)$, cm³/g	CO, μmol/g	CO_2, μmol/g	O, μmol/g	Cg, F/g
GAC	870	0.37	0.29	420	390	1200	110
ACC	1130	0.48	0.39	1370	665	2700	137
ZTC-L	3650	1.54	—	2640	290	3220	335
LY700	950	0.35	0.26	1430	400	2230	143

Local anodic oxidation of graphene using atomic force microscopy was employed for the preparation of graphene/graphene oxide/graphene junctions with tunable semiconducting properties (Masubuchi et al. 2011). However, graphene is a peculiar case regarding electrochemical treatments in comparison with other carbon materials. Electrochemical treatments on this material are mostly focused to the electroreduction of the graphene oxide surface in order to obtain reduced graphene (Shao et al. 2010; Chakrabarti et al. 2013; Pumera 2013). Therefore, instead of graphene oxidation, researchers are very much interested in the selective reduction of oxygen groups to adapt the material to the desired application.

After presenting the background of this topic, we will explain in more detail the electrochemical oxidation of carbon materials to generate SOGs through the review of the work done by our research group in which different experimental conditions, as well as different carbon forms (granular activated carbons [GACs], activated carbon felts, and zeolite-templated carbons [ZTCs], among others), have been studied (Berenguer et al. 2009, 2012, 2013; Tabti et al. 2013; Ruiz-Rosas et al. 2014).

Table 9.1 summarizes the porosity and surface chemistry properties of the pristine porous carbon materials, which will be used in the following sections. Evolved CO, CO_2, and total oxygen content measured by temperature-programmed desorption (TPD) and capacitance obtained in 1 M H_2SO_4 between 0 and 1 V versus Reversible Hydrogen Electrode (RHE) are included. These data show that porous carbons with different porosities, from low surface area and narrow micropore size distribution materials to superporous carbons with specific surface areas above 3000 m²/g, have been used. In addition, different forms have been used, from GAC (sample GAC), activated carbon felt (sample activated carbon cloth [ACC]), powdered templated carbon with a very low particle size (sample ZTC-L), and hierarchical porous carbons (sample LY700).

9.4.2.2.1 Electrochemical treatment of granular activated carbon in a filter press

GACs are very much used in many applications including gas and water purification. The electrochemical treatment of GAC should be done without the addition of any binder in order to get the modified material free of any additive. This means that the design of the electrochemical reactor must fulfill the earlier critical condition. In our case, we used an electrochemical filter-press cell (Berenguer et al. 2009) in which a commercial GAC was subjected to galvanostatic electrolysis (GAC, Waterlink Sutcliffe Carbons, 207A, pHpzc = 9; Mesh: 12 × 20). These kinds of cells constitute a mature technology, being available at pilot plant and industrial scales and opening the door for scaling up this functionalization method.

Figure 9.4 shows a diagram of a lab-scale filter press. It is basically composed by two compartments that can allocate an electrode and that are separated by a membrane. Flow distributors are placed between the membrane and each compartment for feeding them while avoiding preferential paths. Centrifugal pumps connected to the inlet and outlet of the flow distributors are used to force stirring of the electrolyte in the compartments. The activated carbon sample is loosely sandwiched between the electrode and a confinement mesh, resulting in a kind of GAC fluidized bed. A platinized titanium or even cheaper electrodes like stainless steel can be used in the negatively charged compartment, while a dimensional stable anode (DSA) electrode, which is more durable and favors the formation of hydroxyl radicals, is used in the positively charged compartment. Cationic or anionic membranes are selected as separator depending on the electrolyte. For instance, cationic membrane is recommended for an electrolyte containing H_2SO_4, while anionic membranes are employed in electrolytes containing NaOH or NaCl.

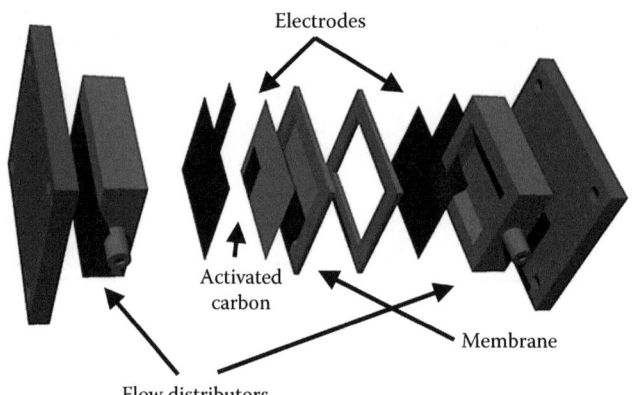

FIGURE 9.4
Scheme of the electrochemical filter-press cell. Geometrical electrode area, 20 cm².

The electrochemical modification of the GAC sample loaded in the compartment of the positively charged DSA electrode can be done at different currents and times. In our case, we used currents between 0.2 and 2.0 A, with a treatment time being 3 h in all the cases (Berenguer et al. 2009). The effect of acid, neutral, and basic electrolytes, being 0.5 M NaOH, 0.5 M H_2SO_4, and 0.34 M NaCl, was also analyzed. The characterization of the SOGs of the GAC after each treatment was performed using TPD. TPD is based upon the different decomposition temperatures of the SOGs when heated at constant rate under inert atmosphere. Decomposition of acidic groups as carboxylic, anhydrides, and lactones proceeds with CO_2 evolution at temperatures between 200°C and 700°C, while neutral and basic groups as phenols, esters, carbonyls, and quinones decompose as CO at temperatures from 500°C to more than 1000°C (Donnet 1968; Román-Martínez et al. 1993; Boehm 2002). This technique is very convenient to characterize the changes occurring in the surface chemistry of the carbon materials in terms of amount and nature of the SOGs.

The results of galvanostatic oxidation show that all the treatments are able to generate SOGs and that an increase in the current provokes an increase in the oxidation degree of the activated carbon, being this effect specially marked when NaCl is chosen as the electrolyte (Berenguer et al. 2009). Interestingly, the nature of the generated SOGs is mainly related to the electrolyte and to the applied current to a lesser extent. In general, the formation of CO-evolving SOGs is obtained at low current. When current is raised, negligible or even negative amounts of CO groups are generated in acid and alkaline mediums, while the amount of CO_2 evolving groups increases. This suggests that, at higher currents, CO-type groups are further oxidized to CO_2-type groups. For these kinds of groups, it can be seen that the formation of carboxylic groups is favored in acid medium, while lactones are the main product in alkaline electrolyte (Berenguer et al. 2009). Notorious differences are observed when NaCl is chosen as the working electrolyte. In this case, a high functionalization degree of the carbon surface occurs even at the lowest current tested, with phenols being the predominantly generated SOG (Berenguer et al. 2009). The greater oxidation degree obtained with this treatment could be due to the electrogeneration of chlorine or chlorine radicals on the DSA electrode, as well as other oxidant species that can be produced in an aqueous Cl_2 solution by disproportionation like hypochlorite/hypochlorous acid species (Cotton et al. 1999). In the other tested electrolytes, the oxidizing agents are produced by the oxidation of water and hydroxyl ions to form hydroxyl radicals (Montilla et al. 2005). These findings can be used for easily tailoring the surface chemistry of activated carbons that could be achieved by selecting the appropriate combination of electrolyte and current.

The porosity of the GAC was also analyzed by means of nitrogen adsorption–desorption isotherms at −196°C and CO_2 adsorption isotherms at 0°C. It was observed that there is a small drop of ca. 10% in surface area and micropore volume after the modification of the surface chemistry of GAC, independently of the applied treatment (Berenguer et al. 2009). This constitutes an important advantage for electrooxidation methods compared to conventional chemical oxidation since, for achieving a similar oxidation degree than after electrooxidation in NaCl, most chemical oxidation methods would have rendered a severe decrease in the porosity of the obtained materials (Moreno-Castilla et al. 1995).

Functionalization of the GAC sample by putting it in contact with the negatively charged electrode was also tested. Some functionalization was observed in this configuration, but in a much lower degree than for the positive configuration. The oxidation was partially related to the presence of oxygen in the electrolyte that can be electroreduced to form hydrogen peroxide, superoxide radicals, and superoxide ions. These latter reactions are suppressed in acid media and favored in alkaline electrolyte, which explains the enhanced electrooxidation in negative configuration when alkaline and neutral electrolytes are utilized (Yeager 1984; Kinoshita 1988). Furthermore, a reduction of the electrogenerated SOGs is observed when the current is increased in this configuration. As electroreduction of carboxylic acids upon negative polarization has been claimed in literature (Cheng and Teng 2003), this result can be explained in terms of the competitive mechanism through indirect oxidation by peroxide- and superoxide-derived species and the direct reduction of SOGs by the electrode potential.

9.4.2.2.2 Some insights about the electrooxidation mechanism: Functionalization of activated carbon felts

GAC remains as a kind of fluidized bed in the filter-press cell, which results in a poor contact with the electrode, and then, it is difficult to extract clear conclusions about the electrooxidation mechanism. For this purpose and because of the flexibility of the experimental setup, the electrochemical functionalization of ACCs was carried out (Tabti et al. 2013). In this case, a good contact with the electrode and an efficient propagation of the applied electric field will occur. For comparison purposes, the same protocol used over ACC was applied to the commercial GAC already used in the previous section.

Both materials were submitted to currents between 0.2 and 1.0 A for 3 h in 0.5 M NaCl electrolyte. Anionic membrane and Ti/Pt and Ti/SnO$_2$-Sb-Pt DSA electrodes for the negative and positive compartment, respectively, were utilized. The carbon samples were placed in the positive compartment for the functionalization, since it seems to proceed more effectively in such arrangement (Berenguer et al. 2009). The main difference between the treatment of GAC and ACC samples was the contact with the electrode, which can affect the performance of the treatment as will be discussed below.

Figure 9.5 shows the TPD analyses of GAC and ACC samples. The electrochemical oxidation of GAC sample proceeds similarly to that observed in 0.34 M NaCl. CO-evolving groups are preferentially produced at low currents, with phenols (which decompose as CO between 600°C and 850°C) being the main contribution. As current increases, more phenols and anhydrides (which decompose as CO and CO$_2$ between 300°C and 600°C) are generated. At the highest tested current (1.0 A), carboxylic acid formation (which decomposes as CO$_2$ between 200°C and 400°C) is observed.

For the ACC sample, extensive oxidation is achieved even at the lowest current (Figure 9.5). Around 6 mmol/g of oxygen (almost 10% wt) is fixed in the ACC surface, in contrast with the 1.4

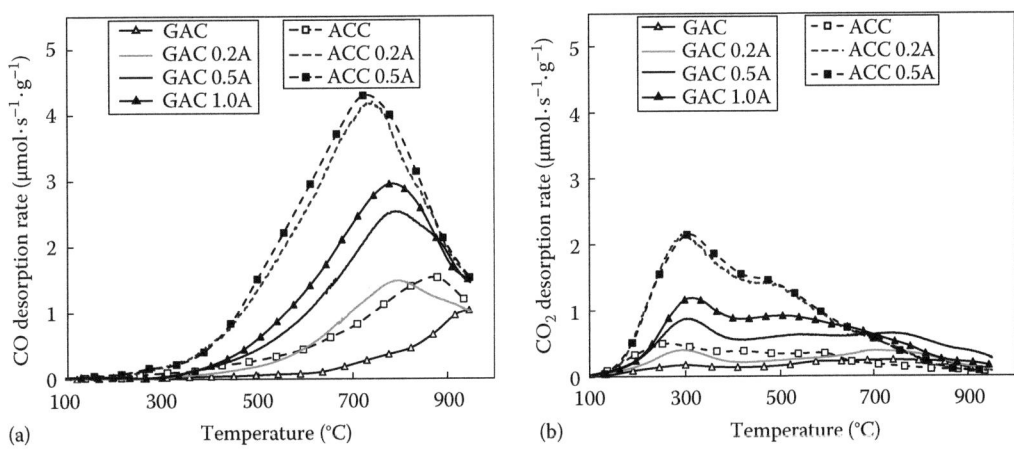

FIGURE 9.5
(a) CO and (b) CO$_2$ TPD profiles for the original and electrooxidized carbon samples.

mmol/g of oxygen generated in the GAC sample using the same current. This large amount of oxygen incorporated in the ACC is close to the maximum oxygen that can be loaded in the surface of this sample even at the highest current used (maximum of 7.1 mmol/g of oxygen at 1.0 A is obtained).

The reasons behind these relevant differences can be deduced considering the oxidation mechanism. Electrooxidation of a positively polarized carbon surface may occur through two mechanisms. The first one is a direct oxidation of the carbon surface that can proceed through the direct polarization of the material giving the formation of phenol-type groups and oxidation to quinones and with carboxylic groups being subsequently formed through the oxidation of CO-type species:

$$C + H_2O \rightarrow C(OH) + H^+ + e^- \tag{9.1}$$

$$C(OH) \rightarrow C(O) + H^+ + e^- \tag{9.2}$$

$$C(O) + H_2O \rightarrow C(OOH) + H^+ + e^- \tag{9.3}$$

A more detailed explanation of direct oxidation mechanism is difficult at present, but it can be tentatively proposed that the anodic polarization may withdraw some electron density from the aromatic rings, promoting their destabilization and the nucleophilic attack by electron-donor species, like water molecules or hydroxyl species.

The second one is *indirect oxidation* that occurs through the formation of oxidizing species over the metal oxide electrodes or even over the carbon surface itself. The nature of these species is different and depends on the electrolyte and the electrode. In chlorine-free media, oxygen evolution reaction (OER) from water oxidation produces oxidizing intermediates such as hydroxyl radicals. In Cl-containing electrolyte, the chlorine evolution reaction (CER) can occur together with oxygen evolution, with the formation of highly oxidized species, such as chlorine, hypochlorous acid, and hypochlorite ions (Berenguer et al. 2013). This kind of oxidation can be compared to the wet oxidation method, and in fact, it may result in rather similar surface functionalization when it is performed using harsh conditions (Berenguer et al. 2013).

Direct oxidation is affected by several properties of the carbon material such as electrical conductivity, crystallinity, edge-to-basal site ratio, wettability, surface area, and porosity. GAC and ACC have a very similar BET surface area and micropore volume (Table 9.1), and they have initially a sufficiently high amount of SOGs (Table 9.1) to improve the wettability of their inner surface. As activated carbons, they have a low degree of crystallinity with small-size, randomly oriented basic structural units and an enhanced edge-to-basal site ratio. This structural disorder results in a relatively poor electrical conductivity compared to carbon materials like graphite, CNTs, or graphene. However, the practical ohmic resistance of the electrode is a function of the intraparticle resistance and the contact or interparticle resistance. The fibrous morphology and knitted structure of the ACC reduces the interparticle contact problems and provides a more intimate and effective contact with the DSA electrode, thus lowering the electrical resistance of the DSA–ACC electrode assembly. In addition, the ACC is in tight contact with the electrode, whereas the GAC remains with a loose contact and like a fluidized bed. Thus, the surface of the ACC will be exposed to a higher polarization than that of GAC making efficient the direct oxidation mechanism.

Regarding indirect oxidation, ACCs have enhanced mass transfer because of their porous structure (Suzuki 1994). Diffusivity coefficients can be considered similar in GAC and ACC, since these porous carbons have a similar BET surface area, micropore volume, and average pore size (1.7 vs. 1.8 nm for GAC and ACC when estimated as $4 \cdot V_{DR}/S_{BET}$). However, the tortuosity of the porosity of a carbon fiber is much lower than that of GAC (Suzuki 1994; Bleda-Martínez et al. 2010), and the diameter of the fibers forming the ACC is around 7 μm, which is 200 times lesser than the particle size of the GAC (~1.5 mm). Consequently, the diffusional limitations that the formed oxidizing species have to overcome in ACC to reach the inner surface are much lower, and the concentration of oxidizing species inside the micropores of ACC will be higher than for GAC. This will result in a higher oxidation degree of the ACC by the indirect oxidation mechanism.

The textural properties of the ACC and GAC samples were studied after the functionalization. It was found that textural properties remained mostly the same for the GAC sample, although some noticeable drop was found in S_{BET} and micropore volume measured by CO_2 and N_2 adsorption in the case of the ACC sample. This decrease was found to be similar in both narrow and wide micropore volumes, which seems to confirm that the oxidation obtained by this method takes place homogeneously in all the porosity. Although some surface area was lost after the functionalization of ACC, their capacitance, which is known to be proportional to the surface area, was determined in 0.5 M H_2SO_4, and it was found to increase after the treatment at 0.2 and 0.5 A. This effect can be explained taking into account that the CO-evolving groups enhance the capacitance through the contribution of redox processes (pseudocapacitance) (Bleda-Martínez et al. 2005). These results highlight the potential application of electrochemical functionalization for enhancing the capacitance of carbon electrodes used in energy storage.

9.4.2.2.3 Electrochemical generation of oxygen-containing groups in zeolite-templated carbons

ZTC constitutes a family of porous carbon materials that are prepared by carbonization of a carbon precursor inside the porosity of a zeolite, which acts as hard template for porosity development (Kyotani et al. 1997). The subsequent removal of the template leaves a carbon material that constitutes a negative replica of the channels of the parent zeolite. The use of different zeolites as hard templates and the appropriate selection of the carbon precursor and conditions for the infiltration or the CVD can lead to different porous textures, from strictly micro to micro-mesoporous ZTC, and also with a 3D structural order and periodicity (Ma et al. 2001; Kyotani et al. 2003; Nishihara et al. 2009). Templated carbons have prospective applications in energy storage devices, thanks to their unique porous structure (Nishihara and Kyotani 2012). In this sense, zeolite-templated superporous carbons, ZTC-L, were prepared by CVD with high electrochemical capacitance and an outstanding rate performance (Itoi et al. 2011). On the other hand, a hierarchical porous carbon, LY700, was prepared from lignin, a biomass by-product that constitutes a renewable precursor, by liquid-phase infiltration of Y zeolite (Ruiz-Rosas et al. 2014). The peculiar morphology of these materials, such as the curvature of their graphene layers induced by the zeolite structure and the high edge-to-basal plane ratio (Nishihara et al. 2009), makes them especially prone to be functionalized.

Thus, ZTC-L and LY700 were submitted to electrochemical oxidation. In the case of ZTC-L, a paste of pure ZTC was tightly attached to a Ti/RuO_2 anode, and the electrode was assembled inside a membrane basket to an intimate contact (Berenguer et al. 2013). The electrooxidation experiments were carried out using a galvanostatic procedure in a three-electrode cell and 0.34 M NaCl, 0.5 M NaOH, and 0.5 M H_2SO_4 were selected as electrolytes. Ag/AgCl/Cl$^-$ (3 M KCl) and Pt wire were used as reference and counter electrodes, respectively. For the LY700 sample, a paste was conformed using 5% of Teflon as binder and 5% of acetylene black as conductivity promoter, which is a composition similar to those employed in the preparation of activated carbon electrodes for commercial supercapacitors. The LY700 electrode was assembled in a stainless steel mesh, and the electrochemical functionalization was performed by cyclic voltammetry, being carried out in a three-electrode cell using 1 M H_2SO_4 as electrolyte and a Pt wire as counter electrode.

Figure 9.6 shows an example of the kind of polarization curves that are obtained at low, medium, and high currents. When a low positive current is applied to the electrode, the electrode potential starts to increase from its equilibrium value at open-circuit potential. The electrode potential during galvanostatic experiments of the carbon material (solid lines) is lower than that observed in the absence of the carbon (dashed lines), which is manifesting that the surface of the carbon material is acting as electrode. The electrode potential increases with time until it reaches a constant potential very close to that obtained in the absence of carbon material. The higher the applied current, the faster the increase of the electrode potential. The reached constant potential corresponds to the electrolyte oxidation in Ti/RuO_2 electrodes, for example, CER in the case of using chlorine-containing electrolyte or OER from water oxidation.

Thus, the first part of the polarization curve is associated to the direct oxidation mechanism, while the second part, corresponding to the indirect oxidation, occurs after depletion of a great part of the reactive sites that form CO-type groups. This allows the accommodation of more ions in the electric double layer and therefore slowly moves the potential into the region where the formation of oxidizing species starts (indirect oxidation mechanism). If a high current is applied, the potential moves faster into the region of OER and CER formation, producing a higher participation of the

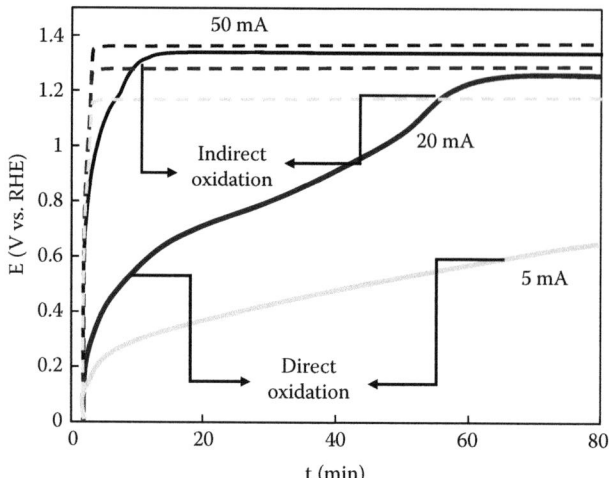

FIGURE 9.6
Examples of polarization curves for the electrode with (solid lines) and without carbon (dashed lines) attained under galvanostatic condition at different current.

indirect oxidation mechanism in the electrooxidation process. As a result, the observed electrooxidation of carbon materials is due to a combination of the direct polarization by the applied positive current and the electrogenerated oxidizing species, that is, the so-called indirect electrochemical oxidation. These two contributions can be easily differentiated by electrochemical techniques and, even, can be chosen to oxidize the carbon surface through a dominant process, which is of interest to tailor the surface chemistry in a more efficient way than the classical chemical oxidation methods.

In this sense, selectivity to the formation of different SOGs can be achieved by adjusting the time, the current, or potential and the electrolyte used in the electrochemical treatment. For example, the use of NaOH in the galvanostatic treatments of ZTC-L favors the formation of CO_2-evolving groups, while H_2SO_4 promotes the formation of CO-type-evolving groups, so that higher CO/CO_2 ratios can be achieved; in NaCl, however, the kind of groups greatly depends on the applied current, which can modulate the CO/CO_2 ratio (Berenguer et al. 2013). These results indicate that depending on the electrolyte, not only different amounts but also different kinds of oxygen functional groups can be introduced. Interestingly, by carefully selecting the electrooxidation conditions, an important amount of oxygen can be incorporated with a less important damage of the structure of the ZTC-L compared to similar oxidation degree by chemical methods (Itoi et al. 2014).

As mentioned previously, the LY700 sample was electrooxidized using a potentiodynamic method, consisting of three cycles in 1 M H_2SO_4 at low scan rate (1 mV/s) in the −0.2 to 1.0 V window (Ruiz-Rosas et al. 2014). The voltammograms obtained for LY700 and a sample that consists of lignin carbonized directly at 700°C (the same temperature utilized in the preparation of LY700 sample), L700 (O content: 7.1 wt.%), are shown in Figure 9.7. During the first positive scan, an irreversible anodic current starts at 0.6 V in LY700. In the next scans, the anodic current above 0.6 V decreases, and a reversible process is clearly observed at 0.42 (forward scan) and 0.30 V (reverse scan). After the third scan, a steady-state voltammogram is obtained. This behavior points out that most of the functionalization of the carbon surface occurs during the first scan. After electrooxidation, the capacitance of LY700 increases significantly from 143 to 195 F/g, thanks to the contribution of a redox couple, as seen in cyclic voltammograms (CVs) recorded between −0.2 and 0.6 V in Figure 9.7. These redox processes are attributed to the presence of SOGs that evolve as CO (Bleda-Martínez et al. 2005). The increase in such kind of groups was confirmed by TPD measurements (Ruiz-Rosas et al. 2014). Interestingly, L700 sample did not show such behavior (Figure 9.7). In this material, the first and subsequent scans performed on this sample between −0.2 and 1.0 V were rather similar, and the voltammogram obtained after the treatment is practically the same as earlier. Such a different behavior can be related to the low concentration of active sites in carbonized lignin when compared to the large amount of reactive sites in the curved and edge site–rich structure generated in the presence of the zeolite template.

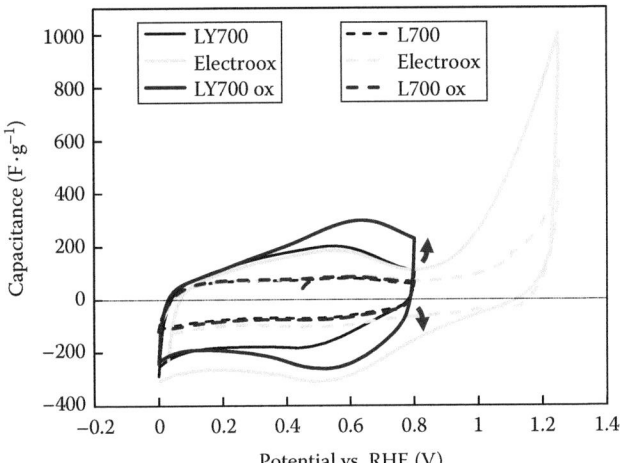

FIGURE 9.7
Potentiodynamic electrooxidation of LY700 and carbonized lignin in 1 M H_2SO_4.

Valuable information can be obtained from the voltammograms about the electrooxidation mechanism. If the double-layer-capacitance current (which is estimated from the capacitance value on the first cycle between −0.2 and 0.6 V) is subtracted from the oxidation current in each scan (the one obtained from 0.6 to 1.0 V), the total charge employed in the functionalization treatment can be estimated to be 3.2 mmol e^-/g in LY700 sample. The total O introduced can be calculated from the difference between the total amount of CO- and CO_2-type groups measured by TPD from pristine and electrooxidized samples, which is 2.6 mmol O/g. From these results, it can be deduced that nearly one electron is required for the insertion of one O atom, which suggests that electrooxidation under these conditions proceeds through the aforementioned direct oxidation mechanism (see the reaction in Equation 9.1). It also highlights the effectiveness of the treatment, since the close values of e^- and O amounts are manifesting that most of the applied charge is employed in the oxidation of the polarized carbon surface, while the side OER is negligible at these conditions.

These results confirm that the electrooxidation treatment can be used for the selective generation of electroactive CO-type groups that can be profited for energy storage in electrodes for supercapacitors, even when the carbon material is already casted into a conductive paste.

9.5 COMPARISON BETWEEN ELECTROCHEMICAL AND CHEMICAL OXIDATION TREATMENTS

As indicated in the previous section, important differences may exist between the surface chemistry of a carbon material when subjected to a wet (chemical) or electrochemical oxidation. This section is dedicated to explain in more detail the differences (and similarities) of both treatments.

For this purpose, GAC and ZTC-L were subjected to wet oxidation with $(NH_4)_2S_2O_8$-saturated 1 M H_2SO_4 solution and 65 wt% HNO_3 solution for comparison with the functionalization attained by the previously presented electrochemical treatment (Berenguer et al. 2012, 2013).

In the case of GAC oxidation, the persulfate wet oxidation was carried out at 2, 5, and 24 h. The amounts of CO-type, CO_2-type oxygen groups, and total O were evaluated from TPD experiments and were compared with those obtained by electrochemical treatment in 0.2A NaCl electrolyte for 1, 3, and 5 h (Figure 9.8). Since functionalization in NaCl proceeds at these conditions by indirect oxidation mechanism, which is fairly similar to wet oxidation, these results allow to discriminate differences in functionalities that only rely on the method and not in the oxidation mechanism. It can be seen that the generation rate of CO- and CO_2-type groups with $(NH4)_2S_2O_8$ is faster within the first 2 h, and then, it decreases until the amount of SOGs levels off. On the contrary, the increase in SOGs with time by the anodic treatment at 0.2 A, that is, the oxidation rate, is approximately constant with time. It suggests that the anodic oxidation process in NaCl allows a much more precise

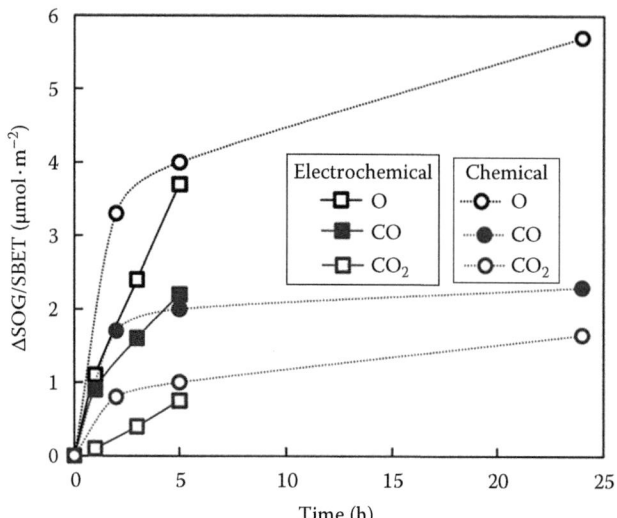

FIGURE 9.8
Generated SOG amount by chemical and electrochemical treatments in GAC.

control of the oxidation level and amount of SOGs. In the electrochemical treatment, the oxidizing species are supplied to the reaction medium at a constant rate determined by the applied current, most probably in the vicinity of GAC particles, whereas during the chemical oxidation, the oxidizing species are in large excess since the beginning, remaining constant during the experiment. The higher generation of CO_2 evolving groups at short times for the chemical treatment with persulfate also indicates that the oxidation of surface-oxidized sites is favored, because of the high concentration of oxidant, in comparison to the anodic treatment, which seems to promote the oxidation of free surface sites. Moreover, provided a proper input current is selected, the anodic treatment in NaCl can produce, during similar treatment times, much higher oxidation degrees in the AC than the conventional chemical treatments with wet oxidation (Berenguer et al. 2012).

If the direct oxidation mechanism is predominant during the electrochemical treatment, for example, at low treatment times at galvanostatic conditions, then the electrochemical treatment permits to create with a high selectivity CO-type groups with a high concentration of phenol groups, which is difficult to achieve by the chemical oxidation.

Another relevant aspect of surface oxidation of porous carbons is that the treatment must not change remarkably the porosity of the pristine material because, otherwise, we will negatively modify the most important property that determines the application of the material. In this sense, since electrochemical oxidation can be done in more controlled conditions than chemical oxidation, it is possible to have a better control of the porosity changes in the final materials. As an example, Table 9.2 includes the surface area of pristine ZTC-L sample and, after different electrochemical treatments, the predominant electrochemical oxidation mechanism and the increase in oxygen content with respect to the pristine sample. For comparison purposes, some data for the same sample after chemical oxidation are included.

In general terms, the higher the amount of introduced oxygen (ΔO), the lower the surface area remaining after the oxidation treatment. However, a strong influence of the oxidation method on these properties can be clearly observed. Thus, comparing samples with similar or higher amounts of introduced oxygen (ΔO), the electrochemically oxidized samples under direct, mixed, or soft-indirect conditions exhibit higher surface areas than those produced by chemical treatments, at least until ΔO around 8000 µmol/g. As demonstrated earlier, electrochemical treatment produces a more controlled reaction. On the contrary, the chemical oxidation is more aggressive because of the higher concentration of reactants. A slower oxidation process could permit a preferential attack of the most reactive sites (defects, edge sites, etc.), thus preserving the nanographene structure. By contrast, the chemical treatments or the electrochemical ones, under marked indirect oxidation conditions, may give rise to a fast and nonselective attack of any carbon atom of the ZTC structure, producing important structural changes.

TABLE 9.2 Functionalization Yield (Increase in Oxygen Content) and BET Surface Area of ZTC after Different Electrooxidation Treatments

Sample	ΔO, µmol/g	S_{BET}, m²/g	Oxidation Mechanism
ZTC	—	3650	—
NaCl, 5 mA, 15 h	3750	2430	Mixed
NaCl, 50 mA, 1 h	3290	2680	Mixed
NaCl, 50 mA, 3 h	6180	1210	Indirect
NaOH, 5 mA, 15 h	4220	1610	Mixed
NaOH, 50 mA, 1 h	2090	2480	Indirect
H_2SO_4, 5 mA, 15 h	5060	2290	Direct
H_2SO_4, 2 mA, 36 h	7880	1860	Direct
HNO_3, 80°C, 15 min	4960	1870	Chemical
HNO_3, 80°C, 1 h	6570	1420	Chemical
HNO_3, 80°C, 2 h	7220	1140	Chemical

9.6 CONCLUSIONS

The use of electrochemically assisted reactions has emerged in the last two decades as a flexible, simple, and effective tool for controlled functionalization of carbon surfaces. Synergic effects rise when novel nanostructured carbon materials of high electrical conductivity are selected as the substrate for their grafting. By fine-tuning of the working parameters, that is, electrode, electrolyte, potential, current intensity, and time, it is possible to introduce selectively functional groups in precise amounts. Most direct outcome is the ability to generate electrochemically different kinds of SOGs by electrooxidation, which are already very useful for a number of applications and can be further profited as the building units for a huge number of functionalization pathways. Moreover, functional groups of valuable heteroatoms for catalytic and energy storage applications, such as nitrogen, phosphorus, and sulfur, or larger molecules, polymers, and enzymes can be bonded to the surface of the material. Knowledge of the reactivity of the carbon surface, which is sometimes missed, is necessary for the development of efficient electrochemical functionalization protocols. Electrochemistry can be used alone for functionalization of carbon material but can be also considered as a complementary tool for the classical surface chemistry techniques.

ACKNOWLEDGMENTS

The authors thank MINECO and FEDER for their financial support (projects 2013-42007-P and CTQ2012-31762). RRR gratefully acknowledges funding from MINECO through *Juan de la Cierva* program (JCI-2012-12664). CGG acknowledges funding from Generalitat Valenciana for the *Santiago Grisolía* grant (GRISOLIA/2013/005).

REFERENCES

Abrahamson, J. 1973. The surface energies of graphite. *Carbon* 11 (4): 337–362.
Ago, H., Kugler, T., Cacialli, F. et al. 1999. Work functions and surface functional groups of multiwall carbon nanotubes. *The Journal of Physical Chemistry B* 103 (38): 8116–8121.
Agüí, L., Yáñez-Sedeño, P., and Pingarrón, J.M. 2008. Role of carbon nanotubes in electroanalytical chemistry: A review. *Analytica Chimica Acta* 622 (1–2): 11–47.
Allardice, D.J. and Walker, P. 1970. The effect of substitutional boron on the kinetics of the carbon-oxygen reaction. *Carbon* 8 (3): 375–385.

Allongue, P., Delamar, M., Desbat, B. et al. 1997. Covalent modification of carbon surfaces by aryl radicals generated from the electrochemical reduction of diazonium salts. *Journal of the American Chemical Society* 7863 (8): 201–207.

Antonucci, P.L., Romeo, F., Minutoli, M. et al. 1988. Electrochemical corrosion behavior of carbon black in phosphoric acid. *Carbon* 26 (2): 197–203.

Ayala, P., Arenal, R., Rümmeli, M. et al. 2010. The doping of carbon nanotubes with nitrogen and their potential applications. *Carbon* 48 (3): 575–586.

Bagotsky, V.S. 2006. *Fundamentals of Electrochemistry*, 2nd edn. Edison, NJ: John Wiley and Sons.

Bahr, J.L. and Tour, J.M. 2001. Highly functionalized carbon nanotubes using in situ generated diazonium compounds. *Chemistry of Materials* 13 (11): 3823–3824.

Bahr, J.L. and Tour, J.M. 2002. Covalent chemistry of single-wall carbon nanotubes. *Journal of Materials Chemistry* 12 (7): 1952–1958.

Bahr, J.L., Yang, J., Kosynkin, D.V. et al. 2001. Functionalization of carbon nanotubes by electrochemical reduction of aryl diazonium salts: A bucky paper electrode. *Journal of the American Chemical Society* 123 (27): 6536–6542.

Bai, H., Xu, Y., Zhao, L. et al. 2009. Non-covalent functionalization of graphene sheets by sulfonated polyaniline. *Chemical Communications (Cambridge, England)* (13) (April): 1667–1669.

Balasubramanian, K. and Burghard, M. 2005. Chemically functionalized carbon nanotubes. *Small (Weinheim an Der Bergstrasse, Germany)* 1 (2): 180–192.

Bandosz, T.J. 2008. Surface chemistry of carbon materials. In *Carbon Materials for Catalysis*, P. Serp and J. L. Figueiredo, eds., pp. 45–92. New York: John Wiley and Sons, Inc.

Bandosz, T.J. and Ania, C.O. 2006. *Activated Carbon Surfaces in Environmental Remediation. Interface Science and Technology*, Vol. 7. Amsterdam, the Netherlands: Elsevier.

Banerjee, S. and Wong, S.S. 2002. Functionalization of carbon nanotubes with a metal-containing molecular complex. *Nano Letters* 2 (1): 49–53.

Bansal, R.C., Vastola, F.J., and Walker, P. 1970. Studies on ultra-clean carbon surfaces—IV. Decomposition of carbon-oxygen surface complexes. *Carbon* 8 (4): 443–448.

Baranton, S. and Bélanger, D. 2005. Electrochemical derivatization of carbon surface by reduction of in situ generated diazonium cations. *The Journal of Physical Chemistry B* 109 (51): 24401–24410.

Bard, A.J.J., Faulkner, L.R.R., Swain, E. et al. 2001. *Electrochemical Methods: Fundamentals and Applications*, 2nd edn. New York: John Wiley and Sons.

Barpanda, P., Fanchini, G., and Amatucci, G.G. 2011. Structure, surface morphology and electrochemical properties of brominated activated carbons. *Carbon* 49 (7): 2538–2548.

Barthos, R., Méhn, D., Demortier, A. et al. 2005. Functionalization of single-walled carbon nanotubes by using alkyl-halides. *Carbon* 43 (2): 321–325.

Barton, S.S., Evans, M.J.B., Halliop, E. et al. 1997. Anodic oxidation of porous carbon. *Langmuir* 13 (5): 1332–1336.

Baskaran, D., Mays, J.W., and Bratcher, M.S. 2005. Noncovalent and nonspecific molecular interactions of polymers with multiwalled carbon nanotubes. *Chemistry of Materials* 17 (13): 3389–3397.

Bedia, J., Ruiz-Rosas, R., Rodríguez-Mirasol, J. et al. 2010. A kinetic study of 2-propanol dehydration on carbon acid catalysts. *Journal of Catalysis* 271 (1): 33–42.

Bélanger, D. and Pinson, J. 2011. Electrografting: A powerful method for surface modification. *Chemical Society Reviews* 40 (7): 3995–4048.

Bellafont, N.P., Mañeru, D.R., and Illas, F. 2014. Identifying atomic sites in N-doped pristine and defective graphene from ab initio core level binding energies. *Carbon*, 76: 155–164.

Berenguer, R., Marco-Lozar, J.P., Quijada, C. et al. 2009. Effect of electrochemical treatments on the surface chemistry of activated carbon. *Carbon* 47 (4): 1018–1027.

Berenguer, R., Marco-Lozar, J.P., Quijada, C. et al. 2012. A comparison between oxidation of activated carbon by electrochemical and chemical treatments. *Carbon* 50 (3): 1123–1134.

Berenguer, R., Nishihara, H., Itoi, H. et al. 2013. Electrochemical generation of oxygen-containing groups in an ordered microporous zeolite-templated carbon. *Carbon* 54 (April): 94–104.

Bhatnagar, A., Hogland, W., Marques, M. et al. 2013. An overview of the modification methods of activated carbon for its water treatment applications. *Chemical Engineering Journal* 219 (March): 499–511.

Bianco, A., Kostarelos, K., and Prato, M. 2005. Applications of carbon nanotubes in drug delivery. *Current Opinion in Chemical Biology* 9 (6): 674–679.

Biniak, S., Szymański, G., Siedlewski, J. et al. 1997. The characterization of activated carbons with oxygen and nitrogen surface groups. *Carbon* 35 (12): 1799–1810.

Bleda-Martínez, M.J., Lozano-Castelló, D., Cazorla-Amorós, D. et al. 2010. Kinetics of double-layer formation: Influence of porous structure and pore size distribution. *Energy and Fuels* 24 (6): 3378–3384.

Bleda-Martínez, M.J., Maciá-Agulló, J.A., Lozano-Castelló, D. et al. 2005. Role of surface chemistry on electric double layer capacitance of carbon materials. *Carbon* 43 (13): 2677–2684.

Bleda-Martínez, M.J., Morallón, E., and Cazorla-Amorós, D. 2007. Polyaniline/porous carbon electrodes by chemical polymerisation: Effect of carbon surface chemistry. *Electrochimica Acta* 52 (15): 4962–4968.

Boehm, H.-P. 1966. Chemical identification of surface groups. *Advances in Catalysis* 16: 179–274.

Boehm, H.-P. 1994. Some aspects of the surface chemistry of carbon blacks and other carbons. *Carbon* 32 (5): 759–769.

Boehm, H.-P. 2002. Surface oxides on carbon and their analysis: A critical assessment. *Carbon* 40 (2): 145–149.

Boehm, H.-P. 2008. *Adsorption by Carbons*. Amsterdam, the Netherlands: Elsevier.

Boukhvalov, D.W. and Katsnelson, M.I. 2008. Chemical functionalization of graphene with defects. *Nano Letters* 8 (12): 4373–4379.

Chakrabarti, M.H., Low, C.T.J., Brandon, N.P. et al. 2013. Progress in the electrochemical modification of graphene-based materials and their applications. *Electrochimica Acta* 107 (September): 425–440.

Chan, C.K., Beechem, T.E., Ohta, T. et al. 2013. Electrochemically driven covalent functionalization of graphene from fluorinated aryl iodonium salts. *The Journal of Physical Chemistry C* 117 (23): 12038–12044.

Chen, D., Feng, H., and Li, J. 2012. Graphene oxide: Preparation, functionalization, and electrochemical applications. *Chemical Reviews* 112 (11): 6027–6053.

Chen, J., Liu, H., Weimer, W.A. et al. 2002. Noncovalent engineering of carbon nanotube surfaces by rigid, functional conjugated polymers. *Journal of the American Chemical Society* 124 (31): 9034–9035.

Chen, J.H., Huang, Z.P., Wang, D.Z. et al. 2001a. Electrochemical synthesis of polypyrrole films over each of well-aligned carbon nanotubes. *Synthetic Metals* 125 (3): 289–294.

Chen, R.J., Bangsaruntip, S., Drouvalakis, K.A. et al. 2003. Noncovalent functionalization of carbon nanotubes for highly specific electronic biosensors. *Proceedings of the National Academy of Sciences of the United States of America* 100 (9): 4984–4989.

Chen, R.J., Zhang, Y., Wang, D. et al. 2001b. Noncovalent sidewall functionalization of single-walled carbon nanotubes for protein immobilization. *Journal of the American Chemical Society* 123 (16): 3838–3839.

Cheng, P.-Z. and Teng, H. 2003. Electrochemical responses from surface oxides present on HNO_3-treated carbons. *Carbon* 41 (11): 2057–2063.

Chung, T.C.M., Jeong, Y., Chen, Q. et al. 2008. Synthesis of microporous boron-substituted carbon (b/c) materials using polymeric precursors for hydrogen physisorption. *Journal of the American Chemical Society* 130 (21): 6668–6669.

Cotton, F.A., Wilkinson, G., Murillo, C.A. et al. 1999. *Advanced Inorganic Chemistry*, 6th edn. New York: John Wiley and Sons.

Dahn, J.R., Reimers, J.N., Sleigh, A.K. et al. 1992. Density of states in graphite from electrochemical measurements on Lix(C1-zBz)6. *Physical Review B* 45 (7): 3773–3777.

Dai, L. 2013. Functionalization of graphene for efficient energy conversion and storage. *Accounts of Chemical Research* 46 (1): 31–42.

Datsyuk, V., Kalyva, M., Papagelis, K. et al. 2008. Chemical oxidation of multiwalled carbon nanotubes. *Carbon* 46 (6): 833–840.

Deinhammer, R.S., Ho, M., Anderegg, J.W. et al. 1994. Electrochemical oxidation of amine-containing compounds: A route to the surface modification of glassy carbon electrodes. *Langmuir* 10 (4): 1306–1313.

De la Puente, G., Pis, J.J., Menéndez, J.A. et al. 1997. Thermal stability of oxygenated functions in activated carbons. *Journal of Analytical and Applied Pyrolysis* 43 (2): 125–138.

Delamar, M., Désarmot, G., Fagebaume, O. et al. 1997. Modification of carbon fiber surfaces by electrochemical reduction of aryl diazonium salts: Application to carbon epoxy composites. *Carbon* 35 (6): 801–807.

Delamar, M., Hitmi, R., Pinson, J. et al. 1992. Covalent modification of carbon surfaces by grafting of functionalized aryl radicals produced from electrochemical reduction of diazonium salts. *Journal of the American Chemical Society* (114): 5883–5884.

Domingo-García, M., López-Garzón, F., and Pérez-Mendoza, M. 2000. Modifications produced by O_2 plasma treatments on a mesoporous glassy carbon. *Carbon* 38 (4): 555–563.

Dongil, A.B., Bachiller-Baeza, B., Guerrero-Ruiz, A. et al. 2011. Surface chemical modifications induced on high surface area graphite and carbon nanofibers using different oxidation and functionalization treatments. *Journal of Colloid and Interface Science* 355 (1): 179–189.

Donnet, J.B. 1968. The chemical reactivity of carbons. *Carbon* 6 (2): 161–176.

Donnet, J.B. and Ehrburger, P. 1977. Carbon fibre in polymer reinforcement. *Carbon* 15 (3): 143–152.

Downard, A.J. 2000. Electrochemically assisted covalent modification of carbon electrodes. *Electroanalysis* 12 (14): 1085–1096.

Elliott, C.M. and Murray, R.W. 1976. Chemically modified carbon electrodes. *Analytical Chemistry* 48 (8): 1247–1254.

Englert, J.M., Dotzer, C., Yang, G. et al. 2011. Covalent bulk functionalization of graphene. *Nature Chemistry* 3 (4): 279–286.

Esumi, K., Nishina, S., Sakurada, S. et al. 1987. Plasma treatment of heat-treated meso-carbon microbeads. *Carbon* 25 (6): 821–825.

Fanning, P.E. and Vannice, M.A. 1993. A DRIFTS study of the formation of surface groups on carbon by oxidation. *Carbon* 31 (5): 721–730.

Figueiredo, J.L. 2013. Functionalization of porous carbons for catalytic applications. *Journal of Materials Chemistry A* 1 (33): 9351.

Figueiredo, J.L., Pereira, M.F.R., Freitas, M.M. et al. 1999. Modification of the surface chemistry of activated carbons. *Carbon* 37 (9): 1379–1389.

Fitzer, E. and Weiss, R. 1987. Effect of surface treatment and sizing of c-fibres on the mechanical properties of CFR thermosetting and thermoplastic polymers. *Carbon* 25 (4): 455–467.

Fryda, M., Herrmann, D., Schäfer, L. et al. 1999. Properties of diamond electrodes for wastewater treatment. *New Diamond and Frontier Carbon Technology* 9 (3): 229–240.

Fujimori, T., Radovic, L.R., Silva-Tapia, A.B. et al. 2012. Structural importance of stone–thrower–wales defects in rolled and flat graphenes from surface-enhanced Raman scattering. *Carbon* 50 (9): 3274–3279.

Fukushima, T. and Aida, T. 2007. Ionic liquids for soft functional materials with carbon nanotubes. *Chemistry (Weinheim an Der Bergstrasse, Germany)* 13 (18): 5048–5058.

Gabriel, S., Jérôme, R., and Jérôme, C. 2010. Cathodic electrografting of acrylics: From fundamentals to functional coatings. *Progress in Polymer Science* 35 (1–2): 113–140.

Gao, M., Huang, S., Dai, L. et al. 2000. Aligned coaxial nanowires of carbon nanotubes sheathed with conducting polymers. *Angewandte Chemie* 112 (20): 3810–3813.

García, A. 1998. Modification of the surface properties of an activated carbon by oxygen plasma treatment. *Fuel* 77 (6): 613–624.

Geneste, F., Cadoret, M., Moinet, C. et al. 2002. Cyclic voltammetry and XPS analyses of graphite felt derivatized by non-Kolbe reactions in aqueous media. *New Journal of Chemistry* 26 (9): 1261–1266.

Geneste, F. and Moinet, C. 2005. Electrochemically linking TEMPO to carbon via amine bridges. *New Journal of Chemistry* 29 (2): 269.

Georgakilas, V., Otyepka, M., Bourlinos, A.B. et al. 2012. Functionalization of graphene: Covalent and non-covalent approaches, derivatives and applications. *Chemical Reviews* 112 (11): 6156–6214.

Godino-Salido, M.L., López-Garzón, R., Gutiérrez-Valero, M.D. et al. 2014. Effect of the surface chemical groups of activated carbons on their surface adsorptivity to aromatic adsorbates based on Π-Π interactions. *Materials Chemistry and Physics* 143 (3): 1489–1499.

Gorgulho, H.F., Mesquita, J.P., Gonçalves, F. et al. 2008. Characterization of the surface chemistry of carbon materials by potentiometric titrations and temperature-programmed desorption. *Carbon* 46 (12): 1544–1555.

Gupta, V., Mathur, R.B., Bahl, O.P. et al. 1995. Structural and transport properties of bromine intercalated carbon fibers. *Carbon* 33 (11): 1633–1639.

Han, W., Bando, Y., Kurashima, K. et al. 1999. Boron-doped carbon nanotubes prepared through a substitution reaction. *Chemical Physics Letters* 299 (5): 368–373.

Hao, G.-P., Li, W.-C., Qian, D. et al. 2010. Rapid synthesis of nitrogen-doped porous carbon monolith for CO_2 capture. *Advanced Materials (Deerfield Beach, FL)* 22 (7): 853–857.

Hill, A. and Marsh, H. 1968. A study of the adsorption of iodine and acetic acid from aqueous solutions on characterized porous carbons. *Carbon* 6 (1): 31–39.

Hirsch, A. 2002. Functionalization of single-walled carbon nanotubes. *Angewandte Chemie International Edition* 41 (11): 1853.

Hirsch, A., Englert, J.M., and Hauke, F. 2013. Wet chemical functionalization of graphene. *Accounts of Chemical Research* 46 (1): 87–96.

Holm, A.H., Vase, K.H., Winther-Jensen, B. et al. 2007. Evaluation of various strategies to formation of pH responsive hydroquinone-terminated films on carbon electrodes. *Electrochimica Acta* 53 (4): 1680–1688.

Hu, C.Y., Xu, Y.J., Duo, S. et al. 2009. Non covalent functionalization of carbon nanotubes with surfactants and polymers. *Journal of the Chinese Chemical Society* 56 (2): 234–239.

Hutson, N.D., Attwood, B.C., and Scheckel, K.G. 2007. XAS and XPS characterization of mercury binding on brominated activated carbon. *Environmental Science and Technology* 41 (5): 1747–1752.

Itoi, H., Nishihara, H., Ishii, T. et al. 2014. Large pseudocapacitance in quinone-functionalized zeolite-templated carbon. *Bulletin of the Chemical Society of Japan* 87 (2): 250–257.

Itoi, H., Nishihara, H., Kogure, T. et al. 2011. Three-dimensionally arrayed and mutually connected 1.2-nm nanopores for high-performance electric double layer capacitor. *Journal of the American Chemical Society* 133 (5): 1165–1167.

Jagtoyen, M. and Derbyshire, F. 1998. Activated carbons from yellow poplar and white oak by H_3PO_4 activation. *Carbon* 36 (7–8): 1085–1097.

Jansen, R.J.J. and van Bekkum, H. 1994. Amination and ammoxidation of activated carbons. *Carbon* 32 (8): 1507–1516.

Johns, J.E. and Hersam, M.C. 2013. Atomic covalent functionalization of graphene. *Accounts of Chemical Research* 46 (1): 77–86.

Jousselme, B., Bidan, G., Billon, M. et al. 2008. One-step electrochemical modification of carbon nanotubes by ruthenium complexes via new diazonium salts. *Journal of Electroanalytical Chemistry* 621 (2): 277–285.

Jung, M.-J., Jeong, E., Kim, S. et al. 2011. Fluorination effect of activated carbon electrodes on the electrochemical performance of electric double layer capacitors. *Journal of Fluorine Chemistry* 132 (12): 1127–1133.

Karousis, N., Tagmatarchis, N., and Tasis, D. 2010. Current progress on the chemical modification of carbon nanotubes. *Chemical Reviews* 110 (9): 5366–5397.

Kiciński, W., Szala, M., and Bystrzejewski, M. 2014. Sulfur-doped porous carbons: Synthesis and applications. *Carbon* 68 (March): 1–32.

Kinoshita, K. 1988. *Carbon: Electrochemical and Physicochemical Properties*. New York: John Wiley and Sons.

Kinoshita, K. and Bett, J. 1973a. Electrochemical oxidation of carbon black in concentrated phosphoric acid at 135°C. *Carbon* 11 (3): 237–247.

Kinoshita, K. and Bett, J. 1973b. Potentiodynamic analysis of surface oxides on carbon blacks. *Carbon* 11 (4): 403–411.

Kuila, T., Bose, S., Mishra, A.K. et al. 2012. Chemical functionalization of graphene and its applications. *Progress in Materials Science* 57 (7): 1061–1105.

Kundu, S., Xia, W., Busser, W. et al. 2010. The formation of nitrogen-containing functional groups on carbon nanotube surfaces: A quantitative XPS and TPD study. *Physical Chemistry Chemical Physics* 12 (17): 4351–4359.

Kuzmany, H., Kukovecz, A., Simon, F. et al. 2004. Functionalization of carbon nanotubes. *Synthetic Metals* 141 (1–2): 113–122.

Kuznetsova, A., Popova, I., Yates, J.T. et al. 2001. Oxygen-containing functional groups on single-wall carbon nanotubes: NEXAFS and vibrational spectroscopic studies. *Journal of the American Chemical Society* 123 (43): 10699–10704.

Kyotani, T., Ma, Z., and Tomita, A. 2003. Template synthesis of novel porous carbons using various types of zeolites. *Carbon* 41 (7): 1451–1459.

Kyotani, T., Nagai, T., Inoue, S. et al. 1997. Formation of new type of porous carbon by carbonization in zeolite nanochannels. *Chemistry of Materials* 9 (2): 609–615.

Lehman, J.H., Terrones, M., Mansfield, E. et al. 2011. Evaluating the characteristics of multiwall carbon nanotubes. *Carbon* 49 (8): 2581–2602.

Leon y Leon, C.A. and Radovic, L.R. 1992. Chemistry of carbon surfaces. In *Chemistry and Physics of Carbon*, P.A. Thrower, ed., Vol. 24. New York: Dekker.

Leon y Leon, C.A., Solar, J., Calemma, V. et al. 1992. Evidence for the protonation of basal plane sites on carbon. *Carbon* 30 (5): 797–811.

Li, B., Zhou, L., Wu, D. et al. 2011. Photochemical chlorination of graphene. *ACS Nano* 5 (7): 5957–5961.

Li, J., Vergne, M.J., Mowles, E.D. et al. 2005. Surface functionalization and characterization of graphitic carbon nanofibers (GCNFs). *Carbon* 43 (14): 2883–2893.

Li, X., Wan, Y., and Sun, C. 2004. Covalent modification of a glassy carbon surface by electrochemical oxidation of R-aminobenzene sulfonic acid in aqueous solution. *Journal of Electroanalytical Chemistry* 569 (1): 79–87.

Linares-Solano, A., Salinas-Martinez de Lecea, C., Cazorla-Amorós, D. et al. 1990. Nature and structure of calcium dispersed on carbon. *Energy and Fuels* 4 (5): 467–474.

Liu, J., Cheng, L., Liu, B. et al. 2000. Covalent modification of a glassy carbon surface by 4-aminobenzoic acid and its application in fabrication of a polyoxometalates-consisting monolayer and multilayer films. *Langmuir* 16 (19): 7471–7476.

Liu, Z., Sun, X., Nakayama-Ratchford, N. et al. 2007. Supramolecular chemistry on water-soluble carbon nanotubes for drug loading and delivery. *ACS Nano* 1 (1): 50–56.

Liu, Z.-W., Peng, F., Wang, H.-J. et al. 2011. Phosphorus-doped graphite layers with high electrocatalytic activity for the O_2 reduction in an alkaline medium. *Angewandte Chemie International Edition* 50 (14): 3257–3261.

Lu, W., Soukiassian, P., and Boeckl, J. 2012. Graphene: Fundamentals and functionalities. *MRS Bulletin* 37 (12): 1119–1124.

Ma, P.-C., Siddiqui, N.a., Marom, G. et al. 2010. Dispersion and functionalization of carbon nanotubes for polymer-based nanocomposites: A review. *Composites Part A: Applied Science and Manufacturing* 41 (10): 1345–1367.

Ma, Z., Kyotani, T., Liu, Z. et al. 2001. Very high surface area microporous carbon with a three-dimensional nano-array structure: Synthesis and its molecular structure. *Chemistry of Materials* 13 (12): 4413–4415.

Mabena, L.F., Sinha Ray, S., Mhlanga, S.D. et al. 2011. Nitrogen-doped carbon nanotubes as a metal catalyst support. *Applied Nanoscience* 1 (2): 67–77.

Masubuchi, S., Arai, M., and Machida, T. 2011. Atomic force microscopy based tunable local anodic oxidation of graphene. *Nano Letters* 11 (11): 4542–4546.

McBreen, J., Olender, H., Srinivasan, S. et al. 1981. Carbon supports for phosphoric acid fuel cell electrocatalysts: Alternative materials and methods of evaluation. *Journal of Applied Electrochemistry* 11 (6): 787–796.

McCreery, R.L. 2008. Advanced carbon electrode materials for molecular electrochemistry. *Chemical Reviews* 108 (7): 2646–2687.

McKee, D.W. 1981. The catalyzed gasification of carbon. In *Chemistry and Physics of Carbon*, P.L. Walker, ed., Vol. 16. New York: Marcel Dekker.

McKee, D.W., Spiro, C.L., and Lamby, E.J. 1984a. The inhibition of graphite oxidation by phosphorus additives. *Carbon* 22 (3): 285–290.

McKee, D.W., Spiro, C.L., and Lamby, E.J. 1984b. The effects of boron additives on the oxidation behavior of carbons. *Carbon* 22 (6): 507–511.

Mehta, M.P. and Flora, J.R.V. 1997. Effects of electrochemical treatment of granular activated carbon on surface acid groups and the adsorptive capacity for phenol. *Water Research* 31 (9): 2171–2176.

Metrot, A. and Fischer, J.E. 1981. Charge transfer reactions during anodic oxidation of graphite in H_2SO_4. *Synthetic Metals* 3 (3–4): 201–207.

Mickelson, E.T., Huffman, C.B., Rinzler, A.G. et al. 1998. Fluorination of single-wall carbon nanotubes. *Chemical Physics Letters* 296 (1–2): 188–194.

Montilla, F., Morallón, E., and Vázquez, J.L. 2005. Evaluation of the electrocatalytic activity of antimony-doped tin dioxide anodes toward the oxidation of phenol in aqueous solutions. *Journal of The Electrochemical Society* 152 (10): B421.

Moreno-Castilla, C., Ferro-Garcia, M.A., Joly, J.P. et al. 1995. Activated carbon surface modifications by nitric acid, hydrogen peroxide, and ammonium peroxydisulfate treatments. *Langmuir* 11 (11): 4386–4392.

Moreno-Castilla, C., López-Ramón, M., and Carrasco-Marín, F. 2000. Changes in surface chemistry of activated carbons by wet oxidation. *Carbon* 38 (14): 1995–2001.

Nishihara, H. and Kyotani, T. 2012. Templated nanocarbons for energy storage. *Advanced Materials (Deerfield Beach, FL)* 24 (33): 4473–4498.

Nishihara, H., Yang, Q.-H., Hou, P.-X. et al. 2009. A possible buckybowl-like structure of zeolite templated carbon. *Carbon* 47 (5): 1220–1230.

Niyogi, S., Hamon, M.A., Hu, H. et al. 2002. Chemistry of single-walled carbon nanotubes. *Accounts of Chemical Research* 35 (12): 1105–1113.

Ortiz, B., Saby, C., Champagne, G.Y. et al. 1998. Electrochemical modification of a carbon electrode using aromatic diazonium salts. 2. Electrochemistry of 4-nitrophenyl modified glassy carbon electrodes in aqueous media. *Journal of Electroanalytical Chemistry* 455 (1–2): 75–81.

Osborn, J.A., Ianniello, R.M., Wieck, H.J. et al. 1982. Use of chemically modified activated carbon as a support for immobilized enzymes. *Biotechnology and Bioengineering* 24 (7): 1653–1669.

Osuna, S., Torrent-Sucarrat, M., Solà, M. et al. 2010. Reaction mechanisms for graphene and carbon nanotube fluorination. *The Journal of Physical Chemistry C* 114 (8): 3340–3345.

Otake, Y. and Jenkins, R.G. 1993. Characterization of oxygen-containing surface complexes created on a microporous carbon by air and nitric acid treatment. *Carbon* 31 (1): 109–121.

Panchakarla, L.S., Subrahmanyam, K.S., Saha, S.K. et al. 2009. Synthesis, structure, and properties of boron- and nitrogen-doped graphene. *Advanced Materials*, 21 (46): 4726–4730.

Pantelides, S.T., Puzyrev, Y., Tsetseris, L. et al. 2012. Defects and doping and their role in functionalizing graphene. *MRS Bulletin* 37 (12): 1187–1194.

Paredes, J.I., Martínez-Alonso, A., and Tascón, J.M.D. 2000. Comparative study of the air and oxygen plasma oxidation of highly oriented pyrolytic graphite: A scanning tunneling and atomic force microscopy investigation. *Carbon* 38 (8): 1183–1197.

Pels, J.R., Kapteijn, F., Moulijn, J.A. et al. 1995. Evolution of nitrogen functionalities in carbonaceous materials during pyrolysis. *Carbon* 33 (11): 1641–1653.

Peng, X. and Wong, S.S. 2009. Functional covalent chemistry of carbon nanotube surfaces. *Advanced Materials* 21 (6): 625–642.

Petrov, P., Lou, X., Pagnoulle, C. et al. 2004. Functionalization of multi-walled carbon nanotubes by electrografting of polyacrylonitrile. *Macromolecular Rapid Communications* 25 (10): 987–990.

Pletcher, D., Greff, R., Peat, R. et al. 2010. *Instrumental Methods in Electrochemistry*. Cambridge, U.K.: Woodhead Publishing Limited.

Poh, H.L., Šimek, P., Sofer, Z. et al. 2013. Halogenation of graphene with chlorine, bromine, or iodine by exfoliation in a halogen atmosphere. *Chemistry* 19 (8): 2655–2662.

Polovina, M., Babić, B., Kaluderović, B. et al. 1997. Surface characterization of oxidized activated carbon cloth. *Carbon* 35 (8): 1047–1052.

Price, B.K., Hudson, J.L., and Tour, J.M. 2005. Green chemical functionalization of single-walled carbon nanotubes in ionic liquids. *Journal of the American Chemical Society* 127 (42): 14867–14870.

Price, B.K. and Tour, J.M. 2006. Functionalization of single-walled carbon nanotubes "on water." *Journal of the American Chemical Society* 128 (39): 12899–12904.

Pumera, M. 2013. Electrochemistry of graphene, graphene oxide and other graphenoids: Review. *Electrochemistry Communications* 36 (November): 14–18.

Punckt, C., Pope, M.A., Liu, J. et al. 2010. Electrochemical performance of graphene as effected by electrode porosity and graphene functionalization. *Electroanalysis* 22 (23): 2834–2841.

Puziy, A., Poddubnaya, O., Martínez-Alonso, A. et al. 2002. Synthetic carbons activated with phosphoric acid. *Carbon* 40 (9): 1493–1505.

Quintana, M., Spyrou, K., Grzelczak, M. et al. 2010. Functionalization of graphene via 1,3-dipolar cycloaddition. *ACS Nano* 4 (6): 3527–3533.

Quintana, M., Vazquez, E., and Prato, M. 2013. Organic functionalization of graphene in dispersions. *Accounts of Chemical Research* 46 (1): 138–148.

Radovic, L.R. 2009. Active sites in graphene and the mechanism of CO_2 formation in carbon oxidation. *Journal of the American Chemical Society* 131 (47): 17166–17175.

Radovic, L.R. 2010. Surface chemical and electrochemical properties of carbons. In *Carbons for Electrochemical Energy Storage and Conversion Systems*, F. Beguin and E. Frackowiak, eds., pp. 163–219. Boca Raton, FL: Taylor & Francis Group.

Radovic, L.R. and Bockrath, B. 2005. On the chemical nature of graphene edges: Origin of stability and potential for magnetism in carbon materials. *Journal of the American Chemical Society* 127 (16): 5917–5927.

Radovic, L.R., Silva, I.F., Ume, J.I. et al. 1997. An experimental and theoretical study of the adsorption of aromatics possessing electron-withdrawing and electron-donating functional groups by chemically modified activated carbons. *Carbon* 35 (9): 1339–1348.

Radovic, L.R., Silva-Tapia, A.B., and Vallejos-Burgos, F. 2011. Oxygen migration on the graphene surface. 1. Origin of epoxide groups. *Carbon* 49 (13): 4218–4225.

Rafailov, P.M., Thomsen, C., Dettlaff-Weglikowska, U. et al. 2008. High levels of electrochemical doping of carbon nanotubes: Evidence for a transition from double-layer charging to intercalation and functionalization. *The Journal of Physical Chemistry B* 112 (17): 5368–5373.

Rai Puri, B. and Bansal, R.C. 1966. Studies in surface chemistry of carbon blacks part III. Interaction of carbon blacks and aqueous bromine. *Carbon* 3 (4): 533–539.

Ramanathan, T., Fisher, F.T., Ruoff, R.S. et al. 2005. Amino-functionalized carbon nanotubes for binding to polymers and biological systems. *Chemistry of Materials* 17 (6): 1290–1295.

Raymundo-Piñero, E., Cazorla-Amorós, D., and Linares-Solano, A. 2003. The role of different nitrogen functional groups on the removal of SO_2 from flue gases by N-doped activated carbon powders and fibres. *Carbon* 41 (10): 1925–1932.

Raymundo-Piñero, E., Cazorla-Amorós, D., Linares-Solano, A. et al. 2002. Structural characterization of N-containing activated carbon fibers prepared from a low softening point petroleum pitch and a melamine resin. *Carbon* 40 (4): 597–608.

Robinson, J.T., Burgess, J.S., Junkermeier, C.E. et al. 2010. Properties of fluorinated graphene films. *Nano Letters* 10 (8): 3001–3005.

Román-Martínez, M.C., Cazorla-Amorós, D., Linares-Solano, A. et al. 1993. TPD and TPR characterization of carbonaceous supports and Pt/C catalysts. *Carbon* 31 (6): 895–902.

Rosas, J.M., Bedia, J., Rodríguez-Mirasol, J. et al. 2009. HEMP-derived activated carbon fibers by chemical activation with phosphoric acid. *Fuel* 88 (1): 19–26.

Rosas, J.M., Bedia, J., Rodríguez-Mirasol, J. et al. 2010. On the preparation and characterization of chars and activated carbons from orange skin. *Fuel Processing Technology* 91 (10): 1345–1354.

Rosas, J.M., Ruiz-Rosas, R., Rodríguez-Mirasol, J. et al. 2012. Kinetic study of the oxidation resistance of phosphorus-containing activated carbons. *Carbon* 50 (4): 1523–1537.

Ruiz-Rosas, R., Valero-Romero, M.J., Salinas-Torres, D. et al. 2014. Electrochemical performance of hierarchical porous carbon materials obtained from the infiltration of lignin into zeolite templates. *ChemSusChem*, 7 (5): 1458–1467.

Saby, C., Ortiz, B., Champagne, G.Y. et al. 1997. Electrochemical modification of glassy carbon electrode using aromatic diazonium salts. 1. Blocking effect of 4-nitrophenyl and 4-carboxyphenyl groups. *Langmuir* 7463 (14): 6805–6813.

Salame, I.I. and Bandosz, T.J. 2001. Surface chemistry of activated carbons: Combining the results of temperature-programmed desorption, Boehm, and potentiometric titrations. *Journal of Colloid and Interface Science* 240 (1): 252–258.

Salavagione, H.J., Martínez, G., and Ellis, G. 2011. Recent advances in the covalent modification of graphene with polymers. *Macromolecular Rapid Communications* 32 (22): 1771–1789.

Salinas-Torres, D., Sieben, J.M., Lozano-Castelló, D. et al. 2012. Characterization of activated carbon fiber/polyaniline materials by position-resolved microbeam small-angle x-ray scattering. *Carbon* 50 (3): 1051–1056.

Salinas-Torres, D., Sieben, J.M., Lozano-Castelló, D. et al. 2013. Asymmetric hybrid capacitors based on activated carbon and activated carbon fibre–PANI electrodes. *Electrochimica Acta* 89 (February): 326–333.

Santos, L.M., Ghilane, J., Fave, C. et al. 2008. Electrografting polyaniline on carbon through the electroreduction of diazonium salts and the electrochemical polymerization of aniline. *The Journal of Physical Chemistry C* 112 (41): 16103–16109.

Shao, Y., Wang, J., Engelhard, M. et al. 2010. Facile and controllable electrochemical reduction of graphene oxide and its applications. *Journal of Materials Chemistry* 20 (4): 743.

Shen, W. and Fan, W. 2013. Nitrogen-containing porous carbons: Synthesis and application. *Journal of Materials Chemistry A* 1 (4): 999.

Shi, Q., Peng, F., Liao, S. et al. 2013. Sulfur and nitrogen co-doped carbon nanotubes for enhancing electrochemical oxygen reduction activity in acidic and alkaline media. *Journal of Materials Chemistry A* 1 (47): 14853–14857.

Shim, M., Shi Kam, N.W., Chen, R.J. et al. 2002. Functionalization of carbon nanotubes for biocompatibility and biomolecular recognition. *Nano Letters* 2 (4): 285–288.

Silva, A., Martins, M., Freitas, M.M. et al. 2002. Immobilisation of amine-functionalised nickel(II) Schiff Base complexes onto activated carbon treated with thionyl chloride. *Microporous and Mesoporous Materials* 55 (3): 275–284.

Silva-Tapia, A.B., García-Carmona, X., and Radovic, L.R. 2012. Similarities and differences in O_2 chemisorption on graphene nanoribbon vs. carbon nanotube. *Carbon* 50 (3): 1152–1162.

Simmons, T.J., Bult, J., Hashim, D.P. et al. 2009. Noncovalent functionalization as an alternative to oxidative acid treatment of single wall carbon nanotubes with applications for polymer composites. *ACS Nano* 3 (4): 865–870.

Sorokina, N.E., Maksimova, N.V., and Avdeev, V.V. 2001. Anodic oxidation of graphite in 10 to 98% HNO_3. *Inorganic Materials* 37 (4): 360–365.

Spitsyn, B.V., Bouilov, L.L., and Derjaguin, B.V. 1981. Vapor growth of diamond on diamond and other surfaces. *Journal of Crystal Growth* 52 (April): 219–226.

Stacy, W., Imperial, G., and Walker, P. 1965. 117. Fixation of gaseous chlorine by anthracite. *Carbon* 3 (3): 358.

Stańczyk, K., Dziembaj, R., Piwowarska, Z. et al. 1995. Transformation of nitrogen structures in carbonization of model compounds determined by XPS. *Carbon* 33 (10): 1383–1392.

Stein, A., Wang, Z., and Fierke, M.A. 2009. Functionalization of porous carbon materials with designed pore architecture. *Advanced Materials* 21 (3): 265–293.

Stöhr, B., Boehm, H.-P., and Schlögl, R. 1991. Enhancement of the catalytic activity of activated carbons in oxidation reactions by thermal treatment with ammonia or hydrogen cyanide and observation of a superoxide species as a possible intermediate. *Carbon* 29 (6): 707–720.

Stonehart, P. 1984. Carbon substrates for phosphoric acid fuel cell cathodes. *Carbon* 22 (4–5): 423–431.

Strelko Jr., V., Streat, M., and Kozynchenko, O. 1999. Preparation, characterisation and sorptive properties of polymer based phosphorus-containing carbon. *Reactive and Functional Polymers* 41 (1–3): 245–253.

Su, Y., Zhang, Y., Zhuang, X. et al. 2013. Low-temperature synthesis of nitrogen/sulfur co-doped three-dimensional graphene frameworks as efficient metal-free electrocatalyst for oxygen reduction reaction. *Carbon* 62 (October): 296–301.

Suzuki, M. 1994. Activated carbon fiber: Fundamentals and applications. *Carbon* 32 (4): 577–586.

Tabti, Z., Berenguer, R., Ruiz-Rosas, R. et al. 2013. Electrooxidation methods to produce pseudocapacitance-containing porous carbons. *Electrochemistry* 81 (10): 833–839.

Tagmatarchis, N. and Prato, M. 2004. Functionalization of carbon nanotubes via 1,3-dipolar cycloadditions. *Journal of Materials Chemistry* 14 (4): 437.

Tanada, S., Kawasaki, N., Nakamura, T. et al. 1999. Removal of formaldehyde by activated carbons containing amino groups. *Journal of Colloid and Interface Science* 214 (1): 106–108.

Tang, S., Lu, N., Wang, J.K. et al. 2007. Novel effects of surface modification on activated carbon fibers using a low pressure plasma treatment. *Journal of Physical Chemistry C* 111 (4): 1820–1829.

Tobias, H. and Soffer, A. 1985. Chemisorption of halogen on carbons—I stepwise chlorination and exchange of C-Cl with C-H bonds. *Carbon* 23 (3): 281–289.

Touhara, H. and Okino, F. 2000. Property control of carbon materials by fluorination. *Carbon* 38 (2): 241–267.

Tripathi, B.P., Schieda, M., Shahi, V.K. et al. 2011. Nanostructured membranes and electrodes with sulfonic acid functionalized carbon nanotubes. *Journal of Power Sources* 196 (3): 911–919.

Tsubokawa, N. 1992. Functionalization of carbon black by surface grafting of polymers. *Progress in Polymer Science* 17 (3): 417–470.

Van Dommele, S., Romero-Izquirdo, A., Brydson, R. et al. 2008. Tuning nitrogen functionalities in catalytically grown nitrogen-containing carbon nanotubes. *Carbon* 46 (1): 138–148.

Vastola, F.J., Hart, P.J., and Walker, P. 1964. A study of carbon-oxygen surface complexes using O18 as a tracer. *Carbon* 2 (1): 65–71.

Verma, P., Maire, P., and Novák, P. 2011. Concatenation of electrochemical grafting with chemical or electrochemical modification for preparing electrodes with specific surface functionality. *Electrochimica Acta* 56 (10): 3555–3561.

Vinke, P., van der Eijk, M., Verbree, M. et al. 1994. Modification of the surfaces of a gas-activated carbon and a chemically activated carbon with nitric acid, hypochlorite, and ammonia. *Carbon* 32 (4): 675–686.

Vix-Guterl, C., Couzi, M., Dentzer, J. et al. 2004. Surface characterizations of carbon multiwall nanotubes: Comparison between surface active sites and Raman spectroscopy. *The Journal of Physical Chemistry B* 108 (50): 19361–19367.

Walker, P., Taylor, R.L., and Ranish, J.M. 1991. An update on the carbon-oxygen reaction. *Carbon* 29 (3): 411–421.

Wang, R., Higgins, D.C., Hoque, M.A. et al. 2013. Controlled growth of platinum nanowire arrays on sulfur doped graphene as high performance electrocatalyst. *Scientific Reports* 3 (January): 2431.

Wang, Y., Iqbal, Z., and Malhotra, S.V. 2005. Functionalization of carbon nanotubes with amines and enzymes. *Chemical Physics Letters* 402 (1–3): 96–101.

Wang, Z.-M., Yamashita, N., Wang, Z.-X. et al. 2004. Air oxidation effects on microporosity, surface property, and CH_4 adsorptivity of pitch-based activated carbon fibers. *Journal of Colloid and Interface Science* 276 (1): 143–150.

White, R.J., Antonietti, M., and Titirici, M.-M. 2009. Naturally inspired nitrogen doped porous carbon. *Journal of Materials Chemistry* 19 (45): 8645.

Willman, K.W., Rocklin, R.D., Nowak, R. et al. 1980. Electronic and photoelectron spectral studies of electroactive species attached to silanized carbon and platinum electrodes. *Journal of the American Chemical Society* 102 (26): 7629–7634.

Wohlgemuth, S.-A., White, R.J., Willinger, M.-G. et al. 2012. A one-pot hydrothermal synthesis of sulfur and nitrogen doped carbon aerogels with enhanced electrocatalytic activity in the oxygen reduction reaction. *Green Chemistry* 14 (5): 1515–1523.

Wood, K.N., O'Hayre, R., and Pylypenko, S. 2014. Recent progress on nitrogen/carbon structures designed for use in energy and sustainability applications. *Energy and Environmental Science* 7 (4): 1212.

Wu, G., More, K.L., Johnston, C.M. et al. 2011. High-performance electrocatalysts for oxygen reduction derived from polyaniline, iron, and cobalt. *Science (New York, N.Y.)* 332 (6028): 443–447.

Wu, M., Ren, Y., Guo, N. et al. 2012. Hydrothermal co-doping of boron and phosphorus into porous carbons prepared from petroleum coke to improve oxidation resistance. *Materials Letters* 82 (September): 124–126.

Wu, X. and Radovic, L.R. 2006. Inhibition of catalytic oxidation of carbon/carbon composites by phosphorus. *Carbon* 44 (1): 141–151.

Yamada, Y., Kim, J., Matsuo, S. et al. 2014. Nitrogen-containing graphene analyzed by x-ray photoelectron spectroscopy. *Carbon* 70 (April): 59–74.

Yang, D.-S., Bhattacharjya, D., Inamdar, S. et al. 2012a. Phosphorus-doped ordered mesoporous carbons with different lengths as efficient metal-free electrocatalysts for oxygen reduction reaction in alkaline media. *Journal of the American Chemical Society* 134 (39): 16127–16130.

Yang, G., Liu, B., and Dong, S. 2005. Covalent modification of glassy carbon electrode during electrochemical oxidation process of 4-aminobenzylphosphonic acid in aqueous solution. *Journal of Electroanalytical Chemistry* 585 (2): 301–305.

Yang, G., Shen, Y., Wang, M. et al. 2006. Copper hexacyanoferrate multilayer films on glassy carbon electrode modified with 4-aminobenzoic acid in aqueous solution. *Talanta* 68 (3): 741–747.

Yang, L., Jiang, S., Zhao, Y. et al. 2011. Boron-doped carbon nanotubes as metal-free electrocatalysts for the oxygen reduction reaction. *Angewandte Chemie (International Ed. in English)* 50 (31): 7132–7135.

Yang, Z., Yao, Z., Li, G. et al. 2012b. Sulfur-doped graphene as an efficient metal-free cathode catalyst for oxygen reduction. *ACS Nano* 6 (1): 205–211.

Yao, N., Lordi, V., Ma, S.X.C. et al. 2011. Structure and oxidation patterns of carbon nanotubes. *Journal of Materials Research* 13 (09): 2432–2437.

Yao, Z., Nie, H., Yang, Z. et al. 2012. Catalyst-free synthesis of iodine-doped graphene via a facile thermal annealing process and its use for electrocatalytic oxygen reduction in an alkaline medium. *Chemical Communications (Cambridge, England)* 48 (7): 1027–1029.

Ye, X.R., Chen, L.H., Wang, C. et al. 2006. Electrochemical modification of vertically aligned carbon nanotube arrays. *The Journal of Physical Chemistry. B* 110 (26): 12938–12942.

Yeager, E. 1984. Electrocatalysts for O_2 reduction. *Electrochimica Acta* 29 (11): 1527–1537.

Ying, Y., Saini, R.K., Liang, F. et al. 2003. Functionalization of carbon nanotubes by free radicals. *Organic Letters* 5 (9): 1471–1473.

You, C., Liao, S., Li, H. et al. 2014. Uniform nitrogen and sulfur co-doped carbon nanospheres as catalysts for the oxygen reduction reaction. *Carbon* 69 (April): 294–301.

Yue, Z.R., Jiang, W., Wang, L. et al. 1999. Surface characterization of electrochemically oxidized carbon fibers. *Carbon* 37 (11): 1785–1796.

Zeng, L. 2008. *Single-Walled Carbon Nanotubes: Functionalization, Characterization and Application*. Houston, TX: Rice University.

Zhang, J., Liu, X., Blume, R. et al. 2008. Surface-modified carbon nanotubes catalyze oxidative dehydrogenation of N-butane. *Science (New York)* 322 (5898): 73–77.

Zhang, R. and Ritter, J.A. 1997. New approximate model for nonlinear adsorption and diffusion in a single particle. *Chemical Engineering Science* 52 (18): 3161–3172.

Zhao, J., Park, H., Han, J. et al. 2004. Electronic properties of carbon nanotubes with covalent sidewall functionalization. *The Journal of Physical Chemistry B* 108 (14): 4227–4230.

Zhao, Y., Yang, L., Chen, S. et al. 2013. Can boron and nitrogen co-doping improve oxygen reduction reaction activity of carbon nanotubes? *Journal of the American Chemical Society* 135 (4): 1201–1204.

Zhou, J., Shan, X., Ma, J. et al. 2014. Facile synthesis of P-doped carbon quantum dots with highly efficient photoluminescence. *RSC Advances* 4 (11): 5465.

Zhou, J.-H., Sui, Z.-J., Zhu, J. et al. 2007. Characterization of surface oxygen complexes on carbon nanofibers by TPD, XPS and FT-IR. *Carbon* 45 (4): 785–796.

Zhou, Y., Zhu, Y., Lin, S. et al. 2013. Synthesis of sulfur-doped carbon nanotubes by liquid precursor. *Materials Focus* 2 (1): 44–47.

Zielke, U., Hüttinger, K.J., and Hoffman, W.P. 1996. Surface-oxidized carbon fibers: I. Surface structure and chemistry. *Carbon* 34 (8): 983–998.

CHAPTER 10

CONTENTS

Surface Oxygen Functionalities and Microstructure of Activated Carbons Derived from Lignocellulosic Resources

10

P. González-García

10.1 INTRODUCTION

Activated carbon (AC) has been described through the years as a unique and versatile material. Its disordered micro-/nanostructure develops a diverse porous structure providing high surface area in small volumes. Aside from these textural parameters, the presence of surface functionalities is a distinctive feature of AC, since it provides the chemical polarity, ion interchange capacity, and surface charge. All these characteristics contribute, or not, to the performance of the carbons for specific applications.

The development of energy-sustainable and energy-efficient economy depends on the ability to produce low-cost and high-performance materials for electrical energy storage devices. The current years have been characterized by an exponential increase in use of lightweight, compact, and portable electronic devices that require power sources with high energy and power density. The electrochemical double-layer capacitors (EDLCs) with AC electrodes from lignocellulosic precursors have attracted considerable attention due to their great cycle stability, combined with moderate cost and attractive overall performance [1,2]. The key factors that dictate the selection of a carbon material for this purpose are chemical stability, accessibility, low cost, simple manufacture, and nontoxicity (lack of heavy metals such as Cd or Pb), as well as the ability to charge/discharge quickly, long cycle life, high efficiency in the charge/discharge cycle [3], adaptable porosity [4,5], and surface functionality [6–8]. Carbon electrodes are well polarizable, chemically stable in different electrolytes (acidic, basic, and aprotic), and in a wide range of temperatures. Here, the amphoteric character of ACs, both electron donor/acceptor and the simultaneous presence of acidic/basic surface groups, allows the electrochemical properties of materials based on this element to be extensively varied [9].

On the other hand, a high efficiency of ACs as adsorbents for diverse types of pollutants has been reported in several works, since AC has been found efficient for removing both organic and inorganic pollutants [10–14]. Once again, the presence of oxygen functional groups on the AC surface is considered to be responsible for the uptake of pollutants. Hence, efforts are ongoing to substantially improve the potential of carbon surface by using different chemicals or suitable treatment methods that will enable AC to improve its efficiency for the removal of specific contaminants from aqueous phase. The physical and chemical structure of carbon could be changed by various methods, that is, activation conditions (different agents, temperature, and time of the process), precursor, and additives; from here, different methods have been reported in the literature to modify AC surface [15–18].

The global demand for AC was 12,045,000 tons in 2012 and is expected to be significantly increased each year. Mercury and Air Toxics Standards by the U.S. Environmental Protection Agency are expected to be one of the key factors driving the growth of the AC market. In addition, the growing stringent water purification standards across the globe are a key factor affecting the demand for AC products. However, volatility in prices of raw material due to scarcity is expected to hamper the market [19]. Hence, researches have proposed for several years the use of renewable lignocellulosic precursors as an alternative to decrease prices in the production of activation carbons.

10.2 ACTIVATED CARBONS DERIVED FROM LIGNOCELLULOSIC PRECURSORS

AC is a common term used to describe carbon-based materials that have developed high surface area, an internal pore structure, consisting of pores having diverse size distribution as well as a wide spectrum of oxygenated functional groups present on the surface of AC [20]. These characteristics make AC versatile materials that have numerous applications in many areas including water and air treatments, catalyst applications, metal extraction and purification, and gas storage and recently energy-related applications.

For decades, AC has been produced from a variety of carbonaceous-rich materials such as anthracite [21, 22] and bituminous coal [23], lignite [24], peat [25], several types of wood [18,26], and coconut shell [27,28]. Wood and coconut shells are the most common precursors for the large-scale synthesis of AC, yielding a world production of more than 300,000 tons/year [29]. Nowadays, utilization of lignocellulosic biomass (from natural resources and wastes) to produce ACs has become an important and worldwide approach during the last decades. The properties of each produced AC depend of the initial carbonaceous material selected as precursor. Nevertheless, in order to improve the AC performance, the surface area and pore network must be developed during AC preparation, by using adequate physical or chemical treatments. Physical activation is normally made by carbonization of the raw material followed by activation using steam or CO_2. During carbonization, the material is pyrolyzed to remove noncarbon elements, and then activation occurs, at temperatures ranging from 700°C to 1100°C, using gases that open and develop the porosity of the carbonized material [30].

Chemical activation is a one-step method used for the preparation of AC. Here, the raw material is impregnated using different chemical agents such as $ZnCl_2$, H_3PO_4, HNO_3, H_2SO_4, NaOH, and KOH [20,30,31]. Although not common, K_2CO_3 [32] or formamide [33] has been also used as an activating agent. Then, precursor–agent mixture is carbonized at a maximum of around 750°C and finally washed to eliminate the activating agent. The application of a gaseous stream such as air, nitrogen, or argon is a common practice during pyrolysis that generates a better development of the material porosity. The chemical agents help to develop the AC porosity by means of dehydration and degradation of the biomass structure. The use of a lower temperature, compared to physical activation, is compensated by the interaction between the chemicals and the carbon skeleton. The major advantages of chemical activation are the higher yield, lower temperature of activation (less energy costs), less activation time, and, generally, higher development of porosity; among the disadvantages are the activating agent costs and the need to perform an additional washing stage to remove the chemical agent [14,20,30].

The obtained ACs have developed, as the aforementioned, surface area ranging from 80 to 2500 m²/g, pore size distribution from 0.5 nm to a few micrometers, pore volume from 0.1 to 2.5 cm³/g, and density below the graphite one (2.2 cm³/g). Additionally, depending on the type of activation agent used, the AC surface exhibits numerous functional groups, mainly acidic, which favor the specific interactions that allow it to act as an ionic interchanger with the different kinds of adsorbates [34].

10.2.1 Lignocellulosic Resources as Low-Cost Precursors

ACs are commonly considered as expensive materials because of the chemical and physical treatments used for their synthesis, the relative low yield, the high energy consumption for their production, or the thermal treatments used for their regeneration and the losses generated meanwhile. However, their efficient performance, compared to other porous materials, compensates the cost of production. AC production costs can be reduced by either choosing a cheap raw material or applying a proper production method. With the goal of diminishing the cost of producing AC, contemporary research is taking a turn toward industrial or vegetable (lignocellulosic) wastes to be used as raw materials as one alternative to decrease the production cost [10,12,14,18,26,30].

One of the advantages of this new generation of carbon precursors frequently termed as *low cost* lies in the possibility to get benefits from waste materials since, at the same time, this contributes to decreased costs of waste disposal and the negative impact on the environment. Selection of the precursor for the development of low-cost ACs depends upon many factors. The precursor should be freely available, inexpensive, and nonhazardous in nature. Moreover, for good development of the surface, structural, and textural characteristics, high contents of fixed carbon and low amount of ash are desirable [5,10,13,14,18]. In this sense, most of the researches mentioned that it is still a challenge to prepare AC with very specific characteristics, such as a given pore size distribution, using low-cost raw materials processed at low temperature (less energy costs) [35]. Therefore, it is of extreme relevance to find suitable low-cost raw materials that are economically attractive and at the same time present similar or even better characteristics than the conventional ones.

Nowadays, there are many studies on search of low-cost precursors and the development of the also termed *low-cost adsorbents*, so called because they are prepared using waste materials. A great variety of lignocellulosic residues as precursors of AC are reported: wood from several tree species (as cited before), diverse kinds of nuts and shells [36–39], seeds [32,40], stones [41], or different parts of the same precursor such as the root, stem, brunches, leaves, seeds, husks, or peels [42–45]. In addition, some other domestic wastes [46], agricultural and forest crops and wastes [14], agroindustrial by-products, and several sludges are also reported [47,48]. Table 10.1 summarizes some of these lignocellulosic resources used as AC precursors and their most common physicochemical properties.

TABLE 10.1 Lignocellulosic Resources Used as AC Precursors and Their Physicochemical Composition (%)

Precursor	Moisture	Ash	Volatile Matter	Fixed Carbon	Cellulose	Hemicellulose	Lignin
Silver fir [49]	14.4	0.4	78.7	6.5	52.1	15.0	29.9
Holm oak [49]	9.5	2.3	80.8	7.4	39.7	25.9	27.8
Stone pine [49]	9.8	0.7	82.1	7.4	41.0	20.1	31.2
Pyrenean oak [49]	11.1	2.4	80.5	6.0	33.9	25.5	31.2
Bambusa vulgaris [50]	9.1	1.5	73.9	24.7	43.12	31.54	24–56
Jatropha curcas [51]	1.0	6.0	55.0	37.0	56.0	18.0	24.0
Bagasse [52]	52.2	4.5	N/A	53.0	30.27	56.73	13.02
Coconut [52]	27.1	5.1	N/A	51.5	32.6	7.95	59.4
Grape stalk [53]	7.00	8.70	63.13	21.17	N/A	N/A	N/A
Almond sell [54]	6.55	6.85	N/A	N/A	21.72	27.74	36.12
Tomato stems [54]	3.58	10.65	N/A	N/A	27.03	21.08	16.01
Jute fibers [55]	N/A	0.62	N/A	N/A	60.5	21.1	13.5
Waste tea [56]	5.8	4.2	N/A	N/A	17.1	19.9	47.1
Peach stone [57]	9.3	1.1	71.7	17.0	46.0	14.0	33.0
Guadua amplexifolia [58]	8.0	2.0	87.0	11.0	48.0	37.0	5.0
Carya illinoinensis [59]	8.7	1.7	84.4	5.2	30.0	26.0	41.0
Siriguela seeds [60]	9.1	8.7	N/A	N/A	18.8	9.0	16.8
Cocoa shell [60]	9.4	2.0	N/A	N/A	13.2	10.8	13.2
Jackfruit peel [61]	10.0	4.0	50.0	36.0	N/A	N/A	N/A
Barley husks [62]	8.8	6.7	75.0	18.1	N/A	N/A	N/A

10.2.1.1 Chemical structures of the lignocellulosic biomass

In terms of weight, the main constitutive fractions of the lignocellulosic materials are hemicellulose, cellulose, and lignin (see Table 10.1), the thermal decomposition of which forms the carbonaceous structure of the carbon precursor and the subsequent AC. Lignocellulosic materials also contain extractives such as chlorophylls, ash, and waxes. Generally, the three main components have high molecular weights and contribute much mass, while the extractives are of small molecular size and available in little quantities. In general, lignocellulosics have been included in the term biomass, but this term has broader implications than that denoted by lignocellulosics. Lignocellulosic materials have also been called photomass because they are a result of photosynthesis [13].

Cellulose is a polysaccharide that consists of a linear chain of D-glucose linked by β-(1,4)-glycosidic bonds. The cellulose strains are associated together to make cellulose fibrils. Cellulose fibers are linked by a number of intra- and intermolecular hydrogen bonds [63]. Therefore, cellulose is insoluble in water and most organic solvents. Cellulose is the principal component in most of natural fibers such as cotton or linen, which is present even over 90%. However, most of the biomass consists of 35%–60% cellulose (see Table 10.1).

Hemicelluloses, located in secondary cell walls, are heterogeneous branched biopolymers containing pentoses (β-D-xylose, α-L-arabinose), hexoses (β-D-mannose, β-D-glucose, α-D-galactose), and/or organic acids (α-D-glucuronic, α-D-4-O-methyl-galacturonic, and α-D-galacturonic acids) [64]. They are relatively easy to hydrolyze because of their disordered and branched structure (with short lateral chain) as well as their lower molecular weight [63]. Hemicelluloses are relatively sensitive to operation conditions; therefore, parameters such as temperature and retention time must be controlled [65].

Lignin is a natural polymer that together with hemicelluloses acts as a cementing agent matrix of cellulose fibers in the structures of plants. Their functions are to provide structural strength, to provide sealing of water conducting system that links roots with leaves, and to protect plants against degradation [13]. Lignin is a macromolecule, which consists of alkylphenols and has a complex 3D structure, and is covalently linked with xylans in the case of hardwoods and with galactoglucomannans in softwoods [66]. The basic chemical phenylpropane units of lignin (primarily syringyl, guaiacyl, and p-hydroxyphenol) are bonded together by a set of linkages to form a very complex matrix. This matrix comprises a variety of functional groups, such as hydroxyl, methoxyl, and carbonyl, which impart a high polarity to the lignin macromolecule [67].

Among the three fractions of the lignocellulosic materials, lignin has been identified as the main component in lignocellulosic biomass responsible for the adsorption process [68]. Lignin based biomass is the most abundant renewable carbon resource on earth after cellulose, with a worldwide production of 40–50 million tons/year [66]. Due to the rich carbon content of lignin, lignocellulosic biomass is a good option to be used as precursor for producing AC.

Finally, an important fraction of the lignocellulosic precursors are the extractives that can provide by-products. These extractives include organic substances with low molecular weight and are soluble in neutral solvents. These include resins (a combination of the components terpenes, lignans, and other aromatics), fats, waxes, fatty acids and alcohols, terpenes, tannins, and flavonoids. Also chlorophylls are considered as extractives [69].

10.2.2 Activation Process of Carbons

The transformation of the lignocellulosic biomass into the disordered/ordered structure of the AC is achieved by several chemical methods evolving high-temperature treatments. However, the resulting carbonaceous material does not always develop the desirable physicochemical, microstructure, or porous structure. Hence, the activation process is a significant step to form the surface functionalities, enlarge the surface area, and increase the porous networks.

10.2.2.1 Carbonization

The carbonization is a process sometimes used prior to activation, where the raw lignocellulosic material undergoes a thermal treatment to enrich carbon content in the carbonaceous material.

Hence, low-molecular-weight volatiles are first released, followed by light aromatics and finally hydrogen gas [10]. The resultant product obtained is in the form of a fixed carbonaceous skeleton [70]. In this process, an initial porosity is formed even though it is still comparatively low. The pores formed during carbonization are filled with tarry pyrolysis residues and require activation in order to develop the specific characteristics of the carbon. Careful selection of carbonization parameters is important because this process leaves a significant effect on the final product [71]. In this process, the carbonization temperature has the most significant effect, followed by heating rate, nitrogen flow rate, and finally residence time [14,72]. Normally, higher carbonization temperatures (600°C–700°C) result in reduced yield of char while increasing the liquid and gas release rates [14]. Higher temperature will also increase ash and fixed carbon content and lower amount of volatile matter. Thus, high temperatures result in better-quality char but also decrease yield.

10.2.2.2 Chemical activation

There are two widespread methods to activate carbonaceous materials, commonly known as chemical and physical activations. In chemical activation, most of the times, the raw lignocellulosic precursor is directly impregnated with certain chemical agent such as H_3PO_4, H_2SO_4, HNO_3, $ZnCl_2$, K_2CO_3, NaOH, and KOH. The resulting precursor–agent mixture is thermally treated at temperatures ranging from 400°C to 1000°C under a controlled atmosphere. Besides, the activation temperature, time, and heating rate are important preparation variables for obtaining AC with specific characteristics [25,28,29,39,60,61]. In other occasions, the raw lignocellulosic precursor is preconditioned by a carbonization step; later on, the obtained carbon is the impregnated and undergoes higher thermal treatments in inert atmospheres [50,73].

During the impregnation stage, the chemical agent–precursor mass ratio, impregnation agent, temperature time, and stirring are strictly controlled, since the function of the dehydrating agents is to inhibit the formation of tar and other undesired products during the carbonization process. Also, the pore size distribution and surface area are determined by the ratio between the mass of the chemical agent and the raw material [4,31]. For instance, during phosphoric acid activation, concentration of acid modifies the micropores volume. High amount of acid inhibits the development of microporosity promoting the formation of meso- and macropores [74]. Zinc chloride activation produces a template effect and takes place the micropore formation, which is the reason why efficient elimination of chloride is significant. In addition, the pore size increases with concentration of chloride [20]. In contrast, activation with alkalis does not promote the development of meso- and micropores; however, the disintegration of the raw material has been also observed [73].

10.2.2.3 Physical activation

In a physical activation process, the lignocellulosic precursor is carbonized under an inert atmosphere, and the resulting carbon is subjected to a partial and controlled gasification at high temperature. The activation atmosphere is a high oxidizing agent such as CO_2, water steam, O_2, or a mixture of them at elevated temperatures. The reactions that may occur under such condition to extract carbon atoms from the char structure are summarized as follows [20]:

$$C + H_2O \rightarrow CO + H_2, \Delta H = 117 \text{ kJ/mol} \tag{10.1}$$

$$C + 2H_2O \rightarrow CO_2 + 2H_2, \Delta H = 75 \text{ kJ/mol} \tag{10.2}$$

$$C + CO_2 \rightarrow CO, \Delta H = 159 \text{ kJ/mol} \tag{10.3}$$

All of these reactions are endothermic; therefore, external energy is required to supply the high activation temperature that is almost above 800°C [75]. As explained, during the activation process, steam or CO_2 reacts with carbon to produce CO, CO_2, H_2, or CH_4 [76]. Physical activation of lignocellulosic chars with steam or CO_2 causes different effects on the development of microporosity. In early stages of activation process, CO_2 develops narrow micropores, while steam widens the initial micropores of the char. At high degrees of burn-off, steam generates activated chars that exhibit larger meso- and macropore volume than those prepared by CO_2. As a result, CO_2 creates activated chars with larger micropore volume and narrower micropore size distribution than those activated by steam [36].

10.3 PRESENCE OF THE SURFACE OXYGEN GROUPS ON CARBONS

The adsorption behavior of ACs is affected to a considerable extent by the chemical state of their surfaces, which is also of great practical importance in many other applications of carbon materials such as for catalysts and catalyst supports and carbon–polymer composites [77]. Additionally, as most of the surface groups are electrochemically active, they are likely to impact on the electrochemical performance of carbon electrodes.

It is well known that the average structure of the AC is formed by disordered arrangements of graphene-like layers, which are responsible for the physicochemical properties of carbons. Two distinct sites can be distinguished in the graphene layers: basal and edge carbon atoms [37]. The disordered fraction of carbon materials contains a large number of imperfections and defects such as structural carbon vacancies or nonaromatic rings (formed by five or seven or more carbon atoms). These sites along with the edges of the carbon layers, called *dangling bonds* or active sites, are associated with higher densities of unpaired electrons and therefore show a strong tendency to chemisorb external atoms, chemically known as heteroatoms [20,37,77].

ACs contain diverse heteroatoms, such as hydrogen, oxygen, nitrogen, and sulfur, in the main component. Sometimes, calcium or silica is present, depending on the precursor. Hydrogen is directly bonded to the edge atoms of the graphene-like layers of the AC; however, oxygen, nitrogen, and sulfur can be bonded both at the edge of the layers and also in the ring within them. All these heteroatoms contribute to the properties of the AC in several ways. Nevertheless, the greatest contribution comes from the presence of oxygen, in particular the edge-bonded oxygen [20].

The several types of oxygen functionalities to be found on the AC surfaces, due to the electronegativity of the oxygen atoms, posses dipole moments, and their presence has a marked effect on the shape of the adsorption isotherms of polar adsorbates. This effect is of singular importance for systems using adsorption from solutions, in particular adsorption from aqueous solutions where the water molecules are competitively adsorbed at the sites of the oxygen surface complexes [78]. Therefore, surfaces of ACs must be analyzed in terms of amounts of oxygen present and equally in terms of the chemical composition of the several forms of oxygen that are present as surface oxygen groups (SOGs).

10.3.1 Formation of SOG during the Activation Process

All carbon allotropic forms are metastable against exposure to oxygen-containing gases and against oxidizing agents such as nitric or sulphuric acids and hydrogen peroxide. These reagents and gases like dioxygen, ozone, nitric oxide, CO_2, and water steam can all be used to create oxygen functional groups [20,34,77,79]. Exposure of clean carbon surfaces to any of these reagents will also adventitiously create oxygenated functional groups [80]. All oxidation reactions occur in two steps [81]. One of these is the reductive activation of the reagent producing an oxygen dianion. This can occur either by chemisorption and reductive dissociation at the basal planes of sp² carbon [82] or by direct activation of a radical center formed by a dangling bond of a sp²/sp³ carbon atom at the surface. As these centers are very reactive, they will only be present after aggressive activation of carbon in extremely inert environments (ultrahigh vacuum, ultrapure inert gases) or during in situ gasification and during formation of solid carbon from atomic carbon sources [77].

The second reaction is the diffusion of the activated oxygen dianion to a site of covalent bond formation that must be a prismatic edge/defect site in sp² carbons. In strongly bent carbons of fullerenoid structures, this site may also be a localized double bond at the curved graphene layers where an epoxide ring structure is formed [83]. When the initial C–O bond is formed, there exist several possibilities of further reactions that are controlled by the boundary conditions and the reaction kinetics: (1) at too high temperature, the C–O complex will desorbs as CO; (2) at too high temperatures and under excess of activated oxygen, a C–O_2 unit will form and desorb as carbon dioxide; (3) at intermediate temperature, a carbon oxygen group with low disturbance of the graphene will form; and (4) at very low temperature, a complex carbon oxygen group with strong disturbance of the graphene layer will form [84]. Therefore, it is clear to notice that two different types of oxygenated surfaces are formed, depending on the reaction conditions, during the impregnation and/or activation steps.

10.3.1.1 Basic surfaces

Surfaces with basic character are formed during cooling, or even at room temperature, in a carbonaceous material that was previously outgassed at high temperatures. Here, the oxygen atoms present in dry oxygen, CO_2, or water steam atmospheres, or even aqueous acids, are chemisorbed to the carbon material [85]. The reaction is quite fast in the beginning, but slows down gradually. Much more oxygen is bonded on the surface in a slow reaction in the presence of moist air. This phenomenon termed as *aging* occurs gradually and was described for the first time in 1966 by Puri [86]. The aging process takes several months at room temperature to become easily measurable. It can be followed easily within a few weeks when the reaction occurs mildly. However, the presence of water vapor accelerates this reaction significantly [87–89].

The aging process causes changes in the properties of carbon materials. The surface becomes more and more hydrophilic, and the adsorption capacity of ACs for noxious gases or methyl iodide is greatly reduced [77,87]. With porous carbons, the surface oxidation begins at the outer surface of the particles but progresses very slowly into their interior due to very slow diffusion of oxygen in narrow pores [90]. In consequence, the exterior and interior surfaces of ACs can differ significantly in their adsorption properties. An often used way to reduce the aging and oxidize the surface of carbons is by treatment with oxidizing aqueous solutions, for example, of hydrogen peroxide, ammonium peroxodisulfate, or sodium hypochlorite. Nitric acid is very frequently used because its oxidizing effect can be easily controlled by the concentration, reaction temperature, and time. One disadvantage of nitric acid is, however, that the pore structure of the carbon is considerably changed. The micropores become wider and the micropore volume is reduced [91], and also some nitrogen is bound on the surface [91,92]. Oxidation with $(NH_4)_2S_2O_8$ has little effect on the pore structure, and more relatively strongly acidic groups are produced than in the reaction with HNO_3 [93].

10.3.1.2 Acidic surfaces

The chemistry of acidic surface is more complex. Differences between acidic and basic surface oxides were first introduced by Steenberg in 1944 [34], and so far, the acidic or basic nature of surface functionalities of carbon materials has received much attention by many researchers [34,90,94]. There is general consensus about the type of surface functionalities that contribute to the acidic character of a carbon material (i.e., carboxyl groups, anhydrides, hydroxyls, lactones, and lactol groups). Nevertheless, the issue of carbon basic surfaces is still controversial and remains under research [90,94,95]. Figure 10.1 shows a variety of possible acidic structures. Carbonyl groups might exist free or in the form of their anhydrides. They may also react with hydroxyl groups to form lactones or with carbonyls to produce lactols. Phenolic-type hydroxyl groups are known to be weakly acidic. Carbonyl groups may be isolated or combined into quinonic-type structures. Even when ether or xanthenes groups are not acidic, their presence to neighboring carbonyl groups induces an acidic character.

10.3.2 Effect of Posttreatments

It is well known that the physical and chemical structures of carbon could be changed by various methods, that is, activation conditions (different agents, temperature and time of the process), precursor, and additives [96]. For this reason, different methods have been reported in the literature [15–17,97] to modify AC surface that throw some light on the chemistry and mechanism behind treatment methods that enable AC surface for higher uptake of specific pollutants. In general, the

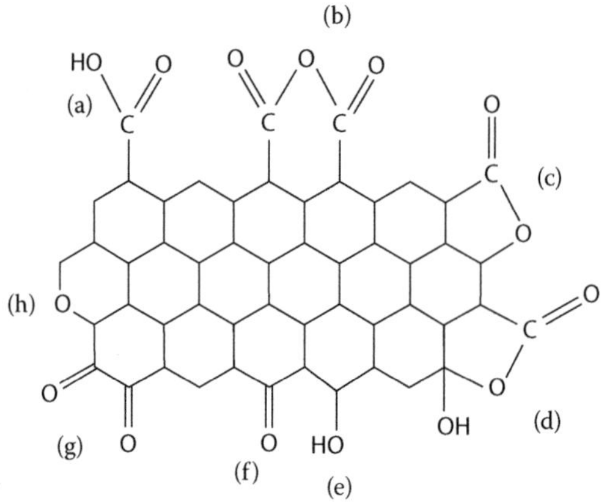

FIGURE 10.1
Schematic representation of the possible SOG in acidic surfaces. Carboxyl groups (a), anhydrides (b), lactones (c), lactols (d), phenols (e), isolated carbonyls (f), quinones (g), and ethers (h).

surface modification of AC is carried out after the activation step. The modification can be classified into three categories: (1) chemical modification, (2) physical modification, and (3) biological modification. Furthermore, oxidative [98] and nonoxidative [99] methods of surface treatments of AC have been reported in the literature.

The incorporation of acidic oxygen functional groups into AC by HNO_3 oxidation and the potential of modified carbon toward cadmium removal were reported a few years ago [91]. It was observed that carboxylic acid groups were the major surface species incorporated and phenol and quinone groups were introduced during the oxidation process. The formation of lactone groups during heat treatment was also suggested. The functional groups had a range of thermal stabilities with carboxylic acid groups being the least stable.

Alkaline treatment of AC promotes positive surface charge, which in turn is helpful to adsorb negatively charged species in higher amounts. The easiest way of producing porous carbons with basic surface properties is to treat it at high temperature in inert, hydrogen, or ammonia atmosphere [100–102]. The treatment of AC with NH_3 at 400°C–900°C leads to the formation of basic nitrogen functionalities [103,104]. Amides, aromatic amines, and protonated amides are produced at 400°C–600°C. Furthermore, pyridine-type structures occur at higher temperatures, which enhance the basicity of the carbon surface [105]. The nitrogen-containing groups generally provide the basic property, which could enhance the interaction between porous carbon and acid molecules, such as dipole–dipole interaction, hydrogen bonding, and covalent bonding. Furthermore, under alkaline basic solutions, it is expected that hydroxyl ions react with the surface functional groups of AC. The alkaline treatment of AC is beneficial in enhancing the adsorption of especially organic species from water [97].

It was reported that treatment of AC by urea supported the formation of the basic groups and carbonyls. The presence of surface functional groups affected the adsorption capacity of the produced samples for the removal of phenols. Urea-treated samples with a basic character and high nitrogen content presented the highest phenol uptake capacity; nitric acid–treated carbons and oxygen-gasified samples presented an acidic surface functionality and a low phenol adsorption capacity. The beneficial role of nitrogen on phenol adsorption was attributed to adsorbate–adsorbent interactions [106].

In recent years, modification of AC by means of microwave radiation is gaining wide attention due to its capacity in heating at molecular level leading to homogenous and quick thermal reactions [97,107–109]. In comparison to conventional heating, microwave heating offers many advantages: microwave energy heats the material from inside out; there is no need for heat convection through a fluid; microwave energy provides rapid heating; there is no direct contact between the microwave heating source and the heated material; there is ease in the heating process control; it has high-temperature capabilities; it allows time and energy savings; and it results in an increase of chemical reactivity [110]. Moreover, microwave processing systems are also relatively compact, portable, maintainable, and cost-effective. Liu et al. [111] prepared modified bamboo-based AC

TABLE 10.2 Selected Studies of Lignocellulosic-Derived AC Modifications and Their Effect on the Surface Chemistry

Precursor	Posttreatment	Chemical Modifications
Almond shells [116]	Hydrogen peroxide oxidation	Carboxyl, ketone, ether groups, and probable carboxyl–carbonate were formed.
Coconut shell [117]	Hydrogen peroxide oxidation	CO_2-evolving groups might be largely induced at oxidation temperature <100°C.
Almond shells [116]	Nitric acid oxidation	Carboxyl groups and probable carboxyl–carbonate structures as well as nitro and nitrate groups were detected.
Coconut shell [117]	Nitric acid oxidation	Oxygen, nitrogen, and hydrogen content increased and carbon content decreased.
Corncob [118]	Nitric acid oxidation	Increased density of acidic oxygen groups at carbon surface.
Olive stones [119]	Nitric acid oxidation	CO_2-evolving oxygen groups strongly increased.
Coconut shell [120]	Ozone	Increase in surface oxygen functional groups, especially in the phenolic and carboxylic categories.
Cherry stones [121]	Ozone	Fixation of acidic groups and the removal of basic ones. As a result, the PZC was considerably lower in the oxidized carbon.
Olive palm brunches [102]	Ammonia	High presence of nitrogen-containing species.
Eucalyptus wood [122]	Ammonia	Increase in the amount of phenolic and carboxylic groups. Presence of C–N bonding.
Wood [123]	Amination	Formation of amides and carboxylic acids.

in a microwave oven under N_2 atmosphere. A gradual decrease in the surface acidic groups was observed during the modification, while the surface basicity was enhanced to some extent, which gave rise to an increase in the point of zero charge (PZC). An increase in the micropores was observed in the beginning, and micropores were then extended into larger ones, resulting in an increase in the pore volume and average pore size.

One of the most widely used oxidants is ozone for the depuration of toxic organic compounds present in water. The simultaneous use of ozone and AC has recently been proposed as an alternative to ozonation [112]. Initially, the high adsorption capacity of AC was considered responsible for the higher efficacy of the combined system. However, some researchers [113,114] reported that other phenomena besides adsorption increase the efficiency of the combined system, such as (1) reaction between the ozone and the organic matter adsorbed on the AC and (2) the generation of free radicals from the reaction between the dissolved ozone and the AC, which can produce mineralization of the organic matter.

The porous structure of the ozone-treated carbons remained practically unchanged in relation to the virgin granular AC. However, important modifications of the chemical surface and hydrophobicity were observed from IR spectroscopy, pH titrations, and determination of the PZC. The ozone treatment gave rise to acidic SOGs. At room temperature, primarily carboxylic groups were formed, while a more homogeneous distribution of carboxylic, lactonic, hydroxyl, and carbonyl groups was obtained at 100°C [115]. Table 10.2 presents a brief summary about changes in the SOG on ACs, as a result of the most common posttreatments.

10.3.3 SOG Identification and Quantification

Traditional methods and modern instrumentation have facilitated the chemical analyses of surface oxygen complexes, since utilization of both as complementary techniques provides a better description and quantification of the oxygen, nitrogen, and other atoms bonded to the carbon surface. SOG identification has been usually performed by infrared spectroscopy, whereas their quantification has been obtained by several years from the Boehm titration method. The development of advanced techniques such as x-ray photoelectron spectroscopy (XPS) or electron energy loss spectroscopy (EELS) has permitted to obtain quantitative information about the local chemical composition, bonding type, and hybridization state.

10.3.3.1 Infrared spectroscopy

The IR is the spectroscopy that deals with the infrared region of the electromagnetic spectrum, which is light with a longer wavelength and lower frequency than visible light. The infrared portion of the electromagnetic spectrum is usually divided into three regions: The far region (400–10 cm^{-1}) provides information from the rotational transitions, vibrational modes of the crystal lattice vibrations, and the skeleton of the molecule. The mid region (4,000–400 cm^{-1}) is used to study the fundamental vibrations and the vibrational–rotational structure, and the third is the near region (14,000–4,000 cm^{-1}) where most of contributions come from the overtones or harmonic vibrations associated to hydrogen tensions [124].

The major types of molecular vibrations are stretching and bending. Infrared radiation is absorbed and the associated energy is converted into these types of motions. Since infrared spectroscopy is based on the fact that the molecules rotate and vibrate as a result of interaction with infrared radiation, IR absorption information is generally presented in the form of a spectrum with wavelength or wave number as the *x*-axis and absorption intensity or percent transmittance as the *y*-axis. Hence, position and intensity of the formed transmittance bands are related to the presence of oxygen (hydrogen, nitrogen, silicon, phosphorous, or sulfur) bonded to the carbon atoms as functionalities [20,34,77,85,89]. Nowadays, these data can be easily found in several websites and manuscripts. As an example, Table 10.3 is included as a guide.

10.3.3.2 X-ray photoelectron spectroscopy

XPS or electron spectroscopy for chemical analysis is a technique for the characterization of solids and powder samples due to its ability to measure binding-energy variations resulting from their chemical environment. From this technique, it is possible to obtain the chemical composition of various material surfaces up to 1 nm in depth. Hence, it is possible to find out if the material is superficially oxidized, if it contains iron or sulfur and other elements [34].

XPS uses x-rays to excite electrons, and then their kinetic energy is measured. X-ray photons with hv energy (characteristic energy of a X-ray photon from the excitation source) are absorbed by

TABLE 10.3 Infrared Adsorption Bands on Carbon Materials and Their Corresponding Assignments to Oxygen Functionalities

Group or Functionality	Assignment Regions (cm^{-1})		
	1000–1500	1500–2050	2050–3700
C–O stretch of ethers	1000–1300		
Ether bridge between rings	1230–1250		
Cyclic ethers containing COCOC groups	1025–1141		
Alcohols	1049–1276		3200–3640
Phenolic groups			
C–O stretch	1000–1220		
O–H bend/stretch	1160–1200		2500–3620
Carbonates, carboxyl–carbonate	1000–1500	1590–1600	
Aromatic C=C stretching		1585–1600	
Quinones		1550–1680	
Carboxylic acids	1120–1200	1665–1760	2500–3300
Lactones	1160–1370	1675–1790	
Anhydrides	980–1300	1740–1880	
Ketenes (C=C=O)			2080–2200
C–H stretch			2600–3000

the atoms of the solid, leading to ionization and the emission of an inner-shell electron. By absorbing a photon, atoms gain the energy of the photon and release an electron to get back to the original stable energy state. The ejected electron retains all the energy from the striking photon, and it can then escape from the atom with a kinetic energy that depends on its binding energy, the incident photon energy, and a correction factor called the work function. Since the kinetic energy of the ejected electrons is dependent on their binding energy, different chemical species can be identified. For each and every element, there will be a characteristic set of peaks in the photoelectron spectrum at kinetic energies determined by the photon energy and the respective binding energies. Furthermore, the intensity of the peaks can be related to the concentration of the element within the sampled region, so that quantitative analysis of the surface composition can be obtained [34,125].

10.3.3.3 Electron energy loss spectroscopy

The EELS is a technique associated to the transmission electron microscope. During the electron microscopy observations, the electron bean is focused on the specimen and the spectrometer measures the loss of kinetic energy of electrons inelastically that have interacted with the sample [126]. This energy loss is a characteristic of the structure of each of the elements present in a region under study; therefore, it can be detected and subsequently analyzed [127]. Most of time, EELS is considered as a complementary technique to the X-ray Energy Dispersive Spectroscopy (EDS) analysis for the study of crystalline materials. However, since it is more sensitive for light elements ($Z < 10$), it can be used for carbon, nitrogen, and oxygen quantifications, while EDS does not provide accuracy for these elements. As EELS is very sensitive to the chemical environment of the atoms, a little change of the oxygen content would be reflected in the fine structure of the spectrum.

In a typical oxygen Electron Energy-Loss (EEL) spectrum, the ionization oxygen-K edge reflects the electronic transitions from the oxygen 1s core level to the unoccupied final state of the 2p hybridized levels. These transitions occur at ≈532 and 540 eV, respectively. Beside the oxygen-K edge, the carbon-K edge must be also present in the collected spectra, since the presence of both indicates existence of bonding between them [127]. For a disordered carbon material, the corresponding spectrum has two characteristic features: the 1s → π* transition peak at ≈284 eV and the 1s → σ* transition peak occurring at ≈291 eV [128].

There are a few mathematical methods to quantify the local chemical composition obtained from EELS. The most common are the windows-based methods that permit to obtain two important features from the integrated areas of the transition peaks for each element [129]. First, the percent of each element present in the material is quantified by the integration of all the peaks emerged along the whole spectrum. Second, the relative bonding ratio content of each element is detected in the spectrum. From here, it is possible to know accurately the oxygen content and its hybridization type. However, the simplest form to obtain the chemical composition is from the DigitalMicrograph software. A typical EEL spectrum from an AC is depicted in Figure 10.2.

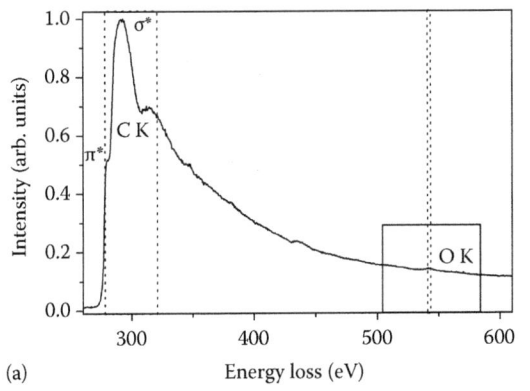
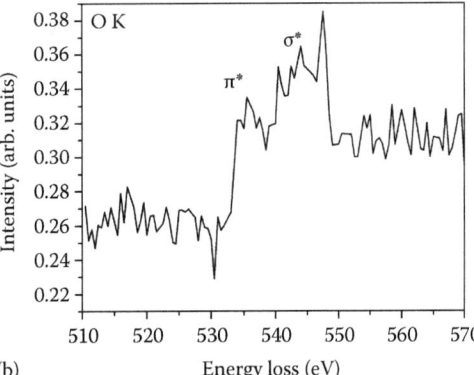

FIGURE 10.2
Electron energy loss spectrum for an AC (a). The dashed windows are used for the quantification of the amounts of C and O present in the sample. An enlargement of the squared area allows to see the transition peaks at 532 and 540 eV characteristics for oxygen (b).

10.3.3.4 Boehm titration method

By several decades, the Boehm titration method, developed in 1966 by Boehm, has been the most used chemical analyses to determine the type and the amount of functional groups on the carbon surfaces [130]. This method is based on the selective neutralization by equilibration with a series of bases of different strengths. Free carboxyl groups are the strongest acidic type. They are sufficiently strong to liberate acetic acid from calcium acetate or carbon dioxide from sodium bicarbonate. Therefore, they can be easily neutralized by titration. Carboxylic groups condensed into lactones or lactols present reduced acidity. They can be neutralized (with ring opening) by sodium carbonate, but not by bicarbonate. Phenolic groups are weakly acidic and they are completely neutralized by sodium hydroxide, but not to a significant extent by dilute sodium carbonate. Still higher uptakes than with NaOH were found with sodium ethoxide in ethanol. This difference corresponds to quinonic and xanthene groups [85].

The Boehm titration is a reliable method to evaluate the general trends in surface acidity. It is simple and fast and usually gives good reproducibility. Its big deficiency is that all groups are classified as oxygen-containing acids. Although the Boehm titration is widely used, it has not been standardized. Explicit methodology has not been specified and certain titration steps vary between research groups. The lack of standardization makes it difficult to compare literature results. Recently, a standardization of the Boehm titration method has been reported. Some of the factors taken into account were the ratio of carbon to reaction base, the length of time that samples are shaken or stirred, the method of CO_2 expulsion from the solutions, and the method of titration endpoint determination [131,132].

Authors found that dissolved CO_2 in the basic solutions has a significant effect on the amount of surface functionalities determined by the Boehm titration, and therefore it must be completely removed. CO_2 removal methods including the use of heat (heating or refluxing) are insufficient to fully remove the effect of the dissolved CO_2 and will result in extremely large variances. Since it was found that it is very difficult to sufficiently remove CO_2 from the solutions used in a direct titration, the acidified aliquots should be titrated immediately after degassing for 2 h with N_2 or Ar, and degassing should continue during the titration, to ensure sufficient removal of CO_2 [131]. In addition, the best method of agitation was found to be shaking, as originally suggested by Boehm. Stirring and sonication resulted in a change in the macroscopic surface of the carbon particles and should be avoided in terms of sample agitation. Duration of shaking is dependent of the surface area of the carbon, and high-surface-area carbons (i.e., many small pores) would require a longer agitation time as diffusion into the pores is important in the Boehm titration [132].

10.3.4 Nitrogen, Sulfur, and Phosphorous as Functionalities

It is common to think that nitrogen functionalities can be easily formed on the surface of AC such as the oxygen groups, since both are present in the air. However, the amount of nitrogen in carbon materials is almost negligible and the carbon–nitrogen complexes are insignificant [34]. The presence of nitrogen in the final structure of the ACs comes from the precursor, PAN fibers most of the times, and also the activation or modification procedures. Here, nitrogen is incorporated to dope, functionalize, or enrich the AC structure for a desirable application. Most of the chemical methods used to introduce nitrogen in the AC include strong oxidation with HNO_3 or NO_x and thermal treatments with chemical agents containing nitrogen such as ammonia, urea, or melamine [91,100,106]. From here, existence of several nitrogen functionalities is possible, as observed in Figure 10.3. In case of lignocellulosic-derived ACs, their lignocellulosic nature does not provide the enough amount of nitrogen to form nitrogen surface groups. However, they can be formed by the aforementioned posttreatments.

Sulfur is naturally present in carbon materials in many forms such as elemental sulfur, inorganic species, and organosulfur compounds. Sulfur content is usually higher than nitrogen, varying from 0% to 5%. Carbon–sulfur complexes are extremely stable even at high temperatures [34]. It has been described that maximum adsorption of sulfur on carbon takes place at 873 K and that the amount of sulfur fixed depends on the carbon nature. Carbon reacts with sulfur-containing compounds such as H_2S, CS_2, or SO_2 at high and low temperatures, resulting in the fixation of

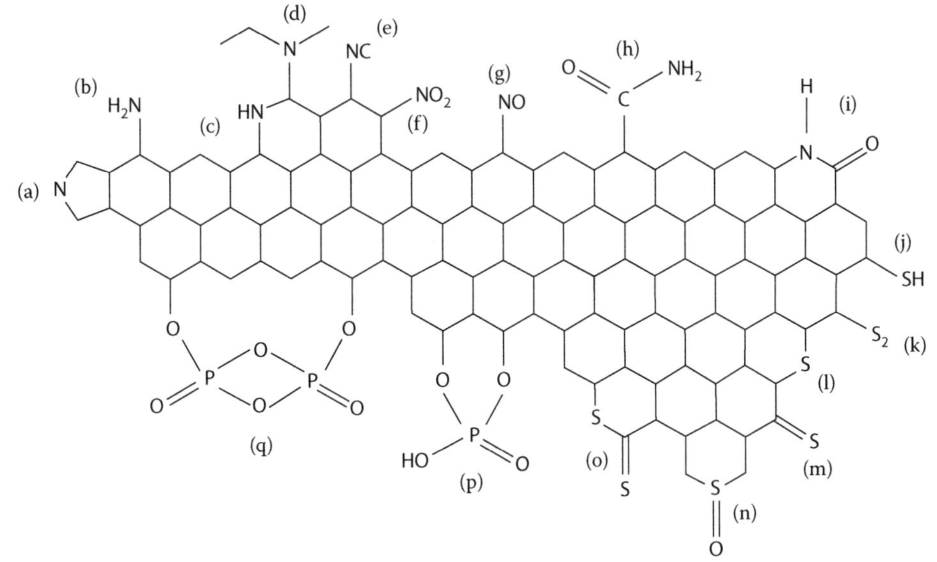

FIGURE 10.3
Schematic representation of the possible nitrogen, sulfur, and phosphorous functionalities. Pyrrole (a); primary (b), secondary (c), and tertiary (d) amines; nitrile (e), nitroso (f), and nitro (g) groups; amide (h); lactam (i); thiophenol (j); disulfide (k); sulfide (l); thymoquinone (m); sulfoxide (n); thiolactone (o); phosphocarbonaceous esters (p); and pyrophosphate (q).

sulfur-containing groups [133]. Sulfur fixation is thought to proceed via capillary condensation, adsorption, chemisorption, and solution in the carbon structure [34]. In several works, it has been proved that sulfur can be fixed by interaction with the oxygenated acidic groups of carbon or by addition to unsaturated sites [16,134–136]. These substitution reactions involve quinone and phenolic groups giving rise to thioquinone and thiophenol structures, whereas the addition to certain active sites results in the formation of sulfide and sulfoxide groups [134,137] (see Figure 10.3).

Phosphorus interactions with AC precursors are well known since phosphoric acid has been widely used as chemical activation agent. Phosphorous promotes bond cleavage reactions and the formation of phosphate and polyphosphate bridges that expand and cross-link the carbon matrix, driving to an accessible pore structure after removal of the acid [138]. The introduction of phosphorus into the carbon matrix is usually achieved by polymerization of a carbonaceous material in the presence of phosphorus compounds that provide multifunctionality for cross-linking the materials. Some evidences have been provided indicating that the phosphorus is fixed in the carbon matrix as red phosphorus and as some chemically bonded forms, mainly as C–O–P bonds [139], as depicted in Figure 10.3.

10.4 STRUCTURAL PROPERTIES OF ACTIVATED CARBONS

After several decades of intense research concerning to the synthesis, characterization, and applications of ACs, a detailed microstructure and the adsorption properties of these materials still cannot be deduced, even from those studies performed from advance experimental techniques such as x-ray diffraction or high-resolution transmission electron microscopy (HRTEM). As a consequence, suitable atomistic models have been developed trying to describe their atomic structure to understand their different properties and to help in tailoring them for specific purposes.

10.4.1 Average Microstructure

The structure of ACs has been described by several authors as a mixture of graphite-like crystallites and disordered domains formed by complex aromatic–aliphatic forms [20,34,140]. Nowadays, from x-ray diffraction and Raman spectroscopy, it has been deduced that these graphitic crystallites actually are local domains formed by a few, about three, parallel graphene-like layers of variable length. These domains are randomly oriented, extensively interconnected, and imbibed into a disordered carbon matrix; such can be clearly observed by HRTEM [73,141]. The array of carbon bonds, almost 100% sp^2, depending on the surface acidic groups on the surface of these domains, is disrupted during the activation process, yielding relative free valences. The anisotropic graphene-like misalignment is associated to the presence of voids. During the activation, the spaces between the domains become cleared of less organized carbonaceous matter, and at the same time, some of the carbon is removed from the graphenic–graphitic structure. The resulting *channels* through the graphitic regions and the interstices between the crystallites of the AC, together with fissures inside and parallel to the graphenic planes, make up the porous structure that usually has a large surface area [142]. These ranges over several sizes are therefore classified into the three groups according to the IUPAC classification [143].

10.4.2 Surface Area and Porous Texture

ACs are identified by two distinctive features: their surface oxygenated acidic groups and their high surface area and porous texture. These parameters can be controlled by several factors: carbonization atmosphere, activation agent, precursor, time and temperature of thermal treatment, use of templates to synthesize the precursor, particle size, and chemical treatment. Surface and structural properties of nanoporous solids can be studied directly by employing advance characterization techniques such as atomic force microscopy, electron microscopy, x-ray analysis, and various spectroscopic methods suitable for materials characterization and surface imaging [4,140]. In addition, these properties can be obtained by indirect methods such as adsorption [144,145], chromatography [146], and thermal analysis [147]. The quantities evaluated from adsorption, chromatography, and thermodesorption data provide information about the whole adsorbent–adsorbate system. These data can be used mainly to extract information about surface heterogeneity and porosity of the carbons.

The low-temperature (77 K) nitrogen adsorption has become a standard and probably the most used method to determine the surface area and pore size distribution of nanoporous adsorbents [20,34,144,145,148,149]. Here, the amount of gas adsorbed (n) on the surface of the solid is proportional to the mass of the adsorbent, and it depends on the pressure (P), the temperature (T), and the nature of both the adsorbent and the adsorptive. The representation of the amount adsorbed versus equilibrium absolute pressure (P) or relative pressure (P/P_0) gives the adsorption isotherm. If the temperature is below the critical temperature of the gas, P_0 is the liquid vapor pressure at the adsorption temperature. The different types of adsorbents (carbon materials, clays, zeolites, etc.) and the large family of possible adsorptives give physical adsorption isotherms with different shapes. The IUPAC has classified the physical adsorption isotherms into six principal groups according to the type of porosity developed by the solid. Their representation and description can be easily found in detail in several manuscripts and books [20,34,144,145,148,149].

Due to the disordered structure of the AC, an appropriate characterization of their textural parameters must be used. The most popular method for determining the surface area of porous solids is based on the Brunauer–Emmett–Teller (BET) theory, developed by Brunauer et al. [150]. The obtained surface area value is usually expressed as S_{BET}. This model is based on the kinetic model of monolayer adsorption proposed by Langmuir and extended to describe the multilayer adsorption. Additionally, use of standard isotherms (normalized forms of the isotherm data for a unique adsorbate–adsorbent system) is widely used, principally those calculated from the t and α_s methods [145,151]. However, it has been reported that the classical analysis of the low-temperature nitrogen adsorption–desorption isotherms by the BET theory often overestimates the surface area with respect to other determinations, especially when the carbon develops pores below 0.9 nm [152].

The pore diameters, from the classification given by the IUPAC, are macropores (>50 nm), mesopores (2–5 nm), and micropores (<2 nm); additionally, micropores can be divided into two classifications: wide (0.7–2 nm) and narrow (<0.7 nm) [143]. The accurate pore size analysis of porous carbons with a combination of micropores and small mesopores is a challenge, particularly when the shapes of these pores at the micro and meso levels are different; fortunately, reliable methods for the characterization of porosity in carbon materials have been developed. Classical mathematical procedures based on the Kevin equation have been proposed for the calculation of mesoporosity from nitrogen adsorption isotherm. The Barrett–Joyner–Halenda and Dellimore–Heal methods are the most frequently used. The micropore size distribution has been described by the Horvath–Kawazoe and Saito–Foley methods, considering the existence of slit-like shape geometry for the pores, described as two graphitic walls separated by certain distance [143,149].

Phenomenological models of adsorption based on the Dubinin theory of volume filling of micropores have found also a great utility. The Dubinin–Radushkevich, Dubinin–Astakhov, and Dubinin–Stoeckli approaches have become widely accepted methods for calculating the micropore volume, the characteristic energy of adsorption, and the distribution of micropore sizes [145,153,154]. In the last years, the use of models based on the density functional theory (DFT) for determining the PSD seems to provide a more precise description of the adsorption into the pores. The local DFT, the nonlocal DFT, and recently the quenched solid-state DFT approaches are based on the behavior of the nonhomogeneous fluids at a solid interface [155–158]. Although these models depart from the slit pore model, most of these do not provide a realistic description of the highly disordered nature of these carbons. Reconstruction methods, in which a 3D structural model is built, offer the most promising route to realistic models of such carbons at the present time. Reverse Monte Carlo method is one such reconstruction method in which an atomistic model is built that matches experimental structure factor data from x-ray or neutron diffraction [158,159].

As these textural parameters depend from the experimental conditions, including the precursor, the activation step becomes the most significant for the development of surface area and porous texture. The activation agent, temperature, and time are the most analyzed parameters on the textural properties. Several works on the use of statistical methods, such as the response surface, to optimize them have been reported [43,58,160–162]. In addition, there is an intense search of novel lignocellulosic precursors to achieve these optimal characteristics and their accurate description. Nevertheless, the S_{BET} value is the most used as indicative of the surface area. As evidence, Figure 10.4 shows a comparative result of the reported surface area values for some ACs from different lignocellulosic precursor and activation methods.

Notice the strong influences of the impregnation agent on the development of the reported the S_{BET} values. For instance, carbons obtained from corncob or artichoke leaves with phosphoric acid impregnation ratio reach S_{BET} as high as 2081 and 2038 m^2/g, respectively [163,164]. All the carbons prepared by chemical activation with ZnCl$_2$ reach S_{BET} higher than 750 m^2/g, reaching also S_{BET} as high as 2100 m^2/g [165]. The preparation of AC by chemical activation with KOH and NaOH allows to obtain AC with high specific surface areas more than 1000 m^2/g. However, KOH and NaOH are corrosive and deleterious chemicals. For this reason, recent studies have proposed the preparation

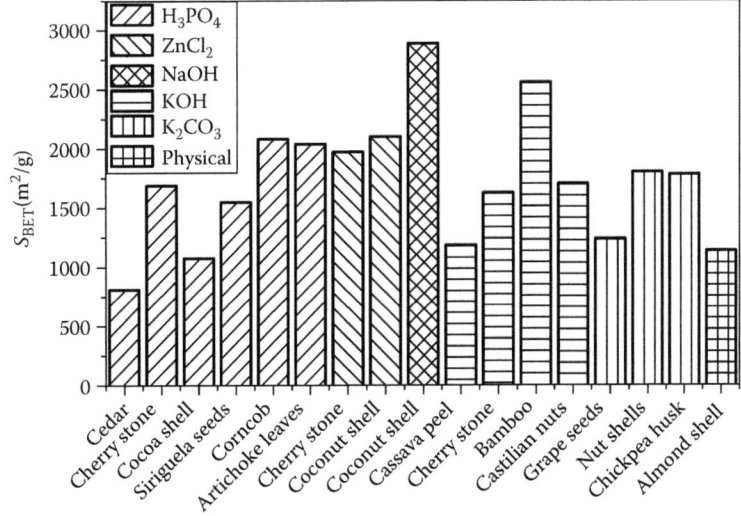

FIGURE 10.4
Comparison of the surface area values for ACs from different lignocellulosic precursor and activation methods.

of ACs by chemical activation with K_2CO_3 in one step. Also, the activation with K_2CO_3 renders carbons with a competitive S_{BET} (between 1200 and 1800 m²/g) compared with those obtained by activation with KOH or NaOH [166]. For the ACs obtained by physical activation, those obtained from CO_2 show a higher specific surface area than those obtained by activation with steam.

10.5 APPLICATION OF ACTIVATED CARBONS AS A RESULT OF THEIR SURFACE CHEMISTRY

10.5.1 Adsorption of Hazardous Materials

Adsorption of both organic and inorganic compounds in aqueous solutions has been a very important application of AC. In fact, it is known that around 80% of the world production of AC is used in liquid–phase applications [167]. Also, the treatment of wastewater and contaminated groundwater using AC is increasing throughout the world as a result of the limited sources of water supply [168]. When using AC, the adsorption process results from interactions between the carbon surface and the adsorbate. These interactions can be electrostatic or nonelectrostatic. When the adsorbate is an electrolyte that dissociates in aqueous solution, electrostatic interactions occur. These interactions can be attractive or repulsive depending on the (1) charge density of the carbon surface, (2) chemical characteristics of the adsorbate, and (3) ionic strength of the solution. Nonelectrostatic interactions are always attractive and can include (1) van der Waals forces, (2) hydrophobic interactions, and (3) hydrogen bonding [18].

It has been reported that the properties of the adsorbate that mainly influence the adsorption process in AC are (1) molecular size, (2) solubility, (3) pKa values, and (4) nature of the substituents (in the case of aromatic adsorbates). When the AC is in contact with an aqueous solution, an electric charge is generated. This charge results from either the dissociation of the surface functional groups of the carbon or the adsorption of ions from the solution and strongly depends on the solution pH and on the surface characteristics of the adsorbent [169].

It has been reported that carboxyl groups were proven to be directly responsible for heavy metal sorption onto lignocellulosic-derived adsorbents [170–172], whereas phenolic groups are believed to be accountable for the formation of complexes with heavy metals [173]. In addition, carbonyl and carboxyl groups increase the adsorption capacity via formation of complexes and/or chelates on the surface of the AC, even at high alkaline pH values [174].

Heavy metals are considered to be one of the most hazardous water contaminants. According to the World Health Organization (2011), among the most toxic metals are arsenic, cadmium, chromium, copper, lead, mercury, zinc, and nickel [175]. Up to date, many studies report that the removal of heavy metals by AC is successfully, economically favorable and technically easy; for these reasons, ACs are widely used to treat waters contaminated with heavy metals. Additional information regarding the adsorption of heavy metals from aqueous solutions can be found in several manuscripts and some reviews [4,16,59,91,99,133,135,174,176–179].

Adsorption of metallic ions from aqueous solution is far from being a straightforward process. Metallic species have small size, being frequently charged in solution; therefore, the predominant interactions in their adsorption process on AC are of electrostatic nature [180]. The factors that mainly control the extent of adsorption on AC are (1) the chemistry of the metal ion (speciation) or metal ion complex, (2) the solution pH and the PZC of the surface, (3) the surface area and porosity (narrow and wider microporosity), (4) the surface composition (oxygen functionality), and (5) the size of the adsorbing species (hydrated ions in the range 1.0–1.8 nm), mainly for carbons with significant volumes of narrow microporosity [16]. Figure 10.5 depicts the effect of the total content of SOG on the heavy metal adsorption capacity of some ACs. Notice that in most cases, high concentrations of SOG decrease the performance of the ACs. The only exception seems to be adsorption of Cr(IV) on coconut-derived ACs.

On the other hand, discharge of dyes in the environment is a matter of concern for both toxicological and esthetical reasons. It is estimated that more than 100,000 commercially available dyes with over 7×10^5 tonnes of dyestuff are produced annually [181]. Adsorption of dyes onto active carbons depends of the surface charge in the carbon in the water source. ACs demonstrate a high capacity for

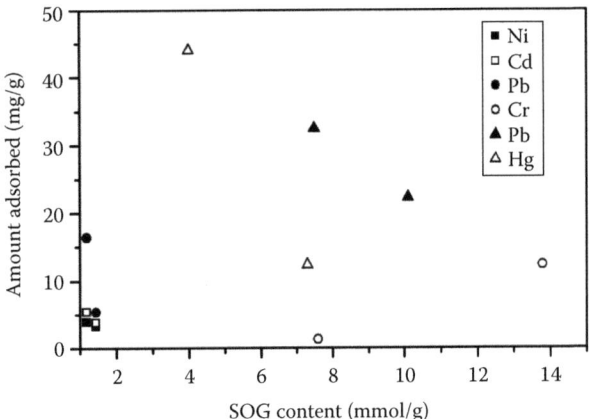

FIGURE 10.5
Relationship between the total content of SOG and the adsorption capacity of some heavy metals.

both acid and basic dyes [13]. The effects of various experimental parameters of dye adsorption are initial pH, dye concentration, sorbent dosage, ion strength, and residence time [67,182].

10.5.2 Electrodes for Electrochemical Double-Layer Capacitors

ACs with oxygen-containing functional groups contribute to the capacitance of EDLCs, which is called pseudocapacitance generally credited to Faradaic reaction of these groups with electrolyte ions [6–9,110]. The oxygen-containing surface groups on ACs can be divided into three types, depending on the nature of the C–O bonds: (1) The first type is the chemically fixed groups (e.g., carbonyl), which are degassed as CO, only upon heating the carbon to above 800°C. These groups are believed to be electrochemically inactive, and their main effect is an influence of the PZC. (2) Surface groups (e.g., carboxylic), which provide surface acidity, are degassed at above 400°C, mostly as in CO_2. (3) And the third type is the surface groups with electrochemical redox activity such as quinine/hydroquinone moieties, which can also be degassed between 400°C and 800°C [183]. Such surface groups are easily recognized by the electrochemical response of carbon electrodes (e.g., a couple of reversible peaks in cyclic voltammetric measurements). It should be noted that these redox sites are active commonly only under acidic conditions [184], as observed in Figure 10.6.

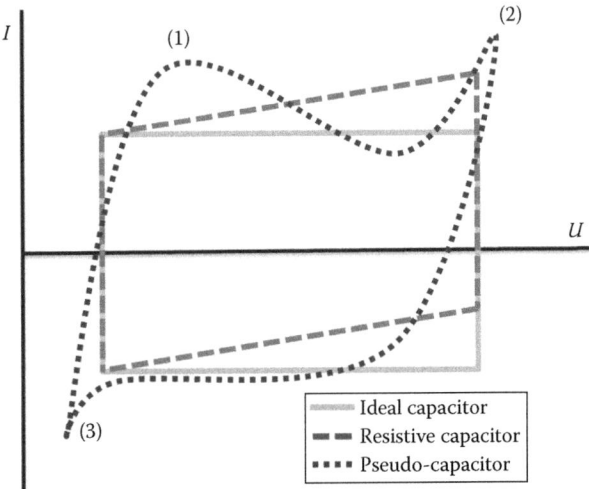

FIGURE 10.6
Representation of the different types of cyclic voltammograms. In the curve for a pseudocapacitor, the numbered maxima correspond to the contributions from carboxylic groups (1), phenolic groups (2), and quinones (3).

Many oxygen-containing redox-active functional groups increase both the capacitance and high-rate performance in aqueous electrolyte solutions without significant penalties. The wetting of the hydrophobic carbon surface in aqueous electrolyte solution can be poor. Therefore, the presence of oxygen-containing polar functional groups (hydroxyl, carboxyl, carbonyl, quinine, etc.) greatly increases the hydrophilicity and surface area accessibility to aqueous electrolytes [185,186]. In addition to improved wetting, pseudocapacitance from selected functional groups may contribute to over 22% of the total capacitance values [187]. However, certain functional groups may also lead to the leakage and degradation of EDLCs in strong acidic and basic aqueous electrolytes, similar to their negative effects in organic electrolytes. Oxygen species on the surface are commonly disliked for the carbon materials used in nonaqueous electrolytes, because they were found to be unstable and have adverse effects on the reliability of capacitors, causing self-discharge, leakage current, degradation, and other negative contributions [188].

10.6 SUMMARY AND CONCLUSIONS

Even when fullerenes, carbon nanotubes, and grapheme seem to cover most of the scientific contributions about carbon materials, ACs receive special interest for their always inspected properties. It is clear that the presence of SOG on ACs and their high surface area are the most important features for an AC. Their formation mechanism is still under research; however, it is possible to describe how both parameters define the favorable performance of these now called *low-cost* materials. Worldwide researches make important contributions to find novel precursors, develop novel activation procedures, and control the final properties of these interesting materials. In addition, use of lignocellulosic resources for the production of AC decreases the cost of production and the environmental impact of the agricultural and agroindustrial wastes.

ACKNOWLEDGMENTS

The author would like to acknowledge CONACTY, México, for the support given through a postdoctoral grant.

REFERENCES

1. Wei, L. Yushin, G. 2012. Nanostructured activated carbons from natural precursors for electrical double layer capacitors. *Nanoenergy* 1: 552–565.
2. Kalyani, P. Anitha, A. 2013. Biomass carbon and its prospects in electrochemical energy systems. *Int J Hydrogen Energy* 38: 4034–4045.
3. Inagaki, M. Kang, F. 2014. Engineering and applications of carbon materials. In *Materials Science and Engineering of Carbon: Fundamentals*, eds. M. Inagaki, F. Kang, M. Toyoda, H. Konno, pp. 237–265. Philadelphia, PA: Elsevier.
4. Inagaki, M. Tascon, J.M.D. 2006. Pore formation and control in carbon materials. In *Activated Carbon Surfaces in Environmental Remediation*, ed. T.J. Bandosz, pp. 49–105. Philadelphia, PA: Elsevier Ltd.
5. Román, S. Valente Nabais, J.M. Ledesma, B. et al. 2013. Production of low-cost adsorbents with tunable surface chemistry by conjunction of hydrothermal carbonization and activation processes. *Micropor Mesopor Mater* 165: 127–133.

6. Centeno, T.A. Stoeckli, F. 2006. The role of textural characteristics and oxygen–containing surface groups in the supercapacitor performances of activated carbons. *Electrochim Acta* 52: 560–566.
7. Inagaki, M. Konno, H. Tanaike, O. Carbon materials for electrochemical capacitors. *J Power Sources* 195: 7880–7903.
8. Béguin, F. Raymundo-Piñero, E. Frackowiak, E. 2010. Electrical double-layer capacitors and pseudocapacitors. In *Carbons for Electrochemical Energy Storage and Conversion Systems*, eds. F. Béguin, E. Frackowiak, pp. 329–346. Boca Raton, FL: CRC Press.
9. Béguin, F. Frackowiak, E. 2006. Nanotextured carbons for electrochemical energy storage. In *Nanomaterials Handbook*, ed. Y. Gogotsi, pp. 713–723. Boca Raton, FL: CRC Press.
10. Ali, I. Asim, M. Khan, T.A. 2013. Low cost adsorbents for the removal of organic pollutants from wastewater. *J Environ Manage* 113: 170–183.
11. Pintor, A.M.A. Ferreira, C.I.A. Pereira, J.C. et al. 2012. Use of cork powder and granules for the adsorption of pollutants: A review. *Water Res* 46: 3152–3166.
12. Robinson, T. McMullan, G. Marchant, R. Nigam, P. 2001. Remediation of dyes in textile effluent: A critical review on current treatment technologies with a proposed alternative. *Biores Technol* 77: 247–255.
13. Demirbas, A. 2009. Agricultural based activated carbons for the removal of dyes from aqueous solutions: A review. *J Hazard Mater* 167: 1–9.
14. Ioannidou, O. Zabaniotou, A. 2007. Agricultural residues as precursors for activated carbon production—A review. *Renew Sustain Energy Rev* 11: 1705–1766.
15. Yin, C.Y. Aroua, M.K. Daud, W.M.A.W. 2007. Review of modifications of activated carbon for enhancing contaminant uptakes from aqueous solutions. *Separ Purif Technol* 52: 403–415.
16. Rivera-Utrilla, J. Sánchez-Polo, M. Gómez-Serrano, V. et al. 2011. Activated carbon modifications to enhance its water treatment applications: An overview. *J Hazard Mater* 187: 1–23.
17. Shafeeyan, M.S. Daud, W.M.A.W. Houshmand, A. Shamiri, A. 2010. A review on surface modification of activated carbon for carbon dioxide adsorption. *J Anal Appl Pyrol* 89: 143–151.
18. Dias, J.M. Alvim-Ferraz, M.C.M. Almeida, M.F. Rivera-Utrilla, J. Sánchez-Polo, M. 2007. Waste materials for activated carbon preparation and its use in aqueous-phase treatment: A review, *J Environ Manage* 85: 833–846.
19. Transparency Market Research. 2013. *Activated Carbon Market (Powdered, Granular) for Liquid Phase and Gas Phase Applications in Water Treatment, Food and Beverage Processing, Pharmaceutical and Medical, Automotive and Air purification—Global Industry Analysis, Size, Share, Growth, Trends and Forecast, 2013-2019.* Albany, NY: Transparency Market Research.
20. Marsh, H. Rodriguez-Reinoso, F. 2006. *Activated Carbon.* Oxford, U.K.: Elsevier Ltd.
21. Lozano-Castelló, D. Lillo-Ródemas, M.A. Carzola-Amorós, D. Linares-Solano, A. 2001. Preparation of activated carbons from Spanish anthracite: I. Activation by KOH. *Carbon* 39: 741–749.
22. Lillo-Ródemas, M.A. Lozano-Castelló, D. Carzola-Amorós, D. Linares-Solano, A. 2001. Preparation of activated carbons from Spanish anthracite: II. Activation by NaOH. *Carbon* 39: 751–759.

23. El Qada, E.N. Allen, S.J. Walker, G.M. 2006. Adsorption of methylene blue onto activated carbon produced from steam activated bituminous coal: A study of equilibrium adsorption isotherm. *Chem Eng J* 124: 103–110.

24. Carrott, P.J.M. Cansado, I.P.P. Mourão, P.A.M. et al. 2012. On the use of ethanol for evaluating microporosity of activated carbons prepared from Polish lignite. *Fuel Proc Technol* 103: 34–38.

25. Donald, J. Ohtsuka, Y. Xu, C. 2011. Effects of activation agents and intrinsic minerals on pore development in activated carbons derived from a Canadian peat. *Mater Lett* 65: 744–747.

26. Díaz-Diéz, M.A. Gómez-Serrano, V. Fernández-González, C. Cuerda-Correa, E.M. Macías-García, A. 2004. Porous texture of activated carbons prepared by phosphoric acid activation of woods. *Appl Surf Sci* 238: 309–313.

27. Cazetta, A.L. Vargas, A.M.M. Nogami, E.M. et al. 2011. NaOH-activated carbon of high surface area produced from coconut shell: Kinetics and equilibrium studies from the methylene blue adsorption. *Chem Eng J* 174: 117–125.

28. Foo, K.Y. Hameed, B.H. 2012. Coconut husk derived activated carbon via microwave induced activation: Effects of activation agents, preparation parameters and adsorption performance. *Chem Eng J* 184: 57–65.

29. Mourão, P.A.M. Laginhas, C. Custódio, F. Nabais, J.M.V. Carrott, P. Ribeiro-Carrott, M.M.L. 2001. Influence of oxidation process on the adsorption capacity of activated carbons from lignocellulosic precursors. *Fuel Proc Technol* 92: 241–246.

30. Rodríguez-Reinoso, F. Molina-Sabio, M. 1992. Activated carbons from lignocellulosic materials by chemical and/or physical activation: An overview. *Carbon* 30: 1111–1118.

31. Mohamed, A.R. Mohammadi, M. Darzi, G.N. 2010. Preparation of carbon molecular sieve from lignocellulosic biomass: A review. *Renew Suitan Energy Rev* 14: 1591–1599.

32. Okman, I. Karagöz, S. Tay, T. Redem, M. 2014. Activated carbons from grape seeds by chemical activation with potassium carbonate and potassium hydroxide. *Appl Surf Sci* 293: 138–142.

33. Cossarutto, L. Zimny, T. Kaczmarczyk, J. et al. 2001. Transport and sorption of water vapour in activated carbons. *Carbon* 39: 2339–2346.

34. Bandosz, T.J. Ania, C.O. 2006. Surface chemistry of activated carbon and its characterization. In *Activated Carbon Surfaces in Environmental Remediation*, ed. T.J. Bandosz, pp. 159–230. Philadelphia, PA: Elsevier Ltd.

35. Sudaryanto, Y. Hartono, S.B. Irawaty, W. Hindarso, H. Ismadji, S. 2006. High surface area activated carbon prepared from cassava peel by chemical activation. *Biores Technol* 97: 734–739.

36. Hayashi, J. Horikawa, T. Takeda, I. Muroyama, K. Ani, F.N. 2002. Preparing activated carbon from various nutshells by chemical activation with K_2CO_3. *Carbon* 40: 2381–2386.

37. Foo, K.Y. Hameed, B.H. 2011. Preparation and characterization of activated carbon from pistachio nut shells via microwave-induced chemical activation. *Biomass Bioenergy* 35: 3257–3261.

38. Nabais, J.M.V. Laginhas, C.E.C. Carrott, P.J.M. Ribeiro-Carrott, M.M.L. 2011. Production of activated carbons from almond shell. *Fuel Proc Technol* 92: 234–240.

39. Martínez de Yuso, A. Rubio, B. Izquierdo, M.T. 2014. Influence of activation atmosphere used in the chemical activation of almond shell on the characteristics and adsorption performance of activated carbons. *Fuel Proc Technol* 119: 74–80.

40. Jimenez-Cordero, D. Heras, F. Alonso-Morales, N. Gilarranz, M.A. Rodríguez, J.J. 2014. Preparation of granular activated carbons from grape seeds by cycles of liquid phase oxidation and thermal desorption. *Fuel Proc Technol* 118: 148–155.

41. Olivares-Marín, M. Fernández, J.A. Lázaro, M.J. et al. 2009. Cherry stones as precursor of activated carbons for supercapacitors. *Mater Chem Phys* 114: 323–327.

42. Arami-Niya, A. Daud, W.M.A.W. Mjalli, F.S. 2011. Comparative study of the textural characteristics of oil palm shell activated carbon produced by chemical and physical activation for methane adsorption. *Chem Eng Res Des* 89: 657–664.

43. Hameed, B.H. Tan, I.A.W. Ahmad, A.L. 2008. Optimization of basic dye removal by oil pal fibre-based activated carbon using response surface methodology. *J Hazard Mater* 158: 324–332.

44. Alam, M.Z. Ameem, E.S. Muyibi, S.A. Kabbashi, N.A. 2009. The factors affecting the performance of activated carbon prepared from oil palm empty fruit branches for adsorption of phenol. *Chem Eng J* 155: 191–198.

45. Nerm, A.E. Khaled, A. Abdelwahab, O. El-Sikaily, A. 2008. Treatment of wastewater containing toxic chromium using new activated carbon developed from date palm seed. *J Hazard Mater* 152: 263–275.

46. Miranda, R. Bustos-Martínez, D. Sosa-Blanco, C. Gutiérrez-Villareal, M.H. Rodríguez-Cantú, M.E. 2009. Pyrolysis of sweet orange (*Citrus sinensis*) dry peel. *J Anal Appl Pyrol* 86: 245–251.

47. Ibrahim, S. Fatimah, I. Ang, H.M. Wang, S. 2010. Adsorption of anionic dyes in aqueous solution using chemically modified barley straw. *Water Sci Technol* 62: 1177–1182.

48. Sun, D. Zhang, X. Wu, Y. Liu, X. 2010. Adsorption of anionic dyes from aqueous solution on fly ash. *J Hazard Mater* 181: 335–342.

49. López, F.A. Centeno, T.A. García-Díaz. I. Alguacil, F.J. 2013. Textural and fuel characteristics of the chars produced by the pyrolysis waste wood, and the properties of activated carbons prepared from them. *J Anal Appl Pyrol* 104: 551–558.

50. González, P.G. Pliego-Cuervo, Y.B. 2013. Physicochemical and microtextural characterization of activated carbons produced from water steam activation of three bamboo species. *J Anal Appl Pyrol* 99: 32–39.

51. Tongpoothorn, W. Sriuttha, M. Homchan, P. Chanthai, S. Ruangviriyachai, C. 2011. Preparation of activated carbon derived from *Jatropha curcas* fruit shell by simple thermochemical activation and characterization of their physico-chemical properties. *Chem Eng Res Des* 89: 335–340.

52. Jústiz-Smith, N.G. Virgo, G.J. Buchanan, V.E. 2008. Potential of Jamaican banana, coconut coir and bagasse fibres as composite material. *Mater Charact* 59: 1273–1278.

53. Ozdemir, I. Sahin, M. Orhan, R. Erdem, M. 2014. Preparation and characterization of activated carbon from grape stalk by zinc chloride activation. *Fuel Proc Technol* 25: 200–206.

54. Tiryaki, B. Yangmur, E. Banford, A. Aktas, Z. 2014. Comparison of activated carbon produced from natural biomass and equivalent chemical compositions. *J Anal Appl Pyrol* 105: 276–283.

55. Phan, N.H. Rio, S. Faur, C. et al. 2006. Production of fibrous activated carbons from natural cellulose (jute, coconut) fibers for water treatment applications. *Carbon* 44: 2569–2577.

56. Gurten, I.I. Ozmak, M. Yagmur, E. Aktas, Z. 2012. Preparation and characterization of activated carbon from waste tea using K_2CO_3. *Biomass Bioenergy* 37: 73–81.

57. Uysal, T. Duman, G. Onal, Y. Yasa, I. Yanik, J. 2014. Production of activated carbon and fungicidal oil from peach stone by two-stage process. *J Anal Appl Pyrol* 108: 47–55.

58. González, P.G. Hernández-Quiroz, T. García-González, L. 2014. The use of experimental design and response surface methodologies for the synthesis of chemically activated carbons produced from bamboo. *Fuel Proc Technol* 127: 133–139.

59. Hernández-Montoya, V. Mendoza-Castillo, D.I. Bonilla-Petriciolet, A. Montes-Morán, M.A. Pérez-Cruz, M.A. 2011. Role of the pericarp of *Carya illinoinensis* as biosorbent and as precursor of activated carbon for the removal of lead and acid blue 25 in aqueous solutions. *J Anal Appl Pyrol* 92: 143–151.

60. Gonçalves Pereira, R. Martins Veloso, C. Mendes da Silva, M. et al. 2014. Preparation of activated carbons from cocoa shells and siriguela seeds using H_3PO_4 and $ZnCl_2$ as activating agents for BSA and α-lactalbumin adsorption. *Fuel Proc Technol* 26: 476–486.

61. Prahas, D. Kartika, Y. Indraswati, N. Ismadji. S. 2008. Activated carbon from jackfruit peel waste by H_3PO_4 chemical activation: Pore structure and surface chemistry characterization. *Chem Eng J* 40: 32–42.

62. Loredo-Cancino, M. Soto-Regalado, E. Cerino-Córdova, F.J. et al. 2013. Determining optimal conditions to produce activated carbon from barley husks using single or dual optimization. *J Environ Manage* 125: 117–125.

63. Li, M.F. Fan, Y.M. Xu, F. Sun, R.C. Zhang, X.L. 2010. Cold sodium hydroxide/urea based pretreatment of bamboo for bioethanol production: Characterization of the cellulose rich fraction. *Ind Crops Prod* 32: 551–559.

64. Girio, F.M. Fonseca, C. Carvalheiro, F. et al. 2010. Hemicelluloses for fuel ethanol: A review. *Biores Technol* 101: 4775–4800.

65. Mood, S.H. Golfeshan, A.H. Tabatabaei, M. Lignocellulosic biomass to bioethanol, a comprehensive review with a focus on pretreatment. *Renew Sutain Energy Rev* 27: 77–93.

66. Suhas, P.J.M. Carrott, M.M.L. Carrott, R. 2007. Lignin—From natural adsorbent to activated carbon: A review. *Biores Technol* 98: 2301–2312.

67. Hashem, A. Akasha, R.A. Ghith, A. Hussein, D.A. 2007. Adsorbent based on agricultural wastes for heavy metal and dye removal: A review. *Energy Edu Sci Technol* 19: 69–86.

68. Nor, M.M. Chung, L.L. Teong, L.K. Mohamed, A.R. 2013. Synthesis of activated carbon from lignocellulosic biomass and its applications in air pollution control—A review. *J Environ Chem Eng* 1: 658–666.

69. Demirbas, A. 2008. The sustainability of combustible renewable. *Energy Source Part A* 30: 1114–1119.

70. Lewis, I.C. 1982. Chemistry of carbonization. *Carbon* 20: 519–529.

71. Daud, W.M.A.W. Ali, W.S.W. Sulaiman, M.Z. 2000. The effects of carbonization temperature on pore development in palm-shell-based activated carbon. *Carbon* 38: 1925–1932.

72. Lua, A.C. Lau, F.Y. Guo, J. 2006. Influence of pyrolysis conditions on pore development of oil-palm-shell activated carbons. *J Anal Appl Pyrol* 76: 96–102.

73. González-García, P. Centeno, T.A. Urones-Garrote, E. Ávila-Brande, D. Otero.Díaz, L.C. 2013. Microstructure and surface properties of lignocellulosic-based activated carbons. *Appl Surf Sci* 256: 731–737.

74. Zuo, S. Yang, J. Liu, J. 2010. Effects of the heating history of impregnated lignocellulosic material on pore development during phosphoric acid activation. *Carbon* 48: 3293–3295.

75. Hayashi, J. Kazehaya, A. Muroyama, K. Watkinson, A. 2000. Preparation of activated carbon from lignin by chemical activation. *Carbon* 38: 1873–1878.

76. Reed, A.R. Williams, P.T. 2004. Thermal processing of biomass natural fiber wastes by pyrolysis. *Int J Energy Res* 28: 131–145.

77. Boehm, H.P. 2008. Surface chemical characterization of carbons from adsorption studies. In *Adsorption by Carbons*, eds. E.J. Bottani, J.M.D. Tascón, pp. 301–327. Amsterdam, the Netherlands: Elsevier Ltd.

78. Newcombe, G. 2008. Adsorption from aqueous solutions: Water purification. In *Adsorption by Carbons*, eds. E.J. Bottani, J.M.D. Tascón, pp. 679–709. Amsterdam, the Netherlands: Elsevier Ltd.

79. Delennay, F. Tysoe, W.T. Heinemann, H. Somorjai, G.A. 1984. The role of KOH in the steam gasification of graphite: Identification of the reaction steps. *Carbon* 22: 401–407.

80. Kelemen, S.R. Freund, H. Model CO_2 gasification reactions on uncatalyzed and potassium catalyzed glassy carbon surfaces. *J Catal* 102: 80–91.

81. Schlögl, R. Loose, G. Wesemann, M. 1990. On the mechanism of the oxidation of graphite by molecular oxygen. *Solid State Ionics* 43: 183–192.

82. Henschke, B. Schubert, H. Blöcker, J. Atammy, F. Schlögl, R. 1994. Mechanistic aspects of the reaction between carbon and oxygen. *Thermochim Acta* 234: 53–83.

83. Werner, H. Schedel-Niedrig, Th. Wohlers, D. et al. 1994. Reaction of molecular oxygen with C_{60}: Spectroscopic studies. *J Chem Soc Faraday Trans* 90: 403–409.

84. Yang, J. Mestl, G. Herein, D. Schlögl, R. Find, J. 2000. Reaction of NO with carbonaceous materials 1. Reaction and adsorption of NO on ashless carbon black. *Carbon* 38: 715–727.

85. Boehm, H.P. 1990. Surface oxides on carbon. *High-Temp High-Press* 22: 275–288.

86. Puri, B.R. 1966. Chemisorbed oxygen evolved as carbon dioxide and its influence on surface reactivity of carbons. *Carbon* 4: 391–400.

87. Billinge, B.H.M. Docherty, J.B. Bevan, M.J. 1984. The desorption of chemisorbed oxygen from activated carbons and its relationship to ageing and methyl iodide retention efficiency. *Carbon* 22: 83–89.

88. Adams, L.B. Hull, C.R. Holmes, R.J. Newton, R.A. 1988. An examination of how exposure to humid air can result in changes in the adsorption properties of activated carbons. *Carbon* 26: 451–459.

89. Boehm, H.P. 1994. Some aspects of the surface chemistry of carbon blacks and other carbons. *Carbon* 32: 759–769.

90. Menendez, J.A. Illin-Gomez, M.J. Leon y Leon, C.A. Radovic, L.R. 1995. On the difference between the isoelectric point and the point of zero charge of carbon. *Carbon*, 33: 1655–1657.
91. Jia, Y.F. Thomas, K.M. 2000. Adsorption of cadmium ions on oxygen surface sites in activated carbon. *Langmuir* 16: 1114–1122.
92. Strelko, V. Malik, D.J. Streat, M. 2002. Characterisation of the surface of oxidised carbon adsorbents. *Carbon* 40: 95–104.
93. Moreno-Catilla, C. Carrasco-Marin, F. Mueden, A. 1997. The creation of acidic carbon surfaces by treatment with (NH4)$_2$S$_2$Os. *Carbon* 35: 1619–1626.
94. Leon y Leon, C.A. Solar, J.M. Calemma, V. Radovic, L.R. 1992. Evidence for the protonation of basal plane sites on carbon. *Carbon* 30: 797–811.
95. Montes-Morán, M.A. Suárez, D. Menéndez, J.A. Fuente, E. 2004. On the nature of basic sites on carbon surface: An overview. *Carbon* 42: 1219–1225.
96. Pietrowski, P. Ludwiczak, I. Tyczkowski, J. 2012. Activated carbons modified by Ar and CO plasmas—Acetone and cyclohexane adsorption. *Mater Sci (MEDZIAGOTYRA)* 18: 158–162.
97. Baatnagar, A. Hogland, W. Marques, M. Sillanpää, M. 2013. An overview of the modification methods of activated carbon for its water treatment applications. *Chem Eng J* 219: 499–511.
98. Santiago, M. Stüber, F. Fortuny, A. Fabregat, A. Font, J. 2005. Modified activated carbons for catalytic wet air oxidation of phenol. *Carbon* 43: 2134–2145.
99. Sato, S. Yoshihara, K. Moriyama, K. Machida, M. Tatsumoto, H. 2007. Influence of activated carbon surface acidity on adsorption of heavy metal ions and aromatics from aqueous solution. *Appl Surf Sci* 253: 8554–8559.
100. Menendez, J.A. Phillips, J. Xia, B. Radovic, L.R. 1996. On the modification and characterization of chemical surface properties of activated carbon: In the search of carbons with stable basic properties. *Langmuir* 12: 4404–4410.
101. Biniak, S. Szymanski, G. Siedlewski, J. Swiatkoski, A. 1997. The characterization of activated carbons with oxygen and nitrogen surface groups. *Carbon* 35: 1799–1810.
102. Shaarani, F.W. Hameed, B. 2011. Ammonia-modified activated carbon for the adsorption of 2,4-dichlorophenol. *Chem Eng J* 169: 180–185.
103. Mangun, C.L. Benak, K.R. Economy, J. Foster, K.L. 2001. Surface chemistry, pore sizes and adsorption properties of activated carbon fibers and precursors treated with ammonia. *Carbon* 39: 1809–1820.
104. Jansen, R.J.J. Bekkum, H.V. 1995. XPS of nitrogen-containing functional groups on activated carbon. *Carbon* 33: 1021–1027.
105. Raymundo-Pinero, E. Cazorla-Amoros, D. Linares-Solano, A. 2003. The role of different nitrogen functional groups on the removal of SO$_2$ from flue gases by N-doped activated carbon powders and fibres. *Carbon* 41: 1925–1932.
106. Stavropoulos, G.G. Samaras, P. Sakellaropoulos, G.P. 2008. Effect of activated carbons modification on porosity, surface structure and phenol adsorption *J Hazard Mater* 151: 414–421.
107. Ania, C.O. Parra, J.B. Menéndez, J.A. Pis, J.J. 2005. Effect of microwave and conventional regeneration on the microporous and mesoporous network and on the adsorptive capacity of activated carbons. *Micropor Mesopor Mater* 85: 7–15.
108. Yuen, F.K. Hameed, B.H. 2009. Recent developments in the preparation and regeneration of activated carbons by microwaves. *Adv Colloid Interface Sci* 149: 19–27.

109. Hesas, R.H. Daud, W.M.A.W. Sahu, N.J. Arami-Niya, A. 2013. The effects of a microwave heating method on the production of activated carbon from agricultural waste: A review. *J Anal Appl Pyrol* 100: 1–11.
110. Dehdashti, A. Khavanin, A. Rezaee, A. Assilian, H. Motalebi, M. 2011. Application of microwave irradiation for the treatment of adsorbed volatile organic compounds on granular activated carbon. *Iran J Environ Health Sci Eng* 8: 85–94.
111. Liu, Q.-S. Zheng, T. Li, N. Wang, P. Abulikemu, G. 2010. Modification of bamboo based activated carbon using microwave radiation and its effects on the adsorption of methylene blue. *Appl Surf Sci* 256: 3309–3315.
112. Zaror, C.A. 1997. Enhanced oxidation of toxic effluents using simultaneous ozonation and activated carbon treatment. *J Chem Technol Biotechnol* 70: 21–28.
113. Jans, U. Hoigné, J. 1998. Activated carbon and carbon black catalyzed transformation of aqueous ozone into OH-radicals. *Ozone Sci Eng* 20: 67–90.
114. Rivera-Utrilla, J. Sanchez-Polo, M. 2002. Ozonation of 1,3,6-naphthalenetrisulphonic acid catalysed by activated carbon in aqueous phase. *Appl Catal B: Environ* 39: 319–329.
115. Álvarez, P.M. García-Araya, J.F. Beltrán, F.J. Masa, F.J. Medina, F. 2005. Ozonation of activated carbons: Effect on the adsorption of selected phenolic compounds from aqueous solutions. *J Colloid Interface Sci* 283: 503–512.
116. Moreno-Castilla, C. Ferro-García, M.A. Joly, J.P. Bautista-Toledo, I. Carrasco-Marín, F. Rivera-Utrilla, J. 1995. AC surface modifications by nitric acid, hydrogen peroxide, and ammonium peroxydisulfate treatments. *Langmuir* 11: 4386–4392.
117. Qiao, W. Korai, Y. Mochida, I. Hori, Y. Maeda, T. 2002. Preparation of an AC artifact: Oxidative modification of coconut shell-based carbon to improve the strength. *Carbon* 40: 351–358.
118. El-Hendawy, A.N.A. 2003. Influence of HNO_3 oxidation on the structure and adsorptive properties of corncob-based AC. *Carbon* 41: 713–722.
119. Haydar, S. Ferro-García, M.A. Rivera-Utrilla, J. Joly, J.P. 2003. Adsorption of p-nitrophenol on an AC with different oxidations. *Carbon* 41: 387–395.
120. Chiang, H.L. Huang, C.P. Chiang, P.C. 2002. The surface characteristics of AC as affected by ozone and alkaline treatment. *Chemosphere* 47: 257–265.
121. Jaramillo, J. Gómez-Serrano, V. Álvarez, P.M. 2009. Enhanced adsorption of metal ions onto functionalized granular ACs prepared from cherry stones. *J Hazard Mat* 161: 670–676.
122. Heidari, A. Younesi, H. Rashidi, A. Ghoreyshi, A.A. 2014. Evaluation of CO_2 adsorption with eucalyptus wood bases activated carbon modified by ammonia solution through heat treatment. *Chem Eng J* 254: 503–513.
123. Deliyanni, E. Bandosz, T.J. 2010. Effect of carbon surface modification with dimethylamine on reactive adsorption of NOx. *Langmuir* 27: 1837–1843.
124. Stuart, B. 2004. *Infrared Spectroscopy: Fundamentals and Applications*. England, U.K.: Wiley and Sons Ltd.
125. Heide, P. 2011. *X-Ray Photoelectron Spectroscopy: An Introduction to Principles and Practices*. Hoboken, NJ: Wiley and Sons Ltd.
126. Williams, D.B. Carter, C.B. 2009. *Transmission Electron Microscopy: A Textbook for Materials Science*. New York: Plenum Publishing Co.
127. Egerton, R.F. 1996. *Electron Energy-Loss in the Electron Microscope*. New York: Plenum Publishing Co.
128. Yuan, J. Brown, L.M. 2000. Investigation of atomic structures of diamond-like amorphous carbon by electron energy loss spectroscopy. *Micron* 31: 515–525.

129. Bernier, N. Bocquet, F. Allouche, A. et al. 2008. A methodology to optimize the quantification of sp^2 carbon fraction from K edge EELS spectra. *J Electron Spectrosc Relat Phenom* 164: 34–43.

130. Boehm, H.P. 1966. Chemical identification of surface groups. In *Advances in Catalysis*, eds. D.D. Eley, H. Pines, P.B. Weisz, 179–274. New York: Academic Press.

131. Goertzan, S.L. Thériault, K.D. Oickle, A.M. Tarasuk, A.C. Andreas, H.A. 2010. Standardization of the Boehm titration. Part I. CO_2 expulsion and endpoint determination. *Carbon* 48: 1252–1261.

132. Oickle, A.M. Goertzan, S.L. Hopper, K.R. Abdalla, Y.O. Andreas, H.A. 2010. Standardization of the Boehm titration: Part II. Method of agitation, effect of filtering and dilute titrant. *Carbon* 48: 3313–3322.

133. Macías-García, A. Valenzuela-Calahorro, C. Gómez-Serrano, V. Espinosa-Mansilla, A. 1993. Adsorption of Pb^{2+} by heat-treated and sulfurized activated carbon. *Carbon* 31: 1249–1255.

134. Gryglewicz, G. Rutkowski, P. Yperman, J. 2004. Characterization of sulfur functionalities of supercritical extracts from coals of different rank, using reductive pyrolysis. *Energy Fuels* 18: 1595–1602.

135. Krishnan, K.A. Anirudhan, T.S. 2003. Removal of cadmium (II) from aqueous solutions by steam activated sulphurised carbon prepared from sugar-cane bagasse pith: Kinetics and equilibrium studies. *Water* 29: 147–156.

136. Le Leuch, L.M. Subrenat, A. Le Cloirec, P. 2003. Hydrogen sulfide adsorption and oxidation onto activated carbon cloths: Applications to odorous gaseous emission treatments. *Langmuir* 19: 10869–10877.

137. Wang, J. Deng, B. Wang, X. Zheng, J. 2009. Adsorption of aqueous Hg(II) by sulfur impregnated active carbon. *Environ Eng Sci* 26: 1693–1699.

138. Jagtoyen, M. Derbyshire, F. 1998. Activated carbons from yellow poplar and white oak by H_3PO_4 activation. *Carbon* 36: 1085–1097.

139. Puziy, A.M. Poddubnaya, O.I. Martínez-Alonso, A. Suárez-García, F. Tascón, J.M.D. 2005. Surface chemistry of phosphorous-containing carbons of lignocellulosic origin. *Carbon* 43: 2857–2868.

140. Inaganki, M. 2010. Structure and texture of carbon materials. In *Carbons for Electrochemical Energy Storage and Conversion Systems*, eds. F. Béguin, E. Frackowiak, pp. 37–76. Boca Raton, FL: CRC Press.

141. Harris, P.J.F. Lui, Z. Suenaga, K. 2008. Imaging the atomic structure of activated carbon. *J Phys Condens Matter* 20: 362201–362205.

142. Menéndez-Díaz, J.A. Martín-Gullón, I. 2006. Surface chemistry of activated carbon and its characterization. In *Activated Carbon Surfaces in Environmental Remediation*, ed. T.J. Bandosz, pp. 1–48. Oxford, U.K.: Elsevier Ltd.

143. Rouquerol, J. Avnir, D. Fairbridge, C.W. et al. 1994. Recommendations for the characterization of porous solids. *Pure Appl Chem* 66: 1739–1758.

144. Lozano-Castelló, D. Suárez-García, F. Carzola-Amorós, D. Linares-Solano, A. 2010. Porous texture of carbons. In *Carbons for Electrochemical Energy Storage and Conversion Systems*, eds. F. Béguin, E. Frackowiak, pp. 115–118. Boca Raton, FL: CRC Press.

145. Gregg, S.J. Sing, K.S.W. 1982. *Adsorption, Surface Area and Porosity*. London, U.K.: Academic Press.

146. Paryjczak, T. 1986. *Gas Chromatography in Adsorption and Catalysis*. Chichester, U.K.: Horwood Ltd.

147. Wunderlich, B. 1990. *Thermal Analysis*. New York: Academic Press.

148. Sing, K. 2001. The use of nitrogen adsorption for the characterization of porous materials. *Colloid Surf A* 31: 3–9.

149. Rouquerol, F. Rouquerol, J. Sing, K.S.W. 1999. Adsorption by powders and porous solids. London, U.K.: Academic Press.

150. Brunauer, S. Emmett, P.H. Teller, E. 1938. Adsorption of gases in multimolecular layers. *J Am Chem Soc* 60: 309–319.

151. Lippens, B.C. de Boer, J.H. 1965. Studies on pore systems in catalysts: V. The t method. *J Catal* 4: 319–323.

152. Centeno, T.A. Stoeckli, F. 2010. The assessment of surface areas in porous carbons by two model-independent techniques, the DR equation and DFT. *Carbon* 48: 2478–2486.

153. Gil, A. Grange, P. 1996. Application of the Dubinin-Radushkevich and Dubinin-Astakhov equations in the characterization of microporous solids. *Colloid Suf A* 113: 39–50.

154. Stoeckli, F. López-Ramón, M.V. Hugi-Cleary, D. Guillot, A. 2001. Micropores sizes in activated carbons determined from the Dubinin-Radushkevich equation. *Carbon* 39: 1115–1116.

155. Olivier, J.P. 1995. Modeling physical adsorption on porous a nonporous solids using density functional theory. *J Porous Mater* 2: 9–17.

156. Lastokie, C. Gubbins, K.E. Quirke, N. 1993. Pore size distribution analysis of microporous carbons: A density functional theory approach. *J Phys Chem* 97: 4786–4796.

157. Neimark, A.V. Lin, Y. Ravikovitch, P.I. Thommes, M. 2009. Quenched solid density functional theory and pore size analysis of micro-mesoporous carbons. *Carbon* 47: 1617–1628.

158. Thomson, T.K. Gubbins, K.E. 2000. Modeling structural morphology of microporous carbons by reverse Monte Carlo. *Langmuir* 16: 5761–5773.

159. Palmer, J.C. Brennan, J.K. Hurley, M.M. Balboa, A. Gubbins, K.E. 2009. Detailed structural models for activated carbons from molecular simulation. *Carbon* 47: 2904–2913.

160. Hesas, R.H. Arami-Niya, A. Daud, W.M.A.W. Sahu, J.N. 2013. Preparation of granular activated carbon from oil palm shell by microwave-induced chemical activation: Optimisation using surface response methodology. *Chem Eng Res Des* 91: 2447–2456.

161. Gil, M.V. Martínez, M. García, S. Rubiera, F. Pis, J.J. Pevida, C. Response surface methodology as an efficient tool for optimizing carbon adsorbents for CO_2 capture. *Fuel Proc Technol* 106: 55–61.

162. Garba, Z.N. Rahim, A.A. 2014. Process optimization of $K_2C_2O_4$-activated carbon from *Prosopis africana* seed hulls using response surface methodology. *J Anal Appl Pyrol* 107: 306–312.

163. Sych, N.V. Trofymenko, S.I. Poddubnaya, O.I. 2012. Porous structure and surface chemistry of phosphoric acid activated carbon from corncob. *Appl Surf Sci* 261: 75–82.

164. Benadjemia, M. Milliere, L. Reinert, L. Benderdouche, N. Duclaux, L. 2011. Preparation, characterization and methylene blue adsorption of phosphoric acid activated carbons from globe artichoke leaves. *Fuel Proc Technol* 92: 1203–1212.

165. Azevedo, D.C.S. Araújo, J.C.S. Bastos-Neto, M. et al. 2007. Microporous activated carbon prepared from coconut shells using chemical activation with zinc chloride. *Micropor Mesopor Mater* 100: 361–364.

166. Hayashi, J. Horikawa, T. Moruyama, K. Gomes, V.G. 2002. Activated carbon from chickpea husk by chemical activation with K_2CO_3: Preparation and characterization. *Micropor Mesopor Mater* 55: 63–68.

167. Moreno-Castilla, C. Rivera-Utrilla, J. 2001. Carbon materials as adsorbents for the removal of pollutants from the aqueous phase. *Mater Res Soc Bull* 26: 890–894.

168. Meidl, J.A. 1997. Responding to changing conditions: How powdered activated carbon systems can provide the operational flexibility necessary to treat contaminated groundwater and industrial wastes. *Carbon* 35: 1207–1216.

169. Moreno-Castilla, C. 2004. Adsorption of organic molecules from aqueous solutions on carbon materials. *Carbon* 42: 83–94.

170. Romero-González, M. Williams, C.J. Gardiner, P.H.E. 2001. Study of the mechanisms of cadmium biosorption by dealginated seaweed waste. *Environ Sci Technol* 35: 3025–3030.

171. Yun, Y.S. Park, D. Park, J.M. Volesky, B. 2001. Biosorption of trivalent chromium on the brown seaweed biomass. *Environ Sci Technol* 35: 4353–4358.

172. Min, S.H. Han, J.S.E.W. Park, J.K. 2004. Improvement of cadmium ion removal by base treatment of juniper fiber. *Water Res* 38: 1289–1295.

173. Bailey, S.E. Olin, T.J. Bricka, M. Adrian, D.D. 1999. A review of potentially low-cost sorbents for heavy metals. *Water Res* 33: 2469–2479.

174. González, P.G. Pliego-Cuervo, Y.B. 2014. Adsorption of Cd(II), Hg(II) and Zn(II) from aqueous solution using mesoporous activated carbon produced from Bambusa vulgaris striata. *Chem Eng Res Des* 92: 2715–2724.

175. W.H.O. 2011. Guidelines for drinking-water quality. Geneva, Switzerland: World Health Organization Press.

176. Kong, J. Yue, Q. Sun, S. et al. 2014. Adsorption of Pb(II) from aqueous solution using keratin waste-hide waste: Equilibrium, kinetic and thermodynamic modeling studies. *Chem Eng J* 241: 393–400.

177. Momcilvic, M. Purenovic, M. Bojic, A. Zarubica, A. Randelovic, M. Re, oval of lead(II) ions from aqueous solutions by adsorption into pine cone activated carbon. *Desalination* 276: 53–59.

178. Bulut, Y. Tez, Z. 2007. Adsorption studies on ground shells of hazelnut and almond. *J Hazard Mater* 149: 35–41.

179. Liu, S.X. Chen, X. Che, X.Y. Liu, Z.F. Wang, H.L. 2007. Activated carbon with excellent chromium(VI) adsorption performance prepared by acid-base surface modification. *J Hazard Mater* 141: 315–319.

180. López-Ramón, V. Moreno-Castilla, C. Rivera-Utrilla, J. Radovic, L.R. 2002. Ionic strength effects in aqueous phase adsorption of metal ions on activated carbons. *Carbon* 4: 2020–2022.

181. Rafatullah, M. Sulaiman, O. Hashim, R. Ahmad, A. Adsorption of methylene blue on low-cost adsorbents: A review. *J Hazard Mater* 177: 70–80.

182. Crini, G. 2006. Non-conventional low-cost adsorbents for dye removal: A review. *Biores Technol* 97: 1061–1085.

183. Conway, B.E. Birss, V. Wojtowisz, J. 1997. The role and utilization of pseudocapacitance for energy storage by supercapacitors. *J Power Source* 66: 1–14.

184. Noked, M. Soffer, A. Aurbach, D. 2011. The electrochemistry of activated carbonaceous materials: Past, present, and future. *J Solid State Electrochem* 15: 1563–1578.

185. Bleda-Martínez, M.J. Maciá-Agulló, J.A. Lozano-Castelló, D. et al. 2005. Role of the surface chemistry on the electric double layer capacitance of carbon materials. *Carbon* 43: 2677–2684.

186. Bleda-Martínez, M.J. Lozano-Castelló, D. Morallón, E. Cazorla-Amorós, D. Linares-Solano, A. 2006. Chemical and electrochemical characterization of porous carbon materials. *Carbon* 44: 2642–2651.
187. Qu, D. 2002. Studies of the activated carbons used in double-layer supercapacitors. *J Power Source* 109: 403–411.
188. Lazzari, M. Soavi, F. Mastragostino, M. 2009. Dynamic pulse power and energy of ionic-liquid-based supercapacitor for HEV application batteries and energy storage. *J Electrochem Soc* 156: A661–A666.

CHAPTER 11

CONTENTS

Fluorescence Sensing by Functionalized Carbon Dots Nanoparticles

11

João M.M. Leitão, Eliana F.C. Simões,
and Joaquim C.G. Esteves da Silva

11.1 INTRODUCTION

Fluorescent carbon dots (CDs) nanoparticles are usually referred as the natural and sustainable alternative to the semiconductor quantum dots (QDs). Indeed, CDs can easily be synthesized from renewable biomass. Compared to QDs, they have a higher solubility in water, have higher resistance to photo degradation, show a nonblinking luminescence, and have lower toxicity and generally good biocompatibility. The color-tuning properties in the visible wavelength range and quantum yields (QYs) of CDs are usually referred as less favorable properties when compared with QDs, but recent research advances are markedly improving these characteristics of CDs (Esteves da Silva 2011, 2012; Zhang et al. 2013).

Since their accidental discovery in 2004 in arc-discharged soot (Xu et al. 2004), other techniques have been used in CDs synthesis, like top-down physical methods and bottom-up chemical methodologies, and their multiple applications in diverse fields of science (Baker and Baker 2010; Esteves da Silva et al. 2011; Li et al. 2012; Ruedas-Rama et al. 2012; Luo et al. 2013). Some of the synthesized CDs are intrinsically fluorescent, but they also can be modified to alter their luminescence properties (Zhang et al. 2013; Ding et al. 2014). The carbon surface of bare CDs can be easily modified by oxidation, passivation, and/or functionalization with different types of compounds that modified the intrinsically fluorescent properties. Indeed, the properties of bare CDs can be changed only by oxidation; passivation of their surface; functionalization, by linking some species to the surface of bare CDs; or yet functionalization of oxidized or passivated CDs. The passivation and/or functionalization of bare CDs is done in order to obtain fluorescence, increasing the QY, to achieve sensitive and/or selective sensing of some compounds, or for the delivery of some therapeutic drug.

The characteristic feature of CDs is the intrinsically fluorescent properties of the synthesized bare or functionalized CDs that confers them a particularly useful role in sensing and/or bioimaging (Esteves da Silva et al. 2011; Li et al. 2012; Ruedas-Rama et al. 2012; Luo et al. 2013; Zhang et al. 2013; Ding et al. 2014). The increasing trend in the relatively high QY and recent advances in the tunable fluorescent properties open new areas for CDs applications that could be yet enlarge as a consequence of their functionalization. In preliminary applications of CDs, the nonfluorescent raw CDs, were surface passivated with diamine-terminated poly(ethylene glycol), with an average molecular weight of 1500 (PEG1500N), and poly(propionylethyleneimine-*co*-ethyleneimine), an ethyleneimine copolymer (PPEI-EI), was then used for bioimaging (Sun et al. 2006; Cao et al. 2007). Since these first applications, the CDs have been used for drug delivery, biological labeling, biomedicine, inks, solar cells patterning, and optoelectronic devices in electronic, photonic, energy, catalysis, and medicine fields.

Beside the former top-down methods of synthesis, several bottom-up methods as thermal decomposition, ultrasound, microwave, autoclave, and acid dehydration and similar top-down methods of synthesis were developed for the synthesis of raw CDs and/or posterior passivation and/or functionalization (Esteves da Silva et al. 2011; Li et al. 2012; Zhang et al. 2013). All the bottom-up chemical methods are based on thermal or acid digestion of some CDs precursor. Several different types of CDs precursors have been used. Based on these methods of synthesis, the functionalized CDs could be obtained in one or more steps of synthesis.

The focus of this chapter is the fluorescence sensing by functionalized CDs, including the surface oxidation and passivation. This chapter presents an overview of the main functionalized CDs designed as nanosensors for selected analytical species. The CDs fluorescence sensing will be discussed as well as the methods of synthesis of the functionalized CDs, the functionalization labels, and the target species.

11.2 CDs FLUORESCENCE MECHANISM

CDs belong to the family of carbon nanomaterials together with nanodiamonds, graphene, nanotubes, and fullerenes. CDs are mainly composed of graphitic carbon, that is, the predominate carbon atoms have a sp^2 hybridization and show high percentages of oxygen functional groups. These structural characteristics contrast, for example, with nanodiamonds, where carbon has a sp^3 hybridization, but are similar to nanographene (viz., the oxidized form, graphene oxide).

Both CDs and nanographene (particularly graphene oxide) nanoparticles are fluorescent nanomaterials that have the characteristic feature of multicolor emission when different excitation wavelengths are used (Liu et al. 2007, 2009; Peng and Travas-Sejdic 2009; Zhu et al. 2009; Li et al. 2010; Mao et al. 2010, 2014; Kozák et al. 2013; Cushing et al. 2014). This property is a major advantage of these fluorescent nanomaterials because it allows tuning of the emission spectra simply by changing the excitation wavelength (keeping the size of the nanoparticle). Indeed, inorganic semiconductors like the QDs show the *normal* excitation-independent emission spectra, and, because they have a strong quantum confinement effect, the emission can be tuned by varying the size of the nanoparticle (Esteves da Silva 2011, 2012).

The justification for the existence of fluorescence in graphite-like nanomaterials is still not clearly understood, but it has long been proposed that it should be due to surface defects, which originates energy levels that justifies the light emission. These levels would allow radiative recombination of excitons (Peng and Travas-Sejdic 2009; Zhu et al. 2009). The existence of surface defect, and the corresponding energy states/bands, was proposed based on the observations that the fluorescence appears and/or the QY increased when the nanoparticles are subjected to polymer passivation and nitric acid treatment (Liu et al. 2007, 2009; Peng and Travas-Sejdic 2009; Zhu et al. 2009; Li et al. 2010; Mao et al. 2010).

Recently, the origin of the strong excitation dependence of the emission spectra of graphene oxide has been attributed to *giant red-edge effect* (Cushing et al. 2014). Taking into consideration the structural and chemical similarities of graphene oxide and CDs, the existence of a similar *edge effect* in CDs is expected (the previously discussed surface defects), and the resulting edge states/ bands are responsible for the existence of unique fluorescence properties.

11.3 CDs FLUORESCENCE SENSING

The fluorescence sensing by functionalized CDs nanoparticles is being done in the analysis of different analytical ionic or molecular species using CDs synthesized by different methods of different carbon precursors. In sensor development, the main objective of the functionalization is to increase the sensitivity and selectivity. The main fluorescence strategies used and the kind of analytical species evaluated by functionalized CDs are presented in Figure 11.1.

Most of the analytical applications have been done in the quantification of ionic species, namely, cations. Also different types of biological molecules, some of them small molecules, were the focus of the fluorescence sensing by functionalized CDs. Besides the fluorescence sensing based on quenching or fluorescence enhancement of the intrinsic CDs fluorescence, also, ratiometric analytical, turn off/on, turn on/off, and fluorescence resonance energy transfer (FRET) fluorescence sensing strategies have been proposed.

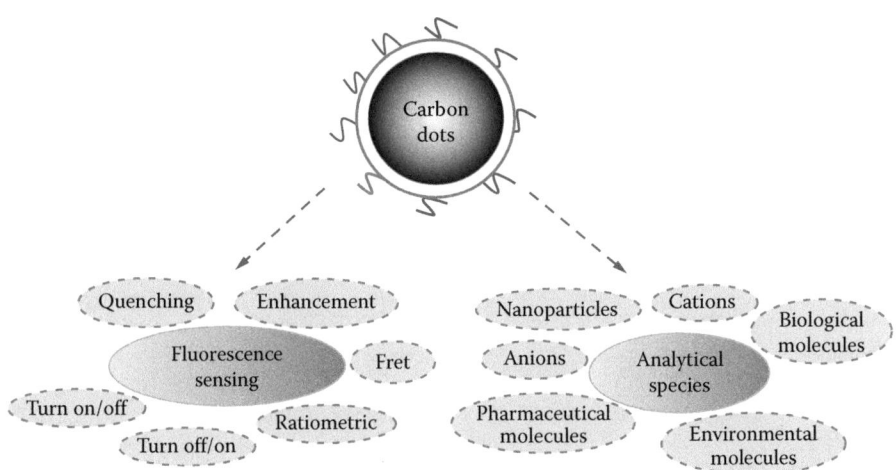

FIGURE 11.1
Fluorescence sensing strategies and analytical species evaluated by functionalized CDs.

11.3.1 Ionic Species

Several types of ionic species have been sensed by functionalized CDs. Tables 11.1 through 11.4 present the cationic species and Table 11.5 the anionic species analyzed using fluorescence sensing by functionalized CDs, the functionalization label used, the methods of synthesis of the functionalized CDs, and the observed limit of detection (LOD). Examples of the fluorescence sensing strategies for the ionic species evaluation by functionalized CDs are presented in Figure 11.2.

The highest number of analytical applications of functionalized CDs was in the fluorescence sensing of cationic species. Fewer applications were observed in the fluorescence sensing of anionic species. Different functionalization labels were used for the fluorescence sensing by the functionalized CDs of the ionic species. In most of the applications, CDs are functionalized with some amine. Other less specific functionalization/passivation labels for the ionic sensing of CDs are some types

TABLE 11.1 Cu^{2+} Fluorescence Sensing by Functionalized CDs Nanoparticles

References	Functionalization Label	CDs Synthesis	CDs Functionalization	LOD (μM)
Liu et al. (2014a)	AEAPMS/Si nanoparticles doped with rhodamine B	Thermic	Mixing/two steps	0.0352
Zhu et al. (2014)	HNO_3/AE-TPEA	Electrochemical	1-Ethyl-3-(3-dimethylaminopropyl) carbodiimide hydrochloride (EDC)–NHS/two steps	—
Zhang et al. (2014)	HNO_3	Thermic	Thermic/one step	—
Liu et al. (2014b)	Branched polyethylenimine (BPEI)	Autoclave	Mixing/two steps	0.115
Salinas-Castillo et al. (2013)	PEI	Microwave	Microwave/one step	0.09/0.12
Dong et al. (2012)	BPEI	Thermic	Thermic/one step	0.006
Liu et al. (2012)	PEG200/bovine serum albumin (BSA)/lysine	Microwave	EDC–NHS/mixing/three steps	—
Zhu et al. (2012)	CdSe-ZnS/AE-TPEA	Electrochemical	Mixing/EDC–NHS/three steps	1
Qu et al. (2012)	AE-TPEA	Electrochemical	EDC–NHS/two steps	0.010
Gonçalves et al. (2010)	HNO_3/PEG200N/NAC	Laser ablation	Thermic refluxing/three steps	—

TABLE 11.2 Hg^{2+} Fluorescence Sensing by Functionalized CDs Nanoparticles

References	Functionalization Label	CDs Synthesis	CDs Functionalization	LOD (μM)
Yuan et al. (2014)	4,7,10-Trioxa-1,13-tridecanediamine (TTDDA)/ bis(dithiocarbamato) copper (II) complex (CuDTC$_2$)	Thermic	Mixing/two steps	0.05
Yan et al. (2014)	1,2-Ethylenediamine (EDA)	Autoclave	Autoclave/one step	0.226
—	AEAPMS	Thermic	Thermic/one step	0.845
Zhang and Chen (2014)	Ethylene glycol (EG)	Autoclave	Autoclave/one step	0.23
Gonçalves et al. (2010a,b, 2012)	HNO$_3$/PEG200N/NAC	Laser ablation	Thermic refluxing/ three steps	0.1/0.1/0.01
Li et al. (2011b)	FAM/ssDNA	Thermic	Adsorption/two steps	0.01

TABLE 11.3 H^+ Fluorescence Sensing by Functionalized CDs Nanoparticles

References	Functionalization Label	CDs Synthesis	CDs Functionalization	LOD (μM)
Du et al. (2013a)	EDA/fluorescein isothiocyanate (FITC)	Microwave	Mixing/two steps	—
Nie et al. (2014)	Diethylamine (DEA)/FITC	Thermic refluxing	Mixing/two steps	—
Kong et al. (2014)	AEAPMS	Thermic	Thermic/one step	—
Qu et al. (2013)	Urea	Microwave	Microwave/one step	—
Shen et al. (2013)	HNO$_3$/PEI	Thermic refluxing	Thermic refluxing/ one step	—
Shi et al. (2012)	TTDDA/FITC/rhodamine B isothiocyanate (RBITC)	Thermic	Mixing/two steps	—
Kong et al. (2012)	4'-(Aminomethylphenyl)-2,2':6', 2''-terpyridine (AE-TPY)	Electrochemical	EDC–NHS/two steps	—
Gonçalves et al. (2010)	HNO$_3$/PEG200N/NAC	Laser ablation	Thermic refluxing/ three steps	—
Gonçalves and Esteves da Silva (2010)	HNO$_3$/PEG200N/ mercaptosuccinic acid (MSA)	Laser ablation	Thermic refluxing/ three steps	—

of poly(ethylene glycol) or oxidation by a strong acid. Most CDs are synthesized by a bottom-up method of synthesis based on a thermal digestion of a carbon precursor. Usually, the CDs functionalization is done in one step by thermal treatment of the carbon precursor and the functionalization label. Another common method of CDs functionalization is the carbodiimide cross-linking chemistry that allows covalent bonding of a carboxyl group in the CDs surface with an amine group of the functionalization label.

The principal cationic species analyzed by fluorescence sensing of functionalized CDs were Cu^{2+} (Gonçalves et al. 2010b; Dong et al. 2012; Liu et al. 2012, 2014a,b; Qu et al. 2012; Zhu et al. 2012; Salinas-Castillo et al. 2013; Zhu et al. 2014; Zhang et al. 2014), Hg^{2+} (Gonçalves et al. 2010a,b, 2012; Li et al. 2011b; Yan et al. 2014; Yuan et al. 2014; Zhang and Chen 2014), and pH (Gonçalves and Esteves da Silva 2010; Gonçalves et al. 2010b; Kong et al. 2012; Shi et al. 2012; Du et al. 2013a; Shen et al. 2013; Qu et al. 2013; Kong et al. 2014; Nie et al. 2014). Besides these cations, Fe^{3+}, K^+, Ca^{2+}, Ag^+, and Be^{2+} and the organometallic cation CH_3Hg^+ were also synthesized by functionalized CDs for their fluorescence sensing (Sun et al. 2010; Li et al. 2011a; Wei et al. 2012; Krishna et al. 2013; Qu et al. 2013; Zhu et al. 2013; Algarra et al. 2014; Costas-Mora et al. 2014; Li et al. 2014; Song et al. 2014).

TABLE 11.4 Fluorescence Sensing of Other Cationic Species by Functionalized CDs Nanoparticles

References	Cationic Species	Functionalization Label	CDs Synthesis	CDs Functionalization	LOD (μM)
Song et al. (2014)	Fe^{3+}	EDA	Autoclave	Autoclave/one step	—
Zhu et al. (2013)	—	EDA	Autoclave	Autoclave/one step	17.91
Qu et al. (2013)	—	Urea	Microwave	Microwave/one step	0.04
Sun et al. (2010)	—	2-(2-Aminoethoxy)-ethanol	Autoclave	Autoclave/one step	11.2
Li et al. (2014)	—	Urea/polyvinyl pyrrolidone (PVP), polyacrylamide (PAM), or polyacrylic acid (PAA)	Autoclave	Autoclave/two steps	—
Wei et al. (2012)	K^+	EDA	Thermic refluxing	EDC–NHS/two steps	10
Krishna et al. (2013)	Ca^{2+}	PEG1500N/glutamic acid and hyaluronic acid	Thermic	EDC/two steps	—
Algarra et al. (2014)	Ag^+	H_2SO_4/MSA	Ultrasonic	Mixing/two steps	0.386
Li et al. (2011a)	—	ROX dye labeled ssDNA	Ultrasonic	Mixing/two steps	0.0005
Li et al. (2014)	Be^{2+}	Urea/PVP, PAM, or PAA	Autoclave	Autoclave/two steps	23
Costas-Mora et al. (2014)	CH_3Hg^+	PEG200	Ultrasonic	Ultrasonic/one step	0.0059

TABLE 11.5 Fluorescence Sensing of Anionic Species by Functionalized CDs Nanoparticles

References	Anionic Species	Functionalization Label	CDs Synthesis	CDs Functionalization	LOD (μM)
Liu et al. (2013)	F^-	PEG200/$Zr(H_2O)_2$EDTA	Microwave	Mixing/two steps	0.031
Du et al. (2013b)	I^-	EDA	Microwave	Microwave/one step	0.430
Gonçalves and Esteves da Silva (2010)	—	HNO_3/PEG200N/ MSA	Laser ablation	Thermic refluxing/ two steps	—
Hou et al. (2013)	S^{2-}	EDA/N,N'-bis(2-(ethylthio)ethyl) propane-1,3-diamine	Microwave	EDC–NHS/two steps	0.78
Zhang et al. (2014)	$C_2O_4^{2-}$	HNO_3	Thermic	Thermic/one step	1.0
Zhao et al. (2011)	PO_4^{3-}	11-Aminoundecanoic acid	Thermic refluxing	Thermic/one step	0.051

The sensing of these cationic species is quite important in environmental analysis where some of these cationic species are environmental pollutants, like the inorganic mercury ion (Hg^{2+}) and methylmercury (CH_3Hg^+). Heavy metal ions contamination, including Cu^{2+}, is monitored due to their critical role in some physiological and pathological events. Cu^{2+} ion sensing is based on the quenching of the fluorescence of CDs at the maximum excitation and emission wavelengths. For the polyethylenimine (PEI)-CDs, due to their upconversion properties, and in order to eliminate the Fe^{3+} interference, the Cu^{2+} fluorescence sensing was done at a different wavelength (850 nm) (Salinas-Castillo et al. 2013). Two of the developed CDs sensors are ratiometric and are based on the calculation of a fluorescence ratio of the quenching provoked by Cu^{2+} concentration and a constant fluorescence reference signal, in order to correct some instrumental variations or sample effects

(Liu et al. 2014a). In one of the sensors, rhodamine B-doped Si nanoparticles were coated with functionalized CDs, and the quenching effect at 467 nm of the functionalized CDs was compared with the fluorescence emission of the rhodamine B at 585 nm (Liu et al. 2014a). In the other sensor, a nanocomposite of CDs and CdSe/ZnS QD were functionalized with *N*-(2-aminoethyl)-*N*,*N*,*N*′-tris(pyridin-2-ylmethyl)ethane-1,2-diamine (AE-TPEA). The quenching effect of the nanocomposite CdSe/ZnS-C-TPEA at 485 nm was compared with the fluorescence emission at 644 nm of CdSe/ZnS embedded in silica shells (Zhu et al. 2012).

Most of the Hg^{2+} fluorescence sensing by the functionalized CDs is also based on the fluorescence dynamic quenching by the Hg^{2+} (Gonçalves et al. 2010a,b, 2012; Yan et al. 2014; Zhang and Chen 2014). An optical sensor was developed based on the dynamic fluorescence extinction of the *N*-acetyl-ʟ-cysteine (NAC)–amine poly(ethylene glycol)-200 (PEG200N)–CDs after immobilization of the functionalized CDs sensor in the tip of the optical fiber by a sol-gel (Gonçalves et al. 2010a) and by a layer-by-layer deposition technique (Gonçalves et al. 2012). Also, the optical fiber detection of Cu^{2+} with a lower sensitivity, considered as interfering on the Hg^{2+} sensing, was evaluated (Gonçalves et al. 2010a, 2012). Based on the enhancement of fluorescence, two different turn off/on sensors were developed for the fluorescence sensing of Hg^{2+} (Li et al. 2011b; Yuan et al. 2014): the recovery of fluorescence by the Hg^{2+} was achieved after the quenching of the CDs induced by the linked $CuDTC_2$ complex (Yuan et al. 2014) and the quenching of the fluorescence of linked fluorescein amidite (FAM)–ssDNA induced by the CDs (Li et al. 2011b).

An extensive work has been done in order to develop functionalized CDs as fluorescence sensor for pH. A tunable fluorescent emission functionalized CD was synthesized for the pH fluorescence sensing (Nie et al. 2014). The developed sensors are based on the decrease (Shi et al. 2012; Du et al. 2013a; Qu et al. 2013; Kong et al. 2014; Nie et al. 2014) or in the increase (Gonçalves and Esteves da Silva 2010; Gonçalves et al. 2010b; Kong et al. 2012; Shen et al. 2013) of the fluorescence intensity of a pH-sensitive label with

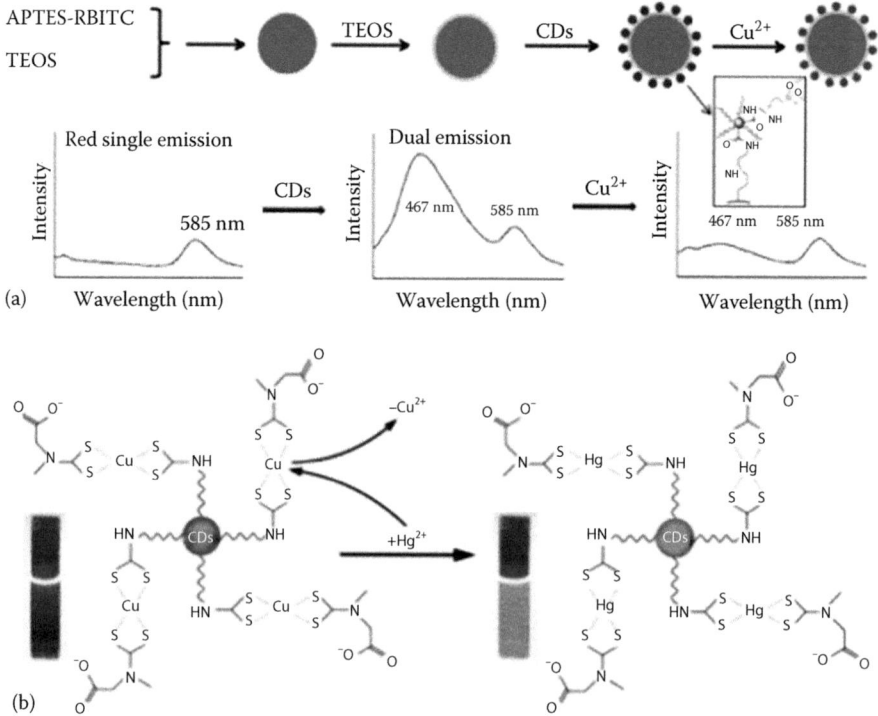

FIGURE 11.2
Examples of fluorescence sensing strategies for the ionic species evaluation by functionalized CDs nanoparticles. (a) Cu^{2+} ratiometric sensor. (Reprinted with permission from Liu, X., Zhang, N., Bing, T., and Shangguan, D., Carbon dots based dual-emission silica nanoparticles as a ratiometric nanosensor for Cu(2+), *Anal. Chem.*, 86(5), 2289. Copyright 2014a American Chemical Society.) (b) Hg^{2+} turn off/on sensor. (Reprinted with permission from Yuan, C., Liu, B., Liu, F., Han, M. Y., and Zhang, Z., Fluorescence "turn on" detection of mercuric ion based on bis(dithiocarbamato) copper(II) complex functionalized carbon nanodots. *Anal. Chem.*, 86(2), 1123. Copyright 2014 American Chemical Society.) *(Continued)*

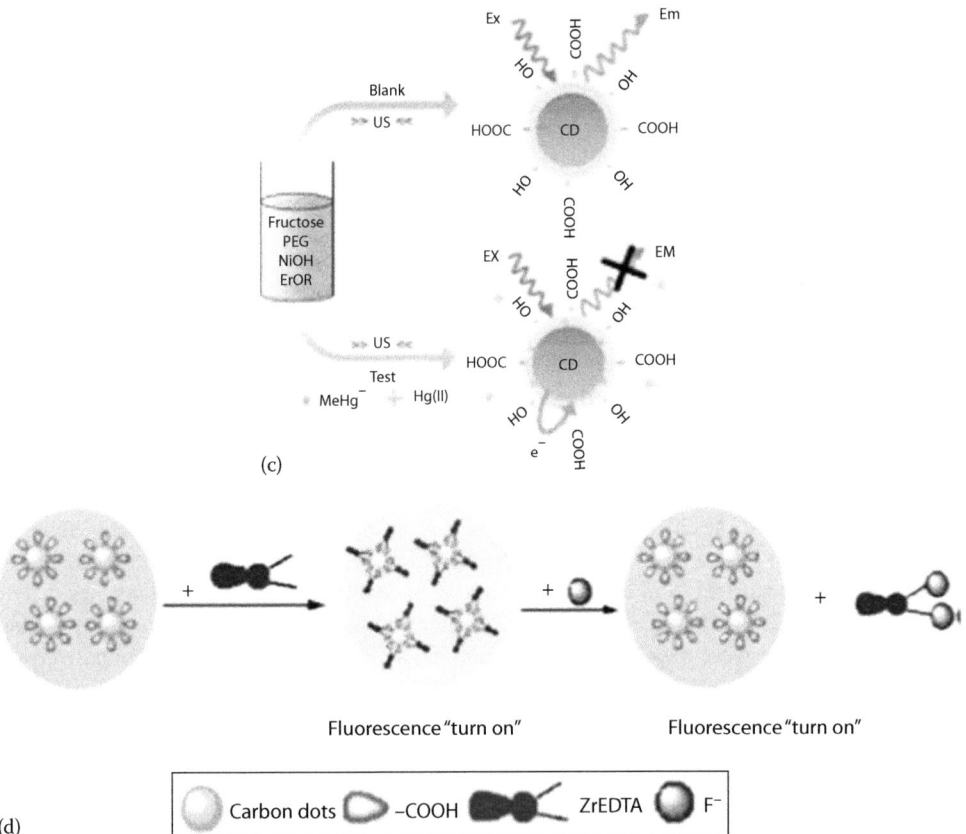

FIGURE 11.2 (Continued)
Examples of fluorescence sensing strategies for the ionic species evaluation by functionalized CDs nanoparticles. (c) CH_3Hg^+ in situ fluorescence quenching sensor. (Reprinted with permission from Costas-Mora, I., Romero, V., Lavilla, I., and Bendicho, C., In situ building of a nanoprobe based on fluorescent carbon dots for methylmercury detection, *Anal. Chem.*, 86(9), 4536. Copyright 2014 American Chemical Society.) (d) F^- turn on/off sensor. (Liu, J. M., Lin, L. P., Wang, X. X. et al., $Zr(H_2O)2EDTA$ modulated luminescent carbon dots as fluorescent probes for fluoride detection, *Analyst*, 138(1), 278, 2013. Reproduced by permission of The Royal Society of Chemistry.)

the increase of the H^+ concentration. The FICT is one of the preferentially used pH-sensitive labels. Some of the pH sensors are ratiometric sensors (Shi et al. 2012; Du et al. 2013a; Nie et al. 2014). In all of these ratiometric sensors, the calculated ratio decreases with the proton concentration. In two of these sensors, the ratio of the fluorescence intensity of a pH-sensitive label and of the CDs (Du et al. 2013a; Nie et al. 2014) is calculated, and in the other, the ratio of the fluorescence intensity of two pH-sensitive labels used for the functionalization (Shi et al. 2012) is used. One of the ratiometric sensors developed is also a FRET-based sensor where the CD also acts as a donor and the FICT, used as pH-sensitive label, also acts as an acceptor of a FRET mechanism (Du et al. 2013a).

Besides these species (Cu^{2+}, Hg^{2+}, and pH), the majority of the fluorescence sensing of other cations studied by fluorescence functionalized CDs are based on the fluorescence quenching induced by them (Sun et al. 2010; Krishna et al. 2013; Qu et al. 2013; Zhu et al. 2013; Algarra et al. 2014; Costas-Mora et al. 2014; Li et al. 2014; Song et al. 2014). Also, based on the fluorescence quenching, CDs were in situ synthesized in order to sense CH_3Hg^+ (Costas-Mora et al. 2014), and, based on the fluorescence quenching at two wavelengths (455 and 520 nm), a ratiometric sensor was developed for the Fe^{3+} sensing (Qu et al. 2013). Based on the enhancement of fluorescence intensity, the sensing of K^+, by a FRET pair of fluorescence functionalized CDs and graphene and an ion-selective crown ether (Wei et al. 2012), and the sensing of Ag^+, by a dye-labeled single-stranded DNA (ssDNA) (Li et al. 2011a), were done.

The anionic fluorescence sensing by functionalized CDs nanoparticles was done for the F^-, I^-, S^{2-}, $C_2O_4^{2-}$, and PO_4^{3-} anions (Table 11.5). The main strategy used for the fluorescence anion sensing is based on a previous fluorescence enhancement or quenching followed by a fluorescence quenching or enhancement, respectively, by the anion under analysis (Zhao et al. 2011; Du et al.

2013b; Hou et al. 2013; Liu et al. 2013; Zhang et al. 2014). Fluoride sensing was done after a previous enhancement of CDs fluorescence by the $Zr(H_2O)_2EDTA$ (Liu et al. 2013). Also, the sensing of S^{2-} and $C_2O_4^{2-}$ was done after the CDs fluorescence quenching by the Cu^{2+} (Hou et al. 2013; Zhang et al. 2014), I^-, and PO_4^{3-}, after the CDs fluorescence quenching by the Hg^{2+} (Du et al. 2013b) and Eu^{3+} (Zhao et al. 2011) cations.

11.3.2 Molecular Species

Several types of molecular species had been fluorescence sensed by functionalized CDs. The biological, pharmaceutical, and the environmental molecular species evaluated by fluorescence sensing, the functionalization label used, the methods of synthesis of the functionalized CDs, and the LOD found are presented in Tables 11.6 through 11.8. Examples of the fluorescence sensing strategies for the molecular species evaluation by functionalized CDs are presented in Figure 11.3.

The main applications of functionalized CDs were for the fluorescence sensing of molecular biological species (Li et al. 2011c; Liu et al. 2011; Mao et al. 2012; Xu et al. 2012; Baruah et al. 2013; Wang et al. 2013; Yeh et al. 2013; Yu et al. 2013a,b; Shan et al. 2014; Shi et al. 2014; Song et al. 2014; Wang et al. 2014; Zhu et al. 2014). Fewer examples were found in the fluorescence sensing of molecular pharmaceutical (Sun et al. 2010; Yang et al. 2014) or environmental species (Cayuela et al. 2013; Bu et al. 2014; Huang et al. 2014). A greater number of functionalization labels specific for a certain molecule were used for the molecular fluorescence sensing by functionalized CDs. Except for one top-down electrochemical CDs synthesis, all the CDs are also synthesized by a bottom-up method of synthesis based in a thermal digestion of a carbon precursor. As for the ionic species, the CDs functionalization was also mainly done in one-step thermal treatment of both the carbon precursor and the functionalization label. The CDs functionalization based on the carbodiimide cross-linking chemistry was also frequently used for the CDs functionalization.

Glucose is one of the chemical species most frequently screened in clinical analysis, and it was also the focus of the fluorescence sensing by functionalized CDs. Both of the developed sensors

TABLE 11.6 Fluorescence Sensing of Biological Molecules by Functionalized CDs Nanoparticles

References	Molecular Species	Functionalization Label	CDs Synthesis	CDs Functionalization	LOD (μM)
Shan et al. (2014)	Glucose	BBr_3	Autoclave	Autoclave/one step	8
Yeh et al. (2013)	—	RGO	Autoclave	Autoclave/one step	0.140/0.230
Wang et al. (2013)	Acetylcholine	RGO	Autoclave	Autoclave/one step	3×10^{-5}
Mao et al. (2012)	Dopamine	AEAPMS/APTES/ TEOS	Thermic	Mixing/two steps	0.017
Song et al. (2014)	Hemin	EDA	Autoclave	Autoclave/one step	—
Baruah et al. (2013)	—	β-cyclodextrin/lactose monohydrate/ sucrose	Microwave	Mixing/two steps	—
Liu et al. (2011)	Thrombin	FAM-labeled thrombin aptamer	Thermic	Mixing/two steps	0.020
Xu et al. (2012)	—	Thrombin aptamers (TBA_{15}/TBA_{29})	Microwave	EDC–NHS/two steps	0.001
Wang et al. (2014)	Insulin	HNO_3	Thermic	Thermic/one step	—
Li et al. (2011c)	dsDNA	FAM-labeled ssDNA	Thermic	Mixing/two steps	—
Shi et al. (2014)	GSH	2,20-(Ethylene-dioxy) bis(ethylamine)	Microwave	Microwave/one step	0.050

TABLE 11.7 Fluorescence Sensing of Small Biological Molecules by Functionalized CDs Nanoparticles

References	Molecular Species	Functionalization Label	CDs Synthesis	CDs Functionalization	LOD (μM)
Yu et al. (2013b)	NO	EDA/naphthalimide with an o-phenylenediamine group	Microwave	EDC–N-hydroxy sulfosuccinimide (sulfo-NHS)/two steps	0.003
Shan et al. (2014)	H_2O_2	BBr_3	Autoclave	Autoclave/one step	—
Song et al. (2014)	—	EDA	Autoclave	Autoclave/one step	0.0265
Yeh et al. (2013)	—	RGO	Autoclave	Autoclave/one step	—
Wang et al. (2013)	—	RGO	Autoclave	Autoclave/one step	3×10^{-5}
Zhu et al. (2014)	H_2S	HNO_3/AE-TPEA-Cu^{2+}	Electrochemical	EDC–NHS/two steps	0.7
Yu et al. (2013a)	—	EDA/naphthalimide azide	Microwave	EDC–sulfo-NHS/two steps	0.01

TABLE 11.8 Fluorescence Sensing of Pharmaceutical, Environmental, and Biological Molecules by Functionalized CDs Nanoparticles

References	Molecular Species	Functionalization Label	CDs Synthesis	CDs Functionalization	LOD (μM)
Pharmaceutical					
Sun et al. (2010)	Ferrous succinate	2-(2-Aminoethoxy)-ethanol	Autoclave	Autoclave/one step	11.2
Yang et al. (2014)	Tetracyclines	P_2O_5	Mixing	Mixing/one step	0.042–0.075
Environmental					
Huang et al. (2014)	HClO/ClO$^-$	H_3PO_4	Microwave	Mixing/two steps	0.015
Bu et al. (2014)	PBB15	PEG-200/antigen PBB15	Microwave	Microwave/one step	0.125
Cayuela et al. (2013)	DNP/DMQ	H_2SO_4/HNO_3/acetone	Thermic refluxing	Thermic/two steps	2.176/5.676

were also H_2O_2 sensors, and the glucose fluorescence sensing was obtained indirectly through sensing of H_2O_2 after the oxidation of glucose by the glucose oxidase in gluconic acid and H_2O_2 (Yeh et al. 2013; Shan et al. 2014). In one of the developed glucose sensors, the H_2O_2 induces a decrease of fluorescence of boron-doped CDs (Shan et al. 2014), and in the other, the H_2O_2 induces an increase of fluorescence of CDs attached to reduced graphene oxide (RGO) by reaction of the H_2O_2 with reactive oxygen species that quenched the fluorescence of the RGO-CDs (Yeh et al. 2013).

Using the same mechanism for glucose sensing (Yeh et al. 2013), the fluorescence sensing of the acetylcholine neurotransmitter by RGO-CDs was successfully achieved using the acetylcholinesterase and the choline oxidase enzymes (Wang et al. 2013). However, in this last work, a different fluorescence sensing strategy is based on the fluorescence quenching by the reactive oxygen species generated by the H_2O_2. Acetylcholinesterase converts the acetylcholine to choline, which is then oxidized by the choline oxidase in betaine and H_2O_2, which generates the reactive oxygen species that induce the fluorescence quenching of the FGO-CDs (Wang et al. 2013). Besides the acetylcholine neurotransmitter, also, the fluorescence sensing of the most important catecholamine neurotransmitter dopamine was done by functionalized CDs. Based on the fluorescence quenching, the dopamine sensing was performed with N-(β-aminoethyl)-γ-aminopropyl methyldimethoxysilane

(AEAPMS)-CDs encapsulated with a silica imprinted sol-gel film obtained with tetraethoxysilane (TEOS) and 3-aminopropyltriethoxysilane (APTES) (Mao et al. 2012).

The fluorescence quenching of functionalized CDs by the Fe^{3+} ion is the bases of several molecular sensors (Sun et al. 2010; Baruah et al. 2013; Song et al. 2014). The fluorescence quenching of Fe^{3+} due to iron-containing porphyrin is the bases of a hemin sensor and, due to its role on the Fenton reaction, of a H_2O_2 sensor (Song et al. 2014). Based on the same principle, another hemin sensor was developed (Baruah et al. 2013). However, the sensitivity of the fluorescence quenching of the β-cyclodextrin, lactose, or sucrose-CDs by the hemin is higher than that observed with Fe^{3+} or Fe^{2+}, and the fluorescence quenching by amine depends on the number of hydroxyl groups of the functionalization label (Baruah et al. 2013). Also, in other work, the ferrous succinate was indirectly quantified after the conversion of Fe^{2+} to Fe^{3+} by the H_2O_2 and consequent fluorescence quenching of the fluorescence functionalized CDs by the Fe^{3+} (Sun et al. 2010).

Artificial DNA oligonucleotides known as aptamers were used for the CDs fluorescence sensing of thrombin, insulin, and double-stranded DNA (dsDNA) (Li et al. 2011c; Liu et al. 2011; Xu et al. 2012; Wang et al. 2014). Different fluorescence strategies were used for the fluorescence sensing of

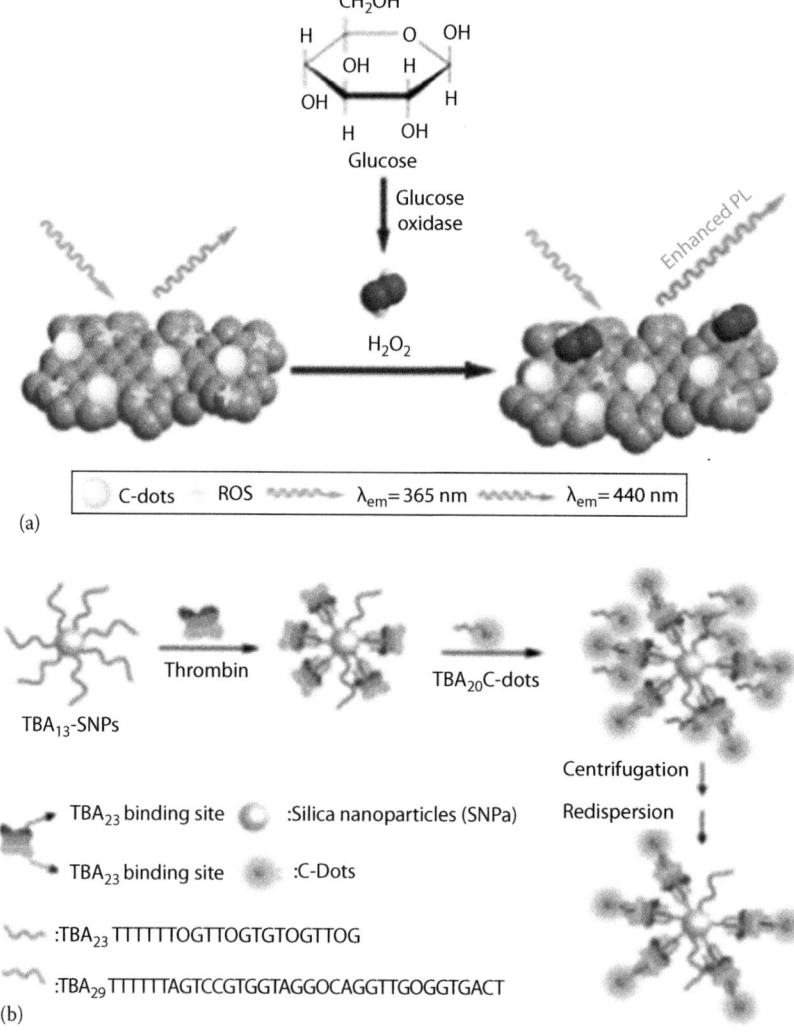

(a)

(b)

FIGURE 11.3

Examples of fluorescence sensing strategies for the molecular species evaluation by functionalized CDs nanoparticles: (a) Glucose turn off/on sensor. (Reprinted from *Talanta*, 115, Yeh, T., Wang, C., and Chang, H. T., Photoluminescent C-dots@RGO for sensitive detection of hydrogen peroxide and glucose, 718–723, Copyright 2013, with permission from Elsevier.) (b) Thrombin aptamer sandwich sensor. (Xu, B., Zhao, C., Wei, W. et al., Aptamer carbon nanodot sandwich used for fluorescent detection of protein, *Analyst*, 137(23), 5483, 2012. Reproduced by permission of The Royal Society of Chemistry.)

(Continued)

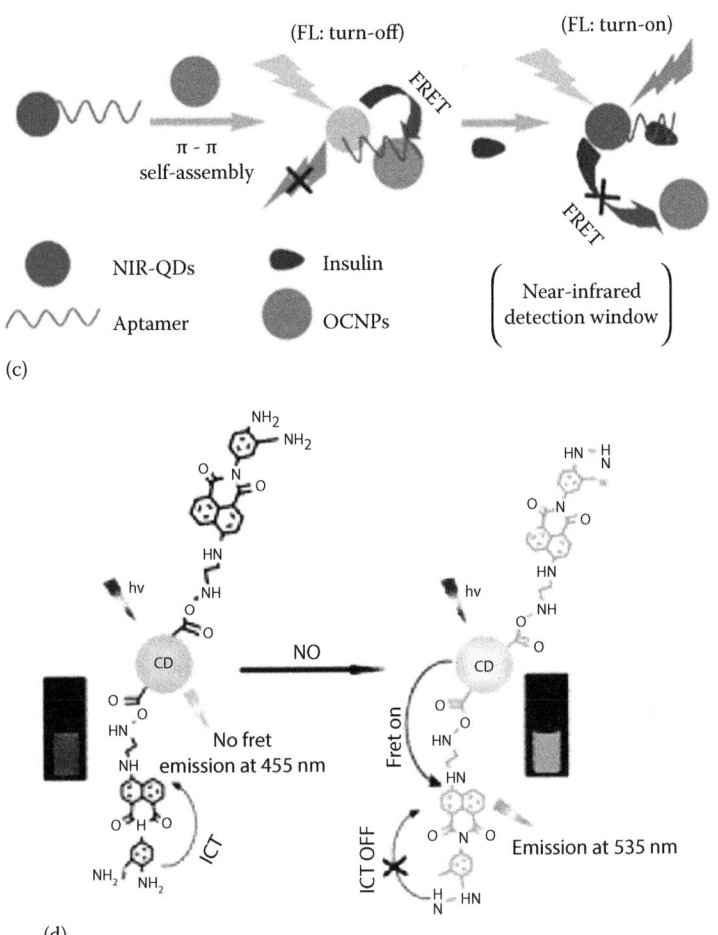

(c)

(d)

FIGURE 11.3 (*Continued*)
Examples of fluorescence sensing strategies for the molecular species evaluation by functionalized CDs nanoparticles: (c) Insulin turn off/on–based FRET aptamer sensor. (Wang, Y., Gao, D.,
Zhang P. et al., A near infrared fluorescence resonance energy transfer based aptamer biosensor
for insulin detection in human plasma, *Chem. Commun. (Camb.)*, 50(7), 811, 2014. Reproduced by
permission of The Royal Society of Chemistry.) (d) NO ratiometric sensor. (Yu, C., Wu, Y., Zeng, F.,
and Wu, S., A fluorescent ratiometric nanosensor for detecting NO in aqueous media and imaging
exogenous and endogenous NO in live cells, *J. Mater. Chem. B*, 1(33), 4152, 2013b. Reproduced by
permission of The Royal Society of Chemistry.)

these biological compounds. The sensing of thrombin was performed by linking thrombin aptamers
to the CDs (Liu et al. 2011; Xu et al. 2012). In one of the developed sensors, the thrombin fluorescence sensing was achieved by the recovery of the fluorescence of a FAM-labeled thrombin aptamer
after the fluorescence quenching of the FAM fluorophore by the CDs due to the lowest affinity of
the thrombin aptamer complex to the CDs (Liu et al. 2011). In the other, TBA_{15} and TBA_{29} aptamers, which binds at different sites of thrombin, were, respectively, linked to Si nanoparticles and
CDs. An enhancement of fluorescence was observed with the thrombin due to the successively
sandwich structure formed by the thrombin with the TBA_{29}–CDs and TBA_{15}–Si nanoparticles (Xu
et al. 2012). The sensing of insulin was done by a FRET mechanism between an insulin aptamer
functionalized near infrared (NIR) QDs as a donor and an oxidized CDs as an acceptor was developed for the fluorescence sensing of insulin. In this insulin fluorescence sensing strategy, the CDs
quenched the NIR fluorescence that is recovered by higher affinity of the insulin to the aptamer
(Wang et al. 2014). As for the thrombin fluorescence sensing based on the recovery of the FAM-
labeled thrombin aptamer after the fluorescence quenching of the FAM fluorophore by the CDs,
also, the dsDNA fluorescence sensing was done by the same principle. In the dsDNA-developed
CDs fluorescent sensor, the recovery of fluorescence was achieved through the hybridization of the
ssDNA forming the dsDNA and consequent desorption from the CDs surface (Li et al. 2011c).

The sensing of the glutathione (GSH) thiol tripeptide was done by a FRET mechanism between the fluorescent functionalized CD as a donor and Au nanoparticle as an acceptor and posterior recover of fluorescence by GSH. Besides the GSH sensing by this fluorescent turn off/on sensor, also, the spectrophotometric GSH sensing was done using the absorbance ratio at the two maximum wavelengths of Au nanoparticles. Good results were found with this dual nanosensor. The possibility of the GSH sensing in the presence of cysteine and homocysteine-related compounds was also established (Shi et al. 2014).

Besides the previously described hydrogen peroxide sensors, used as sensors of glucose and acetylcholine, some functionalized CDs were also developed for the fluorescence sensing of two important small biological molecules such as NO (Yu et al. 2013b) and H_2S (Yu et al. 2013a; Zhu et al. 2014) in different physiological and pathological processes. With the same approach, a ratiometric fluorescent sensor based on a FRET mechanism of a naphthalimide-derived functionalized CDs was developed for NO (Yu et al. 2013b) and for H_2S (Yu et al. 2013a) fluorescence sensing. These two ratiometric fluorescence sensors were used for NO in a naphthalimide with an o-phenylenediamine group and for H_2S in a naphthalimide azide functionalized CDs. With both developed sensors, a reaction of NO and H_2S with the naphthalimide-derived CDs allows a FRET mechanism, between the CDs and the product of the reaction of the derived naphthalimide linked to CDs, which induces a fluorescence decrease of the naphthalimide-derived functionalized CDs at 455 and 425 nm and an increase at 535 and 526 nm, respectively, for NO and H_2S. Both the calculated ratios increase with NO and H_2S concentrations. Also, for H_2S, a two-photon turn on/off sensor based on a previous Cu^{2+} fluorescence quenching of an AE-TPEA functionalized CDs and posterior recovery of fluorescence was developed (Zhu et al. 2014).

For the sensing of pharmaceutical molecules, and besides the previously described fluorescence sensing of ferrous succinate (Sun et al. 2010), only for the tetracycline was developed a fluorescent sensor based on functionalized CDs. The tetracycline fluorescent sensing is based in the fluorescence quenching of the functionalized CDs by the tetracyclines (Yang et al. 2014).

Environmental molecules subject to fluorescence sensing by functionalized CDs were hypochlorous acid (HClO) (Huang et al. 2014), 4,4′-dibrominated biphenyl (PBB15) (Bu et al. 2014), and 2,4-dinitrophenol (DNP) and 2-amino-3,4,8-trimethyl-3H-imidazo[4,5-f] quinoxaline (DMQ) (Cayuela et al. 2013). The HClO fluorescence sensing was done with tunable fluorescent emission CDs by quenching induced by HClO (Huang et al. 2014). The fluorescence sensing by functionalized CDs of the three environmental organic pollutants (PBB15, DNP, and DMQ) is done for PBB15 by an immunosensor based in a FRET mechanism (Bu et al. 2014) and for the DNP and DMQ through the fluorescence quenching of acetone-passivated CDs (Cayuela et al. 2013). The PBB15 immunosensor is based in the FRET between the antigen PBB15–poly(ethylene glycol)-200 (PEG200)–CDs as a donor and the modified anti-PBB15 antibody Au nanoparticles as the acceptor. By reaction of these two modified nanoparticles, a decrease of fluorescence intensity of the functionalized CDs is observed due to FRET. The recovery of the fluorescence of functionalized CDs is obtained due to the competitive immunoreactions with PBB15 (Bu et al. 2014).

11.3.3 Other Species

Besides the fluorescence sensing of ionic or molecular species, an application for Au nanoparticles sensing using a CD-based sensor was developed (Cayuela et al. 2014). The developed Au nanoparticle CDs sensor was synthesized in a three-step synthesis by thermal refluxing multiwalled carbon nanotubes (MWCNTs) in an acid medium, passivation with acetone, and posterior functionalization with cysteamine by N,N'-diisopropylcarbodiimide (DIC)–N-hydroxysuccinimide (NHS). The fluorescence sensing is based in the static quenching of fluorescence of the functionalized cysteamine CDs by the Au nanoparticles. An LOD of 0.0002 μM was obtained (Cayuela et al. 2014).

11.3.4 Conclusions/Perspectives

Several fluorescence sensing strategies have been exploited for the sensitive and selective detection of different ionic and molecular analytical species by functionalized CDs. A great effort has been done in order to design different methods of synthesis/functionalization of CDs with higher QY to obtain higher sensitivity. The majority of the CD-based analytical methodologies are selective. Even so an intense research in new functionalization labels and functionalization methodologies is still needed in order to achieve the required selectivity and in order to enlarge the application field of the functionalized CDs to new ionic and molecular analytical species.

Some research work was already done in the synthesis of excitation wavelength–dependent tunable emission CDs. Most of the tunable synthesized CDs were applied in bioimaging. A research trend line using CDs excitation wavelength-dependent emission tunability with similar fluorescence intensities could also be explored in order to the sensitivity and selective fluorescence sensing by functionalized CDs.

REFERENCES

Algarra, M., B. B. Campos, K. Radotić et al. 2014. Luminescent carbon nanoparticles: Effects of chemical functionalization, and evaluation of Ag⁺ sensing properties. *Journal of Materials Chemistry A* 2 (22):8342–8351.

Baker, S. N. and G. A. Baker. 2010. Luminescent carbon nanodots: Emergent nanolights. *Angewandte Chemie (International ed. in English)* 49 (38):6726–6744.

Baruah, U., N. Gogoi, G. Majumdar, and D. Chowdhury. 2013. Capped fluorescent carbon dots for detection of Hemin: Role of number of –OH groups of capping agent in fluorescence quenching. *The Scientific World Journal* 2013:1–9.

Bu, D., H. Zhuang, G. Yang, and X. Ping. 2014. An immunosensor designed for polybrominated biphenyl detection based on fluorescence resonance energy transfer (FRET) between carbon dots and gold nanoparticles. *Sensors and Actuators B: Chemical* 195:540–548.

Cao, L., X. Wang, M. J. Meziani et al. 2007. Carbon dots for multiphoton bioimaging. *Journal of American Chemical Society* 129:11318–11319.

Cayuela, A., M. L. Soriano, M. C. Carrion, and M. Valcarcel. 2014. Functionalized carbon dots as sensors for gold nanoparticles in spiked samples: Formation of nanohybrids. *Analytica Chimica Acta* 820:133–138.

Cayuela, A., M. L. Soriano, and M. Valcarcel. 2013. Strong luminescence of carbon dots induced by acetone passivation: Efficient sensor for a rapid analysis of two different pollutants. *Analytica Chimica Acta* 804:246–251.

Costas-Mora, I., V. Romero, I. Lavilla, and C. Bendicho. 2014. In situ building of a nanoprobe based on fluorescent carbon dots for methylmercury detection. *Analytical Chemistry* 86 (9):4536–4543.

Cushing, S. K., M. Li, F. Huang, and N. Wu. 2014. Origin of strong excitation wavelength dependent fluorescence of graphene oxide. *ACS Nano* 8 (1):1002–1013.

Ding, C., A. Zhu, and Y. Tian. 2014. Functional surface engineering of C-dots for fluorescent biosensing and in vivo bioimaging. *Accounts of Chemical Research* 47 (1):20–30.

Dong, Y., R. Wang, G. Li et al. 2012. Polyamine-functionalized carbon quantum dots as fluorescent probes for selective and sensitive detection of copper ions. *Analytical Chemistry* 84 (14):6220–6224.

Du, F., Y. Ming, F. Zeng, C. Yu, and S. Wu. 2013a. A low cytotoxic and ratiometric fluorescent nanosensor based on carbon-dots for intracellular pH sensing and mapping. *Nanotechnology* 24 (36):365101.

Du, F., F. Zeng, Y. Ming, and S. Wu. 2013b. Carbon dots-based fluorescent probes for sensitive and selective detection of iodide. *Microchimica Acta* 180 (5–6):453–460.

Esteves da Silva, J. C. G. 2011. Carbon and silicon fluorescent nanomaterials. In *Nanomaterials*, ed. M. M. Rahman. In-Tech, Rijeka, Croatia.

Esteves da Silva, J. C. G. 2012. Localized surface plasmons effect in the fluorescence quantum dots. In *Quantum Dots: Applications, Synthesis an Characterization*, ed. D. O. Ciftja. Nova Publishers, Hauppauge, NY.

Esteves da Silva, J. C. G. and H. M. R. Gonçalves. 2011. Analytical and bioanalytical applications of carbon dots. *TrAC Trends in Analytical Chemistry* 30 (8):1327–1336.

Gonçalves, H. M., A. J. Duarte, F. Davis, S. P. Higson, and J. C. Esteves da Silva. 2012. Layer-by-layer immobilization of carbon dots fluorescent nanomaterials on single optical fiber. *Analytica Chimica Acta* 735:90–95.

Gonçalves, H. M., A. J. Duarte, and J. C. Esteves da Silva. 2010a. Optical fiber sensor for Hg(II) based on carbon dots. *Biosensors and Bioelectronics* 26 (4):1302–1306.

Gonçalves, H. and J. C. Esteves da Silva. 2010. Fluorescent carbon dots capped with PEG200 and mercaptosuccinic acid. *Journal of Fluorescence* 20 (5):1023–1028.

Gonçalves, H., P. A. S. Jorge, J. R. A. Fernandes, and J. C. G. Esteves da Silva. 2010b. Hg(II) sensing based on functionalized carbon dots obtained by direct laser ablation. *Sensors and Actuators B: Chemical* 145 (2):702–707.

Hou, X., F. Zeng, F. Du, and S. Wu. 2013. Carbon-dot-based fluorescent turn-on sensor for selectively detecting sulfide anions in totally aqueous media and imaging inside live cells. *Nanotechnology* 24 (33):335502.

Huang, Z., F. Lin, M. Hu et al. 2014. Carbon dots with tunable emission, controllable size and their application for sensing hypochlorous acid. *Journal of Luminescence* 151:100–105.

Kong, B., A. Zhu, C. Ding et al. 2012. Carbon dot-based inorganic-organic nanosystem for two-photon imaging and biosensing of pH variation in living cells and tissues. *Advanced Materials* 24 (43):5844–5848.

Kong, W., H. Wu, Z. Ye et al. 2014. Optical properties of pH-sensitive carbon-dots with different modifications. *Journal of Luminescence* 148:238–242.

Kozák, O., K. K. R. Datta, M. Greplová et al. 2013. Surfactant-derived amphiphilic carbon dots with tunable photoluminescence. *The Journal of Physical Chemistry C* 117 (47):24991–24996.

Krishna, A. S., C. Radhakumary, and K. Sreenivasan. 2013. In vitro detection of calcium in bone by modified carbon dots. *Analyst* 138 (23):7107–7111.

Li, H., Z. Kang, Y. Liu, and S. Lee. 2012. Carbon nanodots: Synthesis, properties and applications. *Journal of Materials Chemistry* 22 (46):24230–24253.

Li, H., J. Zhai, and X. Sun. 2011a. Sensitive and selective detection of silver(I) ion in aqueous solution using carbon nanoparticles as a cheap, effective fluorescent sensing platform. *Langmuir* 27 (8):4305–4308.

Li, H., J. Zhai, J. Tian, Y. Luo, and X. Sun. 2011b. Carbon nanoparticle for highly sensitive and selective fluorescent detection of mercury(II) ion in aqueous solution. *Biosensors and Bioelectronics* 26 (12):4656–4660.

Li, H., Y. Zhang, L. Wang, J. Tian, and X. Sun. 2011c. Nucleic acid detection using carbon nanoparticles as a fluorescent sensing platform. *Chemical Communications (Cambridge)* 47 (3):961–963.

Li, Q., T. Y. Ohulchanskyy, R. Liu et al. 2010. Photoluminescent carbon dots as biocompatible nanoprobes for targeting cancer cells in vitro. *The Journal of Physical Chemistry C* 114 (28):12062–12068.

Li, X., S. Zhang, S. A. Kulinich, Y. Liu, and H. Zeng. 2014. Engineering surface states of carbon dots to achieve controllable luminescence for solid-luminescent composites and sensitive Be^{2+} detection. *Scientific Reports* 4:4976–4983.

Liu, H., T. Ye, and C. Mao. 2007. Fluorescent carbon nanoparticles derived from candle soot. *Angewandte Chemie (International ed. in English)* 46 (34):6473–6475.

Liu, J., J. Li, Y. Jiang et al. 2011. Combination of pi-pi stacking and electrostatic repulsion between carboxylic carbon nanoparticles and fluorescent oligonucleotides for rapid and sensitive detection of thrombin. *Chemical Communications (Cambridge)* 47 (40):11321–11323.

Liu, J. M., L. P. Lin, X. X. Wang et al. 2012. Highly selective and sensitive detection of Cu^{2+} with lysine enhancing bovine serum albumin modified-carbon dots fluorescent probe. *Analyst* 137 (11):2637–2642.

Liu, J. M., L. P. Lin, X. X. Wang et al. 2013. $Zr(H_2O)2EDTA$ modulated luminescent carbon dots as fluorescent probes for fluoride detection. *Analyst* 138 (1):278–283.

Liu, R., D. Wu, S. Liu et al. 2009. An aqueous route to multicolor photoluminescent carbon dots using silica spheres as carriers. *Angewandte Chemie (International ed. in English)* 48 (25):4598–4601.

Liu, X., N. Zhang, T. Bing, and D. Shangguan. 2014a. Carbon dots based dual-emission silica nanoparticles as a ratiometric nanosensor for Cu(2+). *Analytical Chemistry* 86 (5):2289–2296.

Liu, Y., Y. Zhao, and Y. Zhang. 2014b. One-step green synthesized fluorescent carbon nanodots from bamboo leaves for copper(II) ion detection. *Sensors and Actuators B: Chemical* 196:647–652.

Luo, P. G., S. Sahu, S. Yang et al. 2013. Carbon "quantum" dots for optical bioimaging. *Journal of Materials Chemistry B* 1 (16):2116–2127.

Mao, L., W. Tang, Z. Deng et al. 2014. Facile access to white fluorescent carbon dots toward light-emitting devices. *Industrial and Engineering Chemistry Research* 53 (15):6417–6425.

Mao, X. J., H. Z. Zheng, Y. J. Long et al. 2010. Study on the fluorescence characteristics of carbon dots. *Spectrochimica Acta. Part A, Molecular and Biomolecular Spectroscopy* 75 (2):553–557.

Mao, Y., Y. Bao, D. Han, F. Li, and L. Niu. 2012. Efficient one-pot synthesis of molecularly imprinted silica nanospheres embedded carbon dots for fluorescent dopamine optosensing. *Biosensors and Bioelectronics* 38 (1):55–60.

Nie, H., M. Li, Q. Li et al. 2014. Carbon dots with continuously tunable full-color emission and their application in ratiometric pH sensing. *Chemistry of Materials* 26 (10):3104–3112.

Peng, H. and J. Travas-Sejdic. 2009. Simple aqueous solution route to luminescent carbogenic dots from carbohydrates. *Chemistry of Materials* 21 (23):5563–5565.

Qu, Q., A. Zhu, X. Shao, G. Shi, and Y. Tian. 2012. Development of a carbon quantum dots-based fluorescent Cu^{2+} probe suitable for living cell imaging. *Chemical Communications (Cambridge)* 48 (44):5473–5475.

Qu, S., H. Chen, X. Zheng, J. Cao, and X. Liu. 2013. Ratiometric fluorescent nanosensor based on water soluble carbon nanodots with multiple sensing capacities. *Nanoscale* 5 (12):5514–5518.

Ruedas-Rama, M. J., J. D. Walters, A. Orte, and E. A. Hall. 2012. Fluorescent nanoparticles for intracellular sensing: A review. *Analytica Chimica Acta* 751:1–23.

Salinas-Castillo, A., M. Ariza-Avidad, C. Pritz et al. 2013. Carbon dots for copper detection with down and upconversion fluorescent properties as excitation sources. *Chemical Communications (Cambridge)* 49 (11):1103–1105.

Shan, X., L. Chai, J. Ma et al. 2014. B-doped carbon quantum dots as a sensitive fluorescence probe for hydrogen peroxide and glucose detection. *Analyst* 139 (10):2322–2325.

Shen, L., L. Zhang, M. Chen, X. Chen, and J. Wang. 2013. The production of pH-sensitive photoluminescent carbon nanoparticles by the carbonization of polyethylenimine and their use for bioimaging. *Carbon* 55:343–349.

Shi, W., X. Li, and H. Ma. 2012. A tunable ratiometric pH sensor based on carbon nanodots for the quantitative measurement of the intracellular pH of whole cells. *Angewandte Chemie (International ed. in English)* 51 (26):6432–6435.

Shi, Y., Y. Pan, H. Zhang et al. 2014. A dual-mode nanosensor based on carbon quantum dots and gold nanoparticles for discriminative detection of glutathione in human plasma. *Biosensors and Bioelectronics* 56:39–45.

Song, Y., S. Zhu, S. Xiang et al. 2014. Investigation into the fluorescence quenching behaviors and applications of carbon dots. *Nanoscale* 6 (9):4676–4682.

Sun, W., Y. Du, and Y. Wang. 2010. Study on fluorescence properties of carbogenic nanoparticles and their application for the determination of ferrous succinate. *Journal of Luminescence* 130 (8):1463–1469.

Sun, Y., B. Zhou, Y. Lin et al. 2006. Quantum-sized carbon dots for bright and colorful photoluminescence. *Journal of the American Chemical Society* 128 (24):7756–7757.

Wang, C. I., A. P. Periasamy, and H. T. Chang. 2013. Photoluminescent C-dots@RGO probe for sensitive and selective detection of acetylcholine. *Analytical Chemistry* 85 (6):3263–3270.

Wang, Y., D. Gao, P. Zhang et al. 2014. A near infrared fluorescence resonance energy transfer based aptamer biosensor for insulin detection in human plasma. *Chemical Communications (Cambridge)* 50 (7):811–813.

Wei, W., C. Xu, J. Ren, B. Xu, and X. Qu. 2012. Sensing metal ions with ion selectivity of a crown ether and fluorescence resonance energy transfer between carbon dots and graphene. *Chemical Communications (Cambridge)* 48 (9):1284–1286.

Xu, B., C. Zhao, W. Wei et al. 2012. Aptamer carbon nanodot sandwich used for fluorescent detection of protein. *Analyst* 137 (23):5483–5486.

Xu, X., R. Ray, Y. Gu et al. 2004. Electrophoretic analysis and purification of fluorescent single-walled carbon nanotube fragments. *Journal of the American Chemical Society* 126 (40):12736–12737.

Yan, F., Y. Zou, M. Wang et al. 2014. Highly photoluminescent carbon dots-based fluorescent chemosensors for sensitive and selective detection of mercury ions and application of imaging in living cells. *Sensors and Actuators B: Chemical* 192:488–495.

Yang, X., Y. Luo, S. Zhu et al. 2014. One-pot synthesis of high fluorescent carbon nanoparticles and their applications as probes for detection of tetracyclines. *Biosensors and Bioelectronics* 56:6–11.

Yeh, T., C. Wang, and H. T. Chang. 2013. Photoluminescent C-dots@RGO for sensitive detection of hydrogen peroxide and glucose. *Talanta* 115:718–723.

Yu, C., X. Li, F. Zeng, F. Zheng, and S. Wu. 2013a. Carbon-dot-based ratiometric fluorescent sensor for detecting hydrogen sulfide in aqueous media and inside live cells. *Chemical Communications (Cambridge)* 49 (4):403–405.

Yu, C., Y. Wu, F. Zeng, and S. Wu. 2013b. A fluorescent ratiometric nanosensor for detecting NO in aqueous media and imaging exogenous and endogenous NO in live cells. *Journal of Materials Chemistry B* 1 (33):4152–4159.

Yuan, C., B. Liu, F. Liu, M. Y. Han, and Z. Zhang. 2014. Fluorescence "turn on" detection of mercuric ion based on bis(dithiocarbamato)copper(II) complex functionalized carbon nanodots. *Analytical Chemistry* 86 (2):1123–1130.

Zhang, L., H. Wei, F. Gao et al. 2013. Controllable synthesis of fluorescent carbon dots and their detection application as nanoprobes. *Nano-Micro Letters* 5 (4):247–259.

Zhang, R. and W. Chen. 2014. Nitrogen-doped carbon quantum dots: Facile synthesis and application as a "turn-off" fluorescent probe for detection of Hg^{2+} ions. *Biosensors and Bioelectronics* 55:83–90.

Zhang, S., Q. Wang, G. Tian, and H. Ge. 2014. A fluorescent turn-off/on method for detection of Cu^{2+} and oxalate using carbon dots as fluorescent probes in aqueous solution. *Materials Letters* 115:233–236.

Zhao, H. X., L. Q. Liu, Z. D. Liu et al. 2011. Highly selective detection of phosphate in very complicated matrixes with an off-on fluorescent probe of europium-adjusted carbon dots. *Chemical Communications (Cambridge)* 47 (9):2604–2606.

Zhu, A., Z. Luo, C. Ding et al. 2014. A two-photon "turn-on" fluorescent probe based on carbon nanodots for imaging and selective biosensing of hydrogen sulfide in live cells and tissues. *Analyst* 139 (8):1945–1952.

Zhu, A., Q. Qu, X. Shao, B. Kong, and Y. Tian. 2012. Carbon-dot-based dual-emission nanohybrid produces a ratiometric fluorescent sensor for in vivo imaging of cellular copper ions. *Angewandte Chemie (International ed. in English)* 51 (29):7185–7189.

Zhu, H., X. Wang, Y. Li et al. 2009. Microwave synthesis of fluorescent carbon nanoparticles with electrochemiluminescence properties. *Chemical Communications (Cambridge)* (34):5118–5120.

Zhu, S., Q. Meng, L. Wang et al. 2013. Highly photoluminescent carbon dots for multicolor patterning, sensors, and bioimaging. *Angewandte Chemie (International ed. in English)* 52 (14):3953–3957.

CHAPTER 12

CONTENTS

Aligned Carbon Nanotube–Based Sensing Materials

12

Xuemei Sun, Xin Lu, Xin Fang, and Huisheng Peng

12.1 INTRODUCTION

The alignment of carbon nanotubes (CNTs) represents a general and efficient strategy to extend their remarkable electrical, mechanical, thermal, and optical properties to a macroscopic scale that is required for many practical applications (Dai et al. 2003, Liu et al. 2011, Sun et al. 2012a). Over the past decade, significant efforts have been made to align CNTs by various in situ and ex situ techniques. The direct growth of CNT array is the most effective way to obtain neat aligned CNTs. Other macroscopic CNT assemblies, for example, CNT sheets and CNT fibers, can be further spun from the arrays and be easily engineered under various conditions. Although great progress have been made in aligned arrays of both multiwalled carbon nanotube (MWCNT) (Fan et al. 1999) and single-walled carbon nanotube (SWCNT) (Hata et al. 2004), all spinnable arrays for aligned sheets and fibers are still on the basis of MWCNTs synthesized by chemical vapor deposition (CVD). Due to the highly aligned and porous structure, aligned CNTs can be incorporated with a lot of polymers to fabricate composite materials without alignment destruction, and the resulting composites exhibit more dramatically enhanced physical properties than traditional randomly dispersed CNTs (Huang et al. 2011, Li et al. 2011, Sun et al. 2012a, 2013b). In addition, the synergetic interactions between aligned CNTs and polymers have been found to create some new sensing properties different from any individual component. In this chapter, the synthesis of aligned CNT materials will be first summarized. The properties of aligned CNTs are then introduced with a highlight of electromechanical properties of CNT fiber for artificial muscles. The incorporation of functional polymers into the aligned CNTs is finally emphasized to develop novel chromatic and deformable composite materials. The challenge and application of aligned CNT-based sensing materials will be also described.

12.2 SYNTHESIS OF ALIGNED CNT MATERIALS

12.2.1 Aligned CNT Array

First reported in 1996, aligned CNT bundles containing 70% of SWCNTs exhibited locally oriented ropes, but still randomly dispersed in bulk materials (Thess et al. 1996). In the same year, a macroscopically aligned MWCNT array was synthesized by pyrolysis of acetylene through a thermal CVD process, which was catalyzed by iron nanoparticles embedded in mesoporous silica (Li et al. 1996). After that, the synthesis of aligned CNT arrays was optimized and expanded with the development of various methods. To date, large-scale CNT arrays with controllable diameters and lengths were grown on nickel-coated glass by plasma-enhanced hot filament CVD (Ren et al. 1998). Densely packed long SWCNT arrays were synthesized by water-assisted CVD process (Hata et al. 2004). Continuous aligned CNTs were massively produced by fluidized bed catalytic CVD (Zhang et al. 2010). More importantly, superaligned CNT arrays were grown using a thermal CVD method, with the morphology and structure of the resulting CNT array being controlled by the catalyst, growth temperature, reaction time, flow rate, and carbon sources and partial pressure (Andrews et al. 2002, Li et al. 2006).

In a typical CVD procedure, a catalyst layer is first deposited on a silicon substrate through physical deposition methods like electron-beam evaporation, ion beam–assisted deposition, and magnetron sputtering or solution coating methods such as spin coating, spray coating, dip coating, and microcontact printing (Qiu et al. 2012). Iron, cobalt, nickel, and their alloys are widely used as catalysts. Additionally, Al_2O_3, as a buffer layer, is previously deposited on the silicon substrate to prevent reaction between the silicon and metal catalyst and preserve activity of catalyst (Qiu et al. 2012). The aligned CNTs are grown in a tube furnace filled with a constant flow of reacting gas. The carbon source is provided by hydrocarbons such as acetylene, ethylene, xylene, and methane and usually carried by nitrogen or argon mixed with hydrogen. The temperature is programmed during the process. In the heating process at 500°C, the catalyst layer, driven by the surface energy minimization, is crack into individual nanoparticles densely distributed over the substrate. With temperature rising to 740°C, the hydrocarbon is decomposed into carbonaceous clusters and dissolved in the catalytic nanoparticles. Subsequently, the carbonaceous species are saturated and ejected to nucleate at the surface of the catalytic nanoparticles, which forms the initial cap of the CNT. Then, the CNT grows continuously until the growth is stopped by cutting off the carbon source, cooling the tube furnace or deactivating the catalyst. The synthesis of the aligned CNT array is shown in Figure 12.1. Due to the strong van der Waals interaction among the neighboring CNTs, the resulting CNTs are inclined to vertically align on the substrate by crowding effect (Figure 12.2a through c, Fan et al. 1999). The morphology and structure of the CNT array strongly depend on the catalyst such as the thickness and roughness and synthesis parameters such as the temperature program, growing time, gas components, and flowing rates. Currently, CNT arrays can be synthesized with various heights ranging from micrometers to millimeters.

The macroscopic assemblies of aligned CNTs, CNT sheets and fibers, can be prepared from spinnable CNT arrays through a simple dry-drawing technique. The spinnability of a CNT array is benefited from the strong interaction and dense arrangement of CNTs. In fact, the CNTs in the spinnable array are prone to aggregate into CNT bundles, which usually contain CNTs intercalating the neighboring bundle. Therefore, the spinnability of a CNT array necessitates a clean CNT surface, narrow distribution of CNT diameters, and appropriate length and nucleation density, which can be delicately controlled during synthesis. A clean surface can ensure a strong van der Waals interaction

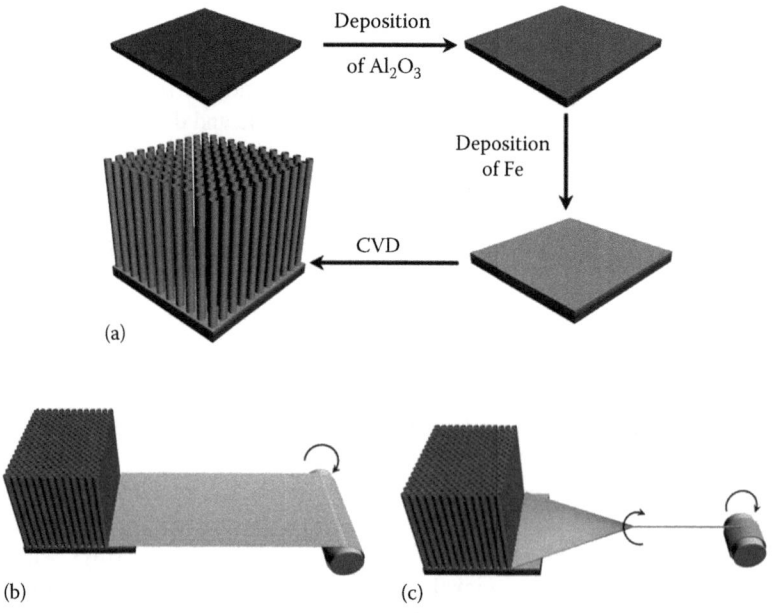

FIGURE 12.1
Schematic illustration of the synthesis of aligned CNT materials. (a) Synthesis of aligned CNT arrays through the CVD process. (b) and (c) Synthesis of CNT sheets and fibers from the CNT array by dry spinning, respectively.

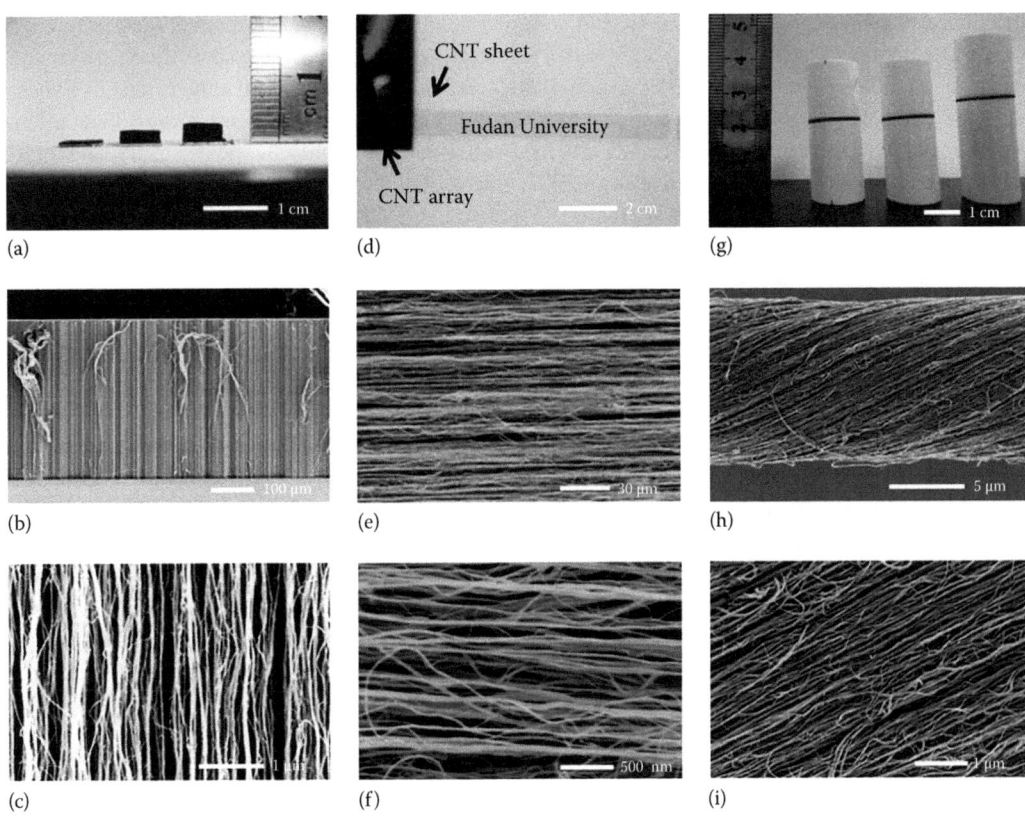

FIGURE 12.2
Photographs and SEM images of aligned CNT materials. (a through c) CNT arrays, (d through f) CNT sheets, and (g through i) CNT fibers at different magnifications, respectively. (Reprinted with permission from Yang, Z. et al., *Adv. Mater.*, 23, 5436, 2011a; Yang, Z. et al., *J. Mater. Chem.*, 21, 13772, 2011b.)

among CNTs, which may be undermined by the amorphous carbon (Zhang et al. 2006, Liu et al. 2008). The length of CNT bundle is another factor affecting the spinnability. Too short CNT bundle cannot provide sufficient interaction to pull out the continuous bundles. If the CNT is too long, however, the interaction is too strong to form a top-end connection. The heights of spinnable CNT arrays (e.g., the length of CNTs) are usually in the range of 100 μm to 1.5 mm, providing an appropriate interaction and interconnection (Jiang et al. 2002, Li et al. 2006, Liu et al. 2008). Of the utmost importance, the spinnability benefits from the superior alignment and appropriate density, which require a uniform size distribution of catalytic particles (Zhang et al. 2009). It is calculated that, when diameters of CNTs are ~10 nm, the good spinnable CNT array should have an area density of ~10^{11} cm^{-2} enabling CNTs to form interconnections. In contrast, the arrays with area density of less than 10^{10} cm^{-2} are not spinnable (Zhang and Baughman 2011).

12.2.2 Aligned CNT Sheet

Continuous CNT yarn was firstly drawn out from superaligned MWCNT arrays in 2002, and this preparation process has been widely employed ever since (Jiang et al. 2002). Drawing can be initiated using a tweezer or a blade by cutting into the array to contact CNTs. Once the CNTs at the edge of the array are pulled out, the neighboring CNTs are consecutively followed forming a continuous, long yarn or sheet (Figure 12.1b). Figure 12.2d through f further shows SEM images of CNT sheet in which CNTs are highly aligned along the drawing direction. Generally, the sizes of the resulting CNT sheet are tuned by the height and width of CNT arrays. Technically, a 1 cm × 1 cm CNT array with a height of 200 μm can produce a 1 m long sheet with width of 1 cm. These as-produced sheets have advantages in structure including good alignment, high CNT volume fraction, and tunable thickness and density (Figure 12.2d). The thickness of the dry-spun CNT sheet is increased with the increase of array height (Zhang et al. 2005). The typical volumetric density of the CNT sheet is ~0.0015 g cm^{-3} and can be improved to 0.5 g cm^{-3} simply by solvent treatment, along with the sheet thickness changing from ~20 μm to 50 nm (Zhang et al. 2005). Benefited from the structure advantage, the CNT sheet has high conductivity, optical transmittance, lightweight, and flexibility, enabling promising applications in solar cells, electrochemical capacitors, field emission transistors, chemical sensors, and high-performance structure materials (Jiang et al. 2002, Zhang et al. 2006, Wei et al. 2008).

12.2.3 Aligned CNT Fiber

Two approaches can convert a 2D CNT sheet into a 1D CNT fiber. One is pulling the CNT sheet across a volatile solvent. The interaction of solvent and CNTs makes the CNT sheets shrink into a fiber. The as-prepared CNT fiber has a rough surface and irregular morphology. The CNTs in the fiber are highly aligned along the fiber axis, but rarely entangled. The second more efficient method is twisting the CNT sheet during dry spinning process (Figure 12.1c), which can be collected continuously and scaled up easily for industrial production. The fiber can be further densified by solvent treatment. The introduction of twisting can increase the interaction among CNTs and prevent the sliding of CNTs, thus resulting in higher performance of the CNT fiber. Figure 12.2g through i shows that CNTs in the fiber are highly aligned and densely twisted. The angle between the aligned CNT and the fiber axis is defined as the helical angle, which will affect the diameter of the fiber as well as its package density. The helical angle can be tailored by controlling the speed ratio of the drawing and twisting. The diameter of fibers can be controlled in a wide range by the initial width of sheet typically from several to tens of micrometers. Twisted fibers with diameters of 2 and ~15 μm result from 200 μm and 1 cm wide sheets, respectively. The density of the CNT fibers is usually in the range of 0.2–0.8 g cm^{-3} (Zhang et al. 2004, 2007a). Multi-ply CNT fibers can be prepared by twisting CNT fibers together, which can be further knitted and knotted to form hierarchical structures (Zhang et al. 2004).

Apart from aligned CNT arrays, CNT fibers can also be prepared by CVD spinning and wet spinning process. CVD spinning is a high-yield method to continuously produce CNT fibers. In this process, a mixture of carbon source (e.g., ethanol, hexane, and acetone) and catalyst (e.g., ferrocene) is injected into a heated furnace assisted with a hydrogen flow. The carbon source decomposes into CNTs, which then assemble into CNT aerogels by van der Waals interactions in the gas flow. The continuous CNT fiber is prepared by densifying the resulting CNT aerogels with solvent. The as-produced CNT fiber exhibits a macroscopically aligned structure and delivers a high strength (~1 GPa) and electrical conductivity (~1500 S cm^{-1}). This method can continuously produce CNT fibers as long as several kilometers, which is favorable for industrial production (Li et al. 2004, Koziol et al. 2007, Zhong et al. 2010).

The dry spinning and CVD spinning method are currently applicable to MWCNTs, since the spinnable SWCNT array is not available yet. Wet spinning is especially applicable to prepare SWCNT fiber (Vigolo et al. 2000). In a typical process, SWCNTs are dispersed in organic, acid, or aqueous solvent, and then the dispersion is extruded through a die into a coagulation bath, forming continuous SWCNT fibers (Vigolo et al. 2000). Randomly dispersed in solvent, the SWCNTs are prone to aggregate, and the addition of surfactants will cause impurity residue thus decreasing their electrical and thermal properties. Several strategies have been made to solve the problem, such as dissolving the SWCNTs in strong acid and passing through a water coagulation bath and tuning the pH of SWCNT dispersion, resulting in improved electrical (5000 S cm^{-1}) and thermal conductivity (21 W m^{-1}k^{-1}) (Ericson et al. 2004, Kozlov et al. 2005, Davis et al. 2009).

12.3 PROPERTIES OF ALIGNED CNT MATERIALS

12.3.1 Mechanical Properties

The individual CNT has a tensile strength of 150 GPa and a high modulus of 1 TPa (Yu et al. 2000, Demczyk et al. 2002). However, when CNTs are assembled into sheet or fiber, their mechanical performances could probably disappoint us. The reported tensile strength of the CNT sheet is usually below 100 MPa, and 10^2–10^3 MPa for CNT fibers. The slump in strength should be ascribed to the different failure mechanisms. The CNTs in sheets and fibers interact with each other primarily by van der Waals force, rendering a ready slippage when stretched, rather than a fracture of CNT (Figure 12.3a and b) (Vilatela et al. 2011). Thus the interaction and the connection among CNTs are very important for the strength. Generally, the mechanical behavior of a CNT fiber is affected by the CNT structure as well as the fiber geometry. CNT structures include diameter, length, wall thickness, alignment, and defect. Normally, fibers containing longer CNTs with smaller diameters and fewer walls possess higher tensile strengths, because longer CNTs tend to tangle and interconnect more efficiently with each other, and fewer-walled CNTs with smaller diameters provide larger intertube contact areas. For instance, CNT fibers from arrays with lengths of 300, 500 and 650 μm have tensile strengths of 0.32, 0.56 and 0.85 GPa, respectively; CNT fibers with 6-walled and 8–10-walled CNTs have 0.86 and 0.69 GPa, respectively (Zhang et al. 2007b, Jia et al. 2011).

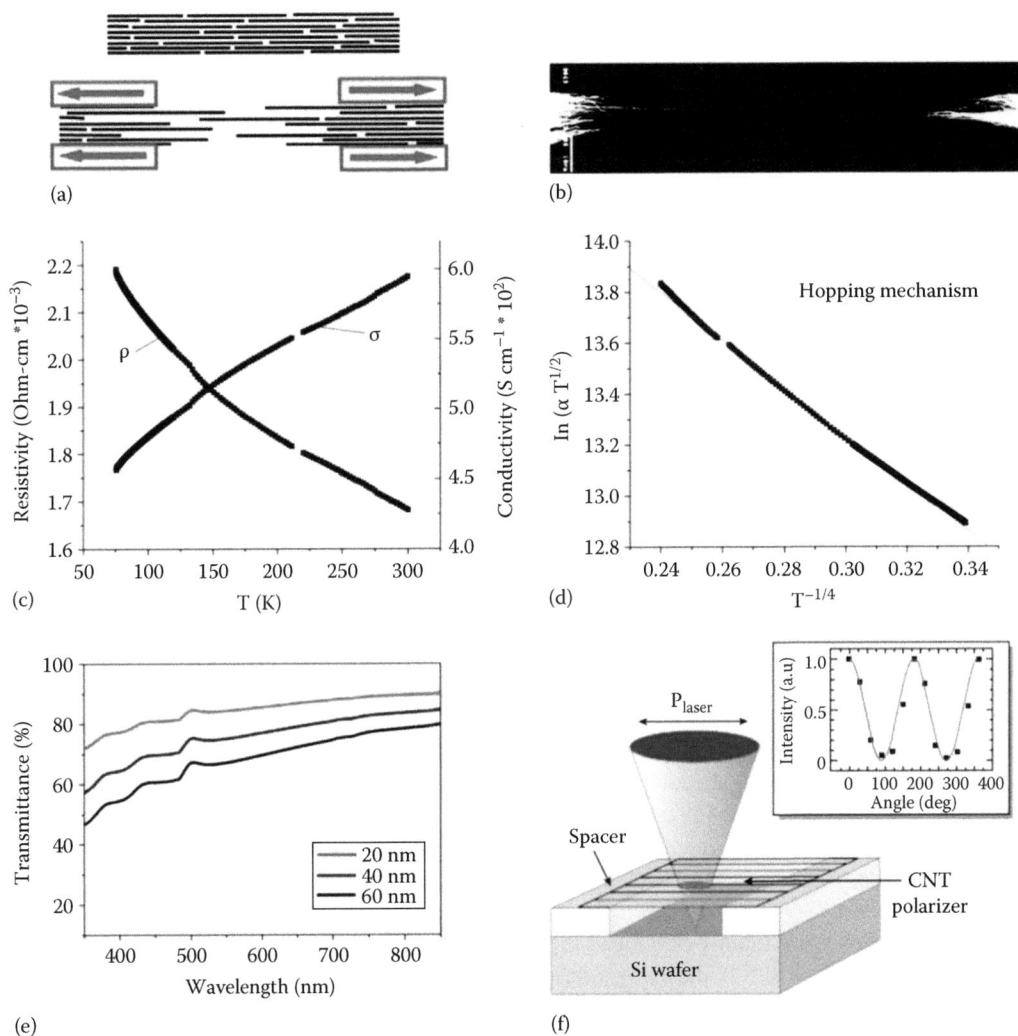

FIGURE 12.3
Mechanical, electrical, and optical properties of aligned CNT materials. (a) Schematic illustration of the fracture mechanism of a CNT fiber. (b) SEM image of a fractured CNT fiber (scale bar, 10 μm). (c) Temperature dependence of the resistance (ρ) and conductivity (σ) of a CNT fiber. (d) Fitting of the conductivity of a CNT fiber based on the variable range hopping mechanism. (e) Transmittances of CNT sheets of different thicknesses measured by UV–vis spectroscopy. (f) Schematic illustration of a CNT polarizer. The inserted is the dependence of the normalized intensity against the polarization angle. (Reprinted with permission from Vilatela, J.J. et al., *ACS Nano*, 5, 1921, 2011; Li, Q.W. et al., *Adv. Mater.*, 19, 3358, 2007; Sun, H. et al., *J. Mater. Chem. A*, 1, 12422, 2013a; Jiang, K.L. et al., *Nature*, 419, 801, 2002.)

Moreover, the fibers from aligned CNT arrays exhibit a 4.5 times higher tensile strength and Young's modulus compared with the fibers from wavy arrays (Zheng et al. 2010).

Parameters of the CNT fiber geometries, including helix angle and fiber diameter, also exert a significant influence. The twist decreases voids among CNTs and densifies the fiber, creating more contacting interfaces and resulting more efficient load transfer in the CNT fibers. As a result, the strength and modulus are increased with the increasing angle until the peak is reached. Beyond the optimal angle, they will be decreased because the load compresses the fibers rather than transports along the fiber axis. Jiang et al. found that both tensile strength and Young's modulus of CNT fibers were decreased monotonously with increasing helical angles from 10° to 40° (Liu et al. 2010a). Li et al. further found that it showed a double-peak behavior between helical angle and strength (Zhao et al. 2010). The first peak (15°–20°) in strength was attributed to the balance between load transport and load compress in fibers. The second peak (~30°) in strength came from the hollow structural collapse of the CNT fibers that made stronger CNT bundles and smaller cross-sectional areas. In addition, it was found that the CNT fiber with diameter of ~10 μm exhibited the highest tensile strength under similar conditions, due to appropriate radial pressure and defects in the fiber (Liu et al. 2010a, Zhao et al. 2010).

The mechanical performance can be further improved through posttreatment and compositing with polymers (Dalton et al. 2003, Kozlov et al. 2005, Koziol et al. 2007, Ma et al. 2009). The strength of the CNT fiber was improved from 0.85 to 1.91 GPa after post-spin twisting (Zhang et al. 2007b). The solvent densification can enhance interactions, preventing the slippage among individual CNTs and increasing strengths from 0.63 to 1.10 GPa (Liu et al. 2010a). CNT fibers have been found to exhibit the highest specific strength after annealing at 300°C both in air and argon because the defects were healed (Yang et al. 2011b). The fiber treated by using concentrated nitric acid had a tensile strength of 1.52 GPa, increased by 52% compared with the untreated one, due to the enhanced interfacial interaction (Meng et al. 2012). The treatment of a chemical agent aided by ultraviolet (UV) irradiation resulted in an increase of 100% in specific strength and 300% in toughness of the CNT fibers by increasing the anchor points (Boncel et al. 2011). The tensile strengths of CNT fiber were improved from 0.32 to 0.42 GPa and from 0.21 to 0.8 GPa by incorporation of silica and polyacrylic acid, respectively (Guo et al. 2012a). Up to now, the optimized tensile strength of CNT fibers can reach as high as 8.8 GPa measured at a small gauge length and obtained from gas-phase-CVD-grown CNTs (Koziol et al. 2007).

12.3.2 Electrical Properties

The electron transport in aligned CNT fiber is different from that in individual CNTs. It is reported to possess a 3D hopping mechanism according to the temperature dependence of conductivity (Figure 12.3c and d). Although the electrical conductivity of aligned CNTs is as high as 10^3 S cm^{-1}, it is still two orders of magnitude lower than that of the individual CNT. This decrease mainly arises from the nonuniformity of the CNTs including the length, diameter, alignment, and impurity, which inevitably increases scattering and interfacial contact resistances among CNTs. Therefore, a lot of efforts are made to decrease the contact resistance in order to improve the electrical conductivity of aligned CNTs.

Two approaches are mainly developed for improving the conductivity of CNT fibers. One is to modify the structure of CNTs during synthesis involving diameter, wall, length, crystallinity, and alignment. The other one is to reduce the impurities through posttreatments like solvent densifying, annealing, and chemical doping. The electrical conductivity of CNT sheets can be improved by stacking CNT sheets together into multilayers with the introduction of chemical treatment (Di et al. 2012). The electrical conductivities of CNT fibers are increased from 595 to 818 and 969 S cm^{-1} when oxidized in air and HNO$_3$, respectively, while decreased to 29 S cm^{-1} when annealed in hydrogen atmosphere (Li et al. 2007). It was found that the π-conjugated system is critical to maintain the conductivity. The oxidation dopes the CNT without interrupting the π-conjugated system. The conductivity of CNT fibers can be further improved to be 13,000 S cm^{-1} by using KAuBr$_4$ doping solution (Alvarenga et al. 2010). Conversely, annealing in hydrogen atmosphere gives rise to the formation of sp^3 carbon that destroys the π-conjugated structure of CNTs, resulting in a dramatic decrease of conductivity.

12.3.3 Thermal Properties

An individual CNT has longitudinal thermal conductivity of more than 3000 W m^{-1} K^{-1} (Kim et al. 2001). An idealized CNT comprises atomically perfect sp^2 carbon networks, which is conducive for phonon transport. However, similar to their strength and conductivity, the thermal conductivities of CNT assemblies, that is, arrays, sheets, and fibers, are several orders lower, ranging from 0.5 to 60 W m^{-1} K^{-1} (Badaire et al. 2004, Ericson et al. 2004, Aliev et al. 2007, Koziol et al. 2007, Behabtu et al. 2013). The dramatic decrease is mainly attributed to the morphology difference including nanotube defects, local disorder, and interconnections among CNT bundles. The extremely high surface area also leads to an excessive radial heat radiation that hinders the thermal performance of CNTs. Density is an important factor to the thermal conductivity of aligned CNTs. As reported, the thermal conductivities were increased from 2.5 to 50 and 60 W m^{-1} K^{-1} with the density of aligned CNTs rising from 0.039 to 0.6 and 0.9 g cm^{-3} (Aliev et al. 2007, Jakubinek et al. 2012, Pöhls et al. 2012). It should be noted that the aligned CNTs exhibit an anisotropic thermal conductivity. Generally, the thermal conductivity was 50 W m^{-1} K^{-1} in longitudinal direction, while ~2.1 W m^{-1} K^{-1} in the transverse direction of CNT orientation (Aliev et al. 2007).

The thermal conductivity of aligned CNTs can be improved by increasing quality of CNTs and doping with other components. It was reported that water-assisted CVD could produce a clean and uniform CNT array, which led to a less defective CNT fiber with a higher thermal conductivity of 80 W m^{-1} K^{-1} (Jakubinek et al. 2012). Moreover, the CNT fibers doped with chlorosulfonic acid showed a thermal conductivity of 380 W m^{-1} K^{-1}, which was 10 times higher than the nontreated fibers. The additional doping in iodine further improved the thermal conductivity to 635 W m^{-1} K^{-1} (Behabtu et al. 2013).

12.3.4 Optical Properties

A single-layered CNT sheet exhibits high transmittance (>85%) for visible (vis) light and thus can be used as a transparent electrode for flexible optoelectronic devices (Figure 12.3e) (Peng 2007, Yang et al. 2011a, Sun et al. 2013a). As previously reported, SWCNTs and nanowires have a polarized light absorption along their axes (Li et al. 2001, Wang et al. 2001). Due to the similar anisotropic structure, with an average diameter of 10 nm, aligned CNTs exhibit a polarization effect for light absorption. When a beam of light is incident on the aligned CNTs, photons with a polarization direction parallel to the axis of the CNTs will be absorbed, while those perpendicularly will pass through (Figure 12.3f). The degree of polarization of the aligned CNT polarizer, $P = (I_{max} - I_{min})/(I_{max} + I_{min})$, reaches up to 0.92 (Jiang et al. 2002). Therefore, highly aligned CNT sheets can be used as polarized infrared thermal detectors, simultaneously realizing the detection of power intensity and polarization in a simple integrating method (Xiao et al. 2011). Furthermore, the aligned CNT bundles, as lamp filaments, can emit an axially polarized light with a polarization degree of 0.33 (Li et al. 2003b).

12.3.5 Electromechanical Properties

Due to high electrical conductivities, mechanical strengths, and electrochemical activities, CNTs can be electromechanically actuated deriving from the contraction and expansion caused by the charge injection and extraction (Figure 12.4a) (Baughman et al. 1999). CNTs can also be pneumatically actuated with a 300% strain, driving by the electrochemically generated gas confined within the CNT scaffolds (Spinks et al. 2002). Aligned CNTs with anisotropic properties is favorable for charge accumulation and transport, making them eligible as electromechanical actuators with large and rapid actuating responses. In fact, it has been observed that the aligned CNT fibers could generate a stress one order of magnitude greater than that achieved with unaligned assemblies of CNTs (Viry et al. 2010). Technically, many electromechanical responses of aligned CNTs result from the charge ingress and outflow, so they are usually investigated in liquid electrolytes.

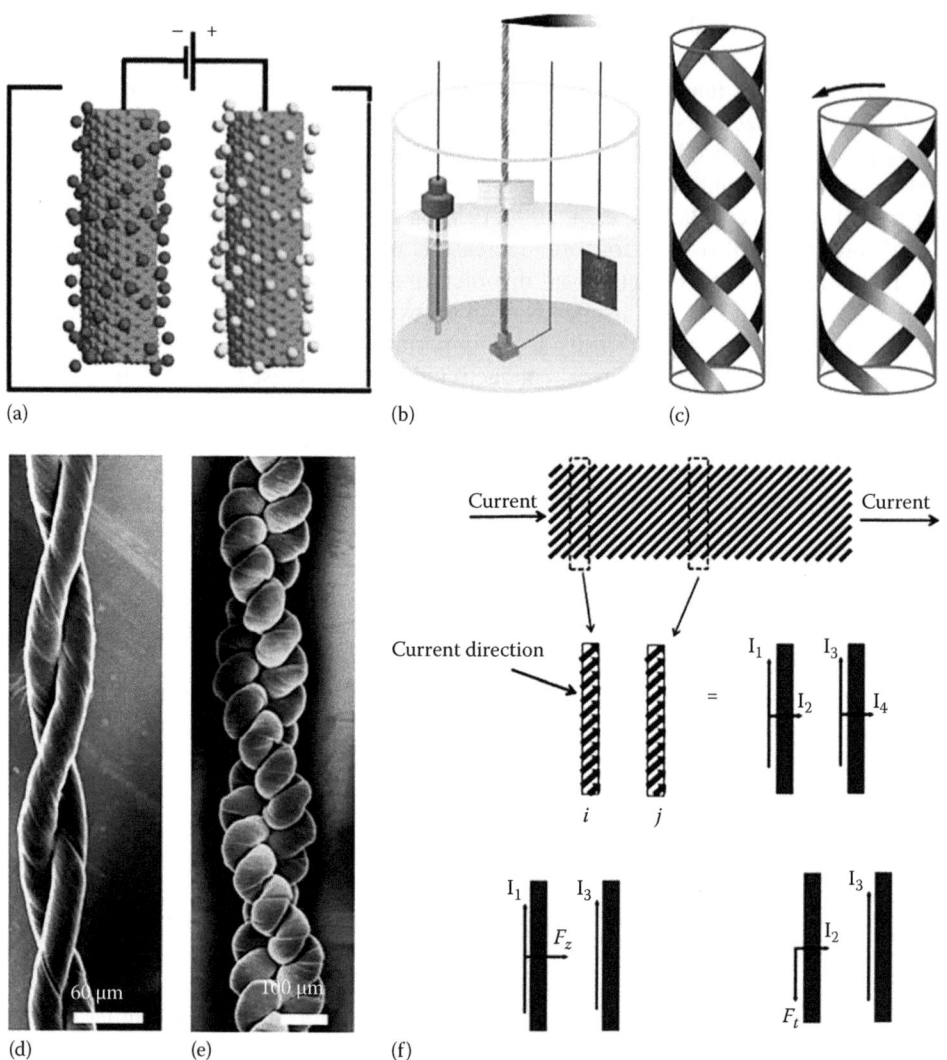

FIGURE 12.4
The actuation performance of the CNT fiber. (a) The generated electrical layer with opposite charges along two CNT electrodes in an electrolyte when voltage is applied. (b) Schematic illustration of an electrochemical cell used for actuation tests with a three-electrode system in electrolyte. (c) The volume expansion of the CNT fiber during charge injection. (d) and (e) SEM images of twisted and overtwisted two-ply CNT fibers for torsional and tensile actuations, respectively. (f) The mechanism for the contraction and rotation of a CNT fiber according to Ampere's law among helically aligned CNTs. (Reprinted with permission from Mirfakhrai, T. et al., *Smart Mater. Struct.*, 16, S243, 2007; Foroughi, J. et al., *Science*, 334, 494, 2011; Lee, J.A. et al., *Nano Lett.*, 14, 2664, 2014; Guo, W. et al., *J. Mater. Chem.*, 22, 903, 2012a; Guo, W. et al., *Adv. Mater.*, 24, 5379, 2012b.)

Various aligned CNT assemblies, array, sheet, and fiber, can provide an electromechanical response. For example, a square CNT array tower, with a width of 1 mm and height of 4 mm, was peeled off from the substrate and connected with an affixed electrode in a 2 M NaCl solution (Yun et al. 2006). A three-electrode electrochemical system was used to study its electrochemical actuation, with the CNT array tower as the working electrode, the Ag/AgCl electrode as the reference electrode, and the platinum plate as the counter electrode. The CNT array showed a unidirectional strain of 0.15% at 2 V and was actuated up to 10 Hz with small decrease in strain.

CNT sheet shows an anisotropic mechanical strength. In a transverse direction, they can be stretched by up to 300%, which is beneficial for large-strain actuation (Aliev et al. 2009). The CNT sheet actuator was developed with a CNT sheet as a working electrode and a distant ground plane as a counter electrode. Upon applying a positive voltage of 5 kV, the CNT sheet expanded in the width direction and showed a giant deformation of 220%. The average actuation rate was $3.7 \times 10^4\%$ s^{-1} at operating temperatures from 80 to 1900 K, indicating a rapid response. However, the CNT sheets required high voltage of kilovolts to achieve such giant strains and had low tensile strengths.

CNT fibers, densely packed and twisted, have a higher tensile strength that can provide better performance of actuation. The mechanisms of the actuation including contraction and rotation are attributed to charge/ion injection in electrolyte and electromagnetic force. The CNT fiber is actuated to contract in a liquid electrolyte with a three-electrode system. CNT fiber, aqueous Ag/AgCl, and polypyrrole (PPy) sheet were used as working, reference, and counter electrodes, respectively (Mirfakhrai et al. 2007). The CNT fiber was found to longitudinally contract when charged and recovered when discharged. The maximal actuating strain was 0.5% at 2.5 V. The mechanism of this contraction was related to the structure changes within CNTs and a radial swelling due to its helical structure upon the insertion of ions in the fiber.

Apart from contraction, CNT fiber can also be actuated to rotate, which is favorable as an artificial muscle. Generally, the CNT fiber is assembled into a three-electrode electrochemical system and immersed in liquid electrolyte (Figure 12.4b). The CNT fiber provided a reversible rotation of 15,000°C and speed of 590 revolutions min^{-1} (Foroughi et al. 2011). The rotation was actuated by a hydrostatic pressure generated from the ion influx and release during electrochemical charge/discharge. In this process, the fiber volume was changed resulting simultaneous lengthwise contraction and torsional rotation (Figure 12.4c).

Recently, all-solid-state CNT fiber torsional and tensile actuators are developed, without the use of the conventional three-electrode system. The actuation of the CNT fiber could conduct in the absence of aqueous circumstances by adopting a solid gel electrolyte (Lee et al. 2014). Unlike the previously discussed actuators, which commonly consist of single CNT fibers, this actuator displays a unique architecture. The twisted, uncoiled CNT fibers were first prepared and infiltrated with solid gel electrolyte. Two CNT fibers were plied together using an opposite twist direction than for the single fiber twist and used as anode and cathode for torsional muscles (Figure 12.4d). The two-ply CNT fiber showed reversible rotation of 53° mm^{-1} at 5 V. The two twisted and coiled CNT fibers infiltrated with solid gel electrolyte were plied together for the tensile actuation (Figure 12.4e). Contractions of 0.52% and 1.3% were obtained at 1 and 2.5 V upon loads of 11 and 10.1 MPa, respectively. This load lifting capability was ~25 times higher than that of human skeletal muscle, enabling promising applications in artificial muscle.

The CNT fiber actuation is not exclusively conducted in an electrolyte. The electromechanical contraction and torsion of CNT fibers can be realized in a broad environmental media, including air, water, and organic solvents without electrolytes. Moreover, the actuation of CNT fibers can be performed by directly passing a low electric current along them, without the use of electrolytes and a relatively complex three-electrode system (Guo et al. 2012b). The CNT fiber was directly connected to an external current source in air. When passing through a direct current of several microamperes, the fiber contracted along the axial direction, which was accompanied by a rotation and self-twisting. The contraction strain and rotation angle were higher than 2% and 360°, respectively. After removal of the current, the CNT fiber was fully recovered to its original state, indicating a decent reversibility. The CNT fiber with a length of 5 mm generated a stress of 6 MPa under a current of 4 mA. The similar phenomena were observed in various liquids such as water, organic solvents, and electrolyte solutions. A different mechanism, Ampere's law among helically aligned CNTs, was proposed to explain the phenomena (Figure 12.4f). When a current flows in the CNT fiber, the electromagnetic forces are produced from parallel arranged CNTs. The collective force generated by more than a million CNTs in the fiber is high enough to induce macroscopic electromechanical actuation of longitudinal contraction and torsional rotation.

In addition to the electromechanical actuation of CNT fibers based on charge injection and electromagnetic force, the electrothermal heating, direct heating, light, and chemicals can also trigger the torsional and tensile actuation induced by expansion and structural transformation of guest-filled CNT fibers such as wax, bismaleimide, and poly(vinyl alcohol) (Lima et al. 2012, Meng et al. 2014).

12.4 ELECTROCHROMATIC POLYMER COMPOSITES BASED ON ALIGNED CNTs

The remarkable properties of the CNTs, including mechanical strength, electrical and thermal conductivity, and chemical stability, make their aligned assemblies uniquely promising for a broad range of applications, especially as building blocks for composites (Sun et al. 2012a). Many functional materials, such as inorganic materials, small organic molecules, and polymers, can be incorporated to produce high-performance composite films and fibers. Particularly, the high electrical properties of the aligned CNTs enable the composite a high electrochromic sensitivity. For instance, aligned CNTs were incorporated with tungsten oxide to obtain an electrochromic composite showing color change from yellowish to blue. The aligned CNTs served as the scaffold and provided an anisotropic, continuous pathway for electron transport, and tungsten oxide acted as an electrochromic component (Yao et al. 2012). The aligned CNTs could also composite with liquid crystal molecules to prepare a prototype liquid crystal display cell, which showed dark and bright when switching a voltage on and off (Fu et al. 2010). In addition, the synergetic interactions between aligned CNTs and functional polymers have been found to create some new electrochromatic properties (Sun et al. 2012a).

12.4.1 Electrochromatic Polydiacetylene/ CNT Composite Fiber

Polydiacetylene (PDA) changes colors in response to various external stimuli such as temperature, pH, mechanical rubbing, chemical reagent, light, and magnetic field and has been widely explored as a material for chromatic sensors (Sun et al. 2010a,b, Chen et al. 2011b). This color transition, typically from blue to red, is identified as the conformation change of PDA backbone induced by stimuli. However, most of the observed chromatic phenomena of PDA are irreversible under limited stimuli, which hinder its practical applications. Therefore, massive efforts have been made to improve the reversibility and sensitivity and increase the scope of stimuli (Peng et al. 2005, 2006), although electrochromism is rarely realized due to the very low conductivity of bare PDA. To this end, CNT fibers with high conductivity and electron transport following a 3D hopping mechanism are promising candidates to incorporate PDA to realize the desired electrochromism.

PDA/CNT composite fibers are prepared by dipping bare CNT fibers into a solution of diacetylene monomers, followed by topochemical polymerization under UV light (Peng et al. 2009). CNTs remain highly aligned and PDA is uniformly distributed in the resulting PDA/CNT composite fiber. The composite fiber exhibits a high strength of 1.1 GPa and conductivity of 350 S cm^{-1}, respectively. When passing with a direct current of 10–30 mA, the PDA/CNT composite fiber showed a color change from blue to red and recovered to blue when the current was cut off (Figure 12.5). The chromatic transition was completed within a second, indicating a rapid responsive capability. The aligned CNT plays a critical role in the electrochromatic behavior. When an electric current passed through the PDA/CNT composite fiber, the polar end groups of the side chain in PDA are inclined to reorganize in response to the generated electric fields among aligned CNTs, which further induces the conformation change of PDA backbones and finally leads to a chromatic transition. Benefited from its high flexibility, strength, and conductivity, the PDA/CNT composite fiber could be particularly useful for nondestructive detection and structure monitoring in a broad spectrum of fields from aircraft to small electronic facilities.

12.4.2 Electrothermal Chromatic Polymer/CNT Composite

Another sensing polyacetylene derivative, poly[1-phenyl-2-(p-trimethylsilyl)phenylacetylene] (PTP), exhibits a reversible change in fluorescent intensity induced by heat and solvent, which results from the molecular perturbation and exciton deconfinement within the PTP backbone (Kwak et al. 2006, 2008). Particularly, the PTP has a higher thermal stability than PDA due to a rigid

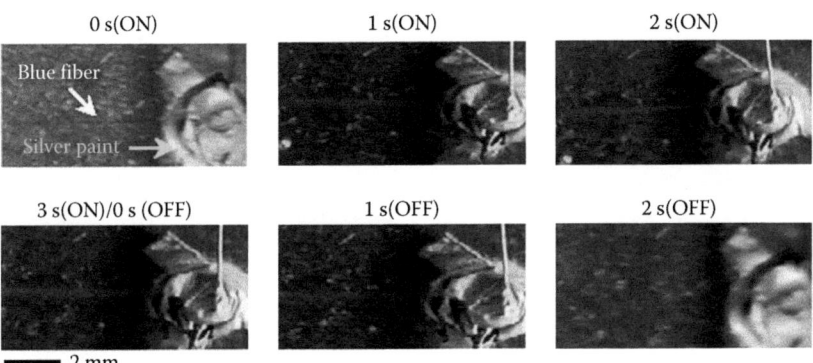

FIGURE 12.5
Chromatic transitions of a PDA/CNT composite fiber in response to an electric current. (Reprinted with permission from Peng, H. et al., *Nat. Nanotechnol.*, 4, 738, 2009.)

jacket, which is formed through the intra- and interchain molecular interactions of benzene and silicon moieties in the side chains and shields the conjugated PTP backbone from the thermal attack. In fact, the decomposition temperature of PTP is measured up to 400°C. Therefore, a new family of electrochromatic composites with high thermal stability and reversibility can be prepared by incorporating aligned CNTs into PTP.

The PTP/CNT composite fibers are prepared by coating PTP solution onto CNT fibers, followed by evaporation of solvent (Sun et al. 2013c). The resulted composite fibers exhibit a rapid change in fluorescent intensity as well as color with electric currents ranging from 5 to 40 mA. The color change can repeat for more than a thousand cycles without any fatigue in structure and sensitivity. After incorporating aligned CNTs, the PTP/CNT composites have a decomposition temperature of higher than 500°C, tensile strength of ~500 MPa, and electrical conductivity of ~370 S cm^{-1}. Moreover, the PTP/CNT composite film can also be prepared by coating the PTP solution onto CNT sheets (Figure 12.6a and b). Similarly, the composite film showed a rapid and reversible fluorescent quenching and color change when currents were applied, but with a much wider range of 20–240 mA. A gradual color change from yellowish green to dark green was clearly observed (Figure 12.6c), and the chromatic transition finished within a second with current on and off. As previously declared, the aligned CNTs function as continuous pathways for electron and phonon transport and induce the reorganization of polymer chains under the electric current, leading to a color change. The electrochromatic PTP/CNT composite films can be made at a large scale with flexibility and are promising as electric current indicator similar to the pH test paper.

The polymer for electrothermal chromatic composite can be expanded to heterochain polymers, polyurethane (PU), for example. PU shows a shape memory or switchable transparency in response to temperatures. Normally, PU has hard segments (e.g., 4,4′-diphenylmethane diisocyanate and 1,4-butanediol) and soft segments (e.g., aliphatic polyester polyol) in its backbone. The soft segment of large molecules intends to crystallize into spherulites, so PU is opaque due to the scattering of crystal grains and boundaries at room temperature. When heated, the soft segments melt into amorphous structures and PU becomes transparent due to the refraction, reflection, and transition of light. The transmittance change was simply a result of phase transition of the soft segments under heating, which, however, is very slow due to the limited heat transport. Meanwhile, there are some places inconvenient to be provided with heat, which limited the application of PU. Considering the remarkable electrical and thermal conductivity of the aligned CNTs, CNT sheet is introduced into PU film to realize the electrothermal transmittance change and improve the switchability.

The PU/CNT composite films are prepared by sandwiching a layer of CNT sheet between two PU films through a simple solution process (Figure 12.6d) (Meng et al. 2011). The obtained PU/CNT composite film exhibited a rapid transition from opacity to transparence in a few seconds under an electric current and slowly reversed after the current was cut off (Figure 12.6e). The opaque-to-transparent transition results from the heat generated from the current, which induces the phase transition of the soft segments in PU. One layer of CNT sheet provides an electrically conductive pathway but with a relatively high resistance. Moreover, the spatial

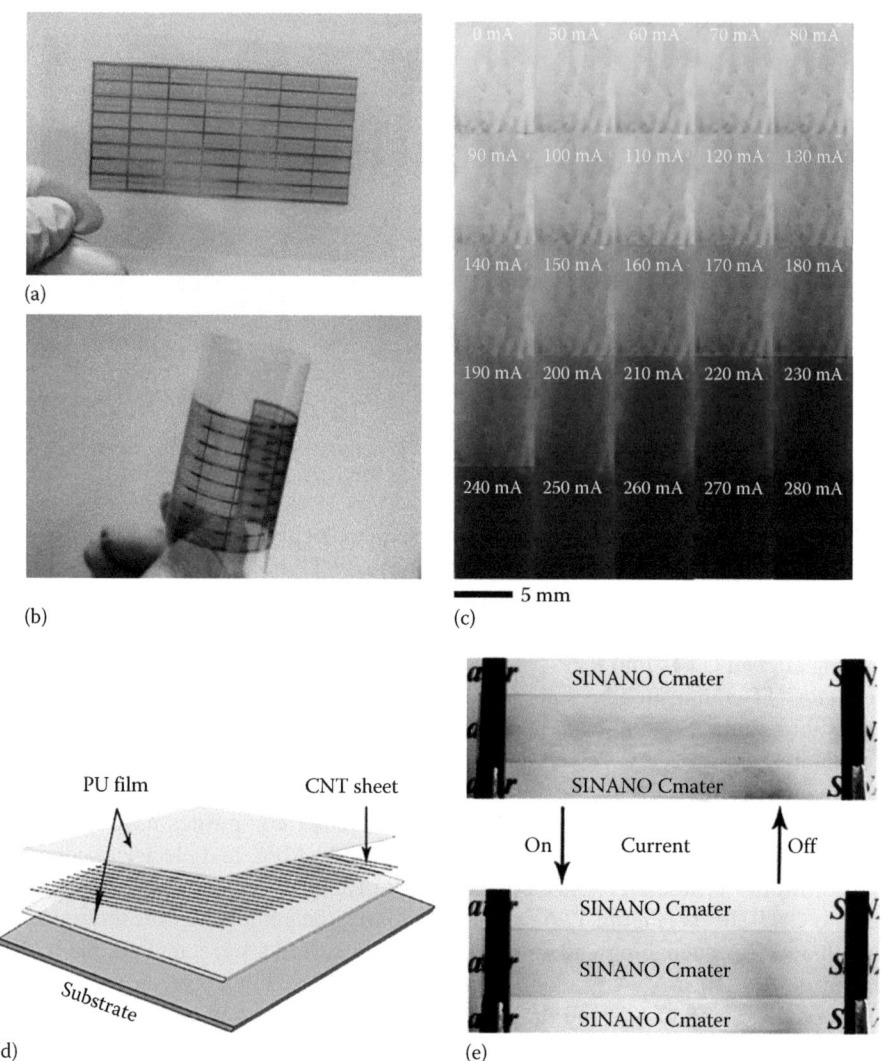

FIGURE 12.6
(a) and (b) Photographs of a flexible PTP/CNT composite film on a poly(ethylene terephthalate) substrate. (c) Photographs of a PTP/CNT composite film under different currents from 0 to 280 mA. (d) The structure of a PU/CNT composite film. (e) The reversible change of transmittance of a PU/CNT composite film with an electric current off and on. (Reprinted with permission from Sun, H. et al., *J. Mater. Chem. A*, 1, 12422, 2013a; Sun, X. et al., *Adv. Mater.*, 25, 5153, 2013b; Meng, F. et al., *ACS Appl. Mater. Inter.*, 3, 658, 2011.)

confinement of aligned CNTs and interaction between aligned CNTs and PU induce the formation of small and uniform spherulites along the aligned direction leading to a decreased melting point of soft segments from 52°C to 47°C and faster transition. In fact, for a 1.4 cm long PU/CNT film, it took 5 s for transition at a voltage higher than 30 V, while 15–20 s for a bare PU film to be transparent on a hot stage with a temperature of 60°C. This method was easy to operate with low energy cost and can be used at a large scale for various other polymers applied in optically switchable windows.

12.4.3 Electrochemical Chromatic Polymer/CNT Composite

When chromatic polymers meet energy devices, it is expected to create new ideas to develop smart devices. For example, by integrating the electrochromic polymer with a supercapacitor, one is able to monitor the charge state of a supercapacitor by displaying different colors. Polyaniline (PANI), a widely used active material in supercapacitor, exhibits a typical electrochromatic behavior at different oxidation states (Argun et al. 2004). PANI is green in its intrinsic state and turns to blue at oxidized state and light yellow at reduced state. Aligned CNTs can be introduced to enable high electrical conductivities, electrochemical and thermal properties of PANI, and improve the rate and contrast of electrochromism. Therefore, a smart supercapacitor can be fabricated by compositing PANI with aligned CNTs as electrodes (Lin et al. 2013).

The PANI/CNT composite film is prepared by first paving a layer of CNT sheet onto a flexible and transparent substrate followed by depositing PANI through electropolymerization of aniline. The supercapacitor was finally fabricated by assembling two PANI/CNT composite films together coated with electrolyte (Figure 12.7a) (Chen et al. 2014). The resulting PANI/CNT composite films are explored as both electrochromic display and electrodes of supercapacitor. CNTs have a high conductivity for charge transfer and accumulation, while the PANI has a chromatic behavior and dominates the capacitance of the device. Specifically, the device displays light yellow, green, and blue when charged from 0 to 1 V and reverses when a negative voltage is applied, as shown in Figure 12.7b. The electrochromism, indicating different redox states of PANI, makes the charge state of the device obviously detected. Moreover, the unique properties of aligned CNTs and the special fabrication process enable high stretchability and flexibility for supercapacitors, which paints an encouraging picture for smart devices.

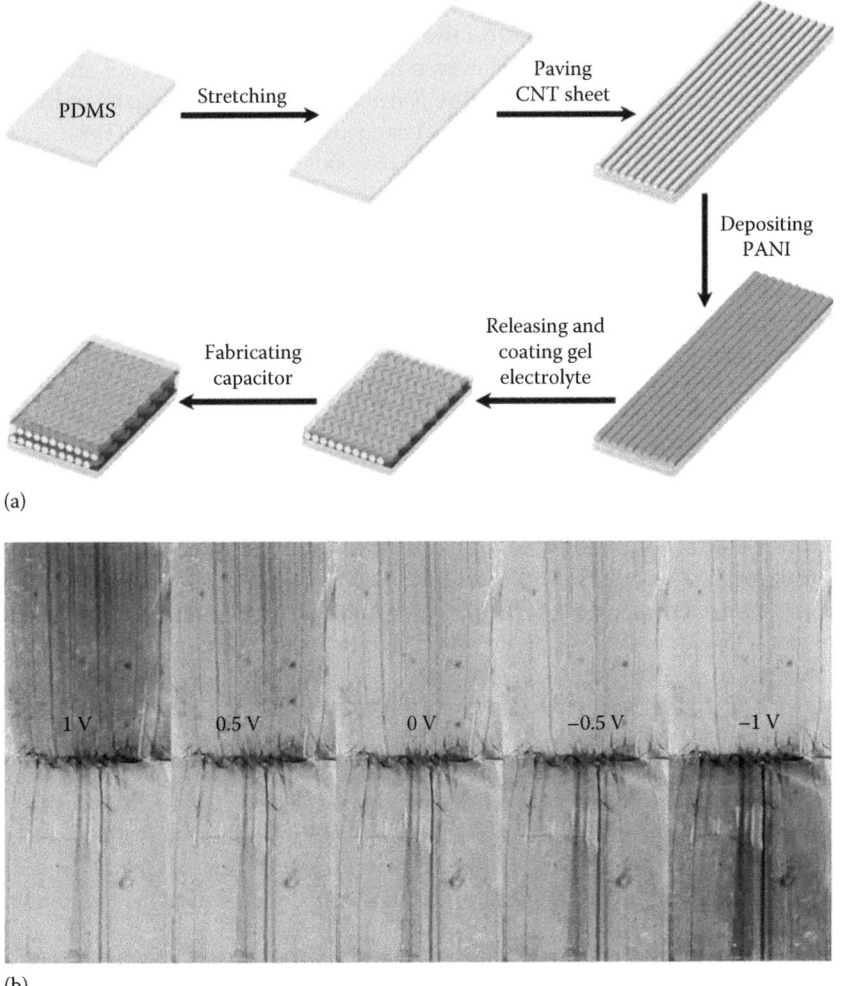

(a)

(b)

FIGURE 12.7
A chromatic supercapacitor based on the PANI/CNT composite film. (a) Schematic illustration of the preparation process. (b) The chromatic transition during a charging–discharging process. (Reprinted with permission from Chen, X. et al., *Adv. Mater.*, 26, 4444, 2014, doi: 10.1002/adma.201400842.)

12.5 DEFORMABLE POLYMER COMPOSITES BASED ON ALIGNED CNTs

In many cases, the sensitive polymers are expected to provide a direction-selective actuation in response to external stimuli, such as electric field, light, and chemical species. Amorphous polymers, however, are reluctant to generate an anisotropic deformation under external stimuli because of its disordered arrangement in assemblies and generate a low degree of deformation (Ikeda et al. 2007). The orientation of polymer chains indeed contributes to an anisotropic behavior in deformation. Aligned CNT arrays and sheets have shown a unique anisotropic structure and excellent properties, which are promising to induce the orientation of polymer chains by mutual interaction. Therefore, the introduction of aligned CNTs is conducive to achieve the anisotropic actuation with higher degree.

12.5.1 Electroactive Polymer/CNT Composite

The ionic electroactive polymer is a branch of sensitive polymers and shows a unique electrome-chanical actuation behavior. It can be directly integrated with microelectronic controlling circuits to perform complex actuations (Brochu and Pei 2010). Among them, conducting polymers, such as PPy, have the advantages of low operating voltage, large deformation, and high mechanical strength (Madden et al. 2000). They exhibit electroactive deformations due to the uptake of counter ions during electrochemical redox processes. However, the conducting polymers are annoyed by some inevitable problems. For instance, the durability is deteriorated by stress relaxation like creep; the electromechanical response is slow due to the low conductivity. The aligned CNTs provide an effective approach to overcome the inherent limitation of conducting polymers. As we declared, the aligned CNTs possess a high conductivity and mechanical strength, which can be introduced as second component to enhance its electromechanical response.

Similar to PANI, the PPy/CNT composite film is prepared by electropolymerizing pyrrole on the CNT sheet, and multilayer composites can be obtained by alternating PPy/CNT films and bare PPy layers (Zheng et al. 2011). The orientation and weight percentage of CNTs can be simply controlled in this process. The resulted PPy/CNT composite films have a several fold enhanced strength and modulus than bare PPy film, and the electrical conductivity is 50–60% improved compared with bare PPy film. The PPy/CNT composite film exhibits an anisotropic deformation, with elongation of 7.5% in transverse direction while less than 1% in longitudinal direction, which arises from the anisotropic property of CNT sheets, compared to a 5.7% deformation for the bare PPy (Figure 12.8a). In longitudinal direction, the CNT sheets have a higher strength and modulus than the transverse direction, leading to a high stiffness and low electrochemical strain (0.2%). Moreover, the incorporation of CNT sheets also dramatically reduces the creep during cyclic deformation (Figure 12.8b). For example, less than 0.2% creep strain occurred in the PPy/CNT composite in 35 min under a load of 6 MPa, while the bare PPy film exhibited more than 2% creep strain in just 15 min under a load of only 4 MPa.

Apart from PPy, Nafion, the trade name for a sulfonated tetrafluoroethylene–based fluoropoly-mer copolymer, is another ionic electroactive polymer, which possesses high ionic conductivity and excellent mechanical properties (Mauritz and Moore 2004). Commonly, the actuation of conductive polymer is accomplished in electrolyte solutions. Nafion, however, is luckily exempt from that concern, because it can absorb ionic liquids as electrolytes, enabling an in-air actuation. The electrome-chanical performance of Nafion is derived from the ion transport in materials under a certain voltage. The concentration and mobility of ions and the modulus of materials determine the responding rate, strain, and actuation efficiency. Aligned CNTs show high volume fraction, conductivity, and modulus and are expected to incorporate and improve the electromechanical performance of Nafion.

The composite actuator is made in a sandwich structure, two Nafion/CNT composite films as electrodes separated by a Nafion ionomer, which is absorbed with imidazolium ionic liquids as electrolytes (Figure 12.8c) (Liu et al. 2010b). When applying a voltage, the ions accumulate at one Nafion/CNT composite electrode while depleted at the other one. The ion transfer leads to an opposite volume strain: one expands while the other contracts. As a result, the composite film shows a bending behavior actuated by voltage (Figure 12.8d). The remarkable properties of aligned CNTs

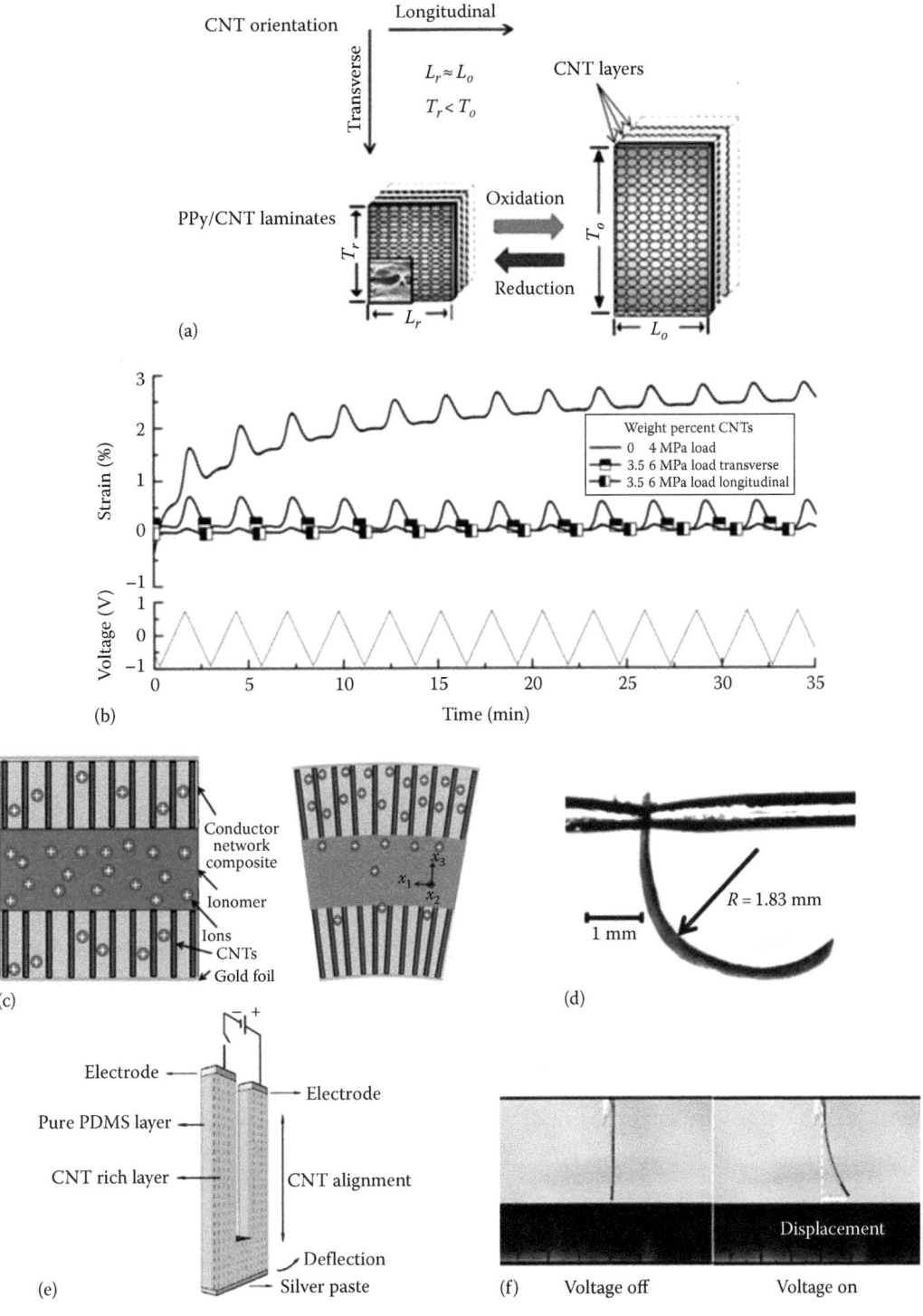

FIGURE 12.8

The electroactive deformation of aligned CNT-based composites. (a) Schematic illustration of the actuation performance of PPy/CNT laminates. (b) The actuation behavior of the bare PPy and PPy/CNT laminates in response to pulse voltage. (c) Schematic illustration of a three-layer bimorph actuator with Nafion/CNT composite films as electrodes with voltage off and on. (d) An optical image of the bending behavior of a Nafion/CNT composite actuator under a voltage of 4 V. (e) The structure of a PDMS/CNT composite actuator. (f) The bending behavior of a PDMS/CNT composite actuator with a voltage of 40 V on and off. (Reprinted with permission from Zheng, W. et al., *Adv. Mater.*, 23, 2966, 2011; Liu, S. et al., *Adv. Funct. Mater.*, 20, 3266, 2010b; Chen, L. et al., *ACS Nano*, 5, 1588, 2011a.)

benefit the actuation performance of the composite film. High surface area provides abundant space for ion accumulation, resulting in a high ion concentration and large strain. High electrical conductivity provides the effective path for fast ion transport, ensuring a rapid actuation response. The alignment of CNTs facilitates the actuation amplitude due to anisotropy of mechanical properties. Therefore, the composite exhibited a strain of over 8% actuated by a 4 V pulse voltage with 0.5 Hz in frequency, and this actuation lasted more than 10 min without degradation.

In addition to electrochemical actuation, electrothermal actuation is an important branch of actuation behavior that is induced by electric current. Specifically, the actuation is derived from the mismatch in thermal properties like thermal conductivity and coefficients of thermal expansion. Therefore, an electrothermal actuator can be fabricated by assembling materials with different thermal properties. Aligned CNTs are electrically conductive and have high thermal conductivity, while different coefficients of thermal expansion from polymer, thus, can be used for electrothermal actuator via compositing with flexible polymers, such as polydimethylsiloxane (PDMS) (Chen et al. 2008).

A U-shaped PDMS/CNT composite film is prepared by embedding CNT sheets within the PDMS surface against the bare PDMS (Figure 12.8e) (Chen et al. 2011a). The large interfaces between aligned CNTs and PDMS restrict the thermal expansion of PDMS. Therefore, the longitudinal coefficient of thermal expansion of the CNT-rich PDMS layer (6×10^{-6} K^{-1}) is two orders of magnitude lower than that of bare PDMS (3.1×10^{-4} K^{-1}). Meanwhile, conductive CNT sheets inside can convert electrical energy to thermal energy and heat up the entire composite. When a 40 V voltage (current of 78.5 mA) was applied for 5 s, a free-end PDMS/CNT composite film (30 mm × 6 mm × 0.77 mm) bent immediately, with a displacement of 9.5 mm, and recovered to the original state slowly, typically 60 s, after the voltage was removed (Figure 12.8f). The bending actuation can be repeated for 100 cycles with a negligible change in displacement.

12.5.2 Light-Driven Liquid-Crystalline Polymer/CNT Composite

Light-driven polymers have an advantage over the aforementioned electroactive polymers, that is, they can realize a remote and wireless control, which is competitive as artificial muscles and photo mobile soft actuators due to easy fabrication and lightweight (Yamada et al. 2008). Liquid-crystalline (LC) polymers containing photoactive moieties such as azobenzene are intensively explored. The photoactive LC polymer exhibits a bending deformation arising from a change in the orientation of mesogens triggered by the *trans–cis* photoisomerization in azobenzene (Li et al. 2003a, Yu et al. 2003). The orientation of LC mesogens dominates the anisotropic deformation; thus how to effectively orient the LC mesogens is of great importance (Kondo et al. 2006). Traditionally, LC mesogens were oriented through an aligned polyimide layer with parallel grooves produced by mechanical rubbing. But this method brings about several problems such as the production of broken debris and structure damage and electrostatic charge accumulation on the surface. To this end, aligned CNT materials with parallel nanosized grooves are ideal templates to orient LC mesogens.

An azobenzene-containing liquid-crystalline polymer (ALCP) was investigated as a demonstration (Wang et al. 2012). The ALCP/CNT composite film was prepared by injecting the mixture of monomers, cross-linker, and initiator into the CNT sheet through a melting process followed by photopolymerization. The ALCP mesogens were effectively oriented along the CNT length (Figure 12.9a). The resulting composite film underwent a reversible bending deformation by alternate irradiations of UV and vis lights (Figure 12.9b). The reversible deformation can be stably repeated for more than 100 cycles without obvious fatigue.

Moreover, the orientation of mesogens can be further tuned by the fabrication process. When the precursors are introduced through a solution process, the mesogens prefer to orient perpendicularly to the aligned CNTs (Sun et al. 2012b). The resulting ALCP/CNT composite films bent against the incident direction of UV light and reversed by opposite irradiation, that is, solely UV light can induce this reversible deformation (Figure 12.9c and d). Although the bending behavior can be effectively repeated and attributed to the interactions between CNTs and ALCP, more efforts are still required to clearly understand the mechanism. The introduction of aligned CNTs provided the ALCP composites with excellent properties such as high mechanical strength up to ~1 GPa and

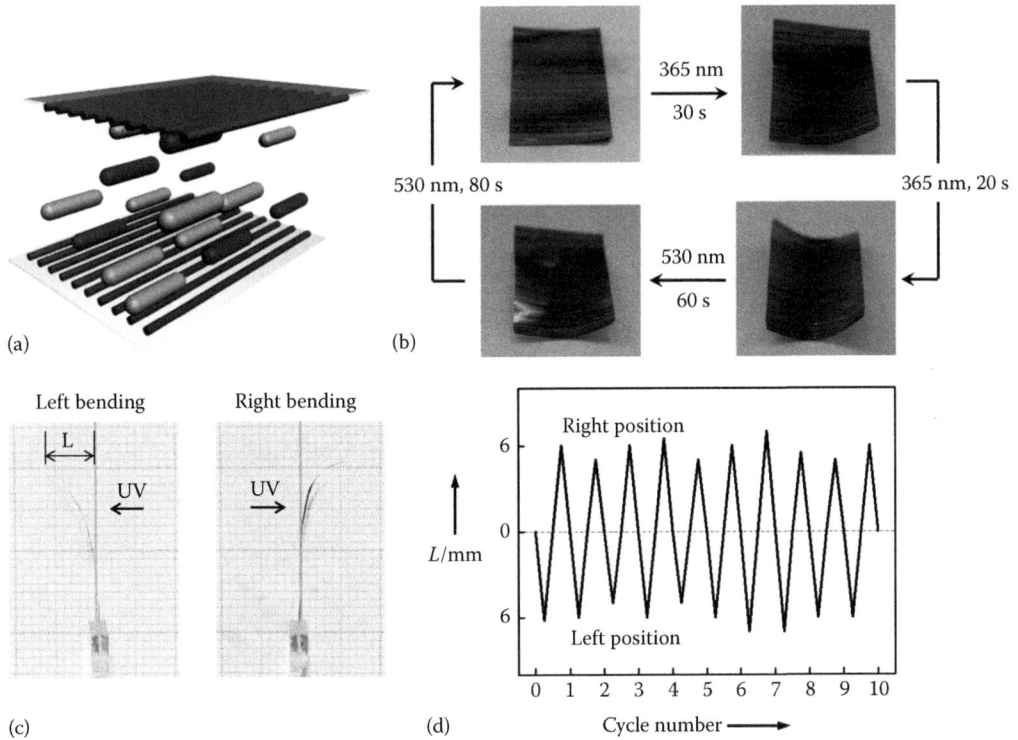

FIGURE 12.9
(a) Schematic illustration of the orientation of ALCP induced by aligned CNTs. (b) The reversible deformation of an ALCP/CNT composite film under alternate irradiation of UV and vis light. (c) The deformation behavior of an ALCP/CNT composite strip irradiated by UV light in opposite directions. (d) The displacement in (c) recorded at different cycles. (Reprinted with permission from Wang, W. et al., *Angew. Chem. Int. Ed.*, 51, 4644, 2012; Sun, X. et al., *Angew. Chem. Int. Ed.*, 51, 8520, 2012b.)

electrical conductivity of 10^2–10^3 S cm^{-1}. These properties enable the composite film some unique applications, such as remote control electric switch.

12.5.3 Solvent-Driven Deformable Polymer/CNT Composite

Besides LC polymers with mesogens in the side chains, aligned CNTs can also induce the orientation of polymers with conjugated backbones to achieve the anisotropic actuation. PTP, which shows chromatic behaviors as previously discussed, has rigid conjugated backbones with benzene and silicon moieties as side chains, resulting in a high free volume. The interchain distance can be changed upon absorption and evaporation of organic solvent (Kwak et al. 2008, Lee et al. 2012), which is expected to exhibit a remarkable actuation if properly designed.

The PTP/CNT composite films are prepared by spin coating a PTP solution onto the CNT sheets paved on a PDMS substrate, followed by evaporation of the solvent at room temperature (Lu et al. 2014). The resulted PTP/CNT composite film exhibited a reversible deformation, for example, rolling into a tube and recovering to the flat format triggered by organic solvents such as ethanol (Figure 12.10a). The bending and unbending can be completed in ~0.6 s and less than 1 s for recovery. In particular, the anisotropic deformations are produced along the direction perpendicular to the CNT length. It probably derives from the preferred orientation of conjugated polymers induced by aligned CNTs and reversible reorganization of polymer backbones with different interchain distances driven by the solvent. The removal of the solvent made the backbones closely stacked, while the diffusion of the solvent reversed the process (Figure 12.10b). The PTP/CNT composite film maintains intact even after repeating for 300 cycles, suggesting a considerable deformation stability. The generated stress during actuation is up to 15 MPa, which is 43 times of the strongest natural muscle. Therefore, these composite films can be widely used as a new family of artificial materials, such as the use of a miniature crane to lift up the object and an artificial flower to mimic the leaf of the Venus flytrap.

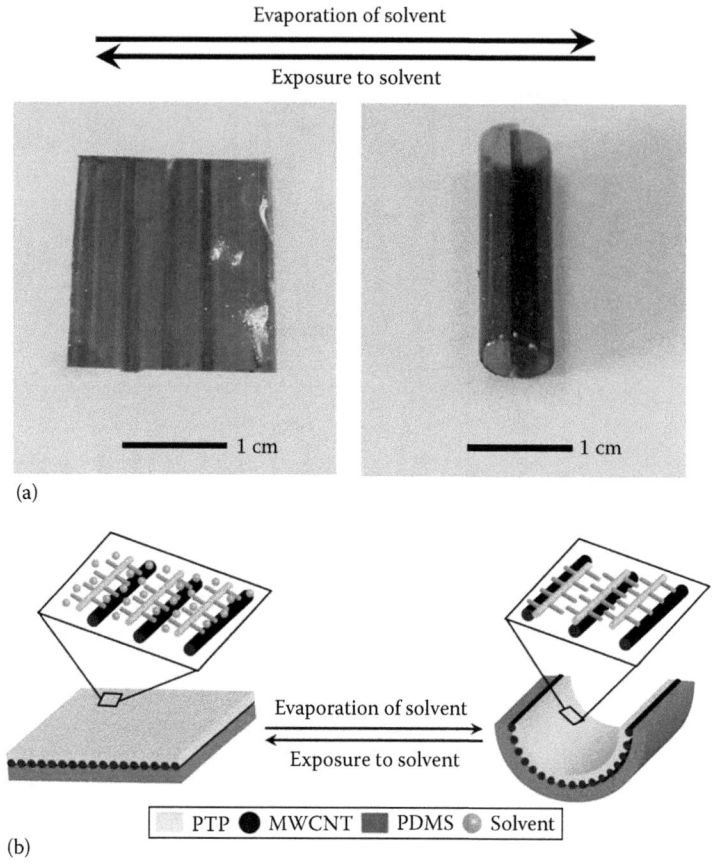

FIGURE 12.10
(a) The reversible deformation of a PTP/CNT composite film induced by a solvent. (b) Schematic illustration of the mechanism of the reversible deformation of (a). (Reprinted from Lu, X. et al., *J. Mater. Chem. A*, 2, 17272, 2014. With permission.)

12.6 CONCLUSION AND PERSPECTIVE

Individual CNTs have remarkable properties, and aligned CNTs provide a very efficient approach for assembling CNTs together to make macroscopic materials and take full advantage of them. Since the successful synthesis of aligned CNT array and the fabrication of aligned CNT sheet, yarn, and fiber from arrays, diverse efforts have been made to optimize the growth condition, drawing process, and posttreatment to improve the quality and mechanical, electrical, and thermal properties of aligned CNT materials. However, even the strongest CNT fibers up to now are still much weaker than individual CNTs. Significant research is required to optimize the alignment and intertube contact, such as through crosslinking the CNTs by chemical bonds, in order to obtain higher performance. The unique structure makes aligned CNT films and fibers promising for use as artificial muscles. Although the electromechanical actuations with large stroke, high rate, and stress have been achieved for aligned CNTs and explained by electrochemical charge injection and electromagnetic force, the mechanism of actuation should be further studied for clear understanding. The CNT fibers are expected to show more complex actuation by designing hierarchical structures such as overtwisting, multi-plying, knitting, and combination under stimuli of solvent, heat, and light apart from electricity. On these bases, a multidisciplinary research of materials, physics, and chemistry and a combination of experimental, theoretical, and engineering studies are necessary and helpful for the effective mass production and practical application of aligned CNTs.

Several sensing polymer/CNT composites are developed with new chromatic and deformable properties, including electrochromism based on molecular perturbation, ion and heat, and deformation under stimuli of electricity, light, and solvent. Further research should be made to find out the intrinsic mechanism of the interaction between aligned CNTs and polymers, for example, the orientation and crystallization of polymer induced by aligned CNTs and the transfer of external stimuli in composite. At the same time, the sensitivity and stability of sensing composite must be improved to meet the large-scale practical application. By introducing proper responsive polymers and appropriate constructions, the polymer/CNT composites are expected to realize deformation and chromism simultaneously, which can be used more widely. The chromism and deformation of sensing polymer/aligned CNT composites are also promising for application in smart energy devices, such as showing the work state of devices by color change and producing portable devices by self-folding.

REFERENCES

Aliev, A. E., Guthy, C., Zhang, M. et al. 2007. Thermal transport in MWCNT sheets and yarns. *Carbon* 45:2880–2888.

Aliev, A. E., Oh, J., Kozlov, M. E. et al. 2009. Giant-stroke, superelastic carbon nanotube aerogel muscles. *Science* 323:1575–1578.

Alvarenga, J., Jarosz, P. R., Schauerman, C. M. et al. 2010. High conductivity carbon nanotube wires from radial densification and ionic doping. *Appl. Phys. Lett.* 97:182106.

Andrews, R., Jacques, D., Qian, D. L., Rantell, T. 2002. Multiwall carbon nanotubes: Synthesis and application. *Acc. Chem. Res.* 35:1008–1017.

Argun, A. A., Aubert, P.-H., Thompson, B. C. et al. 2004. Multicolored electrochromism in polymers: Structures and devices. *Chem. Mater.* 16:4401–4412.

Badaire, S., Pichot, V., Zakri, C. et al. 2004. Correlation of properties with preferred orientation in coagulated and stretch-aligned single-wall carbon nanotubes. *J. Appl. Phys.* 96:7509.

Baughman, R. H., Cui, C., Zakhidov, A. A. et al. 1999. Carbon nanotube actuators. *Science* 284:1340–1344.

Behabtu, N., Young, C. C., Tsentalovich, D. E. et al. 2013. Strong, light, multifunctional fibers of carbon nanotubes with ultrahigh conductivity. *Science* 339:182–186.

Boncel, S., Sundaram, R. M., Windle, A. H., Koziol, K. K. K. 2011. Enhancement of the mechanical properties of directly spun CNT fibers by chemical treatment. *ACS Nano* 5:9339–9344.

Brochu, P., Pei, Q. 2010. Advances in dielectric elastomers for actuators and artificial muscles. *Macromol. Rapid. Commun.* 31:10–36.

Chen, L., Liu, C., Liu, K. et al. 2011a. High-performance, low-voltage, and easy-operable bending actuator based on aligned carbon nanotube/polymer composites. *ACS Nano* 5:1588–1593.

Chen, L. Z., Liu, C. H., Hu, C. H., Fan, S. S. 2008. Electrothermal actuation based on carbon nanotube network in silicone elastomer. *Appl. Phys. Lett.* 92:263104.

Chen, X., Li, L., Sun, X. et al. 2011b. Magnetochromatic polydiacetylene by incorporation of Fe_3O_4 nanoparticles. *Angew. Chem. Int. Ed.* 50:5486–5489.

Chen, X., Lin, H., Chen, P., Guan, G., Deng, J., Peng, H. 2014. Smart, stretchable supercapacitors. *Adv. Mater.* 26:4444–4449, doi: 10.1002/adma.201400842.

Dai, L., Patil, A., Gong, X. et al. 2003. Aligned nanotubes. *Chemphyschem* 4:1150–1169.

Dalton, A. B., Collins, S., Munoz, E. et al. 2003. Super-tough carbon-nanotube fibres. *Nature* 423:703.

Davis, V. A., Parra-Vasquez, A. N. G., Green, M. J. et al. 2009. True solutions of single-walled carbon nanotubes for assembly into macroscopic materials. *Nat. Nanotechnol.* 4:830–834.

Demczyk, B. G., Wang, Y. M., Cumings, J. et al. 2002. Direct mechanical measurement of the tensile strength and elastic modulus of multiwalled carbon nanotubes. *Mater. Sci. Eng. A* 334:173–178.

Di, J., Hu, D., Chen, H. et al. 2012. Ultrastrong, foldable, and highly conductive carbon nanotube film. *ACS Nano* 6:5457–5464.

Ericson, L. M., Fan, H., Peng, H. et al. 2004. Macroscopic, neat, single-walled carbon nanotube fibers. *Science* 305:1447–1450.

Fan, S., Chapline, M. G., Franklin, N. R., Tombler, T. W., Cassell, A. M., Dai, H. 1999. Self-oriented regular arrays of carbon nanotubes and their field emission properties. *Science* 283:512–514.

Foroughi, J., Spinks, G. M., Wallace, G. G. et al. 2011. Torsional carbon nanotube artificial muscles. *Science* 334:494–497.

Fu, W., Liu, L., Jiang, K., Li, Q., Fan, S. 2010. Super-aligned carbon nanotube films as aligning layers and transparent electrodes for liquid crystal displays. *Carbon* 48:1876–1879.

Guo, W., Liu, C., Sun, X., Yang, Z., Kia, H. G., Peng, H. 2012a. Aligned carbon nanotube/polymer composite fibers with improved mechanical strength and electrical conductivity. *J. Mater. Chem.* 22:903–908.

Guo, W., Liu, C., Zhao, F. et al. 2012b. A novel electromechanical actuation mechanism of a carbon nanotube fiber. *Adv. Mater.* 24:5379–5384.

Hata, K., Futaba, D. N., Mizuno, K., Namai, T., Yumura, M., Iijima, S. 2004. Water-assisted highly efficient synthesis of impurity-free single-walled carbon nanotubes. *Science* 306:1362–1364.

Huang, S., Li, L., Yang, Z. et al. 2011. A new and general fabrication of an aligned carbon nanotube/polymer film for electrode applications. *Adv. Mater.* 23:4707–4710.

Ikeda, T., Mamiya, J.-i., Yu, Y. 2007. Photomechanics of liquid-crystalline elastomers and other polymers. *Angew. Chem. Int. Ed.* 46:506–528.

Jakubinek, M. B., Johnson, M. B., White, M. A. et al. 2012. Thermal and electrical conductivity of array-spun multi-walled carbon nanotube yarns. *Carbon* 50:244–248.

Jia, J., Zhao, J., Xu, G. et al. 2011. A comparison of the mechanical properties of fibers spun from different carbon nanotubes. *Carbon* 49:1333–1339.

Jiang, K. L., Li, Q. Q., Fan, S. S. 2002. Nanotechnology: Spinning continuous carbon nanotube yarns—Carbon nanotubes weave their way into a range of imaginative macroscopic applications. *Nature* 419:801.

Kim, P., Shi, L., Majumdar, A., McEuen, P. L. 2001. Thermal transport measurements of individual multiwalled nanotubes. *Phys. Rev. Lett.* 87:215502.

Kondo, M., Yu, Y., Ikeda, T. 2006. How does the initial alignment of mesogens affect the photoinduced bending behavior of liquid-crystalline elastomers? *Angew. Chem. Int. Ed.* 118:1406–1410.

Koziol, K., Vilatela, J., Moisala, A. et al. 2007. High-performance carbon nanotube fiber. *Science* 318:1892–1895.

Kozlov, M. E., Capps, R. C., Sampson, W. M., Ebron, V. H., Ferraris, J. P., Baughman, R. H. 2005. Spinning solid and hollow polymer-free carbon nanotube fibers. *Adv. Mater.* 17:614–617.

Kwak, G., Fukao, S., Fujiki, M., Sakaguchi, T., Masuda, T. 2006. Temperature-dependent, static, and dynamic fluorescence properties of disubstituted acetylene polymer films. *Chem. Mater.* 18:2081–2085.

Kwak, G., Lee, W.-E., Jeong, H., Sakaguchi, T., Fujiki, M. 2008. Swelling-induced emission enhancement in substituted acetylene polymer film with large fractional free volume: Fluorescence response to organic solvent stimuli. *Macromolecules* 42:20–24.

Lee, J. A., Kim, Y. T., Spinks, G. M. et al. 2014. All-solid-state carbon nanotube torsional and tensile artificial muscles. *Nano Lett.* 14:2664–2669.

Lee, W.-E., Jin, Y.-J., Park, L.-S., Kwak, G. 2012. Fluorescent actuator based on microporous conjugated polymer with intramolecular stack structure. *Adv. Mater.* 24:5604–5609.

Li, L., Yang, Z., Gao, H. et al. 2011. Vertically aligned and penetrated carbon nanotube/polymer composite film and promising electronic applications. *Adv. Mater.* 23:3730–3735.

Li, M. H., Keller, P., Li, B., Wang, X., Brunet, M. 2003a. Light-driven side-on nematic elastomer actuators. *Adv. Mater.* 15:569–572.

Li, P., Jiang, K., Liu, M., Li, Q., Fan, S., Sun, J. 2003b. Polarized incandescent light emission from carbon nanotubes. *Appl. Phys. Lett.* 82:1763–1765.

Li, Q. W., Li, Y., Zhang, X. F. et al. 2007. Structure-dependent electrical properties of carbon nanotube fibers. *Adv. Mater.* 19:3358–3363.

Li, Q. W., Zhang, X. F., DePaula, R. F. et al. 2006. Sustained growth of ultralong carbon nanotube arrays for fiber spinning. *Adv. Mater.* 18:3160–3163.

Li, W. Z., Xie, S. S., Qian, L. X. et al. 1996. Large-scale synthesis of aligned carbon nanotubes. *Science* 274:1701–1703.

Li, Y. L., Kinloch, I. A., Windle, A. H. 2004. Direct spinning of carbon nanotube fibers from chemical vapor deposition synthesis. *Science* 304:276–278.

Li, Z., Tang, Z., Liu, H. et al. 2001. Polarized absorption spectra of single-walled 4 Å carbon nanotubes aligned in channels of an AlPO4–5 single crystal. *Phys. Rev. Lett.* 87:127401.

Lima, M. D., Li, N., Jung de Andrade, M. et al. 2012. Electrically, chemically, and photonically powered torsional and tensile actuation of hybrid carbon nanotube yarn muscles. *Science* 338:928–932.

Lin, H., Li, L., Ren, J. et al. 2013. Conducting polymer composite film incorporated with aligned carbon nanotubes for transparent, flexible and efficient supercapacitor. *Sci. Rep.* 3:1353.

Liu, K., Sun, Y., Chen, L. et al. 2008. Controlled growth of super-aligned carbon nanotube arrays for spinning continuous unidirectional sheets with tunable physical properties. *Nano Lett.* 8:700–705.

Liu, K., Sun, Y., Zhou, R. et al. 2010a. Carbon nanotube yarns with high tensile strength made by a twisting and shrinking method. *Nanotechnology* 21:045708.

Liu, L., Ma, W., Zhang, Z. 2011. Macroscopic carbon nanotube assemblies: Preparation, properties, and potential Applications. *Small* 7:1504–1520.

Liu, S., Liu, Y., Cebeci, H. et al. 2010b. High electromechanical response of ionic polymer actuators with controlled-morphology aligned carbon nanotube/Nafion nanocomposite electrodes. *Adv. Funct. Mater.* 20:3266–3271.

Lu, X., Zhang, Z., Li, H., Sun, X., Peng, H. 2014. Conjugated polymer composite artificial muscle with solvent-induced anisotropic mechanical actuation. *J. Mater. Chem. A* 2:17272–17280.

Ma, W. J., Liu, L. Q., Zhang, Z. et al. 2009. High-strength composite fibers: Realizing true potential of carbon nanotubes in polymer matrix through continuous reticulate architecture and molecular level couplings. *Nano Lett.* 9:2855–2861.

Madden, J. D., Cush, R. A., Kanigan, T. S., Hunter, I. W. 2000. Fast contracting polypyrrole actuators. *Synth. Met.* 113:185–192.

Mauritz, K. A., Moore, R. B. 2004. State of understanding of Nafion. *Chem. Rev.* 104:4535–4586.

Meng, F., Zhang, X., Li, R. et al. 2014. Electro-induced mechanical and thermal responses of carbon nanotube fibers. *Adv. Mater.* 26:2480–2485.

Meng, F., Zhang, X., Xu, G. et al. 2011. Carbon nanotube composite films with switchable transparency. *ACS Appl. Mater. Inter.* 3:658–661.

Meng, F., Zhao, J., Ye, Y., Zhang, X., Li, Q. 2012. Carbon nanotube fibers for electrochemical applications: Effect of enhanced interfaces by an acid treatment. *Nanoscale* 4:7464–7468.

Mirfakhrai, T., Oh, J., Kozlov, M. et al. 2007. Electrochemical actuation of carbon nanotube yarns. *Smart Mater. Struct.* 16:S243.

Peng, H. 2007. Aligned carbon nanotube/polymer composite films with robust flexibility, high transparency, and excellent conductivity. *J. Am. Chem. Soc.* 130:42–43.

Peng, H., Sun, X., Cai, F. et al. 2009. Electrochromatic carbon nanotube/polydiacetylene nanocomposite fibres. *Nat. Nanotechnol.* 4:738–741.

Peng, H., Tang, J., Pang, J. et al. 2005. Polydiacetylene/silica nanocomposites with tunable mesostructure and thermochromatism from diacetylenic assembling molecules. *J. Am. Chem. Soc.* 127:12782–12783.

Peng, H., Tang, J., Yang, L. et al. 2006. Responsive periodic mesoporous polydiacetylene/silica nanocomposites. *J. Am. Chem. Soc.* 128:5304–5305.

Pöhls, J.-H., Johnson, M. B., White, M. A. et al. 2012. Physical properties of carbon nanotube sheets drawn from nanotube arrays. *Carbon* 50:4175–4183.

Qiu, L., Sun, X., Yang, Z., Guo, W., Peng, H. 2012. Preparation and application of aligned carbon nanotube/polymer composite material. *Acta Chim. Sinica* 70:1523.

Ren, Z. F., Huang, Z. P., Xu, J. W. et al. 1998. Synthesis of large arrays of well-aligned carbon nanotubes on glass. *Science* 282:1105–1107.

Spinks, G. M., Wallace, G. G., Fifield, L. S. et al. 2002. Pneumatic carbon nanotube actuators. *Adv. Mater.* 14:1728–1732.

Sun, H., You, X., Yang, Z., Deng, J., Peng, H. 2013a. Winding ultrathin, transparent, and electrically conductive carbon nanotube sheets into high-performance fiber-shaped dye-sensitized solar cells. *J. Mater. Chem. A* 1:12422–12425.

Sun, X., Chen, T., Huang, S. et al. 2010a. UV-induced chromatism of polydiacetylenic assemblies. *J. Phys. Chem. B* 114:2379–2382.

Sun, X., Chen, T., Huang, S., Li, L., Peng, H. 2010b. Chromatic polydiacetylene with novel sensitivity. *Chem. Soc. Rev.* 39:4244–4257.

Sun, X., Chen, T., Yang, Z., Peng, H. 2012a. The alignment of carbon nanotubes: An effective route to extend their excellent properties to macroscopic scale. *Acc. Chem. Res.* 46:539–549.

Sun, X., Sun, H., Li, H., Peng, H. 2013b. Developing polymer composite materials: Carbon nanotubes or graphene? *Adv. Mater.* 25:5153–5176.

Sun, X., Wang, W., Qiu, L., Guo, W., Yu, Y., Peng, H. 2012b. Unusual reversible photomechanical actuation in polymer/nanotube composites. *Angew. Chem. Int. Ed.* 51:8520–8524.

Sun, X., Zhang, Z., Lu, X., Guan, G., Li, H., Peng, H. 2013c. Electric current test paper based on conjugated polymers and aligned carbon nanotubes. *Angew. Chem. Int. Ed.* 52:7776–7780.

Thess, A., Lee, R., Nikolaev, P. et al. 1996. Crystalline ropes of metallic carbon nanotubes. *Science* 273:483–487.

Vigolo, B., Pénicaud, A., Coulon, C. et al. 2000. Macroscopic fibers and ribbons of oriented carbon nanotubes. *Science* 290:1331–1334.

Vilatela, J. J., Elliott, J. A., Windle, A. H. 2011. A model for the strength of yarn-like carbon nanotube fibers. *ACS Nano* 5:1921–1927.

Viry, L., Mercader, C., Miaudet, P. et al. 2010. Nanotube fibers for electromechanical and shape memory actuators. *J. Mater. Chem.* 20:3487–3495.

Wang, J., Gudiksen, M. S., Duan, X., Cui, Y., Lieber, C. M. 2001. Highly polarized photoluminescence and photodetection from single indium phosphide nanowires. *Science* 293:1455–1457.

Wang, W., Sun, X., Wu, W., Peng, H., Yu, Y. 2012. Photoinduced deformation of cross-linked liquid-crystalline polymer film oriented by a highly aligned carbon nanotube sheet. *Angew. Chem. Int. Ed.* 51:4644–4647.

Wei, Y., Liu, L., Liu, P., Xiao, L., Jiang, K., Fan, S. 2008. Scaled fabrication of single-nanotube-tipped ends from carbon nanotube micro-yarns and their field emission applications. *Nanotechnology* 19:475707.

Xiao, L., Zhang, Y., Wang, Y. et al. 2011. A polarized infrared thermal detector made from super-aligned multiwalled carbon nanotube films. *Nanotechnology* 22:025502.

Yamada, M., Kondo, M., Mamiya, J.-i. et al. 2008. Photomobile polymer materials: Towards light-driven plastic motors. *Angew. Chem. Int. Ed.* 47:4986–4988.

Yang, Z., Chen, T., He, R. et al. 2011a. Aligned carbon nanotube sheets for the electrodes of organic solar cells. *Adv. Mater.* 23:5436–5439.

Yang, Z., Sun, X., Chen, X. et al. 2011b. Dependence of structures and properties of carbon nanotube fibers on heating treatment. *J. Mater. Chem.* 21:13772–13775.

Yao, Z., Di, J., Yong, Z., Zhao, Z., Li, Q. 2012. Aligned coaxial tungsten oxide-carbon nanotube sheet: A flexible and gradient electrochromic film. *Chem. Commun.* 48:8252–8254.

Yu, M.-F., Files, B. S., Arepalli, S., Ruoff, R. S. 2000. Tensile loading of ropes of single wall carbon nanotubes and their mechanical properties. *Phys. Rev. Lett.* 84:5552–5555.

Yu, Y., Nakano, M., Ikeda, T. 2003. Photomechanics: Directed bending of a polymer film by light. *Nature* 425:145.

Yun, Y., Shanov, V., Tu, Y. et al. 2006. A multi-wall carbon nanotube tower electrochemical actuator. *Nano Lett.* 6:689–693.

Zhang, M., Atkinson, K. R., Baughman, R. H. 2004. Multifunctional carbon nanotube yarns by downsizing an ancient technology. *Science* 306:1358–1361.

Zhang, M., Baughman, R. 2011. Assembly of carbon nanotube sheets. In *Electronic Properties of Carbon Nanotubes*, ed. Marulanda J. M., pp. 3–20. InTech, Rijeka, Croatia.

Zhang, M., Fang, S., Zakhidov, A. A. et al. 2005. Strong, transparent, multifunctional, carbon nanotube sheets. *Science* 309:1215–1219.

Zhang, Q., Zhao, M. Q., Huang, J. Q., Nie, J. Q., Wei, F. 2010. Mass production of aligned carbon nanotube arrays by fluidized bed catalytic chemical vapor deposition. *Carbon* 48:1196–1209.

Zhang, X., Jiang, K., Feng, C. et al. 2006. Spinning and processing continuous yarns from 4-inch wafer scale super-aligned carbon nanotube arrays. *Adv. Mater.* 18:1505–1510.

Zhang, X., Li, Q., Holesinger, T. G. et al. 2007a. Ultrastrong, stiff, and lightweight carbon-nanotube fibers. *Adv. Mater.* 19:4198–4201.

Zhang, X., Li, Q., Tu, Y. et al. 2007b. Strong carbon-nanotube fibers spun from long carbon-nanotube arrays. *Small* 3:244–248.

Zhang, Y., Zou, G., Doorn, S. K. et al. 2009. Tailoring the morphology of carbon nanotube arrays: From spinnable forests to undulating foams. *ACS Nano* 3:2157–2162.

Zhao, J., Zhang, X., Di, J. et al. 2010. Double-peak mechanical properties of carbon-nanotube fibers. *Small* 6:2612–2617.

Zheng, L., Sun, G., Zhan, Z. 2010. Tuning array morphology for high-strength carbon-nanotube fibers. *Small* 6:132–137.

Zheng, W., Razal, J. M., Whitten, P. G. et al. 2011. Artificial muscles based on polypyrrole/carbon nanotube laminates. *Adv. Mater.* 23:2966–2970.

Zhong, X. H., Li, Y. L., Liu, Y. K. et al. 2010. Continuous multilayered carbon nanotube yarns. *Adv. Mater.* 22:692–696.

CHAPTER 13

CONTENTS

Density Functional Investigation on $M_3N@C_{80}$ and $M_2C_2@C_{82}$ (M = Sc, Y)

13

Debesh Ranjan Roy and Esha V. Shah

13.1 INTRODUCTION

Endohedral metallofullerenes ($M@C_{2n}$) are known as the fullerenes containing metal (one, two, or three) compounds inside the cage [1]. These metallofullerenes have gained lots of attention to the research community in the recent days for their characteristic structural and electrical properties, which have potential applications as nanodevices in spintronics [2–4]. Until now, investigations are under progress over 30 metal elements inside the *atomic park* [5], mainly chosen from the second and third group in the periodic table. In this line, metal nitride– or metal carbide–encapsulated fullerenes have gained immense focus due to their characteristic functionality toward potential applications. Trimetallic nitride–encapsulated fullerenes, for example, $Sc_3N@C_{68}$, $Sc_3N@C_{80}$, and $Er_xScN@C_{68}$ ($x = 1$–3), have been synthesized recently [6]. Also, metal carbide–encapsulated fullerenes, for example, $Sc_2C_2@C_{84}$, $Sc_3C_2@C_{80}$, and $Y_2C_2@C_{82}$ [7–9], are synthesized and their structures are investigated. One of the most important goals of such work is to understand and determine the precise position of metal and its compound inside the fullerene cage. This information is very important for understanding the physical properties and growth mechanism of these endohedral fullerene families. Although experimental techniques like XRD [10] have been tendered for such purpose, it is hardly possible to avail detailed structural information due to multiple-disorder and/or giant motion of metal atoms. In this regard, a standard theoretical investigation on the specific structural information on the metal nitride– and metal carbide–encapsulated endohedral fullerenes becomes relevant [11–14]. Although a number of theoretical studies have been performed to resolve structure and property issues, the field is lacking with a lot more standard theoretical investigations for precise information and properties of these metallofullerenes.

The purpose of the present work is to perform a detailed theoretical study on the structure, electronic properties, and aromaticity of metal nitride– and metal carbide–encapsulated endohedral fullerenes, viz., $M_3N@C_{80}$ and $M_2C_2@C_{82}$ with group III metal elements M = Sc, Y. The complete work is carried out under the density functional theory (DFT) [15–17].

13.2 THEORY AND COMPUTATION

The ionization potential (*I.P.*) and electron affinity (*E.A.*) can be expressed in terms of the highest occupied (ϵ_{HOMO}) and the lowest unoccupied (ϵ_{LUMO}) molecular orbital energies using Koopmans' approximation [17] as

$$IP \approx -\epsilon_{HOMO}; \quad EA \approx -\epsilon_{LUMO} \tag{13.1}$$

Reprinted as a part from *Physica E*, 70, Shah, E.V. and Roy, D.R., Structure, electronic properties, aromaticity and dynamics of $M_3N@C_{80}$ and $M_2C_2@C_{82}$ (M = Sc, Y): A density functional study, 157–164, Copyright (2015), with permission from Elsevier.

To account for the stability of a molecule, Pearson [19] has introduced an important parameter, viz., *chemical hardness*. For an N-electron system, the second derivative of energy with respect to N, keeping external potential $v(\vec{r})$ fixed, is considered to be a measure of the chemical hardness [20]:

$$\eta = \frac{1}{2}\left(\frac{\partial^2 E}{\partial N^2}\right)_{v(\vec{r})} \tag{13.2}$$

The hardness can be expressed in terms of ϵ_{HOMO} and ϵ_{LUMO} as follows [17]:

$$\eta \approx \frac{IP - EA}{2} \approx \frac{\epsilon_{LUMO} - \epsilon_{HOMO}}{2} \tag{13.3}$$

Electronegativity (χ) [21] (negative chemical potential [μ] [22]) for an N-electron system is defined as the first-order number derivatives of total energy (E) as follows:

$$\chi = -\mu = \left(\frac{\partial E}{\partial N}\right)_{v(\vec{r})} \tag{13.4}$$

where $v(\vec{r})$ is the external potential.

Using a finite difference approximation, χ may be expressed as [17]

$$\chi = \frac{I + A}{2} \tag{13.5}$$

The nucleus-independent chemical shift (NICS) has been defined as [23] "absolute magnetic shieldings computed at ring centers (nonweighted mean of the heavy atom coordinates)." The NICS is considered as the negative value of the magnetic shielding at a chosen point in a molecule. In general, a small value (negative or positive) of NICS identifies a system as nonaromatic, whereas a significant negative NICS value indicates the system as aromatic and a significant positive NICS value classifies a system as antiaromatic.

The theoretical investigations are carried out using a gradient-corrected approach within the DFT framework [15–17]. A molecular orbital approach using a linear combination of atomic orbitals is applied to the probe electronic structure. The wave functions for the cluster were expressed as a linear combination of Gaussian-type orbitals (GTOs) situated at the atomic positions in the cluster. The actual calculation is performed using the implementation in the Gaussian 09 program [24]. For exchange and correlation functions, we have used a very successful and widely used hybrid functional, Becke's three-parameter exchange functional with the Lee–Yang–Parr correlation (B3LYP) [25]. For the basis set we have used all-electron STO-3G (three primitive Gaussian orbitals are fitted to a single Slater-type orbital) [26] for all the considered compounds. All structures were fully optimized without imposing any symmetry constrains, to allow for full variational freedom. To locate the lowest energy configuration for each cluster, all possible initial guess is considered.

13.3 RESULTS AND DISCUSSION

13.3.1 Metal Nitride–Encapsulated Fullerenes: $M_3N@C_{80}$

The optimized geometries of Sc_3N and Y_3N clusters are presented in Figure 13.1. Sc_3N shows a planar structure with perfect D_{3h} symmetry with Sc–N distance as 2.10 Å. On the other hand, energy-minimized Y_3N cluster implies a pyramidal structure with C_{3v} symmetry. Bond lengths of Y–N and Y–Y in Y_3N are found to be 2.20 and 3.35 Å, respectively. Figure 13.2 provides optimized geometries of C_{80} (D_{5h}) and M_3N-encapsulated C_{80}, viz., $Sc_3N@C_{80}$ and $Y_3N@C_{80}$ clusters. It may be noted that both the Sc_3N and Y_3N are perfectly centered inside the C_{80} cage, as expected from the D_{5h} symmetry of C_{80}, making a perfect core–cage system. Although Sc_3N retains its planar structure during the core (Sc_3N)–cage (C_{80}) interaction, the pyramidal conformation of Y_3N gets flattened from δ (Y–Y–Y–N) = 46.8° to 24.0°. It is found that nitride encapsulation enhances the dimension of C_{80} (8.31 Å) to 8.53 Å (for $Sc_3N@C_{80}$) and 8.57 Å (for $Y_3N@C_{80}$).

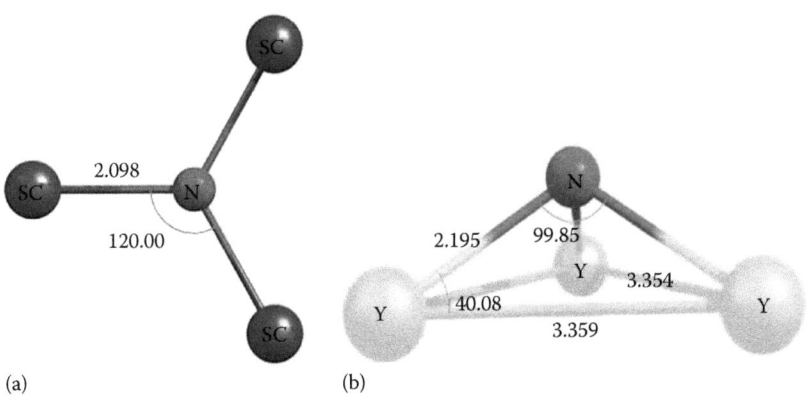

FIGURE 13.1
Optimized geometries of (a) Sc₃N and (b) Y₃N clusters.

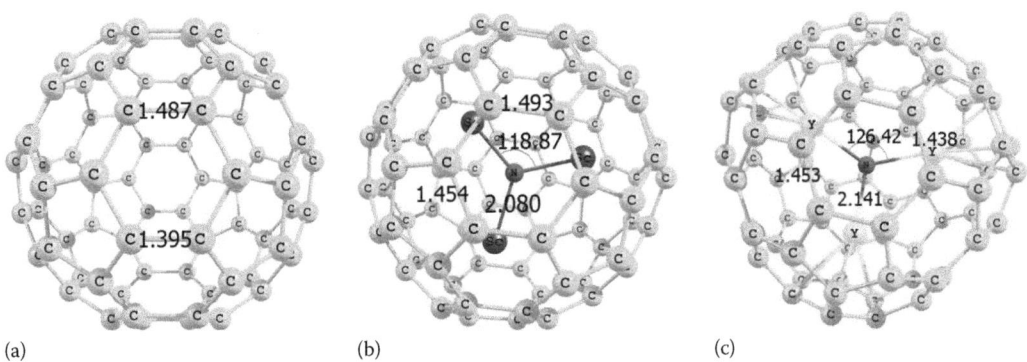

FIGURE 13.2
Optimized geometries of (a) C₈₀, (b) Sc₃N@C₈₀, and (c) Y₃N@C₈₀ clusters.

| | | | TABLE 13.1 | HOMO–LUMO Energy Gap (HLG, eV), Ionization Potential (*I.P.*, eV), Electron Affinity (*E.A.*, eV), Chemical Hardness (η, eV), and Electronegativity (χ) Values of C₈₀, Sc₃N, Y₃N, Sc₃N@C₈₀, and Y₃N@C₈₀ Clusters, along with the NICS Values at the Respective Cage Centers | | |

TABLE 13.1 HOMO–LUMO Energy Gap (HLG, eV), Ionization Potential (*I.P.*, eV), Electron Affinity (*E.A.*, eV), Chemical Hardness (η, eV), and Electronegativity (χ) Values of C₈₀, Sc₃N, Y₃N, Sc₃N@C₈₀, and Y₃N@C₈₀ Clusters, along with the NICS Values at the Respective Cage Centers

Clusters	HLG	*I.P.*	*E.A.*	η	χ	NICS
C₈₀	0.66	3.33	2.67	0.33	3.00	23.90
Sc₃N	3.71	1.88	−1.80	1.85	0.03	—
Y₃N	1.63	0.84	−0.80	0.81	0.02	—
Sc₃N@C₈₀	2.47	3.15	0.69	1.23	1.92	−144.20
Y₃N@C₈₀	2.44	3.28	0.83	1.22	2.06	4.45

Table 13.1 provides the HOMO–LUMO energy gap (HLG), *I.P.*, *E.A.*, chemical hardness (η), and electronegativity (χ) of C₈₀, Sc₃N, Y₃N, Sc₃N@C₈₀, and Y₃N@C₈₀ clusters along with the NICS values at the respective cage centers. It may be noticed that nitride encapsulation provides a significant enhancement of HLG for C₈₀ (0.66 eV) to 2.47 eV (for Sc₃N@C₈₀) and 2.44 eV (for Y₃N@C₈₀), implying their possibility toward novel semiconductor applications. Also, increase of η due to complexation implies that the encapsulated complex becomes harder and stable. The large *I.P.* (>3eV) values of the complexed system also support their stability. Further, it may be noted that *E.A.* and χ of C₈₀ significantly decrease due to complexation indicating an electrostatic interaction between the C₈₀ cage and M₃N core.

The reaction energies (E_R) for both the nitride encapsulations are as follows:

$$E(Sc_3N@C_{80}) - \{E(C_{80}) + E(Sc_3N)\} = -13.38\,eV \tag{13.6}$$

$$E(Y_3N@C_{80}) - \{E(C_{80}) + E(Y_3N)\} = -18.78\,eV \tag{13.7}$$

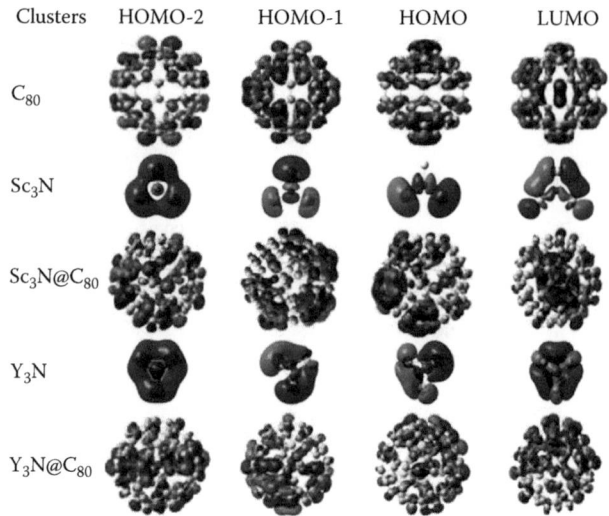

FIGURE 13.3
FMOs for C_{80}, Sc_3N, $Sc_3N@C_{80}$, Y_3N, and $Y_3N@C_{80}$ clusters.

The negative reaction energy for both the complexations implies that Sc and Y nitride encapsulations are thermodynamically favorable.

To gain further insight into these nitride endohedral fullerenes, we have predicted important representative frontier molecular orbitals (FMOs) and NICS values. Figure 13.3 represents some important FMOs (HOMO-2, HOMO-1, HOMO, and LUMO) for C_{80}, Sc_3N, $Sc_3N@C_{80}$, Y_3N, and $Y_3N@C_{80}$ clusters. It may be noticed for all the occupied energy levels of the metal-encapsulated systems ($M_3N@C_{80}$), orbital distribution mostly lies at the cage, implying the existence of electrostatic interaction and possible electron transfer between the anionic cage (C_{80}) and cationic core (M_3N). This feature indicates an important aspect of these encapsulated fullerenes toward spintronics applications. The NICS value (+23.9 ppm) for C_{80} (D_{5h}) identifies it as an antiaromatic cluster. Although Y_3N encapsulation of C_{80} converts the system toward a nonaromatic (NICS = +4.45 ppm) one, in case of Sc_3N encapsulation, the complex ($Sc_3N@C_{80}$) becomes highly aromatic (NICS = −144.2 ppm). The aromatic feature of $Sc_3N@C_{80}$ may be understood from its occupied MOs, where a number of π-overlapping may be noticed on the hexa- and pentagonal carbon planes on the cage surface.

13.3.2 Metal Carbide–Encapsulated Fullerenes: $M_2C_2@C_{82}$

The optimized geometries of Sc_2C_2 and Y_2C_2 clusters are presented in Figure 13.4. Sc_2C_2 shows a planar rhombus structure with perfect D_{2h} symmetry with Sc–C distance of 2.16 Å. On the other hand, the Y_2C_2 cluster is found to be a planar structure with C_{2h} symmetry. Bond lengths of Y–C in Y_2C_2 are found to be 2.90 and 3.31 Å and for C–C as 1.29 Å. Figure 13.5 provides optimized geometries of C_{82} (C_{2v}) and M_2C_2-encapsulated C_{82}, viz., $Sc_2C_2@C_{82}$ and $Y_2C_2@C_{82}$ clusters. It may be noted that both the Sc_2C_2 and Y_2C_2 are perfectly centered inside the C_{82} cage, as expected from the C_{2v} symmetry of C_{82}, making a perfect cage–core system. Although Sc_2C_2 retains its planar structure during the core (Sc_2C_2)–cage (C_{82}) interaction with a C–C bond squeezing of 0.36 Å, the planar C_{2h} conformation of Y_2C_2 (in its bare state) forms a bent conformation with δ (Y–C–C–Y) = 125.15°. Also, Sc and Y are found to be coordinated with one and four carbon atoms, respectively, residing on the surface of these cage complexes. It is found that carbide encapsulation enhances the dimension of C_{82} (8.60 Å) to 8.87 Å (for $Sc_2C_2@C_{82}$) and 8.79 Å (for $Y_2C_2@C_{82}$). The bent conformation of Y_2C_2 inside the C_{82} cage is expected, as the average bond length of Y–C in the present study is found to be 2.40 Å, which implies that the required dimension of the cage should be at least four times of the Y–C distance to accommodate a planar Y_2C_2 cluster. The bent conformation of Y_2C_2 in the $Y_2C_2@C_{82}$ complex is also confirmed by experimental studies, for example, NMR [27] and XRD/MEM [10].

Table 13.2 provides the HLG, *I.P.*, *E.A.*, chemical hardness (η), and electronegativity (χ) of C_{82}, Sc_2C_2, Y_2C_2, $Sc_2C_2@C_{82}$, and $Y_2C_2@C_{82}$ clusters along with the NICS values at the respective cage centers. It may be noticed that carbide encapsulation increases the HLG of C_{82} (1.02 eV) to 1.29 eV

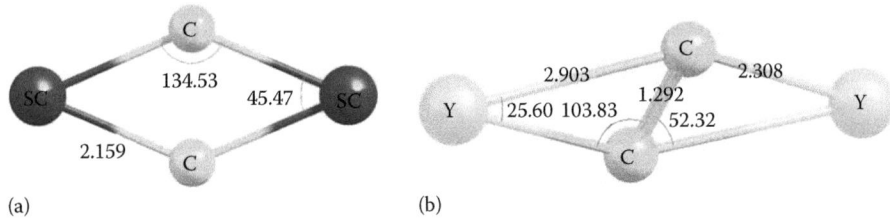

FIGURE 13.4
Optimized geometries of (a) Sc_2C_2 and (b) Y_2C_2 clusters.

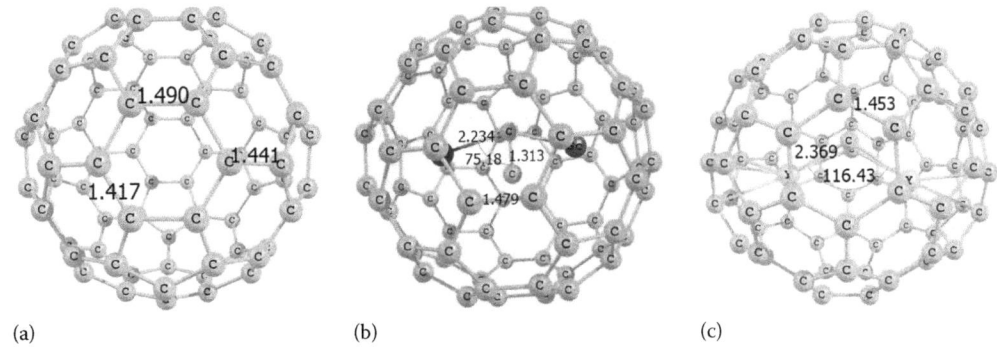

FIGURE 13.5
Optimized geometries of (a) C_{82}, (b) $Sc_2C_2@C_{82}$, and (c) $Y_2C_2@C_{82}$ clusters.

(for $Sc_2C_2@C_{82}$) and 1.18 eV (for $Y_2C_2@C_{82}$), implying their stability toward complexation. Also, increase of η due to complexation implies that the encapsulated complex becomes harder and stable. The large *I.P.* (~3 eV) values of the complexed system imply their robustness. Further, it may be noted that *E.A.* and χ of C_{82} significantly decrease due to complexation indicating an existence of electrostatic interaction between the C_{82} cage and M_2C_2 core.

The reaction energies (E_R) for both the carbide encapsulations are as follows:

$$E(Sc_2C_2@C_{82}) - \{E(C_{82}) + E(Sc_2C_2)\} = -11.06\,\text{eV} \tag{13.8}$$

$$E(Y_2C_2@C_{82}) - \{E(C_{82}) + E(Y_2C_2)\} = -13.44\,\text{eV} \tag{13.9}$$

The negative reaction energy for both the complexations implies that Sc and Y carbide encapsulations are thermodynamically favorable.

To gain further insight into these carbide endohedral fullerenes, we have predicted important representative FMOs and NICS values. Figure 13.6 represents some important FMOs (HOMO-2, HOMO-1, HOMO, and LUMO) for C_{82}, Sc_2C_2, $Sc_2C_2@C_{82}$, Y_2C_2, and $Y_2C_2@C_{82}$ clusters. It may be noticed that for all the occupied energy levels of the metal-encapsulated systems ($M_2C_2@C_{82}$), orbital distribution mostly lies at the cage, implying the existence of electrostatic interaction and possible electron transfer between the anionic cage (C_{82}) and cationic core (M_2C_2). This feature indicates an

TABLE 13.2 HOMO–LUMO Energy Gap (HLG, eV), Ionization Potential (*I.P.*, eV), Electron Affinity (*E.A.*, eV), Chemical Hardness (η, eV), and Electronegativity (χ) Values of C_{82}, Sc_2C_2, Y_2C_2, $Sc_2C_2@C_{82}$, and $Y_2C_2@C_{82}$ Clusters, along with the NICS Values at the Respective Cage Centers

Clusters	HLG	I.P.	E.A.	η	χ	NICS
C_{82}	1.02	3.54	2.52	0.51	3.03	3.678
Sc_2C_2	1.34	1.97	0.62	0.67	1.29	—
Y_2C_2	2.56	1.63	−0.9	1.28	0.35	—
$Sc_2C_2@C_{82}$	1.29	3.08	1.79	0.65	2.44	−39.74
$Y_2C_2@C_{82}$	1.18	2.93	1.75	0.59	2.34	−45.84

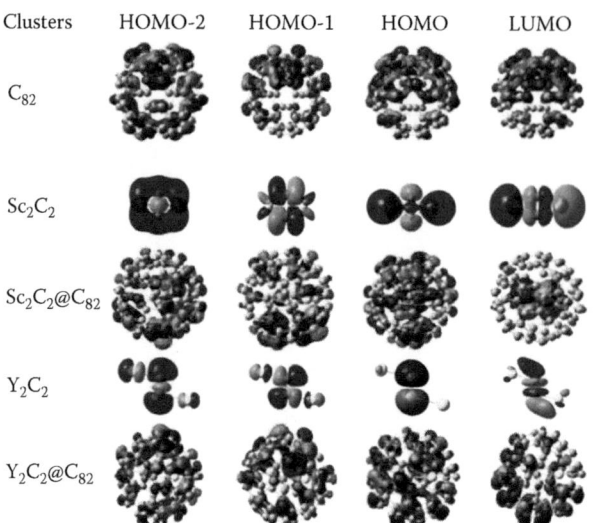

FIGURE 13.6
FMOs for C_{82}, Sc_2C_2, $Sc_2C_2@C_{82}$, Y_2C_2, and $Y_2C_2@C_{82}$ clusters.

important aspect of these encapsulated fullerenes toward spintronics applications. The NICS value (+3.68 ppm) for C_{82} (C_{3v}) identifies it as a nonaromatic cluster. It is heartening to note that both the carbide encapsulations in C_{82} convert the system into aromatic systems (NICS = −39.7 ppm for $Sc_2C_2@C_{82}$ and −45.8 ppm for $Y_2C_2@C_{82}$). The aromatic feature of both the $M_2C_2@C_{82}$ may be understood from its occupied MOs, where a number of π-overlapping may be noticed on the hexa- and pentagonal carbon planes on the cage surface along with the π-overlapping on the core carbide M_2C_2.

13.4 CONCLUDING REMARKS

A detailed investigation on the structure, electronic properties, and aromaticity is performed on metal nitride– and metal carbide–encapsulated endohedral fullerenes, viz., $M_3N@C_{80}$ and $M_2C_2@C_{82}$ (M = Sc, Y). It is noticed that both the nitrides and carbides are stabilized at the center of C_{80} (D_{5h}) and C_{82} (C_{2v}) cages making them a perfect cage–core system. All the nitride and carbides retain their lowest energy conformation (planar or pyramidal) inside the fullerene cages except for Y_2C_2, whose bent conformation is found inside the C_{82} cage, as expected from the C_{82} cage dimension (8.60 Å) that cannot occupy a planar Y_2C_2 with an average bond length of Y–C (2.40 Å). The same aspect is also reflected by reports from different experimental researchers. It is noticed that Sc atoms are coordinated with a single carbon atom, whereas Y atoms are quadruply coordinated with carbon atoms on the surface of the cages. It is found that nitride and carbide encapsulation in the fullerene cages enhances the HOMO–LUMO gap (HLG) and chemical hardness (η) of the C_{80} and C_{82} cages, implying that system gains more stability during complexation. On the other hand, decrease of *E.A.* and electronegativity (χ) of these bare cages during nitride and carbide encapsulation provides a clear existence of electrostatic interaction and possible electron transfer between cationic cores (M_3N and M_2C_2) and anionic cages (C_{80} and C_{82}), which is one important aspect for spintronics applications. The negative reaction energies for all the nitride and carbide encapsulation reactions imply that formation of all these endohedral fullerenes is thermodynamically favorable. It is heartening to note that nitride and carbide encapsulation converts antiaromatic C_{80} and nonaromatic C_{82} cages into nonaromatic ($Y_3N@C_{80}$) and aromatic systems ($Sc_3N@C_{80}$, $Sc_2C_2@C_{82}$, and $Y_2C_2@C_{82}$).

ACKNOWLEDGMENTS

DRR is thankful to the Science and Engineering Research Board (SERB), Department of Science and Technology (DST), New Delhi, Government of India, for financial support by awarding the FastTrack project grant (D.O. SR/FTP/PS-199/2011). EVS is thankful to Sardar Vallabhbhai National Institute of Technology (SVNIT), Surat, for the institute research fellowship.

REFERENCES

1. Chai, Y., Guo, T., Jin, C., Haufler, R. E., Chibante, L. P. F., Fure, J., Wang, L., Alford, J. M., and Smalley, R. E. *J. Phys. Chem.* 1991, *95*, 7564.
2. Shinohara, H. *Rep. Prog. Phys.* 2000, *63*, 843.
3. Lee, J., Kim, H., Kahng S. J., Kim, G., Son, Y. W., Ihm, J., Kato, H., Wang, Z. W., Okazaki, T., Shinohara, H., and Kuk, Y. *Nature(London)* 2002, *415*, 1005.
4. Hirahara, K., Suenaga, K., Bandow, S., Kato, H., Okazaki, T., Shinohara, H., and Iijima, S. *Phys. Rev. Lett.* 2000, *85*, 5384.
5. Koltover, V. K. *J. Mol. Liq.* 2006, *127*, 139.
6. Akasaka, N. and Nagase, S., *Endofullerenes: A New Family of Carbon Clusters*, Kluwer Academic Publishers, Dordrecht, the Netherlands, 2003.
7. Wang, C.-R., Kai, T., Tomiyama, T., Yoshida, T., Kobayashi, Y., Nishibori, E., Takata, M., Sakata, M., and Shinohara, H. *Angew. Chem. Int. Ed.* 2001, *40*, 397.
8. Iiduka, Y., Wakahara, T., Nakahodo, T., Tsuchiya, T., Sakuraba, A., Maeda, Y., Akasaka, T. et al. *J. Am. Chem. Soc.* 2005, *127*, 12500.
9. Nishibori, E., Narioka, S., Takata, M., Sakata, M., Inoue, T., and Shinohara, H. *Chem. Phys. Chem.* 2006, *7*, 345.
10. Nishibori, E., Ishihara, M., Takata, M., Sakata, M., Ito, Y., Inoue, T., and Shinohara, H. *Chem. Phys. Lett.* 2006, *433*, 120–124.
11. Wu, J., Wang, T., Shu, C., Lu, X., and Wang, C. *Chin. J. Chem.* 2012, *30*, 765.
12. Yang, H., Lu, C., Liu, Z., Jin, H., Che, Y., Olmstead, M. M., and Balch, A. L. *J. Am. Chem. Soc.* 2008, *130*, 17296.
13. Cao, B., Hasegawa, M., Okada, K., Tomiyama, T., Okazaki, T., Suenaga, K., and Shinohara, H. *J. Am. Chem. Soc.* 2001, *123*, 9679.
14. Akiyama, K., Sueki, K., Kodama, T., Kikuchi, K., Takigawa, Y., Nakahara, H., Ikemoto, I., and Katada, M. *Chem. Phys. Lett.* 2000, *317*, 490.
15. Hohenberg, P. and Kohn, W. *Phys. Rev. B* 1964, *136*, B864.
16. Kohn, W. and Sham, L. J. *Phys. Rev. A* 1965, *140*, A1133.
17. Parr, R. G. and Yang, W. *Density Functional Theory of Atoms and Molecules*, Oxford University Press, New York, 1989.
18. Shah, E. V. and Roy, D. R. *Physica E* 2015, *70*, 157–164.
19. Pearson, R. G. *Hard and Soft Acids and Bases*, Dowden, Hutchinson and Ross, Stroudsburg, PA, 1973.
20. Parr, R. G. and Pearson, R. G. *J. Am. Chem. Soc.* 1983, *105*, 7512.
21. Sen, K. D. and Jorgenson, C. K. *Structure and Bonding*, Vol. 66: *Electronegativity*; Springer, Berlin, Germany, 1987.
22. Parr, R. G., Donnelly, R. A., Levy, M., and Palke, W. E. *Chem. Phys.* 1978, *68*, 3801.
23. Schleyer, P. v. R., Maerker, C., Dransfeld, A., Jiao, H., and Hommes, N. J. R. v. E. *J. Am. Chem. Soc.* 1996, *118*, 6317.
24. Frisch, M. J., Trucks, G. W., Schlegel, H. B., Scuseria, G. E., Robb, M. A., Cheeseman, J. R., Scalmani, G., Barone, V., Mennucci, B., Petersson, G. A. et al., GAUSSIAN 09, Revision D.01; Gaussian, Inc., Pittsburgh PA, 2009.
25. Becke, A. D. *J. Chem. Phys.* 1993, *98*, 5648.
26. Hehre, W. J., Stewart, R. F., and Pople, J. A. *J. Chem. Phys.* 1969, *51*, 2657.
27. Inoue, T., Tomiyama, T., Sugai, T., and Shinohara, H. *Chem. Phys. Lett.* 2003, *382*, 226.

CHAPTER 14

CONTENTS

Modification of Polymer Membranes Using Carbon Nanotubes

14

A.G. Tkachev, A.E. Burakov, I.V. Romantsova, and A.E. Kucherova

14.1 MEMBRANE APPLICATIONS

14.1.1 Introduction

Industrial applications are divided into six main subgroups: reverse osmosis (RO), ultrafiltration (UF), microfiltration (MF), gas separation, pervaporation, and electrodialysis. Medical applications comprise artificial kidneys, blood oxygenators, and controlled-release pharmaceuticals. Relatively few companies are actually involved in more than one industry subgroup (Baker 2004).

Membranes have gained an important place in chemical technology and are used in a broad range of applications. The key property exploited is the ability of a membrane to control permeation rates of chemical species. In controlled drug delivery systems, the goal is to moderate the permeation rate of a drug from a reservoir to the body. In separation applications, the purpose is to allow one component of a mixture to pass freely through the membrane while hindering the permeation of other components.

14.1.2 Historical Development of Membranes

Systematic studies of membrane phenomena can be traced to the eighteenth-century scientists. For example, in 1748, Abbé Nolet coined the word *osmosis* to describe permeation of water through a diaphragm. Through the nineteenth and early twentieth centuries, membranes had no industrial or commercial uses but were employed as laboratory tools to develop physical/chemical theories. At about the same time, the concept of a perfectly selective semipermeable membrane was used by Maxwell and others in developing the kinetic theory of gases (Baker 2004).

Early investigators experimented with every type of diaphragm available to them, such as bladders of pigs, cattle, or fish and sausage casings made of animal gut. Later, nitrocellulose membranes were preferred, because they could be manufactured reproducibly. In 1907, Bechhold devised a technique to prepare nitrocellulose membranes of graded pore size, which were characterized by a bubble test method. Other early workers, particularly Elford (1937) and Ferry (1936), improved Bechhold's technique, and by the early 1930s, microporous membranes were commercially available. During the next 20 years, this early MF membrane technology was expanded to other polymers, notably CA. Membranes found their first significant application in the testing of drinking water at the end of World War II.

By 1960, the elements of modern membrane science had been developed, but membranes were employed in only a few laboratory and small, specialized industrial applications. They suffered from four problems that prohibited their widespread use in separation processes: they were too unreliable, too slow, too unselective, and too expensive.

In the early 1960s, Loeb and Sourirajan developed a method for making defect-free, high-flux, anisotropic RO membranes (Loeb and Sourirajan 1963)—the seminal discovery that transformed membrane separation from a laboratory to an industrial process. These membranes consist of an ultrathin, selective surface film supported by a much thicker but much more permeable microporous substrate to provide mechanical strength. The flux of the first Loeb–Sourirajan RO membrane was 10 times higher than that of any membrane then available and made RO a potentially practical method of water desalination. The work of Loeb and Sourirajan and the timely infusion of large sums of research and development dollars from the US Department of the Interior and Office of Saline Water (OSW) resulted in the commercialization of RO, and it was a major factor in the development of UF and MF. The studies on electrodialysis processes were also funded by the OSW.

Concurrently with the development of these industrial applications was the independent development of membranes for medical separation processes, in particular, artificial kidneys. In 1945, in the Netherlands, Kolf and Berk (1944) demonstrated the first successful artificial kidney. It took almost 20 years to refine the technology for use on a large scale, but these developments had been completed by the early 1960s.

The period from 1960 to 1980 produced a significant change in the status of membrane technology. Based on the original Loeb–Sourirajan technique, other membrane formation processes, including interfacial polymerization and multilayer composite casting and coating, were developed for making high-performance membranes. Using these processes, membranes with selective layers as thin as 0.1 μm or less are now being produced by a number of companies. Methods of packaging membranes into large-membrane-area spiral wound, hollow fine fiber, capillary, and plate-and-frame modules were also developed, and advances were made in improving membrane stability. By 1980, MF, UF, RO, and electrodialysis had all been established processes with large plants installed worldwide.

14.1.3 Membrane Types

Water treatment processes employ several types of membranes—MF, UF, RO, and nanofiltration (NF) (Amjad 1993, Perry and Green 1997). MF membranes have the largest pore size and typically reject large particles and various microorganisms. UF membranes have smaller pores than MF membranes, and therefore, in addition to large particles and microorganisms, they can reject bacteria and soluble macromolecules such as proteins. RO membranes are effectively nonporous and therefore exclude particles and even many low-molecular-mass species such as salt ions and organics (Perry and Green 1997).

NF membranes are relatively new and are sometimes called *loose* RO membranes. They are porous membranes, but since the pores are on the order of ten angstroms or less, they exhibit performance between that of RO and UF membranes (Baker 2004).

The features and applications of the membrane processes are presented in Table 14.1.

14.1.4 Membrane Characteristics

Membranes can be classified as follows:

1. Construction materials
   a. Organic
      i. Natural (modified cellulose natural products [CA, cellulose nitrate, regenerated cellulose, etc.])
      ii. Synthetic (polyamide, polyethylene, polypropylene [PP], polyolefin, polysulfide, etc.)
   b. Organic (metallic oxide deposits [aluminum, titanium, zircon, etc.] on metallic or ceramic tubes)
2. Structure
   a. Isotropic (or symmetric): Symmetric structure with channels; not used
   b. Anisotropic (or asymmetric): 0.1–0.5 μm layer on 150–200 μm support
   c. Composite (thin film on asymmetric membranes with polymers not otherwise used)
3. Porosity (mean pore size of membranes determined with statistical methods)
   a. Rejection
   b. Molecular weight cutoff (MWCO)

TABLE 14.1 Features and Applications of the Membrane Processes

	Reverse Osmosis	Nanofiltration	Ultrafiltration	Microfiltration
Membrane	Asymmetric	Asymmetric	Asymmetric	Symmetric Asymmetric
Range of use	Angstrom	Angstrom	Angstrom	μm
Rejection of:	Low- and high-molecular-weight compounds, sodium chloride, glucose, amino acids. Sectors: desalination, industrial wastes, process water and civil wastewater reuse, sugar concentration	High-molecular-weight compounds, polyvalent negative ions, mono-, bi-, and oligosaccharides. Sectors: drinkable water (softening), textile (color removal), galvanic, paper, disinfection	Bacteria, macromolecules, proteins, polysaccharides, viruses. Sectors: food, textile, painting, paper	Microorganisms (bacteria), suspended solids, oily emulsions, sterilization of liquids, clay, beverage clarification. Does not remove viruses and macromolecules
Membrane material	Cellulose acetate (CA) Thin film	CA Thin film	Ceramic polysiphone (PSO), polyvinylidene fluoride (PVDF), CA Thin film	Ceramic polypropylene (PP), PSO, PVDF
Membrane module	Tubular, spiral wound, plane	Tubular, spiral wound, plane	Tubular, hollow fiber, spiral wound, plane	Tubular, hollow fiber
Operating pressure (bar)	10–60	5–40	1–10	<2

14.1.4.1 Isotropic membranes

A microporous membrane is very similar in structure and function to a conventional filter. It has a rigid, highly voided structure with randomly distributed, interconnected pores. However, these pores differ from those in a conventional filter by being extremely small, on the order of 0.01–10 μm in diameter. All particles larger than the largest pores are completely rejected by the membrane (Baker 2004).

Particles smaller than the largest pores but larger than the smallest pores are partially rejected, according to the pore size distribution of the membrane. Particles much smaller than the smallest pores will pass through the membrane. Thus, separation of solutes by this membrane type is mainly a function of molecular size and pore size distribution. In general, only molecules that differ considerably in size can be separated effectively, for example, in UF and MF.

Nonporous, dense membranes consist of a dense film, through which permeates are transported by diffusion under the driving force of a pressure, concentration, or electrical potential gradient. The separation of various components of a mixture is related directly to their relative transport rate within the membrane, which is determined by their diffusivity and solubility in the membrane material. Thus, nonporous, dense membranes can separate permeates of similar size if their concentration in the membrane material (i.e., their solubility) differs significantly. Most gas separation, pervaporation, and RO membranes use dense membranes to perform the separation. Usually, these membranes have an anisotropic structure to improve the flux.

Electrically charged membranes can be dense or microporous but are most commonly very finely microporous, with pore walls carrying fixed positively or negatively charged ions. A membrane with fixed positively charged ions is referred to as an anion-exchange membrane, because it binds anions in the surrounding fluid. Similarly, a membrane containing fixed negatively charged ions is called a cation-exchange membrane. Separation with charged membranes is achieved mainly by the exclusion of ions of the same charge as fixed ions of the membrane structure and to a much lesser extent by the pore size. The separation is affected by the charge and concentration of ions in the solution. For example, monovalent ions are excluded less effectively than divalent ions, and selectivity decreases in solutions of high ionic strength. Electrically charged membranes are used for processing electrolyte solutions in electrodialysis (Baker 2004).

14.1.4.2 Anisotropic membranes

The transport rate of a species through a membrane is inversely proportional to the membrane thickness. High transport rates are desirable in membrane separation processes for economic reasons; therefore, the membrane should be as thin as possible. Conventional film fabrication technologies limit the manufacture of mechanically strong, defect-free films to about 20 μm thickness. The development of novel membrane fabrication techniques to produce anisotropic membrane structures was one of the major breakthroughs in the membrane technology during the past 30 years. Anisotropic membranes consist of an extremely thin surface layer supported on a much thicker, porous substructure. The surface layer and its substructure may be formed in a single operation or separately. In composite membranes, layers are usually made of different polymers. Separation properties and permeation rates of the membrane are determined exclusively by the surface layer; the substructure acts as mechanical support. The advantages of higher fluxes provided by anisotropic membranes are so great that almost all commercial processes use such membranes.

14.1.4.3 Ceramic, metal, and liquid membranes

The discussion so far implies that membrane materials are organic polymers, and in fact, the vast majority of membranes used commercially are polymer based. However, in recent years, interest in membranes formed from less conventional materials has increased (Baker 2004). Ceramic membranes, a special class of microporous membranes, are being used in UF and MF applications, for which solvent resistance and thermal stability are required. Dense metal membranes, particularly palladium membranes, are being considered for the separation of hydrogen from gas mixtures, and supported liquid films are being developed for carrier-facilitated transport processes.

14.1.5 Theory of Membrane Transport

The most important property of membranes is their ability to control the permeation rate of different species. Two models used to describe the permeation mechanism are presented in Figure 14.1. The first one is the solution-diffusion model, where permeates dissolve in the membrane material and then diffuse through the membrane down a concentration gradient. The permeates are separated because of the differences in solubilities of the materials in the membrane and the differences in rates, at which the materials diffuse through the membrane (Hillis 2000). The other one is the pore-flow model, where permeates are transported by pressure-driven convective flow through tiny pores. Separation occurs because one of the permeates is excluded (filtered) from some of the pores in the membrane, through which the other permeates move. Among these models proposed in the nineteenth century, the pore-flow model was closer to normal physical experience, being more popular until the mid-1940s. However, during the 1940s, the solution-diffusion model was used to explain the transport of gases through polymer films. This implementation of the solution-diffusion model was relatively uncontroversial, but the transport mechanism in RO membranes was a hotly debated issue in the 1960s and early 1970s (Meares 1966, Paul 1974, 1976, Paul and Ebra-Lima 1970, Sourirajan 1970, Yasuda and Peterlin 1973). However, by 1980, the proponents of the solution-diffusion model had carried the day; currently, only a few die-hard pore-flow modelers employ this approach to rationalize RO.

Diffusion, the basis of the solution-diffusion model, is the process of matter transport from one part of a system to another by a concentration gradient. Individual molecules in a membrane medium are in constant random molecular motion, whereas in an isotropic medium, they have no

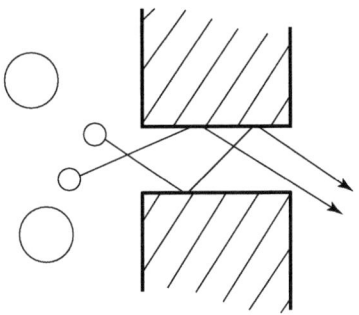

Microporous membranes
separate by molecular
filtration

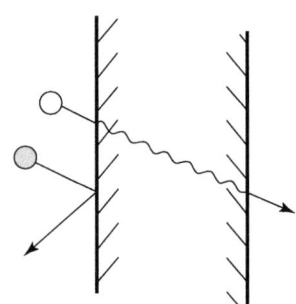

Dense solution-diffusion
membranes separate because
of differences in the solubility
and mobility of permeants
in the membrane material

FIGURE 14.1
Molecular transport through membranes can be described by a flow through permanent pores or by the solution-diffusion mechanism.

preferred direction of motion. Although the mean displacement of a molecule from its starting point can be calculated after a period of time, nothing can be said about the direction, in which any individual molecule would move. However, if a concentration gradient of permeate molecules is formed in the medium, simple statistics show that a net transport of matter will occur from high-concentration regions to low-concentration ones.

For example, when two adjacent volume elements with slightly different permeate concentrations are separated by an interface, more molecules would rather move from the concentrated interface side to the less concentrated one than in the opposite direction, simply because of the difference in the number of molecules for each volume element. This concept was first recognized theoretically and experimentally by Fick (1855), who presented his results in the form of the equation now known as Fick's law of diffusion:

$$J_i = -D_i \frac{dc_i}{dx}, \tag{14.1}$$

where
   J_i is the rate of transfer of component i or flux (kg/m^2·s)
   dc_i/dx is the concentration gradient of component i
   D_i is the diffusion coefficient (m^2/s) representing and a mobility measure for individual molecules

The *minus* sign shows that the direction of diffusion is down the concentration gradient. Diffusion is an inherently slow process. In practical diffusion-controlled separation processes, useful fluxes across membranes are achieved by making them very thin and creating large concentration gradients (Schippers et al. 2004).

Pressure-driven convective flow, the basis of the pore-flow model, is most commonly used to describe flows in a capillary or porous medium. The basic equation covering this type of transport represents Darcy's law:

$$J_i = K'c_i \frac{dp}{dx}, \tag{14.2}$$

where
   dp/dx is the pressure gradient in the porous medium
   c_i is the concentration of component i in the medium
   K' is the coefficient related to the nature of the medium

In general, convective pressure–driven membrane fluxes are high compared with those obtained by simple diffusion.

The difference between the solution-diffusion and pore-flow mechanisms lies in the relative size and permanence of pores. For membranes, across which the transport is best described by the solution-diffusion model and Fick's law, the free volume elements (pores) are tiny spaces between polymer chains caused by the thermal motion of polymer molecules. These volume elements appear and disappear on about the same time scale as the motion of permeates traversing the membrane. On the other hand, for membranes, across which the transport is best described by the pore-flow model and Darcy's law, the free volume elements (pores) are relatively large and fixed, and they do not fluctuate in position or volume on a time scale of the permeate motion, being connected to one another. The larger the individual free volume elements (pores), the more likely they are to be present long enough to produce pore-flow characteristics in the membrane. As a rough rule of thumb, the transition between transient (solution-diffusion) and permanent (pore-flow) pores is in the range 5–10 Å in diameter.

The mean pore diameter in a membrane is difficult to measure directly and must often be inferred from the size of the molecules that permeate the membrane or by some other indirect technique. With this caveat in mind, membranes can be organized into three general groups, as shown in Figure 14.2:

1. UF, MF, and microporous Knudsen flow gas separation membranes are all clearly microporous, and transport occurs by pore flow.
2. RO, pervaporation, and polymer gas separation membranes have a dense polymer layer with no visible pores, in which the separation occurs. These membranes show different transport rates for molecules as small as 2–5 Å in diameter. The fluxes of permeates through these membranes are also much lower than through the microporous membranes. Transport is best described by the solution-diffusion model. The spaces between the polymer chains in these membranes are less than 5 Å in diameter and so are within the normal range of thermal motion of the polymer chains that make up the membrane matrix.

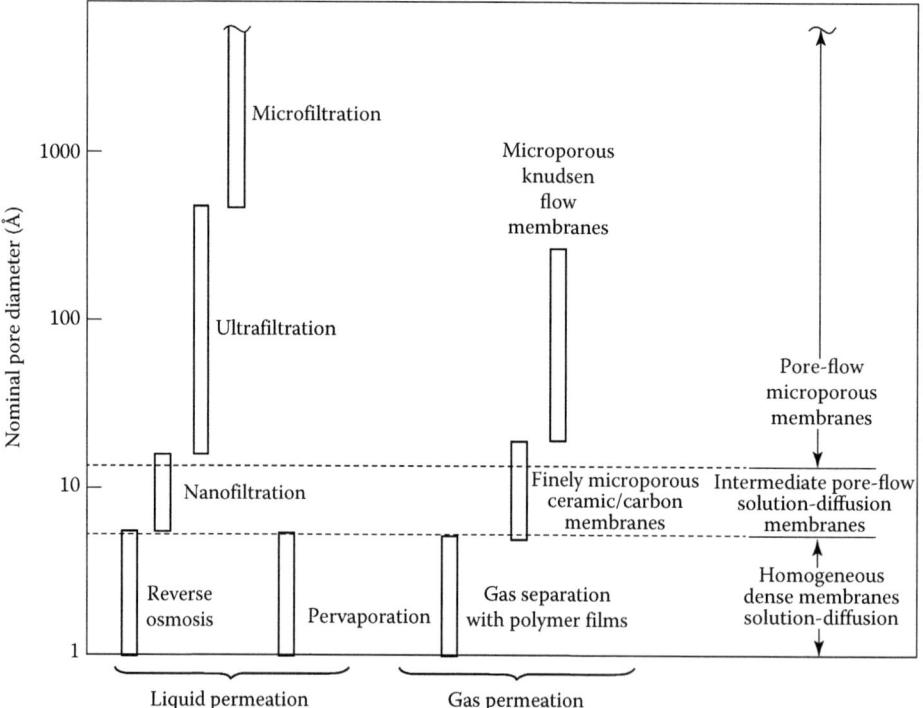

FIGURE 14.2
Schematic representation of the nominal pore size and best theoretical model for the principal membrane separation processes.

Molecules permeate the membrane through free volume elements between the polymer chains that are transient on the time scale of the diffusion processes occurring.

3. Membranes in the third group contain pores with diameters between 5 and 10 Å and are intermediate between truly microporous and truly solution-diffusion membranes.

Transport through dense, nonporous membranes (this includes permeation in RO, pervaporation, and gas separation membranes) occurs by molecular diffusion and is described by the solution-diffusion model. The predictions of this model are in good agreement with experimental data, and a number of simple equations that usefully rationalize the properties of these membranes result. Transport through microporous UF and MF membranes occurs by convective flow with some form of sieving mechanism producing the separation. However, the ability of theory to rationalize transport in these membranes is poor (Baker 2004).

A number of factors concurrently affect permeation, so a simple quantitative description of the process is not possible.

14.1.6 Problems with Membranes

14.1.6.1 Polarization effect

In membrane separation processes, a gas or liquid mixture contacts the feed side of the membrane, and a permeate enriched in one of the components of the mixture is withdrawn from the downstream side of the membrane. Because the feed mixture components permeate the membrane at different rates, there is a gradual buildup in the concentration of nonpermeating or slowly permeating components in the feed as the more permeable components pass through the membrane.

A layer is formed near the surface of the membrane, whereby the solution immediately adjacent to the membrane surface becomes depleted in the permeating solute on the feed side of the membrane, and its concentration is lower than that in the bulk fluid. On the other hand, the concentration of the nonpermeating component increases at the membrane surface. A concentration gradient is formed in the fluid adjacent to the membrane surface. This phenomenon is known as concentration polarization, and it serves to reduce the permeating component's concentration difference across the membrane, thereby lowering its flux and the membrane selectivity (Baker 2004).

The importance of concentration polarization depends on the membrane separation process. Concentration polarization can significantly affect membrane performance in RO, but it is usually well controlled in industrial systems. On the other hand, membrane performance in UF, electrodialysis, and some pervaporation processes is seriously affected by concentration polarization. To minimize concentration polarization,

- Select a membrane well suited to the process.
- Keep the concentration low if this is practical.
- Keep the pressure differential low if this is practical.
- Maintain flow across the membrane to reduce the film thickness and to scour deposits.

14.1.6.2 Membrane fouling

Membrane fouling is a process where solute or particles deposit onto a membrane surface or into membrane pores in a way that degrades the membrane's performance. It is a major obstacle to the widespread use of this technology. Membrane fouling can cause severe flux decline and affect the quality of the water produced. Severe fouling may require intense chemical cleaning or membrane replacement. This increases the operating costs of a treatment plant. There are various types of foulants: colloidal (clays, flocs), biological (bacteria, fungi), organic (oils, polyelectrolytes, humics), and scaling (mineral precipitates) (Baker 2004).

Fouling can be divided into reversible and irreversible fouling based on the attachment strength of particles to the membrane surface. Reversible fouling can be removed by a strong sheer force of backwashing. Formation of a strong matrix of fouling layer with the solute during a continuous filtration process will result in reversible fouling being transformed into an irreversible fouling layer. Irreversible fouling is the strong attachment of particles that cannot be removed by physical cleaning (Choi et al. 2005).

Factors that affect membrane fouling are as follows:

- Membrane properties such as pore size, hydrophobicity, pore size distribution, and membrane material
- Solution properties such as concentration, the nature of the components, and particle size distribution
- Operating conditions such as pH, temperature, flow rate, and pressure

Flux and transmembrane pressure (TMP) are the best indicators of membrane fouling. Under constant flux operation, TMP increases to compensate for the fouling. On the other hand, under constant pressure operation, flux declines due to membrane fouling.

Even though membrane fouling is an inevitable phenomenon during membrane filtration, it can be minimized by strategies such as cleaning, appropriate membrane selection, and choice of operating conditions.

Membranes can be cleaned physically, biologically, or chemically. Physical cleaning includes sponges, water jets, or backflushing using a permeate. Biological cleaning uses biocides to remove all viable microorganisms, whereas chemical cleaning involves the use of acids and bases to remove foulants and impurities.

Another strategy to minimize membrane fouling is the use of the appropriate membrane for a specific operation. The nature of the feed water must first be known; then a membrane that is less prone to fouling with that solution is chosen. For aqueous filtration, a hydrophilic membrane is preferred.

Operating conditions during membrane filtration are also vital, as they may affect fouling conditions during filtration. For instance, cross-flow filtration is always preferred to dead-end filtration, because turbulence generated during the filtration entails a thinner deposit layer and therefore minimizes fouling (Baker 2004).

14.2 METHODS OF MEMBRANE MODIFICATION

Since the early 1960s, synthetic membranes have been used successfully in a wide variety of large-scale industrial applications (Cabasso 1987). The rapid adoption of membranes in industry resulted from breakthrough developments in membrane materials, membrane structures, and large-scale membrane production methods. This overview provides a brief introduction to some of the most common membrane formation and modification methods used for production of commercial membranes.

The properties of a membrane are controlled by the membrane material and the membrane structure. To be useful in an industrial separation process, a membrane must exhibit at least the following characteristics (Pinnau 1994):

- High flux
- High selectivity (rejection)
- Mechanical stability
- Tolerance to all feed stream components (fouling resistance)
- Tolerance to temperature variations
- Manufacturing reproducibility
- Low manufacturing cost
- Ability to be packaged into high surface area modules

Of the aforementioned requirements, flux and selectivity (rejection) determine the selective mass transport properties of a membrane. The higher the flux of a membrane at a given driving force, the lower is the membrane area required for a given feed flow rate, and therefore, the lower are the capital costs of a membrane system (Strathmann 1985a,b). Selectivity determines the

TABLE 14.2   Membrane Modification Methods

Modification Method	Goal	Application
Annealing		RO, gas separation (GS), UF
Heat treatment	Elimination of membrane defects	
Solvent treatment	Control of pore size	
Solvent exchange	Elimination of membrane defects	GS, UF
Surface coating	Elimination of membrane defects	GS
	Improvement of fouling resistance	RO, NF, UF
Chemical treatment		
Fluorination	Improvement of flux and selectivity	GS
Cross-linking	Improvement of chemical resistance	UF
Pyrolysis	Improvement of flux and selectivity	RO, GS, pervaporation (PV)

extent of separation. Membranes with higher selectivity are desirable because higher product purity can be achieved in a separation process.

The development of high-performance membranes involves the selection of a suitable membrane material and the formation of this material into a desired membrane structure. However, it is often necessary to modify the membrane material or the structure to enhance the overall performance of the membrane.

The objectives for modification of preformed membranes are increasing flux and/or selectivity and increasing chemical resistance (solvent or fouling resistance). In general, two kinds of modifications could be effected to improve membrane performance (Strathmann 1985a,b). They are

1. Modification of the polymer, then membrane formation with modified material
2. Modification of the membrane surface

Some of the most commonly practiced membrane modification methods are listed in Table 14.2.

The first reported membrane modification method involved annealing of porous membranes by heat treatment. Zsigmondy and Bachmann demonstrated that the pore size of a preformed nitrocellulose membrane could be decreased with a hot water or steam treatment (Zsigmondy and Bachmann 1922). Loeb and Sourirajan used the same method to improve the salt rejection of integrally skinned asymmetric CA RO membranes (Loeb and Sourirajan 1962). The properties of gas separation membranes can also be improved by annealing with a heat treatment (Hoehn 1974, Kusuki et al. 1992). Heat treatment of gas separation membranes typically leads to an increase in selectivity, because of elimination of microdefects in the thin separating layer. An alternative method of reducing the number of defects in the separating layer in a membrane is based on a solvent swelling technique (Pinnau and Wind 1991, Rezac et al. 1994). In this method, microdefects in a membrane can be eliminated by swelling the thin separating layer with a vapor or a liquid. As a result of swelling, the modulus of the polymer decreases sharply. It has been suggested that capillary forces can pull the swollen polymer layer together, thereby eliminating membrane defects (Rezac et al. 1994).

During the development of integrally skinned asymmetric CA gas separation membranes, it was found that water-wet membranes collapse and form an essentially dense film upon drying. This collapse occurs because of the strong capillary forces within the finely porous structure during the drying process. This phenomenon can be described by the well-known Young–Laplace relationship. Hence, the capillary pressure is directly proportional to the surface tension of a liquid but inversely proportional to the pore radius. If the modulus of the membrane material (in the swollen state) is lower than the capillary force of the liquid in the pore space, the pores will collapse and form a dense polymer film. Because water has a very high surface tension, it is often difficult to dry water-wet membranes without collapsing the membrane structure. An exchange of water with liquids having lower surface tension, such as alcohols or aliphatic hydrocarbons, results in maintaining the original membrane structure upon drying. Typical solvent exchange methods involve replacing water first with isopropanol and then with w-hexane (Admassu 1989, Manos 1978, Rowley and Slowig 1971). Other methods of eliminating the collapse of finely porous membrane structures include freeze drying and addition of surfactants to the water prior to drying of the wet membranes (Gantzel and Merten 1970, Hayes 1991, Jensvold et al. 1992).

Surface modification of flat sheet UF membranes, polyethersulfone (PES), was investigated to improve the hydrophilicity of the membrane surface, thereby reducing the adsorption of the proteins onto the membrane. Grafting of hydrophilic polymers onto UV/ozone-treated PES was used to improve the hydrophilicity of the commercial PES membranes. Hydrophilic polymers, that is, polyvinyl alcohol (PVA), polyethylene glycol (PEG), and chitosan, were employed to graft onto PES membrane surfaces because of their excellent hydrophilic property. It is concluded that grafting of PVA, PEG, or chitosan onto UV/ozone-treated PES membranes increases hydrophilicity and lowers protein adsorption by 20%–60% compared to the virgin PES membrane (Dworecki et al. 2004, Kochkodan et al. 2006, Muller 2005, Raghavan et al. 1996, Tsarenko et al. 2007).

Pyrolysis of polymer precursors is a relatively new modification method that can lead to significantly improved separation performance of synthetic membranes. Specifically, molecular sieve membranes made from pyrolyzed polyacrylonitrile and polyimide (Jones and Koros 1994, Koresh and Soffer 1987, Kusuki et al. 1997) as well as selective surface flow membranes made from polyvinylidene chloride–acrylate terpolymer (Rao 1996) have significantly better separation performance than polymer membranes in gas separation applications. However, the use of these membranes in industrial applications is currently limited by their poor mechanical stability, susceptibility to fouling, and very high production costs. If these problems can be solved in future work, inorganic membranes made from polymer precursors will present a new generation of high-performance membranes for the next millennium.

14.3 CARBON NANOTUBES

14.3.1 General Information

Carbon nanotubes (CNTs) consist of multiple rolled-up graphite layers. These carbon molecules have novel properties, making them potentially useful in many applications in nanotechnology, electronics, optics, water and gas purification, and other fields of materials science, as well as potential uses in architectural fields (O'Connell 2006). They may also have applications in the construction of body armor. They exhibit extraordinary strength and unique electrical properties and are efficient thermal conductors (Ajayan 1999, Reich et al. 2004).

14.3.2 Geometry

There are three unique geometries of CNTs. The three different geometries are also referred to as flavors. The three flavors are armchair, zigzag, and chiral [e.g., zigzag (n, 0); armchair (n, n); and chiral (n, m)].

Applied quantum chemistry, specifically, orbital hybridization best describes chemical bonding in nanotubes. The chemical bonding of nanotubes is composed entirely of sp^2 bonds, similar to those of graphite. These bonds, which are stronger than the sp^3 bonds found in alkanes, provide nanotubules with their unique strength. Moreover, nanotubes naturally align themselves into *ropes* held together by van der Waals' forces.

Nanotubes are categorized as single-walled nanotubes (SWNTs) and multiwalled nanotubes (MWNTs).

MWNTs consist of multiple rolled layers (concentric tubes) of graphite. There are two models that can be used to describe the structures of MWNTs. In the Russian doll model, sheets of graphite are arranged in concentric cylinders. In the Parchment model, a single sheet of graphite is rolled in around itself, resembling a scroll of parchment or a rolled newspaper. The interlayer distance in MWNTs is close to the distance between graphene layers in graphite, approximately 3.4 Å. The outer diameter of MWNTs may range from 1 to 50 nm, while the inner diameter is usually of several nm (Harris 1999).

14.3.3 Preparation and Purification

Techniques have been developed to produce nanotubes in sizeable quantities, including arc discharge, laser ablation, high-pressure carbon monoxide (HiPco), and chemical vapor deposition (CVD) (Ebbesen and Ajayan 1992). Most of these processes take place in vacuum or with process gases. CVD growth of CNTs can occur in vacuum or at atmospheric pressure. During CVD, a substrate is prepared with a layer of metal catalyst particles, most commonly nickel, cobalt (Inami et al. 2007), iron, or a combination (Ishigami et al. 2008). The metal nanoparticles can also be produced by other ways, including reduction of oxides or oxide solid solutions. The diameters of the nanotubes that are to be grown are related to the size of the metal particles (Ebbesen 1996). This can be controlled by patterned (or masked) deposition of the metal, annealing, or by plasma etching of a metal layer. The substrate is heated to approximately 700°C. To initiate the growth of nanotubes, two gases are bled into the reactor: a process gas (such as ammonia, nitrogen, or hydrogen) and a carbon-containing gas (such as acetylene, ethylene, ethanol, or methane). Nanotubes grow at the sites of the metal catalyst; the carbon-containing gas is broken apart at the surface of the catalyst particle, and the carbon is transported to the edges of the particle, where it forms the nanotubes. This mechanism is still being studied. The catalyst particles can stay at the tips of the growing nanotube during the growth process or remain at the nanotube base, depending on the adhesion between the catalyst particle and the substrate. Thermal catalytic decomposition of hydrocarbon has become an active area of research and can be a promising route for the bulk production of CNTs. Fluidized bed reactor is the most widely used reactor for CNT preparation. Scale-up of the reactor is the major challenge (Ionimos Makris et al. 2006, Muradov 2001).

CVD is a common method for the commercial production of CNTs. For this purpose, the metal nanoparticles are mixed with a catalyst support such as MgO or Al_2O_3 to increase the surface area for higher yield of the catalytic reaction of the carbon feedstock with the metal particles. One issue in this synthesis route is the removal of the catalyst support via an acid treatment, which sometimes could destroy the original structure of the CNTs. However, alternative catalyst supports that are soluble in water have proven effective for nanotube growth.

Of the various means for nanotube synthesis, CVD shows the most promise for industrial-scale deposition, because of its price/unit ratio and because CVD is capable of growing nanotubes directly on a desired substrate, whereas the nanotubes must be collected in the other growth techniques.

Purification of CNTs generally refers to the separation of CNTs from other entities, such as carbon nanoparticles, amorphous carbon, residual catalyst, and other unwanted species. The classic chemical techniques for purification have been tried, but they have not been found to be effective in removing the undesirable impurities. Three basic methods have been used with varying degrees of success, namely, gas-phase, liquid-phase, and intercalation methods (Ebbesen 1994).

It is now possible to cut CNTs into smaller segments, by extended sonication in concentrated acid mixtures or by using an extrusion system. The resulting CNTs form a colloidal suspension in solvents. They can be deposited on substrates, or further manipulated in solution, and can have many different functional groups attached to the ends and sides of the CNTs.

14.3.4 Functionalization

Pristine CNTs are unfortunately insoluble in many liquids such as water, polymer resins, and most solvents. Thus, they are difficult to evenly disperse in a liquid matrix such as epoxies and other polymers. This complicates efforts to utilize the CNT's outstanding physical properties in the manufacture of composite materials as well as in other practical applications that require preparation of uniform mixtures of CNTs with many different organic, inorganic, and polymer materials. To make CNTs more easily dispersible in liquids, it is necessary to physically or chemically attach certain molecules, or functional groups, to their smooth sidewalls without significantly changing the CNT's desirable properties. This process is called functionalization. The production of robust composite materials requires strong covalent chemical bonding between the filler particles and the polymer matrix, rather than the much weaker van der Waals' physical bonds that occur if the CNTs are not properly functionalized (Harris 1999).

14.3.5 Applications

The structure of CNTs influences their properties—including electrical and thermal conductivity, density, and lattice structure.

CNTs are the best known field emitters of any material. This is understandable, given their high electrical conductivity and the incredible sharpness of their tip. The smaller the tip's radius of curvature, the more concentrated the electric field will be, leading to increased field emission. The sharpness of the tip also means that they emit at especially low voltage, an important fact for building low-power electrical devices that utilize this feature (Saito 1998).

CNTs have the intrinsic characteristics desired in material used as electrodes in batteries and capacitors, two technologies of rapidly increasing importance. CNTs have a tremendously high surface area and good electrical conductivity, and very importantly, their linear geometry makes their surface highly accessible to the electrolyte.

The record-setting anisotropic thermal conductivity of CNTs is enabling many applications where heat needs to move from one place to another. Such an application is found in electronics, particularly heat sinks for chips used in advanced computing, where uncooled chips now routinely reach over 100°C. The technology for creating aligned structures and ribbons of CNTs (Walters et al. 2001) is a step toward realizing incredibly efficient heat conduits. In addition, composites with CNTs have been shown to dramatically increase their bulk thermal conductivity, even at very small loadings.

Fibers spun of pure CNTs have recently been demonstrated (Baughman 2000) and are undergoing rapid development, along with CNT composite fibers. Such superstrong fibers will have many applications including body and vehicle armor, transmission line cables, woven fabrics, and textiles.

The exploration of CNTs in biomedical applications is just underway but has significant potential. Since a large part of the human body consists of carbon, it is generally thought of as a very biocompatible material. Cells have been shown to grow on CNTs, so they appear to have no toxic effect. The cells also do not adhere to the CNTs, potentially giving rise to applications such as coatings for prosthetics and surgical implants.

The large surface area and high absorbency of CNTs make them ideal candidates for use in air, gas, and water filtration. A lot of research is being done in replacing activated charcoal with CNTs in certain ultrahigh purity applications (Morinobu et al. 2004).

Many researchers and corporations have already developed CNTs based on air and water filtration devices. It has been reported that these filters can not only block the smallest particles but also kill most bacteria. This is another area where CNTs have already been commercialized, and products are on the market now. Someday CNTs may be used to filter other liquids such as fuels and lubricants as well.

A lot of research is being done in the development of CNTs based air and gas filtration. Filtration has been shown to be another area where it is cost-effective to use CNTs already.

In closing, CNTs have many unique and desirable properties. Although many applications may take significant investments of time and money to develop to reach commercial viability, there are plenty of applications today in which CNTs add significant benefits to existing products with relatively low implementation costs. Most of these applications are in the polymer, composite materials, batteries, paints, plastics, ceramics, automotive, textiles industries, etc.

14.4 EXPERIMENTAL INVESTIGATION

The technology of polymer membrane modification consists of the following stages:

- Synthesis of CNTs
- Preparation of steady-state colloidal solutions by means of ultrasound (the application of surfactant species as stabilizers is supposed)
- Formation of CNT porous structures on membrane surfaces

14.4.1 Synthesis of Carbon Nanotubes

The presence of catalysts is the key factor and a powerful instrument determining the quality of CNTs obtained. There are numerous methods of catalyst preparation: for instance, precipitation and thermal decomposition. However, the best homogeneity of a catalyst is provided by solgel technology due to the uniform distribution of components in a starting solution.

The citrate solgel method includes several stages (Figure 14.3).

First, metal salts react with citric acid to form low-molecular-weight oligomers containing metal chelating groups. The heating of these oligomers with polyatomic alcohols results in their etherification. Next, the etherified oligomers form a gel-like metal-containing polymer at higher temperatures, involving the entire volume of the reaction mixture. The mixed metal oxide catalyst is obtained by pyrolysis of the polymer.

The Ni–Co/MgO catalyst was prepared according to the aforementioned procedure. $Ni(NO_3)_2 \cdot 6H_2O$ was added to an appropriate amount of distilled water under stirring until a transparent solution was obtained. After that, $Mg(NO_3)_2 \cdot 6H_2O$ and $Co(NO_3)_2 \cdot 6H_2O$ were added to the solution under stirring followed by introducing ethylene glycol $(CH_2OH)_2$ and citric acid $(HOC(CH_2COOH)_2 \cdot COOH)$. The solution was heated at 80°C for 3 h until the formation of green foamy material and then at 140°C for 1 h. Next, this material was ground into powder.

The powder was placed into a muffle furnace and heated at 550°C for 1 h to remove any organic compounds. To grow CNTs, the catalyst powder was sprayed on a metal disk located in the middle section of a CVD reactor. Then a gas mixture consisting of propane and butane was introduced into this reactor. The temperature inside the reactor vessel was kept at 650°C. The process time was 30 min.

The catalysts and as-grown CNTs were characterized by scanning electron microscopy (SEM) (Carl Zeiss, Neon 40), transmission electron microscopy (TEM) (JEOL, JEM–200CX), and high-resolution transmission electron microscopy (HRTEM) (JEOL, JEM–2010) (Figure 14.4).

The particle size of the catalyst was in the range of 10–25 nm. The high specific area support (MgO) with a high loading of Co^{2+} nanoparticles was the main reason for an extremely efficient growth of CNTs. In the present work, structural and textural parameters of the catalyst were studied. Surface area, pore volume, and pore size distribution are the most valuable characteristics of the catalyst. These parameters were measured using a SORBTOMETR-M surface analyzer, and the following values were obtained: specific surface area 133 m²/g and mesopore surface area 138 m²/g. The particle size distribution in the catalyst was determined with a MicroSizer 201C laser analyzer. The analysis of the data obtained shows that the predominant size of the catalyst particles was in the range of 37.7–43.3 μm.

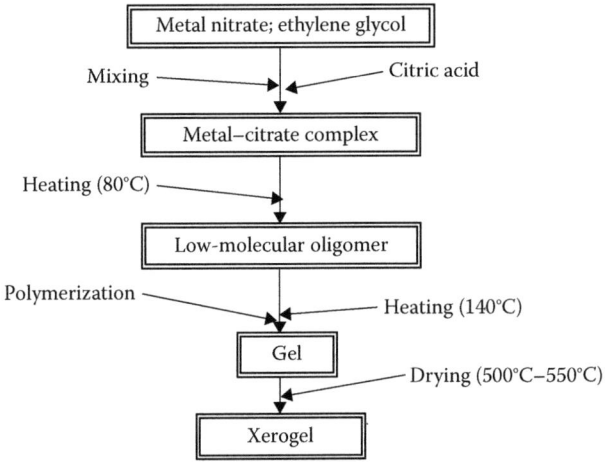

FIGURE 14.3
Scheme of the solgel process for catalyst preparation.

FIGURE 14.4
SEM images of the catalyst structure.

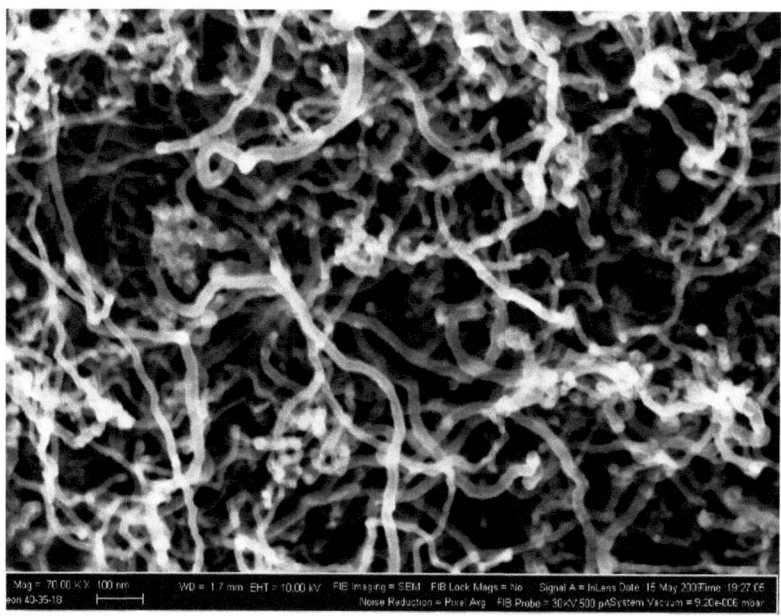

FIGURE 14.5
CNTs grown on the Ni–Co/MgO catalyst.

Figure 14.5 presents the SEM images of the MWNTs synthesized using propane–butane as carbon source and Ni–Co/MgO as catalyst. The diameter of the nanotubes was in the range of 5–30 nm.

HRTEM observations disclosed the CNT's well-graphitized structure with a conical wall (Figure 14.6). The conical nanotubes were most subject to connection of various functional groups and showed increased adsorption characteristics.

However, the CNTs, after the CVD procedure, represented agglomerates of various sizes (50–500 μm), thereby being unsuitable for such a kind of modification.

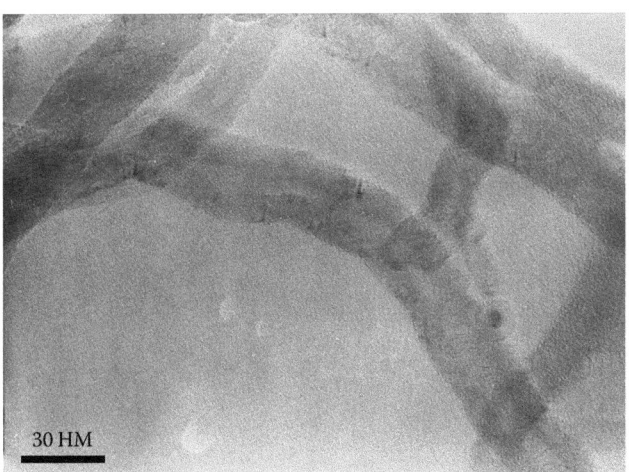

FIGURE 14.6
HRTEM image of an individual MWNT.

14.4.2 Preparation of Colloidal Solutions

The preparation of a water suspension, in which CNTs are uniformly distributed, is necessary for subsequent formation of CNT layers on the surface of polymer membranes. There are a few ways to considerably reduce the size of CNT aggregates.

In the present work, the method ultrasonic dispersion was used. Usually, the term *dispersion* means attrition of a solid in a liquid medium. Ultrasonic machining of nanotubes in the liquid medium makes it possible to obtain superfine, dispersed, homogeneous, and chemically pure mixes.

Based on the analysis of references (modifying additives to various construction and functional materials) and the results of experimental researches, it was revealed that the effective amount of the nanotubes additives is <0.01 wt.%. At such concentration, it was possible to obtain stable suspensions and enough homogeneity nanotube layers on the membrane surface. Stable dispersion will last for days or weeks.

14.4.3 Formation of Carbon Nanotube Porous Structures on Membrane Surfaces

In the present work, the process of CNT layer formation was carried out on a laboratory facility containing a two-compartment partition cell (Figure 14.7).

The system operates as follows. The working solution is injected from day tank 1 into partition cell 4 through high-pressure valves, plunger pump 2, and receiver 3. After the partition, the solution returns to day tank 1, and its flow is controlled by flow-type flowmeters 5. Receiver 3 is used to smooth out pressure pulsations in the flow at 30%–40% of the working pressure. It is a welded cylindrical vessel filled with precompressed air from high-pressure compressor 7. The pressure in the system is controlled by an exemplary pressure gauge mounted on the cell. The flow is adjusted with plunger pump 2. The liquid that passed through the separation membrane is collected in permeate collection tank 6. The pressure can be reset by a needle valve. The pressure and flow fluctuations in the described system do not exceed 5% of the established value.

The main element of the facility is the plate-type two-compartment partition cell, which directly drives the membrane separation process.

14.4.3.1 Comparative analysis of the productivity of polymer microfiltration membranes

In this chapter, the modification effect of CNT layers on the productivity of MF fluoroplastic composite membranes (MFFK series) was estimated. The choice of these membranes was based

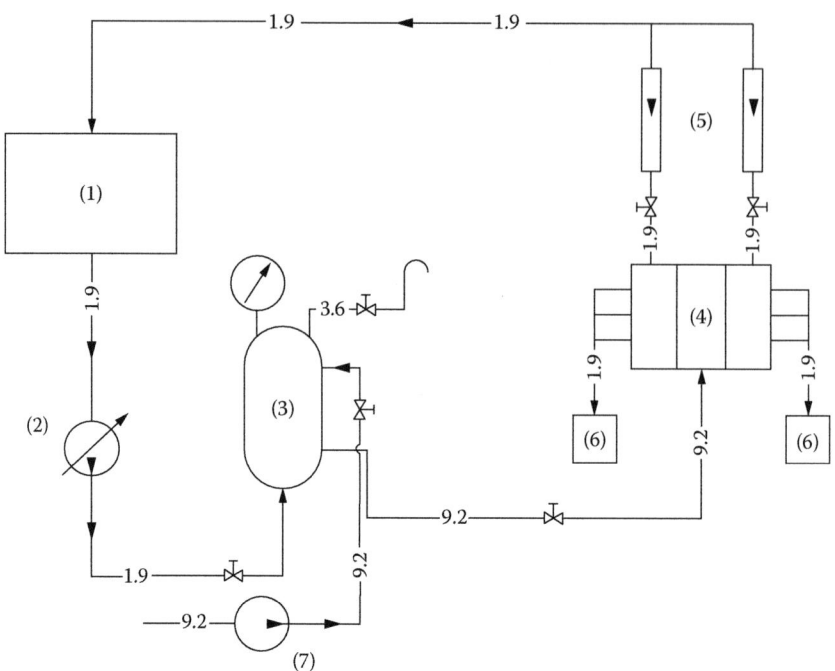

FIGURE 14.7
Scheme of the experimental laboratory facility: (1) Day tank, (2) plunger pump, (3) receiver, (4) partition cell, (5) float-type flowmeters, (6) permeate collection tank, (7) high-pressure compressor.

on the fact that their structure provides the best conditions for sludge formation. The main characteristic is the membrane productivity. MF membranes with a pore size of 0.1–1.0 μm retain fine suspended and colloidal particles. They are commonly used when there is a need for rough water purification or pretreatment before deeper cleaning.

The experiment was performed as follows. A sample concentrate (0.5% magnesium sulfate solution) was introduced at a given pressure into the laboratory facility containing a standard separation membrane. The permeate sample volume was measured at regular time intervals (5 min) to evaluate changes in the volume of the solution that passed through the membrane.

The duration of each experiment was 180 min. The operating pressure in the cell was 0.2 MPa. After the study was conducted with a model, the procedure was repeated with the nanomodified membrane.

The results of the experiments were plotted as a graph of the permeate sample volume against the separation time for the MF membranes of MFFK-1 and MFFK-4 (Figures 14.8 and 14.9, respectively).

The productivity was determined as follows. Distilled water was passed through the modified membrane at a pressure of 0.4 atm (working pressure for MF membranes). The permeate sample volume was measured at regular time intervals (5 min). The experiment was continued until the permeate sample volume was constant.

From these graphs (Figures 14.8 and 14.9), it is possible to conclude that the mean productivity of the modified membrane was significantly higher than that of the standard sample. Moreover, for both types of membranes, a mean increase of 1.5 times can be observed.

14.4.3.2 Comparative analysis of the selectivity of polymer nanofiltration and reverse osmosis membranes

The quantitative analysis of the modification effect of CNTs on the selectivity of polymer membrane samples was performed with RO (MGA-95, porous polymer semitransparent film on the basis of CA on a substrate; energy-saving polyamide [ESPA]) and NF (OPMN-P) membranes using standard salt solutions.

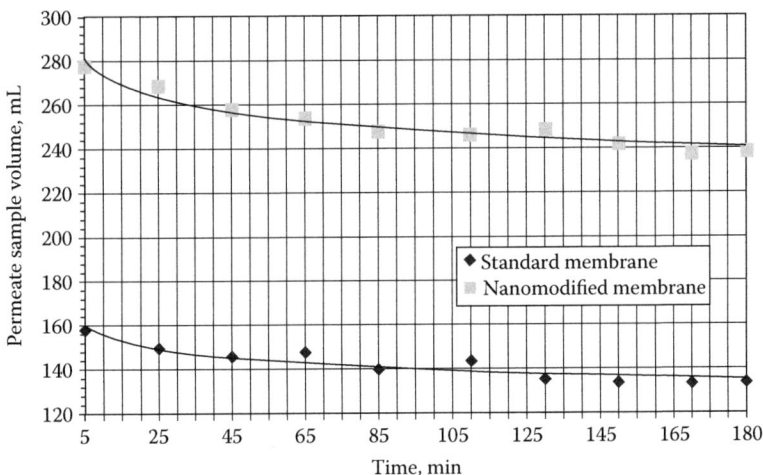

FIGURE 14.8
Productivity of standard and nanomodified membranes (MFFK-1).

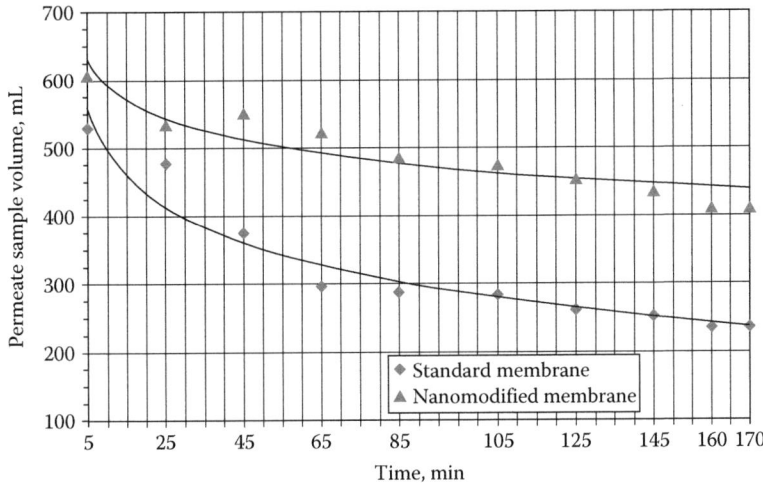

FIGURE 14.9
Productivity of the standard and nanomodified membranes (MFFK-4).

Separation mechanisms for RO and NF membranes are almost the same. These membranes present conditionally nonporous partitioning not prone to the formation of precipitates. To work with two-stage NF and RO membrane systems, it is important to improve the quality of separation while reducing energy consumption for the process.

The experiment was carried out as follows. A standard sample concentrate (5% magnesium sulfate solution) was introduced at a given pressure into the laboratory facility. Samples were taken at intervals of 120 min. The operating pressure in the cell was 1.4 MPa.

The results of the experiment are shown in Figure 14.10.

From this figure, one can see that the selectivity of the nanomodified membrane was significantly higher than that of the standard sample. Moreover, the presence of impurities in the magnesium sulfate nanomodified membrane permeate was decreased by an average of eight times compared to the standard membrane.

Besides, the experiments were conducted with magnesium and calcium chloride solutions (0.2% solution mixture) at an operating pressure of 10 MPa.

The results are presented in Figure 14.11.

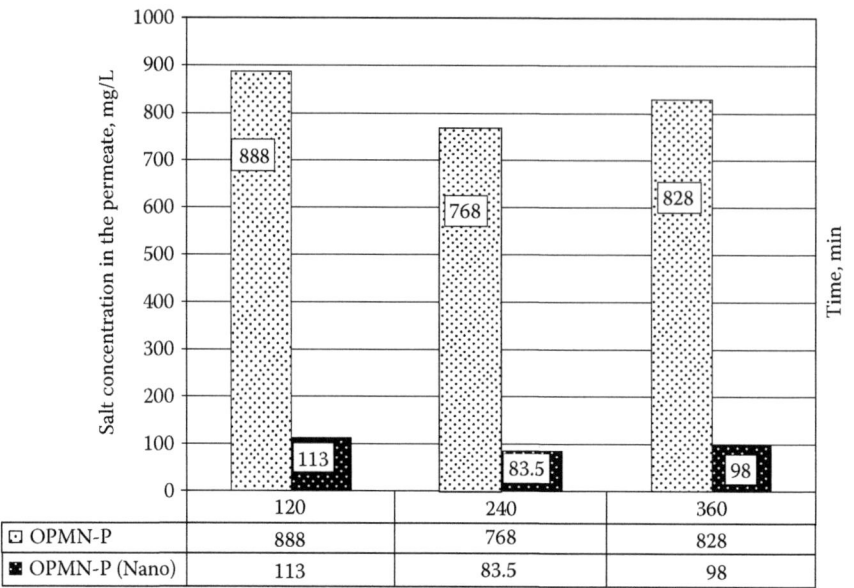

FIGURE 14.10
Selectivity of the standard and nanomodified membranes (OPMN-P) for magnesium sulfate.

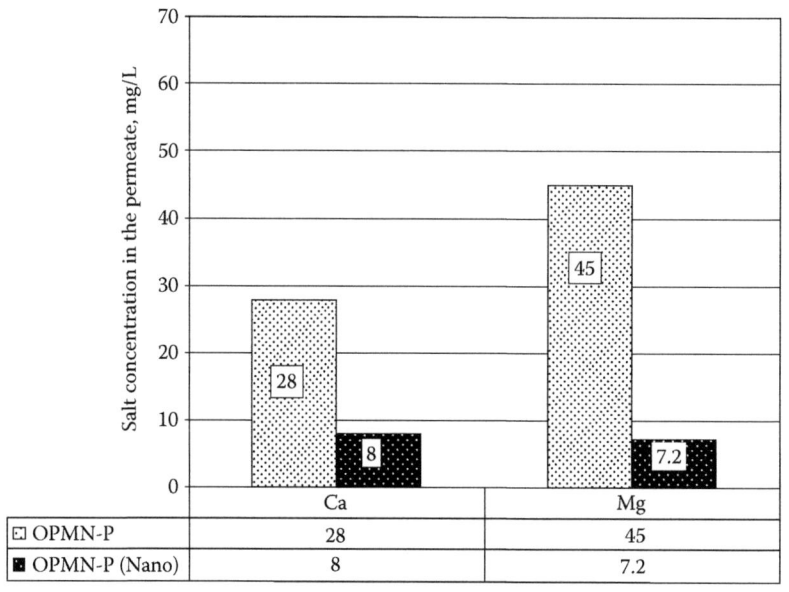

FIGURE 14.11
Selectivity of the OPMN-P membrane for the solution mixture of calcium and magnesium chlorides.

This figure demonstrates that the selectivity of the modified membrane was much higher than that of the standard sample. Moreover, the concentration of calcium and magnesium ions in the permeate after the modified membrane separation process has increased more than 3.5 times and more than 6 times, respectively. The obtained results give a clear indication of the effectiveness of using the CNTs to improve the bandwidth of NF membranes.

The description of the RO membrane used in the experiments is given in the following section:

ESPA membranes offer high performance and the ability to operate at high specific flow rates without changing standards for high selectivity. They are used for treating brackish water, manufacturing bottled water, and in some *light* industrial processes.

ESPA elements are employed when the cost of energy is high enough or source water has a low temperature and/or when the source water total salt content requires the use of high pressures.

MGA membranes are translucent or white porous polymer films on the basis of CA substrates (nonwoven PP, woven and nonwoven polyester).

They are used for desalination of brackish and saline waters with salt contents up to 20 g/L, decontamination of sewage and industrial effluents, and isolation of valuable substances from aqueous solutions.

Their features and benefits are as follows: high hydrophilicity and stiffness promoting long-term use of membranes, low toxicity, and safety at work.

The results of the comparative analysis of the quality and standard purification of the modified RO membrane samples are shown in Figures 14.12 and 14.13.

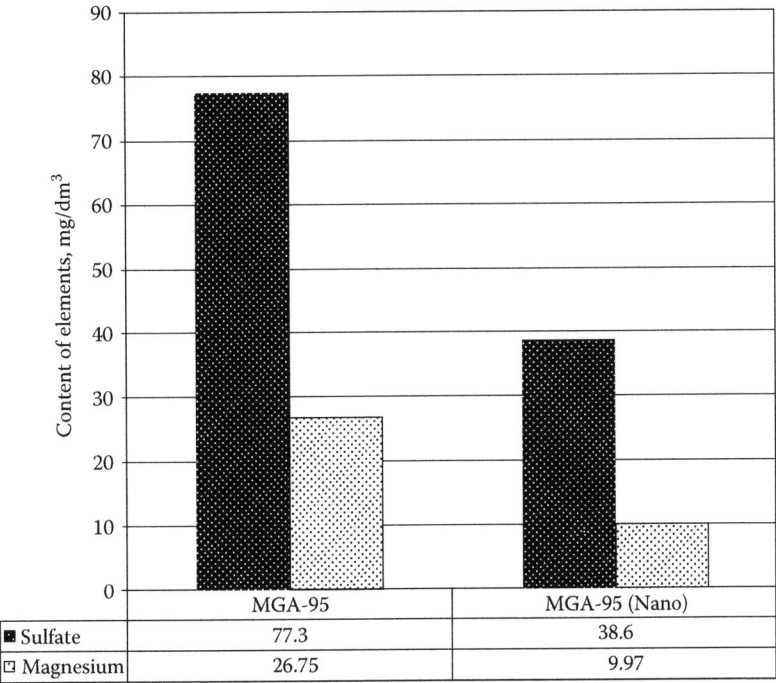

	MGA-95	MGA-95 (Nano)
■ Sulfate	77.3	38.6
▯ Magnesium	26.75	9.97

FIGURE 14.12
Comparative diagrams of selectivity for the standard and modified membranes (MGA-95).

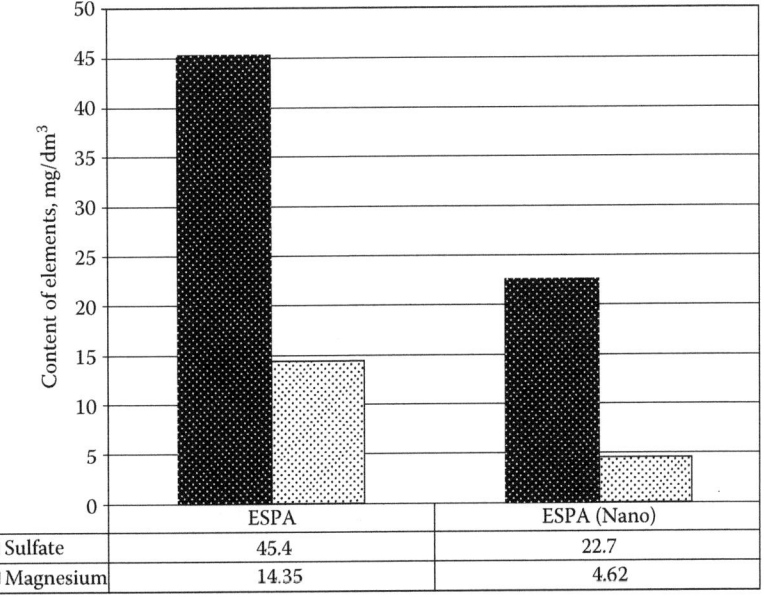

	ESPA	ESPA (Nano)
■ Sulfate	45.4	22.7
▯ Magnesium	14.35	4.62

FIGURE 14.13
Comparative diagrams of selectivity for the standard and modified membranes (ESPA).

It can be observed that the impurity content in the membrane element permeate after the modification was decreased by an average of more than 2.2 times.

The results of the experiments confirmed the effectiveness and feasibility of amending polymer membranes based on a combination of the unique properties of CNTs and innovative modification of membrane elements. The studies showed that there is a real possibility of applying the modification process on an industrial scale with an appropriate technical and economic effect due to a marked increase in indicators of quality cleaning.

14.5 CONCLUSIONS

In the present research, a novel and effective method for the surface modification of polymer membranes with CNTs was presented. The basic stages of the membrane modification process were considered aiming to decrease impurity contents in the permeates and increase the membrane productivity.

The CNT synthesis on the Ni–Co/MgO catalyst with an extremely high yield and purity was described. The structural and dispersion characteristics of this catalyst were studied. The morphology and structure of the CNTs and catalyst were characterized by SEM and HRTEM.

Thus, it was shown that the modification with nanotubes results in increased selectivity and productivity of the studied membranes.

It should be noted that the research requires performing significant optimality and specific testing, which will be carried out afterward. Nevertheless, the appropriateness and importance of the studied issue do not cause any doubt.

REFERENCES

Admassu, W. 1989. Process for drying water–wet polycarbonate membranes. U.S. Patent 4,843,733.

Ajayan, P.M. 1999. Nanotubes from carbon. *Chem. Rev.* 99:1787.

Amjad, Z. 1993. *Reverse Osmosis: Membrane Technology, Water Chemistry, and Industrial Applications.* New York: Van Nostrand Reinhold.

Baker, R.W. 2004. *Membrane Technology and Applications.* San Francisco, CA: John Wiley and Sons.

Baughman, R.H. 2000. Materials science—Putting a new spin on carbon nanotubes. *Science* 290:1310–1311.

Cabasso, I. 1987. *Encyclopedia of Polymer Science and Engineering.* Vol. 9, pp. 509–579. New York: John Wiley and Sons.

Choi, H., Zhang, K., Dionysiou, D.D., and Sorial, G.A. 2005. Effect of permeate flux and tangential flow on membrane fouling for wastewater treatment. *J. Separ. Purif. Technol.* 45:68–78.

Dworecki, K., Drabik, M., Hasegawa, T., and Wsik, S. 2004. Modification of polymer membranes by ion implantation. *Nucl. Instrum. Methods Phys. Res. Sect. B* 225:483–488.

Ebbesen, T.W. 1994. Carbon nanotubes. *Annu. Rev. Mater. Sci.* 24:235.

Ebbesen, T.W. 1996. *Carbon Nanotubes—Preparation and Properties.* Boca Raton, FL: CRC Press.

Ebbesen, T.W. and Ajayan, P.M. 1992. Large-scale synthesis of carbon nanotubes. *Nature* 358:220–236.

Elford, W.J. 1937. Principles governing the preparation of membranes having graded porosities. The properties of "Gradocol" membranes as ultrafilters. *Trans. Faraday Soc.* 33:1094–2011.

Ferry, J.D. 1936. Ultrafilter membranes and ultrafiltration. *Chem. Rev.* 18:373–394.

Fick, A. 1855. Uber Diffusion, Poggendorff's Annal. *Phys. Chem.* 94:59–86.

Gantzel, P.K. and Merten, U. 1970. Gas separation with high-flux cellulose acetate membranes. *Ind. Eng. Chem. Proc. Des. Dev.* 9:331–347.

Harris, P.F. 1999. *Carbon Nanotubes and Related Structures: New Materials for the Twenty-First Century.* Cambridge, U.K.: Cambridge University Press.

Hayes, R.A. 1991. Surfactant treatment of polyaramide gas separation membranes. U.S. Patent 5,032,149.

Hillis, P. 2000. *Membrane Technology in Water and Wastewater Treatment.* Cambridge, U.K.: Royal Society of Chemistry.

Hoehn, H. 1974. Heat treatment of membranes of selected polyimides, polyesters and polyamides. U.S. Patent 3,822,202.

Inami, N. et al. 2007. Synthesis-condition dependence of carbon nanotube growth by alcohol catalytic chemical vapor deposition method. *Sci. Technol. Adv. Mater.* 8:292–307.

Ionimos Makris, Th., Giorgi, L., Giorgi, R., Lisi, N., Salernitano, E., Alvisi, M., and Rizzo, A. 2006. CVD synthesis of carbon nanotubes on different substrates. *Carbon Nanotubes Math. Phys. Chem.* 222:59–60.

Ishigami, N., Ago, H., Imamoto, K., Tsuji, M., Lakoubovskii, K., and Minami, N. 2008. Crystal plane dependent growth of aligned single-walled carbon nanotubes on sapphire. *J. Am. Chem. Soc.* 130:9918–9924.

Jensvold, J.A., Cheng, T., and Schmidt, D.L. 1992. Polycarbonate, polyester, and polyestercarbonate semi-permeable gas separation membranes possessing improved gas selectivity and recovery, and processes for making and using the same. U.S. Patent 5,141,530.

Jones, C.W. and Koros, W.J. 1994. Carbon molecular sieve gas separation membranes—I. Preparation and characterization based on polyimide precursors. *Carbon* 324:1419–1425.

Kochkodan, V.M., Hilal, N., Goncharuk, V.V., Al-Khatib, L., and Levadna, T.I. 2006. Effect of the surface modification of polymer membranes on their microbiological fouling. *Colloid J.* 68:267–273.

Kolf, W.J. and Berk, H.T. 1944. The artificial kidney: A dialyzer with great area. *Acta Med. Scand.* 117:121–144.

Koresh, J.E. and Soffer, A. 1987. The carbon molecular sieve membrane. General properties and the permeability of CH_4/H_2 mixture. *Sep. Sci. Technol.* 22:973–985.

Kusuki, Y., Shimazaki, H., Tanihara, N., Nakanishi, S., and Yoshinaga, T. 1997. Gas permeation properties and characterization of asymmetric carbon membranes prepared by pyrolyzing asymmetric polyimide hollow fiber membrane. *J. Membr. Sci.* 134:245–253.

Kusuki, Y., Yoshinaga, T., and Shimazaki, H. 1992. Aromatic polyimide double layered hollow filamentary membrane and process for producing same. U.S. Patent 5,141,642.

Loeb, S. and Sourirajan, S. 1962. Seawater demineralisation by means of an osmotic membrane. *Adv. Chem. Ser.* 38:117–132.

Loeb, S. and Sourirajan, S. 1963. Sea water demineralization by means of an osmotic membrane, in saline water conversion-II. *Adv. Chem. Ser.* 28:117–132.

Manos, P. 1978. Solvent exchange drying of membranes for gas separation. U.S. Patent 4,120,098.

Meares, P. 1966. On the mechanism of desalination by reversed osmotic flow through cellulose acetate membrane. *Eur. Polym. J.* 2:241–263.

Morinobu, E. et al. 2004. Applications of carbon nanotubes in the twenty-first century. *Phil. Trans. R. Soc. Lond. A* 362:2223–2238.

Muller, H.-J. 2005. Modified membranes. U.S. Patent 6,884,350.

Muradov, N. 2001. Hydrogen via methane decomposition: An application for decarbonization of fossil fuels. *Int. J. Hydrogen Energy* 26:1165–1175.

O'Connell, M.J. 2006. *Carbon Nanotubes: Properties and Applications: A Book.* Claremont, CA: Taylor & Francis Group.

Paul, D.R. 1974. Diffusive transport in swollen polymer membranes, in *Permeability of Plastic Films and Coatings*, ed. H.B. Hopfenberg, pp. 35–48. New York: Plenum Press.

Paul, D.R. 1976. The solution–diffusion model for swollen membranes. *Sep. Purif. Methods* 5:33–51.

Paul, D.R. and Ebra-Lima, O.M. 1970. Pressure–induced diffusion of organic liquids through highly swollen polymer membranes. *J. Appl. Polym. Sci.* 14:2201–2214.

Perry, R.H. and Green, D.W. 1997. *Perry's Chemical Engineers' Handbook.* New York: McGraw-Hill.

Pinnau, I. 1994. Recent advances in the formation of ultrathin polymeric membranes for gas separations. *Polym. Adv. Technol.* 5:733–744.

Pinnau, I. and Wind, J. 1991. Process for increasing the selectivity of asymmetric membranes. U.S. Patent 5,007,944.

Raghavan, S., Jan, D., and Chilkunda, R. 1996. Modification of polyvinylidene fluoride membrane and method of filtering. U.S. Patent 5,531,900.

Rao, M.B., Sircar, S., and Golden, T.C. 1996. Composite porous carbonaceous membranes. U.S. Patent 5,507,860.

Reich, S. et al. 2004. *Carbon Nanotubes: Basic Concepts and Physical Properties.* Berlin, Germany: Wiley-VCH.

Rezac, M.E., Le Roux J.D., Chen, H., Paul, D.R., and Koros, W.J. 1994. Effect of mild solvent post-treatments on the gas transport properties of glassy polymer membranes. *J. Membr. Sci.* 90:213–221.

Rowley, M.E. and Slowig, W.D. 1971. Dry stabilized, rewettable semipermeable cellulose ester and ether membranes and their preparation. U.S. Patent 3,592,672.

Saito, R. 1998. *Physical Properties of Carbon Nanotubes.* Singapore: World Scientific Publishing.

Schippers, J.C., Kruithof, J., and Nederl, M. 2004. *Integrated Membrane Systems.* Denver, CO: American Water Works Association.

Sourirajan, S. 1970. *Reverse Osmosis.* New York: Academic Press.

Strathmann, H. 1985a. Production of microporous media by phase inversion processes. In *Materials Science of Synthetic Membranes*, ed. D.R. Lloyd, pp. 165–195. Washington, DC: ACS Symp.

Strathmann, H. 1985b. Synthetic membranes and their preparation. In *Handbook of Industrial Membrane Technology*, ed. M.C. Porter, pp. 1–60. Park Ridge, NJ: Noyes Publications.

Tsarenko, S.A., Kochkodan, V.M., Potapchenko, N.G., Kosinova, V.N., and Goncharuk, V.V. 2007. Use of titanium dioxide for surface modification of polymeric membranes to diminish their biological contamination. *Russ. J. Appl. Chem.* 80:600–604.

Walters, D. et al. 2001. In-plane-aligned membranes of carbon nanotubes. *Chem. Phys. Lett.* 338:14–20.

Yasuda, H. and Peterlin, A. 1973. Diffusive and bulk flow transport in polymers. *J. Appl. Polym. Sci.* 17:433–456.

Zsigmondy, R. and Bachmann, W. 1922. Filter and method of producing same. U.S. Patent 1,421,341.

CHAPTER 15

CONTENTS

Structures, Fabrication, Physical Properties, and Applications of Graphene Oxide and Hydrogenated Graphene

15

Lizhao Liu, Haili Gao, and Jijun Zhao

15.1 INTRODUCTION

As a superstar nanomaterial, graphene holds great promise for many technological applications and attracts significant attentions. However, graphene is a semimetal with zero gap. Oxidation and hydrogenation are the two common ways to convert sp^2 hybridization state into sp^3 configuration and then open a tunable bandgap in the graphene-based materials. The other physical properties, such as mechanical strength and work function, can be also tailored by the ratio and species of functional groups. Moreover, the existence of surface functional groups will facilitate metal adsorption on the graphene basal plane, resulting in applications in the fields of energy and environmental science. In this chapter, we will summarize recent researches on the structures and physical properties of graphene oxide (GO) and hydrogenated graphene (HG). The relevant experimental progresses will also be briefly reviewed.

GO is a monolayer of graphite oxide. Graphite oxide was successfully fabricated in the laboratory more than one and a half centuries ago [1]. Generally, there are three main methods to synthesize graphite oxide, which will be described in this chapter. By exfoliating graphite oxide into monolayered structures, GO can be obtained. After the experimental synthesis, one important issue is to determine the structure of GO. However, since the oxygen-containing groups and their arrangements across the carbon network vary with different synthesis conditions, the structure of GO is still ambiguous hitherto. Generally, two kinds of techniques have been employed to determine the structure of GO, that is, spectroscopic and microscopic techniques [2]. The former provides deep insights into the types of oxygenated functional groups in GO and their distributions, while the latter directly shows the atomic structure and lattice of GO. Besides characterization, structural modeling of GO is very helpful to elucidate its geometry from atomistic scale. Starting from the structural models of GO, one can further explain its various physical and chemical properties observed in experiments or exploit its application in potential fields. To model the structure of GO, computational simulation is an important approach. So far, numerous computations have been carried out to study the structure of GO, focusing on the effects of coverage, ratio, and arrangement of the oxygen-containing groups. Generally, a structural model with epoxies and hydroxyls orderly arranged in a chained form is considered to be thermodynamically stable [3].

On the other hand, the fundamental properties of GO are closely related to the coverage, ratio, and arrangement of the functional groups. The electronic property of GO is tunable, mainly depending on the coverage and ratio of the epoxies and hydroxyls [4,5]. The tunable bandgap also leads to the tunable optical property of GO. Moreover, the mechanical property of GO mainly relies on the coverage and arrangement of the functional groups, as well as the thickness of GO sheet. The abundant physical and chemical properties enable GO-promising applications in numerous fields. First of all, since GO is oxidized graphene, one fundamental usage of GO is to serve as material to further fabricate graphene by reduction. In addition, the oxygen-containing groups of GO can be used as the active sites for adsorbing and detecting gases, which is very useful in energy storage and environmental science.

HG is another important functionalized graphene derivative by partially or fully transforming the sp^2 hybridization of carbon atoms on graphene sheet to sp^3 by hydrogen chemisorption, which has attracted tremendous research interests for potential applications in graphene-based electronic devices, hydrogen storage, and so on [6–8]. Nowadays, various synthesis and surface analysis techniques are successfully developed for hydrogenation of graphene, such as exposing to hydrogen plasma and atomic hydrogen beams [9–12]. These approaches provide a route to engineer HG with local functionalization, meanwhile with the troubles in controlling the accurate amount of induced atomic defects and the exact hydrogen concentrate. Electrical transport experiments and angle-resolved photoemission spectroscopy (ARPES) are usually utilized to measure the bandgap of HG [13,14]. The most stable graphane with fully hydrogenation on both sides is an insulator with a big energy gap of about 5.4 eV calculated with GW method [15] and has also been demonstrated in the reversible hydrogenation process of graphene by electrical transport measurement [9].

Depending on hydrogen coverage and configurations, bandgap of HG with paired H vacancies can be tailored continuously from 0 to 4.66 eV. Such tunable gaps between midultraviolet and near-infrared regions (NIR) (ca. 0.5–5 eV) in the HG may lead to potential applications in future electronics and photonics (e.g., solar cells) [16–18]. Furthermore, H vacancy clusters of graphane-like structures formed in hydrogenation process may be an ideal support material for noble metal catalysts with lower-energy barrier of chemical reactions [19]. Commonly, since that it is almost impossible to synthesize perfect HG without defects, it is important to choose the most appropriate method for fabrication and characterization of HG [6], whereas the theoretical studies can provide vital atomistic insight into the physical properties in regard with the detailed conformations.

15.2 GRAPHENE OXIDE

15.2.1 Fabrication

Usually, GO can be obtained by exfoliating graphite oxide into monolayered sheets through a variety of thermal and mechanical methods. One common approach is sonicating graphite oxide in water or polar organic media, which can completely exfoliate the graphite oxide [20–23]. Another common approach is mechanical stirring of graphite oxide in water [24,25]. Also, sonicating and mechanical stirring can be combined to exfoliate graphite oxide [26]. Compared with the mechanical stirring, sonicating in water or polar organic media is much faster, but it has a great disadvantage in causing substantial damage to the GO platelets [23].

The experimental fabrication of graphite oxide can go back to more than one and a half centuries ago. By treating graphite with the mixture of potassium chlorate ($KClO_3$) and nitric acid (HNO_3), Brodie reported the synthesis of graphite oxide during investigating the structure of graphite [1]. The ratio of C:H:O in the product was measured to be 61.04:1.85:37.11 with a molecular formula of $C_{2.19}H_{0.80}O_{1.00}$. After heating to a temperature of 220°C, the ratio of C:H:O of this material changed to 80.13:0.58:19.29, with a molecular formula of $C_{5.51}H_{0.48}O_{1.00}$, coupled with a loss of carbonic acid and carbonic oxide. Since the material was dispersible in pure or basic water, but not in acidic media, Brodie termed the material "graphic acid." [26]

Later in 1898, Staudenmaier [27] improved Brodie's approach by adding the chlorate (such as $KClO_3$ or $NaClO_3$) in multiple aliquots during the reaction, as well as adding the concentrated sulfuric acid (H_2SO_4) to increase the acidity of the mixture, rather than in a single addition as Brodie did. Then the graphite was treated with the chlorate, H_2SO_4, and HNO_3. This slight change in the procedure resulted in an overall extent of oxidation similar to Brodie's multiple oxidation approach (C:O ~ 2:1) but performed more practically in a single reaction vessel.

TABLE 15.1 Methods to Fabricate Graphite Oxide from Oxidizing Graphite

Method	Brodie	Staudenmaier	Hummers	Modified Hummers	
Year	1859	1898	1958	1999	2004
Oxidants	$KClO_3$,HNO_3	$KClO_3$ (or $NaClO_3$), HNO_3, H_2SO_4	$NaNO_3$, $KMnO_4$, H_2SO_4	Pre-ox: $K_2S_2O_8$, P_2O_5, H_2SO_4 Ox: $KMnO_4$, H_2SO_4	$NaNO_3$, $KMnO_4$, H_2SO_4
C:O ratio	2.16 [1]; 2.28 [30]	N/A [27]; 1.85 [30]	2.25 [28]; 0.17 [30]	1.3 [31]	1.8 [32]
Reaction time (days)	3–4 [1]; 5/12 [30]	1–2 [27]; 10 [30]	~1/12 [28]; ~5/12 [30]	1/4 pre-ox + 1/12 ox [31]	~5 [32]
Intersheet spacing (Å)	5.95 [30]	6.23 [30]	6.67 [30]	6.9 [31]	8.3 [32]

Source: Compton, O.C., Nguyen, S.T.: Graphene oxide, highly reduced graphene oxide, and graphene: Versatile building blocks for carbon-based materials. *Small.* 2010. 6. 711–723. Copyright Wiley-VCH Verlag GmbH & Co. KGaA. Reproduced with permission.

In 1958, Hummers and Offeman [28] developed an alternate oxidation method by reacting graphite with a mixture of potassium permanganate ($KMnO_4$) and concentrated H_2SO_4, which also achieved similar levels of oxidation. Graphite oxide was prepared by mixing ultrapure graphite powder and sodium nitrate ($NaNO_3$) in H_2SO_4. Then, $KMnO_4$ was added to catalyze the reaction through the oxidant of diamanganese heptoxide (Mn_2O_7). The Mn_2O_7 is able to selectively oxidize unsaturated aliphatic double bonds over aromatic double bonds, which may have important implications for the structure of graphite and reaction pathway(s) during the oxidation.

Generally speaking, Brodie, Staudenmaier, and Hummers methods are the three major ways to produce graphite oxide from graphite. Though others have developed slightly modified versions, these three methods remain the primary routes. Table 15.1 [29] summarizes the experimental parameters of different methods for fabricating graphite oxide.

However, both Brodie and Staudenmaier methods generate ClO_2 gas, which must be handled with caution due to its high toxicity and tendency to decompose in air to produce explosions. Fortunately, this drawback can be avoided in the Hummers method, which has relatively shorter reaction time and is absent of hazardous ClO_2. Nowadays, the Hummers method has been widely used. However, the Hummers method still has one shortcoming, which is potential contamination by excess permanganate ions. The contamination can be removed by H_2O_2 treatment, followed by washing and thorough dialysis [33]. On the other hand, it has been demonstrated that the products of graphite oxide synthesis reactions show strong variance, depending on not only the particular oxidants used but also the graphite source and reaction conditions.

15.2.2 Structural Characterization

The experimentally fabricated GO powders are usually water dispersible, insulating, and light brown in color. However, since the oxygen-containing groups and their arrangements across the carbon network vary much more in different synthesis conditions, the structure of GO is still ambiguous hitherto. In general, two kinds of techniques have been employed to determine the structure of GO. The spectroscopic technique provides deep insights into the types and distributions of oxygenated functional groups on GO, while the microscopic means directly shows the atomic structure and lattice of GO.

15.2.2.1 Spectroscopic techniques

The solid-state nuclear magnetic resonance (NMR) spectra of different GO samples exhibit similar resonance patterns featuring three peaks (60, 70, and 130 ppm), and their relative intensities do not change significantly upon oxidation. Early in 1997 and 1998, Lerf et al. assigned the peak around

60 ppm to epoxy (C–O–C), the peak around 70 ppm to hydroxyl (C–OH), and the peak around 130 ppm to nonaromatic carbon double bonds (>C=C<) [34,35]. Afterward, high-resolution NMR using magic angle spinning by Cai et al. [36] confirmed that the peak around 60 ppm corresponds to epoxy, the peak around 70 ppm corresponds to hydroxyl, and the peak around 130 ppm corresponds to sp^2 carbon, as shown in Figure 15.1a. Now, the three major chemical-shift peaks around 60, 70, and 130 ppm are commonly accepted and assigned to epoxy, hydroxyl, and graphitic sp^2

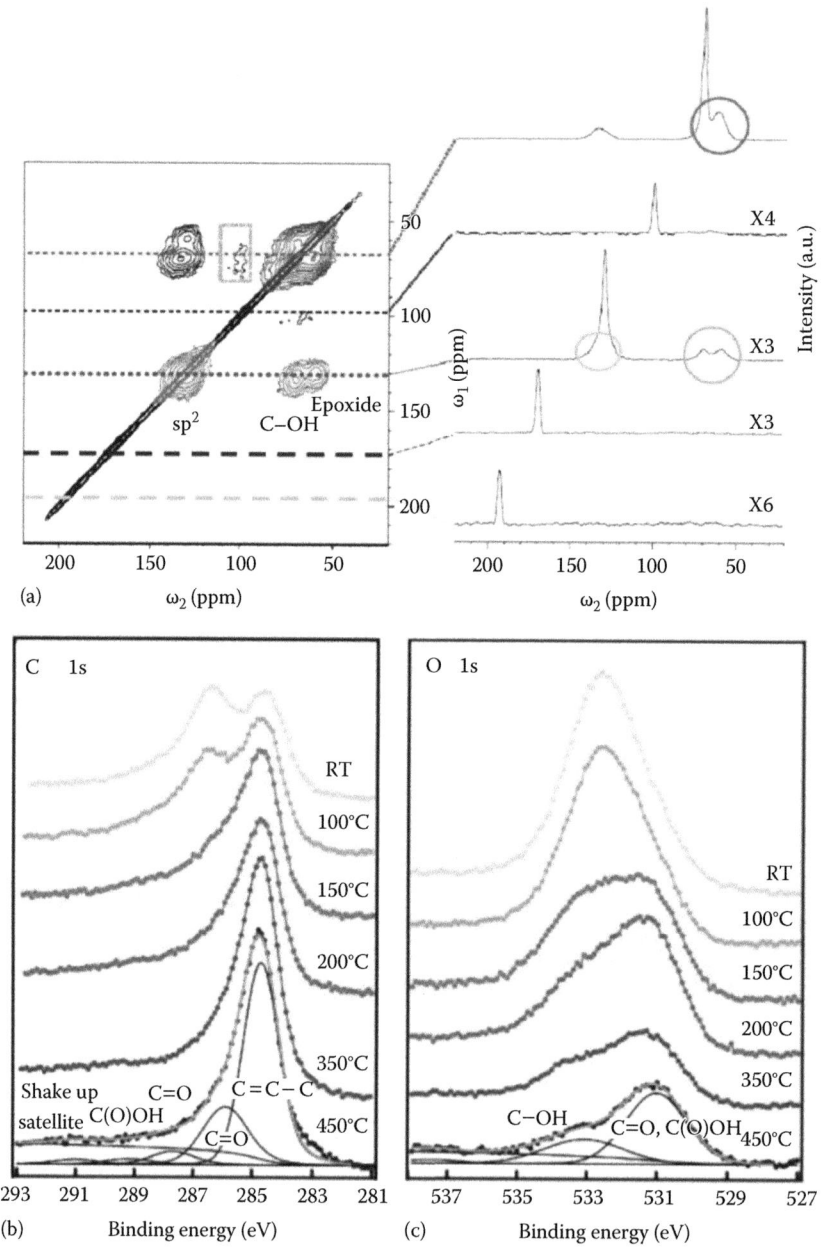

FIGURE 15.1

(a) 2D ^{13}C/^{13}C chemical-shift correlation solid-state NMR spectra of GO and slices selected from the 2D spectrum at the indicated positions (70, 101, 130, 169, and 193 ppm) in the ω_1 dimension. (Cai, W.W., Piner, R.D., Stadermann, F.J., Park, S., Shaibat, M.A., Ishii, Y., Yang, D.X. et al., Synthesis and solid-state NMR structural characterization of ^{13}C-labeled graphite oxide. *Science*, 321, 1815–1817, 2008. Reprinted with permission of AAAS.) (b,c) are high-resolution XPS spectra of the C 1s and O 1s signal in GO, respectively. (Mattevi, C., Eda, G., Agnoli, S., Miller, S., Mkhoyan, K.A., Celik, O., Mastrogiovanni, D., Granozzi, G., Garfunkel, E., Chhowalla, M.: Evolution of electrical, chemical, and structural properties of transparent and conducting chemically derived graphene thin films. *Adv. Funct. Mater.* 2009. 19. 2577–2583. Copyright Wiley-VCH Verlag GmbH & Co. KGaA. Reproduced with permission.)

carbon, respectively. Epoxy and hydroxyl were determined to be two major functional groups across the basal plane in GO. To gain information about the distribution of major functional groups, two- and multidimensional NMR spectra from Ruoff's group [36,37] revealed that epoxy and hydroxyl were close to each other, with some tiny islands of pure epoxies or hydroxyls. The major peaks in the NMR mentioned earlier were related to the carbon atoms singly bonded to oxygen atoms. In addition, in the high-resolution ^{13}C NMR spectra [36,38,39], three other minor peaks were also found at about 101, 167, and 191 ppm, which were tentatively assigned to lactol, the ester carbonyl, and the ketone groups, respectively.

On the other hand, x-ray photoelectron spectroscopy (XPS) can further unveil the nature of the carbon atoms in different chemical environments. The high-resolution XPS spectra demonstrated that in the C 1s signal of pristine GO, there are five different chemically shifted components at 284.5, 285.86, 286.55, 287.5, and 289.2 eV, respectively, which can be assigned to sp^2 carbons in aromatic rings (284.5 eV) and C atoms bonded to hydroxyl (C–OH, 285.86 eV), epoxy (C–O–C, 286.55 eV), carbonyl (>C=O, 287.5 eV), and carboxyl groups (COOH, 289.2 eV) accordingly [40–43], as shown in Figure 15.1b. However, the presence of carbonyl (>C=O) groups is still ambiguous. Some reports [35,44] only considered four feature components of the deconvolution of the C 1s spectra by ignoring the presence of the >C=O groups, that is, sp^2 carbons, C–OH, C–O–C, and COOH. Further information provided by the O 1s spectra can complement the information from the C 1s spectra. Deconvolution of the O 1s spectra indicates three main peaks around 531.08, 532.03, and 533.43 eV, which are assigned to C=O (oxygen doubly bonded to aromatic carbon) [40,45], C–O (oxygen singly bonded to aliphatic carbon) [46,47], and phenolic (oxygen singly bonded to aromatic carbon) [46,47] groups, respectively, as shown in Figure 15.1c. Moreover, the pristine GO shows an additional peak at a higher binding energy (534.7 eV) [42], which can be related to the chemisorbed/intercalated water molecules.

15.2.2.2 Microscopic techniques

Transmission electron microscopy (TEM) is a useful microscopic technique to directly image the lattice atoms and topological defects in GO [48–50]. Using high-resolution TEM (HRTEM), Erickson et al. [50] demonstrated that the specific atomistic features of GO show three major regions, which are holes, graphitic regions, and high-contrast disordered regions with approximate area percentages of 2%, 16%, and 82%, respectively, as shown in Figure 15.2a. The holes in GO were proposed to be formed by releasing CO and CO_2 during the aggressive oxidation and sheet exfoliation. The graphitic regions were suggested to be resulted from incomplete oxidation of the basal plane, which preserves the honeycomb structure of graphene. The disordered regions of the basal plane were originated from abundant oxygen-containing groups aggregated in these regions, including hydroxyls, epoxies, and carbonyls. Then using aberration-corrected HRTEM, Gómez-Navarro et al. [49] further unraveled the topological defects in GO. It was found that the dominant clustered pentagons and heptagons as well as the in-plane distortions and strain in the surrounding lattice exist in GO.

Another important microscopic technique is scanning tunneling microscopy (STM). Previously, several groups have used STM to study the surface of GO and observed highly defective regions [51–54]. According to Gómez-Navarro et al.'s measurement [51], it is distinguishable from pristine graphene and oxidized regions through the bright spots, where the oxidized regions are marked by green contours, as shown in Figure 15.2b. By estimating the ratio of oxidized regions, the degree of functionalization can be obtained. Then, Kudin et al. [52] compared the STM images of highly oriented pyrolytic graphite (HOPG) and GO. The STM image of HOPG is in a highly crystalline order, while the STM image of GO appears rough, featuring a peak-to-peak topography of 1 nm. This roughness is due to functional groups and defects. Fourier transformation of the STM image shows a clear signature of a graphitic backbone where the hexagonal symmetry is highlighted by manually added lines. This indicates reemergence of graphitic order during the reduction process. Furthermore, Pandey et al. [54] examined the oxidized regions of GO and surprisingly observed a periodic arrangement of oxygen atoms spanned over a few nanometers, as shown in Figure 15.2c. This periodic arrangement can be understood by a structural model with oxygen atoms arranged in a rectangular lattice, suggesting a series of epoxy groups present in strips.

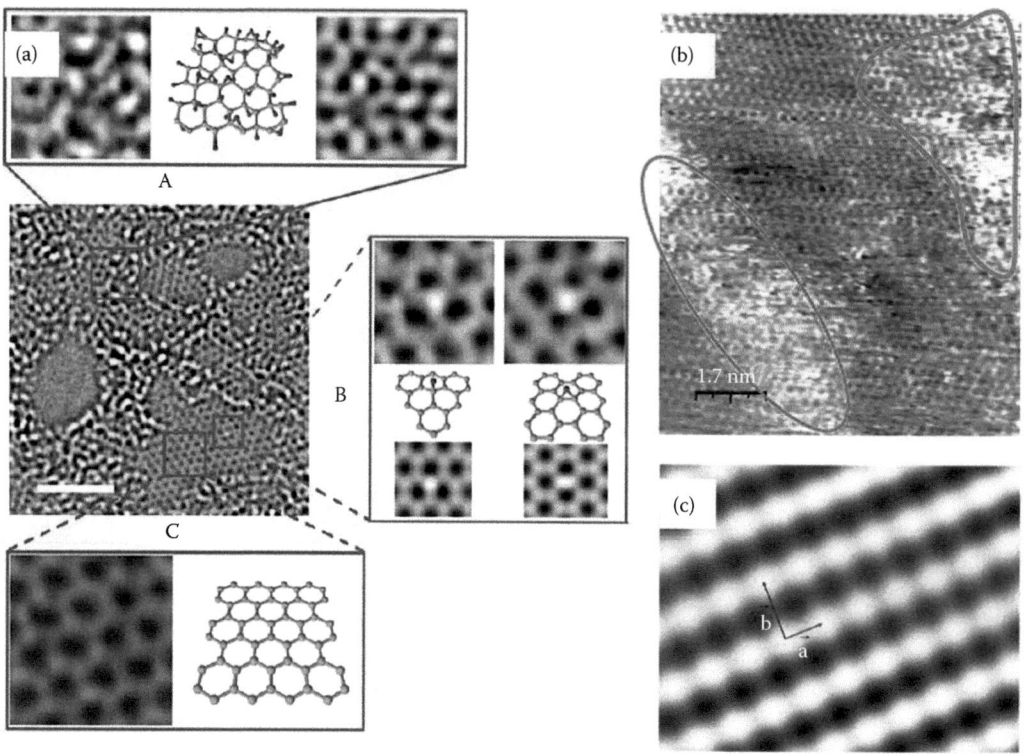

FIGURE 15.2
(a) Aberration-corrected TEM image of a single sheet of suspended GO with three typical regions: (A) the oxidized region of the GO; (B) the functional group residual region; (C) reconstruction of the graphitic region. (Erickson, K., Erni, R., Lee, Z., Alem, N., Gannett, W., Zettl, A.: Determination of the local chemical structure of graphene oxide and reduced graphene oxide. *Adv. Mater.* 2010. 22. 4467–4472. Copyright Wiley-VCH Verlag GmbH & Co. KGaA. Reproduced with permission.) (b) STM image of a GO monolayer on a HOPG substrate where oxidized regions are marked by contours. (Reprinted with the permission from Gómez-Navarro, C., Weitz, R.T., Bittner, A.M., Scolari, M., Mews, A., Burghard, M., Kern, K., Electronic transport properties of individual chemically reduced graphene oxide sheets. *Nano Lett.*, 7, 3499–3503. Copyright 2007 American Chemical Society.) (c) High-resolution STM image of the oxidized regions of GO revealing a rectangular lattice of oxygen atoms. (Reprinted from *Surf. Sci.*, 602, Pandey, D., Reifenberger, R., Piner, R., Scanning probe microscopy study of exfoliated oxidized graphene sheets, 1607–1613, Copyright 2008, with permission from Elsevier.)

15.2.3 Structural Modeling

15.2.3.1 Sketching models

Early in 1939, Hofmann and Holst originally proposed the structure of GO with only epoxy groups [55]. They supposed that the oxygen was bound to the carbon atoms of the hexagon layer planes by epoxy linkages with an ideal formula of C_2O, as shown in Figure 15.3a. Then in 1947, considering the hydrogen content of GO, Ruess suggested a structural model incorporated with hydroxyl groups [56], as shown in Figure 15.3b. This model also indicates that the basal plane structure of GO is in an sp^3 hybridization form rather than the sp^2-hybridized model of Hofmann and Holst [55]. Scholz and Boehm reconsidered the stoichiometric ratio and revised Ruess's model, suggesting a model that consists of ribbons of conjugated carbon backbone and regular quinoidal species but complete removal of the epoxy groups [57], as depicted in Figure 15.3c. Based on the spectroscopic characterization of GO, Szabó et al. [38] revived the model as proposed by Scholz and Boehm and presented a model with a corrugated carbon network including a ribbonlike arrangement of flat carbon hexagons connected by C=C double bonds. Therefore, the resulting carbon skeleton is a mixture of the Ruess and Scholz–Boehm skeleton, including a random distribution of two kinds of domains: the translinked cyclohexane chairs and the corrugated hexagon ribbons [38], as displayed

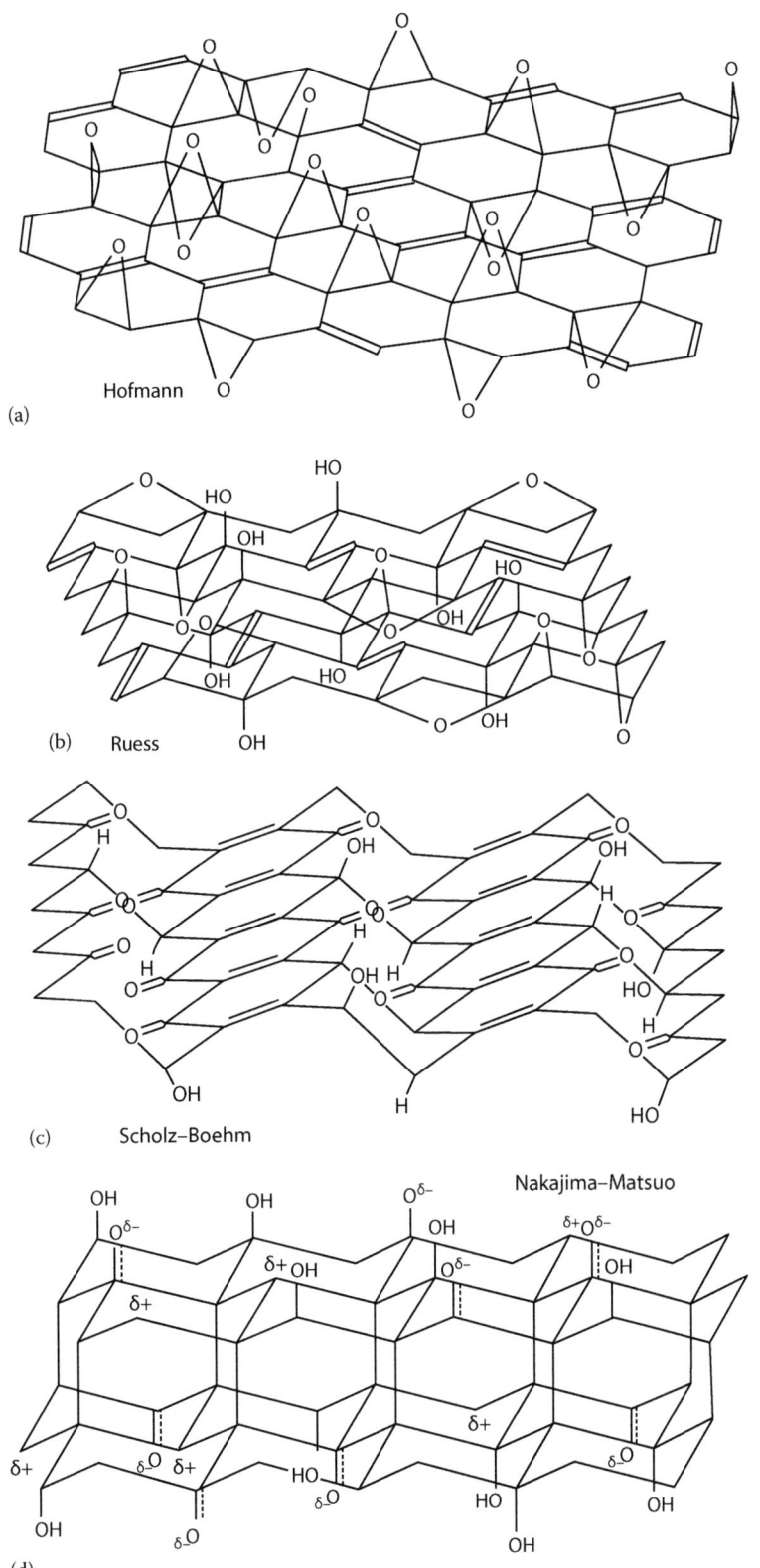

FIGURE 15.3

Sketching models for GO proposed in early stage. (a) Hofmann–Host's model, (b) Ruess' model, (c) Scholz–Boehm's model, and (d) Nakajima–Matsuo's model. (Szabó, T., Berkesi, O., Forgó, P., Josepovits, K., Sanakis, Y., Petridis, D., Dékány, I., Evolution of surface functional groups in a series of progressively oxidized graphite oxides, *Chem. Mater.*, 18, 2740–2749, 2006. Reproduced by permission of The Royal Society of Chemistry.) *(Continued)*

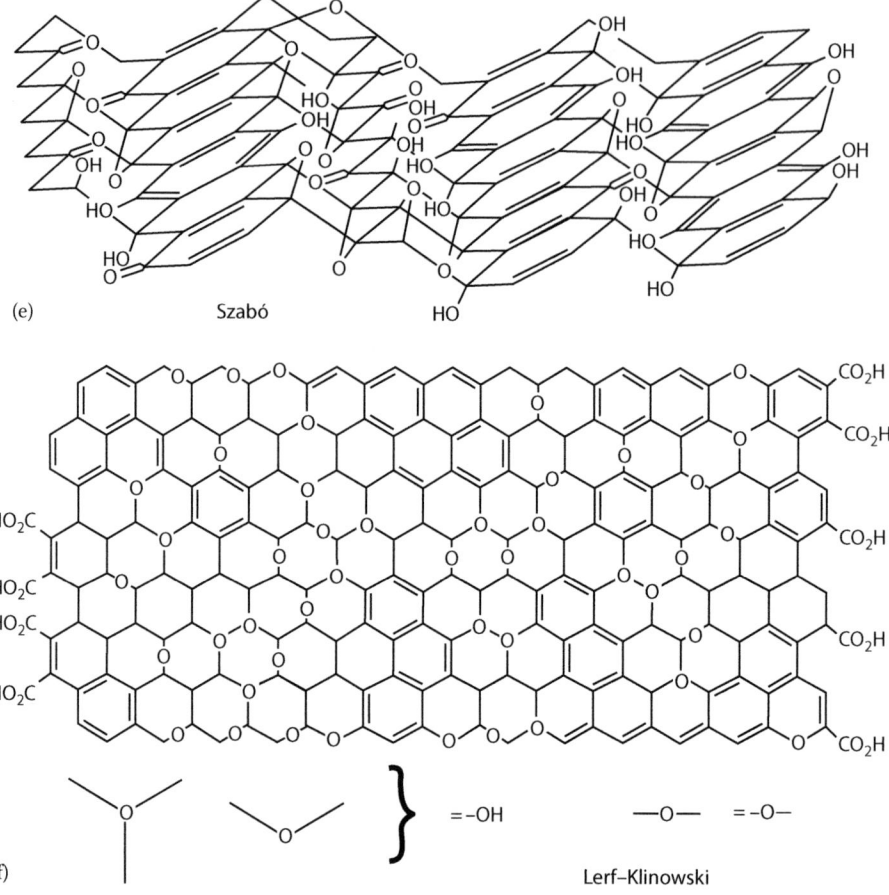

FIGURE 15.3 (*Continued*)
Sketching models for GO proposed in early stage. (e) Szabó et al.'s model and (f) Lerf–Klinowski's model. (Lerf, A., He, H., Forster, M., Klinowski, J., Structure of graphite oxide revisited. *J. Phys. Chem. B*, 102, 4477–4482, 1998. Reproduced by permission of The Royal Society of Chemistry.)

in Figure 15.3d. On the other hand, Nakajima et al. [58,59] proposed a poly-like $(C_2F)_n$ model by fluorination of GO, as shown in Figure 15.3e. In this model, two carbon layers linked to each other by sp^3 carbon–carbon bonds perpendicular to the layers, where carbonyl and hydroxyl groups were present in relative amounts depending on the level of hydration. Afterward, Leaf et al. [34,35,60] replaced the previous models by a structural model with randomly distributed flat aromatic and wrinkled regions. The flat aromatic regions consisted of unoxidized benzene rings and the wrinkled regions be a red C=C, C–OH, and ether groups, as presented in Figure 15.3f.

15.2.3.2 Simulation predictions

Recently, it has been experimentally demonstrated that epoxy and hydroxyl groups are the two major functional groups on the basal plane of GO [36,61]. Based on this, theoretical simulations were carried out to study the atomistic structure of GO. Employing the SIESTA package based on density functional theory (DFT), Boukhvalov and Katsnelson studied the GO with epoxies only, hydroxyls only, and both epoxy and hydroxyl groups [62]. They found that the oxygen-containing groups prefer to site on both sides of the graphene. Particularly, the hydroxyls are energetically favorable to site at neighboring carbon atoms from opposite sides of the graphene. Moreover, GO with both epoxy and hydroxyl groups is more stable than the one with only epoxy or hydroxyl groups. Afterward, Lahaye et al. considered the epoxy in details [63]. They pointed out that the epoxy should be the 1,2-ether oxygen, while the 1,3-ether oxygen in GO is not energetically to be formed. Then they proposed a structure of GO with the 1,2-ether oxygen dominated and closely arranged. But at the reverse side of the carbon plane, the hydroxyl groups are located. This arrangement repeats along the carbon network with subtle variations, leading to a random pattern when the oxidation over a macroscopic region emerges. In addition, Yan et al. [3,4] studied the arrangement of epoxy and hydroxyl groups on graphene by first-principles calculations and found that the epoxy and hydroxyl groups are preferable to aggregate together, forming specific oxygen-containing group chains with sp^2 carbon regions in between. This model was further confirmed by other groups [64,65]. Taking both the thermodynamic and kinetic factors into account, Lu et al. [65] suggested that the hydroxyl chain is a very stable structure. Wang et al. [66] systematically investigated the structure of GO and constructed a structural phase diagram of the GO with respect to the chemical potentials of oxygen and hydrogen, as shown in Figure 15.4a. It was found that the fully covered GO without any sp^2 carbon is thermodynamically stable, such as the hydroxyl only, epoxy only, and mixed hydroxyl and epoxy phases. However, the fully covered GO only exists under stringent experimental conditions due to competition with the formation of water. Considering the kinetic factors, GO with both functional groups and sp^2 carbons is a kinetically hindered metastable phase, as routinely observed in experiments. Using the genetic algorithm combined with DFT, Xiang et al. performed global search of the lowest-energy structure of GO [67]. It was found that the phase separation between bare graphene and fully oxidized graphene is thermodynamically favorable in GO.

On the other hand, some efforts have been devoted to the amorphous structural models of GO since experimental characterizations indicated that GO is amorphous. Using the Monte Carlo method, Paci et al. randomly placed the epoxy and hydroxyl groups on either side of a graphene basal plane [68]. Hydrogen bonds were formed between the functional groups, such as hydroxyl–hydroxyl and hydroxyl–epoxy hydrogen bonds. During evolution, various defects were observed. For example, small holes due to break of C–C bonds can be formed, where the resulting dangling bond carbons can further form carbonyl and alcohol groups. Other molecules such as peroxide and water can also be generated. Employing first-principles calculations, Liu et al. studied the amorphous GO with randomly distributed epoxy and hydroxyl groups [5]. They suggested that for the amorphous GO, the energetically preferable structure always contains some local ordered motifs, as shown in Figure 15.4b. This can be explained by the formation of hydrogen bonds in the ordered

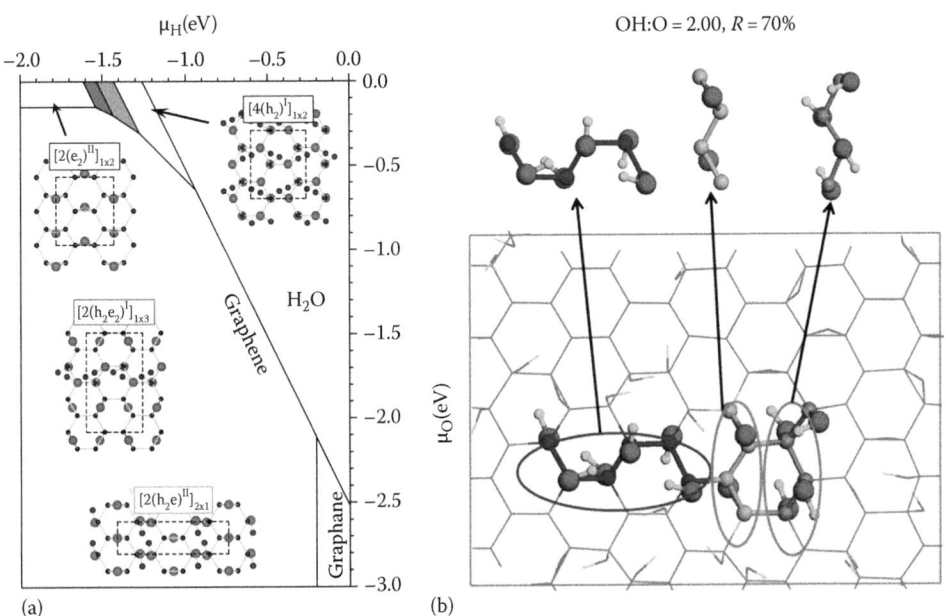

(a)   (b)

FIGURE 15.4
(a) Thermodynamic stability diagram of the GO phases with respect to the chemical potentials of oxygen and hydrogen. Insets show the atomic structures of the corresponding GO phases. (Reprinted with the permission from Wang, L., Sun, Y.Y., Lee, K., West, D., Chen, Z.F., Zhao, J.J., Zhang, S.B., Stability of graphene oxide phases from first-principles calculations, *Phys. Rev. B*, 82, 161406(R), 2010. Copyright 2010 by the American Physical Society.) (b) Thermodynamic stable amorphous GO with OH:O = 2.00 and R = 70%. The highlighted parts are the local ordered motifs. (Reprinted from *Carbon*, 50, Liu, L., Wang, L., Gao, J., Zhao, J., Gao, X., Chen, Z., Amorphous structural models for graphene oxides, 1690–1698, Copyright 2012, with permission from Elsevier.)

motifs, which will reduce the total energy of the system. Moreover, in these locally ordered motifs, clusters completely formed by epoxy and (or) hydroxyl groups were observed, consistent with the experimental observations.

15.2.4 Fundamental Properties

15.2.4.1 Thermodynamic stability

Usually, GO is thermally unstable and slowly decomposes above 60°C–80°C [5,69]. Though a reduction of GO from high coverage to low coverage, such as from 75% to 6.25%, is relatively easy, further reduction seems to be rather difficult [62]. Also, the thermodynamic stability of GO relies on its detailed geometric structure and chemical stoichiometry. Due to significant local distortion, a single functional group adsorbed on graphene has an adsorption energy of −4.72 eV for epoxy and −9.34 eV for hydroxyl, respectively [4]. Further attachment of the functional groups to form pairs on both sides of the graphene can stabilize the GO. Especially, it is energetically favorable for the hydroxyl and epoxy groups to aggregate together and to form specific types of chains with sp^2 carbon regions in between [3,4]. Moreover, the oxidation of graphene into GO is an exothermic process with negative formation energy [5]. As the coverage of functional groups increases, more and more portion of graphene will be oxidized, leading to lower formation energy. Therefore, the fully oxidized GO is thermodynamic stable, as indicated by Wang et al. [66]. While, taking the kinetic factors into account, GO with both functional groups and sp^2 carbons is a kinetically hindered metastable phase. Meanwhile, increase of the OH:O ratio will also lead to enhanced stability of GO since large population of hydroxyls will form more hydrogen bonds [5]. In addition, the ordered GO is more stable than the amorphous one in a certain chemical stoichiometry. As the coverage of functional groups increases, the energy difference between the ordered and amorphous GO enlarges accordingly. Within the coverage less than 5%, their energy difference becomes negligible and the amorphous GO can be as stable as the ordered one [5].

15.2.4.2 Electronic property

The experimental fabricated GO is usually insulating with a large bandgap. Fortunately, the electronic property of the GO can be tuned by varying the coverage and ratio of functional groups. Typically, GO has a sheet resistance value of $\sim 10^{12}$ Ω sq^{-1} or higher due to large population of sp^3-hybridized carbons bonded with the oxygen-containing groups [24]. However, after reduction, the sheet resistance can be degraded by several orders of magnitude and hence transform the insulating GO into a semiconductor or even into a graphene-like semimetal [2]. On the other hand, analysis of electron density of states for GO shows that the electronic gap reduces from 2.8 to 1.8 eV as the coverage of functional groups drops from 75% to 50%. Further reduction will make the GO conducting [62]. Employing the LDA calculation, Yan et al. indicated that bandgap of the GO can be increased from 0.2 to 4.2 eV by enhancing the coverage of oxygen-containing groups [3,4]. Other studies also demonstrated the tunable electronic property by changing the coverage of functional groups [5,63]. In addition, it was found that the electronic property of GO can also be affected by the ratio of hydroxyl to epoxy (OH:O). As the OH:O ratio increases, energy gap of the electron density of states increases accordingly due to enhanced degree of sp^3 hybridization [5]. In contrast to the clean gap of the ordered GO, there exist some defective states in the gap region [5].

15.2.4.3 Mechanical property

Similar to the electronic property, GO also possesses tunable mechanical property. First of all, for the graphite oxide, the reported Young's modulus and intrinsic strength values show a wide distribution, ranging from 6 to 42 GPa and from 76 to 293 MPa, respectively [70]. Moreover, the mechanical properties of graphite oxide can be tailored by doping or compositing. By doping a small amount (less than 1 wt%) of Mg^{2+} and Ca^{2+} ions, significant enhancement in mechanical stiffness (10%–200%) and fracture strength (~50%) of the graphite oxide can be achieved [71]. Introducing glutaraldehyde or water molecules into the gallery regions will effectively tailor the

TABLE 15.2 Young's Modulus (E) and Intrinsic Strength (τ_c) of GO with OH:O = 2.00 but Different Coverages (R) and Arrangements of the Functional Groups

R (%)	E (GPa)		τ_c (GPa)	
	Ordered GO	Amorphous GO	Ordered GO	Amorphous GO
0	495		47.8	
10	468.6	430.9	46.3	40.9
20	453.6	395.3	44.4	37.5
40	420.9	367.4	40.0	33.1
50	407.7	324.7	38.6	27.9

Source: Liu, L., Zhang, J., Zhao, J., Liu, F., Mechanical properties of graphene oxides, *Nanoscale*, 4, 5910–5916, 2012. Reproduced by permission of The Royal Society of Chemistry.

Note: Notice that the E and τ_c for the case of R = 0% are the scaled values of graphene with a vdW distance of 7 Å.

interlayer adhesions of the graphite oxide. Both the tensile modulus and strength show significant improvements for the glutaraldehyde-treated graphite oxide, but reduced mechanical properties are observed for the H_2O-treated graphite oxide [70]. Park et al. also demonstrated that graphite oxide chemically cross-linked by polyallylamine exhibit enhanced mechanical stiffness and strength [72].

As mentioned earlier, the graphite oxide has a Young's modulus of 6–42 GPa [70]. However, the Young's modulus of graphite oxide closely depends on the thickness, owing to the weak interlayer interaction. When the thickness is reduced down to a few layers, its Young's modulus increases dramatically to about 200 GPa [73,74]. Especially, GO monolayer has a much larger Young's modulus than that of thick graphite oxide. A measurement with atomic force microscopy (AFM) suggests that GO monolayer has a mean Young's modulus of 250 ± 150 GPa [75]. Taking a distance of 7 Å for van der Waals interaction of the GO, Young's modulus of 207.6 ± 23.4 GPa and elastic constant of 145.3 ± 16.4 N/m were reported by Suk et al. using AFM measurement combined with finite element analysis [74]. Other factors such as coverage, arrangement, and ratio of the functional groups contribute to the mechanical property of GO. Zheng et al. [76] studied the mechanical properties of graphene with different functional groups, such as –OH and –COOH, and pointed out that Young's modulus of the functionalized graphene reduces with the increasing coverage of surface functional groups. Besides, Liu et al. systematically investigated the mechanical property of GO [77]. It was found that the Young's modulus and intrinsic strength of GO mainly depend on the coverage and arrangement of the epoxy and hydroxyl groups, both of which decrease with increasing coverage. At the same coverage, GO with orderly arranged functional groups possess larger Young's modulus and intrinsic strength than those of the randomly arranged one. Meanwhile, the Young's modulus of GO only slightly fluctuates with the OH:O ratio due to change of thickness. Particularly, GO exhibits significant electromechanical effect. As a GO sheet is uniaxially elongated, C–O hybridization becomes weaker and more electrons are released, resulting in a reduction of bandgap. For an ordered GO with coverage of 50% and OH:O = 2.00, when it undergoes a tensile strain from 0% to 10%, the bandgap shrinks from 1.41 to 0.61 eV with a reduction extent of ~57%. For an amorphous GO with the same stoichiometry, under a tensile strain from 0% to 8%, the bandgap shrinks from 1.03 to 0.78 eV with a reduction extent of ~24%. Table 15.2 gives the Young's modulus and intrinsic strength of GO with different coverages and arrangements of the functional groups.

15.2.4.4 Optical property

Generally, a suspension of GO film in water is dark brown to light yellow, depending on the concentration. However, the optical transmittance of GO films can be continuously tuned by varying the film thickness or the extent of reduction [78]. It was found that reduced thin GO films (with a thickness less than 30 nm) is semitransparent [79], and the atomically thin GO can be highly transparent in the visible spectrum [80]. Therefore, GO has been widely investigated for transparent conductor applications [81,82]. On the other hand, an important optical feature of GO is the fluorescence in NIR regions, visible and ultraviolet regions [83–86], which is the most notably different from graphene since no fluorescence in the zero-gap graphene [87]. For instance, laterally nanosized GO aqueous suspensions show low-energy fluorescence in the red to NIR regions [88,89]. Meanwhile, weak blue to ultraviolet fluorescence was observed in as-synthesized GO thin films (centered around 390 nm) and solutions (centered around 440 nm) under ultraviolet radiation [85,90,91]. In addition, the blue fluorescence was found in a graphene quantum dot, which is pH dependent [83]. Similar blue fluorescence was observed in water-soluble GO fragments produced by ionic-liquid-assisted electrochemical exfoliation of graphite [92].

15.2.5 Applications

Since GO is oxidized graphene, a fundamental application of GO is to serve as raw material to further fabricate graphene by chemical reduction. On the other hand, due to the oxygen-containing groups, GO is also promising in the fields of energy and environmental science.

15.2.5.1 Energy storage

Due to the existence of functional groups, GO is able to adsorb atoms or molecules, which is useful in energy storage. Wang et al. proposed that titanium-anchored GO is an ideal material for hydrogen storage [64]. The titanium atoms can be steadily bonded with the hydroxyl groups on the GO surface without clustering. Each titanium atom can adsorb several H_2 molecules with moderate binding energies of 14–41 kJ mol^{-1}. As a result, the theoretical capacity of such a metal–GO hybrid reaches 4.9 wt% or 64 g L^{-1}. The oxygen-containing groups can also bind lithium amidoborane (LiAB) with GO to form a hybrid GO_3–LiAB [93]. The hybrid GO_3–LiAB complex has a great dehydrogenation performance for chemical hydrogen storage, which can further store up to 5.0 wt% H_2 via physisorption with binding energies close to the ideal range. Besides, by using the well-known reaction between boronic acids and hydroxy groups, GO layers can be linked together to form a new layered structure, that is, graphene oxide framework (GOF) [94]. Such GOF structures have tunable pore widths, volumes, and binding sites depending on the linkers chosen, which are promising in gas sorption. Grand canonical Monte Carlo simulations at 77 K and 1 bar for several representative GOF structures show that the absolute hydrogen adsorption increases with the size of GOF, which can reach a limit of 10 wt%.

In addition to hydrogen storage, GO-based materials are also promising in cells and batteries. It was demonstrated that the 3D mixture of reduced GO (rGO) and Ni particles can be used for direct ethanol fuel cells [95]. Upon electrochemical cycling of this Ni/rGO-coated electrode in alkaline media, the Ni nanoparticles undergo a surface oxidation, leading to a fast and reversible $Ni(OH)_2$/NiOOH redox transition. During ethanol oxidation, the redox centers showed a high catalytic activity toward the ethanol oxidation with a high peak current density of about 6 mA cm^{-2} at a potential of 0.66 V, showing great potential for direct application in ethanol fuel cells. Meanwhile, the usage of rGO in solar cell was also illustrated. Using N-doped rGO sponge coated on top of doped tin oxide glass as the counter electrode, a power conversion efficiency as high as 7.07% can be achieved, which is much better than that with undoped rGO sponge (4.84%) and comparable to the 7.44% efficiency achieved with a platinum counter electrode [96]. Moreover, GO-based materials show potential in lithium batteries. By combining Al-rGO and SnS_2 as the electrode material for lithium batteries, excellent rate capability can be obtained [97]. At a low current density of 100 mA g^{-1}, Al-RGO/SnS_2 has a capacity of 540 mAh g^{-1} after five cycles, which is much higher than 420 mAh g^{-1} of GO/SnS_2. At a high current density of 1000 mA g^{-1}, the capacity of Al-RGO/SnS_2 reaches 420 mAh g^{-1}, and the value of GO/SnS_2 is only 242 mAh g^{-1}. From 540 mAh g^{-1} at 100 mA g^{-1} to 420 mAh g^{-1} at 1000 mA g^{-1}, the capacity retention of Al-RGO/SnS_2 remains as high as 77.8%.

Furthermore, the large surface area also enables GO useful in electrochemical double-layer capacitors. Usually, the freestanding GO shows low capacitance due to its low conductivity. However, the capacitance of GO-based capacitors can be enhanced via some optimization processes, such as doping and compositing. For example, GO mixed with MnO_2 achieved a large capacitance of 197 F g^{-1}, which was significantly higher than that of GO (10.9 F g^{-1}) and bulk MnO_2 (6.8 F g^{-1}) [98].

Then, using the electrode composed of in situ grown rGO sponge and nickel foam, Shibing et al. achieved a high specific capacitance of 366 F g^{-1} at 2 A g^{-1}. In addition, N-doped GO containing ~18 wt% nitrogen shows outstanding performance as a supercapacitor electrode material, which achieves a specific capacitance as high as 461 F g^{-1} at 5 mV s^{-1} [99].

15.2.5.2 Environmental science

The extensive functional groups on GO can serve as an efficient site for reduction events, leading to adsorption and detection of gases [100]. Garcia-Gallastegui et al. suggested that the lightweight and charge complementary GO can interact effectively with the layered double hydroxides (LDHs), which in turn enhance the CO_2 uptake capacity and multicycle stability of the assembly [101]. As a consequence, the absolute capacity of the LDHs reduces by 62% using only 7 wt% GO as a support. Meanwhile, Zhao et al. reported the CO_2 capture by aminated GO [102]. They demonstrated that GO with 50 wt% ethylenediamine has the largest adsorption capacity of 46.55 mg CO_2/g sample. Furthermore, GO is useful for gas detection. Due to high specific surface area and improved conductivity, the rGO–Cu_2O mesocrystals presented a higher sensitivity toward NO_2 at room temperature over the constituent counterparts, attaining an unprecedented detection limit of 64 ppb [103]. Also, Li et al. illustrated that the palladium-decorated rGO shows highly sensitive, recoverable, and reliable detection of NO gas ranging from 2 to 420 ppb with response time of several hundred seconds at room temperature [104].

GO has also been demonstrated to be promising in photocatalyst [105–109]. In 2010, Yeh et al. firstly studied the photocatalytic H_2 evolution activity of GO with a bandgap of 2.4–4.3 eV [105]. They found that GO exhibited stable H_2 generation from an aqueous methanol solution or pure water, even in the absence of Pt cocatalyst under mercury light irradiation. Then, they further investigated the photocatalytic activity of GO with various oxidation levels and established an inverse relationship between the amount of H_2 evolution and the population of the oxygen-containing groups on the GO sheets [107]. It can be concluded that the GO with higher oxidation degree has a larger bandgap and limited absorption of light, exhibiting a lower photocatalytic activity than the GO with lower oxidation degree in turn. To explain the mechanism of GO as promising photocatalyst material and search the optimal composition of GO for higher photocatalytic activity, Jiang et al. have studied the electronic properties of GO systems responsible for photocatalytic water splitting using density functional theory (DFT) calculations, especially the effect of epoxy and hydroxyl groups on the work function, bandgap, CBM/VBM position, and optical absorption spectra [110]. During varying the coverage and relative ratio of the epoxy and hydroxyl groups, both the bandgap and work function of the GO can meet the requirements of a photocatalyst. Particularly, they found that the electronic structures of GO materials with 40%–50% (33%–67%) coverage and OH:O ratio of 2:1 (1:1) are suitable for both reduction and oxidation reactions for water splitting. Among the structures studied, the GO with 50% coverage and OH:O (1:1) ratio is the most promising material for visible light–driven photocatalyst [110].

15.3 HYDROGENATED GRAPHENE

Depending on hydrogen coverage, HG holds a large tunable bandgap through adsorbing atomic hydrogen on one or two sides of graphene, which extends the scope of graphene-based electronic and optical devices [9,111–113]. However, it is difficult to estimate the coverage and adsorption structures of hydrogen atoms in hydrogenation process through exposing the graphene to hydrogen plasma or atomic hydrogen beams and to determine the accurate value of bandgap observed in the electrical transport and optical spectroscopy measurements. Ab initio calculations can overcome such shortages through investigating physical properties of HG with well-defined configurations, which are very helpful to understand the evolution of the geometrical and electronic structures of HG [114–116]. Here, we focus on the theoretical investigations of HG in this section from two folds: fabrication and characterization of HG and structures and physical properties of epitaxial and freestanding HG.

15.3.1 Fabrication and Characterization of HG

15.3.1.1 Fabrication

Hydrogenation of graphene has attracted tremendous research interests from the two aspects: tuning bandgap to remove the obstacles for graphene for semiconductor-based electronic devices and possible devotions to hydrogen storage [117–121]. Nowadays, various hydrogenation techniques of graphene are successfully developed for different special destinations [122,123]. For example, remote hydrogen plasma, atomic hydrogen beams, electron-induced dissociation of hydrogen silsesquioxane (HSQ), and plasma-enhanced chemical vapor deposition (CVD) are used to synthesize locally functionalized HG with an opening bandgap in the semimetallic graphene sheet and Birch reduction technique for the applications of graphene in hydrogen storage [112,124–128].

The remote hydrogen plasma technique provides a possible approach to obtain large H coverage when the graphene deposited on substrates with only one single side exposed to atomic hydrogen, although it brings obstacles in controlling the amount of induced atomic defects and hydrogen concentrate [12]. The hydrogenation process using atomic hydrogen beams demands the graphene grown previously on oriented transition metal substrates to achieve the high degree of hydrogenation, which also suffers the shortcomings in regulating hydrogen coverage and atomic hydrogen defects [7]. Breaking Si–H bonds of spin-on-graphene HSQ is another effective hydrogenation method to achieve the controlled local functionalization, which may allow the coexisting of conductive graphene and semiconducting HG in one circuit [125,129]. In the structures of graphene hydrogenated by such method, there are many disorder-induced states with subgap energies, which block the direct measure of the bandgap in transistor structures. Plasma-enhanced CVD can bring out the interconvertibility between graphane-like structures and graphene through dehydrogenation of graphane-like films deposited on Cu/Ti-coated SiO$_2$–Si (the schematic illustration shown in Figure 15.5) [126]. This method provides a possible fabrication route to vary the conductivity in a single deposition system, which satisfies the requirements of developing whole-graphene electronics in the future.

The aforementioned methods are more appropriate to engineer HG with local functionalization. In contrast, the wet chemistry methods such as Birch reduction are more attractive in applications of graphene for hydrogen storage. Birth reduction technique transfers the hybridization of carbon atoms from sp^2 to sp^3 by introducing a covalent C–H bond, but the low values of binding energy (0.7 eV) and chemisorption barrier (0.3 eV) of hydrogen atoms on graphene result in the dehydrogenation phenomenon even at moderate temperatures [7]. Noteworthy, each graphene-hydrogenated method has its own advantages and disadvantages; thus, one has to carefully choose the suitable method based on the purpose of applications.

FIGURE 15.5
Schematic illustration of the in situ "grow-and-pattern" process through incorporating the remote discharged plasma beam source and a laser. (Reprinted with permission from Wang, Y., Xu, X., Lu, J., Lin, M., Bao, Q., Ozyilmaz, B., Loh, K.P., Toward high throughput interconvertible graphane-to-graphene growth and patterning, *ACS Nano*, 4, 6146–6152. Copyright 2010 American Chemical Society.)

15.3.1.2 Characterization

After the fabrication process, HG can be characterized by various surface analysis techniques, like Raman spectroscopy, STM, ARPES, and so on [130–132]. Raman spectra of pristine, hydrogenated and dehydrogenated graphene are very different, providing a convenient route to probe electronic structures and structural evolution of HG. Commonly, the intensity ratio (I_D/I_G) of D band to G band can indirectly reflect the extent of hydrogenation in graphitic materials containing sp^2 and sp^3 hybridization, and the excitation energies of D and 2D bands can be used to explore the electronic structures of graphene after hydrogenation [132]. STM images recorded on graphene layers exposed to increasing doses of atomic hydrogen yield the hydrogen coverage and adsorption configurations of HG [130]. Other methods like x-ray photoemission and x-ray absorption spectroscopy are also used to study the saturated hydrogen coverage and the resulting variations of graphene-substrate chemical bonds [133]. Electrical transport experiments and ARPES are usually utilized to detect the bandgap of HG [9,130]. Due to the presence of disorder-induced states with subgap energies, the bandgap measured by electrical transport is in contrast with the ARPES results in some cases [7]. Limited by the disordered defects in HG and characterization techniques, it is still difficult to synthesize the perfect graphane or give unambiguous information of electronic properties with regard to the structures and hydrogen concentration at present.

15.3.1.3 Theoretical studies

As discussed previously, it is nearly impossible to observe the detailed structural evolution from graphene to HG by experimental approaches. Either as the hydrogen receptor for hydrogen storage or as a semiconductor material for graphene-based electronics, both kinds of applications demand the knowledge of the adsorption and diffusion of hydrogen atoms on graphene. Thus, it becomes necessary to investigate hydrogenation process of graphene by theoretical simulations. Yu et al. have examined the formation of hydrogen clusters on graphene by ab initio method, which provides a perspective of a phase nucleation process for hydrogenation of graphene [134]. According to their calculations, reducing the number of unpaired π electrons increases the binding energy, and two adjacent carbon atoms hydrogenated from two opposite sides is energetically favorable compared with the ones with two hydrogen atoms attached on the same side. Based on these two points, it is convenient to explain the energy increment ΔE_n due to nth atom sorption versus its number. Adding an odd number of H atoms yields one unpaired π electron destroying graphene's aromaticity, which leads to the oscillations of energy increments ΔE_n for the oddth adsorbed H less than those for the eventh one. The configurations with completely hydrogenated six-rings are more stable than the incomplete ones, and the binding energy per H atom becomes higher as increasing the extent of hydrogenation, as shown in Figure 15.6a.

Further investigation of the nucleation barrier shows that chemisorption of a single hydrogen atom is too weak to overweigh the bond strength of molecular H$_2$, and the possible nucleation centers should be lattice defects or the metal particle itself [134]. Abhishek et al. also demonstrated the computed energy barrier of the motion of an H from the catalyst to the HG is small (0.7 eV) and can be overcome at operational temperatures [135]. Doping and chemical modification of graphene are proven to determine the binding energy versus the diffusion barrier of hydrogen atoms on graphene through DFT calculations by Angela and his coauthors [136]. They reported both boron substitutionally doped graphene and hydroxylated graphene have the potential to simultaneously meet thermodynamic and kinetic constraints for reversible room-temperature hydrogenation.

Since the presence of H frustrated domains during the hydrogenation of graphene, it is still unlikely to form large domains of perfect graphane-like structures with the assistance of nucleation centers. Flores et al. indicated that there is a significant probability for the appearance of H frustration, if considering H atoms are randomly incorporated during plasma exposure [137]. Figure 15.6b displays the scheme of the formation of H frustration between different hydrogenated domains. In the separated domains, H atoms bind on the graphene with the sequence of alternating up and down. As increasing the size of perfect hydrogenated domain, the alternating up/down sequence H atoms is no longer possible in the area between the two domains. Molecular dynamics simulations provide a direct observation of the formation of H frustration from graphene to graphane. The triangle path shows that the sequence of up and down H atoms is broken during the hydrogenation of graphene

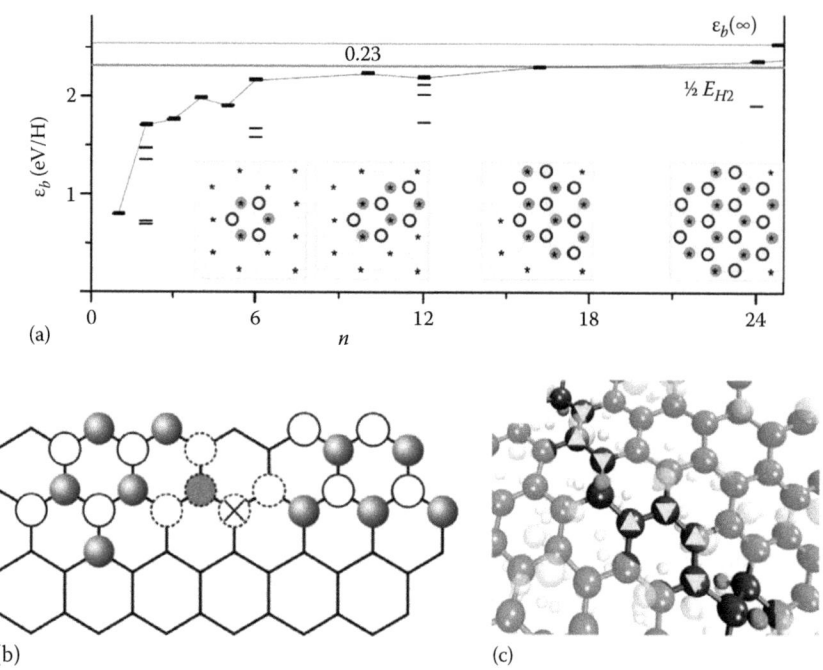

(a)

(b)   (c)

FIGURE 15.6

(a) Binding energy per H for all clusters considered as a function of their size *n*. (Reprinted with permission from Lin, Y., Ding, F., Yakobson, B.I., Hydrogen storage by spillover on graphene as a phase nucleation process, *Phys. Rev. B*, 78, 041402, 2008. Copyright 2008 by the American Physical Society.) (b) Scheme of the formation of H frustrated domains. Closed circles refer to up hydrogen atoms, open circles refer to down ones, and an open cross refer to H frustrated site. (c) Zoomed region indicating H frustrated domains formed. The triangle path shows that a sequence of up and down H atoms is no longer possible. (b, c) (From Flores, M.Z., Autreto, P.A., Legoas, S.B., Galvao, D.S., Graphene to graphane: A theoretical study, *Nanotechnology*, 20, 465704, 2009.)

(see Figure 15.6c). The existence of H frustration during hydrogenation of graphene has also been demonstrated by several groups in experiments yet. In order to get a perfect HG with a clean bandgap, there remain various challenges in the fabrication and characterization of HG.

15.3.2 Structures and Physical Properties of HG

Due to different geometric structures, epitaxial and suspended HG possess distinctive physical properties. In the hydrogenation process of epitaxial graphene, adsorption and distribution of H atoms are affected by interaction between the graphene and substrate, which does not exist in the freestanding HG. Theoretical studies indicate that the most stable configuration of suspended HG has the stoichiometry of CH with a chair-like structure in which hydrogen atoms distribute alternately on both sides of the graphene sheet (i.e., graphane) and display insulating behavior with a bandgap of 5.4 eV (calculated by GW method) [15,138]. However, for epitaxial HG, constraining hydrogen atoms to one side results in an expected H coverage of only 25% and smaller bandgap [139]. Furthermore, the physics behind the gap tuning mechanism is different for the epitaxial and suspended HG. Hence, in the following, we discuss the structures and physical properties of epitaxial HG and freestanding HG separately.

15.3.2.1 Epitaxial HG

Various potential applications of graphene such as electronic devices and transparent electrodes demand modification of the intrinsic properties of pristine graphene through chemical functionalization [112]. Hydrogenation of epitaxial graphene can apply strain to the graphene sheet, as well as preactivation of the basal plane of graphene with hydrogen, both of which are helpful for further functionalization of graphene [140–142]. Therefore, elucidating the adsorption structures of hydrogen atoms on supported graphene is important for exploring graphene-based semiconductor materials. When graphene grows on a substrate-like metal or semiconductor surface, diffusion of hydrogen atoms through the graphene to form two-sided hydrogenation is somehow hindered; thus, partial or full hydrogenation from a single surface is more possible [143,144]. In the hydrogenation process of epitaxial graphene, various materials can be used as substrates, that is, transition metal such as Ir or Pt and semiconducting materials such as SiO_2 or SiC [114,145–148].

15.3.2.1.1 Transition metal supported HG

The lattice mismatch between graphene lattice and metal surface leads to formation of the super-lattice structures, which has an important influence on the adsorption position of hydrogen atoms. For transition metal surfaces such as Ir(111) and Ni(111), due to lattice mismatch, there are distinct regions defined as atop, face-centered cubic (fcc) regions, and hexagonal close-packed (hcp) regions in the Moiré superlattice of graphene on the substrate [112]. The binding between carbon and metal atoms in the latter two regions is slightly stronger than that in atop regions by DFT calculations. Figure 15.7 depicts the different parts of the moiré supercell with graphene on Ir(111). In the fcc and hcp regions, every other C atoms binding to a hydrogen atoms on top and every other to an Ir atom on Ir(111) surface may generate the graphane-like structures, which has more favorable energy

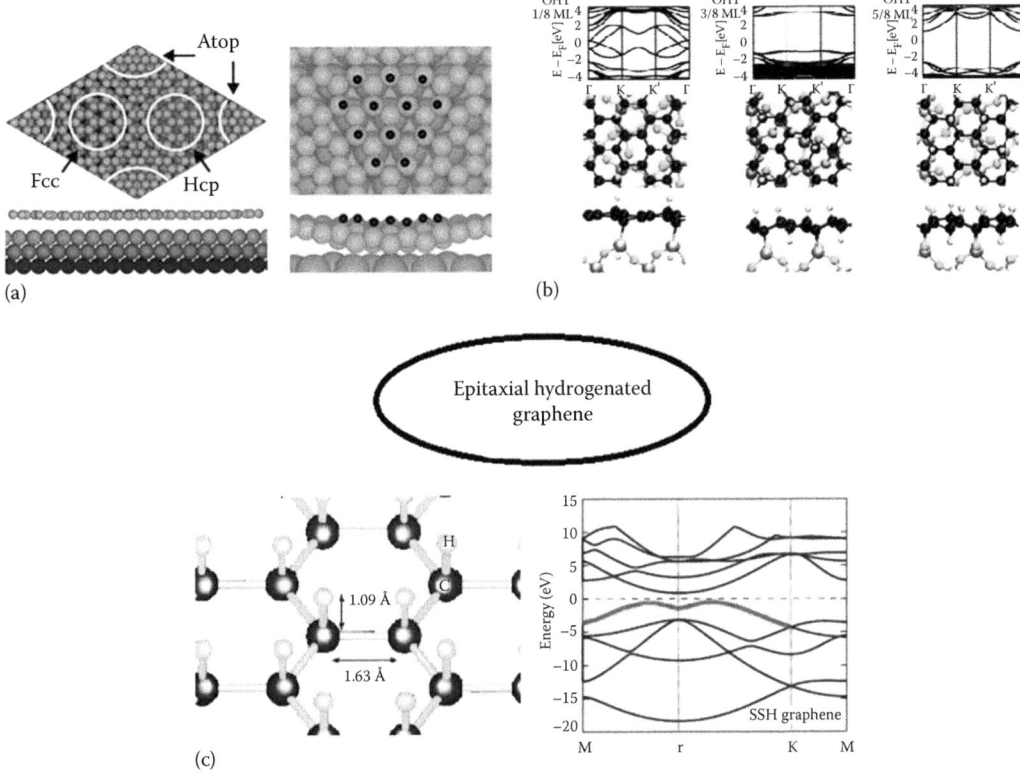

FIGURE 15.7
Three typical configurations of epitaxial HG: (a) HG on Ir(111) surface or (b) HG on SiO_2 surface and (c) fully single-sided HG as well as its band structures. (a) (Reprinted with permission from Balog, R., Andersen, M., Jørgensen, B., Sljivancanin, Z., Hammer, B., Baraldi, A., Larciprete, R., Hofmann, P., Hornekær, L., Lizzit, S., Controlling hydrogenation of graphene on Ir (111), *ACS Nano*, 7, 3823–3832. Copyright 2013 American Chemical Society.) (b) (Reprinted with permission from Havu, P., Ijäs, M., Harju, A., Hydrogenated graphene on silicon dioxide surfaces, *Phys. Rev. B*, 84, 205423, 2011. Copyright 2011 by the American Physical Society.) (c) (Reprinted with permission from Pujari, B.S., Gusarov, S., Brett, M., Kovalenko, A., Single-side-hydrogenated graphene: Density functional theory predictions, *Phys. Rev. B*, 84, 041402, 2011. Copyright 2011 by the American Physical Society.)

compared to other hydrogen adsorbate structures. Such preferential adsorption structures have been proved by recent experiments through STM images. When Ir(111) supported graphene is exposed to increasing doses of atomic hydrogen, H atoms are located at the bright parts of the moiré pattern where every C atom is placed above a surface Ir atom [130]. The interaction between Ir atom and C atoms results in rehybridization from sp^2 to sp^3 of C atoms, which may lower the adsorption barrier of atomic hydrogen. Otherwise, different interaction between graphene and metal substrates will lead to distinctive adsorption behavior of hydrogen atoms on graphene [147]. For a weakly interacting substrate (e.g., Pt), the binding energy of H clusters increases with the cluster size. In contrast, for a strongly interacting substrate (e.g., Ni), the binding energy is almost constant with the cluster size. For these two substrates, the stable graphane-like structures are formed.

Such graphane-like structures exhibit novel electronic properties such as opening of a large bandgap in semimetal graphene surface. Balog et al. reported the existence of a bandgap at least 450 meV for patterned hydrogen adsorption graphene on Ir(111), as detected by ARPES [130]. Their density functional based tight binding (DFTB) calculations demonstrated that the origin of such gap opening can be ascribed to the confinement effect in the residual bare graphene regions, which is independent of the local disorder and formation of graphane-like islands. Such confinement-induced bandgap opening has also been observed by other groups. In the measurements of Roberto et al., the bandgap width of graphene on an iridium substrate is eventually up to 1 eV [111]. In addition, HG deposited on Ni(111) surface with intercalation of Au atoms still retains a large bandgap of 1 eV under the hydrogen coverage of 8% [13].

15.3.2.1.2 Semiconductor supported HG

Besides the metal surfaces, semiconducting materials such as SiO_2 and SiC are commonly used to fabricate the hydrogenated epitaxial graphene as well [150–154]. For the graphene supported on 6H-SiC(0001) substrate, hydrogenation is a useful approach to tailor the atomic geometry and electronic band structures of graphene. Bora et al. reported two essential types of carbon atoms on the graphene buffer: threefold C atoms without bonding with the substrate and fourfold C atoms bonding above Si atoms [148]. The most stable adsorption configuration for a single H atom is on the top of a threefold C atom on the graphene surface, because such site allows C atoms becoming sp^3-like with H adsorption. In the case of low hydrogen concentrations ($<1.43 \times 10^{15}$ cm^{-2}), the hydrogenation energy on the graphene nearly remains constant values. But when H concentration exceed 7.15×10^{14} cm^{-2}, binding of H atoms in the interface of graphene and SiC surface is higher than adsorbing on the graphene buffer. Meanwhile, the band structure of the detached graphene by intercalating atomic hydrogen is identical to the freestanding graphene, which provides an effective technique to produce freestanding HG [150].

Compared to SiC substrate, different terminations of SiO_2 surface are more likely to affect the energetics and extent of graphene hydrogenation [123,155]. Using first-principles calculations, Havu et al. indicated that the substrate should be oxygenated and hydrogen passivated first in order to facilitate HG on SiO_2 [114]. On the hydrogen-saturated oxygen terminated SiO_2 surfaces, the lowest-energy configuration of HG resembles the chair-like freestanding graphane, like the configurations shown in Figure 15.7. In the 1/8-ML coverage, only the upper surface of graphene is attached to H atoms, which cannot open a bandgap. With increasing adsorption amount of H atoms to 3/8- and 5/8-ML, the presence of hydrogen atoms above and below the graphene sheet facilitates the band-gap opening of epitaxial graphene. Such H coverage dependence of bandgap was also found in the freestanding HG [16], which will be discussed in detail later.

15.3.2.1.3 Single-sided HG

Since the interaction between graphene and the substrates is weak, hydrogenation of graphene is usually less affected by supported surfaces, allowing scientists to investigate one-sided HG by ab initio calculations without worrying about the errors arising from neglecting the substrate [149,156,157]. For example, Bhalchandra et al. reported a novel single-sided HG (SSHG) with all carbon atoms saturated by hydrogen atoms on one side, which is a semiconductor with an indirect bandgap of 1.35 eV [149]. The origin of indirect bandgap is that the band arising in graphene from the p_z orbitals on carbon atoms shifts upward in energy, and this variation comes from enhanced repulsion among hydrogen atoms on one single surface. In addition, the hydrogenated energy in SSHG is 5.90 eV/atom and comparable to acetylene, which implies such HG may be synthesized in experiments. When hydrogen atoms only bind in a sublattice of graphene from one side, the resulting semi-HG (referred to as "graphone") becomes a ferromagnetic semiconductor with a small indirect gap of 0.46 eV, which is derived from the localized and unpaired electrons in the unhydrogenated carbon atoms [158]. Such ferromagnetism has been proved in partially hydrogenated epitaxial graphene grown on 4H-SiC(0001) surface at room temperature [159].

15.3.2.2 Freestanding HG

The intercalation of hydrogen atoms between graphene and the substrates eventually results in the complete delamination of the graphene from the substrate and becomes quasi-freestanding graphene [13,160–162]. On the other hand, exposing the suspended graphene to atomic hydrogen can directly fabricate freestanding HG [9,129]. Thus, investigations of freestanding HG comprise an important part of the research of graphene-based materials. For freestanding HG, there are two kinds of typical structures with distinctive properties: H atoms adsorbed graphene (with only one or two H atoms) and graphane, a fully HG.

15.3.2.2.1 H atoms adsorbed graphene

Chemisorption of a single hydrogen atom on graphitic monolayer transforms the sp^2 hybridization state of a carbon atom to sp^3, which induces new states in the pristine sp^2 lattice [163–165]. Even before the discovery of stable graphene films in experiments, Elizabeth et al. have investigated the hallmark of one H adsorbed graphene with or without Stone–Wales defects by DFT calculations yet [166]. The adsorption of atomic hydrogen opens a substantial gap of 1.25 eV between occupied and unoccupied graphene bands and is accompanied by a spin-polarized gap state, as shown in Figure 15.8a, which is induced by the potential of H$^+$ ionic core. In addition, incorporation of five- or seven-membered rings has a strong effect on the electronic structures of one hydrogen-doped graphene and brings the ground state back to metallic and spin-paired throng quenching the spin. Commonly, the chemisorption of a single hydrogen atom on graphene leads to the appearance of magnetic moments in the system, because that opening double C=C bond creates one unpaired electron without H saturation. In contrast to one single H case, the adsorption of hydrogen pairs could generate nonmagnetic sp^3 bonding with more favorable binding energies. Boukhvalov et al. reported the configurations of H pairs adsorbed on carbon atoms from different sublattices are more stable than other cases, and for all types of hydrogen pairs, the chemisorption of hydrogen atoms is more favorable in energy when the distance of hydrogenated carbon atoms is less than 5 Å [167]. The STM investigation of atomic hydrogen adsorbate structures on graphene also revealed that at low coverage hydrogen dimers preferentially form on the graphene–SiC surface, and at higher coverage hydrogen atoms randomly adsorb into large hydrogen clusters [11].

15.3.2.2.2 Graphane

Another important type of HG is graphane, with the two favorable conformations: a chair-like conformer with the hydrogen atoms alternating on both sides of the plane (shown in Figure 15.8b) and a boat-like conformer with the hydrogen atoms alternating in pairs [138]. Both these conformations have the same stoichiometry with C:H ratio of 1:1 and similar binding energy of hydrogen atoms with 6.56 eV/atom in the chair-like one and 6.53 eV/atom in the boat-like one. Such configurations are very stable with even lower binding energy than other hydrocarbons like benzene and acetylene. As mentioned previously, the formation of atomic defects occurs inevitably during hydrogenation process, and thus it is almost impossible to synthesize pure graphane with perfect chair-like or boat-like lattices. For a significantly broader variation of lattice spacing observed in experiments, Duminda et al. believed such broad distribution of lattice spacing can be attributed to the existence of locally stable twist-boat membranes other than the chair form [168]. In contrast to pristine graphene, all isomers of graphane with the chair-like, boat-like, and twist-boat-chair configurations have very similar band structures. They have a direct bandgap at the Γ point with 3.5 eV for the chair structure, 3.7 eV for the boat structure, and 3.8 eV for the twist-boat-chair structure [138,168].

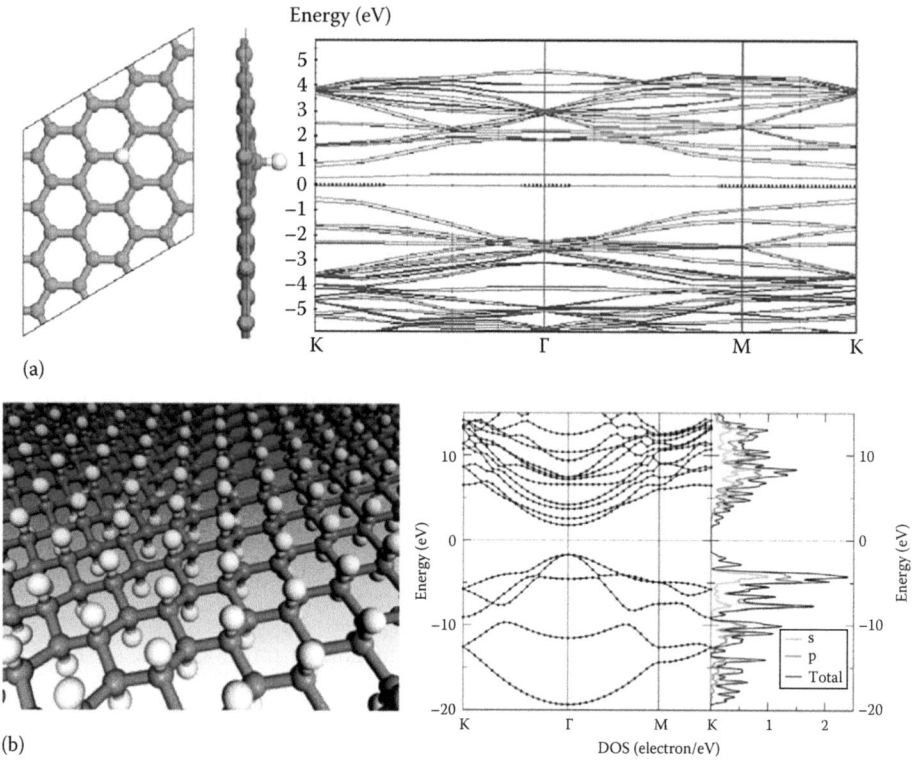

FIGURE 15.8
(a) Geometrical and band structures of one hydrogen-doped graphene. (Reprinted with permission from Duplock, E.J., Scheffler, M., Lindan, P.J., Hallmark of perfect graphene, *Phys. Rev. Lett.*, 92, 225502, 2004. Copyright 2004 by the American Physical Society.) (b) Geometrical and band structures of chair-like graphane. (Reprinted with permission from Sofo, J.O., Chaudhari, A.S., Barber, G.D., Graphane: A two-dimensional hydrocarbon, *Phys. Rev. B*, 75, 153401, 2007. Copyright 2007 by the American Physical Society.)

Instead of the generalized gradient approximation (GGA), more accurate GW calculations give a larger bandgap with 5.4 eV in the ground state chair-like configuration and 4.9 eV in the metastable boat-like configuration, which implies the graphane being an insulator [15]. The transformation from a semimetal graphene into an insulating graphane in experiments is a landmark on the applications of graphene due to its opening of a sizeable bandgap, which is the basic requirement for applications in the semiconductor microelectronics [9,169].

15.3.2.2.3 Partially HG

The zero bandgap of graphene and the large bandgap of graphane imply that it is possible to tune the bandgap in a wide range by controlling the hydrogen coverage. Through DFT calculations, Gao et al. have demonstrated that the bandgap of HG can be continuously tuned from 0 to 4.66 eV by random dehydrogenation of pure graphane, which may lead to potential applications in future electronics and photonics [16]. Starting from the chair-like graphane represented by a large supercell containing 240 carbon atoms and 240 hydrogen atoms, we have investigated three types of HG configurations: (I) randomly removing H pairs from fully HG (shown in Figure 15.9a), (II) randomly removing individual H atoms from fully HG, and (III) creating paired H vacancies according to some ordered pattern. First, HG with paired H vacancies has larger hydrogen-binding energy than the unpaired configurations. On a perfect graphane of sp^3-bonding network, a single H vacancy creates a dangling bond and results in some local strain as well, each of which raises the total energy. In the condition of paired dehydrogenation, both unsaturated C atoms are turned to sp^2-hybridized state, and thus these two adjacent C atoms form a C=C double bond, through which the two π electrons pair together and the vacancy-induced local strain can be partially released.

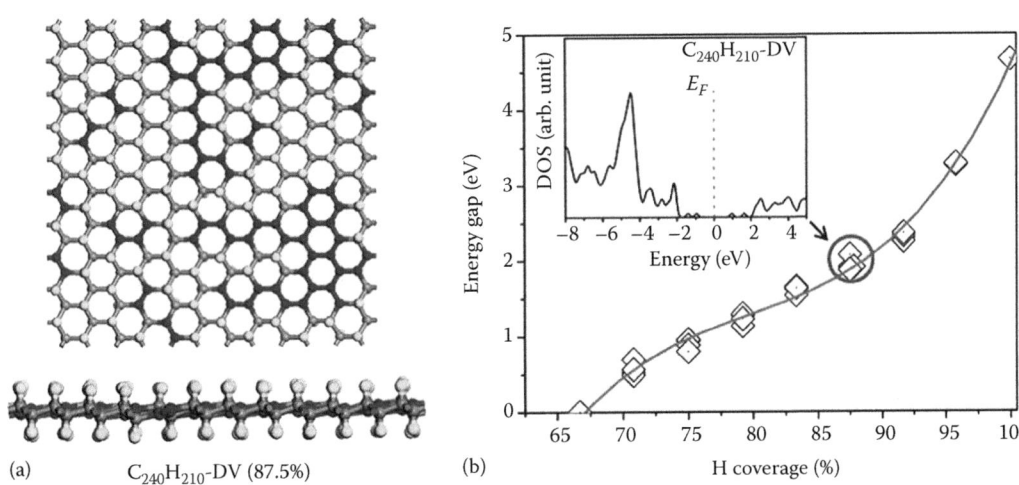

(a) $C_{240}H_{210}$-DV (87.5%)   (b)   H coverage (%)

FIGURE 15.9
One configuration of partially hydrogenated graphane with 75% H coverage ($C_{240}H_{160}$ in supercell) (a) and energy gap of HG as a function of H coverage (b). The inset figure is DOS of $C_{240}H_{210}$ with paired H vacancies. The gray balls stand for sp^3 C atoms saturated with H atom, dark gray balls highlight unsaturated sp^2 C atoms, and light-gray balls represent H atoms. (Reprinted with permission from Gao, H., Wang, L., Zhao, J., Ding, F., Lu, J., Band gap tuning of hydrogenated graphene: H coverage and configuration dependence, *J. Phys. Chem. C*, 115, 3236–3242. Copyright 2011 American Chemical Society.)

In addition, the theoretical calculations by Gao et al. revealed the bandgap of randomly HG with paired vacancies highly dependent on the H coverage (bandgap of HG versus H coverage shown in Figure 15.9b). The bandgap of partial HG configurations of type I reduces with decreasing H coverage. The theoretical GGA gap drops from 4.66 eV for a fully HG to zero for partially HG with 66.7% coverage. Further comparisons of the electron density of state for perfect graphane, dehydrogenated graphane with paired and unpaired H vacancies, and dehydrogenated graphane with paired H vacancies indicate that the middle and edge impurity states are associated with the unpaired and paired H vacancies, respectively. Due to the presence of midstates induced by unpaired π electrons, it seems impossible for the HG configurations with unpaired vacancies to create a clean bandgap. Under low H coverage, there are enough H vacancies in graphene configurations, which may result in the contact of different H-vacancy clusters and make it impossible to confine the π electrons in pristine graphene into chair-like graphane lattice with sp^3 hybridization. Thus, at low H coverage, it becomes necessary to constrain hydrogen atoms adsorption in patterns for tailoring the bandgap of HG. For example, cutting the graphane sheet into nanoribbons can tune the bandgap of graphane from 3.58 to 3.82 eV [170], and creating a vacancy H cluster in graphane can reduce the bandgap as the vacancy size increases [171]. Through adsorbing H atoms in lines on graphene, one can even open a big bandgap at low H coverage, such as 1.69 eV for $C_{240}H_{90}$ (37.5%) and 1.58 eV for $C_{240}H_{30}$ (12.5%), respectively [16].

Well-structured partially HG derivatives holding very interesting magnetic, metallic, and semiconducting properties are still a challenge for fabrication and characterization [142]. The randomly dehydrogenated graphane with paired H vacancies has attracted the extensive interests in experimental and theoretical studies since proposed. Due to individual H vacancies corresponding to a high-energy state in graphane, a practical way of creating such HG structures is through a kinetic process at low temperature, such as electron/ion irradiation [131]. Wojtaszek et al. [172] reported that the argon-hydrogen plasma produced in a reactive ion-etching system is a passway to partially hydrogenating graphene. Under the chosen plasma conditions, the hydrogenation process does not introduce considerable damages to the graphene sheet. Birch reduction of graphite or graphite oxides is also an effective approach to synthesize the partially HG. For example, Yang et al. have observed both edge and interior HG with a large bandgap of 4 eV detected by using ultraviolet-visible spectroscopy [127], and Alex et al. even found a weak ferromagnetism in HG during the whole temperature range up to room temperature [173]. Further functionalization of HG through removing hydrogen atoms will form an H vacancy cluster, which can support the noble metal catalysts.

The energy barrier is only 0.47 eV for the Langmuir–Hinshelwood oxidation process for a CO coadsorbed on Au_{16} gold cage with an O_2 molecule [19].

15.4 CONCLUSIONS

Two kinds of superstar carbon nanomaterials, GO and HG, are discussed in this chapter, from their experimental fabrication and characterization to fundamental properties, and finally the potential applications. GO contains abundant oxygen-containing groups. Change of coverage, arrangements, types, and relative ratios of the oxygen-containing groups results in different electronic, mechanical, and optical properties of GO. Moreover, the tunable properties of GO enable its application in various fields, such as thin-film transistors, energy storage devices, and sensors in chemical, environmental, and biotechnological sciences. On the other hand, HG with partially or fully sp^3-hybridized carbons induced by hydrogen chemisorption shows great potential in bandgap engineering of graphene. Generally, bandgap of HG with paired H vacancies can be tailored continuously from 0 to 4.66 eV through controlling hydrogen coverage and configurations. Such a big tunable gap regime lies between midultraviolet and NIR regions, leading to potential applications of HG in electronics and photonics. In addition, since the fabricated HG always contains defects like H vacancies, noble metal catalysts could be adsorbed on the vacancies, making HG an ideal support material for catalyzing chemical reactions.

As mentioned earlier, though GO is widely used in many fields, its atomistic structure is still under debate due to the complex stoichiometry and numerous arrangements of the functional groups. Therefore, how to fabricate GO with controllable stoichiometry and geometry is a big challenge. Also, more theoretical works should be carried out to study the atomistic structure of GO, its oxidation and reduction mechanism, and multilayered GO. As for HG, theoretical studies considering the defective HG with detailed conformations should be carried forward.

ACKNOWLEDGMENTS

This work was supported by the National Natural Science Foundation of China (No. 11304030, No. 11134005). L. Liu also acknowledges support from the Fundamental Research Funds for the Central Universities of China (No. DUT14RC(3)114).

REFERENCES

1. Brodie, B.C.: On the atomic weight of graphite. *Philos. Trans. R. Soc. Lond.* 149, 249–259 (1859).
2. Chen, D., Feng, H., Li, J.: Graphene oxide: Preparation, functionalization, and electrochemical applications. *Chem. Rev.* 112, 6027–6053 (2012).
3. Yan, J.A., Xian, L.D., Chou, M.Y.: Structural and electronic properties of oxidized graphene. *Phys. Rev. Lett.* 103, 086802 (2009).
4. Yan, J.A., Chou, M.Y.: Oxidation functional groups on graphene: Structural and electronic properties. *Phys. Rev. B* 82, 125403 (2010).
5. Liu, L., Wang, L., Gao, J., Zhao, J., Gao, X., Chen, Z.: Amorphous structural models for graphene oxides. *Carbon* 50, 1690–1698 (2012).
6. Robinson, J.T., Perkins, F.K., Snow, E.S., Wei, Z., Sheehan, P.E.: Reduced graphene oxide molecular sensors. *Nano Lett.* 8, 3137–3140 (2008).
7. Robinson, J.A., Snow, E.S., Badescu, S.C., Reinecke, T.L., Perkins, F.K.: Role of defects in single-walled carbon nanotube chemical sensors. *Nano Lett.* 6, 1747–1751 (2006).

8. Lin, Y.-M., Avouris, P.: Strong suppression of electrical noise in bilayer graphene nanodevices. *Nano Lett.* 8, 2119–2125 (2008).
9. Elias, D., Nair, R., Mohiuddin, T., Morozov, S., Blake, P., Halsall, M., Ferrari, A., Boukhvalov, D., Katsnelson, M., Geim, A.: Control of graphene's properties by reversible hydrogenation: Evidence for graphane. *Science* 323, 610–613 (2009).
10. Castellanos-Gomez, A., Wojtaszek, M., Tombros, N., van Wees, B.J.: Reversible hydrogenation and bandgap opening of graphene and graphite surfaces probed by scanning tunneling spectroscopy. *Small* 8, 1607–1613 (2012).
11. Balog, R., Jørgensen, B., Wells, J., Lægsgaard, E., Hofmann, P., Besenbacher, F., Hornekær, L.: Atomic hydrogen adsorbate structures on graphene. *J. Am. Chem. Soc.* 131, 8744–8745 (2009).
12. Haberer, D., Petaccia, L., Farjam, M., Taioli, S., Jafari, S., Nefedov, A., Zhang, W., Calliari, L., Scarduelli, G., Dora, B.: Direct observation of a dispersionless impurity band in hydrogenated graphene. *Phys. Rev. B* 83, 165433 (2011).
13. Haberer, D., Vyalikh, D., Taioli, S., Dora, B., Farjam, M., Fink, J., Marchenko, D., Pichler, T., Ziegler, K., Simonucci, S.: Tunable band gap in hydrogenated quasi-freestanding graphene. *Nano Lett.* 10, 3360–3366 (2010).
14. Whitener Jr, K.E., Lee, W.K., Campbell, P.M., Robinson, J.T., Sheehan, P.E.: Chemical hydrogenation of single-layer graphene enables completely reversible removal of electrical conductivity. *Carbon* 72, 348–353 (2014).
15. Lebegue, S., Klintenberg, M., Eriksson, O., Katsnelson, M.: Accurate electronic band gap of pure and functionalized graphane from GW calculations. *Phys. Rev. B* 79, 245117 (2009).
16. Gao, H., Wang, L., Zhao, J., Ding, F., Lu, J.: Band gap tuning of hydrogenated graphene: H coverage and configuration dependence. *J. Phys. Chem. C* 115, 3236–3242 (2011).
17. Tachikawa, H., Iyama, T., Kawabata, H.: Effect of hydrogenation on the band gap of graphene nano-flakes. *Thin Solid Films* 554, 199–203 (2013).
18. Matis, B.R., Burgess, J.S., Bulat, F.A., Friedman, A.L., Houston, B.H., Baldwin, J.W.: Surface doping and band gap tunability in hydrogenated graphene. *ACS Nano* 6, 17–22 (2012).
19. Chen, G., Li, S., Su, Y., Wang, V., Mizuseki, H., Kawazoe, Y.: Improved stability and catalytic properties of au16 cluster supported on graphane. *J. Phys. Chem. C* 115, 20168–20174 (2011).
20. Stankovich, S., Piner, R.D., Nguyen, S.T., Ruoff, R.S.: Synthesis and exfoliation of isocyanate-treated graphene oxide nanoplatelets. *Carbon* 44, 3342–3347 (2006).
21. Stankovich, S., Piner, R.D., Chen, X., Wu, N., Nguyen, S.T., Ruoff, R.S.: Stable aqueous dispersions of graphitic nanoplatelets via the reduction of exfoliated graphite oxide in the presence of poly(sodium 4-styrenesulfonate). *J. Mater. Chem.* 16, 155–158 (2006).
22. Stankovich, S., Dikin, D.A., Dommett, G.H.B., Kohlhaas, K.M., Zimney, E.J., Stach, E.A., Piner, R.D., Nguyen, S.T., Ruoff, R.S.: Graphene-based composite materials. *Nature* 442, 282–286 (2006).
23. Paredes, J.I., Villar-Rodil, S., Martínez-Alonso, A., Tascón, J.M.D.: Graphene oxide dispersions in organic solvents. *Langmuir* 24, 10560–10564 (2008).
24. Becerril, H.A., Mao, J., Liu, Z., Stoltenberg, R.M., Bao, Z., Chen, Y.: Evaluation of solution-processed reduced graphene oxide films as transparent conductors. *ACS Nano* 2, 463–470 (2008).
25. Zhu, Y., Stoller, M.D., Cai, W., Velamakanni, A., Piner, R.D., Chen, D., Ruoff, R.S.: Exfoliation of graphite oxide in propylene carbonate and thermal reduction of the resulting graphene oxide platelets. *ACS Nano* 4, 1227–1233 (2010).

26. Dreyer, D.R., Park, S., Bielawski, C.W., Ruoff, R.S.: The chemistry of graphene oxide. *Chem. Soc. Rev.* 39, 228–240 (2010).
27. Staudenmaier, L.: Verfahren zur darstellung der graphitsäure. *Ber. Dtsch. Chem. Ges.* 31, 1481–1487 (1898).
28. Hummers, W.S., Offeman, R.E.: Preparation of graphitic oxide. *J. Am. Chem. Soc.* 80, 1339 (1958).
29. Compton, O.C., Nguyen, S.T.: Graphene oxide, highly reduced graphene oxide, and graphene: Versatile building blocks for carbon-based materials. *Small* 6, 711–723 (2010).
30. Scholz, W., Boehm, H.P.: Untersuchungen am Graphitoxid. VI. Betrachtungen zur Struktur des Graphitoxids. *Z. Anorg. Allg. Chem.* 369, 327–340 (1969).
31. Kovtyukhova, N.I., Ollivier, P.J., Martin, B.R., Mallouk, T.E., Chizhik, S.A., Buzaneva, E.V., Gorchinskiy, A.D.: Layer-by-layer assembly of ultrathin composite films from micron-sized graphite oxide sheets and polycations. *Chem. Mater.* 11, 771–778 (1999).
32. Hirata, M., Gotou, T., Horiuchi, S., Fujiwara, M., Ohba, M.: Thin-film particles of graphite oxide 1: High-yield synthesis and flexibility of the particles. *Carbon* 42, 2929–2937 (2004).
33. Johnson, J.A., Benmore, C.J., Stankovich, S., Ruoff, R.S.: A neutron diffraction study of nano-crystalline graphite oxide. *Carbon* 47, 2239–2243 (2009).
34. Lerf, A., He, H., Riedl, T., Forster, M., Klinowski, J.: ^{13}C and ^1H MAS NMR studies of graphite oxide and its chemically modified derivatives. *Solid State Ion* 101–103, 857–862 (1997).
35. Lerf, A., He, H., Forster, M., Klinowski, J.: Structure of graphite oxide revisited. *J. Phys. Chem. B* 102, 4477–4482 (1998).
36. Cai, W.W., Piner, R.D., Stadermann, F.J., Park, S., Shaibat, M.A., Ishii, Y., Yang, D.X. et al.: Synthesis and solid-state NMR structural characterization of ^{13}C-labeled graphite oxide. *Science* 321, 1815–1817 (2008).
37. Casabianca, L.B., Shaibat, M.A., Cai, W.W., Park, S., Piner, R., Ruoff, R.S., Ishii, Y.: NMR-based structural modeling of graphite oxide using multidimensional ^{13}C solid-state NMR and ab initio chemical shift calculations. *J. Am. Chem. Soc.* 132, 5672–5676 (2010).
38. Szabó, T., Berkesi, O., Forgó, P., Josepovits, K., Sanakis, Y., Petridis, D., Dékány, I.: Evolution of surface functional groups in a series of progressively oxidized graphite oxides. *Chem. Mater.* 18, 2740–2749 (2006).
39. Stankovich, S., Dikin, D.A., Piner, R.D., Kohlhaas, K.A., Kleinhammes, A., Jia, Y., Wu, Y., Nguyen, S.T., Ruoff, R.S.: Synthesis of graphene-based nanosheets via chemical reduction of exfoliated graphite oxide. *Carbon* 45, 1558–1565 (2007).
40. Mattevi, C., Eda, G., Agnoli, S., Miller, S., Mkhoyan, K.A., Celik, O., Mastrogiovanni, D., Granozzi, G., Garfunkel, E., Chhowalla, M.: Evolution of electrical, chemical, and structural properties of transparent and conducting chemically derived graphene thin films. *Adv. Funct. Mater.* 19, 2577–2583 (2009).
41. Yang, D., Velamakanni, A., Bozoklu, G., Park, S., Stoller, M., Piner, R.D., Stankovich, S. et al.: Chemical analysis of graphene oxide films after heat and chemical treatments by X-ray photoelectron and micro-Raman spectroscopy. *Carbon* 47, 145–152 (2009).
42. Akhavan, O.: The effect of heat treatment on formation of graphene thin films from graphene oxide nanosheets. *Carbon* 48, 509–519 (2010).
43. Ganguly, A., Sharma, S., Papakonstantinou, P., Hamilton, J.: Probing the thermal deoxygenation of graphene oxide using high-resolution in situ X-ray-based spectroscopies. *J. Phys. Chem. C* 115, 17009–17019 (2011).

44. Lahaye, R.J.W.E., Jeong, H.K., Park, C.Y., Lee, Y.H.: Density functional theory study of graphite oxide for different oxidation levels. *Phys. Rev. B* 79, 125435 (2009).
45. Bagri, A., Mattevi, C., Acik, M., Chabal, Y.J., Chhowalla, M., Shenoy, V.B.: Structural evolution during the reduction of chemically derived graphene oxide. *Nat. Chem.* 2, 581–587 (2010).
46. Hontoria-Lucas, C., López-Peinado, A.J., López-González, J.d.D., Rojas-Cervantes, M.L., Martín-Aranda, R.M.: Study of oxygen-containing groups in a series of graphite oxides: Physical and chemical characterization. *Carbon* 33, 1585–1592 (1995).
47. Schniepp, H.C., Li, J.-L., McAllister, M.J., Sai, H., Herrera-Alonso, M., Adamson, D.H., Prud'homme, R.K., Car, R., Saville, D.A., Aksay, I.A.: Functionalized single graphene sheets derived from splitting graphite oxide. *J. Phys. Chem. B* 110, 8535–8539 (2006).
48. Pacilé, D., Meyer, J.C., Fraile Rodríguez, A., Papagno, M., Gomez-Navarro, C., Sundaram, R.S., Burghard, M., Kern, K., Carbone, C., Kaiser, U.: Electronic properties and atomic structure of graphene oxide membranes. *Carbon* 49, 966–972 (2011).
49. Gómez-Navarro, C., Meyer, J.C., Sundaram, R.S., Chuvilin, A., Kurasch, S., Burghard, M., Kern, K., Kaiser, U.: Atomic structure of reduced graphene oxide. *Nano Lett.* 10, 1144–1148 (2010).
50. Erickson, K., Erni, R., Lee, Z., Alem, N., Gannett, W., Zettl, A.: Determination of the local chemical structure of graphene oxide and reduced graphene oxide. *Adv. Mater.* 22, 4467–4472 (2010).
51. Gómez-Navarro, C., Weitz, R.T., Bittner, A.M., Scolari, M., Mews, A., Burghard, M., Kern, K.: Electronic transport properties of individual chemically reduced graphene oxide sheets. *Nano Lett.* 7, 3499–3503 (2007).
52. Kudin, K.N., Ozbas, B., Schniepp, H.C., Prud'homme, R.K., Aksay, I.A., Car, R.: Raman spectra of graphite oxide and functionalized graphene sheets. *Nano Lett.* 8, 36–41 (2008).
53. Paredes, J.I., Villar-Rodil, S., Solis-Fernandez, P., Martinez-Alonso, A., Tascon, J.M.D.: Atomic force and scanning tunneling microscopy imaging of graphene nanosheets derived from graphite oxide. *Langmuir* 25, 5957–5968 (2009).
54. Pandey, D., Reifenberger, R., Piner, R.: Scanning probe microscopy study of exfoliated oxidized graphene sheets. *Surf. Sci.* 602, 1607–1613 (2008).
55. Hofmann, U., Holst, R.: Über die Säurenatur und die Methylierung von Graphitoxyd. *Ber. Dtsch. Chem. Ges. B* 72, 754–771 (1939).
56. Ruess, G.: Über das graphitoxyhydroxyd (graphitoxyd). *Monatsh. Chem.* 76, 381–417 (1947).
57. Scholz, W., Boehm, H.P.: Untersuchungen am graphitoxid. VI. Betrachtungen zur struktur des graphitoxids. *Z. Anorg. Allg. Chem.* 369, 327–340 (1969).
58. Nakajima, T., Mabuchi, A., Hagiwara, R.: A new structure model of graphite oxide. *Carbon* 26, 357–361 (1988).
59. Nakajima, T., Matsuo, Y.: Formation process and structure of graphite oxide. *Carbon* 32, 469–475 (1994).
60. He, H., Riedl, T., Lerf, A., Klinowski, J.: Solid-state NMR studies of the structure of graphite oxide. *J. Phys. Chem.* 100, 19954–19958 (1996).
61. Gao, W., Alemany, L.B., Ci, L.J., Ajayan, P.M.: New insights into the structure and reduction of graphite oxide. *Nat. Chem.* 1, 403–408 (2009).
62. Boukhvalov, D.W., Katsnelson, M.I.: Modeling of graphite oxide. *J. Am. Chem. Soc.* 130, 10697–10701 (2008).

63. Lahaye, R., Jeong, H.K., Park, C.Y., Lee, Y.H.: Density functional theory study of graphite oxide for different oxidation levels. *Phys. Rev. B* 79, 125435 (2009).
64. Wang, L., Lee, K., Sun, Y.Y., Lucking, M., Chen, Z.F., Zhao, J.J., Zhang, S.B.B.: Graphene oxide as an ideal substrate for hydrogen storage. *ACS Nano* 3, 2995–3000 (2009).
65. Lu, N., Yin, D., Li, Z.Y., Yang, J.L.: Structure of graphene oxide: Thermodynamics versus kinetics. *J. Phys. Chem. C* 115(24), 11991–11995 (2011).
66. Wang, L., Sun, Y.Y., Lee, K., West, D., Chen, Z.F., Zhao, J.J., Zhang, S.B.: Stability of graphene oxide phases from first-principles calculations. *Phys. Rev. B* 82, 161406(R) (2010).
67. Xiang, H.J., Wei, S.H., Gong, X.G.: Structural motifs in oxidized graphene: A genetic algorithm study based on density functional theory. *Phys. Rev. B* 82, 035416 (2010).
68. Paci, J.T., Belytschko, T., Schatz, G.C.: Computational studies of the structure, behavior upon heating, and mechanical properties of graphite oxide. *J. Phys. Chem. C* 111, 18099–18111 (2007).
69. Gao, X., Jiang, D.-E., Zhao, Y., Nagase, S., Zhang, S., Chen, Z.: Theoretical insights into the structures of graphene oxide and its chemical conversions between graphene. *J. Comput. Theor. Nanosci.* 8, 2406–2422 (2011).
70. Gao, Y., Liu, L.-Q., Zu, S.-Z., Peng, K., Zhou, D., Han, B.-H., Zhang, Z.: The effect of interlayer adhesion on the mechanical behaviors of macroscopic graphene oxide papers. *ACS Nano* 5, 2134–2141 (2011).
71. Park, S., Lee, K.-S., Bozoklu, G., Cai, W., Nguyen, S.T., Ruoff, R.S.: Graphene oxide papers modified by divalent ions–enhancing mechanical properties via chemical cross-linking. *ACS Nano* 2, 572–578 (2008).
72. Park, S., Dikin, D.A., Nguyen, S.T., Ruoff, R.S.: Graphene oxide sheets chemically cross-linked by polyallylamine. *J. Phys. Chem. C* 113, 15801–15804 (2009).
73. Robinson, J.T., Zalalutdinov, M., Baldwin, J.W., Snow, E.S., Wei, Z., Sheehan, P., Houston, B.H.: Wafer-scale reduced graphene oxide films for nanomechanical devices. *Nano Lett.* 8, 3441–3445 (2008).
74. Suk, J.W., Piner, R.D., An, J., Ruoff, R.S.: Mechanical properties of monolayer graphene oxide. *ACS Nano* 4, 6557–6564 (2010).
75. Gómez-Navarro, C., Burghard, M., Kern, K.: Elastic properties of chemically derived single graphene sheets. *Nano Lett.* 8, 2045–2049 (2008).
76. Zheng, Q., Geng, Y., Wang, S., Li, Z., Kim, J.-K.: Effects of functional groups on the mechanical and wrinkling properties of graphene sheets. *Carbon* 48, 4315–4322 (2010).
77. Liu, L., Zhang, J., Zhao, J., Liu, F.: Mechanical properties of graphene oxides. *Nanoscale* 4, 5910–5916 (2012).
78. Eda, G., Fanchini, G., Chhowalla, M.: Large-area ultrathin films of reduced graphene oxide as a transparent and flexible electronic material. *Nat. Nanotechnol.* 3, 270–274 (2008).
79. Eda, G., Chhowalla, M.: Chemically derived graphene oxide: Towards large-area thin-film electronics and optoelectronics. *Adv. Mater.* 22, 2392–2415 (2010).
80. Loh, K.P., Bao, Q.L., Eda, G., Chhowalla, M.: Graphene oxide as a chemically tunable platform for optical applications. *Nat. Chem.* 2, 1015–1024 (2010).
81. Eda, G., Lin, Y.-Y., Miller, S., Chen, C.-W., Su, W.-F., Chhowalla, M.: Transparent and conducting electrodes for organic electronics from reduced graphene oxide. *Appl. Phys. Lett.* 92, 233305 (2008).
82. Wassei, J.K., Kaner, R.B.: Graphene, a promising transparent conductor. *Mater. Today* 13, 52–59 (2010).

83. Pan, D., Zhang, J., Li, Z., Wu, M.: Hydrothermal route for cutting graphene sheets into blue-luminescent graphene quantum dots. *Adv. Mater.* 22, 734–738 (2010).

84. Luo, Z., Vora, P.M., Mele, E.J., Johnson, A.T.C., Kikkawa, J.M.: Photoluminescence and band gap modulation in graphene oxide. *Appl. Phys. Lett.* 94, 111909 (2009).

85. Eda, G., Lin, Y.Y., Mattevi, C., Yamaguchi, H., Chen, H.A., Chen, I., Chen, C.W., Chhowalla, M.: Blue photoluminescence from chemically derived graphene oxide. *Adv. Mater.* 22, 505–509 (2010).

86. Cuong, T.V., Pham, V.H., Tran, Q.T., Hahn, S.H., Chung, J.S., Shin, E.W., Kim, E.J.: Photoluminescence and Raman studies of graphene thin films prepared by reduction of graphene oxide. *Mater. Lett.* 64, 399–401 (2010).

87. Essig, S., Marquardt, C.W., Vijayaraghavan, A., Ganzhorn, M., Dehm, S., Hennrich, F., Ou, F., Green, A.A., Sciascia, C., Bonaccorso, F.: Phonon-assisted electroluminescence from metallic carbon nanotubes and graphene. *Nano Lett.* 10, 1589–1594 (2010).

88. Sun, X., Liu, Z., Welsher, K., Robinson, J.T., Goodwin, A., Zaric, S., Dai, H.: Nanographene oxide for cellular imaging and drug delivery. *Nano Res.* 1, 203–212 (2008).

89. Liu, Z., Robinson, J.T., Sun, X., Dai, H.: PEGylated nanographene oxide for delivery of water-insoluble cancer drugs. *J. Am. Chem. Soc.* 130, 10876–10877 (2008).

90. Subrahmanyam, K.S., Kumar, P., Nag, A., Rao, C.N.R.: Blue light emitting graphene-based materials and their use in generating white light. *Solid State Commun.* 150, 1774–1777 (2010).

91. Chen, J.-L., Yan, X.-P.: A dehydration and stabilizer-free approach to production of stable water dispersions of graphene nanosheets. *J. Mater. Chem.* 20, 4328–4332 (2010).

92. Lu, J., Yang, J.-X., Wang, J., Lim, A., Wang, S., Loh, K.P.: One-pot synthesis of fluorescent carbon nanoribbons, nanoparticles, and graphene by the exfoliation of graphite in ionic liquids. *ACS Nano* 3, 2367–2375 (2009).

93. Li, F., Gao, J., Zhang, J., Xu, F., Zhao, J., Sun, L.: Graphene oxide and lithium amidoborane: A new way to bridge chemical and physical approaches for hydrogen storage. *J. Mater. Chem. A* 1, 8016–8022 (2013).

94. Burress, J.W., Gadipelli, S., Ford, J., Simmons, J.M., Zhou, W., Yildirim, T.: Graphene oxide framework materials: Theoretical predictions and experimental results. *Angew. Chem. Int. Ed.* 49, 8902–8904 (2010).

95. Ren, L., Hui, K.S., Hui, K.N.: Self-assembled freestanding three-dimensional nickel nanoparticle/graphene aerogel for direct ethanol fuel cells. *J. Mater. Chem. A* 1, 5689–5694 (2013).

96. Xue, Y., Liu, J., Chen, H., Wang, R., Li, D., Qu, J., Dai, L.: Nitrogen-doped graphene foams as metal-free counter electrodes in high-performance dye-sensitized solar cells. *Angew. Chem. Int. Ed.* 51, 12124–12127 (2012).

97. Wan, D., Yang, C., Lin, T., Tang, Y., Zhou, M., Zhong, Y., Huang, F., Lin, J.: Low-temperature aluminum reduction of graphene oxide, electrical properties, surface wettability, and energy storage applications. *ACS Nano* 6, 9068–9078 (2012).

98. Chen, S., Zhu, J., Wu, X., Han, Q., Wang, X.: Graphene oxide–MnO_2 nanocomposites for supercapacitors. *ACS Nano* 4, 2822–2830 (2010).

99. Gopalakrishnan, K., Govindaraj, A., Rao, C.N.R.: Extraordinary supercapacitor performance of heavily nitrogenated graphene oxide obtained by microwave synthesis. *J. Mater. Chem. A* 1, 7563–7565 (2013).

100. Chabot, V., Higgins, D., Yu, A., Xiao, X., Chen, Z., Zhang, J.: A review of graphene and graphene oxide sponge: Material synthesis and applications to energy and the environment. *Energy Environ. Sci.* 7, 1564–1596 (2014).

101. Garcia-Gallastegui, A., Iruretagoyena, D., Gouvea, V., Mokhtar, M., Asiri, A.M., Basahel, S.N., Al-Thabaiti, S.A., Alyoubi, A.O., Chadwick, D., Shaffer, M.S.P.: Graphene oxide as support for layered double hydroxides: Enhancing the CO_2 adsorption capacity. *Chem. Mater.* 24, 4531–4539 (2012).

102. Zhao, Y., Ding, H., Zhong, Q.: Preparation and characterization of aminated graphite oxide for CO_2 capture. *Appl. Surf. Sci.* 258, 4301–4307 (2012).

103. Deng, S., Tjoa, V., Fan, H.M., Tan, H.R., Sayle, D.C., Olivo, M., Mhaisalkar, S., Wei, J., Sow, C.H.: Reduced graphene oxide conjugated Cu_2O nanowire mesocrystals for high-performance NO_2 gas sensor. *J. Am. Chem. Soc.* 134, 4905–4917 (2012).

104. Li, W., Geng, X., Guo, Y., Rong, J., Gong, Y., Wu, L., Zhang, X., Li, P., Xu, J., Cheng, G.: Reduced graphene oxide electrically contacted graphene sensor for highly sensitive nitric oxide detection. *ACS Nano* 5, 6955–6961 (2011).

105. Yeh, T.F., Syu, J.M., Cheng, C., Chang, T.H., Teng, H.: Graphite oxide as a photocatalyst for hydrogen production from water. *Adv. Funct. Mater.* 20, 2255–2262 (2010).

106. Yeh, T.-F., Chen, S.-J., Yeh, C.-S., Teng, H.: Tuning the electronic structure of graphite oxide through ammonia treatment for photocatalytic generation of H_2 and O_2 from water splitting. *J. Phys. Chem. C* 117, 6516–6524 (2013).

107. Yeh, T.-F., Chan, F.-F., Hsieh, C.-T., Teng, H.: Graphite oxide with different oxygenated levels for hydrogen and oxygen production from water under illumination: The band positions of graphite oxide. *J. Phys. Chem. C* 115, 22587–22597 (2011).

108. Yeh, T.-F., Teng, H.: Graphite oxide with different oxygen contents as photocatalysts for hydrogen and oxygen evolution from water. *ECS Trans.* 41, 7–26 (2012).

109. Yeh, T.F., Teng, C.Y., Chen, S.J., Teng, H.: Nitrogen-doped graphene oxide quantum dots as photocatalysts for overall water-splitting under visible light illumination. *Adv. Mater.* 26, 3297–3303 (2014).

110. Jiang, X., Nisar, J., Pathak, B., Zhao, J., Ahuja, R.: Graphene oxide as a chemically tunable 2-D material for visible-light photocatalyst applications. *J. Catal.* 299, 204–209 (2013).

111. Grassi, R., Low, T., Lundstrom, M.: Scaling of the energy gap in pattern-hydrogenated graphene. *Nano Lett.* 11, 4574–4578 (2011).

112. Balog, R., Andersen, M., Jørgensen, B., Sljivancanin, Z., Hammer, B., Baraldi, A., Larciprete, R., Hofmann, P., Hornekær, L., Lizzit, S.: Controlling hydrogenation of graphene on Ir (111). *ACS Nano* 7, 3823–3832 (2013).

113. Robinson, J.A., Hollander, M., LaBella III, M., Trumbull, K.A., Cavalero, R., Snyder, D.W.: Epitaxial graphene transistors: Enhancing performance via hydrogen intercalation. *Nano Lett.* 11, 3875–3880 (2011).

114. Havu, P., Ijäs, M., Harju, A.: Hydrogenated graphene on silicon dioxide surfaces. *Phys. Rev. B* 84, 205423 (2011).

115. Li, Y., Chen, Z.: Patterned partially hydrogenated graphene (C4H) and its one-dimensional analogues: A computational study. *J. Phys. Chem. C* 116, 4526–4534 (2012).

116. Fokin, A.A., Gerbig, D., Schreiner, P.R.: σ/σ-and π/π-interactions are equally important: multilayered graphanes. *J. Am. Chem. Soc.* 133, 20036–20039 (2011).

117. Seah, T.H., Poh, H.L., Chua, C.K., Sofer, Z., Pumera, M.: Towards graphane applications in security: The electrochemical detection of trinitrotoluene in seawater on hydrogenated graphene. *Electroanalysis* 26, 62–68 (2014).

118. Eng, A.Y.S., Sofer, Z., Šimek, P., Kosina, J., Pumera, M.: Highly hydrogenated graphene through microwave exfoliation of graphite oxide in hydrogen plasma: Towards electrochemical applications. *Chem. Eur. J.* 19, 15583–15592 (2013).

119. Zhu, S., Li, T.: Hydrogenation-assisted graphene origami and its application in programmable molecular mass uptake, storage, and release. *ACS Nano* 8, 2864–2872 (2014).

120. Tozzini, V., Pellegrini, V.: Prospects for hydrogen storage in graphene. *Phys. Chem. Chem. Phys.* 15, 80–89 (2013).

121. Hussain, T., De Sarkar, A., Ahuja, R.: Functionalization of hydrogenated graphene by polylithiated species for efficient hydrogen storage. *Int. J. Hydrogen Energy* 39, 2560–2566 (2014).

122. Craciun, M., Khrapach, I., Barnes, M., Russo, S.: Properties and applications of chemically functionalized graphene. *J. Phys.: Condens. Matter* 25, 423201 (2013).

123. Jones, J., Mahajan, K., Williams, W., Ecton, P., Mo, Y., Perez, J.: Formation of graphane and partially hydrogenated graphene by electron irradiation of adsorbates on graphene. *Carbon* 48, 2335–2340 (2010).

124. Diankov, G., Neumann, M., Goldhaber-Gordon, D.: Extreme monolayer-selectivity of hydrogen-plasma reactions with graphene. *ACS Nano* 7, 1324–1332 (2013).

125. Balakrishnan, J., Koon, G.K.W., Jaiswal, M., Neto, A.C., Özyilmaz, B.: Colossal enhancement of spin-orbit coupling in weakly hydrogenated graphene. *Nat. Phys.* 9, 284–287 (2013).

126. Wang, Y., Xu, X., Lu, J., Lin, M., Bao, Q., Ozyilmaz, B., Loh, K.P.: Toward high throughput interconvertible graphane-to-graphene growth and patterning. *ACS Nano* 4, 6146–6152 (2010).

127. Yang, Z., Sun, Y., Alemany, L.B., Narayanan, T.N., Billups, W.: Birch reduction of graphite. Edge and interior functionalization by hydrogen. *J. Am. Chem. Soc.* 134, 18689–18694 (2012).

128. Subrahmanyam, K., Kumar, P., Maitra, U., Govindaraj, A., Hembram, K., Waghmare, U.V., Rao, C.: Chemical storage of hydrogen in few-layer graphene. *Proc. Natl. Acad. Sci. USA.* 108, 2674–2677 (2011).

129. Ryu, S., Han, M.Y., Maultzsch, J., Heinz, T.F., Kim, P., Steigerwald, M.L., Brus, L.E.: Reversible basal plane hydrogenation of graphene. *Nano Lett.* 8, 4597–4602 (2008).

130. Balog, R., Jørgensen, B., Nilsson, L., Andersen, M., Rienks, E., Bianchi, M., Fanetti, M., Lægsgaard, E., Baraldi, A., Lizzit, S.: Bandgap opening in graphene induced by patterned hydrogen adsorption. *Nat. Mater.* 9, 315–319 (2010).

131. Sessi, P., Guest, J.R., Bode, M., Guisinger, N.P.: Patterning graphene at the nanometer scale via hydrogen desorption. *Nano Lett.* 9, 4343–4347 (2009).

132. Luo, Z., Yu, T., Ni, Z., Lim, S., Hu, H., Shang, J., Liu, L., Shen, Z., Lin, J.: Electronic structures and structural evolution of hydrogenated graphene probed by Raman spectroscopy. *J. Phys. Chem. C* 115, 1422–1427 (2011).

133. Ng, M.L., Balog, R., Hornekær, L., Preobrajenski, A., Vinogradov, N.A., Mårtensson, N., Schulte, K.: Controlling hydrogenation of graphene on transition metals. *J. Phys. Chem. C* 114, 18559–18565 (2010).

134. Lin, Y., Ding, F., Yakobson, B.I.: Hydrogen storage by spillover on graphene as a phase nucleation process. *Phys. Rev. B* 78, 041402 (2008).

135. Singh, A.K., Ribas, M.A., Yakobson, B.I.: H-spillover through the catalyst saturation: An ab initio thermodynamics study. *ACS Nano* 3, 1657–1662 (2009).

136. Lueking, A.D., Psofogiannakis, G., Froudakis, G.E.: Atomic hydrogen diffusion on doped and chemically modified graphene. *J. Phys. Chem. C* 117, 6312–6319 (2013).

137. Flores, M.Z., Autreto, P.A., Legoas, S.B., Galvao, D.S.: Graphene to graphane: A theoretical study. *Nanotechnology* 20, 465704 (2009).

138. Sofo, J.O., Chaudhari, A.S., Barber, G.D.: Graphane: A two-dimensional hydrocarbon. *Phys. Rev. B* 75, 153401 (2007).

139. Boukhvalov, D., Katsnelson, M.: Tuning the gap in bilayer graphene using chemical functionalization: Density functional calculations. *Phys. Rev. B* 78, 085413 (2008).

140. Boukhvalov, D.W., Son, Y.W.: Covalent functionalization of strained graphene. *ChemPhysChem* 13, 1463–1469 (2012).

141. Sun, Z., Pint, C.L., Marcano, D.C., Zhang, C., Yao, J., Ruan, G., Yan, Z., Zhu, Y., Hauge, R.H., Tour, J.M.: Towards hybrid superlattices in graphene. *Nat. Commun.* 2, 559 (2011).

142. Georgakilas, V., Otyepka, M., Bourlinos, A.B., Chandra, V., Kim, N., Kemp, K.C., Hobza, P., Zboril, R., Kim, K.S.: Functionalization of graphene: Covalent and noncovalent approaches, derivatives and applications. *Chem. Rev.* 112, 6156–6214 (2012).

143. Guisinger, N.P., Rutter, G.M., Crain, J.N., First, P.N., Stroscio, J.A.: Exposure of epitaxial graphene on SiC (0001) to atomic hydrogen. *Nano Lett.* 9, 1462–1466 (2009).

144. Riedl, C., Coletti, C., Starke, U.: Structural and electronic properties of epitaxial graphene on SiC (0 0 0 1): A review of growth, characterization, transfer doping and hydrogen intercalation. *J. Phys. D: Appl. Phys.* 43, 374009 (2010).

145. Ulstrup, S., Nilsson, L., Miwa, J.A., Balog, R., Bianchi, M., Hornekær, L., Hofmann, P.: Electronic structure of graphene on a reconstructed Pt (100) surface: Hydrogen adsorption, doping, and band gaps. *Phys. Rev. B* 88, 125425 (2013).

146. Busse, C., Lazić, P., Djemour, R., Coraux, J., Gerber, T., Atodiresei, N., Caciuc, V., Brako, R., Blügel, S., Zegenhagen, J.: Graphene on Ir (111): Physisorption with chemical modulation. *Phys. Rev. Lett.* 107, 036101 (2011).

147. Andersen, M., Hornekær, L., Hammer, B.: Graphene on metal surfaces and its hydrogen adsorption: A meta-GGA functional study. *Phys. Rev. B* 86, 085405 (2012).

148. Lee, B., Han, S., Kim, Y.-S.: First-principles study of preferential sites of hydrogen incorporated in epitaxial graphene on 6 H-SiC (0001). *Phys. Rev. B* 81, 075432 (2010).

149. Pujari, B.S., Gusarov, S., Brett, M., Kovalenko, A.: Single-side-hydrogenated graphene: Density functional theory predictions. *Phys. Rev. B* 84, 041402 (2011).

150. Watcharinyanon, S., Virojanadara, C., Osiecki, J., Zakharov, A., Yakimova, R., Uhrberg, R., Johansson, L.I.: Hydrogen intercalation of graphene grown on 6H-SiC (0001). *Surf. Sci.* 605, 1662–1668 (2011).

151. Giesbers, A. J. M., Uhlířová K., Konečný M., Peters E. C., Burghard M., Aarts J., Flipse C. F. J.: Interface-induced room-temperature ferromagnetism in hydrogenated epitaxial graphene. *Phys. Rev. Lett.* 111, 166101 (2013).

152. Tokarczyk, M., Kowalski, G., Możdżonek, M., Borysiuk, J., Stępniewski, R., Strupiński, W., Ciepielewski, P., Baranowski, J.: Structural investigations of hydrogenated epitaxial graphene grown on 4H-SiC (0001). *Appl. Phys. Lett.* 103, 241915 (2013).

153. Eren, B., Glatzel, T., Kisiel, M., Fu, W., Pawlak, R., Gysin, U., Nef, C., Marot, L., Calame, M., Schönenberger, C.: Hydrogen plasma microlithography of graphene supported on a Si/SiO$_2$ substrate. *Appl. Phys. Lett.* 102, 071602 (2013).

154. Fessler, G., Eren, B., Gysin, U., Glatzel, T., Meyer, E.: Friction force microscopy studies on SiO$_2$ supported pristine and hydrogenated graphene. *Appl. Phys. Lett.* 104, 041910 (2014).

155. Wojtaszek, M., Vera-Marun, I., Maassen, T., van Wees, B.: Enhancement of spin relaxation time in hydrogenated graphene spin-valve devices. *Phys. Rev. B* 87, 081402 (2013).

156. Xiang, H., Kan, E., Wei, S.-H., Gong, X., Whangbo, M.-H.: Thermodynamically stable single-side hydrogenated graphene. *Phys. Rev. B* 82, 165425 (2010).

157. Openov, L., Podlivaev, A.: Insulator band gap in single-side-hydrogenated graphene nanoribbons. *Semiconductors* 46, 199–202 (2012).

158. Zhou, J., Wang, Q., Sun, Q., Chen, X., Kawazoe, Y., Jena, P.: Ferromagnetism in semihydrogenated graphene sheet. *Nano Lett.* 9, 3867–3870 (2009).

159. Xie, L., Wang, X., Lu, J., Ni, Z., Luo, Z., Mao, H., Wang, R., Wang, Y., Huang, H., Qi, D.: Room temperature ferromagnetism in partially hydrogenated epitaxial graphene. *Appl. Phys. Lett.* 98, 193113 (2011).

160. Riedl, C., Coletti, C., Iwasaki, T., Zakharov, A., Starke, U.: Quasi-freestanding epitaxial graphene on SiC obtained by hydrogen intercalation. *Phys. Rev. Lett.* 103, 246804 (2009).

161. Speck, F., Jobst, J., Fromm, F., Ostler, M., Waldmann, D., Hundhausen, M., Weber, H., Seyller, T.: The quasi-free-standing nature of graphene on H-saturated SiC (0001). *Appl. Phys. Lett.* 99, 122106 (2011).

162. Bocquet, F.C., Bisson, R., Themlin, J.-M., Layet, J.-M., Angot, T.: Reversible hydrogenation of deuterium-intercalated quasi-free-standing graphene on SiC (0001). *Phys. Rev. B* 85, 201401 (2012).

163. Casolo, S., Løvvik, O.M., Martinazzo, R., Tantardini, G.F.: Understanding adsorption of hydrogen atoms on graphene. *J. Chem. Phys.* 130, 054704 (2009).

164. Sluiter, M.H., Kawazoe, Y.: Cluster expansion method for adsorption: Application to hydrogen chemisorption on graphene. *Phys. Rev. B* 68, 085410 (2003).

165. Ito, A., Nakamura, H., Takayama, A.: Molecular dynamics simulation of the chemical interaction between hydrogen atom and graphene. *J. Phys. Soc. Jpn.* 77, (2008).

166. Duplock, E.J., Scheffler, M., Lindan, P.J.: Hallmark of perfect graphene. *Phys. Rev. Lett.* 92, 225502 (2004).

167. Boukhvalov, D., Katsnelson, M., Lichtenstein, A.: Hydrogen on graphene: Electronic structure, total energy, structural distortions and magnetism from first-principles calculations. *Phys. Rev. B* 77, 035427 (2008).

168. Samarakoon, D.K., Wang, X.-Q.: Chair and twist-boat membranes in hydrogenated graphene. *ACS Nano* 3, 4017–4022 (2009).

169. Savchenko, A.: Transforming graphene. *Science* 323, 589–590 (2009).

170. Li, Y., Zhou, Z., Shen, P., Chen, Z.: Structural and electronic properties of graphane nanoribbons. *J. Phys. Chem. C* 113, 15043–15045 (2009).

171. Singh, A.K., Penev, E.S., Yakobson, B.I.: Vacancy clusters in graphane as quantum dots. *ACS Nano* 4, 3510–3514 (2010).

172. Wojtaszek, M., Tombros, N., Caretta, A., Van Loosdrecht, P., Van Wees, B.: A road to hydrogenating graphene by a reactive ion etching plasma. *J. Appl. Phys.* 110, 063715 (2011).

173. Eng, A.Y.S., Poh, H.L., Sanek, F., Maryško, M., Matejkova, S., Sofer, Z.k., Pumera, M.: Searching for magnetism in hydrogenated graphene: Using highly hydrogenated graphene prepared via birch reduction of graphite oxides. *ACS Nano* 7, 5930–5939 (2013).

CHAPTER 16

CONTENTS

Click Functionalization of Carbon Nanotubes and Graphene with Polymers

16

Horacio J. Salavagione, Marián A. Gómez-Fatou,
Gerardo Martínez, Carlos Marco, and Gary J. Ellis

16.1 INTRODUCTION

While the unique properties of graphene and carbon nanotubes (CNTs) make them two of the most highly promising nanomaterials available, one of the fundamental requirements to effectively translate their properties to real applications is integration into devices or multicomponent systems. At the frontiers of materials technology, one of the most successful approaches to reach specific goals with the greatest property efficiency and cost effectiveness is the concept of synergy, achieved by adequately combining various types of materials. In this respect, the covalent combination of carbon nanostructures (CNs) with polymers provides a versatile route to the generation of materials with wide applications, through the union of two families of materials.

In graphene-based or CNT-based polymer nanocomposites, as is the case for polymer nanocomposites in general, optimum improvements in final properties are only effective when the nanoparticles are homogeneously dispersed throughout the matrix, and strong filler/polymer interfacial interactions are developed. However, the inert nature of these CNs makes them difficult to disperse within the majority of polymers, since they can only achieve moderate interaction with a limited group of polymers, typically those containing aromatic moieties. In addition, low solubility of pristine graphene or CNTs also limits materials design. In order to make these nanoparticles dispersible in—or compatible with—a wider range of polymer matrices, as well as to maximize the interfacial interactions, some type of chemical modification is generally required to introduce functional groups that confer reactivity to the pristine material.

CNTs can undergo different chemical reactions and their chemical functionalization has been widely discussed in several reviews [1–6]. The main approaches for the modification of CNTs are (1) the covalent attachment of chemical groups through reactions onto the π-conjugated skeleton of CNT, (2) the noncovalent adsorption or wrapping of various functional molecules, and (3) the endohedral filling of their inner empty cavity [4]. Covalent modification can be performed at the end of the tubes or in their sidewalls. Direct covalent sidewall functionalization can be made by reaction with some molecules of a high chemical reactivity, such as fluorine, carbene, and nitrene, whereas defect functionalization takes advantage of the chemical transformation of defect sites [1,4]. Defects can be created by an oxidative process with strong acids or strong oxidants, and the functional groups generated (COOH, OH) can be used as precursors for further chemical reactions. Finally, the covalent modification of CNTs with polymers is especially interesting for the development of polymer/CNT nanocomposites using *grafting-to* or *grafting-from* approaches [7,8].

The chemistry of graphene-like materials has mainly focused on the chemistry of graphite oxide (GO) or in coupling diazonium salts to pristine graphene sheets. Indeed, the former gives access to many synthetic protocols that can provide graphene with a wide range of functionality, since GO, prepared by strong oxidation of graphite, has different oxygen-containing species on the surface, such as epoxides, hydroxyls, ketones, carboxylic acids, and lactones, each with their own specific chemistry. However, to prepare GO the initial oxidation process leads inevitably to a highly defective material with a damaged sp^2 network and, consequently, far inferior properties (mechanical, thermal, electrical, etc.) than pristine graphene [9–11]. The latter approach via diazonium chemistry occurs directly on the graphene surface, requiring no previous treatment; the degree of functionalization can be controlled and the damage to the sp^2 network is reduced. This type of reaction is very popular in other carbonaceous materials like CNTs, graphite, and glassy carbon. However, diazonium chemistry is restricted to only a handful of aromatic amines, limiting the range of functionalities that can be introduced to the graphene surface.

Consequently, the search for well-defined strategies to combine CNs with polymers has generated recent interest. In order to provide pristine graphene with a specific functional group of interest, the development of a general chemical strategy is required, and in this respect *click* chemistry approaches represent a very attractive option. This chapter will focus on the covalent modification of graphene and CNTs with polymers via click chemistry reactions including Cu-catalyzed alkyne–azide additions and thiol–radical attack to double (thiol–ene) and triple bonds (thiol–yne). The possibilities of these CNs in click reactions will be contrasted, the state of the art of click chemistry for the incorporation of graphene and CNTs into polymers will be revised and balanced, and future perspectives of this exciting field will be discussed.

16.2 CLICK CHEMISTRY

The click chemistry concept, based on the coupling of small molecules by means of carbon- and heteroatom links, was developed by Sharpless and coworkers who defined a set of stringent criteria that a process must meet in order to be classified as *click* [12]. They specified that the reaction must be modular, be broad in scope, be stereospecific, generate very high yields with only inoffensive and easily removable by-products, be simple, be insensitive to oxygen and water, use readily available starting materials and reagents, and use no solvents (or only benign solvents), with simple product isolation avoiding chromatographic methods. Subsequently, many click reactions have been reported, and their main impact has been in the biological chemistry and pharmaceutical areas [13,14].

16.2.1 Cu(I)-Catalyzed Azide–Alkyne Cycloaddition Reaction

Among the collection of click reactions currently available, the Huisgen 1,3-dipolar azide–alkyne cycloaddition, producing 1,2,3 triazole, is probably the most widely investigated [15]. This chemical transformation is characterized by high reliability and broad tolerance to diverse reaction conditions and functional groups. However, it was not until the discovery of the Cu (I)-catalyzed variant of this reaction (Scheme 16.1) (a) by the groups of Meldal [16] and Fokin and Sharpless [17] that the real potential of this coupling technology was demonstrated. In practice, the copper (I) catalyst can be generated in situ using copper (II) sulfate and sodium ascorbate as reducing agent. Alternatively, a copper (I) halide can be used together with a stabilizing ligand. The mechanism of copper-catalyzed azide–alkyne cycloaddition reactions has been studied in detail [18,19] and is beyond the scope of this chapter; however, the triazole formed is essentially chemically inert to reactive conditions, for example, oxidation, reduction, and hydrolysis, and has an intermediate polarity. For a more detailed discussion on this reaction, the reader is directed to a comprehensive review [20].

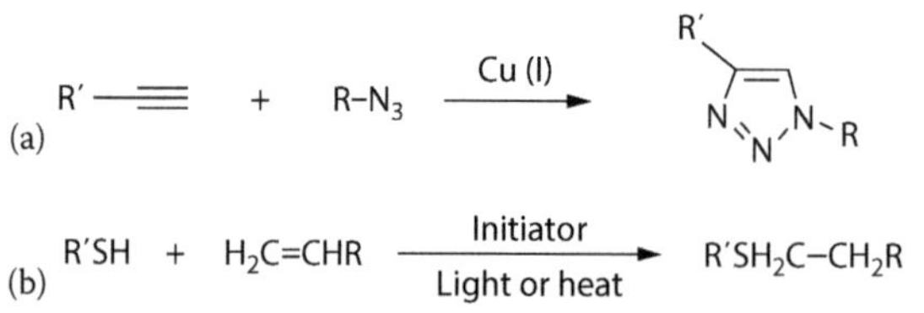

SCHEME 16.1
Mechanism of click reactions employed to connect carbon nanostructures and polymers: (a) the copper (I)-catalyzed azide–alkyne cycloaddition (CuAAC) reaction; (b) thiol–radical reactions.

16.2.2 Thiol–Radical Click Reactions

The most important drawback of the Cu(I)-catalyzed azide–alkyne cycloaddition (CuAAC) is the toxicity of the copper catalyst. In order to avoid the use of any toxic metal catalysts, other metal-free click reactions have been developed [21]. For example, reactions based on the addition of the very reactive thiol–radical to unsaturated carbon–carbon bonds probably represent the most promising metal-free click reactions (Scheme 16.1). Radical–thiol click reactions are based on the formation of thiol–radicals that attack double (thiol–ene) or triple (thiol–yne) bonds via a typical chain process, with initiation, propagation, and termination steps [22–24]. These reactions possess all the desirable features of a click reaction, such as simplicity, high efficiency, no by-products, and high yield, and also present an additional advantage since they can be externally triggered by thermal or optical stimuli [22]. Indeed, thiol-based click reactions are most frequently photoinitiated, which generates unique capabilities with regard to spatial and temporal control of the click reaction [25]. When compared to the thiol–ene reaction, an important distinguishing feature of the thiol–yne reaction is the ability to react with two thiol equivalents, that is, to form a double addition product with 1,2-regioselectivity while, mechanistically, the two thiol addition reaction steps each proceed in a similar manner to the radical thiol–ene reaction (Scheme 16.1).

16.3 CLICK CHEMISTRY IN POLYMER SCIENCE

The impact of the discovery of click chemistry on polymer science has been tremendous, since often the main limitations in polymer preparation and modification are related to experimental factors where design protocols and purification procedures that are standard in organic chemistry generally cannot be efficiently applied. In other words, the characteristics of click chemistry including efficiency, absence of side products, and easy purification open new possibilities for the preparation of new, otherwise unattainable functionalized polymers [26]. However, the direct extension of the click concept to macromolecular chemistry is not strictly correct, and adapted definitions in the context of polymer chemistry with a set of requirements have been described [27]. Figure 16.1 shows the most important criteria for click reaction in macromolecular synthesis. Following the Sharpless

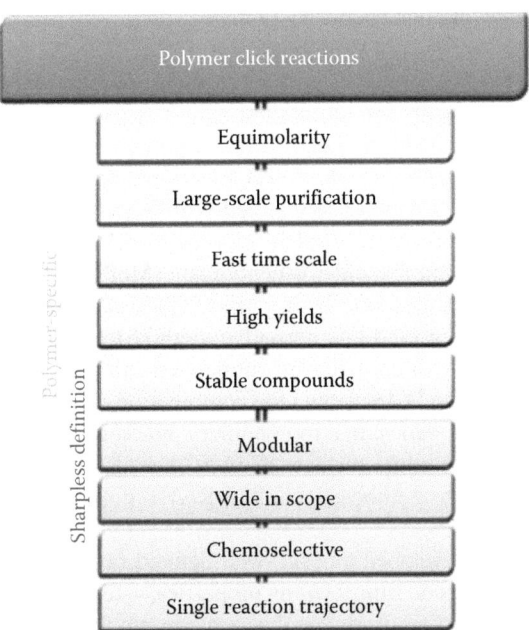

FIGURE 16.1
List of general and specific requirements for click reactions with polymers. (Barner-Kowollik, C., Du Prez, F.E., and Espeel, P.: "Clicking" polymers or just efficient linking: What is the difference. Angewandte Chemie International Edition. 2011. 50. 60–62. Copyright Wiley-VCH Verlag GmbH & Co. KGaA. Reproduced with permission.)

definition (blue boxes), as previously indicated, a click reaction must be modular, wide in scope, and chemoselective/orthogonal, among other characteristics. The green boxes show a series of specific requirements of click reactions in polymer chemistry. The blue-green boxes indicate requirements that are also part of the original definition but have major consequences in the polymer field. As can be seen from the figure, some critical requirements in polymer science are related with equimolarity and large-scale purification steps.

During the past 15 years, the number of novel synthetic methods that allow control over the composition, molecular structure, and the number and placement of functional groups in polymeric materials has grown almost explosively. Living polymerization and, especially, controlled radical polymerization (CRP) [28] methods have become very popular due to the relatively simple experimental setup and applicability to a wide range of monomers. The most important CRP methods include atom transfer radical polymerization (ATRP) [29,30], stable free radical (mostly) nitroxide-mediated polymerization (NMP) [31], and degenerative transfer polymerization, particularly reversible addition–fragmentation chain transfer (RAFT) polymerization [32]. Due to the CRP mechanism, polymer chains prepared by such techniques are end-capped by a *dormant* unit (a halogen atom with ATRP, an alkoxyamine moiety with NMP, a dithioester moiety in RAFT), which can be transformed into diverse functional groups after polymerization. ATRP is probably the most practical technique, because the terminal alkyl halide can be used for standard nucleophilic substitutions or elimination reactions [33].

On the other hand, click chemistry, in particular the CuAAC reaction, is a powerful new synthetic tool in polymer chemistry and materials science [34–38] very practical for dendrimer synthesis, coupling of preformed polymer segments [35], preparation of new polymers [36], or postfunctionalization of polymer backbones [37,38]. Some reviews on click chemistry in polymers are available [39–41].

16.3.1 CuAAC in Polymer Science

Success of the CuAAC in the engineering of polymer architectures stems, in part, from the possibility of introducing the required azide and alkyne functionalities at determined locations in macromolecular building blocks and is a result of advances in controlled polymerization techniques combined with the efficiency of click chemistry [42].

16.3.1.1 End functionalization of well-defined polymers

Lutz et al. [43] first reported the end functionalization of polystyrenes (PSs) prepared using ATRP. The copper (I)-catalyzed cycloaddition of several functional acetylenes with azido-PS allowed the quantitative formation of PS end functionalized with a primary alcohol, a carboxylic acid, or a vinyl group. The copolymers were used in a second click reaction (1,3-dipolar cycloaddition) with poly (ethylene oxide) methyl ether 4-pentynoate (MePEO-P), resulting in the formation via *grafting-to* of polymeric brushes with hydrophilic PEO side chains with a moderate grafting density. Haddleton et al. [44] reported an alternative method for clicking ATRP-generated macromolecules by polymerizing with an azide-functionalized initiator. These initial reports led to rapid proliferation of other methods describing the combination of ATRP and CuAAC to facilitate the synthesis of functional telechelics, macromonomers, macrocycles, stars, networks, bioconjugates, bionanoparticles, and functional surfaces. CuAAC has also been effectively combined with other controlled polymerization methods, including RAFT polymerization, NMP, and cationic polymerization.

Fleichchmann et al. [45] developed a novel, well-defined random terpolymer, synthesized by NMP, composed of styrene for good film-forming properties, 4-(ethynyl) styrene as a carrier for the indispensable alkyne moiety, and a glycidyl methacrylate group for anchoring to silicon substrates. The films were modified in a heterogeneous reaction that efficiently and quantitatively converted the alkyne groups of the surface by click chemistry with azides.

An important number of publications combining RAFT with azide/alkyne click reactions have been described. For example, Gondi et al. [46] described the polymerization of styrene and

N,N-dimethylacrylamide with azide-functionalized chain transfer agents (CTAs), used to mediate RAFT. The resulting azide-terminated polymers were reacted with various acetylene species, demonstrating the ability to prepare a range of telechelics and functional CTAs. Li et al. [47] polymerized 2-azidoethyl methacrylate (AzMA) via RAFT polymerization, and postfunctionalization of polyAzMA was successfully performed via CuAAC with phenyl acetylene with minimal variation in the molecular weight distribution. The strategy of combining RAFT with click chemistry can be applied to prepare a wide range of functional polymers, particularly when the pendant moiety may interfere with the polymerization reaction.

16.3.1.2 Side-chain functionalization

Side-chain modified polymers with pendant azide or acetylenic moieties for the generation of graft polymers have been reported. Matyjaszewsky et al. [48] used N_3Na to open epoxide rings in glycidyl methacrylate and methyl methacrylate copolymers prepared by ATRP to yield azide monomer units. Ladmiral et al. [49] also used click methods to incorporate appropriate sugar azides to poly(methacrylates) bearing terminal alkyne functionalities.

16.3.2 Thiol–Radical Reactions in Polymer Science

Thiol–radical addition to unsaturated CC bonds has been extended to the polymer field for post-modification or polymer network formation. However, the case of coupling two polymeric chains is more complex since thiol–ene chemistry generates significant side reactions, in direct contradiction to the click concept [50]. However, despite not meeting the stringent criteria to be considered a click reaction with polymers, it has been successfully employed for polymer modification. The facility with which thiol or alkene end groups can be incorporated into polymers has allowed the synthesis of a variety of end-functional macromolecules. One particularly straightforward manner of obtaining thiol-terminated polymers is by reduction of thiocarbonyl thiol-end groups resulting from the RAFT process. Indeed, the fact that sulfur-containing end groups are inherent to RAFT has given rise to a variety of applications that capitalize on the versatile chemical reactivity of sulfhydryl groups [41].

The groups of Hoyle and Lowe [51] recently describe a facile one-pot reaction for the convergent synthesis of star polymers. A homopolymer of *N,N*-diethylacrylamide was prepared under standard RAFT conditions employing 1-cyano-1-methylethyl dithiobenzoate as the RAFT agent in conjunction with 2,2′-azobis(isobutyronitrile) (AIBN). Reduction with hexylamine in the presence of dimethylphenylphosphine and trimethylolpropane triacrylate results in a 3-arm star formation via a macromolecular thiol–ene click reaction.

As well as addition to alkenes, thiols can efficiently add to alkynes by a radical-based *thiol–yne* process. For example, Lowe et al. [52] employed a combination of thiol–ene and thiol–yne chemistries to prepare end-functional poly(*N*-isopropylacrylamide) (PNIPAM).

16.4 REACTIVITY OF CARBON NANOSTRUCTURES

The possibilities for the participation of CNTs and graphene in these reactions are naturally related to their chemical reactivity, and in order to rationalize the differences in reactivity between these CNs, several variables must be considered, that is, (1) the available local sites for reaction (edge or basal planes), (2) strain generated by the geometry of the CNs, and (3) the π-orbital alignments of the carbon atoms [53]. Ideally, graphene is a 2D flat monolayer of carbon atoms in a honeycomb-like structure having basal and edge planes that can produce fullerenes and CNTs when appropriately closed. Thus, fullerenes and CNTs can be viewed as 0D and 1D structures of a 2D graphene sheet, respectively.

In single-walled carbon nanotube (SWCNT) sidewalls, the π-orbital misalignment between adjacent carbon atoms plays a large influence in determining overall reactivity (Figure 16.2).

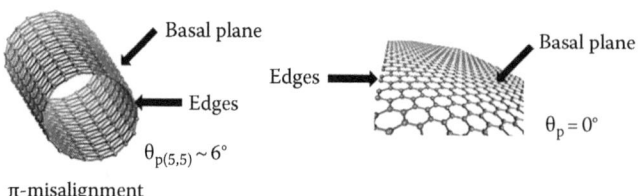

FIGURE 16.2
Structures of carbon nanotubes and graphene.

This misalignment, associated with bonds forming an angle to the tube circumference (i.e., bonds that are neither parallel nor perpendicular to the tube axis), is the origin of torsional strain in nanotubes, and the relief of this strain controls the extent to which addition reactions with nanotubes can occur [6]. Since π-orbital misalignment, as well as pyramidization, scales inversely with the tube diameter [54], nanotubes with smaller diameters are expected to be more reactive than larger ones. Thus, multiwalled carbon nanotubes (MWCNTs) with an outer tube diameter in the range of 5–50 nm generally exhibit a much lower tendency toward sidewall addition reactions than SWCNT with diameter distributions in the range of 0.7–2 nm.

Graphene is seen a priori as the least reactive of this family of CNs [55]. However, different chemical reactions at both edge and basal planes of graphene have been successfully attempted. Edge sites are considered the most reactive [55,56], since edge carbons can adopt tetrahedral geometries more freely (without causing extra strain) than carbon atoms in the basal plane. Therefore, edge carbon atoms are preferred in covalent addition reactions, where zigzag edges are particularly reactive. With regard to the basal reactivity of graphene, it must be taken into account that graphene is chemically unsaturated, composed of sp^2 carbon atoms protected by a π-conjugation system with constrained motion because of the adjacent carbon atoms. Therefore, basal plane covalent addition usually encounters large energy barriers, and reactive chemical groups, such as atomic hydrogen, fluorine, and precursors of other strong chemical radicals, are usually needed in the reactants. To date, the chemical modification of graphene with these reactive chemical groups cannot be fully controlled and can also take place on graphene edges [55]. However, internal carbons of graphene have been modified recently by cycloaddition reactions [57–59]. Here, enhanced reactivity may result from the influence of structural distortion, where the origin is spontaneous rippling of ca. 1 nm [60], causing some local strain [61] as predicted [62].

In summary, the feasibility of reactions occurring at the edges of graphitic planes decreases in passing from graphene to CNT as a consequence of the reduced concentration of active sites, while the possibility of reactions on basal planes can be said to be higher in CNTs than in graphene. In summary, the reactivity of these CNs is dominated by reactions on the basal planes and is higher in nonplanar structures than in planar 2D laminates [63].

16.5 CLICK FUNCTIONALIZATION OF CARBON NANOSTRUCTURES WITH POLYMERS

The click connection between CNs and polymers generates a new type of nanocomposite where the polymer and the filler are covalently linked. In this type of nanocomposites, the concept of interface changes from a traditional view of polymer composites where molecular interactions like van der Waals, hydrogen bonding, halogen bonding, and π–π occur between components at the polymer–filler interface to the single compound concept where the CN forms an integral part of the macromolecular chain or polymer brushes emerge from a graphene surface, for example [64]. A complete elimination of the *interface barrier* would be expected to lead to a full transfer of the properties of the CNs to the matrix and, consequently, to maximum improvements in the final properties. In other words, ideally the strategy of covalent functionalization of graphene and CNs with polymers would allow homogeneous dispersions of the CNs and adequate control of the microstructure of the nanocomposites to be achieved.

16.5.1 Click Functionalization of Carbon Nanotubes with Polymers

CNTs are one of the most attractive nanomaterials due to their unique structure and exceptional thermal, electrical, magnetic, and mechanical properties [65,66]. These outstanding properties make them ideal candidates for many potential applications such as molecular electronics, energy conversion and storage, sensors, biosensors and biomedicine, and advanced composite materials [67–71]. The development of polymer/CNT nanocomposites for structural and functional applications has been a very active area of research over the last decade. The two main factors that control the properties of these materials are an efficient dispersion of individual nanotubes and the establishment of a strong chemical affinity (covalent or noncovalent) with the polymer matrix. However, CNTs usually form bundles due to van der Waals interactions and are extremely difficult to disperse in a polymer matrix. Moreover, the carbon atoms on CNT walls are chemically stable because of the aromatic nature of the bond and interact with the polymer matrix mainly through van der Waals interactions that generally do not provide an efficient load transfer across the CNT/matrix interface. An enormous effort has been made to overcome these two problems, and several recent critical reviews discuss the different approaches [7,8,72]. Noncovalent functionalization of nanotubes using surfactants, biomacromolecules, or wrapping with polymers and covalent functionalization with polymers via *grafting-to* or *grafting-from* routes are techniques used to modify the surface properties of CNTs and to achieve good CNT dispersion and interactions with the polymer matrix [7,8,72]. Click chemistry, on the other hand, is a simple and efficient strategy for the introduction of a large variety of molecules onto the surface of CNTs [73], and it has been demonstrated that functionalization with polymers via click chemistry can enhance their dispersion and interaction with the matrix in nanocomposites [7,73,74]. Likewise, taking advantage of the easy introduction of azide and alkyne groups in polymers and the merits of click chemistry, such as the toleration of other functional groups, short reaction times, and regiospecificity, the functionalization of CNTs with polymers using this approach can open new opportunities to tailor target functionalities for specific applications.

As described in previous sections, the most common click reaction is CuAAC. In order to perform CuAAC reactions with CNT, they should be furnished with either alkyne or azide moieties. Although both azide- and alkyne-modified CNTs are environmentally stable and can be easily clicked to chemical entities having the appropriate counterpart moiety, CNTs have been mainly employed as the alkyne-bearing part in click reactions. However, although to a lesser extent they have also been provided with azide moieties, principally by amidation with amine-azide compounds. Probably the reason why azide-modified CNTs are not preferred is the need to use previously oxidized CNTs for amidation, as this oxidation step is detrimental to the sp^2 network of CNTs, degrading their electrical properties. However, coupling with the diazonium salt of an aromatic amine–azide has also been used for providing CNTs with this click-reagent group.

The synthetic approaches for functionalizing nanotube surfaces with alkyne or azide moieties are limited to a few reactions. In fact, two main strategies are involved in most of the reported work: (1) the addition of diazonium salts to the sidewalls of CNT and (2) the amidation or esterification with carboxylic groups located at edges planes and/or defects created after purification of the tubes by oxidation. Although the former was preferred for alkyne modification of sidewalls of CNTs, recently it has been employed in an attempt to endow SWCNTs with azide groups [75]. However, it must be pointed out that the carbon atoms in the walls of the tubes display certain strain and π-orbital misalignment that make them somewhat reactive for addition of single molecules.

The most significant studies on polymer functionalization of CNTs via click chemistry undertaken to date are described in this section, in most cases employing a Cu(I)-catalyzed azide/alkyne reaction where the CNTs bear alkyne moieties. However, recent examples of the use of thiol–ene reaction to modify CNTs with polymers are also described.

Andronov and coworkers were the first to report the functionalization of CNTs with polymers using click chemistry via the CuAAC reaction in 2005 [76]. They prepared alkyne-functionalized SWCNTs using a phenyl diazonium generated in situ from *p*-aminophenyl propargyl ether and isoamyl nitrite following the solvent-free procedure described by Tours et al. [77]. In parallel,

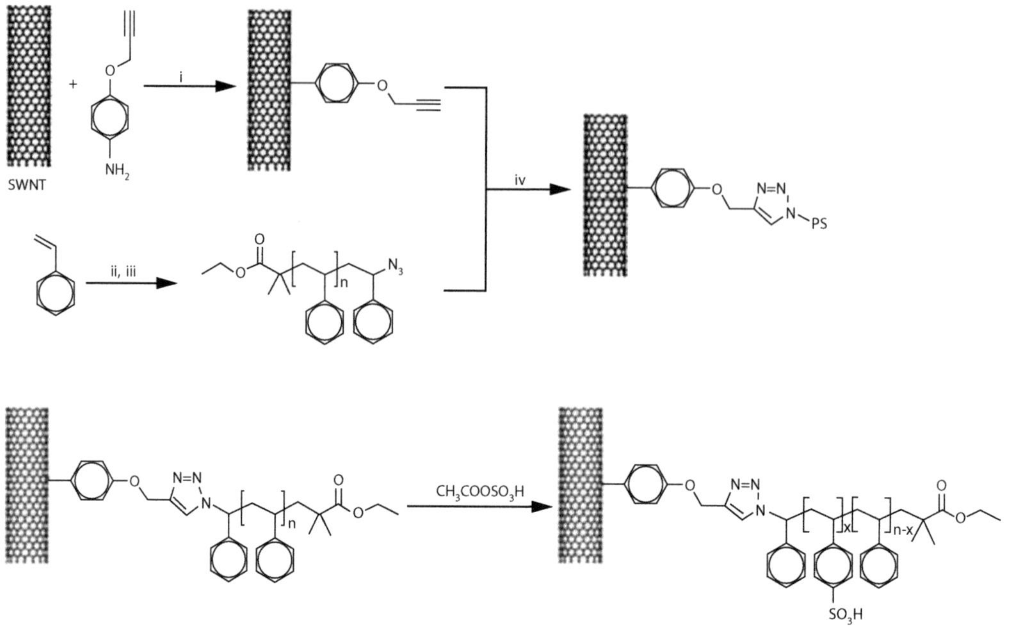

SCHEME 16.2
SWCNT functionalization with PS. (Adapted with permission from Li, H., Cheng, F., Duft, A.M.,
Adronov, A., Functionalization of single-walled carbon nanotubes with well-defined polystyrene
by "click" coupling, *J. Am. Chem. Soc.*, 127, 14518–14524. Copyright 2005 American Chemical
Society; Reprinted from *Carbon*, 45, Li, H. and Adronov, A., Water-soluble SWCNTs from sulfonation
of nanotube-bound polystyrene, 984–990, Copyright 2007, with permission from Elsevier.)

a series of PSs with different molecular weights were synthesized by ATRP. The bromo-terminated
PSs were transformed to azide-terminated PS with NaN_3 in dimethylformamide (DMF). Alkyne-
functionalized SWCNTs and azide-modified PS were coupled by the Cu(I)-catalyzed formation of
1,2,3-triazoles as shown in Scheme 16.2. The reaction very efficiently produced polymer–nanotube
conjugates even under mild coupling conditions (low reaction temperatures and short reaction times).
The grafting density determined by thermogravimetric analysis (TGA) was one polymer chain for
every 200–700 nanotube carbons, depending on molecular weight, leading to a material with a composition of approximately 45% of polymer. These materials showed high solubility in organic solvents
such as THF, $CHCl_3$ and CH_2Cl_2, and debundled polymer-coated SWCNTs were clearly observed by
transmission electron microscopy (TEM) and atomic force microscopy. Some years later, the same
authors employed the same reaction to prepare water-soluble SWCNT-PS hybrids via sulfonation of
the grafted PS chains (see Scheme 16.2) [78]. By controlling the degree of sulfonation from 20 to 34
mol.% and varying the concentration of acetyl sulfate, they prepared SWCNT derivatives with different degrees of solubility in water. Moreover, the sulfonated PS-functionalized SWCNTs exhibited pH
responsivity, being highly soluble between pH 3 and 13 and completely insoluble outside this range.

Yadav et al. [79] also used the CuAAC click reaction to functionalize MWCNTs with
poly(styrene-*b*-(ethylene-*co*-butylene)-*b*-styrene) (SEBS) triblock copolymer. However, in this case
the azide groups were not located at the polymer ends but were introduced chemically on the styrene
units of the copolymer. Coupling of the azide moiety of the block copolymer (SEBS-CH_2N_3) with
alkyne-functionalized MWCNTs was carried out via Cu(I)-catalyzed click chemistry, and different
compositions of MWCNT derivatives were prepared by varying the alkyne–MWCNT ratio. These
materials showed a very important enhancement in mechanical properties, dielectric constant, electrical conductivity, and thermal stability, as shown in Figure 16.3. The covalent attachment of SEBS
to MWCNTs resulted in excellent dispersion of the CNTs in the polymer matrix and accordingly in
significant improvements in their properties.

Other polymers such as polyurethane (PU) [80] or poly(ε-caprolactone) (PCL) [81] have also
been used to prepare nanocomposites with grafted CNTs via click chemistry. Cho et al. synthesized PU-grafted SWCNTs by coupling alkyne-modified nanotubes with azide-containing PU using
CuAAC [80]. Previously, several functionalized PU with different azide contents were prepared

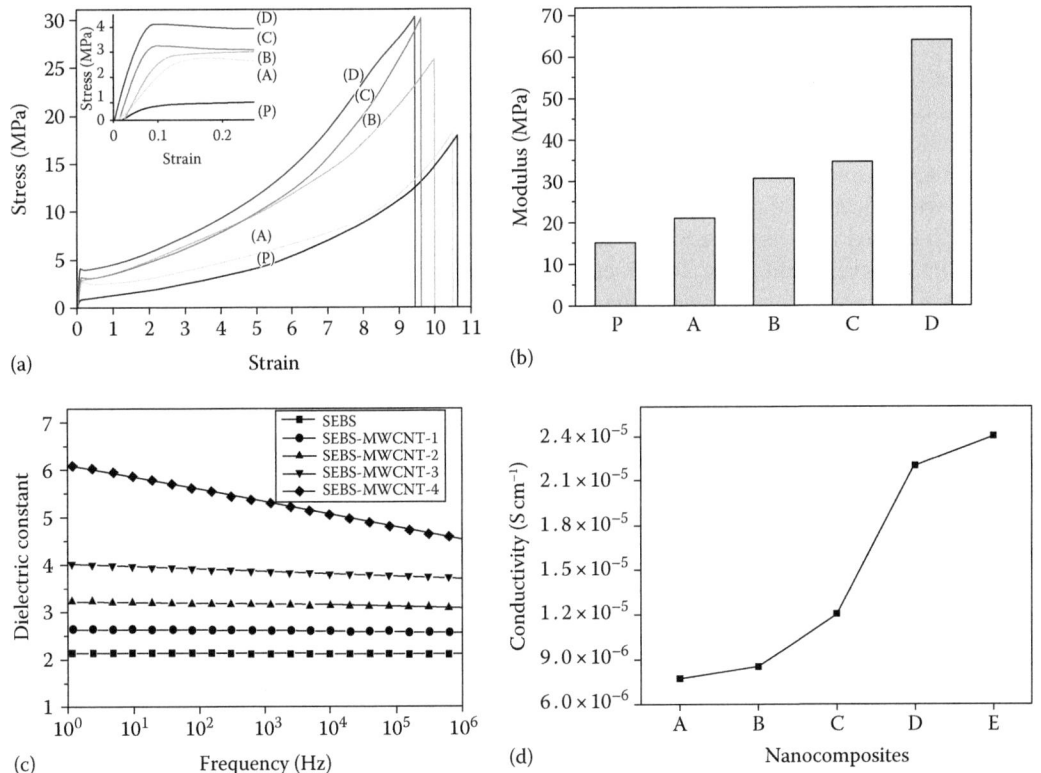

FIGURE 16.3

(a) Stress–strain curves, (b) modulus, (c) dielectric constants, and (d) electrical conductivity of SEBS (p), SEBS-MWCNT-1 (A), SEBS-MWCNT-2 (B), SEBS-MWCNT-3 (C), and SEBS-MWCNT-4 (D). (Reprinted with permission from Yadav, S.K., Mahapatra, S.S., Cho, J.W., and Lee, J.L., Functionalization of multiwalled carbon nanotubes with poly(styrene-b-(ethylene-co-butylene)-b-styrene) by click coupling, *J. Phys. Chem. C*, 114, 11395–11400. Copyright 2010 American Chemical Society.)

from the reaction of poly(ε-caprolactone)diol with azide moieties, 4,4′-methylenebis(phenylisocya nate) and 1,4-butanediol. The content of grafted polymer in the PU-*g*-SWCNTs increased from 26 to 71% with increasing azide functionality, as determined by TGA. The nanocomposites showed higher thermal stability than PU due to uniform dispersion of SWCNTs and enhanced thermal conductivity and good solubility in organic solvents. The same group also used this strategy to functionalize PCL with MWNTs [81]. For this purpose, a chlorine moiety containing PCL was synthesized by the copolymerization of α-chloro-ε-caprolactone with the ε-caprolactone monomer using ring-opening polymerization, and subsequently converted to azide, as in the earlier PU work.

Lee et al. used a different approach to graft PCL to MWCNTs via CuAAC [82]. Instead of a *grafting-to* strategy as in all the examples described up to now, they used a *grafting-from* approach. MWCNTs were modified with hydroxyl groups and used as coinitiators to polymerize PCL or poly(α-chloro-ε-caprolactone) (PαClCL) by surface-initiated ring-opening polymerization. The chlorides were converted into azides by the reaction with sodium azides, and several types of terminal alkynes were reacted with the azides by CuAAC. Subsequent characterization by several methods clearly demonstrated the success of the grafting process. All these results show that click chemistry is a very valuable tool for the efficient grafting of a fully biocompatible and biodegradable polymer such as a PCL onto CNTs, of great interest for potential biomedical applications especially in tissue engineering and neural growth and regeneration [83,84].

On the other hand, Gao et al. [85] demonstrated the versatility of the use of ATRP together with CuAAC click chemistry for the surface modification of CNTs with amphiphilic (Janus) polymer brushes. They developed a novel strategy based on a clickable macroinitiator and combined conventional *grafting-to* and *grafting-from* techniques to grow diverse multifunctional polymer brushes. First, a clickable macroinitiator possessing azide groups for the click reaction and bromo groups for initiating ATRP was synthesized. This macroinitiator was clicked onto alkyne-containing

CNTs resulting in a CNT-based clickable macroinitiator with only a small part of the azide groups consumed. The bromo and residual azide groups were used to initiate ATRP (*grafting-from*) and to click alkyne-terminated polymers (*grafting-to*), respectively. As an example, poly(3-azido-2-(2-bromo-2-methylpropanoyloxy)propylmethacrylate) (polyBrAzPMA) was used as macroinitiator. Both alkyne-functionalized MWCNTs (MWCNTs-Alk) and alkyne-functionalized SWCNTs (SWCNTs-Alk) were used to prepare the CNT-based macroinitiator. The MWCNTs-alk was prepared by esterification of propargyl alcohol with the carboxylic groups of MWCNT-COOH activated with thionyl chloride. The SWCNTs-alk were prepared via the synthesis of SWCNT-OH by the 1,3-dipolar cycloaddition of 2-azidoethanol on the sidewall and the subsequent esterification with a carboxylic acid containing a propargyl group. In the MWCNT-based macroinitiator, hydrophobic poly(*n*-butyl methacrylate) (P*n*BMA) brushes were first grown by in situ ATRP followed by grafting hydrophilic monoalkyne-terminated poly(ethylene glycol) (PEG-Alk) brushes via click chemistry. In the SWCNT-based macroinitiator, PEG-Alk was first grafted via click chemistry and then PS was introduced by in situ ATRP. The structures and morphologies of the functionalized CNTs were exhaustively characterized and confirmed by X-ray photoelectron spectroscopy (XPS), fourier transform infrared (FTIR), Raman spectroscopy, scanning electron microscopy (SEM), TEM, and TGA. The results showed that this strategy could be used as an efficient route to tailor the surface functionalities of CNs, the design of the macroinitiator being the key factor. A year later, the same group combined the layer-by-layer (LbL) technique with click chemistry to functionalize MWCNTs [86]. Two clickable polymers, poly(2-azidoethyl methacrylate) (poly-AzEMA) and poly(propargyl methacrylate) (polyPgMA), were synthesized by ATRP and RAFT, respectively. Then polyAzEMA was coated on alkyne-modified MWCNTs using the CuAAC click reaction as the first polymeric layer. Subsequently, polyPgMA and polyAzEMA were coated via click chemistry as the second and third polymeric layers. SEM, TEM, and TGA demonstrated that in the core-shell structure the layers of polymer were uniformly dispersed and that the thickness and amount of grafted polymer on the MWCNTs was controlled by the LbL cycles. XPS and FTIR showed the presence of residual azide groups on the surface of MWCNTs that could be further functionalized via click chemistry. As a proof of concept, the authors carried out the MWCNT reaction via click coupling with alkyne-modified rhodamine B to develop fluorescent CNTs or monoalkyne-terminated PS to prepare CNT-based PS brushes. It is important to note that the LbL click chemistry approach is a very valuable and versatile tool to functionalize CNTs that provide reactive nanoplatforms that can be used to build more complex and functional architectures.

Liu et al. [87] reported a novel strategy based on the formation of reactive micelles for the covalent functionalization of alkyne-modified MWCNTs with an azide-derivatized thermo-responsive diblock copolymer, poly (*N,N*-dimethylacrylamide) (PDMA)–PNIPAM, via click chemistry. The alkyne-functionalized MWCNTs were prepared from MWCNT-OH with toluene diisocyanate to produce isocyanate-functionalized MWCNTs and then with propargyl alcohol. Different molecular weight azide-terminated copolymers were synthesized by RAFT. Taking into account that these copolymers can form reversible micelles in aqueous solution as a function of temperature, which locate all the active groups in the exterior, this characteristic was used to favor the introduction of high molecular weight copolymers onto the surface of the MWCNTs with a high degree of functionality. In this regard, the CuAAC reaction was carried out in water with PDMA–PNIPAM micelles with the azide groups on the outer shell. TGA results showed an enhancement of grafting efficiency for the copolymers with high molecular weight with around 45% polymer content on the functionalized CNTs. Very recently, MWCNTs have been covalently functionalized with PNIPAM via the CuAAC click reaction but only a low grafting density was obtained [88]. The functionalized CNTs had good solubility in water but no temperature-response behavior was observed.

These click coupling strategies have also been used to develop novel CNT-supported nanoparticle heterostructures that are of great interest in areas such as catalysis, energy conversion, electronic nanodevices, and sensors. Chergui et al. prepared poly(glycidyl methacrylate) (PGMA) via ATRP and its epoxy groups were further acid-hydrolyzed (POH) prior to end functionalization by sodium azide (POH-N$_3$) [89]. The alkyne-modified CNTs were prepared with diazonium salts. The azide-functionalized polymer was clicked to the modified CNTs generating CNT-POH hybrids that were subsequently oxidized to carboxylated polymer-modified CNTs (CNT-PCOOH) (see Scheme 16.3). These CNT derivatives were used as efficient platforms for the in situ synthesis

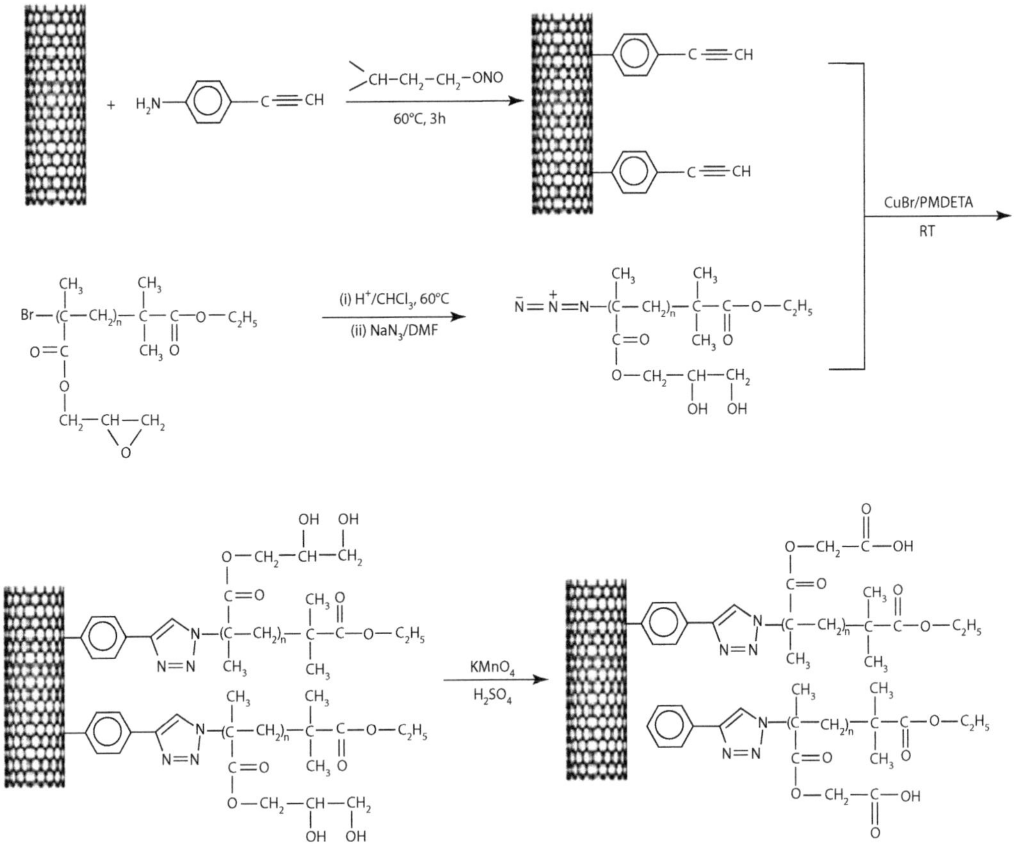

SCHEME 16.3
Synthetic route for the in situ grafting of alkylated aryl groups and click reaction of the ATRP-prepared PGMA on CNTs. (Reprinted with permission from Chergui, S.M., Ledebt, A., Mammeri, F. et al., Hairy carbon nanotube@nano-Pd heterostructures: Design, characterization and application in Suzuki C-C coupling reactions, *Langmuir*, 26, 16115–16121. Copyright 2010 American Chemical Society.)

and massive loading of palladium nanoparticles. The characterization of the resulting heterostructures and their precursors showed important evidence of surface attachment of Pd nanoparticles (Figure 16.4), and these heterostructures were evaluated as catalysts for the well-known Suzuki C–C coupling reaction between bromobenzene and phenylboronic acid, showing good stability and an efficient catalytic effect.

Another example of hybrid materials based on CNTs and metal nanoparticles prepared via click chemistry was reported by Gao et al. [90] who developed a methodology to synthesize magnetic nanohybrids from magnetic nanoparticles (Fe_3O_4) and polymer-coated MWCNTs. They initially prepared Fe_3O_4 nanoparticles of controlled size that they subsequently functionalized with azide and alkyne moieties. Alkyne MWCNTS were modified with polymers containing abundant azide and alkyne groups, and the functionalized Fe_3O_4 nanoparticles were coupled with polymer-coated nanotubes to give the magnetic nanohybrids. It is important to remark that the soft polymer interlayer was crucial for the surface click reaction between nanoparticles because attempts to prepare the nanohybrids directly from alkyne MWCNTs and azide-modified Fe_3O_4 were unsuccessful. The same strategy can be extended to other nanomaterials and, in fact, these authors also prepared magnetic nanohybrids replacing CNTs with silica spheres [90].

Recently, Campidelli et al. used the *grafting-from* approach based on polymerization of monomers on SWNTs via click chemistry to develop hyperbranched structures [91]. They synthesized two porphyrin AB and AB3 monomers containing either one azide (A) and one triple bond (B) (porphyrin 1) or one azide (A) and three triple bonds (B) (porphyrin 2). In the presence of SWCNTs modified with phenylacetylene groups, these monomers were able to react by CuAAC leading to linear or hyperbranched polymers on the nanotube surfaces. These new photoactive nanotube–porphyrin polymer

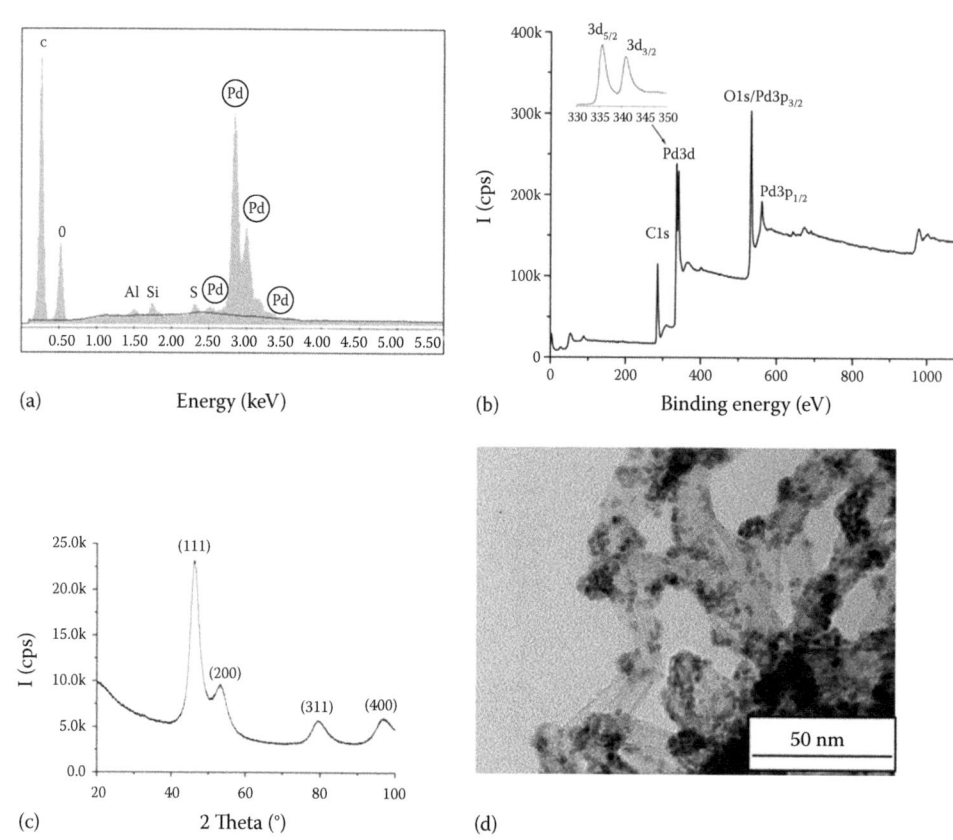

FIGURE 16.4

Analysis of CNT-PCOOH-Pd nanoparticle heterostructures by (a) Energy dispersive X-ray analyses (EDX) and (b) XPS. The inset in (b) shows the high-resolution Pd 3d doublet (c) XRD pattern and (d) TEM image. (Reprinted with permission from Chergui, S.M., Ledebt, A., Mammeri, F. et al., Hairy carbon nanotube@nano-Pd heterostructures: Design, characterization and application in Suzuki C-C coupling reactions, *Langmuir*, 26, 16115–16121. Copyright 2010 American Chemical Society.)

hybrids present good optical and electrochemical properties. They also demonstrated that the number of chromophores on the nanotubes significantly increased when hyperbranched structures were built on the nanotubes.

Finally, Temel et al. used thiol–ene click chemistry to modify MWCNTs with PS via the thermal initiation method [92]. Thiol end-functionalized linear PS (PS-SH) was prepared by the nucleophilic substitution reaction of bromide end groups of PS obtained by ATRP, while the MWCNTs acted as the *ene* component. A thermally induced thiol–ene reaction was applied at 80°C using AIBN to graft PS onto the MWCNTs. This strategy can be used with any polymer with thiol-end groups to prepare CNT nanocomposites.

In summary, click chemistry presents an efficient route to functionalize CNTs with polymers and to develop nanocomposites with enhanced dispersion and properties. The combination of CRP with click chemistry opens new perspectives to tailor complex hybrid materials based on polymer/CNTs with innovative and functional architectures for different areas of application from energy, catalysis, or sensing platforms to diverse electronic nanodevices, among others (Table 16.1).

16.5.2 Click Functionalization of Graphene with Polymers

Based on the combination of reliability and experimental simplicity for nonspecialists of click reactions and the variety of chemical structures of polymers with their subsequent compendium of properties, click chemistry approaches for graphene to develop new graphene–polymer conjugates with enhanced properties mediated by the chemical and physical characteristics of the grafted polymer are of some considerable interest.

TABLE 16.1   Summary List of the Different Experimental Conditions Used to Click Polymers and CNTs

CNT Type	Clickable Group on CNT	Method to Provide CNT with Clickable Entities	Clicked Polymer (Synthesis)/Approach	Click Reaction Conditions	Remarks on the Final Materials	References
Cul-catalyzed azide–alkyne cycloaddition						
SWCNT	–C≡C–	*p*-aminophenyl propargyl ether, isoamyl nitrite, 60°C	PS (ATRP)/*grafting-to*	(PPh3)3 CuBr/DMF, 60°C–100°C Cul/DBU,DMF, 20–90°C	Good solubility in THF, CH₂Cl₂, CHCl₃,	[76]
SWCNT	–C≡C–	*p*-aminophenyl propargyl ether, isoamyl nitrite, 60°C	S-PS (ATRP)/*grafting-to*	(PPh3)3 CuBr/DMF, 60°C–100°C Cul/DBU, DMF, 20–90°C	Good solubility in H₂O pH = 3–13	[78]
MWCNT	–C≡C–	*p*-aminophenyl propargyl ether, isoamyl nitrite, 60°C	SEBS/*grafting-to*	Cul/DBU, DMF, 60°C	Excellent dispersion and enhancement of properties of nanocomposites	[79]
SWCNT	–C≡C–	*p*-aminophenyl propargyl ether, isoamyl nitrite, 60°C	PU (ring-opening polymerization)/*grafting-to*	Cul/DBU, DMF, 60°C	Good dispersion and enhancement of thermal stability of nanocomposites Good solubility in organic solvents	[80]
MWCNT	–C≡C–	*p*-aminophenyl propargyl ether, isoamyl nitrite, 60°C	PCL (ring-opening polymerization)/*grafting-to*	Cul/DBU,DMF, 60°C	Good solubility in THF, CHCl₃	[81]
MWCNT	–C≡C–		PCL (ring-opening polymerization)/*grafting-from*	Cul/phenylboronic acid (PBA), triethyl amine, 60°C	Good solubility in THF, CHCl₃	[82]
MWCNT	–C≡C–	Propargyl alcohol, SOCl₂, CHCl₃, TEA	polyBrAzPMABr macroinitiator/*grafting-to*	CuBr, PMDETA, r.t.	Backbone to synthesize amphiphilic polymer brushes on CNT	[85]
SWCNT	–C≡C–	(i) azidoethanol, NMP, N₂, 160°C (ii) 4-oxo-4-(prop-2-ynyloxy)butanoic acid, CH2Cl₂, SOCl₂, CHCl₃, TEA	polyBrAzPMABr macroinitiator/*grafting-to*	CuBr, PMDETA, r.t.	Backbone to synthesize amphiphilic polymer brushes on CNT	[85]

(Continued)

TABLE 16.1 (Continued)   Summary List of the Different Experimental Conditions Used to Click Polymers and CNTs

CNT Type	Clickable Group on CNT	Method to Provide CNT with Clickable Entities	Clicked Polymer (Synthesis)/Approach	Click Reaction Conditions	Remarks on the Final Materials	References
MWCNT	–C≡C–	Propargyl alcohol, SOCl$_2$, CHCl$_3$, TEA	polyAzEMA (ATRP)/graft to polyPgMA (RAFT)/grafting-to	CuBr, PMDETA, r.t.	LbL reactive nanoplatforms	[86]
MWCNT	–C≡C–	(i) Toluene diisocyanate, N$_2$, 80°C (ii) propargyl alcohol, toluene	PDMA–PNIPAM (RAFT) Reactive micelles/grafting-to	Cu(II) sulfate pentahydrate, H$_2$O sodium ascorbate, 55°C	High degree of aqueous solubility	[87]
MWCNT	–C≡C–	Propargyl alcohol, SOCl$_2$, CHCl$_3$, TEA	PNIPAM (RAFT)/grafting-to	CuBr, PMDETA, 70°C	Good solubility in water and no temperature response	[88]
MWCNT	–C≡C–	Ethynylaniline, isoamyl nitrile, 60°C	PGMA (ATRP)/grafting-to	CuBr, DMF, PMDETA, r.t.	Supported nano-Pd heterostructures	[89]
MWCNT	–C≡C–	Propargyl alcohol, SOCl$_2$, CHCl$_3$, TEA	polyAzEMA (ATRP)/grafting-to	CuBr, DMF, PMDETA, r.t.	Magnetic nanohybrids	[90]
SWCNT	–C≡C–		Porphyrin polymer/grafting-from	Cu(MeCN)$_4$PF$_6$, NMP, 2,6-lutidine, THPTA, r.t.	Hyperbranched structures	[91]
Thiol–radical click reactions						
MWCNT	–C≡C–	—	PS-SH (ATRP)/grafting-to	AIBN, DMF, 80°C	Good dispersion in THF	[92]

The methods employed to click graphene and polymers can be divided in two main families that employ either *grafting-from* or *grafting-to* approaches. In parallel with these strategies, for CNTs *grafting-from* is based on the growth of polymer chains from the surface of graphene, which can be considered a macroinitiator. Here the initiator-modified graphene is a 2D platform from which polymer chains are grown, and the method is principally related to the click chemistry of graphene with discrete organic molecules rather than chemical reactions of graphene with polymeric matrices. Therefore, the background on the click chemistry of graphene with simple molecules [93–96] is fundamental to the design of the synthetic protocols for anchoring the appropriate initiators to the 2D sheets. The design of these experiments has the advantage that many strategies have been previously employed with other carbon materials, so the experience gained on the click chemistry of CNTs, for example, paves the way for the synthetic design of graphene macroinitiators. The development of polymer-grafted CNTs through surface-initiated polymerization, via acid-defect group chemistry and sidewall modification of CNTs [1,4,7,97], can serve as a guide to conduct similar experiments on graphene. In addition, GO is significantly more versatile than both graphene and CNTs since a wide range of chemical reactions can be undertaken on its surface due to the abundance of oxygenated species present, each with their specific chemical reactivity [98]. This is an additional advantage, as well as an important potential tool for chemists for the exploration of new routes for the immobilization of initiator molecules, an area yet to be studied in any detail.

Regarding the real possibilities of this approach, from a chemical standpoint, it is widely viable because the reaction of graphene laminates with simple molecules is easier and can be better controlled than in the case of modification with polymeric macromolecules. In addition, from the point of view of the polymerization method, despite the huge dimensions of graphene or GO laminates, in principle no steric hindrance is expected with *grafting-from* since graphene is the 2D platform and the polymer chains grow from it. In other words, as the polymer grows from the graphene, and hence away from the surface, the steric effect is minimized, which is a very different situation to that occurring when graphene links to the main chain. Therefore, the success of using graphene as 2D platforms by growing polymer brushes is only governed by the advantages and drawbacks of the polymerization process itself in terms of the type of monomer, initiator, catalyst, solvent, additives (if used), and experimental conditions as pressure and temperature.

The *grafting-from* method depends fundamentally on the feasibility of immobilizing specific initiators on the graphitic sheets. However, sometimes this may not be possible, and the covalent linkage between the presynthesized polymer and graphene is the only alternative, which on the other hand can expand the type of polymers that can be bound to graphene. *Grafting-to* methods allow the incorporation of graphene either at the end (terminal) or in the middle of the macromolecular chain. In the latter case graphene can be regarded as forming part of the main chain, and as such, a greater influence on the final properties of the material is expected than in the case of *grafting-from*, where the graphene can only be terminal. Further, in the *grafting-from* method, the amount of graphene in the final product is usually low and depends on both the molecular weight and polydispersity of the polymer. Whereas, in *grafting-to*, the degree of modification does not strongly depend on these factors, since the linkage sites are distributed along the polymeric chain. However, due to the huge size of the graphene sheets, steric factors may determine the degree of modification in this case.

While the number of initiators that can be linked to the graphitic surface in *grafting-from* is limited to a few, the wide chemistry covered by diverse families of polymers makes possible the development of many synthetic strategies to link it with graphene in the case of *grafting-to*. This section is subdivided to consider the specific chemistries addressed to connect the polymers with graphene.

16.5.2.1 Clicking graphene and polymers by CuAAC

Since graphene is totally composed of sp^2 hybridized carbon atoms, to take part in this kind of click reaction, it must to be provided with either alkyne or azide moieties. These groups can be introduced on pristine graphite, as well as on its chemically richer cousin GO produced by the strong oxidation of graphite, a harsh method that produces a highly defective material since disruption to the extended 2D π-conjugation of the graphene sheets occurs, leading to nanoscale graphitic sp^2 islands surrounded by a disordered, highly oxidized sp^3 sea, as well as other defects such as carbon vacancies [99]. While this treatment provokes insulating behavior of the resulting GO sheets, conductivity may be partially restored by convenient thermal or chemical reduction treatment, producing chemically modified graphene sheets.

In general, the approaches to provide graphene with clickable groups are limited to the few used for CNTs. In fact, alkyne groups have been incorporated to the 2D laminates by using well-known procedures: (1) the esterification or amidation of carboxylic groups at the edges of GO with certain alcohols or amines under typical catalytic conditions [100–102] and (2) the coupling of reactive aryl radicals generated from arene-diazonium salts [94,96,103,104]. In the case of the azide, the most employed procedures are the typical nucleophilic attack with sodium azide to the epoxide groups in GO producing the ring opening [105], acylation [106], or amidation [101].

Regarding the specific reaction conditions, the CuAAC reaction with graphene derivatives has been addressed using the same protocols already developed for other CNs thus allowing the incorporation of different moieties to graphene sheets [73,107,108]. PS was one of the first polymers used to prepare graphene-based polymer nanocomposites [109] and also the first to be clicked to graphene by CuAAC [102]. Here, almost monodisperse PS (polydispersity index, PDI = 1.04) bearing a terminal azide group was prepared by ATRP, with controlled molecular weight (in this case 4600 g/mol). Also, GO was provided with acetylenic moieties by acylation with propargyl alcohol. Finally both materials GO and PS were reacted using CuBr and N,N,N',N'',N''–pentamethyldiethylenetriamine (PMDETA) as the catalyst in DMF, giving products with good dispersibility and full exfoliation in common organic solvents. It is very important to note that this study combines click chemistry and a type of CRP that constitutes a powerful tool for preparing well-defined nanomaterials with desired functionalities [110]. Here ATRP has two important advantages: firstly, the terminal alkyl halide can be easily converted to azide moiety for click reactions and, secondly, the amount of initiator in the process determines the final molecular weight of the polymer at full monomer conversion allowing control of the chain length.

In another example including PS, the roles were inverted and the PS derivative carrying alkyne functionalities was clicked to azide-modified GO [106] prepared by the acylation of the hydroxyl groups on GO with 2-bromoisobutyryl bromide followed by nucleophilic substitution of the pendant bromine with NaN$_3$. Acetylene-terminated PS was prepared by ATRP of styrene using propargyl 2-bromoisobutyrate as initiator, and the click reaction was conducted under the same conditions as before generating products with similar solubility properties.

PS and other polymers like poly(methylmethacrylate) (PMMA), poly(methacrylic acid) (PMAA), poly(4-vynilpyridine) (P4VP), and poly((dimethylamino)ethylmethacrylate) (PDMA) were grown from, or attached to, graphene by combining RAFT polymerization and click chemistry [111]. In this excellent study, a comparison between grafting approaches (Scheme 16.4) was made and their influence on parameters like grafting density, polymer molecular weight, or polymer-modified graphene solubility was established. Here, both *grafting-from* and *grafting-to* methods were considered separately. Regarding the former approach, PMMA was used as a model to determine the optimum synthetic conditions (time, catalytic system, reagent ratios, etc.) as shown in Table 16.2. For both methods, the preferred click conditions employed CuI and 1,8-diazabicyclo[5.4.0] undec-7-ene (DBU). The effect of chain length was also studied and the authors concluded that the higher the molecular weight, the lower the grafting density (Figure 16.5). This result was explained by a decrease in the polymer reactivity to the terminal azide due to increased hindrance as the polymer adopted a more random coil structure. In addition the decrease in grafting density resulted in a decrease in solubility. In other words, although long-chain polymer arms emerging from graphene were better for improving the solubility, the low degree of functionalization (low grafting density) achieved was detrimental to solubility, the latter having the greatest influence on this parameter. This issue could be resolved by using a *grafting-from* protocol, where the grafting density was tailored by controlling the concentration of the initiator linked to the graphene surface. Once the desired grafting density was achieved, the polymer molecular weight could be increased via longer polymerization times and/or varying the monomer/initiator ratio, resulting in improved solubility. The only problem with this procedure appeared to be the broadening of the polydispersity index (PDI).

Elastomeric polymers like SEBS triblock copolymers have also been grafted to GO employing the same synthetic protocols as in the earlier examples (CuBr/PMDETA) [100]. The SEBS copolymer carried the azide groups in the styrene segments while the acetylene moieties were located in the GO. The product was soluble in a variety of organic solvents that allowed their homogeneous

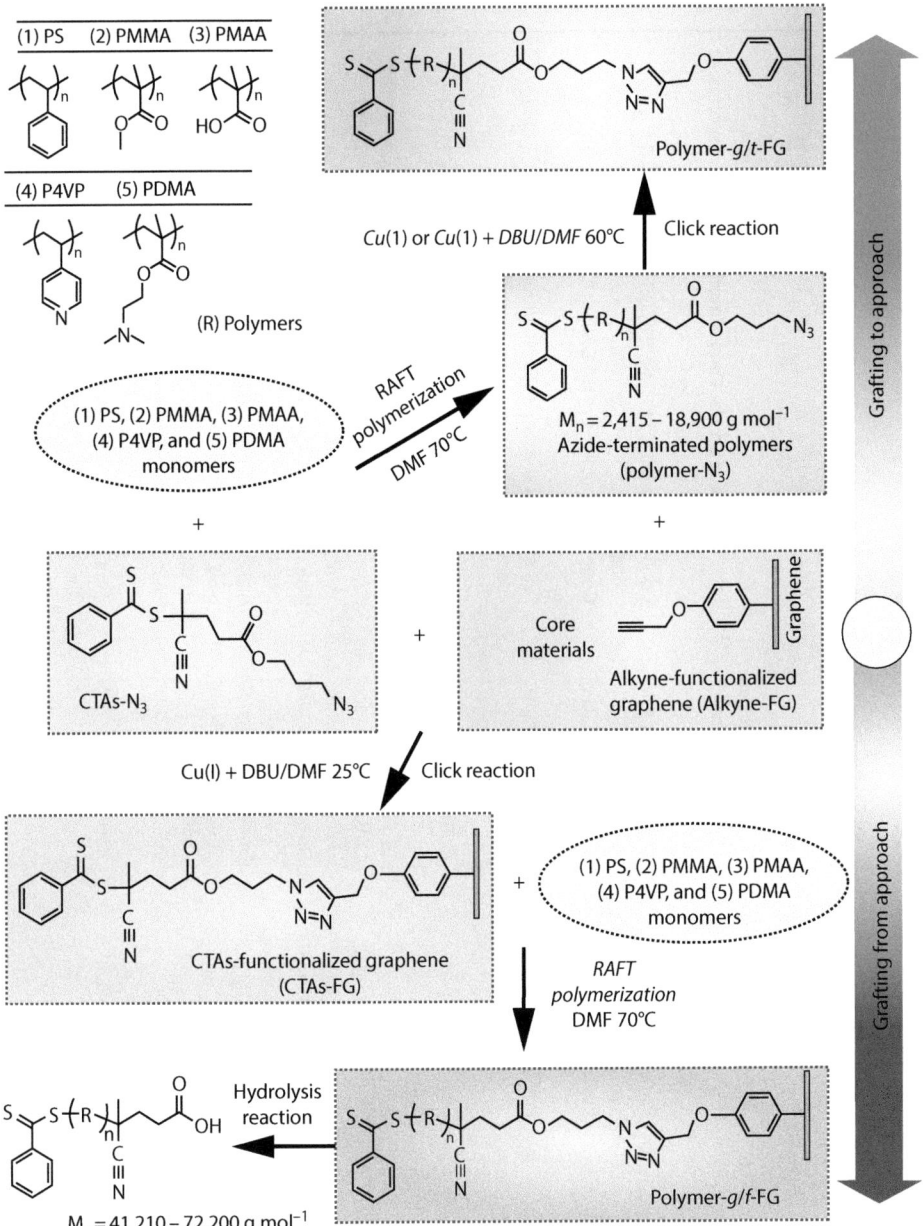

SCHEME 16.4
Synthetic procedures employed in the two types of grafting reactions addressed to click polymers and graphene. (Reprinted with permission from Ye, Y.S., Chen, Y.N., Wang, J.S. et al., Versatile grafting approaches to functionalizing individually dispersed graphene nanosheets using RAFT polymerization and click chemistry, *Chem. Mater.*, 24, 2987–2997. Copyright 2012 American Chemical Society.)

incorporation into polymer matrices. The authors investigated the compatibility and reinforcing effect of the clicked material in polymer composites, using PS as a host polymer, and observed remarkable improvements in the mechanical properties and thermal stability of the resulting composite films.

A thermoresponsive PNIPAM has also been clicked to modified GO [101]. In this study the azide-terminated PNIPAM was clicked with alkyne-modified GO obtained by an amidation reaction with propargylamine. The most outstanding finding of this work was the potential biomedical applications of the graphene–polymer conjugate that is an amphiphilic material composed of a hydrophobic graphene core and hydrophilic PNIPAM arms. The graphene–PNIPAM conjugate was used as an effective vehicle to internalize a water-insoluble anticancer drug,

TABLE 16.2 Summary List of the Different Experimental Conditions Used to Click Polymers and Graphene

Graphene Source	Clickable Group on Graphene	Method to Provide Graphene with Clickable Entities	Clicked Polymer/ Approach	Click Reaction Conditions	Remarks on the Final Materials	References
CuI-catalyzed azide-alkyne cycloaddition						
GO	$-C\equiv C-$	(i) SOCl$_2$, 70°C; (ii) propargyl alcohol, TEA, Cl$_3$CH, r.t.	PS/*grafting-to*	CuBr/PMDETA, DMF, r.t.	Good dispersibility in DMF, THF, CH$_2$Cl$_2$, toluene	[102]
GO	$-N_3$	(i) 2-bromoisobutyryl bromide, TEA, DMF, 0°C; (ii) NaN$_3$, r.t.	PS/*grafting-to*	CuBr/PMDETA, DMF, r.t.	Good dispersibility in THF, DMF CHCl$_3$	[106]
r-GO	$-C\equiv C-$	P-aminophenyl propargyl ether, isoamyl nitrite, 80°C	PMMA, PMAA, PS, P4VP; PDMA/*grafting-to* and *grafting-from*	CuI/DBU, DMF, r.t	Full study on the synthetic conditions and parameters	[111]
GO	$-C\equiv C-$	(i) SOCl$_2$, 70°C; (ii) propargyl alcohol, TEA, CHCl$_3$, r.t.	SEBS/*grafting-to*	CuBr/PMDETA, DMF, r.t	Excellent compatibility with PS. Improved mechanical properties and thermal stability of nanocomposites	[100]
GO	$-C\equiv C-$	Propargylamine, DCC/DMAP, DMF, r.t.	PNIPAM/*grafting-to*	CuBr/PMDETA/sodium L-ascorbate, DMF, r.t.	Potential biomedical: drug delivery and killing cancer cell	[101]
r-GO	$-C\equiv C-$	(i) Propargyl P-aminobenzoate, NO$_2$BF$_4$,CH$_3$CN, −30°C; (ii) r-GO, r.t.	PNIPAM/*grafting-from*	CuBr/PMDETA, DMF, r.t	Water solubility, lower critical solution temperature at 33.2°C	[112]
Pristine graphene	$-C\equiv C-$	Propargyloxybenzenediazonium Tetrafluoroborate, water, 45°C	PEG/*grafting-to*	Sodium bicarbonate/ CuSO$_4$/THPTA/sodium ascorbate	Higher reactivity on CVD-grown samples. Edges and defects are preferential sites of reactions.	[103]
Graphene	$-C\equiv C-$	4-ethynylaniline, isoamyl nitrite, 80°C	PE/*grafting-to*	CuI/DBU, DMF, 60°C	Filler for PE matrices	[113]
GO	$-C\equiv C-$	Propargylamine, DCC/DMAP, DMF	PCL/*grafting-to*	CuBr/PMDETA, DMF, r.t.	Enhancement of the mechanical properties, thermal stability, thermal conductivity, and thermoresponsive shape memory properties	[114]

(Continued)

TABLE 16.2 (Continued) Summary List of the Different Experimental Conditions Used to Click Polymers and Graphene

Graphene Source	Clickable Group on Graphene	Method to Provide Graphene with Clickable Entities	Clicked Polymer/ Approach	Click Reaction Conditions	Remarks on the Final Materials	References
r-GO	–C≡C–	Route I: (i) SOCl$_2$, o-DCB, (ii) propargylamine, DCC/DMAP, o-DCB, r.t. Route II: 4-ethynylaniline, isoamyl nitrite, 80°C	PFA/grafting-to	CuBr/PMDETA, DMF, r.t	Improved solubility. Solvent-dependent electronic properties	[104]
GO	–N$_3$	APTMS, DMF	P3HT/grafting-to	CuI/DIPEA, DMF	Red-shifted optical absorption maximum	[115]
GO	–C≡C–	(i) SOCl$_2$, DMF, 70°C; (ii) propargyl alcohol, TEA, 60°C	PMPCS/grafting-to	CuBr/PMDETA, DMF, 90°C	Improved mechanical properties	[118]
GO	–C≡C–	Ethynylaniline, DCC, DMF, 90°C, N$_2$, 2 h	PEDOT:PSS/grafting-to	CuSO$_4$, sodium ascorbate, r.t. 48 h	Good electrical conductivity	[117]
Thiol–radical click reactions						
r-GO	–C≡C–	4-ethynylaniline, isoamyl nitrite, 60°C	PE/grafting-to	AIBN, o-DCB, 70°C	Better thermal, mechanical, and electrical properties	[113,119]
r-GO	–C=C–, –C≡C–	—	PMMA/grafting-to	AIBN, sodium methoxide, Cl$_2$CH$_2$, 60°C, inert atmosphere	Better electrical conductivity	[122]

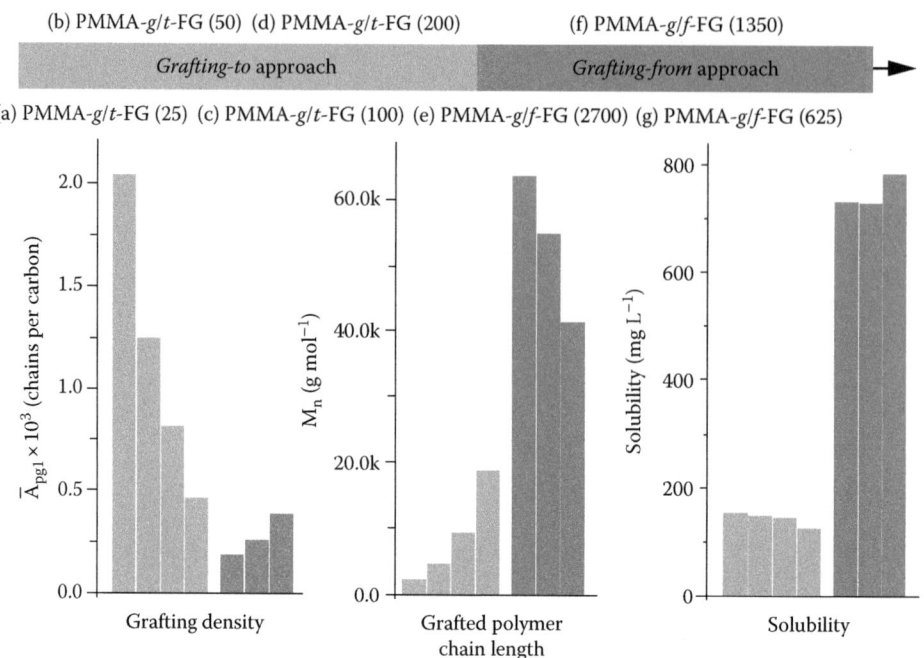

FIGURE 16.5

Characterization of PMMA/graphene materials prepared by (a) *grafting-to* and (b) *grafting-from* approaches. (Reproduced with permission from Ye, Y.S., Chen, Y.N., Wang, J.S. et al., Versatile grafting approaches to functionalizing individually dispersed graphene nanosheets using RAFT polymerization and click chemistry, *Chem. Mater.*, 24, 2987–2997. Copyright 2012 American Chemical Society.)

camptothecin (CPT), into the cells, where the hydrophobic graphene absorbs the aromatic water-insoluble drug, while the water-soluble PNIPAM transports it into the cells. Interestingly, the graphene–PNIPAM conjugate was able to load CPT with a superior loading capacity of 15.6 wt.%, with low cytotoxicity. Thus, the PNIPAM–graphene–CPT complex appears as an excellent system with a high potency of killing A-5RT3 cancer cells in vitro. Brushes of the same polymer, PNIPAM, have also been grafted from the surface of r-GO by combining the CuAAC reaction and RAFT polymerization [112]. The reduced graphene oxide (r-GO) was reacted with propargyl *p*-aminobenzoate by diazonium coupling. Then the acetylene-modified compound was clicked to the RAFT agent *S*-1-dodecyl-*S*′-(α,α′-dimethyl-α″-3-azido-1-propyl acetate)trithiocarbonate followed by the NIPAM RAFT polymerization with AIBN at 60°C.

Another water-soluble polymer poly(ethylene glycol) (PEG), which is one of the most employed polymers for biomedical applications for cell internalization, has been also clicked to graphene [103]. The click reaction was studied in bulk graphene dispersions as well as on CVD-grown graphene. The graphene was furnished with alkyne groups by diazotization in the presence of sodium dodecylsulfate (SDS) surfactant, and the CuAAC reaction with azide-terminated short-chain PEG was conducted in the presence of sodium bicarbonate, $CuSO_4$, tris(3-hydroxypropyltriazolylmethyl) amine (THPTA), and sodium ascorbate. It was demonstrated that the CVD graphene is more reactive than solution-dispersed graphene and that edges and defects preferentially react. In addition, this is one of the few cases where neither GO nor r-GO is employed as the graphene source, but highly ordered pyrolytic graphite (HOPG) and CVD-grown graphene were used.

Very recently the most abundant thermoplastic polymer, polyethylene (PE), was clicked to graphene using several click chemistry approaches. In the specific case of the CuAAC reaction, short-chain PE brushes were coupled to alkyne-modified graphene prepared using CuI/DBU catalyst in DMF at 60°C [113]. The PE brushes were also linked to the graphene by thiol–radical click reactions such as thiol–ene and thiol–yne, and a more detailed discussion will be given in the next section.

Another polymer that has been clicked to graphene is PCL [114]. Alkynyl-modified GO obtained by esterification of GO with propargylamine was clicked with PCL bearing azide side groups under typical synthetic conditions (Table 16.2). The click products displayed good interaction with a PU matrix, leading to remarkable improvements in the mechanical properties, thermal stability, thermal conductivity, and thermoresponsive shape memory properties in the resulting nanocomposites. For instance, the breaking stress, Young's modulus, elongation at break, and thermal stability of nanocomposites containing 2% of the click material showed improvements of 109%, 158%, 28%, and 71°C, respectively.

The CuAAC protocol has also been used to prepare hybrids of graphene and conjugated polymers [104,115–117]. In the first example, an azide-polyfluorene derivative was clicked to alkynyl-modified r-GO under the typical catalytic conditions [104]. The r-GO functionalization was addressed by using two protocols, that is, amidation and diazotization, the latter being the most efficient. The clicked products were soluble in organic solvents like DMF, NMP, and 1,2-dichlorobenzene(o-DCB). But striking differences were observed in the absorption and emission spectra of the graphene–polyfluorene conjugate depending on the solvent, due to different effects on the electronic behavior of polyfluorene in the presence of the electron-acceptor graphene. While the photoluminescence of polyfluorene was slightly affected in NMP and o-DCB, it was totally inhibited in DMF. These changes were not related to the polarity of the solvents (for instance, DMF and NMP have similar polarities) but rather to their affinity to graphene in terms of surface energy.

Poly(3-hexylthiophene) (P3HT) brushes have been attached to GO surfaces [115]. The GO was first furnished with azide groups by a silylation reaction with 3-azidopropyltrimethoxysilane (APTMS), and the click reaction with the ethynyl-terminated P3HT was carried out in the presence of CuI and N,N-diisopropylethylamine (DIPEA) in DMF. The clicked products were compared with materials prepared by the amidation reaction between GO and P3HT. The grafting efficiency of the click approach is much higher because of the greater quantity of hydroxyl and epoxide groups than the carboxylic acid groups on GO sheets. In addition the electronic properties changed with respect to P3HT due to some ordering/crowding of the P3HT grafted on GO. The most famous polythiophene, poly(3,4-ethylene-dioxythiophene)–polystyrene sulfonate (PEDOT:PSS), has also been clicked to r-GO [117]. Terminal alkyne-modified r-GO sheets were prepared by amidation of GO with 4-ethynylaniline followed by chemical reduction (Figure 16.6). The click reaction with azide-functionalized PEDOT:PSS (PEDOT-N$_3$:PSS) was carried out in aqueous solution of CuSO$_4$ and sodium ascorbate using the PEDOT-N$_3$:PSS aqueous dispersion as the solvent. It was observed that clicked composites were smoother than those of unclicked composites suggesting better dispersion of graphene because of interfacial interaction enhancement, which also results in an increase in electrical conductivity (Figure 16.6).

In the last example of this type of click reaction, a thermotropic liquid crystalline polymer has been linked to graphene [118]. The azido-functionalized liquid crystalline polymer, poly(2,5-bis[(4-methoxyphenyl)oxycarbonyl]styrene) (PMPCS), prepared by ATRP was coupled to alkyne-modified GO using the common catalyst (CuBr/PMEDTA) at a higher than usual temperature of 90°C, using PMPCS with two different molecular weights. The lower molecular weight PMPCS remained amorphous after heating, while the higher molecular weight polymer manifested a thermotropic liquid crystalline transition on heating. In agreement with the results mentioned earlier [111], this study shows that the lower the molecular weight, the higher the grafting density. It was demonstrated by Raman and XPS measurements that the higher molecular weight sample with lower grafting density had stronger interactions with the GO sheets due to the increased aromaticity resulting in a higher density of π–π interactions and a more intimate contact between PMPCS and GO. However, when using the click product as a nanofiller for a neat PMPCS matrix, rheological measurements from solution-blended samples showed that higher grafting density improved reinforcement due to better compatibility with the PMPCS matrix.

16.5.2.2 Clicking graphene and polymers by thiol–radical addition

Despite the advantages of thiol–radical click reactions and their demonstrated utility in polymer chemistry [23], the interest of applications in CNTs and graphene has been only recently initiated. The first study on this topic involves the covalent modification of graphene with PE [119], perhaps one of the polymers that has demonstrated most difficulties when preparing graphene/PE nanocomposites. This study employed a short-chain PE linked to pristine graphene by a thiol–ene

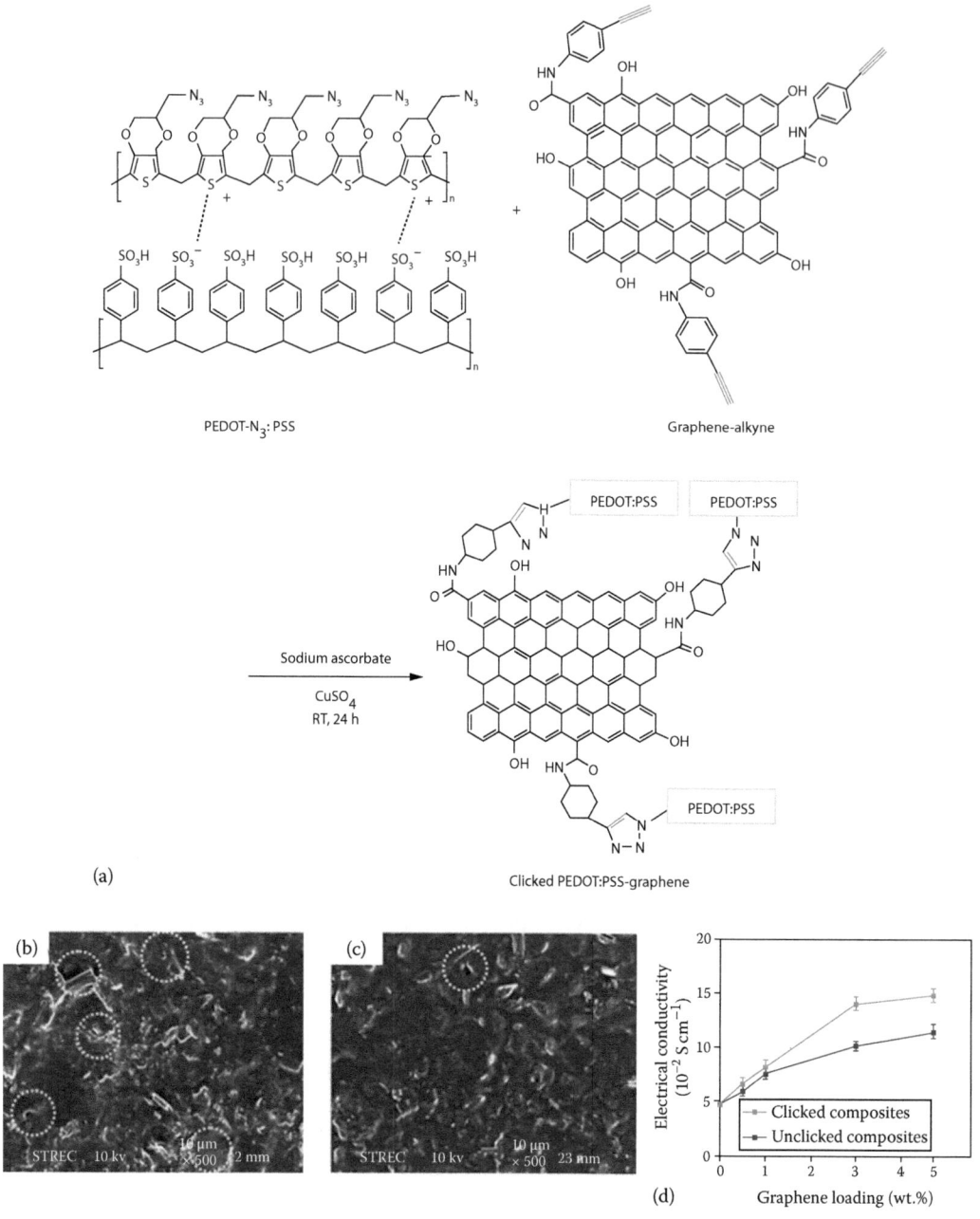

FIGURE 16.6

(a) CuAAC click reaction between PEDOT-N$_3$:PSS and graphene alkyne, SEM images of the surface of (b) unclicked PEDOT-N$_3$:PSS/graphene composite at 1 wt.% graphene loading and (c) clicked PEDOT:PSS/graphene composite at 1 wt.% graphene loading, and (d) comparison of the electrical conductivities of unclicked and clicked PEDOT:PSS/graphene composites. (Reprinted from *Composition of Science and Technology*, 93, Deetuam, C., Samthong, C., Thongyai, S., Praserthdam, P., and Somwangthanaroj, A., Synthesis of well dispersed graphene in conjugated poly(3,4-ethylenedioxythiophene): Polystyrene sulfonate via click chemistry, 1–8, Copyright 2014, with permission from Elsevier.)

reaction and to alkyne-modified graphene by thiol–yne and CuAAC click reactions. For thiol–ene and thiol–yne reactions, the SH-terminated PE was reacted with graphene and alkyne-modified graphene, respectively, in the presence of a thermal initiator AIBN at 70°C. The case of PE is one of the best examples to highlight the importance of the interface in graphene-based polymer nanocomposites. As PE has a somewhat low melting point, it offers the possibility for rapid sample preparation of nanocomposites by blending in the melt. However, melting mixing does not assure an efficient polymer/filler interface and the final properties of the nanocomposites are quite similar to those of the neat polymer. Therefore, other strategies oriented to tailor the interface properties are needed. The strategy of functionalizing graphene with short-chain polymers and their subsequent use, via *masterbatch*, as a filler for a higher molecular weight polymer can be extended to the case of PE. In the case of thiol–ene-modified graphene, it has been observed that the properties of high density polyethylene (HDPE) were improved using a two-step nanocomposite preparation method. The first step consisted of mixing short-chain PE-modified graphene with short-chain PE, while the second uses this mixture as a filler for HDPE. Under these conditions a gradient interface system is created (Figure 16.7), and relevant changes in the mechanical and thermal and, especially, electrical properties of HDPE were observed. The electrical conductivity showed typical percolation behavior with a percolation threshold of between 0.5 to 0.8 wt.% graphene and conductivity values in the order of 1 S cm^{-1} with less than 5 wt% of graphene (Figure 16.7), almost 3 orders of magnitude higher than that reported for graphene-PE nanocomposites employing diverse blending methodologies [120,121]. However, this result was not achieved when graphene was modified with PE brushes by using the other two click approaches, which appears to be reasonable since the graphene source is graphene itself and no new defects are generated by premodification steps. However, the mechanical properties varied in a different manner. For instance, the elongation at break of all clicked-reinforced samples was slightly higher than that of the neat polymer, the value of the material prepared by thiol–ene being the highest. In addition, Young's modulus of the samples using thiol–yne and CuAAC modified graphene was remarkably lower than that of the neat polymer, while that for the thiol–ene-modified graphene, similar values were achieved.

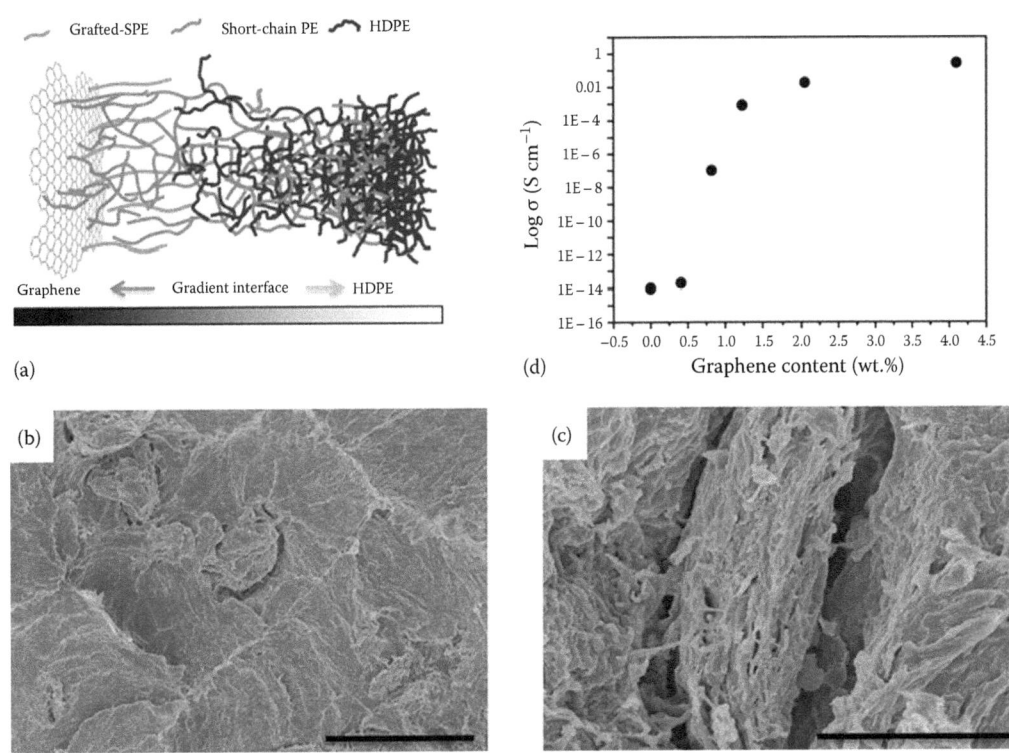

FIGURE 16.7
(a) Schematic describing the gradient interface effect in HDPE/graphene nanocomposites, SEM images of the nanocomposites at different magnification with scale bars corresponding to (b) 20 and (c) 5 μm, and (d) plot of the electric conductivity of the nanocomposites as a function of the graphene content. (b, c—Adapted with permission from Castelaín, M., Martínez, G., Ellis, G., Marco, C., and Salavagione, H.J., The effect of the chemical approaches to modify graphene on the electrical, thermal and mechanical properties of graphene/polyethylene nanocomposites, *Macromolecules*, 46, 8980–8987. Copyright 2013 American Chemical Society; a, d—Castelaín, M., Martínez, G., Ellis, G., and Salavagione, H.J., Versatile chemical tool for the preparation of conductive graphene-based polymer nanocomposites, *Chem. Commun.*, 49, 8967–8969, 2013. Reproduced by permission of The Royal Society of Chemistry.)

The click thiol–ene reaction has been extended to acrylic polymers and copolymers [122]. Contrary to the case for a PMMA homopolymer with a thiol-end group, in the copolymers, the thiol moieties were located along the polymer chain, allowing multiple bonding sites to the graphene surface. This comparative study showed that the copolymers formed a thin layer covering the graphene surface, contrary to the case for typical polymer brushes obtained with thiol-terminated PMMA. The modification of graphene with polymer brushes renders it easier to disperse than in the case of the copolymer. However, when thin layer–modified graphene is used as filler for pure PMMA, the PMMA/graphene interface is much stronger leading to nanocomposites with higher conductivity than in the case of brush-modified graphene.

16.6 CONCLUSIONS AND PERSPECTIVES

The aim of this chapter was to provide a general view on click chemistry strategies developed to incorporate CNT and graphene into polymer matrices. Materials science in the twenty-first century will depend on the development of tailored materials that can be targeted to specific, high-value applications. Consequently, the demand for highly efficient reaction chemistries that yield high-purity, readily accessible products is significant, and its importance is likely to increase in the future. Click chemistry and its extension to the combination of macromolecular chemistry and CNs is an emerging field, where a number of important features can provide powerful tools to the materials chemist, such as effective covalent strategies that allow incorporation of CNs to almost any polymer matrix. The preparation of advanced materials through click approaches is only limited by the imagination of scientists in designing innovative synthetic strategies to furnish the CNs or the polymers with specific click functionalities.

ACKNOWLEDGMENTS

The authors wish to thank the MINECO, Spain, for financial support under the grant MAT2013-47898-C2-2-R and MAT2010-21070-C02-01. HJS acknowledges the MINECO for a Ramón y Cajal Senior Research Fellowship.

REFERENCES

1. Hirsch, A. 2002. Functionalization of single-walled carbon nanotubes. *Angew. Chem. Int. Ed.* 41:1853–1859.
2. Dyke, C.A., Tour, J.M. 2004. Covalent functionalization of single-walled carbon nanotubes for materials applications. *J. Phys. Chem. A* 108:11151–11159.
3. Banerjee, S., Hemraj-Benny, T., Wong, S.S. 2005. Covalent surface chemistry of single walled carbon nanotubes. *Adv. Mater.* 17:17–29.
4. Tasis, D., Tagmatarchis, N., Bianco, A., Prato, M. 2006 Chemistry of carbon nanotubes. *Chem. Rev.* 106:1105–1136.
5. Singh, P., Campidelli, S., Giordani, S., Bonifazi, D., Bianco, A., Prato, M. 2009. Organic functionalisation and characterisation of single-walled carbon nanotubes. *Chem. Soc. Rev.* 38:2214–2230.
6. Peng, X., Wong, S.S. 2009. Functional covalent chemistry of carbon nanotube surfaces. *Adv. Mater.* 21:625–642.
7. Sahoo, N.G., Rana, S., Cho, J.W. et al. 2010. Polymer nanocomposites based on functionalized carbon nanotubes. *Prog. Polym. Sci.* 35:837–867.

8. Ma, P., Siddiqui, N.A., Marom, G. et al. 2010. Dispersion and functionalization of carbon nanotubes for polymer-based nanocomposites: A review. *Comp. Part A* 41:1345–1367.

9. Mattevi, C., Eda, G., Agnoli, S. et al. 2009. Evolution of electrical, chemical, and structural properties of transparent and conducting chemically derived graphene thin films. *Adv. Funct. Mater.* 19:2577–2583.

10. Acik, M., Mattei, C., Gong, C. et al. 2010. The role of intercalated water in multilayered graphene oxide. *ACS Nano* 4:5961–5968.

11. Jung, I., Field, D.A., Clark, N.J. et al. 2009. Reduction kinetics of graphene oxide determined by electrical transport measurements and temperature programmed desorption. *J. Phys. Chem. C* 113:18480–18486.

12. Kolb, H.C., Finn, M.G., Sharpless, K.B. 2001. Click chemistry: Diverse chemical function from a few good reactions. *Angew. Chem. Int. Ed.* 40:2004–2021.

13. Avti, P.K., Mayysinger, D., Kakkar, A. 2013. Alkyne-azide "Click" chemistry in designing nanocarriers for applications in biology. *Molecules* 18:9531–9549.

14. Iha, R.K., Wooley, K.L., Nystrom, A.M., Burke, D.J., Kade, M.J., Hawker, C.J. 2009. Applications of orthogonal "Click" chemistries in the synthesis of functional soft materials. *Chem. Rev.* 109:5620–5686.

15. Huisgen, R. 1963. 1,3-Dipolar cycloadditions. Past and future. *Angew. Chem. Int. Ed.* 2:565–598.

16. Tornøe, C.W., Christensen, C., Meldal, M. 2002. Peptidotriazoles on solid phase: [1,2,3]-triazoles by regiospecific copper (I)-catalyzed 1,3-dipolar cycloadditions of terminal alkynes to azides. *J. Org. Chem.* 67:3057–3064.

17. Rostovtsev, V.V., Green, L.G., Fokin, V.V., Sharpless, K.B. 2002. A stepwise Huisgen cycloaddition process: Copper (I)-catalyzed regioselective ligation of azides and terminal alkynes. *Angew. Chem. Int. Ed.* 41:2596–2599.

18. Rodionov, V.O., Fokin, V.V., Finn, M.G. 2005. Mechanism of the ligand-free CuI-catalyzed azide–alkyne cycloaddition reaction. *Angew. Chem. Int. Ed.* 44:2210–2215.

19. Himo, F., Lovell, T., Hilgraf, R., Rostovtsev, V.V., Noodleman, L., Sharpless, K.B., Fokin, V.V. 2005. Copper (I)-catalyzed synthesis of azoles. DFT study predicts unprecedented reactivity and intermediates. *J. Am. Chem. Soc.* 127:210–216.

20. Meldal, M., Tornøe, C.W. 2008. Cu-catalyzed azide-alkyne cycloaddition. *Chem. Rev.* 108:2952–3015.

21. Becer, C.R., Hoogenboom, R., Schubert, U.S. 2009. Click chemistry beyond metal-catalyzed cycloaddition. *Angew. Chem. Int. Ed.* 48:4900–4908.

22. Hoyle, C.H., Bowman, C.N. 2010. Thiol-ene click chemistry. *Angew. Chem. Int. Ed.* 49:1540–1573.

23. Hoyle, C.E., Lowe, A.B., Bowman, C.N. 2010. Thiol-click chemistry: A multifaceted toolbox for small molecule and polymer synthesis. *Chem. Soc. Rev.* 39:1355–1387.

24. Lowe, A.B. 2010. Thiol-ene "click" reactions and recent applications in polymer and materials synthesis. *Polym. Chem.* 1:17–36.

25. Hensarling, R.M., Doughty, V.A. Chan, J.W., Patton, D.L. 2009. "Clicking" polymer brushes with thiol-yne chemistry: Indoors and out. *J. Am. Chem. Soc.* 131:14673–14675.

26. Barner-Kowollik, C., Inglis, A.J. 2009. Has click chemistry lead to a paradigm shift in polymer material design? *Macromol. Chem. Phys.* 210:987–992.

27. Barner-Kowollik, C., Du Prez, F.E., Espeel, P. 2011. "Clicking" polymers or just efficient linking: What is the difference? *Angew. Chem. Int. Ed.* 50:60–62.

28. Matyjaszewski, K. 2005. Macromolecular engineering: From rational design through precise macromolecular synthesis and processing to targeted macroscopic material properties *Prog. Polym. Sci.* 30:858–875.

29. Kamigaito, M., Ando, T., Sawamoto, M. 2001. Metal-catalyzed living radical polymerization. *Chem. Rev.* 101:3689–3745.

30. Matyjaszewski, K., Xia, J. 2001. Atom transfer radical polymerization. *Chem. Rev.* 101:2921–2990.

31. Hawker, C.J., Bosman, A.W., Harth, E. 2001. New polymer synthesis by nitroxide mediated living radical polymerizations. *Chem. Rev.* 101:3661–3688.

32. Chiefari, J., Rizzardo, E. 2002. Control of free-radical polymerization by chain transfer methods. In *Handbook of Radical Polymerization*, eds. K. Matyjaszewski, and T.D. Davis, pp. 629–690. Hoboken, NJ: Wiley Interscience.

33. Coessens, V., Pintauer, T., Matyjaszewski, K. 2001. Functional polymers by atom transfer radical polymerization. *Prog. Polym. Sci.* 26:337–377.

34. Wu, P., Feldman, A.K., Nugent, A.K. et al. 2004. Efficiency and fidelity in a click-chemistry route to triazole dendrimers by the copper (I)-catalyzed ligation of azides and alkynes. *Angew. Chem. Int. Ed.* 43:3928–3932.

35. Opsteen J.A., van Hest, J.C.M. 2005. Modular synthesis of block copolymers via cycloaddition of terminal azide and alkyne functionalized polymers. *Chem. Commun.* 7:57–59.

36. Díaz, D.D., Punna, S., Holzer, P., McPherson, A.K., Sharpless, K.B., Fokin, V.V., Finn, M.G. 2004. Click chemistry in materials synthesis. 1. Adhesive polymers from copper-catalyzed azide-alkyne cycloaddition. *J. Polym. Sci. Polym. Chem.* 42:4392–4403.

37. Helms, B., Mynar, J.L., Hawkersumpter, C.J., Frechet, J.M.J. 2004. Dendronized linear polymers via "click chemistry". *J. Am. Chem. Soc.* 126:15020–15021.

38. Tsarevsky, N.V., Bernaerts, K.V., Dufour, B., Du Perez, F.E., Matyjaszewski, K. 2004. Well-defined (co)polymers with 5-vinyltetrazole units via combination of atom transfer radical (co)polymerization of acrylonitrile and "click chemistry"-type post-polymerization modification. *Macromolecules* 37:9308–9313.

39. Fournier, D., Hoogenboom, R., Schubert, U.S. 2007. Clicking polymers: A straight-forward approach to novel macromolecular architectures. *Chem. Soc. Rev.* 36:1369–1380.

40. Binder, W.H., Schsenhofer, R. 2008. 'Click' chemistry in polymer and material science: An update. *Macromol. Rapid Commun.* 29:952–981.

41. Sumerlin, B.S., Vogt, A.P. 2010. Macromolecular engineering through click chemistry and other efficient transformations. *Macromolecules* 43:1–13.

42. Matyjaszewski, K., Nakagawa, Y., Gaynor, S.G. 1997. Synthesis of well-defined azido and amino end functionalized polystyrene by atom transfer radical polymerization. *Macromol. Rapid Commun.* 18:1057–1066.

43. Lutz, J.F., Börner, H.G., Weichenhan, K. 2005. Combining atom transfer radical polymerization and click chemistry: A versatile method for the preparation of end-functional polymers. *Macromol. Rapid. Commun.* 26:514–518.

44. Mantovani, G., Ladmiral, V., Tao, L., Haddleton, D.M. 2005. One-pot tandem living radical polymerisation–Huisgens cycloaddition process ("click") catalysed by N-alkyl-2-pyridylmethanimine/Cu(I)Br complexes. *Chem. Commun.* 5:2089–2091.

45. Fleischmann, S., Hinrichs, K., Oertel, U., Reichelt, S., Eichhorn, K.J., Voit, B. 2008. Modification of polymer surfaces by click chemistry. *Macromol. Rapid Commun.* 29:1177–1185.

46. Gondi, S.R., Vogt, A.P., Sumerlin, B.S. 2007. Versatile pathway to functional telechelics via RAFT polymerization and click chemistry. *Macromolecules* 40:474–481.

47. Li, Y., Yang, J., Benicewicz, B.C. 2007. Well-controlled polymerization of 2-azidoethyl methacrylate at near room temperature and click functionalization. *J. Polym. Sci. Polym. Chem.* 45:4300–4308.

48. Tsarevsky, N.V., Bencherif, S.A., Matyjaszewski, K. 2007. Graft copolymers by a combination of ATRP and two different consecutive click reactions. *Macromolecules* 40:4439–4445.

49. Ladmiral, V., Mantovani, G., Clarkson, G.J., Cauet, S., Irwin, J.L., Haddleton, D.M. 2006. Synthesis of neoglycopolymers by a combination of "click chemistry" and living radical polymerization. *J. Am. Chem. Soc.* 128:4823–4830.

50. Koo, S.P.S., Stamenovic, M.M., Prasath, A. et al. 2010. Limitations of radical thiol-ene reactions for polymer–polymer conjugation. *J. Polym. Sci. A: Polym. Chem.* 48:1699–1713.

51. Chan, J.W., Yu, B., Hoyle, C.E., Lowe, A.B. 2008. Convergent synthesis of 3-arm star polymers from RAFT-prepared poly(*N,N*-diethylacrylamide) via a thiol–ene click reaction. *Chem. Commun.* 40:4959–4961.

52. Yu, B., Chan, J.W., Hoyle, C.E., Lowe, A.B. 2009. Sequential thiol-ene/thiol-ene and thiol-ene/thiol-yne reactions as a route to well-defined mono and bis end-functionalized poly(*N*-isopropylacrylamide). *J. Polym. Sci. Polym. Chem.* 47:3544–3557.

53. Niyogi, S., Hamon, M.A., Hu, H. et al. 2002. Chemistry of single-walled carbon nanotubes. *Acc. Chem. Res.* 35:1105–1113.

54. Chen, Z., Thiel, W., Hirsch, A. 2003. Reactivity of the convex and concave surfaces of single-walled carbon nanotubes (SWCNT) towards addition reactions: Dependence on the carbon-atom pyramidalization. *Chem. Phys. Chem.* 4:93–97.

55. Yan, L., Zheng, Y.B., Zhao, F. et al. 2012. Chemistry and physics of a single atomic layer: Strategies and challenges for functionalization of graphene and graphene-based materials. *Chem. Soc. Rev.* 41:97–114.

56. Jiang, D., Sumpter, B.G., Dai, S. 2007. Unique chemical reactivity of a graphene nanoribbon's zigzag edge. *J. Chem. Phys.* 126:134701.

57. Quintana, M., Spyrou, K., Grzelczak, M., Browne, W.R., Rudolf, P., Prato, M. 2010. Functionalization of graphene via 1,3-dipolar cycloaddition. *ACS Nano* 4:3527–3533.

58. Zhong, X., Jin, J., Li, S. et al. 2010. Aryne cycloaddition: Highly efficient chemical modification of graphene. *Chem. Commun.* 46:7340–7342.

59. Zhang, X., Hou, L., Cnossen, A. et al. 2011. One-pot functionalization of graphene with porphyrin through cycloaddition reactions. *Chem. Eur. J.* 17:8957–8964.

60. Meyer, J.C., Geim, A.K., Katsnelson, M.I., Novoselov, K.S., Booth, T.J., Roth, S. 2007. The structure of suspended graphene sheets. *Nature* 446:60–63.

61. Ryu, S., Han, M.Y., Maultzsch, J. et al. 2008. Reversible basal plane hydrogenation of graphene. *Nano Lett.* 8:4597–4602.

62. Fasolino, A., Los, J.H., Katsnelson, M.I. 2007. Intrinsic ripples in graphene. *Nat. Mater.* 6:858–861.

63. Lin, T., Zhang, W.D., Huang, J., He, C. 2005. A DFT study of the amination of fullerenes and carbon nanotubes: Reactivity and curvature. *J. Phys. Chem. B* 109:13755–13760.

64. Salavagione, H.J., Martínez, G., Ellis, G. 2011. Recent advances in the covalent modification of graphene with polymers. *Macromol. Rapid Commun.* 32:1771–1789.
65. Ijima, S. 1991. Helical microtubules of graphitic carbon. *Nature.* 354:56–58.
66. Dresselhaus, M.S., Dresselhaus, G., Avourios, P. 2001. *Carbon Nanotubes, Topics in Applied Physics*, Vol. 80, pp. 1–9. Springer-Verlag: Berlin, Germany.
67. Avouris, P. 2002. Molecular electronics with carbon nanotubes. *Acc. Chem. Res.* 35:1026–1034.
68. Arico, A.S., Bruce, P., Scrosati, B. et al. 2005 Nanostructured materials for advanced energy conversion and storage devices. *Nat. Mater.* 4:366–377.
69. Pandey, P., Datta, M., Malhotra, B.D. 2008. Prospects of nanomaterials in biosensors. *Anal. Lett.* 41:159–209.
70. Martin, C.R., Kohli, P. 2003.The emerging field of nanotube biotechnology. *Nat. Rev. Drug Discov.* 2:29–37.
71. Moniruzzaman, M., Winey, K.I. 2006. Polymer nanocomposites containing carbon nanotubes. *Macromolecules* 39:5194–5205.
72. Spitalsky, Z., Tasis, D., Papagelis, K. et al. 2010. Carbon nanotube-polymer composites: Chemistry, processing, mechanical and electrical properties. *Prog. Polym. Sci.* 35:357–401.
73. Clave, G., Campidelli, S. 2011. Efficient covalent functionalization of carbon nanotubes: The use of "click chemistry". *Chem. Sci.* 18:1887–1896.
74. Castelain, M., Salavagione, H.J., Martinez, G. 2011. Funcionalización de nanotubos de carbono y grafeno con polímeros mediante química click. *Rev. Iberoam. Polím.* 12:239–254.
75. Tuci, G., Vinattieri, C., Luconi, L. et al. 2012. Click on tubes: A versatile approach towards multimodal functionalization of SWCNT. *Chem. Eur. J.* 18:8454–8463.
76. Li, H., Cheng, F., Duft, A.M., Adronov, A. 2005. Functionalization of single-walled carbon nanotubes with well-defined polystyrene by "click" coupling. *J. Am. Chem. Soc.* 127:14518–14524.
77. Bahr, J., Tour, J.M. 2001. Highly functionalized carbon nanotubes using in situ generated diazonium compounds. *Chem. Mater.* 13:3823–3824.
78. Li, H., Adronov, A. 2007. Water-soluble SWCNTs from sulfonation of nanotube-bound polystyrene. *Carbon* 45:984–990.
79. Yadav, S.K., Mahapatra, S.S., Cho, J.W., Lee, J.L. 2010. Functionalization of multiwalled carbon nanotubes with poly(styrene-b-(ethylene-co-butylene)-b-styrene) by click coupling. *J. Phys. Chem. C* 114:11395–11400.
80. Rana, S., Cho, J.W., Kumar, I. 2010. Synthesis and characterization of polyurethane-grafted single-walled carbon nanotubes via click chemistry. *J. Nanosci. Nanotechnol.* 10:5700–5707.
81. Rana, S., Yoo, H.J., Cho, J.W. et al. 2011. Functionalization of multi-walled carbon nanotubes with poly(ε-caprolactone) using click chemistry. *J. Appl. Polym. Sci.* 119:31–37.
82. Lee, R.S., Ch, W.H., Lin, J.H. 2011. Polymer-grafted multi-walled carbon nanotubes through surface-initiated ring-opening polymerization and click reaction. *Polymer* 52:2180–2188.
83. Harrinson, B.S., Atala, A. 2007. Carbon nanotube applications for tissue engineering. *Biomaterials* 28:344–353.
84. Voge, C.M., Stegemann, J.P. 2011. Carbon nanotubes in neural interfacing applications. *J. Neural Eng.* 8:011001.

85. Zhang, Y., He, H., Gao, C. 2008. Clickable macroinitiator strategy to build amphiphilic polymer brushes on carbon nanotubes. *Macromolecules* 41:9581–9594.

86. Zhang, Y., He, H., Gao, C. et al. 2009. Covalent layer-by-layer functionalization of multiwalled carbon nanotubes by click chemistry. *Langmuir* 23:5814–5824.

87. Liu, J., Nie, Z., Gao, Y., Adronov, A., Li, H. 2008. "Click" coupling between alkyne-decorated multiwalled carbon nanotubes and reactive PDMA-PNIPAM micelles. *J. Polym. Sci. A: Polym. Chem.* 46:7187–7199.

88. Su, S., Shuai, Y., Guo, Z. et al. 2013. Functionalization of multi-walled carbon nanotubes with thermo-responsive azide terminated poly(*N*-isopropylacrylamide) via click reactions. *Molecules* 18:4599–4612.

89. Chergui, S.M., Ledebt, A., Mammeri, F. et al. 2010. Hairy carbon nanotube@ nano-Pd heterostructures: Design, characterization and application in Suzuki C-C coupling reactions. *Langmuir* 26:16115–16121.

90. He, H., Zhang, Y., Gao, Ch. et al. 2009. "Clicked" magnetic nanohybrids with a soft polymer interlayer. *Chem. Commun.* 2009:1655–1657.

91. Hijazi, I., Jousselme, B., Jégou, P. et al. 2012. Formation of linear and hyperbranched porphyrin polymers on carbon nanotubes via a CuAAC "grafting from" approach. *J. Mater. Chem.* 22:20936–20942.

92. Temel, G., Uygun, M., Arsu, N. 2013. Modification of multiwall carbon nanotube by thiol-ene click chemistry. *Polym. Bull.* 70:3563–3574.

93. Rodríguez-Pérez, L., Herranz, M.A., Martín, N. 2013. The chemistry of pristine graphene. *Chem. Commun.* 49:3721–3735.

94. Castelain, M., Salavagione, H.J., Segura, J.L. 2012. "Click"-functionalization of [60]fullerene and graphene with an unsymmetrically functionalized diketopyrrolopyrrole (DPP) derivative. *Org. Lett.* 14:2798–2801.

95. Wang, Z., Ge, Z., Zheng, X. 2012. Polyvalent DNA–graphene nanosheets "click" conjugates. *Nanoscale* 4:394–399.

96. Wang, H.X., Zhou, K.G., Xie, Y.L. 2011. Photoactive graphene sheets prepared by "click" chemistry. *Chem. Commun.* 47:5747–5749.

97. Baskaran, D., Mays, J.W., Bratcher, M.S. 2004. Polymer-grafted multiwalled carbon nanotubes through surface-initiated polymerization. *Angew. Chem. Int. Ed.* 43:2138–2142.

98. Dreyer, D.R., Park, S., Bielawski, C.W., Ruoff, R.S. 2010. The chemistry of graphene oxide. *Chem. Soc. Rev.* 39:228–240.

99. Erickson, K., Erni, R., Lee, Z., Alem, N., Gannett, W., Zettl, A. 2010. Determination of the local chemical structure of graphene oxide and reduced graphene oxide. *Adv. Mater.* 22:4467–4472.

100. Cao, Y., Lai, Z., Feng, J. 2011. Graphene oxide sheets covalently functionalized with block copolymers via click chemistry as reinforcing fillers. *J. Mater. Chem.* 21:9271–9278.

101. Pan, Y., Bao, H., Sahoo, N.G., Wu, T., Li, L. 2011. Water-soluble poly(*N*-isopropylacrylamide)–graphene sheets synthesized via click chemistry for drug delivery. *Adv. Funct. Mater.* 21:2754–2763.

102. Sun, S., Cao, Y., Feng, J. 2010. Click chemistry as a route for the immobilization of well-defined polystyrene onto graphene sheets. *J. Mater. Chem.* 20:5605–5607.

103. Jin, Z., McNicholas, T.P., Shih, C.J. et al. 2011. Click chemistry on solution-dispersed graphene and monolayer CVD graphene. *Chem. Mater.* 23:3362–3370.

104. Castelaín, M., Martínez, G., Merino, P. et al. 2012. Graphene functionalisation with a conjugated poly(fluorene) by click coupling: Striking electronic properties in solution. *Chem. Eur. J.* 18:4965–4973.

105. Salvio, R., Krabbenborg, S., Naber, W.J.M., Velders, A.H., Reinhoudt, D.N., van de Wiel, W.G. 2009. The formation of large-area conducting graphene-like platelets. *Chem. Eur. J.* 15:8235–8240.

106. Yang, X., Ma, L., Wang, S., Li, Y., Tu, Y., Zhu, X. 2011. "Clicking" graphite oxide sheets with well-defined polystyrenes: A new strategy to control the layer thickness. *Polymer* 52:3046–3052.

107. Segura, J.L., Castelaín, M., Salavagione, H. 2013. Synthesis and properties of [60]fullerene derivatives functionalized through copper catalyzed Huisgen cycloaddition reactions. *Curr. Org. Synth.* 10:724–736.

108. Segura, J.L., Salavagione, H.J. 2013. Graphene in copper catalyzed azide-alkyne cycloaddition reactions: Evolution from [60] fullerene and carbon nanotubes strategies. *Curr. Org. Chem.* 17:1680–1693.

109. Stankovich, S., Dikin, D.A., Dommett, G.H.B. 2006. Graphene-based composite materials. *Nature* 442:282–286.

110. Fu, R., Fu, G.D. 2011. Polymeric nanomaterials from combined click chemistry and controlled radical polymerization. *Polym. Chem.* 2:465–475.

111. Ye, Y.S., Chen, Y.N., Wang, J.S. et al. 2012. Versatile grafting approaches to functionalizing individually dispersed graphene nanosheets using RAFT polymerization and click chemistry. *Chem. Mater.* 24:2987–2997.

112. Yang, Y., Song, X., Yuan, L. et al. 2012. Synthesis of PNIPAM polymer brushes on reduced graphene oxide based on click chemistry and RAFT polymerization. *J. Polym. Sci. A: Polym. Chem.* 50:329–337.

113. Castelaín, M., Martínez, G., Ellis, G., Marco, C., Salavagione, H.J. 2013. The effect of the chemical approaches to modify graphene on the electrical, thermal and mechanical properties of graphene/polyethylene nanocomposites. *Macromolecules* 46:8980–8987.

114. Yadav, S.K., Yoo, H.Y., Cho, J.W. 2013. Click coupled graphene for fabrication of high-performance polymer nanocomposites. *J. Polym. Sci., Polym. Phys.* 51:39–47.

115. Meng, D., Sun, J., Jiang, S. et al. 2012. Grafting P3HT brushes on GO sheets: Distinctive properties of the GO/P3HT composites due to different grafting approaches. *J. Mater. Chem.* 22:21583–21591.

116. Wang, H.X., Wang, Q., Zhou, K.G., Zhang, H.L. 2013. Graphene in light: Design, synthesis and applications of photo-active graphene and graphene-like materials. *Small* 9:1266–1283.

117. Deetuam, C., Samthong, C., Thongyai, S., Praserthdam, P., Somwangthanaroj, A. 2014. Synthesis of well dispersed graphene in conjugated poly(3,4-ethylenedioxythiophene): Polystyrene sulfonate via click chemistry. *Compos. Sci. Technol.* 93:1–8.

118. Jing, Y., Tang, H., Yu, G., Wu, P. 2013. Chemical modification of graphene with a thermotropic liquid crystalline polymer and its reinforcement effect in the polymer matrix. *Polym. Chem.* 4:2598–2607.

119. Castelaín, M., Martínez, G., Ellis, G., Salavagione, H.J. 2013. Versatile chemical tool for the preparation of conductive graphene-based polymer nanocomposites. *Chem. Commun.* 49:8967–8969.

120. Du, J., Zhao, L., Zeng, Y. et al. 2011. Comparison of electrical properties between multi-walled carbon nanotube and graphene nanosheet/high density polyethylene composites with a segregated network structure. *Carbon* 49:1094–1100.

121. Fim, F.C., Basso, N.R.S., Graebin, A.P., Azambuja, D.S., Galland, G.B. 2013. Thermal, electrical, and mechanical properties of polyethylene-graphene nanocomposites obtained by in situ polymerization. *J. Appl. Polym. Sci.* 128:2630–2637.

122. Liras, M., García, O., Quijada-Garrido, I., Ellis, G., Salavagione, H.J. 2014. Homogenous thin layer coated graphene via one pot reaction with multidentate thiolated PMMAs. *J. Mater. Chem. C* 2:1723–1729.

CHAPTER 17

CONTENTS

Chemical Functionalization as an Approach for the Creation of Arrays of Graphene Quantum Dots Embedded in Dielectric Matrix

17

I.V. Antonova, N.A. Nebogatikova, and V.Ya. Prinz

17.1 INTRODUCTION

Graphene is a material that is one atom thick and has a high conductivity, high mobility of charge carriers, a high thermal conductivity, and a large number of other attractive properties (Geim 2009, Novoselov et al. 2012). Moreover, from this material, many graphene-based dielectrics, such as oxidized, hydrogenated, and fluorinated graphene, can be obtained (Bera et al. 2010, Soldano et al. 2010, Bimberg and Pohl 2011). Fabrication of new graphene-based materials that involve quantum structures and are obtained by means of graphene local functionalization is widening considerably the potential applications targeted by graphene (Craciun et al. 2013). Quantum dot (QD) systems offer materials that have the potential for many applications in physical, chemical, and biological areas, such as optoelectronic, photovoltaic, and memory devices, sensors, bioimaging systems, and medicine (e.g., Shao et al. 2012, Shen et al. 2012, Li et al. 2013). Currently, significant progress has been made in the creation of graphene quantum dots (G-QDs) and the investigation of their properties (Ponomarenko et al. 2008, Güttinger et al. 2012).

A QD is a nanometer-sized object in which excitons are confined in all three spatial dimensions. G-QDs are graphene fragments that are small enough to cause exciton confinement and quantum-size effects. Excitons in graphene have an infinite Bohr diameter. Thus, graphene fragments of any size will show quantum-confinement effects (Li and Yan 2010, Novoselov et al. 2012, Bacon et al. 2014). Generally, graphene or few-layer graphene (FLG) QDs are defined as graphene sheets that are less than ~100 nm in their lateral dimensions and are arranged in one or a few (<10) QD layers at room temperature; such QDs demonstrate size-, shape-, and edge-dependent properties (Zarenia et al. 2011, Zhou et al. 2012). In principle, the bandgap width of the G-QDs can be tuned, by varying the QD sizes, from 0 eV to the bandgap energy of benzene (from 4.4 eV [Yan et al. 2010] to 7.2 eV [Neaton et al. 2006]). Expectedly, G-QDs will impart QD systems that have novel chemical and physical properties that cannot be obtained in systems that are based on semiconductors (Li et al. 2013).

Suspension with small G-QDs (<10 nm) is a new class of zero-dimensional fluorescent carbon material. Such QDs demonstrate much promise for application properties: low toxicity, excellent solubility, high-chemical stability in various solvents, and good surface activity (Xu et al. 2013). Graphene or graphene oxide flakes in suspension are then cut into QDs, which then have sizes of a few nanometers (1–7 nm) during the hydrothermal (Pan et al. 2012, Feng et al. 2013, Hu et al. 2013), microwave hydrothermal (Tang et al. 2013, Xie et al. 2013), and solvothermal process (Zhu et al. 2012, Liu et al. 2013). Blue or green fluorescence is observed for these QD systems. The hydrothermal process is widely known as a method for breaking down carbon nanotubes into nanoribbons (Kosynkin et al. 2009). G-QD properties could be varied by means of being equipped with different functional groups at the edges of the particles. These properties make them much more desirable for many applications in comparison with inorganic semiconductor QDs (Xu et al. 2013, Bacon et al. 2014). A method for obtaining suspensions with QDs is the chemical or ultrasonic disintegration of graphite in nanoflakes (e.g., Zhuo et al. 2012). A specific feature of the graphene flakes that are obtained by this method is the decoration of the flake edges with various functional groups. With functional groups of several types involved, the stoichiometric composition could show variations depending on specific synthesis conditions of QDs. For example, the synthesis of oxyfluorinated graphene flakes has been demonstrated after high-temperature treatment of graphene oxide in a hydrofluoric acid (HF) (Wang 2012).

The influence of functional groups on the properties of G-QDs was discussed in Lee et al. (2012), where QDs were obtained with the help of nanolithography and block copolymers, which were used as a mask. In their study, the authors have obtained and examined arrays of QDs that have identical sizes of 10 or 20 nm, and they found that there was no correlation between the spectral positions of the peaks in the photoluminescence spectra of their samples and the QD sizes. It was established that the oxidation of the lattice defects with oxygen was a determining factor for the emergence of photoluminescence in the examined structures. The functional groups that form and surround the QDs can therefore be regarded as an important factor for defining the properties of the QDs.

The second approach that is most frequently used for obtaining graphene-based QDs is the creation of QDs by nanolithography. In various experiments, the lateral sizes of the QDs that were formed by this method ranged from 50 to 250 nm (Ponomarenko et al. 2008, Güttinger et al. 2012). While studying the conductivity in the transistor structures that have one QD at low (helium) temperatures, the authors were also able to investigate various quantum-dimensional effects: electron confinement in G-QDs is observed by measuring Coulomb blockade and transport through excited states. Measurements in a magnetic field that is perpendicular to the sample plane allowed them to identify the regime with only a few charge carriers in the dot (electron–hole transition) and the crossover to the formation of the graphene-specific zero-energy Landau level at high fields. Nonetheless, the latter approach is not meant for obtaining QD arrays and for examining the interactions between the QDs.

Arrays of QDs that are embedded in a nonconducting matrix are interesting objects for the study of energy quantization–induced quantum-dimensional effects, interaction between QDs, and charge transport in QD arrays (Bera et al. 2010, Bimberg and Pohl 2011) and have a great potential for numerous applications. Recently, some studies of incomplete chemical functionalization, which presumably leads to the formation of high-resistivity graphene derivatives, such as hydrogenated or oxidized graphene, in the formation of QD arrays, were reported (Chuang et al. 2012, Shen et al. 2012). It should be noted here that hydrogenated or oxidized graphene have low stability and reproducibility of their properties due to a low temperature at which they reduce to graphene ~200°C–300°C (Elias et al. 2009, Luo et al. 2009, Acik et al. 2011).

FLG shares all of the attractive properties that are demonstrated by graphene except for its linear dispersion spectrum, and simultaneously, it provides additional possibilities in the fabrication of multilayered graphene-based heterostructures (Antonova et al. 2011a, Kotin et al. 2013). The possibility of controlling the modification of properties of graphene and FLG films while creating derivatives of those materials, which would have varying degrees of functionalization and locality, has stimulated interest in studying structures such as QDs. In this chapter, we have demonstrated conditions for creating arrays of graphene and FLG QDs that are self-formed during chemical functionalization of graphene films in an aqueous solution of HF along with the results of their property investigation. This fluorination process was introduced in Nebogatikova et al. (2013, 2014a). The simplicity of the functionalization procedure, the involvement of noncorrosive media and high temperatures, and also the possibility of achieving controllable variation in the QD sizes were the advantageous features of the method that was used for obtaining the QD arrays. As a result, we obtained graphene or FLG-conducting islands, or fluorographene (FG)-embedded graphene or FLG QDs, which were 70–10 nm in diameter and available in densities that ranged from 1×10^9 to 6×10^{10} cm^{-2} (Nebogatikova et al. 2014b). In this investigation, charge transient spectroscopy of the QD systems (size 50–70 nm) allowed us to examine the QD energy spectra and the timing of the carrier emission (or charge relaxation) from the QDs as a function of the film thickness (Antonova et al. 2014a). It was found that the characteristic time of the carrier emission from the QDs decreased markedly (by about four orders of magnitude) on increasing the QD thickness from one graphene monolayer to 3 nm. Daylight-assisted measurements also demonstrate a strong decrease in the carrier emission time (Antonova et al. 2014b). In addition, the high stability of FG matrix ensures, first, a high stability of the obtained films with QD arrays and, second, stable functionalization of the QD edges.

17.2 CONDITIONS FOR THE CREATION OF GRAPHENE QDs IN A FLUOROGRAPHENE MATRIX

17.2.1 Functionalization of Graphene in Aqueous Solution of Hydrofluoric Acid

The graphene and FLG films that were used in this study were prepared by electrostatic exfoliation of graphene from highly ordered pyrolytic graphite (HOPG) (Sidorov et al. 2007) or graphene grown at 1000°C by means of chemical vapor deposition. In both cases, graphene and FLG flakes were polycrystalline and were obtained on top of silicon substrates that were covered by 300 nm of SiO$_2$. It is known (Blake et al. 2007) that when films of different thicknesses are placed on SiO$_2$/Si substrates with an oxide thickness of 300 nm, a high contrast image is observed. Films of various thicknesses were easily distinguishable against one another using even the optical microscope. The precise values of the thickness of the selected films were determined using atomic force microscopy (AFM).

Synthesis of FG is generally performed using two methods. In the first method, mechanical exfoliation of graphene flakes from fluorinated graphite is used. Fluorinated graphite is normally created by exposing graphite to a fluorinating agent, F$_2$ or XeF$_2$, which is admixed to some ambient medium; this process, which involves holding at temperatures of 120°C–400°C, normally takes from 10 h up to 1 month (Cheng et al. 2010, Nair et al. 2010, Withersy et al. 2010). Graphite fluoride can also be obtained at room temperature using a reaction of graphite with a gaseous mixture of F$_2$, HF, or MF$_n$ (the volatile fluorides MF$_n$ used were IF$_5$, BF$_3$, and ClF$_3$), with subsequent treatment of the resulting material in a fluorine-containing ambient at temperatures of 100°C–600°C (Boukhvalov and Katsnelson 2009). Mechanical cleaving of individual FG atomic planes from commercially available fluorinated graphite was found to be a surprisingly difficult task: FG sheets were found to be extremely fragile and prone to rupture due to many structural defects that result from the harsh fluorination conditions that are used to obtain bulk fluorinated graphite (Nair et al. 2010). In the second method, FG is created by mechanical exfoliation of graphene flakes from graphite with subsequent fluorination of the flakes by a procedure that is quite similar to that normally used to achieve graphite fluorination (Robinson et al. 2010). Due to the small sizes of the graphene flakes, it is possible to obtain fully fluorinated graphene flakes by a XeF$_2$ agent using treatments that last for approximately 2 weeks (Cheng et al. 2010). Because XeF$_2$ rapidly etches Si and readily diffuses even through amorphous SiO$_2$, it appears impossible to employ SiO$_2$/Si substrates in the fluorination procedure; using a chemically inert support is therefore required. The need for an inert support makes the whole process a rather complicated procedure.

HF is widely used in silicon technology. It is known that graphene or carbon tubes that are treated in solutions of HF in water show an enhanced stability when stressed with electric voltages in comparison with nonmodified tubes. Graphene edge passivation and defect decoration in graphene are considered to be responsible for the resulting enhanced stability (Boukhvalov and Katsnelson 2009). Treatments in HF aqueous solutions are also used in fabricating suspended graphene films (Bolotin et al. 2008). Synthesis of oxyfluorinated graphene has been demonstrated in Bruna et al. (2011) and Wang et al. (2012) with the use of treatment in aqueous solution of HF at an enhanced temperature (150°C–180°C) in an autoclave. A similar, relatively simple method for synthesizing fluorinated graphene oxide by the fluorination of graphene oxide in anhydrous hydrogen fluoride was suggested in Pu et al. (2013). In that case, it is possible to obtain a relatively weak degree of fluorination.

In this chapter, graphene or FLG samples were subjected to treatment in aqueous solutions of HF at room temperature. The functionalization of graphene and FLG films was achieved using two methods, either by treating samples in aqueous solution of HF or by exposing the samples to the vapor of the same solution. After treatment in HF, the samples were rinsed with deionized water for 5–10 min. These treatments typically lasted for several minutes. This method of fluorination is very simple and can be performed in any laboratory. Depending on the treatment time, films that have a different degree of graphene functionalization and different properties were obtained (Nebogatikova et al. 2013, 2014a,b).

A first-principles study of fluorinated graphene with different coverage of fluorine based on the density functional theory demonstrated that the properties of graphene can be strongly modified by absorbing different amounts of fluorine (Liu et al. 2012): a precise adsorption of fluorine enables a tuning of the bandgap from 0 to ~3.13 eV as well as a transformation from nonmagnetic semimetal to magnetic/nonmagnetic material.

For further discussion of the conditions of QD formation, we must introduce the threshold time of the sample treatment in HF, t_{th}. The time t_{th} is the time of an HF treatment of graphene or FLG films, which leads to an abrupt change in many properties of the samples, in particular, to a transition of the films from a conducting state into an insulating state. This transition is presented in Figure 17.1a; the resistance increases by seven or eight orders of magnitude, reaching a value of 10^{11}–10^{12} ohm/sq. The measured resistance shows a good correlation with the available resistance for FG obtained from graphene sheets by a traditional method, the treatment of graphene sheets in XeF_2, or in other fluorine-containing gases at elevated temperatures (Cheng et al. 2010, Nair et al. 2010, Withersy et al. 2010). The abrupt growth of the resistance is due to the formation of a connected network of fluorinated regions. The dependences of the time t_{th} on the film thickness and on the solution concentration are shown in Figure 17.1a and b. Evidently, for relatively thick films, a longer time is required for such films to be transferred into an isolating state. For thin films (<7 nm), the time t_{th} depends almost linearly on the film thickness; this dependence presumably results from successive (layer by layer) fluorination of all of the layers. The highest fluorination rate was attained in solutions that contained 1.5%–3% of HF (see Figure 17.1c). Additionally, it was observed that the shortest times required for reaching the resistance jump in the films were attained in a 2.5% solution of HF in water. Variation of the solution temperature clearly demonstrated that the fluorination of graphene and FLG films is a thermally activated process, with energy E_{act} ~ 1.4 eV.

It should be noted here that an increase in the thickness of the FLG films in excess of 10–15 nm resulted in either a pronounced deceleration or a complete termination of the fluorination process.

As shown earlier, changes in the properties of the films occur only upon the treatment of samples in a narrow range of concentrations of the HF solution. The properties of the HF solution depend strongly on the concentration of HF. It is known that hydrogen bonds bind molecules into associates $(HF)_n$ when $n \gg 2$, which have decreasing chemical reactivity of the acid. This effect is especially strong at relatively high concentrations of HF. In the range of concentrations in which fluorination is observed, the maximum number of molecules in the associates corresponds to n ~ 2 ($2HF \leftrightarrow HF_2^- + H^+$), and their linear size of 0.23 nm does not exceed the interlayer distance for graphite and FLG, which makes it possible for them to penetrate into the interlayer space (Karapetyants and Drakin 1994). Therefore, it is possible to explain such a strong dependence of the experimental results on the concentration of HF while accounting for the influence of the hydrogen bonds.

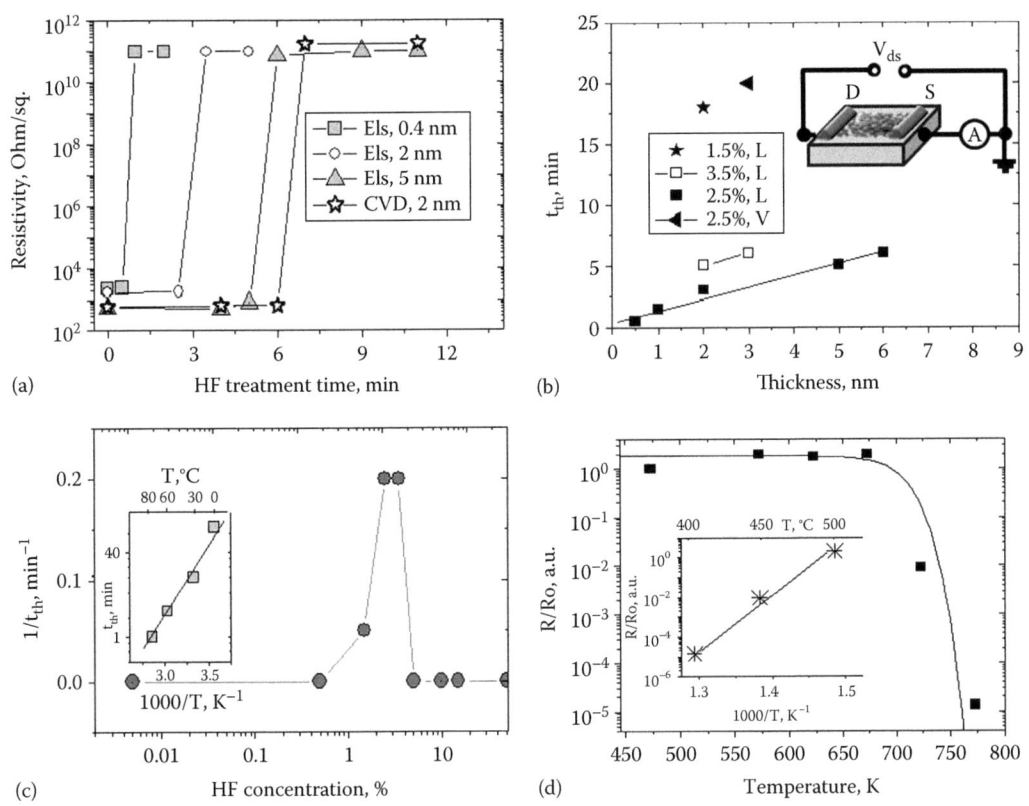

FIGURE 17.1
(a) The specific resistances of the samples depending on the time of treatment in a 2.5% aqueous solution of HF for films that were created by different methods with different thicknesses. The time of the resistivity step is mentioned as the threshold time t_{th}. Els, electrostatic exfoliation; CVD, chemical vapor deposition of graphene. (b) Threshold treatment duration required for reaching the conductor–insulator transition t_{th} versus the film thickness for different compositions of the aqueous solutions of HF(L) or their vapors (V). The insert in (b) demonstrates the configuration of the contacts for the measurement of the current–voltage characteristics on treated films. (c) Reverse time t_{th} versus the solution concentration for 5 nm thick samples at room temperature. The insert in (c) demonstrates the dependence of t_{th} on the temperature of a 2.5% aqueous solution of HF for FLG with a thickness of 5 nm. (d) The resistance of HF-treated samples versus the isochronal annealing temperature. The annealing time was 30 min. Insertion gives the Arrhenius dependence for the determination of the activation energy of the annealing. (From Nebogatikova, N.A. et al., *Physica E.*, 52, 106, 2013; Nebogatikova, N.A. et al., *Nanotechnol. Russia*, 9, 51, 2014a; Nebogatikova, N.A. et al., *Carbon*, 2014b.)

17.2.2 Evidence of the Graphene Fluorination

Now let us consider the main evidence for the fluorination process of graphene or FLG, which caused in many cases partial fluorination with a tunable amount of fluorine:

1. FG is known to be a high-quality insulator (resistivity $\sim 10^{12}$ ohm/sq.) with an approximate 3 eV optical bandgap; this material is stable up to a temperature of 400°C–450°C (Cheng et al. 2010, Nair et al. 2010). The strong increase in the film resistivity due to the treatment in aqueous solution of HF is the first argument toward the fluorination process.

   A series of isochronal anneals was performed for chemically modified samples, for which the time of treatment exceeded t_{th} and the specific resistance increased to $\sim 10^{12}$ ohm/sq. Samples were annealed in the atmosphere of argon in the range of temperatures of 200°C–500°C. It was shown that the resistance decreased by more than five orders of magnitude as a result of annealing at a temperature of ~ 450°C (Figure 17.1d). The recovery of the sample resistance means that the functionalization reaction is reversible, which must be upon the fluorination. It was assumed that the resistance of the samples depended on the

amount of unannealed complexes, the formation of which provided graphene in the nonconductive state (C–F bonds) and could be described by the following formula (Komarov 2004):

$$r(T,t) = R_0 \exp\left(-\beta t \exp\left(\frac{E}{kT}\right)\right),\qquad(17.1)$$

where

      $r(T, t)$ is the resistance of the sample after annealing at the temperature T and for the
        time t
      β is the frequency factor of the decomposition reaction
      E is the energy that is necessary for the bond to break
      k is the Boltzmann constant
      R_0 is the resistance of the modified sample before annealing

It was established in fitting the parameters that the energy of annealing activation E was ~2 eV
(see insertion in Figure 17.1d), while the parameter β was ~10^{11} s^{-1}. The line in Figure 17.1d
was obtained using Formula (17.1) and the parameters E and β defined earlier. The resistivity
recovery temperature was ~450°C and the energy of annealing activation E was ~2 eV, which
coincide with parameters that are known for FG (Cheng et al. 2010, Nair et al. 2010).

2. A comparison of the structural properties of pristine FLG films and the structural properties
of films modified in aqueous solution of HF was studied using Raman spectroscopy under
ambient conditions. Figure 17.2a and b shows changes in the Raman spectra of films that
were treated in aqueous solution of HF for different amounts of time. The most prominent
peak G at ~1580 cm^{-1} is known to be due to in-plane vibrations of sp^2-hybridized carbon
atoms. The peak D at 1350 cm^{-1} is an indicator of some intrinsic defects that are presented in
graphene sheets or a basal-plane chemical reaction breaking the π-bonds and converting the
sp^2 hybridization of the carbon atoms into sp^3 hybridization. The peak 2D is the second-order

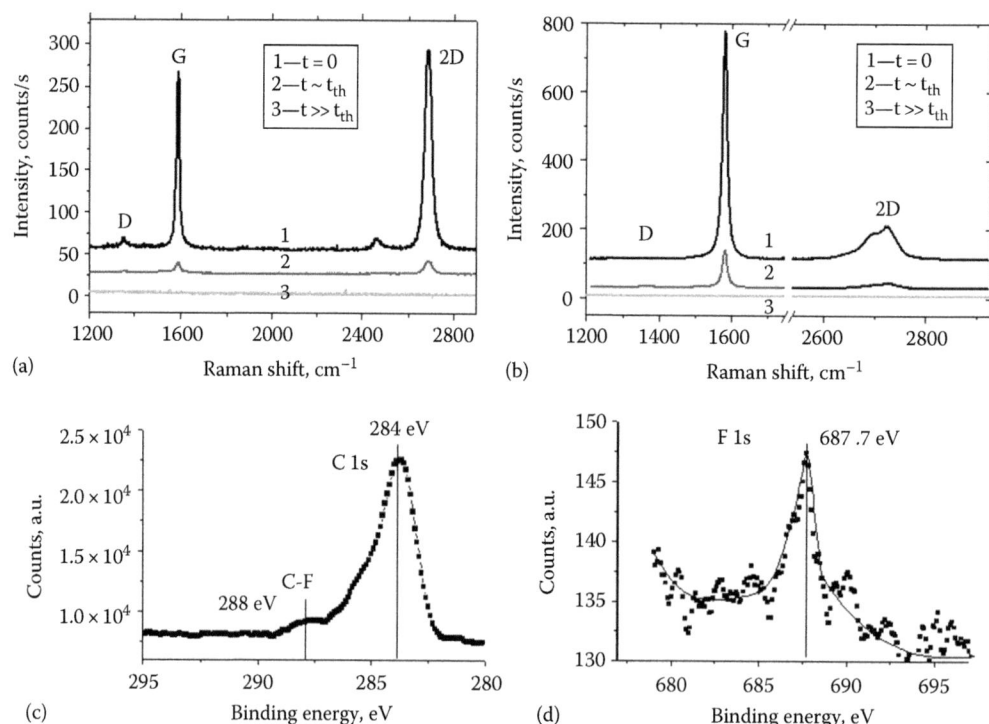

FIGURE 17.2
Raman spectra for pristine films and after different times of functionalization in aqueous solution
of HF for (a) graphene and (b) FLG. XPS spectrum in the area of signals from (c) carbon atoms and
(d) fluorine for FLG with a thickness of 7 nm treated in an aqueous solution of HF for 12 min (t ~ t_{th}).
(From Nebogatikova, N.A. et al., *Physica E.*, 52, 106, 2013.)

scattering band. The dynamics of Raman spectrum changes for HF-treated films includes an initial increase in the D-peak intensity and a subsequent decrease in the intensity of all of the peaks in the spectra. The position of the excitation laser spot on the film was checked using an optical camera. Changes in the Raman spectra are similar to the dynamics of Raman spectrum changes observed for FG (e.g., Cheng et al. 2010, Nair et al. 2010). Such transformations in the Raman spectra occurred because the opening of bandgaps in FG (~3.0) and also because the energy of laser-emitted photons (2.41 eV) were insufficient for band-to-band excitation of the material that caused the output of the resonance conditions. We therefore assume that treatments of graphene films in the aqueous solution of HF for different times led to the formation of FG or few-layer FG films with different degrees of fluorination.

Dramatic changes in the Raman spectra show that graphene functionalization in an aqueous solution of HF occurs in all monolayers of the used FLG. Intercalation of HF into the FLG interlayer space during a few minutes of treatment is impossible. Thus, we must suggest the penetration of fluorination agents to the underlayers through the defects at the grain boundaries (our films are polycrystalline films).

In samples that were treated for the time t_{th}, the Raman spectra still exhibit G and 2D peaks, whose intensity amounts to ~0.1–0.07 of the initial intensity of the peaks for both graphene and FLG (see Figure 17.2a and b). The latter observation implies that the film still contains nonfluorinated regions.

3. X-ray photoelectron spectroscopy (XPS) is an informative tool for investigating the chemical composition and structure of the initial and modified films. The XPS spectra of untreated films and films treated for time $t < t_{th}$ demonstrated peaks that are associated with carbon (C 1s, 284 eV), oxygen (533 eV), and a silicon–oxygen bond (103.5 eV). The first peak belongs to the samples that are under investigation, while the two other peaks belong to the SiO_2 substrate. The decrease in the intensity of the signal that is associated with carbon was detected upon the increase in the time of functionalization of FLG, which is most probably due to the formation of a buckled surface (we will discuss the surface morphology in the following text). The spectra of the samples that were treated in an HF solution for time $t \sim t_{th}$ are given in Figure 17.2c and d. A signal from fluorine F1s was detected in the energy region of 687.7 eV. Such a position of the peak corresponds to sp^3-hybridized chemically bound fluorine ions (Nair et al. 2010, Sherpa et al. 2012), and the bond has a covalent character (Plank et al. 2003). In comparison, the position of the line of fluorine at 685.5 eV means that fluorine ions are bound with carbon by an ionic bond and carbon is sp^2-hybridized. *Free* chemically unbound fluorine (intercalated between the layers of graphene or formed due to the cleavage of a C–F bond) gives a peak that is close to the energy of 684 eV, while bonds with the energy of 691–693 eV correspond to the complexes CF_2 and CF_3 (Nair et al. 2010). The peak for sp^2-hybridized carbon (284 eV) and the peak that corresponds to the C–F bond (288 eV) are very distinguishable on the spectra (Jung et al. 2009). The detection of the peaks of fluorine in the energy regions of 687.7 and 288 eV is direct evidence for the fluorination reaction; however, a low amplitude of the F1s signal is an unsettled question about the degree of graphene fluorination.

17.3 SURFACE MORPHOLOGY OF THE FILMS TREATED IN AQUEOUS SOLUTION OF HF

Consider the present structural transformations that proceed in the films during fluorination. Because our films were polycrystalline films, it is the defects and domain boundaries that were fluorinated first in the films. At this stage, the formation of fluorinated regions in our films was studied using scanning electron microscopy (SEM) and AFM. A SEM image of a partially fluorinated sample (i.e., prior to the occurrence of the transition into an insulating state) is shown in Figure 17.3a and b. As a result of fluorination, domain boundaries (a network of dark lines) have become distinctly observed. It is significant that in the vicinity of domain boundaries, there appear buckled regions in the image that appear as light regions that stretch along domain boundaries. A schematic representation of fluorinated domain boundaries is shown in Figure 17.3c.

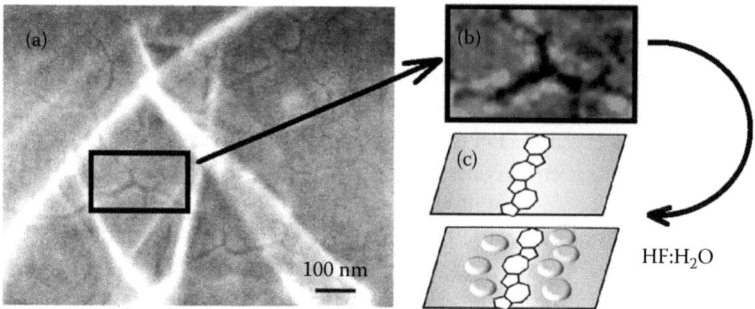

FIGURE 17.3
(a) SEM image of the surface of a 4 nm thick FLG film treated in a solution of HF in water. In the image, large light folds that have formed during the transfer of films onto the silicon dioxide substrate are observed. (b) Enlarged fragment of image (a). Functionalization of film regions in the vicinity of domain boundaries and buckling of fluorinated regions lead to the formation of buckled regions that appear as light zones that stretch along the domain boundaries. (c) Schematic representation of a domain boundary prior to and after a fluorination procedure. In the nonfluorinated film, the film surface was a smooth surface that exhibited no corrugations. (From Nebogatikova, N.A. et al., *Carbon*, 2014b.)

It is a well-known fact that domain boundaries are regions that are rich in lattice defects, such as five- and seven-membered rings of carbon (Figure 17.3c) (An et al. 2011, Kim et al. 2011). Such regions normally show a greater chemical reactivity in comparison with inner domain regions. On the other hand, vacancy defects that are present at domain boundaries are capable of providing access for the solution to underlying film layers, especially in the case of a buckled upper layer. The observed transition into an insulating state proves that fluorination has occurred throughout the whole film thickness, that is, in all layers of the film. However, because of the difficult access of fluorinating complexes to inner film layers, we can expect that, on increasing the film thickness, a longer time will be required for a larger amount of fluorination of the film. Such dependence of t_{th} on the film thickness was in fact observed (see Figure 17.1). In addition, a lower degree of fluorination in the inner monolayers can be expected. As a result, the characteristic size of the QDs and also the degree of fluorination of the barriers that isolate the QDs could appear different in different film layers.

The lattice constant d of graphene is $\approx$2.46 A. According to Cheng et al. (2010) and Nair et al. (2010), FG has a unit cell that is approximately 1% larger than that of graphene, which is d $\approx$ 2.48 A. An increase in d is expected because fluorination leads to sp^3-type bonding, which features a larger interatomic distance in comparison with sp^2 bonding. However, the observed increase is small in comparison with fluorographite, where d was reported to be 2.8%–4.5% larger than in graphite (Ryu et al. 2008, Ribas et al. 2011). The smaller lattice constant d in the FG is probably due to the possibility that 2D FG sheets could undergo corrugation if out-of-plane displacements of carbon atoms in FG are not restricted by the surrounding 3D matrix. For FLG, we have an intermediate situation. Local fluorination leads to the production of local mechanical stain (Euler instability), which leads to corrugation of the whole FLG layer. Corrugation of HF-treated samples allows us to assume that functionalization of the FLG structure with HF initially starts with the top graphene layer. It can be expected that fluorination of the top graphene layer will lead to the formation of a strained top graphene monolayer that is weakly connected with lower-laying graphene underlayers.

The strong modification of the film surface morphology was revealed due to functionalization of graphene with HF. AFM images of FLG films that were treated in aqueous solution of HF for different amounts of time are shown in Figure 17.4a and b; these images are to be compared to the image of the surface of a pristine sample without any relief on the surface. Treatment in aqueous solution of HF for time t of approximately t_{th} leads to the formation of an irregular surface morphology (network, Figure 17.4a). Later, for more prolonged treatment times, the morphology transforms into a regular swell-like periodic morphology (corrugation) of the surface (Figure 17.4b), which was observed. An increase in the duration of HF treatment for $t > t_{th}$ induces no substantial changes in the size and height of the corrugation. We have found that the formation of the corrugated surface corresponds to the formation of the system G-QDs in the insulated matrix of FG, and the properties of this system are considered in more detail in the next sections.

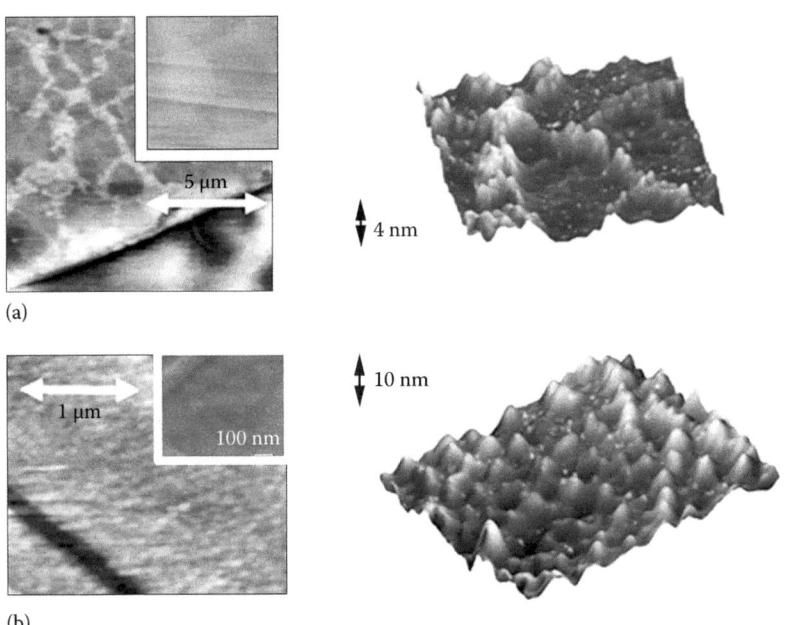

FIGURE 17.4
AFM 2D and 3D images of the surface of pristine samples and samples treated in 2.5% aqueous solution of HF. (a, b) A *network* is formed on the surface after treatment for the time of approximately t_{th}. The height of the *walls* of the grid is 4–10 nm. Insertion in figure (a): the image of the surface of an untreated sample. (b) The AFM image of the surface of samples after a more long-term treatment in HF ($t > t_{th}$), which leads to the formation of corrugations. The characteristic sizes of the formed buckling: a period is approximately 100–200 nm, and the height is 6–10 nm. The insert in (b) gives the SEM image of a corrugated surface. (From Nebogatikova, N.A. et al., *Nanotechnol. Russia*, 9, 51, 2014a.)

17.4 STRUCTURAL PROPERTIES AND SIZE OF QDs

AFM images of the film surface that was given an HF treatment for the fluorination time $t \sim t_{th}$ with different magnifications are shown in Figure 17.5a and b. As was noted earlier, the fluorination process in the film material first begins at the domain boundaries, and then, it penetrates into defectless regions. The latter regions are less reactive and more stable to the action of HF. However, under the action of mechanical stresses and strains and the redistribution of

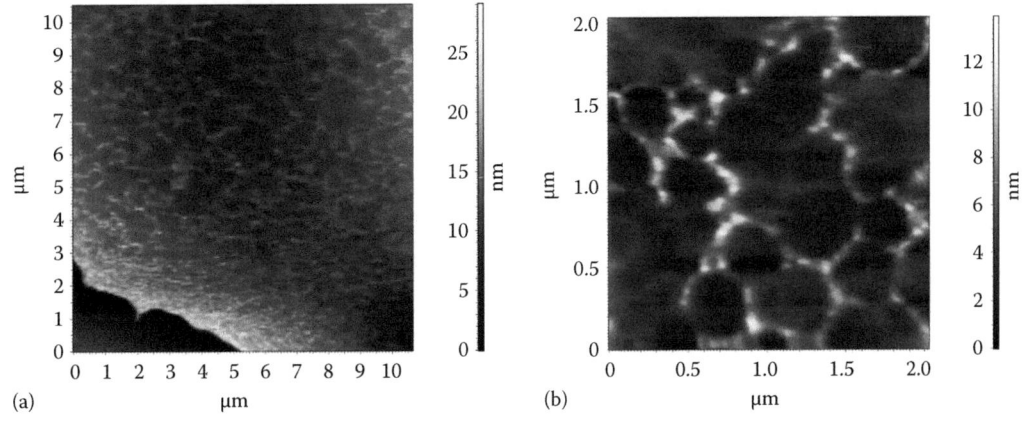

FIGURE 17.5
(a) Image of the surface of a 7 nm thick FLG film treated in a solution of HF in water for 5 min, that is, for a treatment duration in the region of the transition to insulating state (fluorination time of approximately t_{th}). (b) Enlarged fragment of image (a). (From Nebogatikova, N.A. et al., *Carbon*, 2014b.)

electron density in the neighboring fluorinated regions, chemical functionalization processes subsequently become initiated in defectless regions also (Gupta et al. 2006, Boukhvalov and Katsnelson 2009). The formation of networks with graphene islands of 200–500 nm size can be seen clearly in these images.

To study the transformations that proceeded in graphene and FLG films during fluorination, AFM measurements were performed not only in the surface-relief scan mode but also in the lateral-force mode. The latter measurements have allowed us to visualize fluorinated material parts on the film surface and make a comparison between images that were taken from the surface of the samples with different degrees of fluorination. AFM images that were taken during one pass and similar passes of the scanner over the surface of an FLG sample treated in the solution of HF in water during a time $t \geq t_{th}$ are shown in Figure 17.6a and b. It can be seen that as a result of the treatment, a relief with a ~100 nm period and 3–5 nm height formed on the surface of the sample. Performing lateral-force measurements, we obtained an image of a *network* that was composed of fluorinated regions (Figure 17.6b). The fluorinated regions in the figure have a lighter tint. The dark areas are graphene regions, which still remained nonfluorinated. The density of such regions on the unit surface was $(4–6) \times 10^{10}\,cm^{-2}$, and their sizes were 20–70 nm. Simultaneous registration of the same surface region in various scan modes has allowed us to correlate the various surface features that were observed in the different images that were taken.

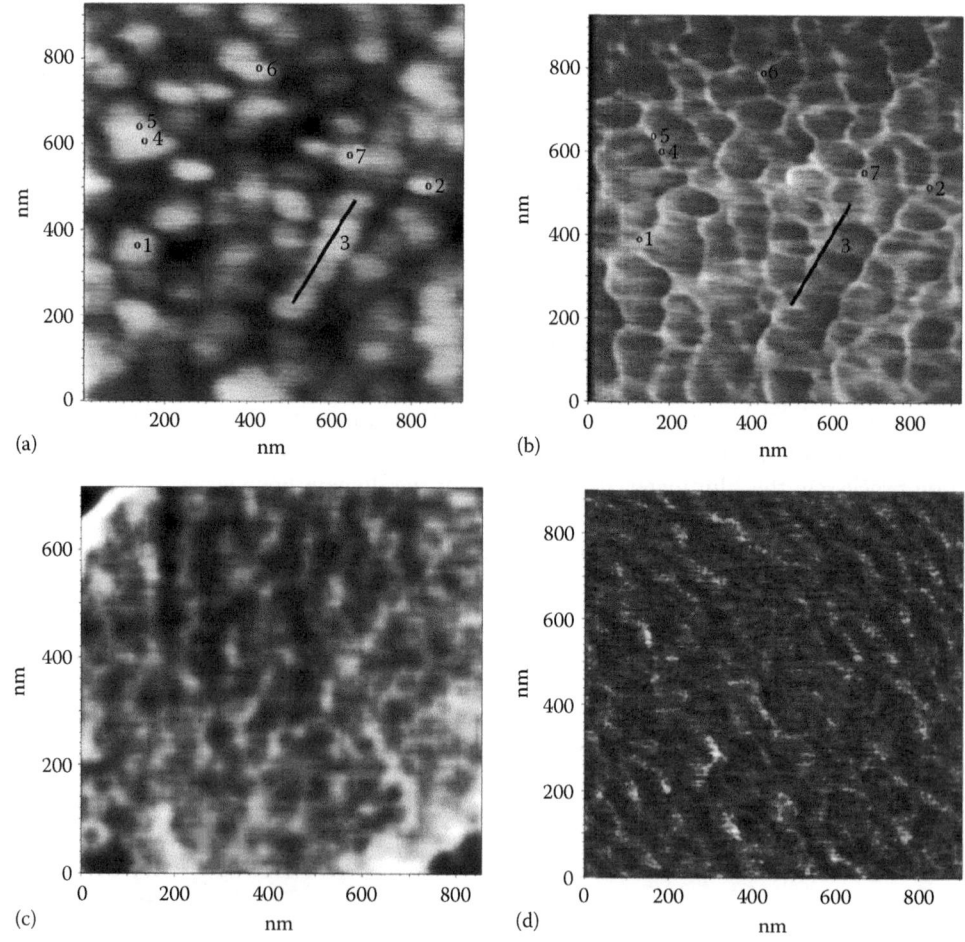

FIGURE 17.6
Surface morphology of an FLG film versus the duration of functionalization of the film in a 2.5% solution of HF in water. (a) Image of the surface of a 1.5 nm thick film treated in the solution for 4 min ($t \geq t_{th}$); the image was obtained in the surface-relief scan mode. (b) Image of the same surface simultaneously registered in the lateral-force mode. (c, d) Image of the surface of the same sample registered in AFM lateral-force measurements after 10 and 20 min functionalization treatments. (From Nebogatikova, N.A. et al., *Carbon*, 2014b.)

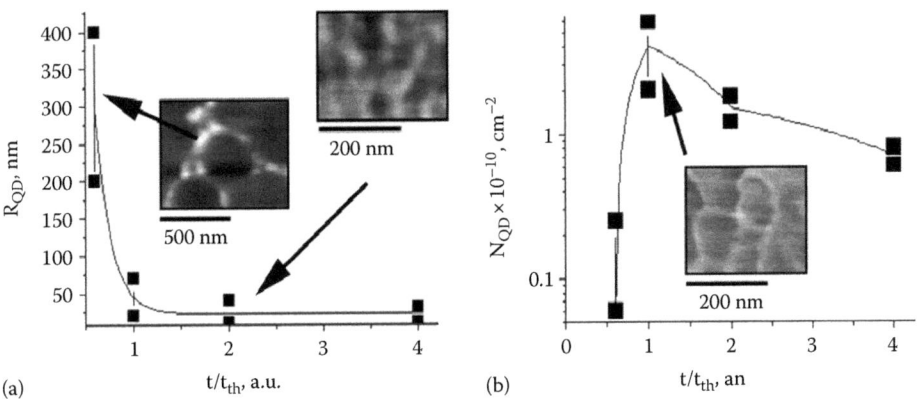

FIGURE 17.7
(a) Sizes of graphene islands in insulating fluorographene matrix versus the duration of treatment of samples in a solution of HF in water. (b) The density of graphene islands in insulating fluorographene matrix per unit surface area of sample versus the HF treatment duration. The insets explain which experimental data were used in evaluating the sizes and concentrations of the QDs. (From Nebogatikova, N.A. et al., *Carbon*, 2014b.)

It can be seen that points at the boundaries of the fluorinated network in Figure 17.6b refer to the regions that have the highest buckling height in Figure 17.6a.

The graphene islands (darker surface regions in Figure 17.6b) are localized at the *bottom* of the corrugated surface shown in Figure 17.6a. It is precisely such a structure that had to be expected on the basis of the mechanism that underlies the formation of the films—buckling of fluorinated regions accompanied by the relaxation of the generated stresses. When increasing the duration of the graphene and the FLG film treatment in aqueous solution of HF, both the sizes and the total number of graphene islands per unit surface area should decrease, and the width of the fluorinated part should increase. Data that confirm the latter prediction are shown in Figures 17.6c and d and 17.7a and b. The QD concentration varies within the range of 6×10^{10}–1×10^{10} cm^{-2}; with an increase in the time of the HF treatment to 20 min, the size of these QDs is decreased from 70 to 10–20 nm. The decrease in the number of graphene islands per unit surface area that was observed by prolonging the duration of the HF treatment can be attributed to the fact that, by that time, some of graphene islands had already undergone more complete fluorination.

Usually, as a result of fluorination, arrays of QDs formed in our samples, and the sizes and density of the QDs showed a decrease on prolonging the duration of the HF treatment. In the latter situation, the separation between the QDs or the widths of the fluorinated regions increases, and we have the possibility of directionally varying the QD sizes and the widths of the FG barriers between the QDs.

17.5 ELECTRICAL PROPERTIES OF ARRAYS OF QDs

In films that have the minimal width of FG barriers ($t \sim t_{th}$), measurements of current–voltage characteristics under the application of high voltages were performed with the aim to organize the conditions for initiating a high electric conductivity achieved through a film with potential barriers in between graphene islands (the voltage was up to 13 V, and the separation between the planar contacts was ~80 μm). As a result, we have found that, indeed, an emergence of conductivity in structure with potential barriers occurred in the films at voltages of more than 7–8 V (Figure 17.8a and c). When there was a decrease in the temperature at which the measurements were performed, from 350 to 80 K, no change in the voltage was required for restoration of conductivity to be observed; however, the density of the current through the film showed a distinct decrease. During the reverse sweep of the voltage, a pronounced hysteresis (3–4 V) was observed; this hysteresis was presumably due to the charge that was trapped either at the graphene islands or at the states in the FG barriers.

A pronounced field effect and hysteresis were also found while performing measurements on transistor configurations, with the silicon substrate being used as the gate (Figure 17.8b).

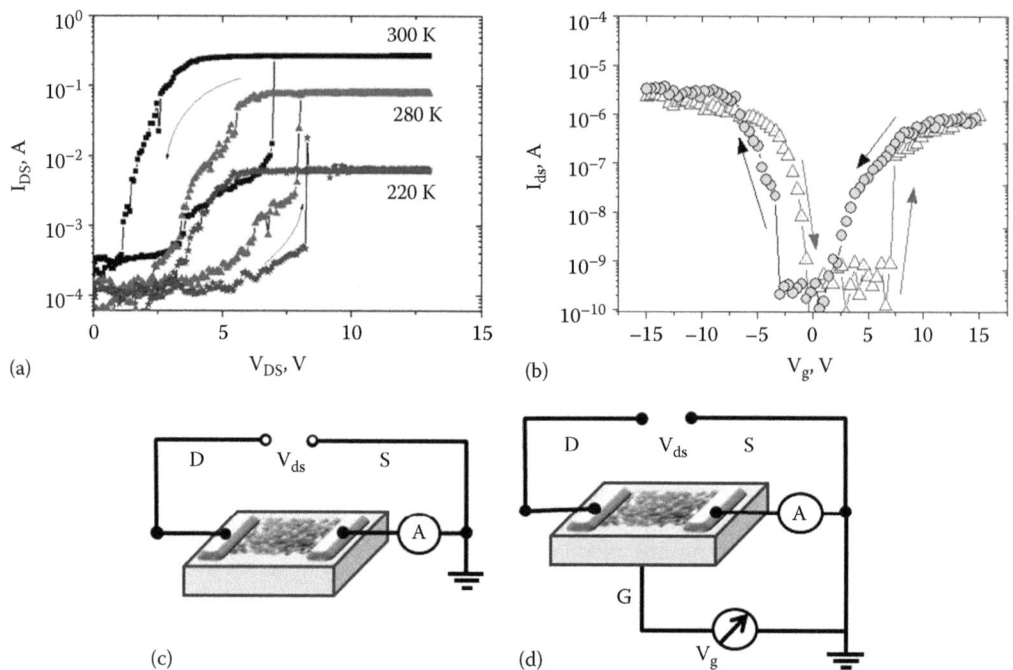

FIGURE 17.8

(a) Source–drain current–voltage characteristics versus the voltage applied between the planar contacts for a sample whose degree of fluorination was at the transition to insulator. Data for various temperatures are shown. (b) Transistor characteristics of the same sample, or the curves of the source–drain current versus the gate voltage, measured at room temperature with the silicon substrate used as the gate. Parts (c) and (d) of the figure show the sketches of the electric circuits that were used to perform measurements (a) and (b), respectively. In cases (b) and (d), the drain–source voltage V_{ds} was 0.2 V. For all of the measurements, the rate of voltage sweep was 0.5 V/s. (From Nebogatikova, N.A. et al., *Carbon*, 2014b.)

The measuring circuit that was used in the latter case is also shown in Figure 17.8d. Here, the neutrality point in the current–voltage characteristic or in the curve of source–drain current versus silicon-substrate potential was found to exhibit a shift toward positive voltages; this direction of the shift was indicative of *p*-type conductivity of graphene/FLG with QDs (sweeping toward positive voltages). The sheet density of the charge carriers was evaluated from the voltage shift ΔV_g of the neutrality point, p = 7.2 × 10^{10} ΔV_g, which was at (2–5) × 10^{11} cm^{-2}. The energy position of the Fermi level in the QDs, as evaluated by the expression $E_F = \hbar\nu_F(\pi p)^{0.5}$ (Güttinger et al. 2012) for several of the films treated in aqueous HF solution during the times t ~ t_{th}, proved to be 0.05–0.08 eV below the Dirac point. When sweeping the voltage in the opposite direction, the current–voltage characteristic turned out to be shifted toward zero voltage; such a shift pointed to a reduced doping level due to electron capturing in the system.

Films that had a higher fluorination degree—featuring, as will be shown in the following text, a greater width of FG barriers between the QDs, when biased with voltages of up to 15 V—no longer initiated conductivity.

Thus, measurements of current–voltage characteristics performed on samples after the transition to the insulating state confirm the presence of conducting graphene islands in the FG matrix and demonstrate the trapping of charges in such films. Recall that the amplitude of the main Raman peaks in such samples amounted to approximately 0.1 of the initial amplitude; this finding is also indicative of only partial fluorination of the films.

The curve of the saturation current at the positive bias voltages (see Figure 17.8a) versus the reciprocal temperature for a film that was functionalized to the state of conductor–insulator transition is shown in Figure 17.9. From the slopes of the curve, the activation energies were determined by I = I_0 exp(−E/kT), from which the numbers proved to be E1 = 0.12 ± 0.02 eV and E2 = 0.35 ± 0.02 eV. The inset to the figure shows the assumed band diagram of the film with G-QDs in the FG matrix. The electron work function Φ_G in the graphene and FLG films is known to be

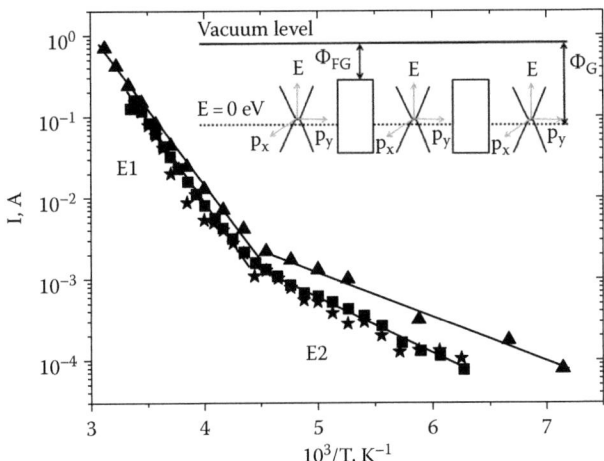

FIGURE 17.9
The saturation current at positive bias voltages (see Figure 17.8a) versus the reciprocal temperature for a 1.5 nm thick film whose degree of fluorination corresponded to the occurrence of a conductor–insulator transition in the film. The various symbols refer to measurements that were performed repeatedly on the same structure. E1 and E2 are the activation energies that were extracted from the linear parts of the I(1/T) curves. The inset shows the inferred band diagram of the film with graphene QDs in the fluorographene matrix. (From Nebogatikova, N.A. et al., *Carbon*, 2014b.)

approximately 4.5–4.7 eV (Giovannetti et al. 2008, Lee et al. 2011), while the detailed band alignment between the graphene and FG is not currently known. Theoretical studies predict a high value of the electron work function Φ_{FG} in the FG (approximately 6.0–7.3 eV) (Lee et al. 2011, Markevich et al. 2011), whereas experiments prove the existence of a barrier for the flow of charge carriers in the graphene/FG/graphene structure (Moon et al. 2013). The values of the activation energies that were determined from the temperature dependence of the electric current (Figure 17.9) also confirm the existence of barriers in the system *graphene/FG with embedded QDs*. The change in the barrier height at the temperature of 230 K is most likely due to the activation of charge carriers from different dimensional quantization levels in the QDs to an energy level at which optimal tunneling of charge carriers across the barriers proceeds. Measurements of the temperature dependence of the current in initial and weakly fluorinated films (i.e., before the conductor–insulator transition) have shown that the current was either independent of the temperature or showed a weakly decreasing behavior with the temperature, thus exhibiting no activation slope on the curve of the current versus the temperature.

The authors (Neek-Amal and Peeters 2011, Natsuki et al. 2012) have theoretically studied graphene buckling under a local change in the lattice constant and a generation of radial mechanical stresses. In Neek-Amal and Peeters (2011), it was shown that graphene buckling can occur under rather moderate strains of 0.4%, with the critical size of the buckled regions defined as the minimal size for the occurrence of buckling, which is approximately 6 nm. According to the calculations of Natsuki et al. (2012), an increase in the number of layers in an FLG film can be expected to have a pronounced influence on the formation of corrugations due to the interaction between the layers, which is induced by van der Waals forces. As a result, a change in the corrugation period can be expected to occur in samples upon increasing the film thickness. However, it was observed experimentally that the corrugation period remained roughly unchanged when the film thickness increased from one monolayer to 8 nm. This experimental finding calls for further study.

17.6 CHARGE SPECTROSCOPY OF QD ARRAYS

We discuss next, the additional arguments in favor of our statement that graphene or FLG QDs are in fact formed in insulating FG matrix. We have applied charge deep-level transient spectroscopy (Q-DLTS) to observe the size quantization levels in QDs (Antonova et al. 2014a).

Q-DLTS, which measures the charge (Q), enables the characterization of traps (quantum-confinement levels) in complicated structures that involve insulating layers and quantum wells or dots (Antonova et al. 2001, 2013). The widely used conventional capacitance modification of the DLTS technique (C-DLTS) was developed for uniform semiconductor structures without dielectric layers. In contrast, the Q-DLTS technique, which measures the charge (Q), enables the characterization of traps (quantum-confinement levels) in complicated structures that have insulated layers and quantum wells or QD (Antonova et al. 2001, 2013). This method uses voltage pulses to fill the graphene islands with charge carriers and, in this way, provides suitable conditions for subsequent observation of the charge emissions to be analyzed. The measurements were performed by varying the time window $\tau_m = (t_2 - t_1)/\ln(t_2/t_1)$ while keeping the temperature unchanged (t_1 and t_2 are the times at which the signal due to the relaxation of the trapped charge $\Delta Q = Q(t_2) - Q(t_1)$ was observed after the end of the filling pulse). The magnitude of the filling pulse was equal to 9–11 V. The analysis of the Q-DLTS spectrum provides the parameters of the localized states in our samples: the carrier emission time, the activation energy(s), and the density of the states. The Q-DLTS measurements are performed in a closed holder and in a holder that has an open window for daylight (~10^{17} photons/cm^{2}s). The emission rate of the holes (electrons) $e_{p(n)}$ from the QDs in a bulk semiconductor material can be written as follows:

$$\tau_m^{-1} = \sigma v_F N_C \exp\left(\frac{-E_a}{kT}\right), \tag{17.2}$$

$$N_C = 4\frac{E}{2\pi\hbar^2 v_F^2} = 4\frac{\sqrt{n_s}}{2\sqrt{\pi}},$$

where
  σ is the carrier capture cross section
  T is the temperature
  E_a is the activation energy
  N_C is the concentration of the states in the conduction band
  $\hbar$ is the reduced Planck's constant
  k is the Boltzmann constant
  v_F is the Fermi velocity around K points ($v_F = 10^6$ m/s)

Thus, the activation energies E_a are extracted from Arrhenius plots in the coordinates $\ln(\tau_m^{-1})$ versus the reversed temperature.

Q-DLTS spectra were obtained for a set of graphene films of different thicknesses (one monolayer to 3 nm) before and after the treatment of samples in 2.5% solution of HF in water for different amounts of time. It was found that the peaks in the Q-DLTS spectra could be observed only in samples that were treated in HF for the time t ~ t_{th} or for slightly longer times (Table 17.1). The latter means that, first, the QD-related peaks in Q-DLTS spectra were only found in films that indeed contained QDs and, second, the potential barriers (fluorinated graphene regions) that impeded the charge transport between neighboring QDs were tunnelable barriers. Typical measured Q-DLTS spectra and typical Arrhenius plots that were obtained for one of the examined samples are shown in Figure 17.10a. The values of the activation energies that were extracted from the plots are listed in Table 17.1. One to three activation energies were identified in each film given a fluorination treatment in aqueous solution of HF for the time t ~ t_{th}.

The position of the maximum of a specific peak in measured Q-DLTS spectra allows us to evaluate the time of the carrier emission from QDs at different temperatures. The temperature dependences of the carrier emission times for all of the films are summarized in Figure 17.10b. Evidently, an increase in the film thickness leads to a strong decrease in the emission time. The ratio between the carrier emission times for G-QDs and 3 nm thick FLG QD is approximately 10^4.

As was mentioned earlier, the fluorination process is a self-organized process, which leads to the formation of a corrugated (buckled) surface due to the ~1% volume difference between the unit cells of graphene and FG. It is widely recognized now that the buckling in the layered systems is driven by strongly modulated strain fields. We hypothesize that, in our samples, the buckling of the top graphene sheet in combination with the structural defects that are present at the grain

TABLE 17.1 Description of the Used Samples and Activation Energies Extracted from DLTS

Sample	Description	DLTS Level Energies
FLG-1	CVD, d = 2.5 nm, $t/t_{cr} < 1$	No peaks
FLG-2	CVD, d = 2 nm, $t/t_{cr} \sim 1$	E2 = 0.14 eV
FLG-3	CVD, d = 3 nm, $t/t_{cr} > 1$	E1 = 0.06 eV
		E2 = 0.10 eV
FLG-4	CVD, d = 3 nm, $t/t_{cr} > 1$	E1 = 0.07 eV
		E2 = 0.22 eV
		E3 = 0.34 eV
FLG-5	CVD, d = 2 nm, $t/t_{cr} > 1$	No peaks
G-1	Electr. exfoliation, bigraphene, $t/t_{cr} < 1$	No peaks
	FWHM = 40 cm^{-1}, d = 0.8 nm	
G-2	Electr. exfoliation, bigraphene, $t/t_{cr} \sim 1$	E2 = 0.15 eV
	FWHM = 40 cm^{-1}, d = 0.8 nm	E3 = 0.29 eV
G-3	CVD, graphene, $t/t_{cr} > 1$	
	FWHM = 38 cm^{-1}, d = 0.5 nm	E3 = 0.29 eV
G-4	CVD, bigraphene, $t/t_{cr} > 1$	E1 = 0.09 eV
	FWHM = 43 cm^{-1}, d = 0.9 nm	E2 = 0.18 eV
		E3 = 0.33 eV

Notes: The fabrication method (electrostatic exfoliation of highly ordered pyrolytic graphite or low-pressure chemical vapor deposition (CVD) growth of graphene sheets on Cu substrates at 1000°C followed by the subsequent transfer of graphene sheets onto a 300 nm SiO$_2$/Si substrate, the full width at half magnitude (FWHM) of the 2D Raman peak, the film thickness *d* as determined by atomic force microscopy, the duration of the treatment of the samples in the 2.5% solution of HF in water expressed in the units of resistivity step time t_{cr}), and the activation energies E1, E2, and E3 of the emission process of the charge carriers from the QDs as extracted from Arrhenius plots.

boundaries could provide access for HF:H$_2$O solution to the underlying graphene sheets. Generally, the buckling of an FLG film must lead to a reduction in the lateral dimensions of the buckled regions with the depth (see the inset to Figure 17.10b). In the latter situation, it can be expected that the QDs created by our method in FLG films would be shaped as truncated pyramids, and the fluorinated graphene regions would be shaped as *cones* (see the inset to Figure 17.10b).

Consider now the strong dependence of the time of the carrier emission from localized states in the QDs on the thickness of the film in which the QDs were created. We assume that the effect is due to the tapering shape of the FG regions that present tunnel barriers for charge carriers that are trapped at QDs and appear to be shaped as triangles in the cross-sectional view of the film (see the inset to Figure 17.10b). The fluorination proceeds successively, layer by layer, and this process leads to a reduction in the width of the fluorinated region in the deep film layers. The charge-carrier tunneling probability depends exponentially on the tunnel-barrier width; this relationship leads to a strong dependence of the carrier emission time on the film thickness or on the QD thickness, which is the same as the film thickness. On the assumption that the ratio between the tunneling probabilities of the charge carriers at different depths amounts to 10^3 (this value is taken to be equal to the ratio between the carrier emission times in bigraphene and in a 3 nm thick FLG film), one can readily obtain that the barrier width for the 3 nm thick FLG film is expected to be approximately 2 nm smaller than that in bigraphene. Such an estimate looks quite reasonable. It should be noted here that similar measurements that are performed for Si and Ge QDs in SiO$_2$ and Al$_2$O$_3$ dielectric layers normally yield a carrier emission time of several milliseconds at room temperature, irrespective of the QD concentration and QD creation method (Antonova et al. 2011b, 2013). Moreover, it was impossible to alter that time by varying the method. In those cases, the emission time value was defined by the technological rule for the thickness of reliable dielectric layers (2–3 nm). In the present study, we show that it is possible to control the time of the carrier emission from the QDs. Choosing a film of a desired thickness and using the proposed method for creating QDs in this film, one can predefine the carrier emission time in a broad range of values from micro- to milliseconds, which is a possibility that is of extreme significance for developing memory devices.

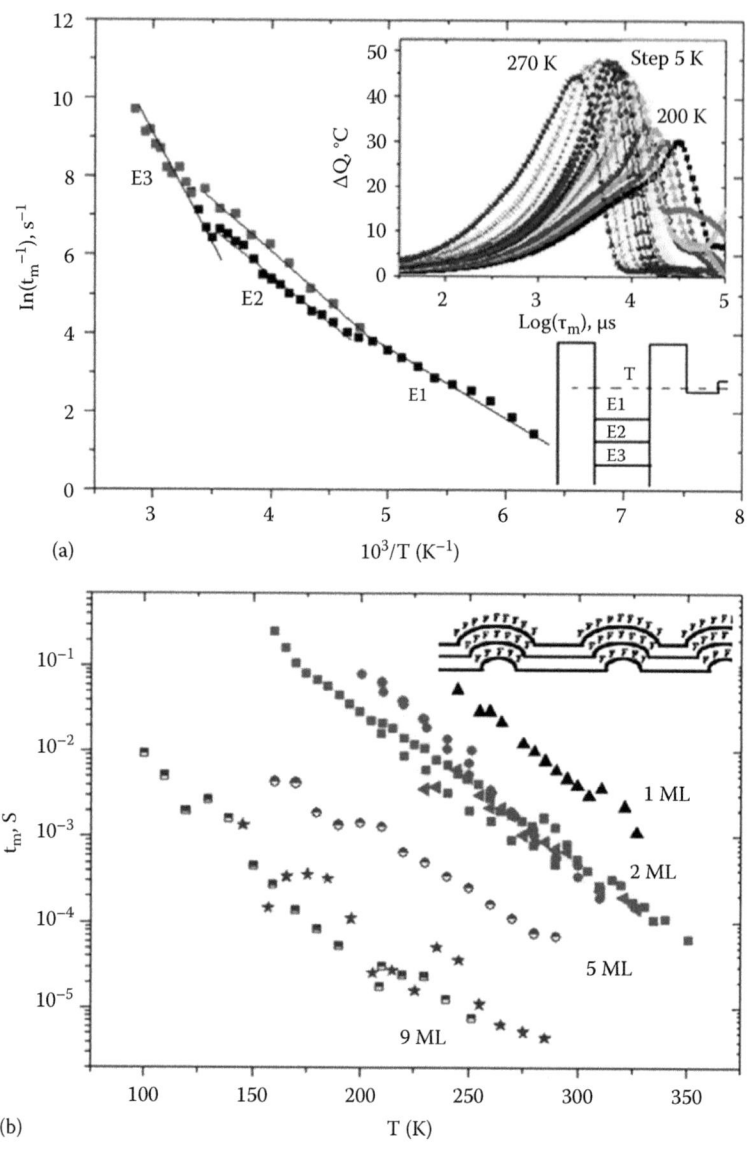

FIGURE 17.10

(a) Arrhenius plots for a bigraphene film with QDs (sample G–4) and (b) the emission time of charge carriers as a function of the temperature for all of the investigated samples. The Q-DLTS measurements were performed by varying the time window $\tau_m = (t_2 - t_1)/\ln(t_2/t_1)$ while keeping the temperature unchanged (t_1 and t_2 are the times at which the signal due to the relaxation of the trapped charge $\Delta Q = Q(t_2) - Q(t_1)$ was observed after the end of the filling pulse). The filling pulse magnitude was in the interval of 9–11 V. The thickness of the measured films is given in (b) as a parameter in the monolayers (ML). The insets in (a) show the Q-DLTS spectra measured for a graphene film with QDs in different temperature ranges and the suggested interpretation of the activation energies E1–E3. The inset in (b) shows a sketch that illustrates the shape of the self-organized QDs and the insulating matrix of the corrugated morphology formed by chemical functionalization of the graphene layers. (From Antonova, I.V. et al., *J. Appl. Phys.*, 2014b.)

Even if there exists no strict lateral correlation in the in-depth propagation of the fluorination process, and nontriangular (conic) cross-sectional barriers thus form in the film, the addition of new layers to the film can be expected to lead to an increased number of potential hopping paths for charge carriers that migrate between QDs and, hence, to a reduced value of the critical hopping distance for such migration; as a result, the time of carrier emission from the QDs would also show a decrease. The latter mechanism offers an alternative explanation for the strong dependence of the carrier emission time on the film thickness. In spite of the absence of electrical conduction in our films, the percolation/hopping conduction theory could prove useful in describing transient processes in examined films (Stalinga 2011). The probability P of carrier hopping between QDs spaced by a distance R with the energy separation of hole levels in QDs E is P = exp (−aR − E/kT), where a is a constant and k is the Boltzmann constant (Sklovskii and Efros 1984). A decrease in R will lead to an increased value of P.

Simultaneously, the activation energies for carrier emission as deduced from experimental data differ quite moderately between QD systems of different thicknesses. Data in Table 17.1 demonstrate only a weak increase in the activation energy with an increase in the film thickness. The latter finding does not contradict our assumption about a conic shape of FG barriers and the truncated pyramidal shape of the QDs in our films because the dependence of the size quantization levels in the QDs on the QD size is normally weaker and obeys a root law.

We mentioned earlier the *p*-type of doping of our QDs and estimated the energy position of the Fermi level as 0.05–0.08 eV below the Dirac point. We assume here that the experimentally observed energies E1, E2, and E3 can be interpreted as energies that are required for holes to be activated from QD size quantization levels to an optimum energy level at which the tunneling of holes through the FG barriers proceeds at the highest possible rate (level T in the inset to Figure 17.10a). Realization of an optimal energy level for the tunneling transport of the charge carriers between the QDs is intimately related to specific features of the implemented QD self-formation process: because of the considerable scatter in the width of the tunnel barriers, in certain directions, there could arise narrow *conducting* channels or regions that have a low degree of fluorination of the material and, hence, the barrier height could exhibit local spatial variations. Alternatively, defect centers that have levels that are situated inside the barriers might exist. Nevertheless, the observed reproducibility of the activation energies that are obtained in repeatedly performed Q-DLTS measurements and also the nearly identical values of activation energies that are obtained in graphene films of different thicknesses point to reproducibility of the energy level for the optimal tunneling of charge carriers. As follows from the inset to Figure 17.4a, the difference E3–E2 between the energy levels of the charge carriers in the QDs must be roughly equal to the QD bandgap width. The value of the bandgap energy in G-QDs embedded in an FG matrix was theoretically analyzed by Ribas et al. (2011), who showed that for G-QDs with armchair (AC)- and zigzag (ZZ)-type edges, reasonable estimates of the G-QD bandgap energy can be obtained using the following formulas:

$$E_g = \frac{14.1}{\frac{1}{n^2} - 0.01} \text{ (eV)} \quad \text{for QDs with AC edges,} \tag{17.3}$$

$$E_g = \frac{19.4}{\frac{1}{n^2} + 0.14} \text{ (eV)} \quad \text{for QDs with ZZ edges.} \tag{17.4}$$

In (17.3) and (17.4), *n* is the total number of atoms in the QD. Accounting for the fact that the energy difference E3–E2 is roughly equal to 0.12–0.14 eV, we can evaluate the sizes of the QDs that contributed to the Q-DLTS signal in our samples. We obtain n = (1.4 − 1.2) × 10⁴ for G-QDs with AC edges and n = (3.0 − 2.2) × 10³ for G-QDs with ZZ edges. Such numbers of atoms refer to QD sizes ~23–20 nm for AC and ~9–8 nm for ZZ edges. The bandgap energy can also be estimated using the following formula for nanoribbons, which is also widely used for QDs[5]:

$$E_g = \frac{2\pi\hbar\upsilon_F}{W} \text{ (eV)} \tag{17.5}$$

In (17.4), W is the width of the nanoribbon. Using formula (17.3), we obtain that the energies 0.14–0.12 eV in the forbidden gap correspond to the QD sizes 30–34 nm. All of the obtained estimates correlate well with the values of the QD sizes in our samples as revealed by lateral-force AFM measurements, specifically, 20–70 nm.

From the amplitudes of the Q-DLTS peaks, we were also able to evaluate the density of the states that cause the emergence of Q-DLTS peaks in our samples. The density-of-states values that are obtained in different structures was found to vary moderately, within the range N = (0.7 − 2) × 10¹¹ cm⁻². Using the known film-thickness values for our films (20–40 μm), the

TABLE 17.2 Description of the Used Samples

Sample	Description	Dark	Daylight
G-3	CVD, graphene, FWHM = 38 cm^{-1}, d = 0.5 nm		E0* = 0.00 eV
			E1* = 0.02 eV
			E2* = 0.08 eV
		E3 = 0.29 eV	E3* = 0.17 eV
G-4	CVD, bigraphene, FWHM = 43 cm^{-1}, d = 0.9 nm		E0* = 0.00 eV
		E1 = 0.09 eV	
		E2 = 0.18 eV	
		E3 = 0.33 eV	E3* = 0.16 eV

Source: Antonova, I.V. et al., *J. Appl. Phys.*, 2014b.

Notes: The method of fabrication and FWHM of the 2D peak extracted from Raman spectra, thickness of the film d extracted from AFM measurements, and time of treatment in the 2.5% HF aqueous solutions in comparison with the time of resistivity step t_{cr} and the activated energies E1, E2, and E3 for carrier emission from QDs, as extracted from Arrhenius plots.

corrugation-period values (100–150 nm) and the density-of-states values as evaluated from Q-DLTS data, we can draw a conclusion that only 5%–10% of all QDs contributed to the Q-DLTS signal in our samples. The latter implies that the conditions for QD recharging were realized under the adopted experimental conditions only for an insignificant fraction of QDs.

The Q-DLTS spectra for the samples measured in the closed holder and in the holder with an open window for the daylight demonstrate peaks in the Q-DLTS spectra in the entire temperature range of 350–80 K, whereas in the case when the closed holder is used, the peaks are observed only in the temperature range of 350 to 170–250 K. For temperatures that are lower than the 170–250 K, the peaks were observed in the Q-DLTS spectra only in the light-assisted case, and the peak position was not changed when the temperature was decreased (to zero activation energy). If we had closed the window in the holder, then the peak in the spectra would have vanished completely. The activation energy values that were extracted from the Arrhenius plots are listed in Table 17.2. From one to three activated energies are found for the samples. The suggested origin of E1–E3 are the energies that are required for carrier activation from a quantum-confinement level in QDs. A comparison of peak magnitudes that were measured with and without the daylight demonstrates that approximately the same number of states recharged during the Q-DLTS measurements in both cases. The daylight during the Q-DLTS measurements leads to a strong decrease in the carrier emission time from the QDs. The relation between the carrier emission time for the G-QDs and bilayer graphene (BG)-QDs in a closed holder and in a holder with an open window for daylight is approximately 10^2–10^4 times.

A set of activation energies that were extracted from Q-DLTS are connected by overcoming the barrier between the levels in the QDs and the additional minimum. The carrier transport from the additional minimum is suggested to be a migration that is most likely described by the variable range hopping theory. The light assistance during the Q-DLTS measurements leads to a carrier excitation due to the light absorption, which results in a possibility of the carrier migration from QDs without activation. Small energies from 0 to 0.17 eV (E0* – E4*) are suggested to correspond with the energy separation during the hopping migration of the carriers to contacts. The last suggestion follows from the very small emission times that are observed for the light-assisted Q-DLTS measurements.

An alternative explanation is based on a suggestion about a charging component in the barrier. In this case, the photon absorption must lead to a decrease in the immobile charge in the barrier, a decrease in the barrier height, and a decrease in the activation energies.

17.7 CONCLUSION, REMARKS, AND OUTLOOK

Chemical functionalization of graphene and FLG is a powerful approach when the physical properties can be engineered to fit a specific requirement for both fundamental science and future applications. Treatments in aqueous solution of HF were found to result in dramatic changes in the structural and electrical properties of graphene and FLG; those changes include an increase in the resistivity ($\geq 10^{11}$–10^{12} ohm/sq.), the vanishing of all of the peaks from the Raman spectra of the samples, the formation of a periodic nanoswell relief on the surface of the films (step approximately 100–200 nm and height ~4–10 nm), and the appearance of the F-related peaks in x-ray photoelectron spectra (687.7 and 288 eV). Reversible behavior of HF-functionalized samples due to additional anneals at 400°C–450°C was also demonstrated. Graphene fluorination is assumed to be under the HF treatment. The conditions and parameters of the fluorination process were determined. The treatments typically lasted for several minutes. Formation of G-QDs was revealed to occur after the steplike increase in film resistance. Creation of an array of conducting graphene islands, or G-QDs, in the films, the QDs being isolated from each other by FG potential barriers, is a self-organized process. Formation of C–F bonds in local areas due to partial fluorination leads to a local increase in the lattice constant and to the development of in-plane Euler instability, and relaxation of the strain leads to the formation of a self-organized corrugation of the film. The QD sizes were estimated to range from 70–20 nm (at a QD density of 6×10^{10} cm^{-2}) to 10–20 nm (at a QD density of 1×10^{10} cm^{-2}). The heights of the graphene/FG barriers were estimated as E1 = 0.12 ± 0.02 eV at room temperature and E2 = 0.35 ± 0.02 eV for T < 230 K. One to three energy levels were extracted from the charge spectroscopy measurements on samples that had graphene and FLG QDs (the QD size was ~50–70 nm). The bandgap energy in the QDs as evaluated from the energy position of the quantum-confinement levels is suggested to equal 0.12–0.14 eV. The most surprising finding was a strong (approximately four orders of magnitude) decrease in the carrier emission time (or charge relaxation time) that was observed on increasing the QD thickness. The latter time is a parameter that is important for different applications, for example, for memory devices. This experimental finding provides a useful instrument for governing the emission of carriers from QDs in FLG films by choosing an appropriate thickness for the QDs. The time of daylight-assisted carrier emission from the QDs was found to also strongly decrease down to ~$(0.8$–$4) \times 10^{-5}$ s; this finding suggests that the barriers for emission include a charge-dependent component.

Such films with QDs are very attractive for fundamental study and different applications. For example, the films that demonstrate the recovery of conductivity and large hysteresis are propitious for memory applications. Moreover, a strong dependence of the carrier emission time on the film thickness provides a useful instrument for governing the charge relaxation on the QDs in the FLG films by choosing an appropriate film thickness. The further study of this system with QDs most likely would demonstrate other surprising and promising findings with respect to application properties. It is necessary also to mention here that the method of partial fluorination and the formation of the layers with QDs are very simple and, in practice, feasible. Making the systems with graphene ODs would enable the development of novel devices that cannot be fabricated using the conventional materials. It was also found that the functionalization of graphene with HF can be controlled by a preliminary treatment of graphene in isopropyl alcohol (Nebogatikova et al. 2014a). The latter treatment suppresses the effects due to the subsequent functionalization of graphene or FLG in HF-containing solutions. The combination of the two treatments gives us a key for the nanodesign of graphene-based devices.

The ability to dial up a specific physical property in graphene-based materials simply by selecting chemical treatment or chemical species that functionalizes graphene extends dramatically the horizons of future electronic applications targeted by graphene. For example, the ability to engineer electrical properties in functionalized FLG might enable a new class of graphene-based flexible and transparent memories.

REFERENCES

Acik, M., Lee, G., Mattevi, C. et al. 2011. The role of oxygen during thermal reduction of graphene oxide studied by infrared absorption spectroscopy. *J. Phys. Chem. C.* 115: 19761–19781.

An, J., Voelkl, E., Suk, J.W. et al. 2011. Domain (grain) boundaries and evidence of "twinlike" structures in chemically vapor deposited grown graphene. *ACS Nano* 5: 2433–2439.

Antonova, I.V., Mutilin, S., Seleznev, V. et al. 2011a. Extremely high response of electrostatically exfoliated few-layered graphene flakes to ammonia adsorption. *Nanotechnology* 22: 285502.

Antonova, I.V., Naumova, O.V., Stano, J. et al. 2001. Traps at bonded interface in SOI structures. *Appl. Phys. Lett.* 79: 4539–4541.

Antonova, I.V., Nebogatikova, N.A., and V.Y. Prinz. 2014a. Self-organized arrays of graphene and few-layer graphene quantum dots in fluorographene matrix: Charge transient spectroscopy. *Appl. Phys. Lett.* 104: 193108.

Antonova, I.V., Nebogatikova, N.A., Prinz, V.Y. et al. 2014b. Light-assisted recharging of graphene quantum dots in fluorographene matrix. *J. Appl. Phys.* 116: 134310.

Antonova, I.V., Popov, V.I., Smagulova, S.A. et al. 2013. Charge deep-level transient spectroscopy of SiO_2 and Al_2O_3 layers with embedded Ge nanocrystals. *J. Appl. Phys.* 113: 084308.

Antonova, I.V., Smagulova, S.A., Neustroev, E.P. et al. 2011b. Charge spectroscopy of SiO_2 layers with embedded silicon nanocrystals modified by irradiation with high-energy ions *Semiconductors* 45: 582–586.

Bacon, M., Bradley, S.J., and T. Nann. 2014. Graphene quantum dots. *Part Part. Syst. Charact.* 31: 415–428.

Bera, D., Qian, L., Tseng, T.K. et al. 2010. Quantum dots and their multimodal applications: A review. *Material* 3: 2260–2345.

Bimberg, D. and U.W. Pohl. 2011. Quantum dots: Promises and accomplishments. *Mater. Today* 14: 388–397.

Blake, P., Hill, E.W., Neto, A.H.C. et al. 2007. Making graphene visible. *Appl. Phys. Lett.* 91: 063124.

Bolotin, K.I., Sikes, K.J., Jiang, Z. et al. 2008. Ultrahigh electron mobility in suspended graphene. *Solid State Commun.* 146: 351–355.

Boukhvalov, D.W. and M.I. Katsnelson. 2009. Chemical functionalization of graphene. *J. Phys. D Condens. Matter.* 21: 344205.

Bruna, B., Massessi, C., Cassiago, A. et al. 2011. Synthesis and properties of monolayer graphene oxyfluoride. *J. Mater. Chem.* 21: 18730–18737.

Cheng, S.-H., Zou, K., Okino, F. et al. 2010. Reversible fluorination of graphene: Evidence of a two-dimensional wide bandgap semiconductor. *Phys. Rev. B* 81: 20543.

Chuang, C., Puddy, R.K., Connolly, M.R. et al. 2012. Evidence for formation of multiquantum dots in hydrogenated graphene. *Nano Res. Lett.* 7: 459.

Craciun, M.F., Khrapach, I., Barnes, M.D. et al. 2013. Properties and applications of chemically functionalized graphene. *J. Phys. Condens. Matter* 25: 423201.

Elias, D.C., Nair, R.R., Mohiuddin, T.M.G. et al. 2009. Control of graphene's properties by reversible hydrogenation: Evidence for graphane. *Science* 323: 610–613.

Feng, Q., Cao, Q.Q., Li, M. et al. 2013. Synthesis and photoluminescence of fluorinated graphene quantum dots. *Appl. Phys. Lett.* 102: 013111.

Geim, A.K. 2009. Graphene: Status and prospects. *Science* 324: 1530–1534.

Giovannetti, G., Khomyakov, P.A., Brocks, G. et al. 2008. Doping graphene with metal contacts. *Phys. Rev. Lett.* 101: 026803.

Gupta, A., Chen, G., Joshi, P. et al. 2006. Raman scattering from high-frequency phonons in supported n-graphene layer films. *Nano Lett.* 6: 2667–2673.

Güttinger, J., Molitor, F., Stampfer, C. et al. 2012. Transport through graphene quantum dots. *Rep. Prog. Phys.* 75: 126502.

Hu, C.F., Liu, Y.L., Yang, Y.H. et al. 2013. One-step preparation of nitrogen-doped graphene quantum dots from oxidized debris of graphene oxide. *J. Mater. Chem. B* 1: 39–42.

Jung, N., Kim, N., Jockusch, S. et al. 2009. Charge transfer chemical doping of few layer graphenes: Charge distribution and band gap formation. *Nano Lett.* 9: 4133–4137.

Karapet'yants, M.K. and S.I. Drakin. 1994. *General and Inorganic Chemistry*. Khimiya, Moscow, Russia [in Russian].

Kim, K., Lee, Z., Regan, W. et al. 2011. Grain boundary mapping in polycrystalline graphene. *ACS Nano* 5: 2142–2146.

Komarov, B.A. 2004. Special features of radiation-defect annealing in silicon p-n structures: The role of Fe impurity atoms. *Semiconductors* 38: 1041–1046.

Kosynkin, D.V., Higginbotham, A.L., Sinitskii, A. et al. 2009. Longitudinal unzipping of carbon nanotubes to form graphene nanoribbons. *Nature* 458: 872–876.

Kotin, I.A., Antonova, I.V., Komonov, A.I. et al. 2013. High carrier mobility in graphene on atomically flat high-resistivity layer. *J. Phys. D Appl. Phys.* 46: 285303.

Lee, J., Kim, K., and W.I. Park. 2012. Uniform graphene quantum dots patterned from self-assembled silica nanodots. *Nano Lett.* 12: 6078–6083.

Lee, W.K., Robinson, J.T., Gunlycke, D. et al. 2011. Chemically isolated graphene nanoribbons reversibly formed in fluorographene using polymer nanowire masks. *Nano Lett.* 11: 5461.

Li, L., Wu, G., Yang, G. et al. 2013. Focusing on luminescent graphene quantum dots: Current status and future perspectives. *Nanoscale* 5: 4015–4039.

Li, L.-S. and X. Yan. 2010. Colloidal graphene quantum dots. *J. Phys. Chem. Lett.* 1: 2572–2576.

Liu, H.Y., Hou, Z.F., Hu, C.H. et al. 2012. Electronic and magnetic properties of fluorinated graphene with different coverage of fluorine. *J. Phys. Chem. C* 116: 18193–18201.

Liu, Q., Guo, B.D., Rao, Z.Y. et al. 2013. Strong two-photon-induced fluorescence from photostable, biocompatible nitrogen-doped graphene quantum dots for cellular and deep-tissue imaging. *Nano Lett.* 13: 2436–2441.

Luo, Z., Yu, T., Kim, K. et al. 2009. Thickness-dependent reversible hydrogenation of graphene layers. *ACS Nano* 3: 1781–1788.

Markevich, A., Jones, R., and P.R. Briddon. 2011. Doping of fluorographene by surface adsorbates. *Phys. Rev. B* 84: 115439.

Moon, J.S., Seo, H.C., Stratan, F. et al. 2013. Lateral graphene heterostructure field-effect transistor. *IEEE Electron Device Lett.* 34: 1190.

Nair, R.R., Ren, W., Jalil, R. et al. 2010. Fluorographene: A two-dimensional counterpart of Teflon. *Small* 6: 2877–2884.

Natsuki, T., Shi, J.X., and Q.Q. Ni. 2012. Buckling instability of circular double-layered graphene sheets. *J. Phys. Condens. Matter* 24: 135004.

Neaton, J.B., Hybertsen, M.S., and S.G. Louie. 2006. Renormalization of molecular electronic levels at metal-molecule interfaces. *Phys. Rev. Lett.* 97: 216405.

Nebogatikova, N.A., Antonova, I.V., Prinz, V.Y. et al. 2014a. Functionalization of graphene and few-layer graphene films in hydrofluoric acid aqueous solution. *Nanotechnol. Russia* 9: 51–59.

Nebogatikova, N.A., Antonova, I.V., Prinz, V.Y. et al. 2014b. Graphene quantum dots in fluorographene matrix formed by means of chemical functionalization. *Carbon* 77: 1095–1103.

Nebogatikova, N.A., Antonova, I.V., Volodin, V.A. et al. 2013. Functionalization of graphene and few-layer graphene with aqueous solution of hydrofluoric acid. *Physica E* 52: 106–111.

Neek-Amal, M. and F.M. Peeters. 2011. Buckled circular monolayer graphene: A graphene nano-bowl. *J. Phys. Condens. Matter* 23: 0945002.

Novoselov, K.S., Falko, V.I., Colombo, L. et al. 2012. A roadmap for graphene. *Nature* 498: 192–200.

Pan, D.Y., Guo, L., Zhang, J.C. et al. 2012. Cutting sp^2 clusters in graphene sheets into colloidal graphene quantum dots with strong green fluorescence. *J. Mater. Chem.* 22: 3314–3318.

Plank, N.O., Jiang, L., and R. Cheung. 2003. Fluorination of carbon nanotubes in CF4 plasma. *Appl. Phys. Lett.* 83: 2426–2428.

Ponomarenko, L.A., Schedin, F., Katsnelson, M.I. et al. 2008. Chaotic dirac billiard in graphene quantum dots. *Science* 320: 356–358.

Pu, L., Ma, Y., Zhang, W. et al. 2013. Simple method for the fluorinated functionalization of graphene oxide. *RSC Adv.* 3: 3881–3884.

Ribas, M.A., Singh, A.K., Sorokin, P.B. et al. 2011. Pattering nanoroads and quantum dots on fluorinated graphene. *Nano Res.* 4: 143–152.

Robinson, J.T., Burgess, J.S., Junkermeier, C.E. et al. 2010. Properties of fluorinated films. *Nano Lett.* 10: 3001–3005.

Ryu, S., Han, M.Y., Maultzsch, J. et al. 2008. Reversible basal plane hydrogenation of graphene *Nano Lett.* 8: 4597–4602.

Shao, D., Sawyer, S., Hu, T. et al. 2012. Photoconductive enhancement effects of graphene quantum dots on zno nanoparticle photodetectors. *Proceedings of Lester Eastman Conference on High Performance Devices (LEC),* IEEE Press, pp. 1–4.

Shen, J., Zhu, Y., Yang, X. et al. 2012. One-pot hydrothermal synthesis of graphene quantum dots surface-passivated by polyethylene glycol and their photoelectric conversion under near-infrared light. *New J. Chem.* Singapore, 36: 97–101.

Sherpa, S.D., Levitin, G., and D.W. Hess. 2012. Effect of the polarity of carbon-fluorine bonds on the work function of plasma-fluorinated epitaxial graphene. *Appl. Phys. Lett.* 101: 111602.

Sidorov, A.N., Yazdanpanah, M.M., Jalilian, R. et al. 2007. Electrostatic deposition of graphene. *Nanotechnology* 18: 135301.

Sklovskii, B.I. and A.L. Efros. 1984. *Electronic Properties of Doped Semiconductors.* Springer, Heidelberg, Germany.

Soldano, C., Mahmood, A., and E. Dujardin. 2010. Production, properties and potential of graphene. *Carbon* 48: 2127–2150.

Stalinga, P. 2011. Electronic transport in organic materials: Comparison of band theory with percolation/(variable range) hopping theory. *Adv. Mater.* 23: 3356–3362.

Tang, L.B., Ji, R.B., Li, X.M. et al. Size-dependent structural and optical characteristics of glucose-derived graphene quantum dots. *Part. Part. Syst. Charact.* 30: 523–531.

Walter, A.L., Jeon, K.-J., Speck, A.B. et al. 2011. Highly p-doped graphene obtained by fluorine intercalation. *Appl. Phys. Lett.* 98: 184102.

Wang, Z., Wang, J., Li, Z. et al. 2012. Synthesis of fluorinated graphene with tunable degree of fluorination. *Carbon* 50: 5403–5410.

Withers, F., Dubois, M., and A.K. Savchenko. 2010. Electron properties of fluorinated single-layer graphene transistors. *Phys. Rev. B.* 82: 1–4.

Xie, M.M., Su, Y.S., Lu, X.N. et al. 2013. Blue and green photoluminescence graphene quantum dots synthesized from carbon fibers. *Mater. Lett.* 93: 161–164.

Xu, M., Li, Z., Zhu, X. et al. 2013. Hydrothermal/solvothermal synthesis of graphene quantum dots and their biological applications. *Nano Biomed. Eng.* 5: 65–71.

Yan, X., Cui, X., Li, L.B. et al. 2010. Large, solution-processable graphene quantum dots as light absorbers for photovoltaics. *Nano Lett.* 10: 1869–1873.

Zarenia, M., Chaves, A., Farias, G.A. et al. 2011. Energy levels of triangular and hexagonal graphene quantum dots: A comparative study between the tight-binding and Dirac equation approach. *Phys. Rev. B* 84: 245403.

Zhou, X., Zhang, Y., Wang, C. et al. 2012. Photo-fenton reaction of graphene oxide: A new strategy to prepare graphene quantum dots for DNA cleavage. *ACS Nano* 6: 6592–6599.

Zhu, S.J., Zhang, J.H., Liu, X. et al. 2012. Graphene quantum dots with controllable surface oxidation, tunable fluorescence and up-conversion emission. *RSC Adv.* 2: 2717–2720.

Zhuo, S., Shao, M., Lee, S.T. 2012. Up-conversion and down-conversion fluorescent graphene quantum dots: ultrasonic preparation and photocatalysis. *ACS Nano* 6: 1059–1064.

CHAPTER 18

CONTENTS

Sonochemical Preparation of Graphene Nanocomposites

Process Intensification

18

B.A. Bhanvase and M.P. Deosarkar

18.1 INTRODUCTION

Graphene is a single-atom-thick sheet containing sp^2-bonded carbon atoms in a hexagonal lattice. Although graphene is the origin of 0D fullerene, 1D carbon nanotubes (CNTs), and 3D graphite that have been widely studied for decades, not too much attention was given to graphene previously (Geim and Novoselov 2007). Graphene is an exclusive material that has the benefit of being both conducting and transparent. It is gaining more attention for their applications in condensed-matter physics, chemical, electronics, and materials science (Geim and Novoselov 2007) and due to its remarkable electrical, thermal, optical, and mechanical properties (Peigney et al. 2001; Zhang et al. 2005; Ghosh et al. 2008; Lee et al. 2008; Lu et al. 2009). Graphene-based materials have been used in various fields such as nanoelectronic devices, biomaterials, intercalation materials, drug delivery, and catalysis (Stankovich et al. 2006; Gilje et al. 2007; Robinson et al. 2008; Arsat et al. 2009; Pasricha et al. 2009; Seger and Kamat 2009; Liang et al. 2010; Akhavan et al. 2011). Currently, several methods have been presented in the literature for the preparation of graphene nanosheets that includes micromechanical exfoliation (Novoselov et al. 2004), chemical vapor deposition (Reina et al. 2009), epitaxial growth (Sutter et al. 2008), scotch tape method (Novoselov et al. 2004), electrostatic deposition of graphene (Sidorov et al. 2007), thermal or chemical decomposition of graphitic materials (Dujardin et al. 1998), and thermal (Hannes et al. 2006) and substrate-free gas-phase synthesis (Dato et al. 2008). Further, for bulk synthesis of graphene, the common process is the chemical exfoliation of graphite (Dresselhaus and Dresselhaus 2002), in which graphite at first has been oxidized to graphite oxide by oxidation with oxidizing agents such as $KMnO_4$ (Hummers and Offeman 1958), $K_2Cr_2O_7$ (Chandra et al. 2010), and $KClO_3$ (McAllister et al. 2007). These methods have been used for the preparation of a graphene dispersion; however, its critical disadvantages are low yield (1%–10%), a significant amount of solvent requirements, and poor stability of the graphene dispersion making many applications unworkable (Hamilton et al. 2009; Wang et al. 2009).

The excellent properties like Young's modulus (~1.0 TPa) (Lee et al. 2008), large specific surface area (2630 $m^2\ g^{-1}$) (Stoller et al. 2008), outstanding thermal conductivity (~5000 W $m^{-1}\ K^{-1}$) (Balandin et al. 2008), high mobility of charge carriers (200,000 $cm^2\ V^{-1}\ s^{-1}$) (Bolotin et al. 2008), and optical transmittance (~97.7%) (Nair et al. 2008) support graphene as ideal building blocks in nanocomposites. Nanocomposites are multiphase materials, in which nanosize material is dispersed in a second phase, that is, a matrix that is called continuous phase, resulting in a combination of the individual properties of the component materials. A significant attention of a researcher has been attached to a new type of nanocomposites like CNT and graphene nanocomposites in the last decade due to the combination of the intrinsic properties of CNTs/graphene and dispersed materials resulting into exceptional performance in various applications (Peng et al. 2009; Eder 2010).

Over the last decades, graphene nanocomposites have been widely developed and established to exhibit a range of exclusive and valuable properties, which are attracting more and more attention from researchers.

The problem of agglomeration of dispersed phase on graphene nanosheets during the preparation of graphene nanocomposite with the use of the different conventional method of synthesis is a critical issue. This type of agglomeration can be reduced with the use of the ultrasound-assisted method of synthesis of graphene nanocomposites, which result in the formation of finely dispersed composite. Several researchers have focused on the use of ultrasonic irradiations for improving the dispersion process and thereby to achieve the uniform loading of metal or metal oxide nanoparticles on graphene nanosheets and the surface properties of the final synthesized product. Large numbers of microbubbles form, grow, and collapse in very short time when ultrasonic waves pass through a liquid medium, resulting in the cavitational effects. These cavitational effects lead to generation of extreme pressure (>500 atm) and temperature (>10,000 K) with cooling rates of >10^{10} K s^{-1} conditions that lead to intense turbulence, liquid circulation currents, and also the formation of the microjets (Bhanvase and Sonawane 2010; Bhanvase et al. 2011, 2012a,b, 2013; Deosarkar et al. 2013; Patel et al. 2013). Moreover, ultrasound-assisted method is useful for the preparation of nano-sized particles and finely dispersed nanocomposite. The reasons are the improved solute transfer and nucleation rate in aqueous suspension observed due to intense micromixing, which lead to the formation of nanometer size particles.

This chapter focuses on the use of ultrasound-assisted processes for the preparation of graphene nanocomposite and their potential applications. The chemical and physical effects of cavitation play an important role in the preparation of different size, shape, and structure of graphene nanocomposites. The use of ultrasonic irradiations during the synthesis of graphene nanocomposites leads to uniform and fine loading of metal and metal oxide nanoparticles on graphene nanosheets compared to conventional preparation methods. Also, case studies for the preparation of SnO_2–graphene and graphene–Fe_3O_4 nanocomposites with the use of ultrasonic irradiation and their potential application are presented.

18.2 SONOCHEMICAL PREPARATION OF GRAPHENE NANOCOMPOSITES

18.2.1 Graphene/Metal Nanocomposites

Graphene–metal nanocomposites are an exceptionally attractive research area due to their outstanding catalytic activity, drug delivery, and electronic and thermal conductivity. The deposition of different metal nanoparticles on graphene sheets, like Pt, Au, Ag, Pd, TiO_2, Co_3O_4, and Fe_3O_4 (Muszynski et al. 2008; Nath et al. 2008; Si and Samulski 2008; Williams et al. 2008; Xu et al. 2008a,b; Goncalves et al. 2009; Jasuja and Berry 2009; Paek et al. 2009; Pasricha et al. 2009; Xu and Wang 2009; Yoo et al. 2009; Cong et al. 2010; Shen et al. 2010), has been reported on the enhanced properties of graphene sheets. Due to the large surface area as well as huge chemical and mechanical stability of graphene and also due to the strong interaction between the metal nanoparticles with its carbonaceous support, those nanocomposites are fairly stable, demonstrate higher catalytic activity, and are easily dispersible in water (Planeix et al. 1994; Liu and Han 2009). Further, the limited surface area, high cost of production, and difficulty in the purification of CNTs supported with metal or metal oxide nanoparticles restrict its utilization for large-scale application as compared to graphene–metal or metal oxide nanocomposites. Hence, graphene is a very promising material for new carbonaceous supports.

Chandra et al. (2011) have successfully synthesized highly water-dispersible rhodium–graphene nanocomposite by the simple reduction of Rh^{3+} salt on poly(ethylene oxide)/poly(propylene oxide)/poly(ethylene oxide) triblock copolymer or pluronic-stabilized graphene oxide (GO) nanosheets with borohydride in the presence of ultrasound. Rhodium nanoparticles with an average particle size of 1–3 nm have been homogeneously and uniformly loaded on the graphene sheets. Further, some porous structures of graphene sheets have also been reported after the reduction of pluronic-stabilized GO in the presence of metal ions. The formation of these types of porous structures is possibly due to the generation of some unstable particles with stable ones. Those unstable particles might be easily removed during the sonochemical reduction of the GO–rhodium nanocomposite. The reported surface area of the resultant graphene–metal nanocomposite is 285 m² g⁻¹, with a pore volume of 0.164 cm³ g⁻¹. This is attributed to the use of ultrasound during the preparation of rhodium–graphene nanocomposite. This material is very efficient for hydrogenation of arenes, particularly for benzene as the substrate material at the room temperature and 5 atm pressure of hydrogen. Further, Vinodgopal et al. (2012) have for the first time used a combination of two ultrasound frequencies at 20 and 211 kHz for the preparation of layered reduced graphene oxide (RGO)–Pt composites. It has been reported that this type of unique dual-frequency arrangement, operating in tandem, yields large exfoliated graphene sheets with platinum nanoparticles uniformly deposited on them. The reported transmission electron and atomic force microscopes confirm the uniform morphology of resulting assemblies to be bi- and single-layered sheets. Further, using an appropriate and novel combination of ultrasound frequencies (Vinodgopal et al. 2012), these graphene assemblies can be obtained with desired morphologies. It has been also reported that these prepared composites show good electrocatalytic activity toward methanol oxidation. Shi et al. (2011) have also used sonoelectrochemical method for the successful fabrication of alloy–graphene nanocomposites. This technique not only offers an easy way for the preparation of alloy nanoparticles but also shows a general approach for the preparation of graphene-based nanostructures with anticipated properties. According to their reported method, Pd was co-electrodeposited with Pt at different atomic ratios, and this formed alloy has been anchored with RGO simultaneously in the presence of poly(diallyl dimethylammonium) (PDDA) chloride. The fabricated PDDA–RGO–PdPt nanocomposite morphologies and structures have been comprehensively studied by transmission electron microscopy, scanning electron microscopy (SEM), and x-ray diffraction (XRD). It has been observed that the ultrasonic irradiations play an important role in the dispersion of graphene sheets, regenerating the electrode by vibrating off the electrodeposited alloy nanoparticles and direct assembling of the nanoparticles onto RGO. It has been also reported that as-prepared PDDA–RGO–PdPt nanocomposites have unique properties of 2D graphene with alloying effects and have exhibited considerably improved catalytic activity and stability toward ethanol electrooxidation, signifying their potential use as efficient electrocatalysts for direct alcohol fuel cells in alkaline conditions. Further, Shi and Zhu (2011) have fabricated a new electrochemical sensor for chlorophenols by using the Pd–graphene nanocomposite and ion liquid. In their report, they have explained the preparation of Pd–graphene nanocomposite with the use of sonoelectrochemical technique, and also they have proposed the possible formation mechanism. In their report, the experimental results showed that Pd nanospheres made of small Pd nanoparticles were uniformly deposited on graphene sheets. This might be due to the improved solute transfer and nucleation rate in the reaction medium due to intense micromixing, which leads to the formation of nanometer size particles, which get deposited on graphene sheets. Further, the electrocatalytic properties also have been investigated by cyclic voltammetry and differential pulse voltammetry, which is an indication of high activity of Pd–graphene nanocomposite for chlorophenol oxidation. In their report, 2-chlorophenol has been selected as the model molecules for oxidation reaction, and the results showed that graphene played an important role in the fabrication of the chlorophenol sensor. The nanocomposite with high electrochemically active surface leads to the outstanding electrocatalytic activity, and ionic liquid further improved the catalytic activity of Pd–graphene for chlorophenols. Recently, Zhu et al. (2014) have used ultrasonic irradiations in the initial stage during the preparation of SnCo–graphene composites through a facile one-pot coprecipitation method. The reported results indicate that SnCo nanoparticles are uniformly loaded on the surface of graphene. It is attributed to the use of ultrasound that improves the dispersion of SnCo on graphene nanosheets. Further, due to the use of ultrasonic irradiation, the composite exhibits elevated reversible capacity with exceptional cyclic stability and rate capability.

GO nanosheets were employed as a template, and hydrazine hydrate was used by Peng et al. (2014) as reductant for GO nanosheets and cupric ion. As per their report, highly dispersed 2D copper/RGO nanosheet (Cu/RGOS) nanocomposites have been successfully prepared by ultrasound-assisted electroless copper plating process. It has been reported that sandwich-like 2D Cu/RGOS nanocomposites consisting of a uniform Cu layer on both sides of centric RGOS have been formed. It is attributed to the use of ultrasound during the preparation of 2D Cu/RGOS nanocomposites. Further, the Cu layer with thickness of about 60 nm reveals approximately single crystalline with preferred crystalline direction and has tight binding with RGOS. The possible reasons are (1) accelerating deposition rate, (2) improving interfacial bonding, and (3) avoiding 2D Cu/RGOS nanocomposites from aggregating. Anandan et al. (2014) have synthesized Sn nanoparticle–stabilized RGO nanodisks by a sonochemical method using $SnCl_2$ and GO nanosheets as precursors in a polyol medium. A very interesting observation has been reported, that is, the difference in RGO morphology in the absence and presence of $SnCl_2$ in the reaction medium. It has been seen that sonication of pure GO in EG leads to the formation of RGO nanosheets. These RGO nanosheets are of few micrometers in size with somewhat scrolled on the edges, which is an indication that the use of ultrasonic procedure efficiently exfoliated the GO. Use of sonication generally leads to the formation of mainly single-layered (>70%) sheets with a lateral width of 21 nm. Conversely, in the presence of $SnCl_2$ in the solution, RGO nanodisks have been formed, and the particle size distribution of GO nanodisks is around 80–100 nm in diameter. Neppolian et al. (2014) have prepared GO-supported Ag and Au monometallic and Au–Ag bimetallic catalysts using a sonochemical technique. Bimetallic catalysts containing varying weight ratios of Au and Ag have been deposited onto GO using a low-frequency horn-type ultrasonicator. Also in their report, the use of high-frequency ultrasonication has been reported efficiently to reduce Ag(I) and Au(III) ions in the presence of polyethylene glycol and 2-propanol. The prepared catalysts have been compared using 4-nitrophenol reduction reaction. Further, it has been found from the reported results that Au–Ag–GO bimetallic catalysts have an elevated activity for the conversion of 4-nitrophenol compared to their monometallic counterparts. It has been also observed that applying dual-frequency ultrasonication in an extremely efficient way for the preparation of bimetallic catalysts requires comparatively low levels of added chemicals and produces bimetallic catalysts with GO with improved catalytic efficiency.

18.2.2 Graphene/Metal Oxide Nanocomposites

TiO_2 has paid much attention because of its high photoconversion efficiency (Gaya and Abdullah 2008). TiO_2 is able to generate photo-induced electron–hole pairs under the irradiation of ultraviolet (UV) light. In view of photoconversion efficiency, the photocatalytic properties of TiO_2 can be more improved by avoiding the recombination of the photo-induced electron–hole pairs efficiently. This can be achieved by loading TiO_2 on graphene, thus forming graphene–TiO_2 composites, which is a promising photocatalytic material, as graphene can act as an electron transfer path that decreases the recombination of the photo-generated electron holes and leads to enhanced photoconversion efficiency (Wang et al. 2009; Zhang et al. 2009). Further, it has been reported that a combination of commercially pure TiO_2 powder and graphene showed a comparatively elevated photocatalytic property (Zhang et al. 2010). But the particle size of commercially pure TiO_2 powder is comparatively large and it is in micrometers. Also, it can agglomerate when it is used alone in the photocatalytic degradation application. Therefore, the photocatalytic property further can be enhanced by reducing the size of TiO_2 to nanometers. This is possible by loading TiO_2 nanoparticles on graphene nanosheets where agglomeration is not significant, and the photocatalytic activity of TiO_2 can be effectively used when TiO_2 size is reduced down to nanometers for liquid-phase oxidation. In view of these benefits, Guo et al. (2011) have succeeded in a controlled loading of TiO_2 nanoparticles with its size around 4–5 nm on the graphene layers uniformly in the presence of ultrasonic irradiation without the use of any surfactant. This is due to the use of ultrasonic irradiation for the pyrolysis and condensation of the dissolved $TiCl_4$ into TiO_2. It has been reported that the photocatalytic activity of the prepared graphene–TiO_2 nanocomposites containing 25 wt.% TiO_2 was superior than that of commercially pure TiO_2. The reason of this enhancement is the extremely small size of the TiO_2 nanoparticles and uniform dispersion of crystalline TiO_2 on graphene nanosheets. Further, the graphene present in the nanocomposite is in a very good contact with the TiO_2 nanoparticles, and that improves the photoelectron conversion of TiO_2 by decreasing the recombination of photo-generated electron–hole pairs. Min et al. (2012) have used ultrasonic irradiations to carry out chemical reactions between GO and titanium alkoxide. As per their report, in the first step, they have activated the GO using poly(acrylic acid) in order to increase carboxylic groups on a graphene surface. Then in the second step, a hydrolysis product of $Ti(OH)_x$ from tetrabutyl titanate has been assembled on the activated GO surface with the use of sonochemical reaction (100 W, 50 Hz). Further, annealing has been reported to carry out at 400°C in Ar that results in a high-quality TiO_2/graphene nanocomposite with the great visible response were obtained. The possible reported reasons are (1) ultrasonic irradiation enhances the degree of dispersion of TiO_2 on GO sheets and (2) accelerates chemical reaction resulted in the partial formation of a Ti–C and Ti–O–C bonds. It has been reported that the results of both the RhB degradation and photocurrent tests indicate that the photocatalytic and photoelectrochemical activities of the TiO_2/graphene nanocomposite are greatly superior to those of commercial P25. Lei et al. (2012) have also fabricated a series of graphene–TiO_2 composites from GO and titanium n-butoxide by an ultrasonic-assisted method. The reported average particle size of the TiO_2 nanoparticles deposited on graphene nanosheets was controlled at around 10–15 nm without using surfactant, and this is due to the pyrolysis and condensation of dissolved titanium n-butoxide into TiO_2 by ultrasonic irradiation. The graphene–TiO_2 nanocomposites have a higher specific surface area, and that increases the decolorization rate for RhB solution. The reported reason is that the graphene and TiO_2 nanoparticles in the nanocomposites interact strongly, which improves the photoelectric conversion of TiO_2 by reducing the recombination of photo-generated electron–hole pairs in the presence of ultrasonic irradiation. Recently, Neppolian et al. (2012) have used an ultrasound-assisted method for the preparation of nanosized Pt–GO–TiO_2 photocatalyst. In their report, they have carried out the photocatalytic and the sonophotocatalytic degradation of a normally used anionic surfactant, dodecyl benzene sulfonate, in aqueous solution using Pt–GO–TiO_2 nanoparticles in order to investigate the photocatalytic efficiency. Also, they have carried out the sonolytic degradation of dodecyl benzene sulfonate for comparison purpose. It has been observed that the Pt–GO–TiO_2 catalyst degrades dodecyl benzene sulfonate faster than P–25 (TiO_2), prepared TiO_2, or GO–TiO_2 photocatalysts. The mineralization of dodecyl benzene sulfonate has been reported to be enhanced by the degree of 3 using Pt–GO–TiO_2 compared to the P–25 (TiO_2). Further, in the presence of GO, an improvement in dodecyl benzene sulfonate oxidation has been reported, and when doped with platinum, mineralization of dodecyl benzene sulfonate was further enhanced. Also as per their report, the Pt–GO–TiO_2 catalyst showed a substantial amount of degradation of dodecyl benzene sulfonate under visible light irradiation. The capability of GO to provide as a solid support to loaded platinum nanoparticles on GO–TiO_2 is useful in developing new photocatalysts.

Metal oxides are promising as Li insertion substitutes to graphite anode materials including their high theoretical capacity, low cost, low toxicity, and widespread availability (Xiong and Xia 2007; Choi et al. 2008; Li et al. 2011). Reducing the size of these types of materials to the nanoscale level is supposed to enable an increase in the number of active reaction sites and accelerate the electron–Li ion transport (Zhang et al. 2006). The controlled preparation of metal oxide nanoparticles is one of the most attractive objectives in nanoscience due to their shape and size-dependent physical and chemical properties in photoelectric devices, catalysts, gas sensors, and even biological areas (Ghosh et al. 2005; Liu et al. 2006). Further, cuprous oxide (Cu_2O) is one of the vital p-type semiconductors with a narrow bandgap and low toxicity, which is eco-friendly and attractive in large-scale applications, like solar cells (Brsikman 1992), gas sensing (Zhang et al. 2006), biosensors (Zhu et al. 2009), and lithium-ion batteries (Poizot et al. 2000). Recently, the photocatalytic properties of Cu_2O have been widely studied. Cu_2O can efficiently adsorb molecular oxygen, which can scavenge photo-generated electrons to restrain the recombination of electron–hole pairs and subsequently enhance the photocatalytic efficiency (Li et al. 2004; Zheng et al. 2009; Yao et al. 2010). Further, Cu_2O can be prepared with different morphologies or various exposed facets and is showing an improvement in the photocatalytic activity during degradation of organic molecules (Ho and Huang 2009; Yang and Liu 2010; Dong et al. 2011). Several conventional methods like conventional mixing, hydrothermal, and solvothermal methods have been used for the preparation of Cu_2O-based graphene composites for use as anode materials in lithium-ion batteries (Xu et al. 2009), electrocatalysts (Yan et al. 2012), photocatalysts (Gao et al. 2012), and the degradation of dyes, supercapacitors (Li et al. 2011), and sensors (Zhang et al. 2011; Deng et al. 2012). However, the use of ultrasound for the preparation of Cu_2O–graphene nanocomposite may lead to its improved properties. Zhang et al. (2012) have demonstrated a facile ultrasound aqueous method to synthesize Cu_2O nanospheres supported on graphene nanosheets. This reported material has a high specific capacity of over 400 mAh g^{-1} at a rate of 100 mA g^{-1} after 20 cycles that is much higher than bare Cu_2O nanospheres (less than 100 mAh g^{-1}). In their report, they have illustrated that the protection of graphene nanosheets is very efficient and powerful for increasing both the capacity and stability of anode material. It has been also reported that use of the ultrasonic method has opened a new window for green and ultrafast synthesis of high-quality novel hybrid functional nanomaterials. Abulizi et al. (2014) have also developed an easy, one-step synthesis of Cu_2O–RGO composites using a simple sonochemical route without any surfactants or templates. The reported results indicated that the Cu_2O sphere is approximately 200 nm in diameter and composed of small Cu_2O particles roughly 20 nm in diameter. It is also possible that the morphology and composition of the Cu_2O–RGO nanocomposites could be well controlled by simply changing the mole ratio of the reactants under ultrasonic irradiation. The prepared Cu_2O–RGO nanocomposites showed improved photocatalytic performance for the degradation of methyl orange than pure Cu_2O spheres, which may have potential applications in water treatment, sensors, and energy storage. Further, Yan et al. (2013) have used a facile way to synthesize Cu_2O/RGO nanocomposites with octahedron-like morphology in aqueous solution without any surfactant. The reported morphology of Cu_2O/RGO shows that the Cu_2O particles and the RGO distribute hierarchically and the primary Cu_2O particles have been encapsulated well in the graphene nanosheets. Further, it has been reported that the Cu_2O/RGO nanocomposites maintain a reversible capacity of 348.4 mAh g^{-1} after 50 cycles at a current density of 100 mA g^{-1}. This may be due to the use of ultrasound for the preparation of Cu_2O/RGO nanocomposites, which allows fine distribution of Cu_2O nanoparticles on reducing GO nanosheets. Zhou et al. (2014) have successfully prepared CuO nanoleaves on the graphene sheets (CuO/GNS) nanocomposite by a facile microwave-assisted method using $Cu(NO_3)_2 \cdot 3H_2O$, NaOH, and ammonia as raw materials and absolute ethanol as solvent. It has been observed that the leaf-shaped rhombic CuO was uniformly dispersed on GNS with a width of 40 nm in the middle and a length of 100–140 nm. As an anode material for Li-ion battery, the CuO/GNS composites have been reported to show enhanced electrochemical performance with high lithium storage capacity, satisfactory cyclic durability, and rate capacity than pristine CuO. The reversible capacity of the CuO/GNS composites was found to be 600 mAh g^{-1} after 50 cycles at 100 mA g^{-1}. The outstanding electrochemical performance was attributed to the synergy effect of the homogeneously loaded CuO nanoleaves and graphene sheets due to ultrasonic irradiations.

Zhuo et al. (2012) have used a facile ultrasonic route for the fabrication of graphene quantum dots in which upconverted emission is presented. As per their report, the prepared graphene quantum dots demonstrate an excitation-independent downconversion and upconversion photoluminescent behavior, and the complex photocatalysts (rutile TiO_2/graphene quantum dot and anatase TiO_2/graphene quantum dot systems) have been designed by them to harness the visible spectrum of sunlight. It has been interesting to see that the photocatalytic rate of the rutile TiO_2/graphene quantum dot complex system is significantly larger than that of the anatase TiO_2/graphene quantum dot complex under visible light ($\lambda > 420$ nm) irradiation in the degradation of methylene blue. For harvesting the energy and being eco-friendly, these prepared graphene quantum dots may be useful for a broad range of applications, including biosensors, bioimaging, laser, and light-emitting diodes. Deosarkar et al. (2013) have reported the preparation of SnO_2–graphene nanocomposite by using a novel ultrasound-assisted solution-based chemical synthesis method. The preparation of GO by the modified Hummers–Offeman method in the presence of ultrasonic irradiation has been reported by them. The deposition of SnO_2 nanoparticles on GO has been also reported with an ultrasound-assisted solution-based synthesis route. The transmission electron microscopic analysis reported by them, the uniform and fine loading of SnO_2 nanoparticles with particle size 3–5 nm on graphene nanosheets by ultrasonically prepared graphene–SnO_2 nanocomposite was observed. It can be concluded in their report that the agglomeration of graphene–SnO_2 nanocomposite has been substantially reduced when ultrasonic irradiation is used. The reason is the cavitational effects created because of the ultrasonic irradiations during the preparation of graphene–SnO_2 nanocomposite that enhance the fine and uniform loading of SnO_2 nanoparticles on graphene nanosheets by oxidation–reduction reaction compared to conventional synthesis methods. Zhu et al. (2013) have further prepared Fe_3O_4 nanoparticles of 30–40 nm with the use of a sonochemical method, and these nanoparticles have been homogeneously deposited on the RGO sheets, that is, Fe_3O_4/RGO nanocomposite. It is also found from their report that the superparamagnetic property of Fe_3O_4/RGO nanocomposite with saturated magnetization of 30 emu g^{-1} is attributed to the ultrasonic irradiation, which leads to the successful and uniform loading of Fe_3O_4 nanoparticles on RGO nanosheets, resulting in the formation of Fe_3O_4/RGO nanocomposite.

18.3 COMPARATIVE STUDY ON CONVENTIONAL AND ULTRASOUND-ASSISTED METHODS FOR THE PREPARATION OF GRAPHENE NANOCOMPOSITE

18.3.1 Graphene–SnO$_2$ Nanocomposite

As reported in earlier section, it has been found that the use of ultrasound plays an important role during the preparation of graphene–metal and graphene–metal oxide nanocomposite, which leads to the formation of uniformly distributed graphene nanocomposites. The problem of agglomeration is significantly eliminated with the use of ultrasonic irradiation, which is generally present in conventional method of synthesis. This finely and homogeneously distributed metal or metal oxide–graphene nanocomposite has a significantly improved property profile.

Deosarkar et al. (2013) have prepared GO from graphite power by using modified Hummers method (Hummers and Offeman 1958) in the presence of ultrasonic irradiations using an ultrasonic horn (Hielscher Ultrasonics GmbH, 22 kHz frequency, 240 W) at 50% amplitude. According to this method, they have added graphite powder (1 g) and NaNO$_3$ (1 g) to 46 mL H$_2$SO$_4$ solution (98 wt.%) in an ice bath. The prepared mixture has been ultrasonicated for 10 min, and gradual addition of 5 g KMnO$_4$ to the as-prepared mixture has been accomplished, and again, it was ultrasonicated for 30 min at 35°C. As per their report to the prepared solution, 100 mL deionized water has been slowly added and again ultrasonicated for 5 min after which 8 mL H$_2$O$_2$ has been added to the resulting solution. The obtained graphite oxide has been ultrasonicated for 30 min, which results in the formation of GO layers. Further, they have been using the prepared 25 mL GO solution, which was diluted with 450 mL deionized water followed by the addition of 8.5 mL of aqueous HCl solution (35 wt.%) under ultrasonic irradiation (5 min). Then they have added 3 g SnCl$_2$·2H$_2$O into the resulted GO–HCl aqueous solution and further ultrasonicated for 30 min at room temperature (Hummers and Offeman 1958). SnO$_2$–graphene nanocomposite was formed during the reduction of GO and, after its formation, was washed with deionized water. SnO$_2$–graphene nanocomposite was finally dried in oven at 60°C for 2 h.

The formation mechanism of SnO$_2$–graphene nanocomposite by the innovative in situ ultrasound-assisted oxidation–reduction of GO with SnCl$_2$·2H$_2$O is shown in Figure 18.1. During this formation process, graphite powder has been oxidized in the presence of ultrasonic irradiation with strong oxidizing agent KMnO$_4$ in the presence of H$_2$SO$_4$, and this leads to the formation of GO. Further, in this reaction, Sn^{2+} ions get hydrolyzed to form SnO$_2$ nanoparticles at room temperature, and the GO nanosheets get simultaneously reduced to graphene nanosheets in the presence of ultrasonic irradiation resulting in the formation of uniformly and finely distributed SnO$_2$–graphene nanocomposite.

Further, Deosarkar et al. (2013) have studied the structural properties of the sonochemically and conventionally prepared SnO$_2$–graphene nanocomposite and compared to the GO by XRD analysis (Figure 18.2). The GO diffraction peaks have been reported at 12.5° and 23.29°, which are attributed to short-range order in stacked graphene sheets. The diffraction pattern of sonochemically and conventionally prepared SnO$_2$–graphene nanocomposite is shown in Figure 18.2. The characteristic peaks of SnO$_2$ at 24.9°, 37.6°, and 52° corresponding to the (1 1 0), (2 0 0), and (2 1 1) planes, respectively, of a tetragonal rutile-like phase of SnO$_2$ (JCPDS Card No. 41-1445) have been observed (Yao et al. 2009; Kim et al. 2010). The peaks at various crystal planes of SnO$_2$–graphene nanocomposite prepared by an ultrasound-assisted method are similar to that of conventionally prepared SnO$_2$–graphene nanocomposite. It has been also reported that conventionally synthesized SnO$_2$–graphene nanocomposite shows higher crystallinity than sonochemically synthesized SnO$_2$–graphene nanocomposite. The reason behind this is the intense environments generated by ultrasonic irradiation that facilitates faster reaction, not allowing the nucleation and crystal growth to occur fully (Patel et al. 2013). It has been reported that the cavitational effects generated because of ultrasonic irradiations produce high local energy and micromixing that acts as a reaction aid, which increases the randomness of the Brownian motion of the SnO$_2$ molecules not allowing regular

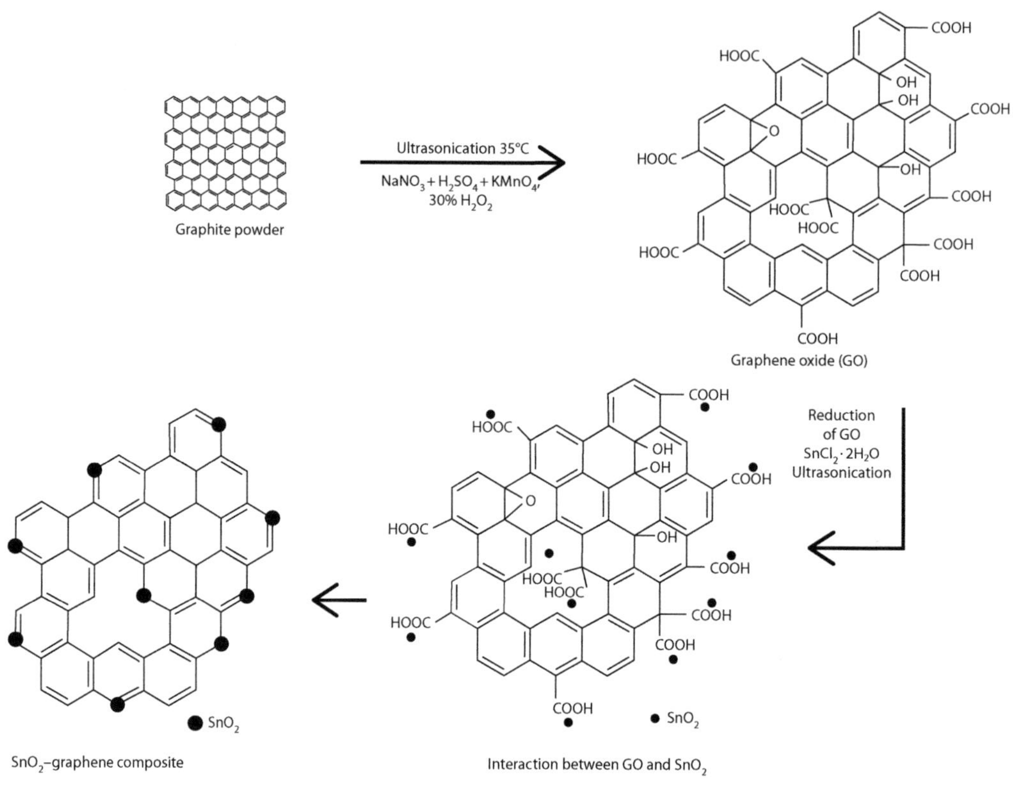

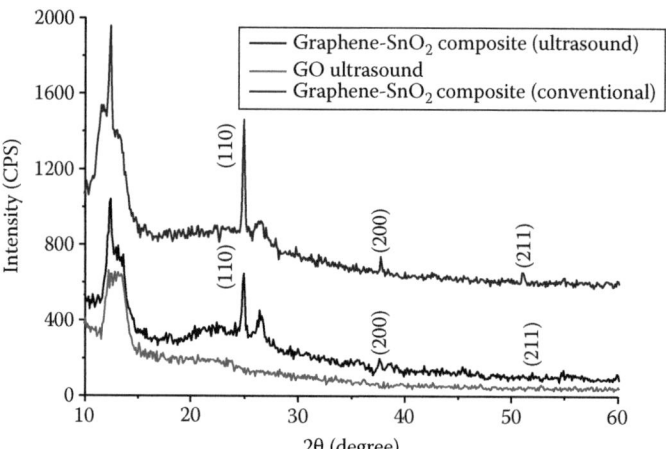

FIGURE 18.1
Diagram of the formation process of GO and SnO_2–graphene nanocomposite. (From Deosarkar, M.P. et al., *Chem. Eng. Process. Process. Intens.*, 70, 48, 2013.)

FIGURE 18.2
XRD pattern of GO prepared by ultrasound method and sonochemically and conventionally prepared SnO_2–graphene nanocomposite. (From Deosarkar, M.P. et al., *Chem. Eng. Process. Process. Intens.*, 70, 48, 2013.)

crystal formation. This will decrease the crystallinity of synthesized product due to unstable position of molecules in the lattice. Further, it has been reported that the peak intensity in case of conventionally synthesized SnO_2–graphene nanocomposite is larger than that of sonochemically synthesized SnO_2–graphene nanocomposite, which is an indication of the loading of larger-sized SnO_2 nanoparticles on graphene nanosheets for conventionally prepared SnO_2–graphene nanocomposite. Moreover, the diffraction peak attributed to graphite has not been reported indicating that the bundle of graphene sheets in the SnO_2–graphene nanocomposite is completely exfoliated in the presence of ultrasonic irradiations.

Deosarkar et al. (2013) have also depicted the TEM images of ultrasonically and conventionally synthesized SnO_2–graphene nanocomposites (Figure 18.3). As per their report, a clear outline with a much sharper contrast than that of the pristine GO nanosheets was observed, demonstrating a layer of attachments was adhered to the graphene nanosheets. Further, TEM images shown in Figure 18.3a and b reveal more fine and homogeneous loading of the SnO_2 nanoparticles of size 3–5 nm on the graphene nanosheets. The reason behind the fine and homogeneous loading of SnO_2 nanoparticles on graphene nanosheets (Figure 18.3a and b) is attributed to the cavitational effect induced due to ultrasonic irradiations, which are responsible for the complete exfoliation of GO nanosheets and are available for the reduction reaction with $SnCl_2 \cdot 2H_2O$. The agglomerated morphology of SnO_2–graphene nanocomposites synthesized by conventional method has been observed from their report, which is depicted in Figure 18.3c. This nonuniform loading of SnO_2 nanoparticles with agglomerated morphology on graphene nanosheets has been reported in the case of conventionally prepared graphene–SnO_2 nanocomposite. Further, it has been also reported that the particle size is larger than that of ultrasonically prepared SnO_2–graphene nanocomposite. The reduction in the size of the SnO_2–graphene nanocomposite is attributed to cavitational effects generated due to

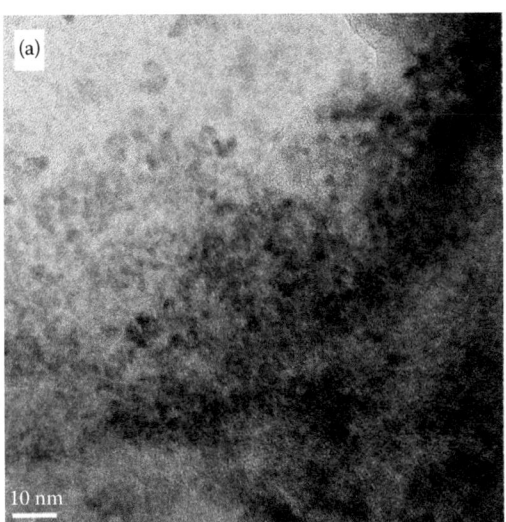 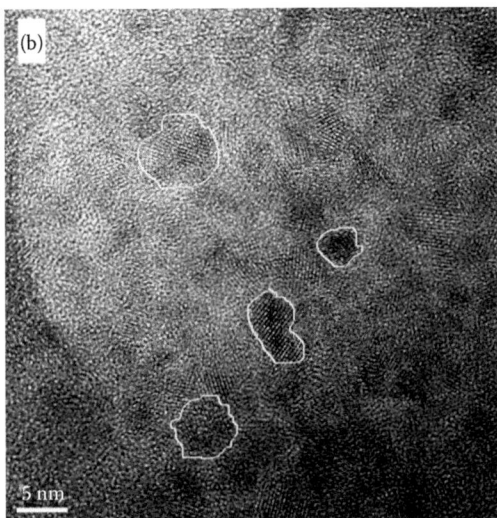

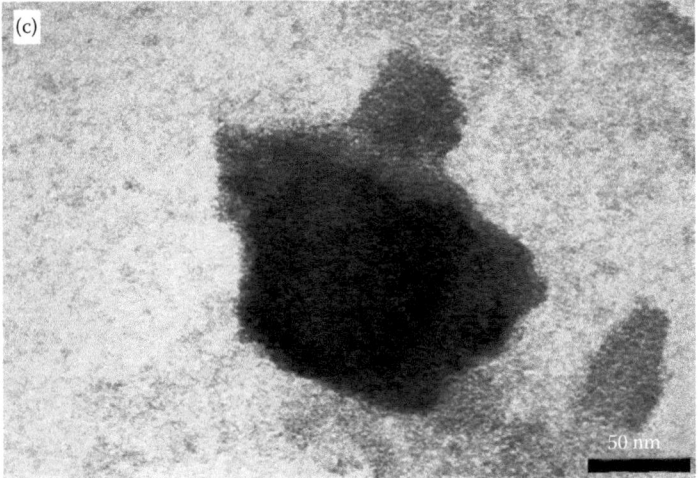

FIGURE 18.3
TEM image of SnO_2–graphene nanocomposite prepared by sonochemical method at (a) 10 and (b) 5 nm magnification and conventional method at (c) 50 nm magnification. (From Deosarkar, M.P. et al., *Chem. Eng. Process. Process. Intens.*, 70, 48, 2013.)

ultrasonic irradiations. Thus, it is clear that the ultrasonic waves play an important role to intensify the fine and homogeneous loading of SnO_2 nanoparticles on graphene nanosheets.

Finally, Deosarkar et al. (2013) have concluded with the successful formation of SnO_2–graphene nanocomposite through a novel and efficient ultrasound-assisted solution-based chemical synthesis route. The morphological analysis shows the homogeneous and fine distribution of SnO_2 in graphene nanosheets. It has been further concluded that the cavitational effects generated due to the ultrasonic irradiations have been shown to intensify the fine and homogeneous distribution of SnO_2 on graphene nanosheets during an oxidation–reduction reaction between GO and $SnCl_2 \cdot 2H_2O$. The intensified fine and homogeneous distribution of SnO_2 nanoparticles with particle size 3–5 nm on reduced graphene nanosheets is attributed to the enhanced nucleation rate and solute transfer rate because of cavitational effect produced by ultrasonic irradiations.

18.3.2 Graphene–Fe$_3$O$_4$ Nanocomposite

Deosarkar et al. (2014) have reported the preparation of GO in the presence of ultrasound as per the method reported in the previous section. Further, they have reported the preparation of Fe_3O_4–graphene nanocomposite using an ultrasound-assisted method as follows: Initially, they have prepared GO solution by adding 40 mg GO to 40 mL of deionized water. The addition of 110 mg $FeCl_3$ and 43 mg $FeCl_2$ to 50 mL deionized water has been accomplished by them and is well mixed with a prepared GO solution at the temperature of the mixture of 85°C and pH equal to 10 in the presence of ultrasound. Finally, they have separated the formed precipitate of graphene–Fe_3O_4 nanocomposite by centrifugation, washed, and dried in oven at 60°C for 2 h.

The synthesis of Fe_3O_4–graphene nanocomposite has been reported to be carried out by using in situ ultrasound-assisted deposition method in the presence of GO as carriers (Figure 18.4). As per the report of Deosarkar et al. (2014), graphite has been oxidized by $KMnO_4$ initially in the presence of H_2SO_4 using ultrasonic irradiations, which results in the formation of GO. At the same time, the

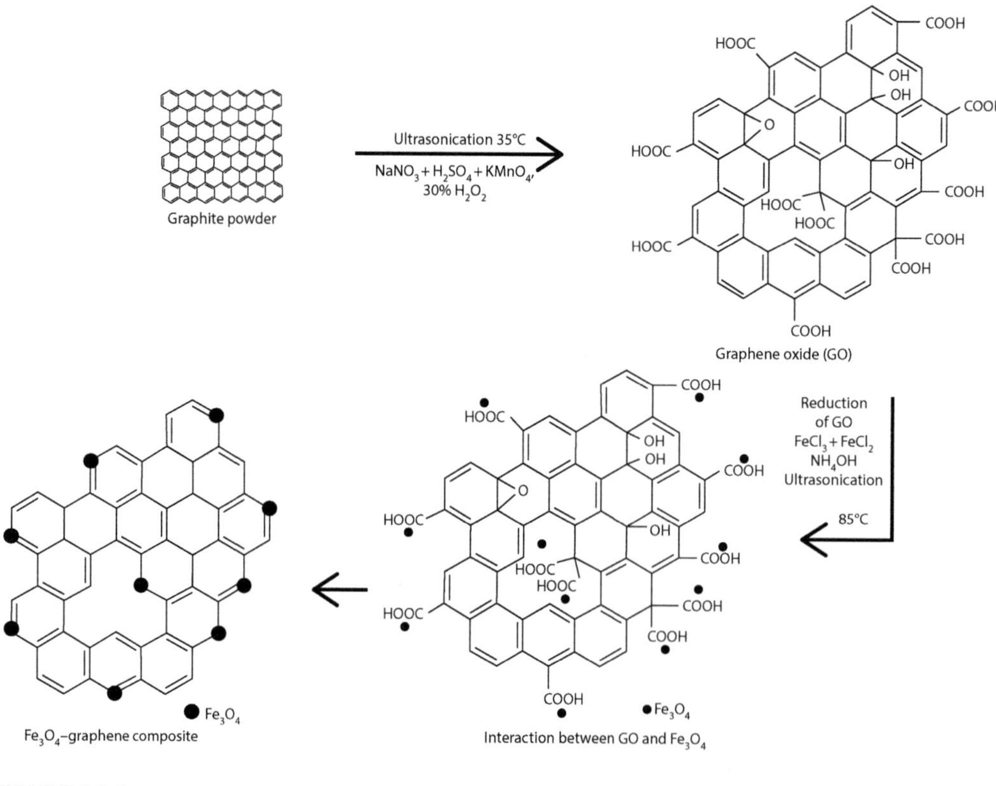

FIGURE 18.4
Mechanism of the formation process of GO and Fe_3O_4–graphene nanocomposites. (From Deosarkar, M.P. et al., *Chem. Eng. Process. Process. Intens.*, 83, 49, 2014.)

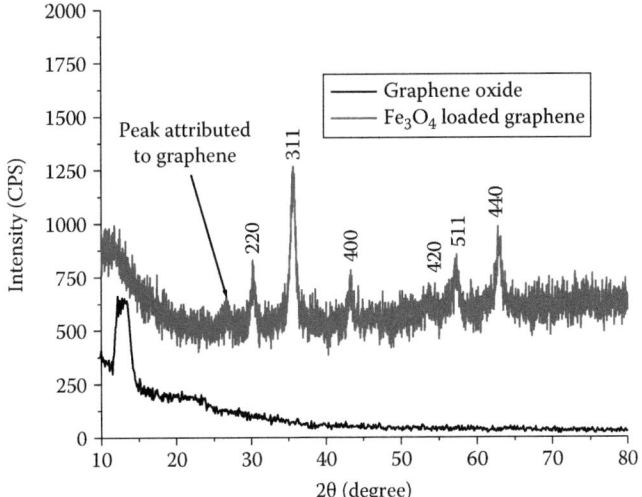

FIGURE 18.5
XRD patterns of GO and Fe_3O_4–graphene nanocomposites. (From Deosarkar, M.P. et al., *Chem. Eng. Process. Process. Intens.*, 83, 49, 2014.)

reduction of GO takes place with Fe^{3+}/Fe^{2+} ions in the presence of ultrasonic irradiation. As per their report, these ions were captured by carboxylate anions on the GO by coordination leading to the formation of graphene nanosheets, and Fe_3O_4 nanoparticles get distributed on graphene nanosheets. Further, it has been reported that the loading of Fe_3O_4 nanoparticles on the GO is dependent on the carboxylic acid groups present on GO. Uniform distribution of chemically deposited Fe_3O_4 nanoparticles on graphene nanosheets is also expected in the presence of ultrasonic irradiation, and this fine and homogeneous loading is due to the micromixing and turbulence created because of the cavitational effect of ultrasonic irradiation.

Deosarkar et al. (2014) also compared the structural properties of the prepared Fe_3O_4–graphene nanocomposite and GO with the help of XRD analysis (Figure 18.5). The diffraction peaks of GO observed at 12.5° and 23.29° indicate a short-range order in stacked graphene sheets. It has been reported that the XRD pattern of Fe_3O_4–graphene nanocomposite has found to be similar to that of magnetite (JCPDS Card No. 88-0315). The reported characteristic peaks of Fe_3O_4 nanoparticles at 30.2°, 35.3°, 43.5°, 49.2°, and 56.7° correspond to the (2 2 0), (3 1 1), (4 0 0), (4 2 2), and (5 1 1) planes, respectively, indicating the magnetite spinel structure of Fe_3O_4 nanoparticles (Berger et al. 1999; Cui et al. 2009). XRD pattern of Fe_3O_4–graphene nanocomposite depicts a low-intensity diffraction peak that indicates cubic-shaped Fe_3O_4 nanoparticles. Further, the reported broad XRD peaks indicate the smaller-size Fe_3O_4 nanoparticles in the presence of ultrasound, which is attributed to the extreme environments generated with the help of sonication that assists a faster reaction, not allowing the nucleation and crystal growth to occur fully (Patel et al. 2013). Further, the diffraction hump present in the range of 24°–28° (2θ value) is attributed to the formation of graphene (Zhou et al. 2010) in graphene–Fe_3O_4 nanocomposite by the reduction of GO in the presence of Fe_3O_4 nanoparticles. Additionally, the diffraction peak of graphite is not being observed, indicating that the bundle of graphene sheets in the Fe_3O_4–graphene nanocomposite is well exfoliated and the ultrasound-assisted synthesis is responsible for the better exfoliation of graphene sheets that leads to homogeneous chemical deposition of Fe_3O_4 nanoparticles on the graphene sheets.

Further, Deosarkar et al. (2014) have analyzed the morphology of ultrasonically prepared Fe_3O_4–graphene nanocomposite using the TEM, which is depicted in Figure 18.6. The observed morphology of ultrasonically prepared Fe_3O_4–graphene nanocomposite indicates that the spheres like particles are loaded homogeneously on the surface and edges of the graphene nanosheets. The reported size of Fe_3O_4 nanoparticles distributed on graphene nanosheets is ranging between 20 and 30 nm with a narrow size distribution. However, a slight agglomeration of Fe_3O_4 nanoparticles has been also observed with a large amount of Fe_3O_4 nanoparticles gets immobilized onto the graphene nanosheets. The reason behind is the cavitational effect of ultrasonic irradiations, which is

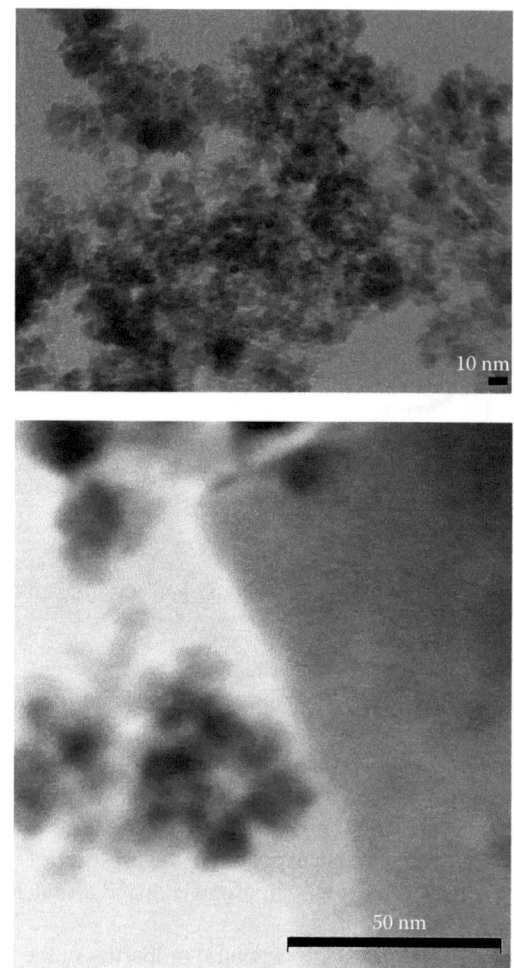

FIGURE 18.6
TEM images of Fe_3O_4–graphene nanocomposites at different magnifications. (From Deosarkar, M.P. et al., *Chem. Eng. Process. Process. Intens.*, 83, 49, 2014.)

responsible for the complete exfoliation of GO nanosheets, and large surface is available for the reduction reaction with Fe^{3+}/Fe^{2+} ions. Further, the distribution of Fe_3O_4 nanoparticles on both sides of the graphene nanosheets has been established with the functional groups such as hydroxyl and carboxylic groups present onto both sides of the GO nanosheets as indicated in the formation mechanism of Fe_3O_4–graphene nanocomposite (Figure 18.4). Further, the sonochemically prepared Fe_3O_4–graphene nanocomposite is compared with the conventionally prepared Fe_3O_4–graphene nanocomposite by Yang et al. (2009) and Lian et al. (2010). As per the reports of Yang et al. (2009) and Lian et al. (2010), nonuniform distribution and aggregation of Fe_3O_4 nanoparticles on the graphene nanosheets have been observed, and use of conventional method cannot remove aggregation of Fe_3O_4 nanoparticles completely. The more homogeneous distribution of Fe_3O_4 nanoparticles on graphene nanosheets has been observed in the work reported by Deosarkar et al. (2014), which may be attributed to cavitational effects induced by ultrasonic irradiations that eliminate the nanoparticle aggregation to the maximum extent (Patel et al. 2013). Further, the morphological investigation has been also reported by Deosarkar et al. (2014) using SEM and EDX analyses of Fe_3O_4–graphene nanocomposite, which are shown in Figure 18.7. It has been found that the Fe_3O_4 nanoparticles are distributed between the layers of the graphene sheets that lead to the formation of porous Fe_3O_4–graphene nanocomposite with large amount of void spaces, and this will lead to the enhancement in the cyclic performance of Fe_3O_4–graphene nanocomposite as an anode material for lithium-ion batteries (Chou et al. 2010; Lian et al. 2010). Moreover, Fe_3O_4 nanoparticles homogeneously distributed between the graphene sheets can decrease the aggregation of Fe_3O_4 nanoparticles to the

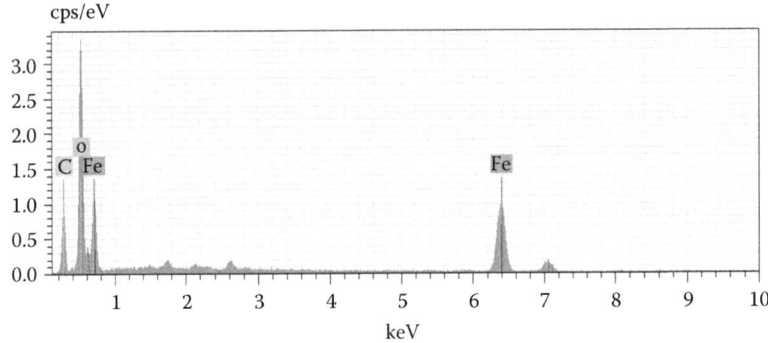

FIGURE 18.7
SEM image and EDX of Fe_3O_4–graphene nanocomposites. (From Deosarkar, M.P. et al., *Chem. Eng. Process. Process. Intens.*, 83, 49, 2014.)

maximum extent in the presence of ultrasonic irradiation, which can be of great benefit to cycle life. Also the presence of Fe, O, and C in EDX analysis confirms the formation of Fe_3O_4–graphene nanocomposite.

18.4 CONCLUSIONS

The current development of graphene nanocomposites prepared by ultrasound-assisted method has been reviewed and reported in this chapter. Synthesis of exfoliated graphene nanocomposite has been successfully reported to be carried out in the presence of ultrasonic irradiations. The reported results of the several researchers established the exfoliation of the stacked nature of graphene with homogeneously distributed metal or metal oxide nanoparticles. This is attributed to the extreme environment created by applying acoustic cavitation. The cavitational effects generated because of the use of ultrasonic irradiations have been shown to intensify the fine and homogeneous loading of metal or metal oxide nanoparticles on graphene nanosheets. Further, homogeneous distribution of metal or metal oxide on graphene nanosheets is attributed to the enhanced nucleation and solute transfer rate due to cavitational effect induced by ultrasonic irradiations. This uniform distribution of metal or metal oxide is leading to a superior property profile of the prepared graphene nanocomposites.

REFERENCES

Abulizi, A., G. Yang, and J. Zhu. 2014. One-step simple sonochemical fabrication and photo-catalytic properties of Cu_2O–rGO composites. *Ultrasonics Sonochemistry* 21:129–135.

Akhavan, O., E. Ghaderi, and A. Esfandiar. 2011. Wrapping bacteria by graphene nanosheets for isolation from environment, reactivation by sonication, and inactivation by near-infrared irradiation. *Journal of Physical Chemistry B* 115:6279–6288.

Anandan, S., A. M. Asiri, and M. Ashokkumar. 2014. Ultrasound assisted synthesis of Sn nanoparticles-stabilized reduced graphene oxide nanodiscs. *Ultrasonics Sonochemistry* 21:920–923.

Arsat, R., M. Breedon, M. Shafiei, P. G. Spizziri, S. Gilje, R. B. Kaner, K. Kalantar-zadeh, and W. Wlodarski. 2009. Graphene-like nano-sheets for surface acoustic wave gas sensor applications. *Chemical Physics Letters* 467:344–347.

Balandin, A. A., S. Ghosh, W. Z. Bao, I. Calizo, D. Teweldebrhan, F. Miao, and C. N. Lau. 2008. Superior thermal conductivity of single-layer graphene. *Nano Letters* 8:902–907.

Berger, P., N. B. Adelman, K. J. Beckman, D. J. Campbell, A. B. Ellis, and G. C. Lisensky. 1999. Preparation and properties of an aqueous ferrofluid. *Journal of Chemical Education* 76:943–948.

Bhanvase, B. A., Y. Kutbuddin, R. N. Borse, N. Selokar, D. V. Pinjari, S. H. Sonawane, and A. B. Pandit. 2013. Ultrasound assisted intensification of calcium zinc phosphate pigment synthesis and its nanocontainer for active anticorrosion coatings. *Chemical Engineering Journal* 231:345–354.

Bhanvase, B. A., D. V. Pinjari, P. R. Gogate, S. H. Sonawane, and A. B. Pandit. 2011. Process intensification of encapsulation of functionalized $CaCO_3$ nanoparticles using ultrasound assisted emulsion polymerization. *Chemical Engineering and Processing: Process Intensification* 50:1160–1168.

Bhanvase, B. A., D. V. Pinjari, P. R. Gogate, S. H. Sonawane, and A. B. Pandit. 2012b. Synthesis of exfoliated poly(styrene–co–methyl methacrylate)/montmorillonite nanocomposite using ultrasound assisted in-situ emulsion copolymerization. *Chemical Engineering Journal* 181–182:770–778.

Bhanvase, B. A., D. V. Pinjari, S. H. Sonawane, P. R. Gogate, and A. B. Pandit. 2012a. Analysis of semibatch emulsion polymerization: Role of ultrasound and initiator. *Ultrasonics Sonochemistry* 19:97–103.

Bhanvase, B. A. and S. H. Sonawane. 2010. New approach for simultaneous enhancement of anticorrosive and mechanical properties of coatings: Application of water repellent nano $CaCO_3$-PANI emulsion nanocomposite in alkyd resin. *Chemical Engineering Journal* 156:177–183.

Bolotin, K. I., K. J. Sikes, Z. Jiang, M. Klima, G. Fudenberg, J. Hone, P. Kim, and H. L. Stormer. 2008. Ultrahigh electron mobility in suspended graphene. *Solid State Communications* 146:351–355.

Brsikman, R. N. 1992. A study of electrodeposited cuprous oxide photovoltaic cells. *Solar Energy Materials and Solar Cells* 27:361–368.

Chandra, S., S. Bag, R. Bhar, and P. Pramanik. 2011. Sonochemical synthesis and application of rhodium–graphene nanocomposite. *Journal of Nanoparticle Research* 13:2769–2777.

Chandra, S., S. Sahu, and P. Pramanik. 2010. A novel synthesis of graphene by dichromate oxidation. *Materials Science and Engineering B* 167:133–136

Choi, S., K. M. Nam, B. K. Park, W. S. Seo, and J. T. Park. 2008. Preparation and optical properties of colloidal, monodisperse, and highly crystalline ITO nanoparticles. *Advanced Materials* 20:2609–2611.

Chou, S. L., J. Z. Wang, M. Choucair, H. K. Liu, J. A. Stride, and S. X. Dou. 2010. Enhanced reversible lithium storage in a nanosize silicon/graphene composite. *Electrochemistry Communications* 12:303–306.

Cong, H. P., J. J. He, Y. Lu, and S. H. Yu. 2010. Water-soluble magnetic functionalized reduced graphene oxide sheets: In situ synthesis and magnetic resonance imaging applications. *Small* 6:169–173.

Cui, Z. M., L. Y. Jiang, W. G. Song, and Y. G. Guo. 2009. High-yield gas–liquid interfacial synthesis of highly dispersed Fe_3O_4 nanocrystals and their application in lithium-ion batteries. *Chemistry of Materials* 21:1162–1166.

Dato, A., V. Radmilovic, Z. Lee, J. Phillips, and M. Frenklach. 2008. Substrate-free gas-phase synthesis of graphene sheets. *Nano Letters* 8:2012–2016.

Deng, S. Z., V. Tjoa, H. M. Fan, H. R. Tan, D. C. Sayle, M. Olivo, S. Mhaisalkar, J. Wei, and C. H. Sow. 2012. Reduced graphene oxide conjugated Cu_2O nanowire mesocrystals for high-performance NO_2 gas sensor. *Journal of the American Chemical Society* 134:4905–4917.

Deosarkar, M. P., S. M. Pawar, and B. A. Bhanvase. 2014. In-Situ sonochemical synthesis of Fe_3O_4-graphene nanocomposite for lithium rechargeable batteries. *Chemical Engineering and Processing: Process Intensification* 83:49–55.

Deosarkar, M. P., S. M. Pawar, S. H. Sonawane, and B. A. Bhanvase. 2013. Process intensification of uniform loading of SnO_2 nanoparticles on graphene oxide nanosheets using a novel ultrasound assisted in situ chemical precipitation method. *Chemical Engineering and Processing: Process Intensification* 70:48–54.

Dong, C. S., M. L. Zhong, T. Huang, M. X. Ma, D. Wortmann, M. Brajdic, and I. Kelbassa. 2011. Photodegradation of methyl orange under visible light by micro-nano hierarchical Cu_2O structure fabricated by hybrid laser processing and chemical dealloying. *ACS Applied Materials and Interfaces* 3:4332–4338.

Dresselhaus, M. S. and G. Dresselhaus. 2002. Intercalation compounds of graphite. *Advances in Physics* 51:1–186.

Dujardin, E., T. W. Ebbesen, A. Krishnan, and M. M. Treacy. 1998. Wetting of single shell carbon nanotubes. *Advanced Materials* 10:1472–1475.

Eder, D. 2010. Carbon nanotube-inorganic hybrids. *Chemical Reviews* 110:1348–1385.

Gao, Z. Y., J. L. Liu, F. Xu, D. P. Wu, Z. L. Wu, and K. Jiang. 2012. One-pot synthesis of graphene–cuprous oxide composite with enhanced photocatalytic activity. *Solid State Sciences* 14:276–280.

Gaya, U. I. and A. H. Abdullah. 2008. Heterogeneous photocatalytic degradation of organic contaminants over titanium dioxide: A review of fundamentals, progress and problems. *Journal of Photochemistry and Photobiology C: Photochemistry Reviews* 9:1–12.

Geim, K. and K. S. Novoselov. 2007. The rise of graphene. *Nature Materials* 6:183–191.

Ghosh, M., E. V. Sampathkumaran, and C. N. R. Rao. 2005. Synthesis and magnetic properties of CoO nanoparticles. *Chemistry of Materials* 17:2348–2352.

Ghosh, S., I. Calizo, D. Teweldebrhan, E. P. Pokatilov, D. L. Nika, A. A. Balandin, W. Bao, F. Miao, and C. N. Lau. 2008. Extremely high thermal conductivity of graphene: Prospects for thermal management applications in nanoelectronic circuits. *Applied Physics Letters* 92:1–3.

Gilje, S., S. Han, M. Wang, K. L. Wang, and R. B. Kaner. 2007. A chemical route to graphene for device applications. *Nano Letters* 7:3394–3398.

Goncalves, G., P. A. A. P. Marques, C. M. Granadeiro, H. I. S. Nogueira, M. K. Singh, and J. Gracio. 2009. Surface modification of graphene nanosheets with gold nanoparticles: The role of oxygen moieties at graphene surface on gold nucleation and growth. *Chemistry of Materials* 21:4796–4802.

Guo, J., S. Zhu, Z. Chen, Y. Li, Z. Yu, Q. Liu, J. Li, C. Feng, and D. Zhang. 2011. Sonochemical synthesis of TiO_2 nanoparticles on graphene for use as photocatalyst. *Ultrasonics Sonochemistry* 18:1082–1090.

Hamilton, C. E., J. R. Lomeda, Z. Z. Sun, J. M. Tour, and A. R. Barron. 2009. High-yield organic dispersions of unfunctionalized graphene. *Nano Letters* 9:3460–3642.

Hannes, C. S., J. L. Li, M. J. McAllister, H. Sai, M. Herrera-Alonso, H. A. Douglas, R. K. Prud'homme, R. Car, D. A. Saville, and I. A. Aksay. 2006. Functionalized single graphene sheets derived from splitting graphite oxide. *Journal of Physical Chemistry B* 110:8535–8539.

Ho, J. Y. and M. H. Huang. 2009. Synthesis of submicrometer-sized Cu_2O crystals with morphological evolution from cubic to hexapod structures and their comparative photocatalytic activity. *Journal of Physical Chemistry C* 113:14159–14164.

Hummers, W. S. and R. E. Offeman. 1958. Preparation of graphitic oxide. *Journal of American Chemical Society* 80:1339.

Jasuja, K. and V. Berry. 2009. Implantation and growth of dendritic gold nanostructures on graphene derivatives: Electrical property tailoring and Raman enhancement. *ACS Nano* 3:2358–2366.

Kim, H., S. W. Kim, Y. U. Park, H. Gwon, D. H. Seo, Y. Kim, and K. Kang. 2010. SnO_2/graphene composite with high lithium storage capability for lithium rechargeable batteries. *Nano Research* 3:813–821.

Lee, C., X. Wei, J. W. Kysar, and J. Hone. 2008. Measurement of the elastic properties and intrinsic strength of monolayer graphene. *Science* 321:385–388.

Lei, Z., T. Ghosh, C. Y. Park, M. Ze-Da, and O. Won-Chun. 2012. Enhanced sonocatalytic degradation of Rhodamine B by graphene-TiO_2 composites synthesized by an ultrasonic-assisted method. *Chinese Journal of Catalysis* 33:1276–1283.

Li, B. J., H. Q. Cao, G. Yin, Y. X. Lu, and J. F. Yin. 2011. Cu_2O@reduced graphene oxide composite for removal of contaminants from water and supercapacitors. *Journal of Materials Chemistry* 21:10645–10648.

Li, J. L., L. Liu, Y. Yu, Y. W. Tang, H. L. Li, and F. P. Du. 2004. Preparation of highly photocatalytic active nano-size TiO_2–Cu_2O particle composites with a novel electrochemical method. *Electrochemistry Communications* 6:940–943.

Li, Y., X. Fan, J. Qi, J. Ji, S. Wang, G. Zhang, and F. Zhang. 2010. Palladium nanoparticle–graphene hybrids as active catalysts for the Suzuki reaction. *Nano Research* 3:429–437.

Lian, P., X. Zhu, H. Xiang, Z. Li, W. Yang, and H. Wang. 2010. Enhanced cycling performance of Fe_3O_4–graphene nanocomposite as an anode material for lithium-ion batteries. *Electrochimica Acta* 56:834–840.

Liang, Y. Y., H. L. Wang, H. S. Casalongue, Z. Chen, and H. J. Dai. 2010. TiO_2 nanocrystals grown on graphene as advanced photocatalytic hybrid materials. *Nano Research* 3:701–705.

Liu, D. P., G. D. Li, Y. Su, and J. S. Chen. 2006. Highly luminescent ZnO nanocrystals stabilized by ionic-liquid components. *Angewandte Chemie International Edition* 45:7370–7373.

Liu, Z. M. and B. X. Han. 2009. Synthesis of carbon-nanotube composites using supercritical fluids and their potential applications. *Advanced Materials* 21:825–829.

Lu, Y. H., W. Chen, Y. P. Feng, and P. M. He. 2009. Tuning the electronic structure of graphene by an organic molecule. *Journal of Physical Chemistry B* 113:2–5.

McAllister, M. J., J. L. Li, D. H. Adamson, H. C. Schniepp, A. A. Abdala, J. Liu, M. Herrera-Alonso et al. 2007. Single sheet functionalized graphene by oxidation and thermal expansion of graphite. *Chemistry of Materials* 19:4396–4404.

Min, Y., K. Zhang, L. Chen, Y. Chen, and Y. Zhang. 2012. Sonochemical assisted synthesis of a novel TiO$_2$/graphene composite for solar energy conversion. *Synthetic Metals* 162:827–833.

Muszynski, R., B. Seger, and P. V. Kamat. 2008. Decorating graphene sheets with gold nanoparticles. *Journal of Physical Chemistry C* 112:5263–5266.

Nair, R. R., P. Blake, A. N. Grigorenko, K. S. Novoselov, T. J. Booth, T. Stauber, N. M. R. Peres, and A. K. Geim. 2008. Fine structure constant defines visual transparency of graphene. *Science* 320:1308.

Nath, S., C. Kaittanis, A. Tinkham, and J. M. Perez. 2008. Dextran coated gold nanoparticles for the assessment of antimicrobial susceptibility. *Analytical Chemistry* 80:1033–1038.

Neppolian, B., A. Bruno, C. L. Bianchi, and M. Ashokkumar. 2012. Graphene oxide based Pt–TiO$_2$ photocatalyst: Ultrasound assisted synthesis, characterization and catalytic efficiency. *Ultrasonics Sonochemistry* 19:9–15.

Neppolian, B., C. Wang, and M. Ashokkumar. 2014. Sonochemically synthesized mono and bimetallic Au–Ag reduced graphene oxide based nanocomposites with enhanced catalytic activity. *Ultrasonics Sonochemistry* 21:1948–1953.

Novoselov, K. S., A. K. Geim, S. V. Morozov, D. Jiang, Y. Zhang, S. V. Dubonos, I. V. Grigorieva, and A. A. Firsov. 2004. Electric field effect in atomically thin carbon films. *Science* 306:666–669.

Paek, S. M., E. Yoo, and I. Honma. 2009. Enhanced cyclic performance and lithium storage capacity of SnO$_2$/graphene nanoporous electrodes with three-dimensionally delaminated flexible structure. *Nano Letters* 9:72–75.

Pasricha, R., S. Gupta, and A. K. Srivastava. 2009. A facile and novel synthesis of Ag–graphene-based nanocomposites. *Small* 5:2253–2259.

Patel, M. A., B. A. Bhanvase, and S. H. Sonawane. 2013. Production of cerium zinc molybdate nano pigment by innovative ultrasound-assisted approach. *Ultrasonics Sonochemistry* 20:906–913.

Peigney, A., C. H. Laurent, E. Flahaut, R. R. Bacsa, and A. Rousset. 2001. Specific surface area of carbon nanotubes and bundles of carbon nanotubes. *Carbon* 39:507–514.

Peng, X. H., J. Y. Chen, J. A. Misewich, and S. S. Wong. 2009. Carbon nanotube-nanocrystal heterostructures. *Chemical Society Reviews* 38:1076–1098.

Peng, Y., Y. Hu, L. Han, and C. Ren. 2014. Ultrasound-assisted fabrication of dispersed two-dimensional copper/reduced graphene oxide nanosheets nanocomposites. *Composites: Part B* 58:473–477.

Planeix, J. M., N. Coustel, B. Coq, V. Brotons, P. S. Kumbhar, R. Dutartre, P. Geneste, P. Bernier, and P. M. Ajayan. 1994. Application of carbon nanotubes as supports in heterogeneous catalysis. *Journal of American Chemical Society* 116:7935–7936.

Poizot, P., S. Laruelle, S. Grugeon, L. Dupont, and J.-M. Taracon. 2000. Nano-sized transition-metal oxides as negative-electrode materials for lithium-ion batteries. *Nature* 407:496–499.

Reina, A., X. T. Jia, J. Ho, D. Nezich, H. B. Son, V. Bulovic, M. S. Dresselhaus, and J. Kong. 2009. Large area, few-layer graphene films on arbitrary substrates by chemical vapor deposition. *Nano Letters* 9:30–35.

Robinson, J. T., F. K. Perkins, E. S. Snow, Z. Q. Wei, and P. E. Sheehan. 2008. Reduced graphene oxide molecular sensors. *Nano Letters* 8:3137–3140.

Seger, B. and P. V. Kamat. 2009. Electrocatalytically active graphene-platinum nano composites. Role of 2-D carbon support in PEM fuel cells. *Journal of Physical Chemistry C* 113:7990–7995.

Shen, J., M. Shi, N. Li, B. Yan, H. Ma, Y. Hu, and M. Ye. 2010. Facile synthesis and application of Ag-chemically converted graphene nanocomposite. *Nano Research* 3:339–349.

Shi, J., G. Yang, and J. Zhu. 2011. Sonoelectrochemical fabrication of PDDA-RGO-PdPt nanocomposites as electrocatalyst for DAFCs. *Journal of Materials Chemistry* 21:7343–7349.

Shi, J. and J. Zhu. 2011. Sonoelectrochemical fabrication of Pd-graphene nanocomposite and its application in the determination of chlorophenols. *Electrochimica Acta* 56:6008–6013.

Si, Y. C. and E. T. Samulski. 2008. Exfoliated graphene separated by platinum nanoparticles. *Chemistry of Materials* 20:6792–6797.

Sidorov, A. N., M. M. Yazdanpanah, R. Jalilian, P. J. Ouseph, R. W. Cohn, and G. U. Sumanasekera. 2007. Electrostatic deposition of graphene. *Nanotechnology* 18:135301.

Stankovich, S., D. A. Dikin, G. H. B. Dommett, K. M. Kohlhaas, E. J. Zimney, E. A. Stach, R. D. Piner, S. T. Nguyen, and R. S. Ruoff. 2006. Graphene-based composite materials. *Nature* 442:282–286.

Stoller, M. D., S. J. Park, Y. W. Zhu, J. H. An, and R. S. Ruoff. 2008. Graphene-based ultracapacitors. *Nano Letters*. 8:3498–3502.

Sutter, P. W., J. I. Flege, and E. A. Sutter. 2008. Epitaxial graphene on ruthenium. *Nature Materials* 7:406–411.

Vinodgopal, K., B. Neppolian, N. Salleh, I. V. Lightcap, F. Grieser, M. Ashokkumar, T. T. Ding, and P. V. Kamat. 2012. Dual-frequency ultrasound for designing two dimensional catalyst surface: Reduced graphene oxide–Pt composite. *Colloids and Surfaces A: Physicochemical and Engineering Aspects* 409:81–87.

Wang, D. H., D. W. Choi, J. Li, Z. G. Yang, Z. M. Nie, R. Kou, D. H. Hu et al. 2009. Self-assembled TiO_2-graphene hybrid nanostructures for enhanced Li-Ion insertion. *ACS Nano* 3:907–914.

Wang, H. L., J. T. Robinson, X. L. Li, and H. J. Dai. 2009. Solvothermal reduction of chemically exfoliated graphene sheets. *Journal of American Chemical Society* 131:9910–9911.

Williams, G., B. Seger, and P. V. Kamat. 2008. TiO_2-graphene nanocomposites. UV-assisted photocatalytic reduction of graphene oxide. *ACS Nano* 2:1487–1491.

Xiong, Y. J. and Y. N. Xia. 2007. Shape-controlled synthesis of metal nanostructures: The case of palladium. *Advanced Materials* 19:3385–3391.

Xu, C. and X. Wang. 2009. Fabrication of flexible metal-nanoparticle films using graphene oxide sheets as substrates. *Small* 5:2212–2217.

Xu, C., X. Wang, L. C. Yang, and Y. P. Wu. 2009. Fabrication of a graphene–cuprous oxide composite. *Journal of Solid State Chemistry* 182:2486–2490.

Xu, C., X. Wang, and J. Zhu. 2008a. Graphene–metal particle nanocomposites. *Journal of Physical Chemistry C* 112:19841–19845.

Xu, C., X. Wang, J. W. Zhu, X. J. Yang, and L. D. Lu. 2008b. Deposition of Co_3O_4 nanoparticles onto exfoliated graphite oxide sheets. *Journal of Material Chemistry* 18:5625–5629.

Yan, G., X. Li, Z. Wang, H. Guo, Q. Zhang, and W. Peng. 2013. Synthesis of Cu_2O/reduced graphene oxide composites as anode materials for lithium ion batteries. *Transactions of Nonferrous Metals Society of China* 23:3691–3696.

Yan, X. Y., X. L. Tong, Y. F. Zhang, X. D. Han, Y. Y. Wang, G. Q. Jin, Y. Qin, and X. Y. Guo. 2012. Cuprous oxide nanoparticles dispersed on reduced graphene oxide as an efficient electrocatalyst for oxygen reduction reaction. *Chemical Communications* 48:1892–1894.

Yang, H. and Z. H. Liu. 2010. Facile synthesis, shape evolution, and photocatalytic activity of truncated cuprous oxide octahedron microcrystals with hollows. *Crystal Growth and Design* 10:2064–2067.

Yang, X., X. Zhang, Y. Ma, Y. Huang, Y. Wang, and Y. Chen. 2009. Superparamagnetic graphene oxide–Fe_3O_4 nanoparticles hybrid for controlled targeted drug carriers. *Journal of Materials Chemistry* 19:2710–2714.

Yao, J., X. Shen, B. Wang, H. Liu, and G. Wang. 2009. In situ chemical synthesis of SnO₂–graphene nanocomposite as anode materials for lithium-ion batteries. *Electrochemistry Communications* 11:1849–1852.

Yao, K. X., X. M. Yin, T. H. Wang, and H. C. Zeng. 2010. Synthesis, self-assembly, disassembly, and reassembly of two types of Cu₂O nanocrystals unifaceted with {001} or {110} planes. *Journal of the American Chemical Society* 132:6131–6144.

Yoo, E., T. Okata, T. Akita, M. Kohyama, J. Nakamura, and I. Honma. 2009. Enhanced electrocatalytic activity of Pt subnanoclusters on graphene nanosheet surface. *Nano Letters* 9:2255–2259.

Zhang, F. Y., Y. J. Li, Y. E. Gu, Z. H. Wang, and C. M. Wang. 2011. One-pot solvothermal synthesis of a Cu₂O/graphene nanocomposite and its application in an electrochemical sensor for dopamine. *Microchimica Acta* 173:103–109.

Zhang, H., X. J. Lv, Y. M. Li, Y. Wang, and J. H. Li. 2010. P25–graphene composite as a high performance photocatalyst. *ACS Nano* 4:380–386.

Zhang, J. T., J. F. Liu, Q. Peng, X. Wang, and Y. D. Li. 2006. Nearly monodisperse Cu₂O and CuO nanospheres: Preparation and applications for sensitive gas sensors. *Chemistry of Materials* 18:867–871.

Zhang, X. Y., H. P. Li, and X. L. Cui. 2009. Preparation and photocatalytic activity for hydrogen evolution of TiO₂/graphene sheets composite. *Chinese Journal of Inorganic Chemistry* 25:1903–1907.

Zhang, Y., Y. W. Tan, H. L. Stormer, and P. Kim. 2005. Experimental observation of the quantum Hall effect and Berry's phase in graphene. *Nature* 438:201–204.

Zhang, Y., X. Wang, L. Zeng, S. Song, and D. Liu. 2012. Green and controlled synthesis of Cu₂O–graphene hierarchical nanohybrids as high-performance anode materials for lithium-ion batteries via an ultrasound assisted approach. *Dalton Transactions* 41:4316–4319.

Zheng, Z. K., B. B. Huang, Z. Y. Wang, M. Guo, X. Y. Qin, X. Y. Zhang, P. Wang, and Y. Dai. 2009. Crystal faces of Cu₂O and their stabilities in photocatalytic reactions. *Journal of Physical Chemistry C* 113:14448–14453.

Zhou, G., D. W. Wang, F. Li, L. Zhang, N. Li, Z. S. Wu, L. Wen, G. Q. Lu, and H. M. Cheng. 2010. Graphene-wrapped Fe₃O₄ anode material with improved reversible capacity and cyclic stability for lithium ion batteries. *Chemistry of Materials* 22:5306–5313.

Zhou, X., J. Zhang, Q. Su, J. Shi, Y. Liu, and G. Du. 2014. Nanoleaf-on-sheet CuO/graphene composites: Microwave-assisted assemble and excellent electrochemical performances for lithium ion batteries. *Electrochimica Acta* 125:615–621.

Zhu, H. T., J. X. Wang, and G. Y. Xu. 2009. Fast synthesis of Cu₂O hollow microspheres and their application in DNA biosensor of hepatitis B virus. *Crystal Growth and Design* 9:633–638.

Zhu, J., D. Wang, T. Liu, and C. Guo. 2014. Preparation of Sn-Co-graphene composites with superior lithium storage capability. *Electrochimica Acta* 125:347–353.

Zhu, L., T. Ghosh, C. Park, Z. D. Meng, and W. C. Oh. 2012. Enhanced sonocatalytic degradation of Rhodamine B by graphene-TiO₂ composites synthesized by an ultrasonic-assisted method, *Chinese Journal of Catalysis* 33:1276–1283.

Zhu, S., J. Guo, J. Dong, Z. Cui, T. Lu, C. Zhu, D. Zhang, and J. Ma. 2013. Sonochemical fabrication of Fe₃O₄ nanoparticles on reduced graphene oxide for biosensors. *Ultrasonics Sonochemistry* 20:872–880.

Zhuo, S., M. Shao, and S. Lee. 2012. Upconversion and downconversion fluorescent graphene quantum dots: Ultrasonic preparation and photocatalysis. *ACS Nano* 6:1059–1064.

CHAPTER 19

CONTENTS

Graphene Plasmonics

19

Deyang Du, Xiaoguang Luo, and Teng Qiu

19.1 INTRODUCTION OF GRAPHENE PLASMONICS

Since being exfoliated successfully from graphite by Novoselov and Geim in 2004, graphene has attracted tremendous attention in many fields due to the novel and unique electrical, mechanical, thermal, and optical properties,[1,2] of which the good optical properties and high carrier mobility of two-dimensional (2D) electrons develop many potential applications in the fields of electromagnetisms and optics, especially in plasmonics.[3]

Because of the linear and cone-like electronic structure without bandgap, a graphene monolayer can have a constant absorption coefficient of 2.3% over a wide spectral range from the visible to the infrared.[4] The single atom layer thickness and low photoconductivity in the visible region of the spectrum, results in that graphene interacts with white light very inefficiently.[5] The way to enhance the interaction between light and graphene was always the key problem for further researches. It is known that two effective ways can be used to increase light absorption. One is nanopatterning, and the other is decoration by noble metal nanostructures. Both of them ascribe to the coupling between surface plasmon (SP) and light. The former refers to the inherent SP of graphene, while the latter is related to SP of the graphene/metal composite structure. For example, gold plasmonic nanoantennas fabricated onto graphene have been shown to substantially enhance its light-harvesting efficiency.[6] It has also been reported that Ag–graphene plasmonic composites can improve light harvesting and photoluminescence quenching at the same time and graphene here also acts as an electron acceptor.[7] Due to remarkable properties, including extreme confinement, beneficial tunability, and low losses, graphene has become a promising plasmonic material for new-generation plasmonic nanoscaled devices, such as high-sensitive graphene-based sensors and detectors.[8]

Unfortunately, regarded as a zero-bandgap semiconductor, monolayer graphene cannot be applied in optical devices flexibly.[9] The lack of a bandgap stops graphene from luminescencing.[10] Thus, many efforts have been made to open the bandgap of graphene through controlling the size (graphene quantum dots), surface chemistry of graphene (graphene oxide),[11] and physical structure (carbon nanotube).[12] Researches show that these treatments make the graphene performing better in photoluminescence[13] and electroluminescence.[14] For example, graphene oxide has shown interesting steady-state photoluminescence emission, ranging from near-infrared to blue fluorescence.[15] In addition, similar to enhancement of light absorption, the SP can also be employed to enhance light emission.

The conductivity of graphene $\sigma(\omega, E_F, \tau, T)$ is an important and complex parameter,[16] which is strongly dependent on the radian frequency ω, Fermi level E_F, electron relaxation time τ, and even the temperature T. Experimental and theoretical results show that the conductivity and permittivity of graphene satisfy the relationship of $\varepsilon = \varepsilon_0 + i\sigma/\omega\Delta$, where Δ denotes the graphene thickness. Therefore, the electromagnetic parameters of graphene can be effectively controlled through adjusting Fermi level, for example, by electrostatic gating or chemical doping. On the basis of this gate-voltage-dependent properties, optical splitters,[17] spatial switches,[18] and ultracompact Mach–Zehnder interferometers[19] have also been designed with graphene and exhibit better tunability than devices consisting of traditional plasmonics materials.[20]

Combined with the relatively high confinement and low loss, graphene can be regarded as a good alternative for new plasmonic materials. The manipulation of light can be more effective somewhat based on graphene. Moreover, composited with traditional noble metals or other plasmonic materials, graphene will perform versatile in plasmonics in the future. In the following text, with a brief introduction of applications of graphene plasmonics, these excellent properties will be discussed in detail.

19.1.1 Electronic Structure of Graphene

Single-layer graphene, a gapless semiconductor, is a monolayer of carbon atoms (with thickness ≈ 0.34 nm) packed in a 2D honeycomb lattice with the lattice constant $a \approx 0.142$ nm. Three sp^2 hybridized orbitals are oriented in the x–y plane and have mutual 120 angles which causes the honeycomb formation consisting of six covalent σ bonds. The remaining unhybridized $2p_z$ orbital is perpendicular to the x–y plane and form π bonds,[21] as the atomic structure shown in Figure 19.1a. Since each $2p_z$ orbital has one extra electron, the π band is half filled. Nevertheless, the half-filled bands in transition elements have played an important role in the physics of strongly correlated systems because of their strong tight-binding character and the large Coulomb energies.[22]

From the tight-binding approach when $\hbar = 1$, the energy bands of a single-layer pure graphene from the π electrons can be expressed as[22]

$$E_{\pm}\left(\mathbf{k}\right) = t\sqrt{3 + f\left(\mathbf{k}\right)} - t'f\left(\mathbf{k}\right), \tag{19.1}$$

where

    $\mathbf{k}$ is the wave vector

    t and t' are the nearest-neighbor and the next nearest-neighbor hopping energy, respectively, and

$$f\left(\mathbf{k}\right) = 2\cos\left(\sqrt{3}k_y a\right) + 4\cos\left(\sqrt{3}k_y a/2\right)\cos\left(3k_x a/2\right). \tag{19.2}$$

Based on this equation, it is found that the obtained two bands are symmetric around zero energy if $t' = 0$, while for $t' \neq 0$ the electron–hole symmetry is broken, as shown in Figure 19.1b. Each carbon atom contributes one π electron, causing the lower band E_- (named as π band) completely filled and upper band E_+ (known as π^* band) completely empty. The two bands touch each other at the Dirac point at each corner of the graphene Brillouin zone, and the band structure close to the Dirac point is cone-like, where the dispersion can be approximately regarded as linear relationship at small wave vector, as shown in the enlarge view of Figure 19.1b.

For pristine graphene, Fermi level E_F is equal to the energy at Dirac point. There is only one kind of electron–hole excitation (interband transition) at low electron hopping energy because of the empty π^* band (conduction band) and the completely filled π band (valence band). But for n/p-doped graphene, the Fermi level E_F is away from the Dirac point which causes the other kind of electron–hole excitation: intraband transition. When the graphene is n-doped, the Fermi level E_F will be higher than the Dirac point, and electrons will be found in the π^* band (conduction band). Electrons both at the bottom of the conduction band and at the top of the valence band can be excited after absorbing a certain amount of energy and momentum. The electron–hole continuum or single-particle excitation region then will be formed in $(\mathbf{q}, \omega)$ space for the wave

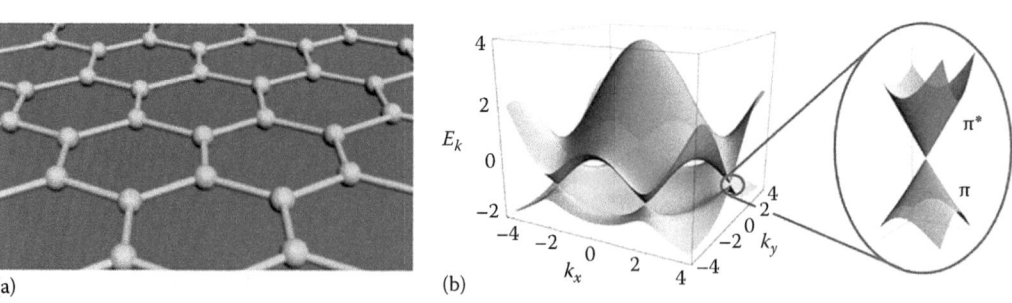

(a)   (b)

FIGURE 19.1
(a) The molecule schematic of graphene. (b) The conical band structure with a zero-energy bandgap at the Dirac point, where $t = 2.7$ eV and $t' = -0.2t$. (From Castro Neto, A.H. et al., *Rev. Mod. Phys.*, 81(1), 109, 2009.)

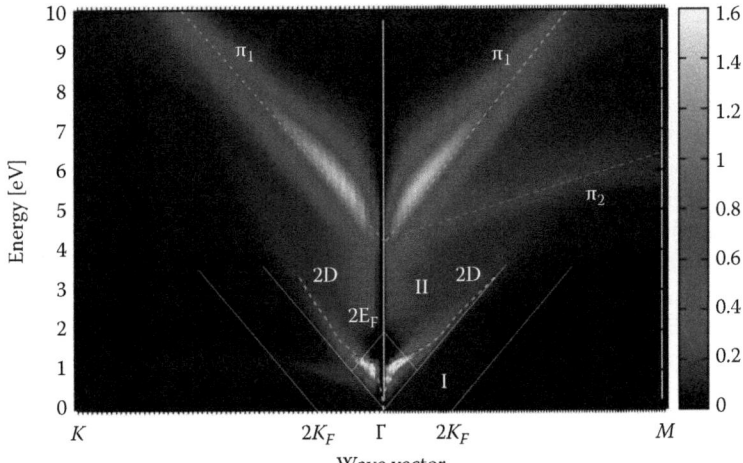

FIGURE 19.2
Intensities of electronic excitations which also show the dispersion relations of 2D plasmon and π plasmons for $E_F = 1$ eV. (From Despoja, V. et al., *Phys. Rev. B*, 87(7), 075447, 2013.)

vector $\mathbf{q} = \mathbf{k} - \mathbf{K}$, where $\mathbf{K}$ is the Dirac point in momentum space. And the intraband (I) and interband (II) transitions (shown in Figure 19.2) of electrons in n-doped graphene have distinct boundary in the single-particle excitation region.

19.1.2 Surface Plasmons in Graphene

The plasmon was firstly reported by Wood in 1902,[23] with the result of uneven distribution of light in a diffraction grating spectrum. With series of experimental and theoretical contributions by Fano, Crookes, Langmuir, Tonks, Pines, Bohm, etc., the field of plasmon was developing with each passing day. Finally in 1960, Stern and Ferrell studied the plasma oscillations of the degenerate electron gas related to the material surface and firstly named them as surface plasmons (SPs).[24]

The concept of SPs is the collective oscillations of charges at the surface of plasmonic materials. Many reports show that, with heavy energy loss, plasmon inside the materials evanesces severely, but it can propagate quite a long distance along the surface. With the development of this field, researchers have found that, SPs can be excited or coupled with the different quantized energies or particles, that is, photons, electrons, and phonons.[25] Coupled with photons, SPs can then form the composite particles of surface plasmon polaritons (SPPs) or induce the localized surface plasmon resonances (LSPRs). SPPs are often transported along the smooth surface of plasmonic materials although with considerable evanesces, while LSPRs can only be localized in the roughness surface accompanying with a strong near field.

The dispersion relationship between the frequency and wave vector for SPPs propagating along the interface of semi-infinite medium and dielectric can be obtained by surface mode solutions of Maxwell's equations under appropriate boundary conditions.[26] It should be noted that SPPs cannot be excited by light in an ideal semi-infinite medium. To excited SPPs, the light must match with the wave vector of SPs by exploiting some ways, or using special structures such as prism, topological defect, and periodic corrugation.[27] The nonradiation solution is the SPs dispersion $k_{SP} = k_0 \sqrt{\varepsilon_m \varepsilon_d / (\varepsilon_m + \varepsilon_d)}$, where ε_m and ε_d are the relative permittivities of medium and dielectric, and k_0 is the wave vector of light in free space. From the dispersion relation, k_{SP} can be complex with the positive real part standing for propagation and imaginary part determining the internal absorption which mainly rely on ε_m. For metallic materials, ε_m can be derived from the Drude model with the result of $\varepsilon_m = 1 - \omega_0^2 / (\omega^2 + i\tau^{-1}\omega)$.[28] Coupled with electrons, photons, or phonons, SPs have appeared promising potential for applications in engineering and applied sciences.[29] Meanwhile, nanophotonics,[30] metamaterials,[31] photovoltaic devices,[32] sensors,[33] and so on have gotten great development ascribing to plasmonics. With strong confinement and great control of electromagnetic (EM) field at subwavelength,[27] SPs have been exploited in many plasmonic materials such as Au,

Ag, Cu, Cr, Al, Mg, etc., which were regarded as the best plasmonic materials in the past. However, these conventional metal materials suffer large energy losses (e.g., Ohmic loss and radiative loss), and SPs have relatively bad tunability in the fixed structure or device.[34] Such shortcomings limit the further development of plasmonics and it is necessary to find new plasmonic materials which will do better in these aspects.

Graphene has proved itself to be a promising alternative plasmonic material from excellent properties, such as high tunability, strong confinement, and low energy loss with efficient wave localization up to mid-infrared frequencies.[35] The low loss property may be due to the gapless cone-like band structure and massless Dirac-Fermions (electrons in graphene), causing the carriers (both electrons and holes) in graphene having ultra-high-mobility and long mean free path.[22] More importantly, SPs excited in graphene are confined much more strongly than those in conventional noble metals, owing to its 2D nature of the collective excitations. And the most exciting advantage of graphene should be the tunability of SPs, as the carriers, densities in the graphene can be controlled by electrical gating and doping.[36]

Figure 19.2 shows that two kinds of SPs exist in graphene, including low-energy 2D SPs and higher-energy π SPs. The latter, generally is difficult to employ in experiments because that the unavoidable damage will happen to graphene after SPs excitation. Instead, 2D SPs have attracted great interest, where the loss in interband transition region (region II) is remarkable, and in the following content SPs in graphene are regarded as 2D SPs. From the results of random-phase approximation, the SPs dispersion in the long-wavelength limit ($q \to 0$) can be expressed as[38]

$$\omega_{SP}(q \to 0) = 2e\sqrt{E_F q / 2\kappa} \qquad (19.3)$$

where κ denotes the background lattice dielectric constant of the system. It should be noted that $q = k$ refer to the Dirac point. Equation 19.3 also implies that $\omega_{SP} \propto n^{1/4}$ in the long-wavelength limit because of $E_F \propto \sqrt{n}$, where n is carrier concentration.[39] Generally, Drude model can express plasmonic properties of metallic materials very well. As a semimetal, graphene can also be described by Drude model once being highly doped or gated. The excitation of SPs in graphene often refers to the terahertz (THz) and infrared frequency regions due to the unique band structure, in which region the intraband contribution is dominating. Then, the conductivity $\sigma(\omega, E_F, \tau, T)$ can be simplified as

$$\sigma_{intra} = \frac{e^2 E_F}{\pi \hbar^2} \frac{i}{\omega + i\tau^{-1}} \qquad (19.4)$$

Using nonretarded approximation or electrostatic limit, one can obtain the analytical expression of SP dispersion relation of graphene on a substrate (with permittivity ε_r) as

$$\lambda \approx \lambda_0 \alpha \frac{4E_F}{\varepsilon_r + 1} \frac{1}{\hbar(\omega + i\tau^{-1})} \qquad (19.5)$$

where $\alpha \approx 1/137$ is the fine structure constant.

Consequently, graphene can be a promising material in further applications in the field of terahertz (THz) material.[39] Besides, devices such as flexible plasmonic waveguide,[40] transformation optical devices,[41] and so on can also be exploited by using graphene. Those achievements manifest much more advantages of graphene in the control of EM wave compared to the conventional metal materials. By the way, SPs in graphene can be coupled with photons, electrons, and phonons. It will form SPPs with photons[42] which have already been observed in a standing wave mode and will composite "plasmaron" particles with electrons.[43] SPs can definitely enhance the light absorption of graphene.[44] Due to the collective effect of graphene SPs, light can even be completely absorbed by the graphene nanodisk array for the featured incident angle.[45] Moreover, graphene can help to turn the SPs in conventional plasmonic material (such as Au),[46] which really shows the abundant further applications of graphene.

19.2 PLASMONS FOR STRONG INTERACTION OF GRAPHENE WITH LIGHT

19.2.1 Localized Surface Plasmon Resonances in Graphene

Single-atom-layer graphene suffers from very bad optical absorption, 2.3% of normal incidence light for a wide frequency region, because of the cone-like electron energy band. To utilize the excellent properties (e.g., ultrahigh electron mobility) in electrophotonic applications, higher absorption of light by graphene is desired. An alternative approach for enhancing optical absorption in graphene is to exploit the relatively stronger absorption properties of LSPRs. Direct optical excitation and facile electrical tunability are both possible, benefiting from plasmons of nanostructured graphene. The use of subwavelength structures enables one to overcome the momentum and energy mismatch between photons and plasmons, causing the coupling of them to enhance.

It is known that the structure of graphene is not always a perfect plane of hexatomic carbon, inhomogeneities such as corrugations, edges, overlaps, defects, impurities, etc. may be caused intentionally or unintentionally in practical applications. SPs in inhomogeneous graphene could be easily localized in the form of LSPRs. Being significantly less loss than traditional plasmonic materials (e.g., coinage metals), graphene plasmons can exhibit large Q-factors, resulting in narrow absorption profiles. By annular dark-field imaging, structural and chemical information around point defects in graphene can be detected, and these point defects can lead to LSPRs. Taking silicon and nitrogen atoms point defects as examples, obvious field enhancement around those defects is observed by electron energy loss spectroscopy imaging.[47] Absolutely, this experiment is a direct proof of LSPR at the single atom level, and this strategy may be a potential way to enhance optical absorption.

Besides, direct patterning of graphene may be more promising means to enhance the optical absorption. For example, achieved by using electron-beam lithography and subsequent lift-off, graphene nanostructures of nanoribbons, nanodisks, and nanorings can be easily produced, and all of them are capable of supporting SPs, especially LSPRs. By stacking these graphene nanostructures (e.g., graphene/insulator disks[48]), the optical absorption can be further increased. Since the Fermi level of graphene is readily controlled by applying a voltage (electrostatic doping), a wide range of plasmon tunability in the IR is possible in those tailored graphene devices.

The mostly investigated graphene structures are micro/nano-ribbons (GMRs/GNRs). By reducing the degree of freedom, SPPs in these ribbons can propagate along the ribbons, while LSPRs are enhanced due to confinement in other directions. The LSPR in GMR can also enhance the optical absorption. To strengthen the resonance and increase the absorption area, GMRs arrays are the best choice. For incident EM wave polarized perpendicular to GMRs, the prominent room-temperature optical absorption peaks can be obtained in THz region, as shown in Figure 19.3, and the resonances can also be tuned by electrical doping, incident angle, and the array scales.[39] By suppressing transmission in these ribbons, light passing through GMRs arrays can even be completely absorbed. Although being approximately proportional to the coverage of graphene, the absorption in GMRs arrays is still stronger than that in the continuous graphene sheet due to the sufficiently high relaxation time.

Different structures proved to be helpful for enhancing the absorption of the graphene. To reduce the dimensionality further, graphene with structures of micro/nano-disks, -antidots, and -rings are studied. Like metal nanoparticles, the EM field in graphene disk behaves like a dipole. Calculated by random-phase approximation and similar to GNRs, finite-size effects cause substantial plasmons broadening compared to the homogeneous graphene when the disk diameter is below ~20 nm.[54] Owing to the zero-dimensional nature, LSPR in disk structure becomes very strong and causes a strong enhanced electrical field. Therefore, the disk structure of graphene might be promising alternative to the metal materials for LSPRs. However, combined with a nearby quantum emitter (e.g., a quantum dot or a molecule), the nonlinear optical response of graphene nanodisks can be obtained. Because of the energy transfer and plasmon–plasmon blockage, the optical response of graphene disks can be easily tuned by doping.

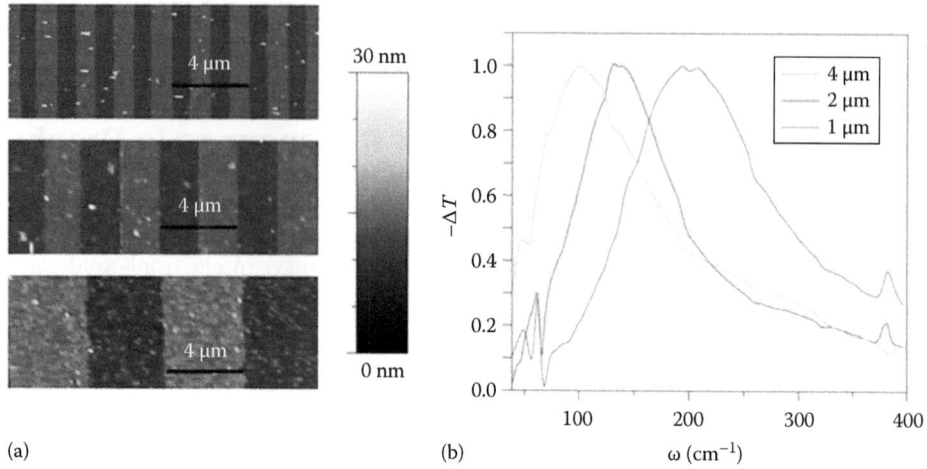

(a)   (b)

FIGURE 19.3
Plasmonic THz metamaterials made by graphene microribbons, the structure is shown by the atomic force microscope imaging (a), the absorption peak (b) is observed relative absorbance spectrum which is impacted by the structure size. (From Ju, L. et al., *Nat. Nanotechnol.*, 6(10), 630, 2011.)

The EM field in graphene antidots can also be regarded as a dipole, while plasmons in graphene ring can be treated as the plasmons hybridization (symmetric and antisymmetric ways) from a graphene disk and a smaller diameter antidot. As shown in Figure 19.4, for similar sizes, the energy of the plasmons hybridization for different graphene structures from low to high are: ring (symmetric coupling), disk or antidot, and ring (antisymmetric coupling), respectively; and the hybridization can be effectively tuned by the size.[55,36] Generally, low hybridization energy is beneficial to the plasmons resonance and also the near-field enhancement. The enhancement factor of EM field by plasmons in graphene ring (symmetric coupling) can reach as large as 10^3 times in THz region, which is almost 20 times larger than similar structure made by gold. Furthermore, the relatively high relaxation times of the charge carriers in these graphene structures enhance the coupling between plasmons and other quasi-particles, and lead to enhanced absorption and suppressed transmission.

Large-scale patterns with graphene micro/nano-structures are necessary for practical applications. However, the coupling of SPs between graphene nanostructures on the same plane is still relatively weak, so an alternative way to enhance the coupling is to stack those graphene nanostructures, for example, the graphene/insulator stack structures.

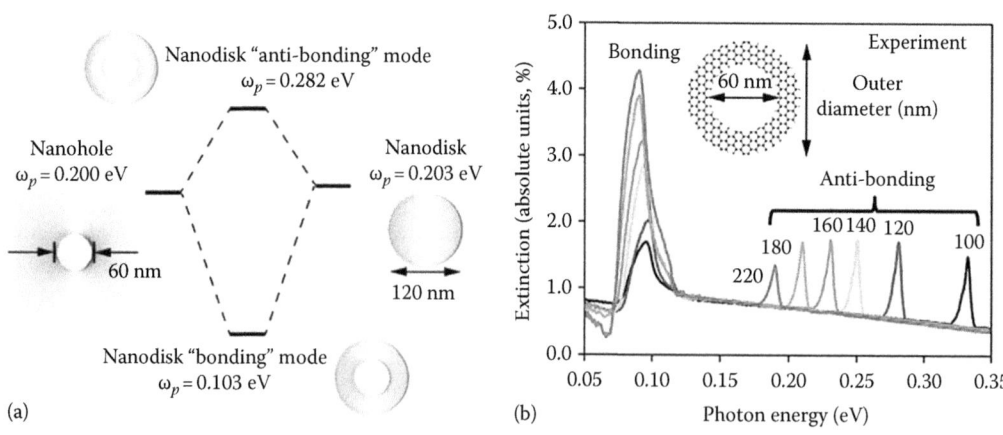

(a)   (b)

FIGURE 19.4
(a) Calculated energy level of the plasmon hybridization for graphene ring, disk, and antidot (hole). (b) Measured extinction spectra of graphene ring for fixed inner diameter (60 nm) and different outer diameters. (From Fang, Z. et al., *ACS Nano*, 6(11), 10222, 2012.)

19.2.2 Surface Plasmon Polaritons in Graphene

As mentioned earlier, graphene is capable of supporting localized and propagated SPs modes. The latter mode often propagates along the graphene plane in the form of SPPs, which is also formed by the coupling between SPs and photons. Therefore, wave guides can be realized by using graphene nanostructure, especially graphene ribbons. Theoretically, two modes, including waveguide and edge modes, can be found in the THz frequency range when SPPs propagate along GMR, and are separated from each other by a gap in wave numbers. Moreover, higher frequency or wider ribbon can increase the number of SPPs modes, and the propagation length, rather than wave vector, is strongly sensitive to the relaxation time of charge carriers.[56]

In experiment, electron energy loss spectroscopy and other spectroscopic studies have revealed the existence of SPs in graphene and the interaction between SPs and low-energy electrons or photons, but the direct visualization of propagating and localized graphene SPs is still highly desired. The convenient way to visualized SPs in graphene is to probe the corresponding SPPs. Similar to the traditional metals, SPs in graphene also face the mismatch of energy and momentum with those of light in free space. Thus, prism, topological defects, and periodic corrugations were adopted to solve this problem. Recently, Chen et al.[57] and Fei et al.[58] obtained the scanning near-field infrared light nanoscopy of SPPs in gated graphene on 6H-SiC and SiO_2 substrate, respectively. They illuminated the sharp tip of an atomic force microscope with a focused infrared beam (with λ_0 around 10 μm) to let the wave vectors of light match those of plasmons. At the fixed frequency of incident infrared light, the SPPs excited by the illuminated tip can propagate along the sheet, and they will be reflected, interfered, and damped at the graphene edges, defects, and at the boundary between different layers of graphene. According to those wave properties, standing waves can be formed when the SPPs evanesce incompletely, as shown in Figure 19.5. The wavelength of SPPs can be conveniently measured from these standing waves rather than the propagating ones, which is ~200 nm, agree with the theoretical

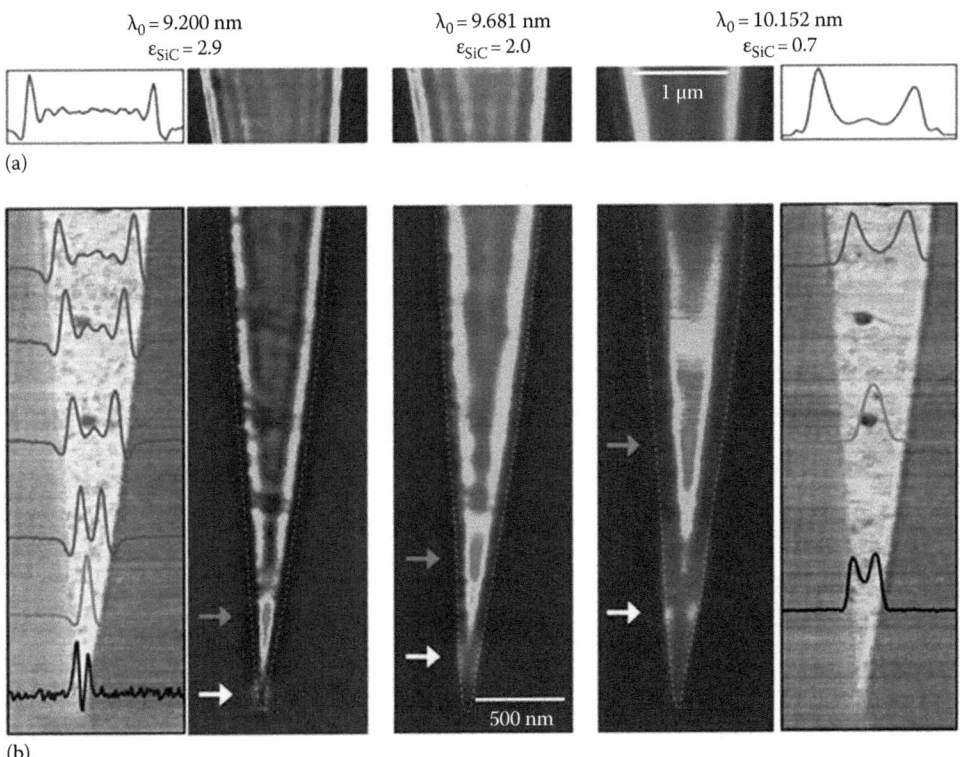

(a)

(b)

FIGURE 19.5
Optical nanoimaging of SPs in tapered graphene ribbon on the carbon-terminated surface of 6H-SiC. Fringes SPs formed by standing wave can be tuned by the incident light and the dielectric constant of substrate. (a) Images of a graphene ribbon, revealing strong dependence of the fringe spacing, and thus plasmon wavelength, on the excitation wavelength. (b) images of a tapered grapheme ribbon. (From Chen, J. et al., *Nature*, 487, 77, 2012.)

prediction of Equation 19.5. Moreover, both amplitude and wavelength of SPPs in graphene can be successfully tuned by different wavelengths of incident light, dielectric constants of substrate and gate voltages. However, the intensities of fringes in the middle of the ribbon are very weak, which implies that the energy loss cannot be ignored in graphene SPPs, even though it is at the weak damping intraband region. It is a big deal of confirming the existence of SPs directly in graphene, which also provides confidence to researchers for developing subwavelength devices with graphene in the future.

19.2.3 Properties of Surface Plasmons in Graphene

The plasmonics referred to metals have been studied for centuries and different applications have been proposed or realized. The possibility of graphene to be the alternative material to the conventional noble metals (such as Au, Ag, Cu, Al, etc.) in plasmonics relies on its unique properties, which will be discussed in detail in the following paragraphs.

For the reason of wide application, we focus on the plasmons related to photons, that is, SPPs. Generally, the existence of SPs depends on a negative real part of dielectric constant, that is, $\varepsilon' < 0$ if $\varepsilon = \varepsilon' + i\varepsilon''$. SPs are well pronounced as resonances when $\varepsilon'' \ll -\varepsilon'$ and the losses are very small.[60] These two conditions are also the criterions for the choosing of good plasmonic materials. SPPs propagate along the surface of the plasmonic materials and attenuate both in the direction parallel and perpendicular to the surface. The wavelength (λ_{SPP}) and propagation distance (δ_{SPP}) can be derived from the complex dispersion relation by taking the real and imaginary parts of the wave vector (k_{SPP}), that is, $\lambda_{SPP} = 2\pi/\mathrm{Re}[k_{SPP}]$ and $\delta_{SPP} = 1/2\mathrm{Im}[k_{SPP}]$, respectively.[61] Besides, the penetration depth in the medium-dielectric semi-infinite system can be obtained by $\delta_i = 1/\mathrm{Im}[k_z]$, where i denotes penetrating into mediums (m) or dielectric (d), and k_z is the complex wave vector perpendicular to the interface, for example, $k_z = k_0\sqrt{\varepsilon_m^2/(\varepsilon_m + \varepsilon_d)}$ inside the mediums and $k_z = k_0\sqrt{\varepsilon_d^2/(\varepsilon_m + \varepsilon_d)}$ inside the dielectric.

When the mediums are metals, all the parameters can be expressed analytically, as shown in Table 19.1, where $\varepsilon_m = 1 - \omega_0^2/(\omega^2 + i\tau^{-1}\omega)$ from Drude model. It is clearly shown that $\delta_{SPP} > \lambda_{SPP}$ for metal plasmonic materials. When $|\varepsilon_m'| \gg |\varepsilon_d|$ and the metal is at low loss, the propagation distance can be approximated by $\delta_{SPP} \approx \lambda_0\left(\varepsilon_m'\right)^2/\left(2\pi\varepsilon_m''\right)$, from which one can see that a large (negative) real part and a small imaginary part of dielectric constant of metals are beneficial for SPPs.[61] For graphene, the dielectric constant is hard to be expressed directly. The common method is to deal with the conductivity from the Kubo formula. Fortunately, for the highly doped case, the dispersion has a simple Drude-like form (Equation 19.5) which can be used to describe the properties of SPPs in graphene. From Table 19.1, it is found that long relaxation time, high Fermi energy, and low dielectric constant of substrate benefit the propagation of SPPs in graphene.

19.2.3.1 Relatively low loss of surface plasmons in graphene

In the visible and infrared parts of spectrum, $|\varepsilon'|$ and ε'' of noble metals meet criterions for good plasmonic materials, of which Au and Ag are the best choices. Tassin et al.[62] have made the comparison of the plasmonic properties between metals and graphene. As shown in Figure 19.6a (the frequency

---

TABLE 19.1 The Expressions of Key Parameters of SPP for Metals and Highly Doped Graphene

Parameter	Metal	Graphene
Wavelength (λ_{SPP})	$\lambda_0\left[\left(\varepsilon_d + \varepsilon_m'\right)/\varepsilon_d\varepsilon_m'\right]^{1/2}$	$\lambda_0(4\alpha E_F)/\hbar\omega(\varepsilon_r + 1)$
Propagation distance (δ_{SPP})	$\lambda_0\left(\varepsilon_m'\right)^2\left[\left(\varepsilon_d + \varepsilon_m'\right)/\varepsilon_d\varepsilon_m'\right]^{3/2}\Big/2\pi\varepsilon_m''$	$\lambda_0(\tau\alpha E_F)/\pi\hbar(\varepsilon_r + 1)$
Penetration depth δ_i	$\delta_d = \lambda_0\left\|\left(\varepsilon_d + \varepsilon_m'\right)/\varepsilon_d^2\right\|^{1/2}\Big/2\pi$ $\delta_m = \lambda_0\left\|\left(\varepsilon_d + \varepsilon_m'\right)/\left(\varepsilon_m'\right)^2\right\|^{1/2}\Big/2\pi$	$\lambda_{SPP}/2\pi$

Source: Barnes, W.L., *J. Opt. A Pure Appl. Opt.*, 8(4), S87, 2006.
Notes: $\alpha = 1/137$ is the fine structure constant; ε_r is the dielectric constant of the substrate.

nanoparticles, where the LSPRs of metal nanoparticles can be tuned. In the hybrid graphene-gold nanorod structure, plasmon resonances at optical frequencies can be controlled and modulated by tuning the interband transitions in graphene through electrical gating. The plasmon resonance will be redshifted or blueshifted depending on the gate voltage. Without doubt, SPs in other noble metals, such as Ag, can also be affected by graphene. With the near-field enhancement of LSPRs, these composite structures can be used for tunable SERS in visible frequency.

19.3 THE APPLICATION OF GRAPHENE PLASMONICS

19.3.1 Graphene Plasmonic Metamaterials

Plasmons play important roles in metamaterials, which have been rising for many years. Owing to its peculiar properties, graphene can also be used as plasmonic metamaterials, see Figure 19.7. As shown in Figure 19.4, absorption peaks in THz region is observed in graphene microribbons array. With an ion-gel top gate, this metamaterial can be controlled electrically. In addition, the width of the ribbon is also an important parameter for graphene microribbons array.[39] The local EM resonances (i.e., LSPRs) of GMRs will be enhanced with appropriate values of width smaller than the wavelength of the input light. As metamaterials, some stacked structures of graphene, shapes and antishapes,[63] can enhance the resonance further, and improve the transmittance. However, with different structures, graphene can be regarded as versatile metamaterial for transmission, absorption, modulator, polarizer, or even for the mysterious cloaking.[64]

Incorporating with graphene, properties of conventional metamaterials will get a prominent improvement. It is found that the coverage of graphene on Fano rings, see Figure 19.7a, can lead to ~250% enhancement of transmission of the metamaterials at the resonant frequency.[46] And this experimental result was interpreted by the renormalization of the plasmonic modes and the frequency shift of the trapped-mode transmission resonance in the presence of graphene. There are other experimental results that show the transmittance of the composited structure can be higher than that of the metamaterials, and that of the graphene. Moreover, the absorbance and reflectance of the metamaterials can also be improved at the specific frequency due to the presence of graphene. This change can happen in other part of the spectrum with some specific structures, such as a layer of hexagonal metallic meta-atoms covered by monolayer graphene sheet, see Figure 19.7b. This nanostructures can do work in THz region. Combined with the tunability of graphene, the properties of metamaterials can be further controlled by graphene.

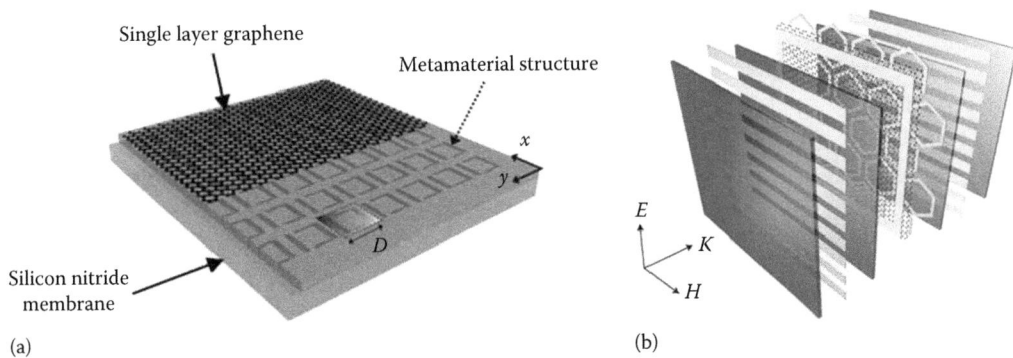

FIGURE 19.7
(a) Asymmetrically split ring made by Au based on Si_3N_4 membrane whose top is covered by a monolayer graphene. (From Papasimakis, N. et al., *Opt. Express*, 18(8), 8353, 2010.) (b) Schematic of a gate-controlled active graphene metamaterial composed of a monolayer graphene on a layer of hexagonal metallic meta-atoms contacted with extraordinary optical transmission electrodes. (From Lee, S.H. et al., *Nat. Mater.*, 11(11), 36, 2012.)

19.3.2 Plasmonic Light Harvesting in Graphene

A graphene monolayer can have a constant absorption coefficient of 2.3% over a wide spectral range from the visible to the infrared which has been mentioned earlier. The relatively poor light absorbance of graphene has limited its applications in photodetector and solar cell (photovoltage). Plasmons can be used to solve this problem, where the LSPR can efficiently enhance the optical absorption at resonant frequency. Many discontinuous graphene structures, such as nanodisk, nanoribbons, and so on exhibit high efficiency for light absorption due to LSPR. Taking micro/nano-ribbons as an example, the THz absorption peaks are found for the input light polarized perpendicular to the ribbon length.[39] The size and the carrier concentration impact dramatically the absorbance because of the changes of plasmon resonances. Generally, smaller size structures lead to higher frequency absorption peak. Theoretical calculations have also proved that resonances appear in the frequency ranges from infrared to microwave mostly.[66] Therefore, by adjusting the ribbons' size, the substrate and the chemical potential, the transmission, and reflection can be completely suppressed while the absorbance can even reach 100%. However, owing to its 1D nature, the absorption efficiency of the ribbons is not good enough for the input light polarized in other directions.

For the sake of absorption of input light with different polarizations, the zero-dimensional structures of antidote and disk have been studied.[67] Similar to metal nanoparticles, LSPRs are stronger in these graphene structures. For the discontinuous graphene sheet with periodic antidote array, see Figure 19.8a and b; LSPRs enhance absorption and suppresses transmission more efficiently because of the higher relaxation times of charge carriers. These rich absorption peaks still appear

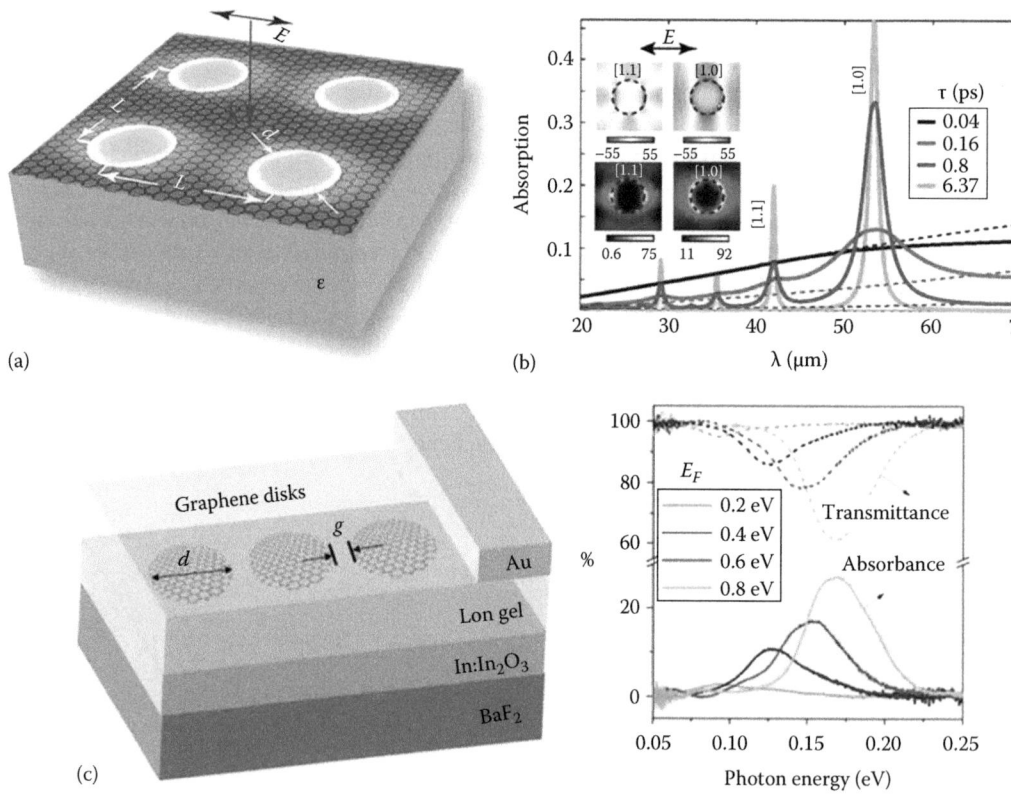

(a)

(b)

(c)

FIGURE 19.8
(a) and (b) Enhanced absorption in the graphene sheet with periodic antidots array, the dashed lines indicate the spectra for graphene monolayer, where $\varepsilon = 1$, $\mu = 0.2$ eV, $L = 2d = 5$ μm; the insets of (b) show the modulus of the spatial distribution of the electric field (below) and the real part on the direction of the incident wave electric field shown by an arrow (above), respectively. (From Nikitin, A. et al., *Appl. Phys. Lett.*, 101(15), 151119, 2012.) (c) A completely optical absorber made by the graphene with periodic nanodisks array. For continuous frequencies input light, there exit a absorption peak for the fixed structure and input angle, and the lower two panels show the absorption peaks for the different structures and different input angles of s-polarized and p-polarized light. (From Thongrattanasiri, S. et al., *Phys. Rev. Lett.*, 108(4), 047401, 2012.)

in THz frequencies, however, they will blueshift with the decrease of the antidot size. The converse structure of micro/nano-disk can also have similar properties. Figure 19.8c shows the complete optical absorption of periodic graphene nanodisks array, which is valid for input light polarized in all directions. With the effect of LSPR, many absorption peaks will appear in a fixed structure. The maximum absorbance can reach 100% which is dependent on the incidence angle of the input light. As calculated, smaller size of the disk is beneficial for higher maximum absorbance and the range of incidence angles to fulfill the maximum absorbance is allowed to be larger.

Some other graphene structures can exhibit efficient absorption, such as graphene sheet. Owing to its fine electrical properties and convenient fabrication, graphene sheet gets more attention. The LSPR in graphene can also be changed and enhanced locally by adjusting the backgrounds, such as substrates, supporter, electrical gate, and so on.

Because of the properties of plasmons, light absorption in visible frequencies is still hard to be realized in graphene. Therefore, composite materials are adopted to solve this problem. For instance, the structure of graphene on 1D photonic crystal has been proposed to improve the light absorption, where the enhancement indeed happen in the visible frequencies for both TM and TE modes of light.[69] Plasmons in metals can resonate in visible frequencies, and in the hybrid structure the resonant frequency can be easily tuned by graphene to some degrees. Above all, graphene can be used in photoelectrics and photovoltaics due to the enhancement of light absorption.

19.3.3 Surface-Enhanced Raman Scattering with Graphene

As a powerful technique to provide detailed information of molecular structures, Raman spectroscopy plays an important role in the analyzation of graphene. From the Raman spectra, many intrinsic information can be implied, such as structure, defect, and so on.[70] Most importantly, the microanalysis of other molecule can be carried out with the enhancement effect from graphene substrate, which is named as graphene-enhanced Raman scattering (GERS).[71]

The origin of GERS has proved to be a chemical mechanism, as plasmons in graphene are dominating in THz and infrared frequencies. However, the enhancement factor of GERS is not strong enough which limits graphene as a sensitive surface enhancement substrate. This problem can be solved by combining graphene and metal nanoparticles. Metal particles have an enhancement effect for Raman scattering which has been proved to be a physical mechanism. The SERS effect can be greatly improved, which is attributed to the electromagnetic interactions between metals and graphene. Therefore, composites of graphene with other metal plasmonic structure are often used for SERS studies. About 100 times of enhancement factor is observed for graphene covered by Ag nanoisland[72]; tens of times is found in graphene covered by Au-nanoparticle array[73] being located under graphene. In addition, for the composite of graphene and Au film or nanoparticles, graphene can suppress the photo-luminescence background so as to get clearer results.

Polarized plasmonic enhancement can also be realized in the composite graphene–metal particle structures. Form the obtained experimental results, it is better to put graphene sheet on the top of metal plasmonic structures. In the composition of graphene sheet and plasmonic strucutres, the electromagnetic field of the metal plasmonic structures can pass through the ultrathin graphene sheet, and then Raman signals will get an enhancement by this chemical and physical mechanism. And that makes graphene/metal substract to be a good choice for SERS. The flat surface and stability of graphene would also help for the uniform, stable, clean, and reproducible signals.[74]

19.3.4 Plasmonic Detectors and Sensors with Graphene

Graphene can be used as high-performance detectors and sensors for room-temperature THz radiation,[75] biomacromolecule,[76] visible light,[77] or even individual gas molecules.[78] With high efficient optical absorption in THz and infrared, graphene could be used as plasmonic detectors for light in those frequencies. With enhancement effect of SPs, the photodetectors and sensors hybridized graphene with other plasmonic structures are exploited for detecting light and optical sensing of macromolecules or biomolecules. Owing to its other advantages, such as tunability, ultrathin thickness, and biocompatibility, graphene proved to have a good application in these detectors and sensors.

Photovoltage and photodetection based on graphene can also be enhanced by combining with plasmonic materials,[79] with the enhancement of produced photocurrent. On the structures of Au nanoparticles array supported by back-gate graphene transistors, see Figure 19.9a, different independent variable characteristics can be obtained depending on the size of Au nanoparticles and the frequencies of the incident light. Moreover, the enhancements of the photocurrent can reach up to 1500% as the thickness of the Au film increases to 12 nm. Generally, the enhancement peaks will redshift with increase of Au nanoparticles sizes.

In addition to the monolayer, multicolor photodetection can also be realized in multilayer graphene-based plasmonic structures. As shown in Figure 19.9b, Au heptamer antenna structure is sandwiched between two monolayer graphene sheets. This structure can be used as multicolor photodetector by adjusting the size of Au heptamer array. The enhancement of the photocurrent in this sandwich structure can reach up to 800% for visible and near-infrared light, which was attributed to the transfer of hot electrons and SPs.

The quality of a sensor or detector can be evaluated by four parameters: sensitivity, selectivity, adsorption efficiency, and signal-to-noise ratio. The SPs resonant sensors exhibit promising potentials because of their very high sensitivity to the sensing medium, where the change in the refractive

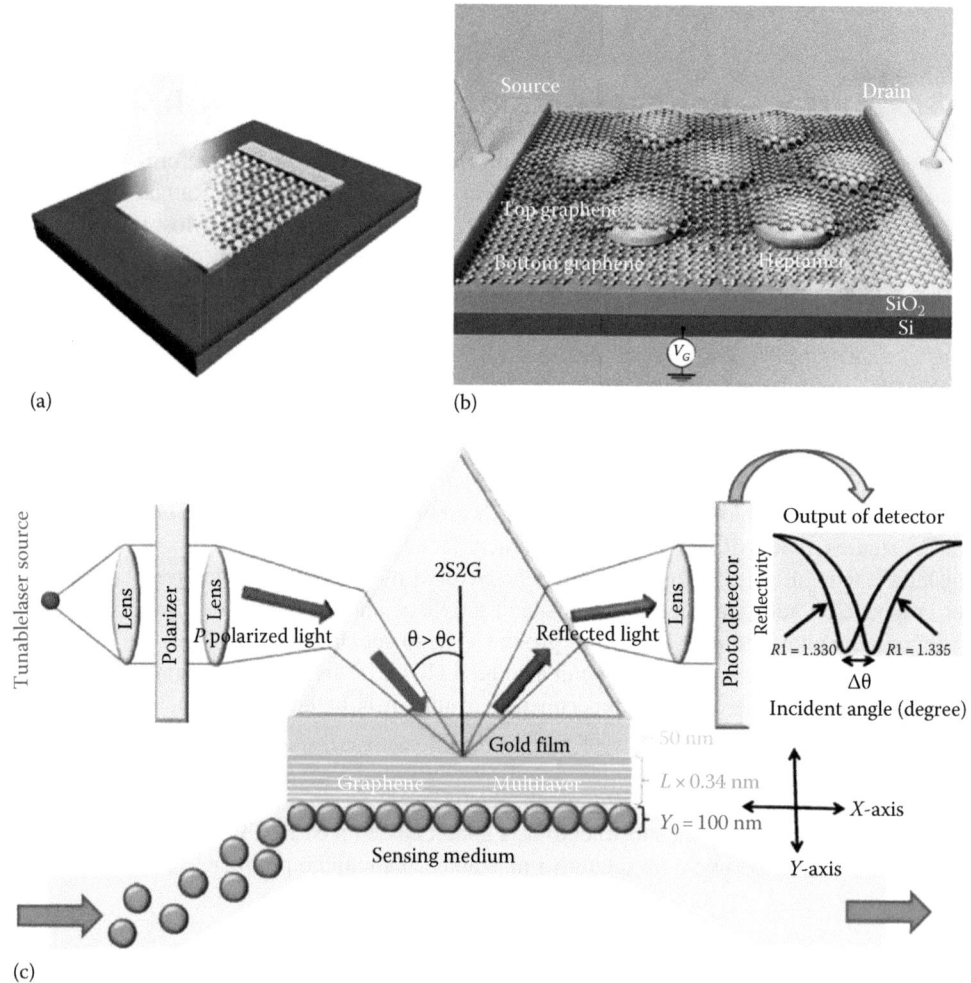

(a)

(b)

(c)

FIGURE 19.9
(a) Schematic of graphene photodetector, with a laser scanning across the graphene–metal junction. (From Liu, Y. et al., *Nat. Commun.*, 2, 579, 2011.) (b) Schematic of one single gold heptamer sandwiched between two monolayer graphene sheets, where the bias is used to electrically dope the graphene. (From Fang, Z. et al., *Nano Lett.*, 12(7), 3808, 2012.) (c) Schematic of experimental setup for plasmonic biosensor with chalcogenide prism, gold, and graphene multilayers. (From Maharana, P.K. and Rajan, J., *Sens. Actuat. B Chem.*, 169, 161, 2012.)

index of the sensed medium induces specific alternations in the characteristics of SPs resonances.[80] As a conventional structure, pure metal films are not the ideal SPs resonant sensors, because of its inactivation and the bad adsorption capacity. Metal film can be covered by graphene layers, where the adsorption for organic molecule or biomolecule can also be improved by $\pi-\pi$ stacking interaction because of the aromatic structure and the signal-to-noise ratio can be greatly enhanced. Moreover, graphene can also protect the SPs in metal from the inactivation and the SPs will not be weakened due to the ultrathin thickness of graphene. A typical schematic of graphene-based SPs resonant sensors is shown in Figure 19.9c.

The metal film can also use some conventional plasmonic materials, such as Au, Ag, Al, etc. and the optimized film thickness (usually about 40–70 nm) is depending on the incident light and the background of the structure. According to theoretical calculation, Ag film is the best due to the sharpest peaks of reflectance and sensitivity with respect to the incidence angle; Au film is the second best and then the others. The number of layer will also impact the sensitivity of the sensors, and generally fewer layers are beneficial to higher sensitivity. In order to get higher sensitivity, the structure can be further optimized; for instance, replacing the silica prism by chalcogenide prism, adding a silicon layer or a silica-doped B_2O_3 between metal and graphene layers, and so on. Different plasmonic materials work in different frequencies, so the selectivity is the tender spot of the graphene-based SPs resonant sensor especially for the mixed sensing medium. Fixing acceptor on graphene for sensing-specific molecular might be the sally port for better selectivity similar to aptasensors.

19.3.5 Tunable Terahertz Surface Plasmons for Amplifier, Laser, and Antenna

The extraordinary properties of SPs in graphene, plus its good flexibility, stability, and good biocompatibility make it a good candidate for various applications, including electronics, optics, THz technology, energy storage, biotechnology, medical sciences, and so on. In the following paragraphs, we will introduce some meaningful applications[84] of graphene plasmonics.

It is known that many matters emit THz EM radiations, such as H_2O, CO_2, N_2, O_2, CO molecules, and so on. THz radiations can be used in detections to unveil intrinsic physical and chemical characters of the matters, such as security check. Moreover, low-energy THz radiations (<41.3 meV) can also be used in nondestructive testing for biological detection. The applications of graphene plasmonic in THz technology will be a revolution for telecommunication. Unfortunately, the lack of effective THz sources and detectors limits its applications. With linear band structures near Dirac point and the possibility of opening a bandgap, graphene proved to be a promising material for THz application. With graphene, the application of THz radiations could get a good improvement.[85]

The THz plasmons in highly doped graphene are mainly related to the intraband transitions of electrons. In such case, the Landau damping is weak while the TM modes are dominating. In order to excite THz plasmons, the incident light should match the dispersions of plasmons. In theory, elementary dipole or quantum emitter can excite THz SPs on graphene surfaces. SPs can dominate the response along the suspended graphene sheet and exhibits a strong tunable excitation peak in the THz region. For supported graphene, the interaction is dependent on the dielectric support layer, for example, the excitation peak will decrease and redshift with poor field confinement on SiO2, while the field confinement will be enhanced for Si. The evanescent wave of incident light can also excite SPs. For instance, with a high-index coupling prism, SPs in doped monolayer graphene can be excited with the frequencies up to about 10 THz, and higher frequencies for few-layer graphene. Due to the enhancement effect, THz SPs in graphene can be used in many fields, such as THz laser (Figure 19.10), THz plasmonic antenna, THz metamaterials, and so on.

The conductivity of graphene is complex with the imaginary part implies which mode is supported, while the real part indicates emission or absorption depending on the negative or positive situation, respectively. As calculated, by choosing appropriate substrate and heavily doping, the real part can be negative in THz and infrared frequencies, so the optical radiation will get an increment. In addition, the carrier relaxation process in graphene is ultrafast, for example, picosecond time

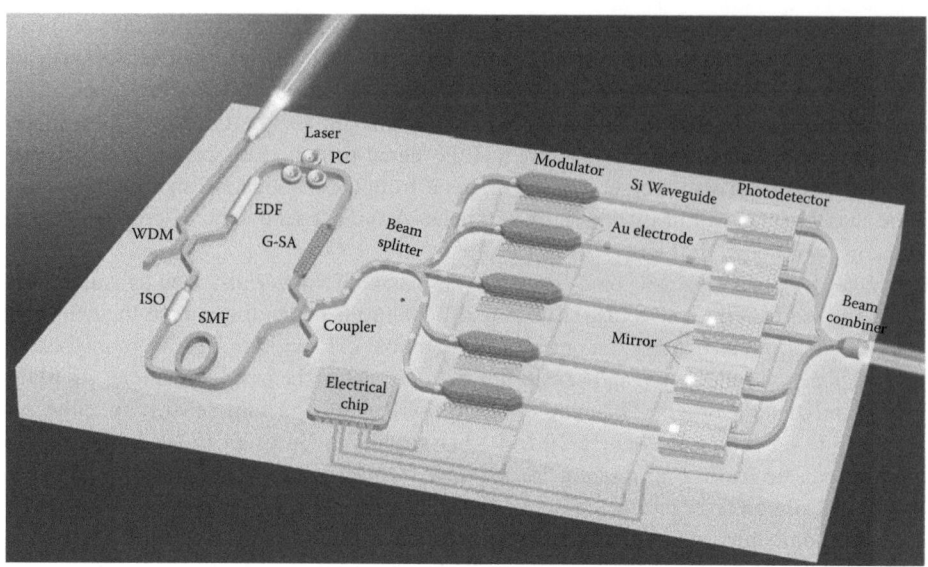

FIGURE 19.10
Schematic illustration of integrated graphene silicon hybrid photonic circuit, which consists of a graphene modelocked laser with a ring cavity (left), graphene optical modulator (middle), and microcavity-enhanced graphene photodetector (right). (From Bao, Q. and Kian, P.L., *ACS Nano*, 6(5), 3677, 2012.)

scale is obtained in epitaxial graphene layers grown on SiC wafers by using optical-pump THz-probe spectroscopy. As a result, it is possible to make ultrafast THz laser based on graphene.

In order to make a laser, the interband population inversion need be achieved first. In doped or undoped graphene, femtosecond population inversion can be easily realized by carrier injection, plasmons, or optical pumping. Taking plasmons as an example, the population inversion in THz region can be achieved by absorbing plasmons. During the population inversion, the coupling of the plasmons to interband electron–hole transitions in graphene can lead to plasmon amplification through stimulated emission. Similar to optical gain, plasmon gain is also an important parameter for a plasmon laser. Owing to the slow group velocity and strong confinement of graphene plasmons, graphene oscillator exhibits as a THz plasmonic amplifer with the plasmon gain values being much larger than the typical gain values in semiconductor interband lasers, and most importantly, it can work in THz frequencies. Besides the interband gain, the intraband absorption will lessen the gain. Nevertheless, the optimized working environment can still be found with high gain through adjusting the parameters, for example, doping, substrate, and so on. After defined as the difference between interband gain and intraband loss, the net plasmon gain will increase at a fixed frequency. And the frequency range of the plasmon gain will be broadened when the carrier density increases. Similar results can also be obtained in optically pumped graphene structures, where the absorbance can exhibit negative value. In addition, the number of graphene layers can also impact the gain. Generally, the plasmon amplification weakens with the increase of number of graphene layers.

Nevertheless, a strong dephasing of the plasmon mode is caused by the large plasmon gain which will prevent THz lasing. In addition, the strong coupling between the plasmons and EM radiation will also hinder the lasing from nonequilibrium plasmons. For solving this problem, Popov et al.[85] predicted a planar array of graphene resonant micro/nano-cavities for THz laser. The cavities have the properties of high confinement of the plasmons and the superradiant nature of EM emission, so the amplification of THz waves at the plasmon resonance frequency will be enhanced than that away from the resonances. Furthermore, the plasmons coherence restore in the graphene micro/nano-cavities and strongly couple to the THz radiation at the balance between the plasmon gain and plasmon radiative damping.

Due to the tunability and the strong confinement of the SPs in graphene, other tunable plasmonic devices can also be realized. For example, THz plasmonic antennas can be made by combining the SPs and THz EM wave. Moreover, Tamagnone et al.[86] proposed a dipole-like plasmonic resonant antenna in a graphene/Al2O3/graphene sandwich structure. By tuning electrical field, this structure can exhibit different properties. It is noted that the silicon lens in structure will lead to higher directivity but has negligible impact on the antenna input impedance.

The integration of all-optical and electro-optical devices requires strategies for controlling and manipulating the phase, direction, polarization, and amplitude of optical beams. Graphene has potential to control light intensity, as in optical limiting and optical switching, which is important in optical communications and optical computing.[83] Methods to control the amplitude and pulse width of light are important in energy modulation. Owing to its unique properties, graphene can be used in wide-ranging optical control schemes. The fabrication of silicon-based integrated optical circuits, which are designed to assume multiple functions of light creation, routing, modulation, computing, and detection, can be enabled by the broadband optical opacity and tunable dynamical conductivity of graphene. Hopefully, breakthroughs in the direct deposition of graphene on silicon can pave the way for the integration of graphene in hybrid silicon-photonic circuits, as schematically illustrated in Figure 19.10.[83]

ACKNOWLEDGMENTS

This work was jointly supported by the National Natural Science Foundation of China under Grant No. 51271057, the Natural Science Foundation of Jiangsu Province, China, under Grant No. BK2012757, and the Program for New Century Excellent Talents in University of Ministry of Education of China under Grant No. NCET-11-0096.

REFERENCES

1. Geim, A. K. and K. S. Novoselov. 2007. The rise of graphene. *Nature Materials* 6 (3):183–191.
2. Novoselov, K. S., A. K. Geim, S. V. Morozov et al. 2004. Electric field effect in atomically thin carbon films. *Science* 306 (5696):666–669.
3. Luo, X., Q. Teng, L. Weibing et al. 2013. Plasmons in graphene: Recent progress and applications. *Materials Science and Engineering: R: Reports* 74 (11):351–376.
4. Mak, K. F., M. Y. Sfeir, Y. Wu et al. 2008. Measurement of the optical conductivity of graphene. *Physical Review Letters* 101:196405–196419.
5. Fang, Z., Y. Wang, A. E. Schather et al. 2014. Active tunable 002. Absorption enhancement with graphene nanodisk arrays. *Nano Letters* 14 (1):299–304.
6. Naik, G. V., V. M. Shalaev, and A. Boltasseva. 2013. Alternative plasmonic materials: Beyond gold and silver. *Advanced Materials* 25 (24):3264–3294.
7. Ran, C., M. Wang et al. 2014. Employing the plasmonic effect of the Ag-graphene composite for enhancing light harvesting and photoluminescence quenching efficiency of poly[2-methoxy-5-(2-ethylhexyloxy)1,4-phenylene-vinylene]. *Physical Chemistry Chemical Physics* 16 (10):4561–4568.
8. Al-Mashat, L., K. Shin, K. Kalantar-zadeh et al. 2010. Graphene/polyaniline nanocomposite for hydrogen sensing. *Journal of Physical Chemistry C* 114 (39):16168–16173.
9. Peng, J., W. Gao, B. K. Gupta et al. 2012. Graphene quantum dots derived from carbon fibers. *Nano Letters* 12 (2):844–849.
10. Eda, G., Y.-Y. Lin, C. Mattevi et al. 2010. Blue photoluminescence from chemically derived graphene oxide. *Advanced Materials* 22 (4):505–509.
11. Zhu, S., S. Tang, J. Zhang et al. 2012. Control the size and surface chemistry of graphene for the rising fluorescent materials. *Chemical Communications* 48 (38):4527–4539.

12. Hong, G., S. M. Tabakman, K. Welsher et al. 2010. Metal-enhanced fluorescence of carbon nanotubes. *Journal of the American Chemical Society* 132 (45):15920–15923.

13. Cuong, T. V., V. H. Pham, Q. T. Tran et al. 2010. Photoluminescence and Raman studies of graphene thin films prepared by reduction of graphene oxide. *Materials Letters* 64 (3):399–401.

14. Kim, B.-J., C. Lee, Y. Jung et al. 2011. Large-area transparent conductive few-layer graphene electrode in GaN-based ultra-violet light-emitting diodes. *Applied Physics Letters* 99:14310114.

15. Sun, X., Z. Liu, K. Welsher et al. 2008. Nano-graphene oxide for cellular imaging and drug delivery. *Nano Research* 1 (3):203–212.

16. Gusynin, V. P., S. G. Sharapov, and J. P. Carbotte. 2007. Magneto-optical conductivity in graphene. *Journal of Physics: Condensed Matter* 19:0262222.

17. Lin, H., M. F. Pantoja, L. D. Angul et al. 2012. FDTD modeling of graphene devices using complex conjugate dispersion material model. *IEEE Microwave and Wireless Components Letters* 22 (12):612–614.

18. Li, H., L. Wang, Z. Huang et al. 2013. Mid-infrared, plasmonic switches and directional couplers induced by graphene sheets coupling system. *Europhysics Letters* 104 (370013).

19. Wang, B., X. Zhang, X. Yuan et al. 2012. Optical coupling of surface plasmons between graphene sheets. *Applied Physics Letters* 100:13111113.

20. Li, H.-J., L.-L. Wang, H. Zhang et al. 2014. Graphene-based mid-infrared, tunable, electrically controlled plasmonic filter. *Applied Physics Express* 7:0243012.

21. Goerbig, M. O. 2011. Electronic properties of graphene in a strong magnetic field. *Reviews of Modern Physics* 83 (4):1193–1243.

22. Castro Neto, A. H., F. Guinea, N. M. R. Peres et al. 2009. The electronic properties of graphene. *Reviews of Modern Physics* 81 (1):109–162.

23. Wood, R. W. 1902. XLII. On a remarkable case of uneven distribution of light in a diffraction grating spectrum. *The London, Edinburgh, and Dublin Philosophical Magazine and Journal of Science* 4 (21):396–402.

24. Stern, E. A. and R. A. Ferrell. 1960. Surface plasma oscillations of a degenerate electron gas. *Physical Review* 120 (1):130.

25. Ritchie, R. H. and R. E. Wilems. 1969. Photon-plasmon interaction in a nonuniform electron gas. I. *Physical Review* 178 (1):372.

26. Raether, H. 1988. *Surface Plasmons on Smooth Surfaces*. Springer: Berlin, Germany.

27. Barnes, W. L., Alain, D., and Thomas, W. E. 2003. Surface plasmon subwavelength optics. *Nature* 424 (6950):824–830.

28. Kreibig, U. and M. Vollmer. 1995. *Optical Properties of Metal Clusters*. Springer: Berlin, Germany.

29. Sarid, D. and W. Challener. 2010. *Modern Introduction to Surface Plasmons: Theory, Mathematica Modeling, and Applications*. Cambridge University Press, Cambridge, U.K.

30. Pendry, J. B. 2000. Negative refraction makes a perfect lens. *Physical Review Letters* 85 (18):3966.

31. Zheludev, N. I. and Y. S. Kivshar. 2012. From metamaterials to metadevices. *Nature Materials* 11 (11):917–924.

32. Atwater, H. A. and A. Polman. 2010. Plasmonics for improved photovoltaic devices. *Nature Materials* 9 (3):205–213.

33. Anker, J. N., W. Paige Hall, O. Lyandres et al. 2008. Biosensing with plasmonic nanosensors. *Nature Materials* 7 (6):442–453.

34. Krasavin, A. V. and A. V. Zayats. 2007. Passive photonic elements based on dielectric-loaded surface plasmon polariton waveguides. *Applied Physics Letters* 90 (21):211101.
35. Jablan, M., H. Buljan, and M. Soljačić. 2009. Plasmonics in graphene at infrared frequencies. *Physical Review B* 80 (24):245435.
36. Fang, Z., Y. Wang, Z. Liu et al. 2012. Plasmon-induced doping of graphene. *ACS Nano* 6 (11):10222–10228.
37. Despoja, V., D. Novko, K. Dekanić et al. 2013. Two-dimensional and π plasmon spectra in pristine and doped graphene. *Physical Review B* 87 (7):075447.
38. Hwang, E. H. and S. Das Sarma. 2007. Dielectric function, screening, and plasmons in two-dimensional graphene. *Physical Review B* 75 (20):205418.
39. Ju, L., B. Geng, J. Horng et al. 2011. Graphene plasmonics for tunable terahertz metamaterials. *Nature Nanotechnology* 6 (10):630–634.
40. Kim, J. T. and S.-Y. Choi. 2011. Graphene-based plasmonic waveguides for photonic integrated circuits. *Optics Express* 19 (24):24557–24562.
41. Vakil, A. and N. Engheta. 2011. Transformation optics using graphene. *Science* 332 (6035):1291–1294.
42. Koppens, F. H. L., D. E. Chang, and F. J. G. De Abajo. 2011. Graphene plasmonics: A platform for strong light–matter interactions. *Nano Letters* 11 (8):3370–3377.
43. Bostwick, A., F. Speck, T. Seyller et al. 2010. Observation of plasmarons in quasi-freestanding doped graphene. *Science* 328 (5981):999–1002.
44. Nikitin, A. Yu, F. Guinea, Francisco J. Garcia-Vidal et al. 2012. Surface plasmon enhanced absorption and suppressed transmission in periodic arrays of graphene ribbons. *Physical Review B* 85 (8):081405.
45. Thongrattanasiri, S., F. H. L. Koppens, and F. J. G. de Abajo. 2012. Complete optical absorption in periodically patterned graphene. *Physical Review Letters* 108 (4):047401.
46. Papasimakis, N., Z. Luo, Z. X. Shen et al. 2010. Graphene in a photonic metamaterial. *Optics Express* 18 (8):8353–8359.
47. Zhou, W., J. Lee, J. Nanda et al. 2012. Atomically localized plasmon enhancement in monolayer graphene. *Nature Nanotechnology* 7 (3):161–165.
48. Yan, H., X. Li, B. Chandra et al. 2012. Tunable infrared plasmonic devices using graphene/insulator stacks. *Nature Nanotechnology* 7 (5):330–334.
49. Soukoulis, C. M. and M. Wegener. 2010. Optical metamaterials–more bulky and less lossy. *Science* 330 (6011):1633–1634.
50. Li, Z. Q., Eric, A. Henriksen, Z. Jiang et al. 2008. Dirac charge dynamics in graphene by infrared spectroscopy. *Nature Physics* 4 (7):532–535.
51. Peres, N. M. R., R. M. Ribeiro, and A. H. Castro Neto. 2010. Excitonic effects in the optical conductivity of gated graphene. *Physical Review Letters* 105 (5):055501.
52. Azad, A. K. and Weili Zhang. 2005. Resonant terahertz transmission in sub-wavelength metallic hole arrays of sub-skin-depth thickness. *Optics Letters* 30 (21):2945–2947.
53. Lu, W. B., W. Zhu, H. J. Xu et al. 2013. Flexible transformation plasmonics using graphene. *Optics Express* 21 (9):10475–10482.
54. Thongrattanasiri, S., A. Manjavacas, and F. J. G. De Abajo. 2012. Quantum finite-size effects in graphene plasmons. *ACS Nano* 6 (2):1766–1775.
55. Yan, H., F. Xia, Z. Li et al. 2012. Plasmonics of coupled graphene micro-structures. *New Journal of Physics* 14 (12):125001.

56. Christensen, J., A. Manjavacas, S. Thongrattanasiri et al. 2011. Graphene plasmon waveguiding and hybridization in individual and paired nanoribbons. *ACS Nano* 6 (1):431–440.

57. Chen, J.-H., C. Jang, S. Xiao et al. 2008. Intrinsic and extrinsic performance limits of graphene devices on SiO_2. *Nature Nanotechnology* 3 (4):206–209.

58. Fei, Z., A. S. Rodin, G. O. Andreev et al. 2012. Gate-tuning of graphene plasmons revealed by infrared nano-imaging. *Nature* 487:82–85.

59. Chen, J., M. Badioli, P. Alonso-González et al. 2012. Optical nano-imaging of gate-tunable graphene plasmons. *Nature* 487:77–81.

60. Stockman, M. I. 2011. Nanoplasmonics: Past, present, and glimpse into future. *Optics Express* 19 (22):22029–22106.

61. Barnes, W. L. 2006. Surface plasmon–polariton length scales: A route to sub-wavelength optics. *Journal of Optics A: Pure and Applied Optics* 8 (4):S87.

62. Tassin, P., T. Koschny, M. Kafesaki et al. 2012. A comparison of graphene, superconductors and metals as conductors for metamaterials and plasmonics. *Nature Photonics* 6 (4):259–264.

63. Fallahi, A. and J. Perruisseau-Carrier. 2012. Design of tunable biperiodic graphene metasurfaces. *Physical Review B* 86 (19):195408.

64. Chen, P.-Y. and A. Alù. 2011. Atomically thin surface cloak using graphene monolayers. *ACS Nano* 5 (7):5855–5863.

65. Lee, S. H., M. Choi, T.-T. Kim et al. 2012. Switching terahertz waves with gate-controlled active graphene metamaterials. *Nature Materials* 11 (11):936–941.

66. Alaee, R., M. Farhat, C. Rockstuhl et al. 2012. A perfect absorber made of a graphene micro-ribbon metamaterial. *Optics Express* 20 (27):28017–28024.

67. Begliarbekov, M., O. Sul, J. Santanello et al. 2011. Localized states and resultant band bending in graphene antidot superlattices. *Nano Letters* 11 (3):1254–1258.

68. Nikitin, A. Yu, F. Guinea, and L. Martin-Moreno. 2012. Resonant plasmonic effects in periodic graphene antidot arrays. *Applied Physics Letters* 101 (15):151119.

69. Liu, J.-T., N.-H. Liu, J. Li et al. 2012. Enhanced absorption of graphene with one-dimensional photonic crystal. *Applied Physics Letters* 101 (5):052104.

70. Narula, R., R. Panknin, and S. Reich. 2010. Absolute Raman matrix elements of graphene and graphite. *Physical Review B* 82 (4):045418.

71. Ling, X., L. Xie, Y. Fang et al. 2009. Can graphene be used as a substrate for Raman enhancement? *Nano Letters* 10 (2):553–561.

72. Urich, A., A. Pospischil, M. M. Furchi et al. 2012. Silver nanoisland enhanced Raman interaction in graphene. *Applied Physics Letters* 101 (15):153113.

73. Schedin, F., E. Lidorikis, A. Lombardo et al. 2010. Surface-enhanced Raman spectroscopy of graphene. *ACS Nano* 4 (10):5617–5626.

74. Xu, W., X. Ling, J. Xiao et al. 2012. Surface enhanced Raman spectroscopy on a flat graphene surface. *Proceedings of the National Academy of Sciences of the United States of America* 109 (24):9281–9286.

75. Vicarelli, L., M. S. Vitiello, D. Coquillat et al. 2012. Graphene field-effect transistors as room-temperature terahertz detectors. *Nature Materials* 11 (10):865–871.

76. Liu, Y., X. Dong, and P. Chen. 2012. Biological and chemical sensors based on graphene materials. *Chemical Society Reviews* 41 (6):2283–2307.

77. Liu, Y., R. Cheng, L. Liao et al. 2011. Plasmon resonance enhanced multicolour photodetection by graphene. *Nature Communications* 2:579.

78. Echtermeyer, T. J., L. Britnell, P. K. Jasnos et al. 2011. Strong plasmonic enhancement of photovoltage in graphene. *Nature Communications* 2:458.

79. Schedin, F., A. K. Geim, S. V. Morozov et al. 2007. Detection of individual gas molecules adsorbed on graphene. *Nature Materials* 6 (9):652–655.
80. Szunerits, S., N. Maalouli, E. Wijaya et al. 2013. Recent advances in the development of graphene-based surface plasmon resonance (SPR) interfaces. *Analytical and Bioanalytical Chemistry* 405 (5):1435–1443.
81. Fang, Z., Z. Liu, Y. Wang et al. 2012. Graphene-antenna sandwich photodetector. *Nano Letters* 12 (7):3808–3813.
82. Maharana, P. K. and R. Jha. 2012. Chalcogenide prism and graphene multilayer based surface plasmon resonance affinity biosensor for high performance. *Sensors and Actuators B: Chemical* 169:161–166.
83. Bao, Q. and K. P. Loh. 2012. Graphene photonics, plasmonics, and broadband optoelectronic devices. *ACS Nano* 6 (5):3677–3694.
84. Docherty, C. J. and M. B. Johnston. 2012. Terahertz properties of graphene. *Journal of Infrared, Millimeter, and Terahertz Waves* 33 (8):797–815.
85. Popov, V. V., O. V. Polischuk, A. R. Davoyan et al. 2012. Plasmonic terahertz lasing in an array of graphene nanocavities. *Physical Review B* 86 (19):195437.
86. Tamagnone, M., J. S. Gomez-Diaz, J. R. Mosig et al. 2012. Analysis and design of terahertz antennas based on plasmonic resonant graphene sheets. *Journal of Applied Physics* 112 (11):114915.

CHAPTER 20

CONTENTS

Quantum Spin Hall Effect in Edge-Functionalized Graphene Nanoribbon

20

Jun-Won Rhim and Kyungsun Moon

20.1 INTRODUCTION

Graphene, an allotrope of carbon made of the 2D sheet of carbon atoms placed on the honeycomb lattice, has attracted a lot of theoretical and experimental attentions due to superb physical properties such as high mobility, strong elastic, and excellent thermal properties [1–3].

Especially, the exotic electronic structure of graphene described by the massless Dirac fermions and the nontrivial Berry phase was found to be the origin of various interesting phenomena such as unconventional transport properties and the half-integer quantum Hall effect [3].

For application point of view, graphene has been widely employed for the fabrication of various sensors and other new devices due to their distinct properties and broad performance range. Functionalized graphene particularly appears to be exceptionally promising for chemical and biological sensor applications. One can functionalize graphene using several different methods. Surface modification of graphene will be one way to fabricate functionalized graphene such as graphene oxide by chemical method.

Physically pulling graphene apart and manipulating the resulting edge geometry will be another clever way to achieve functionalization of graphene. Graphene has been widely employed for the fabrication of various sensors and other new devices because of their distinct properties and high-performance range. When graphene is cut into 1D stripe, we call them as graphene nanoribbon (GNR). The GNR has been initially proposed theoretically to generate the finite bandgap for graphene, which has been one of the main obstacles to the semiconductor applications of graphene. It has been reported that very narrow GNRs with smooth edges can be fabricated by ripping apart carbon nanotubes [4,5]. Remarkably, recent STM measurement has clearly demonstrated the experimental evidence for various correlated edge structures such as zigzag edges and Stone–Wales reconstructed edges. In order to manipulate the edge structure of graphene, diverse experimental detection and fabrication techniques have been developed [6–8].

Recently, the edge state of zigzag GNR (ZGNR) has been of great interest due to its peculiar dispersion relation with the almost flat edge bands near the zero energy [9,10]. When the Coulomb interactions are taken into account, the existence of the flat edge bands may lead to the edge magnetism for various kinds of ribbon edges [9,10]. It has been shown that the edge state of ZGNR is ferromagnetic along each edge, while antiferromagnetic between the two different edges due to the bipartite nature of the lattice structure. Furthermore, it has been theoretically demonstrated that application of lateral electric field to ZGNRs may induce half metallicity [11]. Hence, they can be potentially useful for future spintronics application, which has been theoretically confirmed both for ZGNR and bilayer ZGNR as well [10,12].

Here, we will investigate the role of spin–orbit coupling (SOC) on the edge-functionalized ZGNR in addition to the Coulomb interaction between electrons, which has attracted much attention as one of the model systems to realize the new quantum state of matter called the quantum spin Hall effect(QSHE) [13–15].

In a superconductor, supercurrent can flow without energy dissipation even in a disordered system, which has been one of the most intriguing features in condensed matter physics. In a nutshell, superconductivity can be understood as a macroscopic phase coherence of bosonic Cooper pairs formed by distinct gluing mechanisms. This has led to a bright future for electric applications, which we are still expecting to be realized. The quantum Hall effect is another interesting phenomenon, which carries a dissipationless current [16]. In the semiconductor heterostructures made of GaAs or Si such as MOSFET, 2D electron gas resides in the inversion layer, where electrons are strongly correlated. Upon applying the strong magnetic field, incompressible liquid phases are stabilized between Landau levels, which exhibit a quantized Hall resistance (h/ne^2) with n being an integer. The quantum Hall liquid manifests as a plateau of the Hall resistance, where the longitudinal resistance vanishes implying a flow of dissipationless current. The very existence of dissipationless current can be understood in terms of broken time reversal invariance due to magnetic field. In a rectangular Hall bar geometry, while the bulk excitation has a gap due to discrete Landau levels, gapless chiral edge states exist on the boundary. Since the chiral edge state flows only in one direction, it strongly suppresses a backscattering of electrons due to impurities leading to dissipationless current. The stability of the quantized Hall resistance has a topological origin: Thouless–Kohmoto–Nightingale–de Nijs or the Chern number of the quantum Hall insulator [17,18]. The topological invariants are typically classified by integer numbers, which describe the global topological properties of the quantum state. Hence, the local perturbations cannot change the topological invariant, and this is the reason why the quantum Hall effect is extremely robust against disorder.

Is the time reversal symmetry (TRS) breaking a necessary condition to have a topologically nontrivial quantum state? To answer the question, Kane and Mele considered a time reversal pair of Hamiltonians proposed by Haldane [13,14]. Haldane demonstrated that, in the honeycomb lattice, one can have a nonzero Chern number without uniform magnetic field or Landau levels by allowing the imaginary next-nearest hopping processes in addition to the minimal nearest-neighboring tight-binding model [19]. The imaginary hopping parameters are obtained from the Berry phases corresponding to an artificial periodic local magnetic field he imagined. Since the Haldane model also has a topological order, we have chiral edge states at the boundaries. Kane and Mele noticed that graphene with the intrinsic SOC can be described by a time reversal–symmetrized Haldane model that yields imaginary hopping terms of the form $i\xi_1/3\sqrt{3}\nu_{ij}s^z_{\alpha\beta}c^\dagger_{i\alpha}c_{j\beta}$ where ξ_1 is the strength of the intrinsic SOC, $\nu_{ij} = \pm 1$ is the circulation index, and $s^z_{\alpha\beta}$ is the Pauli matrix for the real spin. The Kane–Mele (KM) model is just a conventional insulator in the point of view of the Chern number, because we have counterflowing chiral edge channels to respect the TRS so that the Chern number becomes zero. However, they demonstrated by using Laughlin's gauge threading argument that this system accumulates spins at the edge, so that we have a nonzero spin Hall conductance $G^s_{xy} = e/h(\langle S_z\rangle_L - \langle S_z\rangle_R)$. This is so-called quantum spin Hall effect. The quantum spin Hall insulator is clearly distinguished from the conventional insulator, and the counterpropagating edge modes, so-called helical edge states, are dissipationless in the sense that they cannot be backscattered without any time reversal breaking perturbations. Kane and Mele also introduced a topological invariant, Z_2 index, to classify various types of insulators under the time reversal symmetric environment. This invariant can be calculated from the number of zeros of $P(\vec{k}) = \mathrm{Pf}\left[\left\langle u_i(\vec{k})\left|\Theta\right|u_j(\vec{k})\right\rangle\right]$ in the Brillouin zone where Pf represents the Pfaffian, Θ is the time reversal operator, and $u_i(\vec{k})$ is the Bloch wave function below the Fermi level. Triggered by their work, the role of intrinsic SOC in graphene and graphene multilayer systems has attracted a lot of attention as one of the model systems to realize the new quantum state of matter [20–44]. Furthermore, this interesting role of the SOC in graphene has stimulated studies on a novel kind of material called the topological insulator (TI) [45–50].

However, there are several practical obstacles to observe the QSHE in graphene. First, the SOC strength of graphene is too tiny due to the small mass of the carbon atom [22–24]. Furthermore, based on the group-theoretical and perturbative arguments, it has been elucidated that the SOC term for the nearest-neighbor hopping process vanishes at the Dirac point by the lattice symmetry of graphene, and hence, the leading SOC term originates from the next-nearest-neighboring hopping processes that leads to additional reduction of the strength of the SOC [22,23,51]. As a result, the size of the spin–orbit gap is beyond the conventional resolution of various spectroscopy techniques to detect the edge states inside the gap. Several authors have indicated that one can obtain much more enhanced intrinsic SOC such as the new hopping processes in the bilayer graphene system, which arise due to the different lattice symmetries around each carbon atom [52,53]. Second, the dangling bonds at the edges of graphene are not stable and easily saturated by organic elements such as the hydrogen, nitrogen, and oxygen [54–57]. While we hope to detect dissipationless edge states in graphene, the passivation cannot be regarded as a perturbation since every p-orbital at the edge are macroscopically involved to this and the orbitals of the passivated elements closely overlap with the edge-localized states. In case of the hydrogen passivation, the s-orbital of the hydrogen atom has vanishing overlap with the p_z orbital of the neighboring carbon atom and just makes a covalent bond with p_x and p_y orbitals so that this system can be described by the same tight-binding model for the p_z orbital of graphene with an open boundary condition. As a result, most researchers assume the hydrogen passivation in studying electronic structures of GNRsand the KM model is also included in this category. One of the interesting results from the hydrogen-passivated graphene is that we have almost flat bands at the Fermi level for the case of the zigzag edge. As mentioned earlier, the high instability of this diverging density of states results in the magnetic ground state with the sublattice ferromagnetism when the Coulomb interaction is included [58]. Since the carbon atoms at the opposite edges of the ZGNR always belong to the different sublattices, when the transverse in-plane electric field is applied, we have a half-metallic property in this system [11]. All those intriguing properties, however, appear when the dangling bonds at the edges are saturated by hydrogens, which is not easy to be realized in real experimental situations. When the oxygen is passivated, it was found out that its p-orbitals actively contribute to the edge-localized states near the Fermi level and yield quite dispersive band dispersions so that there is no such instability like the hydrogen passivation case [55].

We have investigated the effect of the edge geometry on the low-energy physics of the intrinsic SOC both in the zigzag and armchair nanoribbons based on the tight-binding Hamiltonian. In order to describe the SOC, we have included the s, p_x, p_y, and p_z orbitals of carbon atoms instead of using the effective KM term containing only the p_z orbital. For the pristine zigzag nanoribbon, we have demonstrated that one of the σ edge bands, made of the s, p_x, and p_y orbitals located near $E = 0$, lifts up the energy of the spin-filtered chiral edge states at the zone boundary, and hence, the QSHE does not occur at half filling. By increasing the carrier density within a certain range, the system exhibits the QSHE with spin-filtered chiral edge states. By further increasing the carrier density above a certain critical value, the QSHE disappears again by adding an extra pair of edge states leading to even number of edge states at each side of the edges. We propose that the aforementioned characteristic behavior provides a new example of TI belonging to the Class AII with the broken particle-hole symmetry, following the standard classification of the TIs [59]. We have also studied the role of hydrogen passivation on the edges, whose orbitals hybridize with the σ edge bands located near $E = 0$, and then the two edge bands are repelled from each other by creating large energy gaps. Remarkably, we have noticed that the original feature of the QSHE revives with hydrogen passivation. For the armchair GNR, we have noticed that the QSHE is mostly quite stable with or without passivation. However, the edge state of the armchair nanoribbon is too widely spread from the edge of the order of $(\gamma_0/\xi_1)a \cong 1$ mm, where γ_0 represents the nearest-neighbor hopping amplitude of the π electrons, ξ_1 is the SOC-induced next-nearest-neighbor hopping amplitude, and a is the lattice spacing.

In Section 20.2, we have introduced the tight-binding Hamiltonian, which describes the GNR system. The Hamiltonian includes the p_z and s, p_x, and p_y orbitals of carbon atom, the atomic SOC term, and the edge passivation term. In Section 20.3, we have investigated the band structure of the pristine ZGNR, and then the band structure of the armchair GNR is subsequently calculated. In Section 20.4, the effect of hydrogen passivation on the band structure of the ZGNR nanoribbon is studied. In Section 20.5, we have constructed the real-space effective Hamiltonian for the ZGNR. The summary will follow in Section 20.4.

20.2 THE TIGHT-BINDING HAMILTONIAN

The π and σ bands of the ZGNR can be obtained by the following tight-binding Hamiltonian:

$$H = H_\pi + H_\sigma + H_{SO} \tag{20.1}$$

where

H_π represents the nearest-neighbor hopping processes between p_z orbitals
H_σ is the matrix elements among s, p_x, and p_y orbitals
H_{SO} is the on-site atomic SOC term, which connects the two Hilbert spaces together

The Hamiltonian H_π for the p_z orbitals of the ZGNR is given as follows [3]:

$$H_\pi = \gamma_0 \sum_{\langle i,j,m,n \rangle} [c^\dagger_{p_z,B,j,n} c_{p_z,A,i,m} + h.c.] \tag{20.2}$$

where the indices $A(B)$, $i(j)$, and $m(n)$ represent the sublattices, dimer lines from 1 to N, and unit cells along the direction with the discrete translational symmetry, respectively, as shown in Figure 20.1. The hopping parameter between the nearest-neighboring p_z orbitals is $\gamma_0 = V_{pp\pi} = -3.03$ eV. The width of the ribbon N is defined by the number of the dimer lines along x axis, which is half the number of the atoms in the unit cell of the ribbon. The angular bracket under the summation means that only the hopping processes between the nearest-neighbor pairs are considered. The Hamiltonian H_σ can be written as

$$H_\sigma = \sum_{\alpha,i,m} \mathbb{C}^\dagger_{\alpha,i,m} \mathbb{E}_0 \mathbb{C}_{\alpha,i,m} + \sum_{\langle i,j,m,n \rangle} [\mathbb{C}^\dagger_{B,j,n} \Sigma_l \mathbb{C}_{A,i,m} + h.c.] \tag{20.3}$$

where the index α represents the sublattices A and B. The on-site SOC term is given by $H_{SO} = \xi_0/2 \sum \vec{L} \cdot \vec{s}$. While the SOC induces a complexity by introducing the spin-flipping processes, one can note that our system is still block diagonalizable, because the SOC does not mix between two groups $\{p_z\uparrow, p_x\downarrow, p_y\downarrow, s\downarrow\}$ and $\{p_z\downarrow, p_x\uparrow, p_y\uparrow, s\uparrow\}$. Simultaneously, p_z orbital has vanishing overlap with other orbitals by symmetry unless any mirror symmetry breaking term along z direction is included. We have omitted the summation over the spin indices in H_σ, since the Hamiltonian is invariant over spin. $\mathbb{C}^\dagger_{\alpha,i,m}$ is a column vector of a form $[c_{p_y}\ c_{p_x}\ c_s]^T$ at each site $\{\alpha, i, n\}$, while $\mathbb{E}_0$ and Σ_l represent the on-site and hopping matrices, respectively, with $l = \{1 + i-j\}\{m-n + (3-(-1)^i)/2\}$, which have the following matrix elements for the case of zigzag termination of the edges:

$$\mathbb{E}_0 = \begin{pmatrix} 0 & 0 & 0 \\ 0 & 0 & 0 \\ 0 & 0 & \varepsilon_s \end{pmatrix}, \tag{20.4}$$

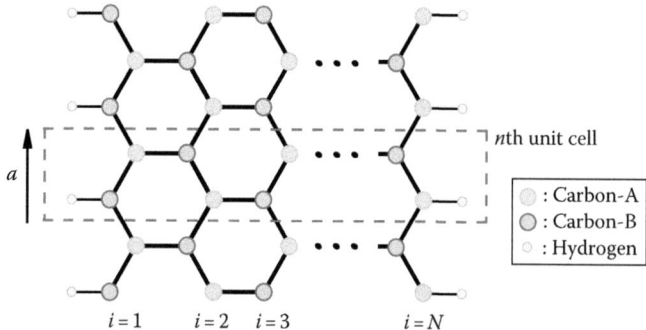

$i=1$ $i=2$ $i=3$ $i=N$

FIGURE 20.1

Hydrogen-passivated zigzag graphene nanoribbon with width N. Two sublattices of the honeycomb lattice are represented by *carbon-A* and *carbon-B*.

$$\Sigma_0 = \begin{pmatrix} V_{pp\pi} & 0 & 0 \\ 0 & -V_{pp\sigma} & V_{sp\sigma} \\ 0 & -V_{sp\sigma} & V_{ss\sigma} \end{pmatrix}, \tag{20.5}$$

$$\Sigma_1 = \begin{pmatrix} (V_{pp\pi} - 3V_{pp\sigma})/4 & \sqrt{3}(V_{pp\pi} + V_{pp\sigma})/3 & \sqrt{3}V_{sp\sigma}/2 \\ \sqrt{3}(V_{pp\pi} + V_{pp\sigma})/3 & (3V_{pp\pi} - V_{pp\sigma})/4 & -V_{sp\sigma}/2 \\ -\sqrt{3}V_{sp\sigma}/2 & V_{sp\sigma}/2 & V_{ss\sigma} \end{pmatrix}, \tag{20.6}$$

$$\Sigma_2 = \begin{pmatrix} (V_{pp\pi} - 3V_{pp\sigma})/4 & -\sqrt{3}(V_{pp\pi} + V_{pp\sigma})/3 & -\sqrt{3}V_{sp\sigma}/2 \\ -\sqrt{3}(V_{pp\pi} + V_{pp\sigma})/3 & (3V_{pp\pi} - V_{pp\sigma})/4 & -V_{sp\sigma}/2 \\ \sqrt{3}V_{sp\sigma}/2 & V_{sp\sigma}/2 & V_{ss\sigma} \end{pmatrix}. \tag{20.7}$$

Here, $s = -8.87$ eV is the on-site energy of the s-orbital relative to that of the p-orbital and the various hopping parameters are chosen to be $V_{pp\pi} = -3.03$, $V_{pp\sigma} = -5.04$, $V_{sp\sigma} = -5.58$, and $V_{ss\sigma} = -6.77$ in eV [60]. $V_{pp\pi}(V_{pp\sigma})$ is the overlap matrix between p-orbitals with $\pi(\sigma)$ bonding. $V_{sp\sigma}$ is the overlap matrix between s and p-orbital where the s-orbital is on the axis of rotational symmetry of p-orbital. Finally, $V_{ss\sigma}$ is the overlap element between s-orbitals. In the bulk 2D case, one can obtain the intrinsic spin–orbit interaction strength $\xi_1 \approx 2\xi_0^2 \varepsilon_s / 9V_{sp\sigma}^2 \sim 10^{-3}$ meV at Dirac points using the aforementioned parameters with $\xi_0 \approx 4$ meV [24]. We will also include the Hamiltonian H_P to take into account the passivation of the dangling orbitals at the edges of the GNR. The specific form of H_P will be given in Section 20.4. For the pristine GNRs, that is, the nonpassivated ZGNR in which the dangling bonds at the edges are kept intact, we will apply the open boundary condition. Other extrinsic SOC terms such as the Rashba interaction are not considered here.

20.3 PRISTINE ZIGZAG GRAPHENE NANORIBBON

The effects of the SOC on the electronic properties of the Dirac particles in graphene have been extensively studied recently. Since the energy levels of σ bands are well separated from the Dirac points, one can obtain the low-energy effective Hamiltonian projected to the Hilbert space of only the p_z orbital by integrating out the high-energy processes involving the σ bands [61]: $H_{eff} \simeq H_\pi - H_{SO} H_\sigma^{-1} H_{SO}$. It has been shown that the nearest-neighbor hopping amplitude induced by the SOC is canceled by the lattice symmetry, and thus the leading contribution due to the intrinsic SOC results from the effective next-nearest-neighbor hopping processes of the following form: $i\xi_1 / 3\sqrt{3} (\vec{d}_{ik} \times \vec{d}_{kj}) c_i^\dagger s^z c_j$, where ξ_1 is on the order of 0.05 K. This effective Hamiltonian, the so-called KM Hamiltonian, has been generally used to study the edge states of GNR, which led to the QSHE. However, we want to point out that since the lattice symmetry is broken at the graphene edges, one should, in principle, use the full Hamiltonian instead of the truncated low-energy effective KM Hamiltonian. In doing so, it is generally observed that the characteristic behavior of the QSHE in the ZGNR depends largely on the edge geometry and the passivation.

In Figure 20.2a, we have shown the calculated band structure of the pristine ZGNR with width $N = 150$. From the two degenerate uncoupled compositions of the spin and orbit $\{p_z\uparrow, p_x\downarrow, p_y\downarrow, s\downarrow\}$ and $\{p_z\downarrow, p_x\uparrow, p_y\uparrow, s\uparrow\}$, we have plotted the former one in Figure 20.2a. Here, we have taken a relatively large value of the SOC, $\xi_0 = 0.1$ eV, for the sake of clarity, since we have noticed that the magnitude of the SOC hardly affects the qualitative features of the band structures. In order to investigate the edge states made of the p_z orbital in detail, we focus on the π edge bands within the dashed box of Figure 20.2a, which are magnified in Figure 20.2b. We have also calculated the band structure based on the effective KM Hamiltonian, which is shown in Figure 20.2c, and compared to the result from our model. We have chosen $\xi_1 \approx 2\xi_0^2 \varepsilon_s / 9V_{sp\sigma}^2 = 6.34 \times 10^{-4}$ eV for the KM Hamiltonian, which corresponds to $\xi_0 = 0.1$ eV for our Hamiltonian.

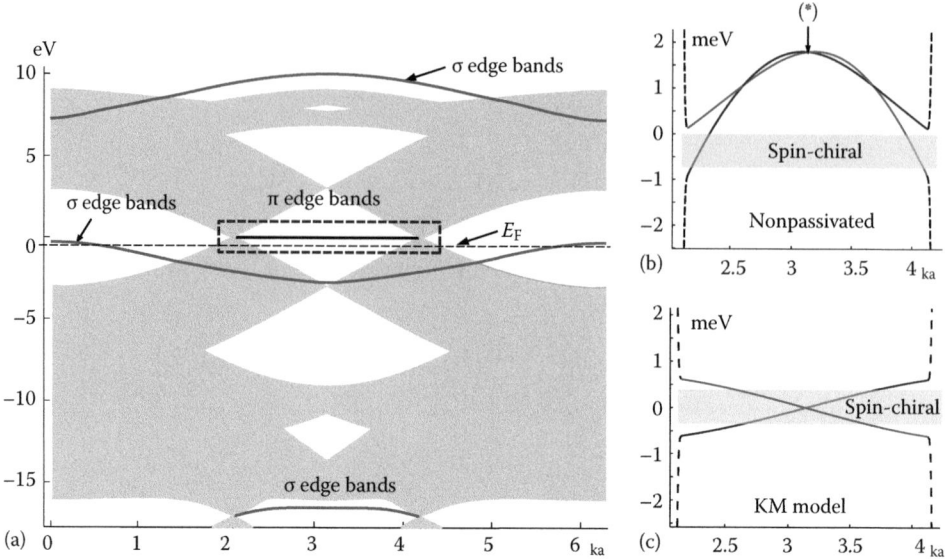

FIGURE 20.2
(a) Band structure of the pristine zigzag graphene nanoribbon with width $N = 150$ and the atomic spin–orbit coupling strength $\xi_0 = 0.1$. Here, the x axis represents the dimensionless wave number ka and the y axis is the energy in eV. The light gray curves represent the σ edge bands made of s, p_x, and p_y orbitals. In the dashed box at the band center, we have the π edge bands, which look almost flat at this energy scale. The bulk bands are shaded in gray. (b) The dashed box in Figure 20.2a is magnified. (c) The band structure near the band center obtained by the KM model with $\xi_1 = 6.34 \times 10^{-4}$. In (b) and (c), the light gray (dark gray) bands include edge states confined to the right (left) side of the ribbon.

By the simple band counting, one can expect to have $(2N - 2) \times 4$ bulk bands and the $2 \times 4 = 8$ edge bands composed of the four atomic orbitals, all of which are twofold spin degenerate due to the time reversal and the inversion symmetry. The most part of the band structure can be understood as a confinement effect on the bulk 2D graphene. The almost flat bands within the black dashed box and the red bands are the newly introduced states, which are absent in the bulk graphene. They represent the edge-localized states, where the former ones (π edge bands) are mostly made of p_z orbital and the latter ones (σ edge bands) mainly consists of s, p_x, and p_y orbitals. These features of the band structures of the ZGNR are consistent with the previous first-principles calculations [62]. It has also been widely known that the gapless flat bands of p_z orbital are formed at $E = 0$ within the finite region of $2\pi/3 < k < \pi$ in the absence of the SOC. It is shown in Figure 20.2a that there exist six σ edge bands with twofold spin degeneracy, and three pairs of them are almost degenerate so that it seems only three σ bands exist. For the stronger SOC, they will split into distinct six nondegenerate bands. While two pairs of them are well separated from the π edge bands, a pair of the σ edge bands appears quite close to the π edge ones. Since two pairs of edge σ bands lie below the π edge bands, the Fermi level of the undoped ZGNR is located below the π edge ones as shown in Figure 20.2a. This can be proved as follows. The Fermi energy for the pristine ZGNR as a function of the ribbon width N is given by $E_F \approx E_B - 2\gamma_0/\sqrt{N}$, where E_B represents the energy at the bottom of the spin-filtered edge states shown in Figure 20.2b. Since the four out of six σ edge bands are located below the π edge bands, the two spin-filtered edge states should be unoccupied at half filling. This formula yields $E_F \approx -0.47$ eV for the $N = 150$ ZGNR, which is consistent with our numerical result shown in Figure 20.2a.

At this charge-neutral point, we have the edge-localized states coming from the two σ edge bands, which become degenerate at the zone boundary. At a fixed value of k away from the zone boundary, the two almost degenerate σ bands are shown to be localized at different edges leading to the even number of pairs of edge states at each edge. Hence, the QSHE does not occur. In contrast to our results, the KM model exhibits the spin-filtered edge states within the energy window of width $2\xi_1$ around the band-crossing point at the zone boundary, and thus one can expect the QSHE to occur, as shown in Figure 20.2c.

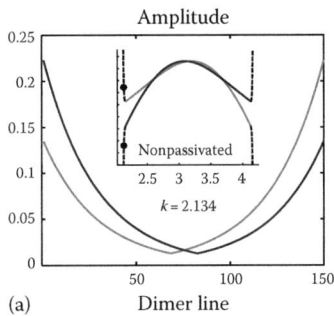
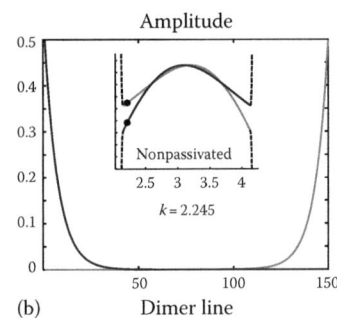
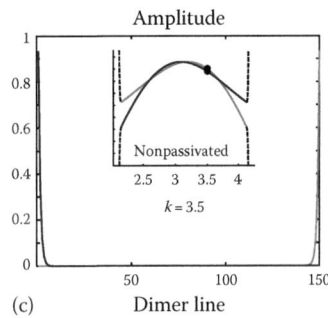

FIGURE 20.3
Eigenvectors of the π edge states at several values of k. The x axis represents the index of the dimer lines of the ribbon and the y axis is the absolute value of the amplitude of the eigenvector at each dimer line. (a) The eigenvector at $k = 2.134$, (b) the eigenvector at $k = 2.245$, and (c) the eigenvector at $k = 3.5$. The dark gray (light gray)-colored eigenvectors correspond to the dark gray (light gray) bands. Here, the dark gray (light gray) bands consist of the edge states localized at the left (right) side.

The π edge bands are shown in Figure 20.2b and c, where the red and blue bands consist of the edge states confined to the right and left edge of the ribbon, respectively. The black dashed bands represent the states whose amplitudes are spread over both edges. For another set of the spin–orbit composition $\{p_z\downarrow, p_x\uparrow, p_y\uparrow, s\uparrow\}$, one can simply interchange the red and blue colors of the bands. In Figure 20.3, we have shown that the states within the π edge bands ($k = 2.245$ and 3.5) are strongly confined to one of the ribbon edges with the localization length $\xi \cong a/\ln(-2\cos ka/2)$, which vanishes as k approaches to the zone boundary ($k = \pi$). In contrast, the states at $k = 2.134$ have a bulk feature, which have finite amplitudes along the width direction.

By increasing the carrier density, one can adjust the Fermi energy into the region, where spin-filtered chiral edge states made of the π orbitals do exist within the energy window of width $2\xi_1$ and the QSHE appears. Here, the red and blue bands are monotonic and move in opposite directions to each other. With a further increase of the carrier density, one can raise the Fermi energy to the band-crossing point at the zone boundary. At this time reversal–invariant point, denoted by an asterisk in Figure 20.2b, the edges states are most strongly confined to the ribbon edges for both the KM model and ours. While the KM model yields the QSHE near this point with a single pair of edge states at each edge as shown in Figure 20.2c, our model demonstrates that two chiral spin bands are nonmonotonic and they cross the Fermi level several times, as shown in Figure 20.2b. There exist edge states propagating in both directions at each edge of the ribbon. Although the number of bands crossing the Fermi level is odd, one of them (black dashed one) is always dispersed at both edges manifesting the feature of quantum confined bulk band. This means that there exist two pairs of edge states at each edge, and hence, the QSHE is not feasible in this energy range.

The main difference between our model and the KM Hamiltonian for graphene is whether the particle-hole symmetry exists or not. We notice that the KM model possesses both the TRS and the particle-hole symmetry at half filling, while our model obeys only the TRS. According to the classification of TIs, the particle-hole symmetry is not required to realize the TIs belonging to the Class AII. Hence, our model clearly demonstrates the existence of the QSHE in the absence of the particle-hole symmetry, although the π edge bands are warped and the Fermi level is shifted due to the broken particle-hole symmetry.

We have also studied the effect of the SOC on the band structure of the armchair GNR. In comparison to the ZGNR, the Dirac cone is located at the Γ point ($k = 0$), which becomes split in energy upon the inclusion of the SOC. It is well known that in the absence of the SOC, the gapless edge bands exist, which cross $E = 0$ at $k = 0$ for $N + 1$ being an integer multiple of three. Hence, the main difference between the ZGNR and the armchair GNR lies in the fact that while there exists a finite range of gapless flat bands for the ZGNR, there is a single gapless point at $k = 0$. In Figure 20.4, the band structure of the armchair GNR with width $N = 152$ is plotted. One can clearly see that the π edge bands are formed around $k = 0$ within the energy range of $2\xi_1$, and hence, the QSHE is expected to occur. The amplitudes of the eigenvectors at a fixed value of $k = 0.05$ are plotted as a function of dimer line index for three different values of $\xi_1 = 6.34 \times 10^{-4}$,

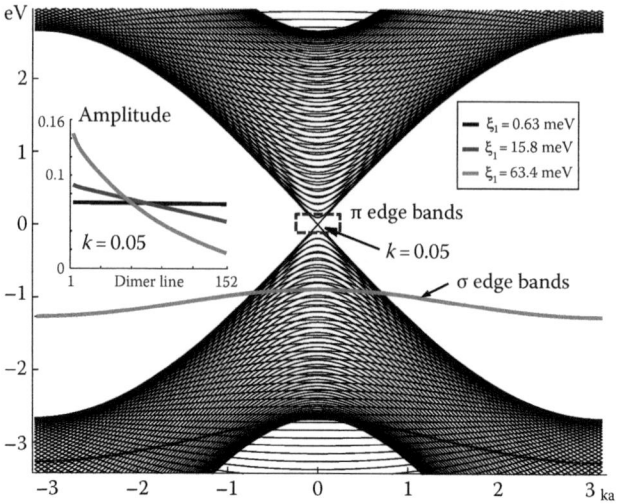

FIGURE 20.4
Band structure of the nonpassivated armchair graphene nanoribbon with width $N = 152$. The red bands represent the σ edge bands, which are almost doubly degenerate. Within the dashed box, we have the edge-localized π bands. In the inset, the amplitudes for the eigenvectors at $k = 0.05$ are plotted as a function of dimer line index for three different values of $\xi_1 = 6.34 \times 10^{-4}$, 1.58×10^{-2}, and 6.34×10^{-2} in eV. One can notice that the edge states are quite broadly dispersed rather than localized at the edge.

1.58×10^{-2}, and 6.34×10^{-2} in eV. One can notice that in contrast to the case of ZGNR, the π edge states for the armchair GNR are quite widely spread. The localization length of the edge states for the armchair GNR can be approximately given by $(\gamma_0/\xi_1)a$ with a being the lattice spacing.

20.4 HYDROGEN-PASSIVATED ZIGZAG GRAPHENE NANORIBBON

In this section, we will investigate the effect of the hydrogen passivation on the edge dangling bonds of the ZGNR. Since the hydrogen atom has a single s-orbital, only the p_x and s-orbitals of the adjacent carbon atom will have a finite overlap with the hydrogen atom so that the Hamiltonian H_P for the edge passivation can be written as follows:

$$H_P = \sum_n (\widetilde{V}_{sp} c^\dagger_{p_x,A,N,n} h_{N,n} + \widetilde{V}_{ss} c^\dagger_{s,A,N,n} h_{N,n} - \widetilde{V}_{sp} c^\dagger_{p_x,B,1,n} h_{1,n} + \widetilde{V}_{ss} c^\dagger_{s,B,1,n} h_{1,n} + h.c.)$$

$$+ \sum_n \varepsilon_h (h^\dagger_{1,n} h_{1,n} + h^\dagger_{N,n} h_{N,n}) \tag{20.8}$$

where the two hopping and on-site parameters between the carbon and hydrogen atom are taken to be $\widetilde{V}_{sp} = -4.5$, $\widetilde{V}_{ss} = -4.2$, and $\varepsilon_h = -2.7$ in eV [63]. The operator $h^\dagger_{i,n}(h_{i,n})$ represents the creation (annihilation) operator of an electron at hydrogen atom bonded to the ith carbon dimer line in the nth unit cell.

The band structure of the hydrogen-passivated ZGNR is shown in Figure 20.5a. There exist eight σ edge bands (red lines), which are composed of four pairs of almost doubly degenerate bands. In comparison to the nonpassivated ZGNR, we have an additional pair of σ edge bands, since the hydrogen s-orbitals at both edges have been coupled to the original s and p_x orbitals in the edge carbon atoms. Since the on-site energy $\varepsilon_h = -2.7$ eV of hydrogen atom is quite close to the band bottom of the σ edge band located in the middle for the nonpassivated ZGNR, they strongly interact with each other and then repel as shown in Figure 20.5a. This makes two significant effects on the edge-state characteristics. Focusing on the π edge bands shown in Figure 20.5b, one can notice that

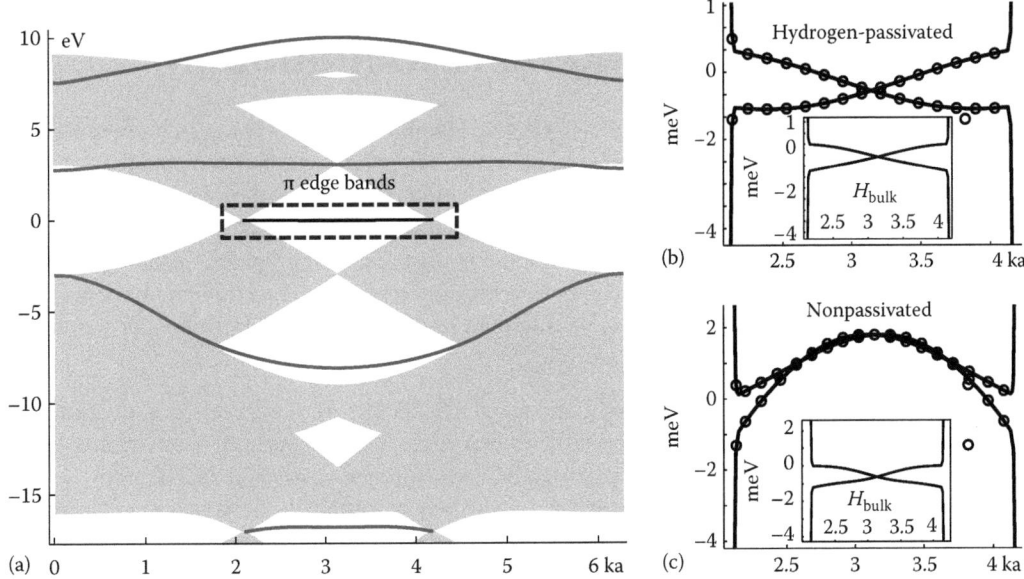

FIGURE 20.5

Band structure of the zigzag graphene nanoribbon with width $N = 150$ (a) with hydrogen pas-
sivation. The passivation parameters are taken to be $\tilde{V}_{sp} = -4.5$, $\tilde{V}_{ss} = -4.2$, and $\varepsilon_h = -2.7$ in eV.
The red bands represent the σ edge states. The bulk bands are shaded in gray. (b) The solid lines
represent the magnified view of the π edge bands in the blue dashed box. The open circles correspond to the result obtained from the perturbation method. (c) The solid lines represent the π edge
bands without passivation. The open circles correspond to the result obtained from the perturbation method for the nonpassivated zigzag graphene nanoribbon. The insets of (b) and (c) show the
results obtained from H_{bulk}.

the general feature of the KM model is recovered upon hydrogen passivation. In addition, a newly
introduced pair of σ edge bands is placed above the π edge bands. This balances the number of
energy bands above and below the π edge bands so that the Fermi level at half filling is placed on
the π edge bands.

By comparing the positions of the σ edge bands in Figures 20.2a and 20.5a, one may presume
that the σ edge bands located close to $E = 0$ mainly affect the energy dispersion of the π edge band.
In order to confirm this scenario, we have studied the effects of the σ edge bands by using the per-
turbation method, where the low-energy effective Hamiltonian is given by $H_{\text{eff}} \simeq H_\pi - H_{\text{SO}}H_\sigma^{-1}H_{\text{SO}}$.
The Hamiltonian H_{eff} can be decomposed into two terms: $H_{\text{eff}} = H_{\text{bulk}} + H_{\text{edge}}$, where H_{bulk} and H_{edge}
for a given k can be written as

$$H_{\text{bulk}} = H_\pi - \sum_{i \in \text{bulk}} H_{\text{SO}}|v_{\sigma i}\rangle E_{\sigma i}^{-1}(k)\langle v_{\sigma i}|H_{\text{SO}},$$

$$H_{\text{edge}} = -\sum_{i \in \text{edge}} H_{\text{SO}}|v_{\sigma i}\rangle E_{\sigma i}^{-1}(k)\langle v_{\sigma i}|H_{\text{SO}}.$$

(20.9)

Here, $|v_{\sigma i}\rangle$ represents the eigenstate of the Hamiltonian $\widetilde{H}_\sigma = H_\sigma + H_P$ in the ith σ band, and
$\sum_{i \in \text{bulk}} \left(\sum_{i \in \text{edge}} \right)$ stands for the sum over the bulk (edge) eigenstates. We expect that H_{bulk} will
reproduce the KM Hamiltonian and thus will always produce spin chiral edge bands at the band
center, which is clearly demonstrated in the insets of both Figure 20.5b and c.

By including the σ edge band contribution H_{edge} to H_{bulk}, we have obtained the results denoted
by the open circles in Figure 20.5b and c. The solid lines represent the exact numerical results of
the full Hamiltonian, which give an excellent agreement with those from the perturbation method.
Hence, we have clearly demonstrated that the hydrogen passivation can change the general features
of the π edge band by modifying the σ edge band profile.

20.5 EFFECTIVE REAL-SPACE HAMILTONIAN FOR THE ZIGZAG GRAPHENE NANORIBBON

In the previous section, we have obtained an effective Hamiltonian $H_{eff}(\vec{k})$ projected to the Hilbert space spanned by the p_z orbitals in the momentum space using the perturbation method.

Here, we will obtain the real-space effective Hamiltonian by applying the inverse Fourier transformation (IFT) to $H_{eff}(\vec{k})$ and analyze the spatial dependence of the on-site energy and hopping parameters. For instance, if one applies the IFT to the $H_{eff}(\vec{k})$ of the 2D graphene including the SOC term, one can obtain the next-nearest-neighboring hopping terms as a leading imaginary hopping process in addition to the original Dirac Hamiltonian leading to the KM Hamiltonian written as

$$H_{KM} = \gamma_0 \sum_{\langle ij \rangle} c_{i\alpha}^{\dagger} c_{j\alpha} + \sum_{\langle\langle ij \rangle\rangle} i \xi_1 / 3\sqrt{3} \nu_{ij} s_{\alpha\beta}^z c_{i\alpha}^{\dagger} c_{j\beta}.$$

By investigating the real-space Hamiltonian for the ZGNR, we have been able to study the effect of the broken translation symmetry at the ribbon edges and the hydrogen passivation as well. The IFTs of the $H_{eff}(\vec{k})$ for both the nonpassivated and hydrogen-passivated ZGNR have yielded as the leading orders the spatially dependent on-site potential and the imaginary next-nearest-neighboring hopping amplitude as shown in Figure 20.6a through d. We have also checked the

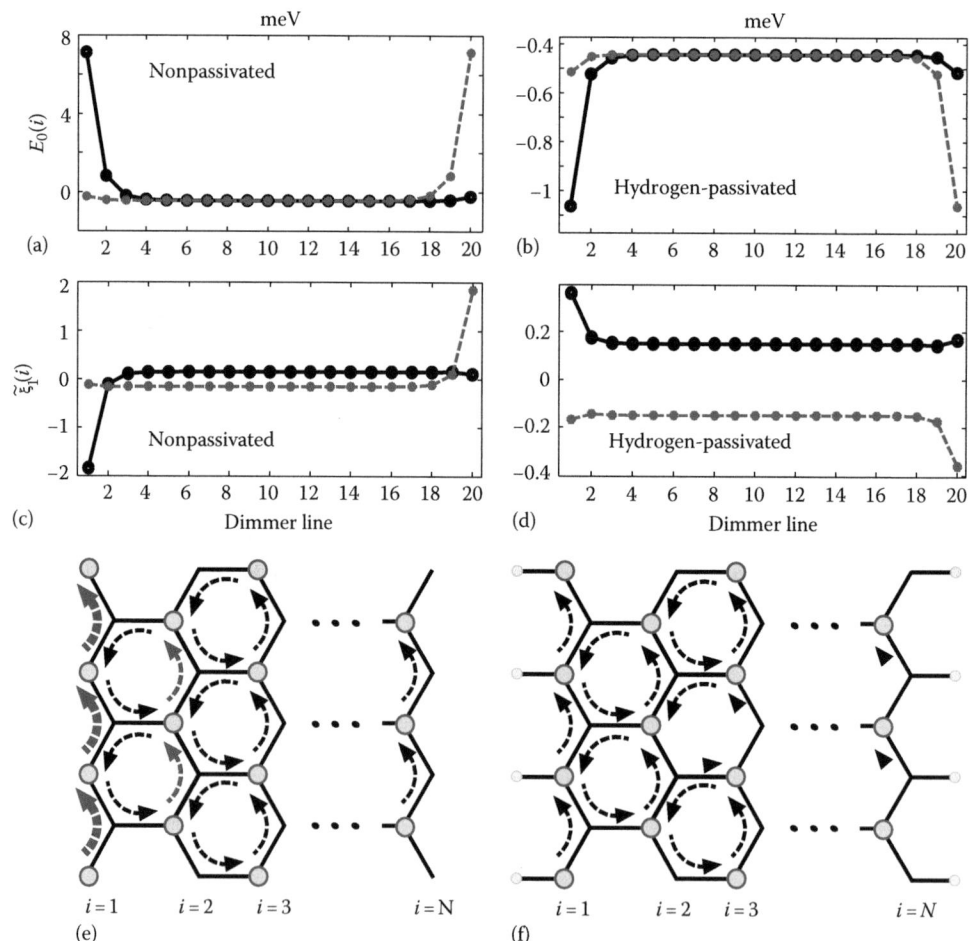

FIGURE 20.6

The spatial dependence of the on-site energy $E_0(i)$ and the effective next-nearest-neighboring hopping parameter $\xi_1(j)$ in eV for the nonpassivated and hydrogen-passivated zigzag graphene nanoribbon with width $N = 20$ and $\tilde{\xi}_1(i) = \xi_1(j)/3\sqrt{3}$. $E_0(i)$ and $\tilde{\xi}_1(i)$ are plotted in (a) and (c) for the nonpassivated zigzag graphene nanoribbon, while in (b) and (d) for the hydrogen-passivated one. The dashed red and solid black curves are plotted for the A and B sublattices, respectively. In (e) and (f), we depict schematically the signs and magnitudes of $\tilde{\xi}_1(i)$ for the counterclockwise hopping processes. The black (red) arrows mean that the sign of $\tilde{\xi}_1(i)$ is + (−), and their thickness represents the magnitude of $\tilde{\xi}_1(i)$.

additional nearest-neighbor hopping term induced by the SOC, which is absent in the 2D graphene due to the bulk lattice symmetry. We note that it is finite but much smaller than that of the next-nearest-neighboring hopping processes at the edges and exponentially decreases away from the edges approaching zero, which is its asymptotic limit. Based on the aforementioned parameters, we have constructed the following real-space effective Hamiltonian to describe the ZGNR:

$$H_{\text{eff}} = \gamma_0 \sum_{\text{n.n.}} c_{\alpha,i,n}^\dagger c_{\beta,j,n} + i \sum_{\text{n.n.n.}} \xi_1(j)/3\sqrt{3} v_{in,jm} s_{\alpha\beta}^z c_{\alpha,i,n}^\dagger c_{\alpha,j,m} + \sum_{\alpha,i,n} E_0(i) c_{\alpha,i,n}^\dagger c_{\alpha,i,m} \qquad (20.10)$$

where the indices $\alpha(\beta)$, $i(j)$, and $m(n)$ stand for sublattices, dimer lines, and unit cells along the longitudinal direction, respectively. The first term represents the noninteracting Dirac Hamiltonian, and $\xi_1(j)$ and $E_0(i)$ stand for the imaginary next-nearest-neighboring hopping amplitude and the on-site potential energy, which depend on the dimer line index. In Figure 20.6a through d, we plot $E_0(i)$ and $\widetilde{\xi}_1(j) = \xi_1(j)/3\sqrt{3}$ as a function of the dimer line index both for nonpassivated and hydrogen-passivated ZGNR with a ribbon width $N = 20$. We have also performed the similar calculations for ZGNRs with much larger width and obtained essentially the same curves for the tight-binding parameters.

First, we compare the spatial dependence of $E_0(i)$ for the nonpassivated and hydrogen-passivated ZGNRs as shown in Figure 20.6a and b, respectively. For both cases, $E_0(i)$ has shown a steep increase as one approaches to one side of the ZGNR for each sublattice. The rate of increase for the nonpassivated ZGNR is much higher than that of the hydrogen-passivated one. We notice that the bending of the π edge bands for the nonpassivated ZGNR originates from the steep confinement potential. Concerning the π edge bands, the edge states near the zone boundary ($k = \pi$) are much more strongly localized than the other states. Since the on-site energy shows a steep rise at the edges, the states localized tightly at the edge will be more strongly influenced and will gain an upward energy shift. This explains the fact that the edge bands near the zone boundary are more warped than the other regions. For the case of the hydrogen-passivated ZGNR, however, the effect of the on-site potential is not noticeable, since the on-site potential difference between the edge and the middle of the ribbon is one order of magnitude smaller than that of the nonpassivated ZGNR. In Figure 20.6c and d, the imaginary next-nearest-neighboring hopping amplitudes $\widetilde{\xi}_1(i)$ are plotted, which demonstrate the strongly enhanced values near the edges of both the nonpassivated and hydrogen-passivated ZGNRs. For the nonpassivated ZGNR, we have found that concerning the direction of the hopping, the sign of the imaginary next-nearest-neighboring hopping parameter near the ribbon edges is opposite to that inside the ribbon, as shown in Figure 20.6e. For the hydrogen-passivated one, the sign is shown to be identical all over the ribbon, just like the KM Hamiltonian, as shown in Figure 20.6f. Interestingly, we observe that whether the edge states are localized to one side or the other can be manipulated by controlling both the sign and the magnitude of the imaginary next-nearest-neighboring hopping parameter near the edge.

20.6 CONCLUSIONS

We have studied the effect of the edge termination on the low-energy physics of the ZGNR and armchair GNR by directly solving the tight-binding Hamiltonian, which includes all the hopping processes between s, p_x, p_y, and p_z in addition to the intrinsic SOC. We have obtained the warped π edge bands for the nonpassivated ZGNR and also noticed that the Fermi level lies below the π edge bands, which crosses the σ edge bands. Hence, at the charge-neutral point, we do not expect the QSHE to occur. We have shown that by electron doping, one can raise the Fermi level into the region where the QSHE can occur. Interestingly, we have demonstrated that the hydrogen passivation at the edges of the ZGNR can recover the standard features of the π edge bands suggested by the KM model. Hence, our observation implies that the ZGNR is a nice example that demonstrates the importance of the interplay between the topological classification based on the bulk property and the edge geometry. We have pointed out that in comparison to the KM Hamiltonian the various features of our model suggest a new example of TI belonging to the Class AII with the broken particle-hole symmetry.

We have also shown that the warping of the π edge bands is due to the strong influence from the σ edge bands located close to the π edge bands, which has been confirmed by the systematic perturbation analysis. Following the IFT, we have been able to obtain the real-space effective Hamiltonian.

Based on the Hamiltonian thus obtained, one can see that the on-site energy and the effective SOC strength are strongly enhanced as one approaches to the ribbon edges. The steep rise of the confinement potential leading to the strong effective lateral electric field can also explain the warping of the π edge bands as well.

ACKNOWLEDGMENTS

This research was supported by Basic Science Research Program through the National Research Foundation of Korea (NRF) funded by the Ministry of Education, Science and Technology(NRF-2012R1A1A2006927) and is partially reproduced with permission from K. Moon, *Phys. Rev. B*, 84, 035402. Copyright © 2011 by the American Physical Society.

REFERENCES

1. A. K. Geim and K. S. Novoselov, *Nat. Mater.* 6, 183 (2007).
2. K. S. Novoselov, V. I. Fal'ko, L. Colombo, P. R. Gellert, M. G. Schwab, and K. Kim, *Nature.* 490, 192 (2012).
3. A. H. Castro Neto, F. Guinea, N. M. R. Peres, K. S. Novoselov, and A. K. Geim, *Rev. Mod. Phys.* 81, 109 (2009).
4. L. Jiao, L. Zhang, X. Wang, G. Diankov, and H. Dai, *Nature.* 458, 877 (2009).
5. D. V. Kosynkin, A. L. Higginbotham, A. Sinitskii, J. R. Lomeda, A. Dimiev, B. K. Price, and J. M. Tour, *Nature.* 458, 872 (2009).
6. Z. Klusek, W. Kozlowski, Z. Waqar, S. Datta, J. S. Burnell-Gray, I. V. Makarenko, N. R. Gall, E. V. Rutkov, A. Y. Tontegode, and A. N. Titkov, *Appl. Surf. Sci.* 252, 1221 (2005).
7. C. O. Girit, J. C. Meyer, R. Erni, M. D. Rossell, C. Kisielowski, L. Yang, C.-H. Park et al., *Science.* 27, 1705 (2009).
8. G. Xie, Z. Shi, R. Yang, D. Liu, W. Yang, M. Cheng, D. Wang, D. Shi, and G. Zhang, *Nano Lett.* 12, 4642–4646 (2012).
9. Y.-W. Son, M. L. Cohen, and S. G. Louie, *Phys. Rev. Lett.* 97, 216803 (2006).
10. J.-W. Rhim and K. Moon, *Phys. Rev. B.* 80, 155441 (2009).
11. Y.-W. Son, M. L. Cohen, and S. G. Louie, *Nature.* 444, 347 (2006).
12. J.-W. Rhim and K. Moon, *J. Phys. Condens. Matter.* 20, 365202 (2008).
13. C. L. Kane and E. J. Mele, *Phys. Rev. Lett.* 95, 146802 (2005).
14. C. L. Kane and E. J. Mele, *Phys. Rev. Lett.* 95, 226801 (2005).
15. J.-W. Rhim and K. Moon, *Phys. Rev. B.* 84, 035402 (2011).
16. R. E. Prange and S. M. Girvin, *The Quantum Hall Effect* (Springer, Heidelberg, Germany, 1990).
17. D. J. Thouless, M. Kohmoto, M. P. Nightingale, and M. den Nijs, *Phys. Rev. Lett.* 49, 405 (1982); Y. Hatsugai, *Phys. Rev. Lett.* 71, 3697 (1993).
18. T. Chakraborty and P. Pietilainen, *The Quantum Hall Effects* (Springer, Heidelberg, Germany, 1995).
19. F. D. M. Haldane, *Phys. Rev. Lett.* 61, 2015 (1988).
20. N. A. Sinitsyn, J. E. Hill, H. Min, J. Sinova, and A. H. MacDonald, *Phys. Rev. Lett.* 97, 106804 (2006).
21. V. K. Dugaev, V. I. Litvinov, and J. Barnas, *Phys. Rev. B.* 74, 224438 (2006).
22. D. Huertas-Hernando, F. Guinea, and A. Brataas, *Phys. Rev. B.* 74, 155426 (2006).
23. H. Min, J. E. Hill, N. A. Sinitsyn, B. R. Sahu, L. Kleinman, and A. H. MacDonald, *Phys. Rev. B.* 74, 165310 (2006).
24. Y. Yao, F. Ye, X.-L. Qi, S.-C. Zhang, and Z. Fang, *Phys. Rev. B.* 75, 041401(R) (2007).

25. X.-F. Wang and T. Chakraborty, *Phys. Rev. B.* 75, 033408 (2007).
26. J. C. Boettger and S. B. Trickey, *Phys. Rev. B.* 75, 121402(R) (2007).
27. A. M. Essin and J. E. Moore, *Phys. Rev. B.* 76, 165307 (2007).
28. M. Zarea and N. Sandler, *Phys. Rev. Lett.* 99, 256804 (2007).
29. S. Onari, Y. Ishikawa, H. Kontani, and J.-I. Inoue, *Phys. Rev. B.* 78, 121403(R) (2008).
30. M. Zarea, C. Busser, and N. Sandler, *Phys. Rev. Lett.* 101, 196804 (2008).
31. P. K. Pyatkovskiy, *J. Phys. Condens. Matter.* 21, 025506 (2009).
32. A. H. Castro Neto and F. Guinea, *Phys. Rev. Lett.* 103, 026804 (2009).
33. Z. Wang, N. Hao, and P. Zhang, *Phys. Rev. B.* 80, 115420 (2009).
34. M. Gmitra, S. Konschuh, C. Ertler, C. Ambrosch-Draxl, and J. Fabian, *Phys. Rev. B.* 80, 235431 (2009).
35. P. Ingenhoven, J. Z. Bernad, U. Zülicke, and R. Egger, *Phys. Rev. B.* 81, 035421 (2010).
36. R. van Gelderen and C. Morais Smith, *Phys. Rev. B.* 81, 125435 (2010).
37. M. J. Schmidt and D. Loss, *Phys. Rev. B.* 81, 165439 (2010).
38. E. McCann and M. Koshino, *Phys. Rev. B.* 81, 241409(R) (2010).
39. E. Prada, P. San-Jose, L. Brey, and H. A. Fertig, *Solid State Commun.* 151, 1075 (2011).
40. D. Soriano and J. Fernández-Rossier, *Phys. Rev. B.* 82, 161302(R) (2010).
41. D. Gosálbez-Martìnez, J. J. Palacios, and J. Fernández-Rossier, *Phys. Rev. B.* 83, 115436 (2011).
42. R. Winkler and U. Z¨ulicke, *Phys. Rev. B.* 82, 245313 (2010).
43. S. Konschuh, M. Gmitra, and J. Fabian, *Phys. Rev. B.* 82, 245412 (2010).
44. A. Yamakage, K. I. Imura, J. Cayssol, and Y. Kuramoto, *Europhys. Lett.* 87, 47005 (2009).
45. L. Fu and C. L. Kane, *Phys. Rev. B.* 74, 195312 (2006).
46. J. E. Moore and L. Balents, *Phys. Rev. B.* 75, 121306(R) (2007).
47. B. Andrei Bernevig, T. L. Hughes, and S.-C. Zhang, *Science.* 314, 1757 (2006).
48. M. Konig, S. Wiedmann, C. Brune, A. Roth, H. Buhmann, L. W. Molenkamp, X.-L. Qi, and S.-C. Zhang, *Science.* 318, 766 (2007).
49. D. Hsieh, D. Qian, L. Wray, Y. Xia, Y. S. Hor, R. J. Cava, and M. Z. Hasan, *Nature (London)* 452, 970 (2008).
50. X.-L. Qi, T. L. Hughes, and S.-C. Zhang, *Nat. Phys.* 4, 273 (2008).
51. G. Dresselhaus and M. S. Dresselhaus, *Phys. Rev.* 140, A401 (1965).
52. F. Guinea, *New J. Phys.* 12, 083063 (2010).
53. H.-W. Liu, X. C. Xie, and Q.-F. Sun, e-print arXiv:1004.0881 [cond-mat] (unpublished).
54. A. P. Seitsonen, A. M. Saitta, T. Wassmann, M. Lazzeri, and F. Mauri, *Phys. Rev. B.* 82, 115425 (2010).
55. M. Vanin, J. Gath, K. S. Thygesen, and K. W. Jacobsen, *Phys. Rev. B.* 82, 195411 (2010).
56. P. Koskinen, S. Malola, and H. Hakkinen, *Phys. Rev. Lett.* 101, 115502 (2010).
57. Y. H. Lu, R. Q. Wu, L. Shen, M. Yang, Z. D. Sha, Y. Q. Cai, P. M. He, and Y. P. Feng, *Appl. Phys. Lett.* 94, 122111 (2009).
58. M. Fujita, K. Wakabayashi, K. Nakada, and K. Kusakabe, *J. Phys. Soc. Jpn.* 65, 1920 (1996).
59. A. P. Schnyder, S. Ryu, A. Furusaki, and A. W. W. Ludwig, *Phys. Rev. B.* 78, 195125 (2008).
60. R. Saito, G. Dresselhaus, and M. S. Dresselhaus, *Physical Properties of Carbon Nanotubes* (Imperial College Press, London, U.K., 1998).
61. L. Petersen and P. Hedegard, *Surf. Sci.* 459, 49 (2000).
62. G. Lee and K. Cho, *Phys. Rev. B.* 79, 165440 (2009).
63. M. L. Elert, J. W. Mintmire, and C. T. White, *J. Phys. Colloq.* 44, C3-451 (1983).

CHAPTER 21

CONTENTS

Functionalization Methods of Graphene

21

*Daniela Iannazzo, Alessandro Pistone,
and Signorino Galvagno*

21.1 INTRODUCTION

Graphene is considered today the most amazing and versatile substance available to mankind, and since its discovery in 2004, it is without any doubt the most intensively studied material [1,2]. From a chemical point of view, it is constituted by a single atomic layer of sp²-hybridized carbon atoms, with a molecule bond length of 0.142 nm tightly packed in a 2D honeycomb lattice (Figure 21.1). Because of its chemical structure, this material is the thinnest compound known. Although it is just one atom thick, graphene possesses outstanding mechanical, electronic, optical, thermal, and chemical properties (Table 21.1) [3–9]. Graphene is mechanically very strong and flexible. The breaking strength of graphene is more than 100 times greater than that of steel [4]. Besides its extraordinary strength, graphene is also very light, with 0.77 mg/m². Its crystal structure does not break even after being stretched up to 20% [5]. Moreover, this material has shown to be the best conductor of heat and electricity [6,7]. From an electrical aspect, the carrier mobility of graphene at room temperature is more than 100 times higher than that of silicon [6]. Its thermal conductivity, measured at room temperature, is much higher than the value observed in all the other carbon structures as carbon nanotubes, graphite, and diamond and is 10 times higher than Cu [7]. Furthermore, graphene is an ultra-wideband optical material that interacts strongly with light of a wide range of wavelengths; it absorbs ~2.3% of light in the visible to infrared region, and this absorption coefficient is one to three orders of magnitude higher than those of conventional semiconductor materials [9].

Owing to these extraordinary properties, graphene and its derivatives are envisaged by scientific community as new nanomaterials able to revolutionize multiple industries from flexible, wearable, and transparent electronics to high-performance computing and spintronics [10–15]. As an example, great interest was devoted to the graphene-based compounds as versatile building material for fabrication of electrochemical devices, and recently, it has been receiving extensive research interest in electrochemical energy conversion and storage [16]. Moreover, this nanomaterial has been investigated for the development of fluorescence resonance energy transfer (FRET) biosensors due to its quenching capability toward various organic dyes and quantum dots [17], as well as fast DNA sequencing [18].

As in other fields, the research on biomedical applications of graphene has seen dramatic progress and is expanding rapidly; once functionalized with biomolecules such as proteins, sugars, and nucleic acids, the graphene-based nanostructures may open a gateway to new fields in biotechnology [19–21]. Potential applications in biomedicine and in particular in drug delivery, cancer therapy, and biological imaging have been reported [22–25]. Moreover, the research on graphene and graphene oxide (GO)–based materials for cell culture is of particular interest. The study in this field demonstrated that graphene and GO are able to accelerate the growth, differentiation, and proliferation of stem cells and therefore hold great promise in tissue engineering, regenerative medicine, and other biomedical fields [26,27].

While scientists had theorized about graphene for decades, it was first produced in the lab in 2004 [1]. Since the first experimental evidence of the electronic properties of graphene, obtaining pure and highly ordered graphene has been a challenge. The commonly applied methods include the micromechanical or chemical exfoliation of graphite [28]. At present, there is an increasing

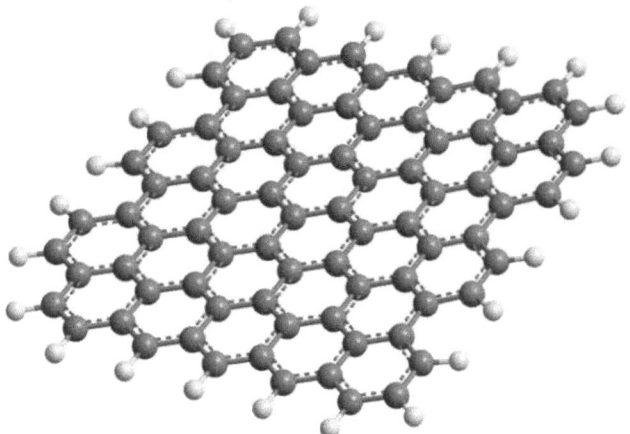

FIGURE 21.1
Idealized structure of a graphene fragment.

TABLE 21.1 Main Chemical and Physical Properties of Graphene

Property	Value	Comparison with Other Materials	Reference
Breaking strength	42 N m^{-1}	More than 100 times greater than steel	[4]
Elastic limit	~20%		[5]
Carrier mobility at room temperature	200,000 cm^2 V^{-1} s^{-1}	More than 100 times higher than Si	[6]
Thermal conductivity	~5000 Wm^{-1} K^{-1}	More than 10 times higher than Cu	[7]
Maximum current density	>108 A cm^{-1}	~100 times larger than Cu	[8]
Optical absorption coefficient	2.3%	~50 times higher than GaAs	[9]

concern in the chemical community about the starting material quality, and recent efforts are directed to wet chemical approaches toward high-quality graphene flakes that encompass the use of graphite as initial material. Thus, graphene has been synthesized by chemical vapor deposition (CVD) and plasma-enhanced CVD growth, thermal decomposition on SiC and other substrates, and chemical, electrochemical, thermal, or photocatalytic reduction of GO and fluorographene [29–33]. Since the different synthetic ways and experimental conditions to produce graphene may significantly affect the graphene properties and in particular its applications in electronic field, a lot of research is still focused on the development of new synthetic routes enabling an effective production of well-defined sheets.

Chemical functionalization of graphene is one of the key topics in graphene research and in the studies related to its applications in electronics, in the nanocomposite synthesis, and in biomedical field. Despite the remarkable electronic and transport properties of graphene and its great potential of replacing silicon for next-generation electronics and optoelectronics, pristine graphene is a semi-metal with zero bandgap; the local density of states at the Fermi level is zero, and conduction can only occur by the thermal excitation of electrons [34]. This lack of an electronic bandgap limits the utilization of graphene in nanoelectronic and photonic devices. However, since the graphene band structure is sensitive to lattice symmetry, chemical modifications may break this symmetry and, by opening the bandgap, can transform graphene into a semiconductor, thus enhancing the potential practical electronic applications [35]. In biomedical field, the functionalization of graphene allows the development of more efficient graphene or GO-based nanoplatform for drug delivery, biosensing, and other applications such as tissue engineering; for these reasons, a huge part of literature reports the functionalization of graphene and its derivatives with organic and inorganic molecules, by covalent and noncovalent interactions [36–39].

The chemical reactivity of graphene is often associated with those of the parent sp^2-hybridized carbon allotropes: fullerenes, carbon nanotubes, and graphite. Graphene may be considered as a 2D building material for those carbon nanomaterials: the 0D buckyball fullerenes can be seen as spherical wrapped graphene sheets, the 1D nanotubes are structurally rolled-up graphene sheets,

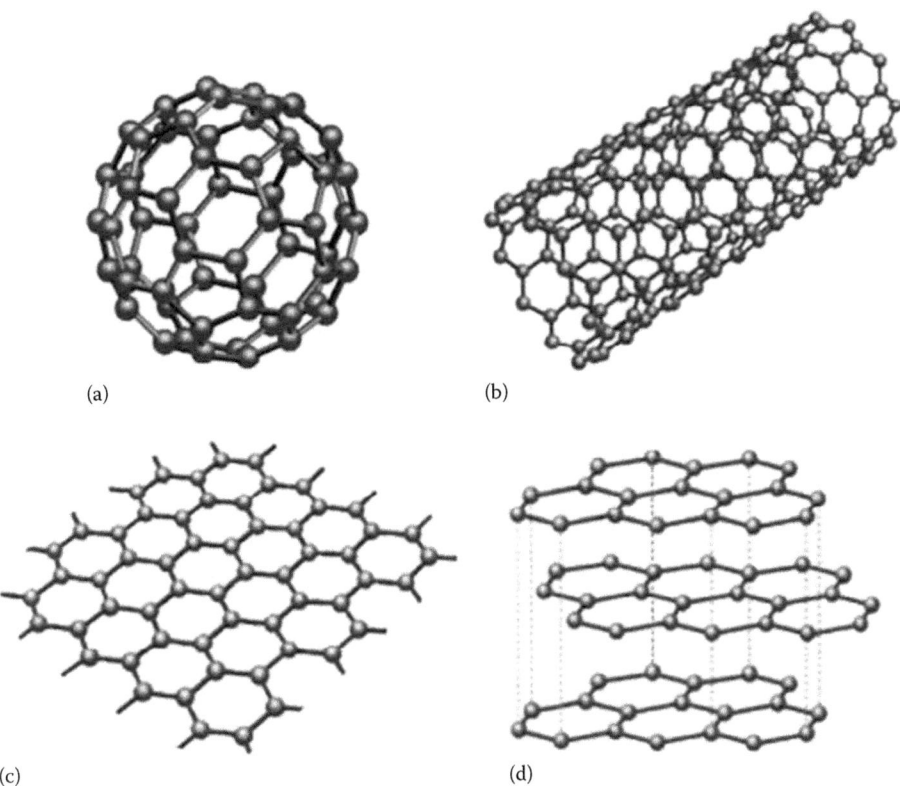

FIGURE 21.2
Examples of sp^2-hybridized carbon allotropes: (a) 0D fullerene (C$_{60}$), (b) 1D carbon nanotubes, (c) 2D graphene, and (d) 3D graphite.

and the 3D graphite consists of multilayer of graphene. However, despite these similarities, the chemical reactivity of 2D graphene differs from that of closed fullerenes and is closer to that of carbon nanotubes [40]; however, with difference to fullerenes and nanotubes that have a curved π surface, graphene is planar and hence less reactive (Figure 21.2).

The chemistry of graphene is considered similar to that of graphite even if in the case of the 1D graphene, chemical reactions may occur at both faces, thus affording chemical species that would be unstable if only one face would be exposed [41]. Similarly to carbon nanotubes, graphene presents defects along the carbon–carbon double-bond network, whose number mainly depends on the method used for graphene synthesis. These imperfections are typically due to the presence of atoms with sp^3 hybridization; these carbon atoms surrounding the defects present a different electronic structure and are chemically activated for further chemical reactions. The chemical functionalization of graphene may occur on the π surface or on the edges even if edge carbon atoms are more reactive than those located in the inner π surface [42,43]. This effect can be explained by considering that the covalent derivatization of the graphene layer, which generally involves a rehybridization process from sp^2 to sp^3, leads to the saturation of a double bond and therefore to a perturbation of the π system; the so-formed tetrahedral sp^3 carbon atoms result in a strong strain, which is significantly lower when the tetrahedral geometry is formed on the carbon atoms located at the edge. Moreover, the honeycomb π system is responsible for the complexation of graphene with a variety of chemical species, mainly through π–π interactions or electron transfer processes, thus opening the way to supramolecular chemistry in solution or on solid surfaces [44,45]. In the following paragraphs, a detailed review on the advances of chemical functionalization of graphene is presented. The chemical functionalization of graphene will be analyzed by the formation of new covalent bonds between the atoms native to GO or to reduced graphene oxide (RGO) and the guest functional groups or by noncovalent π–π interaction between guest molecules and graphene surfaces, which occurs mainly as a physical interaction. In both cases, the surface modification of graphene by covalent and noncovalent modification techniques is very effective in the preparation of processable graphene.

21.2 FUNCTIONALIZATION OF GRAPHENE BY COVALENT BONDING

The covalent approaches to graphene functionalization have been mainly focused on graphite oxide and its reduced form, namely, reduced graphite oxide or chemically converted graphene. In this chapter, several chemical routes to afford graphene derivatives by incorporating a large number of different atoms/organic groups into pristine graphene, radical additions, electrophilic substitution, and cycloaddition reactions will be reported. In Table 21.2, a summary of the recent chemical functionalization of pristine graphene with organic groups is described. Moreover, the oxidation methods to produce GO, the hydrogenation reactions to afford graphane, and the synthesis of graphene fluoride will be briefly discussed.

The functionalization of pristine graphene with organic functional groups has been developed for several purposes. As an example, the attachment of suitable organic groups allows the obtainment of graphene sheets dispersible in common organic solvents, which is of pivotal importance for the formation of nanocomposite materials, while organic functional groups such as chromophores offer new properties that could be combined with the properties of graphene such as conductivity [46]. In general, the functionalization of graphene by organic covalent bonding may occur in the following two different routes: (1) by the formation of covalent bonds between free radicals, electrophiles or dienophiles, and the carbon–carbon double bonds of pristine graphene or (2) by the formation of covalent bonds between the oxygen groups of GO, inserted after the oxidation process on pristine graphene and different organic functional groups. The derivatization of pristine graphene or GO allows the development of more efficient graphene or GO-based nanoplatform for drug delivery, biosensing, and other applications such as tissue engineering.

21.2.1 Radical Addition to sp² Carbon Atoms of Graphene

Considering the low reactivity of pristine graphene, the radical addition of phenyl species generated in situ upon heating of aryl diazonium salts is particularly suitable for the transformation of the carbon atom framework from sp² to sp³ hybridization, thus allowing the formation of new covalent bonds at the honeycomb lattice. The aryl diazonium salt covalent modification, which was firstly proposed by Tour and coworkers on carbon nanotubes and graphene starting from nitrophenyls [47,48] and after developed by other research groups, allows the preparation of self-tailored functional materials with improved dispersion capability in organic solvents and water, by simply changing the substituents in the benzene ring (Figure 21.3) [49–60].

This functionalization method, involving a change in hybridization, introduces an electron flow barrier by opening a bandgap and allows the formation of insulating and semiconducting regions on the graphene surface; this modification has shown to deeply modify the electronic and transport properties of epitaxial graphene decreasing, in a controlled manner, the conductivity of the system from near metallic to semiconducting [50]. The same research group reported another example of diazonium salt addition on bulk graphite, able to induce a steric repulsion between modified graphene layers, thus avoiding reaggregation and providing solubility in common organic media; the in situ reaction of 4-bromophenyl diazonium salts with thermally expanded graphite under mild sonication in N,N'-dimethylformamide (DMF) afforded the exfoliation of graphene layers from the bulk graphite. Noteworthy, after this covalent functionalization, no stabilizer additives are needed to prevent the reaggregation of graphene, and more than 70% of those graphene flakes presented less than five layers. Moreover, the reaction showed a selective edge reactivity compared to the basal plane, as confirmed by electron energy loss spectroscopy (EELS); the authors explained this result considering the low exposure of the thermally expanded graphite interior planes to the reagents [51]. Starting again from pristine graphite, the aryl diazonium salt modification was used by Hirsch et al. to develop a wet chemical bulk functionalization route that does not require initial oxidative damage of the graphene basal planes. Through effective reductive activation using a sodium–potassium alloy, the in situ reaction of graphite with 4-tert-butylphenyldiazonium or 4-sulfonylphenyldiazonium tetrafluoroborate, noticed by the increasing temperature and nitrogen formation, allowed the large insertion of highly reactive aryl radicals on the graphene

TABLE 21.2 Chemical Functionalization of Pristine Graphene with Organic Groups

Functionalization Process	Reaction	Chemical Group Inserted	References
Radical additions	Diazonium coupling	$R = NO_2$	[47–50,53]
		$R = Br$	[51,54]
		$R = t\text{-}Bu$	[52]
		$R = SO_3H$	[52]
		$R = CH_2CH_2OH$	[55,56]
		$R = OCH_2C{\equiv}CH$	[58]
		$R = C{\equiv}CH$	[59,60]
	Photochemical radical addition	$R = H$	[61,62]
Electrophilic substitution	Friedel–Crafts acylation	$R_1 = R_2 = R_4 = R_5 = H$	[68]
Cycloaddition	[3+2]-cycloaddition with azomethine ylides	$R_3 = $ (O-CH₂CH₂-O-CH₂CH₂-NH-Boc chain)	[74]
		$R_1 = R_4 = R_5 = H$ $R_2 = $ (catechol, OH, OH) $R_3 = CH_3$	[75]
		$R_1 = R_4 = R_5 = H$ $R_2 = Ph\text{-}COOH$ $R_3 = CH_3$	[76]
	[2+1]-cycloaddition with nitrenes	$R = $ (pentafluorophenyl ester –O–(CH₂CH₂O)₃–OH)	[77]
		$R = $ (phenylalanine NHBoc, HOOC)	[78]
		$R = $ hexyl, dodecyl, hydroxyl-undecanyl, carboxy-undecanyl	[79]
		$R = $ hydroxyl, carboxyl, amino, bromine, alkyl chain, polyethylene glycol, polystyrene	[80]
		$R = (CH_3)_3Si$	[81]
	[2+1]-cyclopropanation	$R = $ (bis-tetrathiafulvalene/dithiole anthracene)	[83]
	[2+2]-aryne cycloaddition	$R = H, F, CH_3$ $R_1 = H, F$	[84]
	[4+2] Diels–Alder reaction	Dienes: tetracyanoethylene, maleic anhydride Dienophiles: 2,3-dimethoxybutadiene, 9-methylanthracene	[85,86]

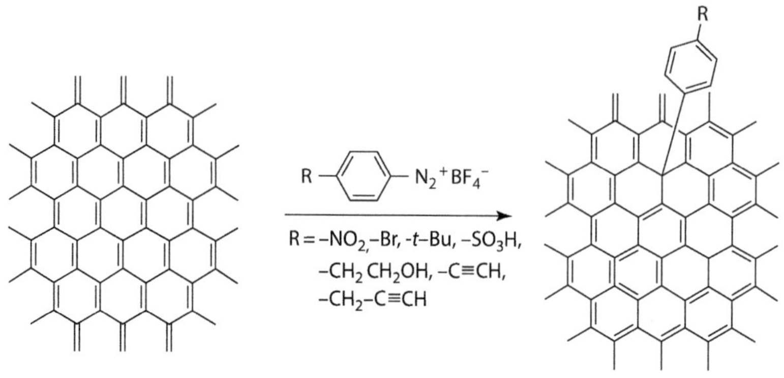

FIGURE 21.3
Aryl diazonium salt additions on graphene.

surface, thus avoiding the intermolecular or intramolecular π–π stacking and preventing the reaggregation of graphene sheets [52]. Single-layer, double-layer, or multilayer graphene shows different chemical reactivity toward electron transfer chemistries with aryl diazonium salts as well as different reactivity of edges and basal planes of graphene. A study by Strano and coworkers, using aryl radical 4-nitrobenzenediazonium tetrafluoroborate, found that single graphene sheets were almost 10 times more reactive than bi- or multilayers of graphene according to the relative disorder (D) peak in the Raman spectrum, examined before and after chemical reaction in water. Moreover, it was found that the reactivity of edges was at least two times higher than the reactivity of the bulk single graphene sheet [53]. In a similar study, the site-dependent and spontaneous functionalization of 4-bromobenzene diazonium tetrafluoroborate and its doping effect on mechanically exfoliated graphene have been investigated by Lim et al. [54]. In this study, the authors reported that the diazonium salt addition is noncovalent for the defect-free mechanically exfoliated graphene basal planes and covalent only in the edges, and thus, the intrinsic electrical properties of the modified graphene are modulated in this study by the noncovalent doping of the graphene surface.

The radical addition to sp^2 carbon atoms of graphene via diazonium salt functionalization is an efficient method for the further functionalizations since it allows the insertion of valuable intermediates to which it is possible to bind other functional groups and thus address the synthesis of this modified graphene sheet for specific purposes. As an example, hydroxylated aryl groups grafted covalently on graphene by the diazonium addition reaction can act as initiators for the polymerization of styrene via the atomic transfer radical polymerization (ATRP) method; as a consequence, the polymeric chains are covalently grafted on the graphene surface offering possibilities for optimizing the processing properties and interface structure of graphene–polymer nanocomposites [55]. Long-chain aromatic amines, chemically similar to the curing agent, have been covalently bonded on the surface of graphene sheets after diazonium addition promoting the graphene exfoliation and molecular level dispersion in the matrix, serving as a linker between graphene and epoxy networks for improved load transfer and modulating the stoichiometric ratio around graphene for the construction of a hierarchical structure able to dissipate more strain energy during fracture [56]. Among the different mechanisms to further functionalize the intermediates inserted via diazonium salt addition, one of the more useful and investigated reactions is the Cu(I)-catalyzed Huisgen [3+2] cycloaddition reaction between azide and alkyne moieties forming a 1,4-substituted 1,2,3-triazole [57]. By means of this so-called click reaction, it is possible to obtain, with high regioselectivity, easy reaction conditions, good reliability, and good yields, a rapid and efficient attachment of functional groups to various materials, including graphene. Figure 21.4 reports some examples of the insertion of organic groups by click chemistry. A short polyethylene glycol chain with terminal carboxylic end group (PEG–COOH) on graphene prepared either from thermally expanded graphite or CVD and functionalized with 4-propargyloxy benzenediazonium tetrafluoroborate was successfully performed by Jin et al., allowing the creation of water-soluble graphene. Moreover, the creation of a triazole ring on graphene sheets modified by phenylacetylene moieties able, by conjugation, to maintain the electronic transport into the graphene sheets has been used to study the photocurrent responses of various photoactive functional molecules introduced on graphene via *click* chemistry [59]. For a similar purpose, Salavagione et al. synthesized

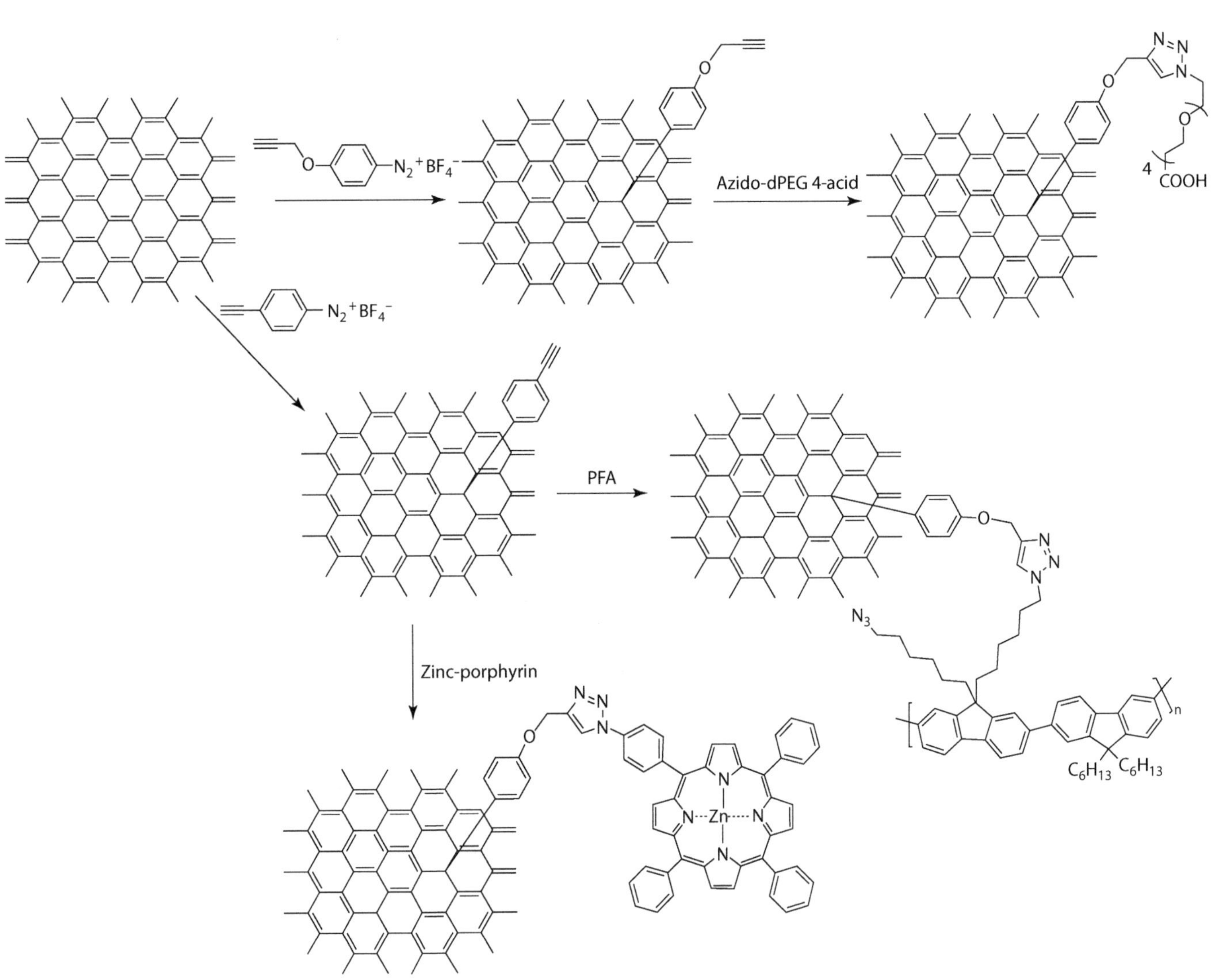

FIGURE 21.4
Examples of *click chemistry* on graphene.

graphene flakes covalently modified with a conjugated polymer, poly[(9,9-dihexylfluorene)-co-alt-(9,9-*bis*-(6-azidohexyl)fluorene)] (PFA), by the Cu-catalyzed Huisgen 1,3-dipolar cycloaddition between alkyne-modified graphene and the azide-functionalized polymer. The resulting materials exhibit absorption and emission spectra with a high dependence on the solvent used. Variations on the graphene solvation can induce different polymer structures depending on the surrounding environment; a high graphene solvation minimizes its influence on the polymer structure, slightly affecting the polymer properties [60].

An alternative free radical addition method includes the reaction of benzoyl peroxide with graphene sheets. Liu et al. used benzoyl peroxide as radical precursor under laser irradiation in the reaction with mechanically exfoliated graphite. The attachment of phenyl groups was confirmed by the appearance of a strong D band at 1343 cm^{-1} in the Raman spectra, due to the formation of sp^3 carbon atoms in the basal plane of graphene after the covalent attachment of phenyl groups and by the increase in the hole doping attributed to the physisorbed benzoyl peroxide that produced a transient benzoyl radical anion [61]. This excited species decomposed to produce the phenyl radical, which is responsible for introducing sp^3 centers onto the graphene basal planes.

A similar reaction was performed in homogeneous dispersions of graphene nanosheets in orthodichlorobenzene (*o*-DCB) obtained from various graphite materials. In this work, the thermal decomposition of benzoyl peroxide was used to initiate radical addition of alkyl iodides to graphene in *o*-DCB dispersions [62].

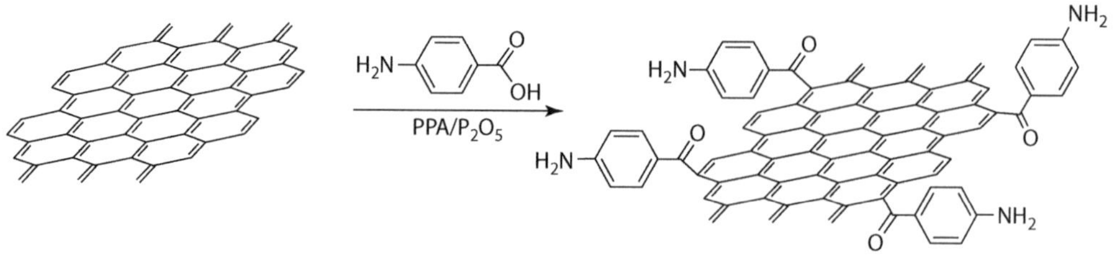

FIGURE 21.5
Friedel–Crafts reaction on graphene.

21.2.2 Friedel–Crafts Acylation

Since the pioneering work of Baek et al. [63] which explored the direct Friedel–Crafts acylation reaction to inherent defective sp^2 C–H sites of CNT, several research groups have expanded this reaction to all types of carbon-based nanomaterials such as fullerenes, carbon nanofibers, nanodiamonds, and graphene [64–67]. The Friedel–Crafts acylation reaction has several advantages such as nondestructive reaction nature, sufficient modification level, utilization of diverse materials, and suitability for mass production; it represents a good alternative to the aryl diazonium salt covalent modification of graphene edges and thus for the exfoliation of graphite [68]. In a recent paper, Choi et al. reported the electrophilic substitution reaction of graphite defects, located principally on the edges, with the organic molecular wedge 4-aminobenzoic acid in poly(phosphoric acid) (PPA)–phosphorous pentoxide (P_2O_5) medium [68] (Figure 21.5). The reaction produced exfoliated graphene and graphene-like sheets. This functionalization improved the graphite solubility and thus allows further penetration of the viscous medium, leading to the final isolation of exfoliated graphene.

21.2.3 Addition of Dienophiles to Carbon–Carbon Bonds

Among the different dienophiles able to react with sp^2 carbons of graphene, azomethine ylides represent the most common species that have been successfully applied in the functionalization of carbon nanostructures such as fullerenes, nanotubes, onions, nanohorns, and graphene [69–72]. This reaction, also called *Prato reaction*, is a 1,3-dipolar cycloaddition reaction of azomethine ylides generated in situ by thermal condensation of aldehydes and α-amino acids [69]. This reaction allowed the preparation of a variety of organic derivatives, which show interesting applications in several areas, such as polymer nanocomposites, biotechnology, nanoelectronic devices, drug delivery, and solar cells [73,74]. With difference in carbon nanotubes and fullerenes, graphene does not present a curvature, and as encountered with other organic functionalizations, the cycloaddition reactions mainly occur at the sheet edges, providing highly functionalized materials. Prato et al. reported the functionalization of exfoliated graphene with *N*-methylpyrrolidone using the 1,3-dipolar functionalization of azomethine ylides. These amino groups were further bound to gold nanorods, which were introduced as contrast markers for the identification of the graphene reactive sites; the uniform distribution of gold nanorods on the graphene surface, with a higher density localized next to the defects, demonstrated that this reaction has taken place not just at the edges but also at the internal carbon–carbon bond network (Figure 21.6).

Graphene sheets were also decorated with dihydroxyphenyl groups by pyrrolidine rings that were formed perpendicular to the graphene surface by the addition of azomethine ylide precursors, formed by the condensation of 3,4-dihydroxybenzaldeyde and sarcosine. The hydroxyl groups introduced onto a graphene sheet increase its dispersibility in polar solvents such as ethanol and DMF [75].

The further functionalization with molecules able to maintain the electronic transport into the graphene sheets makes graphene a very useful platform for novel solar cell applications. Thus, Ragoussi et al. reported the 1,3-dipolar cycloaddition of azomethine ylides obtained by reaction of sarcosine and 4-carboxybenzaldehyde on basal plane of few-layer graphene and the subsequent covalent coupling with light-harvesting and electron-donating phthalocyanine [76]. Physicochemical

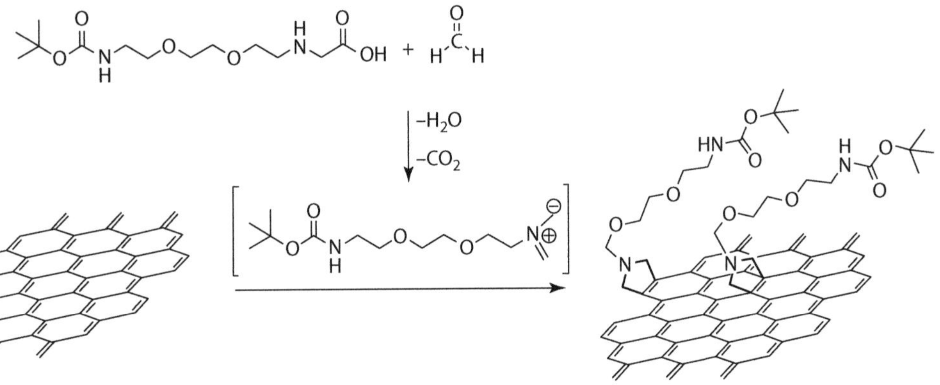

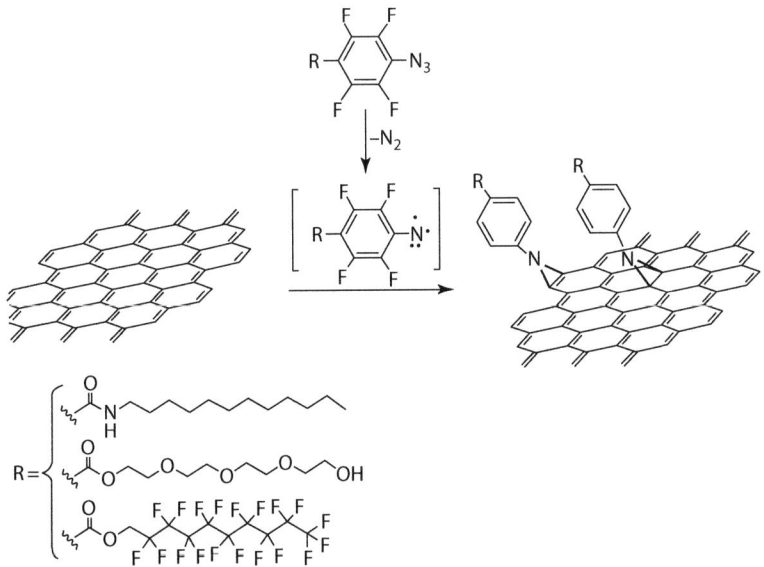

FIGURE 21.6
Prato reaction on graphene.

characterizations reveal an ultrafast charge separation from the photoexcited phthalocyanine to few-layer graphene followed by a slower charge recombination. Raman and TGA analyses indicate a relatively low degree of graphene functionalization, thus allowing the retention of the inherent properties of graphene; moreover, the morphology of the hybrid materials is not affected by the cycloaddition functionalization processes. The fluorescence or phosphorescence quenching of the chromophores and the bleaching of the singlet excited state species in transient absorption measurements suggest excited state energy/electron transfer between graphene and the covalently attached photoactive molecules.

Phenyl and alkyl azides have shown to react with the C–C bonds of graphene by the formation of the reactive intermediate nitrene; the reaction involves a [2+1] cycloaddition to the double bonds of graphene, forming an aziridino-ring linkage by thermal decomposition at high temperatures [77–81]. Liu et al. reported the graphene functionalization by nitrene chemistry using different perfluorophenyl azides (PFPA), which, upon photochemical or thermal activation, were converted into the highly reactive perfluorophenylnitrenes that readily react with the sp[2] carbon network of graphene to form aziridine adducts through a [2+1] cycloaddition (Figure 21.7). Graphene synthesized with these defined chemical functionalities resulted soluble in organic solvents or water depending on the nature of the functional group on PFPA [77].

Pristine few-layer graphene, obtained from exfoliated graphite in *o*-DCB, was used in the covalent nitrene addition of azidophenylalanine [78]. The graphene sheets were reacted with

FIGURE 21.7
Nitrene cycloaddition using different perfluorophenyl azides.

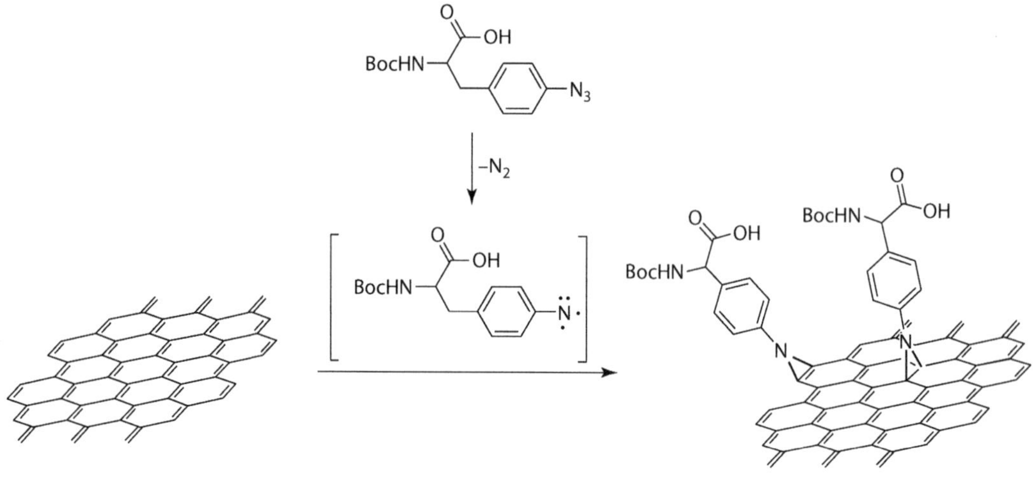

FIGURE 21.8
Nitrene addition of azidophenylalanine.

Boc-protected azidophenylalanine in o-DCB (Figure 21.8); the authors reported a degree of functionalization of 1 phenylalanine substituent per 13 carbons.

Vadukumpully et al. described a facile and simple approach for the covalent functionalization of surfactant-wrapped graphene sheets. Thus, by reaction with the corresponding alkyl azides, alkyl groups such as hexyl, dodecyl, hydroxyl-undecanyl, and carboxy-undecanyl have been inserted into graphene sheets [79]. The free carboxylic acid groups were bound to gold nanoparticles, which were introduced as markers for the reactive sites. The gold nanoparticle–graphene composite was characterized by transmission electron microscopy and atomic force microscopy, demonstrating the uniform distribution of gold nanoparticles all over the surface.

Various functional groups and polymeric chains have been inserted onto graphene sheets by He and Gao via nitrene cycloaddition resulting in functional graphene sheets and 2D macromolecular brushes, respectively. The so-functionalized graphene sheets show enhanced chemical and thermal stabilities compared with GO, are electrically conductive, and display excellent dispersibility and processability in solvents [80]. The formation of covalent bonds between thermally generated nitrene and epitaxial graphene has been demonstrated by Choi et al. [81]. The bonding nature between nitrene, generated from azidotrimethylsilane, and graphene was confirmed to be covalent by high-resolution photoemission spectroscopy (HRPES). Nitrene radicals obtained by thermal treatment are covalently linked to the surface with a minor amount of physisorbed nitrene species, as observed in the XPS analysis by the bonding nature of N 1s peaks, where two distinct N peaks (bonded and physisorbed) were clearly distinguished in the spectra. Also, the formed bandgap of the functionalized epitaxial graphene can be controlled by the amount of the added nitrene in the reaction.

Another cycloaddition reaction used for the functionalization of graphene is the Bingel–Hirsch reaction [82], a nucleophilic [2+1] cyclopropanation of a double bond, usually used on fullerene, by the reaction of a bromomalonate in the presence of a base. For pristine graphene, the microwave irradiation plays an important role as an activating method for the Bingel reaction [83]. Thus, Economopoulos et al. reported chemically modified exfoliated graphene obtained by sonication in benzylamine, following the Bingel reaction conditions, with the aid of microwave irradiation, producing highly functionalized graphene-based hybrid materials. The resulting hybrid materials, possessing cyclopropanated malonate units, bearing with the electroactive π-extended tetrathiafulvalene (ex-TTF) derivative, covalently grafted onto the graphene skeleton, formed stable suspensions for several days in a variety of organic solvents (Figure 21.9). Moreover, the authors observed that no reaction occurs under microwave irradiation when graphite flakes and the reagents for the cyclopropanation reaction were mixed without previous exfoliation. Although no photophysical details were provided for these donor–acceptor nanohybrids, the formation of a radical ion pair was proposed by the authors, based on the estimated electrochemical bandgap (1.23 eV).

Zhong et al. described the aryne cycloaddition to the graphene surface; this [2+2] cycloaddition was carried out by the reaction of 2-(trimethylsilyl)aryl triflates as active benzyne precursors, which

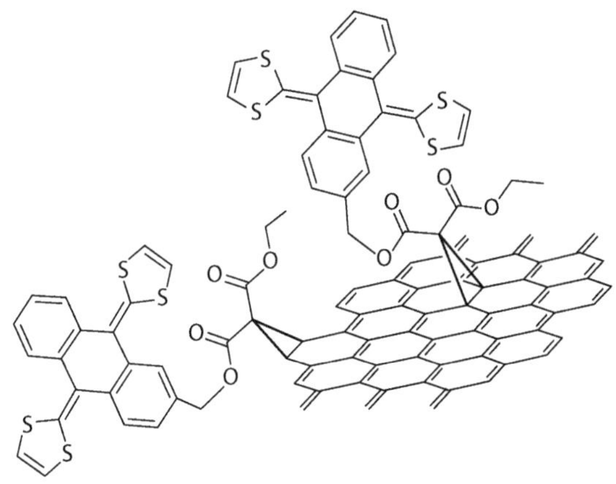

FIGURE 21.9
Functionalized graphene by Bingel–Hirsch reaction.

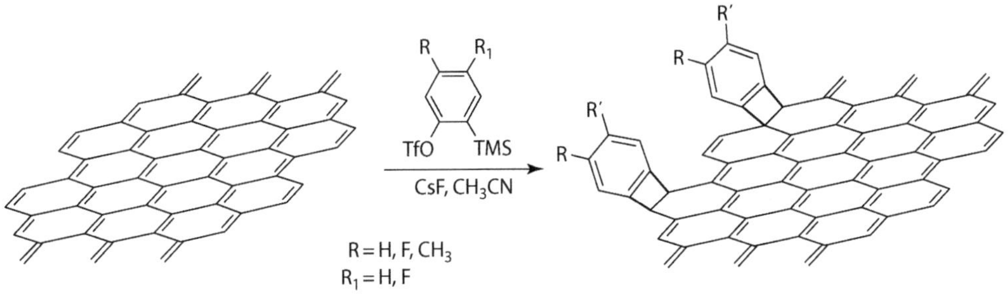

R = H, F, CH$_3$
R$_1$ = H, F

FIGURE 21.10
Aryne cycloaddition on graphene.

were reacted with graphene sheets under mild and neutral conditions, owing to the high activity fluoride-induced decomposition methodology for benzyne generation [84]. The reaction resulted in the formation of a four-membered ring between the aromatic ring and the graphene surface (Figure 21.10). The aryl rings used in this reaction can be substituted by different organic groups, and the synthesized materials are easily dispersible in o-DCB, ethanol, chloroform, and water.

Most of the interesting physical properties of graphene result from the singular electronic band structure at the so-called Dirac point, where the conduction and valence bands cross in momentum space. Although graphene is very stable thermodynamically, the electronic structure at the Dirac point facilitates basal plane chemistry including pericyclic reactions such as the Diels–Alder reaction [85]. Haddon and coworkers discovered a series of facile Diels–Alder reactions in which graphene can act either as a diene when paired with tetracyanoethylene and maleic anhydride or as a dienophile when paired with 2,3-dimethoxybutadiene and 9-methylanthracene [86]. To prove this reaction, the authors used different types of graphene (epitaxial graphene, exfoliated graphene, and highly oriented pyrolytic graphite [HOPG]), under very mild conditions (Figure 21.11).

The reaction conditions are strongly affected by the substrate used and the type of graphene, achieving different functionalization degrees. The authors also reported the reverse processes at higher temperatures than those of the direct reaction. The simultaneous transformation of a pair of neighboring sp^2 carbons in the graphene honeycomb lattice into 1,2-sp^3-hybridized centers suggests the effectiveness of Diels–Alder chemistry in opening a bandgap in graphene and thus the capability to modify its electronic and optical applications.

21.2.4 Hydrogenation of Pristine Graphene

The decoration of the graphene surface with hydrogen atoms to create graphane was the first covalent reaction performed on pristine graphene [87]. This chemical functionalization has attracted extensive attentions for controlling physical properties of graphene since the hydrogenation, by changing

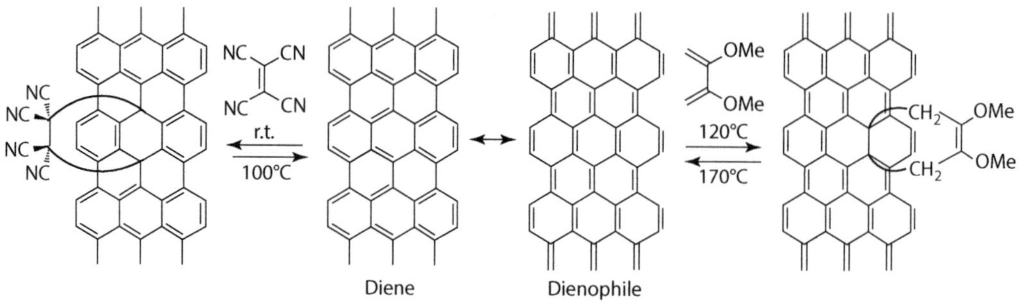

FIGURE 21.11
[4+2]- Diels–Alder reactions on graphene.

the hybridization from sp^2 to sp^3, may create local geometric distortions. Thus, significant progresses and methodologies have been performed on the controlled hydrogenation of graphene from the perspective of hydrogenation style and size [88,89]. This transformation converts the chemically inert and highly conductive zero overlap semimetal graphene into an insulator [89]. Graphene layers on SiO$_2$/Si substrate have been chemically decorated by radio-frequency (RF) hydrogen plasma [90]. The reaction was carried out under a low-pressure hydrogen/argon mixture and reached the saturation after 2 h. The electron diffraction of graphane showed hexagonal symmetry and thus the same crystallinity of graphene; an important decrease in the lattice constant, however, indicated its chemical modification. The reverse reaction was achieved under annealing conditions, recovering almost all the original graphene properties. Both the hydrogenation and dehydrogenation processes of the graphene layers are controlled by the corresponding energy barriers, which show significant dependence on the number of layers. Even if the cold hydrogen plasma treatment also created a large number of undesirable defects on the 2D surface structure, the extent of decorated carbon atoms in graphene layers can be manipulated reversibly up to the saturation coverage, which facilitates engineering of chemically decorated graphene with various functional groups via plasma techniques. Zheng et al. developed a facile method for graphene hydrogenation using a radio-frequency catalytic CVD (rf-cCVD) method to hydrogenate the graphene films [91]. Bulk amounts of catalysts (Ni/Al$_2$O$_3$) were placed upstream of the graphene films and heated at 820°C for 3 h under hydrogen flow. The increased hydrophobicity of the obtained material confirmed the successful introduction of C–H groups on the 2D lattice. Monolayer graphene synthesized by CVD was subjected to controlled and sequential hydrogenation using RF plasma while monitoring its electrical properties in situ. Low-temperature transport properties, namely, electrical resistance, thermopower, Hall mobility, and magnetoresistance, were measured for each sample and correlated with ex situ Raman scattering and x-ray photoemission characteristics. The authors reported that for weak hydrogenation, the transport is seen to be governed by electron diffusion, and low-temperature transport properties show metallic behavior; for strong hydrogenation, the transport is found to be describable by variable range hopping, and the low conductance showed insulating behavior. A clear transition to strong localization is evident with the emergence of pronounced negative magnetoresistance for strongly hydrogenated graphene. [92].

Several research groups have investigated the wet chemical approaches based on Birch reduction for the preparation of hydrogenated graphene [93,94]. The presence of water as the hydrogen source seems to slow down the process and to favor hydrogenation of graphene over the competing H$_2$ formation. Subrahmanyam et al. reported the Birch reduction of few-layer graphene samples giving hydrogenated samples containing up to 5 wt% of hydrogen. The material decomposed readily on heating to 500°C or on irradiation with UV or laser radiation releasing all the hydrogen, thereby demonstrating the possible use of few-layer graphene for chemical storage of hydrogen [93].

The electronic, magnetic, mechanical, and thermal conductivity properties of graphene are determined to a large extent by its unique 2D graphene morphology. Thus, great efforts have been devoted to manipulate the graphene morphology. In this context, Zhang et al. reported a programmable hydrogenation of graphene for novel nanocages. The authors performed molecular mechanics simulations to study the effect of doping patterns on the graphene folding and comprehensively investigated the fundamental mechanism governing the graphene folding with precise control from the energetic viewpoint. Molecular dynamics simulation has been performed to create a cross-shaped cubic graphene nanocage encapsulating a biomolecule by warping the top graphene layer downward and the bottom graphene layer upward to mimic the drug delivery vehicle. Such a paradigm, programmable enabled graphene nanocage, opens up a new avenue to control the 3D architecture of folded graphene and therefore provides a feasible way to exploit and fabricate the graphene-based unconventional nanomaterials and nanodevices for drug delivery [95].

21.2.5 Halogenated Graphenes

Graphite fluoride is a 3D covalent derivative of graphite with interesting electrochemical properties and potential application in hydrogen storage [96]. A stoichiometric derivative of graphene with a fluorine atom attached to each carbon can be considered as the 2D counterpart of the graphite fluoride. Graphene fluoride was prepared from pristine graphite fluoride by mechanical exfoliation, in an analogous approach used for the preparation of graphene [97,98]. Nair et al. reported the preparation of fluorographene by exposure of pristine graphene to atomic fluorine formed by decomposition of xenon difluoride (XeF_2). Fluorination of graphene dramatically changes its structure and electronic and optical properties; the zero bandgap of pristine graphene opens, and fluorographene behaves like an insulator with a resistivity higher than 10^{12} Ω [98]. The chemical exfoliation of graphite fluoride, prepared by fluorination of highly ordered pyrolytic graphite at 600°C and 1 atm of fluorine, provides graphene fluoride with stoichiometry $C_{0.7}F_1$ [99]. However, high-temperature synthesis leads to highly defective graphene fluoride. Thus, Zboril et al. obtained stoichiometric graphene fluoride monolayers in a single step by the liquid-phase exfoliation of graphite fluoride with sulfolane. The author reported comparative quantum mechanical calculations revealing that graphene fluoride is the most thermodynamically stable of the five studied hypothetical graphene derivatives: graphane, graphene fluoride, bromide, chloride, and iodide. The graphene fluoride is transformed into graphene via graphene iodide, a spontaneously decomposing intermediate. Recently, a gas-phase photochemical chlorination of graphene produced partially chlorinated graphene of uniform 8% coverage with stoichiometry $C_{16}Cl$. The resistance increases over four orders of magnitude, and a bandgap appears upon photochlorination, confirmed by electrical measurements. Moreover, localized photochlorination of graphene can facilitate chemical patterning, which may offer a feasible approach to the realization of all-graphene circuits [100]. Moreover, the preparation of few-layered graphene chlorinated up to 56 wt.% by irradiation with UV light in a liquid chlorine medium has been reported. The chlorinated sample decomposes on heating or on laser irradiation releasing all the chlorine. Similar results have been obtained with the bromination of few-layer graphene [101].

21.2.6 Oxidation of Graphene and GO Derivatives

Despite the relative recent discovery of the extraordinary properties of graphene, GO has a history that extends back many decades to some of the earliest studies involving the chemistry of graphite. Brodie first demonstrated the synthesis of GO in 1859 by adding a portion of potassium chlorate to a slurry of graphite in fuming nitric acid [102]. The resulting material, composed of carbon, hydrogen, and oxygen, results in an increase in the overall mass of the flake graphite. Brodie found the material to be dispersible in pure or basic water, but not in acidic media, which prompted him to term the material *graphic acid*. Nearly 60 years after Staudenmaier, Hummers, and Hoffman developed alternative oxidation methods by reacting graphite with a mixture of potassium permanganate ($KMnO_4$) and concentrated sulfuric acid (H_2SO_4), again, achieving similar levels of oxidation. In 1898, Staudenmaier improved this protocol by using mixture of concentrated H_2SO_4 and fuming nitric acid followed by gradual addition of chlorate to the reaction mixture. This small change in the procedure provided a simple protocol for the production of highly oxidized GO [103]. In 1958, Hummers reported an alternative method for the synthesis of GO by using $KMnO_4$ and $NaNO_3$ in concentrated H_2SO_4 [104]. GO prepared by this method could be used for the preparation of large graphitic film. Noteworthy, it has since been demonstrated that the products of these reactions show strong variance, depending not only on the particular oxidants used but also on the graphite source and reaction conditions [105]. Although the exact structure of GO is difficult to determine, it is evident that after the oxidative process, the contiguous aromatic lattice of graphene becomes interrupted by epoxides, alcohols, ketone carbonyls,

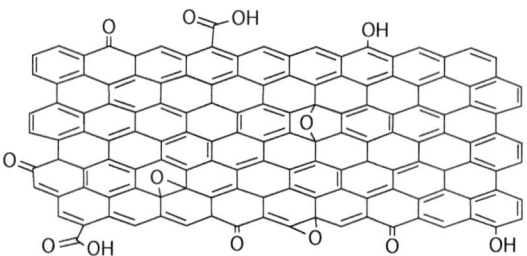

FIGURE 21.12
GO from graphite flake oxidation.

and carboxylic groups [106] (Figure 21.12). This disruption is reflected in an increase in interlayer spacing from 0.335 nm for graphite to more than 0.625 nm for GO [107]. For expanded graphite, the most common reagent used is the Jones' reagent (H_2CrO_4/H_2SO_4).

The most common source of graphite used for chemical reactions, including its oxidation, is flake graphite, which is a naturally occurring mineral, purified to remove heteroatomic contamination. The numerous localized defects in its π-structure may serve as seed points for the oxidation process, but the complexity of flake graphite and the defects that are inherent as a result of its natural source make the elucidation of precise oxidation mechanisms very challenging. GO prepared from flake graphite can be readily dispersed in water and has been used on a large scale for the preparation of graphitic films, as a binder for carbon products and as a component of the cathode of lithium batteries. Moreover, the hydrophilicity of GO allowed its uniform deposition on different substrates as thin films, necessary for applications in electronics [108]. The transformation of GO into a conductive material may require a partial restoration of the graphitic structure, and indeed, either in thin films or in bulk, it can be accomplished by chemical reduction to chemically converted graphene. When colloidally dispersed, a variety of chemical means may be used to reduce GO; the most common and one of the first to be reported was hydrazine monohydrate [109]. With difference in most strong reductants, this reagent does not react with water, making it an attractive option for reducing aqueous dispersions of GO. The most straightforward goal of any reduction protocol is to produce graphene-like materials similar to the pristine graphene achieved from direct mechanical exfoliation of individual layers of graphite. However, the hydrazine (N_2H_4) reduction mainly involves the epoxy groups, and thus, after this reduction, graphitic structure is not fully restored, significant defects are introduced, and the desired properties are not always obtained [110] (Figure 21.13).

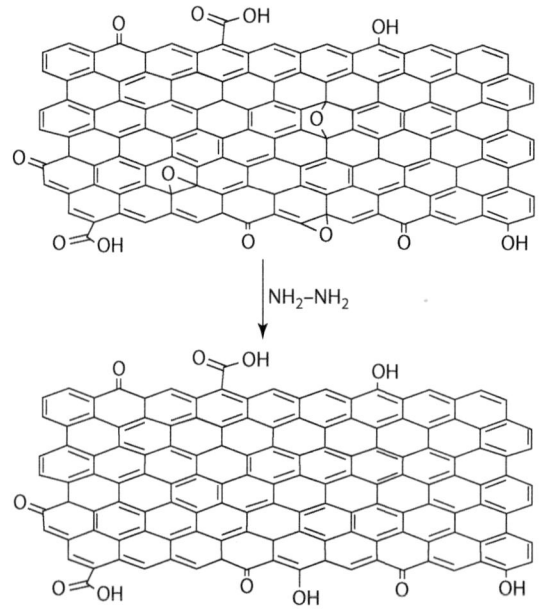

FIGURE 21.13
Chemical reduction of GO by hydrazine.

It was found that the sheet resistance of graphite oxide film, reduced using sodium borohydride ($NaBH_4$), is much lower than that of films reduced using (N_2H_4). This is attributed to the formation of C–N groups in the N_2H_4 case, which may act as donors, compensating the hole carriers in reduced graphite oxide. In the case of $NaBH_4$ reduction, the interlayer distance is first slightly expanded by the formation of intermediate boron oxide complexes and then contracted by the gradual removal of carbonyl and hydroxyl groups along with the boron oxide complexes. Shin et al. produced conducting film comprising a $NaBH_4$-reduced graphite oxide, which revealed a sheet resistance comparable to that of dispersed graphene [111]. In a very recent study, Bo et al. report the caffeic acid (CA) as a new, green, and efficient reducing agent for GO reduction. The CA-reduced GO (CA-rGO) shows a high C/O ratio (7.15). Electronic gas sensors and supercapacitors have been fabricated with the CA-rGO and show good performance, which demonstrates the potential of CA-rGO for sensing and energy storage applications [112].

Marcano et al. reported the scalable preparation of graphene oxide nanoribbons (GONRs) from multiwalled carbon nanotubes by treatment with KMnO4 and concentrated H_2SO_4 [113]; the authors reported that the addition of phosphoric acid (H_3PO_4) to this reaction-produced GONRs with more intact graphitic basal planes [114]. The same research group recently used the same oxidation procedure starting with graphite flakes, to prepare improved GO with fewer defects in the basal plane as compared to GO prepared by the Hummers' method. The method provides a greater amount of hydrophilic oxidized graphite material relative to the other two procedures with a more regular structure and a greater amount of basal plane framework retained. Even though the GO produced is more oxidized than that prepared by Hummers' method, when both are reduced in the same chamber with hydrazine, chemically converted graphene produced from this new method resulted equivalent in its electrical conductivity [115]. Recently, Huang et al. reported a simple room-temperature preparation of high-yield large-area GO. The 100% conversion reported by the authors was achieved using a simplified Hummers' method from large graphite flakes [116].

Functionalization of GO is one of the key topics in graphene research. The addition of further chemical groups to GO may be achieved using various chemical reactions by covalent or noncovalent approaches. The covalent approach is realized through the formation of new covalent bonds between the atoms native to GO or to RGO and the guest functional groups; in contrast, noncovalent functionalization is mainly based on π interaction between countermolecules and RGO/GO, that is, mainly a physical interaction. Such approaches that add functionality to groups already present on the GO render GO/graphite oxide a more versatile precursor for a wide range of applications. Various chemical routes have been proposed, which successfully incorporate a large number of different atoms/ organic groups into GO, including amidation, silanization, esterization, and substitution [117–128]. Such functionalized graphene finds more extensive application in polymer science electronics and in biomedical field.

21.3 NONCOVALENT FUNCTIONALIZATION OF GRAPHENE

Noncovalent functionalization of graphene is very helpful to preserve the intrinsic electronic structure of graphene sheets; from this point of view, the linkage of countermolecules to graphene planes by noncovalent approach is very attractive in order to induce additional selective properties to graphene without loss of its intrinsic chemical and physical characteristics. The noncovalent scientific approach to functionalize graphene is deeply based on the highly π-electron delocalized chemical structure of graphene surface; nowadays, all the scientific approaches to optimize the noncovalent functionalization of graphene are focused to tune properly the π interactions between graphene and countermolecules to design innovative nanomaterials and nanodevices.

Noncovalent interactions between π electron clouds of benzene-like rings of graphene and countermolecules can be included in large families such as rare gas–π, H–π, π–π, cation–π, anion–π systems, and the main energetic factors involved regard often electrostatics, dispersion, induction or exchange repulsion terms, and combination of these, where one or more of these terms prevail on the remaining ones by changing the countermolecules involved in the functionalization approach [129–131]. In particular, for the graphene systems characterized by a very large charge delocalization in the aromatic network, the dispersion energy represents often the driving force in the functionalization with rare gas, hydrogen, or aromatic countermolecules [132–136] where the π–π interactions play a crucial role. Also for the anion–π interactions between graphene and anionic-like countermolecules, the main energetic terms are related to dispersion energy high enough to overcome the repulsive contributions [137–139]. The cation–π interactions represent the basis for the functionalization of graphene with metal cation or a positively charged aromatic countermolecules; in these cases, electrostatic and induction terms are the main energetic terms involved, and the polarizability and charge dispersion ability of the π-electron grid of graphene play a key role in governing such kind of functionalization approaches [140–143].

Among the noncovalent functionalizations of graphene, pyrene moiety has attracted great attentions because of its known strong affinity toward the graphite basal planes [144]. Xu et al. functionalized graphene with 1-pyrenebutyrate to increase the aqueous dispersibility of graphene in aqueous medium [145]. Wang et al. employed pyrenebutanoic acid succinimidyl ester (PBSA) in the π–π noncovalent functionalization of graphene, obtaining improved power conversion efficiency of the graphene-based photovoltaic devices [146]. An et al. found unique optical and sensing properties of graphene noncovalently functionalized with 1-pyrenecarboxylic acid by π–π interactions while preserving the intrinsic conductive properties of graphene [147]. Kodali et al. have demonstrated that the noncovalent functionalization of graphene with PBSA has no destructive effects on the electronic structure of graphene planes [148].

Kinetic studies carried out by Mann et al. reported a tripodal motif that binds multivalently to graphene through three pyrene moieties and projects easily varied functionality away from the surface [149]. The authors demonstrated that with difference in individual pyrene units that readily desorb from graphene in organic solvents, this tripodal receptor forms stable monolayers that withstand infinite dilution conditions for hours. The thermodynamic and kinetic binding parameters of the tripod bearing a redox-active Co(II)*bis*-terpyridine complex were investigated electrochemically, and the formation of monolayers with a saturation coverage of 73.9 ± 0.2 pmol cm^{-2}, which corresponds to a 2.3 nm^2 molecular footprint, was achieved (Figure 21.14).

In a further study, the same research group investigated the surface diffusion of the tripodal system by scanning electrochemical microscopy (SECM) on single-layer graphene and the basal plane of HOPG [150]; the electrocatalytic activity of the tripod was investigated in two electrochemical processes: the oxygen reduction reaction and the oxidation of ferrocyanide. The authors reported that, in both cases, the reactions proceed faster than with a bare graphene electrode; this result is of particular interest for the development of dynamic electrode surfaces for sensing or in electrocatalysis.

Other similar scientific investigations regard the noncovalent functionalization of graphene with perylene-based countermolecules showing increased aqueous dispersion or conductivity or solar energy conversion [151–154]. Perylene bisimides (PBIs) have been successfully used for π–π stacking interactions, which, besides being well suited for these interactions, are also powerful dye-type chromophores. Hirsch et al. have recently reported the binding and electronic interaction of graphene with an organic PBI molecule in homogeneous solution [155] (Figure 21.15). The authors demonstrated that the π–π bonding promotes electronic interactions between graphene and a π-conjugated perylene in a liquid dispersion. Thus, the noncovalent functionalization of graphene with small polycyclic aromatic hydrocarbons (PAHs) in organic solvents could be a useful approach for the systematic tuning of the electronic properties of graphenes.

A variety of organic and inorganic countermolecules have been investigated in the noncovalent functionalization of graphene through π–π, electrostatic, H bond interactions, and so on. Several investigations were performed in the noncovalent functionalization of graphene with metal-based porphyrin [156–158], or with polymeric countermolecules for the preparation of graphene-based nanocomposites [159–162], or with various electron donor or acceptor molecules [163,164], or with ionic liquid polymers [165], or with amphiphilic copolymer [166] in order to combine the unique properties of graphene, unaffected in the noncovalent approach, with the characteristics of the countermolecule adopted, such as solubility in polar or nonpolar solvent, interaction with polymeric matrix in nanocomposite materials, and electronic or conducting properties in sensing or photovoltaic devices.

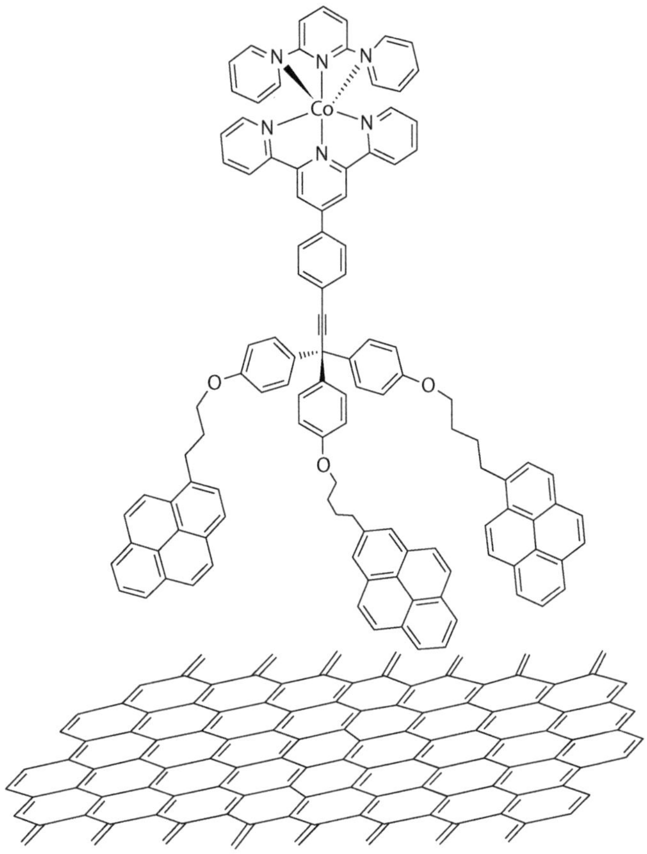

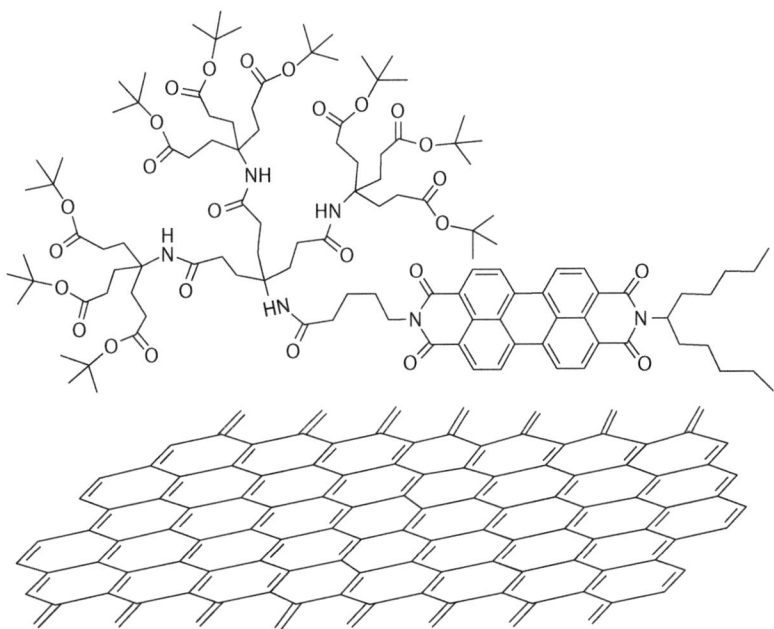

FIGURE 21.14
Adsorption on graphene of a tripodal pyrene derivative bearing a redox-active Co(tpy) complex.

FIGURE 21.15
Dendronized perylene bisimide–graphene hybrid system.

21.4 CONCLUSIONS

Since graphene discovery in 2004, the scientific community was immediately fascinated from the unique chemical and physical properties of this new nanomaterial, that is, able to revolutionize multiple industries from flexible, wearable, and transparent electronics, to high-performance computing and spintronics. In order to improve the outstanding properties and thus the fields of application, pristine graphene requires previous covalent and/or noncovalent chemical functionalization. Furthermore, the selective insertion of specific functional groups on the graphene surface allows the development of high-performance graphene-based nanomaterials, opening the way to innovative applications in many technological fields, such as electronic, sensing, polymeric-based nanocomposites, and in biomedical field.

On the basis of the chemical reactivity of the parent sp^2-hybridized fullerenes and carbon nanotubes, well-known covalent and noncovalent reactions for these nanomaterials have been applied also on this flat form of carbon, despite its lower chemical reactivity when compared with the parent curved carbon nanoforms. In this regard, the most successful reactions involving covalent and/or noncovalent bonding to pristine graphene have been discussed in this chapter.

REFERENCES

1. Novoselov, K. S., Geim, A. K., Morozov, S. V., Jiang, D., Zhang, Y, Dubonos, S. V. et al. 2004. Electric field effect in atomically thin carbon films. *Science* 306:666–669.
2. Singh, V., Joung, D., Zhai, L., Das, S., Khondaker, S. I., Seal, S. 2011. Graphene based materials: Past, present and future. *Prog. Mater. Sci.* 56:1178–1271.
3. Hibino, H. 2013. Graphene Research at NTT. *NTT Tech. Rev* 11:1–5.
4. Lee, C., Wei, X., Kysar, J. W., Hone, J. 2008. Measurement of the elastic properties and intrinsic strength of monolayer graphene. *Science* 321:385–388.
5. Kim, K. S., Zhao, Y., Jang, H., Lee, S. Y., Kim, J. M., Kim, K. S. et al. 2009. Large-scale pattern growth of graphene films for stretchable transparent electrodes. *Nature* 457:706–710.
6. Chen, J.-H., Jang, C., Xiao, S., Ishigami, M., Fuhrer, M. S. 2008. Intrinsic and extrinsic performance limits of graphene devices on SiO_2. *Nat. Nanotechnol.* 3:206–209.
7. Ghosh, S., Bao, W., Nika, D. L., Subrina, S., Pokatilov, E. P., Lau, C. N. et al. 2010. Dimensional crossover of thermal transport in few-layer graphene. *Nat. Mater.* 9:555–558.
8. Murali, R., Yang, Y., Brenner, K., Beck, T., Meindl, J. D. 2009. Breakdown current density of graphene nanoribbons. *Appl. Phys. Lett.* 94(24):243114.
9. Nair, R. R., Blake, P, Grigorenko, A. N., Novoselov, K. S., Booth, T. J., Stauber, T., R. 2008. Fine structure constant defines visual transparency of graphene. *Science* 320(5881):1308.
10. Xuan, Y., Wu, Y. Q., Shen, T., Qi, M., Capano, M. A., Cooper, J. A. et al. 2008. Atomic-layer-deposited nanostructures for graphene-based nanoelectronics. *Appl. Phys. Lett.* 92(1):13101–13103.
11. Liu, C., Alwarappan, S., Chen, Z. F., Kong, X., Li, C. Z. 2010. Membrane less enzymatic biofuel cells based on graphene nanosheets. *Biosens. Bioelectron.* 25:1829–1833.
12. Stoller, M. D., Park, S., Zhu, Y. W., An, J., Ruoff, R. S. 2008. Graphene-based ultra-capacitors. *Nano Lett.* 8(10):3498–3502.

13. Wang, L., Lee, K., Sun, Y. Y., Lucking, M., Chen, Z., Zhao, J. J. et al. 2009. Graphene oxide as an ideal substrate for hydrogen storage. *ACS Nano* 3(10):2995–3000.
14. Lu, C. H., Yang, H. H., Zhu C. L., Chen, X., Chen, G. N. 2009. A graphene platform for sensing biomolecules. *Angew. Chem. Int. Ed.* 121(26):4879–4881.
15. Loh, K. P., Bao, Q. L., Eda, G., Chhowalla, M. 2010. Graphene oxide as a chemically tunable platform for optical applications. *Nat. Chem.* 2:1015–1024.
16. Hu, Y., Sun, X. 2013. Chemically functionalized graphene and their applications in electrochemical energy conversion and storage. Nanotechnology and nanomaterials. In *Advances in Graphene Science*, Chapter 7, ed. M. Aliofkhazraei. Intech Publisher, Rijeka, Croatia.
17. Chen, D., Tang, L. H., Li, J. H. 2010. Graphene-based materials in electrochemistry. *Chem. Soc. Rev.* 39:3157–3180.
18. Min, S. K., Kim, W. Y., Cho, Y., Kim, K. S. 2011. Fast DNA sequencing with a graphene-based nanochannel device. *Nat. Nanotechnol.* 6:162–165.
19. Feng, L. Z, Liu, Z. 2011. Graphene in biomedicine: Opportunities and challenges. *Nanomedicine* 6(2):317–324.
20. Shen, H., Zhang, L., Liu, M., Zhang, Z. 2012. Biomedical applications of graphene. *Theranostics* 2(3):283–294.
21. Jiang, H. J. 2011. Chemical preparation of graphene-based nano-materials and their applications in chemical and biological sensors. *Small* 7(17):2413–2427.
22. Sun, X. M., Liu, Z., Welsher, K., Robinson, J. T., Goodwin, A., Zaric, S. 2008. Nanographene oxide for cellular imaging and drug delivery. *Nano Res.* 1(3):203–212.
23. Liu, Z., Robinson, J. T., Sun, X. M., Dai, H. 2008. PEGylated nanographene oxide for delivery of water-insoluble cancer drugs. *J. Am. Chem. Soc.* 130(33):10876–10877.
24. Depan, D., Shah, J., Misra, R. D. K. 2011. Controlled release of drug from folate-decorated and graphene mediated drug delivery system: Synthesis, loading efficiency, and drug release response. *Mater. Sci. Eng. C* 31(7):1305–1312.
25. Zhang, L. M., Xia, J. G., Zhao, Q. H., Liu, L., Zhang, Z. 2010. Functional graphene oxide as a nanocarrier for controlled loading and targeted delivery of mixed anticancer drugs. *Small* 6(4):537–544.
26. Nayak, T. R., Andersen, H., Makam, V. S., Khaw, C., Bae, S., Xu, X. 2011. Graphene for con-trolled and accelerated osteogenic differentiation of human mesenchymal stem cells. *ACS Nano* 5(6):4670–4678.
27. Kalbacova, M., Broz, A., Kong, J., Kalbac, M. 2010. Graphene substrates promote adherence of human osteoblasts and mesenchymal stromal cells. *Carbon* 48(15):4323–4329.
28. Lotya, M., Hernandez, Y., King, P. J., Smith, R. J., Nicolosi, V., Karlsson, L. S. 2009. Liquid phase production of graphene by exfoliation of graphite in surfactant/water solutions. *J. Am. Chem. Soc.* 131:3611–3620.
29. Lee, S., Lee, K., Zhong, Z. 2010.Wafer Scale homogeneous bilayer graphene films by chemical vapor deposition. *Nano Lett.* 10:4702–4707.
30. Vitchev, R., Malesevic, A., Petrov, R. H., Kemps, R., Mertens, M., Vanhulsel, A. 2010. Initial stages of few-layer graphene growth by microwave plasma-enhanced chemical vapour deposition. *Nanotechnology* 21:095602.
31. Forbeaux, I., Themlin, J. M., Debever, J. M. 1998. Heteroepitaxial graphite on 6H–SiC(0001): Interface formation through conduction band electronic structure. *Phys. Rev. B* 58:16396.
32. Akhavan, O., Ghaderi, E. 2009. Photocatalytic reduction of graphene oxide nanosheets on TiO_2 thin film for photoinactivation of bacteria in solar light irradiation. *J. Phys. Chem. C* 113:20214.

33. Zhu, Y., Stoller, M. D., Cai, W., Velamakanni, A., Piner, R. D., Chen, D. 2010. Exfoliation of graphite oxide in propylene carbonate and thermal reduction of the resulting graphene oxide platelets. *ACS Nano* 4:1227–1233.

34. Zhang, Y., Tan, Y. W., Stormer, H. L., Kim, P. 2005. Experimental observation of the quantum Hall effect and Berry's phase in graphene. *Nature* 438:201–204.

35. Yavari, F., Kritzinger, C., Gaire, C., Song, L., Gullapalli, H., Borca-Tasciuc, T. 2010. Tunable bandgap in graphene by the controlled adsorption of water molecules. *Small* 6(22):2535–2538.

36. Chung, C., Kim, Y.-K., Shin, D., Ryoo, S.-R., Hong, B.-H., Min, D.-H. 2013. Biomedical applications of graphene and graphene oxide. *Acc. Chem. Res.* 46(10):2211–2224.

37. Kuila, T., Bose, S., Mishra, A. K., Khanra, P., Kim, N. H., Lee, J. H. 2012. Chemical functionalization of graphene and its applications. *Prog. Mater. Sci.* 57(7):1061–1105.

38. Nguyen, L. H., Nguyen, T. D., Tran, V. H., Dang, T. T. H, Tran, D. L. 2014. Functionalization of reduced graphene oxide by electroactive polymer for biosensing applications. *Adv. Nat. Sci: Nanosci. Nanotechnol.*, 5:035005 (5pp).

39. Georgakilas, V., Otyepka, M., Bourlinos, A. B., Chandra, V., Kim, N., Kemp, K. C. 2012. Functionalization of graphene: Covalent and non-covalent approaches, derivatives and applications. *Chem. Rev.* 112(11):6156–6214.

40. Park, S., Srivastava, D., Cho, K. 2003. Generalized chemical reactivity of curved surfaces: Carbon nanotubes. *Nano Lett.* 3(9):1273–1277.

41. Rodríguez-Pérez, L., Herranz, M. A., Martín, N. 2013. The chemistry of pristine graphene. *Chem. Commun.* 49:3721–3735.

42. Rao, C. N. R., Subrahmanyam, A. K. S., Govindaraj, A. 2009. Graphene: A new two-dimensional nanomaterial. *Angew. Chem. Int. Ed.* 48:7752–7777.

43. Allen, M. J., Tung, V. C., Kaner, R. B. 2010. Honeycomb carbon: A review of graphene. *Chem. Rev.* 110(1):132–145.

44. Sun, H., Wu, L., Wei, W., Qu, X. 2013. Recent advances in graphene quantum dots for sensing. *Mater. Today* 16(11):433–442.

45. He, B., Tang, Q., Wang, M., Ma, C., Yuan, S. 2014. Complexation of polyaniline and graphene for efficient counter electrodes in dye-sensitized solar cells: Enhanced charge transfer ability. *J. Power Sources* 256:8–13.

46. Niyogi, S., Bekyarova, E., Itkis, M. E., Zhang, H., Shepperd, K., Hicks, J. et al. 2010. Spectroscopy of Covalently Functionalized Graphene. *Nano Lett.* 10:4061–4066.

47. Kosynkin, D. V., Higginbotham, A. L., Sinitskii, A., Lomeda, J. R., Dimiev, A., Price, B. K. et al. 2009. Longitudinal unzipping of carbon nanotubes to form graphene nanoribbons. *Nature* 458:872–876.

48. Sinitskii, A., Dimiev, A., Corley, D. A., Fursina, A. A., Kosynkin, D. V., Tour, J. M. 2010. Kinetics of diazonium functionalization of chemically converted graphene nanoribbons. *ACS Nano* 4:1949–1954.

49. Hong, J., Niyogi, S., Bekyarova, E., Itkis, M. E., Ramesh, P., Amos N. et al. 2011. Effect of nitrophenyl functionalization on the magnetic properties of epitaxial graphene. *Small* 7(9):1175–1180.

50. Bekyarova, E., Itkis, M., Ramesh, P., Berger, C., Sprinkle, M., deHeer, W. A. et al. 2009. Chemical modification of epitaxial graphene: Spontaneous grafting of aryl groups. *J. Am. Chem. Soc.* 131(4):1336–1337.

51. Sun, Z., Kohama, S., Zhang, Z., Lomeda, J. R., Tour, J. M. 2010. Soluble graphene through edge-selective functionalization. *Nano Res.* 3:117–125.

52. Englert, J. M., Dotzer, C., Yang, G., Schmid, M., Papp, C., Gottfried, J. M. et al. 2011. Covalent bulk functionalization of graphene. *Nat. Chem.* 3:279–286.

53. Sharma, R., Baik, J. H., Perera, C. J., Strano, M. S. 2010. Anomalously large reactivity of single graphene layers and edges toward electron transfer chemistries. *Nano Lett.* 10:398–405.

54. Lim, H., Lee, J. S., Shin, H. J., Shin, H. S., Choi, H. C. 2010. Spatially resolved spontaneous reactivity of diazonium salt on edge and basal plane of graphene without surfactant and its doping effect. *Langmuir.* 26:12278–12284.

55. Fang, M., Wang, K., Lu, H., Yang, Y., Nutt, S. 2009. Covalent polymer functionalization of graphene nanosheets and mechanical properties of composites. *J. Mater. Chem.* 19:7098–7105.

56. Fang, M., Zhang, Z., Li, J. 2010. Constructing hierarchically structured interphases for strong and tough epoxy nanocomposites by amine-rich graphene surface. *J. Mater. Chem.* 20(43):9635–9643.

57. Johnson, J. A., Finn, M. G., Koberstein, J. T., Turro, N. J. 2008. Construction of linear polymers, dendrimers, networks, and other polymeric architectures by copper-catalyzed azide-alkyne cycloaddition "Click" chemistry. *Macromol. Rapid Commun.* 29:1052–1072.

58. Jin, Z., McNicholas, T. P., Shih, C. J., Wang, Q. H., Paulus, G. L. C., Hilmer, A. J. et al. 2011. Click chemistry on solution-dispersed graphene and monolayer CVD graphene. *Chem. Mater.* 23:3362–3370.

59. Wang, H. X., Zhou, K. G., Xie, Y. L., Zeng, J., Chai, N. N., Li, J. et al. 2011. Photoactive graphene sheets prepared by "click" chemistry. *Chem. Commun.* 47:5747–5749.

60. Castelaín, M., Martínez, G., Merino, P., Martín-Gago, J. A., Segura, J. L., Ellis, G. et al. 2012. Graphene functionalisation with a conjugated poly(fluorene) by click coupling:Striking electronic properties in solution. *Chem. Eur. J.* 18(16):4965–4973.

61. Liu, H., Ryu, S., Chen, Z., Steigerwald, M. L., Nuckolls, C., Brus, L. E. 2009. Photochemical reactivity of graphene. *J. Am. Chem. Soc.* 131:17099–17101.

62. Hamilton, C. E., Lomeda, J. R., Sun, Z., Tour, J. M., Barron, A R. 2009. High-yield organic dispersions of unfunctionalized graphene. *Nano Lett.* 9:3460–3462.

63. Baek, J. B., Lyons, C. B., Tan, L. S. 2004. Covalent modification of vapour-grown carbon nanofibers via direct Friedel-Crafts acylation in polyphosphoric acid. *J. Mater. Chem.*14:2052–2056.

64. Han, S. W., Oh, S. J., Tan, L. S., Baek, J. B. 2009. Grafting of 4-(2, 4, 6-Trimethylphenoxy) benzoyl onto single-walled carbon nanotubes in poly (phosphoric acid) via amide function. *Nanoscale Res. Lett.* 4:766–772.

65. Lim, D. H., Lyons, C. B., Tan, L. S., Baek, J. B. 2008. Regioselective chemical modification of fullerene by destructive electrophilic reaction in polyphosphoric acid/phosphorus pentoxide. *J. Phys. Chem. C.* 112:12188–12194.

66. Baek, J. B., Lyons, C. B., Tan, L. S. 2004. Grafting of vapor-grown carbon nanofibers via in-situ polycondensation of 3-phenoxybenzoic acid in poly (phosphoric acid). *Macromolecules* 37:8278–8285.

67. Chang, D. W., Jeon, I. Y., Choi, H. J., Baek, J. B., Bae, I. Y., Lee, S. Y. et al. 2010. Mild and nondestructive chemical modification of carbon nanotubes (CNTs): Direct Friedel-Crafts acylation reaction. Chapter 12. In *Nanotechnology and Nanomaterials, Physical and Chemical Properties of Carbon Nanotubes.* ed. S. Suzuki. Intech publisher, Rijeka, Croatia.

68. Choi, E. K., Jeon, I. Y., Bae, S. Y., Lee, H. J., Shin, H. S., Dai, L. 2010. High-yield exfoliation of three-dimensional graphite into two-dimensional graphene-like sheets. *Chem. Commun.* 46:6320–6322.

69. Maggini, M., Scorrano, G., Prato, M. 1993. Addition of azomethine ylides to C60: Synthesis, characterization, and functionalization of fullerene pyrrolidines. *J. Am. Chem. Soc.* 115:9798–9799.

70. Georgakilas, V., Kordatos, K., Prato, M., Guldi, D. M., Holzinger, M., Hirsch, A. 2002. Organic functionalization of carbon nanotubes. *J. Am. Chem. Soc.* 124(5):760–761.

71. Tasis, D., Tagmatarchis, N., Bianco, A., Prato, M. 2006. Chemistry of carbon nanotubes. *Chem. Rev.* 106:1105–1136.

72. Quintana, M., Spyrou, K., Grzelczak, M., Browne, W. R., Rudolf, P., Prato, M. et al. 2010. Functionalization of graphene via 1,3-dipolar cycloaddition. *ACS Nano* 4:3527–3533.

73. Singh, R., Pantarotto, D., Lacerda, L., Pastorin, G., Klumpp, C., Prato, M. 2006. Tissue biodistribution and blood clearance rates of intravenously administered carbon nanotube radiotracers. *Proc. Natl. Acad. Sci. USA* 103:3357–3362.

74. Kostarelos, K., Lacerda, L., Pastorin, G., Wu, W., Wieckowski, S., Luangsilivay, L. et al. 2007. Cellular uptake of functionalized carbon nanotubes is independent of functional group and cell type. *Nat. Nanotechnol.* 2:108–113.

75. Georgakilas, V., Bourlinos, A. B., Zboril, R., Steriotis, T. A., Dallas, P., Stubos, A. K. et al. 2010. Organic functionalisation of graphenes. *Chem. Commun.* 46(10):1766–1768.

76. Ragoussi, M. E., Malig, J., Katsukis, G., Butz, B., Spiecker, E., De la Torre, G. et al. 2012. Linking photo- and redoxactive phthalocyanines covalently to graphene. *Angew. Chem. Int. Ed.* 51(26):6421–6425.

77. Liu, L. H., Lerner, M. M., Yan, M. 2010. Derivitization of pristine graphene with well-defined chemical functionalities. *Nano Lett.* 10(9):3754–3756.

78. Strom, T. A., Dillon, E. P., Hamilton, C. E., Barron, A. R. 2010. Nitrene addition to exfoliated graphene: A one-step route to highly functionalized grapheme. *Chem. Commun.* 46:4097–4099.

79. Vadukumpully, S., Gupta, J., Zhang, Y., Xu, G. Q., Valiyaveettil, S. 2011. Functionalization of surfactant wrapped graphene nanosheets with alkylazides for enhanced dispersibility. *Nanoscale* 3:303–308.

80. He, H., Gao, G. 2010. General approach to individually dispersed, highly soluble, and conductive graphene nanosheets functionalized by nitrene chemistry. *Chem. Mater.* 22(17):5054–5064.

81. Choi, J., Kim, K., Kim, B., Lee, H., Kim, S. 2009. Covalent functionalization of epitaxial graphene by azidotrimethylsilane. *J. Phys. Chem.* 113(22):9433–9435.

82. Hirsch, A., Brettreich, M. 2005. *Fullerenes—Chemistry and Reactions*. Wiley-VCH, Weinheim, Germany.

83. Economopoulos, S. P., Rotas, G., Miyata, Y., Shinohara, H., Tagmatarchis, N. 2010. Exfoliation and chemical modification using microwave irradiation affording highly functionalized graphene. *ACS Nano* 4:7499–7507.

84. Zhong, X., Jin, J., Li, S., Niu, Z., Hu, W., Li, R. et al. 2010. Aryne cycloaddition: Highly efficient chemical modification of graphene. *Chem. Commun.* 46:7340–7342.

85. Sarkar, S., Bekyarova, E., Haddon, R. C. 2012. Chemistry at the Dirac point: Diels-Alder reactivity of graphene. *Acc. Chem. Res.* 45(4):673–682.

86. Sarkar, S., Bekyarova, E., Niyogi, S., Haddon, R. C. 2011. Diels-Alder chemistry of graphite and graphene: Graphene as diene and dienophile. *J. Am. Chem. Soc.* 133(10):3324–3327.

87. Savchenko, A. 2009. Transforming graphene. *Science* 323:589–590.

88. Sofo, J. O., Chaudhari, A. S., Barber, G. D. 2007. Graphane: A two-dimensional hydrocarbon. *Phys. Rev.* 75:153401.

89. Elias, D. C. 2009. Control of graphene's properties by reversible hydrogenation: Evidence for graphane. *Science* 323:610–613.
90. Luo, Z., Yu, T., Kim, K.-J., Ni, Z., You, Y. S., Shen, Z. et al. 2009. Thickness-dependent reversible hydrogenation of graphene layers. *ACS Nano* 3:1781–1788.
91. Zheng, L., Li, Z., Bourdo, S., Watanabe, F., Ryersonb, C. C., Biris, A. S. 2011. Catalytic hydrogenation of graphene films. *Chem. Commun.* 47:1213–1215.
92. Jayasingha, R., Sherehiy, A., Wu, S. Y., Sumanasekera, G. U. 2013. In situ study of hydrogenation of graphene and new phases of localization between metal–insulator transitions. *Nano Lett.* 13(11):5098–5105.
93. Subrahmanyam, K. S., Kumar, P., Maitra, U., Govindaraj, A., Hembram, K. P. S. S., Waghmare, U. V., Rao, C. N. R. 2011. Chemical storage of hydrogen in few-layer graphene. *Proc. Natl. Acad. Sci. USA* 108:2674–2677.
94. Schäfer, R. A., Englert, J. M., Wehrfritz, P., Bauer, W., Hauke, F., Seyller, T., Hirsch A. 2013. On the way to graphane-pronounced fluorescence of polyhydrogenated graphene. *Angew. Chem. Int. Ed.* 52:754–757.
95. Zhang, L., Zeng, X., Wang, X. 2013. Programmable hydrogenation of graphene for novel nanocages. *Sci. Rep.* 3:3162.
96. Cheng, H. S., Sha, X. W., Chen, L., Cooper, A. C., Foo, M. L., Lau, G. C. et al. 2009. An enhanced hydrogen adsorption enthalpy for fluoride intercalated graphite compounds. *J. Am. Chem. Soc.* 131:17732.
97. Withers, F., Dubois, M., Savchenko, A. K. 2010. Electron properties of fluorinated single-layer graphene transistors. *Phys. Rev. B* 82:073403.
98. Nair, R. R., Ren, W. C., Jalil, R., Riaz, I., Kravets, V. G., Britnell, L. et al. Fluorographene: A two-dimensional counterpart of teflon. *Small* 6(24):2877–2884.
99. Cheng, S. H., Zou, K., Okino, F., Gutierrez, H. R., Gupta, A., Shen, N. et al. 2010. Reversible fluorination of graphene: Evidence of a two-dimensional wide bandgap semiconductor. *Phys. Rev. B* 81:205435.
100. Li, B., Zhou, L., Wu, D., Peng, H., Yan, K., Zhou, Y. et al. 2011. Photochemical chlorination of graphene. *ACS Nano* 5:5957–5961.
101. Gopalakrishnan, K., Subrahmanyam, K. S., Kumar, P., Govindaraj, A., Rao, C. N. R. 2012. Reversible chemical storage of halogens in few-layer graphene. *RSC Adv.* 2:1605–1608.
102. Brodie, B. C. 1859. On the atomic weight of graphite. *Philos. Trans. R. Soc. Lond.* 14:249–259.
103. Staudenmaier, L. 1898. Verfahren zur darstellung der graphitsäure. *Ber. Dtsch. Chem. Ges.* 31:1481–1487.
104. Hummers, W. S., Offeman, R. E. 1958. Preparation of graphitic oxide. *J. Am. Chem. Soc.* 80:1339.
105. Dreyer, D. R., Park, S., Bielawski, C. W., Ruoff, R. S. 2010. The chemistry of graphene oxide. *Chem. Soc. Rev.* 39:228–240.
106. He, H., Klinowski, J., Forster, M. A. 1998. New structural model for graphite oxide. *Chem. Phys. Lett.* 287:53–56.
107. Hontoria-Lucas, C., López-Peinado, A., López-Gonzàlez, J., Rojas-Cervantes, M., Martìn-Aranda, R. 1995. Study of oxygen- containing groups in a series of graphite oxides: Physical and chemical characterization. *Carbon* 33:1585–1592.
108. Mkhoyan, K., Contryman, A., Silcox, J., Stewart, D., Eda, G., Mattevi, C. 2009. Atomic and electronic structure of graphene-oxide. *Nano Lett.* 9:1058–1063.
109. Stankovich, S., Dikin, D. A., Piner, R. D., Kohlhaas, K. A., Kleinhammes, A., Jia, Y. et al. 2007. Synthesis of graphene-based nanosheets via chemical reduction of exfoliated graphite oxide. *Carbon* 45:1558–1565.

110. Gomez-Navarro. C., Meyers, J. C., Sundaram, R. S., Chuvilin, A., Kurash, S., Burghard, M. 2010. Atomic structure of reduced graphene oxide. *Nano Lett.* 10:1144–1148.

111. Shin, H. J., Kim, K. K., Benayad, A., Yoon, S. M., Park, H. K., Jung, I. S. et al. 2009. Efficient reduction of graphite oxide by sodium borohydride and its effect on electrical conductance. *Adv. Funct. Mater.* 19(12):1987–1992.

112. Bo, Z., Shuai, X., Mao, S., Yang, H., Qian, J., Chen, J. et al. 2014. Green preparation of reduced graphene oxide for sensing and energy storage applications. *Sci. Rep.* 4:4684.

113. Marcano, D. C. Kosynkin, D. V., Berlin, J. M., Sinitskii, A., Sun, Z., Slesarev, A. 2010. Improved synthesis of graphene oxide. *ACS Nano* 4(8):4806–4814.

114. Higginbotham, A., Kosynkin, D., Sinitskii, A., Sun, Z., Tour, J. M. 2010. Lower-defect graphene oxide nanoribbons from multiwalled carbon nanotubes. *ACS Nano* 4:2059–2069.

115. Chen, J., Yao, B., Li, C., Shi, G. 2013. An improved Hummers method for eco-friendly synthesis of graphene oxide. *Carbon* 64:225–229.

116. Huang, N. M., Lim, H. N., Chia, C. H., Yarmo, M. A., Muhamad, M. R. 2011. Simple room-temperature preparation of high-yield large-area graphene oxide. *Int. J. Nanomed.* 6:3443–3448.

117. Hou, S., Kasner, M., Su, S. 2010. Highly sensitive and selective dopamine biosensor fabricated with silianized graphene. *J. Phys. Chem.* C 114(35):14915–14921.

118. Melucci, M., Treossi, E., Ortolani, L. 2010. Facile covalent functionalization of graphene oxide using microwaves: Bottom-up development of functional graphitic materials. *J. Mater. Chem.* 20(41):9052–9060.

119. Xu, Y., Liu, Z., Zhang, X. 2009. A graphene hybrid material covalently functionalized with porphyrin: Synthesis and optical limiting property. *Adv. Mater.* 21(12):1275–1279.

120. Liu, Y., Zhou, J., Zhang, X. 2009. Synthesis, characterization and optical limiting property of covalently oligothiophene-functionalized graphene material. *Carbon* 47(13):3113–3121.

121. Zhu, J., Li, Y., Chen, Y. 2011. Graphene oxide covalently functionalized with zinc phthalocyanine for broadband optical limiting. *Carbon* 49(6):1900–1905.

122. Zhang, B., Chen, Y., Zhuang, X. 2010. Poly(N-vinylcarbazole) chemically modified graphene oxide. *J. Polym. Sci. Part A Polym. Chem.* 48(12):2642–2649.

123. Bao, H., Pan, Y., Ping, Y. 2011. Chitosan-functionalized graphene oxide as a nano-carrier for drug and gene delivery. *Small* 7(11):1569–1578.

124. Tang, X., Li, W., Yu, Z. 2011. Enhanced thermal stability in graphene oxide covalently functionalized with 2-amino-4,6-didodecylamino-1,3,5-triazine. *Carbon* 49(4):1258–1265.

125. Yu, D., Yan, Y., Durstock, M. 2010. Soluble P3HT-grafted graphene for efficient bilayer-heterojunction photovoltaic devices. *ACS Nano* 4(10):5633–5640.

126. Pham, T., Kumar, N., Jeong, Y. 2010. Covalent functionalization of graphene oxide with polyglycerol and their use as templates for anchoring magnetic nanoparticles. *Synth. Methods* 160(17–18):2028–2036.

127. Shen, J., Li, N., Shi, M. 2010 Covalent synthesis of organophilic chemically functionalized graphene sheets. *J. Coll. Int. Sci* 348(2):377–383.

128. Pramoda, K., Hussain, H., Koh, H. 2010. Covalent bonded polymer graphene nano-composites. *J. Polym. Sci Part A Polym. Chem.* 48(19):4262–4267.

129. Tarakeshwar, P., Choi, H. S., Kim, K. S. 2000. Molecular clusters of pi-systems: Theoretical studies of structures, spectra, and origin of interaction energies. *Chem. Rev.* 100:4145–4185.

130. Riley, K. E., Pitonak, M., Jurecka, P., Hobza, P. 2010. Stabilization and structure calculations for noncovalent interactions in extended molecular systems based on wave function and density functional theories. *Chem. Rev.* 110:5023–5063.

131. Hong, B. H., Small, J. P., Purewal, M. S., Mullokandov, A., Sfeir, M. Y., Wang, F. et al. 2005. Extracting subnanometer singles hells from ultra long multiwalled carbon nanotubes. *Proc. Natl. Acad. Sci. USA* 102:14155–14158.

132. Tarakeshwar, P., Choi, H. S., Kim, K. S. 2001. Olefinic vs. aromatic π–H interaction: A theoretical investigation of the nature of interaction of first-row hydrides with ethene and benzene. *J. Am. Chem. Soc.* 123:3323–3331.

133. Lee, E. C., Hong, B. H., Lee, J. Y., Kim, J. C., Kim, D., Kim, Y. et al. 2005. Substituent effects on the edge-to-face aromatic interactions. *J. Am. Chem. Soc.* 127:4530–4537.

134. Kwon, J. Y., Singh, N. J., Kim, H. A., Kim, S. K., Kim, K. S., Yoon, J. 2004. Fluorescent GTP-sensing in aqueous solution of physiological pH. *J. Am. Chem. Soc.* 126:8892–8893.

135. Grimme, S. 2004. On the importance of electron correlation effects for the π-π interactions in cyclophane. *Chem. Eur. J.* 10:3423–3429.

136. Lee, E. C., Kim, D., Jurecka, P., Tarakeshwar, P., Hobza, P., Kim, K. S. 2007. Understanding of assembly phenomena by aromatic-aromatic interactions: Benzene dimer and the substituted systems. *J. Phys. Chem. A* 111:3446–3457.

137. Quinonero, D., Garau, C., Rotger, C., Frontera, A., Ballester, P., Costa, A. et al. 2002. M. Anion–π interactions: Do they exist? *Angew. Chem. Int. Ed.* 41:3389–3392.

138. Kim, D., Tarakeshwar, P., Kim, K. S. 2004. Theoretical investigations of anion–π interactions: The role of anions and the nature of π system. *J. Phys. Chem. A* 108:1250–1258.

139. Geronimo, I., Singh, N. J., Kim, K. S. 2011. Can electron-rich π systems bind anions? *J. Chem. Theory Comput.* 7:825–829.

140. Yi, H. B., Lee, H. M., Kim, K. S. J. 2009. Interaction of benzene with transition metal cations: Theoretical study of structures, energies, and IR spectra. *Chem. Theory Comput.* 5:1709–1717.

141. Yi, H. B., Diefenbach, M., Choi, Y. C., Lee, E. C., Lee, H. M., Hong, B. H. et al. 2006. Interactions of neutral and cationic transition metals with the redox system of hydroquinone and quinone: Theoretical characterization of the binding topologies, and implications for the formation of nanomaterials. *Chem. Eur. J.* 12:4885–4892.

142. Ihm, H., Yun, S., Kim, H. G., Kim, J. K., Kim, K. S. 2002. Tripodal nitro-imidazolium receptor for anion binding driven by (C–H)$^+$---X$^-$ hydrogen bonds. *Org. Lett.* 4:2897–2900.

143. Chellappan, K., Singh, N. J., Hwang, I. C., Lee, J. W., Kim, K. S. 2005. A Calix[4] imidazolium[2]pyridine as an anion receptor. *Angew. Chem. Int. Ed.* 44:2899–2903.

144. Jaegfeldt, H., Kuwana, T., Johansson, G. 1983. Electrochemical stability of catechols with a pyrene side chain strongly adsorbed on graphite electrodes for catalytic oxidation of dihydronicotinamide adenine dinucleotide. *J. Am. Chem. Soc.* 105:1805–1814.

145. Xu, Y., Bai, H., Lu, G., Li, C., Shi, G. Q. 2008. Flexible graphene films via the filtration of water-soluble noncovalent functionalized graphene sheets. *J. Am. Chem. Soc.* 130:5856–5857.

146. Wang, Y., Chen, X., Zhong, Y., Zhu, F., Loh, K. P. 2009. Large area, continuous, few-layered graphene as anodes in organic photovoltaic devices. *Appl. Phys. Lett.* 95:063302.

147. An, X., Butler, T. W., Washington, M., Nayak, S. K., Kar, S. 2011. Optical and sensing properties of 1-pyrenecarboxylic acid-functionalized graphene films laminated on polydimethylsiloxane membranes. *ACS Nano* 5:1003–1011.

148. Kodali, V. K., Scrimgeour, J., Kim, S., Hankinson, J. H., Carroll, K. M., de Heer, W. et al. 2011. Nonperturbative chemical modification of graphene for protein micropatterning. *Langmuir* 27:863–865.

149. Mann, J. A., Rodríguez-López J., Abruña, H. D., Dichtel, W. R. 2011. Multivalent binding motifs for the noncovalent functionalization of graphene. *J. Am. Chem. Soc.* 133(44):17614–17617.

150. Rodrìguez-Lòpez, J., Ritzert, N. L., Mann, J. A., Tan, C., Dichtel, W. R., Abruña, D. 2012. Quantification of the surface diffusion of tripodal binding motifs on graphene using scanning electrochemical microscopy. *J. Am. Chem. Soc.* 34:6224–6236.

151. Su, Q., Pang, S. P., Alijani, V., Li, C., Feng, X. L., Mullen, K. 2009. Composites of graphene with large aromatic molecules. *Adv. Mater.* 21:3191–3195.

152. Wang, Q. H., Hersam, M. C. 2009. Room-temperature molecular resolution characterization of self-assembled organic monolayers on epitaxial graphene. *Nat. Chem.* 1:206–211.

153. Cheng, H. C., Shiue, R. J., Tsai, C. C., Wang, W. H., Chen, Y. T. 2011. High-quality graphene p-n junctions via resist-free fabrication and solution-based noncovalent functionalization. *ACS Nano* 5:2051–2059.

154. Wang, X., Tabakman, S. M., Dai, H. 2008. Atomic layer deposition of metal oxides on pristine and functionalized graphene. *J. Am. Chem. Soc.* 130:8152–8153.

155. Kozhemyakina, N. V., Englert, J. M., Yang, G., Spiecker, E., Schmidt, C. D., Hauke, F. et al. 2010. A non-covalent chemistry of graphene: Electronic communication with dendronized perylene bisimides. *Adv. Mater.* 22:5483–5487.

156. Tu, W., Lei, J., Zhang, S., Ju, H. 2010. Characterization, direct electrochemistry, and amperometric biosensing of graphene by noncovalent functionalization with picket-fence porphyrin. *Chem. Eur. J.* 16:10771–10777.

157. Geng, J., Jung, H. T. 2010. Porphyrin functionalized graphene sheets in aqueous suspensions: From the preparation of graphene sheets to highly conductive graphene films. *J. Phys. Chem. C* 114:8227–8234.

158. Zhang, S., Tang, S., Lei, J., Dong, H., Ju, H. J. 2011. Functionalization of graphene nanoribbons with porphyrin for electrocatalysis and amperometric biosensing. *Electroanal. Chem.* 656:285–288.

159. Bai, H., Xu, Y., Zhao, L., Li, C., Shi, G. 2009. Non-covalent functionalization of graphene sheets by sulfonated polyaniline. *Chem. Commun.* 1667–1669.

160. Wu, H., Zhao, W., Hu, H., Chen, G. 2011. One-step in situ ball milling synthesis of polymer-functionalized graphene nanocomposites *J. Mater. Chem.* 21:8626–8632.

161. Choi, B. G., Park, H., Park, T. J., Yang, M. H., Kim, J. S., Jang, S. Y. et al. 2010. Solution chemistry of self-assembled graphene nanohybrids for high-performance flexible biosensors. *ACS Nano* 4:2910–2918.

162. Saxena, A. P., Deepa, M., Joshi, A. G., Bhandari, S., Srivastava, A. K. 2011. Poly(3,4-ethylenedioxythiophene)-ionic liquid functionalized graphene/reduced graphene oxide nanostructures: Improved conduction and electrochromism. *ACS Appl. Mater. Interf.* 3:1115–1126.

163. Ghosh, A., Rao, K. V., Voggu, R., George, S. 2010. Non-covalent functionalization, solubilization of graphene and single-walled carbon nanotubes with aromatic donor and acceptor molecules. *J. Chem. Phys. Lett.* 488:198201.

164. Choi, E. Y., Han, T. H., Hong, J. H., Kim, J. E., Lee, S. H., Kim, H. W. 2010. Noncovalent functionalization of graphene with end-functional polymers. *J. Mater. Chem.* 20:1907–1912.

165. Kim, T. Y., Lee, H. W., Kim, J. E., Suh, K. S. 2010. Synthesis of phase transferable graphene sheets using ionic liquid polymers. *ACS Nano* 4:1612–1618.

166. Qi, X., Pu, K. Y., Li, H., Zhou, X., Wu, S., Fan, Q. L. et al. 2010. Amphiphilic graphene composites. *Angew. Chem. Int. Ed.* 49:9426–9429.

CHAPTER 22

CONTENTS

Functionalization of Carbon Nanotubes and Graphene with Amines and Biopolymers Containing Amino Groups

22

F. Navarro-Pardo, A.L. Martínez-Hernández, and C. Velasco-Santos

22.1 INTRODUCTION

Carbon nanotubes (CNTs) and graphene are the most widely studied carbon-based nanomaterials due to their outstanding structural features and excellent chemical, electrical, and mechanical properties.[1–3] CNTs are rolled-up cylinders of graphene sheets with small diameter and large length/diameter ratio; they are considered as 1D nanostructures and can be classified according to the number of graphene cylinders they possess into single-walled CNTs (SWCNTs) and multiwalled CNTs (MWCNTs).[3] In addition, CNTs can have a variety of individual structures, morphologies, collective arrangements, and properties, all of which are dependent on their synthesis technique.[4,5] Graphene is a 2D one-atom-thick planar sheet of sp^2 bonded carbon atoms.[3] This carbon allotrope can be found as single-, double-, and few-layer graphene; the maximum number of graphene layers is 10, because the electronic structure with more sheets approaches the 3D limit of graphite.[6]

Pristine forms of these nanomaterials are insoluble in water and in most organic solvents; therefore, in order to exploit their exceptional properties, a variety of strategies have been developed to obtain stable and uniform dispersions of them.[7–9] Chemical modification represents an attractive alternative for the attachment of organic moieties on the graphitic surface of these nanomaterials often using the rich chemistry of hydroxyl, carboxyl, and epoxy groups, of the oxidated carbon nanomaterials.[10–13] A wide variety of aliphatic and aromatic amines, amino acids, amine-terminated biomolecules, ionic liquids (ILs), oligomers, and silane compounds have been successfully used in the preparation of functionalized carbon nanomaterials.[14–17]

Primo et al.[18] have done a critical discussion of the different strategies for fabricating electrochemical biosensors from MWCNTs in polymer matrices, where affinity properties play an important role in the recognition of biomolecules. Tiwari et al.[19] have provided a review of several polymers employed for the fabrication of biosensors; among them, a variety of polymers containing amino groups have been used due to the affinity with biomolecules. Choi et al.[1] have also summarized a variety of approaches regarding the application of graphene in biosensors. There is a variety of reviews showing outstanding advances in carbon nanomaterial–based electrochemical sensors and biosensors.[20–22] Pavlidis and coworkers[23] have examined a variety of methods to couple enzymes on graphene sheets for their use as nanobiocatalytic systems. Battigelli and coworkers[24] have provided an overview on advances in chemistry for bioapplications of CNTs focusing on the functionalization of them with bioactive molecules and they have also summarized the strategies for the insertion of molecules in the inner core of CNTs for drug delivery and imaging applications.

Modification of graphene-based materials with amines or molecules containing amino groups has been commonly used for the fabrication of luminescent graphene quantum dots due to the decrease of surface defects for fluorescence enhancement.[15] These carbon nanomaterials are versatile and effective for drug delivery carriers for many small molecule drugs; the procedure for attaching these drugs often involves amino modification of the graphitic surface.[25–27] Graphene-functionalized composites with biocompatible and biodegradable polymers have been developed by many research groups; their results have shown to have a dual function: directing stem cell differentiation and improving physical properties of scaffolds.[28]

Moreover, the use of carbon nanomaterials as reinforcing fillers represents another widely studied application. Controlled dispersion and stabilization of carbon nanomaterials in a solvent or a polymer matrix remains a challenge due to the strong van der Waals' binding energies associated with the nanomaterial agglomerates.[8] Chemical functionalization is the most promising approach for enhancing dispersibility and processability of carbon nanomaterials.[29] Covalent modification of carbon nanomaterials enables efficient tailoring of polymer/carbon nanomaterial interface through reactive coupling between the functional moieties on the nanofiller surface with the available functional groups of the polymer.[7] Taking advantage of this strategy, amine-modified carbon nanomaterials have led to the improvement on structural properties of nanocomposites, for example, epoxies, polyamides, and biopolymers have resulted in increased thermal stability and mechanical properties when incorporating these nanofillers.[2,7,10–12]

Amino groups have also been employed to obtain adsorbent materials based on graphene or nanotubes resulting in high removal efficiency of contaminants from aqueous solutions.[30–32] This has been possible because of the ability to control the electric interactions between the contaminants and the adsorbents.[32]

In this chapter, we examine the recent advances regarding the different strategies for chemical modification of CNTs and graphene with a variety of amines and biopolymers, where special attention is paid to functionalization with biopolymers such as chitosan and keratin. As a start, this chapter describes generally the different covalent modification methods for functionalization with a variety of molecules containing amino moieties. The following sections include the strategies employed for CNTs and graphene according to the type of functionalizing molecules. To conclude, the improvements in the dispersion and solubility of the nanomaterials are highlighted.

22.2 METHODS OF CHEMICAL AMINO MODIFICATION

Covalent modification of carbon nanomaterials can be achieved by different reaction mechanisms, including nucleophilic substitution,[33–40] electrophilic addition,[15,41–44] condensation,[45,46] and cycloaddition.[47–53] Kuila et al.[14] have provided a review detailing the mechanisms of a variety of covalent functionalization approaches of GO. Singh et al.[54] have done a survey about the commonly used methods for amino-functionalized CNTs by either using acid chlorides as intermediates or carbodiimide-activated coupling. Liu et al.[44] have presented a variety of methods for surface modification of graphene where agents with amino functionalities are often employed. Servant and coworkers[55] have summarized different strategies for the conjugation of graphene with therapeutically active molecules.

Oxidation of carbon nanomaterials is highly effective to increase the surface area and to attach organic functional groups on carbon surface. There is a wide variety of oxidation methods for CNTs or graphene, which have been outlined by different authors.[37,56] Covalent functionalization of carbon nanomaterials is usually preceded by an oxidation process.[57] Hydroxyl, carboxyl, and epoxy groups on the nanomaterial surface serve as precursors for the anchoring of organic molecules.[13,57,58] Alternatively, carboxylic groups can be converted to acyl chlorides, formed by reaction with oxalyl chloride ($COCl_2$) or thionyl chloride ($SOCl_2$).[24]

Amino moieties play an important role because of the features they confer, including biocompatibility, cell affinity, and peptide synthesis, among others.[59,60] These functional groups are very useful as anchoring sites for binding other materials by means of a variety of bonds.[61] Carboxylic acids or carboxylic chlorides can be further converted into amide groups by reacting them with molecules containing amine moieties such as aliphatic amines, aryl amines, amino acid derivatives, peptides, amino-group-substituted dendrimers, or any other nucleophiles.[23,27,55,62–65]

Therefore, most of these strategies involve multistep methods to modify carbon nanomaterials with amine groups.[66–68] Preactivation of the carboxylic acids is necessary to ensure covalent bonding; carbodiimide coupling enhances the reaction efficiency of covalent links between carboxylic acids and amines.[54,69,70] These coupling agents are known for playing a key role in condensation reactions.[41] Our research group has used this strategy to amine modification of CNTs and graphene using 1-ethyl-3-(dimethylaminopropyl)carbodiimide hydrochloride (EDAC) and dimethylamine.[10,12] Figure 22.1 shows the pristine nanotubes characterized by a larger diameter when compared to the functionalized ones; after functionalization, the nanotubes exhibit open ends due to the oxidation prior to amination and also rougher surface on the walls due to the defects imposed by the amine treatment. Furthermore, the characteristic peaks of amines are found in the FTIR spectrum of these nanomaterials. The scheme of reaction of graphene and also the FTIR spectrum of this 2D carbon nanomaterial are displayed in Figure 22.2, showing similarities in the peaks found in the 1D carbon nanomaterials. These amino-functionalized carbon nanomaterials have been incorporated into a polymer matrix compatible with these groups such as polyamide 6,6. The behavior obtained in the mechanical response of nanocomposites obtained using electrospinning and injection molding is in agreement, where the 1D nanomaterials provide the highest reinforcement at the lowest content, whereas the 2D nanomaterials enhanced the storage modulus as the content of these nanofillers was increased.[10,12] Ma et al.[70] have employed dicyclohexylcarbodiimide (DCC) and dimethylaminopyridine. DCC has also been used for amine functionalization by several groups.[71–73]

As described earlier, modification with amines has been widely applied in works where biomolecules are involved; the surface chemistry of nanomaterials has a great effect on the compatibility of these biomaterials. Pavlidis et al.[74] found that amino functionalization affects to a great extent the catalytic behavior of lipases. Enzymes have also been attached by using the acylation approach to amino functionalization and increase the compatibility of the graphitic surface with these biomolecules, which can be employed to fabricate devices for drug delivery and biosensor applications.[23,75] Amino acids such as L-lysine, phenylalanine, and tyrosine have also been employed due to their

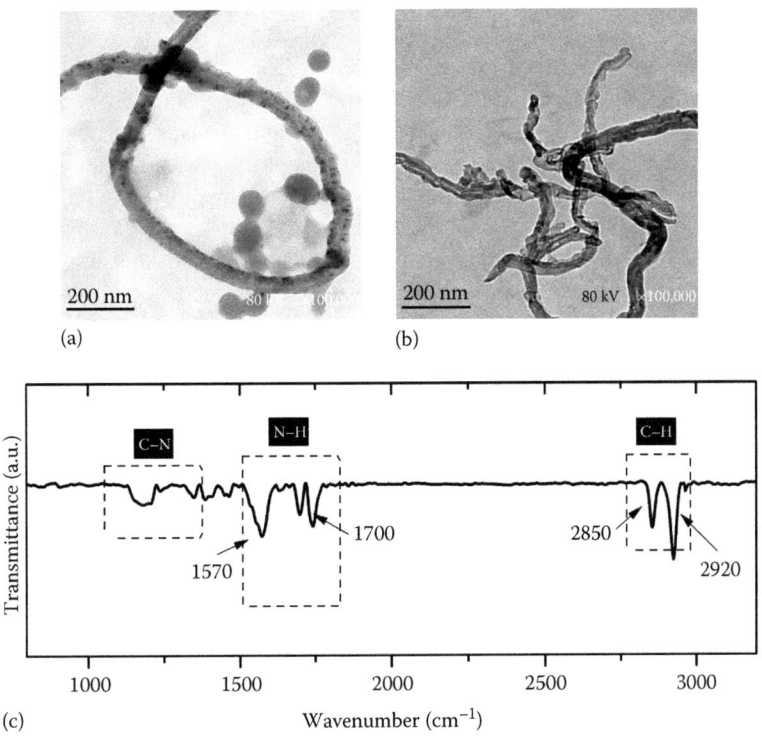

FIGURE 22.1
Transmission electron micrographs of (a) pristine MWCNTs and (b) amino-functionalized MWCNTs; inset (c) shows the FTIR spectrum of the aminated MWCNTs.

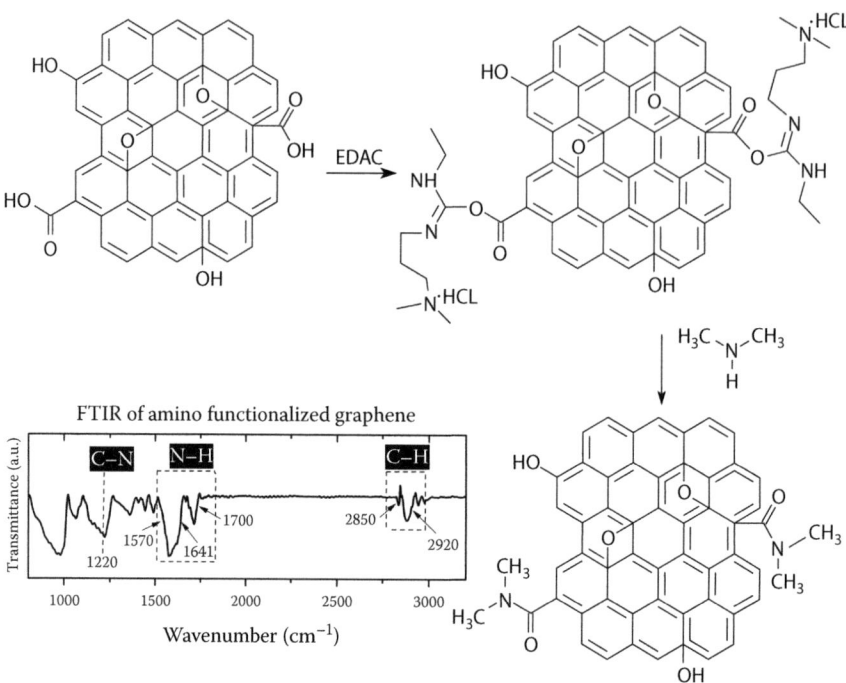

FIGURE 22.2
Scheme of reaction of amino functionalization of graphene sheets and FTIR spectrum of the resulting nanomaterial.

biocompatibility, plentiful active functional groups, relatively good solubility in water, and chirality.[30,40] Graphitic surface of the nanomaterials studied in this chapter has shown unique interactions with DNA and RNA, which make them attractive in sensing and delivery of both molecules.[28] These biomolecules won't be further discussed in this chapter; the reader is encouraged to find more information in the variety of reviews regarding this topic.[21,28,76–79]

Other molecules, such as polymers or oligomers, can also be covalently bonded to carbon nanomaterials by attaching these macromolecules by the *grafting to* or *grafting from* approaches, depending on the building of polymer chains.[13] The former involves the bonding of preformed polymer chains reacting with the surface of the carbon nanoform, and the latter relies on the immobilization of the initiators on the graphitic surface followed by the polymer chain growth from the carbon surface via *in situ* polymerization.[58,80]

Polymers are also useful for the fabrication of biosensors.[19] Chitosan is a natural glucosamine and a random copolymer made of β-(1→4)-*N*-acetyl-D-glucosamine and α-(1→4)-D-glucosamine.[81] The carboxyl and amine side groups on its structure allow the direct binding of enzymes for immobilization or the covalent attachment of functional groups for the modification of polymer properties such as hydrophobicity that can achieve a hydrophilic character by the protonation of its amine side groups at pH values below approximately 5.0.[82] Additionally, nanometric carbon/chitosan nanocomposites have special pH-responsive association behavior.[26] Chitosan has been used as a biocompatible coating for a range of nanoparticle (NP) core materials due to its amine functionality for subsequent attachment of biofunctional ligands.[69,83] Keratin is the term referred to the broad category of insoluble proteins that associate as intermediate filaments and form the bulk of hair, wool, horns, hooves, and nails.[84] Keratin-based materials have shown promise in diverse applications due to their intrinsic biocompatibility, biodegradability, mechanical durability, and natural abundance.[84,85] This fibrous protein contains a large number of disulfide bonds and hydrophobic amino acids with balancing hydrophilic amino acids that confer chemical stability.[86] Carbon nanomaterials grafted with poly(amidoamine) (PAMAM) have been proved to have antibacterial properties.[87] Polyethylenimine (PEI) has been used as a chemical scissor to unzip GO sheets and as a stabilizing agent to prevent graphene quantum dots from layer stacking and lateral aggregation.[88] Functionalization of graphene sheets with this polymer has been also used for the self-assembly with negatively charged nanomaterials (oxidized CNTs).[89] Electrochemical behavior of the polyaniline (PANI) backbone has been exploited for fabricating chemical sensors when grafted to carbon nanomaterials.[78,90] Amine functionalization is commonly used when processing epoxy nanocomposites.[65,91] Furthermore, amines have been used for curing of epoxies because of their autocatalytic character owned by hydroxyl groups formed during the reaction.[91]

Other alternative approaches involving different species containing amino moieties have been developed. Diverse studies have employed ammonia (NH_3) as a precursor to introduce amine functionality onto carbon nanomaterials, enhancing their hydrophilicity and biocompatibility.[59,38,92] NH_3 or a mixture of N_2 and H_2 is commonly used for plasma treatments; the latter components are safer and nontoxic.[93,94] A range of nitrogen groups are introduced to the surface of carbon nanomaterials when using NH_3 plasma treatment including imine, nitrile, and amide, among other groups; however, this method favors the formation of primary amines.[94,95] Allylamine or heptylamine has been also used when employing this technique; the former is more toxic and unstable.[94] 3-Aminopropyl trimethoxysilane (APTES) has been used for treating SiO_2 NPs and forming hybrids with carbon nanostructures.[95,96] APTES has also been employed to modify graphitic surfaces for generating free amino groups for bacteria or enzyme attachment, which can be employed as biosensors.[97-102] This approach is of considerable interest because of the mass production ease to obtain reagentless amperometric biosensors.[98] This molecule is also known for forming stable chelates with metal ions, making it suitable for removal of species containing them.[30] APTES has also been used for improving interfacial bonding in nanocomposites.[91,103,104]

22.3 CARBON NANOTUBES

CNTs are usually purified by acids in order to separate the nanotubes from amorphous carbon and metallic catalyst impurities; this procedure yields CNTs containing –COOH and –OH groups bonded to the end cap and/or sidewall defect sites.[61] Oxidized nanotubes will be denoted as OSWCNTs or OMWCNTs. Figure 22.3 shows two different approaches employed by Gabriel and coworkers for the functionalization of SWCNTs with different amines; they oxidized these nanotubes in hot air, and these nanotubes were also chemically oxidized followed by acylation and amination of the SWCNTs. Similar procedures based on these reactions have been followed by different research groups.[62,54,105-109] However, this approach requires long reaction times, usually taking several days.[75,108,110] Table 22.1 shows a variety of amines that have been employed for the functionalization of CNTs, and in the following sections, a diversity of treatments is examined according to the type of amines or molecules with amino groups employed.

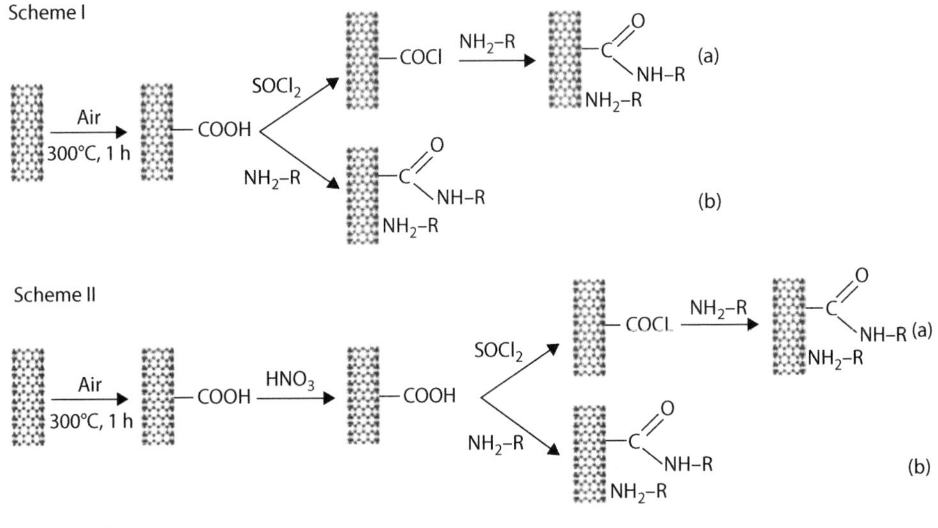

FIGURE 22.3
Schemes illustrating the different strategies used to functionalize SWCNTs. (Reprinted from *Carbon*, 44, Gabriel, G., Sauthier, G., Fraxedas, J. et al., Preparation and characterisation of single-walled carbon nanotubes functionalised with amines, 1891–1897, Copyright 2006, with permission from Elsevier.)

TABLE 22.1 Functionalizing Molecules for CNT Modification

Classification	Name	References
Aliphatic amines	Octadecylamine	[61]
	Ethylenediamine	[10,12,63,65,70,109,111,112,147]
	Hexamethylenediamine	[108,109,112]
	Hexamethylenetetramine	[106]
Single ring aromatic amines	1,4-Phenylenediamine	[112]
	1,3-Phenylenediamine	[115]
	2,4-Dinitroaniline	[75]
	4,4'-Oxydianiline	[104]
	2,6-Dinitroaniline	[75]
	Aniline	[67,90]
	N-Decyl-2,4,6-trinitroaniline	[75]
Polymers	Polyamidoamine	[116,134]
	Polyaniline	[81,131–133]
	Polyethyleneimine	[33,89]
	Poly(thiourea-amide)	[45]
	Amine modified polyethylene glycol	[43]
	Monoamine-terminated polyethylene oxide	[83,120]
Others	1-Aminoanthracene	[51]
	2-Aminoanthracene	[61]
	2,20-Ethylenedioxy-diethylamine	[105]
	4-Aminobenzoic acid	[131,132]
	4,4'-Diaminodiphenylmethane	[108]
	3-(Aminopropyl)triethoxysilane	[135–137]
	Aminomethyltriethoxysilane	[138]
	Aziridine	[49]
	Bis(p-aminocyclohexyl)methane diamine	[66]
	N-propylamide	[35]
	$N_2 + H_2$ plasma	[141]
	NH_3 plasma	[142]
	Perfluorinated amines	[61]

22.3.1 Aliphatic Amines

Syrgiannis et al.[35] attached *n*-propylamide (*in situ* generated) onto SWCNTs by a direct nucleophilic addition reaction. The negatively charged intermediate was reoxidized by air leading to SWCNTs with amino functionalities. Jimeno and coworkers[62] compared the functionalization degree of three amines with different molecular structure; they found amines with low molecular weights, for example, triethylenetetramine (TETA), achieving the highest functionalization degree. This research also showed that the steric effects have an influence on the degree of functionalization. A diazonium reaction method that has been commonly used for the attachment of para-substituted benzene rings to CNTs was used to functionalize SWCNTs with ethylenediamine (EDA).[111] Amino functionalization was achieved by using a microwave-assisted treatment with EDTA and hexamethylenediamine (HMTA) in presence of sodium nitrite to initiate a semistable diazonium ion, which then resulted in a radical reaction with CNTs.[112,113] According to Amiri and coworkers,[112] EDA treatment functionalized the nanotubes more effectively when using this method. Davarpanah et al.[42] also functionalized MWCNTs with three amines, monoethanolamine (MEA), diethanolamine (DEA), and triethanolamine (TEA), by a rapid microwave-assisted method. This group proposed the reaction mechanisms based on previous studies where amine functionality favored functionalization via amidation and esterification mechanisms, while alcohols are known to be effective via electrophilic addition reaction. This suggested that the amine and the alcoholic functionalities on the ethanolamines could lead to all the reactions proposed, as seen in Figure 22.4. Le and coworkers attached amine functionalities by reacting dodecylamine (DDA) in the presence of H_2SO_4, which was used as catalyst; they found that the content of DDA was 34.8%.[97]

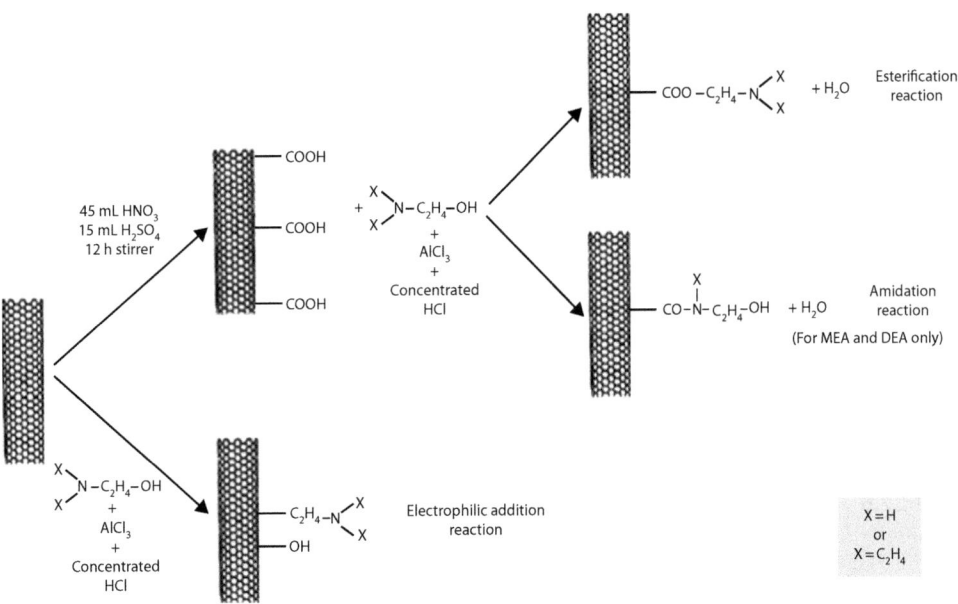

FIGURE 22.4
Anticipated mechanisms for functionalization of MWCNTs with MEA, DEA, and TEA. X stands for –H, or –C$_2$H$_4$OH, wherever applicable. It should be noted that AlCl$_3$ as a Lewis acid and also water-absorbent can catalyze all three reactions. (With kind permission from Springer Science+Business Media: *Appl. Phys. A: Mater. Sci. Process*, Microwave-induced high surface functionalization of multi-walled carbon nanotubes for long-term dispersion in water, 115, 2014, 167–175, Davarpanah, M., Maghrebi, M., and Hosseinipour, E., Copyright 2014.)

22.3.2 Aromatic Amines

Ellison et al.[64] developed a single-step technique to functionalize SWCNTs with 1,4-benzenediamine, resulting in SWCNTs containing phenyl NH$_2$ groups, which were able to undergo additional reactions, such as with the heterofunctional cross-linker succinimidyl-4-(*N*-maleimidomethyl) cyclohexane-1-carboxylate. Basiuk and coworkers[114] developed a solvent-free method based on thermal activation for functionalizing MWCNTs with three aliphatic diamines and an aromatic diamine; this method showed the presence of cross-linking molecules at the sites of close contact between nanotubes and a trend to self-organization into parallel structures. SWCNTs were acyl chloride functionalized with COCl$_2$ prior to aniline functionalization where polymerization of aniline was avoided.[67] OMWCNTs have been functionalized with phenylenediamine (PDA); the proposed procedure and mechanism is shown in Figure 22.5. However, this method required 4 days to complete the amino functionalization.[115] Sun and coworkers[51] found that semiconducting SWCNTs have higher reactivity with 1-aminoanthracene toward the Diels–Alder cycloaddition reaction than metallic SWCNTs.[51]

22.3.3 Polymers

22.3.3.1 Grafting to

Liao et al. employed two techniques to graft PEI to MWCNTs for obtaining a nanomaterial with polyvalent reactive amine groups. In the first route, the carboxyl groups of OMWCNTs were activated by using SOCl$_2$ and aminated with PEI; however, this method required several steps and also degraded the nanotubes. A second alternative method was considered that involved treatment of pristine MWCNTs with PEI in *N,N*-dimethylformamide (DMF); the functionalized nanotubes consisted of 6–8 wt.% PEI chains grafted. They also found that the amine groups of the grafted species could be converted into amides using an activated octadecanoic acid derivative.[33] Mao et al. attached poly(acrylic acid) (PAA) onto the surface of aminated MWCNTs, and then the carboxylic

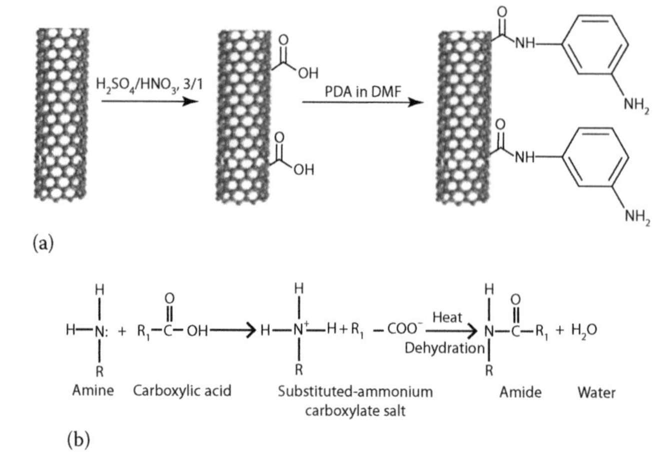

(a)

(b)

FIGURE 22.5
Procedure for carboxylic and amine functionalization of MWCNTs (a) and mechanism of amine functionalization of MWCNTs (b). (Reprinted from *Desalination*, 286, Rahimpour, A., Jahanshahi, M., Khalili, S. et al., Novel functionalized carbon nanotubes for improving the surface properties and performance of polyethersulfone (PES) membrane, 99–107, Copyright 2012, with permission from Elsevier.)

groups on PAA moiety were used to react with octadecylamine (ODA) through amidation, resulting in the generation of multiple pendant ODA groups in the side chains of PAA at the surface of the MWCNTs.[105] Amine functionalization of OCNTs has also been performed with dendrimers such as PAMAM.[116] Nanoscale ionic materials (NIMs) have been obtained via the neutralization or hydrogen bonding of hydroxyl or carbonyl groups on the surface of MWCNTs and amine groups of organic modifiers, as poly(ethylene) glycol (PEG)-substituted tertiary amines.[117,118]

Biopolymers such as keratin and chitosan have also been grafted to CNTs. Acyl chloride–functionalized MWCNTs were used to graft low-molecular-weight chitosan (LMC) by a nucleophilic substitution reaction; according to Ke et al.,[119] ~58 wt.%, corresponding to four molecular chains of the LMC, was attached to 1000 carbon atoms of the nanotube sidewalls. Acyl chlorination has also been used to functionalize OMWCNTs with polymers such as monoamine-terminated poly(ethylene oxide) and chitosan.[83,120] Covalently bounded chitosan/MWCNTs were obtained by grafting chitosan onto nanotubes in the presence of EDC and *N*-hydroxysuccinimide (NHS) as coupling agents; the reaction yielded an amide bond by activation of carboxylic groups on surfaces of MWCNTs, followed by aminolysis of the *o*-isoacylurea intermediates by the amino groups of chitosan.[121] Castillo et al.[122] solubilized SWCNTs in chitosan and prepared a folic acid (FA) chitosan–SWCNTs conjugate by a reaction of FA with chitosan–SWCNTs in the presence of EDC. Liu et al.[123] also prepared chitosan and FA conjugates by a covalently cross-linking reaction using EDC; then chitosan/FA/MWCNT hybrids were obtained by a liquid–gel transition of chitosan–FA conjugates upon the ionic interaction with polyanions (tripolyphosphate). Nitrogen plasma technique was applied for grafting chitosan onto MWCNTs; in this process, the active nitrogen species react with MWCNTs to form active carbon site. This active sites react with the functional groups of chitosan C–O–C and C–NH–C bonds between polymer and MWCNTs.[32] Magnetic chitosan/MWCNTs were obtained via a suspension cross-linking method by using Fe_3O_4 NPs.[124] Li et al.[125] reported an *in situ* synthesis method for obtaining chitosan NPs/MWCNTs by using an ionotropic gelation process that involves the mixture of two aqueous solutions at room temperature. Dai et al.[126] grafted chitosan using the scheme of reaction displayed in Figure 22.6, where as a prior step to this functionalization, chitosan was treated to prepare a Schiff base chitosan in order to increase its chemical stability in acid solutions and improve its metal-ion-adsorbing properties. Shariatzadeh et al.[127] employed a microwave-assisted method to synthesize MWCNTs grafted with chitosan and MgO NPs. MWCNTs have also been decorated with magnetite and have been incorporated into a chitosan matrix.[128] Storage modulus of the films decreased as the concentration of decorated MWCNT increased, indicating dependence of storage modulus on aggregate size; the storage modulus increased in the temperature range of 150–250°C, and this behavior was more noticeable at 1:3 ratios of MWCNT/magnetite NPs due to the stronger interfacial adhesion between NPs and the polymeric matrix.

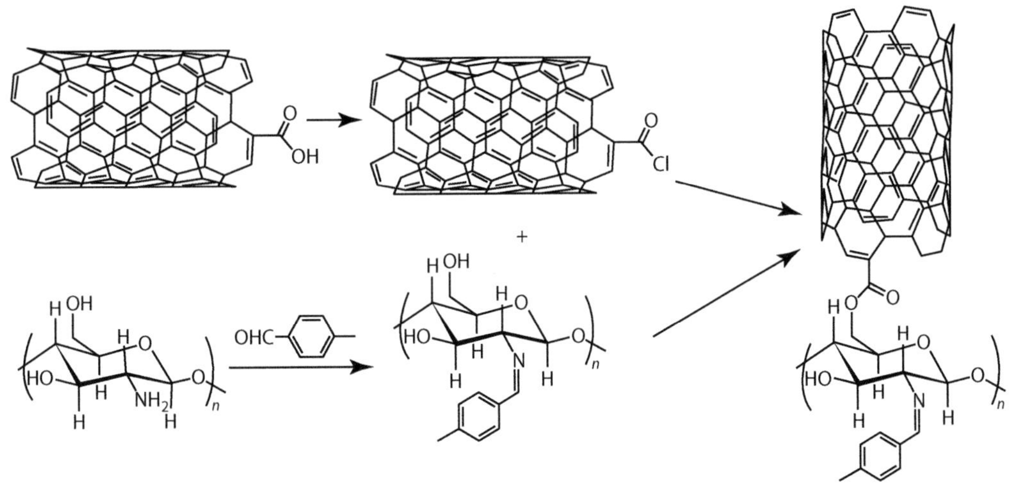

FIGURE 22.6
Synthesis and structure of S-CS-MWCNTs. (Reprinted from *J. Hazard. Mater.*, 219–220, Dai, B., Cao, M., Fang, G. et al., Schiff base-chitosan grafted multiwalled carbon nanotubes as a novel solid-phase extraction adsorbent for determination of heavy metal by ICP-MS. 103–110, Copyright 2012, with permission from Elsevier.)

Keratin is a fibril protein consisting of polypeptide chains formed by the condensation of different amino acids in which functional groups can be exploited for grafting carbon nanomaterials.[85] However, to date, reports concerning functionalization routes related to this biopolymer are limited. A patent has claimed CNTs can be used to voluminize hair; this patent also proposed to graft keratin onto CNTs in order to increase the affinity of these nanotubes with keratin in the hair.[129] Keratin was grafted to OMWCNTs by a redox reaction; in this route, malic acid/$KMnO_4$ generates macroradicals by the attack of the species to primary radicals of the peptide bonds in keratin; pendant functional groups such as –NH_2, –COOH, –SH, and –OH can form covalent bonds with functional groups –COOH and COO– present in nanotube walls.[130] Figure 22.7 shows the FTIR of the MWCNT grafted with dialyzed keratin at different conditions, at 75°C with OMWCNTs/keratin 1:1 wt./wt. (CK7511), 65°C with OMWCNTs/keratin 1:1 wt./wt. (CK6511), and 55°C with OMWCNTs/keratin 1:3 wt./wt. (CK5513).

22.3.3.2 Grafting from

Guo et al.[90] preconcentrated OSWCNTs into a solution of $HAuCl_4$ and 4-dodecylaniline to produce nanocomposites in an *in situ* one-pot fashion. Baek and coworkers have developed protocols to graft PANI onto MWCNTs by functionalizing the nanotubes with 4-ABA in polyphosphoric acid/phosphorous pentoxide as a *direct* Friedel–Crafts acylation reaction medium, followed by their reaction with aniline using ammonium persulfate/aqueous hydrochloric acid to promote a chemical oxidative polymerization[131] or by the polymerization of aniline via an *in situ* static interfacial polymerization in H_2O/CH_2Cl_2 in the presence of the aforementioned functionalized nanotubes.[132] Figure 22.8 shows that as static interfacial polymerization was progressed, the color of top aqueous layer turned darker and that of bottom organic layer containing 4-aminobenzoyl-functionalized MWCNTs (AF-MWCNTs) turned lighter. The amount of AF-MWCNTs incorporated into the reaction was estimated by residual amount of AF-MWNT on the bottom of vial (white arrow), which was 45%; hence, AF-MWCNTs incorporated into the reaction and migrated to top aqueous layer was 55%. OMWCNTs have also been coated with PANI via polycondensation of aniline.[133] Hyperbranched PAMAM has been grafted from OMWCNTs via a multistep Michael addition reaction. The route employed consisted of obtaining amino-functionalized nanotubes by reacting OMWCNTs in excess 1-(2-aminoethyl)piperazine (AEPZ) using EDC; then the amino-modified MWCNTs reacted with excess *N,N'*-methylenebisacrylamide to obtain vinyl-functionalized MWCNTs and subsequently treated with excess AEPZ. The last two reactions were repeated for two cycles to finally terminate the vinyl terminal groups with AEPZ.[134] A similar procedure was employed by Yuan et al. to obtain a nanohybrid containing silver NPs.[87] Poly(thiourea-amide)/epoxy/CNT hybrids were obtained by *in situ* polymerization using both pristine and oxidized MWCNTs.[45]

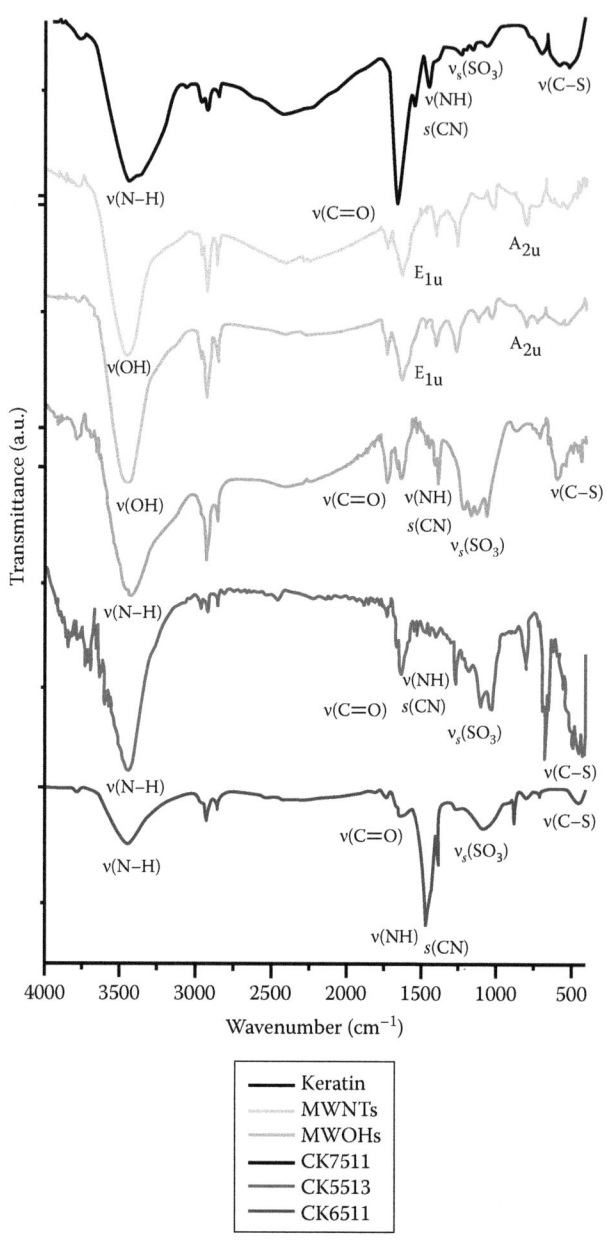

FIGURE 22.7
FTIR spectra of the samples of pristine nanotubes (MWNTs), oxidized nanotubes (MWOHs), and the keratin-grafted nanotubes obtained at different reaction temperatures. (Reprinted from Estévez-Martínez, Y. et al., *J. Nanomater.*, 702157, Copyright 2013. With permission from Hindawi.)

22.3.4 Other Methods

Pristine and oxidized CNTs are commonly functionalized with APTES where peptide bond is formed as a result of carboxyl group reacting with the amine group from the APTES; this aminosilanization protocol has been used by other groups.[135–137] Kang and coworkers[138] have developed a route for treating SWCNTs using tetraethyl orthosilicate and aminomethyltriethoxysilane with aluminum-tri-sec-butoxide in a glove box filled with nitrogen. This leads to the formation of aluminosilicate nanotubes with up to 15% covalently bonded primary amine moieties on their inner wall. Amine functionalization has been accomplished by reaction of azomethine ylides onto the nanotube sidewalls, in order to further functionalize these CNTs with oligonucleotides.[139] Jiang et al.[50] developed a method to produce amino CNTs by cycloaddition of 11-azido-3,6,9-trioxaundecan-1-amine.

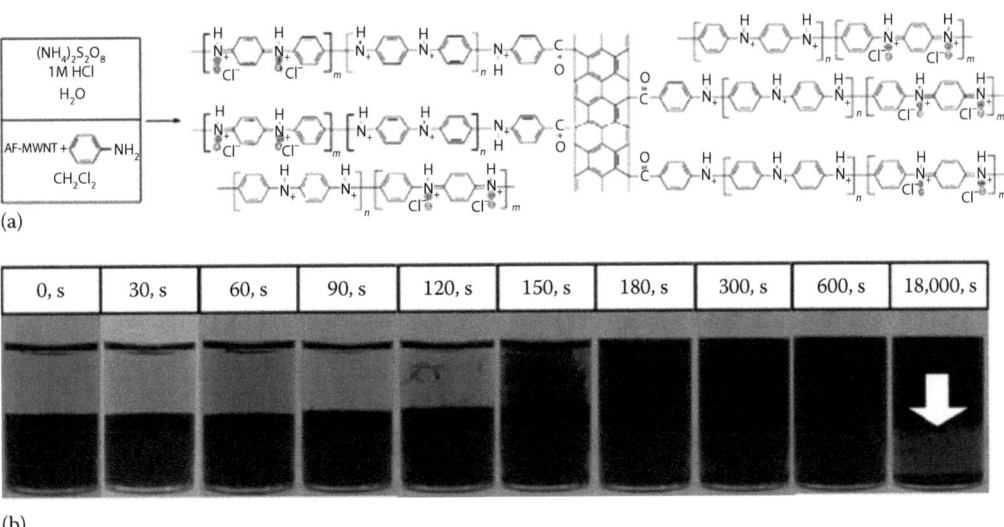

(a)

(b)

FIGURE 22.8
(a) Interfacial polymerization of aniline: top aqueous 1 M hydrochloric acid layer contains ammonium persulfate; bottom organic layer contains aniline and AF-MWCNTs; (b) photographs of reaction vials as function of reaction time. (From Jeon, I.N.-Y., Loon-Seng, T. A.N., and Baek, J.-B.: Synthesis and electrical properties of polyaniline/polyaniline grafted multiwalled carbon nanotube mixture via *in situ* static interfacial polymerization. *J. Polym. Sci. Part A Polym. Chem.* 2010. 48. 1962–1972. Copyright Wiley-VCH Verlag GmbH & Co. KGaA. Reproduced with permission.)

Supercritical water oxidation (SCWO) was used to amino-functionalized MWCNTs in an aqueous dispersion containing NH_3.[140]

Plasma treatment is a convenient method for functionalizing the surface and adjacent areas of materials; the excited species, radicals, electrons, ions, and UV light generated in this technique strongly interact with the graphitic surface breaking the sp^2-hybridized carbon (C=C) bonds and creating active sites for binding of functional groups.[59,94] Plasma treatments for modification of nanomaterials with amine groups have used nitrogen-containing gases (NH_3 or NH_3 mixed with other gases, $N_2, N_2/H_2$)[34,93,141–143] or have grafted a polymer onto the surface, which has been activated by a plasma pretreatment.[94] Microwave-excited surface-wave plasma (MW-SWP) is another technique employed where the high electron energy fractionizes NH_3, forming metastable ions of NH_2, NH, N, and H, as well as radicals. Chen et al. introduced amino groups onto MWCNTs by using NH_3/Ar MW-SWP treatment, resulting in a rough surface and enhanced defects of MWCNTs with no significant damage.[59] Yook et al.[95] found that when maintaining a minimum level of plasma power, the properties of CNTs were preserved, and the partially decomposed species are more favorable for amino functionalization than the atomic species generated from the full decomposition of NH_3. Chen and coworkers[94] achieved a level of primary amines of 2.3%, higher than previous report using the nitrogen-containing gas plasma treatment by using a mixture of N_2 and H_2, which was found to be preferable to using NH_3. They also coated MWCNTs with a thin layer of plasma-polymerized heptylamine and obtained even higher levels (3.5%) of primary amines. Cao et al.[34] studied the reaction behavior of the chemical modification of boron nitride nanotubes with NH_3 plasmas. They found that the NH_2* radical is more favorably bonded to B atoms than to the N atoms. Friedrich et al.[36] found that C–Br bonds from plasma chemical processing of graphitic materials were well-suited for grafting of organic molecules by nucleophilic substitution with 1,6-diaminohexane and APTES. Brunetti et al.[49] reported a microwave-assisted method to cycloaddition of azomethine ylides to pristine SWCNTs in solvent-free conditions.

22.4 GRAPHENE

GO obtained from the exfoliation of graphite oxide through thermal or mechanical methods is a commonly used route to obtain graphene sheets.[144,145] Carboxylic acid groups at the edges of graphene sheets often require activation of this functional group by using $SOCl_2$ or the previously

TABLE 22.2 Functionalizing Molecules for Graphene Modification

Classification	Name	References
Aliphatic amines	Methylamine	[156,159]
	Ethylenediamine	[31,71–73,146,148,151,157]
	Butylamine	[153,155,156,158,159]
	Hexylamine	[37]
	Octylamine	[155]
	Decylamine	[154]
	Dodecylamine	[155]
	Octadecylamine	[154,183]
	Hexamethylenediamine	[74,150]
	1,8-Diaminooctane	[114]
	1,10-Diaminodecane	[114]
	1,12-Diaminododecane	[114]
Single ring aromatic amines	4-Aminobenzylamine	[162]
	Phenylenediamine	[149]
	p-Xylylenediamine	[163]
	Phenylalanine	[30]
Polymers	Poly(2-(diethylamino)ethyl methacrylate)	[174]
	Polyaniline	[72,73]
	Polyethyleneimine	[88,164]
	Polypyrrole	[71,73]
Others	3-(Aminopropyl)triethoxysilane	[184]
	4-Aminobenzoic acid	[160]
	4,4'-Oxydianiline	[183]
	4,4'-Methylenedianiline	[161]
	4,4'-Diaminodiphenylsulfone	[183]
	Ionic liquid	[178–180]
	L-lysine	[40]
	N-aminoethylpiperazine	[175]
	N,N-dimethylformamide	[177]
	Tetraphenylporphyrin	[47]
	NH_3	[181]

mentioned carbodiimides; addition of nucleophilic species like amines produces covalently attached functional groups on GO via the formation of amides.[145–151] On the other hand, epoxy groups can go through ring-opening reactions, reaction of these groups with amines yielding nucleophilic attack at the α-carbon by the amine.[145,152,153] Table 22.2 shows a number of molecules containing amino groups that have been used for graphene functionalization, and in the following sections, we review these routes and many other several techniques employed for the amino functionalization of graphene sheets.

22.4.1 Aliphatic Amines

Ryu et al.[154] have employed the commonly used acyl chlorination reaction by $SOCl_2$ adopted in the midstep of the amine functionalization for grafting graphene sheets with varying chain length alkylamines, as displayed in Figure 22.9. Compton et al.[37] functionalized GO paper with hexylamine by flowing a methanol solution of the amine through the as-prepared wet paper; the long alkyl chain of hexylamine intercalated between the sheets as *spacers*, expanding the intersheet gallery and reducing the interactions between the hydrophilic groups on adjacent sheets. Furthermore, this hexylamine-modified GO paper was reduced to obtain modified GO conductive alkylated graphene paper where the hexylamine served as a structure-stabilizing agent. GO sheets were decorated with $MnFe_2O_4$ NPs by using oleylamine as an intermediary spacer, resulting in water-soluble nanocomposites.[43] Stankovich and coworkers[155] modified graphene with different chain alkylamines

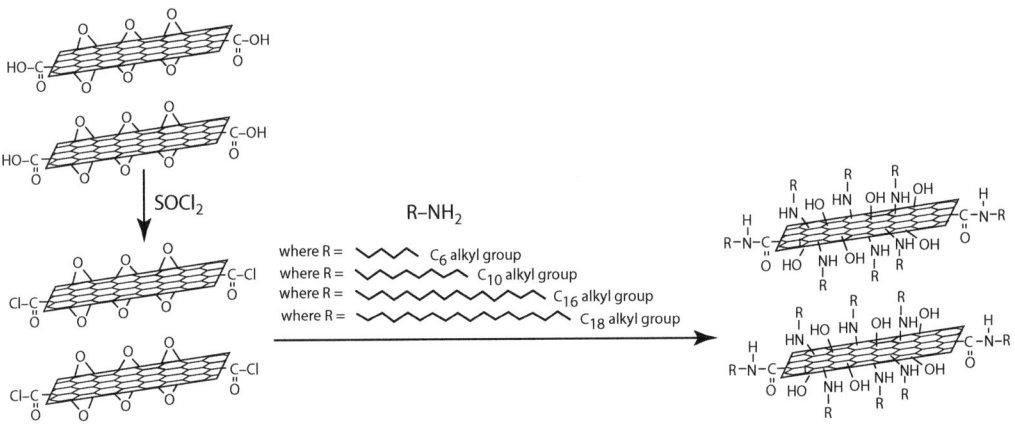

FIGURE 22.9
Possible interaction mechanism between GO and long-chain alkylamines. (Reprinted from *Chem. Eng. J.*, 244, Ryu, S.H. and Shanmugharaj, A.M., Influence of long-chain alkylamine-modified graphene oxide on the crystallization, mechanical and electrical properties of isotactic polypropylene nanocomposites, 552–560, Copyright 2014, with permission from Elsevier.)

via solution or vapor-phase intercalation; they observed a direct correlation between the number of carbon atoms in the alkyl chain of the chemisorbed amine and the d-spacing of the graphene paper in the solution-based method. On the other hand, the absence of solvent during vapor-phase intercalation lowered the rate of reaction between the amine and the GO surface. Guan et al.[156] synthesized methylamine-modified graphene and *n*-butyl amine–modified graphene based on a solvothermal route. Stine and coworkers[157] developed a method to activate graphene sheets using fluorination followed by reaction with EDA. Fluorinated graphene obtained via plasma-assisted decomposition of CF_4 also served to functionalize with butylamine using an ultrasonication probe.[158] Ikeda and coworkers[159] prepared amine hydrochloride solutions from alkyl amines and hydrochloric acid in methanol in order to amino functionalize GO at room temperature.

22.4.2 Aromatic Amines

Chua and Pumera[160] have employed a Friedel–Crafts acylation method to amino functionalize graphene with 4-ABA by using polyphosphoric acid/phosphorus pentoxide. Figure 22.10 shows the scheme of this functionalization and the preparation of nanocomposites based on polynorbornene dicarboximide (PND) matrix. Fang and coworkers[161] covalently bonded long-chain aromatic amines on the surface of graphene nanosheets by diazonium addition. Lee et al.[162] developed a simple procedure of amino functionalization, which consisted of stirring GO sheets in 4-aminobenzylamine at room temperature for 24 h. Mallakpour et al.[46] functionalized GO with aromatic–aliphatic amines such as phenylalanine and tyrosine by condensation and nucleophilic addition reactions between –NH_2 groups of the amino acids and carboxylic acid or epoxy groups on the GO sheets; tyrosine-functionalized GO exhibited a scrolled structure attributed to the π–π stacking interaction between the aromatic ring of tyrosine and graphene and also H-bonding interactions between the phenolic –OH groups of guest molecules and O-containing groups of GO layers. P-xylylenediamine was covalently intercalated between GO sheets for the fabrication of pillared-type graphene-based materials with a tailored interlayer spacing.[163]

22.4.3 Polymers

22.4.3.1 Grafting to

Amino-functionalized PEG was grafted on high-quality superparamagnetic NPs/GO nanocomposites through carbodiimide chemistry; the resulting nanocomposites enhanced the colloidal stability and improved the nanocomposite biocompatibility.[43] PEI molecules were grafted *in situ* resulting in

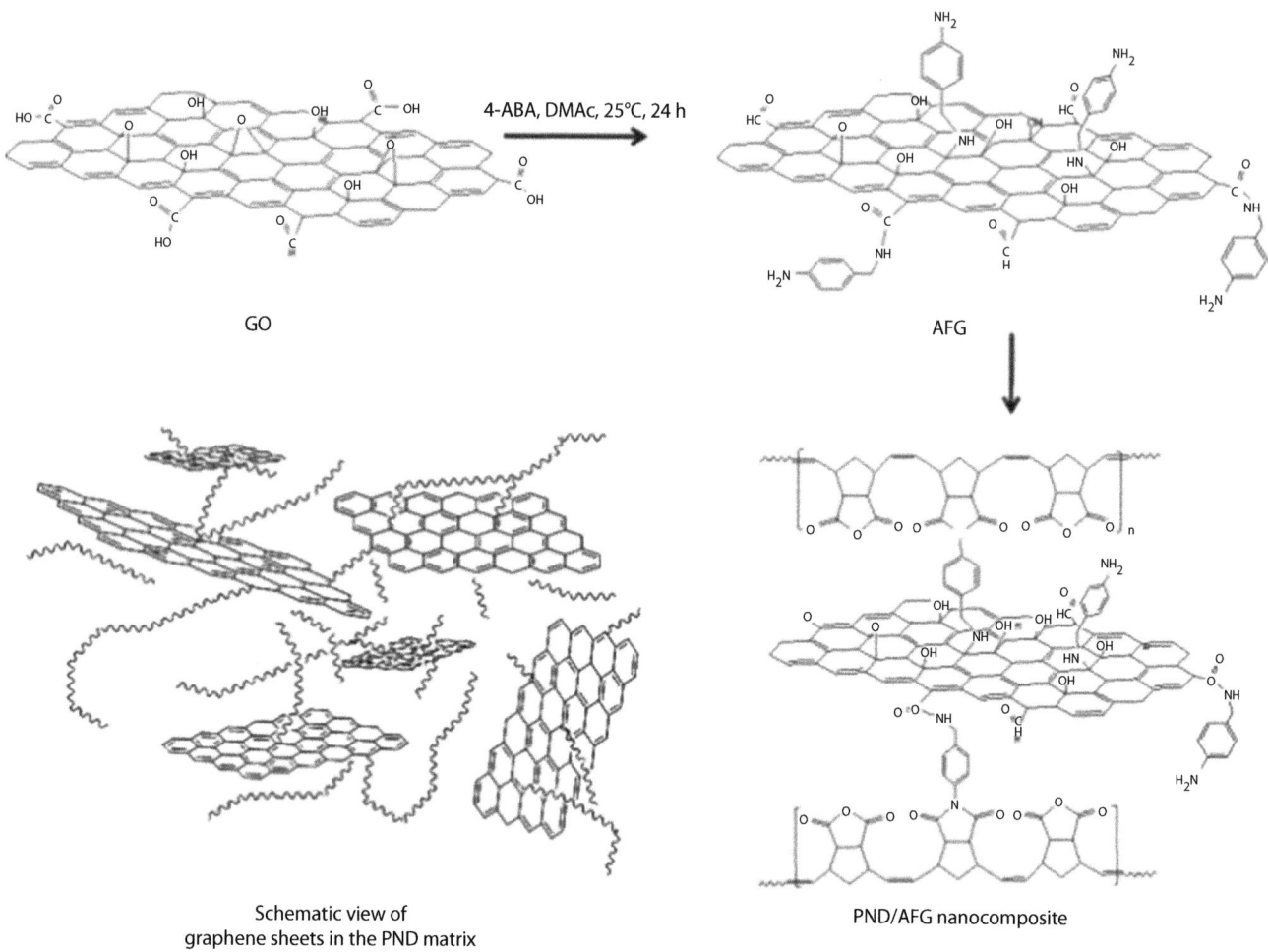

FIGURE 22.10
Preparation of amino-functionalized graphene (AFG) and PND/AFG hybrids and the schematic view of graphene sheets in the PND matrix. (From Chua, C.K. and Pumera, M.: Friedel-crafts acylation on graphene. *Chem. Asian J.* 2012. 7. 1009–1012. Copyright Wiley-VCH Verlag GmbH & Co. KGaA. Reproduced with permission.)

a hydrothermal reduction and cutting of graphene sheets through the formation of amide–carbonyl bonds based on the Lewis acid–alkaline reaction between carboxylic groups of GO and amine moieties in PEI.[88] PEI was also grafted onto graphene sheets via the nucleophilic ring-opening reaction between the amine groups in PEI and epoxy groups of GO; the GO/PEI species provided a suitable environment for constructing a multilayer film using a layer-by-layer complexation method.[164] Bao and coworkers[26] used a route to functionalize GO with chitosan where EDC initiated the amide linkage between GO and chitosan by forming an active intermediate, which was further stabilized by the addition of NHS. Alkynyl groups on GO were generated by the reaction of GO and propargylamine in the presence of DCC and 4-dimethylaminopyridine; this material served for coupling chitosan containing azide moieties by a Cu-catalyzed Huisgen cycloaddition click reaction.[53] Yang and coworkers[165] covalently grafted hydroxypropyl cellulose or chitosan onto the activated GO sheets via esterification reactions, and the functionalized GO reduced hydrazine. Figure 22.11 shows the scheme of reaction of this method starting from the two-stage oxidation of graphite (step 1); then the carboxyl groups of GO were partially converted into acyl chloride (step 2); these activated groups served to graft LMC or hydroxypropyl cellulose (step 3). A similar procedure was employed by Zeng et al.[166] for grafting phthaloyl chitosan onto graphene sheets, which were further decorated with Pd NPs. Shao et al.[167] prepared chitosan/GO via the solution casting method, and the films obtained were treated at elevated temperature; the epoxy groups in GO reacted with the amino groups in chitosan resulting in a cross-linking reaction. Bustos-Ramírez et al.[168] reported a route for grafting chitosan onto GO via redox reaction; the temperature of reaction was found to play an

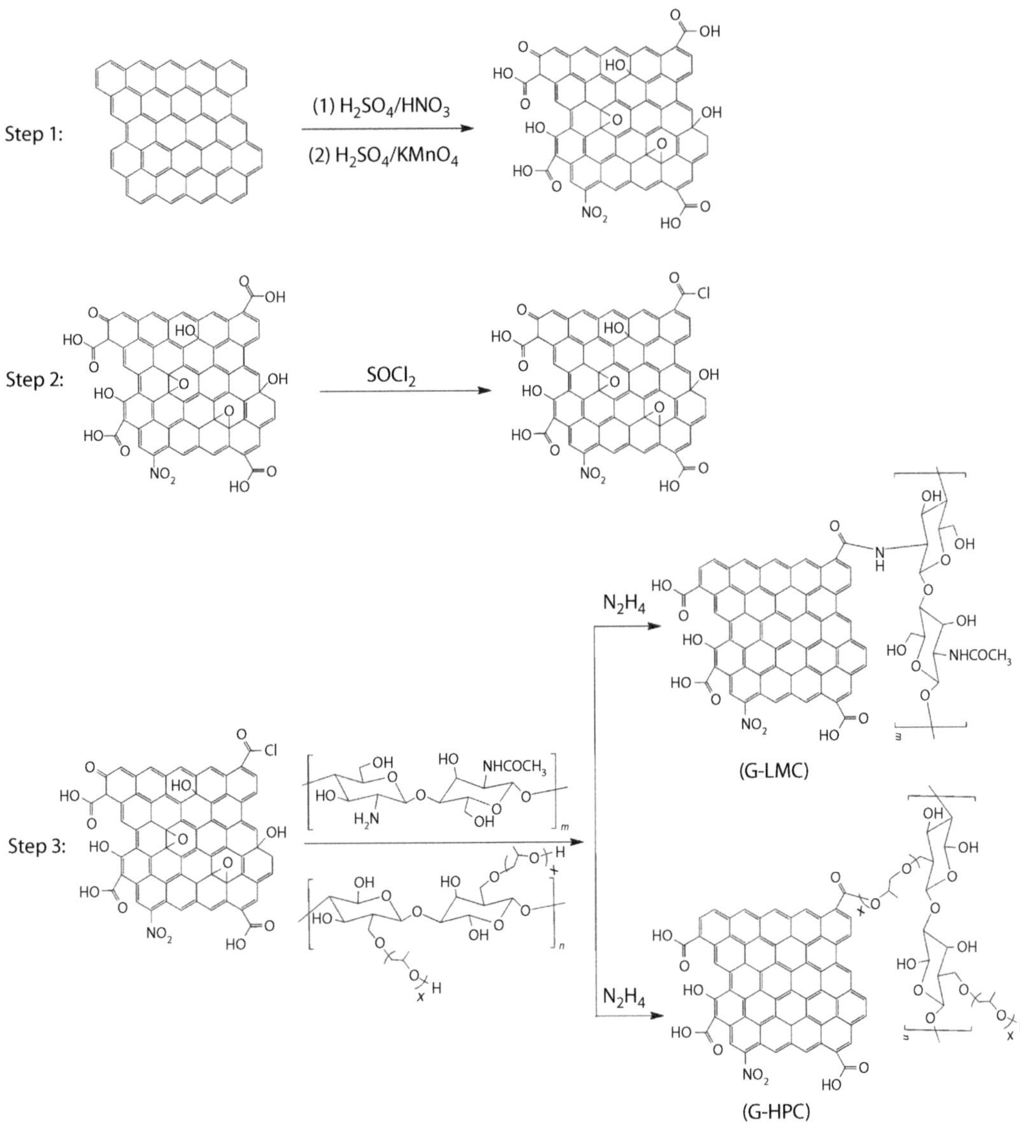

FIGURE 22.11
Preparation routes of polysaccharide-functionalized graphene. (Reprinted with permission from Yang, Q., Pan, X., Clarke, K. et al., Covalent functionalization of graphene with polysaccharides, *Ind. Eng. Chem. Res.*, 51, 310–317. Copyright 2012 American Chemical Society.)

important role in the morphology of the grafted sheets and the solubility properties; the scheme of reaction is presented in Figure 22.12. Feng and coworkers[169] simultaneously reduced GO sheets and functionalized them with chitosan for their incorporation into a polyvinyl alcohol matrix; this provided improved dispersion of the nanofillers and loading transfer due to the hydrogen bonding between polymer matrix and chitosan. Carboxylic groups of GO reacted with amino functionalities of chitosan under continual microwave radicalization in DMF medium, achieving grafting of graphene sheets with chitosan chains; this process was followed by hydrazine hydrate treatment to reduce GO sheets.[170] Rana and coworkers[171] covalently bonded chitosan onto GO sheets via an amide bond using DMF in the presence of pyridine. Pendant groups of keratin are sensitive to react under a redox system of malic acid/KMnO$_4$ in H$_2$SO$_4$ and generate free radicals involved in the formation of the graft; this favored the grafting of keratin onto GO.[172,173] These keratin-grafted graphene sheets have been used to incorporate them into a chitosan starch matrix, showing improved mechanical response as displayed in Figure 22.13; however, when compared to the nanocomposites reinforced with GO, the storage modulus gradually decreased at higher temperatures suggesting that the high content of grafted nanofiller may modify the interfacial interactions due to the hydrophobic nature of keratin.[173]

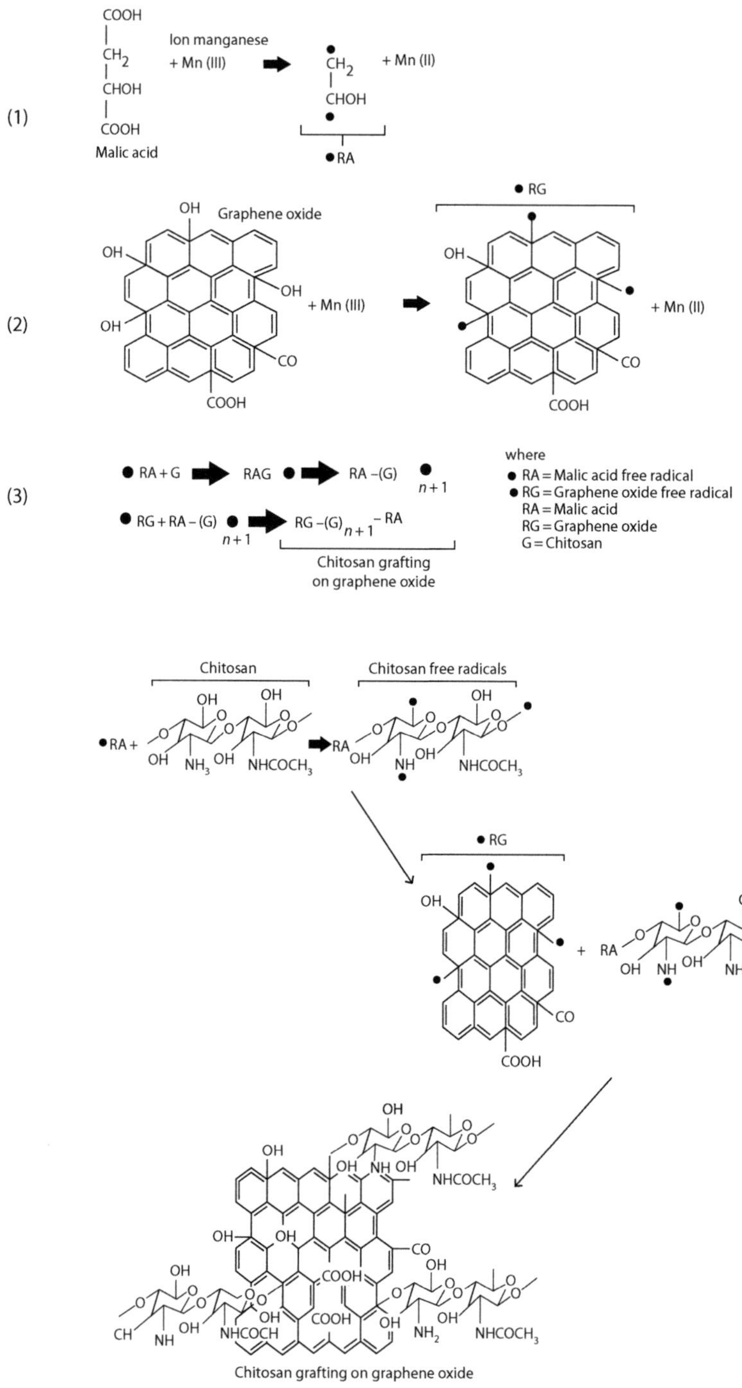

FIGURE 22.12
Reaction scheme of chitosan grafting onto GO. (Reprinted from Bustos-Ramírez, K. et al., *Materials*, 6, 911, Copyright 2013. With permission from MDPI.)

22.4.3.2 Grafting from

Surface-initiated *in situ* atom transfer radical polymerization (ATRP) was used to functionalize GO with poly(2-(diethylamino)ethyl methacrylate) (PDEA). The procedure used was the following: the carboxylic acids on GO were used to react NHS and 1,3-diaminopropane. The ATRP initiator molecules were coupled to the GO via reaction of 2-bromo-2-methylpropionylbromide with the hydroxyl and amine groups. Finally, the surface-initiated ATRP of DEA served to graft PDEA chains on graphene.[174] Graphene sheets were functionalized with phenylene diamine followed by

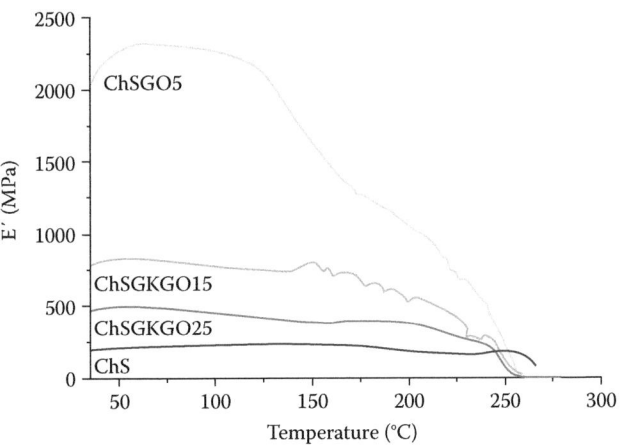

FIGURE 22.13
Storage modulus of neat chitosan/starch film and the nanocomposite containing 5 wt.% GO (Ch/S/GO) and the films containing 5 wt.% keratin-grafted GO at two different conditions (Ch/S/GKGO15 and Ch/S/GKGO25). (Reprinted with permission from Rodríguez-González, C., Kharissova, O.V., Martínez-Hernández, A.L. et al., Graphene oxide sheets covalently grafted with keratin obtained from chicken feathers, *Dig. J. Nanomater. Biostruct.*, 8. Copyright 2012 American Chemical Society.)

the *in situ* polymerization of aniline.[149] Polyamide 6 was grafted to graphene sheets by *in situ* polycondensation of caprolactam using GO, which was previously functionalized with NH_3 and hexamethylenediamine.[150] GO was functionalized with *N*-aminoethylpiperazine (AEPZ) using DCC to proceed with an *in situ* polymerization among AEPZ, di(acryloyloxyethyl) benzenephosphonate, and AEPZ-modified GO, as displayed in Figure 22.14.[175] Devi and coworkers,[176] electropolymerized

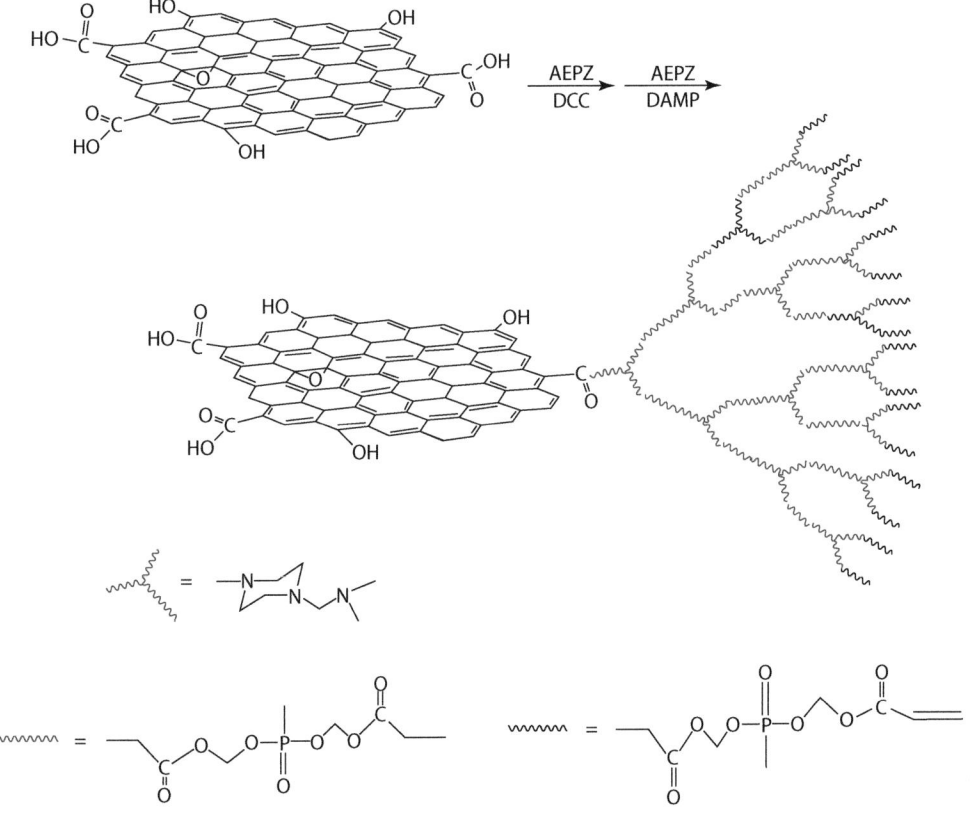

FIGURE 22.14
Schematic of the fabrication of hyperbranched functionalized GO. (From Devi, R. et al., *Sens. Actuat. B Chem.*, 186, 17, 2013.)

pyrrole in the presence of GO onto an Au electrode, to subsequently electropolymerize chitosan and aniline, which resulted in an improved amperometric oxalate biosensor.

22.4.4 Other Methods

Bosch-Navarro et al.[177] employed DMF for the reduction of GO; this procedure not only promoted reduction of the oxygenated groups but also introduced nitrogen functionalities across the graphene-like lattice from their reaction with the dimethylamine molecules generated *in situ* from the decomposition of the solvent; the release of dimethylamine can be controlled by temperature and reaction time, whose parameters can be used to control efficiency of the subsequent functionalization of GO to generate *N*-functionalities across the graphene layers. Furthermore, they attached Au NPs because of the affinity between these functional groups and gold, as seen in Figure 22.15. Similar results were obtained in our research group when reducing GO with HMTA, where a few functional groups related to amine moieties were found by FTIR spectroscopy.[11,12] Mo et al. functionalized GO with L-lysine and divalent copper ions (Lys–Cu–Lys), where the terminal amino acids of Lys–Cu–Lys were linked to the residual epoxide groups of GO by nucleophilic substitution; the copper ion was eliminated using ethylenediaminetetraacetic acid disodium salt.[40] Bayazit and coworkers[52] have studied theoretically and experimentally the functionalization of SWCNTs with pyridinium ylide generated *in situ* via the addition of triethylamine to the Kröhnke salts *N*-(4-methyl sodium benzene sulfonate)-pyridinium bromide and *N*-(4-nitrobenzyl)-pyridinium bromide under microwave conditions. GO was functionalized with *N*-methylpyrrolidone via a 1,3-dipolar cycloaddition of azomethine ylides; this resulted in a degree of functionalization of approximately 1 functional group in 128 carbon atoms.[48]

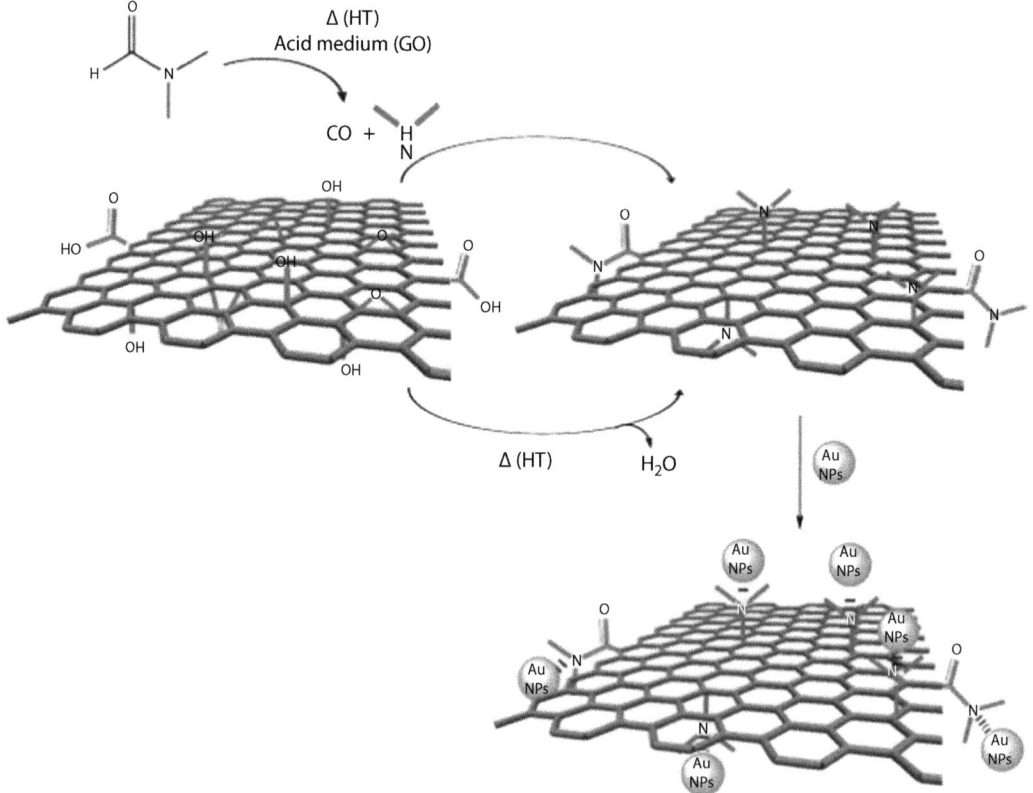

FIGURE 22.15
Functionalization/reduction process to obtain reduced graphene oxide and Au NPs/graphene sheets. (Reprinted from *Carbon*, 54, Bosch-Navarro, C., Coronado, E., and Martí-Gastaldo, C., Controllable coverage of chemically modified graphene sheets with gold nanoparticles by thermal treatment of graphite oxide with *N,N*-dimethylformamide, 201–207, Copyright 2013, with permission from Elsevier.)

Yang et al.[178] attached an amine-terminated IL on the surface of graphene sheets by a nucleophilic ring-opening reaction between the epoxy groups of GO and the amine groups of the IL using potassium hydroxide as a catalyst; the cations of the IL were introduced to the graphene sheets, contributing to a stabilization of graphene dispersions via electrostatic repulsion. Functionalized GO with characteristics similar to those of room-temperature ILs has been obtained by grafting of chloroacetic acid and the subsequent carboxyl neutralization with mono-functionalized polyetheramine (Jeffamine 2070).[39] Similar nanomaterials were obtained by the reaction between GO and 2-azidoethanol via nitrene chemistry, followed by the thermal reduction of the functionalized GO and the silanization of these functionalized graphene sheets (G–OH–SIT). Finally, the NIMs were obtained by titrating G–OH–SIT via acid–base reaction with Jeffamine 2070.[179] Li and coworkers[180] used 1,3-di(4-amino-1-pyridinium)propane tetrafluoroborate IL to modify GO through covalent binding of amino groups and epoxy groups in an alkaline solution. This functionalization allowed the fabrication of a biosensor; hemoglobin was immobilized biocompatible hybrid nanomaterial, which provided a microenvironment around the protein to retain the enzymatic bioactivity.

APTES attachment has been performed on various substrates containing surface hydroxyl groups, which can react with alkoxysilanes.[102,103] Addiego et al. functionalized graphene with APTES for bacteria attachment combined with GaN to build a biosensor that operated as high-electron-mobility transitor.[99] Luo et al.[30] prepared an APTES oligomer (PAS) for functionalizing GO sheets. This group also directly functionalized GO with APTES and tested their efficiency in Pb (II) adsorption; they found that PAS–GO possessed a higher amount of amino groups than the APTES–GO and the oligomer-based material formed a 3D network with multiarm PAS bridges.

GO was subjected to mild amino hydrothermal treatment at 70–150°C using NH_3 solution, followed by thermal annealing at 100°C in order to extract sp^2 domains without deconstructing their graphitic structure. In this method, the NH_3 reacts with epoxy groups to form a primary amine and alcohols by nucleophilic substitution.[38] Lai and coworkers[181] reduced GO and amine modified the resulting graphene via a one-pot solvothermal method using ethylene glycol as solvent and NH_3 water as nitrogen precursor; the reaction proceeds by the nucleophilic substitution of –COOH and C–O–C groups of GO by the NH_3 radicals. Guo et al.[182] also employed this method for amino functionalization of GO. Amine-functionalized silica particles (NH_2–SiO_2) formed a uniform coating with GO via homogenous dispersion of colloids in the solutions, derived from the electrostatic interactions between the strongly negatively charged groups (–COOH, –O–, –OH) of GO and positively charged groups (–NH_2) of NH_2–SiO_2.[95]

Kim et al.[146] used acid chloride–functionalized GO and amino-functionalized MWCNTs to obtain hybrids resembling graphene sheets scrolled on nanotubes as depicted in Figure 22.16. The proposed formation process was the following. The carbonyl chloride groups at the edge of GO reacted with amine groups on MWCNTs, anchoring the GO sheets on the surface of MWCNTs via amide bonds. Secondly, the anchored GO sheets curved along MWCNT's surface due to shear stress from vigorous stirring and continuous perturbation of the surrounding solvent. Finally, the curved GO sheets formed scrolls around the aminated MWCNTs. Functionalization of graphene with PEI was used for sequential self-assembly with OMWCNT to fabricate hybrid carbon films.[89]

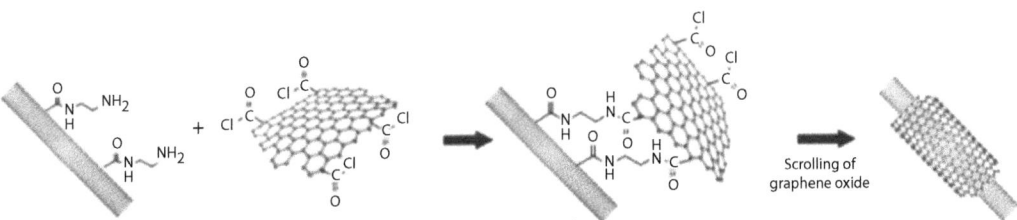

FIGURE 22.16
Scheme showing GO scroll formation around an MWCNT template by covalent conjugation. (Reprinted from *Carbon*, 48, Kim, Y.-K. and Min, D.-H., Preparation of scrolled graphene oxides with multi-walled carbon nanotube templates, 4283–4288, Copyright 2010, with permission from Elsevier.)

22.5 SOLUBILITY AND DISPERSIBILITY

As explained earlier, covalent functionalization in nanotubes can proceed by oxidative processes involving strong acids or oxidants, which result in opening of nanotube ends or at defect sites on the nanotube sidewalls or structural damage of graphene sheets.[16,22,70,144,153] Alternative routes have been researched, such as the mild nondestructive method developed by Chang and coworkers to yield a Friedel–Crafts acylation of CNTs resulted in greatly improved dispersibility and compatibility of the functionalized nanotubes while keeping their intrinsic properties.[29] Further functionalization has allowed important changes in the solubility and dispersibility properties of carbon nanomaterials.[16,17,23] Amine groups create a distribution of positive charges on the carbon nanomaterial surface and greatly increase their hydrophilicity.[143] The amino functionalization technique used by Syrgiannis and coworkers leads to the formation of negatively charged intermediates of SWCNTs that were individualized and effectively solubilized in organic solvents. The subsequent reoxidation of these intermediates yielded neutral derivatives (uncharged material), which exhibited a high degree of solubility due to the n-propylamine addends.[35] Gabriel et al.[61] found that CNTs modified with aliphatic amines are soluble in nonpolar solvents (ether, chloroform, etc.) and the ones modified with perfluorinated amines are soluble in polar solvents (ethanol, DMF, etc.). ODA-functionalized GO showed better dispersion in H_2O when compared to EDA and 4,4′-diaminodiphenylsulfone-functionalized GO; this was attributed to the higher amino functionalization of ODA-GO than the other two. The same behavior occurred in acetone; on the other hand, EDA-modified GO showed best dispersion in methyl alcohol. Furthermore, this research showed that size effect of graphenes has an effect on dispersion, as consolidated by UV–Vis spectroscopy.[183] Mao et al.[105] obtained dispersions of ODA-functionalized MWCNTs in chloroform that were stable for over 1 year; however, these nanotubes were insoluble in water due to the attachment of this amine, which is insoluble in polar solvents.

Solubility of aniline-functionalized SWCNTs was found to be very effective in CH_2Cl_2, $CHCl_3$, and CS_2 and completely insoluble in water. Solvents like toluene and DMF resulted in solubility inferior to 0.05 mg/mL. However, a very stable suspension could be obtained for a longer period. These results indicated the remarkable degree of covalent functionalization after microwave solubilization.[67] Kakade and coworkers[67] found that upon addition of aniline hydrochloride to amine-functionalized SWCNTs, the solubility of the functionalized nanotubes was suppressed due to the formation of zwitterionic linkages. 4-Aminobenzylamine-functionalized graphene was especially well dispersed in polar solvents such as *N,N*-dimethylacetamide. After sonication in this solvent, a small amount of the functionalized graphene precipitated within the first few days, and the homogeneous colloidal suspension was separated from the mixture after 2 weeks. This remaining suspension was stable for up to 3 months without any further precipitation graphene sheets.[162]

Grafting of polymers is expected to have greater influence on the carbon nanomaterial properties and their affinity to polymer matrices as compared to the addition of low-molecular-weight functionalities.[80] Amine-modified MWCNTs were dispersed more homogenously in a polymer matrix than the acid-modified MWCNTs; however, the amino-functionalized MWCNTs possessed entangled structures.[104] An effective method to increase the dispersion of carbon nanomaterials in the polymer matrix is usually by preparing composites from solution; critical to this method is the separation of carbon nanomaterial aggregates in the solution.[170] Surface modification of RGO by chitosan was a decisive factor for the stable dispersion of the nanofillers in a PVA matrix.[169]

Apart from improving the chemical affinity of carbon nanomaterials with polymer matrices, the diverse modification routes also assist the effective solubility in organic solvents. Liao and coworkers[33] found that the cationic ammonium groups expected to be present in the PEI grafts on MWNCT–NH–PEI at pH < 7 facilitated dispersion of MWNT–NH–PEI in water. MWCNT–NH–PEI dispersion in water (pH < 7) remained stable for over 1 year. On the other hand, when dispersing the nanotubes at pH 12 (0.01 M NaOH), the MWNT–NH–PEI material began to settle after only 1 day. Amino-functionalized MWCNTs were effectively complexed with monoamine-terminated PEO, which enhanced the dispersibility of the complex in ethanol solution.[120] PDEA chains grafted onto the GO rendered this hybrid an exceptional pH-responsive property with high solubility and good stability in physiological solutions.[174] Chitosan-grafted GO sheets showed high solubility and stability in acidic and neutral aqueous solutions and formed stable dispersions for a long term. Furthermore, due to the conjugated structure of graphene basal planes, chitosan-modified GO can attach and absorb aromatic, water-insoluble drugs via van der Waals' interactions making it an important composite for biomedical applications.[26] Chitosan NP/MWCNT hybrids showed good dispersibility and stability in aqueous solution for up to 30 days, which can be favorable for the immobilization of biomolecules.[125] Amino functionalization enhances dispersion and interfacial interactions of CNTs and epoxy resins.[70]

The polymer grafting strategy is not only an effective way to solubilize carbon nanomaterials in water, but this technique plays a key role for the preparation of polymeric carbon packaging, metal-ion adsorption, novel drug delivery, and gene composites.[171] Table 22.3 shows a variety of works concerning carbon nanomaterial modification with chitosan and the application of these materials. Chitosan provided stabilization against van der Waals' attraction, which is essentially due to steric hindrance of the polymer chains that enabled to disperse graphene in aqueous solution.[170] Decreasing chitosan degree of acetylation under 50% by partial deacetylation of chitin results in a higher solubility in aqueous media, which facilitates also carbon nanomaterial dispersion.[81] Due to the hydrophobic and hydrophilic properties of chitosan, stable polymer/MWCNT solutions yielded a homogeneous distribution of nanotubes during the thermally induced phase separation process to obtain nanocomposite scaffolds.[82] LMC/MWCNTs were found to be insoluble in pure water due to the increase in crystallinity, whereas they were easily soluble in DMF, dimethyl sulfoxide (DMSO), dimethylacetamide, and acetic acid aqueous solution.[119] Magnetic chitosan/MWCNTs were effectively dispersed and formed a very stable suspension in aqueous solution. Furthermore, these nanotubes displayed the advantages of recycling from aqueous solution and high adsorption for effective removal of dye molecules and other pollutants from aqueous solution.[124]

Amine-functionalized MWCNT dispersions obtained by the SCWO method showed to be highly soluble (~84 mg/L) and stable in water for 2 weeks; contact angle test proved that the amine functionalization increased the surface energy on the MWCNTs.[140] GO modified with amino IL allowed the good dispersion of the graphene sheets into water, DMF, and DMSO at various concentrations, forming long-term, stable, and homogeneous dispersions after ultrasonic treatment; the unfunctionalized chemically converted graphene (u-CCG) sheets were also prepared, and a poor dispersibility of u-CCG was clearly observed.[178] Carbon-based NIMs possess excellent amphiphilic behavior.[39,179] Amino IL–functionalized graphene sheets (GO–IL) were well dispersed in water, DMF, and DMSO, due to an enhanced solubility and electrostatic intersheet repulsion provided by the IL units. Furthermore, these dispersions were stable for more than 3 months.[178] A combination of continuous wave and pulsed plasma mode was found to be effective for producing surfaces with a higher selectivity for primary amines and higher levels of functional groups for improved interfacial bonding and dispersion in epoxy nanocomposites.[94]

TABLE 22.3 Research Involving of Chitosan-Modified Carbon Nanomaterials

Carbon Nanomaterial	Purpose	References
Carbon nanotubes	Hierarchically structured conductive nanocomposite transducers	[81]
	Scaffolds for bone graft substitutes	[82,83]
	Nanocomposite for biochemical or electrochemical applications	[121]
	Gene delivery vectors	[123]
	Potential applications in catalysis and environmental protection	[119]
	Conjugated systems for the treatment of infectious diseases and cancer cells	[122]
	Removal of pollutants by adsorption	[124]
	Immobilization of therapeutic biomolecules	[125]
	Solid-phase extraction adsorbent for determination of heavy metals	[126]
Graphene	Biocomposite containing two antimicrobial fragments	[127]
	Nanocarrier for an anticancer drug	[26]
	Construction of a glucose biosensor	[166]
	Biocompatible nanocomposite membranes	[167]
	Nanofiller for nanocomposites	[53,168]
	Fire-resistant nanocomposites	[169]
	Nanocomposites for biological applications	[170]
	Drug release	[171]
	Biosensor to measure oxalate level in urine and plasma	[176]

22.6 CONCLUSION

Chemical functionalization of carbon nanomaterials such as CNTs and graphene has been extensively studied by different research groups in order to obtain materials that can be employed in a diverse applications. A variety of works have been summarized in this chapter, highlighting that covalent modification enables effective interaction between molecules containing amino groups and CNTs or graphene. Furthermore, this route significantly improves the physicochemical properties of these nanomaterials. Covalent modification covers diverse reaction mechanisms including nucleophilic substitution, electrophilic addition, condensation, and cycloaddition. A number of compounds containing amine moieties have been studied due to the rich chemistry they offer and because they represent an efficient pathway to confer biocompatibility, cell affinity, peptide synthesis, and functionality for binding other molecules by means of a variety of bonds. Modification of CNTs and graphene with amines or molecules containing this functional group offers essential advantages for expanding the applications of these carbon nanostructures due to the tailored affinity with biomolecules or polymers and improved stability and dispersibility.

REFERENCES

1. Choi, W., Lahiri, I., Seelaboyina, R. et al. 2010. Synthesis of graphene and its applications: A review. *Critical Reviews in Solid State and Materials Sciences*, 35(1), 52–71.
2. Kingston, C., Zepp, R., Andrady, A. et al. 2014. Release characteristics of selected carbon nanotube polymer composites. *Carbon*, 68, 33–57.
3. Liu, W.-W., Chai, S.-P., Mohamed, A.R. et al. 2014. Synthesis and characterization of graphene and carbon nanotubes: A review on the past and recent developments. *Journal of Industrial and Engineering Chemistry*, 20(4), 1171–1185.
4. Koziol, K., Boskovic B.O., and Yahya, N. 2011. Synthesis of carbon nanostructures by CVD method. In *Carbon and Oxide Nanostructures*, ed. N. Yahya, pp. 23–49, Berlin, Germany: Springer-Verlag.
5. Joselevich, E., Dai, H., Liu, J. et al. 2008. Carbon nanotube synthesis and organization. In *Carbon Nanotubes*, eds. A. Jorio, G. Dresselhaus, and M.S. Dresselhaus, pp. 101–164, Berlin, Germany: Springer-Verlag,
6. Partoens, B. and Peeters, F.M. 2006. From graphene to graphite: Electronic structure around the K point. *Physical Review B: Condensed Matter and Materials Physics*, 74(7), 075404.
7. Bose, S., Khare, R.A., and Moldenaers, P. 2010. Assessing the strengths and weaknesses of various types of pre-treatments of carbon nanotubes on the properties of polymer/carbon nanotubes composites: A critical review. *Polymer*, 51(5), 975–993.
8. Huang, Y.Y. and Terentjev, E.M. 2012. Dispersion of carbon nanotubes: Mixing, sonication, stabilization, and composite properties. *Polymers*, 4(1), 275–295.
9. Ma, P.-C., Siddiqui, N.A., Marom, G. et al. 2010. Dispersion and functionalization of carbon nanotubes for polymer-based nanocomposites: A review. *Composites Part A: Applied Science and Manufacturing*, 41(10), 1345–1367.
10. Navarro-Pardo, F., Martínez-Barrera, G., Martínez-Hernández, A.L. et al. 2012. Nylon 6,6 electrospun fibres reinforced by amino functionalised 1D and 2D carbon. *IOP Conference Series: Materials Science and Engineering*, 40(1), Article ID: 012023.
11. Navarro-Pardo, F., Martínez-Barrera, G., Martínez-Hernández, A.L. et al. 2013. Effects on the thermo-mechanical and crystallinity properties of nylon 6,6 electrospun fibres reinforced with one dimensional (1D) and two dimensional (2D) carbon. *Materials*, 6(8), 3494–3513.

12. Navarro-Pardo, F., Martínez-Barrera, G., Martínez-Hernández, A.L. et al. 2014. Influence of 1D and 2D carbon fillers and their functionalisation on crystallisation and thermo-mechanical properties of injection moulded nylon 6,6 nanocomposites. *Journal of Nanomaterials*, 2014 (2014), Article ID: 670261.

13. Layek, R.K. and Nandi, A.K. 2013. A review on synthesis and properties of polymer functionalized graphene. *Polymer*, 54(19), 5087–5103.

14. Kuila, T., Bose, S., Mishra, A.K. et al. 2012. Chemical functionalization of graphene and its applications. *Progress in Materials Science*, 57(7), 1061–1105.

15. Lin, L., Rong, M., Luo, F. et al. 2014. Luminescent graphene quantum dots as new fluorescent materials for environmental and biological applications. *TrAC—Trends in Analytical Chemistry*, 54, 83–102.

16. Martínez-Hernández, A.L., Velasco-Santos, C., and Castaño, V.M. 2010. Carbon nanotubes composites: Processing, grafting and mechanical and thermal properties. *Current Nanoscience*, 6(1), 12–39.

17. Velasco-Santos, C., Martínez-Hernández, A.L., and Castaño, V.M. 2011. Silanization of carbon nanotubes: Surface modification and polymer nanocomposites. In *Carbon Nanotubes—Polymer Nanocomposites*, ed. S. Yellampadi, Rijeka, Croatia: In Tech.

18. Primo, E.N., Gutierrez, F.A., Luque, G.L. et al. 2013. Comparative study of the electrochemical behavior and analytical applications of (bio)sensing platforms based on the use of multi-walled carbon nanotubes dispersed in different polymers. *Analytica Chimica Acta*, 805, 19–35.

19. Tiwari, I., Singh, K.P., and Singh, M. 2009. An insight review on the application of polymer-carbon nanotubes based composite material in sensor technology. *Russian Journal of General Chemistry*, 79(12), 2685–2694.

20. Jacobs, C.B., Peairs, M.J., and Venton, B.J. 2010. Review: Carbon nanotube based electrochemical sensors for biomolecules. *Analytica Chimica Acta*, 662(2), 105–127.

21. Shao, Y., Wang, J., Wu, H. et al. 2010. Graphene based electrochemical sensors and biosensors: A review. *Electroanalysis*, 22(10), 1027–1036.

22. Wang, Y., Li, Z., Wang, J. et al. 2011. Graphene and graphene oxide: Biofunctionalization and applications in biotechnology. *Trends in Biotechnology*, 29(5), 205–212.

23. Pavlidis, I.V., Patila, M., Bornscheuer, U.T. et al. 2014. Graphene-based nanobiocatalytic systems: Recent advances and future prospects. *Trends in Biotechnology*.

24. Battigelli, A., Ménard-Moyon, C., Da Ros, T. et al. 2013. Endowing carbon nanotubes with biological and biomedical properties by chemical modifications. *Advanced Drug Delivery Reviews*, 65(15), 1899–1920.

25. Wong, B.S., Yoong, S.L., Jagusiak, A. et al. 2013. Carbon nanotubes for delivery of small molecule drugs. *Advanced Drug Delivery Reviews*, 65(15), 1964–2015.

26. Bao, H., Pan, Y., Ping, Y. et al. 2011. Chitosan-functionalized graphene oxide as a nanocarrier for drug and gene delivery. *Small*, 7(11), 1569–1578.

27. Hwang, J.-Y., Shin, U.S., Jang, W.-C. et al. 2013. Biofunctionalized carbon nanotubes in neural regeneration: A mini-review. *Nanoscale*, 5(2), 487–497.

28. Goenka, S., Sant, V., and Sant, S. 2014. Graphene-based nanomaterials for drug delivery and tissue engineering. *Journal of Controlled Release*, 173(1), 75–88.

29. Chang, D.W., Jeon, I.Y., Choi H.J. et al. 2013. Mild and nondestructive chemical modification of carbon nanotubes (CNTs): Direct Friedel-Crafts acylation reaction. In *Physical and Chemical Properties of Carbon Nanotubes*, ed. S. Suzuki, In Tech. http://dx.doi.org/10.5772/50805.

30. Luo, S., Xu, X., Zhou, G. et al. 2014. Amino siloxane oligomer-linked graphene oxide as an efficient adsorbent for removal of Pb(II) from wastewater. *Journal of Hazardous Materials*, 274, 145–155.

31. Zhang, Y., Ma, H.-L., Peng, J. et al. 2013. Cr(VI) removal from aqueous solution using chemically reduced and functionalized graphene oxide. *Journal of Materials Science*, 48(5), 1883–1889.

32. Shao, D.D., Hu, J., Wang, X.K. et al. 2011. Plasma induced grafting multiwall carbon nanotubes with chitosan for 4,4′-dichlorobiphenyl removal from aqueous solution. *Chemical Engineering Journal*, 170(2–3), 498–504.

33. Liao, K.-L., Wan, A., Batteas, J.D. et al. 2008. Superhydrophobic surfaces formed using layer-by-layer self-assembly with aminated multiwall carbon nanotubes. *Langmuir*, 24(8), 4245–4253.

34. Cao, F., Ren, W., Ji, Y.-M. et al. 2009. The structural and electronic properties of amine-functionalized boron nitride nanotubes via ammonia plasmas: A density functional theory study. *Nanotechnology*, 20(14), Article ID:145703.

35. Syrgiannis, Z., Hauke, F., Röhrl, J. et al. 2008. Covalent sidewall functionalization of SWNTs by nucleophilic addition of lithium amides. *European Journal of Organic Chemistry*, 2008(15), 2544–2550.

36. Friedrich, J.F., Wettmarshausen, S., Hanelt, S. et al. 2010. Plasma-chemical bromination of graphitic materials and its use for subsequent functionalization and grafting of organic molecules. *Carbon*, 48(13), 3884–3894.

37. Compton, O.C., Dikin, D.A., Putz, K.W. et al. 2010. Electrically conductive "alkylated" graphene paper via chemical reduction of amine-functionalized graphene oxide paper. *Advanced Materials*, 22(8), 892–896.

38. Tetsuka, H., Asahi, R., Nagoya, A. et al. 2012. Optically tunable amino-functionalized graphene quantum dots. *Advanced Materials*, 24(39), 5333–5338.

39. Liu, X., Zeng, C., Tang, Z. et al. 2014. Liquefied graphene oxide with excellent amphiphilicity. *Chemistry Letters*, 43(2), 222–224.

40. Mo, Z., Gou, H., He, J. et al. 2012. Controllable synthesis of functional nanocomposites: Covalently functionalize graphene sheets with biocompatible L-lysine. *Applied Surface Science*, 258(22), 8623–8628.

41. Nakayama, K., Takada, T., Abe, S. et al. 2012. Amide bond formation between carboxylated multi-walled carbon nanotubes and glass surface by using carbodiimide condensing agent and triazole derivatives. *Molecular Crystals and Liquid Crystals*, 568(1), 38–45.

42. Davarpanah, M., Maghrebi, M., and Hosseinipour, E. 2014. Microwave-induced high surface functionalization of multi-walled carbon nanotubes for long-term dispersion in water. *Applied Physics A: Materials Science and Processing*, 115(1), 167–175.

43. Peng, E., Choo, E.S.G., Chandrasekharan, P. et al. 2012. Synthesis of manganese ferrite/graphene oxide nanocomposites for biomedical applications. *Small*, 8(23), 3620–3630.

44. Liu, J., Cui, L., and Losic, D. 2013. Graphene and graphene oxide as new nanocarriers for drug delivery applications. *Acta Biomaterialia*, 9(12), 9243–9257.

45. Kausar, A., Iqbal, A., and Hussain, S.T. 2013. Novel hybrids derived from poly(thiourea-amide)/Epoxy and carbon nanotubes. *Polymer—Plastics Technology and Engineering*, 52(11), 1169–1174.

46. Mallakpour, S., Abdolmaleki, A., and Borandeh, S. 2014. Covalently functionalized graphene sheets with biocompatible natural amino acids. *Applied Surface Science*, 307, 533–542.

47. Zhang, X., Hou, L., Cnossen, A. et al. 2011. One-pot functionalization of graphene with porphyrin through cycloaddition reactions. *Chemistry—A European Journal*, 17(32), 8957–8964.

48. Quintana, M., Spyrou, K., Grzelczak, M. et al. 2010. Functionalization of graphene via 1,3-dipolar cycloaddition. *ACS Nano*, 4(6), 3527–3533.

49. Brunetti, F.G., Herrero, M.A., Muñoz, J.D.M. et al. 2007. Reversible microwave-assisted cycloaddition of aziridines to carbon nanotubes. *Journal of the American Chemical Society*, 129(47), 14580–14581.

50. Jiang, Y., Jin, C., Yang, F. et al. 2011. A new approach to produce amino-carbon nanotubes as plasmid transfection vector by [2 + 1] cycloaddition of nitrenes. *Journal of Nanoparticle Research*, 13(1), 33–38.

51. Sun, J.-T., Zhao, L.-Y., Hong, C.-Y. et al. 2011. Selective Diels-Alder cycloaddition on semiconducting single-walled carbon nanotubes for potential separation application. *Chemical Communications*, 47(38), 10704–10706.

52. Bayazit, M.K., Celebi, N., and Coleman, K.S. 2014. A theoretical and experimental exploration of the mechanism of microwave assisted 1,3-dipolar cycloaddition of pyridinium ylides to single walled carbon nanotubes. *Materials Chemistry and Physics*, 145(1–2), 99–107.

53. Ryu, H.J., Mahapatra, S.S., Yadav, S.K. et al. 2013. Synthesis of click-coupled graphene sheet with chitosan: Effective exfoliation and enhanced properties of their nanocomposites. *European Polymer Journal*, 49(9), 2627–2634.

54. Singh, P., Campidelli, S., Giordani, S. et al. 2009. Organic functionalisation and characterisation of single-walled carbon nanotubes. *Chemical Society Reviews*, 38(8), 2214–2230.

55. Servant, A., Bianco, A., Prato, M. et al. 2014. Graphene for multi-functional synthetic biology: The last 'zeitgeist' in nanomedicine. *Bioorganic and Medicinal Chemistry Letters*, 24(7), 1638–1649.

56. Osswald, S. and Etzold, B.J.M. 2013. Oxidation and purification of carbon nanostructures. In *Carbon Nanomaterials*, eds. Y. Gogotsi and V. Presser, CRC. http://dx.doi.org/10.1201/b15591.

57. Marques, P.A.A.P., Gonçalves, G., Cruz, S. et al. 2011. Functionalized graphene nanocomposites. In *Nanocomposite Technology*, ed. A. Hashim, In Tech. http://dx.doi.org/10.5772/18209.

58. Zheng, D., Vashist, S.K., Dykas, M.M. et al. 2013. Graphene versus multi-walled carbon nanotubes for electrochemical glucose biosensing. *Materials*, 6(3), 1011–1027.

59. Chen, C., Liang, B., Lu, D. et al. 2010. Amino group introduction onto multiwall carbon nanotubes by NH_3/Ar plasma treatment. *Carbon*, 48(4), 939–948.

60. Singh, S., Singh, M.K., Kulkarni, P.P. et al. 2012. Amine-modified graphene: Thrombo-protective safer alternative to graphene oxide for biomedical applications. *ACS Nano*, 6(3), 2731–2740.

61. Gabriel, G., Sauthier, G., Fraxedas, J. et al. 2006. Preparation and characterisation of single-walled carbon nanotubes functionalised with amines. *Carbon*, 44(10), 1891–1897.

62. Jimeno, A., Goyanes, S., Eceiza, A. et al. 2009. Effects of amine molecular structure on carbon nanotubes functionalization. *Journal of Nanoscience and Nanotechnology*, 9(10), 6222–6227.

63. Zhou, H., Wang, T., and Duan, Y.Y. 2013. A simple method for amino-functionalization of carbon nanotubes and electrodeposition to modify neural microelectrodes. *Journal of Electroanalytical Chemistry*, 688, 69–75,

64. Ellison, M.D. and Gasda, P.J. 2008. Functionalization of single-walled carbon nanotubes with 1,4-benzenediamine using a diazonium reaction. *Journal of Physical Chemistry C*, 112(3), 738–740.

65. Cividanes, L.S., Brunelli, D.D., Antunes, E.F. et al. 2013. Cure study of epoxy resin reinforced with multiwalled carbon nanotubes by Raman and luminescence spectroscopy. *Journal of Applied Polymer Science*, 127(1), 544–553.

66. Davis, D.C., Wilkerson, J.W., Zhu, J. et al. 2011. A strategy for improving mechanical properties of a fiber reinforced epoxy composite using functionalized carbon nanotubes. *Composites Science and Technology*, 71(8), 1089–1097.

67. Kakade, B.A. and Pillai, V.K. 2008. An efficient route towards the covalent functionalization of single walled carbon nanotubes. *Applied Surface Science*, 254(16), 4936–4943.

68. Yang, K., Gu, M., Han, H. et al. 2008. Influence of chemical processing on the morphology, crystalline content and thermal stability of multi-walled carbon nanotubes. *Materials Chemistry and Physics*, 112(2), 387–392.

69. Thanh, N.T.K. and Green L.A.W. 2010. Functionalisation of nanoparticles for biomedical applications. *Nano Today*, 5(3), 213–230.

70. Ma, P.-C., Mo, S.-Y., Tang, B.-Z. et al. 2010. Dispersion, interfacial interaction and re-agglomeration of functionalized carbon nanotubes in epoxy composites. *Carbon*, 48(6), 1824–1834.

71. Sahoo, S., Nayak, G.C., and Das, C.K. 2012. Synthesis and electrochemical characterization of modified graphene/polypyrrole nanocomposites. *Macromolecular Symposia*, 315(1), 177–187.

72. Sahoo, S., Karthikeyan, G., Nayak, G.C. et al. 2012. Modified graphene/polyaniline nanocomposites for supercapacitor application. *Macromolecular Research*, 20(4), 415–421.

73. Sahoo, S., Bhattacharya, P., Hatui, G. et al. 2013. Sonochemical synthesis and characterization of amine-modified graphene/conducting polymer nanocomposites. *Journal of Applied Polymer Science*, 128(3), 1476–1483.

74. Pavlidis, I.V., Vorhaben, T., Gournis, D. et al. 2012. Regulation of catalytic behaviour of hydrolases through interactions with functionalized carbon-based nanomaterials. *Journal of Nanoparticle Research*, 14, 842.

75. Wang, Y., Iqbal, Z., and Malhotra, S.V. 2005. Functionalization of carbon nanotubes with amines and enzymes. *Chemical Physics Letters*, 402(1–3), 96–101.

76. Daniel, S., Rao, T.P., Rao, K.S. et al. 2007. A review of DNA functionalized/grafted carbon nanotubes and their characterization. *Sensors and Actuators B: Chemical*, 122(2), 672–682.

77. Makarucha, A.J., Todorova, N., and Yarovsky, I. 2011. Nanomaterials in biological environment: A review of computer modelling studies. *European Biophysics Journal*, 40(2), 103–115.

78. Ali, S.R., Parajuli, R.R., Balogun, Y. et al. 2008. A non oxidative electrochemical sensor based on a self-doped polyaniline/carbon nanotube composite for sensitive and selective detection of the neurotransmitter dopamine: A review. *Sensors*, 8(12), 8423–8452.

79. Gao, H. and Kong, Y. 2004. Simulation of DNA-nanotube interactions. *Annual Review of Materials Research*, 34, 123–150.

80. Spitalsky, Z., Tasis, D., Papagelis, K. et al. 2010. Carbon nanotube-polymer composites: Chemistry, processing, mechanical and electrical properties. *Progress in Polymer Science (Oxford)*, 35(3), 357–401.

81. Kumar, B., Castro, M., and Feller, J.F. 2012. Controlled conductive junction gap for chitosan-carbon nanotube quantum resistive vapour sensors. *Journal of Materials Chemistry*, 22(21), 10656–10664.

82. Lau, C., Cooney, M.J., and Atanassov, P. 2008. Conductive macroporous composite chitosan-carbon nanotube scaffolds. *Langmuir*, 24(13), 7004–7010.

83. Venkatesan, J., Qian, Z.-J., Ryu, B. et al. 2011. Preparation and characterization of carbon nanotube-grafted-chitosan—Natural hydroxyapatite composite for bone tissue engineering. *Carbohydrate Polymers*, 83(2), 569–577.

84. Rouse, J.G. and Van Dyke, M.E. 2010. A review of keratin-based biomaterials for biomedical applications. *Materials*, 3(2), 999–1014.

85. Martínez-Hernández, A.L. and Velasco-Santos, C. 2012. Keratin fibers from chicken feathers: Structure and advances in polymer composites. In *Keratin Structure, Properties and Applications*, R. Dullaart and J. Mousquès (Eds.), pp. 149–21, Nova Science Publishers. http://www.novapublishers.com/catalog/product_info.php?products_id=32840.

86. Martínez-Hernández, A.L., Velasco-Santos, C., De Icaza, M. et al. 2005. Microstructural characterisation of keratin fibres from chicken feathers. *International Journal of Environment and Pollution*, 23(2), 162–178.

87. Yuan, W., Jiang, G., Che, J. et al. 2008. Deposition of silver nanoparticles on multi-walled carbon nanotubes grafted with hyperbranched poly(amidoamine) and their antimicrobial effects. *Journal of Physical Chemistry C*, 112(48), 18754–18759.

88. Xue, Q., Huang, H., Wang, L. et al. 2013. Nearly monodisperse graphene quantum dots fabricated by amine-assisted cutting and ultrafiltration. *Nanoscale*, 5(24), 12098–12103.

89. Yu, D. and Dai, L. 2010. Self-assembled graphene/carbon nanotube hybrid films for supercapacitors. *Journal of Physical Chemistry Letters*, 1(2), 467–470.

90. Guo, L. and Peng, Z. 2008. One-pot synthesis of carbon nanotube-polyaniline-gold nanoparticle and carbon nanotube-gold nanoparticle composites by using aromatic amine chemistry. *Langmuir*, 24(16), 8971–8975.

91. Park, J.K. and Kim, D.S. 2014. Effects of an aminosilane and a tetra-functional epoxy on the physical properties of di-functional epoxy/graphene nanoplatelets nanocomposites. *Polymer Engineering and Science*, 54(4), 969–976.

92. Inagaki, N., Narushima, K., Hashimoto, H. et al. 2007. Implantation of amino functionality into amorphous carbon sheet surfaces by NH_3 plasma. *Carbon*, 45(4), 797–804.

93. Sainsbury, T., Ikuno, T., Okawa, D. et al. 2007. Self-assembly of gold nanoparticles at the surface of amine- and thiol-functionalized boron nitride nanotubes. *Journal of Physical Chemistry C*, 111(35), 12992–12999.

94. Chen, Z., Dai, X.J., Lamb, P.R. et al. 2012. Practical amine functionalization of multi-walled carbon nanotubes for effective interfacial bonding. *Plasma Processes and Polymers*, 9(7), 733–741.

95. Yook, J.Y., Jun, J. and Kwak, S. 2010. Amino functionalization of carbon nanotube surfaces with NH3 plasma treatment. *Applied Surface Science*, 256, 23, 6941–6944.

96. Lee, J.-S., Yoon, J.-C., and Jang, J.-H. 2013. A route towards superhydrophobic graphene surfaces: Surface-treated reduced graphene oxide spheres. *Journal of Materials Chemistry A*, 1(25), 7312–7315.

97. Shin, S.R., Bae, H., Cha, J.M. et al. 2012. Carbon nanotube reinforced hybrid microgels as scaffold materials for cell encapsulation. *ACS Nano*, 6(1), 362–372.

98. Luong, J.H.T., Hrapovic, S., Wang, D. et al. 2004. Solubilization of multiwall carbon nanotubes by 3-aminopropyltriethoxysilane towards the fabrication of electrochemical biosensors with promoted electron transfer. *Electroanalysis*, 16(1–2), 132–139.

99. Addiego, C. 2013. Fabrication of graphene field effect transistors on boron nitride substrates. *National Nanotechnology Infrastructure Network REU Proceedings*, pp. 210–229. http://www.nnin.org/sites/default/files/2013_reu_ra/2013nninRA_Addiego.pdf.

100. Suehiro, J., Ikeda, N., Ohtsubo, A. et al. 2008. Fabrication of bio/nano interfaces between biological cells and carbon nanotubes using dielectrophoresis. *Microfluidics and Nanofluidics*, 5(6), 741–747.

101. Zheng, D., Vashist, S.K., Dykas, M.M. et al. 2013. Graphene versus multiwalled carbon nanotubes for electrochemical glucose biosensing. *Materials*, 6, 1011–1027.

102. Teixeira, S., Burwell, G., Castaing, A. et al. 2014. Epitaxial graphene immunosensor for human chorionic gonadotropin. *Sensors and Actuators B: Chemical*, 190, 723–729.

103. Li, W., Shi, C., Shan, M. et al. 2013. Influence of silanized low-dimensional carbon nanofillers on mechanical, thermomechanical, and crystallization behaviors of poly(L-lactic acid) composites—A comparative study. *Journal of Applied Polymer Science*, 130(2), 1194–1202.

104. Yuen, S.-M., Ma, C.-C.M., Lin, Y.-Y. et al. 2007. Preparation, morphology and properties of acid and amine modified multiwalled carbon nanotube/polyimide composite. *Composites Science and Technology*, 67(11–12), 2564–2573.

105. Mao, Z., Wu, W., Xie, C. et al. 2011. Lipophilic carbon nanotubes and their phase-separation in SBS. *Polymer Testing*, 30(2), 260–270.

106. Krajcik, R., Jung, A., Hirsch, A. et al. 2008. Functionalization of carbon nanotubes enables non-covalent binding and intracellular delivery of small interfering RNA for efficient knock-down of genes. *Biochemical and Biophysical Research Communications*, 369(2), 595–602.

107. Lee, S.W., Kim, B.-S., Chen, S. et al. 2009. Layer-by-layer assembly of all carbon nanotube ultrathin films for electrochemical applications. *Journal of the American Chemical Society*, 131(2), 671–679.

108. Shen, J., Huang, W., Wu, L. et al. 2007. Study on amino-functionalized multiwalled carbon nanotubes. *Materials Science and Engineering A*, 464(1–2), 151–156.

109. Benlikaya, R., Slobodian, P., and Riha, P. 2014. The enhanced alcohol sensing response of multiwalled carbon nanotube networks induced by alkyl diamine treatment. *Sensors and Actuators B*, 201, 122–130.

110. Shen, J., Huang, W., Wu, L. et al. 2007. The reinforcement role of different amino-functionalized multi-walled carbon nanotubes in epoxy nanocomposites. *Composites Science and Technology*, 67(15–16), 3041–3050.

111. Chidawanyika, W. and Nyokong, T. 2010. Characterization of amine-functionalized single-walled carbon nanotube-low symmetry phthalocyanine conjugates. *Carbon*, 48(10), 2831–2838.

112. Amiri, A., Maghrebi, M., Baniadam, M. et al. 2011. One-pot, efficient functionalization of multi-walled carbon nanotubes with diamines by microwave method. *Applied Surface Science*, 257(23), 10261–10266.

113. Moaseri, E., Baniadam, M., Maghrebi, M. et al. 2013. A simple recoverable titration method for quantitative characterization of amine-functionalized carbon nanotubes. *Chemical Physics Letters*, 555, 164–167.

114. Basiuk, E.V., Basiuk, V.A., Meza-Laguna, V. et al. 2012. Solvent-free covalent functionalization of multi-walled carbon nanotubes and nanodiamond with diamines: Looking for cross-linking effects. *Applied Surface Science*, 259, 465–476.

115. Rahimpour, A., Jahanshahi, M., Khalili, S. et al. 2012. Novel functionalized carbon nanotubes for improving the surface properties and performance of polyethersulfone (PES) membrane. *Desalination*, 286, 99–107.

116. Sager, R.J., Klein, P.J., Davis, D.C. et al. 2011. Interlaminar fracture toughness of woven fabric composite laminates with carbon nanotube/epoxy interleaf films. *Journal of Applied Polymer Science*, 121(4), 2394–2405.

117. Lei, Y., Xiong, C., Guo, H. et al. 2008. Controlled viscoelastic carbon nanotube fluids. *Journal of the American Chemical Society*, 130(11), 3256–3257.
118. Li, Q., Dong, L., Fang, J. et al. 2010. Property-structure relationship of nanoscale ionic materials based on multiwalled carbon nanotubes. *ACS Nano*, 4(10), 5797–5806.
119. Ke, G., Guan, W., Tang, C. et al. 2007. Covalent functionalization of multiwalled carbon nanotubes with a low molecular weight chitosan. *Biomacromolecules*, 8(2), 322–326.
120. An, J.S., Nam, B.-U., Tan, S.H. et al. 2007. Study on the functionalization of multi-walled carbon nanotube with monoamine terminated poly(ethylene oxide). *Macromolecular Symposia*, 249–250, 276–282.
121. Cao, X., Dong, H., Li, C.M. et al. 2009. The enhanced mechanical properties of a covalently bound chitosan-multiwalled carbon nanotube nanocomposite. *Journal of Applied Polymer Science*, 113(1), 466–472.
122. Castillo, J.J., Torres, M.H., Molina, D.R. et al. 2012. Monitoring the functionalization of single-walled carbon nanotubes with chitosan and folic acid by two-dimensional diffusion-ordered NMR spectroscopy. *Carbon*, 50(8), 2691–2697.
123. Liu, X., Zhang, Y., Ma, D. et al. 2013. Biocompatible multi-walled carbon nanotube-chitosan-folic acid nanoparticle hybrids as GFP gene delivery materials. *Colloids and Surfaces B: Biointerfaces*, 111, 224–231.
124. Zhu, H., Fu, Y., Jiang, R. et al. 2013. Preparation, characterization and adsorption properties of chitosan modified magnetic graphitized multi-walled carbon nanotubes for highly effective removal of a carcinogenic dye from aqueous solution. *Applied Surface Science*, 285(PARTB), 865–873.
125. Li, C., Yang, K., Zhang, Y. et al. 2011. Highly biocompatible multi-walled carbon nanotube-chitosan nanoparticle hybrids as protein carriers. *Acta Biomaterialia*, 7(8), 3070–3077.
126. Dai, B., Cao, M., Fang, G. et al. 2012. Schiff base-chitosan grafted multiwalled carbon nanotubes as a novel solid-phase extraction adsorbent for determination of heavy metal by ICP-MS. *Journal of Hazardous Materials*, 219–220, 103–110.
127. Shariatzadeh, B. and Moradi, O. Surface functionalization of multiwalled carbon nanotubes with chitosan and magnesium oxide nanoparticles by microwave-assisted synthesis. *Polymer Composites*, 35, 2050–2055.
128. Castrejón-Parga, K.Y., Camacho-Montes, H., and Rodríguez-González, C.A. Chitosan–starch film reinforced with magnetite-decorated carbon nanotubes. *Journal of Alloys and Compounds*, 615(S1), S505–S510.
129. Giroud, F. and Favreau, V. Cosmetic composition for voluminizing keratin fibers and cosmetic use of nanotubes for voluminizing keratin fibers. US Patent 0115232, filed June 2003 and issued June 2004.
130. Estévez-Martínez, Y., Velasco-Santos, C., Martínez-Hernández, A.L. et al. 2013. Grafting of multiwalled carbon nanotubes with chicken feather keratin. *Journal of Nanomaterials*, 2013, 702157.
131. Jeon, I.N.-Y., Kang, S.-W., Loon-Seng, T.A.N. et al. 2010. Grafting of polyaniline onto the surface of 4-aminobenzoylfunctionalized multiwalled carbon nanotube and its electrochemical properties. *Journal of Polymer Science, Part A: Polymer Chemistry*, 48(14), 3103–3112.
132. Jeon, I.N.-Y., Loon-Seng, T.A.N., and Baek, J.-B. 2010. Synthesis and electrical properties of polyaniline/polyaniline grafted multiwalled carbon nanotube mixture via in situ static interfacial polymerization. *Journal of Polymer Science, Part A: Polymer Chemistry*, 48(9), 1962–1972.

133. Kumar Mishra, A. and Ramaprabhu, S. 2012. Polyaniline/multiwalled carbon nanotubes nanocomposite-an excellent reversible CO_2 capture candidate. *RSC Advances*, 2(5), 1746–1750.

134. You, Y.-Z., Yan, J.-J., Yu, Z.-Q. et al. 2009. Multi-responsive carbon nanotube gel prepared via ultrasound-induced assembly. *Journal of Materials Chemistry*, 19(41), 7656–7660.

135. Gui, M.M., Yap, Y.X., Chai, S.-P. et al. 2013. Multi-walled carbon nanotubes modified with (3-aminopropyl)triethoxysilane for effective carbon dioxide adsorption. *International Journal of Greenhouse Gas Control*, 14, 65–73.

136. Scheibe, B., Borowiak-Palen, E., and Kalenczuk, R.J. 2009. Effect of the silanization processes on the properties of oxidized multiwalled carbon nanotube. *Proceedings of the Third National Conference on Nanotechnology*, 116, 150–155.

137. Sarkar, A. and Daniels-Race, T. 2013. Electrophoretic deposition of carbon nanotubes on 3-amino-propyl-triethoxysilane (APTES) surface functionalized silicon substrates. *Nanomaterials*, 3, 272–288.

138. Kang, D.-Y., Brunelli, N.A., Yucelen, G.I. et al. 2014. Direct synthesis of single-walled aminoaluminosilicate nanotubes with enhanced molecular adsorption selectivity. *Nature Communications*, 5, 3342.

139. Villa, C.H., McDevitt, M.R., Escorcia, F.E. et al. 2008. Synthesis and biodistribution of oligonucleotide-functionalized, tumor-targetable carbon nanotubes. *Nano Letters*, 8(12), 4221–4228.

140. Chun, K.-Y., Moon, I.-K., Han, J.-H. et al. 2013. Highly water-soluble multi-walled carbon nanotubes amine-functionalized by supercritical water oxidation. *Nanoscale*, 5(21), 10171–10174.

141. Dai, X.J., Chen, Y., Chen, Z. et al. 2011. Controlled surface modification of boron nitride nanotubes. *Nanotechnology*, 22(24), Article ID: 245301.

142. Ikuno, T., Sainsbury, T., Okawa, D. et al. 2007. Amine-functionalized boron nitride nanotubes. *Solid State Communications*, 142(11), 643–646.

143. Khodadadei, F., Ghourchian, H., Soltanieh, M. et al. 2014. Rapid and clean amine functionalization of carbon nanotubes in a dielectric barrier discharge reactor for biosensor development. *Electrochimica Acta*, 115, 378–385.

144. Cooper, D.R., D'Anjou, B., Ghattamaneni, N. et al. 2012. Experimental review of graphene, *Condensed Matter Physics*, 2012, 1–50.

145. Dreyer, D.R., Park, S., Bielawski, C.W. et al. 2010. The chemistry of graphene oxide, *Chemical Society Reviews*, 39(1), 228–240.

146. Kim, Y.-K. and Min, D.-H. 2010. Preparation of scrolled graphene oxides with multi-walled carbon nanotube templates. *Carbon*, 48(15), 4283–4288.

147. Byon, H.R., Lee, S.W., Chen, S. et al. 2011. Thin films of carbon nanotubes and chemically reduced graphenes for electrochemical micro-capacitors. *Carbon*, 49(2), 457–467.

148. Park, J.S., Cho, S.M., Kim, W.-J. et al. 2011. Fabrication of graphene thin films based on layer-by-layer self-assembly of functionalized graphene nanosheets. *ACS Applied Materials and Interfaces*, 3(2), 360–368.

149. Remyamol, T., John, H., and Gopinath, P. 2013. Synthesis and nonlinear optical properties of reduced graphene oxide covalently functionalized with polyaniline. *Carbon*, 59, 308–314.

150. Hou, W., Tang, B., Lu, L. et al. 2014. Preparation and physico-mechanical properties of amine-functionalized graphene/polyamide 6 nanocomposite fiber as a high performance material. *RSC Advances*, 4(10), 4848–4855.

151. Matsuo, Y., Miyabe, T., Fukutsuka, T. et al. 2007. Preparation and characterization of alkylamine-intercalated graphite oxides. *Carbon*, 45(5), 1005–1012.

152. Bourlinos, A.B., Gournis, D., Petridis, D. et al. 2003. Graphite oxide: Chemical reduction to graphite and surface modification with primary aliphatic amines and amino acids. *Langmuir*, 19(15), 6050–6055.

153. Compton, O.C., Jain, B., Dikin, D.A. et al. 2011. Chemically active reduced graphene oxide with tunable C/O ratios. *ACS Nano*, 5(6), 4380–4391.

154. Ryu, S.H. and Shanmugharaj, A.M. 2014. Influence of long-chain alkylamine-modified graphene oxide on the crystallization, mechanical and electrical properties of isotactic polypropylene nanocomposites. *Chemical Engineering Journal*, 244, 552–560.

155. Stankovich, S., Dikin, D.A., Compton, O.C. et al. 2010. Systematic post-assembly modification of graphene oxide paper with primary alkylamines. *Chemistry of Materials*, 22(14), 4153–4157.

156. Guan, W., Li, Z., Zhang, H. et al. 2013. Amine modified graphene as reversed-dispersive solid phase extraction materials combined with liquid chromatography-tandem mass spectrometry for pesticide multi-residue analysis in oil crops. *Journal of Chromatography A*, 1286, 1–8.

157. Stine, R., Ciszek, J.W., Barlow, D.E. et al. 2012. High-density amine-terminated monolayers formed on fluorinated CVD-grown graphene. *Langmuir*, 28(21), 7957–7961.

158. Valentini, L., Cardinali, M., Bittolo Bon, S. et al. 2010. Use of butylamine modified graphene sheets in polymer solar cells. *Journal of Materials Chemistry*, 20(5), 995–1000.

159. Ikeda, Y., Karim, M.R., Takehira, H. et al. 2013. Hydrogen generation by graphene oxide-alkylamine hybrids through photocatalytic water splitting. *Chemistry Letters*, 2013, 1–3.

160. Chua, C.K. and Pumera, M. 2012. Friedel-crafts acylation on graphene. *Chemistry—An Asian Journal*, 7(5), 1009–1012.

161. Fang, M., Zhang, Z., Li, J. et al. 2010. Constructing hierarchically structured interphases for strong and tough epoxy nanocomposites by amine-rich graphene surfaces. *Journal of Materials Chemistry*, 20(43), 9635–9643.

162. Lee, D., Choi, M.-C., and Ha, C.-S. 2012. Polynorbornene dicarboximide/amine functionalized graphene hybrids for potential oxygen barrier films. *Journal of Polymer Science, Part A: Polymer Chemistry*, 50(8), 1611–1621.

163. Tsoufis, T., Tuci, G., Caporali, S. et al. 2013. P-xylylenediamine intercalation of graphene oxide for the production of stitched nanostructures with a tailored interlayer spacing. *Carbon*, 59, 100–108.

164. Shan, C., Wang, L., Han, D. et al. 2013. Polyethyleneimine-functionalized graphene and its layer-by-layer assembly with prussian blue. *Thin Solid Films*, 534, 572–576.

165. Yang, Q., Pan, X., Clarke, K. et al. 2012. Covalent functionalization of graphene with polysaccharides. *Industrial and Engineering Chemistry Research*, 51(1), 310–317.

166. Zeng, Q., Cheng, J.-S., Liu, X.-F. et al. 2011. Palladium nanoparticle/chitosan-grafted graphene nanocomposites for construction of a glucose biosensor. *Biosensors and Bioelectronics*, 26(8), 3456–3463.

167. Shao, L., Chang, X., Zhang, Y. et al. 2013. Graphene oxide cross-linked chitosan nanocomposite membrane. *Applied Surface Science*, 280, 989–992.

168. Bustos-Ramírez, K., Martínez-Hernández, A.L., Martínez-Barrera, G. et al. 2013. Covalently bonded chitosan on graphene oxide via redox reaction. *Materials*, 6(3), 911–926.

169. Feng, X., Wang, X., Xing, W. et al. 2013. Simultaneous reduction and surface functionalization of graphene oxide by chitosan and their synergistic reinforcing effects in PVA films. *Industrial and Engineering Chemistry Research*, 52(36), 12906–12914.

170. Hu, H., Wang, X., Wang, J. et al. 2011. Microwave-assisted covalent modification of graphene nanosheets with chitosan and its electrorheological characteristics. *Applied Surface Science*, 257(7), 2637–2642.

171. Rana, V.K., Choi, M.-C., Kong, J.-Y. et al. 2011. Synthesis and drug-delivery behavior of chitosan-functionalized graphene oxide hybrid nanosheets. *Macromolecular Materials and Engineering*, 296(2), 131–140.

172. Rodríguez-González, C., Kharissova, O.V., Martínez-Hernández, A.L. et al. 2012. Graphene oxide sheets covalently grafted with keratin obtained from chicken feathers. *Digest Journal of Nanomaterials and Biostructures*, 8(1), 127–138.

173. Rodríguez-González, C., Martínez-Hernández, A.L., Castaño, V.M. et al. 2012. Polysaccharide nanocomposites Reinforced with graphene oxide and keratin-grafted graphene oxide, *Industrial Engineering and Chemistry Research*, 51(3), 619–629.

174. Kavitha, T., Haider Abdi, S.I., and Park, S.-Y. 2013. PH-sensitive nanocargo based on smart polymer functionalized graphene oxide for site-specific drug delivery. *Physical Chemistry Chemical Physics*, 15(14), 5176–5185.

175. Hu, W., Zhan, J., Wang, X. et al. 2014. Effect of functionalized graphene oxide with hyper-branched flame retardant on flammability and thermal stability of cross-linked polyethylene. *Industrial and Engineering Chemistry Research*, 53(8), 3073–3083.

176. Devi, R., Relhan, S., and Pundir, C.S. 2013. Construction of a chitosan/polyaniline/graphene oxide nanoparticles/polypyrrole/Au electrode for amperometric determination of urinary/plasma oxalate. *Sensors and Actuators B: Chemical*, 186, 17–26.

177. Bosch-Navarro, C., Coronado, E., and Martí-Gastaldo, C. 2013. Controllable coverage of chemically modified graphene sheets with gold nanoparticles by thermal treatment of graphite oxide with *N,N*-dimethylformamide. *Carbon*, 54, 201–207.

178. Yang, H., Shan, C., Li, F. et al. 2009. Covalent functionalization of polydisperse chemically-converted graphene sheets with amine-terminated ionic liquid. *Chemical Communications*, 26, 3880–3882.

179. Wu, L., Zhang, B., Lu, H. et al. 2014. Nanoscale ionic materials based on hydroxyl-functionalized graphene. *Journal of Materials Chemistry A*, 2(5), 1409–1417.

180. Li, R., Liu, C., Ma, M. et al. 2013. Synthesis of 1,3-di(4-amino-1-pyridinium)propane ionic liquid functionalized graphene nanosheets and its application in direct electrochemistry of hemoglobin. *Electrochimica Acta*, 95, 71–79.

181. Lai, L., Chen, L., Zhan, D. et al. 2011. One-step synthesis of NH2-graphene from in situ graphene-oxide reduction and its improved electrochemical properties. *Carbon*, 49(10), 3250–3257.

182. Guo, X., Wei, Q., Du, B. et al. 2013. Removal of metanil yellow from water environment by amino functionalized graphenes (NH$_2$-G)—Influence of surface chemistry of NH 2-G. *Applied Surface Science*, 284, 862–869.

183. Hu, Y., Shen, J., Li, N. et al. 2010. Amino-functionalization of graphene sheets and the fabrication of their nanocomposites. *Polymer Composites*, 31(12), 1987–1994.

184. Zhang, F., Jiang, H., Li, X. et al. 2014. Amine-functionalized GO as an active and reusable acid-base bifunctional catalyst for one-pot cascade reactions. *ACS Catalysis*, 4(2), 394–401.

SECTION II

CHEMICAL FUNCTIONALIZATION OF CARBON NANOMATERIALS

Properties and Applications

CHAPTER 23

CONTENTS

Applications of Functionalized Carbon-Based Nanomaterials

23

J. Sadhik Basha

23.1 INTRODUCTION

Nanotechnology deals with structures of 100 nm and involves developing materials or devices within this range. Modern nanotechnology techniques lead to synthesize the metallic/nonmetallic particles of nanosize dimensions to play a vital role in many different scientific regimes and technological areas, particularly in the energy-related science and technology (Sadhik Basha 2014). The size of the nanoparticle overcomes the difficulties of settling, abrasion, and clogging compared to that of microsized particles (Wickham et al. 2006). The research community is actively questing hundreds of applications of nanotechnology in the prominent domain of bionanotechnology, nanostructured catalysts, and carbon-based nanomaterials. *Carbon* is one of the most environment-friendly and versatile elements profusely found in various states in our earth. The application of carbon materials dates back from the past decades, sighting a speedy scientific importance incited by the innovations in *carbon-based materials*. Carbon-based materials include the elemental carbon in various allotropic forms, carbon composites, and currently those materials play a vital role in the advancement of nanotechnology. The major advantages of carbon-based materials are surface functionalization, unique morphology, and chemical versatility. One of the potential applications of carbon-based materials is the carbon nanotube (CNT).

CNTs possess unique properties due to their specific structure, and hence it is one of the propitious candidates for scientific applications. They vary in sizes or features, ranging from 1 to 100 nm in one or more dimensions, and hence it is one of the core features of emerging technological revolution. The main advantages of CNTs are unique thermal, electronic, material, and biological properties, which are not found in conventional materials. With regard to these unique properties and significant recognition capabilities, CNTs have resulted in systems with improved performance and novel applications in all the research domains including fuel engineering science.

One of the very recent applications of nanotechnology is the incorporation of nanoparticles (such as CNT, alumina, ceria) with diesel fuel to act as a fuel-borne catalyst. Choi et al. (1995) observed that nanofluids exhibit superior thermal and stability properties when compared to that of conventional heat transfer fluids, owing to their enhanced surface-area-to-volume ratio. They also revealed that nanofluids behave like molecules of liquid, since they do not cause any clogging problems in the fluid flow pipelines. Deluca et al. (2005) studied that the surface-area-to-volume ratio of a nanoparticle increases considerably and hence allows the more amount of fuel to contact with the oxidizer (i.e., air). This in turn makes the fuel undergo improved combustion to enhance the performance and reduce the harmful emissions (such as NO_x, HC, CO, smoke, and soot). Further, the investigation of nanoparticles reported by Prasher et al. (2005) revealed that the addition of nanoparticles to a fluid would enhance its physical properties, such as thermal conductivity, mass diffusivity, and radiative heat transfer. Furthermore, the improved heat transfer performance of nanofluids is also due to the enhancement in larger surface area, heat capacity, thermal conductivity, and collisions between the fluid, particles, and flow passages (Wickham et al. 2006).

Yetter et al. (2009) critically reviewed the reports on nanoparticle combustion. They revealed that nanosize metallic powders that possess a high specific surface area (SSA) will have the tendency to store energy, which may in turn lead to high reactivity. Owing to those potential properties of nanoparticles, they observed that nanosize powders will play a vital role in future fuels, propellants, pyrotechnics, explosives, etc. In their detailed report on nanoparticle combustion, they predicted that adding a nanocatalyst to hydrocarbon (HC) fuels (such as diesel) will lead to lower ignition delay and enhanced ignition temperature. In addition, they revealed that adding a fuel-borne nanocatalyst to diesel fuels will also lead to reduced harmful emissions (such as NO_x, smoke, and soot). They concluded that nanoenergetic particles have various potential advantages in the field of fuel engineering science and rocket-based propellants to achieve a better performance in coming years. However, very few efforts have been reported on incorporating the potential nanoparticles with fossil fuels to attain an improved performance and reduce harmful emissions of diesel engines. The nanoparticles possess excellent dispersions in all the base fluids (such as water, oil, and ethylene glycol) to form stable suspensions even after several weeks or months (Eastman et al. 1999).

23.2 APPLICATIONS OF CARBON NANOMATERIALS/CNT

One of the major applications of nanotechnology is the invention of CNTs. The unique properties of CNTs have led to many potential applications ranging from chemical electrodes to composites and also in the fuel engineering fields. CNTs are termed as 1D cylinders of carbon with single layer (SWCNT) or multiple layers (MWCNT) (Iijima and Ichihashi 1993). CNTs have been of great interest for various applications to improve the performance of thermal systems. Very recently, researchers are focusing to incorporate CNTs with fossil fuels to act as a fuel-borne nanoadditive in order to attain better working characteristics of diesel engines. Marquis and Chibante (2005) suspended the CNT in a fluid and observed a long-term chemical stability and substantial enhancement in the thermal properties. They have also revealed that the heat transfer fluids (such as diesel) possess low thermal conductivity, and their thermal properties can be improved by adding CNT. Moy et al. (2002) reported that the CNT could act as a potential nanoadditive for the fuels to perform the following functions:

1. To enhance the burning rate of the fuel (particularly for the low-volatile fuels like diesel) and cetane number (ignition quality of the fuel)
2. To act as an antiknock additive
3. To afford as a fuel conductive
4. To promote clean burning
5. To suppress smoke formation

They have also claimed that the CNTs have the ability to trap the free radicals to function as an antiknock additive. In addition, they have also revealed that the addition of CNT to diesel fuel will lead to enhancement in the cetane number, which may, in turn, lead to shorten the ignition delay. In fuel engineering science, the fuel that has a higher cetane number will undergo shortened ignition delay and vice versa and in turn lead to better combustion characteristics in diesel engines.

In this section, we will discuss some of the potential applications of nanoparticles and carbon-based nanomaterials that the researchers are focusing on. Some recent investigations have been carried out on adding nanomaterials to fossil fuels to enhance the thermal efficiency and to reduce the harmful emission from the internal combustion engines. The notion of adding the additives to the diesel fuel is to attain better fuel properties so as to perform the following specific functions (Nubia et al. 2007):

1. To reduce harmful emissions
2. To improve fluid stability over a wider range of conditions
3. To improve the viscosity index
4. To improve the ignition by reducing its delay time and flash point
5. To reduce the wear with agents that adsorb onto metal surfaces

Yetter et al. (2009) and Dreizin (2000) have critically reviewed the reports on nanoparticle combustion and revealed that due to their high SSA, it could lead to high reactivity. They have also revealed that adding nanocatalyst to HC fuels (such as diesel) will shorten the ignition delay and reduce soot emissions. Several researchers (Roger 2006; Roos et al. 2008; Sabourin et al. 2009) have revealed that adding nanosize particles to fuel will act as a liquid fuel catalyst and thereby enhances the ignition and combustion characteristics of the engine. In general, if the size of the catalyst was reduced in the order of nanometer, there is an existence of actives surfaces, which in turn (1) ameliorates the reaction efficiency (Luisa and Duncan 2007), (2) enhances the dispersion rate, and (3) eliminates the clogging in the fuel injectors (Wickham et al. 2006). Luisa and Duncan (2007) critically reviewed the catalytic properties of the potential nanoparticles and observed that if the size of the catalyst was reduced to nanosize, the active surfaces will be increased and could enhance the reaction efficiency of the concerned thermal systems.

23.3 CARBON NANOMATERIALS/CNT AS AN ADDITIVE WITH LIQUID FUELS

In present-day global context scenario, the automobile revolution focuses on controlling the harmful emissions from diesel engines. For the past three decades, the use of diesel in all sectors like the industry, transportation, and power plant sectors has been increased due to population growth and rapid industrialization. Incomplete combustion of fuel in a diesel engine results to lower power output and increase in hazardous emissions such as NO_x, CO, soot, smoke, HC, and particulate matter (PM). Due to the stringent emission regulations all over the world, the effect of minimizing the deleterious emissions is underway. Low combustion efficiency and high pollutant emissions causing air pollution are the peculiar characteristics of diesel fuel. Many investigations (Cunningham et al. 1990; Ullman et al. 1994) have been focused on the improvement of diesel fuel properties, such as using cetane improver, diesel ignition improvers, and fuel-borne nanoadditives.

The concept and application of nanotechnology is one of the efficient techniques to achieve improved fuel properties in the coming years. One of the major applications of nanotechnology is the invention of potential nanoparticles and CNT. Nanoparticles/CNT can be incorporated in a thermal-related system by dispersing it into a base fluid medium to achieve better thermal properties. Nanofluid is a solid–liquid mixture that consists of nanoparticles and a base liquid, and the preparation of nanofluids is a challenge in the area of thermosciences (Yoo et al. 2007). Henceforth, the research community is focusing on the use of nanoadditives (such as carbon-based material—CNT) to improve the performance and emission attributes of diesel engines.

Recent studies on the use of CNT and metallic nanoparticles as fuel-borne catalysts in diesel engines have been carried out by few researchers (Moy et al. 2002; Sadhik Basha and Anand 2011b, 2013, 2014). Very few efforts have been carried out by adding nanoparticles to the conventional fuel such as diesel. Nanoparticles possess several potential thermophysical properties, which have led to an immediate solution for the thermal science–related environment in order to achieve better efficiency and potential output. The effect of achieving the desired properties of fuels is one of the major applications of nanofluid technology (Tyagi et al. 2008).

Marquis and Chibante's (2005) work on CNT indicated that low thermal conductivity is a primary demerit in the development of energy-efficient heat transfer fluids required in many heat transfer systems. To overcome those limitations, they proposed the idea of suspending nanoparticles/CNT in a base fluid with an attempt to enhance the thermal conductivity and thermal efficiency of the system. They identified the potential advantages using nanofluids in thermal-related systems such as cost-effectiveness, improved thermal properties, minimized clogging, long-term chemical stability, and long-term effective performance life. They found a considerable enhancement of thermal conductivity when dispersed in various heat transfer fluids such as mineral and synthetic oils, water, and water/ethylene glycol mixtures. They concluded that CNTs have the potential in various applications such as engine cooling systems, oil coolers, and heat pumps to ameliorate the thermal and lubricating performance. In general, heat transfer liquids display low thermal conductivity, and in turn, they possess a limited effectiveness.

Sadhik Basha and Anand (2011b) have studied the effects of CNTs on blending with diesel fuel in a diesel engine. CNTs were synthesized by the electric arc method and their SSA was determined by the following equation (Peigney et al. 2001):

$$\text{Specific surface area } (\text{SSA}) = 1315 \times d_e / nd_e - 0.68 \sum_{i=1}^{n} i$$

where
  d_e is the external diameter of CNT
  n is the number of shells of CNT

The same team (Sadhik Basha and Anand 2011b) utilized CNTs and blended with diesel fuel (with the dosage of 50, 100, and 150 ppm/L) using an ultrasonicator, and the prepared CNT-blended fuels were subjected for stability characteristics. The prepared stable CNT-blended fuels were subjected to an experimental investigation in a diesel engine. They have revealed a substantial improvement in the brake thermal efficiency and brake-specific fuel consumption for the CNT-blended diesel fuels compared to that of neat diesel. They have also revealed that the enhancement in the brake thermal efficiency was due to the better combustion characteristics and enhanced surface-area-to-volume ratio of CNT. Owing to CNT's better combustion characteristics and improved degree of mixing with air, they observed considerable reduction in content of NO_x, UBHC, CO, and smoke at all the loads compared to that of neat diesel fuel. Henceforth, it was observed that CNTs have a potential to improve the fuel properties and thereby played a vital role in improving the working characteristics of internal combustion engines.

23.4 PREPARATION OF CARBON NANOMATERIALS/CNT-BLENDED DIESEL AND BIODIESEL FUELS

CNTs are basically solid nanoparticles, and a standard method is adopted to disperse the same in a base fluid. Ultrasonication technique was preferred to prepare the CNT-blended diesel fuels compared to that of mechanical agitation. The ultrasonication method was widely adopted to disperse CNT in a base fluid, since it facilitates the possible agglomerate particles back to nanometer range (Putra et al. 2003).

CNTs were weighed by means of a digital weighing machine (model: Shimadzu AY220, Japan) to a predefined dosage of, say, 25 ppm and dispersed in the neat diesel for 30 min with the aid of an ultrasonicator (model: Lark SB5200, 120 W, 40 kHz) at 30°C as shown in Figure 23.1. Subsequently, the prepared CNT-blended diesel fuels were subjected to stability investigations. The prepared fuels were kept in a 200 mL transparent graduated scale glass test tube under static conditions. It was observed that the CNT-blended diesel fuels were found stable for more than a week. In addition, it was also observed that there was no separation of diesel fuel and CNT. Consequently, the prepared fuels were subjected to experimental analysis in diesel engines.

23.5 PREPARATION OF CNT-BLENDED WATER–DIESEL AND WATER–BIODIESEL EMULSION FUELS

Water–diesel emulsion fuels were prepared by mixing two immiscible fluids (say, diesel and water) in the presence of surfactants. Surfactants were added to the water–diesel mixture, to reduce the interfacial surface tension and to maximize their superficial contact areas to

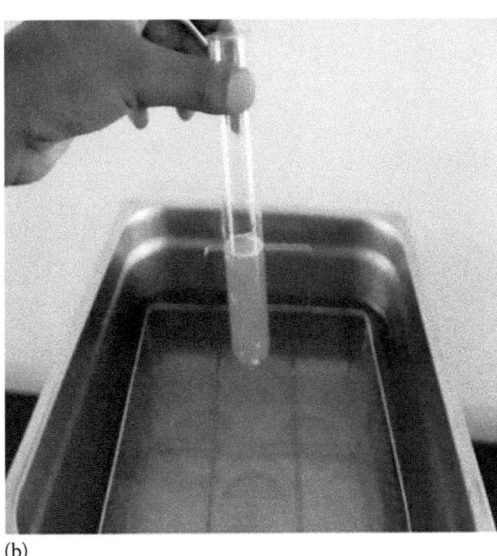

(a)   (b)

FIGURE 23.1
Photograph of (a) CNT and (b) CNT dispersed in diesel after ultrasonication. (From Sadhik Basha, J. and Anand, R.B., *Alexandria Eng. J.*, 53, 259, 2014.)

make stable emulsions (Chiaramonti et al. 2003). There are basically two types of water–diesel emulsion fuels: water-in-diesel emulsion and diesel-in-water emulsion fuel. In the water-in-diesel emulsion fuel type, a very small water droplet is dispersed within the diesel continuous phase, whereas in the case of diesel-in-water emulsion fuel type, very fine diesel droplets are dispersed within the water continuous phase in the presence of potential surfactants. A mixture of two nonionic surfactants, Span80 (hydrophobic) and Tween80 (hydrophilic), was used to produce the water-in-diesel emulsion fuel. The hydrophilic group is polarized and oil repelling, whereas the nature of the hydrophobic group is the opposite. CNT-blended diesel fuels and biodiesel emulsion fuels were prepared systematically using both mechanical homogenizer and ultrasonicator broadly in three steps. In the first step, the CNTs were weighed separately with the aid of a digital weighing machine to a predetermined dosage of, say, 25 ppm and dispersed in distilled water (say, 5% by volume) using the ultrasonicator for 30 min at 30°C. In the second step, the surfactant mixture (Span80 and Tween80) was prepared by mixing the two surfactants with 2% by volume. In the third step, the neat diesel (93% by volume) was mixed with the surfactant mixture by means of a mechanical homogenizer (set at various agitation speeds of 1000, 1500, 2000, and 2500 rpm), and simultaneously, the CNTs dispersed in a distilled water solution, prepared earlier, were added by a metering pump for 15 min at 30°C. This process was carried out for 30 min, and thus the CNT-blended water–diesel emulsion fuel was prepared. Likewise, the highest percentage of water–diesel emulsion fuel (say 10% and 15% of water content by volume) was prepared. The prepared samples of CNT-blended water–diesel emulsion fuels at various agitation speeds had a creamy white color and were kept in the graduated glass test tubes for stability investigations (in accordance with the stability study carried out by Selim and Ghannam [2010]). The percentage of the separated water layer (by volume) with respect to the elapsed time was measured for each sample. In the aforementioned fashion, the CNT-blended biodiesel emulsion fuels were prepared as shown in Figure 23.2. It was observed that the CNT nanoparticles were completely dispersed in distilled water during ultrasonication, and this leads to possible encapsulation of nanoparticles within the water droplet present in the continuous oil layer (diesel fuel). Basically, the emulsion fuels have an inferior ignition quality (Armas et al. 2005; Sadhik Basha and Anand 2014), and hence several researchers tried out various additives to improve the burning characteristics of emulsion fuels. In this regard, CNTs were tried out as a potential additive with emulsion fuels to enhance the ignition quality of emulsion fuels.

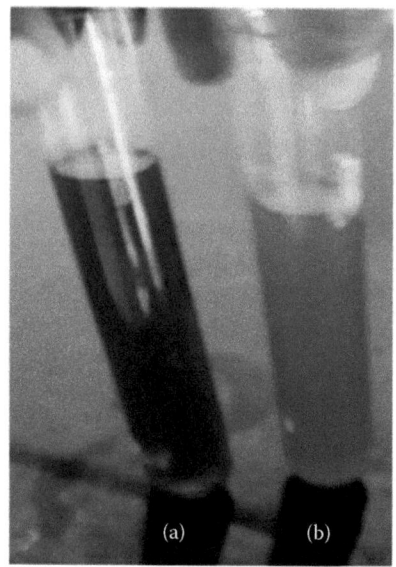

(a)   (b)   (c)

FIGURE 23.2
Photograph of (a) JME, (b) JME emulsion fuel, and (c) CNT-blended JME emulsion fuel. (From Sadhik Basha, J. and Anand, R.B., *Alexandria Eng. J.*, 53, 259, 2014.)

23.6 EXPERIMENTAL OUTCOME ON USING CARBON NANOMATERIALS/CNT-BLENDED FUELS IN ENGINES

The application of carbon-based nanomaterials (such as CNT) has played a critical role in fuel engineering science. The technical community has been confronted with the use of CNTs on blending with fossil fuels and modified alternative fuels (such as water–diesel and water–biodiesel emulsion fuels). Very recently, some researchers have utilized CNT as a fuel-borne additive with the modified form of fossil fuels (such as emulsion fuels) in an attempt to improve fuel qualities. CNTs were incorporated with emulsion fuels systematically (as discussed in the previous section) to achieve the desirable fuel quality characteristics.

The functionalization of carbon-based materials (i.e., CNT) has played a peculiar role in the working characteristics of diesel engine. Due to the addition of CNTs with fuels, the ignition quality of fuels has been improved (Moy et al. 2002; Sadhik Basha and Anand 2010c, 2011b, 2013, 2014, Sadhik Basha 2015). In the fuel engineering field, the ignition quality is indexed by the magnitude of the cetane number. The higher the cetane number of the fuel, the more enhancement in the ignition quality (Heywood 1988). This information indicates that by adding CNT to emulsion fuels, the cetane number of fuels increased and the performance, combustion, and emission attributes of diesel engines improved (Sadhik Basha and Anand 2011b, 2013, 2014).

Sadhik and Anand (2011b) carried out several experimental investigations on using CNT as an additive with both diesel and biodiesel emulsion fuels. Basically, emulsion fuels have inferior ignition quality (Armas et al. 2005; Sadhik Basha and Anand 2010a,b, 2011a,c, 2014) and longer ignition delay, and hence several researchers tried out various additives to improve the burning characteristics of emulsion fuels. In this regard, CNT is considered as a potential additive to enhance the ignition quality of emulsion fuels.

23.6.1 Combustion Characteristics of Diesel Engine on Using CNT-Blended Water–Diesel and Water–Biodiesel Emulsions

Carbon-based materials such as CNT were used as additive with diesel and biodiesel fuels. Sadhik Basha and Anand (2011b) utilized CNT-blended diesel and biodiesel emulsion fuels as alternative fuels in diesel engines. They found an appreciable improvement in the performance, combustion, and emission attributes of diesel engine. With regard to the combustion characteristics of CNT-blended water–diesel emulsion fuels, the problem of longer ignition delay was shortened due to the addition of CNT as shown in Figure 23.3. The figure illustrates that the peak pressure and the heat release rate were lesser for the CNT-blended diesel fuels than that of neat diesel and neat water–diesel emulsion fuel, owing to the quick burning characteristics of CNT (Moy et al. 2002). The presence of CNT in diesel fuel has improved the degree of fuel–air mixing and in turn accelerated combustion when compared to that of neat diesel operation. The effect of CNT was more pronounced on imparting reduced cylinder pressure on the account of shortened ignition delay. In the case of neat diesel, the premixed combustion phase was dominant, and hence the burning characteristics were inferior compared to that of CNT-blended diesel fuels. Due to the quick burning characteristics of CNT-blended diesel fuels, the premixed combustion phase was shortened compared to that of neat diesel operation. As a result, there was an observance of reduced cylinder peak pressure and heat release rate for CNT-blended diesel fuels compared to that of neat diesel at all the loads.

A similar trend of reduced peak cylinder pressure was observed for CNT-blended biodiesel emulsion fuels as shown in Figure 23.4. The addition of CNT to jatropha methyl ester (JME) emulsion fuel (25, 50, and 100 ppm) has exhibited a gradual decrement in the cylinder pressure on the account of shortened premixed burning phase. This was due to the enhancement in the cetane number (refer to Table 23.1) and better combustion characteristics of CNT (Moy et al. 2002; Sadhik Basha and Anand 2011b, 2014).

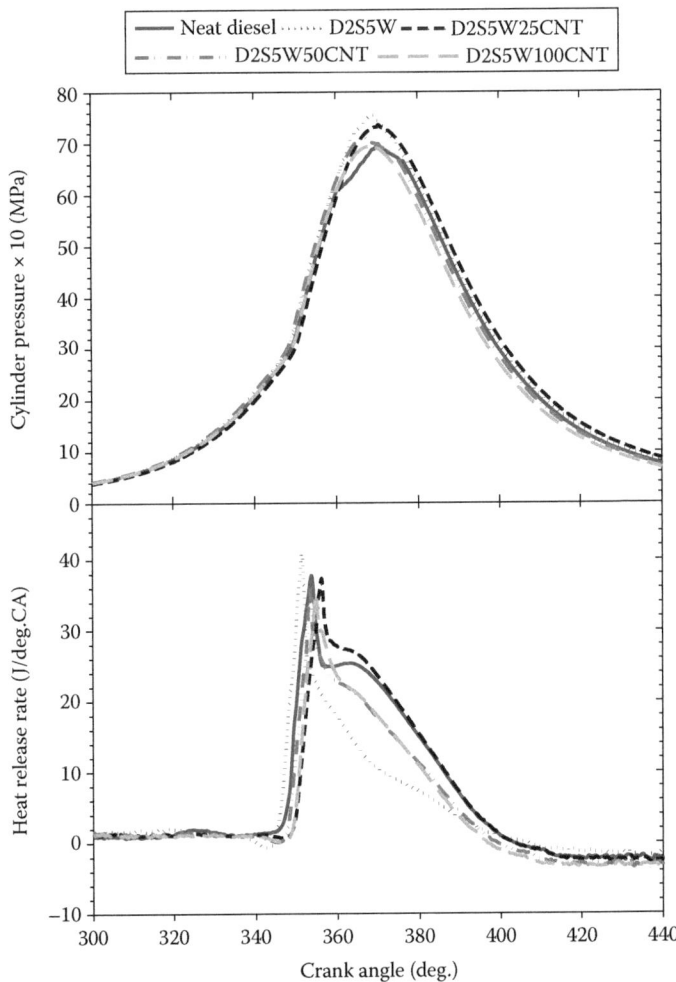

FIGURE 23.3
Variation of cylinder pressure and heat release rate with crank angle at full load for CNT-blended water–diesel emulsion fuels. (From Sadhik Basha, J. and Anand, R.B., *Proc. Inst. Mech. Eng. J. Power Energy*, 225, 279, 2011b.)

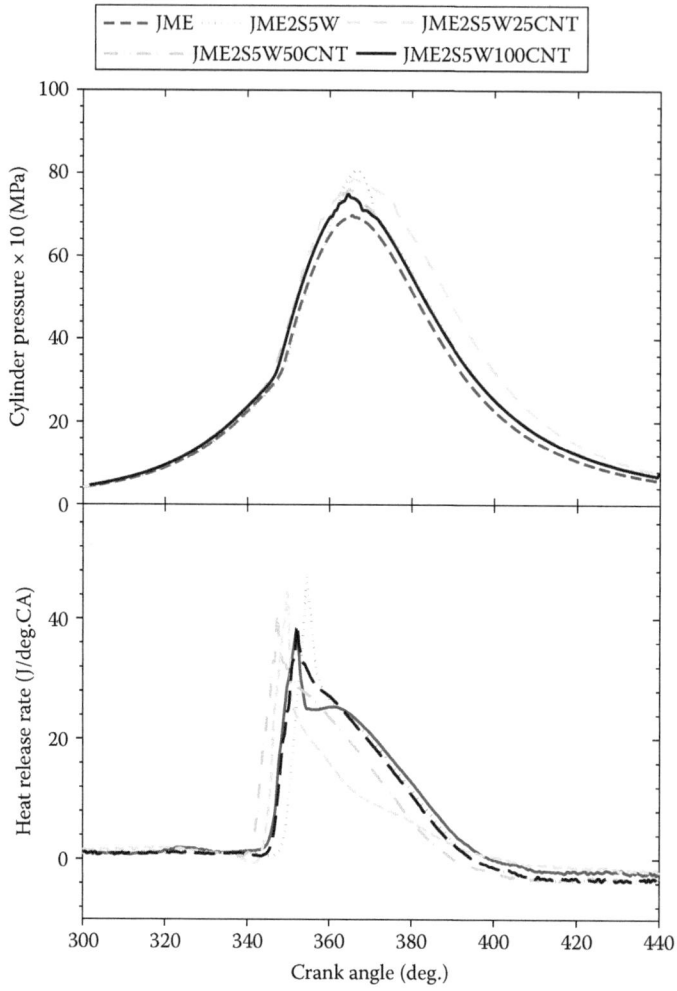

FIGURE 23.4
Variation of cylinder pressure and heat release rate with crank angle at full load for CNT-blended water–biodiesel emulsion fuels. (From Sadhik Basha, J. and Anand, R.B., *Alexandria Eng. J.*, 53, 259, 2014.)

TABLE 23.1 Fuel Properties of CNT-Blended Biodiesel Emulsion Fuel

Properties	JME	JME2S5W	JME2S5W25CNT	JME2S5W50CNT	JME2S5W100CNT
Density at 15°C, kg/m³	895	899.8	897.2	897.8	899.4
Kinematic viscosity at 40°C (×10⁻⁶ m²/s)	5.05	5. 40	5.43	5.76	5.91
Flash point,°C	85	140	130	125	122
Net calorific value, MJ/kg	38.88	37.05	37.28	37.35	37.85
Cetane no.	53	51	54	55	56

Source: Sadhik Basha, J. and Anand, R.B., *Alexandria Eng. J.*, 53, 259, 2014.

23.6.2 Performance Characteristics of Diesel Engine on Using CNT-Blended Water–Diesel and Water–Biodiesel Emulsions

The incorporation of CNT in diesel and biodiesel fuels has improved the brake thermal efficiency and brake-specific fuel consumption of diesel engine (refer to Figure 23.5). This could be probably attributed to the better combustion characteristics of CNT. In general, the nanosize particles possess high surface area and reactive surfaces that contribute to higher chemical reactivity to act as

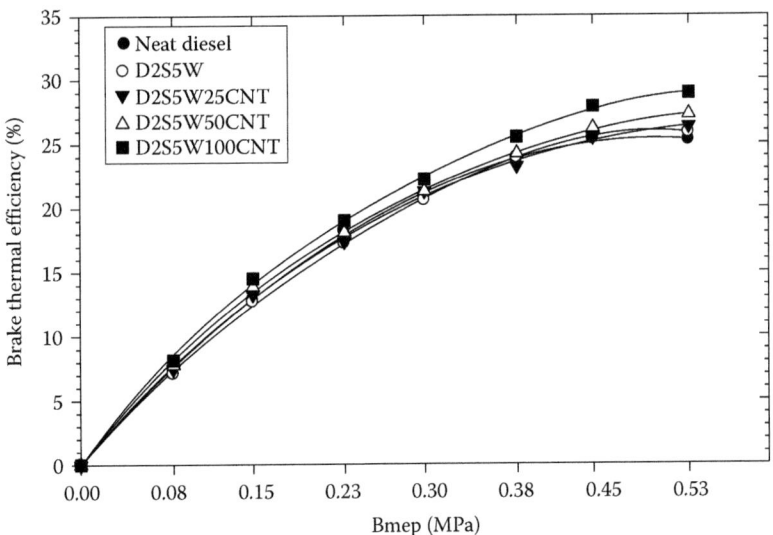

FIGURE 23.5
Variation of brake thermal efficiency for water–diesel emulsion fuels. (From Sadhik Basha, J. and Anand, R.B., *Proc. Inst. Mech. Eng. J. Power Energy*, 225, 279, 2011b.)

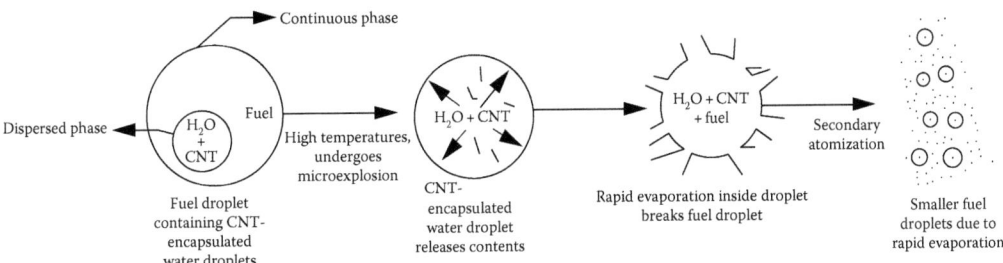

FIGURE 23.6
Schematic of microexplosion phenomenon of CNT-blended emulsion fuels. (From Sadhik Basha, J. and Anand, R.B., *Proc. Inst. Mech. Eng. J. Power Energy*, 225, 279, 2011b.)

a potential catalyst (Luisa and Duncan 2007). In this perspective, the presence of CNT could have improved due to the existence of high surface area and active surfaces. In addition, CNT-blended water–diesel emulsion fuels may also exhibit microexplosion and secondary atomization, similar to that of water–diesel emulsion fuel as shown in Figure 23.6. As the boiling point of water is lower than that of diesel, the encapsulated CNT water droplets in diesel fuel will absorb heat quickly to form water vapor and explode through the surrounding oil layers, through the effect called *microexplosion*. Subsequently, the number of secondary encapsulated CNT droplets of fine size will take place as a result of secondary atomization. Thus, the encapsulated CNT in the water droplets is finely atomized along with the diesel fuel and reacts with air effectively, owing to their enhanced surface-area-to-volume ratio. This phenomenon could have led to enhanced burning of CNT and in turn enhanced the thermal efficiency of diesel engine (Sadhik Basha and Anand 2011b) compared to that of water-diesel emulsion fuel. In the case of neat water–diesel emulsion fuel, the brake thermal efficiency is better than that of neat diesel. However, in comparison with CNT-blended water–diesel emulsion fuels, the brake thermal efficiency of neat water–diesel emulsion fuel is low. This could be due to the adverse effect of water addition, longer ignition delay, and reduced engine temperature of neat water–diesel emulsion fuel (Sadhik Basha and Anand 2011a).

A similar trend of improved performance characteristics was observed for CNT-blended biodiesel emulsion fuels when compared to that of neat biodiesel and biodiesel emulsion fuels as shown in Figure 23.7. It is observed that the brake thermal efficiency of emulsion fuels was higher than JME fuel due to the high heat release rate at all the loads. As the emulsion fuel gets ignited, the amount of heat release was enhanced and caused higher brake thermal efficiency (Sadhik Basha and Anand 2014) when compared to that of JME fuel. Furthermore, due to the

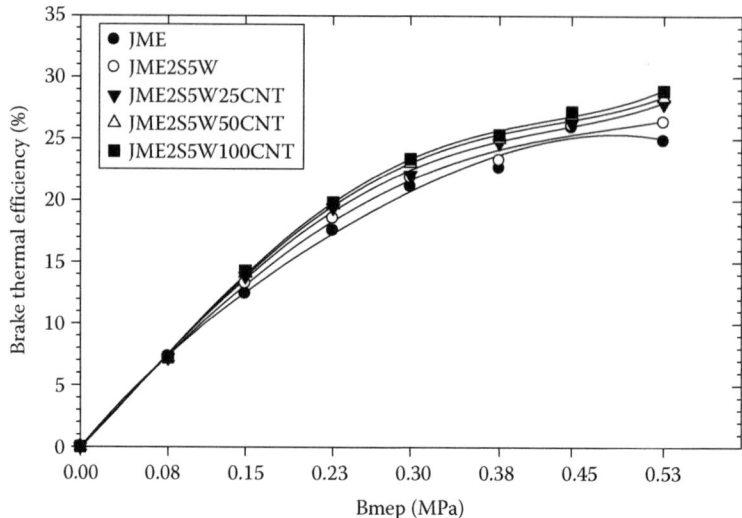

FIGURE 23.7
Variation of brake thermal efficiency for water–biodiesel emulsion fuels. (From Sadhik Basha, J. and Anand, R.B., *Alexandria Eng. J.*, 53, 259, 2014.)

combined effects of microexplosion and secondary atomization phenomenon, the combustion rate was improved (Marwan et al. 2001) for emulsion fuels when compared to that of neat JME fuel operation. On the other hand, CNT-blended JME emulsion fuels have shown further improvement in the brake thermal efficiency when compared to that of neat water–biodiesel emulsion fuel due to the accelerated combustion. Once CNT-blended JME emulsion fuels are subjected to high-pressure and high-temperature environment in the combustion chamber, the water droplets encased in the fuel absorbed the heat quickly (due to the low boiling point of water). As an effect, the spray jet momentum of CNT-blended JME emulsion fuels could have increased, inducing intensive secondary atomization as shown in Figure 23.6. Owing to the factors mentioned earlier, there could be an occurrence of improved homogenization of fuel and air mixing in the presence of CNT causing improved combustion and burning characteristics (Moy et al. 2002) compared to that of neat water–biodiesel emulsion fuel. As a result, the degree of fuel–air mixing in the presence of CNT could have enhanced (Sadhik Basha and Anand 2011b, 2014) for the CNT-blended JME emulsion fuels resulting to higher brake thermal efficiency compared to that of neat water–biodiesel emulsion fuel.

23.6.3 Emission Characteristics of Diesel Engine on Using CNT-Blended Water–Diesel and Water–Biodiesel Emulsion Fuels

Recent information indicates that CNT influences the reduction of harmful pollutants from diesel engine. The addition of CNT to fossil fuels has led to a reduction in harmful pollutants from diesel engine. Owing to the improved combustion characteristics of CNT, the level of the harmful pollutants was decreased considerably. At higher engine loads, greater burning temperature occurs inside the combustion chamber and leads to higher NO_x emissions from diesel engines. The emissions of NO_x for the neat diesel were observed higher due to high combustion temperature (Sadhik and Anand 2012) compared to that of neat water–diesel emulsion fuel and CNT-blended water–diesel emulsion fuels as shown in Figure 23.8. Due to the significant microexplosion phenomenon associated with water–diesel emulsion fuels, water rapidly evaporates during combustion and lowers the cylinder average temperature and thereby induced lower NO_x emissions. On the other hand, the CNT-blended water–diesel emulsion fuels produced further reduction of NO_x emissions due to the quick burning characteristics (Moy et al. 2002) and improved combustion. A similar trend of reduced NO_x emissions was observed for CNT-blended biodiesel emulsion fuels when compared to that of neat biodiesel and biodiesel emulsion fuels as shown in Figure 23.9.

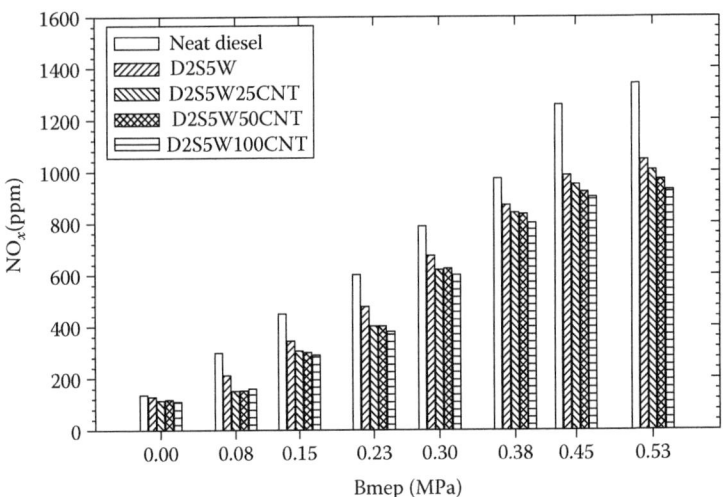

FIGURE 23.8
Variation of NO$_x$ for CNT-blended water–diesel emulsion fuels. (From Sadhik Basha, J. and Anand, R.B., *Proc. Inst. Mech. Eng. J. Power Energy*, 225, 279, 2011b.)

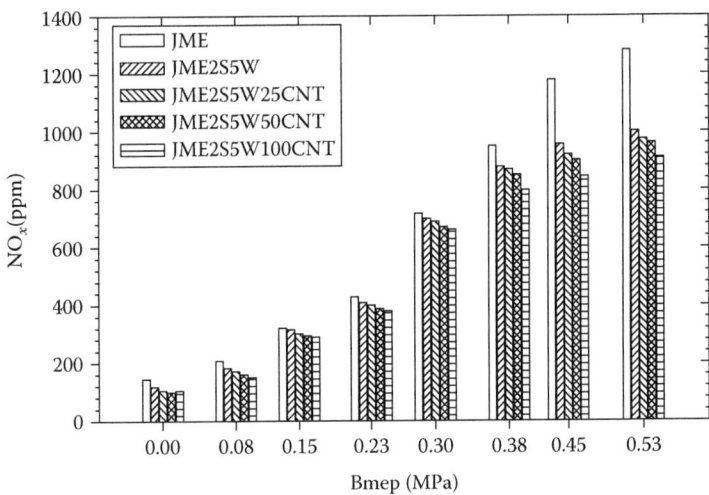

FIGURE 23.9
Variation of NO$_x$ for CNT-blended water–biodiesel emulsion fuels. (From Sadhik Basha, J. and Anand, R.B., *Alexandria Eng. J.*, 53, 259, 2014.)

23.6.4 Evaporation Characteristics of CNT-Blended Water–Diesel and Water–Biodiesel Emulsion Fuels

The evaporation characteristics of fuels are the very important aspects to note to understand their rate of vaporization. In general, if the fuel undergoes rapid evaporation, it may facilitate better combustion characteristics in diesel engine. This is experimentally proved by means of a hot-plate evaporation test, which is an additional investigation to predict the performance of combustion and the rate of vaporization of emulsion fuel droplets during the physical ignition delay period (Sadhik and Anand 2012). The variation of evaporation characteristics of the neat diesel, neat water–diesel emulsion fuel, and the CNT-blended water–diesel emulsion fuels is illustrated in Figure 23.10. It is inferred from the figure that the evaporation time for CNT-blended diesel fuels (D2S5W25CNT, D2S5W50CNT, and D2S5W100CNT) was reduced drastically compared to that of neat diesel and neat water–biodiesel emulsion fuel. This could be due to the enhanced heat transfer characteristics of CNT (Marquis and Chibante 2005) in the emulsion fuels. The CNT-blended water–diesel emulsion fuels produced the

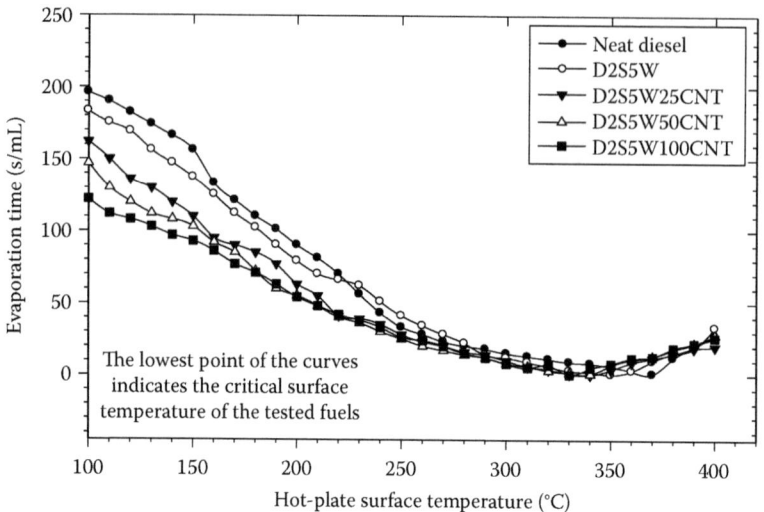

FIGURE 23.10
Variation of evaporation characteristics of CNT-blended water–diesel emulsion fuels. (From Sadhik Basha, J., Impact of nano-additives on the performance, emission and combustion characteristics of direct injection compression ignition engine, PhD dissertation, National Institute of Technology, Trichy, India, 2011.)

advancement in the critical surface temperatures compared to that of neat diesel and D2S5W fuel. The critical surface temperature for neat diesel was 370°C, whereas it was 357°C, 347°C, 338°C, and 332°C for D2S5W, D2S5W25CNT, D2S5W50CNT, and D2S5W100CNT fuels, respectively. The advancement of the critical surface temperature of the CNT-blended water–diesel emulsion fuel clearly indicated the accelerated evaporation effect compared to that of neat diesel and D2S5W fuel. During the preparation of CNT-blended water–diesel emulsion fuels, the CNT was encapsulated within the water droplet. Hence, when the encapsulated CNT water droplet was subjected to a hot environment, it exhibited microexplosion and in turn evaporates rapidly. A similar trend was also observed for CNT-blended biodiesel emulsion fuels as depicted in Figure 23.11. It was observed from the figure that the critical surface temperature for Jatropha Biodiesel (JBD) fuel was 340°C, whereas it was 334°C, 331°C, and 327°C for JBD25CNT, JBD50CNT, and JBD100CNT fuels, respectively. Hence, it was asserted that the addition of CNT to the biodiesel fuel could induce rapid evaporation compared to that of JBD fuel.

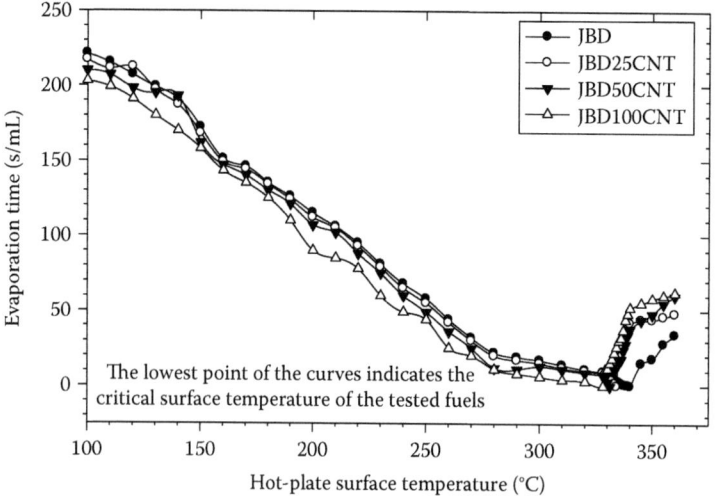

FIGURE 23.11
Variation of evaporation characteristics of CNT-blended water–biodiesel emulsion fuels. (From Sadhik Basha, J., Impact of nano-additives on the performance, emission and combustion characteristics of direct injection compression ignition engine, PhD dissertation, National Institute of Technology, Trichy, India, 2011.)

23.7 SUMMARY

Nanoparticles have numerous and potential properties that are attractive for use in various applications such as in fuel engineering fields and thermal-related systems. Owing to their enhanced surface-area-to-volume ratio and unique thermal properties, they play a critical role in various applications. Recently, carbon-based materials (such as CNT) are utilized as a potential candidate to improve fuel properties. Information indicates that on adding CNT to fossil fuels, the ignition quality is improved, and hence it ameliorates the burning characteristics of emulsion fuels. It is observed that adding CNT in the form of nanosize to a base fluid (such as diesel) will facilitate better radiative and heat/mass transfer properties, and henceforth such an enhancement in thermophysical properties of the fuel will have the potential of reducing the evaporation (ignition) time of droplets within a diesel engine and hence favorably leads to shorten ignition delay. The effect of adding nanoadditives to conventional fuels such as diesel and gasoline is still in the developing stage, and it needs more research to understand the performance and emission characteristics in detail. On the other hand, the effect of adding nanoparticles to diesel fuel should not cause a new mode of air pollution. The degree of quantity and degree of stability of a nanomaterial in a base fluid are to be investigated in detail for better performance and reduced emissions in coming years.

NOMENCLATURE

Bmep	Brake mean effective pressure, bar
bTDC	Before top dead center
CA	Crank angle
CNT	Carbon nanotube
CO	Carbon monoxide
D2S5W	93% Diesel + 2% surfactant + 5% water
D2S5W25CNT	93% Diesel + 2% surfactant + 5% water + 25 ppm of CNT
D2S5W50CNT	93% Diesel + 2% surfactant + 5% water + 50 ppm of CNT
HC	Hydrocarbons
JME	Jatropha methyl esters
JME2S5W	93% Jatropha methyl esters + 2% surfactant + 5% water
JME2S5W25CNT	93% Jatropha methyl esters + 2% surfactant + 5% water + 25 ppm CNT
JME2S5W50CNT	93% Jatropha methyl esters + 2% surfactant + 5% water + 50 ppm CNT
JME2S5W100CNT	93% Jatropha methyl esters + 2% surfactant + 5% water + 100 ppm CNT
NO_x	Nitrogen oxides
PM	Particulate matter

REFERENCES

Armas, O., R. Ballesteros, F.J. Martos et al. 2005. Characterization of light duty diesel engine pollutant emissions using water-emulsified fuel. *Fuel*, 84: 1011–1018.

Chiaramonti, D., M. Bonini, E. Fratini et al. 2003. Development of emulsions from biomass pyrolysis liquid and diesel and their use in engines—Part 1: Emulsion production. *Biomass Bioenergy*, 25: 85–99.

Choi, S.U.S. and J.A. Eastman. 1995. *Enhancing Thermal Conductivity of Fluids with Nanoparticles*, ASME International Mechanical Engineering Congress and Exhibition, San Francisco, CA.

Cunningham, L.J., T.J. Henly, and A.M. Kulinowski. 1990. The effects of diesel ignition improvers in low-sulfur fuels on heavy-duty diesel emissions. SAE 902173. SAE International, doi: 10.4271/902173.

Deluca, L.T., L. Galfetti, and F. Severini. 2005. Combustion of composite solid propellants with nanosized aluminium. *Combustion, Explosion and Shock Waves*, 41: 680–692.

Dreizin, E.L. 2000. Phase changes in metal combustion. *Progress in Energy and Combustion Science*, 26: 57–78.

Eastman, J.A., U.S. Choi, S. Li et al. 1999. Novel thermal properties of nanostructured materials. *Material Science Forum,* 312–314: 629–634.

Heywood, J.B. 1988. *Internal Combustion Engines.* Singapore: McGraw-Hill.

Iijima, S. and T. Ichihashi. 1993. Single-shell carbon nanotubes of 1-nm diameter. *Nature*, 363: 603–605.

Luisa, F. and S. Duncan. 2007. *Applications of Nanotechnology: Environment.* Aarhus, Denmark: Interdisciplinary Nanoscience Centre, pp. 1–14.

Marquis, F.D.S. and L.P.F. Chibante. 2005. Improving the heat transfer of nanofluids and nanolubricants with carbon nanotubes. *Journal of the Minerals, Metals and Materials Society*, 57: 32–43.

Marwan, A.A.N., R. Hobina, and S.A. Wagstaff. 2001. The use of emulsion, water induction and EGR for controlling diesel engine emissions. SAE Technical Paper No. 2001-01-1941. SAE International, doi: 10.4271/2001-01-1941.

Moy, D., C. Niu et al. 2002. Carbon nanotubes in fuels. US Patent No. 6419717.

Nubia, M.R., A.C. Pinto, M.Q. Cristina et al. 2007. The role of additives for diesel and diesel blended (ethanol or biodiesel) fuels: A review. *Energy and Fuels*, 21: 2433–2445.

Peigney, A., C. Laurent, E. Flahaut et al. 2001. Specific surface area of carbon nanotubes and bundles of carbon nanotubes. *Carbon*, 39: 507–514.

Prasher, R.S., P. Bhattacharya, and P.E. Phelan. 2005. Thermal conductivity of nano scale colloidal solutions. *Physical Review Letters*, 94. doi: http://dx.doi.org/10.1103/PhysRevLett.94.025901.

Putra, N., W. Roetzel, and S.K. Das. 2003. Natural convection of nano-fluids. *Heat and Mass Transfer*, 39: 775–784.

Roger, S. 2006. Cerium oxide nanoparticles as fuel additives. US Patent No. US2006/0254130 A1.

Roos, W.J., D. Richardson, and D.J. Claydon. 2008. Diesel fuel additives containing cerium or manganese and detergents. US Patent No. US2008/0066375 A1.

Sabourin, J.L., M.D. Daniel, R.A. Yetter et al. 2009. Functionalized graphene sheet colloids for enhanced fuel/propellant combustion. *ACS Nano*, 3: 3945–3954.

Sadhik Basha, J. 2011. Impact of nano-additives on the performance, emission and combustion characteristics of direct injection compression ignition engine. PhD dissertation, National Institute of Technology, Trichy, India.

Sadhik Basha, J. 2014. An experimental analysis of a diesel engine using alumina nanoparticles blended diesel fuel. SAE Technical Paper 2014-01-1391. SAE International, doi: 10.4271/2014-01-1391.

Sadhik Basha, J. and R.B. Anand. 2010a. Effects of nanoparticle blended water-biodiesel emulsion fuel on working characteristics of a diesel engine. *International Journal of Global Warming*, 2: 330–346.

Sadhik Basha, J. and R.B. Anand. 2010b. Applications of nanoparticles/nanofluid in compression ignition engines—A case study. *International Journal of Applied Engineering and Research*, 4: 697–708.

Sadhik Basha, J. and R.B. Anand. 2010c. Performance and emission characteristics of a DI compression ignition engine using carbon nanotubes blended diesel. *International Journal of Advances in Thermal Science and Engineering*, 1: 67–76.

Sadhik Basha, J. and R.B. Anand. 2011a. An experimental study in a CI engine using nano-additive blended water-diesel emulsion fuel. *International Journal of Green Energy*, 8: 332–348.

Sadhik Basha, J. and R.B. Anand. 2011b. An experimental investigation in a diesel engine using CNT blended water-diesel emulsion fuel. *Proceedings of Institution of Mechanical Engineers, Journal of Power and Energy*, 225: 279–288.

Sadhik Basha, J. and R.B. Anand. 2011c. Role of nano-additive blended biodiesel emulsion fuel on the working characteristics of a diesel engine. *Journal of Renewable and Sustainable Energy*, 3: 1–17.

Sadhik Basha, J. and R.B. Anand. 2012. Effects of nanoparticle additive in the water-diesel emulsion fuel on the performance, emission and combustion characteristics of a diesel engine. *International Journal of Vehicle Design*, 59: 164–181.

Sadhik Basha, J. and R.B. Anand. 2013. The influence of nano additive blended biodiesel fuels on the working characteristics of a diesel engine. *International Journal of the Brazilian Society of Mechanical Sciences and Engineering*, 35: 257–264.

Sadhik Basha, J. and R.B. Anand. 2014. Performance, emission and combustion characteristics of a diesel engine using carbon nanotubes blended jatropha methyl esters emulsions. *Alexandria Engineering Journal*, 53: 259–273.

Sadhik Basha, J. 2015. Preparation of Water-Biodiesel Emulsion Fuels with CNT and Alumina Nano-Additives and their Impact on the Diesel Engine Operation. SAE Technical Paper 2015-01-0904, SAE International, http://papers.sae.org/2015-01-0904/.

Selim, M.Y.E. and M.T. Ghannam. 2010. Combustion study of stabilized water-in-diesel fuel emulsion. *Energy Sources, Part A: Recovery, Utilization, and Environmental Effects*, 32: 256–274.

Tyagi, H., P.E. Phelan, R. Prasher et al. 2008. Increased hot plate ignition probability for nanoparticle-laden diesel fuel. *Nano Letters*, 8: 1410–1416.

Ullman, T.L., K.B. Spreen, and R.L. Mason. 1994. Effects of cetane number, cetane improver, aromatics and oxygenates on heavy-duty diesel emissions. SAE 941020. SAE International, doi: 10.4271/941020.

Wickham, D.T., R.L. Cook, S. De Voss et al. 2006. Soluble nanocatalysts for high performance fuels. *Journal of Russian Laser Research*, 27: 552–561.

Yetter, R.A., G.A. Risha, and S.F. Son. 2009. Metal particle combustion and nanotechnology. *Proceedings of the Combustion Institute*, 32: 1819–1838.

Yoo, D.H., K.S. Hong, and H.S. Yang. 2007. Study of thermal conductivity of nanofluids for the application of heat transfer fluids. *Thermochimica Acta*, 455: 66–69.

CHAPTER 24

CONTENTS

Chemical Functionalization of Carbon Materials and Its Applications

24

Shivani Dhall, Kapil Sood, and A. Rubavathy Jaya Priya

24.1 INTRODUCTION

With the developments in the nanotechnology, all nano-carbon materials viz. graphene, fullerenes, nanohorns, carbon nanotubes (CNTs) have opened up new avenues for exploration and thus giving rise to entirely new area of study known as carbon nanotechnology. Among them CNTs, nano-horns and graphene have received considerable attention from researchers due to their extraordinary mechanical, electrical, and thermal properties. These excellent properties provide thrilling opportunities to fabricate new materials for advanced applications. Owing to their outstanding properties, scientists are interested to investigate and modify these new substance classes. In the last decade, different types of chemical functionalization process have been investigated and are now widely used for the construction of multifunctional architectures with C_{60} as the integral building unit. However, they are available in form of fluffy powder, which is difficult to handle. As-synthesized carbon materials contain a large amount of impurities such as growth metal catalyst (Fe, Ni, and Cu), particle soot, and amorphous carbon as a synthetic residue. The well homogeneous dispersion of these materials into the host media, which can be in the form of liquid or solid materials, is one of the major challenges encountered in the fabrication of nanocomposite and devices used in nano-electronic and interconnected technologies. In addition, these materials are chemically inert and have poor solubility in solvents as well as in water because of the strong carbon–carbon interaction and van der Waals' attractions between them. This limitation often requires controlled chemical modification/functionalization of these materials on surface of the carbon materials in order to improve their solubility and dispersion stability in the solvent including water. Chemical functionalization has been simply dispersed and solubilizes these carbon materials, by attaching hydrophilic species to their hydrophobic structures. In spite of this, functionalization process removes impurities like amorphous/carbonaceous carbon and dissolution of metal catalyst although aggressive functionalization of carbon materials shortens its length and destructs their structure, which strongly affect the electrical properties. These functionalized materials are proving to be valuable in a broad range of areas starting from the preparation of nanocomposites to the fabrication of sensors. Functionalization process also improves their compatibility with the host material.

Functionalization process is generally divided into covalent and noncovalent methods depending on the mechanism of the connection between the functionalized component and the carbon atoms on the surface of carbon materials. Both of these approaches have some advantages and drawbacks, so the selection of functionalization methods depends on the outcomes of the final material. Covalent functionalization creates covalent linkage of chemical bonds or functional groups onto the surface and ends tips using different acids. While noncovalent functionalization can be achieved by forming van der Waals' bonds between planar groups and the carbon or by wrapping molecules helically round the nanotubes. Noncovalent approach avoids disrupting the structure of the carbon materials, enabling their original properties to be retained.

24.2 NONCOVALENT FUNCTIONALIZATION

Noncovalent functionalization of materials is achieved by wrapping of polymer and adsorption of surfactants or small aromatic molecules. Polymers like poly(phenylene vinylene) and polystyrene are widely used to wrap the carbon material specially CNTs [1,2]. The stirring of carbon materials in the presence of polymer can lead to a wrapping process and hence polymers coating onto the surface materials. The physical adsorption primarily depends on the properties of surfactants, medium chemistry, or the polymer matrix. Surfactants wrapping on the surface of carbon materials can effectively decrease their surface tension, which leads to prevention of the formation of aggregates. Surfactants are mainly classified into three types, namely,

1. Nonionic surfactants
2. Ionic surfactants
3. Cationic surfactants

The ionic surfactant like sodium dodecyl sulfate (SDS) and sodium dodecyl benzene sulfonate (SDBS) is widely used for solubilizing the CNTs in the different solvents. Both of them have the same length of the alkyl chain, but the SDBS has a phenyl ring attached between the alkyl chain and the hydrophilic group. Kang et al. [3] found that the presence of the phenyl ring makes SDBS more efficient for solubilization of CNTs as compared to SDS because of aromatic stacking formed between the single-walled carbon nanotubes (SWCNTs) and the phenyl rings of the SDBS within the micelle. SDBS is more effective than SDS surfactant because it contains a benzyl ring in dispersing SWCNTs in water [4]. Generally, about 1% by weight SDS/SDBS is added to the distilled water to prepare the aqueous suspension, and then ultrasonication is mainly used to get the dispersion of the CNTs. These surfactants transfer the charge on the nanotubes and then dispersed them by electrostatic forces. It is found that these surfactant molecules create ordered layers on the nanotube surfaces [5]. Moreover, the diameter-dependent distribution of CNTs in the solvent of SDBS, measured by atomic force microscopy (AFM), showed that even at 20 mg/mL solution of SDBS in H_2O separates, 63.5% of the SWCNT bundles exfoliate in single tubes [1,2].

In spite of the aforesaid surfactants, two different approaches to the noncovalent functionalization of nanotubes are also widely used to get the dispersion of the nanotubes. The first approach established by Hongjie Dai's group in 2001 [6] involves planar group molecules that form van der Waals' bonds and a *tail* connecting to succinimidyl ester group (Figure 24.1), irreversibly adsorb to the surface of nanotubes by π-stacking forces, while the second approach includes wrapping of nanotubes via larger polymeric molecules.

The noncovalent functionalization of nanotubes with pyrene [7,8] proved helpful to attach the magnetic nanoparticles and fullerene to nanotubes [9,10]. Dirk Guldi and their colleagues investigated the interactions between SWCNTs and pyrenes, porphyrins, and other molecules [10,11].

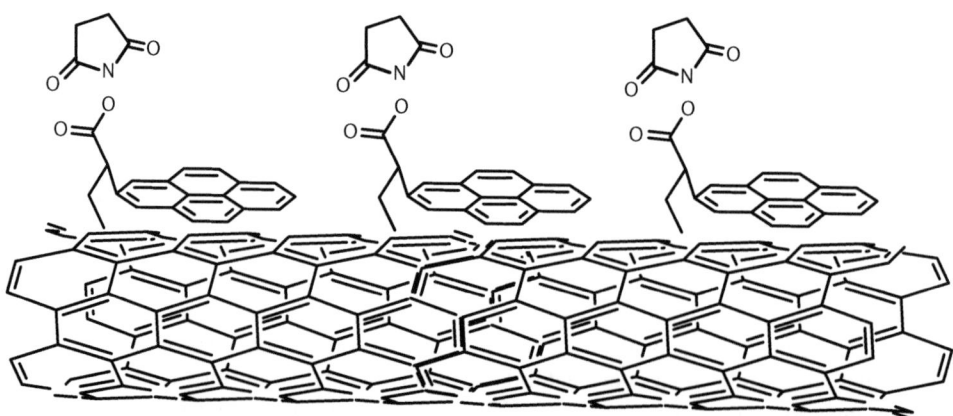

FIGURE 24.1
Noncovalent functionalization of SWCNT with 1-pyrene-butanoic acid and succinimidyl ester.

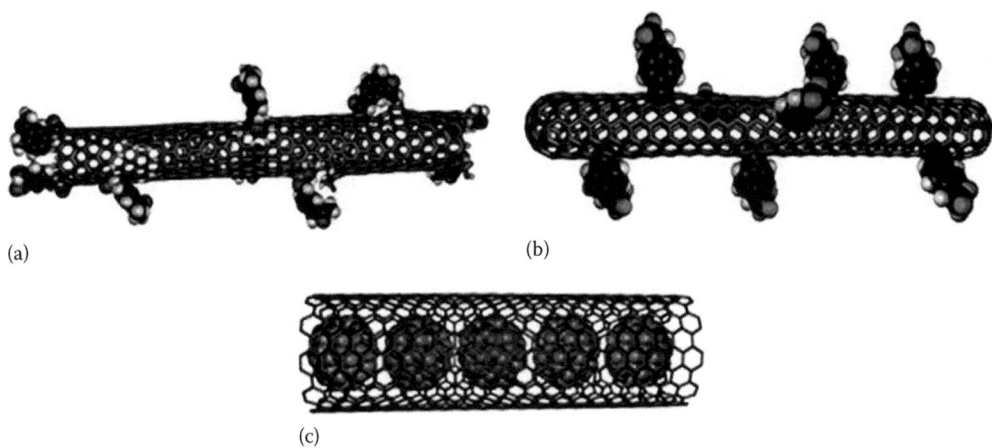

FIGURE 24.2
Different derivation strategies for CNTs. (a) defect functionalization, (b) side wall functionalization, and (c) endohedral functionalization.

24.3 COVALENT FUNCTIONALIZATION

Covalent functionalization process introduces various functional/chemical bonds such as carboxylic and hydroxyl groups on the nanotubes that have rich chemistry. In addition, these functional groups improve the solubility and dispersion stability of the nanotubes in solvents as well as in water, which is important to equalize the strong van der Waals' forces and π–π stacking interaction between nanotubes [12]. This process also enhance the bonding of nanotubes with polymer and nanoparticles for their applications in bio, gas and chemical sensors [13–18]. These functional groups initiate new partially occupied bands in the electronic band structure of nanotubes, and also they behave as electron acceptor/donor facilitating the charge transfer phenomena between the nanotubes [17,19–22]. However, due to specific chemical groups (COOH or OH), various localized kinds of perturbation are the changes in the local electronic structure of the nanotubes. This kind of changes is specific to the type of functional group and is localized about the chemical groups [1,2]. During functionalization, some sp^2 hybridization changes into sp^3, and various defects such as irregular arrangements of pentagon/heptagon, creation of holes in the sidewalls, and oxidization of sites can be possible [12,19,21,23]. This hybridization change creates considerable effect on the charge transport properties of the nanotubes. Furthermore, functionalization process opens the bandgap in the case of metallic nanotubes and a suppression of the optical transitions between the nearest van Hove singularities. In general, CNT covalent functionalization can be divided into different categories as shown in Figure 24.2.

24.4 DEFECT FUNCTIONALIZATION

Defects in CNTs are of enormous importance because they are used as anchor groups for further functionalization and attachments of the nanoparticles. The presence of these defects on the CNTs is promising, opening point for the progress of the covalent chemistry. Although these defects destruct the CNTs backbone, they lead to production of new material properties of CNTs with other substance.

In general, as-produced CNTs contain broad variety of defects and dislocated carbon atoms that are preferentially situated at the ends of the nanotubes where the catalyst/metal nanoparticles were attached [1,2,24]. The type and amount of intrinsic defects in the CNTs change with the production and subsequent purification of the nanotubes and encompass the introduction of pentagon/heptagon defects, resulting in strain and hence bending of the CNTs.

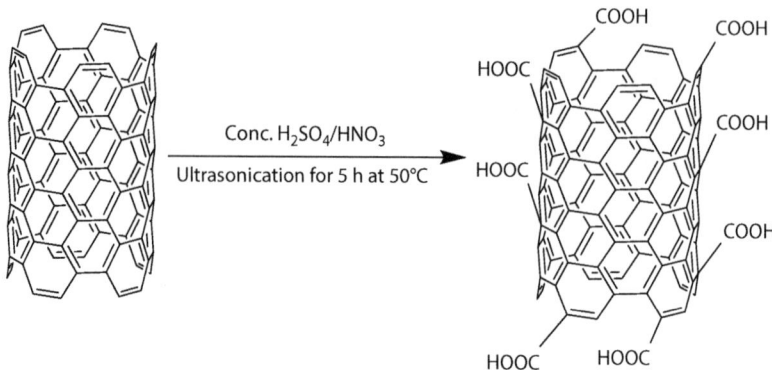

FIGURE 24.3
Carboxylic functionalization of CNTs using concentrated oxidizing acids at elevated temperatures.

The purification methods are based on liquid-phase or gas-phase oxidative treatment procedures. Most common oxidants for CNTs are concentrated nitric (HNO_3), sulphuric (H_2SO_4), hydrogen peroxide (H_2O_2) acids, gaseous oxygen, ozone, and potassium permanganate ($KMnO_4$). Using the mixture of HNO_3 and H_2SO_4 for the functionalization of nanotubes and nanohorns is the most prominent method in the nanotechnology as shown in Figure 24.3. Smalley's group [25] initially reported the covalent functionalization of CNTs using HNO_3 and H_2SO_4 mixture by sonication technique. Afterward, various publications have been reported on the functionalization of the nanotubes and its applications [26–34]. Generally, functionalization occurs at the defect sites/sidewall as well as at the ends of the CNTs.

For instance, it is predicted that carbon atoms adjacent to vacancies in the nanotube framework should be the positions where the introduction of carboxylic acid functionalities takes place. As a consequence, the removal of amorphous carbon and catalyst impurities is accompanied by the introduction of carboxylic acid functionalities and other oxygen-bearing groups at the ends and defect sites of the CNTs framework. Actually, it has turned out that it is not an easy task to develop purification strategies that are capable of removing the impurities only, without touching the structural integrity of the nanotubes. Furthermore, the defects that are present at the sp^2 hybridization of CNT network indicate the locations of higher reactivity toward aggressive chemicals such as oxidizing agents, and therefore, the opening of the CNTs framework at defect sides is the dominant pathway for oxidation-driven functionalization.

The well understanding and control on carboxylic acid functionality content are of utmost importance for defect group functionalization sequences, as these functionalities are the foundation for the attachment of other functional entities—the final functionalization degree is determined by the starting defect density.

Recently, Dhall et al. [22] have reported the covalent functionalization of pristine multi-walled carbon nanotubes (MWCNTs) at the room temperature conditions. They confirmed that the intertubular spacing of nanotubes is increased due to the attachment of the various functional groups at the surface as well as at the their ends by TEM and XRD spectra. They also showed that the value of intensity I_D/I_G ratio for functionalized MWCNTs is found to be increased to 1.14 as compared to 1.09 which is for pristine MWCNTs. This is because of breaking of chemical bonds and insertion of various functional groups on ends as well as on the defect sites/sidewalls of nanotubes. The functional groups on F-MWCNTs walls are considered as defects in their structure. They also found that the crystallite size of functionalized MWCNTs was decreased to from 15.2 to 14.6 nm.

24.5 IN SITU COVALENT FUNCTIONALIZATION

From the above discussion, it is clear that the covalent functionalizations of carboxylic acids or hydroxylic acids on the walls of the CNTs are of great importance due to their practical applications in various fields. But by modifying the CNTs using covalent functionalizations, the electronic and spectroscopic

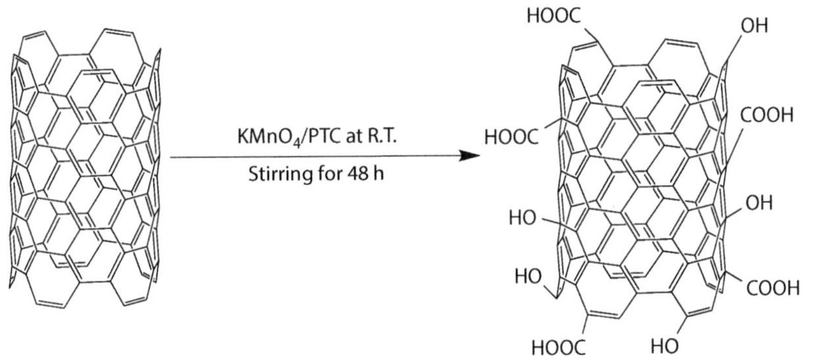

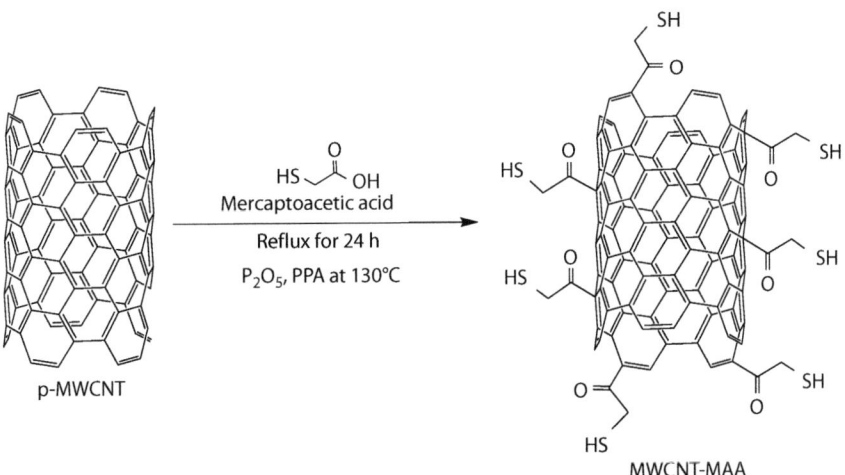

FIGURE 24.4
Carboxylic/hydroxyl functionalization of CNTs using mild oxidizing acids at room temperature.

properties should not get altered. In order to overcome this strategy, a mild oxidizing agent is necessary for the reaction to be carried out at room temperature. This was achieved by functionalization of the hydroxyl/carboxylic groups on the sidewalls of the CNTs using $KMnO_4$ and phase transfer catalyst (TBAB) at room temperature as shown in Figure 24.4. This type of covalent functionalization decreases the defect sites in the walls of the nanotubes and increases the degree of functionalization for the further moiety to be attached for the designed application. The addition of polymer onto the surface of the CNTs to obtain the polymer brushes increases the degree of functionalization and thereby directly increases its application in the field of biosensors, antimicrobial, anticancer activity, nanoprobes, nanotweezers, etc. Similarly, several conducting CNT–polymer composites, namely, CNTs–poly(3-octylthiophene), CNTs–polyaniline, and MWCNTs–sulfonated polyaniline, were prepared [35,36].

Repeated functionalization of nanotubes via carboxylic acids/hydroxyl/amine groups for the attachments of linker groups, namely, polymer matrices/bioconjugates/nanoprobes required for decoration of metal/metal oxide nanoparticles onto the surface of the CNTs, increases the defects onto the surface of CNTs. In order to avoid those problems, Murugan and Rubavathy [37] carried out in a single-step procedure the acid functionalization of MWCNTs by following the Friedel–Crafts acylation (FcAc) reaction using polyphosphoric acid and phosphorus pentoxide as mild oxidizing agents to introduce the –SH (functional group) onto the surface of the CNTs by in situ mechanism. This single-step functionalization of mercaptoacetic acid onto the surface of MWCNTs was performed similar to the earlier reported method [38] (Figure 24.5).

The Raman spectra for pristine MWCNTs and MWCNT–SH were depicted in Figure 24.6a and b, respectively. Manna et al. [39] have reported that the intensity of the D band observed

FIGURE 24.5
Schematic representation for the synthesis of MWCNT–SH.

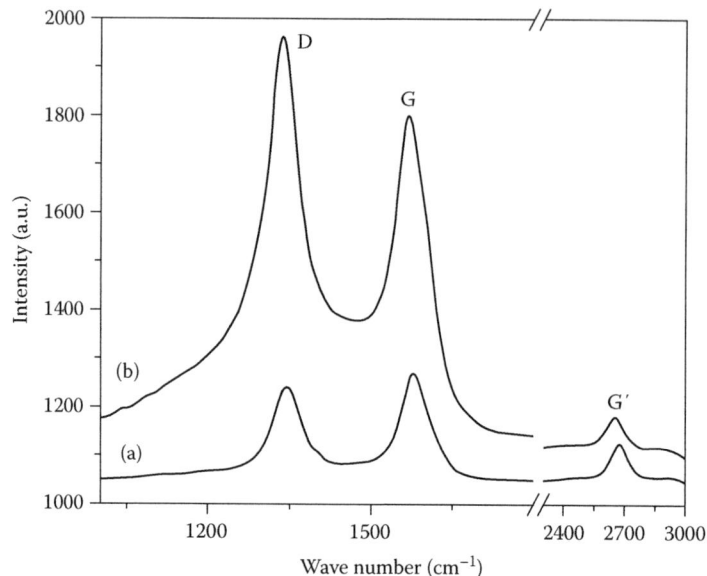

FIGURE 24.6
Raman spectra of (a) p-MWCNTs and (b) MWCNT–SH.

at 1340 cm⁻¹ is dependent on the number of defects in the sidewalls, presence of in-plane sub-stitutional groups, vacancies, and finite-size effects. Filho et al. [40] describe that graphite (G) band exhibits one single Lorentzian peak at 1573 cm⁻¹ due to tangential mode vibration of the carbon atoms. Further, the appearance of peak at 2664 cm⁻¹ confirms the characteristic G′ band from the second-order Raman spectra [41]. The calculated I_D/I_G was found to increase from 0.9 for pristine MWCNTs to 1.1 for MWCNT–SH (Figure 24.6b.) and thus confirms that the MAA was covalently functionalized on p-MWCNTs with minimum defect. The FESEM images of MWCNT–SH were shown in Figure 24.7. From the microscopic images, it is clear that the surface morphology is thickened and loosely packed. Obviously, the debundled nature and the increase in the diameter size of pristine MWCNTs from ~10 to ~30 nm in MWCNT–SH were noticed (Figure 24.7).

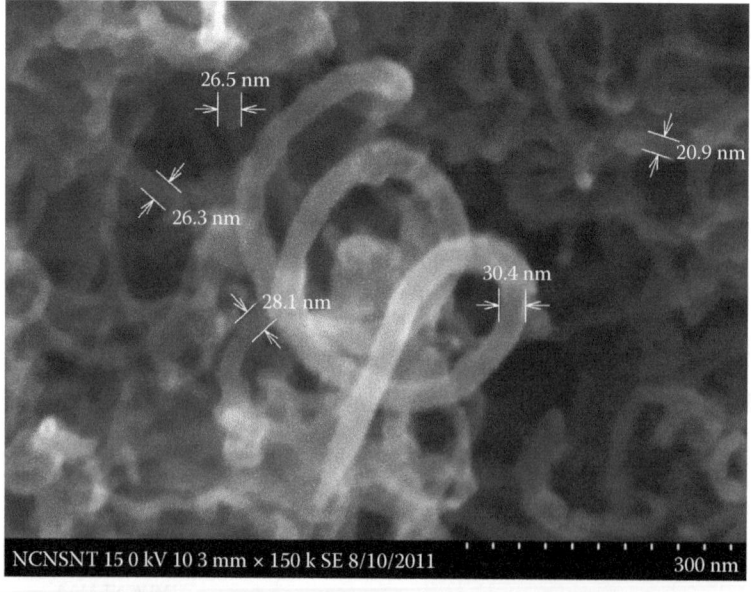

FIGURE 24.7
FESEM image of thiol-functionalized MWCNTs (MWCNT–SH).

24.6 APPLICATIONS OF CHEMICALLY FUNCTIONALIZED CARBON MATERIALS

24.6.1 Gas Sensors

CNTs are the prominent building blocks for the fabrication of the electronic devices and sensors as when compared with other carbon forms, namely, graphite, carbon blacks, glassy carbon, activated carbon, carbon microbeads, carbon fibers, carbon cloths, and carbon aerogels [42–45]. CNTs have the ability to detect small concentrations of molecules with high sensitivity under ambient conditions [46,47]. Nowadays, the fabrication of novel gas sensors for the detection of the odorless, colorless, tasteless, toxic, and hazardous gases (H_2, NH_3, O_2, and NO_2) has attracting wide attention [48–52]. Although a variety of metal and semiconducting metal oxide–based gas sensors are available in the market because of their structural simplicity and low cost, the main drawback of these sensors is their high operating temperature (~200°C) and low sensitivity [53,54]. Therefore, researcher's goal is to fabricate the sensors that can operate at room temperature and consume low power. Carbon materials including CNTs, carbon particles, and horn-based sensors have been recently studied due to their surface area. Among these carbon materials, CNTs can absorb large amount of gases, which make them a probable contender for gas sensor with very high sensitivity and low response time. In addition, CNTs are well known to behave as p-type semiconductors and respond to both oxidizing (electron withdrawing, e.g., NO_2 and O_2) and reducing (electron donating, e.g., NH_3 and H_2) gases under ambient conditions. The fabrication of CNT-based gas sensors opens a new way for multitransducer and multisensor array. There are two types of approaches used for the fabrication of gas sensors, namely, chemiresistor and field-effect gas sensors. Among these, chemiresistive sensors are easy to fabricate as compared to field-effect gas sensors.

Chemiresistor is a type of sensor that measures the change in current with gas concentration, which further reflects the changes in electrical conductance or resistance when they are exposed to different gases (NO_2, H_2S, CO, NH_3, and H_2). A chemiresistor is generally fabricated on the silicon (Si) substrate on which a thin layer of insulting material like silicon oxide (SiO_2) is grown. By using lithography beam, desired patterns, like interdigitated electrodes (IDEs), are fabricated on oxide layer. Network of well-dispersed CNTs, horn, and nanoparticles can be casted on the electrodes. Figure 24.8 shows the fabricated chemiresistor gas sensor containing functionalized CNTs [22]. The inset of Figure 24.8 shows the optical image of fabricated gold IDEs using photolithography. The SEM shows the network of CNTs on the IDEs. The sensitivity of this sensor at room temperature is 0.8% for 0.05% by volume of H_2 gas at room temperature [22]. It shows the complete resistance recovery, good repeatability, and less recovery time.

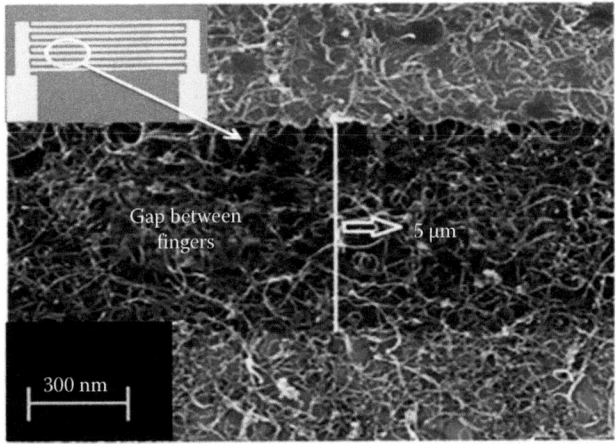

FIGURE 24.8
CNT-based chemiresistor gas sensor.

Dong et al. [33] fabricated acid-functionalized SWCNT-based sensors for the detection of CO and NH_3 gas mixture at 150°C. They found that acid-treated tubes were more sensitive toward these gases. Dhall et al. [22] have reported the MWCNT-based H_2 gas sensors at the room temperature conditions. They found that covalent functionalization of pristine MWCNTs enhance significantly their electrical and sensing response. Randeniya et al. [55] have reported the detection of 20 ppm to 2% H_2 gas in the nitrogen atmosphere using CNT yarns. They have deposited the layers of Pd and Pt on the yarns to enhance the sensitivity for H_2 gas. Kim et al. [56] proposed the detection of H_2 and CO gas using carbon nanoparticles at the room temperature conditions. They found that pristine carbon nanoparticles showed no response to reducing gases at room temperature but a significant response at elevated temperature as compared to functionalized nanoparticles. In addition, they observed that functionalized nanoparticle–based sensor operates at room temperature to measure the low concentrations of CO and H_2 even without any Pd or Pt catalysts, which are mainly used for splitting H_2 molecules into reactive H atoms. Sano et al. [57] have synthesized and fabricated single-walled carbon nanohorns (SWCNHs) produced by a gas-injected arc-in-water method for the detection of NH_3 and O_3 at room temperature. They found that the electrical resistance of the SWCNH film increased with adsorption of NH_3 and it decreases when they are exposed to O_3. It is found that SWCNH-based sensor illustrates higher sensitivity as compared to SWCNTs of the same electric resistance at the present condition with relatively a large interelectrode gap.

24.6.2 Biosensors

CNTs and SWCNHs have been extensively used as biosensors in a variety of ways. Before fabrication of biosensors, the immobilization of biomolecules with specific functionalities on the sensing material is very important. These biomolecules act as anchoring groups to bind the particular species in the testing sample and catalyze the reaction of a specific analyte. The recognition of specific target molecules is the main feature of the biological sensing. On the basis of the structure, biosensor is divided into two categories, namely, single-cell or single-molecule sensors and field-effect transistor (FET) biosensor.

The use of CNTs as single probes to gain great spatial resolutions is the main application of CNTs for biological sensing [58]. Due to the small size of CNT-based probes, it can be inserted into single cell for in situ measurements with excellent sensitivity and minimum disturbance. Murugan and Rubavathy [37] investigated the antibody-based nanobiosensor for the detection of benzo[a] pyrene tetrol (BPT), a biomarker for exposure of human to find the carcinogen benzo[a]pyrene (BaP) via pulling an optical fiber to the range of nanometer at the tip. For this sensor, the distal ends of the nanotubes are covalently coated via silane linkers with anti-BPT antibodies. By adopting similar technique, Gao [59] studied the electrophysiology phenomena in single cell or in single molecule using individual MWCNT nanoelectrode probe. Due to the small diameter, high aspect ratio, and large Young's modulus and mechanical strength, individual CNTs attached to an SPM/AFM tip have attracted wide attention [24]. The functionalization of this tip with different functional group provides more chemical force information. SWCNT tips could resolve the double-helix structure of DNA and provide high resolutions [24]. On the other hand, the bundle of MWCNTs of diameter from 10 to 100 nm attach to the pyramidal tip with an acrylic adhesive.

FET-based protein biosensors investigated by Hongjie Dai's group contained noncovalent functionalization of SWCNTs with a surfactant and then attachment of specific receptors to these nanotubes. For the fabrication of these devices, the SWCNTs were directly grown in situ on quartz substrates. In this sensor, the variation in electrical current with the additions of proteins in the solution is recorded to measure the sensitivity.

To fabricate the mediator-free biosensors, CNTs have been used as molecular wire to understand direct electron transfer phenomena between electrode surfaces and redox enzymes [60–63]. However, most of synthesis methods of CNTs produce some impurities like graphite and metal catalyst, so purification of nanotubes is very necessary. Although SWCNHs are free from impurities, it can be directly used in biosensing applications. SWCNHs firstly were used to fabricate a glucose biosensor by encapsulating glucose oxidize in the nanocomposite of Nafion-SWCNHs. This sensor shows high sensitivity, low detection limit, and good selectivity [64-68]. The high sensitivity of this sensor is due to the large conductivity of SWCNHs, and also Nafion-based nanocomposite film provided a biocompatible microenvironment to maintain enzymatic activity. After that, SWCNHs were widely used as novel biocompatible matrix for fabricating H_2O_2 biosensor. In addition, the noncovalent functionalized SWCNHs are widely used in the fields of biosensor and electrochemistry [69].

24.6.3 Acoustic Sensors

CNTs have also been used to detect the sound in the same manner as ear. In this sensor, the network of the MWCNTs response to the acoustic energy, and by using electrical technique, the motion is converted into measurable electrical signal. The electrical signal is similar to the signal coming from microphone except the fainter sounds, which mostly contained in this sensor. It is found that, as compared to SWCNTs, MWCNTs are naturally directional and bend away from the source of sounds. Generally, directionality derivative from the two ears in normal hearing, however, in the case of nanotechnology, especially in the case of nanotubes, was provided from a single sensor. This novel technique produces the new class of microactuators and microsensors for the real-world applications in liquid/gas atmosphere [2].

24.6.4 Force Sensors

CNTs can be used as force sensors in AFM and STM. Wood et al. [70] have investigated the use of SWCNTs as molecular and microscopic pressure sensors. It was found that when hydrostatic pressure is applied to SWCNTs, then there is a shift of nanotube band in Raman spectroscopy. The direct microscopic pressure in diamond anvil cell allows the manipulation of nanostructure for comparison with the peak shift. They also observed that the shifting in the G band of the Raman spectra occurs at the pressure of 2000 atm. The CNTs are used as strain sensors for getting the strain profile in the matrix of polymer. The strain is transformed from polymer to CNTs and can be measured when the distance between the CNTs is 1 μm from each other [71–73]. To investigate the individual strain/stress component, the Raman band of CNTs in a specific direction must be selected. The Raman technique can also be used to measure the individual stress component.

The rapid and accurate detection of H_2O_2 is of great interest because it is not only catalyzed by many highly selective oxides but it also acts as an essential compound in food, pharmaceutical, clinical, and environmental analyses. There are many methods applied for the detection of H_2O_2, such as fluorometry, chemiluminescence, and electrochemical methods. Usually, enzymes such as horseradish peroxidase, cytochrome, and myoglobin [74] were used on the functionalized electrodes depending upon the immobilization material coated for the detection of H_2O_2. Generally, the electrooxidation or electroreduction of H_2O_2 on bare gold or the carbon electrode, which is extensively used in the electrochemical experiments, requires high overpotential, while common electroactive species will confuse the measurements. Under these circumstances, it is of great interest to fabricate chemically modified electrodes for the detection of H_2O_2 with high efficiency and selectivity. Most of these methods are based on the immobilization of enzymes such as horseradish peroxidase [75], cytochrome c [76], and myoglobin (on the functionalized electrodes). Although the enzyme-based biosensors can acquire remarkable selectivity, they usually suffer from the complicated enzyme immobilization processes and the instability of the immobilized biomolecules. Thus, the nonenzymatic sensor for the detection of H_2O_2 is greatly appreciated to current researchers (Table 24.1).

24.7 POLYMERIC DEVICES

The electronic structure of CNTs makes its use for the fabrication of composite devices. The composite of SWCNTs and polymer poly(3-octylthiophene) shows those properties that are widely used in photovoltaic cells. The electron transfer in CNT-based photovoltaic cell from nanotube to polymer causes to increase short-circuit current and fill factor in the diode. Composite with poly(m-phenylenevinylene-co-2,5-dioctyloxy-p-phenylenevinylene) has been used as effective organic light-emitting diodes [22,92]. In addition, actuators based on SWNT have also been used as

TABLE 24.1 Electrodes in Use for H_2O_2 Gas Sensors

Electrode	Applied Potential (V)	Lower Detection Limit (µm)	Linear Range (mm)	References
Se/Pt nanocomposites	0 (vs. SCE)	3.1	10–15,000	[76]
Pt nanoparticles/ordered mesoporous carbon nanocomposite	−0.1 (vs. Ag/AgCl)	1.2	2–4,212	[77]
Pt nanoparticle-loaded carbon nanofiber electrode	0 (vs. Ag/AgCl)	0.6	1–800	[78]
Pt/polypyrrole hybrid hollow microspheres	−0.1 (vs. Ag/AgCl)	1.2	1,000–8,000	[79]
Gold nanowire	−0.2 (vs. Ag/AgCl)	1.2	1.2–800	[80]
Hemoglobin/laponite/chitosan	−0.25 (vs. SCE)	6.2	6.2–2,550	[81]
Au–graphene–HRP–chitosan biocomposites	−0.3 (vs. Ag/AgCl)	1.7	5–5,130	[82]
Mesoporous platinum microelectrodes	0.6 (vs. SCE)	4.5	20–40,000	[83]
Pt/graphene nanocomposite	−0.20 (vs. SCE)	0.8	2.5–6,650	[84]
MWCNTs/GCE	_____		10–100	[85]
MWCNTs–NF (DDAB/Cat)	_____	0.6	700–4,633	[86]
Cat/SWNTs–CHI/GCE	_____	2.5	5–50	[87]
Cat/NiO/GCE	_____	15.9	1–1,000	[85]
Cat/cysteine/Si solgel/GCE	_____		1–30	[88]
HRP/NAEs	_____	0.42	10–15,000	[89]
PtNP/NAEs	_____	194.6	200–20,000	[90]
Cat/AuNPs/graphene–NH₂/GCE	_____	0.05	0.3–600	[91]

microswitches and artificial muscles [22]. Moreover, SWNT composite actuators show no relaxation with time, and the conductivity required to achieve actuation is less than that of metal. Deflection amounts were found to increase with increasing nanotube concentration.

24.8 SUMMARY

The functionalizations of the nanocarbon materials have been discussed. Functionalizations of these materials are the building block for the attachment of metal/metal oxide and polymers on their surface. With this method, we can disperse/separate the nanoparticles for applications like gas sensors, chemical sensors, pressure sensors, FET devices, and other interdisciplinary areas. The excess covalent functionalization of the carbon materials can destroy their physical and structural properties. On the other hand, noncovalent functionalization does not affect the structural properties of carbon materials. So, using this method, we can separate out CNTs in accordance with their diameter and length. For sensing applications, the covalent functionalization of the carbon materials plays an important role to enhance the sensitivity and make possible to operate at the room temperature conditions.

REFERENCES

1. D.M. Guldi and N. Martín, *Carbon Nanotubes and Related Structures, Synthesis, Characterization, Functionalization, and Applications*, Wiley–VCH GmBH & Co.KGaA, Germany (2010).
2. P.J.F. Harris, *Carbon Nanotube Science Synthesis, Properties and Applications*, Cambridge University Press, Cambridge, U.K. (2011).
3. M. Kang, S.J. Myung, and H.J. Jin, *Polymer*, 47 (2006) 3961.

4. M.F. Islam et al., *Nano Lett.*, 3 (2003) 269.
5. C. Richard et al., *Science*, 300 (2003) 775.
6. R.J. Chen et al., *J. Am. Chem. Soc.*, 123 (2001) 3838.
7. N. Nakashima, Y. Tomonari, and H. Murakami, *Chem. Lett.*, 31 (2002) 638.
8. H. Murakami and N. Nakashima, *J. Nanosci. Nanotechnol.*, 6 (2006) 16.
9. V. Georgakilas et al., *Chem. Mater.*, 17 (2005) 1613.
10. D. M. Guldi, *J. Phys. Chem. B*, 109 (2005) 432.
11. C. Ehli et al., *J. Am. Chem. Soc.*, 128 (2006) 222.
12. C.A. Dyke and J.M. Tour, *Chem. Eur. J.*, 10 (2004) 813.
13. K. Molhave, S.B. Gudnason, A.T. Pedersen, C.H. Clausen, A. Horsewell, and P. Boggild, *Nano Lett.*, 6 (2006) 1663.
14. S. Mathew, U.M. Bhatta, J. Ghatak, B.R. Sekhar, and B.N. Dev, *Carbon*, 45 (2007) 2659.
15. B.P. Singh, D. Singh, R.B. Mathur, and T.L. Dhami, *Nano Res. Lett.*, 3 (2008) 444.
16. A. Amiri, M. Shanbedi, H. Eshghi, S.Z. Heris, and M. Baniadam, *Phys. Chem.*, 116 (2012) 3369.
17. H.C. Lau, R. Cervini, S.R. Clarke, M.G. Markovic, J.G. Matisons, S.C. Hawkins, C.P. Huynh, and G.P. Simon, *J. Nanoparticle. Res.*, 10 (2008) 77.
18. K.Y. Dong, J. Choi, Y.D. Lee, B.H. Kang, Y.Y. Yu, H.H. Choi, and B.K. Ju, *Nano Res. Lett.*, 8 (2013) 12.
19. H. Peng, L.B. Alemany, J.L. Margrave, and V.N. Khabashesku, *J. Am. Chem. Soc.*, 125 (2003) 15174.
20. C. Wang, G. Zhou, H. Liu, J. Wu, Y. Qiu, B. L. Gu, and W. Duan, *Phys. Chem. B*, 110 (2006) 10266.
21. K.A. Worsley, I. Kalinina, E. Bekyarova, and R.C. Haddon, *J. Am. Chem. Soc.*, 131 (2009) 18153.
22. S. Dhall, N. Jaggi, and R. Nathawat, *Sens. Actuators A*, 201 (2013) 321.
23. S. Niyogi, M.A. Hamon, H. Hu, B. Zhao, P. Bhowmik, R. Sen, M.E. Itkis, and R.C. Haddon, *Acc. Chem. Res.*, 35 (2002) 1105.
24. M. Meyyappan, *Carbon Nanotubes Science and Applications*, CRC Press, FL, USA (2005).
25. J. Liu et al., *Science*, 280 (1998) 1253.
26. J.N. Coleman, A.B. Dalton, S. Curran, A. Rubio, A.P. Davey, and A. Drury, *Adv. Mater.*, 12 (2000) 213.
27. R.K. Saini, I.W. Chiang, H. Peng, R.E. Smalley, W.E. Billups, R.H. Hauge, and J.L. Margrave, *J. Am. Chem. Soc.*, 125 (2003) 3617.
28. X.B. Yan, B.K. Tay, and Y. Yang, *J. Phys. Chem. B*, 110 (2006) 25844.
29. A.G. Osorio et al., *Appl. Surf. Sci.*, 255 (2008) 2485.
30. F.A. Abuilaiwi, T. Laoui, M.A. Harthi, and A. Muataz, *Arabian J. Sci. Eng.*, 35 (2010) 1.
31. Z.P. Nhlabatsi et al., Synthesis and characterization of copper containing carbon nanotubes (CNTs) and their use in the removal of impurities in water, (2011). http://www. docstoc.com/docs/74409791.
32. G.M. Neelgund and A. Oki, *J. Nanosci. Nanotechnol.*, 11 (2011) 3621.
33. K.Y. Dong et al., *Nanoscale Res. Lett.*, 8 (2013) 12.
34. A.J. Haider, M.R. Mohammed, E.A.J.A. Mulla, and D.S. Ahmed, *Rend. Fis. Acc. Lincei*, 25 (2014) 403.
35. H.J. Kim et al., *Polym. Adv. Technol.*, 20 (2009) 736.
36. R.R. Kakarla, P.L. Kwang, G.A. Iyengar, K.M. Seok, M.S. Ali, and N.Y. Chang, *J. Polym. Sci., Part A: Polym. Chem.*, 44, (2006) 3355.
37. E. Murugan and A.R.J. Priya, *Int. Rev. Appl. Eng. Res.*, 3 (2013) 5.

38. H. Choi, I. Jeon, D.W. Chang, D. Yu, L. Dai, L. Tan, and J. Baek, *J. Phys. Chem. C*, 115, (2011) 1746.
39. A. Manna, T. Imae, K. Aoi, M. Okada, and T. Yogo, *Chem. Mater.*, 13, (2001) 1674.
40. A.G.S. Filho, A. Jorio, G.G. Samsonidze, G. Dresselhaus, R. Saito, and M.S. Dresselhaus, *Nanotechnology*, 14 (2003) 1130.
41. M.S. Dresselhaus, G. Dresselhaus, R. Saito, and A. Jorio, *Phys. Rep.*, 409 (2005) 47.
42. R.L. McCreery, *Carbon Electrodes: Structural Effects on Electron Transfer Kinetics*, in Electroanalytical Chemistry, (A.J. Bard, eds.), M. Dekker, Inc., New York (1991), p. 221.
43. A.J. Downward, *Electroanalysis*, 12 (2000) 1085–1096.
44. L. Rabinovich and O. Lev, *Electroanalysis*, 13, (2001) 265.
45. S.A. Wring and J.P. Hart, *Analyst*, 117 (1992) 1215.
46. D.H. Oh, N.D. Hoa, and D.J. Kim, *Nanosci. Nanotechnol.*, 11 (2011) 1601.
47. S.K. Arya, S. Krishnan, H. Silva, S. Jean, and S. Bhansali, *Analyst*, 137 (2012) 2743.
48. H. Gu, Z. Wang, and Y. Hu, *Sensors*, 12 (2012) 5517.
49. L. De Luca, A. Donato, S. Santangelo, G. Faggio, G. Messina, N. Donato, and G. Neri, *Int. J. Hydrogen Energy*, 37 (2012) 1842.
50. T. Zhang, S. Mubeen, N.V. Myung, and M.A. Deshusses, *Nanotechnology*, 19 (2008) 332001.
51. S. Srivastava, S.S. Sharma, S. Agrawala, S. Kumara, M. Singha, and Y.K. Vijay, *Synth. Met.*, 160 (2010) 529.
52. B. Philip, J.K. Abraham, A. Chandrasekhar, and V.K. Varadan, *Smart Mater. Struct.*, 12 (2003) 935.
53. C. Wongchoosuk, A. Wisitsoraat, D. Phokharatkul, A. Tuantranont, and T. Kerdcharoen, *Sensors*, 10 (2010) 7705.
54. L.D. Luca, A. Donato, S. Santangelo, G. Faggio, G. Messina, N. Donato, and G. Neri, *Int. J. Hydrogen Energy*, 37 (2012) 1842.
55. L.K. Randeniya, P.J. Martin, and A. Bendavid, *Carbon*, 50 (2012) 1786.
56. D. Kim, P.V. Pikhitsa, H. Yang, and M. Choi, *Nanotechnology*, 22 (2011) 485501.
57. N. Sano, M. Kinugasa, F. Otsuki, and J. Suehiro, *Adv. Powder Technol.*, 18 (2007) 455.
58. B. Mahar, C. Laslu, R. Yip, and Y. Sun, *IEEE Sensors,* 7 (2007) 266.
59. H. Gao, *J. Mech. Phys. Solids*, 1 (1993) 457.
60. J. X. Wang, M. X. Li, Z. J. Shi, N. Q. Li, and Z. N. Gu, *Anal. Chem.*, 74 (2002) 1993.
61. J. Wang, *Electroanalysis*, 17 (2005) 7.
62. J.J. Gooding, R. Wibowo, J. Q. Liu, W. Yang, D. Losic, S. Orbon, F. J. Mearns, J. G. Shapter, and D. B. Hibbert, *J. Am. Chem. Soc.*, 125 (2003) 9006.
63. F. Patolsky, Y. Weizmann, and I. Willner, *Angew. Chem. Int. Ed.*, 43 (2004) 2113.
64. X. Q. Liu, H. J. Li, F. A. Wang, S. Y. Zhu, Y. L. Wang, and G. B. Xu, *Biosens. Bioelectron.*, 25 (2010) 2194.
65. Y. Zou, C. Xiang, L.X. Sun, and F. Xu, *Biosens. Bioelectron.*, 23 (2008) 1010.
66. D.R. Shobha and S.S. Narayanan, *Biosens. Bioelectron.*, 23 (2008) 1404.
67. G.G. Vasilis, S.A. Law, J.C. Ball, R. Andrews, and L.G. Bachas, *Anal. Biochem.*, 329 (2004) 247.
68. C. Deng, J. Chen, X. Chen, C. Xiao, L. Nie, and S. Yao, *Biosens. Bioelectron.*, 23 (2008) 1272.
69. S. Zhu and G. Xu, *Nanoscale*, 2 (2010) 2538.
70. J.R. Wood et al., *J. Phys. Chem. B*, 103 (1999) 10388.
71. Q. Zhao, M.D. Frogley, and H.D. Wagner, *J. Compos. Sci. Technol.*, 62 (2002) 147.

72. J. Hu, Y. Yu, H. Guo, Z. Chen, A. Li, X. Feng, B. Xi, and G. Hu, *J. Mater. Chem.*, 21 (2011) 5352.
73. Y.H. Zhu, H.M. Cao, L.H. Tang, X.L. Yang, and C.Z. Li, *Electrochim. Acta*, 54 (2009) 2823.
74. L. Zhang, *Biosens. Bioelectron.*, 23 (2008) 1610.
75. Y. Li, J.J. Zhang, J. Xuan, L.P. Jiang, and J.J. Zhu, *Electrochem. Commun.*, 12 (2010) 777.
76. X.J. Bo, J.C. Ndamanisha, J. Bai, and L.P. Guo, *Talanta*, 82 (2010) 85.
77. Y. Liu, D.W. Wang, L. Xu, H.Q. Hou, and T.Y. You, *Biosens. Bioelectron.*, 26 (2011) 4585.
78. X.J. Bian, X.F. Lu, E. Jin, L.R. Kong, W.J. Zhang, and C. Wang, *Talanta*, 81 (2010) 813.
79. S.J. Guo, D. Wen, S.J. Dong, and E.K. Wang, *Talanta*, 77 (2009) 1510.
80. D. Shan, E. Han, H.G. Xue, and S. Cosnier, *Biomacromolecules*, 8 (2007) 3041.
81. K.F. Zhou, Y.H. Zhu, X.L. Yang, J. Luo, C.Z. Li, and S.R. Luan, *Electrochim. Acta*, 55 (2010) 3055.
82. S.A.G. Evans, J.M. Elliott, L.M. Andrews, P.N. Bartlett, P.J. Doyle, and G. Denuault, *Anal. Chem.*, 74 (2002) 1322.
83. F. Zhang, Z. Wang, Y. Zhang, Z. Zheng, C. Wang, Y. Du, and W. Ye, *Int. J. Electrochem. Sci.*, 7 (2012) 1968.
84. A. Salimi, A. Noorbakhsh, and M. Ghadermarz, *Anal. Biochem.*, 344 (2005) 16.
85. P.A. Prakash, U. Yogeswaran, and S.M. Chen, *Talanta*, 78 (2009) 1414.
86. H.J. Jiang, H. Yang, and D.L. Akins, *J. Electroanal. Chem.*, 623 (2008) 181.
87. J.W. Di, M. Zhang, K. Yao, and S.P. Bi, *Biosens. Bioelectron.*, 22 (2006) 247.
88. J. Xua, F.J. Shang, J.H.T. Luong, K.M. Razeeb, and J.D. Glennon, *Biosens. Bioelectron.*, 25 (2010) 1313.
89. M. Jamal, J. Xu, and K.M. Razeeb, *Biosens. Bioelectron.*, 26 (2010) 1420.
90. A. Salimi, E. Sharifi, A. Noorbakhsh, and S. Soltanian, *Biophys. Chem.*, 125 (2007) 540.
91. K.J. Huang, D.J. Niu, X. Liu, Z.W. Wu, Y. Fan, Y.F. Chang, and Y.Y. Wu, *Electrochim. Acta*, 56 (2011) 2947.
92. P. Fournet et al., *J. Appl. Phys.*, 90 (2001) 969.

CHAPTER 25

CONTENTS

Carbon Nanotube Field-Effect Transistor

The Application of Carbon Nanotube

25

Sanjeet Kumar Sinha and Saurabh Chaudhury

25.1 INTRODUCTION

Carbon is the 15th most abundant element in the Earth's crust, and it is the 4th most abundant element in the universe by mass after hydrogen, helium, and oxygen. After oxygen, carbon is the second most abundant element by mass (18.5%) in the human body. In the periodic table, carbon is a member of group 14; it is nonmetallic and has four electrons to form covalent bonds with other elements. Carbon has the unique characteristic among all elements to form long chains of its own atoms, a property called catenation. The strength of carbon–carbon bond gives rise to a number of molecular forms, so the study of the carbon element and its compounds is another field. The different carbon atoms can be identified with respect to the number of carbon neighbors.

The property of some chemical elements to exist in two or more different forms is called allotropism. Two well-known allotropes of carbon are diamond and graphite. Allotropes are different structural modifications of an element, and through a different manner, the atoms of the element are bonded together. A free carbon atom has the electronic structure $1s^2 2s^2 2p^2$. In order to form covalent bonds, one of the 2s electrons is shifted to 2p and form sp^2 hybridization. Due to other single electrons in the s orbital, it has semimetal characteristics. The overlap of orbitals on adjacent atoms in a given plane provides the electron bond network, which gives graphite its relatively high electrical conductivity. Graphene is a thick sheet of graphite and has great potential for device application in nanoelectronics because of its properties such as high electrical conductivity and mechanical strength.

Carbon nanotube (CNT) is a one-atom-thick sheet of graphite (called graphene) rolled up into a cylinder with a diameter of the order of a nanometer. A carbon nanotube field-effect transistor (CNTFET) utilizes CNT as the channel material instead of bulk silicon in the traditional metal oxide semiconductor field-effect transistor (MOSFET) structure. The exponential increase of leakage currents in a scaled device of MOSFET is an inevitable consequence.

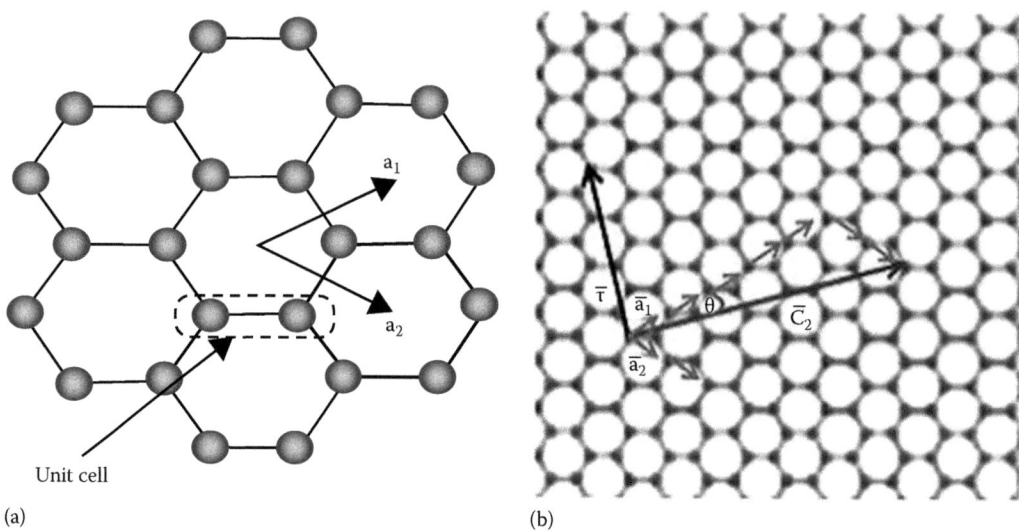

(a)   (b)

FIGURE 25.1
(a) Graphene with vectors a_1 and a_2 in real space; (b) graphene atomic structure with chiral vector.

In MOSFET, constant field scaling reaches a limit. This is due to an ideal constant field scaling impossible for some nonscaling quantities. Several circuit techniques to reduce the leakage power consumption have been proposed in literature. Field-effect transistors based on CNTs have been a focus of active research in recent years. CNTFET devices are the viable alternatives to MOSFET devices in the nanometer regime and emerging to be the technology for future semiconductor devices. CNTFET devices have a number of excellent characteristics in the nanometer regime over MOSFET devices. It has higher mobility, high threshold voltage, low leakage, and reduced propagation delay in deep nanometer range. Research is in progress globally for setting up the technology and mass-scale production of CNTFET. A single-walled CNT (SWCNT) can act as either a conductor or a semiconductor, depending on the angle of the atomic arrangement along the tube. This is referred to as the chirality vector and is represented by the integer pair (m, n).

A CNT can act as either a conductor or a semiconductor, depending on the angle of the atomic arrangement along the tube. This is referred to as the chirality vector and is represented by the integer pair (m, n). The circumference of a CNT can be expressed in terms of a chiral vector, $c = na_1 + ma_2$, which connects two equivalent sites of the 2D graphene sheet, as shown in Figure 25.1a and b, where n and m are the integers and a_1 and a_2 are the unit vectors of the hexagonal honeycomb lattice. A simple method to determine if a CNT is metallic or semiconducting is to consider its indexes (m, n). The nanotube is metallic if m = n or m − n = 3i, where i is an integer, otherwise the tube is semiconducting [1].

Although compared to silicon, CNTs have unique properties such as stiffness, strength, and tenacity [2], it is still experiencing lack of technology for mass production and high production cost. The geometry of the graphene lattice and the chiral vector of the tube shown in Figure 25.2 determines the structural parameters like diameter, unit cell, and its carbon atoms, as well as the size and shape of the Brillouin zone.

There are two types of CNT, single walled and multiwalled. These are typically a few nanometers in diameter and several micrometers to centimeters in length. It has an important role in the context of the nanometer regime, because of their novel chemical and physical properties. They are mechanically very strong and can conduct electricity extremely well. These remarkable properties give CNTs a range of potential applications such as in nanoelectronics.

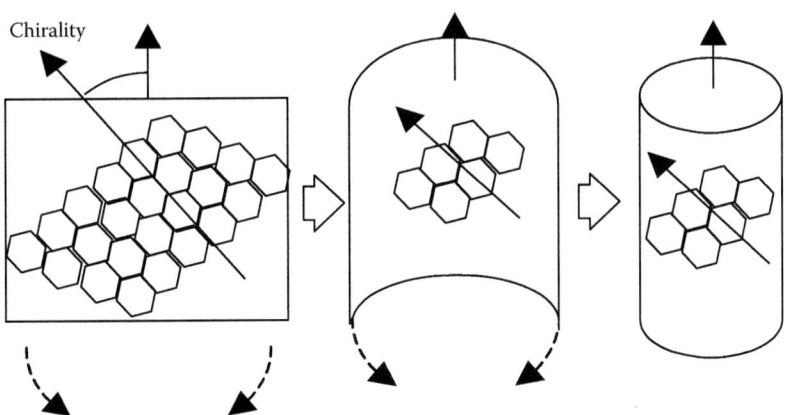

FIGURE 25.2
Schematic animations of a CNT that can be formed by rolling up graphene and the chirality of CNT.

25.2 APPLICATION

CNTFET uses CNT as their semiconducting channels. An SWCNT consists of one cylinder only, and the simple manufacturing process of this device makes it a very promising alternative to MOSFET. Most SWCNTs have a diameter of 1 nm, with a tube length that can be many millions of times longer. The structure of an SWCNT can be conceptualized as a one-atom-thick layer of graphite, called graphene, wrapped into a seamless cylinder. CNT is a promising alternative to conventional silicon technology for future nanoelectronics because of their unique electrical properties. A CNTFET is a transistor that utilizes a single CNT as the channel material instead of bulk silicon in the traditional MOSFET structure. As shown in Figure 25.3a, CNTs are rolled-up sheets of graphene. Graphene is the strongest material ever tested; however, the separating process of it from graphite will require some technological development before it is economical to be used in industrial processes.

The I–V characteristics of the CNTFET are similar to the MOSFET. The threshold voltage is defined as the voltage required to turn on the transistor. The threshold voltage of the intrinsic CNT channel can be approximated to be of first order as the half band gap is an inverse function of the diameter. Similar to the traditional silicon device, the CNTFET has also four terminals. As shown in Figure 25.3b, the undoped semiconducting nanotubes are placed under the gate as the

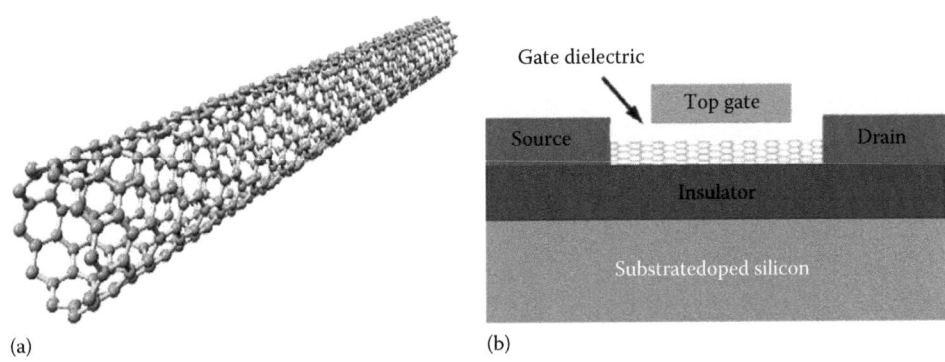

(a)   (b)

FIGURE 25.3
(a) Rolled-up sheet of graphene sheet; (b) schematic of CNTFET device.

channel region, while doped CNTs are placed between the gate and the source/drain to allow for a low series resistance in the ON state. As the gate potential increases, the device is electrostatically turned on or off.

The diameter of the CNT can be calculated following Equation 25.1 as defined in [3]:

$$D_{CNT} = \frac{\sqrt{3}a_0}{\pi} \sqrt{n^2 + m^2 + mn} \tag{25.1}$$

where $a_0 = 0.142$ nm is the interatomic distance between each carbon atom and its neighbor. Clearly, the chiral vector pair (m, n) and the interatomic distance decide the diameter of a CNT. The direction of the chiral vector is measured by the chiral angle Θ. The chiral angle Θ can be calculated as

$$Cos\,\Theta = \frac{(n+m)/2}{\sqrt{n^2 + m^2 + mn}} \tag{25.2}$$

The differences in the chiral angle and the diameter cause the differences in the properties of the CNTs. Threshold voltage of a CNTFET in terms of the diameter of CNT as in [3] can be given as

$$V_{TH} = a(V_{\Pi})/\sqrt{3}\,qD_{CNT} \tag{25.3}$$

where
   q is the electronic charge
   a = 2.49 Å is the lattice constant
   V_{Π} = 3.033 eV is the carbon π to π bond energy

Again, the threshold voltage of the CNT is inversely proportional to its diameter. Thus, CNTFETs provide a unique opportunity to control the threshold voltage by changing the chirality vector or the diameter of the CNT, as shown in Equations 25.1 through 25.3. The gate-to-source voltage that generates the same reference current is taken as the threshold voltage for the transistor that has different chirality.

The chiral vector has an important role in deciding the diameter of the CNT. When the difference of the chiral vector (m, n) is small, for example, (m, n) = (4, 0), the threshold voltage is larger, which is about 0.69 V, but at the same time, when the difference is small and the chiral vector (m, n) is large, for example, (m, n) = (16, 12), then the threshold voltage is very small and is found to be 0.17 V (Figure 25.4)

The gate-to-source voltage that generates the same reference current is taken as the threshold voltage for the transistor that has different chirality. CNTFETs provide a unique opportunity to control the threshold voltage by changing the chirality vector or the diameter of the CNT [4]. The CNTFETs are particularly attractive due to the possibility of near-ballistic channel transport, easy application of high-k gate insulator, and novel device physics. The ac properties are technologically most relevant, although most of the work on CNTFETs has concentrated on their dc properties. It is observed that a short nanotube operating in the ballistic regime, and the quantum capacitance (Qc), should be able to provide a gain in the THz range [5]. The comparison of CNTFET-based logic circuits to CMOS logic circuits is necessary to establish the means of evaluation for performance metrics such as current density, device

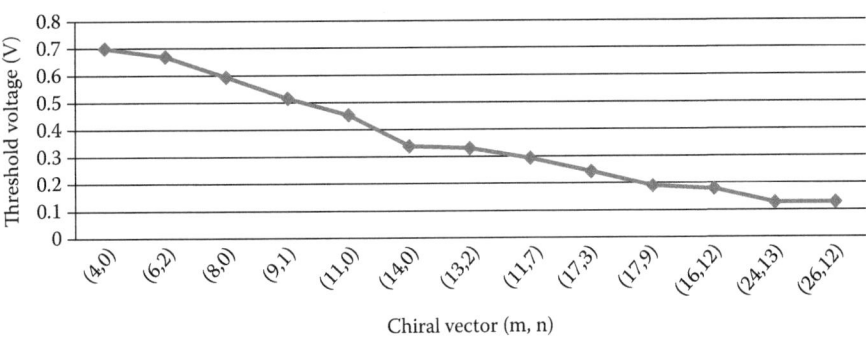

FIGURE 25.4
Threshold voltage at different chiral vectors.

switching speed, propagation delay through the gates, switching energy, operating temperature, and cost. However, the technology is not sufficiently mature to enable meaningful comparisons as the positioning techniques must still evolve to enable high-yield volume manufacturing and contact technology must be improved to reduce the impact of contacts on circuit performance.

25.3 CHARACTERISTICS OF CNTFET

Simulation results of single-gate MOSFET for Qc at different gate voltages and oxide thickness at a constant drain voltage of 1 V are shown in Figure 25.5. It can be observed from the figure that in single-gate MOSFET, as the oxide thickness goes down from 1.5 to 0.7 nm, the Qc increases significantly as the gate voltage increases from 0.5 V and above. This increment in Qc is observed up to a gate voltage of 1 V. Figure 25.5a shows the bar diagram representation of Qc against the gate voltage of MOSFET. It can also be observed from the simulation that at a very low gate voltage such as at 0 and 0.083 V, the value of Qc is the same for all oxide thicknesses considered here.

Qc has an important role in nanoscale devices [6], and it is the property of channel material. Since the density of state is finite in a semiconductor quantum well, but as the charge in the quantum well increases, the Fermi level needs to move up above the conduction band. This movement requires energy and this conceptually corresponds to Qc [7].

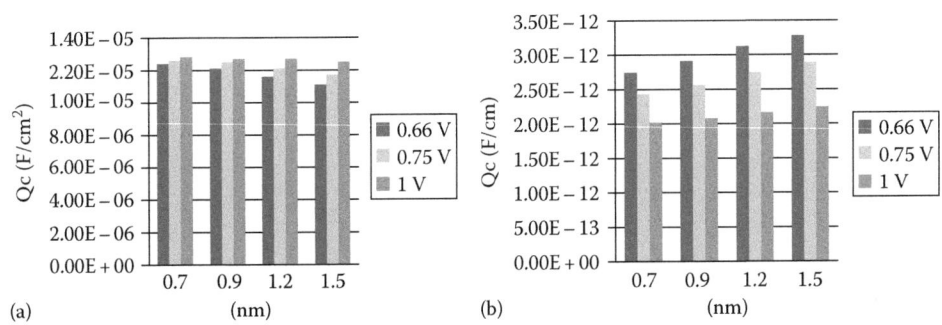

FIGURE 25.5
Quantum capacitance (Qc) vs. gate voltage for different oxide thickness in (a) MOSFET and (b) CNTFET.

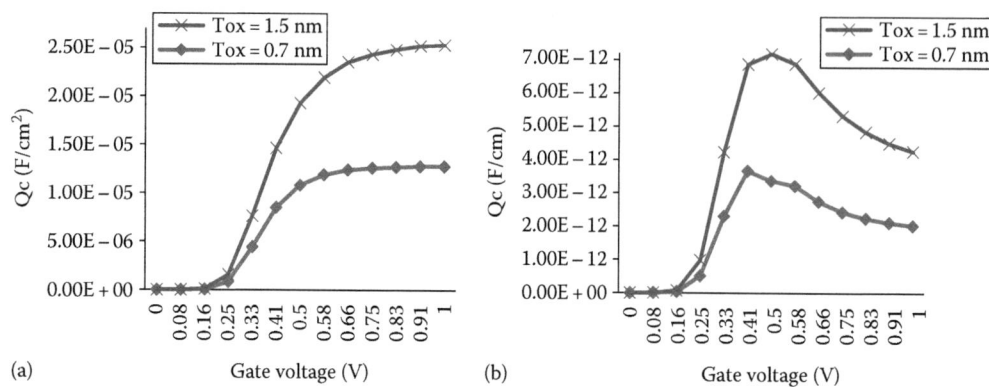

FIGURE 25.6
(a) Qc vs. gate voltage of MOSFET device; (b) Qc vs. gate voltage of CNTFET device.

It can be observed from Figure 25.6a that Qc increases as we decrease the oxide thickness for gate voltages up to 0.5 V. However, as the gate voltage is increased above 0.5 V, it shows a decreasing nature, that is, Qc decreases with decreasing oxide thickness. However, the effect is significant in and around a gate voltage of 1 V.

The plot of Figure 25.6a and b shows that in the nanoscale regime, CNTFET devices has an advantage over MOSFET due to lesser and lesser Qc, which leads to an increase in threshold voltage as explained in the earlier section, while in MOSFET the value of Qc goes on increasing and thereby reducing threshold voltage, which leads to increased leakage and performance degradation [8].

25.4 CONCLUSION

CNTs are rolled-up sheets of graphene, and CNTFET uses CNT as the channel material instead of silicon in conventional MOSFET devices. It has some unique property of decreasing Qc with the reduction in oxide thickness, which is not possible to get in MOSFET. Thereafter, we have analyzed the effect of the variation of the chiral vector on threshold voltage in CNTFET devices by HSPICE simulation and observed that at a low value of chiral vector pair, the threshold voltage is higher, whereas at higher value of chiral vector pair, the threshold voltage is small. From the result shown earlier, we can conclude that in nanoscale regime, CNTFET devices are advantageous over single-gate MOSFET devices due to lesser and lesser Qc.

REFERENCES

1. M. Dresselhaus, G. Dresselhaus, and R. Saito, Carbon fibers based on C60 and their symmetry, *Physical Review B*, 45(11), 6234–6244, 1992.
2. R. H. Baughman, A. A. Zhakidov, and W. A. deHeer, Carbon nanotubes the route toward applications, *Science*, 297, 787–792, 2002.
3. A. Imran and M. Azam, Impact of CNT's diameter variation on the performance of CNTFET dual-X CCII, *International Journal of Computer Applications* (0975–8887), 56(16), 1–6, October 2012.

4. P. Avouris, Z. Chen, and V. Perebeinos, Carbon-based electronics, *Nature Nanotechnology*, 2(10), 605–615, 2007.

5. P. G. Collins and P. Avouris, Nanotubes for electronics, *Scientific American*, 283, 62–69, 2000.

6. C. Dekker, Carbon nanotubes as molecular quantum wires, *Physics Today*, 52(5), 22–28, 1999.

7. S. Luryi, Quantum capacitance devices, *Applied Physics Letters*, 52, 501–503, 1998.

8. S. K. Sinha and S. Chaudhury, Impact of oxide thickness on gate capacitance— A comprehensive analysis on MOSFET, nanowire FET, and CNTFET devices, *IEEE Transactions on Nanotechnology*, 12(6), 958–964, November 2013.

CHAPTER 26

CONTENTS

On the Mechanical Properties of Functionalized CNT Reinforced Polymer

26

Roham Rafiee and Reza Pourazizi

26.1 INTRODUCTION

Possessing exceptional mechanical, thermal, and electrical properties, carbon nanotubes (CNTs) received extreme interests among researchers as a potential reinforcing agent of polymeric composites (Beling and Epron 2005, Lau et al. 2006). The experimentally observed significant growth in the mechanical properties of polymer by adding small portions of CNTs (Schadler et al. 1998, Li et al. 2000, Qian et al. 2000, Zhu et al. 2004, Li et al. 2005) has established a new era in the field of advanced materials.

The efficiency of CNTs in reinforcing polymer depends on the atomic structure of CNTs at nanoscale, while load transfer phenomenon from matrix to CNT through the intermediate phase between CNT and surrounding polymer plays a key role at the scale of micro in this regard. This issue has been discussed in the literature by many researchers (Chen et al. 1998, Mickelson et al. 1998, Bahr et al. 2001, Huang et al. 2002, Lin et al. 2002, Zhang et al. 2013a,b, Zhao et al. 2013).

Interfacial bonding in the interphase region between embedded CNT and its surrounding polymer is a crucial issue for the load transfer and reinforcement phenomena. Naturally, the interaction between CNT and resin takes place through the weak van der Waals' (vdW) and electrostatic interactions (Bahr and Tour 2002, Sinnott 2002). Thanks to chemical functionalization, some researchers tried to improve load transfer mechanism by providing covalent cross-links between the carbon atoms of CNT and the polymer chain of the matrix (Bahr and Tour 2002, Frankland et al. 2002, Sinnott 2002, Bufa et al. 2007). In this method, covalent cross-links will be established between CNT and surrounding polymer through adding small organic groups to nanotube sidewalls (Chen et al. 1998, Mickelson et al. 1998, Bahr and Tour 2001), fluorination of nanotubes (Mickelson et al. 1998), and oxidizing the nanotube (Huang et al. 2002, Lin et al. 2002). In this process, hybridization of carbon atoms is changed from sp^2 to sp^3 in the CNT structure, and covalent cross-links will be constructed between CNT and resin atoms (Georgakilas et al. 2002).

The functionalization procedure has a main drawback by inducing defects in the nanostructure of CNT, which may reduce Young's modulus of CNT. Moreover, almost all produced CNTs accommodate some processing-induced defects occurring during their growth process. Therefore, it is of great importance to study the degree to which these structural defects will affect the mechanical properties of CNT.

In this chapter, first, the influence of structural defects on the longitudinal Young's modulus of isolated CNT is investigated using nanoscale continuum mechanics approach with the aid of finite element (FE) modeling. Then the influence of interphase region on the mechanical property of carbon nanotube reinforced polymer (CNTRP) at microscale is studied. Interfacial bonding in the interphase region between embedded CNT and its surrounding polymer, which plays a crucial role in load transfer issue, is studied in the absence and presence of functionalization.

26.2 MODELING ISOLATED CNT

26.2.1 Nondefected CNTs

Due to the associated challenges with the experimental study at the nanoscale, theoretical modeling techniques play an important role in understanding the mechanical behavior of nanostructures like CNTs. The theoretical studies on the mechanical properties of CNT can be categorized in three main groups as atomistic modeling, continuum mechanics–based approaches, and nanoscale continuum modeling (Rafiee and Maleki Moghadam 2014).

Atomistic modeling techniques are limited to short time and small length scales, preventing them to be applicable to the large number of atoms demanding huge amount of computations. Continuum mechanics–based approaches employ theories of rods, trusses, beams, shells, or curved plates. In this method, the whole lattice structure of a CNT is replaced with a continuum structure. It is necessary to carefully investigate the degree of validity of these theories, especially when they are applied to a lattice structure, which is inherently a discrete one. Thus, nanoscale continuum theories that integrate continuum mechanics theories with the nanoscale molecular structure are preferred. The nanoscale continuum mechanics link the interatomic potentials of atomic structure to the continuum level of materials. In this method, instead of replacing the structure of CNT with a continuum cylinder structure (solid or hollow), the C–C bonds of CNT atomic structure are replaced by continuum beam elements. The advantage of this method is that the discrete structure of CNT is kept in the model. The nanoscale continuum mechanics approach has been originally developed by Li and Chou (2003) taking into account the linear interatomic potentials and successfully applied to CNT. The mechanical properties of beam elements representing C–C bonds are obtained by a correlation between interatomic potential energy in molecular space and strain energy of beam element in structural mechanics space. Therefore, the following formulations can be used for obtaining mechanical and geometrical properties of the beam element representing C–C bonds:

$$\frac{EA}{L} = k_r, \quad \frac{EI}{L} = k_\theta, \quad \frac{GJ}{L} = k_\tau \tag{26.1}$$

where
  k_r, k_θ, and k_τ are the bond stretching force constant, bond angle bending force constant, and torsional resistance, respectively
  E, A, I, G, and J represent Young's modulus, cross-sectional area, moment of inertia, shear modulus, and polar moment of inertia of the beam, respectively
  L is the length of the C–C bond and is chosen as 0.142 nm

The associated molecular space constants are selected using AMBER force field parameters as follows (Cornell et al. 1991):

$$k_r = 6.52 \times 10^{-7} \left[\frac{N}{nm}\right], \quad k_\theta = 8.76 \times 10^{-10} \left[\frac{N \cdot nm}{rad^2}\right], \quad k_\tau = 2.78 \times 10^{-10} \left[\frac{N \cdot nm}{rad^2}\right] \tag{26.2}$$

Each C–C bond was replaced with beam element, and lattice molecular structure was substituted with equivalent discrete frame structure.

CNT can be virtually imagined as a rolled graphene sheet. Thus, the coordinate of CNT atoms is obtained from the coordinate of carbon atoms in a graphene sheet using the following mapping equation (Koloczek et al. 2001):

$$(X, Y, Z) = \left[R\cos\left(\frac{x}{R}\right), R\sin\left(\frac{x}{R}\right), y \right] \tag{26.3}$$

The Cartesian coordinates of carbon atoms are fed into ANSYS as nodal coordinates of FE model. Each node is connected to other three adjacent nodes using built-in BEAM4 element of ANSYS.

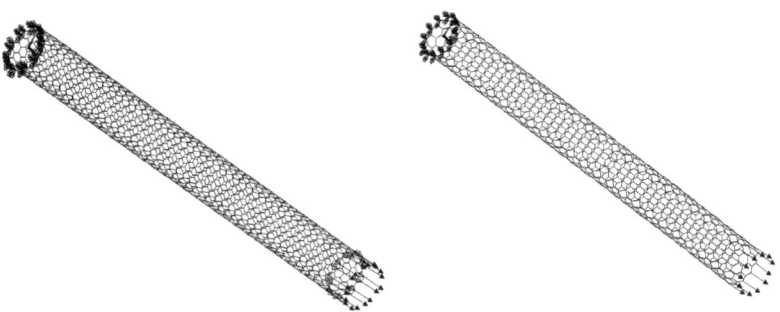

FIGURE 26.1
FE models of nondefected CNTs.

Different FE models of single-walled carbon nanotubes (SWCNTs) with various diameters and chiralities are constructed. The models are subjected to uniform tensile axial displacement at one end, while the other end is restricted from the movement in axial direction and is free to move in radial direction. The length of the CNT along its longitudinal axis is 10 times larger than its radius to prohibit any edge effects. For instance, two different types of CNT, that is, (9, 9) armchair and (12, 0) zigzag, are presented in Figure 26.1.

The elastic modulus of investigated CNT can be obtained using the following equation:

$$E_{CNT} = \frac{\sigma_z}{\varepsilon_z} = \frac{\sum f_z / A}{\Delta L / L} \tag{26.4}$$

where

f_z represents the reaction force at applied constraints to the model and obtained from the output of the FE analysis
A stands for the cross section of CNT calculated as

$$A = \pi \times \left[\left(\frac{R+t}{2} \right)^2 - \left(\frac{R-t}{2} \right)^2 \right] \tag{26.5}$$

where t and R denote the thickness and radius of the CNT, respectively. There is no any consistent value for the wall thickness of CNT in literature. Here, the widely accepted value of 0.34 nm is chosen as interplanar spacing between adjacent graphene sheets in graphite (Robertson et al. 1992, Lu 1997). The radius of CNT is calculated using the following equation:

$$R = \frac{\sqrt{3}L}{2\pi} \sqrt{(n^2 + m^2 + nm)} \tag{26.6}$$

where (n, m) resembles of CNT chiral index.

The obtained results for the Young's modulus of nondefected CNTs in comparison with available analytical solution presented by Shokrieh and Rafiee (2010a) are illustrated in Figure 26.2. It is worth mentioning that the analytical solution presented in Figure 26.2 for the sake of comparison was also developed for a discrete structure of CNT on the basis of nanoscale continuum approach using beam structural member (Shokrieh and Rafiee 2010a).

A very good agreement between the obtained results by FE analysis and the reported results by analytical solution implies on the proper level of accuracy for employed modeling procedure.

26.2.2 Defected CNTs

As it was previously explained, the functionalization procedure will cause structural defects in the nanostructure of functionalized CNT. The defect type is known as vacancy defect leading to

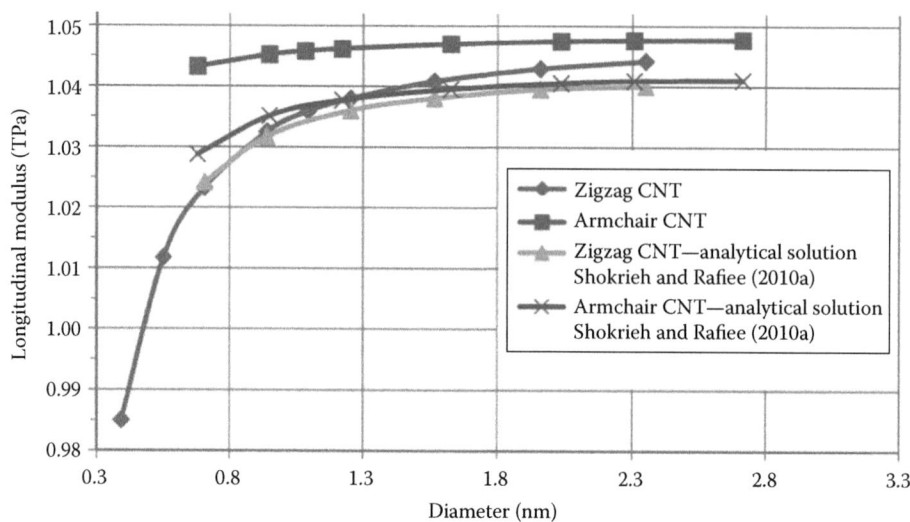

FIGURE 26.2
Young's modulus of nondefected CNT. (From Rafiee, R. and Pourazizi, R., *J. Mater. Res.*, 2014.)

broken C–C bonds. Subsequently, prior to studying the influence of functionalization on the load transfer issue from matrix to CNT, it is required to investigate the influence of vacancy defects on the Young's modulus of the isolated CNT. For this purpose, a computer code was written in ANSYS that randomly removes C–C bonds in the CNT nanostructure on the basis of Monte Carlo technique (Kleiber and Hien 1992) representing vacancy defects.

Number of defects, distribution, and the position of removed C–C bonds along the axial and circumferential directions of the CNT are all selected randomly resembling full stochastic analysis. It means that in addition to the number of defects, both the location and arrangement of broken C–C bonds are also treated as random parameters. The same modeling and computational procedure that has been already explained in the preceding section is followed here to obtain the Young's modulus of defected CNT.

Preventing the unwelcomed effects of end-effect phenomenon in FE analysis, if the removed element is placed at very both ends of the model, the model is ignored and another realization will be produced. Thus, the model is carefully checked prohibiting such a specific located defect. The flowchart of the stochastic code simulating CNT with vacancy defect is mentioned in Figure 26.3.

For quantitative analysis of the results, the defect density parameter is defined to express the amount of incorporated vacancy defects as follows:

$$D = \frac{N_d}{N_t} \tag{26.7}$$

where N_d and N_t are the number of removed bonds and the total bonds of CNT before incorporating defects, respectively.

Due to the stochastic implementation of modeling procedure, the code is repeated sufficiently for each defined number of defects until the standard deviation (SD) of obtained results for the Young's modulus is less than 1% as convergence criterion. Finally, the average of Young's modulus obtained from different realizations is reported as the Young's modulus of the investigated defected CNT.

For instance, it was observed that for a defected CNT with (12, 0) chiral index and 4% defect density, 450 numbers of realizations are required to fulfill convergence criterion, while for 5% defect density, 600 numbers of realization have to be produced.

The written code is executed for different CNTs, and the results are reported in Figure 26.4.

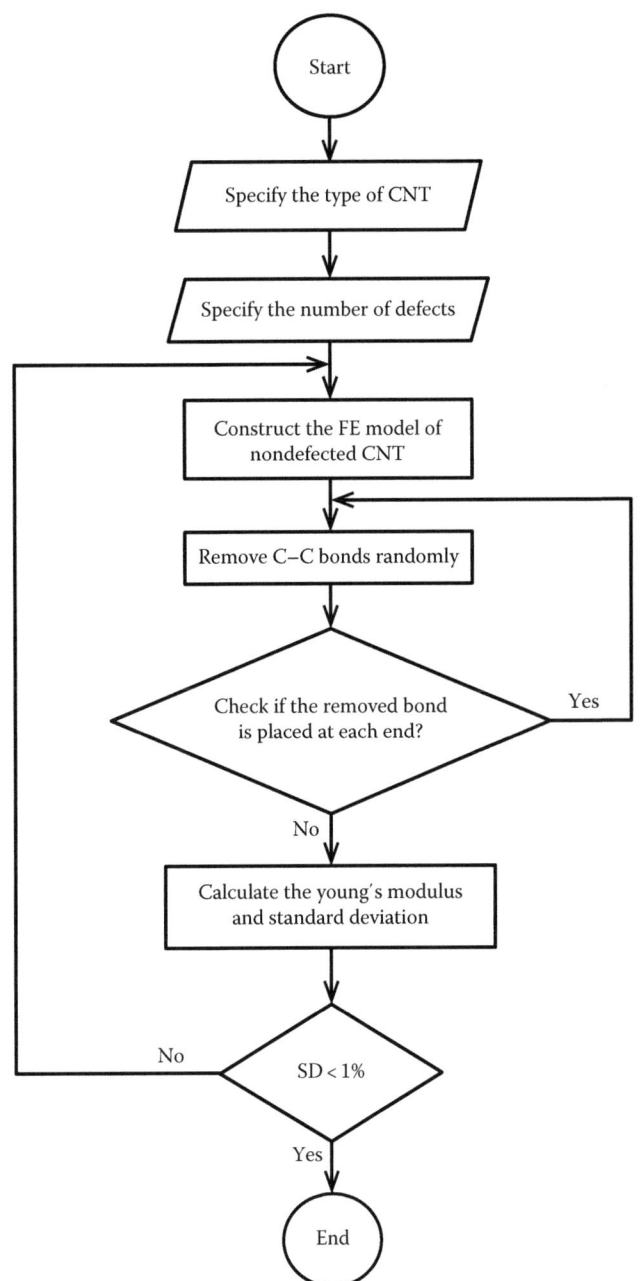

FIGURE 26.3
Flowchart of simulating defected CNT. (From Rafiee, R. and Pourazizi, R., *J. Mater. Res.*, 2014.)

It can be understood from Figure 26.3 that a monotonically linear decreasing trend can be considered with a good level of accuracy for the reduction in CNT's Young's modulus with respect to the defect density. The Young's modulus of defected CNTs decreases about 6% and 12% by considering defect density as 1% and 2%, respectively.

Comparison between the results presented in the published data using molecular dynamics simulation or FE methods (Yuan and Liew 2009, Chen et al. 2010, Saxena and Lal 2012, Sharma et al. 2012) and the results obtained in this study on the basis of nanoscale continuum mechanics approach shows a very good agreement implying on the acceptable accuracy in the modeling.

It is also observed that the level of reduction in CNT's Young's modulus depends on the distribution of the broken bond along the CNT length. Placed at the same CNT length, when all broken bonds are just distributed around the CNT circumference, the minimum Young's modulus will be

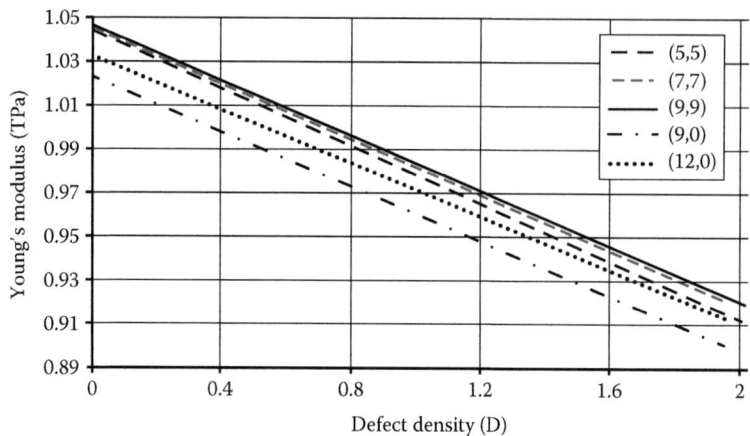

FIGURE 26.4
Young's modulus of defected CNT versus defect density. (From Rafiee, R. and Pourazizi, R., *J. Mater. Res.*, 2014.)

obtained implying on the maximum reduction; however, if all broken bonds are distributed in different stations of CNT lengthwise instead of concentrating at the local position, this will lead to the maximum Young's modulus as a minimum reduction level. In general, concentration of the broken bonds in local region will considerably reduce the Young's modulus of defected CNTs in comparison with nondefected ones. This behavior was also reported in literature by Shofner et al. (2006) using molecular dynamics simulation.

Besides the influence of either locally concentrated or distributed defects on modulus reduction, it was verified that the orientation of broken C–C bonds plays an important role in Young's modulus of CNT. In zigzag CNTs, the axial bonds contribute to load carrying capacity much more than inclined bonds with respect to the axial axis. However, for the case of armchair CNT, the influence of inclined covalent bonds becomes more pronounced than that of perpendicular bonds to the axial axis in load carrying capacity of CNTs. Consequently, removal of axial bonds in zigzag CNTs and inclined bonds in armchair CNTs will reduce CNT's Young's modulus more significantly than the other orientations of bonds.

From statistical point of view, a case study is conducted for a defected CNT with (12, 0) chiral index incorporating 5% defect density. The modulus of defected CNTs varies between 602.7 and 800.5 GPa in comparison with 1.03 GPa for intact CNT. The associated configurations of defected CNTs for the extreme cases are shown in Figure 26.5.

The mode value of 0.75 TPa and the mean value of 0.74 TPa are achieved for the Young's modulus of investigated defected CNT. The histogram of obtained results accompanied with continuous probability density function in the form of Weibull and Gaussian distribution is presented in Figure 26.6.

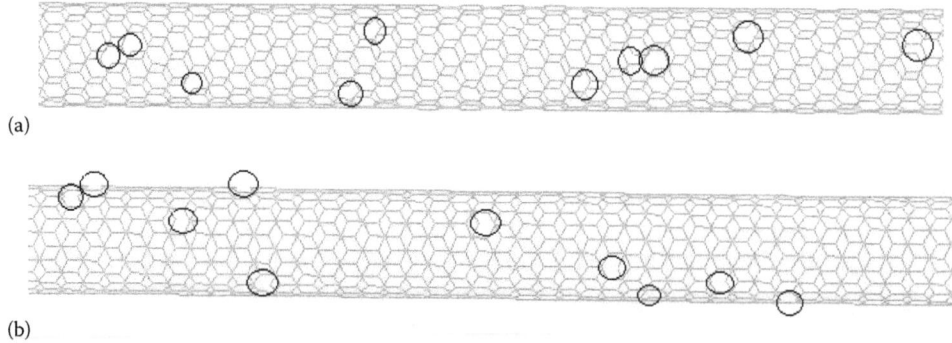

FIGURE 26.5
Defected CNTs with 5% defect density. (a) Minimum modulus and (b) maximum modulus.

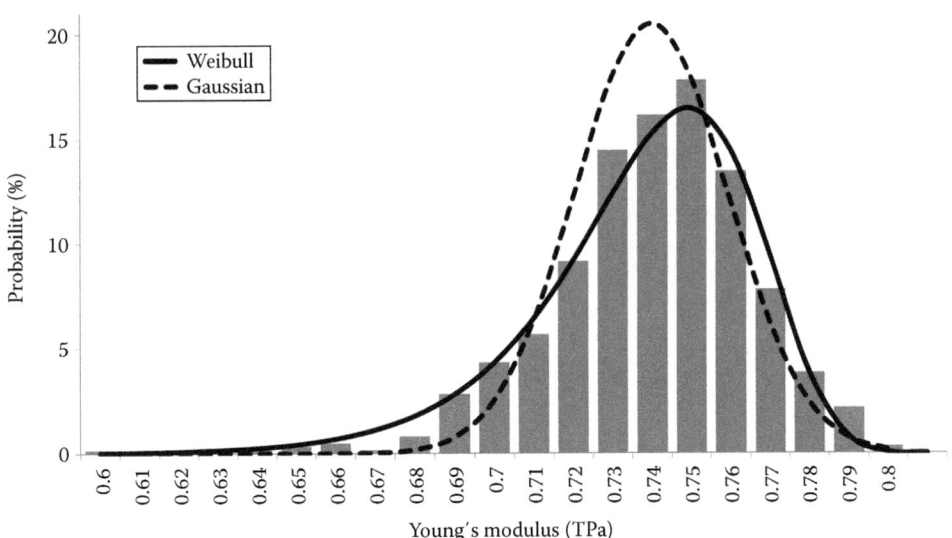

FIGURE 26.6
Statistical plots of the Young's modulus for defected CNT (12, 0) with 5% defect density.

26.3 MODELING CNTRP AT MICROSCALE

The efficiency of a CNT in reinforcing polymer is attributed not only to its own mechanical properties but also to other key factors like load transfer issue from matrix to CNT, nonstraight shapes of CNT, formation of aggregates, and nonuniform dispersion of CNTs. Among the mentioned parameters, the Young's modulus of isolated CNT is pertinent to the nanoscale, which has been already discussed in the preceding section. At the level of micro, the interaction between CNT and surrounding polymer has to be studied through simulating the interphase region. The interphase region accounts for load transferring from the matrix to the CNT. Other parameters like nonstraight shapes, agglomeration, and nonuniform dispersion of CNTs in resin are associated with meso- and macroscales, which are beyond the scope of this study.

CNTs naturally interact with polymer chains of the matrix through weakly nonbonded vdW and electrostatic interactions in the absence of chemical functionalization (Lordi and Yao 2000). Electrostatic interactions can be neglected in comparison with vdW interactions, since vdW contributes more considerably in three higher orders of magnitude than the electrostatic energy (Gou et al. 2004).

Chemical functionalization can improve load transfer mechanism of interphase by providing covalent cross-links between the carbon atoms of CNTs and the molecules of polymer. This procedure induces vacancy defects in the nanostructure of CNT due to formation of sp^3 hybridized sites. The reduction in the Young's modulus of isolated CNTs has been studied in preceding section.

Therefore, the appropriate representative volume element (RVE) at microscale should address interaction between CNT and surrounding polymer accomplished through stress transferring from matrix to CNT in the interphase region. The RVE of CNTRP at microscale consists of three different phases as CNT, interphase region, and surrounding polymer.

In the next sections, the interaction between nondefected CNT and surrounding polymer through solely vdW interactions is firstly studied. Then an interphase region containing combination of vdW and covalent links is simulated representing functionalized CNT wherein defected CNT is taken into account.

Concurrent multiscale FE modeling is used wherein CNT is modeled at nanoscale, while the surrounding polymer is simulated at microscale using continuum medium. The interphase region between CNT and surrounding polymer is simulated using semicontinuum approach. In semicontinuum approach, the interactions between CNT and polymer are established using continuum element instead of replacing the interphase region with a continuum intermediate medium (Rafiee et al. 2013).

26.3.1 Modeling RVE with Nonbonded Interphase Region

The same procedure explained in Section 26.2.1 is employed to construct the CNT model with chiral index of (12, 0) as a case study. The longitudinal modulus of this CNT is obtained as 1032.46 GPa. The surrounding resin is modeled as a continuum environment. The radius of the resin is selected in a manner to reflect 5% as a CNT volume fraction in the RVE. The interphase is modeled by considering only the vdW interactions between CNT and polymer. Generally, the vdW forces are expressed by Lennard-Jones potential as follows (Rafii-Tabar 2004):

$$F_{LJ} = 4\frac{\varepsilon}{r}\left[-12\left(\frac{\sigma}{r}\right)^{12} + 6\left(\frac{\sigma}{r}\right)^{6}\right] \qquad (26.8)$$

where

    r is the distance between the atoms
    ε and σ are the vdW parameters, with values of $\varepsilon = 0.0556$ kcal/mol and 0.34 nm for CNT, respectively

The vdW force is a nonlinear force consisting of two distinct regions of repulsion and attraction (Kalamkarov et al. 2006). Furthermore, the vdW interaction can be neglected when the interatomic distance is equal or greater than 0.85 nm.

Due to the nonlinear nature of the vdW bonds, the nonlinear spring element in ANSYS is used, and its parameters are adjusted accordingly. The center of the carbon atoms in the CNT is placed at the midsection of the tube thickness. It is also assumed that the innermost layer of the resin is located at the same position of the outer surface of the CNT. Consequently, the thickness of interphase is considered as 0.17 nm.

The constructed FE model of investigated RVE is mentioned in Figure 26.7. As it can be seen from the figure, the CNT is extended through the whole span of the RVE; therefore, the RVE accounts for the long CNT.

One end of the whole mode is fully restrained from rotational and translational movements, while a uniform axial displacement is applied to the other end of the model. A nonlinear analysis is performed using full Newton–Raphson iterative method due to the nonlinear nature of the vdW interaction. The longitudinal modulus of investigated RVE is obtained using resultant forces appeared in the constraint after FE analysis. The simulated resin was treated as an isotropic material with Young's modulus of 3–5 GPa and 0.3 as Poisson's ratio, which are the corresponding values for typically available epoxy resins in the market. Thus, the Young's modulus of the RVE is placed between 34.18 and 36.42 GPa implying on negligible influence of resin modulus on the results. This is originated from this fact that the CNT is the main contributor to the stiffness of the RVE.

The stress–strain diagram of investigated RVE containing long CNT compared with pure resin and the rule of mixture (ROM) is presented in Figure 26.8.

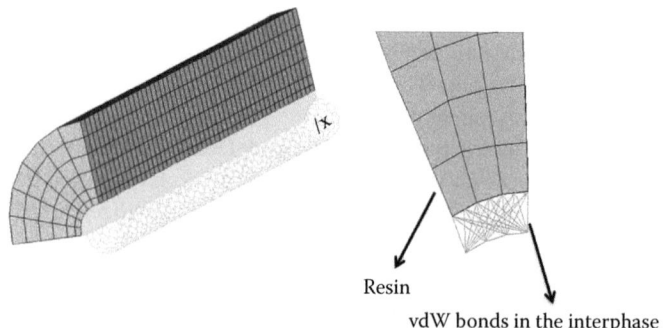

Resin

vdW bonds in the interphase

FIGURE 26.7
FE model of CNT, nonbonded interphase, and resin. (From Rafiee, R. and Pourazizi, R., *Comput. Mater. Sci.*, 2015.)

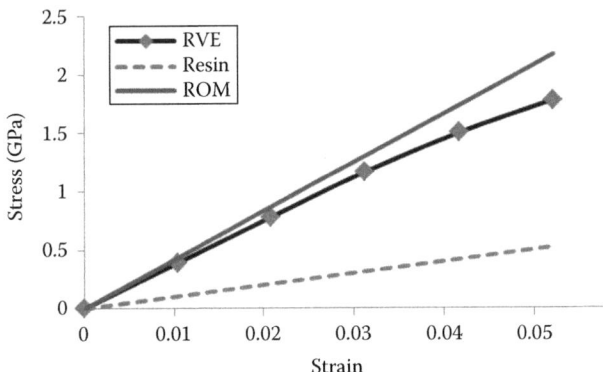

FIGURE 26.8
Stress–strain curves. (From Rafiee, R. and Pourazizi, R., *Comput. Mater. Sci.*, 2015.)

Recalling from Section 26.2.1, due to the linear behavior of CNT arisen from employed linear interatomic potentials as well as the linear behavior of resin, the nonlinear behavior of investigated RVE depicted in Figure 26.8 stems from the nonlinear behavior of vdW interactions in the interphase. The stress–strain curve of the RVE can be considered nearly linear below the 4% of strain. It can be also understood that ROM overestimates the Young's modulus of the RVE. ROM considers perfect bonding between reinforcing agents and surrounding polymer, which is not pertinent to the case of CNTRP. Moreover, ROM simply neglects the lattice structure of the CNT by assuming fibrous reinforcing agent as a continuum medium. The same trend for the stress–strain curve was also reported by Frankland et al. (2003) using molecular dynamics simulations.

26.3.2 Equivalent Continuum Medium for Nonbonded Interphase

The performed modeling procedure is very time consuming due to the numerous nonlinear elements simulating vdW interaction. Consequently, for the sake of simplicity, it is more by investigators to replace the interphase region with a continuum medium. In other words, instead of simulating interphase region on the basis of previously explained semicontinuum approach, the whole interphase is modeled using hollow cylinder. A careful approach should be employed to obtain the Young's modulus of this equivalent continuum medium. Generally, in continuum modeling of the interphase region between CNT and surrounding polymer, the modulus of the interphase is selected between the Young's moduli of the resin and the CNT without any specific theoretical background. There is no any consistent reason for selecting the interphase modulus, and most often, the influence of different chosen values for interphase modulus is investigated on the mechanical behavior of the RVE (see Kulkarni et al. 2010, Needleman et al. 2010, Ayatollahi et al. 2011).

In this section, it is intended to obtain the Young's modulus of the interphase using the performed analysis in Section 26.3.1. For this purpose, the lattice structure of CNT and semicontinuum model of the interphase are replaced with a continuum medium called equivalent fiber. Thus, the investigated RVE will be replaced with the equivalent fiber and surrounding resin. Therefore, the Young's modulus of the equivalent fiber will be derived. Afterward, the developed equivalent fiber is divided into two continuum phases of CNT and interphase region. Subsequently, the Young's modulus of the virtually continuum interphase can be obtained. The explained strategy of mentioned conversions is depicted in Figure 26.9.

26.3.2.1 Young's modulus of equivalent fiber

Following the first step of illustrated strategy of converting CNT and interphase into equivalent fiber, it can be seen that the equivalent fiber is perfectly bonded to the resin. Therefore, it is permissible

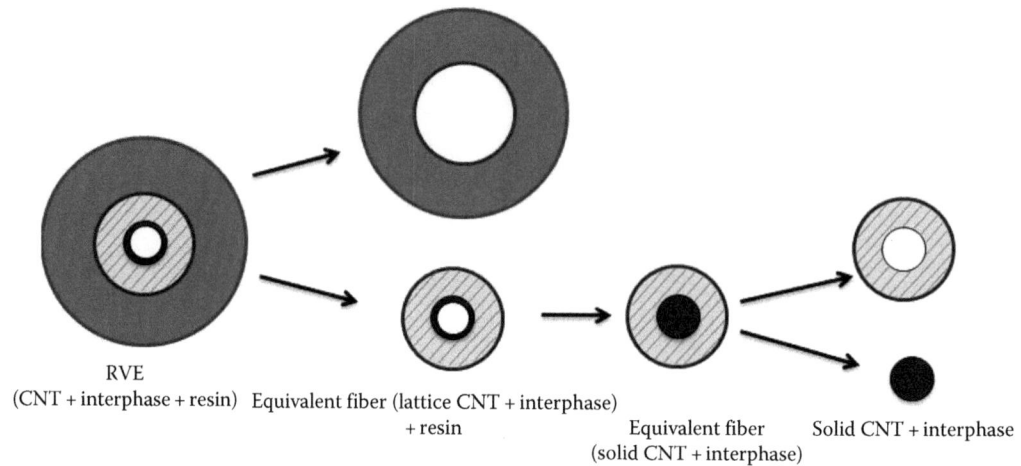

RVE
(CNT + interphase + resin)   Equivalent fiber (lattice CNT + interphase)
+ resin   Equivalent fiber   Solid CNT + interphase
(solid CNT + interphase)

FIGURE 26.9
Schematic representation of continuum modeling of the interphase.

to use micromechanics rules. The ROM is employed inversely to obtain the Young's modulus of the equivalent fiber as follows:

$$E_{EQF} = \frac{E_{RVE}}{V_{EQF}} - \frac{E_M V_M}{V_{EQF}}$$ (26.9)

where

  E_{EQF}, E_{RVE}, and E_M are the longitudinal modulus of equivalent fiber, RVE, and matrix, respectively
  V_M and V_{EQF} are the volume fractions of matrix and equivalent fiber, respectively

Following the aforementioned procedure, the obtained modulus for the equivalent fiber is 642.4 GPa. Shokrieh and Mahdavi (2011) have obtained 642 GPa as the Young's modulus of the equivalent fiber using the analytical method. It was also reported by Shokrieh and Rafiee (2010b), using the adaptive vdW interaction (AVI) technique, that the Young's modulus of the equivalent fiber is 649.12 GPa. In the AVI modeling developed by Shokrieh and Rafiee (2010b), the instantaneous status of vdW links is updated in each and every substep of nonlinear analysis: namely, some vdW interactions are removed as their lengths exceed the cutoff range dictated by Lennard-Jones potential (more than 0.85 nm), and some new vdW interactions are established due to the relative changes between the nodes of CNT and polymer models. This will lead to the very-time-consuming modeling procedure. In contrast to the AVI modeling, in the present study, just deactivation procedure is taken into account without constructing new vdW interactions. Therefore, the modeling time is reduced considerably with very slight reduction in load transfer issue from matrix to resin. It could be said that current compromise in the modeling procedure is a rational compromise as the level of changes in the results is quite negligible. The very good agreement between obtained Young's modulus for the equivalent fiber in this research in comparison with published data (Shokrieh and Rafiee 2010b, Shorkrieh and Mahdavi 2011) established our confident toward accuracy of the modeling procedure.

26.3.2.2 Young's modulus of nonbonded interphase

After obtaining Young's modulus of the equivalent fiber, the Young's modulus of the interphase region can be calculated as well. The lattice structure of CNT is replaced with solid cylinder. Both solid and lattice CNTs are subjected to the same axial force. They are expected to experience the same deformation in the longitudinal direction. Thus, the effective radius of solid cylinder can be obtained by equaling the deformations of solid and lattice CNT structures. The solid CNT deformation is obtained as follows:

$$\Delta L_{eff} = \frac{F_z L}{E_z^{eff} \cdot A_{eff}} \tag{26.10}$$

where
E_z^{eff} and A_{eff} are the longitudinal modulus and cross section of solid CNT, respectively
F_z and L denote the tensile force applied to CNT and the length of CNT in both lattice and solid structures, respectively

The deformation of the lattice CNT can be calculated by the following equation:

$$\delta L = \frac{F_z L}{A_{CNT} \cdot E_{CNT}} \tag{26.11}$$

where A_{CNT} stands for the cross section of lattice CNT and it can be calculated using Equation 26.5. Considering equal longitudinal modulus for both structures, both the cross-sectional areas of lattice CNT and solid CNT have to be the same, that is, $A_{CNT} = A_{eff}$. Finally, the radius of the solid CNT is derived as follows:

$$R_{eff} = \sqrt{2 R_{CNT} t} \tag{26.12}$$

Following the explained procedure, the effective radius of solid CNT with chiral index of (12, 0) becomes 0.565 nm.

On the other hand, the volume fraction of solid CNT in equivalent fiber is obtained using the following equation:

$$V_{CNT} = \frac{A_{CNT}}{A_{EQF}} = \left(\frac{R_{CNT}}{R_{EQF}} \right)^2 \tag{26.13}$$

Substituting R_{CNT} with 0.47 nm and R_{EQF} with 0.64 (0.47 + 0.17) nm, V_{CNT} is obtained as 0.539. This will lead to 0.461 for V_{int}.

Having in hand the effective radius of the CNT in the form of solid cylinder, the Young's modulus of the virtually continuum medium for the interphase region can be obtained by using ROM inversely as follows:

$$E_{int} = \frac{E_{EQF}}{V_{int}} - \frac{E_{CNT} V_{CNT}}{V_{int}} \tag{26.14}$$

Recalling from the very first part of the Section 26.3.1, Young's modulus of investigated CNT with chiral index of (12, 0) is 1032.46 GPa (E_{CNT}), and Young's modulus of the equivalent fiber is obtained as 642.4 GPa in the previous section. Thus, Young's modulus of the nonbonded interphase is obtained as 186.43 GPa.

FE model of the RVE containing solid CNT and continuum interphase was constructed to verify the obtained Young's modulus for the interphase region. In this model, the solid CNT is modeled with Young's modulus of 1032.46 GPa and radius of 0.583 nm. The continuum interphase region is modeled with Young's modulus of 186.43 GPa and outer radius of 0.794 nm. The obtained longitudinal modulus of the equivalent fiber consisting of solid CNT and continuum interphase is reported as 646.4 GPa, which is in a very good agreement with previously obtained 642.4 GPa (less than 1% difference) for the equivalent fiber consisting of lattice CNT and semicontinuum interphase region.

TABLE 26.1 Young's Modulus of Two Different RVEs

CNT Type	Interphase Interaction	CNT Modulus (GPa)	RVE Modulus (GPa)	EQF Modulus (GPa)	Interphase Modulus (GPa)
Nondefected CNT	Fully vdW	1032.46	41.62	642.4	186.34
Defected/ functionalized CNT	vdW and covalent links	945.47–990.46	39.03–40.68	590.54–623.66	180.2–195.2

Source: Rafiee, R. and Pourazizi, R., *Comput. Mater. Sci.*, 2015.

26.3.3 Modeling RVE Incorporating Functionalized CNT

In constant to Section 26.3.1 that interphase region encompasses of only vdW interaction, in this section, the interphase region comprises a combination of vdW links and transverse covalent bonds between CNT and surrounding polymer. Nonlinear spring elements representing nonbonded vdW interactions and linear beam elements simulating covalent cross-links are used for modeling interphase region between functionalized CNT and polymer. The transverse covalent bonds are established between those nodes in the CNT model incorporating defect and adjacent nodes on the resin elements. In accordance with the molecular structures of resin and CNT, the covalent bonds established between the carbon atoms of CNT structure and the carbon atoms of epoxy structure. The properties of this bond are obtained using AMBER force field (Ferguson et al. 1995) as 6.565 MPa for Young's modulus, 1.042 MPa for shear modulus, and 2.15 for Poisson's ratio.

Recalling from Section 26.2.2, it was observed that topology of the defect affects the degree to which Young's modulus of defected CNT reduces in comparison with pristine CNT. Addressing the lowest and the highest reduction in Young's modulus in comparison with nondefected CNTs, two FE models of RVEs including defected CNTs with 1% defect density and 5% CNT volume fraction are constructed. The obtained results for the RVEs of functionalized CNT in comparison with the results of RVE with fully nonbonded interphase are reported in Table 26.1.

The results reveal that the modulus of RVE incorporating functionalized CNT decreases about 2.2%–6.2% when it is compared with the modulus of the RVE containing of nondefected CNT. Kuang and He (2009) have used molecular mechanics method to model the RVE consisting of functionalized CNT and the reported 3.47% reduction in the longitudinal modulus of RVE. The equivalent fiber (EQF) modulus decreases approximately 2.9%–8.1%, which is relatively good agreement with the results reported by Odegard et al. (2005). They used an molecular dynamics (MD) method and reported reduction in the modulus of EQF after adding functionalized CNT to polymer. Based on their research, the longitudinal modulus of EQF containing (10, 10) CNTs is equal to 548.7 GPa, and the modulus of EQF containing (10, 10) CNTs with 1.1% defect is 487.7 that shows 11% reduction in EQFs modulus.

It can be understood from the results that the modulus of the RVE after functionalization of CNT reduces, thanks to incorporating defect in the CNT nanostructure. Although load transfer mechanism in the interphase is improved after functionalization, the effective modulus of the RVE decreases implying on negative side effect of functionalization process at microscale. So it can be concluded that positive effect of functionalization cannot be captured at the scale of micro and it appears at the scale of meso by reducing the level of local concentration of CNTs in the local aggregates. It has been reported in literature that CNTs are more likely to get dispersed when chemical functionalization is applied to CNTs.

26.4 CNT LENGTH ANALYSIS

Following the obtained value for the continuum modeling of the nonbonded interphase region, the influence of CNT length on load transfer capability is studied in this section. In contrast to the previous section where the CNT length was the same as RVE length, in this section, the length of CNT is shorter than that of RVE length. Representing long CNT, it was assumed that the CNT

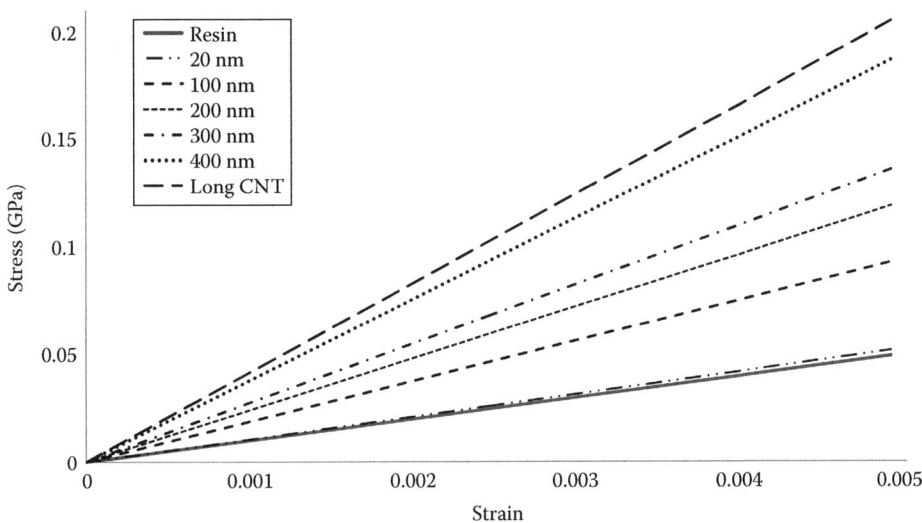

FIGURE 26.10
Stress–strain diagrams for RVE containing short CNT, nonbonded interphase, and resin.

extended throughout the whole span of RVE, but in this section, the influence of CNT length along the RVE span is investigated. Therefore, the RVE that is used in this part is 2D and contains CNT, nonbonded interphase, and resin. The properties of all these phases have been obtained in the preceding sections for the continuum modeling strategy. The only variable parameter is considered the CNT. As a case study, the resin length and the volume fraction of CNT in RVE are considered as 408 nm and 5%, respectively. In this model, the modulus of CNT, resin, and interphase are taken as 1032.46, 10, and 186.34 GPa, respectively. The obtained stress–strain curves for the RVEs with different CNT lengths are presented in Figure 26.10.

It should be pointed out that the stress–strain diagrams have been extracted for strains up to 1% wherein linear region of stress–strain diagram of the RVE can be observed. It could be understood that increasing the length of CNT, load transfer issue is improved. In the RVE containing long CNT, in addition to the interphase region that is responsible for indirectly transferring the applied load from matrix to CNT, a portion of loading is directly applied to the CNT's end. However, for the RVE containing short CNT, the only available mechanism for load transferring from matrix to CNT is the interphase region. Consequently, increasing the length of CNT, the role of interphase in stress transferring becomes more pronounced.

It can be seen that in very small length (20 nm), the stress–strain curve of the RVE coincides with resin's curve. It can be interpreted that smaller CNT length will not contribute in reinforcing the polymer at all and the interphase is not able to transfer any load to the CNT. When the length of short CNT increases, the results approach the bounding value reported by long CNT. Therefore, the main difference between the modulus of RVEs containing long CNT and short CNT can be attributed to the efficiency of load transfer mechanism between CNT and resin.

26.5 CONCLUSION

The influence of chemical functionalized on the Young's modulus of CNTRP at microscale is the subject of this chapter. Thus, CNTRP is simulated at microscale taking into account two different types of interphase region between CNT and surrounding polymer. The simulation is divided into two scales of nano and micro. Young's modulus of defected CNT is obtained and compared with intact CNT to study the influence of vacancy defects on the mechanical property of the SWCNT. Nanoscale continuum modeling using FE modeling is employed for this purpose. The vacancy defect caused due to the chemical functionalization of the CNT is simulated and analyzed. Full stochastic modeling procedure is developed taking into account all numbers, orientations, and locations of the defect as random parameters. Consequently, distribution of defects along the length and circumference of CNTs is considered as random parameters in order to be able to capture the real nature of defect topology. A computer code is written for this purpose to incorporate arbitrary number of defects into the discrete structure of CNT and reports the results.

The results demonstrate that the vacancy defect can reduce the Young's modulus of CNT to about 12%–13% when 2% of C–C bonds are broken in the CNT nanostructure. Defect topology indicates that among all broken bonds, those bonds that are oriented more closely to the longitudinal axis of CNT have a greater influence on the reduction of CNT's Young's modulus. Moreover, local concentration of broken bonds will significantly reduce the Young's modulus in comparison with distributed broken bonds along the length of CNT. Finally, due to the stochastic modeling, a statistical analysis is performed to obtain probability of occurrence associated with different obtained results.

At microscale, an RVE containing single CNT and polymer interacting through interphase region is investigated. A multiscale FE modeling is employed wherein the lattice structure of CNT is modeled at nanoscale as discrete structure and the surrounding polymer is simulated at microscale as continuum medium. The interaction between CNT and polymer occurring through the interphase region is simulated using the semicontinuum approach. Two different types of CNTs, that is, non-defected CNT and functionalized CNT, are simulated. The former interacts with the surrounding polymer through the vdW interactions, while the interphase of the latter contains both nonbonded vdW interactions and strong covalent bonds.

Young's moduli of two RVEs containing nondefected CNT and functionalized CNT are obtained. The results show that functionalization will reduce the Young's modulus of the RVE, as the contribution of the CNT is more pronounced in the Young's modulus of RVE in comparison with interphase region. It could be concluded that the influence of functionalization in improving mechanical properties of CNT-based nanocomposites will appear at the scale of meso wherein the probability of aggregate formations and concentration of CNT in local regions decreases considerably. Chemical functionalization will hinder CNT to interact with neighboring CNT in the resin due to establishing cross covalent bonds between CNT and surrounding polymer. This will lead to a better quality of the CNT dispersion in resin.

The Young's modulus of the fully vdW interphase region is obtained as 186.4 GPa, which can be used in the continuum simulation of CNT-based polymer. Thanks to the obtained value, the influence of CNT length on the Young's modulus of the RVE consisting of short CNT is studied. The results reveal that smaller lengths of CNT (less than 10 nm) will not contribute to the reinforcing polymer. Moreover, increasing the length of short CNT, the CNT reinforcement capability improves significantly until it approaches the bounding value dictated by the long CNT. Thanks to the considerable reduction in the runtime of the model, the efficiency of the continuum modeling as a very good alternative of semicontinuum modeling is shown.

REFERENCES

Ayatollahi, M., Shadlou, S., and Shokrieh, M. 2011. Multiscale modeling for mechanical properties of carbon nanotube reinforced nanocomposites subjected to different types of loading. *Compos. Struct.* 93: 2250–2259.

Bahr, J.L. and Tour, J.M. 2001. Highly functionalized carbon nanotubes using in situ generated diazonium compounds. *Chem. Mater.* 13(11): 3823–3824.

Bahr, J.L. and Tour, J.M. 2002. Covalent chemistry of single-wall carbon nanotubes. *J. Mater. Chem.* 12: 1952–1958.

Belin, T. and Epron, F. 2005. Characterization methods of carbon nanotubes: A review. *Mater. Sci. Eng. B* 119: 105–118.

Buffa, F., Abraham, G.A., Grady, B.P., and Resasco, D. 2007. Effect of nanotube functionalization on the properties of single-walled carbon nanotube/polyurethane composites. *J. Polym. Sci., Part B: Polym. Phys.* 45: 490–501.

Chen, J., Hamon, M.A., Hu, H. et al. 1998. Solutions properties of single-walled carbon nanotubes. *Science* 282(5386): 95–98.

Chen, L., Zhao, Q., Gong, Z., and Zhang, H. 2010. The effects of different defects on the elastic constants of single-walled carbon nanotubes. *Proceedings of the Fifth IEEE International Conference*, Xiamen, China, pp. 777–780.

Cornell, W.D., Cieplak, P., Bayly, C.I. et al. 1991. A second generation force field for the simulation of proteins, nucleic acids, and organic molecules. *J. Am. Chem. Soc.* 117: 5179–5197.

Ferguson, D., Cornell, W., Cieplak, P., Bayly, C., Gould, I., and Merz, K. 1995. A second generation force field for the simulation of proteins, nucleic acids, and organic molecules. *J. Am. Chem. Soc.* 117: 5179–5197.

Frankland, S., Harik, V., Odegard, G., Brenner, D., and Gates, T. 2003. The stress–strain behavior of polymer–nanotube composites from molecular dynamics simulation. *Compos. Sci. Technol.* 63: 1655–1661.

Frankland, S.J.V., Caglar, A., Brenner, D.W., and Griebel, M. 2002. Molecular simulation of the influence of chemical cross-links on the shear strength of carbon nanotube polymer interfaces. *J. Phys. Chem. B* 106: 3046–3048.

Georgakilas, V., Kordatos, K., Prato, M., Guldi, D.M., Holzinger, M., and Hirsch, A. 2002. Organic functionalization of carbon nanotubes. *J. Am. Chem. Soc.* 124(5): 760–761.

Gou, J., Minaei, B., Wang, B., Liang, Z., and Zhang, C. 2004. Computational and experimental study of interfacial bonding of single-walled nanotube reinforced composites. *Comput. Mater. Sci.* 31: 225–236.

Huang, W., Taylor, S., Fu, K. et al. 2002. Attaching proteins to carbon nanotubes via diimide-activated amidation. *Nano Lett.* 2(4): 311–314.

Kalamkarov, A.L., Georgiades, A.V., Rokkam, S.K., Veedu, V.P., and Ghasemi-Nejhad, M.N. 2006. Analytical and numerical techniques to predict carbon nanotubes properties. *Int. J. Solids Struct.* 43: 6832–6854.

Kleiber, M. and Hien, T.D. 1992. *The Stochastic Finite Element Method*. New York: John Wiley Publisher Science.

Koloczek, J., Young-Kyun, K., and Burian, A. 2001. Characterization of spatial correlations in carbon nanotubes-modelling studies. *J. Alloys Compd.* 328: 222–225.

Kuang, Y. and He, X. 2009. Young's moduli of functionalized single-wall carbon nanotubes under tensile loading. *Compos. Sci. Technol.* 69: 169–175.

Kulkarni, M., Carnahan, D., Kulkarni, K., Qian, D., and Abot, J.L. 2010. Elastic response of a carbon nanotube fiber reinforced polymeric composite: A numerical and experimental study. *Compos. Part B* 41: 414–421.

Lau, K.T., Gu, C., and Hui, D. 2006. A critical review on nanotube and nanotube/nanoclay related polymer composite materials. *Compos. Part B* 37: 42536.

Li, C. and Chou, T. 2003. A structural mechanics approach for the analysis of carbon nanotubes. *Int. J. Solids Struct.* 40: 2487–2499.

Li, F., Cheng, H.M., Bai, S., Su, G., and Dresselhaus, M.S. 2000. Tensile strength of single walled carbon nanotubes directly measured from their macroscopic ropes. *Appl. Phys. Lett.* 77(20): 3161–3163.

Li, Y.J., Wang, K.K.L., Wei, J.Q. et al. 2005. Tensile properties of long aligned double-walled carbon nanotube strands. *Carbon* 43: 31–35.

Lin, Y., Rao, A.M., Sadanadan, B., Kenik, E.A., and Sun, Y.P. 2002. Functionalizing multiple-walled carbon nanotubes with aminopolymers. *J. Phys. Chem. B* 106(6): 1294–1298.

Lordi, V. and Yao, N. 2000. Molecular mechanics of binding in carbon-nanotube–polymer composites. *J. Mater. Res.* 15(12): 2770–2779.

Lu, J.P. 1997. Elastic properties of carbon nanotubes and nanoropes. *Phys. Rev. Lett.* 79(7): 1297–1300.

Mickelson, E.T., Huffman, C.B., Rinzler, A.G., Smalley, R.E., Hauge, R.H., and Margrave, J.L. 1998. Fluorination of buckytubes. *Chem. Phys. Lett.* 296(1–2): 188–194.

Needleman, A., Borders, T., Brinson, L., Flores V., and Schadler, L. 2010. Effect of an interphase region on debonding of a CNT reinforced polymer composite. *Compos. Sci. Technol.* 70: 2207–2215.

Odegard, G., Frankland, S., and Gates, T. 2005. Effect of nanotube functionalization on the elastic properties of polyethylene nanotube composites. *AIAA J.* 43(8): 1828–1835.

Qian, D., Dickey, E., Andrews, R., and Rantell, T. 2000. Load transfer and deformation mechanisms in carbon nanotube-polystyrene composites. *Appl. Phys. Lett.* 76(20): 2868–2870.

Rafiee, R. and Maleki Moghadam, R. 2014. On the modeling of carbon nanotubes: A critical review. *Compos. Part B* 56: 435–449.

Rafiee, R. and Pourazizi, R. 2014. Evaluating the influence of defects on the Young's modulus of carbon nanotubes using stochastic modeling, *J. Mater. Res.* 17(3): 758–766.

Rafiee, R. and Pourazizi, R. 2015. Influence of CNT functionalization on the interphase regions between CNT and polymer. *Comput. Mater. Sci.* 96: 573–578.

Rafiee, R., Rabczuk, T., Pourazizi, R., Zhao, J., and Zhang, Y. 2013. Challenges of the modeling methods for investigating the interaction between the CNT and surrounding polymer. *Adv. Mater. Sci. Eng.* Article ID 183026. 10pp. http://dx.doi.org/10.1155/2013/183026.

Rafii-Tabar, H. 2004. Computational modelling of thermo-mechanical and transport properties of carbon nanotubes. *Phys. Rep.* 390: 235–452.

Robertson, D.G., Brenner, D.W., and Mintmire, J.W. 1992. Energies of nanoscale graphitic tubule. *Phys. Rev. B* 45(21): 12592–12595.

Saxena, K.K. and Lal, A. 2012. Comparative molecular dynamics simulation study of mechanical properties of carbon nanotubes with number of stone-wales and vacancy defects. *Proc. Eng.* 38: 2347–2355.

Schadler, L., Giannaris, S.C., and Ajayan, P.P. 1998. Load transfer in carbon nanotube epoxy composites. *Appl. Phys. Lett.* 73(26): 3842–3844.

Sharma, K., Saxena, K., and Shukla, M. 2012. Effect of multiple Stone-Wales and vacancy defects on the mechanical behavior of carbon nanotubes using molecular dynamics. *Proc. Eng.* 38: 3373–3380.

Shofner, M.L., Khabashesku, V.N., and Barrera, E.V. 2006. Processing and mechanical properties of fluorinated single-wall carbon nanotube-polyethylene composites. *Chem. Mater.* 18: 906–913.

Shokrieh, M.M. and Mahdavi, S.M. 2011. Micromechanical model to evaluate the effects of dimensions and interphase region on the elastic modulus of CNT/polymer composites. *Modares Mechanical Engineering* 11(3): 13–25.

Shokrieh, M.M. and Rafiee, R. 2010a. Prediction of Young's modulus of graphene sheets and carbon nanotubes using nanoscale continuum mechanics approach. *Mater. Des.* 2: 790–795.

Shokrieh, M.M. and Rafiee, R. 2010b. On the tensile behavior of an embedded carbon nanotube in polymer matrix with non-bonded interphase region. *Compos. Struct.* 3(92): 647–652.

Sinnott, S.B. 2002. Chemical functionalization of carbon nanotubes. *J. Nanosci. Nanotechnol.* 2: 113–123.

Yuan, J. and Liew, K. 2009. Effects of vacancy defect reconstruction on the elastic properties of carbon nanotubes. *Carbon* 47: 1526–1533.

Zhang, Y., Zhao, J., Jia, Y. et al. 2013a. An analytical solution for large diameter carbon nanotube-reinforced composite with functionally graded variation interphase. *Compos. Struct.* 104: 261–269.

Zhang, Y., Zhao, J., Wei, N., Jiang, J.W., and Rabczuk, T. 2013b. Effects of the dispersion of polymer wrapped two neighbouring single walled carbon nanotubes (SWNTs) on nano engineering load transfer. *Compos. Part B* 45(1): 1714–1721.

Zhao, J., Jiang, J.W., Jia, Y., Guo, W., and Rabczuk, T. 2013. A theoretical analysis of cohesive energy between carbon nanotubes, graphene and substrates. *Carbon* 57: 108–119.

Zhu, J., Peng, H., Rodriguez-Macias, F. et al. 2004. Reinforcing epoxy polymer composites through covalent integration of functionalized nanotubes. *Adv. Funct. Mater.* 14(7): 643–648.

CHAPTER 27

CONTENTS

Drug Delivery Aspects of Carbon Nanotubes

27

Neelesh Kumar Mehra and N.K. Jain

27.1 INTRODUCTION

Nanotechnology is a contemporary discipline that has emerged in the field of cell biology in the form of nanosize range particles including liposomes, nanoparticles, dendrimers, quantum dots, drug conjugates, and carbon nanotubes (CNTs). The main objectives of the nanotechnology are to maximize the current therapeutic strategies while minimizing the side effects of the theranostics (Jain et al. 2007, 2014). The term *nanotechnology* is derived from the *Greek* word *nano*, meaning *dwarf*, and is defined as the technology at molecular and nanometric scale dealing with nanoscale phenomena and processes—nanodevice, -science, -pharmacology, -materials, and -systems. However, the term *nanomedicine* is a combination of nanotechnology and medicine that plays a pivotal role in diverse area from drug delivery to imaging and diagnostics (Mehra and Jain 2014).

The targeted drug delivery system is an interface between the drug and patient and has become an extremely demanding prospect in nanoscience employing new generation drugs, which are unstable in the biological milieu owing to hydrophobic nature, poor transport properties across biological membrane, and low bioavailability. In the current scenario, targeted drug delivery has become a multidisciplinary nanoscience comprising of pharmacokinetics and biopharmaceutics Lacerda et al. 2012; Mehra and Jain 2014; Mehra et al. 2008, 2014a). This book chapter provides the detailed milestone discovery, background, classification and general characteristics, functionalization, cellular trafficking pathways, and drug delivery aspects of functionalized CNTs (*f*-CNTs). The major strength of this book chapter is the exploration of not hitherto discussed drug delivery aspects of CNTs.

27.2 CNTs: A NOVEL NANOVECTOR FOR DRUG DELIVERY

Currently, CNTs have sparked a great deal of excitement in the targeted drug delivery being considered as alternative, safe, and effective drug delivery system. CNTs were first discovered by Bacon in 1960 (Bacon 1960). However, after a long journey, a Japanese Microscopist Sumio Iijima fully explored the tiny nanoneedle tubular structure of CNTs in a transmission electron microscopy (TEM) observation and published in the journal *Nature* entitled *Helical microtubules of graphitic carbon* (Iijima 1991) that gained more than 32,381 citations in the scientific communities (as on May 30, 2014). The milestone discovery of CNTs has opened a new door to the researchers working in pharmaceutical and nanotechnology fields. CNTs are unique, sp^2-hybridized, 3D carbon-based nanoparticles, comprised of thin graphite sheets made by condensed benzene rings rolled up into the seamless hollow and nanoneedle tubular structure. CNTs belong to fullerene family, that is, third allotropic form of carbon. The unique physicochemical properties of CNTs like ultralightweight, high surface area and aspect ratio (length/diameter), photoluminescence, nanoneedle, high-rich surface functional chemistry, nonimmunogenicity, biocompatibility, biliary excretion, a rapid uptake by cells due to anisotropic "needle-like" morphology, and extremely high-drug cargo ability are expected to make them attractive nanovector in drug delivery. Additionally, *f*-CNTs mimic a nanomatrix, wherein drug molecules get entrapped well, and hence release can be controlled temporarily or spatially (Boncel et al. 2013; Brahmachari. et al. 2014; Jain et al. 2014; Kayat et al. 2011; Lacerda et al. 2012; Mehra and Jain 2014; Pastorin et al. 2006; Ren et al. 2012; Singh et al. 2013; Wu et al. 2014).

27.2.1 Classification of CNTs

The CNTs have been classified into following four categories depending upon the presence of number of walls and summarized in Figure 27.1:

1. Single-walled carbon nanotubes (SWCNTs): SWCNTs are characterized by the presence of single graphitic sheet and have diameter and length ranging from 0.4 to 3.0 and 20 to 1000 nm, respectively (Jain et al. 2014; Kesharwani et al. 2012; Lodhi et al. 2013; Lu et al. 2012; Mehra and Jain 2013, 2014; Mehra et al. 2008, 2014a,b; Mody et al. 2014).
2. Double-walled carbon nanotubes (DWCNTs): DWCNTs are another class of CNTs that resemble SWCNTs due to their same morphological features, while they consist of exactly two concentric cylindrical graphite walls (Mehra and Jain 2014).
3. Triple-walled carbon nanotubes (TWCNTs): TWCNTs are characterized by the presence of three cylindrical graphitic walls.
4. Multiwalled carbon nanotubes (MWCNTs): MWCNTs are made up of several concentric graphene layers (2–10 layers) of graphene shells having diameter range from 1.4 to 100 nm and length range from 1 to 50 μm (Cheng et al. 2011; Mehra and Jain 2014).

27.2.2 Advantages and Disadvantages of Functionalized CNTs

f-CNTs possess several unique advantages but negligible and manageable disadvantages, as drug delivery systems depending upon the nature of drug (hydrophobic or hydrophilic) being delivered. A brief overview of both the main advantages and disadvantages of CNTs has been summarized in context of targeted drug delivery by Jain and coworkers (Chopdey et al. 2014; Gupta et al. 2013; Jain et al. 2007, 2009, 2013, 2014; Kesharwani et al. 2012; Lodhi et al. 2013; Mehra and Jain 2013, 2014; Mehra et al. 2008, 2013; Mody et al. 2014).

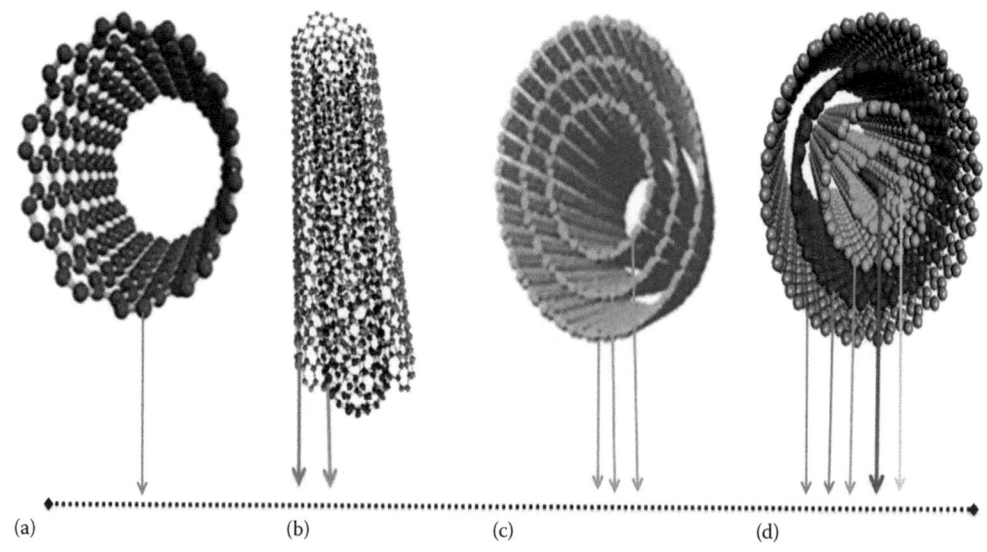

(a)   (b)   (c)   (d)

FIGURE 27.1
Available types of carbon nanotubes: (a) Single-, (b) double-, (c) triple-, and (d) multiwalled carbon nanotubes.

The advantages of *f*-CNTs include the following:

1. Ultralightweight and easy availability.
2. High aspect ratio with uniform ordered structure.
3. Highly elastic nature.
4. Photoluminescence property.
5. Good biocompatibility, nonimmunogenicity, aqueous dispersibility, and low toxicity.
6. Cellular internalization via endocytosis and nanoneedle mechanisms.
7. Biodegradability (?).
8. Minimum cytotoxicity based on in vitro and in vivo studies.
9. Excretion through biliary pathway (urine 96% and remaining 4% by feces).
10. Cell membrane penetration due to its tiny nanoneedle tubular structure.
11. CNTs have open ends on both sides making their inner surface accessible for further incorporation of drug moieties within nanotubes.
12. The biodistribution and pharmacokinetic properties can be easily altered through controlling size, cutting, rich surface chemistry, and degree of functionalization.
13. Longer interior space for endohedral filling relative to diameter, resulting in high drug loading efficiency with controlled drug release.
14. Rare retention of well and highly *f*-CNTs in reticuloendothelial system.

Apart from the various advantages, some drawbacks are also associated with pristine CNTs (*p*-CNTs) as given in the following text:

1. Hydrophobic nature of *p*-CNTs.
2. *p*-CNTs are toxic, hence unsuitable for drug delivery.
3. Nonbiodegradability (?).
4. Bundling/aggregation phenomena.
5. Presence of metallic and amorphous impurities.
6. Chances of accumulation based on the in vivo studies.

Biodegradable or nonbiodegradable nature of CNTs needs to be resolved in future development of a targeted drug delivery system (Kotchey et al. 2013; Mehra and Jain 2014).

27.3 FUNCTIONALIZATION OF CNTs

The *f*-CNTs have emerged as new tool in the field of pharmaceutical and nanotechnology including drug delivery. As-synthesized (first generation) CNTs are hydrophobic and toxic in nature due to the presence of some metallic and amorphous impurities, thus unsuitable for development of pharmaceutical drug products. However, *p*-CNTs can be altered by attaching various chemical functional moieties, rendering them biocompatible and safe for human body. The surface modifications, performed to overcome the major hurdles associated with *p*-CNTs, are referred to as functionalization. Surface functionalization of CNTs plays a pivotal role in the improvement of aqueous solubility that is a critical platform for the design and development of new monohybrid biomaterials. Generally, CNTs have been functionalized using two main strategies (based on both covalent and noncovalent linkages between the surface of CNTs and biochemical functional moieties): (1) covalent and (2) noncovalent functionalization (Lacerda et al. 2012; Mehra and Jain 2014; Mehra et al. 2014a).

27.3.1 Covalent Functionalization

The increased aqueous solubility and biocompatibility of *p*-CNTs are the most important factors in drug delivery and could be achieved through covalent functionalization. Covalent functionalization is an alternative, more reliable, and efficient approach for grafting of chemical functional groups on CNTs. The *ends and defects* and *sidewall* covalent functionalization are the two most widely used functionalization strategies, depending upon the location of surface chemical functional groups (Mehra and Jain 2014; Vardharajula et al. 2012).

The *ends and defects* is more reactive than *sidewall* covalent functionalization strategy. The various oxygen-containing functional groups like carboxylic, hydroxyl, phenolic, lactone, ester, ketone, and alcohol are easily introduced by electrochemical, mechanical, and chemical routes. The oxygen-containing chemical functional groups can also be easily generated using strong oxidizing acid treatment by varying the type of acids, concentration, and reaction conditions such as temperature, sonication, and time via *ends and defects* covalent functionalization of CNTs. The amount and type of oxygen-containing chemical functional groups depend on the treatment strategies (Karousis and Tagmatarchis 2010; Mehra and Jain 2014).

The *sidewall* covalent functionalization is another class of covalent functionalization strategy used for introducing functional groups on the sidewall surfaces of CNTs without loss of Van Hove singularities. The sidewall defects like Stone–Wales may locally enhance the chemical reactivity of the CNTs (Jain et al. 2007; Karousis and Tagmatarchis 2010; Kesharwani et al. 2012).

27.3.1.1 Attachment of polymer

The poor aqueous solubility in various solvents limits the manipulation of CNTs and hampers their possible applications in many promising fields including drug delivery. Thus, surface engineering of CNTs with polymers is a more efficient, reliable, and alternative way to solve this problem that also greatly improves the solubility of CNTs. Generally, two main methodologies *grafting from* and *grafting to* are commonly used for grafting of polymer including linear and multifunctional hyperbranched polymers with the *f*-CNTs (Sun et al. 2011; Tasis et al. 2006). Sun and coworkers exhaustively reviewed and discussed the various methods for functionalization of CNTs with linear and hyperbranched polymers including dendrimers for possible applications (Sun et al. 2011).

27.3.1.1.1 Grafting from approach

The *grafting from* is generally based on covalent immobilization of polymer precursors on surface of CNTs and subsequent propagation by in situ polymerization of monomers, self-condensing vinyl copolymerization, and step-by-step methodology from the attached initiators or reactive sites (Mehra and Jain 2014; Tasis et al. 2006). The high grafting density of polymers including high molecular weight or hyperbranched polymers was achieved in this *grafting from* approach (Sun et al. 2011).

27.3.1.1.2 Grafting to approach

The *grafting to* is more reliable and facile methodology for attachment of the polymer on surface of *f*-CNTs involving reaction between polymers and functional groups of *f*-CNTs. However, it suffers from relatively low grafting density and demand of the reactive functional groups for polymers (Mehra and Jain 2014; Sun et al. 2011). Tao et al. reported the grafting of fourth-generation poly(amidoamine) dendrons to prefunctionalized MWCNTs by esterification (Tao et al. 2006).

It is believed that the *grafting from* approach is somewhat better than *grafting to* for conjugation of various linear or hyperbranched polymers to CNTs with high grafting density and step-by-step methodology, while *grafting to* approach has low grafting density. Currently, several reports have been available on dendrimers conjugation with the *f*-CNTs for biomedical applications (Murugan and Vimala 2011; Qin et al. 2011; Shi et al. 2009).

27.3.2 Noncovalent Functionalization

Another functionalization approach of nanotubes includes noncovalent functionalization where CNTs surface is known to interact with various chemical moieties through noncovalent interactions like hydrophobic, van der Waals force, electrostatic, hydrogen bonding, and π–π stacking interactions. CNTs are successfully functionalized via noncovalent interaction with surfactants, polyethylene glycol (PEG), chitosan (CHI), porphyrin derivatives, fluorophores, polymers, lipids, nucleic acids, proteins, and endohedral functionalization to increase solubility and biocompatibility, thereby making them less toxic (Jain et al. 2007; Karousis and Tagmatarchis 2010; Mehra et al. 2008; Tasis et al. 2006). The various surfactants like sodium dodecyl sulfate, sodium dodecylbenzenesulfonate, cetyltrimethylammonium bromide, Triton X series, and Pluronic (F and E series) enhance the stability and aqueous dispersibility of CNTs (Jain et al. 2007; Mehra et al. 2008). Similarly, PEGylation of CNTs makes them more hydrophilic and improves their pharmacokinetic parameters making them stealth in character (Mehra et al. 2014b).

27.3.3 Characterization of CNTs through Electron Microscope

Microscopy is a major tool for the characterization of CNTs and its conjugates. In drug delivery field, microscopical examinations enable an understanding of a number of variables that govern delivery. These imaging techniques are used to examine the shape, size, and distribution of CNTs as well as their interactions with biological environments. Till date, the number of microscopy techniques is available including the traditional optical microscopy, scanning electron microscopy, TEM, high-resolution TEM, scanning probe microscopy, confocal laser scanning microscopy, and atomic force microscopy (AFM). The AFM is an imaging tool for characterization of the surface topography and other properties of the CNTs. It offers 3D visualization and provides many physical properties like size, morphology, surface texture, and roughness. The AFM is used to characterize the p- and f-CNTs, in particular the length of bundles of the nanotubes. The tapping mode of AFM image of cut and shortened SWCNTs at different time points (2, 6, and 10 h) is represented in Figure 27.2a through c. It is clear from the AFM images that shorten and cut SWCNTs have good dispersibility devoid of aggregation and seem to be nanoneedle tubular structure in nanometric size range.

TEM is a microscopic technique whereby a beam of electrons is transmitted through an ultrathin specimen, interacting with the specimen as it passes through. It is used to investigate the possible morphological changes of CNTs depending on the severity of harsh treatment upon chemical functionalization. The TEM photomicrographs of p-, oxidized, and D-α-tocopheryl polyethylene glycol 1000 succinate (TPGS)–conjugated MWCNTs are shown in Figure 27.3a through c. The TEM images showed that the size are reduced after oxidation, and increase in time of oxidation decreased the size of MWCNTs.

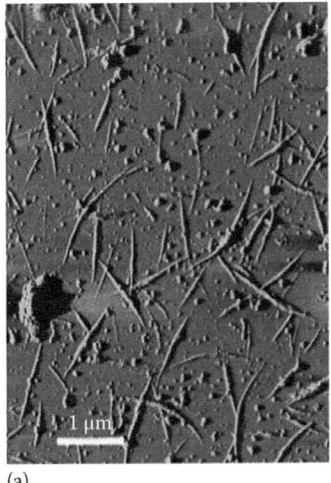 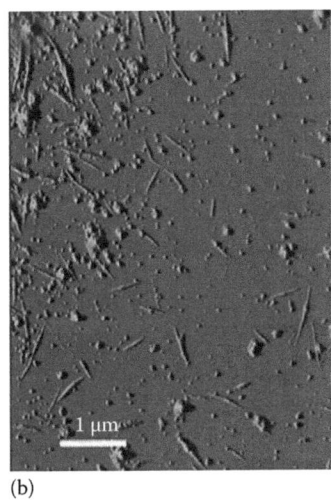 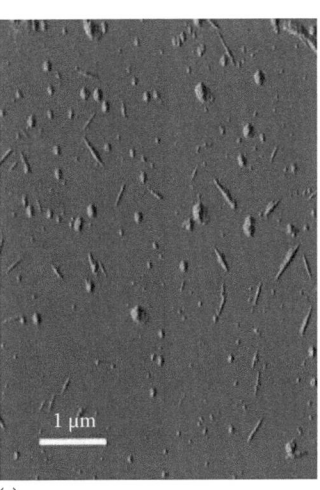

(a) (b) (c)

FIGURE 27.2
Tapping mode AFM images of SWCNTs cut for (a) 2 h, (b) 6 h, and (c) 10 h oxidative treatment. (Reproduced from *Carbon*, 44, Marshall, M.W., Popa-Nita, S., and Shapter, J.G., Measurement of functionalized carbon nanotubes carboxylic acid groups using a simple chemical process, 1137–1141, Copyright 2006, with permission from Elsevier.)

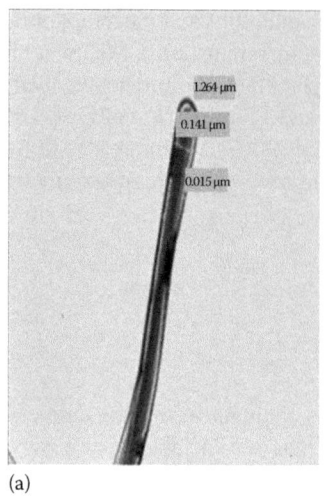
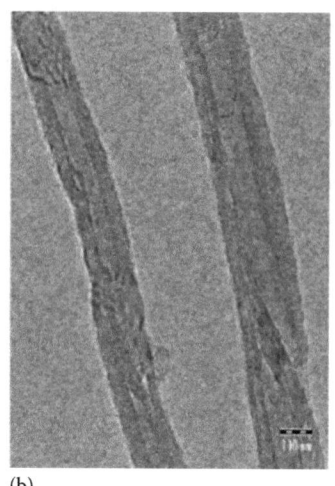
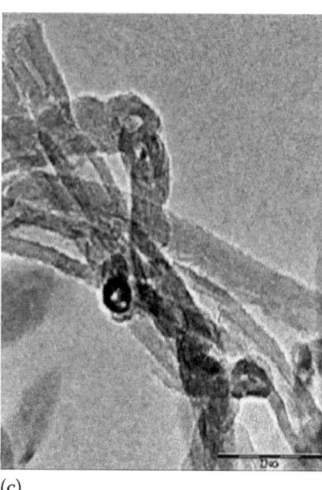

(a)　　　　　　　　　　(b)　　　　　　　　　　(c)

FIGURE 27.3

Transmission electron microscopic images: (a) pristine, (b) oxidized, and (c) TPGS-conjugated MWCNTs. (Reproduced from *Biomaterials*, 35, Mehra, N.K., Verma, A.K., Mishra, P.R., and Jain, N.K., The cancer targeting potential of d-α-tocopheryl polyethylene glycol 1000 succinate tethered multi walled carbon nanotubes, 4573–4588, Copyright 2014b, with permission from Elsevier.)

27.4 DRUG DELIVERY ASPECTS THROUGH FUNCTIONALIZED CNTs

Several drug delivery nanovehicles including polymeric nanoparticles, dendrimers, and liposomes have been continually designed and engineered to evade or overcome the drug extrusion by drug efflux transporter systems, resulting in an increased drug accumulation in the cytosol or nucleus of the infectious cells. Despite these available nanovehicles, *f*-CNTs have been used in the delivery of drugs after proper surface modifications. Recently, we have contributed a book chapter on *functionalized carbon nanotubes and their applications* for interest of researchers (Mehra and Jain 2014).

27.4.1 Delivery of Amphotericin B through Functionalized CNTs

Amphotericin B (AmB) is a first-line amphoteric polyene heptane macrolide antifungal drug used in the treatment of visceral leishmaniasis (VL). AmB is sensitive to heat and light and inactivated at low pH values. It is reported that the AmB is highly toxic to mammalian cells and possible reason is the formation of aggregates as a result of low aqueous solubility. VL is a life-threatening parasitic disease caused by obligate intramacrophage protozoa of the genus *Leishmania*. In humans, VL is always fatal if left untreated, and no alternative treatment option is available (Prajapati et al. 2010, 2012; Pruthi et al. 2012).

The conjugation of AmB to *f*-CNTs has several advantages like increased solubility of the molecules, decrease in the aggregation phenomena, improved antibiotic efficacy due to internalization capacity, and modulation of the antibiotic activity against the mammalian, bacterial, and fungal cells (Wu et al. 2005).

For the very first time, Wu and coworkers explored new strategy for the double functionalization of CNTs and assessed the toxicity and uptake characteristics of the AmB-conjugated CNTs (AmB-CNTs) toward mammalian cells (Wu et al. 2005). Similarly, Benincasa and coworkers explored the antifungal activity of the AmB-CNTs conjugates (Benincasa et al. 2011). The aforementioned studies suggested that the AmB-CNTs conjugates were easily taken up by fungal cells, while preserving the antifungal activity devoid of any toxic effects (Benincasa et al. 2011; Wu et al. 2005).

The efficacy and stability of *f*-CNTs as a delivery system for AmB delivery were studied on a Chinese hamster model following oral administration. The orally administered AmB novel formulation showed 99% inhibition of parasite growth following a 5-day course at 15 mg/kg body weight (Prajapati et al. 2012).

Our laboratory investigated the AmB macrophage-targeted delivery using the mannose-conjugated functionalized MWCNTs (AmBitubes) employing macrophage J774 cell line. The synthesized AmBitubes were able to release AmB in a controlled manner at different pH (4, 7.4, and 10) with enhanced macrophage uptake as well as higher disposition in macrophages-rich organs (Pruthi et al. 2012).

All the aforementioned studies establish that f-CNTs have a significant impact on further development of AmB-based drug delivery system.

27.4.2 Doxorubicin Delivery Using Functionalized CNTs

Doxorubicin (DOX) (anthracycline antibiotic) is a DNA-interacting chemotherapeutic drug most widely used in the treatment of breast, ovarian, prostate, brain, cervix, and lung cancers. However, the clinical application of DOX is limited due to short half-life and acute cardiotoxicity (Mehra and Jain 2014; Mehra et al. 2014b). Integrin receptors are mainly overexpressed on various cancerous cells like melanoma, glioblastoma, breast, and ovarian. Integrin consists of a large N-terminal extracellular domain consisting of 100 and 700 residues for α and β subunits, respectively (Mehra et al. 2013). For the very first time, DOX delivery was reported using cyclic arginine–glycine–aspartic acid peptide–conjugated SWCNTs noncovalently functionalized by phospholipids–PEG (PL-PEG 5000-NH$_2$). The developed conjugate enhanced the DOX delivery to integrin $\alpha_v\beta_3$-positive U87 MG cells, but no noticeable improvement was found on integrin $\alpha_v\beta_3$-negative MCF-7 cell (Liu et al. 2007).

The Pluronic F-127–dispersed DOX-MWCNTs supramolecular complex was developed and showed the enhanced cytotoxicity as compared to free drug on MCF-7 cells (Ali-Boucetta et al. 2008). Folic acid (FA) has a high affinity toward folate receptor protein, which is commonly overexpressed on the surface of cancerous cells and is used as a recognized tumor biomarker. The targeting moiety, FA-conjugated CHI, and alginate (ALG)-functionalized SWCNTs were developed for DOX delivery (DOX-ALG-SWCNTs, DOX-SWCTs, DOX-CHI/ALG-SWCNTs, and DOX-CHI-SWCNTs). The increased cellular uptake of DOX-FA-CHI/ALG-SWCNTs was obtained through receptor-mediated endocytosis (RME) mechanism with high loading efficiency (Zhang et al. 2009).

The dual-targeted drug delivery system based on MWCNTs difunctionalized with folate and iron (FA-MWCNTs@Fe) was developed, employing an external magnetic field. The FA-MWCNTs@Fe showed a prolonged release pattern of DOX with sixfold higher antitumor activity as compared to free DOX due to the biological (active) and magnetic (passive) targeting of difunctionalized CNTs on HeLa cells (Li et al. 2011). Till date, huge research reports are available on DOX delivery employing f-CNTs after conjugating of targeting moieties like hyaluronic acid (Datir et al. 2011), hydroxybenzoic acid (Gu et al. 2011), FA, and CHI (Huang et al. 2011; Ji et al. 2012) for targeting purposes.

A new type of drug delivery system involving CHI- and FA-modified SWCNTs for controlled loading and release of DOX was constructed and characterized employing HCC SMMC-7721 cell lines. The drug delivery system effectively killed and depressed the growth of liver cancer showing superior pharmaceutical efficiency to free DOX (Ji et al. 2012).

Our laboratory has been continuously exploring the drug delivery aspects of f-CNTs. The dexamethasone mesylate anchored MWCNTs (DOX/DEX-MWCNTs) for controlled release of DOX and selective targeting to lung epithelial cancer cells A-549 were developed and characterized. The DOX/DEX-MWCNTs exhibited less hemolytic toxicity and more cytotoxic response as compared to free DOX (Lodhi et al. 2013).

A highly effective drug delivery system was constructed by coating FA-terminated PEG (PEG-FA) on SWCNTs (PEG-FA/SWCNTs) using a facile noncovalent method. The developed DOX-loaded PEG-FA/SWCNTs (DOX/PEG-FA/SWCNTs) exhibited excellent stability under neutral pH conditions such as serum but dramatically released DOX at tumor microenvironment by clathrin-mediated endocytosis mechanism. The DOX/PEG-FA/SWCNTs were found to be more selective and effective than free DOX and nontargeting delivery system (DOX/PEG/SWCNTs) for killing of tumor cells, without showing negligible cytotoxicity to normal 3T3 cells (Niu et al. 2013).

Mehra and Jain demonstrated the targeted delivery of DOX using FA-conjugated DOX-loaded PEG-functionalized MWCNTs (DOX/FA-PEG-MWCNTs) employing human breast cancer MCF-7 cells. The DOX/FA-PEG-MWCNTs showed higher tumor growth suppression efficacy due to its stealth nature, and median survival time was extended up to 30 days as compared to free DOX (Mehra and Jain 2013).

Recently, Jain and coworkers investigated the cancer-targeting potential of DOX-loaded D-α-TGPS-tethered MWCNTs (DOX/TPGS-MWCNTs) and compared it with pristine MWCNTs and free DOX solution. The DOX/TPGS-MWCNTs showed higher entrapment efficiency ascribed to π–π stacking interactions and sustained release pattern at the lysosomal pH (pH 5.3). The DOX/TPGS-MWCNTs showed enhanced cytotoxicity and cellular uptake and were most preferentially taken up by the cancerous cells via endocytosis mechanism as shown in Figure 27.4. The DOX/TPGS-MWCNTs nanoconjugate depicted the significantly longer survival span (44 days, $p < 0.001$) than DOX/MWCNTs (23 days) and free DOX (18 days) (Figure 27.5). The tumor growth inhibition study clearly indicated that inclusion of the pH-responsive characteristics increased the overall pharmaceutical targeting efficiency of the targeted nanotubes formulations (Mehra et al. 2014).

Recently, it was reported that targeted therapeutic semiconducting single-walled carbon nanotubes (sSWCNTs) influence the viscoelasticity of cancer cells employing drug-sensitive OVCAR8 and resistant OVCAR8/ADR to overcome drug resistance. The novel nanoformulation with a cholanic acid–derivatized hyaluronic acid (CAHA) biopolymer wrapped around sSWCNTs and loaded with DOX; CAHA-sSWCNTs-DOX was more effective in killing drug-resistant cancer cells compared with PL-PEG-modified sSWCNTs and free DOX (Bhirde et al. 2014).

Other anticancer agents like daunorubicin (Dau) and epirubicin hydrochloride (EPI) have been successfully delivered using *f*-CNTs. The reversible targeting and controlled release of Dau to cancer cells from sgc8c aptamer-wrapped SWCNTs (Dau-aptamer-SWCNTs complex) were studied employing Molt-4 (target) and U266 (B lymphocyte human nontarget) cells. The Dau-aptamer-SWCNTs complex was able to specify delivery and internalization of Dau to target Molt-4 cells with reduced cytotoxic effects (Taghdisi et al. 2011).

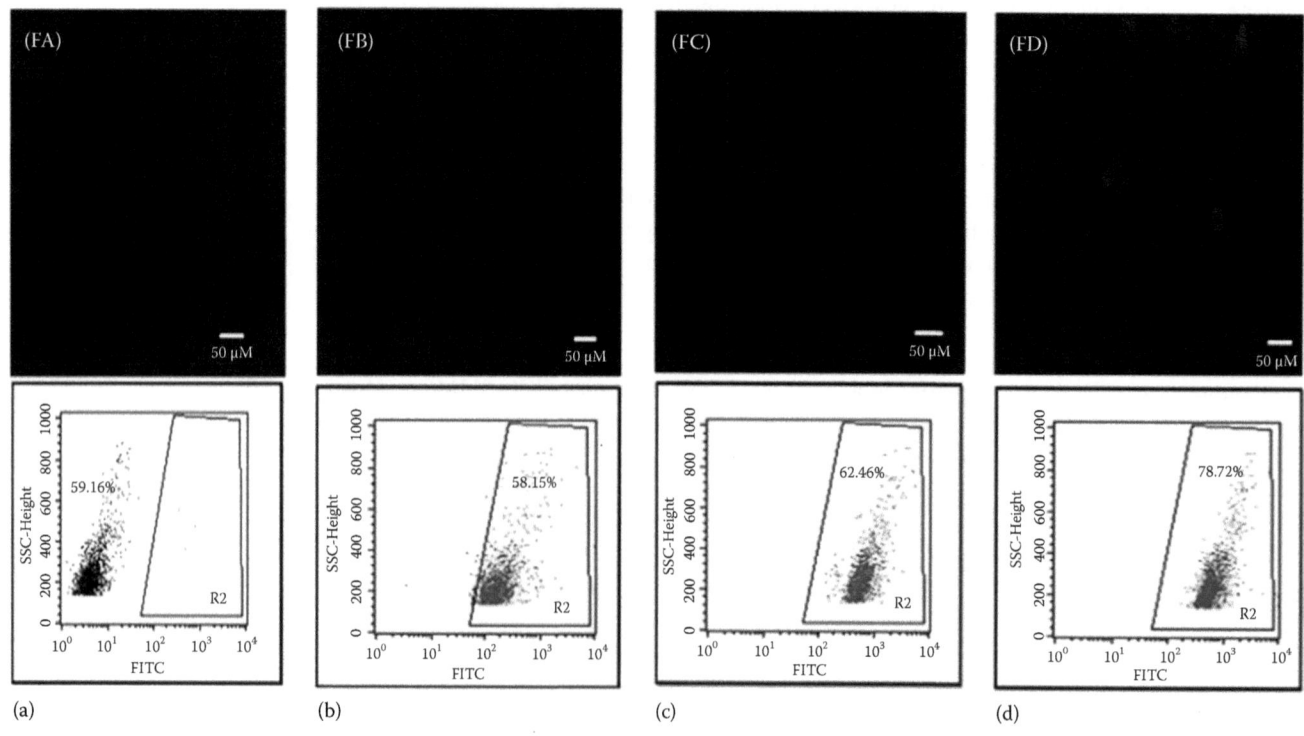

FIGURE 27.4
Qualitative and quantitative cellular uptake of the DOX in MCF-7 cell: (FA and a) Control, (FB and b) Free DOX solution, (FC and c) DOX/MWCNTs, and (FD and d) DOX/TPGS-MWCNTs formulations. (Reproduced from *Biomaterials*, 35, Mehra, N.K., Verma, A.K., Mishra, P.R., and Jain, N.K., The cancer targeting potential of d-α-tocopheryl polyethylene glycol 1000 succinate tethered multi walled carbon nanotubes, 4573–4588, Copyright 2014b, with permission from Elsevier.)

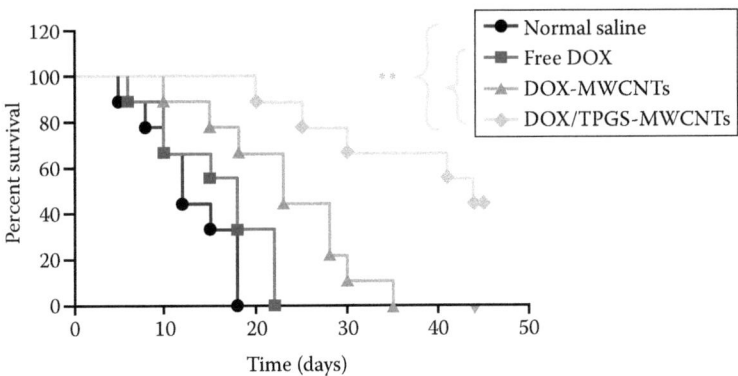

FIGURE 27.5

Kaplan–Meier survival curves of MCF-7-bearing Balb/c mice and analyzed by log-rank (Mantel–Cox) test with normal saline group as control after intravenous administration of free DOX, DOX/MWCNTs, and DOX/TPGS-MWCNTs nanoconjugate (5.0 mg/kg body weight dose). (Reproduced from *Biomaterials*, 35, Mehra, N.K., Verma, A.K., Mishra, P.R., and Jain, N.K., The cancer targeting potential of d-α-tocopheryl polyethylene glycol 1000 succinate tethered multi walled carbon nanotubes, 4573–4588, Copyright 2014b, with permission from Elsevier.)

EPI, an analog of DOX hydrochloride is a highly efficient anthracycline cytostatic antibiotic but causes severe suppression of hematopoiesis and cardiac toxicity. The adsorption behavior, that is, interaction between EPI and carboxylated CNTs (*c*-MWCNTs), was evaluated by Chen and coworkers. The researchers found that the EPI was strongly and rapidly adsorbed on MWCNTs, *c*-MWCNTs, and SWCNTs, mainly attributed to π–π stacking interaction among EPI and graphene surface of MWCNTs. The adsorption capacity of EPI was ranked in the following order: SWCNTs > *c*-MWCNTs > MWCNTs. Due to the high surface area and hydrogen bonding, the adsorption efficiency was found highest and most stable in case of *c*-MWCNTs than MWCNTs. The supramolecular EPI-MWCNTs complex formed through π–π stacking interaction was found to be more effective in the treatment of tumor (Chen et al. 2011). Table 27.1 summarizes the literature on DOX delivery using *f*-CNTs for cancer therapy.

27.4.3 Taxol Derivatives Delivery through Functionalized CNTs

Multifunctional CNTs have been used for efficient delivery of taxol derivatives, that is, paclitaxel (PTX) and docetaxel (DTX), in the treatment of various cancers. In this context, taxol derivative PTX was conjugated on to the surface of *f*-CNTs via cleavable ester or disulfide bond (water-soluble SWCNTs-PTX conjugate) for controlled and sustained delivery. The SWCNTs-PTX conjugate showed higher tumor growth suppression than that of clinical taxol on murine 4T1 breast cancer. The SWCNTs-PTX also showed prolonged blood circulation with onefold higher PTX uptake through EPR devoid of any apparent toxicity to normal organs (Liu et al. 2008).

Chen and coworkers developed a drug delivery system by conjugating biotin (vitamin) to SWCNTs through a cleavable linker for delivery of taxol. The developed biotin-SWCNTs conjugate caused apoptosis and cell death through formation of a stable microtubule-taxoid complex (Chen et al. 2008).

A novel drug delivery system of PTX-conjugated hyperbranched polycitric acid (PCA) functionalized MWCNTs (MWCNTs-*g*-PCA-PTX) was developed for cancer chemotherapy. The MWCNTs-*g*-PCA-PTX showed increased cytotoxicity and was easily taken up by cancerous cells (A-549 and SKOV3 cell line) through endocytosis mechanism, where the cleavable ester bond gets hydrolyzed. Then, released PTX enters into the cytoplasm of cancerous cells (Sobhani et al. 2011).

Another taxol derivative, DTX, was attached to SWCNTs through π–π accumulation linked with Asn-Gly-Arg (NGR) peptide as targeting moiety (SWCNTs-NGR-DTX). The SWCNTs-NGR-DTX showed higher tumor-suppressing growth in cultured PC3 and S180 cell line with minimal

TABLE 27.1 Doxorubicin Delivery Using Surface-Functionalized Carbon Nanotubes for Theranostics Applications

Carbon Nanotubes Conjugates	Dose (Cell Line Used)	Targeting Ligand Used	Inference of the Study	References
Monoclonal antibody-DOX-fluorescein-BSA-SWCNTs	WiDr colon cancer cells	Monoclonal antibody	Enable molecular targeting after interaction of monoclonal antibody.	Heister et al. (2009)
DOX-FA-CHI/ALG-SWCNTs	50 μg/mL on HeLa cell line (cervical carcinoma)	Folic acid	More cytotoxic and selective.	Zhang et al. (2009)
Fluorescein-PL-PEG-SWCNTs-arginine–glycine–aspartic acid (RGD)-DOX	10 μM on MCF-7 (breast cancer) cell line	RGD peptide	Less toxic to MCF-7 cell.	Liu et al. (2007)
DOX-PL-PEG-SWCNTs	10 mg/kg on lymphoma	—	Efficiently treating tumors with less toxic.	Liu et al. (2009)
DOX-Pluronic F127-MWCNTs	10 and 20 μg/mL on MCF-7 (breast cancer) cell line	—	DOX-Pluronic F-127-MWCNTs were found more efficient.	Ali-Boucetta et al. (2008)
DOX-amphiphilic polymers-CNTs	0.5 mg/mL on B16F10 (melanoma) cell line	—	DOX-amphiphilic polymers-CNTs were found more efficient.	Park et al. (2008)
DOX/FA-MWCNTs@Fe	32 μg/mg DOX on FA-MWCNTs@Fe HeLa cells	FA and Fe	It demonstrated both biologically and magnetically targeting capabilities toward HeLa cells in vitro with capacity sixfold higher delivery efficiency of doxorubicin than free doxorubicin.	Li et al. (2011)
DOX/DEX-MWCNTs	A-549 lung epithelial cancer cell	Dexamethasone	DOX/DEX-MWCNTs were found to be less hemolytic and more cytotoxic as compared to free DOX on A-549 lung epithelial cancer cell line.	Lodhi et al. (2013)
DOX/FA-PEG-MWCNTs	MCF-7 (breast cancer) cell line	FA	DOX/FA-PEG-MWCNTs more selectively inhibit the growth of cancer as compared to nontargeted MWCNTs and free DOX through caveolin-mediated endocytosis mechanism.	Mehra and Jain (2013)
^{99m}Tc-MWCNTs-HA-DOX	100 μ Ci(100 μL) A-549 human lung adenocarcinoma cells	Hyaluronic acid (HA)	DOX-loaded HA-MWCNTs exhibited 3.2 times higher cytotoxicity and increased apoptotic activity devoid of any detectable cardiotoxicity, hepatotoxicity, hepatoxicity, or nephrotoxicity.	Datir et al. (2011)
DOX/PEGylated MWCNTs	HeLa, HepG2, KS62 cells	—	It exhibits the efficient anti-MDR effect on liver cancer and leukemia.	Cheng et al. (2011)
DOX-CNT-CHI-FA	—	FA	Showed higher loading efficiency with controlled DOX release at the tumor microenvironment.	Huang et al. (2011)
DOX/FA/CHI/SWCNTs	HCC SMMC-7721 cancer cell line	FA	DOX/FA/CHI/SWCNTs are promising for high tumor treatment efficacy with reduced side effect by showing superior pharmaceutical efficiency than free DOX.	Ji et al. (2012)
DOX-FA-CHI-SWCNTs	5.0 mg/kg body weight on HeLa cells	FA	DOX-FA-CHI-SWCNTs were found to be more safe and effective with lower systemic toxicity than free DOX at equivalent doses.	Meng et al. (2012)
DOX-FA-MN-MWCNTs	U87 cells	FA	DOX-FA-MN-MWCNTs could efficiently be taken up by U87 cells with subsequent intracellular release of DOX, followed by transport of DOX into the nucleus.	Lu et al. (2012)
DOX-hydrazone-CNTs-FA prodrug	HeLa cells	FA	The developed prodrug system killed cancer resulting of photodynamic effects afforded by carrier CNTs.	Fan et al. (2013)
DOX/PEG-FA/SWCNTs	3T3 and HeLa cells	FA	DOX/PEG-FA/SWNTs were found more selective and effective nontargeting DOX/PEG/SWNTs and free DOX to kill tumor cells but showed negligible cytotoxicity to normal 3T3 cells.	Niu et al. (2013)
DOX/TPGS-MWCNTs	MCF-7 cells	TPGS	The developed DOX/TPGS-MWCNTs formulation shows better cancer-targeting potentials on tumor-bearing Balb/c mice.	Mehra et al. (2014b)

side effects than DTX. The targeting efficiency of SWNT-NGR-DTX in tumor tissue was enhanced compared with SWNT-DTX and DTX. Authors suggested that the SWNT drug delivery system may be promising in terms of treatment efficacy for cancer (Wang et al. 2011).

27.4.4 SN-38 Delivery through Functionalized CNTs

The 7-ethyl-10-hydroxycamptothecin (SN-38) is a chemotherapeutic agent and topoisomerase I inhibitor used in the treatment of various cancers including colorectal, lung, and ovarian cancers. Despite the excellent cancer-targeting potential of SN-38, researchers do not recommend it as an anticancer drug directly in humans, due to the poor solubility. SN-38 shows 1000-fold higher cytotoxicity against cancer cells as compared to a potent chemotherapeutic prodrug Irinotecan (CPT-11). The antibody C 225-conjugated SN-38 covalently attached SWCNTs (SWCNTs25/py38) through strong hydrophobic and π–π stacking interactions was developed and characterized for targeted delivery of SN-38 to colorectal cancer cells. The schematic representation of SWCNTs-carrier synthesis is shown in Figure 27.6. The SWCNTs25/py38 was internalized through clathrin-dependent RME into Epidermal Growth Factor Receptor (EGFR)-expressing cells. The SWCNTs25/py38 shows good biocompatibility and efficient cellular uptake with improved targeting ability (Lee et al. 2013).

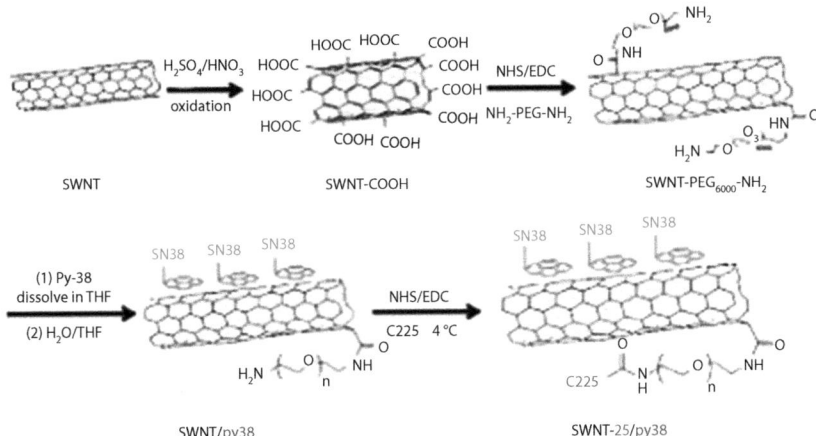

(a) Note that in our notation, the slash "/" represents a π–π stacking association and the dash "–" a covalent association

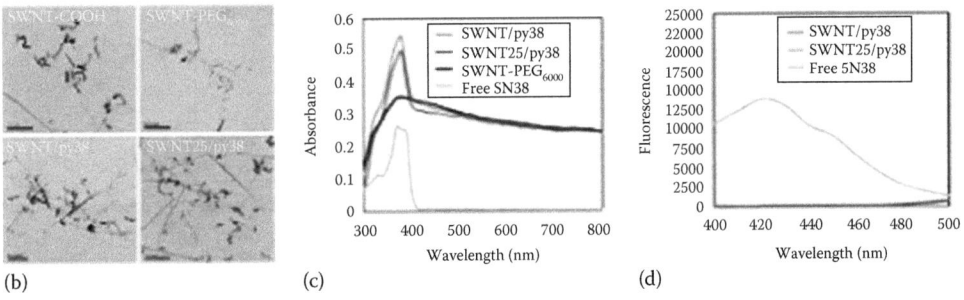

FIGURE 27.6
(a) SWCNTs-carrier synthesis scheme. (b) TEM images showing the size distribution of SWCNTs-COOH, SWCNTs-PEG, SWCNTs/py38, and SWCNTs25/py38. (c) The absorbance spectra of SWCNTs-PEG/py38, SWCNTs-PEG, and SN-38 free drug. (d) The fluorescence spectra of SWCNTs-PEG/py38 and SN-38 free drug. (Reproduced from *Biomaterials*, 34, Lee, P.C., Chiou, Y.C., Wong, J.M., Peng, C.L., and Shieh, M.J., Targeting colorectal cancer cells with single-walled carbon nanotubes conjugated to anticancer agent SN-38 and EGFR antibody, 8756–8765, Copyright 2003, with permission from Elsevier.)

27.4.5 Platinum Analogs Delivery Using Functionalized CNTs

A lot of studies have been reported considering *f*-CNTs as drug delivery system after loading of various drug molecules into the inner cavity and attachment onto the external surface of CNTs through chemical bond interactions. *f*-CNTs are more efficient for transportation of Pt(IV) prodrug across to the cell membrane. The Pt(IV) prodrug (c,c,t-[Pt(NH$_3$)$_2$Cl$_2$(OEt)O$_2$CCH$_2$CH$_2$CO$_2$H)]) through heterobifunctional cross-linker using 1-ethyl-3-(dimethylaminopropyl)carbodiimide hydrochloride and *N*-hydroxysuccinimide onto surface of SWCNTs was conjugated and developed. The CNTs targeted the Pt(IV) prodrug after releasing of active Pt(II) species into cancer cells (Feazell et al. 2007).

Dhar and colleagues developed a longboat delivery system tethering SWCNTs with c,c,t-(Pt[NH$_3$]2Cl$_2$[OEt][O$_2$CCH$_2$CH$_2$CO$_2$H]) through amide bond formation and delivered the platinum-based drug into cells. The longboat structure was easily taken up by the cancerous cells through endocytosis mechanism, followed by the cisplatin release and interaction with the nuclear DNA (Dhar et al. 2008).

Cisplatin (*cis*-diamminedichloroplatinum II, CDDP) is a highly potent and light-sensitive anticancer drug, most widely used in the treatment of testicular, ovarian, breast, and bladder cancer. CDDP can easily be encapsulated inside tip-opened and shortened functionalized SWCNTs (SWCNTs-CDDP) and inhibit the viability of PC3 and DU145 prostate cancer cells in vivo (Tripsciano et al. 2009).

A *CNTs bottle* concept was introduced by Pastorin Georgia and coworkers at National University of Singapore for protection of drug molecules by capping the ends of the tubes with different molecules. Encapsulation of the drug inside the nanotubes seems to be more beneficial than conjugation or attachment on to the external surface or walls of the nanotubes. Additionally, interior of nanotubes has more favorable binding energy for adsorption and encapsulation of drug molecules that interact with CNTs through simple adhesive forces and eliminate the need of forming chemical bonding (Ren and Pastorin 2008; Li et al. 2012). A CNTs bottle structure was proposed for the incorporation, release, and enhanced cytotoxic effect of an FDA-approved chemotherapeutic drug cisplatin (Li et al. 2012).

In another interesting approach, SWCNTs-based cisplatin delivery system was developed for specific destruction of head and neck squamous carcinoma cells and effectively inhibited tumor growth as compared with nontargeted SWCNTs-CDDP (Bhirde et al. 2009). Recently, PEGylated MWCNTs for encapsulation and sustained release of oxaliplatin using nanoextraction strategy on HT 29 cells were reported by Wu and coworkers. The oxaliplatin-encapsulated PEGylated MWCNTs (MWCNTs-PEG-oxaliplatin) showed sustained release pattern with improved cytotoxicity of oxaliplatin on HT-29 cells (Wu et al. 2013).

27.4.6 Miscellaneous Drug Delivery through Functionalized CNTs

Surface chemistry–dependent *switch* regulates the trafficking and therapeutic performance of anticancer drug-loaded CNTs investigated by Das and coworkers. In line with that approach, a series of biofunctionalized MWCNTs decorated with antifouling polymer (PEG), tumor recognition modules (FA/hyaluronic acid/estradiol), and fluorophores (rhodamine B isothiocyanate/Alexa Fluor) were synthesized and loaded separately with methotrexate, DOX, and PTX (Das et al. 2013). In alignment with anticancer-targeted drug delivery using *f*-CNTs, a novel viable strategy combining 2-methoxyestradiol (2-ME)-loaded NGR peptide–conjugated SWCNTs (NGR-SWCNTs-2-ME) for tumor angiogenesis targeting was designed, developed, and characterized on sarcoma (S180) tumor-bearing mice model. The 2-ME is an antiangiogenesis agent that inhibits the growth of various cancer cells. It also inhibits the proangiogenic transcription factor hypoxia-inducible factor 1-alpha. The neovascularity-targeted drug delivery system NGR-SWCNTs-2-ME revealed stronger tumor inhibition effect than 2-ME alone and SWCNTs-2-ME without NGR targeting (Chen et al. 2013).

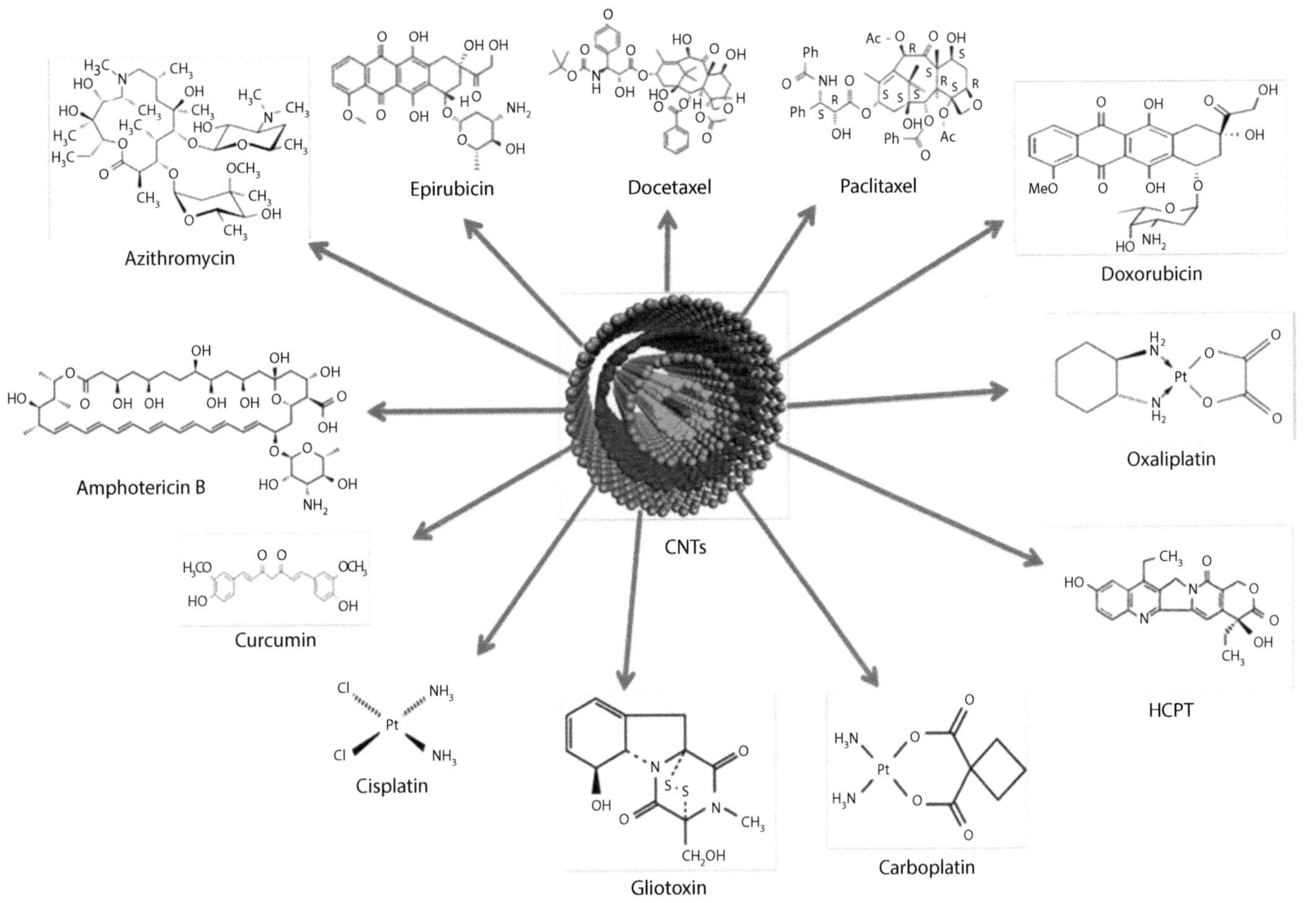

FIGURE 27.7
Drugs delivered through functionalized carbon nanotubes.

Etoposide (ETO) is a semisynthetic derivative of podophyllotoxin that breaks DNA by interacting with the DNA topoisomerase II acting in the late S and early G2 phase of the cell cycle. Mahmood and coworkers and Chen and coworkers reported the improved therapeutic potential of ETO using ligand-decorated *f*-CNTs compared with ETO and CNTs alone (Chen et al. 2012; Mahmood et al. 2009).

Gemcitabine (GEM) is an FDA-approved frontline chemotherapeutic agent for treatment of metastatic pancreatic cancer. It is a cell cycle–dependent (S phase) deoxycytidine analog that inhibits the cellular DNA synthesis. Our group reported the cancer-targeting potential of GEM-loaded FA-conjugated MWCNTs (GEM/FA-MWCNTs) and showed higher entrapment efficiency. The GEM/FA-MWCNTs showed more cytotoxic response on MCF-7 human breast cancer cells. On the basis of in vitro and in vivo studies, surface-engineered CNTs were suggested as alternative, safe, and effective drug delivery system (Singh et al. 2013). Figure 27.7 and Table 27.2 summarize the various drug deliveries through *f*-CNTs.

27.5 CELLULAR TRAFFICKING MECHANISM OF FUNCTIONALIZED CNTs

The *f*-CNTs easily cross the cellular membrane and enter into the cytoplasm for transportation of cargo molecules like drug, protein antigen, antibody, and nucleic acid. In general, it is reported that till date no single, unique cellular uptake mechanism is responsible for cellular

TABLE 27.2 Miscellaneous Drug Delivery Using Surface-Functionalized Carbon Nanotubes for Theranostics Applications

Drug Delivery System	Dose and Drug Used	Cell Line Used	Targeting Ligand Used	Inference of the Study	References
CNTs-amphotericin B (AmB) conjugates	AmB	Human Jurkat lymphoma T cells	—	AmB covalently lined CNTs are taken up by mammalian cells devoid of any specific toxic effect.	Wu et al. (2005)
Gemcitabine (GEM)/ FA-MWCNTs	GEM	MCF-7 human breast cancer cell line	FA	GEM/FA-MWCNTs depict the better cancer-targeting potential employing MCF-7 cell line.	Singh et al. (2013)
EPI-c-MWCNTs	131.3–120.8 mg EPI/gm of epirubicin	—	—	EPI-c-MWCNTs show greater EPI release in acidic medium.	Chen et al. (2011)
CDDP-SWCNTs	100 μg/mL of cisplatin (CDDP)	DU145 and PC3 cells	—	The developed CDDP-SWCNTs show the similar effect on PC3 but less on DU 145 cell.	Tripsciano et al. (2009)
EGF-CDDP-SWCNTs	1.3 μM of cisplatin	Squamous carcinoma	EGF	EGF-CDDP-SWCNTs conjugates were found more efficient on squamous carcinoma.	Bhirde et al. (2009)
Pt(IV)-PL-PEG-SWCNTs	Pt(IV) prodrug	NTera-2 testicular cancer cells	—	Higher toxic to tumorous cells, that is, Ntera-2 testicular cancer cells	Feazell et al. (2007)
Biotin-SWCNTs-taxoid-fluorescein	13.9 μM of taxoid	L1210FR, L1210 and W138 cell lines	Biotin	The biotin-SWCNTs-taxoid-fluorescein was found more efficient.	Chen et al. (2008)
CDDP-CNT	Cisplatin (CDDP)	MCF-7 human breast cancer cell	—	CDDP-CNT exhibits higher drug loading and does not prevent the release of encapsulated drug, while improving its pharmacological effects.	Li et al. (2012)
SWCNTs-PTX	5 mg/kg of paclitaxel	4T1 murine breast cancer cell line	—	SWCNTs-PTX affords higher efficacy by prolonged blood circulation and tenfold higher tumor PTX uptake.	Liu et al. (2008)
f-CNTs-AmB conjugate	—	10 antifungal strains and human Jurkat lymphoma cell	—	f-CNTs-AmB conjugate improves the pharmacological profile of AmB and displays a much reduced toxicity in vitro against mammalian cells.	Benincasa et al. (2011)
MWCNTs-g-PCA-PTX; MWCNTs-g-PCA	Paclitaxel	A-549 and SKOV3 cell lines	—	MWCNTs-g-PCA-PTX shows more cytotoxic effect than the free drug with improved cell penetration.	Sobhani et al. (2011)
HCPT-diaminotriethyleneglycol-MWCNTs	5 mg/kg of HCPT	Hepatic H22 tumor-bearing mice	—	More efficient with pH-triggered drug release.	Wu et al. (2009)
f-CNTs-AmB	15 mg/kg body weight of AmB		—	f-CNTs-AmB showed 99% inhibition of parasite growth following a 5-day course at 15 mg/kg body weight.	Prajapati et al. (2012)
NGR-SWCNTs-2-methoxyestradiol (2-ME)	2-ME	Sarcoma (S180) tumor cell line	NGR peptide	The novel neovasculary NGR-SWCNTs-2-ME could be beneficial as targeted drug delivery system.	Chen et al. (2013)
MWCNTs-PEG-oxaliplatin	Oxaliplatin	HT-29 cancer cell line	—	MWCNTs-PEG-oxaliplatin could be used as sustained release and showed slightly decreased cytotoxic effect.	Wu et al. (2013)
f-SWNTs-COS-GTX p53, f-SWNTs-COS-GTX-lysozyme, f-SWNTs-COS-GTX-FA	Gliotoxin (GTX)	HeLa and MCF-7 cancer cells	Lysozyme, p53, and FA	f-SWNTs-COS-GTX-p53 is the most effective delivery vehicle with a controlled release and enhanced cytotoxicity rendered through apoptosis in human cervical cancer (HeLa) cells.	Bhatnagar et al. (2014)

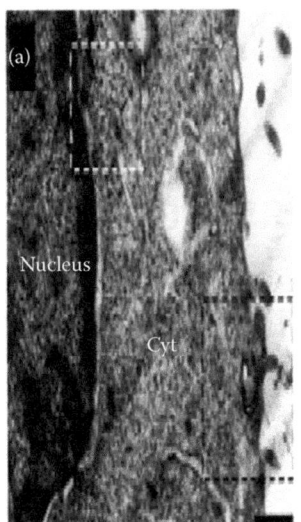 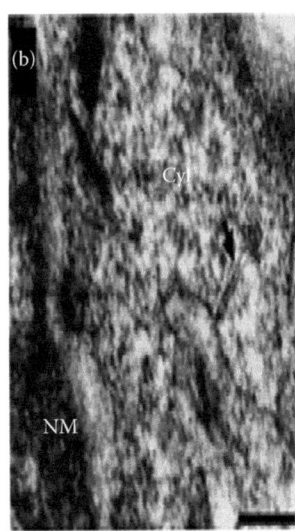 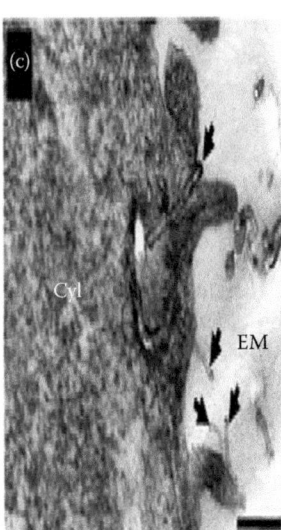

FIGURE 27.8
Intracellular localization of MWCNT-NH$_3^+$ 2 in A549 cells following 60 min incubation period at 4°C. (b) is a higher magnification of the area limited in a with a white-dashed line, and (c) is a higher magnification of the area limited in a with a black-dashed line. Scale bars are 500 nm (a) and 200 nm (b and c). Cyt, cytoplasm; NM, nuclear membrane; EM, extracellular medium. Arrows are pointing to MWCNT- NH$_3^+$. (Reproduced from *Biomaterials*, 33, Lacerda, L., Russier, J., Pastorin, G. et al., Translocation mechanisms of chemically functionalised carbon nanotubes across plasma membranes, 3334 -3343, Copyright 2012, with permission from Elsevier.)

trafficking of *f*-CNTs. Scientist believes that the *f*-CNTs are internalized into the cells mainly by two pathways: (1) energy-dependent internalization, endosomally mediated internalization and (2) direct translocation through the plasma membrane, that is, passive diffusion, which is known as *tiny nanoneedle* mechanism. Due to their tiny nanoneedle tubular structure, CNTs might be able to perforate the cell membrane without any apparent cell damage. It is more interesting that CNTs have unique tiny nanoneedle mechanism for cellular internalization that is well understood. The antibodies, proteins, and nucleic acid–assisted nanotubes have been associated with energy-dependent endocytotic cell uptake. However, macromolecule-assisted CNTs impede direct CNTs–cell membrane interactions, rendering nanotubes unable to directly cross the cell or plasma membrane (Lacerda et al. 2012).

Chen and coworkers successfully developed in vitro CNTs nanoinjector using AFM tip and functionalized MWCNTs by attaching to a model cargo compound through disulfide linker. The prepared AFM-controlled MWCNTs-based nanoinjector is able to penetrate into a cell, where it releases the attached cargo after breaking the disulfide bond (Chen et al. 2007).

Understanding the internalization mechanisms responsible for the entry of *f*-CNTs into live cells is critical from a fundamental point of view and is also considered to be useful for further development of CNT-based drug delivery systems for targeting of therapeutics. Lacerda and coworkers explored the possible mechanisms and studied how *f*-CNTs are able to translocate across cell membranes in phagocytic and nonphagocytic cells. Their finding suggested that the cellular internalization of *f*-CNTs is not solely dependent on one or other previously reported mechanism but is more probably governed by combination of two mechanisms, that is, passive diffusion and RME mechanism (Figure 27.8). Authors also reported that approximately 30%–50% of *f*-CNTs enter into cell through a temperature-insensitive and energy-independent mechanism (Lacerda et al. 2012).

The cellular trafficking pathway using 3D electron tomography imaging was studied in both phagocytotic and nonphagocytic cells. The *f*-CNTs penetrate into phagocytotic and nonphagocytotic cells following three alternative pathways: (1) via direct membrane translocation of individual nanotubes, (2) via membrane wrapping as individual tubes, and (3) in bundles within vesicular compartments (Al-Jamal et al. 2011; Lacerda et al. 2012). We believe and suggest that the internalization of *f*-CNTs is not clear either following solely one mechanism (direct penetration across cell membrane) or mix with energy-dependent mechanism (endocytosis and macropinocytosis).

27.6 CONCLUSION AND FUTURE PERSPECTIVES

In the current paradigm, surface *f*-CNTs have been attracting a great deal of attention, particularly in drug delivery. CNTs are more effective and promising nanocarriers owing to their needlelike shape that enables them to perforate cellular membrane and transport into the cellular compartment through RME or tiny nanoneedle mechanism. The drug molecules can be easily loaded and conjugated to nanotubes through various possible interaction mechanisms. The loading of drug moieties inside the interior chamber of the nanotubes is called endohedral filling. The greater stability and loading capacity of *f*-CNTs make them unique and more exciting advanced biomaterials for drug delivery as compared to existing nanocarriers including liposomes, polymeric nanoparticles, dendrimers, and lipid constructs.

Higher the degree of functionalization, lower is the toxicity and no accumulation in the tissues, because once CNTs are functionalized, they can readily be excreted though renal pathway. Thus, we strongly believe that *f*-CNTs may open a new door in the upcoming years in diverse fields including pharmaceutical. But more reliable and accurate data for *f*-CNTs are needed in their development as safe and effective drug delivery products. Toxicity of *f*-CNTs shall have to be extensively investigated to prove safety of the developed formulation.

ACKNOWLEDGMENT

One of the authors (Neelesh Kumar Mehra) is thankful to the University Grants Commission, New Delhi, India for providing the Senior Research Fellowship.

REFERENCES

Ali-Boucetta, H., Al-Jamal, K.T., McCarthy, D., Prato, M., Bianco, A., Kostarelos, K. 2008. Multiwalled carbon nanotube-doxorubicin supramolecular complexes for cancer therapeutics. *Chem Commun* 8(4): 459–461.

Al-Jamal, K.T., Nerl, H., Muller, K.H. et al. 2011. Cellular uptake mechanisms of functionalised multi-walled carbon nanotubes by 3D electron tomography imaging. *Nanoscale* 3: 2627–2635.

Bacon, R. 1960. Growth, structure, and properties of graphite whiskers. *J Appl Phys* 31: 284–290.

Benincasa, M., Pacor, S., Wu, W., Prato, M., Bianco, A., Gennaro, R. 2011. Antifungal activity of amphotericin B conjugated to carbon nanotubes. *ACS Nano* 5(1): 199–208.

Bhatnagar, I., Jayachandran, V., Se-Kwon, K. 2014. Polymer-functionalized single walled carbon nanotubes mediated drug delivery of Gliotoxin in cancer cells. *J Biomed Nanotechnol* 10(1): 120–130.

Bhirde, A.A., Chikkaveeraiah, B.V., Srivatsan, A. et al. 2014. Targeted therapeutic nanotubes influence the viscoelasticity of cancer cells to overcome drug resistance. *ACS Nano* 8(5): 4177–4189. doi: 10.1021/nn501223q.

Bhirde, A.A., Patel, V., Gavard, J. et al. 2009. Targeted killing of cancer cells in vivo and in vitro with EGF-directed carbon nanotube-based drug delivery. *ACS Nano* 3(2): 307–316.

Brahmachari, S., Ghosh, M., Dutta, S., Das, P.K. 2014. Biotinylated amphiphile-single walled carbon nanotubes conjugate for target-specific delivery to cancer cells. *J Mater Chem* 2: 1160–1173.

Boncel, S., Zajac, P., Koziol, K.K.K. 2013. Liberation of drugs from multi-wall carbon nanotubes carriers. *J Control Release* 169: 126–140.

Chen, C., Xie, X.X., Zhou, Q. et al. 2012. EGF-functionalized single-walled carbon nanotubes for targeting delivery of etoposide. *Nanotechnology* 23: 045104 (1–12).

Chen, C., Zhang, H., Hou, L. et al. 2013. Single-walled carbon nanotubes mediated neovascularity targeted antitumor drug delivery system. *J Pharm Pharmaceut Sci* 16(1): 40–51.

Chen, J., Chen, S., Zhao, X., Kuznetsova, L.V., Wong, S.S., Ojima, I. 2008. Functionalized single-walled carbon nanotubes as rationally designed vehicles for tumor-targeted drug delivery. *J Am Chem Soc* 130: 16778–16785.

Chen, X., Kis, A., Zettl, A., Bertozzi, C.R. 2007 A cell nanoinjector based on carbon nanotubes. *Proc Natl Acad Sci USA* 104: 8218–8222.

Chen, Z., Pierre, D., He, H. et al. 2011. Adsorption behaviour of epirubicin hydrochloride on carboxylated carbon nanotubes. *Int J Pharm* 405: 153–161.

Cheng, J., Meziani, M.J., Sun, Y.P., Cheng, S.H. 2011. Poly(ethylene glycol)-conjugated multi-walled carbon nanotubes as an efficient drug carrier for overcoming multi-drug resistance. *Toxicol Appl Pharmacol* 250(2): 184–189.

Chopdey, P.K., Tekade, R.K., Mehra, N.K., Mody, N., Jain, N.K. 2014. Glycyrrhizin conjugated dendrimer and multi-walled carbon nanotubes for liver specific delivery of doxorubicin. *J Nanosci Nanotechnol* 14: 1–13.

Das, M., Singh, R.P., Datir, S.R., Jain, S. 2013. Surface chemistry dependent "switch" regulates the trafficking and therapeutic performance of drug-loaded carbon nanotubes. *Bioconj Chem* 24: 626–638.

Datir, S.R., Das, M., Singh, R.P., Jain, S. 2011. Hyaluronate tethered smart multi walled carbon nanotubes for tumor-targeted delivery of doxorubicin. *Bioconj Chem* 23(11): 2201–2213.

Dhar, S., Liu, Z., Thomale, J., Dai, H., Lippard, S.J. 2008. Targeted single-wall carbon nanotube-mediated Pt(IV) prodrug delivery using folate as a homing device. *J Am Chem Soc* 130: 11467–11476.

Fan, J., Zeng, F., Xu, J., Wu, S. 2013. Targeted anti-cancer prodrug based on carbon nanotubes with photodynamic therapeutic effect and pH-triggered drug release. *J Nanoparticle Res* 15: 1911.

Feazell, R.P., Nakayama-Ratchford, N., Dai, H., Lippard, S.J. 2007. Soluble single-walled carbon nanotubes as longboat delivery systems for platinum(IV) anticancer drug design. *J Am Chem Soc* 129(27): 8438–8439.

Gu, Y.J., Cheng, J., Jin, J., Cheng, S.H., Wong, W.T. 2011. Development and evaluation of pH-responsive single-walled carbon nanotube-doxorubicin complexes in cancer cells. *Int J Nanomed* 6: 2889–2898.

Gupta, R., Mehra, N.K., Jain, N.K. 2013. Fucosylated multiwalled carbon nanotubes for Kupffer cells targeting for the treatment of cytokine-induced liver damage. *Pharm Res* 31(2): 322–334.

Heister, E., Neves, V., Tilmaci, C. et al. 2009. Triple functionalisation of single-walled carbon nanotubes with doxorubicin, a monoclonal antibody, and a fluorescent marker for targeted cancer therapy. *Carbon* 47: 2152–2160.

Huang, H., Yuan, Q., Shah, J.S., Misra, R.D.K. 2011. A new family of folate-decorated and carbon nanotubes-mediated drug delivery system: Synthesis and drug delivery response. *Adv Drug Deliv Rev* 63:1332–1339.

Iijima, S. 1991. Helical microtubules of graphite carbon. *Nature* 354: 56–58.

Jain, A.K., Dubey, V., Mehra, N.K. et al. 2009. Carbohydrate conjugated multi walled carbon nanotubes: Development and characterization. *Nanomed: Nanotech Biol Med* 5: 432–442.

Jain, A.K., Mehra, N.K., Lodhi, N., Dubey, V., Mishra, D., Jain, N.K. 2007. Carbon nanotubes and their toxicity. *Nanotoxicology* 1(3): 167–197.

Jain, K., Mehra, N.K., Jain, N.K. 2014. Potential and emerging trends in nanopharmacology. *Curr Opin Pharmacol* 15: 97–106.

Jain, N.K., Mishra, V., Mehra, N.K. 2013. Targeted drug delivery to macrophages. *Exp Opin Drug Deliv* 10(3): 353–367.

Ji, Z., Lin, G., Lu, Q. et al. 2012. Targeted therapy of SMMC-7721 liver cancer in vitro and in vivo with carbon nanotubes based drug delivery systems. *J Coll Interf Sci.* 365: 143–149.

Karousis, N., Tagmatarchis, N. 2010. Current progress on the chemical modification of carbon nanotubes. *Chem Rev* 110: 5366–5397.

Kayat, J., Gajbhiye, V., Tekade, R.K., Jain, N.K. 2011. Pulmonary toxicity of carbon nanotubes: A systematic report. *Nanomed: Nanotech Biol Med* 7(1): 40–49.

Kesharwani, P., Ghanghoria, R., Jain, N.K. 2012. Carbon nanotubes exploration in cancer cell lines. *Drug Discov Today* 17(17–18): 1023–1030.

Kotchey, G.P., Zhao, Y., Kagan, V.E., Star, A. 2013. Peroxidase-mediated biodegradation of carbon nanotubes in vitro and in vivo. *Adv Drug Deliv Rev* 65(15): 1921–1932.

Lacerda, L., Russier, J., Pastorin, G. et al. 2012. Translocation mechanisms of chemically functionalised carbon nanotubes across plasma membranes. *Biomaterials* 33: 3334–3343.

Lee, P.C., Chiou, Y.C., Wong, J.M., Peng, C.L., Shieh, M.J. 2013. Targeting colorectal cancer cells with single-walled carbon nanotubes conjugated to anticancer agent SN-38 and EGFR antibody. *Biomaterials* 34: 8756–8765.

Li, J., Yap, S.Q., Yoong, S.L. et al. 2012. Carbon nanotubes bottles for incorporation release and enhanced cytotoxic effect of cisplatin. *Carbon* 50: 1625–1634.

Li, R., Wu, R., Zhao, L. et al. 2011. Folate and iron difunctionalized multiwall carbon nanotubes as dual-targeted drug nanocarrier to cancer cells. *Carbon* 49(5): 1797–1805.

Liu, Z., Chen, K., Davis, C. et al. 2008. Drug delivery with carbon nanotubes for in vivo cancer treatment. *Cancer Res* 68(16): 6652–6660.

Liu, Z., Fan, A.C., Rakhra, K. et al. 2009. Supramolecular stacking of doxorubicin on carbon nanotubes for in vivo cancer therapy. *Angew Chem Int Ed* 48(41): 7668–7672.

Liu, Z., Sun, X., Nakayama-Ratchford, N., Dai, H. 2007. Supramolecular chemistry on water-soluble carbon nanotubes for drug loading and delivery. *ACS Nano* 1(1): 50–56.

Lodhi, N., Mehra, N.K., Jain, N.K. 2013. Development and characterization of dexamethasone mesylate anchored on multi walled carbon nanotubes. *J Drug Target* 21(1): 67–76.

Lu, Y.J., Wei, K.C., Ma, C.C.M., Yang, S.Y., Chen, J.P. 2012. Dual targeted delivery of doxorubicin to cancer cells using folate-conjugated magnetic multi-walled carbon nanotubes. *Coll Surf B Biointerf* 89: 1–9.

Mahmood, M., Karmakar, A., Fejleh, A. et al. 2009. Synergistic enhancement of cancer therapy using a combination of carbon nanotubes and anti-tumor drug. *Nanomedicine* 4(8): 883–893.

Marshall, M.W., Popa-Nita, S., Shapter, J.G. 2006. Measurement of functionalized carbon nanotubes carboxylic acid groups using a simple chemical process. *Carbon* 44: 1137–1141.

Mehra, N.K., Jain, A.K., Lodhi, N., Dubey, V., Mishra, D., Jain, N.K. 2008. Challenges in the use of carbon nanotubes for biomedical application. *Crit Rev Ther Drug Car Syst* 25(2): 169–206.

Mehra, N.K., Jain, N.K. 2013. Development, characterization and cancer targeting potential of surface engineered carbon nanotubes. *J Drug Target* 21(8): 745–758.

Mehra, N.K., Jain, N.K. 2014. *Functionalized Carbon Nanotubes and Their Drug Delivery Applications*, Vol. 4, Ed. B.S. Bhoop. , pp. 327–369. Studium Press, LLC, London, U.K.

Mehra, N.K., Mishra, V., Jain, N.K. 2013. Receptor based therapeutic targeting. *Ther Deliv* 4(3): 369–394.

Mehra, N.K., Mishra, V., Jain, N.K. 2014a. A review of ligand tethered surface engineered carbon nanotubes. *Biomaterials* 35(4): 1267–1283.

Mehra, N.K., Verma, A.K., Mishra, P.R., Jain, N.K. 2014b. The cancer targeting potential of D-α-tocopheryl polyethylene glycol 1000 succinate tethered multi walled carbon nanotubes. *Biomaterials* 35: 4573–4588.

Meng, L., Zhang, X., Lu, Q., Fei, Z., Dyson, P.J. 2012. Single walled carbon nanotubes as drug delivery vehicles: Targeting doxorubicin to tumors. *Biomaterials* 33: 1689–1698.

Mody, N., Tekade, R.K., Mehra, N.K., Chopdey, P., Jain, N.K. 2014. Dendrimers, liposomes, carbon nanotubes and PLGA nanoparticles: One platform assessment of drug delivery potential. *AAPS PharmSciTech* 15(2): 388–399.

Murugan, E., Vimala, G. 2011. Effective functionalization of multiwalled carbon nanotubes with amphiphilic poly(propyleneimine) dendrimer carrying silver nanoparticles for better dispersibility and antimicrobial activity. *J Coll Interf Sci* 357: 354–365.

Niu, L., Meng, L., Lu, Q. 2013. Folate-conjugated PEG on single walled carbon nanotubes for targeting delivery of doxorubicin to cancer cells. *Macromol Biosci* 13(6): 735–744. doi:10.1002/mabi.201200475.

Park, S., Yang, H.S., Kim, D., Jo, K., Jon, S. 2008. Rational design of amphiphilic polymers to make carbon nanotubes water-dispersible, anti-biofouling, and functionalizable. *Chem Commun* 25: 2876–2878.

Pastorin, G., Wu, W., Wieckowski, S. et al. 2006. Double functionalization of carbon nanotubes for multimodal drug delivery. *Chem Commun* 11: 1182–1184.

Prajapati, V.K., Awasthi, K., Gautam, S. et al. 2010. Targeted killing of *Leishmania donovani* in vivo and in vitro with amphotericin B attached to functionalized carbon nanotubes. *J Antimicrob Chemother* 66: 874–879.

Prajapati, V.K., Awasthi, K., Yadav, T.P., Rai, M., Srivastava, O.N., Sundar, S. 2012. An oral formulation of amphotericin B attached to functionalized carbon nanotubes is an effective treatment for experimental visceral leishmaniasis. *J Infect Dis* 205: 333–336.

Pruthi, J., Mehra, N.K., Jain, N.K. 2012. Macrophages targeting of amphotericin B through mannosylated multi walled carbon nanotubes. *J Drug Target* 20(7): 593–604.

Qin, W., Yang, K., Tang, H. et al. 2011. Improved GFP gene transfection mediated by polyamidoamine dendrimer-functionalized multi-walled carbon nanotubes with high biocompatibility. *Coll Surf B Biointerf* 84: 206–213.

Ren, J., Shen, S., Wang, D. et al. 2012. The targeted delivery of anticancer drugs to brain glioma by PEGylated oxidized multi-walled carbon nanotubes modified with angiopep-2. *Biomaterials* 33: 3324–3333.

Ren, Y., Pastorin, G. 2008. Incorporation of hexamethylmelamine inside capped carbon nanotubes. *Adv Mater* 20(11): 2031–2036.

Shi, X., Wang, S.H., Shen, M. et al. 2009. Multifunctional dendrimer-modified multi walled carbon nanotubes: Synthesis, characterization and in vitro cancer cell targeting and imaging. *Biomacromolecules* 10: 1744–1750.

Singh, R., Mehra, N.K., Jain, V., Jain, N.K. 2013. Gemcitabine-loaded smart carbon nanotubes for effective targeting to cancer cell. *J Drug Target* 21(6): 581–592.

Sobhani, Z., Dinarvand, R., Atyabi, F., Ghahremani, M., Adeli, M. 2011. Increased paclitaxel cytotoxicity against cancer cell lines using a novel functionalized carbon nanotube. *Int J Nanomed* 6: 705–719.

Sun, J.T., Hong, C.Y., Pan, C.Y. 2011. Surface modification of carbon nanotubes with dendrimers or hyperbranched polymers. *Polymer Chem* 2: 998–1007.

Taghdisi, S.M., Lavaee, P., Ramezani, M., Abnous, K. 2011. Reversible targeting and controlled release delivery of daunorubicin to cancer cells by aptamer wrapped carbon nanotubes. *Euro J Pharm Biopharm* 77(2): 200–206.

Tao, L., Chen, G., Mantovani, G., York, S., Haddleton, D.M. 2006. Modifications of multi-wall carbon nanotubes surfaces with poly(amidoamine) dendrons: Synthesis and metal templating. *Chem Commun* 47: 4949–4995.

Tasis, D., Tagmatarchis, N., Bianco, A., Prato, M. 2006. Chemistry of carbon nanotubes. *Chem Rev* 106(3): 1105–1136.

Tripisciano, C., Kraemer, K., Taylor, A., Borowiak-Palen, E. 2009. Single-wall carbon nanotubes based anticancer drug delivery system. *Chem Phys Lett* 478(4–6): 200–205.

Vardharajula, S., Ali, S.Z., Tiwari, P.M. et al. 2012. Functionalized carbon nanotubes: Biomedical applications. *Int J Nanomed* 7: 5361–5374.

Wang, L., Zhang, M., Zhang, N. et al. 2011. Synergistic enhancement of cancer therapy using a combination of docetaxel and photothermal ablation induced by single-walled carbon nanotubes. *Int J Nanomed* 6: 2641–2652.

Wu, H., Shi, H., Zhang, H. et al. 2014. Prostate stem cell antigen antibody-conjugated multiwalled carbon nanotubes for targeted ultrasound imaging and drug delivery. *Biomaterials* 35(20): 5369–5380.

Wu, L., Man, C., Wang, H. et al. 2013. PEGylated multi-walled carbon nanotubes for encapsulation and sustained release of oxaliplatin. *Pharm Res* 30: 412–423.

Wu, W., Wieckowski, S., Pastorin, G. et al. 2005. Targeted delivery of amphotericin B to cells by using functionalized carbon nanotubes. *Angew Chem Int Ed Engl* 44: 6358–6362.

Wu, W., Li, R., Bian, X. et al. 2009. Covalently combining carbon nanotubes with anticancer agent: preparation and antitumor activity. *ACS Nano* 3(9): 2740–2750.

Zhang, X., Meng, L., Lu, Q., Fei, Z., Dyson, P.J. 2009. Targeted delivery and controlled release of doxorubicin to cancer cells using modified single wall carbon nanotubes. *Biomaterials* 30(30): 6041–6047.

CHAPTER 28

CONTENTS

Surface Functionalization of Nanodiamond for Biomedical Applications

Polyglycerol Grafting and Further Derivatization

28

Li Zhao and Naoki Komatsu

28.1 INTRODUCTION

Nanodiamond (ND) is a member of nanocarbons that possess great potential for a variety of applications. In particular, ND has been extensively investigated in the fields of biology and medicine over the last decade by utilizing its nontoxic property, chemical stability, large specific surface area, amenability to various chemical functionalization, and nonbleaching and nonblinking fluorescence from nitrogen-vacancy (N-V) centers.[1] These characteristics make ND one of the ideal platforms in the in vivo applications such as an imaging agent for diagnosis[2] and a drug carrier for therapy.[3] Since biomedical in vivo applications of nanoparticles often require multiple functions such as good dispersibility in physiological environment, stealth nature to avoid nonspecific uptake, and active targeting, surface functionalization of ND has been extensively investigated for these purposes.[4] To date, ND has been functionalized noncovalently by protein[5] and silica[6] and covalently by polyethylene glycol (PEG)[7] and polyglycerol (PG)[8] through a variety of surface functional groups such as hydroxyl and carboxyl ones. Covalent functionalization that yields ND with defined chemical structure and better colloidal stability facilitates biomedical applications. Although PEG is the most frequently used hydrophilic polymer for covalent functionalization of nanomaterials, only one terminal in a PEG string can be functionalizable, limiting a number of further chemical derivatization. In contrast, PG possesses a large number of hydroxyl groups on the PEG backbone (Scheme 28.1), which significantly increase not only the hydrophilicity but also sources for further derivatization.[9] In addition, PG is reported to be as biocompatible as[10] and more resistant to protein adsorption than[11] PEG. Thus, PG can be concluded as a better alternative to PEG in view of biomedical applications.

In this chapter, we describe the PG functionalization of ND and its derivatization for the applications as an imaging agent and a drug carrier on the basis of our recent publications on functionalized ND-PG for fluorescence labeling, magnetic resonance imaging, and targeted drug delivery.

28.2 PG FUNCTIONALIZATION OF ND

The ND-PG was synthesized through ring-opening multibranching polymerization of glycidol at high temperature as shown in Scheme 28.1.[8] When the polymerization is initiated at the surface functional groups such as hydroxyl and carboxyl groups, the ND surface is connected covalently to and covered fully with PG. The optimized reaction is conducted at 140°C under neutral condition to graft ND with more and longer polymer trees and to retard the side reaction, that is, the self-initiating

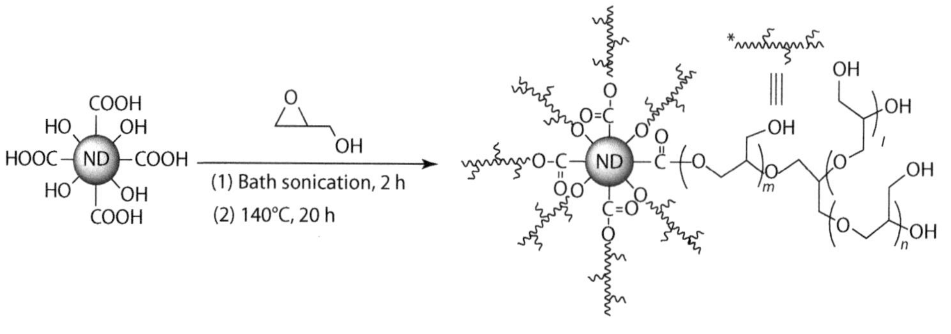

SCHEME 28.1
Synthesis of ND-PG through surface-initiated ring-opening polymerization of glycidol.

ring-opening polymerization of glycidol giving PG without ND core (free PG). The amount of grafted PG is dependent on the particle size or specific surface area of the ND core. In the case of high-pressure high-temperature ND with an average size of 30 nm (ND30), the PG layer is about 40 wt.% as quantified by thermogravimetric analysis. When detonation ND (dND) with 4 nm size is used as a starting material, the PG layer increases to 78 wt.% owing to more growth of polymer trees. Such a large amount of PG greatly enhances the aqueous solubility of ND-PG. For example, the solubility of ND30-PG in phosphate buffered saline (PBS) is not less than 16 mg/mL (Figure 28.1). In sharp contrast, the pristine ND30 completely precipitated in PBS, and the PEGylated ND30 showed very low solubility of 0.04 mg/mL.[7] The hydrosols of ND30-PG are very stable; no precipitates and no significant change in the diameter distribution are observed for more than 3 months. Such high solubility in buffer even enabled chromatographic purification and size sorting of the ND30-PG.[8]

Apart from the aqueous solubility mentioned earlier, the other significant difference between ND and ND-PG is susceptibility to aggregation. As revealed by dynamic light scattering, ND30 displayed a very broad size distribution in water with a mean hydrodynamic diameter of 75.3 nm, indicating that large aggregates were included in the dispersion (Figure 28.2a). In the case of ND30-PG, the size distribution was much narrower, and its mean hydrodynamic diameter was as small as 46.9 nm (Figure 28.2b). Scanning transmission electron microscopy (STEM) images showed that ND30 particles were prone to aggregate, while ND30-PG particles were individually dispersed (Figure 28.2c and d). These results suggest that the grafted PG layer efficiently protects ND-PG particles from aggregation by steric shielding.

(a) (b) (c)

FIGURE 28.1
PBS dispersions of (a) ND30 (completely precipitated), (b) ND30-PEG (0.04 mg/mL), and (c) ND30-PG (>16 mg/mL), after more than 1 month. The ND30-PEG dispersion was irradiated with green laser light to show Tyndall effect.

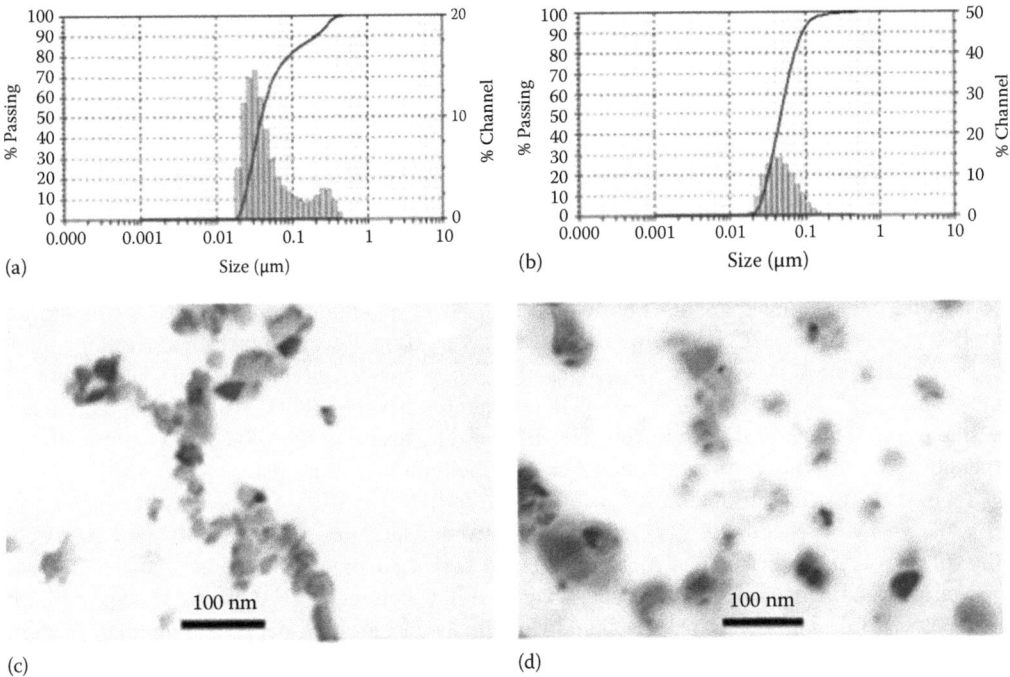

FIGURE 28.2
Hydrodynamic size distribution (volume) and STEM images of the nanoparticles, (a, c) ND30, (b, d) ND30-PG.

28.3 CHEMICAL FUNCTIONALIZATION OF ND-PG FOR BIOMEDICAL APPLICATIONS

Scheme 28.2 illustrates the surface derivatization of ND-PG for biomedical applications. First, some of the hydroxyl groups on the ND-PG surface were converted to more reactive groups such as azido, amino, and carboxyl groups through stepwise organic transformations. Then, a variety of functional moieties including imaging tags, targeting ligands, and therapeutic drugs were immobilized on the

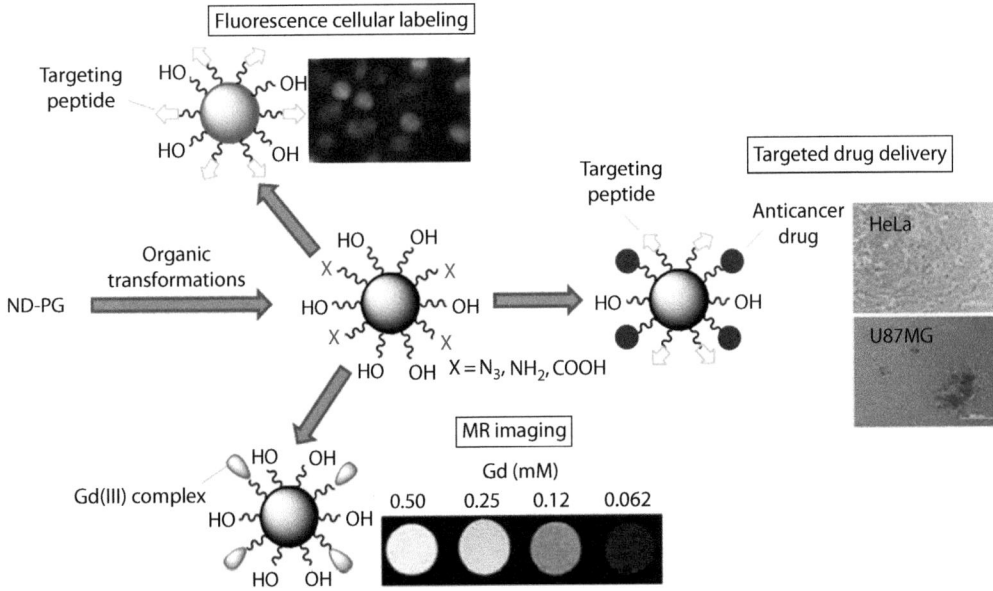

SCHEME 28.2
Surface derivatization of ND-PG for various biomedical applications.

surface of ND-PG through covalent conjugation. The multifunctional ND-PG derivatives with good solubility and stability in physiological media can be used as imaging or therapeutic agents in various biomedical applications.[12–15]

28.3.1 Functionalization of Fluorescent ND-PG with RGD Peptide for Selective Cellular Labeling

The intrinsic fluorescence of ND originated from N-V centers provides an ideal tool for cellular labeling and intracellular tracking (Figure 28.3).[16,17] Therefore, fluorescent ND (fND) was grafted with PG, and the resulting fND-PG was conjugated with arginylglycylaspartic acid (RGD) peptide that specifically binds to $\alpha_v\beta_3$ integrin for selective fluorescence cellular labeling.[18] The synthetic route is shown in Scheme 28.3 (–OH → –OTs [tosylate] → –N$_3$ → –RGD), using click reaction as a key step to immobilize RGD peptide on fND-PG.[12,19] The resulting fND-PG-RGD exhibited good dispersibility and stability in water and cell culture medium (>1.0 mg/mL).

To investigate the cell uptake, fND, fND-PG, and fND-PG-RGD were first incubated with HeLa and U87MG cancer cells. The difference between HeLa and U87MG cells is that U87MG cell expresses $\alpha_v\beta_3$ integrin much more than HeLa cell. Cell uptake of these nanoparticles was quantitatively compared using fluorescence-activated flow cytometry (FACS). As shown in Figure 28.4, fND was taken up by both U87MG and HeLa in a concentration-dependent manner. In sharp contrast, neither U87MG nor HeLa displayed any significant fND-PG uptake even at the highest concentration, indicating stealth effect of the PG coating. In the case of fND-PG-RGD, however,

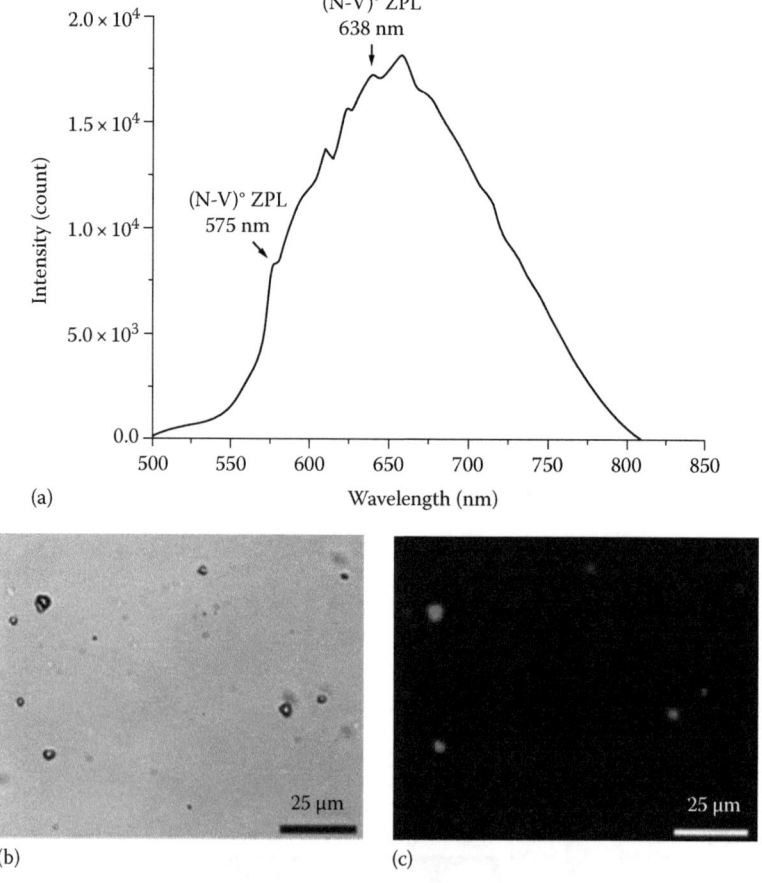

FIGURE 28.3

(a) Photoluminescence spectrum of fND powder measured at room temperature using a 488 nm laser as excitation source (HORIBA LabRAM), (b) bright-field image of fND powder, and (c) fluorescence image under cy3 mode.

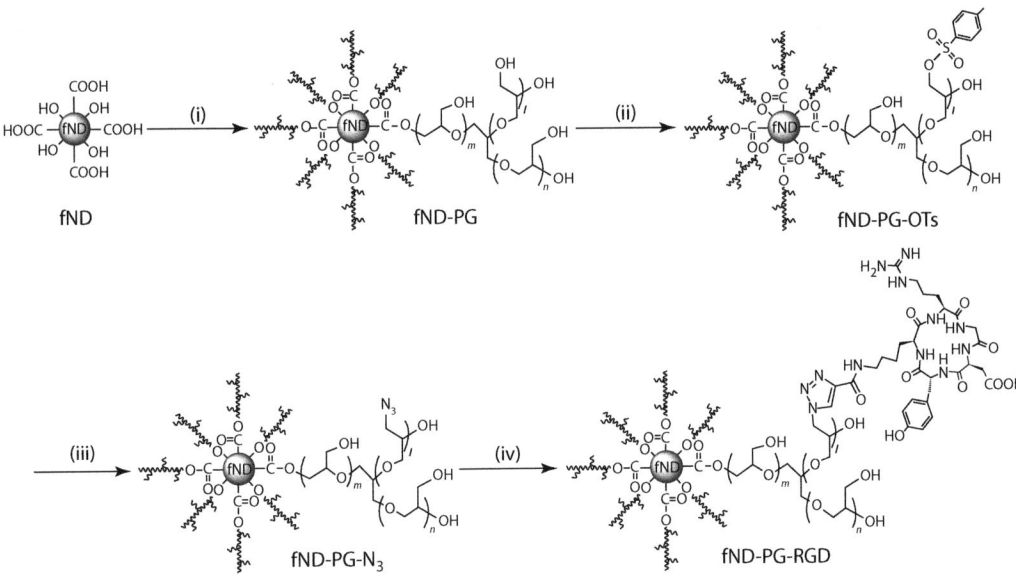

SCHEME 28.3
Synthesis of fND-PG-RGD through stepwise organic reactions: (i) glycidol, 140°C, 20 h; (ii) *p*-TsCl, pyridine, 0°C ~ r. t., overnight; (iii) NaN$_3$, 90°C, overnight; and (iv) RGD propargyl amide, copper(II) sulfate pentahydrate, sodium ascorbate, r. t., 48 h.

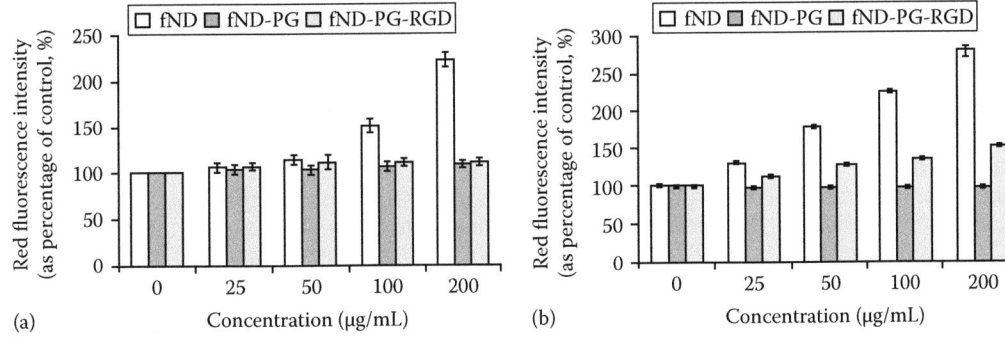

FIGURE 28.4
FACS analysis of uptake of fND, fND-PG, and fND-PG-RGD in separately cultured (a) HeLa and (b) U87MG cells after 24 h treatment. Data were acquired on arithmetic scale, and arithmetic means were used for quantification (*n* = 3).

U87MG and HeLa behaved differently. fND-PG-RGD uptake was dose dependent in U87MG, but was not observed in HeLa even at the maximal concentration. Therefore, the following conclusions can be drawn from the aforementioned results: (1) PG coating has shielding effect against cell uptake and (2) RGD exerts targeting effect. To demonstrate the selective cell uptake better, fND-PG-RGD was further incubated with cocultured U87MG and HeLa cells. Fluorescence microphotographs clearly showed the selective uptake of fND-PG-RGD only in U87MG (Figure 28.5).

Following internalization in U87MG cells, subcellular localization of fND-PG-RGD was further explored using fluorescence microscopy. As shown in Figure 28.6, the red fluorescence originating from the nanoparticles is superimposed on the blue dye staining lysosomes, suggesting that fND-PG-RGD mainly deposited in the lysosomes of U87MG cells.[20]

28.3.2 Functionalization of ND-PG with Basic Polypeptides for Gene Delivery

Basic polypeptide (BPP) (Arg$_8$, Lys$_8$, or His$_8$) conjugated ND-PG (ND-PG-BPP) was designed and synthesized as a gene carrier to load plasmid DNA (pDNA) through electrostatic interaction.[13]

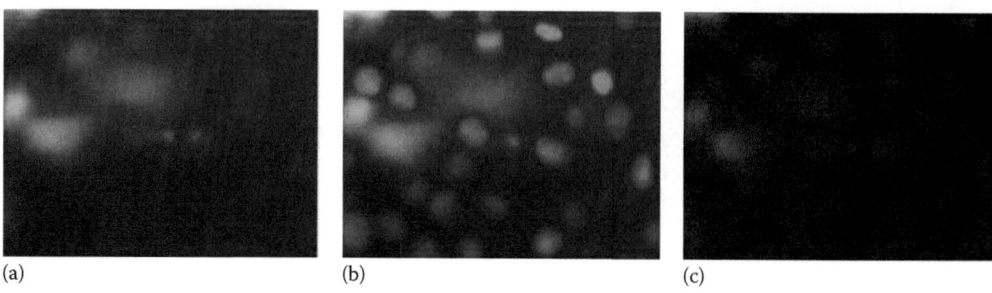

(a)   (b)   (c)

FIGURE 28.5
Fluorescent microphotographs of cocultured U87MG and HeLa cells after 24 h treatment of fND-PG-RGD (200 µg/mL). U87MG cells were first labeled with 5(6)-carboxyfluorescein diacetate succinimidyl ester (CFDA-SE), a green fluorescent supravital dye for cell tracking, and then grown together with HeLa cells. Three images of the same field were taken using different filters: (a) green fluorescence was from U87MG stained with CFDA-SE, (b) blue fluorescence was from the nuclei of both U87MG and HeLa cells, and (c) punctuated red fluorescence was from internalized fND-PG-RGD.

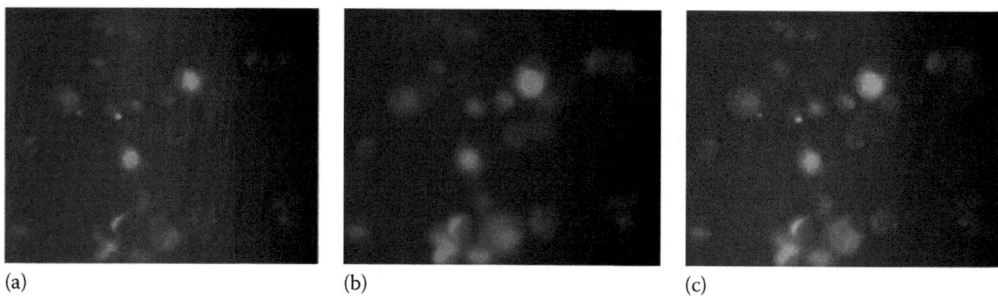

(a)   (b)   (c)

FIGURE 28.6
Subcellular localization of internalized fND-PG-RGD in U87MG cells. Lysosomes were staining by LysoTracker® Blue. Three images of the same field were taken using different filters: (a) fluorescence was from internalized fND-PG-RGD, (b) blue fluorescence was from lysosomes, and (c) merged image of (a, b).

The preparation of ND-PG-BPP is shown in Scheme 28.4. Nonfluorescent ND with 50 nm size (ND50) was chosen as the core, and the ND50-PG-N$_3$ was prepared in a similar method to that of fND-PG-N$_3$ shown in Scheme 28.3. The BPP (Arg$_8$, Lys$_8$, or His$_8$) terminated with propargyl glycine was employed for the click conjugation to give ND-PG-BPP. The synthesized ND-PG-BPP exhibited good solubility in water (>1.0 mg/mL).

Since DNA is immobilized on the surface of nanoparticles mostly by electrostatic attraction between the negative charge of DNA and the positive charge on the surface of the nanoparticle,[21] the zeta potential of ND50, ND50-PG, and ND-PG-BPP was measured at neutral pH in Milli-Q water. The zeta potential of ND50 was −46.7 ± 3.6 mV. The relatively large negative potential can be attributed to a large number of carboxyl groups on the surface of ND. The zeta potential of ND50-PG became less negative (–36.8 ± 1.7 mV) through PG functionalization, probably because some of the carboxyl groups of ND50 were converted to ester by initiation of the ring-opening polymerization of glycidol. The immobilization of polypeptide (Arg$_8$, Lys$_8$, and His$_8$) turned the zeta potentials into positive (+44.1, 38.7, and 14.2 mV, respectively) due to the protonation to the basic groups in the peptides: imidazole, amine, and guanidine. These zeta potentials of the ND-PG-BPP are roughly proportional to the pK_a values of the side chains in these basic amino acids: His (6.0), Lys (10.5), and Arg (12.5).

The positive surface charge of nanoparticles enables complexation with negatively charged DNA through electrostatic interaction. To evaluate the DNA complexation capability of the ND-PG-BPP, an agarose gel retardation assay was performed. The result of the electrophoresis is shown in Figure 28.7. ND-PG-Arg$_8$ and ND-PG-Lys$_8$ with higher positive zeta potential formed complexes with the pDNA, which can be proved by fade and disappearance of the white band corresponding to the pDNA. In particular, ND-PG-Arg$_8$ with the highest positive zeta potential completely retarded

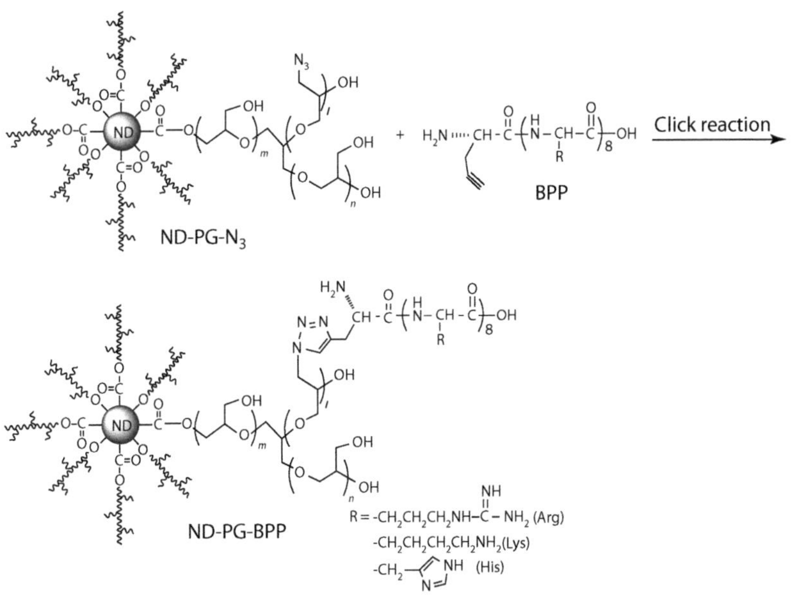

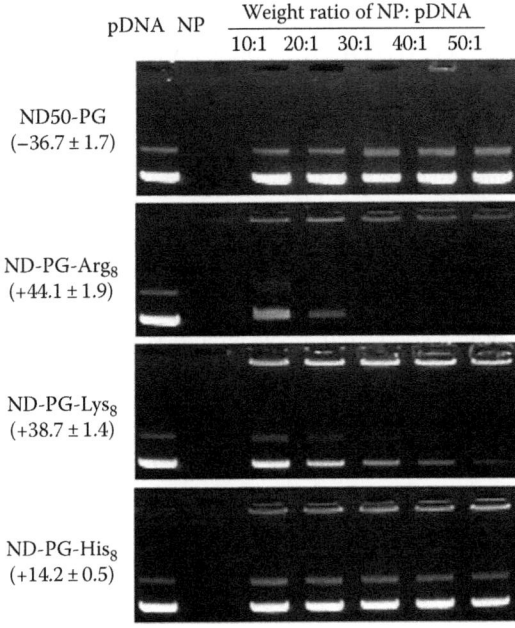

SCHEME 28.4
Synthesis of ND-PG-BBP through click reaction.

FIGURE 28.7
Electrophoretic migration of pDNA, NP (ND50-PG or ND-PG-BPP), and NP/pDNA mixtures at various weight ratios. The zeta potential of the nanoparticles is in the parentheses.

the pDNA at a relatively low NP/pDNA weight ratio (30:1). In contrast, ND50-PG and ND-PG-His$_8$ with negative and less positive potentials, respectively, were not able to form complex with the pDNA even at the highest NP/pDNA weight ratio (50:1).

28.3.3 Functionalization of dND-PG with Gadolinium (III) Complex for MRI

For ND-PG-based imaging probe, gadolinium (Gd[III]) complex was immobilized on the surface of ND-PG. Due to the paramagnetic property of Gd and the extremely high solubility of dND-PG (>80 mg/mL), the resulting dND-PG-Gd(III) worked as a magnetic resonance imaging (MRI) contrast agent.[15]

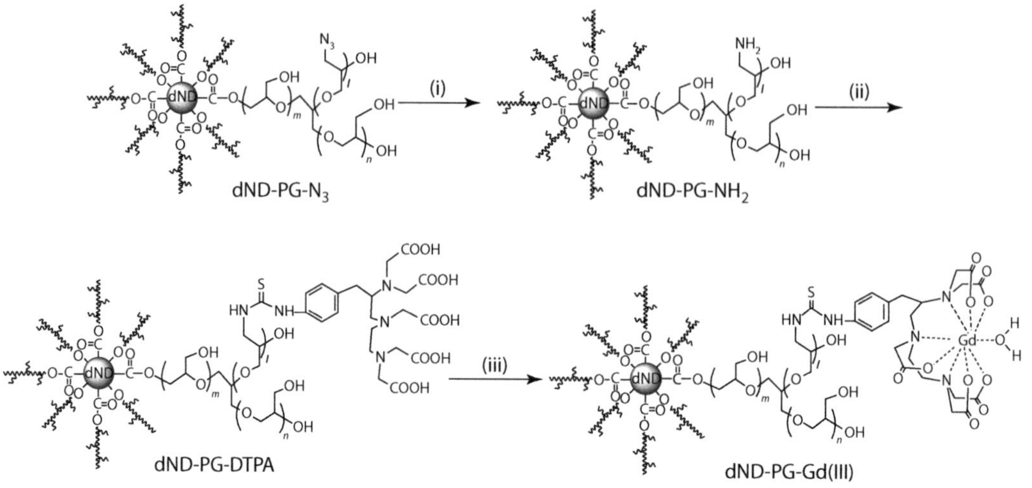

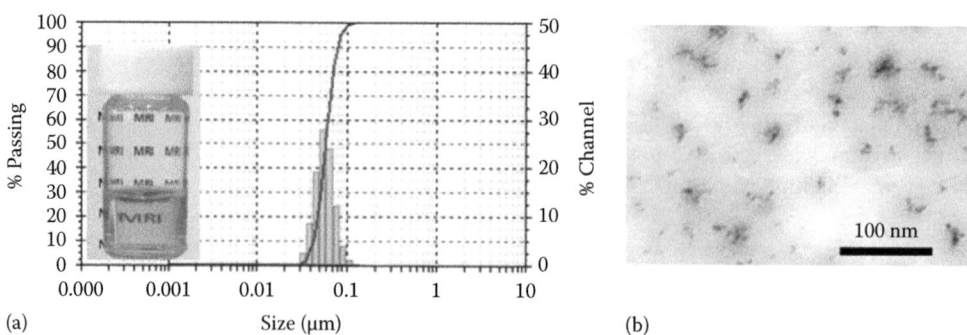

SCHEME 28.5
Synthesis of dND-PG-Gd(III): (i) triphenyl phosphate, *N,N*-dimethylformamide (DMF), ammonia, 60°C, 18 h; (ii) *p*-SCN-Bn-DTPA, pH 10, DMF/H₂O, 60°C, 12 h; and (iii) GdCl₃, pH 6–7, r. t., 5 h.

The preparation of dND-PG conjugated with the Gd(III) complex is shown in Scheme 28.5. From dND-PG-N₃ synthesized according to the process shown in Scheme 28.3, the amine (dND-PG-NH₂) was prepared through reduction of the azido group with triphenylphosphine. The ligand for Gd(III), diethylenetriaminepentaacetic acid (DTPA), was covalently connected through thiourea by reacting isothiocyanate with the amino group on the dND-PG. The resulting dND-PG-DTPA was complexed with Gd(III) ions under mild acidic condition to give dND-PG-Gd(III). dND-PG-Gd(III) showed good solubility in PBS (>4.5 mg/mL) as shown in Figure 28.8a (inset). The PBS solution of dND-PG-Gd(III) was very stable; no precipitates and no significant change in the diameter distribution were observed for more than 3 months. The hydrodynamic diameter of dND-PG-Gd(III) in PBS was measured to be 50.3 ± 14.0 nm, which is slightly larger than that of dND-PG in PBS (49.4 ± 15.6 nm). In addition, no large aggregates were found in the STEM image of dND-PG-Gd(III) (Figure 28.8b).

Since Gd(III) complexes reduce the proton longitudinal relaxation time (T_1) of adjacent water molecules, they provide bright contrast in T_1-weighted MR images.[22] As expected, dND-PG-Gd(III) significantly brightened the MR images of its solutions in contrast to pure water at 3.0 T (Figure 28.9). In particular, the MR images of dND-PG-Gd(III) solutions were brighter than those of Magnevist® solutions at the same Gd concentrations, indicating better contrast efficiency of dND-PG-Gd(III). The longitudinal relaxation rate (r_1) of the nanoparticle and Magnevist® was measured to be 16.7 and 3.5 mM⁻¹ s⁻¹, respectively. dND-PG-Gd(III) has much larger r_1 than Magnevist® because of the restriction in the motion of the Gd(III) complex moiety by connecting to the dND-PG platform.[23,24] The relatively high T_1 relaxivity makes dND-PG-Gd(III) a promising contrast agent for in vivo MRI.

FIGURE 28.8
(a) Size distribution of dND-PG-Gd(III) in PBS. Inset: photograph of dND-PG-Gd(III) well dispersed in PBS (4.5 mg/mL); (b) STEM image of dND-PG-Gd(III).

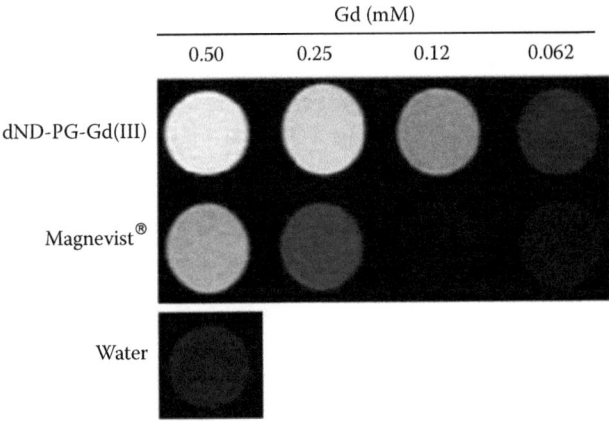

FIGURE 28.9
T_1-weighted MRI images of dND-PG-Gd(III) and Magnevist® at 3.0 T.

28.3.4 Functionalization of ND-PG for Targeted Drug Delivery

As mentioned earlier, the PG coating on ND has shielding effect against cell uptake and the conjugated targeting ligand enables specific cell uptake. Therefore, ND-PG can work as an ideal vehicle for targeted drug delivery, if a drug is loaded on it. For the purpose, we have designed and prepared multifunctional ND-PG conjugated with both RGD peptide and anticancer drugs. Platinum (Pt)-based anticancer drug and doxorubicin (DOX) were loaded on the surface of ND-PG through different synthetic approaches.

28.3.4.1 Functionalization of ND-PG with targeting peptide and Pt-based drug

To immobilize a targeting peptide and a Pt-based anticancer drug, ND50-PG was engineered as shown in Scheme 28.6. ND-PG-N$_3$ was prepared according to the method shown in Scheme 28.3. Carboxyl groups were introduced on ND-PG-N$_3$ as ligands for Pt by nucleophilic ring opening

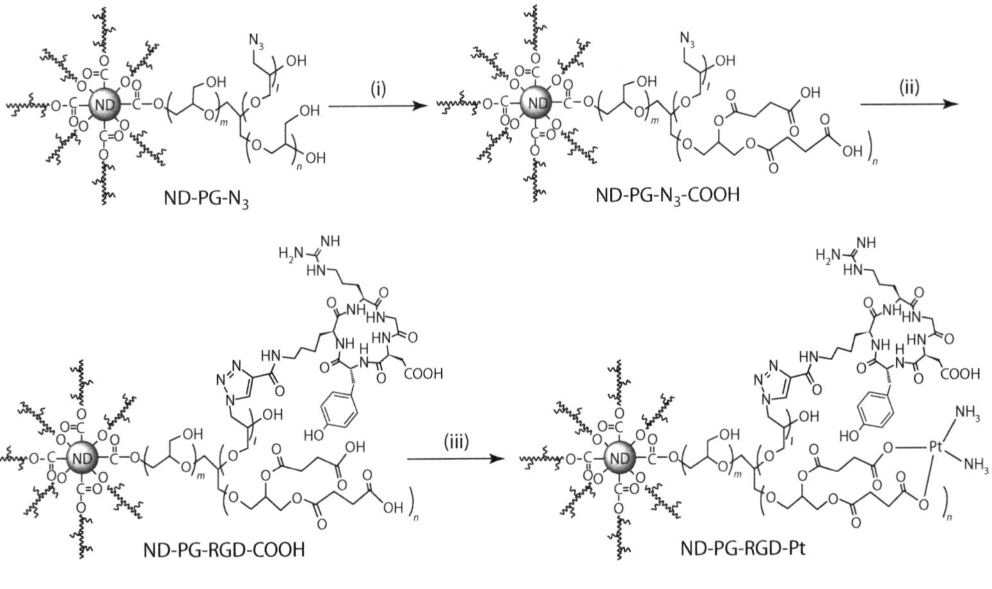

ND-PG-N$_3$ (i)   ND-PG-N$_3$-COOH   (ii)

ND-PG-RGD-COOH   (iii)   ND-PG-RGD-Pt

SCHEME 28.6
Synthesis of ND-PG-RGD-Pt: (i) succinic anhydride, 4-(dimethylamino)pyridine (DMAP), pyridine, 0°C ~ r. t., overnight; (ii) RGD propargyl amide, copper(II) sulfate pentahydrate, sodium ascorbate, r. t., 48 h; and (iii) cisplatin, 0.5 M NaOH, pH 8.0, r. t., 72 h.

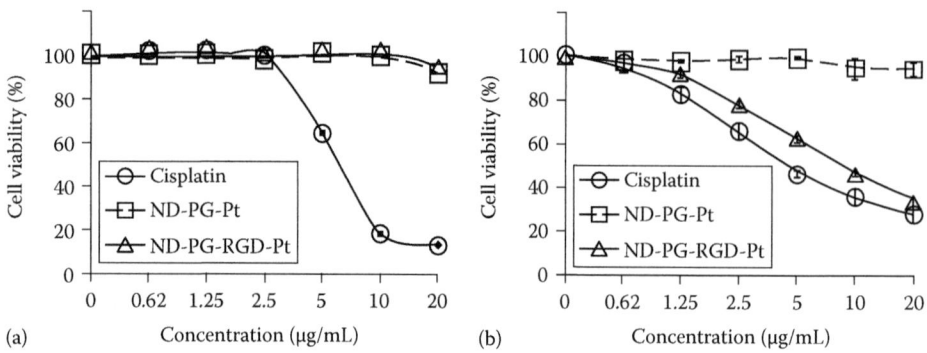

FIGURE 28.10
In vitro therapeutic effect of cisplatin, ND-PG-Pt, and ND-PG-RGD-Pt on (a) HeLa and (b) U87MG cells. Concentrations were normalized to platinum up to 20 μg/mL. Treatment duration was 24 h ($n = 3$).

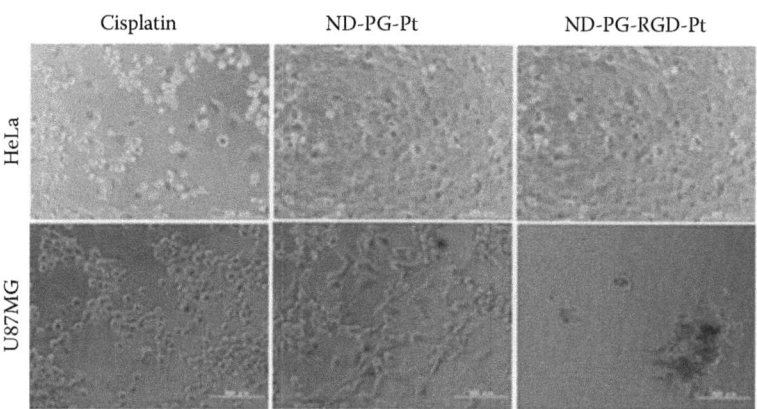

FIGURE 28.11
Microphotographs of HeLa and U87MG cells after incubation with cisplatin, ND-PG-Pt, and ND-PG-RGD-Pt for 24 h. Platinum concentration was normalized to 10 μg/mL.

of succinic anhydride at the hydroxyl groups.[25] The resulting ND-PG-N$_3$-COOH was conjugated with RGD peptide through click chemistry of the azido groups with the propiolic amide of RGD peptide, yielding ND-PG-RGD-COOH.[26] Finally, *cis*-(NH$_3$)$_2$Pt(II) moiety was immobilized on the ND-PG-RGD-COOH through ligand exchange of Cl$^-$ at the cisplatin with –COO– under a slightly basic condition, affording ND-PG-RGD-Pt.[25] ND-PG-Pt without RGD peptide was also prepared as a control from ND-PG-COOH.

ND-PG-Pt and ND-PG-RGD-Pt are readily dispersed in water and Dulbecco's modified Eagle's medium with concentration of >1.0 mg/mL. The content of Pt in ND-PG-Pt and ND-PG-RGD-Pt were up to 17 and 14 wt.%, respectively, based on the measurement of inductively coupled plasma with atomic emission spectroscopy.

To study the therapeutic effect in vitro, separately grown U87MG and HeLa cells were treated with cisplatin, ND-PG-Pt, and ND-PG-RGD-Pt. The results of the WST-8 assay for cell viability are shown in Figure 28.10. Cisplatin is toxic to both U87MG and HeLa, while ND-PG-Pt is almost nontoxic to both cells. ND-PG-RGD-Pt exhibits little toxicity in HeLa but similar toxicity to cisplatin in U87MG. Microphotographs also provided direct visual evidence (Figure 28.11). The nontoxicity of ND-PG-Pt to both cells and the selective toxicity of ND-PG-RGD-Pt to U87MG are well correlated with the cell uptake behavior of fND-PG and fND-PG-RGD mentioned earlier.

28.3.4.2 Functionalization of dND-PG with targeting peptide and doxorubicin

To deliver hydrophobic DOX, dND-PG with extremely high aqueous solubility was used as a vehicle.[14] The surface functionalization of dND-PG is shown in Scheme 28.7. To immobilize DOX on

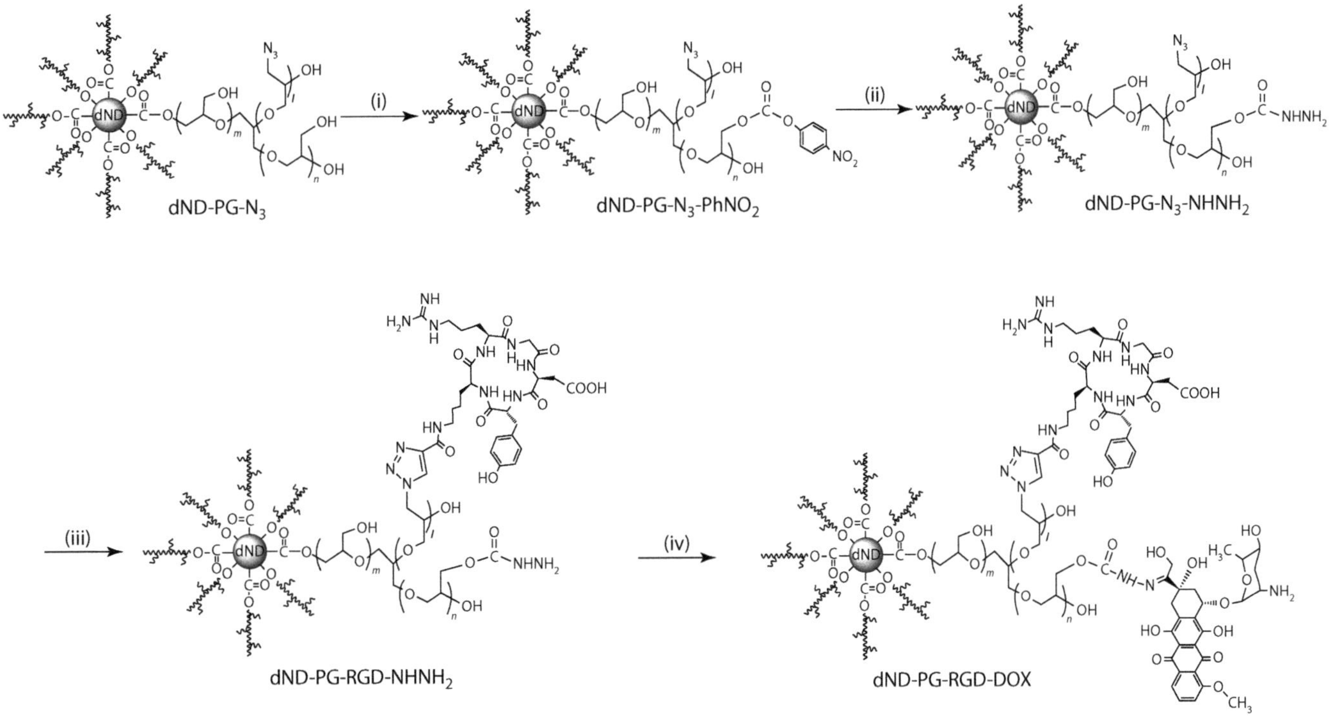

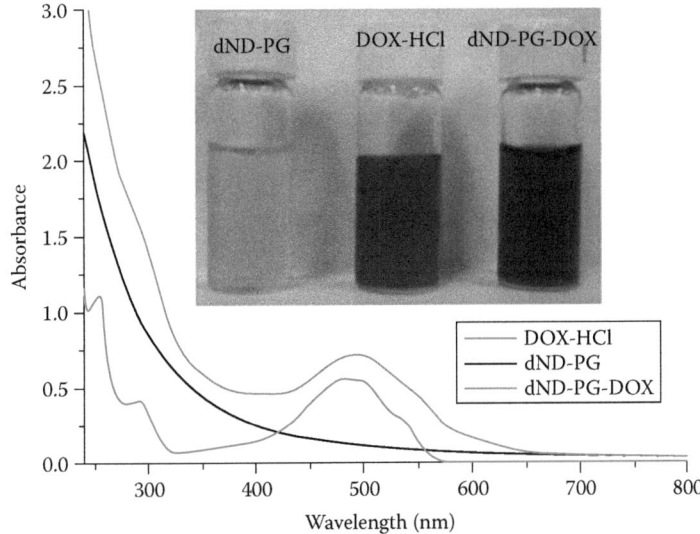

SCHEME 28.7
Synthesis of dND-PG-RGD-DOX: (i) bis(4-nitrophenyl) carbonate, triethylamine, r. t., 24 h; (ii) hydrazine monohydrate, 90°C, overnight; (iii) RGD propargyl amide, copper (II) sulfate pentahydrate, sodium ascorbate, r. t., 48 h; and (iv) doxorubicin hydrochloride, pH 7, 50°C, 24 h.

dND-PG-RGD, hydrazine moiety was introduced through substitution of the p-nitrophenyl group in the p-nitrobenzoate (dND-PG-N_3-PhNO$_2$) to give dND-PG-N_3-NHNH$_2$. After click conjugation of RGD peptide (dND-PG-RGD-NHNH$_2$), DOX was immobilized through acid-labile hydrazone linkage to yield dND-PG-RGD-DOX.[27] dND-PG-DOX without RGD peptide was also prepared as a control from dND-PG-NHNH$_2$. Both dND-PG-DOX and dND-PG-RGD-DOX showed good solubility in water and PBS with concentration of >1.0 mg/mL (Figure 28.12). The content of DOX

FIGURE 28.12
UV-visible absorption of DOX-HCl, dND-PG, and dND-PG-DOX in water. The inset shows photographs of water solutions of DOX-HCl, dND-PG, and dND-PG-DOX.

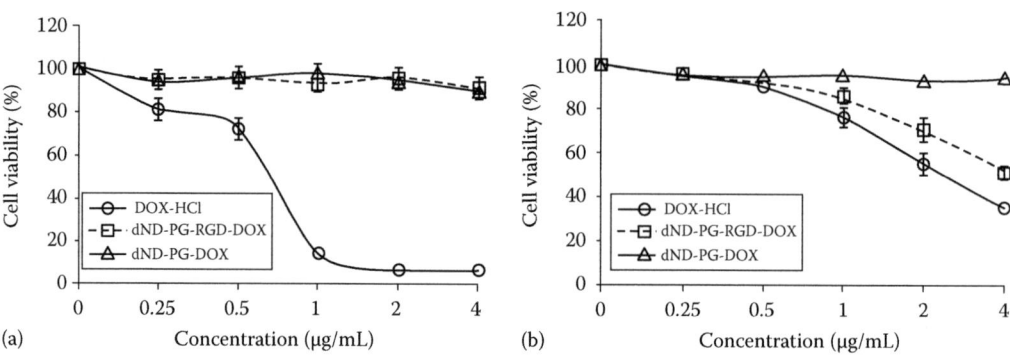

FIGURE 28.13
In vitro therapeutic effect of DOX-HCl, dND-PG, and dND-PG-DOX on (a) U937 macrophage cells and (b) A549 cells. Concentrations were normalized to DOX up to 4 µg/mL. Treatment duration was 48 h ($n = 3$).

in dND-PG-DOX and dND-PG-RGD-DOX was quantified to be 5.6 and 12.2 wt.%, respectively, by measuring their absorbance at 480 nm.

U937 macrophage cells (low expression of $\alpha_v\beta_3$ integrin) and A549 cancer cells (high expression of $\alpha_v\beta_3$ integrin) were employed for in vitro evaluation of the anticancer drugs. As shown in Figure 28.13, DOX showed nonselective toxicity to both A549 cells and U937 macrophages, though U937 macrophages were clearly much more sensitive to DOX than A549 cells. In sharp contrast, dND-PG-DOX appeared almost nontoxic to both types of cells even after 48 h of treatment. dND-PG-RGD-DOX displayed significant toxicity in A549 cells but little toxicity in U937 macrophages. These results provide further concrete evidence for stealth effect of dND-PG against nonspecific cell uptake as well as the targeting efficacy of RGD peptide.

28.4 CONCLUSIONS AND PERSPECTIVES

In summary, we have developed new surface functionalization methodology of ND including PG grafting and further derivatization in view of various biomedical applications. The grafted PG demonstrated several advantages over frequently used PEG, such as better hydrophilicity and ease of chemical derivatization. A variety of functional moieties have been immobilized on ND-PG, enabling cellular targeting, imaging, and therapy. The in vivo application of the multifunctional ND is ongoing in our laboratory.

It is worth mentioning that the strategy of PG grafting and further functionalization is applied not only to ND but also to other nanocarbon materials (CNT and graphene) as well as metal oxide nanoparticles (iron oxide and zinc oxide),[26,28] indicating that this is a general and practical method for surface functionalization of nanomaterials. We expect that this methodology can be applied to a wide variety of nanomaterials to provide them with sufficient properties for their biomedical applications, such as aqueous dispersibility and stability in a dispersion, biocompatibility, and stealth effect.

REFERENCES

1. Mochalin, V. N.; Shenderova, O.; Ho, D.; Gogotsi, Y. *Nat. Nanotechnol.* **2012**, *7*, 11.
2. Mohan, N.; Chen, C. S.; Hsieh, H. H.; Wu, Y. C.; Chang, H. C. *Nano Lett.* **2010**, *10*, 3692.
3. Chow, E. K.; Zhang, X.-Q.; Chen, M.; Lam, R.; Robinson, E.; Huang, H.; Schaffer, D.; Osawa, E.; Goga, A.; Ho, D. *Sci. Transl. Med.* **2011**, *3*, 73ra21.
4. Krueger, A.; Lang, D. *Adv. Funct. Mater.* **2012**, *22*, 890.

5. Tzeng, Y.-K.; Faklaris, O.; Chang, B.-M.; Kuo, Y.; Hsu, J.-H.; Chang, H.-C. *Angew. Chem. Int. Ed.* **2011**, *50*, 2262.
6. Bumb, A.; Sarkar, S. K.; Billington, N.; Brechbiel, M. W.; Neuman, K. C. *J. Am. Chem. Soc.* **2013**, *135*, 7815.
7. Takimoto, T.; Chano, T.; Shimizu, S.; Okabe, H.; Ito, M.; Morita, M.; Kimura, T.; Inubushi, T.; Komatsu, N. *Chem. Mater.* **2010**, *22*, 3462.
8. Zhao, L.; Takimoto, T.; Ito, M.; Kitagawa, N.; Kimura, T.; Komatsu, N. *Angew. Chem. Int. Ed.* **2011**, *50*, 1388.
9. Thomas, A.; Müller, S. S.; Frey, H. *Biomacromolecules* **2014**, *15*, 1935.
10. Sisson, A. L.; Steinhilber, D.; Rossow, T.; Welker, P.; Licha, K.; Haag, R. *Angew. Chem. Int. Ed.* **2009**, *48*, 7540.
11. Kainthan, R. K.; Zou, Y.; Chiao, M.; Kizhakkedathu, J. N. *Langmuir* **2008**, *24*, 4907.
12. Zhao, L.; Xu, Y.-H.; Qin, H.; Abe, S.; Akasaka, T.; Chano, T.; Watari, F.; Kimura, T.; Komatsu, N.; Chen, X. *Adv. Funct. Mater.* **2014**, *24*, 5349.
13. Zhao, L.; Nakae, Y.; Qin, H.; Ito, T.; Kimura, T.; Kojima, H.; Chan, L.; Komatsu, N. *Beilstein J. Org. Chem.* **2014**, *10*, 707.
14. Zhao, L.; Xu, Y.-H.; Akasaka, T.; Abe, S.; Komatsu, N.; Watari, F.; Chen, X. *Biomaterials* **2014**, *35*, 5393.
15. Zhao, L.; Shiino, A.; Qin, H.; Kimura, T.; Komatsu, N. *J. Nanosci. Nanotechnol.*, **2015,** *15*, 1076.
16. Yu, S. J.; Kang, M. W.; Chang, H. C.; Chen, K. M.; Yu, Y. C. *J. Am. Chem. Soc.* **2005**, *127*, 17604.
17. Wu, T.-J.; Tzeng, Y.-K.; Chang, W.-W.; Cheng, C.-A.; Kuo, Y.; Chien, C.-H.; Chang, H.-C.; Yu, J. *Nat. Nanotechnol.* **2013**, *8*, 682.
18. Welsher, K.; Liu, Z.; Sherlock, S. P.; Robinson, J. T.; Chen, Z.; Daranciang, D.; Dai, H. *Nat. Nanotechnol.* **2009**, *4*, 773.
19. Meinhardt, T.; Lang, D.; Dill, H.; Krueger, A. *Adv. Funct. Mater.* **2011**, *21*, 494.
20. Faklaris, O.; Joshi, V.; Irinopoulou, T.; Tauc, P.; Sennour, M.; Girard, H.; Gesset, C. et al. *ACS Nano* **2009**, *3*, 3955.
21. Maeda-Mamiya, R.; Noiri, E.; Isobe, H.; Nakanishi, W.; Okamoto, K.; Doi, K.; Sugaya, T.; Izumi, T.; Homma, T.; Nakamura, E. *Proc. Natl. Acad. Sci. USA.* **2010**, *107*, 5339.
22. Caravan, P.; Ellison, J. J.; McMurry, T. J.; Lauffer, R. B. *Chem. Rev.* **1999**, *99*, 2293.
23. Zhou, Z.; Lu, Z.-R. *Wiley Interdiscip. Rev. Nanomed. Nanobiotechnol.* **2013**, *5*, 1.
24. Manus, L. M.; Mastarone, D. J.; Waters, E. A.; Zhang, X. Q.; Schultz-Sikma, E. A.; Macrenaris, K. W.; Ho, D.; Meade, T. J. *Nano Lett.* **2010**, *10*, 484.
25. Ye, L.; Letchford, K.; Heller, M.; Liggins, R.; Guan, D.; Kizhakkedathu, J. N.; Brooks, D. E.; Jackson, J. K.; Burt, H. M. *Biomacromolecules* **2010**, *12*, 145.
26. Zhao, L.; Chano, T.; Morikawa, S.; Saito, Y.; Shiino, A.; Shimizu, S.; Maeda, T. et al. *Adv. Funct. Mater.* **2012**, *22*, 5107.
27. Hu, X.; Liu, S.; Huang, Y.; Chen, X.; Jing, X. *Biomacromolecules* **2010**, *11*, 2094.
28. Zhao, L.; Takimoto, T.; Kimura, T.; Komatsu, N. *J. Indian Chem. Soc.* **2011**, *88*, 1787.

CHAPTER 29

C O N T E N T S

Applications of Functionalized Carbon-Based Nanomaterials in Analytical Chemistry

29

Victoria F. Samanidou and Antigoni E. Koletti

29.1 INTRODUCTION

Research in nanomaterials is currently an area of intense scientific interest, due to the wide range of potential applications in various fields, such as analytical chemistry and biochemistry. These materials not only have unique thermal, chemical, and mechanical properties, but they also have a high surface-to-volume ratio. However, besides their unique and exceptional properties, in some cases it is preferable to functionalize them for improving their characteristics. The functionalization of nanomaterials can be achieved with covalent and noncovalent modification techniques. Different carbon-based nanomaterials used for chemical analysis are shown in Figure 29.1 (Zhang et al., 2013).

Graphenes are constituted of sp^2-bond carbon atoms packed in a honeycomb crystal lattice. They can be described as a one-atom-thick layer of graphite and they are the basic structural element of other carbon allotropes such as graphite, carbon nanotubes, and fullerenes. Functionalization processes prevent the accumulation of graphenes and form stable dispersions. As a result, graphenes can be used in solvent-assisted techniques (Kuila et al., 2012).

Another form of nanomaterial are nanotubes. Carbon-based nanotubes (CNTs) are macromolecules with a cylindrical shape, a few nanometers radius, and a length sometimes longer than 20 cm (Balasubramanian and Burghard, 2005). They can be classified based on the layers of graphene sheets, into single-walled carbon nanotubes (SWCNT), double-walled carbon nanotubes (DWCNT), and multiwalled carbon nanotubes (MWCNT) (Chen et al., 2011). MWCNTs were discovered in 1991 and SWCNTs in 1993. The main advantages of these structures are the high length-to-diameter ratio, the exceptional electrochemical properties, and the low density (Balasubramanian and Burghard, 2005). On the other hand, their insolubility in many common solvents limited their use in various applications. This disadvantage could be overcome by modifying CNTs with suitable functional groups (Huang et al., 2013). The modification of CNTs surface can be achieved through covalent and noncovalent approaches. In noncovalent approaches, different kinds of molecules are linked to CNTs surface through hydrogen bounding, π–π sacking, electrostatic forces, van der Waals forces, or hydrophobic/hydrophobic forces. Covalent modification occurs either with direct binding of functional groups on CNTs surface or with the use of carboxylic groups (Huang et al., 2013; Sun et al., 2002). The final structure and properties are dependent on the method of modification (Sun et al., 2002).

Due to their thermal and pH stability and their high absorption capacity, CNTs are used in a wide range of application in analytical science, including solid-phase extraction and microextraction, stationary-phase chromatographic techniques, and as matrices in matrix-assisted laser

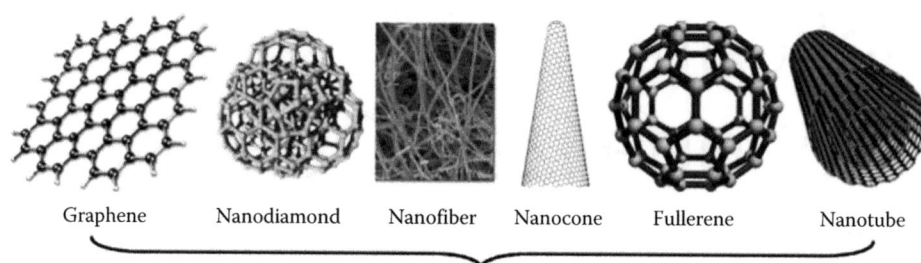

Graphene   Nanodiamond   Nanofiber   Nanocone   Fullerene   Nanotube

Carbon-based nanomaterials in sample preparation

FIGURE 29.1
Different carbon-based nanomaterials used in chemical analyses. (Reprinted from *Analytica Chimica Acta*, 784, Zhang, B.-T., Zheng, X., Li, H.-F., Lin, J.-M., Application of carbon-based nanomaterials in sample preparation: A review, 1–17, Copyright 2013, with permission from Elsevier.)

desorption ionization techniques. Moreover, CNTs can be functionalized easily, which enhances their physicochemical properties and makes them suitable candidates for numerous applications (Samanidou and Karageorgou, 2012).

Fullerenes are another form of carbon-based nanomaterials, which were discovered in 1985 (Kroto et al., 1985). They are closed-cage structures in which carbon atoms are connected in pentagonal and hexagonal rings. The most studied fullerene is C_{60}, which has high symmetry. The functionalized forms of these carbon allotropes increase their solubility in organic solvents and enhance their selectivity (Zhang et al., 2013).

Functionalized graphene, fullerene, carbon nanotubes, and other carbon-based nanostructures have a wide spectrum of applications, including in biochemistry, in drug delivery systems, as biosensors, and many others. This chapter focuses on their application in analytical chemistry. Modified carbon nanostructures have been employed as sorbent materials in extraction techniques, stationary-phase separation techniques, and as modifiers on electrode surfaces. We report the developments in the application of functionalized carbon-based nanomaterials in analytical chemistry during the last decade.

In recent years, many effective methods have been developed for the functionalization of carbon-based nanomaterials. Mainly these can be classified into two wide categories including covalent modification and noncovalent processes. In addition, CNTs have also functionalized with chemical or hydromechanical methods (Chen et al., 2011). In noncovalent processes, functionalization groups are anchored on CNTs surface under various noncovalent interactions such as van der Waals forces, hydrogen bonds, electrostatic, and π–π stacking interactions. Usually, this modification was achieved with harsh oxidation. CNTs were treated with strong acids under refluxing and thus defects are generated on the surface of nanotubes, in which functional groups could be absorbed (Chen et al., 2011; Meng et al., 2009). The main advantage of noncovalent approaches is that the electronic properties of nanotubes are retained due to the preservation of their sp^2 structure (Chen et al., 2011).

The possibilities for the functionalization of SWCNTs (1) π–π interaction, (2) defect group functionalization, and (3) noncovalent functionalization with polymers are shown in Figure 29.2 (Choudhary and Gupta, 2011).

On the other hand, in covalent approaches, functional substitutes are linked to CNTs through covalent linkages. There are two main categories in covalent functionalization. In the first approach, the substitute is attached directly on the CNT's sidewall, which is achieved through the hybridization to sp^3 and consequent loss of conjugation. In the second approach, functional groups are attached indirectly through carboxylic groups. However, covalent process may affect the CNTs' structure, changing their properties (Meng et al., 2009).

Graphenes have also been functionalized with covalent and noncovalent techniques, enhancing their solubility and averting their agglomeration. Regardless of the functionalization approach used, graphenes were first oxidized and then reduced. Even though modification processes improve their solubility, graphenes lose their electrical conductivity and surface area (Kuila et al., 2012).

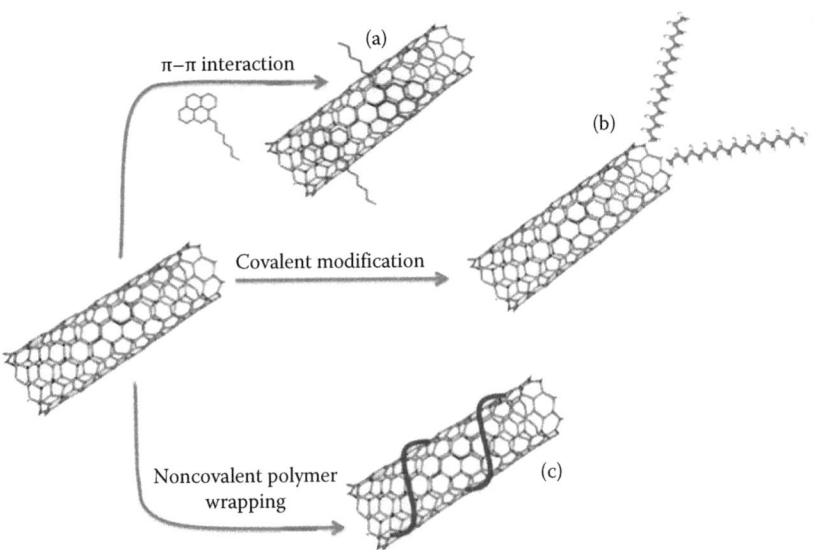

FIGURE 29.2
Possibilities for the functionalization of SWCNTs (a) π–π interaction, (b) defect group functionalization, and (c) noncovalent functionalization with polymers. (Reprinted from Choudhary, V. and Gupta, A., Polymer/carbon nanotube nanocomposites, in *Carbon Nanotubes-Polymer Nanocomposites*, S. Yellampalli (Ed.). Copyright 2011. With permission.)

29.2 SAMPLE PREPARATION

In analytical processes, sample preparation is one of the most crucial steps. Preparation techniques might help to isolate the target analyte from complex matrices or to enrich the final sample, if the initial concentration is low. Carbon-based nanostructures not only have unique physicochemical properties but also a high surface-to-volume ratio. Moreover, their structure enhances the interaction with various organic molecules through noncovalent forces. All these properties render them good candidates for fabricating sorbent materials. Conventional carbon-based nanomaterials and their functionalized forms have been used as sorbent materials in solid-phase extraction and microextraction and as matrices in matrix-assisted laser desorption ionization techniques. Modified carbon nanotubes have extensively been studied as sorbent materials. Their large surface area combined with the functional groups makes them great capacitors for various types of organic and inorganic molecules (Parodi et al., 2011).

Functionalized carbon-based nanotubes have been investigated as sorbent materials for the extraction of metal ions from different matrices. For solid-phase extraction of cadmium from solid environmental samples, they were compared to unmodified CNT, oxidized CNT, and L-alanine CNT. Oxidized CNT exhibited the highest absorption capacity, which was 130 μmol g^{-1}. For the determination of cadmium, oxidized CNT coupled with on-line ultrasonic nebulization inductively coupled plasma optical emission spectrometry was used. The detection limit was 1.03 μg L^{-1} and the limit of quantification was 3.42 μg L^{-1}. Moreover, the column can be regenerated after use, without decreasing its capacity (Parodi et al., 2011). A solid-phase extraction method for Cu^{2+} and Ni^{2+}, in which MWCNT was functionalized with hexahistidine-target protein was used as a sorbent material, was developed by Liu et al. The experiments were carried out in on-line microcolumn, which was packed with modified sorbent material with a pH range of 3–4.5 for copper cations and 4.5–6 for nickel cations. Absorbed ions were recovered with 0.2 mol L^{-1} imidazole-hydrogen chloride solution and they were determined using on-line flame atomic absorption spectrometry. This modification enhances the tolerable concentration of Cu^{2+} and Ni^{2+} up to 20,000 and 1,800, compared with unfunctionalized MWCNTs, respectively (Liu et al., 2009b). The simultaneous preconcentration of Cr(III), Fe(III), and Pb(II) from biological and natural water using MWCNTs modified with ethylenediamine as sorbent material has been studied. For the synthesis of ethylenediamin-MWCNTs (EDA-MWCNTs), MWCNTs were carboxylated with HNO_3 followed by suspension in EDA for obtaining the final functionalized nanotubes. According to the authors, these sorbent materials have an excellent selectivity of Cr(III), Fe(III), and Pb(II) and can be prepared easily and rapidly. For the detection of metal ions, inductively coupled plasma optical emission spectrometry was used and their limits of detection were 0.24, 0.19, and 0.33 ng mL^{-1} for Cr(III), Fe(III), and Pb(II), respectively (Zang et al., 2009).

A new sorbent has been synthesized using gallium (III) ion and MWCNTs by surface imprinting techniques. First, MWCNTs were functionalized with vinyl and phenyl groups, then the Ga(III) ion complex of 8-hydroxyquinoline was polymerized selectively on the surface of CNTs. Finally, the template (8-hydroxyquinoline) was removed with ethanol and acetic acid. The synthesized material was applied for the enrichment of gallium ion of fly ash lixivium and for sample detection using flame atomic absorption spectrometry. The detection limit was 3.03 ng mL^{-1} and the linearity ranged from 1.5 to 150 ng mL^{-1} (Zhang et al., 2010).

A polypyrrole-graphene composite was fabricated for coating a fiber, which was used in solid-phase microextraction of five phenols, namely, phenol, o-cresol, m-cresol, 2,4-dichlorophenol, and p-bromophenol, from water. According to the authors, this modified fiber offers better extraction efficiency compared to fibers coated with polypyrrole or polypyrrole-graphene oxide. Furthermore, polypyrrole-graphene fiber has better or comparable extraction efficiency than various commercial fibers. For preparing the coated fiber, pyrrole and graphene were polymerized on a stainless steel wire. It is important that the fiber is chemically and mechanically stable. It can be used more than 50 times without losing its extraction ability, but strong acid or base pH decreases its ability, so the pH value must be adjusted within the range 6–9. The determination of phenols was carried out by gas chromatography with limits of detection at 45, 0.9, 0.3, 4.1, and 3.4 µg L^{-1} for phenol, o-cresol, m-cresol, 2,4-dichlorophenol, and p-bromophenol, respectively (Zou et al., 2011).

An immobilized carboxylated CNT minicolumn was fabricated for preconcentrating nonsteroidal anti-inflammatory drugs (NSAIDs) in urine. Functionalized CNTs incorporated with porous glass were used for fabricating the sorption material. The authors report that the immobilization boosts the interaction between CNTs and the target analyte, because of the orientation of CNTs on glass surface and the decrease in the interaction between the CNTs. After solid-phase extraction, samples (keteprofen, tolmetin, and indomethacine) were analyzed by capillary electrophoresis coupled with mass spectrometry and their detection limits were determined at 1.9, 1.6, and 2.6 µg L^{-1} for keteprofen, tolmetin, and indomethacine, respectively (Suárez et al., 2007a).

Nanocellulose, carboxylated CNTs, and ionic liquids were used for direct immersion single drop microextraction. This technique was used for the extraction and preconcentration of the mutagenic 2-amino-3,8-dimethylimidazo[4,5-f]quinoxaline from fried food. The combination of these materials offers a stable droplet with selectivity in the target heterocyclic amine. The detection of the target compound was carried out using a capillary electrophoresis-diode array detector system. The linear range of concentration was 0.1–10 mg L^{-1} and the detection limit was 0.29 mg L^{-1} (Ruiz-Palomero et al., 2014).

Magnetic MWCNTs functionalized with polyethylene glycol were used for the isolation of puerarin from rat plasma followed by high-performance liquid chromatography (HPLC) for the determination of the target analyte. Magnetite immobilized on the surface of nanotubes improved their solubility in water, which is important for ensuring sufficient contact between the sorbent material and the target compound. The magnetic nanotube was modified noncovalently with polyethylene glycol. The advantages of these sorbents are the large surface area and their capability to interact intensively. Moreover, this material was easily isolated from the matrix, by means of a magnet, and puerarin was recovered using acetonitrile. The limit of detection was determined at 0.005 µg mL^{-1} (Yu et al., 2014).

A new absorbent for the selective isolation of hemoglobin from whole human blood has been fabricated by immobilizing graphene oxide sheets on the surface of SiO_2 nanoparticles. The absorption was achieved by adding 3 mg of nanocomposite in 1 mL solution at pH 7, and for the recovery of hemoglobin, Tris-HCL buffer was used at pH 8.9. Hemoglobin in real samples has been determined using sodium dodecyl sulfate polyacrylamide gel electrophoresis (Liu et al., 2011a).

Graphene oxides coated with hydrophilic polyamide organic membrane have been used as sorbent material in micro-solid-phase extraction. This micro-SPE method was developed for the enrichment of parabens from water and vinegar samples followed by gas chromatography mass spectrometry. Graphene oxide was chosen as the sorbent material, because of its high extraction efficiency of parabens, as a result of the large specific area and the interactions with target compounds. The linear range was determined at 0.1–100 ng mL^{-1} for water samples and 0.5–100 ng mL^{-1} for vinegar samples, and the detection limits were in the range of 0.005–0.01 ng mL^{-1} for water sample and 0.01–0.05 ng mL^{-1} for vinegar samples (Wang et al., 2014).

Oxidized carbon nanofibers were used as sorbent material in a micro-column in which trace rare earth elements were pre-concentrated from biological samples prior to their determination by inductively coupled plasma mass spectrometry (ICP-MS). The modified absorbent exhibited high adsorption capacity against lanthanum (18.1 mg g^{-1}), cerium (19.3 mg g^{-1}), samarium (23.6 mg g^{-1}), europium (17.6 mg g^{-1}), dysprosium (22.3 mg g^{-1}), and yttrium (19.5 mg g^{-1}). An eluent solution was used, 2 mL of 0.5 mol L^{-1} nitric acid, which efficiently eliminated the carry over effect. The limits of detection were 1, 1.2, 0.4, 0.6, 0.2, and 0.6 pg mL^{-1} for La, Ce, Sm, Eu, Dy, and Y, respectively (Chen et al., 2007).

MWCNTs functionalized covalently with poly(ethylene glycol) have been developed to be used as coating material for an SPME fiber. This modified fiber was used in headspace SPME procedure for selective extraction of benzene, toluene, ethylbenzene, and o-xylene for water samples (tap water, mineral water, well water, and wastewater). Highly porous surface, preparations simplicity, stability in high temperature, and reuse more than 200 extractions are the main advantages of PEG-g-MWCNTs sol-gel fibers. The extractions were analyzed by gas chromatography with a flame ionization detector. The methods' limits of detection ranged from 0.6 to 3 pg mL^{-1} and the limits of quantification from 2 to 10 pg mL^{-1} (Sarafraz-Yazdi et al., 2011). In addition, a polypropylene hollow fiber modified with MWCNTs has been developed for the preconcentration of benzene, toluene, ethylbenzene, and xylenes from wastewater and human hair. A sol-gel solution containing acidified MWCNTs was injected into hollow fiber segments (2 cm). The fiber was immersed into aqueous sample under stirring for 35 min and then transferred to methanol under ultrasonic agitation for recovering analytes. The determination of target analytes was carried out by gas chromatography flame ionization detector system. The limits of detection were in the range of 0.49–0.7 ng L^{-1} (Es'haghi et al., 2011).

Functionalized carbon allotropes, such as fullerenes and CNTs, have also been applied as matrix materials in matrix-assisted laser desorption/ionization mass spectrometry (MALDI-MS). Functionalized CNTs have been fabricated as matrix for the detection of low-weight metabolites with matrix-assisted laser desorption/ionization mass spectrometry. Carboxylic MWCNTs were coated with polyaniline, which enhanced their dispersibility in water, which in turn is important in the analysis of water-soluble molecules. Six metabolites were determined by MALDI-MS and the results showed that CNTs coated with polyaniloine could be used as matrix material for the analysis of low-weight metabolites (Meng et al., 2011). Shi and coworkers have also developed magnetic graphene (Fe$_3$O$_4$-graphene) as matrix of MALDI-TOF-MS for the detection of small molecules. As target analytes, active compounds of traditional Chinese medicine (berberine hydrochloride, curcumin, wogonin, scutellarin, luteoloside, and chlorogenic acid) and nicotine metabolites (cotinine, cotinine nitrogen oxides, and nicotine nitrogen oxides), which were detected at both ion modes, have been chosen. Moreover, the large surface area combined with the hydrophobicity of graphenes favors the interactions with the benzene ring; consequently, they can be used as a pre-enrichment material (Shi et al., 2012). Graphene and graphene oxides have also been investigated as matrix material in MALDI-TOF-MS for the detection and enrichment of long-chain fatty acids. The target compounds were n-dodecanoic acid (C12), n-tetradecanoic acid (C14), n-hexadecanoic acid (C16), n-octadecanoic acid (C18), and n-eicosanoic acid (C20). The applicability of this method was tested in biological samples, serum, and urine metabolites, from which the precipitated proteins and other solid particles had been removed prior to detection. The advantages of this method are simple preparation, enhancement of enrichment efficiency and the detection limit, as well as minimization of background ion interferences (Liu et al., 2011b). The present applications of functionalized carbon-based nanomaterials for sample preparation are presented in Table 29.1.

TABLE 29.1 Applications of Functionalized Carbon-Based Nanostructures in Preparation Techniques

Nanostructure	Analytes	Matrix	Preparation Technique	Analytical Technique	LOD	Reference
Oxidized-CNTs	Cd	Solid environmental samples	SPE	On-line USN-ICPOES	1.03 µg L^{-1}	Parodi et al. (2011)
6His-tagget protein/MWCNT	Cu^{2+}, Ni^{2+}	CRF: GBW07405 (soils), GBW08513 (tea leaf), GBW09101 (human hair), and GBW10016 (tea)	On-line SPE	On-line FAAS	0.31 (Cu^{2+}) 0.63 gL^{-1} (Ni^{2+})	Liu et al. (2009b)
EDA-MWCNTs	Cr(III), Fe(III), Pb(II)	Biological and natural water	SPE	ICPOES	0.19–0.33 ng mL^{-1}	Zang et al. (2009)
Ga(III) imprinted MWCNTs	Ga	Fly ash	SPE	FAAS	3.03 ng mL^{-1}	Zhang et al. (2010)
Polypyrrole-graphene	Five phenols	Natural water	SPME	GC-FID	0.3–45 µg L^{-1}	Zou et al. (2011)
c-SWCNT immobilized on glass particles	NSAIDs	Urine	SPE	CE-MS	1.6–2.6 µg L^{-1}	Suárez et al. (2007a)
Nanocellulose, c-CNTs, and ionic liquids	Heterocyclic amine	Fried food	SDME	CE-DAD	0.29 mg L^{-1}	Ruiz-Palomero et al. (2014)
PEGlated-MWCNTs@Fe$_3$O$_4$	Puerarin	Rat plasma	SPE	HPLC-DAD	0.005 µg mL^{-1}	Yu et al. (2014)
GO immobilized on SiO$_2$	Hemoglobin	Human blood	SPE	Gel electrophoresis		Liu et al. (2011a)
GO coated with hydrophilic polyamide membrane	Parabens	Water and vinegar	µ-SPE	GC-MS	0.005–0.01 ng mL^{-1} (water), 0.01–0.05 ng mL^{-1} (vinegar)	Wang et al. (2014)
Oxidized carbon nanofibers	Trace rare elements	CRM: GBW 07601 (human hair)	µ-SPE	ICP-MS	0.2–1.2 pg mL^{-1}	Chen et al. (2007)
PEG-g-MWCNTs	BETX	Water	SPME	GC-FID	0.6–3 pg mL^{-1}	Sarafraz-Yazdi et al. (2011)
PP-MWCNTs	BETX	Water and human hair	SPME	GC-FID	0.49–0.7 ng L^{-1}	Es'haghi et al. (2011)
Polyaniline-MWCNTs	Six low-weight metabolites		MALDI	MS		Meng et al. (2011)
Fe$_3$O$_4$G	Active pharmaceutical compound and nicotine metabolites	Water	MALDI	TOF-MS		Shi et al. (2012)
G and GO	Long-chain fatty acids	Serum and urine metabolites	MALDI	TOF-MS	16–380 fmol (GO) and 74–715 fmol (G)	Liu et al. (2011b)

29.3 SEPARATION

Chromatographic techniques are widely used since they offer many advantages, the most important of which is the effective separation of complex mixtures. The heart of the separation system is the analytical column in which the separation is carried out. The kind, the structure, and the size of packing material affect the absorption of analytes in the stationary phase, therefore they are crucial parameters for separation. Functionalized forms of graphene allotropes have been developed as stationary phase in gas and liquid chromatography and in capillary electrochromatography. The kind of the functional group linked with the carbon nanostructure strongly affects its interaction with target compounds, and as a result modification is an important step for improving the resolution (Merli et al., 2010).

Functionalized carbon nanotubes have been studied as stationary phase in gas chromatography. For separating mixtures of aliphatic C1–C5 alcohols and esters, functionalized MWCNTs have been developed. The derivatization of MWCNTs with 22′-(ethylenedioxy)-diethylamine makes them a suitable stationary phase for separating these isomeric alcohols, esters, and other by-products of alcoholic fermentation with good resolution and perfect peak shape (Merli et al., 2010). Speltini et al. (2010) developed stationary phases for a gas chromatographic column based on MWCNTs. Derivatized MWCNTs were packed in glass columns that were applied for the separation of alkanes and aromatic compounds. The results showed that MWCNTs-CO_2H is a suitable stationary phase for separating linear and branched C3–C5 alkanes and MWCNTs-$CONH_2$ for aromatic hydrocarbons. However, temperatures more than 200°C and pH value affect the functional groups, which could be a limitation for specific separation processes (Speltini et al., 2010).

Shell-core beats were developed as stationary phase for ion chromatography. The core of the beats consists of polystyrene-divinylbenzene and the shell of functionalized MWCNTs. Acidified MWCNTs are bound with a branching polymer, which consists of methylamine and 1,4-butanediol diglycidyl ether, for obtaining quaternary ammonium groups and a positive charge. The core beats were sulfonated for charging negatively, as a result, there were electrostatic interactions between opposite charged components. The new material was applied as stationary phase in anion chromatography for the separation of inorganic anions, namely, fluoride, chloride, nitrite, bromide, and nitrate. The packing material was stable in high pH values and exhibited excellent permeability (Huang et al., 2013).

C_{60} fullerenes coated with surfactants such as sodium dodecyl sulfate (SDS) were investigated as pseudostationary phase, in electrokinetic chromatography, for separating β-lactams antibiotics, NSAIDs, and amphenicols. The surfactant coat amplifies the interaction between fullerenes and target compounds and enhances the stability of baseline. Moreover, this pseudostationary phase offers better resolution, in optimum conditions, compared with MWCNTs and SWCNTs coated with surfactant. The analysis of pharmaceuticals and urine samples shows that fullerenes-SDS can achieve separation of target analytes without matrix interferences (Moliner-Martínez et al., 2007). Another modification of the pseudostationary phase in electrokinetic chromatography using functionalized MWCNTs was described by Cao et al. (2011). They developed MWCNTs coated with ionic liquids for the determination of flavonoids (calycosin-7-o-β-D-glucoside, formononetin, dihydroquercetin, ononin), phenolic acids (rosmarinic acid, danshensu, salvianolic acid B, protocatechuic acid), and saponinis (ginsenoside Rg_1, ginsenoside Rb_1) simultaneously. The modified surfaces of the nanotubes interact selectively with target compounds resulting in improved separation. The optimum conditions established that offered a separation within 11 min are as follows: 28 kV separation voltages, 10 mM borate buffer (pH 9.0) with 100 mM surfactant (SDS), 6% propanol, and 4 g mL^{-1} ILs-MWNTs. Detection limits ranged between 1.01 and 76.32 μg mL^{-1} (Cao et al., 2011). An alternative pseudostationary phase in which SWCNTs were coated with surfactant was proposed by Suárez et al. (2007b). The surfactant coat enhanced the solubility of nanotubes in water without affecting their absorption capabilities. This modified pseudostationary phase applied in micellar nanoparticle dispersed electrokinetic chromatography (MiNDEKC) equipped with diode array detector for the separation of analytes used to study the effect of SWNTs on electrochromatographic separation of chlorophenols (2,6-dichlorophenol, 2,3-dichlorophenol, 2,5-dichlorophenol, 3,4-dichlorophenol,

3,5-dichlorophenol, 2,4,5-trichlorophenol, 2,3,6-trichlorophenol, 2,3,5-trichlorophenol, and pentachlorophenol), NSAIDs (tolmetin, indomethacin, acetylsalicylic acid, fenbufen, ibuprofen, ketoprofen), and penicillins (penicillin-G sodium salt, ampicillin, oxacillin, penicillin V, amoxicillin, cloxacillin). Even though the increase of CNTs concentration improves the separation, because they interact with aromatic groups, an increase over a maximum amount of CNTs was not compatible with the detector. To surpass this limitation, a partial filling was placed just before sampling (Suárez et al., 2007b). Alternatively, in capillary electrochromatography carboxylic MWCNTs were used as stationary phase. In this case, carboxylic MWCNTs were immobilized in fused silica capillary, enhancing the resolution. The modified capillary has been applied for the separation of three pharmaceuticals compounds, namely, amoxicillin, ketoprofen, and chloramphenicol. The proposed modification obtained good reproducibility, improving the separation, and the capillary could be used more than 6 months if it regenerated properly (Sombra et al., 2008).

Graphene oxides have also been used as stationary phase in open-tubular capillary electrochromatography. Fused silica capillary was modified by introducing amino groups, after which graphene oxides passed through the capillary and immobilized onto the inner surface. The applicability of the new stationary phase was studied with NSAIDs (indoprofen, ketoprofen, suprofen, fenoprofen, flurbiprofen, and naproxen). The determination was achieved at a voltage of 25 kV with 50 mmol L^{-1} borate buffer (pH 10) in 40 cm effective capillary length with the diode array detector set at 214 nm. Modified stationary phase possesses excellent resolution and the results were reproducible with limits of detection ranging at 1–1.8 μg mL^{-1} (Wang et al., 2013).

An effective separation of purine and pyrimidine bases was achieved by using functionalized MWCNTs in capillary zone electrophoresis. For improving the separation of target compounds, namely, adenine, hypoxanthine, 8-azaadenine, thymine, cytosine, uracil, and guanine, researchers added carboxylic MWCNTs in running solution buffer. The optimum separation voltage defined was at +8 kV and the buffer solution was 23 mM tetraborate with 8×10^5 g mL^{-1} functionalized nanotubes. Under these conditions, the separation was completed within 16 min. This method was applied for the determination of analytes in yeast RNA (Xiong et al., 2006).

SWCNTs were bonded with 3-amino-propyl silica gel for studying their potential use as stationary phase in HPLC. The developed material was packed in HPLC columns and applied for the separation of seven polycyclic aromatic hydrocarbons (benzene, diphenyl, naphthalene, fluorene, anthracene, phenanthrene, and benzanthracene). Chromatographic separation was carried out in an HPLC system equipped with a UV detector, set at 254 nm. Different eluent systems were tested, and the best resolution was achieved with acetonitrile/water 6:4 v/v (Chang et al., 2007).

A sandwich structure C_8 modified graphene/mesoporous silica composites have been fabricated for the enrichment of endogenous peptides. The new stationary phase was synthesized by depositing *n*-octyltriethoxysilane (C_8)/CTAB (surfactant) composites on both sides of a hydrophilic graphene sheet followed by removing surfactant. This new nanocomposite is dispersible in aqueous media owing to hydrophilic surfaces; their pore structure is highly open with a uniform size (2.8 nm) and a large surface area (632 m^2 g^{-1}). The determination of target peptides was achieved with a UPLC system equipped with a mass spectrometry detector. Mouse brain extracts were used for investigating the efficiency of this new stationary phase. As a result, 89 endogenous peptides were identified without large molecules (Yin et al., 2012).

Carbon nanoparticles have been recently proposed to be used in TLC. Fang and Olesik (2014) worked with MWCNTs and edge-plane carbon nanorods (EPCNs) and compared their separation ability when they were electrospun with different kinds of polymers, including PAN and polystyrene.

According to the authors, these new materials due to their improved mechanical, thermal, and electrical properties are suitable for UTLC phase due to changes in the surface properties. This first report of a UTLC phase that contains carbon nanoparticles has shown that the composite materials provide increased and improved selectivity for aromatic compounds. Table 29.2 presents the recent applications of functionalized forms of carbon-based nanomaterials in separation.

TABLE 29.2 Recent Applications of Functionalized Nanomaterials in Separation

Nanostructure	Analyte	Matrix	Analytical Technique	LOD	Reference
MWCNTs—22'-(ethylenedioxy) diethylamine	C1–C5 alcohols and esters	Grappa	GC-FID		Merli et al. (2010)
MWCNTs-CO₂H MWCNTs-CONH₂	Linear and branched C3–C5 alkanes aromatic hydrocarbons		GC		Speltini et al. (2010)
Polymer-MWCNTs	Inorganic anions		Anion chromatography		Huang et al. (2013)
C₆₀-SDS	β-lactams antibiotics, NSAIDs, and amphenicols	Pharmaceuticals and urine	CE	0.34–1.02 mg L^{-1}	Moliner-Martínez et al. (2007)
C₆₀-Ionic liquid	Flavonoids, phenolic acids, and saponinis	Qishenyiqi dropping pills	CE	1.01–76.32 μg mL^{-1}	Cao, Li and Yi (2011)
SWCNTs-surfactant	Chlorophenols, NSAIDs, and penicillins		MiNDEKC-DAD		Suárez et al. (2007b)
c-MWCNTs immobilized on fused silica	Pharmaceutical compounds	Pharmaceutical formulations	CE	0.05–0.3 mg L^{-1}	Sombra et al. (2008)
GO immobilized on fused silica	NSAIDs		Open-tubular CE	1–1.8 μg mL^{-1}	Wang et al. (2013)
c-MWCNTS in running buffer	Purine and pyrimidine bases	Yeast RNA	CZE	0.9–3 μg mL^{-1}	Xiong et al. (2006)
SWCNTS-3-amino-propyl silica gel	Polycyclic aromatic hydrocarbons		HPLC-UV		Chang et al. (2007)
C8 modified G/mesoporous silica	89 endogenous peptides	Mouse brain	UPLC-MS		Yin et al. (2012)

29.4 DETECTION

Functionalized carbon-based nanomaterials have been extensively studied as detectors for numerous organic and inorganic molecules. Their structure combined with exceptional electrochemical properties makes them suitable for sensing applications.

Cooper et al. studied the chemiresistor performance of functionalized carbon nanotubes, graphenes, and gold nanoparticles for detecting petroleum hydrocarbons (cyclohexane, naphthalene, benzene, toluene, ethylbenzene, and three xylene isomers) in deionized water. Researchers functionalized SWCNT, MWCNT, and reduced graphene oxide nanosheets with octadecyl-1-amine and gold nanoparticles with 1-hexanethiol. This functional group was chosen due to the fact that strait chain alkanes provide stable dispersions of nanostructures and thus a suitable intrinsic electrical resistance. The results showed that the sensor coated with gold particles assembly was the most sensitive against all target hydrocarbons with detection limits ranging from 0.2 to 0.6 ppm. The least sensitive chemiresistor was one which was modified with MWCNTs. Moreover, graphene-modified sensors exhibited remarkable stability over a 26-day period without changing their electrical resistance (Cooper et al., 2014).

Yaxiong and coworkers developed a modified glassy carbon electrode with DNA functionalized SWCNT for the detection of 2,4,6-trinitrotoluene. This modification offered an abundance of π-electrons and hydrogen bond binding sites enhancing the selectivity of the sensor and reducing overpotentials. Furthermore, the electrode responded rapidly completing the detection within 15 s. The limit of detection was obtained at 0.5 μg L^{-1}. Modified electrode was tested by spiking ground waters (Liu Yaxiong et al., 2009a). For the detection of trinitrotoluene in water samples, Wei et al. developed a sensitive chemiresistor using noncovalently functionalized SWCNTs. 1-Pyrenemethylamine was linked on the surface of the nanotube through π–π stacking. The amino substitutes could interact with the target compound, creating complexes negatively charged, which affected the electrical conductance of the nanotubes. For the detection of the target molecule, a network of functionalized SWCNTs was applied between electrodes. The advantages of the new sensor are sensitive detection at the ppt levels of the analyte, fast response, and selectivity over interferences (Wei et al., 2014).

A gas chemiresistor based on MWCNT functionalized with nanoclusters of metal Au and Pt has been fabricated for the detection of NO_2 and NH_3. Modification with novel metals enhances gas sensitivity and the sensor exhibits rapid response to electrical conductivity changes. The detection limit was established under the ppm level and the repeatability was satisfactory (Penza et al., 2007). Functionalized graphene oxides, linked with histidine (His)-tagged acetylcholinesterase have been investigated as paraoxon biosensors. Although biosensors exhibited good short- and long-term stability and sensitivity, the lack of selectivity was the main drawback (Zhang et al., 2014). An amperometric sensor for hydrogen peroxide based on thionine functionalized MWCNTs has been developed. CNTs were carboxylated with nitric acid, followed by linking of thionine and were then transferred onto a paraffin-impregnated graphite electrode coated with nafion. This modified sensor was stable on a wide pH range, had an excellent sensitivity and a low detection limit, and responded rapidly. It can be used as an amperometric detector in chromatography and flow injection analysis (Jeykumari et al., 2007). Liu et al. proposed, for the detection of hydrogen peroxide, a cytochrome C immobilized ionic liquid functionalized MWCNTs modified glassy carbon electrode. Derivatization with the ionic liquid achieved a large surface area for immobilizing enzymes and a good environment for the Cytochrome C to remain bioelectrocatalytically active. The applicability of modified electrode was examined with amperommetry. This sensor features sensitivity, stability, low detection limit ($1.3 \ 10^{-8}$ M), and long-term stability (Liu et al., 2013).

A modified electrode with palladium nanoparticles on the surface of functionalized graphene oxide has been fabricated for the detection of nitro aromatic compounds. The reduced graphene oxide was functionalized with 1,3,6,8-pyrene tetra sulfonic acid sodium salt, as a result, the solubility and dispensability of graphenes in aqueous media was enhanced providing more binding sites for nanoparticles, and subsequent binding palladium nanoparticles with sulfonic acid groups. The modified electrode exhibited high electrocatalytic activity and sensitivity for nitrobenzene detection with a detection limit of 0.62 ppb (Zhou et al., 2014b).

A chemoresistive sensor has been developed for the detection of formaldehyde vapor using SWCNTs network-modified with solid organic acid tetrafluorohydroquinone (TFQ) by Shi et al. (2013). For the fabrication of this sensor, first the nanotube network was set between the electrodes followed by treating with solid organic acid TFQ. The interaction between the target analyte and tetrafluorohydroquinone increased the conductance of SWCNTs and the sensors achieved rapid response (less than 1 min) in ppb level concentrations (Shi et al., 2013).

The detection of volatile organic compounds (VOCs) was achieved using functionalized GOs. The chemoresistive sensor modified with an assembly of magnetic nanoparticles-decorated reduced graphene oxide with poly(3,4-ethylene dioxythiophene) and poly(ionic liquid). According to the authors, this sensing material offered more sensitive, selective, and rapid detection in comparison with its structural elements. The results showed that its signal was stable for polar (ethanol, methanol, acetone, water) and nonpolar (chloroform, styrene, dichlorobenzene, toluene) volatile organic compounds at room temperature (Tung et al., 2014).

For the detection of cancer biomarkers, Yang and coworkers developed a sandwich-type immunosensor, immobilizing the primary antibody on graphene sheets and using quantum dots functionalized graphene sheets as labels for the secondary antibody. This immunosensor was applied for the detection of prostate-specific antigen and exhibited linear response between 0.005 and 10 ng mL^{-1}, limit of detection at pg levels (3 pg mL^{-1}), good selectivity, stability, and reproducibility. The authors point out that the low detection limit derives from the high amount of quantum dots, which was immobilized on graphenes, the high amount of primary antibodies immobilized on graphene sheets, and graphene's good conductivity. The developed sensor was used for detecting target cancer biomarkers from patient's serum with effective results (Yang et al., 2011).

A glassy carbon electrode modified with polyethyleneimine functionalized graphene was developed for the sensitive detection of gallic acid (3,4,5-trihydroxybenzoic acid). The new electrode exhibited better performance compared with bare glassy carbon electrode and graphene oxide glassy carbon electrode. The modification with polyethyleneimine enhances the accumulation of gallic acid on the electrodes surface through π–π stacking and electrostatic and hydrogen bond forces. Furthermore, the large surface of modified electrode would increase the absorption capacity of the target analyte, thus enhancing electrochemical response and sensitivity. The limit of detection was established at 0.07 mg L^{-1} (Luo et al., 2013).

A composition based on functionalized graphene oxide sheets was applied on a glassy carbon electrode for the determination of quercetin with square wave voltammetry. *P*-aminothiophenol was functionalized with graphene mixed with gold nanoparticles and the final aqueous dispersion was applied on the electrode's surface followed by solvents evaporation under an infrared heat lamp. The modified electrode was applied for the detection of quercetin in pharmaceuticals with square wave voltammetry (Yola et al., 2013).

The simultaneous determination of hydroquinone and catechol was achieved using a modified carbon ionic liquid electrode. A poly(crystal violet)-functionalized graphene was electrodeposited on the surface of an electrode enhancing its electrochemical performance. The detection of target compounds was carried out by cyclic voltammetry. The separation was achieved with a potential of 112 mV and pH 2 (phosphate buffer solution). The detection of hydroquinone and catechol in artificial water was satisfactory (Sun et al., 2013). Another modified glassy carbon electrode was proposed for the simultaneous detection of dihydroxybenzene isomers (hydroquinone and catechol) by Zhou et al. (2014a). In this case, the electrode was modified with graphene–graphene oxide composition. It is important that neither stabilizing compound nor organic solvent is used. The applicability of this nanocomposite was investigated by cyclic and differential pulse voltammetry. The linear range of oxidation peaks was determined to be 0.5–300 μM and the detection limits were 0.16 μM for hydroquinone and 0.2 μM for catechol (Zhou et al., 2014a). For the simultaneous detection of these isomers, a modified glassy carbon electrode has been proposed with a three-dimensional functionalized graphene. This modified electrode can distinguish between hydroquinone and catechol, with higher oxidation currents compared to bare glass carbon electrode and two-dimensional graphene modified electrodes, showing enhanced catalytic activity and sensitivity. The results of the differential pulse voltammetry show that the isomers can be detected sensitively and selectively with a peak-to-peak separation of about 100 mV. The detection limits were 1.0×10^{-7} and 8.0×10^{-8} M for hydroquinone and catechol, respectively (Du et al., 2014).

TABLE 29.3 Applications of Functionalized Nanomaterials in Detection Techniques

Nanostructure	Analytes	Matrix	Detection Technique	LOD	Reference
Octadecyl-1-amine-CNTs Octadecyl-1-amine-RGON	Petroleum hydrocarbons	Deionized water			Cooper et al. (2014)
DNA-SWCNTs	TNT	Spiked water		$0.5\ \mu g\ L^{-1}$	Liu et al. (2009a)
1-pyrenemethylamine-SWCNTs	TNT	Water		ppt levels	Wei et al. (2014)
Au/Pt-MWCNTs	NO_2 and NH_3				Penza et al. (2007)
His-tagged acetylcholinesterase -GO	Paraoxon		Cyclic voltammetry amperometry	0.65 nM	Zhang et al. (2014)
Thionine-MWCNTs	Hydrogen peroxide		Amperometry	2.9×10^{-8} M	Jeykumari et al. (2007)
Cytochrome C immobilized IL-MWCNTs	Hydrogen peroxide		Amperometry	1.3×10^{-8} M	Liu et al. (2013)
Pd-GO	Nitrobenzene		Cyclic and differential pulse voltammetry	0.62 ppb	Zhou et al. (2014b)
TFQ-SWCNTs	Formaldehyde vapor			Ppb levels	Shi et al. (2013)
Magnetic NP-reduced graphene oxide with poly(3,4-ethylene dioxythiophene and poly(ionic liquid)	VOCs			Ppm level	Tung et al. (2014)
Primary antibody on GS and quantum dots-GS as labels	Prostate antigen	Patient's serum	Immunosensor	$3\ pg\ mL^{-1}$	Yang et al. (2011)
Polyethyleneimine-G	Gallic acid	Green and black tea	Cyclic voltammetry	$0.07\ mg\ L^{-1}$	Luo et al. (2013)
P-aminothiophenol-G mixed Au NP	Quercetin	Pharmaceutical products	Square wave voltammetry	3×10^{-13} M	Yola et al. (2013)
Poly(crystal violet)-G	Hydroquinone and catecholo	Artificial water	Cyclic voltammetry	0.033 and $0.097\ mol\ L^{-1}$	Sun et al. (2013)
G-GO	Hydroquinone and catecholo	Tap water	Differential pulse voltammetry	0.16 and $0.2\ \mu M$	Zhou et al. (2014a)
3-dimensional G	Hydroquinone and catecholo	Lake and tap water	Differential pulse voltammetry	1.0×10^{-7} and 8.0×10^{-8} M	Du et al. (2014)
Poly(diallyldimethylammonium chloride)-GO anchored Pb-Pt	Dopamine, ascorbic acid, and uric acid	Human urine and blood serum	Differential pulse voltammetry	0.61, 0.04, and $0.10\ \mu M$	Yan et al. (2013)
4,9,10-perylene tetracarboxylic acid-GS/MWCNTs/IL	Dopamine		Differential pulse voltammetry		Niu et al. (2013)
C_{60}-CNTs/IL	Catecholamines	Serum and urine	Voltammetry	15–22 ± 2 nM	Mazloum-Ardakani and Khoshroo (2014)
GO-chitosan	Microcystin-LR	Tap and river water	Immunosensor	$0.016\ \mu g\ L^{-1}$	Zhao et al. (2013)
Poly(allylamine hydrochloride)-CNTs	Cholesterol	Serum	Cyclic voltammetry	0.02 mM	Cai et al. (2013)
Carboxylic MWCNTs	6-MP	Rabbit blood	Amperometric detector	2.0×10^{-7}	Cao et al. (2003)
$[Co(phen)_3]^{3+}$-MWCNTs	6-MP	Rabbit blood	Voltammetry		Lu et al. (2008)

Poly(diallyldimethylammonium chloride)-functionalized graphene oxide sheets in which Pb-Pt nanoparticles were anchored were developed for the simultaneous detection of dopamine, ascorbic acid, and uric acid in human urine and blood serum. The modified electrode was used in differential pulse voltammetry with satisfactory results. The limit of detection (based on S/N = 3) was 0.61, 0.04, and 0.10 μM for ascorbic acid, dopamine, and uric acid, respectively (Yan et al., 2013). In addition, Niu et al. have proposed a glass carbon electrode modified with graphene sheets, MWCNTs, and ionic liquid functionalized with 3,4,9,10–perylene tetracarboxylic acid for the determination of dopamine. The detection of the target analyte was performed by differential pulse voltammetry (Niu et al., 2013). A glassy carbon electrode was modified with a film that consisted of fullerenes (C_{60}) functionalized CNTs and ionic liquid. This new electrode was applied in voltammetric methods for the determination of catecholamines (norepinephrine, isoprenaline, and dopamine) in serum and urine samples. The detection limits were established at 18 ± 2, 22 ± 2, and 15 ± 2 nM for norepinephrine, isoprenaline, and dopamine, respectively (Mazloum-Ardakani and Khoshroo, 2014).

An electrochemical immunosensor was developed for the determination of microcystin-LR. On the glassy carbon electrode surface, graphene oxide sheets and chitosan were dropped, which were used as immobilization material. Signal amplification was achieved using horseradish peroxidase-carbon nanospheres-antibody system. The linear detection range was determined between 0.05 and 15 μg L^{-1} and the limit of detection at 0.016 μg L^{-1}, under optimum conditions (Zhao et al., 2013).

Functionalized CNTs with cationic poly(allylamine hydrochloride)-gold nanoparticles composite were developed as a base for a bienzyme sensor. The positively charged nanocomposites interacted, through electrostatic interactions, with positively charged enzymes of horseradish peroxide (HRP) and cholesterol oxidase (ChOx) fabricating the biosensor with layer-by-layer assembly technique. This new composition was applied on the surface of a gold electrode for the detection of cholesterol. The bienzyme biosensor obtained the linear range from 0.18 to 11 mM, with a limit of detection of 0.02 mM (Cai et al., 2013).

A chemically modified electrode (CME) with carboxylic MWCNTs was used as the working electrode in liquid chromatography for the amperometric detection of 6-mercaptopurinein. In comparison with glassy carbon electrode (GCE), the modified one exhibited better sensitivity and lower detection limit, 1.3×10^{-6} for GCE and 2.0×10^{-7} for the CME. The developed method was used for the determination of 6-mercaptopurine in rabbit blood with microdialysis sampling together with liquid chromatography chemically modified electrode. It is notable that the modified electrode possessed excellent stability for more than a 1-month period storage (Cao et al., 2003). A new electrode was fabricated by immobilizing $[Co(phen)_3]^{3+}$ on the surface of MWCNTs for the detection of 6-mercaptopurine (6-MP). The functional groups interact with target compound, forming a free $[Co(phen)_3]^{3+}$ or the $[Co(phen)_3]^{3+}$–6-MP complex, and as a result the reductive current value was changed and the determination achieved. The voltametric determination of target analyte was selective even though rabbit blood was present (Lu et al., 2008). Table 29.3 presents the applications in detection.

29.5 CONCLUSIONS

Carbon-based nanomaterials present excellent thermal, chemical, and mechanical stability; high surface-to-volume ratio; and exceptional electrochemical properties. All these unique properties make them ideal candidates for numerous performances, but their hydrophobicity may limit their applicability. To overcome this limitation, the nanostructures could be modified with suitable functional groups. The functionalization processes improve their properties thus making them appealing molecules for further study. Till now they have been used for a great number of methods in analytical chemistry, but future developments are expected to expand the range of applications.

ACKNOWLEDGMENT

The authors thank Elsevier for permission to reproduce the image shown in Figure 29.1.

REFERENCES

Balasubramanian, K. and M. Burghard. 2005. Chemically functionalized carbon nanotubes. *Small* 2: 180–192.

Cai, X., Gao, X., Wang, L., Wu, Q., and Lin, X. 2013. A layer-by-layer assembled and carbon nanotubes/gold nanoparticles-based bienzyme biosensor for cholesterol detection. *Sensors and Actuators B* 181: 575–583.

Cao, J., Li, P., and Yi, L. 2011. Ionic liquids coated multi-walled carbon nanotubes as a novel pseudostationary phase in electrokinetic chromatography. *Journal of Chromatography A* 1218: 9428–9434.

Cao, X.-N., Lin, L., Zhou, Y.-Y. et al. 2003. Amperometric determination of 6-mercaptopurineon functionalized multi-wall carbon nanotubes modified electrode by liquid chromatography coupled with microdialysis and its application to pharmacokinetics in rabbit. *Talanta* 60: 1063–1070.

Chang, Y.X., Zhou, L.L., Li, G.X., Li, L., and Yuan, L.M. 2007. Single-wall carbon nanotubes used as stationary phase in HPLC. *Journal of Liquid Chromatography and Related Technologies* 30: 2953–2958.

Chen, L., Xie, H., and Yu, W. 2011. Functionalization methods of carbon nanotubes and its applications. In *Carbon Nanotubes Applications on Electron Devices*, J.M. Marulanda (Ed.). InTech. doi: 10.5772/18547.

Chen, S., Xiao, M., Lu, D., and Zhan, X. 2007. Use of microcolumn packed with modified carbon nanofibers coupled with inductively coupled plasma mass spectrometry for simultaneous on-line preconcentration and determination of trace rare earth elements in biological samples. *Rapid Communication Mass Spectrometry* 21: 2524–2528.

Choudhary, V. and Gupta, A. 2011. Polymer/carbon nanotube nanocomposites. In *Carbon Nanotubes-Polymer Nanocomposites*, S. Yellampalli (Ed).

Cooper, J.S., Myers, M., Chow, E. et al. 2014. Performance of graphene, carbon nanotube and gold nanoparticle chemiresistor sensor for the detection of petroleum hydrocarbons in water. *Journal of Nanoparticle Research* 16: 1–13. doi: 10.1007/s11051-013-2173-5.

Du, J., Ma, L., Shan, D. et al. 2014. An electrochemical sensor based on the three-dimensional functionalized graphene for simultaneous determination of hydroquinone and catechol. *Journal of Electroanalytical Chemistry* 722–723: 38–45.

Es'haghi, Z., Ebrahimi, M., and Hosseini, M.-S. 2011. Optimization of a novel method for determination benzene, toluene, ethylbenzene and xylenes in hair and waste water sample by carbon nanotubes reinforced sol-gel based hollow fiber solid phase microextraction and gas chromatography using factorial experimental design. *Journal of Chromatography A* 1218: 3400–3406.

Fang, X. and Olesic, S.V. 2014. Carbon nanotube and carbon nanorod-filler polyacrylonitrile electrospun stationary phase for ultrathin layer chromatography. *Analytica Chemica Acta* 830: 1–10.

Huang, Z., Xi, L., Subhani, Q., Yan, W., Guo, W., and Zhu, Y. 2013. Covalent functionalization of multi-walled carbon nanotubes with quaternary ammonium groups and its application in ion chromatography. *Carbon* 62: 127–134.

Jeykumari, D.R., Ramaprabhu, S., and Narayanan, S.S. 2007. A thionine functionalized MWCNT modified electrode for the determination of hydrogen peroxide. *Carbon* 45: 1340–1353.

Kroto, H.W., Heath, J.R., O'Brien, S.C., Curl, R.F., and Smalley, R.E. 1985. C60: Buckminsterfullerene. *Nature* 318: 162–163.

Kuila, T., Bose, S., Mishara, A.K., Khanra, P., Kim, N.H., and Lee, J.H. 2012. Chemical functionalization of graphene and its applications. *Progress in Material Science* 57: 1061–1105.

Liu, J.-W., Zhang, Q., Chen, X.-W., and Wang, J.-H. 2011a. Surface assembly of graphene oxide nanosheets on SiO_2 particles for the selective isolation of hemoglobin. *Chemistry-A European Journal* 17: 4864–4870.

Liu, X., Bu, C., Nan, Z., Zheng, L., Qui, Y., and Lu, X. 2013. Enzymes immobilized on amine-terminated ionic liquid-functionalized carbon nanotube for hydrogen peroxide determination. *Talanta* 105: 63–68.

Liu, Y., Lan, D., and Wei, W. 2009a. Layer-by-layer assembled DNA-functionalized single-walled carbon nanotube hybrids-modified electrodes for 2,4,6-trinitrotoluene detection. *Journal of Electroanalytical Chemistry* 637: 1–5.

Liu, Y., Li, Y., Wu, Z.-Q., and Yan, X.-P. 2009b. Fabrication and characterization of hexa-histidine-tagged protein functionalized multi-walled carbon nanotubes for selective solid-phase extraction of Cu^{2+} and Ni^{2+}. *Talanta* 79: 1464–1471.

Liu, Y., Liu, J., Deng, C., and Zhang, X. 2011b. Graphene and graphene oxides: two ideal choices for the enrichment and ionization of long-chain fatty acids free from matrix assisted laser desorption/ionization matrix interferences. *Rapid Communication Mass Spectrometry* 25: 3223–3234.

Lu, B.-Y., Li, H., Deng, H., Xu, Z., Li, W.-S., and Chen, H.-Y. 2008. Voltammetric determination of 6-mercaptopurine using $[Co(Phen)_3]^{3+}$/MWCNT modified graphite electrode. *Journal of Electroanalytical Chemistry* 621: 97–102.

Luo, J.H., Li, B.L., Li, L.B., and Luo, H.Q. 2013. Sensitive detection of gallic acid based on polyethyleneimine-functionalized graphene modified glassy carbon electrode. *Sensors and Actuators B* 186: 84–89.

Mazloum-Ardakani, M. and Khoshroo, A. 2014. High performance electrochemical sensor based on fullerene-functionalized carbon nanotubes/ionic liquid: Determination of some chatecholamines. *Electrochemistry Communications* 42: 9–12.

Meng, J., Shi, C., and Deng, C. 2011. Facile synthesis of water-soluble multi walled carbon nanotubes and polyaniline composites and their application in detection of small metabolites by matrix assisted laser desorption/ionization mass spectrometry. *Chemical Communications* 47: 11017–11019.

Meng, L., Fu, C., and Lu, Q. 2009. Advanced technology for functionalization of carbon nanotubes. *Progress in Natural Science* 19: 801–810.

Merli, D., Speltini, A., Ravelli, D., Quartarone, E., Costa, L., and Profumo, A. 2010. Multi-walled carbon nanotubes as the gas chromatography stationary phase: Role of their functionalization in the analysis of aliphatic alcohols and esters. *Journal of Chromatography A* 1217: 7275–7281.

Moliner-Martínez, Y., Cárdenas, S., and Valcárcel, M. 2007. Surfactant coated fullerenes C_{60} as pseudostationary phase in electrokinetic chromatography. *Journal of Chromatography A* 1167: 210–216.

Niu, X., Yang, W., Guo, H., Ren, J., and Gao, J. 2013. Highly sensitive and selective dopamine biosensor based on 3,4,9,10-perylene tetracarboxylic acid functionalized graphene sheets/multi-wall carbon nanotubes/ionic liquid composite film modified electrode. *Biosensor and Bioelectronics* 41: 225–231.

Parodi, B., Savio, M., Martinez, L.D., Gil, R.A., and Smichowski, P. 2011. Study of carbon nanotubes and functionalized-carbon nanotubes as substrates for flow injection solid phase extraction associated to inductively coupled plasma with ultrasonic nebulization. Application to Cd monitoring in solid environmental samples. *Microchemical Journal* 98: 225–230.

Penza, M., Cassano, G., Rossi, R. et al. 2007. Enhancement of sensitivity in gas chemiresistor based on carbon nanotube surface functionalized with noble metals (Au, Pt) nanoclusters. *Applied Physics Letters* 90: 173123. doi: 10.1063/1.2722207.

Ruiz-Palomero, C., Soriano, M.L., and Valcárcel, M. 2014. Ternary composites of nanocellulose, carbon nanotubes and ionic liquids as new extractants for direct immersion single drop microextraction. *Talanta* 125: 72–77.

Samanidou, V.F. and Karageorgou, E.G. 2012. Carbon nanotubes in sample preparation. *Current Organic Chemistry* 16: 1645–1669.

Sarafraz-Yazdi, A., Amiri, A., Rounaghi, G., and Hosseini. E.H. 2011. A novel solid phase microextraction using coated fiber based sol-gel technique using poly(ethylene glycol) grafted multi-walled carbon nanotubes for determination of benzene, toluene, ethylbenzene and o-xylene in water samples with gas chromatography-flam ionization spectrometry. *Journal of Chromatography A* 1218: 5757–5764.

Shi, C., Meng, J., and Deng, C. 2012. Enrichment and detection of small molecules using magnetic graphene as an absorbent and a novel matrix of MALDI-TOF-MS. *Chemical Communications* 48: 2418–2420.

Shi, D., Wei, L., Wang, J., Zhao, J., Chen, C., and Xu, D. 2013. Solid organic acid tetrafluorohydroquinone functionalized single-walled carbon nanotube chemiresistive sensors for highly sensitive and selective formaldehyde detection. *Sensors and Actuators B* 177: 370–375.

Sombra, L., Moliner-Martínez, Y., Cárdenas, S., and Valcárcel, M. 2008. Carboxylic multi-walled carbon nanotubes as immobilized stationary phase in capillary electrochromatography. *Electrophoresis* 29: 3850–3857.

Speltini, A., Merli, D., Quartarone, E., and Profumo, A. 2010. Separation of alkanes and aromatic compounds by packed column gas chromatography using functionalized multi-walled carbon nanotubes as stationary phase. *Journal of Chromatography A* 1217: 2918–2924.

Suárez, B., Simonet, B.M., Cárdenas, S., and Valcárcel, M. 2007a. Determination of non-steroidal anti-inflammatory drugs in urine by combing an immobilized carboxylated carbon nanotubes minicolumn for solid phase extraction with capillary electrophoresis-mass spectrometry. *Journal of Chromatography A* 1159: 203–207.

Suárez, B., Simonet, B.M., Cárdenas, S., and Valcárcel, M. 2007b. Surfactant-coated single-walled carbon nanotubes as a novel pseudostationary phase in capillary EKC. *Electrophoresis* 29: 1714–1722.

Sun, W., Wang, Y., Lu, Y., Hu, A., Shi, F., and Sun, Z. 2013. High sensitive simultaneously electrochemical detection of hydroquinone and catechol with apoly(crystal violet) functionalized graphene modified carbon ionic liquid detection. *Sensors and Actuators B* 188: 564–570.

Sun, Y.P., Fu, K., Lin, Y., and Huang, W. 2002. Functionalized carbon nanotubes: Properties and applications. *Accounts of Chemical Research* 35: 1096–1104.

Tung, T.T., Castro, M., Pillin, I., Kim, T.Y., Suh, K.S., and Feller, J.-F. 2014. Graphene-Fe$_3$O$_4$/PIL-PEDOT for the design of sensitive and stable quantum chemo-resistive VOC sensor. *Carbon* 74: 104–112.

Wang, C., de Rooy, S., Lu, C.-F. et al. 2013. An immobilized graphene oxide stationary phase for open-tubular capillary electrochromatography. *Electrophoresis* 34: 1197–1202.

Wang, L., Zang, X., Wang, C., and Wang, Z. 2014. Graphene oxide as a micro-solid-phase extraction sorbent for the enrichment of parabens from water and vinegar samples. *Journal of Separation Science* 37: 1656–1662. doi: 10.1002/jssc.201400028.

Wei, L., Lu, D., Wang, J. et al. 2014. Highly sensitive detection of trinitrotoluene in water by chemiresistive sensor based on noncovalently amino functionalized single-walled carbon nanotube. *Sensors and Actuators B* 190: 529–534.

Xiong, X., Ouyang, J., Baeyens, W.R.G., Delanghe, J.R., Shen, X., and Yang, Y. 2006. Enhanced separation of purine and pyrimidine bases using carboxylic MWCNTs additive in capillary zone electrophoresis. *Electrophoresis* 27: 3243–3253.

Yan, J., Liu, S., Zhang, Z. et al. 2013. Simultaneous determination of ascorbic acid, dopamine and uric acid based on graphene anchored with Pd-Pt nanoparticles. *Colloids and Surfaces B: Biointerfaces* 111: 392–397.

Yang, M., Javadi, A., and Gong, S. 2011. Sensitive electrochemical immunosensor for the detection of cancer biomarker using quantum dot functionalized graphene sheets as labels. *Sensors and Actuators B* 155: 357–360.

Yin, P., Wang, Y., Li, Y., Deng, C., Zhang, X., and Yang, P. 2012. Preparation of sandwich-structured graphene/mesoporous silica composites with C8 modified pore wall for highly efficient selective enrichment of endogenous peptides for mass spectrometry analysis. *Proteomics* 12: 2784–2791.

Yola, M.L., Atar, N., Üstündağ, Z., and Solak, A.O. 2013. A novel voltametric sensor based on *p*-aminothiophenol functionalized graphene oxide/gold nanoparticles for determining quercetin in presence of ascorbic acid. *Journal of Electroanalytical Chemistry* 698: 9–16.

Yu, P., Wang, Q., Ma, H., Wu, J., and Shen, S. 2014. Determination of puerarin in rat plasma using PEGylated magnetic carbon nanotubes by HPLC. *Journal of Chromatography B* 959: 55–61.

Zang, Z., Hu, Z., Li, Z., He, Q., and Chang, X. 2009. Synthesis, characterization and application of ethylenediamine-modified multiwalled carbon nanotubes for selective solid-phase extraction and preconcentration of metal ions. *Journal of Hazardous Materials* 172: 958–963.

Zhang, B.-T., Zheng, X., Li, H.-F., and Lin, J.-M. 2013. Application of carbon-based nano-materials in sample preparation: A review. *Analytica Chemica Acta* 784: 1–17.

Zhang, H., Li, Z.-F., Snyder, A., Xie, J., and Stanciu, L. 2014. Functionalized graphene oxide for the fabrication of paraoxon biosensors. *Analitica Chimica Acta* 827: 86–94. http://dx.doi.org/10.1016/j.aca.2014.04.014.

Zhang, Z., Zhang, H., Hu, Y., Yang, X., and Yao, S. 2010. Novel surface molecularly imprinted material modified multi-walled carbon nanotubes as solid-phase extraction sorbent for selective extraction gallium ion from fly ash. *Talanta* 82: 304–311.

Zhao, H., Tian, J., and Quan, X. 2013. A graphene and multienzyme functionalized carbon nanosphere-based electrochemical immunosensor for microcystin-LR detection. *Colloids and Surfaces B: Biointerfaces* 103: 38–44.

Zhou, X., He, Z., Lian, Q., Li, Z., Jiang, H., and Lu, X. 2014a. Simultaneous determination of dihydroxybenzene isomers based on graphene-graphene oxide nanocomposite modified glassy carbon electrode. *Sensors and Actuators B* 193: 198–204.

Zhou, X., Yuan, C., Qin, D. et al. 2014b. Pd nanoparticles on functionalized graphene for excellent detection of nitro aromatic compounds. *Electrochimica Acta* 119: 243–250.

Zou, J., Song, X., Ji, J. et al. 2011. Polypyrrole/graphene composite-coated fiber for the solid-phase microextraction of phenols. *Journal of Separation Science* 34: 2765–2772.

CHAPTER 30

CONTENTS

Surface Modification of Carbon Nanotubes and Its Application in Environmental Protection

30

*Omid Moradi, Hamidreza Sadegh,
and Ramin Shahryari-ghoshekandi*

30.1 INTRODUCTION

Industrial wastes and wastewaters such as those produced by petrochemistry, metal plating, fertilizers, and pesticides were the main sources for discharging the toxic ions in the environment [1,2].

The most common toxic ions in wastewaters responsible for particular problems were heavy metal ions, azo dyes, etc. [3–5]. Despite the fact that the human body needs small dose such as Zn^{2+}, an excess of it may cause eminent health problems such as depression, lethargy, neurological signs, and increased thirst. Exposure to such toxic ions can cause health problems such as liver or kidney damage, Wilson disease, insomnia, cancer, diarrhea, nausea, vomiting, dermatitis, chronic asthma, coughing, and headache [6–11].

Removal of these toxic ions from wastewater is necessary for many health and environmental considerations. Conventional methods such as reduction, precipitation, adsorption, oxidation, and ion exchange have been used to remove these ions. However, adsorption process was the most suitable method because of its high efficiency and economic consideration [12]. Some adsorbents such as activated carbon (AC), zeolites, biomaterials, nanoparticles, and polymers have been extensively used for adsorption of toxic heavy metal ions [13–25]. However, the adsorption efficiency of these adsorbents was very low. Therefore, it has become the center of attention of different research groups to find more efficient adsorbents.

Since the discovery of carbon nanotubes (CNTs) [26], it has become the favorite adsorbent among carbon nanomaterials (CNMs) because of its unique physical and chemical properties [27–31]. Note that there are two main types of CNTs: single-walled (SWCNTs) and multiwalled (MWCNTs) as shown in Figure 30.1.

In the last decade, CNTs have found a great potential application for removing heavy metals and azo dyes [3–5,14–22,32–46]. In this review, we highlight the past and present attempts at using CNTs as adsorbents to removing heavy metal ions and azo dyes. Furthermore, the effect of surface modification of CNTs by removing heavy metal ions and azo dyes is in the core of this review.

30.2 SURFACE MODIFICATION OF CNTs

CNTs surfaces are by nature highly hydrophobic. To overcome this problem in raw CNTs, different methods of surface modification are being pursued by noncovalent and covalent functionalization strategies [47].

Noncovalent functionalization strategies do not have any effect on the physical properties of the CNTs because they keep the structure of intrinsic sp^2-hybridized orbitals unchanged.

(a)   (b)

FIGURE 30.1
Schematics of (a) SWCNTs and (b) MWCNTs.

This can be done by taking advantage of the π–π interaction between conjugated molecules and the graphitic sidewall of CNTs [48–50]. Another method such as noncovalent hydrophobic interactions (amphiphilic molecules) with aromatic surface of CNTs in aqueous media has been explored, which can reduce the hydrophobic interface between the CNTs and their polar environment [51–59].

Functionalization in covalent method depends on the chemical bonds between carbon atoms of CNTs, chemical reactions, and conjugation of hydrophilic organic molecules on the surface of CNTs [28]. According to the location of the functional groups, the strategies to covalently functionalized CNTs can be classified into two main categories: defect functionalization and sidewall functionalization. The covalent functionalization of CNTs is more robust and better controllable compared to functionalization based on noncovalent methods [60–65].

The defect functionalization of CNTs is based on the conversion of carboxylic groups and other oxygenated sites formed through oxidative purification. The carboxylic groups at the end of the CNTs can be coupled with other functional groups. The oxidized CNTs usually react with thionyl chloride to activate the carboxylic group for a later reaction with amines or alcohols as shown in Figure 30.2 [66–70].

The sidewall functionalization is based on covalent linkage of functional groups onto the sidewall of CNTs. The covalent sidewall functionalization is associated with the change of hybridization from sp^2 to sp^3 and a simultaneous loss of conjugation.

In contrast to the well-developed chemistry of fullerenes, the covalent functionalization chemistry of CNTs has only been achieved recently [71]. This might be either related to the lack of availability of SWCNTs in the 1990s or to the significantly lower reactivity of the commercially available CNTs [72] compared to fullerenes. Not surprisingly, most of the reported sidewall functionalization reactions of CNTs require very reactive reagents. Nevertheless, within the last few years, a diverse chemistry was developed to covalently modify the surface of CNTs (Figure 30.3) [73].

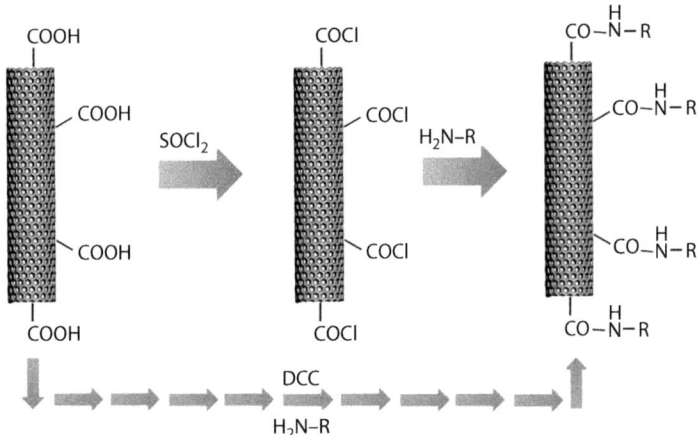

FIGURE 30.2
Schematic representation of the oxidative treatment of CNTs followed by the treatment with thionyl chloride and subsequent amidation.

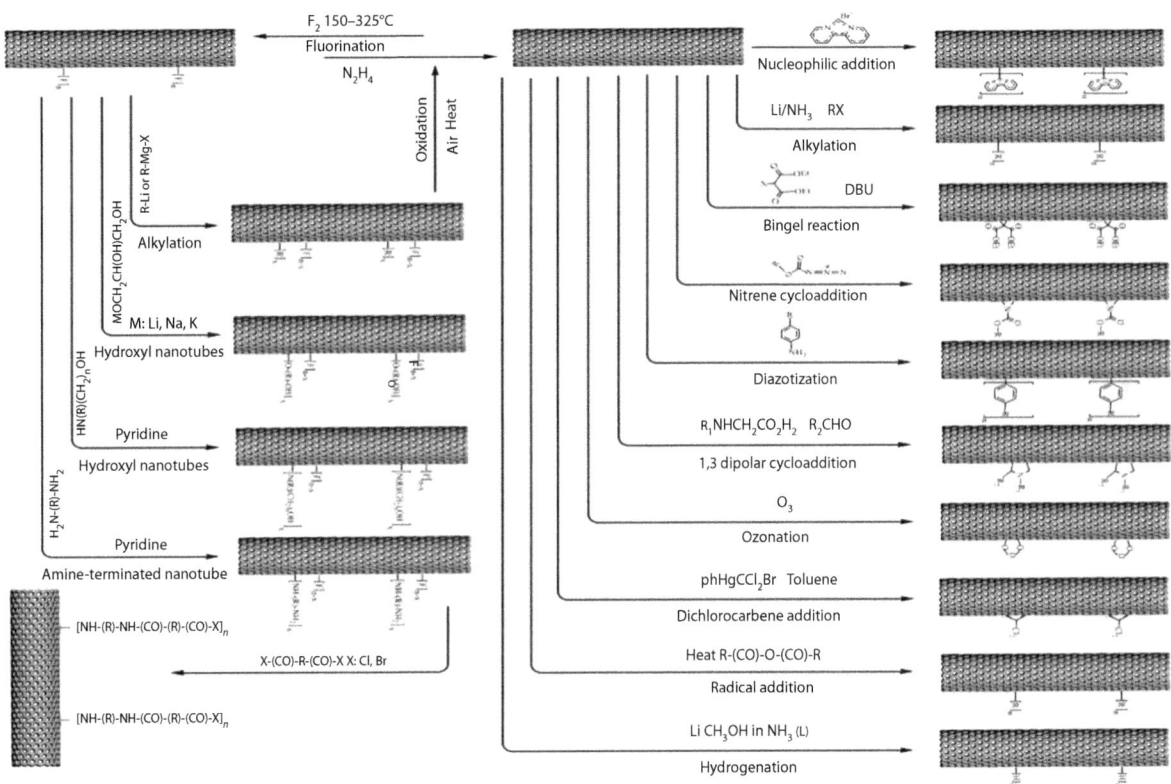

FIGURE 30.3
Schematic describing various covalent sidewall functionalization reactions of CNT.

The pioneer study on covalent sidewall functionalization of CNTs was achieved by Mickelson et al. [108] in 1998, through the treatment of CNTs with elemental fluorine. The degree of addition ranges from 0.1% to the complete combustion of the CNTs under these drastic conditions. Subsequent treatment of the fluoronanotubes with N_2H_4 or $LiBH_4/LiAlH_4$ leads to the restoration of the CNTs [108]. Fluorinated CNTs are now commercially available and therefore give rise to a widespread chemistry by using them as starting material to carry out subsequent derivatization reactions. Thus, sidewall alkylation of the nanotubes was achieved by the nucleophilic substitution with Grignard reagents or the reaction with alkyl-lithium precursors [74]. Oxidation in air or heat treatment leads to the removal of the alkyl groups and therefore the functionalization of the SWCNTs. Electrochemical addition of aryl radicals to the CNTs has also been reported by Bahr et al. [75]. The addition of radicals on CNTs was first introduced by Holzinger et al. [110] who used heptadecafluorooctyl iodide to photoinduce the generation of the perfluorinated radicals followed by their addition to the SWCNTs. Other studies used organic peroxides as precursors to achieve covalent sidewall functionalization [76].

Chen et al. reported the addition of carbenes by the reaction of SWCNTs with dichlorocarbene [77]. The carbene was first generated from chloroform with potassium hydroxide [106] and later from phenyl (bromodichloromethyl) mercury [78]. The addition of nucleophilic carbenes on the electrophilic SWCNTs resulted in the formation of a zwitterionic poly adduct [110]. The covalently bound imidazolidene addend bears a positive charge, whereas the nanotube bears one negative charge for each addend attached to the sidewall causing the n-doping of the CNTs [79]. This offered a new approach for the controlled modification of the electronic properties of SWCNTs. The addition of nitrenes on the sidewall of CNTs was achieved via reactive alkyl oxycarbonyl nitrenes obtained from alkoxycarbonyl azides [80,81]. Prato et al. succeeded in the development of one of the most powerful techniques for the functionalization of CNTs using the 1,3-dipolar cycloaddition of azomethine ylides [82]. The treatment of pristine SWCNTs with an aldehyde and an N-substituted glycine derivative resulted in the formation of substituted pyrrolidine moieties on the SWCNT surface [83].

30.3 APPLICATION OF CNTs IN ENVIRONMENTAL PROTECTION

30.3.1 Removal of Heavy Metal Ions

Zinc is one of the most common heavy metals occurring in wastewater. In water, Zn ion may associate or react with neutral or ionic compounds to form inorganic salts, stable organic complexes or inorganic or organic colloids. The quantity of Zn ion available in water from each of these forms is dependent upon the solubility of these forms, the pH and temperature, the total amount of the Zn form present in water, and the presence of other metal ions, organic compounds, and inorganic compounds. Zn ion exhibits high toxicity to aquatic organisms, and it may cause high chronic toxicity in some cases. For example, Zn ion is highly toxic to aquatic organisms and has a high potential to bioaccumulate. Therefore, it is important to remove Zn from wastewater during the treatment. Lu and Chiu [84] in their study explained the trend of the graph obtained from the adsorption capacity of Zn^{2+} onto the CNTs, where adsorption increased when the pH is adjusted from 1 to 8, achieved maximum between pH 8 and 11 and decreased at pH 12. The time taken for the adsorption to reach the equilibrium for the SWCNTs and MWCNTs are 1 h, while for the powdered activated carbon (PAC) it took twice that long or 2 h.

A comparison on the adsorption of Zn^{2+} between the CNTs and commercial PAC was also carried out. With the initial Zn^{2+} concentration range of 10–80 mg/L, the maximum adsorption capacities of Zn^{2+} calculated by the Langmuir model for the SWCNTs, MWCNTs, and PAC were 43.66, 32.68, and 13.04 mg/g, respectively. The time taken to reach equilibrium was short, and they suggested that in order to obtain high adsorption capacity, SWCNTs and MWCNTs are preferably used for Zn^{2+} removal from water. In another case study, Lu et al. [85] reported that the amount of Zn^{2+} sorbed onto the CNTs increased with a rise in temperature. Using the same conditions, the Zn^{2+} sorption capacity of the CNTs was much greater than that of the commercially available PAC, reflecting that the SWCNTs and MWCNTs are effective sorbents. The thermodynamic analysis revealed that the sorption of Zn^{2+} onto the CNTs is endothermic and spontaneous. The Zn^{2+} ions could be easily removed from the surface site of the SWCNTs and MWCNTs by a 0.1 mol/L nitric acid solution, and the sorption capacity was maintained after 10 cycles of the sorption/desorption process. This suggests that both CNTs can be reused through many cycles of water treatment and regeneration.

Not only the Zn ions could be removed by the CNTs adsorption, other studies show that some other trace elements such as Ni, Pb, Cd, Cr, and Cu can also be absorbed by this powerful CNT agent.

Yang et al. [86], in their study on adsorption of Ni^{2+} on oxidized MWCNTs, found that the CNT adsorption capacity increases with the increase of pH in the pH range of 2–9 from 0% to ~99%. They found that oxidized MWCNTs were the most suitable material for the solidification and pre-concentration of Ni^{2+} from aqueous solutions. In another article, Kandah and Meunier [87] reported on their achievement in adsorption of some heavy metals using functionalized CNTs. With a large adsorption capacity, they also reported on the ability of the CNTs to remove some organic dyes from water. In their study, for both produced and oxidized CNTs, the Langmuir model determined 18.08 and 49.26 mg/g, respectively, as the maximum Ni^{2+} adsorption uptake. These two sets of isotherm models fit well with the experimental data. Therefore we conclude that the experiment was successful where the CNTs were found to be the most effective nickel ion absorbent based on the high adsorption capacity as well as the short adsorption time [88].

Chromium at low-level exposure can irritate the skin and cause ulceration. Long-term exposure can cause kidney and liver damages as well as damaging the circulatory and nerve tissues. Once again, it is important to eliminate such traces in our wastewater by the aid of the CNTs. Atieh [89] has reported using CNTs supported by AC to remove Cr^{6+} ions from polluted water. The highest adsorption capacity by using AC-CNT coated adsorbent obtained from batch adsorption experiments was 9.0 mg/g. Therefore, it seems that AC coated with CNTs is most effective for the removal of chromium ions. Di et al. [90] reported the removal of Cr^{6+} from drinking water using ceria nanoparticles supported on aligned CNTs (CeO_2/ACNTs).

The largest adsorption capacity of CeO_2/ACNTs reaches 30.2 mg/g at an equilibrium Cr^{6+} concentration of 35.3 mg/L at pH 7.0. They also concluded that the adsorption capacity of the CeO_2/ACNTs is 1.5 times higher than that of the AC, 2.0 times higher than that of Al_2O_3, and 1.8 times higher than that of ball-milled ACNTs. The high adsorption capacity and the wide range of pH values make CeO_2/ACNTs a good candidate material for removal of Cr^{6+} from drinking water. Hu et al. [91] reported the removal of Cr^{6+} from aqueous solution using oxidized MWCNTs. The maximum removal of Cr^{6+} was found at low pH, and the adsorption kinetics of Cr^{6+} was suitable for pseudo-second-order models. The removal of chromium mainly depends on the occurrence of redox reaction of adsorbed Cr^{6+} on the surface of oxidized MWCNTs to the formation of Cr^{3+} and subsequent sorption of Cr^{3+} on MWCNTs. This Cr^{3+} sorption appears to be the leading mechanism for chromium uptake to MWCNTs. To remove the anionic chromate (CrO_4^{2-}) from wastewater, functionalized CNTs were found to be the most suitable compared to unmodified CNTs in terms of the adsorption capacity [92]. The authors explained the reason for the CNTs showing excellent adsorption capability of anionic chromate, because of the interaction of CrO_4^{2-} with the surface oxygen-containing functional groups on the modified CNTs. Hence, functionalized CNTs once again can be potentially employed in adsorbing heavy metal from wastewater. Lu and Su [93] reported that MWCNTs were thermally treated to remove natural organic matter (NOM) in aqueous solution. The amount of adsorbed NOM onto CNTs increased with a rise in initial NOM concentration and solution ionic strength but decreased with a rise in solution pH. Also concluded that the performance of adsorption using raw CNTs and treated CNTs and granular AC. Results proved that treated CNTs have the best NOM adsorption performance compared to others.

Besides Zn and Cr, Pb is also one of the famous heavy metals occurring in wastewater. Pb compounds are generally soluble in soft and slightly acidic water. Pb waterworks were often applied in past days, and these may still be present in old buildings. The Pb from pipes may partially dissolve in the water flowing through. It occurs in almost all water resources as well as wastewater. Despite its toxicity, the presence of lead in the water may affect the human and marine lives' health. Therefore, again, the discussion on the CNTs to absorb lead from wastewater is still going on. Wang et al. [94] reported that the role of functional groups in the adsorption of Pb^{2+} to create a chemical complex was critical for efficient adsorption. 75.3% of Pb^{2+} adsorption capacity was achieved. Pb^{2+} is in the form of PbO and $Pb(OH)_2$, and $PbCO_3$ is adsorbed on the surface of the acidified MWCNTs, which is only 3.4% of the total Pb^{2+} adsorption capacity.

They concluded that the Pb^{2+} species adsorbed acidified MWCNTs on the ends and at the defects sites. Wang et al. [95] reported that Mn oxidecoated CNTs (MnO_2/CNTs) were used to remove Pb^{2+} from aqueous solution. The Pb^{2+} removal capacity of MnO_2/CNTs decreased with the decrease of pH. From the Langmuir isotherms, maximum adsorption capacity was 78.74 mg/g, comparing with CNTs, significant improvement of Pb^{2+} adsorption shows MnO_2/CNTs can be good Pb^{2+} absorbers. The adsorption of Pb^{2+} by MnO_2/CNTs occurred during the first 15 min of contact time, and full equilibrium was reached in 2 h. The kinetic adsorption described was a pseudo-second-order rate equation.

Wang et al. [96], in the study on the adsorption of acidified MWCNTs to Pb^{2+}, found that the experiment met the Langmuir model. Results demonstrated a high adsorption capacity of acidified MWCNTs for Pb^{2+}. The reason stated was the formation of oxygenous functional groups on the surface of the CNTs. Salt formation or complex deposited occurred on the surface of CNTs when acidified MWCNTs react with Pb^{2+}. The adsorption contact time to reach equilibrium is about 20 min, which proved to be shorter than AC (reported to be 120 min). Besides the discussion on adsorption contact time, authors also stated that Pb^{2+} can be easily regenerated from the acidified MWCNTs by adjusting the solution to pH 2.

In another study, Kabbashi et al. [97] discussed the removal of Pb^{2+}, which has reached the maximum value of 85% or 83% at pH 5 or 40 mg/L of the CNTs, respectively. Higher correlation coefficients from the Langmuir isotherm model indicates the strong adsorptions of Pb^{2+} on the surface of CNTs (adsorption capacity Xm = 102 mg/g). The results indicate that the highest percentage removal of Pb^{2+} (96.03%) can be achieved at pH 5, 40 mg/L of CNTs, contact time 80 min and agitation speed 50 rpm.

The effectiveness of an adsorbent depends on the adsorptive properties of its surface. Adsorption takes place when a solid surface is in contact with a solution and tends to accumulate a surface layer of solute molecules created by the imbalanced surface forces. Moradi et al. [98] explained the interaction between some heavy metal ions such as Pb^{2+}, Cd^{2+}, and Cu^{2+} ions from aqueous solution adsorbed by SWCNTs and carboxylate group functionalized single-walled carbon nanotube (SWCNT–COOH) surfaces. The maximum adsorption capacities (qm) for Pb^{2+}, Cu^{2+}, and Cd^{2+} ions onto the SWCNT–COOH obtained are 96.02, 77.00, and 55.89 mg/g, respectively, and by the SWCNTs are 33.55, 24.29, and 24.07 mg/g, respectively. The thermodynamic parameters values showed that the adsorption of ions on the SWCNT–COOH and SWCNTs at 283–313 K is spontaneous and endothermic.

One of the most hazardous heavy metals is cadmium (Cd). In humans, long-term exposure to Cd is associated with severe renal dysfunction. Therefore, it is essential to remove Cd in wastewater treatment. Li et al. [99] discussed the adsorption of Cd^{2+} by CNTs. The adsorption capacity is highly pH dependent where H_2O_2 oxidizes the CNTs to reach equilibrium state with a high adsorption capacity at lower pH, and it is more obvious compared to the HNO_3- and $KMnO_4$-oxidized CNTs. The experiments of the CNT dosage effect on Cd^{2+} adsorption reflect that the adsorption capacity for $KMnO_4$-oxidized CNTs has a sharper increase at the CNT dosage from 0.03 to 0.08 g/100 mL than the as-grown H_2O_2 and HNO_3-oxidized CNTs and its removal efficiency almost reaches 100% at the CNT dosage of 0.08 g/100 mL. This shows that CNTs are able to adsorb Cd^{2+} with high adsorption capacity.

To prove the efficiency of CNTs in adsorbing heavy metals, Gao et al. [100] studied the adsorption of Ni, Cu, Zn, and Cd from aqueous solutions on CNTs oxidized with concentrated nitric acid. The adsorption was carried out in single, binary, ternary, and quaternary systems and multicomponent solutions. They studied the effect of process parameters such as the pH value, initial concentration of ions, the surface chemistry of the adsorbent, the kind and number of components in the adsorption system, and the ratio of the metal ion species in solution. For the single and binary systems, it turned out that the amount adsorbed on oxidized CNTs followed the order of Cu^{2+}(aq) > Ni^{2+}(aq) > Cd^{2+}(aq) > Zn^{2+}(aq). The good correlation between the amount adsorbed and standard electrode potential indicated that the redox process may serve a role in the mechanism of adsorption. Meanwhile, for the ternary and quaternary systems, the adsorption was more complex. The order of amount adsorbed at the same concentration was Cu^{2+}(aq) > Cd^{2+}(aq) > Zn^{2+}(aq) > Ni^{2+}(aq). Stafiej and Pyrzynska [101] reported that adsorption characteristics of some divalent metal ions (Cu, Co, Cd, Zn, Mn, Pb) was very pH depending. The effect of metal ion removal at pH 9 was Cu^{2+} > Pb^{2+} > Co^{2+} > Zn^{2+} > Mn^{2+}.

Li et al. [102] studied different heavy metals (Pb^{2+}, Cu^{2+}, and Cd^{2+}) adsorption using oxidized CNTs. The maximum adsorption capacities of Pb^{2+}, Cu^{2+}, and Cd^{2+} ions by oxidized CNTs, calculated from Langmuir isotherm, were 63.29, 23.89, and 11.01 mg/g, respectively, in the following order Pb^{2+} > Cu^{2+} > Cd^{2+}. The kinetic models of first, pseudo-second-, and second-order rate model fit well the experimental data. For initial metal concentration of 30 mg/L, the adsorption rates of Pb^{2+}, Cu^{2+}, and Cd^{2+} are 0.033, 0.049, and 0.096 mg/g·min, respectively, in the order of $Cd^{2+} > Cu^{2+} > Pb^{2+}$. Rao et al. [103] reported that the sorption capacities of metal ions to different CNTs follow roughly the order

$$Pb^{2+} > Ni^{2+} > Zn^{2+} > Cu^{2+} > Cd^{2+}$$

The sorption capacities of metal ions by raw CNTs are very low but significantly increased after oxidation of CNT surfaces by HNO_3, NaOCl, and $KMnO_4$ solutions. The surface-oxidized CNTs show great potential as superior sorbents for environmental protection applications. The sorption mechanism appears mainly attributable to chemical interaction between the metal ions and the surface functional groups.

The order of heavy metal ions removed from aqueous solutions by CNTs mentioned earlier shows that their adsorption does not depend clearly on the ionic radius of metal ions, but depend on properties of CNTs such as ionic strength, pH, foreign ions, CNT mass, contact time, initial metal ion concentration, and temperature. However, synthesis conditions have major influence on the nature of the CNTs product formed. Reaction conditions including temperature, gas composition, and the nature and composition of metallic catalysts leading to CNT formation, which, in turn, affect the properties of the final product, still need to be explored in order to understand these influences.

Copper is an essential substance to human life, but in high doses, it can cause anemia, liver, and kidney damages as well as stomach and intestinal irritation. This is the reason why copper should not be present in wastewater. Li et al. [104] reported that environmental friendly adsorbents that were CNTs immobilized by calcium alginate (CNTs/CA) were designed for copper adsorption. CNTs/CA copper adsorption properties were investigated via equilibrium studies. Experimental results showed that copper removal efficiency of the CNTs/CA is high and reaches 69.9% even at a low pH of 2.1. The CNTs/CA copper adsorption capacity can attain 67.9 mg/g at the copper equilibrium concentration of 5 mg/L.

In other cases, Li et al. [105] also studied fluoride adsorption on alumina-based CNTs. They reported that the fluoride adsorption isotherms were high at a pH range of 5.0–9.0 with the fluoride adsorption capacity for Al_2O_3/CNTs of about 13.5 times higher than that of the AC-300 carbon and 4 times higher than that of the c–Al_2O_3 at equilibrium fluoride concentration of 12 mg/L. The broad range of pH values and high adsorption capacity of the Al_2O_3/CNTs indicates its suitability to be applied in fluoride removal from water.

Ruparelia et al. [106] reported the removal of heavy metal cations such as cadmium, lead, nickel, and zinc by CNMs, the adsorption capacity of nanocarbon, nanoporous carbon (NPC), and activation were tested. NPC was the best among other adsorbents due to its ion exchange with acidic oxygen-containing functional groups introduced by activation during the posttreatment process as well as due to its unique nanoporous structure. They concluded that NPC has good potential in water treatment applications. Many CNT applications require handling in solution phase; however, CNTs have proven difficult to disperse in solvents [107]. Chemical modification of SWCNTs is often required for more versatile suspension capabilities and enablement of certain applications. This has encouraged greater exploitation of their intrinsic properties, as well as the capability to modify these properties. In particular, the functionalization of CNTs is required for their aqueous suspensions and to allow for molecular interactions with biological systems. Native CNTs adsorb surfactants that have been found to associate with CNTs via van der Waals interactions through their hydrophobic chains, rendering CNTs hydrophilic and able to disperse in aqueous environments. SWCNTs can be isolated from aggregated bundles [108], allowing for spectroscopic probing of individual SWCNTs. Surfactant additives prevented nonspecific interactions between SWCNTs and proteins. Also, surfactants with modified head groups have also been used to link SWCNTs to specific molecules [109]. Kim et al. [110] reported that lipids comprise a class of molecules that interact with CNTs similarly to surfactants. This class of molecule offers control of its interactions with CNTs through modification of their hydrophobic chains, while also providing versatility of the functionality that they impart through the modification of their head groups. Among these lipids, chemical interaction between the heavy metal ions and the surface functional groups of CNTs is the major adsorption mechanism. Table 30.1 illustrates the brief summary of the removal of heavy metals using nonmodified and functionalized CNTs.

TABLE 30.1 Data of Maximum Adsorption Capacity of Heavy Metal Ions with CNTs

Adsorbent	q_m							SA	RP	Conditions	References
	Zn^{2+}	Cu^{2+}	Ni^{2+}	Cd^{2+}	Pb^{2+}	Cr^{2+}	Co^{2+}				
SWCNTs	43.66							423	NA	pH: 8–11, Contact time: 60 min, C_0: 10–80	[62]
MWCNTs	32.68							297			
PAC	13.04							852			
As-produced CNTs		8.25						82.2	NA	pH: 2–9, Ionic strength: 0.01 M, Temperature: 280–320 K, CNT dosage: 0.5 g/L, C_0: 43.1	[63]
As-produced CNTs			18.083					134	NA	pH: 2–8, Contact time: 20–50 min, CNTs dosage: 20 mg, C_0: 1–200	[64]
MWCNTs/iron oxides			9.18					88.53	87	pH: 6.5, CNTs dosage: 2 g, Temperature: 298 K, C_0: 6	[65]
As-grown CNTs				1.1–3.5				122	NA	pH: 2–12, CNTs dosage: 0.05–0.20 g, Contact time: 4 h, C_0: 1.1–9.50	[66]
MWCNTs					1			134	NA	pH: 5, CNTs dosage: 0.05 g, Temperature: 280–321 K, C_0: 10–80	[60]
MWCNTs AC					4 18			162.16 1124.58	NA	pH: 5, Temperature: 298–323 K, CNTs dosage: 1 g, Contact time: 20–120 min, C_0: 10–60	[68]
Normal AC, AC coated with CNTs							8.25 9.0	123 755.8	66.0 72.0	pH: 2–4, Contact time: 60 min, Agitation speed: 100–200 rpm, C_0: 0.5	[69]
CNTs		27.03		9.30	62.50			98.69	NA	pH: 5, CNTs dosage: 0.05 g, Temperature: 298 K, C_0: 5–60	[70]
HNO$_3$ modified		13.87						64.3	NA	pH: 2–9, Ionic strength: 0.01 M, Temperature: 280–320 K, CNTs dosage: 0.5 g/L, C_0: 43.1	[63]
NaOCl modified		47.39						94.5			

(Continued)

TABLE 30.1 (Continued) Data of Maximum Adsorption Capacity of Heavy Metal Ions with CNTs

Adsorbent	q_m							SA	RP	Conditions	References
	Zn^{2+}	Cu^{2+}	Ni^{2+}	Cd^{2+}	Pb^{2+}	Cr^{2+}	Co^{2+}				
Oxidized CNTs	58	50.37		75.84	104.05		69.63	NA	66, 41, 63, 84,58	pH: 7, Temperature: 298 K, C_0: 100–1200	[73]
SWCNTs-COOH		65.21		50.31	78.73			NA	NA	pH: 5–6, Contact time: 60–70 min, Temperature: 293 K, C_0: 10–60	[2]
MWCNTs (oxidized)			NA					197	99	pH: 2–9, Contact time: 2 h, C_0: 2–24	[74]
Oxidized CNTs	NA	NA	NA	NA				77	96, 98, 90, 93	pH: 2–12, Reaction time: 4 h, CNTs dosage: 0.15 g, C_0: 1–5	[75]
Oxidized CNTs			49.261					145	NA	pH: 2–8, Contact time: 20–50 min, CNTs dosage: 1 g, C_0: 10–200	[64]
Oxidized SWCNTs MWCNTs			47.85 38.46					397 307	47 36	pH: 1–8, CNTs dosage: 0.05 g, Agitation speed: 180 rpm, C_0: 10–80	[76]
Oxidized SWCNTs MWCNTs			47.86 38.46					380 323	93.4 90.6	pH: 7, CNTs dosage: 1 g, Contact time: 12 h, C_0:10–80	[77]
MWCNTs SWCTNs	14.4		13.05					397 307	NA	pH: 7, CNTs dosage: 1 g, Contact time: 12 h, C_0: 10–80	[78]
Oxidized CNTs			49.261					145	NA	pH: 2–8, Contact time: 20–50 min, CNTs dosage: 20 mg, C_0: 10–200	[64]
Oxidized CNTs				34.36				NA	19.14	pH: 7, CNTs dosage: 10 mg, Contact time: 120 min, Agitation speed: 150 rpm, Temperature: 298 K, C_0: 1	[79]
Oxidized MWCNTs and ethylenediamine functionalized MWCNTs (e-MWCNT)				23.32 25.7				78.49 101.24	93 99	pH: 5.5–6.5, Temperature: 318 K, CNTs dosage: 1 mg, Contact time: 5–120 min, C_0: 0.1–5	[80]

(Continued)

TABLE 30.1 (Continued) Data of Maximum Adsorption Capacity of Heavy Metal Ions with CNTs

Adsorbent	q_m							SA	RP	Conditions	References
	Zn²⁺	Cu²⁺	Ni²⁺	Cd²⁺	Pb²⁺	Cr²⁺	Co²⁺				
MWCNTs					15.6			145	87.8	pH: 5, CNTs dosage: 0.05 g, Temperature: 280–321 K, C₀: 10–80	[60]
Oxidized MWCNTs		28.49		10.86	97.08			NA	23.7–100, 1.3–75.4, 56.1–100	pH: 6–11, CNT dosage: 0.05–0.3 g, C₀: 5–30, 2–15, 10–60	[81]
Oxidized MWCNTs					59			237.33	85	pH: 5, Temperature: 298–323 K, CNTs dosage: 1 g, C₀: 10–60	[68]
Oxidized MWCNTs					91			195.49	75.3	Temperature: 298 K, Oxidation: 1–6 h, CNTs dosage: 0.025 g, C₀: 10–100	[82]
Pristine Annealed					7.2			162.1	2.4		
					21			253.8	24.7		
Oxidized MWCNTs					2.6–11.70			197	90	pH: 7, Temperature: 298 K, Contact time: 36 h, CNTs dosage: 3 g, C₀: 1.0	[83]
Oxidized CNTs					28			153	100	pH: 5, Contact time: 10 min, Temperature: 298 K, CNTs dosage: 0.05 g, C₀: 10–80	[84]

NA, not available; q_m, maximum adsorption capacity (mg/g); SA, surface area (m²/g); RP, removal percentage (%); C₀, initial concentration (mg/L).

30.3.2 Removal of Organic Dyes

Organic dyes are one of the most hazardous materials in industrial effluents that are discharged from various industries (e.g., textiles, leather, cosmetics, and paper) and act as contaminants to the environment in general and water sources in particular. Many organic dyes have high levels of biotoxicity that cause potential mutagenic and carcinogenic effects in humans. For example, dyes such as Sudan red I, II, III, and VI, whose use in food is prohibited as a result of their toxicity or carcinogenicity at even low concentrations, are widely used in other industries. Most dye compounds contain complex aromatic structures (Figure 30.4) that make them highly resistant to biodegradation and recalcitrant to conventional biological and physical oxidation treatments. Therefore, the targeted removal of such compounds has attracted a growing amount of attention [111].

A wide range of materials has been used for the removal of organic dyes from wastewaters, including AC, zeolite, clay, and polymers, to name but a few. The current priority is to develop novel adsorbent materials with high adsorption capacities and removal efficiencies to realize effective control of these environmental pollutants. CNTs could be one of the most promising absorbents for

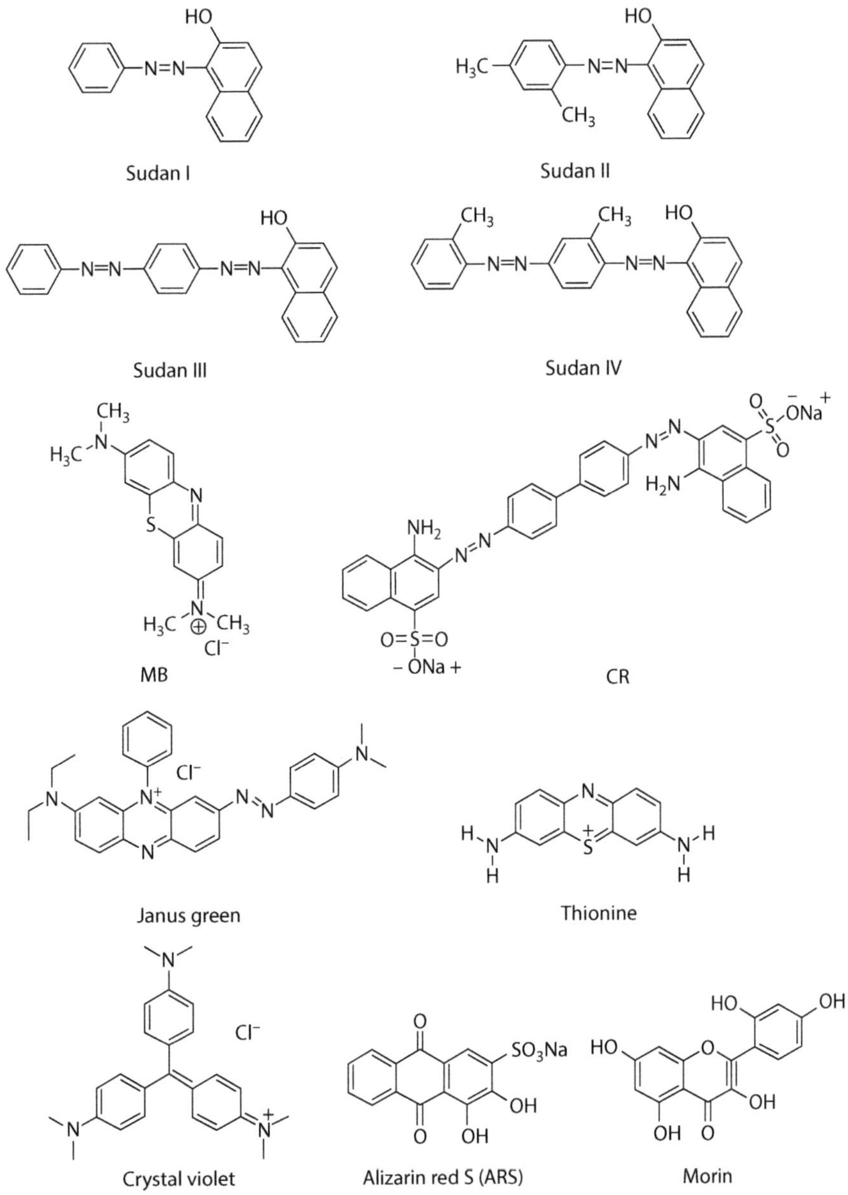

FIGURE 30.4
Chemical structures of organic dyes targeted for removal by CNTs.

this purpose because of their large adsorption capacity for organic dyes. Indeed, MWCNTs have been shown to outperform cadmium hydroxide nanowire–loaded AC (Cd(OH)$_2$–NW–AC) with respect to their efficient removal of safranine O (SO) from wastewater [112]. However, only a few reports on the application of CNTs for dye removal from aqueous solution have been published until now [113–119] and the CNTs were typically directly used without further treatment [114–116,118] (Table 30.2).

Functionalization of CNTs has been undertaken because the introduction of various functional groups can provide new adsorption sites for organic dyes. Among such modifications, oxidation is an easy method of introducing hydroxyl and carbonyl groups to the sidewalls of CNTs. Oxidized

TABLE 30.2 Removal of Different Organic Dyes Using CNTs

Type of CNT	Modification Method	Type of Dye Removed	Adsorption Effect	References
MWCNTs	Refluxed pristine MWCNTs in concentrated HNO$_3$/H$_2$SO$_4$ mixture for 4 h	Toluidine blue, MB, methyl green (MG), and bromopyrogallol red	Not provided	[113]
MWCNTs	Untreated	ECR	73.18 mg/g	[115]
MWCNTs	Untreated	Arsenazo(III)	Not provided	[116]
MWCNTs	Untreated	Alizarin red S (ARS), morin	ARS: 161.290 mg/g; morin: 26.247 mg/g	[114]
MWCNTs	Untreated	Reactive red M-2BE (RRM) textile dye	335.7 mg/g	[118]
MWCNTs	Oxidized using concentrated nitric acid	MO	Not provided	[117]
SWCNTs	Untreated	RB29	496 mg/g	[128,125]
MWCNTs	Untreated	CR	Not provided	[129]
MWCNTs	Untreated	MO	Not provided	[130]
MWCNTs	Alkali activated	MB, MO	MO: 149 mg/g; MB: 399 mg/g	[131]
MWCNTs	Untreated	Acid red 18 (Azo-Dye)	166.67 mg/g	[132]
MWCNTs	Untreated	MB, acid dye (acid red 183, AR183)	MB: 59.7 mg/g; AR183: 5.2 mg/g	[133]
MWCNTs	Untreated	Acid blue 161 (AB 161)	91.68%	[134]
MWCNTs	Untreated	Reactive blue 4 (RB4) and acid red 183 (AR183)	RB4: 69 mg/g; AR183: 45 mg/g	[135]
MWCNTs	Fabricated magnetic MWCNTs by Fenton's reagent method (M-MWCNTs)	MO	28 mg/g	[136]
SWCNTs	Pristine and oxidized	Basic red 46 (BR 46)	SWCNTs: 38.35 mg/g; oxidized: 49.45 mg/g	[137]
MWCNTs	Oxidized	5-(4-Dimethyl benzylidene amino) rhodanine (DMBAR)	15.52 mg/g	[138]
MWCNTs	Produced by Ni nanoparticle catalyzed pyrolysis of methane in a hydrogen and nitrogen flow at 650°C	MB	188.68 mg/g	[139]
MWCNTs	Untreated	Triclosan	153.1 mg/g	[140]
SWCNTs	Untreated	Reactive red 120 (RR-120)	426.49 mg/g	[141]

MWCNTs have been shown to be effective in the removal of MR [120] and methylene blue (MB) from aqueous solutions [121]. Another work has focused on the development of CNT-impregnated chitosan hydrogel beads (CSBs) for the removal of Congo red (CR). In Langmuir adsorption modeling, CSBs demonstrated a higher maximum adsorption capacity than normal chitosan CSBs (450.4 vs. 200.0 mg/g; [122]). A new generation of CSBs prepared by using sodium dodecyl sulfate and MWCNTs to improve upon their mechanical properties has also demonstrated a high maximum adsorption capacity for CR (375.94 mg/g; [123]).

Compared to MWCNTs and hybrid CNTs (HCNTs), SWCNTs can demonstrate better adsorption properties for organic contaminants because of their higher specific surface area. Indeed, SWCNTs are more efficient at removing benzene and toluene, and has shown maximum adsorption capacities of 9.98 and 9.96 mg/g, respectively [124]. A maximum adsorption capacity of 496 mg/g was achieved when reactive blue 29 (RB29) was removed from aqueous solution using SWCNTs [125].

Recently a novel self-assembled cylindrical graphene-CNT (G-CNT) hybrid was developed, and it achieved a maximum adsorption capacity of 81.97 mg/g for the removal of MB from aqueous solution and the removal efficiency reached 97% for low (10 mg/L) initial MB concentrations [126]. Finally, Zeng et al. [127] proposed a new concept of using entangled CNTs as porous frameworks to enhance the adsorption of organic dyes. The composites obtained through polymerization with polyaniline (PANI) possessed large surface areas. At an initial malachite green concentration of 16 mg/L, the CNT/PANI composites exhibited a 15% higher equilibrium adsorption capacity of 13.95 mg/g compared to neat PANI. The research on the removal and adsorption of organic dyes using CNTs is summarized in Table 30.2 [140–143].

30.4 CONCLUSIONS

The presence of heavy metal ions and dyes in wastewater is a major concern for environment conservation and human health. Until now, the process of removal of these ions has not reached the optimum conditions. It is evident from the literature survey of 141 articles that adsorption was the most frequently studied for the treatment of wastewater. CNTs have been widely used to remove heavy metal and dyes from wastewater. Adsorption was process by functionalized CNTs and low-cost adsorbents, effective, and economic for wastewater treatment.

REFERENCES

1. N. K. Srivastava, C. B. Majumder, Novel biofiltration methods for the treatment of heavy metals from industrial wastewater, *J. Hazard. Mater.* 151 (2008) 1–8.
2. O. Moradi, K. Zare, M. Monajjemi, M. Yari, H. Aghaie, The studies of equilibrium and thermodynamic adsorption of Pb(II), Cd(II) and Cu(II) Ions from aqueous solution onto SWCNTs and SWCNT–COOH surfaces, *Fullerenes, Nanotubes, Carbon Nanostruct.* 18 (2010) 285–302.
3. Y. S. Al-Degsa, M. I. E-Barghouthia, A. Issaa, M. A. Khraishehb, G. M. Walker, Sorption of Zn(II), Pb(II), and Co(II) using natural sorbents: Equilibrium and kinetic studies, *Water Res.* 40 (2006) 2645–2658.
4. R. Molinari, T. Poerio, R. Cassano, N. Picci, P. Argurio, Copper (II) removal from wastewaters by a new synthesized selective extractant and SLM viability, *Ind. Eng. Chem. Res.* 43 (2004) 623–628.
5. O. Moradi, K. Zare, Adsorption of Pb(II), Cd(II) and Cu(II) ions in aqueous solution on SWCNTs and SWCNT –COOH surfaces: Kinetics studies, *Fullerenes, Nanotubes, Carbon Nanostruct.* 19 (2011) 628–652.

6. S. Babel, T. A. Kurniawan, Low-cost adsorbents for heavy metals uptake from contaminated water: A review, *J. Hazard. Mater.* 97.1 (2003) 219–243.
7. American Water Works Association (AWWA), *Water Quality and Treatment: A Handbook of Community Water Supplies*, McGraw-Hill, New York, 2003.
8. R. C. Bansal, M. Goyal, *Activated Carbon Adsorption*, CRC Press, Boca Raton, FL, 2005.
9. D. F. Flick, H. F. Kraybill, J. M. Dlmitroff, Toxic effects of cadmium: A review, *Environ. Res.* 4 (1971) 71–85.
10. R. Khlifi, A. Hamza-Chaffai, Head and neck cancer due to heavy metal exposure via tobacco smoking and professional exposure: A review, *Toxicol. Appl. Pharm.* 248 (2010) 71–88.
11. C. B. Ernhart, A critical review of low-level prenatal lead exposure in the human: Effects on the fetus and newborn, *Reprod. Toxicol.* 6 (1992) 9–19.
12. O. Moradi, B. Mirza, M. Norouzi, A. Fakhri, Removal of Co(II), Cu(II) and Pb(II) ions by Polymer based 2-hydroxyethyl methacrylate: Adsorption and desorption, *Iran. J. Environ. Health Sci. Eng.* 9 (2012) 31.
13. H. Mahmoodian, O. Moradi, B. Shariatzadeh, Grafting chitosan and polyHEMA on carbon nanotubes surfaces: "Grafting to" and "Grafting from" methods, *Int. J. Biol Macromol.* 63 (2014) 92–97.
14. J. Ayala, F. Blanco, P. Garcia, P. Rodriguez, Asturian fly ash as a heavy metals removal material, J. Sancho, *Fuel* 77 (1998) 1147–1154.
15. O. Moradi, The removal of ions by functionalized carbon nanotube: Equilibrium, isotherms and thermodynamic studies, *Chem. Biochem. Eng. Q.* 25.2 (2011) 229–240.
16. S. C. Pan, C. C. Lin, D. Hwa, Reusing sewage sludge ash as adsorbent for copper removal from waste water Resources, *Resour. Conserv. Recycl.* 39 (2003) 79–90.
17. M. M. Rao, A. Ramesh, G. P. C. Rao, K. Seshaiah, Removal of copper and cadmium from the aqueous solutions by activated carbon derived from ceibapentandra hulls, *J. Hazard. Mater. B* 129 (2006) 123–129.
18. B. Biskup, B. Subotic, Removal of heavy metal ions from solutions using zeolites (III) influence of sodium ion concentration in the liquid phase on the kinetics of exchange processes between cadmium ions from solution and sodium ions from zeolite A, *Sep. Sci. Technol.* 39 (2004) 925–940.
19. Q. Li, S. Wu, G. Liu, X. Liao, X. Deng, D. Sun, Y. Hu, Y. Huang, Simultaneous biosorption of cadmium (II) and lead (II) ions by pretreated biomass of phanerochaete chrysosporium, *Sep. Purif. Technol.* 34 (2004) 135–142.
20. F. Ekmekyapar, A. Aslan, Y. K. Bayhan, A. Cakici, Biosorption of copper (II) by nonliving lichen biomass of *Cladonia rangiformis* Hoffm, *J. Hazard. Mater.* 137 (2006) 293–298.
21. O. Moradi, M. Aghaie, K. Zare, M. Monajjemi, H. Aghaie, The study of adsorption characteristics Cu^{2+} and Pb^{2+} ions onto PHEMA and P(MMA-HEMA) surfaces from aqueous single solution, *J. Hazard. Mater.* 170 (2009) 673–679.
22. W. Chu, Lead metal removal by recycled alum sludge, *Water Res.* 33 (1999) 3019–3025.
23. R. Sublet, M. O. Simonnot, A. Boireau, M. Sardin, Selection of an adsorbent for lead removal from drinking water by a point-of-use treatment device, *Water Res.* 37 (2003) 4904–4912.
24. S. M. Hoseyni, O. Moradi, S. Tahmacebi, Removal of COD from dairy wastewater by MWCNTs: Adsorption isotherm and modeling, *Fullerenes, Nanotubes, Carbon Nanostruct.* 21 (2013) 794–803.

25. M. Arias, M. T. Barral, J. C. Mejuto, Enhancement of copper and cadmium adsorption on kaolin by the presence of humic acids, *Chemosphere* 48 (2002) 1081–1088.
26. S. Iijima, Helical microtubules of graphitic carbon, *Nature* 354.6348 (1991) 56–58.
27. M. Ouyang, J. L. Huang, C. M. Lieber, One-dimensional energy dispersion of single-walled carbon nanotubes by resonant electron scattering, *Phys. Rev. Lett.* 88 (2002) 6.
28. O. Moradi, M. Yari, K. Zare, B. Mirza, F. Najafi, Carbon nanotubes: Chemistry principles and reactions: Review, *Fullerenes, Nanotubes, Carbon Nanostruct.* 20 (2012) 138–151.
29. E. T. Thostenson, Z. F. Ren, T. W. Chou, Advances in the science and technology of carbon nanotubes and their composites: A review, *Compos. Sci. Technol.* 61.13 (2001) 1899–1912.
30. H. E. Troiani, M. Miki-Yoshida, G. A. Camacho-Bragado, M. A. L. Marques, A. Rubio, J. A. Ascencio, M. Jose-Yacaman, Direct observation of the mechanical properties of single-walled carbon nanotubes and their junctions at the atomic level, *Nano Lett.* 3.6 (2003) 751–755.
31. X. G. Wan, J. M. Dong, D. Y. Xing, Optical properties of carbon nanotubes, *Phys. Rev. B* 58.11 (1998) 6756–6759.
32. C. Lu, H. Chiu, Adsorption of zinc (II) from water with purified carbon nanotubes, *Chem. Eng. Sci.* 61 (2006) 1138–1145.
33. C. Lu, H. Chiu, C. Liu, Removal of zinc (II) from aqueous solution by purified carbon nanotubes: Kinetics and equilibrium studies, *Ind. Eng. Chem. Res.* 45.8 (2006) 2850–2855.
34. O. Moradi, K. Zare, M. Yari, Interaction of some heavy metal ions with single walled carbon nanotube, *Int. J. Nano Dim.* 1.3 (2011) 203–220.
35. A. Stafiej, K. Pyrzynska, Adsorption of heavy metal ions with carbon nanotubes, *Sep. Purif. Technol.* 58.1 (2007) 49–52.
36. O. Moradi, A. Fakhri, SA. Adami, SE. Adami, Isotherm, thermodynamic, kinetics and adsorption mechanism studies of ethidium bromide by single-walled carbon nanotube and carboxylate group functionalized single-walled carbon nanotube, *J. Colloid. Interf. Sci.* 395 (2013) 224–229.
37. Y. H. Li, Y. Zhu, Y. Zhao, D. Wu, Z. Luan, Different morphologies of carbon nanotubes effect on the lead removal from aqueous solution, *Diamond. Relat. Mater.* 15 (2006) 90–94.
38. X. Peng, Z. Luan, Z. Di, Z. Zhang, C. Zhu, Carbon nanotubes–ironoxides magnetic composites as adsorbent for removal of Pb(II) and Cu(II) from water, *Carbon* 43 (2005) 880–883.
39. D. Robati, Pseudo-second-order kinetic equations for modeling adsorption systems for removal of lead ions using multi-walled carbon nanotube, *J. Nanostruct. Chem.* 3 (2013) 1–6.
40. Z. Saadi, R. Saadi, R. Fazaeli, Fixed-bed adsorption dynamics of Pb(II) adsorption from aqueous solution using nanostructured γ-alumina, *J. Nanostruct. Chem.* 3 (2013) 48.
41. C. Chen, X. Wang, Adsorption of Ni(II) from aqueous solution using oxidized multi-wall carbon nanotubes, *Ind. Eng. Chem. Res.* 45 (2006) 9144–9149.
42. H. Y. Li, Z. Luan, X. Xiao, X. Zhou, C. Xu, D. Wu, B. Wei, Removal Cu^{2+} ions from aqueous solutions by carbon nanotubes, *Adsorp. Sci. Technol.* 21 (2003) 475–485.

43. A. A Farghali, M. Bahgat, W. M. A. El Rouby, M. H. Khedr, Decoration of multi-walled carbon nanotubes (MWCNTs) with different ferrite nanoparticles and its use as an adsorbent, *J. Nanostruct. Chem.* 3 (2013) 1–12.

44. O. Moradi, M. Yari, P. Moaveni, M. Norouzi, Removal of p-nitrophenol and naphthalene from petrochemical wastewater using SWCNTs and SWCNT-COOH surfaces, *Fullerenes, Nanotubes, Carbon Nanostruct.* 20 (2012) 85–98.

45. C. Chen, X. Li, D. Zhao, X. Tan, X. Wang, Adsorption kinetic, thermodynamic and desorption studies of Th(IV) on oxidized multi-wall carbon nanotubes, *Colloid Surf. A: Physicochem. Eng. Asp.* 302 (2007) 449–454.

46. M. Tuzen, M. Soylak, Multiwalled carbon nanotubes for speciation of chromium in environmental samples, *J. Hazard. Mater.* 147 (2007) 219–225.

47. S. Anna, K. Pyrzynska, Adsorption of heavy metal ions with carbon nanotubes, *Sep. Purif. Technol.* 58 (2007) 49–52.

48. J. Zhang, J. K. L. Lee, Y. Wu, R. W. Murray, Photoluminescence and electronic interaction of anthracene derivatives adsorbed on sidewalls of single-walled carbon nanotubes, *Nano Lett.* 3 (2003) 403–407.

49. S. A. Curran, J. Cech, D. Zhang, J. L. Dewald, A. Avadhanula, M. Kandadai, S. Roth, Thiolation of carbon nanotubes and sidewall functionalization, *J. Mater. Res.* 21 (2006) 1012–1018.

50. O. Moradi, Adsorption behavior of basic Red 46 by single walled carbon nanotubes surfaces, *Fullerenes, Nanotubes, Carbon Nanostruct.* 21 (2013) 286–301.

51. O. Moradi, H. Sadegh, R. Shahryari-ghoshekandi, *Adsorption and Desorption in Carbon Nanotubes, Discovery and Evolution*, Lambert Academic Publishing, Saarbrücken, Germany, 2014.

52. O. Moradi, M. S. Maleki, S. Tahmasebi, Comparison between kinetics studies of protein adsorption by single walled carbon nanotube and gold nanoparticles surfaces, *Fullerenes, Nanotubes, Carbon Nanostruct.* 21 (2013) 733–748.

53. Y. Yao, F. Xu, M. Chen, Z. Xu, Z. Zhu, Adsorption behavior of methylene blue on carbon nanotubes, *Bioresour. Technol.* 101 (2010) 3040–3046.

54. S. T. Yang, S. Chen, Y. Chang, A. Cao, Y. Liu, H. Wang, Removal of methylene blue from aqueous solution by graphene oxide, *J. Colloid Interf. Sci.* 359 (2011) 24–29.

55. O. Moradi, K. Zare, Adsorption of ammonium ion by multi-walled carbon nanotube: Kinetics and thermodynamic studies, *Fullerenes, Nanotubes, Carbon Nanostruct.* 21 (2013) 449–459.

56. O. Moradi, M. Norouzi, A. Fakhri, K. Naddafi, Interaction of removal ethidium bromide with carbon nanotube: Equilibrium and isotherm studies, *J. Environ. Health Sci. Eng.* 12.1 (2014) 17.

57. J. V. Pearce, M. A. Adams, O. E. Vilches, M. R. Johnson, H. R. Glyde, One-dimensional and two-dimensional quantum systems on carbon nanotube bundles, *Phys. Rev. Lett.* 95 (2005) 185302.

58. M. Bienfait, P. Zeppenfeld, N. Dupont-Pavlovsky, M. Muris, M. R. Johnson, T. Wilson, M. DePies, O. E. Vilches, Thermodynamics and structure of hydrogen, methane, argon, oxygen, and carbon dioxide adsorbed on single-wall carbon nanotube bundles, *Phys. Rev. B* 70 (2004) 035410.

59. E. B. Mackie, R. A. Wolfson, L. M. Arnold, K. Lafdi, A. D. Migone, Adsorption studies of methane films on catalytic car bon nanotubes and on carbon filaments, *Langmuir* 13 (1997) 7197–7201.

60. Y. H. Li, S. Wang, J. Wei, X. Zhang, C. Xu, Z. Luan, D. Wu, B. Wei, Lead adsorption on carbon nanotubes, *Chem. Phys. Lett.* 357 (2002) 263–266.

61. J. W. Shim, S. J. Park, S. K. Ryu, Effect of modification with HNO_3 and NaOH on metal adsorption by pitch-based activated carbon fibers, *Carbon* 39 (2001) 1635–1642.

62. Q. R. Long, R. T. Yang, Carbon nanotubes as superior sorbent for dioxin removal, *J. Am. Chem. Soc.* 123 (2001) 2058–2059.

63. C. H. Wu, Studies of the equilibrium and thermodynamics of the adsorption of Cu^{2+} onto as-produced and modified carbon nanotubes, *J. Colloid Interface Sci.* 311 (2007) 338–346.

64. M. I. Kandah, J. L. Meunier, Removal of nickel ions from water by multi-walled carbon nanotubes, *J. Hazard. Mater.* 146 (2007) 283–288.

65. C. L. Chen, J. Hu, D. D. Shao, J. X. Li, X. K. Wang, Adsorption behavior of multiwall carbon nanotube/iron oxide magnetic composites for Ni(II) and Sr(II), *J. Hazard. Mater.* 164 (2009) 923–928.

66. Y. H. Li, S. W. Wang, Z. L. Luan, J. D. Ding, C. X. Xu, D. Wu, Adsorption of cadmium (II) from aqueous solution by surface oxidized carbon nanotubes, *Carbon* 41 (2003) 1057–1062.

67. H. J. Wang, P. A. Murphy, Isoflavone composition of American and Japanese soybeans in Iowa: Effects of variety, crop year, and location, *J. Agricult. Food Chem.* 42 (1994) 1674–1677.

68. H. J. Wang, A. L. Zhou, F. Peng, H. Yu, L. F. Chen, Adsorption characteristic of acidified carbon nanotubes for heavy metal Pb(II) in aqueous solution, *Mater. Sci. Eng. A* 466 (2007) 201–206.

69. M. A. Atieh, Removal of chromium (VI) from polluted water using carbon nanotubes supported with activated carbon, *Proc. Environ. Sci.* 4 (2011) 281–293.

70. S. H. Hsieh, J. J. Horng, C. K. Tsai, Growth of carbon nanotube on micro-sized Al_2O_3 particle and its application to adsorption of metal ions, *J. Mater. Res.* 21 (2006) 1269–1273.

71. D. H. Lin, B. S. Xing, Adsorption of phenolic compounds by carbon nanotubes: Role of aromaticity and substitution of hydroxyl groups, *Environ. Sci. Technol.* 42 (2008) 7254–7259.

72. X. J. Peng, Y. H. Li, Z. K. Luan, Z. C. Di, H. Y. Wang, B. H. Tian, Adsorption of 1,2-dichlorobenzene from water to carbon nanotubes, *Chem. Phys. Lett.* 376 (2003) 154–158.

73. M. A. Tofighy, T. Mohammadi, Adsorption of divalent heavy metal ions from water using carbon nanotube sheets, *J. Hazard. Mater.* 185 (2011) 140–147.

74. S. Yang, J. Li, D. Shao, J. Hu, X. Wang, Adsorption of Ni(II) on oxidized multi-walled carbon nanotubes: Effect of contact time, pH, foreign ions and PAA, *J. Hazard. Mater.* 166 (2009) 109–116.

75. Z. M. Gao, T. J. Bandosz, Z. B. Zhao, M. Han, J. S. Qiu, Investigation of factors affecting adsorption of transition metals on oxidized carbon nanotubes, *J. Hazard. Mater.* 167 (2009) 357–365.

76. C. Lu, C. Liu, Removal of nickel(II) from aqueous solution by purified carbon nanotubes, *J. Chem. Technol. Biotechnol.* 81 (2006) 1932–1940.

77. C. Y. Lu, C. Liu, G. P. Rao, Comparisons of sorbent cost for the removal of Ni^{2+} from aqueous solution by carbon nanotubes and granular activated carbon, *J. Hazard. Mater.* 151 (2008) 239–246.

78. C. Lu, C. Liu, F. Su, Sorption kinetics, thermodynamics and competition of Ni^{2+} from aqueous solutions onto surface oxidized carbon nanotubes, *Desalination* 249 (2009) 18–23.

79. N. A. Kabbashi, J. I. Daoud, Y. Q. Isam, M. E. S. Mirghami, N. F. Rosli, F. R. Nurhasni, Statistical analysis for removal of cadmium from aqueous solution at high pH, *J. Basic Appl. Sci.* 5.6 (2011) 440–446.

80. G. D. Vukovic, A. D. Marinkovic, M. Colic, M. D. Ristic, R. Aleksic, A. A. P. Grujic, P. S. Uskokovic, Removal of cadmium from aqueous solutions by oxidized and ethylenediamine-functionalized multi-walled carbon nanotubes, *Chem. Eng. J.* 57 (2010) 238–324.

81. Y. H. Li, J. Ding, Z. Lun, Z. Di, Y. Zhu, C. Xu, D. Wu, B. Wei, Competitive adsorption of Pb^{2+}, Cu^{2+} and Cd^{2+} ions from aqueous solutions by multiwalled carbon nanotubes, *Carbon* 41 (2003) 2787–2792.

82. H. Wang, A. Zhou, F. Peng, H. Yu, J. Yang, Mechanism study on adsorption of acidified multiwalled carbon nanotubes to Pb(II), *J. Colloid Interface Sci.* 316 (2007) 277–283.

83. D. Xu, X. Tan, C. Chen, X. Wang, Removal of Pb(II) from aqueous solution by oxidized multiwalled carbon nanotube, *J. Hazard. Mater.* 154 (2008) 407–416.

84. Y. H. Li, Z. Di, J. Ding, D. Wu, Z. Luan, Y. Zhu, Adsorption thermodynamic, kinetic and desorption studies of Pb^{2+} on carbon nanotubes, *Water Res.* 39 (2005) 605–609.

85. B. Pan, B. S. Xing, Adsorption mechanisms of organic chemicals on carbon nanotubes, *Environ. Sci. Technol.* 42 (2008) 9005–9013.

86. H. H. Cho, K. Wepasnick, B. A. Smith, F. K. Bangash, D. H. Fairbrother, W. P. Ball, Sorption of aqueous Zn[II] and Cd[II] by multiwall carbon nanotubes: The relative roles of oxygen-containing functional groups and graphenic carbon, *Langmuir* 26 (2010) 967–981.

87. H. Khani, O. Moradi, Influence of surface oxidation on the morphological and crystallographic structure of multi-walled carbon nanotubes via different oxidants, *J. Nanostruct. Chem.* 3 (2013) 73.

88. S. Rosenzweig, G. A. Sorial, E. Sahle-Demessie, J. Mack, Effect of acid and alcohol network forces within functionalized multiwall carbon nanotubes bundles on adsorption of copper(II) species, *Chemosphere* 90 (2013) 395–402.

89. J. L. Bahr, E. T. Mickelson, M. J. Bronikowski, R. E. Smalley, J. M. Tour, Dissolution of small diameter single-wall carbon nanotubes in organic solvents? *Chem. Commun.* 2 (2001) 193–194.

90. R. J. Chen, Y. Zhang, D. Wang, H. Dai, Noncovalent sidewall functionalization of single-walled carbon nanotubes for protein immobilization, *J. Am. Chem. Soc.* 123 (2001) 3838–3839.

91. M. F. Islam, E. Rojas, D. M. Bergey, A. T. Johnson, A. G. Yodh, High weight fraction surfactant solubilization of single-wall carbon nanotubes in water, *Nano Lett.* 3 (2003) 269–273.

92. C. Richard, F. Balavoine, P. Schultz, T. W. Ebbesen, C. Mioskowski, Supramolecular self-assembly of lipid derivatives on carbon nanotubes, *Science* 300 (2003) 775–778.

93. M. J. O'Connell, S. M. Bachilo, C. B. Huffman, V. C. Moore, M. S. Strano, E. H. Haroz, K. L. Rialon et al., Band gap fluorescence from individual single-walled carbon nanotubes, *Science* 297 (2002) 593–596.

94. V. C. Moore, M. S. Strano, E. H. Haroz, R. H. Hauge, R. E. Smalley, Individually suspended single-walled carbon nanotubes in various surfactants, *Nano Lett.* 3 (2003) 1379–1382.

95. S. M. Bachilo, M. S. Strano, C. Kittrell, R. H. Hauge, R. E. Smalley, R. B. Weisman, Structureassigned optical spectra of single-walled carbon nanotubes, *Science* 298 (2002) 2361–2366.

96. A. Hagen, T. Hertel, Quantitative analysis of optical spectra from individual single-wall carbon nanotubes, *Nano Lett.* 3 (2003) 383–388.

97. V. Zorbas, A. Ortiz-Acevedo, A. B. Dalton, M. M. Yoshida, G. R. Dieckmann, R. K. Draper, R. H. Baughman, M. Jose-Yacaman, I. H. Musselman, Preparation and characterization of individual peptide-wrapped single-walled carbon nanotubes, *J. Am. Chem. Soc.* 126 (2004) 7222–7227.

98. V. Zorbas, A. L. Smith, H. Xie, A. Ortiz-Acevedo, A. B. Dalton, G. R. Dieckmann, R. K. Draper, R. H. Baughman, I. H. Musselman, Importance of aromatic content for peptide/single-walled carbon nanotube interactions, *J. Am. Chem. Soc.* 127 (2005) 12323–12328.

99. N. W. S. Kam, H. Dai, Carbon nanotubes as intracellular protein transporters: Generality and biological functionality, *J. Am. Chem. Soc.* 127 (2005) 6021–6026.

100. M. Zheng, A. Jagota, M. S. Strano, A. P. Santos, P. Barone, S. G. Chou, B. A. Diner et al., Structure-based carbon nanotube sorting by sequence-dependent DNA assembly, *Science* 302 (2003) 1545–1548.

101. J. Chen, M. A. Hamon, H. Hu, Y. Chen, A. M. Rao, P. C. Eklund, R. C. Haddon, Solution properties of single-walled carbon nanotubes, *Science* 282 (1998) 95–98.

102. J. Liu, A. G. Rinzler, H. Dai, J. H. Hafner, R. K. Bradley, P. J. Boul, R. E. Smalley, Fullerene pipes, *Science* 280 (1998) 1253–1256.

103. F. Pompeo, D. E. Resasco, Water solubilization of single-walled carbon nanotubes by functionalization with glucosamine, *Nano Lett.* 2 (2002) 369–373.

104. P. W. Chiu, G. S. Duesberg, U. Dettlaff-Weglikowska, S. Roth, Interconnection of carbon nanotubes by chemical functionalization, *Appl. Phys. Lett.* 80 (2002) 3811–3813.

105. J. L. Bahr, J. M. Tour, Covalent chemistry of single-wall carbon nanotubes, *J. Mater. Chem.* 12 (2002) 1952–1958.

106. Y. Chen, R. C. Haddon, S. Fang, A. M. Rao, P. C. Eklund, W. H. Lee, R. E. Smalley, Chemical attachment of organic functional groups to single-walled carbon nanotube material, *J. Mater. Res.* 13 (1998) 2423–2431.

107. G. P. Rao, C. Lu, F. Su, Sorption of divalent metal ions from aqueous solution by carbon nanotubes: A review, *Sep. Purif. Technol.* 58 (2007) 224–231.

108. E. T. Mickelson, I. W. Chiang, J. L. Zimmerman, P. J. Boul, J. Lozano, J. Liu, J. L. Margrave, Solvation of fluorinated single-wall carbon nanotubes in alcohol solvents, *J. Phys. Chem. B* 103 (1999) 4318–4322.

109. M. A. Herrero, M. Prato, Recent advances in the covalent functionalization of carbon nanotubes, *Mol. Crystal Liquid Crystal* 438 (2008) 21–32.

110. M. Holzinger, O. Vostrowsky, A. Hirsch, F. Hennrich, M. Kappes, R. Weiss, F. Jellen, Sidewall functionalization of carbon nanotubes, *Angew. Chem. Int. Ed.* 40 (2001) 4002–4005.

111. W. J. Yang, P. Ding, L. Zhou, J. G. Yu, X. Q. Chen, F. P. Jiao, Preparation of diamine modified mesoporous silica on multi-walled carbon nanotubes for the adsorption of heavy metals in aqueous solution, *Appl. Surf. Sci.* (2013) 38–45.

112. M. Ghaedi, S. Haghdoust, S. N. Kokhdan, A. Mihandoost, R. Sahraie, A. Daneshfar, Comparison of activated carbon, multiwalled carbon nanotubes, and cadmium hydroxide nanowire loaded on activated carbon as adsorbents for kinetic and equilibrium study of removal of safranine O. *Spectrosc. Lett.* 45 (2012) 500–510.

113. M. Bahgat, A. A. Farghali, W. M. A. El Rouby, M. H. Khedr, Synthesis and modification of multiwalled carbon nano-tubes (MWCNTs) for water treatment applications, *J. Anal. Appl. Pyrol.* 92 (2011) 307–313.

114. M. Ghaedi, A. Hassanzadeh, S. N. Kokhdan, Multiwalled carbon nanotubes as adsorbents for the kinetic and equilibrium study of the removal of alizarin red S and morin, *J. Chem. Eng. Data* 56 (2011) 2511–2520.

115. M. Ghaedi, A. Shokrollah, H. Hossainian, S. N. Kokhdan, Comparison of activated carbon and multiwalled carbon nanotubes for efficient removal of eriochrome cyanine R (ECR): Kinetic, isotherm, and thermodynamic study of the removal process, *J. Chem. Eng. Data* 56 (2011) 3227–3235.

116. M. Ghaedi, A. Shokrollahi, H. Tavallali, F. Shojaiepoor, B. Keshavarz, H. Hossainian et al., Activated carbon and multiwalled carbon nanotubes as efficient adsorbents for removal of arsenazo(III) and methyl red from waste water, *Toxicol. Environ. Chem.* 93 (2011) 438–449.

117. S. W. Hu, W. J. Li, Z. D. Chang, H. Y. Wang, H. C. Guo, J. H. Zhang et al., Removal of methyl orange from aqueous solution by magnetic carbon nanotubes, *Spectrosc. Spect. Anal.* 31 (2011) 205–209.

118. F. M. Machado, C. P. Bergmann, T. H. M. Fernandes, E. C. Lima, B. Royer, T. Calvete et al., Adsorption of reactive red M-2BE dye from water solutions by multi-walled carbon nanotubes and activated carbon, *J. Hazard. Mater.* 192 (2011) 1122–1131.

119. H. Sadegh, R. Shahryari-ghoshekandi, S. Agarwal, I. Tyagi, M. Asif, V. K. Gupta, Microwave-assisted removal of malachite green by carboxylate functionalized multi-walled carbon nanotubes: Kinetics and equilibrium study, *J. Mol. Liq.* 206 (2015) 151–158.

120. M. Ghaedi, S. N. Kokhdan, Oxidized multi walled carbon nanotubes for the removal of methyl red (MR): Kinetics and equilibrium study, *Desalin. Water Treat.* 49 (2012) 317–325.

121. M. Ghaedi, H. Khajehsharifi, A. H. Yadkuri, M. Roosta, A. Asghari, Oxidized multiwalled carbon nanotubes as efficient adsorbent for bromothymol blue, *Toxicol. Environ. Chem.* 94 (2012) 873–883.

122. S. Chatterjee, M. W. Lee, S. H. Woo, Adsorption of Congo red by chitosan hydrogel beads impregnated with carbon nanotubes, *Bioresour. Technol.* 101 (2010) 1800–1806.

123. S. Chatterjee, T. Chatterjee, S. R. Lim, S. H. Woo, Effect of the addition mode of carbon nanotubes for the production of chitosan hydrogel core–shell beads on adsorption of Congo red from aqueous solution, *Bioresour. Technol.* 102 (2011) 4402–4409.

124. B. Bina, M. M. Amin, A. Rashidi, H. Pourzamani, Benzene and toluene removal by carbon nanotubes from aqueous solution, *Arch. Environ. Prot.* 38 (2012) 3–25.

125. K. Nadafi, A. Mesdaghinia, R. Nabizadeh, M. Younesian, M. J. Rad, The combination and optimization study on RB29 dye removal from water by peroxy acid and singlewall carbon nanotubes, *Desalin. Water Treat.* 27 (2011) 237–242.

126. L. H. Ai, J. Jiang, Removal of methylene blue from aqueous solution with self-assembled cylindrical graphene-carbon nanotube hybrid, *Chem. Eng. J.* 192 (2012) 156–163.

127. Y. Zeng, L. J. Zhao, W. D. Wu, G. X. Lu, F. Xu, Y. Tong et al., Enhanced adsorption of malachite green onto carbon nanotube/polyaniline composites, *J. Appl. Polym. Sci.* 127 (2013) 2475–2482.

128. M. Jahangiri-Rad, K. Nadafi, A. Mesdaghinia, R. Nabizadeh, M. Younesian, M. Rafiee, Sequential study on reactive blue 29 dye removal from aqueous solution by peroxy acid and single wall carbon nanotubes: Experiment and theory, *Iran J. Environ. Health* (2013) 10.

129. S. Ramazani, M. Ghaedi, K. Mortazavi, Multiwalled carbon nanotubes as efficient adsorbent for the removal of Congo red, *Fresenius Environ. Bull.* 20 (2011) 2514–2520.

130. Y. J. Yao, B. He, F. F. Xu, X. F. Chen, Equilibrium and kinetic studies of methyl orange adsorption on multiwalled carbon nanotubes, *Chem. Eng. J.* 170 (2011) 82–89.

131. J. Ma, F. Yu, L. Zhou, L. Jin, M. X. Yang, J. S. Luan et al., Enhanced adsorptive removal of methyl orange and methylene blue from aqueous solution by alkali-activated multiwalled carbon nanotubes, *ACS Appl. Mater. Inter.* 4 (2012) 5749–5760.

132. M. Shirmardi, A. Mesdaghinia, A. H. Mahvi, S. Nasseri, R. Nabizadeh, Kinetics and equilibrium studies on adsorption of acid red 18 (azo-dye) using multiwall carbon nanotubes (MWCNTs) from aqueous solution, *E-J. Chem.* 9 (2012) 2371–2383.

133. S. B. Wang, C. W. Ng, W. T. Wang, Q. Li, Z. P. Hao, Synergistic and competitive adsorption of organic dyes on multiwalled carbon nanotubes, *Chem. Eng. J.* 197 (2012) 34–40.

134. F. Geyikci, Adsorption of acid blue 161 (AB 161) dye from water by multi-walled carbon nanotubes, *Fullerenes, Nanotubes, Carbon Nanostruct.* 21 (2013) 579–593.

135. S. B. Wang, C. W. Ng, W. T. Wang, Q. Li, L. Q. Li, A comparative study on the adsorption of acid and reactive dyes on multiwall carbon nanotubes in single and binary dye systems, *J. Chem. Eng. Data* 57 (2012) 1563–1569.

136. F. Yu, J. H. Chen, L. Chen, J. Huai, W. Y. Gong, Z. W. Yuan et al., Magnetic carbon nanotubes synthesis by Fenton's reagent method and their potential application for removal of azo dye from aqueous solution, *J. Colloid Interface Sci.* 378 (2012) 175–183.

137. O. Moradi, Adsorption behavior of basic red 46 by single-walled carbon nanotubes surfaces, *Fullerenes, Nanotubes, Carbon Nanostruct.* 21 (2013) 286–301.

138. M. Ghaedi, P. Ghobadzadeh, S. N. Kokhdan, M. Soylak, Oxidized multiwalled carbon nanotubes as adsorbents for kinetic and equilibrium study of removal of 5-(4-dimethyl amino benzylidene)rhodanine, *Arab. J. Sci. Eng.* 38 (2013) 1691–1699.

139. Y. H. Li, Q. J. Du, T. H. Liu, X. J. Peng, J. J. Wang, J. K. Sun et al., Comparative study of methylene blue dye adsorption onto activated carbon, graphene oxide, and carbon nanotubes, *Chem. Eng. Res. Des.* 91 (2013) 361–368.

140. S. Q. Zhou, Y. S. Shao, N. Y. Gao, J. Deng, C. Q. Tan, Equilibrium, kinetic, and thermodynamic studies on the adsorption of triclosan onto multi-walled carbon nanotubes, *Clean Soil Air Water* 41 (2013) 539–547.

141. E. Bazrafshan, F. K. Mostafapour, A. R. Hosseini, A. R. Khorshid, A. H. Mahvi, Decolorisation of reactive red 120 dye by using single-walled carbon nanotubes in aqueous solutions, *J. Chem.* (2013) 938374.

142. H. Mahmoodian, O. Moradi, Rapid microwave synthesis and surface modification of chitosan-MWCNTs nanocomposites using HEMA monomers, accepted for publication, *Int. J. Biol. Macromol.* 63 (2014) 92–97.

143. H. Mahmoodian, O. Moradi, Preparation and characterization of 2-hydroxyethyl methacrylate–chitosan functionalized multiwall carbon nanotubes nanocomposites, *Polym. Compos.* 35 (3) (2014) 495–500.

CHAPTER 31

CONTENTS

The Use of Carbon Nanotubes in the Treatment of Water and Wastewater

31

Geoffrey S. Simate

31.1 INTRODUCTION

Water is one of the most important substances on earth. All plants and animals need water to survive. If there was no water, then there would be no life on earth. However, in the recent past, the demand for water has exceeded the supply of water in many places (UNEP, 2012). The demand for water has been growing rapidly as a result of increasing population coupled with rapid urbanization and industrialization (Liu et al., 2013). On the other hand, the shortage of water resources or existing freshwater is exacerbated by global warming, which is making ice to melt, sea level to rise, submergence of freshwater, and increased water evaporation (Das et al., 2014). Furthermore, pollutants such as heavy metals and distillates from various anthropogenic activities are also entering water resources (Liu et al., 2013).

To address the indisputable need of pure water, numerous water treatment technologies have been proposed and applied at experimental and field levels (Das et al., 2014). However, traditional water and wastewater treatment technologies and infrastructure are reaching their limit for providing adequate water quality to meet human and environmental requirements (Qu et al., 2013). Therefore, drinking water treatment processes should take a new dimension where the treatment plants have to accommodate the additional need for the removal of complex chemical contaminants originating from anthropogenic sources (Upadhyayula et al., 2009).

Fortunately, in recent years, advances in nanotechnology have given outstanding opportunities to develop next-generation water treatment processes. In particular, this chapter presents various potential applications of carbon nanotubes (CNTs) in the treatment of water and wastewater. The chapter is divided into six themes covering (1) properties of CNTs, (2) CNTs as adsorbents, (3) CNTs as heterogeneous coagulants and/or flocculants, (4) CNTs in membrane filtration processes, (5) CNTs for water disinfection and microbial control, and (6) concluding remarks.

31.2 PROPERTIES OF CARBON NANOTUBES

The CNTs are a form of carbon, similar to graphite found in pencils, and their name, CNT, originates from their nanometer-scale size. The structure of CNTs can be visualized as plane sheets of graphite that have been rolled into tubes (Baddour and Briens, 2005) as shown in Figure 31.1. Since their rediscovery (Monthioux and Kuznetsov, 2006) by Iijima (1991), CNTs have attracted a lot of interesting research due to their outstanding properties that have potential impact on broad areas of science and technology (Donaldson et al., 2013).

Comprehensive analyses of the properties of CNTs have been extensively discussed in various literature (Dai, 2002; Popov, 2004; Baddour and Briens, 2005; Paradise and Goswami, 2007).

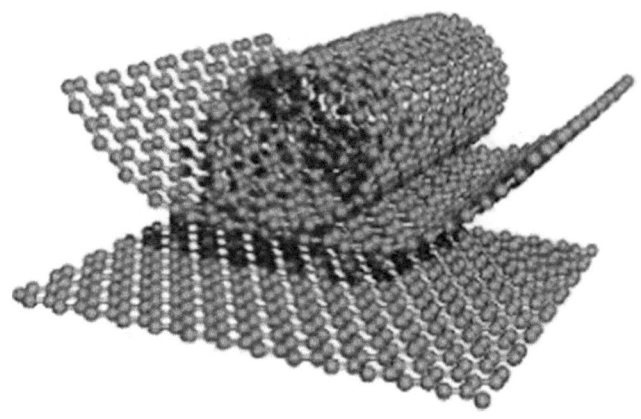

FIGURE 31.1
Schematic diagram of an individual layer of honeycomb-like carbon called graphene, and how this could be rolled in order to form a carbon nanotube. (From Endo, M. et al., *Pure Appl. Chem.*, 78(9), 1703, 2006.)

TABLE 31.1 Mechanical Properties of Carbon Nanotubes

Material	Young's Modulus (GPa)	Tensile Strength (GPa)	Density (g/cm³)
Single-wall nanotube	900–1700	75	
Multiwall nanotube	1800 average, 690–1870	150	2.6
Steel	208	0.4	7.8
Epoxy	3.5	0.005	1.25
Wood	15	0.008	0.6

Sources: Paradise, M. and Goswami, T., *Mater. Des.*, 28, 1477, 2007; Yamabe, T., *Synth. Metals*, 70(1–3), 1511, 1995.

However, this section will cite the most important properties, which make them superior to most traditional materials. It must be noted that it is the arrangement of carbon atoms that imparts great strength to the CNTs, and the cylindrical form also boasts many other remarkable properties.

Many studies have been performed on the mechanical properties of CNTs including those conducted by Treacy et al. (1996), Cornwall and Wille (1997), Lu (1997), Krishnan et al. (1998), Baddour and Briens (2005), etc. The nanotubes are far lighter than steel and are also between 10 and 100 times stronger (Rosso, 2001). They have been described as the strongest fibers known to man (Rosso, 2001). Experimental and theoretical results have shown an elastic modulus greater than 1 TPa (that of diamond is 1.2 TPa) (Treacy et al., 1996; Krishnan et al., 1998; Popov, 2004); with the elastic modulus of multiwalled CNTs (MWCNTs) being higher than that of single-walled CNTs (SWCNTs) (Yamabe, 1995; Rosso, 2001; Paradise and Goswami, 2007). It has been predicted that CNTs have the highest Young's modulus of all different types of composite tubes such as BN, BC_3, BC_2N, C_3N_4, and CN (Table 31.1) (Delmotte and Rubio, 2002; Paradise and Goswami, 2007). Due to high in-plane tensile strength of graphite, both single and multiwall nanotubes are expected to have large bending constants since they mostly depend on Young's modulus (Paradise and Goswami, 2007). In other words, due to the extremely high strength of CNTs, they can bend to very large angles without breaking up. The nanotube has also been found to be very flexible; it can be elongated, twisted, flattened, or bent into circles before fracturing (Iijima et al., 1996; Paradise and Goswami, 2007).

The CNTs have proved to be especially unique with capabilities of acting as either a metallic or semiconductor, which depends on tubule diameter and chiral angle (Paradise and Goswami, 2007). The theoretical and experimental results show that CNT have superior electrical properties compared to conventional materials (Paradise and Goswami, 2007). They can produce electric current carrying capacity 1000 times higher than copper wires (Collins and Avouris, 2000). Depending on their structure, CNTs can be almost perfect 1D conductors in which various phenomena have been observed at low temperatures (Popov, 2004; Baddour and Briens, 2005):

- Single-electron charging
- Resonant tunneling through discrete energy levels
- Proximity-induced superconductivity

The electronic capabilities possessed by CNTs are seen to arise predominately from interlayer interactions, rather than from interlayer interactions between multilayers within a single CNT or between different nanotubes (Dresselhaus et al., 1995).

Theoretical calculations using various models have also predicted CNTs to have very high thermal conductivity (Sinha et al., 2005). In fact, molecular dynamic simulations (MDS) have calculated the thermal conductivity of an isolated SWCNT to be 6.6×10^4 W/m-K, while phonon spectrum analysis of SWCNTs has found the thermal conductivity to be 6×10^4 W/m-K (Sinha et al., 2005). Nanotubes are extremely stable at high temperatures and can withstand 2800°C in a vacuum and up to 750°C at normal atmospheric pressures (Rosso, 2001). It is these thermal characteristics as well as other factors that make nanotubes so well suited to serve as electrical conductors.

Apart from the mechanical, electrical, and thermal properties, CNTs also exhibit excellent chemical properties. These interesting chemical properties make CNTs very attractive as adsorbents, heterogeneous coagulants and/or flocculants, nanofilters, additives to membranes, and disinfectants. Some of these applications utilize the smoothly scalable size-dependent properties of CNTs that relate to the high specific surface area, such as high reactivity, and strong sorption (Qu et al., 2013). However, it is difficult to synthesize CNTs with surface characteristics required for water and wastewater treatment. Therefore, surface modification and interfacial engineering are essential in making advanced CNTs of good bulk and surface properties (Lin et al., 2003). The chemistry of CNTs covering both the covalent and noncovalent reactions at the tips, outerwalls, and innerwalls of SWCNTs and MWCNTs has been extensively documented elsewhere (O'Connell et al., 2001; Dai et al., 2003; Lin et al., 2003; Matarrendona, 2003; Tasis at al., 2006; Trojanowicz, 2006; Liu, 2008), thus it is not discussed in this chapter.

31.3 CARBON NANOTUBES AS ADSORBENTS

Among water and wastewater treatment processes, adsorption has been recognized as an efficient and economical method (Rao et al., 2007). The process is commonly employed as a polishing step to remove organic and inorganic contaminants in water and wastewater treatment (Qu et al., 2013). Activated carbon has been the most widely used adsorbent in adsorption processes, but due to the depletion of coal-based activated carbons, there has been an increase in its cost (Fu and Wang, 2010). Therefore, in recent past, several cheaper adsorbents such as beer yeast (Han et al., 2006), fly ash (Weng and Huang, 1994), green algae (Malkoc and Nuhoglu, 2003), neem saw dust (Vinodhini and Das, 2010), rice husks (Luo et al., 2011), and cassava peel waste (Ndlovu et al., 2013; Simate and Ndlovu, 2015) have been investigated. However, these adsorbents have not been as efficient as activated carbon. Efficiency of the conventional and the recent studied adsorbents is usually limited by the surface area or active sites, the lack of selectivity, and the adsorption kinetics (Qu et al., 2013). On the other hand, nanoadsorbents such as CNTs offer significant improvement with their extremely high specific surface area and associated sorption sites, short intraparticle diffusion distance, and tunable pore size and surface chemistry (Qu et al., 2013).

The CNTs have been proven to possess great potential as superior adsorbents for removing many kinds of organic and inorganic pollutants such as dioxin (Long and Yang, 2001), volatile organic compounds (Agnihotri et al., 2005; Gauden et al., 2006) from air stream or fluoride (Li et al., 2003a), 1,2-dichlorobenzene (Peng et al., 2003), trihalomethanes (Lu et al., 2005), soil organic matters (Yang et al., 2006), and various divalent metal ions from aqueous solution (Li et al., 2002). Actually, several studies show that CNTs are better adsorbents than activated carbon for heavy metals (Li et al., 2003b; Lu et al., 2006), and the adsorption kinetics is fast on CNTs due to the highly accessible adsorption sites and the short intraparticle diffusion distance (Qu et al., 2013).

TABLE 31.2 Maximum Sorption Capacities of Various Divalent Metal Ions with Carbon Nanotubes

Adsorbent	Q_m				
	Cd^{2+}	Cu^{2+}	Ni^{2+}	Pb^{2+}	Zn^{2+}
CNTs	5.1			1.00	
CNTs (HNO_3)				49.95	
SWCNTs			9.22		11.23
SWCNTs (NaOCl)			47.85		43.66
MWCNTs			7.53		10.21
MWCNTs (NaOCl)			38.46		32.68
MWCNTs (HNO_3)	7.2	24.49	9.80	97.08	
CNTs	1.1				
CNTs (H_2O_2)	2.6				
CNTs ($KMnO_4$)	11.0				

Source: Rao, G.P. et al., *Sep. Purif. Technol.*, 58, 224, 2007.
Note: Q_m, maximum sorption capacity (mg/g).

TABLE 31.3 Maximum Sorption Capacities of Various Divalent Metal Ions with Other Adsorbents

Adsorbent	Q_m				
	Cd^{2+}	Cu^{2+}	Ni^{2+}	Pb^{2+}	Zn^{2+}
Fly ash	8.00	8.10			
Inactivated lichen		7.69			
Granulated activated carbon		20.55			
Powdered activated carbon					13.50
Crab shell	198.97	62.28		267.29	
Green macroalgae	4.70	5.57		28.72	2.66
Palm shell activated carbon				95.20	
Kaolinite		11.04	2.79		
Iron slag		88.50		95.24	
Modified chitosan	38.50	109.00	9.60		
Granular biomass	60.00	55.00	26.00	255.00	
Sugar beet pulp	24.39	21.16	11.86	73.76	17.79

Source: Rao, G.P. et al., *Sep. Purif. Technol.*, 58, 224, 2007.
Note: Q_m, maximum sorption capacity (mg/g).

Tables 31.2 and 31.3 show the maximum metal ion sorption capacities of raw and surface oxidized CNTs and other sorbents, respectively, as calculated by the Langmuir equation (Rao et al., 2007).

The CNTs also have better regeneration and reusability capabilities (Li et al., 2005; Lu et al., 2006, 2007). In fact, it has been reported that CNTs can be regenerated and reused up to several hundred times for Zn^{2+} removal while maintaining reasonable adsorption capacity (Lu et al., 2007). Moreover, the adsorption capacity of SWCNT and MWCNT has been found to decrease less than 25% after 10 regeneration and reuse cycles, compared to that of activated carbon that reduced by more than 50% after 1 regeneration (Lu et al., 2006).

31.4 CARBON NANOTUBES AS HETEROGENEOUS COAGULANTS AND/OR FLOCCULANTS

Over the years, coagulation and flocculation have remained the widely used methods for water and wastewater pretreatment (Simate et al., 2012a). At present, inorganic metal salts (e.g., alum and ferric chloride) and organic polymers (e.g., polyacrylamide) have been used extensively in the coagulation and flocculation processes (Simate, 2012; Simate et al., 2012a). However, most of these chemicals have several disadvantages. For example, the metal ions from inorganic flocculants or noxious monomers from polymeric coagulants that remain during the treatment of wastewaters may impinge on human health and have undesirable consequences to the environment (Yang et al., 2011; Simate et al., 2012b). As a result of these disadvantages, the search to find alternative coagulants and/or flocculants is eminent.

Recently, CNTs were evaluated for the heterogeneous coagulation and/or flocculation of brewery wastewater (Simate et al., 2012a). It can be theorized that if positive (or basic) CNTs can adsorb on separate colloidal particles, then the particles can be drawn together; a phenomenon known as bridging flocculation (Simate et al., 2011). Furthermore, the adsorption of CNTs onto particle surfaces can also result in charge neutralization (Simate et al., 2011). Once the surface charge has been neutralized, the ionic cloud dissipates and the electrostatic potential disappears resulting in a near zero net charge, so that the contact among colloidal particles occurs freely (Simate, 2012; Simate et al., 2012a). In this study, turbidity and chemical oxygen demand, including the zeta potential, were used to monitor the progress of the coagulation/flocculation process. The results showed that both pristine and hydrochloric acid functionalized CNTs demonstrated the ability to successfully coagulate colloidal particles in the brewery wastewater. Therefore, this study highlighted the applicability of acid functionalized CNTs as a novel flocculant in the removal of colloidal contaminants from wastewaters. However, ferric chloride was found to be a more effective coagulant than both the pristine and functionalized CNTs. Therefore, a major area for future research is to improve the coagulation and flocculation process and better understand the kinetics. Results also showed that the heterogeneous coagulation of colloidal particles by acid functionalized CNTs in brewery wastewater occurred by the mechanisms of charge neutralization.

The significance of CNTs in this kind of application is that by employing positively charged CNTs to wastewaters, it is expected that soluble pollutants can be adsorbed as shown from Section 31.3; at the same time, colloidal particles can be removed through heterogeneous coagulation, which results from surface charge neutralization between CNTs and colloidal particles (Simate et al., 2012a).

31.5 CARBON NANOTUBES IN MEMBRANE FILTRATION PROCESSES

Membrane technologies constitute vital units of many water treatment systems (Liu et al., 2013). Among the pressure-driven membrane processes, nanofiltration (NF) is the relatively most recent one. The NF process has separation characteristics between ultrafiltration (UF) and reverse osmosis (RO) (Trebouet et al., 2001). Compared to RO membranes, NF membranes have a looser structure and enable higher fluxes and lower operating pressures. Compared to UF membranes, NF membranes have a tighter structure and are, therefore, able to reject small organic molecules having molecular weights as low as 200–300 Da (Trebouet et al., 2001).

Membranes can be made using various types of materials including polymers, ceramics, and metals and have been manufactured using a variety of methods (Mostafavi et al., 2009). Unfortunately, most of the traditional membranes are fragile and nondurable (e.g., polymer membranes), have less throughput (e.g., metal and ceramic membranes), and are also not reusable. Therefore, there is need to fabricate reusable filters that have controlled porosity at the nanoscale and at the same time, can be formed into macroscopic structures with controlled geometric shapes, density, and dimensions (Srivastava et al., 2004). Fortunately, recent years have witnessed impressive breakthroughs toward application of nanostructured materials such as CNTs in membrane filtration. In fact, membrane filtration has been shown to be another area where it is cost effective to use CNTs. Apart from the extremely fast mass-transport properties, additional benefits of membranes based on CNTs are their ability to be functionalized—allowing them to be chemically tuned to suit applications (Majumder and Ajayan, 2010). Furthermore, CNTs have inherent properties, such as supercompressibility and electrical conductivity, which can be exploited for designing active membrane structures. This chapter dwells on two areas of membrane filtration in which CNTs may find widespread application—as nanofilters and as additives to traditional membrane materials.

31.5.1 Carbon Nanotubes as Nanofilters

The inner hollow cavities of CNTs or interstices between vertically oriented CNTs provide a great possibility for filtering water. In fact, using MDS, it has been shown that membranes comprising subnanometer diameter CNTs can provide an efficient means of water desalination when used in RO (Corry, 2008). Moreover, the high aspect ratios, smooth hydrophobic graphitic walls, and inner pore diameter of CNTs give rise to exceptionally efficient transport of water molecules (Kar et al., 2012; Das et al., 2014). In other words, the smooth and hydrophobic inner core of the hollow CNTs allows uninterrupted and spontaneous passage of water molecules with very little absorption (Das et al., 2014). The nanofilters are also advantageous over chemically intensive treatment methods that generate secondary pollutants (Lee and Baik, 2010).

Literature shows that specially aligned CNT filters have been successfully produced. For example, Srivastava et al. (2004) fabricated filtration membranes consisting of hollow cylinders with radially aligned CNT walls and efficiently carried out filtration of heavier hydrocarbon species, C_mH_n (m > 12), from hydrocarbonaceous oil such as petroleum, C_mH_n (n = 2m + 2, m = 1–12). The nanofilter was also able to remove *Escherichia coli* from drinking water and filtered nanometer-sized poliovirus. The produced CNT nanofilters had a uniform nanoporous structure that was favorable for filtration with low blockage. The cylinders also had high mechanical strength and thermal stability. The CNT filter also had a major advantage over conventional membrane filters; it could be cleaned repeatedly by simple ultrasonication and autoclaving. After regeneration, the filter regained its full filtering efficiency.

A membrane filter with both superhydrophobic and superoleophilic properties was developed by synthesizing needlelike, vertically aligned MWCNTs on a stainless steel mesh with microscale pores for the separation of oil and water (Lee and Baik, 2010). The dual-scale structure, nanoscale needlelike tubes on the mesh with microscale pores combined with the low surface energy of carbon amplified both hydrophobicity and oleophilicity. The nanotube filter could separate diesel and water layers and even surfactant-stabilized emulsions. The successful phase separation of the high viscosity lubricating oil and water emulsions was also carried out. Mostafavi et al. (2009) used spray pyrolysis method to fabricate a hollow cylindrical nanofilter from MWCNTs. The fabricated nanofilter was studied for virus separation. The results showed that the fabricated nanofilter had good water permeability, high filtrate flux, and could be used for virus removal with high efficiency. In one study, Holt et al. (2006) microfabricated membranes in which aligned CNTs with diameters of less than 2 nm served as pores. Water permeabilities of these nanotube-based membranes were found to be several orders of magnitude higher than those of commercial polycarbonate membranes, despite having smaller pore sizes.

31.5.2 Carbon Nanotubes as Additives to Membrane Filters

Membrane filtration processes have demonstrated to be very effective in removing a wide range of organic and inorganic materials. However, membrane fouling remains one of the major obstacles for the wider application of membrane processes in wastewater treatment (Xie et al., 2008). In fact, any successful application of membrane technology requires efficient control of membrane fouling (Trebouet et al., 2001). Fouling results from the excessive accumulation of suspended or dissolved contaminants on the membrane surface or within membrane pores, resulting in loss of their hydraulic permeability (Koros et al., 1996; Zhou and Smith, 2002). It depends on the interaction between membrane surface and foulants, which is related to the membrane morphology and chemistry as well as the properties of foulants (Liu et al., 2013). Therefore, modifying the surface chemistry of a membrane is an effective method to control membrane fouling. One way of changing the surface chemistry of a membrane is by adding specially developed nanoparticles (Richards et al., 2012). In fact, embedding nanoparticles to a water-purifying membrane can change its properties, making it hydrophilic or water attracting so that water passes through more easily. More importantly, however, the membrane retains its ability to filter out contaminants.

Studies in the recent past have shown that immobilizing CNTs in different types of membranes alter the solute–membrane interactions, which is one of the major physicochemical factors affecting the permeability and selectivity of a membrane (Gethard et al., 2011). The CNT immobilized in a membrane serves as a sorbent and provides an additional pathway for solute transport (Gethard et al., 2011). Research has shown that CNT-blended polysulfone membrane (Choi et al., 2006) and polyethersulfone membrane (Celik et al., 2011) are more hydrophilic and have an enhanced fouling resistance due to the hydrophilic carboxylic groups of functionalized CNTs. Other functional groups can also be introduced onto CNT surface, such as hydrophilic isophthaloyl chloride groups (Qiu et al., 2009) and amphilic polymer groups with protein-resistant ability (Liu et al., 2010). Besides combating fouling problems, there are several other advantages for blending CNTs with traditional membranes. The superior mechanical strength of CNTs results in improved tensile strength for CNT-blended membranes than the pristine membranes. The tensile strength of the MWCNT/polyacrylonitrile membranes (Majeed et al., 2012) and MWCNT/chitosan composite membranes (Tang et al., 2009) at 2 wt.% MWCNTs loading increased 97% and 90%, respectively, compared to the pristine ones. Maphutha et al. (2013) prepared a CNT-integrated polymer composite membrane with a polyvinyl alcohol barrier layer for the treatment of oil-containing wastewater. Relative to the baseline polymer, an increase of 119% in the tensile strength, 77% in the Young's modulus, and 258% in the toughness were obtained for a 7.5 wt.% MWCNTs in the polymer composite. The permeate through the membrane showed oil concentrations below the acceptable 10 mg/L limit with an excellent throughput and oil rejection of over 95%.

31.6 CARBON NANOTUBES FOR WATER DISINFECTION AND MICROBIAL CONTROL

Water disinfection is usually the final but is considered as an important process in water and wastewater treatment operations. It is a process designed for the removal, deactivation, or killing of pathogenic microorganisms thus preventing the transmission of waterborne diseases (Angeloudis et al., 2014). In order for a material to be used for water disinfection, it must exhibit potent antimicrobial activity while remaining harmless to humans at relevant doses (Mahendra et al., 2009). Currently, disinfection and microbial control practices often rely on chemical oxidants such as free chlorine, chloramines, and ozone (Mahendra et al., 2009). Though these chemicals are effective in killing and inactivating various types of bacteria and viruses, they are ineffective against cyst-forming parasites that cause severe diarrhea, such as *Giardia* and *Cryptosporidium* (Mahendra et al., 2009). Furthermore, these chemicals may produce carcinogenic by-products (Gopal et al., 2007; Mahendra et al., 2009; Simate et al., 2012b). Table 31.4 is a summary of the chlorination

TABLE 31.4 Chlorination By-Products and Their Health Effects

Class of Disinfection By-Products	Compounds	Health Effects
Trihalomethanes	Chloroform	Cancer, liver, kidney, and reproductive effects
	Dibromochloromethane	Nervous system, liver, kidney, and reproductive effects
	Bromodichloromethane	Cancer, liver, kidney, and reproductive effects
	Bromoform	Cancer, liver, kidney, and reproductive effects
Haloacetonitrile	Trichloroacetonitrile	Cancer, mutagenic, and clastogenic effects
Halogenated aldehydes/ketones	Formaldehyde	Mutagenic
Halophenol	2-Chlorophenol	Cancer and tumor promoter
Haloacetic acids	Dichloroacetic	Acid cancer and reproductive and developmental effects
	Trichloroacetic acid	Liver, kidney, spleen, and developmental effects

Source: Gopal, K. et al., *J. Hazard. Mater.*, 140, 1, 2007.

by-products and their health effects (Gopal et al., 2007). As a result of several pitfalls of chlorine and other traditional disinfectants, water scientists and engineers are seeking for alternative disinfectants.

Advances in CNT research suggest that most carbon-based nanomaterials are cytotoxic to bacteria (Fang et al., 2007; Lyon and Alvarez, 2008; Kang et al., 2009), with SWCNTs exhibiting the strongest antimicrobial activity (Kang et al., 2007; Arias and Yang, 2009). The effectiveness of SWNTs as antimicrobial agents is attributed to their unique physicochemical properties such as small diameter (<5 nm), high aspect ratio (Vecitis et al., 2010), and cylinder-like shape (Hossain et al., 2014). In general, the antimicrobial mechanisms of nanomaterials are diverse, including photocatalytic production of reactive oxygen species that inactivate viruses and cleave DNA, disruption of the structural integrity of the bacterial cell envelope resulting in leakage of intracellular components, and interruption of energy transduction (Mahendra et al., 2009). In particular, the antibacterial activity of CNTs is attributed to a physical interaction in which CNTs pierce cells (Li et al., 2008; Mauter and Elimelech, 2008) or oxidative stress that compromise cell membrane integrity (Narayan et al., 2005; Kang et al., 2007, 2008).

The cytotoxicity or antibacterial effects of CNTs could be utilized by coating and immobilizing CNTs onto membrane filter surfaces (Kang et al., 2007). The MWCNTs could also be made into hollow fibers, and bundles of nonaligned SWCNTs or MWCNTs could be applied in a packed column/filter bed to equip them with antibacterial and antiviral properties (Li et al., 2008). Previous studies have shown that including CNTs onto hollow fibers or coating microporous membrane with CNTs could inactivate bacteria and viruses (Srivastava et al., 2004; Brady-Estévez et al., 2008). All these studies clearly show that CNTs may be useful in inhibiting microbial attachment and biofilm formation on surfaces.

Despite clear indications that CNTs may be useful as antimicrobial and antiviral agents, the available evidence shows that CNTs may have adverse effects on human health. Studies have shown that CNTs may be able to enter the human body through the skin, lung, and gastrointestinal tract and can also penetrate epithelial cells and rapidly pass through the blood circulation system and accumulate in the lungs, liver, and bladder (Simate et al., 2012b). However, organized research can increase their benefits and decrease their unfavorable effects (Hossain et al., 2014). In fact, when applied properly, CNTs can act as both adsorbents and disinfectants.

31.7 CONCLUDING REMARKS

This chapter has shown that the unique properties of CNTs have the ability to revolutionize the treatment of water and wastewater. Indeed, the amazing mechanical, electrical, and chemical properties of CNTs are what make them versatile in their applications. The chapter looked at four areas that show most promising in full-scale application now and in the near future. It has shown that CNTs have a strong ability to adsorb many types of chemical and microbial contaminants thus rendering them as nanosorbents. On the other hand, studies have also shown that colloidal particles may be removed from wastewater using CNTs through heterogeneous coagulation. Therefore, by applying positively charged CNTs to wastewater, it is expected that soluble pollutants can be adsorbed, and at the same time, colloidal particles can be removed through heterogeneous coagulation, which results from surface charge neutralization between CNTs and colloidal particles.

Nanofilters comprising subnanometer diameter CNTs can provide an efficient means of filtering water. Although these narrow-pore filters reject contaminants extremely well, they still conduct water at high rates and thus are many times more efficient than existing membranes. Bacteria and viruses can be effectively removed by CNT filters through adsorption and microbe killing. Therefore, the antimicrobial activities of CNTs also make them a promising alternative for water disinfection and microbial control. The incorporation of CNTs into traditional membranes also has the potential to alter membrane surface chemistry. The CNT-immobilized membranes have the ability to resist membrane fouling for extended periods of time, increased hydrophobicity, and enhanced permeate flux.

REFERENCES

Agnihotri, S., Rood, M.J., and Rostam-Abadi, M. (2005). Adsorption equilibrium of organic vapors on single-walled carbon nanotubes. *Carbon* 43: 2379–2388.

Angeloudis, A., Stoesser, T., and Falconer, R.A. (2014). Predicting the disinfection efficiency range in chlorine contact tanks through a CFD-based approach. *Water Research* 60: 118–129.

Arias, L.R. and Yang, L.J. (2009). Inactivation of bacterial pathogens by carbon nanotubes in suspensions. *Langmuir* 25: 3003–3012.

Baddour, C.E. and Briens, C. (2005). Carbon nanotube synthesis: A review. *International Journal of Chemical Reactor Engineering* 3: R3.

Brady-Estévez, A.S., Kang, S., and Elimelech, M. (2008). A single-walled carbon-nanotube filter for removal of viral and bacterial pathogens. *Small* 4(4): 481–484.

Celik, E., Park, H., Choi, H., and Choi, H. (2011). Carbon nanotube blended polyethersulfone membranes for fouling control in water treatment. *Water Research* 45(1): 274–282.

Choi, J.H., Jegal, J., and Kim, W.N. (2006). Fabrication and characterization of multi-walled carbon nanotubes/polymer blend membranes. *Journal of Membrane Science* 284(1–2): 406–415.

Collins, P.G. and Avouris, P. (2000). Nanotubes for electronics. *Scientific American* 283: 62–69.

Cornwell, C.F. and Wille, L.T. (1997). Elastic properties of single-walled carbon nanotubes in compression. *Solid State Communications* 101(8): 555–558.

Corry, B. (2008). Designing carbon nanotube membranes for efficient water desalination. *Journal of Physics Chemistry B* 112: 1427–1434.

Dai, H. (2002). Carbon nanotubes: Opportunities and challenges. *Surface Science* 500: 218–241.

Dai, L., He, P., and Li, S. (2003). Functionalized surfaces based on polymers and carbon nanotubes for some biomedical and optoelectronic applications. *Nanotechnology* 14: 1081–1097.

Das, R., Ali, M. E., Hamid, S. B. A., Ramakrishna, S., and Chowdhury, Z. Z. (2014). Carbon nanotube membranes for water purification: A bright future in water desalination. *Desalination* 336: 97–109.

Delmotte, J.P. and Rubio, A. (2002). Mechanical properties of carbon nanotubes: A fiber digest for beginners. *Carbon* 40(10): 1729–1734.

Donaldson, K., Poland, C.A., Murphy, F.A., MacFarlane, M., Chernova, T., and Schinwald, A. (2013). Pulmonary toxicity of carbon nanotubes and asbestos—Similarities and differences. *Advanced Drug Delivery Reviews* 65(15): 2078–2086.

Dresselhaus, M., Dresselhaus, G., and Saito, R. (1995). Physics of carbon nanotubes. *Carbon* 33(7): 883–891.

Endo, M., Hayashi, T., and Kim, Y.A. (2006). Large-scale production of carbon nanotubes and their applications. *Pure and Applied Chemistry* 78(9): 1703–1713.

Fang, J.S., Lyon, D.Y., Wiesner, M.R., Dong, J.P., and Alvarez, P.J.J. (2007). Effect of a fullerene water suspension on bacterial phospholipids and membrane phase behavior. *Environmental Science and Technology* 41: 2636–2642.

Fu, F. and Wang, Q. (2010). Removal of heavy metal ions from wastewater: A review. *Journal of Environmental Management* 92: 407–418.

Gauden, P.A., Terzyk, A.P., Rychlicki, G., Kowalczyk, P., Lota, K., Raymundo-Pinero, E., Frackowiak, E., and Beguin, F. (2006). Thermodynamic properties of benzene adsorbed in activated carbons and multi-walled carbon nanotubes. *Chemical Physics Letters* 421: 409–414.

Gethard, K., Sae-Khow, O., and Mitra, S. (2011). Water desalination using carbon-nanotube-enhanced membrane distillation. *ACS Applied Materials Interfaces* 3(2): 110–114.

Gopal, K., Tripathy, S.S., Bersillon, J.L., and Dubey, S.P. (2007). Chlorination byproducts, their toxicodynamics and removal from drinking water. *Journal of Hazardous Materials* 140: 1–6.

Han, R., Li, H., Li, Y., Zhang, J., Xiao, H., and Shi, J. (2006). Biosorption of copper and lead ions by waste beer yeast. *Journal of Hazardous Materials* 137(3): 1569–1576.

Holt, J.K., Park, H.G., Wang, Y.M., Stadermann, M., Artyukhin, A.B., Grigoropoulos, C.P., Noy, A., and Bakajin, O. (2006). Fast mass transport through sub-2-nanometer carbon nanotubes. *Science* 312(5776): 1034–1037.

Hossain, F., Perales-Perez, O.J., Hwang, S., and Román, F. (2014). Antimicrobial nanomaterials as water disinfectant: Applications, limitations and future perspectives. *Science of the Total Environment* 466–467: 1047–1059.

Kang, S., Herzberg, M., Rodrigues, D.F., and Elimelech, M. (2008). Antibacterial effects of carbon nanotubes: Size does matter. *Langmuir* 24: 6409–6413.

Kang, S., Mauter, M.S., and Elimelech, M. (2009). Microbial cytotoxicity of carbon-based nanomaterials: Implications for river water and wastewater effluent. *Environmental Science and Technology* 43: 2648–2653.

Kang, S., Pinault, M., Pfefferle, L.D., and Elimelech, M. (2007). Single-walled carbon nanotubes exhibit strong antimicrobial activity. *Langmuir* 23: 8670–8673.

Kar, S., Bindal, R.C., and Tewari, P.K. (2012). Carbon nanotube membranes for desalination and water purification: Challenges and opportunities. *Nano Today* 7: 385–389.

Koros, W.J., Ma, Y.H., and Shimidzu, T. (1996). Terminology for membranes and membrane processes (IUPAC Recommendations 1996). *Journal of Membrane Science* 120: 149–159.

Krishnan, A., Dujardin, E., Ebbesen, T.W., Yianilos, P.N., and Treacy, M.M.J. (1998). Young's modulus of single-walled nanotubes. *Physical Review B* 58(20): 14013–14019.

Iijima, S. (1991). Helical microtubules of graphitic carbon. *Nature* 354: 56–58.

Iijima, S., Brabec, C., Maiti, A., and Bernholc, J. (1996). Structural flexibility of carbon nanotubes. *Journal of Chemical Physics* 104: 2089–2092.

Lee, C. and Baik, S. (2010). Vertically-aligned carbon nano-tube membrane filters with superhydrophobicity and superoleophilicity. *Carbon* 48(8): 2192–2197.

Li, Q., Mahendra, S., Lyon, D.Y., Brunet, L., Liga, M.V., Li, D., and Alvarez, P.J.J. (2008). Antimicrobial nanomaterials for water disinfection and microbial control: Potential applications and implications. *Water Research* 42: 4591–4602.

Li, Y.H., Di, Z.C., Ding, J., Wu, D.H., Luan, Z.K., and Zhu, Y.Q. (2005). Adsorption thermodynamic, kinetic and desorption studies of Pb^{2+} on carbon nanotubes. *Water Research* 39(4): 605–609.

Li, Y.H., Ding, J., Luan, Z.K., Di, Z.C., Zhu, Y.F., Xu, C.L., Wu, D.H., and Wei, B.Q. (2003b). Competitive adsorption of Pb^{2+}, Cu^{2+} and Cd^{2+} ions from aqueous solutions by multiwalled carbon nanotubes. *Carbon* 41(14): 2787–2792.

Li, Y.H., Wang, S., Wei, J., Zhang, X., Xu, C., Luan, Z., Wu, D., and Wei, B. (2002). Lead adsorption on carbon nanotubes. *Chemical Physics Letters* 357: 263–266.

Li, Y.H., Wang, S., Zhang, X., Wei, J., Xu, C., Luan, Z., and Wu, D. (2003a). Adsorption of fluoride from water by aligned carbon nanotubes. *Materials Research Bulletin* 38: 469–476.

Lin, T., Bajpai, V., Ji, T., and Dai, L. (2003). Chemistry of carbon nanotubes. *Australian Journal of Chemistry* 56: 635–651.

Liu, R. 2008. The functionalisation of carbon nanotubes. PhD thesis, University of New South Wales, Sydney, New South Wales, Australia.

Liu, X., Wang, M., Zhang, S., and Pan, B. (2013). Application potential of carbon nanotubes in water treatment: A review. *Journal of Environmental Sciences* 25(7): 1263–1280.

Liu, Y.L., Chang, Y., Chang, Y.H., and Shih, Y.J. (2010). Preparation of amphiphilic polymer-functionalized carbon nanotubes for low-protein-adsorption surfaces and protein-resistant membranes. *ACS Applied Materials and Interfaces* 2(12): 3642–3647.

Long, R.Q. and Yang R.T. (2001). Carbon nanotubes as superior sorbent for dioxin removal. *Journal of American Chemical Society* 123: 2058–2059.

Lu, C., Chiu, H., and Bai, H. (2007). Comparisons of adsorbent cost for the removal of zinc (II) from aqueous solution by carbon nanotubes and activated carbon. *Journal of Nanoscience and Nanotechnology* 7(4–5): 1647–1652.

Lu, C., Chung, Y.L., and Chang, K.F. (2005). Adsorption of trihalomethanes from water with carbon nanotubes. *Water Research* 39: 1183–1189.

Lu, C.S., Chiu, H., and Liu, C.T. (2006). Removal of zinc (II) from aqueous solution by purified carbon nanotubes: Kinetics and equilibrium studies. *Industrial and Engineering Chemistry Research* 45(8): 2850–2855.

Lu, J.P. (1997). Elastic properties of carbon nanotubes and nanoropes. *Physical Review Letters* 79: 1297–1300.

Luo, X., Deng, Z., Lin, X., and Zhang, C. (2011). Fixed-bed column study for Cu^{2+} removal from solution using expanding rice husk. *Journal of Hazardous Materials* 187: 182–189.

Lyon, D.Y. and Alvarez, P.J.J. (2008). Fullerene water suspension (nC(60)) exerts antibacterial effects via ROS-independent protein oxidation. *Environmental Science and Technology* 42: 8127–8132.

Mahendra, S., Li, Q., Lyon, D.Y., Brunet, L., and Alvarez, P. (2009). Nanotechnology-enabled water disinfection and microbial control: Merits and limitations. In: Nora, S., Mamadou, D., Jeremiah, D., Anita, S., and Richard, S. (eds.), *Nanotechnology Applications for Clean Water*, pp. 157–166. William Andrew Publishing, Boston, MA.

Majeed, S., Fierro, D., Buhr, K., Wind, J., Du, B., Boschetti-De-Fierro, A. et al. (2012). Multi-walled carbon nanotubes (MWCNTs) mixed polyacrylonitrile (PAN) ultrafiltration membranes. *Journal of Membrane Science* 403–404: 101–109.

Majumder, M. and Ajayan, P.M. (2010). Carbon nanotube membranes: A new frontier in membrane science. In: Drioli, E. and Giorno, L. (Eds.), *Comprehensive Membrane Science and Engineering*, Vol. 1. Elsevier, Amsterdam, the Netherlands.

Malkoc, E. and Nuhoglu, Y. (2003). The removal of chromium(VI) from synthetic wastewater by *Ulothrix zonata*. *Fresenius Environmental Bulletin* 12(4): 376–381.

Matarredona, O., Rhoads, H., Li, Z., Harwell, J.H., Balzano, L., and Resasco, D.E. (2003). Dispersion of single-walled carbon nanotubes in aqueous solutions of the anionic surfactant NaDDBS. *Journal of Physical Chemistry B* 107: 13357–13367.

Mauter, M.S. and Elimelech, M. (2008). Environmental applications of carbon-based nanomaterials. *Environmental Science Technology* 42(16): 5843–5859.

Maphutha, S., Moothi, K., Meyyappan, M., and Iyuke, S.E. (2013). A carbon nanotube-infused polysulfone membrane with polyvinyl alcohol layer for treating oil-containing waste water. *Scientific Reports* 3: 1509.

Monthioux, M. and Kuznetsov, V.L. (2006). Who should be given the credit for the discovery of carbon nanotubes? *Carbon* 44: 1621–1623.

Mostafavi, S.T., Mehrnia, M.R., and Rashidi, A.M. (2009). Preparation of nanofilter from carbon nanotubes for application in virus removal from water. *Desalination* 238: 271–280.

Narayan, R.J., Berry, C.J., and Brigmon, R.L. (2005). Structural and biological properties of carbon nanotube composite films. *Material Science and Engineering B* 123: 123–129.

Ndlovu, S., Simate, G.S., Seepe, L., Shemi, A., Sibanda, V., and van Dyk, L. (2013). The removal of Co^{2+}, V^{3+} and Cr^{3+} from waste effluents using cassava waste. *South African Journal of Chemical Engineering* 18(1): 1–19.

O'Connell, M.J., Boul, P., Ericson, L.M., Huffman, C., Wang, Y., Haroz, E., Kuper, C., Tour, J., Ausman, K.D., and Smalley, R.E. 2001. Reversible water-solubilisation of single-walled carbon nanotubes by polymers wrapping. *Chemical Physics Letters* 342: 265–271.

Paradise, M. and Goswami, T. (2007). Carbon nanotubes—Production and industrial applications. *Materials and Design* 28: 1477–1489.

Peng, X., Li, Y., Luan, Z., Di, Z., Wang, H., Tian, B., and Jia, Z. (2003). Adsorption of 1,2-dichlorobenzene from water to carbon nanotubes. *Chemical Physics Letters* 376: 154–158.

Popov, V.N. (2004). Carbon nanotubes: Properties and applications. *Materials Science and Engineering Reports* 43: 61–102.

Qiu, S., Wu, L. G., Pan, X. J., Zhang, L., Chen, H.L., and Gao, C.J. (2009). Preparation and properties of functionalized carbon nanotube/PSF blend ultrafiltration membranes. *Journal of Membrane Science* 342(1–2): 165–172.

Qu, X., Alvarez, P.J.J., and Li, Q. (2013). Applications of nanotechnology in water and wastewater treatment. *Water Research* 47: 3931–3946.

Rao, G.P., Lu, C., and Su, F. (2007). Sorption of divalent metal ions from aqueous solution by carbon nanotubes: A review. *Separation and Purification Technology* 58: 224–231.

Richards, H.L., Baker, P.G.L., and Iwuoha, E. (2012). Metal nanoparticle modified polysulfone membranes for use in wastewater treatment: A critical review. *Journal of Surface Engineered Materials and Advanced Technology* 2: 183–193.

Rosso, M.A. (2001). Origins, properties, and applications of carbon nanotubes and fullerenes, IT 283 Advance materials and processes. California State University, Fresno, CA.

Simate, G.S. (2012). The treatment of brewery wastewater using carbon nanotubes synthesized from carbon dioxide carbon source. PhD thesis, University of the Witwatersrand, Braamfontein, Johannesburg.

Simate, G.S., Cluett, J., Iyuke, S.E., Musapatika, E.T., Ndlovu, S., Walubita, L.F., and Alvarez, A.E., 2011. The treatment of brewery wastewater for reuse: State of the art. *Desalination* 273: 235–247.

Simate, G.S., Iyuke, S.E., Ndlovu, S., and Heydenrych, M. (2012a). The heterogeneous coagulation and flocculation of brewery wastewater using carbon nanotubes. *Water Research* 46 (4): 1185–1197.

Simate, G.S., Iyuke, S.E., Ndlovu, S., Heydenrych, M., and Walubita, L.F. (2012b). Human health effects of residual carbon nanotubes and traditional water treatment chemicals in drinking water. *Environment International* 39(1): 38–49.

Simate, G.S. and Ndlovu, S. (2015). The removal of heavy metals in a packed bed column using immobilized cassava peel waste biomass. *Journal of Industrial and Engineering Chemistry* 21: 635–643.

Sinha, S., Barjami, S., Iannacchione, G., Schwab, A., and Muench, G. (2005). Off-axis thermal properties of carbon nanotube films. *Journal of Nanoparticle Research* 7: 651–657.

Srivastava, A., Srivastava, O.N., Talapatra, S., Vajtai, R., and Ajayan, P.M. (2004). Carbon nanotube filters. *Nature Materials* 3: 610–614.

Tang, C. Y., Zhang, Q., Wang, K., Fu, Q., and Zhang, C.L. (2009). Water transport behavior of chitosan porous membranes containing multi-walled carbon nanotubes (MWNTs). *Journal of Membrane Science* 337(1–2): 240–247.

Tasis, D., Tagmatarchis, N., Bianco, A., and Prato, M. (2006).Chemistry of carbon nanotubes. *Chemical Reviews* 106(3): 1105–1136.

Treacy, M.M.J., Ebbesen, T.W., and Gibson, J.M. (1996). Exceptionally high Young's modulus observed for individual carbon nanotubes. *Nature* 381: 678–680.

Trebouet, D., Schlumpf, J.P., Jaouen, P., and Quemeneur, F. (2001). Stabilized landfill leachate treatment by combined physicochemical-nanofiltration processes. *Water Research* 35(12): 2935–2942.

Trojanowicz, M. (2006). Analytical applications of carbon nanotubes: A review. *Trends in Analytical Chemistry* 25(5): 480–489.

UNEP. (2012). The fifth global environmental outlook report, Chapter 4: Water. http://www.unep.org/geo/pdfs/geo5/GEO-5_WATER-small.pdf (accessed May 2012).

Upadhyayula, V.K.K., Deng, S., Mitchell, M.C., and Smith, G.B. (2009). Application of carbon nanotube technology for removal of contaminants in drinking water: A review. *Science of the Total Environment* 408: 1–13.

Vecitis, C.D., Zodrow, K.R., Kang, S., and Elimelech, M. (2010). Electronic-structure-dependent bacterial cytotoxicity of single-walled carbon nanotubes. *American Chemical Society Nano* 4(9): 5471–5479.

Vinodhini, V. and Das, N. (2010). Packed bed column studies on Cr(VI) removal from tannery wastewater by neem sawdust. *Desalination* 264: 9–14.

Weng, C. and Huang, C. (1994). Treatment of metal industrial wastewater by fly ash and cement fixation. *Journal of Environmental Engineering* 120(6): 1470–1487.

Xie, X., Zhou, H., Chong, C., and Holbein, B. (2008). Coagulation assisted membrane filtration to treat high strength wastewater from municipal solid waste anaerobic digesters. *Journal of Environmental Engineering Science* 7: 21–28.

Yamabe, T. (1995). Recent development of carbon nanotube. *Synthetic Metals* 70(1–3): 1511–1518.

Yang, K., Zhu, L., and Xing, B. (2006). Adsorption of polycyclic aromatic hydrocarbons by carbon nanomaterials. *Environmental Science and Technology* 40: 1861–1866.

Yang, Z., Shang, Y., Lu, Y., Chen, Y., Huang, X., Chen, A., Jiang, Y., Gu, W., Qian, X., Yang, H., and Cheng, R. (2011). Flocculation properties of biodegradable amphoteric chitosan-based flocculants. *Chemical Engineering Journal* 172(1): 287–295.

Zhou, H. and Smith, D.W. (2002). Advanced technologies in water and wastewater treatment. *Journal of Environmental Engineering Science* 1: 247–264.

CHAPTER 32

CONTENTS

Application of Functionalized Carbon-Based Nanomaterials in Membrane Separation

Carbon Nanotubes and Graphene

32

Qian Wen Yeang, Abu Bakar Sulong, and Soon Huat Tan

32.1 INTRODUCTION

Membrane separation processes have attracted much attention because of their potential in replacing conventional energy-intensive separation technologies such as distillation and evaporation. Membrane separation offers advantages such as high stability and efficiency, low operating and capital cost, low energy requirement, and ease of operation (Mulder, 1991; Ho and Sirkar, 1992; Baker, 2004). As such, it is a promising candidate in separation processes. Currently, membrane-based separation has been extensively applied in industries for liquid and gas separation.

Nonetheless, the ability of existing membranes in withstanding harsh operating environment and the economic competitiveness of the existing membrane technologies indicated the need to seek for new membrane materials with improved permeability and selectivity. The development of polymeric and inorganic membranes is hindered by the trade-off between permeability and selectivity. Polymeric membranes also suffer from poor solvent, chemical, and thermal resistance (Verkerk et al., 2001). Although inorganic membranes have superior solvent-resistant properties as well as thermal and pore structure stability than polymeric ones (Pietraß, 2006; Li, 2007; Li et al., 2007), the application of inorganic membranes is still being hindered by high membrane production cost, difficulty in handling, and deficient technology to produce continuous and defect-free membranes (Zimmerman et al., 1997; Caro et al., 2000; Ciobanu et al., 2008).

Currently, membrane separation science and technology is experiencing a revolution given the increasing number of studies involving the use of nanotechnology. Carbon-based nanomaterials are promising candidates for improving the efficiency and capability of current membrane separation systems. These nanomaterials have also been extensively applied in the fields of drug delivery, electronics, structural materials, bioimaging, biosensing, and energy conservation (Baughman et al., 2002) because of their unique physical, chemical, and electronic properties, which are due to their hybridization state (structural confirmation) (Ajayan, 1999). Among recently developed carbon-based nanomaterials, carbon nanotubes (CNTs) and graphene have been the most widely studied for next-generation membrane separation technologies in water desalination, pervaporation, and gas separation applications. Generally, carbon-based nanomaterials are either employed directly or functionalized with a specific functional group to improve membrane functionality and efficiency.

This chapter attempts to provide a comprehensive review of functionalized carbon-based nanomaterials with focus on CNTs and graphene as potential materials that can improve membrane separation technologies. Recent development and contributions of functionalized CNTs and graphene in both liquid and gas separation via membrane separation technologies are discussed. Lastly, the challenges and future research direction in the application of these functionalized carbon-based nanomaterials are briefly outlined.

32.2 CARBON NANOTUBES IN MEMBRANE SEPARATION PROCESS

CNTs are nanoscale cylinders made up of rolled-up graphene sheets comprised of carbon atoms and capped at one or both ends with a half fullerene (Iijima, 1991). The outstanding properties of CNTs such as exceptional mechanical strength as well as their high electrical and thermal conductance make them suitable for a wide range of applications in structural materials, energy storage devices, adsorbents, semiconductors, and other electronics (Dresselhaus et al., 1995; Baughman et al., 2002).

The exceptionally smooth and hollowed structure of nanotubes could also facilitate the rapid transport of liquid and gas molecules through the channels, thereby resulting in high flux membrane separation performance. Water transport through CNTs is two to five times higher than the theoretical predictions by the Hagen–Poiseuille equation (Holt et al., 2006; Ahadian and Kawazoe, 2009), whereas gas transport is over an order of magnitude larger than the Knudsen diffusion predictions (Holt et al., 2006). Molecular dynamic simulations show that the exceptionally fast flow rate is due to the atomic smoothness and molecular ordering, in which water molecules are passed through CNTs in a 1D single-file procession (Hummer et al., 2001; Kalra et al., 2003).

Moreover, the pore entrance of CNTs could be functionalized to facilitate selective transport and is often termed as *gate-keeper controlled* separation. Functionalization of CNTs can maximize the possible benefits of CNTs and improve separation performance. Specific physicochemical characteristics can be introduced through the functionalization of CNT tips, which could facilitate the selective removal of targeted molecules based on the physicochemical interaction of target molecules with the functional group attached on the CNT tips. CNTs could also be functionalized with organic moieties that could provide enhanced CNT attachment in the host materials to improve reinforcement in composite materials. Apart from that, the surface and tip of CNTs can be functionalized accordingly to alleviate both hydrophilic and hydrophobic foulings.

The application of functionalized CNTs in membrane-based liquid separation, starting with desalination and water purification followed by pervaporation, will be presented in the next section. Meanwhile, their application in membrane-based gas separation will be discussed in the successive section.

32.2.1 Carbon Nanotubes in Liquid Separation

32.2.1.1 Desalination and water purification

CNTs have been widely exploited in the field of liquid separation with focus on water desalination and purification process. As CNTs are capped with a half fullerene and contain a lot of impurities, pretreatment of CNTs is often needed before application in water purification. On top of that, pristine CNTs form agglomeration easily, which leads to a significant decrease in membrane flux and rejection capacities. After pretreatment, the capped ends of CNTs are unzipped to form open tips, thereby enabling attachment with specific functional groups. CNT functionalization causes the CNT membranes to be selective for a specific pollutant while increasing the water flux through the hollow tubes. Promising results have been reported for functionalized CNT membranes in desalination processes. These results include good water permeability, fouling resistance, pollutant degradation and self-cleaning functions, improved mechanical and thermal properties, and long-term membrane stability.

32.2.1.1.1 Filtration

Filtration is an extremely important water purification process to remove bacteria and other organisms from drinking water. Apart from that, it is also an essential process in the industry for wastewater treatment. Studies have shown that the incorporation of functionalized CNTs in filtration membranes exhibited great improvement in the field of membrane filtration process.

Choi et al. (2006) reported that acid-treated multiwalled CNTs (MWCNTs) with blended polysulfone (PSf) membranes show enhanced flux and rejection compared with PSf membranes without MWCNTs. The increased hydrophilicity of the membrane was attributed to the carboxylic acid functional groups developed by the treatment of MWCNTs surfaces with concentrated nitric (HNO_3) and sulfuric acids (H_2SO_4). Similar observations were reported by Celik et al. (2011) who blended acid-treated MWCNTs with polyethersulfone (PES). Fouling resistance was also exhibited during surface water filtration as the blend membranes demonstrated reduced flux decline compared with pure PES membranes. More recently, Saranya et al. (2014) reported similar findings when applying acid-treated MWCNT/PES blend ultrafiltration membranes in paper mill effluent treatment. Acid-treated MWCNTs were introduced into different polymers for other applications, such as protein separation and oil sand process-affected water treatment; similar trends were also observed (Kumar and Ulbricht, 2013; Kim et al., 2013a).

In a different context, De Lannoy et al. (2013) studied the relationship between the degrees of carboxylation of MWCNTs with the changes in the tensile strength, hydrophilicity, and water flux of MWCNT/PSf ultrafiltration membranes. Although CNTs with a high degree of functionality form homogeneous polymer solutions leading to enhanced membrane characteristics, CNTs have a higher tendency to leave the membrane during membrane cleaning. Thus, establishing a balance between these benefits and shortcomings is very important because CNTs may have negative effects on humans and environmental ecosystems once they are released.

Kim et al. (2012) fabricated a relatively new thin-film nanocomposite membrane via interfacial polymerization of a support layer that consists of acid-modified MWCNTs and a thin-film layer of nanosilver particles. The thin-film nanocomposite membranes exhibited enhanced permeability and antifouling properties because of the acid-modified MWCNTs and nanosilver particles.

Apart from the common flat sheet membranes, acid-treated MWCNTs were also employed in different membrane configurations such as hollow fiber membranes. Compared with conventional flat sheet membranes, hollow fiber membranes are more advantageous in wider applications due to their large membrane surface areas that could be packed per unit volume. The optimized acid-modified MWCNT/PSf mixed matrix hollow fiber membranes synthesized by Yin et al. (2013) demonstrated a remarkable increase in pure water flux of up to 100% without sacrificing the solute rejection capability, and its antifouling properties were also enhanced.

Aside from modification with strong acids, CNTs can be functionalized with different functional groups, polymer chains, and other nanoparticles to increase separation performance. Qiu et al. (2009) grafted the MWCNTs surface with isocyanate and isophthaloyl chloride groups by reacting carboxylated MWCNTs and 5-isocyanato-isophthaloyl chloride to improve MWCNTs and PSf polymer matrix compatibility. In a different approach, Rahimpour et al. (2012) functionalized MWCNTs with amine groups (NH_2) through the chemical treatment of strong acids (H_2SO_4/HNO_3) and 1,3-phenylenediamine to improve the dispersion of MWCNTs in the PES membrane. Both studies reported an enhancement in water flux and a reduction in fouling because of the increased membrane surface hydrophilicity attributed to the addition of functionalized MWCNTs.

Polymer chains such as polycaprolactone (Mansourpanah et al., 2011), hyperbranched poly(amine-ester) (Zhao et al., 2012), and poly(methyl methacrylate) (Shen et al., 2013) can also be grafted to the surface of MWCNTs to enhance the compatibility of MWCNTs with an organic solvent or polymer. Daraei et al. (2013) conducted a comparative study on the effects of introducing four differently functionalized MWCNTs (i.e., acid-functionalized MWCNTs and MWCNTs functionalized with hydrophilic polymer groups such as citric acid, acrylic acid, and acrylamide) on PES membrane performance. Membrane separation performance, such as water flux and fouling resistance, was improved for membranes incorporated with polymer-modified MWCNTs because of the higher compatibility with the polymer matrix and the presence of more functional groups compared with acid-functionalized MWCNTs. In a similar study, Shah and Murthy (2013) reported the effects of different functionalizations of MWCNTs (i.e., oxidized, amide, and azide) in the separation performance of MWCNT/PSf membranes in heavy metal removal. The addition of functionalized MWCNTs increased membrane hydrophilicity and decreased pore size, thereby improving the efficiency of heavy metal removal. The best results were exhibited by amide-functionalized MWCNTs.

Aside from polymer chains, nanoparticles such as titanium dioxide (TiO_2) can also be used to functionalize MWCNTs. Vatanpour et al. (2012) coated oxidized MWCNTs with TiO_2 to form a nanofiltration PES membrane. The resulting PES membrane exhibited excellent antibiofouling property because of the membrane surface hydrophilicity and synergistic photocatalytic activity induced by TiO_2 nanoparticles.

32.2.1.1.2 Reverse osmosis

At present, reverse osmosis (RO) is widely applied in water desalination technologies worldwide. Commercial interest in RO technology is increasing globally due to continuous process improvements, which in turn lead to significant cost reductions. Although commercially available membranes exhibit relatively good performance in many applications, the rapid depletion of water resources has urged the development of membranes with enhanced productivity, selectivity, fouling resistance stability, and availability at a relatively lower cost. Over the past decade, nanotechnology concepts have triggered the synthesis of new water treatment membranes with state-of-the-art performance, including high permeability, selectivity, and fouling resistance. As mentioned earlier, CNTs have become one of the most popular candidates because of their exceptionally high transport of liquid and gas through the smooth and hollowed structure of CNTs as well as their ability to be functionalized.

In an attempt to overcome the problems faced in commercially used polyamide (PA) RO membranes, Chan et al. (2013) incorporated zwitterion-functionalized CNTs into a PA composite membrane. Both simulations and experimental results proved the potential of zwitterion-functionalized CNTs in developing desalination membranes with high efficiency. When the fraction of functionalized CNTs increased, the water flux increased significantly and the ion rejection ratio increased or remained unchanged. This condition suggested that the increased water flux is not attributed to an increase in nonspecific pores in the membrane but rather attributed to an additional transport mechanism caused by the existence of the functionalized CNTs. Molecular simulations also demonstrated that the addition of only two zwitterions per nanotube end induced the complete rejection of ions, while allowing significant water flux for nanotubes with similar diameters utilized in the experiments mainly because of the steric hindrance of zwitterions. Similar observations were reported by Kim et al. (2014) who incorporated acid-modified CNTs into PA RO membranes. The improved membrane performance in terms of water flux and salt rejection as well as the enhanced membrane stability containing acid-modified CNTs may be attributed to the hydrophobic nanochannels of CNTs and interactions between CNTs and PA in the active layers. The transport of water molecules is illustrated in Figure 32.1.

In another research, Zhao et al. (2014a) reported the effectiveness of MWCNTs grafted with abundant long chains of aliphatic acid in facilitating the compatibility of MWCNTs in the PA layer, which led to improved membrane performance in the desalination process. An improvement in flux without sacrificing solute rejection and antifouling property was observed with increasing MWCNT loading. This condition suggested that the incorporation of modified MWCNTs in membranes is effective in enhancing membrane performance.

Vertically aligned CNT membranes also have a greater advantage over conventional RO membranes. Corry (2008) investigated the applicability of vertically aligned CNTs in water desalination through RO by studying water and ion transport through membranes formed from vertically aligned CNTs with diameters ranging from 6 to 11 Å via molecular dynamic simulations at hydrostatic pressure and equilibrium conditions. It was reported that the ions encounter a large energy barrier and are unable to pass through the narrower tubes studied (i.e., (5, 5) and (6, 6) *armchair*-type tubes) but are able to pass through the wider (7, 7) and (8, 8) nanotubes. From this study, CNTs that are incorporated into porous membranes can, in principle, achieve flow rates much higher than those of existing membranes in water desalination through RO. A few years later, Corry (2011) examined water and ion transport through functionalized CNTs via molecular dynamics simulations to study the effect of chemical modification on the performance of the aligned CNTs membranes. The (8, 8)

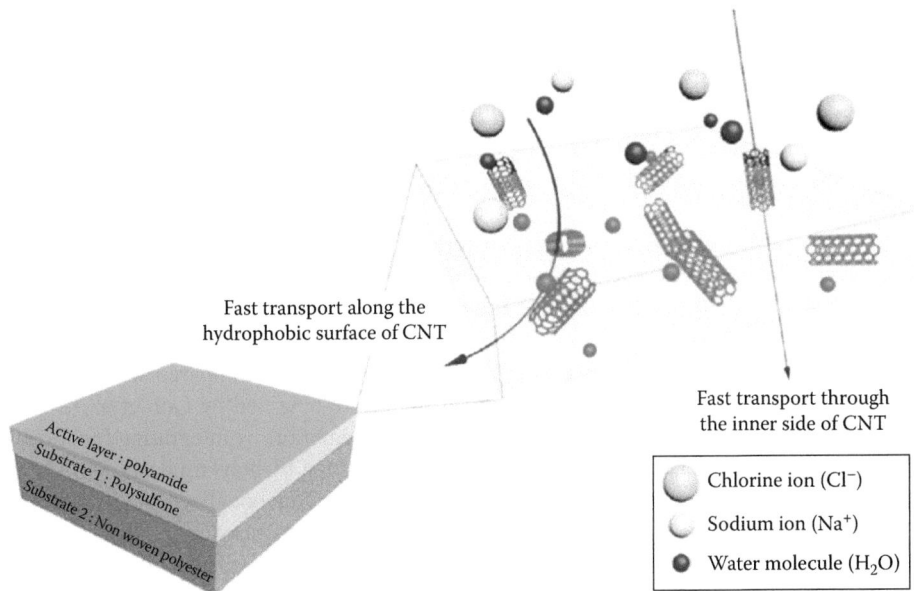

FIGURE 32.1
Schematic illustration of the fast transport of water molecules. (From Kim, H.J. et al., *Appl. Mater. Interf.*, 6, 2819, 2014.)

CNTs with a diameter of 1.1 nm, which were previously reported to be only fairly effective in ion rejection, were modified with a variation in charged and polar functional groups. The functionalization of the pore entrance prevented the ions from passing but also decreased the flow of water through the membrane, which can be attributed to the increased electrostatic interactions between water molecules and CNTs. However, the author concluded that the inclusion of functionalized CNTs in desalination membranes is useful in achieving salt rejection and rapid water flow considering that the performance of these membranes in the simulations is still much better than existing technologies. Although vertically aligned CNTs membranes appeared to be promising for high-performance RO membranes, fabricating vertically aligned CNT membranes with uniform porosity is still very difficult.

32.2.1.1.3 Forward osmosis

Filtration and RO have been widely utilized in water purification, water desalination, and wastewater reclamation. Nevertheless, the aforementioned membrane processes are pressure driven, wherein a high hydraulic pressure is required to separate water from a solution. Forward osmosis (FO) is a low-cost and more environmentally friendly desalination technology that is capable of inducing a net flow of water by employing osmotic pressure as a driving force. FO has been considered as a possible candidate to replace energy-intensive, pressure-driven separation processes. Nonetheless, the development of the FO process is hindered by the lack of high-performance membranes. This scenario has forced the development of new membrane materials by incorporating nanoparticles such as CNTs to enhance water flux without sacrificing salt rejection and membrane fouling resistance.

Using molecular dynamic simulations, Jia et al. (2010) discovered that the incorporation of CNTs in FO membranes showed great potential for seawater desalination. The CNT-incorporated membrane exhibited the capability of breaking the trade-off limit between selectivity and permeability that exists in traditional liquid separation membranes. Thus, considering other properties such as antifouling ability and excellent mechanical property, CNTs have great potential in the development of the FO process.

Research on the incorporation of CNTs in the FO process at the laboratory scale has since been widely reported. Asymmetric membranes with thin selective layers coated on top of a highly porous and hydrophilic support layer are commonly used in FO processes. The internal concentration polarization problem could lead to a significant decrease in the water flux across the membrane.

Thus, Wang et al. (2013) dispersed carboxylated MWCNTs within the PES substrate followed by coating of a PA active layer to form a high-performance FO membrane to solve the previously mentioned crucial problem in FO processes. The incorporation of carboxylated MWCNTs increased membrane porosity, thereby leading to a remarkable decrease in the internal concentration and an increase in osmotic water flux. Notably, the performance of the resulting membrane is also better than that of commercial membranes. The tensile strength of the membrane also improved with the addition of MWCNTs.

Amini et al. (2013) employed a slightly different route by incorporating the MWCNTs into the active layer. The MWCNT surfaces were functionalized with NH_2 to assist their incorporation and dispersion in the PA active layer of the thin-film nanocomposite FO membrane. Surface hydrophilicity was enhanced because of the addition of NH_2-functionalized MWCNTs in the active layer. Flux enhancement and acceptable salt rejection were observed compared with the thin-film composite membrane without MWCNTs. Similar results were obtained by Goh et al. (2013) who fabricated a hollow fiber membrane instead of the flat sheet membrane by integrating functionalized MWCNTs in the poly(ethyleneimine) (PEI) active layer supported by poly(amide-imide).

Instead of modifying the active layer or substrate layer with functionalized MWCNTs, Dumée et al. (2013) fabricated the support layer of a thin-film composite membrane by using buckypaper (BP) made up of hydroxyl functionalized MWCNTs. BPs have been regarded as promising supports for high flux thin composite membranes because of their large porosity, strong chemical resistance, and tunability to enhance water adsorption and transport. Unlike commercial PSf supports, BPs can also be formed in exceptionally thin sheets that can further augment water permeation.

32.2.1.1.4 Membrane distillation

Membrane distillation is considered as a substitute to RO and other desalination methods, especially when the concentration of solutes is high (Lawson and Lloyd, 1997). The membrane distillation technique is practical because of its low energy consumption and high water recovery. Direct contact membrane distillation is defined as a technique for water desalination where a porous membrane is employed as a separation barrier between two liquid streams at different temperatures (Schneider and Vangassel, 1984; Schofield et al., 1987; 1990; El-Bourawi et al., 2006). In this process, highly hydrophobic materials are essential for the separation barrier to stop the processed liquids from wetting and wicking into the membrane pores and forming direct bridges between the two streams of water (Schofield et al., 1990; Cabassud and Wirth, 2003; Cerneaux et al., 2009). Thus, CNT BPs that possess unique properties such as hydrophobicity, high porosity, and specific surface area are considered as potential candidates for membrane distillation applications.

However, CNT BP membranes have several shortcomings when self-supporting CNT BP membranes are applied in the direct contact membrane distillation process (Dumée et al., 2010). The limitations include a significant decrease in water flux with prolonged operation times because of temperature polarization that limit water permeability by creating a stagnant boundary layer on the surface and the destruction of the membrane because of the formation of microcracks at CNT BP membranes, which causes water to be drawn in through capillary forces.

Dumée et al. (2011) countered the said problems by functionalizing the outer walls of CNTs with 3-glycidoxypropyltrimethoxysilane chains via ultraviolet (UV)/ozone treatment to create hydroxyl and carboxylate active sites, which were later substituted with alkoxysilane-based groups. Subsequently, the functionalized CNTs were coated with a thin layer of poly(tetrafluoroethylene). The improvement in the membrane flux was attributed to the enhanced hydrophobicity of material that could increase the surface area of the water meniscus, thereby providing more exchange surface accessible for water evaporation. The membrane lifespan was also prolonged because of the increased membrane stability upon modification. Thus, the modified CNT BP membranes can be considered as promising candidates in the current development of the membrane distillation process.

Aside from CNT BP, carboxylated MWCNTs can be immobilized in prefabricated membranes (referred to as CNT-immobilized membrane [CNIM]), which was demonstrated by Bhadra et al. (2013). In general, CNTs function as sorbents that provide additional passages for solute transport. The significantly more polar carboxylated MWCNTs can increase interactions with the water vapor in CNIM to improve the desalination efficiency in membrane distillation. The encapsulation

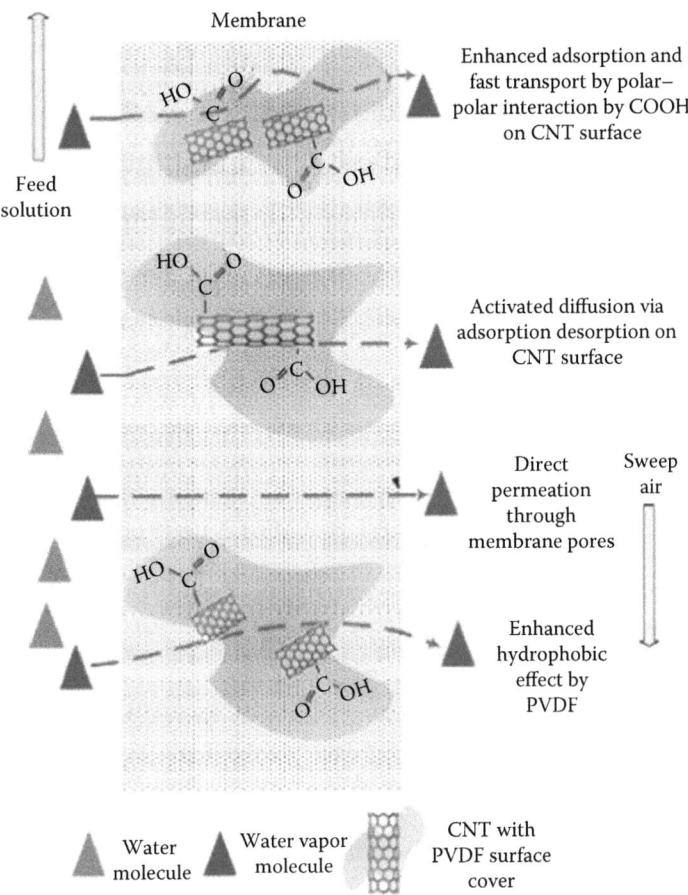

FIGURE 32.2
Mechanism of action on CNIM. (From Bhadra, M. et al., *Sep. Purif. Technol.*, 120, 373, 2013.)

of carboxylated MWCNTs in poly(vinylidene fluoride) (PVDF) hindered the increase in the hydrophilicity of the overall membrane that helps maintain membrane performance. The transport mechanism across the CNIM is shown in Figure 32.2. Desalination performance was enhanced with carboxylated MWCNTs with flux reaching as high as 19.2 kg/m^2·h in a sweep gas membrane distillation mode, which was not observed in their previous study that involved unfunctionalized MWCNTs (Gethard et al., 2011).

32.2.1.2 Pervaporation

Pervaporation is viewed as a potential alternative to conventional distillation due to its ability in separating azeotropic liquid mixtures with low energy consumption, moderate cost, ease of operation, and no entrainers needed. Pervaporation involves the removal of the minor components (usually less than 10 wt.%) of liquid mixtures that decreases the energy consumption of the pervaporation process (Feng and Huang, 1997). Pervaporation has attracted an exceptionally great amount of interest from researchers worldwide and has emerged as one of the most promising membrane technologies in recent years. In general, the applications of pervaporation can be divided into three categories: (1) dehydration of organic solvents (alcohols, esters, ethers, ketones, etc.), (2) recovery of organic compounds from aqueous solutions (recovery of aromatic compound, removal of volatile organic compounds, separation of compounds from fermentation broths in biotechnology, etc.), and (3) organic–organic mixtures separation (ethanol/ethyl *tert*-butyl ether, methanol/methyl *tert*-butyl ether, benzene/cyclohexane, etc.) (Smitha et al., 2004). To date, the pervaporation dehydration of organic solvents is the most developed among these applications. The three essential issues that must be given a great deal of consideration in the development of pervaporation membrane are membrane permeation flux, membrane selectivity, and membrane stability. The objective in developing new pervaporation membranes is either to increase the flux while maintaining the selectivity, achieving a higher selectivity at constant flux, or both (Lipnizki et al., 1999).

The use of CNT-based nanocomposite is one of the prospective technologies for the current membrane separation industry. The transport properties were enhanced significantly due to the smoothness of the interior channel defined by the defect-free CNTs, which results in rapid diffusive transport in CNTs compared to other inorganic fillers. The function of CNT fillers in nanocomposites is similar to that of conventional microporous materials that exhibit large surface areas as well as high chemical and thermal stability. Apart from that, functionalization of CNTs with specific functional groups could increase the compatibility of CNTs with the polymer matrix because of the compatibility between the functional groups attached on CNTs and the functional groups of the polymer itself.

The homogeneous dispersion of CNTs into the polymer matrix is a critical issue in the fabrication process of CNT nanocomposites for pervaporation process. Throughout the years, various functionalization techniques have been explored by researchers worldwide to overcome the problem of CNT agglomeration and improve the dispersion of CNTs into the polymer matrix.

In order to improve the dispersion of MWCNTs in polymer matrix, Choi et al. (2009) treated MWCNTs with strong acid (HNO_3 and H_2SO_4) prior to the addition of MWCNTs into the poly(vinyl alcohol) (PVA) matrix. They investigated the effect of MWCNT addition on the pervaporation performance of the PVA membrane in the dehydration of alcohol by varying the amount of MWCNTs (1–5 wt.%) in PVA membrane. Transmission electron microscopy (TEM) results indicated that the MWCNTs were homogeneously distributed throughout the PVA matrix. In the pervaporation separation of ethanol/water mixtures, the MWCNT/PVA blend membranes exhibited an increase in permeation flux with increasing amount of MWCNTs. On the other hand, the separation factor was maintained up to 1 wt.% MWCNTs followed by a decrease when the amount of MWCNTs was further increased. This phenomenon can be attributed to the inner empty space of the MWCNTs, which could function as a path for the passage of water and alcohol with relatively less resistance.

In another study, Liu et al. (2009) incorporated poly(styrene sulfonic acid)-functionalized CNTs (CNT-PSSA) into the chitosan (CS) matrix in the preparation of CS/CNT nanocomposites (CS/CNT-PSSA). Homogeneous dispersion of CNTs was achieved by the existence of chemical linkages between CS and CNTs that were formed in the nanocomposites via the reaction between the sulfuric acid groups of CNT-PSSA and the amino groups of CS. The CS/CNT-PSSA nanocomposites exhibited superior thermal and mechanical properties, water and solvent uptakes, bond water ratios, and electrical conductivity than the neat CS polymer. In the pervaporation dehydration of 90 wt.% ethanol aqueous solution, CS/CNT-PSSA nanocomposite with 1 wt.% of CNT-PSSA demonstrated higher separation performance with a pervaporation separation index (PSI) of approximately 4.67 folds over that of neat CS.

Similar findings were also reported by Qiu et al. (2010) who incorporated functionalized MWCNTs into the CS membrane pervaporation dehydration of ethanol/water mixtures. In their research, MWCNTs treated with mixed acid were further functionalized with diisobutyryl peroxide. An increase in the permeation flux and a slight decrease in the separation factor were observed with increasing functionalized MWCNT content in the membrane. Unlike the CS homogeneous membrane, the functionalized MWCNT-filled CS membranes displayed higher PSI values and enhanced pervaporation separation property.

Shirazi et al. (2011) also compared the performance of nanocomposite membranes incorporated with HNO_3-functionalized CNTs and pure PVA membranes in the pervaporation dehydration of isopropanol/water mixtures. They reported that the presence of functionalized CNTs in the PVA membranes decreased the degree of swelling of the membranes. The application of the nanocomposite membrane in pervaporation revealed that the water separation factor of the membrane was significantly improved by the incorporation of CNTs into the PVA matrix because of the increased rigidity of the polymer chains.

Employing a different membrane configuration, Hu et al. (2012) successfully prepared composite membranes made up of a separating layer of polyvinylamine incorporated with CNTs supported on a microporous PSf substrate for the pervaporation dehydration of ethylene glycol. An increase in the surface hydrophilicity of the membrane was reported upon the addition of acid-treated CNTs in the membrane. The resulting membrane demonstrated an increase in the permeation flux and separation factor with significant enhancement, especially in separation performance at low feed water concentrations.

Recently, Yee et al. (2014) employed CNT BP, which are commonly used in direct contact membrane distillation for water desalination, in the pervaporation dehydration of a multicomponent etherification reaction mixture. Unlike in previous studies, the fabrication of a novel asymmetric membrane with a preselective layer composed of acid-treated MWCNT-BP followed by a thin layer of PVA to enhance the removal of water from the reaction mixture was emphasized. A remarkable two- and fourfold enhancement of the permeation flux and separation factor, respectively, was observed in MWCNT-BP/PVA asymmetric membranes compared with those of a pure PVA membrane. The hydrophilic group on the oxidized MWCNTs and nanochannels of the preselective layer that facilitate the permeation of water molecules may be the contributing factor to these improvements. An enhancement in the mechanical properties was also observed compared with those of a pure PVA membrane.

In addition to the treatment with acid, CNTs were functionalized with polymers to achieve homogeneous dispersion in the polymer matrix. Sajjan et al. (2013) developed sodium alginate membranes impregnated with CS-wrapped MWCNTs. In the pervaporation dehydration of an isopropanol–water mixture with 10 wt.% water content, the membrane incorporated with 2 wt.% CS-wrapped MWCNTs demonstrated the best separation factor of 6419 accompanied by a flux of $21.76 \times 10^{-2}\,kg/m^2 \cdot h$ at 30°C. They also reported that the total flux and water flux almost overlapped with each other for all CS wrapped-MWCNT incorporated membranes. This result indicated that the fabricated membranes are highly selective toward water.

Aside from homogeneously dispersing CNTs in the polymer matrix, bulk-aligned poly(3-hydroxybutyrate) (PHB)-functionalized MWCNTs into CS were successfully synthesized by Ong et al. (2011) via a simple and practical filtration method. Initially, PHB-functionalized MWCNTs were aligned on a membrane filter template via a filtration process followed by casting the CS solution onto the template to produce PHB-functionalized MWCNTs/CS nanocomposite membranes. The MWCNT amount required to improve the mechanical properties of the nanocomposite membrane was decreased because of the bulk alignment of the PHB-functionalized MWCNTs. The resulting nanocomposite membrane demonstrated a relatively good permeation flux and separation factor toward water in the pervaporation process of 1,4-dioxane dehydration.

In another approach, PVA-functionalized MWCNT was bulk-aligned on the PVDF membrane via a simple filtration method and was further coated with CS to form a novel three-layer nanocomposite membrane (Yeang et al., 2013). Water permeance and selectivity were significantly improved in the pervaporation dehydration of acetone. The three-layer nanocomposite membrane exhibited the best separation performance compared with homogeneously dispersed MWCNT/CS membrane and pure CS membrane. The three-layer nanocomposite membrane can be regarded as a potential solution to the trade-off problem often faced by pervaporation membranes.

Apart from the functionalization methods discussed earlier, carboxyl group–functionalized MWCNTs were encapsulated by a polyelectrolyte–polyelectrolyte complex (PEC) layer and finely dispersed in the bulk PEC nanocomposite film by Zhao et al. (2009). The nanocomposite membrane loaded with 7 wt.% MWCNTs exhibited 2.6 times higher tensile strength and 1.8 times higher modulus compared with those of pristine PEC. Excellent performance was achieved by the PEC/MWCNT nanocomposite membranes in the pervaporation dehydration of isopropanol. The high performance of the nanocomposite membranes remained stable for up to 20 days of operation time.

Amirilargani et al. (2013) synthesized poly(allylamine hydrochloride) (PAH)-wrapped MWCNTs and eventually incorporated them into PVA membranes for the pervaporation dehydration of isopropanol. PAH-wrapped MWCNTs were thoroughly dispersed in the polymer matrix. Water selectivity was also significantly increased because of the rigidification of the polymer chains as a result of the incorporation of modified CNTs into the PVA matrix.

Apart from their application in dehydration processes, functionalized CNTs have also been utilized in organic–organic separation. Peng et al. (2007a) dispersed β-cyclodextrin (β-CD)-functionalized CNTs (β-CD-CNT) in the PVA matrix to prepare β-CD-CNT/PVA hybrid membranes for the pervaporation separation of benzene/cyclohexane mixtures. The β-CD-CNTs were uniformly dispersed in the PVA matrix. A significant improvement in Young's modulus and thermal stability were observed in the β-CD-CNT/PVA hybrid membranes compared with those in pure PVA and β-CD/PVA membranes. Excellent pervaporation properties were also exhibited by the β-CD-CNT/PVA hybrid membranes. In another work by the same group of researchers (Peng et al., 2007b), instead of β-CD, CNTs were wrapped by CS and dispersed into the PVA matrix for the same separation process. The CS-wrapped CNTs were also effectively dispersed into the PVA matrix because of the compatibility of CS with PVA. According to the authors, the simultaneous increase in permeation flux and separation factor observed in this study was attributed to the preferential affinity of CNTs toward benzene and the increased free volume caused by the addition of CS-wrapped CNTs that modified the PVA polymer chain packing.

32.2.2 Carbon Nanotubes in Gas Separation

In addition to the application in membrane-based liquid separation processes discussed in the previous section, functionalized CNTs were also utilized in membrane-based gas separation. Membrane-based gas separation has attracted considerable interest in industries because of its advantages such as having a small footprint, lower capital, and lower operating costs compared with other traditional separation technologies (Rousseau, 1987). The current membrane-based gas separation processes consist of carbon dioxide (CO_2) removal from fuel or flue gas, natural gas separation, hydrogen (H_2) recovery, and oxygen (O_2)–nitrogen (N_2) separation. Given their low cost and ease of processing, polymers are usually employed as membrane materials. Nevertheless, efforts have been concentrated in overcoming the trade-off between the permeability and selectivity of polymeric materials to improve their separation performance. By incorporating inorganic fillers into the polymer matrix to fabricate the polymer–inorganic hybrid membrane, the polymer chain packing is interfered. As such, the free volume increases, which leads to an increase in gas diffusion. Given their porous structure and excellent mechanical properties, CNTs have gained considerable research interests as a type of inorganic filler in the polymer matrix. CNTs have also been proven as ideal candidates for gas adsorption and separation purposes given their superior selectivity and flux for the transport of light gases, which were a few times higher than those of other materials with comparable pore sizes (Skoulidas et al., 2002; Sokhan et al., 2004; Matranga et al., 2006). In this section, the contributions of functionalized CNTs in the CO_2 capture, H_2 recovery, and separation of different gas mixtures will be highlighted.

CO_2 is a greenhouse gas commonly found in natural gas streams, biogas from anaerobic digestion, flue gas from fossil fuel combustion, and coal gasification. CO_2 separation in many industrial processes is important because the presence of CO_2 along with other acid gases decreases the calorific value and causes the gas streams to become acidic and corrosive. This scenario eventually lowers the possibilities of gas compression and transport within transportation systems. The increasing amount of CO_2 emissions has also become a major contributor to the Earth's greenhouse effect. Thus, economic and effective techniques for CO_2 removal and capture are desperately required and have drawn much interest from researchers. One of the most promising solutions is to cultivate a cost-effective and high-performance CO_2 separation membrane. At present, mixed matrix membrane technology has exhibited the most promising results among the many available membrane materials. As mentioned earlier, CNTs emerged as potential inorganic fillers because of their superior gas separation and mechanical properties. Various types of functionalization have been employed to achieve uniform CNT dispersion in a polymer matrix, improve the CNT attachment to the polymer, and induce affinity toward specific molecules.

Aroon et al. (2010) functionalized MWCNTs with CS prior to incorporation in polyimide (PI) to improve the dispersivity of MWCNTs in the PI membrane for its application in CO_2/methane (CH_4) separation. Their TEM results showed open-ended and well-dispersed CS-functionalized MWCNTs. In the context of separation performance, both CO_2/CH_4 selectivity and CO_2 and CH_4 permeabilities were higher for the PI membrane incorporated with CS-functionalized MWCNTs than those for the PI membrane alone. CO_2 permeability increased by 20.48 Barrer, whereas CH_4 permeability increased by 0.71 Barrer with the addition of a mere 1% of CS-functionalized MWCNTs into the casting dope. CO_2/CH_4 selectivity also increased by 51.4% (from 10.9 to 16.5). According to the authors, the enhanced permeabilities were due to the existence of high diffusivity tunnels in the MWCNTs within the PI matrix.

Aside from functionalization with polymer, MWCNTs can also be functionalized with chemical agents such as 3-aminopropyltriethoxylsilane (APTES) to allow the effective dispersion of the tubes in an organic solvent such as *N*-methlypyrrolidone during the preparation of PES/MWCNT-mixed matrix membranes (Ismail et al., 2011). The highest CO_2/CH_4 and O_2/N_2 selectivities were exhibited by the PES membrane incorporated with functionalized MWCNTs, whereas the permeability of gases was almost the same as those of unpurified and unfunctionalized MWCNTs. Notably, the gas permselectivities decreased and the permeabilities increased with increasing MWCNT loading because of the existence of defects and interface voids.

Moreover, the noncovalent functionalization of CNTs with β-CD appears to be a promising way to further develop environmentally friendly nanocomposite materials with enhanced properties (Sanip et al., 2011; Aroon et al., 2013). The characterization results proved that β-CD-functionalized MWCNTs were thoroughly dispersed in the polymer matrix. In terms of gas separation properties, both studies reported an enhanced CO_2/CH_4 selectivity of PI incorporated with β-CD-functionalized MWCNTs mixed with matrix membranes compared with neat PI membrane.

For CO_2/N_2 separation, Cong et al. (2007) utilized brominated poly(2,6-diphenyl-1,4-phenylene oxide) (BPPOdp) membrane embedded with carboxylic acid–functionalized single-walled CNTs (SWCNTs) to improve the mechanical strength and studied the effect of SWCNT addition on gas separation properties. Their results indicated that carboxylic acid–functionalized SWCNTs dispersed more uniformly in BPPOdp compared with pristine SWCNTs. However, their gas separation performances were unchanged.

By contrast, an improved gas separation performance was achieved by Ge et al. (2011) by using PES membrane incorporated with carboxylic acid–functionalized MWCNTs. The improved gas permeability and selectivity were attributed to the free volume of the polymer chains or MWCNT channels. In addition, the carboxyl functional groups have an important function in creating stronger interactions with CO_2, which eventually improved the solubility of the polar gas and restrained the solubility of the nonpolar gas, thereby increasing CO_2/N_2 gas selectivity.

Moreover, Khan et al. (2013) reported an enhanced dispersion and permeability as well as selectivity by using a polymer membrane with intrinsic microporosity and embedded with polyethylene glycol (PEG)–grafted MWCNTs. The PEG chains on MWCNTs interacted with CO_2, which increased the solubility of the polar gas and reduced the solubility of the nonpolar gas. As such, CO_2/N_2 separation was favored. Furthermore, the experimental sorption isotherms of CO_2 and N_2 and the mechanical properties of the mixed matrix membranes were improved.

Recently, promising results were reported by Wang et al. (2014b) who also modified MWCNTs with PEG, whose ethylene oxide unit functioned as an effective group to achieve high CO_2 permeability and CO_2/light gas selectivity, to improve the dispersion in the polyether block amide matrix. They discovered that the incorporation of high-molecular-weight PEG-based polymers favors CO_2/CH_4 separation, but the incorporation of low-molecular-weight PEG-based polymers favors CO_2/N_2 separation. Improved CO_2 permeability was observed in all the hybrid membranes because of the increased amorphous phase of the membrane.

For the CO_2/H_2 separation in syngas production, a highly CO_2-selective membrane capable of withstanding high pressures and high temperatures is necessary. Zhao et al. (2014b) successfully fabricated a membrane with significantly enhanced stability by dispersing acid-treated MWCNTs as reinforcing nanofillers in the PVA matrix containing NH_2. The membrane with 4 wt.% acid-treated MWCNTs exhibited excellent stability without any change in its selectivity and permeability in an evaluation performed at a temperature of 380.15 K and a feed pressure of 1.52 MPa for 264 h (~11 days). The authors also claimed that the synthesized membrane showed remarkable CO_2/H_2 separation and high tolerance of the feed gas at high pressures and high temperatures and therefore could be potentially applied in a stand-alone membrane unit for energy-efficient precombustion carbon capture from coal-derived syngas or in conjunction with water–gas-shift reaction for CO cleanup to produce high-purity H_2 for fuel cells and to simultaneously capture CO_2.

We then focused on the application of functionalized CNTs in H_2 recovery. H_2 is usually recovered via energy-intensive separation processes, such as pressure swing adsorption and cryogenic systems, which are applicable in numerous operating conditions. Membrane processes are considered as one of the most promising techniques for the production of high-purity H_2 because of their cost- and energy-saving properties. Although CNTs have been widely studied for gas separation applications, studies on the application of functionalized CNTs in membranes for H_2 recovery are still limited. Acid-functionalized MWCNTs have been applied in poly(bisphenol A-*co*-4-nitrophthalic anhydride-*co*-1,3-phenylene diamine) membrane for H_2 recovery by Weng et al. (2009). From the results of their study, no obvious increase was observed in the H_2, CO_2, and CH_4 permeabilities at low MWCNT concentrations (1–5 wt.%), whereas the permeabilities and selectivities of H_2, CO_2, and CH_4 were substantially improved at high MWCNT concentrations. This phenomenon was attributed to the gas transport in the nanocomposite membranes via the interstices between the MWCNTs and the polymer chains because of the high MWCNT loading.

In addition, functionalized CNTs can be used in the separation of other gas mixtures. For O_2/N_2 gas separation, a mixed matrix hollow fiber membrane embedded with functionalized MWCNTs instead of a flat sheet mixed matrix membrane was developed by Goh et al. (2012). A facile two-step approach involving dry air oxidation and surfactant (Triton X100) dispersion was conducted to modify the MWCNTs and prevent their agglomeration during dispersion in a PEI matrix. The removal of the disordered amorphous carbons and metal particles via dry air oxidation increased the accessibility of the hollow structure to promote the fast and smooth transport of gas molecules, thereby resulting in excellent O_2/N_2 gas separation. The permeability increased by 60% while maintaining the selectivity of the O_2/N_2 gas pair in mixed matrix hollow fibers embedded with surface-modified MWCNTs.

Hollow fiber membranes based on commercial BTDA-TDI/MDI (P84) co-PI and phenol-functionalized MWCNTs as nanofillers were fabricated by Favvas et al. (2014) for helium (He)/N_2 separation. The excellent dispersion of functionalized MWCNTs in the polymer matrix was observed via scanning electron microscopy. Furthermore, the gas permeance coefficients accordingly increased with MWCNT concentration, thereby indicating an increase in the free volume of the polymer matrix. High permeability values and ideal He–N_2 selectivities from 2.86 to 12.2 were also obtained. In conclusion, the authors claimed that these membranes may be applicable in He enrichment applications and are excellent candidates for carbon molecular sieve membranes.

In addition, functionalized CNTs can be vertically aligned in an anodic aluminum oxide (AAO) template for the separation of hydrogen sulfide (H_2S) from the binary mixture of H_2S/CH_4 (Gilani et al., 2013). To improve the separation of H_2S from the H_2S/CH_4 binary mixture, CNTs were functionalized with dodecylamine. After functionalization with dodecylamine, the selectivity of the CNT/AAO membrane increased about 1.5–2 times. However, wall smoothness of the CNTs was reduced after being functionalized which led to the decrease of the permeability of N_2 gas to approximately half of its permeability exhibited by the nonfunctionalized CNT/AAO membrane. All these findings can be attributed to the polar environment created by the amide functional group, which improved the physical and chemical adsorption of H_2S molecules on the CNT wall. As such, the permeability and selectivity of H_2S were enhanced.

32.3 GRAPHENE IN MEMBRANE SEPARATION PROCESS

In addition to CNTs, graphene is another carbon-based material that has received a substantial amount of attention as a potential candidate for membrane fabrication. Graphene, a single layer of graphite made up of a lattice of hexagonally arranged sp^2-bonded atoms, is a very thin membrane, which is only an atom thick and exhibits high electrical conductivity (Geim, 2009), high breaking strength (Lee et al., 2008), and impermeability to molecules as small as He in its pristine state (Bunch et al., 2008). Since the isolation of graphene from graphite sheets via mechanical exfoliation method (Novoselov et al., 2004), studies on its electrical, optical, and mechanical properties have rapidly grown. Graphene is now widely studied as a selective material for membranes because of its potential in creating ultrathin high flux membranes with size-tunable pores that can function as molecular sieves.

Even though graphene is not permeable, countless efforts have been expended to modify the graphene structure for separation applications. As a result, various types of graphene-based materials with properties deviating from those of pure graphene were synthesized. Porous graphene, a graphene sheet with some holes or pores within the atomic plane, is an example of modified graphene materials. Porous graphene exhibits excellent potential in membranes for applications such as molecular sieves, energy storage components, and nanoelectronics because of its porous structure.

In addition, the functionalization of graphene is one way to modify its structure to achieve desired properties, such as improved solubility in various solvents (Niyogi et al., 2006; Xu et al., 2008), increased dispersibility in matrices (Ramanathan et al., 2008; Quintana et al., 2010), and enhanced processing and manipulation to fabricate various devices (Liu et al., 2008; Li et al., 2008a). Most reported covalent chemical modifications of graphene are in the form of graphene oxide (GO) at the reactive sites of O_2-containing functional groups because the functionalization of defect-free graphene is not easy. Furthermore, GO itself is known as covalent-functionalized graphene with O_2 functional groups.

The application of graphene and its derivatives in the fabrication of membranes can be classified into two categories. The first application involves the direct use of graphene as a separating layer (Sint et al., 2008; Du et al., 2011; Han et al., 2013; Sun et al., 2013), and the second application is through the incorporation of graphene in a polymer matrix to improve membrane performance (Enotiadis et al., 2012; Heo et al., 2013; Zhao et al., 2013b).

In the subsequent section, the development and contribution of graphene-based materials in membrane-based liquid separation, starting with desalination and water purification followed by pervaporation, will be fully discussed. Their application in membrane-based gas separation will also be covered in the succeeding section.

32.3.1 Graphene in Liquid Separation

32.3.1.1 Desalination and water purification

Studies on GO and porous graphene in membranes for desalination and water purification have been steadily growing since the isolation of graphene sheets. Specifically, GO nanosheets, oxygenated graphene sheets that are integrated with carboxyl, hydroxyl, and epoxide functional groups, have emerged as promising candidates for the fabrication of functional nanocomposite materials with high chemical stability, strong hydrophilicity, and exceptional antifouling properties (Dikin et al., 2007; Dreyer et al., 2010; Koinuma et al., 2012), all of which could aid in water treatment processes. The simulation results showed that submicrometer-thick membranes produced from stacked GO sheets, which are impermeable to molecules as small as He, allowed the unhindered permeation of water through the membranes at least 10^{10} times faster than He (Nair et al., 2012). According to the research team, this interesting finding was attributed to the low-friction flow of a monolayer of water through 2D capillaries generated by closely spaced GO sheets. Meanwhile, the diffusion of other molecules was hindered by the reversible narrowing of the capillaries at low humidity and/or by their clogging with water. Hence, the membrane described earlier showed potential as a candidate in the design of filtration and separation membranes for the selective removal of water. In addition to the simulation studies, experimental studies involving the use of graphene-based membranes showed promising results. Detailed studies on the development of graphene-based materials in desalination and water purification processes, such as filtration, RO, and FO, are presented in the following sections.

32.3.1.1.1 Filtration

Filtration efficiency primarily depends on the selection of appropriate membrane materials. Thus, graphene-based sheets are ideal building blocks in preparing graphene-involved functional films because of their unique properties (Dikin et al., 2007; Li et al., 2008b; Xu et al., 2008; Pang et al., 2009; Cote et al., 2010; Yan et al., 2010; Yu and Dai, 2010; Sui et al., 2011).

Ganesh et al. (2013) dispersed GO in PSf to form mixed matrix nanofiltration membranes because of the hydrophilicity and mechanical properties of GO. As a result, the surface hydrophilicity, water flux, and salt rejection property of the membrane were improved. A maximum of 72% sodium sulfate (Na_2SO_4) rejection was achieved when the membrane was incorporated with 2000 ppm GO at an applied pressure of 4 bar.

To obtain a nanoscale stable dispersion, Wang et al. (2012) modified GO sheets with PEI prior to the formation of the membrane. Their dye removal experiment results showed that the membranes demonstrated exceptional dye removal capacity with 99.5% retention of Congo red accompanied by permeance of 8.4 kg/m²·h·MPa. Excellent nanofiltration properties were also demonstrated in the separation of monovalent and divalent ions. Furthermore, the GO addition enhanced the mechanical and thermal stability of the membranes.

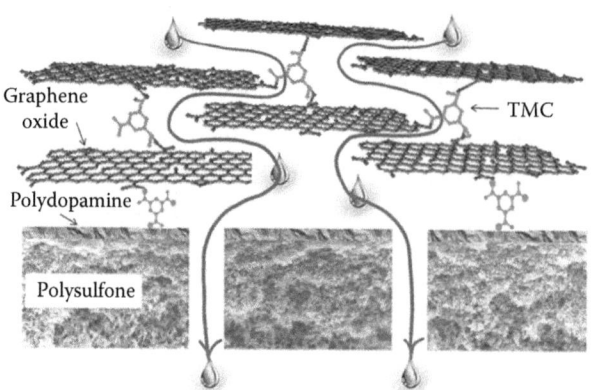

FIGURE 32.3
Transport of water across the membrane. (From Hu, M. and Mi, B., *Environ. Sci. Technol.*, 47, 3715, 2013.)

An enhanced separation was also reported by Hu and Mi (2013) who synthesized a water separation membrane via the layer-by-layer deposition of 1,3,5-benzenetricarbonyl trichloride crosslinked GO nanosheets on a polydopamine-coated PSf support. The objective of this fabrication method was to allow water to flow through the nanochannels between the GO layers. Moreover, this method rejects unwanted solutes through size exclusion and charge effects. The transport of water across the membrane is illustrated in Figure 32.3. The cross-linking of the GO nanosheets improved the stability of the GO nanosheets in a water environment and also fine-tuned the charges, functionality, and spacing of the GO nanosheets. A flux ranging from 80 to 276 L/m²·h·MPa, which is approximately 4–10 times higher than those of most commercial nanofiltration membranes was obtained. Despite exhibiting a relatively low rejection (6%–46%) of monovalent and divalent salts, the GO membrane demonstrated a moderate rejection (46%–66%) of methylene blue and a high rejection (93%–95%) of rhodamine WT.

Although ultrafast water transport through the membranes made from stacked GO sheets has been reported (Nair et al., 2012), the permeation through these GO membranes remains inadequate compared with commercial pressure-driven ultrafiltration membranes because the GO membranes are hindered by their less fluidic channels, including the randomly distributed nanoscale wrinkles (Raidongia and Huang, 2012) and/or spaces strutted by hydroxyl, carboxyl, and epoxy groups (Vandezande et al., 2008; Sun et al., 2013) and also because of the slow diffusion process within. However, Qiu et al. (2011) took advantage of the intrinsic corrugation to form GO membranes with controlled corrugation via hydrothermal treatment. An elevated permeance of 45 L/m²·h·bar caused by the microscopic wrinkles in GO was demonstrated by the resulting membrane during pressure-driven separation. These membranes can be promising materials for nanoscale separation if they are well defined because the nanochannels in the thermally corrugated GO membranes were mainly attributed to the microscopic corrugated GO wrinkles.

However, the formation of nanochannels in the GO membranes proposed by Qiu et al. (2011) is very difficult to control. In an attempt to form a percolated nanochannel network with a well-defined channel size within GO membranes, Huang et al. (2013) synthesized nanostrand-channeled GO ultrafiltration membranes with a network of nanochannels with a narrow size distribution (3–5 nm). The resulting membrane exhibited superior separation performance. A 10-fold enhancement in permeance without compromising the rejection rate was observed compared with the GO membranes. Furthermore, with similar rejection, their permeance is more than 100 times higher than those of commercial ultrafiltration membranes because of the porous structure of the nanostrand-channeled GO membranes.

Graphene-based nanocomposite films with dilated space and channels can be formed via a different approach by integrating TiO_2 nanoparticles between these carbon sheets (Xu et al., 2013) followed by the assembly into GO–TiO_2 films via vacuum filtration. The resulting membranes were used in filtration to remove dye molecules (methyl orange and rhodamine B) from water. These results show that these GO–TiO_2 films capture significant amounts of dye molecules in addition to the adsorption capacities of these dyes.

In addition to the enhanced flux and rejection, an improved fouling mitigation was also observed. Zhao et al. (2013a) reported enhanced antifouling properties in PVDF/GO ultrafiltration membranes compared with pure PVDF membranes because of the enhanced surface hydrophilicity and membrane morphologies. Similar results were also obtained by Zinadini et al. (2014) with a PES-mixed matrix nanofiltration membrane containing GO nanoplates. Upon the addition of GO, the antifouling properties and the water flux remarkably improved because of the increased hydrophilicity of the prepared membranes caused by the hydroxyl and carboxylic acid groups on the GO surface.

Xu et al. (2014) claimed to be the first to functionalize GO with the long polymer chains of APTES to alleviate the fouling behavior of ultrafiltration membranes. Their experimental results showed that the hydrophilicity, water flux, bovine serum albumin flux, and rejection rate of PVDF membranes incorporated with functionalized GO were higher than those of pure PVDF membranes and PVDF membranes incorporated with unfunctionalized GO membranes. The fouling resistance parameters eventually decreased because of the increased hydrophilicity of the hybrid membranes. In addition, the tensile strength was also improved because of the strong interfacial interaction between the functionalized GO and the matrix through the covalent functionalization of GO.

The synergistic effect between 2D GO and 1D oxidized MWCNTs on the fouling resistance and separation was also investigated by Zhang et al. (2013). The hierarchical GO/oxidized MWCNT architecture, which prevented the stacking of GO attributed to the long and tortuous oxidized MWCNTs, sufficiently dispersed the GO/oxidized MWCNTs on the membrane surface and PVDF matrix. As a result, the antifouling properties were improved because of the increased surface hydrophilicity from the addition of GO-MWCNTs. The pure water flux increased by 251.73%, 103.54%, and 85.68% compared with those of PVDF, PVDF/oxidized MWCNTs, and PVDF/GO, respectively, because of the synergistic effects of GO and oxidized MWCNTs.

To produce a multifunctional membrane for concurrent water filtration and photodegradation, Gao et al. (2013) used GO as the cross-linker for individual TiO_2 microspheres and electron acceptor to enhance the photocatalytic activity in the fabrication of GO–TiO_2 microsphere hierarchical membrane. The permeate flux slightly decreased with the addition of GO compared with that of the TiO_2 microsphere membrane because of the reduction of both inter- and intrapore of the GO–TiO_2 membrane, which is favorable for high separation efficiency. Excellent photodegradation activity, no membrane fouling, and a substantial enhancement in GO–TiO_2 membrane strength and flexibility were also reported.

Gao et al. (2014) also adopted GO to expand the light response range of TiO_2 by modifying the surface of the water filtration membranes with TiO_2–GO to improve their photocatalytic activities under both UV radiation and sunlight. TiO_2 nanoparticles and GO nanosheets were sequentially deposited on the PSf base membrane via a layer-by-layer technique. The significant improvement in methylene blue photodegradation kinetics under UV radiation and sunlight as well as the increased hydrophilicity of the TiO_2–GO membrane led to an improved membrane flux.

Shao et al. (2014) reported that apart from its use in water filtration, GO-based membrane also showed excellent potential for practical applications in solvent-resistant nanofiltration. The incorporation of GO in polypyrrole-hydrolyzed polyacrylonitrile composite solvent-resistant nanofiltration membranes improved the solvent permeance without compensating the Rose Bengal rejection. The methanol, ethanol, and isopropanol permeances of the GO-incorporated polypyrrole-hydrolyzed polyacrylonitrile composite membrane were higher by approximately 945%, 635%, and 302%, respectively, compared with those of pure polypyrrole-hydrolyzed polyacrylonitrile composite membranes.

GO was also used as a filler material to synthesize GO-filled PVDF microfiltration tubes via simple decompress filtration process for application in the separation of citral solution and the removal of copper ions (Cu^{2+}) (Ji et al., 2014). The resulting membrane was suitable for the separation of an organic solution with different molecular weights because of the reduced effective mean pore size of the composite microfiltration membrane caused by the significant amount of GO spheres that filled the PVDF channel. Furthermore, outstanding separation or adsorption was shown by a series of citral separation or transition metal adsorption.

Meanwhile, to the best of our knowledge, porous graphene has not been experimentally studied in filtration applications. Recently, using molecular dynamics simulations, He et al. (2013) designed bioinspired nanopores in graphene sheets with the ability to distinguish between sodium ion (Na^+)

and potassium (K^+) ion, which have very alike properties. The selectivity of the three carboxylate-functionalized nanopores was found to be voltage dependent. The Na^+ passage was favored at low voltages because of the selective blockage of the pore by Na^+, whereas preferential K^+ passage was favored at high voltages because Na^+ no longer blocks the pore. Hence, these membranes were considered promising candidates as nanofiltration membranes for Na^+/K^+ separation. However, the feasibility of the proposed porous graphene membrane is yet to be experimentally proven.

32.3.1.1.2 Reverse osmosis

One of the major challenges in RO membrane technology is improving the membrane stability related to fouling and chlorine resistance. To overcome these challenges, Choi et al. (2013) coated GO multilayers on PA thin-film composite membrane surfaces via layer-by-layer deposition of oppositely charged GO nanosheets. The fouling resistance was profoundly improved because of the increased surface hydrophilicity caused by the GO-coated layers. Furthermore, the GO-coated layer functioned as a chlorine barrier for the underlying PA membrane because of the chemically inert nature of GO nanosheets.

Chlorine-resistant thin-film composite RO membranes were also fabricated by Kim et al. (2013c) by incorporating GO and aminated GO (aGO) into sulfonated poly(arylene ether sulfone) material containing amino groups (aPES). Chlorine resistance was enhanced by the aPES/GO/aGO RO membrane compared with the typical PA RO membrane with excellent RO performances. These positive results were attributed to the integration of the GO and aGO layers and hydrophilic aPES active layer.

The water transport through a porous graphene membrane was compared with that through a thin CNT membrane via molecular dynamic simulations (Suk and Aluru, 2010). The water flux through the porous graphene was nearly twice that of CNTs because of the lower energy barrier at the entry regions of the graphene pores. Nevertheless, experimental studies are still needed to verify the unique transport properties of the porous graphene membrane in RO applications.

Cohen-Tanugi and Grossman (2012) also reported that nanoscale pores in single layer free-standing graphene is effective in removing sodium chloride and enabling water flow with permeabilities of several orders of magnitude higher than existing RO membranes. From the simulation results, the desalination performance strongly depended on the pore chemistry and pore diameter and adequately sized pores allowed water passage while rejecting ions. The water flux was doubled when graphene pores bonded with the hydroxyl groups, which was attributed to their hydrophilic nature. By contrast, the water flow of hydrogenated pores with hydrophobic character decreased but had better salt rejection compared with hydroxylated pores. Hence, nanoporous graphene may have an important function in water purification.

32.3.1.1.3 Forward osmosis

As previously mentioned, the internal concentration polarization is detrimental to water permeation flux in FO processes. The internal concentration polarization can be undoubtedly negated because of the one-atom thickness of graphene. Hence, the fabrication of FO membrane consisting of functionalized porous single-layer graphene and the study on its permeation properties via molecular dynamics simulations were conducted (Gai and Gong, 2014). The porous graphene membrane showed excellent results, in which the internal concentration polarization decreased to zero in the FO processes. In addition, the water flux was approximately 1.7×10^3 times higher than that of typical cellulose triacetate FO membranes. In another similar study conducted by the same group of researchers (Gai et al., 2014), an excellent salt rejection and a water flux of 28.1 L/cm^2·h, which is about 1.8×10^4 times higher than that of a typical cellulose triacetate membrane, was reported for the FO system because of the fluorinated porous graphene. The earlier results suggested the possible application of porous graphene in the FO water desalination process. To the best of our knowledge, porous graphene is yet to be realized as a FO membrane despite of its potential application in the FO processes.

32.3.1.2 Pervaporation

In addition to studies on CNTs, studies involving graphene-based nanomaterials in pervaporation have been gradually growing because of their unique separation and mechanical properties.

Different techniques that harness the beneficial properties of GO nanomaterials in pervaporation have been reported.

The ultimate goal in pervaporation separation is to overcome a trade-off in the selectivity and permeance. To enhance the pervaporation dehydration of isopropanol, Suhas et al. (2013) incorporated a highly oxygenated surface-functionalized graphene sheets as the nanofiller into the sodium alginate matrix. The membrane permeance and selectivity improved because of the high surface areas and polar functionalization of the graphene sheets. Nevertheless, the selectivity started to decrease at higher functionalized graphene sheet concentrations because of the increase in particle agglomeration.

Apart from being utilized as a filler in mixed matrix membrane, multilayered GO was coated onto a thin film of nanofibrous composite mat to form a high flux membrane for ethanol dehydration (Yeh et al., 2013). Based on the experimental data, the GO-based thin film of nanofibrous composite membranes exhibited a higher separation factor compared with commercial membranes because of the hindered transport of ethanol through the GO barrier layer. A water permeate flux of 2.2 $kg/m^2 \cdot h$, which is twice higher than that of the commercialized pervaporation membrane, and a separation factor of 308, which is almost four times higher than that of the commercialized membrane, were shown by the membrane with a GO barrier layer thickness of 93 nm.

In another study, highly ordered flexible layers of GO were coated on modified polyacrylonitrile substrates via a pressure-assisted self-assembly method (Hung et al., 2014). The composite membrane showed excellent permeation flux and selectivity in the pervaporation separation of a 70 wt.% isopropanol/water mixture with 99.5 wt.% water in the permeate and 2047 $g/m^2 \cdot h$ permeation flux. The dense GO film consisted of highly ordered and packed laminates led to the high selectivity of the membrane by allowing water to pass through while inhibiting isopropanol molecules. The schematic representation of the mechanism for separating isopropanol and water using a freestanding GO membrane is illustrated in Figure 32.4.

In addition to the examples previously discussed, free-standing GO thin films can also be assembled via ultrafiltration method to form membranes that can be used in the pervaporation dehydration of ethanol (Tang et al., 2014). The resulting GO thin films have high structural stability and excellent mechanical strength caused by the hydrogen bonds, which also promoted the interlocking between GO nanosheets. Preferential water transport with relatively high water permeability and selectivity was observed in the pervaporation dehydration of ethanol because of high hydrophilicity and appropriate interlaminar spacing.

Besides dehydration of organic solvents, the separation of aromatic/aliphatic mixtures is considered as one of the most crucial processes in the chemical industry. With the growing interest on pervaporation as a simple and feasible technique for the separation of aromatic/aliphatic mixtures,

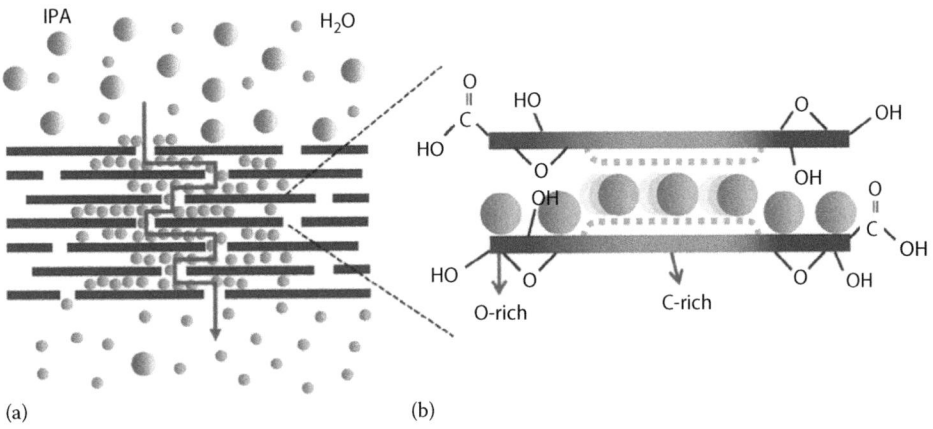

FIGURE 32.4
Schematic representation of the mechanism for separating isopropanol and water using a free-standing GO film. (a) Cross-sectional structure of a pure GO laminate, showing the interlocking layers and the interlayer spacing, which permit water but inhibit isopropanol to pass through. (b) Magnified GO sheets illustrating the hydrophilic O-rich edge, to which water molecules are adsorbed, and the hydrophobic C-rich central region, where there is a low-fiction channel (dashed lines) of monolayer water molecules. (Hung, W.-S. et al., *Carbon*, 68, 670, 2014.)

the development of membrane materials is critical because the stability of the membrane is a major concern. Hence, Wang et al. (2014a) fabricated a *pore-filling* membrane via the dynamic pressure-driven assembly of a PVA–GO nanohybrid layer onto an asymmetric polyacrylonitrile ultrafiltration membrane to enhance the stability of the membrane. The affinity of the membrane toward aromatic compounds was improved by the molecular-level dispersion of GO in PVA, which further enhanced the separation in the pervaporation of toluene/*n*-heptane mixtures. The pervaporation performance of the composite membrane was stable even after immersion of the membrane in 50 wt.% toluene/*n*-heptane solution for 480 h.

32.3.2 Graphene in Gas Separation

Apart from their application in membrane-based liquid separation processes previously presented, graphene-based materials were also studied for application in membrane-based gas separation. The generation of holes within the graphene plane is necessary to allow gas permeability because pure graphene sheets are impermeable to gases caused by the high electron density of their aromatic rings, which prevent any atom or molecule from passing through the graphitic plane (Jiang et al., 2009). Hence, porous graphene has been widely investigated for its use as membrane for gas separation (Blankenburg et al., 2010; Schrier, 2010; Du et al., 2011; Hauser and Schwerdtfeger, 2012).

A theoretical study was conducted by Jiang et al. (2009) to demonstrate the possibility of the use of porous graphene as a membrane for the separation of molecular gases. The permeability and selectivity of the graphene sheets with designed subnanometer pores were studied via first-principle density functional theory calculations. N_2-functionalized pores exhibited high selectivities in the order of 10^8 for H_2/CH_4 with a high H_2 permeance. However, extremely high selectivities in the order of 10^{23} for H_2/CH_4 were observed for all-H_2 passivated pores, which have small widths (at 2.5 Å) that function as barriers for CH_4.

Koenig et al. (2012) carried out an experimental study on porous graphene as a membrane for molecular sieving. Graphene sheets fabricated via mechanical exfoliation of graphite deposited on a silicon substrate were subjected to UV-induced oxidative etching to create pores. Through a pressurized blister test and mechanical resonance, the transport of a variety of gases, including H_2, CO_2, argon (Ar), CH_4, N_2, and sulfur hexafluoride (SF_6), through the pores was measured. The measured gas leak rates were consistent with the theoretical models based on the effusion through angstrom-sized pores (Jiang et al., 2009, Blankenburg et al., 2010).

Shan et al. (2012) utilized molecular dynamic simulations to investigate the effects of chemical functionalization of graphene sheets and pore rims on the gas separation performance of porous graphene membranes for CO_2/N_2 separation. They discovered that the absorption of CO_2 improved through the chemical functionalization of the graphene sheet. In addition, the selectivity of CO_2 over N_2 significantly improved through the chemical functionalization of the pore rim. The porous graphene membrane with all-N modified pores exhibited a higher CO_2 selectivity over N_2 compared with the unmodified graphene membrane because of the enhanced electrostatic interactions.

Meanwhile, the water solubility–driven separation of CO_2 from CO_2/O_2, CO_2/N_2, and CO_2/CH_4 mixtures by using a porous graphene membrane was investigated by Lee and Aluru (2013) via molecular dynamic simulations. After the introduction of a water slab between a gas mixture and a graphene membrane, the separation ratio of the gas mixtures depended on the water-solubility ratio of the gas molecules in the mixture. The results showed that the separation of the gas mixture was controlled by the water slab, and a thick water slab resulted in a high selectivity. Compared with CNTs, the graphene membrane offers a higher selectivity ratio, which is attributed to its single-atom thickness.

Furthermore, Liu et al. (2013a) demonstrated via molecular dynamic simulations that porous graphene with pore sizes of ~3.4 Å selectively separated CO_2 from N_2. High selectivity was achieved because of the increasing higher number of CO_2 that passes through compared with N_2. From the simulation results, permeances of 2.8×10^5 GPU (gas permeation unit) for CO_2 and around 2.9×10^3 GPU for N_2 were obtained. The predicted CO_2/N_2 selectivity is approximately 300. Meanwhile, the H_2 permeance through a porous graphene membrane was also studied by the same group of researchers (Liu et al., 2013b) via classical molecular dynamic simulations. The H_2 permeate flux linearly increased with the pressure driving force. Permeance within a relatively narrow range of 1×10^5 GPU to 4×10^5 GPU was obtained after normalizing the flux through pressure drop.

Jungthawan et al. (2013) applied the first-principle density functional theory and reported that the diffusion rates of H_2, O_2, and CO_2 in porous graphene can be enhanced up to 7, 13, and 20 orders of magnitude, respectively, with the application of a tensile stress. Hence, the authors believed that applying tensile stress is an effective way to control the diffusion rate of gases through porous graphene. Porous graphene can also allow the transport of larger molecules such as O_2 because of the significant improvement in the diffusion rates of porous graphene under strain.

Qin et al. (2013) designed porous graphene with a new line defect consisting of a sequence of octagons and all-H_2 passivated pores as gas separation membranes via first-principle calculations. The resulting membrane displayed extremely high separation capability in favor of H_2 among all studied species, and the selectivity was in the order of 10^{22} for the H_2/CH_4 separation. This phenomenon was attributed to the all-H_2 passivated pore that produces a formidable barrier of 1.5 eV for CH_4, thereby allowing the permeation of H_2.

Lei et al. (2014) recently investigated the effect of the charges around the pore on the separation of H_2S/CH_4 mixture by using a porous graphene membrane via molecular dynamic simulations. Their findings revealed that the reduction in the potential energy of H_2S around the charged pore was caused by electrostatic interactions between H_2S and the membrane. More H_2S molecules eventually accumulated around the pores and led to an enhanced selectivity. However, this accumulation also slightly lowered the permeation ratio of H_2S.

In addition to porous graphene, GO membranes were also investigated in gas separation. Kim et al. (2013b) experimentally studied the gas permeation behavior of GO membranes. Selective gas diffusion was achieved through the GO membranes by controlling gas flow channels and pores via different stacking techniques. Well-interlocked GO membranes exhibited high CO_2/N_2 selectivity at a high relative humidity, which suggested their potential as membranes for postcombustion CO_2 separation processes.

Li et al. (2013) deposited ultrathin GO membranes on AAO via a simple filtration process. Based on the permeation test results, H_2 permeated ~300 times faster through an 18 nm-thick GO membrane than CO_2. Furthermore, the H_2 and He permeances exponentially decreased when the membrane thickness was increased from 1.8 to 20 nm. According to the authors, the exponential dependence of the gas permeances on membranes thickness was attributed to the particular molecular transport pathway via the selective structural defects in the GO membranes. In addition, these membranes were considered promising candidates in H_2 recovery for ammonia production because of the superior molecular-sieving performance of H_2 compared with other gas molecules, as demonstrated by the ultrathin GO membranes.

Regardless of the simulation and experimental efforts previously discussed, significant technical complications still exist in the fabrication of such membranes for practical gas separation applications. More effort is needed to overcome the hurdles by further examining and understanding the molecular transport of gases through the graphene-based membranes and to develop and prepare graphene-based materials via facile and economic methods.

32.4 CONCLUSION AND FUTURE OUTLOOK

This chapter summarizes the recent developments of functionalized carbon-based nanomaterials, particularly CNTs and graphene, in the application of membrane separation process. Owing to the unique properties of CNTs and graphene, significant enhancement in membrane separation properties, mechanical properties, and stability were observed. Furthermore, CNTs and graphene can be functionalized to overcome limitations and to achieve revolutionary performances in specific applications. The application of CNTs and graphene showed tremendous advancements for liquid and gas separations. It can be foreseen that these materials may be able to compete with the existing separation membranes in the near future.

Nevertheless, their realization in membrane separation may be hampered by challenges related to their synthesis and processing. Hence, further studies on the innovative approaches for the synthesis and processing of these carbon-based nanomaterials are crucial to make this material function more effectively in membrane separation applications. In addition, more information on the effects of the different types of functionalization on the potential applications of these carbon-based nanomaterials in membrane separation is also critical for future developments. The applications of CNTs and graphene in membrane technology are also hindered by their costs and operational issues. These setbacks remain a challenge in scaling up the technology from laboratory-scale to industrial level. While most of the studies have focused on the potential advantages of these carbon-based nanomaterials in membrane separation, it is also crucial to investigate the potential human health and environmental risk caused by carbon-based nanomaterials.

To date, numerous experimental studies on the potential application of CNTs in both liquid and gas separation have been reported. However, more scientific and technical inputs are needed to discover more possibilities and potentials. On the other hand, liquid and gas separation based on single layer/multilayers graphene membrane is also considered as an ideal model but the related studies are mostly at the simulation stage. Thus, increased effort is needed to experimentally verify the interesting simulation findings. In a nutshell, with the encouraging progress of carbon-based nanomaterials in membrane separation applications, these materials will play an important role in the future development of membrane separation technology.

ACKNOWLEDGMENTS

Yeang, Q. W. acknowledges the MyPhD fellowship support from the Ministry of Education of Malaysia. This research is also supported by the research grant (FRGS 6071295).

LIST OF ABBREVIATIONS

3-aminopropyltriethoxylsilane	APTES
Aminated graphene oxide	aGO
Amine groups	NH_2
Anodic aluminum oxide	AAO
Argon	Ar
Brominated poly(2,6-diphenyl-1,4-phenylene oxide)	BPPOdp
Buckypaper	BP
Carbon dioxide	CO_2
Carbon nanotube	CNT
Carbon nanotube immobilized membrane	CNIM
Copper ion	Cu^{2+}
Forward osmosis	FO
Gas permeation unit	GPU
Graphene oxide	GO
Helium	He
Hydrogen	H_2
Hydrogen sulfide	H_2S
Methane	CH_4
Multiwalled carbon nanotube	MWCNT
Nitric acid	HNO_3
Nitrogen	N_2
Oxygen	O2
Pervaporation separation index	PSI
Poly(3-hydroxybutyrate)	PHB
Poly(allylamine hydrochloride)	PAH
Poly(ethyleneimine)	PEI
Poly(styrene sulfonic acid)-functionalized CNT CNT-PSSA	
Poly(vinyl alcohol)	PVA
Poly(vinylidene fluoride)	PVDF
Polyamide	PA
Polyelectrolyte–polyelectrolyte complex	PEC

Polyethersulfone	PES
Polyethylene glycol	PEG
Polyimide	PI
Polysulfone	PSf
Potassium ion	K^+
Reverse osmosis	RO
Single-walled carbon nanotube	SWCNT
Sodium ion	Na+
Sodium sulfate	Na_2SO_4
Sulfonated poly(arylene ether sulfone) material containing amino groups	aPES
Sulfur hexafluoride	SF_6
Sulfuric acid	H_2SO_4
Titanium dioxide	TiO_2
Transmission electron microscopy	TEM
Ultraviolet	UV
β-cyclodextrin	β-CD
β-cyclodextrin functionalized carbon nanotubes	β-CD-CNTs

REFERENCES

Ahadian, S. and Kawazoe, Y. 2009. An artificial intelligence approach for modeling and prediction of water diffusion inside a carbon nanotube. *Nanoscale Research Letters*, 4, 1054–1058.

Ajayan, P. M. 1999. Nanotubes from carbon. *Chemical Reviews*, 99, 1787–1800.

Amini, M., Jahanshahi, M., and Rahimpour, A. 2013. Synthesis of novel thin film nano-composite (TFN) forward osmosis membranes using functionalized multi-walled carbon nanotubes. *Journal of Membrane Science*, 435, 233–241.

Amirilargani, M., Ghadimi, A., Tofighy, M. A., and Mohammadi, T. 2013. Effects of poly(allylamine hydrochloride) as a new functionalization agent for preparation of poly vinyl alcohol/multiwalled carbon nanotubes membranes. *Journal of Membrane Science*, 447, 315–324.

Aroon, M. A., Ismail, A. F., and Matsuura, T. 2013. Beta-cyclodextrin functionalized MWCNT: A potential nano-membrane material for mixed matrix gas separation membranes development. *Separation and Purification Technology*, 115, 39–50.

Aroon, M. A., Ismail, A. F., Montazer-Rahmati, M. M., and Matsuura, T. 2010. Effect of chitosan as a functionalization agent on the performance and separation properties of polyimide/multi-walled carbon nanotubes mixed matrix flat sheet membranes. *Journal of Membrane Science*, 364, 309–317.

Baker, R. W. 2004. Overview of membrane science and technology. In: Baker, R. W. (ed.), *Membrane Technology and Applications*, 2nd edn., pp. 1–14. John Wiley and Sons Ltd.

Baughman, R. H., Zakhidov, A. A., and De Heer, W. A. 2002. Carbon nanotubes—The route toward applications. *Science*, 297, 787–792.

Bhadra, M., Roy, S., and Mitra, S. 2013. Enhanced desalination using carboxylated carbon nanotube immobilized membranes. *Separation and Purification Technology*, 120, 373–377.

Blankenburg, S., Bieri, M., Fasel, R. et al. 2010. Porous graphene as an atmospheric nanofilter. *Small*, 6, 2266–2271.

Bunch, J. S., Verbridge, S. S., Alden, J. S. et al. 2008. Impermeable atomic membranes from graphene sheets. *Nano Letters*, 8, 2458–2462.

Cabassud, C. and Wirth, D. 2003. Membrane distillation for water desalination: How to chose an appropriate membrane? In *Conference on Desalination and the Environment—Fresh Water for All*, pp. 307–314. Elsevier Science BV: Malta, Italy.

Caro, J., Noack, M., Kölsch, P., and Schäfer, R. 2000. Zeolite membranes—State of their development and perspective. *Microporous and Mesoporous Materials*, 38, 3–24.

Celik, E., Park, H., Choi, H., and Choi, H. 2011. Carbon nanotube blended polyether-sulfone membranes for fouling control in water treatment. *Water Research*, 45, 274–282.

Cerneaux, S., Strużyńska, I., Kujawski, W. M., Persin, M., and Larbot, A. 2009. Comparison of various membrane distillation methods for desalination using hydrophobic ceramic membranes. *Journal of Membrane Science*, 337, 55–60.

Chan, W.-F., Chen, H.-Y., Surapathi, A. et al. 2013. Zwitterion functionalized carbon nanotube/polyamide nanocomposite membranes for water desalination. *ACS Nano*, 7, 5308–5319.

Choi, J.-H., Jegal, J., and Kim, W.-N. 2006. Fabrication and characterization of multi-walled carbon nanotubes/polymer blend membranes. *Journal of Membrane Science*, 284, 406–415.

Choi, J.-H., Jegal, J., Kim, W. N., and Choi, H. S. 2009. Incorporation of multiwalled carbon nanotubes into poly(vinyl alcohol) membranes for use in the pervaporation of water/ethanol mixtures. *Journal of Applied Polymer Science*, 111, 2186–2193.

Choi, W., Choi, J., Bang, J., and Lee, J.-H. 2013. Layer-by-layer assembly of graphene oxide nanosheets on polyamide membranes for durable reverse-osmosis applications. *ACS Applied Materials and Interfaces*, 5, 12510–12519.

Ciobanu, G., Carja, G., and Ciobanu, O. 2008. Structure of mixed matrix membranes made with SAPO-5 zeolite in polyurethane matrix. *Microporous and Mesoporous Materials*, 115, 61–66.

Cohen-Tanugi, D. and Grossman, J. C. 2012. Water desalination across nanoporous graphene. *Nano Letters*, 12, 3602–3608.

Cong, H., Zhang, J., Radosz, M., and Shen, Y. 2007. Carbon nanotube composite membranes of brominated poly(2,6-diphenyl-1,4-phenylene oxide) for gas separation. *Journal of Membrane Science*, 294, 178–185.

Corry, B. 2008. Designing carbon nanotube membranes for efficient water desalination. *Journal of Physical Chemistry B*, 112, 1427–1434.

Corry, B. 2011. Water and ion transport through functionalised carbon nanotubes: Implications for desalination technology. *Energy and Environmental Science*, 4, 751–759.

Cote, L. J., Kim, J., Zhang, Z., Sun, C., and Huang, J. 2010. Tunable assembly of graphene oxide surfactant sheets: Wrinkles, overlaps and impacts on thin film properties. *Soft Matter*, 6, 6096–6101.

Daraei, P., Madaeni, S. S., Ghaemi, N. et al. 2013. Enhancing antifouling capability of PES membrane via mixing with various types of polymer modified multi-walled carbon nanotube. *Journal of Membrane Science*, 444, 184–191.

De Lannoy, C.-F., Soyer, E., and Wiesner, M. R. 2013. Optimizing carbon nanotube-reinforced polysulfone ultrafiltration membranes through carboxylic acid functionalization. *Journal of Membrane Science*, 447, 395–402.

Dikin, D. A., Stankovich, S., Zimney, E. J. et al. 2007. Preparation and characterization of graphene oxide paper. *Nature*, 448, 457–460.

Dresselhaus, M. S., Dresselhaus, G., and Saito, R. 1995. Physics of carbon nanotubes. *Carbon*, 33, 883–891.

Dreyer, D. R., Park, S., Bielawski, C. W., and Ruoff, R. S. 2010. The chemistry of graphene oxide. *Chemical Society Reviews*, 39, 228–240.

Du, H., Li, J., Zhang, J. et al. 2011. Separation of hydrogen and nitrogen gases with porous graphene membrane. *Journal of Physical Chemistry C*, 115, 23261–23266.

Dumée, L., Germain, V., Sears, K. et al. 2011. Enhanced durability and hydrophobicity of carbon nanotube bucky paper membranes in membrane distillation. *Journal of Membrane Science*, 376, 241–246.

Dumée, L., Lee, J., Sears, K. et al. 2013. Fabrication of thin film composite poly(amide)-carbon-nanotube supported membranes for enhanced performance in osmotically driven desalination systems. *Journal of Membrane Science*, 427, 422–430.

Dumée, L. F., Sears, K., Schütz, J. et al. 2010. Characterization and evaluation of carbon nanotube Bucky-Paper membranes for direct contact membrane distillation. *Journal of Membrane Science*, 351, 36–43.

El-Bourawi, M. S., Ding, Z., Ma, R., and Khayet, M. 2006. A framework for better understanding membrane distillation separation process. *Journal of Membrane Science*, 285, 4–29.

Enotiadis, A., Angjeli, K., Baldino, N., Nicotera, I., and Gournis, D. 2012. Graphene-based Nafion nanocomposite membranes: Enhanced proton transport and water retention by novel organo-functionalized graphene oxide nanosheets. *Small*, 8, 3338–3349.

Favvas, E. P., Nitodas, S. F., Stefopoulos, A. A. et al. 2014. High purity multi-walled carbon nanotubes: Preparation, characterization and performance as filler materials in co-polyimide hollow fiber membranes. *Separation and Purification Technology*, 122, 262–269.

Feng, X. and Huang, R. Y. M. 1997 Liquid separation by membrane pervaporation: A review. *Industrial and Engineering Chemistry Research*, 36, 1048–1066.

Gai, J.-G. and Gong, X.-L. 2014. Zero internal concentration polarization FO membrane: Functionalized graphene. *Journal of Materials Chemistry A*, 2, 425–429.

Gai, J.-G., Gong, X.-L., Wang, W.-W., Zhang, X., and Kang, W.-L. 2014. An ultrafast water transport forward osmosis membrane: Porous graphene. *Journal of Materials Chemistry A*, 2, 4023–4028.

Ganesh, B. M., Isloor, A. M., and Ismail, A. F. 2013. Enhanced hydrophilicity and salt rejection study of graphene oxide-polysulfone mixed matrix membrane. *Desalination*, 313, 199–207.

Gao, P., Liu, Z., Tai, M., Sun, D. D., and Ng, W. 2013. Multifunctional graphene oxide–TiO2 microsphere hierarchical membrane for clean water production. *Applied Catalysis B: Environmental*, 138–139, 17–25.

Gao, Y., Hu, M., and Mi, B. 2014. Membrane surface modification with TiO_2–graphene oxide for enhanced photocatalytic performance. *Journal of Membrane Science*, 455, 349–356.

Ge, L., Zhu, Z., and Rudolph, V. 2011. Enhanced gas permeability by fabricating functionalized multi-walled carbon nanotubes and polyethersulfone nanocomposite membrane. *Separation and Purification Technology*, 78, 76–82.

Geim, A. K. 2009. Graphene: Status and prospects. *Science*, 324, 1530–1534.

Gethard, K., Sae-Khow, O., and Mitra, S. 2011. Water desalination using carbon-nanotube-enhanced membrane distillation. *Applied Materials and Interfaces*, 3, 110–114.

Gilani, N., Towfighi, J., Rashidi, A. et al. 2013. Investigation of H_2S separation from H_2S/CH_4 mixtures using functionalized and non-functionalized vertically aligned carbon nanotube membranes. *Applied Surface Science*, 270, 115–123.

Goh, K., Setiawan, L., Wei, L. et al. 2013. Fabrication of novel functionalized multi-walled carbon nanotube immobilized hollow fiber membranes for enhanced performance in forward osmosis process. *Journal of Membrane Science*, 446, 244–254.

Goh, P. S., Ng, B. C., Ismail, A. F., Aziz, M., and Hayashi, Y. 2012. Pre-treatment of multi-walled carbon nanotubes for polyetherimide mixed matrix hollow fiber membranes. *Journal of Colloid and Interface Science*, 386, 80–87.

Han, Y., Xu, Z., and Gao, C. 2013. Ultrathin graphene nanofiltration membrane for water purification. *Advanced Functional Materials*, 23, 3693–3700.

Hauser, A. W. and Schwerdtfeger, P. 2012. Nanoporous graphene membranes for efficient 3He/4He separation. *The Journal of Physical Chemistry Letters*, 3, 209–213.

He, Z., Zhou, J., Lu, X., and Corry, B. 2013. Bioinspired graphene nanopores with voltage-tunable ion selectivity for Na^+ and K^+. *ACS Nano*, 7, 10148–10157.

Heo, Y., Im, H., and Kim, J. 2013. The effect of sulfonated graphene oxide on sulfonated poly(etheretherketone) membrane for direct methanol fuel cells. *Journal of Membrane Science*, 425–426, 11–22.

Ho, W. S. W. and Sirkar, K. K. 1992. Overview. In: Ho, W. S. W. and Sirkar, K. K. (eds.), *Membrane Handbook*, pp. 3–15. Chapman & Hall: New York.

Holt, J. K., Park, H. G., Wang, Y. M. et al. 2006. Fast mass transport through sub-2-nanometer carbon nanotubes. *Science*, 312, 1034–1037.

Hu, M. and Mi, B. 2013. Enabling graphene oxide nanosheets as water separation membranes. *Environmental Science and Technology*, 47, 3715–3723.

Hu, S. Y., Zhang, Y., Lawless, D., and Feng, X. 2012. Composite membranes comprising of polyvinylamine-poly(vinyl alcohol) incorporated with carbon nanotubes for dehydration of ethylene glycol by pervaporation. *Journal of Membrane Science*, 417–418, 34–44.

Huang, H., Song, Z., Wei, N. et al. 2013. Ultrafast viscous water flow through nanostrand-channelled graphene oxide membranes. *Nature Communications*, 4, Article number: 2979.

Hummer, G., Rasaiah, J. C., and Noworyta, J. P. 2001. Water conduction through the hydrophobic channel of a carbon nanotube. *Nature*, 414, 188–190.

Hung, W.-S., An, Q.-F., De Guzman, M. et al. 2014. Pressure-assisted self-assembly technique for fabricating composite membranes consisting of highly ordered selective laminate layers of amphiphilic graphene oxide. *Carbon*, 68, 670–677.

Iijima, S. 1991. Helical microtubules of graphitic carbon. *Nature*, 354, 56–58.

Ismail, A. F., Rahim, N. H., Mustafa, A. et al. 2011. Gas separation performance of polyethersulfone/multi-walled carbon nanotubes mixed matrix membranes. *Separation and Purification Technology*, 80, 20–31.

Ji, T., Sun, M., Zou, L., and Ma, N. 2014. Fabrication of graphene oxide-filled poly(vinylidene fluoride) microfiltration tubes and the applications in a separation of citral solution and a removal of Cu^{2+} ions. *Materials Letters*, 120, 30–32.

Jia, Y.-X., Li, H.-L., Wang, M., Wu, L.-Y., and Hu, Y.-D. 2010. Carbon nanotube: Possible candidate for forward osmosis. *Separation and Purification Technology*, 75, 55–60.

Jiang, D., Cooper, V. R., and Dai, S. 2009. Porous graphene as the ultimate membrane for gas separation. *Nano Letters*, 9, 4019–4024.

Jungthawan, S., Reunchan, P., and Limpijumnong, S. 2013. Theoretical study of strained porous graphene structures and their gas separation properties. *Carbon*, 54, 359–364.

Kalra, A., Garde, S., and Hummer, G. 2003. Osmotic water transport through carbon nanotube membranes. In: Cozzarelli, N. R. (ed.) *Proceedings of the National Academy of Sciences of the United States of America*. Stanford University's HighWire Press: Stanford, CA.

Khan, M. M., Filiz, V., Bengtson, G. et al. 2013. Enhanced gas permeability by fabricating mixed matrix membranes of functionalized multiwalled carbon nanotubes and polymers of intrinsic microporosity (PIM). *Journal of Membrane Science*, 436, 109–120.

Kim, E.-S., Hwang, G., El-Din, M. G., and Liu, Y. 2012. Development of nanosilver and multi-walled carbon nanotubes thin-film nanocomposite membrane for enhanced water treatment. *Journal of Membrane Science*, 394–395, 37–48.

Kim, E.-S., Liu, Y., and Gamal El-Din, M. 2013a. An in-situ integrated system of carbon nanotubes nanocomposite membrane for oil sands process-affected water treatment. *Journal of Membrane Science*, 429, 418–427.

Kim, H. J., Choi, K., Baek, Y. et al. 2014. High-performance reverse osmosis CNT/polyamide nanocomposite membrane by controlled interfacial interactions. *Applied Materials and Interfaces*, 6, 2819–2829.

Kim, H. W., Yoon, H. W., Yoon, S.-M. et al. 2013b. Selective gas transport through few-layered graphene and graphene oxide membranes. *Science*, 342, 91–95.

Kim, S. G., Hyeon, D. H., Chun, J. H., Chun, B.-H., and Kim, S. H. 2013c. Novel thin nanocomposite RO membranes for chlorine resistance. *Desalination and Water Treatment*, 51, 6338–6345.

Koenig, S. P., Wang, L., Pellegrino, J., and Bunch, J. S. 2012. Selective molecular sieving through porous graphene. *Nature Nanotechnology*, 7, 728–732.

Koinuma, M., Ogata, C., Kamei, Y. et al. 2012. Photochemical engineering of graphene oxide nanosheets. *Journal of Physical Chemistry C*, 116, 19822–19827.

Kumar, M. and Ulbricht, M. 2013. Novel antifouling positively charged hybrid ultrafiltration membranes for protein separation based on blends of carboxylated carbon nanotubes and aminated poly(arylene ether sulfone). *Journal of Membrane Science*, 448, 62–73.

Lawson, K. W. and Lloyd, D. R. 1997. Membrane distillation. *Journal of Membrane Science*, 124, 1–25.

Lee, C., Wei, X., Kysar, J. W., and Hone, J. 2008. Measurement of the elastic properties and intrinsic strength of monolayer graphene. *Science*, 321, 385–388.

Lee, J. and Aluru, N. R. 2013. Water-solubility-driven separation of gases using graphene membrane. *Journal of Membrane Science*, 428, 546–553.

Lei, G., Liu, C., Xie, H., and Song, F. 2014. Separation of the hydrogen sulfide and methane mixture by the porous graphene membrane: Effect of the charges. *Chemical Physics Letters*, 599, 127–132.

Li, H., Song, Z., Zhang, X. et al. 2013. Ultrathin, molecular-sieving graphene oxide membranes for selective hydrogen separation. *Science*, 342, 95–98.

Li, K. 2007. *Ceramic Membrane for Separation and Reaction.* John Wiley and Sons: Chichester, U.K.

Li, X., Wang, X., Zhang, L., Lee, S., and Dai, H. 2008a. Chemically derived, ultrasmooth graphene nanoribbon semiconductors. *Science*, 319, 1229–1232.

Li, X., Zhang, G., Bai, X. et al. 2008b. Highly conducting graphene sheets and Langmuir-Blodgett films. *Nature Nanotechnology*, 3, 538–542.

Li, Y., Zhou, H., Zhu, G., Liu, J., and Yang, W. 2007. Hydrothermal stability of LTA zeolite membranes in pervaporation. *Journal of Membrane Science*, 297, 10–15.

Lipnizki, F., Hausmanns, S., Ten, P.-K., Field, R. W., and Laufenberg, G. 1999. Organophilic pervaporation: Prospects and performance. *Chemical Engineering Journal*, 73, 113–129.

Liu, H., Dai, S., and Jiang, D.-E. 2013a. Insights into CO_2/N_2 separation through nanoporous graphene from molecular dynamics. *Nanoscale*, 5, 9984–9987.

Liu, H., Dai, S., and Jiang, D.-E. 2013b. Permeance of H_2 through porous graphene from molecular dynamics. *Solid State Communications*, 175–176, 101–105.

Liu, Y.-L., Chen, W.-H., and Chang, Y.-H. 2009. Preparation and properties of chitosan/carbon nanotube nanocomposites using poly(styrene sulfonic acid)-modified CNTs. *Carbohydrate Polymers*, 76, 232–238.

Liu, Z., Liu, Q., Huang, Y. et al. 2008. Organic photovoltaic devices based on a novel acceptor material: Graphene. *Advanced Materials*, 20, 3924–3930.

Mansourpanah, Y., Madaeni, S. S., Rahimpour, A. et al. 2011. Fabrication new PES-based mixed matrix nanocomposite membranes using polycaprolactone modified carbon nanotubes as the additive: Property changes and morphological studies. *Desalination*, 277, 171–177.

Matranga, C., Bockrath, B., Chopra, N., Hinds, B. J., and Andrews, R. 2006. Raman spectroscopic investigation of gas interactions with an aligned multiwalled carbon nanotube membrane. *Langmuir*, 22, 1235–1240.

Mulder, M. 1991. Introduction. In: Mulder, M. (ed.), *Basic Principle of Membrane Technology*, pp. 1–15. Kluwer Academic Publisher: Dordrecht, the Netherlands.

Nair, R. R., Wu, H. A., Jayaram, P. N., Grigorieva, I. V., and Geim, A. K. 2012. Unimpeded permeation of water through helium-leak–tight graphene-based membranes. *Science*, 335, 442–444.

Niyogi, S., Bekyarova, E., Itkis, M. E. et al. 2006. Solution properties of graphite and graphene. *Journal of the American Chemical Society*, 128, 7720–7721.

Novoselov, K. S., Geim, A. K., Morozov, S. V. et al. 2004. Electric field effect in atomically thin carbon films. *Science*, 306, 666–669.

Ong, Y. T., Ahmad, A. L., Zein, S. H. S., Sudesh, K., and Tan, S. H. 2011. Poly(3-hydroxybutyrate)-functionalised multi-walled carbon nanotubes/chitosan green nanocomposite membranes and their application in pervaporation. *Separation and Purification Technology*, 76, 419–427.

Pang, S., Tsao, H. N., Feng, X., and Müllen, K. 2009. Patterned graphene electrodes from solution-processed graphite oxide films for organic field-effect transistors. *Advanced Materials*, 21, 3488–3491.

Peng, F., Hu, C., and Jiang, Z. 2007a. Novel poly(vinyl alcohol)/carbon nanotube hybrid membranes for pervaporation separation of benzene/cyclohexane mixtures. *Journal of Membrane Science*, 297, 236–242.

Peng, F., Pan, F., Sun, H., Lu, L., and Jiang, Z. 2007b. Novel nanocomposite pervaporation membranes composed of poly(vinyl alcohol) and chitosan-wrapped carbon nanotube. *Journal of Membrane Science*, 300, 13–19.

Pietraß, T. 2006. Carbon-based membranes. *MRS Bulletin*, 31, 765–769.

Qin, X., Meng, Q., Feng, Y., and Gao, Y. 2013. Graphene with line defect as a membrane for gas separation: Design via a first-principles modeling. *Surface Science*, 607, 153–158.

Qiu, L., Zhang, X., Yang, W. et al. 2011. Controllable corrugation of chemically converted graphene sheets in water and potential application for nanofiltration. *Chemical Communications*, 47, 5810–5812.

Qiu, S., Wu, L., Pan, X. et al. 2009. Preparation and properties of functionalized carbon nanotube/PSF blend ultrafiltration membranes. *Journal of Membrane Science*, 342, 165–172.

Qiu, S., Wu, L., Shi, G. et al. 2010. Preparation and pervaporation property of chitosan membrane with functionalized multiwalled carbon nanotubes. *Industrial and Engineering Chemistry Research*, 49, 11667–11675.

Quintana, M., Spyrou, K., Grzelczak, M. et al. 2010. Functionalization of graphene via 1,3-dipolar cycloaddition. *ACS Nano*, 4, 3527–3533.

Rahimpour, A., Jahanshahi, M., Khalili, S. et al. 2012. Novel functionalized carbon nanotubes for improving the surface properties and performance of polyethersulfone (PES) membrane. *Desalination*, 286, 99–107.

Raidongia, K. and Huang, J. X. 2012. Nanofluidic ion transport through reconstructed layered materials. *Journal of the American Chemical Society*, 134, 16528–16531.

Ramanathan, T., Abdala, A. A., Stankovich, S. et al. 2008. Functionalized graphene sheets for polymer nanocomposites. *Nature Nanotechnology*, 3, 327–331.

Rousseau, R. 1987. *Handbook of Separation Process Technology.* John Wiley and Sons: Paris, France.

Sajjan, A. M., Jeevan Kumar, B. K., Kittur, A. A., and Kariduraganavar, M. Y. 2013. Novel approach for the development of pervaporation membranes using sodium alginate and chitosan-wrapped multiwalled carbon nanotubes for the dehydration of isopropanol. *Journal of Membrane Science*, 425–426, 77–88.

Sanip, S. M., Ismail, A. F., Goh, P. S. et al. 2011. Gas separation properties of functionalized carbon nanotubes mixed matrix membranes. *Separation and Purification Technology*, 78, 208–213.

Saranya, R., Arthanareeswaran, G., and Dionysiou, D. D. 2014. Treatment of paper mill effluent using Polyethersulfone/functionalised multiwalled carbon nanotubes based nanocomposite membranes. *Chemical Engineering Journal*, 236, 369–377.

Schneider, K. and Vangassel, T. J. 1984. Membrane distillation. *Chemie Ingenieur Technik*, 56, 514–521.

Schofield, R. W., Fane, A. G., and Fell, C. J. D. 1987. Heat and mass transfer in membrane distillation. *Journal of Membrane Science*, 33, 299–313.

Schofield, R. W., Fane, A. G., Fell, C. J. D., and Macoun, R. 1990. Factors affecting flux in membrane distillation. *Desalination*, 77, 279–294.

Schrier, J. 2010. Helium separation using porous graphene membranes. *The Journal of Physical Chemistry Letters*, 1, 2284–2287.

Shah, P. and Murthy, C. N. 2013. Studies on the porosity control of MWCNT/polysulfone composite membrane and its effect on metal removal. *Journal of Membrane Science*, 437, 90–98.

Shan, M., Xue, Q., Jing, N. et al. 2012. Influence of chemical functionalization on the CO_2/N_2 separation performance of porous graphene membranes. *Nanoscale*, 4, 5477–5482.

Shao, L., Cheng, X., Wang, Z., Ma, J., and Guo, Z. 2014. Tuning the performance of polypyrrole-based solvent-resistant composite nanofiltration membranes by optimizing polymerization conditions and incorporating graphene oxide. *Journal of Membrane Science*, 452, 82–89.

Shen, J. N., Yu, C. C., Ruan, H. M., Gao, C. J., and Van Der Bruggen, B. 2013. Preparation and characterization of thin-film nanocomposite membranes embedded with poly(methyl methacrylate) hydrophobic modified multiwalled carbon nanotubes by interfacial polymerization. *Journal of Membrane Science*, 442, 18–26.

Shirazi, Y., Tofighy, M. A., and Mohammadi, T. 2011. Synthesis and characterization of carbon nanotubes/poly vinyl alcohol nanocomposite membranes for dehydration of isopropanol. *Journal of Membrane Science*, 378, 551–561.

Sint, K., Wang, B., and Kral, P. 2008. Selective ion passage through functionalized graphene nanopores. *Journal of the American Chemical Society*, 130, 16448–16449.

Skoulidas, A. I., Ackerman, D. M., Johnson, J. K., and Sholl, D. S. 2002. Rapid transport of gases in carbon nanotubes. *Physical Review Letters*, 89, 185901.

Smitha, B., Suhanya, D., Sridhar, S., and Ramakrishna, M. 2004. Separation of organic-organic mixtures by pervaporation-a review. *Journal of Membrane Science*, 241, 1–21.

Sokhan, V. P., Nicholson, D., and Quirke, N. 2004. Transport properties of nitrogen in single walled carbon nanotubes. *Journal of Chemical Physics*, 120, 3855–3863.

Suhas, D. P., Raghu, A. V., Jeong, H. M., and Aminabhavi, T. M. 2013. Graphene-loaded sodium alginate nanocomposite membranes with enhanced isopropanol dehydration performance via a pervaporation technique. *RSC Advances*, 3, 17120–17130.

Sui, D., Huang, Y., Huang, L. et al. 2011. Flexible and transparent electrothermal film heaters based on graphene materials. *Small*, 7, 3186–3192.

Suk, M. E. and Aluru, N. R. 2010. Water transport through ultrathin graphene. *The Journal of Physical Chemistry Letters*, 1, 1590–1594.

Sun, P., Zhu, M., Wang, K. et al. 2013. Selective ion penetration of graphene oxide membranes. *ACS Nano*, 7, 428–437.

Tang, Y. P., Paul, D. R., and Chung, T. S. 2014. Free-standing graphene oxide thin films assembled by a pressurized ultrafiltration method for dehydration of ethanol. *Journal of Membrane Science*, 458, 199–208.

Vandezande, P., Gevers, L. E. M., and Vankelecom, I. F. 2008. Solvent resistant nanofiltration: Separating on a molecular level. *Chemical Society Reviews*, 37, 365–405.

Vatanpour, V., Madaeni, S. S., Moradian, R., Zinadini, S., and Astinchap, B. 2012. Novel antibifouling nanofiltration polyethersulfone membrane fabricated from embedding TiO_2 coated multiwalled carbon nanotubes. *Separation and Purification Technology*, 90, 69–82.

Verkerk, A. W., Van Male, P., Vorstman, M. A. G., and Keurentjes, J. T. F. 2001. Properties of high flux ceramic pervaporation membranes for dehydration of alcohol/water mixtures. *Separation and Purification Technology*, 22–23, 689–695.

Wang, N., Ji, S., Li, J., Zhang, R., and Zhang, G. 2014a. Poly(vinyl alcohol)–graphene oxide nanohybrid "pore-filling" membrane for pervaporation of toluene/n-heptane mixtures. *Journal of Membrane Science*, 455, 113–120.

Wang, N., Ji, S., Zhang, G., Li, J., and Wang, L. 2012. Self-assembly of graphene oxide and polyelectrolyte complex nanohybrid membranes for nanofiltration and pervaporation. *Chemical Engineering Journal*, 213, 318–329.

Wang, S., Liu, Y., Huang, S. et al. 2014b. Pebax–PEG–MWCNT hybrid membranes with enhanced CO_2 capture properties. *Journal of Membrane Science*, 460, 62–70.

Wang, Y., Ou, R., Ge, Q., Wang, H., and Xu, T. 2013. Preparation of polyethersulfone/carbon nanotube substrate for high-performance forward osmosis membrane. *Desalination*, 330, 70–78.

Weng, T.-H., Tseng, H.-H., and Wey, M.-Y. 2009. Preparation and characterization of multi-walled carbon nanotube/PBNPI nanocomposite membrane for H_2/CH_4 separation. *International Journal of Hydrogen Energy*, 34, 8707–8715.

Xu, C., Cui, A., Xu, Y., and Fu, X. 2013. Graphene oxide–TiO_2 composite filtration membranes and their potential application for water purification. *Carbon*, 62, 465–471.

Xu, Y., Bai, H., Lu, G., Li, C., and Shi, G. 2008. Flexible graphene films via the filtration of water-soluble noncovalent functionalized graphene sheets. *Journal of the American Chemical Society*, 130, 5856–5857.

Xu, Z., Zhang, J., Shan, M. et al. 2014. Organosilane-functionalized graphene oxide for enhanced antifouling and mechanical properties of polyvinylidene fluoride ultrafiltration membranes. *Journal of Membrane Science*, 458, 1–13.

Yan, X., Chen, J., Yang, J., Xue, Q., and Miele, P. 2010. Fabrication of free-standing, electrochemically active, and biocompatible graphene oxide–Polyaniline and graphene–Polyaniline hybrid papers. *ACS Applied Materials and Interfaces*, 2, 2521–2529.

Yeang, Q. W., Zein, S. H. S., Sulong, A. B., and Tan, S. H. 2013. Comparison of the pervaporation performance of various types of carbon nanotube-based nanocomposites in the dehydration of acetone. *Separation and Purification Technology*, 107, 252–263.

Yee, K. F., Ong, Y. T., Mohamed, A. R., and Tan, S. H. 2014. Novel MWCNT-buckypaper/ polyvinyl alcohol asymmetric membrane for dehydration of etherification reaction mixture: Fabrication, characterisation and application. *Journal of Membrane Science*, 453, 546–555.

Yeh, T.-M., Wang, Z., Mahajan, D., Hsiao, B. S., and Chu, B. 2013. High flux ethanol dehydration using nanofibrous membranes containing graphene oxide barrier layers. *Journal of Materials Chemistry A*, 1, 12998–13003.

Yin, J., Zhu, G., and Deng, B. 2013. Multi-walled carbon nanotubes (MWNTs)/ polysulfone(PSU) mixed matrix hollow fiber membranes for enhanced water treatment. *Journal of Membrane Science*, 437, 237–248.

Yu, D. and Dai, L. 2010. Self-assembled graphene/carbon nanotube hybrid films for supercapacitors. *The Journal of Physical Chemistry Letters*, 1, 467–470.

Zhang, J., Xu, Z., Shan, M. et al. 2013. Synergetic effects of oxidized carbon nanotubes and graphene oxide on fouling control and anti-fouling mechanism of polyvinylidene fluoride ultrafiltration membranes. *Journal of Membrane Science*, 448, 81–92.

Zhao, C., Xu, X., Chen, J., and Yang, F. 2013a. Effect of graphene oxide concentration on the morphologies and antifouling properties of PVDF ultrafiltration membranes. *Journal of Environmental Chemical Engineering*, 1, 349–354.

Zhao, H., Qiu, S., Wu, L. et al. 2014a. Improving the performance of polyamide reverse osmosis membrane by incorporation of modified multi-walled carbon nanotubes. *Journal of Membrane Science*, 450, 249–256.

Zhao, Q., Qian, J., Zhu, M., and An, Q. 2009. Facile fabrication of polyelectrolyte complex/ carbon nanotube nanocomposites with improved mechanical properties and ultrahigh separation performance. *Journal of Materials Chemistry*, 19, 8732–8740.

Zhao, X., Ma, J., Wang, Z. et al. 2012. Hyperbranched-polymer functionalized multiwalled carbon nanotubes for poly(vinylidene fluoride) membranes: From dispersion to blended fouling-control membrane. *Desalination*, 303, 29–38.

Zhao, Y., Jung, B. T., Ansaloni, L., and Ho, W. S. W. 2014b. Multiwalled carbon nanotube mixed matrix membranes containing amines for high pressure CO_2/H_2 separation. *Journal of Membrane Science*, 459, 233–243.

Zhao, Y., Xu, Z., Shan, M. et al. 2013b. Effect of graphite oxide and multi-walled carbon nanotubes on the microstructure and performance of PVDF membranes. *Separation and Purification Technology*, 103, 78–83.

Zimmerman, C. M., Singh, A., and Koros, W. J. 1997. Tailoring mixed matrix composite membranes for gas separations. *Journal of Membrane Science*, 137, 145–154.

Zinadini, S., Zinatizadeh, A. A., Rahimi, M., Vatanpour, V., and Zangeneh, H. 2014. Preparation of a novel antifouling mixed matrix PES membrane by embedding graphene oxide nanoplates. *Journal of Membrane Science*, 453, 292–301.

CHAPTER 33

CONTENTS

(Kang et al. 2006). In addition, the lengths of CNTs/CNFs range from less than a micrometer to over a millimeter. Therefore, CNTs/CNFs usually have an aspect ratio of more than 1000.

With extremely high aspect ratios, CNTs/CNFs tend to assemble into bundles or ropes and form clumps. Moreover, van der Waals' forces and high surface energies make it hard for CNTs/CNFs to be dispersed uniformly and effectively in aqueous solutions (Xie et al. 2005; Yazdanbakhsh et al. 2012). For example, liquids with a surface tension below 130–170 mN/m should be used to wet bundles of SWCNTs. However, CNTs showed lower dispersion ability in water, which has a surface tension of 72 mN/m at room temperature (Dujardin et al. 1998).

Ideally, the ends of CNTs/CNFs are closed by two caps (semifullerenes), and the sidewall is made of a hexagonal structure of sp^2-hybridized carbon atoms formed highly delocalized π-electrons systems. These delocalized π-electrons can be used to combine with other compounds containing π-electrons to develop noncovalent modified CNTs/CNFs by π–π stacking interaction. In addition, CNTs/CNFs can be solubilized in aqueous solutions by noncovalently associating them with linear polymers, wherein the polymer disrupts the hydrophobic interface with water and the tubes/fibers (Banerjee et al. 2005).

33.2.2 Defects on CNTs/CNFs

In reality, CNTs/CNFs cannot always form perfect six-membered-ring carbon structure but rather contain defects formed during synthesis. Typically around 1%–3% of the carbon atoms of a nanotube/nanofiber are located at defect sites. Some defects stem from the initial fabrication of CNTs/CNFs, such as Stone–Wales defects caused by the presence of five- or seven-membered rings in the carbon framework and sp^3-hybridized defects that include H or OH. The purification processes also introduce defects to CNTs/CNFs, such as strong acid oxidation that damages the carbon framework, which leaves defect sites (vacancies) on the sidewall and open ends of CNTs/CNFs, terminated with carboxyl, hydroxyl, ester, and nitro groups (Hirsch 2002). Such defects are reactive spots to produce functional groups and offer the basis for covalent functionalization, noncovalent modification combining with covalent modification, in situ synthesis of CNTs/CNFs, and grafting on functionalized fibers precisely. Furthermore, pyramidalization and π-orbital misalignment between adjacent carbon atoms produce the chemical reactivity of the surface of CNTs/CNFs and make CNTs/CNFs possible for chemical modification (Banerjee et al. 2005; Haddon 1988).

33.3 METHODS OF THE COVALENT/NONCOVALENT MODIFICATION OF CNTs/CNFs

33.3.1 Methods of the Covalent Modification of CNTs/CNFs

The covalent modification method is based on surface chemical functionalization to form covalent bond onto the backbone of CNTs/CNFs. It can be achieved at different sites such as the ends of CNTs/CNFs as well as at their sidewalls (Balasubramanian and Burghard 2005; Hirsch 2002). Direct covalent sidewall modification of SWCNTs transforms sp^2-hybridized carbon framework into sp^3 hybrid carbons, corresponding with a critical loss of conjugation (Hirsch and Vostrowsky 2005). Covalent bond formation takes place preferentially at the preexisting defect sites due to their high chemical reactivity compared with that of purely hexagonal lattice (Banerjee et al. 2005). Covalent modification of CNTs/CNFs can improve their chemical compatibility with the target medium, improve their wettability, and reduce their tendency to agglomerate. A great deal of work has been investigated and mainly focused on covalent modification of CNTs/CNFs for fabricating polymer-based materials. These include oxidation and subsequent elaboration of CNTs (Datsyuk et al. 2008; Rosca et al. 2005; Zhang et al. 2003), fluorination and further substitution by alkyl groups (Boul et al. 1999; Mickelson et al. 1998), and addition of carbenes, nitrenes, and radicals (Holzinger et al. 2001). More details can be found in the relevant review (Hirsch et al. 2005). However, most of these covalent modification methods for polymer-based materials are forbidden in cement and concrete materials due to the retarded hydration of cement or the introduction of hazardous substance. To date, just oxidation approach is suitable and has been utilized in cement and concrete field. Therefore, this section mainly focuses on oxidation covalent modification of CNTs/CNFs in cement and concrete materials.

33.3.1.1 Mechanism of oxidation covalent modification of CNTs/CNFs

With the purpose to purify and enhance the chemical reactivity of carbon network, the oxidation of CNTs/CNFs either by acid treatment (Saito et al. 2002), plasma stimulation (Wang et al. 2009), ozone treatment (Mawhinney et al. 2000a), or strong oxidants such as $KMnO_4$ (Yu et al. 1998) has gained a lot of attention (Datsyuk et al. 2008). Liu et al. firstly developed a successful oxidation process for SWCNTs involved treatment with sonication in mixtures of sulfuric and nitric acids (Liu et al. 1998). Through such harsh treatments, the ends of SWCNTs are opened as well as the formation of holes in the sidewalls. In addition, the raw materials are cleaned by removal of amorphous and graphitic carbon (Banerjee et al. 2005; Mawhinney et al. 2000b). Figure 33.1 shows an oxidation process. Oxidation of CNTs leads to the ends and sidewalls being decorated with oxygen-containing groups (carboxyl, hydroxyl, ester, and nitro), which could be utilized for further derivatization and cuts and thus shortens the tubes (Qin et al. 2003; Sun et al. 2001; Ziegler et al. 2005). The cutting effect can be determined and demonstrated by atomic force microscopy (AFM) and scanning electron microscope (SEM) (Liu et al. 1998). The presence of carboxylic acid groups and hydroxyl groups bound to SWCNTs can be observed by Fourier transform infrared (FTIR) absorption spectroscopy (Wang et al. 2003; Zhang et al. 2003). Such superficial chemical groups, on the one hand, can make chemical reactions take place and originate strong chemical bonds between CNTs and cement and concrete matrix, thus enhancing the reinforcement efficiency (Musso et al. 2007, 2009). On the other hand, the presence of the carboxylic acid groups and hydrogen atoms on the surface of CNTs/CNFs decreases the contact angle and enhanced the wettability due to a dipole moment that causes the polarity of the water molecules to form additional hydrogen bonds (Kotsalis et al. 2005). Moving forward, introduction of carboxyl groups reduces the van der Waals' interactions between CNTs and CNFs, which separate bundles into individual tubes/fibers or smaller bundles (Kim et al. 2012). The electrostatic repulsion between the carboxylic anions on the CNTs/CNFs by the reason of relatively high zeta potentials of oxidized CNTs/CNFs also plays an important role in enhancing the dispersion of CNTs/CNFs in aqueous solutions (Chen et al. 2005; Sano et al. 2001).

33.3.1.2 Oxidation covalent modification of CNTs/ CNFs in cement and concrete materials

Oxidation treatment of CFs has been proved to have an ameliorating effect on fiber dispersion and increase the bond strength between CFs and cement and concrete materials (Fu et al. 1998). According to this case, Li et al. firstly introduced oxidation covalent modification of CNTs by using a mixture solution of H_2SO_4 and HNO_3 (3:1 by volume) and uniformly dispersed CNTs into cement paste by means of ultrasonic energy. The treated CNTs were characterized by FTIR, which demonstrated that carboxylic and hydroxyl groups attach to the surfaces of CNTs. Due to the presence of carboxylic acid

Pristine CNTs/CNFs   Functionalized CNTs/CNFs

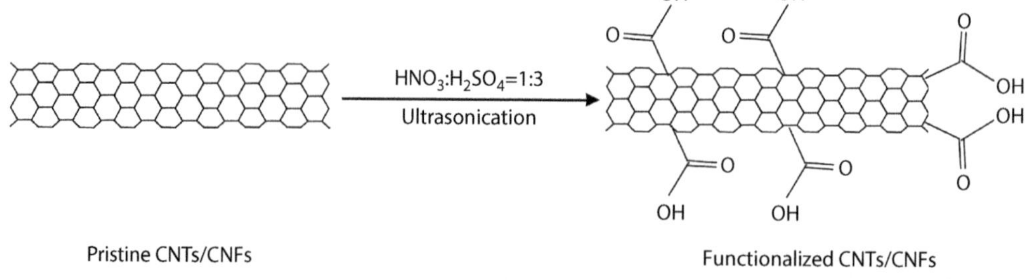

FIGURE 33.1
Schematic presentation of oxidation covalent modification of CNTs/CNFs by a mixture solution of HNO_3 and H_2SO_4 (1:3). The carboxylic acid groups attach to the sidewalls and ends of CNTs/CNFs.

groups, chemical reactions take place between the carboxylic acid and the calcium silicate hydrate (C–S–H) or Ca(OH)$_2$. As a result, the microstructure and mechanical properties are improved (Li et al. 2005). Additionally, Li et al. observed that the contact points of the treated CNTs were much fewer than those of the untreated CNTs, which means the treated CNTs have a better dispersion in cement pastes (Li et al. 2007). Yu et al. found CNTs successfully dispersed in water by treating CNTs with a mixture of sulfuric acid and nitric acid. After attaching functional groups on the CNT surfaces, the functionalized CNTs are less likely to agglomerate. The electrostatic repulsion force between these negative charges can be utilized to disperse CNTs in water without any surfactant (Yu and Kwon 2009). Luo et al. also treated MWCNTs with acid oxidization and found the active groups (such as –COOH group and C–O group) that improved the bonding strength of treated MWCNTs to cement hydration. The tubes presented superior dispersion and interface adhesion within cement matrix (Luo et al. 2011). Sanchez et al. studied the dispersion of CNFs in the cement pastes by using covalent surface modification with nitric acid. The CNFs were found as individual tubes throughout the cement pastes and as entangled networks in pockets (Sanchez 2009; Sanchez et al. 2009).

However, under intense reactive conditions, oxidation covalent modification of CNTs/CNFs might introduce structural defects and disrupt sp^2 hybridization and even destruct CNTs/CNFs. It has considerable effect on the properties of CNTs/CNFs, especially the electrical and mechanical properties (Burghard 2005; Coleman et al. 2006; Garg and Sinnott 1998; Park et al. 2006). Such phenomenon also appears in cement and concrete materials reinforced by CNTs/CNFs. Yu found that the untreated CNTs have more forceful effect on reducing electrical resistivity of cement composites (Li et al. 2007). Sanchez et al. observed failure with a fair level of mechanical integrity for composites containing treated CNFs (Sanchez et al. 2009). Tyson reported that acid-treated CNTs/CNFs had weaker mechanical properties than untreated nanofilaments in that the excessive formation of ettringite caused by the presence of sulfates (Abu Al-Rub et al. 2011). Musso demonstrated that carboxyl-group functionalized MWCNTs worsened the mechanical properties of the composite, while annealed MWCNTs showed an improvement due to the removal of lattice defects from the walls of CNTs (Musso et al. 2009).

33.3.2 Methods of the Noncovalent Modification of CNTs/CNFs

According to the aforementioned analysis, two major drawbacks in covalent modification of CNTs/CNFs are summarized: (1) aggressive processes, especially the oxidation with strong acids, originate structural defects and deteriorate the intrinsic properties of CNTs/CNFs; (2) the inert surface of CNTs/CNFs provides the limited active sites, followed by a low efficiency of functionalization and dispersibility (Geng et al. 2008; Kim et al. 2005; Ma et al. 2009). In contrast with covalent modification, the noncovalent modification of CNTs/CNFs is particularly attractive in that it preserves the structure of CNTs/CNFs; conserves sp^2-conjugated framework and properties, especially the electrical properties; consumes low energy; and is easy to control (Chen et al. 2001; Han et al. 2011a; Kim et al. 2012; Vaisman et al. 2006). The noncovalent surface modification method is based on adsorption or wrapping various species of surfactants or polymers through the van der Waals' interactions, π–π stacking, hydrogen bonding, and hydrophobic bonding between CNTs/CNFs and guest molecules and thus prevents the formation of aggregates, improves dispersibility of CNTs/CNFs, enhances the specific interfacial area, and leads to superior nanocomposites (Cadek et al. 2004; Hilding et al. 2003; Paria and Khilar 2004; Star et al. 2003; Vaisman et al. 2006).

33.3.2.1 Mechanism of noncovalent modification of CNTs/CNFs: Surfactant adsorption

Surfactants usually are amphiphilic compounds. They contain both hydrophobic groups and hydrophilic groups at their tails and heads, respectively. Surfactants usually consist of one or few hydrocarbon chains (Paria and Khilar 2004; Vaisman et al. 2006). Thus, the process of dispersing CNTs/CNFs depends strongly on the types of surfactants (Rausch et al. 2010). According to the charge of their head groups, surfactants can be categorized in several ways: cationic, anionic, nonionic, or zwitterionic.

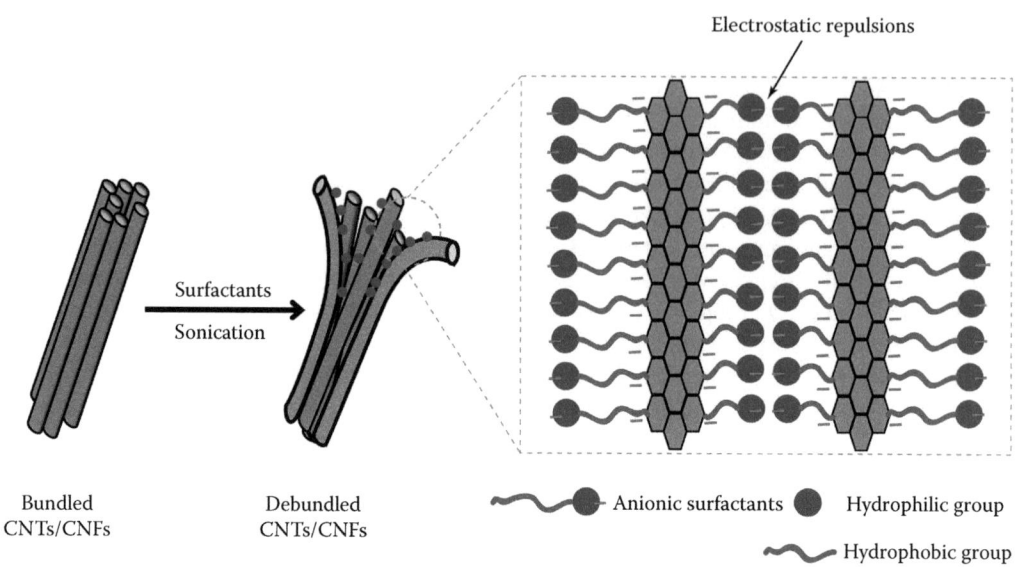

FIGURE 33.2
Schematic presentation of dispersion mechanism of anionic surfactants adsorbing onto the tube/fiber surface. Electrostatic repulsions are formed between the hydrophilic groups of anionic surfactants (negatively charged).

Although both ionic and nonionic surfactants can successfully disperse CNTs/CNFs in water, the stabilization mechanism for these two categories has subtle differences. Generally, the dispersion mechanism of ionic surfactants is electrostatic repulsions, namely, the so-called *unzipping* mechanism (as shown in Figure 33.2) (Strano et al. 2003). In a typical dispersion procedure, the surfactant is adsorbed on the CNT/CNF surface, and the CNT/CNF bundle ends are opened, and then surfactant adsorption accelerates, and the surfactant gets into the small spaces between the bundle and the isolated tube/fiber; finally, CNTs/CNFs are separated into individual tubes/fibers because of electrostatic interactions (Vaisman et al. 2006). In this case, ultrasonication is requisite to facilitate the surfactant adsorption onto the tube/fiber surface. It is worth noting that since the surface charge of CNTs is negative via zeta potential measurements in aqueous solutions, electrostatic interactions are formed between the surfactant's positively charged head group and the negatively charged CNT surface (Barraza et al. 2002). Thus, surfactants with cationic head groups, such as dodecyltrimethylammonium bromide (DTAB) (Luo et al. 2009; Whitsitt and Barron 2003) or cetyltrimethylammonium bromide (CTAB) (Barraza et al. 2002), are more attractive than those with anionic head groups such as sodium dodecyl sulfate (SDS) (Yu et al. 2007) or sodium dodecylbenzene sulfonate (SDBS, also called NaDDBS) (Moore et al. 2003; Tan and Resassco 2005) in the terms of stabilizing CNT/CNF dispersion by neutralizing the filler surface charge. Different from ionic surfactants, the treatment of nonionic surfactants, such as octylphenoxy polyethoxy (Triton X-100, also called as TX100) (Geng et al. 2008), is based on a strong hydrophobic attraction between the CNT/CNF surface and the hydrophilic parts of surfactants, which forms steric repulsion.

Generally, ionic surfactants are preferable for CNTs/CNFs in water-soluble solutions. In organic solvents, alternatively, nonionic surfactants are recommended (Vaisman et al. 2006). The lengths of the hydrophobic regions and the types of hydrophilic groups of surfactants affect the CNT/CNF's solubilizing capabilities (Islam et al. 2003). Surfactants with benzene rings, smaller head-group sizes, and longer chain lengths behave better adsorption resulting in relatively higher dispersive efficiency. For example, SDS has a weaker interaction with the nanotube surface compared to that of SDBS and TX100 because it does not have a benzene ring. SDBS disperses better in water than TX100 and SDS because of its head group and slightly longer alkyl chain (Islam et al. 2003). The efficiency of tube/fiber stabilization also depends on how surfactant molecules adsorbed onto the CNTs/CNFs (Islam et al. 2003). Once the surfactant is adsorbed onto the filler surface, the surfactant molecules are self-assembled into micelles above a critical micelle concentration. Because the larger the micelle size, the stronger the steric repulsive force introduced by micelles, it is expected that the surfactants with a higher concentration can more effectively disentangle large CNTs/CNFs agglomerates (Geng et al. 2008). The manners of surfactant adsorption have been postulated in three ways: CNTs/CNFs encapsulated in a cylindrical surfactant micelle, hemimicellar adsorption, and random adsorption of surfactant molecules on the surface of CNTs/CNFs (Yurekli et al. 2004).

33.3.2.2 Mechanism of noncovalent modification of CNTs/CNFs: Polymer wrapping

Wrapping mechanism usually occurs in the case that CNTs/CNFs disperse in aqueous solution in the presence of polymers, for example, polyvinylpyrrolidone (PVP) (O'Connell et al. 2001), gum arabic (GA) (Bandyopadhyaya et al. 2002), or polyacrylic acid (PAA) (Liu et al. 2006). The polymer wrapping process is achieved through the van der Waals' interactions and π–π stacking between CNTs/CNFs and polymer chains (Gao et al. 2011). It not only helps to improve the CNT/CNF dispersion but also leads to a better interaction between the modified CNTs/CNFs and matrix (Rahmat and Hubert 2011). The reversible wrapping mechanism of CNTs/CNFs with polymers was identified from the view of kinetics (Paria and Khilar 2004; Shvartzman-Cohen et al. 2004) and thermodynamics (O'Connell et al. 2001). Thermodynamically, the wrapping is driven by eliminating the hydrophobic interface between the tubes/fibers and the aqueous medium. Otherwise, based on the view of kinetics, the agglomeration of CNTs/CNFs is prevented due to long-ranged entropic repulsion among polymer-modified tubes/fibers. The dispersion efficiency is partly influenced by the structure of the polymers. Polymers with flexible backbones and aromatic side groups such as polystyrene (PS) tend to wrap around nanotube form an interchain coiling conformation (Tallury and Pasquinelli 2010a). However, stiff polymer chains such as poly(paraphenylene vinylene) (PPV) form a helical configuration rather than an interchain coiling due to the preference for optimizing π–π interactions (as shown in Figure 33.3) (Rahmat et al. 2011; Tallury and Pasquinelli 2010b). In addition, CNTs/CNFs, which are functionalized through the oxidative introduction of carboxyl groups, present useful sites for polymer wrapping. By this method, zwitterionic interactions or hydrogen bonds are formed between the defect sites of tubes/fibers and the functional groups of the polymers (Lee et al. 2008).

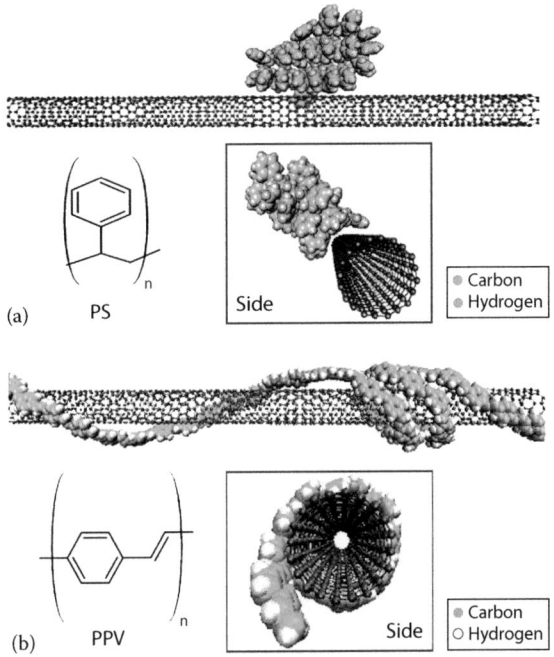

FIGURE 33.3
Schematic presentation of polymers with different structure wrapping: (a) polymers with flexible backbones (PS) wrapping on CNTs/CNFs with an interchain coiling conformation; (b) polymers with stiff backbones (PPV) wrapping on CNTs/CNFs with a helical configuration more tightly (from the side view).

33.3.2.3 Noncovalent modification of CNTs/CNFs in cement and concrete materials

Noncovalent modification of CNTs/CNFs has been widely and most commonly utilized to promote stable dispersions of CNTs/CNFs in cement and concrete materials, including surfactants and polymers. This is mainly because chemical admixtures are time-honoredly used in cement and concrete materials to obtain some characteristics not obtainable from plain cement and concrete materials. Commonly used admixtures include water-reducing admixture, accelerators, retarders, and air entrainments and most of such admixtures are surfactants and polymers. Therefore, it is quite natural to adopt surfactants or polymers to disperse CNTs/CNFs in cement and concrete materials.

Various kinds of surfactants and polymers have been shown to effectively disperse CNTs/CNFs in cement and concrete materials. GA, a natural polymer, is among the earliest choice to employ as a powerful dispersing agent for CNTs. It has the advantage of being compatible with cement and concrete materials and can disperse the nanotubes without impairing the cement hydration (De Ibarra et al. 2006). Later on, investigations have been expanded to anionic, cationic, nonionic, and mixed surfactants. Yu et al. employed an anionic surfactant, SDS, in combination with ultrasonication to achieve an effective dispersion of MWCNTs in water and cement paste (Yu and Kwon 2009). Luo et al., respectively, employed an anionic surfactant CTAB, a nonionic surfactant TX100, and the combination of SDBS and TX100 to disperse MWCNTs in cement matrix (Luo et al. 2009, 2011). Gao et al. used SDS to disperse the nonoxidized CNFs in concrete (Gao et al. 2009). The proper choice of a suitable surfactant needs consideration of its structure, its optimum ratio to nanotubes/nanofibers, thermal stability, and PH value of its solution (Rastogi et al. 2008). Because cement in aqueous solution forms a negatively charged hydration system, anionic surfactants, such as SDBS, have higher dispersion capability than cationic surfactants. Luo et al. compared five surfactants, which, respectively, were SDBS, NaDC, TX100, GA, and CTAB, to enhance solubilization/dispersion of MWCNTs in aqueous solution and cement paste. They found that the solubilization/dispersion capabilities of these five surfactants were distinguishing with a decreasing sequence as SDBS&TX100 > SDBS > NaDC/TX100 > NaDC > AG > TX100 > CTAB. The SDBS and TX100 with a mixing ratio of 3:1 by weight of cement and CNTs exhibit the best solubilization/dispersion capability (Luo et al. 2009). In addition, the PH value of surfactants/water solution affects the solubilization/dispersion capabilities of surfactants. For example, when adding $Ca(OH)_2$, dispersions present a tendency to diminish its absorbance value along time, which can be an indicative of a reagglomeration process (Mendoza et al. 2013).

Although surfactants are very helpful for dispersing CNTs/CNFs in cement and concrete materials, further investigations show that surfactants have a harmful impact on mechanical, electrical, and durable properties of cement and concrete materials, react with the water-reducing admixtures, interfere with the hydration reaction of cement, and usually inhibit cement and concrete setting and hardening (Cwirzen et al. 2008; Han et al. 2009b; Sanchez and Ince 2009; Sobolkina et al. 2012; Yazdanbakhsh et al. 2010). Therefore, weaker surfactants such as water-reducing admixtures (plasticizers and superplasticizers) are used as surfactants of CNTs/CNFs. These typically polycarboxylate-based and specially made surfactants can facilitate dispersion process of CNTs/CNFs and enhance the uniformity and workability of nanocomposites (Konsta-Gdoutos et al. 2010b; Makar and Beaudoin 2004; Shah et al. 2009). Table 33.1 lists noncovalent modification of CNTs/CNFs in cement and concrete materials.

33.3.3 Covalent Modification Combined with Noncovalent Modification of CNTs/CNFs in Cement and Concrete Materials

The covalent surface modification method and the noncovalent surface modification method can be combined to improve CNTs/CNFs in cement and concrete materials. Normally, with the noncovalent surface modification method alone, CNTs/CNFs, which are natural inert, are almost fully

TABLE 33.1 Noncovalent Modification of CNTs/CNFs in Cement and Concrete Materials

Surface Modifier	CNTs/ CNFs	Charge	Modification Mechanism	References
SDS	CNTs	Anionic	Surfactant adsorption	Han et al. (2009), Sobolkina et al. (2012), Yu et al. (2009)
	CNFs			Gao et al. (2009)
SDBS (NaDDBS)	CNTs	Anionic	Surfactant adsorption	Han et al. (2009a,b, 2010, 2012), Luo et al. (2009, 2011, 2013)
CTAB	CNTs	Cationic	Surfactant adsorption	Luo et al. (2009)
MC	CNTs	Nonionic	Surfactant adsorption	Azhari (2008), Azhari and Banthia (2012), Luo et al. (2013), Veedu (2011), Wang et al. (2012)
CMC	CNTs	Nonionic	Surfactant adsorption	Keriene et al. (2013), Yakovlev et al. (2013)
MB	CNTs	Nonionic	Polymer wrapping	Yakovlev et al. (2013)
NaDC	CNTs	Anionic	Surfactant adsorption	Luo et al. (2009)
TX100	CNTs	Nonionic	Surfactant adsorption	Luo et al. (2009, 2011, 2013)
Brij 35	CNTs	Nonionic	Surfactant adsorption	Sobolkina et al. (2012)
GA	CNTs	Nonionic	Polymer wrapping	Cwirzen et al. (2008), De Ibarra et al. (2006), Luo et al. (2009)
SP	CNTs	Zwitterionic	Surfactant adsorption	Abu Al-Rub et al. (2011), Han et al. (2011b), Konsta-Gdoutos et al. (2010b), Mendoza et al. (2013), Petrunin et al. (2013), Tyson et al. (2011), Yazdanbakhsh et al. (2010)
	CNFs			Abu Al-Rub et al. (2011), Azhari and Banthia (2012), Galao et al. (2012), Han et al. (2011b), Howser et al. (2011), Tyson et al. (2011)
PAA	CNTs	Nonionic	Polymer wrapping	Cwirzen et al. (2008)
	CNFs			Peyvandi et al. (2013)
NG	CNTs	NG	Surfactant adsorption	Konsta-Gdoutos et al. (2008, 2010b), Shah et al. (2009)
	CNFs			Metaxa et al. (2013)

Notes: MC, methyl cellulose; CMC, carboxymethyl cellulose; MB, masterbatch; Brij 35, polyoxyethylene(23) lauryl ether; SP, superplasticizers; NG, no given.

covered by surfactants or polymer molecules. This prevents bond formation between matrix materials and carbon nanomaterials, which consequently are usually not able to effectively enhance mechanical properties of nanocomposites (Nasibulina et al. 2012b). However, creating covalent bonds between the matrix and CNTs/CNFs will produce a high bonding strength due to the improvement of chemical activity of CNTs/CNFs and interfacial interactions between surface-modified nanotubes and hydrations of cement by the introduction of functional groups on the CNT/CNF surface. The combination naturally leads to a dual effect of increasing the steric barrier or dispersing CNTs/CNFs in cement and concrete materials. Cwirzen et al. obtained stable and homogenous dispersions of MWCNTs in water by using noncovalent surface functionalization and additional treatment with PAA polymers. The study also showed that the surface covalent functionalization combined with modification using PAA polymers has more superior dispersion effect compared with only using PAA polymers or GA (Cwirzen et al. 2008). Peyvandi et al. used PAA for wrapping the CNFs–COOH in order to improve their dispersion in water and interaction with cement hydrates (Peyvandi et al. 2013). Similarly, Han et al. achieved an effective dispersion of CNTs in cement paste and cement mortar by using carboxylation of MWCNTs and a surfactant, SDBS (Han et al. 2009b).

33.4 METHODS OF IN SITU GROWTH OF CNTs/CNFs ON RAW MATERIALS OF CEMENT AND CONCRETE MATERIALS

Methods of the covalent/noncovalent modification of CNTs/CNFs are multistep and time consuming and have high cost. Recently, researchers proposed an innovative approach to grow CNTs/CNFs directly on the surface of raw materials of cement and concrete materials, which include cement, clinker, silica fume, fly ash, and sand (Cwirzen et al. 2009; Dunens et al. 2009; Hlavacek and Smilauer 2012; Hlavacek et al. 2011; Ludvig et al. 2009, 2011 Mudimela et al. 2009; Nasibulin et al. 2009, 2013; Nasibulina et al. 2010, 2012b; Sanchez et al. 2009; Veedu et al. 2006). Under this approach, CNTs/CNFs attach to raw materials of cement and concrete materials, which provides homogeneous dispersion of the carbon nanomaterials in cement and concrete materials. Meanwhile, the mechanical and electrical properties of cement and concrete composites are improved (Ludvig et al. 2009; Nasibulina et al. 2010, 2012b; Sun et al. 2013). The methods of in situ growth of CNTs/CNFs mainly include chemical vapor deposition (CVD) method and microwave irradiation (MI) method (Sun et al. 2013).

33.4.1 CVD Method

The CVD method is generally used and considered as the most efficient process for high-yield fabrication of CNTs/CNFs (Popov 2004). Nasibulin et al. firstly employed a modified CVD method to grow CNTs/CNFs on the surface of cement particles for synthesizing cement hybrid material (CHM) in which CNTs/CNFs grow on cement particles by two different methods: screw feeder and fluidized bed reactors. The synthesis uses acetylene as the main carbon source and carbon monoxide and dioxide as the additives to enhance the yield. Extra catalysts and inert support substances are free for requirement since cement particles contain Fe_2O_3, SiO_2, MgO, and Al_2O_3 known to be substantive materials for the growth of CNTs/CNFs, which solves the main drawback of the pristine CVD method (Nasibulin et al. 2009, 2010). CNTs/CNFs were successfully grown on silica fume, fly ash, sand, and soil by the CVD method (Dunens et al. 2009; Ludvig et al. 2011; Mudimela et al. 2009; Nasibulin et al. 2013). The setup of the CVD method and the mechanism of in situ growth of CNTs/CNFs on cement particles are illustrated in Figure 33.4.

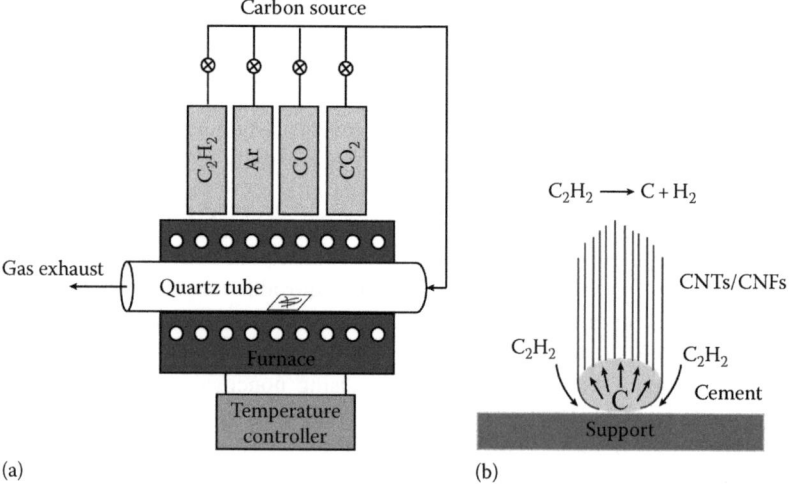

FIGURE 33.4
Schematic presentations: (a) the setup of the CVD method; (b) the mechanism of in situ growth of CNTs/CNFs on cement particles.

TABLE 33.2   In-Situ Growth of CNTs/CNFs on Raw Materials of Cement and Concrete Materials by CVD Method

Substrate	CS	Catalyst	Temperature	Products	Morphology	Reference
Cement	C_2H_2	—	450°C–700°C	CNFs and CNTs	—	Nasibulin et al. (2009)
SF	C_2H_2	Fe	750°C	CNFs and CNTs	~30–60 nm in diameter	Ludvig et al. (2011)
Clinker	C_2H_2	—	550°C	CNFs	~10 μm in length	Nasibulina et al. (2012b)
					~30–50 nm in diameter	
Sand	C_2H_4	—	800°C	CNFs	—	Nasibulin et al. (2013)
Fly ash	C_2H_4	FA	650°C	CNFs and CNTs	~10–40 nm in diameter	Dunens et al. (2009)
					~1 μm in length	
Soil	C_2H_2	—	550°C	CNFs	—	Nasibulin et al. (2013)

Note: CS, carbon source; SF, silica fume; FA, Fe impregnated fly ash.

The amount and morphology of the grown CNTs/CNFs depend on the applied temperatures, the chemical component and flow speed of pristine cement, and the type/amount of carbon resource. Table 33.2 concludes CNTs/CNFs in different experimental conditions of in situ growth of CNTs/CNFs on raw materials of cement and concrete materials by CVD method.

33.4.2 MI Method

MI method is an ultrafast approach for CNT/CNF's growth, which could reach a very high temperature (>1100°C) in only 15–30 s through heating the individual molecules of precursors, such as conducting polymers or other conductive materials. Compared to the CVD method, MI reacts at room temperature in air, without the need for any inert gas protection and additional feed stock gases (Zhang and Liu 2012). Liu et al. firstly utilized MI method to initiate CNTs/CNFs in situ growth on fly ash. The synthesis process is shown in Figure 33.5. In the process, conducting polymers were served as heating precursors, which were the physical mixture of ferrocene powder with

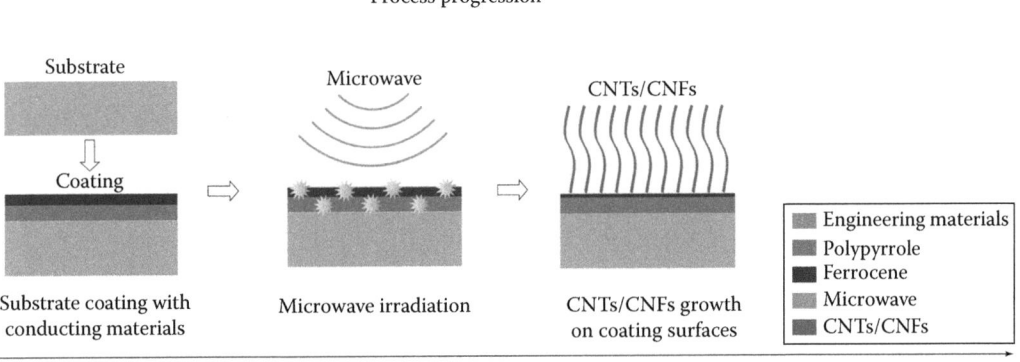

FIGURE 33.5
Schematic presentation of MI method for CNT/CNF's growth on engineering materials.

polypyrrole·Cl powder in the solid state. Rodlike, hollow CNTs were observed with lengths in several micrometers and diameters in the range of 150–200 nm (Liu et al. 2011). By MI method, CNTs/CNFs can directly grow on a wide selection of engineering materials including glass fibers (GFs), CFs, fly ash, and glass microballoons (Liu et al. 2011).

33.5 METHODS OF FABRICATING CNT/CNF–FIBER HIERARCHICAL STRUCTURES

CNTs/CNFs as nanoscale particles have an extremely high aspect ratio, which increases the difficulty for dispersing them uniformly in the matrix. In comparison to CNTs/CNFs, conventional fibers, such as steel fibers (SFs), GFs, silica fibers (SiFs), and CFs, are micrometer scale, have lower aspect ratio, and have consequently easier dispersion. The idea of growing or grafting CNTs/CNFs onto fibers to fabricate hierarchical structures has been put forward recently and considered as a potential technique to provide high loadings of CNTs/CNFs in the composites, while alleviating the dispersion problem as well as enhancing matrix properties (Qian et al. 2010a). The dispersion of CNTs/CNFs in the hierarchical structures is transformed into the dispersion of micrometer-scale fibrous materials (Yamamoto et al. 2009). Meanwhile, a significant decrease in contact angle and a polarity change of fibers are expected after growing or grafting CNTs/CNFs due to increase in surface roughness of the pristine fibers, which means that the wettability of hierarchical structures is improved (Qian et al. 2010b). Grafting or growing CNTs/CNFs is also expected to increase the surface contact area with the matrix and create mechanical interlocking, which may improve stress transfer (Downs and Baker 1995). To date, most work has focused on polymer composites, and exciting results have been achieved in improvements in mechanical, electrical, and thermal properties of the hierarchical polymer composites. Usually, CNFs/CNTs and fibers (especially SFs and CFs) as reinforced materials are mixed in cement and concrete materials separately. However, no literature has reported about applying CNF/CNT–fiber hierarchical structures in cement and concrete materials. It is potentially and readily scalable for application of hierarchical structures in cement and concrete materials. Therefore, this section mainly focuses on the fabrication of CNF/CNT–CF hierarchical structures (as shown in Figure 33.6) to provide guidance for their potential applications in cement and concrete materials. A wide range of other fibrous materials for CNF/CNT's growth or graft, such as SFs, GFs, and silicon fibers, are summarized in Table 33.3, as well as CFs.

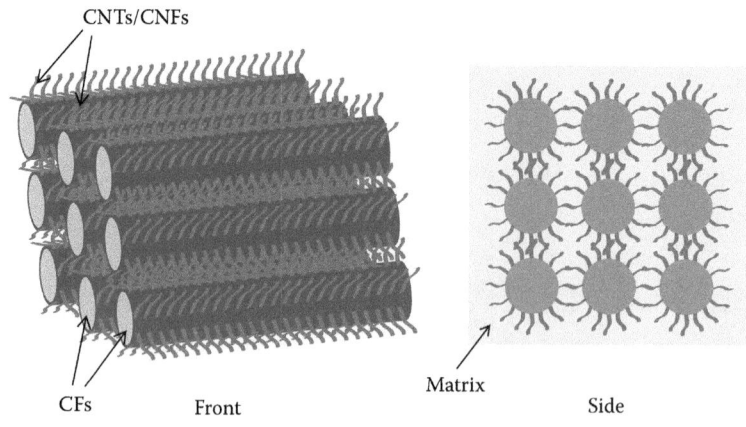

FIGURE 33.6
Schematic presentation of CNT/CNF grafting on the surface of CFs and dispersing them uniformly in the matrix to fabricate hierarchical/multiscale carbon structures.

TABLE 33.3   Various Fibers Used for Fabricating CNT/CNF–Fiber Hierarchical Structures

Fibers	Matrix	Fillers	Method of Fabrication	Catalyst	Way of Attachment	Reference
CFs	Epoxy	CNTs	CVD	Stainless steel 304	Grow	Thostenson et al. (2002)
CFs	—	CNTs	Thermal CVD	Iron nanoparticles	Grow	Zhu et al. (2003)
CFs	Epoxy	CNTs	EPD	—	Graft	Bekyarova et al. (2007)
f-CFs	Epoxy	F-MWCNTs	Chemical reactions	—	Graft	He et al. (2007)
f-CFs	—	F-CNTs	Chemical reactions	—	Graft	Laachachi et al. (2008)
CFs	Phenolic	CNTs	CVD	Ferrocene	Grow	Mathur et al. (2008)
Ceramic fibers	Epoxy	CNTs	Thermal CVD	Fe salt-based	Grow	Yamamoto et al. (2009)
PAMAM–CFs	—	F-MWCNTs	Chemical reactions	—	Graft	Mei et al. (2010)
CFs		CNTs	EPD + ultrasonication	—	Graft	Guo et al. (2012)
SFs		CNTs	Thermal CVD	—	Grow	Park and Lee (2006)
Stainless SFs	—	CNTs	CVD	La$_2$NiO$_4$	Grow	Gao et al. (2008)
Stainless SFs	—	CNTs	Plasma-assisted CVD	—	Grow	Sugimoto et al. (2009)
Quartz fibers	—	CNTs	Floating catalyst method	Iron nanoparticles	Graft	Zhang et al. (2008)
GFs	Epoxy	CNTs	Impregnation	—	Graft	Siddiqui et al. (2009)
SiFs	PMMA	CNTs	Injection CVD	Ferrocene	Grow	Qian et al. (2010a)
SiC fibers	Epoxy	MWCNTs	CVD	Ferrocene	Grow	Veedu et al. (2006)
GFs	Polyester	CNFs	VARTM	—	Graft	Sadeghian et al. (2006)

Note: VARTM, vacuum-assisted resin transfer molding.

Predominantly, the combination of CNTs/CNFs and CFs has been achieved through two different routes: growing CNTs/CNFs onto the fiber surface directly using CVD process (Garcia et al. 2008; Kepple et al. 2008; Mathur et al. 2008; Qian et al. 2008, 2010b; Thostenson et al. 2002; Zhang et al. 2009b) and depositing and grafting premade CNTs/CNFs onto the fiber surface by electrophoresis deposition (EPD) (Bekyarova et al. 2007; Guo et al. 2012) or chemical reactions (He et al. 2007, 2012 ; Laachachi et al. 2008; Liu et al. 2013; Mei et al. 2010; Zhang et al. 2009a). It is necessary to note that the term *graft*, herein, is refined to describe premade CNTs/CNFs that are attached to premade CFs through any method, which has different meaning with the term *grow*.

33.5.1 Growth of CNTs/CNFs Directly on CFs via CVD

Similar to the method of growing CNTs/CNFs on the surface of raw materials of cement and concrete materials, the growth of CNTs/CNFs directly on CFs using CVD process is mostly investigated and reported. However, unlike cement/mineral admixtures, extra catalyst, such as iron, is required by using CFs as substrates that have no substantive materials for the growth of CNTs/CNFs. The deposition of catalyst has been made by using various methods including thermal evaporation

(Kepple et al. 2008; Mathur et al. 2008; Zhang et al. 2009b), electron beam evaporation (Garcia et al. 2008), incipient wetness (Qian et al. 2008, 2010b), or magnetron sputtering (Thostenson et al. 2002). The diameter of the CNTs/CNFs grown on the surface is considered to depend strongly on the size of the employed catalyst. Generally, the diameter of the catalyst particles is consistent with the diameter of the CNTs/CNFs grown from them (Qian et al. 2008, 2010b). Consequently, in order to stabilize catalyst particles and help to obtain uniform deposition of catalyst, it is crucial to pre-treat the surface of CFs before the initial catalyst deposition, particularly via incipient wetness or other solution-based techniques. Heat treatment (Thostenson et al. 2002), polymer coating (Zhang et al. 2009b), and particularly, chemical oxidation (e.g., nitric acid oxidation) (Qian et al. 2010b) have been explored for conducting surface treatment of CFs.

In addition, when CNTs/CNFs grow onto CFs, an appropriate growth density or coverage and morphology are desirable, both to maximize stress transfer and to minimize the potential damage of the primary fiber. Too long and randomly oriented CNTs/CNFs tend to become entangled and disrupted each other, furthermore leading to asymmetric growth of CNTs/CNFs. This demonstrated completely different enhancement of interfacial bonding strength between matrix and hierarchical structures (Qian et al. 2010). The morphology, density, and coverage of CNTs/CNFs on CFs by CVD strongly depend on the following three parameters: growth temperature, flow rate of the carbon source/catalyst mixture, and growth time (Zhang et al. 2009b). For example, Zhang et al. investigated the effect of CNT's different growth conditions on CNT morphology. After CVD, the CNT alignment morphology and density were varied with growth time, growth temperature, and atmosphere flow rate. At 700°C, CNTs on the CF surface appeared to be sparse low density. With the increase in growth temperature from 700°C to 750°C, the density of CNTs increased substantially, and the morphology of CNTs was predominantly aligned and uniform in length. While the temperature and atmosphere flow rate continued to increase, the morphology of CNTs changed into randomly oriented and variable length. They also found that with sufficient growth time, uniform coverage of CNTs on CFs surface can be obtained, which can be selected to tailor the interfacial properties of CNT–CF hierarchical structures (Zhang et al. 2009b).

Although the CVD process is an efficient technique for the growth of CNTs/CNFs on a variety of surfaces, there still appear some remarkable issues. Because active catalysts deposit and react on the surface of CFs, such reactions may damage the primary fibers and cause the degradation of their mechanical properties. Furthermore, the pretreatment of CF surface and high growth temperature during the CVD process also contribute to the damage. The other existing problems are the deposition of catalyst particles on CFs before the growth of CNTs/CNFs and the purge of catalyst particles after growth, taken together with the difficulties in purification of as-achieved CNTs/CNFs (Bekyarova et al. 2007; Qian et al. 2010). It means that the growth of CNTs/CNFs on CFs via CVD is not readily automated and utilized for industrial applications.

33.5.2 Grafting of CNTs/CNFs Directly on CFs via EPD or Chemical Reactions

Grafting of CNTs/CNFs directly on CFs via EPD or chemical reactions is considered to be a simpler and more readily automated technique than using CVD for CNT/CNF's growth (Bekyarova et al. 2007; Laachachi et al. 2008). One advantage of the grafting process is the absence of the effect of catalyst residue at the interface between CNTs/CNFs and CFs because either CNTs/CNFs or CFs have been purified by acid before grafting (He et al. 2012).

EPD is known to be one very promising technique being developed for depositing CNTs/CNFs and allows the formation of uniform films and coatings on objects with complex shapes and rough surfaces (Bekyarova et al. 2007; Boccaccini et al. 2006). Deposition is achieved through the motion and accumulation of charged particles toward an electrode under an applied electric field in solvent (Boccaccini et al. 2006). Grafting CNTs/CNFs on the surface of CFs by the use of EPDDP is based on the negative charge of CNTs/CNFs to respond to an electric field and migrate toward the positive CFs electrode and subsequently, deposited on the fiber surface (Bekyarova et al. 2007). The negative charge is attributed to the introduction of carboxylic acid groups or hydroxyl groups by covalent modification of CNTs/CNFs. Bekyarova et al. observed a homogeneous deposition of both SWCNTs and MWCNTs on the surface of the CF electrode. The morphologies of the deposited SWCNTs and MWCNTs showed differences. The more rigid structure of the MWCNTs was deposited as individual tubes, whereas the SWCNTs tended to form bundles and appeared to be the filmlike deposition associating with the surface of CFs tightly (Bekyarova et al. 2007). However, during the EPD process, water electrolysis always causes bad and sparse deposition. Alternatively, ultrasonically assisted EPD increases deposition sites and subsequently increases the amount and uniformity of CNTs/CNFs (Guo et al. 2012).

Recently, chemical methods are developed to covalently graft CNTs/CNFs onto CFs (He et al. 2007, 2012; Laachachi et al. 2008; Liu et al. 2013; Mei et al. 2010; Zhang et al. 2009a). Both CNTs/CNFs and CFs are functionalized separately to create chemical groups such as carboxyl, hydroxyl, or amine groups. The grafting is carried out via esterification, anhydration, or amidization between functional groups to form bonds between CNTs/CNFs and CFs with different solvents such as acetone, Dimethylformamide (DMF) DMF, or toluene (He et al. 2007; Laachachi et al. 2008). The grafting quality is directly related to not only the chemical solvent nature but also the CNT/CNF dispersion in solvent. Laachachi et al. studied CNTs grafted onto a CF surface with different solvents. They found that CNTs shown better dispersion by using DMF or acetone as a solvent than toluene. Although relatively more polar of DMF than acetone, a lower CNT/CF interaction was observed in DMF, and acetone as a solvent gave the best grafting results. Even so, because of the low chemical active sites on the CFs, the CNT grafting has low density and is inhomogeneous (Laachachi et al. 2008). Mei et al. introduced a new efficient method to graft CNTs onto the CFs by the use of polyamide amine (PAMAM), a kind of dendrimer, to get a uniform grafting structure. The periphery of PAMAM has many amino groups, which can provide more active sites and thus act as a *bridge* to connect the functionalized CNTs/CNFs onto CFs by the chemical reaction (Mei et al. 2010).

However, compared with the CVD method for growing CNTs/CNFs directly on CFs, the grafting methods via EPD or chemical reactions provide little control over the CNT/CNF's orientation, currently. The grafting strength is still much lower than the intrinsic CNT/CNF strength, thus limiting the potential application in hierarchical composites (He et al. 2012; Qian et al. 2010).

33.6 CONCLUSIONS AND FUTURE PERSPECTIVES

CNTs/CNFs play significant roles in improving the performance of cement and concrete materials and promote the development of novel, multifunctional, and sustainable construction materials. The dispersion issue of CNTs/CNFs in cement and concrete materials directly affects the properties of composites and must be addressed. In this chapter, we provide an overview on major progresses and advances of the chemical modification methods of CNTs/CNFs in cement and concrete materials. The reason that CNTs/CNFs can be chemically modified is the specific hexagonal structure with delocalized π-electrons and defects located on the carbon framework, which are inherent or latterly generated by the purification processes. Uniform dispersion of CNTs/CNFs can be achieved with chemical modification methods of CNTs/CNFs including the covalent modification methods, the noncovalent modification methods, the in situ synthesis methods, and methods of fabricating CNT/CNF–fiber hierarchical structures. The first three methods have been utilized in cement and concrete materials. The noncovalent modification methods based on surfactant adsorption or polymer wrapping are considered as the most available methods and have been widely used because they preserve the structure of CNTs/CNFs. The drawback is that they may interfere with the hydration reaction of cement and usually inhibit cement and concrete setting and hardening. The covalent modification method here focuses on oxidation treatment of CNTs/CNFs with strong acids and may result in damage to CNT/CNF structures. The in situ growth methods including CVD method and MI method provide a new approach to disperse CNTs/CNFs into cement and concrete materials. Although the methods of fabricating CNT/CNF–fiber hierarchical structures have never been applied in cement and concrete materials, the application in polymers shows its tremendous possibility and potential in cement and concrete field.

The study on chemical modification of CNTs/CNFs for application in cement and concrete field is still unsystematic, and many fundamental issues have not been fully explored and need to be further addressed including (1) modification mechanisms of each methods, (2) quantitative analysis of the degree of dispersion of CNTs/CNFs in solvents, (3) effect of the structural parameters of CNTs/CNFs on chemical modification processes, (4) optimization of chemical modification processes including cost and efficiency, (5) interfacial bond between modified CNTs/CNFs and cement and concrete materials, (6) compatibility of modified CNTs/CNFs in cement and concrete materials, (7) combination of different chemical modification methods, (8) development of new efficient chemical modification method, and (9) destruction of the environment. Nevertheless, it is obvious that the chemical modification of CNTs/CNFs is the most available and promising method for incorporating CNTs/CNFs into cement and concrete materials, and this will produce the optimum reinforcing/modifying effect of CNTs/CNFs to cement and concrete materials.

ACKNOWLEDGMENTS

The authors thank the funding supports from the National Natural Science Foundation of China (grant no. 51178148), the Program for New Century Excellent Talents in University of China (grant no. NCET-11-0798), the Ministry of Science and Technology of China (grant no. 2011BAK02B01), and the Fundamental Research Funds for the Central Universities of China.

REFERENCES

Abu Al-Rub, R. K., Tyson, B. M., Yazdanbakhsh, A., and Grasley, Z. 2011. Mechanical properties of nanocomposite cement incorporating surface-treated and untreated carbon nanotubes and carbon nanofibers. *Journal of Nanomechanics and Micromechanics* 2(1):1–6.

Aitcin, P. C. 2000. Cements of yesterday and today: Concrete of tomorrow. *Cement and Concrete Research* 30(9):1349–1359.

Azhari, F. 2008. Cement-based sensors for structural health monitoring. Dissertation for the Master Degree of Applied Science. Vancouver, British Columbia, Canada: University of British Columbia.

Azhari, F. and Banthia, N. 2012. Cement-based sensors with carbon fibers and carbon nanotubes for piezoresistive sensing. *Cement and Concrete Composites* 34(7):866–873.

Balasubramanian, K. and Burghard, M. 2005. Chemically functionalized carbon nanotubes. *Small* 1(2):180–192.

Bandyopadhyaya, R., Nativ-Roth, E., Regev, O., and Yerushalmi-Rozen, R. 2002. Stabilization of individual carbon nanotubes in aqueous solutions. *Nano Letters* 2(1):25–28.

Banerjee, S., Hemraj Benny, T., and Wong, S. S. 2005. Covalent surface chemistry of single-walled carbon nanotubes. *Advanced Materials* 17(1):17–29.

Barraza, H. J., Pompeo, F., O'Rea, E. A., and Resasco, D. E. 2002. SWNT-filled thermoplastic and elastomeric composites prepared by miniemulsion polymerization. *Nano Letters* 2(8):797–802.

Baughman, R. H., Zakhidov, A. A., and De Heer, W. A. 2002. Carbon nanotubes-the route toward applications. *Science* 297(5582):787–792.

Bekyarova, E., Thostenson, E. T., and Yu, A. 2007. Multiscale carbon nanotube-carbon fiber reinforcement for advanced epoxy composites. *Langmuir* 23(7):3970–3974.

Boccaccini, A. R., Cho, J., and Roether, J. A. 2006. Electrophoretic deposition of carbon nanotubes. *Carbon* 44(15):3149–3160.

Boul, P. J., Liu, J., and Mickelson, E. T. 1999. Reversible sidewall functionalization of buckytubes. *Chemical Physics Letters* 310(3–4):367–372.

Burghard, M. 2005. Electronic and vibrational properties of chemically modified single-wall carbon nanotubes. *Surface Science Reports* 58(1):1–109.

Cadek, M., Coleman, J. N., and Ryan, K. P. 2004. Reinforcement of polymers with carbon nanotubes: The role of nanotube surface area. *Nano Letters* 4(2):353–356.

Chen, R. J., Zhang, Y., Wang, D., and Dai, H. 2001. Noncovalent sidewall functionalization of single-walled carbon nanotubes for protein immobilization. *Journal of the American Chemical Society* 123(16):3838–3839.

Chen, S., Shen, W., Wu, G., Chen, D., and Jiang, M. 2005. A new approach to the functionalization of single-walled carbon nanotubes with both alkyl and carboxyl groups. *Chemical Physics Letters* 402(4):312–317.

Coleman, J. N., Khan, U., Blau, W. J., and Gunko, Y. K. 2006. Small but strong: A review of the mechanical properties of carbon nanotube-polymer composites. *Carbon* 44(9):1624–1652.

Cwirzen, A., Habermehl-Cwirzen, K., and Penttala, V. 2008. Surface decoration of carbon nanotubes and mechanical properties of cement/carbon nanotube composites. *Advances in Cement Research* 20(2):65–73.

Cwirzen, A., Habermehl-Cwirzen, K., and Shandakov, D. 2009. Properties of high yield synthesised carbon nano fibres/Portland cement composite. *Advances in Cement Research* 21(4):141–146.

Dai, H. 2002. Carbon nanotubes: Opportunities and challenges. *Surface Science* 500(1):218–241.

Datsyuk, V., Kalyva, M., and Papagelis, K. 2008. Chemical oxidation of multiwalled carbon nanotubes. *Carbon* 46(6):833–840.

De Ibarra, Y. S., Gaitero, J. J., Erkizia, E., and Campillo, I. 2006. Atomic force microscopy and nanoindentation of cement pastes with nanotube dispersions. *Physica Status Solidi (a)* 203(6):1076–1081.

Downs, W. B. and Baker, R. 1995. Modification of the surface properties of carbon fibers via the catalytic growth of carbon nanofibers. *Journal of Materials Research* 10(3):625–633.

Dujardin, E., Ebbesen, T. W., Krishnan, A., and Treacy, M. M. 1998. Wetting of single shell carbon nanotubes. *Advanced Materials* 10(17):1472–1475.

Dunens, O. M., MacKenzie, K. J., and Harris, A. T. 2009. Synthesis of multiwalled carbon nanotubes on fly ash derived catalysts. *Environmental Science and Technology* 43(20):7889–7894.

Fu, X., Lu, W., and Chung, D. 1998. Ozone treatment of carbon fiber for reinforcing cement. *Carbon* 36(9):1337–1345.

Galao, O., Zornoza, E., Baeza, F. J., Bernabeu, A., and Garces, P. 2012. Effect of carbon nanofiber addition in the mechanical properties and durability of cementitious materials. *Materials of Construction* 62(307):343–357.

Gao, D., Sturm, M., and Mo, Y. L. 2009. Electrical resistance of carbon-nanofiber concrete. *Smart Materials and Structures* 18(9):95039.

Gao, J., Loi, M. A., De Carvalho, E. J. F., and Dos Santos, M. C. 2011. Selective wrapping and supramolecular structures of polyfluorene-carbon nanotube hybrids. *ACS Nano* 5(5):3993–3999.

Gao, L. Z., Kiwi-Minsker, L., and Renken, A. 2008. Growth of carbon nanotubes and microfibers over stainless steel mesh by cracking of methane. *Surface and Coatings Technology* 202(13):3029–3042.

Garcia, E. J., Hart, A. J., and Wardle, B. L. 2008. Long carbon nanotubes grown on the surface of fibers for hybrid composites. *Aiaa Journal* 46(6):1405–1412.

Garg, A. and Sinnott, S. B. 1998. Effect of chemical functionalization on the mechanical properties of carbon nanotubes. *Chemical Physics Letters* 295(4):273–278.

Geng, Y., Liu, M. Y., Li, J., Shi, X. M., and Kim, J. K. 2008. Effects of surfactant treatment on mechanical and electrical properties of CNT/epoxy nanocomposites. *Composites Part A: Applied Science and Manufacturing* 39(12):1876–1883.

Guo, J., Lu, C., An, F., and He, S. 2012. Preparation and characterization of carbon nanotubes/carbon fiber hybrid material by ultrasonically assisted electrophoretic deposition. *Materials Letters* 66(1):382–384.

Haddon, R. C. 1988. pi-Electrons in three dimensional. *Accounts of Chemical Research* 21(6):243–249.

Han, B., Yu, X., and Kwon, E. 2009a. A self-sensing carbon nanotube/cement composite for traffic monitoring. *Nanotechnology* 20(44):445501.

Han, B., Yu, X., and Ou, J. 2009b. Dispersion of carbon nanotubes in cement-based composites and its influence on the piezoresistivities of composites. In *ASME 2009 Conference on Smart Materials, Adaptive Structures and Intelligent Systems*, pp. 57–62. American Society of Mechanical Engineers, Oxnard, CA.

Han, B., Yu, X., and Ou, J. 2010. Effect of water content on the piezoresistivity of MWNT/cement composites. *Journal of Materials Science* 45(14):3714–3719.

Han, B., Yu, X., and Ou, J. 2011a. Multifunctional and smart carbon nanotube reinforced cement-based materials. In *Nanotechnology in Civil Infrastructure*, eds. G. Kasthurirangan, B. Bjorn, T. Peter, and A. O. Nii, pp. 1–47. Berlin, Germany: Springer.

Han, B., Zhang, K., Yu, X., Kwon, E., and Ou, J. 2011b. Fabrication of piezoresistive CNT/CNF cementitious composites with superplasticizer as dispersant. *Journal of Materials in Civil Engineering* 24(6):658–665.

Han, B., Zhang, K., Yu, X., Kwon, E., and Ou, J. 2012. Electrical characteristics and pressure-sensitive response measurements of carboxyl MWNT/cement composites. *Cement and Concrete Composites* 34(6):794–800.

He, X., Wang, C., and Tong, L. 2012. Direct measurement of grafting strength between an individual carbon nanotube and a carbon fiber. *Carbon* 50(10):3782–3788.

He, X., Zhang, F., Wang, R., and Liu, W. 2007. Preparation of a carbon nanotube/carbon fiber multi-scale reinforcement by grafting multi-walled carbon nanotubes onto the fibers. *Carbon* 45(13):2559–2563.

Hilding, J., Grulke, E. A., George Zhang, Z., and Lockwood, F. 2003. Dispersion of carbon nanotubes in liquids. *Journal of Dispersion Science and Technology* 24(1):1–41.

Hirsch, A. 2002. Functionalization of single-walled carbon nanotubes. *Angewandte Chemie International Edition* 41(11):1853–1859.

Hirsch, A. and Vostrowsky, O. 2005. Functionalization of carbon nanotubes. In *Functional Molecular Nanostructures*. ed. A. D. Schluter, pp. 193–237. Berlin, Germany: Springer.

Hlavacek, P. and Smilauer, V. 2012. Fracture properties of cementitious composites reinforced with carbon nanofibers/nanotubes. *Engineering Mechanics* 211:391–397.

Hlavacek, P., Smilauer, V., Padevet, P., Nasibulina, L., and Nasibulin, A. G. 2011. Cement grains with surface-synthesized carbon nanofibres: Mechanical properties and nanostructure. In *Third International Conference Nanocon 2011*. Proceedings Brno, Czech Republic: Tanger Ltd.

Holzinger, M., Vostrowsky, O., and Hirsch, A. 2001. Sidewall functionalization of carbon nanotubes. *Angewandte Chemie International Edition* 40(21):4002–4005.

Howser, R. N., Dhonde, H. B., and Mo, Y. L. 2011. Self-sensing of carbon nanofiber concrete columns subjected to reversed cyclic loading. *Smart Materials and Structures* 20(8):85031.

Iijima, S. 1991. Helical microtubules of graphitic carbon. *Nature* 354(6348):56–58.

Islam, M. F., Rojas, E., Bergey, D. M., Johnson, A. T., and Yodh, A. G. 2003. High weight fraction surfactant solubilization of single-wall carbon nanotubes in water. *Nano Letters* 3(2):269–273.

Kang, I., Heung, Y. Y., and Kim, J. H. 2006. Introduction to carbon nanotube and nanofiber smart materials. *Composites Part B: Engineering* 37(6):382–394.

Kepple, K. L., Sanborn, G. P., Lacasse, P. A., Gruenberg, K. M., and Ready, W. J. 2008. Improved fracture toughness of carbon fiber composite functionalized with multi walled carbon nanotubes. *Carbon* 46(15):2026–2033.

Keriene, J., Kligys, M., and Laukaitis, A. 2013. The influence of multi-walled carbon nanotubes additive on properties of non-autoclaved and autoclaved aerated concretes. *Construction and Building Materials* 49(0):527–535.

Kim, S. W., Kim, T., and Kim, Y. S. 2012. Surface modifications for the effective dispersion of carbon nanotubes in solvents and polymers. *Carbon* 50(1):3–33.

Kim, Y. J., Shin, T. S., and Choi, H. D. 2005. Electrical conductivity of chemically modified multiwalled carbon nanotube/epoxy composites. *Carbon* 43(1):23–30.

Konsta-Gdoutos, M. S., Metaxa, Z. S., and Shah, S. P. 2008. Nanoimaging of highly dispersed carbon nanotube reinforced cement based materials. In *Seventh International RILEM Symposium on Fiber Reinforced Concrete: Design and Applications*, pp. 125–131. Chennai, India.

Konsta-Gdoutos, M. S., Metaxa, Z. S., and Shah, S. P. 2010a. Multi-scale mechanical and fracture characteristics and early-age strain capacity of high performance carbon nanotube/cement nanocomposites. *Cement and Concrete Composites* 32(2):110–115.

Konsta-Gdoutos, M. S., Metaxa, Z. S., and Shah, S. P. 2010b. Highly dispersed carbon nanotube reinforced cement based materials. *Cement and Concrete Research* 40(7):1052–1059.

Kotsalis, E. M., Demosthenous, E., Walther, J. H., Kassinos, S. C., and Koumoutsakos, P. 2005. Wetting of doped carbon nanotubes by water droplets. *Chemical Physics Letters* 412(4):250–254.

Laachachi, A., Vivet, A., and Nouet, G. 2008. A chemical method to graft carbon nanotubes onto a carbon fiber. *Materials Letters* 62(3):394–397.

Lee, S. H., Park, J. S., Lim, B. K., and Kim, S. O. 2008. Polymer/carbon nanotube nanocomposites via noncovalent grafting with end-functionalized polymers. *Journal of Applied Polymer Science* 110(4):2345–2351.

Li, G., Wang, P., and Zhao, X. 2005. Mechanical behavior and microstructure of cement composites incorporating surface-treated multi-walled carbon nanotubes. *Carbon* 43(6):1239–1245.

Li, G. Y., Wang, P. M., and Zhao, X. 2007. Pressure-sensitive properties and microstructure of carbon nanotube reinforced cement composites. *Cement and Concrete Composites* 29(5):377–382.

Liu, A., Honma, I., Ichihara, M., and Zhou, H. 2006. Poly (acrylic acid)-wrapped multiwalled carbon nanotubes composite solubilization in water: Definitive spectroscopic properties. *Nanotechnology* 17(12):2845.

Liu, J., Rinzler, A. G., and Dai, H. 1998. Fullerene pipes. *Science* 280(5367):1253–1256.

Liu, X., Song, Y., Li, C., and Wang, F. 2013. Effects of carbon nanotubes grafted on a carbon fiber surface on their interfacial properties with the matrix in composites. *New Carbon Materials* 27(6):455–461.

Liu, Z., Wang, J., and Kushvaha, V. 2011. Poptube approach for ultrafast carbon nanotube growth. *Chemical Communications* 47(35):9912–9914.

Lu, K. L., Lago, R. M., and Chen, Y. K. 1996. Mechanical damage of carbon nanotubes by ultrasound. *Carbon* 34(6):814–816.

Ludvig, P., Calixto, J. M., Ladeira, L. O., and Gaspar, I. C. 2011. Using converter dust to produce low cost cementitious composites by in situ carbon nanotube and nanofiber synthesis. *Materials* 4(3):575–584.

Ludvig, P., Ladeira, L. O., Calixto, J. M., Gaspar, I., and Melo, V. S. 2009. In-situ synthesis of multiwall carbon nanotubes on portland cement clinker. In *11th International Conference on Advanced Materials*. Rio de Janeiro, Brazil.

Luo, J., Duan, Z., and Li, H. 2009. The influence of surfactants on the processing of multi-walled carbon nanotubes in reinforced cement matrix composites. *Physica Status Solidi (a)* 206(12):2783–2790.

Luo, J., Duan, Z., Xian, G., Li, Q., and Zhao, T. 2015. Damping performances of carbon nanotube reinforced cement composite. *Mechanics of Advanced Materials and Structures* 22(03):224–232.

Luo, J. L., Duan, Z., Xian, G., Li, Q., and Zhao, T. 2011. Fabrication and fracture toughness properties of carbon nanotube-reinforced cement composite. *The European Physical Journal Applied Physics* 53(03):30402.

Ma, P., Siddiqui, N. A., Marom, G., and Kim, J. 2010. Dispersion and functionalization of carbon nanotubes for polymer-based nanocomposites: A review. *Composites Part A: Applied Science and Manufacturing* 41(10):1345–1367.

Ma, P. C., Wang, S. Q., Kim, J., and Tang, B. Z. 2009. In-situ amino functionalization of carbon nanotubes using ball milling. *Journal of Nanoscience and Nanotechnology* 9(2):749–753.

Makar, J. M. and Beaudoin, J. J. 2004. Carbon nanotubes and their application in the construction industry. *Special Publication—Royal Society of Chemistry* 292:331–342.

Mathur, R. B., Chatterjee, S., and Singh, B. P. 2008. Growth of carbon nanotubes on carbon fibre substrates to produce hybrid/phenolic composites with improved mechanical properties. *Composites Science and Technology* 68(7–8):1608–1615.

Mawhinney, D. B., Naumenko, V., and Kuznetsova, A. 2000a. Infrared spectral evidence for the etching of carbon nanotubes: Ozone oxidation at 298 K. *Journal of the American Chemical Society* 122(10):2383–2384.

Mawhinney, D. B., Naumenko, V., and Kuznetsova, A. 2000b. Surface defect site density on single walled carbon nanotubes by titration. *Chemical Physics Letters* 324(1–3):213–216.

Mei, L., He, X., and Li, Y. 2010. Grafting carbon nanotubes onto carbon fiber by use of dendrimers. *Materials Letters* 64(22):2505–2508.

Mendoza, O., Sierra, G., and Tobon, J. I. 2013. Influence of super plasticizer and Ca(OH)$_2$ on the stability of functionalized multi-walled carbon nanotubes dispersions for cement composites applications. *Construction and Building Materials* 47(0):771–778.

Metaxa, Z. S., Konsta-Gdoutos, M. S., and Shah, S. P. 2013. Carbon nanofiber cementitious composites: Effect of debulking procedure on dispersion and reinforcing efficiency. *Cement and Concrete Composites* 36:25–32.

Mickelson, E. T., Huffman, C. B., and Rinzler, A. G. 1998. Fluorination of single-wall carbon nanotubes. *Chemical Physics Letters* 296(1–2):188–194.

Mo, Y. L. and Roberts, R. H. 2013. Carbon nanofiber concrete for damage detection of infrastructure. In *Advances in Nanofibers*, ed. R. Maguire. InTech, Rijeka, Croatia.

Moore, V. C., Strano, M. S., and Haroz, E. H. 2003. Individually suspended single-walled carbon nanotubes in various surfactants. *Nano Letters* 3(10):1379–1382.

Mudimela, P. R., Nasibulina, L. I., and Nasibulin, A. G. 2009. Synthesis of carbon nanotubes and nanofibers on silica and cement matrix materials. *Journal of Nanomaterials* 2009(29):256128.

Musso, S., Porro, S., and Vinante, M. 2007. Modification of MWNTs obtained by thermal-CVD. *Diamond and Related Materials* 16(4):1183–1187.

Musso, S., Tulliani, J., Ferro, G., and Tagliaferro, A. 2009. Influence of carbon nanotubes structure on the mechanical behavior of cement composites. *Composites Science and Technology* 69(11):1985–1990.

Nasibulin, A. G., Koltsova, T., and Nasibulina, L. I. 2013. A novel approach to composite preparation by direct synthesis of carbon nanomaterial on matrix or filler particles. *Acta Materialia* 61(6):1862–1871.

Nasibulin, A. G., Shandakov, S. D., and Nasibulina, L. I. 2009. A novel cement-based hybrid material. *New Journal of Physics* 11(2):23013.

Nasibulina, L. I., Anoshkin, I. V., and Nasibulin, A. G. 2012a. Effect of carbon nanotube aqueous dispersion quality on mechanical properties of cement composite. *Journal of Nanomaterials* 2012(35):169262.

Nasibulina, L. I., Anoshkin, I. V., and Semencha, A. V. 2012b. Carbon nanofiber/clinker hybrid material as a highly efficient modificator of mortar mechanical properties. *Materials Physics and Mechanics* 13(1):77–84.

Nasibulina, L. I., Anoshkin, I. V., and Shandakov, S. D. 2010. Direct synthesis of carbon nanofibers on cement particles. *Transportation Research Record: Journal of the Transportation Research Board* 2142(1):96–101.

O'Connell, M. J., Boul, P., and Ericson, L. M. 2001. Reversible water-solubilization of single-walled carbon nanotubes by polymer wrapping. *Chemical Physics Letters* 342(3):265–271.

Paria, S. and Khilar, K. C. 2004. A review on experimental studies of surfactant adsorption at the hydrophilic solid—Water interface. *Advances in Colloid and Interface Science* 110(3):75–95.

Park, H., Zhao, J., and Lu, J. P. 2006. Effects of sidewall functionalization on conducting properties of single wall carbon nanotubes. *Nano Letters* 6(5):916–919.

Park, S. J. and Lee, D. G. 2006. Development of CNT-metal-filters by direct growth of carbon nanotubes. *Current Applied Physics* 6:182–186.

Petrunin, S., Vaganov, V., and Sobolev, K. 2013. The effect of functionalized carbon nanotubes on the performance of cement composites. *Nanocon* 10:16–18.

Peyvandi, A., Sbia, L. A., Soroushian, P., and Sobolev, K. 2013. Effect of the cementitious paste density on the performance efficiency of carbon nanofiber in concrete nanocomposite. *Construction and Building Materials* 48:265–269.

Popov, V. N. 2004. Carbon nanotubes: Properties and application. *Materials Science and Engineering: R: Reports* 43(3):61–102.

Qian, H., Bismarck, A., Greenhalgh, E. S., Kalinka, G., and Shaffer, M. S. 2008. Hierarchical composites reinforced with carbon nanotube grafted fibers: The potential assessed at the single fiber level. *Chemistry of Materials* 20(5):1862–1869.

Qian, H., Bismarck, A., Greenhalgh, E. S., and Shaffer, M. S. 2010a. Carbon nanotube grafted silica fibres: Characterising the interface at the single fibre level. *Composites Science and Technology* 70(2):393–399.

Qian, H., Bismarck, A., Greenhalgh, E. S., and Shaffer, M. S. 2010b. Carbon nanotube grafted carbon fibres: A study of wetting and fibre fragmentation. *Composites Part A: Applied Science and Manufacturing* 41(9):1107–1114.

Qian, H., Greenhalgh, E. S., Shaffer, M. S., and Bismarck, A. 2010. Carbon nanotube-based hierarchical composites: A review. *Journal of Materials Chemistry* 20(23):4751–4762.

Qin, Y., Liu, L., and Shi, J. 2003. Large-scale preparation of solubilized carbon nanotubes. *Chemistry of Materials* 15(17):3256–3260.

Rahmat, M. and Hubert, P. 2011. Carbon nanotube-polymer interactions in nanocomposites: A review. *Composites Science and Technology* 72(1):72–84.

Rastogi, R., Kaushal, R., and Tripathi, S. K. 2008. Comparative study of carbon nanotube dispersion using surfactants. *Journal of Colloid and Interface Science* 328(2):421–428.

Rausch, J., Zhuang, R., and Mader, E. 2010. Surfactant assisted dispersion of functionalized multi-walled carbon nanotubes in aqueous media. *Composites Part A: Applied Science and Manufacturing* 41(9):1038–1046.

Rosca, I. D., Watari, F., Uo, M., and Akasaka, T. 2005. Oxidation of multiwalled carbon nanotubes by nitric acid. *Carbon* 43(15):3124–3131.

Sadeghian, R., Gangireddy, S., Minaie, B., and Hsiao, K. 2006. Manufacturing carbon nanofibers toughened polyester/glass fiber composites using vacuum assisted resin transfer molding for enhancing the mode-I delamination resistance. *Composites Part A: Applied Science and Manufacturing* 37(10):1787–1795.

Saito, T., Matsushige, K., and Tanaka, K. 2002. Chemical treatment and modification of multi-walled carbon nanotubes. *Physica B: Condensed Matter* 323(1):280–283.

Sanchez, F. 2009. Carbon nanofibre/cement composites: Challenges and promises as structural materials. *International Journal of Materials and Structural Integrity* 3(2):217–226.

Sanchez, F. and Ince, C. 2009. Microstructure and macroscopic properties of hybrid carbon nanofiber/silica fume cement composites. *Composites Science and Technology* 69(7):1310–1318.

Sanchez, F. and Sobolev, K. 2010. Nanotechnology in concrete-a review. *Construction and Building Materials* 24(11):2060–2071.

Sanchez, F., Zhang, L., and Ince, C. 2009. Multi-scale performance and durability of carbon nanofiber/cement composites. In *Nanotechnology in Construction* 3, eds. B. Zdenek, P. J. M. Bartos, N. Jiri, V. Smilauer, and J. Zeman, pp. 345–350. Springer, Berlin, Germany.

Sano, M., Okamura, J., and Shinkai, S. 2001. Colloidal nature of single-walled carbon nanotubes in electrolyte solution: The Schulze-Hardy rule. *Langmuir* 17(22):7172–7173.

Shah, S. P., Konsta-Gdoutos, M. S., Metaxa, Z. S., and Mondal, P. 2009. Nanoscale modification of cementitious materials. In *Nanotechnology in Construction 3*, eds. B. Zdenek, P. J. M. Bartos, N. Jiri, V. Smilauer, and J. Zeman, pp. 125–130. Springer, Berlin, Germany.

Shvartzman-Cohen, R., Nativ-Roth, E., and Baskaran, E. 2004. Selective dispersion of single-walled carbon nanotubes in the presence of polymers: The role of molecular and colloidal length scales. *Journal of the American Chemical Society* 126(45):14850–14857.

Siddiqui, N. A., Sham, M., Tang, B. Z., Munir, A., and Kim, J. 2009. Tensile strength of glass fibres with carbon nanotube-epoxy nanocomposite coating. *Composites Part A: Applied Science and Manufacturing* 40(10):1606–1614.

Sobolkina, A., Mechtcherine, V., and Khavrus, V. 2012. Dispersion of carbon nanotubes and its influence on the mechanical properties of the cement matrix. *Cement and Concrete Composites* 34(10):1104–1113.

Star, A., Liu, Y., and Grant, K. 2003. Noncovalent side-wall functionalization of single-walled carbon nanotubes. *Macromolecules* 36(3):553–560.

Strano, M. S., Moore, V. C., and Miller, M. K. 2003. The role of surfactant adsorption during ultrasonication in the dispersion of single-walled carbon nanotubes. *Journal of Nanoscience and Nanotechnology* 3(1–2):81–86.

Sugimoto, S., Matsuda, Y., and Mori, H. 2009. Carbon nanotube formation directly on the surface of stainless steel materials by plasma-assisted chemical vapor deposition. *Journal of Plasma and Fusion Research* 8:522–525.

Sun, S., Yu, X., Han, B., and Ou, J. 2013. In situ growth of carbon nanotubes/carbon nanofibers on cement/mineral admixture particles: A review. *Construction and Building Materials* 49:835–840.

Sun, Y., Huang, W., and Lin, Y. 2001. Soluble dendron-functionalized carbon nanotubes: Preparation, characterization, and properties. *Chemistry of Materials* 13(9):2864–2869.

Tallury, S. S. and Pasquinelli, M. A. 2010a. Molecular dynamics simulations of flexible polymer chains wrapping single-walled carbon nanotubes. *The Journal of Physical Chemistry B* 114(12):4122–4129.

Tallury, S. S. and Pasquinelli, M. A. 2010b. Molecular dynamics simulations of polymers with stiff backbones interacting with single-walled carbon nanotubes. *The Journal of Physical Chemistry B* 114(29):9349–9355.

Tan, Y. and Resasco, D. E. 2005. Dispersion of single-walled carbon nanotubes of narrow diameter distribution. *The Journal of Physical Chemistry B* 109(30):14454–14460.

Thostenson, E. T., Li, W. Z., Wang, D. Z., Ren, Z. F., and Chou, T. W. 2002. Carbon nanotube/carbon fiber hybrid multiscale composites. *Journal of Applied Physics* 91(9):6034–6037.

Tyson, B. M., Abu Al-Rub, R. K., Yazdanbakhsh, A., and Grasley, Z. 2011. Carbon nanotubes and carbon nanofibers for enhancing the mechanical properties of nanocomposite cementitious materials. *Journal of Materials in Civil Engineering* 23(7):1028–1035.

Vaisman, L., Wagner, H. D., and Marom, G. 2006. The role of surfactants in dispersion of carbon nanotubes. *Advances in Colloid and Interface Science* 128:37–46.

Veedu, V. P. 2011. Multifunctional cementitious nanocomposite material and methods of making the same. United States Patent: 7666327.

Veedu, V. P., Cao, A., and Li, X. 2006. Multifunctional composites using reinforced laminae with carbon-nanotube forests. *Nature Materials* 5(6):457–462.

Wang, B., Zhang, Y., Guo, Z., Han, Y., and Ma, H. 2012. Dispersion of carbon nanofibers in aqueous solution. *Nano* 7(06):521–528.

Wang, S. C., Chang, K. S., and Yuan, C. J. 2009. Enhancement of electrochemical properties of screen-printed carbon electrodes by oxygen plasma treatment. *Electrochimica Acta* 54(21):4937–4943.

Wang, Z., Luo, G., Chen, J., Xiao, S., and Wang, Y. 2003. Carbon nanotubes as separation carrier in capillary electrophoresis. *Electrophoresis* 24(24):4181–4188.

Whitsitt, E. A. and Barron, A. R. 2003. Silica coated single walled carbon nanotubes. *Nano Letters* 3(6):775–778.

Xie, X., Mai, Y., and Zhou, X. 2005. Dispersion and alignment of carbon nanotubes in polymer matrix: A review. *Materials Science and Engineering: R: Reports* 49(4):89–112.

Yakovlev, G., Pervushin, G., and Maeva, I. 2013. Modification of construction materials with multi-walled carbon nanotubes. *Procedia Engineering* 57:407–413.

Yamamoto, N., John Hart, A., and Garcia, E. J. 2009. High-yield growth and morphology control of aligned carbon nanotubes on ceramic fibers for multifunctional enhancement of structural composites. *Carbon* 47(3):551–560.

Yazdanbakhsh, A., Grasley, Z., Tyson, B., and Abu Al-Rub, R. K. 2010. Distribution of carbon nanofibers and nanotubes in cementitious composites. *Transportation Research Record: Journal of the Transportation Research Board* 2142(1):89–95.

Yazdanbakhsh, A., Grasley, Z., Tyson, B., and Al-Rub, R. A. 2012. Challenges and benefits of utilizing carbon nanofilaments in cementitious materials. *Journal of Nanomaterials* 2012(19):1–8.

Yu, J., Grossiord, N., Koning, C. E., and Loos, J. 2007. Controlling the dispersion of multi-wall carbon nanotubes in aqueous surfactant solution. *Carbon* 45(3):618–623.

Yu, R., Chen, L., and Liu, Q. 1998. Platinum deposition on carbon nanotubes via chemical modification. *Chemistry of Materials* 10(3):718–722.

Yu, X. and Kwon, E. 2009. A carbon nanotube/cement composite with piezoresistive properties. *Smart Materials and Structures* 18(5):55010.

Yurekli, K., Mitchell, C. A., and Krishnamoorti, R. 2004. Small-angle neutron scattering from surfactant-assisted aqueous dispersions of carbon nanotubes. *Journal of the American Chemical Society* 126(32):9902–9903.

Zhang, F., Wang, R., He, X., Wang, C., and Ren, L. 2009a. Interfacial shearing strength and reinforcing mechanisms of an epoxy composite reinforced using a carbon nanotube/carbon fiber hybrid. *Journal of Materials Science* 44(13):3574–3577.

Zhang, J., Zou, H., and Qing, Q. 2003. Effect of chemical oxidation on the structure of single-walled carbon nanotubes. *The Journal of Physical Chemistry B* 107(16):3712–3718.

Zhang, Q., Liu, J., Sager, R., Dai, L., and Baur, J. 2009b. Hierarchical composites of carbon nanotubes on carbon fiber: Influence of growth condition on fiber tensile properties. *Composites Science and Technology* 69(5):594–601.

Zhang, Q., Qian, W., and Xiang, R. 2008. In situ growth of carbon nanotubes on inorganic fibers with different surface properties. *Materials Chemistry and Physics* 107(2):317–321.

Zhang, X. and Liu, Z. 2012. Recent advances in microwave initiated synthesis of nano-carbon materials. *Nanoscale* 4(3):707–714.

Zhu, S., Su, C., Lehoczky, S. L., Muntele, I., and Ila, D. 2003. Carbon nanotube growth on carbon fibers. *Diamond and Related Materials* 12(10):1825–1828.

Zhu, W., Bartos, P. J. M., and Porro, A. 2004. Application of nanotechnology in construction. *Materials and Structures* 37(9):649–658.

Ziegler, K. J., Gu, Z., and Peng, H. 2005. Controlled oxidative cutting of single-walled carbon nanotubes. *Journal of the American Chemical Society* 127(5):1541–1547.

CHAPTER 34

CONTENTS

Emerging Applications of Functionalized Carbon-Based Nanomaterials

34

P.S. Goh, B.C. Ng, and A.F. Ismail

34.1 INTRODUCTION

Undoubtedly, the distinguishing properties of carbon nanomaterials, particularly carbon nanotubes (CNTs) and graphenes, have vividly guaranteed several imminent applications in laboratory scale and market place. Carbon nanomaterials that have been modified by surface functionalization or doping with heteroatoms to create specific tailored properties can be addressed as second generation of nanocarbon structure. This new class of carbon materials open the doors for the possibility to design and build hybrid or hierarchical materials. They also serve as the catalyst to the development of the third generation of carbon-nanomaterials that can be potentially used in nanoarchitectured supramolecular hybrids or composites (Su and Centi, 2013).

Further variations on these carbon nanomaterials include the addition of an almost infinite variety of functionalities ranging from simple hydroxylation to the grafting of biological molecules such as deoxyribonucleic acid (DNA). The functionalization of carbon nanomaterials can extremely extend their applications in a wide range of fields such as biomedical, energy, and environment (Balasubramanian and Burghard, 2005). Driven by the immediate applications upon the virtues rendered by the subsequent functionalization of these carbon nanomaterials, there has been a significant emphasis placed on the development of more versatile and feasible approaches for the modifications (Xu et al., 2011). As such, the establishment of simple and cost-effective chemical methods for covalent functionalization of CNT materials is becoming an area of growing fundamental and industrial importance (Daniel et al., 2007). Currently, much effort is underway to finding cost-effective approaches to functionalize these carbon nanomaterials by attachment of different functional molecules (Xue et al., 2011), which can generally be attached on the carbon structure through two main approaches: covalent functionalization and noncovalent functionalization (Balasubramanian and Burghard, 2005).

Among diverse classes of carbon nanomaterials, CNTs have attracted particular attention due to their unique physicochemical properties. Over the course of the last two decades, CNTs have been used in biomedical field for cellular therapy, drug delivery, and more recently as sensors for detecting specific proteins and other biomolecules in serum (Yang et al., 2010). Despite their many outstanding characteristics, pristine CNTs are difficult to handle due to the issue related to the dispersion. Moreover, the physical dimensions of CNTs, such as length and diameter, also have some bearings on the toxicity of CNTs, where longer and thinner structures tend to inflict greater cytotoxicity (Nagai et al., 2011). These major constrains associated to pristine CNTs could be overcome using functionalization techniques, which have been recognized as powerful ways to increase the affinity for polar solvents and to prevent aggregation in tight bundles (Speltini et al., 2013). According to some previous studies focused on the biomedical application of CNTs, properly functionalized CNTs are also able to achieve prolonged circulation half-life and improved bioavailability by escaping opsonization-induced reticuloendothelial system (RES) clearance (Vitchev et al., 2010). In the recent decade, research has placed more focuses on the application of functionalized CNTs. Variety of functionalized CNTs, that is, oxidized, amidated, acylated, surfactant and biopolymer-assisted, and biomolecules modified have been developed and utilized as effective, safe, nano-sized, and smart systems (Mehra et al., 2014). The organic functional groups that have been attached to the nanotubes assist effectively in unroping the nanotube bundles and improving the solubility, dispersion, processing, and compatibility of CNTs when blended with the particular matrix. While improving the water dispersibility and reducing the cytotoxicity of CNTs, these surface functionalization also provides additional attachment sites for secondary loading of chemical or supramoleculars based on their purposes (Heister et al., 2009). From the mechanical properties point of view, functionalized nanotubes can provide multiple bonding sites to the organic/inorganic polymer matrix so that the load can be transferred to nanotubes and thus inhibit separation between the surfaces of polymer and nanotubes (Khabashesku et al., 2005).

Compared to CNT, graphene possesses important qualities such as low cost, facile fabrication and modification, a higher surface area with two external surfaces and the absence of toxic metal particles (Mendes et al., 2013). The planar structure of graphene offers an excellent capability to immobilize a large number of substances, including metals, drugs, biomolecules, and fluorescent probes and cells (Yang et al., 2013). Therefore, it is not surprising that graphene and its derivatives such as graphene oxide (GO), and reduced graphene oxide (rGO), or more generally simplified as graphene-family nanomaterials (GFNs), have generated great interest and offered promising nano-platform for widespread applications including in nanoelectronics, energy technology, sensors, and biomedical applications (Guo and Mei, 2014). This class of novel nanomaterial is nowadays under intensive study. Worth saying that the utilization of GFNs in biological applications generates a series of advantages with respect to CNTs due to the absence of metal catalysts during the synthesis consequently possess higher purity. Hence, these materials have already been used as carriers in nanomedicine. Suitably modified GFNs possess vast potential in biomedical field where they can serve as an excellent drug delivery platform for anticancer/gene delivery, biosensing, bioimaging, antibacterial applications, cell culture, and tissue engineering (Liu et al., 2011). Furthermore, graphene consists of a one atom thick carbon (sp^2 hybridized) sheet composed of six-member rings (Scida et al., 2011) providing an exposed surface area that is nearly twice as large as that of single-walled CNTs and also the absence of metallic impurities that can affect the accuracy of a sensor to make it attractive for analytical applications (Pumera et al., 2009).

Unquestionably, the advances made in the functionalization of carbon nanomaterials have unfolded avenues to explore the impact of these materials on various fields and has certainly a great impact in their future development. This chapter intends to provide valuable insight into the applications of functionalized carbon nanomaterials in some major fields where the carbon nanostructures have made significant contributions in the recent progress. These fields include biomedical, energy, sensor, and wastewater treatment.

34.2 APPLICATIONS OF FUNCTIONALIZED CARBON NANOMATERIALS

34.2.1 Biomedical

Based on their extraordinary physicochemical properties, carbon nanomaterials have spurred the interest in the domain of biological research since a few years ago (Ménard-Moyon et al., 2010). In terms of biomedical applications, carbon nanomaterials have demonstrated immense potentials and contributions that covering but not limited to drug/gene delivery, cell and tumor imaging, adsorption of enzymes, and cancer therapy (Fabbro et al., 2012). The successful application of carbon nanomaterials in biomedical field requires the consideration of many underlying issues. First and utmost, the biocompatibility and toxicity must be closely monitored and improved before preclinical and clinical studies can be undertaken. Secondly, the pristine carbon nanostructures demonstrated poor interaction with biomolecules due to their pronounced tendency to agglomerate or bundling and lack of solubility in aqueous media. Therefore, carefully tailored functionalizations are required to sort out this hurdle (Ji et al., 2012). It is in well agreement that facile approaches of purification, carboxylation, acylation, and amidation of these carbon nanostructures have facilitated successful conjugation of various biomolecules as depicted in Figure 34.1 for their wide applications in biomedical field.

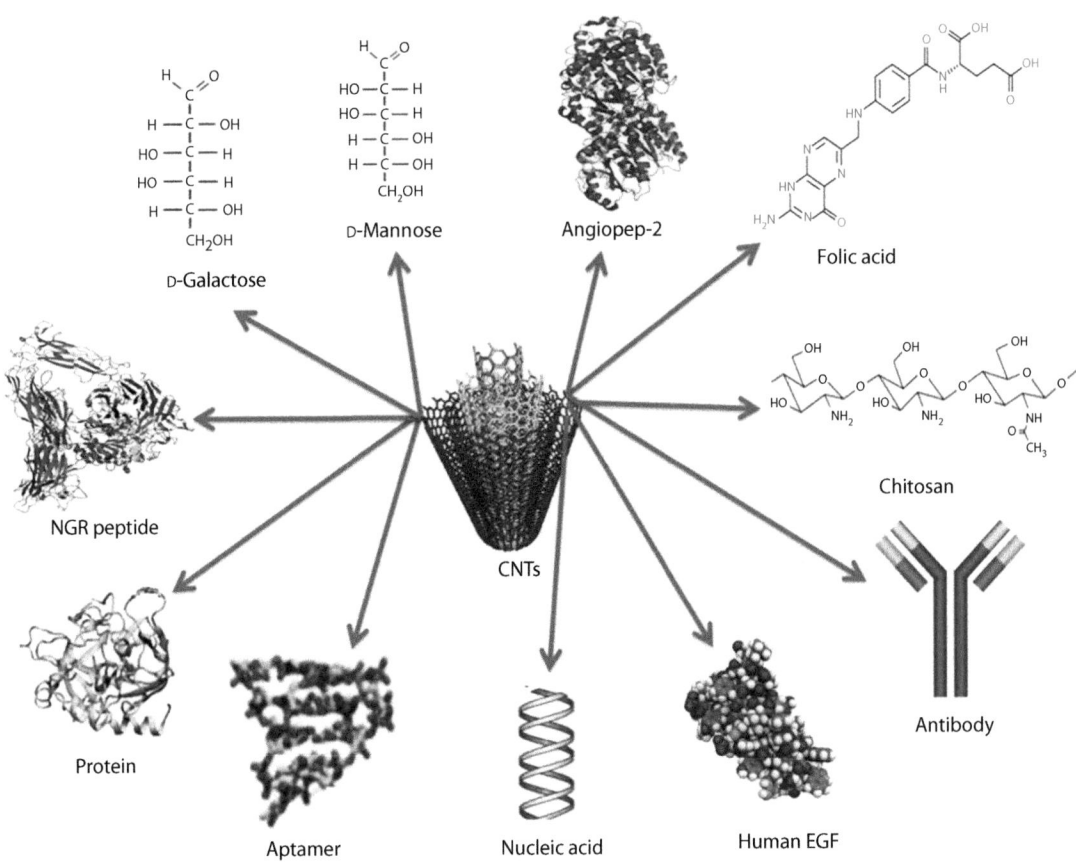

FIGURE 34.1
The conjugation of various biomolecules to CNTs for biomedical application. (From Mehra, N.K. et al., *Biomaterials*, 35, 1267, 2014.)

In the past several years, several review articles have summarized the progresses, advantages, and challenges of using carbon nanomaterials, particularly CNTs in the area of biomedicine (Kostarelos et al., 2009). For instance, Wu et al. (2010) summarized different surface modification strategies to obtain functionalized single-walled CNTs (SWCNTs) for biomedical use. Liu et al. (2009) published a comprehensive review that covers the surface modification, potential toxicity, as well as both in vitro and in vivo applications of CNTs. Recently, Meng et al. (2012) evaluated the performance of SWCNT as drug delivery nanocarrier for targeting DOX to tumors. On the other hand, in the area of nanomedicine, graphene and its derivatives and composites have come into view as new biomaterials that provide exciting opportunities for new generation of nanocarriers for drug delivery, tissue engineering, and probes for cell and biological imaging (Wang et al., 2011). A exponential increase in research efforts in this new emerging field has been evidenced by hundreds of related publications and reviews on this topic (Feng et al., 2013).

34.2.1.1 Drug delivery

Nanoscaled drug carriers have emerged as a bridge linking nanotechnology and advanced drug delivery. Surface functionalized carbon nanomaterials are attracting a great deal of attention in targeted drug delivery. The engineered surface decoration can be made by exploiting the several targeting motifs, which are capable controlling and sustaining drug delivery along with minimizing dose, minimum side effects associated with the free drug and most importantly the improved biocompatibility for patient compliance (Mehra et al., 2014). However, a feasible system must be properly designed to build an efficient nanocarrier with optimized drug-loading capacity as well as to release drugs in a controllable way with optimized dosage at a specific site required for successful therapy (Liu et al., 2013a). Such a system not only improves the efficacy of the drug, but minimizes the systemic toxicity to improve the quality of the patient's life (Meng et al., 2012). Due to the completely different properties from their bulk counterpart, the manipulations and exploitations of biomolecules on the nanoscale are of great importance to the scientists and engineers in which they assure a range of exceptional properties that are could hardly be found from other molecules and polymers (Premkumar and Geckeler, 2012). Due to the presence of reactive sites and relatively large surface area, functionalized carbon nanomaterials can be easily conjugated with numerous biomolecules with ligand–receptor interactions, just to name a few, proteins and amino acid, enzymes, nucleic acid (DNA and siRNA), aptamers, vitamins, monoclonal antibodies, and peptides. Ligand–receptor interactions play a crucial role in targeted drug delivery as these biomolecules may interact with the receptors present on the surface of a cell, tissue, or organ. It may bind with specific receptor site and can be easily internalized inside the cell. The conjugation concept sparked the development of a new generation of nanohybrid materials with potential applications in diagnostics and targeted therapy where they can be designed to dimensionally approach at subcellular scale (Mehra et al., 2014).

Functionalized MWCNTs have been shown as a promising system for the delivery of drugs to intracellular targets in order to improve the immunogenic potential of DCs for therapeutic purposes. Colic et al. (2014) established a promising protocol for the preparation of dendritic cells (DCs) as cancer vaccines based on stimulation of Toll-like receptors (TLRs). DCs are one of the essential components in immune system for the initiation and regulation of the immune response. Therefore, interactions between CNTs and DCs seem to be crucial for understanding the mechanisms by which CNTs influence the immune response (Palomaki et al., 2010). 7-Thia-8-oxoguanosine (7-TOG) was covalently attached to MWCNTs to activate human monocyte-derived (Mo) DCs. The delivery of agonists to endosomal TLRs could increase the intracellular concentrations of the compounds and their prolonged release in order to stimulate stronger immune response. In this context, functionalized CNTs could be advantageous as a carrier system since this nanomaterial hold promising feature of easily phagocytosed by DCs (Wang et al., 2009), hence may improve the current protocols for preparation of DC vaccines that are based on the preparation of immunogenic DCs with the ability to polarize Th1 immune response.

The conjugation of targeting antibodies to the CNTs surface to yield prototype nanoconstructs is a major advancement in the targeted delivery. Generally, nanotubes-based antibody targeting is a promising nanomodality in therapeutic and diagnostic oncology. SWNTs functionalized with antibody specific for tumor cell receptors may be exploited for in vivo specific detection of cancer cells at early stages with the advanced made in deep tissue imaging using spectroscopic techniques. Till date, only a few studies have been reported for conjugation of antibody to functionalized CNTs (McDevitt et al., 2007). Xiao et al. conducted a facile method to covalently conjugate anti-HER2 chicken IgY antibody onto carboxylated HiPco SWCNTs via a microwave-assisted functionalization method. The synthesized HER2 IgY-SWCNTs complexes demonstrated very high specificity for HER2-expressing cancer cell during the selective targeting of cancer cells (Xiao et al., 2009). The resultant complex was successfully used in vitro for both detection and selective destruction of HER2-expressing breast cancer cells. As the dual-function agent of does not IgY-SWCNT complex require internalization by the cancer cells in order to achieve the selective photothermal ablation, thus it has the potential to be extended to detect and treat various cancer types.

The applications of functionalized CNTs in delivery of doxorubicin (DOX) for cancer treatment have been visualized in the last few years. DOX is a chemotherapeutic agent and has been used for the treatment of many human cancers. CNTs were found more efficient in DOX delivery with higher holding capacity, controlled/sustained release and targeting potential along with minimum or no toxicity, hence demonstrating great targeting potential in DOX delivery as compared to the other available nanocarriers like liposomes and dendrimers (Mehra et al., 2014). More importantly, the supramolecular stacking of DOX on CNT were found to be more effective and less toxic as compared to the free DOX in equimolar concentration (Huang et al., 2011). In the attempt made by Zhang et al. (2009), DOX has been successfully loaded on the chitosan and sodium alginate wrapped functionalized SWCNTs and tested on human cervical carcinoma cells. While the mixing of pristine SWCNTs with DOX generally yield nanotubes with very heterogeneous surfaces, uniformly coated nanotubes could be obtained when DOX is loaded onto polysaccharide (chitosan and sodium alginate) modified SWCNTs, The developed DOX-SWCNTs conjugates selectively accumulated in the targeted cancerous tissues and efficiently released DOX in controlled manner. The exceptional properties can be attributed to the complementary fashion rendered by polysaccharides to facilitate further functionalization with a targeting DOX group. Similarly, Heister et al. (2009)

prepared a triple functionalized SWCNT comprising anticancer drug DOX, a monoclonal antibody, and a fluorescent marker at noncompeting binding sites for the recognition of carcinoembryonic antigen (CEA), which is a glycoprotein expressed only in cancer cells, especially adenocarcinoma such as colon cancers. The attachment of the three agents was accomplished by two different chemical approaches noncovalent and covalent binding. While DOX was noncovalently attached, both fluorescein and CEA antibody were covalently attached to the SWCNTs via bovine serum albumin (BSA) as a hydrophilic multifunctional linker. The complex was efficiently taken up by cancer cells with subsequent intracellular release of DOX, which then translocated to the nucleus while the nanotubes remain in the cytoplasm. In this context, nanotubes played an important role to enable molecular targeting via the attachment of monoclonal antibodies that are crucial for the delivery of drugs that are not taken up by cells under normal conditions. Nevertheless, further study needs to be done to address the question whether CNTs are able to enter the nuclei of cells.

Graphene and its derivatives have been extensively explored as some of the most promising biomaterials for biomedical applications due to their unique properties: two-dimensional planar structure, large surface area, chemical and mechanical stability, superb conductivity, and good biocompatibility (Liu et al., 2013a). GFNs, particularly monolayer graphene, have the theoretical maximum surface area owing to the arrangement of every atom that lies on the surface to offer extremely high capacity for drug delivery (Guo and Mei, 2014). Furthermore, graphene sheets as drug carrier are interesting because both sides of a single sheet could be accessible for drug binding. While drugs are mainly loaded onto CNTs via surface and tips, the graphene sheet is expected to load drugs via its two faces and edges (Yang et al., 2008). Generally, GFNs with small size, sharp edges, and rough surfaces easily internalize into the cell as compared to larger and smooth GFNs. It is worth pointing out that, the low cost and large production scale of graphite and GO is unmatched by CNTs (Liu et al., 2008). Particularly for GO, it is more widely used than graphene for biomedical applications as it offers several distinct advantages over other nonviral vectors. Many hydroxyl, carboxylic acid, and other reactive groups are introduced on the surface during the fabrication of GO via oxidation of graphite. These functional groups are amenable to ligand conjugation, cross-linking, and other modifications, rendering GO tailored for a range of biomedical and other applications. GO-functionalized conjugates can be used as effective adsorbents for hydrophobic drug molecules owing to their amphiphilic nature and oxygenated unsubstituted graphene domains (Goenka et al., 2014).

Graphene-based materials have been conjugated with a number of natural biopolymers such as gelatin and chitosan as functionalizing agents for drug delivery applications due to the biocompatibility, biodegradability, and low immunogenicity of these biopolymer to significantly reduce the toxicity of graphene. For instance, when gelatin was used as a reducing and functionalizing agent to load DOX onto graphene nanosheets (gelatin–GS) (Liu et al., 2011a), the complex showed higher drug loading capacity due to large surface area and relatively higher π interactions. The gelatin–GS–DOX complex also exhibited high toxicity toward MCF-7 cells through endocytosis. Recently, development of a dual-targeted (magnetic and biological) drug delivery system has improved efficiency of such system. Triple functionalized GO-DOX, encapsulated by Fe_3O_4, chitosan, and folic acid exhibited high loading efficiency and targeted drug delivery to the tumor area (Wang et al., 2013). Chitosan was used as a bridge to combine folic acid with functionalized GO while improving the whole stability and biocompatibility as well as encapsulate and control release of drug molecules. It was found that the loading and release of DOX depends upon the hydrogen bonding interaction with GO and chitosan-folic acid conjugate and subjected to the pH change. In view of the different releasing behaviors of DOX on the functionalized GO under different pH environment, this multifunctionalized GO can be used as a good candidate material for intelligent drug carrier. In addition, delivery of more than one anticancer drug by GO also has been reported (Zhang et al., 2010). Controlled loading of both DOX and camptothecin onto folic acid-conjugated GO via pep stacking and hydrophobic interactions resulted in both target specificity and much higher cytotoxicity to MCF-7 cells than conjugated GO loaded with either drug alone.

Dai et al. (Liu et al., 2008) investigated the ability of graphene in the attachment and delivery of aromatic water-insoluble drugs. Upon the conjugation of six-armed polyethylene glycol (PEG)-amine stars to the carboxylic acid groups on the nanographene sheets (Figure 34.2a), they found that the PEG conjugation on the functionalized nanographene sheets (NGO-PEG) are biocompatible without obvious toxicity as shown in the in vitro cell toxicity assay illustrated in Figure 34.2b. The NGO-PEG can be attached with various forms of aromatic anticancer drug

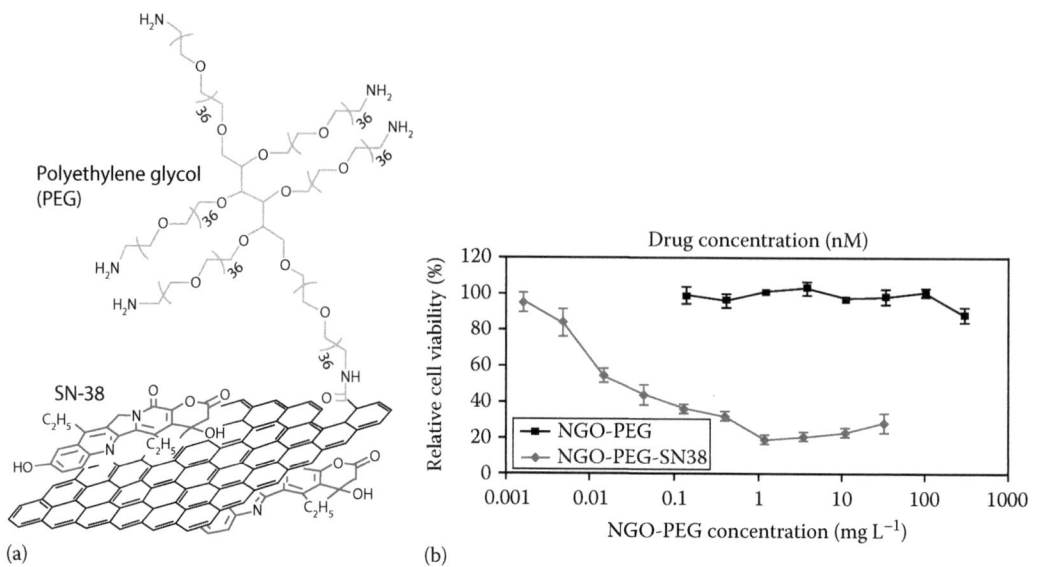

FIGURE 34.2
(a) Schematic drawing of conjugation of six-armed polyethylene glycol (PEG)-amine stars to the carboxylic acid groups on the nanographene sheets; (b) in vitro cell toxicity assay exhibited that no obvious toxicity was measured for various concentrations of NGO–PEG without drug loading. (From Liu, Z. et al., *J. Am. Chem. Soc.*, 130, 10876, 2008.)

such as camptothecin analogues and Iressa (geftinib), a potent epidermal growth factor receptor (EGFR) inhibitor with high efficiency via simple adsorption. Yang et al. showed that the DOX molecules could make a strong bond with the GO surface through π–π stacking interactions with the hydrophobic quainine part of DOX. Besides that, hydrogen bond reaction was also anticipated between carboxyl functional groups of GO and amino groups of DOX (Yang et al., 2008). Based on the established interactions, high loading of 235 mg mL^{-1} could be achieved via simple non-covalent mixing. However, the pH-dependent loading and releasing was observed, which may be due to the hydrogen-bonding interactions between GO and DOX.

34.2.1.2 Gene therapy

Based on the use of gene vectors that protect DNA from nuclease degradation and facilitates the cellular uptake of DNA into cells with improved transfection efficiency, gene therapy is a promising approach to treat various diseases caused by genetic disorders, including cystic fibrosis, Parkinson's disease, and cancer (Liu et al., 2013a). Chen et al. (2011) developed polyethylenimine (PEI)-functionalized GO to serve as a candidate for gene delivery vector. The nanohybrid of PEI-GO that formed by the covalent linking of PEI and GO via an amide bond exhibited an excellent ability to condense plasmid DNA onto the surface of a GO sheet through an electrostatic interaction arising from the cationic PEI at a low mass ratio with a positive potential of 49 mV. The conjugate could effectively deliver plasmid DNA into cells and be localized in the nucleus. Similarly, a PEI-grafted GO nanocarrier was also developed for sequential delivery of siRNA and anticancer drugs by Zhang et al. (2011). The PEI–GO composite showed significantly lower cytotoxicity and substantially higher transfection efficacy of siRNA. Their results demonstrated that delivery of siRNA and DOX by the PEI–GO nanocarrier led to significantly enhanced anticancer efficacy.

Carbon nanohorns (CNHs) possess a spherical shape and larger diameters than CNTs, which could lead to differences in the mechanism of metabolism, degradation, or dissolution, clearance and bioaccumulation. CNHs can be considered as ideal carriers to anchor biologically active molecules as their high surface areas allow the incorporation of molecular entities, such as polyamidoamine (PAMAM) dendrimers. A hybrid system based on PAMAM dendrimers anchored to the CNH surface has been reported (Guerra et al., 2012). The system was aimed to release interfering genetic material diminishing the levels of a house-keeping protein and a protein directly involved in prostate cancer development (Guerra et al., 2012). Synergetic effect was achieved in which PAMAM dendrimers responsible for the electrostatic binding to siRNA and enhance the solubility of CNHs to improve biocompatibility meanwhile CNHs serve as platform for dendrimers. Besides that, this hybrid material is also far less toxic than the corresponding free dendrimer as it did not display any cytotoxicity up to 25 lg mL^{-1} while it is very effective to couple siRNA. The ability to efficiently transfect siRNA suggested that this hybrid system holds bright future for routine gene therapy.

34.2.1.3 Tissue engineering

Major advances in the knowledge of cell and organ transplantation and of material chemistry in recent years have aided in the sustained development of tissue engineering based on carbon nanomaterials (Zhang et al., 2010b). Tissue engineering involves the seeding of cells onto a scaffold, which is then cultured as a whole in vitro and finally implanted into the body as a tissue engineered construct when matured (Pan et al., 2012). In general, there are four areas that these carbon nanostructures can be used in which are relevant for tissue engineering, that is, cell tracking and labeling, sensing cellular behavior, augmenting cellular behavior, and enhancing tissue matrices (Harrison and Atala, 2007). Possible utilization of carbon nanomaterials with various kinds of biopolymer molecules substances in tissue engineering has been extensively reviewed.

In this field, CNTs can be used as additives to reinforce the mechanical strength of tissue scaffolding and conductivity by dispersing a small fraction of CNTs into a biopolymer molecules or to improve the benefits of native extracellular matrix (Wang et al., 2005). Furthermore, CNT-coated 3D scaffold could be applied with high functionalization with many kinds of proteins this could be advantageous for bone regeneration such as osteoblast transplantation. Osteoblast cell growths have been observed on the CNT composite scaffolds, which indicated that functionalized CNT may be easily interacting with cell and increased all cell metabolic function (Venkatesan et al., 2012). Many researches revealed that functionalized CNTs (carboxylated or hydroxylated CNTs) are water-soluble and could eventually be cleared from systemic blood circulation through the renal excretion route (Singh et al., 2006), which means that functionalized CNTs are safe in biomedical application such as scaffold (Abarrategi et al., 2008). Hirata et al. (2011) showed that collagen/SWCNT composite materials have great utility as scaffolds in tissue engineering. Collagen has wide applications in tissue engineering and regenerative medicine because of its regular helical structure, excellent biocompatibility, and moderate immunogenicity. By preparing rat primary osteoblasts (ROBs) cultured on each sponge for 1 day for the transplantation, they reported that after the implantation of the MWCNT-coated collagen sponge honeycomb for 28 days, bone tissues were successfully formed in the pores according to its honeycomb structure. On the other hand, flattened and fibrous bone-like tissue formed in the uncoated sponge, suggesting the uncoated collagen sponge was absorbed before the bone formed solidly. This observation proposed that MWCNTs might contribute to the favorable bone formation and stability against bone remodeling around the MWCNT-coated sponge. Besides that, severe inflammatory responses such as necrosis and degeneration were not observed around SWCNTs after implantation, indicating the good biocompatibility of SWCNTs.

The biocompatibility, biodegradability, and antimicrobial activity of chitosan have promoted its application in bone tissue engineering to mimic all the natural functions of the normal bone such as porosity and cell proliferation. CNT functionalized with chitosan is a promising biomaterial for bone tissue engineering due to high mechanical strength and electrical conductivity of CNT as well as lower toxicity of CNT when it is functionalized and incorporated into chitosan matrix (Zhang et al., 2010b). Venkatesan et al. (2012) grafted chitosan with functionalized MWCNTs and observed that the MWCNT play a structural reinforcement with chitosan as well as imparting novel properties for cell growth. Potential cytotoxic effects associated with CNT can be minimized by chemically functionalizing the surface and combination with the chitosan natural polymer. The main advantage of the bicomponent system of chitosan/f-MWCNT scaffolds is better than chitosan scaffold in terms cell proliferation, alkaline phosphatase activity, and mineralization as checked in.

Graphene-based materials have been explored for wound healing, stem cell engineering, regenerative medicine, and tissue engineering. Very recently, graphene/hydroxyapatite (HA) nanorod composite has been synthesized through in situ growth method to grow HA over the GNS surface to improve the mechanical property and osseointegration ability of HA (Fan et al., 2014). Despite the good bone bonding and bone regeneration through the interactions of osteogenic cells with HA, the poor tensile strength and fracture toughness of HA limit its practical application as bone replacement.

Therefore, integrating GNS and HA nanorods will result in the development of new composites with good mechanical property and excellent biocompatibility. The results obtained suggested that HA nanorods play a vital role in enhancing osteoblastic adhesion and differentiation where the increased number of filopodia enabled the cells to tightly bind to GNS/HA surface. It was found that the synthesized complex with 40 wt.% HA exhibits better biocompatibility and higher bone cellular proliferation than GO and HA do.

The application of carbon nanomaterials for tissue engineering can also be extended to wound healing therapy. Lu et al. (2012) prepared chitosan–PVA nanofibrous scaffolds containing graphene in order to understand wound healing potential of this complex in mice and rabbit. They found that the samples containing graphene healed completely and at a faster rate as compared to others in both mice and rabbit. By performing antibacterial studies using *Escherichia coli*, Agrobacterium and yeast, they confirmed that the presence of free electron in graphene does not affect the multiplication of eukaryotic cells but inhibits the prokaryotic cell multiplication, thereby preventing the growth of microbes. Besides wound healing, graphene-based materials can also be potentially applied in musculoskeletal tissue engineering. Ku and Park (2013) myotube formation on graphene-based nanomaterials, particularly GO and rGO using mouse myoblast C2C12 cell lines. GO showed higher myotube fusion/maturation index and upregulated expression of myogenic genes (MyoD, myogenin, troponin T, and myosin heavy chain) compared to rGO. The enhanced cellular behavior for GO was probably due to the greater surface roughness and higher density of surface oxygen content that influence adsorption of serum proteins. The remarkably enhanced myogenic differentiation on GO, which resulted from serum protein adsorption and nanotopographical cues has demonstrated the ability of GO to stimulate myogenic differentiation, showing a potential for skeletal tissue engineering applications. Engineering substrates to induce desired cell phenotype and genotype is an important strategy of scaffold design for tissue-engineering applications. Wang et al. developed fluorinated graphene sheets as the scaffold for stem cell growth (Wang et al., 2012). The fluorination of graphene was performed by exposing the samples to a fluorinating agent, xenon difluoride. The Fluorinated graphene films were found to be highly supportive of bone marrow derived mesenchymal stem cells (MSCs) growth and the coverage of fluorine has significant effects on cell morphology, cytoskeletal, and nuclear elongation of MSCs. Higher proliferation and stronger polarization of MSCs promoting neuronal differentiation were induced (Wang et al., 2012). This was further enhanced when MSCs were confined into microchannels patterned onto fluorinated graphene in the absence of any other chemical inducers.

34.2.2 Sustainable Energy

Carbon nanomaterials have been playing a significant role in the development of alternative clean and sustainable energy technologies (Candelaria et al., 2012). A number of attractive and unique characteristics of these nanostructured carbon materials render them promising materials for energy applications: (1) high surface area, (2) tuneable pore structure, (3) high electron, phonon and heat transport, (4) easy accessibility for reactant or ions, and (5) new storage mechanism (Su and Centi, 2013). For instance, fullerene-containing p-type semiconducting polymers are one of the key foundations in rapidly advancing organic photovoltaics (Dennler et al., 2009). CNTs and GNFs are emerging classes of carbon nanomaterials that have been explored as promising candidates for the next generation of optically transparent electronically conductive films for solar cells (Hecht et al., 2011). They have also been widely studied for the development of batteries, supercapacitors, and fuel cells (Sun et al., 2011).

34.2.2.1 Fuel cells

Polymer electrolyte membrane fuel cell (PEMFC) has received intensive researches from both alternative energy and environmental consideration owing to their attractive features of high power density, low operating temperature, and converting fuel to water as the only by-product (Muller et al., 2006). Currently, the investigation on suitable materials for the components in fuel cells, especially for bipolar plate which accounts for nearly 38% in a fuel cell stack cost (Dhakate et al., 2007), has become a critical research issue to the commercial manufacturers and research community in order to reduce the fabrication cost while increasing the reliability of fuel cells. Liao et al. reported the fabrication of functionalized CNT-based polymer composite bipolar plates as the substitute for the conventional graphite plates (Liao et al., 2010). The acid chloride-functionalized MWCNTs are covalently modified with poly(oxyalkylene)-amine bearing the diglycidyl ether of bisphenol A epoxy oligomers prior to the preparation of the MWCNT/polypropylene composite. Increased flexural strength of the resultant composite bipolar plate was mainly associated with the stronger functionalized MWCNT–polymer interaction, causing the composite bipolar plate to transfer the load from the host polymer matrix to MWCNTs more efficiently. The maximum current density and power density of the single cell test for the nanocomposite bipolar plate with are 1.32 A cm^{-2} and 0.533 W cm^{-2}, respectively. The overall performance indicated that the functionalized MWCNTs/polypropylene nanocomposite bipolar plates prepared in this study are suitable for PEMFC application.

Proton exchange membrane (PEM) is one of the most essential components of fuel cell that provides proton conduction and prevents direct connection between fuel and oxidant. Nafion has been commonly used for this purpose owing to its physical and chemical stability at moderate temperatures. However, many strategies have been pursued to tackle the shortcoming of this commercial membranes. This includes the fabrication of alternative membrane based on the incorporation of functionalized carbon-based nanomaterials (Asgari et al., 2013). CNTs functionalized with histidine, an imidazole-based amino acid have been used as fillers for the preparation of Nafion polymer nanocomposites. Uniform distribution of the functionalized CNTs within the Nafion matrix decrease the methanol permeability efficiently and higher selectivity value leads to better results in improvement of transport properties and single cell performance. Incorporation of modified CNTs into the Nafion matrix resulted in changes in the water uptake and ion exchange capacity. The nanocomposite membranes showed higher proton conductivity especially at elevated temperatures. This phenomenon can be attributed to the imidazole groups that their basic nitrogen sites can enhance proton transport by Grotthuss-type mechanism.

Despite the intensive research investigating the applications of CNTs as catalyst supporting materials in fuel cell fabrication to exhibit enhanced catalytic activities (Choi et al., 2012), many recent studies have shifted their research target toward graphene since the existence of 2D graphene is reported (Dikin et al., 2007). Graphenes are granted with some outstanding characteristics as compared to their CNTs counterpart, such as higher surface area that are up to theoretical value of 2630 m^2 g^{-1}, excellent conductivity for electrochemical applications, unique graphitized basal plane structure, and, more importantly, they can be produced at a much lower cost and higher purity. Graphene sheets have been identified as a promising candidate for cathode catalyst support in PEMFCs. Recently, graphene received attention as the catalyst support in methanol oxidation for fuel cell application. Kou et al. (2009) functionalized graphene sheets (FGSs) through thermal expansion of GO, followed by the impregnation of Pt nanoparticles with average diameter of 2 nm. The Pt/FGS shows high initial current density and good retention on both the electrochemical surface area and oxygen reduction activity compared with the commercial Pt/C. It has been interestingly found that higher stability of Pt catalyst on FGSs than MWCNT may be attributed to the functionalized surface groups and the higher surface area of FGSs. FGSs, with more π-sites and functional groups, may lead to a strong metal support interaction and resultant resistance of Pt to sintering hence enhanced the durability.

Efforts have also been constantly devoted to the research in cathodic oxygen reduction reaction (ORR) due to its crucial role in electrochemical energy conversion in fuel cells (Gong et al., 2009). GO has been functionalized with a quaternary ammonium salt, tridodecylmethylammonium-chloride TDMAC, to create net positive charge on carbon atoms in a total graphene sheet via intermolecular charge transfer as shown in Figure 34.3a (Ahmed and Jeon, 2012). Successive adsorption of TDMAC made the surface of rGO positively charged, hence leading to good dispersion via electrostatic repulsion among individual graphene sheets. As a result, the metal-free catalyst demonstrated enhanced electrocatalytic activities toward ORR in which the onset potential of the ORR for the TDMAC-rGO electrode shifted positively with a more pronounced increase in current density as displayed in Figure 34.3b. Additionally, the developed TDMAC-rGO catalyst was also found to be inert to cross-over effects of methanol/ethanol in the ORR, where it exhibited excellent selectivity in ORR, with no visible response to methanol/ethanol fuel molecules oxidation in an alkaline medium. The exceptional characteristics of this efficient ORR catalyst provide an opportunity for practical applications in fuel cells.

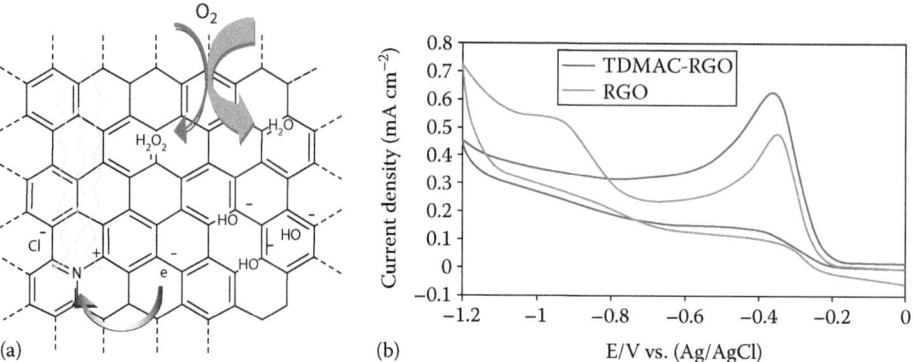

FIGURE 34.3
(a) Schematic figure showing electron receiving by TDMAC from RGO sheet for enhancing the ORR and O_2 reduced to H_2O as a main product and (b) a cathodic reduction peak displays the onset potential of the ORR for the TDMAC-RGO electrode shifted positively with a more pronounced increase in current density. (From Ahmed, M.S. and Jeon, S., *J. Power. Sources*, 218, 168, 2012.)

34.2.2.2 Lithium-ion batteries

Lithium-ion batteries (LIBs) have received significant attention due to high-energy density for applications in electrical vehicles, electronic and portable device, and grid technology. Moreover, high-energy and high-power rechargeable LIB has become a key enabler for vehicle electrification including plug-in hybrid electrical vehicles and full electric vehicles (EV) (Ahn et al., 2012). The key parameters for LIBs are energy and power density (both gravimetric and volumetric), cyclability, rate capability, safety, dependence from temperature, and the cost of production (Kucinskis et al., 2013). Efficient electrode materials with further improved volumetric and gravimetric capacity are critical to enhance the desired properties LIBs.

Scientists have used functionalized CNTs as an anode material for LIBs and found that they had improved electrochemical properties. Ahn et al. (2012) deposited SnO_2 nanoparticles on the functionalized SWCNTs to serve as negative electrodes for the LIB. The SWCNTs functionalized with carboxylic acid groups was used as the buffering agent to suppress the mechanical degradation of SnO_2-based negative electrode and the conducting medium to facilitate electron transport and lithium ion diffusion. The observations indicated that the carboxylic acid group presented on CNTs surface in fact suppress the electrolyte decomposition. Thus, the capacity degradation was not observed in the SnO_2/CNT composite, and it was mitigated by functionalizing the CNTs buffering fillers which inter-connect the Sn cluster. It was also believed that the carboxylic group helped the nucleation of SnO_2 on CNT surface and enhanced the adhesion between them. SnO_2/functionalized SWCNT composite electrodes were shown to have a high specific capacity up to 650 mA h g^{-1} as well as stable cycling performance with 85% capacity retention after 100 cycles, which suggest them a suitable anode candidate for high-power LIBs.

In a recent attempt, ionic liquids (ILs) have been utilized to modify the physicochemical properties of MWCNTs (Bak et al., 2014). ILs adsorbed on the carbon nanomaterials via van der Waals and π–π interactions, leading to the enhancement in their mechanical and chemical properties as well as improving their processability. MWCNT that was functionalized with 1-butyl-3-methyl-imidazolium tetrafluoroborate that served as ILs acted as the free-standing conductive networks to improve electrical conductivity and mechanical integrity of active Fe_2O_3 and to reduce the total mass of LIB anode electrode by excluding the usage of additive and binder during the processing for higher specific capacity. In this hybrid system, the well-defined nanostructure of Fe_2O_3 was constructed on the surface of the functionalized CNT, where ILs played a role of dispersing CNT for large accessible surface area and of providing a favorable interface between CNT and Fe_2O_3. The nanoscale deposition of Fe_2O_3 nanoparticles on the IL functionalized CNT greatly improved the discharge capacity up to 413 mA h g^{-1} due to the redox reaction of Fe_2O_3 nanoparticles. After 50 cycles of charge/discharge, the coulombic efficiency and cycle stability of the composite were estimated to be 98% and 67%, respectively.

Although graphene nanostructures were thought to have significant disorder and defects, which might lower their electrical conductivity, some reports verified highly disordered graphene nanosheets as electrodes with high reversible capacities (794–1054 mA h g^{-1}) and good cyclic stability (Kuchinskis et al., 2013). Recent researches have raised the possibility of using GNF as an electron conducting additive for LIB electrode materials this class of carbon nanomaterial is found to significantly improve electrode electrochemical performance. It has been well agreed that the hydrophilic oxygen-containing functional groups (epoxides, hydroxides, carboxylic groups) on GO can serve as anchor sites and consequently make nanoparticles attach on the surfaces and edges of GO sheets. Furthermore, it has been hypothesized that the surface oxidation of carbon materials could improve their electro-chemical properties. In their recent work, Reddy Channu et al. (2014) confirmed that functionalized GO-coated vanadium oxide electrodes have outperformed discharge capacity than the carbon-coated vanadium oxide electrodes that can be attributed to higher electrical conductivity and larger surface area of functionalized GO than the conventional carbon. The functionalized graphene suppressed the agglomeration and deformation of the vanadium oxide nanostructure and accommodate the large volume variation and maintain good electronic contact during the charge discharge process and improve their rate performance.

34.2.2.3 Hydrogen storage

Hydrogen has been considered the core of an energy economy in its own right that can be conveniently converted into electricity, particularly through the use of fuel cell technology. The carbon-based nanomaterials have gain vast attentions from the scientific community as possible materials that stand to deliver unparalleled performance as the next generation of base materials for hydrogen storage (Yürüm et al., 2009). But, in their pristine form, the carbon nanostructures are chemically too inert to be used for this technological application. Therefore, to improve the reactivity of the carbon-based nanostructures, functionalization of their surfaces or interior is highly required. The modification of these carbon nanomaterials by the addition of foreign atoms or molecules could lead to enhanced interaction between hydrogen and the carbon surface which resulted in higher storage capacity. Further, it has been pointed out that the functionalization of carbon nanomaterials with transition metal atoms itself occupies more weight percentage and also they forms strong metal hydrides while hydrogenation.

In the group of various carbon nanostructures, CNT is most probably the most widely investigated potential hydrogen storage material (Silambarasan et al., 2014), primarily due to their remarkable properties such as hollowness, cylindrical shape, interstitial sites, and nanometer scale diameter and porosity make them as one of the interesting candidates for hydrogen storage. One of the facile modification approaches of SWCNTs for the purpose of hydrogen storage can be made by means of functionalizing them with borane (BH$_3$). The simulation studies based on density functional theory (DFT) have indicated that functionalization of SWCNTs with BH$_3$ could enhance the binding energy of hydrogen molecules and thereby increase the storage capacity (Surya et al., 2011). Silambarasan et al. (Silambarasan et al., 2014) reported the functionalization of SWCNT of BH$_3$ where LiBH$_4$ was used as the precursor for BH3 and hydrogenation and dehydrogenation studies were performed based on the functionalized SWCNTs. It has been deduced that, the interaction between the functionalized SWCNTs and hydrogen molecules is primarily due to the combination of electrostatic, inductive and covalent charge transfer mechanisms. Here, borane acts as a bridge between the hydrogen molecules and SWCNTs. On hydrogenation, it helps to hold the hydrogen molecules onto SWCNTs in the ideal binding energy limits. A maximum storage capacity of 4.77 wt.% was achieved at 50°C, which is close to the USDOE target of 5.5 wt.% for a hydrogen storage medium to be used for on-board applications. The hydrogenated system is stable at room temperature and the entire (100%) stored hydrogen are released in the temperature range of 90°C–125°C. Since the binding energy of hydrogen molecules lies within the ideal range, it is supposed that the stored hydrogen binds sufficiently strong with the host and is desorbed in the temperature range feasible for mobile applications.

The hydrogen storage capacity of graphene, the hydrogenated form of graphene, CH has been reported (Hussain et al., 2014). The advantage of using graphene as hydrogen storage material is its nano-size, large stability, and relatively stronger CH-metal binding. This would allow having uniform distribution of dopants over the surface during the functionalization of graphene. Hussain et al. (2014) substituted H atoms of graphene with Li atoms to investigate the H_2 storage capacity in molecular form. The polar nature of bonds in CeLi leaves a significant amount of charge on Li atoms, which in turn can polarize and adsorb H_2 molecules, thus resulting in a good storage of H_2 molecules. The binding energies of CLi_3 and CLi_4 on CH sheet have been found to be 3.80 and 2.2 eV, respectively, which is large enough to avoid any kind of clustering among these species. The molecular dynamic simulations at 350 K suggested that CLi_3 and CLi_4 on CH sheet, a maximum of 24H2 molecules can be adsorbed with average adsorption energy within the ideal range of 0.25–0.35 eV suitable for practical applications and a very high storage capacity of more than 13 wt.% can be achieved.

34.2.2.4 Solar cell

All organic and hybrid solar cells that highly capable in harnessing solar energy on a large scale are anticipated to play pivotal roles in fulfilling the long-term consumption for sustainable and efficient sources of energy. The novel bulk heterojunction structures of carbon nanomaterials have attracted tremendous interest for both scientific fundamentals and potential applications in various new opto-electronic devices, particularly in the development of photovoltaic solar cells (Zhe et al., 2009). It is a general consensus that the extraordinary properties possessed by the carbon nanostructures, such as light weight, excellent mechanical strength, flexibility, excellent carrier mobility, and outstanding electrocatalytic property, are expected to improve the performance of solar cells. Carbon nanostructures have been widely used in different aspects of photovoltaics to render many advantages, for example, increase the internal quantum efficiency and hence the short circuit current in solar cells as well as to increase the carrier mobility by providing a ballistic pathway for a photoexcited carrier to transport (Li and Liu, 2012).

SWCNTs with outstanding electron-transporting properties and high surface area have been successfully used in organic photovoltaic devices as the electron acceptor material to provide effective carriers transport channels (Yan et al., 2014). Functionalization of the SWCNT with carboxyl acid groups was performed prior to the wrapping of poly(3-hexylthiophene) (P3HT) and [6,6]-phenyl-C 61-butyric acid methyl ester (PCBM) to improve the dispersity and result in a more homogenous photoactive layer. Maximum power conversion efficiency (PCE) of 3.02% with a short-circuit current density of 11.46 mA cm^{-2} was obtained from the photovoltaic cell. The PCE enhancement of P3HT:PCBM:SWCNTs-based solar cells can be mainly ascribed to the role of functionalized SWCNTs to work as electron-acceptors and electron-transporting channels, charge extraction and surface morphology, as well as increasing of the crystallinity of the photoactive layer. SWCNTs fit the need of a wide conductive network and allowed the extension of excitons dissociation area and fastening charge carriers transfer across the active layer, thus favorably led to an increased PCE.

In traditional dye-sensitized solar cell (DSSC), nanocrystalline and porous TiO_2 serve as charge transport medium onto which a monolayer of dye sensitizer is chemically adsorbed. Composites combining carbon-based materials have exhibited synergistic effects with regard to their electrochemical properties and catalytic activities. MWCNTs have been incorporated into TiO_2 nanocrystalline electrodes in a DSSC to enhance charge carriers mobility and to act as electron collector, hence to improve the electronic transport of the solar cell (de Morais et al., 2013). Acid functionalization was carried out to introduce carboxylic groups on to the surface of the CNT, increasing their polarity and reactivity and contributing to a greater interaction with the TiO_2 matrix. The authors found strong correlations between the performance of DSSC based on TiO_2–MWCNT electrodes and the amount of the functionalized MWCNT added, in which the incorporation of lower concentration of 0.02 wt.% MWCNT into the TiO_2 film offered a more efficient electron transport through the nanoparticles. As a result, the MWCNT-TiO_2 electrode provided the best electrical parameters with short circuit photocurrent density (Jsc) of 8.88 mA cm^{-2}, open-circuit voltage (Voc) of 0.65 V, and a conversion efficiency of 3.05%. For similar purpose, Yue et al. (2013) reported the preparation of graphene/PEDOT:PSS (poly(3,4-ethylen-edioxythiophene):polystyrenesulfonate) composite film for counter electrode in DSSC. The graphene/PEDOT:PSS was electrodeposited on fluorine-doped tin oxide

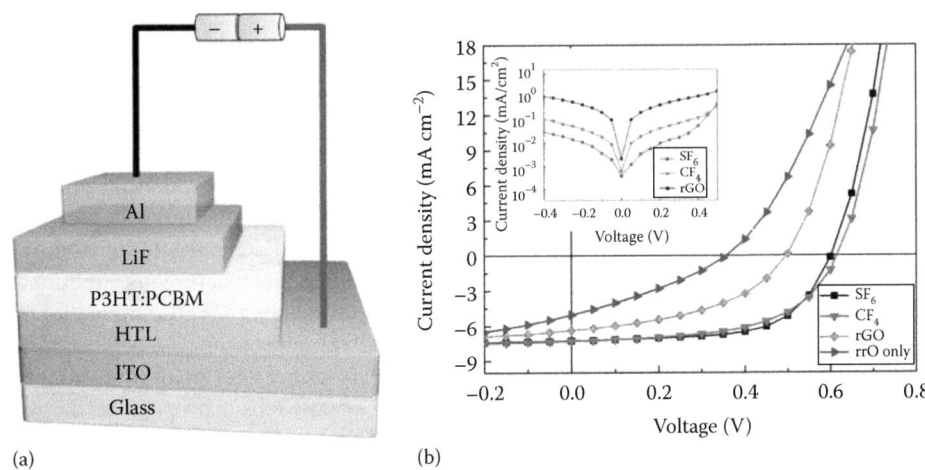

(a)   (b)

FIGURE 34.4
(a) Device structure of the organic solar cell ITO/glass substrate as interfacial layer and (b) J–V curves characteristic based on plasma treated fluorinated rGO. (From Yu, Y.Y. et al., *Appl. Surf. Sci.*, 287, 91, 2013.)

conductive substrate by one-step electrochemical polymerization method. The results obtained indicated that the graphene/PEDOT:PSS composite film has low charge-transfer resistance on the electrolyte/electrode interface. Additionally, the graphene/PEDOT:PSS CE provided a large active surface area enhancing the electrolyte/electrode contact area and offering more electrocatalytic sites at the counter electrode. Under optimal conditions, the DSSC based on graphene/PEDOT:PSS CE showed a high PCE of 7.86%, which is comparable with the performance of the DSSC based on the Pt CE (7.31%).

Yu et al. (2013) investigated the plasma treatment of rGO with sulfur hexafluoride (SF_6) and tetrafluoromethane (CF_4) using reactive ion etching upon the spin-coating of the rGO thin films onto the ITO/glass substrate as interfacial layer in organic solar cell as depicted in Figure 34.4a. The functionalization of fluorine compounds induced dipole on the plasma treated rGO thin films based on the existence of C–F covalent bonds with fluorine on the surface of rGO that formed due to the difference in electronegativity between carbon and fluorine. The work function has been increased by the plasma-treated rGO films hence contributed to a better performance in the solar cells. This functionalization route increased the electrostatic potential energy and, consequently, the PCE of the organic solar cells increased more than the rGO films without plasma treatment. As illustrated in the J–V curves in Figure 34.4b, the cells with SF_6 and CF_4 showed V_{OC} of 0.60 and 0.61 V, J_{SC} of 7.27 and 7.29 mA cm^{-2}, fill factor of 62.33% and 56.36%, and PCE of 2.72% and 2.52%, respectively. These fluorine gas plasma treated rGO films increased J_{SC}, which in turn resulted in a 48% increase in efficiency compared to non-functionalized rGO thin films.

34.2.3 Sensor

Bio and chemical sensor research has been taking one of the leading directions in recent research trends. New frontiers targeting the development of highly specific receptor molecules and novel transducer platforms have been established for their huge potentials in bringing sensor technology to the next level (Fam et al., 2011). Carbon nanomaterials with inherent nanoscale features have potential for becoming attractive components for the next generation of autonomous sensor technology. Their high surface-to-volume ratio is also a definite asset toward the development of biosensing platforms for single-molecule detection (Vashist et al., 2011). This intrinsic coupling of electrical properties and mechanical deformation in these carbon nanomaterials makes them ideal candidates to combine adaptive and sensory capabilities to cater for the increasing needs for simple, sensitive and stable electronic sensors suited for trace detection in a wide spectrum of applications ranging from lab-on-a-chip and in vivo biosensors to environmental monitoring and warfare agent detection (Li et al., 2008). Among the carbon nanomaterials, CNTs stand out as the most extensively studied material as a part of the transducer element as well as a functional receptor element in an electronic device for deployment in electronic sensing platforms due to its superior chemical and electronic properties. On the other hand, the recent efforts undertaken on GFNs have also opened opportunities for using them as sensor for their ultra sensitive sensor response with the lowest detection capability approaching even a single molecule, as well as ultra fast speed and long-term durability (Basu and Bhattacharyya, 2012). The surface of these carbon nanostructures is functionalized for mainly two purposes: First, to solubilize/isolate them and second to render them biocompatible or functional for a specific application. When an analyte approaches a CNT based sensor it interacts with the molecules surrounding the CNT. This direct chemical environment of a nanoscaled object is called the corona. It determines important sensor properties like toxicity or selectivity.

34.2.3.1 Biosensor

Numerous advantages of carbon nanomaterials as electrode materials have been visualized for the assays of diversified chemicals of food quality, clinical, and environmental interest. Carbon nanomaterials have been proven as promising materials for the development of advanced biosensors with improved detection sensitivity and selectivity.

Glucose biosensor has been the most widely studied because of its importance in food and fermentation analysis, in medical diagnosis, and in environmental monitoring. Glucose biosensor based on electrocatalytic oxidation of reduced nicotinamide adenine dinucleotide phosphate (NADPH) generated by the enzymatic reaction has been fabricated through the immobilization of glucose dehydrogenase (GDH) on the surface of SWNTs (Du et al., 2008). SWCNT functionalized with poly(nile blue A) (PNb) acted as a mediator and an enzyme immobilization matrix to electrocatalyze the oxidation of NADPH at a very low potential (ca. −30 mV vs. SCE) and led to a substantial decrease in the overpotential by about 750 mV compared with the bare glassy carbon electrode. These results suggested that PNb and SWNTs have the synergistic electrocatalytic effects on the oxidation of NADPH. The synergistic effects may probably result from the relatively smaller electron-transfer resistance, higher electroactive surface area, and the special three-dimensional structure of PNb–SWNTs. Additionally, the biosensor also exhibited fast response, broad and useful linear range, reproducibility and selectivity to render the application for determine the concentration of glucose on real sample of clinic. In another study for direct electrochemical detection of glucose, MWCNTs were thiol functionalized using dielectric barrier discharge plasma prior to the decoration of MWCNTs with Au-nanoparticles for the immobilization glucose oxidase enzyme as depicted in Figure 34.5a and b (Vesali-Naseh et al., 2013). The oxidation current shows a linear behavior in the range of 0.4–4.0 mM. The detection limit is estimated to be 3.0 mM (based on the signal to noise ratio = 3) with a relatively high sensitivity of 23.1 μA mM^{-1} cm^2. The high sensitivity of the hybrid electrode was attributed to the good electrocatalytic activity of Au-NPs and high surface area of MWCNTs that have improved the accessibility of glucose to the active sites of the enzyme. During the detection of glucose at −0.55 V, the response of the glucose oxidase immobilized Au/MWCNTs electrode showed that upon addition of 2.0 mM glucose and 0.1 mM amino acid (AA) and uric acid (UA), only a noticeable current for glucose oxidation is observed, while no response could be seen after addition of AA and UA as shown in Figure 34.5c, which implied that AA and UA do not interfere with the detection thus the sensor is quite selective.

A major barrier for construction of graphene-based biosensors is the insolubility of graphene in most solvents owing to the high interlayer attraction energies. Zhang et al. developed a graphene multilayer film modified electrode for glucose biosensing based on layer by layer (LbL) assembly of graphene composites functionalized with copper phthalocyanine-3,4′,4″,4‴-tetrasulfonic acid tetrasodium salt (TSCuPc) or alcian blue (AB) pyridine variant (Zhang et al., 2013b). Significantly enhanced the electrocatalytic activity of the modified electrode toward O$_2$ reduction has been observed owing to the synergistic effect of graphene and copper phthalocyanines, the multilayer film facilitated the electron transfer, enhancing the electrochemical response and exhibiting excellent

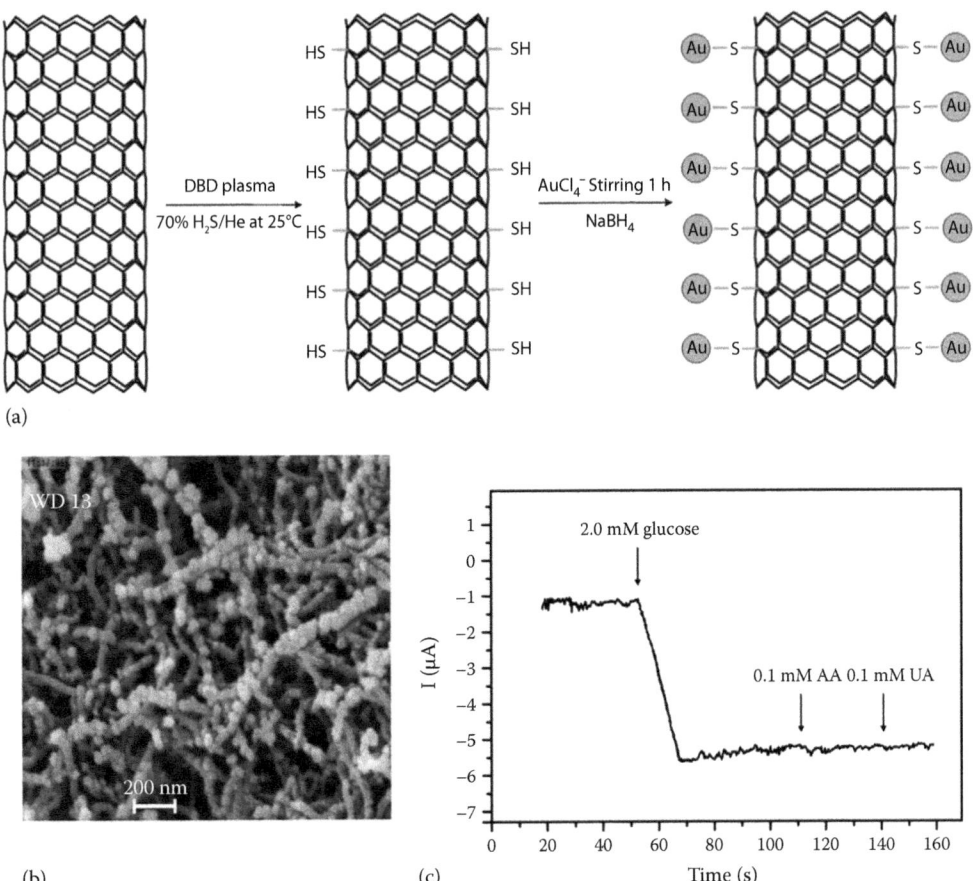

(a)

(b)

(c)

FIGURE 34.5
(a) Schematic of deposition of Au on the thiol (−SH) functionalized MWCNTs, (b) SEM micrograph of the Au/MWCNTs-SH hybrid, and (c) amperometric response of glucose oxidase immobilized hybrid electrode to subsequent additions of 2.0 mM glucose and 0.1 mM UA and AA at a potential of −0.55 V. (From Vesali-Naseh, M. et al., *Sens. Act. B Chem.*, 188, 488, 2013.)

reproducibility, stability, sensitivity, and selectivity. Based on the O_2 consumption during the oxidation process of glucose, the biosensor exhibited a low detection limit of 0.05 mmol L^{-1} with response linear up to 8 mmol L^{-1} glucose concentration.

Hydrogen peroxide (H_2O_2) is one of the by-products of various important oxidase enzymes in countless biological reactions (Vashist et al., 2011). The relationship between H_2O_2 concentration and human health has attracted great attention. Therefore, rapid, accurate and reliable determination of these compounds is of practical importance in food, pharmaceutical, clinical, and environmental analysis. For this purpose, many researches focused on CNTs-enzyme-based biosensor because CNTs can maintain the bioactivity of enzymes and improve the sensitivity of sensors. Liu et al. reported detection of hydrogen peroxide based on cytochrome C (Cyt c) immobilized IL-functionalized MWCNTs glass carbon electrode (Liu et al., 2013b). It is well agreed that, IL-functionalized CNTs exhibit switchable solubility, high charge transfer activity, and high electronic conductivity (Yu et al., 2006). In addition, low interfacial tensions of IL resulted in high nucleation rates, allowing formation of very small particles (Antonietti et al., 2004). In this case, the IL-NH_2 (imidazolium cation) functionalized MWNT with positively charged was in favor of conjugation of negatively charged Cyt c, hence provided an extra ionic affinity between MWNT and Cyt c. Also, the covalent modification of MWCNTs with amine-terminated ILs exhibited a high surface area for the enzyme immobilization and provided a good microenvironment for Cyt c to retain its bioelectrocatalytic activity toward H_2O_2. The biosensor exhibited a wide linear response range nearly four orders of magnitude of H_2O_2 (4.0×10^{-8} M to 1.0×10^{-4} M) with a good linearity (0.9980) and a low detection limit of 1.3×10^{-8} M (based on S/N = 3), while displaying high selectivity, good reproducibility, and long-term stability.

Immobilization of nanotubes with specific recognition biosystems indeed provides ideal miniaturized biosensor (Daniel et al., 2007). Also, the concept of using DNA to direct the assembly of nanotubes into nanoscale devices is attracting attention because of its potential to assemble a multicomponent system in one step by using different base sequence for each component. A H_2O_2 biosensor based on the immobilization of horseradish peroxidase (HRP) on DNA functionalized MWCNTs has been established (Zeng et al., 2009). Under optimized condition, the biosensor exhibited excellent performance in terms of wide linear range of 0.0006–1.8 mM and low limit of detection 0.3 μM. Besides that, the enzyme biosensor also showed excellent selectivity to H_2O_2 in the presence of glucose, ascorbic acid, uric acid, and citric acid. This result indicated that DNA functionalized SWCNTs could provide a favorable microenvironment for HRP to transfer electrons directly with underlying GC electrode. The recovery rate in the range 97%–101% was obtained when tested with disinfector samples, which indicated that the hybrid electrode is feasible to determine H_2O_2 in real sample analysis. In another similar study based on immobilization of HRP for the detection of H_2O_2, a complex film that consist of graphene, CNT, gold-platinum nanoparticles was fabricated prior to the assembling on electrode surface (Sheng et al., 2011). Upon the immobilization of HRP, the deposition of conducting polymer polyaniline (PANI) layer was enzymatically induced on the electrode. It was found that the matrix structure of GE–CNT–Nafion complex film was highly beneficial in maintaining a high surface area on the electrode surface since the sheets could not readily collapse. The resultant electrode is endowed with excellent synergistic electrochemical effects for the reduced process of H_2O_2, good reproducibility, and long-term stability. The electronic conductivity between HRP and the electrode was promoted by the synergistic effect of GE–CNT hybrid materials, AuPt nanoparticles and PANI. Also, the method that the highly efficient polymerization of aniline under the enzymatic reaction improved the sensitivity and detection limit for H_2O_2 determination.

To address the shortcomings of high cost, limited lifetime and the critical operating situation limit of most of the enzyme-biosensors, Li et al. (2013b) constructed a novel nonenzymatic sensor based on electrodepositing of Ag nanoparticles on a glassy carbon electrode modified with IL functionalized CNT composite to detect H_2O_2 with low detection limit and a wide linear range. The feasibility of the device for practical applications was carried out by analyzing the 1% human serum. The concentration of hydrogen peroxide in serum sample was determined to be 15.8 ± 0.029 nM. The performance of the proposed sensor was better than that of some other modified electrodes or comparable to some enzyme-biosensors, particularly at low limit of detection. The excellent performance of H_2O_2 sensor was ascribed to the good conductivity and network like structure of IL-MWCNT to provide more sites for loading of Ag nanoparticles (Li et al., 2013b). Nonenzyme-based biosensor with dual sensing platforms for both H_2O_2 and glucose has also been developed (Guo et al., 2011). To achieve this target, Hemin, a well-known natural metalloporphyrin and the active center of the heme-protein family, has been functionalized on grapheme nanosheet to simultaneously catalyze the reduction of H_2O_2 at the modified electrode but also can catalyze H_2O_2 oxidizing peroxidase substrate. The electrode demonstrated a wide linear response to logarithm concentration of H_2O_2 ranging from 0.05 μM to 0.5 mM and the detection limit is as low as 20 nM meanwhile the linear range of glucose was 0.05 μM to 0.5 mM with the detection limit of 30 nM.

34.2.3.2 Gas sensor

The principle of carbon nanomaterials gas sensors is based on the change in electrical properties induced by transfer of charged between gas molecules and the carbon nanostructure as well as with the attached functional groups (Castro et al., 2009). These carbon nanomaterials change their electrical resistance when they are exposed to different explosive and flammable gases such as H_2, NO_2, H_2S, CO, and NH_3, as well as some toxic chlorinated aliphatic hydrocarbons such as CCl_4 at room temperature. The use of pristine carbon nanomaterials as gas sensor has many drawbacks such as lack of specificity to different gases analytes and low-sensitivity toward analyte that have poor affinity. Furthermore, the sensitivity and selectivity of the pristine carbon nanostructures toward inorganic gases is limited due to the weak interaction and minimum charge transfer between the carbon nanomaterials and gas molecules (Hatchett and Josowicz, 2008). Hence, the drawbacks are normally minimized via the functionalization of carbon nanomaterials with acid, nanoparticles, and polymers. It has been well reported that the polar groups on the nanotube surface increase the adsorption affinity of the electron-donor or acceptor gases and consequently offer better response (Ménard-Moyon et al., 2010).

Dhall et al. (2013) reported the detection of 0.05% H_2 gas at room temperature using acid functionalized multiwalled CNTs. The current carrying capacity of functionalized-MWCNTs is found to increase to 35 mA from 49 µA at low sweep voltage, which implied that the electrical transport properties of the functionalized MWCNTs increases to thousand times as compared with the pristine MWCNTs. This is due to the presence of functional groups to increase the number of bands close to the Femi level, hence resulting in electrons transfer between the carbon atoms. The adsorption mechanism of the H_2 molecules during the sensing is depicted in Figure 34.6a. The increased sensing sensitivity of the carboxylic acid-MWCNTs was due the dissociation of H_2 gas molecules to H atoms on the edges of oxidized/functionalized MWCNTs. These functional groups acted as catalytic sites for the dissociation of H_2 gas molecules. Meanwhile in the reverse process when hydrogen is off, these functional groups again help in the desorption of H_2 gas even in the absence of ambient oxygen. As a result, the recovery time of functionalized MWCNTs sensor decreased to 100 s for 0.05% of H_2 gas as compared to 190 s for the pristine MWCNTs. Additionally, as compared to the pristine MWNCT, the sensor response based on the functionalized MWCNTs was fast and reproducible even at low concentration of H_2 gas as exhibited in Figure 34.6b.

Synergetic effect can be achieved through the proper combination of organic functional group and metal oxide in enhancing the response toward a targeted gas. For this purpose, the functionalization of MWCNTs was carried out by ester hydrolysis of p-aminomethyl benzoate to allow the introduction of benzoic acids groups on the MWCNTs, and then followed by the decoration of nanoparticles of Fe_3O_4 on the functionalized surface (Pistone et al., 2013). In such hybrid system, the walls functionalization of MWCNTs was an important step to introduce end carboxyl groups

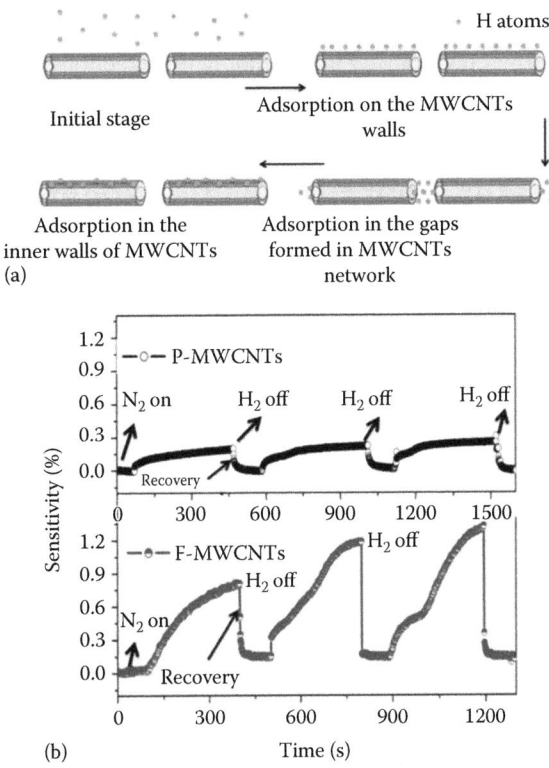

FIGURE 34.6
(a) Schematic representation of the adsorption sites of hydrogen gas molecules during sensing mechanism and (b) sensor sensitivity of ristine MWCNTs and functionalized-MWCNTs film at different concentrations of H_2. (From Dhall, S. et al., *Sens. Act. A Phys.*, 201, 321, 2013.)

for the direct synthesis of metal oxide nanoparticles on the as these oxygen-containing groups could act as nucleation sites for the growth of nanoparticles. During the sensing of ammonia gas, the mechanism likely involved the interaction of ammonia with COOH group, which promoted the extent of ammonia adsorption as well as diffusion into the sensor layer. Next, the adsorbed ammonia could interact via hydrogen bonding with O atoms on the metal oxide surface, leading to higher response of Fe_3O_4/MWCNT PhCOOH hybrid composites. As such, the developed composite sensors have shown much more reversible characteristics, especially a shorter response time, in the presence of UV light that acts accelerating the desorption of adsorbed ammonia on the sensing layer.

Most of the conducting polymers such as PANI exhibit highly reversible redox behavior with a distinguishable chemical memory. Hence, they have been considered as effective materials for fabrication of the chemical sensors. The combination of conducting polymer and CNTs is an attractive route to fabricate a new sensor material having faster response, high sensitivity, and good reproducibility at ambient conditions. A surface-functionalized MWCNT doped PANI sensor has been developed for chloroform sensing (Kar and Choudhury, 2013). The interactions between the surface carboxyl functional groups of CNTs and the functional groups in the polymer chains have promoted the dispersion of the nanotubes in the composites. The nanocomposite sensors exhibited stable baseline and improved responses to chloroform vapor at all concentrations. The carboxyl-functionalized MWCNTs favorably interact with the conjugated PANI chains and create a pathway for strong dipole interaction with the analyte molecules, hence facilitated electron delocalization and charge transport through the polymer chain and consequently decrease the resistance of the exposed nanocomposite sensor. A β-cyclodextrin (CD) functionalized graphene-modified carbon paste electrode was fabricated for the simultaneous detection of toxic chlorophenols isomers: 2-chlorophenols and 3-chlorophenols (Wei et al., 2014). The multifunctional nanocomposite obtained from graphene and β-CD has the advantages of both graphene and β-CD, including the large surface area, high conductivity, and high supramolecular recognition and enrichment capability properties. Under the optimized conditions, the oxidation peak currents displayed a good linear relationship to concentration in the ranges from 0.5 to 40 μM for 2-chlorophenol and 0.4 to 77 μM for 3-chlorophenol, with detection limits of 0.2 and 0.09 μM, respectively.

34.2.4 Wastewater Treatment and Purification

Carbon-based nanotechnologies have found water treatment applications in many aspects, such as sorbents, catalyst, and membranes (Liu et al., 2013c). Owing to their tunable physical, chemical, electrical, and structural properties, they can inspire innovative technologies to address the alarming global issues related to water shortage and water pollution problems. It has been proven that most of these carbon nanomaterials possess potential for the removal of many kinds of pollutants from water because of their ability to establish π–π electrostatic interactions and their large surface areas.

34.2.4.1 Metal ions adsorption

Adsorption is a promising process for the removal of metal ions from water and wastewater because it is a simple, economic, and suitable operation method, and it is significant to exploit simple, novel, and high efficient adsorbents for this approach. For the past decades, CNTs have proven their great potentials to become the promising third generation of carbonaceous adsorbents (Liu et al., 2013c). Theoretically, with their hollow and layered structures, all adsorption sites of CNTs are located on the inner and outer layer surface to facilitate efficient adsorption process. It has been found that surface chemistry is an important factor influencing the CNTs adsorption behavior. Functional groups such as –OH, –C=O and –COOH could be intentionally introduced onto CNT surface to make CNTs more hydrophilic and suitable for the adsorption of relatively low molecular weight and polar contaminants, such as phenol (Lin and Xing, 2008) and 1,2-dichlorobenzene (Peng et al., 2003). Nevertheless, some studies that investigated on the influence of surface oxidation of MWCNT on the adsorption capacity and affinity of organic compounds in water pointed out that surface oxidation of MWCNTs decreased the surface area–normalized adsorption capacity of organic compounds significantly because of the competition of water molecules (Wu et al., 2012). On the other hand, the adsorption affinity of organic chemicals were not greatly influenced as the adsorption interactions, that is, hydrophobic effect, π–π interaction and hydrogen bond remained constant.

Surface functionalization of CNTs is favorable for the uptake of heavy metal ions as the adsorption process depends mainly on the specific complexation between metal ions and the hydrophilic functional groups of CNTs (Rao et al., 2007). For example, it has been reported that (Rosenzweig et al., 2013) the –OH and –COOH functional groups on CNTs play a vital role to provide accessible sites for copper adsorption in which the surface functionalized CNTs exhibited higher adsorption capacity for copper than pristine CNTs. Another significant contribution of surface functionalization of CNTs with hydrophilic groups is the improvement of CNT dispersion in aqueous media. Cho et al. (2010) reported that surface oxidation of CNTs accomplished through nitric acid treatment to enhance the adsorption of zinc and cadmium ions from aqueous solutions. The authors found that both COOH-functionalized carbon surface and graphenic-carbon sites were contributing substantially to the metal ion sorption. Using two-site Langmuir adsorption modeling, they found that the carboxyl-carbon sites of CNTs were over 20 times more energetic for Zn(II) uptake than the unoxidized carbon sites. Besides that, the maximum adsorption capacities obtained for functionalized MWCNTs were found to be slightly higher than those of activated carbon when normalized to surface area.

Due to their high reactivity with many chemical species, it has been suggested that amino groups together with oxygen groups could serve as coordination and electrostatic interaction sites for transition metal sorption (Yu et al., 2007). The possibility of the employment of amino-functionalized MWCNTs as a sorbent for the removal of Cd^{2+} ions from aqueous solutions was examined by Vukovi´c et al. (2010). Direct coupling of ethylenediamine (EDA) with carboxylic groups was performed to introduce amino groups on MWCNT surface. It was found that the functionalization significantly improved the dispersibility of CNTs in water and reduces their cytotoxicity. The adsorptions of Cd^{2+} onto pristine-MWCNT were mainly physisorption processes; meanwhile, both physisorption and chemisorption contributed to the adsorption of Cd^{2+} ions onto the functionalized-MWCNT. It was also found that the metal ion sorption capacity of o-MWCNT was not in direct correlation with their specific surface area, pore specific volume and mean pore diameter but strongly depended on their total surface acidity. In this context, the metal ion sorption capacity of MWCNTs increased with increasing total surface acidity, including carboxyls, lactones, and phenols, present on the surface sites of MWCNTs. Moreover, acidic oxygen–containing groups behave as ion-exchange sites for the retention of Cd^{2+} cations, hence giving rise to the formation of metal ligand surface complexes (Jaramillo et al., 2009). At the surface of amino functionalized-MWCNT, free amino and non-reacted oxygen containing functional groups were present. Hence, additional coordination and electrostatic interactions are possible between Cd^{2+} ions and unprotonated amino groups at pH values higher than 7.

Xu et al. (2011) demonstrated the superior adsorption capability of functionalized CNTs for anionic chromate (CrO_4^{2-}) from aqueous solution, as compared to that of without oxygen containing functional groups. The estimated maximum adsorption capacity per unit surface area on the functionalized CNT was 0.83 μmol m^{-2}, which was about 15 times higher than 0.06 μmol m^{-2} on the CNTs. Based on both the adsorption capacity per unit mass and per unit surface area, the functionalized CNTs exhibited the superior performance for capturing CrO_4^{2-} from aqueous solution. The superior adsorption capability of chromate on the surface oxidized CNTs as compared with the CNTs can be ascribed to the interaction of chromate with the abundant surface oxygen-containing functional groups on the functionalized CNT. The abundant oxygen-containing surface functional groups, particularly the OH groups, has the excellent CrO_4^{2-} adsorption capacity because the adsorption of CrO_4^{2-} occurs mainly via ion exchange with acidic surface OH groups.

GO is found very useful for heavy metal ion adsorption due to the presence of several functional groups on its surface. Nevertheless, functionalization of GO with certain reactive functional groups is found helpful to facilitate considerable affinity towards the metal ions the form of electrostatic interaction. Trioctylamine, a long chain tertiary amine, has been functionalized on GO through lone pair-p interaction as well as hydrogen bonding for the adsorption of chromium (Kumar et al., 2013). At the optimum pH range 2.5–4.0, the adsorption of hexavalent chromium onto the functionalized GO adsorbent was taken place. The hydrogen bonding and electrostatic interaction of $HCrO_4$ with the protonated tertiary amine on GO surface have favored the facile adsorption of chromium(VI) to achieve a high adsorption capacity of 232.55 mg g^{-1}.

Despite the presence of abundant surface functional groups hence high adsorption performance of metal ions shown by GO, its application in water treatment is sometimes limited by the difficulty to be separated from treated water. The functionalization of GO with magnetic organic functional groups with high adsorption capacity and can be easily separated through physical magnetic separation is a promising way to tackle the aforementioned issue. Magnetic β-cyclodextrin–chitosan has been successfully attached to the GO surface through the formation of chemical bond between carboxyl group of GO and amine group of magnetic β-cyclodextrin–chitosan to favorably adsorb and remove of $HCrO^{4-}$ very quickly and efficiently from wastewater at pH below 6.8 (Li et al., 2013a). At low pH, the redox reactions in the aqueous and solid phases were promoted where Cr(VI) was partially reduced to Cr(III) by the reductive surface hydroxyl groups on the functionalized GO. The resulting Cr(III) can be released back into the solution at lower pH in the form of water-soluble Cr(III) species or precipitated on the surface of the CCGO in the form of Cr_2O_3, hence achieving the performance of adsorption chromium. Another major reason that contributed to the enhanced adsorption was the extension of the active surface from 342.3 m^2 g^{-1} for GO to 445.6 m^2 g^{-1} for functionalized GO that allowed more efficient utilization of the respective adsorption sites for the sorption of chromium.

34.2.4.2 Dye compounds adsorption

Textile industries are one of the main sources of water pollution. Wastewater containing dyes present a serious environmental problem because of its high toxicity and possible accumulation in the environment. The conventionally used activated carbon has been widely used as adsorbent for removal of dyes from the aqueous solution, but it presents some disadvantages such as flammable and difficult to regenerate as it needs to be reclaimed (Gao et al., 2013). In this context, functionalized carbon nanomaterials have been considered as good alternatives for extracting the dye compounds from aqueous media discharged from textile industries. The dispersion and versatility of the nanomaterials in aqueous phase can be improved by blending them with organic functional groups or biopolymers in a conjugated system. Functionalized carbon nanomaterials are hydrophilic in nature resulting in the high affinity for the adsorption of cations and anions dye from the aqueous solution due to the presence of oxygen containing functional groups at the surface and hence it is advantageous over activated carbon. Functionalized MWCNTs have been successfully used for the adsorption of three different azoic dyes, namely direct Congo red, reactive green HE4BD, and golden yellow MR dyes. During the adsorption, the mechanisms involved not only the π–π interactions between bulk π-systems on CNT surfaces and organic molecules with C=C double bonds or benzene rings, but also the hydrogen bonds and electrostatic interactions that mainly contributed by the charged functional groups on the MWCNT surfaces. As a result, promising adsorption capacity of 148, 152, and 141 mg g^{-1} was obtained for direct Congo red, reactive green HE4BD, and golden yellow MR dyes, respectively.

Soluble starch-functionalized MWCNT were prepared to improve the hydrophilicity and biocompatibility of MWCNTs where starch acted as a template for the uniform growth of iron oxide nanoparticles to promote efficient adsorption of anionic methyl orange (MO) and cationic methylene blue (MB) dyes (Chang et al., 2011). These magnetic MWCNTs possess the properties of adsorption capacity and magnetic separation and can therefore be used as magnetic adsorbents to remove organic contaminants from aqueous solutions. It was noticed that the hydrophilic property of soluble starch improved the dispersion of MWCNT-starch-iron oxide in the aqueous solution. Besides that, the increased contact surface between magnetic MWCNT and dyes can effectively reduce the aggregates of MWCNTs and facilitated the diffusion of dye molecules to the surface of MWCNTs. The removal of metanil yellow (MY) dye from aqueous solution by amino functionalized graphene was studied by Guo et al. (2013). Three possible mechanisms, including ion exchange, dissolution/precipitation, and surface complexation, have been proposed for the adsorption of MY onto the adsorbent. The mechanisms of the adsorption process of the acid MY dyes on the functionalized graphene are likely to be the ionic interactions of the colored dye ions with the amino groups. The reason for the outstanding adsorption capacity up to 71.62 mg g^{-1} observed at low pH may be attributed to the large number of H$^+$ present at these pH, which protonated the amino groups on the surface of the functionalized graphene.

34.2.4.3 Membrane-based separation

In the recent development, carbon nanomaterials have been potentially used in difference approaches to increase the water treatment and desalination efficiency and capacity through the direct involvement of this material as a free-standing vertically aligned membrane or as a nanofillers incorporated in a polymer membrane. Particularly, functionalized carbon nanomaterials have been identified as a suitable filler to be embedded into a polymer matrix to manufacture polymer nanocomposite membranes granted with rapid water transport for various forms of membrane-based separations ranging from microfiltration, ultrafiltration, nanofiltration to reverse osmosis.

Owing to the several merits: strong antimicrobial activity, higher water flux than other porous materials of comparable size, tunable pore size and surface chemistry, and electrical conductivity, CNTs are promising materials for membrane-based separation. Studies have demonstrated that CNT filters can achieve high water flux at reasonably low pressure (Brady-Estévez et al., 2008). In fact, the extraordinarily fast transport of water in CNTs could be utilized for the production of high-flux nanotube-based filtration membranes, in which the nanotubes serve as a water channel in an impermeable support matrix, hence significantly increase the water fluxes (Joseph and Aluru, 2008). CNT is increasingly investigated, and it has been proposed as a promising option for enhancing the efficiency and capability of currently available desalination system in order to meet the municipal and industrial water supply demands. It has been anticipated that the novel membranes drawing on the unique properties of CNTs may reduce significantly the energy and cost of desalination (Mauter et al., 2008). Chemical functionalization at the entrance to CNT cores affects the selectivity of chemical transport across the membrane structure. Via tip functionalization, CNT membranes can acquire ion exclusion ability. Majumder et al. (2007) modified the tips of CNTs using electrochemical grafting of diazonium salts and suggested the separation coefficient of the CNT membranes can be tuned by changing the voltage apply. Upon the introduction of negatively charged groups by plasma treatment, Fornasiero et al. (2008) performed functionalization of carboxyl groups at the pore entrance of a large diameter CNT array to achieve high salt rejection capability through electrostatic repulsion. With the presence of the functional groups exerts electrostatic repulsion to allow exclusion of salt ions. Salt rejection of nearly 100% has been achieved, particularly for the salt solution with greater anions valence charge such as $K_3Fe(CN)_6$. Similarly, Gong et al. (2010) designed a controllable ion-selective nanopore based on SWCNTs with specially arranged carbonyl oxygen atoms modified inside the nanopores. The different patterns of carbonyl oxygen atoms determined the hydration structure of K$^+$ and Na$^+$ within the nanopores, leading to a tunable ionic selectivity.

It has been pointed out that functionalization of oxygen-containing groups in an appropriate amount was beneficial to the membrane hydrophilicity. Besides that, the increased roughness also served as the main contributor to the increase of efficient filtration area, which can progressively lead to elevation of the water flux. One of the earliest efforts in promoting CNTs as appealing nanofiller for UF membrane was pursued by Choi et al. (2006), who reported the blending of 1.5 wt.% of carboxylated MWCNTs with PSf to enhance the pure water flux of the resultant nanocomposite. De Lannoy et al. (2012) fabricated the nanocomposite UF membrane by pressure filtering a thin layer of PVA crosslinked carboxylated CNT and succinic acid onto a support membrane. The complementary factors contributed by the hydrophilicity from both carboxyl groups of CNTs and hydroxyl-saturated chains of PVA explained the enhanced hydrophilicity of the membrane surface. The desirable porous structure resulted from the rigidity of CNTs that created larger separation distance between the PVA chains has in turn increased the

PWF. Additionally, polyethylene oxide (PEO) rejected as much as 90% can be achieved with a small loading of PVA crosslinked CNTs.

Despite the similar qualitative property as CNTs, the application of graphene in water treatment has not been studied much compared to CNTs. However, recent attentions have been progressively switched to the utilization of graphene for the fabrication of nanocomposite membrane to combat some shortages encountered in CNTs. it has been evidenced that the presence of surface oxidized graphene in polymeric membrane material would induce hydrophilicity, which would facilitate high water permeation and impede fouling owing to the low interfacial energy between a surface and water. In addition, its functional groups would ensure a large negative zeta potential, which could also constrain attachment of bio-foulants and their accumulation on the membrane surface (Krishnan et al., 2008). Zhao et al. (2013) reported the increase in PWF and permeate flux by 79% and 99%, respectively, with the incorporation of 2 wt.% of GO in PVDF UF nanocomposite membrane. The drastically improved water permeation was attributed to the high hydrophilicity due to the presence of abundant oxygen-containing functional groups on the GO surface that attracted water molecules inside the membrane matrix and facilitate the passage of water molecules through the membranes. Furthermore, the adsorption of BSA was largely decreased due to the formation of hydrated layer and steric hindrance on the membrane surface and the slow change of flux ratio indicated better antifouling properties due to the introduction of hydrophilic GO nanosheets.

It is known that by bringing together two nanofillers like CNTs and graphene derivatives, they form a co-supporting network of both fillers like a hybrid net structure to improve some desired properties (Chatterjee et al., 2012). Integration one-dimensional oxidized MWCNTs and two-dimensional GO resulted in a strong synergistic effect between the two materials, consequently leading to a superior ultrafiltration membrane with higher antifouling performance. Zhang et al. (2013a) investigated the permeation and antifouling performance of polyvinylidene fluoride (PVDF) composite membranes incorporated with GO/oxidized-MWCNTs. It was observed that the long and tortuous oxidized MWCNTs that can bridge adjacent GO and inhibit their aggregation. The pure water flux recovery achieved 98.28% for membranes with the ratio of 5:5 (GO/oxidized-MWCNTs), which contributed to the synergistic effect of the hybrid membrane. The excellent antifouling performance of the modified membranes is attributable to the affinity of oxidized low-dimensional carbon nanomaterials with water, creating an energetic barrier to the adsorption of BSA.

34.3 CONCLUDING REMARKS

A broad protocol for covalent and noncovalent functionalization is available for the facile modification of carbon nanomaterials. The feasibility and versatility of functionalized carbon nanomaterials in various applications have been evidenced in the last decade. However, more attention should be given to understand the cost-effectiveness of some of the functionalization procedures as well as the long-term stability of the materials under practical operations. The use of CNTs, coupled with well-defined covalent or noncovalent functionalization, might offer tremendous properties for a wide spectrum of applications. However, the progress in this area is still very limited, owing partly due to its high production cost for large-scale applications. However, it is not really possible yet to grow these structures in a controlled way. In this context, graphene and their derivatives could be a formidable competitor of CNTs since the former also exhibits several unique characteristics, including being easy and relatively inexpensive to make (Vashist et al., 2011). One of the major challenges for the application of CNTs, particularly in clinical and environmental, is the concern of their potential long-term toxicity (Yang and Liu, 2012). CNTs without appropriate surface functionalized have been found to be toxic in vivo, inducing a wide range of toxicity to animals (Poland et al., 2008). There is a need for standardizing the terminology, the fabrication of GFNs, and the validation of toxicological methodologies. Standardization will provide necessary information to researchers for better understanding the physicochemical characteristics and the potential toxicological effects in cells and animals, thus facilitating the practical applications of these promising new nanomaterials in humans (Guo and Mei, 2014).

The recent advances and significant development of carbon nanomaterials in biomedical has opened exciting opportunities for the future and broad use of nanomaterials in real clinical conditions (Liu et al., 2013a). For biomedical application, stability is an important concern in the development of any safe, effective, and clinically meaningful formulation. Despite the rapid development of carbon nanomaterials in the past decade, a comprehensive understanding of the interaction of these nanostructures with living systems and their adverse effects in vitro and in vivo are essential for further development and safe use. An agreement has been gradually reached that the functionalization of pristine graphene and GO significantly improves their biocompatibility, and this step is essential for designing stable and safe drug delivery nanocarriers. In this regard, functionalized nanostructures show greater stability in terms of minimum chances for leaking of drug molecules as compared to liposomal as well as other nanoformulations (Mehra et al., 2014). Intriguingly, despite discrepancies in findings on the clearance mechanism of nanotubes, majority of the studies have suggested that functionalized carbon nanomaterials, when intravenously injected into animals such as mice or rats are prone to accumulate in the RES, for example, liver and spleen, and then gradually excreted, possibly via both fecal and renal routes (Ruggiero et al., 2010). In addition, highly functionalized CNTs with well-known biocompatible moieties (such as PEG) have demonstrated reduced in vivo toxicity after being intravenously injected into animals as compared to their raw, unfunctionalized counterparts (Schipper et al., 2008). However, more toxicity studies using in vivo animal models are required in future to prove their biocompatibility.

The design and fabrication of electrochemical biosensors using carbon nanomaterials remains a vibrant research area. However, due to the fact that the studies on these biosensors have been mainly conducted at the laboratory level, there is still a long journey waiting for scientists to realize the application of these biosensors in real clinical cases. Also, for commercial viability carbon-based sensors must be manufactured with high consistency and cost-effectiveness so that they can compete with other commercially available products. One of the main obstacles is the exorbitant cost that is beyond the means of a small enterprise to introduce a new device to compete with existing technologies. Surface modification of carbon nanomaterials is often an effective way to fully take advantage of their unique physical, chemical, and electrical properties. Despite the euphoria, when they are applied for water treatment purposes, the potential threat of these carbon nanomaterials to the environment and human health should be taken into consideration before large-scale applications. Therefore, before using CNTs in real water treatment devices, the potential risk of CNT leaking should be evaluated prudently since CNTs have been shown to exhibit cytotoxicity (Liu et al., 2013b).

At the current stage, it is still difficult to draw definite conclusions about the successful implementations of carbon nanomaterials in practical application. However, advanced and innovative strategies will find themselves better positioned to combat challenges and boost the current development to stimulate more innovative improvement in both academia and industry with increasing competitive intensity. Though it would be still early to make an absolute impact, it seems certain that the efforts pursued will be critical in determining the future position of carbon nanomaterials for industry applications.

REFERENCES

Abarrategi, A., Gutiérrez, M. C., Moreno-Vicente, C. et al. 2008. Multiwall carbon nanotube scaffolds for tissue engineering purposes. *Biomaterials* 29: 94–102.

Ahmed, M. S., Jeon, S. 2012. New functionalized graphene sheets for enhanced oxygen reduction as metal-free cathode electrocatalysts. *Journal of Power Sources* 218: 168–173.

Ahn, D., Xiao, X., Li, Y. 2012. Applying functionalized carbon nanotubes to enhance electrochemical performances of tin oxide composite electrodes for Li-ion battery. *Journal of Power Sources* 212: 66–72.

Antonietti, M., Kuang, D., Smarsly, B., Zhou, Y. 2004. Ionic liquids for the convenient synthesis of functional nanoparticles and other inorganic nanostructures. *Angewandte Chemie International Edition* 43: 4988–4992.

Asgari, M. S., Nikazar, M., Molla-abbasi, P., Hasani-Sadrabadi, M. M. 2013. Nafion® / histidine functionalized carbon nanotube: High-performance fuel cell membranes. *International Journal of Hydrogen Energy* 38: 5894–5902.

Bak, B. M., Kim, S.-K., Park, H. S. 2014. Binder-free, self-standing films of iron oxide nanoparticles deposited on ionic liquid functionalized carbon nanotubes for lithium-ion battery anodes. *Materials Chemistry and Physics* 144: 396–401.

Balasubramanian, K., Burghard, M. 2005. Chemically functionalized carbon nanotubes, *Small* 1: 180–192.

Basu, S., Bhattacharyya, P. 2012. Recent developments on graphene and graphene oxide based solid state gas sensors. *Sensors and Actuators B: Chemical* 173: 1–21.

Brady-Estévez, A. S., Kang, S., Elimelech, M. 2008. A singlewalled-carbon-nanotube filter for removal of viral and bacterial pathogens. *Small* 4: 481–484.

Candelaria, S. L., Shao, Y., Zhou, W. et al. 2012. Nanostructured carbon for energy storage and conversion. *Nano Energy* 1: 195–220.

Castro, B. I. R., Contes, E. J., Colon, M. L., Meador, M. A., Pomales, G. S., Cabrer, C. R. 2009. Combined electron microscopy and spectroscopy characterization of as-received, acid purified and oxidized HiPCO single-wall carbon nanotubes. *Material Character* 60: 1442–1453.

Chang, P. R., Zheng, P., Liu, B., Anderson, D. P., Yu, J., Ma, X. 2011. Characterization of magnetic soluble starch-functionalized carbon nanotubes and its application for the adsorption of the dyes. *Journal of Hazardous Materials* 186: 2144–2150.

Chatterjee, S., Nafezarefi, F., Tai, N. H., Schlagenhauf, L., Nueesch, F. A., Chu, B. T. T. 2012. Size and synergy effects of nanofiller hybrids including graphene nanoplatelets and carbon nanotubes in mechanical properties of epoxy composites. *Carbon* 50: 5380–5386.

Chen, B. A., Liu, M., Zhang, L. M. et al. 2011. Polyethylenimine-functionalized graphene oxide as an efficient gene delivery vector. *Journal of Material Chemistry* 21: 7736–7741.

Cho, H. H., Wepasnick, K., Smith, B. A., Bangash, F. K., Fairbrother, D. H., Ball, W. P. 2010. Sorption of aqueous Zn(II) and Cd(II) by multiwall carbon nanotubes: The relative roles of oxygen-containing functional groups and graphenic carbon. *Langmuir* 26: 967–981.

Choi, H.-J., Jung, S.-M., Seo, J.-M., Chang, D. W., Dai, L., Baek, J.-B. 2012. Graphene for energy conversion and storage in fuel cells and supercapacitors. *Nano Energy* 1: 534–551.

Choi, J. H., Jegal, J., Kim, W.-N. 2006. Fabrication and characterization of multi-walled carbon nanotubes/polymer blend membranes. *Journal of Membrane Science* 284: 406–415.

Čolić, M., Džopalić, T., Tomić, S. et al. 2014. Immunomodulatory effects of carbon nanotubes functionalized with a Toll-like receptor 7 agonist on human dendritic cells. *Carbon* 67: 273–287.

Daniel, S., Rao, T. P., Rao, K. S. et al. 2007. A review of DNA functionalized/grafted carbon nanotubes and their characterization. *Sensors and Actuators B: Chemical* 122: 672–682.

De Lannoy, C. F., Jassby, D., Davis, D. D., Wiesner, M. R. 2012. A highly electrically conductive polymer-multiwalled carbon nanotube nanocomposite membrane. *Journal of Membrane Science* 415–416: 718–724.

De Morais, A., Loiola, L. M. D., Benedetti, J. E. 2013. Enhancing in the performance of dye-sensitized solar cells by the incorporation of functionalized multi-walled carbon nanotubes into TiO_2 films: The role of MWCNT addition. *Journal of Photochemistry and Photobiology A: Chemistry* 251: 78–84.

Dennler, G., Scharber, M. C., Brabec, C. J. 2009. Polymer-fullerene bulk-heterojunction solar cells. *Advanced Materials* 21: 1323–1338.

Dhakate, S. R., Mathur, R. B., Kakati, B. K., Dhami, T. L. 2007. Properties of graphite-composite bipolar plate prepared by compression molding technique for PEM fuel cell. *International Journal of Hydrogen Energy* 32: 4537–4543.

Dhall, S., Jaggi, N., Nathawat, R. 2013. Functionalized multiwalled carbon nanotubes based hydrogen gas sensor. *Sensors and Actuators A: Physical* 201: 321–327.

Dikin, D. A., Stankovich, S., Zimney, E. J. 2007. Preparation and characterization of graphene oxide paper. *Nature* 448: 457–460.

Du, P., Wu, P., Cai, C. 2008. A glucose biosensor based on electrocatalytic oxidation of NADPH at single-walled carbon nanotubes functionalized with poly(nile blue A). *Journal of Electroanalytical Chemistry* 624: 21–26.

Fabbro, C., Ali-Boucetta, H., Da Ros, T. 2012. Targeting carbon nanotubes against cancer. *Chemical Communications* (*Cambridge*) 48: 3911–3926.

Fam, D. W. H., Palaniappan, Al., Tok, A. I. Y., Liedberg, B., Moochhala, S. M. 2011. A review on technological aspects influencing commercialization of carbon nanotube sensors. *Sensors and Actuators B: Chemical* 157: 1–7.

Fan, Z., Wang, J., Wang, Z. et al. 2014. One-pot synthesis of graphene/hydroxyapatite nanorod composite for tissue engineering. *Carbon* 66: 407–416.

Feng, L., Wu, L., Qu, X. 2013. New horizons for diagnostics and therapeutic applications of graphene and graphene oxide. *Advanced Materials* 25: 168–86.

Fornasiero, F., Park, H. G., Holt, J. K. et al. 2008. Ion exclusion by sub-2-nm carbon nanotube pores. *Proceedings of the National Academy of Sciences of the United States of America* 105: 17250–17255.

Gao, H., Zhao, S., Cheng, X., Wang, X., Zheng, L. 2013. Removal of anionic azo dyes from aqueous solution using magnetic polymer multi-wall carbon nanotube nanocomposite as adsorbent. *Chemical Engineering Journal* 223: 84–90.

Goenka, S., Sant, V., Sant, S. 2014. Graphene-based nanomaterials for drug delivery and tissue engineering. *Journal of Controlled Release* 173: 75–88.

Gong, P., Du, F., Xia, Z. H., Durstock, M., Dai, L. 2009. Nitrogen-doped carbon nanotube arrays with high electrocatalytic activity for oxygen reduction. *Science* 323: 760–764.

Gong, X. J., Li, J. C., Xu, K., Wang, J. F., Yang, H. 2010. A controllable molecular sieve for Na^+ and K^+ ions. *Journal of the American Chemical Society* 132 (6): 1873–1877.

Guerra, J., Herrero, M. A., Carrión, B. et al. 2012. Carbon nanohorns functionalized with polyamidoamine dendrimers as efficient biocarrier materials for gene therapy. *Carbon* 50: 2832–2844.

Guo, X. Mei, N. 2014. Assessment of the toxic potential of graphene family nanomaterials. *Journal of Food and Drug Analysis* 22: 105–115.

Guo, X., Wei, Q., Du, B. et al. 2013. Removal of Metanil Yellow from water environment by amino functionalized graphenes (NH2-G)—Influence of surface chemistry of NH2-G. *Applied Surface Science* 284: 862–869.

Guo, Y., Li, J., Dong, S. 2011. Hemin functionalized graphene nanosheets-based dual biosensor platforms for hydrogen peroxide and glucose. *Sensors and Actuators B: Chemical* 160: 295–300.

Harrison, B. S., Atala, A. 2007. Carbon nanotube applications for tissue engineering. *Biomaterials* 28: 344–353.

Hatchett, D. W., Josowicz, M. 2008. Composites of intrinsically conducting polymers as sensing nanomaterials, *Chemical Reviews* 108: 746–769.

Hecht, D. S., Hu, L. B., Irvin, G. 2011. Emerging transparent electrodes based on thin films of carbon nanotubes, graphene, and metallic nanostructures. *Advanced Materials* 23: 1482–1513.

Heister, E., Neves, V., Tilmaciu, C. 2009. Triple functionalisation of single-walled carbon nanotubes with doxorubicin, a monoclonal antibody, and a fluorescent marker for targeted cancer therapy. *Carbon* 47: 2152–2160.

Hirata, E., Uo, M., Takita, H., Akasaka, T., Watari, F., Yokoyama, A. 2011. Multiwalled carbon nanotube-coating of 3D collagen scaffolds for bone tissue engineering. *Carbon* 49: 3284–3291.

Huang, H., Yuan, Q., Shah, J. S., Misra, R. D. K. 2011. A new family of folate-decorated and carbon nanotubes-mediated drug delivery system: Synthesis and drug delivery response. *Advanced Drug Delivery Reviews* 63: 1332–1339.

Hussain, T., De Sarkar, A., Ahuja, R. 2014. Functionalization of hydrogenated graphene by polylithiated species for efficient hydrogen storage. *International Journal of Hydrogen Energy* 39: 2560–2566.

Jaramillo, J., Gómez-Serrano, V., Álvareza, P. M. 2009. Enhanced adsorption of metal ions onto functionalized granular activated carbons prepared from cherry stones. *Journal of Hazardous Materials* 161: 670–676.

Ji, Z., Lin, G., Lu, Q. et al. 2012. Targeted therapy of SMMC-7721 liver cancer in vitro and in vivo with carbon nanotubes based drug delivery system. *Journal of Colloid and Interface Science* 365: 143–149.

Joseph, S., Aluru, N. R. 2008. Why are carbon nanotubes fast transporters of water? *Nano Letters* 8: 452–458.

Kar, P., Choudhury, A. 2013. Carboxylic acid functionalized multi-walled carbon nanotube doped polyaniline for chloroform sensors. *Sensors and Actuators B: Chemical* 183: 25–33.

Khabashesku, V. N., Margrave, J. L., Barrera, E. V. 2005. Functionalized carbon nanotubes and nanodiamonds for engineering and biomedical applications. *Diamond and Related Materials* 14: 859–866.

Kostarelos, K., Bianco, A., Prato, M. 2009. Promises, facts and challenges for carbon nanotubes in imaging and therapeutics. *Nature Nanotechnology* 4: 627–633.

Kou, R., Shao, Y., Wang, D. 2009. Enhanced activity and stability of Pt catalysts on functionalized grapheme sheets for electrocatalytic oxygen reduction. *Electrochemistry Communications* 11 (5): 954–957.

Krishnan, S., Weinman, C. J., Ober, C. K. 2008. Advances in polymers for anti-biofouling surfaces. *Journal of Material Chemistry* 18: 3405–3413.

Ku, S. H., Park, C. B. 2013. Myoblast differentiation on graphene oxide. *Biomaterials* 34: 2017–2023.

Kucinskis, G., Bajars, G., Kleperis, J. 2013. Graphene in lithium ion battery cathode materials: A review. *Journal of Power Sources* 240: 66–79.

Kumar, A. S. K., Kakan, S. S., Rajesh, N. 2013. A novel amine impregnated graphene oxide adsorbent for the removal of hexavalent chromium. *Chemical Engineering Journal* 230: 328–337.

Li, C., Thostenson, E. T., Chou, T.-W. 2008. Sensors and actuators based on carbon nanotubes and their composites: A review. *Composites Science and Technology* 68: 1227–1249.

Li, G., Liu, L. 2012. Organic solar cells enhanced by carbon nanotubes, Chapter 11. *Nano Optoelectronic Sensors and Devices*, N. Xi and K.W.C. Lai (eds.), pp. 183–198. Elsevier Inc.

Li, L., Fan, L., Sun, M. et al. 2013a. Adsorbent for chromium removal based on graphene oxide functionalized with magnetic cyclodextrin–chitosan. *Colloids and Surfaces B: Biointerfaces* 107: 76–83.

Li, X., Liu, Y., Zheng, L. 2013b. A novel nonenzymatic hydrogen peroxide sensor based on silver nanoparticles and ionic liquid functionalized multiwalled carbon nanotube composite modified electrode. *Electrochimica Acta* 113: 170–175.

Liao, S.-H., Weng, C.-C., Yen, C.-Y. 2010. Preparation and properties of functionalized multiwalled carbon nanotubes/polypropylene nanocomposite bipolar plates for polymer electrolyte membrane fuel cells. *Journal of Power Sources* 195: 263–270.

Lin, D. H., Xing, B. S. 2008. Tannic acid adsorption and its role for stabilizing carbon nanotube suspensions. *Environmental Science and Technology* 42: 5917–5923.

Liu, J., Cui, L., Losic, D. 2013a. Graphene and graphene oxide as new nanocarriers for drug delivery applications. *Acta Biomaterialia* 9: 9243–9257.

Liu, K., Zhang, J.-J., Cheng, F.-F., Zheng, T.-T., Wang, C., Zhu, J.-J. 2011a. Green and facile synthesis of highly biocompatible graphene nanosheets and its application for cellular imaging and drug delivery. *Journal of Material Chemistry* 21: 12034–12040.

Liu, X., Bu, C., Nan, Z., Zheng, L., Qiu, Y., Lu, X. 2013b. Enzymes immobilized on amine-terminated ionic liquid-functionalized carbon nanotube for hydrogen peroxide determination. *Talanta* 105: 63–68.

Liu, X., Wang, M., Zhang, S., Pan, B. 2013c. Application potential of carbon nanotubes in water treatment: A review. *Journal of Environmental Sciences* 25: 1263–1280.

Liu, Z., Robinson, J. T., Sun, X., Dai, H. 2008. PEGylated nanographene oxide for delivery of water-insoluble cancer drugs. *Journal of American Chemical Society* 130: 10876–10877.

Liu, Z., Robinson, J. T., Tabakman, S. M. et al. 2011b. Carbon materials for drug delivery and cancer therapy. *Materials Today* 14: 316–323.

Liu, Z., Tabakman, S., Welsher, K., Dai, H. 2009. Carbon nanotubes in biology and medicine: In vitro and in vivo detection, imaging and drug delivery. *Nano Research* 2: 85–120.

Lu, B., Li, T., Zhao, H. et al. 2012. Graphene-based composite materials beneficial to wound healing. *Nanoscale* 4: 2978–2982.

Majumder, M., Zhan, X., Andrews, R., Hinds, B. J. 2007. Voltage gated carbon nanotube membranes. *Langmuir* 23: 8624–8631.

Mauter, M. S., Elimelech, M. 2008. Environmental applications of carbon-based nanomaterials. *Environmental Science and Technology* 42: 5843–5859.

McDevitt, M. R., Chattopadhyay, D., Kappel, B. J. et al. 2007. Tumor targeting with antibody-functionalized radiolabeled carbon nanotubes. *Journal of Nuclear Medicine* 48: 1180–1189.

Mehra, N. K., Mishra, V., Jain, N. K. 2014. A review of ligand tethered surface engineered carbon nanotubes. *Biomaterials* 35: 1267–1283.

Ménard-Moyon, C., Kostarelos, K., Prato, M. et al. 2010. Functionalized carbon nanotubes for probing and modulating molecular functions. *Chemistry and Biology* 2010: 107–115.

Mendes, R. G., Bachmatiuk, A., Büchner, B. et al. 2013. Carbon nanostructures as multi-functional drug delivery platforms. *Journal of Material Chemistry B* 1: 401–428.

Meng, L., Zhang, X., Lu, Q., Fei, Z., Dyson, P. J. 2012. Single walled carbon nanotubes as drug delivery vehicles: Targeting doxorubicin to tumors. *Biomaterials* 33: 1689–1698.

Muller, A., Kauranen, P., Ganski, A. V., Hell, B. 2006. Injection moulding of graphite composite bipolar plates. *Journal of Power Sources* 154: 467–471.

Nagai, H., Okazaki, Y., Chew, S. H. et al. 2011. Diameter and rigidity of multiwalled carbon nanotubes are critical factors in mesothelial injury and carcinogenesis. *Proceedings of the National Academy of Sciences of the United States of America* 108: E1330–E1338.

Palomaki, J., Karisola, P., Pylkkanen, L., Savolainen, K., Alenius, H. 2010. Engineered nanomaterials cause cytotoxicity and activation on mouse antigen presenting cells. *Toxicology* 267: 125–131.

Pan, L., Pei, X., He, R., Wan, Q., Wang, J. 2012. Multiwall carbon nanotubes/polycaprolactone composites for bone tissue engineering application. *Colloids and Surfaces B: Biointerfaces* 93: 226–234.

Peng, X. J., Li, Y. H., Luan, Z. K. et al. 2003. Adsorption of 1,2-dichlorobenzene from water to carbon nanotubes. *Chemical Physics Letters* 376: 154–158.

Pistone, A., Piperno, A., Iannazzo, D. et al. 2013. Fe_3O_4–MWCNT single bond PhCOOH composites for ammonia resistive sensors. *Sensors and Actuators B: Chemical* 186: 333–342.

Poland, C. A., Duffin, R., Kinloch, I. et al. 2008. Carbon nanotubes introduced into the abdominal cavity of mice show asbestos-like pathogenicity in a pilot study. *Nature Nanotechnology* 3: 423–428.

Premkumar, T., Geckeler, K. E. 2012. Graphene–DNA hybrid materials: Assembly, applications, and prospects. *Progress in Polymer Science* 37: 515–529.

Pumera, M., Ambrosi, A., Bonanni, A. et al. 2009. Graphene for electrochemical sensing and biosensing. *Nanoscale* 1: 260–265.

Rao, G. P., Lu, C., Su, F. S. 2007. Sorption of divalent metal ions from aqueous solution by carbon nanotubes: A review. *Separation and Purification Technology* 58: 224–231.

Reddy Channu, V. S., Ravichandran, D., Rambabu, B., Holze, R. 2014. Carbon and functionalized graphene oxide coated vanadium oxide electrodes for lithium ion batteries. *Applied Surface Science* 305: 596–602.

Rosenzweig, S., Sorial, G. A., Sahle-Demessie, E., Mack, J. 2013. Effect of acid and alcohol network forces within functionalized multiwall carbon nanotubes bundles on adsorption of copper(II) species. *Chemosphere* 90: 395–402.

Ruggiero, A., Villa, C. H., Bander, E. 2010. Paradoxical glomerular filtration of carbon nanotubes. *Proceedings of the National Academy of Sciences of the United States of America* 107: 12369–12374.

Schipper, M. L., Nakayama-Ratchford, N., Davis, C. R. et al. 2008. A pilot toxicology study of single-walled carbon nanotubes in a small sample of mice. *Nature Nanotechnology* 3: 216–221.

Scida, K., Stege, P. W., Haby, G. 2011. Recent applications of carbon-based nanomaterials in analytical chemistry: Critical review. *Analytica Chimica Acta* 691: 6–17.

Sheng, Q., Wang, M., Zheng, J. 2011. A novel hydrogen peroxide biosensor based on enzymatically induced deposition of polyaniline on the functionalized graphene–carbon nanotube hybrid materials. *Sensors and Actuators B: Chemical* 160: 1070–1077.

Silambarasan, D., Vasu, V., Iyakutti, K., Surya, V. J., Ravindran, T. R. 2014. Reversible hydrogen storage in functionalized single-walled carbon nanotubes. *Physica E: Low-Dimensional Systems and Nanostructures* 60: 75–79.

Singh, R,. Pantarotto, D., Lacerda, L. et al. 2006. Tissue biodistribution and blood clearance rates of intravenously administered carbon nanotube radiotracers. *Proceedings of the National Academy of Sciences of the United States of America* 103: 3357–3362.

Speltini, A., Merli, D., Profumo, A. 2013. Analytical application of carbon nanotubes, fullerenes and nanodiamonds in nanomaterials-based chromatographic stationary phases: A review. *Analytica Chimica Acta* 783: 1–16.

Su, D. S., Centi, G. 2013. A perspective on carbon materials for future energy application. *Journal of Energy Chemistry* 22: 151–173.

Sun, Y. Q., Wu, Q. O., Shi, G. Q. 2011. Graphene based new energy materials. *Energy and Environmental Science* 4: 1113–1132.

Surya, V. J., Iyakutti, K., Venkataramanan, N. S., Mizuseki, H., Kawazoe, Y. 2011. Single walled carbon nanotubes functionalized with hydrides as potential hydrogen storage media: A survey of intermolecular interactions. *Physica Status Solidi B* 248: 2147–2158.

Vashist, S. K., Zheng, D., Al-Rubeaan, K., John Luong, H. T. Sheu, F.-S. 2011. Advances in carbon nanotube based electrochemical sensors for bioanalytical applications. *Biotechnology Advances* 29: 169–188.

Venkatesan, J., Ryu, B. M., Sudha, P. N., Kim, S.-K. 2012. Preparation and characterization of chitosan–carbon nanotube scaffolds for bone tissue engineering. *International Journal of Biological Macromolecules* 50: 393–402.

Vesali-Naseh, M., Mortazavi, Y., Khodadadi, A. A., Parsaeian, P., Moosavi-Movahedi, A. A. 2013. Plasma thiol-functionalized carbon nanotubes decorated with gold nanoparticles for glucose biosensor. *Sensors and Actuators B: Chemical* 188: 488–495.

Vitchev, R., Malesevic, A., Petrov, R. H. et al. 2010. Initial stages of few-layer graphene growth by microwave plasma-enhanced chemical vapour deposition. *Nanotechnolgy* 21: 095602.

Vuković, G. D., Marinković, A. D., Čolić, M. et al. 2010. Removal of cadmium from aqueous solutions by oxidized and ethylenediamine-functionalized multi-walled carbon nanotube. *Chemical Engineering Journal* 157: 238–248.

Wang, J., Sun, R. H., Zhang, N. et al. 2009. Multi-walled carbon nanotubes do not impair immune functions of dendritic cells. *Carbon* 47 (7): 1752–1760.

Wang, S. F., Shen, L., Zhang, W.-D., Tong, Y.-J. 2005. Preparation and mechanical properties of chitosan, carbon nanotubes composites. *Biomacromolecules* 6: 3067–3072.

Wang, Y., Lee, W. C., Manga, K. K. et al. 2012. Fluorinated graphene for promoting neuro-induction of stem cells. *Advanced Materials* 24: 4285–4290.

Wang, Y., Li, Z., Wang, J. et al. 2011. Graphene and graphene oxide: Biofunctionalization and applications in biotechnology. *Trends in Biotechnology* 29: 205–212.

Wang, Z., Zhou, C., Xia, J. et al. 2013. Fabrication and characterization of a triple functionalization of graphene oxide with Fe_3O_4, folic acid and doxorubicin as dual-targeted drug nanocarrier. *Colloids Surfaces B: Biointerfaces* 106: 60–65.

Wei, M., Tian, D., Liu, S., Zheng, X., Duan, S., Zhou, C. 2014. β-Cyclodextrin functionalized graphene material: A novel electrochemical sensor for simultaneous determination of 2-chlorophenol and 3-chlorophenol. *Sensors and Actuators B: Chemical* 195: 452–458.

Wu, H.-C., Chang, X., Liu, L., Zhao, F., Zhao, Y. 2010. Chemistry of carbon nanotubes in biomedical applications. *Journal of Material Chemistry* 20: 1036–1052.

Wu, W. H., Chen, W., Lin, D. H., Yang, K. 2012. Influence of surface oxidation of multiwalled carbon nanotubes on the adsorption affinity and capacity of polar and nonpolar organic compounds in aqueous phase. *Environmental Science and Technology* 46: 5446–5454.

Xiao, Y., Gao, X., Taratula, O. et al. 2009. Anti-HER2IgY antibody-functionalized single-walled carbon nanotubes for detection and selective destruction of breast cancer cells. *BMC Cancer* 9: 351–353.

Xu, Y. J., Rosa, A., Liu, X. et al. 2011. Characterization and use of functionalized carbon nanotubes for the adsorption of heavy metal anions. *New Carbon Materials* 26: 57–62.

Xue, Y., Bao, L., Xiao, X. et al. 2011. Noncovalent functionalization of carbon nanotubes with lectin for label-free dynamic monitoring of cell-surface glycan expression. *Analytical Biochemistry* 410: 92–97.

Yan, J., Ni, T., Zou, F. 2014. Towards optimization of functionalized single-walled carbon nanotubes adhering with poly(3-hexylthiophene) for highly efficient polymer solar cells. *Diamond and Related Materials* 41: 79–83.

Yang, K., Feng, L., Shi, X. et al. 2013. Nano-graphene in biomedicine: Theranostic applications. *Chemical Society Review* 42: 530–547.

Yang, K., Liu, Z. 2012. In vivo biodistribution, pharmacokinetics, and toxicology of carbon nanotubes. *Current Drug Metabolism* 13: 1057–1067.

Yang, W. R., Ratinac, K. R., Ringer, S. P. et al. 2010. Carbon nanomaterials in biosensors: Should you use nanotubes or graphene. *Angewandte Chemie International Edition* 49: 2114–2138.

Yang, X. Y., Zhang, X. Y., Liu, Z. F. 2008. High-efficiency loading and controlled release of doxorubicin hydrochloride on graphene oxide. *Journal of Physical Chemistry C* 112: 17554–17558.

Yu, B., Zhou, F., Liu, G., Liang, Y., Huck, W. T. S., Liu, W. 2006. The electrolyte switchable solubility of multi-walled carbon nanotube/ionic liquid (MWCNT/IL) hybrids. *Chemical Communications* 22: 2356–2358.

Yu, J., Tong, M., Sun, X., Li, B. 2007. Cystine-modified biomass for Cd(II) and Pb(II) biosorption. *Journal of Hazardous Materials* 143: 277–284.

Yu, Y.-Y., Kang, B. H., Lee, Y. D., Lee, S. B., Ju, B.-K. 2013. Effect of fluorine plasma treatment with chemically reduced graphene oxide thin films as hole transport layer in organic solar cells. *Applied Surface Science* 287: 91–96.

Yue, G., Wu, J., Xiao, Y. 2013. Functionalized graphene/poly(3,4-ethylenedioxythiophene): polystyrenesulfonate as counter electrode catalyst for dye-sensitized solar cells. *Energy* 54: 315–321.

Yürüm, Y., Taralp, A., Nejat Veziroglu, T. 2009. Storage of hydrogen in nanostructured carbon materials. *International Journal of Hydrogen Energy* 34: 3784–3798.

Zeng, X., Li, X., Liu, X. et al. 2009. A third-generation hydrogen peroxide biosensor based on horseradish peroxidase immobilized on DNA functionalized carbon nanotubes. *Biosensors and Bioelectronics* 25: 896–900.

Zhang, J., Xu, Z., Shan, M. et al. 2013a. Synergetic effects of oxidized carbon nanotubes and graphene oxide on fouling control and anti-fouling mechanism of polyvinylidene fluoride ultrafiltration membranes. *Journal of Membrane Science* 448: 81–92.

Zhang, L., Xia, J., Zhao, Q. et al. 2010a. Functional graphene oxide as a nanocarrier for controlled loading and targeted delivery of mixed anticancer drugs. *Small* 6: 537–544.

Zhang, L. M., Lu, Z. X., Zhao, Q. H. et al. 2011. Enhanced chemotherapy efficacy by sequential delivery of siRNA and anticancer drugs using PEI-grafted graphene oxide. *Small* 7: 460–464.

Zhang, X., Meng, L., Lu, Q., Fei, Z., Dyson, P. J. 2009. Targeted delivery and controlled release of doxorubicin to cancer cells using modified single wall carbon nanotubes. *Biomaterials* 30: 6041–6047.

Zhang, Y., Bai, Y., Yan, B. 2010b. Functionalized carbon nanotubes for potential medicinal applications. *Drug Discovery Today* 15: 428–435.

Zhang, Y.-Q., Fan, Y.-J., Cheng, L. 2013b. A novel glucose biosensor based on the immobilization of glucose oxidase on layer-by-layer assembly film of copper phthalocyanine functionalized graphene. *Electrochimica Acta* 104: 178–184.

Zhao, C., Xu, X., Chen, J., Yang, F. 2013. Effect of graphene oxide concentration on the morphologies and antifouling properties of PVDF ultrafiltration membranes. *Journal of Environmental Chemical Engineering* 1: 349–354.

CHAPTER 35

CONTENTS

Quantum Chemical Prospective of Open-Shell Carbon Nanomaterials for Nonlinear Optical Applications

35

Shabbir Muhammad and Masayoshi Nakano

35.1 INTRODUCTION

In contemporary science, nonlinear optical (NLO) material has gotten tremendous importance because of its several novel applications in hi-tech optoelectronic and photonic devices (Papadopoulos et al., 2006; Muhammad and Nakano, 2013). Conventionally, NLO materials were designed using some basic design rules to guide synthetic explorations. The modern synthetic chemistry in coordination with reliable computational techniques has resulted in many interesting syntheses of several chromophores. There are many types of materials that have explored for NLO applications including organic (Nalwa and Miyata, 1996; Muhammad et al., 2009a), inorganic (Kong et al., 2006), organometallic (Lin et al., 1999), and organic–inorganic hybrid materials (Judeinstein and Sanchez, 1996; Muhammad et al., 2012, 2013c). Among them, organic compounds have been extensively studied (theoretically and experimentally) for NLO applications because of their large NLO properties, ease of fabrication, and relatively low cost (Kawata et al., 2001; Terenziani et al., 2008). In particular, the third-order NLO property (γ) of organic π-conjugated materials has been studied in order to build new photonic devices such as optical switches and 3D memories. The γ is a molecular response at the origin of third-order NLO property. Several structure–NLO property relationships have been studied in organic π-conjugated materials to tune the amplitude of γ, for example, the π-conjugated length, charge states, and changing the strength of donor/acceptor groups (De Melo and Silbey, 1987; De Melo and Fonseca, 1996; Yamada et al., 1997; Nakano et al., 2005). Recently, it has been observed that the open-shell diradical character (Nakano et al., 2002, 2007) and spin multiplicity (Nakano et al., 2004) have significant influence on the amplitude of molecular third-order NLO response (γ). Along this line, we have deduced several structure–property relationships to spotlight this novel functionalization strategy for several function materials. The open-shell intermediate diradical character showed a remarkable effect on the γ amplitudes in a variety of materials ranging from a simple H_2 molecule dissociation model (Nakano et al., 2006b) to *p*-quinodimethane (Nakano et al., 2005), benzene compounds (Nakano et al., 2006a), transition metal systems with metal–metal multiple bonds (Fukui et al., 2011), polyaromatic diphenalenyl diradicaloids (Nakano et al., 2006c; Ohta et al., 2007), graphene nanoflakes (Nagai et al., 2010), and fullerene systems (Muhammad et al., 2013a). The field of NLO material design got a further momentum with the discovery of carbon nanomaterials ranging form graphene (2D) to carbon nanotubes (CNTs) (1D) and fullerenes (0D) as shown in Figure 35.1. Interestingly, due to the extensive π-conjugation, these carbon nanomaterials show excellent NLO properties.

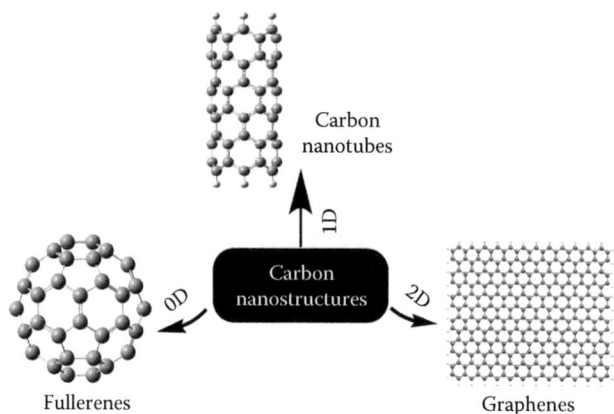

FIGURE 35.1
Representation of three different forms of carbon nanomaterials.

There has been realized several strategies to incorporate and to functionalize carbon nanomaterials for NLO applications. The graphenes and CNTs achieve the open-shell character due to the localization of α and β spins on zigzag edges (Yoneda et al., 2011), while in fullerenes, the spin density are usually localized on terminal edges with highly constrain bonding parameters (Muhammad et al., 2013a). These magnetizations in carbon nanomaterials have been reported even in the absence of transition metal complexes. The proposal about the origin of open-shell character from the zigzag edges of graphenes and CNTs is very important especially for their functionalization. The zigzag edges can have two types of CNTs and graphenes, that is, nonpassivated or hydrogen-passivated (Son et al., 2006). A flat band at the Fermi level is present in both cases of edges, which cause an open-shell ground state with nonzero magnetic moments at the edges. An antiferromagnetic (AFM) coupling can be observed in mutually opposite edges of nanographenes and CNTs. Furthermore, under the strong electric field, the spin-polarized edges of nanographenes and CNTs result in half-metallicity, which is a useful property for the practical application of graphenes and CNTs in spintronic devices (Son et al., 2006). Although the open-shell character for graphenes, CNTs, and fullerenes have extensively studied (Cervantes-Sodi et al., 2008), the research in their open-shell character with reference to their diradical character is still rare and is limited to few nanomaterial systems. In the present chapter, we will spotlight not only the origin of open-shell diradical character in carbon nanomaterials but also check the influence on the third-order NLO properties. An overview of different design strategies for functionalization of nanomaterials will be presented based on several previous reports of magnetic properties of carbon nanomaterials.

35.2 QUANTUM CHEMICAL METHODOLOGY

35.2.1 Diradical Character

The diradical character represents the instability of a chemical bond (Yamaguchi, 1975; Yamaguchi et al., 1990, 1999), and for multiradical systems, there are usually more than one nonzero (y_i) values. Herein, using the LC-UBLYP/6-31G* method, the diradical character y_i, which is associated with the highest occupied natural orbital (HONO)-i and the lowest unoccupied natural orbital (LUNO) + i (where i = 0, 1,...), is defined by the occupation numbers (n_i) of the natural orbitals (NOs) that has been calculated by

$$y_i = n_{\text{LUNO}+i} = 2 - n_{\text{HONO}-i}. \tag{35.1}$$

Here, y_i yields a value varying from 0 to 1 for closed-shell to pure diradical, respectively. So, in spin-restricted (R) single-determinant treatments like RHF or RDFT, the diradical characters that should be all 1 or 0 can take fractional values when using spin-unrestricted (U) schemes (UHF or UDFT). The wave function can be expanded in the form of a limited configuration interaction when considering the different orbitals for different spin, and it leads to fractional occupation numbers other than 0, 1, or 2. For example, in the H_2 molecule stretching model, increasing internuclear distance increases its LUNO occupation numbers from 0 to 1, and relates to the more mixing of the bonding (HONO) and antibonding (LUNO) orbitals in the wave function of singlet ground state (Nakano et al., 2005, 2006b). As the spin contamination effect is smaller in UDFT than UHF, we evaluate spin-projected diradical characters with UDFT methods. According to the MC-SCF theory, the diradical character is defined as twice the weight of the doubly excited configuration in the singlet ground state, and in spin-unrestricted single determinant schemes including UHF and UDFT, the diradical character is formally expressed by Equation 35.1 as given earlier. In our several previous reports, the diradical character has been discussed and explained in connection with the odd electron number and density (Nakano et al., 2011; Yoneda et al., 2012). The diradical character is not observable but provides an index of chemical bond. These measures are useful for obtaining intuitive and pictorial descriptions of the open-shell characters and of their impact on various response properties (Nakano et al., 2007; Kamada et al., 2010).

35.2.2 Third-Order Nonlinear Optical Property (γ)

The light propagating across a macroscopic material interacts with the electromagnetic field of the molecules in the material and induces a polarization. The polarization (P) of the material at macroscopic level can be expressed as

$$P = \chi^{(1)}E + \chi^{(2)}E \cdot E + \chi^{(3)}E \cdot E \cdot E + \cdots, \tag{35.2}$$

where
  E is the applied electronic field intensity
  the coefficient $\chi^{(n)}$ is the nth-order polarizability of the macroscopic material

Generally, a higher order polarizability is much smaller than the lower one, which means $\chi^{(1)} \gg \chi^{(2)} \gg \chi^{(3)}$. When a weak electronic field interacts with a material, the electric polarization of the material is directly proportional to the initial electric field intensity, that is,

$$P = \chi^{(1)}E. \tag{35.3}$$

However, the principles of traditional linear optics cannot explain many newly discovered phenomena of laser. Therefore, when a strong electronic field interacts with a material, the nonlinear coefficients $\chi^{(2)}$ and $\chi^{(3)}$ are large enough and should be considered. For microscopic molecules, the relationship between the molecular electric polarization (P') and the electric field intensity E is expressed as

$$P' = \alpha_{ij}E_j + \beta_{ijk}E_j \cdot E_k + \gamma_{ijkl}E_j \cdot E_k \cdot E_l + \cdots, \tag{35.4}$$

where
  α_{ij} is the ij component of the polarizability tensor (α)
  β_{ijk} is the ijk component of the molecular first hyperpolarizability tensor (β)
  γ_{ijkl} is the $ijkl$ component of the molecular second hyperpolarizability tensor (γ)

Generally, the macroscopic polarization (P) of the material is the statistical average of the microscopic polarization (P'). For quantum chemistry investigations, an important measure to design and exploit a high-performance NLO material is the calculation of the microscopic polarization nonlinear coefficient β or γ of the molecules. The longitudinal electronic γ values (γ_{zzzz}) of these systems are calculated by adopting the finite-field (FF) approach (Cohen and Roothaan, 1965), which consists of the fourth-order differentiation of the energy (E) with respect to the electric field (F).

The perturbation series expansion convention (usually called B convention) is chosen for defining γ. We adopt the following numerical differentiation procedure:

$$\gamma = \{E(3F) - 12E(2F)39E(F) - 56E(0)$$

$$+ 39E(-E) - 12E(-2F) + E(-3F)\}/36F^4. \tag{35.5}$$

Here, $E(F)$ corresponds to the total energy in the presence of field F applied in longitudinal direction of z-axis.

35.3 PERTURBATION THEORY AND SECOND HYPERPOLARIZABILITY

According to the perturbation theory, the summation-over-states (SOS) expression for γ is related to the excitation energies (E_{n0}), transition moments between the states (μ_{mn}), and dipole moments differences between the ground and excitation states ($\mu_{nn} - \mu_{00}$) of a molecule (Morrell and Albrecht, 1979; Willetts et al., 1992).

$$\gamma_{ijkl} = \frac{1}{6} P(i,j,k,l) \left\{ \begin{array}{c} \displaystyle\sum_{n=1} \frac{\mu_{n0}^i \Delta\mu_{nn}^j \Delta\mu_{nn}^k \mu_{n0}^l}{E_{n0}^3} - \sum_{n,m} \frac{\mu_{n0}^i \mu_{n0}^j \mu_{m0}^k \mu_{m0}^l}{E_{n0} E_{m0}^2} + \sum_{\substack{n,m \\ (m \neq n)}} \frac{\mu_{0n}^i \Delta\mu_{nn}^j \mu_{nm}^k \mu_{m0}^l}{E_{n0} E_{m0}} \\ + \sum_{\substack{n,n',m \\ (m \neq n, n' \neq m)}} \frac{\mu_{0n}^i \mu_{nm}^j \mu_{mn'}^k \mu_{n'0}^l}{E_{n0} E_{m0} E_{n'0}} \end{array} \right\} \tag{35.6}$$

where
  subscript "0" represents the ground state
  μ_{n0} is transition moment between the ground state (0) and the nth excited state (n)
  $\Delta\mu_{nn}$ is the difference of dipole moment between the ground state (0) and the nth excited state (n)
  E_{n0} is defined as the transition energy ($E_{n0}, E_n - E_0$)
  $P(i,j,k,l)$ is a permutation operator

The excitation energies are found in the denominators of the SOS expression, while the transition moments and dipole moment differences appear in the numerators. The above expression indicates that small excitation energies and large transition moments as well as large dipole moment differences are required for large third-order NLO properties. On the basis of these guiding principles as well as empirical chemical structural dependences of these quantities, various classes of NLO compounds have so far been designed.

Physically, the introduction of a $\Delta\mu$ term means that as the electrons interact with the oscillating electric field of light, they prefer to shift from one direction relative to the other. Accordingly, molecules for second-order NLO applications were based upon unsymmetrical π-electron systems end capped with electron donor and accepter groups to impart the electronic bias (Marder et al., 1991). Another important factor is ΔE, which is associated with the size of π-conjugation systems or the number of π-electrons (Muhammad et al., 2009b, 2013b; Zhong et al., 2012). The delocalization of electrons in the molecular frame enhances the optical nonlinearity and this has been one of the most successful strategies to improve nonlinearity. In this respect, carbon nanomaterial with extensive π-conjugation can play a vital role to tune the NLO properties.

35.4 CATEGORIES OF CARBON NANOMATERIALS

Carbon atom is known for many years to have the ability to generate a variety of chemical structures with good chemical and physical properties because of its flexible bonding patterns with other atoms. The ability of carbon to make nanomaterials of all carbon atoms, especially those containing networks of sp²-hybridized C=C double bonds, has a key role in the current widespread application of carbon nanomaterials in nanotechnology. Carbon nanomaterials do not exhibit the instabilities of other nanomaterials due to the high activation energy for their structural deformation. Carbon nanomaterial has many types ranging from zero-dimensional fullerenes (Kroto et al., 1985) to one-dimensional CNTs (Iijima, 1991) as well as two-dimensional graphene sheets (Novoselov et al., 2004) as represented in Figure 35.1. The discovery of fullerene in 1985 by Kroto et al. and the subsequent finding of CNTs by Iijima in 1991 were two milestones in history of carbon nanomaterials that have stimulated intense research efforts about nanostructures in general and nanocarbon structures in particular. In carbon nanomaterials, the sp² hybridization of carbon leads to a strong covalent bonding and abundant delocalized π-electrons. In the present prospect, based on earlier stated structure–NLO property relationship, we briefly discuss the different electro-optical and structure–property aspects of carbon nanomaterials.

35.4.1 Open-Shell Nanographenes

The graphene is the basic structure of carbon nanomaterials. Geim et al. have obtained the first monolayer of graphene carbon atoms, that is, a quasi-two-dimensional (2D) material (Geim and Novoselov, 2007). Graphenes have attracted much scientific attention due to their excellent physical properties. Graphenes have also open up new horizons for novel functional material designing due to their unique spin and electronic properties. Most of the theoretical investigations started from acenes and later on included larger molecules like nanographenes. Recently, the field of graphenes got a further momentum from the study of their open-shell diradical character (Sun and Wu, 2012). In early studies, it has been noted that olegocenes prefered triplet states onwards from nonacene. A similar investigation carried out by Houk et al. (2001) at the B3LYP level of theory defined polyacenes as triplet ground-state two parrallel ribbons of polyacetylene with nonvashing bandgap. Unlike the previous studies, Bendikov et al. (2004) found that the wave function of the RB3LYP method become unstable for oligocenes larger than hexacenes. They have found a stable singlet ground state with diradical character for oligocenes larger than hexacenes using the broken symmetry (UB3LYP) approach. The advancement of nanographenes received great advantage from oligocene studies due to the resemblance of their edges. The nanographenes have been prepared by different experimental schemes like top-down and bottom-up schemes (Zhi and Müllen, 2008). A major issue of graphenes is their low stability because of their diradical character. Most of the nanographenes are short lived and have high reactivity, which perhaps hamper their practicle application in advance technology. Theoretical and experimental studies have discovered many strategies to resolve the low stability issue, for example, attaching bulky groups near to radical site, delocalization of diradical over π-conjugated fragments, Clar's sextet formation, etc. Recently, the relatively small size nanographenes have been successfully prepared by bottom-up scheme (Konishi et al., 2013). For example, polycyclic aromatic hydrocarbon with two types of edges, that is, zigzag and armchair, have been recently prepared as shown in Figure 35.2. It is important to investigate the effect of edge type on electronic properties of such nanographenes. In this regard, we have made some pioneering theoretical investigations to check the effect of edge type on open-shell character

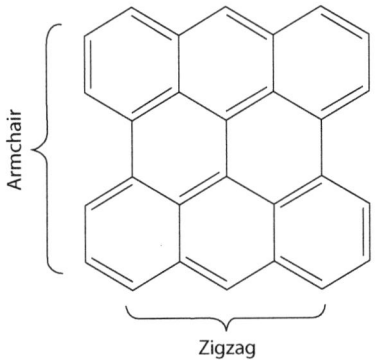

FIGURE 35.2
A short PAH [3,3] graphene layer with zigzag and armchair edges.

TABLE 35.1 The Diradical Character (y_0) and First Hyperpolarizability [$\times 10^3$ a. u.] of Some Nanographenes

Type of Nanographene	Spin State	y_0	γ_L	γ_{xxxx}	References
PAH [3,3]	Singlet	0.510	144.5	34.1	Nakano et al. (2008)
	Triplet	—	35.0	34.8	
PAH [5,5]	Singlet	0.989	405.4	499.5	Nakano et al. (2008)
	Triplet	—	350.5	311.8	
Trigonal	Singlet	0.996	21.4	—	Yoneda et al. (2011)
	Triplet	—	21.0	—	
Rhombic	Singlet	0.418	452.0	—	Yoneda et al. (2011)
	Triplet	—	115.0	—	
Bow tie	Singlet	0.970	199.0	—	Yoneda et al. (2011)
	Triplet	—	191.0	—	

and second hyperpolarizability of different shapes of nanographenes. For example, in two different studies, it has been elucidated that the open-shell diradical character originates from the spin polarization distributions around the zigzag edges. For nanographenes with intermediate diradical character, the diagonal γ_L component along the armchair edges (γ_L) is four times enhanced as compared to γ_{xxxx} component along the zigzag edges, whereas these γ_L components are of similar amplitudes when the system displays nearly pure diradical character (Nakano et al., 2008).

These results are in-line with the diradical character dependence of γ_L predicted in our previous studies on other types of chemical systems. Similar results have been obtained on extended PAHs where multiradical character beyond diradical character arises due to the extra extension of [X,Y] PAHs with $X,Y \geq 7$ singlet systems (Nagai et al., 2010). It has been found that multiradical character also contributes to further exaltation of second hyperpolarizability besides diradical character. Furthermore, in case of three different types of graphene nanoflakes including trigonal, rhombic, and bow-tie shapes, the rhombic illustrated an intermediate diradical character, while trigonal and bow-tie shapes presented pure diradical characters (Yoneda et al., 2011). These results show that the diradical character for GNFs depends not only on the edge shapes, that is, zigzag or armchair, and their π-conjugation lengths, but also on their geometry or shape. It is also important to note here that the triplet states have lower γ values than their corresponding singlet states. For example, γ_L of PAH [3,3] in singlet state is about five times larger than that of triplet ground state (see Table 35.1). A similar trend can also be seen in rhombic graphene where singlet state has larger amplitude of γ_L than its triplet state. Thus, it can be concluded that GNFs constitute a promising class of open-shell NLO materials because of the possibility to control γ by adjusting their shape, the nature of the edge and spin state in addition to the π-conjugation size. Previously, we have used the valence configuration interaction (VCI) method with a simple two-site model A•−B• with two electrons in two orbitals to trace the origin of γ amplitude of symmetric diradical systems with diradical character y.

This two-level model involves three singlet states (g [the ground state], $e1$ [the first excited state], and $e2$ [the second excited state]). There are two main contributions to the static electronic γ expression of symmetric molecules including type II (negative) and type III-2 (positive) terms composed of the products of transition moments ($\mu_{i,j}$) in the numerator and those of excitation energies ($E_{i,g}$) in the denominator as

$$\gamma = \gamma^{II} + \gamma^{III-2} = 4\left[-\frac{(\mu_{g,e1})^4}{E_{e1,g}^3} + \frac{(\mu_{g,e1})^2(\mu_{e1,e2})^2}{(E_{e1,g})^2 E_{e2,g}} \right] \tag{35.7}$$

From the VCI results, it is found that the ground state (g) involves the covalent (diradical) and ionic components with an equal weight at $y = 0$, and as y increases, the weight of diradical component relatively increases, while in the second excited state ($e2$), which also involves these components with an equal weight at $y = 0$ and correlates with the ground state, the ionic component relatively increases. In contrast, the first excited state ($e1$) remains a pure ionic state during the variation of y. These features of each state lead to the variations in excitation energies and transition properties as y increases: (1) the transition moment amplitude between $e2$ and $e1$ ($|\mu_{e1,e2}|$) increases, while that

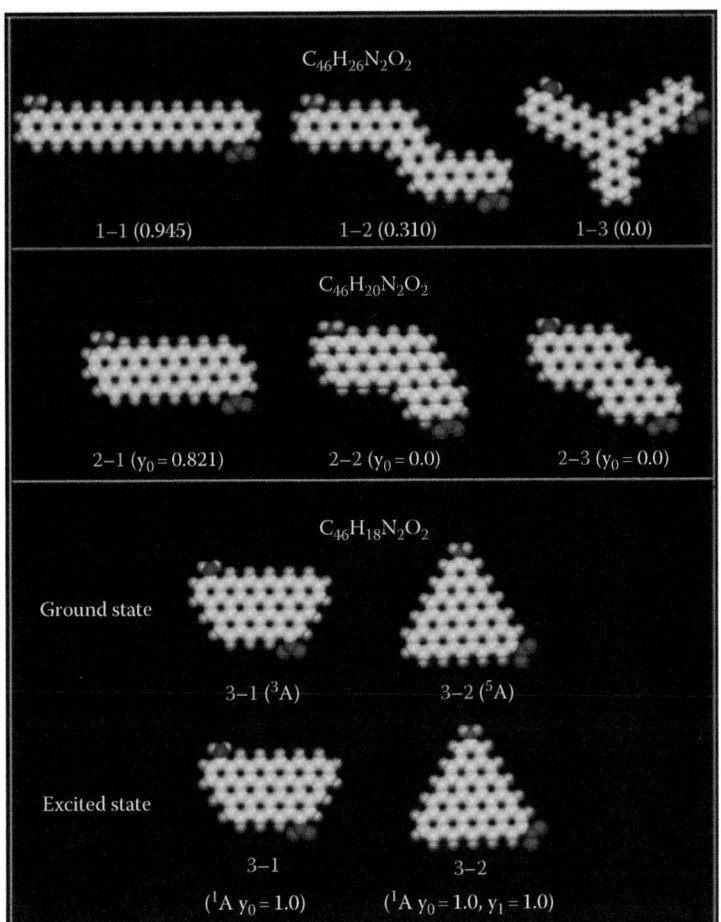

FIGURE 35.3
Structures of disubstituted GQDs. The light green, darkblue, red, and pink balls denote C, N, O, and H atoms, respectively. (Reprinted with permission from Zhou, Z.-J., Liu, Z.-B., Li, Z.-R., Huang, X.-R., and Sun, C.-C., Shape effect of graphene quantum dots on enhancing second-order nonlinear optical response and spin multiplicity in NH_2–GQD–NO_2 systems, *J. Phys. Chem. C*, 115, 2011, 16282. Copyright 2011 American Chemical Society.)

between $e1$ and g ($|\mu_{g,e1}|$) decreases toward zero value at $y = 1$ and (2) the excitation energies of states $e2$ ($E_{e2,g}$) and $e1$ ($E_{e1,g}$) decrease and coincide with each other at the limit of $y = 1$. These variations with increasing y values turns out to cause a bell-shaped variation of γ for y, that is, γ attains a maximum at an intermediate y value, which is dominated by that of type III-2 term in Equation 35.7.

For the first hyperpolarizability, graphenes have been also extensively studied but rarely from the viewpoint of open-shell diradical character and first hyperpolarizability. Recently, Zhou et al. (2011) in this regard have carried out an important study on ultrashort (US) nanographenes or graphene quantum dots (GQDs). They have investigated the shape effect on the first hyperpolarizability of donor-quantum dots-acceptor configurations. A variety of GQDs with different shapes have studied at the spin-unrestricted B3LYP/6-31G* level of theory (see Figure 35.3). In GQDs, the diradical character has been calculated using the spin-projected UHF method proposed by previous studies of Nakano et al. (2008). In UHF theory, the diradical is formally expressed as (Yamaguchi, 1975)

$$y_i = 1 - \frac{2T_i}{1 + T_i^2}. \tag{35.8}$$

Here, T_i represents the orbital overlap between the corresponding orbital pairs and T_i can be expressed in terms of occupation numbers of UHF neutral orbitals:

$$T_i = \frac{(n_{\text{HOMO}-i} - n_{\text{LUMO}+i})}{2}. \tag{35.9}$$

The diradical character GQDs have been obtained using UHF occupation number and the range of diradical character varies from 0 to 1 representing closed-shell and pure singlet diradical character. It has been found that GQDs with same number of carbon atoms (46 carbon atoms) showed different amplitudes of first hyperpolarizability as well as also different spin multiplicities that is useful in designing strategy for tuning first hyperpolarizability and spin properties of these nanosystems. It has also been found that low spin GQDs possess larger amplitudes of first hyperpolarizability.

35.4.2 Open-Shell Fullerenes

Fullerenes are challenging compounds in quantum computational chemistry because of their enormously large sizes, which are difficult to be fully optimized with modern *ab initio* method. Several empty fullerenes have been found unstable due the presence of open-shell diradical character. There are two main reasons, that is, electronic structure and geometrical strains, for this instability or diradical nature of fullerenes (Kovalenko and Khamatgalimov, 2003). Fullerenes having different open-shell diradical characters can be potential candidates for efficient NLO molecules. Both the π-character and substantial σ-character has been found in curved π-conjugations of C_{60} within unique sp^2-hybridized carbon atoms, which show that fullerenes are remarkably different from the planar π-conjugations within graphite and planar polycyclic aromatic hydrocarbons with only π-character (Kovalenko and Khamatgalimov, 2003). Although several fullerenes and its derivatives have been studied for NLO application, unfortunately, open-shell character of fullerenes has been hardly exploited because the most popular C_{60} and C_{70} fullerenes are closed-shell systems. The first fullerene that is found to be open shell was C_{26} due to its highly constrain structure. Similarly, Kovalenko et al. have reported the open-shell structure of C_{74} fullerene and concluded that an empty structure of C_{74} fullerene is unstable because of its biradical character with the presence of two phenalenyl-radical substructures symmetrically disposed in C_{74} cage. On the other hand, the addition of two electrons changes the open-shell configuration of C_{74} cage into closed shell and increased its stability (Kovalenko and Khamatgalimov, 2003). Very recently, we have carried out a detailed investigation about several fullerene systems. (Muhammad et al., 2013a) This study was based on some previously reported fullerenes including fullerenes C_{20}, C_{26}, C_{30}, C_{36}, C_{40}, C_{42}, C_{48}, C_{60}, and C_{70} (Weber et al., 2005; Sackers et al., 2006; Song et al., 2010). We have used spin-unrestricted (U) B3LYP/6-31G*//LC-UBLYP/6-31G* level to clarify the interplay between second hyperpolarizability γ and diradical character. A novel structure–property relationship has been proposed between the third-order NLO polarizability (second hyperpolarizability γ) and diradical character, which is a chemical index of bond nature of open-shell singlet systems. Unlike the previous experimental studies that are about their synthesis and characterization, we have studied their open-shell diradical character. The above-mentioned fullerene systems have been categorized into three groups depending on their diradical character. The first group includes closed-shell fullerenes with diradical character $y_0 \cong 0$. The second group includes the fullerenes that have an intermediate diradical with $0 < y_0 < 1$, while the third group includes nearly pure diradical ($y_0 \cong 1$) fullerenes. The fullerenes C_{20}, C_{60}, and C_{70} are categorized as closed-shell systems with $y_0 \cong 0$. The second category includes open-shell fullerenes with intermediate diradical characters, that is, C_{30}, C_{40}, C_{42}, and C_{48} fullerenes. The third group includes fullerenes having the highest diradical characters consisting of C_{36} ($y_0 = 0.923$) and C_{26} ($y_0 = 0.827$) as shown in Table 35.2.

The spin and odd electron densities (Takatsuka et al., 1978), resonance forms on the Schlegel projections of different fullerenes in the light of Clar's sextet rule, and bond length alternation parameters have been calculated to trace the origin of di-/multi-radical characters of the open-shell fullerenes. Interestingly, it can be seen from Table 35.2 that the γ_{zzzz} values are significantly larger for fullerenes with intermediate diradical characters than for closed-shell and pure open-shell systems, regardless of their different π-conjugation sizes. These findings are in accordance with our previous results on the VCI two-site diradical model (Nakano et al., 2005, 2007).

35.4.3 Open-Shell Carbon Nanotubes

There is a growing interest in CNTs due to their physical, chemical, and medical applications. The two main strategies are mainly used for functionalization of nanotubes, that is, covalent and noncovalent functionalization (Balasubramanian and Burghard, 2005). However, these functionalization

TABLE 35.2 Diradical Character (y_0), Polarizability (α_{zz}), and Second Hyperpolarizability (γ_{zzzz}) [$\times 10^3$ a.u.] of All Examined Fullerene Systems

Fullerene Systems	y_0	α_{zz} [a.u.]	γ_{zzzz} [a.u.]
Fullerene C_{20}	0.078	151	2.619
Fullerene C_{26}	0.826	181	2.911
Fullerene C_{30}	0.435	278	17.34
Fullerene C_{36}	0.923	267	3.917
Fullerene C_{40}	0.681	456	43.44
Fullerene C_{42}	0.611	413	11.05
Fullerene C_{48}	0.701	563	31.43
Fullerene C_{60}	0.000	451	3.695
Fullerene C_{70}	0.000	589	9.723

techniques are far difficult to carry out because of extreme complications in the preparation and separation of native CNTs (Shim et al., 2002). These complications insisted the scientists to use computational approaches to answer several questions about CNTs. Sheka et al. have used the spin-unrestricted Hartree–Fock (UHF) SCF methodology that corresponds to the state with the definite value of the spin projection S_z to investigate the chemical susceptibility of CNTs (Sheka and Chernozatonskii, 2007, 2010). Two classes of CNTs including armchair and zigzag fragments have been studied by the above-mentioned approach. A variety of CNTs has been considered in this study including short and long nanotubes as well as caped and free-end nanotubes. The curvature effect on the magnetism of US zigzag CNTs have shown that for both in finite-length nanotubes and nanographenes, the gradual increase in curvature leads to a magnetic transition from AFM coupling to ferromagnetic coupling between the two edges (Yang et al., 2010). The study indicated the possibility of tuning the magnetization including magnetic coupling as well as magnetic moment in CNTs and graphenes by controlling the curvature. The differences of magnetism in nanotubes with higher and lower curvature has been explained using spin densities of two cases of US-tubes (11, 0) and (9, 0) nanotubes as examples. Figure 35.4a and b shows the spin density of (11, 0) US-tube and (9, 0) US-tube, respectively. For both US-tubes, the spin density is mainly localized at the zigzag edges and decays from the edges to the middle. It has been found that with higher curvature value, the density is mainly from sp^2 dangling bond states, while for lower curvature, it is mixed sp^2–sp^3 rehybridized

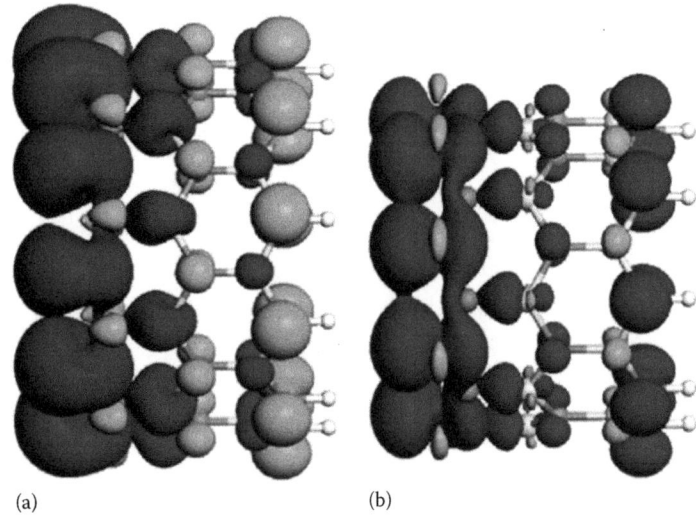

(a)   (b)

FIGURE 35.4
Isosurfaces of spin density for (a) (11, 0) US-tube and (b) (9, 0) US-tube. The isovalue is 0.01 e/Å³. Black and gray represent positive and negative spin densities, respectively. (Reprinted with permission from Yang, Y., Yan, X.H., Shen, X., Zhang, X., and Xiao, Y., Curvature effects on the magnetism of ultrashort zigzag carbon nanotubes and nanographenes, *J. Phys. Chem. C*, 114, 2010, 7553. Copyright 2010 American Chemical Society.)

TABLE 35.3 Second Hyperpolarizabilities (γ_{zzzz} [a.u.] and Diradical Characters (y_0)) and for [N] Cyclacenes and Cyclophenylenes, where $N = 5$–10 at the LC-UBLYP/6-31G* Level

Systems	y_0	γ_{zzzz}
[5] Cyclacene	0.320	4.89×10^3
[6] Cyclacene	0.900	1.81×10^3
[7] Cyclacene	0.565	5.27×10^3
[8] Cyclacene	0.942	2.47×10^3
[9] Cyclacene	0.720	4.75×10^3
[10] Cyclacene	0.965	3.81×10^3
[5] Cyclophenylene	0.000	0.807×10^3
[6] Cyclophenylene	0.000	0.958×10^3
[7] Cyclophenylene	0.000	1.11×10^3
[8] Cyclophenylene	0.000	1.26×10^3
[9] Cyclophenylene	0.000	1.41×10^3
[10] Cyclophenylene	0.000	1.56×10^3

states. Thus, it is possible to manipulate the magnetic coupling and magnetic moment of zigzag carbon edges by adjusting the curvature size, which may be useful in practical applications of CNTs.

Along similar lines, Chen et al. (2007) have studied the open-shell singlet character of cyclacenes as well as US zigzag nanotubes using density functional theory combined with broken symmetry approach. The ground-state geometries have been found to be open-shell singlet for [n] cyclacenes and US zigzag nanotubes. The results have been further supported using both the complete-active-space self-consistent-field (CASSCF) method as well as multireference perturbation theory method for [6] cyclacene. The relative energies of the open-shell (OS) singlet to the closed-shell (CS) singlet for [n] cyclacenes have been compared. These energies fluctuate when n is greater than 8 because of the fact that even [n] cyclacenes are more stable than the odd [n] cyclacenes for the closed-shell singlets.

Recently, we have used spin-unrestricted density functional theory with long-range corrected functional (LC-UBLYP) to study singlet ground states for their open-shell diradical character (y_i) and third-order NLO properties of linear/cyclic [N]acenes and [N]phenylenes, where N is the number

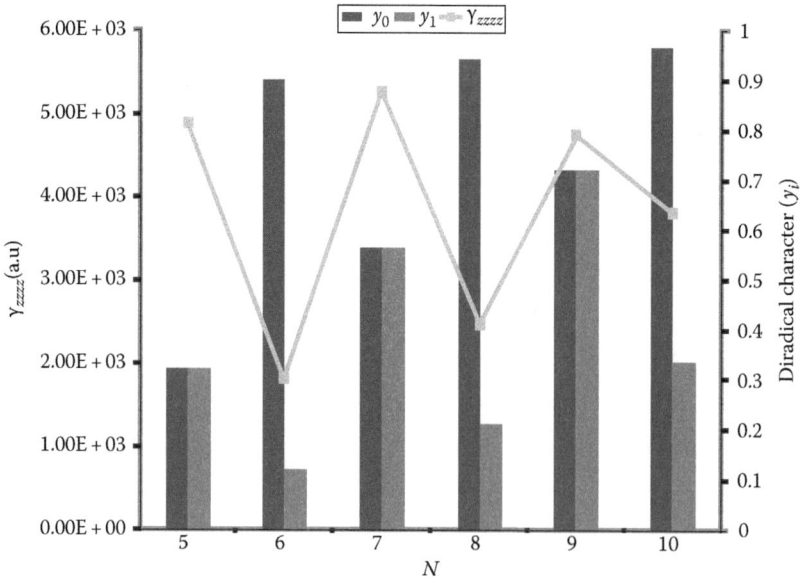

FIGURE 35.5

γ_{zzzz} versus diradical values (y_i) for [N]cyclic acenes with $N = 5$–10. (From Muhammad, S., Minami, T., Fukui, H., Yoneda, K., Minamide, S., Kishi, R., Shigeta, Y., and Nakano, M.: Comparative study of diradical characters and third-order nonlinear optical properties of linear/ cyclic acenes versus phenylenes. *Int. J. Quant. Chem.* 2013b. 113. 592–598. Copyright Wiley-VCH Verlag GmbH & Co. KGaA. Reprinted with permission.)

of benzene rings in each linear and cyclic configuration (Muhammad et al., 2013b). The cyclic and linear acenes have shown singlet diradical character in their ground states, while all phenylenes are found to have closed-shell configurations in their ground states. Furthermore, second hyperpolarizability (γ_{zzzz}) amplitudes also showed the diradical dependence for open-shell linear/cyclic acenes and registered larger amplitudes of hyperpolarizability especially in the intermediate range of diradical character than those of closed-shell phenylene counterparts. For example, an evaluation γ_{zzzz} value of cyclic acenes and phenylenes denotes that cyclic acenes posses larger amplitudes than phenylenes. Similarly, [5]C-Ac has γ_{zzzz} value of 4.89×10^3 a.u. that is about six times larger than 0.807×10^3 a.u of phenylenes (see Table 35.3). As shown in Figure 35.5, the cyclacene systems [5]C-Ac, [7]C-Ac and [9]C-Ac possess intermediate diradical character and larger amplitudes of γ_{zzzz} than the systems [6]C-Ac, [8]C-Ac and [10]C-Ac with high diradical values. These findings are in-line with our previous investigations on nanographenes (Nagai et al., 2010; Yoneda et al., 2011).

35.5 CONCLUSIONS AND PROSPECTIVE

Thus, from the above discussion, it is clear that the open-shell diradical character of several carbon nanomaterials play an essential role to control their functionalization. In case of graphenes, the α and β electrons are found on opposite zigzag edges. For example, in case of cyclacenes, the origin of AFM coupling has also been found from its zigzag edges. It is also important to mention that besides zigzag edge, the shape and size of graphenes are important to tune the open-shell diradical character of graphenes, which has been indicated by studies mentioned earlier. CNTs suppose to form by rolling graphene layer also show significant open-shell diradical character. The open-shell diradical character of CNTs depends on the type of tube, curvature size, tube length as well as the odd and even number of benzene rings in the tube. Unlike the graphenes and CNTs, the open-shell diradical character of fullerenes, that is, having zero-dimensional structures, strongly depends on their shape and size. The C_{36} and C_{26} fullerenes have shown the highest diradical characters due to their smaller size and highly strained structure, which has clarified using odd electron density as well as resonance forms, etc. Interestingly, in all the open-shell diradical carbon nanomaterials, it can be seen that first hyperpolarizability amplitude strongly depends on the range of diradical value where intermediate diradical species have shown the largest amplitude of first hyperpolarizability. Recently, there has been several highlights on future possibilities of new carbon-based open-shell compounds including open-shell graphene aggregates (Nakano et al., 2009; Yoneda et al., 2014) and carbon nanomaterials with intrinsic defects for NLO application (Zhou et al., 2014), quantum dots and nanoribbons for half-mettalicity (Dutta et al., 2009), as well as multilayer graphenes and hybrid carbon–boron–nitrogen nanomaterials for NLO and molecular magnetism, etc. (Ouyang et al., 2014). These compounds show a variety of functional properties, which are in the spotlight of materials designing fields in contemporary science. In future study, it would be worthwhile to address these above-mentioned new open-shell carbon materials for their potential applications as efficient electro-optical materials.

ACKNOWLEDGMENT

Shabbir Muhammad acknowledges the support of Research Center for Advanced Materials Science (RCAMS), King Khalid University, Abha, Saudi Arabia.

REFERENCES

Balasubramanian, K. and Burghard, M., 2005. Chemically functionalized carbon nanotubes. *Small*, 1, 180–192.

Bendikov, M., Duong, H.M., Starkey, K., Houk, K.N., Carter, E.A., and Wudl, F., 2004. Oligoacenes: Theoretical prediction of open-shell singlet diradical ground states. *Journal of the American Chemical Society*, 126, 7416–7417.

Cervantes-Sodi, F., Csanyi, G., Piscanec, S., and Ferrari, A.C., 2008. Edge-functionalized and substitutionally doped graphene nanoribbons: Electronic and spin properties. *Physical Review B*, 77, 165427.

Chen, Z., Jiang, D.-E., Lu, X., Bettinger, H.F., Dai, S., Schleyer, P.V.R., and Houk, K.N., 2007. Open-shell singlet character of cyclacenes and short zigzag nanotubes. *Organic Letters*, 9, 5449–5452.

Cohen, H.D. and Roothaan, C.C.J., 1965. Electric dipole polarizability of atoms by the hartree—Fock method. I. Theory for closed—Shell systems. *The Journal of Chemical Physics*, 43, S34–S39.

De Melo, C.P. and Fonseca, T.L., 1996. Ab initio polarizabilities of polyenic chains with conformational defects. *Chemical Physics Letters*, 261, 28–34.

De Melo, C.P. and Silbey, R., 1987. Non-linear polamzabilities of conjugated chains: Regular polyenes, solitons, and polarons. *Chemical Physics Letters*, 140, 537–541.

Dutta, S., Manna, A.K., and Pati, S.K., 2009. Intrinsic half-metallicity in modified graphene nanoribbons. *Physical Review Letters*, 102, 096601.

Fukui, H., Nakano, M., Shigeta, Y., and Champagne, B., 2011. Origin of the enhancement of the second hyperpolarizabilities in open-shell singlet transition-metal systems with metal–metal multiple bonds. *The Journal of Physical Chemistry Letters*, 2, 2063–2066.

Geim, A.K. and Novoselov, K.S., 2007. The rise of graphene. *Nature Materials*, 6, 183–191.

Houk, K.N., Lee, P.S., and Nendel, M., 2001. Polyacene and cyclacene geometries and electronic structures: Bond equalization, vanishing band gaps, and triplet ground states contrast with polyacetylene. *The Journal of Organic Chemistry*, 66, 5517–5521.

Iijima, S., 1991. Helical microtubules of graphitic carbon. *Nature*, 354, 56–58.

Judeinstein, P. and Sanchez, C., 1996. Hybrid organic–inorganic materials: A land of multidisciplinarity. *Journal of Materials Chemistry*, 6, 511–525.

Kamada, K., Ohta, K., Shimizu, A., Kubo, T., Kishi, R., Takahashi, H., Botek, E., Champagne, B., and Nakano, M., 2010. Singlet diradical character from experiment. *The Journal of Physical Chemistry Letters*, 1, 937–940.

Kawata, S., Sun, H.-B., Tanaka, T., and Takada, K., 2001. Finer features for functional microdevices. *Nature*, 412, 697–698.

Kong, F., Huang, S.-P., Sun, Z.-M., Mao, J.-G., and Cheng, W.-D., 2006. Se2 (B2O7): A new type of second-order NLO material. *Journal of the American Chemical Society*, 128, 7750–7751.

Konishi, A., Hirao, Y., Matsumoto, K., Kurata, H., Kishi, R., Shigeta, Y., Nakano, M., Tokunaga, K., Kamada, K., and Kubo, T., 2013. Synthesis and characterization of quarteranthene: Elucidating the characteristics of the edge state of graphene nanoribbons at the molecular level. *Journal of the American Chemical Society*, 135, 1430–1437.

Kovalenko, V.I. and Khamatgalimov, A.R., 2003. Open-shell fullerene C 74: Phenalenyl-radical substructures. *Chemical Physics Letters*, 377, 263–268.

Kroto, H.W., Heath, J.R., O'brien, S.C., Curl, R.F., and Smalley, R.E., 1985. C 60: Buckminsterfullerene. *Nature*, 318, 162–163.

Lin, W., Wang, Z., and Ma, L., 1999. A novel octupolar metal-organic NLO material based on a chiral 2D coordination network. *Journal of the American Chemical Society*, 121, 11249–11250.

Marder, S.R., Beratan, D.N., and Cheng, L.T., 1991. Approaches for optimizing the first electronic hyperpolarizability of conjugated organic molecules. *Science*, 252, 103–106.

Morrell, J.A. and Albrecht, A.C., 1979. Second-order hyperpolarizability of *p*-nitroaniline calculated from perturbation theory based expression using CNDO/S generated electronic states. *Chemical Physics Letters*, 64, 46–50.

Muhammad, S., Fukuda, K., Minami, T., Kishi, R., Shigeta, Y., and Nakano, M., 2013a. Interplay between the diradical character and third—Order nonlinear optical properties in fullerene systems. *Chemistry—A European Journal*, 19, 1677–1685.

Muhammad, S., Janjua, M.R.S.A., and Su, Z., 2009a. Investigation of dibenzoboroles having π-electrons: Toward a new type of two-dimensional NLO molecular switch? *The Journal of Physical Chemistry C*, 113, 12551–12557.

Muhammad, S., Liu, C., Zhao, L., Wu, S., and Su, Z., 2009b. A theoretical investigation of intermolecular interaction of a phthalimide based "on–off" sensor with different halide ions: Tuning its efficiency and electro-optical properties. *Theoretical Chemistry Accounts*, 122, 77–86.

Muhammad, S., Minami, T., Fukui, H., Yoneda, K., Kishi, R., Shigeta, Y., and Nakano, M., 2012. Halide ion complexes of decaborane (B10H14) and their derivatives: Noncovalent charge transfer effect on second-order nonlinear optical properties. *The Journal of Physical Chemistry A*, 116, 1417–1424.

Muhammad, S., Minami, T., Fukui, H., Yoneda, K., Minamide, S., Kishi, R., Shigeta, Y., and Nakano, M., 2013b. Comparative study of diradical characters and third—Order nonlinear optical properties of linear/cyclic acenes versus phenylenes. *International Journal of Quantum Chemistry*, 113, 592–598.

Muhammad, S. and Nakano, M., 2013. Computational strategies for nonlinear optical properties of carbon nano-systems. *Nanoscience and Computational Chemistry: Research Progress*, 309–332.

Muhammad, S., Xu, H., Su, Z., Fukuda, K., Kishi, R., Shigeta, Y., and Nakano, M., 2013c. A new type of organic–inorganic hybrid NLO-phore with large off-diagonal first hyperpolarizability tensors: A two-dimensional approach. *Dalton Transactions*, 42, 15053–15062.

Nagai, H., Nakano, M., Yoneda, K., Kishi, R., Takahashi, H., Shimizu, A., Kubo, T., Kamada, K., Ohta, K., and Botek, E., 2010. Signature of multiradical character in second hyperpolarizabilities of rectangular graphene nanoflakes. *Chemical Physics Letters*, 489, 212–218.

Nakano, M., Fujita, H., Takahata, M., and Yamaguchi, K., 2002. Theoretical study on second hyperpolarizabilities of phenylacetylene dendrimer: Toward an understanding of structure-property relation in NLO responses of fractal antenna dendrimers. *Journal of the American Chemical Society*, 124, 9648–9655.

Nakano, M., Fukui, H., Nagai, H., Minami, T., Kishi, R., and Takahashi, H., 2009. Third-order nonlinear optical properties of open-shell singlet molecular aggregates composed of diphenalenyl diradicals. *Synthetic Metals*, 159, 2413–2415.

Nakano, M., Kishi, R., Nakagawa, N., Ohta, S., Takahashi, H., Furukawa, S.-I., Kamada, K., Ohta, K., Champagne, B., and Botek, E., 2006a. Second hyperpolarizabilities (γ) of bisimidazole and bistriazole benzenes: Diradical character, charged state, and spin state dependences. *The Journal of Physical Chemistry A*, 110, 4238–4243.

Nakano, M., Kishi, R., Nitta, T., Kubo, T., Nakasuji, K., Kamada, K., Ohta, K., Champagne, B., Botek, E., and Yamaguchi, K., 2005. Second hyperpolarizability (γ) of singlet diradical system: Dependence of γ on the diradical character. *The Journal of Physical Chemistry A*, 109, 885–891.

Nakano, M., Kishi, R., Ohta, S., Takahashi, H., Kubo, T., Kamada, K., Ohta, K., Botek, E., and Champagne, B., 2007. Relationship between third-order nonlinear optical properties and magnetic interactions in open-shell systems: A new paradigm for nonlinear optics. *Physical Review Letters*, 99, 033001.

Nakano, M., Kishi, R., Ohta, S., Takebe, A., Takahashi, H., Furukawa, S.-I., Kubo, T., Morita, Y., Nakasuji, K., and Yamaguchi, K., 2006b. Origin of the enhancement of the second hyperpolarizability of singlet diradical systems with intermediate diradical character. *The Journal of Chemical Physics*, 125, 074113.

Nakano, M., Kubo, T., Kamada, K., Ohta, K., Kishi, R., Ohta, S., Nakagawa, N., Takahashi, H., Furukawa, S.-I., and Morita, Y., 2006c. Second hyperpolarizabilities of polycyclic aromatic hydrocarbons involving phenalenyl radical units. *Chemical Physics Letters*, 418, 142–147.

Nakano, M., Minami, T., Yoneda, K., Muhammad, S., Kishi, R., Shigeta, Y., Kubo, T., Rougier, L., Champagne, B.T., and Kamada, K., 2011. Giant enhancement of the second hyperpolarizabilities of open-shell singlet polyaromatic diphenalenyl diradicaloids by an external electric field and donor–acceptor substitution. *The Journal of Physical Chemistry Letters*, 2, 1094–1098.

Nakano, M., Nagai, H., Fukui, H., Yoneda, K., Kishi, R., Takahashi, H., Shimizu, A., Kubo, T., Kamada, K., and Ohta, K., 2008. Theoretical study of third-order nonlinear optical properties in square nanographenes with open-shell singlet ground states. *Chemical Physics Letters*, 467, 120–125.

Nakano, M., Nitta, T., Yamaguchi, K., Champagne, B., and Botek, E., 2004. Spin multiplicity effects on the second hyperpolarizability of an open-shell neutral π-conjugated system. *The Journal of Physical Chemistry A*, 108, 4105–4111.

Nalwa, H.S. and Miyata, S., 1996. *Nonlinear Optics of Organic Molecules and Polymers*. CRC Press: Boca Raton, FL.

Novoselov, K.S., Geim, A.K., Morozov, S.V., Jiang, D., Zhang, Y., Dubonos, S.V., Grigorieva, I.V., and Firsov, A.A., 2004. Electric field effect in atomically thin carbon films. *Science*, 306, 666–669.

Ohta, S., Nakano, M., Kubo, T., Kamada, K., Ohta, K., Kishi, R., Nakagawa, N., Champagne, B., Botek, E., and Takebe, A., 2007. Theoretical study on the second hyperpolarizabilities of phenalenyl radical systems involving acetylene and vinylene linkers: Diradical character and spin multiplicity dependences. *The Journal of Physical Chemistry A*, 111, 3633–3641.

Ouyang, Q., Xu, Z., Lei, Z., Dong, H., Yu, H., Qi, L., Li, C., and Chen, Y., 2014. Enhanced nonlinear optical and optical limiting properties of graphene/ZnO hybrid organic glasses. *Carbon*, 67, 214–220.

Papadopoulos, M.G., Sadlej, A.J., and Leszczynski, J., 2006. *Non-Linear Optical Properties of Matter*. Springer: Dordrecht, the Netherlands.

Sackers, E., Oßwald, T., Weber, K., Keller, M., Hunkler, D., Wörth, J., Knothe, L., and Prinzbach, H., 2006. Bromination of unsaturated dodecahedranes—En route to C20 fullerene. *Chemistry—A European Journal*, 12, 6242–6254.

Sheka, E.F. and Chernozatonskii, L.A., 2007. Bond length effect on odd-electron behavior in single-walled carbon nanotubes. *The Journal of Physical Chemistry C*, 111, 10771–10779.

Sheka, E.F. and Chernozatonskii, L.A., 2010. Broken symmetry approach and chemical susceptibility of carbon nanotubes. *International Journal of Quantum Chemistry*, 110, 1466–1480.

Shim, M., Shi Kam, N.W., Chen, R.J., Li, Y., and Dai, H., 2002. Functionalization of carbon nanotubes for biocompatibility and biomolecular recognition. *Nano Letters*, 2, 285–288.

Son, Y.-W., Cohen, M.L., and Louie, S.G., 2006. Half-metallic graphene nanoribbons. *Nature*, 444, 347–349.

Song, J., Parker, M., Schoendorff, G., Kus, A., and Vaziri, M., 2010. A study on the electronic and structural properties of fullerene C 30 and azafullerene C 18 N 12. *Journal of Molecular Structure: THEOCHEM*, 942, 71–76.

Sun, Z. and Wu, J., 2012. Open-shell polycyclic aromatic hydrocarbons. *Journal of Materials Chemistry*, 22, 4151–4160.

Takatsuka, K., Fueno, T., and Yamaguchi, K., 1978. Distribution of odd electrons in ground-state molecules. *Theoretica Chimica Acta*, 48, 175–183.

Terenziani, F., Katan, C., Badaeva, E., Tretiak, S., and Blanchard-Desce, M., 2008. Enhanced two-photon absorption of organic chromophores: Theoretical and experimental assessments. *Advanced Materials*, 20, 4641–4678.

Weber, K., Voss, T., Heimbach, D., Weiler, A., Keller, M., Wörth, J., Knothe, L., Exner, K., and Prinzbach, H., 2005. From unsaturated dodecahedranes to C40 cages? *Tetrahedron Letters*, 46, 5471–5474.

Willetts, A., Rice, J.E., Burland, D.M., and Shelton, D.P., 1992. Problems in the comparison of theoretical and experimental hyperpolarizabilities. *The Journal of Chemical Physics*, 97, 7590–7599.

Yamada, S., Nakano, M., Shigemoto, I., Kiribayashi, S., and Yamaguchi, K., 1997. Theoretical study of the third-order nonlinear optical susceptibilities for the β-phase crystal of p-NPNN. *Chemical Physics Letters*, 267, 438–444.

Yamaguchi, K., 1975. The electronic structures of biradicals in the unrestricted Hartree-Fock approximation. *Chemical Physics Letters*, 33, 330–335.

Yamaguchi, K., Carbo, R., and Klobukowski, M., 1990. *Self-Consistent Field: Theory and Applications*. Carbo, R., 727. Elsevier: Amsterdam.

Yamaguchi, K., Yamanaka, S., Nishino, M., Takano, Y., Kitagawa, Y., Nagao, H., and Yoshioka, Y., 1999. Symmetry and broken symmetries in molecular orbital descriptions of unstable molecules II. Alignment, flustration and tunneling of spins in mesoscopic molecular magnets. *Theoretical Chemistry Accounts*, 102, 328–345.

Yang, Y., Yan, X.H., Shen, X., Zhang, X., and Xiao, Y., 2010. Curvature effects on the magnetism of ultrashort zigzag carbon nanotubes and nanographenes. *The Journal of Physical Chemistry C*, 114, 7553–7557.

Yoneda, K., Nakano, M., Fukuda, K., Matsui, H., Takamuku, S., Hirosaki, Y., Kubo, T., Kamada, K., and Champagne, B.T., 2014. Third-order nonlinear optical properties of one-dimensional open-shell molecular aggregates composed of phenalenyl radicals. *Chemistry—A European Journal*, 20, 11129–11136.

Yoneda, K., Nakano, M., Fukui, H., Minami, T., Shigeta, Y., Kubo, T., Botek, E., and Champagne, B., 2011. Open-shell characters and second hyperpolarizabilities of one-dimensional graphene nanoflakes composed of trigonal graphene units. *ChemPhysChem*, 12, 1697–1707.

Yoneda, K., Nakano, M., Inoue, Y., Inui, T., Fukuda, K., Shigeta, Y., Kubo, T., and Champagne, B., 2012. Impact of antidot structure on the multiradical characters, aromaticities, and third-order nonlinear optical properties of hexagonal graphene nanoflakes. *The Journal of Physical Chemistry C*, 116, 17787–17795.

Zhi, L. and Müllen, K., 2008. A bottom-up approach from molecular nanographenes to unconventional carbon materials. *Journal of Materials Chemistry*, 18, 1472–1484.

Zhong, R.L., Xu, H.L., Su, Z.M., Li, Z.R., Sun, S.L., and Qiu, Y.Q., 2012. Spiral intramolecular charge transfer and large first hyperpolarizability in Möbius cyclacenes: New insight into the localized π electrons. *ChemPhysChem*, 13, 2349–2353.

Zhou, Z.-J., Liu, Z.-B., Li, Z.-R., Huang, X.-R., and Sun, C.-C., 2011. Shape effect of graphene quantum dots on enhancing second-order nonlinear optical response and spin multiplicity in NH_2–GQD–NO_2 systems. *The Journal of Physical Chemistry C*, 115, 16282–16286.

Zhou, Z.-J., Yu, G.-T., Ma, F., Huang, X.-R., Wu, Z.-J., and Li, Z.-R., 2014. Theoretical investigation on nonlinear optical properties of carbon nanotubes with Stone–Wales defect rings. *Journal of Materials Chemistry C*, 2, 306–311.

CHAPTER 36

CONTENTS

Functionalization of Carbon Nanotubes with Polymers

36

Ana M. Díez-Pascual

36.1 INTRODUCTION

Carbon nanotubes (CNTs) are one-dimensional carbon-based nanomaterials discovered by Iijima (1991) that possess very large aspect ratio (>1000), high flexibility, low density (~1.8 g/cm^3), and exceptional mechanical, thermal, and electrical properties (Popov 2004; Thostenson et al. 2001), which make them ideal candidates for the fabrication of multifunctional nanocomposites. Their extraordinary high Young's modulus (up to 1.2 TPa) and tensile strength (ca. 50–200 GPa) make them one of the strongest and stiffest materials on earth and, in principal, ideal candidates for polymer reinforcement. There are two main types of CNTs: those consisting of a single graphite sheet wrapped into a cylindrical tube with a diameter in the range of 0.7–3 nm, the single-walled carbon nanotubes (SWCNTs), and those composed of more than two coaxial cylinders, each rolled out of single sheets, with diameters between 2 and 40 nm, the multiwalled carbon nanotubes (MWCNTs). Both types of CNTs can be synthesized using different routes (Dervishi et al. 2009), including high-temperature evaporation using arc-discharge or laser ablation and gas-phase catalytic growth from carbon monoxide or chemical vapour deposition (CVD) from hydrocarbons. CVD materials generally contain residual catalyst particles, while the main contaminants in the products of high-temperature reactions are carbonaceous impurities. The laser process generates CNTs of the highest quality (few defects, high crystallinity, extremely high aspect ratio) compared to other synthesis methods (Díez-Pascual et al. 2009), while nanotubes produced by the CVD technique present a larger number of defects as a result of the lower growth temperature. The purity, quality, aspect ratio, and the nature of impurities, hence the source of the CNTs, can have a strong influence on the final properties of CNT-reinforced composites.

However, two fundamental issues must be taken into account when dealing with this type of nanofiller. First, the strong tendency for agglomeration of the CNTs, in particular SWCNTs, that are generally encountered as bundles of aligned tubes, and second, the persisting impurities that accompany the CNTs, such as metallic catalyst particles and carbonaceous fragments (amorphous and crystalline carbon beads). The most commonly employed purification method consists in the oxidation of the CNTs with strong acids (Tsang et al. 1994) such as nitric, sulfuric, or a mixture of both. Nevertheless, this approach introduces defects into the tubular framework that can have an adverse effect on their mechanical properties and/or disrupt the electronic continuum, reducing surface electrical and thermal conductivities, and can produce significant damage, such as sidewall opening or tube breakage. The successful utilization of CNTs in composite applications depends on their homogenous dispersion throughout the matrix. Another challenge is to achieve a strong CNT–polymer interfacial adhesion, hence an effective load transfer, a prerequisite to obtain materials with enhanced mechanical properties (Gonzalez-Dominguez et al. 2011).

In view of the preceding, several methods have been reported such as mechanical dispersion (i.e., ultrasonication (Ma et al. 2010), ball-milling (Li et al. 1999), plasma treatment (Chen et al. 2013), and chemical modification involving either covalent or noncovalent bonding between nanotubes and polymers) (Figure 36.1, Hirsch 2002). The noncovalent approach consists in the physical adsorption and/or wrapping of polymers to the CNT surface. The graphitic CNT sidewalls can

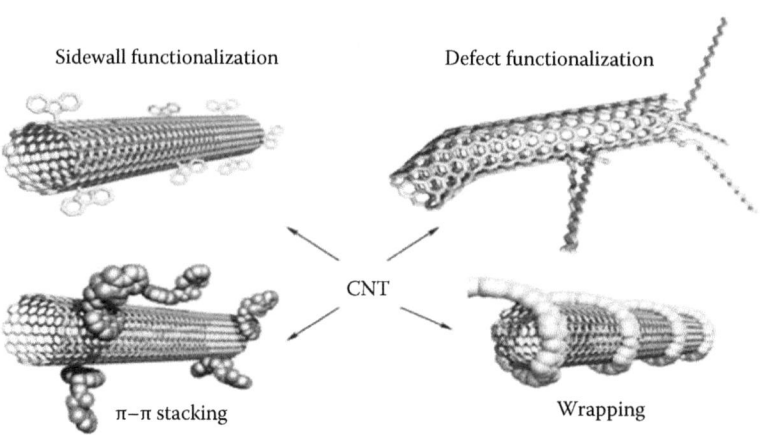

FIGURE 36.1

Schematic representation of the CNT functionalization routes using polymers: covalent functionalization at the CNT sidewalls or defects and noncovalent functionalization via π–π stacking interactions or wrapping. (From Hirsch, A.: Functionalization of carbon nanotubes. *Angew. Chem. Int. Ed.* 2002. 41. 1853–1859. Copyright Wiley-VCH Verlag GmbH & Co. KGaA. Reproduced with permission.)

interact with conjugated polymers via π–π stacking as well as with those containing heteroatoms with free electron pairs. This route preserves the nanotube integrity and properties, since it does not destroy the conjugated system of the CNT sidewalls. The covalent method involves the chemical bonding (grafting) of polymer chains to CNTs, and can be performed via *grafting to* or *grafting from* strategies. The first approach is based on the synthesis of a polymer derivative that can react with the surface of pristine, oxidized, or functionalized CNTs (Díez-Pascual and Naffakh 2012). However, the oxidation/functionalization treatments performed in acid media generally shorten the CNTs and can bring significant damage such as sidewall opening or tube breakage (Huang et al. 2002), introducing defects in the tubular framework that can adversely impact their mechanical properties. Another disadvantage of this method is that the grafted polymer content is restricted due to the low reactivity and high steric hindrance of the polymer chains. In the *grafting from* strategy, the polymer is grown from the CNT surface via in situ polymerization of monomers initiated by chemical species immobilized on the CNT sidewalls and edges. The high reactivity of monomers makes the *grafting from* process efficient and controllable, enabling the preparation of nanocomposites with a high degree of grafting. However, this method requires strict control of the amounts of each reactant and the polymerization conditions. Alternatively, CNT functionalization can be performed by plasma treatment (Chen et al. 2012, 2013), a time saving and environmentally friendly technique for modifying CNTs by directly introducing a high density of functional groups. The CNT surfaces can readily change from hydrophobic to hydrophilic with plasma processing, thus facilitating the dispersion within the polymer matrix. Furthermore, this treatment does not alter the intrinsic mechanical properties of the tubes.

In this chapter, various strategies for modifying CNTs using polymers will be illustrated, including covalent and noncovalent approaches, and the characteristics and advantages of each method will be discussed. Due to the huge number of papers reported to date on this topic, it is impossible to make a comprehensive overview of all aspects of this large subject in the framework of this chapter. Therefore, only the most characteristic and important recent examples are described. Finally, challenges and future perspectives in this exciting research field are provided.

36.2 COVALENT FUNCTIONALIZATION OF CARBON NANOTUBES

As mentioned earlier, the modification of CNTs by polymers may be divided into two categories, involving either noncovalent or covalent bonding of polymer chains to CNTs. In the covalent approach, CNTs are functionalized by not reversible attachment of polymers to their sidewall

(sidewall functionalization) or at defect sites (defect functionalization) usually localized at the tips (Figure 36.1), and can be carried out via *grafting to* or *grafting from* strategies.

36.2.1 Grafting to Strategy

The *grafting to* method typically involves preformed polymer chains reacting with the surface of pristine, oxidized, or functionalized CNTs. The most important approaches used for this type of functionalization are radical or carbanion additions as well as cycloaddition reactions to the double bonds of the CNTs (Li et al. 2005), given that the curvature of the carbon nanomaterials makes them susceptible to various addition reactions. Alternatively, defect sites on the surface of oxidized CNTs such as open-ended nanostructures with terminal carboxylic acid groups allow covalent bonding of oligomers or polymer chains (Tasis et al. 2006). Frequently, amino- or hydroxyl-terminated polymers are anchored via amidation or esterification reactions with the carboxylic acid groups onto the CNT surface. An advantage of the *grafting to* method is that preformed commercial polymers of controlled molecular weight and polydispersity can be employed. The main drawback of the technique is that it typically results in a low grafting density (Spitalsky et al. 2010). In addition, it is limited to polymers containing reactive functional groups. In the following sections, a variety of examples on *grafting to* methodologies reported to date in the literature will be discussed.

36.2.1.1 Coupling and nucleophilic addition reactions

Since 2003, several works have been reported on coupling and nucleophilic addition reactions between polymers and functionalized CNTs. For instance, Xie et al. (2007) incorporated phenyl-alkyne groups onto the surface of SWCNTs via diazonium salt chemistry (Bahr and Tour 2001) followed by reaction with benzyl chlorinated polystyrene-*co*-poly(pchloromethylstyrene) and polystyrene-*co*-poly(*p*-chloromethylstyrene)-b-polystyrene, which were synthesized by free radical polymerization. The coupling reaction between benzyl chloride moieties and alkyne-modified SWCNTs was carried out under mild reaction conditions using Pd as catalyst, leading to 53 and 81 wt% polymer-grafted content, respectively. Likewise, You et al. (2006) anchored thiol-reactive moieties onto MWCNT sidewalls, which were subsequently reacted with a thiol-terminated poly[*N*-(2-hydroxypropyl)methacrylamide] (PHPMA) by a coupling reaction. Results demonstrated that this thiol-coupling reaction is efficient for synthesizing water-soluble polymer-modified MWCNTs under mild conditions. Liu et al. (2004a,b,d) modified oxidized MWCNTs by direct monolithiation with ferrocene using tert-butyllithium, resulting in *p*-chloromethylstyrene-terminated MWCNTs. Subsequently, the modified CNTs were functionalized with polystyryllithium chains, and the polymer content was found to be around 80 wt% according to TGA analysis.

An efficient route for CNT functionalization is the use of an organometallic approach. Recently, it was found that epoxy groups can be anchored onto the surface of CNTs under mild reaction conditions via a one-pot procedure involving two reactive steps (Baudot et al. 2009). The first consist in the nucleophilic addition of an organolithium reactant to double bonds of the hexagonal rings of the CNTs and the formation of nanotube carbanions via transfer of charge. The second stage is the nucleophilic substitution with halogen or hydroxyl oxacylcopropanes, for example, epichloro-hydrin, and involves the nucleophilic attack from the CNT carbanions to the carbon atom bearing the chlorine in the epichlorohydrin via elimination of lithium halides. The epoxy group may be converted into different kinds of functionalities through a ring-opening process, enabling cross-linking reactions with hydroxyl or amine-containing polymers. Using this approach, Diez-Pascual and Naffakh (2012) reported the grafting of an aminated polyphenylene sulphide (PPS-NH2) derivative to epoxy-functionalized CNTs, and the extent of the grafting process was about 25%. Wu et al. (2003) described the nucleophilic substitution reaction of polystyryllithium anions prepared through anionic polymerization with acyl choride-modified MWCNTs (Figure 36.2), and the polymer content in the resulting sample was about 40%. Similar results were obtained by Baskaran et al. (2007) with living polystyryllithium and polybutadienyllithium anions, where grafted fractions up to 25% were reported. It was found that the percentage of polystyrene (PS) grafted onto the MWCNTs increased upon reducing the molecular weight of the polymer precursor. Blake et al. (2004) used butyllithium-functionalized MWCNTs that reacted with chlorinated polypropylene (PP) to yield CNTs covalently bonded to polymer chains, and the amount of polymer content was 31 wt%.

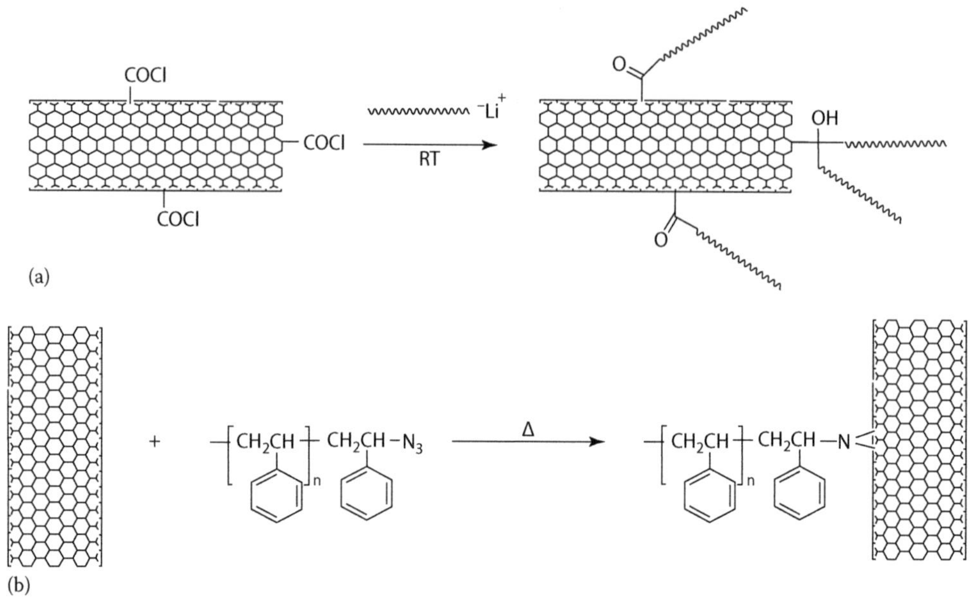

FIGURE 36.2
(a) Nucleophilic substitution reaction of polystyryllithium anions with acyl chloride-modified CNTs.
(b) Cycloaddition reaction of azide-terminated polystyrene onto the CNT surface. (Reprinted from
Prog. Polym. Sci., 35, Spitalsky, Z., Tasis, D., Papagelis, K., and Galiotis, C., Carbon nanotube–
polymer composites: Chemistry, processing, mechanical and electrical properties, 357–401,
Copyright 2010, with permission from Elsevier.)

36.2.1.2 Cycloaddition reactions

Several works have been reported on the cycloaddition reaction of PS derivatives onto functionalized CNTs. Thus, Qin et al. (2004a,b) described the cycloaddition reaction of azide-terminated
PS and SWCNTs under inert atmosphere (Figure 36.2), leading to a functionalization degree of
one polymer chain per 48 CNT carbons. Li et al. (2005) used a similar approach in which azide-
terminated PS and alkyne-functionalized SWCNTs were coupled via [3 + 2] Huisgen cycloaddition
between the alkyne and azide end groups, yielding a 45% amount of grafted polymer. The same
authors carried out a sulfonation reaction to the grafted PS chains (Li and Andronov 2007), and the
sulfonation degree was controlled by the amount of sulfonation reagent employed in the reaction. In
addition to PS, other polymers like poly(1-phenyl-1-alkyne) and poly(diphenylacetylene) derivatives
carrying azido-terminal groups have been used to covalently functionalize SWCNTs via cyclization
reactions (Li et al. 2006), leading to soluble polyene-modified nanotubes that emitted intense visible
luminescence.

36.2.1.3 Esterification and amidation reactions

Side defect functionalization typically occurs via amidation or esterification reactions of carboxylic groups of the CNT surface with polymers or their derivatives. For the esterification approach,
the most commonly employed hydroxyl-terminated polymers are poly(vinyl alcohol) (PVA) and
poly(ethylene glycol) (PEG). Thus, Lin and coworkers (2003) grafted PVA chains onto oxidized
SWCNTs and MWCNTs by carbodiimide-activated esterification reactions, and Riggs et al.
(2000) reported the grafting of poly(vinyl acetate-*co*-vinyl alcohol) (PVAc-VA) via ester linkages to acyl-activated SWCNTs. The esterification reaction has also been used for grafting PEG
chains to acyl chloride-activated SWCNTs (Zhao et al. 2005). Grafting of hydroxyl terminated
poly(methyl methacrylate) (PMMA-OH) and poly[(methyl methacrylate)-*co*-(2-hydroxyethyl
methacrylate)] (PMMAHEMA) with acyl chloride-activated MWCNTs have also been carried
out by Baskaran et al. (2005a,b) in various solvents at different temperatures. Díez-Pascual and
coworkers (2010a,b) grafted a hydroxyl polyetheretherketone derivative (PEEK-OH) to the
surface of oxidized SWNCTs via direct esterification through the activation of the carboxylic

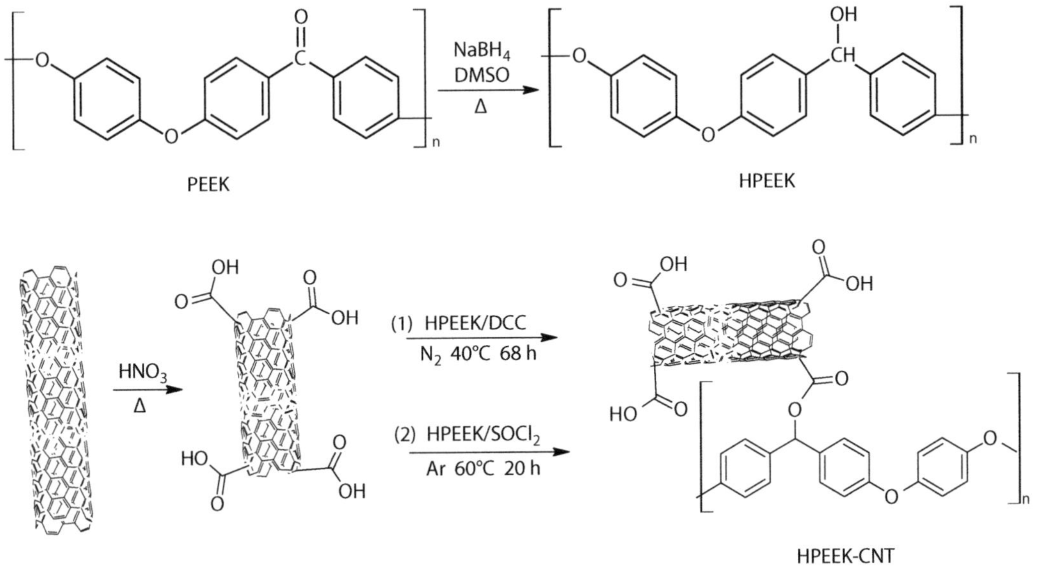

FIGURE 36.3
Schematic representation of the synthesis of a hydroxylated PEEK derivative (HPEEK) and *grafting to* the surface of acid-treated SWCNTs via direct esterification reaction through the activation of carboxylic groups with carbodiimide (1) or via acylation of the carboxylic groups with thionyl chloride (2). (From Díez-Pascual, A.M., Martinez, G., Gonzalez-Dominguez, J.M., Martinez, T., and Gomez-Faou, M.A., Grafting of a hydroxylated PEEK derivative to the surface of single-walled carbon nanotubes, *J. Mater. Chem.*, 20, 8285–8296, 2010a. Reproduced by permission of The Royal Society of Chemistry.)

groups of the CNTs with carbodiimide and via a two-step process through acylation of the CNTs with thionyl chloride (Figure 36.3). The former method was more effective for anchoring the polymer, leading to a nanocomposite with improved thermal and mechanical properties. Both oxidized SWCNT and MWCNT materials have been employed for carbodiimide-activated esterifications with derivatized polyimide (PI)-containing pendant hydroxyl groups (Hill et al. 2005). A similar approach was used for a poly(*N*-vinyl carbazole) copolymer (PVKV) incorporating pendant hydroxyl groups, which was grafted to oxidized SWCNTs via acyl-activation reaction (Wang et al. 2005). Wang and Tseng (2007) prepared polyurethanes (PU) carrying carboxyl groups in the chain extender that underwent esterification reaction with acyl chloride–activated MWCNTs, and found that longer acid treatment of the CNTs resulted in a higher amount of grafted polymer chains.

Polymers containing amino groups such as polyethyleneimine (PEI), polyetherimide, and poly(*N*-isopropylacrylamide) (PNIPAAm) have been employed in amidation reactions with CNTs. For instance, Hu et al. (2005) grafted branched PEI to acyl chloride–modified CNTs, and the CNT adduct as was applied for neurite growth. Ge et al. (2005) grafted oxidized MWCNTs with polyetherimide under inert atmosphere at high temperatures without the need for a catalyst. Similarly, Qu et al. (2004) grafted an amine-terminated PI to SWCNTs and MWCNTs through a carbodiimide-activated reaction. Following a similar method, PNIPAAm was grafted to acid-treated SWCNTs, leading to 8 wt% polymer content. In a different approach, oxidized MWCNTs were anchored onto polyacrylonitrile (PAN) nanoparticles through the reaction of the reduced cyano-groups of the polymer and the carboxylic moieties of CNT surface (Han et al. 2007). Poly(propionylethylenimine-*co*-ethylimine) (PPEI-EI) was anchored to acyl-activated tubes by Lin and coworkers (2002), using either direct heating or carbodiimide-assisted amidation. TGA measurements revealed about 30 wt% polymer content for the acylation–amidation route and 40 wt% for the heating process. These authors also studied three different functionalization methods of diamine-terminated PEG onto SWCNTs (direct thermal heating, acylation–amidation, and carbodiimide-activated coupling) (Huang et al. 2003). The lowest degree of functionalization was attained for direct thermal heating, while the carbodiimide activation led to a higher grafting degree.

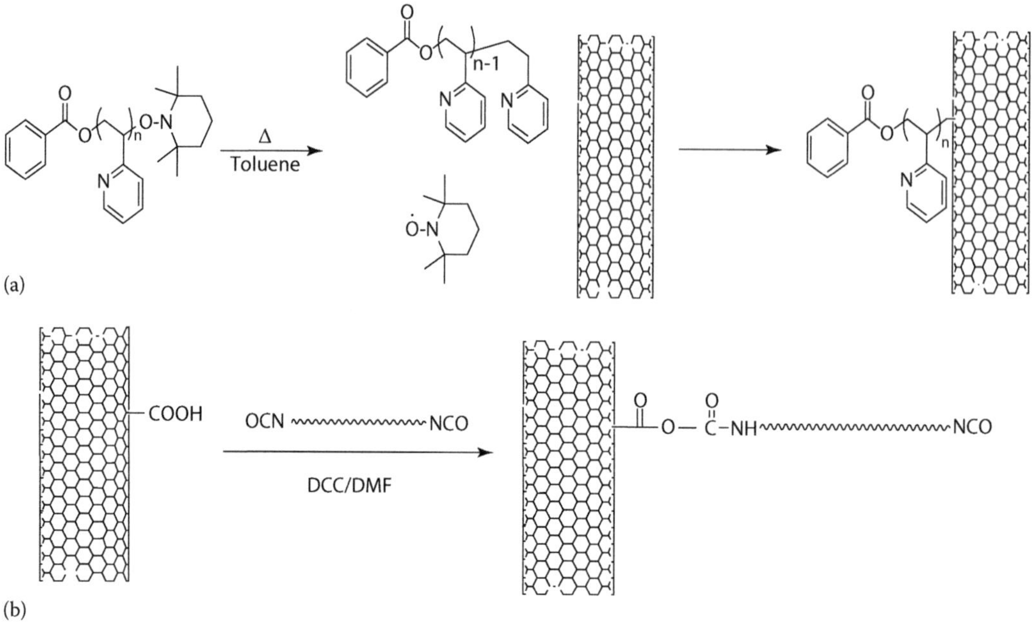

(a)

(b)

FIGURE 36.4
(a) Heating of TEMPO-terminated poly2VP chains to dissociate the TEMPO group, resulting in radical terminated chains that are grafted to MWCNTs. (From Lou, X., Daussin, R., Cuenot, S. et al.: Synthesis of pyrene-containing polymers and noncovalent sidewall functionalization of multiwalled carbon nanotubes. *Chem. Mater.* 2004a. 16. 4005–4011. Copyright Wiley-VCH Verlag GmbH & Co. KGaA. Reproduced with permission.) (b) Covalent boding between isocyanate-terminated PU and amine-functionalized CNTs. (Reprinted from *Compos. Sci. Technol.*, 65, Kuan, H.C., Ma, C.C.M., Chang, W.P., Yuen, S.M., Wu, H.H., and Lee, T.M., Synthesis, thermal, mechanical and rheological properties of multiwall carbon nanotube/waterborne polyurethane nanocomposite, 1703–1710, Copyright 2005, with permission from Elsevier.)

36.2.1.4 Other *grafting to* approaches

Other *grafting to* routes have been reported, such as the grafting by a radical mechanism, condensation, or sonochemical reactions. Regarding the first method, Lou and coworkers (2004a,b) studied the attachment of poly(2-vinylpyridine) (P2VP) of controlled molecular weight end-capped by 2,2,6,6-tetramethylpiperidinyl-1-oxyl (TEMPO) group to MWCNTs (Figure 36.4a). Heating of TEMPO-terminated P2VP chains causes the TEMPO group to dissociate, resulting in radical-terminated chains that were grafted to CNTs. A very similar approach was used by the same authors for grafting PS, polycaprolactone (PCL), and the corresponding block copolymer (PCL-b-PS). On the other hand, condensation reactions have been employed by several authors to anchor PU to oxidized (Jung et al. 2006) or alkoxysilane-functionalized MWCNTs (Wu et al. 2005), or to attach isocyanate-terminated PU to amino-functionalized MWCNTs (Kuan et al. 2005) (Figure 36.4b). Sonochemical reaction was employed by Koshio et al. (2001) to anchor a polymer to SWCNTs via defect site creation on the CNT sidewalls, and the chemical bonding was proved by FTIR spectroscopy.

36.2.2 *Grafting from* Strategy

The *grafting from* approach typically involves the polymerization of monomers from surface-derived initiators on either MWCNTs or SWCNTs, which are covalently anchored through various functionalization reactions developed for small molecules, including acid-defect group chemistry and sidewall functionalization of CNTs. The advantage of this methodology is that the polymer polymerization is not limited by steric hindrance, enabling high-molecular-weight polymers to be efficiently grafted, and, consequently, nanocomposites with relatively high grafting density can be prepared. However, it needs strict control of the amounts of initiator and substrate as well as precise control of the conditions of the polymerization reaction.

36.2.2.1 Atom transfer radical polymerization

The atom transfer radical polymerization (ATRP) technique has been used by several groups for grafting PMMA chains to CNTs (Baskaran et al. 2004; Liu et al. 2007; Matrab et al. 2006; Yao et al. 2003). Yao and coworkers (2003) reported an in situ polymerization process from growing polymers to anchor methylmethacrylate (MMA) chains to CNTs (Figure 36.5). First, SWCNT sidewalls were functionalized with alkyl bromide moieties using a two-step procedure involving first a 1,3-dipolar cycloaddition to introduce phenol functionalities (1 in Figure 36.5), followed by an esterification with 2-bromoisobutyryl bromide. The functionalized CNTs (2 in Figure 36.5) served as macroinitiators for ATRP of MMA, leading to a PMMA-SWCNT hybrid nanocomposite (3 in Figure 36.5). A significant enhancement in the glass transition temperature (T_g) of the nanocomposite was observed, although the hybrid was not soluble in organic solvents. This approach enabled a high grafting density and can be applied to a wide range of monomers. Baskaran et al. (2004) reported ATRP of styrene and MMA from MWCNTs. According to TGA, the amount of PMMA covalently attached was 70 wt%, while in the case of PS it changed from 18 to 33 wt% upon increasing the concentration of the initiator on nanotubes. The authors suggested that the amount of polymer present on the surface, hence the molecular weight of the growing polymer chains, can be tailored by modifying the initiator concentration. Liu et al. (2007) described ATRP polymerization of MMA with SWCNTs, and the amount of polymer was 17 wt%. The grafting of PMMA and PS brushes by ATRP from the surface of aligned MWCNTs was carried out by Matrab et al. (2006), in which the ATRP initiator was grafted to the nanotube surface via electrochemical treatment with diazonium salts.

The ATRP has also been applied to other polymers. Sun et al. (2004) functionalized aligned arrays of CNTs, known as forests, with PNIPAAm. The CNTs were activated with aminopropyltrimethoxysilane (APTES) and subsequently NIPAAm was polymerized by ATRP. A similar approach was reported by Kong et al. (2004) for ATRP of PNIPAAm on nonaligned CNTs. The grafting of styrene-*co*-acrylonitrile (SAN) copolymer, PS, and PAN from MWCNTs was carried out through the introduction of ATRP initiator onto the surface of CNTs (Shanmugharah et al. 2007), and the amount of polymer grafted was reported to be 63.4, 45.5, and 64.5 wt%, respectively. These studies demonstrate that CNTs can influence the progress of the ATRP process, albeit the mechanism is not fully understood yet. It has been postulated that initiating radicals

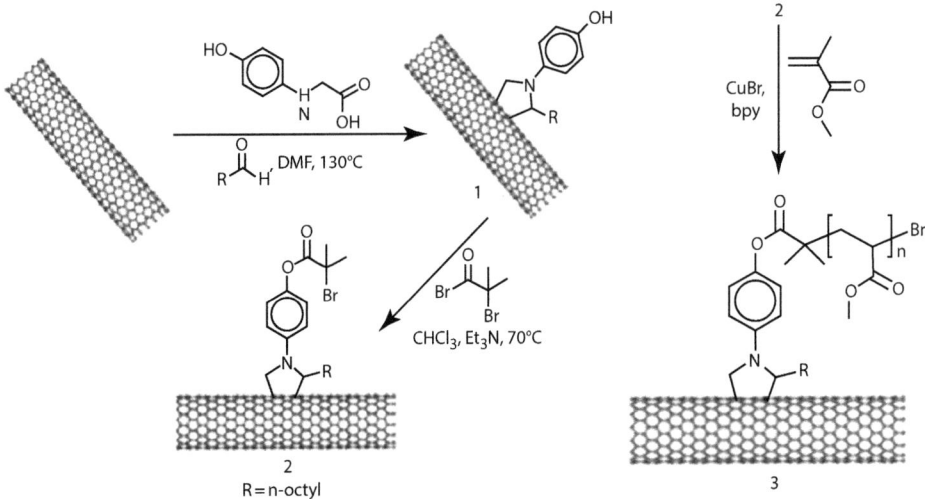

FIGURE 36.5
ATRP *grafting from* approach to prepare PMMA-SWCNT nanocomposites. (Reprinted with permission from Yao, Z., Braidy, N., Botton, G.A., and Adronov, A., Polymerization from the surface of single-walled carbon nanotubes—Preparation and characterization of nanocomposites, *J. Am. Chem. Soc.*, 125, 16015–16024. Copyright 2003 American Chemical Society.)

and propagating polymer chains undergo radical coupling to CNT sidewalls, thus quenching the polymerization and raising the molecular weight distribution.

36.2.2.2 Reversible addition-fragmentation chain transfer

Another way to chemically synthesize polymer-grafted CNTs is reversible addition fragmentation chain-transfer (reversible addition–fragmentation chain transfer [RAFT] polymerization). Using this approach, Cui et al. (2004) grafted PS chains onto MWCNTs. First a thiocarbonylthio RAFT agent was immobilized onto the surface of oxidized nanotubes, and then styrene (St) was polymerized in the presence of the RAFT agent using 2,2-azobisisobutyronitrile (AIBN) as an initiator (Figure 36.6). TEM images of the samples provided direct evidence of the formation of a core-shell nanostructure, in which the MWCNTs were coated with a polymer layer. This methodology could be applied for the preparation of different core-shell polymer nanocomposites. The same procedure was used for grafting PNIPAAm to MWCNTs (Hong et al. 2005) or water-soluble polyacrylamide (PAM) to SWCNTs (Wang et al. 2007), in which the amount of polymer grafted grew with increasing polymerization time. Amphiphilic polymer brushes consisting of an MWCNT hard core and a soft shell of PS-b-PNIPAAM were also synthesized by RAFT polymerization of St followed by NIPAAM (Xu et al. 2007), and the amount of copolymer attached to MWCNTs increased from 56 to 86 wt% when the polymerization time increased from 10 to 36 h.

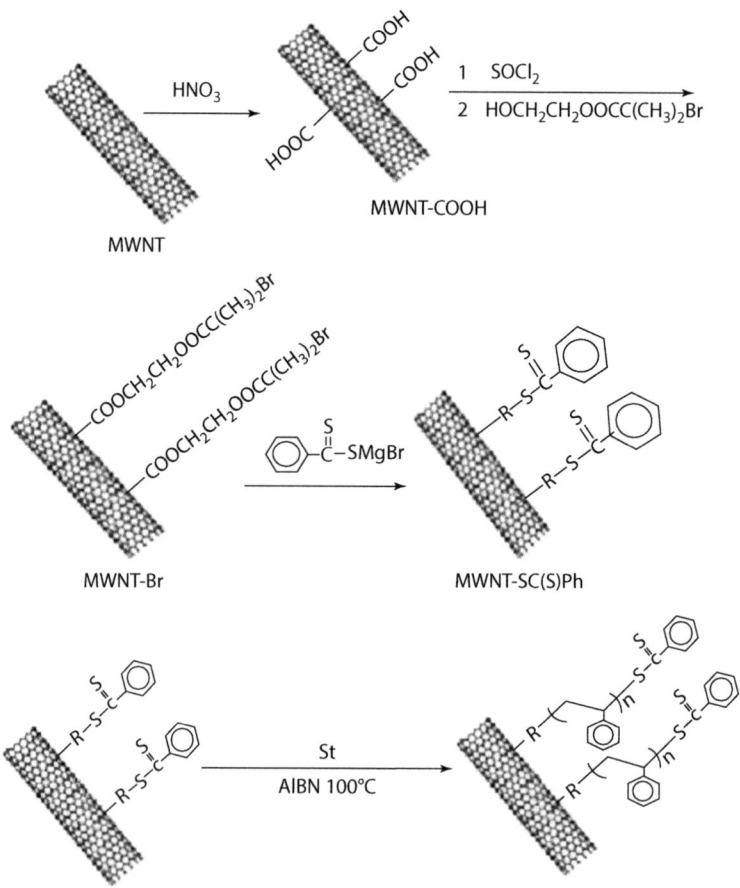

FIGURE 36.6
Synthesis of the RAFT agent and subsequent polymerization of styrene (St) on the nanotube surface. (Reprinted from *Polymer*, 45, Cui, J., Wang, W.P, You, Y.Z., Liu, C., and Wang, P., Functionalization of multiwalled carbon nanotubes by reversible addition fragmentation chain-transfer polymerization, 8717–8721, Copyright 2004, with permission from Elsevier.)

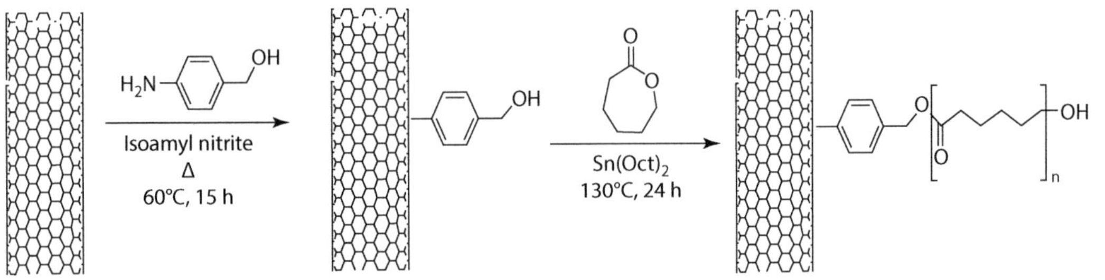

FIGURE 36.7
Nanotube functionalization via the diazonium salt method followed by ROP of ε-caprolactone. (Reprinted with permission from Qu, L., Veca, L.M., Lin, Y. et al., Soluble nylon-functionalized carbon nanotubes from anionic ring opening polymerization from nanotube surface, *Macromolecules*, 38, 10328–10331. Copyright 2005 American Chemical Society.)

36.2.2.3 Ring opening polymerization

The ring opening polymerization (ROP) approach has also been described by several authors for grafting PCL onto modified CNTs. Thus, Buffa et al. functionalized SWCNTs with 4-hydroxy-methylaniline (HMA) via diazonium salt chemistry, and the OH group generated from the functionalization was used to start the ROP of ε-caprolactone (Figure 36.7) (Qu et al. 2005). The PCL-modified nanotubes showed increased suspendibility in chloroform. Alternatively, ROP can be performed by glycol ester-activated MWCNTs in the presence of stannous octoate as catalyst and butyl alcohol as coinitiator (Zeng et al. 2006). In this case, the grafted polymer content was tailored by modifying the ratio of monomer to MWCNTs. Yang et al. (2007) grafted MWCNTs to polyamide chains by attaching caprolactam molecules to isocyanate-modified MWCNTs via anionic ROP. The polymer weight percentage increased with polymerization time, reaching a saturation value of 65% after 6 h. This approach can be applied to other monomers that polymerize via the same mechanism as PCL.

36.2.2.4 Radical and anionic polymerization

In situ free radical polymerization of St or MMA has been frequently employed for CNT functionalization. Following this procedure, Shaffer and Koziol (2002) grafted oxidized MWCNTs with PS. The water dispersion of oxidized MWCNTs was combined with St monomers and a radical initiator. Grafting ratios in the range of 50%–90% were obtained depending on the initial MWCNT content. Yang et al. (2006) synthesized vinyl-modified MWCNTs by the amidation of acyl chloride-functionalized MWCNTs and allylamine, followed by an in situ free radical polymerization of St to generate MWCNT-*g*-PS, and it was found that 1 of every 100 carbon atoms of MWCNTs was functionalized. Qin et al. (2004a,b) reported the grafting of SWCNTs by poly(sodium 4-styrenesulfonate) (PSS) using the same approach (Figure 36.8a), and a 45 wt% of grated PPS was attained. In another work, a PSS/PAA copolymer was grafted to vinyl-functionalized MWCNTs by free radical polymerization of the appropriate monomers (Du et al. 2008). Acid-treated MWCNTs were modified with PMMA chains by the same procedure, and the molecular weight of PMMA was found to rise with MWCNT loading (Jia et al. 1999). Afterward, this route was modified by using the CNTs as macroinitiators (Liu et al. 2004c). First, potassium persulfate was dissolved in hydrogen peroxide solution to adsorb the initiator on the CNT surface. Then, the MMA monomer containing vinyl benzene as the cross-linking agent was emulsified and subsequently added to the solution of activated MWCNTs. Similarly, PMMA chains were grafted onto SWCNTs in a poor solvent for the polymer. Alternatively, Yue et al. (2007) carried out in situ polymerization of MMA in supercritical CO_2. The SWCNTs were first functionalized with amino ethyl methacrylate by amidation of oxidized nanotubes. The supercritical fluid improved the diffusivity of the monomers and favored the growth of PMMA chains.

On the other hand, Viswanathan et al. (2003) performed in situ anionic polymerization of St in the presence of modified SWCNTs. Carbanions were introduced onto the SWCNT surface via

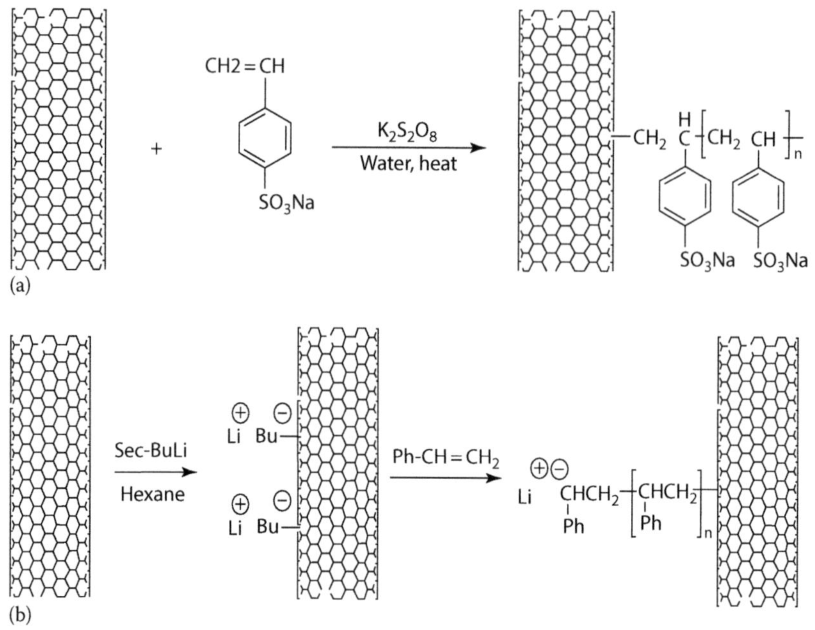

FIGURE 36.8
(a) Grafting of polystyrene derivate by in situ radical polymerization and (b) anionic polymerization of styrene onto carbon nanotubes. (Reprinted from *Prog. Polym. Sci.*, 35, Spitalsky, Z., Tasis, D., Papagelis, K., and Galiotis, C., Carbon nanotube–polymer composites: Chemistry, processing, mechanical and electrical properties, 357–401, Copyright 2010, with permission from Elsevier.)

treatment with sec-butyllithium, a procedure that exfoliated the bundles and provided initiating sites for the polymerization of St (Figure 36.8b). However, the PS content in grafted SWCNTs was only about 10 wt%. A similar method was used by Liang and coworkers (2006) for PMMA, and the PMMA content was around 45 wt%.

36.2.2.5 Other *grafting from* approaches

Water-soluble SWCNTs and MWCNTs can be prepared by grafting polyacrylamide chains from the nanotube surface via redox radical polymerization (Tasis et al. 2007). The first stage consists in the covalent attachment of organic radicals thermally generated from organic peroxides to the CNTs and the second is the polymerization of acrylamide. The advantage of this procedure was that grafting proceeds in aqueous media and at room temperature. Metallocene catalysis polymerization has also been used for the grafting process. For instance, the Ziegler–Natta catalyst ($MgCl_2/TiCl_4$) anchored to the surface of acid-treated CNTs was used for in situ polymerization of ethylene monomer (Liu et al. 2004c), leading to a very high polymer content (94 wt%). Datsyuk et al. (2005) reported an in situ nitroxide-mediated polymerization (NMP) of MMA onto double-walled carbon nanotubes (DWCNTs). First, short chains of PAA or PS were polymerized in situ in the presence of an NMP initiator. Then, the presence of the stable nitroxide radical on the CNT surface enabled reinitiating polymerization of different monomers. The main advantage of this route is that it does not require any CNT pretreatment or functionalization. An identical procedure was applied by these authors to graft PAA-PMA or PAA-PS to MWCNTs and DWCNTs (Datsyuk et al. 2007). A different grafting approach is the endohedral filling, which consists in the encapsulation of small molecules inside hollow CNTs. For example, Liu et al. (2004a,b,d) reported the encapsulation of PS into MWCNTs. For such purposes, the monomer and the radical initiator were placed into the cavities of the CNTs with the aid of supercritical CO_2 and they remained inside the CNTs after removing the CO_2. Subsequently, the monomers were polymerized at a suitable temperature, leading to MWCNT-PS nanocomposites. Steinmetz et al. (2006) filled MWCNTs with PVK and conducting polypyrrole (PPy) prepared inside the nanotubes by free radical polymerization of *N*-vinyl carbazole and oxidation polymerization of pyrrole, respectively, using a super critical fluid. An analogous route was used for the development of MWCNTs filled by polyacetylene (Steinmetz et al. 2007). This procedure with supercritical fluids requires elevated pressures, hence is limited to a few polymers.

36.3 NONCOVALENT FUNCTIONALIZATION OF CARBON NANOTUBES

This type of functionalization offers the advantage of leaving the aromatic structure of the CNTs unperturbed, since the adsorption of polymers occurs without disruption of the extended π-conjugated system but via van der Waals or electrostatic interactions. Therefore, the original electronic and optical properties of the CNTs are preserved, which is a key factor for application. However, the main drawback is that because the nature of the interactions is inherently weak, problems associated with desorption may arise, leaving poorly functionalized carbon nanotubes that tend to reaggregate.

36.3.1 Noncovalent Functionalization through π–π Interactions

Polymers have been demonstrated to be excellent wrapping materials for the noncovalent functionalization of CNTs as a result of the π–π stacking and van der Waals interactions between their chains containing aromatic rings and the CNT surface. They are chosen based on their affinity to the matrix and the fillers, and those with aromatic groups, such as the styrene-based ones, are commonly used to wrap carbon nanomaterials, acting as compatibilizing agents for their integration in polymer matrices. Thus, poly-para-hydroxystyrene has been used to wrap around the CNT surface and increase its solubility in various solvents (Park et al. 2008). Copolymers have also been employed to introduce functional groups on CNT sidewalls. For instance, Ajayan and coworkers (2003) reported the functionalization of SWCNTs with hydrolyzed-poly(styrene-*alt*-maleic anhydride) (h-PSMA). The noncovalently attached layer of h-PSMA contains carboxylic acid groups that were used to covalently attach PEI, forming a cross-linked polymer bilayer. Liu et al. (2006) reported the noncovalent functionalization of MWCNTs with a polystyrene-g-(glycidyl methacrylate-*co*-styrene) (PS-*g*-(GMA-*co*-St)) copolymer. The PS component was attached to MWCNTs, leaving the GMA chains far from the nanotube surfaces. Since GMA is readily soluble in many solvents, the swelling behavior of the nanotubes was significantly improved. Other aromatic polymers such as polyetherimide (PEI), poly(bisphenol-A-ether sulfone) (PSF), and poly(1-4-phenylene ether-ether sulfone) (PEES) have been employed as compatibilizing agents for the integration of SWCNTs in PEEK matrix (Diez-Pascual et al. 2010a).

Another typical approach is the noncovalent bonding of CNTs with polymers containing pendant pyrene groups. Pyrene derivatives are ideal molecules for π–π interactions with carbon nanotubes since their aromatic system is analogous to graphite. In a pioneer work, Petrov and coworkers (2003) reported the synthesis of pyrene-modified MWCNTs where (1-pyrene)methyl 2-methyl-2-propenoate (PyMMP) was synthesized and copolymerized with MMA and 2-(dimethylamino)ethyl methacrylate (DMAEMA) by ATRP. The authors concluded that the polymer to CNTs ratio must be at least two orders of magnitude to have a stable dispersion in organic solvents. The dispersion stability was found to improve with the pyrene content, whereas the molecular weight of the polymer had hardly any effect on this property. Stable dispersions in water were also obtained when PDMAEMA was substituted for PMMA in the copolymerization with PyMMP. The same authors synthesized poly(ethyleneco-butylene)-b-poly(MMA-*co*-PyMMP) diblocks to modify the MWCNT surface in order to enhance their dispersion in various organic solvents

(Lou et al. 2004a). Similar studies were carried out by Wang et al. (2006) using SWCNTs instead of MCWNTs. Meuer and coworkers (2008) performed the synthesis of PMMA by RAFT polymerization end-capped with one α-pyrene unit to be absorbed on the MWCNT surface. They reported that this approach is exceptional for disentangling and dispersing MWCNTs into PMMA or PEG matrices with the potential to align the CNTs. Schopf et al. (2009) demonstrated the importance of using poly[poly(ethylene glycol) acrylate] polymers synthesized by RAFT polymerization and functionalized with pyrene groups. The authors developed a simple method for the assembly of CNTs on silicon surfaces via e-beam lithography, which had the ability of selectively controlling the location of the CNTs into nanoscale patterns, which could be greatly interesting for electronic and biological applications.

Yan et al. (2010) described the synthesis of pyrene-capped polystyrene (PyPS) via anionic polymerization to disperse SWCNTs in organic solvents such as chloroform. The modified SWCNTs were subsequently used as fillers in PS matrices, and the composite materials showed superior optical transmittance and excellent antistatic properties at very low nanotube content. More complex polymers were used for similar studies, as reported by Gorur and coworkers (2011) who synthesized a hexa-armed pyrene end-capped phosphazene dendrimer [N3P3-(Pyr)6] and a star poly(3-caprolactone) polymer [N3P3-(PCL-Pyr)6]. The method used, called a "core first approach," started from a hexa-hydroxyl functional phosphazene derivative (N3P3-(OH)6), which then acted as the initiator for the ROP of 3-caprolactone for the preparation of the star polymer (N3P3-(PCLOH)6). The hydroxyl groups were converted into bromide azide and were attached via click chemistry reactions with 1-ethynyl pyrene. The resulting phosphazene dendrimer and star polymer with the pyrene functional groups were used to noncovalently functionalize MWCNTs.

36.3.2 Noncovalent Functionalization via Ionic Interactions

Ionic interactions have also been employed for noncovalent functionalization of CNTs with polyelectrolytes, which endow CNT surfaces with positively or negatively charged properties, an approach that offers a variety of opportunities for generating CNT-based hybrid nanostructures. Rouse and Lillehei (2003) designed a method of assembling polymer/SWCNTs films via sequential absorption of poly(diallyldimethylammoniun chloride) (PDDA) followed by carbon nanotubes onto a substrate. This method results in the uniform growth of films containing a high CNT concentration. The same approach was used by Wang et al. (2001) for the functionalization of CNTs in order to act as metal-free catalysts for oxygen reduction reaction in fuel cells. Ionic interactions between carboxyl-modified MWCNTs and positively charged polymers were also reported by Yan et al. (2007), who fabricated uniform polyaniline (PANI)/MWCNT nanocomposites by mixing an aqueous colloidal suspension of positively charged PANI nanofiber and an aqueous dispersion of oxidized MWNTs (Figure 36.9). Raman and FT IR spectroscopies revealed a strong electrostatic interaction between $C–N^+$ species of the PANI and the COO^- of MWCNTs. Based on noncovalent ionic interactions, the layer-by-layer (LbL) technique has been employed for the preparation of multilayer composite coatings of CNTs. In a pioneer work, Mamedov et al. (2002), silicon wafers were alternatively dipped into dispersions of SWCNTs and polyelectrolyte solutions. This approach enables the preparation of composites with very high CNT loading (i.e., 50 wt%). Lu et al. (2006) studied the viscoelastic properties of polyelectrolyte multilayer films, prepared by LbL assembly of PDDA and PAA with and without SWNTs. The SWCNT loading was about 4.7% by weight and was enough to increase the Young's modulus and the tensile strength of the polyelectrolyte film by 1.57- and 1.82-fold, respectively. Polyelectrolyte layers adsorbed onto CNTs can be subsequently used as an anchorage for different biomacromolecules. In this regard, Zykwinska and coworkers (2010) functionalized polyelectrolyte modified CNTs with a natural biopolymer. Analogously, Aslan et al. (2012) employed two polypeptides, cationic poly(L-lysine) and anionic poly(L-glutamic acid), and the resulting composites exhibited antimicrobial behavior.

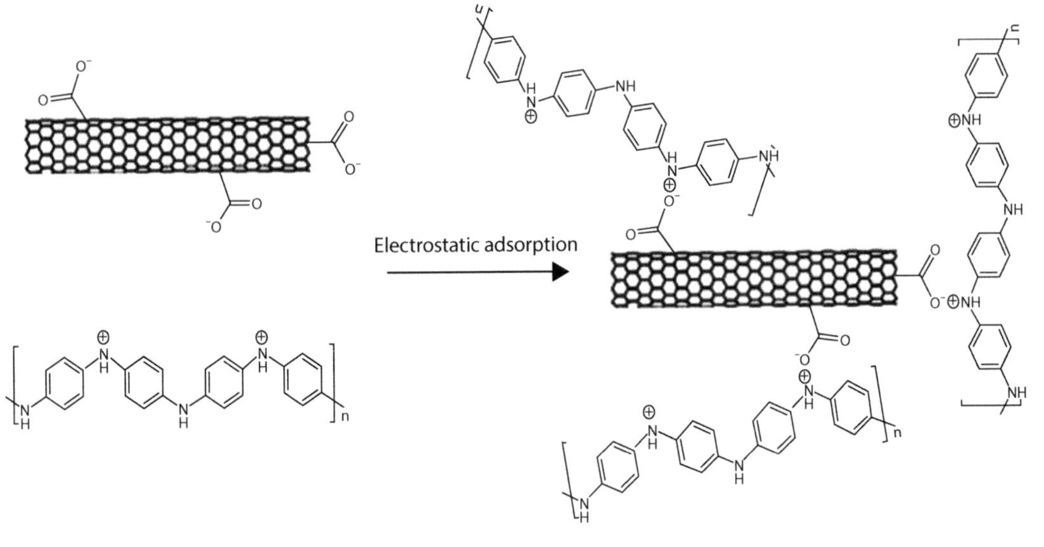

FIGURE 36.9
Schematic representation of the electrostatic adsorption between negatively charged MWCNTs and positively charged PANI molecules. (Reprinted with permission from Yan, X.B., Han, Z.J., Yang, Y., and Tay, B.K., Fabrication of carbon nanotube-polyaniline composites via electrostatic adsorption inaqueous colloids, *J. Phys. Chem. C*, 111, 4125–4131. Copyright 2007 American Chemical Society.)

36.3.3 Noncovalent Functionalization via CH–π and OH–π Interactions

Although the strength of CH–π interactions is only one-tenth of the hydrogen bond, molecules containing many CH linkages can sufficiently interact with CNTs and form complexes. Baskaran et al. (2005a,b) investigated the effect of CH–π interactions between polymers and MWCNTs and found shifts in the IR and Raman frequencies of the composites. The authors concluded that, although the intermolecular CH–π interaction is weak, its effect on the dissolution of MWCNTs in organic solvents is obvious and indicates a general phenomenon occurring between polymers and CNTs. Wang and Chen developed temperature responsive SWCNT dispersions through noncovalent interactions with PNIPAAm (Wang and Chen 2007). These interactions occurred between the hydrocarbon backbone and the isopropyl groups of PNIPAAm and SWCNTs through van der Waals and hydrophobic interactions. Despite the affinity of PNIPAAm to SWCNT sidewalls being relatively low, the suspension can be reversibly changed between the dispersed and aggregated states by modifying the temperature. Naito et al. (2008) investigated the wrapping of poly(dialkylsilane)s with random coiled, flexible, and semiflexible main chains onto SWCNTs. The polymer-wrapping resulted from strong CH–π interactions between the alkyl side chains of the polymers and the CNT surface. Zheng and Xu (2010) compared the morphology and crystallization behavior of polyethylene (PE) and poly(ethylene oxide) (PEO) on SWCNTs with the aid of supercritical CO_2. The interfacial interactions between the polymers and the SWCNTs were investigated by FTIR and Raman spectroscopy. The changes in the vibration frequencies of the composites compared to the pure polymers demonstrated the existence of noncovalent interactions. PE interacted favorably with the SWCNTs, leading to "shish-kebab" structures, while PEO had unfavorable interactions that prevented from forming ordered crystals.

On the other hand, very scarce literature regarding the OH-π noncovalent interactions of CNTs with polymers has been reported, which are mostly related to polysaccharides and their derivatives. Numata and coworkers (2005) demonstrated the existence of a periodically helical wrapping of natural polysaccharides with pristine and shortened SWCNTs. Fu et al. (2007) described the wrapping ability of amylose to form helical superstructures around SWCNTs. Interestingly, amylose/SWCNT complexes had much better biocompatibility than raw SWCNTs. The same authors

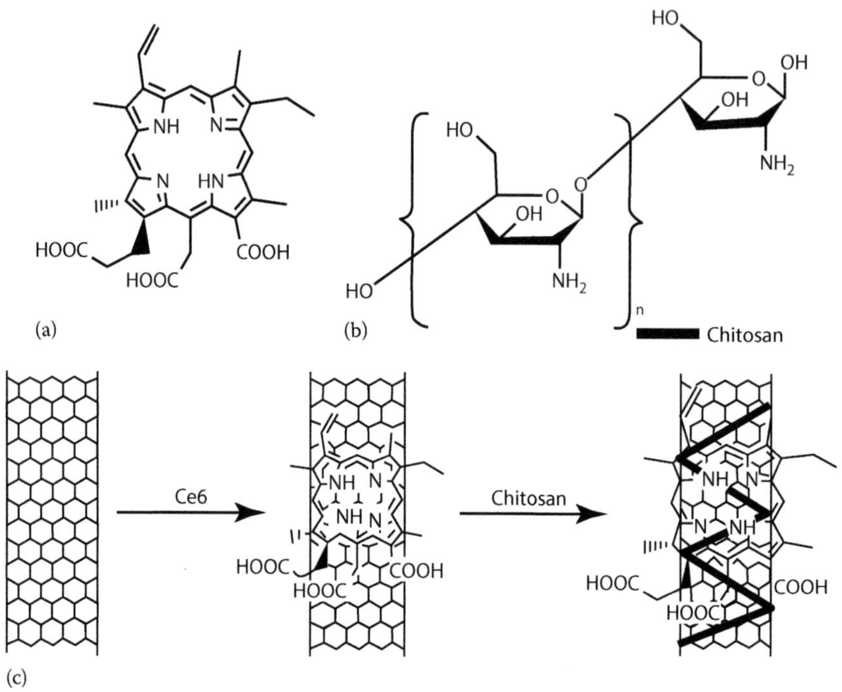

FIGURE 36.10
Schematic representation of the molecular structure of: (a) Ce6, (b) chitosan, and (c) preparation of chitosan–Ce6–SWCNTs complexes. (Reprinted from *Carbon*, 50, Xiao, H., Zhu, B., Wang, D. et al., Photodynamic effects of chlorine e6 attached to single wall carbon nanotubes through noncovalent interactions, 1681–1689, Copyright 2012, with permission from Elsevier.)

reported the modification of SWCNTs with other natural polysaccharides, such as alginate sodium and chitosan by noncovalent interactions (Zhang et al. 2009). They proved that compared to purified and oxidized SWCNTs, the polysaccharide-wrapped ones can successfully mimic nanofibrous extracellular matrix and enhance cell adhesion and proliferation. Moreover, chlorine e6–SWCNTs complexes have been wrapped by chitosan to improve aqueous solubility and biocompatibility (Figure 36.10, Xiao et al. 2012). Chlorin e6 (Ce6) is a promising photosensitizer for photodynamic therapy and was loaded on SWCNTs by noncovalent π–π interactions. Biological tests demonstrated that chitosan-wrapped Ce6–SWCNTs complexes exhibited higher anticancer effect against HeLa cells than free Ce6.

36.4 CONCLUSIONS AND FUTURE PERSPECTIVES

CNTs are new allotropes of carbon greatly interesting to many disciplines including chemistry, physics, biology, material science, and engineering owed to their unique properties and potential applications, ranging from optoelectronics to biotechnology. Nevertheless, their strong tendency to aggregate and insolubility in common solvents limit their wide accessibility for many applications. Various approaches have been adopted to address these issues, including functionalization of CNTs with polymers. These strategies are classified as "covalent" or "noncovalent" depending on whether chemical modification occurs with rehybridization of the carbon atoms from sp^2 to sp^3 (via formation of covalent bonds) or through ionic interactions or van der Waals forces that do not disrupt the sp^2 aromatic structure. While the former method enables the formation of stable modified CNTs, it adversely affects their properties as a result of the partial damage of the π-conjugation framework. In contrast, the noncovalent route preserves their intrinsic mechanical and electronic properties, although it results in complexes with low stability that can lead to desorption of the as-introduced functionalities. The choice criteria for one approach over the other is dependent on several factors, the most important one being the synthetic strategy for the postmodification steps and the application purpose of the resulting hybrid materials. For instance, selective CNT functionalization can be achieved via click chemistry by preparing azide-functionalized polymers. However, this approach has limited practical use owing to the need for multistep reactions for azide and alkyl groups to apply click chemistry. Therefore, although several milestones have been achieved, ongoing research efforts are necessary to design new selective, cost-effective, environmentally friendly, and scalable functionalization routes that could be applied at an industrial level. A large number of parameters including the solvent type, polymer molecular weight, concentration, and viscosity should be considered when screening for new polymers for selective CNT functionalization. All these parameters need to be optimized in order to attain CNT systems with versatile architectures, new functionalities, and superior properties. In summary, the proper selection and careful design of the CNT functionalization method is the key challenge to attain high-performance composites with optimal CNT dispersion and adhesion to the polymer matrix.

REFERENCES

Aslan, S., Deneufchatel, M., Hashmi, S. et al. 2012. Carbon nanotube-based antimicrobial biomaterials formed via layer-by-layer assembly with polypeptides. *J. Colloid Interface Sci.* 388:268–273.

Bahr, J. L. and Tour, J. M. 2001. Highly functionalized carbon nanotubes using in situ generated diazonium compounds. *Chem. Mater.* 13:3823–3824.

Baskaran, D., Dunlap, J. R., Mays, J. W., and Bratcher, M. S. 2005a. Grafting efficiency of hydroxy-terminated poly(methyl methacrylate) with multiwalled carbon nanotubes. *Macromol. Rapid. Commun.* 26:481–486.

Baskaran, D., Mays, J. W., and Bratcher, M. S. 2004. Polymer-grafted multiwalled carbon nanotubes through surface-initiated polymerization. *Angew. Chem. Int. Ed.* 43:2138–2142.

Baskaran, D., Mays, J. W., and Bratcher, M. S. 2005b. Noncovalent and nonspecific molecular interactions of polymers with multiwalled carbon nanotubes. *Chem. Mater.* 17:3389–3397.

Baskaran, D., Sakellariou, G., Mays, J. W., and Bratcher, M. S. 2007. Grafting reactions of living macroanions with multi-walled carbon nanotubes. *J. Nanosci. Nanotechnol.* 7:1560–1567.

Baudot, C., Volpe, M. V., Kong, J. C., and Tan, C. M. 2009. Epoxy functionalized carbon nanotubes and methods of forming the same. US Patent 299,082.

Blake, R., Gun'ko, Y. K., Coleman, J. et al. 2004. A generic organometallic approach toward ultra-strong carbon nanotube polymer composites. *J. Am. Chem. Soc.* 126:10226–10227.

Chen, Z., Dai, X. J., Lamb, P. R. et al. 2012. Practical amine functionalization of multiwalled carbon nanotubes for effective interfacial bonding. *Plasma Process. Polym.* 9:733–741.

Chen, Z., Dai, X. J., Magniez, K. et al. 2013. Improving the mechanical properties of epoxy using multiwalled carbon nanotubes functionalized by a novel plasma treatment. *Compos. Part A Appl. Sci. Manuf.* 45:145–152.

Cui, J., Wang, W. P., You, Y. Z., Liu, C., and Wang, P. 2004. Functionalization of multiwalled carbon nanotubes by reversible addition fragmentation chain-transfer polymerization. *Polymer* 45:8717–8721.

Datsyuk, V., Billon, L., Guerret-Piecourt, C. et al. 2007. in situ nitroxide-mediated polymerized poly(acrylic acid) as a stabilizer/compatibilizer carbon nanotube/polymer composites. *J. Nanomater.* 2007:74769–74781.

Datsyuk, V., Guerret-Piecourt, C., Dagreou, S. et al. 2005. Double walled carbon nanotube/polymer composites via in-situ nitroxide mediated polymerisation of amphiphilic block copolymers. *Carbon* 43:873–876.

Dervishi, E., Li, Z., Xu, Y. et al. 2009. Carbon nanotubes: Synthesis, properties, and applications. *Particul. Sci. Technol.* 27:107–125.

Díez-Pascual, A. M., Martinez, G., Gonzalez-Dominguez, J. M., Martinez, T., and Gomez-Faou, M. A. 2010a. Grafting of a hydroxylated PEEK derivative to the surface of single-walled carbon nanotubes. *J. Mater. Chem.* 20:8285–8296.

Díez-Pascual, A. M. and Naffakh, M. 2012. Grafting of an aminated poly(phenylene sulphide) derivative to functionalized single-walled carbon nanotubes. *Carbon* 50:857–868.

Díez-Pascual, A. M., Naffakh, M., Gómez, M. A. et al. 2009. Development and characterization of PEEK/carbon nanotube composites. *Carbon* 47:3079–3090.

Díez-Pascual, A. M., Naffakh, M., González-Domínguez, J. M. et al. 2010b. High performance PEEK/carbon nanotube composites compatibilized with polysulfones-I. Structure and thermal properties. *Carbon* 48:3485–3499.

Du, F., Wu, K., Yang, Y., Liu, L., Gan, T., and Xie, X. 2008. Synthesis and electrochemical probing of water-soluble poly(sodium 4-styrenesulfonate-coacrylicacid)-grafted multiwalled carbon nanotubes. *Nanotechnology* 19:85716–85724.

Fu, C., Meng, L., Lu, Q., Zhang, X., and Gao, C. 2007. Large-scale production of homogeneous helical amylose/SWNTs complexes with good biocompatibility. *Macromol. Rapid Commun.* 28:2180–2184.

Ge, J. J., Zhang, D., Li, Q. et al. 2005. Multiwalled carbon nanotubes with chemically grafted polyetherimides. *J. Am. Chem. Soc.* 127:9984–9985.

González-Domínguez, J. M., Anson-Casaos, A., Díez-Pascual, A. M. et al. 2011. Solvent-free preparation of high-toughness epoxy-SWNT composite materials. *ACS Appl. Mater. Interf.* 3:1441–1450.

Gorur, M., Yilmaz, F., Kilic, A., Sahin, Z. M., and Demirci, A. 2011. Synthesis of pyrene end-capped A6 dendrimer and star polymer with phosphazene core via "click chemistry." *J. Polym. Sci. Part A Polym. Chem.* 49:3193–3206.

Han, S. J., Kim, B., and Suh, K. D. 2007. Electrical properties of a composite film of poly(acrylonitrile) nanoparticles coated with carbon nanotubes. *Macromol. Chem. Phys.* 208:377–383.

Hill, D., Lin, Y., Qu, L. et al. 2005. Functionalization of carbon nanotubes with derivatized polyimide. *Macromolecules* 38:7670–7675.

Hirsch, A. 2002. Functionalization of carbon nanotubes. *Angew. Chem. Int. Ed.* 41:1853–1859.

Hong, C. Y., You, Y. Z., and Pan, C. Y. 2005. Synthesis of water-soluble multiwalled carbon nanotubes with grafted temperature-responsive shells by surface RAFT polymerization. *Chem. Mater.* 17:2247–2254.

Hu, H., Ni, N., Mandal, S. K. et al. 2005. Polyethyleneimine functionalized single-walled carbon nanotubes as a substrate for neuronal growth. *J. Phys. Chem. B* 109:4285–4389.

Huang, W., Fernando, S., Allard, L. F., and Sun, Y. P. 2003. Solubilization of single walled carbon nanotubes with diamine-terminated oligomeric poly(ethylene glycol) in different functionalization reactions. *Nano Lett.* 3:565–568.

Huang, W., Lin, Y., Taylor, S., Gaillard, J., Rao, A. M., and Sun, Y. P. 2002. Sonication-assisted functionalization and solubilization of carbon nanotubes. *Nano Lett.* 2:231–234.

Iijima, S. 1991. Helical microtubules of graphitic carbon. *Nature* 354:56–58.

Jia, Z., Wang, Z., Xu, C. et al. 1999. Study on poly(methyl methacrylate): Carbon nanotube composites. *Mater. Sci. Eng. A* 271:395–400.

Jung, Y. C., Sahoo, N. G., and Cho, J. W. 2006. Polymeric nanocomposites of polyurethane block copolymers and functionalized multi-walled carbon nanotubes as crosslinkers. *Macromol. Rapid. Commun.* 27:126–131.

Kong, H., Li, W., Gao, C. et al. 2004. Poly(Nisopropylacrylamide)-coated carbon nanotubes: Temperature sensitive molecular nanohybrids in water. *Macromolecules* 37:6683–6686.

Koshio, A., Yudasaka, A. M., Zhang, M., and Iijima, S. 2001. A simple way to chemically react single-wall carbon nanotubes with organic materials using ultrasonication. *Nano Lett.* 1:361–363.

Kuan, H. C., Ma, C. C. M., Chang, W. P., Yuen, S. M., Wu, H. H., and Lee, T. M. 2005. Synthesis, thermal, mechanical and rheological properties of multiwall carbon nanotube/waterborne polyurethane nanocomposite. *Compos. Sci. Technol.* 65:1703–1710.

Lahiff, E., Ryu, C. Y., Curran, S., Minett, A. L., Blau, W. J., and Ajayan, P. M. 2003. Selective positioning and density control of nanotubes within a polymer thin film. *Nano Lett.* 3:1333–1337.

Li, H. and Adronov, A. 2007. Water-soluble SWCNTs from sulfonation of nanotube-bound polystyrene. *Carbon* 45:984–990.

Li, H., Cheng, F., Duft, A. M., and Adronov, A. 2005. Functionalization of single walled carbon nanotubes with well-defined polystyrene by "click" coupling. *J. Am. Chem. Soc.* 127:14518–14524.

Li, Y. B., Wei, B. Q., Liang, J., Yu, Q., and Wu, D. H. 1999. Transformation of carbon nanotubes to nanoparticles by ball milling process. *Carbon* 37:493–497.

Li, Z., Dong, Y., Haussler, M. et al. 2006. Synthesis of, light emission from, and optical power limiting in soluble single-walled carbon nanotubes functionalized by disubstituted polyacetylenes. *J. Phys. Chem. B* 110:2302–2309.

Liang, F., Beach, J. M., Kobashi, K. et al. 2006. in situ polymerization initiated by single-walled carbon nanotube salts. *Chem. Mater.* 18:4764–4767.

Lin, Y., Rao, A. M., Sadanadan, B., Kenik, E. A., and Sun, Y. P. 2002. Functionalizing multiple-walled carbon nanotubes with aminopolymers. *J. Phys. Chem. B* 106:1294–1298.

Lin, Y., Zhou, B., Fernando, K. A. S., Liu, P., Allard, L. F., and Sun, Y. P. 2003. Polymeric carbon nanocomposites from carbon nanotubes functionalized with matrix polymer. *Macromolecules* 36:7199–7204.

Liu, C., Cheng, H. M., Zhao, H., Yang, F., and Zhang, X. 2004a. Surface modification of single-walled carbon nanotubes with polyethylene via in situ Ziegler–Natta polymerization. *J. Appl. Polym. Sci.* 92:3697–3700.

Liu, I. C., Huang, H. M., Chang, C. Y., Tsai, H. C., Hsu, C. H., and Tsiang, R. C. C. 2004c. Preparing a styrenic polymer composite containing well-dispersed carbon nanotubes: Anionic polymerization of a nanotube-bound p-methylstyrene. *Macromolecules* 37:283–287.

Liu, M., Yang, Y., Zhu, T., and Liu, Z. 2007. A general approach to chemical modification of single-walled carbon nanotubes with peroxy organic acids and its application in polymer grafting. *J. Phys. Chem. C* 111:2379–2385.

Liu, Y., Tang, J., and Xin, J. H. 2004d. Fabrication of nanowires with polymer shellsusing treated carbon nanotube bundles as macro-initiators. *Chem. Commun.* 2004:2828–2829.

Liu, Y. T., Zhao, W., Huang, Z. Y. et al. 2006. Non-covalent surface modification of carbon nanotubes for solubility in organic solvents. *Carbon* 44:1613–1616.

Liu, Z., Dai, X., Xu, J. et al. 2004b. Encapsulation of polystyrene within carbon nanotubes with the aid of supercritical CO_2. *Carbon* 42:458–460.

Lou, X., Daussin, R., Cuenot, S. et al. 2004a. Synthesis of pyrene-containing polymers and noncovalent sidewall functionalization of multiwalled carbon nanotubes. *Chem. Mater.* 16:4005–4011.

Lou, X., Detrembleur, C., Pagnoulle, C. et al. 2004b. Surface modification of multiwalled carbon nanotubes by poly(2-vinylpyridine): Dispersion, selective deposition, and decoration of nanotubes. *Adv. Mater.* 16:2123–2127.

Lu, H., Huang, G., Wang, B., Mamedov, A., and Gupta, S. 2006. Characterization of the linear viscoelastic behavior of single-wall carbon nanotube/polyelectrolyte multilayer nanocomposite film using nanoindentation. *Thin Solid Films* 500:197–202.

Ma, P. C., Siddiqui, N. A., Marom, G., and Kim, J. K. 2010. Dispersion and functionalization of carbon nanotubes for polymer-based nanocomposites: A review. *Compos. Part A Appl. Sci. Manuf.* 41:1345–1367.

Mamedov, A. A., Kotov, N. A., Prato, M., Guldi, D. M., Wicksted, J. P., and Hirsch, A. 2002. Molecular design of strong single-wall carbon nanotube/polyelectrolyte multilayer composites. *Nat. Mater.* 1:190–194.

Matrab, T., Chancolon, J., L'hermite, M. M. et al. 2006. Atom transfer radical polymerization (ATRP) initiated by aryl diazonium salts: A new route for surface modification of multiwalled carbon nanotubes by tethered polymer chains. *Colloids Surf. Physicochem. Eng. Asp.* 287:217–221.

Meuer, S., Braun, L., and Zentel, R. 2008. Solubilisation of multi walled carbon nanotubes by α-pyrene functionalised PMMA and their liquid crystalline self-organisation. *Chem. Commun.* 2008:3166–3168.

Naito, M., Nobusawa, K., Onouchi, H. et al. 2008. Stiffness- and conformation-dependent polymer wrapping onto single-walled carbon nanotubes. *J. Am. Chem. Soc.* 130:16697–16703.

Numata, M., Asai, M., Kaneko, K. et al. 2005. Inclusion of cut and as-grown single-walled carbon nanotubes in the helical superstructure of schizophyllan and curdlan (β-1,3-Glucans). *J. Am. Chem. Soc.* 127:5875–5884.

Park, S., Huh, J. O., Kim, N. G. et al. 2008. Photophysical properties of noncovalently functionalized multi-walled carbon nanotubes with poly-*para*-hydroxystyrene. *Carbon* 46:706–720.

Petrov, P., Stassin, F., Pagnoulle, C., and Jerome, R. 2003. Noncovalent functionalization of multi-walled carbon nanotubes by pyrene containing polymers. *Chem. Commun.* 2003:2904–2905.

Popov, V. N. 2004. Carbon nanotubes: Properties and application. *Mater. Sci. Eng. R* 43:61–102.

Qin, S., Qin, D., Ford, W. T., Resasco, D. E., and Herrera, J. E. 2004a. Functionalization of single-walled carbon nanotubes with polystyrene via grafting to and grafting from methods. *Macromolecules* 37:752–757.

Qin, S., Qin, D., Ford, W. T. et al. 2004b. Solubilization and purification of single-wall carbon nanotubes in water by in situ radical polymerization of sodium 4-styrenesulfonate. *Macromolecules* 37:3965–3967.

Qu, L., Lin, Y., Hill, D. E. et al. 2004. Polyimide functionalized carbon nanotubes: Synthesis and dispersion in nanocomposite films. *Macromolecules* 37:6055–6060.

Zheng, X. and Xu, Q. 2010. Comparison study of morphology and crystallization behavior of polyethylene and poly(ethylene oxide) on single-walled carbon nanotubes. *J. Phys. Chem. B* 114:9435–9444.

Zykwinska, A., Radji-Taleb, S., and Cuenot, S. 2010. Layer-by-layer functionalization of carbon nanotubes with synthetic and natural polyelectrolytes. *Langmuir* 26:2779–2784.

CHAPTER 37

CONTENTS

Graphene

Functionalization with Nanoparticles and Applications

37

Shabi Abbas Zaidi and Jae Ho Shin

37.1 INTRODUCTION

Carbon, from the Latin word *carbo* (coal), is a unique element that has an indispensible part in the entire organic chemistry. It shows an outstanding ability to make numerous members containing chains and ring networks by utilizing single, double, and triple bond formations. Carbon forms few allotropes such as diamond, graphite, fullerenes, amorphous carbon, and carbon nanotubes (CNTs); however, diamond and graphite are two usually obtained forms at ambient conditions. It has been delineated that diamond contains sp^3-bonded carbon atoms arranged in a tetrahedral lattice with sigma bonds resulting to one of the hardest of all natural materials. On the contrary, graphite is a very soft and delicate material composed of sp^2-hybridized carbon atoms in planar hexagonal rings (Mukhopadhyay and Gupta 2013).

The building block of all the carbonaceous materials including CNTs, graphene, and fullerene is a single graphitic layer that is covalently functionalized and densely packed in a hexagonal honeycomb crystal lattice. The different forms of carbon are obtained by the rolling mechanism of single graphite layer. For example, 0D fullerene is formed when a single graphitic layer is rolled into a sphere, whereas 1D CNT is obtained after a graphitic layer is rolled with respect to its axis as shown in Figure 37.1 (Geim and Novoselov 2007).

Graphene, which consists of a single atomic sheet in planar 2D structure form, may be designated as single-layer, bilayer, trilayer, few-layer and, multilayer graphene based on its formation with single or stacking of two, three, 5–10, and ~20–30 layers of graphite, respectively. These layers are stacked with strong in-plane bonds and weak van der Waals–like forces. Furthermore, it also has been established successfully that different dimensions in carbon material exhibit dramatically different chemical properties.

In the last two decades, nanostructured materials have received surge of interest owing to their remarkable properties. Among various nanostructured nanomaterials, the carbonaceous materials such as CNTs and graphene are at the forefront of advanced materials research.

Ever since the discovery of graphene by Novoselov et al. in 2004 at the University of Manchester, United Kingdom, it has been heralded as the most profound and next-generation material dominating its other forms (i.e., especially CNTs). The overwhelming interest in the graphene chemistry is due to its enormous potential and impact in material synthesis at nanoscale. Few fascinating physicochemical properties of graphene are theoretically high specific surface area of a single graphene sheet (~2630 m^2 g^{-1}), which is quite higher than that of activated carbons and CNTs, high electronic capabilities, excellent thermal and electrical conductivity, and strong mechanical strength (Enoki and Ando 2013).

Normally, graphene exhibits two oxidation states: graphene oxide (GO) and reduced graphene oxide (RGO). Generally, their solubility and electrical conductivity vary from one state to another. For instance, GO shows fair solubility in water and less electrical conductivity, whereas RGO is sparingly dispersible in many solvents, particularly water, and offers excellent electrical conductivity. There are numerous work reports validating the fact that the existence of rich oxygen-containing and hydrophilic groups, such as hydroxyl, epoxide, carboxyl, and carboxylic groups over planar

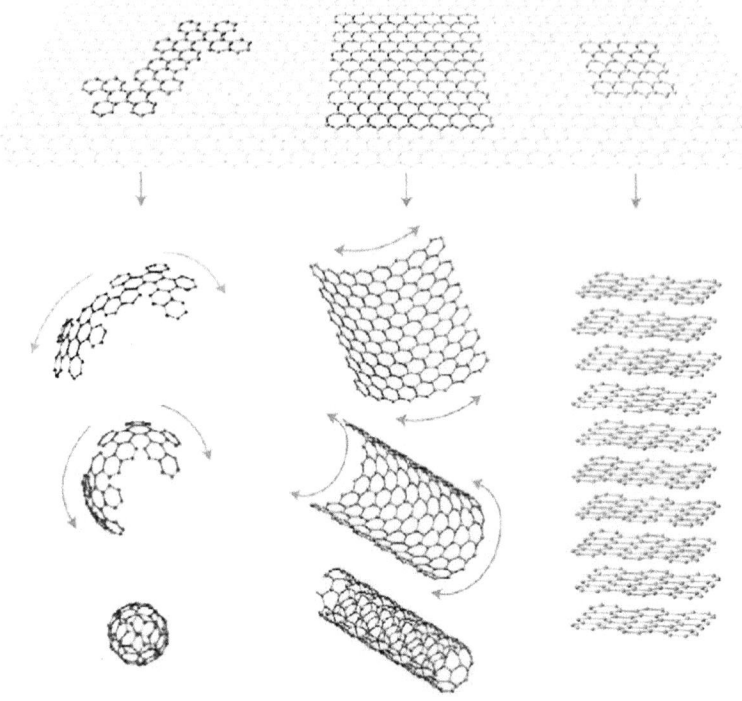

FIGURE 37.1
Schematic illustrations of graphene and graphene-based 0D-, 1D-, and 3D-structured carbon.
(Reprinted by permission from Macmillan Publishers Ltd. *Nat. Mater.*, Geim, A.K. and Novoselov,
K.S., The rise of grapheme, 6(3), 183–191, copyright 2007.)

graphene sheet, is primarily responsible for good water solubility of GO. Upon successful reduction
of GO, most of the oxygen-containing groups, in particular the hydroxyl, epoxide, and carboxyl,
will be completely eliminated, thus resulting into π-conjugation-rich graphene, that is, RGO. The
π-conjugation in GO (now RGO) substantiates the conductivity of graphene tremendously. However,
the excellent conductivity is generated after sacrificing its solubility in water and other organic solvents. The decline in solubility may inevitably reduce the processability of graphene and therefore
limit its applications. Furthermore, some physical properties including optical transmittance and
resistance strictly depend upon the number of stacked graphite layers. Li et al. (2009) reported that
monolayer graphene exhibits ~97% optical resistance with ~2.2 KΩ/sq sheet resistance, but optical
transmittance decreased to ~95%, ~92%, and ~89% with corresponding sheet resistance of 1, ~700,
and ~400 KΩ/sq for 2-layer, 3-layer, and 4-layer graphene sheet stacking, respectively. Hence, the
careful, appropriate fine-tuning and control synthesis are required in order to obtain successful GO/
RGO materials (Choi and Lee 2012).

37.2 SYNTHESIS OF GRAPHENE

Upon reviewing a vast literature available on graphene synthesis, it has been concluded that numerous techniques have been applied for graphene synthesis. Das in 2013 categorized and represented
different graphene synthesis methods into two broad sections: top-down and bottom-up processes.

37.2.1 Top-Down Approaches

The top-down approaches involve the utilization of a large quantity of graphite for oxidation and
exfoliation and finally proceed to obtain reduced graphene (RGO).

Although, it was Lang (1975) who studied and synthesized few-layer graphite material on a
single crystal platinum surface via chemical decomposition process, in 2004 for the first time,

Novoselov et al. reported the successful synthesis of monolayer graphene using *adhesive tape* (scotch tape) peeling method for mechanically splitting strongly stacked layers into individual sheets on a SiO_2/Si substrate. This graphene synthesis method was simple and cost-effective, but industrial production of graphene cannot be realized using this method.

In 1999, Lu et al. proposed atomic force microscopy (AFM) tip approach for production of graphene from few layers down to monoatomic single graphene layer followed by a tipless AFM cantilever method for improved production of graphene at large scale. However, these techniques could provide graphene of a thickness 10 nm (~30 stacked layers of graphene) and were utilized for fabrication of field effect transistors. Zhang et al. (2005) worked for the improvement of large-scale graphene production procedure over AFM cantilever.

Another established method for graphene synthesis is chemical exfoliation, which is realized by intercalating alkali metals (Viculis et al. 2005). After the first report by Viculis et al. (2003) on the production of chemically exfoliated graphite using potassium (K), this synthesis process has been progressed gradually for achieving high monolayer yield of large-sized conductive graphene sheets under mild solution-phase conditions and avoided use of hazardous exfoliating compounds (i.e., oleum and hydrazine).

The chemical synthesis has been one of the most commonly sought methods for production of high-quality graphene. This process involves the synthesis of GO by different oxidation ways of graphite, dispersing the flakes by sonication and reducing it back to graphene by various means. Although, this synthesis route was first demonstrated by Boehm et al. (1962), however, the recent revolutionary work of Rouff and coworkers revived it again (Stankovich et al. 2006b,c).

After reviewing majority of research accomplished on graphene, it can be realized that three oxidation methods named after their inventors known as Brodie (1860), Staudenmaier (1898), and Hummers and Hoffeman (1958) are prevailing. Particularly, the method of Hummers and Hoffeman is more popular and endorsed among the three. These methods involve various strong acids and oxidizing agents (i.e., nitric acid, sulfuric acid sodium nitrite, and potassium permanganate) for oxidation, and the degree of oxidation varies by the reaction parameters including pressure and temperature, stoichiometry, and type of precursor graphite material. The synthesis of GO is followed by exfoliation of few-layer stacked GO using sonication. Here, the conditions of sonication (time, sonication power, and solvents) need strict attention for successful exfoliation results. Finally, the reduction of GO was carried out with hydrazine to obtain RGO.

37.2.2 Bottom-Up Approaches

The bottom-up approaches focus on the utilization of laboratory grade appropriate material as starting material followed by pyrolyzation to synthesize graphene sheets. These methods are known as solvothermal methods.

Choucair et al. (2009) proposed the pyrolysis method where sodium and ethanol were utilized to prepare sodium ethoxide. The sodium ethoxide was pyrolyzed resulting into stacked layers of graphene, which were simply dispersed by sonication to achieve monoatomic graphene sheet. This method was considered highly cost-effective and compatible for mass-scale production of graphene but provided defective graphene.

Another novel graphene synthesis method is called unzipping or unwrapping of CNTs, which leads into a long and thin graphene referred as *graphene nanoribbon*. It is usually carried out by chemical or plasma-etched methods and sometimes intercalation of metals or simpler chemicals (Chen et al. 2007; Cano-Marquez et al. 2009; Jiao et al. 2009, 2010).

When a single crystalline film is grown over single crystalline substrate, it is called epitaxial approach. It can be classified as homoepitaxial and heteroepitaxial approaches based on growth of a layer of the same material as the substrate (Si on Si) and growth of a layer of a different material than the substrate (GaAs on Si), respectively. The epitaxial method requires high-temperature treatment of a precursor material. In 1975, Van Bommel et al. proposed epitaxial growth method for graphene deposition over SiC substrate (heteroepitaxial).

Although, the epitaxial growth approach could help enable large-sized and single-domain graphene production in a controlled manner, however, being only one-atom thick, graphene is vulnerable to perturbations from its supporting substrate. To access the intrinsic electronic properties of graphene, a substrate that does not disturb its electronic structure is highly desired. Recently, Yang et al. (2013) successfully applied epitaxial growth mechanism of single-domain graphene on hexagonal boron nitride and proposed that their approach can be used widely for graphene band engineering through epitaxy on different substrates.

Another widely utilized method for graphene synthesis is thermal chemical vapor deposition (CVD) where thermally decomposed precursor materials are deposited over substrate surface resulting into desired materials.

In 1975 and 1979, two separate groups of Lang and Eizenberg et al., respectively, reported on the synthesis of thermal CVD. However, this technique has not found due attention until Somani et al. (2006) attempted to produce graphene on Ni foil using CVD. Since then, there are numerous work reports on CVD by applying various accessible precursors (Obraztsov et al. 2007; Yu et al. 2008; Obraztsov 2009; Kim et al. 2009; Ismach et al. 2010; Verma et al. 2010). Furthermore, CVD is also used to fabricate various nanomaterials other than carbonaceous materials. The advantage of CVD is that the quality (i.e., size, shape, and morphology) of desired material can be tuned by controlling the parameters of CVD process.

Some groups have worked on plasma-enhanced CVD (PE-CVD) technique. In this method, thermally (usually at low temperature) decomposed gaseous materials react with in situ microwave or ac (RF) frequency–generated plasma, which deposits a thin layer of desired product over substrate. Unlike conventional CVD, PE-CVD has an advantage to achieve vertically aligned graphene deposition at low temperature and low pressure by using reactive species generated in the plasma, at low cost, and contains few defects in the final product (Shang et al. 2008; Chan et al. 2013).

Another group (Wang et al. 2004) utilized varying concentrations of methane gas in order to deposit mono to few layers of graphene on various substrates (i.e., Cu, Ta, Nb, Zr, Si, SiO_2, Mo, and Al_2O_3) with RF frequency. They successfully deposited graphene within 5–40 min based on different experimental parameters.

37.3 CHARACTERIZATION

The characterization of graphene is a very vital factor as it provides the supporting evidences for successful synthesis of graphene. It has been viewed from various articles that the underdiscussed material did not satisfy the criteria of graphene but acknowledged as graphene by mistake. Hence, thorough and careful characterization of graphene is highly desired.

37.3.1 Microscopy Analysis

37.3.1.1 Optical analysis

In 2008, Nair et al. observed that graphene is an optically active material as it absorbs ~2.3% of visible light, and hence, it can be characterized using various optical methods. In 2009, Akcöltekin et al. also supported the Nair et al. observation when they showed that graphene sheets can be optically identified on various substrates (i.e., Si/SiO_2, TiO_2, CaF_2, $SrTiO_3$, and Al_2O_3).

37.3.1.2 Atomic force microscopy analysis

AFM was the first technique utilized for (Novoselov et al. 2004; Young et al. 2012) providing information on number of graphene layers in sample and indicated that some graphene layers had a thickness of 0.4 nm, referred as the signature of single graphene layer. Furthermore, AFM also sheds light on imaging and topography of the material.

37.3.1.3 Scanning tunneling microscopy and scanning electron microscopy analysis

Scanning tunneling microscopy (STM) and scanning electron microscopy (SEM) were also utilized to characterize graphene samples where graphene was grown over electrically conductive substrates.

Under special conditions, STM can provide information with an accuracy of atomic resolution. These techniques also reveal growth mechanism pattern (Gao et al. 2010; Yu et al. 2011).

37.3.1.4 Transmission electron microscopy analysis

Among different transmission electron microscopy (TEM) techniques, state-of-the-art aberration-corrected (Cs) TEM imaging and nanoprobe diffraction analysis in TEM have been proved particularly useful methods for graphene structural studies (i.e., lattice and electron diffraction patterns) (Meyer et al. 2007, 2008; Kotakoski et al. 2011).

37.3.2 Spectroscopy Analysis

37.3.2.1 Raman spectroscopy analysis

Ferrari et al. (2006) showed that graphene exhibits strong Raman scattering resonance, which leads to study of single-layer graphene material. Moreover, Raman spectroscopy provides many useful data including estimation of number of layers, defects, edge states (i.e., armchair and zigzag edges), strain and doping level in graphene due to the variation in 2D (originating from double resonance processes), and G modes (only first-order phonon mode available in graphite). The results proved that Raman spectroscopy is a powerful tool for graphene characterization (Lee et al. 2008; Malard et al. 2009).

37.3.2.2 Auger spectroscopy analysis

Xu et al. (2010) studied the thickness of graphene film using Auger electron spectroscopy (AES). AES is another established technique based on the analysis of energetic electrons emitted from first few atomic layers of energized material. In case of graphene, AES is particularly useful as it avoids and is not affected from coupling between graphene and substrates, which leads to softening of π-bonds in graphene and undetectable spectra (Berger et al. 2004).

37.3.2.3 X-ray photoemission spectroscopy analysis

X-ray photoemission spectroscopy (XPS) is also utilized to acquire information of the chemical compositions of the sample surface as shown by Riedl et al. (2010) and Englert groups (Englert et al. 2011).

37.3.2.4 X-ray diffraction spectroscopy analysis

Another very useful technique is X-ray diffraction (XRD). Rao et al. (2009) determined that due to the presence of sharp Bragg reflection (002, 2θ ~26°), which becomes broader with the decrease in number of graphene layers, it was possible to determine the number of layers using Scherrer formula. Other reports have also successfully used XRD to characterize graphene (Wang et al. 2008; Meng and Park 2012).

37.4 FUNCTIONALIZATION

Graphene has made its niche in catalysts, sensors, and energy storing and composite materials research areas based on its standout properties. Despite the great application potential, it is worth mentioning that graphene itself possesses zero bandgap as well as inertness to reaction, which weakens the competitive strength of graphene in the field of semiconductors and sensors. This is one of the reasons and important motivation for the huge increase in the number of research projects aimed at the functionalization of graphene including reactions of graphene (and its derivatives) with organic and inorganic molecules, chemical modification of the large graphene surface, and the general description of various covalent and noncovalent interactions with graphene. Bandgap opening of graphene by doping, intercalation, and striping would be useful for functionalization and can be deployed in the fabrication of nanoelectronic devices. In addition, it has been well established that graphene has no appreciable solubility in most solvents. This problem was circumvented with functionalization of graphene material via van der Waals forces (noncovalent π–π stacking) or C–C covalent coupling mechanism. Unlike covalent functionalization, which creates defects in graphene, noncovalent functionalization proved to be popular due to its extended π-conjugation properties on the graphene surface (Stankovich et al. 2006a; Dreyer et al. 2010; Loh et al. 2010; Zhu et al. 2010; Chen et al. 2012; Kuila et al. 2012; Craciun et al. 2013; Economopoulos and Tagmatarchis 2013).

Due to the increased interest in graphene arising from its exceptional electrical and mechanical properties, the immobilization of metallic and other nanoparticles on graphene sheets has become the focus of many researchers. A typical pristine graphene sheet can be characterized as an ideal substrate for the dispersion of nanoparticles due to its large active surface area per unit mass, in comparison with CNTs, amorphous carbon, or graphite, which have a lower active surface area because only the external surface is active (Liu et al. 2010a; Young et al. 2012).

The topic of functionalization is very wide; therefore, this chapter will restrict its discussion only on the functionalization of graphene with various types of nanoparticle systems comprehensively and their potential applications.

37.4.1 Applications of Nanoparticle Functionalized Graphene

Recently, a flood of research work utilized different types of nanoparticle systems because of their extraordinary properties. The application of nanoparticles over high surface area of GO and RGO revolutionizes the field of electrochemistry. It heralded to be an unexpected growth in addition to realization of nanoscale composite surfaces useful for various areas including electrochemical sensing, biosensing, catalysis, energy conversion, and storage (Muszynski et al. 2008; Georgakilas et al. 2012; Walcarius et al. 2013). Although there are numerous reports on the utilization of nanoparticles with graphene, this chapter will focus on recently published reports pertaining to the topics discussed in the following sections.

37.4.1.1 Fabrication of electrochemical sensors

The utilization of gold nanoparticles (AuNPs) has been prominent for the fabrication of electrochemical sensors. For example, Fang et al. (2010) proposed self-assembly of graphene/AuNP heterostructure or graphene/Au@Pd hybrid NPs to enhance the electrochemical catalytic ability toward hydrogen peroxide (H_2O_2) sensing. The zoomed TEM images confirmed the uniform distribution and high loading of AuNPs with size of AuNPs ranging from 11 to 18 nm over graphene surface as shown in Figure 37.2.

Another approach reported by Zhou et al. (2013) demonstrated the synergistic effect of AuNPs and ionic liquid (IL)-functionalized GO for ultrasensitive detection of mercury ions (Hg^{2+}). The characterization of deposited material clearly showed the uniform AuNPs of about 100 nm in average diameter formed randomly on the surface of glassy carbon electrode (GCE).

Lu et al. (2011a) fabricated highly dispersed and stable silver nanoparticle (AgNP)-functionalized graphene using $NaBH_4$ as reducing agent for electrochemical detection of nitroaromatic compounds (2011). The size of AgNPs was estimated to be 10 nm according to Scherrer's formula and TEM images. The author successfully detected 2,4-dinitrotoluene and 2,4,6-trinitrotoluene with detection limits of 0.21 and 0.45 ppm, respectively.

The enzymeless detection of H_2O_2 was carried out based on AgNP-decorated graphene nanosheets (Liu et al. 2010b). The AgNPs–graphene nanocomposites were synthesized with two different routes, namely, direct absorption of AgNPs and chemical reduction of silver salts, leading to different loading amount and particle size of AgNP formation over graphene sheets as depicted in TEM images. It was further observed that high loading and large particle size of AgNPs exhibited better electrocatalytic effect toward H_2O_2.

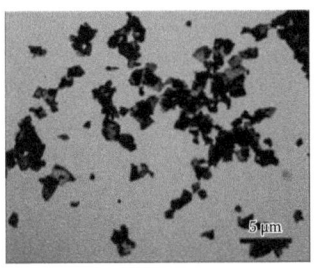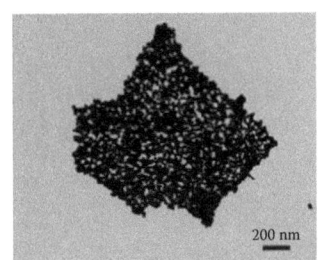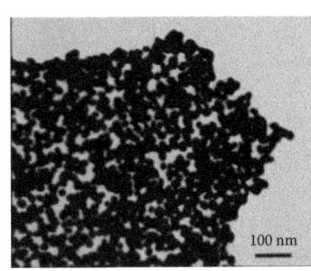

FIGURE 37.2
Typical TEM images of GN/AuNPs at different magnifications. (Reprinted with permission from Fang, Y., Guo, S., Zhu, C., Zhai, Y., and Wang, E., Self-assembly of cationic polyelectrolyte-functionalized graphene nanosheets and gold nanoparticles: A two-dimensional heterostructure for hydrogen peroxide sensing, *Langmuir*, 26(13), 11277. Copyright 2010 American Chemical Society.)

In an interesting approach, Yang et al. (2014) synthesized AgNPs/poly(methylene blue)-functionalized graphene composite for rutin determination. It was found that the synthesized material contains AgNPs of about 20–30 nm in size integrated uniformly with the graphene composite surface. The electrochemical study demonstrated lower charge transfer and enhanced electrocatalytic effect after AgNP integration in the material.

For electrochemical sensing of immunoglobulin E, streptavidin-functionalized AgNP–graphene hybrid was prepared by Song et al. (2013).The XRD pattern and UV–vis spectrum and TEM image characterization reports supported the synthesis of AgNP–graphene composite and uniform covering of graphene with AgNPs.

Lu et al. (2011b) published a facile approach on the utilization of CdSe nanoparticles (CdSe-NPs) over graphene for the sensitive detection of esculetin. The precursor solutions used for the synthesis of CdSe-NPs were prepared by obtaining Na_2SeSO_3 aqueous solution dissolving Se powder in Na_2SO_3 solution and $CdCl_2$ solution followed by the rest of the synthesis process. Owing to high graphene surface area, a large number of CdSe-NPs were distributed with a size ranging from 3 to 9 nm, which was well supported by XRD results. It was concluded that the synergistic effect of functionalized graphene and CdSe-NPs was responsible for the high electrochemical activity of sensor toward esculetin.

A simultaneous detection system was developed (Wu et al. 2014b) for small biomolecules using amino-group-functionalized mesoporous Fe_3O_4 NPs over graphene sheets. The TEM images of stepwise modifications have been shown in Figure 37.3. It was observed from SEM and TEM results that the size of Fe_3O_4 NPs was ~45 nm with pore size of 10 nm providing evidence for its mesoporosity. The incorporation of Fe_3O_4 NPs in graphene greatly improved the electrooxidation and peak separation for ascorbic acid, dopamine, and uric acid.

37.4.1.2 Fabrication of biosensors

This section will focus on the application of various nanoparticles in different types of functionalized graphene for biosensor strategies. Different types of GO or RGO nanocomposite were synthesized by means of various physical and chemical functionalization processes such as electrochemical synthetic reduction, self-assembly method, chemical reduction approach, physical desorption, and ultrasonic methods. A range of nanoparticle systems including simple Au, magnetic AuNPs, Ag, Ag@Au hybrid NPs, CdSe, Fe@Au hybrid NPs, core–shell Fe_2O_3, lanthanide-doped (Yb and Er) upconversion nanoparticles (UCNPs), and platinum nanoparticles (PtNPs) have been applied for the preparation of graphene-based biosensor. For example, Dong et al. (2011b) proposed a triple signal amplification approach for DNA detection by applying ssDNA–streptavidin (SA) horseradish peroxidase (HRP) coupling with AuNPs and carbon sphere as signal tag over RGO (2012). The electrochemical impedance spectroscopy (EIS) showed that the electron-transfer resistance of ssDNA–AuNP-assembled RGO reduced to 800 Ω as compared to 2400 Ω for GO-modified GCE, indicating the generation of high conductivity due to incorporation of AuNPs.

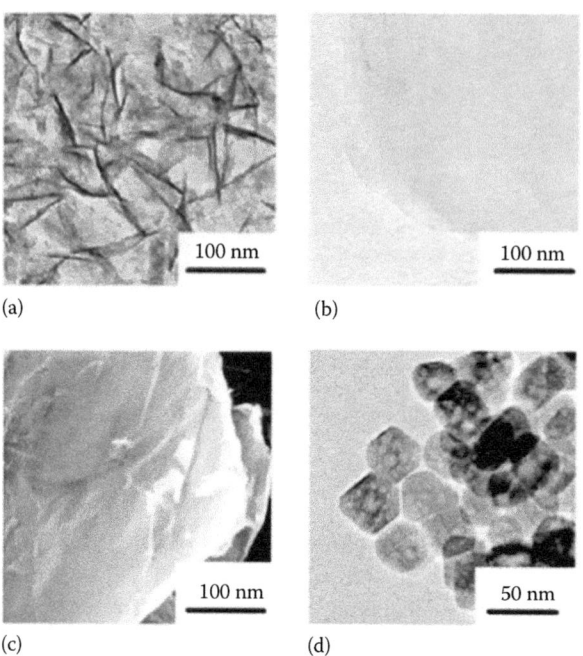

FIGURE 37.3
(a) TEM image of GO, (b) TEM image of GS, (c) SEM image of GS, and (d) TEM image of Fe_3O_4–NH_2. (Reprinted from *Electrochim. Acta*, 116, Wu, D., Li, Y., Zhang, Y., Wang, P., Wei, Q., and Du, B., Sensitive electrochemical sensor for simultaneous determination of dopamine, ascorbic acid, and uric acid enhanced by amino-group functionalized mesoporous Fe3O4@graphene sheets, 244–249, Copyright 2014b, with permission from Elsevier.)

By using Nafion-functionalized thermally exfoliated graphene as an excellent support for AuNPs and cholesterol oxidase, Aravind et al. (2011) reported the fabrication of cholesterol biosensor. The characterization data obtained from FE-SEM, TEM, energy-dispersive X-ray spectroscopy (EDX), and FT-IR revealed the formation of graphene nanoplatelets possessing high surface area suitable for uniform and high loading of AuNPs. The sensitivity of resulting biosensor was found to be 314 nA/μM for cholesterol.

Hong et al. (2010) used facile self-assembly approach for decoration of AuNPs over 1-pyrene butyric acid functionalized graphene by mixing their aqueous dispersions. The diameters of synthesized AuNPs were in the range of 2–6 nm. The sensor showed the high electrocatalytic behavior for UA due to the presence of AuNPs.

The various immunosensors have been reported recently based on AuNPs/magnetic AuNPs and graphene together with other composite materials. For example, recently, Li et al. (2013) fabricated an ultrasensitive immunosensor for the detection of diethylstilbestrol by utilizing AuNPs on mesoporous silica (Au@SBA-15) and amine-functionalized graphene. It was observed that amino-functionalized mesoporous silica (NH_2-SBA-15) offered more binding sites for loading antibody (Ab) and AuNPs due to the large surface area of the mesoporous structure. The obtained results were in agreement with the results provided by various characterization techniques. Furthermore, the synergistic effect of AuNPs and NH_2-SBA-15 increased the electrochemical signal for diethylstilbestrol.

Few reports have been published on the utilization of Fe@AuNP-decorated GO for electrochemical DNA biosensor and acetylcholinesterase-immobilized boronic acid-functionalized Fe@Au magnetic nanoparticles for the determination of acetylthiocholine chloride amperometric approach. It was concluded in both strategies that the incorporation of Fe@AuNP hybrid system had a pivotal role in enhancing the electrochemical responses greatly (Dong et al. 2012a; Yolaa et al. 2014).

Liu et al. (2013) developed a novel and interesting nonenzymatic ultrasensitive electrochemiluminescence (ECL) sandwich-type immunosensor approach based on core–shell Fe_3O_4–Au magnetic nanoparticles over quantum dot–functionalized graphene sheet (FGS) as labels for carbohydrate antigen 125 (CA125), known as one of the markers for cancer. The easy manipulation of magnetic

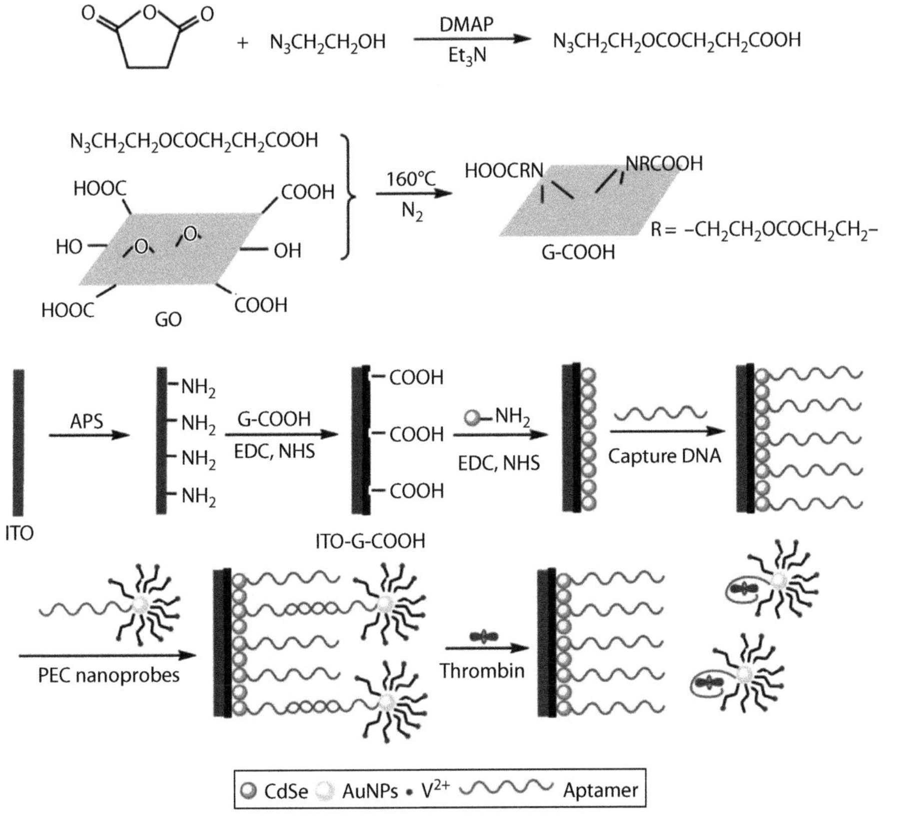

FIGURE 37.4
Schematic diagrams for preparation of carboxyl-functionalized graphene and the fabrication of amplified signal-on thrombin biosensor. (Zhang, X., Xu, Y., Yang Y. et al., A new signal-on photo-electrochemical biosensor based on a graphene/quantum-dot nanocomposite amplified by the dual-quenched effect of bipyridinium relay and AuNPs. *Chem. Eur. J.* 2012. 18(51). 16411–16418. Copyright Wiley-VCH Verlag GmbH & Co. KGaA. Reproduced with permission.)

nanoparticles and the unique biocompatibility of AuNPs encouraged them to utilize the aforementioned system. The resultant immunosensor exhibited a low detection limit of 1.2 mU mL^{-1} and applied to analyze real serum samples successfully.

Zhang et al. (2012) utilized the CdSe quantum dots and carboxyl-functionalized graphene system by the dual-quenched effect of bipyridinium relay and AuNPs for the fabrication of a photo-electrochemical biosensor for thrombin. Figure 37.4 depicted the preparation scheme for thrombin biosensor. This work also focused and studied the effects of various types of functionalized graphene (i.e., carboxyl-functionalized graphene [COOH-G] and amine-functionalized graphene [NH$_2$-G]) and AuNPs with varying sizes (5, 18, and 45 nm). It was concluded that coupling of CdSe-NPs and COOH-G offered improved sensitivity due to better solubility compared to the sensitivity obtained with CdSe-NPs and NH$_2$-G. In addition, regularity and homogenous distribution of smaller-size AuNPs (5 nm) provided excellent selectivity for the determination of thrombin.

The other popular nanoparticle system is AgNPs, which has been proved to be an excellent electrocatalytic system due to their easy dissolution and oxidization of silver without any poisonous reagent and enhanced signal amplification properties compared to other nanoparticles. Hence, the hybrid composite of AgNPs and graphene showed some highly desired applications for the fabrication of biosensors.

For instance, Ren et al. (2011) utilized the synergistic effects of GO and AgNP hybrid system based on surface-enhanced Raman scattering (SERS) substrates for folic acid detection. It was found from zoomed TEM images that AgNPs were loaded on both sides of GO with an average size of 35 nm. The SERS spectra of different concentrations of folic acid in water and serum have been assembled in Figure 37.5. The resultant hybrid material exhibited ultrasensitive SERS efficiency for folic acid as low as 9 nM due to strong enrichment of folic acid enabled by electrostatic interaction and the self-assembled AgNPs.

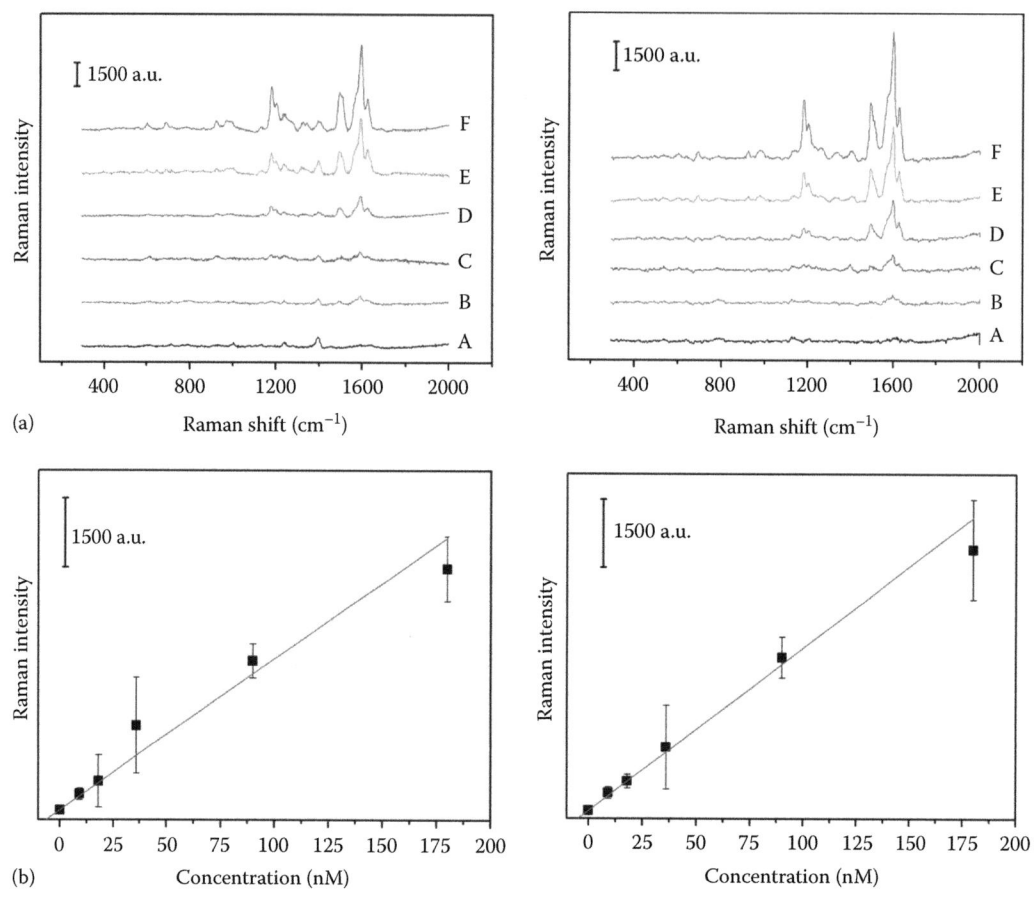

FIGURE 37.5

(a) SERS spectra of different concentrations of folic acid in water (left side figures) and diluted serum (right side figures), blank (A), 9 nM (B), 18 nM (C), 36 nM (D), 90 nM (E), and 180 nM (F), and (b) SERS dilution series of folic acid in water based on the peak located at 1595 cm^{-1}. (Reprinted with permission from Ren, W., Fang, Y., and Wang, E., A binary functional substrate for enrichment and ultrasensitive SERS spectroscopic detection of folic acid using graphene oxide/ Ag nanoparticle hybrids, *ACS Nano*, 5(8), 6425–6433. Copyright 2011 American Chemical Society.)

Another interesting strategy was reported by Gupta et al. (2013) by synthesizing a glucose biosensor based on Ag@AuNP-modified GO nanocomposite (diameter of particle in the range of 10–20 nm as estimated from TEM) and its application as SERS substrates.

Some research groups have utilized the high conductivity of PtNPs and loaded them onto graphene for the fabrication of composite material. A novel ultrasensitive ECL immunoassay based on PtNP-decorated graphene–CNT composite and carbon dot–functionalized mesoporous Pt/Fe for carcinoembryonic antigen detection has been reported (Deng et al. 2014). The EIS studies supported the decrease electron-transfer resistance after GCE was coated by the Pt/GR–CNT composite.

Wang et al. (2013) also prepared a H_2O_2 biosensor with platinum/graphene functionalized by poly(diallyldimethylammonium chloride) (PDDA) as a novel enzyme (HRP) carrier using microwave. For the fabrication of biosensor, PtNPs were synthesized in in situ reduction of $PtCl_6^{2-}$ in the presence of PDDA on the surface of graphene (GO was reduced with hydrazine). The author claimed that the microwave preparation provided a good microenvironment for the rapid preparation of PtNPs resulting into high-quality Pt/GR hybrid with controllable density, small size (in the range of 2–5 nm, mean diameter of 3.3 ± 0.7 nm), and uniform distribution. The EIS and other CV studies showed the excellent electron-transfer properties of Pt/GR between the electrode surface and the electrolyte solution.

Recently, a fluorescence resonance energy transfer (FRET) assay and a highly efficient energy transfer acceptor–based biosensor for human immunodeficiency virus (HIV type 1) antibodies in human serum were proposed (Wu et al. 2014a). A novel approach referred as peptide-functionalized lanthanide-doped (Yb and Er) *UCNPs* to GO was applied due to their outstanding features such

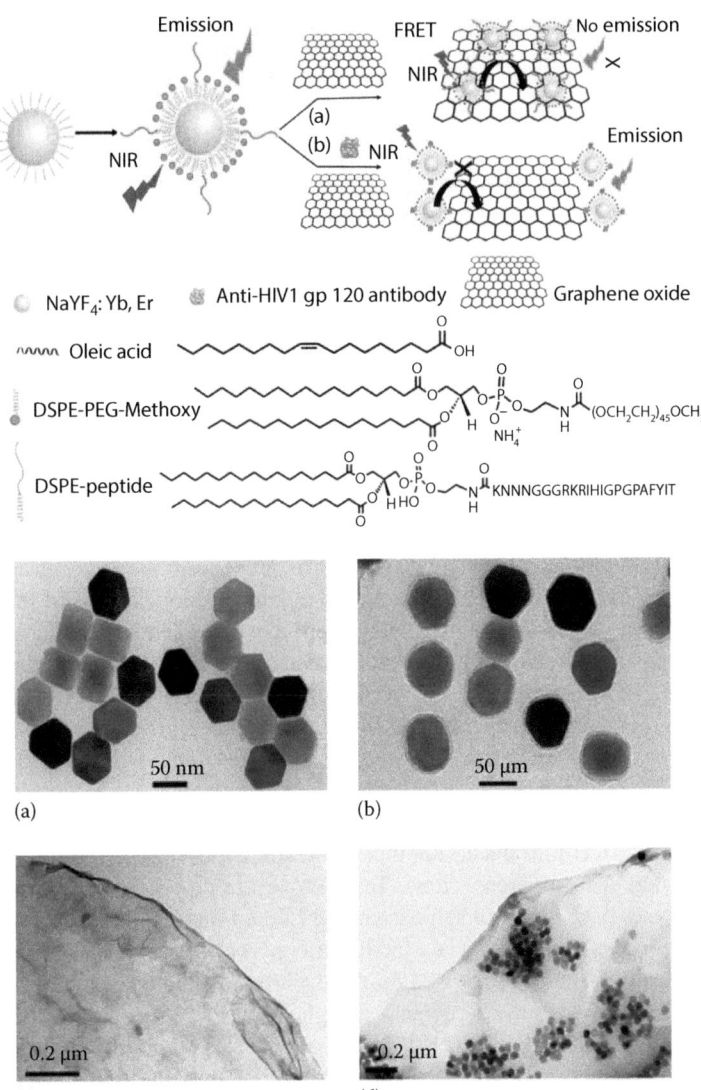

FIGURE 37.6
Schematic illustration of the upconversion FRET-based biosensor for the detection of anti-HIV-1 gp120 antibody. TEM images of the (a) OA-coated UCNPs in cyclohexane, (b) peptide-functionalized UCNPs in water, (c) as-prepared GO, and (d) peptide-functionalized UCNP–GO complexes in water. (Wu, Y.M., Cen, Y., Huang, L.J., Yu, R.Q., and Chu, X., Upconversion fluorescence resonance energy transfer biosensor for sensitive detection of human immunodeficiency virus antibodies in human serum, *Chem. Commun.*, 50(36), 4759–4762, 2014a. Reproduced by permission of The Royal Society of Chemistry.)

as improved quantum yield, tunable multicolor emission, minimal photobleaching, greater light penetration depth in tissue, low toxicity, and strong visible fluorescence under the excitation of near-infrared light. Figure 37.6 shows the fabrication scheme of FRET-based biosensor and TEM images of UCNPs at various stages.

A facile one-step approach was used to prepare water-dispersible and peptide-functionalized UCNPs through the self-assembly of phospholipid–peptide conjugation onto the hydrophobic UCNP surface. The mean diameter for UCNPs varies from 60 to 62 nm (core–shell structure with a uniform, approximately 2 nm thick, hydrophobic oleic acid–lipid layer around the surface), before and after functionalization with phospholipid–peptide conjugation, respectively, as depicted by TEM, XRD, and dynamic light scattering results. In comparison with oleic UCNPs dispersed in cyclohexane (ca. 65 nm), the mean hydrodynamic diameter in water increased to approximately 90 nm. This showed the remarkable fluorescence quenching of peptide-functionalized UCNPs to GO surface.

The developed biosensor was successfully utilized for the determination of HIV type 1 antibodies in human serum with negligible interferences from complex biological samples.

37.4.1.3 Fabrication of energy conversion and storage materials

The ever-increasing demand for energy creates a tremendous amount of attraction from scientific communities to develop reliable, alternative energy storage and conversion devices. The research reports show a clear shift for electrochemical energy storage systems such as much-higher-performance batteries, supercapacitors, and fuel cells based on nanosystems including-nanoparticles and graphene hybrid composites (Choi and Park 2012). Some groups clearly reviewed that inorganic–nanocarbon hybrid materials are the main focus for the development of energy storage devices (Dai 2013; Wang and Dai 2013).

Choi et al. (2012) focused their attention on tailoring the electrode surface design in order to tune structure and morphology for convenient ion and electron pathways needed for high-performance energy storage devices such as lithium-ion (Li-ion) batteries. However, the strong aggregation problem of constituting components was responsible for poor cell performance due to inferior migration of Li^+. The author successfully circumvented it by utilizing 3D hierarchical porous structures of Co_3O_4 nanoparticles/RGO composite films through deposition processes and tuned the pore structures and size by using polystyrene spheres through replication. The SEM images showed the well-interconnected porous RGO structure without any obvious shrinkage, whereas the incorporation of Co_3O_4 nanoparticles within RGO was confirmed by high-angle annular dark-field scanning TEM, HR-TEM, and EDX mapping. It was worth noticing that bare Co_3O_4 NPs (an average size of 35.4 ±12.5 nm with broad NP size distribution from 2 to 47 nm) aggregated without RGO; however, the average diameter of 5.87 ± 2.1 nm for Co_3O_4 NPs was observed with RGO. Finally, the fabricated material exhibited a high rate capability of 71% retention at 1000 mA g^{-1} and demonstrated excellent cycling durability of 90.6% retention.

Chidembo et al. (2014) introduced another interesting strategy for the fabrication of energy storage device (asymmetric supercapacitor). To improve the charge storage capacity of devices, they used ternary synergistic effect of self-assembled RGO, functionalized multiwalled CNTs, and nickel oxide nanoparticles (NiO NPs) via a facile spray pyrolysis of GO-based liquid crystals. The SEM images supported the model proposed for the evolution NiO NPs from nickel nitrite. In addition, the distribution of NiO NPs over graphene, morphology of ternary system, and electrochemical characterization were studied by TEM, SEM, CV, and EIS. The proposed asymmetric supercapacitor system demonstrated excellent conductive network providing a maximum capacitance of 97 F g^{-1}, remarkable cycle life (96.78% capacitance retention over 2000 cycles), and energy density of 23 Wh kg^{-1}.

Recently, Mishra and Ramaprabhu (2011) formulated a supercapacitor electrode using various metal oxides such as RuO_2, TiO_2, and Fe_3O_4 nanoparticles and polyaniline (PANI)-decorated hydrogen-exfoliated hydrophilic-functionalized graphene by using different routes including chemical route, sol–gel, and chemical techniques with precursor materials, respectively. The as-synthesized nanocomposites were characterized using TEM, FT-IR, XRD, and Raman techniques, whereas electrochemical studies were carried out for capacitance measurements successfully. Table 37.1 depicted the specific capacitance of various synthesized nanocomposites at different scan rates in CV. It can be deduced easily from Table 37.1 that RuO_2 NP nanocomposite and PANI–graphene exhibited constant high values of capacitance at all the scan rates. These results suggested the viable commercial potential in these systems.

A novel and interesting nanocomposite was proposed for highly improved rechargeable lithium–sulfur (Li-S) batteries (Cao et al. 2011). The nanocomposite was prepared as a sandwich-type architecture containing FGSs/stacks and a layer of sulfur nanoparticles referred as graphene sheet-sulfur nanocomposite (FGSS). The thermal expansion approach was utilized to prepare FGSs from graphite oxide. To prepare FGSS nanocomposite sandwich-type structure, firstly, appropriate experimental conditions and steps were carried out by mixing sulfur, CS_2, graphene, and so on followed by Nafion coating (Nf@FGSS) for high reversible capacity (high charge/discharge rates). From characterization of the material confirmed by energy-dispersive spectroscopy, it was concluded that each sulfur nanoparticle layer is ~5 nm thick and NPs have been loaded on both sides by 3–6 nm thick graphene stacks. The electrochemical characterization and battery

TABLE 37.1 Specific Capacitance of HEG, f-HEG, RuO₂-f-HEG, TiO₂-f-HEG, Fe₃O₄-f-HEG, and PANI-f-HEG Nanocomposites Using Cyclic Voltammetry with Different Voltage Sweep Rates and Using the Galvanostatic Charge–Discharge Technique

Sample	Voltage Scan Rate (10 mV/s)	Voltage Scan Rate (20 mV/s)	Voltage Scan Rate (50 mV/s)	Voltage Scan Rate (100 mV/s)	Specific Capacitance (F/g) with Current Densities (Galvanostatic)
HEG	80	77	70	65	62 (8 A/g)
f-HEG	125	122	112	105	110 (10 A/g)
RuO₂-f-HEG	265	245	225	215	220 (10 A/g)
TiO₂-f-HEG	60	58	53	40	38 (6 A/g)
Fe₃O₄-f-HEG	180	173	165	160	140 (10 A/g)
Fe₃O₄-f-HEG (1 M Na₂SO₄)	65	62	58	55	45 (4A/g)
PANI-f-HEG	375	360	340	320	355 (10 A/g)

Source: Mishra, A.K. and Ramaprabhu, S., Functionalized graphene-based nanocomposites for supercapacitor application, *J. Phys. Chem. C*, 115(29), 14006. Copyright 2011 American Chemical Society.

performance of the FGSS nanocomposite offered good cycling stability of 84.3% capacity retention over 100 cycles at 1680 mA g^{-1}. The high performance of it can be attributed to novel and excellent nanocomposite structure, the good conductivity of graphene, the high redox activity of sulfur, and the cationic exchange properties of the Nafion coating that mitigates the migration of polysulfide anions.

Choi et al. (2012) studied the preparation of electrodes favorable for energy storage by utilizing various in situ–generated nanoparticles (Au, Pt, Pd, Ru, and RuO₂) and directly deposited them over IL-functionalized RGO. The size, amount, and crystalline structure of NPs were optimized in order to enhance their pseudocapacitance. The GOs were prepared by modified Hummer method followed by chemical reduction. The RGO was functionalized through IL dispersion in aqueous GO solution. The existence of anionic sites of ILs provided active sites for NP growth (NP precursors were nucleated via ultrasound irradiation) and prevented RGO aggregation. The functionalization and morphology characterization data were obtained using FT-IR, TEM, XRD, and XPS techniques. The results were in good agreement with previous reported values. The generality of proposed strategy was confirmed by applying it to various NP systems. The morphologies and sizes of the NPs were dependent on the type of NPs, because of the interactions between NPs and IL-RGOs and the growth kinetics of NPs. Although Pt/IL-RGO demonstrated excellent electrocatalytic activity toward electrooxidation of methanol, RuO₂/IL-RGO hybrids were selected as electrode materials for electrochemical energy storage, as both RuO₂ and RGOs are currently regarded as potential materials for pseudocapacitors and electrical double-layer capacitors, respectively (Chen et al. 2010; Wu et al. 2010). RuO₂/IL-RGO facilitated higher specific capacitance and cycle stability (retained 97% of maximum specific capacitance even after 1000 cycles of charge–discharge cycles) compared to RuO₂/RGO and IL-RGO.

37.4.1.4 Fabrication of catalysis materials

Catalysis is an indispensible technique used in chemical and fuel processing. The study of catalysis by robust metallic nanoparticle–supported graphene system is undoubtedly the most studied due to the exposure of high surface area of graphene capable to support large numbers of nanoparticle systems. Few very interesting and discerning review articles have been published on the topic of graphene in catalysis and catalysis by metallic nanoparticles (Shang et al. 2014). It has been deduced from various research reports that the activity of NPs is strongly dependent on the size of the nanoparticles and types of support system as discussed by Hvolaek et al. (2007) on the role of AuNPs in catalysis by mentioning that only catalysts with Au particles below 5 nm show catalytic activity.

Recently, Li et al. (2014a) reported on a facile and low-cost AuNP-based catalyst for 4-nitrophenol reduction through AuNPs anchored on the IL of 3,4,9,10-perylenetetracarboxylic acid noncovalent-functionalized graphene (Au/PDIL-GR). The successful stepwise synthesis of nanocomposite was determined by various characterization techniques providing wide-size random distribution (the size of NPs greater than 10 nm) of AuNPs over graphene surface.

Another interesting one-step synthesis of hybrid AuNP–GO nanosheets through electrostatic self-assembly was realized (Choi et al. 2011) for improved electrocatalytic reduction of several nitroarenes. They showed effective control of AuNP concentration and their stability over GO. The TEM and AFM images demonstrated the successful covering of graphene surface (edges and basal planes) with AuNPs of 6.0 nm in diameter. The synthesized hybrid material depicted remarkable efficiency toward nitroarene reduction due to the synergistic catalytic effect of AuNPs and GO nanosheets.

In a recent report, a ternary composite system composed of AgNPs with core–shell structure, silica gel, and RGO (SG/RGO/Ag) by electrostatic self-assembly was developed (Zhang et al. 2014). As discussed in the previous sections, the catalytic activity of AgNPs is closely related to its size. The smaller the size of the AgNPs, the higher the catalytic activity of AgNPs (~10–15 nm). Under optimized experimental conditions, SG/RGO/Ag offered remarkable electrocatalytic activity for rhodamine B, and its degradation was observed by change in color that gradually became paler owing to the presence of the large number of active sites in the form of AgNP and synergy interaction between the RGO and AgNPs.

A different and facile synthesis for the fabrication of AgNPs over GO was shown by Jeon et al. (2013) who utilized the mussel-inspired green (bioinspired approach) synthesis approach. Firstly, GO-dopa was synthesized via mussel-inspired chemical motif of dopa over GO surface, which leads to the spontaneous in situ formation of AgNPs by simple mixing with aqueous $AgNO_3$ solution at room temperature for 3 h without any additional reducing or stabilizing agents. The characterization data of as-fabricated composite have been presented in Figure 37.7. This approach realized the formation of well-dispersed AgNPs over GO with an average particle size of 7.71 ± 1.34 nm as shown by TEM. The Ag/GO-dopa composite exhibited superior catalytic activity for nitroarenes compared to GO or GO-dopa alone. Furthermore, this method showed versatility and generality when applied for the formation of other NPs (Cu and Au) based on the same protocol.

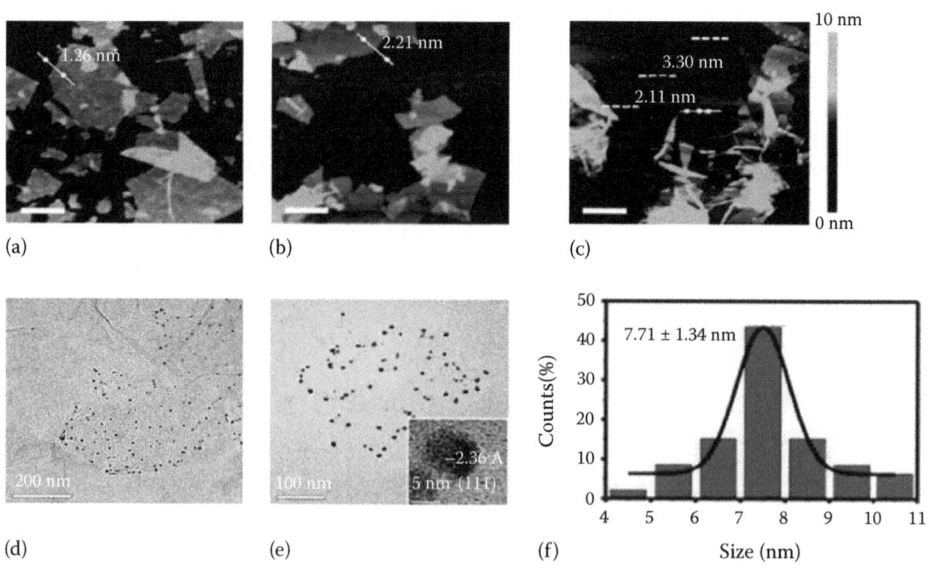

(a) (b) (c)

(d) (e) (f)

FIGURE 37.7

Representative AFM and TEM images of Ag/GO-dopa: (a) GO, (b) GO-dopa, (c–e) Ag/GO-dopa, with an inset image of the AgNPs. (f) Size distribution histogram of the AgNPs. Scale bar in AFM images is 2.5 mm. (Jeon, E.K., Seo, E., Lee, E., Lee, W., Um, M.K., and Kim, B.S., Mussel-inspired green synthesis of silver nanoparticles on graphene oxide nanosheets for enhanced catalytic applications, *Chem. Commun.*, 49(33), 3392–3394, 2013. Reproduced by permission of The Royal Society of Chemistry.)

Rajesh et al. (2014) developed a method for the stabilization of AgNP and AuNP on GO functionalized with poly(amidoamine) (PAMAM) dendrimers. The stepwise synthesis, modifications, and morphology of synthesized material were confirmed by using FT-IR, SEM, HR-TEM, surface analyzer, and TGA techniques. It was observed that the regular structure and versatile chemistry of PAMAM dendrimers were responsible for stabilization of Ag/AuNPs. The size of the AgNPs has been estimated to be ~20 nm comparable to the size of AuNPs. The composite material performed quite well toward the degradation of azo dyes as confirmed by ^{1}H-NMR, FT-IR, and UV–vis kinetics spectra.

There have been reports on successful PdNP decoration on GO/RGO surface for the design and novel preparation of GO-supported electrocatalysts (Zhang et al. 2013). For example, Pd nanoparticles on 9-amino-1-azabenzanthrone-functionalized graphene-like carbon surface and in situ synthesis of palladium nanoparticle on FGSs were carried out for the determination of methanol and ethanol oxidation in alkaline media (Li et al. 2014b). In another report, the exposed functional surfaces of GO and chemically derived graphene were exploited as a support for palladium clusters and nanoparticles, and their efficiency was investigated using Suzuki–Miyaura reaction (C–C coupling reaction) (Scheuermann et al. 2009). It was established that unlike conventional Pd/C catalyst, GO and graphene-based heterogeneous catalysts demonstrated superior activities with turnover frequencies exceeding 39,000 h^{-1}, accompanied by very low palladium leaching (<1 ppm). The high activity of the material was ascribed to the shape and size of the nanoparticles and their size distribution (i.e., the Pd dispersion) and high surface area.

Some attractive strategies for the preparation of hybrid and bimetallic NP–supported graphene systems have been described. For instance, utilizing the palladium and palladium–yttrium (Pd/Pd-Y)-decorated graphene, Seo et al. (2013) developed an efficient electrocatalyst for oxygen reduction and ethanol reduction. Shaabani and Mahyari (2013) employed a facile, efficient, and environmentally friendly procedure for PdCo bimetallic nanoparticles supported on polypropylenimine dendrimer–grafted graphene synthesize heterogeneous catalyst for successful determination of Sonogashira reaction under copper- and solvent-free conditions using ultrasound irradiation at room temperature. The characterization results offer evidence for uniform distribution of PdCo NPs of size 2–3 nm.

For the efficient oxidation of methanol, graphene or functionalized graphene has been extensively utilized as a support for the dispersion of highly active PtNPs. The corresponding electrocatalysts facilitate excellent oxidation of methanol (Luo et al. 2012; Mayavan et al. 2012; Zhong et al. 2013).

In addition to PtNPs, hybrids of Pt–AuNPs and Pt–NiNPs with graphene have been shown as feasible and excellent electrocatalysts toward formic acid oxidation and methanol oxidation (Wang et al. 2011; Luo et al. 2013).

For visible-light photocatalytic degradation of methyl orange, one-step hydrothermal route was employed for the fabrication of CdS nanoparticle/FGS (CdS NP/FGR) nanocomposites as reported by Yan et al. (2013). The successful synthesis of electrocatalyst was confirmed by several characterization techniques, and resulting nanocomposite provided high efficiency compared to pure CdS NPs, owing to strong redox ability and smaller size of CdS in the nanocomposite. Furthermore, the high specific surface area for the absorption of a large number of methyl orange molecules and reduction in electron–hole pair recombination with the introduction of FGR enhanced the degradation efficiency of studied electrocatalysts.

A novel RuNP-decorated graphene nanoelectrocatalyst was developed through dry synthesis (Gopiraman et al. 2013). TEM images and Raman spectroscopy confirmed the good attachment of RuNPs over the graphene surface. XRD and XPS showed the nature of RuNPs (i.e., metallic and nanocrytalline), whereas heterogeneity of catalyst structure was supported by ICP-MS analysis. The versatility and feasible dry synthesis support the easy development of Ru-based catalyst for aerial oxidation of alcohols and transfer hydrogenation of ketones.

Devadoss et al. (2014) investigated the role of AuNP–tungsten oxide (WO$_3$) nanoparticle–supported graphene membrane for improved photoelectrochemical glucose oxidase–mediated enzymatic glucose oxidation. TEM results demonstrated the high degree of monodisperse distribution of electrostatically assembled colloidal AuNPs with a mean diameter of ~5 nm on graphene membrane containing uniformly distributed WO$_3$ NPs of ~20–30 nm size. The proposed catalyst offered superior activity compared to other systems without conducting channels and can be a viable alternative to TiO$_2$ nanomaterials in photoelectrochemical-based applications.

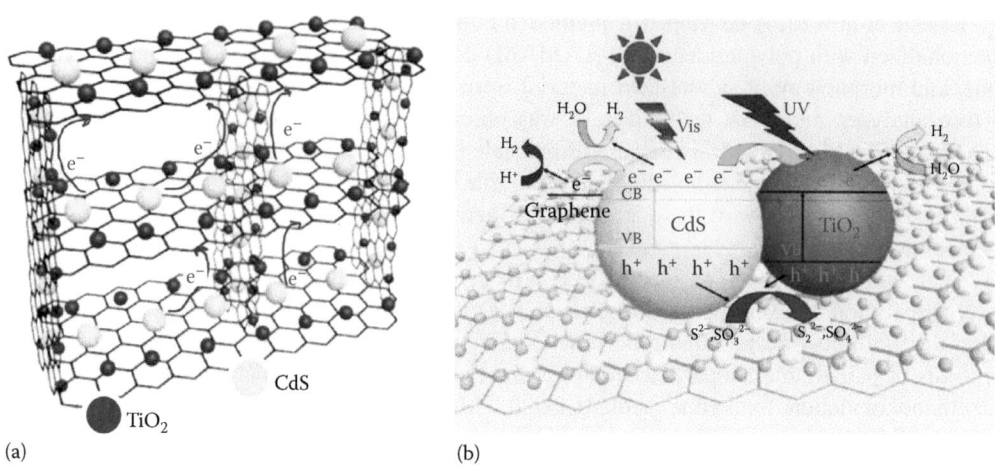

(a)   (b)

FIGURE 37.8
(a) Schematic structure of 3D CdS/P25/graphene aerogel networks. (b) Illustration of the proposed reaction mechanism for hydrogen production over the CdS/P25/graphene aerogel under irradiation of simulated sunlight. (Reprinted with permission from Han, W., Ren, L., Gong, L. et al., Self-assembled three-dimensional graphene-based aerogel with embedded multifarious functional nanoparticles and its excellent photoelectrochemical activities, *ACS Sustain. Chem. Eng.*, 2(4), 741–748. Copyright 2014 American Chemical Society.)

Another excellent electrocatalyst for photoelectrochemical application was fabricated by Han et al. (2014). They synthesized a 3D cylindrical architecture aerogel composed of dual types of nanoparticles (TiO_2 from P25 and CdS) loaded on graphene sheet surface by a very facile one-pot hydrothermal method. The authors observed outstanding photoelectrochemical hydrogen production from water reduction under sunlight. The schematic structure of as-prepared aerogel networks and illustration of the proposed reaction mechanism are shown in Figure 37.8. The achievement of excellent photocatalytic properties including light absorption, improved photocurrent, extremely efficient charge separation properties, and superior durability was ascribed to synergistic effect generated from ternary graphene-based nanocomposite.

Guo and Sun (2012) demonstrated robust and facile solution-phase self-assembly approach for the fabrication of bimetallic FePt NPs over graphene surface to achieve an outstanding catalyst for oxygen reduction reaction. The size of FePt NPs was ~7 nm as evaluated by TEM analysis. The as-synthesized nanocatalyst exhibited better activity and durability compared to carbon supported by the same NPs or commercially available PtNPs.

37.5 CONCLUDING REMARKS AND FUTURE PROSPECTS

This review chapter focused on the systematic efforts taken for the development of various graphene nanoparticle–based nanocomposites in recent years. The synergism effect of graphene and nanoparticles demonstrated excellent efficiency and innovative applications in electrochemical sensors, biosensors, energy conversion and storage, and catalysis. We discussed based on our understanding that the size of nanoparticles plays important roles in efficiency enhancement. Certainly, it can be deduced that smaller-size nanoparticles contribute more compared to nanoparticles with bigger sizes. Furthermore, at some instances, a threshold size of nanoparticles has to be confirmed in order to achieve an outstanding performance from the material. For example, it was found that AuNPs less than ~5 nm in size do not offer good results. In addition, there is always a need to look into more facile, robust, and efficient preparation method of nanoparticles, GO and graphene, and their nanocomposites.

Based on some other reports and in our own experience, it has been realized that the complete understanding of graphene structure still needed exhaustive studies and research practices, owing to the great chemistry that can be plausible from GO and graphene. Another important aspect that needs to be addressed is to look for better synthetic approaches that demonstrate better reproducible and controlled methods. There is also a need for more controlled and facile methods or nanoparticle-based graphene composites.

Recently, graphene is in the news due to its potential to be stretched across glass surfaces of phones or tablets to make them into touch screens, stronger, and more flexible than the current technology. Hence, it's ideal for futuristic gadgets like bendable smart watches or tablets that fold up into smartphones. Finally, as we move into the next generation of smart and better-engineered nanocomposites, we can imagine a more potential expansion on the already studied synthetic procedures and can visualize a new realm of opportunities for this material.

ACKNOWLEDGMENTS

We sincerely appreciate and express our gratitude to Kwangwoon University for financial support in 2014 for this work. The authors also appreciate the helpful input from Dr. Akbar Nawab at the University of Florida, Gainesville, United States.

REFERENCES

Akcöltekin, S., M. ElKharrazi, B. Köhler, A. Lorke, and M. Schleberger. 2009. Graphene on insulating crystalline substrates. *Nanotechnology* 20 (15): 155601.

Aravind, S. S. J., T. T. Baby, T. Arockiadoss, R. B. Rakhi, and S. Ramaprabhu. 2011. A cholesterol biosensor based on gold nanoparticles decorated functionalized grapheme nanoplatelets. *Thin Solid Films* 519 (16): 5667–5672.

Berger, C., Z. Song, T. Li et al. 2004. Ultrathin epitaxial graphite: 2D electron gas properties and a route toward graphene-based nanoelectronics. *Journal of Physical Chemistry B* 108 (52): 19912–19916.

Boehm, H. P., A. Clauss, G. O. Fischer, and U. D. Hofmann. 1962. Adsorptions verhalten sehr dunner Kohlenstoff-Folien. *Zeitschrift für anorganische und allgemeine Chemie* 316: 119–127.

Brodie, B. C. 1860. Sur le poids atomique du graphite. *Annales des Chimie et des physique* 59: 466.

Cano-Marquez, A. G., F. J. Rodriguez-Macias, J. Campos-Delgado et al. 2009. Ex-MWNTs: Graphene sheets and ribbons produced by lithium intercalation and exfoliation of carbon nanotubes. *Nano Letters* 9 (4): 1527–1533.

Cao, Y., X. Li, I. A. Aksay et al. 2011. Sandwich-type functionalized graphene sheet-sulfur nanocomposite for rechargeable lithium batteries. *Physical Chemistry Chemical Physics* 13 (17): 7660–7665.

Chan S. H., S. H. Chen, W. T. Lin, M. C. Li, U. C. Lin, and C. C. Kuo. 2013. Low-temperature synthesis of graphene on Cu using plasma-assisted thermal chemical vapor deposition. *Nanoscale Research Letters* 8: 285.

Chen, D., H. Feng, and J. Li. 2012. Graphene oxide: Preparation, functionalization, and electrochemical applications. *Chemical Reviews* 112 (11): 6027–6053.

Chen, D., L. Tang, and J. Li. 2010. Graphene-based materials in electrochemistry. *Chemical Society Reviews* 39 (8): 3157–3180.

Chen, Z. H., Y. M. Lin, M. J. Rooks, and P. Avouris. 2007. Graphene nano-ribbon electronics. *Physica E-Low-Dimensional Systems and Nanostructures* 40 (2): 228–232.

Chidembo, A. T., S. H. Aboutalebi, K. Konstantinov, D. Wexler, H. K. Liu, and S. X. Dou. 2014. Liquid crystalline dispersions of graphene-oxide-based hybrids: A practical approach towards the next generation of 3D isotropic architectures for energy storage applications. *Particle and Particle Systems Characterization* 31 (4): 465–473.

Choi, B. G., S. J. Chang, Y. B. Lee, J. S. Bae, H. J. Kim, and Y. S. Huh. 2012. 3D hetero-structured architectures of Co_3O_4 nanoparticles deposited on porous grapheme surfaces for high performance of lithium ion batteries. *Nanoscale* 4 (19): 5924–5930.

Choi, B. G. and H. S. Park. 2012. Controlling size, amount and crystalline structures of nanoparticles deposited on graphenes for highly efficient energy conversion and storage. *ChemSusChem* 5 (4): 709–715.

Choi, W. and J.-W. Lee. 2012. *Graphene: Synthesis and Applications.* Boca Raton, FL: Taylor & Francis.

Choi, Y., H. S. Bae, E. Seo, S. Jang, K. H. Park, and B. S. Kim. 2011. Hybrid gold nanoparticle-reduced graphene oxide nanosheets as active catalysts for highly efficient reduction of nitroarenes. *Journal of Materials Chemistry* 21 (39): 15431–15436.

Choucair, M., P. Thordarson, and J. A. Stride. 2009. Gram-scale production of graphene based on solvothermal synthesis and sonication. *Nature Nanotechnology* 4 (1): 30–33.

Craciun, M. F., I. Khrapach, M. D. Barnes, and S. Russo. 2013. Properties and applications of chemically functionalized graphene. *Journal of Physics: Condensed Matter* 25 (42): 423201.

Dai, L. 2013. Functionalization of graphene for efficient energy conversion and storage. *Accounts of Chemical Research* 46 (1): 31–42.

Deng, W., F. Liu, S. Ge, J. Yu, M. Yan, and X. Song. 2014. A dual amplification strategy for ultrasensitive electrochemiluminescence immunoassay based on a Pt nanoparticles dotted graphene–carbon nanotubes composite and carbon dots functionalized mesoporous Pt/Fe. *Analyst* 139 (7): 1713–1720.

Devadoss, A., P. Sudhagar, S. Das et al. 2014. Synergistic metal-metal oxide nanoparticles supported electro-catalytic graphene for improved photo-electrochemical glucose oxidation. *ACS Applied Materials and Interfaces* 6 (7): 4864–4871.

Dong, H., Z. Zhu, H. Ju, and F. Yan. 2012b. Triplex signal amplification for electrochemical DNA biosensing by coupling probe-gold nanoparticles–graphene modified electrode with enzyme functionalized carbon sphere as tracer. *Biosensors and Bioelectronics* 33 (1): 228–232.

Dong, J., T. Liu, X. Meng et al. 2012a. Amperometric biosensor based on immobilization of acetylcholinesterase via specific binding on biocompatible boronic acid-functionalized Fe@Au magnetic nanoparticles. *Journal of Solid State Electrochemistry* 16 (12): 3783–3790.

Dreyer, D. R., S. Park, C. W. Bielawski, and R. S. Ruoff. 2010. The chemistry of grapheme oxide. *Chemical Society Reviews* 39 (1): 228–240.

Economopoulos, S. P. and N. Tagmatarchis. 2013. Chemical functionalization of exfoliated graphene. *Chemistry: A European Journal* 19 (39): 12930–12936.

Eizenberg, M. and J. M. Blakely. 1979. Carbon monolayer phase condensation on Ni(111). *Surface Science* 82 (1): 228–236.

Englert, J. M., C. Dotzer, G. Yang et al. 2011. Covalent bulk functionalization of graphene. *Nature Chemistry* 3 (4): 279–286.

Enoki, T. and T. Ando. 2013. *Physics and Chemistry of Graphene: Graphene to Nanographene.* Boca Raton, FL: Pan Stanford Publishing.

Fang, Y., S. Guo, C. Zhu, Y. Zhai, and E. Wang. 2010. Self-assembly of cationic polyelectrolyte-functionalized graphene nanosheets and gold nanoparticles: A two-dimensional heterostructure for hydrogen peroxide sensing. *Langmuir* 26 (13): 11277–11282.

Ferrari, A. C., J. C. Mayer, V. Scardaci et al. 2006. Raman spectrum of graphene and graphene layers. *Physical Review Letters* 97: 187401.

Gao, L., J. R. Guest, and N. P. Guisinger. 2010. Epitaxial graphene on Cu(111). *Nano Letters* 10 (9): 3512–3516.

Geim, A. K. and K. S. Novoselov. 2007. The rise of graphene. *Nature Materials* 6(3): 183–191.

Georgakilas, V., M. Otyepka, A. B. Bourlinos et al. 2012. Functionalization of graphene: Covalent and non-covalent approaches, derivatives and applications. *Chemical Reviews* 112 (11): 6156–6214.

Gopiraman, M., S. G. Babu, Z. Khatri et al. 2013. Dry synthesis of easily tunable nano ruthenium supported on graphene: Novel nanocatalysts for aerial oxidation of alcohols and transfer hydrogenation of ketones. *Journal of Physical Chemistry C* 117 (45): 23582–23596.

Guo, S. and S. Sun. 2012. FePt nanoparticles assembled on graphene as enhanced catalyst for oxygen reduction reaction. *Journal of American Chemical Society* 134 (5): 2492–2495.

Gupta, V. K., N. Atar, and M. L. Yola et al. 2013. A novel glucose biosensor platform based on Ag@AuNPs modified grapheme oxide nanocomposite and SERS application. *Journal of Colloid and Interface Science* 406: 231–237.

Han, W., L. Ren, L. Gong et al. 2014. Self-assembled three-dimensional graphene-based aerogel with embedded multifarious functional nanoparticles and its excellent photoelectrochemical activities. *ACS Sustainable Chemistry and Engineering* 2 (4): 741–748.

Hong, W., H. Bai, Y. Xu, Z. Yao, Z. Gu, and G. Shi. 2010. Preparation of gold nanoparticle/grapheme composites with controlled weight contents and their application in biosensors. *Journal of Physical Chemistry C* 114 (4): 1822–1826.

Hummers, W. S. and R. E. Hoffeman. 1958. Preparation of graphite oxide. *Journal of American Chemical Society.* 80 (6): 1339.

Hvolaek, B., T. V. W. Janssens, B. S. Clausen, H. Falsig, C. H. Christensen, and J. K. Norskov. 2007. Catalytic activity of Au nanoparticles. *Nanotoday* 2 (4): 14–18.

Ismach, A., C. Druzgalski, S. Penwell et al. 2010. Direct chemical vapor deposition of graphene on dielectric surfaces. *Nano Letter* 10(5): 1542–1548.

Jeon, E. K., E. Seo, E. Lee, W. Lee, M. K. Um, and B. S. Kim. 2013. Mussel-inspired green synthesis of silver nanoparticles on graphene oxide nanosheets for enhanced catalytic applications. *Chemical Communications* 49 (33): 3392–3394.

Jiao, L. Y., X. R. Wang, G. Diankov, H. L. Wang, and H. J. Dai. 2010. Facile synthesis of high-quality graphene nanoribbons. *Nature Nanotechnology* 5 (5): 321–325.

Jiao, L. Y., L. Zhang, X. R. Wang, G. Diankov, and H. J. Dai. 2009. Narrow graphene nano-ribbon from carbon nanotubes. *Nature* 458 (7240): 877–880.

Kim, K. S., Y. Zhao, H. Jang et al. 2009. Large-scale pattern growth of graphene films for stretchable transparent electrodes. *Nature* 457 (7230): 706–710.

Kotakoski, J., A. V. Krasheninnikov, U. Kaiser, and J. C. Meyer. 2011. From point defects in graphene to two-dimensional amorphous carbon. *Physical Review Letters* 106: 105505.

Kuila, T., S. Bose, A. K. Mishra, P. Khanra, N. H. Kim, and J. H. Lee. 2012. Chemical functionalization of graphene and its applications. *Progress in Materials Science* 57 (7): 1061–1105.

Lang, B. 1975. A LEED study of the deposition of carbon on platinum crystal surfaces. *Surface Science* 53 (1): 317–329.

Lee, D. S., C. Riedl, B. Krauss, K. von Klitzing, U. Starke, and J. H. Smet. 2008. Raman spectra of epitaxial graphene on SiC and of epitaxial graphene transferred to SiO_2. *Nano Letters* 8 (12): 4320–4325.

Li, R., H. Yu, Y. Li et al. 2013. Ultrasensitive label-free immunoassay for diethylstilbestrol based on Au nanoparticles on mesoporous silica and amino-functionalized graphene. *Analytical Methods* 5 (20): 5534–5540.

Li, S., S. Guo, H. Yang et al. 2014a. Enhancing catalytic performance of Au catalysts by noncovalent functionalized grapheme using functional ionic liquids. *Journal of Hazardous Materials* 270: 11–17.

Li, X., Y. Zhao, H. Wang, H. Tang, J. Tian, and Y. Fan. 2014b. Highly dispersed Pd nanoparticles on 9-amino-1-azabenzanthrone functionalized graphene-like carbon surface for methanol electro-oxidation in alkaline medium. *Materials Chemistry and Physics* 144 (1–2): 107–113.

Li, X. S., Y. W. Zhu, W. W. Cai et al. 2009. Transfer of large area graphene films for high performance transparent conductive electrodes. *Nano Letters* 9 (12): 4359–4363.

Liu, L. H., M. M. Lerner, and M. Yan. 2010a. Derivitization of pristine graphene with well-defined chemical functionalities. *Nano Letters* 10 (9): 3754–3756.

Liu, S., J. Tian, L. Wang, H. Li, Y. Zhang, and X. Sun. 2010b. Stable aqueous dispersion of graphene nanosheets: Noncovalent functionalization by a polymeric reducing agent and their subsequent decoration with Ag nanoparticles for enzymeless hydrogen peroxide detection. *Macromolecules* 43 (23): 10078–10083.

Liu, W., Y. Zhang, S. Ge et al. 2013. Core–shell Fe_3O_4–Au magnetic nanoparticles based non-enzymatic ultrasensitive electrochemiluminescence immunosensor using quantum dots functionalized grapheme sheet as labels. *Analytica Chimica Acta* 770: 132–139.

Loh, K. P., Q. Bao, P. K. Ang, and J. Yang. 2010. The chemistry of graphene. *Journal of Materials Chemistry* 20 (12): 2277–2289.

Lu, D., Y. Zhang, S. Lin, L. Wang, and C. Wang. 2011b. Sensitive detection of esculetin based on a CdSe nanoparticles-decorated poly(diallyldimethylammonium chloride)-functionalized grapheme nanocomposite film. *Analyst* 136 (21): 4447–4453.

Lu, X., H. Qi, X. Zhang et al. 2011a. Highly dispersive Ag nanoparticles on functionalized graphene for an excellent electrochemical sensor of nitroaromatic compounds. *Chemical Communications* 47 (46): 12494–12496.

Lu, X. K., M. F. Yu, H. Huang, and R. S. Ruoff. 1999. Tailoring graphite with the goal of achieving single sheets. *Nanotechnology* 10 (3): 269–272.

Luo, B., S. Xu, X. Yan, and Q. Xue. 2013. PtNi alloy nanoparticles supported on polyelectrolyte functionalized graphene as effective electrocatalysts for methanol oxidation. *Journal of the Electrochemical Society* 160 (3): F262–F268.

Luo, B., X. Yan, S. Xu, and Q. Xue. 2012. Polyelectrolyte functionalization of graphene nanosheets as support for platinum nanoparticles and their applications to methanol oxidation. *Electrochimica Acta* 59 (1): 429–434.

Malard, L. M., M. A. Pimenta, G. Dresselhaus, and M. S. Dresselhaus. 2009. Raman spectroscopy of graphene. *Physics Reports* 473 (5–6): 51–87.

Mayavan, S., J. B. Sim, and S. M. Choi. 2012. Simultaneous reduction, exfoliation and functionalization of graphite oxide into a graphene-platinum nanoparticle hybrid for methanol oxidation. *Journal of Materials Chemistry* 22 (14): 6953–6958.

Meng, L. Y. and S. J. Park. 2012. Preparation and characterization of reduced graphene nanosheets via pre-exfoliation of graphite flakes. *Bulletin of Korean Chemical Society* 33 (1): 209–214.

Meyer, J. C., A. K. Geim, M. I. Katsnelson, K. S. Novoselov, T. J. Booth, and S. Roth. 2007. The structure of suspended graphene sheets. *Nature* 446 (7131): 60–63.

Meyer, J. C., C. Kisielowski, R. Erni, M. D. Rossell, M. F. Crommie, and A. Zettl. 2008. Direct imaging of lattice atoms and topological defects in graphene membranes. *Nano Letters* 8 (11): 3582–3586.

Mishra, A. K. and S. Ramaprabhu. 2011. Functionalized graphene-based nanocomposites for supercapacitor application. *Journal of Physical Chemistry C* 115 (29): 14006–14013.

Mukhopadhyay, P. and R. K. Gupta. 2013. *Graphite, Graphene and Their Polymer Nanocomposites.* Boca Raton, FL: Taylor & Francis.

Muszynski, R., B. Seger, and P. V. Kamat. 2008. Decorating graphene sheets with gold nanoparticles. *Journal of Physical Chemistry C* 112 (14): 5263–5266.

Nair, R. R., P. Blake, A. N. Grigorenko et al. 2008. Fine structure constant defines visual transparency of graphene. *Science* 320 (5881): 1308.

Novoselov, K. S., A. K. Geim, S. V. Morozov et al. 2004. Electric field effect in atomically thin carbon films. *Science* 306 (5696): 666–669.

Obraztsov, A. N. 2009. Chemical vapor deposition: Making graphene on a large scale. *Nature Nanotechnology* 4 (4): 212–213.

Obraztsov, A. N., E. A. Obraztsova, A. V. Tyurnina, and A. A. Zolotukhin. 2007. Chemical vapor deposition of thin graphite films of nanometer thickness. *Carbon* 45 (10): 2017–2021.

Rajesh, R., S. Senthil Kumar, and R. Venkatesan. 2014. Efficient degradation of azo dyes using Ag and Au nanoparticles stabilized on graphene oxide functionalized with PAMAM dendrimers. *New Journal of Chemistry* 38 (4): 1551–1558.

Rao, C. N. R., K. Biswas, K. S. Subrahmanyam, and A. Govindaraj. 2009. Graphene, the new carbon. *Journal of Materials Chemistry* 19 (17): 2457–2469.

Ren, W., Y. Fang, and E. Wang. 2011. A binary functional substrate for enrichment and ultrasensitive SERS spectroscopic detection of folic acid using graphene oxide/Ag nanoparticle hybrids. *ACS Nano* 5 (8): 6425–6433.

Riedl, C., C. Coletti, and U. Starke. 2010. Structural and electronic properties of epitaxial graphene on SiC(0 0 0 1): A review of growth, characterization, transfer doping and hydrogen intercalation. *Journal of Physics D: Applied Physics* 43 (37): 374009.

Scheuermann, G. M., L. Rumi, P. Steurer, W. Bannwarth, and R. Maulhaupt. 2009. Palladium nanoparticles on graphite oxide and its functionalized graphene derivatives as highly active catalysts for the Suzuki-miyaura coupling reaction. *Journal of American Chemical Society* 131 (23): 8262–8270.

Seo, M. H., S. M. Choi, J. K. Seo, S. H. Noh, W. B. Kim, and B. Han. 2013. The graphene-supported palladium and palladium–yttrium nanoparticles for the oxygen reduction and ethanol oxidation reactions: Experimental measurement and computational validation. *Applied Catalysis B: Environmental* 129: 163–171.

Shaabani, A. and M. Mahyari. 2013. PdCo bimetallic nanoparticles supported on PPI-grafted grapheme as an efficient catalyst for Sonogashira reactions. *Journal of Material Chemistry A* 1 (32): 9303–9311.

Shang, L., T. Bian, B. Zhang et al. 2014. Graphene-supported ultrafine metal nanoparticles encapsulated by mesoporous silica: Robust catalysts for oxidation and reduction reactions. *Angewandte Chemie International Edition* 53 (1): 250–254.

Shang, N. G., P. Papakonstantinou, M. McMullan et al. 2008. Catalyst-free efficient growth, orientation and biosensing properties of multilayer graphene nanoflake films with sharp edge planes. *Advanced Functional Materials* 18(21): 3506–3514.

Somani, P. R., S. P. Somani, and M. Umeno. 2006. Planar nano-graphenes from camphor by CVD. *Chemical Physics Letter* 430 (1–3): 56–59.

Song, W., H. Li, H. Liu, Z. Wu, W. Qiang, and D. Xu. 2013. Fabrication of streptavidin functionalized silver nanoparticle decorated graphene and its application in disposable electrochemical sensor for immunoglobulin E. *Electrochemistry Communications* 31: 16–19.

Stankovich, S., D. A. Dikin, G. H. B. Dommett et al. 2006a. Graphene-based composite materials. *Nature* 442 (20): 282–286.

Stankovich, S., R. D. Piner, X. Chen, N. Wu, S. T. Nguyen, and R. S. Ruoff. 2006b. Stable aqueous dispersions of graphitic nanoplatelets via the reduction of exfoliated graphite oxide in the presence of poly(sodium 4-styrenesulfonate). *Journal of Materials Chemistry* 16(2): 155–158.

Stankovich, S., R. D. Piner, S. T. Nguyen, and R. S. Ruoff. 2006c. Synthesis and exfoliation of isocyanate-treated graphene oxide nanoplatelets. *Carbon* 44 (15): 3342–3347.

Staudenmaier, L. 1898. Verfahren zur Darstellung der Graphitsaure. *Berichte der Deuschen chemischen gesellschaff* 31: 1481–1499.

Van Bommel, A. J., J. E. Crombeen, and A. Van Tooren. 1975. LEED and AUGER electron observation of the SiC(0001) surface. *Surface Science* 48 (2): 463–472.

Verma, V. P., S. Das, I. Lahiri, and W. Choi. 2010. Large-area graphene on polymer film for flexible and transparent anode in field emission device. *Applied Physics Letters* 96 (20): 203108.

Viculis, L. M., J. J. Mack, and R. B. Kaner. 2003. A chemical route to carbon nanoscrolls. *Science* 299 (5611): 1361.

Viculis, L. M., J. J. Mack, O. M. Mayer, H. Thomas, and R. B. Kaner. 2005. Intercalation and exfoliation routes to graphite nanoplatelets. *Journal of Material Chemistry* 15 (9): 974–978.

Walcarius, A., S. D. Minteer, J. Wang, Y. Lin, and A. Merkoci. 2013. Nanomaterials for bio-functionalized electrodes: Recent trends. *Journal of Material Chemistry B* 1 (38): 4878–4908.

Wang, G., J. Yang, J. Park et al. 2008. Facile synthesis and characterization of graphene nanosheets. *Journal of Physical Chemistry C* 112 (22): 8192–8195.

Wang, H. and H. Dai. 2013. Strongly coupled inorganic–nano-carbon hybrid materials for energy storage. *Chemical Society Reviews* 42 (7): 3088–3113.

Wang, J. J., M. Y. Zhu, R. A. Outlaw, X. Zhao, D. M. Manos, and B. C. Holloway. 2004. Synthesis of carbon nanosheets by inductively coupled radio-frequency plasma enhanced chemical vapor deposition. *Carbon* 42 (14): 2867–2872.

Wang, S., X. Wang, and S. P. Jiang. 2011. Self-assembly of mixed Pt and Au nanoparticles on PDDA-functionalized grapheme as effective electrocatalysts for formic acid oxidation of fuel cells. *Physical Chemistry Chemical Physics* 13 (15): 6883–6891.

Wang, Z., J. Xia, X. Guo et al. 2013. Platinum/grapheme functionalized by PDDA as a novel enzyme carrier for hydrogen peroxide biosensor. *Analytical Methods* 5 (2): 483–488.

Wu, D., Y. Li, Y. Zhang, P. Wang, Q. Wei, and B. Du. 2014b. Sensitive electrochemical sensor for simultaneous determination of dopamine, ascorbic acid, and uric acid enhanced by amino-group functionalized mesoporous Fe_3O_4@graphene sheets. *Electrochimica Acta* 116: 244–249.

Wu, Y. M., Y. Cen, L. J. Huang, R. Q. Yu, and X. Chu. 2014a. Upconversion fluorescence resonance energy transfer biosensor for sensitive detection of human immunodeficiency virus antibodies in human serum. *Chemical Communications* 50 (36): 4759–4762.

Wu, Z. S., D. W. Wang, W. Ren et al. 2010. Anchoring hydrous RuO_2 on graphene sheets for high-performance electrochemical capacitors. *Advanced Functional Materials* 20 (20): 3595–3602.

Xu, M., D. Fujita, J. Gao, and N. Hanagata. 2010. Auger electrons spectroscopy: A rational method for determining thickness of graphene films. *ACS Nano* 4 (5): 2937–2945.

Yan, S., B. Wang, Y. Shi et al. 2013. Hydrothermal synthesis of CdS nanoparticle/functionalized graphene sheet nanocomposites for visible-light photocatalytic degradation of methyl orange. *Applied Surface Science* 285: 840–845.

Yang, S., G. Li, J. Zhao, H. Zhu, and L. Qu. 2014. Electrochemical preparation of Ag nanoparticles/poly(methylene blue) functionalized graphene nanocomposite film modified electrode for sensitive determination of rutin. *Journal of Electroanalytical Chemistry* 717–718: 225–230.

Yang, W., G. Chen, Z. Shi et al. 2013. Epitaxial growth of single-domain grapheme on hexagonal boron nitride. *Nature Materials* 12: 792–797.

Yolaa, M. L., T. Erenb, and N. Atar. 2014. A novel and sensitive electrochemical DNA biosensor based on Fe@Au nanoparticles decorated graphene oxide. *Electrochimica Acta* 125: 38–47.

Young, R. J., I. A Kinloch, L. Gong, and K. S. Novoselov. 2012. The mechanics of grapheme nanocomposites: A review. *Composites Science and Technology* 72 (12): 1459–1476.

Yu, Q. K., L. A. Jauregui, W. Wu et al. 2011. Control and characterization of individual grains and grain boundaries in graphene grown by chemical vapor deposition. *Nature Materials* 10 (6): 443–449.

Yu, Q. K., J. Lian, S. Siriponglert, H. Li, Y. P. Chen, and S. S. Pei. 2008. Graphene segregated on Ni surfaces and transferred to insulators. *Applied Physics Letters* 93 (11): 113103.

Zhang, M., J. Xie, Q. Sun et al. 2013. In situ synthesis of palladium nanoparticle on functionalized grapheme sheets at improved performance for ethanol oxidation in alkaline media. *Electrochimica Acta* 111: 855–861.

Zhang, T., X. Li, S. Z. Kang, L. Qin, G. Li, and J. Mu. 2014. Facile assembly of silica gel/reduced graphene oxide/Ag nanoparticle composite with a core-shell structure and its excellent catalytic properties. *Journal of Materials Chemistry A* 2 (9): 2952–2959.

Zhang, X., Y. Xu, Y. Yang et al. 2012. A new signal-on photoelectrochemical biosensor based on a graphene/quantum-dot nanocomposite amplified by the dual-quenched effect of bipyridinium relay and AuNPs. *Chemistry: A European Journal* 18 (51): 16411–16418.

Zhang, Y. B., J. P. Small, W. V. Pontius, and P. Kim. 2005. Fabrication and electric-field-dependent transport measurements of mesoscopic graphite devices. *Applied Physics Letters* 86 (7): 073104.

Zhong, J. P., Y. J. Fan, H. Wang et al. 2013. Highly active Pt nanoparticles on nickel phthalocyanine functionalized grapheme nanosheets for methanol electro oxidation. *Electrochimica Acta* 113: 653–660.

Zhou, N., J. Li, H. Chen, C. Liao, and L. Chen. 2013. A functional graphene oxide-ionic liquid composites-gold nanoparticle sensing platform for ultrasensitive electrochemical detection of Hg^{2+}. *Analyst* 138 (4): 1091–1097.

Zhu, Y., S. Murali, W. Cai et al. 2010. Graphene and graphene oxide: Synthesis, properties, and applications. *Advanced Materials* 22 (35): 3906–3924.

CHAPTER 38

CONTENTS

Recent Trends in Carbon Nanotubes/Graphene Functionalization for Gas/Vapor Sensing

A Review

S.S. Barkade, G.R. Gajare, Satyendra Mishra, J.B. Naik,
P.R. Gogate, D.V. Pinjari, and S.H. Sonawane

38.1 INTRODUCTION

Carbon nanotubes (CNTs) and graphene hold a great promise for the development of miniaturized chemical sensors because of (1) greater adsorptive capacity due to large surface area to volume ratio, (2) better modulation of electrical properties (e.g., capacitance and resistance) upon exposure to analytes due to a greater interaction zone over the cross-sectional area, (3) ability to tune electrical properties of the nanostructure by adjusting the composition and size, and (4) the ease of configuration into different geometries. But these pristine forms of CNTs and graphene have drawbacks like lack of specificity to different gaseous analytes and the low sensitivity toward analytes that have no affinity to them. These limitations are reduced by functionalization of CNTs and graphene. Several research groups are currently working on the functionalization of CNTs and graphene with different materials to alter their chemical nature and enhance their sensing performance. Functionalization is an important aspect of the chemistry of CNTs and graphene and this chemical manipulation is essential for many of the potential applications like energy storage, energy conversion devices, sensors, hydrogen-storage media, nanometer-sized semiconductor devices, probes, and quantum wires. The development of CNTs and graphene-based sensors has attracted intensive interest in the last several years because of their outstanding sensing properties such as high selectivity and prompt response. In this chapter, an extensive overview of functionalized CNTs and graphene-based gas sensing is presented.

38.1.1 Sensor Criteria

Efficient gas sensor should satisfy the following basic criteria: (1) high sensitivity and selectivity, (2) fast response time and recovery time, (3) low gas consumption, (4) low operating temperature and temperature independence, and (5) stable performances.

However, gas sensors based on pristine CNTs and graphene have certain limitations, such as low sensitivity to analytes, lack of selectivity, irreversibility, or long recovery time. To overcome these limitations, functionalization of CNTs and graphene with different materials is necessary to modify their chemical nature and improve their sensing performance.

38.1.2 Why Carbon Nanotubes and Graphene?

The application of CNTs in the next generation of sensors has the potential of revolutionizing the sensor industry due to their inherent properties such as its extremely small size, high strength, better electrical and thermal conductivity, and high specific surface area (Fam et al. 2011). Following properties of CNTs justifies them as one of the best options for gas sensing: (1) well-organized nanostructure at atomic level, (2) large surface to volume ratio, (3) hollow geometry, (4) high electron mobility, (5) interesting electrical and physicochemical properties, and (6) high capability of gas adsorption. Also due to extraordinary electronic properties, graphene presents as an innovative material for gas sensing. The following properties make it an attractive material for gas sensing: (1) high surface area, (2) superconductivity, (3) wide potential window, (4) unusual fractional quantum Hall effect, (5) ballistic electronic transport, (6) higher gas response, and (7) high selectivity.

38.1.3 Role of Functionalization

The high van der Walls attraction between CNTs causes them to remain in bundles. This represents a problem for their application as gas sensors because it results in less adsorption or interaction sites, which translates into less sensitivity. The functionalization breaks the CNT bundles and allows the dispersion in different solvents and also protects them against agglomeration. It involves covalent modification and noncovalent interaction that enables them solubilizing into polar, nonpolar, and aqueous media. For various polymer-functionalized CNTs, there is direct microscopic evidence for the wrapping of individual nanotubes by the polymers. These polymer-wrapped CNTs are among the most soluble samples. Solubility provides an opportunity in the development of CNTs-based chemical sensors and the use of solubilized CNTs as starting materials for further chemical modifications in selective sensing (Ma et al. 2010).

Functionalization of graphene provides large surface area, extraordinary electronic transport behavior, and good compatibility for sensing applications. Since most atoms of the graphene sheet are exposed to the surface, slight changes in the charge environment due to the adsorption of functionalized group molecules provide significant changes in their electrical properties that help in the fabrication of more sensitive sensors. In graphene, the presence of catalytic impurities is negligible and thus the interference caused by these impurities are reduced and offers more reproducible sensing response. Till date, most of the reports suggest that functionalization is the best way to achieve the best sensor performance based on graphene (Zhang et al. 2009)

38.2 FUNCTIONALIZATION STRATEGIES FOR CARBON NANOTUBES

Several approaches have been adopted for the functionalization of CNTs. These include defect functionalization, covalent functionalization of the sidewalls, noncovalent exohedral functionalization, and endohedral functionalization.

38.2.1 Covalent Functionalization of Carbon Nanotubes

Covalent functionalization of CNTs produce defects in the wall structure and this type of surface functionalization is required for the specific requirement of load transfer properties. Many times the choice of the modifications on the nanotubes surface may be a direct result of the properties required from them. It involves making of bonds between carbon frameworks of nanotubes (Hornyak 2009). The modifications can be done by various methods like halogenation, cycloaddition, radical addition, nucleophilic addition, polymer composites, and various other methods. In these methods, various molecules or groups are employed for the functionalization of CNTs, which include halogen group, halogen group, alkyl chains, crown ethers, etc. Polymer composites are

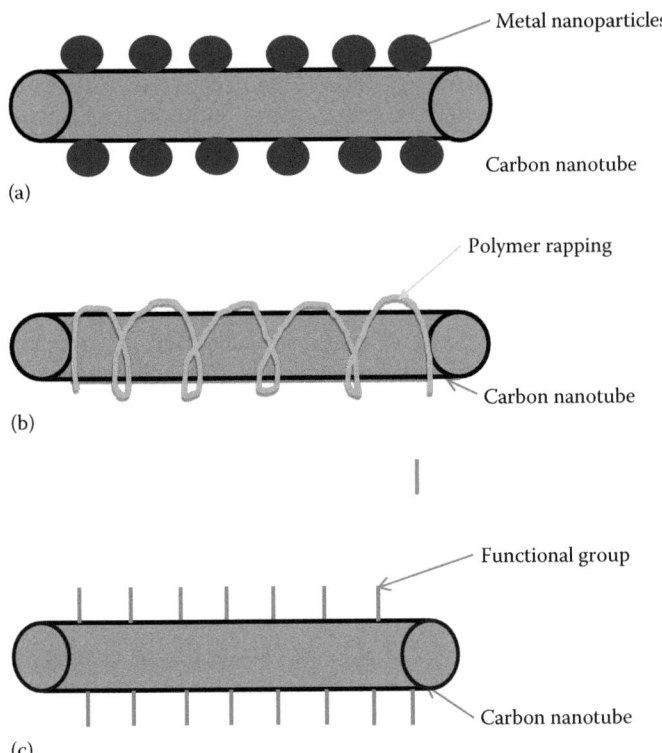

FIGURE 38.1
Carbon nanotube functionalized with (a) metal, (b) metal oxide, and (c) polymeric nanoparticles.

employed by two methods, termed as *grafting to* and *grafting from*. In the former, synthesis of a polymer followed by end-group transformation and attachment of the polymer chain to the surface of CNTs is involved while the latter involves covalent immobilization of the polymer precursors on the surface of the nanotubes and the resulting in situ polymerization (Rao et al. 2011).

38.2.2 Noncovalent Functionalization of Carbon Nanotubes

This method is mainly based on the physical adsorption of surfactants or polymers to obtain homogeneous CNTs suspensions, compelling the advantage of leaving the electronic structure of CNTs intact. This method gains more importance as compared with other functionalization methods because it is a nondestructive process and in this, original electronic structure of CNTs is retained, which is useful for tubes to get solubilized in polar and nonpolar solvents. Functionalization is done using polymer composites, surfactants, and with the help of biomolecules (Gong et al. 2000, Hu et al. 2000, Wang 2005). Compared with the chemical functionalization, noncovalent functionalization has the advantages that it could be operated under relatively mild reaction conditions and the perfect graphitic structure of CNTs could be maintained (Meng et al. 2009). The various conceptual schemes of functionalization for CNTs are shown in Figure 38.1.

38.3 FUNCTIONALIZATION APPROACHES FOR GRAPHENE

Graphene is a 2D monoatomic thick building block of a carbon allotrope and has emerged as an interesting material of the twenty-first century and received worldwide attention due to its remarkable electronic, thermal, optical, and mechanical properties and quantum Hall effect (Singh et al. 2011). Despite the great potential, graphene itself possesses zero band gap as well as inertness to reaction, which weakens the competitive strength of graphene in the field of semiconductors and sensors. Functionalization of graphene can overcome this problem in an efficient way. Many of the researchers demonstrate that functionalization is the best way to achieve the best sensor performance out of graphene (Dan et al. 2009, Zhang et al. 2009). Functionalization increases the sensitivity of the graphene to the adsorption process and reduces nonspecific binding and enhances selectivity (or specificity) for the desired analyte. Similar approaches like CNTs have been employed for the functionalization of graphene (Niyogi et al. 2006, Zhang et al. 2011). However, it is essential to analyze the methods that have been used for functionalization. Many of the chemical functionalization methods employed to date have used covalent bonding that results in the destruction of the sp^2-bonding of graphenes lattice. Methods of noncovalent bonding, however, tend to exploit the extensive opportunities for π bonding on graphenes basal plane and are much more amenable than the harsh methods of covalent functionalization, and, therefore, tend to retain graphene's unique electronic properties for use in sensors.

38.3.1 Functionalization by Covalent Bonding

The main objective of covalent functionalization is to increase the ease of exfoliating graphite to produce modified graphene or to make functionalized graphene for various applications such as polymer nanocomposites. The most widely used chemical production method of graphene uses the oxidative exfoliation of natural graphite and subsequent thermal or chemical reduction. Strong acid treatment of natural graphite generates various oxygen-containing functional groups, such as epoxy, hydroxy, carbonyl, or carboxylic acid, at the basal plane and the edge of graphene, which facilitates the dispersibility in various solvents, including water (Kim et al. 2011). These functional groups on graphene oxide provide endless opportunities for further functionalization to create new hybrid materials for different applications. But due to harsh conditions and loss of transport and chemical properties, it is less preferred for the design of gas sensors.

Covalent functionalization on the surface of graphene nanosheets (GNSs) is one of the strategies in fabricating GNS-based polymer composites and is an effective method for improving the interfacial interactions between GNSs and polymer matrix. Graphene–tetraphenylporphyrin and graphene–palladium tetraphenylporphyrin hybrids were prepared through one-pot cycloaddition reactions and the presence of the covalent linkages between graphene and porphyrin was confirmed by FTIR and Raman spectroscopy (Zhang et al. 2011). Functionalization of EG (exfoliation of graphitic oxide) and DG (conversion of nanodiamond) were carried out through amidation for solubilizing them in nonpolar solvents (Subrahmanyam et al. 2008). The infrared (IR) spectrum of the water-soluble, acid-functionalized graphene shows a prominent band due to carbonyl groups (C=O) in addition to a broad band due to –OH groups. A simple and fast method has been developed to obtain functionalized chemically converted GNSs via covalent functionalization with 3-amino-propyltriethoxysilane (Yang et al. 2009). Covalent functionalization of the charged graphene has been achieved via effective reductive activation by organic diazonium salts (Englert et al. 2011).

38.3.2 Noncovalent Functionalization of Graphene

Noncovalent functionalization of graphene is fascinating, as it does not alter the original electronic structure and planarity of graphene (Rao et al. 2011). Noncovalent functionalization of graphene is based on secondary interactions such as π–π stacking, hydrogen bonding, or charge–charge interactions. It minimizes the damage to the graphitic chemical structure and thus secures the physical and chemical properties of graphitic materials. A very high specific interaction is needed to carry out the functionalization between molecule and graphene (derivatives). The exact choice of functionalizing molecule and the corresponding dispersible solvent system are required for the best performance of functionalized graphene in sensing. In this method, several approaches have been adopted such as wrapping with polyethylene glycol (PEG) and other surfactants and π–π interaction with a pyrene derivative such as 1-pyrenebutanoic acid succinimidyl ester and potassium salt of coronene tetracarboxylic acid (Rao et al. 2011, Zhang et al. 2011). Noncovalent functionalization of

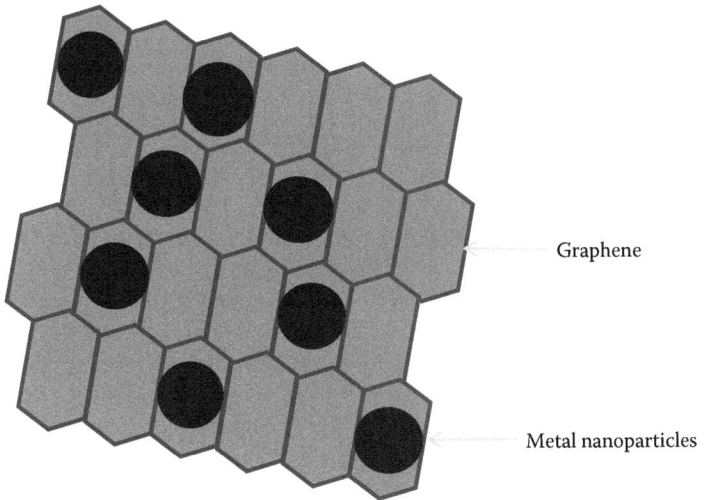

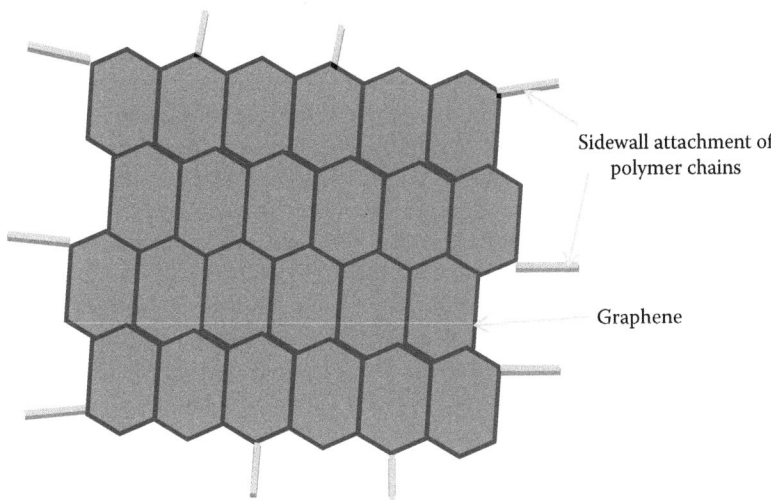

FIGURE 38.2
Graphene functionalized with metal nanoparticles and polymer chains.

graphene by surfactants such as sodium dodecyl sulfate (SDS), cetyltrimethylammonium bromide (CTAB), and polyoxyethylene (40) nonylphenyl ether gives water-soluble graphene (Pati et al. 2011). Noncovalent solubilization of graphene is somewhat easy if π–π interaction is exploited. Different molecules containing π groups can solubilize graphene easily as graphene has a 2D structure with high surface area. Shortcoming of this method is rigorous ultrasonication and sometimes mild heat treatment is required to solubilize CNTs through noncovalent interaction. The various conceptual schemes of functionalization for graphene are shown in Figure 38.2.

38.4 GAS SENSORS BASED ON FUNCTIONALIZED CARBON NANOTUBES

The major drawbacks of using pristine CNTs as sensing materials are the lack of specificity to different gaseous analytes and the low sensitivity toward analytes that have no affinity to CNTs. These limitations can be, at least in part, circumvented by functionalizing the CNTs with analytes-specific entities. Functionalization also sometimes affects the sensor dynamics (Zhang et al. 2008). Functionalization of CNTs with functional groups, metal nanoparticles, metal oxides, and polymers resulted in altered electronic properties, and improving the selectivity and response to specific gases through the interaction of the target molecules with the functional groups or additives is very different (Zhang and Zhang 2009).

38.4.1 CNTs Functionalized via Metal Nanoparticles

Pure CNTs do not show appreciable sensitivity to some gases and the decoration of CNTs by nanoparticles (NPs) shows improved sensing performance and wider area of applications. Nanoparticles of metals like Pd, Al, Pt, Sn, Pd, and Rh have been used to decorate CNTs, allowing selective detection of gases like H_2, NH_3, NO_2, CH_4, H_2S, and CO (Kong et al. 2001, Lu et al. 2004, Star et al. 2006). CNT/NP hybrid structures represent other alternative of nanoscaled gas sensors that can operate at low voltage and power consumption. CNTs decorated with metallic nanoparticles have been widely used to achieve selectivity and improve the sensitivity, response time, and detection limits for a variety of gas detections. Layer by layer, electrodeposition, chemical deposition, electrochemical deposition, and sputtering and other methods used to prepare CNT/NP hybrid structures are discussed in this section.

Gold-modified MWCNT chemiresistors were synthesized for NO_2 gas sensing at higher temperature range of 100°C–250°C by plasma-enhanced chemical vapor deposition (PECVD) technique (Penza et al. 2008). Double wall CNTs has been decorated with a layer of palladium (Pd) nanoparticles of 1, 3, and 6 nm for hydrogen gas sensing (Rumiche et al. 2012). They evaluated the effect of nanotube content and Pd nanoparticle size on H_2-sensing performance at room temperature. Nanoparticle layers with the size of 1 nm did not show reliable sensing response to hydrogen, whereas 3 nm Pd coating exhibited more sensitivity to hydrogen with slightly extended response.

Multiwalled carbon nanotubes (MWCNTs) are functionalized with carboxyl and amide group and are decorated with Mo/Pt nanoparticles for hydrogen sulfide gas sensing (Izadi et al. 2012). Metal-decorated CNTs exhibited better performance as compared to functional groups of carboxyl and amide for H_2S gas monitoring at room temperature. MWCNTs were processed in the form of thin films and decorated with Pd nanoparticles for room temperature hydrogen gas sensing (Zilli et al. 2011). They showed that the Pd-decorated MWCNTs nanocomposite films exhibit a reversible response toward hydrogen at low concentrations. Besides, their sensing capacity is affected by the different stages of purification applied to the CNTs. Palladium nanoparticles have been successfully attached onto the surfaces of single-walled carbon nanotubes (SWCNTs) by dendrimer-mediated synthesis using NH_2-terminated poly(amido amine) (PAMAM) dendrimers for H_2 sensing (Ju et al. 2010). They found much faster response time and better recovery than those prepared without dendrimers. Highly dispersed gold nanoparticles supported on nitrogen-doped MWCNTs were synthesized using an electrochemical method (Sadek et al. 2011). They used simple electrochemical method for modifying MWCNTs and achieved high density decoration of gold nanoparticles onto CNT surface and succeeded in getting high sensitivity, faster response, and good recovery as hydrogen gas sensor.

A nanocomposite plating technique has been demonstrated to fabricate leaf-like CNT/nickel nanostructure film, decorated with Pd nanoparticles to create hydrogen gas sensors on a glass substrate, and showed excellent electrical performance and stability (Lin and Huang 2012). Electronic sensor array based on SWCNTs was developed and decorated by metal nanoparticles (Pt, Pd, Rh, and Sn) in the detection and identification of toxic/combustible gases for personal safety and air pollution monitoring (Star et al. 2006). These small-sized, low power electronic sensor arrays have been reported to be very effective in the detection and identification of toxic/combustible gas. MWCNT have been functionalized in oxygen RF plasma and decorated with Au or Ag nanoclusters as gas-sensitive materials using electron beam evaporation. These hybrids were able to detect as low as 500 ppb of NO_2 even when operated at ambient temperature and similar results were obtained when the operating temperature of the sensors was raised to 150°C (Espinosa et al. 2007).

Hydrogen gas sensor has been fabricated from aligned CNT grown on anodic aluminum oxide template decorated with Pd metal. The sensor was prepared with sound robustness and detected hydrogen gas with good reversible response at room temperature (Ding et al. 2007). MWCNTs have been decorated with rhodium nanoparticles using a colloidal solution in the postdischarge of a radiofrequency (RF) atmospheric plasma of argon (Ar) or argon/oxygen (Ar:O_2) (Leghrib et al. 2011a). The sensing properties of these hybrid materials were investigated toward detection of NO_2 and C_2H_4 at room temperature. They found that the presence of oxygen in the plasma treatment is essential to significantly enhance the gas response of Rh-decorated MWCNTs and to avoid response saturation even at low gas/vapor concentrations. SWCNT-based gas sensors were prepared for NO_2 gas sensing by dielectrophoresis method onto microelectrodes made up of Cr, Pd, or Al. When Al/CNT sensor was exposed to NO_2 gas, the resistance was fast and a large increase in resistance has been noticed, whereas it decreased for other metals/CNT. The Schottky response of the Al/CNT sensor was studied and it was approximately one order of magnitude faster than the CNT response obtained using the other metal electrodes (Suehiro et al. 2006).

Nanostructured platinum (Pt) functionalized MWCNTs have been produced by catalytic CVD and acted as good hydrogen sensor at room temperature with high sensitivity and reversibility (Kumar and Ramaprabhu 2006). The chemical treatment of MWCNTs, synthesized by the pyrolysis of acetylene over $Mm_{0.2}Tb_{0.8}Co_2$-H catalyst that has been obtained using hydrogen decrepitation technique, resulted in enhanced Pt dispersion. Enhanced Pt dispersion is due to functional groups such as –COOH and –OH that acts as fastening sites for Pt and also due to increased hydrophilic nature of chemically treated MWCNTs. Chemiresistor sensor configurations with supported and suspended nanotubes decorated by Ti nanoparticles were used for gas (N_2, Ar, and O_2) sensing at low temperatures. MWCNTs worked as a conductive channel for electrical signal acquisition, a heating element for local heating of attached nanoparticles, and a substrate for nanoparticles deposition for selective gas sensitivity, whereas nanoparticles were employed to provide selectivity to specific gases (Gelamo et al. 2009).

SWCNTs on thin plastic substrates have been decorated with Pd nanoparticles through a simple electrochemical deposition method. CNT/Pd composite nanostructure served as active building blocks for the fabrication of mechanically flexible hydrogen gas sensors. Fabricated sensors showed properties such as fast and sensitive detection under the conditions such as conformal wrapping over curvilinear surfaces, high tolerance toward repeated bending, and mechanical shock resistance. (Sun et al. 2007). A simple fabrication technique has been developed to construct a hydrogen nanosensor by decorating SWCNTs with Pd nanoparticles. They optimized sensing performance by varying the sensors synthesis conditions such as Pd electrodeposition charge, deposition potential, and initial baseline resistance of the SWCNT network. This optimized sensor showed excellent sensing characteristics toward hydrogen. Compared to other functionalization methods, electrodeposition offers a simple, cost-effective, and site-specific way for decorating nanostructured materials (Mubeen et al. 2007).

In recent years, demand for high performance of H_2 as a clean energy source and subsequently hydrogen sensors is high, and it is certainly critical as well as important to monitor leakage of Hydrogen. SWCNT films were made by a highly reproducible filtration process, coated with Pd for the detection of H_2 to the levels of ~10 ppm at room temperature. Sensor showed quick and easy recoverability of 30 s and this property made it suitable for application at room temperature by consumption of low power of ~0.25 mW (Sippel-Oakley et al. 2005). Fabrication of hydrogen sensor was demonstrated by modifying SWCNTs by Pd nanoparticles. The sensor showed high sensitivity, fast response, and better recovery at room temperature (Kong et al. 2001). Nanocomposite of Pd and MWCNTs grafted with 1,6-hexanediamine was prepared by using a chemical reduction method. The effect of carrier gas, temperature, and amount of nanocomposite on the response of methane sensor were investigated. Sensor showed good sensitivity and selectivity toward CH_4 in dry air at ambient temperature conditions along with better response and recovery than other sensors (Li et al. 2009).

CNTs-networked films have been prepared by radiofrequency plasma-enhanced chemical vapor deposition (RF-PECVD) technology onto low-cost alumina substrates, coated with nanosized Co-catalyst, as highly sensitive nanomaterial of a chemiresistor for NO_2, H_2S, and NH_3 sensing at sub-ppm level. Highest gas sensitivity to NO_2, H_2S, and NH_3 has been found by CNTs functionalized with Au loading of 5 nm at 200°C (Penza et al. 2009). SWCNT mesh doped with alkanethiol monolayer-protected gold clusters (MPCs) has been investigated for ultrahigh sensitivity detection of nitrogen dioxide. Detection limit of composite was found to be better (ppb level) than that of pure CNTs (Young et al. 2005). Gas-sensing performance of metal-functionalized CNT gas sensors is summarized in Table 38.1.

TABLE 38.1 CNTs Functionalized Using Metal Nanoparticles

Metal	CNT Type	Functionalization Strategy	Gas	Detection Limit	Response Time (s)	Recovery Time (s)	References
Au	MWCNT	DC-sputtering	NO_2	100 ppb	<600	N/S	Young et al. (2005)
Pd	DWCNT	BOC Edwards Auto 306 vacuum-coating system	H_2	500 ppm	21	~200	Rumiche et al. (2012)
Mo, Pt	MWCNT	Incipient impregnation	H_2S	200 ppm	300	600	Izadi et al. (2012)
Pd	MWCNT	Reduction and filtration	H_2	350 ppm	~150	210	Zilli et al. (2011)
Pd	SWCNT	Dendrimer-mediated synthesis using NH_2-terminated PAMAM dendrimers	H_2	10 ppm	7	N/S	Ju et al. (2010)
Au	MWCNT	Facial electrochemical route	H_2	0.06 vol%	15	20	Sadek et al. (2011)
Pd	CNT–nickel composite	Nanocomposite plating technique	H_2	200 ppm	312	173	Lin and Huang (2012)
Pt, Pd, Sn, Rh	SWCNT	Electrochemical functionalization, electron-beam evaporation	H_2, CH_4, CO, H_2S	N/S	600	N/S	Star et al. (2006)
Au, Ag	MWCNT	Thermal evaporation	NO_2	500 ppb	~1200	N/S	Espinosa et al. (2007)
Pd	ACNT	RF sputtering	H_2	100 ppm	<240	N/S	Ding et al. (2007)
Rh	MWCNT	RF atmospheric plasma of argon (Ar) or argon/oxygen	NO_2, C_2H_4, CO, C_6H_6, moisture	50 ppb NO_2 50 ppb Benzene	~200 <600	N/S	Leghrib et al. (2011a)
Al	SWCNT	Dielectrophoresis	NO_2	<1 ppm	N/S	N/S	Suehiro et al. (2006)
Pt	MWCNT	Catalytic chemical vapor deposition	H_2	N/S	600	900	Kumar and Ramaprabhu (2006)
Ti	MWCNT	Sputtering	N_2, Ar, O_2	N/S	N/S	N/S	Gelamo et al. (2009)
Pd	SWCNT	Electrochemical deposition	H_2	100 ppm	3–60	~300	Sun and Wang (2007)
Pd	SWCNT	Electrodeposition	H_2	100 ppm	600	1200	Mubeen et al. (2007)
Pd	SWCNT	Sputtering, thermal evaporation	H_2	~10 ppm	<600	<30	Sippel-Oakley et al. (2005)
Pd	SWCNT	Electron-beam evaporation	H_2	<40 ppm	5–10	400	Kong et al. (2001)
Pd	MWCNT	Chemical reduction method	CH_4	0.167% v/v	<35	<12	Li et al. (2009)
Au	CNT	Sputtering	NO_2, H_2S, NH_3	100 ppb NO_2	600	N/S	Penza et al. (2009)
Au	SWCNT	Drop casting	NO_2	4.6 ppb	N/S	N/S	Young et al. (2005)

Note: N/S = Not stated.

38.4.2 CNTs Modified by Metal Oxide Nanoparticle

Semiconductor metal oxide sensors are the most promising due to their simple structure and low fabrication cost. Sensors made of metal oxide films have been used for a long time because they provide high sensitivity for the detection of a wide variety of gases. However, their major drawback is their elevated operating temperatures. The development of metal oxide NPs modified CNTs hybrid sensing films has shown advantages like efficient charge transfer, higher surface area, porosity, high catalytic activity, and adsorption capacity. Several studies have reported the excellent sensing properties of CNTs/SnO$_2$ sensor for detection of CO, NO$_2$, and NH$_3$ at low temperatures (Choi et al. 2010, Mehrabi et al. 2010). Thin films of MWCNT-doped WO$_3$ and undoped WO$_3$ were prepared by using electron beam evaporation technique and exposed to 1,000 ppm of H$_2$ at different temperatures (200°C–400°C). They observed that the optimum operating temperature for sensing was 350°C. Further MWCNT-doped WO$_3$ thin films showed higher responses for H$_2$ at any operating temperature when compared to the undoped WO$_3$ thin film (Wongchoosuk et al. 2010).

A compound material of MWCNTs coated with SnO$_2$ was synthesized at ambient conditions. Gas-sensing properties of compound material were studied and found that sensors exhibited good sensing responses to liquefied petroleum gas (LPG), ethanol gas (C$_2$H$_5$OH) with fast response, and recovery within seconds (Liu et al. 2006).

MWCNTs/SnO$_2$ gas sensors were fabricated using ultrasonic-assisted deposition–precipitation method and used for ethanol, acetaldehyde, acetone, toluene, and trichloroethylene at various temperatures. Sensor exhibited significant responses to sub-ppm levels of acetaldehyde and acetone (Feyzabad et al. 2012). Enhanced ethanol sensing properties of CNT/SnO$_2$ core–shell nanostructure have been observed by increasing the reaction time in the synthesis processes. Because of increased reaction time, thicker shell of SnO$_2$ is formed on outer surface of CNT. Core/shell structures exhibit high sensitivity, fast recovery, and good stability toward ethanol gas (Chen et al. 2006).

Different synthesis temperatures for metal oxide NPs modified CNTs hybrid can affect the sensor performance. SnO$_2$/SWCNTs composites were synthesized at different oxidizing temperatures (300°C–600°C) for testing the effect of temperature in their morphology, structure, and gas-sensing properties in the detection of NO$_x$. Sensing study shows that the optimum oxidizing and operating temperatures were 200°C and 400°C, respectively, when the hybrid films were exposed to different concentrations of NO$_x$. Also, the thin films of metal oxide NPs decorated CNTs hybrid showed improved performance for the detection of NO$_x$ under the same conditions (Jang et al. 2012). MWCNTs were treated with nitric acid and used to fabricate MWCNTs-doped SnO$_2$ sensors for the detection of ethanol and LPGs. Thin films of sensors were tested from 100 to 1,000 ppm of ethanol and 1,000 to 10,000 ppm of LPG at different operating temperatures in the range of 10°C–360°C. The detection of ethanol and LPG were improved when the operating temperature was 350°C or lower and it was determined that the optimum operating temperature is 320°C. The hybrid showed better selectivity for LPG than for ethanol. From the results, the selectivity of the MCNTs-doped SnO$_2$ hybrid for LPG can be attributed to the presence of MWCNTs (Van et al. 2010).

Polymer-assisted deposition method was used to prepare the hybrid of cobalt oxide and SWCNTs for No$_x$ (20–100 ppm) and H$_2$ sensing. Polyethyleneimine (PEI) binds the cobalt ions and adjusts the viscosity of the solution during the deposition process in order to get a homogeneous distribution of the particles on the SWCNTs thin film. Hybrid sensor showed proportional increases in response for NO$_x$ as a function of concentration (20–100 ppm), poor recovery at room temperature, and good recovery at 250°C. Higher responses of the Co$_3$O$_4$/SWCNTs composite when compared to pristine CNTs are attributed to the high adsorption power of Co$_3$O$_4$ particles. The composite was also exposed to 4% of H$_2$ in air and showed enhanced responses than pure SWCNTs at room temperature and Co$_3$O$_4$ films at both room temperature and 250°C (Li et al. 2010). Three composites (ZnO/MWCNT, SnO$_2$/MWCNT, and TiO$_2$/MWCNT) were grown simultaneously on silica substrates by catalytic pyrolysis method and used for ethanol sensing.

For 100 ppm ethanol, current–differential voltage (ΔI–V) curves were recorded and better sensitivity was determined when compared to pure MWNT film, ZnO/MWCNTs, and SnO_2/MWCNTs (Liu et al. 2012).

Elements (N, B, and O)-doped CNTs with and without SnO_2 were used to study the effect of functional groups on their gas-sensing properties for 100, 200, 500, and 1,000 ppb of NO_2 at room temperature. All doped-CNTs/SnO_2 hybrids responded better than N-doped, B-doped, and O-doped CNTs. The high sensitivity and improved performance achieved with the B-doped and N-doped-CNTs/SnO_2 hybrids for low concentrations of NO_2 at room temperature are attributed to two main factors: the interaction of the N_2 gas with the n-SnO_2/p-CNTs heterostructure that affects the conduction of the CNTs and the addition of new functionalities (i.e., B and N atoms) to the CNTs surface that affects the conductivity (Leghrib et al. 2011b). Gas-sensing performance of metal oxide nanoparticles functionalized CNT gas sensors is summarized in Table 38.2.

38.4.3 CNTs Functionalized with Polymer

The insolubility of CNTs in solvents or polymers due to strong intertube van der Waals attraction impedes their applications. Now, several studies focus on CNTs functionalized with polymers to take full advantage of their unique properties. Therefore, polymer functionalization of the CNTs may be an important tool to improve the compatibility between organic polymers and CNTs and to provide the nanotube sensors with an ultrahigh sensitivity and selectivity for gas detection. Number of modifications by doping or coating, either partly or completely, has been suggested to further enhance the sensors properties of CNTs (Bekyarova et al. 2004, Ding et al. 2004). CNTs' unique characteristics combined with polymers delocalized bonds, high permeability, and low density have demonstrated that it is possible to detect many different gases with high sensitivity, fast response, and good reproducibility. Polymer/CNTs hybrids used in resistors, surface acoustic wave (SAW) and quartz micro balance (QMB), and other type of sensors are discussed.

Arrays of electrical devices with each comprising multiple SWNTs bridging metal electrodes were obtained by CVD of nanotubes across prefabricated electrode arrays. Polyethyleneimine functionalization with SWCNTs was reported to impart high sensitivity and selectivity to the sensors. Polymer coating affords n-type nanotube devices capable of detecting NO_2 at less than 1 ppb concentrations while being insensitive to other gases (Qi et al. 2003).

A facile fabrication method was used to make chemical gas sensors using SWCNTs electrochemically functionalized with polyaniline (PANI). PANI-SWCNT sensors were used to sense NH_3 gas and exhibited improved sensitivity and reproducibility with low detection limit of 50 ppb (Zhang et al. 2006).

SWCNT/polypyrrole gas sensor was fabricated using a simple chemical polymerization technique followed by spin casting onto prepatterned electrodes. Sensor exhibited greater stability and sensitivity toward NO_2 gas and response of the nanocomposites was about ten times higher than that of polypyrrole. Vertically aligned CNTs have also been modified with polymers for gas sensors (An et al. 2004). A sensor for HCl vapor detection was fabricated by selective growth of aligned CNTs on Si_3N_4/Si substrates patterned by metallic Pt. Poly(o-anisidine) deposition onto the CNTs device was shown to impart higher sensitivity to the sensor (Valentini et al. 2004). A noncovalent functionalization of SWCNTs were carried out by simply submerging nanotube network field effect transistors in an aqueous solution of poly(ethylene imine) (PEI) and starch overnight. PEI-starch polymer-coated SWCNT field effect transistors had n-type characteristics and were used as CO_2 gas sensors (Star et al. 2004).

A compact wireless sensor based on CNT/PMMA composite was developed for the detection of organic vapors. The change in resistance of the CNT/polymer composite film due to the exposure of different gases was utilized as the principle of gas sensor. The resistance changes of the sensor were wirelessly transmitted by a Bluetooth module to a notebook computer for data analysis. The sensor exhibited fast response time (2–5 s) and 10^2–10^3-fold increase in resistance when exposed to saturated dichloromethane, chloroform, and acetone vapors (Abraham et al. 2004). A new approach has been developed based on polyethylene imine noncovalently functionalized SWCNT networks to

TABLE 38.2 CNTs Functionalized Using Metal Oxide Nanoparticles

Metal Oxide	CNT Type	Functionalization Strategy	Gas	Detection Limit	Response Time (s)	Recovery Time (s)	References
SnO$_2$	MWCNT	Sonication-assisted precipitation–deposition	LPG, ethanol gas	10 ppm	N/S	N/S	Liu et al. (2006)
SnO$_2$	MWCNT	Ultrasonic-assisted deposition–precipitation	Ethanol, acetaldehyde, acetone, toluene, trichloroethylene	25 ppm of ethanol	160	118	Feyzabad et al. (2012)
SnO$_2$	MWCNT	Wet chemical method	Ethanol	10 ppm	~1	~10	Chen et al. (2006)

Note: N/S = Not stated.

detect NO in exhaled breath. Interaction of the gas with a functionalized CNT field effect transistor results in a conductivity change that is proportional to the NO gas concentration. Sensor revealed detection up to ppb level and this method offered advantages like low cost, compact size, and simplicity of setup for monitoring NO concentrations while overcoming the limitations of cross-contaminants, etc (Kuzmych et al. 2007).

Conductive polymer composites have been fabricated by in situ polymerization of styrene in the presence of MWCNTs or solution mixing of polystyrene with MWCNTs. Compared with the composites prepared by solution mixing, the ones by polymerization filling had much higher responsivity to organic vapor over a wide MWCNT range. The electrical percolation behaviors of the composites and their resistance responsivities against various organic vapors was investigated (Zhang et al. 2005).A novel approach has been demonstrated toward the development of advanced chemical sensors based on chemically functionalized SWCNTs. SWNCTs were covalently functionalized by poly-(m-amino benzene sulfonic acid) and exhibited improved sensor performance for detection of NH_3 with low detection limit of 5 ppm (Bekyarova et al. 2004). Ammonia, nitrogen dioxide, and water vapor sensor have been developed based on poly-(m-amino benzene sulfonic acid) functionalized single-walled carbon nanotube (SWCNT-PABS) network. SWCNT-PABS sensors exhibited excellent sensitivity with ppb level detection limits (100 ppb for NH_3 and 20 ppb for NO_2) at room temperature. The response time was short and the response was totally reversible (Zhang et al. 2007).

A chemical sensor using a novel nanocomposite material made up of MWCNTs and poly (3-methylthiophene) was studied. Sensor showed change in electrical resistance upon exposure to different chloromethanes and resistance change was proportional to the concentration of the analyte. The sensor was highly selective toward chloromethanes while it was insensitive to methane and other chemicals (Santhanam et al. 2005). A novel amperometric sensor based on MWCNTs-grafted polydiphenylamine (PDPA) has been developed for the determination of carbon monoxide. Sensor exhibited low detection limit of 0.01 ppm with shorter response and recovery time of few seconds (Santhosh et al. 2007). Multifunctional chemical sensors were fabricated by partially coating perpendicularly aligned MWCNTs with polymers such as poly(vinyl acetate), polyisoprene, and then sputtering with gold electrodes. Sensor exhibited rapid and reversible sensing of high concentrations of a variety of volatile organic solvents like ethanol, cyclohexane, and tetrahydrofuran (Wei et al. 2006).

Composite of MWCNTs/polyaniline has been prepared by a method featuring self-assembly and in situ polymerization. The resulting composite showed a very high sensitivity and a low detection limit at the ppb level for the sensing of triethylamine vapor at room temperature (Li et al. 2008). A highly selective gas sensor was constructed by chemical modification of MWCNTs containing carboxyl groups (MWCNT-COOH) with poly ethylene glycol (PEG) in the presence of N,N-dicyclohexylcarbodiimide (DCC). The sensors exhibited high chemical selectivity, fast response, and good reproducibility or long stability to chloroform vapor, which were attributed to the properties of MWNTs-grafted PEG polymers (Niu et al. 2007). Gas sensing performance of polymer functionalized CNT gas sensors is summarized in Table 38.3.

38.4.4 CNTs Decorated with Functional Group

The conductivity of nanotubes changes significantly with the adsorption or the functionalization of chemical reactants on the sidewalls, which leads to the application in sensing. For covalent functionalization of CNTs, several methods have been reported, such as defect site creation and functionalization from the defects, creating carboxylic acids on the end caps of CNTs and subsequent derivatization from the acids, as well as the covalent sidewall functionalization method. For noncovalent functionalization, wrapping of nanotubes with surfactants such as polymers, which could preserve the physical property of nanotubes and also allow their solubilization, have been used. The important features of surfactant, namely adsorption at interface and self-accumulation into supramolecular structures, are effectively used in processing stable colloidal dispersions. CNTs decorated with different functional groups have been used for the development of sensors for detection of VOCs in the environment.

TABLE 38.3 CNTs Functionalized with Polymer

Polymer	CNT Type	Functionalization Strategy	Gas	Detection Limit	Response Time (s)	Recovery Time (s)	References
Nafion, polyethyleneimine	SWCNT	Noncovalently drop coating	NO_2	For NO_2 <1 ppt	~60 to 120	N/S	Qi et al. (2003)
Polyaniline	SWCNT	Electrochemical functionalization	NH_3	50 ppb	N/S	N/S	Zhang et al. (2006)
Polypyrrole	SWCNT	Covalent chemical polymerization	NO_2	N/S	600–1,800	N/S	An et al. (2004)
Poly(o-anisidine)	ACNT	Simple coating	HCl	100 ppm		N/S	Valentini et al. (2004)
Poly(etyleneimine) and starch polymer	CNT	Noncovalently dip coating	CO_2	500 ppm	60	N/S	Star et al. (2004)
Polymethyl methacrylate (PMMA)	MWCNT	Mixed as composite film	Dichloromethane, chloroform, acetone	N/S	2–5	N/S	Abraham et al. (2004)
Polyethylene imine	SWCNT	Noncovalently drop coating	NO	5 ppb	70	N/S	Kuzmych et al. (2007)
Polystyrene	MWCNT	In situ polymerization, solution mixing	Organic vapor	N/S	<240	~60	Zhang et al. (2005)
Poly-(m-aminobenzene sulfonic acid)	SWCNT	Ultrasonication	NH_3	5 ppm	<60	N/S	Bekyarova (2004)
Poly-(m-aminobenzene sulfonic acid)	SWCNT	Ultrasonication	NH_3, NO_2, H_2O vapor	100 ppb NH_3, 20 ppb NO_2	60–600	N/S	Zhang (2007)
Poly(3-methylthiophene)	MWCNT	In situ polymerization	Chloromethane	N/S	60	45	Santhanam (2005)
Polydiphenylamine	MWCNT	Electropolymerization	CO	0.01 ppm	~2	~3	Santhosh et al. (2007)
Poly(vinylacetate), polyisoprene, etc.	ACNT	Drop coating	Cyclohexane, tetrahydrofuran	N/S	<120	<120	Wei et al. (2006)
PANI-TSA	MWCNT	Self-assembly, in situ polymerization	Triethylamine vapor	ppb level	198	600	Li et al. (2008)
PANI-SSA					174	600	
Polyethylene glycol	MWCNT	Direct condensation	Chloroform	N/S	<1	N/S	Niu et al. (2007)

Note: N/S = Not stated.

Aqueous solutions of carboxyl functionalized SWCNTs and MWCNTs were used for resistive nitric oxide gas sensing. Sensor revealed response that was independent of NO concentration with the detection limit of 1 ppm. The sensor showed slow recovery at room temperature after NO exposure that was an indication of a strong interaction between the NO and CNTs (Maklin et al. 2007). A gas sensor for nitrogen dioxide detection was fabricated by conventional photolithography process on an oxidized silicon wafer functionalized with 3-aminopropyltriethoxysilane (APTES). Sensor exhibited high sensitivity and fast response time (a few seconds) compared with the sensor fabricated on a nonsilanized surface (Tran et al. 2008).

Different ammonia-sensing materials based on a zinc oxide (ZnO) layer was overlapped by Pd-doped carboxyl groups functionalized MWCNTs (Pd-COOH-MWCNTs) or by blocks of vertically aligned MWCNTs or by graphite as such and functionalized with fluorinated or nitrogenous functional groups. These sensors showed insensitivity to humidity, while all of them exhibited a good response in NH_3 atmosphere (Tulliani et al. 2011). SWCNTs were modified with porphyrin units and applied as sensing materials in quartz microbalance (QMB) sensors with an integrated preconcentration unit. The application of CNT-porphyrin composite, both at the preconcentrator adsorbing phase and QMB coating, showed a substantial sensitivity improvement to measure volatile organic compounds, such as 1-butanol in ppb level (Lvova et al. 2012). The performance of functionalized CNT-based ethanol vapor sensors were evaluated with the COOH-MWNTs and OH-MWNTs employed as sensing elements. Chemical oxidation method was used to graft functional groups such as COOH and OH on MWCNTs. Sensor performance using COOH/MWCNTs and OH/MWCNTs were compared in terms of various aspects, such as current–voltage curve, response, time constant, etc. COOH/MWCNTs-based sensors exhibited more sensitivity toward ethanol vapor as compared to OH/MWCNTs-based sensors with detection as low as 25 ppt (Ouyang et al. 2010).

The electrical detection of a low concentration of CO can be achieved at room temperature by side wall functionalization of CNTs with COOH group. Sensor revealed high selectivity toward carbon monoxide with the detection of concentration as low as 1 ppm (Fu et al. 2008). Alcohol sensor was fabricated by forming bundles of chemically functionalized MWCNTs across Au electrodes on SiO_2/Si substrates using an AC electrophoretic technique. It was found that the sensors were selective with respect to flow from air, water vapor, and alcohol vapor. The sensor exhibited linear response of ~1 s for alcohol vapor concentrations from 1 to 21 ppm with a detection limit of 0.9 ppm (Sin et al. 2007). SWCNT/porphyrin hybrid–based chemiresistive nanosensor arrays were fabricated for monitoring VOCs in the air. Porphyrins have unique and interesting physicochemical properties and its attachments to CNTs improved selectivity of sensors. Sensor was used for sensing methanol, ethanol, methyl ethyl ketone, and acetone and exhibited good response (Shirsat et al. 2012). Gas sensing performance of CNT functionalization via functional group is summarized in Table 38.4.

38.5 GAS SENSORS BASED ON FUNCTIONALIZED GRAPHENE

38.5.1 Graphene Modified with Metal Nanoparticles

Initially, in many sensing applications, graphene is decorated with metal nanoparticles to increase the sensitivity, selectivity, limit of detection, or a combination of these properties. In most cases the modification is performed by electrochemical reduction of metal salts with the help of external power sources for the reduction of metal ions using graphene flakes obtained from graphene oxide. In other cases, the deposition of nanoparticles was achieved by the chemical reduction of metal salts by the addition of a reducing agent followed by adsorption of the formed nanoparticles in solution (Li et al. 2010, 2011). Catalytically active noble metals have been widely used to increase the sensitivity of graphene-based chemical sensors to various gases.

TABLE 38.4 CNTs Functionalized with Functional Group

Functional Group	CNT Type	Functionalization Strategy	Gas/Vapor	Detection Limit	Response Time (s)	Recovery Time (s)	References
COOH	SWCNT MWCNT	Drop casting	Nitric oxide (NO)	1 ppm	N/S	300–900	Maklin et al. (2007)
Amine	SWNCNT	Photolithography process	NO$_2$	10 ppm	Few seconds	N/S	Tran et al. (2008)
COOH	MWCNT	Screen printing	NH$_3$	50 ppm	350	510	Tulliani et al. (2011)
COOH	SWCNT	Electropolymerization	1-Butanol	20 ppb	N/S	N/S	Tulliani et al. (2011)
COOH, OH	MWCNT	Chemical oxidation	Ethanol vapor	25 ppth	N/S	N/S	Ouyang and Li (2010)
COOH	SWCNT	Drop cast method	CO	1 ppm	N/S	N/S	Fu et al. (2008)
COOH	MWCNT	Chemical functionalization via oxidation	Alcohol	0.9 ppm	~1	N/S	Sin et al. (2007)
Porphyrin	SWCNT	Solvent casting technique	Methanol, MEK, ethanol, acetone	50% saturated	452	N/S	Shirsat et al. (2012)

Note: N/S = Not stated.

Nanosensors were fabricated for hydrogen sensing based on holey-reduced graphene oxide (hRGO). When decorated with Pt nanoparticles, hRGO exhibited a large and selective electronic response toward hydrogen gas with detection limit of 60 ppm (Vedala et al. 2011).

Gas-sensing behavior of graphene with Pt nanoparticles on graphene films was investigated. The sensor showed a better sensitivity when the Pt nanoparticles are deposited on graphene surface as compared to graphene without Pt particles with ppm level detection (Gautam et al. 2011). Palladium-functionalized multilayer graphene nanoribbon (MLGN) networks have been prepared for hydrogen gas sensing. These MLGN networks exhibited high sensitivity to hydrogen at room temperature with a fast response and recovery time and good repeatability (Johnson et al. 2010). Graphene has been decorated with different metal nanoparticles (Pd, Pt, Ag, and Au) by depositing on copper-catalyzed graphene for H_2S sensing. Sensor exhibited moderate response with low detection limit of 2 ppm (Gutes et al. 2012). The multilayered graphene was grown by CVD on a Si-polar 4H-SiC substrate for the detection of hydrogen gas. Current–voltage comparisons showed that Pt acted as a dopant and increased the conductance of graphene (Chu et al. 2011). GNSs decorated with Pt nanoparticles were fabricated on SiC substrates for hydrogen gas sensing. Sensor was exposed to hydrogen gas at various concentrations ranging from 0.06% to 1% and revealed good response toward hydrogen gas (Shafiei et al. 2009).

Graphene–palladium nanocomposite was prepared via layer-by-layer deposition on gold electrodes for hydrogen gas sensing. Sensor formed from composite material exhibited sensing behavior toward hydrogen gas at levels from 0.5% to 1% in synthetic air. Pure graphene was poorly sensitive to hydrogen, but incorporation of Pd nanoparticles increased its sensitivity by more than an order of magnitude (Lange et al. 2011). Platinum-functionalized graphene sheets and decorated MWCNTs were synthesized by a simple drop casting technique and employed as hydrogen sensors. It was found that Pt-decorated graphene sheets have greater sensitivity than that of Pt-decorated MWCNTs and sensors were stable over repeated cycles of hydrogenation and dehydrogenation (Kaniyoor et al. 2009). Carbon monoxide sensing properties of graphene can be enhanced by doping aluminum into the graphene. The enhanced CO sensitivity in the Al-doped graphene is determined by a large electrical conductivity change after the adsorption of CO molecules on Al-doped graphene. By detecting the conductivity change of the Al-doped graphene systems before and after the adsorption of CO, the presence of this toxic molecule was detected sensitively (Ao et al. 2008). Gas sensing performance of metal nanoparticle–functionalized graphene gas sensors is summarized in Table 38.5.

38.5.2 Graphene Decorated with Metal Oxide Nanoparticles

Recently, the development of metal oxide sensors that can operate at room temperatures with high sensitivity and low production cost has attracted much attention. When gas molecules get adsorbed on oxide-functionalized graphene (derivatives) sensor film, metal oxide nanoparticles act as sensing and transducer element and graphene oxide acts like a high-conducting mesh. It amplifies the transduction resulting in large change in conductance as compared to previous results reported for chemically derived graphene-based sensors (Fowler et al. 2009). A solution-processed gas sensor was fabricated based on vertically aligned ZnO nanorods on chemically converted graphene film. Sensor exhibited high sensitivity toward H_2S gas with effective detection of low concentration as low as 2 ppm of H_2S at room temperature (Cuong et al. 2011).

Highly aligned SnO_2 nanorods on graphene 3D array structures were synthesized by a straightforward nanocrystal-seeds-directing hydrothermal method. These 3D array structures were exploited as gas sensors and exhibited improved sensing performances toward H_2S gas. Sensor revealed detection of H_2S with a concentration as low as 1 ppm and exhibited H_2S sensing with fast response and recovery time of a few seconds (Zhang et al. 2011).

Graphene sheets decorated with SnO_2 nanoparticles were prepared through a facile hydrothermal-assisted in situ synthesis route for propanal sensing. Sensor showed high efficiency with detection limit of 0.3 $\mu g\ mL^{-1}$ and very fast recovery and response of a few seconds (Song et al. 2011). Al_2O_3/graphene nanocomposites were produced from graphene oxide solution by a

TABLE 38.5 Graphene Functionalized with Metal Nanoparticles

Metal Used	Form of Graphene	Functionalization Strategy	Gas Detected	Detection Limit	Response Time (s)	Recovery Time (s)	References
Pt	HRGO	Electrodeposition	H_2	60 ppm	N/S	N/S	Vedala et al. (2011)
Pt	Graphene	Photolithography technique and lift off process	NH_3	75 ppm	N/S	N/S	Gautam and Jayatissa (2011)
Pd	Multilayer graphene nanoribbon	Electron beam evaporation	H_2	40 ppm	6–21	23–44	Johnson et al. (2010)
Au, Pt, Pd	Graphene	Electroless deposition	H_2S	2 ppm	N/S	N/S	Gut es et al. (2012)
Pt	Multilayered graphene	Electron beam evaporation	H_2	N/S	~300	~350	Chu et al. (2011)
Pt	Graphene sheets	Electron beam evaporation	H_2	0.06 vol%	N/S	N/S	Shafiei et al. (2009)
Pd	Graphene	Layer-by-layer deposition	H_2, NO_2	H_2 0.5%–1% in synthetic air, 38 ppm for NO_2	N/S	N/S	Lange et al. (2011)
Pt	Graphene sheets	Drop casting	H_2	N/S	~540	N/S	Kaniyoor et al. 2009
Al	Graphene	Doping	CO	N/S	N/S	N/S	Ao et al. (2008)

Note: N/S = Not stated.

one-step, green, low-cost supercritical CO_2 method for ethanol sensing. Sensor exhibited high sensitivity and high selectivity to ethanol gas with short response (10 s) and less recovery time (100 s) (Jiang et al. 2011). Tungsten oxide nanorods/graphene nanocomposite was developed for NO_2 gas sensing. Sensor exhibited remarkably enhanced gas sensing properties with ppb level detection limit (An et al. 2012). Zinc oxide–decorated graphene nanocomposite sensor was prepared for detecting industrial toxins like CO, NH_3, and NO for concentrations as low as 1 ppm at room temperature. Sensor exhibited good response and quick recovery time with a detection limit of ~1 ppm (Singh et al. 2012). A new type of gas sensor has been developed using a hybrid vertically grown ZnO nanorods and free-standing graphene/metal sheets. Sensor demonstrated the ppm level detection of ethanol vapors with high sensitivity toward ethanol (Yi et al. 2011). Functionalized single-layer graphene via photoactive TiO_2 films–based sensor exhibited reversible and linear electrical sensitivity toward oxygen with a minimum detection limit of 0.01% oxygen (Wang et al. 2011). Gas sensing performance of metal oxide nanoparticle–functionalized graphene gas sensors is summarized in Table 38.6.

38.5.3 Graphene Functionalized with Polymer

In polymethyl methacrylate–functionalized graphene sensor, the PMMA acted as an adsorbent layer for concentrating the nonanal molecules on the surface of graphene and exhibits a strong response at ppm level (Dan et al. 2009). When compared with pristine graphene, the response decreases sharply. GO/polypyrene composite sensor exhibited an excellent performance in the selective sensing of toluene, having a fast, linear, and reversible response with a high sensitivity. The high sensitivity of this sensor is attributed to the unique microstructure of GO/PPr hybrid. The incorporation of GO resulted in the formation of a continuous and porous hybrid that led to the generation of uninterrupted conducting paths. Hence, PPr layer on GO sheets can adsorb toluene vapor and increase the conductance of the hybrid (Zhang et al. 2012). Graphene oxide/polypyrrole composite hydrogels were prepared by in situ chemical for ammonia sensing. Among them graphene oxide/polypyrrole composite hydrogel–based sensor exhibited high sensitivity toward ammonia gas (Bai et al. 2011). Hydrogen gas sensor has been made via the incorporation of graphene into polyaniline matrices to produce novel nanocomposite. Graphene/PANI nanocomposite exhibited sensitivity much higher than the sensitivities of sensors based on only graphene sheets and PANI nanofibers (Mashat et al. 2010). Similar to this, graphene oxide/polyaniline nanocomposite was prepared with in situ polymerization of aniline in the presence of GO suspension for hydrogen gas sensing.

Graphene/polyaniline nanocomposite–based sensitivity was 16.57% toward 1% of H_2 gas, which is much larger than the sensitivities of sensors based on only graphene sheets and polyaniline nanofibers (Zheng et al. 2012). Chemical sensors were developed based on chemically converted graphene using spin coating of hydrazine dispersions on interdigitated planar electrode arrays. It was used for the detection of gases like NO_2 and ammonia and revealed a ppm level detection of ammonia (Fowler et al. 2009). Gas-sensing performance of graphene functionalization via polymer is summarized in Table 38.7.

38.6 SELECTION OF RATIONAL CARBON NANOTUBES FUNCTIONALIZATION STRATEGY FOR SENSING

Various researchers have implemented different functionalization strategies to improve the sensing performance (sensitivity, selectivity, and response time) of CNTs with different methods (covalent and noncovalent) and with different materials (metals, metal oxides, and polymers). Molecular specificity can be obtained through rational chemical and/or physical modification of CNTs. In the case of metal nanoparticles–functionalized CNTs, the properties of metal

TABLE 38.6 Graphene Functionalized with Metal Oxide Nanoparticles

Metal Oxide Used	Form of Graphene	Functionalization Strategy	Gas Detected	Detection Limit	Response Time (s)	Recovery Time (s)	References
ZnO	Graphene	Solution-based method	H_2S	2 ppm	N/S	N/S	Cuong et al. (2011)
SnO_2	Graphene sheets	Nanocrystal-seeds-directing hydrothermal method	H_2S	1 ppm	5	10	Zhang et al. (2011)
SnO_2	Graphene sheets	Hydrothermal-assisted in situ synthesis route	Propanal gas	0.3 μg/mL	5	30	Song et al. (2011)
Al_2O_3	Graphene	Supercritical CO_2 method	Ethanol	1.5 μg/mL	10	<100	Jiang et al. (2011)
WO_3	Graphene	Hydrothermal method	NO_2	3.17 ppb	N/S	N/S	An et al. (2012)
ZnO	Graphene oxide	Spin coating	CO, NH_3, and NO	1 ppm	360 for NH_3, 1,500 for NO	120–180 for NH_3	Singh et al. (2012)
ZnO	Graphene	Integration via lifting	Ethanol	~9–10 ppm	N/S	N/S	Yi et al. (2011)
TiO_2	Single-layer graphenes	Coating	O_2	0.01%	130	260	Wang et al. (2011)

Note: N/S = Not stated.

TABLE 38.7 Graphene Functionalized with Polymer

Polymer	Form of Graphene	Functionalization Strategy	Gas/Vapor Detected	Detection Limit	Response Time	Recovery Time	References
Polypyrrole	Graphene oxide	In situ chemical polymerization	Ammonia	800 ppm	N/S	N/S	Bai et al. (2011)
Polyaniline	Graphene	Ultrasound-assisted chemical polymerization	H_2	N/S	N/S	N/S	Mashat et al. (2010)
Polyaniline	Graphene oxide	In situ chemical polymerization	NH_3, toluene, formaldehyde, cyclohexane	3 vol%	N/S	N/S	Zheng et al. (2012)
Hydazine	Graphene film	Spin coating	NO_2, NH_3, dinitrotoluene	5 ppm NH_3 NO_2, 28 ppb dinitrotoluene	N/S	N/S	Fowler et al. (2009)

Note: N/S = Not stated.

nanoparticles like electronic, chemical, and physical that are highly sensitive to changes in their chemical environment are beneficial for sensing. Metal nanoparticles are chemically and mechanically robust and stable as compared to polymer; therefore, metal nanoparticles–functionalized sensors can operate at higher temperature and in harsher environment. The problem of high operating temperature range (200°C and 800°C) of metal oxide–based sensor can be addressed with low operating temperature and with improved sensing properties via functionalizing the CNTs with metal oxide nanoparticles. Hence, modified CNTs with metal oxide nanoparticles can provide the best alternative for metal oxide–based sensor (Mubeen et al. 2013).

Several conducting polymer–based sensors have been demonstrated to be functional at room temperature for the detection of a wide variety of gas or vapors. However, their selectivity and environmental stability are poor. Therefore, polymer functionalization with CNTs is used to overcome these issues for imparting the high sensitivity and selectivity to the sensors. Addition of different functional groups at the ends and side walls of the CNTs provide reactive sites for interacting with target molecules for better sensing. Functional group–attached CNT gas sensor has resulted in high sensitivity and fast response time as compared to pristine CNT. To overcome the limitations posed by pristine CNT sensors, the rational functionalization technique with the proper choice of functional materials provides most exciting building blocks for miniaturized gas sensors and high density gas sensor arrays. By implementing the different aforementioned functionalization strategies via controlled strategy, a customizable CNT sensor with suitable selective receptors may become accessible in the sensing technology.

38.7 SELECTION OF RATIONAL GRAPHENE FUNCTIONALIZATION STRATEGY FOR SENSING

In the early stages, researchers realized that the utility of carbon-based new structures can be enhanced if they are dispersed in solution and huge improvement can be achieved in sensing via covalent modification of CNTs (Bahr and Tour 2002). But the addition of new functional groups disturbs the extended π system of the carbon nanotubes, which was not favorable. Therefore, it affects the electronic properties and in turn the sensing behavior of the sensor.

Initially, the reactions that are effective with CNTs are modified and implemented easily to the graphitic material via covalent functionalization strategy. But due to the known disruption behavior of π system of the CNTs, very few research studies investigated the result of these modifications on electronic properties. Different synthetic methodologies include direct addition of atoms into the graphene lattice, modification of residual functionalities (from a graphene oxide starting material), and direct modifications to the surface, which disrupt the π structure (Liu et al. 2011). Chemical exfoliation of graphite is widely employed to synthesize graphene oxide and it results in the covalent functionalization of graphene. The strong oxidation of graphite produces graphene oxide that results in the addition of oxygen-containing groups such as carboxyl, hydroxyl, carbonyl, epoxide, etc. In nanocomposite formation, the residual functional groups and the wrinkles in basal planes of graphene oxide after rapid heating provides a strong adhesion to polymer due to the introduction of new vacancies and changes in bonding (Liu et al. 2009, Ramanathan et al. 2008, Schniepp et al. 2006).

Covalent functionalization between graphene and benzoyl proxiode were carried out via photochemical reaction that resulted in poorer electronic performance for use in sensors. It involved a focused laser spot to induce the reaction resulting in localized defect areas (Liu et al. 2009). Therefore, many of the chemical functionalization methods used have covalent bonding, and thereby destroy the sp^2 bonding of graphene lattice. Till date, most of the functionalization approaches effectively change the chemical and transport properties of the graphene (derivatives) but they are not suitable to design electrical sensors (Bai et al. 2009). Considering this fact, an alternative strategy is required to modify the graphene for use in gas sensor. In this case, noncovalent modifications should be a superior option. Methods of noncovalent modifications take the advantage of the graphene π structure rather than disrupting it. Also, they are not harsh and retain unique electronic properties of graphene for use in sensors. But with covalent functionalization, there is a formation of many sp^3 bonds and loss of many original properties of graphene.

Three different types of noncovalent strategies have emerged so far. In the first type, π bonding occurs between the π orbitals of graphene basal plane and those of aromatic functional molecules. It is used to functionalize reduced graphene oxide with 1-pyrenebutyrate (Xu et al. 2008) and assemble monolayers of perylene-3,4,9,10-tetracarboxylic dianhydride on epitaxial graphene (Wang and Hersam 2009). In the second method, graphene is coated with polymer, to remarkably enhance its sensitivity for electrical detection of gas molecules (Dan et al. 2009). It might also enhance the selectivity of gas sensor. Metallic nanoparticles have been reduced via electrochemistry on graphene derivatives in the third noncovalent strategy. Decoration of expanded graphite nanoplatelets have been carried out with Pt or Pd nanoparticles through a microwave-heated polyol process (Lu et al. 2008). Still there is a need to check the limits of sensitivity and selectivity imparted to the graphene-based gas sensors by noncovalent methods.

In the current scenario of sensing, instead of designing the single gas sensor, it is better to design an array of sensors (electronic nose) where each component in the array is functionalized differently (Capone et al. 2003, Bondavalli et al. 2009). The basic theme is to take the advantages of cross sensitivity rather than eliminate it. When array is exposed to a mixture of gases, it will produce a matrix of responses. Here each sensor's response contains disparate contributions from various gases due to their particular functionalization. Then there is a characteristic pattern or *fingerprint* for individual gas using multivariate data analysis. Finally, it can be compared with a database of standard patterns for gases and vapors. For instance, an array of CNT field effect sensors, each of which was decorated with one of four different types of metallic nanoparticles to uniquely identify H_2, H_2S, NH_3, and NO_2 gases (Star et al. 2006).

Therefore, this sensor array approach has proven to be effective in a variety of gas sensors made with different transducer materials and with different types of functionalization. Another example involved a tin oxide semiconductor array, with differences in functionalization achieved by varied levels of Pd doping and different electrode geometries. The mixtures of CO, CH_4, and H_2O were measured using eight-element array (Capone et al. 2001). A lot of similarities has been found between the basic functionalization approaches used in the above examples of sensor arrays and the routes already used to functionalize graphene (or its derivatives) nanoparticles. Therefore, sensor array approach could be used, with equal or greater success, on graphene-based gas sensors. Due to easy integration of graphene into lithographic processes for microfabrication or nanofabrication, sensor array approach may be highly fruitful. An array of individual sensor elements can be created from a single, large sheet of graphene, such as that grown by CVD, by means of lithographic patterning and plasma etching. The progress and achievement of these kinds of strategies will permit to tailor the size, sensitivity, and selectivity of graphene to produce a new generation of gas sensor arrays that are compact and that can analyze the mixture of gases and vapors at parts per billion level.

38.8 SUMMARY AND PERSPECTIVES

CNTs and graphene are most prominent nanomaterials for sensing a broad variety of gases/vapors with high response because of their unique structure and electronic properties, and their rich configurations as well. But for practical application, gas sensors are required to possess high selectivity, low operating temperature, and quick response and recovery. The important lacunae of CNTs and graphene-based gas sensors are potential interference from relative humidity at room temperature, slow recovery, and poor selectivity. To overcome these limitations posed by pristine CNTs and graphene-based gas sensors, researchers have directed their efforts toward the functionalization of these materials with various methods and with a wide spectrum of materials such as metal, metal oxide, polymeric nanoparticles, and functional groups. The various functionalization techniques together with appropriate choice of functional materials will make functionalized CNTs and graphene one of the most exciting monolith geometries for miniaturized gas/vapor sensors and high density gas sensor arrays. It is, therefore, necessary to make a suitable choice of the method and the extent of functionalization for gas/vapor sensing applications. Compared to covalent functionalization techniques, noncovalent functionalization avoids many of the pitfalls, but yet there may be limitations due to the presence of a coating around the nanostructure. Functionalization of graphene as compared to CNTs is still not explored and is an attractive field in future through the emergence of novel and task-specific functionalization methods with improved performance. With the help of these functionalization techniques, commercial success of CNTs and graphene-based gas sensing will be possible when dealing with the challenges in the gas sensing field.

REFERENCES

Abraham, J.K., Philip, B., Witchurch, A. et al. 2004. A compact wireless gas sensor using a carbon nanotube/PMMA thin film chemiresistor. *Smart Mater. Struct.* 13: 1045–1049.

An, K.H., Jeong, S.Y., Hwang, H.R. et al. 2004. Enhanced sensitivity of a gas sensor incorporating single-walled carbon nanotube–polypyrrole nanocomposites. *J. Adv. Mater.* 16: 1005–1009.

An, X., Yu, J.C., Wang, Y. et al. 2012. WO_3 nanorods/graphene nanocomposites for high-efficiency visible-light-driven photocatalysis and NO_2 gas sensing. *J. Mater. Chem.* 22: 8525–8531.

Ao, Z.M., Yang, J., Li, S. et al. 2008. Enhancement of CO detection in Al doped graphene. *Chem. Phys. Lett.* 461: 276–279.

Bahr, J.L., Tour, J.M. 2002. Covalent chemistry of single-wall carbon nanotubes. *J. Mater. Chem.* 12: 1952.

Bai, H., Sheng, K., Zhang, P. 2011. Graphene oxide/conducting polymer composite hydrogels. *J. Mater. Chem.* 21: 18653–18658.

Bai, H., Xu, Y.X., Zhao, L. et al. 2009. Non-covalent functionalization of graphene sheets by sulfonated polyaniline. *Chem. Commun.* 13: 1667–1669.

Bekyarova, E., Davis, M., Burch, T. et al. 2004. Chemically functionalized single-walled carbon nanotubes as ammonia sensors. *J. Phys. Chem. B* 108: 19717–19720.

Bondavalli, P., Legagneux, P., Pribat, D. 2009. Carbon nanotubes based transistors as gas sensors: State of the art and critical review. *Sens. Actuators, B* 140: 304–318.

Capone, S., Forleo, A., Francioso L. et al. 2003. Solid-state gas sensors: State of the art and future activities. *J. Optoelect. Adv. Mater.* 5: 1335–1348.

Capone, S., Siciliano, P., Barsan, N. 2001. Analysis of CO and CH_4 gas mixtures by using a micromachined sensor array. *Sens. Actuators, B* 78: 40–48.

Chen, Y., Zhu, C., Wang T. 2006. The enhanced ethanol sensing properties of multi-walled carbon nanotubes/SnO_2 core/shell nanostructures. *Nanotechnology* 17: 3012–3017.

Choi, K.Y., Park, J.S., Park, K.B. et al. 2010. Low power micro gas sensors using mixed SnO_2 nanoparticles and MWCNTs to detect NO_2, NH_3, and xylene gases for ubiquitous sensor network applications. *Sens. Actuators, B* 150: 65–72.

Chu, B.H., Loa, C.F., Nicolosia, J. et al. 2011. Hydrogen detection using platinum coated graphene grown on SiC. *Sens. Actuators, B* 157: 500–503.

Cuong, T.V., Pham, V.H., Chung, J.S. et al. 2011. Solution-processed ZnO-chemically converted graphene gas sensor. *Mater. Lett.* 64: 2479–2482.

Dan, Y.P., Lu, Y., Kybert, N.J. et al. 2009. Intrinsic response of graphene vapor sensors. *Nano Lett.* 9: 1472–1475.

Ding, D., Chen, Z., Rajaputra, S. et al. 2007. Hydrogen sensors based on aligned carbon nanotubes in an anodic aluminum oxide template with palladium as a top electrode. *Sens. Actuators, B* 124: 12–17.

Ding, X.M., Fu, R.W., Zhang, C.Q. et al. 2004. Electrical resistance response of carbon black filled amorphous polymer composite sensors to organic vapors at low vapor concentrations. *Carbon* 42: 2551–2559.

Englert, J.M., Dotzer, C., Yang, G. et al. 2011. Covalent bulk functionalization of graphene. *Nat. Chem.* 3: 279–286.

Espinosa, E., Ionescu, R., Bittencourt, C. et al. 2007. Metal-decorated multi-wall carbon nanotubes for low temperature gas sensing. *Thin Solid Films* 515: 8322–8327.

Fam, D.W.H., Palaniappan, Al., Tok, A.I.Y. et al. 2011. A review on technological aspects influencing commercialization of carbon nanotube sensors. *Sens. Actuators, B* 157: 1–7.

Feyzabad, S.A., Khodadadi, A.A., Naseh, M.V. et al. 2012. Highly sensitive and selective sensors to volatile organic compounds using MWCNTs/SnO_2. *Sens. Actuators, B* 166: 150–155.

Fowler, J.D, Allen, M.J., Tung, V.C. et al. 2009. Practical chemical sensors from chemically derived graphene. *ACS Nano* 3: 301–306.

Fu, D., Lim, H., Shi, Y. et al. 2008. Differentiation of gas molecules using flexible and all-carbon nanotube devices. *J. Phys. Chem. C* 112: 650–653.

Gautam, M., Jayatissa, A.H. 2011. Ammonia sensor device using graphene modified with platinum. *ISDRS*, Maryland, USA, pp. 259–260.

Gelamo, R.V., Rouxinol, F.P., Verissimoa, C. et al. 2009. Low-temperature gas and pressure sensor based on multi-wall carbon nanotubes decorated with Ti nanoparticles. *Chem. Phys. Lett.* 482: 302–306.

Gong, X., Liu, J., Baskaran, S. et al. 2000. Surfactant-assisted processing of carbon nanotube/polymer composites. *Chem. Mater.* 12: 1049–1052.

Gutes, A., Hsia, B., Sussman, A. et al. 2012. Graphene decoration with metal nanoparticles: Towards easy integration for sensing applications. *Nanoscale* 4: 438–440.

Hornyak, G.L. 2009. *Introduction to Nanoscience and Nanotechnology*. Boca Raton, FL: CRC Press.

Hu, C.Y., Xu, Y.J., Duo, S.W. et al. 2000. Non-covalent functionalization of carbon nanotubes with surfactants and polymers. *J. Chin. Chem. Soc.* 56: 234–239.

Izadi, N., Rashidi, A.M., Golzardi, S. 2012. Hydrogen sulfide sensing properties of multi walled carbon nanotubes. *Ceram. Int.* 38: 65–75.

Jang, D.M., Jung, H., Hoa, N.D. et al. 2012. Tin oxide-carbon nanotube composite for NO_x sensing. *J. Nanosci. Nanotechnol.* 12: 1425–1428.

Jiang, Z., Wang, J., Meng, L. et al. 2011. A highly efficient chemical sensor material for ethanol: Al_2O_3/graphene nanocomposites fabricated from graphene oxide. *Chem. Commun.* 47: 6350–6352.

Johnson, J.L., Behnam, A., Pearton, S.J. et al. 2010. Hydrogen sensing using pd-functionalized multi-layer graphene nanoribbon networks. *Adv. Mater.* 22: 4877–4880.

Ju, S., Lee, J.M., Jung, Y. 2010. Highly sensitive hydrogen gas sensors using single-walled carbon nanotubes grafted with Pd nanoparticles. *Sens. Actuators, B* 146: 122–128.

Kaniyoor, A., Jafri, R.I., Arockiadoss, T. et al. 2009. Nanostructured Pt decorated graphene and multi walled carbon nanotube based room temperature hydrogen gas sensor. *Nanoscale* 1: 382–386.

Kim, J.E., Han, T.H., Lee, S.H. et al. 2011. Graphene oxide liquid crystals. *Angew. Chem. Int. Ed. Engl.* 50: 3043–3047.

Kong, J., Chapline, M.G., Dai, H. 2001. Functionalized carbon nanotubes for molecular hydrogen sensors. *Adv. Mater.* 13: 1384–1386.

Kumar, M.K., Ramaprabhu, S. 2006. Nanostructured Pt functionalized multiwalled carbon nanotube based hydrogen sensor. *J. Phys. Chem. B* 110: 11291–11298.

Kuzmych, O., Allen, B.L., Star, A. 2007. Carbon nanotube sensors for exhaled breath components. *Nanotechnology* 18: 375502–375509.

Lange, U., Hirsch, T., Mirsky, V.M. et al. 2011. Hydrogen sensor based on a graphene – palladium nanocomposite. *Electrochim. Acta* 56: 3707–3712.

Leghrib, R., Dufour, T., Demoisson, F. et al. 2011a. Gas sensing properties of multiwall carbon nanotubes decorated with rhodium nanoparticles. *Sens. Actuators, B* 160: 974–980.

Leghrib, R., Felten, A., Pireaux, J.J. et al. 2011b. Gas sensors based on doped-CNT/ SnO$_2$ composites for NO$_2$ detection at room temperature. *Thin Solid Films* 520: 966–970.

Li, W., Jung, H., Hoa, N.D. et al. 2010. Nanocomposite of cobalt oxide nanocrystals and single-walled carbon nanotubes for a gas sensor application. *Sens. Actuators, B* 150: 160–166.

Li, Y., Fan, X., Qi, J. et al. 2010b. Palladium nanoparticle–graphene hybrids as active catalysts for the suzuki reaction. *Nano Res.* 3: 429–437.

Li, Y., Wang, H., Cao, X. et al. 2008. A composite of polyelectrolyte-grafted multi-walled carbon nanotubes and in situ polymerized polyaniline for the detection of low concentration triethylamine vapor. *Nanotechnology* 19: 015503–015507.

Li, Z., Li, J., Wu, X. et al. 2009. Methane sensor based on nanocomposite of palladium/ multi-walled carbon nanotubes grafted with 1,6-hexanediamine. *Sens. Actuators, B* 139: 453–459.

Li, Z., Zhang, P., Wang, K. et al. 2011. Graphene buffered galvanic synthesis of graphene–metal hybrids. *J. Mater. Chem.* 21: 13241–13246.

Lin, T.C., Huang, B.R. 2012. Palladium nanoparticles modified carbon nanotube/ nickel composite rods (Pd/CNT/Ni) for hydrogen sensing. *Sens. Actuators, B* 162: 108–113.

Liu, H., Liu, Y., Zhu, D. 2011. Chemical doping of graphene. *J. Mater. Chem.* 21: 3335–3345.

Liu, H., Ma, H., Zhou, W. et al. 2012. Synthesis and gas sensing characteristic based on metal oxide modification multi wall carbon nanotube composites. *Appl. Surf Sci.* 258: 1991–1994.

Liu, H., Ryu, S., Chen, Z. et al. 2009. Photochemical reactivity of graphene. *J. Am. Chem. Soc.* 131: 17099–17101.

Liu, Y.L., Yang, H.F., Yang, Y. et al. 2006. Gas sensing properties of tin dioxide coated onto multi-walled carbon nanotubes. *Thin Solid Films* 497: 355–360.

Lu, J., Do, I., Drzal, L.T. 2008. Nanometal decorated exfoliated graphite nanoplatelet based glucose biosensors with high sensitivity and fast response. *ACS Nano* 2: 1825–1832.

Lu, Y., Li, J., Han, J. et al. 2004. Room temperature methane detection using palladium loaded single-walled carbon nanotube sensors. *Chem. Phys. Lett.* 391: 344–348.

Lvova, L., Mastroianni, M., Pomarico, G. et al. 2012. Carbon nanotubes modified with porphyrin units for gaseous phase chemical sensing. *Sens. Actuators, B* 170: 163–171.

Ma, P.-C., Siddiqui, N.A., Marom, G. et al. 2010. Dispersion and functionalization of carbon nanotubes for polymer-based nanocomposites: A review. *Composites: Part A* 41: 1345–1367.

Maklin, J., Mustonen, T., Kordas, K. et al. 2007. Nitric oxide gas sensors with functionalized carbon nanotubes. *Phys. Status Solidi B* 244: 4298–4302.

Mashat, L.A., Shin, K., Zadeh, K.K. et al. 2010. Graphene/polyaniline nanocomposite for hydrogen sensing. *J. Phys. Chem. C* 114: 16168–16173.

Mehrabi, B., Mortazavi, Y., Khodadadi, A.A. et al. 2010. Alkaline and template-free hydrothermal synthesis of stable SnO_2 nanoparticles and nanorods for CO and ethanol gas sensing. *Sens. Actuators, B* 151: 140–145.

Meng, L., Fu, C., Lu, Q. 2009. Advanced technology for functionalization of carbon nanotubes. *Prog. Nat. Sci.* 19: 801–810.

Mubeen, S., Lai, M., Zhang, T. et al. 2013. Hybrid tin oxide-SWNT nanostructures based gas sensor. *Electrochim. Acta* 92: 484–490.

Mubeen, S., Zhang, T., Yoo, B. et al. 2007. Palladium nanoparticles decorated single-walled carbon nanotube hydrogen sensor. *J. Phys. Chem. C* 111: 6321–6327.

Niu, L., Luo, Y., Li, Z. 2007. A highly selective chemical gas sensor based on functionalization of multi-walled carbon nanotubes with poly (ethylene glycol). *Sens. Actuators, B* 126: 361–367.

Niyogi, S., Bekyarova, E., Itkis, M.E. et al. 2006. Solution properties of graphite and graphene. *J. Am. Chem. Soc.* 128: 7720–7721.

Oakley, J.S., Wang, H.T., Kang, B.S. 2005. Carbon nanotube films for room temperature hydrogen sensing. *Nanotechnology* 16: 2218–2221.

Ouyang, M., Li, W.J. 2010. Performance of F-cnts sensors towards ethanol vapor using different functional groups. *IEEE-NEMS*, Xiamen, China, pp. 928–931.

Pati, S.K., Enoki, T., Rao, C.N.R. 2011. *Graphene and Its Fascinating Attributes*. Singapore: World Scientific Publishing Co. Pte. Ltd.

Penza, M., Rossi, R., Alvisi, M. et al. 2008. Surface modification of carbon nanotube networked films with Au nanoclusters for enhanced NO_2 gas sensing applications. *J. Sensors* 2: 1–8.

Penza, M., Rossi, R., Alvisi, M. et al. 2009. Functional characterization of carbon nanotube networked films functionalized with tuned loading of Au nanoclusters for gas sensing applications. *Sens. Actuators, B* 140: 176–184.

Qi, P., Vermesh, O., Grecu, M. et al. 2003. Toward large arrays of multiplex functionalized carbon nanotube sensors for highly sensitive and selective molecular detection. *Nano Lett.* 3: 347–351.

Ramanathan, T., Abdala, A.A., Stankovich, S. et al. 2008. Functionalized graphene sheets for polymer nanocomposites. *Nat. Nanotechnol.* 3: 327–331.

Rao, C.N.R., Ghosh, A., Gomathi, A. 2011. Functionalization and solubilization of carbon and inorganic nanostructures. In *Comprehensive Nanoscience and Technology*, D.L. Andrews (ed.), pp. 445–490. London, U.K.: Academic Press.

Rumiche, F., Wang, H.H., Indacochea, J.E. 2012. Development of a fast-response/high-sensitivity double wall carbon nanotube/nanostructured hydrogen sensor. *Sens. Actuators, B* 163: 97–106.

Sadek, A.Z., Bansal, V., McCulloch, D.G. et al. 2011. Facile, size-controlled deposition of highly dispersed gold nanoparticles on nitrogen carbon nanotubes for hydrogen sensing. *Sens. Actuators, B* 160: 1034–1042.

Santhanam, K.S.V., Sangoi, R., Fuller, L. 2005. A chemical sensor for chloromethanes using a nanocomposite of multiwalled carbon nanotubes with poly (3-methylthiophene). *Sens. Actuators B* 106: 766–771.

Santhosh, P., Manesh, K.M., Gopalan, A. et al. 2007. Novel amperometric carbon monoxide sensor based on multi-wall carbon nanotubes grafted with polydiphenylamine—Fabrication and performance. *Sens. Actuators, B* 125: 92–99.

Schniepp, H.C., Li, J.L., McAllister, M.J. et al. 2006. Functionalized single graphene sheets derived from splitting graphite oxide. *J. Phys. Chem. B* 110: 8535–8539.

Shafiei, M., Arsat, R., Yu, J. et al. 2009. Pt/graphene nano-sheet based hydrogen gas sensor. *IEEE Sens. J.* 295–298.

Shirsat, M.D., Sarkar, T., Kakoullis, J. et al. 2012. Porphyrin-functionalized single-walled carbon nanotube chemiresistive sensor arrays for VOCs. *J. Phys. Chem. C* 116: 3845–3850.

Sin, M.L.Y., Chow, G.C.T., Wong, G.M.K. et al. 2007. Ultra low-power alcohol vapor sensors using chemically functionalized multiwalled carbon nanotubes. *Nanotechnology* 6: 571–577.

Singh, G., Choudhary, A., Haranath, D. et al. 2012. ZnO decorated luminescent graphene as a potential gas sensor at room temperature. *Carbon* 50: 385–394.

Singh, V., Joung, D., Zhai, L. et al. 2011. Graphene based materials: Past, present and future. *Prog. Mater. Sci.* 56: 1178–1271.

Song, H., Zhang, L., He, C. et al. 2011. Graphene sheets decorated with SnO_2 nanoparticles: In situ synthesis and highly efficient materials for cataluminescence gas sensors. *J. Mater. Chem.* 21: 5972–5977.

Star, A., Han, T.R., Joshi, V. et al. 2004. Nanoelectronic carbon dioxide sensors. *Adv. Mater.* 16: 2049–2052.

Star, A., Joshi, V., Skarupo, S. et al. 2006. Gas sensor array based on metal-decorated carbon nanotubes. *J. Phys. Chem. B* 110: 21014–21020.

Subrahmanyam, K.S., Ghosh, A., Gomathi, A. et al. 2009. Covalent and noncovalent functionalization and solubilization of graphene. *Nanosci. Nanotechnol. Lett.* 4: 28–31.

Subrahmanyam, K.S., Vivekchand, S.R.C., Govindaraj, A. et al. 2008. A study of graphenes prepared by different methods: Characterization, properties and solubilization. *J. Mater. Chem.* 18: 1517–1523.

Suehiro, J., Imakiire, H., Hidaka, S.I. et al. 2006. Schottky-type response of carbon nanotube NO_2 gas sensor fabricated onto aluminum electrodes by dielectrophoresis. *Sens. Actuators, B* 114: 943–949.

Sun, Y., Wang, H.H. 2007. Electrodeposition of Pd nanoparticles on single-walled carbon nanotubes for flexible hydrogen sensors. *Appl. Phys. Lett.* 90: 213107.

Tran, T.H., Lee, J.W., Lee, K. et al. 2008. The gas sensing properties of single-walled carbon nanotubes deposited on an aminosilane monolayer. *Sens. Actuators, B* 129: 67–71.

Tulliani, J.M., Cavalieri, A., Musso, S. et al. 2011. Room temperature ammonia sensors based on zinc oxide and functionalized graphite and multi-walled carbon nanotubes. *Sens. Actuators, B* 152: 144–154.

Valentini, L., Bavastrello, V., Stura, E. et al. 2004. Sensors for inorganic vapor detection based on carbon nanotubes and poly (o-anisidine) nanocomposite material. *Chem. Phys. Lett.* 383: 617–622.

Van, H.N., Duc, N.A.P., Trung, T. et al. 2010. Gas-sensing properties of tin oxide doped with metal oxides and carbon nanotubes: A competitive sensor for ethanol and liquid petroleum gas. *Sens Actuators, B* 144: 450–456.

Vedala, H., Sorescu, D.C., Kotchey, G.P. et al. 2011. Chemical sensitivity of graphene edges decorated with metal nanoparticles. *Nano Lett.* 11: 2342–2347.

Wang, J. 2005. Carbon-nanotube based electrochemical biosensors: A review. *Electroanalysis* 17: 7–14.

Wang, Q., Guo, X., Cai, L. et al. 2011. TiO_2-decorated graphenes as efficient photo-switches with high oxygen sensitivity. *Chem. Sci.* 2: 1860–1864.

Wang, Q.H., Hersam, M.C. 2009. Room-temperature molecular resolution character-ization of self-assembled organic monolayers on epitaxial graphene. *Nat. Chem.* 1: 206–211.

Wei, C., Dai, L., Roy, A. et al. 2006. Multifunctional chemical vapor sensors of aligned carbon nanotube and polymer composites. *J. Am. Chem. Soc.* 128: 1412–1413.

Wongchoosuk, C., Wisitsoraat, A., Phokharatkul, D. et al. 2010. Multi-walled carbon nanotube-doped tungsten oxide thin films for hydrogen gas sensing. *Sensors* 10: 7705–7715.

Xu, Y.X., Bai, H., Lu, G.W. et al. 2008. Flexible graphene films via the filtration of water-soluble non-covalent functionalized graphene sheets. *J. Am. Chem. Soc.* 130: 5856–5857.

Yang, H., Li, F., Shan, C. et al. 2009. Covalent functionalization of chemically con-verted graphene sheets via silane and its reinforcement. *J. Mater. Chem.* 19: 4632–4638.

Yi, J., Lee, J.M., Park, W. 2011. Vertically aligned ZnO nanorods and graphene hybrid architectures for high-sensitive flexible gas sensors. *Sens. Actuators, B* 155: 264–269.

Young, P., Lu, Y., Terrill, R. et al. 2005. High-sensitivity NO_2 detection with carbon nanotube–gold nanoparticle composite films. *J. Nanosci. Nanotechnol.* 5: 1509–1513.

Zhang, B., Fu, R.W., Zhang, M.Q. et al. 2005. Preparation and characterization of gas-sensitive composites from multi-walled carbon nanotubes/polystyrene. *Sens. Actuators, B* 109: 323–328.

Zhang, L., Li, C., Liu, A.R. et al. 2012. Electrosynthesis of graphene oxide/polypyrene composite films and their applications for sensing organic vapors. *J. Mater. Chem.* 22: 8438–8443.

Zhang, T., Mubeen, S., Bekyarova, E. et al. 2007. Poly (m-aminobenzene sulfonic acid) functionalized single-walled carbon nanotubes based gas sensor. *Nanotechnology* 18: 165504/1-165504/6.

Zhang, T., Mubeen, S., Myung, N.V. et al. 2008. Recent progress in carbon nanotube-based gas sensors. *Nanotechnology* 19: 1–14.

Zhang, T., Nix, M.B., Yoo, B.Y. et al. 2006. Electrochemically functionalized single-walled carbon nanotube gas sensor. *Electroanalysis* 18: 1153–1158.

Zhang, W.D., Zhang, W.H. 2009. Carbon nanotubes as active components for gas sen-sors. *J. Sensors* 2009: 1–16.

Zhang, X., Hou, L., Cnossen, A. et al. 2011. One-pot functionalization of graphene with porphyrin through cycloaddition reaction. *Chem. Eur. J.* 17: 8957–8964.

Zhang, Y.H., Chen, Y.B., Zhou, K.G. et al. 2009. Improving gas sensing properties of gra-phene by introducing dopants and defects: A first-principles study. *Nanotechnology* 20: 185504.

Zhang, Z., Zou, R., Song, G. et al. 2011b. Highly aligned SnO_2 nanorods on graphene sheets for gas sensors. *J. Mater. Chem.* 21: 17360–17365.

Zheng, J., Ma, X., He, X. et al. 2012. Preparation, characterizations, and its potential applications of PANi/graphene oxide nanocomposite. *Procedia Eng.* 27: 1478–1487.

Zilli, D., Bonelli, P.R., Cukierman, A.L. 2011. Room temperature hydrogen gas sensor nanocomposite based on Pd-decorated multi-walled carbon nanotubes thin films. *Sens. Actuators, B* 157: 169–176.

CHAPTER 39

CONTENTS

Chemical Modification of Graphene for Optoelectronic Applications

39

Jian Ru Gong

Graphene is the basic structural element of some carbon allotropes including 0D fullerene, 1D carbon nanotube, and 3D graphite. Its 2D one-atom-thick planar network is composed of the hexagonal crystal lattice structure packed by sp^2-hybridized carbon atoms with the s, p_x, and p_y atomic orbitals on each carbon atom forming three strong σ bonds with other three surrounding atoms.[1,2] Overlap of the remaining p_z orbital on each carbon atom with neighboring carbon atoms produces a filled band of π orbitals called the valence band and an empty band of π* orbitals known as the conduction band. The valence and conduction bands touch at the Brillouin zone corners, thus making graphene a zero-bandgap semiconductor (Figure 39.1). Graphene exhibits many outstanding properties, such as fast room-temperature mobility of charge carriers (200,000 cm^2 V^{-1} s^{-1}), exceptional conductivity (10^6 S cm^{-1}), and large theoretical specific surface area (2,630 m^2 g^{-1}), which promise a technological revolution.[3] However, the practical application of graphene encounters massive problems, for example, lack of economical and reliable ways to produce high-quality graphene and fundamentally unsuitable zero bandgap of graphene for the on–off switching that lies at the heart of digital electronics.

In the quest for applying the excellent properties of graphene into practical applications, our group modifies the structures of the graphene-based nanomaterials on multiscales by systematically adopting three categories of chemical methods, that is, heteroatom doping, molecular covalent bonding, and interfacial noncovalent modification, and modulates the optoelectronic properties by adjusting the electronic movement at the atomic, molecular, and interfacial levels. There is not really another material that has good properties for both optics and electronics like graphene. Therefore, we further explore the application of these nanomaterials in the optoelectronic field. The related research works are summarized as follows based on the different classification of chemical functionalization on the graphene-based nanomaterials.

39.1 HETEROATOM DOPING

The electronic properties of monolayer graphene are unique due to the relativistic nature of charge carriers resulting from the linear dispersion relation, and the mobility of graphene is two orders of magnitude higher than that of silicon. However, the electron devices made from pristine graphene with the intrinsic metallic and Dirac–Fermi characters are difficult to switch off, losing the advantage of the low static power consumption of the current complementary metal oxide semiconductor (CMOS) technology. More importantly, next-generation carbon-based optoelectronics and electronics inevitably require diverse electrical heterojunctions with tunable carrier types (p- or n-type) and Fermi energy levels.[4] Therefore, well-controlled electronic property is a significant challenge for graphene-based electron devices, and we need exact methods to homogenously and reproducibly dope graphene (p- or n-type) for different applications and to control the doping level.

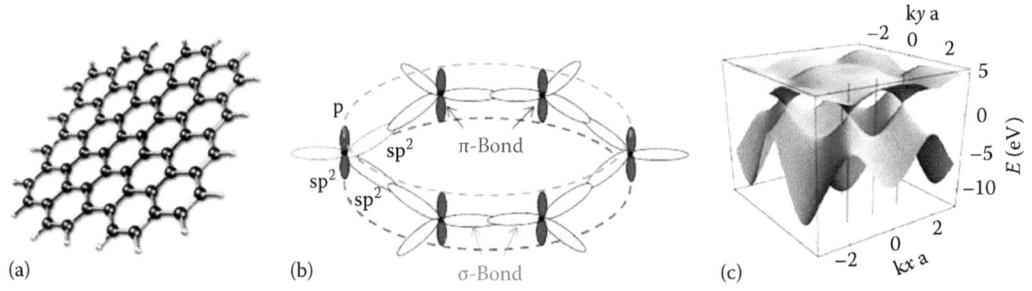

FIGURE 39.1

(a) Two-dimensional graphene plane; (b) basic hexagonal bonding structure for the graphene sheet; carbon nuclei shown as filled black ball, out-of-plane π-bonds represented as delocalized (dotted line), and σ-bonds connect the C nuclei in-plane; (c) electronic dispersion of zero-bandgap graphene.

One of the most feasible methods to control the semiconducting properties of graphene is doping, which is a process intentionally used to tailor the electrical properties of intrinsic semiconductors. The dopant atoms can modify the electronic band structure of graphene and open up an energy gap between the valence and conduction bands. Unfortunately, the inert covalently bonded honeycomb structure of pristine graphene makes it difficult to introduce heteroatoms into the lattice and control the electrical properties of graphene.

Our group demonstrated nitrogen atom (N) doping in the graphene lattice with the graphitic N as the dominant bonding type. It is achieved through high-flux ion bombardment to produce defects (mainly carbon vacancies) in the graphene plane and subsequent annealing in ammonia to fill those vacancies with active N atoms from thermal decomposition of ammonia (Figure 39.2a). Raman spectroscopy can monitor the change of the density of carbon vacancy in graphene with different ion doses for the exact control of the doping level, which is an important parameter of modulating the electronic property of graphene. As the ion dose increases, the ratio of I_D/I_G starts to drop after reaching the highest value, while the ratio of I_G/I_{2D} continues to increase, indicating that the microscopic structure of graphene comes into the stage of transferring from nanocrystalline graphene to low sp^3 amorphous graphene and the formation of amorphous carbon is avoided (Figure 39.2b). After annealing the irradiated graphene in ammonia, the intensity of graphene D peak dramatically decreases, indicating the restoration of defects to the maximal extent. Since it is more useful to get n-type graphene compared to the easily obtained p-typed graphene by adsorbates, we chose N, the natural candidate, because of its similar atomic size as that of C and of its electron donor character for N doping in graphene. When nitrogen atoms are incorporated into the basal plane of graphene, they denote π electrons and fill into the conduction band, then the original charge symmetry is broken, and Fermi level enters the conduction band, leading to n-type doping of graphene and conversion of the linear dispersion of charge carriers to parabolic near zero energy. The graphene-based back-gate field-effect transistors (FETs) were fabricated on a 300 nm $SiO_2/p^{2+}Si$ substrate, and the source/drain electrodes were defined by electron beam lithography and thermal metal deposition of Cr/Au (5 nm/70 nm) (Figure 39.2c and d). The source–drain conductance and back-gate voltage ($G_{sd} - V_g$) curve of the pristine graphene FET shows bipolar transistor effect, and the minimal conduction corresponding to the Dirac point (V_{dirac}) locates at the positive gate voltage indicating the p-type hole doping behavior of pristine graphene due to doping of the physisorbed molecular oxygen (Figure 39.2e). After annealing graphene in ammonia, the FET displays clear n-type behavior with the Dirac point at the negative gate voltage, showing N doping in graphene (Figure 39.2f). The carrier mobility (μ) can be deduced by $μ = (L/WC_gV_{sd})(\Delta I_{sd}/\Delta V_g)$, where C_g is the gate capacitance per unit area (ca. 7 nF cm^{-2}), L and W, channel length and width, are about 2 and 5–10 μm, respectively, and V_{sd} is 30 mV in our experiment. For the pristine graphene, the hole and electron mobilities are about 15,000 and 6,700 cm^2 V^{-1} s^{-1}, respectively. For our N-doped graphene, the hole and electron mobilities are about 6000 cm^2 V^{-1} s^{-1} comparable to those before doping.[5]

In addition, we find that ion irradiation can also controllably dope reduced graphene oxide (RGO) in another fashion. Fourier transform infrared (FTIR) spectroscopy and Raman spectra

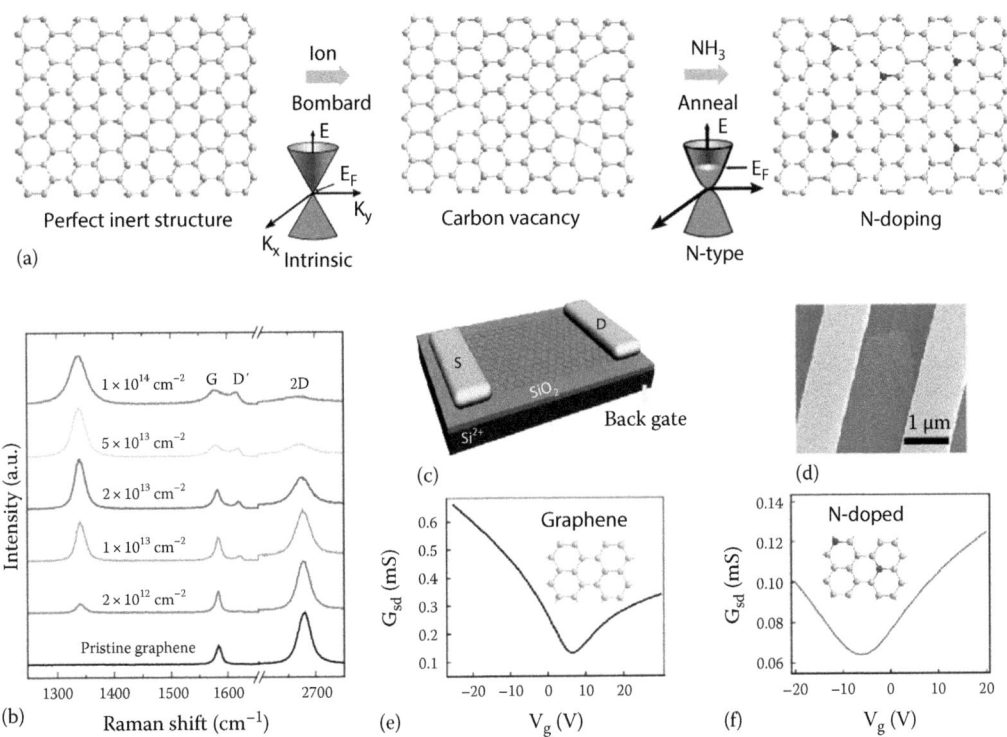

FIGURE 39.2
(a) Process of N atom doping in graphene by ion bombardment and subsequent annealing in ammonia. (b) Raman spectra show the change of the density of carbon vacancy in graphene with different ion doses. (c) Scheme of the graphene-based FET. (d) SEM image of the FET device. (e) G_{sd}–V_g curves of the pristine graphene. (f) G_{sd}–V_g curves of the N-doped graphene. (Reprinted with permission from Guo, B.D., Liu, Q., Chen, E.D., Zhu, H.W., Fang, L., and Gong, J.R., Controllable N-doping of graphene, *Nano Lett.*, 10(12), 4975–4980. Copyright 2010 American Chemical Society.)

results indicate that this difference is attributed to the breaking of the bonds of functional groups of RGO, for example, H–O–H, C–OH, C–H, and the formation of new functional group, such as N–H, on the surface of RGO after ion irradiation. To investigate the effect of doping on the electronic properties, we fabricated RGO-based FET devices with and without N^+ ion irradiation. The threshold voltage of the GFET is more than 30 V due to the existence of oxygen functional groups and the physisorbed molecular oxygen, while the threshold voltage of the FET locates at ca. −20 V after doping at the $1 \times 10^{14}\,cm^{-2}$ flux of irradiation (Figure 39.3). The downshift of the threshold voltage indicates the electron doping effect of the channel. It should be noted that the current was enhanced after N^+ ion irradiation, and the electronic properties of the device were not degraded.[6]

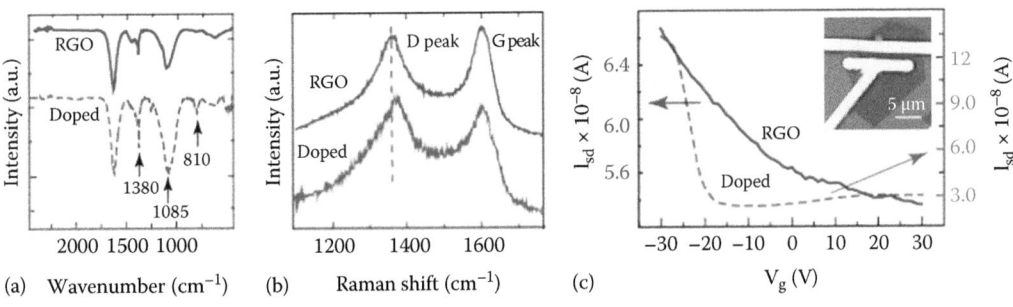

FIGURE 39.3
(a) FTIR, (b) Raman spectra, and (c) G_{sd} – V_g curves of RGO and N-doped RGO. (From Guo, B.D. et al., *Electron. Lett.*, 47(11), 663, 2011.)

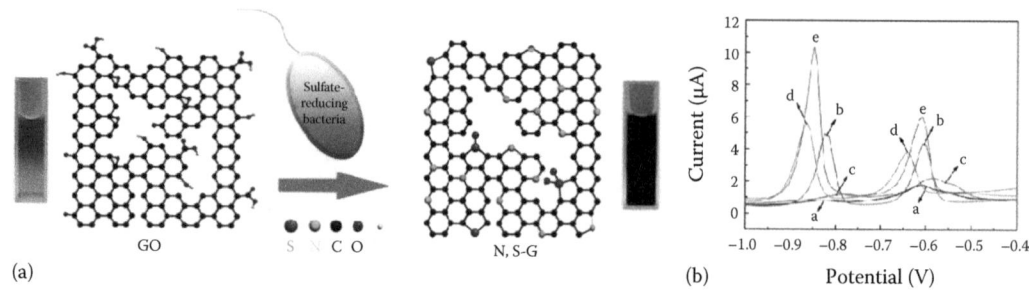

FIGURE 39.4
(a) Simultaneous reduction of GO and doping in RGO by SRB. (b) N, S-G sample shows higher electrocatalytic activity compared to the single-atom-doped graphene. The differential pulse anodic stripping voltammetry of 15 µg L^{-1} of Cd^{2+} and Pb^{2+} on (a) glass carbon electrode (GCE), (b) Bi/GCE, (c) Nafion/GCE, (d) (N, S-G)-Nafion/GCE and (e) Bi/(N, S-G)-Nafion/GCE. (Reprinted with permission from Macmillan Publishers Ltd., *Sci. Rep.*, Guo, P.P., Xiao, F., Liu, Q. et al., One-pot microbial method to synthesize dual-doped graphene and its use as high-performance electrocatalyst, 3, 3499, Copyright 2013.)

Besides single-atom doping, our group also realized multicomponent doping in graphene and demonstrated excellent electrocatalytic performance of the material.[7] It has been recognized that the doping of heteroatoms into graphene is an efficient method to enhance its electrocatalytic performance. Both the theoretical calculations and experimental results have proved that introduction of more electronegative nitrogen (N) atoms into sp^2-hybridized carbon frameworks in graphene is generally effective in modifying their electrical properties and chemical activities.[8] For instance, the spin density and charge distribution of C atom will be influenced by the neighbor N dopants,[9] inducing the activation region on graphene surface, which is expected to directly participate in catalytic reactions and/or provide more nucleation sites to anchor the catalytically active metal nanoparticles.[10] Some recent findings have also shown that doping of sulfur (S) atom into the graphene materials causes the changed spin density of graphene, which results in the excellent catalytic activity, long-term stability, and high tolerance in alkaline media of the S-doped graphene.[11] We report for the first time a novel approach to synthesize N, S-doped graphene (N, S-G) through a one-pot reduction of graphene oxide (GO) by microbial respiration of sulfate-reducing bacteria (SRB) under mild conditions (37°C). In this work, the as-synthesized N, S-G has been applied to fabricate a highly sensitive electrochemical platform to determine trace Cd^{2+} and Pb^{2+} by differential pulse anodic stripping voltammetry. The well-defined reduction peaks with the largest peak currents for target metal ions were obtained for the N, S-G sample due to the synergistic effect of N and S dual doping, which increases the conductivity of the material and improves the electrocatalytic activity of the metal ions compared to the single-atom-doped graphene (Figure 39.4). This method is facile and the reaction temperature is low, which can produce graphene-based electrocatalysts in large scale at low cost.

39.2 MOLECULAR COVALENT BONDING

Based on the general rule that all important photochemical and photophysical processes are the process of transfer and filling of electrons in various molecular orbits, our group can modify graphene quantum dot (GQD) through molecular covalent bonding to modulate the efficiency of charge transfer, coupling strength between the highest occupied molecular orbit (HOMO) and the lowest unoccupied molecular orbit (LUMO), and the magnitude of transition dipole and bandgap, to controllably change the optical property. We will take two-photo bioimaging as an example to introduce our group work on the application of GQD in nanobiomedicine.[12]

Two-photon fluorescence imaging (TPFI) with advantages such as a larger penetration depth, minimized tissue autofluorescence background, and reduced photodamage in biotissues has received much attention for its promising applications in both basic biological research and clinical diagnostics.[13] Advances in two-photon microscope further demonstrate the possibility of TPFI as a powerful technique to probe deep inside various organ tissues of living organisms via a noninvasive way.[14] In the past decades, organic dyes and semiconductor quantum dots (QDs) are widely studied two-photon probes.[15–18] But the rapid photobleaching effect and limited two-photon absorption cross section of organic dyes hamper the imaging depth. The serious toxicity of heavy metals in semiconductor QDs causes concern for in vivo bioimaging, despite their strong two-photon fluorescence. Therefore, the availability of bright fluorescent probes with large two-photon absorption cross sections as well as good biocompatibility is still a critical challenge for deep-tissue TPFI.

The quasi 0D GQD with a single atomic layer gives rise to several advantages over other carbon-based nanomaterials for potential deep-tissue TPFI. First, the bandgap and fluorescence of GQDs can be effectively tuned by doping heteroatoms to the π-conjugated system. A blueshift in fluorescence emission has been reported for GQDs after doped with nitrogen atom through an electrochemical method due to the strong electron-withdrawing effect of the doped nitrogen.[19] It can be expected that doping nitrogen through introducing strong electron-donating groups such as dimethylamido could result in a redshift of fluorescence and achieve longer emission and excitation wavelengths, which are less scattered by biotissues and more practical for bioimaging.[20] Second, GQDs without passivation by any surfactant can exhibit strong fluorescence induced by a pronounced quantum confinement and edge effect,[21] which could also impart a larger two-photon absorption cross section and deeper penetration in turbid tissues.[22] Third, the large rigid π-conjugated electronic structure of GQD can also improve the intramolecular charge-transfer efficiency and therefore enhance the two-photon absorption to achieve larger imaging depth in TPFI.[23]

We developed a facile solvothermal approach for doping nitrogen to GQD using dimethylformamide (DMF) as both solvent and nitrogen sources. Under UV irradiation, the GQD without nitrogen doping showed blue fluorescence rather than the green fluorescence of N-GQD. The fluorescence spectrum of GQD aqueous solution showed a maximum fluorescence peak at 450 nm (Ex 390 nm). Compared with GQD, the maximum emission peak shifted to 520 nm for N-GQD at the same excitation wavelength, indicating that the solvothermal process using DMF could successfully realize the nitrogen doping of GO to form the N-GQD. Two-photon-induced fluorescence of N-GQD was systematically investigated using near-infrared (NIR) femtosecond laser as excitation and applied for efficient two-photon cellular and deep-tissue imaging. A transmission electron microscopy (TEM) image (Figure 39.5a) shows the uniformly dispersed N-GQDs with an average diameter of 3 nm (Figure 39.5b). The obtained N-GQD solution has a high zeta potential (−21 mW in water) and is very stable, exhibiting a transparent homogeneous phase without any precipitation or agglomeration at room temperature for at least 20 months either in water or physiological conditions such as phosphate buffer solution (PBS) and DMEM high glucose culture medium with serum (Figure 39.5b).

Figure 39.5c shows the two-photon fluorescence spectrum of N-GQD solution under the excitation by the femtosecond pulse laser with a wavelength of 800 nm. For the obtained two-photon fluorescence spectrum, the maximum emission wavelength is indistinguishable with that of the one-photon fluorescence spectrum of N-GQD, but the bandwidth is much narrower than that of the one-photon fluorescence spectrum. To explore whether the observed green fluorescence of N-GQD originates from two-photon absorption process with laser excitation in the NIR, the change of the green fluorescence intensity was monitored by adjusting the power of the 800 nm laser. As demonstrated in Figure 39.5d, the obvious quadratic relationship between the fluorescence intensity and the excited laser power suggests that the two-photon excitation is truly responsible for the green fluorescence in nature. For further evaluating the performance of the N-GQD for TPFI, the two-photon absorption cross section (σ_{2PA}) of NGQD was measured using rhodamine B as a reference based on the equation $\sigma_2 = \sigma_1 \times (F_2/F_1) \times (\phi_1/\phi_2) \times (C_1/C_2)$, where F represents the observed fluorescence intensity, φ stands for the quantum yield (QY), C is the concentration, and the subscripts 1 and 2 denote the values for the N-GQD and rhodamine B, respectively.[24] The emission QY of N-GQD aqueous solution was measured to be 0.31. Based on the aforementioned equation, a large σ_{2PA} of 48,000 GM with excitation at 800 nm achieves the highest value ever reported for carbon-based nanomaterials. This σ_{2PA} is two orders of magnitude larger than that of organic molecules and even comparable to that of the high-performance CdSe QDs.[25,26] The large σ_{2PA} is caused by the efficient intramolecular charge transfer, owing to the existence of large π-conjugated systems of N-GQD and the strong electron denoting the effect of the doped dimethylamido. Meanwhile, the increased quantum confinement of the ultrasmall-sized N-GQD with highly symmetric bandgap can contribute to the high σ_{2PA} value. Similar to the chemical structure of polyaromatic compounds, the fluorescence may originate from the π–π electron transition since the N-GQD has a large π-conjugated system and rigid plane. Furthermore, the lone pair electrons from the strong

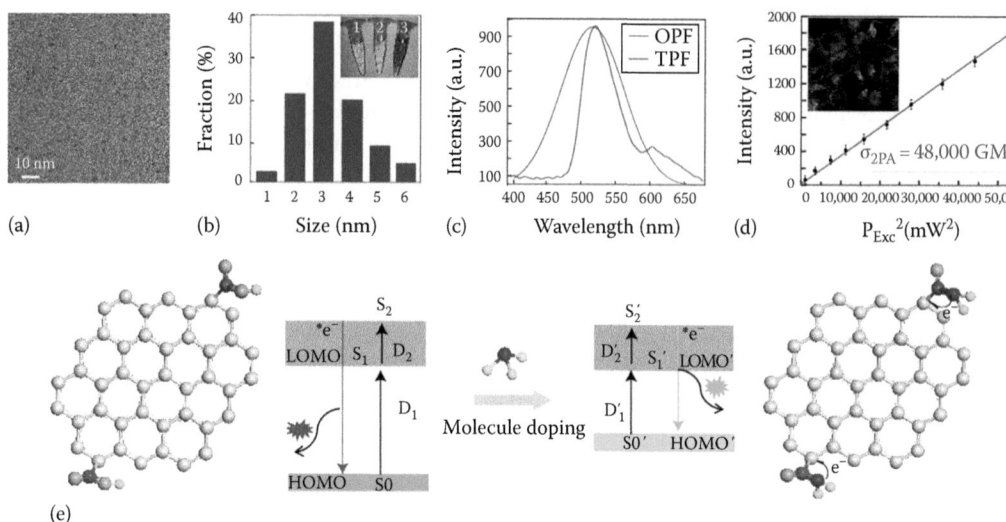

FIGURE 39.5

(a) TEM image of the uniformly dispersed N-GQD. (b) The statistics of the size distribution of N-GQD. Inset: transparent homogeneous phase of N-GQD in various media. (c) Two-photon-induced fluorescence spectrum of N-GQD solution under 800 nm femtosecond laser excitation (red line) and one-photon fluorescence (blue line) at 390 nm laser diode excitation, respectively. (d) Quadratic relationship of the fluorescence intensity of the N-GQD aqueous solution with the different excitation laser powers at 800 nm (P_{Exc}, as measured at the focal plane). Inset: Two-photo cell imaging under 800 nm excitation. The two-photon absorption cross section (σ_{2PA}) is 48,000 GM. (e) Molecular doping modulates the fluorescence of N-GQD. (Reprinted with permission from Liu, Q., Guo, B.D., Rao, Z.Y., Zhang, B.H., and Gong, J.R., Strong two-photon-induced fluorescence from photostable, biocompatible nitrogen-doped graphene quantum dots for cellular and deep-tissue imaging, *Nano Lett.*, 13, 2436–2441. Copyright 2013 American Chemical Society.)

electron-donating group dimethylamido, which is doped to the aromatic ring of N-GQD, can be excited to the aromatic rings to form the p–π conjugation, further enlarging the π-conjugated system.[24] The strong orbital interaction between dimethylamido and π-conjugated system of N-GQD elevates the primary HOMO to a higher energy obit, resulting in a decrease of bandgap and red-shift of fluorescence emission (Figure 39.5e). More importantly, the large π-conjugated system in N-GQD and strong electron-donating effect of dimethylamido can also facilitate the charge-transfer efficiency,[27] enhancing the two-photon absorption and thus imparting strong two-photon-induced fluorescence for NGQD.

To demonstrate the capability of the N-GQD for two-photon bioimaging, we carried out an in vitro bioimaging study using human cervical carcinoma HeLa cells by a multiphoton fluorescence microscope. The results demonstrate that the N-GQD can label both the cell membrane and the cytoplasm of HeLa cells without invading the nucleus in a significant fashion. It should be noted that a low laser power of 1 mW (average power density of 13 W cm⁻²) was sufficient to induce strong fluorescence of the N-GQDs internalized in HeLa cells (Figure 39.5d). Besides the strong two-photon fluorescence and good stability in the physiological conditions, the N-GQD also shows quite low cytotoxicity. For in vivo bioimaging applications, exact knowledge about the maximum tissue penetration depth of N-GQD is required. To explore the potential use of the strongly fluorescent NGQD for deep-tissue TPFI, we investigated the imaging depth of N-GQD in turbid tissue phantom for the first time. Intralipid was chosen as a mock tissue because of its similar scattering properties with the real tissues. The obtained two-photon fluorescence images demonstrate that the N-GQD can be imaged with high resolution and a high signal-to-noise ratio at depths ranging from 0 to 1300 μm in the tissue phantom using NIR laser as excitation source. Even at the depth of 1800 μm, we can easily identify the N-GQDs in the tissue phantom with appreciable two-photon fluorescence signal, though the fluorescence intensity dramatically decreases. In contrast, the one-photon fluorescence imaging (OPFI) shows that the maximum penetration depth is only 400 μm due to the strong scattering and refraction of the visible excitation light in turbid tissue phantom. The large two-photon imaging depth of 1800 μm achieved by the bright N-GQD is far exceeding that of the organic dyes and even larger than that of the semiconductor QDs.[28] The TPFI using N-GQD as a fluorescent probe is particularly suitable for in vivo bioimaging applications where there is considerable interest in investigating biostructures in the 800–1500 μm region.

The excellent photostability of two-photon probes is also of great importance for TPFI, especially for the long-term dynamic bioimaging. Under continued laser excitation at 800 nm, the N-GQD exhibits no attenuation in fluorescence intensity, demonstrating their extraordinary photostability. Meanwhile, concerning the possible negative photothermal effect on living cells with NIR femtosecond laser excitation, we measured the temperature of the N-GQD aqueous solution upon irradiation with 800 nm femtosecond laser at different time. With continuous laser irradiation at the high power density, the temperature of the N-GQD aqueous solution showed little increase. This negligible photothermal effect indicates that N-GQD could be applied for TPFI of living cells and tissues without causing any thermal damage to them. Moreover, the N-GQD was demonstrated to be strongly fluorescent in a wide pH range. The fluorescence emission of N-GQD is strongest in neutral or weakly alkaline conditions (pH = 7–9). Under acidic or strong alkaline conditions, the fluorescence intensity of N-GQD decreases by certain degrees. Even so, the N-GQD exhibits much stronger fluorescence than their N-free counterpart in the acid conditions (pH = 1–6), in which the fluorescence of the N-free GQD will be completely quenched.[29] This result suggests that the fluorescent N-GQD can be used in a wide pH range, such as tumor environment with a relatively low pH value.

In conclusion, N-GQDs were facilely prepared by a one-pot solvothermal approach using DMF as a solvent and nitrogen source. The N-GQD exhibits a two-photon absorption cross section as high as 48,000 GM and is demonstrated as an efficient two-photon fluorescent probe for cellular and deep-tissue imaging. A large imaging penetration depth of 1800 μm achieved by N-GQD in tissue phantom significantly extends the fundamental imaging depth limit of two-photon microscopy. Furthermore, the N-GQD displays little photobleaching and photothermal effects under repeated femtosecond NIR laser irradiation and can emit quite strong fluorescence over a wide range of pH values. It is anticipated that the large imaging depth of N-GQD combining with their excellent biocompatibility and extraordinary photostability will impart potential use for advancing two-photon imaging in virtual applications such as monitoring the biological activity of deep tissues and noninvasively detecting disease of living biosystems.

39.3 INTERFACIAL NONCOVALENT MODIFICATION

The surface/interfacial structures and properties of materials are central to the performance of all kinds of devices. Graphene is an excellent surface/interfacial material for optoelectronic devices due to its unique optical and electronic properties, 2D plane structure, and chemical stability. With the rapid development of graphene-based nanomaterials and nanodevices, study of assembled nanostructures, charge transfer, and optoelectronic properties on the surface/interface of graphene is significant for elucidating the mechanism of optoelectronic devices in depth and exploring the potential applications in various fields. In the following paragraphs, some typical organic and inorganic semiconductor systems are taken as examples to introduce the research work on it.

Self-assembly is a spontaneous process involved in functional materials and living organisms and an effective approach to fabricate a variety of nanostructures as a bottom-up strategy in nanofabrication. The formation of a self-assembly is usually dominated by weaker noncovalent intermolecular actions such as hydrogen bonding, van der Waals' interaction, and π–π interaction. Therefore, understanding and employing the weaker interactions is an important issue in designing assembly and fabricating a molecular device. In addition, this kind of interfacial noncovalent modification has no damage on the structure and property of graphene.

First, we will take a typical system as a sample to demonstrate that we can controllably construct various assembly structures by systematically tuning the functional groups of the molecules. Owing to high conductivity and environmental stability, conjugated polymers are promising candidates in many electronic devices such as light-emitting diodes, thin-film FETs, solid-state lasers, and photovoltaic devices. In particular, polythiophene and its substituted derivatives seem to be of great interest because of its high field-effect mobility and structural organization-dependent performance. We chose four compounds with different thiophene units (3-thiophene acetic acid [TA];

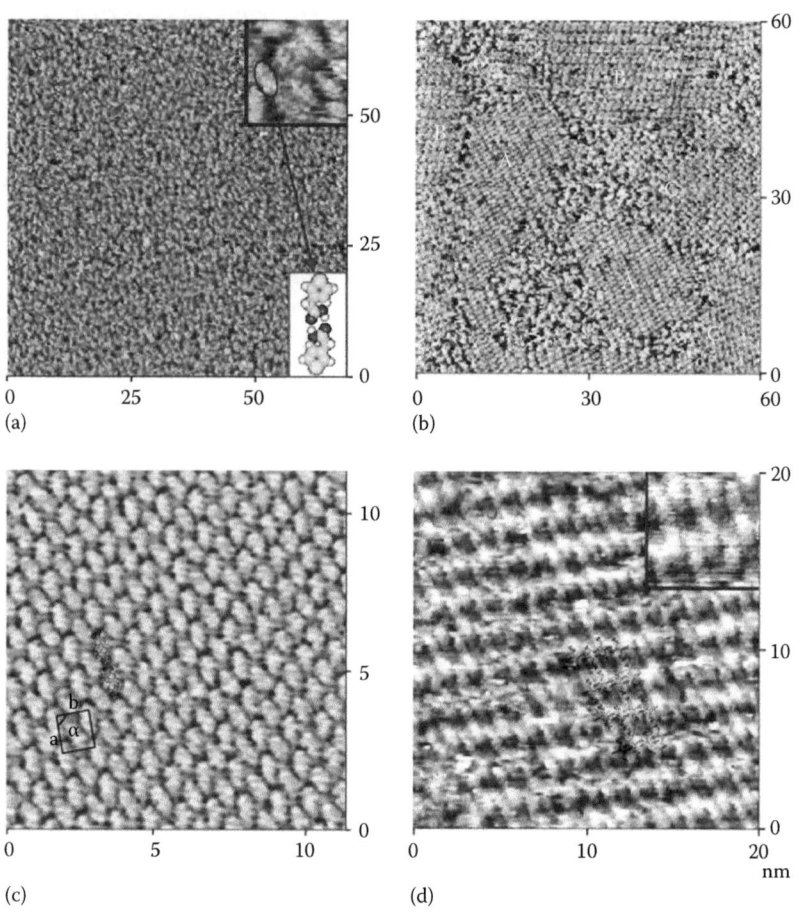

FIGURE 39.6
The molecular assembled structures of (a) TA, (b) DTDA, (c) TTDA, and (d) QTDA. (Reprinted with permission from Xu, L.P., Gong, J.R., Wan, L.J. et al., Molecular architecture of oligothiophene on a highly oriented pyrolytic graphite surface by employing hydrogen bondings, *J. Phys. Chem. B*, 110, 17043–17049. Copyright 2006 American Chemical Society.)

2,2′-bithiophene-5,5′-dicarboxylic acid [DTDA]; 3′-pentyl-5,2′:5′,2″-terthiophene-2,5″-dicarboxylic acid [TTDA]; 4′,3″-dipentyl-5, 2′:5′,2″:5″,2‴-quaterthiophene-2,5‴-dicarboxylic acid [QTDA]) as model molecules to study their assembly on the surface of graphene (Figure 39.6). For TA, two molecules form a dimer by H bonding through the molecular one-end carboxyl group, and the adlayer structure is disordered. The DTDA molecules are strongly hydrogen bonded with the head-to-head well-ordered molecular wire configuration through the carboxyl groups at both ends of DTDA, while TTDA molecules form three different types of ordered packing geometries marked by A, B, and C in the large-scale scanning tunneling microscopy (STM) image. QTDA has a similar molecular structure to TTDA. However, in addition to quaterthiophene rings, there are two alkyl chains forming a balance on the two sides of the molecular skeleton. A stable and ordered 2D QTDA molecular network is formed on the graphene surface. The results show that the symmetric molecular structure favors the stable assembling geometry, and hydrogen bonding plays an essential role in the formation of the ordered assemblies.[30] Likewise, different self-assembled monolayers of three monodendrons, 5-(benzyloxy)-isophthalic acid derivatives, are formed by tuning the number of the molecular alkyl substitutents.[31] The different arrangement configuration of the two conjugated oligo(phenylene ethynylene)s (OPEs) can be constructed by changing the molecular end groups.[32] Two types of molecular arrangements are observed in these self-assemblies with different stripe widths that resulted from the organization of the alkyl chains of a series of banana-shaped liquid crystal molecules, 1,3-phenylene bis[4-(4-n-alkylphenyliminomethyl) benzoates], by varying the length of the molecular alkyl chain. More interestingly, the observation of the bilayer liquid crystal structures is helpful to understand the relative packing orientation of the different molecular layer.[33]

The substrate also has an important effect on the surface assembling structure. Metal-containing macrocycles with high symmetry and precise architecture are a new class of promising supramolecules for future application in nanotechnology because of their exact shape and size, as well as magnetic, photophysical, and electrostatic properties. The adlayer symmetry and molecular arrangement of the metallacyclic rectangle cyclobis[(1,8-bis(transPt(PEt$_3$)$_2$)anthracene) (1,4-bis(4-ethynylpyridyl)benzene](PF$_6$)$_4$ on graphene are dramatically different from that on Au surface. The rectangles spontaneously adsorb on both surfaces and self-organize into well-ordered adlayers. On graphene, the long edge of the rectangle stands on the surface, forming a 2D molecular network. In contrast, the face of the rectangle lays flat on the Au(111) surface, forming linear chains (Figure 39.7a). By imaging the lattice of the substrate simultaneously, the orientation relationship between the adlayer and underlying substrate is revealed that the molecules align along a lattice direction on the graphene surface, and the same behavior is observed on the Au(111) surface (Figure 39.7a). The results show that the molecular self-organization of the supramolecular

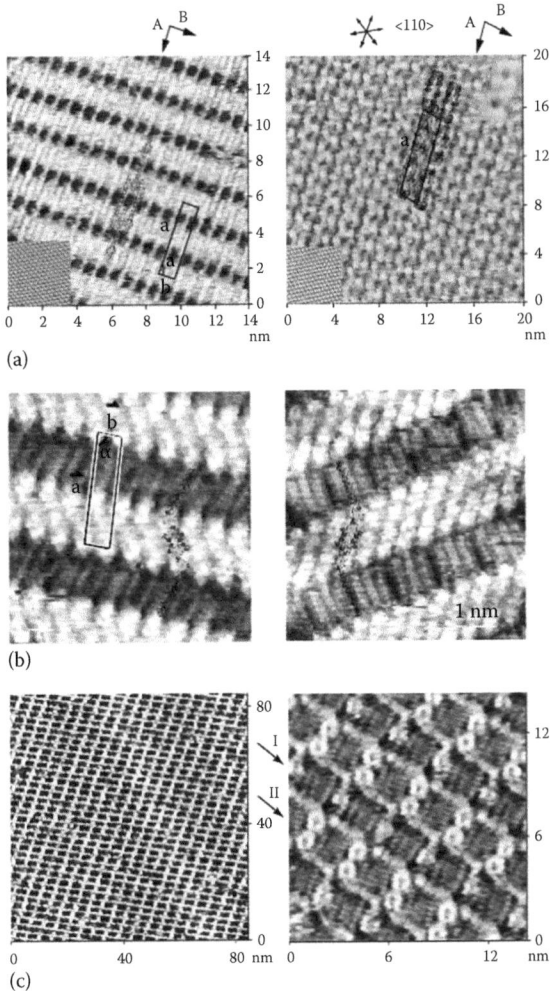

(a)

(b)

(c)

FIGURE 39.7
(a) The molecular arrangement of the same metallacyclic rectangle is different on the different substrate. (b) The substrate can induce the formation of the chiral domain by achiral molecules. (c) The multicomponent molecular arrays. (Reprinted with permission from Gong, J.R., Wan, L.J., Yuan, Q.H. et al., Mesoscopic self-organization of a self-assembled supramolecular rectangle on highly oriented pyrolytic graphite and Au(111) surfaces, *Proc. Natl. Acad. Sci. USA*, 102(4), 971–974; Gong, J.R., Lei, S.B., Pan, G.B., Wan, L.J., Fan, Q.H., and Bai, C.L., Monitoring molecular motion and structure near defect with STM, *Colloids Surf. A Physicochem. Eng. Asp.*, 257–258, 9–13; Gong, J. R.; Yan, H.J., Yuan, Q.H., Xu, L.P., Bo, Z.S., and Wan, L.J., Controllable distribution of single molecules and peptides within oligomer template investigated by STM, *J. Am. Chem. Soc.*, 128(38), 12384–12385. Copyright 2005 and 2006 American Chemical Society.)

rectangle on solid surfaces is governed by intermolecular and molecule–substrate interactions and could be tuned by the appropriate choice of substrate materials.[34] In addition, the graphene substrate can also induce the formation of the chiral domain by achiral molecules as shown by the mirror high-resolution STM image having the same unit cell parameters with its symmetrical packing in Figure 39.7b.[34–36]

Besides single component, the 2D controllable hybrid assembling structures were also studied. The research of hybrid materials provides an opportunity for developing new materials with synergic behavior leading to improved performance or to new useful properties.[32,36] In particular, the controllable arrangement of single molecules is the prerequisite for the development of nanodevices. OPE was chosen as the molecular template because of its well-defined chemical structure, improved solubility and processability, and many applications in organic luminescent field. Taking OPE as the host, the guest molecules ranging from organic and inorganic semiconductor to peptide, various single molecule arrays could be produced just by simply adjusting the molecular molar ratio of different components (Figure 39.7c). H bonding interaction plays an important role for the stability of the complex system. Furthermore, we can monitor the molecular dynamic assembly process in real time. In the case of the monodendron 5-(benzyloxy)-isophthalic acid system, the time-dependent disorder–order transition was observed as displayed in Figure 39.8a.[31,35]

Controlled regulation of the switchable behavior of the supramolecular network is central to the potential application in the molecular-scale nanodevices. We demonstrated that the reversible accommodation of the guest molecules in the nanoporous supramolecular network can be regulated by switching UV and visible light. TCDB/4NN-Macrocycle (*trans,trans,trans,trans*) with photosensitive units is designed to form well-defined nanoporous molecular template on the graphene surface. After UV irradiation, the template can be switched on to encapsulate coronene molecules due to the formation of a new photoisomer (*trans,cis,trans,cis*) and switched off to expel coronene from the inner cavities under visible light (Figure 39.8b). Considering both the substrate effect and the interactions between the adlayer molecules, density functional theory calculations show that the system energy before and after the encapsulation of the guest molecule is approximate, so the transformation can easily take place between the two forms when the sample is irradiated by UV light and visible light. The photoregulated switchable multicomponent supramolecular guest–host network provides a novel strategy for fabricating the functional nanodevices at the molecular scale.[37]

The significance of the study on the nanostructures in real devices was exemplified through exploring the possible effect of the microscopic structural change on organic light-emitting diode (OLED) degradation. Interfacial phenomena are crucial to the performance and stability of OLED, which represent a challenging and important area of OLED science and technology and have been a subject of intensive theoretical and experimental study. An understanding of the phenomena related to the degradation mechanisms of OLED such as the compatibility and thermostability of the host–guest materials in an amorphous state is still of broad interest. Among the various factors that may contribute to the degradation phenomena in OLED, the Joule heating effect on electroluminescent efficiency and lifetime is considered an important factor. Likewise, the detailed understanding of interfacial structural changes with temperature in a nanometer scale is also of interest. In our experiment, thermal treatment was employed to simulate the Joule heating effect. The host–guest mixed layer of 2,2′,2″-(1,3,5-phenylene)tris-[1-phenyl-1Hbenzimidazole] (TPBI) and 1,4-bis(benzothiazole-vinyl) benzene (BT) was found to have good compatibility at room temperature. However, phase separation was observed with increasing temperature, with a concomitant decrease in emission intensity. Furthermore, the scanning tunneling spectroscopy (STS) measurements show that the characteristic bandgap of TPBI and BT remains unchanged with the annealing temperature (Figure 39.8c). This result reveals that ineffective host-to-guest energy transfer is responsible for the decay of OLED due to phase separation.[38] It should be noted that STM was used for molecular imaging, which reflects the electronic state coupling of single-layer molecules and a few layer graphene on a highly oriented pyrolytic graphite (HOPG) surface, and so we can refer to it as the interface between organic molecules and graphene.

Compared to organic molecules, inorganic materials have the advantage of higher thermal and chemical stability when modifying graphene at the interface, especially for photocatalysis. So far, the rapid charge recombination, high cost of noble metals, and the instability of photocatalysts are still big problems that restrict the realization of large-scale industrial application of these materials in this field. Our group greatly improves the photocatalytic performance of inorganic semiconductors

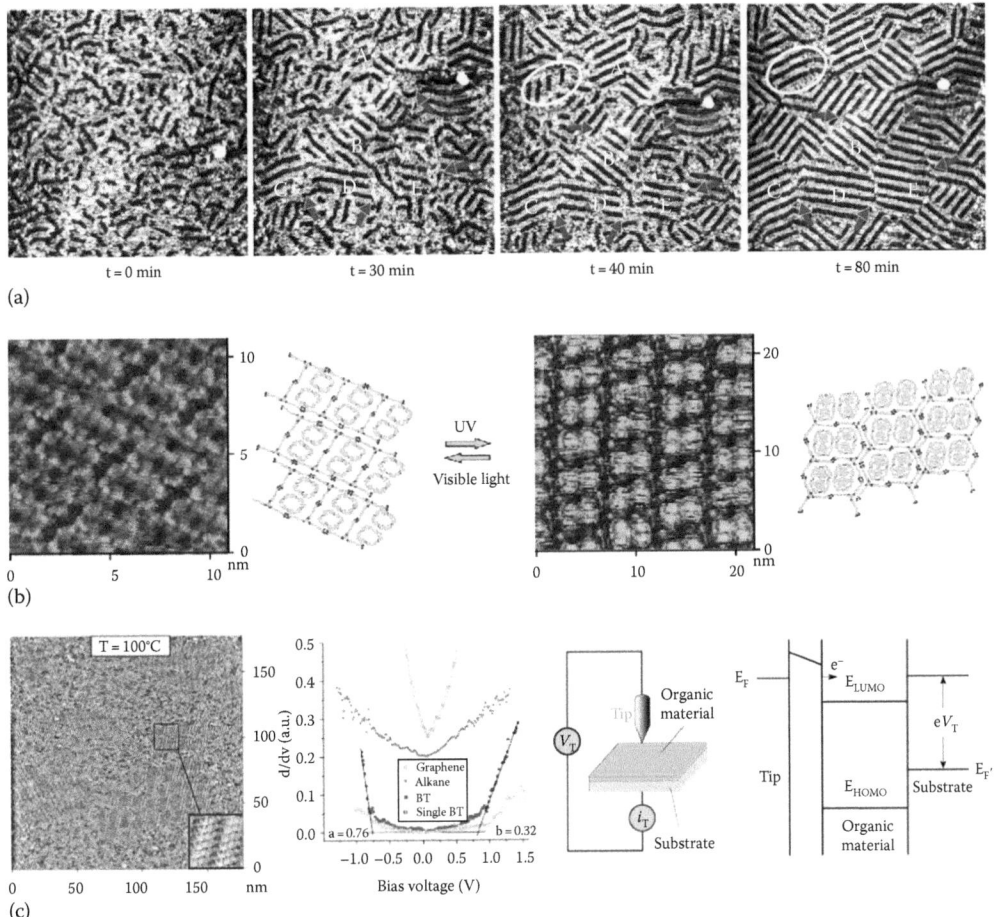

(a) t = 0 min   t = 30 min   t = 40 min   t = 80 min

(b)

(c)

FIGURE 39.8

(a) The real-time observation of the molecular dynamic assembly process.[31] (b) The reversible accommodation of the guest molecules in the nanoporous supramolecular network can be regulated by switching the UV and visible light due to the formation of the new photoisomer. (c) Phase separation occurs at higher temperature; STS results of graphene, alkane, BT, and single BT. A typical STM configuration and energy diagram, in which charge injection into the organic material occurs via tunneling through a vacuum barrier. (Reprinted with permission from Gong, J.R., Lei, S.B., Wan, L.J., Deng, G.J., Fan, Q.H., and Bai, C.L., Structure and dynamic process of two-dimensional monodendron assembly, *Chem. Mater.*, 15(16), 3098–3104; Gong, J.R., Wan, L.J., Lei, S.B., and Bai, C.L., Direct evidence of molecular aggregation and degradation mechanism of organic light-emitting diodes under Joule heating: An STM and photoluminescence study, *J. Phys. Chem. B*, 109, 1675–1682; Li, Q., Guo, B., Yu, J. et al., Highly efficient visible-light-driven photocatalytic hydrogen production of CdS-cluster-decorated graphene nanosheets, *J. Am. Chem. Soc.*, 133, 10878–10884. Copyright 2003, 2005, and 2011 American Chemical Society.)

by combining them with graphene via noncovalent interaction. The interfacial charge-transfer characteristics of the composite are determined by the relative energy levels of the HOMO and LUMO of the catalyst with respect to the Fermi level of graphene. If the HOMO is above the Fermi level of graphene (the Dirac point), electrons will transfer from the catalyst to graphene, whereas if the LUMO is below the Fermi level, electrons will transfer to the dopant (Figure 39.9).

The H_2 production from water splitting over the visible-light responsive chalcogenide system was chosen as an example to introduce the roles of graphene in improving the photocatalytic performance (Figure 39.10). First, graphene can modulate the electronic structure of the photocatalyst. Normally, a higher LUMO energy level possesses an increased driving force for injection of electrons to the H^+/H_2 reduction level, thereby benefitting the photocatalytic performance. And the bandgap narrowing is favorable for light absorption. The experimental results display that the narrowed bandgap and the moderate LUMO level are attributed to the improved performance at the

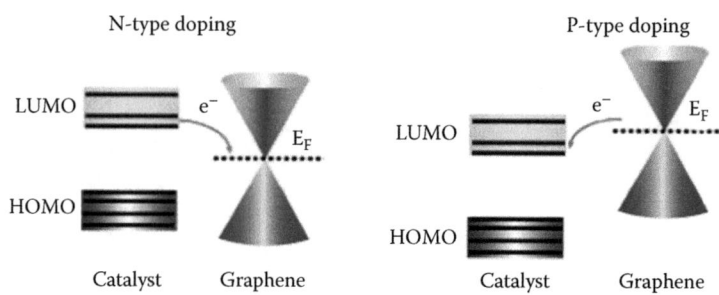

FIGURE 39.9
Scheme of the interfacial charge transfer between graphene and catalyst.

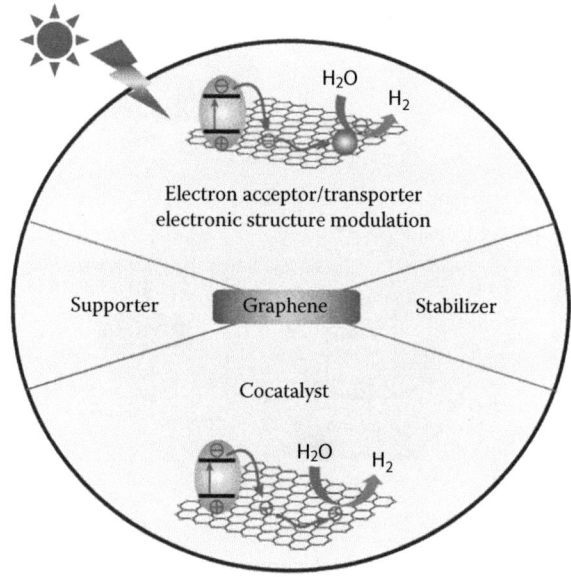

FIGURE 39.10
Various roles of graphene in photocatalysis.

optimal graphene content.[42] Second, when graphene is introduced to the CdS clusters, it can serve as an electron collector and transporter to efficiently separate the photogenerated electron–hole pairs, lengthening the lifetime of the charge carriers. Furthermore, the unique features of graphene allow photocatalytic reactions to take place not only on the surface of CdS but also on the graphene sheet, greatly enlarging the reaction space. The pure CdS nanoparticles show a significant aggregation, while much smaller CdS clusters uniformly and tightly spread on the graphene sheets in the composite, indicating that graphene may interact with CdS nanoparticles and inhibit their aggregation. However, a further increase in the graphene content leads to a deterioration of the catalytic performance. It is reasonable because introduction of a large percentage of black graphene will shield the active sites and the irradiation light. Therefore, a suitable content of graphene is crucial for optimizing the photocatalytic activity of the nanocomposites.[39] Third, graphene can be used as a new cocatalyst for replacing noble metals in hydrogen production. It has been fully demonstrated that noble metals, especially Pt, can function as an efficient H_2-production promoter for many photocatalysts. In the ZnCdS solid solution system, the photocatalytic H_2-production activity of the graphene composite is higher than that of the Pt composite.[40] Fourth, graphene can effectively prohibit the photo corrosion of the metal sulfide and improve the stability of photocatalysts for hydrogen generation. The highest transient photocurrent was obtained for the optimized graphene composite. It is because graphene is an excellent electron acceptor with superior conductivity due to its 2D π-conjugation structure; in the graphene composite, the excited electrons of the graphene composite can transfer from the CB of the solid solution to graphene. Thus, graphene serves as an electron collector and transporter that effectively suppresses the charge recombination in the composite, leaving more charge carriers to form reactive species, which in turn results in the highest photocurrent response and photocatalytic H_2-production rate. Electrochemical impedance spectra (EIS) analysis has become a powerful tool in studying the charge-transfer process occurring in the three-electrode system. The graphene composite shows the smallest semicircle in the middle-frequency region in comparison to the solid solution and the Pt composite electrodes, which indicates the fastest interfacial electron transfer. That is, because of the excellent conductivity, the introduction of graphene can benefit the charge transfer in the graphene composite and thus lowers the charge recombination. Overall, graphene can function as an electron collector and transporter in the composite and inhibit the charge recombination and thus significantly enhance the photocatalytic H_2-production activity.[41,42]

Three categories of chemical methods are introduced for modifying graphene at different levels for various optoelectronic applications in this chapter. Heteroatom doping in lattice has high accuracy and stability, and it is more suitable for fabrication of electronic devices. Molecular covalent bonding is useful for cheap mass-produced solution processing, while interfacial noncovalent modification has advantages for all kinds of optoelectronic devices since the interfacial materials play a crucial role in highly efficient photoelectronic energy conversion. The chemical functionalization approaches and application of graphene-based materials are not limited to the aforementioned. Researchers from different disciplines including our group (http://www.nanoctr.cn/gongjianru) are still working in this field, and more novel methods and meaningful applications will be seen in the near future.

REFERENCES

1. Novoselov, K. S.; Geim, A. K.; Morozov, S. V. et al. 2004. Electric field effect in atomically thin carbon films. *Science* 306: 666–669.
2. Qian, D.; Wagner, G.; Liu, W. K.; Yu, M. F.; Ruoff, R. 2002. Mechanics of carbon nanotubes. *Appl. Mech. Rev.* 55(6): 495–532.
3. Geim, A. K.; Novoselov, K. S. 2007. The rise of graphene. *Nat. Mater.* 6: 183–191.
4. Zhou, C. W.; Kong, J.; Yenilmez, E.; Dai, H. J. 2000. Modulated chemical doping of individual carbon nanotubes. *Science* 290: 1552–1555.
5. Guo, B. D.; Liu, Q.; Chen, E. D.; Zhu, H. W.; Fang, L.; Gong, J. R. 2010. Controllable N-doping of graphene. *Nano Lett.* 10 (12): 4975–4980.
6. Guo, B. D.; Fang, L.; Zhang, B. H.; Gong, J. R. 2011. Doping effect on the shift of threshold voltage of graphene-based field-effect transistors. *Electron. Lett.* 47 (11): 663–664.
7. Guo, P. P.; Xiao, F.; Liu, Q. et al. 2013. One-pot microbial method to synthesize dual-doped graphene and its use as high-performance electrocatalyst. *Sci. Rep.* 3: 3499.
8. Liu, H. T.; Liu, Y. Q.; Zhu, D. B. 2011. Chemical doping of graphene. *J. Mater. Chem.* 21: 3335–3345.
9. Groves, M. N.; Chan, A. S. W.; Malardier-Jugroot, C.; Jugroot, M. 2009. Improving platinum catalyst binding energy to graphene through nitrogen doping. *Chem. Phys. Lett.* 481: 214–219.
10. Bourlinos, A. B.; Gournis, D.; Petridis, D.; Szabo, T.; Szeri, A.; Dekany, I. 2003. Graphite oxide-chemical reduction to graphite and surface modification with primary aliphatic amines and amino acids. *Langmuir* 19: 6050–6055.
11. Yang, Z.; Yao, Z.; Li, G. F. et al. 2012. Sulfur-doped graphene as an efficient metal-free cathode catalyst for oxygen reduction. *ACS Nano* 6: 205–211.
12. Liu, Q.; Guo, B. D.; Rao, Z. Y.; Zhang, B. H.; Gong, J. R. 2013. Strong two-photon-induced fluorescence from photostable, biocompatible nitrogen-doped graphene quantum dots for cellular and deep-tissue imaging. *Nano Lett.* 13: 2436–2441.

13. Helmchen, F.; Denk, W. 2005. Deep tissue two-photon microscopy. *Nat. Methods* 2: 932–940.

14. Jung, W.; Tang, S.; McCormic, D. T. et al. 2008. Miniaturized probe based on a microelectromechanical system mirror for multiphoton microscopy. *Opt. Lett.* 33: 1324–1326.

15. Larson, D. R.; Zipfel, W. R.; Williams, R. M. et al. 2003. Water-soluble quantum dots for multiphoton fluorescence imaging in vivo. *Science* 300: 1434–1436.

16. He, G. S.; Tan, L. S.; Zheng, Q. D.; Prasad, P. N. 2008. Multiphoton absorbing materials: Molecular designs, characterizations, and applications. *Chem. Rev.* 108: 1245–1330.

17. Lee, J. H.; Lim, C. S.; Tian, Y. S.; Han, J. H.; Cho, B. R. 2010. A two-photon fluorescent probe for thiols in live cells and tissues. *J. Am. Chem. Soc.* 132: 1216–1217.

18. Zhu, M. Q.; Zhang, G. F.; Li, C. et al. 2011. Reversible two-photon photoswitching and two-photon imaging of immunofunctionalized nanoparticles targeted to cancer cells. *J. Am. Chem. Soc.* 133: 365–372.

19. Li, Y.; Zhao, Y.; Cheng, H. H. et al. 2012. Nitrogen-doped graphene quantum dots with oxygen-rich functional groups. *J. Am. Chem. Soc.* 134: 15–18.

20. Tetsuka, H.; Asahi, R.; Nagoya, A. et al. 2012. Optically tunable amino-functionalized graphene quantum dots. *Adv. Mater.* 24: 5333–5338.

21. Ponomarenko, L.; Schedin, F.; Katsnelson, M. et al. 2008. Chaotic dirac billiard in graphene quantum dots. *Science* 320: 356–358.

22. Trauzettel, B.; Bulaev, D. V.; Loss, D.; Burkard, G. 2007. Spin qubits in graphene quantum dots. *Nat. Phys.* 3: 192–196.

23. Collini, E. 2012. Cooperative effects to enhance two-photon absorption efficiency: Intra-versus inter-molecular approach. *Phys. Chem. Chem. Phys.* 14: 3725–3736.

24. Xu, C.; Webb, W. W. 1996. Measurement of two-photon excitation cross sections of molecular fluorophores with data from 690 to 1050 nm. *J. Opt. Soc. Am. B* 13: 481–491.

25. Zipfel, W. R.; Williams, R. M.; Webb, W. W. 2003. Nonlinear magic: Multiphoton microscopy in the biosciences. *Nat. Biotechnol.* 21: 1369–1377.

26. Pu, S. C.; Yang, M. J.; Hsu, C. C. et al. 2006. The empirical correlation between size and two-photon absorption cross section of CdSe and CdTe quantum dots. *Small* 2: 1308–1313.

27. Mei, Q. S.; Zhang, K.; Guan, G. J.; Liu, B. H.; Wang, S. H.; Zhang, Z. P. 2010. Highly efficient photoluminescent graphene oxide with tunable surface properties. *Chem. Commun.* 46: 7319–7321.

28. Maestro, L. M.; Ramírez-Hernández, J. E.; Bogdan, N. et al. 2012. Deep tissue bioimaging using two-photon excited CdTe fluorescent quantum dots working within the biological window. *Nanoscale* 4: 298–302.

29. Pan, D. Y.; Zhang, J. C.; Li, Z.; Wu, M. H. 2010. Hydrothermal route for cutting graphene sheets into blue-luminescent graphene quantum dots. *Adv. Mater.* 22: 734–738.

30. Xu, L. P.; Gong, J. R.; Wan, L. J. et al. 2006. Molecular architecture of oligothiophene on a highly oriented pyrolytic graphite surface by employing hydrogen bondings. *J. Phys. Chem. B* 110: 17043–17049.

31. Gong, J. R.; Lei, S. B.; Wan, L. J.; Deng, G. J.; Fan, Q. H.; Bai, C. L. 2003. Structure and dynamic process of two-dimensional monodendron assembly. *Chem. Mater.* 15 (16): 3098–3104.

32. Gong, J. R.; Lei, S. B.; Wan, L. J. et al. 2003. Molecular organization of alkoxy-substituted oligo(phenylene-ethynylene)s studied by STM. *Langmuir* 19 (25): 10128–10131.

33. Gong, J. R.; Wan, L. J. 2005. Two-dimensional assemblies of banana-shaped liquid crystal molecules on HOPG surface. *J. Phys. Chem. B* 109 (40): 18733–18740.

34. Gong, J. R.; Wan, L. J.; Yuan, Q. H. et al. 2005. Mesoscopic self-organization of a self-assembled supramolecular rectangle on highly oriented pyrolytic graphite and Au(111) surfaces. *Proc. Natl. Acad. Sci. USA* 102 (4): 971–974.

35. Gong, J. R.; Lei, S. B.; Pan, G. B.; Wan, L. J.; Fan, Q. H.; Bai, C. L. 2005. Monitoring molecular motion and structure near defect with STM. *Colloids Surf. A Physicochem. Eng. Asp.* 257–258: 9–13.

36. Gong, J. R.; Yan, H. J.; Yuan, Q. H.; Xu, L. P.; Bo, Z. S.; Wan, L. J. 2006. Controllable distribution of single molecules and peptides within oligomer template investigated by STM. *J. Am. Chem. Soc.* 128 (38): 12384–12385.

37. Shen, Y. T.; Deng, K.; Zhang, X. M. et al. 2011. Switchable ternary nanoporous supramolecular network on photo-regulation. *Nano Lett.* 11: 3245–3250.

38. Gong, J. R.; Wan, L. J.; Lei, S. B.; Bai, C. L. 2005. Direct evidence of molecular aggregation and degradation mechanism of organic light-emitting diodes under Joule heating: An STM and photoluminescence study. *J. Phys. Chem. B* 109: 1675–1682.

39. Li, Q.; Guo, B.; Yu, J. et al. 2011. Highly efficient visible-light-driven photocatalytic hydrogen production of CdS-cluster-decorated graphene nanosheets. *J. Am. Chem. Soc.* 133: 10878–10884.

40. Zhang, J.; Yu, J.; Jaroniec, M.; Gong, J. R. 2012. Noble metal-free reduced graphene oxide-ZnxCd1-xS nanocomposite with enhanced solar photocatalytic H2-production performance. *Nano Lett.* 12: 4584–4589.

41. Xie, G.; Zhang, K.; Guo, B.; Liu, Q.; Fang, L.; Gong, J. R. 2013. Graphene-based materials for hydrogen generation from light-driven water splitting. *Adv. Mater.* 25: 3820–3839.

42. Xie, G.; Zhang, K.; Fang, H. et al. 2013. Photoelectrochemical investigation on the synergetic effect between CdS and reduced graphene oxide for solar-energy conversion. *Chem. Asian J.* 8: 2395–2400.

CHAPTER 40

CONTENTS

Chemical Modification of Graphene and Applications for Chemical Sensors

40

Surajit Kumar Hazra and Sukumar Basu

40.1 INTRODUCTION

40.1.1 General Introduction and Literature Review

The electronic configuration of carbon atom is $1s^2 2s^2 2p^2$. The net spin and orbital moment of the unpaired electrons in the p-orbital are "S = 1" and "L = 1," respectively. This yields the ground state with minimum $J = L - S = 0$ ($^3P_0$). In order to satisfy the tetravalent criteria, an "s" electron must be transferred to a vacant 2p-state to yield $2s^1 2p^3$ unsaturated configuration. This is the so-called sp^3-hybridized electron states. Normally, the tendency of atoms or molecules to coalesce in order to form solids leads to reduction of energy due to overlap of electron wave functions, which ultimately result in the formation of energy bands. The energy difference between the free states of atoms and bound states in a solid is normally utilized for the transfer of electron from "s" to "p" orbital. These results in the formation of covalent bonds via sp^3-hybridized orbital and the bonds are directed along the axes of tetrahedron geometry.

The second alternative to this geometry is to form three sp^2-hybridized orbital and dispense one p-orbital electronic charge evenly in the whole structure. So, the three hybridized orbitals form strong covalent bonds, whereas the distributed charge forms a weak π-bond. This yields graphite, another allotrope of carbon.

Graphene is nothing but a honeycomb single-layer graphite lattice having three strong bonds on a plane like joining the corners of a triangle. The π-states are important for stacking the layers, which finally yield bilayers, trilayers, or multilayers of graphene. The π-electrons are highly mobile and are equally distributed on the top and the bottom sides of a graphene sheet. These electrons introduce the concept of valence and conduction bands in graphene by the formation of bonding and antibonding π-orbital. The electronic structure of graphene can be realized as the conical-shaped valence and conduction bands that meet at a single point. This is also called the Dirac crossing energy. So, pure graphene has band structure similar to metals or semimetals. This overlap can be separated by introducing carrier concentration difference between layers. For example, if graphene layers (say two in number) are transferred onto a substrate like SiO_2 or SiC, a possibility of transition from semimetal to semiconductor exists. Basically, this is due to the fact that the electrons are transferred from the substrate to the graphene layer near the interface, which creates a difference in conduction electron density. As a result n-type semiconductivity can be realized due to the presence of excess negative conduction charges. This situation is similar to a very small separation in the overlapped band structure of graphene, giving rise to a small bandgap. The n-type conductivity can also be realized by doping graphene with a material like potassium having single electron in its outer valence shell. So, modulation of charge concentration helps in changing the Fermi state occupancy and the shift in the Fermi level, which eventually changes the mobility energy gap of the charge carriers in graphene. Moreover, the mobility energy gap in graphene can also be accomplished by introducing defects and dopant by applying electric field or by treating with the gases (Balog et al. 2010, Dong et al. 2009, Elias et al. 2009, Ohta et al. 2006, Rudberg et al. 2007, Son et al. 2006, Zhang et al. 2009, Zhou et al. 2008).

The high strength of carbon–carbon sigma bonds imparts stability to the graphene lattice from external perturbations like temperature and pressure. Basically, graphene is a unique material due to its single-layer thickness of ~0.345 nm. Despite such a small thickness, graphene has good mechanical and electrical properties. The fascinating fact about graphene is its electronic mobility, which can be theoretically as high as 200,000 cm^2/(V·s) (Park and Ruoff 2009). Of course, the experimental value of mobility is many times lower in comparison to the theoretical value. Such high mobility implies that electrons in graphene enable ballistic transport with minimum or negligible scattering. However, this transport is possible in only solo graphene layers. When graphene is transferred onto a substrate, the mobility is reduced. For example, the mobility of electrons in graphene on SiO$_2$ substrate is almost one-fifth of the theoretical value. The mechanical strength of graphene is outstanding due to the inherent bonding of carbon–carbon chain with a tensile strength of 130 GPa (Lee et al. 2008). Graphene has light weight, which makes it suitable for versatile applications. Apart from mechanical and electrical attributes, graphene can be optically useful because it can act as a good white light absorber. A single layer of graphene can absorb ~2.3% of sunlight. Multilayer graphene can filter most of the solar radiation that passes through the material.

The easiest method for graphene synthesis is mechanical exfoliation. In this method, exfoliation of highly ordered pyrolytic graphite (HOPG) yields graphene with good electronic properties. However, the graphene, synthesized by this method, lacks good lateral length (maximum of ~10 μm), which limits its use for wafer-scale utilization (Wei and Liu 2010). So, the other methods of graphene formation gain more attractions. Single crystalline SiC was heated at high temperatures to produce graphene (Emtsev et al. 2009). In this method, SiC decomposes and Si evaporates leaving behind free carbon atoms that subsequently arrange into the honeycomb graphene lattice. But the transfer of films onto other substrates is difficult in this method (Reina et al. 2009). So, apart from synthesis of graphene on the large substrate area, the transfer of graphene to other substrates is also a competitive challenge. Moreover, this transfer process can induce cracks and wrinkles thereby deteriorating the quality of graphene. To avoid this transfer limitation, an alternative method is to deposit graphene onto the insulating substrates. However, a thin metal film is deposited onto the insulating substrate beforehand to catalyze and to improve the quality of graphene layer.

Chemical vapor deposition (CVD) technique was employed by Alfonso et al. to deposit graphene on the metal substrates (Sun et al. 2012). The formation of graphene based on the reaction of carbon with different metal substrates may be due to both strong and weak interactions depending on the nature of the metal. The strong interaction is normally with metals like Ni and Co that impart the smallest separation of ~2.1 Å, which is even smaller than the planar separation in graphite. The second category of metals has weak interactions and the separation is little higher (~3.3 Å). Cu, Ag, Au, and Pt are included in the weak interaction category (Batzill 2012). Copper is a good choice for graphene deposition because the solubility of carbon in copper is negligibly small and so monolayer graphene can be easily obtained.

The grain morphology of the substrate also determines the quality of graphene deposited by CVD process (Huang et al. 2011). The metallic films on insulating substrates are polycrystalline in nature with varying grain sizes and abrupt grain boundary junctions. The grain boundaries weaken the strength of the graphene film. Also, the graphene film is likely to get damaged if the film transfer is attempted to other substrate. To obtain large defect-free graphene film, the substrate should have large grains and minimum number of grain boundaries (Li et al. 2010). Apart from the morphology, the thickness of the metal film also plays an important role. Thick metal films can induce thermal strain during the growth process, whereas thin metal films lead to the formation of micro- and macrodefects. Hence, the surface morphology and thickness of the metal films are of utmost importance for the yield of good-quality graphene thin films by CVD method (Kalbac et al. 2012, Tian et al. 2014).

Graphene and graphene oxides (GO) are also prepared by chemical solution methods, but they have limitations for large-scale productions and commercial applications.

Energy is an important concern in today's world. Electrochemical energy conversion and storage using rechargeable batteries, fuel cells, and super capacitors must focus on the improvement of these devices. Graphene is a new material with good electrical characteristics and has the potential to replace graphite, normally used in these devices (Allen et al. 2010, Brownson and Banks 2010, Geim 2009). The special traits of graphene for this purpose are high electronic conductivity, high thermal conductivity, high mobility, and high surface area. The absence of bandgap in graphene limits its electronic device applications. So, GO and reduced graphene oxide (rGO) gain importance. rGO is basically a π-conjugated graphene sheet, which can have equivalent conductivity of pristine graphene. Functionalization can further improve the conductivity and other properties of graphene. For fuel cell applications, catalytic metal and nitrogen-doped graphene matrix play a crucial role for oxygen reduction reaction, and the cell eventually shows the good electrocatalytic activity and operational stability (Jafri et al. 2010, Liang et al. 2012, Zhang et al. 2010). This improvement is due to increased electrical conductivity and carbon–catalyst binding. Sulfur-doped graphene matrix is also sometimes suitable for fuel cells (Yang et al. 2012). In the rechargeable lithium ion batteries, functionalized graphene or N-graphene is used as anode material to increase the charge capacity (Cho et al. 2011, Reddy et al. 2010), and there is a significant improvement over pure graphene. Even the supercapacitors using N-graphene (synthesized by plasma treatment) gave very high capacitance with almost four times that of pure graphene (Jeong et al. 2011). So, nitrogen-doped graphene is an efficient platform for versatile applications.

40.1.2 Scope of Graphene and Chemically Modified Graphene

During the past decade, the potentiality of carbon nanostructures has been established. The fundamental and applied research efforts utilizing carbon nanostructures are directed toward harnessing the field applications. Graphene is an outstanding 2D material comprising of carbon atoms with immense capability for electronic (both high speed and flexible), optoelectronic, spintronics, health and environment, nanocomposites, and sensing applications. The recent success in graphene is due to the cumulative research efforts ever since the discovery of carbon nanotubes (CNTs). Furthermore, the similarity of graphene structure with boron nitride has paved the way to synthesize ternary nanocomposite of boron–carbon–nitrogen. The speculated growth rate for graphene is ~52%, and this implies the market turnover of the order of $986.7 million by the end of 2022 (http://www.bccresearch.com).

Graphene-based capacitor segment and structural materials are expected to have compound annual growth rate of ~ 67% and ~30%, respectively. The highlights of technological innovations will have an explosive effect in the research and development sector of both the developed and the developing countries, although the fast replacement of silicon by graphene in the field of electronics is a mammoth task and is almost impossible. So, slow integration of graphene with silicon electronics and eventually complete replacement of silicon is feasible in the long run for better performance. Therefore, it appears that this excellent thin conductor has tremendous scope in the regime of science, engineering, and technology.

Chemically modified graphene (CMG) is more versatile than pristine graphene. This can easily be synthesized from GO or rGO owing to the presence of oxygen functionalities that facilitates the process of chemical functionalization by attaching other useful molecules.

Presently, a large amount of work is going on the composites based on polymer and graphene (Salavagione et al. 2011, Zheng et al. 2013). Strong bonds between graphene and polymers are formed mainly due to the presence of oxygen functionalities. The dispersion of graphene in polymer is important to have good microstructure. Such composite introduces special attributes like increase in the operating temperature, reduction in moisture contamination, increase in compressive strength, and lightning strike protection. Rollable e-paper is another attractive product of this composite for electronic applications.

Functionalization of graphene is recently adopted to develop chemical gas sensors and biosensors. The existing sensor devices based on other materials can be easily replaced by graphene and functionalized graphene. Some issues like low sensitivity, selectivity, and moisture poisoning of pure graphene can be addressed in respect of chemical functionalization. For example, chemically functionalized graphene yields fast sensor devices with the ability to detect biomolecules like glucose, cholesterol, hemoglobin, and DNA (Kuila et al. 2011).

Due to the large surface, delocalized π-electrons, chemical purity, and easy functionalization, graphene or modified graphene is very useful in bio-related applications like drug delivery. In this application, binding of drug molecules for sufficient distributed loading and subsequent release profiles are important. Since graphene is lipophilic, the concept of hydrophobicity and hydrophilicity can be harnessed to face the challenges in drug delivery. The other interesting applications are tissue engineering and regenerative medicine (Nayak et al. 2011).

The most important aspects for biomedical applications of graphene are biocompatibility and toxicity. Although the properties of graphene improve upon functionalization, the compatibility and toxicity issues need to be addressed prior to any application. More intensive research is required in this area to implement the user-friendly biomedical kits based on chemically functionalized graphene.

40.2 STRUCTURE OF GRAPHENE

40.2.1 Pristine Graphene

Pristine graphene is a perfect 2D monolayer of carbon atoms arranged in honeycomb configuration. It is defect free and is considered as an ideal structure. Normally, pristine graphene has no bandgap, and hence it behaves like a metal. The zero bandgap is the limitation of the use of pristine graphene for electronic applications. Epitaxial growth of graphene on the metallic surfaces, acting as catalysts, is a common technique used for the synthesis of pristine graphene. However, the perfect ideal lattice is sometimes difficult to obtain by this technique, but the electronic properties can be realized from this deviation from the perfect crystal lattice structure. Pristine graphene is hydrophobic in nature, and so it is difficult to disperse it in water. But the solubility can be improved by modifying the surface with the functional groups.

40.2.2 Multilayer Graphene

It is a challenge to obtain a single-layer graphene experimentally. On the other hand, it is relatively easy to obtain few-layer graphene (thickness ~10 nm) or a thick stack of multilayer graphene (thickness ~50 nm). While few-layer graphene can be viewed using a high-resolution microscope, the multilayer graphene of thickness ≥50 nm is visible under normal optical microscope.

The electrical conduction between multilayer graphene and the metals (e.g., palladium, aluminum, gold, titanium, and copper) was studied by Kuroda et al. (2011). While for some metals the conductance of the metal–graphene junctions decreases exponentially, the conductance for other metals gets saturated with the thickness of graphene film. For thick films of graphene, this variation was attributed to the momentum mismatch between the bulk states of the electrode metal and the graphene film. On the other hand, the bonding between the metal and the graphene layer dominates for the thin graphene films and gives rise to the metallic dependence of the conductance and negligible dependence on the number of graphene layers.

40.2.3 Graphene Oxide and Reduced Graphene Oxide

In presence of strong oxidizing chemicals, graphite is oxidized by the addition of oxygen functionalities (Figure 40.1). This makes the material hydrophilic, and as a result, it becomes miscible with water. Then the chemical exfoliation technique in water under sonication yields few-layer GO. The tendency of GO to easily disperse in the solvents (both organic and inorganic) makes it a very useful material to develop composites with ceramic or polymer materials for modulating their mechanical and electrical properties. The sp^2-hybridized bonding is perturbed in GO due to the addition of oxygen. This also destroys the initial honeycomb symmetry of the graphene lattice. As a result unlike graphene, GO has high electrical resistivity. In order to regain the initial conductivity of graphene, GO needs to be modified. GO can form a good bonding network in foreign matrices like

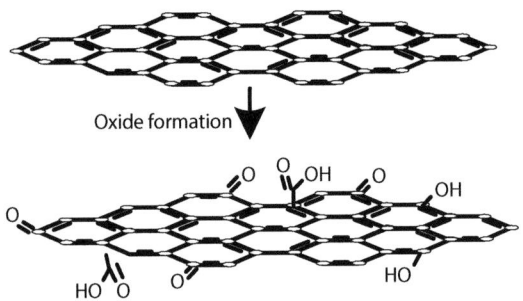

FIGURE 40.1
Oxidation of graphite/graphene to graphene oxide.

ceramics or polymers and show substantial improvement in the intrinsic conductivity of ceramics or polymers. Functionalization of GO using organic groups like porphyrin- and fullerene-based amines can change its properties for possible applications in optoelectronics and biodevices for drug delivery. Porphyrins have drawn considerable interest in molecular electronics due to their excellent electrical (charge transport) and optical properties (absorption and emission) (Otsuki 2010, Tao et al. 1995, Wintjes et al. 2007, Yokoyama et al. 2001). The $\pi-\pi$ loading of especially iron- and zinc-based porphyrins on graphene has influential effect on the electronic properties (Arramel et al. 2013). Scanning tunneling microscopic measurements have revealed semiconducting behavior of porphyrin-functionalized graphene substrates, which clearly implies the existence of an energy gap. The amount of this energy gap can be modulated via the variation of metal core of the organic porphyrin molecules.

Graphene sheets of good quality are necessary for large-scale implementation. The effective way to have large amounts of good-quality graphene is by the reduction of GO to rGO. The process parameters for the conversion of GO to rGO need to be optimized such that the quality of rGO is equivalent to that of pristine graphene obtained by other methods. The reduction can be achieved by employing techniques like treating GO with hydrazine hydrate at 100°C for 24 h or exposing GO to hydrogen plasma or by heating GO to very high temperatures in a furnace. Electrochemical reduction is also a suitable technique to obtain rGO (Zhou et al. 2009). The electrochemical reduction starts in the presence of buffer solution (sodium phosphate, pH ~ 4.0) at 0.6 V. The process is completed by increasing the voltage to 0.87 V, which also results in an increase in the current. The reduction of GO is evident from the color change from brown to black (Zhou et al. 2009). If the carbon to oxygen ratio is very high in rGO, the conductivity will be higher than silver metal. Functionalization of defective rGO is sometimes beneficial to have better properties since it provides the local sites to link chemical functional groups. This is not possible in pristine graphene due to its near perfect lattice. Such surface functionalization of graphene and subsequent incorporation in other material matrices help to widen the applicability of graphene.

40.3 MODIFICATION BY COVALENT AND NONCOVALENT BONDING

Chemical functionality of graphene is necessary to achieve its enhanced performance for different applications. Pristine graphene does not have surface functional groups, and it is a zero bandgap material that is practically unsuitable for electronic applications. So it is necessary to functionalize graphene in order to have new properties along with its pristine characteristics. The main advantages of functionalization of graphene are (1) increase in solubility of graphene in water and other solvents to make the processing steps easier for graphene-based applications and (2) increase of compatibility with other materials like polymers for the synthesis of composites. Mostly, two categories of functionalization are reported for graphene, for example, one with covalent bonds and the other by van der Waals forces. The covalent mode of functionalization allows easy modulation of structural and electrical parameters.

Different density functional codes are used to analyze the covalently modified graphene (Janotti et al. 2001, Marini et al. 2006, Rydberg et al. 2003). For noncovalent functionalization, local density approximation has been adopted (Perdew and Zunger 1981). The third category of functionalization is via ionic bonds. This is effective for intercalated compounds and graphene on metal surfaces (Dresselhaus and Dresselhaus 2002, Enoki et al. 2003, Helveg et al. 2004, Kim et al. 2008, Santos et al. 2008). The ionic bond functionalization is aimed to develop very high conductivity, equivalent to superconductors, in graphene (Weller et al. 2005).

The variation of the energy gap of graphene depends on the type of functionalization. CMG shows low energy gap with some functional groups like NO, NH_2, CN, CCH, and OH, while a wide energy gap of higher than 3 eV is obtained when modified with some other chemical groups like H, F, and Cl as was calculated by Boukhvalov et al. using computation techniques (Boukhvalov and Katsnelson 2009). This difference arises due to the lengthening or shortening of carbon–carbon bond length depending on the repulsive or attractive interactions, respectively, in the graphene functional group assembly. A common example is the experimental difficulty faced to completely functionalize graphite oxide (Boukhvalov and Katsnelson 2008a). This is probably due to the repulsive interaction between functional groups on consecutive carbon atoms of one side of the graphene layer. Due to these repulsive interactions, graphene is generally not a good material for electronic applications due to its stability problems. However, the functional group like fluorine plays an important role to stabilize graphene because it provides a homogeneous coverage. So, graphene functionalized with fluorine can yield wide energy gap and high electron mobility. The high electron mobility is a direct result of wide energy gap due to negligible degree of disorder between layers. However, if graphene has defects, the corrosive nature of fluorine can induce destructive interaction in the graphene lattice at those defect sites after functionalization. This property of fluorine is sometimes used to split nanotubes (Kudin et al. 2001a,b).

Hydrogenation and dehydrogenation of graphene (graphene–hydrogen composite) is an important covalent phenomenon, extremely useful for hydrogen storage application. Basically, hydrogenation of carbon is getting prominence since the use of fullerenes for hydrogen storage applications (Haufler et al. 1990, Henderson and Cahill 1992, 1993, Kroto et al. 1985). Apart from the storage purpose, magnetic behavior of hydrogenated fullerenes ($C_{60}H_x$) is important due to unpaired electrons (Kvyatkovskii et al. 2005). However, the stability of these hydrogenated fullerenes is a concern for magnetic applications if "x" in $C_{60}H_x$ is an even number that destroys the magnetic property (Antonov et al. 2002). Chirality is another important factor for chemical functionalization of the nanostructures (Chang et al. 2008, Gowtham et al. 2008, Tasis et al. 2006). Basically, single-wall CNT can be considered as a scroll of 2D single-layer graphene. So, graphene is a suitable surface for understanding the aspects of chemical functionalization (Lehtinen et al. 2004, Sluiter and Kawazoe 2003, Stojkovic et al. 2003). The existence of similarly charged hydrogen species can result in the repulsive interactions due to which the nanostructures can be distorted, and this can lead to directional hydrogenation properties (Buchs et al. 2007). The hydrogenation of graphene was extensively studied to understand the electronic and the magnetism aspects (Duplock et al. 2004, Lehtinen et al. 2004). The existence of defects like monovacancy and interlayer carbon atom in graphite was also considered to understand these aspects. It can be speculated that the existence of these defects can lead to the creation of energy gap in graphene bilayer by making a difference in carrier density. The hydrogen coverage was also modeled to reveal the most stable configuration in graphene and hence in nanotubes (Boukhvalov et al. 2008, Sluiter and Kawazoe 2003). If the coverage is complete graphane, the hydrogenated graphene is obtained as shown in Figure 40.2.

The magnetism in the functionalized graphene is analyzed on the basis of the interaction of the unpaired electrons at the carbon lattice sites. The structure can be considered as the existence of two sublattices in one lattice. A group of three hydrogen atoms (two in one lattice and one in the other) can be considered to understand the magnetic nature of this carbon–hydrogen system (Yazyev and Helm 2007). The sublattices can be stacked one over the other, which can lead to sp^3 hybridization instead of sp^2 hybridization (Sofo et al. 2007). This transformation is due to the breaking of π-bond, which may happen easily upon chemisorptions of hydrogen. This change may also yield wider energy gap of ~3 eV in the energy spectrum and a small cohesive energy of 0.4 eV. The relatively lower magnitude of cohesive energy may help the hydrogenated system to recover back to graphene efficiently (Boukhvalov and Katsnelson 2008a). The unpaired π-electron charge is now distributed in one sublattice and gives rise to a magnetic moment. Upon chemisorptions of the next hydrogen

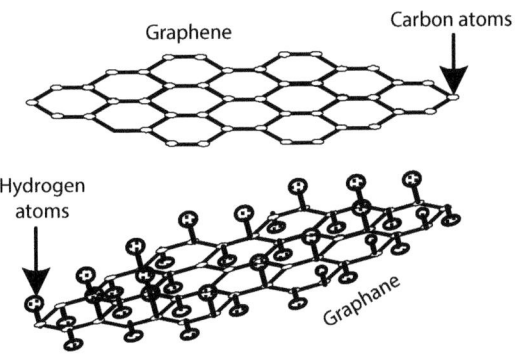

FIGURE 40.2
Structures of graphene and hydrogenated graphene (graphane).

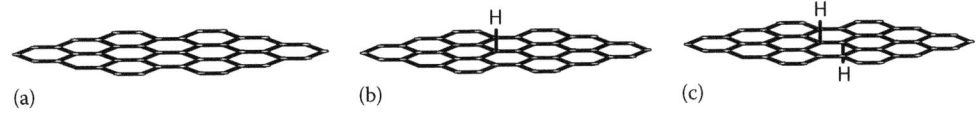

(a)   (b)   (c)

FIGURE 40.3
(a) Pristine graphene. (b) Graphene with one adsorbed hydrogen atom. (c) Graphene with a pair of adsorbed hydrogen atoms.

atom, this distributed charge of the unpaired electron is quenched, and as a result, the magnetic moment disappears. So, this will yield a surface devoid of unpaired electrons or dangling bonds. At the same time, the incorporation of hydrogen in the carbon lattice can create atomic displacements, which can lead to distortion of the crystal lattice. This distortions lead to strong graphene–substrate interactions (Coraux et al. 2008, Du et al. 2008, Preobrajenski et al. 2008, Vazquez et al. 2008). For stability of the chemisorbed hydrogen on the exposed surface of graphene, two hydrogen atoms must bond with the carbon atoms placed at opposite corners of the hexagonal honeycomb lattice (Boukhvalov and Katsnelson 2008b). And for the chemisorbed hydrogen in layered graphene assembly, the distortion will make an energetically stable configuration provided the second hydrogen is attached to the neighboring carbon atom on the other side of the graphene layer in the stacked configuration (Figure 40.3).

As already mentioned, the zero bandgap of graphene is a limitation for its application in electronics. A solution of this problem is the noncovalent functionalization of graphene. The pi-electrons are responsible for the bonding. Basically, the pi-interaction is related to the benzene structure and can be categorized to H–π, metal–π, π–anion, and π–π interactions. π–π interaction is dominant in graphene due to large contact area between two layers. The metal–π interaction is governed by electrostatic forces. The metal ions act as cationic centers and have strong attraction for the negatively charged π-electron. However, the π–anion interaction is repulsive due to the negative charge on both electron and anion.

Functionalization of graphene with transition metals is useful for hydrogen storage (Sofo et al. 2007), for the modification of the electronic structure of bilayer graphene (Robinson et al. 2008), and for the spintronic applications (Roman et al. 2006). The graphene surface can also be functionalized by using aromatic molecules like pyrene (Py) derivatives, which employs π-conjugation (An et al. 2011, Cheng et al. 2011, Jang et al. 2012, Kodali et al. 2010, Lopes et al. 2010, Malig et al. 2012, Parviz et al. 2012, Song et al. 2013, Su et al. 2009, Wang et al. 2009, Zhang et al. 2011). These extra functionalities improve the water solubility of graphene, the efficiency of graphene solar cells, and other graphene-based devices (Cheng et al. 2011, Jang et al. 2012, Su et al. 2009, Wang et al. 2009). Similarly, the chemical functionalization of graphene is possible with other aromatic molecules like coronene carboxylate for nanoelectronic applications (Ghosh et al. 2010), sodium derivative of pyrenene-1-sulfonic acid for solar cell applications (Zhang et al. 2011), 1-pyrenecarboxylic acid for selective ethanol sensing (An et al. 2011), and pyrenebutanoic acid, succinimidyl ester for biosensors (Kodali et al. 2010). The type of conductivity in graphene can also be modulated via noncovalent functionalization. Polyethylenimine and 1-nitropyrene act as n-type and p-type dopant for graphene, respectively (Cheng et al. 2011). Certain functionalities like 1-pyrenebutyric acid can help in the formation of uniformly dispersed polymer composites with graphene (Song et al. 2013). Biocompatibility of graphene can be achieved by heparin functionalization (Lee et al. 2011). Likewise, many other functional groups can be added to graphene to improve its properties for various applications.

40.4 CHEMICAL SENSOR APPLICATIONS

Chemical sensor research with CNTs has demonstrated the tremendous potential for the ppb level detection of various gaseous species (Kong et al. 2000) that are important for health, security, and safety. The unzipped CNT, that is, graphene, is also expected to have good sensor characteristics.

The gas adsorption phenomenon on pristine graphene was theoretically studied for water vapor, ammonia, carbon monoxide, and nitrogen dioxide (Lin and Fang 2013). The maximum adsorption capacity (n) of nitrogen dioxide was found at room temperature following the potential energy curves of the NO_2 molecules on graphene. Again, the experimental value was found to be two orders of magnitude more than the theoretical value ($n = 10^8$ cm^{-2}) for graphene-based devices. This may be attributed to the presence of terminating oxygen atoms on graphene surface, which help in the strong binding of NO_2 molecule on the real experimental surface (binding energy ~ 1 eV) (Lin and Fang 2013). For water vapor, ammonia, and carbon monoxide, theoretical value of "n" is of the order ~10^7 cm^{-2} that is less than NO_2 (Lin and Fang 2013). Similar information on the theoretical and experimental adsorption studies can provide an insight of the gas-sensing mechanism of graphene-based devices. However, it can also be inferred that the presence of oxygen functionalities on the exposed graphene surface is natural, and it can be useful to increase the solid–gas interactions.

Normally, two categories of device configurations can be adopted for gas sensor studies. For a planar-functionalized graphene film on a substrate, two parallel thin electrical contacts can be deposited by physical vapor deposition techniques, for example, e-gun or thermal evaporation (Figure 40.4a). The contacts can be taken either in the form of small-diameter dots or parallel lines. The measurements can be done on the basis of the change of either resistance or current. If the functionalized graphene is interactive with the gas molecules, noncatalytic metal (like gold) can be used for metallic contacts. However, if the graphene matrix is insensitive to gases, the catalytic metal contact either with palladium or platinum may be preferred. Sometimes, the sensing device can be fabricated in the form of a vertical junction between graphene and another semiconductor (Figure 40.4b). In that case, metal contacts are to be taken from both sides of the vertical device.

Normally for vertical devices, one contact is taken from the base of the substrate and other from the top semiconductor. In case, the graphene layer is grown on SiO_2/Si substrate; the insulating SiO_2

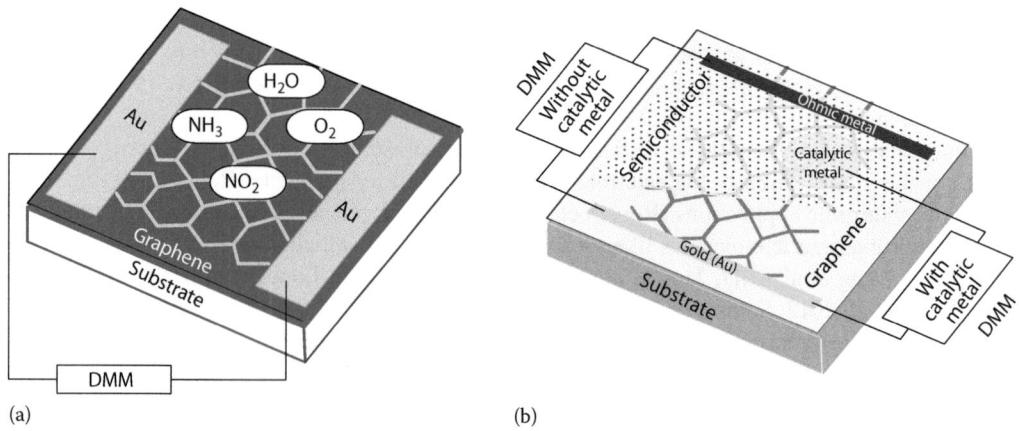

(a)   (b)

FIGURE 40.4

(a) Planar device with parallel electrodes. (b) Vertical heterojunction with or without catalytic contact.

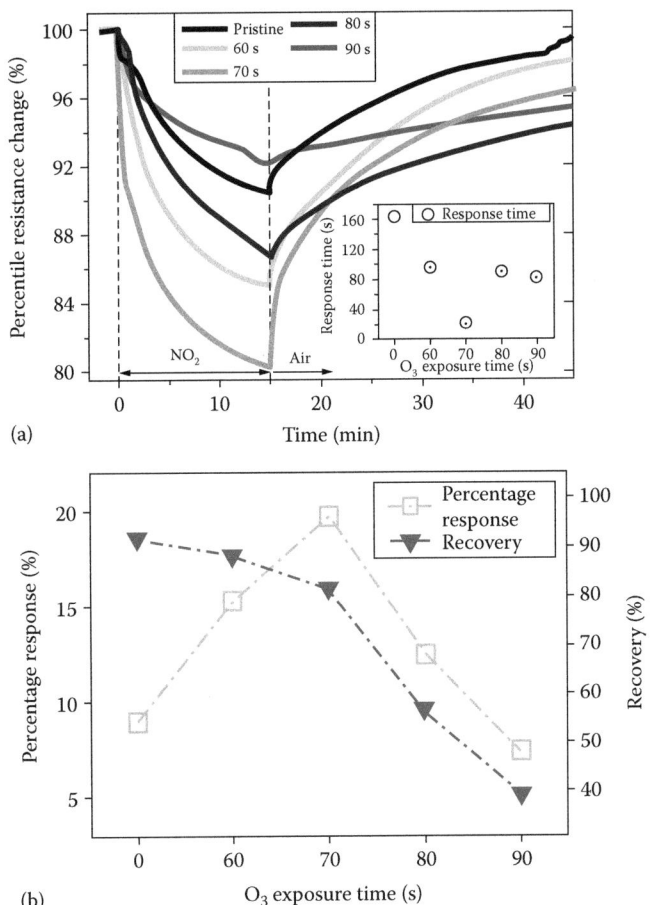

FIGURE 40.5
Sensing performance of graphene with respect to the ozone treatment time. (a) The percentile resistance change of the devices when exposed to 200 ppm NO_2 at room temperature. The NO_2 gas mixed with dry air was injected for 15 min to measure the sensor signal. Dry air was supplied into the tube for 30 min for recovery. The inset depicts the response time with respect to the time of ozone treatment. (b) The percentage response and recovery variations with respect to the ozone treatment time. (Reprinted from *Sens. Actuators B*, 166–167, Chung, M.G., Kim, D.H., Lee, H.M. et al., Highly sensitive NO_2 gas sensor based on ozone treated grapheme, 172–176, Copyright 2012, with permission from Elsevier.)

layer prevents charge carriers to reach the silicon substrate. Therefore, one contact is taken from the top graphene layer that is preserved by partial masking during the growth of the top semiconducting film (Figure 40.4b).

The difference in the NO_2 sensing properties of pristine graphene and surface-modified graphene with ozone was studied by Chung et al. (2012) (Figure 40.5). The ozone treatment time was varied from 0 to 90 s, and this time difference resulted in variation of oxygen species on the graphene surface. The test gas used in this study was NO_2, which has high potential to withdraw electrons from the graphene matrix thereby increasing the hole density and thus a reduction in the device resistance. The sample treated with ozone for 70 s showed the maximum (19.7%) resistance change, which is much higher (almost double) than the resistance change of pristine graphene. This enhancement in the sensor response is probably due to the increase in the affinity of the molecular interaction owing to the transfer of the electronic charges from the sensing matrix. Similarly, the response time was better by a factor of 8 for ozone-treated samples. The recovery characteristics showed a slow trend with increasing ozone treatment time, indicating the slow desorption of NO_2.

The sensing is mainly due to the oxygen groups, which provide ample favorable sites and these sites have very strong affinity toward NO_2 species. Since the sensors were operated at room temperature and a clear difference in response was observed between pristine- and ozone-treated graphene, it could be concluded that ozone treatment increased the adsorption strength of graphene to

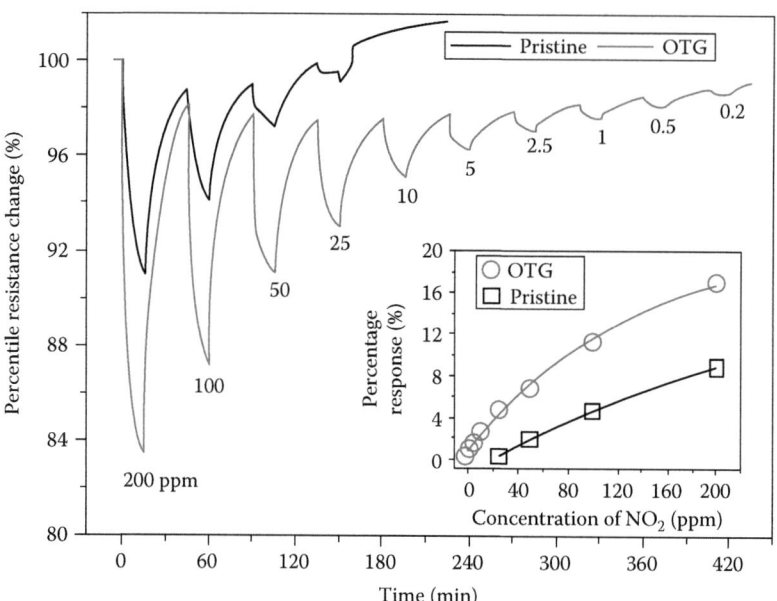

FIGURE 40.6
The percentile resistance changes of the ozone-treated graphene (upto 0.2 ppm) and pristine graphene (upto 25 ppm) sensors. The NO_2 concentration was modulated from 200 ppm to 200 ppb. The inset depicts the correlation between percentage response and concentrations (in agreement with the Langmuir adsorption model). (Reprinted from *Sens. Actuators B*, 166–167, Chung, M.G., Kim, D.H., Lee, H.M. et al., Highly sensitive NO_2 gas sensor based on ozone treated grapheme, 172–176, Copyright 2012, with permission from Elsevier.)

the NO_2 molecules. The deterioration of recovery characteristics is attributed to the hindrance of NO_2 desorption due to the high binding energy between NO_2 molecule and oxygen species on the ozone-treated graphene surface. By exceeding the time of ozone treatment beyond 70 s, the sensor performance (response and response time) was deteriorated due to excessive oxidation of graphene that increased the initial baseline resistance. So, the optimized density of the oxygen-functional group was obtained by treating the graphene surface with ozone for 70 s. The study further revealed a better detection limit of NO_2 for ozone-treated graphene samples (~200 ppb) compared to that for pristine graphene (>10 ppm) (Figure 40.6).

Pure graphene has zero bandgap with excellent electrical conductivity. Therefore, it is not suitable for the development of electronic nose. rGO is almost similar to graphene except that the surface contains oxygen radicals. Sensor studies with vertical CNTs/rGO hybrid films on polyimide substrates showed good sensor response exhibiting an n-type to p-type transition with NO_2 gas (Jeong et al. 2010). The n-type to p-type transition is probably due to the adsorption of electron-withdrawing NO_2 molecules, which makes the matrix an electron-deficient site. However, the transition could also be due to contact barrier modulation between the metal (Au) and rGO, as reported by Lu et al. for NH_3 sensing (Lu et al. 2009).

Apart from rGO, oxygen-functionalized epitaxial graphene (OFEG) sensors are sensitive to polar chemical vapors at room temperature (Nagareddy et al. 2013). An increase in resistance of the resistive sensors based on OFEG is observed upon exposure to polar protic molecules (ethylene glycol and hydrogen peroxide [H_2O_2]), and the opposite phenomenon of decrease in resistance occurs for polar aprotic chemicals (dimethylformamide, dimethylacetamide, n-methyl-2-pyrrolidone, and acetic anhydride). The sensors responded quickly to the test vapor with ~10 s response time. On the other hand, devices based on nonfunctionalized epitaxial graphene have extremely slow recovery characteristics (~1.5 h) and the sensitivity much less than that of OFEG and it may probably be due to the difference in the magnitude of the dipole moment because of the presence of electronegative oxygen.

The adsorption of hydrogen in carbon was investigated for the purpose of hydrogen storage (Darkrim et al. 2002, Dillon et al. 1997, Züttel et al. 2002). Thereafter, studies with graphene–hydrogen interactions were also conducted to understand the potentiality of graphene as a hydrogen storage material (Ao et al. 2014). The theoretical calculations on the storage properties of aluminum-dispersed graphene were supported experimentally. It was found that 10.5 wt% of hydrogen could be adsorbed under ambient conditions. So, if the surface of graphene is modified by hydrogen, it can be speculated that its electrical properties will change. Furthermore, graphene is a single atom thick structure with all its atoms on the surface. This is supposed to increase the solid–gas interaction for good sensor performance. Since pristine graphene does not have much response for hydrogen, the functionalization is a viable alternative to make graphene an efficient sensor material. The graphene equivalent, called rGO with high carbon to oxygen ratio, is also a good material for hydrogen sensor applications (Hafiz et al. 2014, Wang et al. 2013). rGO can be easily produced in large quantities (Zhang et al. 2014). The most efficient hydrogen sensors were prepared by catalytic metal functionalization of graphene.

A single-layer graphene obtained by exfoliation of HOPG was used as a sensor after depositing palladium on the surface (Lim et al. 2013). The interaction between palladium and hydrogen yielded electrons that partially neutralized the holes in the graphene matrix. An asymmetrical I–V was obtained with fairly good response and recovery characteristics. Palladium-doped rGO chemiresistor showed appreciable response toward 50 ppm of hydrogen, and it was a selective sensor to hydrogen because it did not respond to CO, ethanol, and toluene (Pandey et al. 2013).

Similar hydrogen sensor studies were carried out with palladium- and platinum-functionalized graphene/silicon heterojunction in the reverse bias mode (Uddin et al. 2014). This reverse bias diode sensor showed higher sensitivity than graphene chemiresistor due to a large change in current owing to the barrier modulation. The sensors showed the sensitivity to sub-ppm hydrogen concentration.

Hybrid zinc oxide (ZnO)–graphene nanostructures, fabricated by growing ZnO nanowires on single-layer CVD graphene below 100°C, showed high sensitivity toward hydrogen (Liu et al. 2013). The hybrid device performance is governed by the high single crystallinity of ZnO nanowire and the high-charge carrier mobility of graphene.

Functionalization of GO using oxygen-functional groups was also adopted to test the response toward humidity (Phan and Chung 2012). Rapid thermal annealing technique (between 400°C and 1200°C) was used to control the quantity of attached oxygen-functional groups on GO developed over SiO_2/Si substrates by simply spraying GO dispersion solution. Higher annealing temperature (1200°C) removed most of the oxygen functionalities. Also, the films annealed at high temperatures showed lower resistance relative to samples annealed at low temperatures. The sensitivity toward moisture dropped from 35.3% (shown by as deposited GO) to 0.075% upon increasing the annealing temperature to 1200°C. However, the stability and response time of the device are relatively better and poorer, respectively, for high temperature annealed samples. So, a compromise between sensitivity, time of response, and long-term stability is necessary for the field application of oxygen-functionalized GO-based humidity sensors.

Similar humidity sensing studies were carried out with GO films developed from solution having different GO dispersion concentrations (Zhao et al. 2011). The films were annealed at 50°C for 2 h, and an interdigitated capacitor device configuration was used for the study. The devices showed appreciably fast and monotonic response to different concentrations of relative humidity.

Humidity sensors based on the concept of interlayer electronics revealed amplified response (almost twice) relative to the devices based on mono- and bilayer graphene (Rao et al. 2012). This might be due to the tunneling current between the layers of the stacked geometry because of the microscopic material roughness created via occupancy of gas molecules.

GO-based surface acoustic wave devices on quartz substrates were used to monitor moisture vapor pressure in a fixed volume (Balashov et al. 2013). The results were explained from the adsorption kinetics of moisture molecules by GO layer. Metal oxide–graphene variable capacitor also acted as a convenient sensor device to monitor water vapor concentration (Deen et al. 2014). A simple resonant circuit with metal oxide–graphene capacitor and an inductor was used for the sensing experiment. There was a shift in the resonance frequency of this circuit with the change in the relative humidity values in the range from 1% to 97%. The value of the capacitance from the sensing experiment was correlated with that obtained from the separate capacitance–voltage measurements, which established the fact that the sensing mechanism is basically due to the capacitance change in graphene.

Rao et al. reported that unzipped graphene nanoribbon was sensitive to oxygen and it was based on their electron spin resonance studies (Rao et al. 2011b). Upon exposing the ribbons to oxygen gas, the G_c signal that they considered as the carbon signal (G_c) at the spin g-value of 2.0032 from periphery carbon centers diminished, and it again appeared after removing the oxygen gas at room temperature. Further, the cross response was studied by checking the signal, G_c in hydrogen, helium, nitrogen, and argon ambient. The results indicate that oxygen is reversibly physisorbed giving rise to an appreciable difference in the G_c signal. Therefore, the sensing process may be considered as highly selective to oxygen. There are other reports on the CVD-grown monolayer graphene, yielding appreciable response toward different concentrations of oxygen at room temperature (Chen et al. 2011). In these studies, the device resistance/resistivity was monitored following the increase of the hole conductivity upon adsorption of oxygen molecules for the sensing purpose. The minimum detection limit of this oxygen sensor was 1.25% by volume.

The chemical vapor sensing of graphene and CNTs was compared based on the exposure ratio to the sensing agents by Park et al. (2012a). The sensors were fabricated by using HOPG and CNT, respectively. Good sensor response was registered for graphene. In spite of the smaller surface area of 2D graphene compared to CNTs, the experiments revealed a high potential of graphene-based sensors in terms of sensitivity and response time. The reproducibility of graphene-based sensors was checked by Rivera et al. (2010). Monolayer graphene grown by thermal CVD showed good repeatability of sensing for NO_2, NH_3, and ethanol at room temperature.

Reflectance mode optical sensors using GO were developed to sense different concentrations of ethanol in water (Shabaneh et al. 2013). A multimode optical fiber was selected for this study, and its tip was coated with GO by drop-cast technique. The optical reflectance was reduced by 37% in the presence of 5% ethanol in water. The typical response and recovery times of 30 and 45 s, respectively, were obtained. A composite of functionalized multiwall CNTs and graphene also showed high sensitivity toward ethanol (Kumari et al. 2014). Interlayer nanosensors having electrical contacts on different graphene layers showed a resistance change up to ~185% in 0.5% ethanol in nitrogen ambient (Rao et al. 2011a). This value is higher compared to the intralayer sensors, which showed a maximum resistance change of 22.79% under similar conditions. The enhancement is probably due to the change of tunneling current due to vapor adsorption.

The sensor response toward volatile organic compounds (VOCs) could be improved by functionalizing the graphene surface using cobalt metalloporphyrin (Choi et al. 2013). The response toward 10 ppm toluene was enhanced by 200% relative to pristine graphene at room temperature. The substrate flexibility and the transparency were the added advantages from the mechanical and the optical points of view. Surface functionalization of rGO with magnetic Fe_3O_4 and Ag nanoparticles exhibited high sensitivity and selectivity toward VOCs (Tran et al. 2012).

Graphene/indium nitride (InN) heterojunction diode sensor showed good response toward acetone, NO_2, and water vapor (Wilson et al. 2013). The heterojunction was fabricated by transferring the CVD-grown graphene layer onto InN thin film supported on GaN/sapphire substrate. The heterojunction response was relatively better than simple graphene or InN conductometric sensors. Sensor arrays on the silicon substrate using sensing elements like rGO, single-wall CNTs, and copper oxide nanowires were fabricated to detect volatile organic vapors by analyzing the signature response by the sensor assembly (MacNaughton et al. 2011). The resistance of the sensor array was monitored in the presence and in the absence of organic molecules like isopropyl alcohol, acetone, and aromatic compounds. A huge resistance change (~150%) of this electronic nose was observed due to the interaction of the analyte molecules with the sensing materials.

Graphene, GO, or rGO are potential sensing materials for gases like NO_x, NH_3, CF_4, and CO. A composite of single-walled CNTs and graphene is responsive to NO (Liang et al. 2013). The gas adsorption and desorption kinetics follow the Langmuir model that reveals an optimum operating temperature beyond which the desorption dominates over adsorption. Similar studies were carried out with the hybrid films of CNTs and reduced graphene on polyimide substrates to check the response toward NO_2 gas (Jeong et al. 2010). Surface modification of graphene using platinum nanoparticles was adopted to enhance the response toward low concentration of ammonia (15–58 ppm) (Gautam and Jayatissa 2012a). Analysis of adsorption and desorption curves exposed two different adsorption sites for surface-modified graphene, which is probably responsible for the enhanced response. The ammonia response was also checked with graphene-based field effect transistor (FET) in the temperature range 25°C–100°C (Gautam and Jayatissa 2012b). A shift of the Dirac peak was observed upon exposure to ammonia. The device characteristics further revealed two different adsorption sites for ammonia. rGO-based sensors were responsive to ammonia at

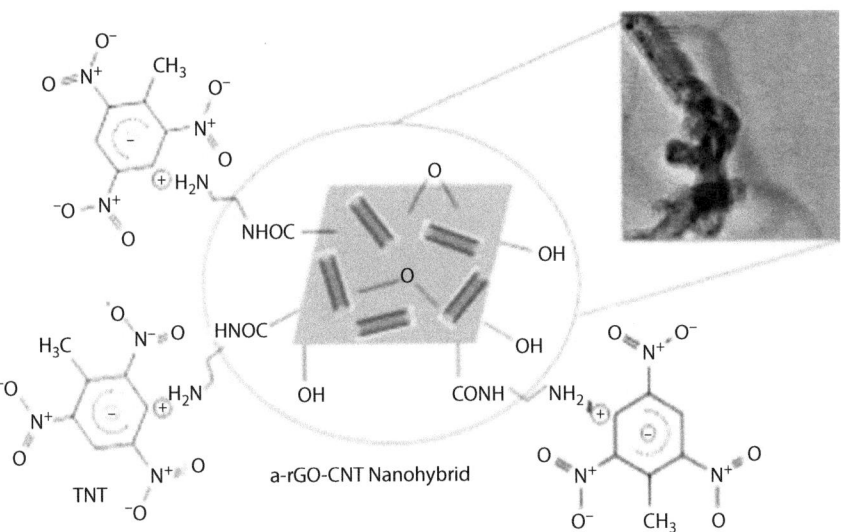

FIGURE 40.7
Amine functionalized reduced graphene oxide and CNT nanocomposite bonded with TNT molecule. (Reprinted from *J. Hazard. Mater.*, 248–249, Sablok, K., Bhalla, V., Sharma, P., Kaushal, R., Chaudhary, S., and Suri, C.R., Amine functionalized graphene oxide/CNT nanocomposite for ultrasensitive electrochemical detection of trinitrotoluene, 322–328, Copyright 2013, with permission from Elsevier.)

room temperature (Hu et al. 2014). Both sensitivity (2.4%) and response time (1.4 s) were good for 1 ppb ammonia. The recovery was amplified with IR illumination. Reproducibility and selectivity were also experimentally observed. Similar ammonia sensors were developed with graphene–silver nanowire composites (Tran et al. 2013) and ink-jet-printed GO (Le et al. 2012). Other gases like CF_4 and CO were detected using graphene-based FETs and pristine graphene, respectively (Lin and Fang 2013, Park et al. 2012b).

Graphene-based sensors can also be useful for detection of explosives for security reasons. Amine-functionalized rGO and CNT nanocomposite chemical sensor can be helpful for detecting TNT (Sablok et al. 2013). The sensing mechanism is based on the bonding of TNT (an electron-deficient molecule) to amine groups (electron rich) on the composite substrate via the formation of Jackson–Meisenheimer complex (Figure 40.7). The detection limit is 0.01 ppb with an excellent repeatability.

Similar explosive sensors based on p-aminothiophenol-functionalized silver nanoparticles dispersed on graphene were tested for the detection of TNT by employing surface-enhanced Raman scattering (Liu and Chen 2013). Electrochemically reduced GO was functionalized with 1-pyrenebutyl-amino-β-cyclodextrin (PyCD-rGO) to selectively sense picric acid, an explosive compound (Huang et al. 2014).

The response curves to different concentrations of picric acid are shown in Figure 40.8.

The selective sensing property of this sensor was tested by using three categories of phenols, viz., p-cresol, 2, 4-xylenol, and 2-nitrophenol. Figure 40.9a shows the response to p-cresol (25–525 µM), 2, 4-xylenol (12.5–562 µM), and 2-nitrophenol (12.5–450 µM). The PyCD-rGO sensor was also responsive to p-cresol, 2, 4-xylenol, and 2-nitrophenol, but the magnitude of response was very low (Figure 40.9b).

So, the PyCD-rGO sensor exhibits excellent selectivity for picric acid. Basically, the rGO was modified with a cyclodextrin (CD) derivative to harness both the hydrophobic and hydrophilic characteristics so that it can act as a good receptor for the organic molecules in aqueous solution. This unique hydrophobic and hydrophilic combination can help to form complex with different hydrophobic organic molecules also. Moreover, the size of β-CD is similar to benzene ring. So, β-CD is preferred for the recognition of picric acid. Also, Py gets loaded on rGO by simple π–π interactions. Molecules of p-cresol, 2, 4-xylenol, and 2-nitrophenol increase the conductance of the PyCD-rGO to a small extent. These molecules have hydrophobic ends that can form complex with β-CD. Also, these molecules have electron-donating –CH_3 group, which reduces the electron-withdrawing

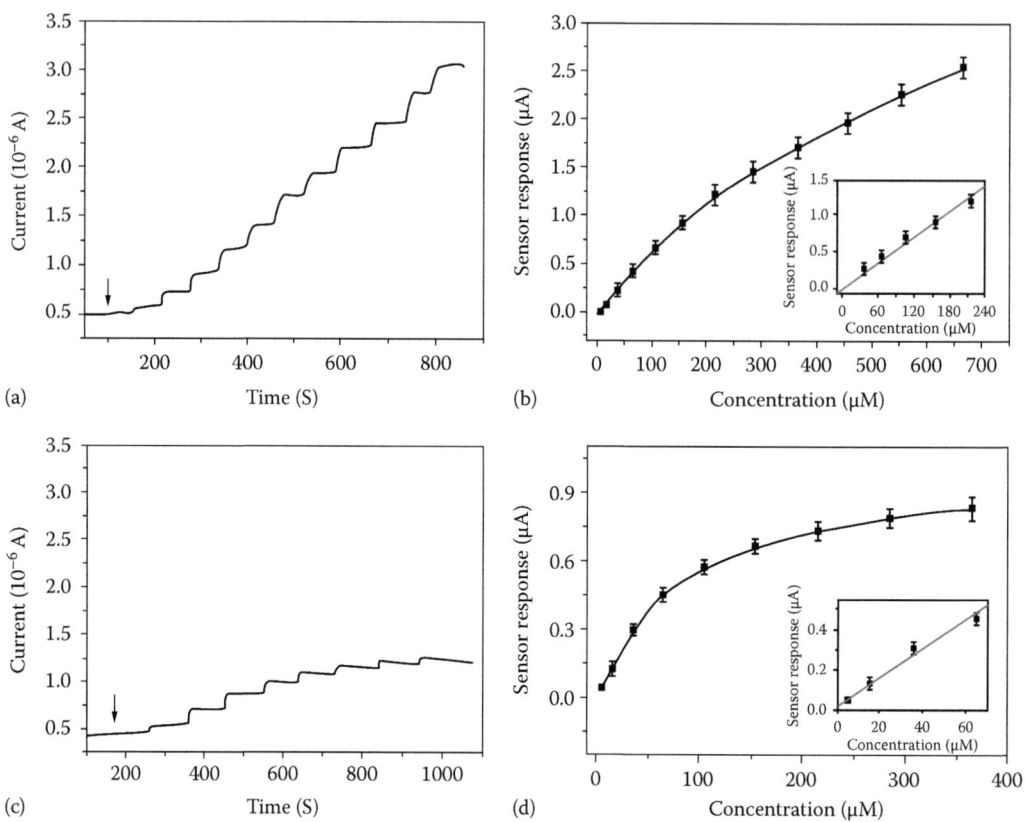

FIGURE 40.8

Real-time response curve and average sensor response are shown for PyCD-rGO sensor (a and b) and for the rGO sensor (c and d) upon exposure to different concentrations of picric acid. The response as measured for three times for each concentration of picric acid. (Reprinted from *Sens. Actuators B*, 196, Huang, J., Wang, L., Shi, C., Dai, Y., Gu, C., and Liu, J., Selective detection of picric acid using functionalized reduced graphene oxide sensor, 567–573, Copyright 2014, with permission from Elsevier.)

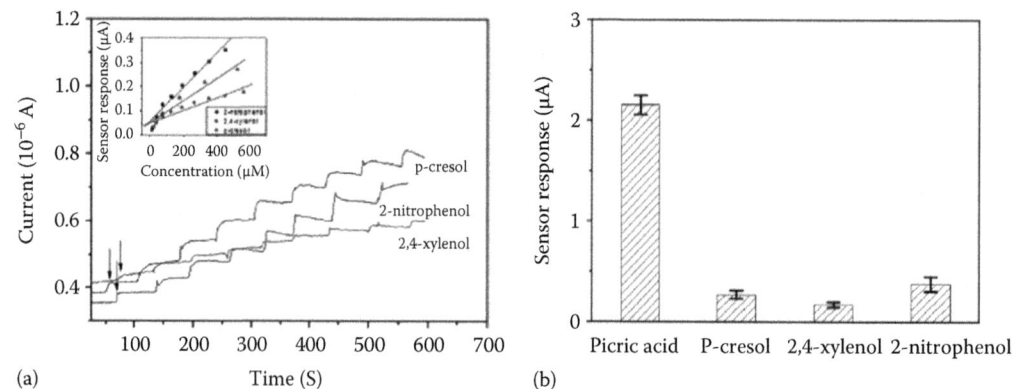

FIGURE 40.9

(a) Real-time response curves and sensor responses (inset) of PyCD-rGO sensor to different concentrations of p-cresol, 2,4-xylenol, and 2-nitrophenol and (b) a comparison of the average responses of PyCD-rGO sensor to 500 µM picric acid, p-cresol, 2,4-xylenol, and 2-nitrophenol. The response for each target was measured for three times. (Reprinted from *Sens. Actuators B*, 196, Huang, J., Wang, L., Shi, C., Dai, Y., Gu, C., and Liu, J., Selective detection of picric acid using functionalized reduced graphene oxide sensor, 567–573, Copyright 2014, with permission from Elsevier.)

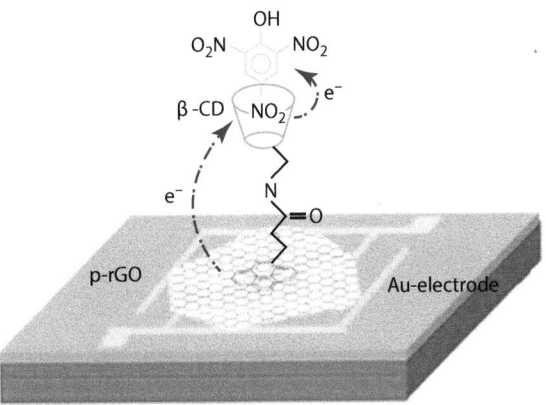

FIGURE 40.10
Schematic view of the sensing layer interacting with picric acid molecule. (Reprinted from *Sens. Actuators B*, 196, Huang, J., Wang, L., Shi, C., Dai, Y., Gu, C., and Liu, J., Selective detection of picric acid using functionalized reduced graphene oxide sensor, 567–573, Copyright 2014, with permission from Elsevier.)

ability of β-CD. The reduction is proportional to the number of available –CH₃ groups. For example, the p-cresol has one –CH₃ group, and 2, 4-xylenol has two –CH₃ groups; so the withdrawing ability of the latter reduces more compared to p-cresol, and hence 2, 4-xylenol has relatively lower sensor response. This is different from the –NO₂ group activity (Figure 40.10). For 2-nitrophenol, the –NO₂ activity is very low. However, for picric acid, the –NO₂ group activity is very high. As a result, PyCD-rGO is very effective to selectively detect picric acid. Also, the reproducibility of these results was checked by testing different other sensor devices based on PyCD-rGO.

The biosensor devices based on functionalized graphene were also fabricated to test glucose, ascorbic acid, H₂O₂, *Escherichia coli*, and cholesterol (Basu et al. 2014, Chang et al. 2013, Li et al. 2014, Pakapongpan et al. 2014, Zhang et al. 2013). For *E. coli* detection, the honeycomb graphene lattice is functionalized with anti-*E. coli* antibodies followed by passivation of the remaining device area by 1 h incubation with 0.1% polysorbate, Tween 20. The antibodies are normally attached with the help of a linking molecule like 1-pyrenebutanoic acid succinimidyl ester. Relatively large conductance change in the device is observed after it is exposed to *E. coli* bacteria at different concentrations. The selectivity is tested by exposing the device to other strains of bacteria. Such devices can also be used to study the glucose-induced metabolic activities of the bacteria.

40.5 CONCLUSIONS

Graphene and its allies like GO and rGO have shown tremendous potential for versatile applications, including chemical sensors. Chemical alteration of the carbon network using different functionalities is quite useful for improving the response, response time, and selectivity toward different test analytes. The surface modification manipulates the carbon–oxygen ratio, which in turn affects the electronic properties of graphene within a single layer or in between the layers for the multilayer graphene. For chemical sensor application, catalytic surface modification is quite useful in the context that additional adsorption sites are available for better response. Also, the focus has shifted toward rGO, an easy synthesizable equivalent to graphene with negligible defects. Furthermore, graphene is compatible to CMOS technology, and hence it is quite useful for fabricating electronic devices with the possibility of subsequent integration. Such devices are expected to function at low operating voltages and with utmost efficiency in sectors like computing, communication, sensors, and energy conversion devices. Continued efforts are being made from both theoretical and experimental approaches to understand more about this unique material and find suitable ways to harness desired applications through chemical surface modifications.

ACKNOWLEDGMENT

Sukumar Basu is thankful to the Coordinator, IC Design and Fabrication Center, Department of Electronics and Telecommunication Engineering, Jadavpur University, Kolkata, India, for the research support.

REFERENCES

Allen, M., Tung, V., and Kaner, B. 2010. Honeycomb carbon: A review of graphene. *Chem. Rev.* 110(1):132–145.

An, X., Butler, T. W., Washington, M., Nayak, S. K., and Kar, S. 2011. Optical and sensing properties of 1-pyrenecarboxylic acid-functionalized graphene films laminated on polydimethylsiloxane membranes. *ACS Nano* 5:1003.

Antonov, V. E., Bashkin, I. O., Khasanov, S. S. et al. 2002. Magnetic ordering in hydrofullerite $C_{60}H_{24}$. *J. Alloy. Compd.* 330–332:365–368.

Ao, Z., Dou, S., Xu, Z., Jiang, Q., and Wang, G. 2014. Hydrogen storage in porous graphene with Al decoration. *Int. J. Hydrogen Energy* 39(28):16244–16251. http://dx.doi.org/10.1016/j.ijhydene.2014.01.044.

Arramel, Castellanos-Gomez, A., and Jan van Wees, B. 2013. Band gap opening of graphene by noncovalent π-π interaction with porphyrins. *Graphene* 2:102–108. http://dx.doi.org/10.4236/graphene.2013.23015.

Balashov, S. M., Balachova, O. V., Braga, A. V. U., Bazetto, M. C. Q., and Pavani Filho, F. November 3–6, 2013. The optimized SAW humidity sensor with nanofilms of graphene oxide. *IEEE Sensors*, Baltimore, MD, pp. 1–4. doi: 10.1109/ICSENS.2013.6688308.

Balog, R., Jørgensen, B., Nilsson, L. et al. 2010. Bandgap opening in graphene induced by patterned hydrogen adsorption. *Nat. Mater.* 9:315–319. doi: 10.1038/nmat2710.

Basu, P. K., Indukuri, D., Keshavan, S. et al. 2014. Graphene based *E. coli* sensor on flexible acetate sheet. *Sens. Actuators B* 190:342–347.

Batzill, M. 2012. The surface science of graphene: Metal interfaces, CVD synthesis, nanoribbons, chemical modifications, and defects. *Surf. Sci. Rep.* 67(3–4):83–115.

Boukhvalov, D. W. and Katsnelson, M. I. 2008a. Modeling of graphite oxide. *J. Am. Chem. Soc.* 130:10697–10701.

Boukhvalov, D. W. and Katsnelson, M. I. 2008b. Tuning the gap in bilayer graphene using chemical functionalization: Density functional calculations. *Phys. Rev. B* 78:085413.

Boukhvalov, D. W. and Katsnelson, M. I. 2009. Chemical functionalization of graphene. *J. Phys. Condens. Matter* 21:344205 (12 pp.). doi: 10.1088/0953-8984/21/34/344205.

Boukhvalov, D. W., Katsnelson, M. I., and Lichtenstein, A. I. 2008. Hydrogen on graphene: Electronic structure, total energy, structural distortions and magnetism from first-principles calculations. *Phys. Rev. B* 77:035427.

Brownson, D. and Banks, C. 2010. Graphene electrochemistry: An overview of potential applications. *Analyst* 135(11):2768–2778.

Buchs, G., Krasheninnikov, A. V., Ruffieux, P., Grönig, P., Foster, A. S., and Nieminen, R. M. 2007. Creation of paired electron states in the gap of semiconducting carbon nanotubes by correlated hydrogen adsorption. *New J. Phys.* 9:275. doi: 10.1088/1367-2630/9/8/275.

Chang, H., Wang, X., Shiu, K.-K. et al. 2013. Layer-by-layer assembly of graphene, Au and poly(toluidine blue O) films sensor for evaluation of oxidative stress of tumor cells elicited by hydrogen peroxide. *Biosens. Bioelectron.* 41:789–794.

Chang, K., Berber, S., and Tománek, D. 2008. Transforming carbon nanotubes by silylation: An Ab initio study. *Phys. Rev. Lett.* 100:236102.

Chen, C. W., Hung, S. C., Yang, M. D. et al. 2011. Oxygen sensors made by monolayer graphene under room temperature. *Appl. Phys. Lett.* 99:243502. http://dx.doi. org/10.1063/1.3668105.

Cheng, H. C., Shiue, R. J., Tsai, C. C., Wang, W. H., and Chen, Y. T. 2011. High-quality graphene p-n junctions via resist-free fabrication and solution-based noncovalent functionalization. *ACS Nano* 5:2051.

Cho, Y., Kim, H., and Im, H. 2011. Nitrogen-doped graphitic layers deposited on silicon nanowires for efficient lithium ion battery anodes. *J. Phys. Chem. C* 115(19):9451–9457.

Choi, J., Pyo, S., Lee, K., Ko, H.-J., and Kim, J. January 20–24, 2013. Transparent and flexible toluene sensor with enhanced sensitivity using adsorption catalyst-functionalized graphene. *Micro Electro Mechanical Systems (MEMS), IEEE 26th International Conference*, Taipei, Republic of China, pp. 512–515. doi: 10.1109/MEMSYS.2013.6474291.

Chung, M. G., Kim, D. H., Lee, H. M. et al. 2012. Highly sensitive NO_2 gas sensor based on ozone treated graphene. *Sens. Actuators B* 166–167:172–176.

Coraux, J. N., Diaye, A. T., Busse, C., and Michely, T. 2008. Structural coherency of graphene on Ir(111). *Nano Lett.* 8:565–570.

Darkrim, F. L., Malbrunot, P., and Tartaglia, G. P. 2002. Review of hydrogen storage by adsorption in carbon nanotubes. *Int. J. Hydrogen Energ.* 27:193–202.

Deen, D. A., Olson, E. J., Ebrish, M. A., and Koester, S. J. 2014. Graphene-based quantum capacitance wireless vapor sensors. *IEEE Sens. J.* 14(5):1459–1466. doi: 10.1109/JSEN.2013.2295302.

Dillon, A. C., Jones, K. M., Bekkedahl, T. A., Kiang, C. H., Bethune, D. S., and Heben, M. J. 1997. Storage of hydrogen in single-walled carbon nanotubes. *Nature* 386:377–379.

Dong, X., Shi, Y., Zhao, Y. et al. 2009. Symmetry breaking of graphene monolayers by molecular decoration. *Phys. Rev. Lett.* 102(13):135501. doi: 10.1103/PhysRevLett.102.135501.

Dresselhaus, M. S. and Dresselhaus, G. 2002. Intercalation compounds of graphite. *Adv. Phys.* 51:1–186. doi: 10.1080/00018730110113644.

Du, X., Skachko, I., Barker, A., and Andrei, E. Y. 2008. Approaching ballistic transport in suspended graphene. *Nat. Nanotechnol.* 3:491–495.

Duplock, E. J., Scheffler, M., and Lindan, P. J. D. 2004. Hallmark of perfect graphene. *Phys. Rev. Lett.* 92:225502.

Elias, D., Nair, R. R., Mohiuddin, T. M. G. et al. 2009. Control of graphene's properties by reversible hydrogenation: Evidence for graphane. *Science* 323(5914):610–613. doi: 10.1126/science.1167130.

Emtsev, K. V., Bostwick, A., Horn, K. et al. 2009. Towards wafer-size graphene layers by atmospheric pressure graphitization of silicon carbide. *Nat. Mater.* 8:203–207.

Enoki, T., Suzuki, M., and Endo, M. 2003. *Graphite Intercalation Compounds and Applications* (New York: Oxford University Press).

Gautam, M. and Jayatissa, A. H. 2012a. Adsorption kinetics of ammonia sensing by graphene films decorated with platinum nanoparticles. *J. Appl. Phys.* 111(9):094317–094319. doi: 10.1063/1.4714552.

Gautam, M. and Jayatissa, A. H. 2012b. Graphene based field effect transistor for the detection of ammonia. *J. Appl. Phys.* 112(6):064304–064307. doi: 10.1063/1.4752272.

Geim, A. 2009. Graphene: Status and prospects. *Science* 324(5934):1530–1534.

Ghosh, A., Rao, K. V., George, S. J., and Rao, C. N. R. 2010. Noncovalent functionalization, exfoliation, and solubilization of graphene in water by employing a fluorescent coronene carboxylate. *Chem. Eur. J.* 16(9):2700–2704.

Gowtham, S., Scheicher, R. H., Pandey, R., Karna, S. P., and Ahuja, R. 2008. First-principles study of physisorption of nucleic acid bases on small-diameter carbon nanotubes. *Nanotechnology* 19(12):125701. doi: 10.1088/0957–4484/19/12/125701.

Graphene: Technologies, Applications and Markets (AVM075B; August 2012) http://www.bccresearch.com/pressroom/avm/global-market-graphene-based-products-reach-$986.7-million-2022; (accessed May, 2014).

Hafiz, S. M., Ritikos, R., Whitcher, T. J. et al. 2014. A practical carbon dioxide gas sensor using room-temperature hydrogen plasma reduced graphene oxide. *Sens. Actuators B* 193:692–700.

Haufler, R. E., Conceicao, J., Chibante, L. P. F. et al. 1990. Efficient production of C_{60} (buckminsterfullerene), $C_{60}H_{36}$, and the solvated buckide ion. *Phys. Chem.* 94:8634–8636.

Helveg, S., López-Cartes, C., Sehested, J. et al. 2004. Atomic-scale imaging of carbon nanofibre growth. *Nature* 427:426. doi: 10.1038/nature02278.

Henderson, C. C. and Cahill, P. A. 1992. Semi-empirical calculations of the isomeric C_{60} dihydrides. *Chem. Phys. Lett.* 198:570–576.

Henderson, C. C. and Cahill, P. A. 1993. $C_{60}H_2$: Synthesis of the simplest C_{60} hydrocarbon derivative. *Science* 259:1885–1887. doi: 10.1126/science.259.5103.1885.

Hu, N., Yang, Z., Wang, Y. et al. 2014. Ultrafast and sensitive room temperature NH_3 gas sensors based on chemically reduced graphene oxide. *Nanotechnology* 25:025502. doi: 10.1088/0957–4484/25/2/025502.

Huang, J., Wang, L., Shi, C., Dai, Y., Gu, C., and Liu, J. 2014. Selective detection of picric acid using functionalized reduced graphene oxide sensor. *Sens. Actuators B* 196:567–573.

Huang, P. Y., Ruiz-Vargas, C. S., van der Zande, A. M. et al. 2011. Grains and grain boundaries in single-layer graphene atomic patchwork quilts. *Nature* 469:389–392.

Jafri, R., Imran, R., and Ramaprabhu, S. 2010. Nitrogen doped graphene as catalyst support for oxygen reduction reaction in proton exchange membrane fuel cell. *J. Mater. Chem.* 20(34):7114–7117.

Jang, A.-R., Jeon, E. K., Kang, D. et al. 2012. Reversibly light-modulated dirac point of graphene functionalized with spiropyran. *ACS Nano* 6:9207–9213.

Janotti, A., Wei, S.-H., and Singh, D. J. 2001. First-principles study of the stability of BN and C. *Phys. Rev. B* 64:174107.

Jeong, H., Lee, J., and Shin, W. 2011. Nitrogen-doped graphene for high-performance ultracapacitors and the importance of nitrogen-doped sites at basal planes. *Nano Lett.* 11(6):2472–2477.

Jeong, H. Y., Lee, D.-S., Choi, H. K. et al. 2010. Flexible room-temperature NO_2 gas sensors based on carbon nanotubes/reduced graphene hybrid films. *Appl. Phys. Lett.* 96:213105. doi: 10.1063/1.3432446.

Kalbac, M., Frank, O., and Kavan, L. 2012. The control of graphene double-layer formation in copper-catalyzed chemical vapour deposition. *Carbon* 50(10):3682–3687.

Kim, G., Jhi, S.-H., Park, N. et al. 2008. Optimization of metal dispersion in doped graphitic materials for hydrogen storage. *Phys. Rev. B* 78:085408.

Kodali, V. K., Scrimgeour, J., Kim, S. et al. 2010. Nonperturbative chemical modification of graphene for protein micropatterning. *Langmuir* 27:863.

Kong, J., Franklin, N. R., Zhou, C. et al. 2000. Nanotube molecular wires as chemical sensors. *Science* 287(5453):622–625.

Kroto, H. W., Heath, J. R., O'Brien, S. C., Curl, R. F, and Smalley, R. E. 1985. C_{60}: Buckminsterfullerene. *Nature* 318:162–163.

Kudin, K. N., Bettinger, H. F., and Scuseria, G. E. 2001a. Fluorinated single-wall carbon nanotubes. *Phys. Rev. B* 63:045413.

Kudin, K. N., Scuseria, G. E., and Yakobson, B. I. 2001b. C_2F, BN, and C nanoshell elasticity from ab initio computations. *Phys. Rev. B* 64:235406.

Kuila, T., Bose, S., Khanra, P. et al. 2011. Recent advances in graphene-based biosensors. *Biosens. Bioelectron.* 26(12):4637–4648. doi.org/10.1016/j.bios.2011.05.039.

Kumari, A., Prasad, N., Kaur, A., Dixit, S. K., Bhatnagar, P. K., and Mathur, P. C. December 17–21, 2014. Effect of surfactant assisted dispersed graphene on sensitivity of MWCNT-alcohol sensor. *AIP Conference Proceedings*, Thapar University, Punjab, India, Vol. 1591, p. 568. http://dx.doi.org/10.1063/1.4872677.

Kuroda, M. A., Tersoff, J., Newns, D. M., and Martyna, G. J. 2011. Conductance through multilayer graphene films. *Nano Lett.* 11(9):3629–3633. doi: 10.1021/nl201436b.

Kvyatkovskii, O. E., Zakharova, I. B., Shelankov, A. L., and Makarova, T. L. 2005. Spin-transfer mechanism of ferromagnetism in polymerized fullerenes: Ab initio calculations. *Phys. Rev. B* 72:214426.

Le, T., Lakafosis, V., Kim, S. et al. October 29–November 1, 2012. A novel graphene-based inkjet-printed WISP-enabled wireless gas sensor. *Microwave Conference (EuMC), 42nd European*, Amsterdam, the Netherlands, pp. 412–415.

Lee, C., Wei, X., Kysar, J. W. et al. 2008. Measurement of the elastic properties and intrinsic strength of monolayer graphene. *Science* 321:385–388.

Lee, D. Y., Khatun, Z., Lee, J.-H., Lee, Y.-K., and In, I. 2011. Blood compatible graphene/heparin conjugate through noncovalent chemistry. *Biomacromolecules* 12:336–341.

Lehtinen, P. O., Foster, A. S., Ma, Y., Krasheninnikov, A. V., and Niemien, R. M. 2004. Irradiation-induced magnetism in graphite: A density functional study. *Phys. Rev. Lett.* 93:187202.

Li, M., Bo, X., Mu, Z., Zhang, Y., and Guo, L. 2014. Electrodeposition of nickel oxide and platinum nanoparticles on electrochemically reduced graphene oxide film as a nonenzymatic glucose sensor. *Sens. Actuators B* 192:261–268.

Li, X., Magnuson, C. W., Venugopal, A. et al. 2010. Graphene films with large domain size by a two-step chemical vapor deposition process. *Nano Lett.* 10:4328–4334.

Liang, S.-Z., Chen, G., Harutyunyan, A. R., Cole, M. W., and Sofo, J. O. 2013. Analysis and optimization of carbon nanotubes and graphene sensors based on adsorption-desorption kinetics. *Appl. Phys. Lett.* 103:233108. http://dx.doi.org/10.1063/1.4841535.

Liang, Y., Wang, H., and Zhou, J. 2012. Covalent hybrid of spinel manganese-cobalt oxide and graphene as advanced oxygen reduction electrocatalysts. *J. Am. Chem. Soc.* 134(7):3517–3523.

Lim, J., Hwang, S., Yoon, H. S., Lee, E., Lee, W., and Jun, S. C. 2013. Asymmetric electron hole distribution in single-layer graphene for use in hydrogen gas detection. *Carbon* 63:3–8.

Lin, X., Ni, J., and Fang, C. 2013. Adsorption capacity of H_2O, NH_3, CO, and NO_2 on the pristine graphene. *J. Appl. Phys.* 113:034306. doi: 10.1063/1.4776239.

Liu, J. W., Wu, J., Ahmad, M. Z., and Wlodarski, W. June 16–20, 2013. Hybrid aligned zinc oxide nanowires array on CVD graphene for hydrogen sensing. *Solid-State Sensors, Actuators and Microsystems (TRANSDUCERS and EUROSENSORS XXVII), Transducers and Eurosensors XXVII: The 17th International Conference*, Barcelona, Spain, pp. 194–197. doi: 10.1109/Transducers.2013.6626735.

Liu, M. and Chen, W. 2013. Graphene nanosheets-supported Ag nanoparticles for ultra-sensitive detection of TNT by surface-enhanced Raman spectroscopy. *Biosens. Bioelectron.* 46:68–73.

Lopes, M., Candini, A., Urdampilleta, M. et al. 2010. Surface-enhanced Raman signal for terbium single-molecule magnets grafted on graphene. *ACS Nano* 4:7531.

Lu, G., Ocola, L. E., and Chen, J. 2009. Reduced graphene oxide for room-temperature gas sensors. *Nanotechnology* 20(44):445502.

MacNaughton, S., Sonkusale, S., Surwade, S., Ammu, S., and Manohar, S. October 28–31, 2011. Electronic nose based on graphene, nanotube and nanowire chemiresistor arrays on silicon. *Sensors IEEE*, Limerick, Ireland, pp. 125–128. doi: 10.1109/ICSENS.2011.6127182.

Malig, J., Ramero-Nieto, C., Jux, N., and Guldi, D. M. 2012. Integrating water-soluble graphene into porphyrin nanohybrids. *Adv. Mater.* 24:800.

Marini, A., Garcia-Gonzáles, P., and Rubio, A. 2006. First-principles description of correlation effects in layered materials. *Phys. Rev. B* 96:136404.

Nagareddy, V. K., Hua, K. C., and Hernandez, S. C. 2013. Improved chemical detection and ultra-fast recovery using oxygen functionalized epitaxial graphene sensors. *IEEE Sens. J.* 13(8):2810–2817. doi: 10.1109/JSEN.2013.2259154.

Nayak, T. R., Andersen, H., Makam, V. S. et al. 2011. Graphene for controlled and accelerated osteogenic differentiation of human mesenchymal stem cells. *ACS Nano* 5(6):4670–4678. doi: 10.1021/nn200500h.

Ohta, T., Bostwick, A., Seyller, T., Horn, K., and Rotenberg, E. 2006. Controlling the electronic structure of bilayer graphene. *Science* 313(5789):951–954. doi: 10.1126/science.1130681.

Otsuki, J. 2010. STM studies on porphyrins. *Coord. Chem. Rev.* 254(19–20):2311–2341. doi: 10.1016/j.ccr.2009.12.038.

Pakapongpan, S., Mensing, J. P., Phokharatkul, D., Lomas, T., and Tuantranont, A. 2014. Highly selective electrochemical sensor for ascorbic acid based on a novel hybrid graphene-copper phthalocyanine-polyaniline nanocomposites. *Electrochim. Acta* 133:294–301.

Pandey, P. A., Wilson, N. R., and Covington, J. A. 2013. Pd-doped reduced graphene oxide sensing films for H_2 detection. *Sens. Actuators B* 183:478–487.

Park, H. G., Hwang, S., Lim, J. et al. 2012a. Comparison of chemical vapor sensing properties between graphene and carbon nanotubes. *Jpn. J. Appl. Phys.* 51:045101.

Park, J., Park, K.-S., Jeong, Y.-S. et al. 2012b. Characteristic variations of graphene field-effect transistors induced by CF_4 gas. *Jpn. J. Appl. Phys.* 51:081301 doi: 10.1143/JJAP.51.081301.

Park, S. and Ruoff, R. S. 2009. Chemical methods for the production of graphenes. *Nat. Nanotechnol.* 4:217–224.

Parviz, D., Das, S., Ahmed, H. S. T., Irin, F., Bhattacharia, S., and Green, M. J. 2012. Dispersions of non-covalently functionalized graphene with minimal stabilizer. *ACS Nano* 6:8857–8867.

Perdew, J. P. and Zunger, A. 1981. Self-interaction correction to density-functional approximations for many-electron systems. *Phys. Rev. B* 23:5048.

Phan, D.-T. and Chung, G.-S. October 28–31, 2012. Effects of oxygen-functional groups on humidity sensor based graphene oxide thin films. *IEEE Sensors*, Taipei, Taiwan, pp. 1–4. doi: 10.1109/ICSENS.2012.6411130.

Preobrajenski, A. B., Ng, M. L., Vinogradov, A. S., and Mårtensson, N. 2008. Controlling graphene corrugation on lattice-mismatched substrates. *Phys. Rev. B* 78:073401.

Rao, F., Almumen, H., Wen, L., and Lixin, D. August 20–23, 2012. Nanosensors based on graphene inter-layer electronic properties: Sensing mechanism and selectivity. *Nanotechnology (IEEE-NANO)*, Birmingham, U.K., pp. 1–4. doi: 10.1109/NANO.2012.6322022.

Rao, F. B., Almumen, H., Dong, L. X., and Li, W. June 5–9, 2011a. Highly sensitive bilayer structured graphene sensor. *Solid-State Sensors, Actuators and Microsystems Conference (TRANSDUCERS), 16th International*, Beijing, Republic of China, pp. 2738–2741. doi: 10.1109/TRANSDUCERS.2011.5969828.

Rao, S. S., Stesmans, A., Keunen, K., Kosynkin, D. V., Higginbotham, A., and Tour, J. M. 2011b. Unzipped graphene nanoribbons as sensitive O_2 sensors: Electron spin resonance probing and dissociation kinetics. *Appl. Phys. Lett.* 98:083116. http://dx.doi.org/10.1063/1.3559229.

Reddy, A., Srivastava, A., and Gowda, S. 2010. Synthesis of nitrogen doped graphene films for lithium ion batteries. *ACS Nano* 4(11):6337–6342.

Reina, A., Jia, X., Ho, J. et al. 2009. Large area, few-layer graphene films on arbitrary substrates by chemical vapor deposition. *Nano Lett.* 9:30–35.

Rivera, I. F., Joshi, R. K., and Jing, W. November 1–4, 2010. Graphene-based ultrasensitive gas sensors. *Sensors IEEE*, Kona, HI, pp. 1534–1537. doi: 10.1109/ICSENS.2010.5690325.

Robinson, J. P., Schomeurs, H., and Oroszlány, F. V. 2008. Adsorbate-limited conductivity of graphene. *Phys. Rev. Lett.* 101:196803.

Roman, T., Dino, W. A., Nakanishi, H., Kasai, H., Sugimoto, T., and Tange, K. 2006. Realizing a carbon-based hydrogen storage material. *Jpn. J. Appl. Phys.* 45:1765–1767.

Rudberg, E., Salek, P., and Luo, Y. 2007. Nonlocal exchange interaction removes half-metallicity in graphene nano-ribbons. *Nano Lett.* 7(8):2211–2213. doi: 10.1021/nl070593c.

Rydberg, H., Dion, M., Jacobson, N. et al. 2003. Van der Waals density functional for layered structures. *Phys. Rev. Lett.* 91:126402.

Sablok, K., Bhalla, V., Sharma, P., Kaushal, R., Chaudhary, S., and Suri, C. R. 2013. Amine functionalized graphene oxide/CNT nanocomposite for ultrasensitive electrochemical detection of trinitrotoluene. *J. Hazard. Mater.* 248–249:322–328.

Salavagione, H. J., Martínez, G., and Ellis, G. 2011. Recent advances in the covalent modification of graphene with polymers. *Macromol. Rapid Commun.* 32:1771–1789.

Santos, E. J. G., Ayuela, A., Fagan, S. B. et al. 2008. Switching on magnetism in Ni-doped graphene: Density functional calculations. *Phys. Rev. B* 78:195420.

Shabaneh, A. A., Arasu, P. T., Girei, S. H. et al. October 28–30, 2013. Reflectance response of optical fiber sensor coated with graphene oxide towards ethanol. *Photonics (ICP), 2013 IEEE 4th International Conference*, Melaka, Malacca, pp. 272–274. doi: 10.1109/ICP.2013.6687136.

Sluiter, M. H. and Kawazoe, Y. 2003. Cluster expansion method for adsorption: Application to hydrogen chemisorptions on graphene. *Phys. Rev. B* 68:085410.

Sofo, J. O., Chaudhari, A. S., and Barber, G. D. 2007. Graphane: A two-dimensional hydrocarbon. *Phys. Rev. B* 75:153401.

Son, Y.-W., Cohen, M. L., and Louie, S. G. 2006. Half-metallic graphene nanoribbons. *Nature* 444(7117):347–349. doi: 10.1038/nature05180.

Song, S. H., Park, K. H., Kim, B. H. et al. 2013. Enhanced thermal conductivity of epoxy–graphene composites by using non-oxidized graphene flakes with non-covalent functionalization. *Adv. Mater.* 25:732.

Stojkovic, D., Zhang, P., Lammert, P. E., and Crespi, V. H. 2003. Collective stabilization of hydrogen chemisorptions on graphenic surfaces. *Phys. Rev. B* 68:195406.

Su, Q., Pang, S., Alijani, V., Li, C., Feng, X., and Mullen, K. 2009. Composites of graphene with large aromatic molecules. *Adv. Mater.* 21(31):3191–3195.

Sun, J., Lindvall, N., Cole, M. T. et al. 2012. Low partial pressure chemical vapor deposition of graphene on copper. *IEEE T. Nanotechnol.* 11:255–260.

Tao, N. J., Cardenas, G., Cunha, F., and Shi, Z. 1995. In situ STM and AFM study of protoporphyrin and iron(III) and zinc(II) protoporphyrins adsorbed on graphite in aqueous solutions. *Langmuir* 11(11):4445–4448. doi: 10.1021/la00011a043.

Tasis, D., Tagmatarchis, N., Bianco, A., and Prato, M. 2006. Chemistry of carbon nanotubes. *Chem. Rev.* 106(3):1105–1136.

Tian, J., Hu, B., Wei, Z. et al. 2014. Surface structure deduced differences of copper foil and film for graphene CVD growth. *Appl. Surf. Sci.* 300:73–79.

Tran, Q. T., Huynh, T. M. H., Tong, D. T., Tran, V. T., and Nguyen, N. D. 2013. Synthesis and application of graphene–silver nanowires composite for ammonia gas sensing, OPEN ACCESS. *Adv. Nat. Sci. Nanosci. Nanotechnol.* 4:045012. doi: 10.1088/2043-6262/4/4/045012.

Tung, T. T., Castro, M., and Feller, J.-F. August 20–23, 2012. Electronic noses for VOCs detection based on the nanoparticles hybridized graphene composites. *Nanotechnology (IEEE-NANO), 2012 12th IEEE Conference*, Birmingham, U.K., pp. 1–5. doi: 10.1109/NANO.2012.6322037.

Uddin, M. A., Singh, A. K., Sudarshan, T. S., and Koley, G. 2014. Functionalized graphene/silicon chemidiode H_2 sensor with tunable sensitivity. *Nanotechnology* 25:125501. doi: 10.1088/0957-4484/25/12/125501.

Vazquez de Parga, A. L., Calleja, F., Borca, B. et al. 2008. Periodically rippled graphene: Growth and spatially resolved electronic structure. *Phys. Rev. Lett.* 100:056807.

Wang, J., Singh, B., Maeng, S., Joh, H.-I., and Kim, G.-H. 2013. Assembly of thermally reduced graphene oxide nanostructures by alternating current dielectrophoresis as hydrogen-gas sensors. *Appl. Phys. Lett.* 103:083112. http://dx.doi.org/10.1063/1.4819378.

Wang, Y., Chen, X., Zhong, Y., Zhu, F., and Loh, K. P. 2009. Large area, continuous, few-layered graphene as anodes in organic photovoltaic devices. *Appl. Phys. Lett.* 95:063302.

Wei, D. and Liu, Y. 2010. Controllable synthesis of graphene and its applications. *Adv. Mater.* 22:3225–3241.

Weller, T. E., Ellerby, M., Saxena, S., Smith, R. P., and Skipper, N. T. 2005. Superconductivity in the intercalated graphite compounds C_6Yb and C_6Ca. *Nat. Phys.* 1:39–41. doi: 10.1038/nphys0010.

Wilson, A., Jahangir, I., Singh, A. K. et al. November 3–6, 2013. Tunable graphene/indium nitride heterostructure diode sensor. *Sensors IEEE*, Baltimore, MD, pp. 1–4. doi: 10.1109/ICSENS.2013.6688451.

Wintjes, N., Bonifazi Dr., D., Cheng Dr., F. et al. 2007. A supramolecular multiposition rotary device. *Angew. Chem. Int. Ed.* 46(22):4089–4092. doi: 10.1002/anie.200700285.

Yang, S., Zhi, L., and Tang, K. 2012. Efficient synthesis of heteroatom (N or S)-doped graphene based on ultrathin graphene oxide-porous silica sheets for oxygen reduction reactions. *Adv. Funct. Mater.* 22(17):3634–3640.

Yazyev, O. V. and Helm, L. 2007. Defect-induced magnetism in graphene. *Phys. Rev. B* 75:125408.

Yokoyama, T., Yokoyama, S., Kamikado, T., Okuno, Y., and Mashiko, S. 2001. Selective assembly on a surface of supramolecular aggregates with controlled size and shape. *Nature* 413(6856):619–621. doi: 10.1038/35098059.

Zhang, L., Liang, X., and Song, W. 2010. Identification of the nitrogen species on N-doped graphene layers and Pt/NG composite catalyst for direct methanol fuel cell. *Phys. Chem. Chem. Phys.* 12(38):12055–12059. doi: 10.1039/C0CP00789G.

Zhang, M., Gao, B., Vanegas, D. C. et al. 2014. Simple approach for large-scale production of reduced graphene oxide films. *Chem. Eng. J.* 243:340–346.

Zhang, M., Yuan, R., Chai, Y., Wang, C., and Wu, X. 2013. Cerium oxide-graphene as the matrix for cholesterol sensor. *Anal. Biochem.* 436(2):69–74.

Zhang, Y., Tang, T.-T., Girit, C. et al. 2009. Direct observation of a widely tunable bandgap in bilayer graphene. *Nature* 459(7248):820–823. doi: 10.1038/nature08105.

Zhang, Z., Huang, H., Yang, X., and Zang, L. 2011. Tailoring electronic properties of graphene by π-π stacking with aromatic molecules. *J. Phys. Chem. Lett.* 2(22):2897–2905. doi: 10.1021/jz201273r.

Zhao, C.-L., Qin, M., and Huang, Q.-A. October 28–31, 2011. Humidity sensing properties of the sensor based on graphene oxide films with different dispersion concentrations. *IEEE Sensors*, Limerick, Ireland, pp. 129–132. doi: 10.1109/ICSENS.2011.6126968.

Zheng, W., Shen, B., and Zhai, W. 2013. Surface functionalization of graphene with polymers for enhanced properties. http://dx.doi.org/10.5772/50490. InTech Publications.

Zhou, M., Wang, Y., Zhai, Y. et al. 2009. Controlled synthesis of large-area and patterned electrochemically reduced graphene oxide films. *Chem. Eur. J.* 15:6116–6120.

Zhou, S. Y., Siegel, D. A., Fedorov, A. V., and Lanzara, A. 2008. Metal to insulator transition in epitaxial graphene induced by molecular doping. *Phys. Rev. Lett.* 101(8):086402. doi: 10.1103/PhysRevLett.101.086402.

Züttel, A., Sudan, P., Mauron, P., Kiyobayashi, T., Emmenegger, C., and Schlapbach, L. 2002. Hydrogen storage in carbon nanostructures. *Int. J. Hydrogen Energ.* 27:203–212.

CHAPTER 41

CONTENTS

Hysteresis in the Resistivity of Graphene Channel

41

M.V. Strikha

41.1 INTRODUCTION: EXPERIMENTAL EVIDENCES FOR HYSTERESIS OF RESISTIVITY IN GRAPHENE CHANNEL

One of the remarkable properties of graphene, which was obtained for the first time in 2004 (Novoselov et al. 2004), consists in a symmetry of its band spectrum with respect to the Dirac point and a resulting symmetry in the dependence of graphene conductivity (resistivity) on the gate voltage (Geim 2009) with respect to the zero bias (when the Fermi level crosses the Dirac point). However, such a perfect symmetry of current–voltage characteristics with respect to the zero bias voltage at the gate is observed rather rarely for graphene channels in real graphene structures. As a rule, a shift of voltage that corresponds to the electroneutrality point from zero is observed, which is induced by a certain chemical doping of graphene specimens in the course of their fabrication. At the same time, a hysteresis in the dependence of specimen resistance (conductance) on the gate voltage was repeatedly observed in experiment for graphene channels on substrates of various natures.

Attention to hysteresis phenomena in the resistance of one- and multilayered graphene channels substantially grew after an intensive study of graphene on ferroelectric substrates has been started (see Strikha 2012a,b and the reference therein). The availability of hysteresis loop in the dependence of ferroelectric polarization on the applied field strength allows neatly distinguishable states with different resistances of graphene channel (states "0" and "1" for logic elements) to be created. For the first time, such a logic element was created on the basis of graphene deposited onto a substrate made of a liquid-crystal ferroelectric polyvinylidene fluoride–trifluoroethylene (PVDF–TtFE) (Zheng et al. 2009). Later, the fabrication technology of such elements was improved (Zheng et al. 2010, Raghavan et al. 2012). However, from the viewpoint of practical applications, their shortcoming consists in a necessity to apply considerable switching voltages to the gate (some tens of volts), which is a consequence of a large coercive field that is necessary to vary the polarization direction in the ferroelectric PVDF–TtFE.

From this viewpoint, the application of a substrate made of ferroelectric ceramics $Pb(Zr_xTi_{1-x})$ O_3 (PZT) turned out to be a fortunate alternative. It is a high-technology substrate characterized by a very high dielectric permittivity (up to 3850 near the morphotropic phase border at $x = 0:52$) (Rouquette 2004). Rather reliable systems with two stable states, which may be taken as a principle of operation for the elements of ferroelectric-based nonvolatile memory of new generation, have already been created on the basis of one- and multilayered graphene on the PZT substrate (Hong et al. 2010, Zheng et al. 2011, Song et al. 2011). An opportunity of using one- and multilayered graphene on the ferroelectric substrate for the creation of effective modulator of radiation in the near- and mid-infrared range for on-chip optical interconnections was also demonstrated (Strikha 2011).

In Strikha (2012b), a model was proposed that takes into account the capture of charge carriers by states at the graphene–ferroelectric interface and the screening of the electric field in the substrate by their charges (with the corresponding reduction of free carrier concentration in graphene). This simple model allowed the behavior of resistance in the graphene channel with the variation of gate voltage, which was observed in the experiment (Hong et al. 2010), to be described qualitatively. In the work of Kurchak et al. (2013a), the finite energy width of the impurity state band was additionally taken into account, which allowed us to explain the increase of memory window followed by its saturation as the switching voltage at the gate grows.

However, the hysteresis phenomena in the graphene channel were observed not only for systems on the ferroelectric substrate, but for usual and the most widespread for today SiO_2 substrates (Sabri et al. 2009, Lafkioti et al. 2010, Wang et al. 2010, Levesque et al. 2011, Veligura et al. 2011). It should also be emphasized that p-doping at the zero gate voltage is often observed in graphene on the quartz substrate. The interpretations of those results in various works are substantially different. Really, in the ideal gate-doped graphene, the electroneutrality point (it takes place when the Fermi level crosses the Dirac point in the band spectrum) corresponds to the zero gate voltage V_g (see, e.g., Das Sarma et al. 2011). However, the majority of graphene-based field-effect transistors usually reveal p-doping (i.e., the electroneutrality point lies at a positive V_g), and, at the same time, the dependences of their channel resistance on V_g have a hysteresis loop.

Moreover, the physical origins of those two phenomena can be different. In particular, p-doping can be induced both by adsorbates located on the free surface of graphene and/or at the graphene–quartz interface (Lafkioti et al. 2010, Wang et al. 2010) and by electrochemical processes with the participation of graphene (Sabri et al. 2009, Levesque et al. 2011). At the same time, two different types of hysteresis can be observed: direct or inverse (Wang et al. 2010) (or, in the terminology of Veligura et al. [2011], negative or positive). The direct hysteresis corresponds to the displacement of electroneutrality point toward negative V_g's, when the gate voltage changes consecutively first to positive and then to negative V_g's. The inverse hysteresis corresponds to the displacement of electroneutrality point toward positive V_g's. In the work of Wang et al. (2010), the direct hysteresis was associated with polarization effects with the participation of adsorbates on the graphene surface, and the inverse one with the capture of charge carriers from graphene by traps at the interface and a subsequent screening of the field in the substrate by that bound charge. Note that the latter mechanism is completely similar to that, the quantitative model of which was developed in the works of Strikha (2012b) and Kurchak et al. (2013a). In particular, in a damp environment, there is a layer of adsorbed water, at room temperature (Lafkioti et al. 2010, Wang et al. 2010), or ice, at low temperatures (Wang et al. 2010), on the graphene surface, with the adsorbed water molecules possessing the dipole moment.

As to traps, this role can be played by the interface states (Wang et al. 2010). Until now, the dependences of hysteresis on such factors as the regimes of graphene fabrication, system heating and annealing, the rate of gate voltage switching, the usage of heavy water instead of the ordinary one, and so forth have been studied experimentally (Veligura et al. 2011). However, no quantitative model, which would allow the transition from the hysteresis of one type to another to be described, has been developed until now. In Kurchak et al. (2013b), such a model was developed taking into account both the adsorbed layer of dipoles on the free graphene surface and the capture of charge carriers from graphene.

41.2 THEORETICAL MODEL FOR DIRECT AND INVERSE HYSTERESIS

In what follows, we assume that there is a layer of adsorbed molecules, for example, of electrolyte, water, or ice on the free graphene surface. The molecules are dipoles, and, owing to some physical and chemical reasons (Wang et al. 2010), they are oriented by their negative charges toward the graphene layer (Figure 41.1a). This means that, provided the zero bias voltage at the gate, there is a 2D concentration of holes in graphene, which is related with the polarization of the adsorbed layer of molecules by the simple relation

$$n_{po} = \frac{P_o}{e} \tag{41.1}$$

We also assume that, up to a certain value of positive voltage at the gate, this polarization remains constant. (This voltage value is connected with a critical electric field near the graphene surface

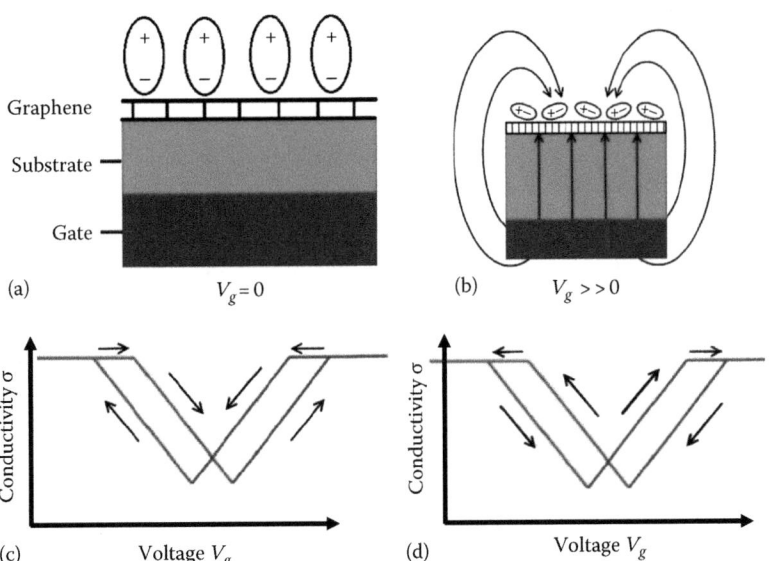

FIGURE 41.1
(a) Dipoles in the adsorbate layer are so oriented that they dope the graphene channel with holes at zero V_g. (b) After the critical value of electric field has been achieved, the polarization in the adsorbate layer becomes destroyed. Dependences of concentration and conductivity on the gate voltage create a closed hysteresis loop: a direct (c) and inverse (d) one.

(Figure 41.1a and b); the field itself is determined by the problem geometry, because, in effect, the graphene channel does not cover the whole substrate surface, and electric field lines pass around its sides to terminate at the charges on the free graphene surface.) Note that this assumption corresponds to the nonlinear dynamics of dipoles bound with the graphene surface (for free dipoles, the orientation of which is described by the Langevin laws, this effect does not take place). Then, the concentration of free carriers in graphene is described as follows:

$$n = \frac{\kappa V_g}{4\pi ed} - n_{po},$$ (41.2)

where
  d is the substrate thickness
  κ the substrate dielectric permittivity

The electroneutrality point ($n = 0$) corresponds to a positive voltage equal to

$$V_{NN} = \frac{-4\pi edn_{po}}{\kappa} = \frac{-4\pi dp_o}{\kappa}$$ (41.3)

At voltages higher than the critical one, the polarization becomes destroyed (see Figure 41.1b); for simplicity, we consider it to equal zero. Then, the concentration is described by the usual formula for the plane capacitor:

$$n = \frac{\kappa V_g}{4\pi ed}$$ (41.4)

If the gate voltage changes rapidly and the polarization has no time to restore spontaneously, the concentration reaches the electroneutrality point at $V_g = 0$. A further increase of negative gate voltages gives rise to that the polarization P_o ultimately becomes restored, and the loop of direct hysteresis gets closed (Figure 41.1c).

There are plenty of models for the critical field in the dipole layer. We use the simplest static model, in which the relative concentration of dipoles in this layer, δ, depends nonlinearly on the applied voltage V (Li et al. 2012), namely,

$$V(\delta) = V_S - \left(\frac{RT}{F}\right)\left(\ln\left(\frac{\delta}{1-\delta}\right) - J(\delta - 0.5)\right) \tag{41.5}$$

The quantity J in Equation 41.5 has the sense of a dimensionless exchange constant, which equals identically zero in the ideal linear case; the relative concentration falls within the interval $0 < \delta < 1$, and the actual concentration is determined by the relation $N = N_0\delta$; R is the universal gas constant; F the Faraday constant; and V_S is a certain equilibrium value corresponding to the steady state of dipoles (for the sake of simplicity, all changes can be reckoned from this state). In the dynamic case, the dependence $V(\delta)$ transforms into the relaxation equation (Kurchak et al. 2013b):

$$-\frac{1}{\tau}\frac{\partial\delta}{\partial t} + \ln\left(\frac{\delta}{1-\delta}\right) - J\left(\delta - \frac{1}{2}\right) = \theta \tag{41.6}$$

where τ is the characteristic time of dipole relaxation. We also introduced the dimensionless voltage $\theta = (V - V_S)F/RT$. It was demonstrated in Kurchak et al. (2013b) that a well-pronounced bistable behavior is observed at positive $J > J_{cr}$, where $J_{cr} = 4$. This fact has an elementary explanation, because

$$\ln\left(\frac{\delta}{1-\delta}\right) \approx 4\left(\delta - \frac{1}{2}\right) + \frac{16}{3}\left(\delta - \frac{1}{2}\right)^3 + \frac{64}{5}\left(\delta - \frac{1}{2}\right)^5$$

and Equation 41.6 has the form

$$-\frac{1}{\tau}\frac{\partial\delta^*}{\partial t} - (J-4)\delta^* + \frac{16}{3}\delta^{*3} = \theta \tag{41.7}$$

in the vicinity of transition point $\delta^* = \delta - (1/2)$. Hence, our assumption can correspond to a situation described by formula (41.5) with a large enough value of exchange constant J. However, since the steady-state energy of dipoles oriented as shown in Figure 41.1a is lower by ΔE than the same energy for the orientation shown in Figure 41.1b, a spontaneous restoration of dipole polarization $P(t) \propto eaN_0\delta^*(t)$, where a is the arm of elementary dipole, will take place starting from the moment when the electric field strength becomes lower than the critical value E_C. The restoration formula reads

$$P(t) = P_o[1 - \exp(-t/\tau)] \tag{41.8}$$

In Equation 41.5, we adopted for simplicity that relaxation is characterized by a unique time τ (this is a natural approximation for a simple dipole layer, but generally speaking, it is rarely realized in real systems) (Glarum 1960). However, in the framework of this simple approximation, the voltage distance between the electroneutrality points in the hysteresis loop (Figure 41.1c) is determined by the relation

$$\left|V^1_{NN} - V^2_{NN}\right| = \frac{4\pi d[P_o - P(t)]}{\kappa} = \frac{4\pi d P_o \exp(-t/\tau)}{\kappa} \tag{41.9}$$

Therefore, if the system's switching time becomes of the same order of magnitude as the relaxations time τ, the hysteresis loop diminishes and ultimately vanishes. For the geometry exhibited in Figure 41.1, $(\varepsilon_S + \varepsilon_l)/2$ should substitute κ in formula (41.9) in the quasi-static case, where ε_S and ε_l are the dielectric permittivities of the substrate and the medium over graphene, respectively.

To describe a phenomenon that is an inversion of hysteresis, let us use the approach developed in the works of Strikha (2012b) and Kurchak et al. (2013a). Here, we consider a monatomic graphene layer doped by means of the gate voltage. The dependence of Fermi energy in graphene on the concentration of free charge carriers n is given by the known relation (see, e.g., Das Sarma et al. 2011)

$$E_F = \hbar v_F (\pi n)^{1/2} \tag{41.10}$$

where $v_F = 10^8$ cm/s. Suppose that there are localized states with the energy E_T and the 2D concentration n_T at the graphene–substrate interface. While V_g increases in the interval for which $E_F < E_T$, the concentration of charge carriers in graphene is determined by expression (41.4). However, at $E_F = E_T$, electrons from graphene start to occupy the interface states, and, as was shown in Kurchak et al. (2013b), this process of state filling takes place in the voltage interval

$$\frac{4\pi e d}{\kappa} \frac{E_T^2}{\pi \hbar^2 v_F^2} \leq V_g < \frac{4\pi e d}{\kappa} \frac{E_T^2}{\pi \hbar^2 v_F^2} + \frac{4\pi e d n_T}{\kappa} \tag{41.11}$$

The negative charge bound at filled interface states screens the field in the substrate. Therefore, the concentration of free charge carriers in graphene remains constant

$$n = \frac{E_T}{\sqrt{\pi} \hbar v_F} \tag{41.12}$$

within range of voltage interval (41.11). At higher voltages, when all interface states are already filled by electrons, the evident relation takes place

$$n(V_g) = \frac{\kappa V_g}{4\pi e d} - n_T \tag{41.13}$$

The next assumption of the works of Strikha (2012b) and Kurchak et al. (2013a) consists in that the lifetime of electrons at the interface states considerably exceeds the time of system switching. Therefore, relation (41.13) remains valid even if the gate voltage diminishes, because the charge carriers captured by the states remain at them, although the Fermi level in graphene becomes lower than those states in the energy scale. Hence, the dependence of concentration n on the voltage V_g has the form shown in Figure 41.1d. Here, the arrows mark the direction of voltage sweep (increase or decrease). If the voltage diminishes after switching, the value of n reaches the electroneutrality point (the Dirac point) at the positive voltage

$$V_{NP} = \frac{4\pi e d n_T}{\kappa} \tag{41.14}$$

determined by the concentration of interface states n_T. Note that the n-values to the left from Dirac points in Figure 41.1d correspond to the hole concentrations. At a certain high negative V_g (the specific value depends on the model of localized states), the captured electrons should ultimately recombine with the holes in the graphene layer. Unfortunately, the data on the microscopic nature of level potential, which would allow a more exact model of this process to be developed, are absent. Therefore, in the work of Kurchak et al. (2013a), an assumption was made that the dependence of free charge carrier concentration n on the gate voltage is symmetric with respect to the voltage $V_{NP}/2$, that is, the level depopulation starts at a certain negative voltage and terminates at a voltage higher by V_{NP}. Then, the dependence of concentration on the voltage has formed (41.4) again. Hence, the hysteresis loop in the dependence of concentration on the gate voltage becomes closed.

It is clear that if the interval of gate voltage variation is narrow, that is, the condition $E_F < E_T$ is satisfied, no hysteresis is observed. The case when the switching voltage V_{sweep} falls in interval (41.11), that is, when the level filling is not complete, is of special interest. In this case, the antihysteresis curve broadens, and the electroneutrality points move away from each other in the voltage scale as V_{sweep} increases. The distance between them is now determined by the relation

$$V_{DP}(V_{sweep}) = \frac{4\pi e d n_T(V_{sweep})}{\kappa} \tag{41.15}$$

where $n_T(V_{sweep})$ is the concentration of interface states filled by electrons before the switching moment.

Hysteresis in the dependence of concentration on the gate voltage should have a counterpart in the similar dependence for the total resistivity (recall that it has the dimension of ohms for 2D structures). Really, the total resistivity of graphene layer is determined by Mattiessen's rule

$$\rho(V_g) = \frac{1}{\sigma(V_g)} + \frac{1}{\sigma_{intr}(T)} + \frac{1}{\sigma_{min}} \tag{41.16}$$

Here, the first summand in the right part of (41.16) corresponds to the conductivity of doped graphene

$$\sigma(V_g) = en\mu \tag{41.17}$$

where
the 2D concentration n is determined from one of V_g-dependences obtained earlier
μ is the charge carrier mobility

Mention that Equation 41.17 with μ independent of n is valid for the case of carriers scattering in graphene channel on the random distributed charged impurities in the substrate only (see Das Sarma et al. 2011); for the other dominant scattering mechanisms, (41.17) is no longer valid. The second summand in the right part of (41.16) describes the intrinsic conductivity of graphene. It has to be taken into account when the Fermi level is located near the electroneutrality point. It should be recalled that the concentration of charge carriers in intrinsic graphene depends on the temperature T as follows:

$$n_{intr}(T) = \frac{2(kT)^2}{\pi(\hbar v_F)^2} \tag{41.18}$$

where k is the Boltzmann constant. According to Equation 41.18, the concentration is $n_{intr} \approx 1.1 \cdot 10^{11}$ cm^{-2} at room temperature. That is why this summand should be taken into account in an interval of about ± 0.1 V around the electroneutrality point if the substrate thickness is of the order of 300 nm and the substrate dielectric permittivity of the order of 400. If the gate voltages fall within this interval, the conductivity has the same order as that described by expression (41.17). At last, the value $\sigma_{min} \approx 4e^2/\hbar$ corresponds to the minimum conductivity near the Dirac point, which has the quantum-mechanical nature (see, e.g., Das Sarma et al. 2011). We should emphasize that, if the both mechanisms (of direct and inverse hysteresis) act simultaneously, then, according to Equation 41.9, the mechanism of direct hysteresis should expectedly dominate at high rates of system switching (the polarization has no time to restore, and the electroneutrality points in Figure 41.1c are located far away from each other). At the same time, at low rates of system switching, the polarization restores, and the electroneutrality points approach each other and ultimately merge together. The phenomenon of inverse hysteresis associated with the capture of charge carriers by the interface states can dominate against this background.

41.3 THEORY VERSUS EXPERIMENT

In the work of Wang et al. (2010), the hysteresis phenomenon in one- and multilayered graphene obtained using the method of chemical deposition from the vapor phase onto the SiO$_2$ substrate was studied. The gate voltage was varied in the interval from -80 to $+80$ V, and the rate of its

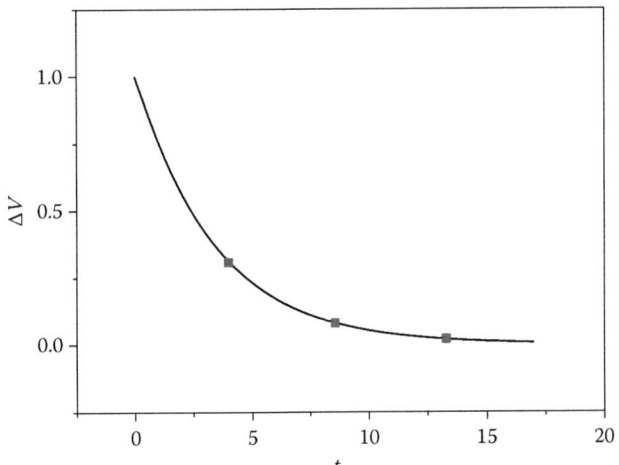

FIGURE 41.2
Dependence of memory window (in volt units) on the time needed for the system to return back from the switching point to the electroneutrality one (in second units): experimental data, reconstructed from Wang et al. (2010) (symbols) and their approximation by formula (41.9) (solid curve).

variation was changed considerably. The dependences of graphene channel conductivity on the voltage at the electrolytic gate (the aqueous solution of KCl was used as an electrolyte) for different voltage sweep rates were obtained. If the rate of voltage sweep decreases, the distance between the electroneutrality points also decreases until they practically merge together at a rate of 0.0625 V/s. Such a behavior is in perfect agreement with the predictions made in the previous section of this work.

In Figure 41.2, the dependence of memory window (the distance between the electroneutrality points, $\Delta V = \left| V_{NN}^1 = V_{NN}^2 \right|$, on the time interval needed for the system to return back from the switching point to the electroneutrality one is depicted by symbols. Here, we used the experimental points reconstructed from the data of Wang et al. (2010). The curve in the same figure shows the approximation of experimental data by formula (41.9) with $\tau = 3.4$ s and the preexponential coefficient approximately equal to 1. (Generally speaking, the preexponential coefficient is more cumbersome in this case than the formula for the plane capacitor (41.4) predicts. However, we cannot derive its exact form due to lack of exhaustive information concerning the problem geometry.) It is evident from Figure 41.2 that the reduction of memory window as the time interval of surface dipole relaxation increases is described well by exponential dependence (41.9). It is important to note that, owing to a strong coupling between negative poles of dipoles with the graphene surface (see, e.g., Veligura et al. 2011), the time τ that is included into this dependence exceeds, by a good many orders of magnitude, the relaxation times characteristic of polarization induced by free dipoles and described by the formula of the Langevin type.

At the same time, the relaxation time for the concentration of charge carriers captured by the interface states is much longer. It equals hours at room temperature and tens of days at the nitrogen one (Hong et al. 2010). Therefore, in the exponential timescale of Wang et al. (2010), this concentration can be regarded as dependent only on the gate voltage changes.

The dependence of graphene channel conductivity on the voltage at the standard Si gate was also studied in Wang et al. (2010) at various rates of this voltage sweep both at high temperatures, when the adsorbed water molecules form a water layer, and at temperatures below the water freezing point, when the covering layer is ice. The chemical nature of those molecules is identical, but the mobility is different, namely, it is substantially higher in the liquid and lower in the crystal. This difference explains a discrepancy between the corresponding experimental dependences. In particular, in the case of ice, the direct hysteresis took place at high rates of voltage variation, and the inverse one at low rates. At the same time, in the case of water, when the time of spontaneous restoration of polarization is very short owing to a high mobility of dipoles, the inverse hysteresis was observed at any rate of voltage variation among those that were used in the experiment.

41.4 CONCLUSIONS

The results of comparison between the experimental data and the theoretical dependences, which was done in the previous section, testify that the model proposed in Kurchak et al. (2013b) can both describe the hysteresis phenomenon in the dependence of conductivity in the graphene channel created on the substrates of various origins on the gate voltage, and provide a comprehension, at the quantitative level, of parameters inherent to the interface states and the polarization of adsorbate bound with the free graphene surface. The model predicts an opportunity to distinguish between two hysteresis types, direct and inverse ones. The former is associated with the repolarization of dipoles bound at the graphene surface. Under experimental conditions, it is water molecules that most often play this role. The latter is connected with the capture of free charge carriers by the interface states. The resolution between the hysteresis types can be made by varying the rate of gate voltage sweep. This becomes possible owing to the hierarchy of spontaneous relaxation times that describe two indicated processes. Namely, these are seconds for the relaxation of polarization induced by adsorbed dipoles, and hours and days for the relaxation of concentration of charge carriers captured at the interface states.

The approximations used in Kurchak et al. (2013) are standard in the graphene physics and the physicists of polarizable media. At the same time, the introduction of parameter τ for the spontaneous relaxation of polarization requires more substantiation, because the approximation applied by us is simplified. The model proposed can be specified on the basis of deeper understanding of the physicochemical nature of coupling between the adsorbed dipoles and the graphene surface, on the one hand, and the model of creation of localized states at the graphene–dielectric substrate interface, on the other hand. A further experimental research of the hysteresis phenomena in the dependence of graphene channel conductivity on the gate voltage can provide us with a valuable material for better understanding of the parameters of this physicochemical structure.

REFERENCES

Das Sarma, S., Adam, Sh., Hwang, E.H., Rossi, E. 2011. Electronic transport in two dimensional graphene. *Rev. Mod. Phys.* 83: 407–470.

Geim, A.K. 2009. Graphene: Status and prospects. *Science* 324: 1530–1539.

Glarum, S.H. 1960. Dielectric relaxation of polar liquids. *J. Chem. Phys.* 33: 1371–1375.

Hong, X., Hoffman, J., Posadas, A. et al. 2010. Unusual resistance hysteresis in *n*-layer graphene field effect transistors fabricated on ferroelectric $Pb(Zr_{0.2}Ti_{0.8})O_3$. *Appl. Phys. Lett.* 97: article number 033114.

Kurchak, A.I., Morozovska, A.N., Strikha, M.V. 2013b. Rival mechanisms of hysteresis in the resistivity of graphene channel. *Ukr. J. Phys.* 58: 472–479.

Kurchak, A.I., Strikha, M.V. 2013a. Antihysteresis of the electrical resistivity of graphene on a $Pb(Zr_xTi_{1-x})O_3$ ferroelectric substrate. *JETP* 116: 112–117.

Lafkioti, N., Krauss, B., Lohmann, T. et al. 2010. Graphene on a hydrophobic substrate: Doping reduction and hysteresis suppression under ambient conditions. *Nano Lett.* 10: 1149–1153.

Levesque, P.L., Sabri, S.S., Aguirre, C.M. et al. 2011. Probing charge transfer at surfaces using graphene transistors. *Nano Lett.* 11: 132–137.

Li, J., Xiao, X., Yang, F., Verbrugge, M.W., Cheng, Y.-T. 2012. Potentiostatic Intermittent titration technique for electrodes governed by diffusion and interfacial reaction. *J. Phys. Chem. C* 116: 1472–1478.

Novoselov, K.S., Geim, A.K., Morozov, S.V. et al. 2004. Electric field effect in atomically thin carbon films. *Science* 306: 666–669.

Raghavan, S., Stolichnov, I., Setter, N. et al. 2012. Long-term retention in organic ferroelectric-graphene memories. *Appl. Phys. Lett.* 100: article number 023507.

Rouquette, J., Haines, J., Bornand, V. et al. 2004. Pressure tuning of the morphotropic phase boundary in piezoelectric lead zirconate titanate. *Phys. Rev. B* 70: article number 014108.

Sabri, S.S., Levesque, P.L., Aguirre, C.M. et al. 2009. Graphene field effect transistors with parylene gate dielectric. *Appl. Phys. Lett.* 95: article number 242104.

Song, E.B., Lian, B., Kim, S.M. et al. 2011. Robust bi-stable memory operation in single-layer graphene ferroelectric memory. *Appl. Phys. Lett.* 99: article number 042109.

Strikha, M.V. 2011. Modulation of a mid-IR radiation by a gated graphene on ferroelectric substrate. *Ukr. J. Phys. Opt.* 12: 162–165.

Strikha, M.V. 2012a. Non-volatile memory and IR radiation modulators based upon graphene-on-ferroelectric substrate. A review. *Ukr. J. Phys. Opt.* 13: S5–S27.

Strikha, M.V. 2012b. Mechanism of the antihysteresis behavior of the resistivity of graphene on a $Pb(Zr_xTi_{1-x})O_3$ ferroelectric substrate. *JETP Lett.* 95: 198–200.

Veligura, A., Zomer, P.J., Vera-Marun, I.J. et al. 2011. Relating hysteresis and electrochemistry in graphene field effect transistors. *J. Appl. Phys.* 110: article number 113708.

Wang, H., Wu, Y., Cong, C. et al. 2010. Hysteresis of electronic transport in graphene transistors. *ACS Nano* 4: 7221–7228.

Zheng, Y., Ni, G.-X., Toh, Z.-T. et al. 2009. Gate-controlled nonvolatile graphene-ferroelectric memory. *Appl. Phys. Lett.* 94: article number 163505.

Zheng, Y., Ni, G.-X., Toh, Z.-T. et al. 2010. Graphene field effect transistors with ferroelectric gating. *Phys. Rev. Lett.* 105: article number 166602.

Zheng, Y., Ni, G.-X., Bae, S. et al. 2011. Wafer-scale graphene/ferroelectric hybrid devices for low voltage electronics. *Europhys. Lett.* 93: article number 17002.

CHAPTER 42

CONTENTS

Electronic and Optical Properties of Boron- and Nitrogen-Functionalized Graphene Nanosheet

42

Suman Chowdhury, Ritwika Das, Palash Nath,
Dirtha Sanyal, and Debnarayan Jana

42.1 INTRODUCTION

The carbon-related nanomaterials in the recent two decades have fostered and played an important key role in the advancement of nanoscience and nanotechnology. Among the various possible allotropes of carbon atoms in low-dimensional systems, graphene, a purely 2D honeycomb-like crystal of carbon atoms (Ando, 2009; Castro Neto et al., 2009; Choi et al., 2010; Loh et al., 2010; Kuila et al., 2012), has been the focus of extensive research studies after its successful synthesis in 2004 (Anton et al.,1998; Novoselov et al., 2004). Two adjacent carbon atoms of the graphene planar structure build up the unit cell of the crystal. Out of the four valence electrons of the carbon atoms, three are used for the planar C–C sp^2 hybridized bonding. Thus, each carbon atom contains single p_z electron. Due to the presence of this single unbound electron, graphene exhibits fascinating electronic and electrochemical properties. One of the key features of graphene is its extremely high conductivity compared to other materials. The reason of this feature lies in the peculiar band structure at some special point known as Dirac points where the conduction band (CB) and the valence band (VB) touch at a single point at the Fermi energy. This indicates that the graphene is a semiconductor with zero bandgap or a semimetal. Note that the origin of Dirac cone is through structural symmetry of graphene. All the carbon atoms in graphene are connected through double bonds. Extremely high mobility along with easy control of the charge carriers is one of the assets of graphene for the next generation of electronics. However, the absence of energy gap in electronic spectra makes it unsuitable for devices such as graphene transistors. When one or more graphene layers are introduced, the inversion symmetric bilayer also possesses a zero bandgap in its pristine form.

Graphyne, another similar structure of graphene, consists of double- and triple-bonded units of carbon atoms. The presence of triple bonds in graphyne can have intriguing features, having other geometry instead of hexagonal as possessed by graphene.

42.2 CHEMICAL FUNCTIONALIZATION

As mentioned, the lack of bandgap is an impediment for any practical device fabrication; methods of tailoring a bandgap into graphene have been the target of the intense study. In this sense, chemical functionalization is one of the techniques for engineering the bandgap. There are in fact many techniques used in graphene for bandgap opening:

1. Application of electric field perpendicular to bilayer of graphene
2. Hydrogenation
3. Doping/adatoms

4. Graphene–substrate interactions (including hexagonal boron nitride)
5. Strain engineering
6. Introduction of Stone–Wales defects (57–57 defect/topological defect)

Applied field can induce a bandgap in spite of the fact that an external power supply is to be maintained. The numbers 2–4 in the aforementioned list fall into the category of chemical functionalization. Substrate chemical functionalization changes the Coulomb potential on different sides of the graphene layer. Hydrogenation (Elias et al., 2009) and fluorination (Cheng et al., 2010, 2012; Robinson et al., 2010; Sheng et al., 2012) are two common types of chemical functionalization in graphene that are used to modify the physical properties of pristine graphene. Even transition metal doping can also modify the electronic [Nakada et al., 2011] and optical [Nath et al., 2015] properties of graphene in compared to pristine graphene. Patterned hydrogen absorption (Balog et al., 2010) can also induce an opening of the bandgap in pristine graphene. Apart from hydrogenation (attaching H) or fluorination (attaching F), chlorine atoms interact with graphene in a characteristic different way (Li et al., 2011; Wu et al., 2011; Gopalakrishnan et al., 2012; Vinogradov et al., 2012; Xu et al., 2013; Zhang et al., 2013). For hydrogenation or fluorination, the interaction between H and F is through the formation of C–H or C–F bonds. However, in the case of chlorine, several types of bonds appear giving rise to quite large bandgap of 1.21 eV for fully chlorinated graphene (Schin and Ciraci, 2012). Recently, Nath et al. (2014b) have studied the effect of boron (B) and nitrogen (N) substitutional codoping in planar graphene network. B and N pair doped in the graphene crystal yields an opening of nonzero bandgap at the Fermi level (E_F) when the N–B pair occupies the different relative sublattice position. However, the linear band crossing without showing any bandgap opening is obtained when the N–B pair occupies the same relative sublattice position. It is also to be noted that when N and B are doped as nearest neighbor, the defect formation energy is smaller than the other possible doping configurations (Nath et al., 2014a). Apart from relative doping sites, the doping concentration also tunes the bandgap value (Nath et al., 2014b). Bandgap value increases (Nath et al. 2014b; Rani et al., 2013a, 2013b, 2014a) with the doping concentration of N, B, and N–B pair in the graphene structure.

Bandgap opening has been noticed in graphene nanoribbons through chemical functionalization (Son et al., 2006; Gorjizadeh and Kawazoe, 2010) and in zigzag triwing graphene nanoribbons by boron and nitrogen doping (Liu and Shen, 2009; Ma et al., 2011; Tachikawa et al., 2011). Bandgap opening has also been controlled through a substrate (Zhou, 2007) in epitaxial graphene and metal–semiconductor–metal nanostructure (Hicks, 2012). The strain can also induce a bandgap opening in pristine graphene (Choi et al., 2010; Jana et al., 2014). Defects and impurities in hexagonal graphene network can significantly alter the electronic density of states (DOS) (Chowdhury et al., 2014) and bandgap of the graphene (Lusk and Carr, 2008; Sahin and Ciraci, 2011; Araujo et al., 2012). The bandgap in graphene can be introduced via noncovalent functionalization with porphyrin molecules such as iron protoporphyrin (FePP) and zinc protoporphyrin (ZnPP). The scanning tunneling spectroscopy (STS) measurements suggested that π–$\pi*$ stacking of FePP on graphene yielded a bandgap of 0.45 eV compared to ZnPP (0.23 eV). The metallic nature of the respective porphyrin is responsible for such a large considerable bandgap in graphene. Both tight-binding model and first-principles electronic structure computations have established (Dvorak et al., 2013) that the presence of bandgap in patterned graphene had a symmetry origin.

42.3 METHODOLOGY USED IN DFT FOR COMPUTATION OF OPTICAL PROPERTIES OF GRAPHENE

Density functional theory (DFT) calculations (Parr et al., 1989; Dreizler et al., 1995; Jana, 2008; Giustino, 2014) enable us to determine the electronic energy $E_n(\vec{k})$ of a given band n as a function of k-vector along some symmetry direction in the Brillouin zone (BZ). Apart from the determination of the bandgaps at various k-points in BZ, one can also calculate the electronic DOS, which plays an important key role in optical properties of the graphene. The generalized DOS at an arbitrary d dimension can be written (Odom et al., 1998, 2000) for a given band n as

$$D_d^n(E) = \left(\frac{L}{2\pi}\right)^d \int \frac{d^d k \delta(k(E_n) - k)}{|\nabla_k(E_n)|}$$

(42.1)

where L denotes the typical size of the system, $E_n(\vec{k})$ describes the dispersion of the given band, and the integral is determined over the BZ. At a typical band edge, $|\nabla_k(E_n)| \rightarrow 0$, and the singularities appear in DOS at some values of E. These singular features of a material known as van Hove singularities (vHSs) are the characteristic features of the dimension of the system. For example, vHS turns out to be kinks in three dimensions, while in two dimensions, there are stepwise discontinuities with increasing values of energy E. However, in 1D system, these vHSs are manifested as spikes/peaks. Any 1D system will eventually exhibit the spike features in their DOS due to the 1D nature of the band structure. It is clear from the expression (Equation 42.1) that the most contribution from the states comes from the situations where the dispersion relation is flat and $|\nabla_k(E_n)|$ is small. The number of electronic states (NES) $\left(= \int_{E_1}^{E_2} D_d^n(E)dE \right)$ between two energy states E_1 and E_2 indicates the typical number of emitting electrons that are responsible for the field emission current. For direct optical transitions between two bands with energy difference $E_{nn'} = E_{n'} - E_n$, the joint density of states (JDOS) is more appropriate, in such a situation, and we can rewrite the aforementioned expression (Equation 42.1) as

$$D_{nn'}(\hbar\omega) = \frac{2}{(2\pi)^3} \int \delta(E_n(k) - E_{n'}(k))d^3k \qquad (42.2)$$

Partial density of states (PDOS) and local density of states (LDOS) represent useful semi-qualitative tools for analyzing the electronic structure.

The optical properties of any system are generally studied by the complex dielectric function defined by $\vec{D}(\omega) = \varepsilon(\omega)\vec{E}(\omega) = [\varepsilon_1(\omega) + i\varepsilon_2(\omega)]\vec{E}(\omega)$. However, $\varepsilon_1(\omega)$ and $\varepsilon_2(\omega)$ are not independent of each other. They are connected by Kramers–Kronig (KK) (Cohen and Chelikowsky, 1988; Mahan, 1990; Yu and Cardona, 1996; Dressel and Grüner, 2002) relation to be discussed later on. In this numerical simulation, the imaginary part of the dielectric function $\varepsilon_2(\omega)$ has been computed by using first-order time-dependent perturbation theory. In the simple dipole approximation used in DFT calculation (Jana et al., 2013), the imaginary part is given by

$$\varepsilon_2\left(q \rightarrow 0_{\vec{u}}, \hbar\omega\right) = \frac{2e^2\pi}{\Omega\varepsilon_0} \sum_{k,v,c} \left|\left\langle \psi_k^c \mid \vec{u} \cdot \vec{r} \mid \psi_k^v \right\rangle\right|^2 \delta\left(E_k^c - E_k^v - E\right) \qquad (42.3)$$

where Ω and ε_0 represent, respectively, the volume of the supercell and the permittivity of the free space. The sum over k is a crucial point in numerical calculation. It actually samples the whole region of BZ in the k space. The other two sums take care of the contribution of the unoccupied CB and occupied VB. Here, $\vec{u}, \vec{r}$, respectively, represent the polarization vector of the incident electric field and position vector. The matrix element of this dot product of these two vectors is computed between the single-electron energy eigenstates. Since the magnetic field effect is weaker by a factor of v/c, the transition matrix elements between the eigenstates of CB and VB have been calculated only due to the electric field. No phonon contribution is taken into account here. Moreover, in this formulation, the local field effect and the excitonic effect have been neglected.

We will notice later that the imaginary part of the dielectric constant of any graphene system (undoped, doped, or defected) is always positive throughout the range of frequency. This can be understood very simply from Equation 42.3 used for this simulation study. The square of the matrix element and the even functional nature of the energy-conserving delta function ensure the positivity of ε_2. This property of ε_2 serves as one of the cross-checks in any numerical computation. To determine the wave functions in Equation 42.3, the first-principles spin-unpolarized DFT is performed using plane-wave pseudopotential methods (Parr et al., 1989; Payne et al., 1992; Dreizler et al., 1995; Milman et al., 2000; Giustino, 2014). Like any *ab initio* calculation, the self-consistent Kohn–Sham (KS) equation (Kohn and Sham, 1965) has been used to compute the eigenfunction here. For the exchange and correlation term, the generalized gradient approximation (GGA) as proposed by Perdew–Burke–Ernzerhof (PBE) (Perdew et al., 1966) is adopted. Compared to the standard local density approximation (LDA) (with appropriate modifications) used mostly in electronic band structure calculation, the optical properties of the system are normally standardized by spin-unpolarized GGA.

The anisotropic behavior in the optical properties can be investigated by taking into account the polarization vector $\vec{u}$ of the electromagnetic field in Equation 42.3. Thus, the dielectric constant can be evaluated for three separate cases: One can choose in the direction of $\vec{u}$ (1) of the electric field vector for the light at normal incidence (polarized); (2) choosing $\vec{u}$ in the direction of propagation of incident light at the normal incidence but the electric field vector is considered as an average over the plane perpendicular to this direction (unpolarized); and (3) choosing $\vec{u}$ not in specified direction while the electric field vectors are taken as full isotropic average (polycrystalline). The directions of the field (k wave vector) have been chosen with respect to the axis of the doped/undoped carbon nanotubes (CNTs). The parallel polarization refers to $k(0, 0, 1)$, while perpendicular one refers to $k(1, 0, 0)$.

The dielectric constant describes the typical causal response, and the real ($\varepsilon_1(\omega)$) and imaginary part ($\varepsilon_2(\omega)$) of it are connected by KK transform (Chelikowsky, 1988; Mahan, 1990; Cohen and Yu and Cardona, 1996; Dressel and Grüner, 2002) given by

$$\varepsilon_1(\omega) - 1 = \frac{2}{\pi} P \int_0^\infty \frac{\omega' \varepsilon_2(\omega') d\omega'}{\omega'^2 - \omega^2} \tag{42.4}$$

$$\varepsilon_2(\omega) = -\frac{2}{\pi\omega} P \int_0^\infty \frac{\omega'^2 (\varepsilon_1(\omega') - 1) d\omega'}{\omega'^2 - \omega^2} \tag{42.5}$$

where P denotes the principal part. If there is no absorption in the entire region of the frequency, then $\varepsilon_2(\omega) = 1$; hence, using the aforementioned relation, we get $\varepsilon_1(\omega) = 1$. Thus, one concludes that there is no frequency dependence of the dielectric constant. Similarly, one can also relate the real part of inverse of the dielectric function to its imaginary counterpart as

$$\text{Re}\left[\frac{1}{\varepsilon(\omega)}\right] - 1 = \frac{1}{\pi} P \int_{-\infty}^{+\infty} \text{Im}\left[\frac{1}{\varepsilon(\omega)}\right] \frac{d\omega'}{\omega' - \omega}$$

$$\text{Im}\left[\frac{1}{\varepsilon(\omega)}\right] = \frac{1}{\pi} P \int_{-\infty}^{+\infty} \left[1 - \text{Re}\left(\frac{1}{\varepsilon(\omega)}\right)\right] \frac{d\omega'}{\omega' - \omega} \tag{42.6}$$

One can also give similar arguments for other optical constants such as refractive index $N = n + ik = \sqrt{\varepsilon(\omega)}$ as

$$n(\omega) - 1 = \frac{2}{\pi} P \int_0^\infty \frac{\omega' k(\omega') d\omega'}{\omega'^2 - \omega^2}$$

$$k(\omega) = -\frac{2}{\pi\omega} P \int_0^\infty \frac{\omega'^2 (n(\omega') - 1) d\omega'}{\omega'^2 - \omega^2} \tag{42.7}$$

This relation is quite useful in the sense that if one is able to measure the only component such as absorption coefficient defined as $\alpha(\omega) = (2k(\omega)\omega)/c$ (c being the speed of light) over a wide range of frequency, then the refractive index $\tilde{n}(\omega)$ can be calculated without separate phase measurements. It is worthy to mention that KK relations are nonlocal in nature. The absorption coefficient α also is related to the imaginary part of the dielectric constant as

$$\alpha = \frac{\varepsilon_2 \omega}{\tilde{n} c} \tag{42.8}$$

where $\tilde{n}$ and c are the refractive index and the speed of light, respectively. The sharp peaks at some particular energy regime are associated with pairwise transitions between vHS present in the electronic energy band spectrum. Besides, this characteristic tool gives us the information regarding the quality of the nanotubes. The purity of the sample is related to the area under the first interband transition peak. For frequencies ω higher than those of the highest absorption (which is described by the imaginary part of the dielectric constant), one can simplify the KK relations as

$$\varepsilon_1(\omega) - 1 = \frac{2}{\pi} P \int_0^\infty \frac{\omega' \varepsilon_2(\omega\omega) d\omega d}{\omega'^2 - \omega^2}$$

$$\varepsilon_2(\omega) = -\frac{2}{\pi\omega} P \int_0^\infty \frac{\omega'^2 (\varepsilon_1(\omega') - 1) d\omega'}{\omega'^2 - \omega^2}$$

(42.9)

The reflectivity $R(\omega)$ of any media at normal incidence is calculated from the refractive indexes via the relation given by

$$R(\omega) = \left(\frac{1 - \sqrt{\varepsilon(\omega)}}{1 + \sqrt{\varepsilon(\omega)}} \right)^2, \quad \varepsilon(\omega) = \varepsilon_1(\omega) + i\varepsilon_2(\omega)$$

(42.10)

The reflectivity $R(\omega)$ of any media at normal incidence also can be rewritten in terms of the refractive indexes $N = \tilde{n} + i\kappa$ as

$$R = \left(\frac{1 - N}{1 + N} \right)^2$$

(42.11)

It is clearly evident from the definition that the reflectivity is always positive in the scheduled range of the frequency and is dimensionless. R is sometimes regarded as the index of refraction as a function of wavelength of light used. It is to be remembered that in this calculation, we are taking into account the long wavelength limit ($q \rightarrow 0$) as the imaginary part of the dielectric constant is evaluated in the long wavelength limit.

Electron beams can be used for probing the electronic structure of solids. In this time-dependent calculation of the ground-state electronic states, the interaction is between the photon and electrons. The transitions between the occupied and unoccupied states are caused by the electric field of the photon. When these excitations are collective in nature, they are termed as plasmons as discussed earlier. The loss function, which is a direct measure of the collective excitations of the systems, is defined as $L(\omega) = \text{Im}[-1/\varepsilon(q,\omega)]$. Since we are taking $q \rightarrow 0$ limit in our calculation, therefore, we are considering the loss function behavior under the long wavelength limit. At this junction, it is to be remembered that the dielectric function enters as $\text{Im}[\varepsilon(q,\omega)]$ into the energy loss by an electromagnetic wave in a solid, while $L(\omega) = \text{Im}[-1/\varepsilon(q,\omega)]$ is responsible for the energy loss by a fast charged particle that passes through a solid. The peak position of this loss function determines the typical energy of the plasmons in the system. Electron energy loss spectroscopy (EELS) actually probes the typical inelastic interactions due to irradiation with beam of electrons. It is highly volume sensitive compared to other spectroscopic measurements. The spectra resulting from these collective excitations can be alternatively understood as a JDOS between VB and CB weighted by appropriate matrix elements. In terms of the real and imaginary dielectric constants, a straightforward algebra reveals that $L(\omega) = \text{Im}\left[-1/\varepsilon(q,\omega)\right] = \dfrac{\varepsilon_2(\omega)}{\varepsilon_1^2(\omega) + \varepsilon_2^2(\omega)}$. At the plasma frequency, this expression attains the higher value when $\varepsilon_1 \rightarrow 0$ and $\varepsilon_2 < 1$. EELS is a direct access to the dielectric properties of a material as it corresponds to the excitations of electronic states close to the Fermi level of the material. EELS actually involves the excitations of valence electrons that define the structure of the bandgap, collective plasma oscillation, and/or interband transition in case of a semiconductor or metal.

42.4 POTENTIAL USE OF GRAPHENE AS VARIOUS SENSOR DEVICES

Chemical functionalization of graphene can be of great importance for fabricating various types of sensors using the various forms of graphene. It has a range of unique potential properties, which could be exploited in different types of sensing application. A sensor is a device that can convert a physical property into an electrical signal, which is to be recorded. A sensor must have some basic properties such as sensitivity, selectivity, resolution, and its response time, that is, stability. Response time is an important timescale such that a sensing system takes to display a change in the physical property. Graphene is now used for the development of optoelectronic devices, super-capacitors, and various types of high-performance sensors. Graphene has large surface-to-volume ratio, unique optical properties, high electrical conductivity, excellent carrier mobility, high carrier density ($\sim 10^{13}$ cm^{-2}), high thermal conductivity (~ 1500–5500 W m^{-1} k^{-1}), room-temperature Hall effect, high signal-to-noise ratio (which is due to low intrinsic noises), and extremely high mechanical strength (200 times greater than that of steel, tensile modulus of ~ 1 TPa) (Castro Neto et al., 2009). The large surface area of 2D graphene has the property to enhance the surface loading of desired biomolecules, either through passive adsorption or by covalent cross-linking to the reactive groups of biomolecules. On the other hand, the excellent conductivity and small bandgap of chemical-functionalized graphene are helpful for the conduction of electrons between the biomolecules and the electrode surface. Graphene has almost twofold higher effective surface area and greater cost-effectiveness than CNTs. Additionally, it has greater homogenous surface that is responsible for functionalization. According to its sensing application, it can be classified in different categories—thermal, mechanical, electrochemical, and biological sensors.

42.5 MODIFICATIONS DUE TO CHEMICAL FUNCTIONALIZATION OF GRAPHENE

42.5.1 Modifications in Optical Properties

Optical properties of a material play a very important role in making optoelectronic and spintronic devices. Two-dimensional graphene or graphene-like nanomaterials have been found to have huge potential to replace the present silicon-based nanotechnology completely. So it is therefore important to study their optical properties in detail. Monolayer and bilayer graphenes show interesting optical properties with and without magnetic field (Ho et al., 2010) by using the relation of the optical spectrum based on Fermi's golden rule at zero temperature (Mahan, 1990). The optical spectra of monolayer and bilayer graphenes are given in Figure 42.1 with varying magnetic field ranging from 0 to 60 T. From the figure, it is evident that the rate of absorption increases monotonically with the photon energy. In the presence of magnetic field, the optical response is mainly governed by the transition between different Landau levels. But there are certain selection rules for the transition between different Landau levels. These rules come from the orthogonality of Hermite polynomials. It says that the transition is possible if the two sublattices in graphene have the same number of nodes. The selection can be written as $\Delta n = n_c - n_v = \pm 1$.

However, in bilayer graphene, the layer–layer interaction becomes very important, which can be seen from Figure 42.1. The optical properties of graphene quantum dot (Sheng et al., 2012; Zielinski et al., 2012) have been explored. In particular, (Sheng et al., 2012) have shown that by controlling the size, shape, and edge of the quantum dots, optical properties can be manipulated significantly. Sometimes it is necessary to increase the distance between the graphene sheets in graphite. It is done by intercalating graphite with lithium (Li), potassium (K), etc. This is equivalent to doping. The optical properties of intercalated graphite have been studied (Blinowski et al., 1980; Hoffman et al., 1985). The optical properties of a new type of semiconductor called graphene monoxide are studied (Yang et al., 2013) under the framework of DFT. All the results of optical properties are illustrated in Figures 42.2 and 42.3. Here, they have calculated real ($\varepsilon_1(\omega)$) and complex ($\varepsilon_2(\omega)$) dielectric functions, EELS ($L(\omega) = \text{Im}[-1/\varepsilon(q,\omega)]$), real ($\eta(\omega)$) and complex ($k(\omega)$) refractive

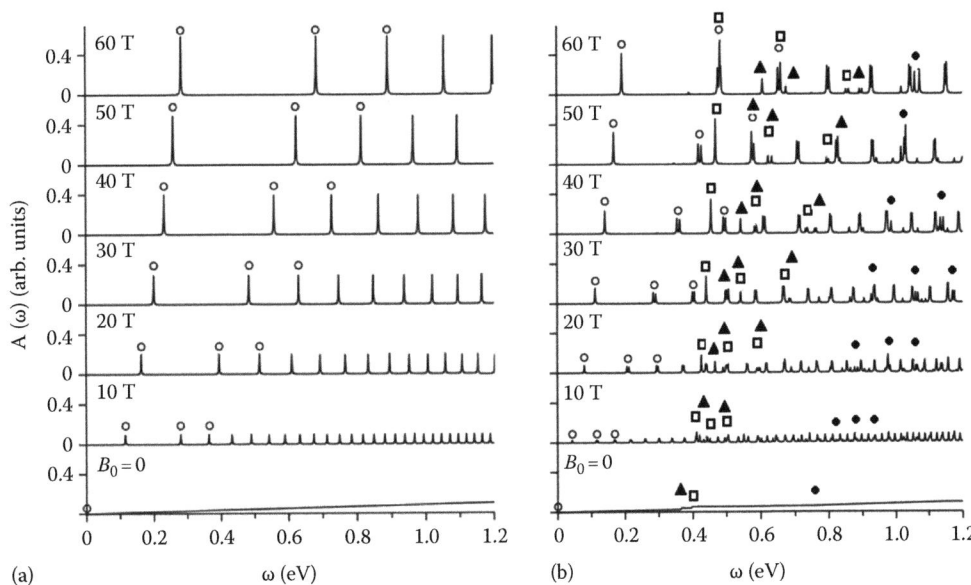

FIGURE 42.1
Optical spectra of (a) monolayer and (b) bilayer graphenes. The magnetic field corresponds to the curves ranging from bottom to top 0–60 T. The hollow black balls, solid black triangles, hollow black squares and solid black balls indicate the first three excitations of $n_1^v \to n_1^c$, $n_2^v \to n_1^c$, $n_1^v \to n_2^c$, and $n_2^c \to n_2^c$, respectively. (Reprinted with permission from Ho, Y.H. et al., *Phil. Trans. R. Soc. A*, 368, 5445, 2010.)

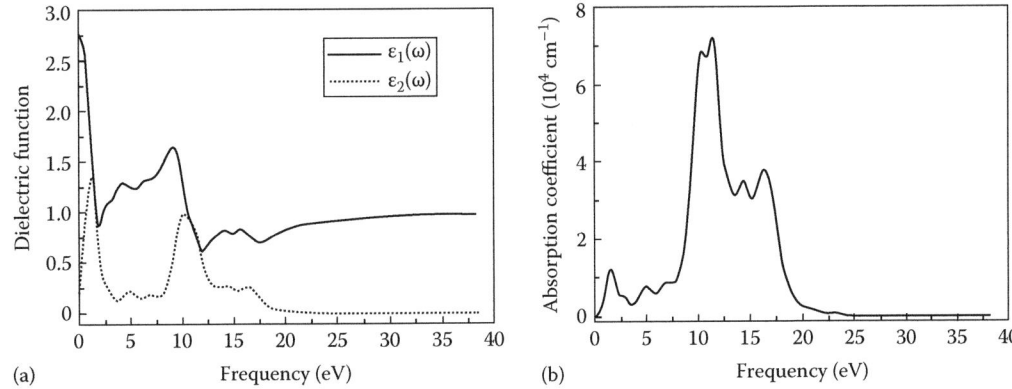

FIGURE 42.2
(a) Dielectric function and (b) absorption coefficient of graphene monoxide. (Reprinted with permission from Sheng, B. et al., *RSC Adv.*, 2, 6761, 2012.)

index, optical conductivity ($\sigma(\omega)$), reflectivity ($R(\omega)$), and absorption coefficient ($\alpha(\omega)$). Cheng et al. (2013) investigated the optical properties of functionalized graphene with hydrogen by using *ab initio* calculation. They have calculated the real and imaginary parts of the optical conductivity for majority, minority, and total spins for different supercell size $\left(\sqrt{3} \times \sqrt{3}, 2 \times 2, 3 \times 3, 4 \times 4, 5 \times 5\right)$. The results are illustrated in Figures 42.4 and 42.5 without considering the spin–orbit coupling effect. Without the spin–orbit coupling, the off-diagonal matrix element of the conductivity tensor is zero. The real part of the conductivity shows completely different behavior from pristine graphene than when it is functionalized with hydrogen. In this study, they have highlighted some important observations: (1) Two spin channels give different optical conductivity. (2) It shows fine structures for band edge transitions. (3) The absorption peak due to vHS around the M point is redshifted than the pristine graphene due to functionalization. (4) There is an enhancement in the optical conductivity at the band edge. Because of the functionalization, apart from the π and π* bands, impurity bands are also formed. In pristine graphene, the transition is only between π and π* bands. But here in the transition process, the impurity bands are also involved. Here, the transition is between the filled impurity band to the π band for the branch involving majority spin and from the π* band to

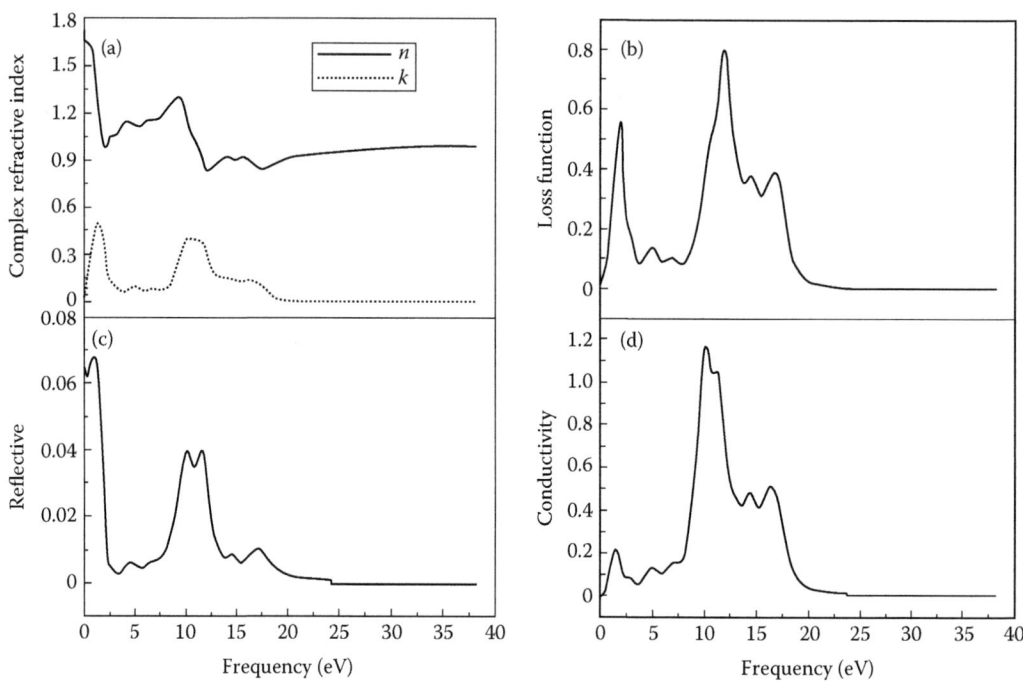

FIGURE 42.3
Optical functions of graphene monoxide. (a) Complex refractive index. (b) Loss function.
(c) Reflective. (d) Conductivity. (Reprinted with permission from Yang, G. et al., *J. Semicond.*, 34(8), 083004-5, 2013.)

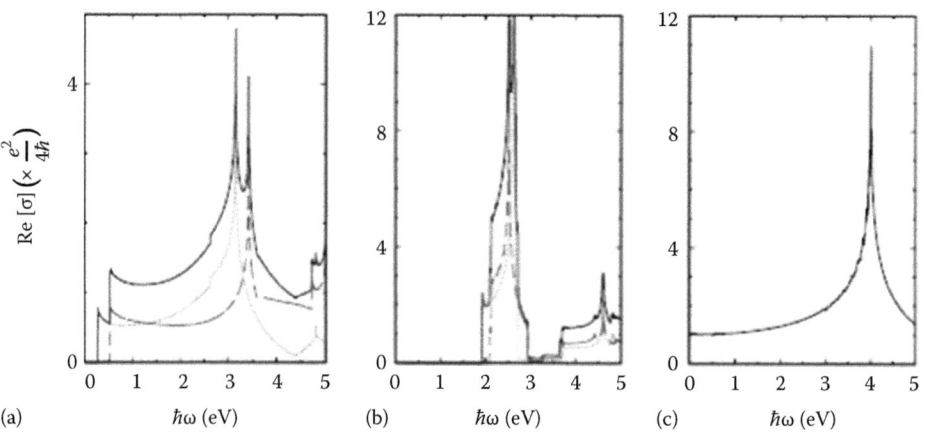

FIGURE 42.4
Real part of the optical conductivity for (a) $\sqrt{3} \times \sqrt{3}$ supercell, (b) 2 × 2 supercell and (c) pristine graphene. Red-dotted (blue-dashed) curves are the contribution from the majority (minority) spin branch; black solid curves are the total spin. (Reprinted with permission from Cheng, J.L. et al., *Phys. Rev. B*, 88, 045438-6, 2013.)

the empty impurity band for the branch involving minority spin. The impurity bands are narrow, so the large electronic DOS enhances the absorption around the band edge. After including the spin–orbit coupling term in the Hamiltonian, due to change of the symmetry in the electronic states, the off-diagonal terms of the conductivity tensor becomes nonzero though the magnitude is about two orders smaller than the diagonal terms. This is represented in Figure 42.6. The most interesting conclusion that can be drawn due to the nonzero value of the off-diagonal terms of the conductivity tensor is the occurrence of Kerr and Faraday effects without any external magnetic field. For functionalized graphene on a substrate, the value of Kerr or Faraday angle is about 10^{-4} rad, where for freestanding functionalized graphene, this value is about 10^{-2} rad.

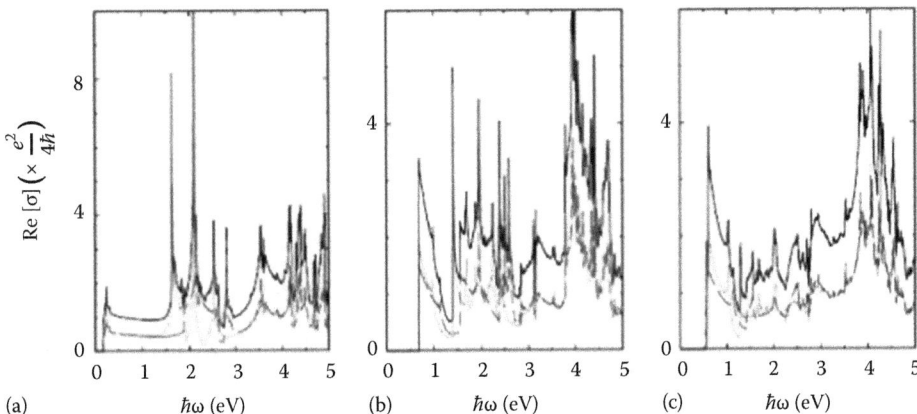

FIGURE 42.5
Optical conductivity for (a) 3 × 3, (b) 4 ×4, and (c) 5 × 5 supercells. (Reprinted with permission from Cheng, J.L. et al., *Phys. Rev. B*, 88, 045438-6, 2013.)

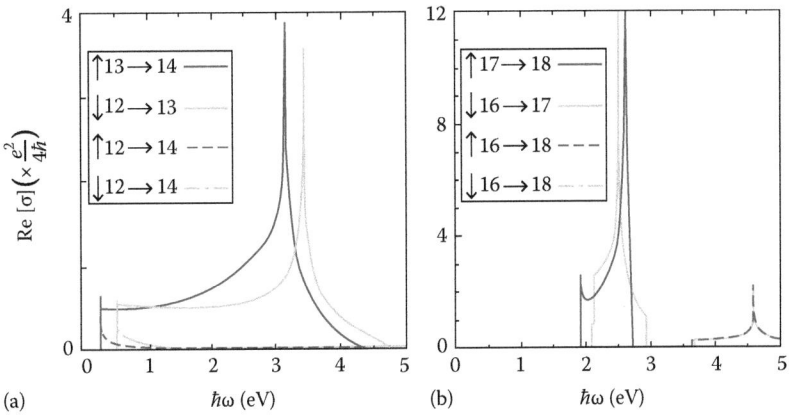

FIGURE 42.6
Band-resolved contribution to the optical conductivity for (a) $\sqrt{3} \times \sqrt{3}$ and (b) 2 × 2 supercells. The impurity bands are labeled by 13 and 17, respectively. We use ↑ (↓) to indicate the majority (minority) bands. (Reprinted with permission from Cheng, J.L. et al., *Phys. Rev. B*, 88, 045438-6, 2013.)

Nath et al. (2014b) studied the optical properties for both parallel ($E_\parallel$) and perpendicular ($E_\perp$) polarization of boron-doped graphene (BG), nitrogen-doped graphene (NG), and boron–nitrogen-codoped graphene (NBG) nanosheet by using *ab initio* computation. In case of pristine graphene, it can be seen that the value of $\varepsilon_1(0)$ is 2.8 for $E_\parallel$ and 1.3 for $E_\perp$. $E_\perp$ does not have any profound effect on $\varepsilon_1(0)$ with increasing concentration of BG, NG, and NBG as evident from Figure 42.7. But $\varepsilon_1(0)$ changes significantly with increasing concentration for $E_\parallel$. The typical spectra of the real part of the dielectric function for pristine, BG, NG, and NBG systems are given in Figure 42.8. The maximum values of $n(\omega)$ also differ for both polarizations. With increasing doping concentration, the peak positions of $n(\omega)$ do not significantly changes for $E_\perp$. But for NG and BG systems, the peak positions shifted toward low-frequency region as depicted in Figure 42.9. The static value of the refractive index ($n(0)$) for parallel polarization for different concentrations of B, N, and BN doping is illustrated in Figure 42.10. The reflectivity data also show some interesting features. It can be noticed that the maximum value of the reflectivity ($R_{max}(\omega)$) for pristine graphene lies in the ultraviolet (UV) region for $E_\parallel$ polarization, but for $E_\perp$, it is in much higher energy range. In around 7.0–10.0 eV region, zero reflectivity can be observed. For $E_\parallel$ polarization, irrespective of doping type and concentration, almost zero reflectivity can be noticed within the region 7.0–8.0 eV. This result is contrary to CNT, where zero reflectivity can be observed only for N system (Jana et al., 2013, 2009). For NG system, the value of the reflectivity is greater than 0.4, which lies near the visible range of the electromagnetic (EM) wave. So it can be inferred that for selective doping with nitrogen, the system can be illuminated with

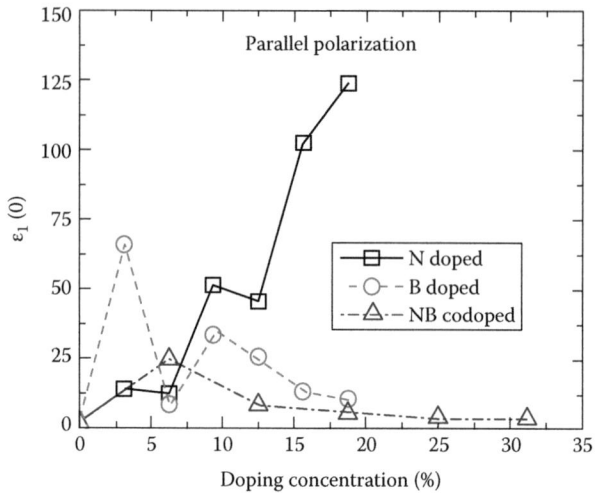

FIGURE 42.7
Variation of static dielectric constant with N, B, and NB doping concentration for $E_{\parallel}$ polarization.
It increases significantly with N concentration. (Reprinted with permission from Nath, P. et al.,
Carbon, 73, 275, 2014.)

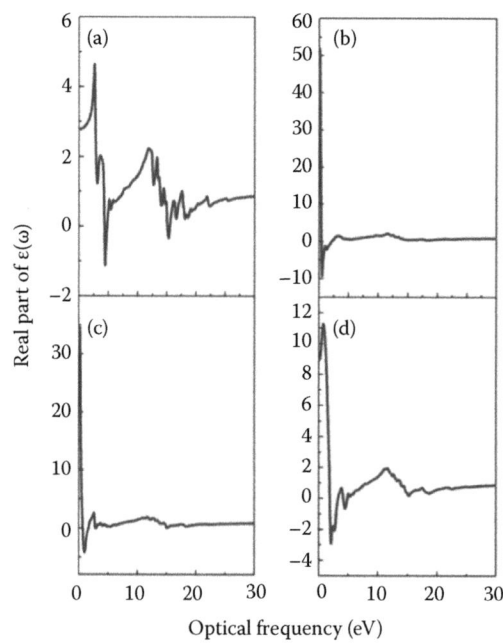

FIGURE 42.8
Representation of the typical spectra of the real part of dielectric function ($\varepsilon(\omega)$) for different doped
systems for parallel polarization. (a) Pristine system, (b) N-doped system, (c) B-doped system, and
(d) NB-codoped system.

visible light. This happens because of the increase in the free charge density in the system for $E_{\parallel}$
polarization, but this effect is totally absent in $E_{\perp}$ polarization. For single-doped system, about 90%
reflectivity is observed for $E_{\parallel}$ polarization. But with increasing B concentration, an overall decreasing
tendency can be observed in the values of $R_{max}(\omega)$ for $E_{\parallel}$ polarization. In case of absorption coefficient,
the maximum value of the absorption coefficient ($\alpha_{max}(\omega)$) is almost the same for both types of polarization in pristine graphene. With increasing concentration of N, B, and NB, the peak height decreases
for both types of polarizations. But for high enough concentration for N doping, the peak height initially decreases, but then it shows an increasing tendency. Here also in the same frequency region,
very small values of $\alpha(\omega)$ can be noticed irrespective of doping type and concentration. Here, it can

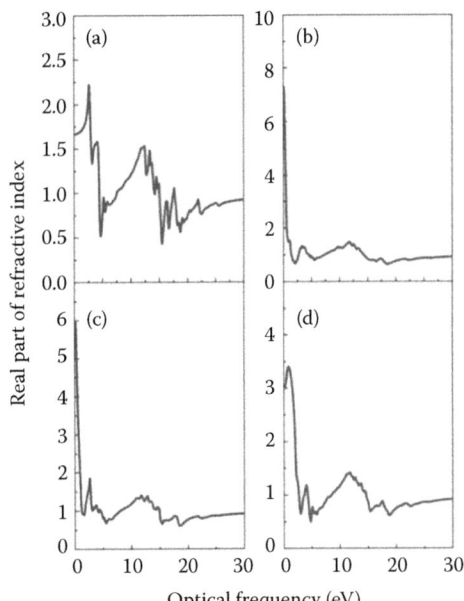

FIGURE 42.9
Representation of the typical spectra of the real part of refractive index ($n(\omega)$) for different doped systems for parallel polarization. (a) Pristine system, (b) N-doped system, (c) B-doped system, and (d) NB-codoped system.

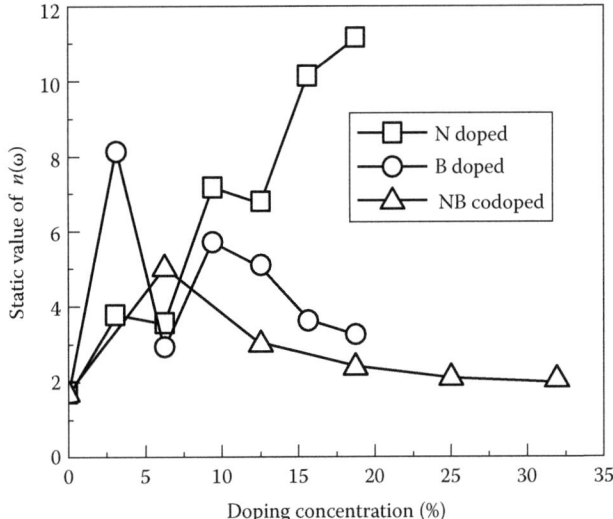

FIGURE 42.10
Representation of the variation of the static value of refractive index ($n(0)$) with respect to doping concentration for parallel polarization.

be inferred that pristine graphene as well as doped graphene nanosheet is highly transparent in this frequency range. Some typical spectra of the absorption coefficient for pristine, NG, BG, and NBG type are illustrated in Figure 42.11. Some prominent EELS peaks can be observed as illustrated in Figure 42.12. The first EELS peak can be seen at 4.9 eV with peak height 1.75. Then three more prominent peaks can be noticed within the frequency range 15.5–19.0 eV for $E_\parallel$ polarization. Only a single prominent peak of height 2.4 can be observed for $E_\perp$ polarization around 15.3 eV. Eberlein et al. (2008) experimentally observed in-plane plasmon excitation in freestanding graphene due to π and ($\pi + \sigma$) plasmon excitations around 4.7 and 14.6 eV, respectively. Out-of-plane plasmon excitation has also been observed by Eberlein et al. (2008) near 15 eV. This peak occurs due to transition from π to π^* bands. These results are in well agreement with the theoretically predicted data for both

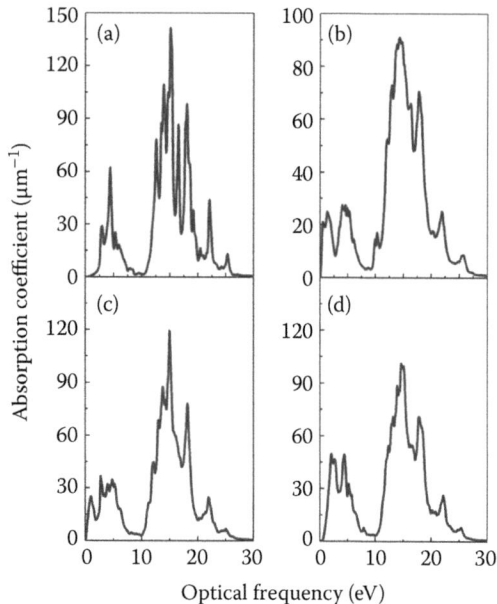

FIGURE 42.11

Representation of the typical spectra of the absorption coefficient ($\alpha(\omega)$) for different doped systems for parallel polarization. (a) Pristine system, (b) N-doped system, (c) B-doped system, and (d) NB-codoped system.

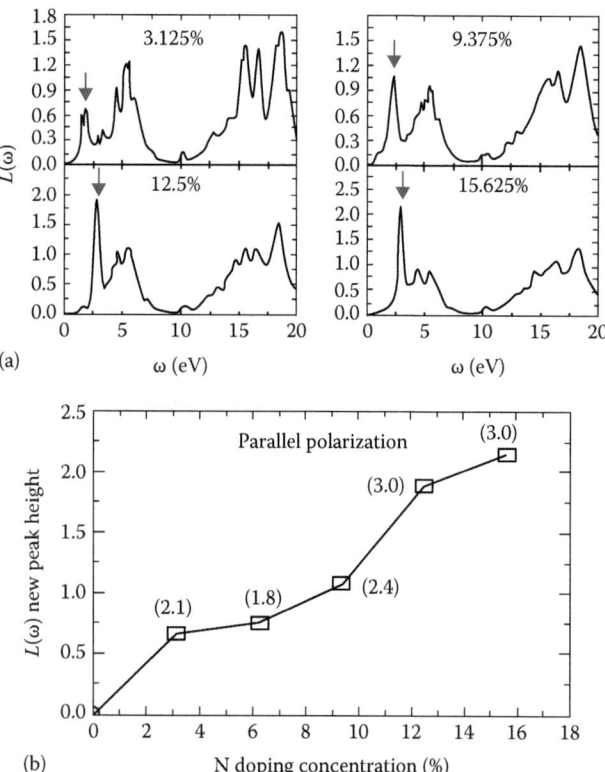

FIGURE 42.12

EELS for NG systems. (a) Depicts the EELS for some NG systems of different concentrations (%) as noted in different graph panels. New EELS peaks are denoted by red downward arrows. (b) New EELS peak height variation with N doping concentration for $E_{\parallel}$ polarization. The numbers corresponding to the data points in this figure denote the EELS peak position in eV unit of the corresponding N doping concentration. (Reprinted with permission from Nath, P. et al., *Carbon*, 73, 275, 2014.)

polarizations. With increasing doping concentration, the $L(\omega)$ peak height at the frequency level 15.3 eV starts decreasing for any type of doping for $E_\perp$ polarization. Broadening of EELS peak starts occurring for $E_\parallel$ polarization around 15.5–19.0 eV, so losing their identity. A new EELS peak can be observed for nitrogen-doped system for $E_\parallel$ polarization around 1.8–3.0 eV. Increasing charge density due to N doping gives rise to this new EELS peak, which may be associated due to the addition of extra π electrons in the system. Now it will be very interesting to study the XES, XAS, and RIXS measurements as reported by Zhang et al. (2012). These spectra can be studied by the variation of dielectric functions. Johari and Shenoy (2011) studied the optical properties of graphene oxide (GO) with different concentration of major functional groups like epoxides, hydroxyls, and carbonyls by using DFT. The structures used are depicted in Figure 42.13. They have mainly studied the EELS spectra for both parallel and perpendicular polarizations to check the sensitivity of π and $\pi + \sigma$ plasmons due to different functional groups. They have found that π plasmon is less sensitive than

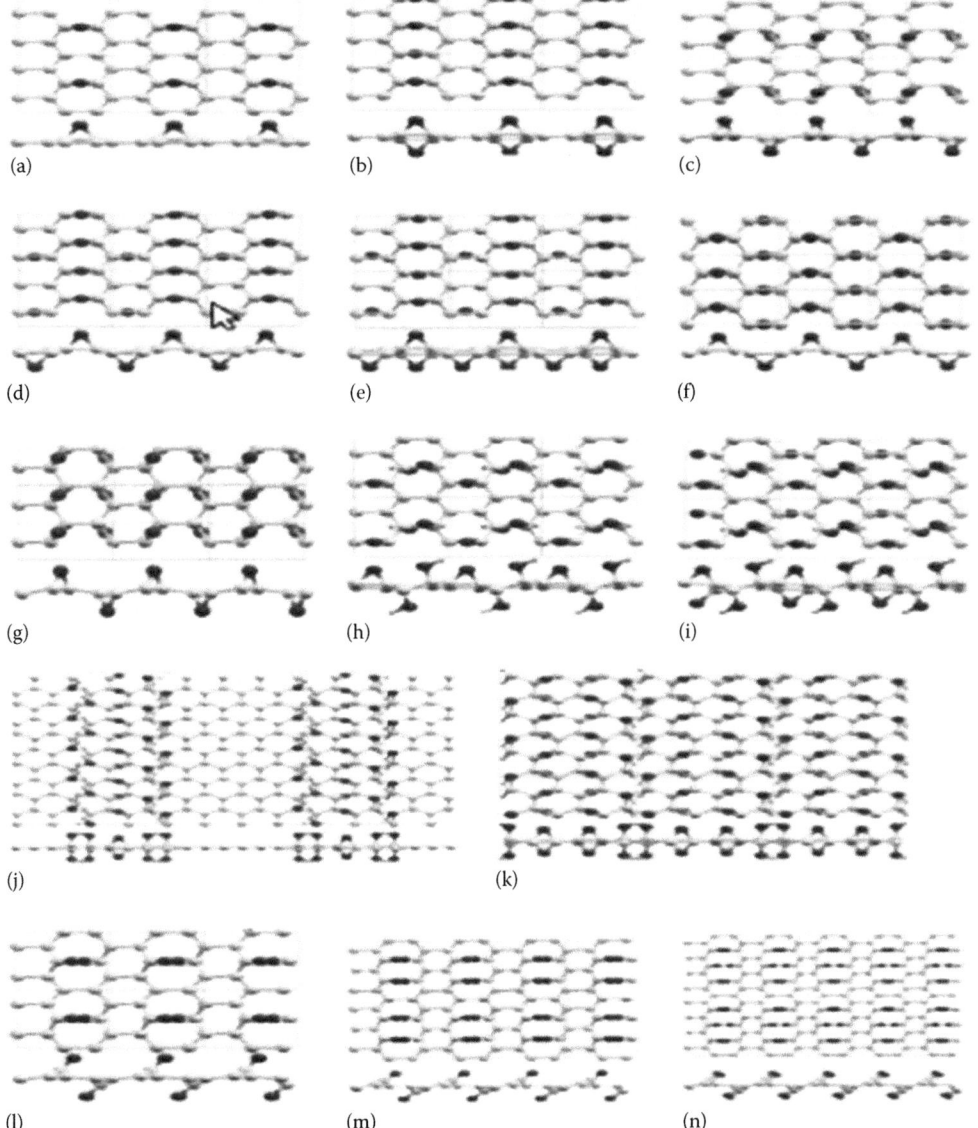

FIGURE 42.13
Top and side views of the various configurations of GO. GO with various coverages of epoxy functional groups is depicted: (a) 25%, (b, c) 50%, (d, e) 75%, and (f, g) 100%. Configurations (h–k) represent 50%, 75%, 50%, and 100%, respectively, coverage of the graphene basal plane by epoxy and hydroxyl functional groups. GO with carbonyl groups is shown in configurations (l, m, and n) with two, four, and six carbonyl groups, respectively. (Reprinted with permission from Johari, P. and Shenoy, V.B., *ACS Nano*, 5, 7640, 2011.)

TABLE 42.1 k-Point Meshes Used for the Relaxation and the Electronic Structure Calculations (vac ≈ 12 Å) and for the Loss Spectra Calculations (vac ≈ 27 Å), for Graphene (Gr) and GO with Various Configurations and Functional Groups Depicted in Figure 42.13

Structure	No. of Atoms in the Unit Cell	sp³-Bonded Carbon Atoms (%)	Functional Groups			K Mesh		E_{form} (eV)
			Epoxy	Hydroxyl	Carbonyl	vac ≈ 12 Å	vac ≈ 27 Å	
Gr	4	0	0	0	0	45 × 45 × 1	45 × 45 × 1	—
a	9	25	1	0	0	35 × 25 × 1	25 × 15 × 1	0.13
b	10	50	2	0	0	35 × 25 × 1	25 × 15 × 1	0.08
c	10	50	2	0	0	35 × 25 × 1	25 × 15 × 1	−0.16
d	11	75	3	0	0	35 × 25 × 1	25 × 15 × 1	−1.09
e	11	75	3	0	0	35 × 25 × 1	25 × 15 × 1	−0.82
f	6	100	2	0	0	45 × 45 × 1	20 × 40 × 1	−0.94
g	6	100	2	0	0	45 × 45 × 1	20 × 40 × 1	−0.76
h	13	50	1	2	0	35 × 25 × 1	25 × 15 × 1	−3.53
i	14	75	2	2	0	35 × 25 × 1	25 × 15 × 1	−4.67
j	42	50	2	8	0	8 × 32 × 1	8 × 32 × 1	−17.04
k	24	100	4	4	0	21 × 42 × 1	10 × 20 × 1	−11.17
l	10	0	0	0	2	35 × 25 × 1	25 × 15 × 1	−0.47
m	16	0	0	0	4	30 × 15 × 1	20 × 10 × 1	−2.24
n	22	0	0	0	6	30 × 15 × 1	12 × 8 × 1	−4.09

Source: Reproduced with permission from Johari, P. and Shenoy, V.B., *ACS Nano*, 5, 7640, 2011.

Note: The number of atoms in the unit cell and formation energies of the system presented in Figure 42.13 are also mentioned.

π + σ plasmon. The π + σ plasmon gets blueshifted up to 3.0 eV due to change in concentration of epoxy and hydroxyl functional groups from 25% to 75%. But due to the addition of carbonyl groups, it gets redshifted because of the generation of holes in the GO system. The structural details are schematically illustrated in Table 42.1. Experimentally, it was confirmed that epoxides and hydroxyls are major functional groups in GO (Lerf et al., 1998; Cai et al., 2008). In this work, they have confirmed this theoretically. The most stable structure is the structure of pristine graphene and oxidized graphene at 50% coverage, which was previously found by Yan et al. (2009). From their work, it can be seen that the π and π + σ plasmon peaks appear around the energy range 4.8 and 14.5 eV (Eberlein et al., 2008; Gass et al., 2008). As mentioned earlier, the epoxide functional group affects the π + σ plasmon peak more than the π plasmon peak compared with the pristine graphene. When both the epoxy and the hydroxyl groups are present, there is a significant blueshift in the π + σ plasmon peak. One important inference can be drawn from the above consideration that, the position of the π + σ plasmon peak, so the optical properties can be tailored over a wide range by controlled deoxidation, whereas the π peaks remain less affected. With 100% coverage, there are no sharp π peaks, as all the carbon atoms become sp³ bonded. As mentioned earlier, due to the addition of carbonyl groups, the π peak gets redshifted, while the π + σ peak almost remains unaffected. The maximum redshift of 1.0 eV is observed when the oxygen-to-carbon ratio is 37.5%. So here the π plasmon can be controlled by varying the coverage of carbonyl group. For perpendicular polarization, the EELS spectra mainly exhibit one peak, which is not sharp but smeared over a wide range of energy. In case of the EELS spectra for perpendicular polarization, also the π + σ plasmon peak remains unaffected. They have also studied the optical gap, which is the lowest π−π* gap in the DOS. The optical gap of pristine graphene and its variation with different coverage and composition is depicted in Figure 42.14. However, DFT calculation is mainly done in independent particle approximation, which is less accurate to account for the excitonic effect. Here, for more accurate results, Bethe–Salpeter equation is suggested. The GGA-PBE exchange correlation functional always underestimates any excited state property, and also any other hybrid functional scheme is not suitable for this type of computational study. But we can rely on these results to get correct trend of the optical properties. Sharma et al. (2013) calculated the optical absorbance of rectangular shape quantum dot of graphene, boron nitride,

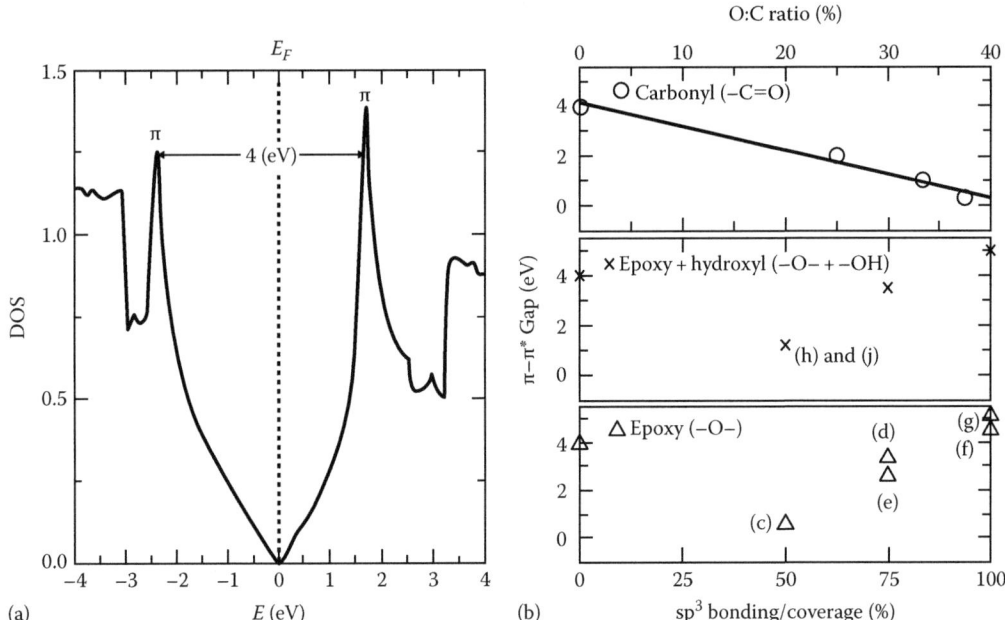

FIGURE 42.14
(a) DOS of pristine graphene, showing the optical gap, i.e., the π–π* gap to be 4 eV. Dotted line depicts the Fermi level. (b) Optical gap of various compositions of GO with respect to its coverage by the functional groups. The captions (c), (d), (e), (f), (g), (h), (j) refer to the various configurations of GO as depicted in Figure 42.13. The optical gap is plotted with respect to the percentage of sp^3-bonded carbon atoms (coverage) for epoxides and hydroxyls (lower and middle plots), while for GO with carbonyl groups, it is plotted with respect to the O:C ratio (upper plot). Solid line shows the linear fit. (Reprinted with permission from Johari, P. and Shenoy, V.B, *ACS Nano*, 5, 7640, 2011.)

and their hybrids by using first-principles calculations. They have shown that all quantum dots absorb in the UV energy region, but only for boron nitride quantum dot, the maximum value of the absorbance lies in the UV energy range. Xu et al. (2010) consider the electron–photon–phonon interaction to compute the optical conductance and transmission coefficient in the presence of a linearly polarized radiation field. In the short wavelength limit, they have obtained the universal value of the optical conductance $\sigma_0 = e^2/(4\hbar)$. Most importantly, they have found an optical window of having wavelength 4–100 μm for optical absorbance induced by different interband and intraband transitions. The width of this window strongly depends on the temperature and free carrier charge density of the system. Prezzi et al. (2008) computed the optoelectronic properties of armchair graphene nanoribbons (A-GNRs) from first-principles calculations. They have incorporated the many-body effect within the calculations. Their calculations show the importance of many-body effects while dealing these types of systems. Here, the optical spectra are mainly dominated due to strongly bound exciton. A-GNRs are proposed to have three distinct classes: N = 3p − 1, N = 3p, and N = 3p + 1, where p is an integer and N is the number of dimer lines across the ribbon width. The optical absorbance shows strong dependence on the family of A-GNRs. The optical response can be tuned by changing the family and also the termination of edge states. Pedersen et al. (2008) obtained the optical properties of graphene antidot lattices by using tight-binding formalism. By antidot lattices of graphene, we mean periodic array of holes within the graphene sheet. This type of artificial nanomaterials shows dipole-allowed direct bandgap, which induces prominent optical absorption edge. In this study, to calculate the optical conductivity, they have used triangle integration method, which is given in the appendix in their paper. For optical conductivity calculation for different lattices, they show that the edge states are clearly visible in the spectra. In experiments, graphene samples are usually placed on a substrate (usually oxidized silicon wafer). So the effect of a substrate on the optical spectra is also very important. They have shown that the thickness of the oxide has also a profound effect on the spectra. The low-energy refractive index has also been studied for different antidot systems. Falkovsky (2008) gives the detailed theoretical study of the optical properties of single-layer and multilayer graphene. Optical conductivity of graphene doped with the career density 10^{10} cm^{-2} is depicted in Figure 42.15.

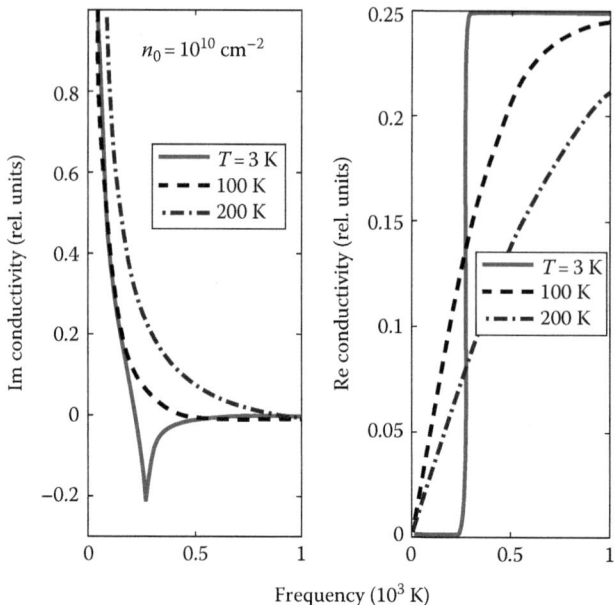

FIGURE 42.15
Imaginary and real parts of optical conductivity in $e^2/\hbar$ for graphene doped with the carrier density 10^{10} cm^{-2}. (Reprinted with permission from Falkovsky, L.A., *J. Phys. Conf. Ser.*, 129, 012004-11, 2008.)

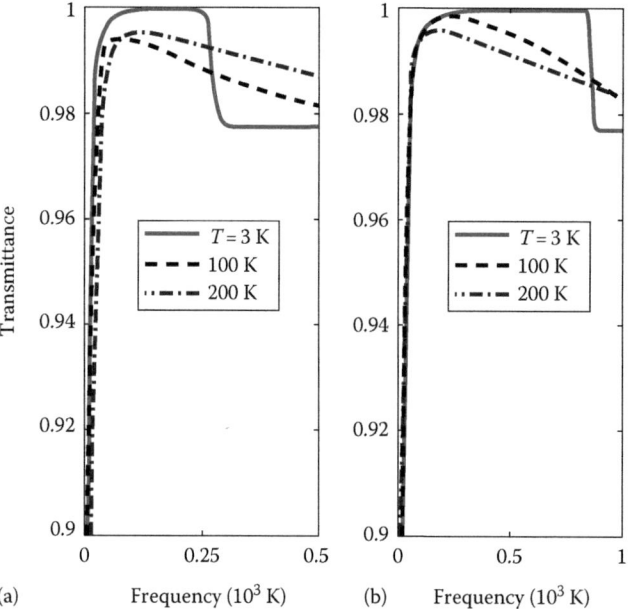

FIGURE 42.16
Transmittance spectrum of graphene with carrier densities (a) $n_0 = 10^{10}$ cm^{-2} and (b) $n_0 = 10^{11}$ cm^{-2} versus frequency; normal incidence. For the carrier density $n_0 = 10^{11}$ cm^{-2}, the chemical potential equals to 428, 389, and 294 at 3, 100, and 200 K, correspondingly. (Reprinted with permission from Falkovsky, L.A., *J. Phys. Conf. Ser.*, 129, 012004-11, 2008.)

Transmittance spectra of single-layer graphene doped with career densities 10^{10} cm^{-2} and 10^{11} cm^{-2} for normal incidence are illustrated in Figure 42.16. The transmittance and reflectance spectra for multi-layer graphene having career density 10^{10} cm^{-2}, layer thickness 3.35 Å, and total plate thickness of 33.5 nm are also depicted in Figure 42.17. Recently, Putz et al. (2014) have studied the complex optical conductivity for graphene supercell of various size. Santoso et al. (2014) have calculated the optical conductivity in single-layer graphene induced by mild oxygen plasma exposure.

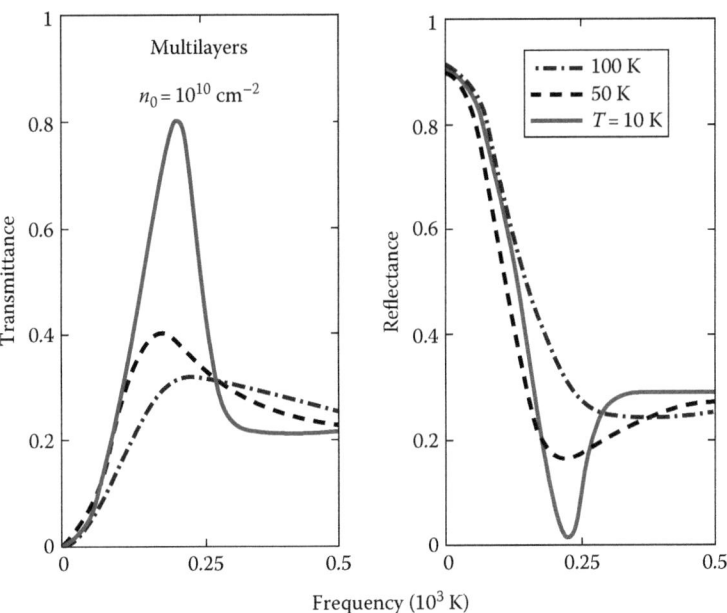

FIGURE 42.17
Transmittance and reflectance at normal incidence for a film with the multilayer graphene and the carrier density $n_0 = 10^{10}$ cm^{-2} in a layer; the distance between the layers is $d = 3.35$ Å; thickness of the plate is $l = 100d$; temperatures are noted at curves. (Reprinted with permission from Falkovsky, L.A., *J. Phys. Conf. Ser.*, 129, 012004-11, 2008.)

42.5.2 Modification in Synthesis of Graphene-Based Sensor

There are a lot of methods that are used to develop graphene such as epitaxial growth, chemical vapor deposition (CVD), micromechanical exfoliation, arc discharge method, intercalation methods in graphite, and unzipping of CNTs. Kumar et al. (2013) reported the preparation of NG by microwave plasma CVD method, which is a very efficient approache for the large-scale synthesis as it is cost-effective and environmentally friendly. Each method has its own advantages and disadvantages. Among all of these methods, chemical method is the efficient as well as profitable method for the production of bulk quantity of graphene toward applications in electrochemical sensors and energy storage devices. A wide range of chemical modification and biomolecular binding strategies (Georgakilas et al., 2012) have been developed to functionalize specific chemical groups (such as carboxyl, hydroxyl, acid chloride, and amine) on graphene and for binding the chemically modified graphene to the biomolecules, respectively. Similarly, several methods have also been developed for preparing graphene nanocomposites by conjugating graphene to nanomaterials.

42.6 POTENTIAL OF SENSOR DEVICE

Recently, various efforts were done to review the structure, preparation, properties, and applications of graphene and its composite materials (Geim, 2009; Rao et al., 2009) in sensor application. Graphene and graphene-based composite electrode materials exhibited high electrocatalytic properties toward sensors, biosensors, solar cells, and pesticides.

42.6.1 Graphene-Based Chemical Sensors

Currently, nanomaterials are used for electrochemical sensor. Depending on the diameter and the degree of helicity, CNTs are used as electrochemical sensor. It was studied that CNTs can act as efficient electron transfer promoter. Graphene and its derivatives are now viable candidates for using as ultrasensitive chemical sensor. So graphene is an electroactive material. For ultrahigh sensitivity detection of various gases existing in environment, graphene is used as chemical sensors. High levels of sensitivity in detection processes are important for different industrial, environmental, and public safety application. Graphene is now used as electrochemical biosensors, which are based on the direct electron transfer between the enzyme and the electrode surface. They have superior analytical performance and excellent antifouling ability (Zheng et al., 2012). Electrochemical sensors offer selectivity and sensitivity with very low detection limits with range from nanomolar to picomolar (Klessen et al., 1994). Electrochemical techniques including cyclic voltammetry (Cui and Zhang, 2012) and differential pulse voltammetry (DPV) were employed to study the electrochemical sensors. Du et al. (2013) reported a nonenzymatic electrochemical uric acid sensor by using graphene-modified carbon fiber electrode. A single-walled carbon nanotubes–graphene hybrid (SWCNT-GNS) film-modified electrode was fabricated for the detection of acetaminophen by DPV.

42.6.2 Graphene-Based Biosensors

Graphene-based biosensors were extensively studied (Chen et al., 2008; Xu et al., 2012) due to the large specific area, good electrical, thermal conductivity, and biocompatibility properties of graphene. Graphene is also suitable for enzyme-based biosensor due to its excellent catalytic behavior toward inorganic molecule like H_2O_2 and nicotinamide adenine dinucleotide (NADH). For the development of biosensor, direct electron communication between electrode and active center of enzyme without any extra molecule is very important. But in most of the cases, active center of redox enzymes is located in hyperbolic cavity of molecule. Due to direct electron transfer in between electrode and enzyme, CNTs are used as biosensors. However, functionalized graphene can be used to promote electron transfer in between electrode and enzyme. NG can show efficient electrocatalytic activity for the reduction of H_2O_2. Besides, NG can perform glucose biosensing in a low concentration of 0.01 N. The CVD method (Wei et al., 2009) can be successfully used to prepare NG. Large area synthesis of N-graphene films can be easily transferred to different substrates for various characterization and measurements. These doped films exhibit remarkable performance for oxygen reduction reaction (ORR) associated with alkaline fuel cell and sensors (Jana et al., 2013). Graphene has been widely used in biosensors (Liu et al., 2012; Xu et al., 2012). It can detect glucose, glutamate, hydrogen peroxide, benzene, xylenes, cyclohexane, NADH, hemoglobin, cholesterol, protein biomarkers like alpha fetoprotein, carcinoembryonic antigen, prostate-specific antigen, human epidermal growth factor receptor 2, epidermal growth factor receptor, immunoglobulin G (IgG) and IgE, saccharides, and cancer cells also. Graphene-based nanoelectronic devices have also been employed for DNA sensors (for detecting single- and double-stranded DNA, nucleobases, and nucleotides). DNA sequencing can be done by graphene. Recently, Xu et al. (2009) have shown that graphene can solve the bottleneck issue of how to achieve single-base resolution of electronic DNA sequencing. Solid-state nanopores act as single-molecule sensors, and they can rapidly sequence DNA molecules. Graphene-based nanopores also can translocate DNA. As DNA molecules move through the pore, the device can simultaneously measure drops in ionic current and changes in local voltage in the transistor, which are used to detect the molecules.

42.6.3 Graphene-Based Gas Sensors

Due to adsorption of gas molecules in graphene surface, it can act as suitable gas sensors. In such a situation, gas molecules act as donors as well as acceptor. Gas sensors are used for detecting hydrogen, carbon monoxide, ammonia, chlorine, nitrogen dioxide, nitric oxide, oxygen, ethanol, water vapors, iodine, methane, hydrogen cyanide, and trimethylamine. All atoms of a single-layer graphene sheet can be considered as surface atoms, and they have the capability of adsorbing gas molecules, providing the largest sensing area per unit volume. Yuan et al. (2013) reported that microscopic sensor produced from graphene can detect individual gas molecules. The adsorbed molecule can change the local carrier concentration in graphene, which is like stepwise change in resistance. This device can show change in electrical resistivity by adsorption of gases. As conductivity is proportional to the product of the number of charge carriers and mobility, the change in conductivity must be due to changes in the number density or mobility of carriers. This implies that gas adsorption can enhance the number of holes if gas acts as an acceptor or can increase the number of electrons if gas acts as a donor. Graphene covered with thin film of platinum showed reduction of resistance in response to exposure of 1% hydrogen at different temperature. Here, basically sensor work is based on splitting of H_2 molecule in the presence of catalytic metal. Individual graphene sheets for sensing can be complex and expensive and can suffer from poor reliability due to contamination and large variability from sample to sample. This problem, however, can be overcome by developing a sensor based on vertically aligned graphene sheets or carbon nanowalls (CNWs).

42.6.4 Graphene-Based Pesticide Sensors

This type of sensor is broadly used in the field of agriculture and to control insecticides and pests. Pesticides were detected using analytical techniques such as gas or liquid chromatography with mass spectrometry and fluorimetry (Diaz and Peinado, 1997). But these techniques often are associated with the main disadvantages like expensivity, complicacy, and requirement of highly skilled labors. Alternatively electrochemical methods such as cyclic voltammetry, chronoamperometry, and DPV offered simple and inexpensive route for the sensitive detection of pesticides at nanomolar concentration.

42.6.5 Graphene-Based Strain Sensors

Strain sensors can detect deformations or structural change occurring in our surrounding infrastructures or in internal activities in human bodies. To make an efficient strain sensor, the quality of the material should be chosen such that it can exhibit large structural change in response to a small strain. Lee et al. (2010) studied the piezoresistance response of graphene and the graphene-based strain sensor with a gauge factor of 6.1. Due to excellent stretchability and transparency (75%–80%) of graphene, it is used for human interface devices for real-life applications. Bae et al. (2013) have shown that graphene strain sensor can measure electrical resistance at the time of tensile deformation due to application of continuous voltage. From Figure 42.18, it is clear that there are two profound regions in the graph: (1) when the applied strain is lower than 1.8%, almost linear relationship between the resistance change and the strain, and (2) when the strain is in the range between 1.8% and 7.1%, there exists a nonlinear relationship. Moreover, when the applied strain is greater than 7.1%, resistance increases abruptly, and the device does not respond properly. In region (1), lower than 1.8%, the gauge actor is about 2.4, which is comparable to metallic strain gauge. In contrast, in region (2), the gauge factor of the strain sensor over 1.8% ranges from 4 to 14. In region (1), change in resistance occurs due to carbon–carbon bond length stretching. In region (2), change in resistance occurs due to temperature, effective mass, doping concentration, and crystal imperfection (Kittel, 1976).

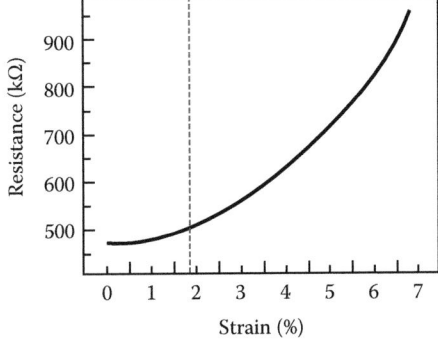

FIGURE 42.18
Representation of the variation of resistance with increasing strain. (Reprinted with permission from Bae, S.H. et al., *Carbon*, 51, 236, 2013.)

42.7 PERSPECTIVE AND FUTURE DIRECTIONS

1. Graphene can easily replace indium-based electrodes in organic light-emitting diode (OLED). These graphene diodes are low of cost, and they are sensitive. So it can be used as sensors in mobile display screens.
2. Graphene sheets can store electrons. So it is used as ultracapacitor. Besides, due to large surface area of graphene, it can provide increase of electric power.
3. Graphene-based sensors have a large surface area. Molecules that are sensitive to a particular disease can attract carbon atoms in graphene. Thus, a sensor can be formed by attaching a fluorescent molecule to the single-stranded DNA. Now these DNA will be attached to graphene. When a large number of single-stranded DNA combines with graphene, it will form double-stranded DNA, which will enhance the fluorescence level, and this gives rise to sensor in application to diagnose disease.
4. Graphene has been shown to be noncytotoxic in cell culture experiments, as it affects the growth of gram-positive and gram-negative bacteria. So graphene-based sensors have significant applications in health care.

42.8 CONCLUSIONS

Chemical functionalization of appropriate forms of graphene can open the door of various applications in optoelectronic and sensor device. Moreover, the electronic structure and the associated optical properties of chemical-functionalized graphene can also serve as benchmark in computational calculations in other 2D materials. Graphene is applicable to a wide range of sensing modalities and offers advantages over more conventional materials (like catalytic sensor, metal oxide sensor, and optical sensor) in nearly all of them. Catalytic sensors are inexpensive but lack in sensitivity. Optical sensors are highly sensitive and expensive. The main advantage of graphene is its mechanical stability and, in particular, its utility for the fabrication of flexible electronic devices particularly for flexible touch screens. Although much work is yet to be done, the future prospect of graphene in sensing devices looks very exciting indeed.

REFERENCES

Ando, T. 2009. The electronic properties of graphene and carbon nanotubes. *NPG Asia Mater.* 1: 17–21.

Anton, L., Heyong, H., Michael, F., and Jacek, K. 1998. Structure of graphite oxide revisited. *J. Phys. Chem. B* 102: 4477–4482.

Araujo, P.T., Terrones, M., and Dresselhaus, M.S. 2012. Defects and impurities in graphene-like materials. *Mater. Today* 15: 98–109.

Arramel, A., Gomez, A.C., and Wees, B.J.V. 2013. Band gap opening of graphene by noncovalent $\pi-\pi^*$ interaction with porphyrins. *Graphene* 2: 102–108.

Bae, S.H., Lee, Y., Sharma, B.K., Lee, H.J., Kim, J.H., and Ahn, J.H. 2013. Graphene-based transparent strain sensor. *Carbon* 51: 236–242.

Balog, R., Jørgensen, B., Nilsson, L., Andersen, M., Rienks, E., Bianchi, M., Fanetti, M., Lægsgaard, E., Baraldi, A., and Lizzit, S. 2010. Band gap opening in graphene induced by patterned hydrogen adsorption. *Nat. Mater.* 9: 315–319.

Blinowski, J., Hau, N.H., Rigaux, C., Vieren, J.P., Toullec, R.L., Furdin, G., Herold, A., and Melin, J. 1980. Band structure model and dynamical dielectric function in lowest stages of graphite acceptor compounds. *J. Phys. (Paris)* 41(1): 667–676.

Liu, Y., Dong, X., and Chen, P. 2012. Biological and chemical sensors based on graphene materials. *Chem. Soc. Rev.* 41: 2283–2307.

Loh, K.P., Bao, Q., Ang, P.K., and Yang, J. 2010. The chemistry of graphene. *J. Mater. Chem.* 20: 2277–2289.

Lusk, M.T. and Carr, L.D. 2008. Nanoengineering defect structures on graphene. *Phys. Rev. Lett.* 100: 175503-07.

Ma, L., Hu, H., Zhu, L., and Wang, J. 2011. Boron and nitrogen doping induced half-metallicity in zigzag triwing graphene nanoribbons. *J. Phys. Chem. C* 175: 6195–6199.

Mahan, G.D. 1990. *Many Particle Physics*. Plenum Press, New York.

Milman, V., Winkler, B., White, J.A., Pickard, C.J., Payne, M.C., and Akhmatskaya, E.V. 2000. Electronic structure, properties, and phase stability of inorganic crystals: A pseudo-potential plane-wave study. *Int. J. Quant. Chem.* 77: 895–910.

Nakada, K. and Ishii, A. 2011. Migration of adatom adsorption on graphene using DFT calculation. *Solid State Commun.* 151: 13–16.

Nath, P., Chowdhury, S., Sanyal, D., and Jana, D. 2014b. Ab-initio calculation of electronic and optical properties of nitrogen and boron doped graphene nanosheets. *Carbon* 73: 275–282.

Nath, P., Sanyal, D., and Jana, D. 2014a. Semi-metallic to semiconducting transition in graphene-nanosheet with site specific co-doping of boron and nitrogen. *Phys. E* 56: 64–68.

Nath, P., Sanyal, D., and Jana, D. 2015. Optical properties of transition metal atom adsorbed graphene: A density functional theoretical calculation. *Physica E* 69: 306–315.

Novoselov, K.S., Geim, A.K., Morozov, S.V. et al. 2004. Electric field effect in atomically thin carbon films. *Science* 306: 666–669.

Odom, T.W., Huang, J.L., Kim, P., and Lieber, C.M. 1998. Atomic structure and electronic properties of single-walled carbon nanotubes. *Nature* 391: 62–64. Structure and electronic properties of carbon nanotubes. *J. Phys. Chem. B* 2000; 104: 2794–2809.

Odom, T.W., Huang, J.L., Kim, P., and Lieber, C.M. 2000. Structure and electronic properties of carbon nanotubes. *J. Phys. Chem. B* 104: 2794–2809.

Parr, R.G. and Yang, W. 1989. *Density Functional Theory of Atoms and Molecules*. Oxford University Press, Oxford, U.K.

Payne, M.C., Teter, M.P., Allan, D.C., Arias, T.A., and Joannopoulos, J.D. 1992. Iterative minimization techniques for *ab initio* total-energy calculations: Molecular dynamics and conjugate gradients. *Rev. Mod. Phys.* 64: 1045–1097.

Pedersen, T.G., Flindt, C., Pedersen, J., Jauho, A., Mortensen, N.A., and Pedersen, K. 2008. Optical properties of graphene antidot lattices. *Phys. Rev. B* 77: 245431-6.

Perdew, J.P., Burke, K., and Enzerhof, M. 1966. Generalized gradient approximation made simple. *Phys. Rev. Lett.* 77: 3865–3868.

Prezzi, D., Varsano, D., Ruini, A., Marini, A.A., and Molinari, E. 2008. Optical properties of graphene nanoribbons: The role of many-body effects. *Phys. Rev. B* 77: 041404(R)-4.

Putz, S., Gmitra, M., and Fabin, J. 2014. Optical conductivity of hydrogenated graphene from first principles. *Phys. Rev. B* 89: 035437-7.

Rani, P., Dubey, G.S., and Jindal, V.K. 2014. DFT study of optical properties of pure and doped graphene. *Phys. E* 62: 28–35.

Rani, P. and Jindal, V.K. 2013a. Designing band gap of graphene by B and N dopant atoms. *RSC Adv.* 3: 802–812.

Rani, P. and Jindal, V.K. 2013b. Stability and electronic properties of isomers of B/N co-doped graphene. *Appl. Nanosci.* 4: 989–996 (doi:10.1007/s 13204-93-0280-3).

Rao, C.N.R., Sood, A.K., Subrahmanyam, K.S., and Govindaraj, A. 2009. Graphene: The new two-dimensional nanomaterial. *Angew. Chem. Int. Ed. Engl.* 48(42): 7752–7777.

Robinson, J.T., Burgess, J.S., Junkermeier, C.E., Badescu, S.C., Reinecke, T.L., Perkins, F.K., Zalalutdniov, M.K., Baldwin, J.W., Culbertson, J.C., and Sheehan, P.E. 2010. Properties of fluorinated graphene films. *Nano Lett.* 10: 3001–3005.

Sahin, H. and Ciraci, S. 2011. Structural, mechanical, and electronic properties of defect patterned graphene nanomeshes from first principles. *Phys. Rev. B* 84: 035452-7.

Santoso, I., Singh, R.S., Gogoi, P.K. et al. 2014. Tunable optical absorption and interactions in graphene via oxygen plasma. *Phys. Rev. B* 89: 075134-9.

Schin, H. and Ciraci, S. 2012. Chlorine adsorption on graphene: Chlorographene. *J. Phys. Chem. C* 116: 24075–24083.

Sharma, Y., Bandyopadhyay, A., and Pati, S.K. 2013. Structural stability, electronic, magnetic, and optical properties of rectangular graphene and boron nitride quantum dots: Effects of size, substitution, and electric field. *J. Phys. Chem. C* 117: 23295–23304.

Sheng, B., Chen, J., Yan, X., and Xue, Q. 2012. Synthesis of fluorine-doped multi-layered graphene sheets by arc-discharge. *RSC Adv.* 2: 6761–6764.

Son, Y.W., Cohen, M.L., and Louie, S.G. 2006. Energy gaps in graphene nanoribbons. *Phys. Rev. Lett.* 97: 216803-07. See also the erratum *Phys. Rev. Lett.* 98: 089901 (2007).

Tachikawa, H., Iyama, T., and Azumi, K. 2011. Density functional theory study of boron- and nitrogen-atom doped graphene chips. *J. Appl. Phys.* 50: 01BJ03-4.

Vinogradov, N.A., Simonov, K.A., Generalov, A.V., Vinogradov, A.S., Vyalikh, D.V., Laubschat, C., Mårtensson, N., and Preobrajenski, A.B. 2012. Controllable P-doping of graphene on Ir(111) by chlorination with $FeCl_3$. *J. Phys. Condens. Matter* 24: 314202-10.

Wei, D., Liu, Y., Wang, Y., Zhang, H., Huang, L., and Yu, G. 2009. Synthesis of N-doped graphene by chemical vapor deposition and its electrical properties. *Nano Lett.* 9: 1752–1758.

Wu, J., Xie, L., Li, Y., Wang, H., Ouyang, Y., Guo, J., and Dai, H. 2011. Controlled chlorine plasma reaction for noninvasive graphene doping. *J. Am. Chem. Soc.* 133: 19668–19671.

Xu, M.S., Fujita, D., and Hanagata, N. 2009. Perspective and challenges of emerging single-molecule DNA sequencing technologies. *Small* 5: 2638–2649.

Xu, M.S., Yan, G., Xi, Y., and Chen, H.Z. 2012. Unique synthesis of graphene-based materials for clean energy and biological sensing applications. *Chin. Sci. Bull.* 57: 3000–3009.

Xu, W., Dong, H.M., Li, L.L., Yao, J.Q., Vasilopoulos, P., and Peeters, F.M. 2010. Optoelectronic properties of graphene in the presence of optical phonon scattering. *Phys. Rev. B* 82: 125304–125309.

Xu, Z., Allen, H., and Yang, H. 2013. Impact of chlorine functionalization on high mobility chemical vapor deposition grown graphene. *ACS Nano* 7: 7262–7270.

Yan, J.A., Xian, L.D., and Chou, M.Y. 2009. Structural and electronic properties of oxidized graphene. *Phys. Rev. Lett.* 103: 086802-06.

Yang, G., Zhang, Y., and Xunwang, Y. 2013. Electronic structure and optical properties of a new type of semiconductor material: Graphene monoxide. *J. Semicond.* 34(8): 083004-5.

Yu, P.Y. and Cardona, M. 1996. *Fundamentals of Semiconductors*. Springer-Verlag, Berlin, Germany.

Yuan, W. and Shi, G. 2013. Graphene-based gas sensors. *J. Mater. Chem.* A 1: 10078–10091.

Zhang, L., Schwertfager, N., Cheiwchanchamnangij, T. et al. 2012. Electronic band structure of graphene from resonant soft x-ray spectroscopy: The role of core-hole effects. *Phys. Rev. B* 86: 245430-5.

Zhang, X., Hsu, A., Wang, H. et al. 2013. Impact of chlorine functionalization on high-mobility chemical vapour deposition grown graphene. *ACS Nano* 7: 7262–7270.

Zheng, D., Vashist, S.K., Al-Rubeaan, K., Luong, J.H., and Sheu, F.S. 2012. Mediatorless amperometric glucose biosensing using 3-aminopropyltriethoxysilane-functionalized graphene. *Talanta* 99: 22–28.

Zhou, S.Y. 2007. Substrate-induced bandgap opening in epitaxial graphene. *Nat. Mater.* 6: 770–775.

Zielinski, M., Potasz, P., Kadantsev, E.S., Voznyy, O., Hawrylak, P., Sheng, W., Korkusinski, M., and Guclu, A.D. 2012. Electronic and optical properties of semiconductor and graphene quantum dots. *Front. Phys.* 7(3): 328–352.

CHAPTER 43

CONTENTS

CVD of Carbon Nanomaterials

From Graphene Sheets to Graphene Quantum Dots

43

Roberto Muñoz and Cristina Gómez-Aleixandre

43.1 INTRODUCTION

The discovery of graphene (single-atom-thick carbon film) by using mechanical exfoliation of parent graphite samples[1] astonished the scientific community that was unaware of the existence of freestanding 2D crystals. Graphene was thought to be a theoretical material, thermodynamically unstable and unable to withstand thermal fluctuations. The mechanical cleavage method allowed isolating graphene layers of micrometer size and suitable quality to demonstrate locally its outstanding physical properties.[2] However, only small-area sheets could be produced and the number of exfoliated layers was not easily controlled. Since then, there has been a growing interest in developing methods to produce large-area graphene samples with suitable lateral size for applications. One method, easily scalable, that perfectly fulfills these requirements is the *bottom-up*[3] growth by chemical vapor deposition (CVD). Currently, the CVD of large-area graphene layers has become a hot topic in material science and technology, and numerous examples of large-area graphene synthesis by CVD can be found in the literature.

Beyond the lateral size, one key question in graphene is the quality of the film that is related to the number of defects and degree of crystallinity. The quality of small *cleavage* samples depends on the defect density and structural order of the parent graphite. A typical parent graphite of high purity and quality is a highly oriented pyrolytic graphite (HOPG). This material is obtained from pyrolysis of organic compounds. Pyrolytic graphite, obtained by heat treatment of pyrolytic carbon or by CVD above 2500 K, exhibits a high degree of preferred crystallographic orientation perpendicular to the surface. Subsequent annealing under compressive stress at approximately 3300 K results in HOPG. Thus, HOPG incorporates a low impurity level of the order of ppm and lateral grain size typically of the order of mm. That's why exfoliated sheets exhibit local properties close to theoretical limits[4]; the sheets can be part of one grain and so isolated single crystals.

In the case of large-area synthesis by CVD, 2D single crystals are highly desirable. However, graphene films exhibit a polycrystalline structure wherein the final film is a result of the coalescence of multiple domains. The standard CVD processes are based on heterogeneous catalysis of organic precursors and performed on metal substrates at much lower temperature and pressure (less than 1350 K and 1 atm) than those mentioned in HOPG. The crystallization mechanism is quite different in both cases. Bulk 3D pyrolytic graphite has a limited degree of crystallization if annealed at low temperature and pressure. Extreme thermodynamic conditions are essential for a large crystal size in bulk carbon materials.[5] As the goal in CVD of graphene is not a 3D, but a 2D, material, the process of heterogeneous catalysis over transition metals has proven to be effective. The metal performs the two different roles of catalyst and substrate. It favors the decomposition of CVD carbon precursors at lower temperature, while 2D pure graphene layer is formed over the entire surface. This process is quite similar to the *catalyst poisoning* in a typical catalytic process and concludes when the metal is saturated or its surface is completely covered. Some reports have already appeared on the synthesis of millimeter-size single crystals by catalytic CVD (CCVD), controlling the thermodynamics and kinetics of the nucleation and growth processes.[6] Even wafer-size single-crystal graphene has been claimed recently.[7]

Currently, CCVD has two major drawbacks: the large time and energy consumption at the typical synthesis temperature and the transference of the film to the desired substrate for usage. The transfer normally induces contamination, wrinkles, or even breakage because of the underlying metal that has to be chemically removed. Regardless, CVD is not expensive for mass production while quality is maintained (Figure 43.1). In this scenario, the game-changing breakthrough would be the development of CVD processes to rapidly deposit high-quality graphene layers on arbitrary substrates, at low temperature. More and more research groups have faced this challenge and recently promising results have been published.

The applications of large-area graphene include transparent electrodes in solar cells, touch screens, flat panel displays, and flexible electronics.[8] Meanwhile, great interest in wafer-size single-crystal samples has emerged in micro- and nanoelectronics with the final objective of integrating graphene into existing silicon-based manufacturing technologies. Depending on the application, it may be acceptable to implement lower-quality graphene with smaller grain size, while maintaining the large-area or also small-area single crystals. Hence, the graphene functional material could exhibit a different degree of perfection. In fact, tailored graphene-based structures, as graphene nanosheets with variable size and number of layers, broaden the way for graphene chemistry and functionalization with straight application in biology, nanoelectronics, photovoltaics, and so on.[9] Even carbon and graphene nanodots under 10 nm size have recently attracted wide attention because of their strong photoluminescence.[10] These tiny dots could be attractive candidates for bioimaging and biosensors.

In this chapter, we will present different CVD processes and technologies to deposit tailored graphene-based *planar* structures, from 2D large-area graphene sheets and single crystals to 0D graphene quantum dots (GQDs). The size and shape of graphene 2D domains as well as the number of layers can be controlled carefully by growth kinetics and thermodynamics involved in the process. The chapter is divided in some sections. In Section 43.2, we introduce the recommended *nomenclature* for graphene 2D materials that has been published recently.[11] We resume this naming system to distinguish the various members of the graphene family. In Section 43.3, the material properties are discussed and also how the size, the number of layers, and the degree of crystallinity affect, if it is so, on these properties. Finally, *we overview the CVD technologies and methods to grow tailored graphene* on metals and dielectric substrates.

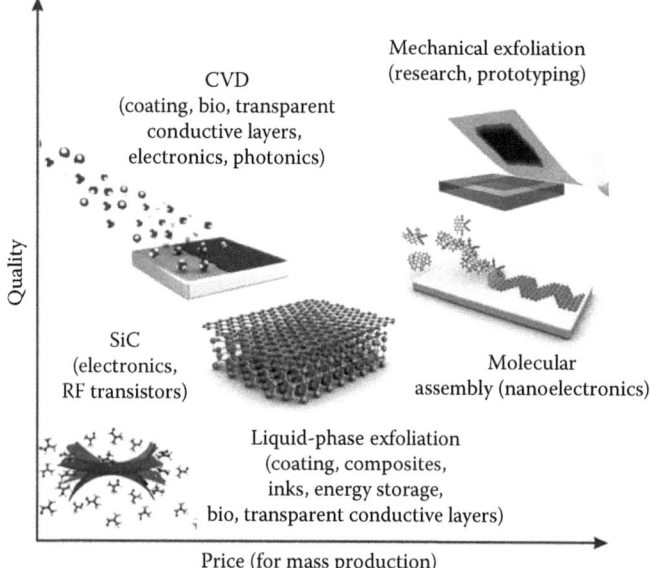

FIGURE 43.1

Several methods of mass production of graphene in terms of size, quality, and price. (Adapted from Novoselov, K.S. et al., *Nature*, 490, 192, 2013.)

43.2 RECOMMENDED NOMENCLATURE FOR 2D GRAPHENE

Interest in 2D carbon forms has expanded beyond monolayer graphene to include related materials with significant variations in lateral dimension, layer number, rotational faulting, and chemical modification. A recommended nomenclature for the new family of 2D carbon materials has appeared recently, which is assumed willingly in this chapter.[11] We resume this naming system to distinguish the various members of the graphene material family but only those that can be obtained by CVD, as either a coating or a sheet adhered on a functional substrate. The most important assumptions that have been taken into account are referred to the lateral dimensions, morphology, and crystallography.

Graphene: A single-atom-thick sheet of hexagonally arranged, sp^2-bonded carbon atoms that is not an integral part of a carbon material but is freely suspended or adhered on a foreign substrate. The lateral dimensions of graphene can vary from several nanometers to the macroscale. The term graphene is usually attributed to Hans Peter Boehm and coworkers who in 1986 defined graphene as a hypothetical final member of infinite size of the polycyclic aromatic hydrocarbon series naphthalene, anthracene, phenanthrene, tetracene, coronene, etc. The substances in this series have the common ending *ene* for organic compounds with carbon–carbon double bonds, and its last member contains *graph* from graphite, a root that derived from the Greek word for drawing, which is the early use of graphite.

   Note that with this definition, other members of the graphene family of 2D materials cannot be simply called *graphene* but must be named using a unique multiword term that distinguishes them from the isolated monolayer (see definitions below).

Graphene layer: A single-atom-thick sheet of hexagonally arranged, sp^2-bonded carbon atoms occurring within a carbon material structure, regardless of whether that material structure has 3D order (graphitic) or not (turbostratic or rotationally faulted).

Turbostratic carbon: 3D sp^2-bonded carbon material in which there is no defined registry of the layers, meaning there is no spatial relationship between the positions of the carbon atoms in one graphene layer with those in adjacent layers. The name derives from *turbo* (rotated) and *strata* (layer) and can also be called rotationally faulted. This is a common structure in carbon materials prepared at low temperature or in *hard carbons* that do not pass through a fluid phase during carbonization.

Multilayer graphene (MLG): 2D (sheetlike) material, either as a freestanding flake or substrate-bound coating, consisting of a small number (between 2 and about 10) of well-defined, countable, stacked graphene layers of extended lateral dimension.

Bilayer graphene, trilayer graphene: 2D (sheetlike) materials, consisting of 2 or 3 well-defined, countable, stacked graphene layers of extended lateral dimension. If the stacking registry is known, it can be specified separately, such as *an AB-stacked bilayer graphene* (see Figure 43.5 b) or *a rotationally faulted trilayer graphene*.

Few-layer graphene (FLG): A subset of MLG (defined as earlier) with layer numbers from 2 to about 5.
   Carbon films containing discontinuous or fragmented graphene layers of very small lateral dimension should be called *carbon thin films* rather than *MLG*, since they do not consist of a defined number of countable graphene layers of extended lateral dimension.

Graphene microsheet: A single-atom-thick sheet of hexagonally arranged, sp^2-bonded carbon atoms that is not an integral part of a carbon material but is freely suspended or adhered on a foreign substrate and has a lateral dimension between 100 nm and 100 μm. This term is recommended over the more general *graphene*, when one wants to emphasize the micrometer scale of the lateral dimension in cases where it is key to properties or behavior.

Graphene nanosheet: A single-atom-thick sheet of hexagonally arranged, sp^2-bonded carbon atoms that has a lateral dimension less than 100 nm.

GQDs: An alternative term for graphene nanosheets or FLG nanosheets, which is used particularly in studies where photoluminescence is the target property. Generally, GQDs have very small lateral dimensions <10 nm (average ~5 nm) at the lower end of the range for graphene nanosheets. Some GQDs may be few-layer materials.

43.3 FUNDAMENTAL (MAIN) PROPERTIES OF 2D GRAPHENE MATERIALS

The amazing properties of graphene justify the research that is being carried out on it. However, some of these characteristics have been measured only for small-area graphene sheets obtained by cleavage because these are single crystals with low impurity level. With this in mind, in this section, we review of the properties of exfoliated graphene trying to compare as far as possible with the material grown by CVD. Because the interest focuses not only in graphene but also in 2D derivatives, we outline the differences with FLG and the importance in the lateral dimensions of the sheets, especially at the nanoscale.

43.3.1 Crystal Structure

Irrespective of its lateral dimensions, graphene exhibits a *nearly* planar hexagonal structure. Each carbon atom is bonded covalently to its three neighbors with a bonding length of 1.42 Å, the average of the typical single (C–C) and double (C=C) covalent σ bonds, and a bonding energy of 5.9 eV.[12] The areal density of carbon atoms in graphene is 3.82×10^{15} cm^{-2}. Figure 43.2 depicts the atomic structure of graphene and its corresponding reciprocal space lattice and Brillouin zone (BZ). The experimental evidence of the structure is shown in Figures 43.3 and 43.4.

The honeycomb lattice is not a Bravais lattice because two neighboring sites are not equivalent and can be decomposed in two sublattices A and B. Both sublattices, however, are triangular Bravais lattices with a two-atom basis (A and B). The lattice vectors of the graphene unit cell and the corresponding reciprocal cell are defined by (see Figure 43.2a)

$$a_1 = \frac{a}{2}\begin{pmatrix} 3 \\ \sqrt{3} \end{pmatrix}, \quad a_2 = \frac{a}{2}\begin{pmatrix} 3 \\ -\sqrt{3} \end{pmatrix}$$

$$b_1 = \frac{2\pi}{3a}\begin{pmatrix} 1 \\ \sqrt{3} \end{pmatrix}, \quad b_2 = \frac{2\pi}{3a}\begin{pmatrix} 1 \\ -\sqrt{3} \end{pmatrix}$$

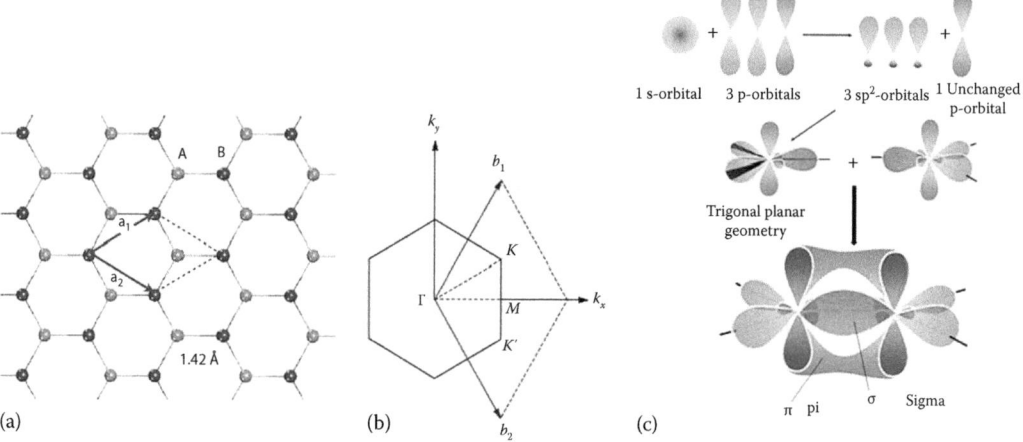

(a)　　　　　　　　　(b)　　　　　　　　　(c)

FIGURE 43.2

(a) Atomic structure of graphene. The carbon atoms are represented by gray spheres, and the dotted blue lines denote the two-atom primitive cell with lattice vectors a_1 and a_2. The lattice can be also represented by two triangular sublattices A (light gray) and B (dark gray). (b) The corresponding reciprocal lattice is denoted by the red dotted lines with lattice vectors b_1 and b_2; the first BZ is denoted by the hexagon with the high symmetry points Γ, *M*, and *K* (Dirac point). (c) Illustration of the carbon valence orbitals. The three in-plane σ orbitals in graphene and the π orbital perpendicular to the sheet. The in-plane σ and the π bonds in the carbon hexagonal network strongly connect the carbon atoms. (Adapted from Pinto, H.P. and Leszczynski, J., Fundamental properties of graphene, in *Handbook of Carbon Nano Materials*, Vol. 5, Chapter 1, World Scientific Series on Carbon Nanoscience, World Scientific, Singapore, 2014; Nottingham Trent University, sp hybridization, Nottingham, U.K. http://www.ntu.ac.uk/cels/molecular_geometry/hybridization/Sp_hybridization/index.html.)

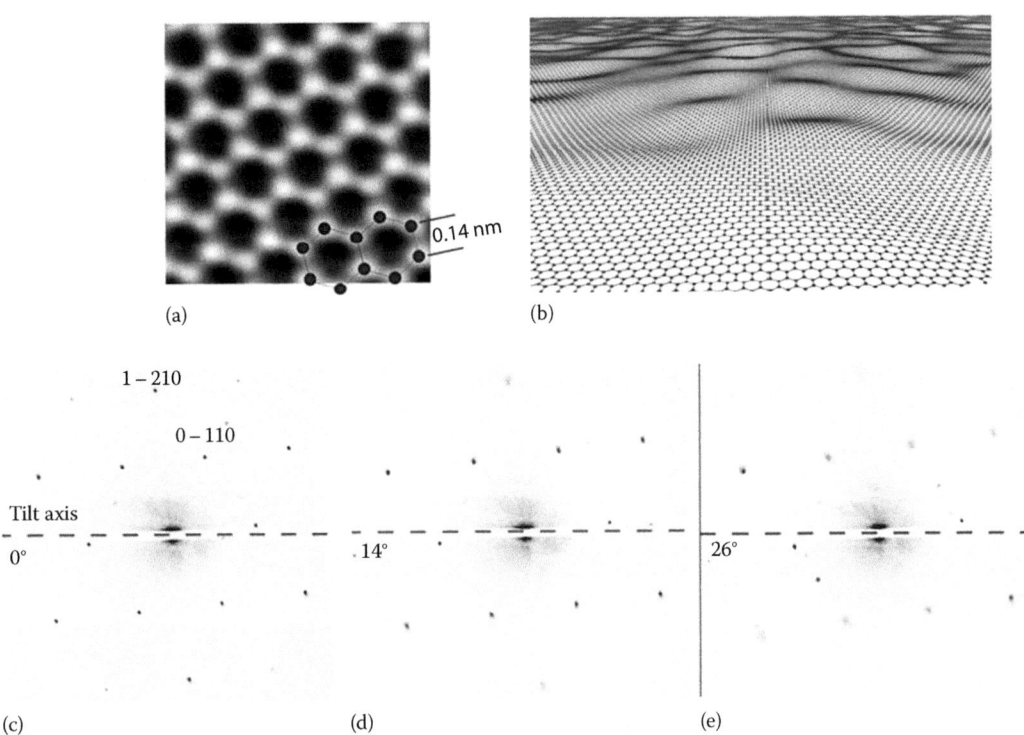

(a)　　　　　　　　　　　　(b)

1–210

0–110

Tilt axis

0°　　　　　　　　14°　　　　　　　　26°

(c)　　　　　　　　　(d)　　　　　　　　(e)

FIGURE 43.3
(a) Atomic TEM image of graphene. (b) The structure that explains deviation from diffraction analysis. The observed corrugations in the third dimension may provide subtle reasons for the stability of 2D crystals. (c–e) Acquired diffraction peaks became broader with increasing tilt angle as is shown. (Adapted from Meyer, J.C. et al., *Nature*, 446, 60, 2007; Bao, W. et al., *Nat. Nanotechnol.*, 4, 562, 2009; http://www.condmat.physics.manchester.ac.uk/publications/; Meyer, J.C. et al., *Nature*, 446, 60, 2007.)

where the C–C distance is $a = 1.42$ Å. The most remarkable feature of graphene lattice is the so-called Dirac points K and K' located at the vertices of the BZ with coordinates (see Figure 43.2b):

$$K = \frac{2\pi}{3a}\begin{pmatrix} 1 \\ \frac{1}{\sqrt{3}} \end{pmatrix} \quad K' = \frac{2\pi}{3a}\begin{pmatrix} 1 \\ -\frac{1}{\sqrt{3}} \end{pmatrix}$$

The formation of this lattice is given by the sp² hybridization of each carbon atom (see Figure 43.2c; carbon has 4 valence electrons distributed in one s orbital and three p orbitals). This leads to the formation of strong covalent σ bonds forming an interbond angle of 120°. The corresponding σ band is a closed shell and gives to the lattice its remarkable strength. In addition, the C–C bonding is enhanced by a fourth bond associated with the overlap of the remaining p orbital oriented perpendicularly to the planar structure. This can interact and form covalent bonds with the adjacent C atoms forming a π band. This band is half-filled since each p orbital has one extra electron.

One of the first transmission electron microscopy (TEM) images of the atomic structure of isolated graphene is shown in Figure 43.3a.[13] Detailed studies by TEM and electron diffraction patterns, acquired as a function of tilt angle, revealed that suspended graphene sheets are not perfectly flat; they exhibit intrinsic microscopic roughening such that the surface normal varies by several degrees and out-of-plane deformations reach 1 nm. The reason that led to this conclusion was that the acquired diffraction peaks became broader with increasing tilt angle as is shown in Figure 43.3c through e. This does not correspond to a flat surface. The only structure that could explain the observed deviation is represented in Figure 43.3b. The observed corrugations in the third dimension may provide subtle reasons for the stability of 2D crystals.[14,15]

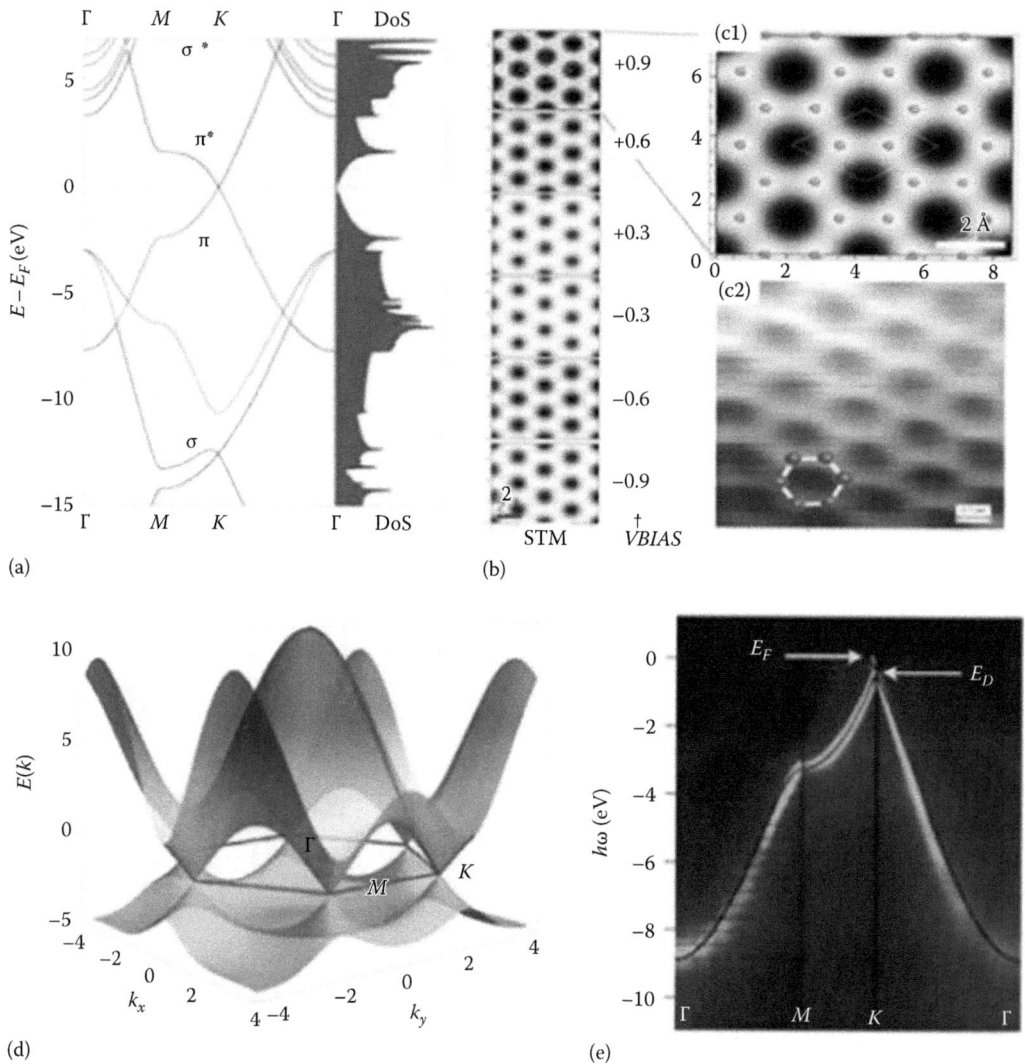

FIGURE 43.4

(a) Band structure along the high symmetry points Γ, M, and K (Dirac point) and density of states (DoS). (b) Simulated constant-current STM images for several Voltage bias or voltage polarization (VBIAS): the bright protrusions are above the C atoms. (c1) Close-up of the STM image for VBIAS = +0.9 V superimposed the graphene lattice. (c2) Experimental constant-current STM image for single-layer graphene with VBIAS = +1 V and tunneling current of 1 nA. (Adapted from Stolyarova, E. et al., *Proc. Nat. Acad. Sci. USA*, 104, 9209, 2007.) Notice that STM with positive (negative) VBIAS is mapping the empty (occupied) states at the Fermi level. (d) Electronic conical dispersion in the vicinity of K points (3D). The red hexagon denotes the first BZ. (e) ARPES measurement of energy distribution along high symmetry points. (Adapted from Reich, S. et al., *Phys. Rev. B*, 66, 035412, 2002; Bostwick, A. et al. *Nat. Phys.*, 3, 36, 2006.)

More atomic resolution real-space images of isolated, single-layer graphene on SiO_2 substrates are available via scanning tunneling microscopy (STM) in Figure 43.4c.[24]

The deviations from the ideal structure of graphene continue in the nanoscale. Different studies performed on graphene nanosheets at $T > 0$ K are in accord with those for large graphene sheets, that is, these structures are not planar but buckle and vibrate. Thus, this feature would appear to be universal for such structures irrespective of their extent.[19] Other factor that one must consider when studying realistic graphene nanosheets is that if the sheet is unterminated, then reconstructions at edges will occur, in order to break aromaticity and to lower the total energy of the flake.[20] As we shall outline in the next section, the edges of nanosheets play a very important role in determining their properties and in particular their electronic and magnetic structure.

43.3.2 Band Structure and Electronic Properties

The electronic properties of graphene depend on the number of layers, the staking order, and the lateral dimensions of the sheets. Furthermore, graphene nanosheets can potentially range in size from molecular to semi-infinite 2D structures, and consequently, their electronic structures can vary from having discrete molecular levels to being band-like as their dimensions are made larger. This leads to the potential of spanning the range of electronic and magnetic properties by using nanosheets of different dimensions.

43.3.2.1 Electronic band structure of graphene

The tight-binding approach was the first approximation used by P. R. Wallace in 1947 to solve the electronic structure of single-atom-thick graphite.[21] In-plane strong covalent σ bonds and perpendicular π bonds form the corresponding bonding σ, π and antibonding σ^*, π^* bands. The σ and σ^* bands are separated by a large energy gap, while the bonding and antibonding π states lie in the vicinity of the Fermi level (E_F). Consequently, the σ bonds are frequently neglected for the prediction of the electronic properties of graphene around the Fermi energy. Thus, the electronic properties of graphene are determined by π and π^* orbitals that form valence and conduction bands (Figure 43.4a).[22]

The electronic structure of graphene can be solved using the aforementioned tight-binding theory or ab initio calculations. Figure 43.4a shows the computed electronic structure of graphene using density-functional theory (DFT)[23,25] that replicates the ab initio calculations.[26] The electronic properties of graphene can also be acceded directly by scanning probe microscopes. The displayed electronic structure is complemented by the computed constant-current STM topologies and compared with the STM images for graphene (Figure 43.4b, c).[24]

The π and π^* bands touch at the corners of the hexagonal BZ (Figure 43.4d). Such corners are called Dirac points and labeled by their momentum vector usually denoted by K and K'. Close to the Fermi energy, the π and π^* bands are quasi-linear, in contrast with the usual quadratic energy momentum relation obeyed by electrons at band edges in conventional semiconductors. With one electron per atom in the π–π^* model, the negative energy band is fully occupied, while the positive branch is totally empty. The energy dispersion $E(k)$ around the K point can also be found using angle-resolved photoemission spectroscopy (ARPES) of a single layer of graphene.[27,28]

Thus, conical distribution of holes and electrons appears in the corners of a 2D BZ whose points touch at the Fermi energy. The quasi-linear electronic band dispersion leads to the massless character for these carriers.

Recent advanced studies claim that the real electronic spectrum near Dirac points is profoundly nonlinear in suspended graphene. The standard approach of the Landau theory, which enabled to map strongly interacting liquid in metals into a noninteracting gas, fails in graphene especially at energy close to the neutrality point, where the density of states vanishes. Graphene's spectrum is filled with electronic states up to the Fermi energy and their coulomb interaction has to be taken into account. In graphene, electron–electron interactions are expected to play a significant role, as the screening length diverges at the charge neutrality point.[29]

As has been noted before, the electronic properties of *FLG* depend on the number of layers and the staking order.[30,31] In the case of *bilayer graphene*, for AA[32] stacking (see figure 43.5 a), each atom is on top of another atom, and in AB or Bernal stacking, the atom on the first layer is on top of the center of the hexagon of the layer beneath. Increasing the number of layers yields more complex ordering. In the bulk case of graphite, there are three common types of staking: (1) AB or Bernal stacking more energetically favorable, (2) the AA staking, and (3) not defined or turbostratic stacking (see Figure 43.5).

The electronic structure of graphene changes markedly with the increasing number of layers being at 10 layers basically the same as graphite.[33] The Dirac point observed in graphene (Figure 43.6a) disappears in bilayer (AB stack) graphene where the linear dispersion is no longer observable and instead parabolic bands with zero value appear at the Fermi level (Figure 43.6b).

Adding a third layer allows the system to recover its linear dispersion, but for additional layers ($N < 10$), the electronic structure has more complexity (Figure 43.6c), suggesting that the overlapping between conduction and valence bands intensifies.[1,33,35] For these reasons, three classes of

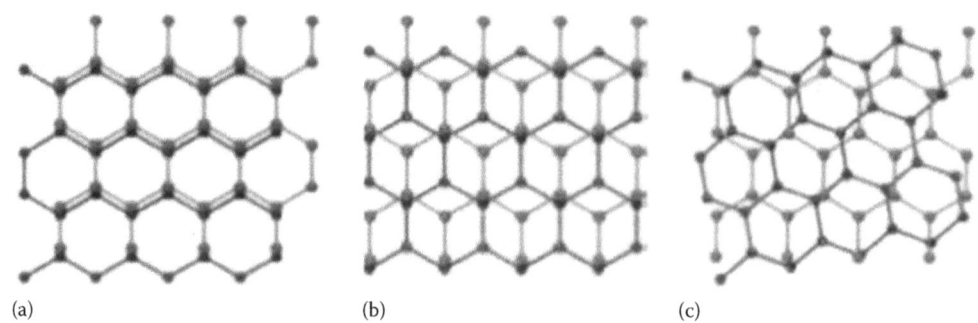

(a)   (b)   (c)

FIGURE 43.5
Staking order showing only the first and second layers of graphite: (a) AA, (b) AB, and
(c) turbostratic stackings.

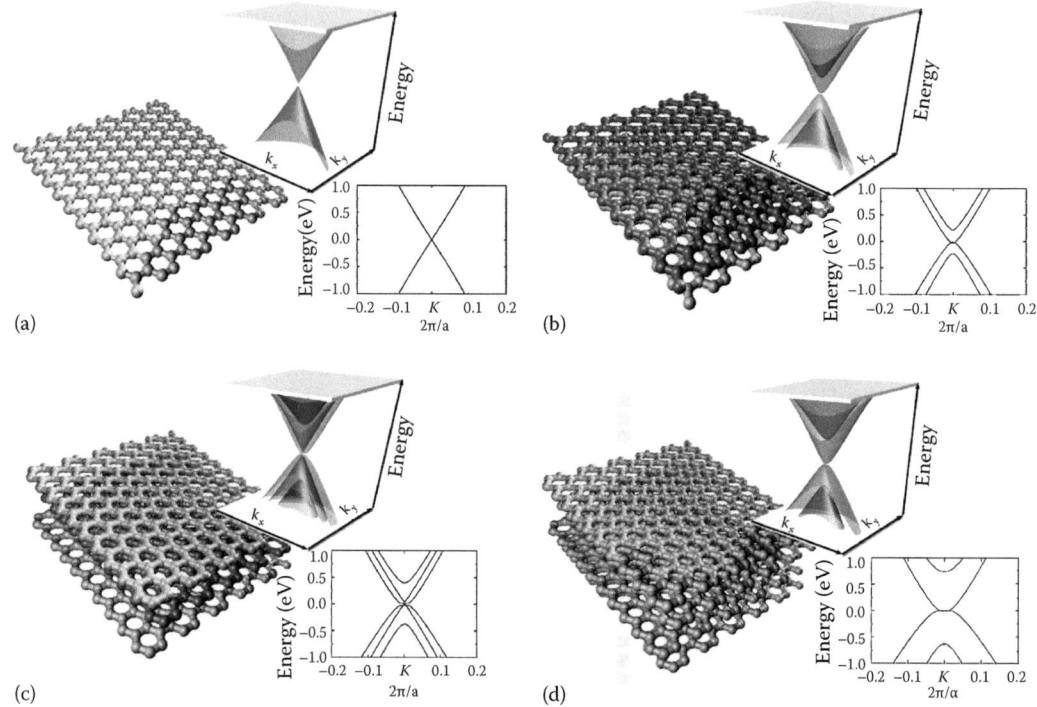

FIGURE 43.6
Effect of stacking on the electronic structure of graphene around the *K* point. (a) Typical dispersion
observed in graphene. (b) The AB-staking double-layer graphene shows a parabolic dispersion.
(c) The AB-staking triple-layer graphene reestablishes the linear dispersion. (d) Band structure for
graphite. (Adapted from Terrones, M. et al., *Nano Today*, 5, 351, 2010.)

layered graphene should be considered: single-layer graphene, double-layer graphene, and FLG
($N < 10$). One way to synthesize bilayer graphene *is via CVD*, which can produce large bilayer
regions that almost exclusively conform a Bernal stack geometry.[36]

    Besides the stacking order and number of layers, *graphene nanosheets* or *dots* show some
very interesting properties and have great potential applications as electronic and magnetic devices.
These potential applications arise because of their intrinsic *nanoscale*, their *edges*, *corners*, and
much larger variety of *shapes*. This great interest is based on the fact that at the *nanoscale* level, the
gap can be opened in monolayer sheets, and even the smallest ones can undergo a transition from
band-like to discrete electronic states. In the limit, the energy of the highest occupied molecular
orbital (HOMO) stands for the Fermi energy (E_F), and the difference in the energies of the HOMO
and that of the lowest unoccupied molecular orbital (LUMO) corresponds to the energy gap (E_g).
In fact, nanosheets only have a continuous band structure when their dimensions are very large.[37]

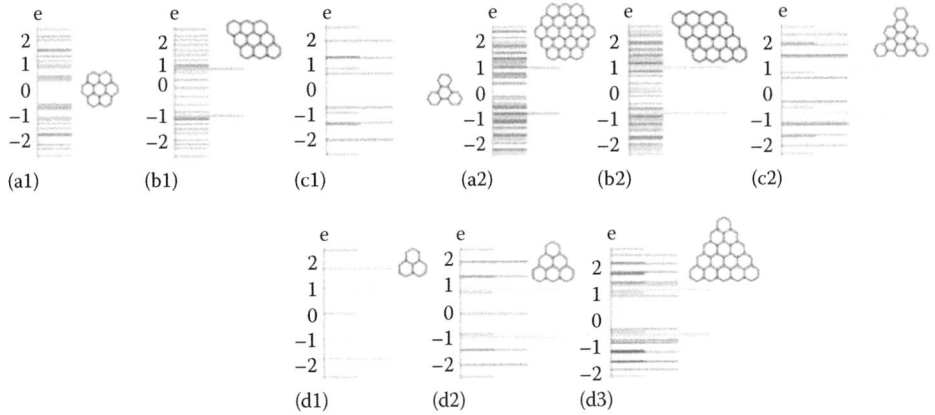

FIGURE 43.7
Discrete electronic structure of graphene nanosheets. (a1,a2) Hexagonal zigzag nanodisks. (b1,b2) Parallelogrammic zigzag nanodisks. (c1,c2) Trigonal armchair nanodisks. (d1,d2,d3) Trigonal zigzag nanodisks. (Adapted from Ezawa, M., *Phys. E*, 40, 1421, 2008.)

The two basic types of edge structures, zigzag (ZZ) and armchair (AC), contain dangling (*uncompleted*) bonds. This introduces a mixture of sp^2 and sp hybridization into the lattice unless the edges are chemically bonded to noncarbon atoms or functional groups. Therefore, both unterminated and terminated edge structures can alter the basic electronic structure of graphene.[38–42] Other types of edge state are less numerous, such as reconstructed ZZs and ACs as well as corner states that are formed where edges meet.[43–46]

Furthermore, the details of this discrete electronic structure can vary strongly with their shape and dimensions, as illustrated in Figure 43.7, and can be heavily influenced by termination, that is, passivation or functionalization.[47–52]

Few-layer nanosheets and few-layer stacks both have been the subject of theoretical and experimental interest. Such structures can have properties that differ from those of a monolayer and of graphite nanocrystals.[42,53] An example of this is that bilayer nanosheets have a different bandgap than single-layer dots and electrons in such structures may be confined with potential barriers in contrast to single-layer nanosheets.[42]

43.3.2.2 Electronic properties

Since the occupied and unoccupied *bands touch at the six zone corners*, the graphene sheet displays a semimetallic or *zero-gap* semiconducting character whose intrinsic Fermi surface is reduced to these six points. Despite having exceptional electronic properties, the gapless character is a severe limitation from the point of view of electronic applications. To overcome this limitation, several methods have been suggested to induce a bandgap.[41,42] Bilayer graphene is one option because the linear dispersion around Dirac points disappears and instead *parabolic bands* turn up. In this case, *a bandgap can be opened* by an external electric field, an interesting property in electronic applications.[54] A gap is also opened intrinsically in monolayer nanosheets and even a transition from band-like to discrete electronic states may occur. Along with this basic change in the nature of the electronic level, there is also a consequent change in electrical character from insulator to semiconductor. The electronic properties of nanosheets may be tailored using their *edge* states by chemically bonding noncarbon atoms of functional groups. Therefore, either unsaturated or saturated edge structures can alter the electronic structure and properties. Corner states are also likely to be invaluable for a variety of purposes. We will discuss later that the discrete electronic states can promote luminescence in semiconducting nanostructures.

Beyond the limitations, many of the electronic properties measured in experiments have exceeded those obtained in any other material, with some characteristics reaching the theoretically predicted limits. Intrinsic electron mobility is a measure of how easily electrons move in a material. The highest room-temperature electron mobility reported for exfoliated samples is more than 200,000 $cm^2 \cdot V^{-1} \cdot s^{-1}$ (intrinsic limit), two orders higher than Si, and 20 times higher than GaAs.[55]

This means that the electrons in graphene move with a negligible scattering level. The corresponding bonding π and antibonding π* bands in graphene form a network that allows the electrons to move without collisions through the structure. This high mobility represents a long mean-free path of ~3 μm that can be considered ballistic transport.[2,4,41,56] High mobility is one way of making semiconductor devices smaller and run faster. Actually, mobility values span a wide range of 2,000 $< \mu <$ 200,000 cm$^2 \cdot$V$^{-1} \cdot$s^{-1}. The weak temperature dependence found for μ indicates that impurity or defect scattering is the dominant scattering mechanism and provides room for future device improvement.[57,58] Meanwhile, CVD promises a scalable method to produce large-area graphene, and recent studies reported CVD graphene samples of comparable quality, $\mu \leq$ 70,000 cm^2/(Vs), to the best reported exfoliated samples. In addition, these samples show low resistivity down to 120 Ω·sq^{-1} at high carrier density (theoretical limit in 30 Ω·sq^{-1}).[59,60] Two important sources of disorder, namely, grain boundaries and processing-induced contamination, are substantially reduced in this case. These results confirm the possibility of achieving high-performance graphene devices based on a scalable synthesis process.

Graphene is able to sustain a high current density of 10^8 A cm^{-2},[61] and therefore the 2D carrier density n_s across the charge neutrality point covers a range of roughly ±10^{13}/cm^2, up to six orders of magnitude higher than Cu. The corresponding resistivity of the graphene sheet would be around 10^{-6} Ω·cm. This is less than the resistivity of silver, the lowest known at room temperature. Although the carrier density approaches zero at the Dirac point, the 2D conductivity remains finite. Such behavior, dubbed the *minimum* conductivity, has been observed experimentally in many graphene devices with resistance ranging from 2 to 7 kΩ at low temperature.[57,62] A slightly larger value from 6 to 9 kΩ is also reported for bilayer graphene.[63] The origin of this finite conductivity remains unclear at the moment.

Recently, graphene has revealed a variety of unusual transport phenomena, such as an anomalous integer quantum Hall effect, a cyclotron mass m_c of massless carriers with an energy E described by $E = m_c c^{*2}$, and Shubnikov–de Haas oscillations that exhibit a phase shift of π due to Berry's phase.

43.3.3 Thermal Properties of Graphene

The strong in-plane bonding, the anisotropic bonding structure, and the low mass of the carbon atoms give graphene materials unique thermal properties. The in-plane thermal conductivity of graphene at room temperature, the highest of any known material, spans a range between 2000 and 5,000 W m^{-1} K^{-1} for freely suspended samples.[64–67] The extremely high value of the thermal conductivity suggests that graphene can outperform other structures as carbon nanotubes in heat conduction and thermal management. For comparison, the thermal conductivity of natural diamond is 2,200 W m^{-1} K^{-1} at room temperature.[68,69] By contrast, it is worthy to indicate that heat flow in the cross plane direction (along the c axis) of graphene and graphite is strongly limited by weak interplane van der Waals interactions. The thermal conductivity along the c axis of pyrolytic graphite is around 6 W m^{-1} K^{-1} at room temperature, which may partially cast a shadow over graphene thermal applications.[70]

The lattice vibrational modes (phonons) of graphene help to explain its thermal properties.[71] The graphene unit cell, marked by dashed lines in Figure 43.8a, leads to the formation of three acoustic (A) and three optical (O) phonon modes, with the dispersions shown in Figure 43.8b. Longitudinal (L) modes correspond to atomic displacements along the wave propagation direction, whereas transverse (T) modes correspond to in-plane displacements perpendicular to the propagation direction. Moreover, the unique 2D nature of graphene allows out-of-plane atomic displacements, also known as flexural (Z) phonons. The group velocities of the transverse acoustic (TA) and longitudinal acoustic (LA) modes 13.6 and 21.3 km s^{-1} are approximately four to six times higher than those in silicon or germanium because of the strong in-plane sp^2 bonds of graphene and the small mass of carbon atoms.[72–75] In contrast, the flexural ZA modes have an approximately quadratic dispersion. An intriguing open question in the theory of phonon transport in graphene is the relative contribution to heat conduction by LA, TA, and ZA. There have been opposite views expressed as to the importance of ZA phonons, from negligible[76–78] to dominant.[79–82] The argument against ZA contributions originates from Klemens' theory, which states that ZA modes have large scattering strength and zero group velocity near the zone center. The argument for the strong

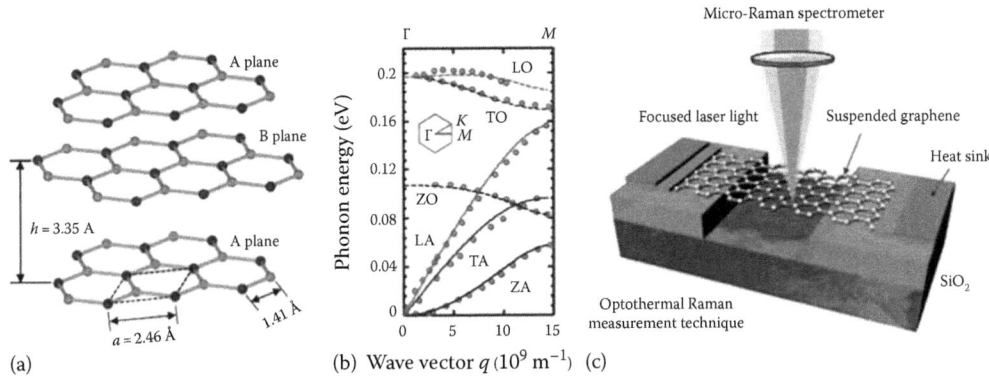

FIGURE 43.8
(a) Atomic arrangement in graphene sheets. Dashed lines in the bottom sheet represent the unit cell. (b) Graphene phonon dispersion along the Γ-to-M crystallographic direction. Lines show numerical calculations; symbols represent experimental data. Note the presence of linear in-plane acoustic modes (LA, TA), as well as flexural out-of-plane acoustic (ZA) modes with a quadratic dispersion. LO, longitudinal optical; TO, transverse optical; ZO, out-of-plane optical. (c) Optothermal Raman technique scheme. (Adapted from Balandin, A.A., *Nat. Mater.*, 10, 569, 2011; Pop, E. et al., *Mrs Bull.*, 37, 1273, 2012.)

contributions of ZA modes is made on the basis of a selection rule in ideal graphene, which restricts the phonon–phonon scattering and increases the lifetime of ZA modes.

The in-plane thermal conductivity of graphene *decreases* significantly when *in contact with a substrate*. At room temperature, the thermal conductivity of graphene supported by SiO_2 was measured as 600 W m^{-1} K^{-1}. This value is still rather high, exceeding the conductivity of Si (145 W m^{-1} K^{-1}) and Cu (400 W m^{-1} K^{-1}).[77,83] For SiO_2-supported graphene, the decrease in thermal conductivity occurs as a result of the coupling and scattering of graphene phonons with substrate vibrational modes. Heat flow perpendicular to a graphene sheet is also limited by weak van der Waals interactions with adjacent substrates, such as SiO_2.[84–86]

Interestingly, the thermal resistance does not change significantly across FLG samples,[85] indicating that the thermal resistance between graphene and its environment dominates that between individual graphene sheets. In the context of nanoscale devices and interconnects,[87,88] dissipation from graphene devices is quite limited by their interfaces, contacts, and surrounding materials, which are often thermal insulators such as SiO_2.

Experimental studies of the thermal conductivity of graphene layers from HOPG and CVD were carried out by the optothermal Raman technique (see Figure 43.8c).[89–95] The G peak in graphene's Raman spectrum exhibits strong temperature dependence. The frequency of the G peak (ω_G) as a function of temperature allows one to convert a Raman spectrometer into an *optical thermometer*. The authors found values exceeding 2,500 W m^{-1} K^{-1} near room temperature for CVD samples, above the bulk graphite limit. Conductivity was as high as ~1,400 W m^{-1} K^{-1} at 500 K and ≈630 W m^{-1} K^{-1} at 600 K. The typical dependence of thermal conductivity with temperature explains these values. Differences in strain distribution in the suspended graphene of various sizes and geometries may also affect the results.

The optothermal Raman study found that thermal conductivity of suspended *FLG* decreases with increasing number of layers *n*, approaching the bulk graphite limit.[92] As *n* in FLG increases, phonon scattering increases leading to a decrease in conductivity.[83,92,96] For *n* > 4, the values can drop below the bulk graphite limit owing to the onset of the phonon-boundary scattering from the top and bottom interfaces; thermal conductivity recovers for sufficiently thick films. A strong decrease as *n* changed from 1 to 2 and a slower decrease for *n* > 2 have also been noted.[97]

The situation is entirely different for *encased graphene* where thermal transport is limited by the acoustic phonon scattering from the top and bottom boundaries and disorder, which is unavoidable when FLG is embedded between two layers of dielectrics. A study conducted with 3ω, another common technique in thermal characterization, found values of thermal conductivity ≈160 W m^{-1} K^{-1} for encased single-layer graphene at 310 K. It increases to 1,000 W m^{-1} K^{-1} for graphite films 8 nm thick.[83] Correspondingly, *conductivity* dependence on *thickness* was similar to other material systems where it is extrinsically limited and scales with it. A similar behavior was observed for ultrathin Diamond-like carbon (DLC) films.[98] The overall values encased DLC films are much smaller than those for encased FLG, as expected for more disordered materials, but the trend is essentially the same.

Finally, when graphene is confined in *nanosheets* that are narrower than the intrinsic phonon mean-free path (λ_0) (width $W \leq \lambda_0$), phonon scattering with boundaries and edge roughness further reduces the thermal conductivity compared to the cases of suspended and supported graphene.[99,100]

43.3.4 Optical Properties

The optical properties of graphene can be analyzed from two different points of view: its opacity or its transparency. The structure of graphene produces an unexpectedly high opacity for an atomic monolayer in vacuum, absorbing $\pi\alpha \approx 2.3\%$ of white light (see Figure 43.9a), where α is the fine-structure constant.[101] For normal incident light, with energy below about 3 eV, the absorption is wavelength independent as a consequence of the electronic structure of monolayer graphene.[102] The nearly linear dispersion of the Dirac electrons implies that for any optical excitation, there will be always an electron–hole pair in resonance. Therefore, this means that its optical transparency >97% can be useful in many optical applications, even more in combination with its high conductivity. Additionally, graphene and bilayer graphene become completely transparent when the optical energy is smaller than double the Fermi level, owing to the Pauli blocking.[41] These properties would suit many controllable photonic devices. The absorption spectrum of graphene is quite flat from 300 to 2500 nm, but a peak arises in the ultraviolet region (~270 nm), due to the exciton-shifted van Hove singularity in the graphene density of states (see two sharp peaks around E_F in DoS included in Figure 43.4a). The linear or wavelength-independent optical absorption can be used to identify graphene on top of a Si/SiO$_2$ substrate by optical image contrast.[1] This scales with the number of layers and is the result of interference, with SiO$_2$ acting as a spacer.

Graphene only reflects <0.1% of the incident light in the visible region,[102] rising to ~2% for 10 layers.[103] Thus, we can take the optical absorption of *FLG* to be proportional to the number of layers, each absorbing ~2.3% over the visible spectrum (inset Figure 43.9b). In a FLG sample, each sheet can be seen as a 2D electron gas, with little perturbation from the adjacent layers, making it optically equivalent to a superposition of almost noninteracting layers.[103] In FLG absorption features can be seen at low energies, associated with interband transitions.[104,105]

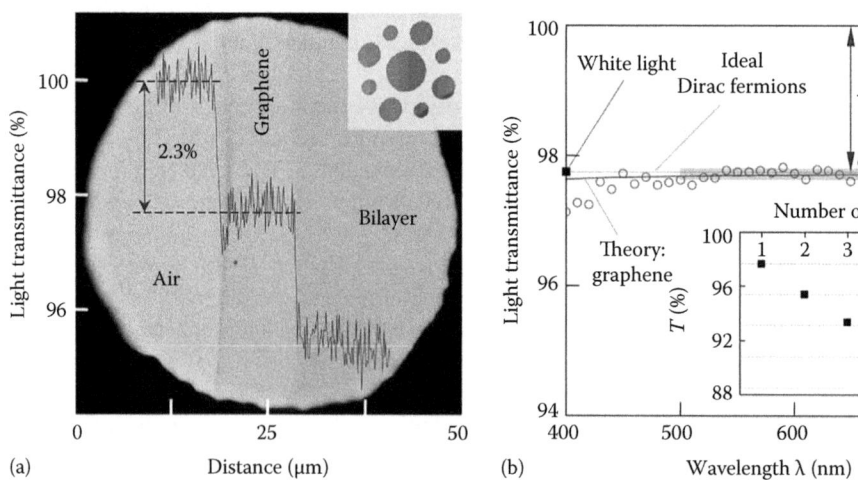

(a)

(b)

FIGURE 43.9

(a) Transmittance for an increasing number of layers. Inset, sample design for the experiment of Reference [102] showing a thick metal support structure with several apertures, on top of which graphene flakes are placed. (b) Transmittance spectrum of single-layer graphene (open circles). The red line is the transmittance $T = (1 + 0.5\pi\alpha)^{-2}$ expected for 2D Dirac fermions, whereas the green curve takes into account a nonlinearity of graphene's electronic spectrum. (Inset) Transmittance of white light as a function of the number of graphene layers (squares). The dashed lines correspond to an intensity reduction by $\pi\alpha$ with each added layer.

Graphene could be made *luminescent* inducing a bandgap by chemical or physical treatments, to reduce the connectivity of the π-electron network. The main exponent is the tailored electronic structure of *graphene nanosheets*.[106] As we have discussed previously, at the nanoscale, a gap can be opened in monolayer sheets, and even a transition can occur from band-like to discrete electronic states. This leads to the interesting possibility of developing graphene nanosheets with optical properties that may be tuned by changing dimensions or by functional groups.[107] Individual graphene flakes can be made brightly luminescent by mild oxygen plasma treatment.[108] Indeed, luminescent graphene-based materials can now be routinely produced that cover the infrared, visible, and blue spectral ranges.[108–112]

An intriguing open question in the theory of photoluminescence remains, whether its origin derives from bandgap emission of electron-confined sp^2 islands[109–111] or arise from oxygen-related states.[108] Whatever the origin, fluorescent organic compounds are of importance to the development of low-cost optoelectronic devices.[113] Blue photoluminescence from aromatic or olefinic molecules is particularly important for display and lighting applications.[114] Luminescent quantum dots are widely used for biolabeling and bioimaging. Fluorescent species in the infrared and near infrared are useful for biological applications, because cells and tissues show little autofluorescence in this region.[115] Even photoluminescent graphene was exploited for live cell imaging in the near infrared with little background.[110]

Broadband nonlinear photoluminescence is also possible following nonequilibrium excitation of graphene layers, as recently reported by several groups.[117–120] Emission occurs throughout the visible spectrum, for energies both higher and lower than the exciting one, in contrast with conventional photoluminescence processes. This broadband nonlinear photoluminescence scales with the number of layers and can be used as a quantitative imaging tool. As for oxygen-induced luminescence, further work is necessary to fully explain this hot luminescence. Electroluminescence was also recently reported in pristine graphene.[116]

43.3.5 Mechanical Properties

The outstanding mechanical properties of graphene is another feature of this material that could find important technological applications.[121–123] The strength observed is a direct consequence of the covalent sp^2 hybridization between carbon atoms. The state-of-the-art measurements confirmed the extreme tensile strength of graphene of 130 GPa, the similar in-plane Young's modulus of graphene and graphite of about 1 TPa, and a spring constant between 1 and 5 N m^{-1} for few-layer material.[124] It is around 200 times stronger than structural steel (A36). It is illustrative to say that this material could sustain a weight of 13 tons suspended in 1 mm^2 surface area. On the other hand, it is very important to outline its intrinsic flexibility.[125,126]

Other mechanical properties include those that indicate the tendency for buckling of graphene. The tension rigidity, at the unstrained equilibrium state for uniaxial stretching, is comparable to 340 GPa·nm[127] value measured recently for graphene by atomic force microscopy (AFM).[128] Using this value, we can derive the flexural rigidity of free graphene to be 3.18 GPa·nm^3. From these values, the critical buckling strain for a sheet of 30 μm can be calculated to be 300 microstrain or 0.03%. This indicates that free graphene could collapse (buckle) at rather small axial compressive strains.

Measuring the intrinsic strength of graphene is challenging. Experimental sophistication of AFM nanoindentation allowed to accurately measure the mechanical properties of graphene.[128] The experimental setup includes a Si/SiO$_2$ epilayer with open holes formed by nanoimprint lithography and reactive ion etching (see Figure 43.10a). The force–displacement response by a AFM tip on monolayer graphene can be described by theoretical models.[128,129] Figure 43.10b shows the experimental stress–strain curve where the maximum stress (or tensile strength) is ~130 GPa. From these results, theory predicts that for small strains ($\varepsilon \lesssim 0.1$), graphene is isotropic with a Young's modulus ~1050 GPa and Poisson's ratio $\nu = 0.186$.[126]

It is important to stress the fact that the reported values of elasticity for graphene apply if no defects are present. For instance, the inclusion of large slits or holes drastically reduces the fracture strength of graphene sheets to 30–40 GPa.[130,131]

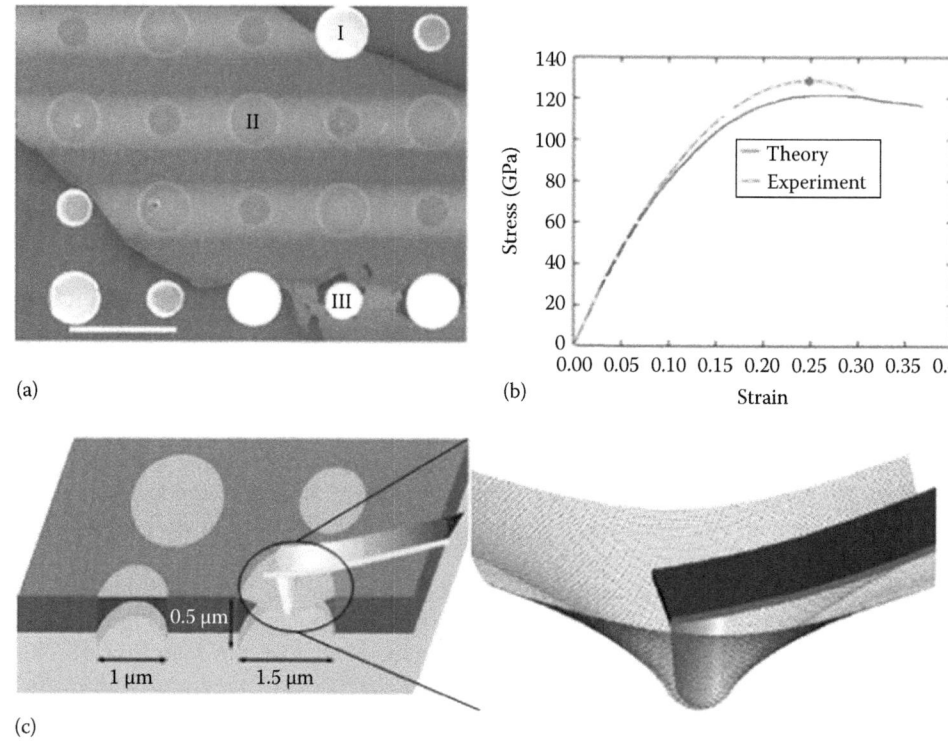

FIGURE 43.10
(a) Scanning electron micrograph of a large graphene flake spanning an array of circular holes 1 and 1.5 μm in diameter. Area I shows a hole partially covered by graphene, area II is fully covered, and area III is fractured from indentation. Scale bar, 3 μm. (b) Nonlinear elastic properties of graphene (dashed line) deduced from analysis of experiments. The diamond point denotes the maximum stress where slope is zero. The stress–strain curve from ab initio calculations is the solid line. (c) AFM nanoindentation scheme of a monolayer graphene membrane deposited onto a Si/SiO₂ substrate with an array of circular holes as in (a); the close-up rightward shows the AFM cantilever with diamond tip. (Adapted from Lee, C. et al., *Science*, 321, 385, 2008; Liu, F. et al., *Phys. Rev. B*, 76, 064120, 2007.)

43.3.6 Chemistry of Graphene

Another property of graphene, already demonstrated, is that it can be readily functionalized.[143,144] The intrinsic chemical reactivity of graphene can be associated to its high specific surface area, 2630 m²·g⁻¹,[132] and its impermeability to any gases due to the electron density of aromatic rings.[133] It is worth mentioning that recently, physicists reported that single-atom-thick sheets of graphene are a hundred times more chemically reactive than thicker sheets.[134] This reactivity explains its moderate burn temperature of around 350°C.[135,136]

In principle, unsaturated graphene can undergo a broad class of organic reactions analogous to unsaturated systems in organic molecules. The detailed reactivity of graphene sheets, in terms of size, shape, and possibility for stoichiometric control, is currently being investigated.[123] Any deviation from the planar structure enhances the reactivity of graphene; covalent sp³ bond on the basal plane enhances the reactivity at the adjacent site, leading to a chain reaction from the point of initial attack.[137] Equally, geometrically strained regions in the graphene lattice are areas of preferential reactivity. Engineering strain in a periodic manner on the surface lattice to control reactivity is a potential way of achieving well-controlled stoichiometric functionalization of graphene. Like a strained plastic, ripples can develop on the graphene surface using simple thermal manipulation.[15] In this context, the edge regions are crucial in the electronic structure and reactivity. In the same way, there is an increased scientific interest in producing by synthetic chemistry (as CVD) graphene nanosheets and nanodots with well-defined shape and size. These open the way for stoichiometric graphene chemistry with straight application in several fields.[9] Graphene nanodots under 10 nm size have recently attracted wide attention because of their strong *photoluminescence*.[138]

These tiny dots could be attractive candidates for bioimaging and biosensors. It can be expected that the ZZ edges will display higher reactivity as compared to the AC edges like in polyaromatic hydrocarbons.[139]

The main purpose of chemical modification of graphene is to overcome its intrinsic limitations or to add functionalities. The *absence of a bandgap* and the *doping* to control the type and increase concentration of charged carriers[140,141] in graphene constitute fundamental problems that can be resolved by functionalization. Modified graphene is *less hydrophobic* and has appreciable solubility.[142]

The graphene sheet can be readily functionalized via noncovalent or covalent coupling reactions. Covalent functionalization of graphene can be performed at the basal plane or at the edges of the sheets. The former modification alters the electronic properties of graphene as new defects are randomly generated after breaking sp^2 configuration. The latter is of higher interest in tailored planar nanostructures. In order to retain unaltered the electronic network graphene, noncovalent methodologies have been also developed.

43.3.6.1 Covalent functionalization

Formation of covalent bonds on the *basal plane* of the graphene sheet often leads to the random formation of sp^3 bonds and produces heterogeneous, nonstoichiometric functionalized products that contain defects and degrade some properties. As a consequence of modification, of course, other properties are modified by the attached chemical groups, and in many cases, this may lead to a new and interesting behavior.

This situation differs from the attachment of chemical species at the *edges* of engineered graphene flakes or dots, in terms of stoichiometric control.

43.3.6.1.1 Functionalization at the basal plane

The *hydrogenation* of graphene by plasma-assisted deposition involves the change of the structure from sp^2 to sp^3. This results in a bandgap opening and a conversion from a metallic graphene to an insulator graphane.[143] However, it is well known in the literature of diamond growth that atomic hydrogen in excess can induce etching producing volatile hydrocarbons[145] Avoidance of exposure to excess atomic hydrogen or energetic hydrogen ions is crucial.

Fluorographene has been also synthesized via reaction with XeF_2, resulting in an insulator with a bandgap of 3 eV.[144,146] Theoretical studies have shown that in fully fluorinated graphene, $(CF)_n$, fluorine is covalently bonded to sp^3-hybridized bonds to carbon atoms, which are elevated above the plane of the original graphene layer,[147,148] much like in graphane. CF_4 plasma has been also used for functionalization[149] (Figure 43.11a). An interesting property of graphite fluorides, CF_x, is their usefulness as lubricants and in the cathodes of primary lithium batteries.[150]

The incorporation of chlorine atoms onto the basal plane of graphene via photochlorination also induces an opening of the bandgap of the graphene layer.[151]

Nitrogen-doped graphene sheets were also prepared when the thermal reduction of graphene oxide took place in the presence of melamine.[152] Moreover, nitrogen-doped and boron-doped graphene sheets and boron-nitride-hybridized graphene heterostructures (h-BNC) have been prepared by a *plasma-enhanced CVD (PECVD)* process tuning their capacitance, electronic, and electrical properties.[153–155]

Besides atomic functionalization, organic molecules can be covalently coupled to the graphene sheet via *carbon–carbon bonding*. The *aryl diazonium salt* chemistry (Figure 43.11b) is another developed route for chemical functionalization, for example, with nitrophenyl groups, which facilitates spontaneous electron transfer from the graphene layer to the diazonium salt resulting in transformation of the electronic structure and transport properties from near metallic to semiconducting. This procedure that can also be applied on graphene as a versatile methodology for tailoring the chemical and electronic properties of graphene.[156–158] As an example, the chemical modification by photoactive chromophore molecules using aryl diazonium salt chemistry is followed by *click chemistry*.[159]

Standard [2 + 2] cycloadditions of in situ generated arynes (Figure 43.11c) have been recently achieved under mild reaction conditions, forming highly functionalized, thermally stable graphene

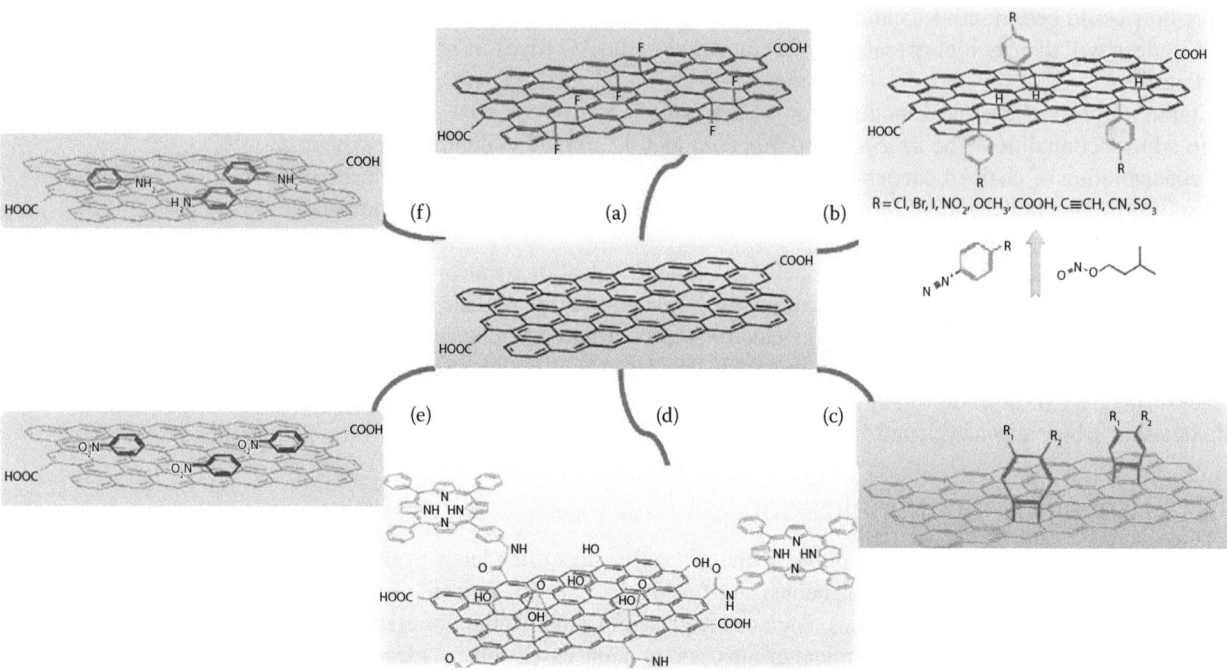

FIGURE 43.11
Covalent functionalization at the basal plane. (a) Fluorination through CF_4 plasma treatment, (b) carbon–carbon bonding via aryl diazonium salt treatment, and (c) standard aryne (2 + 2) cycloaddition. Covalent functionalization at the edges. (d) Porphyrin- and imidazolium-modified graphene by amidation. Noncovalent interactions π–π stacking with (e) electron acceptors and (f) donors.

sheets that can be dispersed in various solvents.[160] *The 1,3-dipolar cycloaddition* of in situ generated azomethine ylides can also be applied in the case of graphene.[161] Traditional organic cycloaddition reactions, such as the Diels–Alder cycloaddition or the Friedel–Crafts acylation reaction, also found use in graphene chemistry.

Azide-based chemistry received intense scientific interest in the field of graphene chemistry, and several reports have emerged over the last years.[162–164] These studies demonstrated a high degree of functionalization in the range of one functional group per 10 carbon atoms using nitrene addition chemistry in microcrystalline graphite. Moreover, carboxyl-terminated alkyl azides were used as a binding site for gold nanoparticles, taking advantage of the well-established nitrene chemistry.

43.3.6.1.2 Functionalization at the edges or vacants

Graphene is hydrophobic and is therefore difficult to solubilize or disperse in most liquids. Thus, chemists have spent considerable effort in functionalizing graphenes at their edges so that they may be more readily solubilized or dispersed, especially in water. Groups such as carboxyl epoxy and hydroxyl are commonly used to solubilize in water, and long alkyl chains make graphene sheets soluble in many organic solvents.

The presence of carboxylic moieties on the edges provides the opportunity for further chemical functionalization via condensation reactions with alcohols or amines with a wide range of small organic molecules and polymers. *Acylation reactions* are among the most common approaches used for linking molecular moieties onto oxygenated groups at the edges where the carboxylic groups can be tethered to amine-functionalized molecules. Extending on this functionalization chemistry, graphene can be functionalized with polyethilenglicol to obtain a biocompatible conjugate that can be used as a platform for high-density loading of hydrophobic drugs, which can bind itself to the graphene via π–π stacking.

The activation of carboxylic groups can take place with thionyl or oxalyl chloride, or a suitable carbodiimide-based molecule, followed by the addition of an appropriate nucleophile. Following this amidation procedure, a variety of biomolecules and moieties[165–168] were prepared and the hybrid materials were evaluated for their photophysical properties. Coupling of graphene sheets with porphyrin and fullerene moieties through *amidation and esterification* reactions (Figure 43.11d), respectively, led to materials with enhanced nonlinear optical performance in the nanosecond regime.[169,170] Besides small organic molecules, the *esterification* reaction of carboxylic groups of graphene oxide by hydroxyl groups of polyvinyl alcohol was also reported.[171]

Besides carboxylic groups located on the edges of graphene, the epoxy groups being on the basal plane are also reactive and can be easily opened by the addition of an amine via *nucleophilic ring opening reactions.*[172]

The interest in graphene nanosheets with well-defined dimensions produced through synthetic chemistry as CVD or molecular chemistry continues to increase.[173,174] As expected, these graphene analogs exhibit potential for a host of applications.[175,176] Due to the nature of obtaining graphene through bottom-up approaches, covalent functionalization on the edges can be inherent in this material, rather than on the basal carbon atoms.[177] Moreover, careful selection of the edge decoration of these quantum dots can allow for precise tuning of the electronic properties, such as the ionization potential (HOMO), the electron affinity (LUMO), and the resulting energy bandgap of the final material.[178] If a graphene nanosheet is covalently bonded to electron-withdrawing oxygen functionalities, p-doping can be induced. Similarly, if it is functionalized by electron-donating nitrogen functionalities, n-doping can be achieved. With precise control of the interface between p-doped and n-doped graphene regions, it is possible to selectively engineer a graphene p–n junction by chemical doping.

The effect of functionalization and doping nanosheets by the H, N, O, F, and V atoms and by groups such as –OH and –CH$_3$ has been also theoretically evaluated, which showed that the electronic and magnetic properties of graphene nanosheets may be readily manipulated by these means.[179–182] This leads to the expectation that they may be extremely useful in areas such as spintronics, sensors, and transistors. This type of atomic functionalization is easily performed *by plasma-assisted CVD* starting from precursor gases as NH$_3$, CF$_4$, H$_2$, O$_2$, and so on.

43.3.6.2 Noncovalent interactions

Unlike covalent functionalization, noncovalent modification hardly disrupts the extended π-conjugation on the graphene surface and retains almost unaltered the electronic network of graphene.

43.3.6.2.1 π–π stacking

The π–π stacking methodology is based on van der Waals forces between aromatic molecules stacking on the graphene plane.[183–185] Donor- and acceptor-type aromatic molecules can be π–π stacked on the graphene surface (Figure 43.11e, f), respectively, to tune electron density of graphene.[186] Recently, this methodology has been used to improve hydrophilic characteristics and compatibility of conductive and transparent CVD graphene films with the spin coating of poly(3,4-ethylenedioxythiophene) needed in photovoltaic device fabrication.[187] Exploiting graphene fluorescence quenching effects and the π–π interactions with nucleobases and aromatic dye compounds, new sensors are developed based on a fluorescence-enhanced detection or photoactivation[188–190] of molecules that are sensitive and selective to the target molecule[191–193] (Figure 43.3b). These experiments indicate that graphene can be used as a platform for selective detection of biomolecules, biomedical applications such as cell imaging, and intracellular drug delivery.

43.3.6.2.2 Electrostatic and ionic interactions

The combination of *electrostatic interactions and π–π stacking* with cationic porphyrin was employed in the fabrication of an optical probe for sensing Cd^{2+} ions.[194,195]

Another class of compounds that have been widely used for the noncovalent modification of reduced graphene oxide is polymers. Amine-terminated polystyrene, carboxylic acid–terminated polystyrene, and hydroxyl-terminated poly(methyl methacrylate) were used for the noncovalent modification of reduced graphene oxide via *ionic* interactions, achieving so stable dispersions.[196]

43.3.6.2.3 Metal nanoparticle immobilization

Graphene can be p-doped *by absorbing metals* with higher electron affinity such as gold.[197] In comparison, alkaline metals are good electron donors owing to their strong ionic bonding and attractive candidates for n-type dopants.[198,199] In addition, the doping of graphene with transition metals such as Ca,[140] K, and Pd[200] can theoretically result in ferromagnetism due to the hybridization between the transition metal and the carbon orbital.

Besides metal nanoparticles, a variety of *metal oxides* were also employed for the decoration of graphene sheets. In this context, Fe_3O_4 magnetite nanoparticles with a size range of 10 nm are formed on the surface of graphene sheets showing high binding capacity for As(III) and As(V) and can be used to remove the heavy and toxic metal from water.[201] Moreover, combination of Fe_3O_4 nanoparticles and modified graphene facilitates the loading of drugs or pollutants onto the basal region of the graphene sheets.[202] In addition, nickel oxide (NiO)[203], Fe_3O_4[204] nanoparticles, and TiO_2 nanospindles,[205] formed and deposited onto graphene sheets, can be used as a novel anode material for high-performance lithium ion batteries. In the same context, an improved electrode for supercapacitors has been proposed, consisting of hydrous ruthenium oxide (RuO_2) nanoparticles deposited between graphene sheets.[206] In addition, graphene decorated with rutile tin oxide (SnO_2) in order to construct nanoporous electrode materials has been also suggested.[207]

Finally, the presence of oxygen functionalities on edges of surface provides the proper reactive sites for the nucleation and growth of several metal components, through both electrostatic and coordinate approaches. In such a way, a nanocatalyst based on palladium nanoparticles was developed.[208,209]

43.4 CVD TECHNOLOGIES AND METHODS FOR GRAPHENE GROWTH

An alternative to the exfoliation method, which resultant graphene layer displays a splendid behavior, may be a bottom-up growth method. Thermal-activated CVD has been widely used to grow single-layer graphene and FLG sheets on metal surfaces, such as Co, Fe, Ni, and Cu.[210–213] Besides the graphene-based 2D sheets, other graphene-related materials have been fabricated/synthesized as well, such as bilayer graphene and FLG,[214,215] 0D GQDs,[216–218] 1D graphene nanoribbons,[219–221] and graphene nanomeshes.[222,223] These materials are expected to possess different electrical and optical properties, due to the variation in size and geometry and the presence of a large amount of edge defects. Usually, an insulating substrate is required for most of graphene applications, and therefore a complex step of transference of the graphene sheet to other substrates is required.[224]

CVD involves the activation of gaseous reactants and the subsequent chemical reaction followed by the formation of a stable solid deposit over a suitable substrate. During the growth process, the precise control of gases flow, temperature, and pressure is essential for successful results. The energy needed for the chemical reaction can be supplied by different sources: *heat*, *light*, or *electric discharges* are used in thermal, laser-assisted, or plasma-assisted CVD, respectively.

Two types of reactions can take place all along the deposition process: homogeneous gas-phase reactions, which occur in the gas phase, and heterogeneous chemical reactions which occur on/near the vicinity of a heated surface, leading to the formation of powders or films, in each case. Two types of reactors may be used for the deposition process depending on the heating area for the process, namely, cold- or hot-wall reactors. In the last case, the gas mixture located in the heated zone reaches the reaction temperature, whereas in the cold-wall reactor, just the substrate and a small zone close to it are heated. In order to finally obtain graphene layers, heterogeneous chemical reactions should be favored and homogeneous chemical reactions avoided during the designed experiments. Figure 43.12 shows a schematic diagram of a typical CVD process.[225]

1. Transport of reactants by forced convection.
2. (a) Thermal or (b) plasma activation. Homogeneous gas reaction with particles and powder production should be avoided in graphene synthesis, controlling the kinetic parameters (P, T, n).
3. Transport of reactants by gas diffusion from the main gas stream through the boundary layer.
4. Adsorption of reactants on the substrate surface.

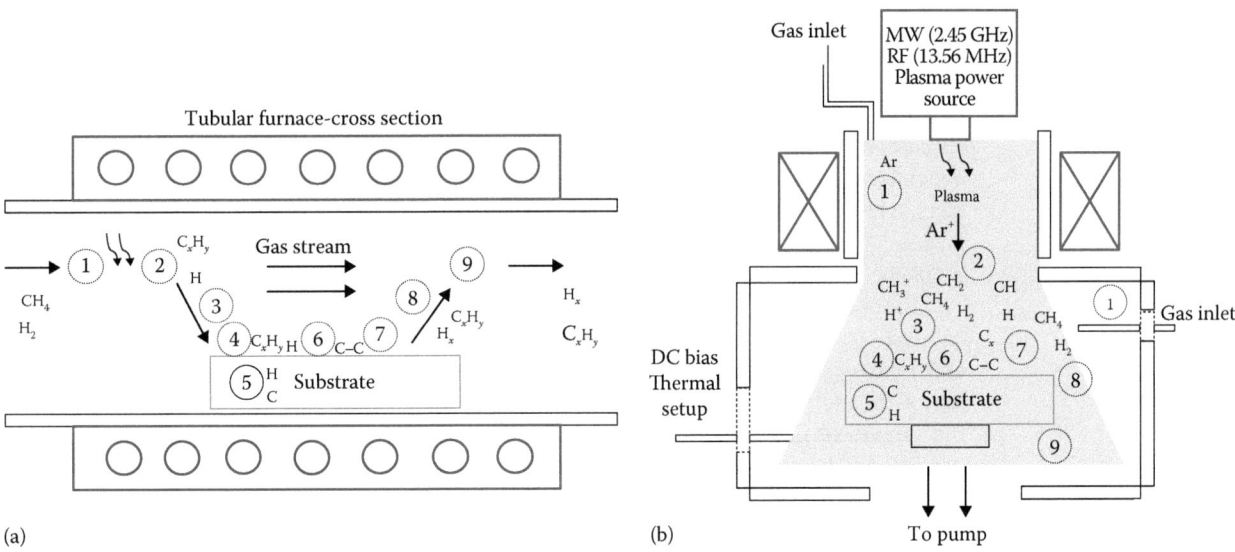

FIGURE 43.12
Schematic diagram of (a) thermal CVD and (b) plasma-assisted CVD processes: case of graphene from CH_4/H_2 mixtures.

5. Dissolution and bulk diffusion of species depending on the solubility and physical properties of the substrate.
6. Thermal activation mediated-surface processes, including chemical decomposition (catalytic), reaction, surface migration to attachment sites (such as atomic-level steps), incorporation, and other heterogeneous surface reactions. Growth of the film.
7. Desorption of by-products from the surface.
8. Transport of by-products by diffusion through the boundary layer and back to the main gas stream.
9. Transport of by-products by forced convection away from the deposition region.

In low-pressure CVD (LPCVD), the usual temperature required for depositing the material from the source gases is excessively high, and therefore it is convenient to include some catalyst (usually metal) in order to favor the reaction and then to obtain the material at a dramatically lower temperature. This is the case of graphene growth that a temperature higher than 1,200°C is necessary as methane is used as precursor gas. However, if the reaction takes place on a copper or nickel surface, the metal catalyzes the hydrocarbon decomposition and the growth may be made at $T < 950°C$. Therefore, in the following, the catalytic preparation of graphene layers will be described.

43.4.1 Thermo-Catalytic CVD of Graphene

CCVD of graphene on metal films has been widely explored. The gas mixture, containing a hydrocarbon, is thermally activated, and also metal substrate acts as catalyst for the hydrocarbon decomposition. Relatively high temperature is required principally for smoothing the metal surface as well as assisting the chemical decomposition reaction. It is important to remark that the catalytic contribution of the metal to the decomposition of the hydrocarbon makes less necessary the excessively high temperature.

Despite the significant progress, CVD graphene is a polycrystalline film made of micrometer- to millimeter-size domains. It was shown that Cu is an excellent candidate for making large-area, uniform-thickness (95%), single-layer graphene films due to the low solubility of C in Cu.[3] Since the chemical decomposition reaction is catalytically activated, the film grown over metal substrate reduces the catalytic activity due to the catalyst poisoning. This should announce the end of the reaction and then the graphene film formation. If the overall process is essentially performed on the surface (adsorption, decomposition, and diffusion of molecules), monolayer graphene is preferentially grown. This is known as *self-limiting* effect and was only observed in Cu to date (and also depending on the process conditions). It was demonstrated that the graphene growth on Cu is somehow surface mediated and self-limiting. On the other hand, on Ni and other common transition metals (Co, Ru, Ir, etc.), it was demonstrated that CVD growth of graphene occurs by carbon bulk diffusion due to the high solubility of carbon and segregation during cooling step. In this latter case, solid solution of a mixture of elements is formed near surface, and the resulting graphene depends on the kinetic parameters selected for the synthesis. Among all the thermodynamic parameters, a fast cooling rate seems to be a critical factor to suppress the formation of multiple graphene layers.[226,227] To date, the graphene films grown on Ni foils or films did not yield uniform monolayer graphene. In most cases, a mixture of monolayer and few layers (multigraphene) was obtained.

From a practical point of view, some critical steps have to be taken to get exposure of the catalyst surface to the gas precursors[225]: (1) heating, (2) annealing, (3) growing, and (4) cooling and final steps. Along these steps, heating up to the process temperature, cleaning and reducing of the metal surface, growing itself, and cooling of the system take place, respectively. For the growth of graphene layers, several hydrocarbons, such as methane, ethane, acetylene, and hexane, can be used as carbon atom source. The hydrocarbon and catalyst chosen for the growth determine the rest of experimental conditions for the process. Methane is the hydrocarbon more widely used, and when graphene is grown on Cu (very low carbon solubility), surface migration process is the main stage controlling the monolayer graphene growth. Differently, as already mentioned, Ni metal can dissolve much more carbon atoms and the graphene growth takes place predominantly from the precipitation during the cooldown of the process. At these conditions, *multigraphene* was often detected. Note that when MLG is preferentially formed, an additional etching step is essential for succeeding in the formation of monolayer graphene.[228,229] Finally, it is worth to mention that good-quality graphene layer has been obtained by CCVD at low as well as atmospheric pressure.[230,224] In general, the CCVD graphene growth requires high temperature, mainly for the treatment of the metal surface, reducing and widening the smooth zones of the surface.

During the annealing step, the catalyst surface is reduced by molecular hydrogen, which subsequently remains close to the bare metal surface; thus, the first step to be considered should be the dissociative chemisorption of H_2 on the metal surface. Under typical conditions of graphene synthesis, this process takes place on Cu and Ni surfaces with different trends. The different kinetic behavior of the CVD growth of graphene on Ni and Cu, which has been clearly pointed out, is noteworthy.[231,232] In the case of Ni, it is more probable for hydrogen to recombine and desorb from the surface, but this is not directly applicable for Cu that exhibits much greater hydrogen solubility.[233-235] In the copper case, saturation would be necessary to desorb molecular hydrogen from the Cu surface. Therefore, before exposure of the catalyst to hydrocarbons, a surface and/or subsurface partially covered with atomic hydrogen could be the starting point.[236] As the hydrocarbon (usually diluted in hydrogen) is put in contact with the metal surface, the dissociative chemisorption of H_2 competes with the physical adsorption and dehydrogenation of CH_4 on available metal surface sites. As a consequence, CH_x species coming from the catalytic decomposition of CH_4 are present over the metal surface. According to theoretical calculations based on DFT[237-239], dehydrogenation reactions probably take place up to $x = 2$, and in particular, the CH monomer dissociation is the rate-limiting step, the last being difficult to complete when copper is used as catalyst. From the monomers covering the Cu surface, the graphene nucleation begins, being dimer formation with simultaneous dehydrogenation a favorable reaction from an energetic point or view.[225] Also it is worth to mention that the formed Carbon-carbon (CC) dimers present a high stability on all sites of the Cu surface.[240] Therefore, the reaction from monocarbon species $CH_x + CH_y \rightarrow C_2 = + (x + y)$ H should be considered as the governing path for carbon deposition on Cu, being the dehydrogenation and the formation of CC bonds with sp^2 hybridization likely completed. Several competing processes have to be considered, namely, (1) dimer formation, (2) dimer migration, (3) back dissociation of dimers into individual atoms, (4) migration of carbon along the surface, and (5) migration of carbon atoms deeper into the bulk. Taking into account that the formation of carbon dimers is exothermic and the migration barrier for the dimer to move on the Cu(111) surface is small, it will cost more energy for the dimer to dissociate than to migrate around the surface. As a consequence and also supported by theoretical calculations, processes (1) and (2) were considered dominant. Riikonen et al.[241] pointed out how migrating dimers could form larger graphitic structures on the Cu(111) facet. It is worthwhile to emphasize that the suitable selection of thermodynamic parameters during the synthesis process is critical, due to the relevance of the fact that the chemical potentials of the surface carbon species are lower than that of carbon in the gas phase.[225] Once the nucleated graphene structure is stable at the surface, further growth can be performed by attachment of carbon species onto graphene edges. Lastly, we would like to emphasize that copper may be considered one of the most proper catalysts for graphene deposition due to its low catalytic activity for methane decomposition together with the also low solubility of carbon into the metal.

Finally, more complex deposition processes result when an extra gas-phase activation (decomposition by plasma or very high temperature [>1,200°C]) is performed. In these cases, the chemical reaction evolves to a mixture of heterogeneous catalysis and decomposition in vapor phase. Therefore, in plasma-assisted processes, the reaction cannot be considered as totally controlled by the catalyst and its study deserves further attention.

43.4.2 Plasma-Enhanced CVD of Graphene

In parallel to the CCVD method at low or atmospheric pressure, plasma-assisted CVD has been also widely developed for the preparation of graphene layers. In this case, low pressure is usually required for the onset of the plasma process. The discharge supplies extra energy and then new species are produced, with particular activation degree. The activation and decomposition of precursor gases, prior to reach the substrate, is effectively performed. Besides, it is noteworthy that after the generation of activated species, the thermal-mediated-surface diffusion of them plays a fundamental role in growth kinetics.

Plasma activation of the gas mixture allows one to work at lower temperature, so the process is highly recommended for industrial applications. Currently, there is a need to develop a reliable and reproducible method for low-temperature synthesis of high-quality graphene for the full exploitation of graphene properties and possible applications. One game-changing breakthrough[8] would be the development of graphene growth on arbitrary surfaces at low temperature with a minimal number of defects. So, plasma-assisted deposition could be a fundamental player in future graphene development.

Radio-frequency (13.56 MHz) PECVD (RF-PECVD)[242] revealed as a very powerful technique for the synthesis of large-scale graphene at relatively low temperature in a short time. Large-area single-layer graphene or MLG of satisfactory quality was synthesized on Ni films deposited on a thermally oxidized Si, at a relatively low temperature (650°C). In the deposition process, trace amounts of CH_4 were introduced into the PECVD chamber only for a short deposition time (30–60 s). Single-layer graphene or MLG was obtained due to that the carbon atoms, coming from the plasma, diffuse into the Ni film and then segregate out its surface. The number of graphene layers increased with longer deposition times for larger CH_4 flows at a cooling rate of about 10°C s^{-1}.

Teresawa et al.[243] investigated MLG grown on Cu foils also by RF-PECVD. The growth of graphene was investigated at various conditions, changing the plasma power, gas pressure, and the substrate temperature (from 500°C to 900°C). At high substrate temperature (~900°C), the growth of the first layer of graphene was affected by the catalytic action of Cu, while at low temperature, the growth of MLG was dominated mostly by radicals generated in the plasma. The main question is that the growth process is no longer self-limiting and it depends on time. For the same plasma conditions, the substrate temperature (T_s) strongly affects the growth rate of graphene. Also outstanding differences in the grain size and number of layers on FLG between 500°C and 900°C were observed. The grain size is much increased at T_s higher than 900°C, and in addition, they detected the decrease in the grain size of graphene with the thickness of the layer. The growth rate of the subsequent layers in MLG was measured approximately five times slower than that of the first layer graphene. In thermal CVD, the difference in the growth rate between the first layer and the second layer was more than 10.[230] This more rapid growth, in comparison with thermal CVD, is one of the features of PECVD. The activated carbon fragment C_2, formed in PECVD, is considered the most important source species for creating graphitic network on the substrate, and the graphene growth occurs even at 500°C.[243] Finally, using a remote radio-frequency (RF) configuration, for eliminating the effect of the plasma electrical field on the orientation of the grown graphene films,

Nandamuri et al.[244] deposited large-area (2.5×2 cm^2) single- and multiple-layer graphene on nickel and polycrystalline nickel films at 650°C.

In addition to RF plasmas, microwave discharges (2.45 GHz) have been also employed for plasma-assisted CVD processes (*microwave plasma CVD* [*MPCVD*]) headed toward graphene growth. As known, the increase in the discharge frequency produces an increase in electron temperature and electron density and thus an improvement in the reaction efficiency. Commonly, as the frequency discharge increases, species with other different activation degrees and also even different species may be produced. Many of these works have been developed applying the microwave plasma and, at the same time, using metal substrates, so that the catalytic decomposition of the hydrocarbon cannot be ruled out. For example, Kim et al.[245] deposited graphene films on polycrystalline nickel foil using a *cold-wall-type microwave* (*MPCVD*) system with a heating stage. A substrate temperature of 450°C–750°C and a total pressure of 20 Torr under various mixing ratios of H$_2$ and CH$_4$ were used. The dependence of monolayer graphene synthesis on temperature was investigated, and it was clearly shown that the higher the temperature, the higher-quality graphene was grown so far. As well, Kumar et al.[246] reported a unique process for rapid synthesis (100 s duration) of FLG films on Cu foil by *MPCVD*. The process can produce films of controllable quality from amorphous to highly crystalline by adjusting plasma conditions during growth and with no supplemental substrate heating (plasma–metal coupling for rapid heating of the foil). The hydrogen plasma was also used to remove the native oxide layer enabling graphene growth on metal Cu. It was suggested that the same process could be used for rapid synthesis of primarily single-layer graphene.

Along the last years, two Japanese groups have developed the *surface wave plasma* (*SWP*)-*CVD* method,[247,248] which exhibits the possibility of synthesizing graphene at low temperature and over substrates other than used in standard processes. At these conditions, the group of Iijima (Tsukuba) did not obtained any monolayer graphene.[247] On the other hand, Kalita et al.[248] reported direct synthesis of graphene nanosheets (very small domain size) on silicon (*n*-Si) and glass substrates by *microwave-assisted SWP* (*MWSWP*) *CVD* at 400°C–560°C. In the MWSWP CVD system (with a hollow quartz plate), the plasma can be started smoothly without any density jump, and plasma density is two times higher than that for a flat plate. High-density SWP can be generated in the meter scale, and therefore large-area deposition can be achieved at a very fast rate. The formation process (C$_2$H$_2$/Ar, 45 Pa) is rapid (70–120 s) and the film can be grown on different substrates. The as-grown deposit consisted of triangular-shaped nanographene domains, 80–100 nm in length, which interconnect among themselves to form a continuous film. In this way, a high density of nanodomains with similar shape is formed, which seems to be particularly adequate for the successful growth of suitable quality graphene nanosheets.

Another high-density microwave plasma method (electron cyclotron resonance plasma–assisted CVD [ECR-CVD]) has been developed at our laboratory for the growth of graphene layers.[249] The ECR discharge is characterized mainly by highly energetic electrons as a consequence of the produced resonance between the microwaves and the electrons. When the natural frequency of the electron gas in the presence of a static magnetic field is coupled to the excitation microwave energy, the electrons are accelerated and their orbit radii increased. Subsequently, the gas injected in the ECR area is excited and ionized by collisions with the accelerated electrons. At these conditions of low pressure, a dense plasma with a relatively high ionization degree, as compared to other types of plasmas, is driven by the divergent magnetic field lines toward the substrate. The main distinctive aspects of this deposition technique are the surprisingly low temperature as well as the high efficiency in the hydrocarbon decomposition; this lasts likely allowing the graphene deposit even on noncatalytic surfaces (dielectric or semiconductor substrates). During the process, small domains of graphene were formed, which remained embedded in an amorphous carbon matrix. Different argon/hydrogen/hydrocarbon mixtures were used and methane (CH$_4$) or acetylene (C$_2$H$_2$) were chosen as carbon atom supplier. We studied the effect of substrate temperature and the gas mixture composition (hydrogen content and type or hydrocarbon). In all the cases, the higher temperature, the higher quality of the grown graphene is. In addition, we detected a positive influence of hydrogen content on the domain size of the deposits. It should be pointed out that the two carbon atoms present in the acetylene molecule account for the larger deposition rate with respect to that reached when methane, instead of acetylene, is introduced in the reactor.

43.4.2.1 Effect of plasma in graphene growth

Similar to the growth mechanism proposed for the CCVD graphene growth, now we will focus on the study of plasma-activated processes, where most of the species are generated within the plasma by electrical activation. As known, when a strong electric field is applied on a low-pressure medium, the breakdown of medium is produced, and as a result, the plasma, formed by ions, electrons, and neutral activated species, is generated. In particular, low-pressure discharges are characterized by an electron temperature (T_e) of 0.5–10 V, and the density of charged species (n) is approximately 10^8–10^{13} cm^{-3}. In these discharges, feedstock gases are broken into positive ions and chemically reactive etchants, deposition precursors, and so on, which then flow toward and physical or chemically interact at the substrate surface. Meanwhile, energy is delivered to the substrate also, for example, in the form of bombarding ions, and the energy flux is there to promote the chemistry at the substrate and not to heat it. The gas pressures for these discharges are often low: $p \approx 1$ mTorr–1 Torr.[250] In such plasmas, the electron temperature is much higher than the gas temperature, and inelastic collisions of the electrons with precursor molecules form chemically active species that participate in the reactions leading to the film growth. Also, surfaces inside the plasma can be bombarded with the active species, such as ions, electrons, and photons, leading to changes in surface chemistry.[251] In general, the use of high-frequency discharges (2.45 GHz) produces a higher electron density as well as a larger ionization degree. In ECR plasmas, the ionization level may reach 10%. Therefore, as microwave or ECR, instead of RF, plasmas are used for the deposition process, the efficiency of the reaction tends to increase, resulting in higher deposition rates.[252-254] However, it must be taken into account that in processes activated at different frequencies, most of the other experimental parameters such as pressure, gas mixture composition, and reactor geometry, required for graphene deposition, also dramatically change from one experiment to the other. Therefore, for the study of processes at other frequencies, the rest of the experimental parameters have to become again adjusted for getting the pursued objective. Therefore, it is difficult to isolate the effect of discharge frequency on the deposition process. Anyway, the higher ionization degree besides the larger efficiency in the dissociation of hydrogen molecules in high-frequency plasmas is undoubtedly a proved fact, which must be kept in mind.

43.4.2.1.1 Analysis of RF and ECR plasmas

Generally, the plasma composition is analyzed by optical emission spectroscopy (OES) and mass spectrometry (MS). Note that only approximately 5% of activated species can be detected by OES, because a little percentage of activated species, generated inside the plasma, deactivates emitting photons in the OES measurement range. In any case, valuable information may be obtained from the study of the influence of different experimental parameters in the plasma composition, which clearly will affect the growth process. We studied methane and acetylene mixtures activated by RF and microwave ECR discharges. The experimental conditions of the processes, which plasma has been analyzed by OES, are shown in the next table (Table 43.1).

In Figure 43.13, the OES spectra of emitting C_2H_2/Ar (with or without hydrogen) plasmas generated by RF and ECR discharges are exhibited at two different scales (Figure 43.13a, b). As can be seen, the emission lines for ECR plasma have to be multiplied by 40 in order to compare them with lines in RF plasma. This means that the intensity of ECR spectra is much less intense than those corresponding to RF-activated plasmas. The difference in the total emission intensity has been explained by the partial pressure of argon, two orders of magnitude lesser in the ECR-activated plasmas. In addition, it is noteworthy to mention that the power in both cases is very different, since a 40 W ECR discharge is applied for the C_2H_2 decomposition, whereas a much higher power of 300 W is required for the C_2H_2 decomposition in the RF plasma. Anyway, as known, for succeeding in graphene growth, very low carbon species concentration must be slowly placed on the substrate surface, and therefore strong acetylene activation is not required. The most significant differences detected between both spectra are (1) the peaks at 358 nm (not assigned) and 388 nm (CH B$^2\Sigma$ → X$^2\Pi$) that appear in the RF spectrum and are not clearly detected in the ECR plasma and (2) the peak at 516 nm, associated with C_2 species, that is actually remarkable in the ECR plasma, whereas it is hardly detected in the RF plasma. Lastly, the emission line at 470 nm, also attributed to the dicarbon species,[33] is strongly intense in the ECR plasma. What does it mean? The shortage

TABLE 43.1 Experimental Conditions of Studied Plasmas

Sample	Plasma	Hydrocarbon	[Hydrocarbon], (sccm)	[H$_2$], (sccm)	[Ar], (sccm)	Power (W)	Pressure (mbar)
A	RF	CH$_4$	2	3	95	300	0.126
B	RF	C$_2$H$_2$	1	4	95	300	0.126
C	ECR (MW)	CH$_4$	4	—	35	100	0.0066
D	ECR (MW)	C$_2$H$_2$	1	—	35	40	0.0061

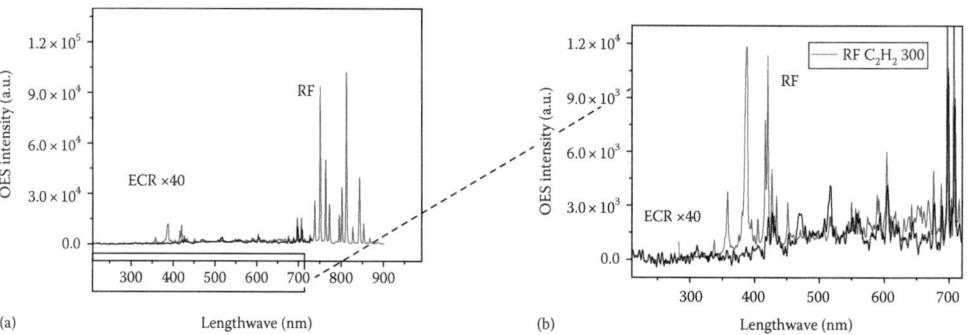

(a)   (b)

FIGURE 43.13
OES spectra of C$_2$H$_2$/Ar plasmas for RF (light color line) and ECR (dark line) discharges.

of CH species (at the B$^2\Sigma$ state) in the ECR plasma coincides with the enrichment of the plasma in C$_2$ species. As known, CH species may be considered as the main precursor species for DLC films growth, and indeed this type of carbon films (DLC), at high deposition rate, is grown when RF plasmas are produced, as experimentally confirmed. However, in ECR discharges, for growing extremely thin films, like graphene layers, a low concentration of carbon-containing species, mainly as dicarbon species, would be particularly appropriate. The contribution of dicarbon species in graphene growth, as previously proposed during catalytic LPCVD graphene formation, would improve the graphene growth at these experimental conditions.

Also, the comparison of the OES spectra corresponding to plasmas generated from different gas precursors for two different frequency discharges (RF and ECR) provides useful information about the processes. As an example, Figure 43.14 displays the OES spectra of plasmas RF and ECR (Figure 43.14a, b) using in each case both CH$_4$ and C$_2$H$_2$ precursor gases. As can be seen, comparing both Figure 43.14a and b, the OES spectra are totally different in intensity and also in the peaks appearing in them at both frequencies. As already mentioned, the differences in intensity are likely due to the exceptionally different experimental conditions (pressure, measuring point, gas flows, etc.) required at every frequency for succeeding in the formation process. On the other hand, it is

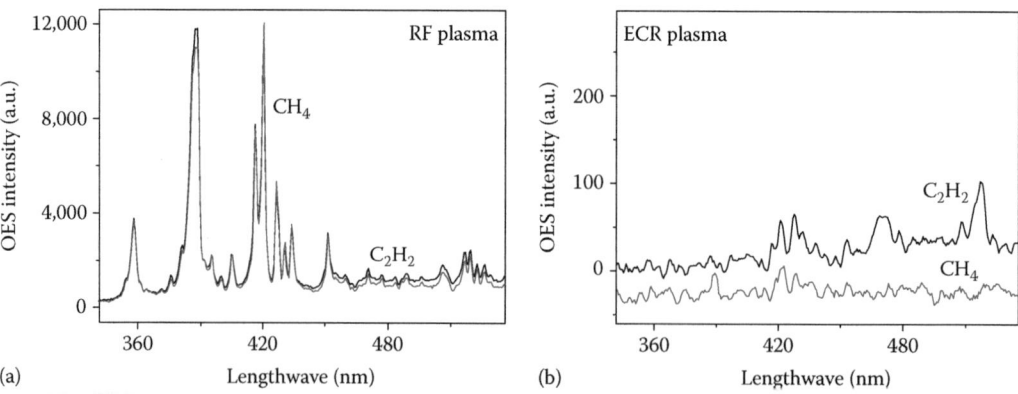

(a)   (b)

FIGURE 43.14
OES spectra of C$_2$H$_2$ and CH$_4$ plasmas activated by (a) RF and (b) ECR discharges.

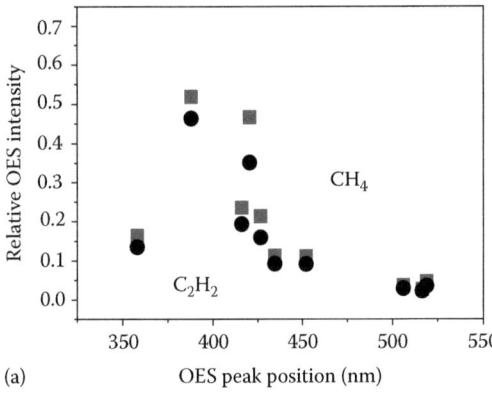
(a)

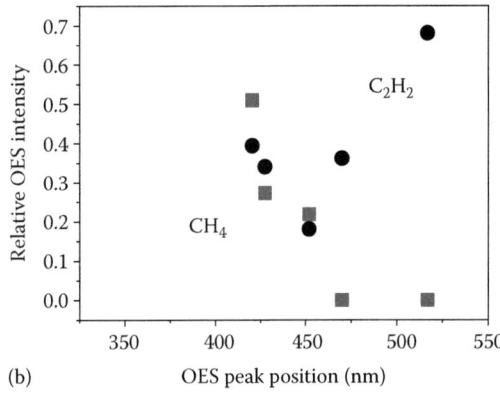
(b)

FIGURE 43.15
Relative intensity of OES lines (with respect to the Ar line at 698 nm) for CH_4 and C_2H_2 plasmas in (a) RF and (b) ECR plasmas.

easy to infer that the energetic conditions are different in RF and ECR plasmas, and so the generated activated species will be also different. It is important to emphasize that the number and type of OES peaks, as well as their relative intensity, are strongly affected by the frequency discharge. For comparing the species present in each case, the intensity of the OES peaks, in the 325–550 nm range, with respect to the argon line at approximately 698 nm, is given in Figure 43.15.

As can be clearly seen, in ECR discharges, no lines have been detected under 400 nm. Having the peak at 388 nm, undoubtedly attributed to the CH radical (transition $B^2\Sigma \rightarrow X^2\Pi$), no detection of this line, together with the weak intensity of the line close to 430 nm (also associated with CH species, transition $A^2\Delta \rightarrow X^2\Pi$), implies that in ECR discharges of methane or acetylene, the CH species is not predominant inside the plasma. Furthermore, in C_2H_2–ECR plasmas, the intensity in the zone around 516 nm (traditionally associated with C_2 species) becomes quite larger (see Figure 43.15b), which along with the appearance of the peak at 470 nm (also related to C_2 species[255]) indicates a high content of these C_2 species during the graphene growth process from acetylene mixtures.

As a conclusion, in RF plasmas, the generated carbon-containing species (plasma composition) is not appreciably affected by the hydrocarbon source. On the contrary, in ECR plasmas, though in no case CH species have been detected, however, whereas C_2 species are not formed in CH_4–ECR, a huge amount of them are produced in C_2H_2–ECR plasmas. The last observation would explain the growth of graphene layers in C_2H_2–ECR plasmas, being dicarbon species the intermediate step. Finally, it is important to remember that, without any doubt, one of the most interesting characteristics of the ECR discharges is its ability to easily produce hydrogen atoms coming from the dissociation of hydrogen molecules. During the formation process, the etching of the growing deposit by atomic hydrogen has been shown to be essential for the growth of graphene layers.

As a supplement of the OES study of plasmas for graphene growth, it is worth emphasizing the results, obtained by MS, on the main species present in the plasma generated by electrical activation. This technique allows detecting all the species with different m/q ratio and so gives a more actual sight of the carbon-containing species during the deposition process. In Table 43.2, the composition of CH_4 and C_2H_2 plasmas generated by an ECR discharge is given. As shown in the table, the CH_4 and C_2H_2 plasmas are made up approximately of 68% and 78% of the source molecule, respectively, and the rest of species are similar though their concentration may appreciably change. For example, a 2.4% of the initial methane molecules decompose, giving finally as a result the formation of acetylene, whereas only approximately a 0.1% of methane molecules are generated in the acetylene plasma. Focusing on the concentration of ethylene, in both plasmas, it is five times higher when acetylene

TABLE 43.2 Composition of CH_4 and C_2H_2 Plasmas Generated by an ECR Discharge

Source Molecule	CH_4	C_2H_2	C_2H_4	C_4H_2	C_4H	C_6H_6	H_2
CH_4	68%	2.4%	1.9%	—	—	Under detection level	25.5%
C_2H_2	~0.1%	78.4%	9.8%	1.5%	0.4%	Under detection level	9.8%

is used as precursor gas, which means that indeed it is much easier to form ethylene from acetylene than from methane. As a result of the given data, the content of dicarbon (multicarbon) species is at least one order of magnitude larger in the acetylene plasmas. This result is in good agreement with the OES measurement conclusions and justifies the growth of graphene layers through the dicarbon species as intermediate step. Furthermore, the large difference in the dicarbon species content in both types of plasma shed light on why 200 W must be applied in methane mixtures for growing graphene, whereas just 40 W is necessary when acetylene is used as precursor gas.

Great advances are being published in this area of graphene synthesis. Monolayer high-quality graphene on metal at low temperature[215] and small domain graphene on dielectric substrates[256-258] have been already published. In the near future, graphene synthesis by plasma-assisted CVD will be an extensive area of research, with the intention to synthesize controllable number of graphene layers on arbitrary substrates by a reliable and reproducible method.

REFERENCES

1. Novoselov, K. S.; Geim, A. K.; Morozov, S. V.; Jiang, D.; Zhang, Y.; Dubonos, S. V.; Grigorieva, I. V.; Firsov, A. A. *Science* 2004, *306*, 666.
2. Morozov, S. V.; Novoselov, K. S.; Katsnelson, M. I; Schedin, F.; Elias, D. C.; Jaszczak, J. A.; Geim A. K. *Phys. Rev. Lett.* 2008, *100*, 016602.
3. Tour, J. M. *Chem. Mater.* 2014, *26*, 163.
4. Mayorov, A. S.; Gorbachev, R. V.; Morozov, S. V.; Britnell, L.; Jalil, R.; Ponomarenko, L. A.; Blake, P. et al. *Nano Lett.* 2011, *11*, 2396.
5. Franklin, R. E. *Proc. R. Soc. Lond.* 1951, *209*, 196.
6. Hao, Y.; Bharati, M. S.; Wang, L.; Liu, Y.; Chen, H.; Nie, S.; Wang, X. et al. *Science* 2013, *342*, 720.
7. Lee, J.-H.; Lee, E. K.; Joo, W.-J.; Jang, Y.; Kim, B.-S.; Lim, J. Y., Choi, S.-H. et al. *Science* 2014, *344*, 286.
8. Novoselov, K. S.; Falko, V. I.; Colombo, L.; Gellert, P. R.; Schwab, M. G.; Kim, K. *Nature* 2013, *490*, 192.
9. Li, H.; Kang, Z.; Liu, Y.; Lee, S.-T. *J. Mater. Chem.* 2012, *22*, 24230.
10. Li, L.; Wu, G.; Yang, G.; Peng, J.; Zhao, J.; Zhu, J.-J. *Nanoscale* 2013, *5*, 4015.
11. Editorial. *Carbon* 2013, *65*, 1.
12. Schabel, M. C.; Martins, J. L., *Phys. Rev. B* 1992, *46*, 7185.
13. Dato, A.; Lee, Z.; Jeon, K.-J.; Erni, R.; Radmilovic, V.; Richardson, T. J.; Frenklach, M. *Chem. Commun.* 2009, *40*, 6095.
14. Meyer, J. C.; Geim, A. K.; Katsnelson, M. I.; Novoselov, K. S.; Booth, T. J.; Rot, S. *Nature* 2007, *446*, 60.
15. Bao, W.; Miao, F.; Chen, Z.; Zhang, H.; Jang, W.; Dames, C.; Lau, C. N. *Nat. Nanotechnol.* 2009, *4*, 562.
16. Pinto, H. P.; Leszczynski, J. Fundamental properties of grapheme. In *Handbook of Carbon Nano Materials*, Vol. 5, Chapter 1, World Scientific Series on Carbon Nanoscience. Singapore: World Scientific, 2014.
17. Nottingham Trent University. sp hybridization, Nottingham, U.K. http://www.ntu.ac.uk/cels/molecular_geometry/hybridization/Sp_hybridization/index.html.
18. Meyer, J. C.; Geim, A. K.; Katsnelson, M. I.; Novoselov, K. S.; Booth, T. J.; Roth, S. *Nature* 2007, *446*, 60.
19. Shi, H.; Lai, L.; Snook, I. K.; Bardnard, A. S. *J. Phys. Chem. C* 2013, *117*, 15375.
20. Barnard, A. S.; Snook, I. K. *J. Chem. Phys.* 2008, *128*, 094707.
21. Wallace, P. R. *Phys. Rev.* 1947, *71*, 622.
22. Kresse, G.; Furthmüller, J. *Comput. Mater. Sci.* 1996, *6*, 15.
23. Kresse, G.; Furthmüller, J. *Phys. Rev. B* 1996, *54*, 11169.

24. Stolyarova, E.; Rim, K. T.; Ryu, S.; Maultzsch, J.; Kim, P.; Brus, L. E.; Heinz, T. F.; Hybertsen, M. S.; Flynn, G. W. *Proc. Nat. Acad. Sci. USA* 2007, *104*, 9209.
25. Perdew, J. P.; Ruzsinszky, A.; Csonka, G. I.; Vydrov, O. A.; Scuseria, G. E.; Constantin, L. A.; Zhou, X.; Burke, K. *Phys. Rev. Lett.* 2008, *100*, 136406.
26. Reich, S.; Maultzsch, J.; Thomsen, C.; Ordejón, P. *Phys. Rev. B* 2002, *66*, 035412.
27. Zhou, S. Y.; Gweon, G. H.; Graf, J.; Fedorov, A. V.; Spataru, C. D.; Diehl, R. D.; Kopelevich, Y.; Lee, D. H.; Louie, S. G.; Lanzara, A. *Nat. Phys.* 2006, *2*, 595.
28. Bostwick, A.; Ohta, T.; Seyller, T.; Horn, K.; Rotenberg, E. *Nat. Phys.* 2006, *3*, 36–40.
29. Elias, D. C.; Gorbachev, R. V.; Mayorov, A. S.; Morozov, S. V.; Zhukov, A. A.; Blake, P.; Ponomarenko, L. A. et al. *Nat. Phys.* 2011, *7*, 701.
30. Luk'yanchuk, I. A.; Kopelevich, Y. *Phys. Rev. Lett.* 2006, *97*, 256801.
31. Latil, S.; Meunier, V.; Henrard, L. *Phys. Rev. B* 2007, *76*, 201402.
32. Liu, Z.; Suenaga, K.; Harris, P. J. F.; Iijima, S. *Phys. Rev. Lett.* 2009, *102*, 015501.
33. Partoens, B.; Peeters, F. M. *Phys. Rev. B* 2006, *74*, 075404.
34. Terrones, M.; Botello-Méndez, A. R.; Campos-Delgado, J.; López-Uras, F.; Vega-Cantú, Y. I.; Rodríguez-Macías, F. J.; Elías, A. L. et al. *Nano Today* 2010, *5*, 351.
35. Morozov, S. V.; Novoselov, K. S.; Schedin, F.; Jiang, D.; Firsov, A. A.; Geim, A. K. *Phys. Rev. B* 2005, *72*, 201401.
36. Brown, L.; Hovden, R.; Huang, P.; Wojcik, M.; Muller, D. A.; Park, J. *Nano Lett.* 2012, *12*, 1609.
37. Chen, F.; Tao, N. *Acc. Chem. Res.* 2009, *42*, 429.
38. Son, Y.; Cohen, M.; Louie, S. *Nature* 2006, *444*, 347.
39. Son, Y.; Cohen, M.; Louie, S. *Phys. Rev. Lett.* 2006, *97*, 216803.
40. Hod, O.; Barone, V.; Peralta, J.; Scuseria, G. *Nano Lett.* 2007, *7*, 2295.
41. Castro Neto, A.; Guinea, F.; Peres, N.; Novoselov, K.; Geim, A. *Rev. Mod. Phys.* 2009, *81*, 109.
42. Abergel, D.; Apalkov, V.; Berashevich, J.; Zieler, K.; Chakraborty, T. *Adv. Phys.* 2010, *59*, 261.
43. Gass, M.; Bangert, U.; Bleloch, A.; Wang, P.; Nair, R.; Geim, A. *Nat. Nanotechnol.* 2008, *3*, 676.
44. Koskinen, P.; Malola, S.; Hakkinen, H. *Phys. Rev. Lett.* 2009, *101*, 115502.
45. Koskinen, P.; Malola, S.; Hakkinen, H. *Phys. Rev. B,* 2009, *80*, 073401.
46. Wu, X.; Zeng, X. *Nano Res.* 2008, *1*, 40.
47. Ezawa, M. *Physica E* 2008, *40*, 1421.
48. Jiang, D.; Sumpte, D.; Dai, S. *J. Chem. Phys.* 2007, *127*, 124703.
49. Jing, D.; Sheng, D. *J. Phys. Chem. A* 2008, *112*, 332.
50. Stein, S.; Brown, R. *J. Am. Chem. Soc.* 1987, *109*, 3721.
51. Banerjee, S.; Bhattacharyya, D. *Comput. Mater. Sci.* 2008, *44*, 41.
52. Ezawa, M. *Phys. Rev. B* 2007, *76*, 245415.
53. Jackel, F.; Watson, M.; Mullen, K.; Rabe, J. *Phys. Rev. B* 2006, *73*, 045423.
54. Zhang, Y.; Tang, T.-T.; Girit, C.; Hao, Z.; Martin, M. C.; Zettl, A.; Crommie, M. F.; Shen, Y. R.; Wang, F. *Nature* 2009, *459*, 820.
55. Walukiewicz, W.; Lagowski, L.; Jastrzebski, L.; Lichtensteiger, M.; Gatos, H. C. *J. Appl. Phys.* 1979, *50*, 899.
56. Chen, J.-H.; Jang, C.; Xiao, S.; Ishigami, M.; Fuhrer, M. S. *Nat. Nanotechnol.* 2008, *3*, 206.
57. Novoselov, K. S.; Geim, A. K.; Morozov, S. V.; Jiang, D.; Katsnelson, M. I.; Girgorieva, I. V.; Dubonos, S. V.; Firsov, A. A. *Nature* 2005, *438*, 197.
58. Berger, C.; Song, Z.; Li, X.; Wu, X.; Brown, N.; Naud, C.; Mayou, D. et al. *Science* 2006, *312*, 1191.

59. Wang, L.; Meric, I.; Huang, P. Y.; Gao, Q.; Gao, Y.; Tran, H.; Taniguchi, T. et al. *Science* 2013, *342*, 614.
60. Petrone, N.; Dean, C. R.; Meric, I.; Van der Zande, A. M.; Huang, P. Y.; Wang, L.; Muller, D.; Shepard, K. L.; Hone, J. *Nano Lett.* 2012, *12*, 2751.
61. Moser, J.; Barreiro, A.; Bachtold, A. *Appl. Phys. Lett.* 2007, *91*, 163513.
62. Zhang, Y.; Tan, Y. W.; Stormer, H. L.; Kim, P. *Nature* 2005, *438*, 201.
63. Novoselov, K. S.; McCann, E.; Morozov, S. V.; Katsnelson, V. I. F. M. I.; Zeitler, U.; Jiang, D.; Schedin, F.; Geim, A. K. *Nat. Phys.* 2006, *2*, 177.
64. Chen, S.; Moore, A. L.; Cai, W.; Suk, J. W.; An, J.; Mishra, C.; Amos, C. et al. *ACS Nano* 2010, *5*, 321.
65. Balandin, A. A. *Nat. Mater.* 2011, *10*, 569.
66. Chen, S.; Wu, Q.; Mishra, C.; Kang, J.; Zhang, H.; Cho, K.; Cai, W.; Balandin, A. A.; Ruoff, R. S. *Nat. Mater.* 2012, *11*, 203.
67. Balandin, A. A.; Ghosh, S.; Bao, W.; Calizo, I.; Teweldebrhan, D.; Miao, F.; Lau, C. N. *Nano Lett. ASAP* 2008, *8*, 902.
68. Anthony, T. R.; Banholzer, W. F.; Fleischer, J. F.; Wei, L. H.; Kuo, P. K.; Thomas, R. L.; Pryor, R. W. *Phys. Rev. B* 1990, *42*, 1104.
69. Berman, R. *Phys. Rev. B* 1992, *45*, 5726.
70. Pierson, H. O. *Handbook of Carbon, Graphite, Diamond and Fullerenes: Properties, Processing and Applications*. Noyes Publications, Park Ridge, NJ, 1993.
71. Pop, E.; Varshney, V.; Roy, A. K. *Mrs Bull.* 2012, *37*, 1273.
72. Mingo, N.; Broido, D. A. *Phys. Rev. Lett.* 2005, *95*, 096105.
73. Nika, D. L.; Pokatilov, E. P.; Askerov, A. S.; Balandin, A. A. *Phys. Rev. B* 2009, *79*, 155413.
74. Popov, V. N. *Phys. Rev. B* 2002, *66*, 153408.
75. Muñoz, E.; Lu, J.; Yakobson, B. I. *Nano Lett.* 2010, *10*, 1652.
76. Reeber, R.; Wang, K. *J. Electron. Mater.* 1996, *25*, 63.
77. Seol, J. H.; Jo, I.; Moore, A. L.; Lindsay, L.; Aitken, Z. H.; Pettes, M. T.; Li, X. S. et al. *Science* 2010, *328*, 213.
78. Xu, Y.; Chen, X. B.; Wang, J. S.; Gu, B. L.; Duan, W. H. *Phys. Rev. B* 2010, *81*, 195425.
79. Li, X.; Maute, K.; Dunn, M. L.; Yang, R. *Phys. Rev. B* 2010, *81*, 245318.
80. Javey, A.; Guo, J.; Paulsson, M.; Wang, Q.; Mann, D.; Lundstrom, M.; Dai, H. *Phys. Rev. Lett.* 2004, *92*, 106804.
81. Wang, J.; Lundstrom, M. *IEEE Trans. Electron Devices* 2003, *50*, 1604.
82. Behnam, A., Lyons, A. S.; Bae, M.-H.; Chow, E. K.; Islam, S.; Neumann, C. M.; Pop, E. *Nano Lett.* 2012, *12*, 4424.
83. Jang, W.; Chen, Z.; Bao, W.; Lau, C. N.; Dames, C. *Nano Lett.* 2010, *10*, 3909.
84. Chen, Z.; Jang, W.; Bao, W.; Lau, C. N.; Dames, C. *Appl. Phys. Lett.* 2009, *95*, 161910.
85. Koh, Y. K.; Bae, M.-H.; Cahill, D. G.; Pop, E. *Nano Lett.* 2010, *10*, 4363.
86. Mak, K. F.; Lui, C. H.; Heinz, T. F. *Appl. Phys. Lett.* 2010, *97*, 221904.
87. Liao, A. D.; Wu, J. Z.; Wang, X. R.; Tahy, K.; Jena, D.; Dai, H. J.; Pop, E. *Phys. Rev. Lett.* 2011, *106*, 256801.
88. Bae, M.-H.; Islam, S.; Dorgan, V. E.; Pop, E. *ACS Nano* 2011, *5*, 7936.
89. Balandin, A. A. *Nano Lett.* 2008, *8*, 902.
90. Ghosh, S.; Calizo. I.; Teweldebrhan, D.; Pokatilov, E. P.; Nika, D. L.; Baladin, A. A.; Bao, W. et al. *Appl. Phys. Lett.* 2008, *92*, 151911.
91. Ghosh, S.; Nika, D. L.; Pokatilov, E. P.; Balandin, A. A. *New J. Phys.* 2009, *11*, 095012.
92. Ghosh, S.; Bao, W.; Nika, D. L.; Subrina, S.; Pokatilov, E. P.; Lau, C. N.; Baladin, A.A. *Nat. Mater.* 2010, *9*, 555.

93. Cai, W.; Moore, A. L.; Zhu, Y.; Li, X.; Chen, S.; Shi, L.; Ruoff, R. S. *Nano Lett.* 2010, *10*, 1645.

94. Faugeras, C.; Faugeras, B.; Orlita, M.; Potemski, M.; Nair, R. R.; Geim, A. K. *ACS Nano* 2010, *4*, 1889.

95. Jauregui, L. A.; Yue, Y.; Sidorov, A. N.; Hu, J.; Yu, Q.; Lopez, G.; Jalilian, R. et al. *ECS Trans.* 2010, *28*, 73.

96. Zhong, W. R.; Zhang, M. P.; Ai, B. Q.; Zheng, D. Q. *Appl. Phys. Lett.* 2011, *98*, 113107.

97. Singh, D.; Murthy, J. Y.; Fisher, T. S. *J. Appl. Phys.* 2011, *110*, 044317.

98. Balandin, A. A.; Shamsa, M.; Liu, W. L.; Casiraghi, C.; Ferrari, A. C. *Appl. Phys. Lett.* 2008, *93*, 043115.

99. Haskins, J.; Kınacı, A.; Sevik, C.; Sevincli, H. L.; Cuniberti, G.; Cağın, T. *ACS Nano* 2011, *5*, 3779.

100. Aksamija, Z.; Knezevic, I. *Appl. Phys. Lett.* 2011, *98*, 141919.

101. Kuzmenko, A. B.; Van Heumen, E.; Carbone, F.; Van Der Marel, D. *Phys. Rev. Lett.* 2008, *100*, 117401.

102. Nair, R. R.; Blake, P.; Grigorenko, A. N.; Novoselov, K. S.; Booth, T. J.; Stauber, T.; Peres, N. M. R.; Geim, A. K. *Science* 2008, *320*, 1308.

103. Casiraghi, C.; Hartschuh, A.; Lidorikis, E.; Qian, H.; Harutyunyan, H.; Gokus, T.; Novoselov, K. S.; Ferrari, A. C. *Nano Lett.* 2007, *7*, 2711.

104. Wang, F.; Zhang, Y.; Chuanshan, T.; Girit, C.; Zettl, A.; Crommiel, M.; Shen, Y. R. *Science* 2008, *320*, 206.

105. Mak, K. F.; Shan, J.; Heinz, T. F. *Phys. Rev. Lett.* 2009, *104*, 176404.

106. Bonaccorso, F.; Sun, Z.; Hasan, T.; Ferrari, A. C. *Nat. Photon.* 2010, *4*, 611.

107. Loh, K.; Bao, Q.; Ang, P.; Yang, J. *J. Mater. Chem.* 2010, *20*, 2277.

108. Gokus, T.; Nair, R. R.; Bonetti, A.; Böhmler, M.; Lombardo, A.; Novoselov, K. S.; Geim, A. K. et al. *ACS Nano* 2009, *3*, 3963.

109. Eda, G.; Lin, Y.-Y.; Mattevi, C.; Yamaguchi, H.; Chen, H.-A.; Chen, I.-S.; Chen, C.-W.; Chhowalla, M *Adv. Mater.* 2010, *22*, 505.

110. Sun, X.; Liu, Z.; Welsher, K.; Robinson, J. T.; Goodwin, A.; Zaric, S.; Dai, H. *Nano Res.* 2008, *1*, 203.

111. Luo, Z.; Vora, P. M.; Mele, E. J.; Johnson, A. T.; Kikkawa, J. M. *Appl. Phys. Lett.* 2009, *94*, 111909.

112. Lu, J.; Yang, J.-X.; Wang, J.; Lim, A.; Wang, S. and Loh, K. P. *ACS Nano* 2009, *3*, 2367.

113. Sheats, J. R.; Antoniadis, H.; Hueschen, M.; Leonard, W.; Miller, J.; Moon, R.; Roitman, D.; Stocking, A. *Science* 1996, *273*, 884.

114. Rothberg, L. J.; Lovinger, A. J. *J. Mater. Res.* 1996, *11*, 3174.

115. Frangioni, J. V. *Curr. Opin. Chem. Biol.* 2003, *7*, 626.

116. Essig, S.; Marquardt, C. W.; Ganzhorn, M.; Dehm, S.; Hennrich, F.; Ou, F.; Green, A. A. et al. *Nano Lett.* 2010, *10*, 1589.

117. Stoehr, R. J.; Kolesov, R.; Pflaum, J.; Wrachtrup, J. *Phys. Rev. B*, 2010, *82*, 121408.

118. Liu, C. H.; Mak, K. F.; Shan, J.; Heinz, T. F. *Phys. Rev. Lett.* 2010, *105*, 127404.

119. Wu, S.; Liu, W.; Schuck, P.J.; Salmeron, M.; Shen, Y. R.; Wang, F. *APS March Meeting*, Portland, OR, 2010.

120. Hartschuh, A. *E-MRS Spring Meeting*, Strasbourg, France, 2010.

121. Huang, Y.; Liang, J.; Chen, Y. *Small* 2012, *8*, 1805.

122. Crespi, V. H.; Benedict, L. X.; Cohen, M. L.; Louie, S. G. *Phys. Rev. B* 1996, *53*, R13303.

123. Geim, A. K. *Science* 2009, *324*, 1530.

124. Frank, I. W.; Tanenbaum, D. M.; Van der Zande, A. M.; McEuen, P. L. *J. Vac. Sci. Technol.* 2007, *25*(6), 2558.

125. Zhao, Q. Z.; Nardelli, M. B.; Bernholc, J. *Phys. Rev. B* 2002, *65*, 144105.
126. Liu, F.; Ming, P. M.; Li, J. *Phys. Rev. B* 2007, *76*, 064120.
127. Tsoukleri, G.; Parthenios, J.; Papagelis, K.; Jalil, R.; Ferrari, A. C.; Geim, A. K.; Novoselov, K. S.; Galiotis, C. *Small* 2009, *5*, 2397.
128. Lee, C.; Wei, X.; Kysar, J. W.; Hone, J. *Science* 2008, *321*, 385.
129. Wan, K.-T.; Guo, S.; Dillard, D. A. *Thin Solid Films* 2003, *425*, 150.
130. Zhao, H.; Aluru, N. R. *J. Appl. Phys.* 2010, *108*, 064321.
131. Khare, R.; Mielke, S. L.; Paci, J. T.; Zhang, S.; Ballarini, R.; Schatz, G. C.; Belytschko, T. *Phys. Rev. B* 2007, *75*, 075412.
132. Zhu, Y.; Murali, S.; Cai, W.; Li, X.; Won Suk, J.; Potts, J. R.; Ruoff, R. S. *Adv. Mater.* 2010, *XX*, 1–19.
133. Bunch, J. S.; Verbridge, S. S.; Alden, J. S.; Van der Zande, A. M.; Parpia, J. M.; Craighead, H. G. and McEuen, P. L. *Nano Lett.* 2008, *8*, 2458.
134. Diankov, G.; Neumann, M.; Goldhaber-Gordon, D. *ACS Nano* 2013, *7*, 1324.
135. Eftekhari, A.; Jafarkhani, P. *J. Phys. Chem. C* 2013, *117*, 22845.
136. Sun, Z.; James, D. K.; Tour, J. M. *J. Phys. Chem. Lett.* 2011, *2*, 2425.
137. Koehler, F. M.; Luechinger, N. A.; Ziegler, D.; Athanassiou, E. K.; Grass, R. N.; Rossi, A.; Hierold, C.; Stemmer, A.; Stark, W. J. *Angew. Chem., Int. Ed.* 2009, *48*, 224.
138. Li, L.; Wu, G.; Yang, G.; Peng, J.; Zhao, J.; Zhu, J.-J. *Nanoscale* 2013, *5*, 4015.
139. Jiang, D. E.; Sumpter, B. G.; Dai, S. *J. Chem. Phys.* 2007, *126*, 134701.
140. Calandra, M.; Mauri, F. *Phys. Rev. B: Condens. Matter Mater. Phys.* 2007, *76*, 161406.
141. Lherbier, A.; Blase, X.; Niquet, Y. M.; Triozon, F.; Roche, S. *Phys. Rev. Lett.* 2008, *101*, 36808.
142. Li, D.; Muller, M. B.; Gilje, S.; Kaner, R. B.; Wallace, G. G. *Nat. Nanotechnol.* 2008, *3*, 101.
143. Elias, D. C.; Nair, R. R.; Mohiuddin, T. M. G.; Morozov, S. V.; Blake, P.; Halsall, M. P.; Ferrari, A. C.; et al. *Science* 2009, *323*, 610.
144. Nair, R. R.; Ren, W.; Jalil, R.; Riaz, I.; Kravets, V. G.; Britnell, L.; Blake, P. et al. *Small* 2010, *6*, 2877.
145. Xie, X. N.; Lim, R.; Li, J.; Li, S. F. Y.; Loh, K. P. *Diamond Relat. Mater.* 2001, *10*, 1218.
146. Robinson, J. T.; Burgess, J. S.; Junkermeier, C. E.; Badescu, S. C.; Reinecke, T. L.; Perkins, F. K.; Zalalutdniov, M. K. et al. *Nano Lett.* 2010, *10*, 3001.
147. Zhou, J.; Wu, M. M.; Zhou, X.; Sun, Q. *Appl. Phys. Lett.* 2009, *95*, 103108.
148. Zhou, J.; Liang, Q.; Dong, J. *Carbon* 2010, *48*, 1405.
149. Bon, S. B.; Valentini, L.; Verdejo, R.; Garcia Fierro, J. L.; Peponi, L.; Lopez-Manchado, M. A.; Kenny, J. M. *Chem. Mater.* 2009, *21*, 3433.
150. Withers, F.; Russo, S.; Dubois, M.; Cracium, M. *Nanoscale Res. Lett.* 2011, *6*, 526.
151. Li, B.; Zhou, L.; Wu, D.; Peng, H.; Yan, K.; Zhou, Y.; Liu, Z. *ACS Nano* 2011, *5*, 5957.
152. Sheng, Z.-H.; Shao, L.; Chen, J.-J.; Bao, W.-J.; Wang, F.-B.; Xia, X.-H. *ACS Nano* 2011, *5*, 4350.
153. Zhang, L. L.; Zhao, X.; Ji, H.; Stoller, M. D.; Lai, L.; Murali, S.; Mcdonnell, S.; Cleveger, B.; Wallace, R. M.; Ruoff, R. S. *Energy Environ. Sci.* 2012, *5*, 9618.
154. Song, L.; Ci, L.; Lu, H.; Sorokin, P. B.; Jin, C.; Ni, J.; Kvashnin, A. G. et al. *Nano Lett.* 2010, *10*, 3209.
155. Muchharla, B.; Pathak, A.; Liu, Z.; Song, L.; Jayasekera, T.; Kar, S.; Vajtai, R. et al. *Nano Lett.* 2013, *13*, 3476.
156. Zhu, Y.; Higginbotham, A. L.; Tour, J. M. *Chem. Mater.* 2009, *21*, 5284.
157. Sun, Z.; Kohama, S.; Zhang, Z.; Lomeda, J.; Tour, J. *Nano Res.* 2010, *3*, 117.

158. Jin, Z.; McNicholas, T. P.; Shih, C.-J.; Wang, Q. H.; Paulus, G. L. C.; Hilmer, A. J.; Shimizu, S.; Strano, M. S. *Chem. Mater.* 2011, *23*, 3362.
159. Wang, H.-X.; Zhou, K.-G.; Xie, Y.-L.; Zeng, J.; Chai, N.-N.; Li, J.; Zhang, H.-L. *Chem. Commun.* 2011, *47*, 5747.
160. Zhong, X.; Jin, J.; Li, S.; Niu, Z.; Hu, W.; Li, R.; Ma, J. *Chem. Commun.* 2010, *46*, 7340.
161. Georgakilas, V.; Bourlinos, A. B.; Zboril, R.; Steriotis, T. A.; Dallas, P.; Stubos, A. K.; Trapalis, C. *Chem. Commun.* 2010, *46*, 1766.
162. Xiao-Yan, F.; Ryo, N.; Li-Chang, Y.; Katsumi, T. *Nanotechnology* 2010, *21*, 475208.
163. Liu, L.-H.; Yan, M. *Nano Lett.* 2009, *9*, 3375.
164. Vadukumpully, S.; Gupta, J.; Zhang, Y.; Xu, G. Q.; Valiyaveettil, S. *Nanoscale* 2011, *3*, 303.
165. Shen, J.; Shi, M.; Yan, B.; Ma, H.; Li, N.; Hu, Y.; Ye, M. *Colloids Surf. B* 2010, *81*, 434.
166. Depan, D.; Girase, B.; Shah, J. S.; Misra, R. D. K. *Acta Biomater.* 2011, *7*, 3432.
167. Karousis, N.; Sandanayaka, A. S. D.; Hasobe, T.; Economopoulos, S. P.; Sarantopoulou, E.; Tagmatarchis, N. *J. Mater. Chem.* 2011, *21*, 109.
168. Zhu, J.; Li, Y.; Chen, Y.; Wang, J.; Zhang, B.; Zhang, J.; Blau, W. J. *Carbon* 2011, *49*, 1900.
169. Liu, Z.-B.; Xu, Y.-F.; Zhang, X.-Y.; Zhang, X.-L.; Chen, Y.-S.; Tian, J.-G. *J. Phys. Chem. B* 2009, *113*, 9681.
170. Zhang, X.; Liu, Z.; Huang, Y.; Wan, X.; Tian, J.; Ma, Y.; Chen, Y. *J. Nanosci. Nanotechnol.* 2009, *9*, 5752.
171. Veca, L. M.; Lu, F.; Meziani, M. J.; Cao, L.; Zhang, P.; Qi, G.; Qu, L.; Shrestha, M.; Sun, Y.-P. *Chem. Commun.* 2009, *18*, 2565–2567.
172. Yang, H.; Shan, C.; Li, F.; Han, D.; Zhang, Q.; Niu, L. *Chem. Commun.* 2009, *26*, 3880–3882.
173. Cai, J.; Ruffieux, P.; Jaafar, R.; Bieri, M.; Braun, T.; Blankenburg, S.; Muoth, M. et al. *Nature* 2010, *466*, 470.
174. Yang, X.; Dou, X.; Rouhanipour, A.; Zhi, L.; Räder, H. J.; Müllen, K. *J. Am. Chem. Soc.* 2008, *130*, 4216.
175. Son, Y.-W.; Cohen, M. L.; Louie, S. G. *Nature* 2006, *444*, 347.
176. Cervantes-Sodi, F.; Csányi, G.; Piscanec, S.; Ferrari, A. C. *Phys. Rev. B* 2008, *77*, 165427.
177. Hamilton, I. P.; Li, B.; Yan, X.; Li, L.-S. *Nano Lett.* 2011, *11*, 1524.
178. Yan, X.; Li, B.; Cui, X.; Wei, Q.; Tajima, K.; Li, L.-S. *J. Phys. Chem. Lett.* 2011, *2*, 1119.
179. Zheng, H.; Duley, W. *Phys. Rev. B* 2008, *78*, 045421.
180. Sahin, H.; Senger, R. *Phys. Rev. B* 2008, *78*, 205423.
181. Berashevich, J.; Chakraborty, T. *Phys. Rev. B* 2009, *80*, 115430.
182. Olivi-Tran, N. *Physica B* 2010, *405*, 2749.
183. Xu, Y.; Zhao, L.; Bai, H.; Hong, W.; Li, C.; Shi, G. *J. Am. Chem. Soc.* 2009, *131*, 5856.
184. Wang, X.; Tabakman, S. M.; Dai, H. *J. Am. Chem. Soc.* 2008, *130*, 8152.
185. Wang, Q. H.; Hersam, M. C. *Nat. Chem.* 2009, *1*, 206.
186. Su, Q.; Pang, S.; Alijani, V.; Li, C.; Feng, X.; Mullen, K. *Adv. Mater.* 2009, *21*, 3191.
187. Wang, Y.; Chen, X.; Zhong, Y.; Zhu, F.; Loh, K. P. *Appl. Phys. Lett.* 2009, *95*, 063302.
188. Zhang, X.; Feng, Y.; Tang, S.; Feng, W. *Carbon* 2010, *48*, 211.
189. Yang, X.; Zhang, X.; Liu, Z.; Ma, Y.; Huang, Y.; Chen, Y. *J. Phys. Chem. C* 2008, *112*, 17554.
190. Depan, D.; Shah, J.; Misra, R. D. K. *Mater. Sci. Eng. C* 2011, *31*, 1305.
191. Li, F.; Bao, Y.; Chai, J.; Zhang, Q.; Han, D.; Niu, L. *Langmuir* 2010, *26*, 12314.
192. Xu, Y.; Malkovskiy, A.; Pang, Y. *Chem. Commun.* 2011, *47*, 6662.

193. Xu, L. Q.; Wang, L.; Zhang, B.; Lim, C. H.; Chen, Y.; Neoh, K.-G.; Kang, E.-T.; Fu, G. D. *Polymer* 2011, *52*, 2376.
194. Xu, Y.; Zhao, L.; Bai, H.; Hong, W.; Li, C.; Shi, G. *J. Am. Chem. Soc.* 2009, *131*, 13490.
195. Wojcik, A.; Kamat, P. V. *ACS Nano* 2010, *4*, 6697.
196. Choi, E.-Y.; Han, T. H.; Hong, J.; Kim, J. E.; Lee, S. H.; Kim, H. W.; Kim, S. O. *J. Mater. Chem.* 2010, *20*, 1907.
197. Gierz, I.; Riedl, C.; Starke, U.; Ast, C. R.; Kern, K. *Nano Lett.* 2008, *8*, 4603.
198. Bostwick, A.; Ohta, T.; Seyller, T.; Horn, K.; Rotenberg, E. *Nat. Phys.* 2007, *3*, 36.
199. Ohta, T.; Bostwick, A.; Seyller, T.; Horn, K.; Rotenberg, E. *Science* 2006, *313*, 951.
200. Uchoa, B.; Lin, C. Y.; Neto, A. H. C. *Phys. Rev. B: Condens. Matter Mater. Phys.* 2008, *77*, 035420.
201. Chandra, V.; Park, J.; Chun, Y.; Lee, J. W.; Hwang, I.-C.; Kim, K. S. *ACS Nano* 2010, *4*, 3979.
202. Zhang, Y.; Chen, B.; Zhang, L.; Huang, J.; Chen, F.; Yang, Z.; Yao, J.; Zhang, Z. *Nanoscale* 2011, *3*, 1446.
203. Lv, W.; Sun, F.; Tang, D.-M.; Fang, H.-T.; Liu, C.; Yang, Q.-H.; Cheng, H.-M. *J. Mater. Chem.* 2011, *21*, 9014.
204. Zhou, G.; Wang, D.-W.; Li, F.; Zhang, L.; Li, N.; Wu, Z.-S.; Wen, L.; Lu, G. Q.; Cheng, H.-M. *Chem. Mater.* 2010, *22*, 5306.
205. Qiu, Y.; Yan, K.; Yang, S.; Jin, L.; Deng, H.; Li, W. *ACS Nano* 2010, *4*, 6515.
206. Wu, Z.-S.; Wang, D.-W.; Ren, W.; Zhao, J.; Zhou, G.; Li, F.; Cheng, H.-M. *Adv. Funct. Mater.* 2010, *20*, 3595.
207. Paek, S.-M.; Yoo, E.; Honma, I. *Nano Lett.* 2008, *9*, 72.
208. Scheuermann, G. M.; Rumi, L.; Steurer, P.; Bannwarth, W.; Mülhaupt, R. *J. Am. Chem. Soc.* 2009, *131*, 8262.
209. Zhang, N.; Qiu, H.; Liu, Y.; Wang, W.; Li, Y.; Wang, X.; Gao, J. *J. Mater. Chem.* 2011, *21*, 11080.
210. Li, X. S.; Cai, W. W.; An, J. H.; Kim, S.; Nah, J.; Yang, D. X.; Piner, R. et al. *Science* 2009, *324*, 1312.
211. Zhang, Y.; Gomez, L.; Ishikawa, F. N.; Madaria, A.; Ryu, K.; Wang, C. A.; Badmaev, A.; Zhou, C. W. *J. Phys. Chem. Lett.* 2010, *1*, 3101.
212. Bhaviripudi, S.; Jia, X. T.; Dresselhaus, M. S.; Kong, J. *Nano Lett.* 2010, *10*, 4128.
213. Reina, A.; Jia, X. T.; Ho, J.; Nezich, D.; Son, H. B.; Bulovic, V.; Dresselhaus, M. S.; Kong, J. *Nano Lett.* 2009, *9*, 3087.
214. Lu, C. C.; Lin, Y.; Liu, Z.; Yeh, C. H.; Suenaga, K.; Chiu, P. W. *ACS Nano* 2013, *7*, 2587.
215. Bi, H.; Sun, S.; Huang, F.; Xie, X.; Jiang, M. *J. Mater. Chem.* 2012, *22*, 411.
216. Yan, X.; Cui, X.; Li, B.; Li, L.-S. *Nano Lett.* 2010, *10*, 1869.
217. Ponomarenko, L. A.; Schedin, F.; Katsnelson, M. I.; Yang, R.; Hill, E. W.; Novoselov, K. S.; Geim, A. K. *Science* 2008, *320*, 356.
218. Ritter, K. A.; Lyding, J. W. *Nat. Mater.* 2009, *8*, 235.
219. Li, X. L.; Wang, X. R.; Zhang, L.; Lee, S. W.; Dai, H. J. *Science* 2008, *319*, 1229.
220. Bai, J.; Duan, X.; Huang, Y. *Nano Lett.* 2009, *9*, 2083.
221. Jiao, L. Y.; Wang, X. R.; Diankov, G.; Wang, H. L.; Dai, H. J. *Nat. Nanotechnol.* 2010, *5*, 321.
222. Bai, J. W.; Zhong, X.; Jiang, S.; Huang, Y.; Duan, X. F. *Nat. Nanotechnol.* 2010, *5*, 190.
223. Liang, X.; Jung, Y.-S.; Wu, S.; Ismach, A.; Olynick, D. L.; Cabrini, S.; Bokor, J. *Nano Lett.* 2010, *10*, 2454.
224. Huang, X.; Yin, Z.; Wu, S.; Qi, X.; He, Q.; Zhang, Q.; Yan, Q.; Boey, F.; Zhang, H. *Small* 2011, *7*, 1876.
225. Muñoz, R.; Gómez-Aleixandre, C. *Chem. Vap. Dep.* 2013, *19*, 297.

226. Reina, A.; Jia, X.; Ho, J.; Nezich, D.; Son, H.; Bulovic, V.; Dresselhaus, M. S.; Kong, J. *Nano Lett.* 2009, *9*, 30–35.
227. Kim, K. S.; Zhao, Y.; Jang, H.; Lee, S. Y.; Kim, J. M.; Kim, K. S.; Ahn, J. H.; Kim, P.; Choi, J. Y.; Hong, B. H. *Nat. Lett.* 2009, *457*, 706.
228. Yao, Y.; Wong, C. P.; *Carbon* 2012, *50*, 5203.
229. Wirtz, C.; Lee, K.; Hallam, T.; Duesberg, G. S. *Chem. Phys. Lett.* 2014, *595–596*, 192.
230. Gao, L.; Ren, W.; Xu, H.; Jin, L.; Wang, Z.; Ma, T.; Zhang, L. P. Z. et al. *Nat. Commun.* 2012, *3*, 699. doi:10.1038/ncomms1702.
231. Li, X.; Cai, W.; Colombo, L.; Ruoff, R. S. *Nanoletters* 2009, *9*, 4268–4272.
232. Kalbac, M.; Frank, O.; Kavan, L. *Carbon* 2012, *50*, 3682–3687.
233. Losurdo, M.; Giangregorio, M. M.; Capezzuto, P.; Bruno, G. *Phys. Chem. Chem. Phys.* 2011, *13*, 20836.
234. Henkelman, G.; Arnaldsson, A.; Jónsson, H. *J. Chem. Phys.* 2006, *124*, 044706.
235. Wang, G.-C.; Nakamura, J. *J. Phys. Chem. Lett.* 2010, *1*, 3053.
236. Greeley, J.; Mavrikakis, M. *J. Phys. Chem. B* 2005, *109*, 3460.
237. Zhang, W.; Wu, P.; Li, Z.; Yang, J. *J. Phys. Chem.* 2011, *115*, 17782.
238. Galea, N. M.; Knapp, D.; Ziegler, T. *J. Catal.* 2007, *247*, 20.
239. Au, C.-T.; Ng, C.-F.; Liao, M.-S. *J. Catal.* 1999, *185*, 12.
240. Chen, H.; Zhu, W.; Zhang, Z. *Phys. Rev. Lett.* 2010, *104*, 186101.
241. Riikonen, S.; Krasheninnikov, A. V.; Halonen, L.; Nieminen, R. M. *J. Phys. Chem. C* 2012, *116*, 5802.
242. Qi, J. L.; Zhang, L. X.; Cao J. et al. *Chin. Sci. Bull.* 2012, *57*, 3040.
243. Terasawa, T.; Saiki, K. *Carbon* 2012, *50*, 869.
244. Nandamuri, G.; Roumimov, S.; Solanki, R. *Appl. Phys. Lett.* 2010, *96*, 154101.
245. Kim, Y.; Song, W.; Lee, S. Y.; Jeon, C.; Jung, W.; Kim, M.; Park, C.-Y. *Appl. Phys. Lett.* 2011, *98*, 263106.
246. Kumar, A.; Voevodin, A. A.; Zemlyanov, D.; Zakharov, D. N.; Fisher, T. S. *Carbon* 2012, *50*, 1546.
247. Kim, J.; Ishihara, M.; Koga, Y.; Tsugawa, K.; Hasegawa, M.; Iijima, S. *Appl. Phys. Lett.* 2011, *98*, 091502.
248. Kalita, G.; Kayastha, M. S.; Uchida, H.; Wakita, K.; Umeno, M. *RSC Adv.* 2012, *2*, 3225.
249. Muñoz, R.; Gómez-Aleixandre, C. *J. Phys. D: Appl. Phys.* 2014, *47*, 045305.
250. Lieberman, M. A.; Lichtenberg, A. J. *Principles of Plasma Discharges and Materials Processing*. John Wiley and Sons, New York, 1994.
251. Alexandrov, S. E.; Hitchman, M. L. *Chem. Vap. Dep.* 2005, *11*, 457.
252. Tzolov, M.; Finger, F.; Carius, R.; Hapke, P. *J. Appl. Phys.* 1997, *81*, 7376.
253. Águas, H.; Silva, V.; Fortunato, E.; Lebib, S.; Roca, I.; Cabarrocas, P.; Ferreira, I.; Guimaraes, L.; Martins, R. *Jpn. J. Appl. Phys.* 2003, *42*, 4935.
254. Amanatides, E.; Mataras, D. *J. Appl. Phys.* 2001, *89*, 1556.
255. Pearse, R. W. B.; Gaydon, A. G. *The Identification of Molecular Spectra*, 2nd edn. Chapman & Hall Ltd, London, U.K., 1950.
256. Wang, J.; Zhu, M.; Outlaw, R. A.; Zhao, X.; Manos, D. M.; Holloway, B. C. *Carbon* 2004, *42*, 2867.
257. Yamada, T.; Kim, J.; Ishihara, M.; Hasegawa, M. *J. Phys. D: Appl. Phys.* 2013, *46*, 063001.
258. Medina, H.; Lin, Y. C.; Jin, C.; Lu, C. C.; Yeh, C. H.; Huang, K. P.; Suenaga, K.; Robertson, J.; Chiu, P. W. *Adv. Funct. Mater.* 2012, *22*, 2123.

CHAPTER 44

CONTENTS

Carbon Nanomaterials for Gas-Sensing Application 44

B.A. Bhanvase and B. Dewangan

44.1 INTRODUCTION

In the past two decades, the rapid progress of nanotechnology, ranging from novel nanoelectronics to molecular assemblies, to nanocomposites, tissue engineering, and biomedicine, has taken place. Nanomaterials, due to their exceptional mechanical, thermal, and electronic properties, have restructured several aspects of current science and engineering. Also, the impact of nanomaterials is increasing in the society, health care, and the environment. Carbon nanomaterials have been used as platforms for ultrasensitive recognition of antibodies (Chen et al. 2003), as nucleic acids sequencers (Wang et al. 2003), and as bioseparators, biocatalysts (Mitchell et al. 2002), and ion channel blockers (Park et al. 2003) for facilitating biochemical reactions and biological processes.

Gas sensing is a very important aspect in industry and gas sensors play a vital role in controlling manufacturing processes. Gas sensors are also being used in a several gas-sensing applications ranging from household gas sensing to gas sensing for environmental safety. Generally, gas sensors for gas-sensing applications are being prepared with the use of sensing materials like semiconducting oxides. However, the major drawback of semiconductor-based gas sensors is its fabrication, and generally, gas sensing is governed by microfabrication techniques (Schuetzle and Hammerle 1986), which impose limitation on the size and geometry of the sensor. These difficulties can be overcome with the use of carbon nanomaterials, which are being investigated for gas-sensing applications. Carbon nanomaterials are a prominent candidate as a gas-sensing material due to their intrinsic properties like small size and good electrical and mechanical properties. An additional key property of carbon nanotubes (CNTs) that favors their use in gas sensing is their high specific surface area (1580 m^2/g) (Cinke et al. 2002). The gas sensing is attributed to the interaction of gas molecules with the surface of carbon nanomaterials, that is, CNTs or graphene. The least amount of carbon nanomaterials can provide abundant sites for interaction of gas due to the high specific surface area of nanotubes or graphene (Adu et al. 2001). Therefore, carbon-based nanomaterials can serve as an economically feasible material for use in gas sensing. Because of these reasons, nanotubes have been explored for gas sensing in a number of applications (Kong et al. 2000; Adu et al. 2001; Varghese et al. 2001).

The change in the electrical property of the sensing material upon exposure to the gas is the basic principle behind gas detection. Generally, with the exposure to different gases, the electrical properties like resistance, thermoelectric power, and dielectric properties of the sensing material get changed (Kong et al. 2000; Sumanasekera et al. 2000; Varghese et al. 2001; Chopra et al. 2002; Savage 2002). These properties of CNTs change substantially in the presence of gases like NH_3, NO_2, LPG, organic vapors, H_2, and CO_2. In this chapter, the use of CNT and graphene-based carbon nanomaterials for different gas-sensing applications has been reported. Their further uses for sensing NH_3, NO_2, organic vapors, and H_2 gases have been explored.

44.2 CARBON NANOMATERIALS AND THEIR GAS-SENSING MECHANISM

Possible gas-sensing mechanisms in CNTs are charge transfer between adsorbed gas molecules and the nanotubes due to the adsorption of gas molecules onto the CNTs or gas-induced changes at the interface between the nanotubes and their metal contacts due to the modifications of the Schottky barriers at the contacts between the CNTs and their metal electrodes. Research on devices made with a single CNT with contacts covered by a protective layer, leaving most of the CNT surface exposed, has provided the evidence that Schottky barrier modulation is the main detection mechanism for gas sensing (Peng et al. 2009). However, the situation became less clear when experiments performed on networks of CNTs suggested that adsorption of gas molecules onto the CNTs was principally responsible for the conductance change (Battie et al. 2011).

A work based on density functional theory argued that NO_2 will adsorb on single tubes within the network, while later work concluded that NO_2 binds more strongly to the interstitial regions between tubes and thus molecules preferentially interact with more than one CNT (Zhao et al. 2002). Furthermore, adsorption can be affected by defects in the tubes or local functionalization of the CNT sidewall, as these sites present additional places for molecular attachment (Robinson et al. 2006). Experimental Fourier transform infrared spectroscopy studies of CNT networks exposed to both NO_2 and NH_3 support the picture of the molecules being trapped in interstitial regions and remaining there well above room temperature (Ellison et al. 2004). However, conflicting experimental work using x-ray photon spectroscopy concluded that above 200 K, NO_2 does not adsorb onto a CNT network at all, arguing against charge transfer as a mechanism for CNT network sensing at room temperature (Larciprete et al. 2007). However, it was not clear whether molecules adsorb onto the CNT network. Even if they were adsorbed it was not clear whether the response can be mainly explained by adsorption on defect-free tubes far from junctions with other tubes, or if other factors (collisions with gas molecules, CNT–CNT junctions, defects) were needed to be considered (Zhong et al. 2010).

Investigation of the electrical response of CNT networks with different densities of nanotubes and CNT–CNT junctions is required to understand how the sensing mechanism changes from high-density networks to single-nanotube devices (Zhang et al. 2006a). When a network has a high-density of nanotube junctions, with a ratio of junction to nanotubes crossing between the electrodes (J/C) higher than 4, they dominate the response. This is because the junctions dominate the device conductance and the electrodes play a minor role. On the other hand, when the number of junctions diminishes, as in the case for low-density networks, with J/C < 2, the electrodes start playing a substantial role in the response and eventually become the main response mechanism for single-CNT devices, where no CNT junctions exist. The gas affects both the Schottky barriers at the CNT/metal interface and the CNT–CNT junctions. The main effect of gas adsorption on the networks is to change the conductance of the CNT–CNT junctions. The CNT–CNT junction density determines which mechanism is dominant. Effects of gas adsorption in regions of the nanotube away from the junctions could not be detected, as demonstrated in single-nanotube devices; therefore, for our chemical vapor deposition (CVD)-grown nanotubes, defects other than the CNT–CNT junctions do not contribute to the response mechanism. These results suggest that gas adsorption occurs preferentially at the CNT–CNT and CNT–electrode junctions (Boyd et al. 2014).

44.3 CARBON NANOMATERIALS AND THEIR APPLICATION FOR SENSING VARIOUS GASES

44.3.1 Sensing of NH_3 and Nitrogen-Based Gases

Ammonia detection and measurement of its concentrations is of interest in industrial and medical areas because of its high toxicity and explosibility. The exposure limit of ammonia is 25 ppm over 8 h and 35 ppm over 15 min exposures. Besides, ammonia is extremely explosive in air at concentrations between 15% and 28% (Liu et al. 2004; Timmer et al. 2005; Zhang et al. 2008; Cui et al. 2012).

Also, the ammonia gas is being widely used in various fields, including the production of fertilizers, plastics, synthetic fibers, dyes, and pharmaceuticals, and it is injurious to human health (Huang et al. 2012). Hence, it is an ecological concern because of severe health risk with the exposure to NH_3, and it is essential to develop ammonia gas sensors to monitor for low concentrations of NH_3 leaks. Several materials have been examined to detect NH_3 including metal nanoparticles (Jiménez-Cadena et al. 2007), metal oxides (Wang et al. 2012), and polymers (Hatchett and Josowicz 2008). Further, single-walled CNTs (SWCNTs) have attracted extensive interest in the sensing application due to their exclusive 1D carbon nanostructure and electrical properties (Kong et al. 2000; Kauffman and Star 2008; Zhang et al. 2008). Battie et al. (2010) has focused on the elaboration of sorted semiconducting SWCNT films (SC-SWCNT films) and the evaluation of a gas microsensor based on these films. During gas sensing, the SC-SWCNT sensor was reported to be exposed to base gas, that is, N_2 gas at 100°C for 15 min in order to eliminate the traces of the water. Further, as per their report, each NH_3 or NO_2 gas exposure was performed for 30 min at room temperature and at 1100 mbar. The desorption of NH_3 was reported for 30 min at 100°C under a N_2 flow. In the case of NO_2, it has been reported that this temperature is not enough to eliminate all adsorbed molecules. For NO_2, the sensor recovery was carried out at 200°C under vacuum (2×10^{-2} mbar). The variation of resistance during the adsorption and desorption of gas for an interelectrode distance equal to 50 μm is reported. The sensitivity to NO_2 is reported to be 10 times larger than the sensitivity to NH_3. As per this report, presently, the sensor is able to detect less than 600 ppb of NH_3 and NO_2. This constraint is due to the experimental setup design limitation, and some new investigations are required to be carried out in order to improve the sensitivity and thus to reduce the response time.

Chopra et al. (2002) have presented the design and development of extremely sensitive and fast-responsive microwave resonant sensors for monitoring the presence of ammonia gas. The designed sensor consists of a circular disk electromagnetic resonant circuit coated with either single- or multiwalled CNTs (MWCNTs) that are highly sensitive to adsorbed gas molecules. It has been reported that upon exposure to ammonia, the electrical resonant frequency of the sensor exhibits a dramatic downshift of 4.375 MHz and the recovery and response times of these sensors are nominally 10 min. It can be concluded that this technology is appropriate for designing remote sensor systems to monitor gases inside sealed opaque packages and environmental conditions that do not allow physical wire connections. Chopra et al. (2003) have also used a circular disk resonator in order to study the gas-sensing properties of CNTs. It has been reported that the presence of gases was detected based on the change in the dielectric constant rather than the electrical conductivity of SWNTs when it is exposed to target gas. Further, noticeable shifts in resonant frequency to both polar (NH_3 and CO) and nonpolar gases (He, Ar, N_2, and O_2) have been reported.

Suehiro et al. (2003) have described a novel approach for fabricating a gas sensor composed of MWCNTs using dielectrophoresis. As per their report, MWCNTs have been dispersed in ethanol and has been trapped and enriched in an interdigitated microelectrode gap under the action of a positive dielectrophoresis force that drove the MWCNTs to a higher electric field region. It has been reported that during the trapping of MWCNTs, the electrode impedance was varied as the number of MWCNTs bridging the electrode gap increased. Further, after the dielectrophoresis process, the ethanol was evaporated and the microelectrode retaining the MWCNTs was exposed to ammonia gas while the electrode impedance was monitored. It has been found that the electrode impedance was altered by ppm levels of ammonia at room temperature. The ammonia exposure decreased the sensor conductance, while the capacitance increased. The sensor showed a reversible response with a time constant of a few minutes. The conductance change was proportional to the ammonia concentration below 10 ppm and then gradually saturated at higher concentrations. The effects of the number of trapped MWCNTs on the sensor response were also discussed by Suehiro et al. (2003). Lucci et al. (2005) have demonstrated NH_3 sensing efficiently in SWCNT ordered by mean of dielectrophoretical process. The reported approach has been used to disperse the nanotubes in $CHCl_3$ and to distribute the suspension between the tracks of multifinger Au electrodes (40 μm spacing) on SiO_2/Si substrates. The controlled arrangement and alignment of the SWCNT bundles has been achieved by applying an alternate voltage (frequency 1 MHz, 10 Vpp) during the solvent evaporation. The sensitivity for NH_3 sensing reported to be strongly improved by the degree of SWCNT alignment between the electrodes. Further, the sensitivity has been improved by increasing the temperature of the devices up to 80°C. Liao et al. (2009) have used mesoporous carbon nanofibers which possesses a high specific surface area 840 m^2/g and a 1.07 eV band gap for NO_2 gas sensor preparation. All mesoporous carbon nanofiber network can operate as the channel material in p-type field-effect transistor (FET) devices with field-effect mobilities over 10 cm^2/V s. Further, it has been reported that the mesoporous carbon nanofiber network demonstrates superior sensitivity and quicker response to NO$_2$ gas than that of CNTs, which makes it a promising candidate as poisonous gas-sensing nanodevices. Recently, Monereo et al. (2013) fabricated and characterized a flexible gas sensor based on carbon nanofibers. The reported sensing device is composed of interdigitated silver electrodes deposited by inkjet printing on Kapton substrates, subsequently coated with carbon nanofibers as sensing element. The fabricated sensor has been characterized in detail by measuring the gas-sensing response to CO, NH$_3$, and humidity. Gas-sensing properties at room temperature with superior selectivity to the tested gases has been reported. It has been further reported that the responsiveness to gases even at room temperature gives significantly reasonable power consumption. The fabricated devices are also reported to exhibit sensitivities to temperature, stress/strain/bending, and light, demonstrating the multifunctionality of the carbon nanofibers and allowing this type of sensors to be used for many applications, as mechanical strain gauges or light detectors.

Further, Kong et al. (2000) established the potential of SWCNT-based gas sensor to NO$_2$ and NH$_3$ at room temperature, which encouraged several researchers to extensively investigate CNTs for gas sensors. SWCNTs have been taken as a promising candidate for gas-sensing applications that can sense the toxic gases like NO$_2$, NH$_3$, O$_2$, H$_2$, CO$_2$, and CO (Fu et al. 2008; Bondavalli et al. 2009; Han et al. 2012). However, recently, in order to enhance the gas-sensing performance of CNT-based gas sensors, many gas-sensing materials like conducting polymers (Bekyarova et al. 2007; Zhang et al. 2007), metals (Penza et al. 2007b), and metal oxides (Bittencourt et al. 2006) have been loaded on the surface of CNTs, which plays a vital role in the enhancement of the sensitivity and selectivity of the resulting gas sensors. Da-Jing et al. (2012) have investigated gas sensors based on CNT and single polyaniline (PANI) nanofiber. The reported CNT gas sensor has been constructed with the use of dielectrophoresis. Single nanofiber has been deposited as nanofiber sensor across two gold electrodes by means of near-field electrospinning without conventional lithography process. The reported nanofiber sensor shows 2.7% reversible resistance change when it is exposed to 1 ppm NH$_3$ gas with a response time of 60 s. Further, it has been reported that the CNT sensor shows good linearity in the concentration range above 20 ppm NH$_3$ gas with response times between 100 and 200 s. Further, from this report, it can be concluded that the size of nanofibers has a significant effect on the response of the gas sensor. This is attributed to the enhanced diffusion mechanism with the smaller diameter of fibers, which results in quicker response to NH$_3$ gas. Therefore, CNT sensor and nanofiber-based sensors could be capable for gas-sensing array and multichemical sensing applications. Jian et al. (2013) have synthesized the composite film with high gas-sensing ability through dielectrophoretic assembly of nanostructured layer of poly(3,4-ethylenedioxythiophene)/poly(styrenesulfonate) (PEDOT/PSS) matrix and O$_2$-plasma-treated SWCNTs. The reported results showed that oxygen-containing functional groups were grafted on the SWCNTs surface by O$_2$ plasma treatment. The functionalized SWCNTs evenly dispersed in the PEDOT/PSS matrix by π–π interaction and electrostatic attraction between PEDOT chains and the nanotubes. The induced morphological and conductive variation by the SWCNTs concentration affects the gas-sensing properties of the composite film. Further, the conductive and gas-sensing characteristics of the composite film have been reported to be enhanced by aligning the functionalized SWCNTs in the polymer and forming the nanostructured film surface using dielectrophoretic manipulation. The gas sensors based on the optimally dielectrophoresis assembled composite film shows highly sensitive, selective, rapid, stable, and reversible responses for sensing 2–300 ppm NH$_3$ and 6–1000 ppb trimethylamine gases at room temperature, suggesting its potential for assessing fish freshness in the fishery chain.

Ghaddab et al. (2010) have reported the development of a hybrid SnO$_2$/SWNT-based gas sensor for the detection of O$_3$ and NH$_3$ gas. They have reported the experimental study of the correlation between the physicochemical properties of CNTs and the sensor's sensitivity. The strong dependence between the characteristics of such nanomaterials and the efficiency of the hybrid layers for gas detection has been reported. It can be concluded that CNTs plays an important role in the adsorption mechanism of target gases and thus on the sensitivity of the chemical gas sensors. Choi et al. (2010) have fabricated NO$_2$, NH$_3$, and xylene gas sensors on microplatforms using mixed SnO$_2$ nanoparticles with 1 wt.% MWCNTs sensing materials. The fabricated gas sensors were reported to be characterized to NO_2, NH_3, and xylene gases, respectively, as a function of concentration at 300°C and temperature from 180 to 380°C at constant concentration. The reported highest sensitivities for the NO_2, NH_3, and xylene were 1.06 at 1.2 ppm and 220°C, 0.19 at 60 ppm and 220°C, and 0.15 at 3.6 ppm and 220°C, respectively. From these reported results, mixed SnO_2 nanoparticles with 1 wt.% MWCNTs showed good sensitivity and selectivity at low-power operation below 30 mW, and fabricated microgas sensors could be used for ubiquitous sensor network applications to monitor environmental pollutants in the air. Ghaddab et al. (2012) have also developed SnO_2/arc-discharge SWNTs hybrid material–based sensor for the detection of NH_3 and O_3 at room temperature. The reported SnO_2/arc-discharge SWNTs hybrid material–based sensor thin films were prepared using solgel and dip-coating techniques. It has been reported that the hybrid material–based sensor was found to have an improved sensitivity as compared to pure SnO_2 or pure SWNT-based sensors. The detection limit at room temperature was reported to be 1 ppm and lower than 20 ppb for NH_3 and O_3, respectively. Further, the hybrid sensor has been reported to exhibit a fast response, a superior sensitivity, and a complete recovery at room temperature. It can be established that the loading of SnO_2 nanoparticles on SWNTs improves the gas-sensing properties of the prepared hybrid nanomaterials.

Penza et al. (2007a) have designed and fabricated a surface acoustic wave sensor using a ZnO guiding layer on 36° Y-cut X-propagation $LiTaO_3$ piezoelectric substrate. Further, as per their report, SWCNT-based nanocomposite coatings have been deposited using the Langmuir–Blodgett (LB) technique. The fabricated sensor has been tested toward H_2, NH_3, and NO_2 gases, in the range of 0.030%–1%, 30–1000 ppm, and 1–10 ppm, respectively. The developed sensor has a high sensitivity, good repeatability, and low detection limit at sub-ppm levels, at room temperature. The experimental results obtained by Penza et al. (2007a) have been compared to a theoretical model, and the surface acoustic wave gas-sensing mechanisms are discussed with the acoustoelectric effect that appears to be dominant in the surface acoustic wave response. Penza et al. (2008) further fabricated a gas chemiresistor onto alumina using MWCNTs networked films grown by radiofrequency plasma–enhanced CVD technology for high-performance gas detection at an operating temperature of 200°C. Modification of MWCNTs tangled bundle films with nominally 5 nm thick Pt and Pd nanoclusters offers higher sensitivity for drastically improved gas detection of NO_2, H_2S, NH_3, CO, up to a lower limit of sub-ppm level. Also, it has been reported that the loading of Pt and Pd nanoclusters on MWCNTs gas sensors exhibits better performances compared to unmodified MWCNTs, making them promising candidates for environmental air-pollutant monitoring. Penza et al. (2009) have also investigated the impact of the customized loading of gold nanoclusters that functionalizes the sidewalls of the CNTs networks on gas-sensing performance of a chemiresistor, operating at a working temperature in the range of 20–250°C. CNTs networked films have been reported to be developed by radiofrequency plasma–enhanced CVD technology onto low-cost alumina substrate, provided with 6 nm nominally thick cobalt growth catalyst. As per this report, nanoclusters of gold have been deposited by sputtering onto CNTs networks with a controlled loading of equivalent thickness of 2.5, 5, and 10 nm. CNTs and gold-modified CNTs reveal a p-type response with a decrease in electrical resistance upon exposure to oxidizing NO_2 gas and an increase in resistance upon exposure to reducing gases like NH_3, CO, N_2O, H_2S, and SO_2. Negligible response has been reported for CNTs and gold-modified CNTs sensors exposed to CO, N_2O, and SO_2. Further, significant improvement in the gas response of NO_2, H_2S, and NH_3, up to a low limit of sub-ppm level, has been reported for gold-modified CNT chemiresistors, and maximum gas sensitivity to NO_2, H_2S, and NH_3 has been reported by CNTs functionalized with gold loading of 5 nm, at 200°C. Intratube modulation and intertube modulation are possible sensing mechanisms modeled for an electrical charge transfer between CNTs networks and adsorbed gas molecules with *p-type* semiconducting characteristics in both the unmodified and gold-modified CNTs sensors. NO_2 gas detection has been reported to be carried out by gold-modified CNTs chemiresistors to serve as very promising chemical nanosensors with high sensitivity, selectivity, reversibility, and very low limit of ppb detection for environmental air-pollutant monitoring.

Yu et al. (2012) have prepared gas sensitive nanostructured hybrid films via microwave assisted solgel process using polyoxyethylene (20) sorbitan monooleate (Tween) and tetraethoxy orthosilicate (TEOS). Tween and TEOS were used as organic and inorganic precursors, respectively, and reported IR spectra of the Tween/TEOS nanohybrid material explored the formation of organic–inorganic networks between Tween and TEOS. The reported response for 1% SWCNT/TTH (Tween80/TEOS nano hybrids) sensor for 30–100 ppb NO concentrations was $R^2 = 0.9958$. From the reported results of Yu et al. (2012), it can be concluded that the hybrid films have a better sensor response and faster response time. Recently, Wang et al. (2014) have prepared noncovalent functionalized SWCNT with *tetra*-α-*iso*-pentyloxyphthalocyanine copper (CuPcTIP) and *tetra*-α-(2,2,4-trimethyl-3-pentyloxy) metal phthalocyanines (CuPcTTMP) hybrid materials. The loading of these CuPc derivatives has been carried out successfully on the surface of nanotubes through π–π stacking interaction. The gas-sensing application of the SWCNT/CuPcTIP and SWCNT/CuPcTTMP hybrid nanomaterials has been investigated with respect to NH_3 at room temperature. The response and recovery properties of the SWCNT sensor with respect to NH_3 gas has been enhanced due to the synergetic behavior between CuPc and SWCNT. Further, the sensing properties of SWCNT/CuPcTIP and SWCNT/CuPcTTMP hybrid nanomaterials with respect to NH_3 gas exhibit excellent reversibility, reproducibility, and selectivity. The improved gas-sensing performance is mainly attributed to the form of the charge transfer conjugate in SWCNT/CuPcTTMP and SWCNT/CuPcTIP hybrid nanomaterials. The achievement of these hybrid sensors with exceptional gas-sensing performance will provide a novel approach for gas-sensing application with the advantages of low cost, low power, and portable properties.

Bekyarova et al. (2004) have demonstrated a new approach toward the development of superior chemical sensors based on chemically modified SWNTs. SWNTs with covalently attached poly(*m*-aminobenzene sulfonic acid), SWNT-PABS, have shown enhanced sensor performance for the detection of NH_3. It has been reported that the devices prepared with SWNT-PABS have shown more than two times higher changes in the resistance compared to pure SWNTs when it was exposed to NH_3. Further, the SWNT-PABS sensors quickly recover their resistance when NH_3 is replaced with nitrogen. It has been also reported that exposure of the prepared sensors to NH_3 induces considerable changes in the electronic structure of SWNT-PABS, which allow the sensing of NH_3 at concentrations as low as 5 ppm. Thin film deposited between interdigitated electrodes was explored as a device configuration for development of gas sensors. Zhang et al. (2006b) have demonstrated a facile fabrication approach to construct chemical gas sensors using SWNT electrochemically functionalized with PANI. The potential benefit of reported method is to enable targeted functionalization with different materials to allow for the creation of high-density individually addressable nanosensor arrays. The reported results demonstrate a higher sensitivity of 2.44% $\Delta R/R$ per ppm_v NH_3. This sensitivity is 60 times higher than pure SWNT-based sensors. The detection limit was as low as 50 ppb_v, with good reproducibility upon repeated exposure to 10 ppm_v NH_3. This enhancement in gas-sensing properties is attributed to the electrochemical functionalization of SWNTs, which provides a promising new approach of creating highly advanced nanosensors with improved sensitivity, detection limit, and reproducibility.

A 2D honeycomb carbon sheet, that is, graphene has gained much interest (Geim 2009; Allen et al. 2010; Han and Gao 2010; Kim et al. 2010; Kou et al. 2010; Zhi et al. 2012) due to its distinctive properties like high Young's modulus (Lee et al. 2008) and specific surface area (Stoller et al. 2008), outstanding thermal conductivity (Balandin et al. 2008), and attractive quantum hall effect. Graphene has been considered as an excellent candidate for gas-sensing applications, mainly due to the following advantages (Ratinac et al. 2010): (1) 2D honeycomb structure in which all carbon atoms are exposed to the environment and offer a large surface area leading to high sensitivity to a variety of gases and (2) its inherently low electric noise due to the high quality of its crystal lattice with its 2D nature that tends to screen charge fluctuations of more than 1D systems such as CNTs. Due to these merits, the graphene sheets prepared by different methods such as exfoliation of graphite (Schedin et al. 2007), CVD (Yu et al. 2011), chemical and thermal reduction of graphene oxide (Robinson et al. 2008; Lu et al. 2009, 2011; Dua et al. 2010), and so on have been extensively studied and it exhibits excellent sensing properties. Further, in order to get better sensing properties of graphene, several sensing materials such as conducting polymers (Al-Mashat et al. 2010), metals (Chu et al. 2011; Lange et al. 2011; Vedala et al. 2011), and metal oxides (Mao et al. 2012) have been loaded on the surface of graphene sheets and that plays a significant role in the enhancement of the sensitivity and selectivity of the resultant gas sensors. Further, graphene is a zero-bandgap semimetal that has amazing electronic (Novoselov et al. 2005; Geim and Novoselov 2007; Bolotin et al. 2008; Chen et al. 2008; Morozov et al. 2008) and mechanical properties (Lee et al. 2008) and is made of a single-layer carbon with every atom on its surface; graphene is a solely 2D material and important candidate for its application for a chemical vapor sensor. It has been reported that the absorption of individual gas molecules onto the surface of a graphene sensor leads to a detectable change in its resistance (Schedin et al. 2007). Further, the intrinsic selectivity of graphene to gaseous vapors can only be estimated with the use of samples where contamination from lithographic processing has been measured. Graphene vapor sensors that are known to be free of chemical contamination should be surface modified to control their chemical sensitivity (Staii et al. 2005; McAlpine et al. 2008).

Huang et al. (2012) have presented a useful ammonia gas sensor based on reduced graphene oxide–PANI hybrids. As per their report, PANI nanoparticles were successfully loaded on the surface of reduced graphene oxide sheets by using reduced graphene oxide–MnO_2 hybrids as both of the templates and oxidants for aniline monomer during the process of polymerization. Further, they have investigated the sensing performance of NH_3 gas sensing of the prepared hybrids and compared with those of the sensors based on bare PANI nanofibers and bare reduced graphene oxide sheets. The sensing performance of prepared hybrids was higher and is attributed to the synergetic behavior between both of the candidates that allowed excellent sensitivity and selectivity to NH_3 gas. The reduced graphene oxide–PANI hybrid device has much superior response, that is, 3.4 and 10.4 times, respectively, with the concentration of NH_3 gas at 50 ppm to NH_3 gas than those of the neat PANI nanofiber and neat graphene-based sensor device. Such type of combinations reported by Huang et al. (2012) facilitates the improvement in the sensing properties and opens a new opportunity for the researcher.

Zhang et al. (2009) have investigated the interactions between four different graphenes and small gas molecules such as CO, NO, NO_2, and NH_3 by using density functional computations to utilize their potential applications as gas sensors. The structural and electronic properties of the graphene-molecule adsorption are mainly dependent on the graphene structure and the molecular adsorption configuration. The doping of dopant in graphene nanosheets, that is, the defective graphene, shows the highest adsorption energy and the tightest binding with different gas molecules and the strong interactions between the adsorbed molecules and the doped graphenes induce remarkable changes to graphene's electronic properties. Because of these properties, the sensitivity of graphene-based chemical gas sensors could be significantly enhanced by adding suitable dopant or defect.

Gautam and Jayatissa (2012a) have studied the ammonia sensing behavior of graphene, which was prepared by a CVD on a copper substrate using a methane and hydrogen gas mixture. It has been reported that the sensitivity and the recovery time of the prepared device is enhanced by the loading of platinum nanoparticles on the surface of graphene. They have reported 80%–85% enhancement of sensor response for platinum-loaded surface compared with the neat graphene surface throughout the measured temperature range for ammonia concentrations of 15–58 ppm. Gautam and Jayatissa (2012b) have also investigated the ammonia gas-sensing behavior of graphene prepared by CVD on copper substrate using a methane and hydrogen gas mixture. The deposition of gold nanoparticles on the surface of graphene films leads to enhancement in the sensitivity and the recovery time of the device. The reported adsorption and desorption data have been analyzed using Langmuir kinetic theory and Freundlich isotherm for the adsorption of ammonia gas, and the activation energy and the heat of adsorption were reported to be around 38 and 41 meV, respectively, for NH_3 gas concentration of 58 ppm at room temperature.

Yavari et al. (2012) have prepared graphene films by CVD that enables sensing of the traces of nitrogen dioxide (NO_2) and ammonia (NH_3) in air at room temperature and atmospheric pressure. The gas species are reported to be detected by monitoring changes in electrical resistance of the graphene film due to gas adsorption. Further, it has been reported that the sensor response time was inversely proportional to the gas concentration. With an increase in the temperature of the film, the removal of chemisorbed molecules from the graphene surface takes place that enables reversible operation. Also, Li et al. (2011), Petit et al. (2010), and Sun et al. (2012) have prepared palladium-decorated reduced graphene oxide for sensing NO gas, Cu-based MOF/graphene composites for NH_3 sensing, and graphene oxide for sensing NH_3 gas.

44.3.2 Organic Vapor Sensing

Highly sensitive chemical sensors development is an attractive research area due to their widespread applications in the chemical and food industry, clinic, agriculture, and environment. CNTs have a high aspect ratio and excellent chemical and environmental stability, which make them ideal candidates for detecting harmful gases. The use of CNTs as a chemical sensor (Collins et al. 2000; Kong et al. 2000) has been attracting a great deal of attention of the several researchers. For sensing applications, semiconducting nanotubes play a significant role and therefore FET geometry is very convenient (Tans et al. 1998; Avouris 2002). Several researchers have reported that CNT transistors are responsive to many gaseous agents (Kong et al. 2000). Further, the conductance of CNTs is known to be sensitive to ambient environments, especially to oxygen and/or oxygen-containing gaseous species (Martel et al. 2001; Derycke et al. 2002). For its practical application, it is further required to examine and to understand the characteristics like reversibility, reproducibility, sensitivity, and selectivity to various gaseous analytes. Also, chlorinated aliphatic hydrocarbons like CCl_4, $CHCl_3$, CH_2Cl_2, and C_2Cl_4 are a group of key environmental pollutants and that creates the problem to the environment (Pilchowski et al. 1992; Yashin and Yashin 1999). These pollutants are released to the environment from different sources, mostly from their use as solvents in paints, adhesives, degreasers, color removers, or related products (Zhang and Li 1996). These organic compounds can lead to severe health problems such as irritation, headache, lung congestion, kidney and liver damage, effects on the brain, convulsion, and cancer. Several techniques like laser-induced fluorescence (Jeffries et al. 1992), gas chromatography (Plumacher and Renner 1993), and solid phase microextraction (Luksbetlej and Bodzek 2000) have been generally used for determination of these type of hydrocarbons. Conversely, these methods are complicated, costly, and difficult. Hence, development of efficient gas-sensing device like CNT-based gas-sensing device with low detection limit is essential for the selective detection of chemical vapors under ambient condition.

Kar and Choudhury (2013) have prepared PANI nanocomposites doped with carboxylic acid modified MWCNTs (c-MWCNT) by in situ chemical oxidation polymerization of aniline monomer using ammonium persulfate in the presence of c-MWCNT. Conducting polymers like PANI exhibit highly reversible redox behavior, and therefore they have been considered as efficient materials for development of the chemical sensors for sensing organic compounds specially, alcohols, ethers, acetone, and inorganic gases like ammonia, Cl_2, CO_2, NO_2, SO_2, and H_2S (Athawale and Kulkarni 2000; Kiattibutr et al. 2002; Jain et al. 2005; Virji et al. 2005; Irimia-Vladu and Fergus 2006; Yan et al. 2007; Choudhury 2009; Choudhury et al. 2009). PANI has attained extensive significance due to its easy doping, exclusive conduction mechanism, and high environmental stability (Heeger 2002; MacDiarmid 2002; Debarnot and Epaillard 2003). The response of the prepared PANI/c-MWCNT nanocomposite to different chlorinated methane vapor reported to be examined and compared with that of the pure PANI by Kar and Choudhury (2013). The prepared nanocomposites reported the better response to chloroform vapor as compared to pure PANI. Zhang et al. (2005) fabricated a new type of conductive polymer composites by in situ polymerization of styrene in the presence of MWCNTs and solution mixing of polystyrene with MWCNTs, respectively. In their report, MWCNTs have been used rather than SWCNTs have been selected as conductive fillers due to the following consideration. (1) MWCNT is much cheaper than SWCNT and, hence, the MWCNT/polymer composites may be more suitable than SWCNT/polymer composites in industrial application. (2) MWCNTs are easy to disperse in the polymer matrix compared with SWCNTs. That makes the composite more suitable for sensing applications. The electrical percolation behaviors of the composites and their resistance responsivities against various organic vapors have been investigated by them. It has been found from this report that the in situ polymerization method is more advantageous to enhance the dispersion of MWCNTs in polystyrene matrix, and resultant composites have higher sensitivity and rate of response for the vapors of good solvents of polystyrene at filler range from 5 to 15 wt.%. It can be further concluded that the MWCNT/polystyrene composites developed in their work are promising candidates for gas sensors to detect, distinguish, and quantify organic vapors. Further, the measurements of the conductivity changes of conducting polymers through charge transfer with certain chemical vapors or nonconducting polymers mixed with conductive fillers via polymer swelling by the gas absorption around the percolation threshold provided the basis for the development of polymer-based chemical vapor sensors (Koul et al. 2001; Dai 2004; Chen et al. 2005). Conversely, the poor environmental stability associated with most conducting polymers limits the scope of their use for practical applications (Dai 2004). Further, the uncertainty on the precise location of the percolation threshold in random dispersion systems remained as one of the main obstacles toward high-performance conductively filled polymer sensors. Wei et al. (2006) have produced the aligned multiwall CNTs by pyrolysis of iron (II) phthalocyanine, which partially covers a polymer coating top-down along their tube length by depositing a droplet of polymer solution like poly(vinyl acetate), polyisoprene, onto the nanotube film. The prepared flexible thin-film device has been reported to be used for chemical vapor sensing through monitoring conductivity changes caused by the charge transfer interaction with gas molecules and/or the intertube distance change induced by polymer swelling via gas absorption. Also, the effect of CNT grafting with poly(ε-caprolactone) on vapor-sensing properties has been investigated by Castro et al. (2009) for a series of conductive polymer composite transducers, which has been developed by layer by layer spray from poly(ε-caprolactone)–CNT solutions. Loading of ε-caprolactone on the CNT surface through in situ ring opening polymerization has been demonstrated in this report by nuclear magnetic resonance after solvent extraction of ungrafted chains. The prepared conductive polymer composite sensors have been reported to be used for measuring the chemoelectrical properties by exposing it to different vapors such as water, methanol, toluene, tetrahydrofuran, and chloroform and it has been analyzed in terms of signal sensitivity, selectivity, reproducibility, and stability. The higher response of the sensor has been reported when poly(ε-caprolactone) is loaded on CNT, which results in the formation of poly(ε-caprolactone)–CNT composites. The use of conductive polymer composite makes it possible to design the sensors with enhanced relative resistance responses amplitude by controlling different parameters like morphology of polymer phases and their crystallinity, combining different fillers, and temperature. CNT filled with a conductive polymer composite has been proved to be capable to detect, quantify, and distinguish various vapors, biomolecules, and temperature variations. Kumar et al. (2010) have further investigated the chemoelectrical properties of chitosan–CNT conductive biopolymer nanocomposites transducers prepared by spray layer-by-layer approach. It has been reported that chitosan provides the transducer with high sensitivity toward not only polar vapors like water and methanol but also to a lesser extent toluene. Further, they have fitted the Langmuir–Henry–Clustering model for quantitative responses that allows to link electrical signal to vapor content. Chitosan–CNT transducers selectivity has been also correlated by them with an exponential law to the inverse of Flory–Huggins interaction parameter χ_{12}. These reported properties make chitosan–CNT a good transducer to be implemented in an electronic nose (e-nose). Further, the morphology of chitosan–CNT suggests a chemical nano-switching mechanism promoting tunneling conduction and originating macroscopic vapor sensing.

Someya et al. (2003) have measured conductance of single-walled semiconducting CNTs in FET geometry and investigated the device response to alcoholic vapors. The significant change has been observed in FET drain current when the device is exposed to various kinds of alcoholic vapors such as methanol, ethanol, 1-propanol, 2-propanol, 1-butanol, tertiary butanol, 1-pentanol, and 1-octanol. These reported responses are reversible and reproducible over several cycles of vapor exposure. They have carried out the measurements in a dry nitrogen environment with and without the addition of saturated or diluted alcoholic vapor. Further, they have reported mechanisms for the change in the FET characteristics and two important findings related to this are as follows: (1) the nanotube FET can be instantly regenerated by removing the applied potentials with discontinuing alcohol vapor exposure. (2) Response to the alcohol vapor in various partial pressure can be scaled by the vapor exposure (pressure × exposure time) for low partial pressure of alcohol. Liu et al. (2010) have manufactured plasma-modified MWCNTs by microwave plasma–enhanced CVD and developed a new gas sensor material. It has been observed that the MWCNT-based gas sensors are a p-type response with resistance enrichment upon exposure to 50–500 ppm ethanol at room temperature. It has been also reported that oxygen plasma modification can enhance the sensor response from 1.03 to 1.16 on process duration of 30 s due to the apparent elimination of amorphous carbon, as established by Raman results, but oxygen plasma modification has no useful support in decreasing the response and recovery time. Also, by applying fluorine plasma modification, the sensor response increases from 1.03 to 1.13 on process duration of 60 s and the response, and recovery time can reduce slightly from 225 to 95 s and 452 to 227 s due to the existence of abundant fluorine-included functional groups, as demonstrated. The sensitivity is also reported to be enhanced three more times (from 0.0003 to 0.0011) and the linear range of measurement can also extend. From these results, the plasma-modified MWCNTs can raise the sensitivity and reactivity for room temperature ethanol sensing, especially fluorinated MWCNTs. Zhai et al. (2012) have prepared carbon-doped ZnO microspheres through a facile hydrothermal process. Carbon-doped ZnO under 500°C calcination showed the better UV-activated room-temperature gas-sensing activity for the detection of ethanol. The transient photovoltage results suggest the presence of sp^2 carbon-type structures that could enhance the separation extent and control the recombination of the photoinduced electron–hole pairs and increase the number of photoinduced oxygen ions on the surface of carbon-doped ZnO take place that enhances the gas-sensing activity.

Staii et al. (2005) have demonstrated a novel, versatile class of nanoscale chemical sensors based on single-stranded DNA as the chemical recognition site and SWCN FETs as the electronic readout component. SwCN-FETs with a nanoscale coating of single-stranded DNA reacts to gas odors that do not cause a noticeable conductivity change in bare devices. Responses of single-stranded DNA/swCN-FETs differ in sign and magnitude for different gases and can be tuned by choosing the *base sequence* of the single-stranded DNA. Single-stranded DNA/swCN-FET sensors sense a range of odors, with quick response and recovery times on the scale of seconds. This significant set of attributes makes sensors based on ssDNA loaded on nanotubes capable for *electronic nose* and *electronic tongue* applications ranging from homeland security to disease diagnosis.

A thin, strongly adhering films of SWNT bundles on flexible substrates like poly(ethylene-terephthalate) were used for vapor sensing (hexane, toluene, acetone, chloroform, acetonitrile, methanol, water, etc.) by Parikh et al. (2006). In this, they have coated poly(ethyleneterephthalate) patterns with films of electronically conductive SWNT bundles by dip coating in aqueous surfactant-supported dispersions and mounted in glass chambers equipped for vapor sensing. The saturated vapor conditions in air showed sensor responses that correlated well with solvent polarity. These types of sensors are very flexible, for example, they can be bent to diameters as small as 10 mm without drastically compromising sensor function. Wang et al. (2010) also fabricated flexible gas sensors by assembling SWNT thin films to sense the nerve agent stimulant dimethyl methylphosphonate vapors at room temperature and studied their response characteristics. The reported sensors have outstanding mechanical flexibility with only a 5.6% deviation in the amount of response when bent in a glass tube with a radius of about 5 mm. It can sense dimethyl methylphosphonate vapor with as low a concentration as 1 ppm, with higher than 3.6% resistance change. Further, it has been reported that the magnitude of response to dimethyl methylphosphonate vapor is 0.232 ± 0.007 ppm^{-1} and the sensor linearity of the responses in the range of 1–40 ppm is about 0.996. These results support that the SWNT flexible sensors are promising in portable on-site detection. Snow et al. (2005) has shown that the capacitance of SWNTs is highly sensitive to a broad class of chemical vapors and that this transduction mechanism can form the basis for a fast, low-power sorption-based chemical sensor. It has been also reported that in the presence of a dilute chemical vapor, molecular adsorbates are polarized by the fringing electric fields radiating from the surface of a SWNT electrode that causes an increase in its capacitance.

Sin et al. (2007) have fabricated alcohol sensors by preparing bundles of chemically modified MWCNTs (f-CNTs) across Au electrodes on SiO$_2$/Si substrates using an AC electrophoretic technique. The developed alcohol vapor sensor uses an ultralow input power of ~0.01–1 µW, which is significantly lower than the power required for most commercially available alcohol sensors. The chemical modification with the COOH groups by oxidation of MWCNTs has been reported. It has been reported that the sensors are selective with respect to flow from air, water vapor, and alcohol vapor, and its response is linear for alcohol vapor concentrations from 1 to 21 ppm with a detection limit of 0.9 ppm. The transient response of these sensors is reported to be ~1 s. Further, it has been demonstrated that the response of the sensors can be increased by one order of magnitude after adding the functional group COOH onto the nanotubes, that is, from ~0.9% of a bare MWCNTs sensor to ~9.6% of an f-CNTs sensor with a dose of 21 ppm alcohol vapor.

The gas ionization induces a positive charge on accumulation and can lead to the field electron emission from CNTs, partially covered with ethocel thin film and ZnO nanorods (Zhan et al. 2013). This effect can result into a new ionization gas-sensing mechanism that can be utilized to recognize the gaseous chemical composition based on the characteristic resistivity of the gases without strong field effect of field ionization and gaseous breakdown. Zhan et al. (2013) have suggested that the hypothesis for this illustration is as follows: (1) the cosmic ray ionization frequency increases 108–1010 times because of the metastable population that resulted from the interaction between the gases and the CNTs and (2) the flux of positive charges is converged because of the ZnO nanorods. The reported experimental results have been shown to be in consistent with the theoretical expectations, qualitatively. This type of device does not experience the degradation effect in the gaseous breakdown processes and exhibits improved repeatability. Further, metal oxide quasi-1D nanostructures have a better gas-sensing performance because of their large surface area and porous structures with a less agglomerated configuration. On the other hand, the well-designed fragile nanostructures could be effortlessly destroyed during the conventional fabrication process of gas sensors. A new materials-sensor integration fabrication strategy on the basis of screen printing technology and calcination, microinjecting have been introduced by Yi et al. (2013) into the fabrication process of sensors to obtain In_2O_3 nanowire–like network directly on the surface of the coplanar sensors array by structure replication from sacrificial CNTs. It has been reported that the obtained In_2O_3 nanowire–like network exhibits an outstanding response (electrical resistance ratio Ra/Rg), about 63.5, for 100 ppm formaldehyde at 300°C, which was about 30 times larger than that of compact In_2O_3 nanoparticles film (nonnetwork film). The improved gas-sensing properties have been essentially attributed to the high surface-to-volume ratio and the nanoscopic structural properties of materials.

The metal oxides like SnO_2 have been extensively investigated for their use in manufacturing gas sensors. As per Choi and Jang (2010), several research articles have been published within the last 30 years, which are focused on SnO_2. It is due to the predominance of SnO_2 in the form of ceramic resistors, thin films, and SnO_2-based nanotubes, belts, and ribbons, which are sensitive to several gases. Also, SnO_2 has outstanding parameters that include detection limit at near-room temperatures, sensitivity, response time, and long-term stability. Further, TiO_2 and SnO_2 are the renowned sensing materials with a good thermal stability of the former and a high sensitivity of the latter. CNTs have also gas-sensing ability at room temperature. CNT-included SnO_2/TiO_2 material is a novel investigation to combine the advantages of three kinds of materials for gas-sensing property. Duy et al. (2008) have prepared a uniform SnO_2/TiO_2 solution by the solgel process with the ratio 3:7 in mole and the CNTs with contents in the range of 0.001–0.5 wt.% have been dispersed in a mixed SnO_2/TiO_2 matrix by using an immersion-probe ultrasonic. The SnO_2–TiO_2 and the CNT-included SnO_2–TiO_2 thin films have been fabricated by the solgel spin-coating method over Pt-interdigitated electrode for gas-sensor device fabrication and they were heat-treated at 500°C for 30 min. Aroutiounian et al. (2012) have also prepared nanocomposite based on SnO_2/multiwall CNTs by the solgel method. Further, ceramic and thick film gas sensors have been developed with the use of the nanocomposite multiwall CNT/SnO_2/Pd structure and investigated. After the sensitization in the Ru(OH)Cl$_3$ water solution, these sensors have a quite high response to isobutane already at 120°C. The effect of changes in the sensitizing solution concentration and the work body temperature on the sensor response has been studied. This type of ceramic sensors has the drawback of slow recovering of the system after the gas supply is stopped. Further, to prepare adequately thick samples, polypropylene powder as a binder has been used during the preparation of sensors made of the SnO_2/MWCNT nanocomposite, and therefore, prepared sensors have been responsive to isobutene at its concentration of 200 ppm. The dependency of the response to isobutene concentration is a nearly linear character for the ceramic sensors made of the MWCNT/SnO_2/Pd nanostructure. These sensors also recovered well and times of the response and recovery are nearly identical and equal ~30 s. Recently, Mendoza et al. (2014) have fabricated chemical sensors based on SnO_2–CNT composite films by hot filament CVD method. The composite films consist of SnO_2 nanoparticles highly and uniformly distributed on the CNTs surface due to which the resistivity of this sensor is extremely sensitive to the presence of adsorbates that gets attached or detached easily at room temperature and ambient pressure depending on their gas phase concentration of ethanol, methanol, and H_2S. It has been shown that the SnO_2–CNT composite films can detect ethanol, methanol, and H_2S down to ppm levels below OSHA's permissible exposure limits at room temperature and ambient pressure. Furthermore, these types of gas sensors get self-recovered within 1 min without requiring any heating or energy source.

Graphene is a zero-bandgap semimetal with amazing electronic (Novoselov et al. 2005; Geim and Novoselov 2007; Bolotin et al. 2008; Chen et al. 2008; Morozov et al. 2008) and mechanical properties (Lee et al. 2008). Graphene is an entirely 2D material made up of a single layer of carbon with every atom on its surface and is an ideal material for the use as a chemical vapor sensor. It has been reported that the deposition of an individual gas molecules onto the surface of a graphene sensor leads to a measurable change in its electrical resistance (Schedin et al. 2007). The intrinsic sensitivity of graphene to different gaseous vapors can be measured during the use of samples where contamination from lithographic processing has been measured. Graphene vapor sensors that are known to be free of chemical contamination should then be amenable to (bio)molecular surface modification to control their chemical sensitivity, as has been done for CNTs (Staii et al. 2005) and semiconductor nanowires (McAlpine et al. 2008). Dan et al. (2009) have shown the contamination layer both degrades the electronic properties of the graphene and masks graphene's intrinsic sensor responses. The contamination layer chemically dopes the graphene, improves carrier scattering, and acts as an absorbent layer that concentrates analyte molecules at the graphene surface, thereby improving the sensor response. Further, they have demonstrated a cleaning process that verifiably eliminates the contamination on the device structure and allows the intrinsic chemical responses of graphene to be measured. Dua et al. (2010) have described a flexible and lightweight chemiresistor made of a thin film composed of overlapped and reduced graphene oxide platelets, which were printed onto flexible plastic surfaces by using inkjet techniques. The reduced graphene oxide films can reversibly and selectively sense chemically aggressive vapors such as NO_2 and Cl_2. Gas sensing has been achieved, without the help of a vapor concentrator, at room temperature, using an air sample containing vapor concentrations ranging from 100 ppm to 500 ppb. Inkjet printing of reduced graphene oxide platelets is carried out for the first time by Dua et al. (2010) with the use of aqueous surfactant-supported dispersions of reduced graphene oxide powder synthesized by the reduction of exfoliated graphite oxide, by using ascorbic acid as a mild and green reducing agent. The prepared film has electrical conductivity properties (15 S/cm) and fewer defects compared to reduced graphene oxide films obtained by using hydrazine reduction have been reported. Han et al. (2011) have observed oxidative etching of graphene flakes to initiate from edges and the occasional defect sites in the basal plane, leading to reduced lateral size and a small number of etch pits. On the contrary, etching of extremely defective graphene oxide and its reduced form resulted in rapid homogeneous fracturing of the sheets into smaller pieces. On the basis of these reported observations, a slow and more controllable etching route has been designed to manufacture nanoporous reduced graphene oxide sheets by hydrothermal steaming at 200°C. The amount of etching and the associated porosity can be suitably tuned by etching time. The steamed nanoporous graphene oxide has exhibited nearly two orders of magnitude increase in the sensitivity and improved recovery time compared to the nonporous reduced graphene oxide annealed at the same temperature, when it has been used as a chemiresistor sensor platform for NO_2 detection. Further, Hu et al. (2012) have presented a useful gas sensor based on chemically reduced graphene oxide by drop-drying method to generate conductive networks between interdigitated electrode arrays. Chemically reduced graphene oxide has been prepared by the reduction of graphene oxide using p-phenylenediamine, which has been used as an outstanding sensing material. Further, they have reported that its well-organized dispersion in organic solvents (i.e., ethanol) benefits the formation of conductive circuits between electrode arrays through drop-drying method. The results reported by Hu et al. (2012) have been presented on the detection of dimethyl methylphosphonate using this simple and scalable fabrication method for practical devices and suggest that p-phenylenediamine reduced chemically reduced graphene oxide exhibits much better (5.7 times with the concentration of dimethyl methylphosphonate at 30 ppm) response to dimethyl methylphosphonate than that of chemically reduced graphene oxide reduced from hydrazine. Also, as per their report, this novel gas sensor based on chemically reduced graphene oxide reduced from p-phenylenediamine shows excellent responsive repeatability to dimethyl methylphosphonate. It has been concluded that the efficient dispersibility of chemically reduced graphene oxide reduced from p-phenylenediamine in organic solvents facilitates the device fabrication through drop-drying method; the resultant chemically reduced graphene oxide-based sensing devices, with miniature, low-cost, portable characteristics, as well as outstanding sensing performances, can ensure its potential application in gas-sensing fields. Recently, Yoon et al. (2013) have carried out cell-based electrochemical sensing of extracellular H_2O_2 with a nanocomposite-modified indium tin oxide electrode to estimate the cytotoxicity of nanomaterials. The indium tin oxide working electrode has been functionalized with a nanocomposite film of Nafion and chemically driven graphene using the electrospray method. The cyclic voltammetric measurements of H_2O_2 molecules using the nanocomposite-modified electrode carried out by Yoon et al. (2013) showed a high sensitivity (82.6 μA/mM/cm^2) due to the improvement of the electrochemical signal by the nanocomposite. Further, the results reported showed that the increased tendency in measured H_2O_2 concentration as the concentration of the chemically driven graphene nanoflakes increased was consistent with the results of the toxicity analysis data obtained by optical bioassays. These toxicity analyses by cell-based electrochemical sensors are of great interest for the toxicity assessment of nanomaterials having various biomedical and environmental applications.

44.3.3 Volatile Organic Compounds Sensing

Volatile organic compounds (VOCs) are organic chemicals that have higher vapor pressure and that can easily convert to vapors at normal temperature and pressure. VOCs are generally found in diverse environments like industrial and residential sites. These compounds have severe effects on the environment through early degradation of the surrounding area and health hazards to people living around the contaminated areas. The VOCs are also considered as the main reason for allergic pathologies, skin, and lung diseases. Because of these reasons, the detection of VOCs has been studied by several researchers and is of a great interest due to stringent environmental standards and regulations in many countries of the world. Further, CNTs, particularly MWCNT, are interesting nanomaterials due to their large surface area–volume ratio, high electron mobility, and high capability of gas adsorption (Iijima 1991). Also, the electronic properties of MWCNTs suggest the transfer of the charges efficiently when it is exposed to analytes (Zhang et al. 2008). Therefore, MWCNTs are highly sensitive to the different gases such as NO_X, NH_3, and SO_2 with a low detection limit and short response time. CNT is a thin-film p-type semiconductor with the advantage of the sensing properties of the individual nanotubes or their bundles/ropes incorporated in the network (Penza et al. 2007c). The increase in the surface coverage with CNTs leads to the formation of *spaghetti-like* structure, which results in increased electron conducting paths between the electrodes (Battie et al. 2011). This effect of CNTs in the sensing films has been explained by percolation theory and recent literature indicates that the percolation effect must be considered for sensor applications (Bondavalli et al. 2009; Battie et al. 2012). Penza et al. (2005) have used LB films consisting of tangled bundles of SWCNTs as sensing nanomaterials onto three different types of sensory systems using complementary transducing principles as surface acoustic waves, quartz crystal microbalance, and standard silica optical fiber for VOCs detection, at room temperature. As per their report, highly sensitive, repeatable, and reversible responses of the SWCNTs sensors indicate feasible VOCs detection in a wide mm Hg vapor pressures range at room temperature. The development of CNTs sensors based on complementary transducing principles is exceptionally capable for sensing applications of multitransducer and multisensor array by using pattern recognition techniques for efficient VOCs chemical analysis, at room temperature. Further, Penza et al. (2010) have also investigated the effect of the surface functionalization of CNT networked films with a metalloporphyrins layer for gas sensing. As per their report, the surface modifications of CNTs networked films with surface layers of metalloporphyrins of zinc and manganese tetraphenylporphyrin result in sensors with improved sensitivity and differentiated selectivity. The modified film is reported to be exhibiting an enhanced sensitivity of the electrical resistance toward the concentrations of common volatile compounds like alcohols, amines, aromatics, and ketones, at room temperature. The reason for the improved sensing properties is adsorption properties of functional units, which are transferred to the sensor signal resulting in improved sensing properties of the sensor. Further, as per their principal component analysis, the functionalization provides adequate selectivity change to turn a triplicate of the same CNT film into an efficient sensor array capable of the compounds recognition. Sayago et al. (2012) have investigated surface acoustic wave gas sensors based on polymers and CNT composites as sensitive layers for sensing the low concentrations of VOCs as octane and toluene. A number of nanocomposites based on polyepichlorohydrin and polyetherurethane with different percentages of MWCNT have been investigated by them to study the effect of MWCNTs in the response of sensors. The reported sensor is selective to volatile gases such as toluene and octane.

Further, Badhulika et al. (2014) have investigated fabrication, characterization, and evaluation of PEDOT/PSS-coated SWNTs sensors for detecting analytes of interest in industrial manufacturing. The effect of the conducting polymer's synthesis conditions in terms of charge-controlled electropolymerization of the monomer EDOT in the presence of the dopant PSS on the sensing performance of the PEDOT:PSS-functionalized SWNT sensors has been systematically studied by them. The optimized sensors have been reported to exhibit sensing properties over a wide dynamic range of concentrations toward saturated vapors of VOCs like methanol, ethanol, and methyl ethyl ketone at room temperature. The reported limit of detection of this sensor is 1.3%, 5.95%, and 3% for saturated vapors of methanol, ethanol, and methyl ethyl ketone, respectively. Tasaltin and Basarir (2014) have prepared novel flexible VOC sensor by the loading of gold nanoparticle and amine modified MWCNTs ($MWCNT-NH_2$) on a polyimide substrate via electrospraying technique. As per their report, they have analyzed chemical sensing behaviors of the sensors against polar (water, propanol, and ethanol) and nonpolar (hexane, toluene, trichloroethylene, and chloroform) VOCs. Further, the effect of the gold nanoparticle/MWCNT-NH$_2$ ratio on the conductivity and sensing has been also reported. The reported sensing mechanism for the sensors occurs in two stages: (1) gold nanoparticles that are believed to wrap the MWCNT-NH$_2$ behave as the filters and determine the gases that will reach to the surface of MWCNT-NH$_2$ and (2) interactions of the VOC's organic functional group on MWCNT-NH$_2$ determine the sensing behavior. Further, as per their report, the gold nanoparticles are assumed to adsorb hexane vapor and do not allow it to reach to MWCNT-NH$_2$, thus very small resistance change. Also, increasing the polarity of the VOCs leads to an increase in resistance change, which is applicable for toluene, trichloroethylene, and chloroform. However, for the polar analytes with higher dielectric constant (water, ethanol, propanol), almost no sensor response was detected because gold nanoparticles do not allow the passage of the said analytes.

Graphene is the focus of current research because of its exceptional properties and found uses in various applications including gas sensors (Schedin et al. 2007; Stoller et al. 2008). Further, many efforts have been made to functionalize graphene with molecules or nanoparticles to get novel properties by a synergistic effect between them (Bai et al. 2011; Huang et al. 2011). Tung et al. (2014) have synthesized quantum chemoresistive vapor sensors from the assembly of magnetic nanoparticles–decorated reduced graphene oxide (Fe$_3$O$_4$-RGO) with PEDOT and poly(ionic liquid) (PIL). As per their report, the prepared new hybrid sensing material confirmed improved sensitivity, selectivity, and signal-to-noise ratio and reduced response time compared to its elementary constituents, which are also sensitive. This suggests that a positive synergy of properties has been achieved during the structuring of the conducting architecture by spray layer by layer. The Fe$_3$O$_4$-RGO/PIL-PEDOT sensor reported to have steady and reproducible signals at room temperature for both polar (ethanol, methanol, acetone, water) and nonpolar (chloroform, styrene, dichlorobenzene, toluene) VOCs, considered as food degradation biomarkers.

44.3.4 H$_2$ Gas Sensing

Hydrogen sensors are in increasing demand due to their numerous applications in chemical and petroleum refining, neon and xenon production, rocket fuels for spacecraft, fuel cells, semiconductor processing for microelectronics, and biomedical applications. Hydrogen in the breath is an excellent indicator for different diseases that include lactose intolerance, fructose malabsorption, fibromyalgia, and neonatal necrotizing enterocolitis (Grimes et al. 2003). Due to these applications, a hydrogen sensor with superior sensitivity, low detection limit, quick response, and recovery time, good specificity with low false positive, and long-term stability with wide temperature operation range is essential. Several researchers have investigated various methods to develop hydrogen sensors. SWNTs could be ideal building blocks for making gas sensors due to their higher surface area. Also, loading of suitable metal or metal oxides on SWNTs leads to improvement of sensitivity toward hydrogen.

Suehiro et al. (2007) have used the dielectrophoresis method for the fabrication of the interfaces between CNT and catalytic palladium (Pd) to realize a CNT-based H$_2$ gas sensor. In their report, two interfaces of CNT/Pd have been tested; those were (1) CNTs have been trapped onto a microelectrode made of Pd, so that the CNT/Pd interfaces were formed at both ends of CNTs lying over the Pd electrode surface, and (2) gas sensor has been fabricated simultaneously by trapping CNTs and Pd nanoparticles under the action of the positive dielectrophoresis. Further, it has been reported that both types of the CNT/Pd gas sensors can detect hydrogen. According to their report, for the hydrogen gas diluted with dry air, the electrical resistance of the CNT/Pd sensors is found increased at the moment of hydrogen exposure and then decreased in the later stage. This is attributed to the reduction of CNTs by H atoms, which were produced by dissociative adsorption of H$_2$ molecules on the catalytic Pd surface. In the later stage, the dissociated H atoms might react with dissociated oxygen atoms to generate H$_2$O molecules accompanying heat generation. Mubeen et al. (2007) have also developed a simple and cost-effective fabrication method to prepare a hydrogen nanosensor by decorating SWCNTs with Pd nanoparticles. According to their findings, the sensor's synthesis conditions like Pd electrodeposition charge, deposition potential, and initial baseline resistance of the SWNT network has a significant effect on the sensing performance. The reported optimized sensor showed exceptional sensing properties toward hydrogen (ΔR/R of 0.42%/ppm) with a lower detection limit of 100 ppm and a linear response up to 1000 ppm. Further, Zhang et al. (2012) have prepared carbon nanofibrous mat (nano-felt) surface attached with palladium nanoparticles (Pd NPs) from electrospun polyacrylonitrile nano-felt surface functionalized with amidoxime groups for hydrogen-sensing application. The prepared material reported to be consisted of comparatively uniform and randomly overlaid carbon nanofibers with diameters of ~300 nm and loaded Pd nanoparticles of size ranging from a few to tens of nanometers. The reported electrospun carbon nano-felt is mechanically flexible/resilient and the resistance of the prepared material found varied when it is exposed to H_2 at room temperature. As per their study, the electrospun carbon nano-felts surface attached with metal nanoparticles could be an alternative material for the fabrication of gas and/or biosensors, and the amidoxime-functionalization of electrospun polyacrylonitrile nano-felt could be a common method for the preparation of different carbon nano-felts surface attached with various metal nanoparticles. Recently, Lin and Huang (2012) have developed a leaflike CNT/nickel (CNT/Ni) composite film modified with Pd NPs for H_2 gas-sensing applications. According to them, a Pd/CNT/Ni film treated by the sensitization and activation process shows an improvement in the hydrogen-sensing properties. The H_2 gas-sensing properties of the Pd/CNT/Ni were reported to have a superior response (7.3%) and quicker response time (312 s) at 200 ppm H_2 gas than those prepared without Pd NPs modified (0.7%, 1092 s).

Wongchoosuk et al. (2010) have also fabricated hydrogen gas sensors based on undoped and 1 wt.% MWCNT-doped tungsten oxide (WO_3) thin films by means of the powder mixing and electron beam evaporation method. They have investigated hydrogen-sensing properties of the thin films at different operating temperatures and gas concentrations ranging from 100 to 50,000 ppm. The reported results indicate that the MWCNT-doped WO_3 thin film have higher sensitivity and selectivity to hydrogen. Thus, as per their report, MWCNT doping based on electron beam coevaporation has shown to be an efficient way of preparation of hydrogen gas sensors with improved sensing and reduced operating temperatures. Further, they have also proposed creation of nanochannels and formation of p–n heterojunctions as the sensing mechanism underlying the improved hydrogen sensitivity of this hybridized gas sensor. De Luca et al. (2012) have prepared platinum-doped titanium dioxide/MWCNTs (Pt/TiO_2/MWCNTs) composites by a solgel technique. Regardless of the nominal C/Ti molar ratio (0.3–17.0), only the anatase phase of TiO_2 has been reported to be detected. However, as per their report, on the composite with the highest C/Ti molar ratio, the formation of a more structurally disordered and nonstoichiometric anatase phase seemed to be favored. Further, they have analyzed electrical characteristics and hydrogen-sensing properties of Pt/TiO_2/MWCNTs composite films deposited on interdigitated ceramic substrates in the temperature interval from room temperature to 100°C. The electrical conductivity of the composite films has been reported to be several orders of magnitude higher than that of pure titania, allowing electrical measurements at room temperature. As per their report, Pt/TiO_2/MWCNTs composite films showed a response to hydrogen concentration, up to 100%, in nitrogen even at room temperature. They have reported *spillover* mechanism, in which hydrogen molecules are first chemisorbed and dissociated on platinum, and finally spill out of the Pt surface, diffusing into the TiO_2 surface layer, with MWCNTs providing a preferential pathway to the current flow to explain the hydrogen-sensing mechanism on these sensors. CNTs are exceptionally responsive to environmental gases and sensing of H_2 gas at room temperature with quick response and recovery time is still a challenge. Considering this, Dhall et al. (2013) have prepared acid-functionalized MWCNTs to detect 0.05% H_2 gas at room temperature. Acids functionalized MWCNTs showed quicker response to H_2 gas as compared to pristine MWCNTs. Further, they have investigated the effect of functionalization on the pristine-MWCNTs structure and their electrical properties using different techniques. The reduced crystallite size of the acids-treated nanotubes and attachments of functional groups on the nanotubes improved the sensing properties.

44.4 CONCLUSIONS AND OUTLOOK

The exceptional electronic structures and properties of some carbon nanomaterials like carbon nanofibers, CNTs, graphene, and reduced graphene oxide are outstanding compared to other well-established nanomaterials in potential applications of electrical sensors. However, for the commercial production of carbon nanomaterials-based gas sensors, fabrication is one of the main challenges and more research is required to find cost-effective, scalable production methods that preserve the necessary properties of such materials. It has been reported that single-molecule adsorption/desorption events are evident using graphene-based devices and the electrical detection of gas adsorption on CNTs or graphene has detection limits at ppb levels, in laboratory conditions. Modification of the carbon nanomaterial surface like decorating with metal or metal oxide nanoparticles or by grafting functional groups is an efficient way to enhance the sensitivity, minimize unwanted effects, and tune selectivity. Also, the selectivity is a key problem as the literature shows the complexity of making carbon nanomaterials absolutely selective and functionalization of carbon nanomaterials might be a solution for the same. Functionalization of carbon nanomaterials with suitable material should be targeted at balancing the strength of adsorption of analytes that is required for high sensitivity against the reversibility of the detection process, which is fundamental for gas sensors to perform continuous measurements. Further, graphene offers good sensing properties toward the target molecules than that of large-diameter CNTs and similar flexibility for functionalization of its surface with notably lower noise levels, which would be preferable for sensors with improved lower detection limits.

REFERENCES

Adu, C. K. W., G. U. Sumanasekera, B. K. Pradhan, H. E. Romero, and P. C. Eklund. 2001. Carbon nanotubes: A thermoelectric nano-nose. *Chemical Physics Letters* 337:31–35.

Allen, M. J., V. C. Tung, and R. B. Kaner. 2010. Honeycomb carbon: A review of graphene. *Chemical Reviews* 110:132–145.

Al-Mashat, L., K. Shin, K. Kalantar-zadeh, J. D. Plessis, S. H. Han, R. W. Kojima, R. B. Kaner, D. Li, X. Gou, S. J. Ippolito, and W. Wlodarski. 2010. Graphene/polyaniline nanocomposite for hydrogen sensing. *Journal of Physical Chemistry C* 114:16168–16173.

Aroutiounian, V. M., V. M. Arakelyan, E. A. Khachaturyan, G. E. Shahnazaryan, M. S. Aleksanyan, L. Forro, A. Magrez, K. Hernadi, and Z. Nemeth. 2012. Manufacturing and investigations of i-butane sensor made of SnO_2/multiwall-carbon-nanotube nanocomposite. *Sensors and Actuators B: Chemical* 173:890–896.

Athawale, A. A. and M. V. Kulkarni. 2000. Polyaniline and its substituted derivatives as sensor for aliphatic alcohols. *Sensors and Actuators B: Chemical* 67:173–177.

Avouris, P. 2002. Carbon nanotube electronics. *Chemical Physics* 281:429–445.

Badhulika, S., N. V. Myung, and A. Mulchandani. 2014. Conducting polymer coated single-walled carbon nanotube gas sensors for the detection of volatile organic compounds. *Talanta* 123:109–114.

Bai, H., C. Li, and G. Shi. 2011. Functional composite materials based on chemically converted graphene. *Advanced Materials* 23:1089–1115.

Balandin, A. A., S. Ghosh, W. Bao, I. Calizo, D. Teweldebrhan, F. Miao, and C. N. Lau. 2008. Superior thermal conductivity of single-layer graphene. *Nano Letters* 8:902–907.

Battie, Y., O. Ducloux, P. Thobois, Y. Coffinier, and A. Loiseau. 2010. Evaluation of sorted semi-conducting carbon nanotube films for gas sensing applications. *C. R. Physique* 11:397–404.

Battie, Y., O. Ducloux, P. Thobois, N. Dorval, J. S. Lauret, and B. Attal-Tretout. 2011. Gas sensors based on thick films of semiconducting single walled carbon nanotubes. *Carbon* 49:3544–3552.

Battie, Y., L. Gorintin, O. Ducloux, P. Thobois, P. Bondavalli, G. Feugnet, and A. Loiseau. 2012. Thickness dependent sensing mechanism in sorted semi-conducting single walled nanotube based sensors. *Analyst* 137:2151–2157.

Bekyarova, E., M. Davis, T. Burch, M. E. Itkis, B. Zhao, S. Sunshine, and R. C. Haddon. 2004. Chemically functionalized single-walled carbon nanotubes as ammonia sensors. *Journal of Physical Chemistry B* 108:19717–19720.

Bekyarova, E., I. Kalinina, E. I. Mikhail, L. Beer, N. Cabrara, and R. C. Haddon. 2007. Mechanism of ammonia detection by chemically functionalized single-walled carbon nanotubes: In situ electrical and optical study of gas analyte detection. *Journal of the American Chemical Society* 129:10700–10706.

Bittencourt, C., A. Felten, E. H. Espinosa, R. Ionescu, E. Llobet, X. Correig, and J. J. Pireaux. 2006. WO$_3$ films modified with functionalised multi-wall carbon nanotubes: Morphological, compositional and gas response studies. *Sensors and Actuators B: Chemical* 115:33–41.

Bolotin, K. I., K. J. Sikes, Zd Jiang, M. Klima, G. Fudenberg, J. Hone, P. Kim, and H. L. Stormer. 2008. Ultrahigh electron mobility in suspended graphene. *Solid State Communications* 146:351–355.

Bondavalli, P., P. Legagneux, and D. Pribat. 2009. Carbon nanotubes based transistors as gas sensors: State of the art and critical review. *Sensors and Actuators B: Chemical* 140:304–318.

Boyd, A., I. Dube, G. Fedorov, M. Paranjape, and P. Barbara. 2014. Gas sensing mechanism of carbon nanotubes: From single tubes to high-density networks. *Carbon* 69:417–423.

Castro, M., J. Lu, S. Bruzaud, B. Kumar, and J. Feller. 2009. Carbon nanotubes/poly(e-caprolactone) composite vapour sensors. *Carbon* 47:1930–1942.

Chen, J. H., C. Jang, S. D. Xiao, M. Ishigami, and M. S. Fuhrer. 2008. Intrinsic and extrinsic performance limits of graphene devices on SiO$_2$. *Nature Nanotechnology* 3:206–209.

Chen, R. J., S. Bangsaruntip, K. A. Drouvalakis, N. W. S. Kam, M. Shim, Y. Li, W. Kim, P. J. Utz, and H. J. Dai. 2003. Noncovalent functionalization of carbon nanotubes for highly specific electronic biosensors. *Proceedings of the National Academy of Sciences* 100:4984–4989.

Chen, S. G., J. W. Hu, M. Q. Zhang, and M. Z. Rong. 2005. Effects of temperature and vapor pressure on the gas sensing behavior of carbon black filled polyurethane composites. *Sensors and Actuators B: Chemical* 105:187–193.

Choi, K. J. and H. W. Jang. 2010. One-dimensional oxide nanostructures as gas-sensing materials: Review and issues. *Sensors* 10:4083–4099.

Choi, K. Y., J. S. Park, K. B. Park, H. J. Kim, H. D. Park, and S. D. Kim. 2010. Low power micro-gas sensors using mixed SnO$_2$ nanoparticles and MWCNTs to detect NO$_2$, NH$_3$, and xylene gases for ubiquitous sensor network applications. *Sensors and Actuators B: Chemical* 150:65–72.

Chopra, S., K. McGuire, N. Gothard, A. M. Rao, and A. Pham. 2003. Selective gas detection using a carbon nanotube sensor. *Applied Physics Letters* 83:2280–2282.

Chopra, S., A. Pham, J. Gaillard, A. Parker, and A. M. Rao. 2002. Carbon-nanotube-based resonant-circuit sensor for ammonia. *Applied Physics Letters* 80:4632–4634.

Choudhury, A. 2009. Polyaniline/silver nanocomposites: Dielectric properties and ethanol vapour sensitivity. *Sensors and Actuators B: Chemical* 138:318–325.

Choudhury, A., P. Kar, M. Mukherjee, and B. Adhikari. 2009. Polyaniline/silver nanocomposite based acetone vapour sensor. *Sensor Letters* 7:592–598.

Chu, B. H., J. Nicolosi, C. F. Lo, W. Strupinski, S. J. Pearton, and F. Ren. 2011. Effect of coated platinum thickness on hydrogen detection sensitivity of graphene-based sensors. *Electrochemical and Solid-State Letters* 14:K43–K45.

Cinke, M., J. Li, B. Chen, A. Cassell, L. Delzeit, J. Han, and M. Meyyappan. 2002. Pore structure of raw and purified HiPco single-walled carbon nanotubes. *Chemical Physics Letters* 365:69–74.

Collins, P. G., K. Bradley, M. Ishigami, and A. Zettl. 2000. Extreme oxygen sensitivity of electronic properties of carbon nanotubes. *Science* 287:1801–1804.

Cui, S. M., H. H. Pu, G. H. Lu, Z. H. Wen, E. C. Mattson, C. Hirschmugl, M. Gajdardziska-Josifovska, M. Weinert, and J. H. Chen. 2012. Fast and selective room temperature ammonia sensors using silver nanocrystal-functionalized carbon nanotubes. *ACS Applied Materials and Interfaces* 4:4898–4904.

Dai, L. 2004. *Intelligent Macromolecules for Smart Devices: From Materials Synthesis to Device Applications*. Springer-Verlag: New York.

Da-Jing, C., L. Sheng, W. Ren-Hui, P. Min, and C. Yu-Quan. 2012. Dielectrophoresis carbon nanotube and conductive polyaniline nanofiber NH_3 gas sensor. *Chinese Journal of Analytical Chemistry* 40:145–149.

Dan, Y., Y. Lu, N. J. Kybert, Z. Luo, and A. C. Johnson. 2009. Intrinsic response of graphene vapor sensors. *Nano Letters* 9:1472–1475.

De Luca, L., A. Donato, S. Santangelo, G. Faggio, G. Messina, N. Donato, and G. Neri. 2012. Hydrogen sensing characteristics of $Pt/TiO_2/MWCNTs$ composites. *International Journal of Hydrogen Energy* 37:1842–1851.

Debarnot, D. N. and F. P. Epaillard. 2003. Polyaniline as a new sensitive layer for gas sensors. *Annali di Chimica* 475:1–15.

Derycke, V., R. Martel, J. Appenzeller, and P. Avouris. 2002. Controlling doping and carrier injection in carbon nanotube transistors. *Applied Physics Letters* 80:2773–2775.

Dhall, S., N. Jaggi, and R. Nathawat. 2013. Functionalized multiwalled carbon nanotubes based hydrogen gas sensor. *Sensors and Actuators A: Physical* 201:321–327.

Dua, V., S. P. Surwade, S. Ammu, S. R. Agnihotra, S. Jain, K. E. Roberts, S. Park, R. S. Ruoff, and S. K. Manohar. 2010. All-organic vapor sensor using inkjet-printed reduced graphene oxide. *Angewandte Chemie International Edition* 49:2154–2157.

Duy, N. V., N. V. Hieu, P. T. Huy, N. D. Chien, M. Thamilselvan, and J. Yi. 2008. Mixed SnO_2/TiO_2 included with carbon nanotubes for gas-sensing application. *Physica E* 41:258–263.

Ellison, M. D., M. J. Crotty, D. Koh, R. L. Spray, and E. Tate. 2004. Adsorption of NH_3 and NO_2 molecules on single-walled carbon nanotubes. *Journal of Physical Chemistry B* 108:7938–7943.

Fu, D. L., H. L. Lim, Y. M. Shi, X. C. Dong, S. G. Mhaisalkar, Y. Chen, S. Moochhala, and L. J. Li. 2008. Differentiation of gas molecules using flexible and all-carbon nanotubes devices. *Journal of Physical Chemistry C* 112:650–653.

Gautam, M. and A. H. Jayatissa. 2012a. Adsorption kinetics of ammonia sensing by graphene films decorated with platinum nanoparticles. *Journal of Applied Physics* 111:094317.

Gautam, M. and A. H. Jayatissa. 2012b. Ammonia gas sensing behavior of graphene surface decorated with gold nanoparticles. *Solid-State Electronics* 78:159–165.

Geim, A. K. 2009. Graphene: Status and prospects. *Science* 324:1530–1534.

Geim, A. K. and K. S. Novoselov. 2007. The rise of graphene. *Nature Materials* 6:183–191.

Ghaddab, B., F. Berger, J. B. Sanchez, and C. Mavon. 2010. Detection of O_3 and NH_3 using tin dioxide/carbon nanotubes based sensors: Influence of carbon nanotubes properties onto sensor's sensitivity. *Procedia Engineering* 5:115–118.

Ghaddab, B., J. B. Sanchez, C. Mavon, M. Paillet, R. Parret, A. A. Zahab, J.-L. Bantignies, V. Flaud, E. Beche, and F. Berger. 2012. Detection of O_3 and NH_3 using hybrid tin dioxide/carbon nanotubes sensors: Influence of materials and processing on sensor's sensitivity. *Sensors and Actuators B: Chemical* 170:67–74.

Grimes, C. A., K. G. Ong, O. K. Varghese, X. Yang, G. Mor, M. Paulose, E. C. Dickey, C. Ruan, M. V. Pishko, J. W. Kendig, and A. J. Mason. 2003. A sentinel sensor network for hydrogen sensing. *Sensors* 3:69–82.

Han, J. and C. Gao. 2010. Functionalization of carbon nanotubes and other nanocarbons by azide chemistry. *Nano-Micro Letters* 2:213–226.

Han, T., Y. Huang, A. T. L. Tan, V. P. Dravid, and J. Huang. 2011. Steam etched porous graphene oxide network for chemical sensing. *Journal of American Chemical Society* 133:15264–15267.

Han, Z. J., H. Mehdipour, X. G. Li, J. Shen, L. Randeniya, H. Y. Yang, and K. Ostrikov. 2012. SWCNT networks on nanoporous silica catalyst support: Morphological and connectivity control for nanoelectronic, gas-sensing, and biosensing devices. *ACS Nano* 6:5809–5819.

Hatchett, D. W. and M. Josowicz. 2008. Composites of intrinsically conducting polymers as sensing nanomaterials. *Chemical Reviews* 108:746–769.

Heeger, A. J. 2002. Semiconducting and metallic polymers: The fourth generation of polymeric materials. *Synthetic Metals* 125:23–42.

Hu, N., Y. Wang, J. Chai, R. Gao, Z. Yang, E. Kong, and Y. Zhang. 2012. Gas sensor based on p-phenylenediamine reduced graphene oxide. *Sensors and Actuators B: Chemical* 163:107–114.

Huang, X., Z. Yin, S. Wu, X. Qi, Q. He, Q. Zhang, Q. Yan, F. Boey, and H. Zhang. 2011. Graphene-based materials: Synthesis, characterization, properties, and applications. *Small* 7:1876–1902.

Huang, X. L., N. T. Hu, R. G. Gao, Y. Yu, Y. Y. Wang, Z. Yang, E. Siu-Wai Kong, H. Wei, and Y. F. Zhang. 2012. Reduced graphene oxide–polyaniline hybrid: Preparation, characterization and its applications for ammonia gas sensing. *Journal of Material Chemistry* 22:22488–22495.

Iijima, S. 1991. Helical microtubules of graphitic carbon. *Nature* 354:56–58.

Irimia-Vladu, M. and J. W. Fergus. 2006. Suitability of emeraldine base polyaniline-PVA composite film for carbon dioxide sensing. *Synthetic Metals* 156:1401–1407.

Jain, S., A. B. Samui, M. Patri, V. R. Hande, and S. V. Bhoraskar. 2005. FEP/polyaniline based multilayered chlorine sensor. *Sensors and Actuators B: Chemical* 106:609–613.

Jeffries, J. B., G. A. Raiche, and L. E. Jusinski. 1992. Detection of chlorinated hydrocarbons via laser-atomization laser-induced fluorescence. *Applied Physics B: Photophysics and Laser Chemistry* 55:76–83.

Jian, J., X. Guo, L. Lin, Q. Cai, J. Cheng, and J. Li. 2013. Gas-sensing characteristics of dielectrophoretically assembled composite film of oxygen plasma-treated SWCNTs and PEDOT/PSS polymer. *Sensors and Actuators B: Chemical* 178:279–288.

Jiménez-Cadena, G., J. Riu, and F. X. Rius. 2007. Gas sensors based on nanostructured materials. *Analyst* 132:1083–1099.

Kar, P. and A. Choudhury. 2013. Carboxylic acid functionalized multi-walled carbon nanotube doped polyaniline for chloroform sensors. *Sensors and Actuators B: Chemical* 183:25–33.

Kauffman, D. R. and A. Star. 2008. Carbon nanotube gas and vapor sensors. *Angewandte Chemie International Edition* 47:6550–6570.

Kiattibutr, P., L. Tarachiwin, L. Ruangchuay, A. Sirivat, and J. Schwank. 2002. Electrical conductivity responses of polyaniline films to SO_2–N_2 mixtures: Effect of dopant type and doping level. *Reactive and Functional Polymers* 53:29–37.

Kim, J., F. Kim, and J. Huang. 2010. Seeing graphene based sheets. *Materials Today* 13:28–38.

Kong, J., N. R. Franklin, C. Zhou, M. G. Chapline, S. Peng, K. Cho, and H. Dai. 2000. Nanotube molecular wires as chemical sensors. *Science* 287:622–625.

Kou, L., H. He, and C. Gao. 2010. Click chemistry approach to functionalize two-dimensional macromolecules of graphene oxide nanosheets. *Nano-Micro Letters* 2:177–183.

Koul, S., R. Chandra, and S. K. Dhawan. 2001. Conducting polyaniline composite: A reusable sensor material for aqueous ammonia. *Sensors and Actuators B: Chemical* 75:151–159.

Kumar, B., J. Feller, M. Castro, and J. Lu. 2010. Conductive bio-polymer nano-composites (CPC): Chitosan-carbon nanotubes transducers assembled via spray layer-by-layer for volatile organic compound sensing. *Talanta* 81:908–915.

Lange, U., T. Hirsch, V. M. Mirsky, and O. S. Wolfbeis. 2011. Hydrogen sensor based on a graphene–palladium nanocomposite. *Electrochimica Acta* 56:3707–3712.

Larciprete, R., L. Petaccia, S. Lizzit, and A. Goldoni. 2007. The role of metal contact in the sensitivity of single-walled carbon nanotube of NO_2. *Journal of Physical Chemistry B* 111:12169–12174.

Lee, C., X. D. Wei, J. W. Kysar, and J. Hone. 2008. Measurement of the elastic properties and intrinsic strength of monolayer graphene. *Science* 321:385–388.

Li, W., X. Geng, Y. Guo, J. Rong, Y. Gong, L. Wu, X. Zhang et al. 2011. Reduced graphene oxide electrically contacted graphene sensor for highly sensitive nitric oxide detection. *ACS Nano* 5:6955–6961.

Liao, L., M. Zheng, Z. Zhang, B. Yan, X. Chang, G. Ji, Z. Shen et al. 2009. The characterization and application of p-type semiconducting mesoporous carbon nanofibers. *Carbon* 47:1841–1845.

Lin, T. and B. Huang. 2012. Palladium nanoparticles modified carbon nanotube/nickel composite rods (Pd/CNT/Ni) for hydrogen sensing. *Sensors and Actuators B: Chemical* 162:108–113.

Liu, C., J. Wu, and H. C. Shih. 2010. Application of plasma modified multi-wall carbon nanotubes to ethanol vapor detection. *Sensors and Actuators B: Chemical* 150:641–648.

Liu, H., J. Kameoka, D. A. Czaplewski, and H. Craighead. 2004. Polymeric nanowire chemical sensor. *Nano Letters* 4:671–675.

Lu, G., S. Park, K. Yu, R. S. Ruoff, L. E. Ocola, D. Rosenmann, and J. Chen. 2011. Toward practical gas sensing with highly reduced graphene oxide: A new signal processing method to circumvent run-to-run and device-to-device variations. *ACS Nano* 5:1154–1164.

Lu, G. H., L. E. Ocola, and J. H. Chen. 2009. Gas detection using low-temperature reduced graphene oxide sheets. *Applied Physics Letters* 94:083111.

Lucci, M., P. Regoliosi, A. Reale, A. Di Carlo, S. Orlanducci, E. Tamburri, M. L. Terranova, P. Lugli, C. Di Natale, A. D'Amico, and R. Paolesse. 2005. Gas sensing using single wall carbon nanotubes ordered with dielectrophoresis. *Sensors and Actuators B: Chemical* 111–112:181–186.

Luksbetlej, K. and D. Bodzek. 2000. Determination of tri-chloromethane, tetrachloromethane and trichloroethane by using microextraction with GC-ECD detection. *Chemia Analityczna* 45:45–51.

MacDiarmid, A. G. 2002. Synthetic metals: A novel role for organic polymers. *Synthetic Metals* 125:11–22.

Mao, S., S. Cui, G. Lu, K. Yu, Z. Wen, and J. Chen. 2012. Tuning gas-sensing properties of reduced graphene oxide using tin oxide nanocrystals. *Journal of Materials Chemistry* 22:11009–11013.

Martel, R., V. Derycke, C. Lavoie, J. Appenzeller, K. Chan, J. Tersoff, and P. Avouris. 2001. Ambipolar electrical transport in semiconducting single-wall carbon nanotubes. *Physical Review Letters* 87:256805.

McAlpine, M. C., H. D. Agnew, R. D. Rohde, M. Blanco, H. Ahmad, A. D. Stuparu, W. A. Goddard Iii, and J. R. Heath. 2008. Peptide-nanowire hybrid materials for selective sensing of small molecules. *Journal of the American Chemical Society* 130:9583–9589.

Mendoza, F., D. M. Hernández, V. Makarov, E. Febus, B. R. Weiner, and G. Morel. 2014. Room temperature gas sensor based on tin dioxide-carbon nanotubes composite films. *Sensors and Actuators B: Chemical* 190:227–233.

Mitchell, D. T., S. B. Lee, L. Trofin, N. Li, T. K. Nevanen, H. Soderlund, and C. R. Martin. 2002. Smart nanotubes for bioseparations and biocatalysis. *Journal of American Chemical Society* 124:11864–11865.

Monereo, O., S. Claramunt, M. Martınez deMarigorta, M. Boix, R. Leghrib, J. D. Prades, A. Cornet, P. Merino, C. Merino, and A. Cirera. 2013. Flexible sensor based on carbon nanofibers with multifunctional sensing features. *Talanta* 107:239–247.

Morozov, S. V., K. S. Novoselov, M. I. Katsnelson, F. Schedin, D. C. Elias, J. A. Jaszczak, and A. K. Geim. 2008. Giant intrinsic carrier mobilities in graphene and its bilayer. *Physical Review Letters* 100:016602.

Mubeen, S., T. Zhang, B. Yoo, M. A. Deshusses, and N. V. Myung. 2007. Palladium nanoparticles decorated single-walled carbon nanotube hydrogen sensor. *Journal of Physical Chemistry* C 111:6321–6327.

Novoselov, K. S. A., A. K. Geim, S. V. Morozov, D. Jiang, M. I. Katsnelson, I. V. Grigorieva, S. V. Dubonos, and A. A. Firsov. 2005. Two-dimensional gas of massless Dirac fermions in graphene. *Nature* 438:197–200.

Parikh, K., K. Cattanach, R. Rao, D. Suh, A. Wu, and S. K. Manohar. 2006. Flexible vapour sensors using single walled carbon nanotubes. *Sensors and Actuators B: Chemical* 113:55–63.

Park, K. H., M. Chhowalla, Z. Iqbal, and F. Sesti. 2003. Single-walled carbon nanotubes are a new class of ion channel blockers. *Journal of Biological Chemistry* 278:50212–50216.

Peng, N., Q. Zhang, C. L. Chow, O. K. Tan, and N. Marzari. 2009. Sensing mechanisms for carbon nanotube based NH_3 gas detection. *Nano Letters* 9:1626–1630.

Penza, M., P. Aversa, G. Cassano, W. Wlodarski, and K. Kalantar-Zadeh. 2007b. Layered SAW gas sensor with single-walled carbon nanotube-based nanocomposite coating. *Sensors and Actuators B: Chemical* 127:168–178.

Penza, M., G. Cassano, P. Aversa, F. Antolini, A. Cusano, M. Consales, M. Giordano, and L. Nicolais. 2005. Carbon nanotubes-coated multi-transducing sensors for VOCs detection. *Sensors and Actuators B: Chemical* 111–112:171–180.

Penza, M., G. Cassano, R. Rossi, M. Alvisi, A. Rizzo, M. A. Signore, T. Dikonimos, E. Serra, and R. Giorgi. 2007a. Enhancement of sensitivity in gas chemiresistors based on carbon nanotube surface functionalized with noble metal (Au, Pt) nanoclusters. *Applied Physics Letters* 90:173123.

Penza, M., G. Cassano, R. Rossi, A. Rizzo, M. A. Signore, M. Alvisi, N. Lisi, E. Serra, and R. Giorgi. 2007c. Effect of growth catalysts on gas sensitivity in carbon nanotube film based chemiresistive sensors. *Applied Physics Letters* 90:103101–103103.

Penza, M., R. Rossi, M. Alvisi, G. Cassano, and E. Serra. 2009. Functional characterization of carbon nanotube networked films functionalized with tuned loading of Au nanoclusters for gas sensing applications. *Sensors and Actuators B: Chemical* 140:176–184.

Penza, M., R. Rossi, M. Alvisi, G. Cassano, M. A. Signore, E. Serra, and R. Giorgi. 2008. Pt- and Pd-nanoclusters functionalized carbon nanotubes networked films for sub-ppm gas sensors. *Sensors and Actuators B: Chemical* 135:289–297.

Penza, M., R. Rossi, M. Alvisi, M. A. Signore, E. Serra, R. Paolesse, A. D'Amico, and C. Di Natale. 2010. Metalloporphyrins-modified carbon nanotubes networked films-based chemical sensors for enhanced gas sensitivity. *Sensors and Actuators B: Chemical* 144:387–394.

Petit, C., B. Mendoza, and T. J. Bandosz. 2010. Reactive adsorption of ammonia on Cu-based MOF/graphene composites. *Langmuir* 26:15302–15309.

Pilchowski, K., C. Averkiou, and B. Solotuschien. 1992. Adsorption of vinyl-chloride on activated carbons, adsorber polymers and zeolites. 2. Kinetics of adsorption. *Journal für Praktische Chemie: Chemiker-Zeitung* 334:681–684.

Plumacher, J. and I. Renner. 1993. Determination of volatile chlorinated hydrocarbons and trichloroacetic-acid in conifer needles by headspace gas-chromatography. *Fresenius Journal of Analytical Chemistry* 347:129–135.

Ratinac, K. R., W. Yang, S. P. Ringer, and F. Breat. 2010. Toward ubiquitous environmental gas sensors-capitalizing on the promise of graphene. *Environmental Science and Technology* 44:1167–1176.

Robinson, J. A., E. S. Snow, S. C. Badescu, T. L. Reinecke, and K. F. Perkins. 2006. Role of defects in single-walled carbon nanotube chemical sensors. *Nano Letters* 6:1747–1751.

Robinson, J. T., F. K. Perkins, E. S. Snow, Z. Q. Wei, and P. E. Sheehan. 2008. Reduced graphene oxide molecular sensors. *Nano Letters* 8:3137–3140.

Savage, T. W. 2002. Effects of oxygen adsorption and nitrogen doping on the thermoelectric power (TEP) in multi-walled carbon nanotubes. MS thesis, Clemson University, Clemson, SC.

Sayago, I., M. J. Fernández, J. L. Fontech, M. C. Horrillo, C. Vera, I. Obieta, and I. Bustero. 2012. New sensitive layers for surface acoustic wave gas sensors based on polymer and carbon nanotube composites. *Sensors and Actuators B: Chemical* 175:67–72.

Schedin, F., A. K. Geim, S. V. Morozov, E. W. Hill, P. Blake, M. I. Katsnelson, and K. S. Novoselov. 2007. Detection of individual gas molecules adsorbed on graphene. *Nature Materials* 6:652–655.

Schuetzle, D. and R. Hammerle. 1986. *Fundamentals and Applications of Chemical Sensors*. Washington, DC: American Chemical Society.

Sin, M. L. Y., G. C. T. Chow, G. M. K. Wong, W. J. Li, P. H. W. Leong, and K. W. Wong. 2007. Ultra-low-power alcohol vapor sensors using chemically functionalized multi-walled carbon nanotubes. *IEEE Transactions of Nanotechnology* 6:571.

Snow, E. S., F. K. Perkins, E. J. Houser, S. C. Badescu, and T. L. Reinecke. 2005. Single-walled carbon nanotube capacitor. *Science* 307:1942–1945.

Someya, T., J. Small, P. Kim, C. Nuckolls, and J. T. Yardley. 2003. Alcohol vapor sensors based on single-walled carbon nanotube field effect transistors. *Nano Letters* 3:877–881.

Staii, C., M. Chen, A. Gelperin, and A. T. Johnson. 2005. DNA-decorated carbon nanotubes for chemical sensing. *Nano Letters* 5:1774–1778.

Stoller, M. D., S. Park, Y. Zhu, J. An, and R. S. Ruoff. 2008. Graphene-based ultracapacitors. *Nano Letters* 8:3498–3502.

Suehiro, J., S. Hidaka, S. Yamane, and K. Imasaka. 2007. Fabrication of interfaces between carbon nanotubes and catalytic palladium using dielectrophoresis and its application to hydrogen gas sensor. *Sensors and Actuators B: Chemical* 127:505–511.

Suehiro, J., G. Zhou, and M. Hara. 2003. Fabrication of a carbon nanotube-based gas sensor using dielectrophoresis and its application for ammonia detection by impedance spectroscopy. *Journal of Physics D: Applied Physics* 36:L109.

Sumanasekera, G. U., C. K. W. Adu, S. Fang, and P. C. Eklund. 2000. Effect of gas adsorption and collisions on electrical transport in single walled carbon nanotubes. *Physical Review Letters* 85:1096–1099.

Sun, Y., S. Liu, and H. Li. 2012. Synthesis of graphene oxide and its gas sensing properties to NH_3. *Journal of Functional Materials* 43:712–714.

Tans, S. J., A. R. M. Verschueren, and C. Dekker, 1998. Room-temperature transistor based on a single carbon nanotube. *Nature* 393:49–52.

Tasaltin, C. and F. Basarir. 2014. Preparation of flexible VOC sensor based on carbon nanotubes and gold nanoparticles. *Sensors and Actuators B: Chemical* 194:173–179.

Timmer, B., W. Olthuis, and A. van den Berg. 2005. Ammonia sensors and their applications—A review. *Sensors and Actuators B: Chemical* 107:666–677.

Tung, T. T., M. Castro, I. Pillin, T. Y. Kim, K. S. Suh, and J.-F. Feller. 2014. Graphene–Fe_3O_4/PIL-PEDOT for the design of sensitive and stable quantum chemo-resistive VOC sensors. *Carbon* 74:104–112.

Varghese, O. K., P. D. Kichambre, D. Gong, K. G. Ong, E. C. Dickey, and C. A. Grimes. 2001. Gas sensing characteristics of multi-wall carbon nanotubes. *Sensors and Actuators B: Chemical* 81:32–41.

Vedala, H., D. C. Sorescu, G. P. Kotchey, and A. Star. 2011. Chemical sensitivity of graphene edges decorated with metal nanoparticles. *Nano Letters* 11:2342–2347.

Virji, S., J. D. Fowler, C. O. Baker, J. Huang, R. B. Kaner, and B. H. Weiller. 2005. Polyaniline nanofiber composites with metal salts: Chemical sensors for hydrogen sulphide. *Small* 1:624–627.

Wang, B., Y. Wu, X. Wang, Z. Chen, and C. He. 2014. Copper phthalocyanine noncovalent functionalized single-walled carbon nanotube with enhanced NH_3 sensing performance. *Sensors and Actuators B: Chemical* 190:157–164.

Wang, J., G. Liu, M. R. Jan, and Q. Zhu. 2003. Electrochemical detection of DNA hybridization based on carbon-nanotubes loaded with CdS tags. *Electrochemistry Communications* 5:1000–1004.

Wang, J., L. M. Wei, L. Y. Zhang, J. Zhang, H. Wei, C. H. Jiang, and Y. F. Zhang. 2012. Zinc doped nickel oxide dendritic crystals with fast response and self-recovery for ammonia detection at room temperature. *Journal of Material Chemistry* 22:20038–20047.

Wang, Y., Z. Yang, Z. Hou, D. Xu, L. Wei, E. Kong, and Y. Zhang. 2010. Flexible gas sensors with assembled carbon nanotube thin films for DMMP vapor detection. *Sensors and Actuators B: Chemical* 150:708–714.

Wei, C., L. Dai, A. Roy, and T. B. Tolle. 2006. Multifunctional chemical vapor sensors of aligned carbon nanotube and polymer composites. *Journal of American Chemical Society* 128:1412–1413.

Wongchoosuk, C., A. Wisitsoraat, D. Phokharatkul, A. Tuantranont, and T. Kerdcharoen. 2010. Multi-walled carbon nanotube-doped tungsten oxide thin films for hydrogen gas sensing. *Sensors* 10:7705–7715.

Yan, X. B., Z. J. Han, Y. Yang, and B. K. Tay. 2007. NO_2 gas sensing with polyaniline nanofibers synthesized by a facile aqueous/organic interfacial polymerization. *Sensors and Actuators B: Chemical* 123:107–113.

Yashin, Y. I. and A. Y. Yashin. 1999. New potentialities of chromatography in the determination of pollutants in potable water. *Journal of Analytical Chemistry* 54:843–849.

Yavari, F., E. Castillo, H. Gullapalli, P. M. Ajayan, and N. Koratkar. 2012. High sensitivity detection of NO_2 and NH_3 in air using chemical vapor deposition grown graphene. *Applied Physics Letters* 100:203120.

Yi, S., S. Tian, D. Zeng, K. Xu, S. Zhang, and C. Xie. 2013. An In_2O_3 nanowire-like network fabricated on coplanar sensor surface by sacrificial CNTs for enhanced gas sensing performance. *Sensors and Actuators B: Chemical* 185:345–353.

Yoon, O., C. Kim, I. Sohn, and N. Lee. 2013. Toxicity analysis of graphene nanoflakes by cell-based electrochemical sensing using an electrode modified with nanocomposite of graphene and Nafion. *Sensors and Actuators B: Chemical* 188:454–461.

Yu, K. H., P. X. Wang, G. H. Lu, K. H. Chen, Z. Bo, and J. H. Chen. 2011. Patterning vertically oriented graphene sheets for nanodevice applications. *Journal of Physical Chemistry Letters* 2:537–542.

Yu, M., G. Suyambrakasam, R. Wu, and M. Chavali. 2012. Preparation of organic–inorganic (SWCNT/TWEEN–TEOS) nano hybrids and their NO gas sensing properties. *Sensors and Actuators B: Chemical* 161:938–947.

Zhai, J., L. Wang, D. Wang, Y. Lin, D. He, and T. Xie. 2012. UV-illumination room-temperature gas sensing activity of carbon-doped ZnO microspheres. *Sensors and Actuators B: Chemical* 161:292–297.

Zhan, C., Y. Pan, Z. Wang, Y. Wang, H. He, and Z. Hou. 2013. Ionization gas sensing mechanism of a hybrid nanostructure with carbon nanotubes and ZnO nanorods. *Sensors and Actuators B: Chemical* 183:81–86.

Zhang, B., R. W. Fu, M. Q. Zhang, X. M. Dong, P. L. Lan, and J. S. Qiu. 2005. Preparation and characterization of gas-sensitive composites from multi-walled carbon nanotubes/polystyrene. *Sensors and Actuators B: Chemical* 109:323–328.

Zhang, J., A. Boyd, A. Tselev, M. Paranjape, and P. Barbara. 2006a. Mechanism of NO_2 detection in carbon nanotube field effect transistor chemical sensors. *Applied Physics Letters* 88:123112-1–123112-3.

Zhang, L., X. Wang, Y. Zhao, Z. Zhu, and H. Fong. 2012. Electrospun carbon nano-felt surface-attached with Pd nanoparticles for hydrogen sensing application. *Materials Letters* 68:133–136.

Zhang, S. and S. F. Y. Li. 1996. Detection of organic solvent vapours and studies of thermodynamic parameters using quartz crystal microbalance sensors modified with siloxane polymers. *Analyst* 121:1721–1726.

Zhang, T., S. Mubeen, E. Bekyarova, B. Y. Yoo, R. C. Haddon, N. V. Myung, and M. A. Deshusses. 2007. Poly (m-aminobenzene sulfonic acid) functionalized single-walled carbon nanotubes based gas sensor. *Nanotechnology* 18:165504–165509.

Zhang, T., S. Mubeen, and N. V. Myung. 2008. Recent progress in carbon nanotube-based gas sensors. *Nanotechnology* 19:332001–332014.

Zhang, T., M. B. Nix, B. Y. Yoo, M. A. Deshusses, and N. V. Myung. 2006b. Electrochemically functionalized single-walled carbon nanotube gas sensor. *Electroanalysis* 18:1153–1158.

Zhang, Y., Y. Chen, K. Zhou, C. Liu, J. Zeng, H. Zhang, and Y. Peng. 2009. Improving gas sensing properties of graphene by introducing dopants and defects: A first-principles study. *Nanotechnology* 20:185504.

Zhao, J., A. Buldum, J. Han, and J. P. Lu. 2002. Gas molecule adsorption in carbon nanotubes and nanotube bundles. *Nanotechnology* 13:195–200.

Zhi, Y., R. Gao, N. Hu, J. Chai, Y. Cheng, L. Zhang, H. Wei, E. S. W. Kong, and Y. Zhang. 2012. The prospective 2D graphene nanosheets: Preparation, functionalization and applications. *Nano-Micro Letters* 4:1–9.

Zhong, J., J. Guo, B. Gao, J. W. Chioi, J. Li, and W. Chu. 2010. Understanding the scattering mechanism of single-walled carbon nanotube based gas sensors. *Carbon* 48:1970–1976.

Nanocarbon Electron Emitters

Advances and Applications

Matthew T. Cole, Clare Collins, Richard Parmee, Chi Li, and William I. Milne

45

45.1 INTRODUCTION

Electron emission is a ubiquitous technology. It is central to electron beam lithography (EBL), high-frequency amplifiers, and X-ray sources. This chapter is concerned with the synthesis, functionalization, and ultimate application of carbon nanotubes (CNTs), carbon nanofibers (CNFs), and graphene in electron emission applications.

Very few other materials in nature enjoy the allotropic diversity of carbon. The history of the use of graphitic carbon has been intimately related to our ability to isolate its various forms, albeit synthetically derived or naturally sourced. One of the earliest recorded examples of the use of engineered carbon, specifically filamentous carbon, was by the pioneer Edison who successfully integrated pyrolysed carbon filaments into early incandescent light bulbs. Unfortunately, their fame was short-lived. They were soon superseded by the more robust tungsten filaments. Nevertheless, during this period, there were various examples of the use of carbon in a myriad of emerging technologies. Soon thereafter, research into the engineering application of carbon stagnated, and for some years. Any ongoing research progressed at a somewhat reserved pace. The demand for lightweight, stiff, and strong composites by the automobile and aeronautics industries stimulated the field once again in the early 1950s. Throughout history, the development of the various nanocarbons has been largely stimulated by industries' desire to develop new devices requiring materials with unique optoelectronic and mechanical properties. Indeed, it was this that motivated the first seminal investigations into the growth of carbon fibers, specifically through the use of polymer-precursor chemical vapor deposition (CVD), which at this time was already a well-established means of depositing thin film materials. During CVD synthesis, the then well-known carbon fibers were, very occasionally, combined with a small number of submicron fibers. In the mid-1970s, Baker et al. [1–4] conjectured that the underlying growth mechanism of these CNFs was a combination of hydrocarbon precursor pyrolysis on the transition metal catalysts, carbon diffusion, and precipitation. The catalysts were deemed essential to CNF growth. The first fibers produced consisted of graphitic walls aligned in an assortment of directions relative to the CNFs' long axis. Baker et al. were not the first, by any means, to observe such submicron fibers. Some very early and largely unnoticed works go back to 1952 by Radushkevich et al. [5] and Tesner et al. [6]. The synthesis of the first sub-50 nm diameter fibers must really be attributed to M. Endo, A. Oberlin, and T. Koyama [7] who, in 1976, reported hollow tubes, with diameters ranging from 2 to 50 nm formed from concentrically nested graphitic planes aligned parallel to the tubes' long axis. These so-called 1D *filamentous carbons*, now known as CNTs, unbeknownst to them at the time, were to be an incredibly rich source of fascinating science with largely unprecedented technological relevance over the next three decades. Unfortunately, the lack of repeatable and macroscale synthesis processes inhibited further progress for some time, and the nanotubes and the other 2D (graphene) and 0D (fullerenes) graphitic carbon allotropes did not return to prominence for another three decades.

At the start of the 1990s, the field of nanotechnology was enlivened by the independently published papers of Iijima et al. [8] and Bethune et al. [9], who, like many at the time, were interested in diamond-like carbon thin films. The discovery of the fullerenes (self-terminated spheres formed from 20–100 trigonally bonded carbon atoms—the most famous of which being the C_{60} Buckminster fullerene, a 60-atom truncated isocahedron 7.1 Å in diameter), 6 years earlier by Kroto et al. [10], set the scene for the nanotubes and later, graphene. The field of carbon nanomaterials was increasing in size, though it nevertheless remained incomplete. The 0D fullerenes had been isolated and even grown, as had the 1D nanotubes. Graphite accounted for the 3D form and was readily available from nature. However, no 2D allotrope had yet been isolated despite the fact that its properties had been understood and widely employed in theoretical studies for more than 70 years. Isolation of the graphitic 2D crystals, termed graphene, had been deemed a natural uncertainty by the seminal theorists Landau and Peierls (1934) and later, Mermin (1965). They argued that it would be impossible to isolate truly 2D single-atomic-layer crystals due to their intrinsic lattice instabilities [11–13]. For some time, as a result, graphene remained a purely academic tool [14]. A material of no real practical consequence, it would be more than a decade until graphene would come to the fore as a useful, rather than a purely hypothetical, engineering material, at which time the range of potential applications seemed almost limitless. During the intervening years, nanotube research was emerging at an ever-increasing rate with industrial applications seemingly increasingly viable.

The emergence of graphene has been somewhat more turbulent than that of the CNTs and the fullerenes. Graphene was not successfully isolated until 2004. As a result, mechanical exfoliation has been widely adopted as this was the approach used in this seminal work [16,17]. Contrary to popular belief, this was not the first demonstration of graphene isolation, merely the most publicized as with the work of Iijima on the CNTs earlier. Initial experiments in the 1950s [18–23] on graphite chemical exfoliation using intercalates—compounds developed to swell and separate graphite crystals—had already isolated graphene, though only in small quantities. Since the work of Geim and Novoselov in 2004 [16], various approaches to graphene synthesis have been developed, with CVD being one of the most promising.

45.2 NANOTUBE AND GRAPHENE SYNTHESIS

The various nanostructures reported to date can be synthesized by either top-down (the successive removal of material) or bottom-up (the self-assembly of single atomic or molecular units) processes. What follows summarizes these two processes.

45.2.1 Top-Down

Graphene has been isolated by many researchers using mechanical exfoliation or cleavage from large graphite crystals (Figure 45.1a). Exfoliation approaches are essentially the repeated removal of graphene planes from larger graphite crystals, continued, ad infinitum, until a single atomic layer remains. Such approaches have proven extremely successful in producing very high-quality flakes, though only over small areas, typically < 10 μm in diameter [16]. Liquid phase chemical exfoliation [24] (Figure 45.1b) offers a commercial means of fabricating disordered graphene thin films using established coating techniques such as Langmuir–Blodgett [25] and drop casting [26]. These films are necessarily microscopically disordered and are formed from graphene flakes coated with deleterious surfactants that smear out the graphene's desirable electronic properties. The aggressive acid treatments and high-power ultrasonication limit the usefulness of these wet chemistry approaches. The graphene flakes are typically rather small (<2 μm). Nonetheless, the approach is simple and, perhaps most importantly, low cost. Such approaches are well suited for integration into flexible transparent conductors, though they are less appealing for active and vacuum electronics. Unlike graphene, CNTs have not yet been synthesized by top-down techniques.

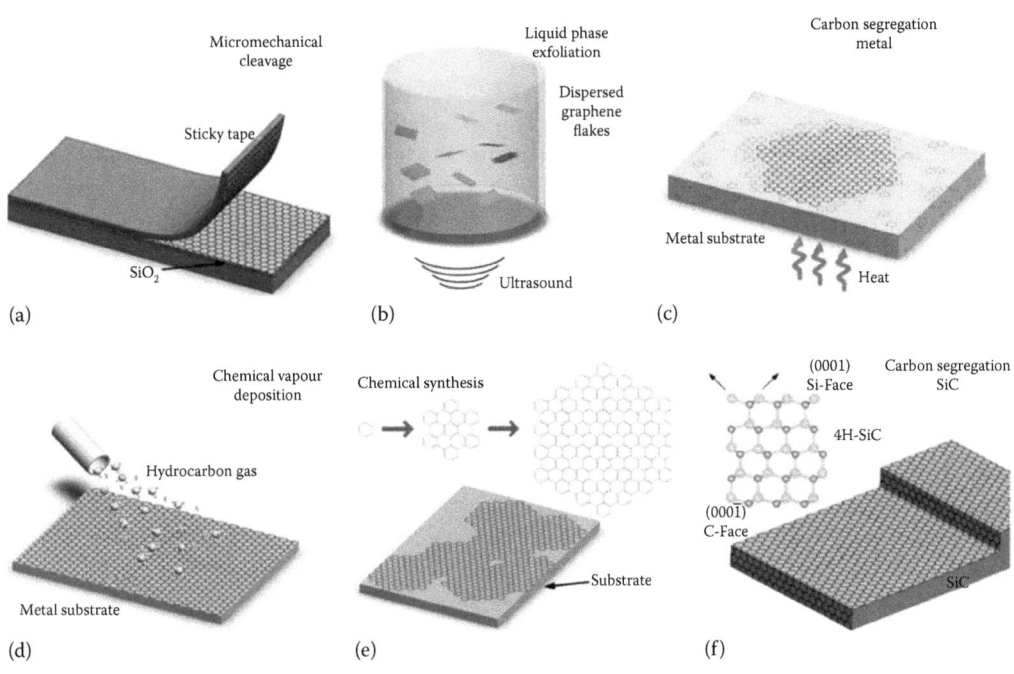

FIGURE 45.1
Common graphene isolation techniques, including (a) scotch-tape/mechanical exfoliation, (b) liquid phase exfoliation, (c) carbon segregation on a metal substrate, (d) CVD, (e) chemical derivation from benzene units, and (f) epitaxial growth on SiC. (Reprinted from *Mater. Today*, 15, Bonaccorso, F., Lombardo, A., Hasan, T., Sun, Z., Colombo, L., and Ferrari, A.C., 12, Copyright 2012, with permission from Elsevier.)

45.2.2 Bottom-Up

In 1970, Huffman et al. [27] reported the first bottom-up synthesis of nanographite. The team's *graphite smoke* was derived from a rudimentary arc discharge system using two high-voltage graphite electrodes. The approach was further developed by Krätschmer et al. [28] who increased the process yields and, in doing so, developed the first means to grow viable amounts of Buckminster fullerenes, C_{60}. It would be more than 20 years until the use of such process were engaged to produce CNTs and graphene. Nonetheless, an important step was made toward a generalized catalytic process: heating a carbon source under an appropriate atmosphere and in the presence of an appropriate catalyst could result in the formation of crystallographically ordered graphitic nanostructures.

The development of graphene growth procedures have, rather unexpectedly, lagged that of nanotubes, despite graphene's central role in our understanding of the growth and exploitation of CNTs and fullerenes. Nevertheless, since 2006, progress has been rapid. Alongside CVD, epitaxial growth on silicon carbide (SiC) is one promising approach to achieving high-quality, large area graphene (Figure 45.1f). Pioneered by De Heer et al. [29] in the mid-1990s, costly, chemically–mechanically polished SiC substrates are heated under ultrahigh vacuum conditions ($<10^{-10}$ mbar) to temperatures in excess of 1300°C. This stimulates Si sublimation and the remaining carbon reforms leaving a surface-bound graphene layer. Technologically, this approach is significant as it produces RF-grade material [30], in part due to the fact that SiC is an insulator and has the ability to deposit graphene directly in a dry process onto dielectrics, opening up a number of important electronics applications. However, the required substrates are expensive and require lengthy thermal processing. CVD (Figure 45.1d), on the other hand, is significantly faster and comparatively inexpensive. Graphene can also be synthesized, with varying yields and degrees of crystallinity, by flame synthesis [31], nanotube plasma unzipping [32,33], carbon segregation on carbon-implanted metal supports, and organic self-assembly of benzene derivatives (Figure 45.1e). However, at present, such approaches remain confined to the laboratory benchtop and are only applicable over small scales (< a few cm) using systems that are at present uncontrollable. CVD, a well-established and industrially familiar process, is one of the few truly viable techniques to mediate widespread adoption of nanocarbon technologies due to its potential for high-quality, large area (>300 mm diameter wafers) synthesis.

Nanotubes and graphene can be grown by laser ablation [34–36], electric arc discharge [37–43], and catalytic CVD [44–49]. Arc discharge and laser ablation are classified as high-temperature (>3000°C) processes with reaction timescales of the order of a few μs to ms. CVD techniques require temperatures of the order of 300°C –1200°C over extended time frames, typically minutes to hours. Many other less common bottom-up synthesis approaches exist, including, but not limited to, ball milling [50] and flame synthesis [51]. These, and other more esoteric synthesis routes, lack scalability and growth selectivity and as such have failed to achieve comparable academic traction and commercial interest as laser ablation, arc discharge, and CVD. Here, we provide a general overview of these more common synthesis techniques.

A typical laser ablation, sometimes termed laser evaporation or vaporization, reactor is illustrated in Figure 45.2a. In 1995, Smalley et al. [52] were one of the first to develop this approach using an inert atmosphere with a continuous wave, high-power laser incident on a solid (or occasionally a liquid) C-, Ni-, Co-, and Fe-containing target. The target composite vaporizes as it is heated rapidly to temperatures of >1000°C. The ablated material sublimes, forming plasma, which is attracted to a cooler counter electrode, where it precipitates to form

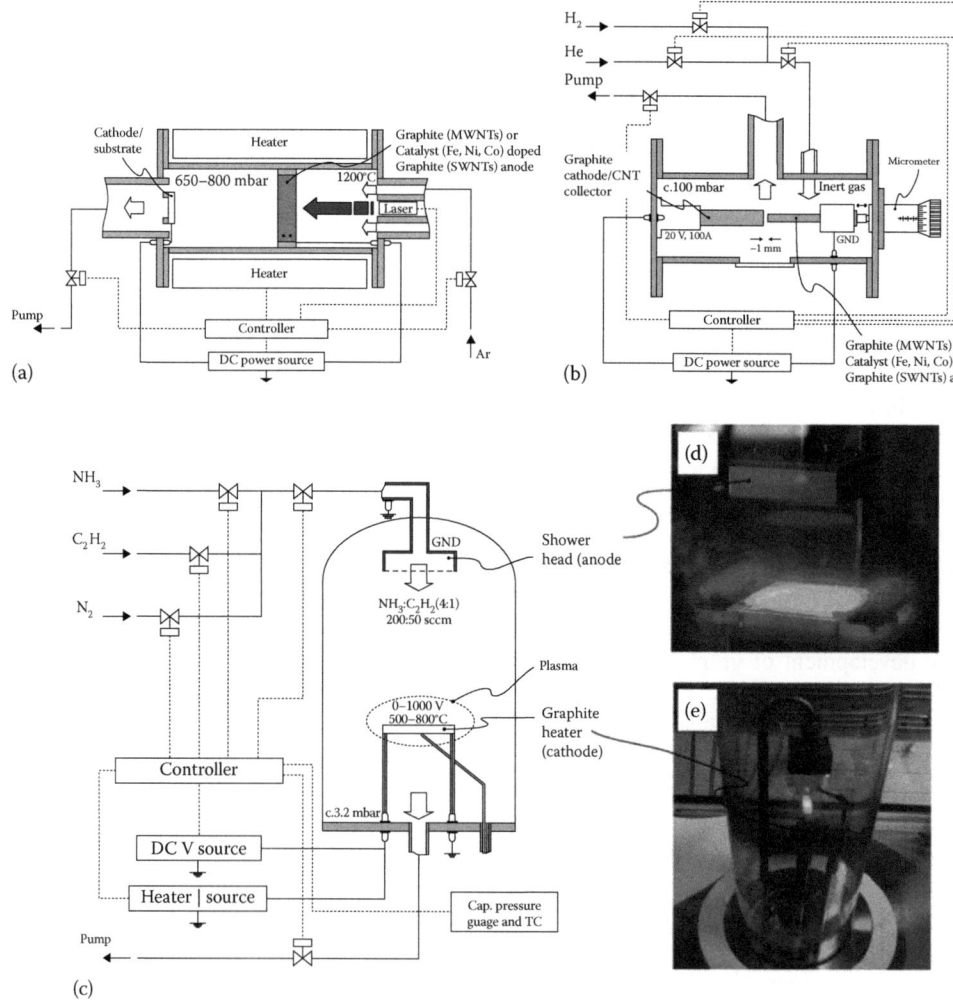

FIGURE 45.2

Common CNT and graphene growth techniques. Schematic depictions of a (a) laser ablation and (b) vacuum arc system. (c) Schematic depiction of a PE-CVD system. (d, e) Photographs of a PE-CVD system during operation. (Adapted from Cole, M.T., *Dry-Transfer of Chemical Vapour Deposited Nanocarbon Thins Films for Functionally Enhanced Flexible Transparent Electronics Applications*, University of Cambridge, Cambridge, U.K., 2011.)

various nanocarbons including fullerenes [53], nanotubes [42], and (only recently) graphene [36]. Though the synthesized nanocarbons are highly graphitic with few lattice defects, such growth is accompanied by a wide range of undesirable carbonaceous species, principal among which is *a*-C; to be of any technological use, this must latterly be removed, which is often achieved by deleterious chemical purification.

Laser ablation and arc discharge (Figure 45.2b) are similar insofar as they are transient sublimation processes. In an arc discharge system, a high potential, typically of the order of a few thousand volts, is applied between two closely spaced graphite electrodes in a vacuum, controlled gaseous atmosphere (such as He, H_2, and N_2 [39,40,42]) or aqueous environment (such as liquid nitrogen or water [41,54]). By controlling the potential and electrode separation, a breakdown event is stimulated resulting in a current of 50–200 A, which, within less than 1s, heats the anode to >1000°C. The electrodes sublime as in the laser ablation case, and the free carbon liberated from the anode precipitates on the cooler anode, ultimately condensing to form nanotubes [42], fullerenes [55], or graphene [56]. As with laser ablation, either pure graphite or catalyst-doped graphite can be used as the electrodes, allowing for approximate control over the predominant material synthesized. In 1993, Iijima et al. [8] and Bethune et al. [9] were some of the first to demonstrate CNT deposition by arc discharge. In an arc discharge system, the extremely high temperatures result in high-quality nanostructure growth, though the resultant soot often has low nanomaterial yields. In the case of a typical nanotube growth, 70–80 wt% of the material is typically unwanted, and after purification, only an extremely low (<1 wt%) yield of CNTs remains [57].

CVD fundamentally differs from the laser ablation and arc discharge. CVD is highly controllable. It allows for high yields and growth orientation, selectivity in the material type, and ready control over the material quality. A typical CVD reactor is illustrated in Figure 45.2c. CVD systems can be hot or cold walled. Hot-walled systems are heated globally; the sample and chamber walls (typically quartz) are ohmically heated. Such systems are inexpensive and are often based on low-cost and readily available tube furnaces. In contrast, cold-walled systems have smaller thermally isolated heaters residing within a larger chamber. Here, the chamber walls remain at significantly lower temperatures, near room temperature in most cases. To date, high-quality graphene [58], nanotubes [59], and nanofibers [60] have all been successfully synthesized via CVD. During CVD, a carbon-containing feedstock gas is thermally and/or plasma decomposed. The carbon-containing gas (such as acetylene, C_2H_2 [61]; methane, CH_4 [62]; ethylene, C_2H_4 [63]; and carbon monoxide, CO [64]) is mixed with a carrier gas, such as NH_3 and H_2, which decomposes to an etching species to inhibit *a*-C deposition. CVD has been used to much success. An extremely wide range of partial pressures (10^{-5}–10^2 mbar), temperatures (270–1200°C), and gas recipes have been used. In contrast to the CVD of CNTs and CNFs, very little carbon is necessary to grow graphene. Extremely low partial pressures, ~10^{-2} mbar [65], employing CH_4 and C_2H_4 are commonly used for graphene synthesis. Such feedstocks are kinetically stable and undergo minimal thermal decomposition at high temperatures (900°C). Despite the wide range of growth techniques available, CVD has gained perhaps the greatest traction due to its large area compatibility, simple automation, industrial familiarity, high controllability, and comparative cheapness.

Accurate control over nanotube alignment and growth temperature reduction can be facilitated by plasma-enhanced CVD (PE-CVD). Here, an applied electric field controls the growth direction and assists in the catalytic kinetic processes. Various plasma types have been studied, including direct current (dc) [67–69], hot filament [70–72] aided dc, RF [73] and magnetically enhanced RF [74], microwave [75–77], and inductively coupled [78,79] approaches. In all cases, the plasma plays a number of important roles: it heats the substrate, typically to a few hundred degrees [80], augments the gaseous atmosphere, modifies the catalyst layer [81], assists in nanotube alignment, and etches *a*-C [82]. To date, PE-CVD is widely adopted for nanotube growth, though recent efforts are ongoing to apply the approach to graphene growth [83,84]. Notwithstanding, the aggressive plasma makes the formation of highly crystalline graphene somewhat challenging.

At the time of publication, the lowest-reported nanofiber growth temperature was 270°C [85] and 300°C for graphene-like films, both of which are highly defective [86]. The chief merit of CVD approaches is their intrinsic controllability. With continued maturity, CVD is widely believed to be the most promising technique for compatible silicon-processing temperatures (<300°C), and a clear understanding of the catalysis and growth kinetics therein will be central in achieving this.

45.3 NANOFIBER, NANOTUBE, AND GRAPHENE CVD

Figure 45.3 illustrates the key stages in nanotube growth. In the case of CNTs, the catalyst, typically deposited through physical vapor deposition (PVD) means such as electron beam evaporation, thermal evaporation, or sputtering, is thermally annealed [60] and/or plasma treated [85] resulting in the formation of nanoparticles or nanoislands, which template the CNT/CNF growth [87–89]. There have been many catalysts considered to date, each with varying yields and areal densities. These include [90] the highly catalytically active transition metals (Fe, Ni, Co), the noble metals (Au [91], Ag [91], Cu [91]), and poor metals (Pb [92], In [93]), as well as semiconductors (Ga [93], Ge [94]), oxides (SiO$_2$ [95,96], ZrO$_2$ [97]), and carbides (SiC [98]). Various metal/metal oxides including MgO [92], Al$_2$O$_3$ [99], and TiO$_2$ [92] nanoparticles have also shown some potential. Binary and tertiary catalyst structures, such as Al$_2$O$_3$/Fe [100], Mo/Co [101,102], and Mo/Al$_2$O$_3$/Fe [103], increase the yields and growth rates while reducing diameter and chiral variations. Various iron-based alloys have also been demonstrated including 304 stainless steel [104], Kanthal [105], Iconel [106], and Nichrome/Chromel [107,108]. In the case of graphene, a wide variety of metallic thin films and foils have been considered, including Cu, Ni, Co, Pt, Ru, and Ir [109–111], as well as numerous alloys [58,112] and compounds like AlN [113] and ZnS [114]. More recently, oxides [115,116] and dielectrics including quartz and sapphire, nitrides [112], and semiconductors [117] have also been shown to be catalytically active, though to a much lesser extent than the widely adopted Ni and Cu alternatives. Researchers continue to experiment with ever more exotic and varied catalyst materials. Nevertheless, there is not, as yet, any clear mechanism as to why oxides catalyze nanocarbon growth or a clear unified theory of nanocarbon growth by CVD across the various allotropes. Evidently, many materials catalyze nanocarbons with varied degrees of activity and material quality, and for the most part, Fe and Ni are frequently employed for nanotube growth and Cu and Ni for graphene growth.

In the case of nanotube and nanofiber growth, catalyst restructuring is of paramount importance. The restructured catalyst films template the nanotube growth and offer a possible means of controlling the nanotube diameter and chirality, which ultimately dictate the CNTs' electronic characteristics. Various approaches have been considered to accurately and reproducibly define nanoislands, including nanoparticle wet deposition, [118], plasma etching, or thermal coalescence, via Oswald ripening, of PVD thin films. Soluble metal salts (such as acetates and nitrates [119]), bicarbonates, biologically derived catalysts (ferritin), organometallic and metallocene compounds

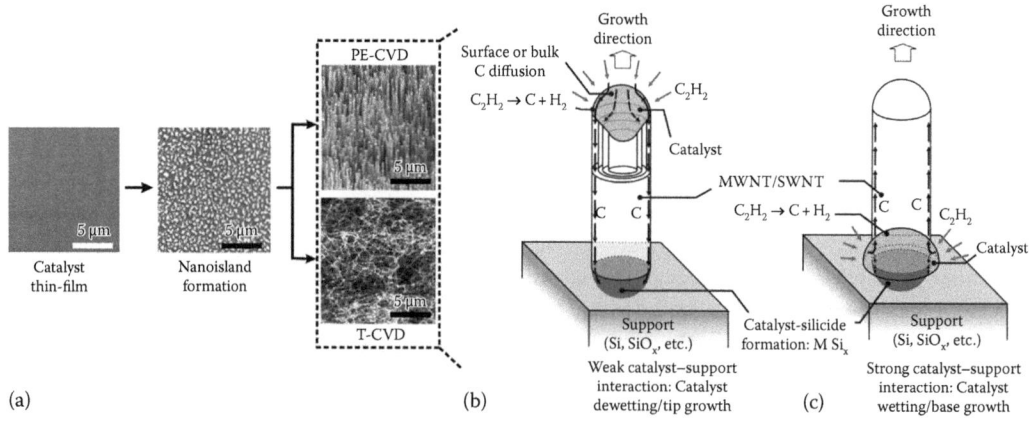

(a)   (b)   (c)

FIGURE 45.3

(a) Central processes to CVD CNT including thin film catalyst deposition, catalyst bed restructuring/nanoisland formation, and typical examples of CNT thin film morphology following PE-CVD and T-CVD. Schemes for (b) tip and (c) route growth. (Adapted from Cole, M.T., *Dry-Transfer of Chemical Vapour Deposited Nanocarbon Thins Films for Functionally Enhanced Flexible Transparent Electronics Applications*, University of Cambridge, Cambridge, U.K., 2011.)

(ferrocene) [89], and metallic colloids are all examples of wet catalyst types that can be deposited by dipping, pipetting, spray coating, casting, electrochemical deposition, ink-jet printing, or microcontact printing. Though offering tight control over the catalyst diameter, in most cases, calcination at 200–500°C is necessary to fully oxidize these catalysts that only become activated following their complete thermal reduction. Such processes are dramatically increasing the process complexity of the subsequent CVD.

Thermal coalescence of sub-10 nm thick PVD thin films is perhaps the most common approach to catalyst nanoisland formation to date. It necessitates an understanding of the catalyst restructuring and free surface energy minimization. During thermal pretreatment, the catalyst dewets and clusters due to surface stresses, as depicted in Figure 45.3a. The exact state of the catalyst is still hotly debated; in situ high-resolution electron microscopy has evidenced a nonliquid state [120], though others have evidence growth even during full catalyst liquefaction [121]. Nonetheless, nanotubes and graphene have been successfully synthesized, at high graphitic quality, at temperatures far below both the catalysts melting and carbide eutectic temperatures. Regardless of the exact state, the catalyst is certainly mobile upon nanoisland formation and, in the case of graphene growth, grain restructuring and growth. For CNTs, the islands agglomerate and form dynamic morphologies, which template the nanocarbon growth. It is this degree of mobility that dictates the exact growth mode adopted—tip or route (as depicted in Figure 45.3b and c). It is also the strength of the catalyst–support interaction that dictates the level of catalyst mobility and the growth mode, which ultimately dominates. For example, Ni on SiO_2 [122] is characterized by a small wetting contact angle giving a correspondingly weak catalyst interaction; this favors tip growth where the catalyst resides at the tip of the nanotube/fiber. Contrastingly, a hydrophilic interface, such as Fe on Si [123], gives a large wetting contact angle and thusly favors root growth; the catalyst particle remains at the substrate interface during carbon extrusion. Different growth modes favor different applications.

A vast number of temperature-dependent interactions occur at the substrate–catalyst interface; catalysts can be consumed due to uncontrolled diffusion, while silicide and carbide alloying inhibit growth, both of which significantly reduce yields. Various diffusion barriers such as ITO, SiO_2, Al_2O_x, and TiN were subsequently developed [124,125]. Less obvious catalyst restructuring occurs in the case of graphene CVD. For common Ni- and Cu-based graphene growth, grains typically grow during annealing, with anneal times often approaching hundreds of minutes in duration. These films also tend to retain their planar form as they are comparatively thick (>100 nm).

To accurately control graphene and nanotube/fiber growth by CVD, insight into the underlying catalysis is necessary. No general theory of graphitic carbon catalysis, which spans the various nanoscale graphitic allotropes, has been put forth to date, and the validity of the generalized vapor–liquid–solid [126–128] and vapor–solid–solid [129] models, which are typically applied to more conventional systems such as Si nanowires, is questionable in the case of the nanocarbons. Hydrocarbon dissociation and subsequent adsorption of atomic carbon on and into the catalytic surfaces are central to heterogeneous catalysis [130]. Much work is ongoing at present to investigate such processes and how they dictate graphene growth. Indeed, this may well result in a generalized catalytic model correlating nanotube, nanofiber, and graphene catalysis. Certainly, such a model is absent from the literature, though many various and often contradicting growth models have been reported [46,131–133]. Nevertheless, it is clear that a great many energy barriers comprise the CVD process, as illustrated in Figure 45.4a and b. Figure 45.4c shows a general scheme, consistent for all nanocarbon growth, comprising of carbon precursor adsorption, hydrocarbon dissociation, and atomic carbon diffusion on and throughout the catalyst [131,133–135]. Figure 45.4d plots the variation in measured growth rate, R, as a function of growth temperature for CNFs, CNTs, and graphene. Through an Arrhenius-like relation, the gradient defines the activation energy of the growth-limiting catalytic process. We note that the activation energy for graphene is almost always >1.0 eV, while for CNTs and CNFs, it is typically ≤1.0 eV, suggesting graphene is likely carbon adatom lattice integration limited, while CNFs and CNTs are surface diffusion limited [136]. Detailed studies of the underlying catalysis are ongoing [131,133,135]. An improved understanding of the catalysis will likely lead to reductions in synthesis temperatures, possibly opening up the potential for direct growth of high-quality, low-defect-density nanocarbons on more exotic dielectric substrates, perhaps possibly even facilitating direct growth on CMOS.

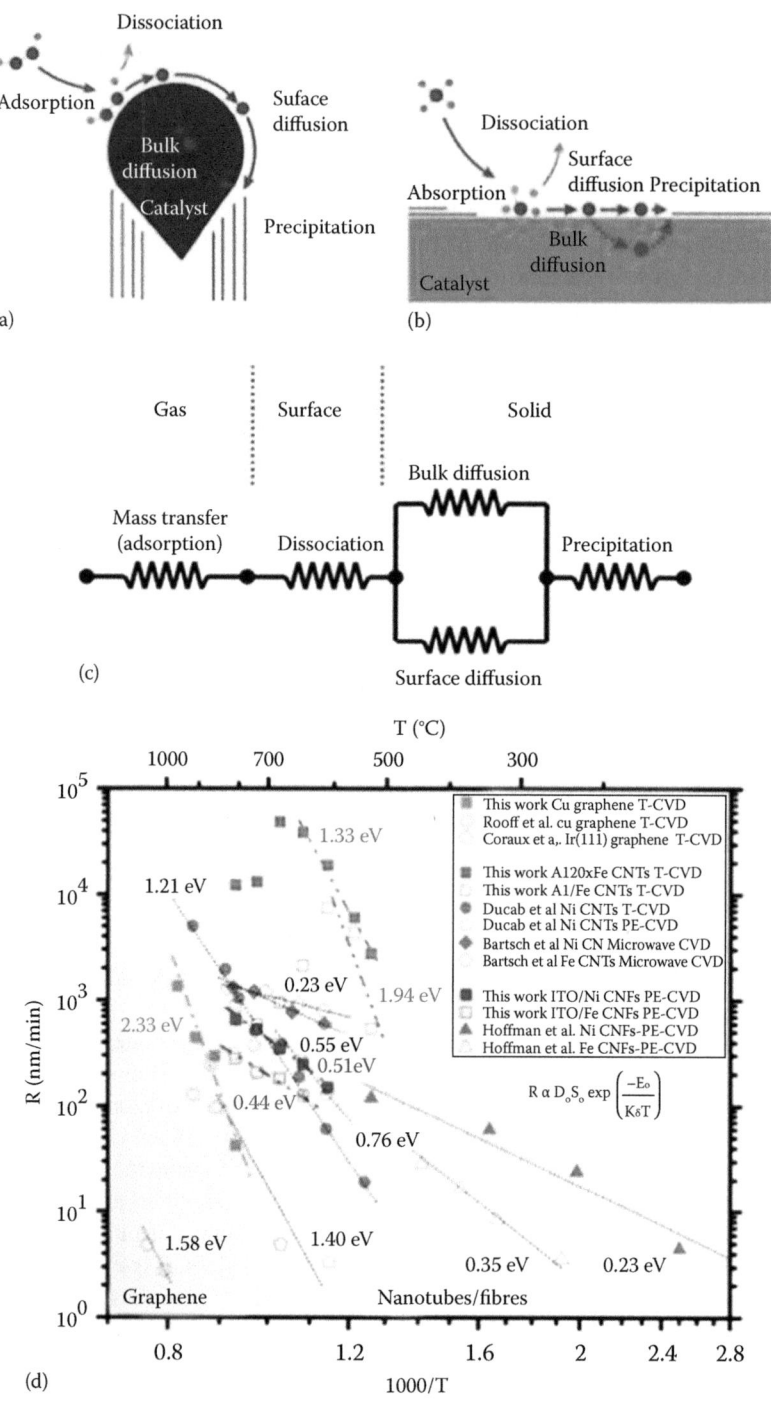

FIGURE 45.4

Catalytic pathways for (a) CNTs/nanofibers and (b) graphene. (c) Conceptual pathways relating to the gas, bulk, and sold activation energies occurring during generalized nanocarbon CVD catalysis. (d) Measured activation energies of various CNTs, CNFs, and graphene. (Adapted from Cole, M.T., *Dry-Transfer of Chemical Vapour Deposited Nanocarbon Thins Films for Functionally Enhanced Flexible Transparent Electronics Applications*, University of Cambridge, Cambridge, U.K., 2011.)

45.4 NANOCARBONS AND FIELD EMISSION

Field emission (FE) is a quantum mechanical process where, as illustrated in Figure 45.5a, electrons tunnel from an electron dense medium through a narrowed surface potential barrier and emit into a vacuum during the application of a high electric field. FE, often termed *cold cathode emission*, is a room temperature process and is significantly more efficient than the widely adopted thermionic emission source—now an ubiquity in electron emission applications including electron microscopy, microwave amplifiers, and X-ray sources.

FE is morphologically enhanced, as described in further detail elsewhere [137], if the emitting material is whisker-like and has a high aspect ratio (Figure 45.5b). As a result, CNTs have had a rich history in FE applications [47,48,138], while graphene is more recently coming to the fore due to its novel electronic character [49,139]. The nanocarbons are near ideal cold cathode emitters; they have a virtually instantaneous time response allowing for extremely high direct electron beam modulation; they are chemically inert, are highly electrically and thermally conductive, and have potentially extremely high aspect ratios, which allow reduction in the driving voltages. Graphene shares many of these properties, though its use in FE applications is, at the time of publication, still in its infancy. The field is gaining traction rapidly, however.

Figure 45.6a summarizes the various FE materials used to date. Figure 45.6a considers an intentionally diverse variety of emitter geometries and morphologies to assess the dependency on emitter work function (Φ) and dimensionality relative to emitter morphology. The respective emission spectra have been normalized to their maximum current density for comparison purposes. E_{on} is defined as the turn-on electric field necessary to derive 10% of the maximum normalized current density, where this $J' = J/J_{max}$, while the threshold field, E_{thr}, is defined at 0.3 on this scale. J_{max} defines the maximum current density and $0 \leq J' \leq 1$. The materials considered include various metals, semimetals, and semiconductors, in addition to 1D, 2D, and bulk emitters. CNTs, graphene, and the graphitic nanocarbons are evidently competitive; they show extremely low-mean E_{on} and E_{thr} (of the order of $5.1_{graphene}/3.7_{CNT}$ V/μm and $6.0_{graphene}/4.3_{CNT}$ V/μm, respectively) as well as high-mean J_{max}, of the order of $26.7_{graphene}/8.8_{CNT}$ mA/cm². Indeed, only a few materials outperform the nanocarbons in this respect, and of those that do, very few such materials can be grown, aligned en masse, and engineered at the submicron scale as is capable with the nanocarbons, and CNTs in particular. Indeed, CNTs couple ultraprecision engineering with highly parallelized fabrication allowing for highly functional electron emitters to be realized.

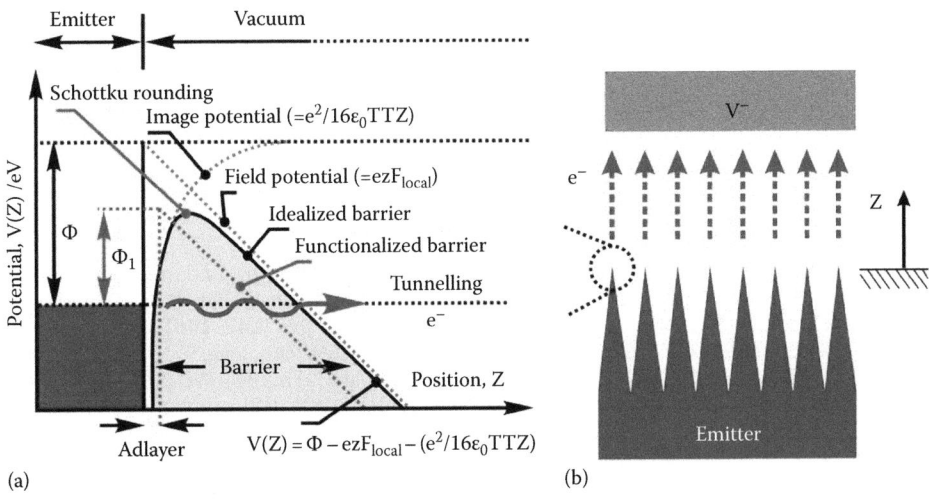

(a)   (b)

FIGURE 45.5
(a) Band diagram of a typical FE electron source. The high aspect, whisker-like morphology of the emitter (gray) enhances the local electric field narrowing the field potential to form a barrier a few nm across through which electrons can efficiently tunnel into the ultrahigh vacuum cavity. (b) Depiction of a whisker-like tip field emitter capable of enhanced FE due to its high aspect ratio; the higher the aspect ratio, the lower the necessary drive voltages.

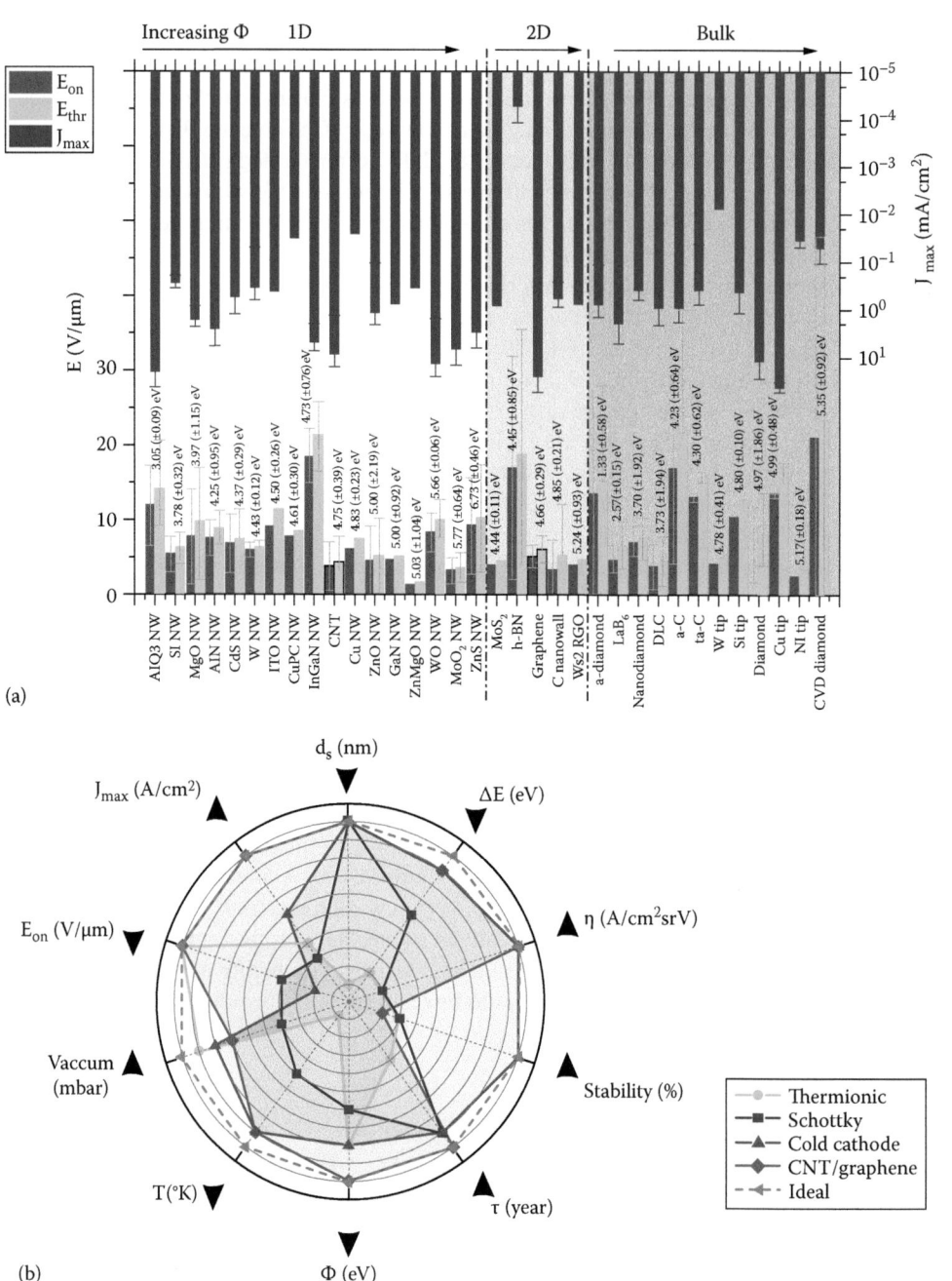

FIGURE 45.6

(a) Overview of the on and threshold electric fields (E_{on} and E_{thr}, respectively) and maximum current density, J_{max}, for various materials used for FE to date, in order of dimensionality (1D, 2D, and bulk) and increasing work function (Φ), including 1D nanowires, AlQ$_3$ [140,141], Si [142–144], MgO [145,146], AlN [147–150], CdS [151–154], W [155–157], ITO [158], CuPC [159,160], InGaN [161–163], CNTs [164–168], Cu [169–171], ZnO [172–177], GaN [178,179], ZnMgO [175,180], WO [181–184], MoO$_2$ [185–187], and ZnS [142,188]; 2D platelets, MoS$_2$ [189], h-BN [190–193], graphene [194–202], C nanowall [203–205], and WS$_2$ RGO [206]; and bulk materials, a-diamond [176,207], LaB$_6$ [208–211], nanodiamond [212,213], DLC [214,215], a-C [216–218], ta-C [219–221], W [222], Si [223–225], diamond [226–230], Cu [170,171,231], Ni [232–235], and CVD diamond [217,236–238]. (b) Polar plot for the various commonplace electron emitters. CNT- and graphene-based field emitters outperform such sources across most metrics, where J_{max} is the current density, E_{on} is the turn-on electric field, *vacuum* denotes the operating vacuum, T is the typical operating temperature, Φ is the emitter work function, τ is the lifetime, *stability* is the temporal stability, η is the electron-optical brightness, ΔE is the energy spread of the emitted electrons, and d_s is the virtual source size. (Adapted from Cole, M.T., *Dry-Transfer of Chemical Vapour Deposited Nanocarbon Thins Films for Functionally Enhanced Flexible Transparent Electronics Applications*, University of Cambridge, Cambridge, U.K., 2011.)

A critical point, often overlooked, is highlighted in Figure 45.6a. Of the wide range of materials considered, there is only a modest enhancement in the emission performance (increased J_{max}, reduced E_{on}, reduced E_{thr}) with decreasing emitter work function, as one would perhaps expect from band arguments. The morphology of the emitter smears out this anticipated trend to a great extent. The emitter morphology and nanoengineered geometry (viz., the degree of alignment and patterning) evidently play a more central role in the emission performance compared to the electronic character of any given emitting material, in all cases. Indeed, the ability to pattern and shape the emitter at the nanoscale is perhaps the most significant benefit of using CNTs over other high-aspect-ratio nanowires, 2D crystals, and nanoplatelets, which are often fabricated through wet chemistry approaches. It is the coupling of CVD techniques with high-resolution lithographic processes that provides the greatest technological benefit and subsequent commercial advantage. Cumulatively, this allows for accurate control of the positioning, alignment, and length of the CNTs, facilitating synthesis of individual high aspect emitters with well-controlled vertically aligned morphologies in an entirely dry process. No single wet chemistry route allows for such fine control over the emitter geometry. Indeed, for comparison purposes, this can be readily quantified across various known CNT deposition techniques.

Wet chemistry techniques offer equivalent control over the CNT length and aspect ratio as they often employ CVD-grown CNTs. As such, we neglect to use CNT length as a metric. However, considering the minimum spacing, or pitch, between CNTs, in the case of a patterned single-walled CNT, of diameter 1 nm, we define the parameter $\alpha = (-1/10)*$ exponent of the minimum intentionally defined feature size. In this instance, we get $\alpha = 0.9$. Now, in the case of screen-printed CNT films coupled to microlithography, α will typically be of the order of 0.6. Clearly, screen-printed CNT thin films, where the CNTs lie at uncontrolled angles with many lying adjacent to the substrate, could potentially be patterned with high-resolution optical or EBL, just as in the CVD case; however, this precludes CNT alignment, which is known to be central to efficient electron sources. Thus, let us now consider the mean variation in individual CNT alignment relative to the perpendicular to the emitter substrate, in a metric termed β. Here, we define $\beta = 0$ for variations in alignment of $90°$, such that there is a wide range of angles within the thin film, with some CNTs lying on the emitter surface and with others vertically orientated at $90°$ to the emitter surface. Let us also define $\beta = 1$ such that there is only a minor variation in alignment of $0°$ from the perpendicular; that is, all CNTs are aligned perpendicular to the emitter surface. Thus, for CVD processes, the maximum attainable $\beta = 0.9$ (vertical CNTs with a mean and conservative angular variation of $<10°$), while $\beta \sim 0.1$ for wet chemistry processed CNT films— indeed, this is also true for many graphene and wet chemistry–deposited nanowire films and exemplifies the underlying reason why such films require mechanical activation with adhesive tapping. Thus, if we consider the simple binary metric $\alpha.\beta$, we find that CVD processes give 0.81, while for wet chemistry processed, we find 0.06. Notably, in terms of the ability to engineer and control the emitter morphology accurately, and hence engineer the FE properties, CVD processes offer much finer control than is attainable elsewhere. The geometry of the electron source dominates the emission performance, compared to the electronic characteristics, and it is this ability to accurately engineer the emitter morphology that is proving pivotal in the technological development on nanoengineered electron sources [47].

Figure 45.6b compares leading electron emission technologies as a function of the maximum current density (J_{max}), the turn-on electric field (E_{on}), the viable operating vacuum (*vacuum*), the typical operating temperature (T), the emitter work function (Φ), the typical emitter lifetime (τ), the temporal stability (*stability*), the electron-optical brightness (η), the energy spread of the emitted electrons (ΔE), and the virtual source size (d_s). The dashed curve depicts the ideal emitter. We note that the nanocarbon FE sources dramatically outperform thermionic, conventional metallic cold cathode, and Schottky emitters across almost all standardized metrics. Notwithstanding, the ongoing primary barrier to commercial adoption of nanocarbon-based electron sources lies in cost reduction, in improved interface adhesion, and in achieving improved robustness toward compromised vacuum conditions, which characteristically manifests as a degraded temporal stability. Nevertheless, recent advancements in nanocarbon functionalization offer one promising solution. Such adlayers decrease the required drive voltages that decrease the likelihood of local microplasma formation while simultaneously reducing energetic ion bombardment and thermal effects, both of which manifest as a distinct enhancement in temporal and vacuum stability. Adlayers also hermetically seal the emitters enhancing their resilience toward otherwise damaging oxidizing environments.

45.5 ENHANCED FIELD EMISSION VIA FUNCTIONALIZATION

As illustrated in Figure 45.5a, for a fixed geometry, functionalizing the emitter surface with a low-work-function material narrows the effective tunnel barrier, which lowers the turn-on electric field and increases the emission current at a given field. The use of adlayers promises a viable means of coupling the beneficial high aspect ratio and potential to nanoengineered CNTs and graphene with the desirable low work function of nominally planar materials. Much work has been conducted in this way to augment the electron emission performance of CNTs via simple adlayer addition, though less so for graphene emitters. Throughout the literature, the FE characteristics of CNT-based emitters have been measured with coatings of Fe [239], Ni [240], Pt [241], Pd [241], Ru [241,242], Ag [243], Cu [244], Co [245], Ti [246], Ta [247], Ag–Cu [248], Hf [249], Ru [242], Er [250], and Cs [251], in addition to LaB_6 [252] and CeB_6 [253]. Various metal oxides and carbides have also been considered, including RuO_2 [254], IrO_2 [255], TiO_2 [256], SnO_2 [257], Fe_2O_3 [258], In_2O_3 [259], MgO [260,261], NiO [240], ZnO [262–264], StO [265], and ZrC [266]. CsI [267] and Ga [268] have also been reported. Other carbon allotrope adlayers have also been studied, such as DLC [269]. Thin films and nanoparticulate films have been considered, using various deposition techniques ranging from conventional PVD to hydrothermal and other chemie douche approaches. In all cases, these functionalized CNT electron sources typically present lower turn-on fields compared with their pristine counterparts. Here, we quantify this by extracting an amplification factor, $\gamma(E)$, where

$$\gamma(E) = \left(\frac{E_{CNT,n}}{E_{CNT/adlayer,n}} \right) \tag{45.1}$$

where E denotes the electric field and n is the value of an arbitrary electric field for a given current density, J. $\gamma(E)$ in essence quantifies the shift of the FE spectra to lower electric fields. Figure 45.7a shows a summary of the CNT adlayers studied to date, where the x-error denotes one standard deviation in the reported range in work function and the y-error one standard deviation in the J-E shift. The dashed curve is the theoretical variation in $\gamma(E)$ as a function of adlayer work function, Φ. Figure 45.7b shows some example SEM micrographs of functionalized CNTs. An approximate 50% decrease in turn-on field was observed by Chen et al. [255] for an IrO_2 adlayer and by Shrestha et al. [250] using an Er adlayer. Indeed, Figure 45.7a shows a clear correlation with increasing $\gamma(E)$ with decreasing Φ, with adlayers such as ZrC and Er showing perhaps the most significant promise. We stress here that Figure 45.7a is independent of the detailed emitter geometry—aligned or disordered—and simply correlates the amplification effects of adlayer inclusion for a fixed emitter geometry.

Functionalization through the inclusion of adlayers has also been shown to enhance the temporal stability of CNT-based emitters, particularly when using adlayers that hermetically seal the CNTs [264]. Li et al. [264] and Ding et al. [270], for example, showed that ZnO NWs grown directly on vertically aligned CVD-CNT arrays gave an improved time stability from 5% (uncoated) to 0.5% (coated). Such emitter sealing prevents lattice degradation of the CNT backbone during transient local outgassing, which results in deleterious arcing and local plasma formation, both of which have been shown to etch the CNTs, which would otherwise result in fluctuations in the emission current.

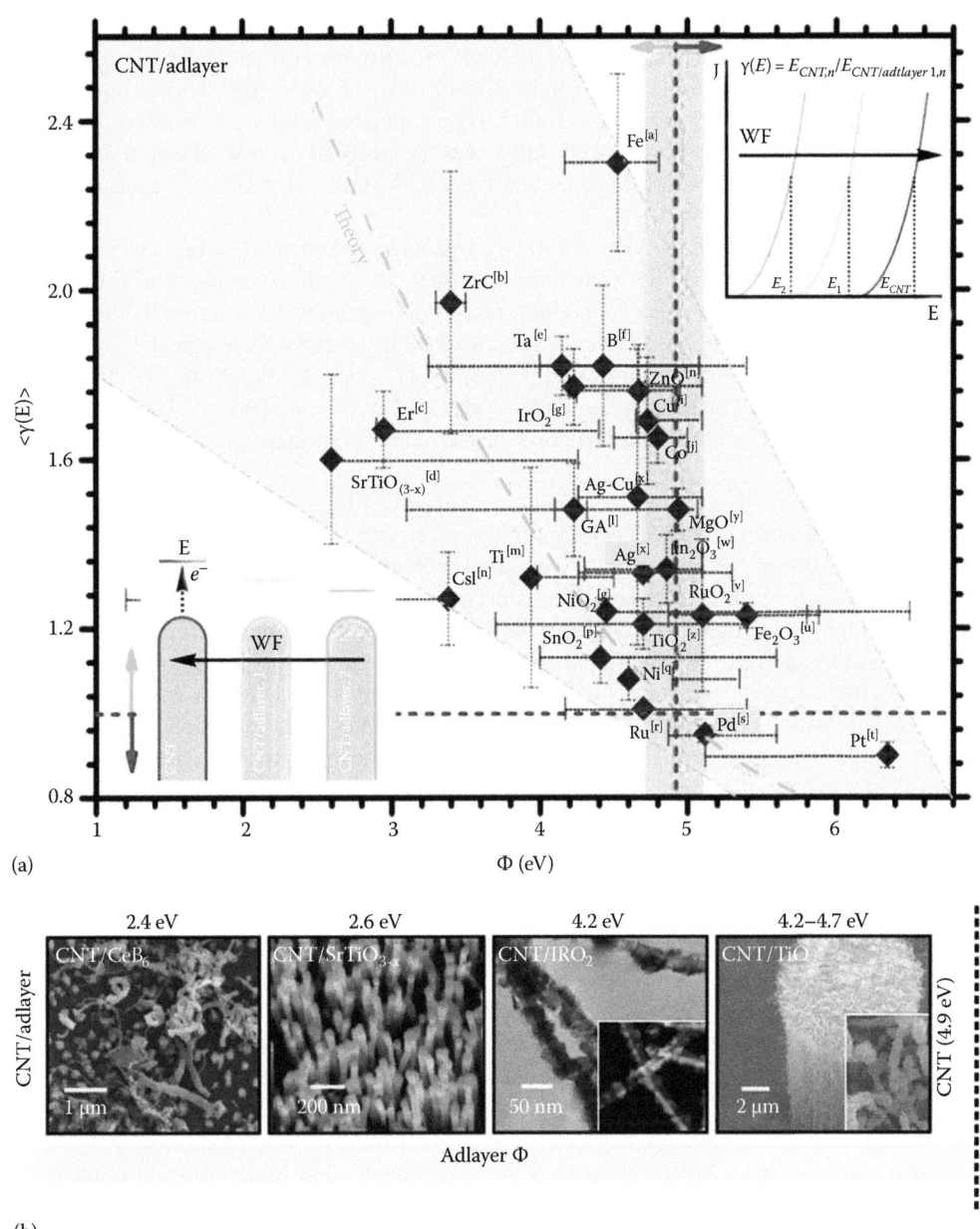

(a)

(b)

FIGURE 45.7

Surface functionalization of CNTs for enhanced FE. (a) Mean amplification factor, <γ>, as a function of adlayer work function (Φ). Note that the lower the adlayer Φ, the greater the γ value. The green curve denotes the theoretical model based on classical Fowler–Nordheim emission. The following adlayers are shown: Pt [241], Pd [241], Ru [241,242], Ni [240], SnO$_2$ [257], Fe$_2$O$_3$ [258], TiO$_2$ [256], NiO [240], RuO$_2$ [254], CsI [267], Ti [246], Ag [243], In$_2$O$_3$ [259], Ga [268], MgO [260,261], Cu [244], Ag–Cu [248], SrTiO$_{3-x}$ (STO) [265], Co [245], IrO$_2$ [255], Er [250], ZnO [262–264], B-doped [271], Ta [247], and ZrC [266]. (b) Sample SEM images of various adlayers in order of increasing work function from left to right, including CeB$_6$ [253] (scale bar, 1 μm), SrTiO$_{3-x}$ [265] (scale bar, 200 nm), IrO$_2$ [255] (scale bar, 50 nm), and TiO$_2$ [256] (scale bar, 2 μm).

45.6 FIELD EMISSION APPLICATIONS

There are a wide range of field electron emission applications. Table 45.1 summarizes the typical emitter performance across some of the most common electron emission applications, principally those shown in Figure 45.8. Much variation in the detailed currents, current densities, and current per tip exists in the literature. Table 45.1 is indicative of these broad ranges. There are undoubtedly devices with operational conditions exterior to those stipulated ranges. Nonetheless, it is clear that an efficient field emitter, with wide market potential, requires an emitting material capable of sourcing the highest values. CNTs, graphene, and the wider family of graphitic nanocarbons, in particular, are one such class of material.

Field emission displays (FEDs) were, in the early 2000s, one of the leading applications of nanostructured field emitters, with Sony championing their development among others. However, the emergence of liquid crystal, plasma, and high-resolution organic light-emitting diode displays saw a decline in FED technologies. Figure 45.8a illustrates the underlying design and function of most FEDs with some examples of fabricated panels, all of which are based on wet chemistry–deposited CNT pixels. For further details, Cole et al. [272] have compiled a detailed historical context of the development, status, and downturn of the field of nanostructured electron emission displays.

Parallel EBL, a CNT-based example of which is shown in Figure 45.8b, aims to exploit the potential nanoscale diameter of the CNT electron sources to realize highly parallelized multibeam devices, typically of the order of a few hundred individually controlled beams, to facilitate extremely rapid nanometer scale patterning and a means of obviating the use of deep and extreme UV photolithography as a route toward further miniaturization of integrated circuits. In the case of EBL, though individual emitters are required to mediate low currents, typically of the order of a few tens of nA, the required current densities are necessarily higher. Nonetheless, individual CNFs can accommodate currents of a few μA before they sublime, and as such, extremely high total currents are attainable using such emitters and the technology has much promise if issues pertaining to multibeam Coulombic repulsion can be resolved.

Field emission lights and lamps (FELs), as illustrated in Figure 45.8c, are another common electron emission application of nanocarbons. FELs are, in essence, single-pixel, high-brightness FEDs. Just as in the case of the displays, light is produced when the excited electrons impinge on a low-voltage phosphor layer. The principal technological challenge lies in achieving homogeneous coverage, and hence uniform electron and optical emission. The so-called Jumbotron lamp, developed by Saito et al., is one such commercially available example [273]. Such FELs are brighter than thermionic light sources, with emission intensities of the order of 10^4 cd/m^2 for beam current densities of 0.25 mA/cm^2 [274]. Indeed, the performance of nanocarbon FELs has proven comparable to conventional commercial fluorescent tubes, which have an emitted intensity of around 1.1×10^4 cd/m^2. Nevertheless, the main barrier to FEL commercialization remains their high power consumption relative to existing high-brightness technologies, such as light-emitting diodes. FELs are nonetheless an exciting alternative to fluorescent tubes as they are mercury-free, turn on almost instantly without flicker, and can be dimmed with ease.

TABLE 45.1 Typical Electron Emitter Performance Criteria across Various Technologies

	Total Emission Current (A)	Typical Current Density (A/cm^2)	Typical Current per Emitter (A)	Beam Energy (eV)
Electron beam lithography	10^{-8}	10^0–10^1	10^{-8}	10^2–10^3
Flat panel display	$>10^{-1}$	10^{-3}	10^{-9}	10^{-1}–10^3
Ion thruster	$>10^{-3}$	10^{-2}	10^{-7}	10^0–10^1
X-ray	$>10^{-3}$	10^{-1}–10^1	10^{-6}–10^{-5}	10^0–10^2
Field emission lamp	$>10^{-2}$	10^{-1}–10^1	10^{-6}–10^{-5}	$>10^2$
Traveling wave tube	$>10^{-2}$	10^0–10^1	10^{-5}	$>10^5$

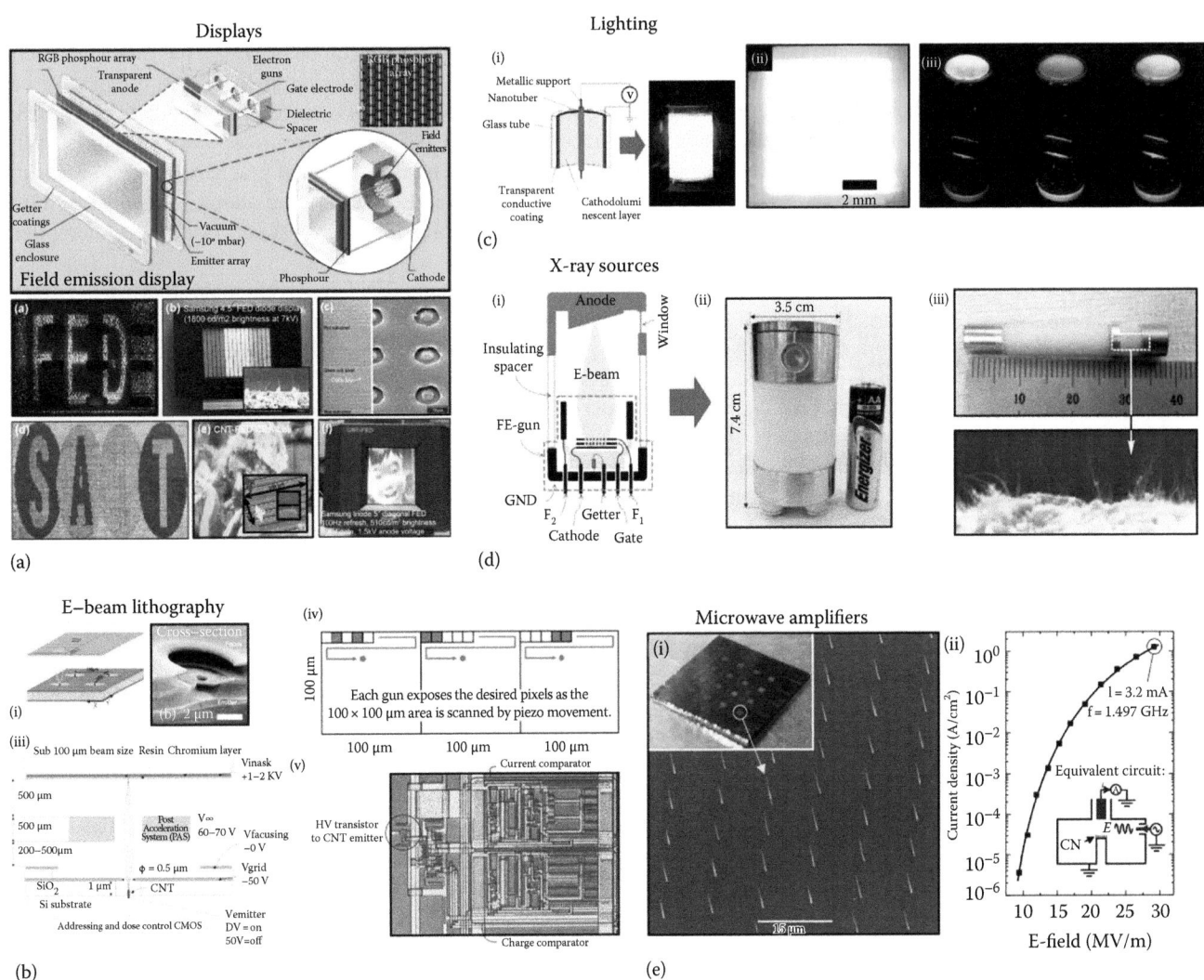

FIGURE 45.8

(a) Various CNT-based electron emission display units. (Adapted from Amaratunga, G., *IEEE Spectrum*, 40, 9, 2003; i: Reprinted from *Development of Field Emission Flat Panel Displays at Motorola*, Talina, A.A., Chalamalaa, B., Colla, B.F., Jaskiea, J.E., Petersena, R., Dworskya, L., Copyright 1998, with permission from Elsevier; Reprinted from *Solid State Electron.*, 45, Talin, A.A., Dean, K.A., Jaskie, J.E., 6, Copyright 2001, with permission from Elsevier, ii: Reprinted with permission from Choi, W.B., Chung, D.S., Kang, J.H., Kim, H.Y., Jin, Y.W., Han, I.T., Lee, Y.H. et al., *Appl. Phys. Lett.*, 75, 20. Copyright 1999, American Institute of Physics; Reprinted with permission from http://phys.org/news86.html 30-inch carbon nanotube based field emission display. Copyright American Institute of Physics, iii: From K.B.K. teo Carbon Nanotube Electron Guns for Displays, http://kennano.com/devices/devices. htm. Copyright Cambridge University, iv: From Talin, A.A. et al., *Solid State Electron.*, 45, 6, 2001; Uemura, S., Carbon nanotube field emission display, in *Perspectives of Fullerene Nanotechnology*, Osawa, E., Ed., Kluwer Academic Publishers, Dordrecht, the Netherlands, 2002, pp. 57–65. Courtesy of J. M Kim, SAIT, v: From P. Foundation http://issuu.com/phantoms_foundation/ docs/e_nano_newsletter_issue20_21/24. Courtesy of CEA-LITEN, vi: Reprinted with permission from Chung, D.S., Park, S.H., Lee, H.W., Choi, J.H., Cha, S.N., Kim, J.W., Jang, J.E. et al., *Appl. Phys. Lett.*, 80, 21. Copyright 2002, American Institute of Physics). (b) A CNT-based parallel EBL system. (Adapted from Milne, W.I., Teo, K.B.K., Amaratunga, G.A.J., Legagneux, P., Gangloff, L., Schnell, J.P., Semet, V., Binh, V.T., Groening, O., *J. Mater. Chem.*, 14, 6, 2004. Reproduced by permission of The Royal Society of Chemistry.) (c) FE lighting devices with integrated screen-printed CNT thin film electron sources. (i: Reprinted from *Carbon*, 40, Bonard, J.-M., Croci, M., Klinke, C., Kurt, R., Noury, O., Weiss, N., 10, Copyright 2002, with permission from Elsevier, ii: Reprinted from *Carbon*, 50, Wu, H.-C., Youh, M.-J., Lin, W.-H., Tseng, C.-L., Juan, Y.-M., Chuang, M.-H., Li, Y.-Y., Sakoda, A., 13, Copyright 2012, with permission from Elsevier, iii: Adapted from Yahachi, S. et al., *Jpn. J. Appl. Phys.*, 37, 3B, Copyright 1998, The Japan Society of Applied Physics.) (d) X-ray sources fabricated with CNT and graphene electron sources. (i, ii: From Jeong, J.-W. et al., *Nanotechnology*, 24, 8, 2013. Copyright IOP Publishing. Reproduced by permission of IOP Publishing, iii: From Kang, J.-T. et al., *ETRI J.*, 35, 6, Copyright 2013, Electronics and Telecommunications Research Institute.) (e) Vertically aligned CNF microwave amplifier. (Reprinted by permission from Macmillan Publishers Ltd., *Nature*, Teo, K.B.K., Minoux, E., Hudanski, L., Peauger, F., Schnell, J.P., Gangloff, L., Legagneux, P., Dieumegard, D., Amaratunga, G.A.J., Milne, W.I., 437, 7061, Copyright 2005.)

X-ray sources are perhaps one of the largest electron emission markets. By 2014, the global X-ray market will be greater than £6 billion per annum, with an estimated 20% growth by 2017 [275]. Central to all commercial X-ray sources is the electron beam. Here, the liberated electrons impinge on a typically Cu or W target resulting in the emission of X-rays. In the past 15 years, there has been significant work on the development of CNT-based X-ray sources, most focusing on the use of screen-printed or vacuum-filtrated CNT electron sources. Some examples of CNT-based sources are shown in Figure 45.8d. Research into CNT- and graphene-based X-ray sources is ongoing with a clear drive toward source miniaturization over added functionality. Further details on the state of the art at the time of publishing can be found in a review compiled by Parmee et al. [276].

High-frequency and microwave power amplifiers, CNT-based examples that are illustrated in Figure 45.8e, are at the heart of most long-range telecommunications systems. With the *Internet of things* becoming increasingly prevalent and with previously isolated devices such as mobile phones, computers, vehicles, and varied consumer products becoming ever more networked, it is clear that the demand for greater data bandwidth and higher channel capacity will continue to increase at an ever more rapid pace. Such data links at present operate at frequencies in excess of 30 GHz. Solid-state electronics have a limited power budget of approximately 1 W at 30 GHz [277]. In practice, more than ten times this is required for efficient wireless data transmission. As such, most data networks based on microwave links employ high-power vacuum electronic devices, termed traveling wave tubes (TWTs). TWTs are based on thermionic electron sources and as such are bulky. Most operate in near Earth orbit and any mass and size savings result in significant financial returns. FE offers a means of aggressive miniaturization allowing for affordable microsatellite networks. In conventional TWTs, data is encoded by direct modulation of a dc electron beam using a circumferential coil. This modulation system accounts for up to 80% of the total device size. One effective means of reducing the device size is to employ FE and directly modulate the electron beam at the source to encode the data, in a triode-like configuration or possibly through the use of integrated pin photodiodes and extremely high-bandwidth microlasers. Spindt-like emitters, with integrated gate electrodes, have been used in the past though the required intergate cathode oxide results in high RC losses, which compromise the attainable bandwidth; nanocarbon-based emitters have no such bottlenecks. In order to be competitive, current densities of the order of 1–10 A/cm^2 are required, and nanocarbons are one of the few available materials, as highlighted previously, that are capable of supporting such current densities without undergoing severe electromigration.

Given the relative youth of nanocarbon FE technologies and the subsequent lack of functional specifications of many of the developed and conceptual FE devices, it is unlikely that many of these devices will make it to market in the very near term. Nevertheless, with increased maturity and assessment of the critical functional parameters—such as lifetime, manufacturing cost, and yield—their competitiveness will dramatically increase. Carbon nanomaterials have a clear proven potential as effective electron emission sources and will likely form the basis of future electron emission–based technologies given suitable technology incubation periods.

45.7 CONCLUSIONS

In this chapter, we have discussed the various approaches to CNT, CNF, and graphene synthesis. The merits of each were highlighted with CVD, seemingly promising perhaps the greatest commercial and technical advantages. The use of the nanocarbons—the CNTs, CNFs, and graphene in FE applications—was then outlined by considering their functional merits relative to other field electron emission materials. The nanocarbons were found to consistently outperform almost all other available materials, and for those materials that showed superior performance over the nanocarbons, it was noted that the ability to nanoengineer, align, and spatially define the nanocarbons with high fidelity and resolution has resulted in them achieving much greater industrial interest than their immediate competitors. Finally, we evidenced that though CNTs do indeed have excellent FE performance, in part due to their extremely high aspect ratio, their turn-on electric fields can be dramatically reduced by the inclusion of surface adlayers. Such nanoscale functionalization was seen to positively enhance the emission spectra with decreasing adlayer work function. Such functionalized systems successfully married the high aspect ratio, high electrical and thermal conductivity, and chemical inertness of the nanocarbons to the low work function of nominally planar materials, resulting in the formation of a novel class of composite emitters capable of meeting the ever-increasing demands of various electron emission applications.

ACKNOWLEDGMENTS

MTC thanks the Oppenheimer Trust, the Technology Strategy Board, and the Engineering and Physical Sciences Research Council for generous financial support. CC thanks the EPSRC Centre for Doctoral Training in Ultra Precision. We express our sincerest apologies to all those individuals who have made contributions to the field that, due to space constraints, we did not mention.

REFERENCES

1. F.S. Baker, J. Williams, A.R. Osborn, *Nature* 239, 5367 (1972).
2. R.T.K. Baker, M.A. Barber, P.S. Harris, F.S. Feates, R.J. Waite, *J. Catal.* 26, 1 (1972).
3. R.T.K. Baker, P.S. Harris, R.B. Thomas, R.J. Waite, *J. Catal.* 30, 1 (1973).
4. R.T.K. Baker, *Carbon* 27, 3 (1989).
5. L.V. Radushkevich, V.M. Lukyanovich, *Zurn Fisic Chim* 26, 88 (1952).
6. P.A. Tesner, A.I. Echeistova, *Doklady Akad. Nauk (USSR)* 87, 1029 (1952).
7. A. Oberlin, M. Endo, T. Koyama, *J. Cryst. Growth* 32, 3 (1976).
8. S. Iijima, *Nature* 354, 6348 (1991).
9. D.S. Bethune, C.H. Kiang, M.S. Devries, G. Gorman, R. Savoy, J. Vazquez, R. Beyers, *Nature* 363, 6430 (1993).
10. H.W. Kroto, J.R. Heath, S.C. Obrien, R.F. Curl, R.E. Smalley, *Nature* 318, 6042 (1985).
11. R.E. Peierls, *Ann. I. H. Poincare* 5, 177 (1935).
12. L.D. Landau, *Phys. Z. Sowjetunion* 11, 26 (1937).
13. N.D. Mermin, *Phys. Rev.* 176, 250–254 (1986).
14. A.K. Geim, K.S. Novoselov, *Nat. Mater.* 6, 3 (2007).
15. F. Bonaccorso, A. Lombardo, T. Hasan, Z. Sun, L. Colombo, A.C. Ferrari, *Mater. Today* 15, 12 (2012).
16. K.S. Novoselov, A.K. Geim, S.V. Morozov, D. Jiang, Y. Zhang, S.V. Dubonos, I.V. Grigorieva, A.A. Firsov, *Science* 306, 5696 (2004).
17. K.S. Novoselov, D. Jiang, F. Schedin, T.J. Booth, V.V. Khotkevich, S.V. Morozov, A.K. Geim, *Proc. Natl. Acad. Sci. USA* 102, 30 (2005).
18. M.S. Dresselhaus, G. Dresselhaus, *Adv. Phys.* 51, 1 (2002).
19. A. Schleede, M. Wellman, *Z. Phys. Chem.* B18, 1 (1932).
20. L.B. Ebert, R.A. Huggins, J.I. Brauman, *Carbon* 12, 2 (1974).
21. H.J. Riley, *Fuel Sci. Pract.* 24, 8–16 (1945).
22. G.R. Hennig, *Progr. Inorg. Chem.* 1, 125–205 (1959).
23. W. Rudorff, *Adv. Inorg. Chem. Radiochem.* 1, 223–266 (1959).
24. H. Wang, J.T. Robinson, X. Li, H. Dai, *J. Am. Chem. Soc.* 131, 29 (2009).
25. X. Li, G. Zhang, X. Bai, X. Sun, X. Wang, E. Wang, H. Dai, *Nat. Nanotechnol.* 3, 9 (2008).
26. H. Choi, H. Kim, S. Hwang, W. Choi, M. Jeon, *Sol. Energ. Mater. Sol, Cell* 95, 1 (2011).
27. K.L. Day, D.R. Huffman, *Nat. Phys. Sci.* 243, (1973).

28. W. Kratschmer, L.D. Lamb, K. Fostiropoulos, D.R. Huffman, *Nature* 347, 354 (1990).
29. C. Berger, Z. Song, T. Li, X. Li, A.Y. Ogbazghi, R. Feng, Z. Dai et al., *J. Phys. Chem. B* 108, 52 (2004).
30. Y.M. Lin, C. Dimitrakopoulos, K.A. Jenkins, D.B. Farmer, H.Y. Chiu, A. Grill, P. Avouris, *Science* 327, 5966 (2010).
31. Z. Li, H. Zhu, D. Xie, K. Wang, A. Cao, J. Wei, X. Li, L. Fan, D. Wu, *Chem. Commun.* 47, 12 (2011).
32. D.V.K.D.V. Kosynkin, A.L. Higginbotham, A. Sinitskii, J.R. Lomeda, A. Dimiev, B.K. Price, J.M. Tour, *Nature* 458, 7240 (2009).
33. L.Y.J.L.Y. Jiao, L. Zhang, X.R. Wang, G. Diankov, H.J. Dai, *Nature* 458, 7240 (2009).
34. M. Yudasaka, R. Yamada, N. Sensui, T. Wilkins, T. Ichihashi, S. Iijima, *J. Phys. Chem. B* 103, 30 (1999).
35. C.D. Scott, S. Arepalli, P. Nikolaev, R.E. Smalley, *Appl. Phys. A* 72, 5 (2001).
36. S. Dhar, A. Roy Barman, G.X. Ni, X. Wang, X.F. Xu, Y. Zheng, S. Tripathy Ariando et al., *AIP Advances* 1, 2 (2011).
37. X. Zhao, M. Ohkohchi, M. Wang, S. Iijima, T. Ichihashi, Y. Ando, *Carbon* 35, 6 (1997).
38. J.T.H. Tsai, A.A. Tseng, *J. Exp. Nanosci.* 4, 1 (2009).
39. Z.J. Shi, Y.F. Lian, X.H. Zhou, Z.N. Gu, Y.G. Zhang, S. Iijima, L.X. Zhou, K.T. Yue, S.L. Zhang, *Carbon* 37, 9 (1999).
40. J.L. Hutchison, N.A. Kiselev, E.P. Krinichnaya, A.V. Krestinin, R.O. Loutfy, A.P. Morawsky, V.E. Muradyan et al., *Carbon* 39, 5 (2001).
41. M.V. Antisari, R. Marazzi, R. Krsmanovic, *Carbon* 41, 12 (2003).
42. Y. Ando, X. Zhao, S. Inoue, T. Suzuki, T. Kadoya, *Diam. Relat. Mater.* 14, 3–7 (2005).
43. N. Li, Z. Wang, K. Zhao, Z. Shi, Z. Gu, S. Xu. *Carbon* 48, 1 (2010).
44. G.F. Zhong, T. Iwasaki, K. Honda, Y. Furukawa, I. Ohdomari, H. Kawarada, *Chem. Vap. Deposition* 11, 3 (2005).
45. C. Zhang, S. Pisana, C.T. Wirth, A. Parvez, C. Ducati, S. Hofmann, J. Robertson, *Diam. Relat. Mater.* 17, 7–10 (2008).
46. R.S. Weatherup, B. Dlubak, S. Hofmann, *ACS Nano* 6, 11 (2012).
47. M.T. Cole, K.B.K. Teo, O. Groening, L. Gangloff, P. Legagneux, W.I. Milne, *Sci. Rep.* 4, 4840 (2014).
48. M.T. Cole, C. Li, Y. Zhang, S.G. Shivareddy, J.S. Barnard, W. Lei, B. Wang, D. Pribat, G.A.J. Amaratunga, W.I. Milne, *ACS Nano* 6, 4 (2012).
49. M.T. Cole, C. Li, W. Lei, K. Qu, K. Ying, Y. Zhang, A.R. Robertson et al., *Adv. Funct. Mater.* 24, 1218–1227 (2013).
50. N. Pierard, A. Fonseca, Z. Konya, I. Willems, G. Van Tendeloo, J.B. Nagy, *Chem. Phys. Lett.* 335, 1–2 (2001).
51. L. Yuan, K. Saito, W. Hu, Z. Chen, *Chem. Phys. Lett.* 346, 1–2 (2001).
52. T. Guo, P. Nikolaev, A.G. Rinzler, D. Tománek, D.T. Colbert, R.E. Smalley, *J. Phys. Chem.* 99, 27 (1995).
53. Z.C. Ying, R.L. Hettich, R.N. Compton, R.E. Haufler, *J. Phys. B Atom. Mol. Opt. Phys.* 29, 21 (1996).
54. N. Sano, H. Wang, I. Alexandrou, M. Chhowalla, K.B.K. Teo, G.A.J. Amaratunga, K. Iimura, *J. Appl. Phys.* 92, 5 (2002).
55. D. Ugarte, *Chem. Phys. Lett.* 198, 6 (1992).
56. K.S. Subrahmanyam, L.S. Panchakarla, A. Govindaraj, C.N.R. Rao, *J. Phys. Chem. C* 113, 11 (2009).

57. Unidym™ Carbon Nanotubes Datasheet, http://www.unidym.com (/files/Unidym_Product_Sheet_SWNT.pdf).
58. A. Reina, X. Jia, J. Ho, D. Nezich, H. Son, V. Bulovic, M.S. Dresselhaus, K. Jing, *Nano Lett.* 9, 1 (2009).
59. C.T. Wirth, C. Zhang, G. Zhong, S. Hofmann, J. Robertson, *ACS Nano* 3, 11 (2009).
60. K.B.K. Teo, M. Chhowalla, G.A.J. Amaratunga, W.I. Milne, G. Pirio, P. Legagneux, F. Wyczisk, D. Pribat, D.G. Hasko, *Appl. Phys. Lett.* 80, 11 (2002).
61. K.B.K. Teo, M. Chhowalla, G.A.J. Amaratunga, W.I. Milne, P. Legagneux, G. Pirio, L. Gangloff, D. Pribat et al., *J. Vac. Sci. Tech. B* 21, 2 (2003).
62. J. Kong, A.M. Cassell, H. Dai, *Chem. Phys. Lett.* 292, 4–6 (1998).
63. C.L. Cheung, A. Kurtz, H. Park, C.M. Lieber, *J. Phys. Chem. B* 106, 10 (2002).
64. S. Huang, M. Woodson, R. Smalley, J. Liu, *Nano Lett.* 4, 6 (2004).
65. J. Sun, N. Lindvall, M.T. Cole, K.T.T. Angel, T. Wang, K.B.K. Teo, D.H.C. Chua, J. Liu, A. Yurgens, *IEEE Trans. Nanotechnol.* 1, 1 (2011).
66. M.T. Cole. *Dry-Transfer of Chemical Vapour Deposited Nanocarbon Thins Films for Functionally Enhanced Flexible Transparent Electronics Applications.* University of Cambridge, Cambridge, U.K., 2011.
67. V.I. Merkulov, D.H. Lowndes, Y.Y. Wei, G. Eres, E. Voelkl, *Appl. Phys. Lett.* 76, 24 (2000).
68. V.I. Merkulov, M.A. Guillorn, D.H. Lowndes, M.L. Simpson, E. Voelkl, *Appl. Phys. Lett.* 79, 8 (2001).
69. Y.Y. Wei, G. Eres, V.I. Merkulov, D.H. Lowndes, *Appl. Phys. Lett.* 78, 10 (2001).
70. S.H. Tsai, C.W. Chao, C.L. Lee, H.C. Shih, *Appl. Phys. Lett.* 74, 23 (1999).
71. K.B.K. Teo, S.B. Lee, M. Chhowalla, V. Semet, V.T. Binh, O. Groening, M. Castignolles et al., *Nanotechnology* 14, 204–211 (2003).
72. O.M. Küttel, O. Groening, C. Emmenegger, L. Schlapbach, *Appl. Phys. Lett.* 73, 15 (1998).
73. G.W. Ho, A.T.S. Wee, J. Lin, W.C. Tjiu, *Thin Solid Films* 388, 1–2 (2001).
74. H. Ishida, N. Satake, G.H. Jeong, Y. Abe, T. Hirata, R. Hatakeyama, K. Tohji, K. Motomiya, *Thin Solid Films* 407, 1–2 (2002).
75. X. Su, C. Stagarescu, G. Xu, D.E. Eastman, I. McNulty, S.P. Frigo, Y. Wang, C.C. Retsch, I.C. Noyan, C.K. Hu, *Appl. Phys. Lett.* 77, 21 (2000).
76. C. Bower, W. Zhu, S.H. Jin, O. Zhou, *Appl. Phys. Lett.* 77, 6 (2000).
77. C. Bower, O. Zhou, W. Zhu, D.J. Werder, S. Jin, *Appl. Phys. Lett.* 77, 17 (2000).
78. L. Delzeit, I. McAninch, B.A. Cruden, D. Hash, B. Chen, J. Han, M. Meyyappan, *J. Appl. Phys.* 91, 9 (2002).
79. K. Matthews, B.A. Gruden, B. Chen, M. Meyyappan, L. Delzeit, *J. Nanosci. Nanotechnol.* 2, 5 (2002).
80. K.B.K. Teo, D.B. Hash, R.G. Lacerda, N.L. Rupesinghe, M.S. Bell, S.H. Dalal, D. Bose et al., *Nano Lett.* 4, 5 (2004).
81. S. Esconjauregui, B.C. Bayer, M. Fouquet, C.T. Wirth, F. Yan, R. Xie, C. Ducati et al., *J. Appl. Phys.* 109, 11 (2011).
82. S. Esconjauregui, M. Fouquet, B. Bayer, J. Robertson, *Phys. Status Solidi B* 247, 11–12 (2010).
83. Y. Woo, D.C. Kim, D.Y. Jeon, H.J. Chung, S.M. Shin, X.S. Li, Y.N. Kwon, D.H. Seo, J. Shin, U.I. Chung, S. Seo, *Electrochem. Soc. Trans.* 19, (2009).
84. G.D. Yuan, W.J. Zhang, Y. Yang, Y.B. Tang, Y.Q. Li, J.X. Wang, X.M. Meng et al., *Chem. Phys. Lett.* 467, 4–6 (2009).
85. S. Hofmann, C. Ducati, J. Robertson, B. Kleinsorge, *Appl. Phys. Lett.* 83, 1 (2003).

86. Z. Li, P. Wu, C. Wang, X. Fan, W. Zhang, X. Zhai, C. Zeng, J. Yang, J. Hou, *ACS Nano* 5, 4 (2011).

87. V.B. Golovko, H.W. Li, B. Kleinsorge, S. Hofmann, J. Geng, M. Cantoro, Z. Yang et al., *Nanotechnology* 16, 9 (2005).

88. R.L. Vander Wal, L.J. Hall, *Carbon* 41, 4 (2003).

89. L. Ci, J. Wei, B. Wei, J. Liang, C. Xu, D. Wu, *Carbon* 39, 3 (2001).

90. M. Hermann Rümmeli, A. Bachmatiuk, F. Börrnert, F. Schäffel, I. Ibrahim, K. Cendrowski, G. Simha-Martynkova et al., *Nanoscale Res. Lett.* 6, 303 (2011).

91. D. Takagi, Y. Homma, H. Hibino, S. Suzuki, Y. Kobayashi, *Nano Lett.* 6, 12 (2006).

92. M.H. Rümmeli, E. Borowiak-Palen, T. Gemming, T. Pichler, M. Knupfer, M. Kalbác, L. Dunsch et al., *Nano Lett.* 5, 7 (2005).

93. R. Rao, K.G. Eyink, B. Maruyama, *Carbon* 48, 13 (2010).

94. T. Uchino, J.L. Hutchison, G.N. Ayre, D.C. Smith, K. De Groot, P. Ashburn, *Jpn. J. Appl. Phys.* 50, 4 (2011).

95. B. Liu, W. Ren, L. Gao, S. Li, S. Pei, C. Liu, C. Jiang, H.M. Cheng, *J. Am. Chem. Soc.* 131, 6 (2009).

96. S. Huang, Q. Cai, J. Chen, Y. Qian, L. Zhang, *J. Am. Chem. Soc.* 131, 6 (2009).

97. S.A. Steiner Iii, T.F. Baumann, B.C. Bayer, R. Blume, M.A. Worsley, W.J. MoberlyChan, E.L. Shaw et al., *J. Am. Chem. Soc.* 131, 34 (2009).

98. M. Kusunoki, M. Rokkaku, T. Suzuki, *Appl. Phys. Lett.* 71, 2620–2622 (1997).

99. H. Liu, D. Takagi, H. Ohno, S. Chiashi, T. Chokan, Y. Homma, *Appl. Phys. Exp.* 1, 1 (2008).

100. P.B. Amama, C.L. Pint, L. McJilton, S.M. Kim, E.A. Stach, P.T. Murray, R.H. Hauge, B. Maruyama, *Nano Lett.* 9, 1 (2009).

101. A. Goyal, L. Simon, Z. Iqbal. *AIChE* 2005, 13993 (2005).

102. L. Ni, K. Kuroda, L.P. Zhou, T. Kizuka, K. Ohta, K. Matsuishi, J. Nakamura, *Carbon* 44, 11 (2006).

103. A.J. Hart, A.H. Slocum, L. Royer, *Carbon* 44, 2 (2006).

104. M.T. Cole, K. Hou, J.H. Warner, J.S. Barnard, K. Ying, Y. Zhang, C. Li, K.B.K. Teo, W.I. Milne, *Diam. Relat. Mater.* 23, 66–71 (2012).

105. A. Rahaman, N. Patra, K.K. Kar, *Fuller. Nanotub. Car. Nanostruct.* 16, 1 (2008).

106. S. Talapatra, S. Kar, S.K. Pal, R. Vajtai, L. Ci, P. Victor, M.M. Shaijumon, S. Kaur, O. Nalamasu, P.M. Ajayan, *Nat. Nanotechnol.* 1, 2 (2006).

107. C. Masarapu, B.Q. Wei, *Langmuir* 23, 17 (2007).

108. T. Chen, L.L. Wang, Y.W. Chen, W.X. Que, Z. Sun, *Appl. Surf. Sci.* 253, 17 (2007).

109. A. Reina, X.T. Jia, J. Ho, D. Nezich, H.B. Son, V. Bulovic, M.S. Dresselhaus, J. Kong, *Nano Lett.* 9, 1 (2009).

110. J. Coraux, A.T. N'Diaye, C. Busse, T. Michely, *Nano Lett.* 8, 2 (2008).

111. A.L.V. de Parga, F. Calleja, B. Borca, M.C.G. Passeggi, J.J. Hinarejos, F. Guinea, R. Miranda, *Phys. Rev. Lett.* 100, 5 (2008).

112. J. Sun, N. Lindvall, M.T. Cole, K.B.K. Teo, A. Yurgens, *Appl. Phys. Lett.* 98, 25 (2011).

113. N. Camara, G. Rius, J.R. Huntzinger, A. Tiberj, L. Magaud, N. Mestres, P. Godignon, J. Camassel, *Appl. Phys. Lett.* 93, 26 (2008).

114. D. Wei, Y. Liu, H. Zhang, L. Huang, B. Wu, J. Chen, G. Yu, *J. Am. Chem. Soc.* 131, 31 (2009).

115. K.-B. Kim, C.-M. Lee, J. Choi, *J. Phys. Chem. C* 115, 30 (2011).

116. J. Sun, N. Lindvall, M.T. Cole, T. Wang, T.J. Booth, P. Boggild, K.B.K. Teo, J. Liu, A. Yurgens, *Appl. Phys. Lett.* 111, 4 (2011).

117. S.A. Ahmed, J. Sun, M.T. Cole, O. Backe, T. Ive, M. Loffler, E. Olsson et al., *IEEE Trans. Semicon. Manu.* 25, 3 (2011).

118. Z. Yu, S.D. Li, P.J. Burke, *Chem. Mater.* 16, 18 (2004).
119. H.E. Unalan, M. Chhowalla, *Nanotechnology* 16, 10 (2005).
120. S. Hofmann, R. Sharma, C. Ducati, G. Du, C. Mattevi, C. Cepek, M. Cantoro et al., *Nano Lett.* 7, 3 (2007).
121. D. Geng, B. Wu, Y. Guo, L. Huang, Y. Xue, J. Chen, G. Yu, L. Jiang, W. Hu, Y. Liu, *Proc. Natl. Acad. Sci. USA* 109, 21 (2012).
122. S. Takenaka, S. Kobayashi, H. Ogihara, K. Otsuka, *J. Catal.* 217, 1 (2003).
123. T. Arcos, F. Vonau, M.G. Garnier, V. Thommen, H.-G. Boyen, P. Oelhafen, M. Düggelin, D. Mathis, R. Guggenheim, *Appl. Phys. Lett.* 80, 13 (2002).
124. J.B.A. Kpetsu, P. Jedrzejowski, C. Côté, A. Sarkissian, P. Mérel, P. Laou, S. Paradis, S. Désilets, H. Liu, X. Sun, *Nanoscale Res. Lett.* 5, 3 (2010).
125. J.M. Simmons, B.M. Nichols, M.S. Marcus, O.M. Castellini, R.J. Homers, M.A. Eriksson, *Small* 2, 7 (2006).
126. E.F. Kukovitsky, S.G. L'Vov, N.A. Sainov, *Chem. Phys. Lett.* 317, 1–2 (2000).
127. K.L. Jiang, C. Feng, K. Liu, S.S. Fan, *J. Nanosci. Nanotechnol.* 7, 4–5 (2007).
128. Y. Saito, *Carbon* 33, 7 (1995).
129. K.W. Kolasinski, *Curr. Opin. Solid State Mater. Sci.* 10, 3–4 (2006).
130. S. Hofmann, G. Csanyi, A.C. Ferrari, M.C. Payne, J. Robertson, *Phys. Rev. Lett.* 95, 0361011–0361014 (2005).
131. K. Celebi, M.T. Cole, J.W. Choi, F. Wyczisk, P. Legagneux, N. Rupesinghe, J. Robertson, K.B. Teo, H.G. Park, *Nano Lett.* 13, 3 (2013).
132. O.V. Yazyev, A. Pasquarello, *Phys. Status Solidi B* 245, 2185–2188 (2008).
133. C. Mattevi, H. Kim, M. Chhowalla, *J. Mater. Chem.* 21, 10 (2011).
134. K. Celebi, M.T. Cole, K.B.K. Teo, H.G. Park, *Electrochem. Solid State Lett.* 15, 1 (2012).
135. H. Kim, C. Mattevi, M.R. Calvo, J.C. Oberg, L. Artiglia, S. Agnoli, C.F. Hirjibehedin, M. Chhowalla, E. Saiz, *ACS Nano* 6, 4 (2012).
136. K. Celebi, M.T. Cole, J.W. Choi, F. Wyczisk, P. Legagneux, N. Rupesinghe, J. Robertson, K.B.K. Teo, H.G. Park, *Nano Lett.* 13, 3 (2013).
137. T. Utsumi, *IEEE Trans. Electron. Dev.* 38, 10 (1991).
138. M.T. Cole, M. Mann, C. Li, K. Hou, Y. Zhang, K. Ying, K.B.K. Teo et al. Novel nanostructured carbon nanotube electron sources. In *International Conference on Materials for Advanced Technologies*, Singapore. Pan Stanford Publishing, Singapore, 2011.
139. M.T. Cole, T. Hallam, W.I. Milne, G.S. Duesberg, *Small* 10, 1, 95–99 (2013).
140. J.J. Chiu, W.S. Wang, C.C. Kei, C.P. Cho, T.P. Perng, P.K. Wei, S.Y. Chiu, *Appl. Phys. Lett.* 83, 22 (2003).
141. C.-P. Cho, T.-P. Perng, *Org. Electron.* 11, 1 (2010).
142. X. Fang, Y. Bando, U.K. Gautam, C. Ye, D. Golberg, *J. Mater. Chem.* 18, 5 (2008).
143. F.C.K. Au, K.W. Wong, Y.H. Tang, Y.F. Zhang, I. Bello, S.T. Lee, *Appl. Phys. Lett.* 75, 12 (1999).
144. Y. Hung, Jr., S.-L. Lee, L.C. Beng, H.-C. Chang, Y.-J. Huang, K.-Y. Lee, Y.-S. Huang, *Thin Solid Films* 556, 146–154 (2014).
145. L.A. Ma, Z.X. Lin, J.Y. Lin, Y.A. Zhang, L.Q. Hu, T.L. Guo, *Phys. E* 41, 8 (2009).
146. H. Tan, N. Xu, S. Deng, *J. Vac. Sci. Technol. B* 28, 2 (2010).
147. J.H. He, R.S. Yang, Y.L. Chueh, L.J. Chou, L.J. Chen, Z.L. Wang, *Adv. Mater.* 18, 5 (2006).
148. X. Ji, P. Chen, J. Deng, W. Zhou, F. Chen, *J. Nanosci. Nanotechnol.* 12, 8 (2012).
149. J. Pelletier, D. Gervais, C. Pomot, *J. Appl. Phys.* 55, 4 (1984).
150. W. Sun, Y. Li, Y. Yang, Y. Li, C. Gu, J. Li, *J. Mater. Chem. C* 2, 13 (2014).

151. P.G. Chavan, M.A. More, D.S. Joag, S.S. Badadhe, I.S. Mulla, Photo-enhanced field emission studies of tapered CdS nanobelts. In *27th International Vacuum Nanoelectronics Conference*, Engelberg, Switzerland, 2014; pp. 83–84.

152. P. Chavan, R. Kashid, S. Badhade, I. Mulla, M. More, D. Joag, *Vacuum* 101, 38–45 (2014).

153. T. Ge, L. Kuai, B. Geng, *J. Alloy Comp.* 509, 39 (2011).

154. Y. Lin, Y. Hsu, S. Lu, S. Kung, *Chem. Commun.* 22, 2391–2393 (2006).

155. Y.H. Lee, C.H. Choi, Y.T. Jang, E.K. Kim, B.K. Ju, N.K. Min, J.H. Ahn, *Appl. Phys. Lett.* 81, 4 (2002).

156. S. Wang, Y. He, X. Fang, J. Zou, Y. Wang, H. Huang, P.M.F.J. Costa et al., *Adv. Mater.* 21, 23 (2009).

157. K.S. Yeong, J.T.L. Thong, *J. Appl. Phys.* 100, 11 (2006).

158. N. Wan, J. Xu, G. Chen, X. Gan, S. Guo, L. Xu, K. Chen, *Acta Mater.* 58, 8 (2010).

159. A.S. Komolov, E.F. Lazneva, S.A. Pshenichnyuk, A.A. Gavrikov, N.S. Chepilko, A.A. Tomilov, N.B. Gerasimova, A.A. Lezov, P.S. Repin, *Semiconductors* 47, 7 (2013).

160. W.Y. Tong, Z.X. Li, A.B. Djurisic, W.K. Chan, S.F. Yu, *Mater. Lett.* 61, 18 (2007).

161. S.-H. Jang, J.-S. Jang, *Electrochem. Solid State Lett.* 13, 12 (2010).

162. T.H. Seo, K.J. Lee, T.S. Oh, Y.S. Lee, H. Jeong, A.H. Park, H. Kim et al., *Appl. Phys. Lett.* 98, 25 (2011).

163. F. Ye, X.M. Cai, X.M. Wang, E.Q. Xie, *J. Cryst. Growth* 304, 2 (2007).

164. J. Lee, Y. Jung, J. Song, J.S. Kim, G.-W. Lee, H.J. Jeong, Y. Jeong, *Carbon* 50, 10 (2012).

165. J.M. Bonard, J.P. Salvetat, T. Stockli, L. Forro, A. Chatelain, *Appl. Phys. A Mater. Sci. Process.* 69, 3 (1999).

166. D.D. Nguyen, Y.-T. Lai, N.-H. Tai, *Diam. Relat. Mater.* 47, 0 (2014).

167. M. Shiraishi, M. Ata, *Carbon* 39, 12 (2001).

168. H. Ago, T. Kugler, F. Cacialli, W.R. Salaneck, M.S.P. Shaffer, A.H. Windle, R.H. Friend, *J. Phys. Chem. B* 103, 38 (1999).

169. K. Zheng, X. Li, X. Mo, G. Chen, Z. Wang, G. Chen, *Appl. Surf. Sci.* 256, 9 (2010).

170. O. Auciello, J.C. Tucek, A.R. Krauss, D.M. Gruen, N. Moldovan, D.C. Mancini, *J. Vac. Sci. Technol. B* 19, 3 (2001).

171. T. Glatzel, H. Steigert, S. Sadewasser, R. Klenk, M.C. Lux-Steiner, *Thin Solid Films* 480, 177–182 (2005).

172. S. Ding, H. Cui, W. Lei, X. Zhang, B. Wang, Stable field emission from ZnO nanowires grown on 3D graphene foam. In *27th International Vacuum Nanoelectronics Conference*, Engelberg, Switzerland, 2014; pp. 178–179.

173. F. Jamali Sheini, D.S. Joag, M.A. More, *Ultramicroscopy* 109, 5 (2009).

174. X. Sun. Designing efficient field emission into ZnO Using morphological and electronic design techniques, low-threshold, high-emission current densities have been obtained with ZnO. [Online], 2006. https://spie.org/x8847.xml. (accessed July 1, 2014)

175. H. Tampo, H. Shibata, K. Maejima, T.W. Chiu, H. Itoh, A. Yamada, K. Matsubara et al., *Appl. Phys. Lett.* 94, 24 (2009).

176. Y.H. Yang, C.X. Wang, B. Wang, N.S. Xu, G.W. Yang, *Chem. Phys. Lett.* 403, 4–6 (2005).

177. Z.P. Zhang, W.Q. Chen, Y.F. Li, J. Chen, Defect-assisted field emission from ZnO nanotrees. In *27th International Vacuum Nanoelectronics Conference*, Engelberg, Switzerland, 2014; pp. 78–79.

178. D.K.T. Ng, M.H. Hong, L.S. Tan, Y.W. Zhu, C.H. Sow, *Nanotechnology* 18, 37 (2007).
179. Y.Q. Wang, R.Z. Wang, M.K. Zhu, B.B. Wang, B. Wang, H. Yan, *Appl. Surf. Sci.* 285, Part B, 0 (2013).
180. R. Yousefi, F.J. Sheini, M.R. Muhamad, M.A. More, *Solid State Sci.* 12, 7 (2010).
181. A. Agiral, J.G.E. Gardeniers, *On-Chip Tungsten Oxide Nanowires Based Electrodes for Charge Injection.* Intech, Croatia, 2010.
182. K. Huang, Q. Pan, F. Yang, S. Ni, D. He, *Mater. Res. Bull.* 43, 4 (2008).
183. Y. Kojima, K. Kasuya, T. Ooi, K. Nagato, K. Takayama, M. Nakao, Jpn. J. Appl. Phys. 46, 9B (2007).
184. S. Yue, H. Pan, Z. Ning, J. Yin, Z. Wang, G. Zhang, *Nanotechnology* 22, 11 (2011).
185. A. Khademi, R. Azimirad, A.A. Zavarian, A.Z. Moshfegh, *J. Phys. Chem. C* 113, 44 (2009).
186. J. Zhou, *Adv. Mater. (Wein.)* 15, 21 (2003).
187. J. Liu, Z. Zhang, C. Pan, Y. Zhao, X. Su, Y. Zhou, D. Yu, *Mater. Lett.* 58, 29 (2004).
188. X. Fang, Y. Bando, G. Shen, C. Ye, U.K. Gautam, P.M.F.J. Costa, C. Zhi, C. Tang, D. Golberg, *Adv. Mater.* 19, 18 (2007).
189. R.V. Kashid, D.J. Late, S.S. Chou, Y.-K. Huang, D. Mrinmoy, D.S. Joag, M.A. More, V.P. Dravid, *Small* 9, 16 (2013).
190. T. Yamada, T. Masuzawa, T. Ebisudani, K. Okano, T. Taniguchi, *Appl. Phys. Lett.* 104, 22 (2014).
191. T. Yamada, T. Ebisudani, K. Okana, T. Masuzawa, Y. Neo, H. Mimura, T. Taniguchi, Field emission characteristics of graphene/h-BN structure. In *27th International Vacuum Nanoelectronics Conference*, Engelberg, Switzerland, 2014; pp. 182–183.
192. A.B. Preobrajenski, A.S. Vinogradov, N. Martensson, *Surf. Sci.* 582, 1–3 (2005).
193. S. Ohtani, T. Yano, S. Kondo, Y. Kohno, Y. Tomita, Y. Maeda, K. Kobayashi, *Thin Solid Films* 546, 53–57 (2013).
194. W. Chen, Y. Su, H. Chen, S. Deng, N. Xu, J. Chen, Temperature dependence of the field emission from monolayer graphene. In *27th International Vacuum Nanoelectronics Conferernce*, Engelberg, Switzerland, 2014; pp. 36–37.
195. T. Hallam, M.T. Cole, W.I. Milne, G.S. Duesberg, *Small* 10, 1 (2014).
196. Z.-S. Wu, S. Pei, W. Ren, D. Tang, L. Gao, B. Liu, F. Li, C. Liu, H.-M. Cheng, *Adv. Mater.* 21, 17 (2009).
197. C. Wu, F. Li, Y. Zhang, T. Guo, *Vacuum* 94, (2013).
198. A. Malesevic, R. Kemps, A. Vanhulsel, M.P. Chowdhury, A. Volodin, C. Van Haesendonck, *J. Appl. Phys.* 104, 8 (2008).
199. U.A. Palnitkar, R.V. Kashid, M.A. More, D.S. Joag, L.S. Panchakarla, C.N.R. Rao, *Appl. Phys. Lett.* 97, 6 (2010).
200. L. Li, W. Sun, S. Tian, X. Xia, J. Li, C. Gu, *Nanoscale* 4, 20 (2012).
201. S. Pandey, P. Rai, S. Patole, F. Gunes, G.-D. Kwon, J.-B. Yoo, P. Nikolaev, S. Arepalli, *Appl. Phys. Lett.* 100, 4 (2012).
202. L. Jiang, T. Yang, F. Liu, J. Dong, Z. Yao, C. Shen, S. Deng, N. Xu, Y. Liu, H.-J. Gao, *Adv. Mater.* 25, 2 (2013).
203. A.T.H. Chuang, J. Robertson, B.O. Boskovic, K.K.K. Koziol, *Appl. Phys. Lett.* 90, 12 (2007).
204. D. Banerjee, S. Mukherjee, K.K. Chattopadhyay, *Appl. Surf. Sci.* 257, 8 (2011).
205. Y.H. Wu, B.J. Yang, B.Y. Zong, H. Sun, Z.X. Shen, Y.P. Feng, *J. Mater. Chem.* 14, 4 (2004).
206. C.S. Rout, P.D. Joshi, R.V. Kashid, D.S. Joag, M.A. More, A.J. Simbeck, M. Washington, S.K. Nayak, D.J. Late, *Sci. Rep.* 3, 1–8 (2013).
207. M.-C. Kan, J.-L. Huang, J.C. Sung, K.-H. Chen, B.-S. Yau, *Carbon* 41, 14 (2003).

208. A. Yutani, A. Kobayashi, A. Kinbara, *Appl. Surf. Sci.* 70–71, Part 2, 0 (1993).
209. Menaka, R. Patra, S. Ghosh, A.K. Ganguli, *RSC Adv.* 2, 20 (2012).
210. M. Jha, R. Patra, S. Ghosh, A.K. Ganguli, *Solid. State Commun.* 153, 1 (2013).
211. J. Xu, G. Hou, H. Li, T. Zhai, B. Dong, H. Yan, Y. Wang, B. Yu, Y. Bando, D. Golberg, *NPG Asia Mater.* 5, 1–9 (2013).
212. S. Raina, W.P. Kang, J.L. Davidson, *Diam. Relat. Mater.* 17, 4–5 (2008).
213. B.-R. Huang, S. Jou, T.-C. Lin, Y.-K. Yang, C.-H. Chou, Y.-M. Wu, *Diam. Relat. Mater.* 20, 3 (2011).
214. H.S. Jung, H.H. Park, S.S. Pang, S.Y. Lee, *Thin Solid Films* 355, 151–156 (1999).
215. H. Zanin, P.W. May, M.H.M.O. Hamanaka, E.J. Corat, *ACS Appl. Mater. Interf.* 5, 23 (2013).
216. D. Banerjee, A. Jha, K.K. Chattopadhyay, *Phys. E* 41, 7 (2009).
217. R.K. Tripathi, O.S. Panwar, A.K. Srivastava, I. Rawal, S. Chockalingam, *Talanta* 125, 276–283 (2014).
218. J. Robertson, *J. Vac. Sci. Technol. B* 17, 2 (1999).
219. A. Hart, B.S. Satyanarayana, W.I. Milne, J. Robertson, *Diam. Relat. Mater.* 8, 2–5 (1999).
220. A. Ilie, A. Hart, A.J. Flewitt, J. Robertson, W.I. Milne, *J. Appl. Phys.* 88, 10 (2000).
221. W.I. Milne, *Appl. Surf. Sci.* 146, 1–4 (1999).
222. M. Setvín, J. Javorský, D. Turčinková, I. Matolínová, P. Sobotík, P. Kocán, I. Ošt'ádal, *Ultramicroscopy* 113, 0 (2012).
223. K.L. Ng, J. Yuan, J.T. Cheung, K.W. Cheah, *Solid State Commun.* 123, 5 (2002).
224. Z.X. Pan, J.C. She, S.Z. Deng, N.S. Xu, Si tip arrays with ultra-narrow nanoscale charge transfer channel. In *27th International Vacuum Nanoelectronics Conference*, Engelberg, Switzerland, 2014; pp. 74–75.
225. S.X. Chen, J.J. Li, C.Z. Gu. In *IEEE, 2007 Seventh IEEE Conference on Nanotechnology*, Hong Kong, Vol. 1–3 (2007).
226. J.W. Glesener, H.B. Lin, A.A. Morrish, *Thin Solid Films* 308, 204–208 (1997).
227. H.-F. Cheng, K.-Y. Teng, H.-C. Chen, G.-C. Tzeng, C.-Y. Tang, I.N. Lin, *Surf. Coat. Technol.* 228, S175–S178 (2013).
228. F. Zhao, D.-D. Zhao, S.-L. Wu, G.-A. Cheng, R.-T. Zheng, *Surf. Coat. Technol.* 228, Supplement 1, 0 (2013).
229. K. Panda, K.J. Sankaran, B.K. Panigrahi, N.-H. Tai, I.N. Lin, *ACS Appl. Mater. Interf.* 6, 11 (2014).
230. P. Abbott, E.D. Sosa, D.E. Golden, *Appl. Phys. Lett.* 79, 17 (2001).
231. X. Song, C.X. Zhao, J.Q. Wu, J. Chen. Field emission from copper micro-cones formed by ion bombardment of copper substrate using an oxide masking. In *25th International Vacuum Nanoelectronics Conference (IVNC)*, Jeju Island, South Korea, 2012; pp. 376–377.
232. J. Wang, L. Wei, L. Zhang, J. Zhang, H. Wei, C. Jiang, Y. Zhang, *Crystengcomm* 15, 7 (2013).
233. E. Le Shim, S.B. Lee, E. You, C.J. Kang, K.W. Lee, Y.J. Choi. Rapid fabrication of wafer scale patterned nickel nanocone arrays for field emission applications. In *25th International Vacuum Nanoelectronics Conference (IVNC)*, Jeju Island, South Korea, 2012; pp. 198–199.
234. S.W. Joo, A.N. Banerjee, *J. Appl. Phys.* 107, 11 (2010).
235. T. Nakane, K. Sano, A. Sakai, A. Magosakon, K. Yanagimoto, T. Sakata, *J. Vac. Sci. Technol. A* 15, 3 (1997).
236. J. Wang, T. Ito, *Diam. Relat. Mater.* 16, 2 (2007).

237. J.B. Cui, J. Ristein, M. Stammler, K. Janischowsky, G. Kleber, L. Ley, *Diam. Relat. Mater.* 9, 3–6 (2000).
238. O. Gröning, *Solid State Electron.* 45, 6 (2001).
239. Y. Zhen-Zhong, G. Jin-Long, W. Zhen-Xia, Z. Zhi-Yuan, H. Jian-Gang, P. Qiang-Yan, *Chin. Phys. Lett.* 24, 1 (2007).
240. C.J. Yang, J.I. Park, Y.R. Cho, *Adv. Eng. Mater.* 9, 1–2 (2007).
241. R.B. Rakhi, A.L. Reddy, M.M. Shaijumon, K. Sethupathi, S. Ramaprabhu, *J. Nanopart. Res.* 10, 1 (2008).
242. C. Liu, K.S. Kim, J. Baek, Y. Cho, S. Han, S.-W. Kim, N.-K. Min, Y. Choi, J.-U. Kim, C.J. Lee, *Carbon* 47, 4 (2009).
243. Z.D. Lin, Y. Sheng-Joue, H. Chih-Hung, C. Shoou-Jinn, C.S. Huang, *IEEE Photon. Tech. Lett.* 25, 11 (2013).
244. Y. Chen, H. Jiang, D. Li, H. Song, Z. Li, X. Sun, G. Miao, H. Zhao, *Nanoscale Res. Lett.* 6, 1 (2011).
245. S.I. Cha, K.T. Kim, S.N. Arshad, C.B. Mo, K.H. Lee, S.H. Hong, *Adv. Mater.* 18, 5 (2006).
246. Y. Zuo, Y. Ren, Z. Wang, X. Han, L. Xi, *Org. Electron.* 14, 9 (2013).
247. Z. Wang, Y. Zuo, Y. Li, X. Han, X. Guo, J. Wang, B. Cao, L. Xi, D. Xue, *Carbon* 73, 0 (2014).
248. S.Y. Lee, W.C. Choi, C. Jeon, C.-Y. Park, J.H. Yang, M.H. Kwon, *Appl. Phys. Lett.* 93, 10 (2008).
249. J. Zhang, C. Yang, Y. Wang, T. Feng, W. Yu, J. Jiang, X. Wang, X. Liu, *Nanotechnology* 17, 1 (2006).
250. S. Shrestha, W.C. Choi, W. Song, Y.T. Kwon, S.P. Shrestha, C.-Y. Park, *Carbon* 48, 1 (2010).
251. A. Wadhawan, R.E. Stallcup, J.M. Perez, *Appl. Phys. Lett.* 78, 1 (2001).
252. M. Kumari, S. Gautam, P.V. Shah, S. Pal, U.S. Ojha, A. Kumar, A.A. Naik et al., *Appl. Phys. Lett.* 101, 12 (2012).
253. R. Patra, S. Ghosh, E. Sheremet, M. Jha, R.D. Rodriguez, D. Lehmann, A.K. Ganguli et al., *J. Appl.* Phys. 115, 9 (2014).
254. S. Youb Lee, C. Jeon, Y. Kim, W. Chel Choi, K. Ihm, T.-H. Kang, Y.-H. Kim, C. Keun Kim, C.-Y. Park, *Appl. Phys. Lett.* 100, 2 (2012).
255. Y.M. Chen, C.A. Chen, Y.S. Huang, K.Y. Lee, K.K. Tiong, *Nanotechnology* 21, 3 (2010).
256. P.H. Chen, Y.S. Huang, W.J. Su, K.Y. Lee, K.K. Tiong, *Mater. Chem. Phys.* 143, 3 (2014).
257. S.K. Pillai, S.C. Motshekga, S.S. Ray, J. Kennedy, *J. Nanomater.* 2012, 1–8 (2012).
258. T.A.J. Loh, D.H.C. Chua, *ECS J. Solid State Sci. Technol.* 3, 4 (2014).
259. J. Lee, T. Park, J. Lee, S. Lee, H. Park, W. Yi, *Carbon* 76, 0 (2014).
260. W.K. Yi, T.W. Jeong, S.G. Yu, J.N. Heo, C.S. Lee, J.H. Lee, W.S. Kim, J.B. Yoo, J.M. Kim, *Adv. Mater.* 14, 20 (2002).
261. S. Chakrabarti, L. Pan, H. Tanaka, S. Hokushin, Y. Nakayama, *Jpn. J. Appl. Phys.* 46, 7R (2007).
262. C. Li, G. Fang, L. Yuan, N. Liu, L. Ai, Q. Xiang, D. Zhao, C. Pan, X. Zhao, *Nanotechnology* 18, 15 (2007).
263. X. Yan, B.-K. Tay, P. Miele, *Carbon* 46, 5 (2008).
264. C. Li, Y. Zhang, M. Mann, P. Hiralal, H.E. Unalan, W. Lei, B.P. Wang et al., *Appl. Phys. Lett.* 96, 14 (2010).
265. A. Pandey, A. Prasad, J.P. Moscatello, M. Engelhard, C. Wang, Y.K. Yap, *ACS Nano* 7, 1 (2012).

266. H. Lei, S.P. Lau, Y.B. Zhang, B.K. Tay, Y.Q. Fu, *Nanotechnology* 15, 5 (2004).

267. M. Shahi, S. Gautam, P.V. Shah, J.S. Rawat, P.K. Chaudhury, H.R.P. Tandon, *J. Nanopart. Res.* 15, 1497 (2013).

268. L. Kun, C. Ming-Ju, L. Hua-Yang, L. Er-Jun, Y. Bin, *J. Inorg. Mater.* 22, 1 (2007).

269. Y. Zou, P.W. May, S.M.C. Vieira, N.A. Fox, *J. Appl. Phys.* 112, 4 (2012).

270. S. Ding, C. Li, W. Lei, Y. Zhang, K. Qasim, H. Cui, X. Zhang, B. Wang, *Thin Solid Films* 524, 0 (2012).

271. J.C. Charlier, M. Terrones, M. Baxendale, V. Meunier, T. Zacharia, N.L. Rupesinghe, W.K. Hsu, N. Grobert, H. Terrones, G.A.J. Amaratunga, *Nano Lett.* 2, 11 (2002).

272. M.T. Cole, W.I. Milne, M. Nakamoto. Field Emission Displays (FED) and Surface-Conduction Electron-Emitter Displays (SED). In *Handbook of Digital Imaging—Field Emission Displays (FED) and Surface-Conduction Electron-Emitter Displays (SED)*. Kriss, M., Ed. John Wiley and Sons Ltd., London, U.K., 2014.

273. Y. Saito, S. Uemura, *Jpn. J. Appl. Phys.* 37, L346 (1998).

274. J.-M. Bonard, M. Croci, C. Klinke, R. Kurt, O. Noury, N. Weiss, *Carbon* 40, 10 (2002).

275. IHS.com Global X-Ray Market, http://press.ihs.com/press-release/design-supply-chain-media/global-x-ray-market-tops-10-billion-2012-and-12-billion-2017 (accessed March 09, 2013).

276. R. Parmee, W.I. Milne, M.T. Cole, *NanoConvergence* 2, 1–27 (2014).

277. W.I. Milne, K.B.K. Teo, G.A.J. Amaratunga, P. Legagneux, L. Gangloff, J.P. Schnell, V. Semet, V.T. Binh, O. Groening, *J. Mater. Chem.* 14, 6 (2004).

278. G. Amaratunga, *IEEE Spectrum* 40, 9 (2003).

279. A.A. Talina, B. Chalamalaa, B.F. Colla, J.E. Jaskiea, R. Petersena, L. Dworskya. In *Development of Field Emission Flat Panel Displays at Motorola*, Materials Research Society, San Francisco, CA 1998.

280. A.A. Talin, K.A. Dean, J.E. Jaskie, *Solid State Electron.* 45, 6 (2001).

281. W.B. Choi, D.S. Chung, J.H. Kang, H.Y. Kim, Y.W. Jin, I.T. Han, Y.H. Lee et al., *Appl. Phys. Lett.* 75, 20 (1999).

282. Phys.Org. 30-inch carbon nanotube based field emission display. http://phys.org/news86.html 30-inch carbon nanotube based field emission display (accessed July 1, 2014).

283. K.B.K. teo Carbon Nanotube Electron Guns for Displays, http://kennano.com/devices/devices.htm. (accessed July 1, 2014)

284. S. Uemura. Carbon nanotube field emission display. In *Perspectives of Fullerene Nanotechnology*, Osawa, E., Ed. Kluwer Academic Publishers, Dordrecht, the Netherlands, 2002; pp. 57–65.

285. P. Foundation http://issuu.com/phantoms_foundation/docs/e_nano_newsletter_issue20_21/24. (accessed July 1, 2014)

286. D.S. Chung, S.H. Park, H.W. Lee, J.H. Choi, S.N. Cha, J.W. Kim, J.E. Jang et al., *Appl. Phys. Lett.* 80, 21 (2002).

287. H.-C. Wu, M.-J. Youh, W.-H. Lin, C.-L. Tseng, Y.-M. Juan, M.-H. Chuang, Y.-Y. Li, A. Sakoda, *Carbon* 50, 13 (2012).

288. S. Yahachi, U. Sashiro, H. Koji, *Jpn. J. Appl. Phys.* 37, 3B (1998).

289. J.-W. Jeong, J.-W. Kim, J.-T. Kang, S. Choi, S. Ahn, Y.-H. Song, *Nanotechnology* 24, 8 (2013).

290. J.-T. Kang, J.-W. Kim, J.W. Jeong, S. Choi, J. Choi, S. Ahn, Y.-H. Song, *ETRI J.* 35, 6 (2013).

291. K.B.K. Teo, E. Minoux, L. Hudanski, F. Peauger, J.P. Schnell, L. Gangloff, P. Legagneux, D. Dieumegard, G.A.J. Amaratunga, W.I. Milne, *Nature* 437, 7061 (2005).

Index